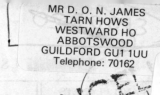

CW00449908

THE NAUTICAL ALMANAC 1998

LIST OF CONTENTS

A2 ALTITUDE CORRECTION TABLES 10°–90°—SUN, STARS, PLANETS

OCT.—MAR. SUN APR.—SEPT.

App. Alt.	Lower Limb	Upper Limb	App. Alt.	Lower Limb	Upper Limb
9 34	+10·8	−21·5	9 39	+10·6	−21·2
9 45	+10·9	−21·4	9 51	+10·7	−21·1
9 56	+11·0	−21·3	10 03	+10·8	−21·0
10 08	+11·1	−21·2	10 15	+10·9	−20·9
10 21	+11·2	−21·1	10 27	+11·0	−20·8
10 34	+11·3	−21·0	10 40	+11·1	−20·7
10 47	+11·4	−20·9	10 54	+11·2	−20·6
11 01	+11·5	−20·8	11 08	+11·3	−20·5
11 15	+11·6	−20·7	11 23	+11·4	−20·4
11 30	+11·7	−20·6	11 38	+11·5	−20·3
11 46	+11·8	−20·5	11 54	+11·6	−20·2
12 02	+11·9	−20·4	12 10	+11·7	−20·1
12 19	+12·0	−20·3	12 28	+11·8	−20·0
12 37	+12·1	−20·2	12 46	+11·9	−19·9
12 55	+12·2	−20·1	13 05	+12·0	−19·8
13 14	+12·3	−20·0	13 24	+12·1	−19·7
13 35	+12·4	−19·9	13 45	+12·2	−19·6
13 56	+12·5	−19·8	14 07	+12·3	−19·5
14 18	+12·6	−19·7	14 30	+12·4	−19·4
14 42	+12·7	−19·6	14 54	+12·5	−19·3
15 06	+12·8	−19·5	15 19	+12·6	−19·2
15 32	+12·9	−19·4	15 46	+12·7	−19·1
15 59	+13·0	−19·3	16 14	+12·8	−19·0
16 28	+13·1	−19·2	16 44	+12·9	−18·9
16 59	+13·2	−19·1	17 15	+13·0	−18·8
17 32	+13·3	−19·0	17 48	+13·1	−18·7
18 06	+13·4	−18·9	18 24	+13·2	−18·6
18 42	+13·5	−18·8	19 01	+13·3	−18·5
19 21	+13·6	−18·7	19 42	+13·4	−18·4
20 03	+13·7	−18·6	20 25	+13·5	−18·3
20 48	+13·8	−18·5	21 11	+13·6	−18·2
21 35	+13·9	−18·4	22 00	+13·7	−18·1
22 26	+14·0	−18·3	22 54	+13·8	−18·0
23 22	+14·1	−18·2	23 51	+13·9	−17·9
24 21	+14·2	−18·1	24 53	+14·0	−17·8
25 26	+14·3	−18·0	26 00	+14·1	−17·7
26 36	+14·4	−17·9	27 13	+14·2	−17·6
27 52	+14·5	−17·8	28 33	+14·3	−17·5
29 15	+14·6	−17·7	30 00	+14·4	−17·4
30 46	+14·7	−17·6	31 35	+14·5	−17·3
32 26	+14·8	−17·5	33 20	+14·6	−17·2
34 17	+14·9	−17·4	35 17	+14·7	−17·1
36 20	+15·0	−17·3	37 26	+14·8	−17·0
38 36	+15·1	−17·2	39 50	+14·9	−16·9
41 08	+15·2	−17·1	42 31	+15·0	−16·8
43 59	+15·3	−17·0	45 31	+15·1	−16·7
47 10	+15·4	−16·9	48 55	+15·2	−16·6
50 46	+15·5	−16·8	52 44	+15·3	−16·5
54 49	+15·6	−16·7	57 02	+15·4	−16·4
59 23	+15·7	−16·6	61 51	+15·5	−16·3
64 30	+15·8	−16·5	67 17	+15·6	−16·2
70 12	+15·9	−16·4	73 16	+15·7	−16·1
76 26	+16·0	−16·3	79 43	+15·8	−16·0
83 05	+16·1	−16·2	86 32	+15·9	−15·9
90 00			90 00		

STARS AND PLANETS

App. Alt.	Corrn
9 56	−5·3
10 08	−5·2
10 20	−5·1
10 33	−5·0
10 46	−4·9
11 00	−4·8
11 14	−4·7
11 29	−4·6
11 45	−4·5
12 01	−4·4
12 18	−4·3
12 35	−4·2
12 54	−4·1
13 13	−4·0
13 33	−3·9
13 54	−3·8
14 16	−3·7
14 40	−3·6
15 04	−3·5
15 30	−3·4
15 57	−3·3
16 26	−3·2
16 56	−3·1
17 28	−3·0
18 02	−2·9
18 38	−2·8
19 17	−2·7
19 58	−2·6
20 42	−2·5
21 28	−2·4
22 19	−2·3
23 13	−2·2
24 11	−2·1
25 14	−2·0
26 22	−1·9
27 36	−1·8
28 56	−1·7
30 24	−1·6
32 00	−1·5
33 45	−1·4
35 40	−1·3
37 48	−1·2
40 08	−1·1
42 44	−1·0
45 36	−0·9
48 47	−0·8
52 18	−0·7
56 11	−0·6
60 28	−0·5
65 08	−0·4
70 11	−0·3
75 34	−0·2
81 13	−0·1
87 03	0·0
90 00	

Additional Corrn

1998

VENUS

Jan. 1–Feb. 5

App. Alt.	Additional Corrn
0	
26	+0·5
46	+0·4
60	+0·3
73	+0·2
84	+0·1

Feb. 6–Feb. 20

App. Alt.	Additional Corrn
0	
29	+0·4
51	+0·3
68	+0·2
83	+0·1

Feb. 21–Mar. 15

App. Alt.	Additional Corrn
0	
34	+0·3
60	+0·2
80	+0·1

Mar. 16–May 5

App. Alt.	Additional Corrn
0	
41	+0·2
76	+0·1

May 6–Dec. 31

App. Alt.	Additional Corrn
0	
60	+0·1

MARS

Jan. 1–Dec. 31

App. Alt.	Additional Corrn
0	
60	+0·1

DIP

Ht. of Eye (m)	Corrn	Ht. of Eye (ft.)
2·4	−2·8	8·0
2·6	−2·9	8·6
2·8	−3·0	9·2
3·0	−3·1	9·8
3·2	−3·2	10·5
3·4	−3·3	11·2
3·6	−3·4	11·9
3·8	−3·5	12·6
4·0	−3·6	13·3
4·3	−3·7	14·1
4·5	−3·8	14·9
4·7	−3·9	15·7
5·0	−4·0	16·5
5·2	−4·1	17·4
5·5	−4·2	18·3
5·8	−4·3	19·1
6·1	−4·4	20·1
6·3	−4·5	21·0
6·6	−4·6	22·0
6·9	−4·7	22·9
7·2	−4·8	23·9
7·5	−4·9	24·9
7·9	−5·0	26·0
8·2	−5·1	27·1
8·5	−5·2	28·1
8·8	−5·3	29·2
9·2	−5·4	30·4
9·5	−5·5	31·5
9·9	−5·6	32·7
10·3	−5·7	33·9
10·6	−5·8	35·1
11·0	−5·9	36·3
11·4	−6·0	37·6
11·8	−6·1	38·9
12·2	−6·2	40·1
12·6	−6·3	41·5
13·0	−6·4	42·8
13·4	−6·5	44·2
13·8	−6·6	45·5
14·2	−6·7	46·9
14·7	−6·8	48·4
15·1	−6·9	49·8
15·5	−7·0	51·3
16·0	−7·1	52·8
16·5	−7·2	54·3
16·9	−7·3	55·8
17·4	−7·4	57·4
17·9	−7·5	59·0
18·4	−7·6	60·5
18·8	−7·7	62·1
19·3	−7·8	63·8
19·8	−7·9	65·4
20·4	−8·0	67·1
20·9	−8·1	68·8
21·4		70·5

Ht. of Eye (m)	Corrn
1·0	−1·8
1·5	−2·2
2·0	−2·5
2·5	−2·8
3·0	−3·0

See table ←

Ht. of Eye (m)	Corrn
20	−7·9
22	−8·3
24	−8·6
26	−9·0
28	−9·3
30	−9·6
32	−10·0
34	−10·3
36	−10·6
38	−10·8
40	−11·1
42	−11·4
44	−11·7
46	−11·9
48	−12·2

Ht. of Eye (ft.)	Corrn
2	−1·4
4	−1·9
6	−2·4
8	−2·7
10	−3·1

See table ←

Ht. of Eye (ft.)	Corrn
70	−8·1
75	−8·4
80	−8·7
85	−8·9
90	−9·2
95	−9·5
100	−9·7
105	−9·9
110	−10·2
115	−10·4
120	−10·6
125	−10·8
130	−11·1
135	−11·3
140	−11·5
145	−11·7
150	−11·9
155	−12·1

App. Alt. = Apparent altitude = Sextant altitude corrected for index error and dip.

ALTITUDE CORRECTION TABLES 0°-10°—SUN,STARS,PLANETS A3

App. Alt.	OCT.—MAR. SUN Lower Limb	Upper Limb	APR.—SEPT. Lower Limb	Upper Limb	STARS PLANETS
° ′	′	′	′	′	′
0 00	−18·2	−50·5	−18·4	−50·2	−34·5
03	17·5	49·8	17·8	49·6	33·8
06	16·9	49·2	17·1	48·9	33·2
09	16·3	48·6	16·5	48·3	32·6
12	15·7	48·0	15·9	47·7	32·0
15	15·1	47·4	15·3	47·1	31·4
0 18	−14·5	−46·8	−14·8	−46·6	−30·8
21	14·0	46·3	14·2	46·0	30·3
24	13·5	45·8	13·7	45·5	29·8
27	12·9	45·2	13·2	45·0	29·2
30	12·4	44·7	12·7	44·5	28·7
33	11·9	44·2	12·2	44·0	28·2
0 36	−11·5	−43·8	−11·7	−43·5	−27·8
39	11·0	43·3	11·2	43·0	27·3
42	10·5	42·8	10·8	42·6	26·8
45	10·1	42·4	10·3	42·1	26·4
48	9·6	41·9	9·9	41·7	25·9
51	9·2	41·5	9·5	41·3	25·5
0 54	−8·8	−41·1	−9·1	−40·9	−25·1
0 57	8·4	40·7	8·7	40·5	24·7
1 00	8·0	40·3	8·3	40·1	24·3
03	7·7	40·0	7·9	39·7	24·0
06	7·3	39·6	7·5	39·3	23·6
09	6·9	39·2	7·2	39·0	23·2
1 12	−6·6	−38·9	−6·8	−38·6	−22·9
15	6·2	38·5	6·5	38·3	22·5
18	5·9	38·2	6·2	38·0	22·2
21	5·6	37·9	5·8	37·6	21·9
24	5·3	37·6	5·5	37·3	21·6
27	4·9	37·2	5·2	37·0	21·2
1 30	−4·6	−36·9	−4·9	−36·7	−20·9
35	4·2	36·5	4·4	36·2	20·5
40	3·7	36·0	4·0	35·8	20·0
45	3·2	35·5	3·5	35·3	19·5
50	2·8	35·1	3·1	34·9	19·1
1 55	2·4	34·7	2·6	34·4	18·7
2 00	−2·0	−34·3	−2·2	−34·0	−18·3
05	1·6	33·9	1·8	33·6	17·9
10	1·2	33·5	1·5	33·3	17·5
15	0·9	33·2	1·1	32·9	17·2
20	0·5	32·8	0·8	32·6	16·8
25	−0·2	32·5	0·4	32·2	16·5
2 30	+0·2	−32·1	−0·1	−31·9	−16·1
35	0·5	31·8	+0·2	31·6	15·8
40	0·8	31·5	0·5	31·3	15·5
45	1·1	31·2	0·8	31·0	15·2
50	1·4	30·9	1·1	30·7	14·9
2 55	1·6	30·7	1·4	30·4	14·7
3 00	+1·9	−30·4	+1·7	−30·1	−14·4
05	2·2	30·1	1·9	29·9	14·1
10	2·4	29·9	2·1	29·7	13·9
15	2·6	29·7	2·4	29·4	13·7
20	2·9	29·4	2·6	29·2	13·4
25	3·1	29·2	2·9	28·9	13·2
3 30	+3·3	−29·0	+3·1	−28·7	−13·0

App. Alt.	OCT.—MAR. SUN Lower Limb	Upper Limb	APR.—SEPT. Lower Limb	Upper Limb	STARS PLANETS
° ′	′	′	′	′	′
3 30	+3·3	−29·0	+3·1	−28·7	−13·0
35	3·6	28·7	3·3	28·5	12·7
40	3·8	28·5	3·5	28·3	12·5
45	4·0	28·3	3·7	28·1	12·3
50	4·2	28·1	3·9	27·9	12·1
3 55	4·4	27·9	4·1	27·7	11·9
4 00	+4·5	−27·8	+4·3	−27·5	−11·8
05	4·7	27·6	4·5	27·3	11·6
10	4·9	27·4	4·6	27·2	11·4
15	5·1	27·2	4·8	27·2	11·2
20	5·2	27·1	5·0	26·8	11·1
25	5·4	26·9	5·1	26·7	10·9
4 30	+5·6	−26·7	+5·3	−26·5	−10·7
35	5·7	26·6	5·5	26·3	10·6
40	5·9	26·4	5·6	26·2	10·4
45	6·0	26·3	5·8	26·0	10·3
50	6·2	26·1	5·9	25·9	10·1
4 55	6·3	26·0	6·0	25·8	10·0
5 00	+6·4	−25·9	+6·2	−25·6	−9·9
05	6·6	25·7	6·3	25·5	9·7
10	6·7	25·6	6·4	25·4	9·6
15	6·8	25·5	6·6	25·2	9·5
20	6·9	25·4	6·7	25·1	9·4
25	7·1	25·2	6·8	25·0	9·2
5 30	+7·2	−25·1	+6·9	−24·9	−9·1
35	7·3	25·0	7·0	24·8	9·0
40	7·4	24·9	7·2	24·6	8·9
45	7·5	24·8	7·3	24·5	8·8
50	7·6	24·7	7·4	24·4	8·7
5 55	7·7	24·6	7·5	24·3	8·6
6 00	+7·8	−24·5	+7·6	−24·2	−8·5
10	8·0	24·3	7·8	24·0	8·3
20	8·2	24·1	8·0	23·8	8·1
30	8·4	23·9	8·1	23·7	7·9
40	8·6	23·7	8·3	23·5	7·7
50	8·7	23·6	8·5	23·3	7·6
7 00	+8·9	−23·4	+8·6	−23·2	−7·4
10	9·1	23·2	8·8	23·0	7·2
20	9·2	23·1	9·0	22·8	7·1
30	9·3	23·0	9·1	22·7	7·0
40	9·5	22·8	9·2	22·6	6·8
7 50	9·6	22·7	9·4	22·4	6·7
8 00	+9·7	−22·6	+9·5	−22·3	−6·6
10	9·9	22·4	9·6	22·2	6·4
20	10·0	22·3	9·7	22·1	6·3
30	10·1	22·2	9·8	22·0	6·2
40	10·2	22·1	10·0	21·8	6·1
8 50	10·3	22·0	10·1	21·7	6·0
9 00	+10·4	−21·9	+10·2	−21·6	−5·9
10	10·5	21·8	10·3	21·5	5·8
20	10·6	21·7	10·4	21·4	5·7
30	10·7	21·6	10·5	21·3	5·6
40	10·8	21·5	10·6	21·2	5·5
9 50	10·9	21·4	10·6	21·2	5·4
10 00	+11·0	−21·3	+10·7	−21·1	−5·3

Additional corrections for temperature and pressure are given on the following page.

For bubble sextant observations ignore dip and use the star corrections for Sun, planets and stars.

ADDITIONAL REFRACTION CORRECTIONS FOR NON-STANDARD CONDITIONS

App. Alt.	A	B	C	D	E	F	G	H	J	K	L	M	N	App. Alt.
0 00	−6.9	−5.7	−4.6	−3.4	−2.3	−1.1	0.0	+1.1	+2.3	+3.4	+4.6	+5.7	+6.9	0 00
0 30	5.2	4.4	3.5	2.6	1.7	0.9	0.0	0.9	1.7	2.6	3.5	4.4	5.2	0 30
1 00	4.3	3.5	2.8	2.1	1.4	0.7	0.0	0.7	1.4	2.1	2.8	3.5	4.3	1 00
1 30	3.5	2.9	2.4	1.8	1.2	0.6	0.0	0.6	1.2	1.8	2.4	2.9	3.5	1 30
2 00	3.0	2.5	2.0	1.5	1.0	0.5	0.0	0.5	1.0	1.5	2.0	2.5	3.0	2 00
2 30	−2.5	−2.1	−1.6	−1.2	−0.8	−0.4	0.0	+0.4	+0.8	+1.2	+1.6	+2.1	+2.5	2 30
3 00	2.2	1.8	1.5	1.1	0.7	0.4	0.0	0.4	0.7	1.1	1.5	1.8	2.2	3 00
3 30	2.0	1.6	1.3	1.0	0.7	0.3	0.0	0.3	0.7	1.0	1.3	1.6	2.0	3 30
4 00	1.8	1.5	1.2	0.9	0.6	0.3	0.0	0.3	0.6	0.9	1.2	1.5	1.8	4 00
4 30	1.6	1.4	1.1	0.8	0.5	0.3	0.0	0.3	0.5	0.8	1.1	1.4	1.6	4 30
5 00	−1.5	−1.3	−1.0	−0.8	−0.5	−0.2	0.0	+0.2	+0.5	+0.8	+1.0	+1.3	+1.5	5 00
6	1.3	1.1	0.9	0.6	0.4	0.2	0.0	0.2	0.4	0.6	0.9	1.1	1.3	6
7	1.1	0.9	0.7	0.6	0.4	0.2	0.0	0.2	0.4	0.6	0.7	0.9	1.1	7
8	1.0	0.8	0.7	0.5	0.3	0.2	0.0	0.2	0.3	0.5	0.7	0.8	1.0	8
9	0.9	0.7	0.6	0.4	0.3	0.1	0.0	0.1	0.3	0.4	0.6	0.7	0.9	9
10 00	−0.8	−0.7	−0.5	−0.4	−0.3	−0.1	0.0	+0.1	+0.3	+0.4	+0.5	+0.7	+0.8	10 00
12	0.7	0.6	0.5	0.3	0.2	0.1	0.0	0.1	0.2	0.3	0.5	0.6	0.7	12
14	0.6	0.5	0.4	0.3	0.2	0.1	0.0	0.1	0.2	0.3	0.4	0.5	0.6	14
16	0.5	0.4	0.3	0.3	0.2	0.1	0.0	0.1	0.2	0.3	0.3	0.4	0.5	16
18	0.4	0.4	0.3	0.2	0.2	0.1	0.0	0.1	0.2	0.2	0.3	0.4	0.4	18
20 00	−0.4	−0.3	−0.3	−0.2	−0.1	−0.1	0.0	+0.1	+0.1	+0.2	+0.3	+0.3	+0.4	20 00
25	0.3	0.3	0.2	0.2	0.1	−0.1	0.0	+0.1	0.1	0.2	0.2	0.3	0.3	25
30	0.3	0.2	0.2	0.1	0.1	0.0	0.0	0.0	0.1	0.1	0.2	0.2	0.3	30
35	0.2	0.2	0.1	0.1	0.1	0.0	0.0	0.0	0.1	0.1	0.1	0.2	0.2	35
40	0.2	0.1	0.1	0.1	−0.1	0.0	0.0	0.0	+0.1	0.1	0.1	0.1	0.2	40
50 00	−0.1	−0.1	−0.1	−0.1	0.0	0.0	0.0	0.0	0.0	+0.1	+0.1	+0.1	+0.1	50 00

The graph is entered with arguments temperature and pressure to find a zone letter; using as arguments this zone letter and apparent altitude (sextant altitude corrected for dip), a correction is taken from the table. This correction is to be applied to the sextant altitude in addition to the corrections for standard conditions (for the Sun, stars and planets from page A2 and for the Moon from pages xxxiv and xxxv).

THE
NAUTICAL
ALMANAC

FOR THE YEAR

1998

LONDON
Issued by
Her Majesty's
Nautical Almanac Office
by order of the
Secretary of State
for
Defence

WASHINGTON
Issued by the
Nautical Almanac Office
United States
Naval Observatory
under the
authority of the
Secretary of the Navy

LONDON: The Stationery Office

1997

ISBN 0 11 772848 9

ISSN 0077-619X

NOTE

Every care is taken to prevent errors in the production of this publication. As a final precaution it is recommended that the sequence of pages in this copy be examined on receipt. If faulty, it should be returned for replacement.

Printed in the United Kingdom for The Stationery Office
Dd303324 2/97 C180 G210055 10170

PREFACE

The British and American editions of *The Nautical Almanac*, which are identical in content, are produced jointly by H.M. Nautical Almanac Office, Royal Greenwich Observatory, under the supervision of A.T. Sinclair and C.Y. Hohenkerk, and by the Nautical Almanac Office, United States Naval Observatory, under the supervision of A.D. Fiala, to the general requirements of the Royal Navy and of the United States Navy. The Almanac is printed separately in the United Kingdom and in the United States of America.

The data in this Almanac can be made available, in a form suitable for direct photographic reproduction, to the appropriate almanac-producing agency in any country; language changes in the headings of the ephemeral pages can be introduced, if desired, during reproduction. Under this arrangement, this Almanac, with minor modifications and changes of language, has been adopted for the Brazilian, Chilean, Danish, Greek, Indian, Indonesian, Mexican and Norwegian almanacs.

J. V. WALL
Director, Royal Greenwich Observatory
Madingley Road
Cambridge, CB3 0EZ
England

KENT W. FOSTER
Captain, U.S. Navy
Superintendent, U.S. Naval Observatory
Washington, D.C. 20392
U.S.A.

November 1996

CALENDAR, 1998

RELIGIOUS CALENDARS

Epiphany	Jan. 6	Low Sunday	Apr. 19
Septuagesima Sunday	Feb. 8	Rogation Sunday	May 17
Quinquagesima Sunday	Feb. 22	Ascension Day—Holy Thursday	May 21
Ash Wednesday	Feb. 25	Whit Sunday—Pentecost	May 31
Quadragesima Sunday	Mar. 1	Trinity Sunday	June 7
Palm Sunday	Apr. 5	Corpus Christi	June 11
Good Friday	Apr. 10	First Sunday in Advent	Nov. 29
Easter Day	Apr. 12	Christmas Day (Friday)	Dec. 25

First Day of Passover (Pesach)	Apr. 11	Day of Atonement (Yom Kippur)	Sept. 30
Feast of Weeks (Shavuot)	May 31	First day of Tabernacles (Succoth)	Oct. 5
Jewish New Year 5759 (Rosh Hashanah)	Sept. 21		

Islamic New Year (1419)	Apr. 28	Ramadân, First day of (tabular)	Dec. 20

The Jewish and Islamic dates above are tabular dates, which begin at sunset on the previous evening and end at sunset on the date tabulated. In practice, the dates of Islamic fasts and festivals are determined by an actual sighting of the appropriate new moon.

CIVIL CALENDAR—UNITED KINGDOM

Accession of Queen Elizabeth II	Feb. 6	Birthday of Prince Philip, Duke of	
St David (Wales)	Mar. 1	Edinburgh	June 10
Commonwealth Day	Mar. 9	The Queen's Official Birthday†	June 13
St Patrick (Ireland)	Mar. 17	Remembrance Sunday	Nov. 8
Birthday of Queen Elizabeth II	Apr. 21	Birthday of the Prince of Wales	Nov. 14
St George (England)	Apr. 23	St Andrew (Scotland)	Nov. 30
Coronation Day	June 2		

PUBLIC HOLIDAYS

England and Wales—Jan. 1†, Apr. 10, Apr. 13, May 4†, May 25, Aug. 31, Dec. 25, Dec. 28†
Northern Ireland—Jan. 1†, Mar. 17, Apr. 10, Apr. 13, May 4†, May 25, July 13†, Aug. 31, Dec. 25, Dec. 28†
Scotland—Jan. 1, Jan. 2†, Apr. 10, May 4, May 25†, Aug. 3, Dec. 25, Dec. 28†

CIVIL CALENDAR—UNITED STATES OF AMERICA

New Year's Day	Jan. 1	Labor Day	Sept. 7
Martin Luther King's Birthday	Jan. 19	Columbus Day	Oct. 12
Lincoln's Birthday	Feb. 12	General Election Day	Nov. 3
Washington's Birthday	Feb. 16	Veterans Day	Nov. 11
Memorial Day	May 25	Thanksgiving Day	Nov. 26
Independence Day	July 4		

†Dates subject to confirmation

PHASES OF THE MOON

New Moon				First Quarter				Full Moon				Last Quarter			
	d	h	m		d	h	m		d	h	m		d	h	m
				Jan.	5	14	18	Jan.	12	17	24	Jan.	20	19	40
Jan.	28	06	01	Feb.	3	22	53	Feb.	11	10	23	Feb.	19	15	27
Feb.	26	17	26	Mar.	5	08	41	Mar.	13	04	34	Mar.	21	07	38
Mar.	28	03	14	Apr.	3	20	18	Apr.	11	22	23	Apr.	19	19	53
Apr.	26	11	41	May	3	10	04	May	11	14	29	May	19	04	35
May	25	19	32	June	2	01	45	June	10	04	18	June	17	10	38
June	24	03	50	July	1	18	43	July	9	16	01	July	16	15	13
July	23	13	44	July	31	12	05	Aug.	8	02	10	Aug.	14	19	48
Aug.	22	02	03	Aug.	30	05	06	Sept.	6	11	21	Sept.	13	01	58
Sept.	20	17	01	Sept.	28	21	11	Oct.	5	20	12	Oct.	12	11	11
Oct.	20	10	09	Oct.	28	11	46	Nov.	4	05	18	Nov.	11	00	28
Nov.	19	04	27	Nov.	27	00	23	Dec.	3	15	19	Dec.	10	17	53
Dec.	18	22	42	Dec.	26	10	46								

Now, what can we do for your tall ship?

Libertad

In London Docklands' West India Dock, we recently accommodated Her Majesty's Yacht Britannia with room to spare.

We'd be happy to do the same for your vessel. In still non-tidal waters – a great advantage over berthing on the tidal River Thames or the Pool of London.

We can offer you 500m of dedicated ships berth, with an additional 200m if required. Dock depth is 9.0m. Locks are available throughout 24 hours, but we may need to lock in deep draughted vessels 2¹/₂ hours before – ¹/₂ hour after high water.

If you are of a gregarious disposition, or an Admiral of the Fleet, we can handle 12 frigate sized vessels together at any one time.

Naturally, electricity, sewage, fresh water, and telephone connections are available. Road transport can drive you right up to your vessel. And your crew on furlough can easily walk to the Docklands Light Railway Station at South Quay to get them to the West End in half an hour.

As for captains – and owners – one of the points that appeals most is the fact that you can take an after-dinner stroll from your berth to Canary Wharf.

While those with more serious business to transact can be in the City of London in 10 minutes.

Call us on VHF channel 68. Write to us at Pierhead Offices, West India Dock, Manchester Road, (Blue Bridge), London E14. Telephone: 44 171 515 1046 or Fax: 44 171 538 5537. We look forward to having you alongside.

London Docklands

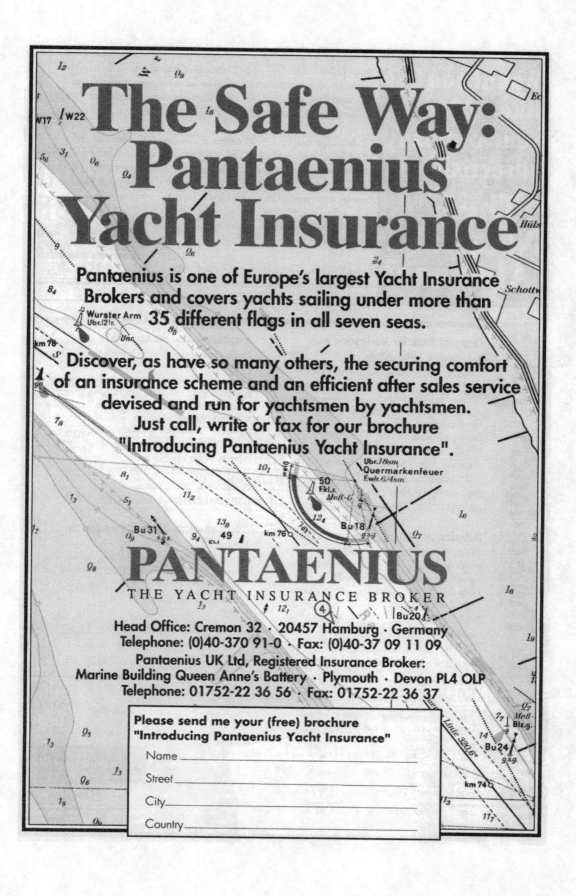

Published by The Stationery Office

Compact Data for Navigation and Astronomy 1996-2000

Compact Data for Navigation and Astronomy provides navigators and astronomers with simple and efficient methods for calculating the positions of the Sun, Moon, navigational planets and stars to a consistent precision with the aid of a personal computer or programmable calculator. Position at sea can therefore be determined without the use of sight reduction tables from observations made with a marine sextant.

The software package, NavPac, is included with this book.
H M Nautical Almanac Office Royal Greenwich Observatory
October 1995   245×188mm   Paperback
ISBN 0 11 772467 X   **£25** (with software package)

Admiralty Manual of Seamanship

This edition of the Admiralty Manual of Seamanship aims to provide the Seaman Specialist with detailed information on all aspects of seamanship concerning the Royal Navy. It is also a source of information on seamanship matters for officers and ratings of other branches.
Ministry of Defence (Navy)
September 1995   297×210mm
ISBN 0 11 772696 6   Hardback   **£80**
(Also available in loose leaf edition 0 11 772695 8   **£90**)

Astronomical Almanac 1997

This publication is prepared jointly by the Nautical Almanac Office, United States Naval Observatory and HM Nautical Almanac Office, Royal Greenwich Observatory and is designed in consultation with astronomers from many countries. It is intended to provide current, accurate astronomical data for use in the making and reduction of observations and for general purposes. A brief description of the use of each ephemeris is given and additional notes are provided in the explanation at the end of the volume.
HM Nautical Almanac Office
November 1996   126×208mm
ISBN 0 11 886505 6   Paperback   Price **£25**

Astronomical Phenomena for the year 1997

Extracts from the Nautical Almanac, including dates and times of planetary and lunar phenomena and other astronomical data of general interest
HM Nautical Almanac Office
255×175
ISBN 0 11 886503 X   Paperback   **£5.95**

Books published by The Stationery Office are available from:
The Stationery Office Bookshops, Agents of The Stationery Office (see Yellow Pages) and through all good booksellers.
Various subject catalogues are available.

Mail order: Stationery Office Publications Centre, P O Box 276, London SW8 5DT

Telephone orders: 0171 873 9090   Fax orders: 0171 873 8200

Enquiries number 0171 873 0011

Internet http://www.the-stationery-office.co.uk/publicat/tsopinf.htm

DAYS OF THE WEEK AND DAYS OF THE YEAR

Day	JAN.		FEB.		MAR.		APR.		MAY		JUNE		JULY		AUG.		SEPT.		OCT.		NOV.		DEC.	
	Wk	Yr	Wk	Yr	Wk	Yr	Wk	Yr	Wk	Yr	Wk	Yr	Wk	Yr	Wk	Yr	Wk	Yr	Wk	Yr	Wk	Yr	Wk	Yr
1	Th.	1	Su.	32	Su.	60	W.	91	F.	121	M.	152	W.	182	Sa.	213	Tu.	244	Th.	274	Su.	305	Tu.	335
2	F.	2	M.	33	M.	61	Th.	92	Sa.	122	Tu.	153	Th.	183	Su.	214	W.	245	F.	275	M.	306	W.	336
3	Sa.	3	Tu.	34	Tu.	62	F.	93	Su.	123	W.	154	F.	184	M.	215	Th.	246	Sa.	276	Tu.	307	Th.	337
4	Su.	4	W.	35	W.	63	Sa.	94	M.	124	Th.	155	Sa.	185	Tu.	216	F.	247	Su.	277	W.	308	F.	338
5	M.	5	Th.	36	Th.	64	Su.	95	Tu.	125	F.	156	Su.	186	W.	217	Sa.	248	M.	278	Th.	309	Sa.	339
6	Tu.	6	F.	37	F.	65	M.	96	W.	126	Sa.	157	M.	187	Th.	218	Su.	249	Tu.	279	F.	310	Su.	340
7	W.	7	Sa.	38	Sa.	66	Tu.	97	Th.	127	Su.	158	Tu.	188	F.	219	M.	250	W.	280	Sa.	311	M.	341
8	Th.	8	Su.	39	Su.	67	W.	98	F.	128	M.	159	W.	189	Sa.	220	Tu.	251	Th.	281	Su.	312	Tu.	342
9	F.	9	M.	40	M.	68	Th.	99	Sa.	129	Tu.	160	Th.	190	Su.	221	W.	252	F.	282	M.	313	W.	343
10	Sa.	10	Tu.	41	Tu.	69	F.	100	Su.	130	W.	161	F.	191	M.	222	Th.	253	Sa.	283	Tu.	314	Th.	344
11	Su.	11	W.	42	W.	70	Sa.	101	M.	131	Th.	162	Sa.	192	Tu.	223	F.	254	Su.	284	W.	315	F.	345
12	M.	12	Th.	43	Th.	71	Su.	102	Tu.	132	F.	163	Su.	193	W.	224	Sa.	255	M.	285	Th.	316	Sa.	346
13	Tu.	13	F.	44	F.	72	M.	103	W.	133	Sa.	164	M.	194	Th.	225	Su.	256	Tu.	286	F.	317	Su.	347
14	W.	14	Sa.	45	Sa.	73	Tu.	104	Th.	134	Su.	165	Tu.	195	F.	226	M.	257	W.	287	Sa.	318	M.	348
15	Th.	15	Su.	46	Su.	74	W.	105	F.	135	M.	166	W.	196	Sa.	227	Tu.	258	Th.	288	Su.	319	Tu.	349
16	F.	16	M.	47	M.	75	Th.	106	Sa.	136	Tu.	167	Th.	197	Su.	228	W.	259	F.	289	M.	320	W.	350
17	Sa.	17	Tu.	48	Tu.	76	F.	107	Su.	137	W.	168	F.	198	M.	229	Th.	260	Sa.	290	Tu.	321	Th.	351
18	Su.	18	W.	49	W.	77	Sa.	108	M.	138	Th.	169	Sa.	199	Tu.	230	F.	261	Su.	291	W.	322	F.	352
19	M.	19	Th.	50	Th.	78	Su.	109	Tu.	139	F.	170	Su.	200	W.	231	Sa.	262	M.	292	Th.	323	Sa.	353
20	Tu.	20	F.	51	F.	79	M.	110	W.	140	Sa.	171	M.	201	Th.	232	Su.	263	Tu.	293	F.	324	Su.	354
21	W.	21	Sa.	52	Sa.	80	Tu.	111	Th.	141	Su.	172	Tu.	202	F.	233	M.	264	W.	294	Sa.	325	M.	355
22	Th.	22	Su.	53	Su.	81	W.	112	F.	142	M.	173	W.	203	Sa.	234	Tu.	265	Th.	295	Su.	326	Tu.	356
23	F.	23	M.	54	M.	82	Th.	113	Sa.	143	Tu.	174	Th.	204	Su.	235	W.	266	F.	296	M.	327	W.	357
24	Sa.	24	Tu.	55	Tu.	83	F.	114	Su.	144	W.	175	F.	205	M.	236	Th.	267	Sa.	297	Tu.	328	Th.	358
25	Su.	25	W.	56	W.	84	Sa.	115	M.	145	Th.	176	Sa.	206	Tu.	237	F.	268	Su.	298	W.	329	F.	359
26	M.	26	Th.	57	Th.	85	Su.	116	Tu.	146	F.	177	Su.	207	W.	238	Sa.	269	M.	299	Th.	330	Sa.	360
27	Tu.	27	F.	58	F.	86	M.	117	W.	147	Sa.	178	M.	208	Th.	239	Su.	270	Tu.	300	F.	331	Su.	361
28	W.	28	Sa.	59	Sa.	87	Tu.	118	Th.	148	Su.	179	Tu.	209	F.	240	M.	271	W.	301	Sa.	332	M.	362
29	Th.	29			Su.	88	W.	119	F.	149	M.	180	W.	210	Sa.	241	Tu.	272	Th.	302	Su.	333	Tu.	363
30	F.	30			M.	89	Th.	120	Sa.	150	Tu.	181	Th.	211	Su.	242	W.	273	F.	303	M.	334	W.	364
31	Sa.	31			Tu.	90			Su.	151			F.	212	M.	243			Sa.	304			Th.	365

ECLIPSES

1. A total eclipse of the Sun on February 26. See map on page 6. The eclipse begins at $14^h 50^m$ and ends at $20^h 06^m$; the total phase begins at $15^h 48^m$ and ends at $19^h 09^m$. The maximum duration of totality is $4^m 13^s$.

2. An annular eclipse of the Sun on August 21–22. See map on page 7. The eclipse begins on August 21 at $23^h 10^m$ and ends on August 22 at $05^h 02^m$; the annular phase begins on August 22 at $00^h 16^m$ and ends on August 22 at $03^h 57^m$. The maximum duration of the annular phase is $3^m 08^s$.

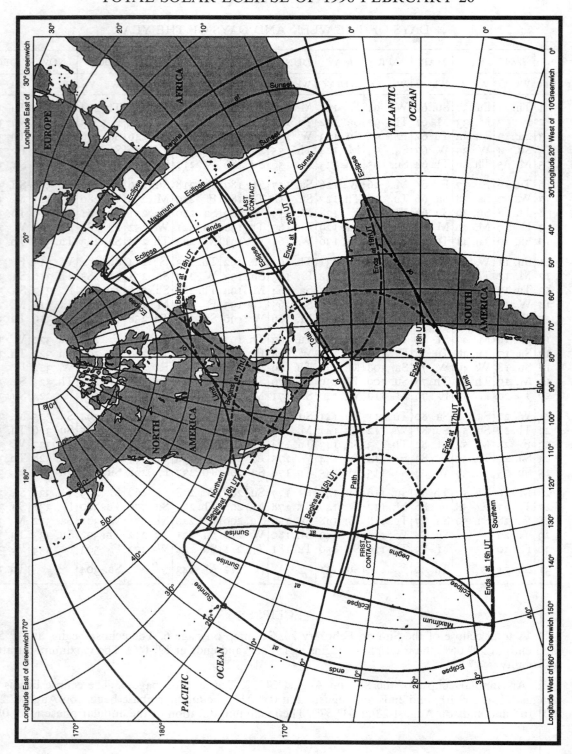

SOLAR ECLIPSE DIAGRAMS

The principal features shown on the above diagrams are: the paths of
total and annular eclipses; the northern and southern limits of partial
eclipse; the sunrise and sunset curves; dashed lines which show the times
of beginning and end of partial eclipse at hourly intervals.

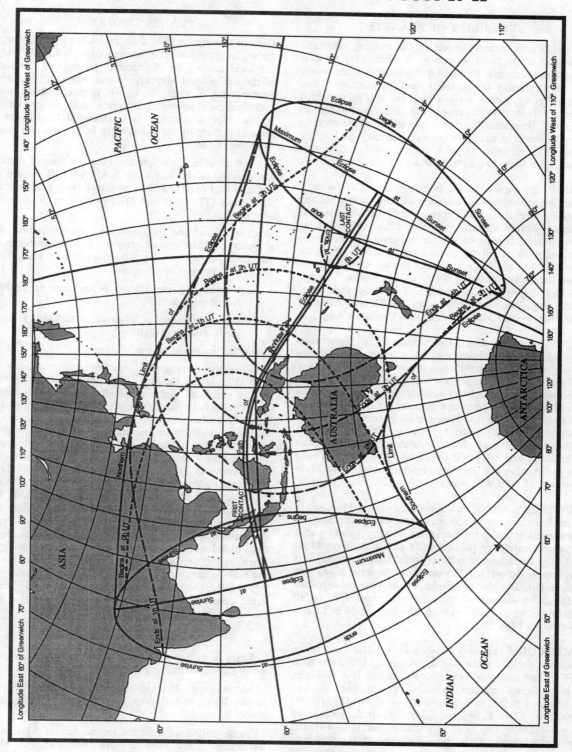

SOLAR ECLIPSE DIAGRAMS

Further details of the paths and times of central eclipse are given in *The Astronomical Almanac.*

VISIBILITY OF PLANETS

VENUS is a brilliant object in the evening sky until mid-way through the second week of January when it becomes too close to the Sun for observation. It reappears at the beginning of the third week of January in the morning sky, where it stays until the end of the third week of September, when it again becomes too close to the Sun; from mid-December it is visible in the evening sky. Venus is in conjunction with Mercury on Jan. 26, Aug. 25 and Sept. 11, with Jupiter on Apr. 23, with Saturn on May 29 and with Mars on August 5.

MARS can be seen in the evening sky in Capricornus, then Aquarius from the last week of January and passing into Pisces at the end of February until just after the first week of March when it becomes too close to the Sun for observation. It reappears in the morning sky during the second week of July in Gemini (passing 6° S of *Pollux* on Aug. 11), moving into Cancer in mid-August, Leo in mid-September (passing 0°9 N of *Regulus* on October 6) and Virgo from the second half of November. Mars is in conjunction with Jupiter on Jan. 21, with Mercury on Mar. 11 and Mar. 30, and with Venus on Aug. 5.

JUPITER can be seen in the evening sky in Capricornus passing into Aquarius at the end of January. It becomes too close to the Sun for observation after the first week of February and reappears in the morning sky during the second week of March. Its westward elongation increases until after mid-June it can be seen for more than half the night, passing into Pisces in early June and back to Aquarius in late August. It is at opposition on Sept. 16 when it is visible all night. Its eastward elongation then decreases, and after mid-December it can only be seen in the evening sky. Jupiter is in conjunction with Mars on Jan. 21, and with Venus on Apr. 23.

SATURN is visible in the evening sky in Pisces until late March when it becomes too close to the Sun for observation. From the beginning of May it can be seen in the morning sky, passing into Cetus in the second half of July and returning to Pisces from mid-September. It is at opposition on Oct. 23 when it can be seen throughout the night. For the remainder of the year its eastward elongation decreases being visible for most of the night. Saturn is in conjunction with Mercury on May 12 and Venus on May 29.

MERCURY can only be seen low in the east before sunrise, or low in the west after sunset (about the time of the beginning or end of civil twilight). It is visible in the mornings between the following approximate dates: January 1 (−0·1) to February 10 (−0·8), April 15 (+2·8) to June 3 (−1·4), August 22 (+2·2) to September 16 (−1·4), December 7 (+1·8) to December 31 (−0·4); the planet is brighter at the end of each period. It is visible in the evenings between the following approximate dates: March 4 (−1·4) to March 30 (+2·3), June 18 (−1·3) to August 6 (+2·7), October 9 (−0·7) to November 26 (+1·8); the planet is brighter at the beginning of each period. The figures in parentheses are the magnitudes.

PLANET DIAGRAM

General Description. The diagram on the opposite page shows, in graphical form for any date during the year, the local mean time of meridian passage of the Sun, of the five planets Mercury, Venus, Mars, Jupiter, and Saturn, and of each 30° of SHA; intermediate lines corresponding to particular stars, may be drawn in by the user if desired. It is intended to provide a general picture of the availability of planets and stars for observation.

On each side of the line marking the time of meridian passage of the Sun a band, 45^m wide, is shaded to indicate that planets and most stars crossing the meridian within 45^m of the Sun are too close to the Sun for observation.

Method of use and interpretation. For any date the diagram provides immediately the local mean times of meridian passage of the Sun, planets and stars, and thus the following information:

(a) whether a planet or star is too close to the Sun for observation;

(b) some indication of its position in the sky, especially during twilight;

(c) the proximity of other planets.

When the meridian passage of an outer planet occurs at midnight the body is in opposition to the Sun and is visible all night; a planet may then be observable during both morning and evening twilights. As the time of meridian passage decreases, the body eventually ceases to be observable in the morning, but its altitude above the eastern horizon at sunset gradually increases; this continues until the body is on the meridian during evening twilight. From then onwards the body is observable above the western horizon and its altitude at sunset gradually decreases; eventually the body becomes too close to the Sun for observation. When the body again becomes visible it is seen low in the east during morning twilight; its altitude at sunrise increases until meridian passage occurs during morning twilight. Then, as the time of meridian passage decreases to 0^h, the body is observable in the west during morning twilight with a gradually decreasing altitude, until it once again reaches opposition.

DO NOT CONFUSE

Mars with Jupiter in the second half of January when Jupiter is the brighter object.

Mercury with Mars in early March, and with Saturn in mid-May; on both occasions Mercury is the brighter object.

Venus with Jupiter in the second half of April, with Saturn from late May to early June and with Mars from late July to mid-August; on all occasions Venus is the brighter object.

Mercury with Venus from late August to mid-September when Venus is the brighter object.

LOCAL MEAN TIME OF MERIDIAN PASSAGE

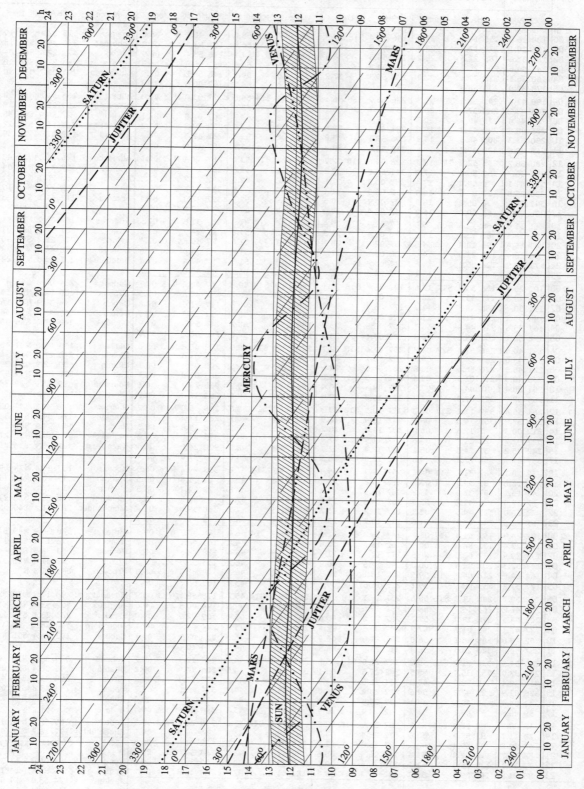

LOCAL MEAN TIME OF MERIDIAN PASSAGE

1998 JANUARY 1, 2, 3 (THURS., FRI., SAT.)

UT	ARIES	VENUS −4.4		MARS +1.2		JUPITER −2.1		SATURN +0.7		STARS		
	GHA	GHA	Dec	GHA	Dec	GHA	Dec	GHA	Dec	Name	SHA	Dec
d h	° ′	° ′	° ′	° ′	° ′	° ′	° ′	° ′	° ′		° ′	° ′
1 00	100 26.6	155 14.8	S17 24.2	146 53.4	S18 39.4	135 32.7	S14 55.0	86 48.6	N 3 06.8	Acamar	315 27.1	S40 19.1
01	115 29.1	170 17.9	23.7	161 53.9	38.8	150 34.6	54.8	101 51.0	06.8	Achernar	335 35.5	S57 15.2
02	130 31.6	185 21.1	23.2	176 54.3	38.3	165 36.6	54.7	116 53.4	06.9	Acrux	173 22.4	S63 04.9
03	145 34.0	200 24.2 ..	22.7	191 54.8 ..	37.7	180 38.5 ..	54.5	131 55.8 ..	06.9	Adhara	255 21.4	S28 58.3
04	160 36.5	215 27.4	22.2	206 55.3	37.2	195 40.5	54.3	146 58.2	07.0	Aldebaran	291 02.7	N16 30.2
05	175 39.0	230 30.6	21.8	221 55.7	36.6	210 42.4	54.2	162 00.6	07.0			
06	190 41.4	245 33.8	S17 21.3	236 56.2	S18 36.1	225 44.4	S14 54.0	177 03.0	N 3 07.0	Alioth	166 31.2	N55 58.0
T 07	205 43.9	260 37.0	20.8	251 56.7	35.5	240 46.3	53.8	192 05.4	07.1	Alkaid	153 08.4	N49 19.2
H 08	220 46.4	275 40.1	20.3	266 57.2	34.9	255 48.3	53.7	207 07.8	07.1	Al Na'ir	27 58.9	S46 58.4
U 09	235 48.8	290 43.3 ..	19.8	281 57.6 ..	34.4	270 50.2 ..	53.5	222 10.2 ..	07.2	Alnilam	275 58.1	S 1 12.4
R 10	250 51.3	305 46.5	19.3	296 58.1	33.8	285 52.2	53.3	237 12.6	07.2	Alphard	218 07.4	S 8 39.1
S 11	265 53.8	320 49.7	18.8	311 58.6	33.3	300 54.2	53.1	252 15.0	07.2			
D 12	280 56.2	335 52.9	S17 18.4	326 59.0	S18 32.7	315 56.1	S14 53.0	267 17.4	N 3 07.3	Alphecca	126 21.3	N26 43.3
A 13	295 58.7	350 56.2	17.9	341 59.5	32.2	330 58.1	52.8	282 19.8	07.3	Alpheratz	357 55.8	N29 04.9
Y 14	311 01.1	5 59.4	17.4	357 00.0	31.6	346 00.0	52.6	297 22.2	07.3	Altair	62 20.1	N 8 51.9
15	326 03.6	21 02.6 ..	16.9	12 00.5 ..	31.1	1 02.0 ..	52.5	312 24.6 ..	07.4	Ankaa	353 27.5	S42 19.3
16	341 06.1	36 05.8	16.4	27 00.9	30.5	16 03.9	52.3	327 27.0	07.4	Antares	112 41.1	S26 25.4
17	356 08.5	51 09.0	16.0	42 01.4	30.0	31 05.9	52.1	342 29.4	07.5			
18	11 11.0	66 12.3	S17 15.5	57 01.9	S18 29.4	46 07.8	S14 51.9	357 31.8	N 3 07.5	Arcturus	146 06.7	N19 11.5
19	26 13.5	81 15.5	15.0	72 02.4	28.8	61 09.8	51.8	12 34.2	07.5	Atria	107 54.1	S69 01.2
20	41 15.9	96 18.8	14.5	87 02.8	28.3	76 11.7	51.6	27 36.6	07.6	Avior	234 22.2	S59 30.2
21	56 18.4	111 22.0 ..	14.0	102 03.3 ..	27.7	91 13.7 ..	51.4	42 39.0 ..	07.6	Bellatrix	278 44.4	N 6 20.7
22	71 20.9	126 25.2	13.6	117 03.8	27.2	106 15.6	51.3	57 41.4	07.7	Betelgeuse	271 13.8	N 7 24.2
23	86 23.3	141 28.5	13.1	132 04.3	26.6	121 17.6	51.1	72 43.8	07.7			
2 00	101 25.8	156 31.8	S17 12.6	147 04.7	S18 26.1	136 19.5	S14 50.9	87 46.2	N 3 07.7	Canopus	264 00.8	S52 41.9
01	116 28.3	171 35.0	12.1	162 05.2	25.5	151 21.5	50.7	102 48.6	07.8	Capella	280 51.5	N45 59.7
02	131 30.7	186 38.3	11.7	177 05.7	24.9	166 23.4	50.6	117 51.0	07.8	Deneb	49 39.9	N45 16.6
03	146 33.2	201 41.6 ..	11.2	192 06.2 ..	24.4	181 25.4 ..	50.4	132 53.4 ..	07.9	Denebola	182 45.7	N14 34.9
04	161 35.6	216 44.8	10.7	207 06.6	23.8	196 27.4	50.2	147 55.8	07.9	Diphda	349 07.8	S18 00.0
05	176 38.1	231 48.1	10.2	222 07.1	23.3	211 29.3	50.0	162 58.2	07.9			
06	191 40.6	246 51.4	S17 09.8	237 07.6	S18 22.7	226 31.3	S14 49.9	178 00.6	N 3 08.0	Dubhe	194 06.0	N61 45.4
07	207 43.0	261 54.7	09.3	252 08.1	22.1	241 33.2	49.7	193 03.0	08.0	Elnath	278 27.2	N28 36.2
F 08	221 45.5	276 58.0	08.8	267 08.5	21.6	256 35.2	49.5	208 05.4	08.1	Eltanin	90 52.2	N51 29.4
R 09	236 48.0	292 01.3 ..	08.4	282 09.0 ..	21.0	271 37.1 ..	49.4	223 07.8 ..	08.1	Enif	33 59.0	N 9 52.0
I 10	251 50.4	307 04.6	07.9	297 09.5	20.4	286 39.1	49.2	238 10.2	08.1	Fomalhaut	15 37.2	S29 38.1
D 11	266 52.9	322 07.9	07.4	312 10.0	19.9	301 41.0	49.0	253 12.6	08.2			
A 12	281 55.4	337 11.2	S17 07.0	327 10.5	S18 19.3	316 43.0	S14 48.8	268 15.0	N 3 08.2	Gacrux	172 14.0	S57 05.8
Y 13	296 57.8	352 14.5	06.5	342 10.9	18.8	331 44.9	48.7	283 17.4	08.3	Gienah	176 04.4	S17 31.7
14	312 00.3	7 17.8	06.0	357 11.4	18.2	346 46.9	48.5	298 19.8	08.3	Hadar	149 04.9	S60 21.4
15	327 02.7	22 21.2 ..	05.6	12 11.9 ..	17.6	1 48.8 ..	48.3	313 22.2 ..	08.3	Hamal	328 14.0	N23 27.2
16	342 05.2	37 24.5	05.1	27 12.4	17.1	16 50.8	48.2	328 24.6	08.4	Kaus Aust.	83 59.9	S34 23.0
17	357 07.7	52 27.8	04.6	42 12.8	16.5	31 52.7	48.0	343 27.0	08.4			
18	12 10.1	67 31.2	S17 04.2	57 13.3	S18 15.9	46 54.7	S14 47.8	358 29.4	N 3 08.5	Kochab	137 20.4	N74 09.7
19	27 12.6	82 34.5	03.7	72 13.8	15.4	61 56.6	47.6	13 31.8	08.5	Markab	13 50.3	N15 11.7
20	42 15.1	97 37.8	03.2	87 14.3	14.8	76 58.6	47.5	28 34.2	08.5	Menkar	314 27.2	N 4 04.8
21	57 17.5	112 41.2 ..	02.8	102 14.8 ..	14.2	92 00.5 ..	47.3	43 36.6 ..	08.6	Menkent	148 21.7	S36 21.3
22	72 20.0	127 44.5	02.3	117 15.2	13.7	107 02.5	47.1	58 39.0	08.6	Miaplacidus	221 41.2	S69 42.4
23	87 22.5	142 47.9	01.8	132 15.7	13.1	122 04.4	46.9	73 41.3	08.7			
3 00	102 24.9	157 51.3	S17 01.4	147 16.2	S18 12.5	137 06.4	S14 46.8	88 43.7	N 3 08.7	Mirfak	308 56.9	N49 51.3
01	117 27.4	172 54.6	00.9	162 16.7	12.0	152 08.3	46.6	103 46.1	08.7	Nunki	76 13.4	S26 17.9
02	132 29.9	187 58.0	00.5	177 17.2	11.4	167 10.3	46.4	118 48.5	08.8	Peacock	53 38.5	S56 44.5
03	147 32.3	203 01.4	17 00.0	192 17.6 ..	10.8	182 12.2 ..	46.2	133 50.9 ..	08.8	Pollux	243 41.8	N28 01.7
04	162 34.8	218 04.8	16 59.5	207 18.1	10.3	197 14.2	46.1	148 53.3	08.9	Procyon	245 11.8	N 5 13.6
05	177 37.2	233 08.1	59.1	222 18.6	09.7	212 16.1	45.9	163 55.7	08.9			
06	192 39.7	248 11.5	S16 58.6	237 19.1	S18 09.1	227 18.1	S14 45.7	178 58.1	N 3 08.9	Rasalhague	96 17.8	N12 33.8
07	207 42.2	263 14.9	58.2	252 19.6	08.6	242 20.0	45.6	194 00.5	09.0	Regulus	207 55.9	N11 58.5
S 08	222 44.6	278 18.3	57.7	267 20.0	08.0	257 22.0	45.4	209 02.9	09.0	Rigel	281 23.1	S 8 12.4
A 09	237 47.1	293 21.7 ..	57.3	282 20.5 ..	07.4	272 23.9 ..	45.2	224 05.3 ..	09.1	Rigil Kent.	140 08.2	S60 49.2
T 10	252 49.6	308 25.1	56.8	297 21.0	06.9	287 25.9	45.0	239 07.7	09.1	Sabik	102 26.4	S15 43.2
U 11	267 52.0	323 28.5	56.4	312 21.5	06.3	302 27.8	44.9	254 10.1	09.2			
R 12	282 54.5	338 31.9	S16 55.9	327 22.0	S18 05.7	317 29.8	S14 44.7	269 12.5	N 3 09.2	Schedar	349 54.0	N56 31.8
D 13	297 57.0	353 35.4	55.5	342 22.5	05.1	332 31.7	44.5	284 14.9	09.2	Shaula	96 38.4	S37 06.0
A 14	312 59.4	8 38.8	55.0	357 22.9	04.6	347 33.7	44.3	299 17.3	09.3	Sirius	258 43.8	S16 43.0
Y 15	328 01.9	23 42.2 ..	54.6	12 23.4 ..	04.0	2 35.6 ..	44.2	314 19.6 ..	09.3	Spica	158 43.8	S11 08.9
16	343 04.4	38 45.6	54.1	27 23.9	03.4	17 37.6	44.0	329 22.0	09.4	Suhail	223 00.7	S43 25.4
17	358 06.8	53 49.1	53.7	42 24.4	02.9	32 39.5	43.8	344 24.4	09.4			
18	13 09.3	68 52.5	S16 53.2	57 24.9	S18 02.3	47 41.5	S14 43.6	359 26.8	N 3 09.5	Vega	80 47.4	N38 47.0
19	28 11.7	83 55.9	52.8	72 25.4	01.7	62 43.4	43.5	14 29.2	09.5	Zuben'ubi	137 18.7	S16 01.8
20	43 14.2	98 59.4	52.3	87 25.8	01.1	77 45.3	43.3	29 31.6	09.5		SHA	Mer.Pass.
21	58 16.7	114 02.8 ..	51.9	102 26.3 ..	00.6	92 47.3 ..	43.1	44 34.0 ..	09.6		° ′	h m
22	73 19.1	129 06.3	51.4	117 26.8	18 00.0	107 49.2	42.9	59 36.4	09.6	Venus	55 06.0	13 31
23	88 21.6	144 09.7	51.0	132 27.3	S17 59.4	122 51.2	42.8	74 38.8	09.7	Mars	45 38.9	14 11
	h m									Jupiter	34 53.8	14 53
Mer.Pass. 17 11.5	v 3.3 d 0.5		v 0.5 d 0.6		v 2.0 d 0.2		v 2.4 d 0.0			Saturn	346 20.4	18 06

UT	SUN GHA	SUN Dec	MOON GHA	v	MOON Dec	d	HP
d h	° ′	° ′	° ′	′	° ′	′	′
1 00	179 10.6	S23 02.0	148 46.5	7.5	S15 07.2	6.8	59.0
01	194 10.3	01.8	163 13.0	7.6	15 00.4	7.0	59.0
02	209 10.0	01.6	177 39.6	7.6	14 53.4	7.0	59.1
03	224 09.7	.. 01.4	192 06.2	7.6	14 46.4	7.1	59.1
04	239 09.4	01.2	206 32.8	7.7	14 39.3	7.2	59.1
05	254 09.1	01.0	220 59.5	7.7	14 32.1	7.3	59.1
06	269 08.8	S23 00.8	235 26.2	7.7	S14 24.8	7.4	59.1
T 07	284 08.5	00.6	249 52.9	7.7	14 17.4	7.5	59.1
H 08	299 08.2	00.4	264 19.6	7.8	14 09.9	7.5	59.1
U 09	314 07.9	.. 00.2	278 46.4	7.8	14 02.4	7.7	59.1
R 10	329 07.6	23 00.0	293 13.2	7.9	13 54.7	7.8	59.2
S 11	344 07.3	22 59.7	307 40.1	7.8	13 46.9	7.8	59.2
D 12	359 07.0	S22 59.5	322 06.9	7.9	S13 39.1	7.9	59.2
A 13	14 06.7	59.3	336 33.8	7.9	13 31.2	8.0	59.2
Y 14	29 06.4	59.1	351 00.7	8.0	13 23.2	8.1	59.2
15	44 06.1	.. 58.9	5 27.7	8.0	13 15.1	8.2	59.2
16	59 05.8	58.7	19 54.7	8.0	13 06.9	8.2	59.2
17	74 05.5	58.5	34 21.7	8.0	12 58.7	8.4	59.2
18	89 05.2	S22 58.3	48 48.7	8.1	S12 50.3	8.4	59.2
19	104 04.9	58.1	63 15.8	8.1	12 41.9	8.5	59.2
20	119 04.6	57.9	77 42.9	8.2	12 33.4	8.6	59.2
21	134 04.4	.. 57.6	92 10.1	8.1	12 24.8	8.6	59.3
22	149 04.1	57.4	106 37.2	8.2	12 16.2	8.8	59.3
23	164 03.8	57.2	121 04.4	8.3	12 07.4	8.8	59.3
2 00	179 03.5	S22 57.0	135 31.7	8.2	S11 58.6	8.9	59.3
01	194 03.2	56.8	149 58.9	8.4	11 49.7	8.9	59.3
02	209 02.9	56.6	164 26.3	8.3	11 40.8	9.0	59.3
03	224 02.6	.. 56.3	178 53.6	8.4	11 31.8	9.1	59.3
04	239 02.3	56.1	193 21.0	8.4	11 22.7	9.2	59.3
05	254 02.0	55.9	207 48.4	8.4	11 13.5	9.2	59.3
06	269 01.7	S22 55.7	222 15.8	8.5	S11 04.3	9.3	59.3
F 07	284 01.4	55.5	236 43.3	8.5	10 55.0	9.4	59.3
R 08	299 01.1	55.2	251 10.8	8.5	10 45.6	9.4	59.3
I 09	314 00.8	.. 55.0	265 38.3	8.6	10 36.2	9.5	59.3
D 10	329 00.5	54.8	280 05.9	8.6	10 26.7	9.6	59.3
A 11	344 00.3	54.6	294 33.5	8.6	10 17.1	9.6	59.3
Y 12	359 00.0	S22 54.3	309 01.1	8.7	S10 07.5	9.7	59.3
13	13 59.7	54.1	323 28.8	8.7	9 57.8	9.8	59.3
14	28 59.4	53.9	337 56.5	8.7	9 48.0	9.8	59.4
15	43 59.1	.. 53.6	352 24.2	8.8	9 38.2	9.9	59.4
16	58 58.8	53.4	6 52.0	8.8	9 28.3	9.9	59.4
17	73 58.5	53.2	21 19.8	8.8	9 18.4	10.0	59.4
18	88 58.2	S22 53.0	35 47.6	8.9	S 9 08.4	10.0	59.4
19	103 57.9	52.7	50 15.5	8.9	8 58.4	10.1	59.4
20	118 57.6	52.5	64 43.4	8.9	8 48.3	10.1	59.4
21	133 57.3	.. 52.3	79 11.3	8.9	8 38.2	10.2	59.4
22	148 57.0	52.0	93 39.2	9.0	8 28.0	10.2	59.4
23	163 56.8	51.8	108 07.2	9.0	8 17.8	10.3	59.4
3 00	178 56.5	S22 51.6	122 35.2	9.1	S 8 07.5	10.4	59.4
01	193 56.2	51.3	137 03.3	9.0	7 57.1	10.4	59.4
02	208 55.9	51.1	151 31.3	9.1	7 46.7	10.4	59.4
03	223 55.6	.. 50.8	165 59.4	9.2	7 36.3	10.5	59.4
04	238 55.3	50.6	180 27.6	9.1	7 25.8	10.5	59.4
05	253 55.0	50.4	194 55.7	9.2	7 15.3	10.5	59.4
06	268 54.7	S22 50.1	209 23.9	9.2	S 7 04.8	10.6	59.4
S 07	283 54.4	49.9	223 52.1	9.3	6 54.2	10.7	59.4
A 08	298 54.2	49.6	238 20.4	9.2	6 43.5	10.7	59.4
T 09	313 53.9	.. 49.4	252 48.6	9.3	6 32.8	10.7	59.4
U 10	328 53.6	49.2	267 16.9	9.4	6 22.1	10.7	59.4
R 11	343 53.3	48.9	281 45.3	9.3	6 11.4	10.8	59.4
D 12	358 53.0	S22 48.7	296 13.6	9.4	S 6 00.6	10.9	59.4
A 13	13 52.7	48.4	310 42.0	9.4	5 49.7	10.8	59.4
Y 14	28 52.4	48.2	325 10.4	9.4	5 38.9	10.9	59.4
15	43 52.1	.. 47.9	339 38.8	9.5	5 28.0	10.9	59.4
16	58 51.9	47.7	354 07.3	9.4	5 17.1	11.0	59.4
17	73 51.6	47.4	8 35.7	9.5	5 06.1	10.9	59.4
18	88 51.3	S22 47.2	23 04.2	9.5	S 4 55.2	11.0	59.4
19	103 51.0	46.9	37 32.7	9.6	4 44.2	11.1	59.4
20	118 50.7	46.7	52 01.3	9.5	4 33.1	11.0	59.4
21	133 50.4	.. 46.4	66 29.8	9.6	4 22.1	11.1	59.4
22	148 50.1	46.2	80 58.4	9.6	4 11.0	11.1	59.4
23	163 49.8	45.9	95 27.0	9.6	S 3 59.9	11.1	59.4
	SD 16.3	d 0.2	SD 16.1		16.2		16.2

Twilight / Sunrise / Moonrise

Lat.	Twilight Naut.	Twilight Civil	Sunrise	Moonrise 1	2	3	4
°	h m	h m	h m	h m	h m	h m	h m
N 72	08 23	10 40	■■■	11 48	11 40	11 33	11 27
N 70	08 04	09 48	■■■	11 21	11 24	11 24	11 24
68	07 49	09 16	■■■	11 01	11 10	11 17	11 22
66	07 37	08 52	10 26	10 45	11 00	11 11	11 21
64	07 26	08 34	09 49	10 31	10 51	11 06	11 19
62	07 17	08 18	09 22	10 20	10 43	11 01	11 18
60	07 09	08 05	09 02	10 10	10 36	10 58	11 17
N 58	07 02	07 54	08 45	10 02	10 30	10 54	11 16
56	06 56	07 44	08 31	09 54	10 25	10 51	11 15
54	06 50	07 35	08 19	09 48	10 20	10 48	11 14
52	06 44	07 28	08 08	09 42	10 16	10 46	11 13
50	06 39	07 20	07 58	09 36	10 12	10 43	11 13
45	06 28	07 05	07 38	09 24	10 03	10 38	11 11
N 40	06 18	06 52	07 22	09 15	09 56	10 34	11 10
35	06 09	06 40	07 08	09 06	09 50	10 30	11 09
30	06 00	06 30	06 56	08 59	09 44	10 27	11 08
20	05 44	06 11	06 35	08 46	09 35	10 22	11 07
N 10	05 28	05 55	06 17	08 35	09 27	10 17	11 05
0	05 12	05 38	06 00	08 24	09 19	10 12	11 04
S 10	04 53	05 20	05 43	08 14	09 11	10 07	11 03
20	04 31	05 00	05 25	08 02	09 03	10 02	11 01
30	04 03	04 36	05 03	07 49	08 53	09 57	11 00
35	03 44	04 21	04 50	07 42	08 48	09 53	10 59
40	03 22	04 03	04 36	07 33	08 41	09 50	10 58
45	02 52	03 41	04 18	07 23	08 34	09 45	10 57
S 50	02 08	03 12	03 56	07 11	08 25	09 40	10 56
52	01 43	02 58	03 46	07 06	08 21	09 38	10 55
54	01 03	02 40	03 34	06 59	08 17	09 35	10 54
56	////	02 19	03 20	06 53	08 12	09 32	10 54
58	////	01 51	03 04	06 45	08 06	09 29	10 53
S 60	////	01 09	02 44	06 36	08 00	09 26	10 52

Sunset / Twilight / Moonset

Lat.	Sunset	Twilight Civil	Twilight Naut.	Moonset 1	2	3	4
°	h m	h m	h m	h m	h m	h m	h m
N 72	■■■	13 28	15 45	17 42	19 44	21 41	23 35
N 70	■■■	14 20	16 04	18 08	19 58	21 47	23 34
68	■■■	14 52	16 19	18 27	20 10	21 52	23 34
66	13 42	15 16	16 31	18 42	20 19	21 57	23 33
64	14 20	15 35	16 42	18 55	20 27	22 00	23 33
62	14 46	15 50	16 51	19 06	20 34	22 03	23 33
60	15 06	16 03	16 59	19 15	20 40	22 06	23 32
N 58	15 23	16 14	17 06	19 22	20 45	22 09	23 32
56	15 37	16 24	17 13	19 29	20 49	22 11	23 32
54	15 49	16 33	17 19	19 35	20 54	22 13	23 32
52	16 00	16 41	17 24	19 41	20 57	22 14	23 31
50	16 10	16 48	17 29	19 46	21 01	22 16	23 31
45	16 30	17 03	17 40	19 57	21 08	22 19	23 31
N 40	16 46	17 17	17 50	20 05	21 14	22 22	23 30
35	17 00	17 28	18 00	20 13	21 19	22 25	23 30
30	17 12	17 38	18 08	20 20	21 23	22 27	23 30
20	17 33	17 57	18 24	20 31	21 31	22 31	23 29
N 10	17 51	18 13	18 40	20 41	21 38	22 34	23 29
0	18 08	18 30	18 56	20 50	21 44	22 37	23 28
S 10	18 25	18 48	19 15	21 00	21 50	22 40	23 28
20	18 43	19 08	19 37	21 09	21 57	22 43	23 28
30	19 05	19 32	20 05	21 20	22 05	22 46	23 27
35	19 18	19 47	20 23	21 27	22 09	22 48	23 27
40	19 32	20 05	20 46	21 34	22 14	22 51	23 26
45	19 50	20 27	21 15	21 42	22 19	22 53	23 26
S 50	20 11	20 55	21 59	21 52	22 26	22 56	23 25
52	20 22	21 10	22 24	21 57	22 29	22 58	23 25
54	20 34	21 27	23 03	22 02	22 32	22 59	23 25
56	20 47	21 48	////	22 08	22 36	23 01	23 24
58	21 03	22 15	////	22 14	22 40	23 03	23 24
S 60	21 23	22 57	////	22 21	22 45	23 05	23 24

SUN / MOON

Day	Eqn. of Time 00ʰ	Eqn. of Time 12ʰ	Mer. Pass.	Mer. Pass. Upper	Mer. Pass. Lower	Age	Phase
d	m s	m s	h m	h m	h m	d	%
1	03 17	03 31	12 04	14 37	02 10	03	10
2	03 46	04 00	12 04	15 32	03 05	04	17
3	04 14	04 27	12 04	16 24	03 58	05	27

1998 JANUARY 4, 5, 6 (SUN., MON., TUES.)

UT	ARIES GHA	VENUS −4.4 GHA	Dec	MARS +1.2 GHA	Dec	JUPITER −2.1 GHA	Dec	SATURN +0.7 GHA	Dec
d h	° ′	° ′	° ′	° ′	° ′	° ′	° ′	° ′	° ′
4 00	103 24.1	159 13.2	S16 50.5	147 27.8	S17 58.8	137 53.1	S14 42.6	89 41.2	N 3 09.7
01	118 26.5	174 16.7	50.1	162 28.3	58.3	152 55.1	42.4	104 43.6	09.8
02	133 29.0	189 20.1	49.6	177 28.7	57.7	167 57.0	42.2	119 45.9	09.8
03	148 31.5	204 23.6 ..	49.2	192 29.2 ..	57.1	182 59.0 ..	42.1	134 48.3 ..	09.8
04	163 33.9	219 27.1	48.8	207 29.7	56.5	198 00.9	41.9	149 50.7	09.9
05	178 36.4	234 30.6	48.3	222 30.2	56.0	213 02.9	41.7	164 53.1	09.9
S 06	193 38.8	249 34.1	S16 47.9	237 30.7	S17 55.4	228 04.8	S14 41.5	179 55.5	N 3 10.0
U 07	208 41.3	264 37.6	47.4	252 31.2	54.8	243 06.8	41.4	194 57.9	10.0
N 08	223 43.8	279 41.1	47.0	267 31.7	54.2	258 08.7	41.2	210 00.3	10.1
D 09	238 46.2	294 44.5 ..	46.6	282 32.1 ..	53.7	273 10.7 ..	41.0	225 02.7 ..	10.1
A 10	253 48.7	309 48.1	46.1	297 32.6	53.1	288 12.6	40.8	240 05.1	10.1
Y 11	268 51.2	324 51.6	45.7	312 33.1	52.5	303 14.5	40.7	255 07.5	10.2
12	283 53.6	339 55.1	S16 45.3	327 33.6	S17 51.9	318 16.5	S14 40.5	270 09.8	N 3 10.2
13	298 56.1	354 58.6	44.8	342 34.1	51.3	333 18.4	40.3	285 12.2	10.3
14	313 58.6	10 02.1	44.4	357 34.6	50.8	348 20.4	40.1	300 14.6	10.3
15	329 01.0	25 05.6 ..	44.0	12 35.1 ..	50.2	3 22.3 ..	40.0	315 17.0 ..	10.4
16	344 03.5	40 09.2	43.5	27 35.5	49.6	18 24.3	39.8	330 19.4	10.4
17	359 06.0	55 12.7	43.1	42 36.0	49.0	33 26.2	39.6	345 21.8	10.4
18	14 08.4	70 16.2	S16 42.7	57 36.5	S17 48.4	48 28.2	S14 39.4	0 24.2	N 3 10.5
19	29 10.9	85 19.8	42.2	72 37.0	47.9	63 30.1	39.3	15 26.6	10.5
20	44 13.3	100 23.3	41.8	87 37.5	47.3	78 32.1	39.1	30 28.9	10.6
21	59 15.8	115 26.8 ..	41.4	102 38.0 ..	46.7	93 34.0 ..	38.9	45 31.3 ..	10.6
22	74 18.3	130 30.4	40.9	117 38.5	46.1	108 35.9	38.7	60 33.7	10.7
23	89 20.7	145 33.9	40.5	132 39.0	45.5	123 37.9	38.6	75 36.1	10.7
5 00	104 23.2	160 37.5	S16 40.1	147 39.5	S17 45.0	138 39.8	S14 38.4	90 38.5	N 3 10.8
01	119 25.7	175 41.1	39.7	162 39.9	44.4	153 41.8	38.2	105 40.9	10.8
02	134 28.1	190 44.6	39.2	177 40.4	43.8	168 43.7	38.0	120 43.3	10.8
03	149 30.6	205 48.2 ..	38.8	192 40.9 ..	43.2	183 45.7 ..	37.9	135 45.6 ..	10.9
04	164 33.1	220 51.8	38.4	207 41.4	42.6	198 47.6	37.7	150 48.0	10.9
05	179 35.5	235 55.4	38.0	222 41.9	42.0	213 49.6	37.5	165 50.4	11.0
M 06	194 38.0	250 58.9	S16 37.5	237 42.4	S17 41.5	228 51.5	S14 37.3	180 52.8	N 3 11.0
O 07	209 40.5	266 02.5	37.1	252 42.9	40.9	243 53.4	37.2	195 55.2	11.1
N 08	224 42.9	281 06.1	36.7	267 43.4	40.3	258 55.4	37.0	210 57.6	11.1
D 09	239 45.4	296 09.7 ..	36.3	282 43.9 ..	39.7	273 57.3 ..	36.8	226 00.0 ..	11.2
A 10	254 47.8	311 13.3	35.8	297 44.4	39.1	288 59.3	36.6	241 02.3	11.2
Y 11	269 50.3	326 16.9	35.4	312 44.8	38.5	304 01.2	36.5	256 04.7	11.2
12	284 52.8	341 20.5	S16 35.0	327 45.3	S17 37.9	319 03.2	S14 36.3	271 07.1	N 3 11.3
13	299 55.2	356 24.1	34.6	342 45.8	37.4	334 05.1	36.1	286 09.5	11.3
14	314 57.7	11 27.7	34.2	357 46.3	36.8	349 07.1	35.9	301 11.9	11.4
15	330 00.2	26 31.3 ..	33.8	12 46.8 ..	36.2	4 09.0 ..	35.7	316 14.3 ..	11.4
16	345 02.6	41 35.0	33.3	27 47.3	35.6	19 10.9	35.6	331 16.6	11.5
17	0 05.1	56 38.6	32.9	42 47.8	35.0	34 12.9	35.4	346 19.0	11.5
18	15 07.6	71 42.2	S16 32.5	57 48.3	S17 34.4	49 14.8	S14 35.2	1 21.4	N 3 11.6
19	30 10.0	86 45.8	32.1	72 48.8	33.8	64 16.8	35.0	16 23.8	11.6
20	45 12.5	101 49.5	31.7	87 49.3	33.2	79 18.7	34.9	31 26.2	11.7
21	60 14.9	116 53.1 ..	31.3	102 49.8 ..	32.7	94 20.6 ..	34.7	46 28.6 ..	11.7
22	75 17.4	131 56.8	30.9	117 50.3	32.1	109 22.6	34.5	61 30.9	11.8
23	90 19.9	147 00.4	30.4	132 50.8	31.5	124 24.5	34.3	76 33.3	11.8
6 00	105 22.3	162 04.1	S16 30.0	147 51.2	S17 30.9	139 26.5	S14 34.2	91 35.7	N 3 11.8
01	120 24.8	177 07.7	29.6	162 51.7	30.3	154 28.4	34.0	106 38.1	11.9
02	135 27.3	192 11.4	29.2	177 52.2	29.7	169 30.4	33.8	121 40.5	11.9
03	150 29.7	207 15.0 ..	28.8	192 52.7 ..	29.1	184 32.3 ..	33.6	136 42.9 ..	12.0
04	165 32.2	222 18.7	28.4	207 53.2	28.5	199 34.2	33.4	151 45.2	12.0
05	180 34.7	237 22.4	27.9	222 53.7	27.9	214 36.2	33.3	166 47.6	12.1
T 06	195 37.1	252 26.0	S16 27.6	237 54.2	S17 27.3	229 38.1	S14 33.1	181 50.0	N 3 12.1
U 07	210 39.6	267 29.7	27.2	252 54.7	26.7	244 40.1	32.9	196 52.4	12.2
E 08	225 42.1	282 33.4	26.8	267 55.2	26.1	259 42.0	32.7	211 54.8	12.2
S 09	240 44.5	297 37.1 ..	26.4	282 55.7 ..	25.6	274 44.0 ..	32.6	226 57.1 ..	12.3
D 10	255 47.0	312 40.8	26.0	297 56.2	25.0	289 45.9	32.4	241 59.5	12.3
A 11	270 49.4	327 44.4	25.6	312 56.7	24.4	304 47.8	32.2	257 01.9	12.4
Y 12	285 51.9	342 48.1	S16 25.2	327 57.2	S17 23.8	319 49.8	S14 32.0	272 04.3	N 3 12.4
13	300 54.4	357 51.8	24.8	342 57.7	23.2	334 51.7	31.8	287 06.7	12.4
14	315 56.8	12 55.5	24.4	357 58.2	22.6	349 53.7	31.7	302 09.1	12.5
15	330 59.3	27 59.2 ..	24.0	12 58.7 ..	22.0	4 55.6 ..	31.5	317 11.4 ..	12.5
16	346 01.8	43 02.9	23.6	27 59.2	21.4	19 57.5	31.3	332 13.8	12.6
17	1 04.2	58 06.7	23.2	42 59.7	20.8	34 59.5	31.1	347 16.2	12.6
18	16 06.7	73 10.4	S16 22.8	58 00.2	S17 20.2	50 01.4	S14 31.0	2 18.6	N 3 12.7
19	31 09.2	88 14.1	22.4	73 00.7	19.6	65 03.4	30.8	17 20.9	12.7
20	46 11.6	103 17.8	22.0	88 01.2	19.0	80 05.3	30.6	32 23.3	12.8
21	61 14.1	118 21.5 ..	21.6	103 01.7 ..	18.4	95 07.2 ..	30.4	47 25.7 ..	12.8
22	76 16.6	133 25.3	21.2	118 02.2	17.8	110 09.2	30.2	62 28.1	12.9
23	91 19.0	148 29.0	20.8	133 02.7	17.2	125 11.1	30.1	77 30.5	12.9
Mer. Pass. 16 59.7		v 3.6	d 0.4	v 0.5	d 0.6	v 1.9	d 0.2	v 2.4	d 0.0

STARS

Name	SHA	Dec
Acamar	315 27.1	S40 19.1
Achernar	335 35.5	S57 15.2
Acrux	173 22.3	S63 05.0
Adhara	255 21.4	S28 58.3
Aldebaran	291 02.7	N16 30.2
Alioth	166 31.1	N55 58.0
Alkaid	153 08.4	N49 19.2
Al Na'ir	27 58.9	S46 58.4
Alnilam	275 58.1	S 1 12.4
Alphard	218 07.4	S 8 39.1
Alphecca	126 21.3	N26 43.3
Alpheratz	357 55.8	N29 04.9
Altair	62 20.1	N 8 51.9
Ankaa	353 27.5	S42 19.3
Antares	112 41.1	S26 25.4
Arcturus	146 06.7	N19 11.5
Atria	107 54.0	S69 01.1
Avior	234 22.1	S59 30.2
Bellatrix	278 44.4	N 6 20.7
Betelgeuse	271 13.8	N 7 24.2
Canopus	264 00.8	S52 41.9
Capella	280 51.5	N45 59.7
Deneb	49 39.9	N45 16.6
Denebola	182 45.6	N14 34.9
Diphda	349 07.8	S18 00.0
Dubhe	194 05.9	N61 45.4
Elnath	278 27.2	N28 36.2
Eltanin	90 52.2	N51 29.4
Enif	33 59.0	N 9 52.0
Fomalhaut	15 37.2	S29 38.1
Gacrux	172 14.0	S57 05.8
Gienah	176 04.4	S17 31.7
Hadar	149 04.8	S60 21.4
Hamal	328 14.0	N23 27.2
Kaus Aust.	83 59.9	S34 23.0
Kochab	137 20.4	N74 09.6
Markab	13 50.3	N15 11.7
Menkar	314 27.2	N 4 04.8
Menkent	148 21.6	S36 21.3
Miaplacidus	221 41.2	S69 42.5
Mirfak	308 57.0	N49 51.3
Nunki	76 13.4	S26 17.8
Peacock	53 38.5	S56 44.5
Pollux	243 41.8	N28 01.7
Procyon	245 11.8	N 5 13.6
Rasalhague	96 17.8	N12 33.8
Regulus	207 55.9	N11 58.5
Rigel	281 23.1	S 8 12.4
Rigil Kent.	140 08.2	S60 49.2
Sabik	102 26.4	S15 43.2
Schedar	349 54.0	N56 31.8
Shaula	96 38.4	S37 05.9
Sirius	258 43.8	S16 43.0
Spica	158 43.8	S11 08.9
Suhail	223 00.7	S43 25.5
Vega	80 47.4	N38 47.0
Zuben'ubi	137 18.7	S16 01.8

	SHA	Mer. Pass.
	° ′	h m
Venus	56 14.3	13 14
Mars	43 16.3	14 09
Jupiter	34 16.6	14 43
Saturn	346 15.3	17 55

UT	SUN GHA	SUN Dec	MOON GHA	v	MOON Dec	d	HP
d h	° ′	° ′	° ′	′	° ′	′	′
4 00	178 49.6	S22 45.7	109 55.6	9.7	S 3 48.8	11.2	59.4
01	193 49.3	45.4	124 24.3	9.7	3 37.6	11.1	59.4
02	208 49.0	45.2	138 53.0	9.6	3 26.5	11.2	59.4
03	223 48.7 ..	44.9	153 21.6	9.7	3 15.3	11.2	59.4
04	238 48.4	44.6	167 50.3	9.7	3 04.1	11.2	59.4
05	253 48.1	44.4	182 19.0	9.8	2 52.9	11.3	59.3
06	268 47.8	S22 44.1	196 47.8	9.7	S 2 41.6	11.2	59.3
07	283 47.6	43.9	211 16.5	9.8	2 30.4	11.2	59.3
S 08	298 47.3	43.6	225 45.3	9.8	2 19.2	11.3	59.3
U 09	313 47.0 ..	43.3	240 14.1	9.8	2 07.9	11.3	59.3
N 10	328 46.7	43.1	254 42.9	9.8	1 56.6	11.3	59.3
D 11	343 46.4	42.8	269 11.7	9.8	1 45.3	11.3	59.3
A 12	358 46.1	S22 42.5	283 40.5	9.8	S 1 34.0	11.3	59.3
Y 13	13 45.9	42.3	298 09.3	9.9	1 22.7	11.3	59.3
14	28 45.6	42.0	312 38.2	9.8	1 11.4	11.3	59.3
15	43 45.3 ..	41.7	327 07.0	9.9	1 00.1	11.3	59.3
16	58 45.0	41.5	341 35.9	9.9	0 48.8	11.3	59.3
17	73 44.7	41.2	356 04.8	9.8	0 37.5	11.3	59.3
18	88 44.4	S22 40.9	10 33.6	9.9	S 0 26.2	11.3	59.3
19	103 44.2	40.7	25 02.5	9.9	0 14.9	11.3	59.3
20	118 43.9	40.4	39 31.4	9.9	S 0 03.6	11.3	59.3
21	133 43.6 ..	40.1	54 00.3	10.0	N 0 07.7	11.4	59.3
22	148 43.3	39.9	68 29.3	9.9	0 19.1	11.3	59.3
23	163 43.0	39.6	82 58.2	9.9	0 30.4	11.3	59.3
5 00	178 42.8	S22 39.3	97 27.1	9.9	N 0 41.7	11.2	59.3
01	193 42.5	39.0	111 56.0	10.0	0 52.9	11.3	59.2
02	208 42.2	38.8	126 25.0	9.9	1 04.2	11.3	59.2
03	223 41.9 ..	38.5	140 53.9	10.0	1 15.5	11.3	59.2
04	238 41.6	38.2	155 22.9	9.9	1 26.8	11.2	59.2
05	253 41.4	37.9	169 51.8	10.0	1 38.0	11.3	59.2
06	268 41.1	S22 37.7	184 20.8	9.9	N 1 49.3	11.2	59.2
07	283 40.8	37.4	198 49.7	10.0	2 00.5	11.2	59.2
M 08	298 40.5	37.1	213 18.7	9.9	2 11.7	11.2	59.2
O 09	313 40.2 ..	36.8	227 47.6	10.0	2 22.9	11.2	59.2
N 10	328 40.0	36.5	242 16.6	9.9	2 34.1	11.2	59.2
D 11	343 39.7	36.3	256 45.5	10.0	2 45.3	11.1	59.2
A 12	358 39.4	S22 36.0	271 14.5	9.9	N 2 56.4	11.1	59.2
Y 13	13 39.1	35.7	285 43.4	10.0	3 07.5	11.1	59.2
14	28 38.8	35.4	300 12.4	9.9	3 18.6	11.1	59.2
15	43 38.6 ..	35.1	314 41.3	10.0	3 29.7	11.1	59.1
16	58 38.3	34.8	329 10.3	9.9	3 40.8	11.0	59.1
17	73 38.0	34.6	343 39.2	9.9	3 51.8	11.0	59.1
18	88 37.7	S22 34.3	358 08.1	10.0	N 4 02.8	11.0	59.1
19	103 37.4	34.0	12 37.1	9.9	4 13.8	11.0	59.1
20	118 37.2	33.7	27 06.0	9.9	4 24.8	10.9	59.1
21	133 36.9 ..	33.4	41 34.9	9.9	4 35.7	10.9	59.1
22	148 36.6	33.1	56 03.8	9.9	4 46.6	10.9	59.1
23	163 36.3	32.8	70 32.7	9.9	4 57.5	10.8	59.1
6 00	178 36.1	S22 32.5	85 01.6	9.8	N 5 08.3	10.8	59.1
01	193 35.8	32.2	99 30.4	9.9	5 19.1	10.8	59.1
02	208 35.5	31.9	113 59.3	9.9	5 29.9	10.7	59.0
03	223 35.2 ..	31.6	128 28.2	9.8	5 40.6	10.7	59.0
04	238 35.0	31.3	142 57.0	9.8	5 51.3	10.7	59.0
05	253 34.7	31.1	157 25.8	9.9	6 02.0	10.6	59.0
06	268 34.4	S22 30.8	171 54.7	9.8	N 6 12.6	10.6	59.1
07	283 34.1	30.5	186 23.5	9.8	6 23.2	10.6	59.0
T 08	298 33.9	30.2	200 52.3	9.8	6 33.8	10.5	59.0
U 09	313 33.6 ..	29.9	215 21.1	9.8	6 44.3	10.4	59.0
E 10	328 33.3	29.6	229 49.8	9.8	6 54.7	10.5	59.0
S 11	343 33.0	29.3	244 18.6	9.7	7 05.2	10.3	59.0
D 12	358 32.8	S22 29.0	258 47.3	9.8	N 7 15.5	10.4	58.9
A 13	13 32.5	28.7	273 16.1	9.7	7 25.9	10.3	58.9
Y 14	28 32.2	28.4	287 44.8	9.7	7 36.2	10.2	58.9
15	43 31.9 ..	28.1	302 13.5	9.6	7 46.4	10.2	58.9
16	58 31.7	27.7	316 42.1	9.7	7 56.6	10.2	58.9
17	73 31.4	27.4	331 10.8	9.6	8 06.8	10.1	58.9
18	88 31.1	S22 27.1	345 39.4	9.7	N 8 16.9	10.0	58.9
19	103 30.9	26.8	0 08.1	9.6	8 26.9	10.0	58.9
20	118 30.6	26.5	14 36.7	9.6	8 36.9	10.0	58.9
21	133 30.3 ..	26.2	29 05.3	9.5	8 46.9	9.8	58.9
22	148 30.0	25.9	43 33.8	9.6	8 56.7	9.9	58.8
23	163 29.8	25.6	58 02.4	9.5	N 9 06.6	9.8	58.8
	SD 16.3 d 0.3		SD 16.2		16.1		16.1

Lat.	Twilight Naut.	Twilight Civil	Sunrise	Moonrise 4	5	6	7
°	h m	h m	h m	h m	h m	h m	h m
N 72	08 20	10 30	■■	11 27	11 21	11 15	11 08
N 70	08 02	09 43	■■	11 24	11 24	11 25	11 26
68	07 47	09 12	11 30	11 22	11 27	11 33	11 40
66	07 35	08 50	10 20	11 21	11 30	11 40	11 52
64	07 25	08 32	09 45	11 19	11 32	11 46	12 01
62	07 16	08 17	09 20	11 18	11 34	11 51	12 10
60	07 08	08 04	09 00	11 17	11 35	11 55	12 17
N 58	07 01	07 53	08 44	11 16	11 37	11 59	12 23
56	06 55	07 43	08 30	11 15	11 38	12 02	12 29
54	06 49	07 35	08 18	11 14	11 39	12 06	12 34
52	06 44	07 27	08 07	11 13	11 41	12 08	12 39
50	06 39	07 20	07 58	11 13	11 42	12 11	12 43
45	06 28	07 05	07 38	11 11	11 44	12 17	12 52
N 40	06 18	06 52	07 22	11 10	11 46	12 22	12 59
35	06 09	06 41	07 09	11 09	11 47	12 26	13 06
30	06 01	06 30	06 57	11 08	11 49	12 29	13 12
20	05 45	06 12	06 36	11 07	11 51	12 36	13 22
N 10	05 30	05 56	06 18	11 05	11 53	12 42	13 31
0	05 13	05 39	06 02	11 04	11 55	12 47	13 39
S 10	04 55	05 22	05 45	11 03	11 58	12 53	13 48
20	04 33	05 02	05 26	11 01	12 00	12 58	13 57
30	04 05	04 38	05 05	11 00	12 03	13 05	14 07
35	03 47	04 23	04 53	10 59	12 04	13 09	14 13
40	03 25	04 06	04 38	10 58	12 06	13 13	14 20
45	02 56	03 44	04 21	10 57	12 08	13 19	14 28
S 50	02 13	03 16	03 59	10 56	12 11	13 25	14 38
52	01 49	03 02	03 49	10 55	12 12	13 28	14 43
54	01 12	02 45	03 37	10 54	12 13	13 31	14 48
56	////	02 24	03 24	10 54	12 14	13 34	14 53
58	////	01 58	03 08	10 53	12 16	13 38	14 59
S 60	////	01 18	02 49	10 52	12 18	13 43	15 06

Lat.	Sunset	Twilight Civil	Twilight Naut.	Moonset 4	5	6	7
°	h m	h m	h m	h m	h m	h m	h m
N 72	■■	13 41	15 52	23 35	25 29	01 29	03 24
N 70	■■	14 28	16 10	23 34	25 21	01 21	03 08
68	12 41	14 59	16 24	23 34	25 15	01 15	02 55
66	13 51	15 21	16 36	23 33	25 09	01 09	02 45
64	14 26	15 40	16 46	23 33	25 05	01 05	02 36
62	14 51	15 54	16 55	23 33	25 01	01 01	02 29
60	15 11	16 07	17 03	23 32	24 58	00 58	02 22
N 58	15 27	16 18	17 10	23 32	24 55	00 55	02 17
56	15 41	16 28	17 16	23 32	24 52	00 52	02 12
54	15 53	16 36	17 22	23 32	24 50	00 50	02 07
52	16 04	16 44	17 27	23 31	24 48	00 48	02 03
50	16 13	16 51	17 32	23 31	24 46	00 46	02 00
45	16 33	17 06	17 43	23 31	24 42	00 42	01 52
N 40	16 49	17 19	17 53	23 30	24 38	00 38	01 45
35	17 02	17 30	18 02	23 30	24 35	00 35	01 40
30	17 14	17 40	18 10	23 30	24 32	00 32	01 35
20	17 35	17 58	18 26	23 29	24 28	00 28	01 26
N 10	17 52	18 15	18 41	23 29	24 24	00 24	01 19
0	18 09	18 31	18 58	23 28	24 20	00 20	01 12
S 10	18 26	18 49	19 16	23 28	24 16	00 16	01 05
20	18 44	19 09	19 38	23 28	24 12	00 12	00 57
30	19 05	19 33	20 05	23 27	24 08	00 08	00 49
35	19 18	19 47	20 23	23 27	24 05	00 05	00 44
40	19 32	20 05	20 45	23 26	24 02	00 02	00 39
45	19 49	20 26	21 14	23 26	23 58	24 32	00 32
S 50	20 11	20 54	21 56	23 25	23 54	24 25	00 25
52	20 21	21 08	22 21	23 25	23 52	24 21	00 21
54	20 33	21 25	22 56	23 25	23 50	24 17	00 17
56	20 46	21 45	////	23 24	23 48	24 13	00 13
58	21 01	22 11	////	23 24	23 45	24 08	00 08
S 60	21 20	22 50	////	23 24	23 43	24 03	00 03

Day	SUN Eqn. of Time 00h	SUN Eqn. of Time 12h	SUN Mer. Pass.	MOON Mer. Pass. Upper	MOON Mer. Pass. Lower	Age	Phase
d	m s	m s	h m	h m	h m	d	%
4	04 41	04 55	12 05	17 16	04 50	06	38
5	05 08	05 22	12 05	18 08	05 42	07	49
6	05 35	05 48	12 06	18 59	06 34	08	60

UT (d h)	ARIES GHA	VENUS −4.3 GHA	VENUS Dec	MARS +1.2 GHA	MARS Dec	JUPITER −2.1 GHA	JUPITER Dec	SATURN +0.7 GHA	SATURN Dec	STARS Name	SHA	Dec
7 00	106 21.5	163 32.7	S16 20.4	148 03.2	S17 16.6	140 13.1	S14 29.9	92 32.8	N 3 13.0	Acamar	315 27.1	S40 19.1
01	121 23.9	178 36.5	20.0	163 03.6	16.0	155 15.0	29.7	107 35.2	13.0	Achernar	335 35.6	S57 15.2
02	136 26.4	193 40.2	19.6	178 04.1	15.4	170 16.9	29.5	122 37.6	13.1	Acrux	173 22.3	S63 05.0
03	151 28.9	208 44.0	.. 19.2	193 04.6	.. 14.8	185 18.9	.. 29.3	137 40.0	.. 13.1	Adhara	255 21.4	S28 58.3
04	166 31.3	223 47.7	18.9	208 05.1	14.2	200 20.8	29.2	152 42.4	13.2	Aldebaran	291 02.7	N16 30.2
05	181 33.8	238 51.5	18.5	223 05.6	13.6	215 22.8	29.0	167 44.7	13.2			
W 06	196 36.3	253 55.2	S16 18.1	238 06.1	S17 13.0	230 24.7	S14 28.8	182 47.1	N 3 13.3	Alioth	166 31.1	N55 58.0
E 07	211 38.7	268 59.0	17.7	253 06.6	12.4	245 26.6	28.6	197 49.5	13.3	Alkaid	153 08.4	N49 19.2
D 08	226 41.2	284 02.7	17.3	268 07.1	11.8	260 28.6	28.5	212 51.9	13.4	Al Na'ir	27 58.9	S46 58.4
N 09	241 43.7	299 06.5	.. 16.9	283 07.6	.. 11.2	275 30.5	.. 28.3	227 54.2	.. 13.4	Alnilam	275 58.1	S 1 12.4
E 10	256 46.1	314 10.3	16.5	298 08.1	10.6	290 32.5	28.1	242 56.6	13.5	Alphard	218 07.4	S 8 39.1
S 11	271 48.6	329 14.0	16.2	313 08.6	10.0	305 34.4	27.9	257 59.0	13.5			
D 12	286 51.0	344 17.8	S16 15.8	328 09.1	S17 09.4	320 36.3	S14 27.7	273 01.4	N 3 13.5	Alphecca	126 21.3	N26 43.3
A 13	301 53.5	359 21.6	15.4	343 09.6	08.8	335 38.3	27.6	288 03.7	13.6	Alpheratz	357 55.8	N29 04.9
Y 14	316 56.0	14 25.4	15.0	358 10.1	08.2	350 40.2	27.4	303 06.1	13.6	Altair	62 20.1	N 8 51.9
15	331 58.4	29 29.2	.. 14.6	13 10.6	.. 07.6	5 42.1	.. 27.2	318 08.5	.. 13.7	Ankaa	353 27.5	S42 19.3
16	347 00.9	44 33.0	14.3	28 11.2	07.0	20 44.1	27.0	333 10.9	13.7	Antares	112 41.1	S26 25.4
17	2 03.4	59 36.7	13.9	43 11.7	06.4	35 46.0	26.8	348 13.3	13.8			
18	17 05.8	74 40.5	S16 13.5	58 12.2	S17 05.8	50 48.0	S14 26.7	3 15.6	N 3 13.8	Arcturus	146 06.6	N19 11.5
19	32 08.3	89 44.3	13.1	73 12.7	05.2	65 49.9	26.5	18 18.0	13.9	Atria	107 54.0	S69 01.1
20	47 10.8	104 48.1	12.7	88 13.2	04.6	80 51.8	26.3	33 20.4	13.9	Avior	234 22.1	S59 30.2
21	62 13.2	119 51.9	.. 12.4	103 13.7	.. 04.0	95 53.8	.. 26.1	48 22.8	.. 14.0	Bellatrix	278 44.4	N 6 20.7
22	77 15.7	134 55.8	12.0	118 14.2	03.4	110 55.7	25.9	63 25.1	14.0	Betelgeuse	271 13.8	N 7 24.2
23	92 18.2	149 59.6	11.6	133 14.7	02.8	125 57.6	25.8	78 27.5	14.1			
8 00	107 20.6	165 03.4	S16 11.2	148 15.2	S17 02.2	140 59.6	S14 25.6	93 29.9	N 3 14.1	Canopus	264 00.8	S52 41.9
01	122 23.1	180 07.2	10.9	163 15.7	01.6	156 01.5	25.4	108 32.3	14.2	Capella	280 51.5	N45 59.7
02	137 25.5	195 11.0	10.5	178 16.2	01.0	171 03.5	25.2	123 34.6	14.2	Deneb	49 39.9	N45 16.6
03	152 28.0	210 14.8	.. 10.1	193 16.7	17 00.4	186 05.4	.. 25.0	138 37.0	.. 14.3	Denebola	182 45.6	N14 34.9
04	167 30.5	225 18.7	09.8	208 17.2	16 59.8	201 07.3	24.9	153 39.4	14.3	Diphda	349 07.8	S18 00.1
05	182 32.9	240 22.5	09.4	223 17.7	59.2	216 09.3	24.7	168 41.7	14.4			
T 06	197 35.4	255 26.3	S16 09.0	238 18.2	S16 58.5	231 11.2	S14 24.5	183 44.1	N 3 14.4	Dubhe	194 05.9	N61 45.4
H 07	212 37.9	270 30.2	08.7	253 18.7	57.9	246 13.1	24.3	198 46.5	14.5	Elnath	278 27.2	N28 36.2
U 08	227 40.3	285 34.0	08.3	268 19.2	57.3	261 15.1	24.1	213 48.9	14.5	Eltanin	90 52.2	N51 29.4
R 09	242 42.8	300 37.9	.. 07.9	283 19.7	.. 56.7	276 17.0	.. 24.0	228 51.2	.. 14.6	Enif	33 59.0	N 9 52.0
S 10	257 45.3	315 41.7	07.6	298 20.2	56.1	291 19.0	23.8	243 53.6	14.6	Fomalhaut	15 37.2	S29 38.1
D 11	272 47.7	330 45.5	07.2	313 20.7	55.5	306 20.9	23.6	258 56.0	14.7			
A 12	287 50.2	345 49.4	S16 06.8	328 21.2	S16 54.9	321 22.8	S14 23.4	273 58.4	N 3 14.7	Gacrux	172 13.9	S57 05.8
Y 13	302 52.7	0 53.3	06.5	343 21.7	54.3	336 24.8	23.2	289 00.7	14.8	Gienah	176 04.4	S17 31.8
14	317 55.1	15 57.1	06.1	358 22.2	53.7	351 26.7	23.1	304 03.1	14.8	Hadar	149 04.8	S60 21.4
15	332 57.6	31 01.0	.. 05.7	13 22.7	.. 53.1	6 28.6	.. 22.9	319 05.5	.. 14.9	Hamal	328 14.0	N23 27.2
16	348 00.0	46 04.8	05.4	28 23.2	52.5	21 30.6	22.7	334 07.8	14.9	Kaus Aust.	83 59.9	S34 23.0
17	3 02.5	61 08.7	05.0	43 23.7	51.8	36 32.5	22.5	349 10.2	15.0			
18	18 05.0	76 12.6	S16 04.7	58 24.3	S16 51.2	51 34.4	S14 22.3	4 12.6	N 3 15.0	Kochab	137 20.3	N74 09.6
19	33 07.4	91 16.4	04.3	73 24.8	50.6	66 36.4	22.2	19 15.0	15.1	Markab	13 50.3	N15 11.7
20	48 09.9	106 20.3	03.9	88 25.3	50.0	81 38.3	22.0	34 17.3	15.1	Menkar	314 27.3	N 4 04.8
21	63 12.4	121 24.2	.. 03.6	103 25.8	.. 49.4	96 40.2	.. 21.8	49 19.7	.. 15.2	Menkent	148 21.6	S36 21.4
22	78 14.8	136 28.1	03.2	118 26.3	48.8	111 42.2	21.6	64 22.1	15.2	Miaplacidus	221 41.2	S69 42.5
23	93 17.3	151 32.0	02.9	133 26.8	48.2	126 44.1	21.4	79 24.4	15.3			
9 00	108 19.8	166 35.8	S16 02.5	148 27.3	S16 47.6	141 46.1	S14 21.3	94 26.8	N 3 15.3	Mirfak	308 57.0	N49 51.3
01	123 22.2	181 39.7	02.2	163 27.8	47.0	156 48.0	21.1	109 29.2	15.4	Nunki	76 13.3	S26 17.8
02	138 24.7	196 43.6	01.8	178 28.3	46.3	171 49.9	20.9	124 31.6	15.4	Peacock	53 38.5	S56 44.5
03	153 27.1	211 47.5	.. 01.5	193 28.8	.. 45.7	186 51.9	.. 20.7	139 33.9	.. 15.5	Pollux	243 41.8	N28 01.7
04	168 29.6	226 51.4	01.1	208 29.3	45.1	201 53.8	20.5	154 36.3	15.5	Procyon	245 11.8	N 5 13.6
05	183 32.1	241 55.3	00.8	223 29.8	44.5	216 55.7	20.3	169 38.7	15.6			
F 06	198 34.5	256 59.2	S16 00.4	238 30.3	S16 43.9	231 57.7	S14 20.2	184 41.0	N 3 15.6	Rasalhague	96 17.7	N12 33.8
R 07	213 37.0	272 03.1	16 00.1	253 30.9	43.3	246 59.6	20.0	199 43.4	15.7	Regulus	207 55.9	N11 58.5
I 08	228 39.5	287 07.0	15 59.7	268 31.4	42.7	262 01.5	19.8	214 45.8	15.8	Rigel	281 23.1	S 8 12.5
D 09	243 41.9	302 10.9	.. 59.4	283 31.9	.. 42.0	277 03.5	.. 19.6	229 48.1	.. 15.8	Rigil Kent.	140 08.1	S60 49.2
A 10	258 44.4	317 14.9	59.0	298 32.4	41.4	292 05.4	19.4	244 50.5	15.9	Sabik	102 26.4	S15 43.2
Y 11	273 46.9	332 18.8	58.7	313 32.9	40.8	307 07.3	19.3	259 52.9	15.9			
12	288 49.3	347 22.7	S15 58.3	328 33.4	S16 40.2	322 09.3	S14 19.1	274 55.2	N 3 16.0	Schedar	349 54.0	N56 31.8
13	303 51.8	2 26.6	58.0	343 33.9	39.6	337 11.2	18.9	289 57.6	16.0	Shaula	96 38.4	S37 05.9
14	318 54.3	17 30.5	57.6	358 34.4	39.0	352 13.1	18.7	305 00.0	16.1	Sirius	258 43.8	S16 43.0
15	333 56.7	32 34.5	.. 57.3	13 34.9	.. 38.3	7 15.1	.. 18.5	320 02.4	.. 16.1	Spica	158 43.7	S11 09.0
16	348 59.2	47 38.4	57.0	28 35.4	37.7	22 17.0	18.3	335 04.7	16.2	Suhail	223 00.7	S43 25.5
17	4 01.6	62 42.3	56.6	43 36.0	37.1	37 18.9	18.2	350 07.1	16.2			
18	19 04.1	77 46.3	S15 56.3	58 36.5	S16 36.5	52 20.9	S14 18.0	5 09.5	N 3 16.3	Vega	80 47.4	N38 47.0
19	34 06.6	92 50.2	55.9	73 37.0	35.9	67 22.8	17.8	20 11.8	16.3	Zuben'ubi	137 18.7	S16 01.8
20	49 09.0	107 54.1	55.6	88 37.5	35.3	82 24.7	17.6	35 14.2	16.4			
21	64 11.5	122 58.1	.. 55.3	103 38.0	.. 34.6	97 26.7	.. 17.4	50 16.6	.. 16.4		SHA	Mer.Pass.
22	79 14.0	138 02.0	54.9	118 38.5	34.0	112 28.6	17.3	65 18.9	16.5	Venus	57 42.8	12 56
23	94 16.4	153 06.0	54.6	133 39.0	33.4	127 30.5	17.1	80 21.3	16.5	Mars	40 54.6	14 07
Mer. Pass.	16 47.9	v 3.9 d 0.4		v 0.5 d 0.6		v 1.9 d 0.2		v 2.4 d 0.1		Jupiter	33 39.0	14 34
										Saturn	346 09.3	17 43

SUN and MOON

UT	SUN GHA	SUN Dec	MOON GHA	v	MOON Dec	d	HP
d h	° ′	° ′	° ′	′	° ′	′	′
7 00	178 29.5	S22 25.3	72 30.9	9.6	N 9 16.4	9.7	58.8
01	193 29.2	25.0	86 59.5	9.4	9 26.1	9.6	58.8
02	208 28.9	24.7	101 27.9	9.5	9 35.7	9.7	58.8
03	223 28.7	.. 24.4	115 56.4	9.5	9 45.4	9.5	58.8
04	238 28.4	24.0	130 24.9	9.4	9 54.9	9.5	58.8
05	253 28.1	23.7	144 53.3	9.4	10 04.4	9.4	58.8
06	268 27.9	S22 23.4	159 21.7	9.4	N10 13.8	9.4	58.7
W 07	283 27.6	23.1	173 50.1	9.4	10 23.2	9.3	58.7
E 08	298 27.3	22.8	188 18.5	9.3	10 32.5	9.2	58.7
D 09	313 27.1	.. 22.5	202 46.8	9.3	10 41.7	9.2	58.7
N 10	328 26.8	22.1	217 15.1	9.3	10 50.9	9.1	58.7
E 11	343 26.5	21.8	231 43.4	9.3	11 00.0	9.0	58.7
S 12	358 26.3	S22 21.5	246 11.7	9.3	N11 09.0	9.0	58.7
D 13	13 26.0	21.2	260 40.0	9.2	11 18.0	8.9	58.7
A 14	28 25.7	20.9	275 08.2	9.2	11 26.9	8.8	58.6
Y 15	43 25.4	.. 20.5	289 36.4	9.2	11 35.7	8.8	58.6
16	58 25.2	20.2	304 04.6	9.2	11 44.5	8.7	58.6
17	73 24.9	19.9	318 32.8	9.1	11 53.2	8.6	58.6
18	88 24.6	S22 19.6	333 00.9	9.1	N12 01.8	8.6	58.6
19	103 24.4	19.2	347 29.0	9.1	12 10.4	8.4	58.6
20	118 24.1	18.9	1 57.1	9.1	12 18.8	8.4	58.6
21	133 23.8	.. 18.6	16 25.2	9.1	12 27.2	8.4	58.6
22	148 23.6	18.3	30 53.3	9.0	12 35.6	8.2	58.5
23	163 23.3	17.9	45 21.3	9.0	12 43.8	8.2	58.5
8 00	178 23.0	S22 17.6	59 49.3	9.0	N12 52.0	8.1	58.5
01	193 22.8	17.3	74 17.3	8.9	13 00.1	8.0	58.5
02	208 22.5	16.9	88 45.2	9.0	13 08.1	7.9	58.5
03	223 22.2	.. 16.6	103 13.2	8.9	13 16.0	7.9	58.5
04	238 22.0	16.3	117 41.1	8.8	13 23.9	7.7	58.5
05	253 21.7	16.0	132 08.9	8.9	13 31.6	7.7	58.4
06	268 21.5	S22 15.6	146 36.8	8.8	N13 39.3	7.6	58.4
T 07	283 21.2	15.3	161 04.6	8.9	13 46.9	7.6	58.4
H 08	298 20.9	14.9	175 32.5	8.8	13 54.5	7.4	58.4
U 09	313 20.7	.. 14.6	190 00.3	8.7	14 01.9	7.3	58.4
R 10	328 20.4	14.3	204 28.0	8.8	14 09.2	7.3	58.4
S 11	343 20.1	13.9	218 55.8	8.7	14 16.5	7.2	58.4
D 12	358 19.9	S22 13.6	233 23.5	8.7	N14 23.7	7.1	58.3
A 13	13 19.6	13.3	247 51.2	8.7	14 30.8	7.0	58.3
Y 14	28 19.3	12.9	262 18.9	8.6	14 37.8	6.9	58.3
15	43 19.1	.. 12.6	276 46.5	8.7	14 44.7	6.8	58.3
16	58 18.8	12.2	291 14.2	8.6	14 51.5	6.8	58.3
17	73 18.6	11.9	305 41.8	8.6	14 58.3	6.6	58.3
18	88 18.3	S22 11.6	320 09.4	8.6	N15 04.9	6.6	58.3
19	103 18.0	11.2	334 37.0	8.5	15 11.5	6.4	58.2
20	118 17.8	10.9	349 04.5	8.5	15 17.9	6.4	58.2
21	133 17.5	.. 10.5	3 32.1	8.5	15 24.3	6.3	58.2
22	148 17.2	10.2	17 59.6	8.5	15 30.6	6.1	58.2
23	163 17.0	09.8	32 27.1	8.5	15 36.7	6.1	58.2
9 00	178 16.7	S22 09.5	46 54.6	8.4	N15 42.8	6.0	58.2
01	193 16.5	09.1	61 22.0	8.5	15 48.8	5.9	58.1
02	208 16.2	08.8	75 49.5	8.4	15 54.7	5.8	58.1
03	223 15.9	.. 08.4	90 16.9	8.4	16 00.5	5.7	58.1
04	238 15.7	08.1	104 44.3	8.4	16 06.2	5.6	58.1
05	253 15.4	07.7	119 11.7	8.4	16 11.8	5.5	58.1
06	268 15.2	S22 07.4	133 39.1	8.4	N16 17.3	5.4	58.1
F 07	283 14.9	07.0	148 06.5	8.3	16 22.7	5.3	58.0
R 08	298 14.7	06.7	162 33.8	8.3	16 28.0	5.2	58.0
I 09	313 14.4	.. 06.3	177 01.1	8.4	16 33.2	5.1	58.0
D 10	328 14.1	06.0	191 28.5	8.3	16 38.3	5.0	58.0
A 11	343 13.9	05.6	205 55.8	8.3	16 43.3	4.9	58.0
Y 12	358 13.6	S22 05.3	220 23.1	8.2	N16 48.2	4.8	58.0
13	13 13.4	04.9	234 50.3	8.3	16 53.0	4.7	57.9
14	28 13.1	04.6	249 17.6	8.3	16 57.7	4.6	57.9
15	43 12.8	.. 04.2	263 44.9	8.2	17 02.3	4.5	57.9
16	58 12.6	03.8	278 12.1	8.3	17 06.8	4.4	57.9
17	73 12.3	03.5	292 39.4	8.2	17 11.2	4.3	57.9
18	88 12.1	S22 03.1	307 06.6	8.2	N17 15.5	4.2	57.9
19	103 11.8	02.8	321 33.8	8.2	17 19.7	4.1	57.8
20	118 11.6	02.4	336 01.0	8.3	17 23.8	3.9	57.8
21	133 11.3	.. 02.0	350 28.3	8.2	17 27.7	3.9	57.8
22	148 11.1	01.7	4 55.5	8.2	17 31.6	3.8	57.8
23	163 10.8	01.3	19 22.7	8.1	N17 35.4	3.6	57.8
	SD 16.3	d 0.3	SD 16.0		15.9		15.8

Moonrise

Lat.	Twilight Naut.	Twilight Civil	Sunrise	Moonrise 7	Moonrise 8	Moonrise 9	Moonrise 10
°	h m	h m	h m	h m	h m	h m	h m
N 72	08 15	10 20	■■■	11 08	11 01	10 51	□
N 70	07 58	09 37	■■■	11 26	11 29	11 37	11 55
68	07 44	09 08	11 10	11 40	11 51	12 07	12 36
66	07 33	08 46	10 13	11 52	12 07	12 30	13 04
64	07 23	08 29	09 40	12 01	12 21	12 48	13 25
62	07 14	08 14	09 16	12 10	12 33	13 03	13 42
60	07 07	08 02	08 57	12 17	12 43	13 15	13 57
N 58	07 00	07 51	08 41	12 23	12 52	13 26	14 09
56	06 54	07 42	08 28	12 29	12 59	13 36	14 19
54	06 48	07 34	08 16	12 34	13 06	13 44	14 29
52	06 43	07 26	08 06	12 39	13 13	13 52	14 37
50	06 38	07 19	07 57	12 43	13 18	13 58	14 45
45	06 28	07 04	07 38	12 52	13 30	14 13	15 01
N 40	06 18	06 52	07 22	12 59	13 40	14 25	15 14
35	06 09	06 41	07 09	13 06	13 49	14 35	15 25
30	06 01	06 31	06 57	13 12	13 57	14 44	15 35
20	05 46	06 13	06 37	13 22	14 10	15 00	15 52
N 10	05 31	05 57	06 19	13 31	14 22	15 14	16 07
0	05 15	05 41	06 03	13 39	14 32	15 27	16 21
S 10	04 57	05 23	05 46	13 48	14 44	15 39	16 35
20	04 35	05 04	05 28	13 57	14 55	15 53	16 50
30	04 08	04 40	05 08	14 07	15 09	16 09	17 07
35	03 50	04 26	04 55	14 13	15 17	16 18	17 17
40	03 28	04 09	04 41	14 20	15 26	16 29	17 29
45	03 00	03 48	04 24	14 28	15 36	16 42	17 42
S 50	02 19	03 20	04 03	14 38	15 49	16 57	17 59
52	01 55	03 06	03 53	14 43	15 55	17 04	18 06
54	01 22	02 50	03 42	14 48	16 02	17 12	18 15
56	////	02 30	03 29	14 53	16 09	17 21	18 25
58	////	02 05	03 14	14 59	16 18	17 31	18 36
S 60	////	01 29	02 55	15 06	16 27	17 42	18 48

Moonset

Lat.	Sunset	Twilight Civil	Twilight Naut.	Moonset 7	Moonset 8	Moonset 9	Moonset 10
°	h m	h m	h m	h m	h m	h m	h m
N 72	■■■	13 54	15 59	03 24	05 21	07 24	□
N 70	■■■	14 37	16 16	03 08	04 54	06 39	08 14
68	13 04	15 06	16 30	02 55	04 34	06 09	07 34
66	14 01	15 28	16 41	02 45	04 18	05 47	07 06
64	14 34	15 45	16 51	02 36	04 05	05 29	06 45
62	14 58	15 59	17 00	02 29	03 54	05 15	06 28
60	15 17	16 12	17 07	02 22	03 45	05 03	06 14
N 58	15 32	16 22	17 14	02 17	03 36	04 52	06 02
56	15 46	16 32	17 20	02 12	03 29	04 43	05 51
54	15 57	16 40	17 25	02 07	03 23	04 35	05 42
52	16 08	16 47	17 31	02 03	03 17	04 28	05 34
50	16 17	16 54	17 35	02 00	03 12	04 22	05 26
45	16 36	17 09	17 46	01 52	03 01	04 08	05 11
N 40	16 52	17 22	17 55	01 45	02 52	03 56	04 58
35	17 05	17 33	18 04	01 40	02 44	03 46	04 47
30	17 17	17 43	18 12	01 35	02 37	03 38	04 37
20	17 36	18 00	18 28	01 26	02 25	03 23	04 21
N 10	17 54	18 17	18 43	01 19	02 14	03 10	04 06
0	18 10	18 33	18 59	01 12	02 05	02 58	03 53
S 10	18 27	18 50	19 17	01 05	01 55	02 46	03 39
20	18 45	19 09	19 38	00 57	01 44	02 33	03 25
30	19 06	19 33	20 05	00 49	01 33	02 19	03 08
35	19 18	19 47	20 23	00 44	01 26	02 10	02 58
40	19 32	20 04	20 45	00 39	01 18	02 01	02 47
45	19 49	20 25	21 13	00 32	01 09	01 49	02 35
S 50	20 10	20 52	21 53	00 25	00 58	01 36	02 19
52	20 20	21 06	22 16	00 21	00 53	01 29	02 12
54	20 31	21 22	22 49	00 17	00 47	01 22	02 03
56	20 44	21 42	////	00 13	00 41	01 14	01 54
58	20 59	22 06	////	00 08	00 34	01 06	01 44
S 60	21 17	22 41	////	00 03	00 27	00 56	01 32

SUN and MOON

	SUN			MOON			
Day	Eqn. of Time 00ʰ	Eqn. of Time 12ʰ	Mer. Pass.	Mer. Pass. Upper	Mer. Pass. Lower	Age	Phase
d	m s	m s	h m	h m	h m	d	%
7	06 01	06 14	12 06	19 52	07 26	09	71
8	06 27	06 40	12 07	20 45	08 18	10	80
9	06 53	07 05	12 07	21 40	09 12	11	88

UT	ARIES GHA	VENUS −4.2 GHA	Dec	MARS +1.2 GHA	Dec	JUPITER −2.0 GHA	Dec	SATURN +0.7 GHA	Dec
10 00	109 18.9	168 09.9	S15 54.3	148 39.5	S16 32.8	142 32.5	S14 16.9	95 23.7	N 3 16.6
01	124 21.4	183 13.9	53.9	163 40.1	32.2	157 34.4	16.7	110 26.0	16.6
02	139 23.8	198 17.8	53.6	178 40.6	31.5	172 36.3	16.5	125 28.4	16.7
03	154 26.3	213 21.8 ..	53.3	193 41.1 ..	30.9	187 38.3 ..	16.3	140 30.8 ..	16.7
04	169 28.8	228 25.8	52.9	208 41.6	30.3	202 40.2	16.2	155 33.1	16.8
05	184 31.2	243 29.7	52.6	223 42.1	29.7	217 42.1	16.0	170 35.5	16.8
06	199 33.7	258 33.7	S15 52.3	238 42.6	S16 29.1	232 44.1	S14 15.8	185 37.9	N 3 16.9
S 07	214 36.1	273 37.6	51.9	253 43.1	28.4	247 46.0	15.6	200 40.2	17.0
A 08	229 38.6	288 41.6	51.6	268 43.6	27.8	262 47.9	15.4	215 42.6	17.0
T 09	244 41.1	303 45.6 ..	51.3	283 44.2 ..	27.2	277 49.9 ..	15.2	230 44.9 ..	17.1
U 10	259 43.5	318 49.6	51.0	298 44.7	26.6	292 51.8	15.1	245 47.3	17.1
R 11	274 46.0	333 53.5	50.6	313 45.2	25.9	307 53.7	14.9	260 49.7	17.2
D 12	289 48.5	348 57.5	S15 50.3	328 45.7	S16 25.3	322 55.7	S14 14.7	275 52.0	N 3 17.2
A 13	304 50.9	4 01.5	50.0	343 46.2	24.7	337 57.6	14.5	290 54.4	17.3
Y 14	319 53.4	19 05.5	49.7	358 46.7	24.1	352 59.5	14.3	305 56.8	17.3
15	334 55.9	34 09.5 ..	49.3	13 47.2 ..	23.4	8 01.4 ..	14.2	320 59.1 ..	17.4
16	349 58.3	49 13.4	49.0	28 47.8	22.8	23 03.4	14.0	336 01.5	17.4
17	5 00.8	64 17.4	48.7	43 48.3	22.2	38 05.3	13.8	351 03.9	17.5
18	20 03.3	79 21.4	S15 48.4	58 48.8	S16 21.6	53 07.2	S14 13.6	6 06.2	N 3 17.5
19	35 05.7	94 25.4	48.1	73 49.3	20.9	68 09.2	13.4	21 08.6	17.6
20	50 08.2	109 29.4	47.8	88 49.8	20.3	83 11.1	13.2	36 11.0	17.6
21	65 10.6	124 33.4 ..	47.4	103 50.3 ..	19.7	98 13.0 ..	13.1	51 13.3 ..	17.7
22	80 13.1	139 37.4	47.1	118 50.9	19.1	113 15.0	12.9	66 15.7	17.8
23	95 15.6	154 41.4	46.8	133 51.4	18.4	128 16.9	12.7	81 18.0	17.8
11 00	110 18.0	169 45.4	S15 46.5	148 51.9	S16 17.8	143 18.8	S14 12.5	96 20.4	N 3 17.9
01	125 20.5	184 49.4	46.2	163 52.4	17.2	158 20.8	12.3	111 22.8	17.9
02	140 23.0	199 53.4	45.9	178 52.9	16.6	173 22.7	12.1	126 25.1	18.0
03	155 25.4	214 57.4 ..	45.6	193 53.4 ..	15.9	188 24.6 ..	12.0	141 27.5 ..	18.0
04	170 27.9	230 01.5	45.2	208 54.0	15.3	203 26.6	11.8	156 29.9	18.1
05	185 30.4	245 05.5	44.9	223 54.5	14.7	218 28.5	11.6	171 32.2	18.1
06	200 32.8	260 09.5	S15 44.6	238 55.0	S16 14.0	233 30.4	S14 11.4	186 34.6	N 3 18.2
07	215 35.3	275 13.5	44.3	253 55.5	13.4	248 32.3	11.2	201 36.9	18.2
08	230 37.8	290 17.5	44.0	268 56.0	12.8	263 34.3	11.0	216 39.3	18.3
S 09	245 40.2	305 21.5 ..	43.7	283 56.6 ..	12.2	278 36.2 ..	10.8	231 41.7 ..	18.4
U 10	260 42.7	320 25.6	43.4	298 57.1	11.5	293 38.1	10.7	246 44.0	18.4
N 11	275 45.1	335 29.6	43.1	313 57.6	10.9	308 40.1	10.5	261 46.4	18.5
D 12	290 47.6	350 33.6	S15 42.8	328 58.1	S16 10.3	323 42.0	S14 10.3	276 48.7	N 3 18.5
A 13	305 50.1	5 37.6	42.5	343 58.6	09.6	338 43.9	10.1	291 51.1	18.6
Y 14	320 52.5	20 41.7	42.2	358 59.2	09.0	353 45.8	09.9	306 53.5	18.6
15	335 55.0	35 45.7 ..	41.9	13 59.7 ..	08.4	8 47.8 ..	09.7	321 55.8 ..	18.7
16	350 57.5	50 49.7	41.6	29 00.2	07.7	23 49.7	09.6	336 58.2	18.7
17	5 59.9	65 53.8	41.3	44 00.7	07.1	38 51.6	09.4	352 00.5	18.8
18	21 02.4	80 57.8	S15 41.0	59 01.2	S16 06.5	53 53.6	S14 09.2	7 02.9	N 3 18.9
19	36 04.9	96 01.9	40.7	74 01.8	05.9	68 55.5	09.0	22 05.3	18.9
20	51 07.3	111 05.9	40.4	89 02.3	05.2	83 57.4	08.8	37 07.6	19.0
21	66 09.8	126 09.9 ..	40.1	104 02.8 ..	04.6	98 59.4 ..	08.6	52 10.0 ..	19.0
22	81 12.3	141 14.0	39.8	119 03.3	04.0	114 01.3	08.5	67 12.3	19.1
23	96 14.7	156 18.0	39.5	134 03.8	03.3	129 03.2	08.3	82 14.7	19.1
12 00	111 17.2	171 22.1	S15 39.2	149 04.4	S16 02.7	144 05.1	S14 08.1	97 17.1	N 3 19.2
01	126 19.6	186 26.1	38.9	164 04.9	02.1	159 07.1	07.9	112 19.4	19.2
02	141 22.1	201 30.2	38.6	179 05.4	01.4	174 09.0	07.7	127 21.8	19.3
03	156 24.6	216 34.2 ..	38.3	194 05.9 ..	00.8	189 10.9 ..	07.5	142 24.1 ..	19.4
04	171 27.0	231 38.3	38.0	209 06.5	16 00.1	204 12.9	07.3	157 26.5	19.4
05	186 29.5	246 42.4	37.7	224 07.0	15 59.5	219 14.8	07.2	172 28.8	19.5
06	201 32.0	261 46.4	S15 37.5	239 07.5	S15 58.9	234 16.7	S14 07.0	187 31.2	N 3 19.5
07	216 34.4	276 50.5	37.2	254 08.0	58.2	249 18.6	06.8	202 33.6	19.6
08	231 36.9	291 54.5	36.9	269 08.5	57.6	264 20.6	06.6	217 35.9	19.6
M 09	246 39.4	306 58.6 ..	36.6	284 09.1 ..	57.0	279 22.5 ..	06.4	232 38.3 ..	19.7
O 10	261 41.8	322 02.7	36.3	299 09.6	56.3	294 24.4	06.2	247 40.6	19.7
N 11	276 44.3	337 06.7	36.0	314 10.1	55.7	309 26.3	06.0	262 43.0	19.8
D 12	291 46.7	352 10.8	S15 35.7	329 10.6	S15 55.1	324 28.3	S14 05.9	277 45.3	N 3 19.9
A 13	306 49.2	7 14.9	35.5	344 11.2	54.4	339 30.2	05.7	292 47.7	19.9
Y 14	321 51.7	22 18.9	35.2	359 11.7	53.8	354 32.1	05.5	307 50.1	20.0
15	336 54.1	37 23.0 ..	34.9	14 12.2 ..	53.1	9 34.1 ..	05.3	322 52.4 ..	20.0
16	351 56.6	52 27.1	34.6	29 12.7	52.5	24 36.0	05.1	337 54.8	20.1
17	6 59.1	67 31.1	34.3	44 13.3	51.9	39 37.9	04.9	352 57.1	20.1
18	22 01.5	82 35.2	S15 34.1	59 13.8	S15 51.2	54 39.8	S14 04.7	7 59.5	N 3 20.2
19	37 04.0	97 39.3	33.8	74 14.3	50.6	69 41.8	04.6	23 01.8	20.3
20	52 06.5	112 43.4	33.5	89 14.8	50.0	84 43.7	04.4	38 04.2	20.3
21	67 08.9	127 47.4 ..	33.2	104 15.4 ..	49.3	99 45.6 ..	04.2	53 06.6 ..	20.4
22	82 11.4	142 51.5	33.0	119 15.9	48.7	114 47.5	04.0	68 08.9	20.4
23	97 13.9	157 55.6	32.7	134 16.4	48.0	129 49.5	03.8	83 11.3	20.5
Mer. Pass. 16 36.1		v 4.0	d 0.3	v 0.5	d 0.6	v 1.9	d 0.2	v 2.4	d 0.1

STARS

Name	SHA	Dec
Acamar	315 27.1	S40 19.1
Achernar	335 35.6	S57 15.2
Acrux	173 22.2	S63 05.0
Adhara	255 21.4	S28 58.4
Aldebaran	291 02.7	N16 30.2
Alioth	166 31.1	N55 58.0
Alkaid	153 08.3	N49 19.2
Al Na'ir	27 58.9	S46 58.4
Alnilam	275 58.1	S 1 12.4
Alphard	218 07.4	S 8 39.1
Alphecca	126 21.2	N26 43.2
Alpheratz	357 55.8	N29 04.9
Altair	62 20.1	N 8 51.9
Ankaa	353 27.5	S42 19.3
Antares	112 41.0	S26 25.5
Arcturus	146 06.6	N19 11.5
Atria	107 54.0	S69 01.1
Avior	234 22.1	S59 30.2
Bellatrix	278 44.4	N 6 20.7
Betelgeuse	271 13.8	N 7 24.2
Canopus	264 00.8	S52 41.9
Capella	280 51.5	N45 59.7
Deneb	49 39.9	N45 16.5
Denebola	182 45.6	N14 34.9
Diphda	349 07.8	S18 00.1
Dubhe	194 05.8	N61 45.4
Elnath	278 27.2	N28 36.2
Eltanin	90 52.1	N51 29.4
Enif	33 59.0	N 9 52.0
Fomalhaut	15 37.2	S29 38.1
Gacrux	172 13.9	S57 05.8
Gienah	176 04.4	S17 31.8
Hadar	149 04.7	S60 21.4
Hamal	328 14.0	N23 27.2
Kaus Aust.	83 59.9	S34 23.0
Kochab	137 20.3	N74 09.6
Markab	13 50.3	N15 11.7
Menkar	314 27.3	N 4 04.8
Menkent	148 21.6	S36 21.4
Miaplacidus	221 41.2	S69 42.5
Mirfak	308 57.0	N49 51.3
Nunki	76 13.3	S26 17.8
Peacock	53 38.4	S56 44.5
Pollux	243 41.8	N28 01.7
Procyon	245 11.7	N 5 13.6
Rasalhague	96 17.7	N12 33.8
Regulus	207 55.8	N11 58.5
Rigel	281 23.1	S 8 12.5
Rigil Kent.	140 08.1	S60 49.2
Sabik	102 26.4	S15 43.2
Schedar	349 54.0	N56 31.8
Shaula	96 38.3	S37 05.9
Sirius	258 43.8	S16 43.0
Spica	158 43.7	S11 09.0
Suhail	223 00.7	S43 25.5
Vega	80 47.4	N38 47.0
Zuben'ubi	137 18.6	S16 01.9

	SHA	Mer. Pass.
Venus	59 27.4	12 38
Mars	38 33.9	14 04
Jupiter	33 00.8	14 25
Saturn	346 02.4	17 32

UT	SUN GHA	SUN Dec	MOON GHA	v	MOON Dec	d	HP
10							
00	178 10.5	S22 00.9	33 49.8	8.2	N17 39.0	3.6	57.7
01	193 10.3	00.6	48 17.0	8.2	17 42.6	3.4	57.7
02	208 10.0	22 00.2	62 44.2	8.2	17 46.0	3.3	57.7
03	223 09.8	21 59.8	77 11.4	8.2	17 49.3	3.3	57.7
04	238 09.5	59.5	91 38.6	8.2	17 52.6	3.1	57.7
05	253 09.3	59.1	106 05.8	8.2	17 55.7	3.0	57.7
06	268 09.0	S21 58.7	120 33.0	8.1	N17 58.7	2.9	57.6
07	283 08.8	58.4	135 00.1	8.2	18 01.6	2.8	57.6
08	298 08.5	58.0	149 27.3	8.2	18 04.4	2.7	57.6
09	313 08.3	.. 57.6	163 54.5	8.2	18 07.1	2.6	57.6
10	328 08.0	57.2	178 21.7	8.2	18 09.7	2.5	57.6
11	343 07.8	56.9	192 48.9	8.2	18 12.2	2.3	57.5
12	358 07.5	S21 56.5	207 16.1	8.2	N18 14.5	2.3	57.5
13	13 07.3	56.1	221 43.3	8.2	18 16.8	2.1	57.5
14	28 07.0	55.8	236 10.5	8.2	18 18.9	2.1	57.5
15	43 06.8	.. 55.4	250 37.7	8.2	18 21.0	1.9	57.5
16	58 06.5	55.0	265 04.9	8.3	18 22.9	1.8	57.4
17	73 06.3	54.6	279 32.2	8.2	18 24.7	1.7	57.4
18	88 06.0	S21 54.2	293 59.4	8.3	N18 26.4	1.6	57.4
19	103 05.8	53.9	308 26.7	8.2	18 28.0	1.5	57.4
20	118 05.5	53.5	322 53.9	8.3	18 29.5	1.4	57.4
21	133 05.3	.. 53.1	337 21.2	8.3	18 30.9	1.3	57.3
22	148 05.0	52.7	351 48.5	8.3	18 32.2	1.2	57.3
23	163 04.8	52.3	6 15.8	8.3	18 33.4	1.0	57.3
11							
00	178 04.5	S21 52.0	20 43.1	8.4	N18 34.4	1.0	57.3
01	193 04.3	51.6	35 10.5	8.3	18 35.4	0.8	57.3
02	208 04.0	51.2	49 37.8	8.4	18 36.2	0.7	57.2
03	223 03.8	.. 50.8	64 05.2	8.4	18 36.9	0.7	57.2
04	238 03.5	50.4	78 32.6	8.4	18 37.6	0.5	57.2
05	253 03.3	50.0	93 00.0	8.4	18 38.1	0.4	57.2
06	268 03.0	S21 49.6	107 27.4	8.5	N18 38.5	0.3	57.2
07	283 02.8	49.3	121 54.9	8.5	18 38.8	0.2	57.1
08	298 02.5	48.9	136 22.4	8.5	18 39.0	0.1	57.1
09	313 02.3	.. 48.5	150 49.9	8.5	18 39.1	0.0	57.1
10	328 02.0	48.1	165 17.4	8.5	18 39.1	0.2	57.1
11	343 01.8	47.7	179 44.9	8.6	18 38.9	0.2	57.1
12	358 01.5	S21 47.3	194 12.5	8.6	N18 38.7	0.4	57.0
13	13 01.3	46.9	208 40.1	8.6	18 38.3	0.4	57.0
14	28 01.1	46.5	223 07.7	8.7	18 37.9	0.6	57.0
15	43 00.8	.. 46.1	237 35.4	8.6	18 37.3	0.6	57.0
16	58 00.6	45.7	252 03.0	8.7	18 36.7	0.8	57.0
17	73 00.3	45.3	266 30.7	8.8	18 35.9	0.9	56.9
18	88 00.1	S21 44.9	280 58.5	8.8	N18 35.0	0.9	56.9
19	102 59.8	44.5	295 26.3	8.8	18 34.1	1.1	56.9
20	117 59.6	44.1	309 54.1	8.8	18 33.0	1.2	56.9
21	132 59.3	.. 43.7	324 21.9	8.9	18 31.8	1.3	56.8
22	147 59.1	43.3	338 49.8	8.9	18 30.5	1.4	56.8
23	162 58.9	42.9	353 17.7	8.9	18 29.1	1.5	56.8
12							
00	177 58.6	S21 42.5	7 45.6	9.0	N18 27.6	1.6	56.8
01	192 58.4	42.1	22 13.6	9.0	18 26.0	1.7	56.8
02	207 58.1	41.7	36 41.6	9.0	18 24.3	1.8	56.7
03	222 57.9	.. 41.3	51 09.6	9.1	18 22.5	1.9	56.7
04	237 57.7	40.9	65 37.7	9.1	18 20.6	2.0	56.7
05	252 57.4	40.5	80 05.8	9.2	18 18.6	2.1	56.7
06	267 57.2	S21 40.1	94 34.0	9.2	N18 16.5	2.2	56.7
07	282 56.9	39.7	109 02.2	9.3	18 14.3	2.3	56.6
08	297 56.7	39.3	123 30.5	9.2	18 12.0	2.4	56.6
09	312 56.4	.. 38.9	137 58.7	9.4	18 09.6	2.5	56.6
10	327 56.2	38.5	152 27.1	9.3	18 07.1	2.6	56.6
11	342 56.0	38.1	166 55.4	9.5	18 04.5	2.7	56.5
12	357 55.7	S21 37.7	181 23.9	9.4	N18 01.8	2.8	56.5
13	12 55.5	37.3	195 52.3	9.5	17 59.0	2.9	56.5
14	27 55.3	36.9	210 20.8	9.6	17 56.1	2.9	56.5
15	42 55.0	.. 36.5	224 49.4	9.6	17 53.2	3.1	56.5
16	57 54.8	36.0	239 18.0	9.6	17 50.1	3.2	56.4
17	72 54.5	35.6	253 46.6	9.7	17 46.9	3.3	56.4
18	87 54.3	S21 35.2	268 15.3	9.7	N17 43.6	3.3	56.4
19	102 54.1	34.8	282 44.0	9.8	17 40.3	3.5	56.4
20	117 53.8	34.4	297 12.8	9.8	17 36.8	3.5	56.3
21	132 53.6	.. 34.0	311 41.6	9.9	17 33.3	3.7	56.3
22	147 53.4	33.6	326 10.5	9.9	17 29.6	3.7	56.3
23	162 53.1	33.1	340 39.4	10.0	N17 25.9	3.8	56.3
	SD 16.3	d 0.4	SD 15.7	15.5	15.4		

SATURDAY (day 10), SUNDAY (day 11), MONDAY (day 12)

Twilight / Sunrise / Moonrise

Lat.	Naut.	Civil	Sunrise	Moonrise 10	11	12	13
N 72	08 10	10 08	■	□	□	12 21	14 37
N 70	07 54	09 29	■	11 55	12 37	13 50	15 21
68	07 40	09 02	10 54	12 36	13 22	14 29	15 50
66	07 29	08 42	10 05	13 04	13 52	14 56	16 12
64	07 20	08 25	09 35	13 25	14 15	15 17	16 29
62	07 12	08 11	09 12	13 42	14 33	15 34	16 43
60	07 05	07 59	08 54	13 57	14 48	15 48	16 55
N 58	06 58	07 49	08 38	14 09	15 00	16 00	17 05
56	06 52	07 40	08 26	14 19	15 11	16 10	17 14
54	06 47	07 32	08 14	14 29	15 21	16 19	17 22
52	06 42	07 25	08 04	14 37	15 29	16 28	17 30
50	06 37	07 18	07 55	14 45	15 37	16 35	17 36
45	06 27	07 04	07 37	15 01	15 54	16 50	17 50
N 40	06 18	06 51	07 21	15 14	16 07	17 03	18 01
35	06 09	06 41	07 08	15 25	16 19	17 14	18 11
30	06 02	06 31	06 57	15 35	16 29	17 24	18 19
20	05 47	06 14	06 37	15 52	16 46	17 40	18 34
N 10	05 32	05 58	06 20	16 07	17 01	17 54	18 47
0	05 16	05 42	06 04	16 21	17 15	18 08	18 58
S 10	04 58	05 25	05 48	16 35	17 29	18 21	19 10
20	04 37	05 06	05 30	16 50	17 44	18 35	19 23
30	04 10	04 43	05 10	17 07	18 02	18 52	19 38
35	03 53	04 29	04 58	17 17	18 12	19 01	19 46
40	03 32	04 12	04 44	17 29	18 23	19 12	19 56
45	03 04	03 52	04 28	17 42	18 37	19 25	20 07
S 50	02 25	03 25	04 07	17 59	18 53	19 40	20 20
52	02 03	03 11	03 57	18 06	19 01	19 48	20 26
54	01 32	02 56	03 46	18 15	19 10	19 56	20 33
56	00 29	02 37	03 34	18 25	19 19	20 05	20 41
58	////	02 13	03 19	18 36	19 30	20 15	20 50
S 60	////	01 40	03 02	18 48	19 43	20 26	21 00

Sunset / Twilight / Moonset

Lat.	Sunset	Civil	Naut.	Moonset 10	11	12	13
N 72	■	14 08	16 07	□	□	11 33	11 04
N 70	■	14 47	16 23	08 14	09 26	10 03	10 19
68	13 23	15 14	16 36	07 34	08 40	09 23	09 49
66	14 11	15 35	16 47	07 06	08 10	08 56	09 27
64	14 42	15 51	16 56	06 45	07 48	08 35	09 09
62	15 04	16 05	17 04	06 28	07 30	08 18	08 54
60	15 23	16 17	17 12	06 14	07 15	08 04	08 42
N 58	15 38	16 27	17 18	06 02	07 02	07 52	08 31
56	15 51	16 36	17 24	05 51	06 51	07 41	08 22
54	16 02	16 44	17 29	05 42	06 41	07 32	08 14
52	16 12	16 51	17 34	05 34	06 33	07 23	08 06
50	16 21	16 58	17 39	05 26	06 25	07 16	07 59
45	16 39	17 12	17 49	05 11	06 09	07 00	07 45
N 40	16 55	17 25	17 58	04 58	05 55	06 47	07 33
35	17 08	17 35	18 07	04 47	05 44	06 36	07 23
30	17 19	17 45	18 14	04 37	05 33	06 26	07 14
20	17 38	18 02	18 29	04 21	05 16	06 09	06 59
N 10	17 56	18 18	18 44	04 06	05 01	05 54	06 45
0	18 12	18 34	19 00	03 53	04 47	05 41	06 32
S 10	18 28	18 51	19 17	03 39	04 33	05 27	06 20
20	18 45	19 09	19 38	03 25	04 18	05 12	06 06
30	19 06	19 32	20 05	03 08	04 00	04 55	05 50
35	19 17	19 46	20 22	02 58	03 50	04 45	05 41
40	19 31	20 03	20 43	02 47	03 39	04 33	05 31
45	19 48	20 24	21 11	02 35	03 25	04 20	05 18
S 50	20 08	20 50	21 50	02 19	03 08	04 04	05 03
52	20 18	21 03	22 11	02 12	03 01	03 56	04 56
54	20 29	21 19	22 41	02 03	02 52	03 47	04 49
56	20 41	21 38	23 37	01 54	02 42	03 38	04 40
58	20 55	22 01	////	01 44	02 31	03 27	04 30
S 60	21 13	22 33	////	01 32	02 18	03 14	04 19

SUN / MOON

Day	SUN Eqn. of Time 00h	12h	Mer. Pass.	MOON Mer. Pass. Upper	Lower	Age	Phase
	m s	m s	h m	h m	h m	d	%
10	07 17	07 29	12 07	22 34	10 07	12	94
11	07 41	07 53	12 08	23 28	11 01	13	98
12	08 05	08 17	12 08	24 20	11 54	14	100 ◯

1998 JANUARY 13, 14, 15 (TUES., WED., THURS.)

18

UT	ARIES	VENUS −4.1		MARS +1.2		JUPITER −2.0		SATURN +0.7		STARS		
d h	GHA	GHA	Dec	GHA	Dec	GHA	Dec	GHA	Dec	Name	SHA	Dec
13 00	112 16.3	172 59.7	S15 32.4	149 16.9	S15 47.4	144 51.4	S14 03.6	98 13.6	N 3 20.5	Acamar	315 27.2	S40 19.1
01	127 18.8	188 03.8	32.1	164 17.5	46.8	159 53.3	03.4	113 16.0	20.6	Achernar	335 35.6	S57 15.2
02	142 21.2	203 07.9	31.9	179 18.0	46.1	174 55.3	03.3	128 18.3	20.7	Acrux	173 22.2	S63 05.0
03	157 23.7	218 11.9	.. 31.6	194 18.5	.. 45.5	189 57.2	.. 03.1	143 20.7	.. 20.7	Adhara	255 21.4	S28 58.4
04	172 26.2	233 16.0	31.3	209 19.0	44.8	204 59.1	02.9	158 23.0	20.8	Aldebaran	291 02.7	N16 30.2
05	187 28.6	248 20.1	31.1	224 19.6	44.2	220 01.0	02.7	173 25.4	20.8			
06	202 31.1	263 24.2	S15 30.8	239 20.1	S15 43.5	235 03.0	S14 02.5	188 27.7	N 3 20.9	Alioth	166 31.0	N55 58.0
07	217 33.6	278 28.3	30.5	254 20.6	42.9	250 04.9	02.3	203 30.1	20.9	Alkaid	153 08.3	N49 19.2
T 08	232 36.0	293 32.4	30.3	269 21.2	42.3	265 06.8	02.1	218 32.5	21.0	Al Na'ir	27 58.9	S46 58.4
U 09	247 38.5	308 36.5	.. 30.0	284 21.7	.. 41.6	280 08.7	.. 02.0	233 34.8	.. 21.1	Alnilam	275 58.1	S 1 12.4
E 10	262 41.0	323 40.6	29.7	299 22.2	41.0	295 10.7	01.8	248 37.2	21.1	Alphard	218 07.4	S 8 39.1
S 11	277 43.4	338 44.7	29.5	314 22.7	40.3	310 12.6	01.6	263 39.5	21.2			
D 12	292 45.9	353 48.8	S15 29.2	329 23.3	S15 39.7	325 14.5	S14 01.4	278 41.9	N 3 21.2	Alphecca	126 21.2	N26 43.2
A 13	307 48.4	8 52.8	28.9	344 23.8	39.0	340 16.4	01.2	293 44.2	21.3	Alpheratz	357 55.8	N29 04.9
Y 14	322 50.8	23 56.9	28.7	359 24.3	38.4	355 18.4	01.0	308 46.6	21.3	Altair	62 20.1	N 8 51.9
15	337 53.3	39 01.0	.. 28.4	14 24.9	.. 37.8	10 20.3	.. 00.8	323 48.9	.. 21.4	Ankaa	353 27.5	S42 19.3
16	352 55.7	54 05.1	28.2	29 25.4	37.1	25 22.2	00.7	338 51.3	21.5	Antares	112 41.0	S26 25.5
17	7 58.2	69 09.2	27.9	44 25.9	36.5	40 24.1	00.5	353 53.6	21.5			
18	23 00.7	84 13.3	S15 27.6	59 26.4	S15 35.8	55 26.1	S14 00.3	8 56.0	N 3 21.6	Arcturus	146 06.6	N19 11.5
19	38 03.1	99 17.4	27.4	74 27.0	35.2	70 28.0	14 00.1	23 58.3	21.6	Atria	107 53.9	S69 01.1
20	53 05.6	114 21.5	27.1	89 27.5	34.5	85 29.9	13 59.9	39 00.7	21.7	Avior	234 22.1	S59 30.3
21	68 08.1	129 25.6	.. 26.9	104 28.0	.. 33.9	100 31.8	.. 59.7	54 03.0	.. 21.8	Bellatrix	278 44.4	N 6 20.7
22	83 10.5	144 29.7	26.6	119 28.6	33.2	115 33.8	59.5	69 05.4	21.8	Betelgeuse	271 13.8	N 7 24.2
23	98 13.0	159 33.8	26.4	134 29.1	32.6	130 35.7	59.3	84 07.7	21.9			
14 00	113 15.5	174 38.0	S15 26.1	149 29.6	S15 31.9	145 37.6	S13 59.2	99 10.1	N 3 21.9	Canopus	264 00.8	S52 41.9
01	128 17.9	189 42.1	25.9	164 30.2	31.3	160 39.5	59.0	114 12.4	22.0	Capella	280 51.5	N45 59.7
02	143 20.4	204 46.2	25.6	179 30.7	30.7	175 41.5	58.8	129 14.8	22.0	Deneb	49 39.9	N45 16.5
03	158 22.9	219 50.3	.. 25.4	194 31.2	.. 30.0	190 43.4	.. 58.6	144 17.1	.. 22.1	Denebola	182 45.6	N14 34.9
04	173 25.3	234 54.4	25.1	209 31.8	29.4	205 45.3	58.4	159 19.5	22.2	Diphda	349 07.8	S18 00.1
05	188 27.8	249 58.5	24.9	224 32.3	28.7	220 47.2	58.2	174 21.8	22.2			
06	203 30.2	265 02.6	S15 24.6	239 32.8	S15 28.1	235 49.2	S13 58.0	189 24.2	N 3 22.3	Dubhe	194 05.8	N61 45.4
W 07	218 32.7	280 06.7	24.4	254 33.4	27.4	250 51.1	57.8	204 26.5	22.3	Elnath	278 27.2	N28 36.2
E 08	233 35.2	295 10.8	24.1	269 33.9	26.8	265 53.0	57.7	219 28.9	22.4	Eltanin	90 52.1	N51 29.4
D 09	248 37.6	310 14.9	.. 23.9	284 34.4	.. 26.1	280 54.9	.. 57.5	234 31.2	.. 22.5	Enif	33 59.0	N 9 52.0
N 10	263 40.1	325 19.0	23.6	299 35.0	25.5	295 56.9	57.3	249 33.6	22.5	Fomalhaut	15 37.2	S29 38.1
E 11	278 42.6	340 23.1	23.4	314 35.5	24.8	310 58.8	57.1	264 35.9	22.6			
S 12	293 45.0	355 27.3	S15 23.1	329 36.0	S15 24.2	326 00.7	S13 56.9	279 38.3	N 3 22.7	Gacrux	172 13.8	S57 05.8
D 13	308 47.5	10 31.4	22.9	344 36.6	23.5	341 02.6	56.7	294 40.6	22.7	Gienah	176 04.3	S17 31.8
A 14	323 50.0	25 35.5	22.7	359 37.1	22.9	356 04.6	56.5	309 43.0	22.8	Hadar	149 04.7	S60 21.4
Y 15	338 52.4	40 39.6	.. 22.4	14 37.6	.. 22.2	11 06.5	.. 56.3	324 45.3	.. 22.8	Hamal	328 14.0	N23 27.2
16	353 54.9	55 43.7	22.2	29 38.2	21.6	26 08.4	56.2	339 47.7	22.9	Kaus Aust.	83 59.9	S34 23.0
17	8 57.3	70 47.8	21.9	44 38.7	20.9	41 10.3	56.0	354 50.0	22.9			
18	23 59.8	85 51.9	S15 21.7	59 39.2	S15 20.3	56 12.2	S13 55.8	9 52.4	N 3 23.0	Kochab	137 20.2	N74 09.6
19	39 02.3	100 56.0	21.5	74 39.8	19.6	71 14.2	55.6	24 54.7	23.1	Markab	13 50.3	N15 11.7
20	54 04.7	116 00.2	21.2	89 40.3	18.9	86 16.1	55.4	39 57.1	23.1	Menkar	314 27.3	N 4 04.8
21	69 07.2	131 04.3	.. 21.0	104 40.8	.. 18.3	101 18.0	.. 55.2	54 59.4	.. 23.2	Menkent	148 21.5	S36 21.4
22	84 09.7	146 08.4	20.8	119 41.4	17.6	116 19.9	55.0	70 01.8	23.2	Miaplacidus	221 41.1	S69 42.5
23	99 12.1	161 12.5	20.5	134 41.9	17.0	131 21.9	54.8	85 04.1	23.3			
15 00	114 14.6	176 16.6	S15 20.3	149 42.4	S15 16.3	146 23.8	S13 54.7	100 06.5	N 3 23.4	Mirfak	308 57.0	N49 51.3
01	129 17.1	191 20.7	20.1	164 43.0	15.7	161 25.7	54.5	115 08.8	23.4	Nunki	76 13.3	S26 17.8
02	144 19.5	206 24.9	19.8	179 43.5	15.0	176 27.6	54.3	130 11.2	23.5	Peacock	53 38.4	S56 44.5
03	159 22.0	221 29.0	.. 19.6	194 44.0	.. 14.4	191 29.5	.. 54.1	145 13.5	.. 23.5	Pollux	243 41.8	N28 01.7
04	174 24.5	236 33.1	19.4	209 44.6	13.7	206 31.5	53.9	160 15.9	23.6	Procyon	245 11.7	N 5 13.6
05	189 26.9	251 37.2	19.2	224 45.1	13.1	221 33.4	53.7	175 18.2	23.7			
06	204 29.4	266 41.3	S15 18.9	239 45.7	S15 12.4	236 35.3	S13 53.5	190 20.5	N 3 23.7	Rasalhague	96 17.7	N12 33.7
07	219 31.8	281 45.4	18.7	254 46.2	11.8	251 37.2	53.3	205 22.9	23.8	Regulus	207 55.8	N11 58.5
T 08	234 34.3	296 49.6	18.5	269 46.7	11.1	266 39.2	53.1	220 25.2	23.8	Rigel	281 23.1	S 8 12.5
H 09	249 36.8	311 53.7	.. 18.3	284 47.3	.. 10.4	281 41.1	.. 53.0	235 27.6	.. 23.9	Rigil Kent.	140 08.0	S60 49.2
U 10	264 39.2	326 57.8	18.0	299 47.8	09.8	296 43.0	52.8	250 29.9	24.0	Sabik	102 26.4	S15 43.2
R 11	279 41.7	342 01.9	17.8	314 48.3	09.1	311 44.9	52.6	265 32.3	24.0			
S 12	294 44.2	357 06.0	S15 17.6	329 48.9	S15 08.5	326 46.8	S13 52.4	280 34.6	N 3 24.1	Schedar	349 54.0	N56 31.8
D 13	309 46.6	12 10.1	17.4	344 49.4	07.8	341 48.8	52.2	295 37.0	24.1	Shaula	96 38.3	S37 05.9
A 14	324 49.1	27 14.3	17.1	359 50.0	07.2	356 50.7	52.0	310 39.3	24.2	Sirius	258 43.8	S16 43.0
Y 15	339 51.6	42 18.4	.. 16.9	14 50.5	.. 06.5	11 52.6	.. 51.8	325 41.7	.. 24.3	Spica	158 43.7	S11 09.0
16	354 54.0	57 22.5	16.7	29 51.0	05.8	26 54.5	51.6	340 44.0	24.3	Suhail	223 00.6	S43 25.5
17	9 56.5	72 26.6	16.5	44 51.6	05.2	41 56.5	51.4	355 46.3	24.4			
18	24 59.0	87 30.7	S15 16.3	59 52.1	S15 04.5	56 58.4	S13 51.3	10 48.7	N 3 24.5	Vega	80 47.4	N38 47.0
19	40 01.4	102 34.8	16.0	74 52.6	03.9	72 00.3	51.1	25 51.0	24.5	Zuben'ubi	137 18.6	S16 01.9
20	55 03.9	117 39.0	15.8	89 53.2	03.2	87 02.2	50.9	40 53.4	24.6		SHA	Mer. Pass.
21	70 06.3	132 43.1	.. 15.6	104 53.7	.. 02.6	102 04.1	.. 50.7	55 55.7	.. 24.6		° '	h m
22	85 08.8	147 47.2	15.4	119 54.3	01.9	117 06.1	50.5	70 58.1	24.7	Venus	61 22.5	12 18
23	100 11.3	162 51.3	15.2	134 54.8	01.2	132 08.0	50.3	86 00.4	24.8	Mars	36 14.2	14 02
	h m									Jupiter	32 22.2	14 16
Mer. Pass. 16 24.3		v 4.1	d 0.2	v 0.5	d 0.7	v 1.9	d 0.2	v 2.3	d 0.1	Saturn	345 54.6	17 21

UT	SUN GHA	SUN Dec	MOON GHA	v	MOON Dec	d	HP
d h	° ′	° ′	° ′	′	° ′	′	′
13 00	177 52.9	S21 32.7	355 08.4	10.0	N17 22.1	3.9	56.3
01	192 52.6	32.3	9 37.4	10.1	17 18.2	4.0	56.2
02	207 52.4	31.9	24 06.5	10.1	17 14.2	4.1	56.2
03	222 52.2	.. 31.5	38 35.6	10.2	17 10.1	4.1	56.2
04	237 51.9	31.0	53 04.8	10.3	17 06.0	4.3	56.2
05	252 51.7	30.6	67 34.1	10.2	17 01.7	4.3	56.1
06	267 51.5	S21 30.2	82 03.3	10.4	N16 57.4	4.5	56.1
07	282 51.2	29.8	96 32.7	10.4	16 52.9	4.5	56.1
T 08	297 51.0	29.4	111 02.1	10.4	16 48.4	4.6	56.1
U 09	312 50.8	.. 28.9	125 31.5	10.5	16 43.8	4.6	56.1
E 10	327 50.5	28.5	140 01.0	10.6	16 39.2	4.8	56.0
S 11	342 50.3	28.1	154 30.6	10.6	16 34.4	4.8	56.0
D 12	357 50.1	S21 27.7	169 00.2	10.6	N16 29.6	5.0	56.0
A 13	12 49.8	27.2	183 29.8	10.7	16 24.6	5.0	56.0
Y 14	27 49.6	26.8	197 59.5	10.8	16 19.6	5.1	55.9
15	42 49.4	.. 26.4	212 29.3	10.8	16 14.5	5.1	55.9
16	57 49.1	25.9	226 59.1	10.9	16 09.4	5.3	55.9
17	72 48.9	25.5	241 29.0	10.9	16 04.1	5.3	55.9
18	87 48.7	S21 25.1	255 58.9	11.0	N15 58.8	5.4	55.9
19	102 48.4	24.6	270 28.9	11.1	15 53.4	5.4	55.8
20	117 48.2	24.2	284 59.0	11.1	15 48.0	5.6	55.8
21	132 48.0	.. 23.8	299 29.1	11.1	15 42.4	5.6	55.8
22	147 47.8	23.3	313 59.2	11.3	15 36.8	5.7	55.8
23	162 47.5	22.9	328 29.5	11.2	15 31.1	5.8	55.7
14 00	177 47.3	S21 22.5	342 59.7	11.3	N15 25.3	5.8	55.7
01	192 47.1	22.0	357 30.0	11.4	15 19.5	5.9	55.7
02	207 46.8	21.6	12 00.4	11.5	15 13.6	6.0	55.7
03	222 46.6	.. 21.2	26 30.9	11.5	15 07.6	6.0	55.7
04	237 46.4	20.7	41 01.4	11.5	15 01.6	6.2	55.6
05	252 46.2	20.3	55 31.9	11.6	14 55.4	6.1	55.6
06	267 45.9	S21 19.9	70 02.5	11.7	N14 49.3	6.3	55.6
W 07	282 45.7	19.4	84 33.2	11.7	14 43.0	6.3	55.6
E 08	297 45.5	19.0	99 03.9	11.8	14 36.7	6.4	55.6
D 09	312 45.2	.. 18.5	113 34.7	11.8	14 30.3	6.5	55.5
N 10	327 45.0	18.1	128 05.5	11.9	14 23.8	6.5	55.5
E 11	342 44.8	17.6	142 36.4	11.9	14 17.3	6.6	55.5
S 12	357 44.6	S21 17.2	157 07.3	12.0	N14 10.7	6.6	55.5
D 13	12 44.3	16.8	171 38.3	12.1	14 04.1	6.7	55.4
A 14	27 44.1	16.3	186 09.4	12.1	13 57.4	6.8	55.4
Y 15	42 43.9	.. 15.9	200 40.5	12.2	13 50.6	6.8	55.4
16	57 43.7	15.4	215 11.7	12.2	13 43.8	6.9	55.4
17	72 43.4	15.0	229 42.9	12.3	13 36.9	6.9	55.4
18	87 43.2	S21 14.5	244 14.2	12.3	N13 30.0	7.0	55.3
19	102 43.0	14.1	258 45.5	12.4	13 23.0	7.1	55.3
20	117 42.8	13.6	273 16.9	12.5	13 15.9	7.1	55.3
21	132 42.5	.. 13.2	287 48.4	12.5	13 08.8	7.2	55.3
22	147 42.3	12.7	302 19.9	12.5	13 01.6	7.2	55.3
23	162 42.1	12.3	316 51.4	12.6	12 54.4	7.3	55.2
15 00	177 41.9	S21 11.8	331 23.0	12.7	N12 47.1	7.3	55.2
01	192 41.7	11.4	345 54.7	12.7	12 39.8	7.4	55.2
02	207 41.4	10.9	0 26.4	12.7	12 32.4	7.5	55.2
03	222 41.2	.. 10.5	14 58.1	12.9	12 24.9	7.5	55.2
04	237 41.0	10.0	29 30.0	12.8	12 17.4	7.5	55.1
05	252 40.8	09.5	44 01.8	13.0	12 09.9	7.6	55.1
06	267 40.5	S21 09.1	58 33.8	12.9	N12 02.3	7.6	55.1
07	282 40.3	08.6	73 05.7	13.1	11 54.7	7.7	55.1
T 08	297 40.1	08.2	87 37.8	13.0	11 47.0	7.8	55.1
H 09	312 39.9	.. 07.7	102 09.8	13.2	11 39.2	7.8	55.1
U 10	327 39.7	07.3	116 42.0	13.2	11 31.4	7.8	55.0
R 11	342 39.4	06.8	131 14.2	13.2	11 23.6	7.9	55.0
S 12	357 39.2	S21 06.3	145 46.4	13.3	N11 15.7	7.9	55.0
D 13	12 39.0	05.9	160 18.7	13.3	11 07.8	8.0	55.0
A 14	27 38.8	05.4	174 51.0	13.4	10 59.8	8.0	55.0
Y 15	42 38.6	.. 05.0	189 23.4	13.4	10 51.8	8.0	54.9
16	57 38.3	04.5	203 55.8	13.5	10 43.8	8.1	54.9
17	72 38.1	04.0	218 28.3	13.5	10 35.7	8.1	54.9
18	87 37.9	S21 03.6	233 00.8	13.6	N10 27.6	8.2	54.9
19	102 37.7	03.1	247 33.4	13.6	10 19.4	8.2	54.9
20	117 37.5	02.6	262 06.0	13.7	10 11.2	8.3	54.9
21	132 37.3	.. 02.2	276 38.7	13.7	10 02.9	8.3	54.8
22	147 37.0	01.7	291 11.4	13.7	9 54.6	8.3	54.8
23	162 36.8	01.2	305 44.1	13.8	N 9 46.3	8.4	54.8
	SD 16.3	d 0.4	SD 15.3		15.1		15.0

Twilight / Sunrise / Moonrise

Lat.	Naut.	Civil	Sunrise	Moonrise 13	14	15	16
°	h m	h m	h m	h m	h m	h m	h m
N 72	08 03	09 56	▓▓	14 37	16 28	18 12	19 50
N 70	07 48	09 21	▓▓	15 21	16 56	18 29	20 00
68	07 36	08 56	10 39	15 50	17 16	18 43	20 08
66	07 26	08 37	09 56	16 12	17 32	18 54	20 15
64	07 17	08 21	09 28	16 29	17 45	19 03	20 20
62	07 09	08 08	09 07	16 43	17 57	19 11	20 25
60	07 02	07 56	08 49	16 55	18 06	19 18	20 30
N 58	06 56	07 46	08 35	17 05	18 14	19 24	20 33
56	06 50	07 38	08 23	17 14	18 21	19 29	20 37
54	06 45	07 30	08 12	17 22	18 28	19 34	20 40
52	06 40	07 23	08 02	17 30	18 34	19 38	20 42
50	06 36	07 17	07 54	17 36	18 39	19 42	20 45
45	06 26	07 03	07 35	17 50	18 50	19 50	20 50
N 40	06 17	06 51	07 20	18 01	18 59	19 57	20 54
35	06 09	06 40	07 08	18 11	19 07	20 03	20 58
30	06 02	06 31	06 57	18 19	19 14	20 09	21 02
20	05 47	06 14	06 38	18 34	19 26	20 18	21 07
N 10	05 33	05 59	06 21	18 47	19 37	20 26	21 12
0	05 17	05 43	06 05	18 58	19 47	20 33	21 17
S 10	05 00	05 27	05 49	19 10	19 57	20 40	21 22
20	04 40	05 08	05 32	19 23	20 07	20 48	21 27
30	04 13	04 46	05 13	19 38	20 19	20 57	21 33
35	03 56	04 32	05 01	19 46	20 26	21 02	21 36
40	03 36	04 16	04 47	19 56	20 34	21 08	21 40
45	03 09	03 56	04 31	20 07	20 43	21 15	21 44
S 50	02 31	03 30	04 12	20 20	20 54	21 23	21 49
52	02 10	03 17	04 02	20 26	20 59	21 27	21 52
54	01 42	03 02	03 51	20 33	21 05	21 31	21 54
56	00 56	02 44	03 39	20 41	21 11	21 36	21 57
58	////	02 21	03 25	20 50	21 18	21 41	22 00
S 60	////	01 51	03 08	21 00	21 26	21 46	22 04

Sunset / Twilight / Moonset

Lat.	Sunset	Civil	Naut.	Moonset 13	14	15	16
°	h m	h m	h m	h m	h m	h m	h m
N 72	▓▓	14 23	16 16	11 04	10 54	10 47	10 41
N 70	▓▓	14 58	16 31	10 19	10 25	10 28	10 29
68	13 40	15 23	16 43	09 49	10 04	10 13	10 29
66	14 22	15 42	16 53	09 27	09 47	10 01	10 12
64	14 50	15 58	17 03	09 09	09 33	09 51	10 05
62	15 12	16 11	17 10	08 54	09 21	09 42	09 59
60	15 29	16 22	17 17	08 42	09 11	09 35	09 54
N 58	15 44	16 32	17 23	08 31	09 03	09 28	09 49
56	15 56	16 41	17 28	08 22	08 55	09 22	09 45
54	16 07	16 48	17 33	08 14	08 48	09 17	09 42
52	16 16	16 56	17 38	08 06	08 42	09 12	09 38
50	16 25	17 02	17 42	07 59	08 36	09 08	09 35
45	16 43	17 16	17 52	07 45	08 24	08 58	09 29
N 40	16 58	17 28	18 01	07 33	08 14	08 50	09 23
35	17 10	17 38	18 09	07 23	08 05	08 44	09 19
30	17 21	17 47	18 17	07 14	07 58	08 38	09 14
20	17 40	18 04	18 31	06 59	07 45	08 27	09 07
N 10	17 57	18 20	18 45	06 45	07 33	08 18	09 01
0	18 13	18 35	19 01	06 32	07 22	08 09	08 55
S 10	18 28	18 51	19 18	06 20	07 11	08 01	08 48
20	18 45	19 10	19 38	06 06	06 59	07 51	08 42
30	19 05	19 32	20 04	05 50	06 46	07 41	08 34
35	19 17	19 46	20 21	05 41	06 38	07 34	08 30
40	19 30	20 02	20 42	05 31	06 29	07 27	08 25
45	19 46	20 22	21 08	05 18	06 19	07 19	08 19
S 50	20 06	20 47	21 45	05 03	06 06	07 09	08 12
52	20 15	21 00	22 06	04 56	06 00	07 04	08 09
54	20 26	21 15	22 33	04 49	05 53	06 59	08 05
56	20 38	21 33	23 17	04 40	05 46	06 54	08 01
58	20 52	21 55	////	04 30	05 38	06 47	07 57
S 60	21 08	22 24	////	04 19	05 28	06 40	07 52

SUN / MOON

Day	Eqn. of Time 00h	Eqn. of Time 12h	Mer. Pass.	Mer. Pass. Upper	Mer. Pass. Lower	Age	Phase
d	m s	m s	h m	h m	h m	d	%
13	08 28	08 39	12 09	00 20	12 46	15	99
14	08 50	09 01	12 09	01 10	13 35	16	97
15	09 12	09 23	12 09	01 58	14 21	17	92

1998 JANUARY 16, 17, 18 (FRI., SAT., SUN.)

UT	ARIES	VENUS −4.1		MARS +1.2		JUPITER −2.0		SATURN +0.7		STARS		
	GHA	GHA	Dec	GHA	Dec	GHA	Dec	GHA	Dec	Name	SHA	Dec
d h	° ′	° ′	° ′	° ′	° ′	° ′	° ′	° ′	° ′		° ′	° ′
16 00	115 13.7	177 55.4	S15 15.0	149 55.3	S15 00.6	147 09.9	S13 50.1	101 02.8	N 3 24.8	Acamar	315 27.2	S40 19.1
01	130 16.2	192 59.5	14.8	164 55.9	14 59.9	162 11.8	49.9	116 05.1	24.9	Achernar	335 35.6	S57 15.2
02	145 18.7	208 03.7	14.6	179 56.4	59.3	177 13.7	49.7	131 07.4	24.9	Acrux	173 22.2	S63 05.0
03	160 21.1	223 07.8	.. 14.3	194 57.0	.. 58.6	192 15.7	.. 49.6	146 09.8	.. 25.0	Adhara	255 21.4	S28 58.4
04	175 23.6	238 11.9	14.1	209 57.5	57.9	207 17.6	49.4	161 12.1	25.1	Aldebaran	291 02.7	N16 30.2
05	190 26.1	253 16.0	13.9	224 58.0	57.3	222 19.5	49.2	176 14.5	25.1			
06	205 28.5	268 20.1	S15 13.7	239 58.6	S14 56.6	237 21.4	S13 49.0	191 16.8	N 3 25.2	Alioth	166 31.0	N55 58.0
07	220 31.0	283 24.2	13.5	254 59.1	56.0	252 23.3	48.8	206 19.2	25.3	Alkaid	153 08.3	N49 19.2
08	235 33.4	298 28.4	13.3	269 59.7	55.3	267 25.3	48.6	221 21.5	25.3	Al Na'ir	27 59.0	S46 58.4
F 09	250 35.9	313 32.5	.. 13.1	285 00.2	.. 54.6	282 27.2	.. 48.4	236 23.8	.. 25.4	Alnilam	275 58.1	S 1 12.4
R 10	265 38.4	328 36.6	12.9	300 00.8	54.0	297 29.1	48.2	251 26.2	25.4	Alphard	218 07.4	S 8 39.1
I 11	280 40.8	343 40.7	12.7	315 01.3	53.3	312 31.0	48.0	266 28.5	25.5			
D 12	295 43.3	358 44.8	S15 12.5	330 01.8	S14 52.6	327 32.9	S13 47.8	281 30.9	N 3 25.6	Alphecca	126 21.2	N26 43.2
A 13	310 45.8	13 48.9	12.3	345 02.4	52.0	342 34.9	47.7	296 33.2	25.6	Alpheratz	357 55.8	N29 04.8
Y 14	325 48.2	28 53.0	12.1	0 02.9	51.3	357 36.8	47.5	311 35.5	25.7	Altair	62 20.1	N 8 51.8
15	340 50.7	43 57.2	.. 11.9	15 03.5	.. 50.6	12 38.7	.. 47.3	326 37.9	.. 25.8	Ankaa	353 27.5	S42 19.3
16	355 53.2	59 01.3	11.7	30 04.0	50.0	27 40.6	47.1	341 40.2	25.8	Antares	112 41.0	S26 25.5
17	10 55.6	74 05.4	11.5	45 04.6	49.3	42 42.5	46.9	356 42.6	25.9			
18	25 58.1	89 09.5	S15 11.3	60 05.1	S14 48.7	57 44.5	S13 46.7	11 44.9	N 3 25.9	Arcturus	146 06.6	N19 11.5
19	41 00.6	104 13.6	11.1	75 05.6	48.0	72 46.4	46.5	26 47.2	26.0	Atria	107 53.9	S69 01.1
20	56 03.0	119 17.7	10.9	90 06.2	47.3	87 48.3	46.3	41 49.6	26.1	Avior	234 22.1	S59 30.3
21	71 05.5	134 21.8	.. 10.7	105 06.7	.. 46.7	102 50.2	.. 46.1	56 51.9	.. 26.1	Bellatrix	278 44.4	N 6 20.7
22	86 07.9	149 25.9	10.5	120 07.3	46.0	117 52.1	45.9	71 54.3	26.2	Betelgeuse	271 13.8	N 7 24.2
23	101 10.4	164 30.0	10.3	135 07.8	45.3	132 54.1	45.8	86 56.6	26.3			
17 00	116 12.9	179 34.1	S15 10.1	150 08.4	S14 44.7	147 56.0	S13 45.6	101 58.9	N 3 26.3	Canopus	264 00.8	S52 41.9
01	131 15.3	194 38.2	09.9	165 08.9	44.0	162 57.9	45.4	117 01.3	26.4	Capella	280 51.5	N45 59.7
02	146 17.8	209 42.4	09.7	180 09.5	43.3	177 59.8	45.2	132 03.6	26.5	Deneb	49 39.9	N45 16.5
03	161 20.3	224 46.5	.. 09.6	195 10.0	.. 42.7	193 01.7	.. 45.0	147 06.0	.. 26.5	Denebola	182 45.5	N14 34.9
04	176 22.7	239 50.6	09.4	210 10.5	42.0	208 03.7	44.8	162 08.3	26.6	Diphda	349 07.8	S18 00.1
05	191 25.2	254 54.7	09.2	225 11.1	41.3	223 05.6	44.6	177 10.6	26.6			
06	206 27.7	269 58.8	S15 09.0	240 11.6	S14 40.7	238 07.5	S13 44.4	192 13.0	N 3 26.7	Dubhe	194 05.8	N61 45.4
07	221 30.1	285 02.9	08.8	255 12.2	40.0	253 09.4	44.2	207 15.3	26.8	Elnath	278 27.2	N28 36.2
S 08	236 32.6	300 07.0	08.6	270 12.7	39.3	268 11.3	44.0	222 17.7	26.8	Eltanin	90 52.1	N51 29.3
A 09	251 35.1	315 11.1	.. 08.4	285 13.3	.. 38.7	283 13.2	.. 43.8	237 20.0	.. 26.9	Enif	33 59.0	N 9 52.0
T 10	266 37.5	330 15.2	08.2	300 13.8	38.0	298 15.2	43.7	252 22.3	27.0	Fomalhaut	15 37.3	S29 38.1
U 11	281 40.0	345 19.3	08.1	315 14.4	37.3	313 17.1	43.5	267 24.7	27.0			
R 12	296 42.4	0 23.4	S15 07.9	330 14.9	S14 36.7	328 19.0	S13 43.3	282 27.0	N 3 27.1	Gacrux	172 13.8	S57 05.8
D 13	311 44.9	15 27.5	07.7	345 15.5	36.0	343 20.9	43.1	297 29.3	27.2	Gienah	176 04.3	S17 31.8
A 14	326 47.4	30 31.6	07.5	0 16.0	35.3	358 22.8	42.9	312 31.7	27.2	Hadar	149 04.7	S60 21.5
Y 15	341 49.8	45 35.7	.. 07.3	15 16.6	.. 34.6	13 24.8	.. 42.7	327 34.0	.. 27.3	Hamal	328 14.0	N23 27.2
16	356 52.3	60 39.8	07.2	30 17.1	34.0	28 26.7	42.5	342 36.4	27.3	Kaus Aust.	83 59.8	S34 23.0
17	11 54.8	75 43.9	07.0	45 17.7	33.3	43 28.6	42.3	357 38.7	27.4			
18	26 57.2	90 48.0	S15 06.8	60 18.2	S14 32.6	58 30.5	S13 42.1	12 41.0	N 3 27.5	Kochab	137 20.1	N74 09.6
19	41 59.7	105 52.0	06.6	75 18.7	32.0	73 32.4	41.9	27 43.4	27.5	Markab	13 50.3	N15 11.7
20	57 02.2	120 56.1	06.4	90 19.3	31.3	88 34.3	41.7	42 45.7	27.6	Menkar	314 27.3	N 4 04.8
21	72 04.6	136 00.2	.. 06.3	105 19.8	.. 30.6	103 36.3	.. 41.5	57 48.0	.. 27.7	Menkent	148 21.5	S36 21.4
22	87 07.1	151 04.3	06.1	120 20.4	30.0	118 38.2	41.4	72 50.4	27.7	Miaplacidus	221 41.1	S69 42.5
23	102 09.5	166 08.4	05.9	135 20.9	29.3	133 40.1	41.2	87 52.7	27.8			
18 00	117 12.0	181 12.5	S15 05.8	150 21.5	S14 28.6	148 42.0	S13 41.0	102 55.1	N 3 27.9	Mirfak	308 57.0	N49 51.3
01	132 14.5	196 16.6	05.6	165 22.0	27.9	163 43.9	40.8	117 57.4	27.9	Nunki	76 13.3	S26 17.8
02	147 16.9	211 20.7	05.4	180 22.6	27.3	178 45.8	40.6	132 59.7	28.0	Peacock	53 38.4	S56 44.5
03	162 19.4	226 24.8	.. 05.2	195 23.1	.. 26.6	193 47.8	.. 40.4	148 02.1	.. 28.1	Pollux	243 41.8	N28 01.7
04	177 21.9	241 28.8	05.1	210 23.7	25.9	208 49.7	40.2	163 04.4	28.1	Procyon	245 11.7	N 5 13.6
05	192 24.3	256 32.9	04.9	225 24.2	25.2	223 51.6	40.0	178 06.7	28.2			
06	207 26.8	271 37.0	S15 04.7	240 24.8	S14 24.6	238 53.5	S13 39.8	193 09.1	N 3 28.2	Rasalhague	96 17.7	N12 33.7
07	222 29.3	286 41.1	04.6	255 25.3	23.9	253 55.4	39.6	208 11.4	28.3	Regulus	207 55.8	N11 58.5
08	237 31.7	301 45.2	04.4	270 25.9	23.2	268 57.4	39.4	223 13.7	28.4	Rigel	281 23.1	S 8 12.5
S 09	252 34.2	316 49.2	.. 04.2	285 26.4	.. 22.5	283 59.3	.. 39.2	238 16.1	.. 28.4	Rigil Kent.	140 08.0	S60 49.2
U 10	267 36.7	331 53.3	04.1	300 27.0	21.9	299 01.2	39.1	253 18.4	28.5	Sabik	102 26.4	S15 43.2
N 11	282 39.1	346 57.4	03.9	315 27.5	21.2	314 03.1	38.9	268 20.7	28.6			
D 12	297 41.6	2 01.5	S15 03.8	330 28.1	S14 20.5	329 05.0	S13 38.7	283 23.1	N 3 28.6	Schedar	349 54.1	N56 31.8
A 13	312 44.0	17 05.5	03.6	345 28.7	19.8	344 06.9	38.5	298 25.4	28.7	Shaula	96 38.3	S37 05.9
Y 14	327 46.5	32 09.6	03.4	0 29.2	19.2	359 08.8	38.3	313 27.7	28.8	Sirius	258 43.8	S16 43.0
15	342 49.0	47 13.7	.. 03.3	15 29.8	.. 18.5	14 10.8	.. 38.1	328 30.1	.. 28.8	Spica	158 43.7	S11 09.0
16	357 51.4	62 17.8	03.1	30 30.3	17.8	29 12.7	37.9	343 32.4	28.9	Suhail	223 00.6	S43 25.5
17	12 53.9	77 21.8	03.0	45 30.9	17.1	44 14.6	37.7	358 34.7	29.0			
18	27 56.4	92 25.9	S15 02.8	60 31.4	S14 16.5	59 16.5	S13 37.5	13 37.1	N 3 29.0	Vega	80 47.4	N38 46.9
19	42 58.8	107 29.9	02.6	75 32.0	15.8	74 18.4	37.3	28 39.4	29.1	Zuben'ubi	137 18.6	S16 01.9
20	58 01.3	122 34.0	02.5	90 32.5	15.1	89 20.3	37.1	43 41.7	29.2		SHA	Mer. Pass.
21	73 03.8	137 38.1	.. 02.3	105 33.1	.. 14.4	104 22.3	.. 36.9	58 44.1	.. 29.2		° ′	h m
22	88 06.2	152 42.1	02.2	120 33.6	13.8	119 24.2	36.7	73 46.4	29.3	Venus	63 21.3	11 58
23	103 08.7	167 46.2	02.0	135 34.2	13.1	134 26.1	36.6	88 48.7	29.4	Mars	33 55.5	13 59
	h m									Jupiter	31 43.1	14 06
Mer. Pass. 16 12.5	v 4.1 d 0.2			v 0.5 d 0.7		v 1.9 d 0.2		v 2.3 d 0.1		Saturn	345 46.1	17 09

UT	SUN GHA	SUN Dec	MOON GHA	v	MOON Dec	d	HP
16 00	177 36.6	S21 00.8	320 16.9	13.9	N 9 37.9	8.4	54.8
01	192 36.4	21 00.3	334 49.8	13.9	9 29.5	8.4	54.8
02	207 36.2	20 59.8	349 22.7	13.9	9 21.1	8.5	54.8
03	222 36.0	.. 59.3	3 55.6	14.0	9 12.6	8.5	54.7
04	237 35.8	58.9	18 28.6	14.0	9 04.1	8.5	54.7
05	252 35.5	58.4	33 01.6	14.1	8 55.6	8.6	54.7
06	267 35.3	S20 57.9	47 34.7	14.1	N 8 47.0	8.6	54.7
07	282 35.1	57.5	62 07.8	14.1	8 38.4	8.7	54.7
08	297 34.9	57.0	76 40.9	14.2	8 29.7	8.6	54.7
F 09	312 34.7	.. 56.5	91 14.1	14.2	8 21.1	8.7	54.7
R 10	327 34.5	56.0	105 47.3	14.3	8 12.4	8.7	54.6
I 11	342 34.3	55.6	120 20.6	14.3	8 03.7	8.8	54.6
D 12	357 34.0	S20 55.1	134 53.9	14.3	N 7 54.9	8.8	54.6
A 13	12 33.8	54.6	149 27.2	14.4	7 46.1	8.8	54.6
Y 14	27 33.6	54.1	164 00.6	14.4	7 37.3	8.8	54.6
15	42 33.4	.. 53.6	178 34.0	14.5	7 28.5	8.9	54.6
16	57 33.2	53.2	193 07.5	14.5	7 19.6	8.9	54.6
17	72 33.0	52.7	207 41.0	14.5	7 10.7	8.9	54.5
18	87 32.8	S20 52.2	222 14.5	14.6	N 7 01.8	8.9	54.5
19	102 32.6	51.7	236 48.1	14.6	6 52.9	9.0	54.5
20	117 32.4	51.2	251 21.7	14.6	6 43.9	9.0	54.5
21	132 32.1	.. 50.7	265 55.3	14.7	6 34.9	9.0	54.5
22	147 31.9	50.3	280 29.0	14.6	6 25.9	9.0	54.5
23	162 31.7	49.8	295 02.6	14.8	6 16.9	9.1	54.5
17 00	177 31.5	S20 49.3	309 36.4	14.7	N 6 07.8	9.1	54.5
01	192 31.3	48.8	324 10.1	14.8	5 58.7	9.1	54.4
02	207 31.1	48.3	338 43.9	14.8	5 49.6	9.1	54.4
03	222 30.9	.. 47.8	353 17.7	14.9	5 40.5	9.1	54.4
04	237 30.7	47.3	7 51.6	14.8	5 31.4	9.1	54.4
05	252 30.5	46.9	22 25.4	14.9	5 22.3	9.2	54.4
06	267 30.3	S20 46.4	36 59.3	15.0	N 5 13.1	9.2	54.4
S 07	282 30.1	45.9	51 33.3	14.9	5 03.9	9.2	54.4
A 08	297 29.9	45.4	66 07.2	15.0	4 54.7	9.2	54.4
T 09	312 29.7	.. 44.9	80 41.2	15.0	4 45.5	9.2	54.4
U 10	327 29.4	44.4	95 15.2	15.0	4 36.3	9.3	54.4
R 11	342 29.2	43.9	109 49.2	15.1	4 27.0	9.2	54.3
D 12	357 29.0	S20 43.4	124 23.3	15.1	N 4 17.8	9.3	54.3
A 13	12 28.8	42.9	138 57.4	15.1	4 08.5	9.3	54.3
Y 14	27 28.6	42.4	153 31.5	15.1	3 59.2	9.3	54.3
15	42 28.4	.. 41.9	168 05.6	15.1	3 49.9	9.3	54.3
16	57 28.2	41.4	182 39.7	15.2	3 40.6	9.3	54.3
17	72 28.0	40.9	197 13.9	15.2	3 31.3	9.4	54.3
18	87 27.8	S20 40.4	211 48.1	15.2	N 3 21.9	9.3	54.3
19	102 27.6	39.9	226 22.3	15.2	3 12.6	9.4	54.3
20	117 27.4	39.4	240 56.5	15.2	3 03.2	9.3	54.3
21	132 27.2	.. 38.9	255 30.7	15.3	2 53.9	9.4	54.3
22	147 27.0	38.4	270 05.0	15.2	2 44.5	9.4	54.3
23	162 26.8	37.9	284 39.2	15.3	2 35.1	9.4	54.3
18 00	177 26.6	S20 37.4	299 13.5	15.3	N 2 25.7	9.3	54.3
01	192 26.4	36.9	313 47.8	15.3	2 16.4	9.4	54.2
02	207 26.2	36.4	328 22.1	15.3	2 07.0	9.4	54.2
03	222 26.0	.. 35.9	342 56.4	15.4	1 57.6	9.5	54.2
04	237 25.8	35.4	357 30.8	15.3	1 48.1	9.4	54.2
05	252 25.6	34.9	12 05.1	15.4	1 38.7	9.4	54.2
06	267 25.4	S20 34.4	26 39.5	15.4	N 1 29.3	9.4	54.2
07	282 25.2	33.9	41 13.9	15.3	1 19.9	9.4	54.2
08	297 25.0	33.4	55 48.2	15.4	1 10.5	9.5	54.2
S 09	312 24.8	.. 32.9	70 22.6	15.4	1 01.0	9.4	54.2
U 10	327 24.6	32.4	84 57.0	15.4	0 51.6	9.4	54.2
N 11	342 24.4	31.9	99 31.4	15.4	0 42.2	9.5	54.2
D 12	357 24.2	S20 31.4	114 05.8	15.5	N 0 32.7	9.4	54.2
A 13	12 24.0	30.8	128 40.3	15.4	0 23.3	9.4	54.2
Y 14	27 23.8	30.3	143 14.7	15.4	0 13.9	9.5	54.2
15	42 23.6	.. 29.8	157 49.1	15.4	N 0 04.4	9.4	54.2
16	57 23.4	29.3	172 23.5	15.5	S 0 05.0	9.4	54.2
17	72 23.2	28.8	186 58.0	15.4	0 14.4	9.4	54.2
18	87 23.0	S20 28.3	201 32.4	15.5	S 0 23.8	9.5	54.2
19	102 22.8	27.8	216 06.9	15.4	0 33.3	9.4	54.2
20	117 22.6	27.2	230 41.3	15.4	0 42.7	9.4	54.2
21	132 22.4	.. 26.7	245 15.7	15.5	0 52.1	9.4	54.2
22	147 22.2	26.2	259 50.2	15.4	1 01.5	9.4	54.2
23	162 22.0	25.7	274 24.6	15.4	S 1 10.9	9.4	54.2
SD	16.3	d 0.5	SD 14.9		14.8		14.8

Lat.	Twilight Naut.	Twilight Civil	Sunrise	Moonrise 16	17	18	19
°	h m	h m	h m	h m	h m	h m	h m
N 72	07 56	09 44	■	19 50	21 25	22 58	24 31
N 70	07 42	09 12	11 45	20 00	21 29	22 56	24 24
68	07 31	08 49	10 24	20 08	21 32	22 55	24 18
66	07 21	08 31	09 47	20 15	21 35	22 54	24 13
64	07 13	08 16	09 21	20 20	21 37	22 53	24 08
62	07 05	08 03	09 01	20 25	21 39	22 52	24 04
60	06 59	07 53	08 45	20 30	21 40	22 51	24 01
N 58	06 53	07 43	08 31	20 33	21 42	22 50	23 58
56	06 48	07 35	08 19	20 37	21 43	22 50	23 56
54	06 43	07 27	08 09	20 40	21 45	22 49	23 53
52	06 39	07 21	08 00	20 42	21 46	22 49	23 51
50	06 34	07 15	07 51	20 45	21 47	22 48	23 49
45	06 25	07 01	07 34	20 50	21 49	22 47	23 45
N 40	06 16	06 50	07 19	20 54	21 51	22 46	23 42
35	06 09	06 40	07 07	20 58	21 52	22 46	23 39
30	06 01	06 31	06 56	21 02	21 54	22 45	23 36
20	05 47	06 14	06 38	21 07	21 56	22 44	23 32
N 10	05 33	05 59	06 22	21 12	21 58	22 43	23 28
0	05 19	05 44	06 06	21 17	22 00	22 42	23 24
S 10	05 02	05 28	05 51	21 22	22 02	22 41	23 21
20	04 42	05 10	05 34	21 27	22 04	22 41	23 17
30	04 16	04 48	05 15	21 33	22 07	22 40	23 13
35	04 00	04 35	05 04	21 36	22 08	22 39	23 10
40	03 40	04 19	04 51	21 40	22 10	22 38	23 08
45	03 14	04 00	04 35	21 44	22 11	22 38	23 04
S 50	02 38	03 35	04 16	21 49	22 14	22 37	23 01
52	02 18	03 22	04 07	21 52	22 15	22 37	22 59
54	01 53	03 08	03 57	21 54	22 16	22 36	22 57
56	01 14	02 51	03 45	21 57	22 17	22 36	22 55
58	////	02 30	03 32	22 00	22 18	22 35	22 52
S 60	////	02 03	03 16	22 04	22 20	22 35	22 50

Lat.	Sunset	Twilight Civil	Twilight Naut.	Moonset 16	17	18	19
°	h m	h m	h m	h m	h m	h m	h m
N 72	■	14 37	16 25	10 41	10 35	10 30	10 24
N 70	12 36	15 09	16 39	10 29	10 29	10 29	10 28
68	13 57	15 32	16 50	10 19	10 24	10 28	10 31
66	14 33	15 50	17 00	10 12	10 20	10 27	10 34
64	14 59	16 05	17 08	10 05	10 16	10 27	10 37
62	15 20	16 17	17 15	09 59	10 13	10 26	10 39
60	15 36	16 28	17 22	09 54	10 10	10 26	10 41
N 58	15 50	16 37	17 28	09 49	10 08	10 25	10 42
56	16 02	16 46	17 33	09 45	10 06	10 25	10 44
54	16 12	16 53	17 38	09 42	10 04	10 25	10 45
52	16 21	17 00	17 42	09 38	10 02	10 24	10 46
50	16 29	17 06	17 46	09 35	10 00	10 24	10 47
45	16 47	17 19	17 56	09 29	09 57	10 24	10 50
N 40	17 01	17 31	18 04	09 23	09 54	10 23	10 52
35	17 13	17 41	18 12	09 19	09 51	10 23	10 54
30	17 24	17 50	18 19	09 14	09 49	10 22	10 55
20	17 42	18 06	18 33	09 07	09 45	10 22	10 58
N 10	17 59	18 21	18 47	09 01	09 41	10 21	11 00
0	18 14	18 36	19 02	08 55	09 38	10 20	11 03
S 10	18 29	18 52	19 18	08 48	09 35	10 20	11 05
20	18 46	19 10	19 38	08 42	09 31	10 19	11 07
30	19 05	19 31	20 03	08 34	09 27	10 18	11 10
35	19 16	19 45	20 20	08 30	09 24	10 18	11 11
40	19 29	20 00	20 39	08 25	09 22	10 17	11 13
45	19 44	20 20	21 05	08 19	09 18	10 17	11 15
S 50	20 03	20 44	21 41	08 12	09 15	10 16	11 18
52	20 12	20 57	22 00	08 09	09 13	10 16	11 19
54	20 22	21 11	22 25	08 05	09 11	10 16	11 20
56	20 34	21 27	23 02	08 01	09 09	10 15	11 21
58	20 47	21 48	////	07 57	09 06	10 15	11 23
S 60	21 03	22 15	////	07 52	09 04	10 14	11 24

Day	SUN Eqn. of Time 00h	SUN Eqn. of Time 12h	Mer. Pass.	MOON Mer. Pass. Upper	MOON Mer. Pass. Lower	Age	Phase
	m s	m s	h m	h m	h m	d	%
16	09 33	09 43	12 10	02 44	15 06	18	87
17	09 54	10 03	12 10	03 28	15 49	19	79
18	10 13	10 23	12 10	04 10	16 31	20	71

1998 JANUARY 19, 20, 21 (MON., TUES., WED.)

UT	ARIES	VENUS −4.2		MARS +1.2		JUPITER −2.0		SATURN +0.7		STARS		
	GHA	GHA	Dec	GHA	Dec	GHA	Dec	GHA	Dec	Name	SHA	Dec
d h	° ′	° ′	° ′	° ′	° ′	° ′	° ′	° ′	° ′		° ′	° ′
19 00	118 11.1	182 50.3	S15 01.9	150 34.7	S14 12.4	149 28.0	S13 36.4	103 51.1	N 3 29.4	Acamar	315 27.2	S40 19.1
01	133 13.6	197 54.3	01.7	165 35.3	11.7	164 29.9	36.2	118 53.4	29.5	Achernar	335 35.7	S57 15.2
02	148 16.1	212 58.4	01.6	180 35.8	11.0	179 31.8	36.0	133 55.7	29.6	Acrux	173 22.1	S63 05.0
03	163 18.5	228 02.4 ..	01.4	195 36.4 ..	10.4	194 33.8 ..	35.8	148 58.1 ..	29.6	Adhara	255 21.4	S28 58.4
04	178 21.0	243 06.5	01.3	210 36.9	09.7	209 35.7	35.6	164 00.4	29.7	Aldebaran	291 02.7	N16 30.2
05	193 23.5	258 10.5	01.1	225 37.5	09.0	224 37.6	35.4	179 02.7	29.8			
06	208 25.9	273 14.6	S15 01.0	240 38.1	S14 08.3	239 39.5	S13 35.2	194 05.1	N 3 29.8	Alioth	166 31.0	N55 58.0
07	223 28.4	288 18.6	00.8	255 38.6	07.7	254 41.4	35.0	209 07.4	29.9	Alkaid	153 08.2	N49 19.2
08	238 30.9	303 22.7	00.7	270 39.2	07.0	269 43.3	34.8	224 09.7	30.0	Al Na'ir	27 59.0	S46 58.4
09	253 33.3	318 26.7 ..	00.6	285 39.7 ..	06.3	284 45.2 ..	34.6	239 12.1 ..	30.0	Alnilam	275 58.1	S 1 12.4
10	268 35.8	333 30.7	00.4	300 40.3	05.6	299 47.2	34.4	254 14.4	30.1	Alphard	218 07.4	S 8 39.1
11	283 38.3	348 34.8	00.3	315 40.8	04.9	314 49.1	34.2	269 16.7	30.2			
12	298 40.7	3 38.8	S15 00.1	330 41.4	S14 04.3	329 51.0	S13 34.0	284 19.0	N 3 30.2	Alphecca	126 21.2	N26 43.2
13	313 43.2	18 42.9	15 00.0	345 41.9	03.6	344 52.9	33.9	299 21.4	30.3	Alpheratz	357 55.8	N29 04.8
14	328 45.6	33 46.9	14 59.9	0 42.5	02.9	359 54.8	33.7	314 23.7	30.4	Altair	62 20.1	N 8 51.8
15	343 48.1	48 50.9 ..	59.7	15 43.1 ..	02.2	14 56.7 ..	33.5	329 26.0 ..	30.4	Ankaa	353 27.5	S42 19.3
16	358 50.6	63 55.0	59.6	30 43.6	01.5	29 58.6	33.3	344 28.4	30.5	Antares	112 41.0	S26 25.5
17	13 53.0	78 59.0	59.4	45 44.2	00.8	45 00.6	33.1	359 30.7	30.6			
18	28 55.5	94 03.0	S14 59.3	60 44.7	S14 00.2	60 02.5	S13 32.9	14 33.0	N 3 30.6	Arcturus	146 06.6	N19 11.5
19	43 58.0	109 07.0	59.2	75 45.3	13 59.5	75 04.4	32.7	29 35.4	30.7	Atria	107 53.8	S69 01.1
20	59 00.4	124 11.1	59.0	90 45.8	58.8	90 06.3	32.5	44 37.7	30.8	Avior	234 22.1	S59 30.3
21	74 02.9	139 15.1 ..	58.9	105 46.4 ..	58.1	105 08.2 ..	32.3	59 40.0 ..	30.8	Bellatrix	278 44.4	N 6 20.7
22	89 05.4	154 19.1	58.8	120 47.0	57.4	120 10.1	32.1	74 42.3	30.9	Betelgeuse	271 13.8	N 7 24.2
23	104 07.8	169 23.1	58.6	135 47.5	56.7	135 12.0	31.9	89 44.7	31.0			
20 00	119 10.3	184 27.1	S14 58.5	150 48.1	S13 56.1	150 14.0	S13 31.7	104 47.0	N 3 31.0	Canopus	264 00.9	S52 42.0
01	134 12.8	199 31.2	58.4	165 48.6	55.4	165 15.9	31.5	119 49.3	31.1	Capella	280 51.5	N45 59.7
02	149 15.2	214 35.2	58.3	180 49.2	54.7	180 17.8	31.3	134 51.7	31.2	Deneb	49 39.9	N45 16.5
03	164 17.7	229 39.2 ..	58.1	195 49.7 ..	54.0	195 19.7 ..	31.1	149 54.0 ..	31.2	Denebola	182 45.5	N14 34.9
04	179 20.1	244 43.2	58.0	210 50.3	53.3	210 21.6	30.9	164 56.3	31.3	Diphda	349 07.8	S18 00.1
05	194 22.6	259 47.2	57.9	225 50.9	52.6	225 23.5	30.8	179 58.6	31.4			
06	209 25.1	274 51.2	S14 57.7	240 51.4	S13 52.0	240 25.4	S13 30.6	195 01.0	N 3 31.4	Dubhe	194 05.7	N61 45.4
07	224 27.5	289 55.2	57.6	255 52.0	51.3	255 27.4	30.4	210 03.3	31.5	Elnath	278 27.2	N28 36.2
08	239 30.0	304 59.2	57.5	270 52.5	50.6	270 29.3	30.2	225 05.6	31.6	Eltanin	90 52.1	N51 29.3
09	254 32.5	320 03.2 ..	57.4	285 53.1 ..	49.9	285 31.2 ..	30.0	240 08.0 ..	31.6	Enif	33 59.0	N 9 52.0
10	269 34.9	335 07.2	57.3	300 53.7	49.2	300 33.1	29.8	255 10.3	31.7	Fomalhaut	15 37.3	S29 38.1
11	284 37.4	350 11.2	57.1	315 54.2	48.5	315 35.0	29.6	270 12.6	31.8			
12	299 39.9	5 15.2	S14 57.0	330 54.8	S13 47.8	330 36.9	S13 29.4	285 14.9	N 3 31.8	Gacrux	172 13.8	S57 05.9
13	314 42.3	20 19.2	56.9	345 55.3	47.2	345 38.8	29.2	300 17.3	31.9	Gienah	176 04.3	S17 31.8
14	329 44.8	35 23.2	56.8	0 55.9	46.5	0 40.7	29.0	315 19.6	32.0	Hadar	149 04.6	S60 21.5
15	344 47.2	50 27.2 ..	56.7	15 56.5 ..	45.8	15 42.7 ..	28.8	330 21.9 ..	32.1	Hamal	328 14.0	N23 27.2
16	359 49.7	65 31.1	56.5	30 57.0	45.1	30 44.6	28.6	345 24.2	32.1	Kaus Aust.	83 59.8	S34 23.0
17	14 52.2	80 35.1	56.4	45 57.6	44.4	45 46.5	28.4	0 26.6	32.2			
18	29 54.6	95 39.1	S14 56.3	60 58.1	S13 43.7	60 48.4	S13 28.2	15 28.9	N 3 32.3	Kochab	137 20.1	N74 09.6
19	44 57.1	110 43.1	56.2	75 58.7	43.0	75 50.3	28.0	30 31.2	32.3	Markab	13 50.3	N15 11.7
20	59 59.6	125 47.0	56.1	90 59.3	42.3	90 52.2	27.8	45 33.5	32.4	Menkar	314 27.3	N 4 04.8
21	75 02.0	140 51.0 ..	56.0	105 59.8 ..	41.7	105 54.1 ..	27.6	60 35.9 ..	32.5	Menkent	148 21.5	S36 21.4
22	90 04.5	155 55.0	55.9	120 00.4	41.0	120 56.1	27.4	75 38.2	32.5	Miaplacidus	221 41.1	S69 42.5
23	105 07.0	170 59.0	55.8	136 01.0	40.3	135 58.0	27.3	90 40.5	32.6			
21 00	120 09.4	186 02.9	S14 55.6	151 01.5	S13 39.6	150 59.9	S13 27.1	105 42.8	N 3 32.7	Mirfak	308 57.0	N49 51.3
01	135 11.9	201 06.9	55.5	166 02.1	38.9	166 01.8	26.9	120 45.2	32.7	Nunki	76 13.3	S26 17.8
02	150 14.4	216 10.8	55.4	181 02.6	38.2	181 03.7	26.7	135 47.5	32.8	Peacock	53 38.4	S56 44.5
03	165 16.8	231 14.8 ..	55.3	196 03.2 ..	37.5	196 05.6 ..	26.5	150 49.8 ..	32.9	Pollux	243 41.8	N28 01.7
04	180 19.3	246 18.8	55.2	211 03.8	36.8	211 07.5	26.3	165 52.1	32.9	Procyon	245 11.7	N 5 13.6
05	195 21.7	261 22.7	55.1	226 04.3	36.1	226 09.4	26.1	180 54.5	33.0			
06	210 24.2	276 26.7	S14 55.0	241 04.9	S13 35.4	241 11.3	S13 25.9	195 56.8	N 3 33.1	Rasalhague	96 17.7	N12 33.7
07	225 26.7	291 30.6	54.9	256 05.5	34.8	256 13.3	25.7	210 59.1	33.2	Regulus	207 55.8	N11 58.5
08	240 29.1	306 34.6	54.8	271 06.0	34.1	271 15.2	25.5	226 01.4	33.2	Rigel	281 23.1	S 8 12.5
09	255 31.6	321 38.5 ..	54.7	286 06.6 ..	33.4	286 17.1 ..	25.3	241 03.8 ..	33.3	Rigil Kent.	140 08.0	S60 49.2
10	270 34.1	336 42.4	54.6	301 07.2	32.7	301 19.0	25.1	256 06.1	33.4	Sabik	102 26.3	S15 43.2
11	285 36.5	351 46.4	54.5	316 07.7	32.0	316 20.9	24.9	271 08.4	33.4			
12	300 39.0	6 50.3	S14 54.4	331 08.3	S13 31.3	331 22.8	S13 24.7	286 10.7	N 3 33.5	Schedar	349 54.1	N56 31.8
13	315 41.5	21 54.3	54.3	346 08.8	30.6	346 24.7	24.5	301 13.1	33.6	Shaula	96 38.3	S37 05.9
14	330 43.9	36 58.2	54.2	1 09.4	29.9	1 26.6	24.3	316 15.4	33.6	Sirius	258 43.8	S16 43.0
15	345 46.4	52 02.1 ..	54.1	16 10.0 ..	29.2	16 28.6 ..	24.1	331 17.7 ..	33.7	Spica	158 43.6	S11 09.0
16	0 48.9	67 06.0	54.0	31 10.5	28.5	31 30.5	23.9	346 20.0	33.8	Suhail	223 00.6	S43 25.5
17	15 51.3	82 10.0	53.9	46 11.1	27.8	46 32.4	23.7	1 22.3	33.9			
18	30 53.8	97 13.9	S14 53.8	61 11.7	S13 27.1	61 34.3	S13 23.5	16 24.7	N 3 33.9	Vega	80 47.3	N38 46.9
19	45 56.2	112 17.8	53.7	76 12.2	26.4	76 36.2	23.3	31 27.0	34.0	Zuben'ubi	137 18.6	S16 01.9
20	60 58.7	127 21.7	53.6	91 12.8	25.8	91 38.1	23.1	46 29.3	34.1		SHA	Mer. Pass.
21	76 01.2	142 25.6 ..	53.5	106 13.4 ..	25.1	106 40.0 ..	23.0	61 31.6 ..	34.1	Venus	° ′ 65 16.9	h m 11 39
22	91 03.6	157 29.5	53.4	121 13.9	24.4	121 41.9	22.8	76 34.0	34.2	Mars	31 37.8	13 56
23	106 06.1	172 33.4	53.4	136 14.5	23.7	136 43.8	22.6	91 36.3	34.3	Jupiter	31 03.7	13 57
Mer. Pass. 16 00.7		v 4.0	d 0.1	v 0.6	d 0.7	v 1.9	d 0.2	v 2.3	d 0.1	Saturn	345 36.7	16 58

UT	SUN GHA	SUN Dec	MOON GHA	v	Dec	d	HP
19 00	177 21.8	S20 25.2	288 59.0	15.5	S 1 20.3	9.4	54.2
01	192 21.7	24.7	303 33.5	15.4	1 29.7	9.4	54.2
02	207 21.5	24.1	318 07.9	15.4	1 39.1	9.4	54.2
03	222 21.3 ..	23.6	332 42.3	15.5	1 48.5	9.4	54.2
04	237 21.1	23.1	347 16.8	15.4	1 57.9	9.4	54.2
05	252 20.9	22.6	1 51.2	15.4	2 07.3	9.3	54.2
06	267 20.7	S20 22.1	16 25.6	15.4	S 2 16.6	9.4	54.2
07	282 20.5	21.5	31 00.0	15.4	2 26.0	9.3	54.2
M 08	297 20.3	21.0	45 34.4	15.3	2 35.3	9.4	54.2
O 09	312 20.1 ..	20.5	60 08.7	15.4	2 44.7	9.3	54.2
N 10	327 19.9	20.0	74 43.1	15.4	2 54.0	9.3	54.2
D 11	342 19.7	19.4	89 17.5	15.3	3 03.3	9.3	54.2
A 12	357 19.5	S20 18.9	103 51.8	15.4	S 3 12.6	9.3	54.2
Y 13	12 19.4	18.4	118 26.2	15.3	3 21.9	9.3	54.2
14	27 19.2	17.9	133 00.5	15.3	3 31.2	9.2	54.2
15	42 19.0 ..	17.3	147 34.8	15.3	3 40.4	9.3	54.2
16	57 18.8	16.8	162 09.1	15.3	3 49.7	9.2	54.2
17	72 18.6	16.3	176 43.4	15.2	3 58.9	9.2	54.3
18	87 18.4	S20 15.7	191 17.6	15.3	S 4 08.1	9.3	54.3
19	102 18.2	15.2	205 51.9	15.2	4 17.4	9.1	54.3
20	117 18.0	14.7	220 26.1	15.2	4 26.5	9.2	54.3
21	132 17.8 ..	14.1	235 00.3	15.2	4 35.7	9.2	54.3
22	147 17.7	13.6	249 34.5	15.2	4 44.9	9.1	54.3
23	162 17.5	13.1	264 08.7	15.2	4 54.0	9.2	54.3
20 00	177 17.3	S20 12.5	278 42.9	15.1	S 5 03.2	9.1	54.3
01	192 17.1	12.0	293 17.0	15.1	5 12.3	9.1	54.3
02	207 16.9	11.5	307 51.1	15.1	5 21.4	9.0	54.3
03	222 16.7 ..	10.9	322 25.2	15.1	5 30.4	9.1	54.3
04	237 16.5	10.4	336 59.3	15.0	5 39.5	9.0	54.3
05	252 16.3	09.9	351 33.3	15.0	5 48.5	9.0	54.3
06	267 16.2	S20 09.3	6 07.3	15.0	S 5 57.5	9.0	54.4
T 07	282 16.0	08.8	20 41.3	15.0	6 06.5	9.0	54.4
U 08	297 15.8	08.2	35 15.3	14.9	6 15.5	8.9	54.4
E 09	312 15.6 ..	07.7	49 49.2	15.0	6 24.4	8.9	54.4
S 10	327 15.4	07.2	64 23.2	14.8	6 33.3	8.9	54.4
D 11	342 15.2	06.6	78 57.0	14.9	6 42.2	8.9	54.4
A 12	357 15.1	S20 06.1	93 30.9	14.8	S 6 51.1	8.9	54.4
Y 13	12 14.9	05.5	108 04.7	14.8	7 00.0	8.8	54.4
14	27 14.7	05.0	122 38.5	14.8	7 08.8	8.8	54.4
15	42 14.5 ..	04.4	137 12.3	14.7	7 17.6	8.8	54.5
16	57 14.3	03.9	151 46.0	14.7	7 26.4	8.7	54.5
17	72 14.2	03.4	166 19.7	14.7	7 35.1	8.7	54.5
18	87 14.0	S20 02.8	180 53.4	14.6	S 7 43.8	8.7	54.5
19	102 13.8	02.3	195 27.0	14.6	7 52.5	8.7	54.5
20	117 13.6	01.7	210 00.6	14.6	8 01.2	8.6	54.5
21	132 13.4 ..	01.2	224 34.2	14.5	8 09.8	8.6	54.5
22	147 13.2	00.6	239 07.7	14.5	8 18.4	8.6	54.6
23	162 13.1	20 00.1	253 41.2	14.4	8 27.0	8.5	54.6
21 00	177 12.9	S19 59.5	268 14.6	14.5	S 8 35.5	8.6	54.6
01	192 12.7	59.0	282 48.1	14.3	8 44.1	8.4	54.6
02	207 12.5	58.4	297 21.4	14.4	8 52.5	8.5	54.6
03	222 12.4 ..	57.9	311 54.8	14.3	9 01.0	8.4	54.6
04	237 12.2	57.3	326 28.1	14.2	9 09.4	8.4	54.6
05	252 12.0	56.8	341 01.3	14.2	9 17.8	8.3	54.7
06	267 11.8	S19 56.2	355 34.5	14.2	S 9 26.1	8.3	54.7
W 07	282 11.6	55.6	10 07.7	14.1	9 34.4	8.3	54.7
E 08	297 11.5	55.1	24 40.8	14.1	9 42.7	8.3	54.7
D 09	312 11.3 ..	54.5	39 13.9	14.0	9 51.0	8.2	54.7
N 10	327 11.1	54.0	53 46.9	14.0	9 59.2	8.1	54.8
E 11	342 10.9	53.4	68 19.9	14.0	10 07.3	8.2	54.8
S 12	357 10.8	S19 52.9	82 52.9	13.9	S10 15.5	8.0	54.8
D 13	12 10.6	52.3	97 25.8	13.8	10 23.5	8.1	54.8
A 14	27 10.4	51.7	111 58.6	13.8	10 31.6	8.0	54.8
Y 15	42 10.2 ..	51.2	126 31.4	13.8	10 39.6	8.0	54.8
16	57 10.1	50.6	141 04.2	13.7	10 47.6	7.9	54.9
17	72 09.9	50.1	155 36.9	13.6	10 55.5	7.9	54.9
18	87 09.7	S19 49.5	170 09.5	13.7	S11 03.4	7.8	54.9
19	102 09.5	48.9	184 42.2	13.5	11 11.2	7.8	54.9
20	117 09.4	48.4	199 14.7	13.5	11 19.0	7.8	55.0
21	132 09.2 ..	47.8	213 47.2	13.5	11 26.8	7.7	55.0
22	147 09.0	47.3	228 19.7	13.4	11 34.5	7.7	55.0
23	162 08.9	46.7	242 52.1	13.3	S11 42.2	7.6	55.0
	SD 16.3	d 0.5	SD 14.8		14.8		14.9

Twilight / Sunrise / Moonrise

Lat.	Naut.	Civil	Sunrise	19	20	21	22
°	h m	h m	h m	h m	h m	h m	h m
N 72	07 48	09 31	■■■	24 31	00 31	02 07	03 48
N 70	07 35	09 03	11 06	24 24	00 24	01 53	03 25
68	07 25	08 41	10 11	24 18	00 18	01 42	03 07
66	07 16	08 24	09 38	24 13	00 13	01 32	02 53
64	07 08	08 10	09 14	24 08	00 08	01 24	02 41
62	07 01	07 59	08 55	24 04	00 04	01 18	02 32
60	06 55	07 48	08 40	24 01	00 01	01 12	02 23
N 58	06 50	07 39	08 26	23 58	25 07	01 07	02 16
56	06 45	07 32	08 15	23 56	25 02	01 02	02 09
54	06 40	07 24	08 05	23 53	24 58	00 58	02 03
52	06 36	07 18	07 56	23 51	24 55	00 55	01 58
50	06 32	07 12	07 48	23 49	24 51	00 51	01 54
45	06 23	06 59	07 32	23 45	24 44	00 44	01 43
N 40	06 15	06 48	07 18	23 42	24 38	00 38	01 35
35	06 08	06 39	07 06	23 39	24 33	00 33	01 28
30	06 01	06 30	06 56	23 36	24 28	00 28	01 22
20	05 47	06 14	06 38	23 32	24 21	00 21	01 11
N 10	05 34	06 00	06 22	23 28	24 14	00 14	01 01
0	05 20	05 45	06 07	23 24	24 08	00 08	00 52
S 10	05 04	05 30	05 52	23 21	24 01	00 01	00 44
20	04 44	05 12	05 36	23 17	23 55	24 34	00 34
30	04 19	04 51	05 18	23 13	23 47	24 24	00 24
35	04 04	04 38	05 07	23 10	23 43	24 18	00 18
40	03 44	04 23	04 54	23 08	23 38	24 11	00 11
45	03 19	04 04	04 39	23 04	23 32	24 03	00 03
S 50	02 45	03 40	04 21	23 01	23 26	23 53	24 25
52	02 26	03 28	04 12	22 59	23 22	23 49	24 19
54	02 03	03 15	04 02	22 57	23 19	23 44	24 13
56	01 30	02 59	03 51	22 55	23 15	23 38	24 06
58	////	02 39	03 38	22 52	23 11	23 32	23 58
S 60	////	02 14	03 23	22 50	23 06	23 26	23 49

Sunset / Twilight / Moonset

Lat.	Sunset	Civil	Naut.	19	20	21	22
°	h m	h m	h m	h m	h m	h m	h m
N 72	■■■	14 52	16 35	10 24	10 18	10 12	10 05
N 70	13 16	15 20	16 48	10 28	10 28	10 28	10 29
68	14 12	15 42	16 58	10 31	10 35	10 41	10 48
66	14 45	15 58	17 07	10 34	10 42	10 51	11 03
64	15 09	16 12	17 15	10 37	10 47	11 00	11 15
62	15 28	16 24	17 21	10 39	10 52	11 07	11 26
60	15 43	16 34	17 27	10 41	10 56	11 14	11 35
N 58	15 56	16 43	17 33	10 42	11 00	11 20	11 43
56	16 07	16 51	17 38	10 44	11 03	11 25	11 50
54	16 17	16 58	17 42	10 45	11 06	11 30	11 56
52	16 26	17 04	17 46	10 46	11 09	11 34	12 02
50	16 34	17 10	17 50	10 47	11 12	11 38	12 07
45	16 51	17 23	17 59	10 50	11 17	11 46	12 18
N 40	17 05	17 34	18 07	10 52	11 22	11 53	12 27
35	17 16	17 44	18 14	10 54	11 25	11 59	12 35
30	17 27	17 52	18 21	10 55	11 29	12 04	12 42
20	17 44	18 08	18 35	10 58	11 35	12 13	12 54
N 10	18 00	18 22	18 48	11 00	11 40	12 21	13 05
0	18 15	18 37	19 02	11 03	11 45	12 29	13 15
S 10	18 29	18 52	19 18	11 05	11 50	12 37	13 25
20	18 45	19 09	19 38	11 07	11 55	12 45	13 35
30	19 04	19 30	20 02	11 10	12 01	12 54	13 48
35	19 15	19 43	20 18	11 11	12 05	12 59	13 55
40	19 27	19 58	20 37	11 13	12 09	13 05	14 03
45	19 42	20 17	21 02	11 15	12 14	13 12	14 12
S 50	20 00	20 41	21 36	11 18	12 19	13 21	14 23
52	20 09	20 52	21 54	11 19	12 22	13 25	14 29
54	20 19	21 06	22 17	11 20	12 24	13 29	14 35
56	20 30	21 22	22 48	11 21	12 28	13 34	14 41
58	20 42	21 41	////	11 23	12 31	13 40	14 48
S 60	20 57	22 05	////	11 24	12 35	13 46	14 57

SUN / MOON

	SUN Eqn. of Time 00h	12h	Mer. Pass.	MOON Mer. Pass. Upper	Lower	Age	Phase
Day	m s	m s	h m	h m	h m	d	%
19	10 32	10 41	12 11	04 52	17 13	21	62
20	10 51	10 59	12 11	05 35	17 56	22	53
21	11 08	11 17	12 11	06 18	18 41	23	44

1998 JANUARY 22, 23, 24 (THURS., FRI., SAT.)

UT	ARIES GHA	VENUS −4.3 GHA	Dec	MARS +1.2 GHA	Dec	JUPITER −2.0 GHA	Dec	SATURN +0.7 GHA	Dec	STARS Name	SHA	Dec
22 00	121 08.6	187 37.4	S14 53.3	151 15.1	S13 23.0	151 45.8	S13 22.4	106 38.6	N 3 34.3	Acamar	315 27.2	S40 19.1
01	136 11.0	202 41.3	53.2	166 15.6	22.3	166 47.7	22.2	121 40.9	34.4	Achernar	335 35.7	S57 15.2
02	151 13.5	217 45.2	53.1	181 16.2	21.6	181 49.6	22.0	136 43.2	34.5	Acrux	173 22.1	S63 05.0
03	166 16.0	232 49.1	.. 53.0	196 16.8	.. 20.9	196 51.5	.. 21.8	151 45.6	.. 34.6	Adhara	255 21.4	S28 58.4
04	181 18.4	247 52.9	52.9	211 17.3	20.2	211 53.4	21.6	166 47.9	34.6	Aldebaran	291 02.7	N16 30.2
05	196 20.9	262 56.8	52.8	226 17.9	19.5	226 55.3	21.4	181 50.2	34.7			
06	211 23.3	278 00.7	S14 52.8	241 18.5	S13 18.8	241 57.2	S13 21.2	196 52.5	N 3 34.8	Alioth	166 30.9	N55 58.0
07	226 25.8	293 04.6	52.7	256 19.0	18.1	256 59.1	21.0	211 54.8	34.8	Alkaid	153 08.2	N49 19.2
T 08	241 28.3	308 08.5	52.6	271 19.6	17.4	272 01.0	20.8	226 57.2	34.9	Al Na'ir	27 59.0	S46 58.4
H 09	256 30.7	323 12.4	.. 52.5	286 20.2	.. 16.7	287 02.9	.. 20.6	241 59.5	.. 35.0	Alnilam	275 58.1	S 1 12.4
U 10	271 33.2	338 16.3	52.4	301 20.8	16.0	302 04.9	20.4	257 01.8	35.1	Alphard	218 07.3	S 8 39.1
R 11	286 35.7	353 20.1	52.3	316 21.3	15.3	317 06.8	20.2	272 04.1	35.1			
S 12	301 38.1	8 24.0	S14 52.3	331 21.9	S13 14.6	332 08.7	S13 20.0	287 06.4	N 3 35.2	Alphecca	126 21.2	N26 43.2
D 13	316 40.6	23 27.9	52.2	346 22.5	13.9	347 10.6	19.8	302 08.8	35.3	Alpheratz	357 55.9	N29 04.8
A 14	331 43.1	38 31.7	52.1	1 23.0	13.2	2 12.5	19.6	317 11.1	35.3	Altair	62 20.1	N 8 51.8
Y 15	346 45.5	53 35.6	.. 52.0	16 23.6	.. 12.5	17 14.4	.. 19.4	332 13.4	.. 35.4	Ankaa	353 27.6	S42 19.3
16	1 48.0	68 39.5	52.0	31 24.2	11.8	32 16.3	19.2	347 15.7	35.5	Antares	112 41.0	S26 25.5
17	16 50.5	83 43.3	51.9	46 24.7	11.1	47 18.2	19.0	2 18.0	35.5			
18	31 52.9	98 47.2	S14 51.8	61 25.3	S13 10.4	62 20.1	S13 18.8	17 20.4	N 3 35.6	Arcturus	146 06.5	N19 11.5
19	46 55.4	113 51.0	51.7	76 25.9	09.7	77 22.0	18.6	32 22.7	35.7	Atria	107 53.8	S69 01.1
20	61 57.8	128 54.9	51.7	91 26.4	09.0	92 24.0	18.4	47 25.0	35.8	Avior	234 22.1	S59 30.3
21	77 00.3	143 58.7	.. 51.6	106 27.0	.. 08.3	107 25.9	.. 18.2	62 27.3	.. 35.8	Bellatrix	278 44.4	N 6 20.7
22	92 02.8	159 02.6	51.5	121 27.6	07.6	122 27.8	18.0	77 29.6	35.9	Betelgeuse	271 13.8	N 7 24.2
23	107 05.2	174 06.4	51.5	136 28.2	06.9	137 29.7	17.8	92 31.9	36.0			
23 00	122 07.7	189 10.2	S14 51.4	151 28.7	S13 06.2	152 31.6	S13 17.7	107 34.3	N 3 36.0	Canopus	264 00.9	S52 42.0
01	137 10.2	204 14.1	51.3	166 29.3	05.5	167 33.5	17.5	122 36.6	36.1	Capella	280 51.5	N45 59.7
02	152 12.6	219 17.9	51.2	181 29.9	04.8	182 35.4	17.3	137 38.9	36.2	Deneb	49 39.9	N45 16.5
03	167 15.1	234 21.7	.. 51.2	196 30.4	.. 04.1	197 37.3	.. 17.1	152 41.2	.. 36.3	Denebola	182 45.5	N14 34.8
04	182 17.6	249 25.5	51.1	211 31.0	03.4	212 39.2	16.9	167 43.5	36.3	Diphda	349 07.8	S18 00.1
05	197 20.0	264 29.4	51.1	226 31.6	02.7	227 41.1	16.7	182 45.8	36.4			
06	212 22.5	279 33.2	S14 51.0	241 32.2	S13 02.0	242 43.0	S13 16.5	197 48.2	N 3 36.5	Dubhe	194 05.7	N61 45.4
07	227 25.0	294 37.0	50.9	256 32.7	01.3	257 45.0	16.3	212 50.5	36.6	Elnath	278 27.2	N28 36.2
08	242 27.4	309 40.8	50.9	271 33.3	13 00.6	272 46.9	16.1	227 52.8	36.6	Eltanin	90 52.1	N51 29.3
F 09	257 29.9	324 44.6	.. 50.8	286 33.9	12 59.9	287 48.8	.. 15.9	242 55.1	.. 36.7	Enif	33 59.0	N 9 52.0
R 10	272 32.3	339 48.4	50.7	301 34.4	59.2	302 50.7	15.7	257 57.4	36.8	Fomalhaut	15 37.3	S29 38.1
I 11	287 34.8	354 52.2	50.7	316 35.0	58.5	317 52.6	15.5	272 59.7	36.8			
D 12	302 37.3	9 56.0	S14 50.6	331 35.6	S12 57.8	332 54.5	S13 15.3	288 02.1	N 3 36.9	Gacrux	172 13.7	S57 05.9
A 13	317 39.7	24 59.8	50.6	346 36.2	57.1	347 56.4	15.1	303 04.4	37.0	Gienah	176 04.3	S17 31.8
Y 14	332 42.2	40 03.6	50.5	1 36.7	56.4	2 58.3	14.9	318 06.7	37.1	Hadar	149 04.6	S60 21.5
15	347 44.7	55 07.4	.. 50.4	16 37.3	.. 55.7	18 00.2	.. 14.7	333 09.0	.. 37.1	Hamal	328 14.1	N23 27.2
16	2 47.1	70 11.2	50.4	31 37.9	55.0	33 02.1	14.5	348 11.3	37.2	Kaus Aust.	83 59.8	S34 23.0
17	17 49.6	85 15.0	50.3	46 38.5	54.3	48 04.0	14.3	3 13.6	37.3			
18	32 52.1	100 18.7	S14 50.3	61 39.0	S12 53.6	63 06.0	S13 14.1	18 16.0	N 3 37.4	Kochab	137 20.0	N74 09.6
19	47 54.5	115 22.5	50.2	76 39.6	52.9	78 07.9	13.9	33 18.3	37.4	Markab	13 50.3	N15 11.7
20	62 57.0	130 26.3	50.2	91 40.2	52.2	93 09.8	13.7	48 20.6	37.5	Menkar	314 27.3	N 4 04.8
21	77 59.5	145 30.1	.. 50.1	106 40.8	.. 51.5	108 11.7	.. 13.5	63 22.9	.. 37.6	Menkent	148 21.5	S36 21.4
22	93 01.9	160 33.8	50.1	121 41.3	50.8	123 13.6	13.3	78 25.2	37.6	Miaplacidus	221 41.1	S69 42.6
23	108 04.4	175 37.6	50.0	136 41.9	50.1	138 15.5	13.1	93 27.5	37.7			
24 00	123 06.8	190 41.3	S14 50.0	151 42.5	S12 49.4	153 17.4	S13 12.9	108 29.8	N 3 37.8	Mirfak	308 57.0	N49 51.3
01	138 09.3	205 45.1	49.9	166 43.1	48.7	168 19.3	12.7	123 32.2	37.9	Nunki	76 13.3	S26 17.8
02	153 11.8	220 48.9	49.9	181 43.6	48.0	183 21.2	12.5	138 34.5	37.9	Peacock	53 38.4	S56 44.5
03	168 14.2	235 52.6	.. 49.8	196 44.2	.. 47.2	198 23.1	.. 12.3	153 36.8	.. 38.0	Pollux	243 41.8	N28 01.7
04	183 16.7	250 56.4	49.8	211 44.8	46.5	213 25.0	12.1	168 39.1	38.1	Procyon	245 11.7	N 5 13.6
05	198 19.2	266 00.1	49.7	226 45.4	45.8	228 26.9	11.9	183 41.4	38.2			
06	213 21.6	281 03.8	S14 49.7	241 45.9	S12 45.1	243 28.8	S13 11.7	198 43.7	N 3 38.2	Rasalhague	96 17.7	N12 33.7
07	228 24.1	296 07.6	49.6	256 46.5	44.4	258 30.8	11.5	213 46.0	38.3	Regulus	207 55.8	N11 58.4
S 08	243 26.6	311 11.3	49.6	271 47.1	43.7	273 32.7	11.3	228 48.4	38.4	Rigel	281 23.1	S 8 12.5
A 09	258 29.0	326 15.0	.. 49.6	286 47.7	.. 43.0	288 34.6	.. 11.1	243 50.7	.. 38.4	Rigil Kent.	140 07.9	S60 49.2
T 10	273 31.5	341 18.8	49.5	301 48.3	42.3	303 36.5	10.9	258 53.0	38.5	Sabik	102 26.3	S15 43.2
U 11	288 33.9	356 22.5	49.5	316 48.8	41.6	318 38.4	10.7	273 55.3	38.6			
R 12	303 36.4	11 26.2	S14 49.4	331 49.4	S12 40.9	333 40.3	S13 10.5	288 57.6	N 3 38.7	Schedar	349 54.1	N56 31.8
D 13	318 38.9	26 29.9	49.4	346 50.0	40.2	348 42.2	10.3	303 59.9	38.7	Shaula	96 38.3	S37 05.9
A 14	333 41.3	41 33.6	49.4	1 50.6	39.5	3 44.1	10.1	319 02.2	38.8	Sirius	258 43.8	S16 43.0
Y 15	348 43.8	56 37.3	.. 49.3	16 51.1	.. 38.8	18 46.0	.. 09.9	334 04.5	.. 38.9	Spica	158 43.6	S11 09.0
16	3 46.3	71 41.0	49.3	31 51.7	38.1	33 47.9	09.7	349 06.9	39.0	Suhail	223 00.6	S43 25.6
17	18 48.7	86 44.7	49.2	46 52.3	37.3	48 49.8	09.5	4 09.2	39.0			
18	33 51.2	101 48.4	S14 49.2	61 52.9	S12 36.6	63 51.7	S13 09.3	19 11.5	N 3 39.1	Vega	80 47.3	N38 46.9
19	48 53.7	116 52.1	49.2	76 53.5	35.9	78 53.6	09.1	34 13.8	39.2	Zuben'ubi	137 18.5	S16 01.9
20	63 56.1	131 55.8	49.1	91 54.0	35.2	93 55.5	08.9	49 16.1	39.3			
21	78 58.6	146 59.5	.. 49.1	106 54.6	.. 34.5	108 57.5	.. 08.7	64 18.4	.. 39.3			
22	94 01.1	162 03.2	49.1	121 55.2	33.8	123 59.4	08.5	79 20.7	39.4			
23	109 03.5	177 06.9	49.0	136 55.8	33.1	139 01.3	08.3	94 23.0	39.5			

	SHA	Mer. Pass.
	° ′	h m
Venus	67 02.5	11 20
Mars	29 21.0	13 54
Jupiter	30 23.9	13 48
Saturn	345 26.6	16 47

Mer. Pass. 15 48.9 | *v* 3.8 *d* 0.1 | *v* 0.6 *d* 0.7 | *v* 1.9 *d* 0.2 | *v* 2.3 *d* 0.1

UT	SUN GHA	SUN Dec	MOON GHA	v	MOON Dec	d	HP
d h	° ′	° ′	° ′	′	° ′	′	′
22 00	177 08.7	S19 46.1	257 24.4	13.3	S11 49.8	7.6	55.0
01	192 08.5	45.6	271 56.7	13.3	11 57.4	7.5	55.1
02	207 08.3	45.0	286 29.0	13.1	12 04.9	7.5	55.1
03	222 08.2	.. 44.4	301 01.1	13.2	12 12.4	7.4	55.1
04	237 08.0	43.9	315 33.3	13.1	12 19.8	7.4	55.1
05	252 07.8	43.3	330 05.4	13.0	12 27.2	7.3	55.2
06	267 07.7	S19 42.7	344 37.4	12.9	S12 34.5	7.3	55.2
T 07	282 07.5	42.1	359 09.3	12.9	12 41.8	7.2	55.2
H 08	297 07.3	41.6	13 41.2	12.9	12 49.0	7.2	55.2
U 09	312 07.2	.. 41.0	28 13.1	12.8	12 56.2	7.1	55.3
R 10	327 07.0	40.4	42 44.9	12.7	13 03.3	7.1	55.3
S 11	342 06.8	39.9	57 16.6	12.7	13 10.4	7.0	55.3
D 12	357 06.7	S19 39.3	71 48.3	12.6	S13 17.4	7.0	55.3
A 13	12 06.5	38.7	86 19.9	12.6	13 24.4	6.9	55.4
Y 14	27 06.3	38.1	100 51.5	12.4	13 31.3	6.8	55.4
15	42 06.2	.. 37.6	115 22.9	12.5	13 38.1	6.8	55.4
16	57 06.0	37.0	129 54.4	12.3	13 44.9	6.8	55.4
17	72 05.8	36.4	144 25.7	12.4	13 51.7	6.7	55.5
18	87 05.7	S19 35.8	158 57.1	12.2	S13 58.4	6.6	55.5
19	102 05.5	35.3	173 28.3	12.2	14 05.0	6.5	55.5
20	117 05.3	34.7	187 59.5	12.1	14 11.5	6.5	55.5
21	132 05.2	.. 34.1	202 30.6	12.1	14 18.0	6.5	55.6
22	147 05.0	33.5	217 01.7	11.9	14 24.5	6.3	55.6
23	162 04.8	32.9	231 32.6	12.0	14 30.8	6.3	55.6
23 00	177 04.7	S19 32.4	246 03.6	11.8	S14 37.1	6.3	55.6
01	192 04.5	31.8	260 34.4	11.8	14 43.4	6.2	55.7
02	207 04.3	31.2	275 05.2	11.8	14 49.6	6.1	55.7
03	222 04.2	.. 30.6	289 36.0	11.6	14 55.7	6.0	55.7
04	237 04.0	30.0	304 06.6	11.7	15 01.7	6.0	55.8
05	252 03.9	29.4	318 37.3	11.5	15 07.7	5.9	55.8
06	267 03.7	S19 28.9	333 07.8	11.5	S15 13.6	5.9	55.8
F 07	282 03.5	28.3	347 38.3	11.4	15 19.5	5.8	55.9
R 08	297 03.4	27.7	2 08.7	11.3	15 25.3	5.7	55.9
I 09	312 03.2	.. 27.1	16 39.0	11.3	15 31.0	5.6	55.9
D 10	327 03.0	26.5	31 09.3	11.2	15 36.6	5.6	55.9
A 11	342 02.9	25.9	45 39.5	11.1	15 42.2	5.5	56.0
Y 12	357 02.7	S19 25.3	60 09.6	11.1	S15 47.7	5.4	56.0
13	12 02.6	24.8	74 39.7	11.0	15 53.1	5.3	56.0
14	27 02.4	24.2	89 09.7	11.0	15 58.4	5.3	56.1
15	42 02.2	.. 23.6	103 39.7	10.8	16 03.7	5.2	56.1
16	57 02.1	23.0	118 09.5	10.8	16 08.9	5.1	56.1
17	72 01.9	22.4	132 39.3	10.8	16 14.0	5.1	56.2
18	87 01.8	S19 21.8	147 09.1	10.6	S16 19.1	4.9	56.2
19	102 01.6	21.2	161 38.7	10.6	16 24.0	4.9	56.2
20	117 01.5	20.6	176 08.3	10.6	16 28.9	4.8	56.3
21	132 01.3	.. 20.0	190 37.9	10.4	16 33.7	4.7	56.3
22	147 01.1	19.4	205 07.3	10.4	16 38.4	4.7	56.3
23	162 01.0	18.8	219 36.7	10.4	16 43.1	4.5	56.4
24 00	177 00.8	S19 18.2	234 06.1	10.2	S16 47.6	4.5	56.4
01	192 00.7	17.6	248 35.3	10.2	16 52.1	4.4	56.4
02	207 00.5	17.0	263 04.5	10.2	16 56.5	4.3	56.5
03	222 00.4	.. 16.4	277 33.7	10.0	17 00.8	4.3	56.5
04	237 00.2	15.8	292 02.7	10.0	17 05.1	4.1	56.5
05	252 00.1	15.2	306 31.7	9.9	17 09.2	4.1	56.6
06	266 59.9	S19 14.6	321 00.6	9.9	S17 13.3	3.9	56.6
S 07	281 59.8	14.0	335 29.5	9.8	17 17.2	3.9	56.6
A 08	296 59.6	13.4	349 58.3	9.7	17 21.1	3.8	56.7
T 09	311 59.4	.. 12.8	4 27.0	9.7	17 24.9	3.7	56.7
U 10	326 59.3	12.2	18 55.7	9.6	17 28.6	3.6	56.7
R 11	341 59.1	11.6	33 24.3	9.5	17 32.2	3.5	56.8
D 12	356 59.0	S19 11.0	47 52.8	9.5	S17 35.7	3.4	56.8
A 13	11 58.8	10.4	62 21.3	9.4	17 39.1	3.4	56.8
Y 14	26 58.7	09.8	76 49.7	9.3	17 42.5	3.2	56.9
15	41 58.5	.. 09.2	91 18.0	9.3	17 45.7	3.2	56.9
16	56 58.4	08.6	105 46.3	9.2	17 48.9	3.0	56.9
17	71 58.2	08.0	120 14.5	9.1	17 51.9	3.0	57.0
18	86 58.1	S19 07.4	134 42.6	9.1	S17 54.9	2.8	57.0
19	101 57.9	06.8	149 10.7	9.0	17 57.7	2.8	57.0
20	116 57.8	06.2	163 38.7	9.0	18 00.5	2.6	57.1
21	131 57.6	.. 05.6	178 06.7	8.8	18 03.1	2.6	57.1
22	146 57.5	05.0	192 34.5	8.9	18 05.7	2.5	57.1
23	161 57.3	04.4	207 02.4	8.7	S18 08.2	2.3	57.2
	SD 16.3	d 0.6	SD 15.1		15.3		15.5

Lat.	Twilight Naut.	Civil	Sunrise	Moonrise 22	23	24	25
°	h m	h m	h m	h m	h m	h m	h m
N 72	07 40	09 19	■	03 48	05 36	07 41	■
N 70	07 28	08 53	10 42	03 25	04 59	06 34	08 01
68	07 18	08 33	09 57	03 07	04 34	05 58	07 14
66	07 10	08 17	09 28	02 53	04 14	05 33	06 44
64	07 03	08 04	09 06	02 41	03 58	05 13	06 21
62	06 57	07 53	08 48	02 32	03 45	04 57	06 03
60	06 51	07 44	08 34	02 23	03 34	04 44	05 48
N 58	06 46	07 35	08 21	02 16	03 25	04 32	05 36
56	06 42	07 28	08 11	02 09	03 16	04 22	05 25
54	06 37	07 21	08 01	02 03	03 09	04 13	05 15
52	06 33	07 15	07 53	01 58	03 02	04 06	05 06
50	06 30	07 09	07 45	01 54	02 56	03 59	04 59
45	06 21	06 57	07 29	01 43	02 43	03 43	04 42
N 40	06 14	06 47	07 16	01 35	02 33	03 31	04 29
35	06 07	06 38	07 05	01 28	02 24	03 21	04 18
30	06 00	06 29	06 55	01 22	02 16	03 11	04 08
20	05 47	06 14	06 38	01 11	02 02	02 56	03 51
N 10	05 35	06 00	06 22	01 01	01 50	02 42	03 36
0	05 21	05 46	06 08	00 52	01 39	02 29	03 22
S 10	05 05	05 31	05 54	00 44	01 28	02 16	03 08
20	04 46	05 15	05 38	00 34	01 17	02 03	02 53
30	04 22	04 54	05 20	00 24	01 03	01 47	02 36
35	04 07	04 42	05 10	00 18	00 56	01 38	02 26
40	03 48	04 27	04 58	00 11	00 47	01 28	02 15
45	03 25	04 09	04 43	00 03	00 37	01 16	02 02
S 50	02 52	03 46	04 26	24 25	00 25	01 02	01 46
52	02 35	03 34	04 17	24 19	00 19	00 55	01 39
54	02 13	03 20	04 08	24 13	00 13	00 48	01 30
56	01 44	03 06	03 58	24 06	00 06	00 39	01 21
58	00 53	02 48	03 45	23 58	24 30	00 30	01 11
S 60	////	02 25	03 31	23 49	24 19	00 19	00 58

Lat.	Sunset	Twilight Civil	Naut.	Moonset 22	23	24	25
°	h m	h m	h m	h m	h m	h m	h m
N 72	■	15 06	16 45	10 05	09 55	09 35	
N 70	13 43	15 32	16 57	10 29	10 33	10 42	11 07
68	14 27	15 51	17 06	10 48	10 59	11 19	11 53
66	14 57	16 07	17 14	11 03	11 20	11 45	12 24
64	15 18	16 20	17 21	11 15	11 36	12 05	12 46
62	15 36	16 31	17 28	11 26	11 49	12 21	13 05
60	15 50	16 41	17 33	11 35	12 01	12 35	13 20
N 58	16 03	16 49	17 43	11 43	12 11	12 47	13 32
56	16 13	16 56	17 43	11 50	12 20	12 57	13 43
54	16 23	17 03	17 47	11 56	12 27	13 06	13 53
52	16 31	17 09	17 51	12 02	12 34	13 14	14 02
50	16 39	17 15	17 54	12 07	12 41	13 21	14 10
45	16 55	17 27	18 03	12 18	12 54	13 37	14 26
N 40	17 08	17 37	18 10	12 27	13 06	13 50	14 40
35	17 19	17 46	18 17	12 35	13 15	14 00	14 51
30	17 29	17 55	18 24	12 42	13 24	14 10	15 01
20	17 46	18 10	18 36	12 54	13 38	14 26	15 19
N 10	18 01	18 23	18 49	13 05	13 51	14 41	15 34
0	18 15	18 37	19 03	13 15	14 03	14 54	15 48
S 10	18 30	18 52	19 18	13 25	14 15	15 07	16 02
20	18 45	19 09	19 37	13 35	14 28	15 21	16 17
30	19 03	19 29	20 01	13 48	14 42	15 38	16 34
35	19 13	19 42	20 16	13 55	14 51	15 48	16 44
40	19 25	19 56	20 34	14 03	15 01	15 59	16 56
45	19 39	20 14	20 58	14 12	15 12	16 12	17 09
S 50	19 57	20 37	21 30	14 23	15 26	16 27	17 26
52	20 05	20 48	21 47	14 29	15 32	16 35	17 33
54	20 14	21 01	22 08	14 35	15 40	16 43	17 42
56	20 25	21 16	22 36	14 41	15 48	16 52	17 51
58	20 37	21 33	23 22	14 48	15 57	17 02	18 02
S 60	20 51	21 56	////	14 57	16 07	17 14	18 15

	SUN Eqn. of Time 00ʰ	12ʰ	Mer. Pass.	MOON Mer. Pass. Upper	Lower	Age	Phase
Day	m s	m s	h m	h m	h m	d %	
22	11 25	11 33	12 12	07 03	19 27	24 34	
23	11 41	11 49	12 12	07 51	20 16	25 25	
24	11 56	12 04	12 12	08 42	21 08	26 17	

1998 JANUARY 25, 26, 27 (SUN., MON., TUES.)

UT (d h)	ARIES GHA	VENUS −4.4 GHA	Dec	MARS +1.2 GHA	Dec	JUPITER −2.0 GHA	Dec	SATURN +0.7 GHA	Dec	STARS Name	SHA	Dec
25 00	124 06.0	192 10.5	S14 49.0	151 56.3	S12 32.4	154 03.2	S13 08.1	109 25.3	N 3 39.6	Acamar	315 27.2	S40 19.1
01	139 08.4	207 14.2	49.0	166 56.9	31.7	169 05.1	07.9	124 27.7	39.6	Achernar	335 35.7	S57 15.2
02	154 10.9	222 17.9	49.0	181 57.5	31.0	184 07.0	07.7	139 30.0	39.7	Acrux	173 22.0	S63 05.0
03	169 13.4	237 21.5	.. 48.9	196 58.1	.. 30.2	199 08.9	.. 07.6	154 32.3	.. 39.8	Adhara	255 21.4	S28 58.4
04	184 15.8	252 25.2	48.9	211 58.7	29.5	214 10.8	07.4	169 34.6	39.9	Aldebaran	291 02.7	N16 30.2
05	199 18.3	267 28.9	48.9	226 59.2	28.8	229 12.7	07.2	184 36.9	39.9			
06	214 20.8	282 32.5	S14 48.8	241 59.8	S12 28.1	244 14.6	S13 07.0	199 39.2	N 3 40.0	Alioth	166 30.9	N55 58.0
07	229 23.2	297 36.2	48.8	257 00.4	27.4	259 16.5	06.8	214 41.5	40.1	Alkaid	153 08.2	N49 19.2
08	244 25.7	312 39.8	48.8	272 01.0	26.7	274 18.4	06.6	229 43.8	40.2	Al Na'ir	27 59.0	S46 58.3
S 09	259 28.2	327 43.5	.. 48.8	287 01.6	.. 26.0	289 20.3	.. 06.4	244 46.1	.. 40.2	Alnilam	275 58.1	S 1 12.4
U 10	274 30.6	342 47.1	48.7	302 02.2	25.3	304 22.2	06.2	259 48.4	40.3	Alphard	218 07.3	S 8 39.2
N 11	289 33.1	357 50.7	48.7	317 02.7	24.5	319 24.1	06.0	274 50.8	40.4			
D 12	304 35.6	12 54.4	S14 48.7	332 03.3	S12 23.8	334 26.0	S13 05.8	289 53.1	N 3 40.5	Alphecca	126 21.1	N26 43.2
A 13	319 38.0	27 58.0	48.7	347 03.9	23.1	349 28.0	05.6	304 55.4	40.5	Alpheratz	357 55.9	N29 04.8
Y 14	334 40.5	43 01.6	48.7	2 04.5	22.4	4 29.9	05.4	319 57.7	40.6	Altair	62 20.0	N 8 51.8
15	349 42.9	58 05.2	.. 48.6	17 05.1	.. 21.7	19 31.8	.. 05.2	335 00.0	.. 40.7	Ankaa	353 27.6	S42 19.3
16	4 45.4	73 08.9	48.6	32 05.6	21.0	34 33.7	05.0	350 02.3	40.8	Antares	112 40.9	S26 25.5
17	19 47.9	88 12.5	48.6	47 06.2	20.3	49 35.6	04.8	5 04.6	40.8			
18	34 50.3	103 16.1	S14 48.6	62 06.8	S12 19.6	64 37.5	S13 04.6	20 06.9	N 3 40.9	Arcturus	146 06.5	N19 11.5
19	49 52.8	118 19.7	48.6	77 07.4	18.8	79 39.4	04.4	35 09.2	41.0	Atria	107 53.7	S69 01.1
20	64 55.3	133 23.3	48.6	92 08.0	18.1	94 41.3	04.2	50 11.5	41.1	Avior	234 22.1	S59 30.3
21	79 57.7	148 26.9	.. 48.5	107 08.6	.. 17.4	109 43.2	.. 04.0	65 13.8	.. 41.1	Bellatrix	278 44.4	N 6 20.7
22	95 00.2	163 30.5	48.5	122 09.1	16.7	124 45.1	03.8	80 16.1	41.2	Betelgeuse	271 13.8	N 7 24.2
23	110 02.7	178 34.1	48.5	137 09.7	16.0	139 47.0	03.6	95 18.4	41.3			
26 00	125 05.1	193 37.7	S14 48.5	152 10.3	S12 15.3	154 48.9	S13 03.4	110 20.8	N 3 41.4	Canopus	264 00.9	S52 42.0
01	140 07.6	208 41.2	48.5	167 10.9	14.6	169 50.8	03.2	125 23.1	41.4	Capella	280 51.5	N45 59.7
02	155 10.1	223 44.8	48.5	182 11.5	13.8	184 52.7	03.0	140 25.4	41.5	Deneb	49 39.9	N45 16.5
03	170 12.5	238 48.4	.. 48.5	197 12.1	.. 13.1	199 54.6	.. 02.8	155 27.7	.. 41.6	Denebola	182 45.5	N14 34.8
04	185 15.0	253 52.0	48.5	212 12.6	12.4	214 56.5	02.6	170 30.0	41.7	Diphda	349 07.9	S18 00.1
05	200 17.4	268 55.5	48.4	227 13.2	11.7	229 58.4	02.4	185 32.3	41.7			
06	215 19.9	283 59.1	S14 48.4	242 13.8	S12 11.0	245 00.3	S13 02.2	200 34.6	N 3 41.8	Dubhe	194 05.7	N61 45.4
07	230 22.4	299 02.7	48.4	257 14.4	10.3	260 02.3	02.0	215 36.9	41.9	Elnath	278 27.2	N28 36.2
08	245 24.8	314 06.2	48.4	272 15.0	09.5	275 04.2	01.8	230 39.2	42.0	Eltanin	90 52.1	N51 29.3
M 09	260 27.3	329 09.8	.. 48.4	287 15.6	.. 08.8	290 06.1	.. 01.6	245 41.5	.. 42.0	Enif	33 59.0	N 9 52.0
O 10	275 29.8	344 13.3	48.4	302 16.2	08.1	305 08.0	01.4	260 43.8	42.1	Fomalhaut	15 37.3	S29 38.1
N 11	290 32.2	359 16.9	48.4	317 16.7	07.4	320 09.9	01.2	275 46.1	42.2			
D 12	305 34.7	14 20.4	S14 48.4	332 17.3	S12 06.7	335 11.8	S13 01.0	290 48.4	N 3 42.3	Gacrux	172 13.7	S57 05.9
A 13	320 37.2	29 23.9	48.4	347 17.9	06.0	350 13.7	00.8	305 50.7	42.4	Gienah	176 04.2	S17 31.8
Y 14	335 39.6	44 27.5	48.4	2 18.5	05.2	5 15.6	00.6	320 53.0	42.4	Hadar	149 04.5	S60 21.5
15	350 42.1	59 31.0	.. 48.4	17 19.1	.. 04.5	20 17.5	.. 00.4	335 55.3	.. 42.5	Hamal	328 14.1	N23 27.2
16	5 44.5	74 34.5	48.4	32 19.7	03.8	35 19.4	00.2	350 57.7	42.6	Kaus Aust.	83 59.8	S34 23.0
17	20 47.0	89 38.0	48.4	47 20.3	03.1	50 21.3	13 00.0	6 00.0	42.7			
18	35 49.5	104 41.6	S14 48.4	62 20.8	S12 02.4	65 23.2	S12 59.8	21 02.3	N 3 42.7	Kochab	137 19.9	N74 09.6
19	50 51.9	119 45.1	48.4	77 21.4	01.6	80 25.1	59.6	36 04.6	42.8	Markab	13 50.3	N15 11.7
20	65 54.4	134 48.6	48.4	92 22.0	00.9	95 27.0	59.4	51 06.9	42.9	Menkar	314 27.3	N 4 04.7
21	80 56.9	149 52.1	.. 48.4	107 22.6	12 00.2	110 28.9	.. 59.2	66 09.2	.. 43.0	Menkent	148 21.4	S36 21.4
22	95 59.3	164 55.6	48.4	122 23.2	11 59.5	125 30.8	59.0	81 11.5	43.0	Miaplacidus	221 41.1	S69 42.6
23	111 01.8	179 59.1	48.4	137 23.8	58.8	140 32.7	58.8	96 13.8	43.1			
27 00	126 04.3	195 02.6	S14 48.4	152 24.4	S11 58.0	155 34.6	S12 58.6	111 16.1	N 3 43.2	Mirfak	308 57.1	N49 51.3
01	141 06.7	210 06.1	48.4	167 25.0	57.3	170 36.5	58.4	126 18.4	43.3	Nunki	76 13.3	S26 17.8
02	156 09.2	225 09.5	48.4	182 25.5	56.6	185 38.4	58.2	141 20.7	43.4	Peacock	53 38.4	S56 44.4
03	171 11.7	240 13.0	.. 48.4	197 26.1	.. 55.9	200 40.3	.. 57.9	156 23.0	.. 43.4	Pollux	243 41.8	N28 01.7
04	186 14.1	255 16.5	48.4	212 26.7	55.2	215 42.2	57.7	171 25.3	43.5	Procyon	245 11.7	N 5 13.6
05	201 16.6	270 20.0	48.4	227 27.3	54.4	230 44.1	57.5	186 27.6	43.6			
06	216 19.0	285 23.4	S14 48.5	242 27.9	S11 53.7	245 46.1	S12 57.3	201 29.9	N 3 43.7	Rasalhague	96 17.6	N12 33.7
07	231 21.5	300 26.9	48.5	257 28.5	53.0	260 48.0	57.1	216 32.2	43.7	Regulus	207 55.7	N11 58.4
08	246 24.0	315 30.4	48.5	272 29.1	52.3	275 49.9	56.9	231 34.5	43.8	Rigel	281 23.1	S 8 12.5
T 09	261 26.4	330 33.8	.. 48.5	287 29.7	.. 51.6	290 51.8	.. 56.7	246 36.8	.. 43.9	Rigil Kent.	140 07.9	S60 49.2
U 10	276 28.9	345 37.3	48.5	302 30.3	50.8	305 53.7	56.5	261 39.1	44.0	Sabik	102 26.3	S15 43.2
E 11	291 31.4	0 40.7	48.5	317 30.8	50.1	320 55.6	56.3	276 41.4	44.1			
S 12	306 33.8	15 44.2	S14 48.5	332 31.4	S11 49.4	335 57.5	S12 56.1	291 43.7	N 3 44.1	Schedar	349 54.1	N56 31.8
D 13	321 36.3	30 47.6	48.5	347 32.0	48.7	350 59.4	55.9	306 46.0	44.2	Shaula	96 38.2	S37 05.9
A 14	336 38.8	45 51.0	48.5	2 32.6	47.9	6 01.3	55.7	321 48.3	44.3	Sirius	258 43.6	S16 43.0
Y 15	351 41.2	60 54.5	.. 48.6	17 33.2	.. 47.2	21 03.2	.. 55.5	336 50.6	.. 44.4	Spica	158 43.6	S11 09.0
16	6 43.7	75 57.9	48.6	32 33.8	46.5	36 05.1	55.3	351 52.9	44.4	Suhail	223 00.6	S43 25.6
17	21 46.2	91 01.3	48.6	47 34.4	45.8	51 07.0	55.1	6 55.2	44.5			
18	36 48.6	106 04.7	S14 48.6	62 35.0	S11 45.1	66 08.9	S12 54.9	21 57.5	N 3 44.6	Vega	80 47.3	N38 46.9
19	51 51.1	121 08.2	48.6	77 35.6	44.3	81 10.8	54.7	36 59.8	44.7	Zuben'ubi	137 18.5	S16 01.9
20	66 53.5	136 11.6	48.6	92 36.2	43.6	96 12.7	54.5	52 02.1	44.8			
21	81 56.0	151 15.0	.. 48.7	107 36.8	.. 42.9	111 14.6	.. 54.3	67 04.4	.. 44.8			
22	96 58.5	166 18.4	48.7	122 37.3	42.2	126 16.5	54.1	82 06.7	44.9			
23	112 00.9	181 21.8	48.7	137 37.9	41.4	141 18.4	53.9	97 09.0	45.0			
Mer. Pass.	h m 15 37.1	v 3.5	d 0.0	v 0.6	d 0.7	v 1.9	d 0.2	v 2.3	d 0.1			

	SHA	Mer. Pass.
	° ′	h m
Venus	68 32.5	11 03
Mars	27 05.2	13 51
Jupiter	29 43.8	13 39
Saturn	345 15.6	16 36

UT	SUN GHA	SUN Dec	MOON GHA	v	MOON Dec	d	HP
d h	° '	° '	° '	'	° '	'	'
25 00	176 57.2	S19 03.8	221 30.1	8.7	S18 10.5	2.3	57.2
01	191 57.1	03.1	235 57.8	8.7	18 12.8	2.2	57.2
02	206 56.9	02.5	250 25.5	8.6	18 15.0	2.0	57.3
03	221 56.8	.. 01.9	264 53.1	8.5	18 17.0	2.0	57.3
04	236 56.6	01.3	279 20.6	8.5	18 19.0	1.8	57.3
05	251 56.5	00.7	293 48.1	8.4	18 20.8	1.8	57.4
06	266 56.3	S19 00.1	308 15.5	8.3	S18 22.6	1.6	57.4
07	281 56.2	18 59.5	322 42.8	8.3	18 24.2	1.6	57.4
08	296 56.0	58.8	337 10.1	8.3	18 25.8	1.4	57.5
S 09	311 55.9	.. 58.2	351 37.4	8.2	18 27.2	1.3	57.5
U 10	326 55.7	57.6	6 04.6	8.1	18 28.5	1.3	57.6
N 11	341 55.6	57.0	20 31.7	8.1	18 29.8	1.1	57.6
D 12	356 55.5	S18 56.4	34 58.8	8.0	S18 30.9	1.0	57.6
A 13	11 55.3	55.8	49 25.8	8.0	18 31.9	0.9	57.7
Y 14	26 55.2	55.1	63 52.8	7.9	18 32.8	0.8	57.7
15	41 55.0	.. 54.5	78 19.7	7.8	18 33.6	0.6	57.7
16	56 54.9	53.9	92 46.5	7.9	18 34.2	0.6	57.8
17	71 54.7	53.3	107 13.4	7.7	18 34.8	0.5	57.8
18	86 54.6	S18 52.7	121 40.1	7.8	S18 35.3	0.3	57.8
19	101 54.5	52.0	136 06.9	7.6	18 35.6	0.2	57.9
20	116 54.3	51.4	150 33.5	7.7	18 35.8	0.2	57.9
21	131 54.2	.. 50.8	165 00.2	7.5	18 36.0	0.0	57.9
22	146 54.0	50.2	179 26.7	7.6	18 36.0	0.1	58.0
23	161 53.9	49.5	193 53.3	7.5	18 35.9	0.3	58.0
26 00	176 53.8	S18 48.9	208 19.8	7.4	S18 35.6	0.3	58.0
01	191 53.6	48.3	222 46.2	7.4	18 35.3	0.4	58.1
02	206 53.5	47.7	237 12.6	7.4	18 34.9	0.6	58.1
03	221 53.3	.. 47.0	251 39.0	7.3	18 34.3	0.7	58.1
04	236 53.2	46.4	266 05.3	7.3	18 33.6	0.8	58.2
05	251 53.1	45.8	280 31.6	7.2	18 32.8	0.9	58.2
06	266 52.9	S18 45.2	294 57.8	7.2	S18 31.9	1.0	58.2
07	281 52.8	44.5	309 24.0	7.2	18 30.9	1.1	58.3
08	296 52.7	43.9	323 50.2	7.2	18 29.8	1.3	58.3
M 09	311 52.5	.. 43.3	338 16.4	7.1	18 28.5	1.4	58.3
O 10	326 52.4	42.6	352 42.5	7.0	18 27.1	1.5	58.4
N 11	341 52.2	42.0	7 08.5	7.1	18 25.6	1.6	58.4
D 12	356 52.1	S18 41.4	21 34.6	7.0	S18 24.0	1.7	58.4
A 13	11 52.0	40.7	36 00.6	6.9	18 22.3	1.8	58.5
Y 14	26 51.8	40.1	50 26.5	7.0	18 20.5	2.0	58.5
15	41 51.7	.. 39.5	64 52.5	6.9	18 18.5	2.1	58.5
16	56 51.6	38.8	79 18.4	6.9	18 16.4	2.2	58.6
17	71 51.4	38.2	93 44.3	6.8	18 14.2	2.3	58.6
18	86 51.3	S18 37.6	108 10.1	6.9	S18 11.9	2.4	58.6
19	101 51.2	36.9	122 36.0	6.8	18 09.5	2.6	58.7
20	116 51.0	36.3	137 01.8	6.8	18 06.9	2.7	58.7
21	131 50.9	.. 35.7	151 27.6	6.8	18 04.2	2.7	58.7
22	146 50.8	35.0	165 53.4	6.7	18 01.5	2.9	58.8
23	161 50.6	34.4	180 19.1	6.7	17 58.6	3.1	58.8
27 00	176 50.5	S18 33.7	194 44.8	6.7	S17 55.5	3.1	58.8
01	191 50.4	33.1	209 10.5	6.7	17 52.4	3.3	58.9
02	206 50.2	32.5	223 36.2	6.7	17 49.1	3.4	58.9
03	221 50.1	.. 31.8	238 01.9	6.7	17 45.7	3.5	58.9
04	236 50.0	31.2	252 27.6	6.6	17 42.2	3.6	59.0
05	251 49.9	30.5	266 53.2	6.6	17 38.6	3.7	59.0
06	266 49.7	S18 29.9	281 18.8	6.6	S17 34.9	3.8	59.0
07	281 49.6	29.3	295 44.4	6.7	17 31.1	4.0	59.0
08	296 49.5	28.6	310 10.1	6.5	17 27.1	4.1	59.1
T 09	311 49.3	.. 28.0	324 35.6	6.6	17 23.0	4.2	59.1
U 10	326 49.2	27.3	339 01.2	6.6	17 18.8	4.3	59.1
E 11	341 49.1	26.7	353 26.8	6.6	17 14.5	4.4	59.2
S 12	356 49.0	S18 26.0	7 52.4	6.6	S17 10.1	4.6	59.2
D 13	11 48.8	25.4	22 18.0	6.5	17 05.5	4.7	59.2
A 14	26 48.7	24.7	36 43.5	6.6	17 00.8	4.7	59.2
Y 15	41 48.6	.. 24.1	51 09.1	6.5	16 56.1	4.9	59.3
16	56 48.5	23.4	65 34.6	6.6	16 51.2	5.0	59.3
17	71 48.3	22.8	80 00.2	6.5	16 46.2	5.2	59.3
18	86 48.2	S18 22.1	94 25.7	6.6	S16 41.0	5.2	59.3
19	101 48.1	21.5	108 51.3	6.5	16 35.8	5.3	59.4
20	116 48.0	20.8	123 16.8	6.6	16 30.5	5.5	59.4
21	131 47.8	.. 20.2	137 42.4	6.5	16 25.0	5.6	59.4
22	146 47.7	19.5	152 07.9	6.6	16 19.4	5.6	59.4
23	161 47.6	18.9	166 33.5	6.6	S16 13.8	5.8	59.5
	SD 16.3	d 0.6	SD 15.7		15.9		16.1

Twilight / Moonrise

Lat.	Naut.	Civil	Sunrise	Moonrise 25	26	27	28
°	h m	h m	h m	h m	h m	h m	h m
N 72	07 30	09 06	11 43	▬	▬	10 25	10 07
N 70	07 20	08 42	10 22	08 01	08 58	09 23	09 33
68	07 11	08 24	09 44	07 14	08 11	08 47	09 08
66	07 04	08 10	09 18	06 44	07 41	08 22	08 49
64	06 57	07 58	08 58	06 21	07 18	08 02	08 33
62	06 52	07 47	08 41	06 03	07 00	07 46	08 20
60	06 47	07 39	08 28	05 48	06 45	07 32	08 09
N 58	06 42	07 31	08 16	05 36	06 32	07 20	08 00
56	06 38	07 24	08 06	05 25	06 21	07 10	07 51
54	06 34	07 17	07 57	05 15	06 12	07 01	07 44
52	06 30	07 12	07 49	05 06	06 03	06 53	07 37
50	06 27	07 06	07 42	04 59	05 55	06 46	07 31
45	06 19	06 55	07 27	04 42	05 39	06 31	07 18
N 40	06 12	06 45	07 14	04 29	05 25	06 18	07 07
35	06 06	06 36	07 03	04 18	05 13	06 07	06 57
30	05 59	06 28	06 54	04 08	05 03	05 58	06 49
20	05 47	06 14	06 37	03 51	04 46	05 41	06 35
N 10	05 35	06 00	06 23	03 36	04 31	05 27	06 23
0	05 22	05 47	06 09	03 22	04 17	05 14	06 11
S 10	05 07	05 33	05 55	03 08	04 03	05 00	05 59
20	04 49	05 17	05 40	02 53	03 48	04 46	05 47
30	04 26	04 57	05 23	02 36	03 31	04 30	05 33
35	04 11	04 45	05 13	02 26	03 20	04 20	05 24
40	03 53	04 31	05 01	02 15	03 09	04 09	05 15
45	03 30	04 14	04 48	02 02	02 56	03 56	05 04
S 50	02 59	03 52	04 31	01 46	02 39	03 41	04 50
52	02 43	03 41	04 23	01 39	02 31	03 34	04 44
54	02 23	03 29	04 14	01 30	02 23	03 26	04 37
56	01 57	03 14	04 04	01 21	02 13	03 16	04 29
58	01 18	02 57	03 53	01 11	02 02	03 06	04 21
S 60	////	02 36	03 39	00 58	01 50	02 54	04 11

Sunset / Twilight / Moonset

Lat.	Sunset	Civil	Naut.	Moonset 25	26	27	28
°	h m	h m	h m	h m	h m	h m	h m
N 72	12 43	15 21	16 56	▬	▬	12 37	14 56
N 70	14 04	15 44	17 06	11 07	12 05	13 39	15 29
68	14 42	16 02	17 15	11 53	12 52	14 14	15 53
66	15 08	16 16	17 22	12 24	13 22	14 40	16 11
64	15 28	16 28	17 29	12 46	13 44	14 59	16 26
62	15 44	16 38	17 34	13 05	14 02	15 15	16 38
60	15 58	16 47	17 39	13 20	14 17	15 28	16 49
N 58	16 10	16 55	17 44	13 32	14 30	15 39	16 58
56	16 20	17 02	17 48	13 43	14 41	15 49	17 06
54	16 29	17 08	17 52	13 53	14 51	15 58	17 13
52	16 37	17 14	17 55	14 02	14 59	16 06	17 19
50	16 44	17 19	17 59	14 10	15 07	16 13	17 25
45	16 59	17 31	18 06	14 26	15 23	16 27	17 37
N 40	17 12	17 41	18 13	14 40	15 37	16 40	17 47
35	17 22	17 49	18 20	14 51	15 48	16 50	17 56
30	17 32	17 57	18 27	15 01	15 58	16 59	18 03
20	17 48	18 11	18 38	15 19	16 15	17 15	18 16
N 10	18 03	18 25	18 50	15 34	16 30	17 28	18 28
0	18 16	18 38	19 03	15 48	16 44	17 41	18 38
S 10	18 30	18 52	19 18	16 02	16 58	17 54	18 49
20	18 45	19 08	19 36	16 17	17 12	18 07	19 00
30	19 02	19 28	19 59	16 34	17 29	18 22	19 12
35	19 12	19 40	20 14	16 44	17 39	18 31	19 19
40	19 23	19 54	20 31	16 56	17 50	18 41	19 28
45	19 37	20 11	20 54	17 09	18 03	18 53	19 37
S 50	19 53	20 32	21 24	17 26	18 19	19 07	19 49
52	20 01	20 43	21 40	17 33	18 27	19 14	19 54
54	20 10	20 55	21 59	17 42	18 35	19 21	20 00
56	20 20	21 09	22 24	17 51	18 44	19 29	20 06
58	20 31	21 26	23 01	18 02	18 55	19 38	20 14
S 60	20 44	21 46	////	18 15	19 07	19 49	20 22

SUN / MOON

Day	Eqn. of Time 00ʰ	12ʰ	Mer. Pass.	Mer. Pass. Upper	Lower	Age	Phase
d	m s	m s	h m	h m	h m	d	%
25	12 11	12 18	12 12	09 35	22 02	27	10	●
26	12 25	12 31	12 13	10 30	22 59	28	4	
27	12 38	12 44	12 13	11 27	23 56	29	1	

1998 JANUARY 28, 29, 30 (WED., THURS., FRI.)

UT	ARIES GHA	VENUS −4.4 GHA	Dec	MARS +1.2 GHA	Dec	JUPITER −2.0 GHA	Dec	SATURN +0.7 GHA	Dec	STARS Name	SHA	Dec
28 00	127 03.4	196 25.2	S14 48.7	152 38.5	S11 40.7	156 20.3	S12 53.7	112 11.3	N 3 45.1	Acamar	315 27.2	S40 19.1
01	142 05.9	211 28.6	48.7	167 39.1	40.0	171 22.2	53.5	127 13.6	45.1	Achernar	335 35.7	S57 15.2
02	157 08.3	226 31.9	48.8	182 39.7	39.3	186 24.1	53.3	142 15.9	45.2	Acrux	173 22.0	S63 05.1
03	172 10.8	241 35.3 ..	48.8	197 40.3 ..	38.5	201 26.0 ..	53.1	157 18.2 ..	45.3	Adhara	255 21.4	S28 58.4
04	187 13.3	256 38.7	48.8	212 40.9	37.8	216 27.9	52.9	172 20.5	45.4	Aldebaran	291 02.7	N16 30.2
05	202 15.7	271 42.1	48.8	227 41.5	37.1	231 29.8	52.7	187 22.8	45.5			
06	217 18.2	286 45.4	S14 48.9	242 42.1	S11 36.4	246 31.7	S12 52.5	202 25.1	N 3 45.5	Alioth	166 30.9	N55 58.0
W 07	232 20.7	301 48.8	48.9	257 42.7	35.6	261 33.6	52.3	217 27.4	45.6	Alkaid	153 08.1	N49 19.2
E 08	247 23.1	316 52.2	48.9	272 43.3	34.9	276 35.5	52.1	232 29.7	45.7	Al Na'ir	27 59.0	S46 58.3
D 09	262 25.6	331 55.5 ..	48.9	287 43.9 ..	34.2	291 37.4 ..	51.9	247 32.0 ..	45.8	Alnilam	275 58.1	S 1 12.4
N 10	277 28.0	346 58.9	49.0	302 44.5	33.5	306 39.3	51.7	262 34.3	45.9	Alphard	218 07.3	S 8 39.2
E 11	292 30.5	2 02.2	49.0	317 45.1	32.7	321 41.2	51.5	277 36.6	45.9			
S 12	307 33.0	17 05.6	S14 49.0	332 45.6	S11 32.0	336 43.1	S12 51.3	292 38.9	N 3 46.0	Alphecca	126 21.1	N26 43.2
D 13	322 35.4	32 08.9	49.1	347 46.2	31.3	351 45.0	51.1	307 41.2	46.1	Alpheratz	357 55.9	N29 04.8
A 14	337 37.9	47 12.2	49.1	2 46.8	30.5	6 46.9	50.9	322 43.5	46.2	Altair	62 20.0	N 8 51.8
Y 15	352 40.4	62 15.6 ..	49.1	17 47.4 ..	29.8	21 48.8 ..	50.7	337 45.8 ..	46.2	Ankaa	353 27.6	S42 19.3
16	7 42.8	77 18.9	49.1	32 48.0	29.1	36 50.8	50.5	352 48.1	46.3	Antares	112 40.9	S26 25.5
17	22 45.3	92 22.2	49.2	47 48.6	28.4	51 52.7	50.3	7 50.4	46.4			
18	37 47.8	107 25.5	S14 49.2	62 49.2	S11 27.6	66 54.6	S12 50.1	22 52.7	N 3 46.5	Arcturus	146 06.5	N19 11.5
19	52 50.2	122 28.9	49.2	77 49.8	26.9	81 56.5	49.9	37 55.0	46.6	Atria	107 53.7	S69 01.1
20	67 52.7	137 32.2	49.3	92 50.4	26.2	96 58.4	49.7	52 57.3	46.6	Avior	234 22.1	S59 30.4
21	82 55.2	152 35.5 ..	49.3	107 51.0 ..	25.4	112 00.3 ..	49.5	67 59.6 ..	46.7	Bellatrix	278 44.4	N 6 20.7
22	97 57.6	167 38.8	49.3	122 51.6	24.7	127 02.2	49.3	83 01.9	46.8	Betelgeuse	271 13.8	N 7 24.2
23	113 00.1	182 42.1	49.4	137 52.2	24.0	142 04.1	49.1	98 04.2	46.9			
29 00	128 02.5	197 45.4	S14 49.4	152 52.8	S11 23.3	157 06.0	S12 48.9	113 06.5	N 3 47.0	Canopus	264 00.9	S52 42.0
01	143 05.0	212 48.6	49.5	167 53.4	22.5	172 07.9	48.7	128 08.8	47.0	Capella	280 51.5	N45 59.7
02	158 07.5	227 51.9	49.5	182 54.0	21.8	187 09.8	48.5	143 11.1	47.1	Deneb	49 39.9	N45 16.5
03	173 09.9	242 55.2 ..	49.5	197 54.6 ..	21.1	202 11.7 ..	48.3	158 13.4 ..	47.2	Denebola	182 45.5	N14 34.8
04	188 12.4	257 58.5	49.6	212 55.2	20.3	217 13.6	48.1	173 15.7	47.3	Diphda	349 07.9	S18 00.1
05	203 14.9	273 01.8	49.6	227 55.8	19.6	232 15.5	47.9	188 18.0	47.4			
06	218 17.3	288 05.0	S14 49.7	242 56.4	S11 18.9	247 17.4	S12 47.7	203 20.3	N 3 47.4	Dubhe	194 05.6	N61 45.4
07	233 19.8	303 08.3	49.7	257 57.0	18.2	262 19.3	47.4	218 22.6	47.5	Elnath	278 27.2	N28 36.2
T 08	248 22.3	318 11.5	49.7	272 57.6	17.4	277 21.2	47.2	233 24.9	47.6	Eltanin	90 52.0	N51 29.3
H 09	263 24.7	333 14.8 ..	49.8	287 58.2 ..	16.7	292 23.1 ..	47.0	248 27.2 ..	47.7	Enif	33 59.0	N 9 52.0
U 10	278 27.2	348 18.0	49.8	302 58.8	16.0	307 25.0	46.8	263 29.5	47.8	Fomalhaut	15 37.3	S29 38.1
R 11	293 29.6	3 21.3	49.9	317 59.4	15.2	322 26.9	46.6	278 31.8	47.8			
S 12	308 32.1	18 24.5	S14 49.9	333 00.0	S11 14.5	337 28.8	S12 46.4	293 34.1	N 3 47.9	Gacrux	172 13.7	S57 05.9
D 13	323 34.6	33 27.8	49.9	348 00.6	13.8	352 30.7	46.2	308 36.4	48.0	Gienah	176 04.2	S17 31.8
A 14	338 37.0	48 31.0	50.0	3 01.2	13.0	7 32.6	46.0	323 38.6	48.1	Hadar	149 04.5	S60 21.5
Y 15	353 39.5	63 34.2 ..	50.0	18 01.7 ..	12.3	22 34.5 ..	45.8	338 40.9 ..	48.2	Hamal	328 14.1	N23 27.2
16	8 42.0	78 37.5	50.1	33 02.3	11.6	37 36.4	45.6	353 43.2	48.2	Kaus Aust.	83 59.8	S34 23.0
17	23 44.4	93 40.7	50.1	48 02.9	10.8	52 38.3	45.4	8 45.5	48.3			
18	38 46.9	108 43.9	S14 50.2	63 03.5	S11 10.1	67 40.2	S12 45.2	23 47.8	N 3 48.4	Kochab	137 19.9	N74 09.6
19	53 49.4	123 47.1	50.2	78 04.1	09.4	82 42.1	45.0	38 50.1	48.5	Markab	13 50.3	N15 11.7
20	68 51.8	138 50.3	50.3	93 04.7	08.6	97 44.0	44.8	53 52.4	48.6	Menkar	314 27.3	N 4 04.7
21	83 54.3	153 53.5 ..	50.3	108 05.3 ..	07.9	112 45.9 ..	44.6	68 54.7 ..	48.6	Menkent	148 21.4	S36 21.4
22	98 56.8	168 56.7	50.4	123 05.9	07.2	127 47.8	44.4	83 57.0	48.7	Miaplacidus	221 41.1	S69 42.6
23	113 59.2	183 59.9	50.4	138 06.5	06.4	142 49.7	44.2	98 59.3	48.8			
30 00	129 01.7	199 03.1	S14 50.5	153 07.1	S11 05.7	157 51.6	S12 44.0	114 01.6	N 3 48.9	Mirfak	308 57.1	N49 51.3
01	144 04.1	214 06.3	50.5	168 07.7	05.0	172 53.5	43.8	129 03.9	49.0	Nunki	76 13.2	S26 17.8
02	159 06.6	229 09.4	50.6	183 08.3	04.2	187 55.4	43.6	144 06.2	49.1	Peacock	53 38.4	S56 44.4
03	174 09.1	244 12.6 ..	50.6	198 08.9 ..	03.5	202 57.3 ..	43.4	159 08.5 ..	49.1	Pollux	243 41.8	N28 01.7
04	189 11.5	259 15.8	50.7	213 09.5	02.8	217 59.2	43.2	174 10.8	49.2	Procyon	245 11.7	N 5 13.6
05	204 14.0	274 19.0	50.7	228 10.1	02.0	233 01.1	43.0	189 13.1	49.3			
06	219 16.5	289 22.1	S14 50.8	243 10.7	S11 01.3	248 03.0	S12 42.8	204 15.3	N 3 49.4	Rasalhague	96 17.6	N12 33.7
07	234 18.9	304 25.3	50.8	258 11.4	11 00.6	263 04.9	42.6	219 17.6	49.5	Regulus	207 55.7	N11 58.4
08	249 21.4	319 28.4	50.9	273 12.0	10 59.8	278 06.8	42.4	234 19.9	49.5	Rigel	281 23.2	S 8 12.5
F 09	264 23.9	334 31.6 ..	51.0	288 12.6 ..	59.1	293 08.7 ..	42.2	249 22.2 ..	49.6	Rigil Kent.	140 07.8	S60 49.2
R 10	279 26.3	349 34.7	51.0	303 13.2	58.4	308 10.6	42.0	264 24.5	49.7	Sabik	102 26.3	S15 43.2
I 11	294 28.8	4 37.9	51.1	318 13.8	57.6	323 12.5	41.8	279 26.8	49.8			
D 12	309 31.2	19 41.0	S14 51.1	333 14.4	S10 56.9	338 14.4	S12 41.6	294 29.1	N 3 49.9	Schedar	349 54.2	N56 31.8
A 13	324 33.7	34 44.1	51.2	348 15.0	56.2	353 16.3	41.3	309 31.4	49.9	Shaula	96 38.2	S37 05.9
Y 14	339 36.2	49 47.3	51.2	3 15.6	55.4	8 18.2	41.1	324 33.7	50.0	Sirius	258 43.8	S16 43.1
15	354 38.6	64 50.4 ..	51.3	18 16.2 ..	54.7	23 20.1 ..	40.9	339 36.0 ..	50.1	Spica	158 43.6	S11 09.0
16	9 41.1	79 53.5	51.4	33 16.8	54.0	38 22.0	40.7	354 38.3	50.2	Suhail	223 00.6	S43 25.6
17	24 43.6	94 56.6	51.4	48 17.4	53.2	53 23.9	40.5	9 40.6	50.3			
18	39 46.0	109 59.7	S14 51.5	63 18.0	S10 52.5	68 25.8	S12 40.3	24 42.9	N 3 50.4	Vega	80 47.3	N38 46.9
19	54 48.5	125 02.8	51.6	78 18.6	51.8	83 27.7	40.1	39 45.1	50.4	Zuben'ubi	137 18.5	S16 01.9
20	69 51.0	140 05.9	51.6	93 19.2	51.0	98 29.6	39.9	54 47.4	50.5		SHA	Mer. Pass.
21	84 53.4	155 09.0 ..	51.7	108 19.8 ..	50.3	113 31.5 ..	39.7	69 49.7 ..	50.6			h m
22	99 55.9	170 12.1	51.7	123 20.4	49.5	128 33.4	39.5	84 52.0	50.7	Venus	69 42.8	10 47
23	114 58.4	185 15.2	51.8	138 21.0	48.8	143 35.3	39.3	99 54.3	50.8	Mars	24 50.2	13 48
Mer. Pass. 15 25.3		v 3.2	d 0.0	v 0.6	d 0.7	v 1.9	d 0.2	v 2.3	d 0.1	Jupiter	29 03.4	13 30
										Saturn	345 04.0	16 25

SUN and MOON

UT	SUN GHA	SUN Dec	MOON GHA	v	MOON Dec	d	HP
28 00	176 47.5	S18 18.2	180 59.1	6.5	S16 08.0	5.9	59.5
01	191 47.3	17.6	195 24.6	6.6	16 02.1	6.1	59.5
02	206 47.2	16.9	209 50.2	6.6	15 56.0	6.1	59.5
03	221 47.1	.. 16.3	224 15.8	6.6	15 49.9	6.2	59.6
04	236 47.0	15.6	238 41.4	6.6	15 43.7	6.4	59.6
05	251 46.9	15.0	253 07.0	6.6	15 37.3	6.4	59.6
W 06	266 46.7	S18 14.3	267 32.6	6.6	S15 30.9	6.6	59.6
E 07	281 46.6	13.6	281 58.2	6.7	15 24.3	6.6	59.7
D 08	296 46.5	13.0	296 23.9	6.6	15 17.7	6.8	59.7
N 09	311 46.4	.. 12.3	310 49.5	6.7	15 10.9	6.8	59.7
E 10	326 46.3	11.7	325 15.2	6.7	15 04.1	7.0	59.7
S 11	341 46.1	11.0	339 40.9	6.7	14 57.1	7.1	59.7
D 12	356 46.0	S18 10.4	354 06.6	6.7	S14 50.0	7.2	59.8
A 13	11 45.9	09.7	8 32.3	6.7	14 42.8	7.2	59.8
Y 14	26 45.8	09.0	22 58.0	6.7	14 35.6	7.4	59.8
15	41 45.7	.. 08.4	37 23.7	6.8	14 28.2	7.5	59.8
16	56 45.5	07.7	51 49.5	6.8	14 20.7	7.6	59.8
17	71 45.4	07.0	66 15.3	6.8	14 13.1	7.6	59.9
18	86 45.3	S18 06.4	80 41.1	6.8	S14 05.5	7.8	59.9
19	101 45.2	05.7	95 06.9	6.8	13 57.7	7.9	59.9
20	116 45.1	05.1	109 32.7	6.9	13 49.8	7.9	59.9
21	131 45.0	.. 04.4	123 58.6	6.9	13 41.9	8.1	59.9
22	146 44.9	03.7	138 24.5	6.9	13 33.8	8.1	59.9
23	161 44.7	03.1	152 50.4	6.9	13 25.7	8.2	60.0
29 00	176 44.6	S18 02.4	167 16.3	7.0	S13 17.5	8.4	60.0
01	191 44.5	01.7	181 42.3	6.9	13 09.1	8.4	60.0
02	206 44.4	01.1	196 08.2	7.1	13 00.7	8.5	60.0
03	221 44.3	18 00.4	210 34.3	7.0	12 52.2	8.6	60.0
04	236 44.2	17 59.7	225 00.3	7.0	12 43.6	8.7	60.0
05	251 44.1	59.1	239 26.3	7.1	12 34.9	8.7	60.1
T 06	266 43.9	S17 58.4	253 52.4	7.1	S12 26.2	8.9	60.1
H 07	281 43.8	57.7	268 18.5	7.1	12 17.3	8.9	60.1
U 08	296 43.7	57.0	282 44.6	7.2	12 08.4	9.0	60.1
R 09	311 43.6	.. 56.4	297 10.8	7.2	11 59.4	9.1	60.1
S 10	326 43.5	55.7	311 37.0	7.2	11 50.3	9.2	60.1
D 11	341 43.4	55.0	326 03.2	7.2	11 41.1	9.2	60.1
A 12	356 43.3	S17 54.4	340 29.4	7.3	S11 31.9	9.4	60.1
Y 13	11 43.2	53.7	354 55.7	7.3	11 22.5	9.4	60.1
14	26 43.1	53.0	9 22.0	7.3	11 13.1	9.5	60.2
15	41 42.9	.. 52.3	23 48.3	7.4	11 03.6	9.5	60.2
16	56 42.8	51.7	38 14.7	7.4	10 54.1	9.6	60.2
17	71 42.7	51.0	52 41.1	7.4	10 44.5	9.7	60.2
18	86 42.6	S17 50.3	67 07.5	7.4	S10 34.8	9.8	60.2
19	101 42.5	49.6	81 33.9	7.5	10 25.0	9.8	60.2
20	116 42.4	48.9	96 00.4	7.5	10 15.2	9.9	60.2
21	131 42.3	.. 48.3	110 26.9	7.6	10 05.3	10.0	60.2
22	146 42.2	47.6	124 53.5	7.5	9 55.3	10.0	60.2
23	161 42.1	46.9	139 20.0	7.6	9 45.3	10.1	60.2
30 00	176 42.0	S17 46.2	153 46.6	7.6	S 9 35.2	10.2	60.2
01	191 41.9	45.5	168 13.2	7.7	9 25.0	10.2	60.2
02	206 41.8	44.9	182 39.9	7.7	9 14.8	10.3	60.2
03	221 41.7	.. 44.2	197 06.6	7.7	9 04.5	10.4	60.2
04	236 41.6	43.5	211 33.3	7.8	8 54.1	10.4	60.3
05	251 41.5	42.8	226 00.1	7.8	8 43.7	10.4	60.3
F 06	266 41.4	S17 42.1	240 26.9	7.8	S 8 33.3	10.5	60.3
R 07	281 41.3	41.5	254 53.7	7.8	8 22.8	10.6	60.3
I 08	296 41.1	40.8	269 20.5	7.9	8 12.2	10.6	60.3
D 09	311 41.0	.. 40.1	283 47.4	7.9	8 01.6	10.7	60.3
A 10	326 40.9	39.4	298 14.3	7.9	7 50.9	10.7	60.3
Y 11	341 40.8	38.7	312 41.2	8.0	7 40.2	10.8	60.3
12	356 40.7	S17 38.0	327 08.2	8.0	S 7 29.4	10.8	60.3
13	11 40.6	37.3	341 35.2	8.0	7 18.6	10.8	60.3
14	26 40.5	36.7	356 02.2	8.1	7 07.8	10.9	60.3
15	41 40.4	.. 36.0	10 29.3	8.1	6 56.9	11.0	60.3
16	56 40.3	35.3	24 56.4	8.1	6 45.9	10.9	60.3
17	71 40.2	34.6	39 23.5	8.2	6 35.0	11.1	60.3
18	86 40.1	S17 33.9	53 50.7	8.1	S 6 23.9	11.0	60.3
19	101 40.0	33.2	68 17.8	8.2	6 12.9	11.1	60.3
20	116 39.9	32.5	82 45.0	8.3	6 01.8	11.2	60.3
21	131 39.8	.. 31.8	97 12.3	8.2	5 50.6	11.1	60.3
22	146 39.7	31.1	111 39.5	8.3	5 39.5	11.2	60.3
23	161 39.6	30.4	126 06.8	8.4	S 5 28.3	11.3	60.3
	SD 16.3	d 0.7	SD 16.3		16.4		16.4

Twilight, Sunrise, Moonrise

Lat.	Naut.	Civil	Sunrise	Moonrise 28	29	30	31
N 72	07 21	08 53	10 57	10 07	09 57	09 50	09 44
N 70	07 11	08 32	10 03	09 33	09 37	09 38	09 39
68	07 04	08 15	09 31	09 08	09 20	09 29	09 35
66	06 57	08 02	09 07	08 49	09 07	09 21	09 31
64	06 51	07 51	08 49	08 33	08 56	09 14	09 28
62	06 46	07 41	08 34	08 20	08 47	09 08	09 26
60	06 42	07 33	08 21	08 09	08 39	09 03	09 24
N 58	06 38	07 26	08 10	08 00	08 32	08 58	09 22
56	06 34	07 19	08 01	07 51	08 25	08 54	09 20
54	06 30	07 13	07 52	07 44	08 20	08 51	09 18
52	06 27	07 08	07 45	07 37	08 14	08 47	09 17
50	06 24	07 03	07 38	07 31	08 10	08 44	09 16
45	06 17	06 52	07 24	07 18	08 00	08 38	09 13
N 40	06 10	06 43	07 11	07 07	07 51	08 32	09 10
35	06 04	06 34	07 01	06 57	07 44	08 27	09 08
30	05 58	06 27	06 52	06 49	07 38	08 23	09 06
20	05 47	06 13	06 36	06 35	07 27	08 16	09 03
N 10	05 35	06 01	06 23	06 23	07 17	08 09	09 00
0	05 23	05 48	06 10	06 11	07 08	08 03	08 58
S 10	05 08	05 34	05 56	05 59	06 59	07 57	08 55
20	04 51	05 19	05 42	05 47	06 49	07 51	08 52
30	04 29	05 00	05 26	05 33	06 38	07 44	08 49
35	04 15	04 48	05 16	05 24	06 31	07 39	08 47
40	03 57	04 35	05 05	05 15	06 24	07 35	08 45
45	03 36	04 18	04 52	05 05	06 15	07 29	08 43
S 50	03 06	03 57	04 36	04 50	06 05	07 22	08 40
52	02 51	03 47	04 29	04 44	06 00	07 19	08 39
54	02 33	03 36	04 20	04 37	05 55	07 16	08 37
56	02 10	03 22	04 11	04 29	05 49	07 12	08 36
58	01 37	03 06	04 00	04 21	05 42	07 08	08 34
S 60	////	02 47	03 47	04 11	05 35	07 03	08 32

Sunset, Twilight, Moonset

Lat.	Sunset	Civil	Naut.	Moonset 28	29	30	31
N 72	13 30	15 35	17 07	14 56	17 04	19 07	21 06
N 70	14 24	15 56	17 16	15 29	17 23	19 16	21 08
68	14 56	16 12	17 24	15 53	17 38	19 24	21 09
66	15 20	16 25	17 30	16 11	17 50	19 31	21 11
64	15 38	16 36	17 36	16 26	18 00	19 36	21 12
62	15 53	16 46	17 41	16 38	18 08	19 41	21 13
60	16 06	16 54	17 45	16 49	18 15	19 45	21 14
N 58	16 17	17 01	17 50	16 58	18 22	19 48	21 15
56	16 26	17 08	17 53	17 06	18 27	19 51	21 15
54	16 34	17 14	17 57	17 13	18 32	19 54	21 16
52	16 42	17 19	18 00	17 19	18 37	19 57	21 16
50	16 49	17 24	18 03	17 25	18 41	19 59	21 17
45	17 03	17 35	18 10	17 37	18 50	20 04	21 18
N 40	17 15	17 44	18 17	17 47	18 57	20 08	21 19
35	17 25	17 52	18 23	17 56	19 04	20 12	21 20
30	17 34	18 00	18 28	18 03	19 09	20 15	21 20
20	17 50	18 13	18 40	18 16	19 18	20 21	21 21
N 10	18 04	18 26	18 51	18 28	19 27	20 25	21 22
0	18 17	18 38	19 04	18 38	19 34	20 29	21 23
S 10	18 30	18 52	19 18	18 49	19 42	20 34	21 24
20	18 44	19 07	19 35	19 00	19 50	20 38	21 25
30	19 00	19 26	19 57	19 12	19 59	20 43	21 26
35	19 10	19 37	20 11	19 19	20 04	20 46	21 27
40	19 21	19 51	20 28	19 28	20 10	20 50	21 27
45	19 33	20 07	20 49	19 37	20 17	20 54	21 28
S 50	19 49	20 28	21 18	19 49	20 25	20 58	21 29
52	19 57	20 38	21 33	19 54	20 29	21 00	21 29
54	20 05	20 49	21 51	20 00	20 33	21 02	21 29
56	20 14	21 02	22 13	20 06	20 38	21 05	21 30
58	20 25	21 18	22 44	20 14	20 43	21 08	21 30
S 60	20 37	21 37	23 44	20 22	20 48	21 11	21 31

SUN and MOON

Day	Eqn. of Time 00h	12h	Mer. Pass.	Mer. Pass. Upper	Lower	Age	Phase
28	12 50	12 56	12 13	12 24	24 53	00	0
29	13 01	13 07	12 13	13 21	00 53	01	2
30	13 12	13 17	12 13	14 16	01 49	02	7

1998 JAN. 31, FEB. 1, 2 (SAT., SUN., MON.)

UT	ARIES	VENUS −4.5		MARS +1.2		JUPITER −2.0		SATURN +0.7		STARS		
	GHA	GHA	Dec	GHA	Dec	GHA	Dec	GHA	Dec	Name	SHA	Dec
d h	° ′	° ′	° ′	° ′	° ′	° ′	° ′	° ′	° ′		° ′	° ′
31 00	130 00.8	200 18.3	S14 51.9	153 21.6	S10 48.1	158 37.2	S12 39.1	114 56.6	N 3 50.8	Acamar	315 27.3	S40 19.1
01	145 03.3	215 21.4	51.9	168 22.2	47.3	173 39.1	38.9	129 58.9	50.9	Achernar	335 35.8	S57 15.2
02	160 05.7	230 24.4	52.0	183 22.8	46.6	188 41.0	38.7	145 01.2	51.0	Acrux	173 22.0	S63 05.1
03	175 08.2	245 27.5 ..	52.1	198 23.4 ..	45.9	203 42.9 ..	38.5	160 03.5 ..	51.1	Adhara	255 21.4	S28 58.4
04	190 10.7	260 30.6	52.1	213 24.0	45.1	218 44.8	38.3	175 05.8	51.2	Aldebaran	291 02.7	N16 30.2
05	205 13.1	275 33.6	52.2	228 24.6	44.4	233 46.7	38.1	190 08.0	51.3			
06	220 15.6	290 36.7	S14 52.3	243 25.2	S10 43.6	248 48.6	S12 37.9	205 10.3	N 3 51.3	Alioth	166 30.8	N55 58.0
07	235 18.1	305 39.7	52.3	258 25.8	42.9	263 50.5	37.7	220 12.6	51.4	Alkaid	153 08.1	N49 19.2
S 08	250 20.5	320 42.8	52.4	273 26.4	42.2	278 52.4	37.5	235 14.9	51.5	Al Na'ir	27 59.0	S46 58.3
A 09	265 23.0	335 45.8 ..	52.5	288 27.0 ..	41.4	293 54.3 ..	37.3	250 17.2 ..	51.6	Alnilam	275 58.1	S 1 12.4
T 10	280 25.5	350 48.9	52.5	303 27.6	40.7	308 56.2	37.1	265 19.5	51.7	Alphard	218 07.3	S 8 39.2
U 11	295 27.9	5 51.9	52.6	318 28.3	39.9	323 58.1	36.9	280 21.8	51.8			
R 12	310 30.4	20 54.9	S14 52.7	333 28.9	S10 39.2	339 00.0	S12 36.6	295 24.1	N 3 51.8	Alphecca	126 21.1	N26 43.2
D 13	325 32.9	35 57.9	52.8	348 29.5	38.5	354 01.9	36.4	310 26.4	51.9	Alpheratz	357 55.9	N29 04.8
A 14	340 35.3	51 01.0	52.8	3 30.1	37.7	9 03.8	36.2	325 28.6	52.0	Altair	62 20.0	N 8 51.8
Y 15	355 37.8	66 04.0 ..	52.9	18 30.7 ..	37.0	24 05.7 ..	36.0	340 30.9 ..	52.1	Ankaa	353 27.6	S42 19.3
16	10 40.2	81 07.0	53.0	33 31.3	36.2	39 07.6	35.8	355 33.2	52.2	Antares	112 40.9	S26 25.5
17	25 42.7	96 10.0	53.0	48 31.9	35.5	54 09.5	35.6	10 35.5	52.2			
18	40 45.2	111 13.0	S14 53.1	63 32.5	S10 34.8	69 11.4	S12 35.4	25 37.8	N 3 52.3	Arcturus	146 06.5	N19 11.4
19	55 47.6	126 16.0	53.2	78 33.1	34.0	84 13.3	35.2	40 40.1	52.4	Atria	107 53.6	S69 01.1
20	70 50.1	141 19.0	53.3	93 33.7	33.3	99 15.2	35.0	55 42.4	52.5	Avior	234 22.1	S59 30.4
21	85 52.6	156 22.0 ..	53.3	108 34.3 ..	32.5	114 17.1 ..	34.8	70 44.7 ..	52.6	Bellatrix	278 44.4	N 6 20.7
22	100 55.0	171 25.0	53.4	123 34.9	31.8	129 19.0	34.6	85 47.0	52.7	Betelgeuse	271 13.8	N 7 24.2
23	115 57.5	186 27.9	53.5	138 35.5	31.1	144 20.9	34.4	100 49.2	52.7			
1 00	131 00.0	201 30.9	S14 53.6	153 36.1	S10 30.3	159 22.8	S12 34.2	115 51.5	N 3 52.8	Canopus	264 00.9	S52 42.0
01	146 02.4	216 33.9	53.6	168 36.8	29.6	174 24.7	34.0	130 53.8	52.9	Capella	280 51.5	N45 59.7
02	161 04.9	231 36.9	53.7	183 37.4	28.8	189 26.6	33.8	145 56.1	53.0	Deneb	49 39.9	N45 16.4
03	176 07.3	246 39.8 ..	53.8	198 38.0 ..	28.1	204 28.5 ..	33.6	160 58.4 ..	53.1	Denebola	182 45.4	N14 34.8
04	191 09.8	261 42.8	53.9	213 38.6	27.4	219 30.4	33.4	176 00.7	53.2	Diphda	349 07.9	S18 00.1
05	206 12.3	276 45.7	54.0	228 39.2	26.6	234 32.3	33.2	191 03.0	53.2			
06	221 14.7	291 48.7	S14 54.1	243 39.8	S10 25.9	249 34.2	S12 33.0	206 05.2	N 3 53.3	Dubhe	194 05.6	N61 45.4
07	236 17.2	306 51.6	54.1	258 40.4	25.1	264 36.1	32.8	221 07.5	53.4	Elnath	278 27.2	N28 36.2
08	251 19.7	321 54.6	54.2	273 41.0	24.4	279 38.0	32.5	236 09.8	53.5	Eltanin	90 52.0	N51 29.3
S 09	266 22.1	336 57.5 ..	54.3	288 41.6 ..	23.6	294 39.9 ..	32.3	251 12.1 ..	53.6	Enif	33 59.0	N 9 52.0
U 10	281 24.6	352 00.4	54.4	303 42.2	22.9	309 41.8	32.1	266 14.4	53.7	Fomalhaut	15 37.3	S29 38.1
N 11	296 27.1	7 03.4	54.4	318 42.8	22.2	324 43.7	31.9	281 16.7	53.8			
D 12	311 29.5	22 06.3	S14 54.5	333 43.5	S10 21.4	339 45.6	S12 31.7	296 19.0	N 3 53.8	Gacrux	172 13.6	S57 05.9
A 13	326 32.0	37 09.2	54.6	348 44.1	20.7	354 47.5	31.5	311 21.2	53.9	Gienah	176 04.2	S17 31.8
Y 14	341 34.5	52 12.1	54.7	3 44.7	19.9	9 49.4	31.3	326 23.5	54.0	Hadar	149 04.5	S60 21.5
15	356 36.9	67 15.0 ..	54.8	18 45.3 ..	19.2	24 51.3 ..	31.1	341 25.8 ..	54.1	Hamal	328 14.1	N23 27.2
16	11 39.4	82 17.9	54.9	33 45.9	18.4	39 53.2	30.9	356 28.1	54.2	Kaus Aust.	83 59.8	S34 23.0
17	26 41.8	97 20.8	54.9	48 46.5	17.7	54 55.1	30.7	11 30.4	54.3			
18	41 44.3	112 23.7	S14 55.0	63 47.1	S10 17.0	69 57.0	S12 30.5	26 32.7	N 3 54.3	Kochab	137 19.8	N74 09.6
19	56 46.8	127 26.6	55.1	78 47.7	16.2	84 58.9	30.3	41 35.0	54.4	Markab	13 50.3	N15 11.7
20	71 49.2	142 29.5	55.2	93 48.3	15.5	100 00.8	30.1	56 37.2	54.5	Menkar	314 27.3	N 4 04.7
21	86 51.7	157 32.4 ..	55.3	108 49.0 ..	14.7	115 02.7 ..	29.9	71 39.5 ..	54.6	Menkent	148 21.4	S36 21.4
22	101 54.2	172 35.2	55.4	123 49.6	14.0	130 04.6	29.7	86 41.8	54.7	Miaplacidus	221 41.1	S69 42.6
23	116 56.6	187 38.1	55.5	138 50.2	13.2	145 06.5	29.5	101 44.1	54.8			
2 00	131 59.1	202 41.0	S14 55.6	153 50.8	S10 12.5	160 08.4	S12 29.3	116 46.4	N 3 54.8	Mirfak	308 57.1	N49 51.3
01	147 01.6	217 43.8	55.6	168 51.4	11.7	175 10.3	29.0	131 48.7	54.9	Nunki	76 13.2	S26 17.8
02	162 04.0	232 46.7	55.7	183 52.0	11.0	190 12.2	28.8	146 50.9	55.0	Peacock	53 38.4	S56 44.4
03	177 06.5	247 49.6 ..	55.8	198 52.6 ..	10.2	205 14.0 ..	28.6	161 53.2 ..	55.1	Pollux	243 41.8	N28 01.7
04	192 08.9	262 52.4	55.9	213 53.2	09.5	220 15.9	28.4	176 55.5	55.2	Procyon	245 11.7	N 5 13.6
05	207 11.4	277 55.2	56.0	228 53.9	08.8	235 17.8	28.2	191 57.8	55.3			
06	222 13.9	292 58.1	S14 56.1	243 54.5	S10 08.0	250 19.7	S12 28.0	207 00.1	N 3 55.4	Rasalhague	96 17.6	N12 33.7
07	237 16.3	308 00.9	56.2	258 55.1	07.3	265 21.6	27.8	222 02.4	55.4	Regulus	207 55.7	N11 58.4
08	252 18.8	323 03.8	56.3	273 55.7	06.5	280 23.5	27.6	237 04.6	55.5	Rigel	281 23.2	S 8 12.5
M 09	267 21.3	338 06.6 ..	56.4	288 56.3 ..	05.8	295 25.4 ..	27.4	252 06.9 ..	55.6	Rigil Kent.	140 07.8	S60 49.3
O 10	282 23.7	353 09.4	56.5	303 56.9	05.0	310 27.3	27.2	267 09.2	55.7	Sabik	102 26.3	S15 43.2
N 11	297 26.2	8 12.2	56.6	318 57.5	04.3	325 29.2	27.0	282 11.5	55.8			
D 12	312 28.7	23 15.0	S14 56.6	333 58.1	S10 03.5	340 31.1	S12 26.8	297 13.8	N 3 55.9	Schedar	349 54.2	N56 31.8
A 13	327 31.1	38 17.8	56.7	348 58.8	02.8	355 33.0	26.6	312 16.1	55.9	Shaula	96 38.2	S37 05.9
Y 14	342 33.6	53 20.7	56.8	3 59.4	02.0	10 34.9	26.4	327 18.3	56.0	Sirius	258 43.8	S16 43.1
15	357 36.1	68 23.5 ..	56.9	19 00.0 ..	01.3	25 36.8 ..	26.2	342 20.6 ..	56.1	Spica	158 43.5	S11 09.0
16	12 38.5	83 26.2	57.0	34 00.6	10 00.5	40 38.7	26.0	357 22.9	56.2	Suhail	223 00.6	S43 25.6
17	27 41.0	98 29.0	57.1	49 01.2	9 59.8	55 40.6	25.7	12 25.2	56.3			
18	42 43.4	113 31.8	S14 57.2	64 01.8	S 9 59.0	70 42.5	S12 25.5	27 27.5	N 3 56.4	Vega	80 47.3	N38 46.9
19	57 45.9	128 34.6	57.3	79 02.5	58.3	85 44.4	25.3	42 29.7	56.5	Zuben'ubi	137 18.5	S16 01.9
20	72 48.4	143 37.4	57.4	94 03.1	57.5	100 46.3	25.1	57 32.0	56.5		SHA	Mer. Pass.
21	87 50.8	158 40.2 ..	57.5	109 03.7 ..	56.8	115 48.2 ..	24.9	72 34.3 ..	56.6		° ′	h m
22	102 53.3	173 42.9	57.6	124 04.3	56.1	130 50.1	24.7	87 36.6	56.7	Venus	70 31.0	10 32
23	117 55.8	188 45.7	57.7	139 04.9	55.3	145 52.0	24.5	102 38.9	56.8	Mars	22 36.2	13 45
	h m									Jupiter	28 22.8	13 21
Mer. Pass. 15 13.5		v 2.9	d 0.1	v 0.6	d 0.7	v 1.9	d 0.2	v 2.3	d 0.1	Saturn	344 51.6	16 14

UT	SUN GHA	SUN Dec	MOON GHA	v	MOON Dec	d	HP
d h	° ′	° ′	° ′	′	° ′	′	′
31 00	176 39.6	S17 29.7	140 34.2	8.3	S 5 17.0	11.2	60.3
01	191 39.5	29.1	155 01.5	8.3	5 05.8	11.3	60.3
02	206 39.4	28.4	169 28.9	8.4	4 54.5	11.4	60.2
03	221 39.3 ..	27.7	183 56.3	8.4	4 43.1	11.3	60.2
04	236 39.2	27.0	198 23.7	8.5	4 31.8	11.4	60.2
05	251 39.1	26.3	212 51.2	8.5	4 20.4	11.4	60.2
06	266 39.0	S17 25.6	227 18.7	8.5	S 4 09.0	11.4	60.2
S 07	281 38.9	24.9	241 46.2	8.5	3 57.6	11.4	60.2
A 08	296 38.8	24.2	256 13.7	8.6	3 46.2	11.5	60.2
T 09	311 38.7 ..	23.5	270 41.3	8.6	3 34.7	11.5	60.2
U 10	326 38.6	22.8	285 08.9	8.6	3 23.2	11.5	60.2
R 11	341 38.5	22.1	299 36.5	8.7	3 11.7	11.5	60.2
D 12	356 38.4	S17 21.4	314 04.2	8.6	S 3 00.2	11.5	60.2
A 13	11 38.3	20.7	328 31.8	8.7	2 48.7	11.6	60.2
Y 14	26 38.2	20.0	342 59.5	8.7	2 37.1	11.5	60.2
15	41 38.1 ..	19.3	357 27.2	8.8	2 25.6	11.6	60.2
16	56 38.0	18.6	11 55.0	8.7	2 14.0	11.5	60.1
17	71 38.0	17.9	26 22.7	8.8	2 02.5	11.6	60.1
18	86 37.9	S17 17.2	40 50.5	8.8	S 1 50.9	11.6	60.1
19	101 37.8	16.5	55 18.3	8.8	1 39.3	11.6	60.1
20	116 37.7	15.8	69 46.1	8.9	1 27.7	11.6	60.1
21	131 37.6 ..	15.1	84 14.0	8.8	1 16.1	11.6	60.1
22	146 37.5	14.4	98 41.8	8.9	1 04.5	11.6	60.1
23	161 37.4	13.7	113 09.7	8.9	0 52.9	11.6	60.1
1 00	176 37.3	S17 13.0	127 37.6	8.9	S 0 41.3	11.6	60.1
01	191 37.2	12.3	142 05.5	9.0	0 29.7	11.6	60.1
02	206 37.2	11.5	156 33.5	8.9	0 18.1	11.6	60.0
03	221 37.1 ..	10.8	171 01.4	9.0	S 0 06.5	11.5	60.0
04	236 37.0	10.1	185 29.4	9.0	N 0 05.0	11.6	60.0
05	251 36.9	09.4	199 57.4	9.0	0 16.6	11.6	60.0
06	266 36.8	S17 08.7	214 25.4	9.1	N 0 28.2	11.5	60.0
S 07	281 36.7	08.0	228 53.5	9.0	0 39.7	11.6	60.0
U 08	296 36.6	07.3	243 21.5	9.1	0 51.3	11.5	60.0
N 09	311 36.5 ..	06.6	257 49.6	9.0	1 02.8	11.6	59.9
D 10	326 36.5	05.9	272 17.6	9.1	1 14.4	11.5	59.9
A 11	341 36.4	05.2	286 45.7	9.1	1 25.9	11.5	59.9
Y 12	356 36.3	S17 04.4	301 13.8	9.1	N 1 37.4	11.4	59.9
13	11 36.2	03.7	315 41.9	9.2	1 48.8	11.5	59.9
14	26 36.1	03.0	330 10.1	9.1	2 00.3	11.5	59.9
15	41 36.0 ..	02.3	344 38.2	9.1	2 11.8	11.4	59.9
16	56 36.0	01.6	359 06.3	9.2	2 23.2	11.4	59.8
17	71 35.9	00.9	13 34.5	9.2	2 34.6	11.4	59.8
18	86 35.8	S17 00.2	28 02.7	9.2	N 2 46.0	11.3	59.8
19	101 35.7	16 59.4	42 30.9	9.1	2 57.3	11.3	59.8
20	116 35.6	58.7	56 59.0	9.2	3 08.6	11.3	59.8
21	131 35.6 ..	58.0	71 27.2	9.3	3 19.9	11.3	59.8
22	146 35.5	57.3	85 55.5	9.2	3 31.2	11.3	59.7
23	161 35.4	56.6	100 23.7	9.2	3 42.5	11.2	59.7
2 00	176 35.3	S16 55.9	114 51.9	9.2	N 3 53.7	11.2	59.7
01	191 35.2	55.1	129 20.1	9.3	4 04.9	11.1	59.7
02	206 35.2	54.4	143 48.4	9.2	4 16.0	11.2	59.7
03	221 35.1 ..	53.7	158 16.6	9.3	4 27.2	11.1	59.6
04	236 35.0	53.0	172 44.9	9.2	4 38.3	11.0	59.6
05	251 34.9	52.3	187 13.1	9.3	4 49.3	11.0	59.6
06	266 34.8	S16 51.5	201 41.4	9.3	N 5 00.3	11.0	59.6
07	281 34.8	50.8	216 09.7	9.2	5 11.3	10.9	59.6
08	296 34.7	50.1	230 37.9	9.3	5 22.3	10.9	59.6
M 09	311 34.6 ..	49.4	245 06.2	9.3	5 33.2	10.8	59.5
O 10	326 34.5	48.7	259 34.5	9.3	5 44.0	10.9	59.5
N 11	341 34.5	47.9	274 02.8	9.2	5 54.9	10.7	59.5
D 12	356 34.4	S16 47.2	288 31.0	9.3	N 6 05.6	10.8	59.5
A 13	11 34.3	46.5	302 59.3	9.3	6 16.4	10.7	59.5
Y 14	26 34.2	45.8	317 27.6	9.3	6 27.1	10.6	59.4
15	41 34.2 ..	45.0	331 55.9	9.3	6 37.7	10.6	59.4
16	56 34.1	44.3	346 24.2	9.3	6 48.3	10.6	59.4
17	71 34.0	43.6	0 52.5	9.3	6 58.9	10.5	59.4
18	86 33.9	S16 42.8	15 20.8	9.2	N 7 09.4	10.4	59.4
19	101 33.9	42.1	29 49.0	9.3	7 19.8	10.4	59.3
20	116 33.8	41.4	44 17.3	9.3	7 30.2	10.4	59.3
21	131 33.7 ..	40.7	58 45.6	9.3	7 40.6	10.3	59.3
22	146 33.6	39.9	73 13.9	9.3	7 50.9	10.2	59.3
23	161 33.6	39.2	87 42.2	9.2	N 8 01.1	10.2	59.2
	SD 16.3	d 0.7	SD 16.4		16.3		16.2

Lat.	Twilight Naut.	Twilight Civil	Sunrise	Moonrise 31	Moonrise 1	Moonrise 2	Moonrise 3
°	h m	h m	h m	h m	h m	h m	h m
N 72	07 10	08 39	10 29	09 44	09 38	09 32	09 26
N 70	07 02	08 21	09 46	09 39	09 39	09 40	09 41
68	06 56	08 06	09 18	09 35	09 40	09 46	09 53
66	06 50	07 54	08 57	09 31	09 41	09 51	10 02
64	06 45	07 43	08 40	09 28	09 42	09 56	10 11
62	06 40	07 35	08 26	09 26	09 43	10 00	10 18
60	06 36	07 27	08 14	09 24	09 43	10 03	10 24
N 58	06 33	07 20	08 04	09 22	09 44	10 06	10 30
56	06 29	07 14	07 55	09 20	09 44	10 09	10 35
54	06 26	07 09	07 47	09 18	09 45	10 11	10 39
52	06 23	07 04	07 40	09 17	09 45	10 13	10 43
50	06 20	06 59	07 34	09 16	09 46	10 16	10 47
45	06 14	06 49	07 20	09 13	09 46	10 20	10 55
N 40	06 08	06 40	07 09	09 10	09 47	10 24	11 01
35	06 02	06 32	06 59	09 08	09 48	10 27	11 07
30	05 57	06 25	06 50	09 06	09 48	10 30	11 12
20	05 46	06 13	06 36	09 03	09 49	10 35	11 21
N 10	05 35	06 01	06 22	09 00	09 50	10 40	11 29
0	05 23	05 48	06 10	08 58	09 51	10 44	11 36
S 10	05 10	05 35	05 58	08 55	09 52	10 48	11 44
20	04 53	05 21	05 44	08 52	09 53	10 53	11 52
30	04 32	05 03	05 28	08 49	09 54	10 58	12 01
35	04 18	04 52	05 19	08 47	09 55	11 01	12 07
40	04 02	04 39	05 09	08 45	09 56	11 05	12 13
45	03 41	04 23	04 56	08 43	09 57	11 09	12 20
S 50	03 14	04 03	04 41	08 40	09 58	11 14	12 28
52	02 59	03 54	04 34	08 39	09 58	11 16	12 32
54	02 43	03 43	04 26	08 37	09 59	11 19	12 37
56	02 22	03 30	04 17	08 36	09 59	11 21	12 42
58	01 54	03 15	04 07	08 34	10 00	11 25	12 47
S 60	01 08	02 58	03 56	08 32	10 01	11 28	12 53

Lat.	Sunset	Twilight Civil	Twilight Naut.	Moonset 31	Moonset 1	Moonset 2	Moonset 3
°	h m	h m	h m	h m	h m	h m	h m
N 72	14 00	15 49	17 19	21 06	23 03	24 59	00 59
N 70	14 42	16 08	17 26	21 08	22 57	24 46	00 46
68	15 10	16 23	17 33	21 09	22 53	24 35	00 35
66	15 31	16 35	17 39	21 11	22 50	24 27	00 27
64	15 48	16 45	17 43	21 12	22 47	24 20	00 20
62	16 02	16 53	17 48	21 13	22 44	24 14	00 14
60	16 14	17 01	17 52	21 14	22 42	24 08	00 08
N 58	16 24	17 08	17 55	21 15	22 40	24 04	00 04
56	16 33	17 14	17 59	21 15	22 38	23 59	25 18
54	16 40	17 19	18 02	21 16	22 37	23 56	25 12
52	16 47	17 24	18 05	21 16	22 35	23 52	25 07
50	16 54	17 29	18 08	21 17	22 34	23 49	25 03
45	17 07	17 39	18 14	21 18	22 31	23 43	24 53
N 40	17 19	17 47	18 20	21 19	22 29	23 37	24 44
35	17 28	17 55	18 25	21 20	22 27	23 32	24 37
30	17 37	18 02	18 31	21 20	22 25	23 28	24 31
20	17 52	18 15	18 41	21 21	22 22	23 21	24 24
N 10	18 05	18 27	18 52	21 22	22 19	23 15	24 11
0	18 17	18 39	19 04	21 23	22 16	23 09	24 02
S 10	18 30	18 52	19 17	21 24	22 14	23 03	23 53
20	18 43	19 06	19 34	21 25	22 11	22 57	23 43
30	18 58	19 24	19 55	21 26	22 08	22 50	23 33
35	19 07	19 35	20 08	21 27	22 06	22 45	23 26
40	19 18	19 48	20 24	21 27	22 04	22 41	23 19
45	19 30	20 03	20 45	21 28	22 01	22 35	23 11
S 50	19 45	20 23	21 12	21 29	21 58	22 29	23 01
52	19 52	20 32	21 26	21 29	21 57	22 26	22 57
54	20 00	20 43	21 42	21 29	21 56	22 22	22 52
56	20 08	20 55	22 02	21 30	21 54	22 19	22 46
58	20 18	21 10	22 29	21 30	21 52	22 15	22 40
S 60	20 30	21 27	23 11	21 31	21 50	22 11	22 33

Day	SUN Eqn. of Time 00h	SUN Eqn. of Time 12h	SUN Mer. Pass.	MOON Mer. Pass. Upper	MOON Mer. Pass. Lower	Age	Phase
d	m s	m s	h m	h m	h m	d	%
31	13 22	13 26	12 13	15 11	02 44	03	14
1	13 31	13 35	12 14	16 04	03 37	04	23
2	13 39	13 42	12 14	16 56	04 30	05	34

UT	ARIES GHA	VENUS −4.5 GHA	Dec	MARS +1.2 GHA	Dec	JUPITER −2.0 GHA	Dec	SATURN +0.7 GHA	Dec
3 00 (TUE)	132 58.2	203 48.4	S14 57.8	154 05.5	S 9 54.6	160 53.9	S12 24.3	117 41.2	N 3 56.9
01	148 00.7	218 51.2	57.9	169 06.1	53.8	175 55.8	24.1	132 43.4	57.0
02	163 03.2	233 54.0	58.0	184 06.8	53.1	190 57.7	23.9	147 45.7	57.1
03	178 05.6	248 56.7 ..	58.1	199 07.4 ..	52.3	205 59.6 ..	23.7	162 48.0 ..	57.1
04	193 08.1	263 59.4	58.2	214 08.0	51.6	221 01.5	23.5	177 50.3	57.2
05	208 10.6	279 02.2	58.3	229 08.6	50.8	236 03.4	23.3	192 52.6	57.3
06	223 13.0	294 04.9	S14 58.4	244 09.2	S 9 50.1	251 05.3	S12 23.1	207 54.8	N 3 57.4
07	238 15.5	309 07.6	58.5	259 09.8	49.3	266 07.2	22.9	222 57.1	57.5
08	253 17.9	324 10.4	58.6	274 10.5	48.6	281 09.1	22.6	237 59.4	57.6
09	268 20.4	339 13.1 ..	58.7	289 11.1 ..	47.8	296 11.0 ..	22.4	253 01.7 ..	57.7
10	283 22.9	354 15.8	58.8	304 11.7	47.1	311 12.9	22.2	268 04.0	57.7
11	298 25.3	9 18.5	58.9	319 12.3	46.3	326 14.8	22.0	283 06.2	57.8
12	313 27.8	24 21.2	S14 59.0	334 12.9	S 9 45.6	341 16.7	S12 21.8	298 08.5	N 3 57.9
13	328 30.3	39 23.9	59.1	349 13.6	44.8	356 18.6	21.6	313 10.8	58.0
14	343 32.7	54 26.6	59.2	4 14.2	44.1	11 20.5	21.4	328 13.1	58.1
15	358 35.2	69 29.3 ..	59.3	19 14.8 ..	43.3	26 22.4 ..	21.2	343 15.3 ..	58.2
16	13 37.7	84 32.0	59.4	34 15.4	42.6	41 24.3	21.0	358 17.6	58.3
17	28 40.1	99 34.7	59.5	49 16.0	41.8	56 26.2	20.8	13 19.9	58.4
18	43 42.6	114 37.3	S14 59.6	64 16.6	S 9 41.1	71 28.0	S12 20.6	28 22.2	N 3 58.4
19	58 45.0	129 40.0	59.7	79 17.3	40.3	86 29.9	20.4	43 24.5	58.5
20	73 47.5	144 42.7	59.8	94 17.9	39.5	101 31.8	20.2	58 26.7	58.6
21	88 50.0	159 45.4	14 59.9	109 18.5 ..	38.8	116 33.7 ..	20.0	73 29.0 ..	58.7
22	103 52.4	174 48.0	15 00.1	124 19.1	38.0	131 35.6	19.7	88 31.3	58.8
23	118 54.9	189 50.7	00.2	139 19.7	37.3	146 37.5	19.5	103 33.6	58.9
4 00 (WED)	133 57.4	204 53.3	S15 00.3	154 20.4	S 9 36.5	161 39.4	S12 19.3	118 35.9	N 3 59.0
01	148 59.8	219 56.0	00.4	169 21.0	35.8	176 41.3	19.1	133 38.1	59.0
02	164 02.3	234 58.6	00.5	184 21.6	35.0	191 43.2	18.9	148 40.4	59.1
03	179 04.8	250 01.3 ..	00.6	199 22.2 ..	34.3	206 45.1 ..	18.7	163 42.7 ..	59.2
04	194 07.2	265 03.9	00.7	214 22.8	33.5	221 47.0	18.5	178 45.0	59.3
05	209 09.7	280 06.5	00.8	229 23.5	32.8	236 48.9	18.3	193 47.2	59.4
06	224 12.2	295 09.1	S15 00.9	244 24.1	S 9 32.0	251 50.8	S12 18.1	208 49.5	N 3 59.5
07	239 14.6	310 11.8	01.0	259 24.7	31.3	266 52.7	17.9	223 51.8	59.6
08	254 17.1	325 14.4	01.1	274 25.3	30.5	281 54.6	17.7	238 54.1	59.7
09	269 19.5	340 17.0 ..	01.2	289 25.9 ..	29.8	296 56.5 ..	17.5	253 56.3 ..	59.7
10	284 22.0	355 19.6	01.4	304 26.6	29.0	311 58.4	17.3	268 58.6	59.8
11	299 24.5	10 22.2	01.5	319 27.2	28.3	327 00.3	17.0	284 00.9	3 59.9
12	314 26.9	25 24.8	S15 01.6	334 27.8	S 9 27.5	342 02.2	S12 16.8	299 03.2	N 4 00.0
13	329 29.4	40 27.4	01.7	349 28.4	26.7	357 04.1	16.6	314 05.5	00.1
14	344 31.9	55 30.0	01.8	4 29.1	26.0	12 06.0	16.4	329 07.7	00.2
15	359 34.3	70 32.6 ..	01.9	19 29.7 ..	25.2	27 07.9 ..	16.2	344 10.0 ..	00.3
16	14 36.8	85 35.1	02.0	34 30.3	24.5	42 09.8	16.0	359 12.3	00.4
17	29 39.3	100 37.7	02.1	49 30.9	23.7	57 11.7	15.8	14 14.6	00.4
18	44 41.7	115 40.3	S15 02.3	64 31.5	S 9 23.0	72 13.6	S12 15.6	29 16.8	N 4 00.5
19	59 44.2	130 42.9	02.4	79 32.2	22.2	87 15.5	15.4	44 19.1	00.6
20	74 46.7	145 45.4	02.5	94 32.8	21.5	102 17.4	15.2	59 21.4	00.7
21	89 49.1	160 48.0 ..	02.6	109 33.4 ..	20.7	117 19.3 ..	15.0	74 23.7 ..	00.8
22	104 51.6	175 50.5	02.7	124 34.0	20.0	132 21.1	14.8	89 25.9	00.9
23	119 54.0	190 53.1	02.8	139 34.7	19.2	147 23.0	14.6	104 28.2	01.0
5 00 (THU)	134 56.5	205 55.6	S15 02.9	154 35.3	S 9 18.4	162 24.9	S12 14.3	119 30.5	N 4 01.1
01	149 59.0	220 58.2	03.1	169 35.9	17.7	177 26.8	14.1	134 32.8	01.1
02	165 01.4	236 00.7	03.2	184 36.5	16.9	192 28.7	13.9	149 35.0	01.2
03	180 03.9	251 03.2 ..	03.3	199 37.2 ..	16.2	207 30.6 ..	13.7	164 37.3 ..	01.3
04	195 06.4	266 05.8	03.4	214 37.8	15.4	222 32.5	13.5	179 39.6	01.4
05	210 08.8	281 08.3	03.5	229 38.4	14.7	237 34.4	13.3	194 41.9	01.5
06	225 11.3	296 10.8	S15 03.6	244 39.0	S 9 13.9	252 36.3	S12 13.1	209 44.1	N 4 01.6
07	240 13.8	311 13.3	03.8	259 39.7	13.1	267 38.2	12.9	224 46.4	01.7
08	255 16.2	326 15.8	03.9	274 40.3	12.4	282 40.1	12.7	239 48.7	01.8
09	270 18.7	341 18.3 ..	04.0	289 40.9 ..	11.6	297 42.0 ..	12.5	254 50.9 ..	01.8
10	285 21.2	356 20.8	04.1	304 41.5	10.9	312 43.9	12.3	269 53.2	01.9
11	300 23.6	11 23.3	04.2	319 42.2	10.1	327 45.8	12.1	284 55.5	02.0
12	315 26.1	26 25.8	S15 04.3	334 42.8	S 9 09.4	342 47.7	S12 11.8	299 57.8	N 4 02.1
13	330 28.5	41 28.3	04.5	349 43.4	08.6	357 49.6	11.6	315 00.0	02.2
14	345 31.0	56 30.8	04.6	4 44.0	07.8	12 51.5	11.4	330 02.3	02.3
15	0 33.5	71 33.3 ..	04.7	19 44.7 ..	07.1	27 53.4 ..	11.2	345 04.6 ..	02.4
16	15 35.9	86 35.7	04.8	34 45.3	06.3	42 55.3	11.0	0 06.9	02.5
17	30 38.4	101 38.2	04.9	49 45.9	05.6	57 57.2	10.8	15 09.1	02.6
18	45 40.9	116 40.7	S15 05.1	64 46.5	S 9 04.8	72 59.1	S12 10.6	30 11.4	N 4 02.6
19	60 43.3	131 43.1	05.2	79 47.2	04.1	88 01.0	10.4	45 13.7	02.7
20	75 45.8	146 45.6	05.3	94 47.8	03.3	103 02.9	10.2	60 15.9	02.8
21	90 48.3	161 48.0 ..	05.4	109 48.4 ..	02.5	118 04.8 ..	10.0	75 18.2 ..	02.9
22	105 50.7	176 50.5	05.5	124 49.0	01.8	133 06.6	09.8	90 20.5	03.0
23	120 53.2	191 52.9	05.7	139 49.7	01.0	148 08.5	09.5	105 22.8	03.1
Mer. Pass. 15 01.7		v 2.6	d 0.1	v 0.6	d 0.8	v 1.9	d 0.2	v 2.3	d 0.1

STARS

Name	SHA	Dec
Acamar	315 27.3	S40 19.1
Achernar	335 35.8	S57 15.2
Acrux	173 21.9	S63 05.1
Adhara	255 21.4	S28 58.5
Aldebaran	291 02.7	N16 30.2
Alioth	166 30.8	N55 58.0
Alkaid	153 08.1	N49 19.2
Al Na'ir	27 59.0	S46 58.3
Alnilam	275 58.1	S 1 12.4
Alphard	218 07.3	S 8 39.2
Alphecca	126 21.1	N26 43.2
Alpheratz	357 55.9	N29 04.8
Altair	62 20.0	N 8 51.8
Ankaa	353 27.6	S42 19.3
Antares	112 40.9	S26 25.5
Arcturus	146 06.4	N19 11.4
Atria	107 53.6	S69 01.1
Avior	234 22.1	S59 30.4
Bellatrix	278 44.4	N 6 20.7
Betelgeuse	271 13.8	N 7 24.2
Canopus	264 00.9	S52 42.0
Capella	280 51.6	N45 59.8
Deneb	49 39.9	N45 16.4
Denebola	182 45.4	N14 34.8
Diphda	349 07.9	S18 00.1
Dubhe	194 05.6	N61 45.5
Elnath	278 27.3	N28 36.2
Eltanin	90 52.0	N51 29.3
Enif	33 59.0	N 9 51.9
Fomalhaut	15 37.3	S29 38.1
Gacrux	172 13.6	S57 05.9
Gienah	176 04.2	S17 31.9
Hadar	149 04.4	S60 21.5
Hamal	328 14.1	N23 27.1
Kaus Aust.	83 59.7	S34 23.0
Kochab	137 19.7	N74 09.6
Markab	13 50.3	N15 11.7
Menkar	314 27.3	N 4 04.7
Menkent	148 21.4	S36 21.4
Miaplacidus	221 41.1	S69 42.6
Mirfak	308 57.1	N49 51.3
Nunki	76 13.2	S26 17.8
Peacock	53 38.4	S56 44.4
Pollux	243 41.8	N28 01.7
Procyon	245 11.7	N 5 13.6
Rasalhague	96 17.6	N12 33.7
Regulus	207 55.7	N11 58.4
Rigel	281 23.2	S 8 12.5
Rigil Kent.	140 07.7	S60 49.3
Sabik	102 26.2	S15 43.2
Schedar	349 54.2	N56 31.7
Shaula	96 38.2	S37 05.9
Sirius	258 43.8	S16 43.1
Spica	158 43.5	S11 09.0
Suhail	223 00.6	S43 25.6
Vega	80 47.3	N38 46.9
Zuben'ubi	137 18.4	S16 01.9

	SHA	Mer. Pass.
Venus	70 56.0	10 19
Mars	20 23.0	13 42
Jupiter	27 42.1	13 12
Saturn	344 38.5	16 03

UT	SUN GHA	SUN Dec	MOON GHA	v	MOON Dec	d	HP
d h	° ′	° ′	° ′	′	° ′	′	′
3 00	176 33.5	S16 38.5	102 10.4	9.3	N 8 11.3	10.1	59.2
01	191 33.4	37.7	116 38.7	9.3	8 21.4	10.1	59.2
02	206 33.4	37.0	131 07.0	9.2	8 31.5	10.0	59.2
03	221 33.3	.. 36.3	145 35.2	9.3	8 41.5	10.0	59.2
04	236 33.2	35.5	160 03.5	9.3	8 51.5	9.9	59.1
05	251 33.2	34.8	174 31.8	9.2	9 01.4	9.8	59.1
06	266 33.1	S16 34.1	189 00.0	9.3	N 9 11.2	9.8	59.1
07	281 33.0	33.3	203 28.3	9.2	9 21.0	9.7	59.1
T 08	296 32.9	32.6	217 56.5	9.2	9 30.7	9.7	59.0
U 09	311 32.9	.. 31.9	232 24.7	9.3	9 40.4	9.6	59.0
E 10	326 32.8	31.1	246 53.0	9.2	9 50.0	9.5	59.0
S 11	341 32.7	30.4	261 21.2	9.2	9 59.5	9.5	59.0
D 12	356 32.7	S16 29.7	275 49.4	9.2	N10 09.0	9.3	59.0
A 13	11 32.6	28.9	290 17.6	9.2	10 18.3	9.4	58.9
Y 14	26 32.5	28.2	304 45.8	9.2	10 27.7	9.2	58.9
15	41 32.5	.. 27.5	319 14.0	9.2	10 36.9	9.2	58.9
16	56 32.4	26.7	333 42.2	9.2	10 46.1	9.1	58.9
17	71 32.4	26.0	348 10.4	9.2	10 55.2	9.1	58.8
18	86 32.3	S16 25.2	2 38.6	9.1	N11 04.3	9.0	58.8
19	101 32.2	24.5	17 06.7	9.2	11 13.3	8.9	58.8
20	116 32.2	23.8	31 34.9	9.1	11 22.2	8.8	58.8
21	131 32.1	.. 23.0	46 03.0	9.2	11 31.0	8.7	58.8
22	146 32.0	22.3	60 31.2	9.1	11 39.7	8.7	58.7
23	161 32.0	21.5	74 59.3	9.1	11 48.4	8.6	58.7
4 00	176 31.9	S16 20.8	89 27.4	9.1	N11 57.0	8.6	58.7
01	191 31.8	20.0	103 55.5	9.1	12 05.6	8.4	58.7
02	206 31.8	19.3	118 23.6	9.1	12 14.0	8.4	58.6
03	221 31.7	.. 18.6	132 51.7	9.1	12 22.4	8.3	58.6
04	236 31.7	17.8	147 19.8	9.1	12 30.7	8.2	58.6
05	251 31.6	17.1	161 47.9	9.1	12 38.9	8.1	58.6
06	266 31.5	S16 16.3	176 16.0	9.0	N12 47.0	8.1	58.5
W 07	281 31.5	15.6	190 44.0	9.1	12 55.1	8.0	58.5
E 08	296 31.4	14.8	205 12.1	9.0	13 03.1	7.9	58.5
D 09	311 31.4	.. 14.1	219 40.1	9.0	13 11.0	7.8	58.5
N 10	326 31.3	13.3	234 08.1	9.1	13 18.8	7.7	58.4
E 11	341 31.2	12.6	248 36.2	9.0	13 26.5	7.6	58.4
S 12	356 31.2	S16 11.8	263 04.2	9.0	N13 34.1	7.6	58.4
D 13	11 31.1	11.1	277 32.2	9.0	13 41.7	7.5	58.4
A 14	26 31.1	10.3	292 00.2	8.9	13 49.2	7.3	58.4
Y 15	41 31.0	.. 09.6	306 28.1	9.0	13 56.5	7.3	58.3
16	56 31.0	08.8	320 56.1	9.0	14 03.8	7.3	58.3
17	71 30.9	08.1	335 24.1	8.9	14 11.1	7.1	58.3
18	86 30.8	S16 07.3	349 52.0	9.0	N14 18.2	7.0	58.3
19	101 30.8	06.6	4 20.0	8.9	14 25.2	7.0	58.2
20	116 30.7	05.8	18 47.9	8.9	14 32.2	6.8	58.2
21	131 30.7	.. 05.1	33 15.8	9.0	14 39.0	6.8	58.2
22	146 30.6	04.3	47 43.8	8.9	14 45.8	6.7	58.2
23	161 30.6	03.6	62 11.7	8.9	14 52.5	6.5	58.1
5 00	176 30.5	S16 02.8	76 39.6	8.9	N14 59.0	6.5	58.1
01	191 30.5	02.1	91 07.5	8.8	15 05.5	6.4	58.1
02	206 30.4	01.3	105 35.3	8.9	15 11.9	6.3	58.1
03	221 30.4	16 00.6	120 03.2	8.9	15 18.2	6.3	58.0
04	236 30.3	15 59.8	134 31.1	8.8	15 24.5	6.1	58.0
05	251 30.3	59.1	148 58.9	8.9	15 30.6	6.0	58.0
06	266 30.2	S15 58.3	163 26.8	8.8	N15 36.6	5.9	58.0
07	281 30.2	57.5	177 54.6	8.9	15 42.5	5.9	58.0
T 08	296 30.1	56.8	192 22.5	8.8	15 48.4	5.7	57.9
H 09	311 30.0	.. 56.0	206 50.3	8.8	15 54.1	5.6	57.9
U 10	326 30.0	55.3	221 18.1	8.9	15 59.7	5.6	57.9
R 11	341 29.9	54.5	235 46.0	8.8	16 05.3	5.4	57.9
S 12	356 29.9	S15 53.7	250 13.8	8.8	N16 10.7	5.4	57.8
D 13	11 29.9	53.0	264 41.6	8.8	16 16.1	5.3	57.8
A 14	26 29.8	52.2	279 09.4	8.8	16 21.4	5.1	57.8
Y 15	41 29.8	.. 51.5	293 37.2	8.8	16 26.5	5.1	57.8
16	56 29.7	50.7	308 05.0	8.8	16 31.6	4.9	57.7
17	71 29.7	49.9	322 32.8	8.8	16 36.5	4.9	57.7
18	86 29.6	S15 49.2	337 00.6	8.8	N16 41.4	4.7	57.7
19	101 29.6	48.4	351 28.4	8.7	16 46.1	4.7	57.7
20	116 29.5	47.6	5 56.1	8.8	16 50.8	4.6	57.6
21	131 29.5	.. 46.9	20 23.9	8.8	16 55.4	4.4	57.6
22	146 29.4	46.1	34 51.7	8.8	16 59.8	4.4	57.6
23	161 29.4	45.4	49 19.5	8.8	N17 04.2	4.2	57.6
SD	16.3	d 0.7	SD 16.1		15.9		15.8

Moonrise

Lat.	Twilight Naut.	Civil	Sunrise	3	4	5	6
°	h m	h m	h m	h m	h m	h m	h m
N 72	06 59	08 26	10 05	09 26	09 19	09 11	08 56
N 70	06 53	08 09	09 30	09 41	09 43	09 50	10 03
68	06 47	07 56	09 05	09 53	10 02	10 16	10 40
66	06 42	07 45	08 46	10 02	10 17	10 37	11 06
64	06 38	07 36	08 31	10 11	10 29	10 53	11 26
62	06 34	07 28	08 18	10 18	10 40	11 07	11 42
60	06 30	07 21	08 07	10 24	10 49	11 19	11 56
N 58	06 27	07 14	07 58	10 30	10 57	11 29	12 08
56	06 24	07 09	07 49	10 35	11 04	11 38	12 18
54	06 22	07 04	07 42	10 39	11 10	11 46	12 27
52	06 19	06 59	07 35	10 43	11 16	11 53	12 35
50	06 16	06 55	07 29	10 47	11 21	11 59	12 43
45	06 11	06 46	07 17	10 55	11 32	12 13	12 58
N 40	06 05	06 37	07 06	11 01	11 41	12 24	13 11
35	06 00	06 30	06 57	11 07	11 49	12 34	13 22
30	05 55	06 24	06 49	11 12	11 56	12 43	13 31
20	05 45	06 12	06 35	11 21	12 08	12 57	13 48
N 10	05 35	06 00	06 22	11 29	12 19	13 10	14 02
0	05 24	05 49	06 10	11 36	12 29	13 23	14 16
S 10	05 11	05 37	05 59	11 44	12 40	13 35	14 30
20	04 55	05 23	05 46	11 52	12 50	13 48	14 44
30	04 35	05 05	05 31	12 01	13 03	14 03	15 01
35	04 22	04 55	05 22	12 07	13 10	14 12	15 11
40	04 06	04 43	05 12	12 13	13 19	14 22	15 22
45	03 47	04 28	05 01	12 20	13 28	14 34	15 35
S 50	03 21	04 09	04 47	12 28	13 40	14 48	15 51
52	03 08	04 00	04 40	12 32	13 46	14 55	15 59
54	02 52	03 50	04 33	12 37	13 52	15 03	16 07
56	02 33	03 38	04 24	12 42	13 59	15 11	16 16
58	02 09	03 24	04 15	12 47	14 06	15 20	16 27
S 60	01 33	03 08	04 04	12 53	14 15	15 31	16 40

Moonset

Lat.	Sunset	Twilight Civil	Naut.	3	4	5	6
°	h m	h m	h m	h m	h m	h m	h m
N 72	14 24	16 03	17 30	00 59	02 55	04 55	07 02
N 70	14 59	16 20	17 37	00 46	02 33	04 17	05 55
68	15 24	16 33	17 42	00 35	02 15	03 51	05 19
66	15 43	16 44	17 47	00 27	02 01	03 32	04 53
64	15 58	16 53	17 51	00 20	01 50	03 16	04 33
62	16 11	17 01	17 55	00 14	01 40	03 02	04 17
60	16 22	17 08	17 58	00 08	01 32	02 51	04 04
N 58	16 31	17 14	18 02	00 04	01 25	02 42	03 52
56	16 39	17 20	18 04	25 18	01 18	02 33	03 42
54	16 46	17 25	18 07	25 12	01 12	02 26	03 34
52	16 53	17 29	18 10	25 07	01 07	02 19	03 26
50	16 59	17 34	18 12	25 03	01 03	02 13	03 19
45	17 12	17 43	18 18	24 53	00 53	02 00	03 03
N 40	17 22	17 51	18 23	24 44	00 44	01 49	02 51
35	17 32	17 58	18 28	24 37	00 37	01 40	02 40
30	17 40	18 04	18 33	24 31	00 31	01 32	02 31
20	17 53	18 16	18 43	24 20	00 20	01 18	02 15
N 10	18 06	18 28	18 53	24 11	00 11	01 06	02 01
0	18 17	18 39	19 04	24 02	00 02	00 55	01 48
S 10	18 29	18 51	19 17	23 53	24 43	00 43	01 35
20	18 42	19 05	19 32	23 43	24 31	00 31	01 21
30	18 56	19 22	19 52	23 33	24 18	00 18	01 05
35	19 05	19 32	20 05	23 26	24 10	00 10	00 56
40	19 15	19 44	20 21	23 19	24 01	00 01	00 46
45	19 26	19 59	20 40	23 11	23 50	24 33	00 33
S 50	19 40	20 17	21 05	23 01	23 37	24 18	00 18
52	19 47	20 26	21 18	22 57	23 32	24 11	00 11
54	19 54	20 36	21 34	22 52	23 25	24 03	00 03
56	20 02	20 48	21 52	22 46	23 18	23 55	24 39
58	20 11	21 01	22 15	22 40	23 10	23 45	24 28
S 60	20 22	21 17	22 49	22 33	23 00	23 34	24 15

	SUN			MOON			
Day	Eqn. of Time 00h	12h	Mer. Pass.	Mer. Pass. Upper	Lower	Age	Phase
d	m s	m s	h m	h m	h m	d	%
3	13 46	13 49	12 14	17 49	05 23	06	45
4	13 52	13 55	12 14	18 42	06 15	07	56
5	13 58	14 00	12 14	19 35	07 09	08	67

34   1998 FEBRUARY 6, 7, 8 (FRI., SAT., SUN.)

UT	ARIES	VENUS −4.6		MARS +1.2		JUPITER −2.0		SATURN +0.7		STARS		
	GHA	GHA	Dec	GHA	Dec	GHA	Dec	GHA	Dec	Name	SHA	Dec
d h	° ′	° ′	° ′	° ′	° ′	° ′	° ′	° ′	° ′		° ′	° ′
6 00	135 55.6	206 55.4	S15 05.8	154 50.3	S 9 00.3	163 10.4	S12 09.3	120 25.0	N 4 03.2	Acamar	315 27.3	S40 19.1
01	150 58.1	221 57.8	05.9	169 50.9	8 59.5	178 12.3	09.1	135 27.3	03.3	Achernar	335 35.8	S57 15.2
02	166 00.6	237 00.2	06.0	184 51.6	58.7	193 14.2	08.9	150 29.6	03.4	Acrux	173 21.9	S63 05.1
03	181 03.0	252 02.7	.. 06.1	199 52.2	.. 58.0	208 16.1	.. 08.7	165 31.8	.. 03.4	Adhara	255 21.4	S28 58.5
04	196 05.5	267 05.1	06.3	214 52.8	57.2	223 18.0	08.5	180 34.1	03.5	Aldebaran	291 02.7	N16 30.2
05	211 08.0	282 07.5	06.4	229 53.4	56.5	238 19.9	08.3	195 36.4	03.6			
06	226 10.4	297 09.9	S15 06.5	244 54.1	S 8 55.7	253 21.8	S12 08.1	210 38.7	N 4 03.7	Alioth	166 30.8	N55 58.0
07	241 12.9	312 12.3	06.6	259 54.7	55.0	268 23.7	07.9	225 40.9	03.8	Alkaid	153 08.0	N49 19.2
08	256 15.4	327 14.7	06.8	274 55.3	54.2	283 25.6	07.7	240 43.2	03.9	Al Na'ir	27 59.0	S46 58.3
F 09	271 17.8	342 17.1	.. 06.9	289 55.9	.. 53.4	298 27.5	.. 07.5	255 45.5	.. 04.0	Alnilam	275 58.1	S 1 12.4
R 10	286 20.3	357 19.5	07.0	304 56.6	52.7	313 29.4	07.2	270 47.7	04.1	Alphard	218 07.3	S 8 39.2
I 11	301 22.8	12 21.9	07.1	319 57.2	51.9	328 31.3	07.0	285 50.0	04.2			
D 12	316 25.2	27 24.3	S15 07.3	334 57.8	S 8 51.1	343 33.2	S12 06.8	300 52.3	N 4 04.3	Alphecca	126 21.0	N26 43.2
A 13	331 27.7	42 26.7	07.4	349 58.5	50.4	358 35.1	06.6	315 54.6	04.3	Alpheratz	357 55.9	N29 04.8
Y 14	346 30.1	57 29.1	07.5	4 59.1	49.6	13 37.0	06.4	330 56.8	04.4	Altair	62 20.0	N 8 51.8
15	1 32.6	72 31.4	.. 07.6	19 59.7	.. 48.9	28 38.9	.. 06.2	345 59.1	.. 04.5	Ankaa	353 27.6	S42 19.3
16	16 35.1	87 33.8	07.8	35 00.4	48.1	43 40.8	06.0	1 01.4	04.6	Antares	112 40.8	S26 25.5
17	31 37.5	102 36.2	07.9	50 01.0	47.3	58 42.7	05.8	16 03.6	04.7			
18	46 40.0	117 38.5	S15 08.0	65 01.6	S 8 46.6	73 44.6	S12 05.6	31 05.9	N 4 04.8	Arcturus	146 06.4	N19 11.4
19	61 42.5	132 40.9	08.1	80 02.2	45.8	88 46.4	05.4	46 08.2	04.9	Atria	107 53.5	S69 01.1
20	76 44.9	147 43.2	08.3	95 02.9	45.1	103 48.3	05.2	61 10.4	05.0	Avior	234 22.1	S59 30.4
21	91 47.4	162 45.6	.. 08.4	110 03.5	.. 44.3	118 50.2	.. 04.9	76 12.7	.. 05.1	Bellatrix	278 44.4	N 6 20.7
22	106 49.9	177 47.9	08.5	125 04.1	43.5	133 52.1	04.7	91 15.0	05.2	Betelgeuse	271 13.8	N 7 24.2
23	121 52.3	192 50.3	08.6	140 04.8	42.8	148 54.0	04.5	106 17.2	05.2			
7 00	136 54.8	207 52.6	S15 08.8	155 05.4	S 8 42.0	163 55.9	S12 04.3	121 19.5	N 4 05.3	Canopus	264 00.9	S52 42.0
01	151 57.3	222 54.9	08.9	170 06.0	41.3	178 57.8	04.1	136 21.8	05.4	Capella	280 51.6	N45 59.8
02	166 59.7	237 57.3	09.0	185 06.7	40.5	193 59.7	03.9	151 24.1	05.5	Deneb	49 39.9	N45 16.4
03	182 02.2	252 59.6	.. 09.1	200 07.3	.. 39.7	209 01.6	.. 03.7	166 26.3	.. 05.6	Denebola	182 45.4	N14 34.8
04	197 04.6	268 01.9	09.3	215 07.9	39.0	224 03.5	03.5	181 28.6	05.7	Diphda	349 07.9	S18 00.1
05	212 07.1	283 04.2	09.4	230 08.6	38.2	239 05.4	03.3	196 30.9	05.8			
06	227 09.6	298 06.5	S15 09.5	245 09.2	S 8 37.4	254 07.3	S12 03.1	211 33.1	N 4 05.9	Dubhe	194 05.6	N61 45.5
07	242 12.0	313 08.8	09.7	260 09.8	36.7	269 09.2	02.8	226 35.4	06.0	Elnath	278 27.3	N28 36.2
S 08	257 14.5	328 11.1	09.8	275 10.5	35.9	284 11.1	02.6	241 37.7	06.1	Eltanin	90 52.0	N51 29.2
A 09	272 17.0	343 13.4	.. 09.9	290 11.1	.. 35.2	299 13.0	.. 02.4	256 39.9	.. 06.1	Enif	33 59.0	N 9 51.9
T 10	287 19.4	358 15.7	10.0	305 11.7	34.4	314 14.9	02.2	271 42.2	06.2	Fomalhaut	15 37.3	S29 38.1
U 11	302 21.9	13 18.0	10.2	320 12.3	33.6	329 16.8	02.0	286 44.5	06.3			
R 12	317 24.4	28 20.3	S15 10.3	335 13.0	S 8 32.9	344 18.7	S12 01.8	301 46.7	N 4 06.4	Gacrux	172 13.6	S57 05.9
D 13	332 26.8	43 22.6	10.4	350 13.6	32.1	359 20.6	01.6	316 49.0	06.5	Gienah	176 04.2	S17 31.9
A 14	347 29.3	58 24.8	10.6	5 14.2	31.3	14 22.4	01.4	331 51.3	06.6	Hadar	149 04.4	S60 21.5
Y 15	2 31.8	73 27.1	.. 10.7	20 14.9	.. 30.6	29 24.3	.. 01.2	346 53.5	.. 06.7	Hamal	328 14.1	N23 27.1
16	17 34.2	88 29.4	10.8	35 15.5	29.8	44 26.2	01.0	1 55.8	06.8	Kaus Aust.	83 59.7	S34 23.0
17	32 36.7	103 31.6	10.9	50 16.1	29.0	59 28.1	00.7	16 58.1	06.9			
18	47 39.1	118 33.9	S15 11.1	65 16.8	S 8 28.3	74 30.0	S12 00.5	32 00.3	N 4 07.0	Kochab	137 19.7	N74 09.6
19	62 41.6	133 36.1	11.2	80 17.4	27.5	89 31.9	00.3	47 02.6	07.1	Markab	13 50.3	N15 11.7
20	77 44.1	148 38.4	11.3	95 18.0	26.7	104 33.8	12 00.1	62 04.9	07.1	Menkar	314 27.4	N 4 04.7
21	92 46.5	163 40.6	.. 11.5	110 18.7	.. 26.0	119 35.7	11 59.9	77 07.1	.. 07.2	Menkent	148 21.3	S36 21.4
22	107 49.0	178 42.9	11.6	125 19.3	25.2	134 37.6	59.7	92 09.4	07.3	Miaplacidus	221 41.1	S69 42.7
23	122 51.5	193 45.1	11.7	140 20.0	24.5	149 39.5	59.5	107 11.7	07.4			
8 00	137 53.9	208 47.3	S15 11.9	155 20.6	S 8 23.7	164 41.4	S11 59.3	122 13.9	N 4 07.5	Mirfak	308 57.1	N49 51.3
01	152 56.4	223 49.6	12.0	170 21.2	22.9	179 43.3	59.1	137 16.2	07.6	Nunki	76 13.2	S26 17.8
02	167 58.9	238 51.8	12.1	185 21.9	22.2	194 45.2	58.9	152 18.5	07.7	Peacock	53 38.3	S56 44.4
03	183 01.3	253 54.0	.. 12.3	200 22.5	.. 21.4	209 47.1	.. 58.6	167 20.7	.. 07.8	Pollux	243 41.8	N28 01.7
04	198 03.8	268 56.2	12.4	215 23.1	20.6	224 49.0	58.4	182 23.0	07.9	Procyon	245 11.7	N 5 13.6
05	213 06.2	283 58.4	12.5	230 23.8	19.9	239 50.9	58.2	197 25.3	08.0			
06	228 08.7	299 00.6	S15 12.7	245 24.4	S 8 19.1	254 52.8	S11 58.0	212 27.5	N 4 08.1	Rasalhague	96 17.6	N12 33.7
07	243 11.2	314 02.8	12.8	260 25.0	18.3	269 54.7	57.8	227 29.8	08.1	Regulus	207 55.7	N11 58.4
08	258 13.6	329 05.0	12.9	275 25.7	17.6	284 56.5	57.6	242 32.0	08.2	Rigel	281 23.2	S 8 12.5
S 09	273 16.1	344 07.2	.. 13.0	290 26.3	.. 16.8	299 58.4	.. 57.4	257 34.3	.. 08.3	Rigil Kent.	140 07.7	S60 49.3
U 10	288 18.6	359 09.4	13.2	305 26.9	16.0	315 00.3	57.2	272 36.6	08.4	Sabik	102 26.2	S15 43.2
N 11	303 21.0	14 11.6	13.3	320 27.6	15.3	330 02.2	57.0	287 38.8	08.5			
D 12	318 23.5	29 13.8	S15 13.4	335 28.2	S 8 14.5	345 04.1	S11 56.7	302 41.1	N 4 08.6	Schedar	349 54.2	N56 31.7
A 13	333 26.0	44 16.0	13.6	350 28.8	13.7	0 06.0	56.5	317 43.4	08.7	Shaula	96 38.1	S37 05.9
Y 14	348 28.4	59 18.1	13.7	5 29.5	13.0	15 07.9	56.3	332 45.6	08.8	Sirius	258 43.8	S16 43.1
15	3 30.9	74 20.3	.. 13.8	20 30.1	.. 12.2	30 09.8	.. 56.1	347 47.9	.. 08.9	Spica	158 43.5	S11 09.1
16	18 33.4	89 22.5	14.0	35 30.8	11.4	45 11.7	55.9	2 50.2	09.0	Suhail	223 00.6	S43 25.6
17	33 35.8	104 24.6	14.1	50 31.4	10.7	60 13.6	55.7	17 52.4	09.1			
18	48 38.3	119 26.8	S15 14.2	65 32.0	S 8 09.9	75 15.5	S11 55.5	32 54.7	N 4 09.2	Vega	80 47.2	N38 46.8
19	63 40.7	134 28.9	14.4	80 32.7	09.1	90 17.4	55.3	47 57.0	09.2	Zuben'ubi	137 18.4	S16 01.9
20	78 43.2	149 31.1	14.5	95 33.3	08.4	105 19.3	55.1	62 59.2	09.3		SHA	Mer. Pass.
21	93 45.7	164 33.2	.. 14.6	110 33.9	.. 07.6	120 21.2	.. 54.9	78 01.5	.. 09.4		° ′	h m
22	108 48.1	179 35.4	14.8	125 34.6	06.8	135 23.1	54.6	93 03.7	09.5	Venus	70 57.8	10 07
23	123 50.6	194 37.5	14.9	140 35.2	06.1	150 25.0	54.4	108 06.0	09.6	Mars	18 10.6	13 39
	h m									Jupiter	27 01.1	13 03
Mer. Pass.	14 49.9	v 2.3	d 0.1	v 0.6	d 0.8	v 1.9	d 0.2	v 2.3	d 0.1	Saturn	344 24.7	15 52

SUN and MOON

UT	SUN GHA	SUN Dec	MOON GHA	v	MOON Dec	d	HP
6 00	176 29.3	S15 44.6	63 47.3	8.7	N17 08.4	4.2	57.6
01	191 29.3	43.8	78 15.0	8.8	17 12.6	4.0	57.5
02	206 29.2	43.1	92 42.8	8.8	17 16.6	4.0	57.5
03	221 29.2	.. 42.3	107 10.6	8.8	17 20.6	3.8	57.5
04	236 29.2	41.5	121 38.4	8.7	17 24.4	3.8	57.5
05	251 29.1	40.8	136 06.1	8.8	17 28.2	3.6	57.4
06	266 29.1	S15 40.0	150 33.9	8.8	N17 31.8	3.6	57.4
07	281 29.0	39.2	165 01.7	8.8	17 35.4	3.4	57.4
08	296 29.0	38.5	179 29.5	8.8	17 38.8	3.3	57.4
F 09	311 28.9	.. 37.7	193 57.3	8.8	17 42.1	3.3	57.3
R 10	326 28.9	36.9	208 25.1	8.8	17 45.4	3.1	57.3
I 11	341 28.9	36.1	222 52.9	8.8	17 48.5	3.0	57.3
D 12	356 28.8	S15 35.4	237 20.7	8.8	N17 51.5	2.9	57.3
A 13	11 28.8	34.6	251 48.5	8.8	17 54.4	2.8	57.3
Y 14	26 28.7	33.8	266 16.3	8.8	17 57.2	2.7	57.2
15	41 28.7	.. 33.1	280 44.1	8.8	17 59.9	2.6	57.2
16	56 28.7	32.3	295 11.9	8.8	18 02.5	2.5	57.2
17	71 28.6	31.5	309 39.7	8.9	18 05.0	2.4	57.2
18	86 28.6	S15 30.7	324 07.6	8.8	N18 07.4	2.3	57.1
19	101 28.5	30.0	338 35.4	8.9	18 09.7	2.2	57.1
20	116 28.5	29.2	353 03.3	8.9	18 11.9	2.1	57.1
21	131 28.5	.. 28.4	7 31.2	8.8	18 14.0	2.0	57.1
22	146 28.4	27.6	21 59.0	8.9	18 16.0	1.9	57.1
23	161 28.4	26.9	36 26.9	8.9	18 17.9	1.7	57.0
7 00	176 28.4	S15 26.1	50 54.8	9.0	N18 19.6	1.7	57.0
01	191 28.3	25.3	65 22.8	8.9	18 21.3	1.5	57.0
02	206 28.3	24.5	79 50.7	8.9	18 22.8	1.5	57.0
03	221 28.3	.. 23.8	94 18.6	9.0	18 24.3	1.3	56.9
04	236 28.2	23.0	108 46.6	8.9	18 25.6	1.3	56.9
05	251 28.2	22.2	123 14.6	8.9	18 26.9	1.1	56.9
06	266 28.1	S15 21.4	137 42.5	9.0	N18 28.0	1.1	56.9
S 07	281 28.1	20.6	152 10.5	9.1	18 29.1	0.9	56.9
A 08	296 28.1	19.9	166 38.6	9.1	18 30.0	0.8	56.8
T 09	311 28.0	.. 19.1	181 06.6	9.1	18 30.8	0.7	56.8
U 10	326 28.0	18.3	195 34.7	9.0	18 31.5	0.7	56.8
R 11	341 28.0	17.5	210 02.7	9.1	18 32.2	0.5	56.8
D 12	356 27.9	S15 16.7	224 30.8	9.1	N18 32.7	0.4	56.8
A 13	11 27.9	15.9	238 58.9	9.2	18 33.1	0.3	56.7
Y 14	26 27.9	15.2	253 27.1	9.1	18 33.4	0.2	56.7
15	41 27.9	.. 14.4	267 55.2	9.2	18 33.6	0.1	56.7
16	56 27.8	13.6	282 23.4	9.2	18 33.7	0.0	56.7
17	71 27.8	12.8	296 51.6	9.2	18 33.7	0.1	56.6
18	86 27.8	S15 12.0	311 19.8	9.3	N18 33.6	0.2	56.6
19	101 27.7	11.2	325 48.1	9.2	18 33.4	0.3	56.6
20	116 27.7	10.5	340 16.3	9.3	18 33.1	0.4	56.6
21	131 27.7	.. 09.7	354 44.6	9.4	18 32.7	0.6	56.6
22	146 27.6	08.9	9 13.0	9.3	18 32.1	0.6	56.5
23	161 27.6	08.1	23 41.3	9.4	18 31.5	0.7	56.5
8 00	176 27.6	S15 07.3	38 09.7	9.4	N18 30.8	0.8	56.5
01	191 27.6	06.5	52 38.1	9.4	18 30.0	0.9	56.5
02	206 27.5	05.7	67 06.5	9.5	18 29.1	1.0	56.5
03	221 27.5	.. 05.0	81 35.0	9.4	18 28.1	1.2	56.4
04	236 27.5	04.2	96 03.4	9.6	18 26.9	1.2	56.4
05	251 27.4	03.4	110 32.0	9.5	18 25.7	1.3	56.4
06	266 27.4	S15 02.6	125 00.5	9.6	N18 24.4	1.4	56.4
07	281 27.4	01.8	139 29.1	9.6	18 23.0	1.5	56.4
08	296 27.4	01.0	153 57.7	9.6	18 21.5	1.7	56.3
S 09	311 27.3	15 00.2	168 26.3	9.7	18 19.8	1.7	56.3
U 10	326 27.3	14 59.4	182 55.0	9.7	18 18.1	1.8	56.3
N 11	341 27.3	58.6	197 23.7	9.8	18 16.3	1.9	56.3
D 12	356 27.3	S14 57.8	211 52.5	9.7	N18 14.4	2.0	56.3
A 13	11 27.2	57.0	226 21.2	9.8	18 12.4	2.1	56.2
Y 14	26 27.2	56.2	240 50.0	9.9	18 10.3	2.2	56.2
15	41 27.2	.. 55.5	255 18.9	9.9	18 08.1	2.3	56.2
16	56 27.2	54.7	269 47.8	9.9	18 05.8	2.4	56.2
17	71 27.2	53.9	284 16.7	9.9	18 03.4	2.5	56.2
18	86 27.1	S14 53.1	298 45.6	10.0	N18 00.9	2.5	56.1
19	101 27.1	52.3	313 14.6	10.0	17 58.4	2.7	56.1
20	116 27.1	51.5	327 43.6	10.1	17 55.7	2.8	56.1
21	131 27.1	.. 50.7	342 12.7	10.1	17 52.9	2.8	56.1
22	146 27.1	49.9	356 41.8	10.1	17 50.1	3.0	56.1
23	161 27.0	49.1	11 10.9	10.2	N17 47.1	3.0	56.0
	SD 16.2	d 0.8	SD 15.6		15.5		15.3

Moonrise

Lat.	Twilight Naut.	Twilight Civil	Sunrise	Moonrise 6	7	8	9
	h m	h m	h m	h m	h m	h m	h m
N 72	06 48	08 12	09 44	08 56	▭	▭	12 05
N 70	06 43	07 58	09 15	10 03	10 34	11 35	12 59
68	06 38	07 46	08 53	10 40	11 19	12 17	13 32
66	06 34	07 36	08 35	11 06	11 48	12 45	13 55
64	06 30	07 28	08 21	11 26	12 10	13 07	14 14
62	06 27	07 20	08 10	11 42	12 28	13 24	14 29
60	06 24	07 14	08 00	11 56	12 43	13 38	14 42
N 58	06 22	07 08	07 51	12 08	12 55	13 51	14 53
56	06 19	07 03	07 43	12 18	13 06	14 01	15 02
54	06 17	06 59	07 37	12 27	13 16	14 11	15 11
52	06 14	06 54	07 30	12 35	13 24	14 19	15 18
50	06 12	06 51	07 25	12 43	13 32	14 26	15 25
45	06 07	06 42	07 13	12 58	13 48	14 42	15 40
N 40	06 02	06 35	07 03	13 11	14 02	14 55	15 52
35	05 58	06 28	06 54	13 22	14 13	15 07	16 02
30	05 53	06 22	06 47	13 31	14 23	15 16	16 11
20	05 44	06 11	06 33	13 48	14 40	15 33	16 26
N 10	05 35	06 00	06 22	14 02	14 55	15 48	16 39
0	05 24	05 49	06 11	14 16	15 09	16 01	16 52
S 10	05 12	05 38	06 00	14 30	15 23	16 15	17 05
20	04 57	05 24	05 47	14 44	15 38	16 30	17 18
30	04 38	05 08	05 34	15 01	15 56	16 46	17 33
35	04 26	04 58	05 25	15 11	16 06	16 56	17 42
40	04 11	04 47	05 16	15 22	16 17	17 07	17 52
45	03 52	04 33	05 05	15 35	16 31	17 20	18 04
S 50	03 28	04 15	04 52	15 51	16 47	17 36	18 18
52	03 16	04 07	04 46	15 59	16 55	17 44	18 25
54	03 01	03 57	04 39	16 07	17 04	17 52	18 32
56	02 44	03 46	04 31	16 16	17 14	18 01	18 40
58	02 22	03 33	04 22	16 27	17 25	18 12	18 50
S 60	01 53	03 18	04 12	16 40	17 37	18 24	19 00

Moonset

Lat.	Sunset	Twilight Civil	Twilight Naut.	Moonset 6	7	8	9
	h m	h m	h m	h m	h m	h m	h m
N 72	14 45	16 18	17 42	07 02	▭		09 20
N 70	15 15	16 32	17 47	05 55	07 15	08 04	08 26
68	15 37	16 44	17 52	05 19	06 31	07 22	07 53
66	15 54	16 54	17 56	04 53	06 02	06 53	07 29
64	16 08	17 02	17 59	04 33	05 40	06 31	07 10
62	16 19	17 09	18 02	04 17	05 22	06 14	06 54
60	16 29	17 15	18 05	04 04	05 07	06 00	06 41
N 58	16 38	17 21	18 08	03 52	04 55	05 47	06 30
56	16 46	17 26	18 10	03 42	04 44	05 36	06 20
54	16 53	17 30	18 12	03 34	04 34	05 27	06 11
52	16 59	17 35	18 15	03 26	04 26	05 19	06 03
50	17 06	17 38	18 17	03 19	04 18	05 11	05 56
45	17 16	17 47	18 22	03 03	04 02	04 55	05 41
N 40	17 26	17 54	18 26	02 51	03 49	04 42	05 29
35	17 35	18 01	18 31	02 40	03 37	04 30	05 18
30	17 42	18 07	18 35	02 31	03 27	04 20	05 09
20	17 55	18 18	18 44	02 15	03 10	04 03	04 53
N 10	18 07	18 28	18 54	02 01	02 56	03 48	04 39
0	18 18	18 39	19 04	01 48	02 42	03 34	04 26
S 10	18 29	18 50	19 16	01 35	02 28	03 20	04 13
20	18 41	19 04	19 31	01 21	02 13	03 05	03 59
30	18 54	19 20	19 50	01 05	01 56	02 48	03 42
35	19 02	19 29	20 02	00 56	01 46	02 38	03 33
40	19 11	19 41	20 16	00 46	01 34	02 27	03 22
45	19 22	19 55	20 35	00 33	01 21	02 13	03 09
S 50	19 35	20 12	20 59	00 18	01 05	01 57	02 54
52	19 41	20 20	21 11	00 11	00 57	01 49	02 46
54	19 48	20 30	21 25	00 03	00 48	01 40	02 38
56	19 56	20 40	21 42	24 39	00 39	01 31	02 29
58	20 04	20 53	22 03	24 28	00 28	01 20	02 19
S 60	20 14	21 07	22 31	24 15	00 15	01 07	02 07

SUN / MOON

	SUN Eqn. of Time 00h	12h	Mer. Pass.	MOON Mer. Pass. Upper	Lower	Age	Phase
Day	m s	m s	h m	h m	h m	d	%
6	14 03	14 05	12 14	20 29	08 02	09	77
7	14 06	14 08	12 14	21 22	08 55	10	85
8	14 10	14 11	12 14	22 14	09 48	11	91

1998 FEBRUARY 9, 10, 11 (MON., TUES., WED.)

UT	ARIES GHA	VENUS −4.6 GHA	Dec	MARS +1.2 GHA	Dec	JUPITER −2.0 GHA	Dec	SATURN +0.7 GHA	Dec
9 00	138 53.1	209 39.6	S15 15.0	155 35.9	S 8 05.3	165 26.9	S11 54.2	123 08.3	N 4 09.7
01	153 55.5	224 41.8	15.2	170 36.5	04.5	180 28.7	54.0	138 10.5	09.8
02	168 58.0	239 43.9	15.3	185 37.1	03.8	195 30.6	53.8	153 12.8	09.9
03	184 00.5	254 46.0 ..	15.4	200 37.8 ..	03.0	210 32.5 ..	53.6	168 15.1 ..	10.0
04	199 02.9	269 48.1	15.6	215 38.4	02.2	225 34.4	53.4	183 17.3	10.1
05	214 05.4	284 50.2	15.7	230 39.1	01.5	240 36.3	53.2	198 19.6	10.2
06	229 07.9	299 52.3	S15 15.8	245 39.7	S 8 00.7	255 38.2	S11 53.0	213 21.8	N 4 10.3
07	244 10.3	314 54.4	16.0	260 40.3	7 59.9	270 40.1	52.7	228 24.1	10.4
08	259 12.8	329 56.5	16.1	275 41.0	59.2	285 42.0	52.5	243 26.4	10.4
09	274 15.2	344 58.6 ..	16.2	290 41.6 ..	58.4	300 43.9 ..	52.3	258 28.6 ..	10.5
10	289 17.7	0 00.7	16.4	305 42.2	57.6	315 45.8	52.1	273 30.9	10.6
11	304 20.2	15 02.8	16.5	320 42.9	56.8	330 47.7	51.9	288 33.2	10.7
12	319 22.6	30 04.9	S15 16.7	335 43.5	S 7 56.1	345 49.6	S11 51.7	303 35.4	N 4 10.8
13	334 25.1	45 07.0	16.8	350 44.2	55.3	0 51.5	51.5	318 37.7	10.9
14	349 27.6	60 09.0	16.9	5 44.8	54.5	15 53.4	51.3	333 39.9	11.0
15	4 30.0	75 11.1 ..	17.1	20 45.4 ..	53.8	30 55.3 ..	51.1	348 42.2 ..	11.1
16	19 32.5	90 13.2	17.2	35 46.1	53.0	45 57.2	50.8	3 44.5	11.2
17	34 35.0	105 15.2	17.3	50 46.7	52.2	60 59.0	50.6	18 46.7	11.3
18	49 37.4	120 17.3	S15 17.5	65 47.4	S 7 51.5	76 00.9	S11 50.4	33 49.0	N 4 11.4
19	64 39.9	135 19.4	17.6	80 48.0	50.7	91 02.8	50.2	48 51.2	11.5
20	79 42.4	150 21.4	17.7	95 48.7	49.9	106 04.7	50.0	63 53.5	11.6
21	94 44.8	165 23.4 ..	17.9	110 49.3 ..	49.1	121 06.6 ..	49.8	78 55.8 ..	11.7
22	109 47.3	180 25.5	18.0	125 49.9	48.4	136 08.5	49.6	93 58.0	11.7
23	124 49.7	195 27.5	18.1	140 50.6	47.6	151 10.4	49.4	109 00.3	11.8
10 00	139 52.2	210 29.6	S15 18.3	155 51.2	S 7 46.8	166 12.3	S11 49.2	124 02.5	N 4 11.9
01	154 54.7	225 31.6	18.4	170 51.9	46.1	181 14.2	48.9	139 04.8	12.0
02	169 57.1	240 33.6	18.5	185 52.5	45.3	196 16.1	48.7	154 07.1	12.1
03	184 59.6	255 35.6 ..	18.7	200 53.1 ..	44.5	211 18.0 ..	48.5	169 09.3 ..	12.2
04	200 02.1	270 37.6	18.8	215 53.8	43.8	226 19.9	48.3	184 11.6	12.3
05	215 04.5	285 39.7	19.0	230 54.4	43.0	241 21.8	48.1	199 13.8	12.4
06	230 07.0	300 41.7	S15 19.1	245 55.1	S 7 42.2	256 23.7	S11 47.9	214 16.1	N 4 12.5
07	245 09.5	315 43.7	19.2	260 55.7	41.4	271 25.6	47.7	229 18.4	12.6
08	260 11.9	330 45.7	19.4	275 56.4	40.7	286 27.4	47.5	244 20.6	12.7
09	275 14.4	345 47.7 ..	19.5	290 57.0 ..	39.9	301 29.3 ..	47.2	259 22.9 ..	12.8
10	290 16.8	0 49.7	19.6	305 57.6	39.1	316 31.2	47.0	274 25.1	12.9
11	305 19.3	15 51.7	19.8	320 58.3	38.4	331 33.1	46.8	289 27.4	13.0
12	320 21.8	30 53.6	S15 19.9	335 58.9	S 7 37.6	346 35.0	S11 46.6	304 29.7	N 4 13.1
13	335 24.2	45 55.6	20.0	350 59.6	36.8	1 36.9	46.4	319 31.9	13.1
14	350 26.7	60 57.6	20.2	6 00.2	36.0	16 38.8	46.2	334 34.2	13.2
15	5 29.2	75 59.6 ..	20.3	21 00.9 ..	35.3	31 40.7 ..	46.0	349 36.4 ..	13.3
16	20 31.6	91 01.5	20.4	36 01.5	34.5	46 42.6	45.8	4 38.7	13.4
17	35 34.1	106 03.5	20.6	51 02.1	33.7	61 44.5	45.6	19 40.9	13.5
18	50 36.6	121 05.5	S15 20.7	66 02.8	S 7 33.0	76 46.4	S11 45.3	34 43.2	N 4 13.6
19	65 39.0	136 07.4	20.9	81 03.4	32.2	91 48.3	45.1	49 45.5	13.7
20	80 41.5	151 09.4	21.0	96 04.1	31.4	106 50.2	44.9	64 47.7	13.8
21	95 44.0	166 11.3 ..	21.1	111 04.7 ..	30.6	121 52.1 ..	44.7	79 50.0 ..	13.9
22	110 46.4	181 13.3	21.3	126 05.4	29.9	136 54.0	44.5	94 52.2	14.0
23	125 48.9	196 15.2	21.4	141 06.0	29.1	151 55.8	44.3	109 54.5	14.1
11 00	140 51.3	211 17.1	S15 21.5	156 06.7	S 7 28.3	166 57.7	S11 44.1	124 56.7	N 4 14.2
01	155 53.8	226 19.1	21.7	171 07.3	27.5	181 59.6	43.9	139 59.0	14.3
02	170 56.3	241 21.0	21.8	186 07.9	26.8	197 01.5	43.6	155 01.3	14.4
03	185 58.7	256 22.9 ..	21.9	201 08.6 ..	26.0	212 03.4 ..	43.4	170 03.5 ..	14.5
04	201 01.2	271 24.8	22.1	216 09.2	25.2	227 05.3	43.2	185 05.8	14.6
05	216 03.7	286 26.8	22.2	231 09.9	24.5	242 07.2	43.0	200 08.0	14.7
06	231 06.1	301 28.7	S15 22.4	246 10.5	S 7 23.7	257 09.1	S11 42.8	215 10.3	N 4 14.7
07	246 08.6	316 30.6	22.5	261 11.2	22.9	272 11.0	42.6	230 12.5	14.8
08	261 11.1	331 32.5	22.6	276 11.8	22.1	287 12.9	42.4	245 14.8	14.9
09	276 13.5	346 34.4 ..	22.8	291 12.5 ..	21.4	302 14.8 ..	42.2	260 17.1 ..	15.0
10	291 16.0	1 36.3	22.9	306 13.1	20.6	317 16.7	41.9	275 19.3	15.1
11	306 18.5	16 38.2	23.0	321 13.8	19.8	332 18.6	41.7	290 21.6	15.2
12	321 20.9	31 40.1	S15 23.2	336 14.4	S 7 19.0	347 20.5	S11 41.5	305 23.8	N 4 15.3
13	336 23.4	46 42.0	23.3	351 15.0	18.3	2 22.3	41.3	320 26.1	15.4
14	351 25.8	61 43.8	23.4	6 15.7	17.5	17 24.2	41.1	335 28.3	15.5
15	6 28.3	76 45.7 ..	23.6	21 16.3 ..	16.7	32 26.1 ..	40.9	350 30.6 ..	15.6
16	21 30.8	91 47.6	23.7	36 17.0	15.9	47 28.0	40.7	5 32.8	15.7
17	36 33.2	106 49.5	23.8	51 17.6	15.2	62 29.9	40.5	20 35.1	15.8
18	51 35.7	121 51.3	S15 24.0	66 18.3	S 7 14.4	77 31.8	S11 40.2	35 37.4	N 4 15.9
19	66 38.2	136 53.2	24.1	81 18.9	13.6	92 33.7	40.0	50 39.6	16.0
20	81 40.6	151 55.0	24.3	96 19.6	12.8	107 35.6	39.8	65 41.9	16.1
21	96 43.1	166 56.9 ..	24.4	111 20.2 ..	12.1	122 37.5 ..	39.6	80 44.1 ..	16.2
22	111 45.6	181 58.8	24.5	126 20.9	11.3	137 39.4	39.4	95 46.4	16.3
23	126 48.0	197 00.6	24.7	141 21.5	10.5	152 41.3	39.2	110 48.6	16.4
Mer. Pass.	14 38.1	v 2.0	d 0.1	v 0.6	d 0.8	v 1.9	d 0.2	v 2.3	d 0.1

STARS

Name	SHA	Dec
Acamar	315 27.3	S40 19.1
Achernar	335 35.8	S57 15.2
Acrux	173 21.9	S63 05.1
Adhara	255 21.4	S28 58.5
Aldebaran	291 02.8	N16 30.2
Alioth	166 30.7	N55 58.0
Alkaid	153 08.0	N49 19.1
Al Na'ir	27 59.0	S46 58.3
Alnilam	275 58.1	S 1 12.4
Alphard	218 07.3	S 8 39.2
Alphecca	126 21.0	N26 43.1
Alpheratz	357 55.9	N29 04.8
Altair	62 20.0	N 8 51.8
Ankaa	353 27.6	S42 19.3
Antares	112 40.8	S26 25.5
Arcturus	146 06.4	N19 11.4
Atria	107 53.4	S69 01.1
Avior	234 22.1	S59 30.4
Bellatrix	278 44.4	N 6 20.7
Betelgeuse	271 13.8	N 7 24.2
Canopus	264 01.0	S52 42.1
Capella	280 51.6	N45 59.8
Deneb	49 39.9	N45 16.4
Denebola	182 45.4	N14 34.8
Diphda	349 07.9	S18 00.1
Dubhe	194 05.5	N61 45.5
Elnath	278 27.3	N28 36.2
Eltanin	90 51.9	N51 29.2
Enif	33 59.0	N 9 51.9
Fomalhaut	15 37.3	S29 38.1
Gacrux	172 13.6	S57 06.0
Gienah	176 04.1	S17 31.9
Hadar	149 04.3	S60 21.5
Hamal	328 14.1	N23 27.1
Kaus Aust.	83 59.7	S34 23.0
Kochab	137 19.6	N74 09.6
Markab	13 50.3	N15 11.6
Menkar	314 27.4	N 4 04.7
Menkent	148 21.3	S36 21.4
Miaplacidus	221 41.1	S69 42.7
Mirfak	308 57.1	N49 51.3
Nunki	76 13.2	S26 17.8
Peacock	53 38.3	S56 44.4
Pollux	243 41.8	N28 01.7
Procyon	245 11.7	N 5 13.6
Rasalhague	96 17.5	N12 33.7
Regulus	207 55.7	N11 58.4
Rigel	281 23.2	S 8 12.5
Rigil Kent.	140 07.7	S60 49.3
Sabik	102 26.2	S15 43.2
Schedar	349 54.2	N56 31.7
Shaula	96 38.1	S37 05.9
Sirius	258 43.8	S16 43.1
Spica	158 43.5	S11 09.1
Suhail	223 00.6	S43 25.7
Vega	80 47.2	N38 46.8
Zuben'ubi	137 18.4	S16 01.9

	SHA	Mer. Pass.
Venus	70 37.4	9 57
Mars	15 59.0	13 36
Jupiter	26 20.1	12 54
Saturn	344 10.3	15 41

Left margin day labels: MONDAY (h 00–23 for day 9), TUESDAY (day 10), WEDNESDAY (day 11).

UT	SUN GHA	Dec	MOON GHA	v	Dec	d	HP
d h	° ′	° ′	° ′	′	° ′	′	′
9 00	176 27.0	S14 48.3	25 40.1	10.2	N17 44.1	3.1	56.0
01	191 27.0	47.5	40 09.3	10.3	17 41.0	3.3	56.0
02	206 27.0	46.7	54 38.6	10.3	17 37.7	3.3	56.0
03	221 27.0 ..	45.9	69 07.9	10.4	17 34.4	3.4	56.0
04	236 26.9	45.1	83 37.3	10.3	17 31.0	3.5	55.9
05	251 26.9	44.3	98 06.6	10.5	17 27.5	3.6	55.9
06	266 26.9	S14 43.5	112 36.1	10.4	N17 23.9	3.6	55.9
M 07	281 26.9	42.7	127 05.5	10.5	17 20.3	3.8	55.9
O 08	296 26.9	41.9	141 35.0	10.6	17 16.5	3.8	55.9
N 09	311 26.8 ..	41.1	156 04.6	10.6	17 12.7	4.0	55.8
D 10	326 26.8	40.3	170 34.2	10.6	17 08.7	4.0	55.8
A 11	341 26.8	39.5	185 03.8	10.7	17 04.7	4.1	55.8
Y 12	356 26.8	S14 38.7	199 33.5	10.8	N17 00.6	4.2	55.8
13	11 26.8	37.9	214 03.3	10.7	16 56.4	4.2	55.8
14	26 26.8	37.1	228 33.0	10.8	16 52.2	4.4	55.7
15	41 26.8 ..	36.3	243 02.8	10.9	16 47.8	4.4	55.7
16	56 26.7	35.5	257 32.7	10.9	16 43.4	4.5	55.7
17	71 26.7	34.7	272 02.6	11.0	16 38.9	4.6	55.7
18	86 26.7	S14 33.9	286 32.6	11.0	N16 34.3	4.7	55.7
19	101 26.7	33.1	301 02.6	11.0	16 29.6	4.8	55.7
20	116 26.7	32.2	315 32.6	11.1	16 24.8	4.8	55.6
21	131 26.7 ..	31.4	330 02.7	11.2	16 20.0	4.9	55.6
22	146 26.7	30.6	344 32.9	11.1	16 15.1	5.0	55.6
23	161 26.7	29.8	359 03.0	11.3	16 10.1	5.1	55.6
10 00	176 26.6	S14 29.0	13 33.3	11.3	N16 05.0	5.2	55.6
01	191 26.6	28.2	28 03.6	11.3	15 59.8	5.2	55.5
02	206 26.6	27.4	42 33.9	11.4	15 54.6	5.3	55.5
03	221 26.6 ..	26.6	57 04.3	11.4	15 49.3	5.4	55.5
04	236 26.6	25.8	71 34.7	11.5	15 43.9	5.4	55.5
05	251 26.6	25.0	86 05.2	11.5	15 38.5	5.6	55.5
06	266 26.6	S14 24.2	100 35.7	11.5	N15 32.9	5.6	55.5
T 07	281 26.6	23.4	115 06.2	11.7	15 27.3	5.6	55.4
U 08	296 26.6	22.5	129 36.9	11.6	15 21.7	5.8	55.4
E 09	311 26.6 ..	21.7	144 07.5	11.7	15 15.9	5.8	55.4
S 10	326 26.5	20.9	158 38.2	11.8	15 10.1	5.9	55.4
D 11	341 26.5	20.1	173 09.0	11.8	15 04.2	5.9	55.4
A 12	356 26.5	S14 19.3	187 39.8	11.8	N14 58.3	6.1	55.4
Y 13	11 26.5	18.5	202 10.6	11.9	14 52.2	6.1	55.3
14	26 26.5	17.7	216 41.5	12.0	14 46.1	6.1	55.3
15	41 26.5 ..	16.9	231 12.5	12.0	14 40.0	6.2	55.3
16	56 26.5	16.0	245 43.5	12.0	14 33.8	6.3	55.3
17	71 26.5	15.2	260 14.5	12.1	14 27.5	6.4	55.3
18	86 26.5	S14 14.4	274 45.6	12.2	N14 21.1	6.4	55.2
19	101 26.5	13.6	289 16.8	12.2	14 14.7	6.5	55.2
20	116 26.5	12.8	303 48.0	12.2	14 08.2	6.6	55.2
21	131 26.5 ..	12.0	318 19.2	12.3	14 01.6	6.6	55.2
22	146 26.5	11.1	332 50.5	12.3	13 55.0	6.6	55.2
23	161 26.5	10.3	347 21.8	12.4	13 48.4	6.8	55.2
11 00	176 26.5	S14 09.5	1 53.2	12.5	N13 41.6	6.8	55.1
01	191 26.5	08.7	16 24.7	12.4	13 34.8	6.8	55.1
02	206 26.5	07.9	30 56.1	12.6	13 28.0	7.0	55.1
03	221 26.5 ..	07.1	45 27.7	12.5	13 21.0	6.9	55.1
04	236 26.5	06.2	59 59.2	12.7	13 14.1	7.1	55.1
05	251 26.4	05.4	74 30.9	12.6	13 07.0	7.0	55.1
06	266 26.4	S14 04.6	89 02.5	12.8	N13 00.0	7.2	55.1
W 07	281 26.4	03.8	103 34.3	12.7	12 52.8	7.2	55.0
E 08	296 26.4	03.0	118 06.0	12.8	12 45.6	7.2	55.0
D 09	311 26.4 ..	02.1	132 37.8	12.9	12 38.4	7.4	55.0
N 10	326 26.4	01.3	147 09.7	12.1	12 31.0	7.3	55.0
E 11	341 26.4	14 00.5	161 41.6	12.9	12 23.7	7.4	55.0
S 12	356 26.4	S13 59.7	176 13.5	13.0	N12 16.3	7.5	55.0
D 13	11 26.4	58.8	190 45.5	13.1	12 08.8	7.5	54.9
A 14	26 26.4	58.0	205 17.6	13.1	12 01.3	7.6	54.9
Y 15	41 26.4 ..	57.2	219 49.7	13.1	11 53.7	7.6	54.9
16	56 26.5	56.4	234 21.8	13.2	11 46.1	7.7	54.9
17	71 26.5	55.5	248 54.0	13.2	11 38.4	7.7	54.9
18	86 26.5	S13 54.7	263 26.2	13.3	N11 30.7	7.7	54.9
19	101 26.5	53.9	277 58.5	13.3	11 23.0	7.9	54.9
20	116 26.5	53.1	292 30.8	13.3	11 15.1	7.8	54.8
21	131 26.5 ..	52.2	307 03.1	13.4	11 07.3	7.9	54.8
22	146 26.5	51.4	321 35.5	13.5	10 59.4	8.0	54.8
23	161 26.5	50.6	336 08.0	13.4	N10 51.4	7.9	54.8
	SD 16.2	d 0.8	SD 15.2		15.1		15.0

Lat.	Twilight Naut.	Civil	Sunrise	Moonrise 9	10	11	12
°	h m	h m	h m	h m	h m	h m	h m
N 72	06 36	07 59	09 25	12 05	13 59	15 44	17 23
N 70	06 32	07 46	08 59	12 59	14 31	16 05	17 36
68	06 29	07 35	08 40	13 32	14 55	16 21	17 46
66	06 25	07 27	08 25	13 55	15 13	16 34	17 55
64	06 23	07 19	08 12	14 14	15 28	16 45	18 02
62	06 20	07 13	08 01	14 29	15 40	16 54	18 08
60	06 18	07 07	07 52	14 42	15 51	17 02	18 13
N 58	06 16	07 02	07 44	14 53	16 00	17 09	18 18
56	06 13	06 57	07 37	15 02	16 08	17 15	18 22
54	06 12	06 53	07 31	15 11	16 15	17 20	18 26
52	06 10	06 49	07 25	15 18	16 21	17 25	18 29
50	06 08	06 46	07 20	15 25	16 27	17 29	18 32
45	06 04	06 38	07 09	15 40	16 39	17 39	18 39
N 40	05 59	06 31	06 59	15 52	16 49	17 47	18 44
35	05 55	06 25	06 51	16 02	16 58	17 54	18 49
30	05 51	06 20	06 44	16 11	17 05	18 00	18 53
20	05 43	06 09	06 32	16 26	17 19	18 10	19 00
N 10	05 34	05 59	06 21	16 39	17 30	18 19	19 07
0	05 25	05 49	06 11	16 52	17 41	18 28	19 12
S 10	05 13	05 39	06 00	17 05	17 51	18 36	19 18
20	04 59	05 26	05 49	17 18	18 03	18 45	19 25
30	04 41	05 11	05 36	17 33	18 16	18 55	19 32
35	04 29	05 02	05 28	17 42	18 23	19 01	19 36
40	04 15	04 51	05 20	17 52	18 32	19 08	19 40
45	03 58	04 37	05 10	18 04	18 42	19 15	19 46
S 50	03 35	04 21	04 57	18 18	18 54	19 25	19 52
52	03 23	04 13	04 51	18 25	18 59	19 29	19 55
54	03 10	04 04	04 45	18 32	19 06	19 34	19 58
56	02 54	03 54	04 38	18 40	19 12	19 39	20 02
58	02 35	03 42	04 30	18 50	19 20	19 45	20 06
S 60	02 10	03 28	04 21	19 00	19 29	19 51	20 10

Lat.	Sunset	Twilight Civil	Naut.	Moonset 9	10	11	12
°	h m	h m	h m	h m	h m	h m	h m
N 72	15 05	16 32	17 54	09 20	09 09	09 02	08 56
N 70	15 31	16 44	17 58	08 26	08 36	08 40	08 42
68	15 50	16 54	18 02	07 53	08 11	08 23	08 30
66	16 05	17 03	18 04	07 29	07 52	08 09	08 21
64	16 18	17 10	18 07	07 10	07 37	07 57	08 13
62	16 28	17 17	18 10	06 54	07 24	07 47	08 06
60	16 37	17 22	18 12	06 41	07 13	07 39	07 59
N 58	16 45	17 27	18 14	06 30	07 04	07 31	07 54
56	16 52	17 32	18 16	06 20	06 55	07 25	07 49
54	16 59	17 36	18 18	06 11	06 48	07 19	07 45
52	17 04	17 40	18 20	06 03	06 41	07 13	07 41
50	17 09	17 43	18 21	05 56	06 35	07 08	07 38
45	17 20	17 51	18 26	05 41	06 22	06 58	07 30
N 40	17 30	17 58	18 30	05 29	06 11	06 49	07 23
35	17 38	18 04	18 34	05 18	06 02	06 42	07 18
30	17 45	18 09	18 37	05 09	05 54	06 35	07 13
20	17 57	18 19	18 45	04 53	05 40	06 23	07 04
N 10	18 07	18 29	18 54	04 39	05 27	06 13	06 57
0	18 18	18 39	19 04	04 26	05 16	06 04	06 49
S 10	18 28	18 50	19 15	04 13	05 04	05 54	06 42
20	18 39	19 02	19 29	03 59	04 51	05 44	06 34
30	18 52	19 17	19 47	03 42	04 37	05 32	06 26
35	18 59	19 26	19 58	03 33	04 29	05 25	06 20
40	19 08	19 37	20 12	03 22	04 19	05 17	06 15
45	19 18	19 50	20 29	03 09	04 08	05 08	06 08
S 50	19 30	20 06	20 52	02 54	03 54	04 57	06 00
52	19 36	20 14	21 03	02 46	03 48	04 51	05 56
54	19 42	20 23	21 16	02 38	03 41	04 46	05 52
56	19 49	20 33	21 32	02 29	03 33	04 39	05 47
58	19 57	20 44	21 50	02 19	03 24	04 32	05 42
S 60	20 06	20 58	22 15	02 07	03 14	04 24	05 36

	SUN			MOON			
Day	Eqn. of Time 00h	12h	Mer. Pass.	Mer. Pass. Upper	Lower	Age	Phase
d	m s	m s	h m	h m	h m	d	%
9	14 12	14 13	12 14	23 04	10 39	12	96
10	14 13	14 14	12 14	23 52	11 28	13	99
11	14 14	14 14	12 14	24 38	12 16	14	100

1998 FEBRUARY 12, 13, 14 (THURS., FRI., SAT.)

UT	ARIES GHA	VENUS −4.6 GHA	VENUS Dec	MARS +1.2 GHA	MARS Dec	JUPITER −2.0 GHA	JUPITER Dec	SATURN +0.7 GHA	SATURN Dec
THURSDAY 12 00	141 50.5	212 02.4	S15 24.8	156 22.2	S 7 09.7	167 43.2	S11 39.0	125 50.9	N 4 16.5
01	156 52.9	227 04.3	24.9	171 22.8	09.0	182 45.1	38.8	140 53.1	16.6
02	171 55.4	242 06.1	25.1	186 23.5	08.2	197 47.0	38.5	155 55.4	16.6
03	186 57.9	257 07.9 ..	25.2	201 24.1 ..	07.4	212 48.9 ..	38.3	170 57.6 ..	16.7
04	202 00.3	272 09.8	25.3	216 24.8	06.6	227 50.7	38.1	185 59.9	16.8
05	217 02.8	287 11.6	25.5	231 25.4	05.9	242 52.6	37.9	201 02.2	16.9
06	232 05.3	302 13.4 S15	25.6	246 26.1 S 7	05.1	257 54.5 S11	37.7	216 04.4 N 4	17.0
07	247 07.7	317 15.2	25.7	261 26.7	04.3	272 56.4	37.5	231 06.7	17.1
08	262 10.2	332 17.0	25.9	276 27.4	03.5	287 58.3	37.3	246 08.9	17.2
09	277 12.7	347 18.9 ..	26.0	291 28.0 ..	02.8	303 00.2 ..	37.1	261 11.2 ..	17.3
10	292 15.1	2 20.7	26.1	306 28.7	02.0	318 02.1	36.8	276 13.4	17.4
11	307 17.6	17 22.5	26.3	321 29.3	01.2	333 04.0	36.6	291 15.7	17.5
12	322 20.1	32 24.3 S15	26.4	336 30.0 S 7	00.4	348 05.9 S11	36.4	306 17.9 N 4	17.6
13	337 22.5	47 26.1	26.5	351 30.6	6 59.7	3 07.8	36.2	321 20.2	17.7
14	352 25.0	62 27.8	26.7	6 31.3	58.9	18 09.7	36.0	336 22.4	17.8
15	7 27.4	77 29.6 ..	26.8	21 31.9 ..	58.1	33 11.6 ..	35.8	351 24.7 ..	17.9
16	22 29.9	92 31.4	26.9	36 32.6	57.3	48 13.5	35.6	6 26.9	18.0
17	37 32.4	107 33.2	27.1	51 33.2	56.5	63 15.3	35.3	21 29.2	18.1
18	52 34.8	122 35.0 S15	27.2	66 33.9 S 6	55.8	78 17.2 S11	35.1	36 31.4 N 4	18.2
19	67 37.3	137 36.7	27.4	81 34.5	55.0	93 19.1	34.9	51 33.7	18.3
20	82 39.8	152 38.5	27.5	96 35.2	54.2	108 21.0	34.7	66 36.0	18.4
21	97 42.2	167 40.3 ..	27.6	111 35.8 ..	53.4	123 22.9 ..	34.5	81 38.2 ..	18.5
22	112 44.7	182 42.0	27.8	126 36.5	52.7	138 24.8	34.3	96 40.5	18.6
23	127 47.2	197 43.8	27.9	141 37.1	51.9	153 26.7	34.1	111 42.7	18.7
FRIDAY 13 00	142 49.6	212 45.5 S15	28.0	156 37.8 S 6	51.1	168 28.6 S11	33.9	126 45.0 N 4	18.8
01	157 52.1	227 47.3	28.2	171 38.4	50.3	183 30.5	33.6	141 47.2	18.8
02	172 54.5	242 49.0	28.3	186 39.1	49.6	198 32.4	33.4	156 49.5	18.9
03	187 57.0	257 50.8 ..	28.4	201 39.7 ..	48.8	213 34.3 ..	33.2	171 51.7 ..	19.0
04	202 59.5	272 52.5	28.6	216 40.4	48.0	228 36.2	33.0	186 54.0	19.1
05	218 01.9	287 54.2	28.7	231 41.0	47.2	243 38.1	32.8	201 56.2	19.2
06	233 04.4	302 56.0 S15	28.8	246 41.7 S 6	46.4	258 40.0 S11	32.6	216 58.5 N 4	19.3
07	248 06.9	317 57.7	29.0	261 42.3	45.7	273 41.8	32.4	232 00.7	19.4
08	263 09.3	332 59.4	29.1	276 43.0	44.9	288 43.7	32.1	247 03.0	19.5
09	278 11.8	348 01.1 ..	29.2	291 43.6 ..	44.1	303 45.6 ..	31.9	262 05.2 ..	19.6
10	293 14.3	3 02.8	29.4	306 44.3	43.3	318 47.5	31.7	277 07.5	19.7
11	308 16.7	18 04.6	29.5	321 44.9	42.5	333 49.4	31.5	292 09.7	19.8
12	323 19.2	33 06.3 S15	29.6	336 45.6 S 6	41.8	348 51.3 S11	31.3	307 12.0 N 4	19.9
13	338 21.7	48 08.0	29.7	351 46.2	41.0	3 53.2	31.1	322 14.2	20.0
14	353 24.1	63 09.7	29.9	6 46.9	40.2	18 55.1	30.9	337 16.5	20.1
15	8 26.6	78 11.4 ..	30.0	21 47.5 ..	39.4	33 57.0 ..	30.7	352 18.7 ..	20.2
16	23 29.0	93 13.1	30.1	36 48.2	38.7	48 58.9	30.4	7 21.0	20.3
17	38 31.5	108 14.7	30.3	51 48.8	37.9	64 00.8	30.2	22 23.2	20.4
18	53 34.0	123 16.4 S15	30.4	66 49.5 S 6	37.1	79 02.7 S11	30.0	37 25.5 N 4	20.5
19	68 36.4	138 18.1	30.5	81 50.2	36.3	94 04.6	29.8	52 27.7	20.6
20	83 38.9	153 19.8	30.7	96 50.8	35.5	109 06.4	29.6	67 30.0	20.7
21	98 41.4	168 21.5 ..	30.8	111 51.5 ..	34.8	124 08.3 ..	29.4	82 32.2 ..	20.8
22	113 43.8	183 23.1	30.9	126 52.1	34.0	139 10.2	29.2	97 34.5	20.9
23	128 46.3	198 24.8	31.1	141 52.8	33.2	154 12.1	28.9	112 36.7	21.0
SATURDAY 14 00	143 48.8	213 26.5 S15	31.2	156 53.4 S 6	32.4	169 14.0 S11	28.7	127 39.0 N 4	21.1
01	158 51.2	228 28.1	31.3	171 54.1	31.6	184 15.9	28.5	142 41.2	21.2
02	173 53.7	243 29.8	31.5	186 54.7	30.9	199 17.8	28.3	157 43.5	21.3
03	188 56.2	258 31.4 ..	31.6	201 55.4 ..	30.1	214 19.7 ..	28.1	172 45.7 ..	21.4
04	203 58.6	273 33.1	31.7	216 56.0	29.3	229 21.6	27.9	187 48.0	21.5
05	219 01.1	288 34.7	31.8	231 56.7	28.5	244 23.5	27.7	202 50.2	21.6
06	234 03.5	303 36.4 S15	32.0	246 57.4 S 6	27.7	259 25.4 S11	27.4	217 52.5 N 4	21.7
07	249 06.0	318 38.0	32.1	261 58.0	27.0	274 27.3	27.2	232 54.7	21.8
08	264 08.5	333 39.6	32.2	276 58.7	26.2	289 29.2	27.0	247 57.0	21.8
09	279 10.9	348 41.3 ..	32.4	291 59.3 ..	25.4	304 31.0 ..	26.8	262 59.2 ..	21.9
10	294 13.4	3 42.9	32.5	307 00.0	24.6	319 32.9	26.6	278 01.4	22.0
11	309 15.9	18 44.5	32.6	322 00.6	23.8	334 34.8	26.4	293 03.7	22.1
12	324 18.3	33 46.2 S15	32.7	337 01.3 S 6	23.1	349 36.7 S11	26.2	308 05.9 N 4	22.2
13	339 20.8	48 47.8	32.9	352 01.9	22.3	4 38.6	25.9	323 08.2	22.3
14	354 23.3	63 49.4	33.0	7 02.6	21.5	19 40.5	25.7	338 10.4	22.4
15	9 25.7	78 51.0 ..	33.1	22 03.3 ..	20.7	34 42.4 ..	25.5	353 12.7 ..	22.5
16	24 28.2	93 52.6	33.3	37 03.9	19.9	49 44.3	25.3	8 14.9	22.6
17	39 30.6	108 54.2	33.4	52 04.6	19.2	64 46.2	25.1	23 17.2	22.7
18	54 33.1	123 55.8 S15	33.5	67 05.2 S 6	18.4	79 48.1 S11	24.9	38 19.4 N 4	22.8
19	69 35.6	138 57.4	33.6	82 05.9	17.6	94 50.0	24.7	53 21.7	22.9
20	84 38.0	153 59.0	33.8	97 06.5	16.8	109 51.9	24.4	68 23.9	23.0
21	99 40.5	169 00.6 ..	33.9	112 07.2 ..	16.0	124 53.8 ..	24.2	83 26.2 ..	23.1
22	114 43.0	184 02.2	34.0	127 07.8	15.3	139 55.6	24.0	98 28.4	23.2
23	129 45.4	199 03.7	34.2	142 08.5	14.5	154 57.5	23.8	113 30.7	23.3
Mer. Pass. 14 26.3		v 1.7	d 0.1	v 0.7	d 0.8	v 1.9	d 0.2	v 2.3	d 0.1

STARS

Name	SHA	Dec
Acamar	315 27.3	S40 19.1
Achernar	335 35.9	S57 15.2
Acrux	173 21.8	S63 05.1
Adhara	255 21.5	S28 58.5
Aldebaran	291 02.8	N16 30.2
Alioth	166 30.7	N55 58.0
Alkaid	153 08.0	N49 19.2
Al Na'ir	27 59.0	S46 58.3
Alnilam	275 58.1	S 1 12.4
Alphard	218 07.3	S 8 39.2
Alphecca	126 21.0	N26 43.1
Alpheratz	357 55.9	N29 04.8
Altair	62 20.0	N 8 51.8
Ankaa	353 27.6	S42 19.3
Antares	112 40.8	S26 25.5
Arcturus	146 06.4	N19 11.4
Atria	107 53.4	S69 01.1
Avior	234 22.1	S59 30.4
Bellatrix	278 44.5	N 6 20.7
Betelgeuse	271 13.8	N 7 24.2
Canopus	264 01.0	S52 42.1
Capella	280 51.6	N45 59.8
Deneb	49 39.9	N45 16.4
Denebola	182 45.4	N14 34.8
Diphda	349 07.9	S18 00.1
Dubhe	194 05.5	N61 45.5
Elnath	278 27.3	N28 36.3
Eltanin	90 51.9	N51 29.2
Enif	33 59.0	N 9 51.9
Fomalhaut	15 37.3	S29 38.1
Gacrux	172 13.5	S57 06.0
Gienah	176 04.1	S17 31.9
Hadar	149 04.3	S60 21.5
Hamal	328 14.1	N23 27.1
Kaus Aust.	83 59.7	S34 23.0
Kochab	137 19.5	N74 09.6
Markab	13 50.3	N15 11.6
Menkar	314 27.4	N 4 04.7
Menkent	148 21.3	S36 21.5
Miaplacidus	221 41.1	S69 42.7
Mirfak	308 57.2	N49 51.3
Nunki	76 13.1	S26 17.8
Peacock	53 38.3	S56 44.4
Pollux	243 41.8	N28 01.7
Procyon	245 11.7	N 5 13.6
Rasalhague	96 17.5	N12 33.7
Regulus	207 55.7	N11 58.4
Rigel	281 23.2	S 8 12.5
Rigil Kent.	140 07.6	S60 49.3
Sabik	102 26.2	S15 43.2
Schedar	349 54.3	N56 31.7
Shaula	96 38.1	S37 05.9
Sirius	258 43.8	S16 43.1
Spica	158 43.5	S11 09.1
Suhail	223 00.6	S43 25.7
Vega	80 47.2	N38 46.8
Zuben'ubi	137 18.4	S16 01.9

	SHA	Mer. Pass.
Venus	69 55.9	9 48
Mars	13 48.1	13 33
Jupiter	25 39.0	12 44
Saturn	343 55.3	15 31

UT	SUN GHA	SUN Dec	MOON GHA	v	MOON Dec	d	HP
d h	° ′	° ′	° ′	′	° ′	′	′
12 00	176 26.5	S13 49.8	350 40.4	13.6	N10 43.5	8.1	54.8
01	191 26.5	48.9	5 13.0	13.5	10 35.4	8.1	54.8
02	206 26.5	48.1	19 45.5	13.6	10 27.3	8.1	54.7
03	221 26.5	.. 47.3	34 18.1	13.7	10 19.2	8.1	54.7
04	236 26.5	46.4	48 50.8	13.7	10 11.1	8.2	54.7
05	251 26.5	45.6	63 23.5	13.7	10 02.9	8.3	54.7
06	266 26.5	S13 44.8	77 56.2	13.8	N 9 54.6	8.2	54.7
07	281 26.5	44.0	92 29.0	13.8	9 46.4	8.4	54.7
08	296 26.5	43.1	107 01.8	13.9	9 38.0	8.3	54.7
09	311 26.5	.. 42.3	121 34.7	13.9	9 29.7	8.4	54.7
10	326 26.5	41.5	136 07.6	13.9	9 21.3	8.4	54.6
11	341 26.5	40.6	150 40.5	14.0	9 12.9	8.5	54.6
12	356 26.6	S13 39.8	165 13.5	14.0	N 9 04.4	8.5	54.6
13	11 26.6	39.0	179 46.5	14.0	8 55.9	8.5	54.6
14	26 26.6	38.1	194 19.5	14.1	8 47.4	8.6	54.6
15	41 26.6	.. 37.3	208 52.6	14.1	8 38.8	8.6	54.6
16	56 26.6	36.5	223 25.7	14.2	8 30.2	8.6	54.6
17	71 26.6	35.6	237 58.9	14.2	8 21.6	8.7	54.6
18	86 26.6	S13 34.8	252 32.1	14.2	N 8 12.9	8.7	54.5
19	101 26.6	34.0	267 05.3	14.3	8 04.2	8.7	54.5
20	116 26.6	33.1	281 38.6	14.3	7 55.5	8.7	54.5
21	131 26.6	.. 32.3	296 11.9	14.3	7 46.8	8.8	54.5
22	146 26.7	31.5	310 45.2	14.4	7 38.0	8.8	54.5
23	161 26.7	30.6	325 18.6	14.4	7 29.2	8.9	54.5
13 00	176 26.7	S13 29.8	339 52.0	14.4	N 7 20.3	8.8	54.5
01	191 26.7	28.9	354 25.4	14.5	7 11.5	8.9	54.5
02	206 26.7	28.1	8 58.9	14.5	7 02.6	8.9	54.4
03	221 26.7	.. 27.3	23 32.4	14.5	6 53.7	9.0	54.4
04	236 26.7	26.4	38 05.9	14.6	6 44.7	8.9	54.4
05	251 26.7	25.6	52 39.5	14.6	6 35.8	9.0	54.4
06	266 26.8	S13 24.8	67 13.1	14.6	N 6 26.8	9.0	54.4
07	281 26.8	23.9	81 46.7	14.6	6 17.8	9.0	54.4
08	296 26.8	23.1	96 20.3	14.7	6 08.8	9.1	54.4
09	311 26.8	.. 22.2	110 54.0	14.7	5 59.7	9.1	54.4
10	326 26.8	21.4	125 27.7	14.8	5 50.6	9.1	54.4
11	341 26.8	20.6	140 01.5	14.7	5 41.5	9.1	54.3
12	356 26.9	S13 19.7	154 35.2	14.8	N 5 32.4	9.1	54.3
13	11 26.9	18.9	169 09.0	14.9	5 23.3	9.2	54.3
14	26 26.9	18.0	183 42.9	14.8	5 14.1	9.1	54.3
15	41 26.9	.. 17.2	198 16.7	14.9	5 05.0	9.2	54.3
16	56 26.9	16.3	212 50.6	14.9	4 55.8	9.2	54.3
17	71 26.9	15.5	227 24.5	14.9	4 46.6	9.2	54.3
18	86 27.0	S13 14.7	241 58.4	14.9	N 4 37.4	9.3	54.3
19	101 27.0	13.8	256 32.3	15.0	4 28.1	9.2	54.3
20	116 27.0	13.0	271 06.3	15.0	4 18.9	9.3	54.3
21	131 27.0	.. 12.1	285 40.3	15.0	4 09.6	9.3	54.3
22	146 27.0	11.3	300 14.3	15.0	4 00.3	9.3	54.2
23	161 27.1	10.4	314 48.3	15.1	3 51.0	9.3	54.2
14 00	176 27.1	S13 09.6	329 22.4	15.0	N 3 41.7	9.3	54.2
01	191 27.1	08.7	343 56.4	15.1	3 32.4	9.3	54.2
02	206 27.1	07.9	358 30.5	15.1	3 23.1	9.3	54.2
03	221 27.1	.. 07.0	13 04.6	15.2	3 13.8	9.4	54.2
04	236 27.2	06.2	27 38.8	15.1	3 04.4	9.3	54.2
05	251 27.2	05.4	42 12.9	15.2	2 55.1	9.4	54.2
06	266 27.2	S13 04.5	56 47.1	15.2	N 2 45.7	9.3	54.2
07	281 27.2	03.7	71 21.3	15.2	2 36.4	9.4	54.2
08	296 27.2	02.8	85 55.5	15.2	2 27.0	9.4	54.2
09	311 27.3	.. 02.0	100 29.7	15.2	2 17.6	9.4	54.2
10	326 27.3	01.1	115 03.9	15.2	2 08.2	9.4	54.2
11	341 27.3	13 00.3	129 38.1	15.3	1 58.8	9.4	54.2
12	356 27.3	S12 59.4	144 12.4	15.3	N 1 49.4	9.4	54.2
13	11 27.4	58.6	158 46.7	15.3	1 40.0	9.4	54.1
14	26 27.4	57.7	173 21.0	15.2	1 30.6	9.4	54.1
15	41 27.4	.. 56.9	187 55.2	15.4	1 21.2	9.4	54.1
16	56 27.4	56.0	202 29.6	15.3	1 11.8	9.5	54.1
17	71 27.5	55.2	217 03.9	15.3	1 02.3	9.4	54.1
18	86 27.5	S12 54.3	231 38.2	15.3	N 0 52.9	9.4	54.1
19	101 27.5	53.4	246 12.5	15.4	0 43.5	9.4	54.1
20	116 27.5	52.6	260 46.9	15.3	0 34.1	9.4	54.1
21	131 27.6	.. 51.7	275 21.2	15.4	0 24.7	9.4	54.1
22	146 27.6	50.9	289 55.6	15.4	0 15.3	9.5	54.1
23	161 27.6	50.0	304 30.0	15.3	N 0 05.8	9.4	54.1
	SD 16.2	d 0.8	SD 14.9		14.8		14.8

Left margin day labels: THURSDAY (12), FRIDAY (13), SATURDAY (14).

Lat.	Twilight Naut.	Twilight Civil	Sunrise	Moonrise 12	13	14	15
°	h m	h m	h m	h m	h m	h m	h m
N 72	06 24	07 45	09 07	17 23	18 59	20 33	22 06
N 70	06 21	07 34	08 44	17 36	19 06	20 34	22 01
68	06 19	07 25	08 27	17 46	19 11	20 34	21 57
66	06 16	07 17	08 14	17 55	19 15	20 34	21 53
64	06 14	07 11	08 02	18 02	19 19	20 35	21 50
62	06 13	07 05	07 52	18 08	19 22	20 35	21 48
60	06 11	07 00	07 44	18 13	19 25	20 35	21 46
N 58	06 09	06 55	07 37	18 18	19 27	20 36	21 44
56	06 08	06 51	07 30	18 22	19 29	20 36	21 42
54	06 06	06 48	07 25	18 26	19 31	20 36	21 40
52	06 05	06 44	07 19	18 29	19 33	20 36	21 39
50	06 03	06 41	07 15	18 32	19 34	20 36	21 38
45	06 00	06 34	07 04	18 39	19 38	20 37	21 35
N 40	05 56	06 28	06 56	18 44	19 41	20 37	21 32
35	05 53	06 22	06 48	18 49	19 43	20 37	21 30
30	05 49	06 17	06 42	18 53	19 46	20 37	21 29
20	05 42	06 08	06 30	19 00	19 49	20 38	21 26
N 10	05 34	05 59	06 20	19 07	19 53	20 38	21 23
0	05 25	05 50	06 11	19 12	19 56	20 38	21 21
S 10	05 14	05 39	06 01	19 18	19 59	20 39	21 18
20	05 01	05 28	05 51	19 25	20 02	20 39	21 15
30	04 44	05 13	05 39	19 32	20 06	20 39	21 13
35	04 33	05 05	05 32	19 36	20 08	20 40	21 11
40	04 20	04 55	05 23	19 40	20 11	20 40	21 09
45	04 03	04 42	05 14	19 46	20 14	20 40	21 07
S 50	03 42	04 27	05 03	19 52	20 17	20 41	21 04
52	03 31	04 19	04 57	19 55	20 19	20 41	21 03
54	03 19	04 11	04 51	19 58	20 20	20 41	21 02
56	03 04	04 02	04 45	20 02	20 22	20 41	21 00
58	02 47	03 51	04 37	20 06	20 24	20 42	20 59
S 60	02 25	03 38	04 29	20 10	20 27	20 42	20 57

Lat.	Sunset	Twilight Civil	Twilight Naut.	Moonset 12	13	14	15
°	h m	h m	h m	h m	h m	h m	h m
N 72	15 23	16 45	18 07	08 56	08 51	08 45	08 40
N 70	15 45	16 56	18 09	08 42	08 42	08 42	08 42
68	16 02	17 05	18 11	08 30	08 36	08 40	08 44
66	16 16	17 13	18 13	08 21	08 30	08 38	08 45
64	16 27	17 19	18 15	08 13	08 25	08 36	08 46
62	16 37	17 25	18 17	08 06	08 21	08 34	08 47
60	16 45	17 30	18 19	07 59	08 17	08 33	08 48
N 58	16 53	17 34	18 20	07 54	08 14	08 32	08 49
56	16 59	17 38	18 22	07 49	08 11	08 31	08 50
54	17 05	17 42	18 23	07 45	08 08	08 30	08 50
52	17 10	17 45	18 25	07 41	08 06	08 29	08 51
50	17 15	17 48	18 26	07 38	08 04	08 28	08 51
45	17 25	17 55	18 30	07 30	07 59	08 26	08 53
N 40	17 33	18 01	18 33	07 23	07 55	08 25	08 54
35	17 40	18 06	18 36	07 18	07 51	08 23	08 54
30	17 47	18 11	18 40	07 13	07 48	08 22	08 55
20	17 58	18 21	18 47	07 04	07 43	08 20	08 56
N 10	18 08	18 30	18 55	06 57	07 38	08 18	08 58
0	18 18	18 39	19 04	06 49	07 33	08 16	08 59
S 10	18 27	18 49	19 14	06 42	07 29	08 15	09 00
20	18 38	19 00	19 27	06 34	07 24	08 13	09 01
30	18 49	19 14	19 44	06 26	07 18	08 11	09 02
35	18 56	19 23	19 55	06 20	07 15	08 09	09 03
40	19 04	19 33	20 08	06 15	07 12	08 08	09 04
45	19 14	19 45	20 24	06 08	07 07	08 06	09 05
S 50	19 25	20 00	20 45	06 00	07 02	08 04	09 06
52	19 30	20 08	20 56	05 56	07 00	08 03	09 06
54	19 36	20 16	21 08	05 52	06 57	08 02	09 07
56	19 42	20 25	21 22	05 47	06 54	08 01	09 07
58	19 50	20 35	21 39	05 42	06 51	08 00	09 08
S 60	19 58	20 48	22 00	05 36	06 47	07 59	09 09

Day	SUN Eqn. of Time 00h	SUN Eqn. of Time 12h	SUN Mer. Pass.	MOON Mer. Pass. Upper	MOON Mer. Pass. Lower	Age	Phase
d	m s	m s	h m	h m	h m	d	%
12	14 14	14 14	12 14	00 38	13 01	15	99
13	14 13	14 13	12 14	01 23	13 45	16	96
14	14 12	14 11	12 14	02 06	14 27	17	92

UT	ARIES	VENUS −4.6		MARS +1.2		JUPITER −2.0		SATURN +0.7		STARS		
	GHA	GHA	Dec	GHA	Dec	GHA	Dec	GHA	Dec	Name	SHA	Dec
d h	° ′	° ′	° ′	° ′	° ′	° ′	° ′	° ′	° ′		° ′	° ′
15 00	144 47.9	214 05.3	S15 34.3	157 09.2	S 6 13.7	169 59.4	S11 23.6	128 32.9	N 4 23.4	Acamar	315 27.4	S40 19.1
01	159 50.4	229 06.9	34.4	172 09.8	12.9	185 01.3	23.4	143 35.2	23.5	Achernar	335 35.9	S57 15.2
02	174 52.8	244 08.5	34.5	187 10.5	12.1	200 03.2	23.2	158 37.4	23.6	Acrux	173 21.8	S63 05.1
03	189 55.3	259 10.0 ..	34.7	202 11.1 ..	11.3	215 05.1 ..	22.9	173 39.6 ..	23.7	Adhara	255 21.5	S28 58.5
04	204 57.8	274 11.6	34.8	217 11.8	10.6	230 07.0	22.7	188 41.9	23.8	Aldebaran	291 02.8	N16 30.2
05	220 00.2	289 13.2	34.9	232 12.4	09.8	245 08.9	22.5	203 44.1	23.9			
06	235 02.7	304 14.7	S15 35.0	247 13.1	S 6 09.0	260 10.8	S11 22.3	218 46.4	N 4 24.0	Alioth	166 30.7	N55 58.0
07	250 05.1	319 16.3	35.2	262 13.8	08.2	275 12.7	22.1	233 48.6	24.1	Alkaid	153 08.0	N49 19.2
08	265 07.6	334 17.8	35.3	277 14.4	07.4	290 14.6	21.9	248 50.9	24.2	Al Na'ir	27 59.0	S46 58.3
S 09	280 10.1	349 19.4 ..	35.4	292 15.1 ..	06.7	305 16.5 ..	21.7	263 53.1 ..	24.3	Alnilam	275 58.1	S 1 12.4
U 10	295 12.5	4 20.9	35.5	307 15.7	05.9	320 18.4	21.4	278 55.4	24.4	Alphard	218 07.3	S 8 39.2
N 11	310 15.0	19 22.5	35.7	322 16.4	05.1	335 20.2	21.2	293 57.6	24.5			
D 12	325 17.5	34 24.0	S15 35.8	337 17.1	S 6 04.3	350 22.1	S11 21.0	308 59.9	N 4 24.6	Alphecca	126 21.0	N26 43.1
A 13	340 19.9	49 25.5	35.9	352 17.7	03.5	5 24.0	20.8	324 02.1	24.7	Alpheratz	357 55.9	N29 04.8
Y 14	355 22.4	64 27.1	36.0	7 18.4	02.7	20 25.9	20.6	339 04.3	24.8	Altair	62 20.0	N 8 51.8
15	10 24.9	79 28.6 ..	36.2	22 19.0 ..	02.0	35 27.8 ..	20.4	354 06.6 ..	24.9	Ankaa	353 27.6	S42 19.2
16	25 27.3	94 30.1	36.3	37 19.7	01.2	50 29.7	20.2	9 08.8	25.0	Antares	112 40.8	S26 25.5
17	40 29.8	109 31.6	36.4	52 20.4	6 00.4	65 31.6	19.9	24 11.1	25.1			
18	55 32.2	124 33.1	S15 36.5	67 21.0	S 5 59.6	80 33.5	S11 19.7	39 13.3	N 4 25.2	Arcturus	146 06.3	N19 11.4
19	70 34.7	139 34.7	36.6	82 21.7	58.8	95 35.4	19.5	54 15.6	25.3	Atria	107 53.3	S69 01.1
20	85 37.2	154 36.2	36.8	97 22.3	58.0	110 37.3	19.3	69 17.8	25.4	Avior	234 22.1	S59 30.5
21	100 39.6	169 37.7 ..	36.9	112 23.0 ..	57.3	125 39.2 ..	19.1	84 20.1 ..	25.5	Bellatrix	278 44.5	N 6 20.7
22	115 42.1	184 39.2	37.0	127 23.6	56.5	140 41.1	18.9	99 22.3	25.6	Betelgeuse	271 13.8	N 7 24.2
23	130 44.6	199 40.7	37.1	142 24.3	55.7	155 43.0	18.7	114 24.5	25.7			
16 00	145 47.0	214 42.2	S15 37.3	157 25.0	S 5 54.9	170 44.8	S11 18.4	129 26.8	N 4 25.8	Canopus	264 01.0	S52 42.1
01	160 49.5	229 43.7	37.4	172 25.6	54.1	185 46.7	18.2	144 29.0	25.9	Capella	280 51.6	N45 59.8
02	175 52.0	244 45.1	37.5	187 26.3	53.3	200 48.6	18.0	159 31.3	26.0	Deneb	49 39.9	N45 16.4
03	190 54.4	259 46.6 ..	37.6	202 26.9 ..	52.6	215 50.5 ..	17.8	174 33.5 ..	26.1	Denebola	182 45.4	N14 34.8
04	205 56.9	274 48.1	37.7	217 27.6	51.8	230 52.4	17.6	189 35.8	26.2	Diphda	349 07.9	S18 00.1
05	220 59.4	289 49.6	37.9	232 28.3	51.0	245 54.3	17.4	204 38.0	26.3			
06	236 01.8	304 51.1	S15 38.0	247 28.9	S 5 50.2	260 56.2	S11 17.2	219 40.3	N 4 26.4	Dubhe	194 05.5	N61 45.5
07	251 04.3	319 52.5	38.1	262 29.6	49.4	275 58.1	16.9	234 42.5	26.5	Elnath	278 27.3	N28 36.3
08	266 06.7	334 54.0	38.2	277 30.3	48.6	291 00.0	16.7	249 44.7	26.6	Eltanin	90 51.9	N51 29.2
M 09	281 09.2	349 55.5 ..	38.3	292 30.9 ..	47.9	306 01.9 ..	16.5	264 47.0 ..	26.7	Enif	33 59.0	N 9 51.9
O 10	296 11.7	4 56.9	38.5	307 31.6	47.1	321 03.8	16.3	279 49.2	26.8	Fomalhaut	15 37.3	S29 38.1
N 11	311 14.1	19 58.4	38.6	322 32.2	46.3	336 05.7	16.1	294 51.5	26.9			
D 12	326 16.6	34 59.8	S15 38.7	337 32.9	S 5 45.5	351 07.5	S11 15.9	309 53.7	N 4 27.0	Gacrux	172 13.5	S57 06.0
A 13	341 19.1	50 01.3	38.8	352 33.6	44.7	6 09.4	15.6	324 55.9	27.1	Gienah	176 04.1	S17 31.9
Y 14	356 21.5	65 02.8	38.9	7 34.2	43.9	21 11.3	15.4	339 58.2	27.2	Hadar	149 04.3	S60 21.5
15	11 24.0	80 04.2 ..	39.1	22 34.9 ..	43.2	36 13.2 ..	15.2	355 00.4 ..	27.3	Hamal	328 14.2	N23 27.1
16	26 26.5	95 05.6	39.2	37 35.5	42.4	51 15.1	15.0	10 02.7	27.4	Kaus Aust.	83 59.6	S34 23.0
17	41 28.9	110 07.1	39.3	52 36.2	41.6	66 17.0	14.8	25 04.9	27.5			
18	56 31.4	125 08.5	S15 39.4	67 36.9	S 5 40.8	81 18.9	S11 14.6	40 07.2	N 4 27.6	Kochab	137 19.5	N74 09.6
19	71 33.8	140 09.9	39.5	82 37.5	40.0	96 20.8	14.4	55 09.4	27.7	Markab	13 50.3	N15 11.6
20	86 36.3	155 11.4	39.6	97 38.2	39.2	111 22.7	14.1	70 11.6	27.8	Menkar	314 27.4	N 4 04.7
21	101 38.8	170 12.8 ..	39.8	112 38.9 ..	38.4	126 24.6 ..	13.9	85 13.9 ..	27.9	Menkent	148 21.3	S36 21.5
22	116 41.2	185 14.2	39.9	127 39.5	37.7	141 26.5	13.7	100 16.1	28.0	Miaplacidus	221 41.1	S69 42.7
23	131 43.7	200 15.6	40.0	142 40.2	36.9	156 28.4	13.5	115 18.4	28.1			
17 00	146 46.2	215 17.1	S15 40.1	157 40.8	S 5 36.1	171 30.3	S11 13.3	130 20.6	N 4 28.2	Mirfak	308 57.2	N49 51.3
01	161 48.6	230 18.5	40.2	172 41.5	35.3	186 32.1	13.1	145 22.8	28.3	Nunki	76 13.1	S26 17.8
02	176 51.1	245 19.9	40.3	187 42.2	34.5	201 34.0	12.8	160 25.1	28.4	Peacock	53 38.3	S56 44.4
03	191 53.6	260 21.3 ..	40.5	202 42.8 ..	33.7	216 35.9 ..	12.6	175 27.3 ..	28.5	Pollux	243 41.8	N28 01.7
04	206 56.0	275 22.7	40.6	217 43.5	32.9	231 37.8	12.4	190 29.6	28.6	Procyon	245 11.7	N 5 13.6
05	221 58.5	290 24.1	40.7	232 44.2	32.2	246 39.7	12.2	205 31.8	28.7			
06	237 01.0	305 25.5	S15 40.8	247 44.8	S 5 31.4	261 41.6	S11 12.0	220 34.1	N 4 28.8	Rasalhague	96 17.5	N12 33.6
07	252 03.4	320 26.9	40.9	262 45.5	30.6	276 43.5	11.8	235 36.3	28.9	Regulus	207 55.7	N11 58.4
T 08	267 05.9	335 28.3	41.0	277 46.1	29.8	291 45.4	11.6	250 38.5	29.0	Rigel	281 23.2	S 8 12.5
U 09	282 08.3	350 29.7 ..	41.1	292 46.8 ..	29.0	306 47.3 ..	11.3	265 40.8 ..	29.1	Rigil Kent.	140 07.6	S60 49.3
E 10	297 10.8	5 31.0	41.2	307 47.5	28.2	321 49.2	11.1	280 43.0	29.2	Sabik	102 26.1	S15 43.2
S 11	312 13.3	20 32.4	41.4	322 48.1	27.4	336 51.1	10.9	295 45.3	29.3			
D 12	327 15.7	35 33.8	S15 41.5	337 48.8	S 5 26.7	351 53.0	S11 10.7	310 47.5	N 4 29.4	Schedar	349 54.3	N56 31.7
A 13	342 18.2	50 35.2	41.6	352 49.5	25.9	6 54.8	10.5	325 49.7	29.5	Shaula	96 38.1	S37 05.9
Y 14	357 20.7	65 36.5	41.7	7 50.1	25.1	21 56.7	10.3	340 52.0	29.6	Sirius	258 43.8	S16 43.1
15	12 23.1	80 37.9 ..	41.8	22 50.8 ..	24.3	36 58.6 ..	10.0	355 54.2 ..	29.7	Spica	158 43.4	S11 09.1
16	27 25.6	95 39.3	41.9	37 51.5	23.5	52 00.5	09.8	10 56.5	29.8	Suhail	223 00.6	S43 25.7
17	42 28.1	110 40.6	42.0	52 52.1	22.7	67 02.4	09.6	25 58.7	29.9			
18	57 30.5	125 42.0	S15 42.1	67 52.8	S 5 21.9	82 04.3	S11 09.4	41 00.9	N 4 30.0	Vega	80 47.2	N38 46.8
19	72 33.0	140 43.3	42.3	82 53.5	21.2	97 06.2	09.2	56 03.2	30.1	Zuben'ubi	137 18.4	S16 02.0
20	87 35.5	155 44.7	42.4	97 54.1	20.4	112 08.1	09.0	71 05.4	30.2		SHA	Mer. Pass.
21	102 37.9	170 46.0 ..	42.5	112 54.8 ..	19.6	127 10.0 ..	08.8	86 07.6 ..	30.3		° ′	h m
22	117 40.4	185 47.4	42.6	127 55.5	18.8	142 11.9	08.5	101 09.9	30.4	Venus	68 55.1	9 40
23	132 42.8	200 48.7	42.7	142 56.1	18.0	157 13.8	08.3	116 12.1	30.5	Mars	11 37.9	13 30
Mer. Pass. 14 14.5		*v* 1.5	*d* 0.1	*v* 0.7	*d* 0.8	*v* 1.9	*d* 0.2	*v* 2.2	*d* 0.1	Jupiter	24 57.8	12 35
										Saturn	343 39.8	15 20

UT	SUN GHA	SUN Dec	MOON GHA	v	MOON Dec	d	HP
d h	° ′	° ′	° ′	′	° ′	′	′
15 00	176 27.6	S12 49.2	319 04.3	15.4	S 0 03.6	9.4	54.1
01	191 27.7	48.3	333 38.7	15.4	0 13.0	9.4	54.1
02	206 27.7	47.5	348 13.1	15.4	0 22.4	9.4	54.1
03	221 27.7	.. 46.6	2 47.5	15.4	0 31.8	9.4	54.1
04	236 27.8	45.8	17 21.9	15.4	0 41.2	9.4	54.1
05	251 27.8	44.9	31 56.3	15.4	0 50.6	9.4	54.1
06	266 27.8	S12 44.0	46 30.7	15.4	S 1 00.0	9.4	54.1
07	281 27.8	43.2	61 05.1	15.4	1 09.4	9.4	54.1
S 08	296 27.9	42.3	75 39.5	15.4	1 18.8	9.4	54.1
U 09	311 27.9	.. 41.5	90 13.9	15.4	1 28.2	9.3	54.1
N 10	326 27.9	40.6	104 48.3	15.4	1 37.5	9.4	54.1
D 11	341 28.0	39.8	119 22.7	15.4	1 46.9	9.3	54.1
A 12	356 28.0	S12 38.9	133 57.1	15.4	S 1 56.2	9.4	54.1
Y 13	11 28.0	38.0	148 31.5	15.4	2 05.6	9.3	54.1
14	26 28.1	37.2	163 05.9	15.4	2 14.9	9.3	54.1
15	41 28.1	.. 36.3	177 40.3	15.4	2 24.2	9.4	54.1
16	56 28.1	35.5	192 14.7	15.3	2 33.6	9.3	54.1
17	71 28.2	34.6	206 49.0	15.4	2 42.9	9.3	54.1
18	86 28.2	S12 33.7	221 23.4	15.4	S 2 52.2	9.2	54.1
19	101 28.2	32.9	235 57.8	15.4	3 01.4	9.3	54.1
20	116 28.3	32.0	250 32.2	15.3	3 10.7	9.3	54.1
21	131 28.3	.. 31.1	265 06.5	15.4	3 20.0	9.2	54.1
22	146 28.3	30.3	279 40.9	15.3	3 29.2	9.2	54.1
23	161 28.4	29.4	294 15.2	15.3	3 38.4	9.2	54.1
16 00	176 28.4	S12 28.6	308 49.5	15.4	S 3 47.6	9.2	54.1
01	191 28.4	27.7	323 23.9	15.3	3 56.8	9.2	54.1
02	206 28.5	26.8	337 58.2	15.3	4 06.0	9.2	54.1
03	221 28.5	.. 26.0	352 32.5	15.3	4 15.2	9.1	54.1
04	236 28.5	25.1	7 06.8	15.3	4 24.3	9.1	54.1
05	251 28.6	24.2	21 41.1	15.2	4 33.4	9.1	54.1
06	266 28.6	S12 23.4	36 15.3	15.3	S 4 42.5	9.1	54.1
07	281 28.7	22.5	50 49.6	15.2	4 51.6	9.1	54.1
M 08	296 28.7	21.6	65 23.8	15.2	5 00.7	9.1	54.1
O 09	311 28.7	.. 20.8	79 58.0	15.2	5 09.8	9.0	54.1
N 10	326 28.8	19.9	94 32.2	15.2	5 18.8	9.0	54.1
D 11	341 28.8	19.0	109 06.4	15.2	5 27.8	9.0	54.1
A 12	356 28.8	S12 18.2	123 40.6	15.2	S 5 36.8	9.0	54.1
Y 13	11 28.9	17.3	138 14.8	15.1	5 45.8	8.9	54.1
14	26 28.9	16.4	152 48.9	15.1	5 54.7	8.9	54.1
15	41 29.0	.. 15.6	167 23.0	15.1	6 03.6	8.9	54.1
16	56 29.0	14.7	181 57.1	15.1	6 12.5	8.9	54.2
17	71 29.0	13.8	196 31.2	15.1	6 21.4	8.9	54.2
18	86 29.1	S12 13.0	211 05.3	15.0	S 6 30.3	8.8	54.2
19	101 29.1	12.1	225 39.3	15.0	6 39.1	8.8	54.2
20	116 29.2	11.2	240 13.3	15.0	6 47.9	8.8	54.2
21	131 29.2	.. 10.3	254 47.3	15.0	6 56.7	8.7	54.2
22	146 29.2	09.5	269 21.3	14.9	7 05.4	8.7	54.2
23	161 29.3	08.6	283 55.2	14.9	7 14.1	8.7	54.2
17 00	176 29.3	S12 07.7	298 29.1	14.9	S 7 22.8	8.7	54.2
01	191 29.4	06.9	313 03.0	14.9	7 31.5	8.6	54.2
02	206 29.4	06.0	327 36.9	14.9	7 40.1	8.6	54.2
03	221 29.5	.. 05.1	342 10.8	14.8	7 48.7	8.6	54.2
04	236 29.5	04.2	356 44.6	14.8	7 57.3	8.6	54.2
05	251 29.5	03.4	11 18.4	14.7	8 05.9	8.5	54.3
06	266 29.6	S12 02.5	25 52.1	14.8	S 8 14.4	8.5	54.3
07	281 29.6	01.6	40 25.9	14.7	8 22.9	8.4	54.3
T 08	296 29.7	12 00.8	54 59.6	14.6	8 31.3	8.4	54.3
U 09	311 29.7	11 59.9	69 33.2	14.7	8 39.7	8.4	54.3
E 10	326 29.8	59.0	84 06.9	14.6	8 48.1	8.4	54.3
S 11	341 29.8	58.1	98 40.5	14.6	8 56.5	8.3	54.3
D 12	356 29.9	S11 57.3	113 14.1	14.5	S 9 04.8	8.3	54.3
A 13	11 29.9	56.4	127 47.6	14.5	9 13.1	8.2	54.3
Y 14	26 30.0	55.5	142 21.1	14.5	9 21.3	8.2	54.4
15	41 30.0	.. 54.6	156 54.6	14.4	9 29.5	8.2	54.4
16	56 30.0	53.7	171 28.0	14.4	9 37.7	8.1	54.4
17	71 30.1	52.9	186 01.4	14.4	9 45.8	8.1	54.4
18	86 30.1	S11 52.0	200 34.8	14.3	S 9 53.9	8.1	54.4
19	101 30.2	51.1	215 08.1	14.3	10 02.0	8.0	54.4
20	116 30.2	50.2	229 41.4	14.3	10 10.0	8.0	54.4
21	131 30.3	.. 49.4	244 14.7	14.2	10 18.0	8.0	54.4
22	146 30.3	48.5	258 47.9	14.2	10 26.0	7.9	54.5
23	161 30.4	47.6	273 21.1	14.1	S10 33.9	7.8	54.5
	SD 16.2	d 0.9	SD 14.7		14.7		14.8

Lat.	Naut.	Civil	Sunrise	Moonrise 15	16	17	18
°	h m	h m	h m	h m	h m	h m	h m
N 72	06 11	07 31	08 49	22 06	23 40	25 17	01 17
N 70	06 10	07 21	08 30	22 01	23 29	24 58	00 58
68	06 08	07 14	08 15	21 57	23 20	24 43	00 43
66	06 07	07 07	08 02	21 53	23 12	24 31	00 31
64	06 06	07 02	07 52	21 50	23 06	24 22	00 22
62	06 05	06 57	07 44	21 48	23 00	24 13	00 13
60	06 04	06 52	07 36	21 46	22 56	24 06	00 06
N 58	06 03	06 48	07 29	21 44	22 52	23 59	25 07
56	06 01	06 45	07 24	21 42	22 48	23 54	25 00
54	06 00	06 42	07 18	21 40	22 45	23 49	24 53
52	05 59	06 39	07 14	21 39	22 42	23 44	24 47
50	05 58	06 36	07 09	21 38	22 39	23 40	24 42
45	05 55	06 30	07 00	21 35	22 33	23 31	24 30
N 40	05 53	06 24	06 52	21 32	22 28	23 24	24 20
35	05 50	06 19	06 45	21 30	22 24	23 18	24 12
30	05 47	06 15	06 39	21 29	22 20	23 12	24 05
20	05 40	06 06	06 29	21 26	22 14	23 03	23 53
N 10	05 33	05 58	06 20	21 23	22 08	22 54	23 42
0	05 25	05 50	06 11	21 21	22 03	22 47	23 32
S 10	05 15	05 40	06 02	21 18	21 58	22 39	23 22
20	05 03	05 29	05 52	21 15	21 52	22 31	23 11
30	04 47	05 16	05 41	21 13	21 46	22 21	22 59
35	04 36	05 08	05 34	21 11	21 43	22 16	22 52
40	04 24	04 58	05 27	21 09	21 39	22 10	22 44
45	04 08	04 47	05 18	21 07	21 34	22 03	22 35
S 50	03 48	04 33	05 08	21 04	21 29	21 55	22 24
52	03 39	04 26	05 03	21 03	21 26	21 51	22 19
54	03 27	04 18	04 57	21 02	21 23	21 47	22 13
56	03 14	04 09	04 51	21 00	21 20	21 42	22 07
58	02 58	03 59	04 45	20 59	21 17	21 37	22 00
S 60	02 39	03 48	04 37	20 57	21 13	21 31	21 52

Lat.	Sunset	Civil	Naut.	Moonset 15	16	17	18
°	h m	h m	h m	h m	h m	h m	h m
N 72	15 41	16 59	18 19	08 40	08 34	08 29	08 22
N 70	16 00	17 08	18 20	08 42	08 42	08 42	08 42
68	16 15	17 16	18 21	08 44	08 47	08 52	08 58
66	16 27	17 22	18 23	08 45	08 52	09 01	09 11
64	16 37	17 28	18 24	08 46	08 57	09 08	09 22
62	16 46	17 33	18 25	08 47	09 00	09 15	09 31
60	16 53	17 37	18 26	08 48	09 03	09 20	09 39
N 58	17 00	17 41	18 27	08 49	09 06	09 25	09 46
56	17 06	17 44	18 28	08 50	09 09	09 29	09 52
54	17 11	17 47	18 29	08 50	09 11	09 33	09 58
52	17 15	17 50	18 30	08 51	09 13	09 37	10 03
50	17 20	17 53	18 31	08 51	09 15	09 40	10 07
45	17 29	17 59	18 33	08 53	09 19	09 47	10 17
N 40	17 37	18 04	18 36	08 54	09 23	09 53	10 26
35	17 43	18 09	18 39	08 54	09 26	09 58	10 33
30	17 49	18 14	18 42	08 55	09 28	10 03	10 39
20	18 00	18 22	18 48	08 56	09 33	10 10	10 50
N 10	18 09	18 30	18 55	08 58	09 37	10 17	10 59
0	18 17	18 39	19 03	08 59	09 41	10 24	11 08
S 10	18 26	18 48	19 13	09 00	09 45	10 30	11 17
20	18 36	18 58	19 25	09 01	09 49	10 37	11 27
30	18 47	19 12	19 41	09 02	09 54	10 45	11 38
35	18 53	19 20	19 51	09 03	09 56	10 50	11 44
40	19 00	19 29	20 03	09 04	09 59	10 55	11 51
45	19 09	19 40	20 18	09 05	10 03	11 01	11 59
S 50	19 19	19 54	20 38	09 06	10 07	11 08	12 10
52	19 24	20 01	20 48	09 06	10 09	11 12	12 14
54	19 29	20 09	20 59	09 07	10 11	11 15	12 20
56	19 35	20 17	21 12	09 07	10 13	11 19	12 25
58	19 42	20 27	21 27	09 08	10 16	11 24	12 32
S 60	19 49	20 38	21 46	09 09	10 19	11 29	12 39

Day	SUN Eqn. of Time 00h	12h	Mer. Pass.	MOON Mer. Pass. Upper	Lower	Age	Phase
d	m s	m s	h m	h m	h m	d	%
15	14 09	14 08	12 14	02 49	15 10	18	86
16	14 06	14 05	12 14	03 31	15 52	19	78
17	14 03	14 01	12 14	04 13	16 35	20	70

UT	ARIES GHA	VENUS −4.6 GHA	Dec	MARS +1.2 GHA	Dec	JUPITER −2.0 GHA	Dec	SATURN +0.7 GHA	Dec
18 00	147 45.3	215 50.1	S15 42.8	157 56.8	S 5 17.2	172 15.7	S11 08.1	131 14.4	N 4 30.6
01	162 47.8	230 51.4	42.9	172 57.4	16.4	187 17.5	07.9	146 16.6	30.7
02	177 50.2	245 52.7	43.0	187 58.1	15.7	202 19.4	07.7	161 18.8	30.8
03	192 52.7	260 54.1 ..	43.1	202 58.8 ..	14.9	217 21.3 ..	07.5	176 21.1 ..	30.9
04	207 55.2	275 55.4	43.2	217 59.4	14.1	232 23.2	07.2	191 23.3	31.0
05	222 57.6	290 56.7	43.3	233 00.1	13.3	247 25.1	07.0	206 25.6	31.1
W 06	238 00.1	305 58.0	S15 43.4	248 00.8	S 5 12.5	262 27.0	S11 06.8	221 27.8	N 4 31.2
E 07	253 02.6	320 59.4	43.6	263 01.4	11.7	277 28.9	06.6	236 30.0	31.3
D 08	268 05.0	336 00.7	43.7	278 02.1	10.9	292 30.8	06.4	251 32.3	31.4
N 09	283 07.5	351 02.0 ..	43.8	293 02.8 ..	10.1	307 32.7 ..	06.2	266 34.5 ..	31.5
E 10	298 09.9	6 03.3	43.9	308 03.4	09.4	322 34.6	05.9	281 36.7	31.6
S 11	313 12.4	21 04.6	44.0	323 04.1	08.6	337 36.5	05.7	296 39.0	31.7
D 12	328 14.9	36 05.9	S15 44.1	338 04.8	S 5 07.8	352 38.4	S11 05.5	311 41.2	N 4 31.8
A 13	343 17.3	51 07.2	44.2	353 05.4	07.0	7 40.3	05.3	326 43.5	31.9
Y 14	358 19.8	66 08.5	44.3	8 06.1	06.2	22 42.1	05.1	341 45.7	32.0
15	13 22.3	81 09.8 ..	44.4	23 06.8 ..	05.4	37 44.0 ..	04.9	356 47.9 ..	32.1
16	28 24.7	96 11.1	44.5	38 07.4	04.6	52 45.9	04.6	11 50.2	32.2
17	43 27.2	111 12.4	44.6	53 08.1	03.8	67 47.8	04.4	26 52.4	32.3
18	58 29.7	126 13.6	S15 44.7	68 08.8	S 5 03.1	82 49.7	S11 04.2	41 54.6	N 4 32.4
19	73 32.1	141 14.9	44.8	83 09.4	02.3	97 51.6	04.0	56 56.9	32.5
20	88 34.6	156 16.2	44.9	98 10.1	01.5	112 53.5	03.8	71 59.1	32.6
21	103 37.1	171 17.5 ..	45.0	113 10.8	5 00.7	127 55.4 ..	03.6	87 01.4 ..	32.7
22	118 39.5	186 18.7	45.1	128 11.4	4 59.9	142 57.3	03.4	102 03.6	32.8
23	133 42.0	201 20.0	45.2	143 12.1	59.1	157 59.2	03.1	117 05.8	32.9
19 00	148 44.4	216 21.3	S15 45.3	158 12.8	S 4 58.3	173 01.1	S11 02.9	132 08.1	N 4 33.0
01	163 46.9	231 22.5	45.4	173 13.5	57.5	188 03.0	02.7	147 10.3	33.1
02	178 49.4	246 23.8	45.5	188 14.1	56.8	203 04.8	02.5	162 12.5	33.2
03	193 51.8	261 25.0 ..	45.6	203 14.8 ..	56.0	218 06.7 ..	02.3	177 14.8 ..	33.3
04	208 54.3	276 26.3	45.7	218 15.5	55.2	233 08.6	02.1	192 17.0	33.4
05	223 56.8	291 27.5	45.8	233 16.1	54.4	248 10.5	01.8	207 19.2	33.5
T 06	238 59.2	306 28.8	S15 45.9	248 16.8	S 4 53.6	263 12.4	S11 01.6	222 21.5	N 4 33.6
H 07	254 01.7	321 30.0	46.0	263 17.5	52.8	278 14.3	01.4	237 23.7	33.7
U 08	269 04.2	336 31.3	46.1	278 18.1	52.0	293 16.2	01.2	252 25.9	33.8
R 09	284 06.6	351 32.5 ..	46.2	293 18.8 ..	51.2	308 18.1 ..	01.0	267 28.2 ..	33.9
S 10	299 09.1	6 33.8	46.3	308 19.5	50.4	323 20.0	00.8	282 30.4	34.0
D 11	314 11.6	21 35.0	46.4	323 20.1	49.7	338 21.9	00.5	297 32.7	34.1
A 12	329 14.0	36 36.2	S15 46.5	338 20.8	S 4 48.9	353 23.8	S11 00.3	312 34.9	N 4 34.2
Y 13	344 16.5	51 37.4	46.6	353 21.5	48.1	8 25.7	11 00.1	327 37.1	34.3
14	359 18.9	66 38.7	46.7	8 22.1	47.3	23 27.6	10 59.9	342 39.4	34.4
15	14 21.4	81 39.9 ..	46.8	23 22.8 ..	46.5	38 29.4 ..	59.7	357 41.6 ..	34.5
16	29 23.9	96 41.1	46.9	38 23.5	45.7	53 31.3	59.5	12 43.8	34.6
17	44 26.3	111 42.3	47.0	53 24.2	44.9	68 33.2	59.2	27 46.1	34.7
18	59 28.8	126 43.5	S15 47.1	68 24.8	S 4 44.1	83 35.1	S10 59.0	42 48.3	N 4 34.8
19	74 31.3	141 44.7	47.2	83 25.5	43.3	98 37.0	58.8	57 50.5	34.9
20	89 33.7	156 45.9	47.3	98 26.2	42.6	113 38.9	58.6	72 52.8	35.0
21	104 36.2	171 47.1 ..	47.3	113 26.8 ..	41.8	128 40.8 ..	58.4	87 55.0 ..	35.1
22	119 38.7	186 48.3	47.4	128 27.5	41.0	143 42.7	58.2	102 57.2	35.2
23	134 41.1	201 49.5	47.5	143 28.2	40.2	158 44.6	57.9	117 59.5	35.3
20 00	149 43.6	216 50.7	S15 47.6	158 28.8	S 4 39.4	173 46.5	S10 57.7	133 01.7	N 4 35.4
01	164 46.0	231 51.9	47.7	173 29.5	38.6	188 48.4	57.5	148 03.9	35.5
02	179 48.5	246 53.1	47.8	188 30.2	37.8	203 50.3	57.3	163 06.2	35.6
03	194 51.0	261 54.3 ..	47.9	203 30.9 ..	37.0	218 52.1 ..	57.1	178 08.4 ..	35.7
04	209 53.4	276 55.5	48.0	218 31.5	36.2	233 54.0	56.9	193 10.6	35.8
05	224 55.9	291 56.7	48.1	233 32.2	35.5	248 55.9	56.6	208 12.9	35.9
F 06	239 58.4	306 57.8	S15 48.2	248 32.9	S 4 34.7	263 57.8	S10 56.4	223 15.1	N 4 36.0
R 07	255 00.8	321 59.0	48.3	263 33.5	33.9	278 59.7	56.2	238 17.3	36.1
I 08	270 03.3	337 00.2	48.3	278 34.2	33.1	294 01.6	56.0	253 19.6	36.2
D 09	285 05.8	352 01.3 ..	48.4	293 34.9 ..	32.3	309 03.5 ..	55.8	268 21.8 ..	36.3
A 10	300 08.2	7 02.5	48.5	308 35.6	31.5	324 05.4	55.6	283 24.0	36.4
Y 11	315 10.7	22 03.7	48.6	323 36.2	30.7	339 07.3	55.3	298 26.3	36.6
12	330 13.2	37 04.8	S15 48.7	338 36.9	S 4 29.9	354 09.2	S10 55.1	313 28.5	N 4 36.7
13	345 15.6	52 06.0	48.8	353 37.6	29.1	9 11.1	54.9	328 30.7	36.8
14	0 18.1	67 07.1	48.9	8 38.2	28.3	24 13.0	54.7	343 33.0	36.9
15	15 20.5	82 08.3 ..	49.0	23 38.9 ..	27.6	39 14.9 ..	54.5	358 35.2 ..	37.0
16	30 23.0	97 09.4	49.0	38 39.6	26.8	54 16.7	54.3	13 37.4	37.1
17	45 25.5	112 10.6	49.1	53 40.3	26.0	69 18.6	54.0	28 39.7	37.2
18	60 27.9	127 11.7	S15 49.2	68 40.9	S 4 25.2	84 20.5	S10 53.8	43 41.9	N 4 37.3
19	75 30.4	142 12.9	49.3	83 41.6	24.4	99 22.4	53.6	58 44.1	37.4
20	90 32.9	157 14.0	49.4	98 42.3	23.6	114 24.3	53.4	73 46.4	37.5
21	105 35.3	172 15.1 ..	49.5	113 42.9 ..	22.8	129 26.2 ..	53.2	88 48.6 ..	37.6
22	120 37.8	187 16.3	49.6	128 43.6	22.0	144 28.1	53.0	103 50.8	37.7
23	135 40.3	202 17.4	49.6	143 44.3	21.2	159 30.0	52.7	118 53.1	37.8
Mer. Pass.	h m 14 02.7	v 1.2	d 0.1	v 0.7	d 0.8	v 1.9	d 0.2	v 2.2	d 0.1

STARS

Name	SHA	Dec
Acamar	315 27.4	S40 19.1
Achernar	335 35.9	S57 15.1
Acrux	173 21.8	S63 05.2
Adhara	255 21.5	S28 58.5
Aldebaran	291 02.8	N16 30.2
Alioth	166 30.6	N55 58.0
Alkaid	153 07.9	N49 19.2
Al Na'ir	27 58.9	S46 58.3
Alnilam	275 58.1	S 1 12.5
Alphard	218 07.3	S 8 39.2
Alphecca	126 20.9	N26 43.1
Alpheratz	357 55.9	N29 04.8
Altair	62 20.0	N 8 51.8
Ankaa	353 27.7	S42 19.2
Antares	112 40.7	S26 25.5
Arcturus	146 06.3	N19 11.4
Atria	107 53.3	S69 01.1
Avior	234 22.2	S59 30.5
Bellatrix	278 44.5	N 6 20.7
Betelgeuse	271 13.8	N 7 24.2
Canopus	264 01.0	S52 42.1
Capella	280 51.6	N45 59.8
Deneb	49 39.9	N45 16.4
Denebola	182 45.3	N14 34.8
Diphda	349 07.9	S18 00.0
Dubhe	194 05.5	N61 45.5
Elnath	278 27.3	N28 36.3
Eltanin	90 51.9	N51 29.2
Enif	33 59.0	N 9 51.9
Fomalhaut	15 37.3	S29 38.1
Gacrux	172 13.5	S57 06.0
Gienah	176 04.1	S17 31.9
Hadar	149 04.2	S60 21.6
Hamal	328 14.2	N23 27.1
Kaus Aust.	83 59.6	S34 23.0
Kochab	137 19.4	N74 09.6
Markab	13 50.3	N15 11.6
Menkar	314 27.4	N 4 04.7
Menkent	148 21.2	S36 21.5
Miaplacidus	221 41.1	S69 42.7
Mirfak	308 57.2	N49 51.3
Nunki	76 13.1	S26 17.8
Peacock	53 38.3	S56 44.3
Pollux	243 41.8	N28 01.7
Procyon	245 11.7	N 5 13.6
Rasalhague	96 17.5	N12 33.6
Regulus	207 55.7	N11 58.4
Rigel	281 23.2	S 8 12.5
Rigil Kent.	140 07.6	S60 49.3
Sabik	102 26.1	S15 43.2
Schedar	349 54.3	N56 31.7
Shaula	96 38.0	S37 05.9
Sirius	258 43.8	S16 43.1
Spica	158 43.4	S11 09.1
Suhail	223 00.6	S43 25.7
Vega	80 47.2	N38 46.8
Zuben'ubi	137 18.3	S16 02.0

	SHA	Mer. Pass.
	o '	h m
Venus	67 36.8	9 34
Mars	9 28.3	13 27
Jupiter	24 16.6	12 26
Saturn	343 23.6	15 09

UT	SUN GHA	SUN Dec	MOON GHA	v	MOON Dec	d	HP
d h	° ′	° ′	° ′	′	° ′	′	′
18 00	176 30.4	S11 46.7	287 54.2	14.1	S10 41.7	7.8	54.5
01	191 30.5	45.8	302 27.3	14.1	10 49.5	7.8	54.5
02	206 30.5	45.0	317 00.4	14.0	10 57.3	7.7	54.5
03	221 30.6 ..	44.1	331 33.4	14.0	11 05.0	7.7	54.5
04	236 30.6	43.2	346 06.4	13.9	11 12.7	7.7	54.6
05	251 30.7	42.3	0 39.3	13.9	11 20.4	7.6	54.6
W 06	266 30.7	S11 41.4	15 12.2	13.9	S11 28.0	7.5	54.6
E 07	281 30.8	40.6	29 45.1	13.8	11 35.5	7.6	54.6
D 08	296 30.8	39.7	44 17.9	13.8	11 43.1	7.4	54.6
N 09	311 30.9 ..	38.8	58 50.7	13.7	11 50.5	7.4	54.6
E 10	326 30.9	37.9	73 23.4	13.7	11 57.9	7.4	54.7
S 11	341 31.0	37.0	87 56.1	13.6	12 05.3	7.3	54.7
D 12	356 31.0	S11 36.1	102 28.7	13.6	S12 12.6	7.3	54.7
A 13	11 31.1	35.3	117 01.3	13.5	12 19.9	7.2	54.7
Y 14	26 31.2	34.4	131 33.8	13.5	12 27.1	7.2	54.7
15	41 31.2 ..	33.5	146 06.3	13.4	12 34.3	7.1	54.8
16	56 31.3	32.6	160 38.7	13.4	12 41.4	7.1	54.8
17	71 31.3	31.7	175 11.1	13.3	12 48.5	7.0	54.8
18	86 31.4	S11 30.8	189 43.4	13.3	S12 55.5	7.0	54.8
19	101 31.4	29.9	204 15.7	13.3	13 02.5	6.9	54.8
20	116 31.5	29.1	218 48.0	13.2	13 09.4	6.8	54.9
21	131 31.5 ..	28.2	233 20.2	13.1	13 16.2	6.8	54.9
22	146 31.6	27.3	247 52.3	13.1	13 23.0	6.8	54.9
23	161 31.6	26.4	262 24.4	13.0	13 29.8	6.7	54.9
19 00	176 31.7	S11 25.5	276 56.4	13.0	S13 36.5	6.6	54.9
01	191 31.8	24.6	291 28.4	12.9	13 43.1	6.6	55.0
02	206 31.8	23.7	306 00.3	12.9	13 49.7	6.5	55.0
03	221 31.9 ..	22.9	320 32.2	12.9	13 56.2	6.5	55.0
04	236 31.9	22.0	335 04.1	12.7	14 02.7	6.4	55.0
05	251 32.0	21.1	349 35.8	12.7	14 09.1	6.3	55.1
T 06	266 32.0	S11 20.2	4 07.5	12.7	S14 15.4	6.3	55.1
H 07	281 32.1	19.3	18 39.2	12.6	14 21.7	6.2	55.1
U 08	296 32.2	18.4	33 10.8	12.6	14 27.9	6.2	55.1
R 09	311 32.2 ..	17.5	47 42.4	12.5	14 34.1	6.1	55.1
S 10	326 32.3	16.6	62 13.9	12.4	14 40.2	6.0	55.2
11	341 32.3	15.7	76 45.3	12.4	14 46.2	6.0	55.2
D 12	356 32.4	S11 14.8	91 16.7	12.3	S14 52.2	5.9	55.2
A 13	11 32.5	14.0	105 48.0	12.3	14 58.1	5.8	55.2
Y 14	26 32.5	13.1	120 19.3	12.2	15 03.9	5.8	55.3
15	41 32.6 ..	12.2	134 50.5	12.1	15 09.7	5.7	55.3
16	56 32.6	11.3	149 21.6	12.1	15 15.4	5.7	55.3
17	71 32.7	10.4	163 52.7	12.1	15 21.1	5.5	55.4
18	86 32.8	S11 09.5	178 23.8	11.9	S15 26.6	5.5	55.4
19	101 32.8	08.6	192 54.7	12.0	15 32.1	5.5	55.4
20	116 32.9	07.7	207 25.7	11.8	15 37.6	5.3	55.4
21	131 32.9 ..	06.8	221 56.5	11.8	15 42.9	5.3	55.5
22	146 33.0	05.9	236 27.3	11.8	15 48.2	5.3	55.5
23	161 33.1	05.0	250 58.1	11.6	15 53.5	5.1	55.5
20 00	176 33.1	S11 04.1	265 28.7	11.7	S15 58.6	5.1	55.5
01	191 33.2	03.2	279 59.4	11.5	16 03.7	5.0	55.6
02	206 33.3	02.3	294 29.9	11.5	16 08.7	4.9	55.6
03	221 33.3 ..	01.4	309 00.4	11.5	16 13.6	4.9	55.6
04	236 33.4	11 00.6	323 30.9	11.3	16 18.5	4.8	55.7
05	251 33.5	10 59.7	338 01.2	11.3	16 23.3	4.7	55.7
F 06	266 33.5	S10 58.8	352 31.5	11.3	S16 28.0	4.6	55.7
R 07	281 33.6	57.9	7 01.8	11.2	16 32.6	4.6	55.8
I 08	296 33.6	57.0	21 32.0	11.1	16 37.2	4.4	55.8
D 09	311 33.7 ..	56.1	36 02.1	11.1	16 41.6	4.4	55.8
A 10	326 33.8	55.2	50 32.2	11.0	16 46.0	4.4	55.8
Y 11	341 33.8	54.3	65 02.2	10.9	16 50.4	4.2	55.9
12	356 33.9	S10 53.4	79 32.1	10.9	S16 54.6	4.2	55.9
13	11 34.0	52.5	94 02.0	10.9	16 58.8	4.0	55.9
14	26 34.0	51.6	108 31.9	10.7	17 02.8	4.0	56.0
15	41 34.1 ..	50.7	123 01.6	10.7	17 06.8	3.9	56.0
16	56 34.2	49.8	137 31.3	10.7	17 10.7	3.8	56.0
17	71 34.2	48.9	152 01.0	10.5	17 14.5	3.8	56.1
18	86 34.3	S10 48.0	166 30.5	10.5	S17 18.3	3.6	56.1
19	101 34.4	47.1	181 00.0	10.5	17 21.9	3.6	56.1
20	116 34.5	46.2	195 29.5	10.4	17 25.5	3.5	56.2
21	131 34.5 ..	45.3	209 58.9	10.3	17 29.0	3.4	56.2
22	146 34.6	44.4	224 28.2	10.3	17 32.4	3.3	56.2
23	161 34.7	43.5	238 57.5	10.2	S17 35.7	3.2	56.3
	SD 16.2	d 0.9	SD 14.9		15.0		15.2

Twilight / Sunrise / Moonrise

Lat.	Naut.	Civil	Sunrise	Moonrise 18	19	20	21
°	h m	h m	h m	h m	h m	h m	h m
N 72	05 58	07 17	08 32	01 17	02 59	04 50	■■■
N 70	05 58	07 09	08 15	00 58	02 30	04 02	05 32
68	05 58	07 03	08 02	00 43	02 08	03 32	04 51
66	05 58	06 57	07 51	00 31	01 51	03 09	04 23
64	05 57	06 52	07 42	00 22	01 37	02 51	04 01
62	05 57	06 48	07 35	00 13	01 26	02 37	03 44
60	05 56	06 45	07 28	00 06	01 16	02 25	03 30
N 58	05 56	06 41	07 22	25 07	01 07	02 14	03 18
56	05 55	06 38	07 17	25 00	01 00	02 05	03 07
54	05 54	06 36	07 12	24 53	00 53	01 57	02 58
52	05 54	06 33	07 08	24 47	00 47	01 49	02 50
50	05 53	06 31	07 04	24 42	00 42	01 43	02 43
45	05 51	06 25	06 55	24 30	00 30	01 29	02 27
N 40	05 49	06 21	06 48	24 20	00 20	01 17	02 14
35	05 47	06 16	06 42	24 12	00 12	01 07	02 03
30	05 44	06 12	06 36	24 05	00 05	00 59	01 53
20	05 39	06 05	06 27	23 53	24 44	00 44	01 37
N 10	05 32	05 57	06 19	23 42	24 31	00 31	01 22
0	05 25	05 49	06 10	23 32	24 19	00 19	01 09
S 10	05 16	05 41	06 02	23 22	24 07	00 07	00 56
20	05 04	05 31	05 53	23 11	23 54	24 42	00 42
30	04 49	05 19	05 43	22 59	23 40	24 25	00 25
35	04 40	05 11	05 37	22 52	23 32	24 16	00 16
40	04 28	05 02	05 31	22 44	23 22	24 05	00 05
45	04 14	04 52	05 23	22 35	23 11	23 53	24 41
S 50	03 55	04 38	05 13	22 24	22 57	23 37	24 25
52	03 46	04 32	05 09	22 19	22 51	23 30	24 17
54	03 35	04 25	05 04	22 13	22 44	23 22	24 09
56	03 23	04 17	04 58	22 07	22 37	23 13	23 59
58	03 09	04 08	04 52	22 00	22 28	23 03	23 48
S 60	02 51	03 57	04 45	21 52	22 18	22 52	23 36

Sunset / Twilight / Moonset

Lat.	Sunset	Civil	Naut.	Moonset 18	19	20	21
°	h m	h m	h m	h m	h m	h m	h m
N 72	15 57	17 13	18 32	08 22	08 15	08 03	■■■
N 70	16 14	17 20	18 32	08 42	08 45	08 51	09 07
68	16 27	17 27	18 32	08 58	09 08	09 23	09 48
66	16 38	17 32	18 32	09 11	09 25	09 46	10 17
64	16 47	17 37	18 32	09 22	09 40	10 04	10 38
62	16 54	17 41	18 32	09 31	09 52	10 19	10 55
60	17 01	17 44	18 33	09 39	10 02	10 32	11 10
N 58	17 07	17 47	18 33	09 46	10 11	10 43	11 22
56	17 12	17 50	18 34	09 52	10 19	10 52	11 33
54	17 17	17 53	18 34	09 58	10 26	11 01	11 42
52	17 21	17 56	18 35	10 03	10 33	11 08	11 51
50	17 25	17 58	18 36	10 07	10 39	11 15	11 58
45	17 33	18 03	18 37	10 17	10 51	11 30	12 14
N 40	17 40	18 08	18 39	10 26	11 01	11 42	12 27
35	17 46	18 12	18 42	10 33	11 10	11 52	12 39
30	17 52	18 16	18 44	10 39	11 18	12 01	12 49
20	18 01	18 23	18 49	10 50	11 31	12 16	13 05
N 10	18 09	18 31	18 55	10 59	11 43	12 30	13 20
0	18 17	18 38	19 04	11 08	11 54	12 43	13 34
S 10	18 25	18 47	19 12	11 17	12 05	12 56	13 48
20	18 34	18 56	19 23	11 27	12 17	13 09	14 02
30	18 44	19 09	19 38	11 38	12 31	13 25	14 19
35	18 50	19 16	19 47	11 44	12 39	13 34	14 29
40	18 56	19 25	19 59	11 51	12 48	13 44	14 40
45	19 04	19 35	20 13	11 59	12 58	13 56	14 54
S 50	19 14	19 48	20 31	12 10	13 11	14 11	15 10
52	19 18	19 54	20 40	12 14	13 17	14 18	15 17
54	19 23	20 01	20 50	12 20	13 23	14 26	15 26
56	19 28	20 09	21 02	12 25	13 31	14 35	15 35
58	19 34	20 18	21 16	12 32	13 39	14 44	15 46
S 60	19 41	20 28	21 33	12 39	13 48	14 56	15 58

SUN / MOON

Day	Eqn. of Time 00ʰ	12ʰ	Mer. Pass.	Mer. Pass. Upper	Lower	Age	Phase
d	m s	m s	h m	h m	h m	d	%
18	13 58	13 56	12 14	04 57	17 20	21	61
19	13 53	13 51	12 14	05 43	18 07	22	52
20	13 48	13 44	12 14	06 31	18 56	23	42

1998 FEBRUARY 21, 22, 23 (SAT., SUN., MON.)

UT	ARIES	VENUS −4.6		MARS +1.2		JUPITER −2.0		SATURN +0.7		STARS		
	GHA	GHA	Dec	GHA	Dec	GHA	Dec	GHA	Dec	Name	SHA	Dec
d h	° ′	° ′	° ′	° ′	° ′	° ′	° ′	° ′	° ′		° ′	° ′
21 00	150 42.7	217 18.5	S15 49.7	158 45.0	S 4 20.4	174 31.9	S10 52.5	133 55.3	N 4 37.9	Acamar	315 27.4	S40 19.1
01	165 45.2	232 19.6	.. 49.8	173 45.6	19.7	189 33.8	52.3	148 57.5	38.0	Achernar	335 35.9	S57 15.1
02	180 47.7	247 20.8	49.9	188 46.3	18.9	204 35.7	52.1	163 59.7	38.1	Acrux	173 21.8	S63 05.2
03	195 50.1	262 21.9	.. 50.0	203 47.0	.. 18.1	219 37.6	.. 51.9	179 02.0	.. 38.2	Adhara	255 21.5	S28 58.5
04	210 52.6	277 23.0	50.0	218 47.7	17.3	234 39.4	51.7	194 04.2	38.3	Aldebaran	291 02.8	N16 30.2
05	225 55.0	292 24.1	50.1	233 48.3	16.5	249 41.3	51.4	209 06.4	38.4			
06	240 57.5	307 25.2	S15 50.2	248 49.0	S 4 15.7	264 43.2	S10 51.2	224 08.7	N 4 38.5	Alioth	166 30.6	N55 58.0
07	256 00.0	322 26.3	50.3	263 49.7	14.9	279 45.1	51.0	239 10.9	38.6	Alkaid	153 07.9	N49 19.2
S 08	271 02.4	337 27.4	50.4	278 50.4	14.1	294 47.0	50.8	254 13.1	38.7	Al Na'ir	27 58.9	S46 58.3
A 09	286 04.9	352 28.5	.. 50.4	293 51.0	.. 13.3	309 48.9	.. 50.6	269 15.4	.. 38.8	Alnilam	275 58.2	S 1 12.5
T 10	301 07.4	7 29.6	50.5	308 51.7	12.5	324 50.8	50.3	284 17.6	38.9	Alphard	218 07.3	S 8 39.2
U 11	316 09.8	22 30.7	50.6	323 52.4	11.7	339 52.7	50.1	299 19.8	39.0			
R 12	331 12.3	37 31.8	S15 50.7	338 53.0	S 4 11.0	354 54.6	S10 49.9	314 22.1	N 4 39.1	Alphecca	126 20.9	N26 43.1
D 13	346 14.8	52 32.9	50.8	353 53.7	10.2	9 56.5	49.7	329 24.3	39.2	Alpheratz	357 55.9	N29 04.8
A 14	1 17.2	67 34.0	50.8	8 54.4	09.4	24 58.4	49.5	344 26.5	39.3	Altair	62 19.9	N 8 51.8
Y 15	16 19.7	82 35.1	.. 50.9	23 55.1	.. 08.6	40 00.3	.. 49.3	359 28.7	.. 39.4	Ankaa	353 27.7	S42 19.2
16	31 22.1	97 36.1	51.0	38 55.7	07.8	55 02.2	49.0	14 31.0	39.5	Antares	112 40.7	S26 25.5
17	46 24.6	112 37.2	51.1	53 56.4	07.0	70 04.0	48.8	29 33.2	39.7			
18	61 27.1	127 38.3	S15 51.1	68 57.1	S 4 06.2	85 05.9	S10 48.6	44 35.4	N 4 39.8	Arcturus	146 06.3	N19 11.4
19	76 29.5	142 39.4	51.2	83 57.8	05.4	100 07.8	48.4	59 37.7	39.9	Atria	107 53.2	S69 01.1
20	91 32.0	157 40.4	51.3	98 58.4	04.6	115 09.7	48.2	74 39.9	40.0	Avior	234 22.2	S59 30.5
21	106 34.5	172 41.5	.. 51.4	113 59.1	.. 03.8	130 11.6	.. 48.0	89 42.1	.. 40.1	Bellatrix	278 44.5	N 6 20.7
22	121 36.9	187 42.6	51.4	128 59.8	03.0	145 13.5	47.7	104 44.3	40.2	Betelgeuse	271 13.9	N 7 24.2
23	136 39.4	202 43.6	51.5	144 00.5	02.2	160 15.4	47.5	119 46.6	40.3			
22 00	151 41.9	217 44.7	S15 51.6	159 01.1	S 4 01.5	175 17.3	S10 47.3	134 48.8	N 4 40.4	Canopus	264 01.0	S52 42.1
01	166 44.3	232 45.7	51.6	174 01.8	4 00.7	190 19.2	47.1	149 51.0	40.5	Capella	280 51.7	N45 59.8
02	181 46.8	247 46.8	51.7	189 02.5	3 59.9	205 21.1	46.9	164 53.3	40.6	Deneb	49 39.9	N45 16.3
03	196 49.3	262 47.8	.. 51.8	204 03.2	.. 59.1	220 23.0	.. 46.7	179 55.5	.. 40.7	Denebola	182 45.3	N14 34.8
04	211 51.7	277 48.9	51.9	219 03.8	58.3	235 24.9	46.4	194 57.7	40.8	Diphda	349 07.9	S18 00.0
05	226 54.2	292 49.9	51.9	234 04.5	57.5	250 26.8	46.2	209 59.9	40.9			
06	241 56.6	307 51.0	S15 52.0	249 05.2	S 3 56.7	265 28.6	S10 46.0	225 02.2	N 4 41.0	Dubhe	194 05.5	N61 45.5
07	256 59.1	322 52.0	52.1	264 05.9	55.9	280 30.5	45.8	240 04.4	41.1	Elnath	278 27.3	N28 36.3
S 08	272 01.6	337 53.1	52.1	279 06.5	55.1	295 32.4	45.6	255 06.6	41.2	Eltanin	90 51.8	N51 29.2
U 09	287 04.0	352 54.1	.. 52.2	294 07.2	.. 54.3	310 34.3	.. 45.3	270 08.9	.. 41.3	Enif	33 59.0	N 9 51.9
N 10	302 06.5	7 55.1	52.3	309 07.9	53.5	325 36.2	45.1	285 11.1	41.4	Fomalhaut	15 37.3	S29 38.1
11	317 09.0	22 56.2	52.3	324 08.6	52.7	340 38.1	44.9	300 13.3	41.5			
D 12	332 11.4	37 57.2	S15 52.4	339 09.3	S 3 52.0	355 40.0	S10 44.7	315 15.5	N 4 41.6	Gacrux	172 13.5	S57 06.0
A 13	347 13.9	52 58.2	52.5	354 09.9	51.2	10 41.9	44.5	330 17.8	41.7	Gienah	176 04.1	S17 31.9
Y 14	2 16.4	67 59.2	52.5	9 10.6	50.4	25 43.8	44.3	345 20.0	41.8	Hadar	149 04.2	S60 21.6
15	17 18.8	83 00.3	.. 52.6	24 11.3	.. 49.6	40 45.7	.. 44.0	0 22.2	.. 41.9	Hamal	328 14.2	N23 27.1
16	32 21.3	98 01.3	52.7	39 12.0	48.8	55 47.6	43.8	15 24.5	42.0	Kaus Aust.	83 59.6	S34 23.0
17	47 23.8	113 02.3	52.7	54 12.6	48.0	70 49.5	43.6	30 26.7	42.1			
18	62 26.2	128 03.3	S15 52.8	69 13.3	S 3 47.2	85 51.4	S10 43.4	45 28.9	N 4 42.2	Kochab	137 19.3	N74 09.6
19	77 28.7	143 04.3	52.9	84 14.0	46.4	100 53.2	43.2	60 31.1	42.4	Markab	13 50.3	N15 11.6
20	92 31.1	158 05.3	52.9	99 14.7	45.6	115 55.1	43.0	75 33.4	42.5	Menkar	314 27.4	N 4 04.7
21	107 33.6	173 06.3	.. 53.0	114 15.3	.. 44.8	130 57.0	.. 42.7	90 35.6	.. 42.6	Menkent	148 21.2	S36 21.5
22	122 36.1	188 07.3	53.0	129 16.0	44.0	145 58.9	42.5	105 37.8	42.7	Miaplacidus	221 41.1	S69 42.8
23	137 38.5	203 08.3	53.1	144 16.7	43.2	161 00.8	42.3	120 40.0	42.8			
23 00	152 41.0	218 09.3	S15 53.2	159 17.4	S 3 42.5	176 02.7	S10 42.1	135 42.3	N 4 42.9	Mirfak	308 57.2	N49 51.3
01	167 43.5	233 10.3	53.2	174 18.1	41.7	191 04.6	41.9	150 44.5	43.0	Nunki	76 13.1	S26 17.8
02	182 45.9	248 11.3	53.3	189 18.7	40.9	206 06.5	41.7	165 46.7	43.1	Peacock	53 38.2	S56 44.3
03	197 48.4	263 12.3	.. 53.3	204 19.4	.. 40.1	221 08.4	.. 41.4	180 49.0	.. 43.2	Pollux	243 41.8	N28 01.7
04	212 50.9	278 13.3	53.4	219 20.1	39.3	236 10.3	41.2	195 51.2	43.3	Procyon	245 11.7	N 5 13.6
05	227 53.3	293 14.3	53.5	234 20.8	38.5	251 12.2	41.0	210 53.4	43.4			
06	242 55.8	308 15.2	S15 53.5	249 21.4	S 3 37.7	266 14.1	S10 40.8	225 55.6	N 4 43.5	Rasalhague	96 17.5	N12 33.6
07	257 58.3	323 16.2	53.6	264 22.1	36.9	281 16.0	40.6	240 57.9	43.6	Regulus	207 55.7	N11 58.4
08	273 00.7	338 17.2	53.6	279 22.8	36.1	296 17.8	40.3	256 00.1	43.7	Rigel	281 23.2	S 8 12.5
M 09	288 03.2	353 18.2	.. 53.7	294 23.5	.. 35.3	311 19.7	.. 40.1	271 02.3	.. 43.8	Rigil Kent.	140 07.5	S60 49.3
O 10	303 05.6	8 19.1	53.7	309 24.2	34.5	326 21.6	39.9	286 04.5	43.9	Sabik	102 26.1	S15 43.3
N 11	318 08.1	23 20.1	53.8	324 24.8	33.7	341 23.5	39.7	301 06.8	44.0			
D 12	333 10.6	38 21.1	S15 53.9	339 25.5	S 3 32.9	356 25.4	S10 39.5	316 09.0	N 4 44.1	Schedar	349 54.3	N56 31.7
A 13	348 13.0	53 22.0	53.9	354 26.2	32.1	11 27.3	39.3	331 11.2	44.2	Shaula	96 38.0	S37 05.9
Y 14	3 15.5	68 23.0	54.0	9 26.9	31.4	26 29.2	39.0	346 13.4	44.3	Sirius	258 43.9	S16 43.1
15	18 18.0	83 23.9	.. 54.0	24 27.5	.. 30.6	41 31.1	.. 38.8	1 15.7	.. 44.5	Spica	158 43.4	S11 09.1
16	33 20.4	98 24.9	54.1	39 28.2	29.8	56 33.0	38.6	16 17.9	44.6	Suhail	223 00.6	S43 25.7
17	48 22.9	113 25.9	54.1	54 28.9	29.0	71 34.9	38.4	31 20.1	44.7			
18	63 25.4	128 26.8	S15 54.2	69 29.6	S 3 28.2	86 36.8	S10 38.2	46 22.3	N 4 44.8	Vega	80 47.1	N38 46.8
19	78 27.8	143 27.8	54.2	84 30.3	27.4	101 38.7	37.9	61 24.6	44.9	Zuben'ubi	137 18.3	S16 02.0
20	93 30.3	158 28.7	54.3	99 30.9	26.6	116 40.6	37.7	76 26.8	45.0			
21	108 32.7	173 29.6	.. 54.3	114 31.6	.. 25.8	131 42.5	.. 37.5	91 29.0	.. 45.1		SHA	Mer. Pass.
22	123 35.2	188 30.6	54.4	129 32.3	25.0	146 44.3	37.3	106 31.2	45.2		° ′	h m
23	138 37.7	203 31.5	54.4	144 33.0	24.2	161 46.2	37.1	121 33.5	45.3	Venus	66 02.8	9 28
	h m									Mars	7 19.3	13 23
Mer. Pass. 13 50.9	v 1.0 d 0.1	v 0.7	d 0.8	v 1.9	d 0.2	v 2.2	d 0.1			Jupiter	23 35.4	12 17
										Saturn	343 06.9	14 59

INDEX TO SELECTED STARS, 1998

Name	No	Mag	SHA	Dec
Acamar	7	3·1	315	S 40
Achernar	5	0·6	336	S 57
Acrux	30	1·1	173	S 63
Adhara	19	1·6	255	S 29
Aldebaran	10	1·1	291	N 17
Alioth	32	1·7	167	N 56
Alkaid	34	1·9	153	N 49
Al Na'ir	55	2·2	28	S 47
Alnilam	15	1·8	276	S 1
Alphard	25	2·2	218	S 9
Alphecca	41	2·3	126	N 27
Alpheratz	1	2·2	358	N 29
Altair	51	0·9	62	N 9
Ankaa	2	2·4	353	S 42
Antares	42	1·2	113	S 26
Arcturus	37	0·2	146	N 19
Atria	43	1·9	108	S 69
Avior	22	1·7	234	S 60
Bellatrix	13	1·7	279	N 6
Betelgeuse	16	Var.*	271	N 7
Canopus	17	−0·9	264	S 53
Capella	12	0·2	281	N 46
Deneb	53	1·3	50	N 45
Denebola	28	2·2	183	N 15
Diphda	4	2·2	349	S 18
Dubhe	27	2·0	194	N 62
Elnath	14	1·8	278	N 29
Eltanin	47	2·4	91	N 51
Enif	54	2·5	34	N 10
Fomalhaut	56	1·3	16	S 30
Gacrux	31	1·6	172	S 57
Gienah	29	2·8	176	S 18
Hadar	35	0·9	149	S 60
Hamal	6	2·2	328	N 23
Kaus Australis	48	2·0	84	S 34
Kochab	40	2·2	137	N 74
Markab	57	2·6	14	N 15
Menkar	8	2·8	314	N 4
Menkent	36	2·3	148	S 36
Miaplacidus	24	1·8	222	S 70
Mirfak	9	1·9	309	N 50
Nunki	50	2·1	76	S 26
Peacock	52	2·1	54	S 57
Pollux	21	1·2	244	N 28
Procyon	20	0·5	245	N 5
Rasalhague	46	2·1	96	N 13
Regulus	26	1·3	208	N 12
Rigel	11	0·3	281	S 8
Rigil Kentaurus	38	0·1	140	S 61
Sabik	44	2·6	102	S 16
Schedar	3	2·5	350	N 57
Shaula	45	1·7	97	S 37
Sirius	18	−1·6	259	S 17
Spica	33	1·2	159	S 11
Suhail	23	2·2	223	S 43
Vega	49	0·1	81	N 39
Zubenelgenubi	39	2·9	137	S 16

No	Name	Mag	SHA	Dec
1	Alpheratz	2·2	358	N 29
2	Ankaa	2·4	353	S 42
3	Schedar	2·5	350	N 57
4	Diphda	2·2	349	S 18
5	Achernar	0·6	336	S 57
6	Hamal	2·2	328	N 23
7	Acamar	3·1	315	S 40
8	Menkar	2·8	314	N 4
9	Mirfak	1·9	309	N 50
10	Aldebaran	1·1	291	N 17
11	Rigel	0·3	281	S 8
12	Capella	0·2	281	N 46
13	Bellatrix	1·7	279	N 6
14	Elnath	1·8	278	N 29
15	Alnilam	1·8	276	S 1
16	Betelgeuse	Var.*	271	N 7
17	Canopus	−0·9	264	S 53
18	Sirius	−1·6	259	S 17
19	Adhara	1·6	255	S 29
20	Procyon	0·5	245	N 5
21	Pollux	1·2	244	N 28
22	Avior	1·7	234	S 60
23	Suhail	2·2	223	S 43
24	Miaplacidus	1·8	222	S 70
25	Alphard	2·2	218	S 9
26	Regulus	1·3	208	N 12
27	Dubhe	2·0	194	N 62
28	Denebola	2·2	183	N 15
29	Gienah	2·8	176	S 18
30	Acrux	1·1	173	S 63
31	Gacrux	1·6	172	S 57
32	Alioth	1·7	167	N 56
33	Spica	1·2	159	S 11
34	Alkaid	1·9	153	N 49
35	Hadar	0·9	149	S 60
36	Menkent	2·3	148	S 36
37	Arcturus	0·2	146	N 19
38	Rigil Kentaurus	0·1	140	S 61
39	Zubenelgenubi	2·9	137	S 16
40	Kochab	2·2	137	N 74
41	Alphecca	2·3	126	N 27
42	Antares	1·2	113	S 26
43	Atria	1·9	108	S 69
44	Sabik	2·6	102	S 16
45	Shaula	1·7	97	S 37
46	Rasalhague	2·1	96	N 13
47	Eltanin	2·4	91	N 51
48	Kaus Australis	2·0	84	S 34
49	Vega	0·1	81	N 39
50	Nunki	2·1	76	S 26
51	Altair	0·9	62	N 9
52	Peacock	2·1	54	S 57
53	Deneb	1·3	50	N 45
54	Enif	2·5	34	N 10
55	Al Na'ir	2·2	28	S 47
56	Fomalhaut	1·3	16	S 30
57	Markab	2·6	14	N 15

*0·1 — 1·2

ALTITUDE CORRECTION TABLES 10°–90°—SUN, STARS, PLANETS

SUN

OCT.—MAR. App. Alt.	Lower Limb	Upper Limb	APR.—SEPT. App. Alt.	Lower Limb	Upper Limb
9 34	+10.8	−21.5	9 39	+10.6	−21.2
9 45	+10.9	−21.4	9 51	+10.7	−21.1
9 56	+11.0	−21.3	10 03	+10.8	−21.0
10 08	+11.1	−21.2	10 15	+10.9	−20.9
10 21	+11.2	−21.1	10 27	+11.0	−20.8
10 34	+11.3	−21.0	10 40	+11.1	−20.7
10 47	+11.4	−20.9	10 54	+11.2	−20.6
11 01	+11.5	−20.8	11 08	+11.3	−20.5
11 15	+11.6	−20.7	11 23	+11.4	−20.4
11 30	+11.7	−20.6	11 38	+11.5	−20.3
11 46	+11.8	−20.5	11 54	+11.6	−20.2
12 02	+11.9	−20.4	12 10	+11.7	−20.1
12 19	+12.0	−20.3	12 28	+11.8	−20.0
12 37	+12.1	−20.2	12 46	+11.9	−19.9
12 55	+12.2	−20.1	13 05	+12.0	−19.8
13 14	+12.3	−20.0	13 24	+12.1	−19.7
13 35	+12.4	−19.9	13 45	+12.2	−19.6
13 56	+12.5	−19.8	14 07	+12.3	−19.5
14 18	+12.6	−19.7	14 30	+12.4	−19.4
14 42	+12.7	−19.6	14 54	+12.5	−19.3
15 06	+12.8	−19.5	15 19	+12.6	−19.2
15 32	+12.9	−19.4	15 46	+12.7	−19.1
15 59	+13.0	−19.3	16 14	+12.8	−19.0
16 28	+13.1	−19.2	16 44	+12.9	−18.9
16 59	+13.2	−19.1	17 15	+13.0	−18.8
17 32	+13.3	−19.0	17 48	+13.1	−18.7
18 06	+13.4	−18.9	18 24	+13.2	−18.6
18 42	+13.5	−18.8	19 01	+13.3	−18.5
19 21	+13.6	−18.7	19 42	+13.4	−18.4
20 03	+13.7	−18.6	20 25	+13.5	−18.3
20 48	+13.8	−18.5	21 11	+13.6	−18.2
21 35	+13.9	−18.4	22 00	+13.7	−18.1
22 26	+14.0	−18.3	22 54	+13.8	−18.0
23 22	+14.1	−18.2	23 51	+13.9	−17.9
24 21	+14.2	−18.1	24 53	+14.0	−17.8
25 26	+14.3	−18.0	26 00	+14.1	−17.7
26 36	+14.4	−17.9	27 13	+14.2	−17.6
27 52	+14.5	−17.8	28 33	+14.3	−17.5
29 15	+14.6	−17.7	30 00	+14.4	−17.4
30 46	+14.7	−17.6	31 35	+14.5	−17.3
32 26	+14.8	−17.5	33 20	+14.6	−17.2
34 17	+14.9	−17.4	35 17	+14.7	−17.1
36 20	+15.0	−17.3	37 26	+14.8	−17.0
38 36	+15.1	−17.2	39 50	+14.9	−16.9
41 08	+15.2	−17.1	42 31	+15.0	−16.8
43 59	+15.3	−17.0	45 31	+15.1	−16.7
47 10	+15.4	−16.9	48 55	+15.2	−16.6
50 46	+15.5	−16.8	52 44	+15.3	−16.5
54 49	+15.6	−16.7	57 02	+15.4	−16.4
59 23	+15.7	−16.6	61 51	+15.5	−16.3
64 30	+15.8	−16.5	67 17	+15.6	−16.2
70 12	+15.9	−16.4	73 16	+15.7	−16.1
76 26	+16.0	−16.3	79 43	+15.8	−16.0
83 05	+16.1	−16.2	86 32	+15.9	−15.9
90 00			90 00		

STARS AND PLANETS

App. Alt.	Corrn
9 56	−5.3
10 08	−5.2
10 20	−5.1
10 33	−5.0
10 46	−4.9
11 00	−4.8
11 14	−4.7
11 29	−4.6
11 45	−4.5
12 01	−4.4
12 18	−4.3
12 35	−4.2
12 54	−4.1
13 13	−4.0
13 33	−3.9
13 54	−3.8
14 16	−3.7
14 40	−3.6
15 04	−3.5
15 30	−3.4
15 57	−3.3
16 26	−3.2
16 56	−3.1
17 28	−3.0
18 02	−2.9
18 38	−2.8
19 17	−2.7
19 58	−2.6
20 42	−2.5
21 28	−2.4
22 19	−2.3
23 13	−2.2
24 11	−2.1
25 14	−2.0
26 22	−1.9
27 36	−1.8
28 56	−1.7
30 24	−1.6
32 00	−1.5
33 45	−1.4
35 40	−1.3
37 48	−1.2
40 08	−1.1
42 44	−1.0
45 36	−0.9
48 47	−0.8
52 18	−0.7
56 11	−0.6
60 28	−0.5
65 08	−0.4
70 11	−0.3
75 34	−0.2
81 13	−0.1
87 03	0.0
90 00	

Additional Corrn

1998

VENUS

Jan. 1–Feb. 5

App. Alt.	Corrn
26	+0.5
46	+0.4
60	+0.3
73	+0.2
84	+0.1

Feb. 6–Feb. 20

App. Alt.	Corrn
29	+0.4
51	+0.3
68	+0.2
83	+0.1

Feb. 21–Mar. 15

App. Alt.	Corrn
34	+0.3
60	+0.2
80	+0.1

Mar. 16–May 5

App. Alt.	Corrn
41	+0.2
76	+0.1

May 6–Dec. 31

App. Alt.	Corrn
60	+0.1

MARS

Jan. 1–Dec. 31

App. Alt.	Corrn
60	+0.1

DIP

Ht. of Eye (m)	Corrn	Ht. of Eye (ft)
2.4	−2.8	8.0
2.6		8.6
2.8	−2.9	9.2
3.0	−3.0	9.8
3.2	−3.1	10.5
3.4	−3.2	11.2
3.6	−3.3	11.9
3.8	−3.4	12.6
4.0	−3.5	13.3
4.3	−3.6	14.1
4.5	−3.7	14.9
4.7	−3.8	15.7
5.0	−3.9	16.5
5.2	−4.0	17.4
5.5	−4.1	18.3
5.8	−4.2	19.1
6.1	−4.3	20.1
6.3	−4.4	21.0
6.6	−4.5	22.0
6.9	−4.6	22.9
7.2	−4.7	23.9
7.5	−4.8	24.9
7.9	−4.9	26.0
8.2	−5.0	27.1
8.5	−5.1	28.1
8.8	−5.2	29.2
9.2	−5.3	30.4
9.5	−5.4	31.5
9.9	−5.5	32.7
10.3	−5.6	33.9
10.6	−5.7	35.1
11.0	−5.8	36.3
11.4	−5.9	37.6
11.8	−6.0	38.9
12.2	−6.1	40.1
12.6	−6.2	41.5
13.0	−6.3	42.8
13.4	−6.4	44.2
13.8	−6.5	45.5
14.2	−6.6	46.9
14.7	−6.7	48.4
15.1	−6.8	49.8
15.5	−6.9	51.3
16.0	−7.0	52.8
16.5	−7.1	54.3
16.9	−7.2	55.8
17.4	−7.3	57.4
17.9	−7.4	58.9
18.4	−7.5	60.5
18.8	−7.6	62.1
19.3	−7.7	63.8
19.8	−7.8	65.4
20.4	−7.9	67.1
20.9	−8.0	68.8
21.4	−8.1	70.5

DIP (additional)

Ht. of Eye	Corrn
m	/
1.0	− 1.8
1.5	− 2.2
2.0	− 2.5
2.5	− 2.8
3.0	− 3.0
See table ←	
m	/
20	− 7.9
22	− 8.3
24	− 8.6
26	− 9.0
28	− 9.3
30	− 9.6
32	− 10.0
34	− 10.3
36	− 10.6
38	− 10.8
40	− 11.1
42	− 11.4
44	− 11.7
46	− 11.9
48	− 12.2
ft.	/
2	− 1.4
4	− 1.9
6	− 2.4
8	− 2.7
10	− 3.1
See table ←	
ft.	/
70	− 8.1
75	− 8.4
80	− 8.7
85	− 8.9
90	− 9.2
95	− 9.5
100	− 9.7
105	− 9.9
110	− 10.2
115	− 10.4
120	− 10.6
125	− 10.8
130	− 11.1
135	− 11.3
140	− 11.5
145	− 11.7
150	− 11.9
155	− 12.1

App. Alt. = Apparent altitude = Sextant altitude corrected for index error and dip.

UT	SUN GHA	SUN Dec	MOON GHA	v	MOON Dec	d	HP
d h	° '	° '	° '	'	° '	'	'
21 00	176 34.7	S10 42.6	253 26.7	10.1	S17 38.9	3.2	56.3
01	191 34.8	41.7	267 55.8	10.1	17 42.1	3.0	56.3
02	206 34.9	40.8	282 24.9	10.0	17 45.1	2.9	56.4
03	221 34.9	.. 39.9	296 53.9	10.0	17 48.0	2.9	56.4
04	236 35.0	39.0	311 22.9	9.9	17 50.9	2.8	56.4
05	251 35.1	38.1	325 51.8	9.8	17 53.7	2.7	56.5
06	266 35.2	S10 37.2	340 20.6	9.8	S17 56.4	2.5	56.5
S 07	281 35.2	36.3	354 49.4	9.7	17 58.9	2.5	56.5
A 08	296 35.3	35.4	9 18.1	9.7	18 01.4	2.4	56.6
T 09	311 35.4	.. 34.4	23 46.8	9.6	18 03.8	2.3	56.6
U 10	326 35.4	33.5	38 15.4	9.5	18 06.1	2.2	56.7
R 11	341 35.5	32.6	52 43.9	9.5	18 08.3	2.1	56.7
D 12	356 35.6	S10 31.7	67 12.4	9.4	S18 10.4	2.0	56.7
A 13	11 35.7	30.8	81 40.8	9.3	18 12.4	1.9	56.8
Y 14	26 35.7	29.9	96 09.1	9.3	18 14.3	1.9	56.8
15	41 35.8	.. 29.0	110 37.4	9.3	18 16.2	1.7	56.8
16	56 35.9	28.1	125 05.7	9.2	18 17.9	1.6	56.9
17	71 36.0	27.2	139 33.9	9.1	18 19.5	1.5	56.9
18	86 36.0	S10 26.3	154 02.0	9.1	S18 21.0	1.4	56.9
19	101 36.1	25.4	168 30.1	9.0	18 22.4	1.3	57.0
20	116 36.2	24.5	182 58.1	8.9	18 23.7	1.3	57.0
21	131 36.3	.. 23.6	197 26.0	8.9	18 25.0	1.1	57.1
22	146 36.3	22.7	211 53.9	8.9	18 26.1	1.0	57.1
23	161 36.4	21.8	226 21.8	8.8	18 27.1	0.9	57.1
22 00	176 36.5	S10 20.9	240 49.6	8.7	S18 28.0	0.8	57.2
01	191 36.6	19.9	255 17.3	8.7	18 28.8	0.7	57.2
02	206 36.6	19.0	269 45.0	8.6	18 29.5	0.6	57.3
03	221 36.7	.. 18.1	284 12.6	8.6	18 30.1	0.4	57.3
04	236 36.8	17.2	298 40.2	8.5	18 30.5	0.4	57.3
05	251 36.9	16.3	313 07.7	8.5	18 30.9	0.3	57.4
06	266 36.9	S10 15.4	327 35.2	8.4	S18 31.2	0.2	57.4
S 07	281 37.0	14.5	342 02.6	8.3	18 31.4	0.0	57.4
U 08	296 37.1	13.6	356 30.0	8.3	18 31.4	0.0	57.5
N 09	311 37.2	.. 12.7	10 57.3	8.3	18 31.4	0.2	57.5
D 10	326 37.3	11.8	25 24.6	8.2	18 31.2	0.3	57.6
A 11	341 37.3	10.8	39 51.8	8.2	18 30.9	0.3	57.6
Y 12	356 37.4	S10 09.9	54 19.0	8.1	S18 30.6	0.5	57.6
13	11 37.5	09.0	68 46.1	8.1	18 30.1	0.6	57.7
14	26 37.6	08.1	83 13.2	8.0	18 29.5	0.7	57.7
15	41 37.7	.. 07.2	97 40.2	8.0	18 28.8	0.8	57.8
16	56 37.7	06.3	112 07.2	7.9	18 28.0	1.0	57.8
17	71 37.8	05.4	126 34.1	7.9	18 27.0	1.0	57.8
18	86 37.9	S10 04.5	141 01.0	7.9	S18 26.0	1.2	57.9
19	101 38.0	03.5	155 27.9	7.8	18 24.8	1.2	57.9
20	116 38.1	02.6	169 54.7	7.8	18 23.6	1.4	58.0
21	131 38.1	.. 01.7	184 21.5	7.7	18 22.2	1.5	58.0
22	146 38.2	10 00.8	198 48.2	7.7	18 20.7	1.6	58.0
23	161 38.3	9 59.9	213 14.9	7.6	18 19.1	1.7	58.1
23 00	176 38.4	S 9 59.0	227 41.5	7.6	S18 17.4	1.9	58.1
01	191 38.5	58.1	242 08.1	7.6	18 15.5	1.9	58.2
02	206 38.6	57.1	256 34.7	7.5	18 13.6	2.1	58.2
03	221 38.6	.. 56.2	271 01.2	7.5	18 11.5	2.2	58.2
04	236 38.7	55.3	285 27.7	7.4	18 09.3	2.3	58.3
05	251 38.8	54.4	299 54.1	7.5	18 07.0	2.4	58.3
06	266 38.9	S 9 53.5	314 20.6	7.3	S18 04.6	2.5	58.4
07	281 39.0	52.6	328 46.9	7.4	18 02.1	2.6	58.4
08	296 39.1	51.6	343 13.3	7.3	17 59.5	2.8	58.4
M 09	311 39.1	.. 50.7	357 39.6	7.3	17 56.7	2.8	58.5
O 10	326 39.2	49.8	12 05.9	7.2	17 53.9	3.0	58.5
N 11	341 39.3	48.9	26 32.1	7.3	17 50.9	3.1	58.5
D 12	356 39.4	S 9 48.0	40 58.4	7.2	S17 47.8	3.2	58.6
A 13	11 39.5	47.0	55 24.6	7.1	17 44.6	3.4	58.6
Y 14	26 39.6	46.1	69 50.7	7.2	17 41.2	3.4	58.7
15	41 39.7	.. 45.2	84 16.9	7.1	17 37.8	3.6	58.7
16	56 39.7	44.3	98 43.0	7.0	17 34.2	3.6	58.7
17	71 39.8	43.4	113 09.0	7.1	17 30.6	3.8	58.8
18	86 39.9	S 9 42.5	127 35.1	7.0	S17 26.8	3.9	58.8
19	101 40.0	41.5	142 01.1	7.0	17 22.9	4.1	58.9
20	116 40.1	40.6	156 27.1	7.0	17 18.8	4.1	58.9
21	131 40.2	.. 39.7	170 53.1	7.0	17 14.7	4.3	58.9
22	146 40.3	38.8	185 19.1	6.9	17 10.4	4.3	59.0
23	161 40.4	37.8	199 45.0	6.9	S17 06.1	4.5	59.0
	SD 16.2	d 0.9	SD 15.5		15.7		16.0

Lat.	Twilight Naut.	Twilight Civil	Sunrise	Moonrise 21	Moonrise 22	Moonrise 23	Moonrise 24
°	h m	h m	h m	h m	h m	h m	h m
N 72	05 44	07 03	08 15	▬	▬	▬	08 25
N 70	05 46	06 56	08 01	05 32	06 43	07 23	07 40
68	05 47	06 51	07 50	04 51	05 56	06 42	07 10
66	05 48	06 47	07 40	04 23	05 26	06 14	06 47
64	05 48	06 43	07 32	04 01	05 03	05 52	06 29
62	05 48	06 40	07 25	03 44	04 45	05 35	06 15
60	05 48	06 37	07 19	03 30	04 30	05 21	06 02
N 58	05 48	06 34	07 14	03 18	04 17	05 08	05 52
56	05 48	06 31	07 09	03 07	04 06	04 58	05 42
54	05 48	06 29	07 05	02 58	03 56	04 48	05 34
52	05 48	06 27	07 01	02 50	03 47	04 40	05 26
50	05 48	06 25	06 58	02 43	03 40	04 32	05 20
45	05 46	06 21	06 50	02 27	03 23	04 16	05 05
N 40	05 45	06 17	06 44	02 14	03 09	04 03	04 53
35	05 43	06 13	06 38	02 03	02 58	03 52	04 43
30	05 41	06 09	06 34	01 53	02 48	03 42	04 34
20	05 37	06 03	06 25	01 37	02 31	03 25	04 19
N 10	05 31	05 56	06 17	01 22	02 16	03 10	04 05
0	05 25	05 49	06 10	01 09	02 02	02 56	03 53
S 10	05 16	05 41	06 03	00 56	01 48	02 43	03 40
20	05 06	05 32	05 55	00 42	01 33	02 28	03 27
30	04 52	05 21	05 46	00 25	01 16	02 11	03 11
35	04 43	05 14	05 40	00 16	01 06	02 01	03 02
40	04 32	05 06	05 34	00 05	00 54	01 50	02 52
45	04 19	04 56	05 27	24 41	00 41	01 37	02 40
S 50	04 01	04 44	05 18	24 25	00 25	01 21	02 25
52	03 53	04 38	05 14	24 17	00 17	01 13	02 18
54	03 43	04 32	05 10	24 09	00 09	01 05	02 11
56	03 32	04 24	05 05	23 59	24 55	00 55	02 02
58	03 19	04 16	04 59	23 48	24 45	00 45	01 53
S 60	03 04	04 06	04 53	23 36	24 32	00 32	01 42

Lat.	Sunset	Twilight Civil	Twilight Naut.	Moonset 21	Moonset 22	Moonset 23	Moonset 24
°	h m	h m	h m	h m	h m	h m	h m
N 72	16 13	17 26	18 45	▬	▬	▬	11 59
N 70	16 28	17 32	18 43	09 07	09 46	11 01	12 43
68	16 39	17 37	18 42	09 48	10 33	11 42	13 12
66	16 48	17 42	18 41	10 17	11 03	12 10	13 34
64	16 56	17 45	18 41	10 38	11 26	12 31	13 51
62	17 03	17 49	18 40	10 55	11 45	12 48	14 05
60	17 09	17 52	18 40	11 10	12 00	13 02	14 17
N 58	17 14	17 54	18 40	11 22	12 12	13 14	14 28
56	17 19	17 57	18 40	11 33	12 23	13 25	14 37
54	17 23	17 59	18 40	11 42	12 33	13 34	14 45
52	17 26	18 01	18 40	11 51	12 42	13 42	14 52
50	17 30	18 03	18 40	11 58	12 50	13 50	14 58
45	17 37	18 07	18 41	12 14	13 06	14 06	15 12
N 40	17 44	18 11	18 43	12 27	13 20	14 18	15 23
35	17 49	18 15	18 44	12 39	13 31	14 29	15 33
30	17 54	18 18	18 46	12 49	13 41	14 39	15 41
20	18 02	18 25	18 50	13 05	13 58	14 55	15 56
N 10	18 10	18 31	18 56	13 20	14 14	15 10	16 08
0	18 17	18 38	19 02	13 34	14 28	15 23	16 20
S 10	18 24	18 45	19 10	13 48	14 42	15 36	16 32
20	18 32	18 54	19 21	14 02	14 57	15 51	16 44
30	18 41	19 05	19 34	14 19	15 14	16 07	16 58
35	18 46	19 12	19 43	14 29	15 24	16 16	17 06
40	18 52	19 20	19 54	14 40	15 35	16 27	17 15
45	18 59	19 30	20 07	14 54	15 48	16 39	17 26
S 50	19 08	19 42	20 24	15 10	16 05	16 55	17 39
52	19 12	19 48	20 32	15 17	16 12	17 02	17 45
54	19 16	19 54	20 42	15 26	16 21	17 09	17 52
56	19 21	20 01	20 53	15 35	16 30	17 18	17 59
58	19 26	20 09	21 05	15 46	16 41	17 28	18 07
S 60	19 32	20 19	21 20	15 58	16 53	17 40	18 17

	SUN Eqn. of Time 00h	SUN Eqn. of Time 12h	SUN Mer. Pass.	MOON Mer. Pass. Upper	MOON Mer. Pass. Lower	MOON Age	MOON Phase
Day							
d	m s	m s	h m	h m	h m	d	%
21	13 41	13 38	12 14	07 21	19 48	24	32
22	13 34	13 30	12 14	08 15	20 42	25	22
23	13 27	13 23	12 13	09 10	21 38	26	14

1998 FEBRUARY 24, 25, 26 (TUES., WED., THURS.)

UT	ARIES GHA	VENUS −4.6 GHA	Dec	MARS +1.2 GHA	Dec	JUPITER −2.0 GHA	Dec	SATURN +0.7 GHA	Dec	STARS Name	SHA	Dec
24 00	153 40.1	218 32.5	S15 54.5	159 33.7	S 3 23.4	176 48.1	S10 36.9	136 35.7	N 4 45.4	Acamar	315 27.4	S40 19.1
01	168 42.6	233 33.4	54.5	174 34.3	22.6	191 50.0	36.6	151 37.9	45.5	Achernar	335 35.9	S57 15.1
02	183 45.1	248 34.3	54.6	189 35.0	21.8	206 51.9	36.4	166 40.1	45.6	Acrux	173 21.7	S63 05.2
03	198 47.5	263 35.3 ..	54.6	204 35.7 ..	21.0	221 53.8 ..	36.2	181 42.4 ..	45.7	Adhara	255 21.5	S28 58.5
04	213 50.0	278 36.2	54.7	219 36.4	20.3	236 55.7	36.0	196 44.6	45.8	Aldebaran	291 02.8	N16 30.2
05	228 52.5	293 37.1	54.7	234 37.1	19.5	251 57.6	35.8	211 46.8	45.9			
T 06	243 54.9	308 38.0	S15 54.8	249 37.7	S 3 18.7	266 59.5	S10 35.5	226 49.0	N 4 46.0	Alioth	166 30.6	N55 58.0
U 07	258 57.4	323 38.9	54.8	264 38.4	17.9	282 01.4	35.3	241 51.3	46.1	Alkaid	153 07.9	N49 19.2
E 08	273 59.9	338 39.9	54.9	279 39.1	17.1	297 03.3	35.1	256 53.5	46.2	Al Na'ir	27 58.9	S46 58.2
S 09	289 02.3	353 40.8 ..	54.9	294 39.8 ..	16.3	312 05.2 ..	34.9	271 55.7 ..	46.3	Alnilam	275 58.2	S 1 12.5
D 10	304 04.8	8 41.7	54.9	309 40.5	15.5	327 07.1	34.7	286 57.9	46.5	Alphard	218 07.3	S 8 39.2
A 11	319 07.2	23 42.6	55.0	324 41.1	14.7	342 09.0	34.5	302 00.1	46.6			
Y 12	334 09.7	38 43.5	S15 55.0	339 41.8	S 3 13.9	357 10.9	S10 34.2	317 02.4	N 4 46.7	Alphecca	126 20.9	N26 43.1
13	349 12.2	53 44.4	55.1	354 42.5	13.1	12 12.7	34.0	332 04.6	46.8	Alpheratz	357 55.9	N29 04.7
14	4 14.6	68 45.3	55.1	9 43.2	12.3	27 14.6	33.8	347 06.8	46.9	Altair	62 19.9	N 8 51.8
15	19 17.1	83 46.2 ..	55.2	24 43.9 ..	11.5	42 16.5 ..	33.6	2 09.0 ..	47.0	Ankaa	353 27.7	S42 19.2
16	34 19.6	98 47.1	55.2	39 44.5	10.7	57 18.4	33.4	17 11.3	47.1	Antares	112 40.7	S26 25.5
17	49 22.0	113 48.0	55.2	54 45.2	09.9	72 20.3	33.1	32 13.5	47.2			
18	64 24.5	128 48.9	S15 55.3	69 45.9	S 3 09.1	87 22.2	S10 32.9	47 15.7	N 4 47.3	Arcturus	146 06.3	N19 11.4
19	79 27.0	143 49.8	55.3	84 46.6	08.4	102 24.1	32.7	62 17.9	47.4	Atria	107 53.1	S69 01.1
20	94 29.4	158 50.7	55.4	99 47.3	07.6	117 26.0	32.5	77 20.1	47.5	Avior	234 22.2	S59 30.5
21	109 31.9	173 51.5 ..	55.4	114 47.9 ..	06.8	132 27.9 ..	32.3	92 22.4 ..	47.6	Bellatrix	278 44.5	N 6 20.7
22	124 34.4	188 52.4	55.4	129 48.6	06.0	147 29.8	32.1	107 24.6	47.7	Betelgeuse	271 13.9	N 7 24.2
23	139 36.8	203 53.3	55.5	144 49.3	05.2	162 31.7	31.8	122 26.8	47.8			
25 00	154 39.3	218 54.2	S15 55.5	159 50.0	S 3 04.4	177 33.6	S10 31.6	137 29.0	N 4 47.9	Canopus	264 01.1	S52 42.1
01	169 41.7	233 55.1	55.5	174 50.7	03.6	192 35.5	31.4	152 31.3	48.0	Capella	280 51.7	N45 59.8
02	184 44.2	248 55.9	55.6	189 51.4	02.8	207 37.4	31.2	167 33.5	48.1	Deneb	49 39.8	N45 16.3
03	199 46.7	263 56.8 ..	55.6	204 52.0 ..	02.0	222 39.2 ..	31.0	182 35.7 ..	48.3	Denebola	182 45.3	N14 34.8
04	214 49.1	278 57.7	55.6	219 52.7	01.2	237 41.1	30.7	197 37.9	48.4	Diphda	349 07.9	S18 00.0
05	229 51.6	293 58.5	55.7	234 53.4	3 00.4	252 43.0	30.5	212 40.1	48.5			
W 06	244 54.1	308 59.4	S15 55.7	249 54.1	S 2 59.6	267 44.9	S10 30.3	227 42.4	N 4 48.6	Dubhe	194 05.4	N61 45.5
E 07	259 56.5	324 00.3	55.7	264 54.8	58.8	282 46.8	30.1	242 44.6	48.7	Elnath	278 27.3	N28 36.3
D 08	274 59.0	339 01.1	55.8	279 55.4	58.0	297 48.7	29.9	257 46.8	48.8	Eltanin	90 51.8	N51 29.2
N 09	290 01.5	354 02.0 ..	55.8	294 56.1 ..	57.2	312 50.6 ..	29.7	272 49.0 ..	48.9	Enif	33 58.9	N 9 51.9
E 10	305 03.9	9 02.8	55.8	309 56.8	56.4	327 52.5	29.4	287 51.3	49.0	Fomalhaut	15 37.3	S29 38.1
S 11	320 06.4	24 03.7	55.9	324 57.5	55.6	342 54.4	29.2	302 53.5	49.1			
D 12	335 08.8	39 04.5	S15 55.9	339 58.2	S 2 54.9	357 56.3	S10 29.0	317 55.7	N 4 49.2	Gacrux	172 13.4	S57 06.0
A 13	350 11.3	54 05.4	55.9	354 58.9	54.1	12 58.2	28.8	332 57.9	49.3	Gienah	176 04.1	S17 31.9
Y 14	5 13.8	69 06.2	56.0	9 59.5	53.3	28 00.1	28.6	348 00.1	49.4	Hadar	149 04.2	S60 21.6
15	20 16.2	84 07.1 ..	56.0	25 00.2 ..	52.5	43 02.0 ..	28.3	3 02.4 ..	49.5	Hamal	328 14.2	N23 27.1
16	35 18.7	99 07.9	56.0	40 00.9	51.7	58 03.9	28.1	18 04.6	49.6	Kaus Aust.	83 59.6	S34 23.0
17	50 21.2	114 08.8	56.0	55 01.6	50.9	73 05.7	27.9	33 06.8	49.7			
18	65 23.6	129 09.6	S15 56.1	70 02.3	S 2 50.1	88 07.6	S10 27.7	48 09.0	N 4 49.8	Kochab	137 19.3	N74 09.6
19	80 26.1	144 10.4	56.1	85 03.0	49.3	103 09.5	27.5	63 11.2	50.0	Markab	13 50.3	N15 11.6
20	95 28.6	159 11.3	56.1	100 03.6	48.5	118 11.4	27.2	78 13.5	50.1	Menkar	314 27.4	N 4 04.7
21	110 31.0	174 12.1 ..	56.1	115 04.3 ..	47.7	133 13.3 ..	27.0	93 15.7 ..	50.2	Menkent	148 21.2	S36 21.5
22	125 33.5	189 12.9	56.2	130 05.0	46.9	148 15.2	26.8	108 17.9	50.3	Miaplacidus	221 41.1	S69 42.8
23	140 36.0	204 13.7	56.2	145 05.7	46.1	163 17.1	26.6	123 20.1	50.4			
26 00	155 38.4	219 14.6	S15 56.2	160 06.4	S 2 45.3	178 19.0	S10 26.4	138 22.3	N 4 50.5	Mirfak	308 57.3	N49 51.3
01	170 40.9	234 15.4	56.2	175 07.1	44.5	193 20.9	26.2	153 24.6	50.6	Nunki	76 13.1	S26 17.8
02	185 43.3	249 16.2	56.3	190 07.7	43.7	208 22.8	25.9	168 26.8	50.7	Peacock	53 38.2	S56 44.3
03	200 45.8	264 17.0 ..	56.3	205 08.4 ..	42.9	223 24.7 ..	25.7	183 29.0 ..	50.8	Pollux	243 41.8	N28 01.7
04	215 48.3	279 17.8	56.3	220 09.1	42.2	238 26.6	25.5	198 31.2	50.9	Procyon	245 11.8	N 5 13.6
05	230 50.7	294 18.6	56.3	235 09.8	41.4	253 28.5	25.3	213 33.4	51.0			
T 06	245 53.2	309 19.5	S15 56.3	250 10.5	S 2 40.6	268 30.4	S10 25.1	228 35.7	N 4 51.1	Rasalhague	96 17.4	N12 33.6
H 07	260 55.7	324 20.3	56.4	265 11.2	39.8	283 32.3	24.8	243 37.9	51.2	Regulus	207 55.7	N11 58.4
U 08	275 58.1	339 21.1	56.4	280 11.8	39.0	298 34.2	24.6	258 40.1	51.3	Rigel	281 23.3	S 8 12.5
R 09	291 00.6	354 21.9 ..	56.4	295 12.5 ..	38.2	313 36.0 ..	24.4	273 42.3 ..	51.5	Rigil Kent.	140 07.5	S60 49.3
S 10	306 03.1	9 22.7	56.4	310 13.2	37.4	328 37.9	24.2	288 44.5	51.6	Sabik	102 26.1	S15 43.3
D 11	321 05.5	24 23.5	56.4	325 13.9	36.6	343 39.8	24.0	303 46.8	51.7			
A 12	336 08.0	39 24.3	S15 56.4	340 14.6	S 2 35.8	358 41.7	S10 23.7	318 49.0	N 4 51.8	Schedar	349 54.3	N56 31.7
Y 13	351 10.5	54 25.1	56.5	355 15.3	35.0	13 43.6	23.5	333 51.2	51.9	Shaula	96 38.0	S37 05.9
14	6 12.9	69 25.8	56.5	10 16.0	34.2	28 45.5	23.3	348 53.4	52.0	Sirius	258 43.9	S16 43.1
15	21 15.4	84 26.6 ..	56.5	25 16.6 ..	33.4	43 47.4 ..	23.1	3 55.6 ..	52.1	Spica	158 43.4	S11 09.1
16	36 17.8	99 27.4	56.5	40 17.3	32.6	58 49.3	22.9	18 57.8	52.2	Suhail	223 00.6	S43 25.7
17	51 20.3	114 28.2	56.5	55 18.0	31.8	73 51.2	22.7	34 00.1	52.3			
18	66 22.8	129 29.0	S15 56.5	70 18.7	S 2 31.0	88 53.1	S10 22.4	49 02.3	N 4 52.4	Vega	80 47.1	N38 46.8
19	81 25.2	144 29.8	56.5	85 19.4	30.2	103 55.0	22.2	64 04.5	52.5	Zuben'ubi	137 18.3	S16 02.0
20	96 27.7	159 30.6	56.6	100 20.1	29.4	118 56.9	22.0	79 06.7	52.6		SHA	Mer. Pass.
21	111 30.2	174 31.3 ..	56.6	115 20.8 ..	28.6	133 58.8 ..	21.8	94 08.9 ..	52.7	Venus	64 14.9	9 24
22	126 32.6	189 32.1	56.6	130 21.4	27.9	149 00.7	21.6	109 11.2	52.8	Mars	5 10.7	13 20
23	141 35.1	204 32.9	56.6	145 22.1	27.1	164 02.6	21.3	124 13.4	53.0	Jupiter	22 54.3	12 08
Mer. Pass. 13 39.1		v 0.8	d 0.0	v 0.7	d 0.8	v 1.9	d 0.2	v 2.2	d 0.1	Saturn	342 49.8	14 48

UT	SUN GHA	SUN Dec	MOON GHA	v	Dec	d	HP
d h	° '	° '	° '	'	° '	'	'
24 00	176 40.4	S 9 36.9	214 10.9	6.9	S17 01.6	4.6	59.0
01	191 40.5	36.0	228 36.8	6.9	16 57.0	4.7	59.1
02	206 40.6	35.1	243 02.7	6.9	16 52.3	4.9	59.1
03	221 40.7	.. 34.2	257 28.6	6.8	16 47.4	4.9	59.2
04	236 40.8	33.2	271 54.4	6.8	16 42.5	5.1	59.2
05	251 40.9	32.3	286 20.2	6.9	16 37.4	5.1	59.2
06	266 41.0	S 9 31.4	300 46.1	6.8	S16 32.3	5.3	59.3
07	281 41.1	30.5	315 11.9	6.7	16 27.0	5.4	59.3
T 08	296 41.2	29.5	329 37.6	6.8	16 21.6	5.5	59.3
U 09	311 41.3	.. 28.6	344 03.4	6.8	16 16.1	5.7	59.4
E 10	326 41.3	27.7	358 29.2	6.7	16 10.4	5.7	59.4
S 11	341 41.4	26.8	12 54.9	6.8	16 04.7	5.8	59.5
D 12	356 41.5	S 9 25.9	27 20.7	6.7	S15 58.9	6.0	59.5
A 13	11 41.6	24.9	41 46.4	6.7	15 52.9	6.1	59.5
Y 14	26 41.7	24.0	56 12.1	6.7	15 46.8	6.1	59.6
15	41 41.8	.. 23.1	70 37.8	6.7	15 40.7	6.3	59.6
16	56 41.9	22.2	85 03.5	6.7	15 34.4	6.4	59.6
17	71 42.0	21.2	99 29.2	6.7	15 28.0	6.5	59.7
18	86 42.1	S 9 20.3	113 54.9	6.7	S15 21.5	6.7	59.7
19	101 42.2	19.4	128 20.6	6.6	15 14.8	6.7	59.7
20	116 42.3	18.5	142 46.2	6.7	15 08.1	6.8	59.8
21	131 42.4	.. 17.5	157 11.9	6.7	15 01.3	6.9	59.8
22	146 42.5	16.6	171 37.6	6.7	14 54.4	7.1	59.8
23	161 42.6	15.7	186 03.3	6.6	14 47.3	7.1	59.9
25 00	176 42.6	S 9 14.7	200 28.9	6.7	S14 40.2	7.3	59.9
01	191 42.7	13.8	214 54.6	6.7	14 32.9	7.3	59.9
02	206 42.8	12.9	229 20.3	6.6	14 25.6	7.5	60.0
03	221 42.9	.. 12.0	243 45.9	6.7	14 18.1	7.6	60.0
04	236 43.0	11.0	258 11.6	6.7	14 10.5	7.6	60.0
05	251 43.1	10.1	272 37.3	6.6	14 02.9	7.8	60.0
06	266 43.2	S 9 09.2	287 02.9	6.7	S13 55.1	7.9	60.1
W 07	281 43.3	08.2	301 28.6	6.7	13 47.2	7.9	60.1
E 08	296 43.4	07.3	315 54.3	6.7	13 39.3	8.1	60.1
D 09	311 43.5	.. 06.4	330 20.0	6.6	13 31.2	8.2	60.2
N 10	326 43.6	05.5	344 45.6	6.7	13 23.0	8.2	60.2
E 11	341 43.7	04.5	359 11.3	6.7	13 14.8	8.4	60.2
S 12	356 43.8	S 9 03.6	13 37.0	6.7	S13 06.4	8.4	60.3
D 13	11 43.9	02.7	28 02.7	6.7	12 58.0	8.6	60.3
A 14	26 44.0	01.7	42 28.4	6.8	12 49.4	8.6	60.3
Y 15	41 44.1	9 00.8	56 54.2	6.7	12 40.8	8.7	60.3
16	56 44.2	8 59.9	71 19.9	6.7	12 32.1	8.8	60.4
17	71 44.3	58.9	85 45.6	6.8	12 23.3	8.9	60.4
18	86 44.4	S 8 58.0	100 11.4	6.7	S12 14.4	9.0	60.4
19	101 44.5	57.1	114 37.1	6.8	12 05.4	9.1	60.4
20	116 44.6	56.1	129 02.9	6.8	11 56.3	9.2	60.5
21	131 44.7	.. 55.2	143 28.7	6.7	11 47.1	9.2	60.5
22	146 44.8	54.3	157 54.4	6.8	11 37.9	9.4	60.5
23	161 44.9	53.4	172 20.2	6.9	11 28.5	9.4	60.5
26 00	176 45.0	S 8 52.4	186 46.1	6.8	S11 19.1	9.5	60.6
01	191 45.1	51.5	201 11.9	6.8	11 09.6	9.6	60.6
02	206 45.2	50.6	215 37.7	6.9	11 00.0	9.7	60.6
03	221 45.3	.. 49.6	230 03.6	6.8	10 50.3	9.7	60.6
04	236 45.4	48.7	244 29.4	6.9	10 40.6	9.9	60.6
05	251 45.5	47.8	258 55.3	6.9	10 30.7	9.9	60.7
06	266 45.6	S 8 46.8	273 21.2	6.9	S10 20.8	9.9	60.7
T 07	281 45.7	45.9	287 47.1	6.9	10 10.9	10.1	60.7
H 08	296 45.8	44.9	302 13.0	6.9	10 00.8	10.1	60.7
U 09	311 45.9	.. 44.0	316 38.9	6.9	9 50.7	10.2	60.7
R 10	326 46.0	43.1	331 04.8	7.0	9 40.5	10.3	60.8
S 11	341 46.1	42.1	345 30.8	7.0	9 30.2	10.3	60.8
D 12	356 46.2	S 8 41.2	359 56.8	7.0	S 9 19.9	10.4	60.8
A 13	11 46.3	40.3	14 22.8	7.0	9 09.5	10.5	60.8
Y 14	26 46.4	39.3	28 48.8	7.0	8 59.0	10.5	60.8
15	41 46.5	.. 38.4	43 14.8	7.0	8 48.5	10.6	60.8
16	56 46.6	37.5	57 40.8	7.1	8 37.9	10.7	60.9
17	71 46.7	36.5	72 06.9	7.1	S 8 27.2	10.7	60.9
18	86 46.8	S 8 35.6					
19	101 47.0	34.7					
20	116 47.1	33.7					
21	131 47.2	.. 32.8					
22	146 47.3	31.8					
23	161 47.4	30.9					
SD 16.2	d 0.9		SD 16.2		16.4		16.6

Lat.	Twilight Naut.	Twilight Civil	Sunrise	Moonrise 24	25	26	27
°	h m	h m	h m	h m	h m	h m	h m
N 72	05 30	06 48	07 59	08 25	08 13	08 05	07 59
N 70	05 33	06 44	07 47	07 40	07 46	07 50	07 51
68	05 36	06 40	07 37	07 10	07 26	07 37	07 44
66	05 37	06 36	07 29	06 47	07 10	07 26	07 39
64	05 39	06 33	07 22	06 29	06 57	07 17	07 34
62	05 40	06 31	07 16	06 15	06 45	07 10	07 30
60	05 40	06 29	07 11	06 02	06 36	07 03	07 26
N 58	05 41	06 26	07 07	05 52	06 27	06 57	07 23
56	05 41	06 24	07 02	05 42	06 20	06 52	07 20
54	05 42	06 23	06 58	05 34	06 13	06 47	07 17
52	05 42	06 21	06 55	05 26	06 07	06 43	07 15
50	05 42	06 19	06 52	05 20	06 02	06 39	07 13
45	05 42	06 16	06 45	05 05	05 50	06 30	07 08
N 40	05 41	06 12	06 40	04 53	05 40	06 23	07 04
35	05 40	06 09	06 35	04 43	05 32	06 17	07 00
30	05 39	06 06	06 31	04 34	05 24	06 12	06 57
20	05 35	06 01	06 23	04 19	05 12	06 03	06 52
N 10	05 30	05 55	06 16	04 05	05 00	05 54	06 48
0	05 24	05 49	06 10	03 53	04 50	05 47	06 43
S 10	05 17	05 42	06 03	03 40	04 39	05 39	06 39
20	05 07	05 34	05 56	03 27	04 28	05 31	06 34
30	04 55	05 23	05 48	03 11	04 15	05 21	06 29
35	04 46	05 17	05 43	03 02	04 08	05 16	06 26
40	04 36	05 10	05 38	02 52	03 59	05 10	06 22
45	04 24	05 01	05 31	02 40	03 49	05 03	06 18
S 50	04 08	04 49	05 23	02 25	03 37	04 54	06 14
52	04 00	04 44	05 20	02 18	03 31	04 50	06 11
54	03 51	04 38	05 16	02 11	03 25	04 46	06 09
56	03 41	04 32	05 12	02 02	03 18	04 41	06 06
58	03 29	04 24	05 07	01 53	03 11	04 35	06 03
S 60	03 15	04 15	05 01	01 42	03 02	04 29	06 00

Lat.	Sunset	Twilight Civil	Twilight Naut.	Moonset 24	25	26	27
°	h m	h m	h m	h m	h m	h m	h m
N 72	16 29	17 40	18 58	11 59	14 10	16 17	18 21
N 70	16 41	17 44	18 55	12 43	14 35	16 31	18 26
68	16 51	17 48	18 53	13 12	14 54	16 42	18 31
66	16 59	17 51	18 51	13 34	15 10	16 51	18 35
64	17 06	17 54	18 49	13 51	15 22	16 59	18 38
62	17 11	17 57	18 48	14 05	15 33	17 05	18 41
60	17 17	17 59	18 47	14 17	15 42	17 11	18 43
N 58	17 21	18 01	18 46	14 28	15 49	17 16	18 45
56	17 25	18 03	18 46	14 37	15 56	17 21	18 47
54	17 29	18 05	18 46	14 45	16 02	17 24	18 49
52	17 32	18 06	18 45	14 52	16 08	17 28	18 50
50	17 35	18 08	18 45	14 58	16 13	17 31	18 52
45	17 42	18 11	18 45	15 12	16 23	17 38	18 55
N 40	17 47	18 14	18 46	15 23	16 32	17 44	18 57
35	17 52	18 17	18 47	15 33	16 40	17 49	18 59
30	17 56	18 20	18 48	15 41	16 46	17 53	19 01
20	18 03	18 26	18 51	15 56	16 58	18 01	19 04
N 10	18 10	18 31	18 56	16 08	17 08	18 08	19 07
0	18 16	18 37	19 02	16 20	17 17	18 14	19 10
S 10	18 23	18 44	19 09	16 32	17 26	18 20	19 12
20	18 30	18 52	19 18	16 44	17 36	18 26	19 15
30	18 38	19 02	19 31	16 58	17 47	18 33	19 18
35	18 43	19 08	19 39	17 06	17 53	18 37	19 20
40	18 48	19 16	19 49	17 15	18 00	18 42	19 22
45	18 54	19 25	20 01	17 26	18 09	18 47	19 24
S 50	19 02	19 36	20 17	17 39	18 19	18 54	19 26
52	19 05	19 41	20 25	17 45	18 23	18 57	19 28
54	19 09	19 46	20 33	17 52	18 28	19 00	19 29
56	19 13	19 53	20 43	17 59	18 34	19 04	19 30
58	19 18	20 00	20 55	18 07	18 40	19 07	19 32
S 60	19 23	20 09	21 08	18 17	18 47	19 12	19 34

A total eclipse of the Sun occurs on this date. See page 5.

Day	SUN Eqn. of Time 00h	SUN Eqn. of Time 12h	SUN Mer. Pass.	MOON Mer. Pass. Upper	MOON Mer. Pass. Lower	Age	Phase
d	m s	m s	h m	h m	h m	d	%
24	13 18	13 14	12 13	10 06	22 35	27	7
25	13 10	13 05	12 13	11 03	23 32	28	2
26	13 00	12 55	12 13	12 00	24 28	29	0

1998 FEB. 27, 28, MAR. 1 (FRI., SAT., SUN.)

UT	ARIES GHA	VENUS −4.6 GHA	Dec	MARS +1.2 GHA	Dec	JUPITER −2.0 GHA	Dec	SATURN +0.7 GHA	Dec
27 00	156 37.6	219 33.6	S15 56.6	160 22.8	S 2 26.3	179 04.5	S10 21.1	139 15.6	N 4 53.1
01	171 40.0	234 34.4	56.6	175 23.5	25.5	194 06.3	20.9	154 17.8	53.2
02	186 42.5	249 35.2	56.6	190 24.2	24.7	209 08.2	20.7	169 20.0	53.3
03	201 44.9	264 35.9	.. 56.6	205 24.9	.. 23.9	224 10.1	.. 20.5	184 22.2	.. 53.4
04	216 47.4	279 36.7	56.6	220 25.5	23.1	239 12.0	20.2	199 24.5	53.5
05	231 49.9	294 37.5	56.6	235 26.2	22.3	254 13.9	20.0	214 26.7	53.6
F 06	246 52.3	309 38.2	S15 56.6	250 26.9	S 2 21.5	269 15.8	S10 19.8	229 28.9	N 4 53.7
R 07	261 54.8	324 39.0	56.7	265 27.6	20.7	284 17.7	19.6	244 31.1	53.8
I 08	276 57.3	339 39.7	56.7	280 28.3	19.9	299 19.6	19.4	259 33.3	53.9
D 09	291 59.7	354 40.5	.. 56.7	295 29.0	.. 19.1	314 21.5	.. 19.2	274 35.6	.. 54.0
A 10	307 02.2	9 41.2	56.7	310 29.7	18.3	329 23.4	18.9	289 37.8	54.1
Y 11	322 04.7	24 42.0	56.7	325 30.4	17.5	344 25.3	18.7	304 40.0	54.2
12	337 07.1	39 42.7	S15 56.7	340 31.0	S 2 16.7	359 27.2	S10 18.5	319 42.2	N 4 54.4
13	352 09.6	54 43.5	56.7	355 31.7	15.9	14 29.1	18.3	334 44.4	54.5
14	7 12.1	69 44.2	56.7	10 32.4	15.1	29 31.0	18.1	349 46.6	54.6
15	22 14.5	84 44.9	.. 56.7	25 33.1	.. 14.3	44 32.9	.. 17.8	4 48.9	.. 54.7
16	37 17.0	99 45.7	56.7	40 33.8	13.6	59 34.8	17.6	19 51.1	54.8
17	52 19.4	114 46.4	56.7	55 34.5	12.8	74 36.6	17.4	34 53.3	54.9
18	67 21.9	129 47.1	S15 56.7	70 35.2	S 2 12.0	89 38.5	S10 17.2	49 55.5	N 4 55.0
19	82 24.4	144 47.9	56.7	85 35.8	11.2	104 40.4	17.0	64 57.7	55.1
20	97 26.8	159 48.6	56.7	100 36.5	10.4	119 42.3	16.7	79 59.9	55.2
21	112 29.3	174 49.3	.. 56.7	115 37.2	.. 09.6	134 44.2	.. 16.5	95 02.1	.. 55.3
22	127 31.8	189 50.0	56.7	130 37.9	08.8	149 46.1	16.3	110 04.4	55.4
23	142 34.2	204 50.8	56.7	145 38.6	08.0	164 48.0	16.1	125 06.6	55.5
28 00	157 36.7	219 51.5	S15 56.6	160 39.3	S 2 07.2	179 49.9	S10 15.9	140 08.8	N 4 55.6
01	172 39.2	234 52.2	56.6	175 40.0	06.4	194 51.8	15.6	155 11.0	55.8
02	187 41.6	249 52.9	56.6	190 40.7	05.6	209 53.7	15.4	170 13.2	55.9
03	202 44.1	264 53.6	.. 56.6	205 41.3	.. 04.8	224 55.6	.. 15.2	185 15.4	.. 56.0
04	217 46.5	279 54.3	56.6	220 42.0	04.0	239 57.5	15.0	200 17.7	56.1
05	232 49.0	294 55.1	56.6	235 42.7	03.2	254 59.4	14.8	215 19.9	56.2
S 06	247 51.5	309 55.8	S15 56.6	250 43.4	S 2 02.4	270 01.3	S10 14.6	230 22.1	N 4 56.3
A 07	262 53.9	324 56.5	56.6	265 44.1	01.6	285 03.2	14.3	245 24.3	56.4
T 08	277 56.4	339 57.2	56.6	280 44.8	00.8	300 05.1	14.1	260 26.5	56.5
U 09	292 58.9	354 57.9	.. 56.6	295 45.5	2 00.0	315 07.0	.. 13.9	275 28.7	.. 56.6
R 10	308 01.3	9 58.6	56.6	310 46.2	1 59.2	330 08.9	13.7	290 30.9	56.7
D 11	323 03.8	24 59.3	56.6	325 46.8	58.5	345 10.7	13.5	305 33.2	56.8
A 12	338 06.3	40 00.0	S15 56.5	340 47.5	S 1 57.7	0 12.6	S10 13.2	320 35.4	N 4 56.9
Y 13	353 08.7	55 00.7	56.5	355 48.2	56.9	15 14.5	13.0	335 37.6	57.1
14	8 11.2	70 01.3	56.5	10 48.9	56.1	30 16.4	12.8	350 39.8	57.2
15	23 13.7	85 02.0	.. 56.5	25 49.6	.. 55.3	45 18.3	.. 12.6	5 42.0	.. 57.3
16	38 16.1	100 02.7	56.5	40 50.3	54.5	60 20.2	12.4	20 44.2	57.4
17	53 18.6	115 03.4	56.5	55 51.0	53.7	75 22.1	12.1	35 46.4	57.5
18	68 21.0	130 04.1	S15 56.5	70 51.7	S 1 52.9	90 24.0	S10 11.9	50 48.7	N 4 57.6
19	83 23.5	145 04.8	56.4	85 52.4	52.1	105 25.9	11.7	65 50.9	57.7
20	98 26.0	160 05.4	56.4	100 53.0	51.3	120 27.8	11.5	80 53.1	57.8
21	113 28.4	175 06.1	.. 56.4	115 53.7	.. 50.5	135 29.7	.. 11.3	95 55.3	.. 57.9
22	128 30.9	190 06.8	56.4	130 54.4	49.7	150 31.6	11.0	110 57.5	58.0
23	143 33.4	205 07.5	56.4	145 55.1	48.9	165 33.5	10.8	125 59.7	58.1
1 00	158 35.8	220 08.1	S15 56.3	160 55.8	S 1 48.1	180 35.4	S10 10.6	141 01.9	N 4 58.2
01	173 38.3	235 08.8	56.3	175 56.5	47.3	195 37.3	10.4	156 04.2	58.4
02	188 40.8	250 09.5	56.3	190 57.2	46.5	210 39.2	10.2	171 06.4	58.5
03	203 43.2	265 10.1	.. 56.3	205 57.9	.. 45.7	225 41.1	.. 09.9	186 08.6	.. 58.6
04	218 45.7	280 10.8	56.3	220 58.6	44.9	240 43.0	09.7	201 10.8	58.7
05	233 48.1	295 11.5	56.2	235 59.2	44.1	255 44.9	09.5	216 13.0	58.8
S 06	248 50.6	310 12.1	S15 56.2	250 59.9	S 1 43.4	270 46.7	S10 09.3	231 15.2	N 4 58.9
U 07	263 53.1	325 12.8	56.2	266 00.6	42.6	285 48.6	09.1	246 17.4	59.0
N 08	278 55.5	340 13.4	56.2	281 01.3	41.8	300 50.5	08.9	261 19.7	59.1
D 09	293 58.0	355 14.1	.. 56.1	296 02.0	.. 41.0	315 52.4	.. 08.6	276 21.9	.. 59.2
A 10	309 00.5	10 14.7	56.1	311 02.7	40.2	330 54.3	08.4	291 24.1	59.3
Y 11	324 02.9	25 15.4	56.1	326 03.4	39.4	345 56.2	08.2	306 26.3	59.4
12	339 05.4	40 16.0	S15 56.1	341 04.1	S 1 38.6	0 58.1	S10 08.0	321 28.5	N 4 59.6
13	354 07.9	55 16.7	56.0	356 04.8	37.8	16 00.0	07.8	336 30.7	59.7
14	9 10.3	70 17.3	56.0	11 05.5	37.0	31 01.9	07.5	351 32.9	59.8
15	24 12.8	85 18.0	.. 56.0	26 06.1	.. 36.2	46 03.8	.. 07.3	6 35.1	4 59.9
16	39 15.3	100 18.6	55.9	41 06.8	35.4	61 05.7	07.1	21 37.4	5 00.0
17	54 17.7	115 19.3	55.9	56 07.5	34.6	76 07.6	06.9	36 39.6	00.1
18	69 20.2	130 19.9	S15 55.9	71 08.2	S 1 33.8	91 09.5	S10 06.7	51 41.8	N 5 00.2
19	84 22.6	145 20.5	55.8	86 08.9	33.0	106 11.4	06.4	66 44.0	00.3
20	99 25.1	160 21.2	55.8	101 09.6	32.2	121 13.3	06.2	81 46.2	00.4
21	114 27.6	175 21.8	.. 55.8	116 10.3	.. 31.4	136 15.2	.. 06.0	96 48.4	.. 00.5
22	129 30.0	190 22.4	55.7	131 11.0	30.6	151 17.1	05.8	111 50.6	00.6
23	144 32.5	205 23.0	55.7	146 11.7	29.8	166 19.0	05.6	126 52.8	00.8
Mer. Pass. 13h 27.3m		*v* 0.7	*d* 0.0	*v* 0.7	*d* 0.8	*v* 1.9	*d* 0.2	*v* 2.2	*d* 0.1

STARS

Name	SHA	Dec
Acamar	315 27.4	S40 19.1
Achernar	335 36.0	S57 15.1
Acrux	173 21.7	S63 05.2
Adhara	255 21.5	S28 58.5
Aldebaran	291 02.8	N16 30.2
Alioth	166 30.6	N55 58.0
Alkaid	153 07.9	N49 19.2
Al Na'ir	27 58.9	S46 58.2
Alnilam	275 58.2	S 1 12.5
Alphard	218 07.3	S 8 39.2
Alphecca	126 20.9	N26 43.1
Alpheratz	357 55.9	N29 04.7
Altair	62 19.9	N 8 51.8
Ankaa	353 27.7	S42 19.2
Antares	112 40.6	S26 25.5
Arcturus	146 06.3	N19 11.4
Atria	107 53.1	S69 01.1
Avior	234 22.2	S59 30.5
Bellatrix	278 44.5	N 6 20.7
Betelgeuse	271 13.9	N 7 24.2
Canopus	264 01.1	S52 42.1
Capella	280 51.7	N45 59.8
Deneb	49 39.8	N45 16.3
Denebola	182 45.3	N14 34.8
Diphda	349 07.9	S18 00.0
Dubhe	194 05.4	N61 45.5
Elnath	278 27.4	N28 36.3
Eltanin	90 51.8	N51 29.2
Enif	33 58.9	N 9 51.9
Fomalhaut	15 37.3	S29 38.0
Gacrux	172 13.4	S57 06.1
Gienah	176 04.1	S17 31.9
Hadar	149 04.1	S60 21.6
Hamal	328 14.2	N23 27.1
Kaus Aust.	83 59.5	S34 23.0
Kochab	137 19.2	N74 09.6
Markab	13 50.3	N15 11.6
Menkar	314 27.4	N 4 04.7
Menkent	148 21.2	S36 21.5
Miaplacidus	221 41.2	S69 42.8
Mirfak	308 57.3	N49 51.3
Nunki	76 13.0	S26 17.8
Peacock	53 38.2	S56 44.3
Pollux	243 41.8	N28 01.7
Procyon	245 11.8	N 5 13.6
Rasalhague	96 17.4	N12 33.6
Regulus	207 55.7	N11 58.4
Rigel	281 23.3	S 8 12.5
Rigil Kent.	140 07.4	S60 49.3
Sabik	102 26.0	S15 43.3
Schedar	349 54.3	N56 31.7
Shaula	96 37.9	S37 05.9
Sirius	258 43.9	S16 43.1
Spica	158 43.4	S11 09.1
Suhail	223 00.6	S43 25.7
Vega	80 47.1	N38 46.8
Zuben'ubi	137 18.3	S16 02.0

	SHA	Mer. Pass.
Venus	62 14.8	9h 20m
Mars	3 02.6	13 17
Jupiter	22 13.2	11 59
Saturn	342 32.1	14 37

SUN and MOON

UT	SUN GHA	SUN Dec	MOON GHA	v	MOON Dec	d	HP
d h	° ′	° ′	° ′	′	° ′	′	′
27 00	176 47.5	S 8 30.0	173 09.7	7.2	S 7 10.9	11.1	61.0
01	191 47.6	29.0	187 35.9	7.2	6 59.8	11.1	61.0
02	206 47.7	28.1	202 02.1	7.2	6 48.7	11.2	61.0
03	221 47.8	.. 27.1	216 28.3	7.3	6 37.5	11.3	61.0
04	236 47.9	26.2	230 54.6	7.2	6 26.2	11.3	61.0
05	251 48.0	25.3	245 20.8	7.3	6 14.9	11.4	61.0
06	266 48.1	S 8 24.3	259 47.1	7.3	S 6 03.5	11.3	61.0
07	281 48.2	23.4	274 13.4	7.3	5 52.2	11.5	61.0
08	296 48.3	22.5	288 39.7	7.3	5 40.7	11.4	61.0
F 09	311 48.5	.. 21.5	303 06.0	7.4	5 29.3	11.6	61.0
R 10	326 48.6	20.6	317 32.4	7.3	5 17.7	11.5	61.0
I 11	341 48.7	19.6	331 58.7	7.4	5 06.2	11.6	61.0
D 12	356 48.8	S 8 18.7	346 25.1	7.4	S 4 54.6	11.6	61.0
A 13	11 48.9	17.7	0 51.5	7.4	4 43.0	11.6	61.0
Y 14	26 49.0	16.8	15 17.9	7.4	4 31.4	11.7	61.0
15	41 49.1	.. 15.9	29 44.3	7.5	4 19.7	11.7	61.1
16	56 49.2	14.9	44 10.8	7.4	4 08.0	11.7	61.1
17	71 49.3	14.0	58 37.2	7.5	3 56.3	11.8	61.1
18	86 49.4	S 8 13.0	73 03.7	7.5	S 3 44.5	11.8	61.1
19	101 49.6	12.1	87 30.2	7.5	3 32.7	11.8	61.1
20	116 49.7	11.2	101 56.7	7.5	3 20.9	11.8	61.1
21	131 49.8	.. 10.2	116 23.2	7.6	3 09.1	11.9	61.1
22	146 49.9	09.3	130 49.8	7.5	2 57.2	11.8	61.1
23	161 50.0	08.3	145 16.3	7.6	2 45.4	11.9	61.1
28 00	176 50.1	S 8 07.4	159 42.9	7.6	S 2 33.5	11.9	61.1
01	191 50.2	06.4	174 09.5	7.6	2 21.6	11.9	61.1
02	206 50.3	05.5	188 36.1	7.6	2 09.7	12.0	61.1
03	221 50.4	.. 04.6	203 02.7	7.6	1 57.7	11.9	61.1
04	236 50.6	03.6	217 29.3	7.7	1 45.8	11.9	61.0
05	251 50.7	02.7	231 56.0	7.6	1 33.9	12.0	61.0
06	266 50.8	S 8 01.7	246 22.6	7.7	S 1 21.9	11.9	61.0
S 07	281 50.9	8 00.8	260 49.3	7.7	1 10.0	12.0	61.0
A 08	296 51.0	7 59.8	275 16.0	7.7	0 58.0	12.0	61.0
T 09	311 51.1	.. 58.9	289 42.7	7.7	0 46.0	11.9	61.0
U 10	326 51.2	57.9	304 09.4	7.7	0 34.1	12.0	61.0
R 11	341 51.4	57.0	318 36.1	7.8	0 22.1	12.0	61.0
D 12	356 51.5	S 7 56.0	333 02.9	7.7	S 0 10.1	11.9	61.0
A 13	11 51.6	55.1	347 29.6	7.8	N 0 01.8	12.0	61.0
Y 14	26 51.7	54.2	1 56.4	7.8	0 13.8	11.9	61.0
15	41 51.8	.. 53.2	16 23.2	7.7	0 25.7	12.0	61.0
16	56 51.9	52.3	30 49.9	7.8	0 37.7	11.9	61.0
17	71 52.1	51.3	45 16.7	7.9	0 49.6	12.0	60.9
18	86 52.2	S 7 50.4	59 43.6	7.8	N 1 01.6	11.9	60.9
19	101 52.3	49.4	74 10.4	7.8	1 13.5	11.9	60.9
20	116 52.4	48.5	88 37.2	7.9	1 25.4	11.9	60.9
21	131 52.5	.. 47.5	103 04.1	7.8	1 37.3	11.8	60.9
22	146 52.6	46.6	117 30.9	7.9	1 49.1	11.9	60.9
23	161 52.8	45.6	131 57.8	7.8	2 01.0	11.8	60.9
1 00	176 52.9	S 7 44.7	146 24.6	7.9	N 2 12.8	11.8	60.9
01	191 53.0	43.7	160 51.5	7.9	2 24.6	11.8	60.8
02	206 53.1	42.8	175 18.4	7.9	2 36.4	11.8	60.8
03	221 53.2	.. 41.8	189 45.3	7.9	2 48.2	11.7	60.8
04	236 53.3	40.9	204 12.2	7.9	2 59.9	11.7	60.8
05	251 53.5	39.9	218 39.1	8.0	3 11.6	11.7	60.8
06	266 53.6	S 7 39.0	233 06.1	7.9	N 3 23.3	11.6	60.8
07	281 53.7	38.0	247 33.0	7.9	3 35.0	11.6	60.7
08	296 53.8	37.1	261 59.9	8.0	3 46.6	11.6	60.7
S 09	311 53.9	.. 36.1	276 26.9	7.9	3 58.2	11.6	60.7
U 10	326 54.1	35.2	290 53.8	8.0	4 09.8	11.5	60.7
N 11	341 54.2	34.2	305 20.8	8.0	4 21.3	11.5	60.7
D 12	356 54.3	S 7 33.3	319 47.8	8.0	N 4 32.8	11.5	60.7
A 13	11 54.4	32.3	334 14.8	7.9	4 44.3	11.4	60.6
Y 14	26 54.5	31.4	348 41.7	8.0	4 55.7	11.4	60.6
15	41 54.7	.. 30.4	3 08.7	8.0	5 07.1	11.3	60.6
16	56 54.8	29.5	17 35.7	8.0	5 18.4	11.3	60.6
17	71 54.9	28.5	32 02.7	8.0	5 29.7	11.3	60.6
18	86 55.0	S 7 27.6	46 29.7	8.0	N 5 41.0	11.2	60.5
19	101 55.2	26.6	60 56.7	8.0	5 52.2	11.1	60.5
20	116 55.3	25.7	75 23.7	8.1	6 03.3	11.1	60.5
21	131 55.4	.. 24.7	89 50.8	8.0	6 14.4	11.0	60.5
22	146 55.5	23.8	104 17.8	8.0	6 25.5	11.0	60.4
23	161 55.6	22.8	118 44.8	8.0	N 6 36.5	11.0	60.4
	SD 16.2	d 0.9	SD 16.6		16.6		16.5

Twilight and Moonrise

Lat.	Twilight Naut.	Twilight Civil	Sunrise	Moonrise 27	Moonrise 28	Moonrise 1	Moonrise 2
°	h m	h m	h m	h m	h m	h m	h m
N 72	05 16	06 34	07 43	07 59	07 53	07 47	07 42
N 70	05 20	06 31	07 33	07 51	07 52	07 52	07 53
68	05 24	06 28	07 25	07 44	07 50	07 56	08 03
66	05 27	06 26	07 18	07 39	07 49	08 00	08 11
64	05 29	06 24	07 12	07 34	07 48	08 02	08 18
62	05 31	06 22	07 07	07 30	07 48	08 05	08 24
60	05 32	06 20	07 02	07 26	07 47	08 07	08 29
N 58	05 33	06 19	06 58	07 23	07 46	08 09	08 33
56	05 34	06 17	06 55	07 20	07 46	08 11	08 37
54	05 35	06 16	06 51	07 17	07 45	08 13	08 41
52	05 36	06 15	06 49	07 15	07 45	08 14	08 44
50	05 36	06 13	06 46	07 13	07 44	08 15	08 47
45	05 37	06 11	06 40	07 08	07 43	08 18	08 54
N 40	05 37	06 08	06 35	07 04	07 43	08 21	08 59
35	05 36	06 06	06 31	07 00	07 42	08 23	09 04
30	05 36	06 03	06 27	06 57	07 41	08 25	09 08
20	05 33	05 59	06 21	06 52	07 40	08 28	09 16
N 10	05 29	05 54	06 15	06 48	07 40	08 31	09 22
0	05 24	05 48	06 09	06 43	07 39	08 34	09 29
S 10	05 18	05 42	06 03	06 39	07 38	08 37	09 35
20	05 09	05 35	05 57	06 34	07 37	08 40	09 42
30	04 57	05 26	05 50	06 29	07 36	08 43	09 49
35	04 49	05 20	05 46	06 26	07 36	08 46	09 54
40	04 40	05 13	05 41	06 22	07 35	08 48	09 59
45	04 29	05 05	05 35	06 18	07 35	08 51	10 05
S 50	04 14	04 55	05 29	06 14	07 34	08 54	10 12
52	04 07	04 50	05 25	06 11	07 34	08 55	10 16
54	03 59	04 45	05 22	06 09	07 33	08 57	10 19
56	03 49	04 39	05 18	06 06	07 33	08 59	10 23
58	03 38	04 32	05 14	06 03	07 32	09 01	10 28
S 60	03 26	04 24	05 09	06 00	07 32	09 03	10 33

Twilight and Moonset

Lat.	Sunset	Twilight Civil	Twilight Naut.	Moonset 27	Moonset 28	Moonset 1	Moonset 2
°	h m	h m	h m	h m	h m	h m	h m
N 72	16 44	17 54	19 12	18 21	20 23	22 23	24 24
N 70	16 54	17 57	19 07	18 26	20 21	22 14	24 06
68	17 02	17 59	19 03	18 31	20 19	22 06	23 51
66	17 09	18 01	19 00	18 35	20 18	22 00	23 39
64	17 15	18 03	18 58	18 38	20 17	21 54	23 29
62	17 20	18 05	18 56	18 41	20 16	21 50	23 21
60	17 24	18 06	18 54	18 43	20 15	21 46	23 14
N 58	17 28	18 08	18 53	18 45	20 14	21 42	23 07
56	17 31	18 09	18 52	18 47	20 13	21 39	23 02
54	17 35	18 10	18 51	18 49	20 13	21 36	22 57
52	17 37	18 11	18 50	18 50	20 12	21 33	22 52
50	17 40	18 13	18 50	18 52	20 12	21 31	22 48
45	17 46	18 15	18 49	18 55	20 11	21 26	22 39
N 40	17 50	18 18	18 49	18 57	20 10	21 22	22 32
35	17 55	18 20	18 49	18 59	20 09	21 18	22 26
30	17 58	18 22	18 50	19 01	20 08	21 15	22 20
20	18 05	18 27	18 52	19 04	20 07	21 09	22 11
N 10	18 10	18 32	18 56	19 07	20 06	21 05	22 02
0	18 16	18 37	19 01	19 10	20 05	21 00	21 55
S 10	18 22	18 43	19 07	19 12	20 04	20 55	21 47
20	18 28	18 50	19 16	19 15	20 03	20 50	21 39
30	18 35	18 59	19 28	19 18	20 01	20 45	21 29
35	18 39	19 05	19 35	19 20	20 01	20 42	21 24
40	18 43	19 11	19 44	19 22	20 00	20 38	21 18
45	18 49	19 19	19 55	19 24	19 59	20 34	21 10
S 50	18 56	19 29	20 10	19 26	19 58	20 29	21 02
52	18 59	19 34	20 17	19 27	19 57	20 27	20 58
54	19 02	19 39	20 25	19 29	19 56	20 24	20 54
56	19 06	19 45	20 34	19 30	19 56	20 21	20 49
58	19 10	19 51	20 44	19 32	19 55	20 18	20 43
S 60	19 15	19 59	20 57	19 34	19 54	20 15	20 37

SUN and MOON

	SUN			MOON			
Day	Eqn. of Time 00ʰ	Eqn. of Time 12ʰ	Mer. Pass.	Mer. Pass. Upper	Mer. Pass. Lower	Age	Phase
d	m s	m s	h m	h m	h m	d	%
27	12 50	12 45	12 13	12 56	00 28	01	1
28	12 40	12 34	12 13	13 52	01 24	02	5
1	12 29	12 23	12 12	14 47	02 19	03	11

UT (d h)	ARIES GHA	VENUS −4.6 GHA	Dec	MARS +1.2 GHA	Dec	JUPITER −2.0 GHA	Dec	SATURN +0.7 GHA	Dec	STARS Name	SHA	Dec
2 00	159 35.0	220 23.7	S15 55.7	161 12.4	S 1 29.1	181 20.9	S10 05.3	141 55.0	N 5 00.9	Acamar	315 27.4	S40 19.1
01	174 37.4	235 24.3	55.6	176 13.1	28.3	196 22.8	05.1	156 57.3	01.0	Achernar	335 36.0	S57 15.1
02	189 39.9	250 24.9	55.6	191 13.7	27.5	211 24.7	04.9	171 59.5	01.1	Acrux	173 21.7	S63 05.2
03	204 42.4	265 25.5	.. 55.6	206 14.4	.. 26.7	226 26.5	.. 04.7	187 01.7	.. 01.2	Adhara	255 21.5	S28 58.5
04	219 44.8	280 26.2	55.5	221 15.1	25.9	241 28.4	04.5	202 03.9	01.3	Aldebaran	291 02.9	N16 30.2
05	234 47.3	295 26.8	55.5	236 15.8	25.1	256 30.3	04.2	217 06.1	01.4			
06	249 49.8	310 27.4	S15 55.4	251 16.5	S 1 24.3	271 32.2	S10 04.0	232 08.3	N 5 01.5	Alioth	166 30.6	N55 58.0
07	264 52.2	325 28.0	55.4	266 17.2	23.5	286 34.1	03.8	247 10.5	01.6	Alkaid	153 07.8	N49 19.2
08	279 54.7	340 28.6	55.4	281 17.9	22.7	301 36.0	03.6	262 12.7	01.7	Al Na'ir	27 58.9	S46 58.2
M 09	294 57.1	355 29.2	.. 55.3	296 18.6	.. 21.9	316 37.9	.. 03.4	277 15.0	.. 01.8	Alnilam	275 58.2	S 1 12.5
O 10	309 59.6	10 29.8	55.3	311 19.3	21.1	331 39.8	03.1	292 17.2	02.0	Alphard	218 07.3	S 8 39.2
N 11	325 02.1	25 30.4	55.2	326 20.0	20.3	346 41.7	02.9	307 19.4	02.1			
D 12	340 04.5	40 31.0	S15 55.2	341 20.7	S 1 19.5	1 43.6	S10 02.7	322 21.6	N 5 02.2	Alphecca	126 20.8	N26 43.1
A 13	355 07.0	55 31.6	55.2	356 21.4	18.7	16 45.5	02.5	337 23.8	02.3	Alpheratz	357 56.0	N29 04.7
Y 14	10 09.5	70 32.2	55.1	11 22.0	17.9	31 47.4	02.3	352 26.0	02.4	Altair	62 19.9	N 8 51.7
15	25 11.9	85 32.8	.. 55.1	26 22.7	.. 17.1	46 49.3	.. 02.0	7 28.2	.. 02.5	Ankaa	353 27.7	S42 19.2
16	40 14.4	100 33.4	55.0	41 23.4	16.3	61 51.2	01.8	22 30.4	02.6	Antares	112 40.6	S26 25.5
17	55 16.9	115 34.0	55.0	56 24.1	15.5	76 53.1	01.6	37 32.6	02.7			
18	70 19.3	130 34.6	S15 54.9	71 24.8	S 1 14.7	91 55.0	S10 01.4	52 34.8	N 5 02.8	Arcturus	146 06.2	N19 11.4
19	85 21.8	145 35.2	54.9	86 25.5	14.0	106 56.9	01.2	67 37.1	02.9	Atria	107 53.0	S69 01.1
20	100 24.2	160 35.8	54.8	101 26.2	13.2	121 58.8	00.9	82 39.3	03.1	Avior	234 22.2	S59 30.5
21	115 26.7	175 36.4	.. 54.8	116 26.9	.. 12.4	137 00.7	.. 00.7	97 41.5	.. 03.2	Bellatrix	278 44.5	N 6 20.7
22	130 29.2	190 37.0	54.7	131 27.6	11.6	152 02.6	00.5	112 43.7	03.3	Betelgeuse	271 13.9	N 7 24.2
23	145 31.6	205 37.5	54.7	146 28.3	10.8	167 04.5	00.3	127 45.9	03.4			
3 00	160 34.1	220 38.1	S15 54.6	161 29.0	S 1 10.0	182 06.4	S10 00.1	142 48.1	N 5 03.5	Canopus	264 01.1	S52 42.1
01	175 36.6	235 38.7	54.6	176 29.7	09.2	197 08.3	9 59.9	157 50.3	03.6	Capella	280 51.7	N45 59.8
02	190 39.0	250 39.3	54.5	191 30.4	08.4	212 10.2	59.6	172 52.5	03.7	Deneb	49 39.8	N45 16.3
03	205 41.5	265 40.0	.. 54.5	206 31.0	.. 07.6	227 12.1	.. 59.4	187 54.7	.. 03.8	Denebola	182 45.3	N14 34.8
04	220 44.0	280 40.4	54.4	221 31.7	06.8	242 13.9	59.2	202 56.9	03.9	Diphda	349 07.9	S18 00.0
05	235 46.4	295 41.0	54.4	236 32.4	06.0	257 15.8	59.0	217 59.2	04.0			
06	250 48.9	310 41.6	S15 54.3	251 33.1	S 1 05.2	272 17.7	S 9 58.8	233 01.4	N 5 04.1	Dubhe	194 05.4	N61 45.6
07	265 51.4	325 42.1	54.2	266 33.8	04.4	287 19.6	58.5	248 03.6	04.3	Elnath	278 27.4	N28 36.3
08	280 53.8	340 42.7	54.2	281 34.5	03.6	302 21.5	58.3	263 05.8	04.4	Eltanin	90 51.7	N51 29.2
T 09	295 56.3	355 43.3	.. 54.1	296 35.2	.. 02.8	317 23.4	.. 58.1	278 08.0	.. 04.5	Enif	33 58.9	N 9 51.9
U 10	310 58.7	10 43.8	54.1	311 35.9	02.0	332 25.3	57.9	293 10.2	04.6	Fomalhaut	15 37.3	S29 38.0
E 11	326 01.2	25 44.4	54.0	326 36.6	01.2	347 27.2	57.7	308 12.4	04.7			
S 12	341 03.7	40 45.0	S15 54.0	341 37.3	S 1 00.5	2 29.1	S 9 57.4	323 14.6	N 5 04.8	Gacrux	172 13.4	S57 06.1
D 13	356 06.1	55 45.5	53.9	356 38.0	0 59.7	17 31.0	57.2	338 16.8	04.9	Gienah	176 04.1	S17 31.9
A 14	11 08.6	70 46.1	53.8	11 38.7	58.9	32 32.9	57.0	353 19.0	05.0	Hadar	149 04.1	S60 21.6
Y 15	26 11.1	85 46.6	.. 53.8	26 39.4	.. 58.1	47 34.8	.. 56.8	8 21.2	.. 05.1	Hamal	328 14.2	N23 27.1
16	41 13.5	100 47.2	53.7	41 40.1	57.3	62 36.7	56.6	23 23.5	05.3	Kaus Aust.	83 59.5	S34 23.0
17	56 16.0	115 47.7	53.6	56 40.8	56.5	77 38.6	56.3	38 25.7	05.4			
18	71 18.5	130 48.3	S15 53.6	71 41.4	S 0 55.7	92 40.5	S 9 56.1	53 27.9	N 5 05.5	Kochab	137 19.2	N74 09.6
19	86 20.9	145 48.8	53.5	86 42.1	54.9	107 42.4	55.9	68 30.1	05.6	Markab	13 50.3	N15 11.6
20	101 23.4	160 49.4	53.5	101 42.8	54.1	122 44.3	55.7	83 32.3	05.7	Menkar	314 27.5	N 4 04.7
21	116 25.9	175 49.9	.. 53.4	116 43.5	.. 53.3	137 46.2	.. 55.5	98 34.5	.. 05.8	Menkent	148 21.1	S36 21.5
22	131 28.3	190 50.5	53.3	131 44.2	52.5	152 48.1	55.2	113 36.7	05.9	Miaplacidus	221 41.2	S69 42.8
23	146 30.8	205 51.0	53.3	146 44.9	51.7	167 50.0	55.0	128 38.9	06.0			
4 00	161 33.2	220 51.5	S15 53.2	161 45.6	S 0 50.9	182 51.9	S 9 54.8	143 41.1	N 5 06.1	Mirfak	308 57.3	N49 51.3
01	176 35.7	235 52.1	53.1	176 46.3	50.1	197 53.8	54.6	158 43.3	06.2	Nunki	76 13.0	S26 17.8
02	191 38.2	250 52.6	53.1	191 47.0	49.3	212 55.7	54.4	173 45.5	06.4	Peacock	53 38.2	S56 44.3
03	206 40.6	265 53.1	.. 53.0	206 47.7	.. 48.5	227 57.6	.. 54.1	188 47.7	.. 06.5	Pollux	243 41.8	N28 01.7
04	221 43.1	280 53.7	52.9	221 48.4	47.7	242 59.5	53.9	203 49.9	06.6	Procyon	245 11.8	N 5 13.6
05	236 45.6	295 54.2	52.8	236 49.1	47.0	258 01.4	53.7	218 52.2	06.7			
06	251 48.0	310 54.7	S15 52.8	251 49.8	S 0 46.2	273 03.3	S 9 53.5	233 54.4	N 5 06.8	Rasalhague	96 17.4	N12 33.6
W 07	266 50.5	325 55.3	52.7	266 50.5	45.4	288 05.2	53.3	248 56.6	06.9	Regulus	207 55.7	N11 58.4
E 08	281 53.0	340 55.8	52.6	281 51.2	44.6	303 07.1	53.0	263 58.8	07.0	Rigel	281 23.3	S 8 12.5
D 09	296 55.4	355 56.3	.. 52.5	296 51.9	.. 43.8	318 09.0	.. 52.8	279 01.0	.. 07.1	Rigil Kent.	140 07.4	S60 49.3
N 10	311 57.9	10 56.8	52.5	311 52.6	43.0	333 10.8	52.6	294 03.2	07.2	Sabik	102 26.0	S15 43.3
E 11	327 00.3	25 57.4	52.4	326 53.3	42.2	348 12.7	52.4	309 05.4	07.3			
S 12	342 02.8	40 57.9	S15 52.3	341 53.9	S 0 41.4	3 14.6	S 9 52.2	324 07.6	N 5 07.5	Schedar	349 54.3	N56 31.6
D 13	357 05.3	55 58.4	52.2	356 54.6	40.6	18 16.5	51.9	339 09.8	07.6	Shaula	96 37.9	S37 05.9
A 14	12 07.7	70 58.9	52.2	11 55.3	39.8	33 18.4	51.7	354 12.0	07.7	Sirius	258 43.9	S16 43.1
Y 15	27 10.2	85 59.4	.. 52.1	26 56.0	.. 39.0	48 20.3	.. 51.5	9 14.2	.. 07.8	Spica	158 43.4	S11 09.1
16	42 12.7	100 59.9	52.0	41 56.7	38.2	63 22.2	51.3	24 16.4	07.9	Suhail	223 00.6	S43 25.8
17	57 15.1	116 00.4	51.9	56 57.4	37.4	78 24.1	51.1	39 18.6	08.0			
18	72 17.6	131 00.9	S15 51.9	71 58.1	S 0 36.6	93 26.0	S 9 50.8	54 20.8	N 5 08.1	Vega	80 47.1	N38 46.8
19	87 20.1	146 01.5	51.8	86 58.8	35.8	108 27.9	50.6	69 23.0	08.2	Zuben'ubi	137 18.2	S16 02.0
20	102 22.5	161 02.0	51.7	101 59.5	35.0	123 29.8	50.4	84 25.3	08.3		SHA	Mer. Pass.
21	117 25.0	176 02.5	.. 51.6	117 00.2	.. 34.3	138 31.7	.. 50.2	99 27.5	.. 08.5			
22	132 27.5	191 03.0	51.5	132 00.9	33.5	153 33.6	50.0	114 29.7	08.6	Venus	60 04.0	9 17
23	147 29.9	206 03.5	51.4	147 01.6	32.7	168 35.5	49.7	129 31.9	08.7	Mars	0 54.9	13 13
Mer. Pass. 13 15.5		v 0.6	d 0.1	v 0.7	d 0.8	v 1.9	d 0.2	v 2.2	d 0.1	Jupiter	21 32.3	11 50
										Saturn	342 14.0	14 27

UT	SUN GHA	SUN Dec	MOON GHA	v	Dec	d	HP
d h	° ′	° ′	° ′	′	° ′	′	′
2 00	176 55.8	S 7 21.9	133 11.8	8.1	N 6 47.5	10.9	60.4
01	191 55.9	20.9	147 38.9	8.0	6 58.4	10.9	60.4
02	206 56.0	20.0	162 05.9	8.1	7 09.3	10.8	60.3
03	221 56.1	.. 19.0	176 33.0	8.0	7 20.1	10.7	60.3
04	236 56.3	18.1	191 00.0	8.0	7 30.8	10.7	60.3
05	251 56.4	17.1	205 27.0	8.1	7 41.5	10.6	60.3
06	266 56.5	S 7 16.2	219 54.1	8.0	N 7 52.1	10.6	60.2
07	281 56.6	15.2	234 21.1	8.1	8 02.7	10.5	60.2
08	296 56.8	14.2	248 48.2	8.1	8 13.2	10.5	60.2
09	311 56.9	.. 13.3	263 15.3	8.0	8 23.7	10.4	60.2
10	326 57.0	12.3	277 42.3	8.1	8 34.1	10.3	60.1
11	341 57.1	11.4	292 09.4	8.0	8 44.4	10.3	60.1
12	356 57.3	S 7 10.4	306 36.4	8.1	N 8 54.7	10.1	60.1
13	11 57.4	09.5	321 03.5	8.1	9 04.8	10.1	60.1
14	26 57.5	08.5	335 30.6	8.0	9 15.0	10.0	60.0
15	41 57.6	.. 07.6	349 57.6	8.1	9 25.0	10.0	60.0
16	56 57.8	06.6	4 24.7	8.1	9 35.0	9.9	60.0
17	71 57.9	05.7	18 51.8	8.0	9 44.9	9.9	60.0
18	86 58.0	S 7 04.7	33 18.8	8.1	N 9 54.8	9.8	59.9
19	101 58.1	03.7	47 45.9	8.1	10 04.6	9.7	59.9
20	116 58.3	02.8	62 13.0	8.1	10 14.3	9.6	59.9
21	131 58.4	.. 01.8	76 40.1	8.0	10 23.9	9.6	59.8
22	146 58.5	7 00.9	91 07.1	8.1	10 33.5	9.5	59.8
23	161 58.7	6 59.9	105 34.2	8.1	10 43.0	9.4	59.8
3 00	176 58.8	S 6 59.0	120 01.3	8.1	N10 52.4	9.3	59.7
01	191 58.9	58.0	134 28.4	8.1	11 01.7	9.2	59.7
02	206 59.0	57.0	148 55.5	8.0	11 10.9	9.2	59.7
03	221 59.2	.. 56.1	163 22.5	8.1	11 20.1	9.1	59.7
04	236 59.3	55.1	177 49.6	8.1	11 29.2	9.0	59.6
05	251 59.4	54.2	192 16.7	8.1	11 38.2	8.9	59.6
06	266 59.6	S 6 53.2	206 43.8	8.1	N11 47.1	8.9	59.6
07	281 59.7	52.3	221 10.9	8.0	11 56.0	8.7	59.5
08	296 59.8	51.3	235 37.9	8.1	12 04.7	8.7	59.5
09	312 00.0	.. 50.3	250 05.0	8.1	12 13.4	8.6	59.5
10	327 00.1	49.4	264 32.1	8.1	12 22.0	8.5	59.4
11	342 00.2	48.4	278 59.2	8.1	12 30.5	8.4	59.4
12	357 00.3	S 6 47.5	293 26.3	8.1	N12 38.9	8.4	59.4
13	12 00.5	46.5	307 53.4	8.1	12 47.3	8.2	59.3
14	27 00.6	45.5	322 20.5	8.1	12 55.5	8.2	59.3
15	42 00.7	.. 44.6	336 47.6	8.1	13 03.7	8.0	59.3
16	57 00.9	43.6	351 14.7	8.1	13 11.7	8.0	59.2
17	72 01.0	42.7	5 41.8	8.1	13 19.7	7.9	59.2
18	87 01.1	S 6 41.7	20 08.9	8.1	N13 27.6	7.8	59.2
19	102 01.3	40.7	34 36.0	8.1	13 35.4	7.7	59.1
20	117 01.4	39.8	49 03.1	8.1	13 43.1	7.6	59.1
21	132 01.5	.. 38.8	63 30.2	8.1	13 50.7	7.5	59.1
22	147 01.7	37.9	77 57.3	8.1	13 58.2	7.4	59.0
23	162 01.8	36.9	92 24.4	8.1	14 05.6	7.4	59.0
4 00	177 01.9	S 6 35.9	106 51.5	8.1	N14 13.0	7.2	59.0
01	192 02.1	35.0	121 18.6	8.2	14 20.2	7.1	58.9
02	207 02.2	34.0	135 45.8	8.1	14 27.3	7.1	58.9
03	222 02.3	.. 33.1	150 12.9	8.1	14 34.4	6.9	58.9
04	237 02.5	32.1	164 40.0	8.2	14 41.3	6.9	58.8
05	252 02.6	31.1	179 07.2	8.1	14 48.2	6.7	58.8
06	267 02.7	S 6 30.2	193 34.3	8.1	N14 54.9	6.7	58.8
07	282 02.9	29.2	208 01.4	8.2	15 01.6	6.5	58.7
08	297 03.0	28.3	222 28.6	8.1	15 08.1	6.5	58.7
09	312 03.1	.. 27.3	236 55.7	8.2	15 14.6	6.3	58.7
10	327 03.3	26.3	251 22.9	8.2	15 20.9	6.3	58.6
11	342 03.4	25.4	265 50.1	8.1	15 27.2	6.1	58.6
12	357 03.6	S 6 24.4	280 17.2	8.2	N15 33.3	6.1	58.6
13	12 03.7	23.4	294 44.4	8.2	15 39.4	5.9	58.5
14	27 03.8	22.5	309 11.6	8.2	15 45.3	5.9	58.5
15	42 04.0	.. 21.5	323 38.8	8.2	15 51.2	5.7	58.5
16	57 04.1	20.6	338 06.0	8.2	15 56.9	5.7	58.4
17	72 04.2	19.6	352 33.2	8.2	16 02.6	5.5	58.4
18	87 04.4	S 6 18.6	7 00.4	8.2	N16 08.1	5.5	58.4
19	102 04.5	17.7	21 27.6	8.3	16 13.6	5.3	58.3
20	117 04.6	16.7	35 54.9	8.2	16 18.9	5.2	58.3
21	132 04.8	.. 15.7	50 22.1	8.2	16 24.1	5.2	58.3
22	147 04.9	14.8	64 49.3	8.3	16 29.3	5.0	58.2
23	162 05.1	13.8	79 16.6	8.3	N16 34.3	4.9	58.2
	SD 16.2	d 1.0	SD 16.4		16.2		16.0

Days of week: 2 = MONDAY, 3 = TUESDAY, 4 = WEDNESDAY

Lat.	Twilight Naut.	Twilight Civil	Sunrise	Moonrise 2	3	4	5
°	h m	h m	h m	h m	h m	h m	h m
N 72	05 01	06 19	07 27	07 42	07 36	07 28	07 18
N 70	05 07	06 17	07 19	07 53	07 56	08 01	08 12
68	05 12	06 16	07 12	08 03	08 12	08 24	08 45
66	05 16	06 15	07 06	08 11	08 25	08 43	09 09
64	05 19	06 14	07 01	08 18	08 35	08 58	09 28
62	05 22	06 13	06 57	08 24	08 45	09 11	09 44
60	05 24	06 12	06 53	08 29	08 53	09 21	09 57
N 58	05 26	06 11	06 50	08 33	09 00	09 31	10 08
56	05 27	06 10	06 47	08 37	09 06	09 39	10 18
54	05 28	06 09	06 44	08 41	09 12	09 46	10 27
52	05 29	06 08	06 42	08 44	09 17	09 53	10 34
50	05 30	06 07	06 40	08 47	09 21	09 59	10 41
45	05 32	06 06	06 35	08 54	09 31	10 12	10 57
N 40	05 32	06 04	06 31	08 59	09 40	10 23	11 09
35	05 33	06 02	06 27	09 04	09 47	10 32	11 20
30	05 32	06 00	06 24	09 08	09 53	10 40	11 29
20	05 31	05 56	06 19	09 16	10 04	10 54	11 45
N 10	05 28	05 52	06 13	09 22	10 14	11 06	11 59
0	05 24	05 48	06 09	09 29	10 23	11 18	12 12
S 10	05 18	05 43	06 04	09 35	10 33	11 30	12 25
20	05 10	05 36	05 58	09 42	10 43	11 42	12 40
30	04 59	05 28	05 52	09 49	10 54	11 56	12 56
35	04 52	05 23	05 49	09 54	11 01	12 05	13 05
40	04 44	05 17	05 44	09 59	11 08	12 14	13 16
45	04 33	05 09	05 39	10 05	11 17	12 26	13 29
S 50	04 20	05 00	05 34	10 12	11 28	12 39	13 45
52	04 13	04 56	05 31	10 16	11 33	12 45	13 52
54	04 06	04 51	05 28	10 19	11 38	12 53	14 00
56	03 57	04 46	05 25	10 23	11 44	13 00	14 09
58	03 48	04 40	05 21	10 28	11 51	13 09	14 20
S 60	03 36	04 33	05 17	10 33	11 59	13 20	14 32

Lat.	Sunset	Twilight Civil	Twilight Naut.	Moonset 2	3	4	5
°	h m	h m	h m	h m	h m	h m	h m
N 72	16 59	18 07	19 26	24 24	00 24	02 26	04 30
N 70	17 07	18 09	19 19	24 06	00 06	01 54	03 37
68	17 14	18 10	19 14	23 51	25 32	01 32	03 04
66	17 19	18 11	19 10	23 39	25 14	01 14	02 41
64	17 24	18 12	19 07	23 29	25 00	01 00	02 22
62	17 28	18 13	19 04	23 21	24 48	00 48	02 07
60	17 32	18 14	19 02	23 14	24 37	00 37	01 54
N 58	17 35	18 14	19 00	23 07	24 28	00 28	01 43
56	17 38	18 15	18 58	23 02	24 21	00 21	01 34
54	17 40	18 16	18 57	22 57	24 14	00 14	01 25
52	17 43	18 17	18 56	22 52	24 08	00 08	01 18
50	17 45	18 17	18 55	22 48	24 02	00 02	01 11
45	17 50	18 19	18 53	22 39	23 50	24 56	00 56
N 40	17 54	18 21	18 52	22 32	23 40	24 44	00 44
35	17 57	18 23	18 52	22 26	23 31	24 34	00 34
30	18 00	18 24	18 52	22 20	23 24	24 25	00 25
20	18 06	18 28	18 53	22 11	23 11	24 10	00 10
N 10	18 11	18 32	18 56	22 02	23 00	23 56	24 52
0	18 15	18 36	19 00	21 55	22 49	23 44	24 38
S 10	18 20	18 41	19 06	21 47	22 39	23 31	24 24
20	18 25	18 48	19 13	21 39	22 28	23 18	24 10
30	18 31	18 56	19 24	21 29	22 15	23 03	23 53
35	18 35	19 01	19 31	21 24	22 08	22 54	23 43
40	18 39	19 06	19 39	21 18	21 59	22 44	23 32
45	18 44	19 14	19 49	21 10	21 49	22 32	23 19
S 50	18 49	19 23	20 03	21 02	21 38	22 18	23 03
52	18 52	19 27	20 09	20 58	21 32	22 11	22 55
54	18 55	19 31	20 16	20 54	21 26	22 04	22 47
56	18 58	19 37	20 25	20 49	21 20	21 55	22 38
58	19 02	19 43	20 34	20 43	21 12	21 46	22 27
S 60	19 06	19 49	20 45	20 37	21 04	21 35	22 15

Day	SUN Eqn. of Time 00h	SUN Eqn. of Time 12h	Mer. Pass.	MOON Mer. Pass. Upper	MOON Mer. Pass. Lower	Age	Phase
d	m s	m s	h m	h m	h m	d	%
2	12 17	12 11	12 12	15 42	03 14	04	20
3	12 05	11 59	12 12	16 36	04 09	05	30
4	11 53	11 46	12 12	17 31	05 04	06	41

1998 MARCH 5, 6, 7 (THURS., FRI., SAT.)

UT	ARIES	VENUS −4.6		MARS +1.2		JUPITER −2.0		SATURN +0.7	
d h	GHA	GHA	Dec	GHA	Dec	GHA	Dec	GHA	Dec
	° ′	° ′	° ′	° ′	° ′	° ′	° ′	° ′	° ′
5 00	162 32.4	221 04.0	S15 51.4	162 02.3	S 0 31.9	183 37.4	S 9 49.5	144 34.1	N 5 08.8
01	177 34.8	236 04.5	51.3	177 03.0	31.1	198 39.3	49.3	159 36.3	08.9
02	192 37.3	251 05.0	51.2	192 03.7	30.3	213 41.2	49.1	174 38.5	09.0
03	207 39.8	266 05.4 ..	51.1	207 04.4 ..	29.5	228 43.1 ..	48.9	189 40.7 ..	09.1
04	222 42.2	281 05.9	51.0	222 05.1	28.7	243 45.0	48.6	204 42.9	09.2
05	237 44.7	296 06.4	50.9	237 05.8	27.9	258 46.9	48.4	219 45.1	09.3
06	252 47.2	311 06.9	S15 50.8	252 06.5	S 0 27.1	273 48.8	S 9 48.2	234 47.3	N 5 09.5
07	267 49.6	326 07.4	50.7	267 07.2	26.3	288 50.7	48.0	249 49.5	09.6
T 08	282 52.1	341 07.9	50.7	282 07.9	25.5	303 52.6	47.8	264 51.7	09.7
H 09	297 54.6	356 08.4 ..	50.6	297 08.6 ..	24.7	318 54.5 ..	47.5	279 53.9 ..	09.8
U 10	312 57.0	11 08.9	50.5	312 09.3	23.9	333 56.4	47.3	294 56.1	09.9
R 11	327 59.5	26 09.3	50.4	327 10.0	23.1	348 58.3	47.1	309 58.3	10.0
S 12	343 02.0	41 09.8	S15 50.3	342 10.6	S 0 22.4	4 00.2	S 9 46.9	325 00.5	N 5 10.1
D 13	358 04.5	56 10.3	50.2	357 11.3	21.6	19 02.1	46.7	340 02.8	10.2
A 14	13 06.9	71 10.8	50.1	12 12.0	20.8	34 04.0	46.4	355 05.0	10.3
Y 15	28 09.3	86 11.2 ..	50.0	27 12.7 ..	20.0	49 05.9 ..	46.2	10 07.2 ..	10.5
16	43 11.8	101 11.7	49.9	42 13.4	19.2	64 07.8	46.0	25 09.4	10.6
17	58 14.3	116 12.2	49.8	57 14.1	18.4	79 09.7	45.8	40 11.6	10.7
18	73 16.7	131 12.7	S15 49.7	72 14.8	S 0 17.6	94 11.6	S 9 45.6	55 13.8	N 5 10.8
19	88 19.2	146 13.1	49.6	87 15.5	16.8	109 13.5	45.3	70 16.0	10.9
20	103 21.7	161 13.6	49.5	102 16.2	16.0	124 15.4	45.1	85 18.2	11.0
21	118 24.1	176 14.1 ..	49.4	117 16.9 ..	15.2	139 17.3 ..	44.9	100 20.4 ..	11.1
22	133 26.6	191 14.5	49.3	132 17.6	14.4	154 19.2	44.7	115 22.6	11.2
23	148 29.1	206 15.0	49.2	147 18.3	13.6	169 21.1	44.5	130 24.8	11.3
6 00	163 31.5	221 15.4	S15 49.1	162 19.0	S 0 12.8	184 23.0	S 9 44.3	145 27.0	N 5 11.5
01	178 34.0	236 15.9	49.0	177 19.7	12.0	199 24.9	44.0	160 29.2	11.6
02	193 36.4	251 16.4	48.9	192 20.4	11.2	214 26.8	43.8	175 31.4	11.7
03	208 38.9	266 16.8 ..	48.8	207 21.1 ..	10.5	229 28.7 ..	43.6	190 33.6 ..	11.8
04	223 41.4	281 17.3	48.7	222 21.8	09.7	244 30.6	43.4	205 35.8	11.9
05	238 43.8	296 17.7	48.6	237 22.5	08.9	259 32.5	43.2	220 38.0	12.0
06	253 46.3	311 18.2	S15 48.5	252 23.2	S 0 08.1	274 34.4	S 9 42.9	235 40.2	N 5 12.1
07	268 48.8	326 18.6	48.4	267 23.9	07.3	289 36.3	42.7	250 42.4	12.2
F 08	283 51.2	341 19.1	48.3	282 24.6	06.5	304 38.2	42.5	265 44.6	12.4
R 09	298 53.7	356 19.5 ..	48.2	297 25.3 ..	05.7	319 40.1 ..	42.3	280 46.8 ..	12.5
I 10	313 56.2	11 20.0	48.1	312 26.0	04.9	334 42.0	42.1	295 49.0	12.6
D 11	328 58.6	26 20.4	47.9	327 26.7	04.1	349 43.8	41.8	310 51.2	12.7
A 12	344 01.1	41 20.9	S15 47.8	342 27.4	S 0 03.3	4 45.7	S 9 41.6	325 53.4	N 5 12.8
Y 13	359 03.6	56 21.3	47.7	357 28.1	02.5	19 47.6	41.4	340 55.6	12.9
14	14 06.0	71 21.7	47.6	12 28.8	01.7	34 49.5	41.2	355 57.9	13.0
15	29 08.5	86 22.2 ..	47.5	27 29.5 ..	00.9	49 51.4 ..	41.0	11 00.1 ..	13.1
16	44 10.9	101 22.6	47.4	42 30.2 S	00.2	64 53.3	40.7	26 02.3	13.2
17	59 13.4	116 23.0	47.3	57 30.9 N	00.6	79 55.2	40.5	41 04.5	13.4
18	74 15.9	131 23.5	S15 47.2	72 31.6	N 0 01.4	94 57.1	S 9 40.3	56 06.7	N 5 13.5
19	89 18.3	146 23.9	47.0	87 32.3	02.2	109 59.0	40.1	71 08.9	13.6
20	104 20.8	161 24.3	46.9	102 33.0	03.0	125 00.9	39.9	86 11.1	13.7
21	119 23.3	176 24.8 ..	46.8	117 33.7 ..	03.8	140 02.8 ..	39.6	101 13.3 ..	13.8
22	134 25.7	191 25.2	46.7	132 34.4	04.6	155 04.7	39.4	116 15.5	13.9
23	149 28.2	206 25.6	46.6	147 35.1	05.4	170 06.6	39.2	131 17.7	14.0
7 00	164 30.7	221 26.0	S15 46.4	162 35.8	N 0 06.2	185 08.5	S 9 39.0	146 19.9	N 5 14.1
01	179 33.1	236 26.5	46.3	177 36.5	07.0	200 10.4	38.8	161 22.1	14.3
02	194 35.6	251 26.9	46.2	192 37.2	07.8	215 12.3	38.5	176 24.3	14.4
03	209 38.1	266 27.3 ..	46.1	207 37.9 ..	08.6	230 14.2 ..	38.3	191 26.5 ..	14.5
04	224 40.5	281 27.7	46.0	222 38.6	09.4	245 16.1	38.1	206 28.7	14.6
05	239 43.0	296 28.1	45.8	237 39.3	10.1	260 18.0	37.9	221 30.9	14.7
06	254 45.4	311 28.5	S15 45.7	252 40.0	N 0 10.9	275 19.9	S 9 37.7	236 33.1	N 5 14.8
07	269 47.9	326 29.0	45.6	267 40.7	11.7	290 21.8	37.4	251 35.3	14.9
S 08	284 50.4	341 29.4	45.5	282 41.4	12.5	305 23.7	37.2	266 37.5	15.0
A 09	299 52.8	356 29.8 ..	45.3	297 42.1 ..	13.3	320 25.6 ..	37.0	281 39.7 ..	15.2
T 10	314 55.3	11 30.2	45.2	312 42.7	14.1	335 27.5	36.8	296 41.9	15.3
U 11	329 57.8	26 30.6	45.1	327 43.4	14.9	350 29.4	36.6	311 44.1	15.4
R 12	345 00.2	41 31.0	S15 45.0	342 44.1	N 0 15.7	5 31.3	S 9 36.3	326 46.3	N 5 15.5
D 13	0 02.7	56 31.4	44.8	357 44.8	16.5	20 33.2	36.1	341 48.5	15.6
A 14	15 05.2	71 31.8	44.7	12 45.5	17.3	35 35.1	35.9	356 50.7	15.7
Y 15	30 07.6	86 32.2 ..	44.6	27 46.2 ..	18.1	50 37.0 ..	35.7	11 52.9 ..	15.8
16	45 10.1	101 32.6	44.4	42 46.9	18.9	65 38.9	35.5	26 55.1	15.9
17	60 12.6	116 33.0	44.3	57 47.6	19.6	80 40.8	35.2	41 57.3	16.0
18	75 15.0	131 33.4	S15 44.2	72 48.3	N 0 20.4	95 42.7	S 9 35.0	56 59.5	N 5 16.2
19	90 17.5	146 33.8	44.0	87 49.0	21.2	110 44.6	34.8	72 01.7	16.3
20	105 19.9	161 34.2	43.9	102 49.7	22.0	125 46.5	34.6	87 03.9	16.4
21	120 22.4	176 34.6 ..	43.8	117 50.4 ..	22.8	140 48.4 ..	34.4	102 06.1 ..	16.5
22	135 24.9	191 35.0	43.6	132 51.1	23.6	155 50.3	34.1	117 08.3	16.6
23	150 27.3	206 35.4	43.5	147 51.8	24.4	170 52.2	33.9	132 10.5	16.7
Mer. Pass.	h m 13 03.8	v 0.4	d 0.1	v 0.7	d 0.8	v 1.9	d 0.2	v 2.2	d 0.1

STARS

Name	SHA	Dec
	° ′	° ′
Acamar	315 27.5	S40 19.1
Achernar	335 36.0	S57 15.1
Acrux	173 21.7	S63 05.2
Adhara	255 21.5	S28 58.5
Aldebaran	291 02.9	N16 30.2
Alioth	166 30.5	N55 58.0
Alkaid	153 07.8	N49 19.2
Al Na'ir	27 58.9	S46 58.2
Alnilam	275 58.2	S 1 12.5
Alphard	218 07.3	S 8 39.2
Alphecca	126 20.8	N26 43.1
Alpheratz	357 55.9	N29 04.7
Altair	62 19.9	N 8 51.7
Ankaa	353 27.7	S42 19.2
Antares	112 40.6	S26 25.5
Arcturus	146 06.2	N19 11.4
Atria	107 53.0	S69 01.1
Avior	234 22.3	S59 30.5
Bellatrix	278 44.5	N 6 20.6
Betelgeuse	271 13.9	N 7 24.2
Canopus	264 01.1	S52 42.1
Capella	280 51.7	N45 59.8
Deneb	49 39.8	N45 16.3
Denebola	182 45.3	N14 34.8
Diphda	349 07.9	S18 00.0
Dubhe	194 05.4	N61 45.6
Elnath	278 27.4	N28 36.3
Eltanin	90 51.7	N51 29.2
Enif	33 58.9	N 9 51.9
Fomalhaut	15 37.3	S29 38.0
Gacrux	172 13.4	S57 06.1
Gienah	176 04.0	S17 32.0
Hadar	149 04.1	S60 21.6
Hamal	328 14.2	N23 27.1
Kaus Aust.	83 59.5	S34 22.9
Kochab	137 19.1	N74 09.6
Markab	13 50.3	N15 11.6
Menkar	314 27.5	N 4 04.7
Menkent	148 21.1	S36 21.5
Miaplacidus	221 41.2	S69 42.8
Mirfak	308 57.3	N49 51.3
Nunki	76 13.0	S26 17.8
Peacock	53 38.1	S56 44.3
Pollux	243 41.8	N28 01.7
Procyon	245 11.8	N 5 13.6
Rasalhague	96 17.4	N12 33.6
Regulus	207 55.7	N11 58.4
Rigel	281 23.3	S 8 12.5
Rigil Kent.	140 07.4	S60 49.4
Sabik	102 26.0	S15 43.3
Schedar	349 54.4	N56 31.6
Shaula	96 37.9	S37 05.9
Sirius	258 43.9	S16 43.1
Spica	158 43.3	S11 09.1
Suhail	223 00.6	S43 25.8
Vega	80 47.1	N38 46.8
Zuben'ubi	137 18.2	S16 02.0

	SHA	Mer. Pass.
	° ′	h m
Venus	57 43.9	9 15
Mars	358 47.5	13 10
Jupiter	20 51.4	11 41
Saturn	341 55.5	14 16

UT	SUN GHA	SUN Dec	MOON GHA	v	Dec	d	HP
d h	° ′	° ′	° ′	′	° ′	′	′
5 00	177 05.2	S 6 12.8	93 43.9	8.3	N16 39.2	4.8	58.2
01	192 05.3	11.9	108 11.2	8.2	16 44.0	4.8	58.1
02	207 05.5	10.9	122 38.4	8.4	16 48.8	4.6	58.1
03	222 05.6 ..	09.9	137 05.8	8.3	16 53.4	4.5	58.1
04	237 05.8	09.0	151 33.1	8.3	16 57.9	4.4	58.0
05	252 05.9	08.0	166 00.4	8.3	17 02.3	4.3	58.0
06	267 06.0	S 6 07.1	180 27.7	8.4	N17 06.6	4.1	58.0
T 07	282 06.2	06.1	194 55.1	8.4	17 10.7	4.1	57.9
H 08	297 06.3	05.1	209 22.5	8.3	17 14.8	4.0	57.9
U 09	312 06.5 ..	04.2	223 49.8	8.4	17 18.8	3.9	57.9
R 10	327 06.6	03.2	238 17.2	8.5	17 22.7	3.7	57.9
S 11	342 06.7	02.2	252 44.7	8.4	17 26.4	3.7	57.8
D 12	357 06.9	S 6 01.3	267 12.1	8.4	N17 30.1	3.5	57.8
A 13	12 07.0	6 00.3	281 39.5	8.5	17 33.6	3.5	57.8
Y 14	27 07.2	5 59.3	296 07.0	8.5	17 37.1	3.3	57.7
15	42 07.3 ..	58.4	310 34.5	8.5	17 40.4	3.3	57.7
16	57 07.4	57.4	325 02.0	8.5	17 43.7	3.1	57.7
17	72 07.6	56.4	339 29.5	8.5	17 46.8	3.0	57.6
18	87 07.7	S 5 55.5	353 57.0	8.6	N17 49.8	2.9	57.6
19	102 07.9	54.5	8 24.6	8.5	17 52.7	2.8	57.6
20	117 08.0	53.5	22 52.1	8.6	17 55.5	2.7	57.5
21	132 08.2 ..	52.6	37 19.7	8.6	17 58.2	2.6	57.5
22	147 08.3	51.6	51 47.3	8.7	18 00.8	2.5	57.5
23	162 08.4	50.6	66 15.0	8.6	18 03.3	2.4	57.4
6 00	177 08.6	S 5 49.7	80 42.6	8.7	N18 05.7	2.2	57.4
01	192 08.7	48.7	95 10.3	8.7	18 07.9	2.2	57.4
02	207 08.9	47.7	109 38.0	8.7	18 10.1	2.0	57.3
03	222 09.0 ..	46.7	124 05.7	8.8	18 12.1	2.0	57.3
04	237 09.2	45.8	138 33.5	8.7	18 14.1	1.8	57.3
05	252 09.3	44.8	153 01.2	8.8	18 15.9	1.8	57.3
06	267 09.4	S 5 43.8	167 29.0	8.8	N18 17.7	1.6	57.2
07	282 09.6	42.9	181 56.8	8.9	18 19.3	1.5	57.2
08	297 09.7	41.9	196 24.7	8.9	18 20.8	1.5	57.2
F 09	312 09.9 ..	40.9	210 52.6	8.9	18 22.3	1.3	57.1
I 10	327 10.0	40.0	225 20.5	8.9	18 23.6	1.2	57.1
11	342 10.2	39.0	239 48.4	8.9	18 24.8	1.1	57.1
D 12	357 10.3	S 5 38.0	254 16.3	9.0	N18 25.9	1.0	57.0
A 13	12 10.5	37.1	268 44.3	9.0	18 26.9	0.9	57.0
Y 14	27 10.6	36.1	283 12.3	9.0	18 27.8	0.8	57.0
15	42 10.8 ..	35.1	297 40.3	9.1	18 28.6	0.7	56.9
16	57 10.9	34.2	312 08.4	9.1	18 29.3	0.6	56.9
17	72 11.0	33.2	326 36.5	9.1	18 29.9	0.4	56.9
18	87 11.2	S 5 32.2	341 04.6	9.2	N18 30.3	0.4	56.9
19	102 11.3	31.2	355 32.8	9.2	18 30.7	0.3	56.8
20	117 11.5	30.3	10 01.0	9.2	18 31.0	0.2	56.8
21	132 11.6 ..	29.3	24 29.2	9.2	18 31.2	0.0	56.8
22	147 11.8	28.3	38 57.4	9.3	18 31.2	0.0	56.7
23	162 11.9	27.4	53 25.7	9.3	18 31.2	0.1	56.7
7 00	177 12.1	S 5 26.4	67 54.0	9.4	N18 31.1	0.3	56.7
01	192 12.2	25.4	82 22.4	9.4	18 30.8	0.3	56.6
02	207 12.4	24.4	96 50.8	9.4	18 30.5	0.5	56.6
03	222 12.5 ..	23.5	111 19.2	9.4	18 30.0	0.5	56.6
04	237 12.7	22.5	125 47.6	9.5	18 29.5	0.6	56.6
05	252 12.8	21.5	140 16.1	9.5	18 28.9	0.8	56.5
06	267 13.0	S 5 20.6	154 44.6	9.6	N18 28.1	0.8	56.5
07	282 13.1	19.6	169 13.2	9.6	18 27.3	1.0	56.5
S 08	297 13.3	18.6	183 41.8	9.6	18 26.3	1.0	56.5
A 09	312 13.4 ..	17.6	198 10.4	9.7	18 25.3	1.2	56.4
T 10	327 13.6	16.7	212 39.1	9.7	18 24.1	1.2	56.4
U 11	342 13.7	15.7	227 07.8	9.8	18 22.9	1.3	56.4
R 12	357 13.9	S 5 14.7	241 36.6	9.7	N18 21.6	1.5	56.3
D 13	12 14.0	13.8	256 05.3	9.9	18 20.1	1.5	56.3
A 14	27 14.2	12.8	270 34.2	9.8	18 18.6	1.6	56.3
Y 15	42 14.3 ..	11.8	285 03.0	9.9	18 17.0	1.8	56.3
16	57 14.5	10.8	299 31.9	10.0	18 15.2	1.8	56.2
17	72 14.6	09.9	314 00.9	10.0	18 13.4	1.9	56.2
18	87 14.8	S 5 08.9	328 29.9	10.0	N18 11.5	2.0	56.2
19	102 14.9	07.9	342 58.9	10.0	18 09.5	2.1	56.2
20	117 15.1	06.9	357 27.9	10.2	18 07.4	2.2	56.1
21	132 15.2 ..	06.0	11 57.1	10.1	18 05.2	2.3	56.1
22	147 15.4	05.0	26 26.2	10.2	18 02.9	2.4	56.1
23	162 15.5	04.0	40 55.4	10.2	N18 00.5	2.5	56.1
	SD 16.1	d 1.0	SD 15.7		15.5		15.4

Lat.	Twilight Naut.	Twilight Civil	Sunrise	Moonrise 5	6	7	8
°	h m	h m	h m	h m	h m	h m	h m
N 72	04 45	06 04	07 12	07 18	▭	▭	09 40
N 70	04 53	06 04	07 05	08 12	08 36	09 27	10 44
68	04 59	06 04	07 00	08 45	09 18	10 10	11 20
66	05 04	06 04	06 55	09 09	09 47	10 39	11 45
64	05 09	06 04	06 51	09 28	10 09	11 01	12 05
62	05 12	06 03	06 48	09 44	10 26	11 19	12 21
60	05 15	06 03	06 45	09 57	10 41	11 33	12 34
N 58	05 17	06 03	06 42	10 08	10 53	11 46	12 46
56	05 19	06 02	06 39	10 18	11 04	11 57	12 56
54	05 21	06 02	06 37	10 27	11 13	12 06	13 04
52	05 23	06 02	06 35	10 34	11 22	12 14	13 12
50	05 24	06 01	06 34	10 41	11 29	12 22	13 19
45	05 26	06 00	06 30	10 57	11 45	12 38	13 34
N 40	05 28	05 59	06 26	11 09	11 59	12 51	13 47
35	05 29	05 58	06 23	11 20	12 10	13 03	13 57
30	05 29	05 57	06 21	11 29	12 20	13 13	14 06
20	05 29	05 54	06 16	11 45	12 37	13 29	14 22
N 10	05 27	05 51	06 12	11 59	12 52	13 44	14 36
0	05 23	05 47	06 08	12 12	13 06	13 58	14 49
S 10	05 18	05 43	06 04	12 25	13 20	14 12	15 02
20	05 11	05 37	05 59	12 40	13 35	14 27	15 16
30	05 02	05 30	05 54	12 56	13 52	14 44	15 31
35	04 55	05 26	05 51	13 05	14 02	14 54	15 41
40	04 48	05 20	05 47	13 16	14 13	15 05	15 51
45	04 38	05 14	05 44	13 29	14 27	15 18	16 03
S 50	04 26	05 06	05 39	13 45	14 43	15 34	16 18
52	04 20	05 02	05 36	13 52	14 51	15 42	16 25
54	04 13	04 57	05 34	14 00	15 00	15 50	16 33
56	04 05	04 53	05 31	14 09	15 10	16 00	16 41
58	03 56	04 47	05 28	14 20	15 21	16 11	16 51
S 60	03 46	04 41	05 24	14 32	15 33	16 23	17 02

Lat.	Sunset	Twilight Civil	Twilight Naut.	Moonset 5	6	7	8
°	h m	h m	h m	h m	h m	h m	h m
N 72	17 13	18 21	19 40	04 30	▭	▭	07 37
N 70	17 19	18 21	19 32	03 37	05 05	06 04	06 33
68	17 25	18 21	19 26	03 04	04 23	05 20	05 57
66	17 29	18 21	19 20	02 41	03 54	04 51	05 31
64	17 33	18 21	19 16	02 22	03 33	04 29	05 11
62	17 36	18 21	19 12	02 07	03 16	04 12	04 55
60	17 39	18 21	19 09	01 54	03 01	03 57	04 42
N 58	17 42	18 21	19 07	01 43	02 49	03 45	04 30
56	17 44	18 21	19 04	01 34	02 38	03 34	04 20
54	17 46	18 22	19 03	01 25	02 29	03 24	04 11
52	17 48	18 22	19 01	01 18	02 21	03 16	04 03
50	17 50	18 22	19 00	01 11	02 13	03 08	03 56
45	17 54	18 23	18 57	00 56	01 57	02 52	03 40
N 40	17 57	18 24	18 55	00 44	01 44	02 39	03 27
35	18 00	18 25	18 54	00 34	01 33	02 27	03 16
30	18 02	18 26	18 54	00 25	01 23	02 17	03 07
20	18 07	18 29	18 54	00 10	01 06	02 00	02 51
N 10	18 11	18 32	18 56	24 52	00 52	01 45	02 36
0	18 15	18 35	18 59	24 38	00 38	01 31	02 23
S 10	18 19	18 40	19 04	24 24	00 24	01 17	02 09
20	18 23	18 45	19 11	24 10	00 10	01 02	01 55
30	18 28	18 52	19 20	23 53	24 45	00 45	01 38
35	18 31	18 56	19 27	23 43	24 35	00 35	01 28
40	18 34	19 02	19 34	23 32	24 23	00 23	01 17
45	18 38	19 08	19 44	23 19	24 10	00 10	01 04
S 50	18 43	19 16	19 56	23 03	23 53	24 48	00 48
52	18 45	19 20	20 01	22 55	23 45	24 41	00 41
54	18 48	19 24	20 08	22 47	23 37	24 33	00 33
56	18 50	19 29	20 16	22 38	23 27	24 23	00 23
58	18 53	19 34	20 24	22 27	23 16	24 13	00 13
S 60	18 57	19 40	20 34	22 15	23 03	24 01	00 01

Day	SUN Eqn. of Time 00ʰ	SUN Eqn. of Time 12ʰ	SUN Mer. Pass.	MOON Mer. Pass. Upper	MOON Mer. Pass. Lower	Age	Phase
d	m s	m s	h m	h m	h m	d %	
5	11 39	11 33	12 12	18 25	05 58	07 52	
6	11 26	11 19	12 11	19 18	06 52	08 62	
7	11 12	11 05	12 11	20 10	07 45	09 72	

UT	ARIES GHA	VENUS −4.5 GHA	VENUS Dec	MARS +1.2 GHA	MARS Dec	JUPITER −2.0 GHA	JUPITER Dec	SATURN +0.7 GHA	SATURN Dec	STARS Name	SHA	Dec
d h												
8 00	165 29.8	221 35.8	S15 43.4	162 52.5	N 0 25.2	185 54.1	S 9 33.7	147 12.7	N 5 16.8	Acamar	315 27.5	S40 19.1
01	180 32.3	236 36.1	43.2	177 53.2	26.0	200 56.0	33.5	162 14.9	16.9	Achernar	335 36.0	S57 15.1
02	195 34.7	251 36.5	43.1	192 53.9	26.8	215 57.9	33.3	177 17.1	17.1	Acrux	173 21.7	S63 05.3
03	210 37.2	266 36.9	.. 42.9	207 54.6	.. 27.6	230 59.8	.. 33.0	192 19.3	.. 17.2	Adhara	255 21.6	S28 58.5
04	225 39.7	281 37.3	42.8	222 55.3	28.4	246 01.7	32.8	207 21.5	17.3	Aldebaran	291 02.9	N16 30.2
05	240 42.1	296 37.7	42.7	237 56.0	29.1	261 03.6	32.6	222 23.7	17.4			
06	255 44.6	311 38.1	S15 42.5	252 56.7	N 0 29.9	276 05.5	S 9 32.4	237 25.9	N 5 17.5	Alioth	166 30.5	N55 58.1
07	270 47.0	326 38.4	42.4	267 57.4	30.7	291 07.4	32.2	252 28.1	17.6	Alkaid	153 07.8	N49 19.2
08	285 49.5	341 38.8	42.2	282 58.1	31.5	306 09.3	31.9	267 30.3	17.7	Al Na'ir	27 58.9	S46 58.2
S 09	300 52.0	356 39.2	.. 42.1	297 58.8	.. 32.3	321 11.2	.. 31.7	282 32.5	.. 17.8	Alnilam	275 58.2	S 1 12.5
U 10	315 54.4	11 39.6	41.9	312 59.5	33.1	336 13.1	31.5	297 34.7	18.0	Alphard	218 07.3	S 8 39.3
N 11	330 56.9	26 39.9	41.8	328 00.2	33.9	351 15.0	31.3	312 36.9	18.1			
D 12	345 59.4	41 40.3	S15 41.7	343 00.9	N 0 34.7	6 16.9	S 9 31.1	327 39.1	N 5 18.2	Alphecca	126 20.8	N26 43.1
A 13	1 01.8	56 40.7	41.5	358 01.6	35.5	21 18.8	30.8	342 41.3	18.3	Alpheratz	357 55.9	N29 04.7
Y 14	16 04.3	71 41.1	41.4	13 02.3	36.3	36 20.7	30.6	357 43.5	18.4	Altair	62 19.8	N 8 51.7
15	31 06.8	86 41.4	.. 41.2	28 03.0	.. 37.1	51 22.6	.. 30.4	12 45.7	.. 18.5	Ankaa	353 27.7	S42 19.2
16	46 09.2	101 41.8	41.1	43 03.7	37.8	66 24.5	30.2	27 47.9	18.6	Antares	112 40.6	S26 25.5
17	61 11.7	116 42.1	40.9	58 04.4	38.6	81 26.4	30.0	42 50.1	18.8			
18	76 14.2	131 42.5	S15 40.8	73 05.1	N 0 39.4	96 28.3	S 9 29.7	57 52.3	N 5 18.9	Arcturus	146 06.2	N19 11.4
19	91 16.6	146 42.9	40.6	88 05.8	40.2	111 30.2	29.5	72 54.5	19.0	Atria	107 52.9	S69 01.1
20	106 19.1	161 43.2	40.5	103 06.6	41.0	126 32.1	29.3	87 56.7	19.1	Avior	234 22.3	S59 30.6
21	121 21.5	176 43.6	.. 40.3	118 07.3	.. 41.8	141 34.0	.. 29.1	102 58.9	.. 19.2	Bellatrix	278 44.6	N 6 20.6
22	136 24.0	191 44.0	40.2	133 08.0	42.6	156 35.9	28.9	118 01.1	19.3	Betelgeuse	271 13.9	N 7 24.2
23	151 26.5	206 44.3	40.0	148 08.7	43.4	171 37.8	28.6	133 03.3	19.4			
9 00	166 28.9	221 44.7	S15 39.8	163 09.4	N 0 44.2	186 39.7	S 9 28.4	148 05.5	N 5 19.5	Canopus	264 01.2	S52 42.1
01	181 31.4	236 45.0	39.7	178 10.1	45.0	201 41.6	28.2	163 07.7	19.7	Capella	280 51.7	N45 59.8
02	196 33.9	251 45.4	39.5	193 10.8	45.7	216 43.5	28.0	178 09.9	19.8	Deneb	49 39.8	N45 16.3
03	211 36.3	266 45.7	.. 39.4	208 11.5	.. 46.5	231 45.4	.. 27.8	193 12.1	.. 19.9	Denebola	182 45.3	N14 34.8
04	226 38.8	281 46.1	39.2	223 12.2	47.3	246 47.3	27.5	208 14.3	20.0	Diphda	349 07.9	S18 00.0
05	241 41.3	296 46.4	39.1	238 12.9	48.1	261 49.2	27.3	223 16.5	20.1			
06	256 43.7	311 46.8	S15 38.9	253 13.6	N 0 48.9	276 51.1	S 9 27.1	238 18.7	N 5 20.2	Dubhe	194 05.4	N61 45.6
07	271 46.2	326 47.1	38.7	268 14.3	49.7	291 53.1	26.9	253 20.9	20.3	Elnath	278 27.4	N28 36.3
08	286 48.7	341 47.5	38.6	283 15.0	50.5	306 55.0	26.7	268 23.1	20.4	Eltanin	90 51.7	N51 29.2
M 09	301 51.1	356 47.8	.. 38.4	298 15.7	.. 51.3	321 56.9	.. 26.4	283 25.3	.. 20.6	Enif	33 58.9	N 9 51.9
O 10	316 53.6	11 48.1	38.2	313 16.4	52.1	336 58.8	26.2	298 27.5	20.7	Fomalhaut	15 37.3	S29 38.0
N 11	331 56.0	26 48.5	38.1	328 17.1	52.9	352 00.7	26.0	313 29.7	20.8			
D 12	346 58.5	41 48.8	S15 37.9	343 17.8	N 0 53.6	7 02.6	S 9 25.8	328 31.9	N 5 20.9	Gacrux	172 13.4	S57 06.1
A 13	2 01.0	56 49.2	37.7	358 18.5	54.4	22 04.5	25.6	343 34.1	21.0	Gienah	176 04.0	S17 32.0
Y 14	17 03.4	71 49.5	37.6	13 19.2	55.2	37 06.4	25.3	358 36.3	21.1	Hadar	149 04.0	S60 21.6
15	32 05.9	86 49.8	.. 37.4	28 19.9	.. 56.0	52 08.3	.. 25.1	13 38.5	.. 21.2	Hamal	328 14.2	N23 27.1
16	47 08.4	101 50.2	37.2	43 20.6	56.8	67 10.2	24.9	28 40.7	21.3	Kaus Aust.	83 59.5	S34 22.9
17	62 10.8	116 50.5	37.1	58 21.3	57.6	82 12.1	24.7	43 42.9	21.5			
18	77 13.3	131 50.8	S15 36.9	73 22.0	N 0 58.4	97 14.0	S 9 24.5	58 45.1	N 5 21.6	Kochab	137 19.1	N74 09.6
19	92 15.8	146 51.2	36.7	88 22.7	0 59.2	112 15.9	24.2	73 47.3	21.7	Markab	13 50.3	N15 11.6
20	107 18.2	161 51.5	36.6	103 23.4	1 00.0	127 17.8	24.0	88 49.5	21.8	Menkar	314 27.5	N 4 04.7
21	122 20.7	176 51.8	.. 36.4	118 24.1	.. 00.7	142 19.7	.. 23.8	103 51.7	.. 21.9	Menkent	148 21.1	S36 21.5
22	137 23.1	191 52.1	36.2	133 24.8	01.5	157 21.6	23.6	118 53.9	22.0	Miaplacidus	221 41.2	S69 42.8
23	152 25.6	206 52.5	36.1	148 25.5	02.3	172 23.5	23.4	133 56.1	22.1			
10 00	167 28.1	221 52.8	S15 35.9	163 26.2	N 1 03.1	187 25.4	S 9 23.1	148 58.3	N 5 22.3	Mirfak	308 57.3	N49 51.3
01	182 30.5	236 53.1	35.7	178 26.9	03.9	202 27.3	22.9	164 00.5	22.4	Nunki	76 13.0	S26 17.8
02	197 33.0	251 53.4	35.5	193 27.6	04.7	217 29.2	22.7	179 02.7	22.5	Peacock	53 38.1	S56 44.3
03	212 35.5	266 53.7	.. 35.4	208 28.3	.. 05.5	232 31.1	.. 22.5	194 04.9	.. 22.6	Pollux	243 41.8	N28 01.7
04	227 37.9	281 54.1	35.2	223 29.0	06.3	247 33.0	22.3	209 07.1	22.7	Procyon	245 11.8	N 5 13.6
05	242 40.4	296 54.4	35.0	238 29.7	07.1	262 34.9	22.0	224 09.3	22.8			
06	257 42.9	311 54.7	S15 34.8	253 30.4	N 1 07.8	277 36.8	S 9 21.8	239 11.5	N 5 22.9	Rasalhague	96 17.4	N12 33.6
07	272 45.3	326 55.0	34.6	268 31.1	08.6	292 38.7	21.6	254 13.7	23.0	Regulus	207 55.7	N11 58.4
08	287 47.8	341 55.3	34.5	283 31.8	09.4	307 40.6	21.4	269 15.8	23.2	Rigel	281 23.3	S 8 12.5
T 09	302 50.3	356 55.6	.. 34.3	298 32.5	.. 10.2	322 42.5	.. 21.2	284 18.0	.. 23.3	Rigil Kent.	140 07.3	S60 49.4
U 10	317 52.7	11 55.9	34.1	313 33.2	11.0	337 44.4	20.9	299 20.2	23.4	Sabik	102 26.0	S15 43.3
E 11	332 55.2	26 56.3	33.9	328 33.9	11.8	352 46.3	20.7	314 22.4	23.5			
S 12	347 57.6	41 56.6	S15 33.7	343 34.6	N 1 12.6	7 48.2	S 9 20.5	329 24.6	N 5 23.6	Schedar	349 54.4	N56 31.6
D 13	3 00.1	56 56.9	33.5	358 35.3	13.4	22 50.1	20.3	344 26.8	23.7	Shaula	96 37.9	S37 05.9
A 14	18 02.6	71 57.2	33.4	13 36.0	14.2	37 52.0	20.1	359 29.0	23.8	Sirius	258 43.9	S16 43.1
Y 15	33 05.0	86 57.5	.. 33.2	28 36.7	.. 14.9	52 53.9	.. 19.8	14 31.2	.. 24.0	Spica	158 43.3	S11 09.1
16	48 07.5	101 57.8	33.0	43 37.4	15.7	67 55.8	19.6	29 33.4	24.1	Suhail	223 00.6	S43 25.8
17	63 10.0	116 58.1	32.8	58 38.1	16.5	82 57.7	19.4	44 35.6	24.2			
18	78 12.4	131 58.4	S15 32.6	73 38.8	N 1 17.3	97 59.6	S 9 19.2	59 37.8	N 5 24.3	Vega	80 47.0	N38 46.8
19	93 14.9	146 58.7	32.4	88 39.5	18.1	113 01.5	19.0	74 40.0	24.4	Zuben'ubi	137 18.2	S16 02.0
20	108 17.4	161 59.0	32.2	103 40.2	18.9	128 03.4	18.7	89 42.2	24.5		SHA	Mer. Pass.
21	123 19.8	176 59.3	.. 32.0	118 40.9	.. 19.7	143 05.3	.. 18.5	104 44.4	.. 24.6		° ′	h m
22	138 22.3	191 59.6	31.9	133 41.7	20.5	158 07.2	18.3	119 46.6	24.8	Venus	55 15.7	9 13
23	153 24.7	206 59.9	31.7	148 42.4	21.2	173 09.1	18.1	134 48.8	24.9	Mars	356 40.4	13 07
	h m									Jupiter	20 10.8	11 32
Mer. Pass.	12 52.0	v 0.3	d 0.2	v 0.7	d 0.8	v 1.9	d 0.2	v 2.2	d 0.1	Saturn	341 36.6	14 06

UT	SUN GHA	SUN Dec	MOON GHA	v	MOON Dec	d	HP
d h	° ′	° ′	° ′	′	° ′	′	′
8 00	177 15.7	S 5 03.0	55 24.6	10.3	N17 58.0	2.5	56.0
01	192 15.8	02.1	69 53.9	10.3	17 55.5	2.7	56.0
02	207 16.0	01.1	84 23.2	10.4	17 52.8	2.8	56.0
03	222 16.1	5 00.1	98 52.6	10.4	17 50.0	2.8	56.0
04	237 16.3	4 59.2	113 22.0	10.4	17 47.2	2.9	55.9
05	252 16.4	58.2	127 51.4	10.5	17 44.3	3.1	55.9
06	267 16.6	S 4 57.2	142 20.9	10.5	N17 41.2	3.1	55.9
07	282 16.7	56.2	156 50.4	10.6	17 38.1	3.2	55.9
S 08	297 16.9	55.3	171 20.0	10.7	17 34.9	3.3	55.9
U 09	312 17.0	54.3	185 49.7	10.6	17 31.6	3.4	55.8
N 10	327 17.2	53.3	200 19.3	10.7	17 28.2	3.4	55.8
D 11	342 17.3	52.3	214 49.0	10.8	17 24.8	3.6	55.8
A 12	357 17.5	S 4 51.4	229 18.8	10.8	N17 21.2	3.6	55.8
Y 13	12 17.7	50.4	243 48.6	10.8	17 17.6	3.7	55.7
14	27 17.8	49.4	258 18.4	10.9	17 13.9	3.8	55.7
15	42 18.0	48.4	272 48.3	11.0	17 10.1	3.9	55.7
16	57 18.1	47.5	287 18.3	11.0	17 06.2	4.0	55.7
17	72 18.3	46.5	301 48.3	11.0	17 02.2	4.1	55.7
18	87 18.4	S 4 45.5	316 18.3	11.1	N16 58.1	4.1	55.6
19	102 18.6	44.5	330 48.4	11.1	16 54.0	4.2	55.6
20	117 18.7	43.6	345 18.5	11.2	16 49.8	4.3	55.6
21	132 18.9	42.6	359 48.7	11.2	16 45.5	4.4	55.6
22	147 19.1	41.6	14 18.9	11.3	16 41.1	4.5	55.5
23	162 19.2	40.6	28 49.2	11.3	16 36.6	4.5	55.5
9 00	177 19.4	S 4 39.6	43 19.5	11.3	N16 32.1	4.6	55.5
01	192 19.5	38.7	57 49.8	11.4	16 27.5	4.7	55.5
02	207 19.7	37.7	72 20.2	11.5	16 22.8	4.8	55.5
03	222 19.8	36.7	86 50.7	11.5	16 18.0	4.9	55.4
04	237 20.0	35.7	101 21.2	11.5	16 13.1	4.9	55.4
05	252 20.1	34.8	115 51.7	11.6	16 08.2	5.0	55.4
06	267 20.3	S 4 33.8	130 22.3	11.7	N16 03.2	5.1	55.4
07	282 20.5	32.8	144 53.0	11.7	15 58.1	5.2	55.3
M 08	297 20.6	31.8	159 23.7	11.7	15 52.9	5.2	55.3
O 09	312 20.8	30.9	173 54.4	11.8	15 47.7	5.3	55.3
N 10	327 20.9	29.9	188 25.2	11.8	15 42.4	5.4	55.3
D 11	342 21.1	28.9	202 56.0	11.9	15 37.0	5.4	55.3
A 12	357 21.2	S 4 27.9	217 26.9	11.9	N15 31.6	5.5	55.3
Y 13	12 21.4	26.9	231 57.8	12.0	15 26.1	5.6	55.2
14	27 21.6	26.0	246 28.8	12.0	15 20.5	5.7	55.2
15	42 21.7	25.0	260 59.8	12.1	15 14.8	5.7	55.2
16	57 21.9	24.0	275 30.9	12.1	15 09.1	5.8	55.2
17	72 22.0	23.0	290 02.0	12.1	15 03.3	5.9	55.2
18	87 22.2	S 4 22.1	304 33.1	12.2	N14 57.4	5.9	55.1
19	102 22.4	21.1	319 04.3	12.3	14 51.5	6.1	55.1
20	117 22.5	20.1	333 35.6	12.3	14 45.4	6.0	55.1
21	132 22.7	19.1	348 06.9	12.3	14 39.4	6.2	55.1
22	147 22.8	18.1	2 38.2	12.4	14 33.2	6.2	55.1
23	162 23.0	17.2	17 09.6	12.5	14 27.0	6.2	55.1
10 00	177 23.2	S 4 16.2	31 41.1	12.4	N14 20.8	6.4	55.0
01	192 23.3	15.2	46 12.5	12.6	14 14.4	6.3	55.0
02	207 23.5	14.2	60 44.1	12.5	14 08.1	6.5	55.0
03	222 23.6	13.2	75 15.6	12.7	14 01.6	6.5	55.0
04	237 23.8	12.3	89 47.3	12.6	13 55.1	6.6	55.0
05	252 24.0	11.3	104 18.9	12.7	13 48.5	6.6	55.0
06	267 24.1	S 4 10.3	118 50.6	12.8	N13 41.9	6.7	54.9
07	282 24.3	09.3	133 22.4	12.8	13 35.2	6.8	54.9
T 08	297 24.4	08.4	147 54.2	12.8	13 28.4	6.8	54.9
U 09	312 24.6	07.4	162 26.0	12.9	13 21.6	6.9	54.9
E 10	327 24.8	06.4	176 57.9	13.0	13 14.7	6.9	54.9
S 11	342 24.9	05.4	191 29.9	12.9	13 07.8	7.0	54.9
D 12	357 25.1	S 4 04.4	206 01.8	13.0	N13 00.8	7.0	54.8
A 13	12 25.2	03.5	220 33.8	13.1	12 53.8	7.1	54.8
Y 14	27 25.4	02.5	235 05.9	13.1	12 46.7	7.2	54.8
15	42 25.6	01.5	249 38.0	13.2	12 39.5	7.2	54.8
16	57 25.7	4 00.5	264 10.2	13.2	12 32.3	7.3	54.8
17	72 25.9	3 59.5	278 42.4	13.2	12 25.0	7.3	54.8
18	87 26.0	S 3 58.6	293 14.6	13.3	N12 17.7	7.3	54.8
19	102 26.2	57.6	307 46.9	13.3	12 10.4	7.5	54.7
20	117 26.4	56.6	322 19.2	13.3	12 02.9	7.4	54.7
21	132 26.5	55.6	336 51.5	13.4	11 55.5	7.5	54.7
22	147 26.7	54.6	351 23.9	13.5	11 48.0	7.6	54.7
23	162 26.9	53.7	5 56.4	13.5	N11 40.4	7.6	54.7
	SD 16.1	d 1.0	SD 15.2		15.1		14.9

Lat.	Twilight Naut.	Twilight Civil	Sunrise	Moonrise 8	9	10	11
°	h m	h m	h m	h m	h m	h m	h m
N 72	04 29	05 49	06 56	09 40	11 36	13 22	15 02
N 70	04 39	05 51	06 51	10 44	12 14	13 46	15 17
68	04 47	05 52	06 47	11 20	12 40	14 05	15 30
66	04 53	05 53	06 44	11 45	13 00	14 19	15 40
64	04 58	05 53	06 41	12 05	13 16	14 31	15 48
62	05 02	05 54	06 38	12 21	13 29	14 42	15 55
60	05 06	05 54	06 36	12 34	13 41	14 50	16 01
N 58	05 09	05 55	06 34	12 46	13 50	14 58	16 07
56	05 12	05 55	06 32	12 56	13 59	15 05	16 11
54	05 14	05 55	06 30	13 04	14 06	15 11	16 16
52	05 16	05 55	06 29	13 12	14 13	15 16	16 20
50	05 18	05 55	06 27	13 19	14 19	15 21	16 23
45	05 21	05 55	06 24	13 34	14 32	15 31	16 31
N 40	05 23	05 55	06 22	13 47	14 43	15 40	16 37
35	05 25	05 54	06 19	13 57	14 52	15 48	16 42
30	05 26	05 53	06 17	14 06	15 00	15 54	16 47
20	05 26	05 52	06 14	14 22	15 14	16 05	16 55
N 10	05 25	05 49	06 10	14 36	15 26	16 15	17 03
0	05 23	05 47	06 07	14 49	15 38	16 24	17 10
S 10	05 18	05 43	06 04	15 02	15 49	16 34	17 16
20	05 12	05 38	06 00	15 16	16 01	16 44	17 24
30	05 04	05 32	05 56	15 31	16 15	16 55	17 32
35	04 58	05 28	05 54	15 41	16 23	17 01	17 36
40	04 51	05 24	05 51	15 51	16 32	17 08	17 42
45	04 43	05 18	05 48	16 03	16 42	17 17	17 48
S 50	04 31	05 11	05 44	16 18	16 55	17 27	17 55
52	04 26	05 07	05 42	16 25	17 01	17 32	17 59
54	04 20	05 04	05 40	16 33	17 08	17 37	18 03
56	04 13	04 59	05 37	16 41	17 15	17 43	18 07
58	04 05	04 55	05 35	16 51	17 23	17 49	18 11
S 60	03 55	04 49	05 32	17 02	17 32	17 57	18 16

Lat.	Sunset	Twilight Civil	Twilight Naut.	Moonset 8	9	10	11
°	h m	h m	h m	h m	h m	h m	h m
N 72	17 27	18 35	19 55	07 37	07 24	07 17	07 11
N 70	17 32	18 33	19 45	06 33	06 46	06 51	06 54
68	17 36	18 31	19 37	05 57	06 19	06 32	06 40
66	17 39	18 30	19 31	05 31	05 58	06 16	06 29
64	17 42	18 30	19 25	05 11	05 42	06 04	06 20
62	17 45	18 29	19 21	04 55	05 28	05 53	06 12
60	17 47	18 28	19 17	04 42	05 16	05 43	06 05
N 58	17 49	18 28	19 14	04 30	05 06	05 35	05 59
56	17 51	18 28	19 11	04 20	04 57	05 28	05 54
54	17 52	18 27	19 08	04 11	04 49	05 22	05 49
52	17 54	18 27	19 06	04 03	04 42	05 16	05 44
50	17 55	18 27	19 04	03 56	04 36	05 10	05 40
45	17 58	18 27	19 01	03 40	04 22	04 59	05 32
N 40	18 00	18 27	18 59	03 27	04 11	04 50	05 24
35	18 02	18 28	18 57	03 16	04 01	04 41	05 18
30	18 04	18 28	18 56	03 07	03 53	04 34	05 13
20	18 08	18 30	18 55	02 51	03 38	04 22	05 03
N 10	18 11	18 32	18 56	02 36	03 25	04 11	04 54
0	18 14	18 35	18 59	02 23	03 13	04 00	04 46
S 10	18 17	18 38	19 04	02 09	03 00	03 50	04 38
20	18 20	18 43	19 08	01 55	02 47	03 39	04 30
30	18 24	18 49	19 17	01 38	02 32	03 26	04 20
35	18 27	18 52	19 22	01 28	02 23	03 19	04 14
40	18 30	18 57	19 29	01 17	02 13	03 10	04 08
45	18 33	19 02	19 38	01 04	02 02	03 01	04 00
S 50	18 37	19 09	19 49	00 48	01 47	02 49	03 51
52	18 38	19 13	19 54	00 41	01 41	02 43	03 46
54	18 40	19 16	20 00	00 33	01 33	02 37	03 42
56	18 42	19 20	20 07	00 23	01 25	02 30	03 36
58	18 45	19 25	20 14	00 13	01 16	02 22	03 31
S 60	18 48	19 31	20 23	00 01	01 05	02 13	03 24

Day	SUN Eqn. of Time 00h	12h	Mer. Pass.	MOON Mer. Pass. Upper	Lower	Age	Phase
d	m s	m s	h m	h m	h m	d	%
8	10 58	10 50	12 11	21 01	08 36	10	81
9	10 43	10 35	12 11	21 49	09 25	11	88
10	10 28	10 20	12 10	22 35	10 13	12	93

1998 MARCH 11, 12, 13 (WED., THURS., FRI.)

UT	ARIES GHA	VENUS −4.5 GHA	Dec	MARS +1.2 GHA	Dec	JUPITER −2.0 GHA	Dec	SATURN +0.7 GHA	Dec	STARS Name	SHA	Dec
d h 11 00	168 27.2	222 00.2	S15 31.5	163 43.1	N 1 22.0	188 11.0	S 9 17.9	149 51.0	N 5 25.0	Acamar	315 27.5	S40 19.1
01	183 29.7	237 00.5	31.3	178 43.8	22.8	203 12.9	17.6	164 53.2	25.1	Achernar	335 36.0	S57 15.1
02	198 32.1	252 00.7	31.1	193 44.5	23.6	218 14.8	17.4	179 55.4	25.2	Acrux	173 21.7	S63 05.3
03	213 34.6	267 01.0 . .	30.9	208 45.2 . .	24.4	233 16.7 . .	17.2	194 57.6 . .	25.3	Adhara	255 21.6	S28 58.6
04	228 37.1	282 01.3	30.7	223 45.9	25.2	248 18.7	17.0	209 59.8	25.4	Aldebaran	291 02.9	N16 30.2
05	243 39.5	297 01.6	30.5	238 46.6	26.0	263 20.6	16.8	225 02.0	25.5			
06	258 42.0	312 01.9	S15 30.3	253 47.3	N 1 26.8	278 22.5	S 9 16.5	240 04.2	N 5 25.7	Alioth	166 30.5	N55 58.1
W 07	273 44.5	327 02.2	30.1	268 48.0	27.6	293 24.4	16.3	255 06.4	25.8	Alkaid	153 07.8	N49 19.2
E 08	288 46.9	342 02.5	29.9	283 48.7	28.3	308 26.3	16.1	270 08.6	25.9	Al Na'ir	27 58.9	S46 58.2
D 09	303 49.4	357 02.7 . .	29.7	298 49.4 . .	29.1	323 28.2 . .	15.9	285 10.7 . .	26.0	Alnilam	275 58.2	S 1 12.5
N 10	318 51.9	12 03.0	29.5	313 50.1	29.9	338 30.1	15.7	300 12.9	26.1	Alphard	218 07.3	S 8 39.3
E 11	333 54.3	27 03.3	29.3	328 50.8	30.7	353 32.0	15.4	315 15.1	26.2			
S 12	348 56.8	42 03.6	S15 29.1	343 51.5	N 1 31.5	8 33.9	S 9 15.2	330 17.3	N 5 26.3	Alphecca	126 20.8	N26 43.1
D 13	3 59.2	57 03.9	28.9	358 52.2	32.3	23 35.8	15.0	345 19.5	26.5	Alpheratz	357 56.0	N29 04.7
A 14	19 01.7	72 04.1	28.7	13 52.9	33.1	38 37.7	14.8	0 21.7	26.6	Altair	62 19.8	N 8 51.7
Y 15	34 04.2	87 04.4 . .	28.5	28 53.6 . .	33.8	53 39.6 . .	14.6	15 23.9 . .	26.7	Ankaa	353 27.7	S42 19.2
16	49 06.6	102 04.7	28.3	43 54.3	34.6	68 41.5	14.4	30 26.1	26.8	Antares	112 40.6	S26 25.5
17	64 09.1	117 05.0	28.1	58 55.0	35.4	83 43.4	14.1	45 28.3	26.9			
18	79 11.6	132 05.2	S15 27.9	73 55.7	N 1 36.2	98 45.3	S 9 13.9	60 30.5	N 5 27.0	Arcturus	146 06.2	N19 11.4
19	94 14.0	147 05.5	27.7	88 56.4	37.0	113 47.2	13.7	75 32.7	27.1	Atria	107 52.3	S69 01.1
20	109 16.5	162 05.8	27.4	103 57.1	37.8	128 49.1	13.5	90 34.9	27.3	Avior	234 22.3	S59 30.6
21	124 19.0	177 06.0 . .	27.2	118 57.8 . .	38.6	143 51.0 . .	13.3	105 37.1 . .	27.4	Bellatrix	278 44.6	N 6 20.6
22	139 21.4	192 06.3	27.0	133 58.5	39.4	158 52.9	13.0	120 39.3	27.5	Betelgeuse	271 13.9	N 7 24.2
23	154 23.9	207 06.6	26.8	148 59.2	40.1	173 54.8	12.8	135 41.5	27.6			
12 00	169 26.4	222 06.8	S15 26.6	163 59.9	N 1 40.9	188 56.7	S 9 12.6	150 43.7	N 5 27.7	Canopus	264 01.2	S52 42.1
01	184 28.8	237 07.1	26.4	179 00.6	41.7	203 58.6	12.4	165 45.9	27.8	Capella	280 51.8	N45 59.8
02	199 31.3	252 07.4	26.2	194 01.4	42.5	219 00.5	12.2	180 48.1	27.9	Deneb	49 39.8	N45 16.3
03	214 33.7	267 07.6 . .	26.0	209 02.1 . .	43.3	234 02.4 . .	11.9	195 50.3 . .	28.1	Denebola	182 45.3	N14 34.8
04	229 36.2	282 07.9	25.8	224 02.8	44.1	249 04.3	11.7	210 52.4	28.2	Diphda	349 08.0	S18 00.0
05	244 38.7	297 08.1	25.5	239 03.5	44.9	264 06.2	11.5	225 54.6	28.3			
06	259 41.1	312 08.4	S15 25.3	254 04.2	N 1 45.6	279 08.2	S 9 11.3	240 56.8	N 5 28.4	Dubhe	194 05.4	N61 45.6
T 07	274 43.6	327 08.6	25.1	269 04.9	46.4	294 10.1	11.1	255 59.0	28.5	Elnath	278 27.4	N28 36.3
H 08	289 46.1	342 08.9	24.9	284 05.6	47.2	309 12.0	10.8	271 01.2	28.6	Eltanin	90 51.7	N51 29.2
U 09	304 48.5	357 09.2 . .	24.7	299 06.3 . .	48.0	324 13.9 . .	10.6	286 03.4 . .	28.7	Enif	33 58.9	N 9 51.9
R 10	319 51.0	12 09.4	24.4	314 07.0	48.8	339 15.8	10.4	301 05.6	28.9	Fomalhaut	15 37.3	S29 38.0
S 11	334 53.5	27 09.7	24.2	329 07.7	49.6	354 17.7	10.2	316 07.8	29.0			
D 12	349 55.9	42 09.9	S15 24.0	344 08.4	N 1 50.4	9 19.6	S 9 10.0	331 10.0	N 5 29.1	Gacrux	172 13.4	S57 06.1
A 13	4 58.4	57 10.2	23.8	359 09.1	51.1	24 21.5	09.7	346 12.2	29.2	Gienah	176 04.0	S17 32.0
Y 14	20 00.8	72 10.4	23.6	14 09.8	51.9	39 23.4	09.5	1 14.4	29.3	Hadar	149 04.0	S60 21.6
15	35 03.3	87 10.7 . .	23.3	29 10.5 . .	52.7	54 25.3 . .	09.3	16 16.6 . .	29.4	Hamal	328 14.2	N23 27.1
16	50 05.8	102 10.9	23.1	44 11.2	53.5	69 27.2	09.1	31 18.8	29.5	Kaus Aust.	83 59.4	S34 22.9
17	65 08.2	117 11.1	22.9	59 11.9	54.3	84 29.1	08.9	46 21.0	29.7			
18	80 10.7	132 11.4	S15 22.7	74 12.6	N 1 55.1	99 31.0	S 9 08.6	61 23.2	N 5 29.8	Kochab	137 19.0	N74 09.6
19	95 13.2	147 11.6	22.4	89 13.3	55.9	114 32.9	08.4	76 25.4	29.9	Markab	13 50.3	N15 11.6
20	110 15.6	162 11.9	22.2	104 14.0	56.6	129 34.8	08.2	91 27.5	30.0	Menkar	314 27.5	N 4 04.7
21	125 18.1	177 12.1 . .	22.0	119 14.7 . .	57.4	144 36.7 . .	08.0	106 29.7 . .	30.1	Menkent	148 21.1	S36 21.6
22	140 20.6	192 12.4	21.7	134 15.4	58.2	159 38.6	07.8	121 31.9	30.2	Miaplacidus	221 41.3	S69 42.9
23	155 23.0	207 12.6	21.5	149 16.1	59.0	174 40.5	07.5	136 34.1	30.3			
13 00	170 25.5	222 12.8	S15 21.3	164 16.8	N 1 59.8	189 42.4	S 9 07.3	151 36.3	N 5 30.5	Mirfak	308 57.4	N49 51.3
01	185 28.0	237 13.1	21.1	179 17.6	2 00.6	204 44.3	07.1	166 38.5	30.6	Nunki	76 13.0	S26 17.8
02	200 30.4	252 13.3	20.8	194 18.3	01.4	219 46.2	06.9	181 40.7	30.7	Peacock	53 38.1	S56 44.3
03	215 32.9	267 13.5 . .	20.6	209 19.0 . .	02.1	234 48.2 . .	06.7	196 42.9 . .	30.8	Pollux	243 41.8	N28 01.7
04	230 35.3	282 13.8	20.4	224 19.7	02.9	249 50.1	06.4	211 45.1	30.9	Procyon	245 11.8	N 5 13.6
05	245 37.8	297 14.0	20.1	239 20.4	03.7	264 52.0	06.2	226 47.3	31.0			
06	260 40.3	312 14.2	S15 19.9	254 21.1	N 2 04.5	279 53.9	S 9 06.0	241 49.5	N 5 31.1	Rasalhague	96 17.3	N12 33.6
07	275 42.7	327 14.5	19.7	269 21.8	05.3	294 55.8	05.8	256 51.7	31.3	Regulus	207 55.7	N11 58.4
08	290 45.2	342 14.7	19.4	284 22.5	06.1	309 57.7	05.6	271 53.9	31.4	Rigel	281 23.3	S 8 12.5
F 09	305 47.7	357 14.9 . .	19.2	299 23.2 . .	06.8	324 59.6 . .	05.3	286 56.0 . .	31.5	Rigil Kent.	140 07.3	S60 49.4
R 10	320 50.1	12 15.1	18.9	314 23.9	07.6	340 01.5	05.1	301 58.2	31.6	Sabik	102 26.0	S15 43.3
I 11	335 52.6	27 15.4	18.7	329 24.6	08.4	355 03.4	04.9	317 00.4	31.7			
D 12	350 55.1	42 15.6	S15 18.5	344 25.3	N 2 09.2	10 05.3	S 9 04.7	332 02.6	N 5 31.8	Schedar	349 54.4	N56 31.6
A 13	5 57.5	57 15.8	18.2	359 26.0	10.0	25 07.2	04.5	347 04.8	31.9	Shaula	96 37.8	S37 05.9
Y 14	21 00.0	72 16.0	18.0	14 26.7	10.8	40 09.1	04.2	2 07.0	32.1	Sirius	258 43.9	S16 43.1
15	36 02.4	87 16.2 . .	17.7	29 27.4 . .	11.5	55 11.0 . .	04.0	17 09.2 . .	32.2	Spica	158 43.3	S11 09.1
16	51 04.9	102 16.5	17.5	44 28.1	12.3	70 12.9	03.8	32 11.4	32.3	Suhail	223 00.7	S43 25.8
17	66 07.4	117 16.7	17.2	59 28.8	13.1	85 14.8	03.6	47 13.6	32.4			
18	81 09.8	132 16.9	S15 16.9	74 29.5	N 2 13.9	100 16.7	S 9 03.4	62 15.8	N 5 32.5	Vega	80 47.0	N38 46.7
19	96 12.3	147 17.1	16.8	89 30.2	14.7	115 18.6	03.2	77 18.0	32.6	Zuben'ubi	137 18.2	S16 02.0
20	111 14.8	162 17.3	16.5	104 31.0	15.5	130 20.5	02.9	92 20.2	32.8		SHA	Mer. Pass.
21	126 17.2	177 17.5 . .	16.3	119 31.7 . .	16.3	145 22.5 . .	02.7	107 22.4 . .	32.9		° '	h m
22	141 19.7	192 17.8	16.0	134 32.4	17.0	160 24.4	02.5	122 24.5	33.0	Venus	52 40.5	9 11
23	156 22.2	207 18.0	15.8	149 33.1	17.8	175 26.3	02.3	137 26.7	33.1	Mars	354 33.6	13 03
	h m									Jupiter	19 30.4	11 23
Mer. Pass. 12 40.2		v 0.2	d 0.2	v 0.7	d 0.8	v 1.9	d 0.2	v 2.2	d 0.1	Saturn	341 17.3	13 55

UT	SUN GHA	SUN Dec	MOON GHA	v	MOON Dec	d	HP
d h	o ′	o ′	o ′	′	o ′	′	′
11 00	177 27.0	S 3 52.7	20 28.9	13.5	N11 32.8	7.7	54.7
01	192 27.2	51.7	35 01.4	13.6	11 25.1	7.7	54.7
02	207 27.4	50.7	49 34.0	13.6	11 17.4	7.8	54.6
03	222 27.5 ..	49.7	64 06.6	13.6	11 09.6	7.8	54.6
04	237 27.7	48.7	78 39.2	13.7	11 01.8	7.8	54.6
05	252 27.8	47.8	93 11.9	13.7	10 54.0	7.9	54.6
06	267 28.0	S 3 46.8	107 44.6	13.8	N10 46.1	7.9	54.6
W 07	282 28.2	45.8	122 17.4	13.8	10 38.2	8.0	54.6
E 08	297 28.3	44.8	136 50.2	13.8	10 30.2	8.0	54.6
D 09	312 28.5 ..	43.8	151 23.0	13.9	10 22.2	8.1	54.6
N 10	327 28.7	42.9	165 55.9	13.9	10 14.1	8.1	54.5
E 11	342 28.8	41.9	180 28.8	13.9	10 06.0	8.2	54.5
S 12	357 29.0	S 3 40.9	195 01.7	14.0	N 9 57.8	8.1	54.5
D 13	12 29.2	39.9	209 34.7	14.0	9 49.7	8.3	54.5
A 14	27 29.3	38.9	224 07.7	14.1	9 41.4	8.2	54.5
Y 15	42 29.5 ..	38.0	238 40.8	14.0	9 33.2	8.3	54.5
16	57 29.7	37.0	253 13.8	14.2	9 24.9	8.4	54.5
17	72 29.8	36.0	267 47.0	14.1	9 16.5	8.4	54.5
18	87 30.0	S 3 35.0	282 20.1	14.2	N 9 08.1	8.4	54.4
19	102 30.2	34.0	296 53.3	14.2	8 59.7	8.4	54.4
20	117 30.3	33.0	311 26.5	14.3	8 51.3	8.5	54.4
21	132 30.5 ..	32.1	325 59.8	14.3	8 42.8	8.5	54.4
22	147 30.7	31.1	340 33.1	14.3	8 34.3	8.6	54.4
23	162 30.8	30.1	355 06.4	14.4	8 25.7	8.5	54.4
12 00	177 31.0	S 3 29.1	9 39.8	14.3	N 8 17.2	8.7	54.4
01	192 31.2	28.1	24 13.1	14.5	8 08.5	8.6	54.4
02	207 31.3	27.1	38 46.6	14.4	7 59.9	8.7	54.4
03	222 31.5 ..	26.2	53 20.0	14.5	7 51.2	8.7	54.3
04	237 31.7	25.2	67 53.5	14.5	7 42.5	8.7	54.3
05	252 31.8	24.2	82 27.0	14.5	7 33.8	8.8	54.3
06	267 32.0	S 3 23.2	97 00.5	14.6	N 7 25.0	8.8	54.3
T 07	282 32.2	22.2	111 34.1	14.6	7 16.2	8.8	54.3
H 08	297 32.3	21.2	126 07.7	14.6	7 07.4	8.9	54.3
U 09	312 32.5 ..	20.3	140 41.3	14.6	6 58.5	8.8	54.3
R 10	327 32.7	19.3	155 14.9	14.7	6 49.7	9.0	54.3
S 11	342 32.8	18.3	169 48.6	14.7	6 40.7	8.9	54.3
D 12	357 33.0	S 3 17.3	184 22.3	14.7	N 6 31.8	8.9	54.3
A 13	12 33.2	16.3	198 56.0	14.8	6 22.9	9.0	54.3
Y 14	27 33.3	15.3	213 29.8	14.8	6 13.9	9.0	54.2
15	42 33.5 ..	14.4	228 03.6	14.8	6 04.9	9.0	54.2
16	57 33.7	13.4	242 37.4	14.8	5 55.9	9.1	54.2
17	72 33.8	12.4	257 11.2	14.8	5 46.8	9.0	54.2
18	87 34.0	S 3 11.4	271 45.0	14.9	N 5 37.8	9.1	54.2
19	102 34.2	10.4	286 18.9	14.9	5 28.7	9.1	54.2
20	117 34.3	09.4	300 52.8	14.9	5 19.6	9.2	54.2
21	132 34.5 ..	08.5	315 26.7	15.0	5 10.4	9.1	54.2
22	147 34.7	07.5	330 00.7	14.9	5 01.3	9.2	54.2
23	162 34.9	06.5	344 34.6	15.0	4 52.1	9.1	54.2
13 00	177 35.0	S 3 05.5	359 08.6	15.0	N 4 43.0	9.2	54.2
01	192 35.2	04.5	13 42.6	15.0	4 33.8	9.3	54.2
02	207 35.4	03.5	28 16.6	15.1	4 24.5	9.2	54.2
03	222 35.5 ..	02.6	42 50.7	15.0	4 15.3	9.2	54.1
04	237 35.7	01.6	57 24.7	15.1	4 06.1	9.3	54.1
05	252 35.9	3 00.6	71 58.8	15.1	3 56.8	9.3	54.1
06	267 36.0	S 2 59.6	86 32.9	15.1	N 3 47.5	9.3	54.1
07	282 36.2	58.6	101 07.0	15.1	3 38.2	9.3	54.1
08	297 36.4	57.6	115 41.1	15.2	3 28.9	9.3	54.1
F 09	312 36.6 ..	56.6	130 15.3	15.1	3 19.6	9.3	54.1
R 10	327 36.7	55.7	144 49.4	15.2	3 10.3	9.3	54.1
I 11	342 36.9	54.7	159 23.6	15.2	3 01.0	9.4	54.1
D 12	357 37.1	S 2 53.7	173 57.8	15.2	N 2 51.6	9.3	54.1
A 13	12 37.2	52.7	188 32.0	15.2	2 42.3	9.4	54.1
Y 14	27 37.4	51.7	203 06.2	15.2	2 32.9	9.4	54.1
15	42 37.6 ..	50.7	217 40.4	15.2	2 23.5	9.3	54.1
16	57 37.8	49.8	232 14.6	15.3	2 14.2	9.4	54.1
17	72 37.9	48.8	246 48.9	15.3	2 04.8	9.4	54.1
18	87 38.1	S 2 47.8	261 23.2	15.2	N 1 55.4	9.4	54.1
19	102 38.3	46.8	275 57.4	15.3	1 46.0	9.4	54.1
20	117 38.4	45.8	290 32.7	15.3	1 36.6	9.4	54.0
21	132 38.6 ..	44.8	305 06.0	15.3	1 27.2	9.4	54.0
22	147 38.8	43.8	319 40.3	15.3	1 17.8	9.5	54.0
23	162 39.0	42.9	334 14.6	15.3	N 1 08.3	9.4	54.0
	SD 16.1	d 1.0	SD 14.9		14.8		14.7

Lat.	Twilight Naut.	Twilight Civil	Sunrise	Moonrise 11	Moonrise 12	Moonrise 13	Moonrise 14
o	h m	h m	h m	h m	h m	h m	h m
N 72	04 12	05 34	06 41	15 02	16 39	18 13	19 46
N 70	04 24	05 37	06 37	15 17	16 47	18 15	19 43
68	04 33	05 39	06 34	15 30	16 54	18 17	19 40
66	04 41	05 41	06 32	15 40	17 00	18 19	19 38
64	04 47	05 43	06 30	15 48	17 04	18 21	19 36
62	04 52	05 44	06 28	15 55	17 09	18 22	19 35
60	04 57	05 45	06 27	16 01	17 12	18 23	19 33
N 58	05 01	05 46	06 25	16 07	17 15	18 24	19 32
56	05 04	05 47	06 24	16 11	17 18	18 25	19 31
54	05 07	05 48	06 23	16 16	17 21	18 26	19 30
52	05 09	05 48	06 22	16 20	17 23	18 26	19 29
50	05 11	05 49	06 21	16 23	17 25	18 27	19 28
45	05 15	05 49	06 19	16 31	17 30	18 28	19 27
N 40	05 19	05 50	06 17	16 37	17 34	18 30	19 25
35	05 21	05 50	06 15	16 42	17 37	18 31	19 24
30	05 22	05 50	06 14	16 47	17 40	18 32	19 23
20	05 24	05 49	06 11	16 55	17 45	18 33	19 21
N 10	05 23	05 48	06 09	17 03	17 49	18 35	19 20
0	05 22	05 46	06 07	17 10	17 53	18 36	19 18
S 10	05 19	05 43	06 04	17 16	17 57	18 37	19 17
20	05 14	05 39	06 01	17 24	18 02	18 39	19 15
30	05 06	05 34	05 58	17 32	18 07	18 40	19 13
35	05 01	05 31	05 56	17 36	18 09	18 41	19 12
40	04 55	05 27	05 54	17 42	18 13	18 42	19 11
45	04 47	05 22	05 52	17 48	18 16	18 43	19 10
S 50	04 37	05 16	05 48	17 55	18 21	18 45	19 08
52	04 32	05 13	05 47	17 59	18 23	18 46	19 08
54	04 26	05 10	05 45	18 03	18 25	18 46	19 07
56	04 20	05 06	05 44	18 07	18 28	18 47	19 06
58	04 13	05 02	05 42	18 11	18 30	18 48	19 05
S 60	04 05	04 57	05 40	18 16	18 33	18 49	19 04

Lat.	Sunset	Twilight Civil	Twilight Naut.	Moonset 11	Moonset 12	Moonset 13	Moonset 14
o	h m	h m	h m	h m	h m	h m	h m
N 72	17 41	18 48	20 11	07 11	07 05	06 59	06 54
N 70	17 44	18 45	19 59	06 54	06 55	06 55	06 54
68	17 47	18 42	19 49	06 40	06 46	06 51	06 55
66	17 49	18 40	19 41	06 29	06 39	06 47	06 55
64	17 51	18 38	19 35	06 20	06 33	06 45	06 55
62	17 53	18 37	19 29	06 12	06 28	06 42	06 55
60	17 55	18 36	19 25	06 05	06 24	06 40	06 55
N 58	17 56	18 35	19 21	05 59	06 20	06 38	06 55
56	17 57	18 34	19 17	05 54	06 16	06 36	06 55
54	17 58	18 33	19 14	05 49	06 13	06 35	06 55
52	17 59	18 32	19 12	05 44	06 10	06 33	06 55
50	18 00	18 32	19 10	05 40	06 07	06 32	06 56
45	18 02	18 31	19 05	05 32	06 01	06 29	06 56
N 40	18 03	18 30	19 02	05 24	05 56	06 27	06 56
35	18 05	18 30	18 59	05 18	05 52	06 25	06 56
30	18 06	18 30	18 58	05 13	05 49	06 23	06 56
20	18 09	18 31	18 56	05 03	05 42	06 19	06 56
N 10	18 11	18 32	18 56	04 54	05 36	06 17	06 56
0	18 13	18 34	18 58	04 46	05 31	06 14	06 56
S 10	18 15	18 36	19 01	04 38	05 25	06 11	06 56
20	18 18	18 40	19 06	04 30	05 19	06 08	06 56
30	18 21	18 45	19 13	04 20	05 13	06 05	06 56
35	18 23	18 48	19 18	04 14	05 09	06 03	06 56
40	18 25	18 52	19 24	04 08	05 04	06 01	06 57
45	18 27	18 57	19 32	04 00	04 59	05 58	06 57
S 50	18 30	19 03	19 41	03 51	04 53	05 55	06 57
52	18 31	19 06	19 46	03 46	04 50	05 53	06 57
54	18 33	19 09	19 52	03 42	04 47	05 52	06 57
56	18 35	19 12	19 58	03 36	04 43	05 50	06 57
58	18 36	19 16	20 05	03 31	04 39	05 48	06 57
S 60	18 39	19 21	20 13	03 24	04 35	05 46	06 57

	SUN Eqn. of Time 00h	SUN Eqn. of Time 12h	SUN Mer. Pass.	MOON Mer. Pass. Upper	MOON Mer. Pass. Lower	Age	Phase
Day							
d	m s	m s	h m	h m	h m	d	%
11	10 12	10 04	12 10	23 20	10 58	13	97
12	09 56	09 48	12 10	24 04	11 42	14	100
13	09 40	09 32	12 10	00 04	12 25	15	100

1998 MARCH 14, 15, 16 (SAT., SUN., MON.)

UT d h	ARIES GHA	VENUS −4.5 GHA	Dec	MARS +1.2 GHA	Dec	JUPITER −2.0 GHA	Dec	SATURN +0.7 GHA	Dec	Name	SHA	Dec
14 00	171 24.6	222 18.2	S15 15.5	164 33.8	N 2 18.6	190 28.2	S 9 02.1	152 28.9	N 5 33.2	Acamar	315 27.5	S40 19.1
01	186 27.1	237 18.4	15.3	179 34.5	19.4	205 30.1	01.8	167 31.1	33.3	Achernar	335 36.0	S57 15.1
02	201 29.6	252 18.6	15.0	194 35.2	20.2	220 32.0	01.6	182 33.3	33.4	Acrux	173 21.7	S63 05.3
03	216 32.0	267 18.8	.. 14.8	209 35.9	.. 21.0	235 33.9	.. 01.4	197 35.5	.. 33.6	Adhara	255 21.6	S28 58.6
04	231 34.5	282 19.0	14.5	224 36.6	21.7	250 35.8	01.2	212 37.7	33.7	Aldebaran	291 02.9	N16 30.2
05	246 36.9	297 19.2	14.2	239 37.3	22.5	265 37.7	01.0	227 39.9	33.8			
06	261 39.4	312 19.4	S15 14.0	254 38.0	N 2 23.3	280 39.6	S 9 00.7	242 42.1	N 5 33.9	Alioth	166 30.5	N55 58.1
07	276 41.9	327 19.6	13.7	269 38.7	24.1	295 41.5	00.5	257 44.3	34.0	Alkaid	153 07.8	N49 19.2
S 08	291 44.3	342 19.8	13.5	284 39.4	24.9	310 43.4	00.3	272 46.5	34.1	Al Na'ir	27 58.9	S46 58.2
A 09	306 46.8	357 20.0	.. 13.2	299 40.1	.. 25.6	325 45.3	9 00.1	287 48.6	.. 34.2	Alnilam	275 58.3	S 1 12.5
T 10	321 49.3	12 20.2	13.0	314 40.8	26.4	340 47.2	8 59.9	302 50.8	34.4	Alphard	218 07.3	S 8 39.3
U 11	336 51.7	27 20.4	12.7	329 41.5	27.2	355 49.1	59.6	317 53.0	34.5			
R 12	351 54.2	42 20.6	S15 12.4	344 42.2	N 2 28.0	10 51.1	S 8 59.4	332 55.2	N 5 34.6	Alphecca	126 20.8	N26 43.1
D 13	6 56.7	57 20.8	12.2	359 42.9	28.8	25 53.0	59.2	347 57.4	34.7	Alpheratz	357 56.0	N29 04.7
A 14	21 59.1	72 21.0	11.9	14 43.7	29.6	40 54.9	59.0	2 59.6	34.8	Altair	62 19.8	N 8 51.7
Y 15	37 01.6	87 21.2	.. 11.7	29 44.4	.. 30.3	55 56.8	.. 58.8	18 01.8	.. 34.9	Ankaa	353 27.7	S42 19.2
16	52 04.0	102 21.4	11.4	44 45.1	31.1	70 58.7	58.5	33 04.0	35.0	Antares	112 40.5	S26 25.5
17	67 06.5	117 21.6	11.1	59 45.8	31.9	86 00.6	58.3	48 06.2	35.2			
18	82 09.0	132 21.8	S15 10.9	74 46.5	N 2 32.7	101 02.5	S 8 58.1	63 08.4	N 5 35.3	Arcturus	146 06.2	N19 11.4
19	97 11.4	147 22.0	10.6	89 47.2	33.5	116 04.4	57.9	78 10.6	35.4	Atria	107 52.8	S69 01.1
20	112 13.9	162 22.2	10.3	104 47.9	34.3	131 06.3	57.7	93 12.7	35.5	Avior	234 22.3	S59 30.6
21	127 16.4	177 22.4	.. 10.1	119 48.6	.. 35.0	146 08.2	.. 57.4	108 14.9	.. 35.6	Bellatrix	278 44.6	N 6 20.6
22	142 18.8	192 22.6	09.8	134 49.3	35.8	161 10.1	57.2	123 17.1	35.7	Betelgeuse	271 13.9	N 7 24.2
23	157 21.3	207 22.7	09.5	149 50.0	36.6	176 12.0	57.0	138 19.3	35.9			
15 00	172 23.8	222 22.9	S15 09.3	164 50.7	N 2 37.4	191 13.9	S 8 56.8	153 21.5	N 5 36.0	Canopus	264 01.2	S52 42.1
01	187 26.2	237 23.1	09.0	179 51.4	38.2	206 15.8	56.6	168 23.7	36.1	Capella	280 51.8	N45 59.8
02	202 28.7	252 23.3	08.7	194 52.1	38.9	221 17.8	56.3	183 25.9	36.2	Deneb	49 39.7	N45 16.3
03	217 31.2	267 23.5	.. 08.4	209 52.8	.. 39.7	236 19.7	.. 56.1	198 28.1	.. 36.3	Denebola	182 45.3	N14 34.8
04	232 33.6	282 23.7	08.2	224 53.5	40.5	251 21.6	55.9	213 30.3	36.4	Diphda	349 08.0	S18 00.0
05	247 36.1	297 23.8	07.9	239 54.2	41.3	266 23.5	55.7	228 32.5	36.5			
06	262 38.5	312 24.0	S15 07.6	254 54.9	N 2 42.1	281 25.4	S 8 55.5	243 34.6	N 5 36.7	Dubhe	194 05.4	N61 45.6
07	277 41.0	327 24.2	07.4	269 55.7	42.8	296 27.3	55.3	258 36.8	36.8	Elnath	278 27.4	N28 36.3
S 08	292 43.5	342 24.4	07.1	284 56.4	43.6	311 29.2	55.0	273 39.0	36.9	Eltanin	90 51.6	N51 29.2
U 09	307 45.9	357 24.6	.. 06.8	299 57.1	.. 44.4	326 31.1	.. 54.8	288 41.2	.. 37.0	Enif	33 58.9	N 9 51.9
N 10	322 48.4	12 24.7	06.5	314 57.8	45.2	341 33.0	54.6	303 43.4	37.1	Fomalhaut	15 37.2	S29 38.0
11	337 50.9	27 24.9	06.2	329 58.5	46.0	356 34.9	54.4	318 45.6	37.2			
D 12	352 53.3	42 25.1	S15 06.0	344 59.2	N 2 46.8	11 36.8	S 8 54.2	333 47.8	N 5 37.4	Gacrux	172 13.4	S57 06.1
A 13	7 55.8	57 25.3	05.7	359 59.9	47.5	26 38.7	53.9	348 50.0	37.5	Gienah	176 04.0	S17 32.0
Y 14	22 58.3	72 25.4	05.4	15 00.6	48.3	41 40.6	53.7	3 52.2	37.6	Hadar	149 04.0	S60 21.7
15	38 00.7	87 25.6	.. 05.1	30 01.3	.. 49.1	56 42.6	.. 53.5	18 54.3	.. 37.7	Hamal	328 14.2	N23 27.1
16	53 03.2	102 25.8	04.8	45 02.0	49.9	71 44.5	53.3	33 56.5	37.8	Kaus Aust.	83 59.4	S34 22.9
17	68 05.6	117 25.9	04.6	60 02.7	50.7	86 46.4	53.1	48 58.7	37.9			
18	83 08.1	132 26.1	S15 04.3	75 03.4	N 2 51.4	101 48.3	S 8 52.8	64 00.9	N 5 38.0	Kochab	137 18.9	N74 09.6
19	98 10.6	147 26.3	04.0	90 04.1	52.2	116 50.2	52.6	79 03.1	38.2	Markab	13 50.3	N15 11.6
20	113 13.0	162 26.4	03.7	105 04.8	53.0	131 52.1	52.4	94 05.3	38.3	Menkar	314 27.5	N 4 04.7
21	128 15.5	177 26.6	.. 03.4	120 05.5	.. 53.8	146 54.0	.. 52.2	109 07.5	.. 38.4	Menkent	148 21.1	S36 21.6
22	143 18.0	192 26.8	03.1	135 06.3	54.6	161 55.9	52.0	124 09.7	38.5	Miaplacidus	221 41.3	S69 42.9
23	158 20.4	207 26.9	02.8	150 07.0	55.3	176 57.8	51.7	139 11.9	38.6			
16 00	173 22.9	222 27.1	S15 02.5	165 07.7	N 2 56.1	191 59.7	S 8 51.5	154 14.0	N 5 38.7	Mirfak	308 57.4	N49 51.3
01	188 25.4	237 27.3	02.3	180 08.4	56.9	207 01.6	51.3	169 16.2	38.9	Nunki	76 12.9	S26 17.8
02	203 27.8	252 27.4	02.0	195 09.1	57.7	222 03.5	51.1	184 18.4	39.0	Peacock	53 38.1	S56 44.3
03	218 30.3	267 27.6	.. 01.7	210 09.8	.. 58.5	237 05.5	.. 50.9	199 20.6	.. 39.1	Pollux	243 41.9	N28 01.7
04	233 32.8	282 27.7	01.4	225 10.5	2 59.2	252 07.4	50.6	214 22.8	39.2	Procyon	245 11.8	N 5 13.6
05	248 35.2	297 27.9	01.1	240 11.2	3 00.0	267 09.3	50.4	229 25.0	39.3			
06	263 37.7	312 28.1	S15 00.8	255 11.9	N 3 00.8	282 11.2	S 8 50.2	244 27.2	N 5 39.4	Rasalhague	96 17.3	N12 33.6
07	278 40.1	327 28.2	00.5	270 12.6	01.6	297 13.1	50.0	259 29.4	39.5	Regulus	207 55.7	N11 58.4
08	293 42.6	342 28.4	15 00.2	285 13.3	02.3	312 15.0	49.8	274 31.6	39.7	Rigel	281 23.3	S 8 12.5
M 09	308 45.1	357 28.5	14 59.9	300 14.0	.. 03.1	327 16.9	.. 49.6	289 33.7	.. 39.8	Rigil Kent.	140 07.3	S60 49.4
O 10	323 47.5	12 28.7	59.6	315 14.7	03.9	342 18.8	49.3	304 35.9	39.9	Sabik	102 25.9	S15 43.3
N 11	338 50.0	27 28.8	59.3	330 15.4	04.7	357 20.7	49.1	319 38.1	40.0			
D 12	353 52.5	42 29.0	S14 59.0	345 16.1	N 3 05.5	12 22.6	S 8 48.9	334 40.3	N 5 40.1	Schedar	349 54.4	N56 31.6
A 13	8 54.9	57 29.1	58.7	0 16.9	06.2	27 24.5	48.7	349 42.5	40.2	Shaula	96 37.8	S37 05.9
Y 14	23 57.4	72 29.3	58.4	15 17.6	07.0	42 26.5	48.5	4 44.7	40.4	Sirius	258 43.9	S16 43.1
15	38 59.9	87 29.4	.. 58.1	30 18.3	.. 07.8	57 28.4	.. 48.2	19 46.9	.. 40.5	Spica	158 43.3	S11 09.1
16	54 02.3	102 29.6	57.8	45 19.0	08.6	72 30.3	48.0	34 49.1	40.6	Suhail	223 00.7	S43 25.8
17	69 04.8	117 29.7	57.5	60 19.7	09.4	87 32.2	47.8	49 51.2	40.7			
18	84 07.3	132 29.9	S14 57.2	75 20.4	N 3 10.1	102 34.1	S 8 47.6	64 53.4	N 5 40.8	Vega	80 47.0	N38 46.7
19	99 09.7	147 30.0	56.9	90 21.1	10.9	117 36.0	47.4	79 55.6	40.9	Zuben'ubi	137 18.2	S16 02.0
20	114 12.2	162 30.2	56.6	105 21.8	11.7	132 37.9	47.1	94 57.8	41.1			
21	129 14.6	177 30.3	.. 56.3	120 22.5	.. 12.5	147 39.8	.. 46.9	110 00.0	.. 41.2			
22	144 17.1	192 30.4	56.0	135 23.2	13.2	162 41.7	46.7	125 02.2	41.3			
23	159 19.6	207 30.6	55.7	150 23.9	14.0	177 43.6	46.5	140 04.4	41.4			

	SHA	Mer. Pass.
	° ′	h m
Venus	49 59.2	9 10
Mars	352 27.0	13 00
Jupiter	18 50.2	11 14
Saturn	340 57.7	13 45

Mer. Pass.	12 28.4	*v* 0.2 *d* 0.3	*v* 0.7 *d* 0.8	*v* 1.9 *d* 0.2	*v* 2.2 *d* 0.1

UT	SUN GHA	SUN Dec	MOON GHA	v	MOON Dec	d	HP
d h	° '	° '	° '	'	° '	'	'
14 00	177 39.1	S 2 41.9	348 48.9	15.4	N 0 58.9	9.4	54.0
01	192 39.3	40.9	3 23.3	15.3	0 49.5	9.4	54.0
02	207 39.5	39.9	17 57.6	15.3	0 40.1	9.4	54.0
03	222 39.6 ..	38.9	32 31.9	15.4	0 30.7	9.5	54.0
04	237 39.8	37.9	47 06.3	15.3	0 21.2	9.4	54.0
05	252 40.0	36.9	61 40.6	15.4	0 11.8	9.4	54.0
06	267 40.2	S 2 36.0	76 15.0	15.3	N 0 02.4	9.4	54.0
07	282 40.3	35.0	90 49.3	15.3	S 0 07.0	9.5	54.0
08	297 40.5	34.0	105 23.7	15.4	0 16.5	9.4	54.0
09	312 40.7 ..	33.0	119 58.1	15.3	0 25.9	9.4	54.0
10	327 40.9	32.0	134 32.4	15.4	0 35.3	9.4	54.0
11	342 41.0	31.0	149 06.8	15.4	0 44.7	9.4	54.0
12	357 41.2	S 2 30.0	163 41.2	15.3	S 0 54.1	9.4	54.0
13	12 41.4	29.1	178 15.5	15.4	1 03.5	9.4	54.0
14	27 41.6	28.1	192 49.9	15.4	1 12.9	9.4	54.0
15	42 41.7 ..	27.1	207 24.3	15.4	1 22.3	9.4	54.0
16	57 41.9	26.1	221 58.7	15.3	1 31.7	9.4	54.0
17	72 42.1	25.1	236 33.0	15.4	1 41.1	9.4	54.0
18	87 42.2	S 2 24.1	251 07.4	15.4	S 1 50.5	9.3	54.0
19	102 42.4	23.1	265 41.8	15.3	1 59.8	9.4	54.0
20	117 42.6	22.2	280 16.1	15.4	2 09.2	9.3	54.0
21	132 42.8 ..	21.2	294 50.5	15.3	2 18.5	9.4	54.0
22	147 42.9	20.2	309 24.8	15.4	2 27.9	9.3	54.0
23	162 43.1	19.2	323 59.2	15.3	2 37.2	9.3	54.0
15 00	177 43.3	S 2 18.2	338 33.5	15.4	S 2 46.5	9.3	54.0
01	192 43.5	17.2	353 07.9	15.3	2 55.8	9.3	54.0
02	207 43.6	16.2	7 42.2	15.4	3 05.1	9.3	54.0
03	222 43.8 ..	15.2	22 16.6	15.3	3 14.4	9.3	54.0
04	237 44.0	14.3	36 50.9	15.3	3 23.7	9.2	54.0
05	252 44.2	13.3	51 25.2	15.3	3 32.9	9.3	54.0
06	267 44.4	S 2 12.3	65 59.5	15.3	S 3 42.2	9.2	54.0
07	282 44.5	11.3	80 33.8	15.3	3 51.4	9.2	54.0
08	297 44.7	10.3	95 08.1	15.3	4 00.6	9.2	54.0
09	312 44.9 ..	09.3	109 42.4	15.2	4 09.8	9.2	54.0
10	327 45.0	08.3	124 16.6	15.3	4 19.0	9.1	54.0
11	342 45.2	07.4	138 50.9	15.2	4 28.1	9.2	54.0
12	357 45.4	S 2 06.4	153 25.1	15.3	S 4 37.3	9.1	54.0
13	12 45.6	05.4	167 59.4	15.2	4 46.4	9.1	54.0
14	27 45.8	04.4	182 33.6	15.2	4 55.5	9.1	54.0
15	42 45.9 ..	03.4	197 07.8	15.2	5 04.6	9.0	54.0
16	57 46.1	02.4	211 42.0	15.1	5 13.6	9.1	54.0
17	72 46.3	01.4	226 16.1	15.2	5 22.7	9.0	54.0
18	87 46.5	S 2 00.4	240 50.3	15.2	S 5 31.7	9.0	54.0
19	102 46.6	1 59.5	255 24.5	15.1	5 40.7	9.0	54.0
20	117 46.8	58.5	269 58.6	15.1	5 49.7	8.9	54.0
21	132 47.0 ..	57.5	284 32.7	15.1	5 58.6	8.9	54.0
22	147 47.2	56.5	299 06.8	15.1	6 07.5	8.9	54.0
23	162 47.4	55.5	313 40.9	15.0	6 16.4	8.9	54.0
16 00	177 47.5	S 1 54.5	328 14.9	15.1	S 6 25.3	8.9	54.0
01	192 47.7	53.5	342 49.0	15.0	6 34.2	8.8	54.0
02	207 47.9	52.5	357 23.0	15.0	6 43.0	8.8	54.0
03	222 48.1 ..	51.6	11 57.0	15.0	6 51.8	8.8	54.0
04	237 48.2	50.6	26 31.0	15.0	7 00.6	8.7	54.0
05	252 48.4	49.6	41 05.0	14.9	7 09.3	8.7	54.1
06	267 48.6	S 1 48.6	55 38.9	14.9	S 7 18.0	8.7	54.1
07	282 48.8	47.6	70 12.8	14.9	7 26.7	8.7	54.1
08	297 48.9	46.6	84 46.7	14.9	7 35.4	8.6	54.1
09	312 49.1 ..	45.6	99 20.6	14.9	7 44.0	8.6	54.1
10	327 49.3	44.6	113 54.5	14.8	7 52.6	8.6	54.1
11	342 49.5	43.7	128 28.3	14.8	8 01.2	8.5	54.1
12	357 49.7	S 1 42.7	143 02.1	14.8	S 8 09.7	8.5	54.1
13	12 49.8	41.7	157 35.9	14.7	8 18.2	8.5	54.1
14	27 50.0	40.7	172 09.6	14.7	8 26.7	8.5	54.1
15	42 50.2 ..	39.7	186 43.3	14.8	8 35.2	8.4	54.1
16	57 50.4	38.7	201 17.1	14.6	8 43.6	8.3	54.1
17	72 50.6	37.7	215 50.7	14.7	8 51.9	8.4	54.1
18	87 50.7	S 1 36.7	230 24.4	14.6	S 9 00.3	8.3	54.1
19	102 50.9	35.8	244 58.0	14.6	9 08.6	8.2	54.1
20	117 51.1	34.8	259 31.6	14.6	9 16.8	8.3	54.2
21	132 51.3 ..	33.8	274 05.2	14.5	9 25.1	8.2	54.2
22	147 51.5	32.8	288 38.7	14.5	9 33.3	8.1	54.2
23	162 51.6	31.8	303 12.2	14.5	S 9 41.4	8.1	54.2
SD 16.1	*d* 1.0		SD 14.7	14.7			14.7

Day columns: SATURDAY (14), SUNDAY (15), MONDAY (16)

Lat.	Twilight Naut.	Twilight Civil	Sunrise	Moonrise 14	15	16	17
°	h m	h m	h m	h m	h m	h m	h m
N 72	03 55	05 18	06 25	19 46	21 19	22 55	24 35
N 70	04 09	05 23	06 23	19 43	21 10	22 39	24 09
68	04 20	05 27	06 22	19 40	21 03	22 26	23 50
66	04 29	05 30	06 21	19 38	20 57	22 16	23 35
64	04 36	05 32	06 19	19 36	20 52	22 07	23 23
62	04 42	05 35	06 18	19 35	20 47	22 00	23 12
60	04 47	05 36	06 18	19 33	20 44	21 54	23 03
N 58	04 52	05 38	06 17	19 32	20 40	21 48	22 56
56	04 56	05 39	06 16	19 31	20 37	21 43	22 49
54	04 59	05 40	06 15	19 30	20 34	21 39	22 43
52	05 02	05 41	06 15	19 29	20 32	21 35	22 37
50	05 05	05 42	06 14	19 28	20 30	21 31	22 32
45	05 10	05 44	06 13	19 27	20 25	21 23	22 22
N 40	05 14	05 45	06 12	19 25	20 21	21 17	22 13
35	05 17	05 46	06 11	19 24	20 18	21 11	22 05
30	05 19	05 46	06 10	19 23	20 15	21 06	21 59
20	05 21	05 47	06 09	19 21	20 09	20 58	21 47
N 10	05 22	05 46	06 07	19 20	20 05	20 50	21 37
0	05 21	05 45	06 06	19 18	20 00	20 44	21 28
S 10	05 19	05 43	06 04	19 17	19 56	20 37	21 19
20	05 15	05 40	06 02	19 15	19 52	20 29	21 09
30	05 08	05 36	06 00	19 13	19 47	20 21	20 58
35	05 04	05 33	05 59	19 12	19 44	20 16	20 51
40	04 58	05 30	05 57	19 11	19 41	20 11	20 44
45	04 51	05 26	05 55	19 10	19 37	20 05	20 35
S 50	04 42	05 21	05 53	19 08	19 32	19 58	20 25
52	04 38	05 18	05 52	19 08	19 30	19 54	20 21
54	04 33	05 16	05 51	19 07	19 28	19 50	20 15
56	04 27	05 12	05 50	19 06	19 25	19 46	20 10
58	04 21	05 09	05 49	19 05	19 23	19 42	20 03
S 60	04 13	05 05	05 47	19 04	19 20	19 37	19 56

Lat.	Sunset	Twilight Civil	Twilight Naut.	Moonset 14	15	16	17
°	h m	h m	h m	h m	h m	h m	h m
N 72	17 55	19 02	20 27	06 54	06 49	06 43	06 37
N 70	17 57	18 57	20 13	06 54	06 54	06 54	06 54
68	17 58	18 53	20 01	06 55	06 58	07 03	07 08
66	17 59	18 50	19 52	06 55	07 02	07 10	07 20
64	18 00	18 47	19 44	06 55	07 05	07 16	07 29
62	18 01	18 45	19 38	06 55	07 08	07 22	07 37
60	18 02	18 43	19 32	06 55	07 10	07 26	07 44
N 58	18 02	18 41	19 28	06 55	07 12	07 31	07 51
56	18 03	18 40	19 24	06 55	07 14	07 34	07 56
54	18 04	18 39	19 20	06 55	07 16	07 38	08 01
52	18 04	18 38	19 17	06 55	07 18	07 41	08 06
50	18 05	18 37	19 14	06 56	07 19	07 43	08 10
45	18 06	18 35	19 09	06 56	07 22	07 50	08 19
N 40	18 07	18 34	19 05	06 56	07 25	07 55	08 26
35	18 07	18 33	19 02	06 56	07 27	07 59	08 33
30	18 08	18 32	19 00	06 56	07 29	08 03	08 38
20	18 10	18 32	18 57	06 56	07 33	08 10	08 48
N 10	18 11	18 32	18 56	06 56	07 36	08 16	08 57
0	18 12	18 33	18 56	06 56	07 38	08 21	09 05
S 10	18 14	18 35	18 59	06 56	07 41	08 27	09 13
20	18 15	18 37	19 03	06 56	07 44	08 33	09 22
30	18 17	18 41	19 09	06 56	07 48	08 39	09 31
35	18 19	18 44	19 14	06 56	07 50	08 43	09 37
40	18 20	18 47	19 19	06 57	07 52	08 48	09 44
45	18 22	18 51	19 26	06 57	07 55	08 53	09 51
S 50	18 24	18 56	19 34	06 57	07 58	08 59	10 00
52	18 25	18 58	19 39	06 57	07 59	09 02	10 05
54	18 26	19 01	19 44	06 57	08 01	09 05	10 09
56	18 27	19 04	19 49	06 57	08 03	09 09	10 15
58	18 28	19 08	19 55	06 57	08 05	09 13	10 20
S 60	18 29	19 11	20 03	06 57	08 07	09 17	10 27

	SUN			MOON			
Day	Eqn. of Time 00ʰ	12ʰ	Mer. Pass.	Mer. Pass. Upper	Lower	Age	Phase
d	m s	m s	h m	h m	h m	d	%
14	09 24	09 16	12 09	00 46	13 07	16	98
15	09 07	08 59	12 09	01 28	13 49	17	95
16	08 50	08 42	12 09	02 11	14 32	18	90

UT	ARIES	VENUS −4.5		MARS +1.2		JUPITER −2.0		SATURN +0.6		STARS		
	GHA	GHA	Dec	GHA	Dec	GHA	Dec	GHA	Dec	Name	SHA	Dec
d h	° ′	° ′	° ′	° ′	° ′	° ′	° ′	° ′	° ′		° ′	° ′
17 00	174 22.0	222 30.7	S14 55.4	165 24.6	N 3 14.8	192 45.6	S 8 46.3	155 06.6	N 5 41.5	Acamar	315 27.5	S40 19.1
01	189 24.5	237 30.9	55.1	180 25.3	15.6	207 47.5	46.1	170 08.7	41.6	Achernar	335 36.1	S57 15.0
02	204 27.0	252 31.0	54.7	195 26.0	16.4	222 49.4	45.8	185 10.9	41.7	Acrux	173 21.6	S63 05.3
03	219 29.4	267 31.1	.. 54.4	210 26.8	.. 17.1	237 51.3	.. 45.6	200 13.1	.. 41.9	Adhara	255 21.6	S28 58.6
04	234 31.9	282 31.3	54.1	225 27.5	17.9	252 53.2	45.4	215 15.3	42.0	Aldebaran	291 02.9	N16 30.2
05	249 34.4	297 31.4	53.8	240 28.2	18.7	267 55.1	45.2	230 17.5	42.1			
06	264 36.8	312 31.5	S14 53.5	255 28.9	N 3 19.5	282 57.0	S 8 45.0	245 19.7	N 5 42.2	Alioth	166 30.5	N55 58.1
07	279 39.3	327 31.7	53.2	270 29.6	20.2	297 58.9	44.7	260 21.9	42.3	Alkaid	153 07.7	N49 19.2
T 08	294 41.7	342 31.8	52.9	285 30.3	21.0	313 00.8	44.5	275 24.0	42.4	Al Na'ir	27 58.8	S46 58.2
U 09	309 44.2	357 31.9	.. 52.5	300 31.0	.. 21.8	328 02.7	.. 44.3	290 26.2	.. 42.6	Alnilam	275 58.3	S 1 12.5
E 10	324 46.7	12 32.1	52.2	315 31.7	22.6	343 04.7	44.1	305 28.4	42.7	Alphard	218 07.3	S 8 39.3
S 11	339 49.1	27 32.2	51.9	330 32.4	23.4	358 06.6	43.9	320 30.6	42.8			
D 12	354 51.6	42 32.3	S14 51.6	345 33.1	N 3 24.1	13 08.5	S 8 43.6	335 32.8	N 5 42.9	Alphecca	126 20.7	N26 43.1
A 13	9 54.1	57 32.5	51.3	0 33.8	24.9	28 10.4	43.4	350 35.0	43.0	Alpheratz	357 56.0	N29 04.7
Y 14	24 56.5	72 32.6	51.0	15 34.5	25.7	43 12.3	43.2	5 37.2	43.1	Altair	62 19.8	N 8 51.7
15	39 59.0	87 32.7	.. 50.6	30 35.2	.. 26.5	58 14.2	.. 43.0	20 39.4	.. 43.3	Ankaa	353 27.7	S42 19.1
16	55 01.5	102 32.8	50.3	45 35.9	27.2	73 16.1	42.8	35 41.5	43.4	Antares	112 40.5	S26 25.5
17	70 03.9	117 33.0	50.0	60 36.7	28.0	88 18.0	42.5	50 43.7	43.5			
18	85 06.4	132 33.1	S14 49.7	75 37.4	N 3 28.8	103 19.9	S 8 42.3	65 45.9	N 5 43.6	Arcturus	146 06.2	N19 11.4
19	100 08.9	147 33.2	49.3	90 38.1	29.6	118 21.9	42.1	80 48.1	43.7	Atria	107 52.8	S69 01.1
20	115 11.3	162 33.3	49.0	105 38.8	30.3	133 23.8	41.9	95 50.3	43.8	Avior	234 22.4	S59 30.6
21	130 13.8	177 33.5	.. 48.7	120 39.5	.. 31.1	148 25.7	.. 41.7	110 52.5	.. 43.9	Bellatrix	278 44.6	N 6 20.6
22	145 16.2	192 33.6	48.4	135 40.2	31.9	163 27.6	41.5	125 54.7	44.1	Betelgeuse	271 14.0	N 7 24.2
23	160 18.7	207 33.7	48.0	150 40.9	32.7	178 29.5	41.2	140 56.8	44.2			
18 00	175 21.2	222 33.8	S14 47.7	165 41.6	N 3 33.4	193 31.4	S 8 41.0	155 59.0	N 5 44.3	Canopus	264 01.3	S52 42.1
01	190 23.6	237 33.9	47.4	180 42.3	34.2	208 33.3	40.8	171 01.2	44.4	Capella	280 51.8	N45 59.8
02	205 26.1	252 34.1	47.0	195 43.0	35.0	223 35.2	40.6	186 03.4	44.5	Deneb	49 39.7	N45 16.3
03	220 28.6	267 34.2	.. 46.7	210 43.7	.. 35.8	238 37.1	.. 40.4	201 05.6	.. 44.6	Denebola	182 45.3	N14 34.8
04	235 31.0	282 34.3	46.4	225 44.4	36.5	253 39.1	40.1	216 07.8	44.8	Diphda	349 08.0	S18 00.0
05	250 33.5	297 34.4	46.0	240 45.1	37.3	268 41.0	39.9	231 10.0	44.9			
06	265 36.0	312 34.5	S14 45.7	255 45.8	N 3 38.1	283 42.9	S 8 39.7	246 12.1	N 5 45.0	Dubhe	194 05.4	N61 45.6
W 07	280 38.4	327 34.6	45.4	270 46.6	38.9	298 44.8	39.5	261 14.3	45.1	Elnath	278 27.5	N28 36.3
E 08	295 40.9	342 34.8	45.0	285 47.3	39.6	313 46.7	39.3	276 16.5	45.2	Eltanin	90 51.6	N51 29.2
D 09	310 43.3	357 34.9	.. 44.7	300 48.0	.. 40.4	328 48.6	.. 39.0	291 18.7	.. 45.3	Enif	33 58.9	N 9 51.9
N 10	325 45.8	12 35.0	44.4	315 48.7	41.2	343 50.5	38.8	306 20.9	45.5	Fomalhaut	15 37.2	S29 38.0
E 11	340 48.3	27 35.1	44.0	330 49.4	42.0	358 52.4	38.6	321 23.1	45.6			
S 12	355 50.7	42 35.2	S14 43.7	345 50.1	N 3 42.7	13 54.3	S 8 38.4	336 25.3	N 5 45.7	Gacrux	172 13.3	S57 06.2
D 13	10 53.2	57 35.3	43.4	0 50.8	43.5	28 56.3	38.2	351 27.4	45.8	Gienah	176 04.0	S17 32.0
A 14	25 55.7	72 35.4	43.0	15 51.5	44.3	43 58.2	38.0	6 29.6	45.9	Hadar	149 04.0	S60 21.7
Y 15	40 58.1	87 35.5	.. 42.7	30 52.2	.. 45.1	59 00.1	.. 37.7	21 31.8	.. 46.0	Hamal	328 14.3	N23 27.1
16	56 00.6	102 35.6	42.3	45 52.9	45.8	74 02.0	37.5	36 34.0	46.2	Kaus Aust.	83 59.4	S34 22.9
17	71 03.1	117 35.7	42.0	60 53.6	46.6	89 03.9	37.3	51 36.2	46.3			
18	86 05.5	132 35.8	S14 41.7	75 54.3	N 3 47.4	104 05.8	S 8 37.1	66 38.4	N 5 46.4	Kochab	137 18.9	N74 09.6
19	101 08.0	147 35.9	41.3	90 55.0	48.2	119 07.7	36.9	81 40.6	46.5	Markab	13 50.3	N15 11.6
20	116 10.5	162 36.0	41.0	105 55.7	48.9	134 09.6	36.6	96 42.7	46.6	Menkar	314 27.5	N 4 04.7
21	131 12.9	177 36.1	.. 40.6	120 56.5	.. 49.7	149 11.6	.. 36.4	111 44.9	.. 46.7	Menkent	148 21.1	S36 21.6
22	146 15.4	192 36.3	40.3	135 57.2	50.5	164 13.5	36.2	126 47.1	46.9	Miaplacidus	221 41.3	S69 42.9
23	161 17.8	207 36.4	39.9	150 57.9	51.3	179 15.4	36.0	141 49.3	47.0			
19 00	176 20.3	222 36.5	S14 39.6	165 58.6	N 3 52.0	194 17.3	S 8 35.8	156 51.5	N 5 47.1	Mirfak	308 57.4	N49 51.3
01	191 22.8	237 36.6	39.2	180 59.3	52.8	209 19.2	35.6	171 53.7	47.2	Nunki	76 12.9	S26 17.8
02	206 25.2	252 36.6	38.9	196 00.0	53.6	224 21.1	35.3	186 55.8	47.3	Peacock	53 38.0	S56 44.2
03	221 27.7	267 36.7	.. 38.5	211 00.7	.. 54.4	239 23.0	.. 35.1	201 58.0	.. 47.4	Pollux	243 41.9	N28 01.7
04	236 30.2	282 36.8	38.2	226 01.4	55.1	254 24.9	34.9	217 00.2	47.6	Procyon	245 11.8	N 5 13.6
05	251 32.6	297 36.9	37.8	241 02.1	55.9	269 26.9	34.7	232 02.4	47.7			
06	266 35.1	312 37.0	S14 37.5	256 02.8	N 3 56.7	284 28.8	S 8 34.5	247 04.6	N 5 47.8	Rasalhague	96 17.3	N12 33.6
07	281 37.6	327 37.1	37.1	271 03.5	57.4	299 30.7	34.2	262 06.8	47.9	Regulus	207 55.7	N11 58.4
T 08	296 40.0	342 37.2	36.8	286 04.2	58.2	314 32.6	34.0	277 09.0	48.0	Rigel	281 23.4	S 8 12.5
H 09	311 42.5	357 37.3	.. 36.4	301 04.9	.. 59.0	329 34.5	.. 33.8	292 11.1	.. 48.1	Rigil Kent.	140 07.3	S60 49.4
U 10	326 45.0	12 37.4	36.0	316 05.7	3 59.8	344 36.4	33.6	307 13.3	48.2	Sabik	102 25.9	S15 43.3
R 11	341 47.4	27 37.5	35.7	331 06.4	4 00.5	359 38.3	33.4	322 15.5	48.4			
S 12	356 49.9	42 37.6	S14 35.3	346 07.1	N 4 01.3	14 40.2	S 8 33.1	337 17.7	N 5 48.5	Schedar	349 54.4	N56 31.6
D 13	11 52.3	57 37.7	35.0	1 07.8	02.1	29 42.2	32.9	352 19.9	48.6	Shaula	96 37.8	S37 05.9
A 14	26 54.8	72 37.8	34.6	16 08.5	02.9	44 44.1	32.7	7 22.1	48.7	Sirius	258 44.0	S16 43.1
Y 15	41 57.3	87 37.9	.. 34.3	31 09.2	.. 03.6	59 46.0	.. 32.5	22 24.2	.. 48.8	Spica	158 43.3	S11 09.1
16	56 59.7	102 37.9	33.9	46 09.9	04.4	74 47.9	32.3	37 26.4	48.9	Suhail	223 00.7	S43 25.8
17	72 02.2	117 38.0	33.5	61 10.6	05.2	89 49.8	32.1	52 28.6	49.1			
18	87 04.7	132 38.1	S14 33.2	76 11.3	N 4 05.9	104 51.7	S 8 31.8	67 30.8	N 5 49.2	Vega	80 47.0	N38 46.7
19	102 07.1	147 38.2	32.8	91 12.0	06.7	119 53.6	31.6	82 33.0	49.3	Zuben'ubi	137 18.2	S16 02.0
20	117 09.6	162 38.3	32.4	106 12.7	07.5	134 55.6	31.4	97 35.2	49.4		SHA	Mer.Pass.
21	132 12.1	177 38.4	.. 32.1	121 13.4	.. 08.3	149 57.5	.. 31.2	112 37.3	.. 49.5	Venus	47 12.7	9 10
22	147 14.5	192 38.5	31.7	136 14.1	09.0	164 59.4	31.0	127 39.5	49.6	Mars	350 20.4	12 57
23	162 17.0	207 38.5	31.3	151 14.9	09.8	180 01.3	30.7	142 41.7	49.8	Jupiter	18 10.2	11 04
Mer.Pass. 12 16.6		v 0.1	d 0.3	v 0.7	d 0.8	v 1.9	d 0.2	v 2.2	d 0.1	Saturn	340 37.9	13 34

UT	SUN GHA	SUN Dec	MOON GHA	v	Dec	d	HP
d h	° ′	° ′	° ′	′	° ′	′	′
17 00	177 51.8	S 1 30.8	317 45.7	14.4	S 9 49.5	8.1	54.2
01	192 52.0	29.8	332 19.1	14.4	9 57.6	8.0	54.2
02	207 52.2	28.8	346 52.5	14.4	10 05.6	8.0	54.2
03	222 52.3	.. 27.9	1 25.9	14.3	10 13.6	8.0	54.2
04	237 52.5	26.9	15 59.2	14.4	10 21.6	7.9	54.2
05	252 52.7	25.9	30 32.6	14.2	10 29.5	7.9	54.2
06	267 52.9	S 1 24.9	45 05.8	14.3	S10 37.4	7.8	54.2
07	282 53.1	23.9	59 39.1	14.2	10 45.2	7.8	54.3
08	297 53.2	22.9	74 12.3	14.2	10 53.0	7.7	54.3
09	312 53.4	.. 21.9	88 45.5	14.1	11 00.7	7.7	54.3
10	327 53.6	20.9	103 18.6	14.2	11 08.4	7.7	54.3
11	342 53.8	19.9	117 51.8	14.0	11 16.1	7.6	54.3
12	357 54.0	S 1 19.0	132 24.8	14.1	S11 23.7	7.5	54.3
13	12 54.2	18.0	146 57.9	14.0	11 31.2	7.5	54.3
14	27 54.3	17.0	161 30.9	13.9	11 38.7	7.5	54.3
15	42 54.5	.. 16.0	176 03.8	14.0	11 46.2	7.4	54.3
16	57 54.7	15.0	190 36.8	13.9	11 53.6	7.4	54.4
17	72 54.9	14.0	205 09.7	13.8	12 01.0	7.3	54.4
18	87 55.1	S 1 13.0	219 42.5	13.8	S12 08.3	7.3	54.4
19	102 55.2	12.0	234 15.3	13.8	12 15.6	7.2	54.4
20	117 55.4	11.1	248 48.1	13.7	12 22.8	7.2	54.4
21	132 55.6	.. 10.1	263 20.8	13.7	12 30.0	7.1	54.4
22	147 55.8	09.1	277 53.5	13.7	12 37.1	7.1	54.4
23	162 56.0	08.1	292 26.2	13.6	12 44.2	7.0	54.5
18 00	177 56.1	S 1 07.1	306 58.8	13.6	S12 51.2	7.0	54.5
01	192 56.3	06.1	321 31.4	13.5	12 58.2	6.9	54.5
02	207 56.5	05.1	336 03.9	13.5	13 05.1	6.8	54.5
03	222 56.7	.. 04.1	350 36.4	13.4	13 11.9	6.8	54.5
04	237 56.9	03.2	5 08.8	13.5	13 18.7	6.7	54.5
05	252 57.0	02.2	19 41.3	13.3	13 25.5	6.7	54.5
06	267 57.2	S 1 01.2	34 13.6	13.3	S13 32.2	6.6	54.6
07	282 57.4	1 00.2	48 45.9	13.3	13 38.8	6.6	54.6
08	297 57.6	0 59.2	63 18.2	13.3	13 45.4	6.5	54.6
09	312 57.8	.. 58.2	77 50.5	13.1	13 51.9	6.5	54.6
10	327 58.0	57.2	92 22.6	13.2	13 58.3	6.4	54.6
11	342 58.1	56.2	106 54.8	13.1	14 04.7	6.4	54.6
12	357 58.3	S 0 55.2	121 26.9	13.0	S14 11.1	6.2	54.7
13	12 58.5	54.3	135 58.9	13.1	14 17.3	6.3	54.7
14	27 58.7	53.3	150 31.0	12.9	14 23.6	6.1	54.7
15	42 58.9	.. 52.3	165 02.9	12.9	14 29.7	6.1	54.7
16	57 59.1	51.3	179 34.8	12.9	14 35.8	6.1	54.7
17	72 59.2	50.3	194 06.7	12.8	14 41.9	5.9	54.7
18	87 59.4	S 0 49.3	208 38.5	12.8	S14 47.8	5.9	54.8
19	102 59.6	48.3	223 10.3	12.7	14 53.7	5.9	54.8
20	117 59.8	47.3	237 42.0	12.7	14 59.6	5.7	54.8
21	133 00.0	.. 46.3	252 13.7	12.7	15 05.3	5.8	54.8
22	148 00.1	45.4	266 45.4	12.6	15 11.1	5.6	54.8
23	163 00.3	44.4	281 17.0	12.5	15 16.7	5.6	54.9
19 00	178 00.5	S 0 43.4	295 48.5	12.5	S15 22.3	5.5	54.9
01	193 00.7	42.4	310 20.0	12.4	15 27.8	5.4	54.9
02	208 00.9	41.4	324 51.4	12.4	15 33.2	5.4	54.9
03	223 01.1	.. 40.4	339 22.8	12.4	15 38.6	5.3	54.9
04	238 01.2	39.4	353 54.2	12.3	15 43.9	5.2	55.0
05	253 01.4	38.4	8 25.5	12.2	15 49.1	5.2	55.0
06	268 01.6	S 0 37.5	22 56.7	12.2	S15 54.3	5.1	55.0
07	283 01.8	36.5	37 27.9	12.2	15 59.4	5.0	55.0
08	298 02.0	35.5	51 59.1	12.1	16 04.4	5.0	55.1
09	313 02.2	.. 34.5	66 30.2	12.0	16 09.4	4.8	55.1
10	328 02.3	33.5	81 01.2	12.0	16 14.2	4.8	55.1
11	343 02.5	32.5	95 32.2	12.0	16 19.0	4.8	55.1
12	358 02.7	S 0 31.5	110 03.2	11.9	S16 23.8	4.6	55.1
13	13 02.9	30.5	124 34.1	11.9	16 28.4	4.6	55.2
14	28 03.1	29.5	139 05.0	11.8	16 33.0	4.5	55.2
15	43 03.3	.. 28.6	153 35.8	11.7	16 37.5	4.4	55.2
16	58 03.4	27.6	168 06.5	11.7	16 41.9	4.4	55.2
17	73 03.6	26.6	182 37.2	11.7	16 46.3	4.2	55.3
18	88 03.8	S 0 25.6	197 07.9	11.6	S16 50.5	4.2	55.3
19	103 04.0	24.6	211 38.5	11.5	16 54.7	4.2	55.3
20	118 04.2	23.6	226 09.0	11.5	16 58.9	4.0	55.3
21	133 04.4	.. 22.6	240 39.5	11.4	17 02.9	3.9	55.4
22	148 04.6	21.6	255 09.9	11.4	17 06.8	3.9	55.4
23	163 04.7	20.7	269 40.3	11.4	S17 10.7	3.8	55.5
	SD 16.1	d 1.0	SD 14.8		14.9		15.0

Lat.	Twilight Naut.	Twilight Civil	Sunrise	Moonrise 17	18	19	20
°	h m	h m	h m	h m	h m	h m	h m
N 72	03 36	05 03	06 10	24 35	00 35	02 20	04 23
N 70	03 53	05 09	06 10	24 09	00 09	01 41	03 11
68	04 06	05 14	06 09	23 50	25 14	01 14	02 34
66	04 16	05 18	06 09	23 35	24 53	00 53	02 08
64	04 25	05 22	06 09	23 23	24 37	00 37	01 48
62	04 32	05 25	06 09	23 12	24 23	00 23	01 31
60	04 38	05 27	06 09	23 03	24 12	00 12	01 18
N 58	04 43	05 29	06 08	22 56	24 02	00 02	01 06
56	04 47	05 31	06 08	22 49	23 54	24 56	00 56
54	04 51	05 33	06 08	22 43	23 46	24 47	00 47
52	04 55	05 34	06 08	22 37	23 39	24 39	00 39
50	04 58	05 36	06 08	22 32	23 33	24 32	00 32
45	05 04	05 38	06 08	22 22	23 20	24 17	00 17
N 40	05 09	05 40	06 07	22 13	23 09	24 04	00 04
35	05 12	05 42	06 07	22 05	22 59	23 54	24 48
30	05 15	05 43	06 07	21 59	22 51	23 45	24 38
20	05 19	05 44	06 06	21 47	22 37	23 29	24 21
N 10	05 20	05 45	06 06	21 37	22 25	23 15	24 06
0	05 20	05 44	06 05	21 28	22 14	23 02	23 52
S 10	05 19	05 43	06 04	21 19	22 02	22 49	23 38
20	05 16	05 41	06 03	21 09	21 50	22 35	23 23
30	05 10	05 38	06 02	20 58	21 37	22 19	23 06
35	05 06	05 36	06 01	20 51	21 29	22 10	22 57
40	05 02	05 33	06 00	20 44	21 20	22 00	22 45
45	04 55	05 30	05 59	20 35	21 09	21 48	22 32
S 50	04 47	05 26	05 58	20 25	20 57	21 33	22 16
52	04 44	05 24	05 58	20 21	20 51	21 26	22 09
54	04 39	05 21	05 57	20 15	20 44	21 19	22 00
56	04 34	05 19	05 56	20 10	20 37	21 10	21 51
58	04 28	05 16	05 55	20 03	20 29	21 01	21 41
S 60	04 22	05 13	05 54	19 56	20 20	20 50	21 28

Lat.	Sunset	Twilight Civil	Twilight Naut.	Moonset 17	18	19	20
°	h m	h m	h m	h m	h m	h m	h m
N 72	18 09	19 17	20 45	06 37	06 30	06 20	05 59
N 70	18 09	19 10	20 27	06 54	06 56	07 00	07 11
68	18 09	19 05	20 14	07 08	07 16	07 28	07 48
66	18 09	19 00	20 03	07 20	07 32	07 49	08 15
64	18 09	18 56	19 54	07 29	07 45	08 06	08 35
62	18 09	18 53	19 47	07 37	07 56	08 20	08 52
60	18 09	18 51	19 40	07 44	08 06	08 32	09 06
N 58	18 09	18 48	19 35	07 51	08 14	08 42	09 17
56	18 09	18 46	19 30	07 56	08 21	08 51	09 28
54	18 09	18 44	19 26	08 01	08 28	08 59	09 37
52	18 09	18 43	19 23	08 06	08 34	09 06	09 45
50	18 09	18 42	19 20	08 10	08 39	09 13	09 52
45	18 09	18 39	19 13	08 19	08 51	09 27	10 08
N 40	18 10	18 37	19 08	08 26	09 00	09 38	10 21
35	18 10	18 35	19 05	08 33	09 09	09 48	10 32
30	18 10	18 34	19 02	08 38	09 16	09 57	10 41
20	18 10	18 32	18 58	08 48	09 28	10 11	10 58
N 10	18 11	18 32	18 56	08 57	09 39	10 24	11 12
0	18 11	18 32	18 56	09 05	09 50	10 37	11 26
S 10	18 12	18 33	18 57	09 13	10 00	10 49	11 39
20	18 13	18 35	19 00	09 22	10 11	11 02	11 54
30	18 14	18 38	19 06	09 31	10 24	11 17	12 10
35	18 14	18 40	19 09	09 37	10 31	11 26	12 20
40	18 15	18 42	19 14	09 44	10 40	11 35	12 31
45	18 16	18 45	19 20	09 51	10 49	11 47	12 44
S 50	18 17	18 49	19 28	10 00	11 01	12 01	12 59
52	18 18	18 51	19 31	10 05	11 07	12 08	13 07
54	18 18	18 54	19 36	10 09	11 13	12 15	13 15
56	18 19	18 56	19 40	10 15	11 20	12 23	13 24
58	18 20	18 59	19 46	10 20	11 27	12 33	13 34
S 60	18 20	19 02	19 52	10 27	11 36	12 43	13 46

Day	SUN Eqn. of Time 00h	12h	SUN Mer. Pass.	MOON Mer. Pass. Upper	Lower	Age	Phase
d	m s	m s	h m	h m	h m	d	%
17	08 33	08 24	12 08	02 54	15 16	19	84
18	08 16	08 07	12 08	03 39	16 02	20	77
19	07 58	07 50	12 08	04 25	16 49	21	68

UT	ARIES	VENUS −4.4		MARS +1.3		JUPITER −2.0		SATURN +0.6		STARS		
d h	GHA	GHA	Dec	GHA	Dec	GHA	Dec	GHA	Dec	Name	SHA	Dec
	° ′	° ′	° ′	° ′	° ′	° ′	° ′	° ′	° ′		° ′	° ′
20 00	177 19.4	222 38.6	S14 31.0	166 15.6	N 4 10.6	195 03.2	S 8 30.5	157 43.9	N 5 49.9	Acamar	315 27.5	S40 19.1
01	192 21.9	237 38.7	30.6	181 16.3	11.3	210 05.1	30.3	172 46.1	50.0	Achernar	335 36.1	S57 15.0
02	207 24.4	252 38.8	30.2	196 17.0	12.1	225 07.0	30.1	187 48.3	50.1	Acrux	173 21.6	S63 05.3
03	222 26.8	267 38.9	.. 29.9	211 17.7	.. 12.9	240 08.9	.. 29.9	202 50.4	.. 50.2	Adhara	255 21.6	S28 58.6
04	237 29.3	282 38.9	29.5	226 18.4	13.7	255 10.9	29.7	217 52.6	50.3	Aldebaran	291 02.9	N16 30.2
05	252 31.8	297 39.0	29.1	241 19.1	14.4	270 12.8	29.4	232 54.8	50.5			
06	267 34.2	312 39.1	S14 28.7	256 19.8	N 4 15.2	285 14.7	S 8 29.2	247 57.0	N 5 50.6	Alioth	166 30.5	N55 58.1
07	282 36.7	327 39.2	28.4	271 20.5	16.0	300 16.6	29.0	262 59.2	50.7	Alkaid	153 07.7	N49 19.2
F 08	297 39.2	342 39.2	28.0	286 21.2	16.7	315 18.5	28.8	278 01.4	50.8	Al Na'ir	27 58.8	S46 58.1
R 09	312 41.6	357 39.3	.. 27.6	301 21.9	.. 17.5	330 20.4	.. 28.6	293 03.5	.. 50.9	Alnilam	275 58.3	S 1 12.5
I 10	327 44.1	12 39.4	27.2	316 22.6	18.3	345 22.3	28.3	308 05.7	51.0	Alphard	218 07.3	S 8 39.3
11	342 46.6	27 39.5	26.9	331 23.3	19.0	0 24.3	28.1	323 07.9	51.2			
D 12	357 49.0	42 39.5	S14 26.5	346 24.1	N 4 19.8	15 26.2	S 8 27.9	338 10.1	N 5 51.3	Alphecca	126 20.7	N26 43.1
A 13	12 51.5	57 39.6	26.1	1 24.8	20.6	30 28.1	27.7	353 12.3	51.4	Alpheratz	357 55.9	N29 04.7
Y 14	27 53.9	72 39.7	25.7	16 25.5	21.4	45 30.0	27.5	8 14.5	51.5	Altair	62 19.8	N 8 51.7
15	42 56.4	87 39.7	.. 25.4	31 26.2	.. 22.1	60 31.9	.. 27.3	23 16.6	.. 51.6	Ankaa	353 27.7	S42 19.1
16	57 58.9	102 39.8	25.0	46 26.9	22.9	75 33.8	27.0	38 18.8	51.7	Antares	112 40.5	S26 25.5
17	73 01.3	117 39.9	24.6	61 27.6	23.7	90 35.8	26.8	53 21.0	51.9			
18	88 03.8	132 40.0	S14 24.2	76 28.3	N 4 24.4	105 37.7	S 8 26.6	68 23.2	N 5 52.0	Arcturus	146 06.1	N19 11.4
19	103 06.3	147 40.0	23.8	91 29.0	25.2	120 39.6	26.4	83 25.4	52.1	Atria	107 52.7	S69 01.1
20	118 08.7	162 40.1	23.4	106 29.7	26.0	135 41.5	26.2	98 27.6	52.2	Avior	234 22.4	S59 30.6
21	133 11.2	177 40.2	.. 23.1	121 30.4	.. 26.7	150 43.4	.. 25.9	113 29.7	.. 52.3	Bellatrix	278 44.6	N 6 20.6
22	148 13.7	192 40.2	22.7	136 31.1	27.5	165 45.3	25.7	128 31.9	52.4	Betelgeuse	271 14.0	N 7 24.2
23	163 16.1	207 40.3	22.3	151 31.8	28.3	180 47.2	25.5	143 34.1	52.6			
21 00	178 18.6	222 40.3	S14 21.9	166 32.5	N 4 29.0	195 49.2	S 8 25.3	158 36.3	N 5 52.7	Canopus	264 01.3	S52 42.1
01	193 21.1	237 40.4	21.5	181 33.3	29.8	210 51.1	25.1	173 38.5	52.8	Capella	280 51.8	N45 59.8
02	208 23.5	252 40.5	21.1	196 34.0	30.6	225 53.0	24.9	188 40.6	52.9	Deneb	49 39.7	N45 16.2
03	223 26.0	267 40.5	.. 20.7	211 34.7	.. 31.4	240 54.9	.. 24.6	203 42.8	.. 53.0	Denebola	182 45.3	N14 34.8
04	238 28.4	282 40.6	20.3	226 35.4	32.1	255 56.8	24.4	218 45.0	53.2	Diphda	349 08.0	S18 00.0
05	253 30.9	297 40.7	19.9	241 36.1	32.9	270 58.7	24.2	233 47.2	53.3			
06	268 33.4	312 40.7	S14 19.6	256 36.8	N 4 33.7	286 00.6	S 8 24.0	248 49.4	N 5 53.4	Dubhe	194 05.4	N61 45.6
07	283 35.8	327 40.8	19.2	271 37.5	34.4	301 02.6	23.8	263 51.6	53.5	Elnath	278 27.5	N28 36.3
S 08	298 38.3	342 40.8	18.8	286 38.2	35.2	316 04.5	23.5	278 53.7	53.6	Eltanin	90 51.6	N51 29.2
A 09	313 40.8	357 40.9	.. 18.4	301 38.9	.. 36.0	331 06.4	.. 23.3	293 55.9	.. 53.7	Enif	33 58.9	N 9 51.9
T 10	328 43.2	12 40.9	18.0	316 39.6	36.7	346 08.3	23.1	308 58.1	53.9	Fomalhaut	15 37.2	S29 38.0
U 11	343 45.7	27 41.0	17.6	331 40.3	37.5	1 10.2	22.9	324 00.3	54.0			
R 12	358 48.2	42 41.1	S14 17.2	346 41.0	N 4 38.3	16 12.1	S 8 22.7	339 02.5	N 5 54.1	Gacrux	172 13.3	S57 06.2
D 13	13 50.6	57 41.1	16.8	1 41.7	39.0	31 14.1	22.5	354 04.7	54.2	Gienah	176 04.0	S17 32.0
A 14	28 53.1	72 41.2	16.4	16 42.5	39.8	46 16.0	22.2	9 06.8	54.3	Hadar	149 03.9	S60 21.7
Y 15	43 55.6	87 41.2	.. 16.0	31 43.2	.. 40.6	61 17.9	.. 22.0	24 09.0	.. 54.4	Hamal	328 14.3	N23 27.1
16	58 58.0	102 41.3	15.6	46 43.9	41.3	76 19.8	21.8	39 11.2	54.6	Kaus Aust.	83 59.4	S34 22.9
17	74 00.5	117 41.3	15.2	61 44.6	42.1	91 21.7	21.6	54 13.4	54.7			
18	89 02.9	132 41.4	S14 14.8	76 45.3	N 4 42.9	106 23.6	S 8 21.4	69 15.6	N 5 54.8	Kochab	137 18.9	N74 09.6
19	104 05.4	147 41.4	14.4	91 46.0	43.6	121 25.6	21.2	84 17.7	54.9	Markab	13 50.3	N15 11.6
20	119 07.9	162 41.5	14.0	106 46.7	44.4	136 27.5	20.9	99 19.9	55.0	Menkar	314 27.5	N 4 04.7
21	134 10.3	177 41.5	.. 13.6	121 47.4	.. 45.2	151 29.4	.. 20.7	114 22.1	.. 55.1	Menkent	148 21.0	S36 21.6
22	149 12.8	192 41.6	13.2	136 48.1	45.9	166 31.3	20.5	129 24.3	55.3	Miaplacidus	221 41.4	S69 42.9
23	164 15.3	207 41.6	12.8	151 48.8	46.7	181 33.2	20.3	144 26.5	55.4			
22 00	179 17.7	222 41.7	S14 12.3	166 49.5	N 4 47.5	196 35.1	S 8 20.1	159 28.6	N 5 55.5	Mirfak	308 57.4	N49 51.3
01	194 20.2	237 41.7	11.9	181 50.2	48.2	211 37.1	19.8	174 30.8	55.6	Nunki	76 12.9	S26 17.8
02	209 22.7	252 41.8	11.5	196 50.9	49.0	226 39.0	19.6	189 33.0	55.7	Peacock	53 38.0	S56 44.2
03	224 25.1	267 41.8	.. 11.1	211 51.7	.. 49.8	241 40.9	.. 19.4	204 35.2	.. 55.8	Pollux	243 41.9	N28 01.7
04	239 27.6	282 41.9	10.7	226 52.4	50.5	256 42.8	19.2	219 37.4	56.0	Procyon	245 11.8	N 5 13.6
05	254 30.0	297 41.9	10.3	241 53.1	51.3	271 44.7	19.0	234 39.6	56.1			
06	269 32.5	312 41.9	S14 09.9	256 53.8	N 4 52.1	286 46.6	S 8 18.8	249 41.7	N 5 56.2	Rasalhague	96 17.3	N12 33.6
07	284 35.0	327 42.0	09.5	271 54.5	52.8	301 48.6	18.5	264 43.9	56.3	Regulus	207 55.7	N11 58.4
S 08	299 37.4	342 42.0	09.1	286 55.2	53.6	316 50.5	18.3	279 46.1	56.4	Rigel	281 23.4	S 8 12.5
U 09	314 39.9	357 42.1	.. 08.6	301 55.9	.. 54.4	331 52.4	.. 18.1	294 48.3	.. 56.5	Rigil Kent.	140 07.2	S60 49.4
N 10	329 42.4	12 42.1	08.2	316 56.6	55.1	346 54.3	17.9	309 50.5	56.7	Sabik	102 25.9	S15 43.3
11	344 44.8	27 42.2	07.8	331 57.3	55.9	1 56.2	17.7	324 52.6	56.8			
D 12	359 47.3	42 42.2	S14 07.4	346 58.0	N 4 56.7	16 58.1	S 8 17.5	339 54.8	N 5 56.9	Schedar	349 54.4	N56 31.6
A 13	14 49.8	57 42.2	07.0	1 58.7	57.4	32 00.1	17.2	354 57.0	57.0	Shaula	96 37.8	S37 05.9
Y 14	29 52.2	72 42.3	06.6	16 59.4	58.2	47 02.0	17.0	9 59.2	57.1	Sirius	258 44.0	S16 43.1
15	44 54.7	87 42.3	.. 06.1	32 00.1	.. 58.9	62 03.9	.. 16.8	25 01.4	.. 57.2	Spica	158 43.3	S11 09.1
16	59 57.2	102 42.3	05.7	47 00.9	4 59.7	77 05.8	16.6	40 03.5	57.4	Suhail	223 00.7	S43 25.8
17	74 59.6	117 42.4	05.3	62 01.6	5 00.5	92 07.7	16.4	55 05.7	57.5			
18	90 02.1	132 42.4	S14 04.9	77 02.3	N 5 01.2	107 09.7	S 8 16.2	70 07.9	N 5 57.6	Vega	80 46.9	N38 46.7
19	105 04.5	147 42.5	04.5	92 03.0	02.0	122 11.6	15.9	85 10.1	57.7	Zuben'ubi	137 18.1	S16 02.0
20	120 07.0	162 42.5	04.0	107 03.7	02.8	137 13.5	15.7	100 12.3	57.8		SHA	Mer. Pass.
21	135 09.5	177 42.5	.. 03.6	122 04.4	.. 03.5	152 15.4	.. 15.5	115 14.4	.. 57.9		° ′	h m
22	150 11.9	192 42.6	03.2	137 05.1	04.3	167 17.3	15.3	130 16.6	58.1	Venus	44 21.8	9 09
23	165 14.4	207 42.6	02.8	152 05.8	05.1	182 19.2	15.1	145 18.8	58.2	Mars	348 14.0	12 53
	h m									Jupiter	17 30.6	10 55
Mer. Pass. 12 04.8		v 0.1	d 0.4	v 0.7	d 0.8	v 1.9	d 0.2	v 2.2	d 0.1	Saturn	340 17.7	13 24

UT	SUN GHA	SUN Dec	MOON GHA	v	MOON Dec	d	HP
d h	° ′	° ′	° ′	′	° ′	′	′
20 00	178 04.9	S 0 19.7	284 10.7	11.3	S17 14.5	3.7	55.4
01	193 05.1	18.7	298 41.0	11.2	17 18.2	3.7	55.4
02	208 05.3	17.7	313 11.2	11.2	17 21.9	3.5	55.5
03	223 05.5	.. 16.7	327 41.4	11.2	17 25.4	3.5	55.5
04	238 05.7	15.7	342 11.6	11.1	17 28.9	3.3	55.6
05	253 05.8	14.7	356 41.7	11.0	17 32.2	3.3	55.6
06	268 06.0	S 0 13.7	11 11.7	11.0	S17 35.5	3.3	55.6
07	283 06.2	12.7	25 41.7	10.9	17 38.8	3.1	55.6
08	298 06.4	11.8	40 11.6	10.9	17 41.9	3.0	55.7
F 09	313 06.6	.. 10.8	54 41.5	10.9	17 44.9	3.0	55.7
R 10	328 06.8	09.8	69 11.4	10.7	17 47.9	2.8	55.7
I 11	343 07.0	08.8	83 41.1	10.8	17 50.7	2.8	55.7
D 12	358 07.1	S 0 07.8	98 10.9	10.7	S17 53.5	2.7	55.8
A 13	13 07.3	06.8	112 40.6	10.6	17 56.2	2.6	55.8
Y 14	28 07.5	05.8	127 10.2	10.6	17 58.8	2.5	55.8
15	43 07.7	.. 04.8	141 39.8	10.5	18 01.3	2.4	55.9
16	58 07.9	03.9	156 09.3	10.5	18 03.7	2.4	55.9
17	73 08.1	02.9	170 38.8	10.5	18 06.1	2.2	55.9
18	88 08.2	S 0 01.9	185 08.3	10.3	S18 08.3	2.1	56.0
19	103 08.4 S	00.9	199 37.6	10.4	18 10.4	2.1	56.0
20	118 08.6 N	00.1	214 07.0	10.3	18 12.5	2.0	56.0
21	133 08.8	.. 01.1	228 36.3	10.2	18 14.5	1.8	56.0
22	148 09.0	02.1	243 05.5	10.2	18 16.3	1.8	56.1
23	163 09.2	03.1	257 34.7	10.2	18 18.1	1.7	56.1
21 00	178 09.4	N 0 04.0	272 03.9	10.0	S18 19.8	1.6	56.1
01	193 09.5	05.0	286 32.9	10.1	18 21.4	1.5	56.2
02	208 09.7	06.0	301 02.0	10.0	18 22.9	1.4	56.2
03	223 09.9	.. 07.0	315 31.0	10.0	18 24.3	1.3	56.2
04	238 10.1	08.0	330 00.0	9.9	18 25.6	1.2	56.3
05	253 10.3	09.0	344 28.9	9.8	18 26.8	1.1	56.3
06	268 10.5	N 0 10.0	358 57.7	9.8	S18 27.9	1.0	56.3
S 07	283 10.7	11.0	13 26.5	9.8	18 28.9	0.9	56.4
A 08	298 10.8	11.9	27 55.3	9.7	18 29.8	0.8	56.4
T 09	313 11.0	.. 12.9	42 24.0	9.7	18 30.6	0.7	56.4
U 10	328 11.2	13.9	56 52.7	9.6	18 31.3	0.6	56.5
R 11	343 11.4	14.9	71 21.3	9.6	18 31.9	0.5	56.5
D 12	358 11.6	N 0 15.9	85 49.9	9.6	S18 32.4	0.4	56.5
A 13	13 11.8	16.9	100 18.5	9.4	18 32.8	0.3	56.6
Y 14	28 12.0	17.9	114 46.9	9.5	18 33.1	0.3	56.6
15	43 12.1	.. 18.9	129 15.4	9.4	18 33.4	0.1	56.6
16	58 12.3	19.8	143 43.8	9.4	18 33.5	0.0	56.7
17	73 12.5	20.8	158 12.2	9.3	18 33.5	0.1	56.7
18	88 12.7	N 0 21.8	172 40.5	9.3	S18 33.4	0.2	56.8
19	103 12.9	22.8	187 08.8	9.2	18 33.2	0.3	56.8
20	118 13.1	23.8	201 37.0	9.2	18 32.9	0.4	56.8
21	133 13.3	.. 24.8	216 05.2	9.2	18 32.5	0.5	56.9
22	148 13.5	25.8	230 33.4	9.1	18 32.0	0.6	56.9
23	163 13.6	26.8	245 01.5	9.1	18 31.4	0.8	56.9
22 00	178 13.8	N 0 27.7	259 29.6	9.0	S18 30.6	0.8	57.0
01	193 14.0	28.7	273 57.6	9.0	18 29.8	0.9	57.0
02	208 14.2	29.7	288 25.6	8.9	18 28.9	1.0	57.0
03	223 14.4	.. 30.7	302 53.5	9.0	18 27.9	1.2	57.1
04	238 14.6	31.7	317 21.5	8.8	18 26.7	1.2	57.1
05	253 14.8	32.7	331 49.3	8.9	18 25.5	1.4	57.2
06	268 14.9	N 0 33.7	346 17.2	8.8	S18 24.1	1.4	57.2
07	283 15.1	34.7	0 45.0	8.7	18 22.7	1.6	57.2
08	298 15.3	35.6	15 12.7	8.8	18 21.1	1.6	57.3
S 09	313 15.5	.. 36.6	29 40.5	8.7	18 19.5	1.8	57.3
U 10	328 15.7	37.6	44 08.2	8.6	18 17.7	1.9	57.3
N 11	343 15.9	38.6	58 35.8	8.7	18 15.8	2.0	57.4
D 12	358 16.1	N 0 39.6	73 03.5	8.5	S18 13.8	2.1	57.4
A 13	13 16.3	40.6	87 31.0	8.6	18 11.7	2.2	57.5
Y 14	28 16.4	41.6	101 58.6	8.5	18 09.5	2.3	57.5
15	43 16.6	.. 42.5	116 26.1	8.5	18 07.2	2.4	57.5
16	58 16.8	43.5	130 53.6	8.5	18 04.8	2.6	57.6
17	73 17.0	44.5	145 21.1	8.4	18 02.2	2.6	57.6
18	88 17.2	N 0 45.5	159 48.5	8.4	S17 59.6	2.8	57.7
19	103 17.4	46.5	174 15.9	8.4	17 56.8	2.8	57.7
20	118 17.6	47.5	188 43.3	8.3	17 54.0	3.0	57.7
21	133 17.7	.. 48.5	203 10.6	8.3	17 51.0	3.1	57.8
22	148 17.9	49.5	217 37.9	8.3	17 47.9	3.1	57.8
23	163 18.1	50.4	232 05.2	8.2	S17 44.8	3.3	57.9
	SD 16.1	d 1.0	SD 15.2		15.4		15.6

Lat.	Twilight Naut.	Civil	Sunrise	Moonrise 20	21	22	23
°	h m	h m	h m	h m	h m	h m	h m
N 72	03 16	04 46	05 54	04 23	▬	▬	06 48
N 70	03 36	04 54	05 56	03 11	04 30	05 22	05 46
68	03 51	05 01	05 57	02 34	03 44	04 37	05 11
66	04 03	05 06	05 58	02 08	03 14	04 07	04 45
64	04 13	05 11	05 58	01 48	02 51	03 44	04 25
62	04 21	05 15	05 59	01 31	02 33	03 26	04 09
60	04 28	05 18	05 59	01 18	02 18	03 11	03 56
N 58	04 34	05 21	06 00	01 06	02 06	02 59	03 44
56	04 39	05 23	06 00	00 56	01 55	02 48	03 34
54	04 44	05 25	06 01	00 47	01 45	02 38	03 25
52	04 48	05 27	06 01	00 39	01 37	02 30	03 17
50	04 51	05 29	06 01	00 32	01 29	02 22	03 10
45	04 58	05 33	06 02	00 17	01 12	02 05	02 55
N 40	05 04	05 35	06 02	00 04	00 59	01 52	02 42
35	05 08	05 38	06 03	24 48	00 48	01 40	02 31
30	05 11	05 39	06 03	24 38	00 38	01 30	02 22
20	05 16	05 42	06 04	24 21	00 21	01 13	02 05
N 10	05 18	05 43	06 04	24 06	00 06	00 58	01 51
0	05 19	05 43	06 04	23 52	24 44	00 44	01 38
S 10	05 19	05 43	06 04	23 38	24 30	00 30	01 25
20	05 16	05 42	06 04	23 23	24 15	00 15	01 10
30	05 12	05 40	06 04	23 06	23 58	24 54	00 54
35	05 09	05 38	06 04	22 57	23 48	24 45	00 45
40	05 05	05 36	06 04	22 45	23 37	24 34	00 34
45	05 00	05 34	06 03	22 32	23 23	24 21	00 21
S 50	04 53	05 31	06 03	22 16	23 07	24 06	00 06
52	04 49	05 29	06 03	22 09	22 59	23 58	25 06
54	04 45	05 27	06 03	22 00	22 51	23 50	24 59
56	04 41	05 25	06 02	21 51	22 41	23 41	24 51
58	04 36	05 23	06 02	21 41	22 30	23 31	24 42
S 60	04 30	05 20	06 02	21 28	22 18	23 19	24 32

Lat.	Sunset	Twilight Civil	Naut.	Moonset 20	21	22	23
°	h m	h m	h m	h m	h m	h m	h m
N 72	18 23	19 31	21 03	05 59	▬	▬	09 02
N 70	18 21	19 23	20 43	07 11	07 37	08 34	10 03
68	18 20	19 16	20 27	07 48	08 23	09 19	10 38
66	18 19	19 10	20 14	08 15	08 53	09 49	11 03
64	18 18	19 06	20 04	08 35	09 16	10 12	11 23
62	18 17	19 02	19 56	08 52	09 34	10 29	11 38
60	18 16	18 58	19 48	09 06	09 49	10 44	11 51
N 58	18 16	18 55	19 42	09 17	10 02	10 57	12 03
56	18 15	18 53	19 37	09 28	10 13	11 08	12 13
54	18 15	18 50	19 32	09 37	10 22	11 17	12 21
52	18 14	18 48	19 28	09 45	10 31	11 26	12 29
50	18 14	18 46	19 25	09 52	10 39	11 33	12 36
45	18 13	18 43	19 17	10 08	10 55	11 50	12 51
N 40	18 13	18 40	19 11	10 21	11 09	12 03	13 03
35	18 12	18 38	19 07	10 32	11 20	12 14	13 13
30	18 12	18 36	19 04	10 41	11 30	12 24	13 22
20	18 11	18 33	18 59	10 58	11 48	12 41	13 38
N 10	18 11	18 32	18 56	11 12	12 03	12 56	13 52
0	18 10	18 31	18 55	11 26	12 17	13 10	14 04
S 10	18 10	18 31	18 56	11 39	12 31	13 23	14 17
20	18 10	18 32	18 58	11 54	12 46	13 38	14 30
30	18 10	18 34	19 02	12 10	13 03	13 55	14 45
35	18 10	18 35	19 05	12 20	13 13	14 05	14 54
40	18 10	18 37	19 09	12 31	13 24	14 16	15 04
45	18 10	18 40	19 14	12 44	13 38	14 29	15 16
S 50	18 11	18 43	19 21	12 59	13 54	14 45	15 30
52	18 11	18 44	19 24	13 07	14 02	14 52	15 37
54	18 11	18 46	19 28	13 15	14 10	15 00	15 44
56	18 11	18 48	19 32	13 24	14 20	15 10	15 52
58	18 11	18 50	19 37	13 34	14 31	15 20	16 01
S 60	18 11	18 53	19 42	13 46	14 43	15 32	16 12

Day	SUN Eqn. of Time 00h	12h	Mer. Pass.	MOON Mer. Pass. Upper	Lower	Age	Phase
d	m s	m s	h m	h m	h m	d	%
20	07 41	07 32	12 08	05 14	17 39	22	58
21	07 23	07 14	12 07	06 04	18 30	23	48
22	07 05	06 56	12 07	06 57	19 24	24	38

1998 MARCH 23, 24, 25 (MON., TUES., WED.)

UT	ARIES GHA	VENUS −4.4 GHA	Dec	MARS +1.3 GHA	Dec	JUPITER −2.0 GHA	Dec	SATURN +0.6 GHA	Dec
23 00	180 16.9	222 42.6	S14 02.3	167 06.5	N 5 05.8	197 21.2	S 8 14.8	160 21.0	N 5 58.3
01	195 19.3	237 42.7	01.9	182 07.2	06.6	212 23.1	14.6	175 23.2	58.4
02	210 21.8	252 42.7	01.5	197 07.9	07.3	227 25.0	14.4	190 25.3	58.5
03	225 24.3	267 42.7 ..	01.0	212 08.6 ..	08.1	242 26.9 ..	14.2	205 27.5 ..	58.7
04	240 26.7	282 42.7	00.6	227 09.3	08.9	257 28.8	14.0	220 29.7	58.8
05	255 29.2	297 42.8	14 00.2	242 10.1	09.6	272 30.8	13.8	235 31.9	58.9
M 06	270 31.7	312 42.8	S13 59.8	257 10.8	N 5 10.4	287 32.7	S 8 13.5	250 34.1	N 5 59.0
O 07	285 34.1	327 42.8	59.3	272 11.5	11.2	302 34.6	13.3	265 36.2	59.1
N 08	300 36.6	342 42.9	58.9	287 12.2	11.9	317 36.5	13.1	280 38.4	59.2
D 09	315 39.0	357 42.9 ..	58.5	302 12.9 ..	12.7	332 38.4 ..	12.9	295 40.6 ..	59.4
A 10	330 41.5	12 42.9	58.0	317 13.6	13.5	347 40.3	12.7	310 42.8	59.5
Y 11	345 44.0	27 42.9	57.6	332 14.3	14.2	2 42.3	12.5	325 45.0	59.6
12	0 46.4	42 43.0	S13 57.1	347 15.0	N 5 15.0	17 44.2	S 8 12.2	340 47.1	N 5 59.7
13	15 48.9	57 43.0	56.7	2 15.7	15.7	32 46.1	12.0	355 49.3	59.8
14	30 51.4	72 43.0	56.3	17 16.4	16.5	47 48.0	11.8	10 51.5	5 59.9
15	45 53.8	87 43.0 ..	55.8	32 17.1 ..	17.3	62 49.9 ..	11.6	25 53.7	6 00.1
16	60 56.3	102 43.1	55.4	47 17.8	18.0	77 51.9	11.4	40 55.9	00.2
17	75 58.8	117 43.1	55.0	62 18.5	18.8	92 53.8	11.2	55 58.0	00.3
18	91 01.2	132 43.1	S13 54.5	77 19.2	N 5 19.5	107 55.7	S 8 10.9	71 00.2	N 6 00.4
19	106 03.7	147 43.1	54.1	92 20.0	20.3	122 57.6	10.7	86 02.4	00.5
20	121 06.1	162 43.1	53.6	107 20.7	21.1	137 59.5	10.5	101 04.6	00.6
21	136 08.6	177 43.2 ..	53.2	122 21.4 ..	21.8	153 01.5 ..	10.3	116 06.8 ..	00.8
22	151 11.1	192 43.2	52.7	137 22.1	22.6	168 03.4	10.1	131 08.9	00.9
23	166 13.5	207 43.2	52.3	152 22.8	23.4	183 05.3	09.9	146 11.1	01.0
24 00	181 16.0	222 43.2	S13 51.8	167 23.5	N 5 24.1	198 07.2	S 8 09.6	161 13.3	N 6 01.1
01	196 18.5	237 43.2	51.4	182 24.2	24.9	213 09.1	09.4	176 15.5	01.2
02	211 20.9	252 43.3	51.0	197 24.9	25.6	228 11.1	09.2	191 17.6	01.4
03	226 23.4	267 43.3 ..	50.5	212 25.6 ..	26.4	243 13.0 ..	09.0	206 19.8 ..	01.5
04	241 25.9	282 43.3	50.1	227 26.3	27.2	258 14.9	08.8	221 22.0	01.6
05	256 28.3	297 43.3	49.6	242 27.0	27.9	273 16.8	08.6	236 24.2	01.7
T 06	271 30.8	312 43.3	S13 49.2	257 27.7	N 5 28.7	288 18.7	S 8 08.3	251 26.4	N 6 01.8
U 07	286 33.3	327 43.3	48.7	272 28.4	29.4	303 20.7	08.1	266 28.5	01.9
E 08	301 35.7	342 43.3	48.2	287 29.1	30.2	318 22.6	07.9	281 30.7	02.1
S 09	316 38.2	357 43.4 ..	47.8	302 29.9 ..	31.0	333 24.5 ..	07.7	296 32.9 ..	02.2
D 10	331 40.6	12 43.4	47.3	317 30.6	31.7	348 26.4	07.5	311 35.1	02.3
A 11	346 43.1	27 43.4	46.9	332 31.3	32.5	3 28.3	07.3	326 37.3	02.4
Y 12	1 45.6	42 43.4	S13 46.4	347 32.0	N 5 33.2	18 30.3	S 8 07.0	341 39.4	N 6 02.5
13	16 48.0	57 43.4	46.0	2 32.7	34.0	33 32.2	06.8	356 41.6	02.6
14	31 50.5	72 43.4	45.5	17 33.4	34.8	48 34.1	06.6	11 43.8	02.8
15	46 53.0	87 43.4 ..	45.1	32 34.1 ..	35.5	63 36.0 ..	06.4	26 46.0 ..	02.9
16	61 55.4	102 43.4	44.6	47 34.8	36.3	78 37.9	06.2	41 48.2	03.0
17	76 57.9	117 43.4	44.1	62 35.5	37.0	93 39.9	06.0	56 50.3	03.1
18	92 00.4	132 43.4	S13 43.7	77 36.2	N 5 37.8	108 41.8	S 8 05.7	71 52.5	N 6 03.2
19	107 02.8	147 43.5	43.2	92 36.9	38.5	123 43.7	05.5	86 54.7	03.3
20	122 05.3	162 43.5	42.8	107 37.6	39.3	138 45.6	05.3	101 56.9	03.5
21	137 07.8	177 43.5 ..	42.3	122 38.3 ..	40.1	153 47.5 ..	05.1	116 59.0 ..	03.6
22	152 10.2	192 43.5	41.8	137 39.1	40.8	168 49.5	04.9	132 01.2	03.7
23	167 12.7	207 43.5	41.4	152 39.8	41.6	183 51.4	04.7	147 03.4	03.8
25 00	182 15.1	222 43.5	S13 40.9	167 40.5	N 5 42.3	198 53.3	S 8 04.4	162 05.6	N 6 03.9
01	197 17.6	237 43.5	40.4	182 41.2	43.1	213 55.2	04.2	177 07.8	04.1
02	212 20.1	252 43.5	40.0	197 41.9	43.9	228 57.2	04.0	192 09.9	04.2
03	227 22.5	267 43.5 ..	39.5	212 42.6 ..	44.6	243 59.1 ..	03.8	207 12.1 ..	04.3
04	242 25.0	282 43.5	39.0	227 43.3	45.4	259 01.0	03.6	222 14.3	04.4
05	257 27.5	297 43.5	38.6	242 44.0	46.1	274 02.9	03.4	237 16.5	04.5
W 06	272 29.9	312 43.5	S13 38.1	257 44.7	N 5 46.9	289 04.8	S 8 03.1	252 18.6	N 6 04.6
E 07	287 32.4	327 43.5	37.6	272 45.4	47.6	304 06.8	02.9	267 20.8	04.8
D 08	302 34.9	342 43.5	37.1	287 46.1	48.4	319 08.7	02.7	282 23.0	04.9
N 09	317 37.3	357 43.5 ..	36.7	302 46.8 ..	49.2	334 10.6 ..	02.5	297 25.2 ..	05.0
E 10	332 39.8	12 43.5	36.2	317 47.5	49.9	349 12.5	02.3	312 27.4	05.1
S 11	347 42.2	27 43.5	35.7	332 48.2	50.7	4 14.5	02.1	327 29.5	05.2
D 12	2 44.7	42 43.5	S13 35.2	347 48.9	N 5 51.4	19 16.4	S 8 01.8	342 31.7	N 6 05.3
A 13	17 47.2	57 43.5	34.8	2 49.7	52.2	34 18.3	01.6	357 33.9	05.5
Y 14	32 49.6	72 43.5	34.3	17 50.4	52.9	49 20.2	01.4	12 36.1	05.6
15	47 52.1	87 43.5 ..	33.8	32 51.1 ..	53.7	64 22.1 ..	01.2	27 38.2 ..	05.7
16	62 54.6	102 43.5	33.3	47 51.8	54.4	79 24.1	01.0	42 40.4	05.8
17	77 57.0	117 43.5	32.9	62 52.5	55.2	94 26.0	00.8	57 42.6	05.9
18	92 59.5	132 43.5	S13 32.4	77 53.2	N 5 56.0	109 27.9	S 8 00.5	72 44.8	N 6 06.1
19	108 02.0	147 43.5	31.9	92 53.9	56.7	124 29.8	00.3	87 47.0	06.2
20	123 04.4	162 43.5	31.4	107 54.6	57.5	139 31.8	8 00.1	102 49.1	06.3
21	138 06.9	177 43.5 ..	30.9	122 55.3 ..	58.2	154 33.7	7 59.9	117 51.3 ..	06.4
22	153 09.4	192 43.5	30.5	137 56.0	59.0	169 35.6	59.7	132 53.5	06.5
23	168 11.8	207 43.4	30.0	152 56.7	59.7	184 37.5	59.5	147 55.7	06.6
Mer. Pass. 11 53.0		v 0.0	d 0.5	v 0.7	d 0.8	v 1.9	d 0.2	v 2.2	d 0.1

STARS

Name	SHA	Dec
Acamar	315 27.5	S40 19.1
Achernar	335 36.1	S57 15.0
Acrux	173 21.6	S63 05.4
Adhara	255 21.6	S28 58.6
Aldebaran	291 02.9	N16 30.2
Alioth	166 30.4	N55 58.1
Alkaid	153 07.7	N49 19.2
Al Na'ir	27 58.8	S46 58.1
Alnilam	275 58.3	S 1 12.5
Alphard	218 07.3	S 8 39.3
Alphecca	126 20.7	N26 43.1
Alpheratz	357 55.9	N29 04.7
Altair	62 19.8	N 8 51.7
Ankaa	353 27.7	S42 19.1
Antares	112 40.5	S26 25.5
Arcturus	146 06.1	N19 11.4
Atria	107 52.6	S69 01.1
Avior	234 22.4	S59 30.6
Bellatrix	278 44.6	N 6 20.6
Betelgeuse	271 14.0	N 7 24.2
Canopus	264 01.3	S52 42.1
Capella	280 51.8	N45 59.8
Deneb	49 39.7	N45 16.2
Denebola	182 45.3	N14 34.8
Diphda	349 08.0	S18 00.0
Dubhe	194 05.4	N61 45.7
Elnath	278 27.5	N28 36.3
Eltanin	90 51.5	N51 29.2
Enif	33 58.8	N 9 51.9
Fomalhaut	15 37.2	S29 38.0
Gacrux	172 13.3	S57 06.2
Gienah	176 04.0	S17 32.0
Hadar	149 03.9	S60 21.7
Hamal	328 14.3	N23 27.1
Kaus Aust.	83 59.3	S34 22.9
Kochab	137 18.8	N74 09.7
Markab	13 50.3	N15 11.6
Menkar	314 27.5	N 4 04.7
Menkent	148 21.0	S36 21.6
Miaplacidus	221 41.4	S69 42.9
Mirfak	308 57.4	N49 51.3
Nunki	76 12.9	S26 17.8
Peacock	53 38.0	S56 44.2
Pollux	243 41.9	N28 01.7
Procyon	245 11.9	N 5 13.6
Rasalhague	96 17.3	N12 33.6
Regulus	207 55.7	N11 58.4
Rigel	281 23.4	S 8 12.5
Rigil Kent.	140 07.2	S60 49.4
Sabik	102 25.9	S15 43.3
Schedar	349 54.4	N56 31.6
Shaula	96 37.7	S37 05.9
Sirius	258 44.0	S16 43.1
Spica	158 43.3	S11 09.2
Suhail	223 00.7	S43 25.8
Vega	80 46.9	N38 46.7
Zuben'ubi	137 18.1	S16 02.0

	SHA	Mer. Pass.
	° ′	h m
Venus	41 27.2	9 09
Mars	346 07.5	12 50
Jupiter	16 51.2	10 46
Saturn	339 57.3	13 13

UT	SUN GHA	SUN Dec	MOON GHA	v	MOON Dec	d	HP
23							
00	178 18.3	N 0 51.4	246 32.4	8.2	S17 41.5	3.4	57.9
01	193 18.5	52.4	260 59.6	8.2	17 38.1	3.6	57.9
02	208 18.7	53.4	275 26.8	8.2	17 34.5	3.6	58.0
03	223 18.9	.. 54.4	289 54.0	8.1	17 30.9	3.7	58.0
04	238 19.1	55.4	304 21.1	8.1	17 27.2	3.9	58.1
05	253 19.2	56.4	318 48.2	8.1	17 23.3	3.9	58.1
06	268 19.4	N 0 57.3	333 15.3	8.1	S17 19.4	4.1	58.1
07	283 19.6	58.3	347 42.4	8.0	17 15.3	4.1	58.2
M 08	298 19.8	0 59.3	2 09.4	8.0	17 11.2	4.3	58.2
O 09	313 20.0	1 00.3	16 36.4	8.0	17 06.9	4.4	58.3
N 10	328 20.2	01.3	31 03.4	8.0	17 02.5	4.5	58.3
D 11	343 20.4	02.3	45 30.4	7.9	16 58.0	4.6	58.3
A 12	358 20.6	N 1 03.3	59 57.3	8.0	S16 53.4	4.7	58.4
Y 13	13 20.7	04.2	74 24.3	7.9	16 48.7	4.9	58.4
14	28 20.9	05.2	88 51.2	7.8	16 43.8	4.9	58.5
15	43 21.1	.. 06.2	103 18.0	7.9	16 38.9	5.0	58.5
16	58 21.3	07.2	117 44.9	7.8	16 33.9	5.2	58.5
17	73 21.5	08.2	132 11.7	7.9	16 28.7	5.2	58.6
18	88 21.7	N 1 09.2	146 38.6	7.8	S16 23.5	5.4	58.6
19	103 21.9	10.2	161 05.4	7.7	16 18.1	5.5	58.7
20	118 22.1	11.1	175 32.1	7.8	16 12.6	5.6	58.7
21	133 22.2	.. 12.1	189 58.9	7.8	16 07.0	5.7	58.7
22	148 22.4	13.1	204 25.7	7.7	16 01.3	5.7	58.8
23	163 22.6	14.1	218 52.4	7.7	15 55.6	5.9	58.8
24							
00	178 22.8	N 1 15.1	233 19.1	7.7	S15 49.7	6.1	58.9
01	193 23.0	16.1	247 45.8	7.7	15 43.6	6.1	58.9
02	208 23.2	17.1	262 12.5	7.7	15 37.5	6.2	58.9
03	223 23.4	.. 18.0	276 39.2	7.6	15 31.3	6.3	59.0
04	238 23.6	19.0	291 05.8	7.7	15 25.0	6.4	59.0
05	253 23.7	20.0	305 32.5	7.6	15 18.6	6.6	59.1
06	268 23.9	N 1 21.0	319 59.1	7.6	S15 12.0	6.6	59.1
07	283 24.1	22.0	334 25.7	7.6	15 05.4	6.7	59.1
T 08	298 24.3	23.0	348 52.3	7.6	14 58.7	6.9	59.2
U 09	313 24.5	.. 23.9	3 18.9	7.6	14 51.8	6.9	59.2
E 10	328 24.7	24.9	17 45.5	7.5	14 44.9	7.1	59.2
S 11	343 24.9	25.9	32 12.0	7.6	14 37.8	7.1	59.3
D 12	358 25.1	N 1 26.9	46 38.6	7.5	S14 30.7	7.3	59.3
A 13	13 25.2	27.9	61 05.1	7.6	14 23.4	7.3	59.4
Y 14	28 25.4	28.9	75 31.7	7.5	14 16.1	7.5	59.4
15	43 25.6	.. 29.9	89 58.2	7.5	14 08.6	7.5	59.4
16	58 25.8	30.8	104 24.7	7.5	14 01.1	7.7	59.5
17	73 26.0	31.8	118 51.2	7.5	13 53.4	7.7	59.5
18	88 26.2	N 1 32.8	133 17.7	7.5	S13 45.7	7.9	59.6
19	103 26.4	33.8	147 44.2	7.5	13 37.8	7.9	59.6
20	118 26.6	34.8	162 10.7	7.5	13 29.9	8.1	59.6
21	133 26.8	.. 35.8	176 37.2	7.5	13 21.8	8.1	59.7
22	148 26.9	36.7	191 03.7	7.4	13 13.7	8.2	59.7
23	163 27.1	37.7	205 30.1	7.5	13 05.5	8.3	59.7
25							
00	178 27.3	N 1 38.7	219 56.6	7.4	S12 57.2	8.5	59.8
01	193 27.5	39.7	234 23.0	7.5	12 48.7	8.5	59.8
02	208 27.7	40.7	248 49.5	7.4	12 40.2	8.6	59.8
03	223 27.9	.. 41.7	263 15.9	7.5	12 31.6	8.7	59.9
04	238 28.1	42.6	277 42.4	7.4	12 22.9	8.7	59.9
05	253 28.3	43.6	292 08.8	7.4	12 14.2	8.9	60.0
06	268 28.4	N 1 44.6	306 35.2	7.4	S12 05.3	9.0	60.0
W 07	283 28.6	45.6	321 01.6	7.4	11 56.3	9.0	60.0
E 08	298 28.8	46.6	335 28.0	7.5	11 47.3	9.1	60.1
D 09	313 29.0	.. 47.5	349 54.5	7.4	11 38.2	9.3	60.1
N 10	328 29.2	48.5	4 20.9	7.4	11 28.9	9.3	60.1
E 11	343 29.4	49.5	18 47.3	7.4	11 19.6	9.4	60.2
S 12	358 29.6	N 1 50.5	33 13.7	7.4	S11 10.2	9.4	60.2
D 13	13 29.8	51.5	47 40.1	7.4	11 00.8	9.6	60.2
A 14	28 30.0	52.5	62 06.5	7.4	10 51.2	9.6	60.3
Y 15	43 30.1	.. 53.5	76 32.9	7.3	10 41.6	9.7	60.3
16	58 30.3	54.4	90 59.2	7.4	10 31.9	9.8	60.3
17	73 30.5	55.4	105 25.6	7.4	10 22.1	9.9	60.4
18	88 30.7	N 1 56.4	119 52.0	7.4	S10 12.2	9.9	60.4
19	103 30.9	57.4	134 18.4	7.4	10 02.3	10.1	60.4
20	118 31.1	58.4	148 44.8	7.4	9 52.2	10.1	60.4
21	133 31.3	1 59.3	163 11.2	7.3	9 42.1	10.1	60.5
22	148 31.5	2 00.3	177 37.5	7.4	9 32.0	10.3	60.5
23	163 31.6	N 2 01.3	192 03.9	7.4	S 9 21.7	10.3	60.5
	SD 16.1	d 1.0	SD 15.9		16.2		16.4

Lat.	Twilight Naut.	Twilight Civil	Sunrise	Moonrise 23	24	25	26
°	h m	h m	h m	h m	h m	h m	h m
N 72	02 55	04 30	05 39	06 48	06 30	06 21	06 14
N 70	03 18	04 40	05 42	05 46	05 56	06 00	06 02
68	03 36	04 48	05 44	05 11	05 31	05 44	05 52
66	03 49	04 54	05 46	04 45	05 12	05 30	05 44
64	04 01	05 00	05 48	04 25	04 56	05 19	05 37
62	04 10	05 05	05 49	04 09	04 43	05 09	05 31
60	04 18	05 09	05 50	03 56	04 32	05 01	05 26
N 58	04 25	05 12	05 51	03 44	04 22	04 54	05 21
56	04 31	05 15	05 52	03 34	04 14	04 47	05 17
54	04 36	05 18	05 53	03 25	04 06	04 42	05 13
52	04 40	05 20	05 53	03 17	03 59	04 36	05 10
50	04 44	05 22	05 55	03 10	03 53	04 32	05 07
45	04 52	05 27	05 56	02 55	03 40	04 21	05 00
N 40	04 59	05 30	05 58	02 42	03 29	04 13	04 54
35	05 04	05 33	05 59	02 31	03 20	04 06	04 49
30	05 08	05 36	05 59	02 22	03 11	03 59	04 45
20	05 13	05 39	06 01	02 05	02 57	03 48	04 37
N 10	05 17	05 41	06 02	01 51	02 45	03 38	04 31
0	05 18	05 42	06 03	01 38	02 33	03 29	04 25
S 10	05 19	05 43	06 04	01 25	02 21	03 19	04 18
20	05 17	05 43	06 05	01 10	02 09	03 09	04 12
30	05 14	05 42	06 06	00 54	01 54	02 58	04 04
35	05 11	05 41	06 06	00 45	01 46	02 52	04 00
40	05 08	05 40	06 07	00 34	01 37	02 44	03 55
45	05 04	05 38	06 07	00 21	01 26	02 35	03 49
S 50	04 58	05 35	06 08	00 06	01 12	02 25	03 42
52	04 55	05 34	06 08	25 06	01 06	02 20	03 39
54	04 51	05 33	06 08	24 59	00 59	02 15	03 35
56	04 48	05 31	06 08	24 51	00 51	02 09	03 31
58	04 43	05 30	06 09	24 42	00 42	02 02	03 27
S 60	04 38	05 28	06 09	24 32	00 32	01 54	03 22

Lat.	Sunset	Twilight Civil	Twilight Naut.	Moonset 23	24	25	26
°	h m	h m	h m	h m	h m	h m	h m
N 72	18 36	19 46	21 24	09 02	11 15	13 21	15 25
N 70	18 33	19 36	20 59	10 03	11 49	13 41	15 35
68	18 31	19 27	20 41	10 38	12 13	13 56	15 43
66	18 28	19 21	20 26	11 03	12 31	14 08	15 50
64	18 27	19 15	20 15	11 23	12 46	14 18	15 56
62	18 25	19 10	20 05	11 38	12 59	14 27	16 01
60	18 24	19 06	19 57	11 51	13 09	14 35	16 05
N 58	18 23	19 02	19 50	12 03	13 18	14 41	16 08
56	18 21	18 59	19 44	12 13	13 26	14 47	16 12
54	18 20	18 56	19 38	12 21	13 33	14 52	16 15
52	18 20	18 54	19 34	12 29	13 40	14 57	16 17
50	18 19	18 51	19 30	12 36	13 46	15 01	16 20
45	18 17	18 47	19 21	12 51	13 58	15 10	16 25
N 40	18 16	18 43	19 15	13 03	14 08	15 17	16 29
35	18 15	18 40	19 10	13 13	14 17	15 24	16 33
30	18 14	18 38	19 06	13 22	14 24	15 29	16 36
20	18 12	18 34	19 00	13 38	14 37	15 39	16 42
N 10	18 11	18 32	18 56	13 52	14 49	15 47	16 47
0	18 10	18 30	18 54	14 04	14 59	15 55	16 51
S 10	18 09	18 30	18 54	14 17	15 10	16 03	16 56
20	18 08	18 30	18 55	14 30	15 21	16 11	17 00
30	18 07	18 30	18 58	14 45	15 34	16 21	17 06
35	18 06	18 31	19 01	14 54	15 41	16 26	17 09
40	18 05	18 32	19 04	15 04	15 50	16 32	17 12
45	18 05	18 34	19 08	15 16	15 59	16 39	17 16
S 50	18 04	18 36	19 14	15 30	16 11	16 47	17 21
52	18 04	18 37	19 17	15 37	16 16	16 51	17 23
54	18 03	18 39	19 20	15 44	16 22	16 55	17 25
56	18 03	18 40	19 24	15 52	16 29	17 00	17 28
58	18 03	18 42	19 28	16 01	16 36	17 05	17 31
S 60	18 02	18 44	19 33	16 12	16 44	17 11	17 34

	SUN Eqn. of Time 00h	SUN Eqn. of Time 12h	SUN Mer. Pass.	MOON Mer. Pass. Upper	MOON Mer. Pass. Lower	Age	Phase
Day	m s	m s	h m	h m	h m	d	%
23	06 47	06 38	12 07	07 51	20 19	25	28
24	06 29	06 20	12 06	08 46	21 14	26	18
25	06 11	06 02	12 06	09 42	22 10	27	10

1998 MARCH 26, 27, 28 (THURS., FRI., SAT.)

UT	ARIES GHA	VENUS −4.4 GHA	VENUS Dec	MARS +1.3 GHA	MARS Dec	JUPITER −2.0 GHA	JUPITER Dec	SATURN +0.6 GHA	SATURN Dec
26 THURSDAY									
00	183 14.3	222 43.4	S13 29.5	167 57.4	N 6 00.5	199 39.4	S 7 59.2	162 57.8	N 6 06.8
01	198 16.7	237 43.4	29.0	182 58.1	01.2	214 41.4	59.0	178 00.0	06.9
02	213 19.2	252 43.4	28.5	197 58.8	02.0	229 43.3	58.8	193 02.2	07.0
03	228 21.7	267 43.4 ..	28.0	212 59.5 ..	02.8	244 45.2 ..	58.6	208 04.4 ..	07.1
04	243 24.1	282 43.4	27.5	228 00.3	03.5	259 47.1	58.4	223 06.5	07.2
05	258 26.6	297 43.4	27.0	243 01.0	04.3	274 49.1	58.2	238 08.7	07.3
06	273 29.1	312 43.4	S13 26.6	258 01.7	N 6 05.0	289 51.0	S 7 57.9	253 10.9	N 6 07.5
07	288 31.5	327 43.4	26.1	273 02.4	05.8	304 52.9	57.7	268 13.1	07.6
08	303 34.0	342 43.4	25.6	288 03.1	06.5	319 54.8	57.5	283 15.3	07.7
09	318 36.5	357 43.3 ..	25.1	303 03.8 ..	07.3	334 56.8 ..	57.3	298 17.4 ..	07.8
10	333 38.9	12 43.3	24.6	318 04.5	08.0	349 58.7	57.1	313 19.6	07.9
11	348 41.4	27 43.3	24.1	333 05.2	08.8	5 00.6	56.9	328 21.8	08.1
12	3 43.8	42 43.3	S13 23.6	348 05.9	N 6 09.5	20 02.5	S 7 56.7	343 24.0	N 6 08.2
13	18 46.3	57 43.3	23.1	3 06.6	10.3	35 04.5	56.4	358 26.1	08.3
14	33 48.8	72 43.3	22.6	18 07.3	11.0	50 06.4	56.2	13 28.3	08.4
15	48 51.2	87 43.3 ..	22.1	33 08.0 ..	11.8	65 08.3 ..	56.0	28 30.5 ..	08.5
16	63 53.7	102 43.3	21.6	48 08.7	12.6	80 10.2	55.8	43 32.7	08.6
17	78 56.2	117 43.2	21.1	63 09.4	13.3	95 12.1	55.6	58 34.8	08.8
18	93 58.6	132 43.2	S13 20.6	78 10.1	N 6 14.1	110 14.1	S 7 55.4	73 37.0	N 6 08.9
19	109 01.1	147 43.2	20.1	93 10.9	14.8	125 16.0	55.1	88 39.2	09.0
20	124 03.6	162 43.2	19.6	108 11.6	15.6	140 17.9	54.9	103 41.4	09.1
21	139 06.0	177 43.2 ..	19.1	123 12.3 ..	16.3	155 19.8 ..	54.7	118 43.5 ..	09.2
22	154 08.5	192 43.1	18.6	138 13.0	17.1	170 21.8	54.5	133 45.7	09.3
23	169 11.0	207 43.1	18.1	153 13.7	17.8	185 23.7	54.3	148 47.9	09.5
27 FRIDAY									
00	184 13.4	222 43.1	S13 17.6	168 14.4	N 6 18.6	200 25.6	S 7 54.1	163 50.1	N 6 09.6
01	199 15.9	237 43.1	17.1	183 15.1	19.3	215 27.5	53.8	178 52.3	09.7
02	214 18.3	252 43.1	16.6	198 15.8	20.1	230 29.5	53.6	193 54.4	09.8
03	229 20.8	267 43.1 ..	16.1	213 16.5 ..	20.8	245 31.4 ..	53.4	208 56.6 ..	09.9
04	244 23.3	282 43.0	15.6	228 17.2	21.6	260 33.3	53.2	223 58.8	10.1
05	259 25.7	297 43.0	15.1	243 17.9	22.3	275 35.2	53.0	239 01.0	10.2
06	274 28.2	312 43.0	S13 14.6	258 18.6	N 6 23.1	290 37.2	S 7 52.8	254 03.1	N 6 10.3
07	289 30.7	327 43.0	14.1	273 19.3	23.8	305 39.1	52.6	269 05.3	10.4
08	304 33.1	342 42.9	13.6	288 20.0	24.6	320 41.0	52.3	284 07.5	10.5
09	319 35.6	357 42.9 ..	13.0	303 20.7 ..	25.3	335 42.9 ..	52.1	299 09.7 ..	10.6
10	334 38.1	12 42.9	12.5	318 21.4	26.1	350 44.9	51.9	314 11.8	10.8
11	349 40.5	27 42.9	12.0	333 22.2	26.8	5 46.8	51.7	329 14.0	10.9
12	4 43.0	42 42.8	S13 11.5	348 22.9	N 6 27.6	20 48.7	S 7 51.5	344 16.2	N 6 11.0
13	19 45.5	57 42.8	11.0	3 23.6	28.3	35 50.6	51.3	359 18.4	11.1
14	34 47.9	72 42.8	10.5	18 24.3	29.1	50 52.6	51.0	14 20.5	11.2
15	49 50.4	87 42.8 ..	10.0	33 25.0 ..	29.8	65 54.5 ..	50.8	29 22.7 ..	11.4
16	64 52.8	102 42.7	09.4	48 25.7	30.6	80 56.4	50.6	44 24.9	11.5
17	79 55.3	117 42.7	08.9	63 26.4	31.3	95 58.4	50.4	59 27.1	11.6
18	94 57.8	132 42.7	S13 08.4	78 27.1	N 6 32.1	111 00.3	S 7 50.2	74 29.2	N 6 11.7
19	110 00.2	147 42.7	07.9	93 27.8	32.8	126 02.2	50.0	89 31.4	11.8
20	125 02.7	162 42.6	07.4	108 28.5	33.6	141 04.1	49.8	104 33.6	11.9
21	140 05.2	177 42.6 ..	06.9	123 29.2 ..	34.3	156 06.1 ..	49.5	119 35.8 ..	12.1
22	155 07.6	192 42.6	06.3	138 29.9	35.1	171 08.0	49.3	134 37.9	12.2
23	170 10.1	207 42.5	05.8	153 30.6	35.8	186 09.9	49.1	149 40.1	12.3
28 SATURDAY									
00	185 12.6	222 42.5	S13 05.3	168 31.3	N 6 36.6	201 11.8	S 7 48.9	164 42.3	N 6 12.4
01	200 15.0	237 42.5	04.8	183 32.0	37.3	216 13.8	48.7	179 44.5	12.5
02	215 17.5	252 42.5	04.2	198 32.7	38.1	231 15.7	48.5	194 46.6	12.6
03	230 19.9	267 42.4 ..	03.7	213 33.4 ..	38.8	246 17.6 ..	48.3	209 48.8 ..	12.8
04	245 22.4	282 42.4	03.2	228 34.2	39.6	261 19.5	48.0	224 51.0	12.9
05	260 24.9	297 42.4	02.7	243 34.9	40.3	276 21.5	47.8	239 53.2	13.0
06	275 27.3	312 42.3	S13 02.1	258 35.6	N 6 41.1	291 23.4	S 7 47.6	254 55.3	N 6 13.1
07	290 29.8	327 42.3	01.6	273 36.3	41.8	306 25.3	47.4	269 57.5	13.2
08	305 32.3	342 42.3	01.1	288 37.0	42.6	321 27.3	47.2	284 59.7	13.4
09	320 34.7	357 42.2 ..	00.5	303 37.7 ..	43.3	336 29.2 ..	47.0	300 01.9 ..	13.5
10	335 37.2	12 42.2	13 00.0	318 38.4	44.0	351 31.1	46.7	315 04.0	13.6
11	350 39.7	27 42.2	12 59.5	333 39.1	44.8	6 33.0	46.5	330 06.2	13.7
12	5 42.1	42 42.1	S12 59.0	348 39.8	N 6 45.5	21 35.0	S 7 46.3	345 08.4	N 6 13.8
13	20 44.6	57 42.1	58.4	3 40.5	46.3	36 36.9	46.1	0 10.6	13.9
14	35 47.1	72 42.1	57.9	18 41.2	47.0	51 38.8	45.9	15 12.7	14.1
15	50 49.5	87 42.0 ..	57.4	33 41.9 ..	47.8	66 40.7 ..	45.7	30 14.9 ..	14.2
16	65 52.0	102 42.0	56.8	48 42.6	48.5	81 42.7	45.5	45 17.1	14.3
17	80 54.4	117 41.9	56.3	63 43.3	49.3	96 44.6	45.2	60 19.3	14.4
18	95 56.9	132 41.9	S12 55.7	78 44.0	N 6 50.0	111 46.5	S 7 45.0	75 21.4	N 6 14.5
19	110 59.4	147 41.9	55.2	93 44.7	50.8	126 48.5	44.8	90 23.6	14.7
20	126 01.8	162 41.8	54.7	108 45.4	51.5	141 50.4	44.6	105 25.8	14.8
21	141 04.3	177 41.8 ..	54.1	123 46.2 ..	52.3	156 52.3 ..	44.4	120 28.0 ..	14.9
22	156 06.8	192 41.8	53.6	138 46.9	53.0	171 54.2	44.2	135 30.1	15.0
23	171 09.2	207 41.7	53.1	153 47.6	53.7	186 56.2	44.0	150 32.3	15.1
Mer. Pass.	11 41.2	v 0.0	d 0.5	v 0.7	d 0.8	v 1.9	d 0.2	v 2.2	d 0.1

STARS

Name	SHA	Dec
Acamar	315 27.6	S40 19.1
Achernar	335 36.1	S57 15.0
Acrux	173 21.6	S63 05.4
Adhara	255 21.7	S28 58.6
Aldebaran	291 03.0	N16 30.2
Alioth	166 30.4	N55 58.1
Alkaid	153 07.7	N49 19.3
Al Na'ir	27 58.8	S46 58.1
Alnilam	275 58.3	S 1 12.5
Alphard	218 07.3	S 8 39.3
Alphecca	126 20.7	N26 43.1
Alpheratz	357 55.9	N29 04.7
Altair	62 19.7	N 8 51.7
Ankaa	353 27.7	S42 19.1
Antares	112 40.4	S26 25.6
Arcturus	146 06.1	N19 11.4
Atria	107 52.6	S69 01.1
Avior	234 22.4	S59 30.6
Bellatrix	278 44.6	N 6 20.7
Betelgeuse	271 14.0	N 7 24.2
Canopus	264 01.3	S52 42.1
Capella	280 51.9	N45 59.8
Deneb	49 39.7	N45 16.2
Denebola	182 45.3	N14 34.8
Diphda	349 08.0	S18 00.0
Dubhe	194 05.4	N61 45.7
Elnath	278 27.5	N28 36.3
Eltanin	90 51.5	N51 29.2
Enif	33 58.8	N 9 51.9
Fomalhaut	15 37.2	S29 38.0
Gacrux	172 13.3	S57 06.2
Gienah	176 04.0	S17 32.0
Hadar	149 03.9	S60 21.7
Hamal	328 14.3	N23 27.1
Kaus Aust.	83 59.3	S34 22.9
Kochab	137 18.8	N74 09.7
Markab	13 50.3	N15 11.6
Menkar	314 27.5	N 4 04.7
Menkent	148 21.0	S36 21.6
Miaplacidus	221 41.4	S69 42.9
Mirfak	308 57.4	N49 51.3
Nunki	76 12.8	S26 17.8
Peacock	53 37.9	S56 44.2
Pollux	243 41.9	N28 01.8
Procyon	245 11.9	N 5 13.6
Rasalhague	96 17.2	N12 33.6
Regulus	207 55.7	N11 58.4
Rigel	281 23.4	S 8 12.5
Rigil Kent.	140 07.2	S60 49.5
Sabik	102 25.8	S15 43.3
Schedar	349 54.4	N56 31.5
Shaula	96 37.7	S37 05.9
Sirius	258 44.0	S16 43.1
Spica	158 43.3	S11 09.2
Suhail	223 00.7	S43 25.8
Vega	80 46.9	N38 46.7
Zuben'ubi	137 18.1	S16 02.0

	SHA	Mer. Pass.
Venus	38 29.7	9 09
Mars	344 01.0	12 46
Jupiter	16 12.2	10 37
Saturn	339 36.7	13 03

UT	SUN GHA	Dec	MOON GHA	v	Dec	d	HP
26							
00	178 31.8	N 2 02.3	206 30.3	7.3	S 9 11.4	10.4	60.6
01	193 32.0	03.3	220 56.6	7.4	9 01.0	10.5	60.6
02	208 32.2	04.3	235 23.0	7.4	8 50.5	10.5	60.6
03	223 32.4	.. 05.2	249 49.4	7.4	8 40.0	10.6	60.7
04	238 32.6	06.2	264 15.8	7.3	8 29.4	10.6	60.7
05	253 32.8	07.2	278 42.1	7.4	8 18.8	10.8	60.7
06	268 33.0	N 2 08.2	293 08.5	7.3	S 8 08.0	10.7	60.7
07	283 33.2	09.2	307 34.8	7.4	7 57.3	10.9	60.8
08	298 33.3	10.1	322 01.2	7.4	7 46.4	10.9	60.8
09	313 33.5	.. 11.1	336 27.6	7.3	7 35.5	11.0	60.8
10	328 33.7	12.1	350 53.9	7.4	7 24.5	11.0	60.8
11	343 33.9	13.1	5 20.3	7.3	7 13.5	11.1	60.9
12	358 34.1	N 2 14.1	19 46.6	7.4	S 7 02.4	11.1	60.9
13	13 34.3	15.1	34 13.0	7.4	6 51.3	11.2	60.9
14	28 34.5	16.0	48 39.4	7.3	6 40.1	11.3	60.9
15	43 34.7	.. 17.0	63 05.7	7.3	6 28.8	11.3	61.0
16	58 34.8	18.0	77 32.1	7.3	6 17.5	11.3	61.0
17	73 35.0	19.0	91 58.4	7.4	6 06.2	11.4	61.0
18	88 35.2	N 2 20.0	106 24.8	7.3	S 5 54.8	11.5	61.0
19	103 35.4	20.9	120 51.1	7.4	5 43.3	11.4	61.0
20	118 35.6	21.9	135 17.5	7.3	5 31.9	11.6	61.1
21	133 35.8	.. 22.9	149 43.8	7.3	5 20.3	11.6	61.1
22	148 36.0	23.9	164 10.1	7.4	5 08.7	11.6	61.1
23	163 36.2	24.9	178 36.5	7.3	4 57.1	11.6	61.1
27							
00	178 36.4	N 2 25.8	193 02.8	7.4	S 4 45.5	11.7	61.1
01	193 36.5	26.8	207 29.2	7.3	4 33.8	11.8	61.2
02	208 36.7	27.8	221 55.5	7.3	4 22.0	11.7	61.2
03	223 36.9	.. 28.8	236 21.8	7.4	4 10.3	11.9	61.2
04	238 37.1	29.8	250 48.2	7.3	3 58.4	11.8	61.2
05	253 37.3	30.7	265 14.5	7.3	3 46.6	11.9	61.2
06	268 37.5	N 2 31.7	279 40.8	7.3	S 3 34.7	11.9	61.2
07	283 37.7	32.7	294 07.1	7.4	3 22.8	11.9	61.2
08	298 37.9	33.7	308 33.5	7.3	3 10.9	11.9	61.3
09	313 38.0	.. 34.7	322 59.8	7.3	2 59.0	12.0	61.3
10	328 38.2	35.6	337 26.1	7.3	2 47.0	12.0	61.3
11	343 38.4	36.6	351 52.4	7.3	2 35.0	12.1	61.3
12	358 38.6	N 2 37.6	6 18.7	7.3	S 2 22.9	12.0	61.3
13	13 38.8	38.6	20 45.0	7.3	2 10.9	12.1	61.3
14	28 39.0	39.5	35 11.3	7.3	1 58.8	12.1	61.3
15	43 39.2	.. 40.5	49 37.6	7.3	1 46.7	12.1	61.3
16	58 39.4	41.5	64 03.9	7.3	1 34.6	12.1	61.4
17	73 39.5	42.5	78 30.2	7.3	1 22.5	12.1	61.4
18	88 39.7	N 2 43.5	92 56.5	7.3	S 1 10.4	12.1	61.4
19	103 39.9	44.4	107 22.8	7.3	0 58.3	12.2	61.4
20	118 40.1	45.4	121 49.1	7.2	0 46.1	12.2	61.4
21	133 40.3	.. 46.4	136 15.3	7.3	0 33.9	12.1	61.4
22	148 40.5	47.4	150 41.6	7.3	0 21.8	12.2	61.4
23	163 40.7	48.4	165 07.9	7.2	S 0 09.6	12.2	61.4
28							
00	178 40.9	N 2 49.3	179 34.1	7.3	N 0 02.6	12.1	61.4
01	193 41.1	50.3	194 00.4	7.2	0 14.7	12.2	61.4
02	208 41.2	51.3	208 26.6	7.3	0 26.9	12.2	61.4
03	223 41.4	.. 52.3	222 52.9	7.2	0 39.1	12.1	61.4
04	238 41.6	53.2	237 19.1	7.2	0 51.2	12.2	61.4
05	253 41.8	54.2	251 45.3	7.3	1 03.4	12.2	61.4
06	268 42.0	N 2 55.2	266 11.6	7.2	N 1 15.6	12.1	61.4
07	283 42.2	56.2	280 37.8	7.2	1 27.7	12.2	61.4
08	298 42.4	57.1	295 04.0	7.2	1 39.9	12.1	61.4
09	313 42.6	.. 58.1	309 30.2	7.2	1 52.0	12.1	61.4
10	328 42.7	2 59.1	323 56.4	7.2	2 04.1	12.1	61.4
11	343 42.9	3 00.1	338 22.6	7.2	2 16.2	12.1	61.4
12	358 43.1	N 3 01.1	352 48.8	7.1	N 2 28.3	12.1	61.4
13	13 43.3	02.0	7 14.9	7.2	2 40.4	12.0	61.4
14	28 43.5	03.0	21 41.1	7.2	2 52.4	12.0	61.4
15	43 43.7	.. 04.0	36 07.3	7.1	3 04.4	12.0	61.4
16	58 43.9	05.0	50 33.4	7.2	3 16.4	12.0	61.4
17	73 44.1	05.9	64 59.6	7.1	3 28.4	12.0	61.4
18	88 44.2	N 3 06.9	79 25.7	7.2	N 3 40.4	11.9	61.4
19	103 44.4	07.9	93 51.9	7.1	3 52.3	11.9	61.4
20	118 44.6	08.9	108 18.0	7.1	4 04.2	11.9	61.4
21	133 44.8	.. 09.8	122 44.1	7.1	4 16.1	11.8	61.4
22	148 45.0	10.8	137 10.2	7.1	4 27.9	11.8	61.4
23	163 45.2	11.8	151 36.3	7.1	N 4 39.7	11.8	61.3
	SD 16.1	d 1.0	SD 16.6		16.7		16.7

Lat.	Naut.	Civil	Sunrise	Moonrise 26	27	28	29
N 72	02 31	04 13	05 23	06 14	06 08	06 02	05 56
N 70	02 59	04 25	05 28	06 02	06 03	06 03	06 04
68	03 19	04 34	05 32	05 52	05 59	06 05	06 11
66	03 35	04 42	05 35	05 44	05 55	06 06	06 17
64	03 48	04 49	05 37	05 37	05 52	06 07	06 21
62	03 59	04 54	05 39	05 31	05 50	06 07	06 26
60	04 08	04 59	05 41	05 26	05 47	06 08	06 29
N 58	04 15	05 03	05 43	05 21	05 45	06 09	06 32
56	04 22	05 07	05 45	05 17	05 44	06 09	06 35
54	04 28	05 10	05 46	05 13	05 42	06 10	06 38
52	04 33	05 13	05 47	05 10	05 40	06 10	06 40
50	04 37	05 16	05 48	05 07	05 39	06 11	06 43
45	04 46	05 21	05 51	05 00	05 36	06 11	06 47
N 40	04 54	05 26	05 53	04 54	05 34	06 12	06 51
35	04 59	05 29	05 54	04 49	05 31	06 13	06 55
30	05 04	05 32	05 56	04 45	05 29	06 14	06 58
20	05 11	05 36	05 58	04 37	05 26	06 15	07 03
N 10	05 15	05 39	06 00	04 31	05 23	06 16	07 08
0	05 17	05 41	06 02	04 25	05 21	06 17	07 13
S 10	05 19	05 43	06 04	04 18	05 18	06 18	07 17
20	05 18	05 44	06 06	04 12	05 15	06 19	07 22
30	05 16	05 44	06 08	04 04	05 12	06 20	07 28
35	05 14	05 43	06 09	04 00	05 10	06 21	07 32
40	05 11	05 43	06 10	03 55	05 08	06 21	07 35
45	05 08	05 42	06 11	03 49	05 05	06 22	07 40
S 50	05 03	05 40	06 12	03 42	05 02	06 24	07 45
52	05 00	05 39	06 13	03 39	05 01	06 24	07 47
54	04 57	05 39	06 14	03 35	04 59	06 25	07 50
56	04 54	05 38	06 15	03 31	04 58	06 25	07 53
58	04 50	05 36	06 15	03 27	04 56	06 26	07 56
S 60	04 46	05 35	06 16	03 22	04 54	06 27	08 00

Lat.	Sunset	Civil	Naut.	Moonset 26	27	28	29
N 72	18 50	20 02	21 47	15 25	17 29	19 33	21 38
N 70	18 45	19 49	21 17	15 35	17 31	19 27	21 24
68	18 42	19 39	20 55	15 43	17 33	19 23	21 12
66	18 38	19 31	20 25	15 50	17 34	19 19	21 03
64	18 35	19 24	20 25	15 56	17 35	19 16	20 55
62	18 33	19 18	20 14	16 01	17 36	19 13	20 49
60	18 31	19 13	20 05	16 05	17 37	19 11	20 43
N 58	18 29	19 09	19 57	16 08	17 38	19 09	20 38
56	18 28	19 05	19 51	16 12	17 39	19 07	20 34
54	18 26	19 02	19 45	16 15	17 39	19 05	20 30
52	18 25	18 59	19 40	16 17	17 40	19 03	20 26
50	18 24	18 56	19 35	16 20	17 40	19 02	20 23
45	18 21	18 51	19 25	16 25	17 42	18 59	20 16
N 40	18 19	18 46	19 18	16 29	17 43	18 56	20 10
35	18 17	18 42	19 12	16 33	17 43	18 54	20 05
30	18 16	18 39	19 08	16 36	17 44	18 52	20 00
20	18 13	18 35	19 01	16 42	17 45	18 49	19 53
N 10	18 11	18 32	18 56	16 47	17 46	18 46	19 46
0	18 09	18 29	18 53	16 51	17 47	18 43	19 40
S 10	18 07	18 28	18 52	16 56	17 48	18 40	19 33
20	18 05	18 27	18 52	17 00	17 49	18 37	19 27
30	18 03	18 27	18 54	17 06	17 50	18 34	19 19
35	18 02	18 27	18 56	17 09	17 50	18 32	19 15
40	18 01	18 28	18 59	17 12	17 51	18 30	19 10
45	17 59	18 28	19 02	17 16	17 52	18 27	19 04
S 50	17 58	18 30	19 07	17 21	17 53	18 24	18 57
52	17 57	18 30	19 10	17 23	17 53	18 23	18 54
54	17 56	18 31	19 12	17 25	17 53	18 21	18 50
56	17 55	18 32	19 15	17 28	17 54	18 20	18 47
58	17 54	18 33	19 19	17 31	17 54	18 18	18 42
S 60	17 53	18 34	19 23	17 34	17 55	18 16	18 38

Day	SUN Eqn. of Time 00h	12h	Mer. Pass.	MOON Mer. Pass. Upper	Lower	Age	Phase
	m s	m s	h m	h m	h m	d	%
26	05 53	05 44	12 06	10 38	23 06	28	4
27	05 35	05 26	12 05	11 34	24 02	29	1
28	05 17	05 08	12 05	12 30	00 02	00	0 ●

1998 MARCH 29, 30, 31 (SUN., MON., TUES.)

UT	ARIES	VENUS −4.4		MARS +1.3		JUPITER −2.0		SATURN +0.6	
d h	GHA	GHA	Dec	GHA	Dec	GHA	Dec	GHA	Dec
29 00	186 11.7	222 41.7 S12 52.5		168 48.3 N 6 54.5		201 58.1 S 7 43.7		165 34.5 N 6 15.2	
01	201 14.2	237 41.6	52.0	183 49.0	55.2	217 00.0	43.5	180 36.7	15.4
02	216 16.6	252 41.6	51.4	198 49.7	56.0	232 02.0	43.3	195 38.8	15.5
03	231 19.1	267 41.6 ..	50.9	213 50.4 ..	56.7	247 03.9 ..	43.1	210 41.0 ..	15.6
04	246 21.5	282 41.5	50.3	228 51.1	57.5	262 05.8	42.9	225 43.2	15.7
05	261 24.0	297 41.5	49.8	243 51.8	58.2	277 07.7	42.7	240 45.4	15.8
06	276 26.5	312 41.4 S12 49.2		258 52.5 N 6 59.0		292 09.7 S 7 42.5		255 47.5 N 6 15.9	
07	291 28.9	327 41.4	48.7	273 53.2 6 59.7		307 11.6	42.2	270 49.7	16.1
08	306 31.4	342 41.4	48.2	288 53.9 7 00.4		322 13.5	42.0	285 51.9	16.2
S 09	321 33.9	357 41.3 ..	47.6	303 54.6 ..	01.2	337 15.5 ..	41.8	300 54.1 ..	16.3
U 10	336 36.3	12 41.3	47.1	318 55.3	01.9	352 17.4	41.6	315 56.2	16.4
N 11	351 38.8	27 41.2	46.5	333 56.0	02.7	7 19.3	41.4	330 58.4	16.5
D 12	6 41.3	42 41.2 S12 46.0		348 56.7 N 7 03.4		22 21.2 S 7 41.2		346 00.6 N 6 16.7	
A 13	21 43.7	57 41.1	45.4	3 57.4	04.2	37 23.2	41.0	1 02.8	16.8
Y 14	36 46.2	72 41.1	44.8	18 58.1	04.9	52 25.1	40.7	16 04.9	16.9
15	51 48.7	87 41.0 ..	44.3	33 58.8 ..	05.7	67 27.0 ..	40.5	31 07.1 ..	17.0
16	66 51.1	102 41.0	43.7	48 59.5	06.4	82 29.0	40.3	46 09.3	17.1
17	81 53.6	117 40.9	43.2	64 00.3	07.1	97 30.9	40.1	61 11.5	17.2
18	96 56.0	132 40.9 S12 42.6		79 01.0 N 7 07.9		112 32.8 S 7 39.9		76 13.6 N 6 17.4	
19	111 58.5	147 40.9	42.1	94 01.7	08.6	127 34.8	39.7	91 15.8	17.5
20	127 01.0	162 40.8	41.5	109 02.4	09.4	142 36.7	39.5	106 18.0	17.6
21	142 03.4	177 40.8 ..	41.0	124 03.1 ..	10.1	157 38.6 ..	39.2	121 20.1 ..	17.7
22	157 05.9	192 40.7	40.4	139 03.8	10.9	172 40.5	39.0	136 22.3	17.8
23	172 08.4	207 40.7	39.8	154 04.5	11.6	187 42.5	38.8	151 24.5	18.0
30 00	187 10.8	222 40.6 S12 39.3		169 05.2 N 7 12.3		202 44.4 S 7 38.6		166 26.7 N 6 18.1	
01	202 13.3	237 40.6	38.7	184 05.9	13.1	217 46.3	38.4	181 28.8	18.2
02	217 15.8	252 40.5	38.2	199 06.6	13.8	232 48.3	38.2	196 31.0	18.3
03	232 18.2	267 40.5 ..	37.6	214 07.3 ..	14.6	247 50.2 ..	38.0	211 33.2 ..	18.4
04	247 20.7	282 40.4	37.0	229 08.0	15.3	262 52.1	37.7	226 35.4	18.5
05	262 23.1	297 40.4	36.5	244 08.7	16.0	277 54.1	37.5	241 37.5	18.7
06	277 25.6	312 40.3 S12 35.9		259 09.4 N 7 16.8		292 56.0 S 7 37.3		256 39.7 N 6 18.8	
07	292 28.1	327 40.3	35.3	274 10.1	17.5	307 57.9	37.1	271 41.9	18.9
08	307 30.5	342 40.2	34.8	289 10.8	18.3	322 59.8	36.9	286 44.1	19.0
M 09	322 33.0	357 40.2 ..	34.2	304 11.5 ..	19.0	338 01.8 ..	36.7	301 46.2 ..	19.1
O 10	337 35.5	12 40.1	33.6	319 12.2	19.7	353 03.7	36.5	316 48.4	19.3
N 11	352 37.9	27 40.1	33.1	334 12.9	20.5	8 05.6	36.2	331 50.6	19.4
D 12	7 40.4	42 40.0 S12 32.5		349 13.6 N 7 21.2		23 07.6 S 7 36.0		346 52.8 N 6 19.5	
A 13	22 42.9	57 40.0	31.9	4 14.3	22.0	38 09.5	35.8	1 54.9	19.6
Y 14	37 45.3	72 39.9	31.4	19 15.1	22.7	53 11.4	35.6	16 57.1	19.7
15	52 47.8	87 39.8 ..	30.8	34 15.8 ..	23.4	68 13.4 ..	35.4	31 59.3 ..	19.8
16	67 50.3	102 39.8	30.2	49 16.5	24.2	83 15.3	35.2	47 01.4	20.0
17	82 52.7	117 39.7	29.7	64 17.2	24.9	98 17.2	35.0	62 03.6	20.1
18	97 55.2	132 39.7 S12 29.1		79 17.9 N 7 25.7		113 19.2 S 7 34.7		77 05.8 N 6 20.2	
19	112 57.6	147 39.6	28.5	94 18.6	26.4	128 21.1	34.5	92 08.0	20.3
20	128 00.1	162 39.6	27.9	109 19.3	27.1	143 23.0	34.3	107 10.1	20.4
21	143 02.6	177 39.5 ..	27.4	124 20.0 ..	27.9	158 25.0 ..	34.1	122 12.3 ..	20.5
22	158 05.0	192 39.5	26.8	139 20.7	28.6	173 26.9	33.9	137 14.5	20.7
23	173 07.5	207 39.4	26.2	154 21.4	29.4	188 28.8	33.7	152 16.7	20.8
31 00	188 10.0	222 39.3 S12 25.6		169 22.1 N 7 30.1		203 30.8 S 7 33.5		167 18.8 N 6 20.9	
01	203 12.4	237 39.3	25.0	184 22.8	30.8	218 32.7	33.3	182 21.0	21.0
02	218 14.9	252 39.2	24.5	199 23.5	31.6	233 34.6	33.0	197 23.2	21.1
03	233 17.4	267 39.2 ..	23.9	214 24.2 ..	32.3	248 36.6 ..	32.8	212 25.4 ..	21.3
04	248 19.8	282 39.1	23.3	229 24.9	33.0	263 38.5	32.6	227 27.5	21.4
05	263 22.3	297 39.1	22.7	244 25.6	33.8	278 40.4	32.4	242 29.7	21.5
06	278 24.8	312 39.0 S12 22.1		259 26.3 N 7 34.5		293 42.3 S 7 32.2		257 31.9 N 6 21.6	
07	293 27.2	327 38.9	21.6	274 27.0	35.3	308 44.3	32.0	272 34.0	21.7
08	308 29.7	342 38.9	21.0	289 27.7	36.0	323 46.2	31.8	287 36.2	21.8
T 09	323 32.1	357 38.8 ..	20.4	304 28.4 ..	36.7	338 48.1 ..	31.5	302 38.4 ..	22.0
U 10	338 34.6	12 38.8	19.8	319 29.1	37.5	353 50.1	31.3	317 40.6	22.1
E 11	353 37.1	27 38.7	19.2	334 29.8	38.2	8 52.0	31.1	332 42.7	22.2
S 12	8 39.5	42 38.6 S12 18.6		349 30.5 N 7 38.9		23 53.9 S 7 30.9		347 44.9 N 6 22.3	
D 13	23 42.0	57 38.6	18.0	4 31.3	39.7	38 55.9	30.7	2 47.1	22.4
A 14	38 44.5	72 38.5	17.4	19 32.0	40.4	53 57.8	30.5	17 49.3	22.6
Y 15	53 46.9	87 38.5 ..	16.9	34 32.7 ..	41.1	68 59.7 ..	30.3	32 51.4 ..	22.7
16	68 49.4	102 38.4	16.3	49 33.4	41.9	84 01.7	30.1	47 53.6	22.8
17	83 51.9	117 38.3	15.7	64 34.1	42.6	99 03.6	29.8	62 55.8	22.9
18	98 54.3	132 38.3 S12 15.1		79 34.8 N 7 43.3		114 05.5 S 7 29.6		77 57.9 N 6 23.0	
19	113 56.8	147 38.2	14.5	94 35.5	44.1	129 07.5	29.4	93 00.1	23.1
20	128 59.2	162 38.1	13.9	109 36.2	44.8	144 09.4	29.2	108 02.3	23.3
21	144 01.7	177 38.1 ..	13.3	124 36.9 ..	45.6	159 11.4 ..	29.0	123 04.5 ..	23.4
22	159 04.2	192 38.0	12.7	139 37.6	46.3	174 13.3	28.8	138 06.6	23.5
23	174 06.6	207 37.9	12.1	154 38.3	47.0	189 15.2	28.6	153 08.8	23.6
Mer. Pass.	h m 11 29.4	v −0.1	d 0.6	v 0.7	d 0.7	v 1.9	d 0.2	v 2.2	d 0.1

STARS

Name	SHA	Dec
Acamar	315 27.6	S40 19.0
Achernar	335 36.1	S57 15.0
Acrux	173 21.6	S63 05.4
Adhara	255 21.7	S28 58.6
Aldebaran	291 03.0	N16 30.2
Alioth	166 30.4	N55 58.1
Alkaid	153 07.7	N49 19.3
Al Na'ir	27 58.8	S46 58.1
Alnilam	275 58.3	S 1 12.5
Alphard	218 07.4	S 8 39.3
Alphecca	126 20.7	N26 43.1
Alpheratz	357 55.9	N29 04.7
Altair	62 19.7	N 8 51.7
Ankaa	353 27.7	S42 19.1
Antares	112 40.4	S26 25.6
Arcturus	146 06.1	N19 11.4
Atria	107 52.5	S69 01.1
Avior	234 22.5	S59 30.6
Bellatrix	278 44.7	N 6 20.6
Betelgeuse	271 14.0	N 7 24.2
Canopus	264 01.4	S52 42.1
Capella	280 51.9	N45 59.8
Deneb	49 39.6	N45 16.2
Denebola	182 45.3	N14 34.8
Diphda	349 08.0	S18 00.0
Dubhe	194 05.5	N61 45.7
Elnath	278 27.5	N28 36.2
Eltanin	90 51.5	N51 29.2
Enif	33 58.8	N 9 51.9
Fomalhaut	15 37.2	S29 38.0
Gacrux	172 13.3	S57 06.2
Gienah	176 04.0	S17 32.0
Hadar	149 03.9	S60 21.7
Hamal	328 14.3	N23 27.1
Kaus Aust.	83 59.3	S34 22.9
Kochab	137 18.7	N74 09.7
Markab	13 50.3	N15 11.6
Menkar	314 27.5	N 4 04.7
Menkent	148 21.0	S36 21.6
Miaplacidus	221 41.5	S69 42.9
Mirfak	308 57.5	N49 51.2
Nunki	76 12.8	S26 17.8
Peacock	53 37.9	S56 44.2
Pollux	243 41.9	N28 01.8
Procyon	245 11.9	N 5 13.6
Rasalhague	96 17.2	N12 33.6
Regulus	207 55.7	N11 58.4
Rigel	281 23.4	S 8 12.5
Rigil Kent.	140 07.2	S60 49.5
Sabik	102 25.8	S15 43.3
Schedar	349 54.4	N56 31.5
Shaula	96 37.7	S37 05.9
Sirius	258 44.0	S16 43.1
Spica	158 43.3	S11 09.2
Suhail	223 00.7	S43 25.9
Vega	80 46.9	N38 46.7
Zuben'ubi	137 18.1	S16 02.0

	SHA	Mer. Pass.
	o '	h m
Venus	35 29.8	9 09
Mars	341 54.4	12 43
Jupiter	15 33.6	10 28
Saturn	339 15.8	12 52

UT	SUN GHA	SUN Dec	MOON GHA	v	MOON Dec	d	HP
29 00	178 45.4	N 3 12.8	166 02.4	7.1	N 4 51.5	11.7	61.3
01	193 45.6	13.7	180 28.5	7.1	5 03.2	11.7	61.3
02	208 45.8	14.7	194 54.6	7.1	5 14.9	11.7	61.3
03	223 45.9	.. 15.7	209 20.7	7.0	5 26.6	11.6	61.3
04	238 46.1	16.7	223 46.7	7.1	5 38.2	11.6	61.3
05	253 46.3	17.6	238 12.8	7.0	5 49.8	11.5	61.3
S 06	268 46.5	N 3 18.6	252 38.8	7.0	N 6 01.3	11.5	61.3
U 07	283 46.7	19.6	267 04.8	7.1	6 12.8	11.4	61.2
N 08	298 46.9	20.6	281 30.9	7.0	6 24.2	11.4	61.2
D 09	313 47.1	.. 21.5	295 56.9	7.0	6 35.6	11.3	61.2
A 10	328 47.3	22.5	310 22.9	7.0	6 46.9	11.3	61.2
Y 11	343 47.4	23.5	324 48.9	7.0	6 58.2	11.2	61.2
12	358 47.6	N 3 24.5	339 14.9	7.0	N 7 09.4	11.2	61.2
13	13 47.8	25.4	353 40.9	7.0	7 20.6	11.2	61.1
14	28 48.0	26.4	8 06.9	6.9	7 31.8	11.0	61.1
15	43 48.2	.. 27.4	22 32.8	7.0	7 42.8	11.0	61.1
16	58 48.4	28.4	36 58.8	6.9	7 53.8	11.0	61.1
17	73 48.6	29.3	51 24.7	7.0	8 04.8	10.9	61.1
18	88 48.8	N 3 30.3	65 50.7	6.9	N 8 15.7	10.8	61.1
19	103 48.9	31.3	80 16.6	6.9	8 26.5	10.7	61.0
20	118 49.1	32.3	94 42.5	7.0	8 37.2	10.7	61.0
21	133 49.3	.. 33.2	109 08.5	6.9	8 47.9	10.7	61.0
22	148 49.5	34.2	123 34.4	6.9	8 58.6	10.5	61.0
23	163 49.7	35.2	138 00.3	6.9	9 09.1	10.5	61.0
30 00	178 49.9	N 3 36.1	152 26.2	6.9	N 9 19.6	10.4	60.9
01	193 50.1	37.1	166 52.1	6.9	9 30.0	10.4	60.9
02	208 50.3	38.1	181 18.0	6.8	9 40.4	10.3	60.9
03	223 50.4	.. 39.1	195 43.8	6.9	9 50.7	10.2	60.9
04	238 50.6	40.0	210 09.7	6.9	10 00.9	10.1	60.8
05	253 50.8	41.0	224 35.6	6.8	10 11.0	10.1	60.8
M 06	268 51.0	N 3 42.0	239 01.4	6.9	N10 21.1	9.9	60.8
O 07	283 51.2	42.9	253 27.3	6.8	10 31.0	9.9	60.8
N 08	298 51.4	43.9	267 53.1	6.8	10 40.9	9.8	60.7
D 09	313 51.6	.. 44.9	282 18.9	6.9	10 50.7	9.8	60.7
A 10	328 51.7	45.9	296 44.8	6.8	11 00.5	9.6	60.7
Y 11	343 51.9	46.8	311 10.6	6.8	11 10.1	9.6	60.7
12	358 52.1	N 3 47.8	325 36.4	6.8	N11 19.7	9.5	60.6
13	13 52.3	48.8	340 02.2	6.9	11 29.2	9.4	60.6
14	28 52.5	49.7	354 28.1	6.8	11 38.6	9.3	60.6
15	43 52.7	.. 50.7	8 53.9	6.8	11 47.9	9.2	60.6
16	58 52.9	51.7	23 19.7	6.8	11 57.1	9.2	60.5
17	73 53.1	52.7	37 45.5	6.8	12 06.3	9.0	60.5
18	88 53.2	N 3 53.6	52 11.3	6.8	N12 15.3	9.0	60.5
19	103 53.4	54.6	66 37.1	6.8	12 24.3	8.8	60.4
20	118 53.6	55.6	81 02.9	6.7	12 33.1	8.8	60.4
21	133 53.8	.. 56.5	95 28.6	6.8	12 41.9	8.7	60.4
22	148 54.0	57.5	109 54.4	6.8	12 50.6	8.6	60.3
23	163 54.2	58.5	124 20.2	6.8	12 59.2	8.5	60.3
31 00	178 54.4	N 3 59.5	138 46.0	6.8	N13 07.7	8.4	60.3
01	193 54.6	N 4 00.4	153 11.8	6.8	13 16.1	8.2	60.2
02	208 54.7	01.4	167 37.6	6.8	13 24.3	8.2	60.2
03	223 54.9	.. 02.4	182 03.4	6.8	13 32.5	8.1	60.2
04	238 55.1	03.3	196 29.2	6.7	13 40.6	8.1	60.1
05	253 55.3	04.3	210 54.9	6.8	13 48.7	7.9	60.1
T 06	268 55.5	N 4 05.3	225 20.7	6.8	N13 56.6	7.8	60.1
U 07	283 55.7	06.2	239 46.5	6.8	14 04.4	7.7	60.0
E 08	298 55.9	07.2	254 12.3	6.8	14 12.1	7.6	60.0
S 09	313 56.0	.. 08.2	268 38.1	6.8	14 19.7	7.4	60.0
D 10	328 56.2	09.1	283 03.9	6.8	14 27.1	7.4	59.9
A 11	343 56.4	10.1	297 29.7	6.8	14 34.5	7.3	59.9
Y 12	358 56.6	N 4 11.1	311 55.5	6.9	N14 41.8	7.2	59.9
13	13 56.8	12.0	326 21.4	6.8	14 49.0	7.1	59.8
14	28 57.0	13.0	340 47.2	6.8	14 56.1	7.0	59.8
15	43 57.2	.. 14.0	355 13.0	6.8	15 03.1	6.8	59.8
16	58 57.3	14.9	9 38.8	6.9	15 09.9	6.8	59.7
17	73 57.5	15.9	24 04.7	6.8	15 16.7	6.6	59.7
18	88 57.7	N 4 16.9	38 30.5	6.9	N15 23.3	6.6	59.7
19	103 57.9	17.9	52 56.4	6.9	15 29.9	6.4	59.6
20	118 58.1	18.8	67 22.3	6.8	15 36.3	6.3	59.6
21	133 58.3	.. 19.8	81 48.1	6.9	15 42.6	6.3	59.5
22	148 58.5	20.8	96 14.0	6.9	15 48.9	6.1	59.5
23	163 58.6	21.7	110 39.9	7.0	N15 55.0	6.0	59.5
	SD 16.0	d 1.0	SD 16.7		16.5		16.3

Lat.	Twilight Naut.	Civil	Sunrise	Moonrise 29	30	31	1
N 72	02 03	03 55	05 08	05 56	05 50	05 43	05 33
N 70	02 38	04 09	05 14	06 04	06 06	06 09	06 17
68	03 02	04 21	05 19	06 11	06 19	06 29	06 47
66	03 21	04 30	05 23	06 17	06 29	06 46	07 09
64	03 35	04 38	05 27	06 21	06 38	06 59	07 26
62	03 47	04 44	05 30	06 26	06 46	07 10	07 41
60	03 57	04 50	05 32	06 29	06 52	07 20	07 53
N 58	04 06	04 55	05 35	06 32	06 58	07 28	08 04
56	04 13	04 59	05 37	06 35	07 04	07 36	08 13
54	04 19	05 03	05 39	06 38	07 08	07 42	08 22
52	04 25	05 06	05 40	06 40	07 13	07 48	08 29
50	04 30	05 09	05 42	06 43	07 17	07 54	08 36
45	04 40	05 16	05 45	06 47	07 25	08 06	08 50
N 40	04 49	05 21	05 48	06 51	07 32	08 15	09 02
35	04 55	05 25	05 50	06 55	07 38	08 24	09 12
30	05 00	05 28	05 52	06 58	07 44	08 31	09 21
20	05 08	05 34	05 56	07 03	07 53	08 44	09 37
N 10	05 13	05 38	05 59	07 08	08 02	08 56	09 50
0	05 17	05 41	06 01	07 13	08 10	09 06	10 03
S 10	05 18	05 43	06 04	07 17	08 17	09 17	10 16
20	05 19	05 44	06 06	07 22	08 26	09 29	10 30
30	05 18	05 45	06 09	07 28	08 36	09 42	10 45
35	05 16	05 46	06 11	07 32	08 42	09 50	10 55
40	05 14	05 46	06 13	07 35	08 48	09 59	11 05
45	05 11	05 45	06 15	07 40	08 56	10 09	11 17
S 50	05 07	05 45	06 18	07 45	09 05	10 21	11 32
52	05 05	05 45	06 18	07 47	09 09	10 27	11 40
54	05 03	05 44	06 19	07 50	09 14	10 34	11 47
56	05 00	05 44	06 21	07 53	09 19	10 41	11 56
58	04 57	05 43	06 22	07 56	09 25	10 49	12 06
S 60	04 54	05 42	06 24	08 00	09 32	10 58	12 18

Lat.	Sunset	Twilight Civil	Naut.	Moonset 29	30	31	1
N 72	19 04	20 18	22 15	21 38	23 44	25 53	01 53
N 70	18 58	20 03	21 37	21 24	23 19	25 10	01 10
68	18 52	19 51	21 11	21 12	23 00	24 41	00 41
66	18 48	19 42	20 52	21 03	22 45	24 19	00 19
64	18 44	19 34	20 37	20 55	22 32	24 02	00 02
62	18 41	19 27	20 24	20 49	22 22	23 48	25 04
60	18 38	19 21	20 14	20 43	22 13	23 36	24 50
N 58	18 36	19 16	20 05	20 38	22 05	23 26	24 38
56	18 34	19 12	19 58	20 34	21 58	23 17	24 28
54	18 32	19 08	19 51	20 30	21 52	23 09	24 19
52	18 30	19 04	19 45	20 26	21 46	23 02	24 11
50	18 28	19 01	19 40	20 23	21 41	22 56	24 03
45	18 25	18 54	19 30	20 16	21 31	22 42	23 48
N 40	18 22	18 49	19 21	20 10	21 22	22 31	23 35
35	18 19	18 45	19 15	20 05	21 14	22 21	23 24
30	18 17	18 41	19 10	20 00	21 08	22 12	23 14
20	18 14	18 36	19 02	19 53	20 56	21 58	22 58
N 10	18 11	18 32	18 56	19 46	20 45	21 45	22 43
0	18 08	18 29	18 53	19 40	20 37	21 33	22 30
S 10	18 05	18 26	18 50	19 33	20 27	21 22	22 16
20	18 02	18 24	18 50	19 27	20 17	21 09	22 02
30	17 59	18 23	18 51	19 19	20 06	20 54	21 46
35	17 58	18 23	18 52	19 15	19 59	20 46	21 36
40	17 54	18 23	18 54	19 10	19 52	20 37	21 25
45	17 54	18 23	18 57	19 04	19 43	20 25	21 12
S 50	17 51	18 23	19 01	18 57	19 32	20 12	20 57
52	17 50	18 23	19 03	18 54	19 28	20 06	20 49
54	17 49	18 24	19 05	18 50	19 22	19 59	20 41
56	17 47	18 24	19 07	18 47	19 16	19 51	20 32
58	17 46	18 25	19 10	18 42	19 10	19 43	20 22
S 60	17 44	18 25	19 14	18 38	19 03	19 33	20 10

Day	SUN Eqn. of Time 00h	12h	Mer. Pass.	MOON Mer. Pass. Upper	Lower	Age	Phase
d	m s	m s	h m	h m	h m	d	%
29	04 59	04 50	12 05	13 26	00 58	01	3
30	04 41	04 32	12 05	14 23	01 55	02	8
31	04 23	04 14	12 04	15 20	02 51	03	16

1998 APRIL 1, 2, 3 (WED., THURS., FRI.)

UT	ARIES GHA	VENUS −4.3 GHA	Dec	MARS +1.3 GHA	Dec	JUPITER −2.0 GHA	Dec	SATURN +0.6 GHA	Dec
d h	° ′	° ′	° ′	° ′	° ′	° ′	° ′	° ′	° ′
1 00	189 09.1	222 37.9	S12 11.5	169 39.0	N 7 47.8	204 17.2	S 7 28.3	168 11.0	N 6 23.7
01	204 11.6	237 37.8	10.9	184 39.7	48.5	219 19.1	28.1	183 13.1	23.9
02	219 14.0	252 37.7	10.3	199 40.4	49.2	234 21.0	27.9	198 15.3	24.0
03	234 16.5	267 37.7	.. 09.7	214 41.1	.. 50.0	249 23.0	.. 27.7	213 17.5	.. 24.1
04	249 19.0	282 37.6	09.1	229 41.8	50.7	264 24.9	27.5	228 19.7	24.2
05	264 21.4	297 37.6	08.5	244 42.5	51.4	279 26.8	27.3	243 21.8	24.3
W 06	279 23.9	312 37.5	S12 07.9	259 43.2	N 7 52.2	294 28.8	S 7 27.1	258 24.0	N 6 24.4
E 07	294 26.4	327 37.4	07.3	274 43.9	52.9	309 30.7	26.9	273 26.2	24.6
D 08	309 28.8	342 37.4	06.7	289 44.6	53.6	324 32.6	26.6	288 28.4	24.7
N 09	324 31.3	357 37.3	.. 06.1	304 45.3	.. 54.4	339 34.6	.. 26.4	303 30.5	.. 24.8
E 10	339 33.7	12 37.2	05.5	319 46.0	55.1	354 36.5	26.2	318 32.7	24.9
S 11	354 36.2	27 37.1	04.9	334 46.7	55.8	9 38.4	26.0	333 34.9	25.0
D 12	9 38.7	42 37.1	S12 04.3	349 47.4	N 7 56.6	24 40.4	S 7 25.8	348 37.0	N 6 25.2
A 13	24 41.1	57 37.0	03.7	4 48.1	57.3	39 42.3	25.6	3 39.2	25.3
Y 14	39 43.6	72 36.9	03.1	19 48.8	58.0	54 44.2	25.4	18 41.4	25.4
15	54 46.1	87 36.9	.. 02.5	34 49.5	.. 58.7	69 46.2	.. 25.2	33 43.6	.. 25.5
16	69 48.5	102 36.8	01.9	49 50.2	7 59.5	84 48.1	24.9	48 45.7	25.6
17	84 51.0	117 36.7	01.3	64 50.9	8 00.2	99 50.0	24.7	63 47.9	25.7
18	99 53.5	132 36.7	S12 00.7	79 51.7	N 8 01.0	114 52.0	S 7 24.5	78 50.1	N 6 25.9
19	114 55.9	147 36.6	12 00.1	94 52.4	01.7	129 53.9	24.3	93 52.2	26.0
20	129 58.4	162 36.5	11 59.4	109 53.1	02.4	144 55.9	24.1	108 54.4	26.1
21	145 00.9	177 36.5	.. 58.8	124 53.8	.. 03.1	159 57.8	.. 23.9	123 56.6	.. 26.2
22	160 03.3	192 36.4	58.2	139 54.5	03.9	174 59.7	23.7	138 58.8	26.3
23	175 05.8	207 36.3	57.6	154 55.2	04.6	190 01.7	23.5	154 00.9	26.5
2 00	190 08.2	222 36.2	S11 57.0	169 55.9	N 8 05.3	205 03.6	S 7 23.2	169 03.1	N 6 26.6
01	205 10.7	237 36.2	56.4	184 56.6	06.1	220 05.5	23.0	184 05.3	26.7
02	220 13.2	252 36.1	55.8	199 57.3	06.8	235 07.5	22.8	199 07.4	26.8
03	235 15.6	267 36.0	.. 55.1	214 58.0	.. 07.5	250 09.4	.. 22.6	214 09.6	.. 26.9
04	250 18.1	282 35.9	54.5	229 58.7	08.3	265 11.3	22.4	229 11.8	27.0
05	265 20.6	297 35.9	53.9	244 59.4	09.0	280 13.3	22.2	244 14.0	27.2
T 06	280 23.0	312 35.8	S11 53.3	260 00.1	N 8 09.7	295 15.2	S 7 22.0	259 16.1	N 6 27.3
H 07	295 25.5	327 35.7	52.7	275 00.8	10.4	310 17.2	21.8	274 18.3	27.4
U 08	310 28.0	342 35.6	52.1	290 01.5	11.2	325 19.1	21.5	289 20.5	27.5
R 09	325 30.4	357 35.6	.. 51.4	305 02.2	.. 11.9	340 21.0	.. 21.3	304 22.6	.. 27.6
S 10	340 32.9	12 35.5	50.8	320 02.9	12.6	355 23.0	21.1	319 24.8	27.7
D 11	355 35.3	27 35.4	50.2	335 03.6	13.4	10 24.9	20.9	334 27.0	27.9
A 12	10 37.8	42 35.3	S11 49.6	350 04.3	N 8 14.1	25 26.8	S 7 20.7	349 29.2	N 6 28.0
Y 13	25 40.3	57 35.3	48.9	5 05.0	14.8	40 28.8	20.5	4 31.3	28.1
14	40 42.7	72 35.2	48.3	20 05.7	15.5	55 30.7	20.3	19 33.5	28.2
15	55 45.2	87 35.1	.. 47.7	35 06.4	.. 16.3	70 32.7	.. 20.1	34 35.7	.. 28.3
16	70 47.7	102 35.0	47.1	50 07.1	17.0	85 34.6	19.9	49 37.8	28.5
17	85 50.1	117 35.0	46.4	65 07.8	17.7	100 36.5	19.6	64 40.0	28.6
18	100 52.6	132 34.9	S11 45.8	80 08.5	N 8 18.5	115 38.5	S 7 19.4	79 42.2	N 6 28.7
19	115 55.1	147 34.8	45.2	95 09.2	19.2	130 40.4	19.2	94 44.4	28.8
20	130 57.5	162 34.7	44.6	110 09.9	19.9	145 42.3	19.0	109 46.5	28.9
21	146 00.0	177 34.7	.. 43.9	125 10.6	.. 20.6	160 44.3	.. 18.8	124 48.7	.. 29.0
22	161 02.5	192 34.6	43.3	140 11.3	21.4	175 46.2	18.6	139 50.9	29.2
23	176 04.9	207 34.5	42.7	155 12.0	22.1	190 48.2	18.4	154 53.0	29.3
3 00	191 07.4	222 34.4	S11 42.0	170 12.7	N 8 22.8	205 50.1	S 7 18.2	169 55.2	N 6 29.4
01	206 09.8	237 34.3	41.4	185 13.4	23.5	220 52.0	17.9	184 57.4	29.5
02	221 12.3	252 34.3	40.8	200 14.1	24.3	235 54.0	17.7	199 59.6	29.6
03	236 14.8	267 34.2	.. 40.1	215 14.8	.. 25.0	250 55.9	.. 17.5	215 01.7	.. 29.8
04	251 17.2	282 34.1	39.5	230 15.5	25.7	265 57.9	17.3	230 03.9	29.9
05	266 19.7	297 34.0	38.9	245 16.2	26.5	280 59.8	17.1	245 06.1	30.0
F 06	281 22.2	312 33.9	S11 38.2	260 16.9	N 8 27.2	296 01.7	S 7 16.9	260 08.2	N 6 30.1
R 07	296 24.6	327 33.9	37.6	275 17.6	27.9	311 03.7	16.7	275 10.4	30.2
I 08	311 27.1	342 33.8	37.0	290 18.3	28.6	326 05.6	16.5	290 12.6	30.3
D 09	326 29.6	357 33.7	.. 36.3	305 19.0	.. 29.4	341 07.5	.. 16.3	305 14.8	.. 30.5
A 10	341 32.0	12 33.6	35.7	320 19.8	30.1	356 09.5	16.0	320 16.9	30.6
Y 11	356 34.5	27 33.5	35.0	335 20.5	30.8	11 11.4	15.8	335 19.1	30.7
12	11 37.0	42 33.5	S11 34.4	350 21.2	N 8 31.5	26 13.4	S 7 15.6	350 21.3	N 6 30.8
13	26 39.4	57 33.4	33.8	5 21.9	32.3	41 15.3	15.4	5 23.4	30.9
14	41 41.9	72 33.3	33.1	20 22.6	33.0	56 17.2	15.2	20 25.6	31.1
15	56 44.3	87 33.2	.. 32.5	35 23.3	.. 33.7	71 19.2	.. 15.0	35 27.8	.. 31.2
16	71 46.8	102 33.1	31.8	50 24.0	34.4	86 21.1	14.8	50 30.0	31.3
17	86 49.3	117 33.0	31.2	65 24.7	35.1	101 23.1	14.6	65 32.1	31.4
18	101 51.7	132 33.0	S11 30.5	80 25.4	N 8 35.9	116 25.0	S 7 14.4	80 34.3	N 6 31.5
19	116 54.2	147 32.9	29.9	95 26.1	36.6	131 26.9	14.1	95 36.5	31.6
20	131 56.7	162 32.8	29.3	110 26.8	37.3	146 28.9	13.9	110 38.6	31.8
21	146 59.1	177 32.7	.. 28.6	125 27.5	.. 38.0	161 30.8	.. 13.7	125 40.8	.. 31.9
22	162 01.6	192 32.6	28.0	140 28.2	38.8	176 32.8	13.5	140 43.0	32.0
23	177 04.1	207 32.5	27.3	155 28.9	39.5	191 34.7	13.3	155 45.2	32.1
Mer. Pass.	h m 11 17.6	v −0.1	d 0.6	v 0.7	d 0.7	v 1.9	d 0.2	v 2.2	d 0.1

STARS

Name	SHA	Dec
Acamar	315 27.6	S40 19.0
Achernar	335 36.1	S57 15.0
Acrux	173 21.6	S63 05.4
Adhara	255 21.7	S28 58.6
Aldebaran	291 03.0	N16 30.2
Alioth	166 30.4	N55 58.2
Alkaid	153 07.7	N49 19.3
Al Na'ir	27 58.7	S46 58.1
Alnilam	275 58.3	S 1 12.5
Alphard	218 07.4	S 8 39.3
Alphecca	126 20.7	N26 43.2
Alpheratz	357 55.9	N29 04.7
Altair	62 19.7	N 8 51.7
Ankaa	353 27.7	S42 19.1
Antares	112 40.4	S26 25.6
Arcturus	146 06.1	N19 11.4
Atria	107 52.5	S69 01.1
Avior	234 22.5	S59 30.6
Bellatrix	278 44.7	N 6 20.6
Betelgeuse	271 14.0	N 7 24.2
Canopus	264 01.4	S52 42.1
Capella	280 51.9	N45 59.8
Deneb	49 39.6	N45 16.2
Denebola	182 45.3	N14 34.8
Diphda	349 07.9	S18 00.0
Dubhe	194 05.5	N61 45.7
Elnath	278 27.5	N28 36.2
Eltanin	90 51.4	N51 29.2
Enif	33 58.8	N 9 51.9
Fomalhaut	15 37.2	S29 37.9
Gacrux	172 13.3	S57 06.2
Gienah	176 04.0	S17 32.8
Hadar	149 03.9	S60 21.7
Hamal	328 14.3	N23 27.1
Kaus Aust.	83 59.3	S34 22.9
Kochab	137 18.7	N74 09.7
Markab	13 50.3	N15 11.6
Menkar	314 27.5	N 4 04.7
Menkent	148 21.0	S36 21.6
Miaplacidus	221 41.5	S69 42.9
Mirfak	308 57.5	N49 51.2
Nunki	76 12.8	S26 17.8
Peacock	53 37.9	S56 44.2
Pollux	243 41.9	N28 01.8
Procyon	245 11.9	N 5 13.6
Rasalhague	96 17.2	N12 33.6
Regulus	207 55.7	N11 58.4
Rigel	281 23.4	S 8 12.5
Rigil Kent.	140 07.1	S60 49.5
Sabik	102 25.8	S15 43.3
Schedar	349 54.4	N56 31.5
Shaula	96 37.6	S37 05.9
Sirius	258 44.0	S16 43.1
Spica	158 43.3	S11 09.2
Suhail	223 00.8	S43 25.9
Vega	80 46.8	N38 46.7
Zuben'ubi	137 18.1	S16 02.0

	SHA	Mer. Pass.
	° ′	h m
Venus	32 28.0	9 10
Mars	339 47.6	12 40
Jupiter	14 55.4	10 18
Saturn	338 54.9	12 42

UT	SUN GHA	SUN Dec	MOON GHA	v	MOON Dec	d	HP	Lat.	Twilight Naut.	Twilight Civil	Sunrise	Moonrise 1	Moonrise 2	Moonrise 3	Moonrise 4
d h	° '	° '	° '	'	° '	'	'	°	h m	h m	h m	h m	h m	h m	h m
1 00	178 58.8	N 4 22.7	125 05.9	6.9	N16 01.0	5.8	59.4	N 72	01 28	03 37	04 52	05 33	05 02	▭	06 59
01	193 59.0	23.7	139 31.8	6.9	16 06.8	5.8	59.4	N 70	02 15	03 53	05 00	06 17	06 35	07 16	08 28
02	208 59.2	24.6	153 57.7	7.0	16 12.6	5.7	59.4	68	02 44	04 07	05 06	06 47	07 15	08 01	09 07
03	223 59.4 ..	25.6	168 23.7	6.9	16 18.3	5.5	59.3	66	03 05	04 17	05 11	07 09	07 43	08 31	09 34
04	238 59.6	26.5	182 49.6	7.0	16 23.8	5.5	59.3	64	03 22	04 26	05 16	07 26	08 04	08 53	09 55
05	253 59.8	27.5	197 15.6	7.0	16 29.3	5.3	59.2	62	03 35	04 34	05 20	07 41	08 21	09 11	10 12
06	269 00.0	N 4 28.5	211 41.6	7.0	N16 34.6	5.2	59.2	60	03 47	04 40	05 23	07 53	08 35	09 26	10 26
W 07	284 00.1	29.4	226 07.6	7.1	16 39.8	5.1	59.2	N 58	03 56	04 46	05 26	08 04	08 47	09 39	10 38
E 08	299 00.3	30.4	240 33.7	7.0	16 44.9	5.0	59.1	56	04 04	04 51	05 29	08 13	08 58	09 50	10 48
D 09	314 00.5 ..	31.4	254 59.7	7.1	16 49.9	4.8	59.1	54	04 11	04 55	05 31	08 22	09 07	09 59	10 57
N 10	329 00.7	32.3	269 25.8	7.1	16 54.7	4.8	59.1	52	04 17	04 59	05 33	08 29	09 16	10 08	11 05
E 11	344 00.9	33.3	283 51.9	7.1	16 59.5	4.7	59.0	50	04 23	05 02	05 35	08 36	09 23	10 16	11 12
S 12	359 01.1	N 4 34.3	298 18.0	7.1	N17 04.2	4.5	59.0	45	04 34	05 10	05 39	08 50	09 39	10 32	11 28
D 13	14 01.2	35.2	312 44.1	7.1	17 08.7	4.4	58.9	N 40	04 43	05 16	05 43	09 02	09 52	10 45	11 41
A 14	29 01.4	36.2	327 10.2	7.2	17 13.1	4.3	58.9	35	04 51	05 20	05 46	09 12	10 03	10 57	11 52
Y 15	44 01.6 ..	37.2	341 36.4	7.2	17 17.4	4.2	58.9	30	04 56	05 25	05 49	09 21	10 13	11 07	12 01
16	59 01.8	38.1	356 02.6	7.2	17 21.6	4.1	58.8	20	05 05	05 31	05 53	09 37	10 30	11 24	12 18
17	74 02.0	39.1	10 28.8	7.2	17 25.7	3.9	58.8	N 10	05 11	05 36	05 57	09 50	10 45	11 39	12 32
18	89 02.2	N 4 40.1	24 55.0	7.3	N17 29.6	3.9	58.7	0	05 16	05 40	06 00	10 03	10 59	11 53	12 45
19	104 02.4	41.0	39 21.3	7.3	17 33.5	3.7	58.7	S 10	05 18	05 43	06 04	10 16	11 13	12 07	12 59
20	119 02.5	42.0	53 47.6	7.3	17 37.2	3.6	58.7	20	05 20	05 45	06 07	10 30	11 28	12 22	13 13
21	134 02.7 ..	42.9	68 13.9	7.3	17 40.8	3.5	58.6	30	05 19	05 47	06 11	10 45	11 45	12 40	13 29
22	149 02.9	43.9	82 40.2	7.4	17 44.3	3.4	58.6	35	05 19	05 48	06 13	10 55	11 55	12 50	13 39
23	164 03.1	44.9	97 06.6	7.4	17 47.7	3.3	58.5	40	05 17	05 49	06 16	11 05	12 06	13 01	13 50
2 00	179 03.3	N 4 45.8	111 33.0	7.4	N17 51.0	3.1	58.5	45	05 15	05 49	06 18	11 17	12 20	13 15	14 02
01	194 03.5	46.8	125 59.4	7.5	17 54.1	3.1	58.5	S 50	05 12	05 50	06 22	11 32	12 36	13 31	14 18
02	209 03.7	47.8	140 25.9	7.4	17 57.2	2.9	58.4	52	05 10	05 50	06 23	11 40	12 44	13 39	14 25
03	224 03.8 ..	48.7	154 52.3	7.6	18 00.1	2.8	58.4	54	05 09	05 50	06 25	11 47	12 52	13 48	14 33
04	239 04.0	49.7	169 18.9	7.5	18 02.9	2.7	58.4	56	05 06	05 50	06 27	11 56	13 02	13 57	14 42
05	254 04.2	50.7	183 45.4	7.6	18 05.6	2.6	58.3	58	05 04	05 50	06 29	12 06	13 13	14 08	14 52
06	269 04.4	N 4 51.6	198 12.0	7.6	N18 08.2	2.4	58.3	S 60	05 01	05 49	06 31	12 18	13 26	14 21	15 04
07	284 04.6	52.6	212 38.6	7.6	18 10.6	2.4	58.2								
T 08	299 04.8	53.5	227 05.2	7.7	18 13.0	2.2	58.2	Lat.	Sunset	Twilight Civil	Twilight Naut.	Moonset 1	Moonset 2	Moonset 3	Moonset 4
H 09	314 04.9 ..	54.5	241 31.9	7.7	18 15.2	2.2	58.2								
U 10	329 05.1	55.5	255 58.6	7.8	18 17.4	2.0	58.1	°	h m	h m	h m	h m	h m	h m	h m
R 11	344 05.3	56.4	270 25.4	7.7	18 19.4	1.9	58.1	N 72	19 18	20 35	22 52	01 53	04 22		06 10
S 12	359 05.5	N 4 57.4	284 52.1	7.8	N18 21.3	1.8	58.0	N 70	19 10	20 17	21 59	01 10	02 49	04 02	04 41
D 13	14 05.7	58.3	299 18.9	7.9	18 23.1	1.6	58.0	68	19 03	20 04	21 28	00 41	02 09	03 17	04 01
A 14	29 05.9	4 59.3	313 45.8	7.9	18 24.7	1.6	58.0	66	18 58	19 53	21 06	00 19	01 42	02 47	03 34
Y 15	44 06.1	5 00.3	328 12.7	7.9	18 26.3	1.5	57.9	64	18 53	19 43	20 48	00 02	01 21	02 25	03 13
16	59 06.2	01.2	342 39.6	8.0	18 27.8	1.3	57.9	62	18 49	19 36	20 35	25 04	01 04	02 07	02 56
17	74 06.4	02.2	357 06.6	8.0	18 29.1	1.2	57.8	60	18 46	19 29	20 23	24 50	00 50	01 52	02 41
18	89 06.6	N 5 03.1	11 33.6	8.0	N18 30.3	1.1	57.8	N 58	18 43	19 23	20 13	24 38	00 38	01 40	02 29
19	104 06.8	04.1	26 00.6	8.1	18 31.4	1.0	57.8	56	18 40	19 18	20 05	24 28	00 28	01 29	02 19
20	119 07.0	05.1	40 27.7	8.2	18 32.4	0.9	57.7	54	18 37	19 14	19 58	24 19	00 19	01 19	02 10
21	134 07.2 ..	06.0	54 54.9	8.1	18 33.3	0.8	57.7	52	18 35	19 10	19 51	24 11	00 11	01 11	02 01
22	149 07.3	07.0	69 22.0	8.3	18 34.1	0.7	57.7	50	18 33	19 06	19 46	24 03	00 03	01 03	01 54
23	164 07.5	07.9	83 49.3	8.2	18 34.8	0.5	57.6	45	18 29	18 58	19 34	23 48	24 46	00 46	01 38
3 00	179 07.7	N 5 08.9	98 16.5	8.3	N18 35.3	0.5	57.6	N 40	18 25	18 52	19 25	23 35	24 33	00 33	01 25
01	194 07.9	09.9	112 43.8	8.4	18 35.8	0.3	57.5	35	18 22	18 47	19 17	23 24	24 21	00 21	01 14
02	209 08.1	10.8	127 11.2	8.4	18 36.1	0.3	57.5	30	18 19	18 43	19 12	23 14	24 11	00 11	01 04
03	224 08.3 ..	11.8	141 38.6	8.4	18 36.4	0.1	57.5	20	18 15	18 37	19 03	22 58	23 54	24 47	00 47
04	239 08.4	12.7	156 06.0	8.5	18 36.5	0.0	57.4	N 10	18 11	18 32	18 56	22 43	23 39	24 32	00 32
05	254 08.6	13.7	170 33.5	8.5	18 36.5	0.1	57.4	0	18 07	18 28	18 52	22 30	23 25	24 18	00 18
06	269 08.8	N 5 14.7	185 01.0	8.6	N18 36.4	0.2	57.4	S 10	18 03	18 24	18 49	22 16	23 11	24 05	00 05
07	284 09.0	15.6	199 28.6	8.6	18 36.2	0.3	57.3	20	18 00	18 22	18 47	22 02	22 56	23 50	24 43
08	299 09.2	16.6	213 56.2	8.7	18 35.9	0.4	57.3	30	17 56	18 20	18 47	21 46	22 39	23 33	24 27
F 09	314 09.4 ..	17.5	228 23.9	8.7	18 35.5	0.5	57.2	35	17 53	18 19	18 48	21 36	22 28	23 23	24 18
R 10	329 09.5	18.5	242 51.6	8.7	18 35.0	0.6	57.2	40	17 51	18 18	18 49	21 25	22 17	23 11	24 08
I 11	344 09.7	19.4	257 19.3	8.9	18 34.4	0.7	57.2	45	17 48	18 17	18 51	21 12	22 03	22 58	23 55
D 12	359 09.9	N 5 20.4	271 47.2	8.8	N18 33.7	0.9	57.1	S 50	17 45	18 17	18 54	20 57	21 47	22 42	23 40
A 13	14 10.1	21.4	286 15.0	8.9	18 32.8	0.9	57.1	52	17 43	18 17	18 56	20 49	21 39	22 34	23 33
Y 14	29 10.3	22.3	300 42.9	9.0	18 31.9	1.0	57.1	54	17 41	18 17	18 57	20 41	21 30	22 25	23 26
15	44 10.5 ..	23.3	315 10.9	9.0	18 30.9	1.2	57.0	56	17 40	18 17	19 00	20 32	21 21	22 16	23 17
16	59 10.6	24.2	329 38.9	9.1	18 29.7	1.2	57.0	58	17 37	18 16	19 02	20 22	21 09	22 05	23 07
17	74 10.8	25.2	344 07.0	9.1	18 28.5	1.3	57.0	S 60	17 35	18 17	19 05	20 10	20 57	21 52	22 56
18	89 11.0	N 5 26.1	358 35.1	9.2	N18 27.2	1.5	56.9								
19	104 11.2	27.1	13 03.3	9.2	18 25.7	1.5	56.9								
20	119 11.4	28.1	27 31.5	9.3	18 24.2	1.7	56.9								
21	134 11.6 ..	29.0	41 59.8	9.3	18 22.5	1.7	56.8								
22	149 11.7	30.0	56 28.1	9.4	18 20.8	1.9	56.8								
23	164 11.9	30.9	70 56.5	9.4	N18 18.9	1.9	56.7								

	SUN		MOON			
Day	Eqn. of Time 00h	Eqn. of Time 12h	Mer. Pass.	Mer. Pass. Upper	Mer. Pass. Lower	Age Phase
d	m s	m s	h m	h m	h m	d %
1	04 05	03 56	12 04	16 16	03 48	04 26
2	03 47	03 38	12 04	17 12	04 44	05 36
3	03 30	03 21	12 03	18 06	05 39	06 46

SD 16.0   d 1.0   SD 16.1   15.8   15.6

1998 APRIL 4, 5, 6 (SAT., SUN., MON.)

UT (d h)	ARIES GHA	VENUS GHA	VENUS Dec	MARS GHA	MARS Dec	JUPITER GHA	JUPITER Dec	SATURN GHA	SATURN Dec
4 00	192 06.5	222 32.5	S11 26.7	170 29.6	N 8 40.2	206 36.6	S 7 13.1	170 47.3	N 6 32.2
01	207 09.0	237 32.4	26.0	185 30.3	40.9	221 38.6	12.9	185 49.5	32.4
02	222 11.5	252 32.3	25.4	200 31.0	41.7	236 40.5	12.7	200 51.7	32.5
03	237 13.9	267 32.2 ..	24.7	215 31.7 ..	42.4	251 42.5 ..	12.5	215 53.8 ..	32.6
04	252 16.4	282 32.1	24.1	230 32.4	43.1	266 44.4	12.2	230 56.0	32.7
05	267 18.8	297 32.0	23.4	245 33.1	43.8	281 46.3	12.0	245 58.2	32.8
S 06	282 21.3	312 31.9	S11 22.8	260 33.8	N 8 44.5	296 48.3	S 7 11.8	261 00.3	N 6 32.9
A 07	297 23.8	327 31.9	22.1	275 34.5	45.3	311 50.2	11.6	276 02.5	33.1
T 08	312 26.2	342 31.8	21.4	290 35.2	46.0	326 52.2	11.4	291 04.7	33.2
U 09	327 28.7	357 31.7 ..	20.8	305 35.9 ..	46.7	341 54.1 ..	11.2	306 06.9 ..	33.3
R 10	342 31.2	12 31.6	20.1	320 36.6	47.4	356 56.1	11.0	321 09.0	33.4
D 11	357 33.6	27 31.5	19.5	335 37.3	48.1	11 58.0	10.8	336 11.2	33.5
A 12	12 36.1	42 31.4	S11 18.8	350 38.0	N 8 48.9	26 59.9	S 7 10.6	351 13.4	N 6 33.6
Y 13	27 38.6	57 31.3	18.2	5 38.7	49.6	42 01.9	10.3	6 15.5	33.8
14	42 41.0	72 31.2	17.5	20 39.4	50.3	57 03.8	10.1	21 17.7	33.9
15	57 43.5	87 31.2 ..	16.8	35 40.1 ..	51.0	72 05.8 ..	09.9	36 19.9 ..	34.0
16	72 45.9	102 31.1	16.2	50 40.8	51.8	87 07.7	09.7	51 22.0	34.1
17	87 48.4	117 31.0	15.5	65 41.5	52.5	102 09.6	09.5	66 24.2	34.2
18	102 50.9	132 30.9	S11 14.9	80 42.2	N 8 53.2	117 11.6	S 7 09.3	81 26.4	N 6 34.4
19	117 53.3	147 30.8	14.2	95 42.9	53.9	132 13.5	09.1	96 28.6	34.5
20	132 55.8	162 30.7	13.5	110 43.6	54.6	147 15.5	08.9	111 30.7	34.6
21	147 58.3	177 30.6 ..	12.9	125 44.3 ..	55.3	162 17.4 ..	08.7	126 32.9 ..	34.7
22	163 00.7	192 30.5	12.2	140 45.0	56.1	177 19.4	08.5	141 35.1	34.8
23	178 03.2	207 30.4	11.5	155 45.7	56.8	192 21.3	08.2	156 37.2	34.9
5 00	193 05.7	222 30.4	S11 10.9	170 46.4	N 8 57.5	207 23.2	S 7 08.0	171 39.4	N 6 35.1
01	208 08.1	237 30.3	10.2	185 47.1	58.2	222 25.2	07.8	186 41.6	35.2
02	223 10.6	252 30.2	09.5	200 47.8	58.9	237 27.1	07.6	201 43.8	35.3
03	238 13.1	267 30.1 ..	08.9	215 48.5	8 59.7	252 29.1 ..	07.4	216 45.9 ..	35.4
04	253 15.5	282 30.0	08.2	230 49.2	9 00.4	267 31.0	07.2	231 48.1	35.5
05	268 18.0	297 29.9	07.5	245 49.9	01.1	282 33.0	07.0	246 50.3	35.7
S 06	283 20.4	312 29.8	S11 06.9	260 50.6	N 9 01.8	297 34.9	S 7 06.8	261 52.4	N 6 35.8
U 07	298 22.9	327 29.7	06.2	275 51.3	02.5	312 36.8	06.6	276 54.6	35.9
N 08	313 25.4	342 29.6	05.5	290 52.0	03.2	327 38.8	06.4	291 56.8	36.0
D 09	328 27.8	357 29.5 ..	04.9	305 52.7 ..	04.0	342 40.7 ..	06.1	306 58.9 ..	36.1
A 10	343 30.3	12 29.4	04.2	320 53.4	04.7	357 42.7	05.9	322 01.1	36.2
Y 11	358 32.8	27 29.3	03.5	335 54.1	05.4	12 44.6	05.7	337 03.3	36.4
12	13 35.2	42 29.2	S11 02.8	350 54.8	N 9 06.1	27 46.6	S 7 05.5	352 05.5	N 6 36.5
13	28 37.7	57 29.2	02.2	5 55.5	06.8	42 48.5	05.3	7 07.6	36.6
14	43 40.2	72 29.1	01.5	20 56.2	07.5	57 50.4	05.1	22 09.8	36.7
15	58 42.6	87 29.0 ..	00.8	35 56.9 ..	08.3	72 52.4 ..	04.9	37 12.0 ..	36.8
16	73 45.1	102 28.9	11 00.1	50 57.6	09.0	87 54.3	04.7	52 14.1	36.9
17	88 47.6	117 28.8	10 59.5	65 58.3	09.7	102 56.3	04.5	67 16.3	37.1
18	103 50.0	132 28.7	S10 58.8	80 59.0	N 9 10.4	117 58.2	S 7 04.3	82 18.5	N 6 37.2
19	118 52.5	147 28.6	58.1	95 59.7	11.1	133 00.2	04.0	97 20.6	37.3
20	133 54.9	162 28.5	57.4	111 00.4	11.8	148 02.1	03.8	112 22.8	37.4
21	148 57.4	177 28.4 ..	56.7	126 01.1 ..	12.5	163 04.1 ..	03.6	127 25.0 ..	37.5
22	163 59.9	192 28.3	56.1	141 01.8	13.3	178 06.0	03.4	142 27.2	37.7
23	179 02.3	207 28.2	55.4	156 02.5	14.0	193 07.9	03.2	157 29.3	37.8
6 00	194 04.8	222 28.1	S10 54.7	171 03.2	N 9 14.7	208 09.9	S 7 03.0	172 31.5	N 6 37.9
01	209 07.3	237 28.0	54.0	186 03.9	15.4	223 11.8	02.8	187 33.7	38.0
02	224 09.7	252 27.9	53.3	201 04.6	16.1	238 13.8	02.6	202 35.8	38.1
03	239 12.2	267 27.8 ..	52.6	216 05.3 ..	16.8	253 15.7 ..	02.4	217 38.0 ..	38.2
04	254 14.7	282 27.7	52.0	231 06.0	17.5	268 17.7	02.2	232 40.2	38.4
05	269 17.1	297 27.6	51.3	246 06.7	18.3	283 19.6	02.0	247 42.3	38.5
M 06	284 19.6	312 27.5	S10 50.6	261 07.4	N 9 19.0	298 21.6	S 7 01.7	262 44.5	N 6 38.6
O 07	299 22.0	327 27.4	49.9	276 08.1	19.7	313 23.5	01.5	277 46.7	38.7
N 08	314 24.5	342 27.3	49.2	291 08.8	20.4	328 25.5	01.3	292 48.8	38.8
D 09	329 27.0	357 27.2 ..	48.5	306 09.5 ..	21.1	343 27.4 ..	01.1	307 51.0 ..	39.0
A 10	344 29.4	12 27.1	47.8	321 10.2	21.8	358 29.3	00.9	322 53.2	39.1
Y 11	359 31.9	27 27.0	47.1	336 10.9	22.5	13 31.3	00.7	337 55.4	39.2
12	14 34.4	42 26.9	S10 46.4	351 11.6	N 9 23.2	28 33.2	S 7 00.5	352 57.5	N 6 39.3
13	29 36.8	57 26.8	45.8	6 12.3	24.0	43 35.2	00.3	7 59.7	39.4
14	44 39.3	72 26.7	45.1	21 13.0	24.7	58 37.1	7 00.1	23 01.9	39.5
15	59 41.8	87 26.6 ..	44.4	36 13.7 ..	25.4	73 39.1	6 59.9	38 04.0 ..	39.7
16	74 44.2	102 26.6	43.7	51 14.4	26.1	88 41.0	59.7	53 06.2	39.8
17	89 46.7	117 26.5	43.0	66 15.1	26.8	103 43.0	59.4	68 08.4	39.9
18	104 49.2	132 26.4	S10 42.3	81 15.8	N 9 27.5	118 44.9	S 6 59.2	83 10.5	N 6 40.0
19	119 51.6	147 26.3	41.6	96 16.5	28.2	133 46.9	59.0	98 12.7	40.1
20	134 54.1	162 26.2	40.9	111 17.2	28.9	148 48.8	58.8	113 14.9	40.2
21	149 56.5	177 26.1 ..	40.2	126 17.9 ..	29.6	163 50.8 ..	58.6	128 17.1 ..	40.4
22	164 59.0	192 26.0	39.5	141 18.6	30.4	178 52.7	58.4	143 19.2	40.5
23	180 01.5	207 25.9	38.8	156 19.3	31.1	193 54.6	58.2	158 21.4	40.6
Mer. Pass. 11 05.8		v −0.1	d 0.7	v 0.7	d 0.7	v 1.9	d 0.2	v 2.2	d 0.1

STARS

Name	SHA	Dec
Acamar	315 27.6	S40 19.0
Achernar	335 36.1	S57 14.9
Acrux	173 21.6	S63 05.4
Adhara	255 21.7	S28 58.6
Aldebaran	291 03.0	N16 30.2
Alioth	166 30.4	N55 58.2
Alkaid	153 07.7	N49 19.3
Al Na'ir	27 58.7	S46 58.1
Alnilam	275 58.3	S 1 12.5
Alphard	218 07.4	S 8 39.3
Alphecca	126 20.6	N26 43.2
Alpheratz	357 55.9	N29 04.6
Altair	62 19.7	N 8 51.7
Ankaa	353 27.7	S42 19.1
Antares	112 40.4	S26 25.6
Arcturus	146 06.1	N19 11.4
Atria	107 52.4	S69 01.1
Avior	234 22.5	S59 30.6
Bellatrix	278 44.7	N 6 20.7
Betelgeuse	271 14.0	N 7 24.2
Canopus	264 01.4	S52 42.1
Capella	280 51.9	N45 59.8
Deneb	49 39.6	N45 16.2
Denebola	182 45.3	N14 34.8
Diphda	349 07.9	S18 00.0
Dubhe	194 05.5	N61 45.7
Elnath	278 27.5	N28 36.2
Eltanin	90 51.4	N51 29.2
Enif	33 58.8	N 9 51.9
Fomalhaut	15 37.2	S29 37.9
Gacrux	172 13.3	S57 06.3
Gienah	176 04.0	S17 32.0
Hadar	149 03.8	S60 21.8
Hamal	328 14.3	N23 27.1
Kaus Aust.	83 59.2	S34 22.9
Kochab	137 18.7	N74 09.7
Markab	13 50.2	N15 11.6
Menkar	314 27.5	N 4 04.7
Menkent	148 21.0	S36 21.6
Miaplacidus	221 41.5	S69 43.0
Mirfak	308 57.5	N49 51.2
Nunki	76 12.8	S26 17.8
Peacock	53 37.8	S56 44.2
Pollux	243 42.0	N28 01.8
Procyon	245 11.9	N 5 13.6
Rasalhague	96 17.2	N12 33.6
Regulus	207 55.7	N11 58.4
Rigel	281 23.4	S 8 12.5
Rigil Kent.	140 07.1	S60 49.5
Sabik	102 25.8	S15 43.3
Schedar	349 54.4	N56 31.5
Shaula	96 37.6	S37 05.9
Sirius	258 44.1	S16 43.1
Spica	158 43.2	S11 09.2
Suhail	223 00.8	S43 25.9
Vega	80 46.8	N38 46.7
Zuben'ubi	137 18.1	S16 02.0

	SHA	Mer. Pass.
	° ′	h m
Venus	29 24.7	9 10
Mars	337 40.7	12 36
Jupiter	14 17.6	10 09
Saturn	338 33.7	12 32

SUN / MOON

UT		SUN GHA	SUN Dec	MOON GHA	MOON v	MOON Dec	MOON d	MOON HP
d h		° ′	° ′	° ′	′	° ′	′	′
4	00	179 12.1	N 5 31.9	85 24.9	9.5	N18 17.0	2.0	56.7
	01	194 12.3	32.8	99 53.4	9.6	18 15.0	2.2	56.7
	02	209 12.5	33.8	114 22.0	9.6	18 12.8	2.2	56.6
	03	224 12.7	.. 34.7	128 50.6	9.6	18 10.6	2.3	56.6
	04	239 12.8	35.7	143 19.2	9.7	18 08.3	2.5	56.6
	05	254 13.0	36.7	157 47.9	9.8	18 05.8	2.5	56.5
S	06	269 13.2	N 5 37.6	172 16.7	9.8	N18 03.3	2.6	56.5
A	07	284 13.4	38.6	186 45.5	9.9	18 00.7	2.7	56.5
T	08	299 13.6	39.5	201 14.4	9.9	17 58.0	2.8	56.4
U	09	314 13.7	.. 40.5	215 43.3	10.0	17 55.2	2.9	56.4
R	10	329 13.9	41.4	230 12.3	10.0	17 52.3	3.0	56.4
D	11	344 14.1	42.4	244 41.3	10.1	17 49.3	3.0	56.4
A	12	359 14.3	N 5 43.3	259 10.4	10.2	N17 46.3	3.2	56.3
Y	13	14 14.5	44.3	273 39.6	10.2	17 43.1	3.3	56.3
	14	29 14.7	45.2	288 08.8	10.2	17 39.8	3.3	56.3
	15	44 14.8	.. 46.2	302 38.0	10.4	17 36.5	3.4	56.2
	16	59 15.0	47.1	317 07.4	10.3	17 33.1	3.6	56.2
	17	74 15.2	48.1	331 36.7	10.5	17 29.5	3.6	56.2
	18	89 15.4	N 5 49.0	346 06.2	10.5	N17 25.9	3.7	56.1
	19	104 15.6	50.0	0 35.7	10.5	17 22.2	3.7	56.1
	20	119 15.7	50.9	15 05.2	10.6	17 18.5	3.9	56.1
	21	134 15.9	.. 51.9	29 34.8	10.7	17 14.6	4.0	56.0
	22	149 16.1	52.9	44 04.5	10.7	17 10.6	4.0	56.0
	23	164 16.3	53.8	58 34.2	10.8	17 06.6	4.1	56.0
5	00	179 16.5	N 5 54.8	73 04.0	10.8	N17 02.5	4.2	56.0
	01	194 16.6	55.7	87 33.8	10.9	16 58.3	4.3	55.9
	02	209 16.8	56.7	102 03.7	11.0	16 54.0	4.4	55.9
	03	224 17.0	.. 57.6	116 33.7	11.0	16 49.6	4.4	55.9
	04	239 17.2	58.6	131 03.7	11.1	16 45.2	4.5	55.8
	05	254 17.4	5 59.5	145 33.8	11.1	16 40.7	4.6	55.8
	06	269 17.6	N 6 00.5	160 03.9	11.2	N16 36.1	4.7	55.8
	07	284 17.7	01.4	174 34.1	11.2	16 31.4	4.8	55.8
S	08	299 17.9	02.4	189 04.3	11.3	16 26.6	4.8	55.7
U	09	314 18.1	.. 03.3	203 34.6	11.3	16 21.8	5.0	55.7
N	10	329 18.3	04.3	218 04.9	11.4	16 16.8	4.9	55.7
D	11	344 18.5	05.2	232 35.3	11.5	16 11.9	5.1	55.6
A	12	359 18.6	N 6 06.2	247 05.8	11.5	N16 06.8	5.2	55.6
Y	13	14 18.8	07.1	261 36.3	11.6	16 01.6	5.2	55.6
	14	29 19.0	08.0	276 06.9	11.6	15 56.4	5.3	55.6
	15	44 19.2	.. 09.0	290 37.5	11.7	15 51.1	5.3	55.5
	16	59 19.4	09.9	305 08.2	11.8	15 45.8	5.5	55.5
	17	74 19.5	10.9	319 39.0	11.8	15 40.3	5.5	55.5
	18	89 19.7	N 6 11.8	334 09.8	11.8	N15 34.8	5.5	55.5
	19	104 19.9	12.8	348 40.6	11.9	15 29.3	5.7	55.4
	20	119 20.1	13.7	3 11.5	12.0	15 23.6	5.7	55.4
	21	134 20.2	.. 14.7	17 42.5	12.0	15 17.9	5.8	55.4
	22	149 20.4	15.6	32 13.5	12.1	15 12.1	5.9	55.4
	23	164 20.6	16.6	46 44.6	12.1	15 06.2	5.9	55.3
6	00	179 20.8	N 6 17.5	61 15.7	12.2	N15 00.3	6.0	55.3
	01	194 21.0	18.5	75 46.9	12.2	14 54.3	6.0	55.3
	02	209 21.1	19.4	90 18.1	12.3	14 48.3	6.1	55.3
	03	224 21.3	.. 20.4	104 49.4	12.4	14 42.2	6.2	55.2
	04	239 21.5	21.3	119 20.8	12.4	14 36.0	6.3	55.2
	05	254 21.7	22.3	133 52.2	12.4	14 29.7	6.3	55.2
	06	269 21.9	N 6 23.2	148 23.6	12.5	N14 23.4	6.4	55.2
	07	284 22.0	24.1	162 55.1	12.6	14 17.1	6.5	55.1
M	08	299 22.2	25.1	177 26.7	12.6	14 10.6	6.5	55.1
O	09	314 22.4	.. 26.0	191 58.3	12.6	14 04.1	6.5	55.1
N	10	329 22.6	27.0	206 29.9	12.7	13 57.6	6.7	55.1
D	11	344 22.8	27.9	221 01.6	12.8	13 50.9	6.6	55.1
A	12	359 22.9	N 6 28.9	235 33.4	12.8	N13 44.3	6.8	55.0
Y	13	14 23.1	29.8	250 05.2	12.9	13 37.5	6.8	55.0
	14	29 23.3	30.8	264 37.1	12.9	13 30.7	6.8	55.0
	15	44 23.5	.. 31.7	279 09.0	12.9	13 23.9	6.9	55.0
	16	59 23.6	32.6	293 40.9	13.1	13 17.0	7.0	55.0
	17	74 23.8	33.6	308 13.0	13.0	13 10.0	7.0	54.9
	18	89 24.0	N 6 34.5	322 45.0	13.1	N13 03.0	7.1	54.9
	19	104 24.2	35.5	337 17.1	13.2	12 55.9	7.1	54.9
	20	119 24.4	36.4	351 49.3	13.2	12 48.8	7.2	54.9
	21	134 24.5	.. 37.4	6 21.5	13.2	12 41.6	7.2	54.9
	22	149 24.7	38.3	20 53.7	13.3	12 34.4	7.3	54.8
	23	164 24.9	39.2	35 26.0	13.4	N12 27.1	7.3	54.8
	SD	16.0	d 0.9	SD 15.3		15.2		15.0

Twilight / Sunrise / Moonrise

Lat.	Twilight Naut.	Twilight Civil	Sunrise	Moonrise 4	Moonrise 5	Moonrise 6	Moonrise 7
°	h m	h m	h m	h m	h m	h m	h m
N 72	00 25	03 17	04 36	06 59	09 11	11 01	12 43
N 70	01 48	03 37	04 46	08 28	09 56	11 29	13 01
68	02 24	03 52	04 53	09 07	10 26	11 50	13 15
66	02 49	04 05	05 00	09 34	10 48	12 06	13 26
64	03 08	04 15	05 05	09 55	11 05	12 20	13 36
62	03 23	04 23	05 10	10 12	11 19	12 31	13 44
60	03 36	04 31	05 14	10 26	11 31	12 40	13 51
N 58	03 46	04 37	05 18	10 38	11 42	12 49	13 57
56	03 55	04 43	05 21	10 48	11 51	12 56	14 02
54	04 03	04 47	05 24	10 57	11 59	13 03	14 07
52	04 10	04 52	05 26	11 05	12 06	13 08	14 12
50	04 16	04 56	05 29	11 12	12 12	13 14	14 16
45	04 28	05 04	05 34	11 28	12 26	13 25	14 24
N 40	04 38	05 11	05 38	11 41	12 38	13 35	14 31
35	04 46	05 16	05 42	11 52	12 47	13 43	14 38
30	04 53	05 21	05 45	12 01	12 56	13 50	14 43
20	05 02	05 28	05 50	12 18	13 10	14 02	14 52
N 10	05 09	05 34	05 55	12 32	13 23	14 13	15 00
0	05 15	05 39	05 59	12 45	13 35	14 23	15 08
S 10	05 18	05 43	06 04	12 59	13 47	14 33	15 16
20	05 20	05 46	06 08	13 13	14 00	14 43	15 24
30	05 21	05 49	06 13	13 29	14 14	14 55	15 33
35	05 21	05 50	06 16	13 39	14 23	15 02	15 38
40	05 20	05 52	06 19	13 50	14 32	15 10	15 44
45	05 19	05 53	06 22	14 02	14 44	15 20	15 51
S 50	05 17	05 54	06 26	14 18	14 57	15 31	16 00
52	05 15	05 55	06 28	14 25	15 04	15 36	16 04
54	05 14	05 55	06 30	14 33	15 11	15 41	16 08
56	05 12	05 56	06 33	14 42	15 18	15 48	16 12
58	05 10	05 56	06 35	14 52	15 27	15 55	16 18
S 60	05 08	05 56	06 38	15 04	15 37	16 03	16 23

Sunset / Twilight / Moonset

Lat.	Sunset	Twilight Civil	Twilight Naut.	Moonset 4	Moonset 5	Moonset 6	Moonset 7
°	h m	h m	h m	h m	h m	h m	h m
N 72	19 33	20 53	////	06 10	05 43	05 33	05 26
N 70	19 23	20 32	22 26	04 41	04 57	05 04	05 07
68	19 15	20 16	21 47	04 01	04 27	04 42	04 51
66	19 08	20 04	21 21	03 34	04 04	04 25	04 39
64	19 02	19 53	21 01	03 13	03 47	04 11	04 29
62	18 57	19 44	20 45	02 56	03 32	03 59	04 20
60	18 53	19 37	20 32	02 41	03 19	03 49	04 12
N 58	18 49	19 30	20 22	02 29	03 09	03 40	04 05
56	18 46	19 25	20 12	02 19	02 59	03 32	03 59
54	18 43	19 19	20 04	02 10	02 51	03 25	03 54
52	18 40	19 15	19 57	02 01	02 44	03 19	03 49
50	18 38	19 11	19 51	01 54	02 37	03 13	03 44
45	18 32	19 02	19 38	01 38	02 22	03 01	03 35
N 40	18 28	18 56	19 28	01 25	02 11	02 51	03 27
35	18 24	18 50	19 20	01 14	02 00	02 42	03 20
30	18 21	18 45	19 14	01 04	01 51	02 34	03 14
20	18 15	18 38	19 03	00 47	01 36	02 21	03 03
N 10	18 10	18 32	18 56	00 32	01 22	02 09	02 54
0	18 06	18 27	18 51	00 18	01 10	01 58	02 45
S 10	18 02	18 23	18 47	00 05	00 57	01 47	02 36
20	17 57	18 19	18 45	24 43	00 43	01 35	02 26
30	17 52	18 16	18 44	24 27	00 27	01 22	02 15
35	17 49	18 15	18 44	24 18	00 18	01 14	02 09
40	17 46	18 13	18 45	24 08	00 08	01 05	02 02
45	17 43	18 12	18 46	23 55	24 54	00 54	01 54
S 50	17 38	18 11	18 48	23 40	24 41	00 41	01 43
52	17 36	18 10	18 49	23 33	24 35	00 35	01 39
54	17 34	18 09	18 50	23 26	24 29	00 29	01 33
56	17 32	18 09	18 52	23 17	24 21	00 21	01 28
58	17 29	18 08	18 54	23 07	24 13	00 13	01 21
S 60	17 26	18 08	18 56	22 56	24 03	00 03	01 14

SUN / MOON

	SUN			MOON			
Day	Eqn. of Time 00h	Eqn. of Time 12h	Mer. Pass.	Mer. Pass. Upper	Mer. Pass. Lower	Age	Phase
d	m s	m s	h m	h m	h m	d	%
4	03 12	03 03	12 03	18 58	06 32	07	57
5	02 54	02 46	12 03	19 47	07 22	08	67
6	02 37	02 29	12 02	20 34	08 11	09	76

1998 APRIL 7, 8, 9 (TUES., WED., THURS.)

UT	ARIES GHA	VENUS −4.3 GHA	VENUS Dec	MARS +1.3 GHA	MARS Dec	JUPITER −2.1 GHA	JUPITER Dec	SATURN +0.6 GHA	SATURN Dec
7 00	195 03.9	222 25.7	S10 38.1	171 20.0	N 9 31.8	208 56.6	S 6 58.0	173 23.6	N 6 40.7
01	210 06.4	237 25.6	37.4	186 20.7	32.5	223 58.5	57.8	188 25.7	40.8
02	225 08.9	252 25.5	36.7	201 21.4	33.2	239 00.5	57.6	203 27.9	41.0
03	240 11.3	267 25.4 ..	36.0	216 22.1 ..	33.9	254 02.4 ..	57.4	218 30.1 ..	41.1
04	255 13.8	282 25.3	35.3	231 22.8	34.6	269 04.4	57.1	233 32.2	41.2
05	270 16.3	297 25.2	34.6	246 23.5	35.3	284 06.3	56.9	248 34.4	41.3
06	285 18.7	312 25.1	S10 33.9	261 24.2	N 9 36.0	299 08.3	S 6 56.7	263 36.6	N 6 41.4
07	300 21.2	327 25.0	33.2	276 24.9	36.7	314 10.2	56.5	278 38.7	41.5
08	315 23.7	342 24.9	32.5	291 25.6	37.5	329 12.2	56.3	293 40.9	41.7
09	330 26.1	357 24.8 ..	31.8	306 26.3 ..	38.2	344 14.1 ..	56.1	308 43.1 ..	41.8
10	345 28.6	12 24.7	31.1	321 27.0	38.9	359 16.1	55.9	323 45.3	41.9
11	0 31.0	27 24.6	30.4	336 27.7	39.6	14 18.0	55.7	338 47.4	42.0
12	15 33.5	42 24.5	S10 29.7	351 28.4	N 9 40.3	29 20.0	S 6 55.5	353 49.6	N 6 42.1
13	30 36.0	57 24.4	28.9	6 29.1	41.0	44 21.9	55.3	8 51.8	42.2
14	45 38.4	72 24.3	28.2	21 29.8	41.7	59 23.9	55.1	23 53.9	42.4
15	60 40.9	87 24.2 ..	27.5	36 30.4 ..	42.4	74 25.8 ..	54.9	38 56.1 ..	42.5
16	75 43.4	102 24.1	26.8	51 31.1	43.1	89 27.8	54.6	53 58.3	42.6
17	90 45.8	117 24.0	26.1	66 31.8	43.8	104 29.7	54.4	69 00.4	42.7
18	105 48.3	132 23.9	S10 25.4	81 32.5	N 9 44.5	119 31.7	S 6 54.2	84 02.6	N 6 42.8
19	120 50.8	147 23.8	24.7	96 33.2	45.2	134 33.6	54.0	99 04.8	43.0
20	135 53.2	162 23.7	24.0	111 33.9	45.9	149 35.6	53.8	114 06.9	43.1
21	150 55.7	177 23.6 ..	23.3	126 34.6 ..	46.6	164 37.5 ..	53.6	129 09.1 ..	43.2
22	165 58.1	192 23.5	22.5	141 35.3	47.3	179 39.5	53.4	144 11.3	43.3
23	181 00.6	207 23.4	21.8	156 36.0	48.1	194 41.4	53.2	159 13.4	43.4
8 00	196 03.1	222 23.3	S10 21.1	171 36.7	N 9 48.8	209 43.4	S 6 53.0	174 15.6	N 6 43.5
01	211 05.5	237 23.2	20.4	186 37.4	49.5	224 45.3	52.8	189 17.8	43.7
02	226 08.0	252 23.1	19.7	201 38.1	50.2	239 47.3	52.6	204 20.0	43.8
03	241 10.5	267 23.0 ..	19.0	216 38.8 ..	50.9	254 49.2 ..	52.4	219 22.1 ..	43.9
04	256 12.9	282 22.9	18.2	231 39.5	51.6	269 51.2	52.1	234 24.3	44.0
05	271 15.4	297 22.7	17.5	246 40.2	52.3	284 53.1	51.9	249 26.5	44.1
06	286 17.9	312 22.6	S10 16.8	261 40.9	N 9 53.0	299 55.1	S 6 51.7	264 28.6	N 6 44.2
07	301 20.3	327 22.5	16.1	276 41.6	53.7	314 57.0	51.5	279 30.8	44.4
08	316 22.8	342 22.4	15.4	291 42.3	54.4	329 59.0	51.3	294 33.0	44.5
09	331 25.3	357 22.3 ..	14.6	306 43.0 ..	55.1	345 00.9 ..	51.1	309 35.1 ..	44.6
10	346 27.7	12 22.2	13.9	321 43.7	55.8	0 02.9	50.9	324 37.3	44.7
11	1 30.2	27 22.1	13.2	336 44.4	56.5	15 04.8	50.7	339 39.5	44.8
12	16 32.6	42 22.0	S10 12.5	351 45.1	N 9 57.2	30 06.8	S 6 50.5	354 41.6	N 6 44.9
13	31 35.1	57 21.9	11.8	6 45.8	57.9	45 08.7	50.3	9 43.8	45.1
14	46 37.6	72 21.8	11.0	21 46.5	58.6	60 10.7	50.1	24 46.0	45.2
15	61 40.0	87 21.7 ..	10.3	36 47.2	9 59.3	75 12.6 ..	49.9	39 48.2 ..	45.3
16	76 42.5	102 21.6	09.6	51 47.9	10 00.0	90 14.6	49.7	54 50.3	45.4
17	91 45.0	117 21.5	08.9	66 48.6	00.7	105 16.5	49.5	69 52.5	45.5
18	106 47.4	132 21.3	S10 08.1	81 49.3	N10 01.4	120 18.5	S 6 49.2	84 54.7	N 6 45.7
19	121 49.9	147 21.2	07.4	96 50.0	02.1	135 20.4	49.0	99 56.8	45.8
20	136 52.4	162 21.1	06.7	111 50.7	02.8	150 22.4	48.8	114 59.0	45.9
21	151 54.8	177 21.0 ..	05.9	126 51.4 ..	03.5	165 24.3 ..	48.6	130 01.2 ..	46.0
22	166 57.3	192 20.9	05.2	141 52.1	04.2	180 26.3	48.4	145 03.3	46.1
23	181 59.7	207 20.8	04.5	156 52.8	04.9	195 28.2	48.2	160 05.5	46.2
9 00	197 02.2	222 20.7	S10 03.7	171 53.5	N10 05.6	210 30.2	S 6 48.0	175 07.7	N 6 46.4
01	212 04.7	237 20.6	03.0	186 54.2	06.3	225 32.1	47.8	190 09.8	46.5
02	227 07.1	252 20.5	02.3	201 54.8	07.0	240 34.1	47.6	205 12.0	46.6
03	242 09.6	267 20.4 ..	01.5	216 55.5 ..	07.7	255 36.0 ..	47.4	220 14.2 ..	46.7
04	257 12.1	282 20.3	00.8	231 56.2	08.4	270 38.0	47.2	235 16.3	46.8
05	272 14.5	297 20.1	10 00.1	246 56.9	09.1	285 39.9	47.0	250 18.5	46.9
06	287 17.0	312 20.0	S 9 59.3	261 57.6	N10 09.8	300 41.9	S 6 46.8	265 20.7	N 6 47.1
07	302 19.5	327 19.9	58.6	276 58.3	10.5	315 43.8	46.6	280 22.9	47.2
08	317 21.9	342 19.8	57.9	291 59.0	11.2	330 45.8	46.3	295 25.0	47.3
09	332 24.4	357 19.7 ..	57.1	306 59.7 ..	11.9	345 47.7 ..	46.1	310 27.2 ..	47.4
10	347 26.9	12 19.6	56.4	322 00.4	12.6	0 49.7	45.9	325 29.4	47.5
11	2 29.3	27 19.5	55.7	337 01.1	13.3	15 51.6	45.7	340 31.5	47.6
12	17 31.8	42 19.4	S 9 54.9	352 01.8	N10 14.0	30 53.6	S 6 45.5	355 33.7	N 6 47.8
13	32 34.2	57 19.3	54.2	7 02.5	14.7	45 55.6	45.3	10 35.9	47.9
14	47 36.7	72 19.1	53.4	22 03.2	15.4	60 57.5	45.1	25 38.0	48.0
15	62 39.2	87 19.0 ..	52.7	37 03.9 ..	16.1	75 59.5 ..	44.9	40 40.2 ..	48.1
16	77 41.6	102 18.9	52.0	52 04.6	16.8	91 01.4	44.7	55 42.4	48.2
17	92 44.1	117 18.8	51.2	67 05.3	17.5	106 03.4	44.5	70 44.5	48.4
18	107 46.6	132 18.7	S 9 50.5	82 06.0	N10 18.2	121 05.3	S 6 44.3	85 46.7	N 6 48.5
19	122 49.0	147 18.6	49.7	97 06.7	18.9	136 07.3	44.1	100 48.9	48.6
20	137 51.5	162 18.5	49.0	112 07.4	19.6	151 09.2	43.9	115 51.0	48.7
21	152 54.0	177 18.4 ..	48.2	127 08.1 ..	20.3	166 11.2 ..	43.7	130 53.2 ..	48.8
22	167 56.4	192 18.2	47.5	142 08.8	21.0	181 13.1	43.5	145 55.4	48.9
23	182 58.9	207 18.1	46.7	157 09.5	21.7	196 15.1	43.2	160 57.5	49.1
Mer. Pass. 10 54.0		v −0.1 d 0.7		v 0.7 d 0.7		v 2.0 d 0.2		v 2.2 d 0.1	

Left margin day labels: **T U E S D A Y** (April 7), **W E D N E S D A Y** (April 8), **T H U R S D A Y** (April 9).

STARS

Name	SHA	Dec
Acamar	315 27.6	S40 19.0
Achernar	335 36.1	S57 14.9
Acrux	173 21.6	S63 05.4
Adhara	255 21.7	S28 58.6
Aldebaran	291 03.0	N16 30.2
Alioth	166 30.4	N55 58.2
Alkaid	153 07.7	N49 19.3
Al Na'ir	27 58.7	S46 58.1
Alnilam	275 58.4	S 1 12.5
Alphard	218 07.4	S 8 39.3
Alphecca	126 20.6	N26 43.2
Alpheratz	357 55.9	N29 04.6
Altair	62 19.7	N 8 51.7
Ankaa	353 27.7	S42 19.0
Antares	112 40.4	S26 25.6
Arcturus	146 06.1	N19 11.4
Atria	107 52.4	S69 01.2
Avior	234 22.6	S59 30.6
Bellatrix	278 44.7	N 6 20.7
Betelgeuse	271 14.1	N 7 24.2
Canopus	264 01.4	S52 42.1
Capella	280 51.9	N45 59.8
Deneb	49 39.6	N45 16.2
Denebola	182 45.3	N14 34.8
Diphda	349 07.9	S17 59.9
Dubhe	194 05.5	N61 45.7
Elnath	278 27.6	N28 36.2
Eltanin	90 51.4	N51 29.2
Enif	33 58.8	N 9 51.9
Fomalhaut	15 37.2	S29 37.9
Gacrux	172 13.3	S57 06.3
Gienah	176 04.0	S17 32.0
Hadar	149 03.8	S60 21.8
Hamal	328 14.3	N23 27.1
Kaus Aust.	83 59.2	S34 22.9
Kochab	137 18.6	N74 09.7
Markab	13 50.2	N15 11.6
Menkar	314 27.6	N 4 04.7
Menkent	148 21.0	S36 21.6
Miaplacidus	221 41.6	S69 43.0
Mirfak	308 57.5	N49 51.2
Nunki	76 12.7	S26 17.8
Peacock	53 37.8	S56 44.2
Pollux	243 42.0	N28 01.8
Procyon	245 11.9	N 5 13.6
Rasalhague	96 17.1	N12 33.6
Regulus	207 55.7	N11 58.4
Rigel	281 23.4	S 8 12.5
Rigil Kent.	140 07.1	S60 49.5
Sabik	102 25.8	S15 43.3
Schedar	349 54.4	N56 31.5
Shaula	96 37.6	S37 06.0
Sirius	258 44.1	S16 43.1
Spica	158 43.2	S11 09.2
Suhail	223 00.8	S43 25.9
Vega	80 46.8	N38 46.7
Zuben'ubi	137 18.0	S16 02.0

	SHA	Mer. Pass.
Venus	26 20.2	9 11
Mars	335 33.7	12 33
Jupiter	13 40.3	10 00
Saturn	338 12.5	12 21

UT	SUN GHA	SUN Dec	MOON GHA	v	MOON Dec	d	HP
7 d h	° '	° '	° '	'	° '	'	'
00	179 25.1	N 6 40.2	49 58.4	13.3	N12 19.8	7.4	54.8
01	194 25.2	41.1	64 30.7	13.5	12 12.4	7.5	54.8
02	209 25.4	42.1	79 03.2	13.5	12 04.9	7.4	54.8
03	224 25.6	.. 43.0	93 35.7	13.5	11 57.5	7.6	54.7
04	239 25.8	44.0	108 08.2	13.5	11 49.9	7.6	54.7
05	254 25.9	44.9	122 40.7	13.7	11 42.3	7.6	54.7
06	269 26.1	N 6 45.8	137 13.4	13.6	N11 34.7	7.7	54.7
07	284 26.3	46.8	151 46.0	13.7	11 27.0	7.7	54.7
08	299 26.5	47.7	166 18.7	13.8	11 19.3	7.8	54.7
09	314 26.7	.. 48.7	180 51.5	13.7	11 11.5	7.8	54.6
10	329 26.8	49.6	195 24.2	13.9	11 03.7	7.9	54.6
11	344 27.0	50.5	209 57.1	13.8	10 55.8	7.9	54.6
12	359 27.2	N 6 51.5	224 29.9	13.9	N10 47.9	7.9	54.6
13	14 27.4	52.4	239 02.8	14.0	10 40.0	8.0	54.6
14	29 27.5	53.4	253 35.8	14.0	10 32.0	8.0	54.6
15	44 27.7	.. 54.3	268 08.8	14.0	10 24.0	8.1	54.6
16	59 27.9	55.2	282 41.8	14.0	10 15.9	8.1	54.5
17	74 28.1	56.2	297 14.8	14.1	10 07.8	8.2	54.5
18	89 28.2	N 6 57.1	311 47.9	14.2	N 9 59.6	8.2	54.5
19	104 28.4	58.0	326 21.1	14.2	9 51.4	8.2	54.5
20	119 28.6	59.0	340 54.3	14.2	9 43.2	8.3	54.5
21	134 28.8	6 59.9	355 27.5	14.2	9 34.9	8.3	54.5
22	149 28.9	7 00.9	10 00.7	14.3	9 26.6	8.3	54.5
23	164 29.1	01.8	24 34.0	14.3	9 18.3	8.4	54.4
8 00	179 29.3	N 7 02.7	39 07.3	14.4	N 9 09.9	8.4	54.4
01	194 29.5	03.7	53 40.7	14.4	9 01.5	8.5	54.4
02	209 29.6	04.6	68 14.1	14.4	8 53.0	8.5	54.4
03	224 29.8	.. 05.5	82 47.5	14.4	8 44.5	8.5	54.4
04	239 30.0	06.5	97 20.9	14.5	8 36.0	8.5	54.4
05	254 30.2	07.4	111 54.4	14.6	8 27.5	8.6	54.4
06	269 30.3	N 7 08.3	126 28.0	14.6	N 8 18.9	8.6	54.3
07	284 30.5	09.3	141 01.5	14.6	8 10.3	8.7	54.3
08	299 30.7	10.2	155 35.1	14.6	8 01.6	8.7	54.3
09	314 30.9	.. 11.2	170 08.7	14.6	7 52.9	8.7	54.3
10	329 31.0	12.1	184 42.3	14.7	7 44.2	8.7	54.3
11	344 31.2	13.0	199 16.0	14.7	7 35.5	8.8	54.3
12	359 31.4	N 7 14.0	213 49.7	14.7	N 7 26.7	8.8	54.3
13	14 31.6	14.9	228 23.4	14.8	7 17.9	8.8	54.3
14	29 31.7	15.8	242 57.2	14.8	7 09.1	8.8	54.3
15	44 31.9	.. 16.8	257 31.0	14.8	7 00.3	8.9	54.2
16	59 32.1	17.7	272 04.8	14.8	6 51.4	8.9	54.2
17	74 32.3	18.6	286 38.6	14.9	6 42.5	9.0	54.2
18	89 32.4	N 7 19.6	301 12.5	14.9	N 6 33.5	8.9	54.2
19	104 32.6	20.5	315 46.4	14.9	6 24.6	9.0	54.2
20	119 32.8	21.4	330 20.3	14.9	6 15.6	9.0	54.2
21	134 32.9	.. 22.4	344 54.2	15.0	6 06.6	9.0	54.2
22	149 33.1	23.3	359 28.2	15.0	5 57.6	9.1	54.2
23	164 33.3	24.2	14 02.2	15.0	5 48.5	9.0	54.2
9 00	179 33.5	N 7 25.2	28 36.2	15.0	N 5 39.5	9.1	54.2
01	194 33.6	26.1	43 10.2	15.0	5 30.4	9.1	54.2
02	209 33.8	27.0	57 44.2	15.1	5 21.3	9.2	54.1
03	224 34.0	.. 27.9	72 18.3	15.1	5 12.1	9.1	54.1
04	239 34.2	28.9	86 52.4	15.1	5 03.0	9.2	54.1
05	254 34.3	29.8	101 26.5	15.1	4 53.8	9.2	54.1
06	269 34.5	N 7 30.7	116 00.6	15.2	N 4 44.6	9.2	54.1
07	284 34.7	31.7	130 34.8	15.1	4 35.4	9.2	54.1
08	299 34.8	32.6	145 08.9	15.2	4 26.2	9.2	54.1
09	314 35.0	.. 33.5	159 43.1	15.2	4 17.0	9.3	54.1
10	329 35.2	34.5	174 17.3	15.2	4 07.7	9.3	54.1
11	344 35.4	35.4	188 51.5	15.2	3 58.4	9.3	54.1
12	359 35.5	N 7 36.3	203 25.7	15.3	N 3 49.1	9.3	54.1
13	14 35.7	37.2	218 00.0	15.2	3 39.8	9.3	54.1
14	29 35.9	38.2	232 34.2	15.3	3 30.5	9.3	54.1
15	44 36.0	.. 39.1	247 08.5	15.3	3 21.2	9.3	54.1
16	59 36.2	40.0	261 42.8	15.3	3 11.9	9.4	54.1
17	74 36.4	41.0	276 17.1	15.3	3 02.5	9.4	54.0
18	89 36.6	N 7 41.9	290 51.4	15.3	N 2 53.2	9.4	54.0
19	104 36.7	42.8	305 25.7	15.3	2 43.8	9.4	54.0
20	119 36.9	43.7	320 00.0	15.3	2 34.4	9.4	54.0
21	134 37.1	.. 44.7	334 34.3	15.4	2 25.0	9.4	54.0
22	149 37.2	45.6	349 08.7	15.4	2 15.6	9.4	54.0
23	164 37.4	46.5	3 43.1	15.3	N 2 06.2	9.4	54.0
	SD 16.0	d 0.9	SD 14.9		14.8		14.7

Rows marked with days: **7** = TUESDAY, **8** = WEDNESDAY, **9** = THURSDAY.

Lat.	Twilight Naut.	Twilight Civil	Sunrise	Moonrise 7	8	9	10
°	h m	h m	h m	h m	h m	h m	h m
N 72	////	02 57	04 20	12 43	14 20	15 54	17 28
N 70	01 13	03 20	04 31	13 01	14 30	15 59	17 26
68	02 02	03 38	04 40	13 15	14 39	16 02	17 25
66	02 32	03 52	04 48	13 26	14 46	16 05	17 25
64	02 53	04 03	04 55	13 36	14 52	16 08	17 24
62	03 11	04 13	05 00	13 44	14 57	16 10	17 23
60	03 24	04 21	05 05	13 51	15 02	16 12	17 23
N 58	03 36	04 28	05 09	13 57	15 06	16 14	17 22
56	03 46	04 34	05 13	14 02	15 09	16 15	17 22
54	03 55	04 40	05 17	14 07	15 12	16 17	17 21
52	04 02	04 45	05 20	14 12	15 15	16 18	17 21
50	04 09	04 49	05 22	14 16	15 18	16 19	17 21
45	04 22	04 58	05 28	14 24	15 23	16 22	17 20
N 40	04 33	05 06	05 33	14 31	15 28	16 24	17 19
35	04 42	05 12	05 38	14 38	15 32	16 26	17 19
30	04 49	05 17	05 42	14 43	15 35	16 27	17 19
20	05 00	05 26	05 48	14 52	15 41	16 30	17 18
N 10	05 08	05 32	05 54	15 00	15 47	16 32	17 17
0	05 14	05 38	05 59	15 08	15 52	16 34	17 17
S 10	05 18	05 43	06 04	15 16	15 57	16 37	17 16
20	05 21	05 47	06 09	15 24	16 02	16 39	17 16
30	05 23	05 51	06 15	15 33	16 08	16 42	17 15
35	05 23	05 53	06 18	15 38	16 12	16 43	17 15
40	05 23	05 55	06 22	15 44	16 16	16 45	17 14
45	05 23	05 57	06 26	15 51	16 20	16 47	17 14
S 50	05 21	05 59	06 31	16 00	16 26	16 50	17 13
52	05 20	06 00	06 33	16 04	16 28	16 51	17 13
54	05 19	06 00	06 36	16 08	16 31	16 52	17 13
56	05 18	06 01	06 39	16 12	16 34	16 53	17 12
58	05 17	06 02	06 42	16 18	16 37	16 55	17 12
S 60	05 15	06 03	06 45	16 23	16 41	16 57	17 12

Lat.	Sunset	Twilight Civil	Twilight Naut.	Moonset 7	8	9	10
°	h m	h m	h m	h m	h m	h m	h m
N 72	19 47	21 13	////	05 26	05 20	05 14	05 08
N 70	19 35	20 48	23 06	05 07	05 08	05 08	05 07
68	19 26	20 30	22 09	04 51	04 58	05 02	05 06
66	19 18	20 15	21 37	04 39	04 49	04 58	05 05
64	19 11	20 03	21 14	04 29	04 42	04 54	05 04
62	19 05	19 53	20 56	04 20	04 36	04 50	05 03
60	19 00	19 45	20 42	04 12	04 31	04 48	05 03
N 58	18 56	19 37	20 30	04 05	04 26	04 45	05 02
56	18 52	19 31	20 20	03 59	04 22	04 43	05 02
54	18 48	19 25	20 11	03 54	04 18	04 40	05 01
52	18 45	19 20	20 03	03 49	04 15	04 38	05 01
50	18 42	19 16	19 57	03 44	04 12	04 37	05 00
45	18 36	19 06	19 43	03 35	04 05	04 33	05 00
N 40	18 31	18 59	19 32	03 27	03 59	04 30	04 59
35	18 27	18 52	19 23	03 20	03 54	04 27	04 58
30	18 23	18 47	19 16	03 14	03 50	04 24	04 58
20	18 16	18 38	19 04	03 03	03 42	04 20	04 57
N 10	18 10	18 32	18 56	02 54	03 36	04 16	04 56
0	18 05	18 26	18 50	02 45	03 29	04 13	04 55
S 10	18 00	18 21	18 46	02 36	03 23	04 09	04 54
20	17 55	18 17	18 42	02 26	03 16	04 05	04 53
30	17 49	18 13	18 40	02 15	03 08	04 00	04 52
35	17 45	18 11	18 40	02 09	03 04	03 58	04 51
40	17 41	18 09	18 40	02 02	02 59	03 55	04 51
45	17 37	18 07	18 40	01 54	02 53	03 51	04 50
S 50	17 32	18 04	18 42	01 43	02 45	03 47	04 49
52	17 30	18 03	18 42	01 39	02 42	03 45	04 48
54	17 27	18 02	18 43	01 33	02 38	03 43	04 48
56	17 24	18 01	18 44	01 28	02 34	03 41	04 47
58	17 21	18 00	18 46	01 21	02 30	03 38	04 47
S 60	17 17	17 59	18 47	01 14	02 25	03 35	04 46

	SUN Eqn. of Time 00h	12h	Mer. Pass.	MOON Mer. Pass. Upper	Lower	Age	Phase
d	m s	m s	h m	h m	h m	d	%
7	02 20	02 12	12 02	21 19	08 56	10	83
8	02 03	01 55	12 02	22 02	09 41	11	90
9	01 47	01 38	12 02	22 45	10 24	12	95

1998 APRIL 10, 11, 12 (FRI., SAT., SUN.)

UT	ARIES GHA	VENUS −4.3 GHA	Dec	MARS +1.3 GHA	Dec	JUPITER −2.1 GHA	Dec	SATURN +0.5 GHA	Dec	STARS Name	SHA	Dec
10 00	198 01.3	222 18.0	S 9 46.0	172 10.2	N10 22.4	211 17.0	S 6 43.0	175 59.7	N 6 49.2	Acamar	315 27.6	S40 19.0
01	213 03.8	237 17.9	45.3	187 10.8	23.1	226 19.0	42.8	191 01.9	49.3	Achernar	335 36.1	S57 14.9
02	228 06.3	252 17.8	44.5	202 11.5	23.8	241 20.9	42.6	206 04.1	49.4	Acrux	173 21.6	S63 05.4
03	243 08.7	267 17.7	.. 43.8	217 12.2	.. 24.5	256 22.9	.. 42.4	221 06.2	.. 49.5	Adhara	255 21.7	S28 58.6
04	258 11.2	282 17.6	43.0	232 12.9	25.2	271 24.9	42.2	236 08.4	49.6	Aldebaran	291 03.0	N16 30.2
05	273 13.7	297 17.4	42.3	247 13.6	25.9	286 26.8	42.0	251 10.6	49.8			
F 06	288 16.1	312 17.3	S 9 41.5	262 14.3	N10 26.6	301 28.8	S 6 41.8	266 12.7	N 6 49.9	Alioth	166 30.4	N55 58.2
R 07	303 18.6	327 17.2	40.8	277 15.0	27.3	316 30.7	41.6	281 14.9	50.0	Alkaid	153 07.7	N49 19.3
I 08	318 21.1	342 17.1	40.0	292 15.7	28.0	331 32.7	41.4	296 17.1	50.1	Al Na'ir	27 58.7	S46 58.1
D 09	333 23.5	357 17.0	.. 39.2	307 16.4	.. 28.6	346 34.6	.. 41.2	311 19.2	.. 50.2	Alnilam	275 58.4	S 1 12.4
A 10	348 26.0	12 16.9	38.5	322 17.1	29.3	1 36.6	41.0	326 21.4	50.3	Alphard	218 07.4	S 8 39.3
Y 11	3 28.5	27 16.8	37.7	337 17.8	30.0	16 38.5	40.8	341 23.6	50.5			
12	18 30.9	42 16.6	S 9 37.0	352 18.5	N10 30.7	31 40.5	S 6 40.6	356 25.7	N 6 50.6	Alphecca	126 20.6	N26 43.2
13	33 33.4	57 16.5	36.2	7 19.2	31.4	46 42.5	40.4	11 27.9	50.7	Alpheratz	357 55.9	N29 04.6
14	48 35.8	72 16.4	35.5	22 19.9	32.1	61 44.4	40.2	26 30.1	50.8	Altair	62 19.6	N 8 51.7
15	63 38.3	87 16.3	.. 34.7	37 20.6	.. 32.8	76 46.4	.. 39.9	41 32.2	.. 50.9	Ankaa	353 27.7	S42 19.0
16	78 40.8	102 16.2	34.0	52 21.3	33.5	91 48.3	39.7	56 34.4	51.0	Antares	112 40.3	S26 25.6
17	93 43.2	117 16.1	33.2	67 22.0	34.2	106 50.3	39.5	71 36.6	51.2			
18	108 45.7	132 15.9	S 9 32.4	82 22.7	N10 34.9	121 52.2	S 6 39.3	86 38.7	N 6 51.3	Arcturus	146 06.1	N19 11.5
19	123 48.2	147 15.8	31.7	97 23.4	35.6	136 54.2	39.1	101 40.9	51.4	Atria	107 52.3	S69 01.2
20	138 50.6	162 15.7	30.9	112 24.0	36.3	151 56.1	38.9	116 43.1	51.5	Avior	234 22.6	S59 30.7
21	153 53.1	177 15.6	.. 30.2	127 24.7	.. 37.0	166 58.1	.. 38.7	131 45.2	.. 51.6	Bellatrix	278 44.7	N 6 20.7
22	168 55.6	192 15.5	29.4	142 25.4	37.7	182 00.1	38.5	146 47.4	51.7	Betelgeuse	271 14.1	N 7 24.2
23	183 58.0	207 15.4	28.6	157 26.1	38.3	197 02.0	38.3	161 49.6	51.9			
11 00	199 00.5	222 15.2	S 9 27.9	172 26.8	N10 39.0	212 04.0	S 6 38.1	176 51.8	N 6 52.0	Canopus	264 01.5	S52 42.1
01	214 03.0	237 15.1	27.1	187 27.5	39.7	227 05.9	37.9	191 53.9	52.1	Capella	280 52.0	N45 59.7
02	229 05.4	252 15.0	26.4	202 28.2	40.4	242 07.9	37.7	206 56.1	52.2	Deneb	49 39.5	N45 16.2
03	244 07.9	267 14.9	.. 25.6	217 28.9	.. 41.1	257 09.8	.. 37.5	221 58.3	.. 52.3	Denebola	182 45.3	N14 34.9
04	259 10.3	282 14.8	24.8	232 29.6	41.8	272 11.8	37.3	237 00.4	52.5	Diphda	349 07.9	S17 59.9
05	274 12.8	297 14.7	24.1	247 30.3	42.5	287 13.8	37.1	252 02.6	52.6			
S 06	289 15.3	312 14.5	S 9 23.3	262 31.0	N10 43.2	302 15.7	S 6 36.9	267 04.8	N 6 52.7	Dubhe	194 05.5	N61 45.7
A 07	304 17.7	327 14.4	22.5	277 31.7	43.9	317 17.7	36.7	282 06.9	52.8	Elnath	278 27.6	N28 36.2
T 08	319 20.2	342 14.3	21.8	292 32.4	44.6	332 19.6	36.5	297 09.1	52.9	Eltanin	90 51.4	N51 29.2
U 09	334 22.7	357 14.2	.. 21.0	307 33.1	.. 45.3	347 21.6	.. 36.3	312 11.3	.. 53.0	Enif	33 58.7	N 9 51.9
R 10	349 25.1	12 14.1	20.2	322 33.8	45.9	2 23.5	36.0	327 13.4	53.2	Fomalhaut	15 37.1	S29 37.9
D 11	4 27.6	27 13.9	19.5	337 34.5	46.6	17 25.5	35.8	342 15.6	53.3			
A 12	19 30.1	42 13.8	S 9 18.7	352 35.1	N10 47.3	32 27.5	S 6 35.6	357 17.8	N 6 53.4	Gacrux	172 13.3	S57 06.3
Y 13	34 32.5	57 13.7	17.9	7 35.8	48.0	47 29.4	35.4	12 19.9	53.5	Gienah	176 04.0	S17 32.0
14	49 35.0	72 13.6	17.1	22 36.5	48.7	62 31.4	35.2	27 22.1	53.6	Hadar	149 03.8	S60 21.8
15	64 37.4	87 13.5	.. 16.4	37 37.2	.. 49.4	77 33.3	.. 35.0	42 24.3	.. 53.7	Hamal	328 14.3	N23 27.1
16	79 39.9	102 13.3	15.6	52 37.9	50.1	92 35.3	34.8	57 26.4	53.9	Kaus Aust.	83 59.2	S34 22.9
17	94 42.4	117 13.2	14.8	67 38.6	50.8	107 37.2	34.6	72 28.6	54.0			
18	109 44.8	132 13.1	S 9 14.1	82 39.3	N10 51.4	122 39.2	S 6 34.4	87 30.8	N 6 54.1	Kochab	137 18.6	N74 09.7
19	124 47.3	147 13.0	13.3	97 40.0	52.1	137 41.2	34.2	102 32.9	54.2	Markab	13 50.2	N15 11.6
20	139 49.8	162 12.9	12.5	112 40.7	52.8	152 43.1	34.0	117 35.1	54.3	Menkar	314 27.6	N 4 04.7
21	154 52.2	177 12.7	.. 11.7	127 41.4	.. 53.5	167 45.1	.. 33.8	132 37.3	.. 54.4	Menkent	148 21.0	S36 21.7
22	169 54.7	192 12.6	11.0	142 42.1	54.2	182 47.0	33.6	147 39.4	54.6	Miaplacidus	221 41.6	S69 43.0
23	184 57.2	207 12.5	10.2	157 42.8	54.9	197 49.0	33.4	162 41.6	54.7			
12 00	199 59.6	222 12.4	S 9 09.4	172 43.5	N10 55.6	212 51.0	S 6 33.2	177 43.8	N 6 54.8	Mirfak	308 57.5	N49 51.2
01	215 02.1	237 12.3	08.6	187 44.2	56.3	227 52.9	33.0	192 46.0	54.9	Nunki	76 12.7	S26 17.8
02	230 04.6	252 12.1	07.8	202 44.8	56.9	242 54.9	32.8	207 48.1	55.0	Peacock	53 37.8	S56 44.2
03	245 07.0	267 12.0	.. 07.1	217 45.5	.. 57.6	257 56.8	.. 32.6	222 50.3	.. 55.1	Pollux	243 42.0	N28 01.8
04	260 09.5	282 11.9	06.3	232 46.2	58.3	272 58.8	32.4	237 52.5	55.3	Procyon	245 11.9	N 5 13.6
05	275 11.9	297 11.8	05.5	247 46.9	59.0	288 00.8	32.2	252 54.6	55.4			
S 06	290 14.4	312 11.7	S 9 04.7	262 47.6	N10 59.7	303 02.7	S 6 32.0	267 56.8	N 6 55.5	Rasalhague	96 17.1	N12 33.6
U 07	305 16.9	327 11.5	03.9	277 48.3	11 00.4	318 04.7	31.7	282 59.0	55.6	Regulus	207 55.7	N11 58.4
N 08	320 19.3	342 11.4	03.2	292 49.0	01.1	333 06.6	31.5	298 01.1	55.7	Rigel	281 23.5	S 8 12.5
D 09	335 21.8	357 11.3	.. 02.4	307 49.7	.. 01.7	348 08.6	.. 31.3	313 03.3	.. 55.8	Rigil Kent.	140 07.1	S60 49.5
A 10	350 24.3	12 11.2	01.6	322 50.4	02.4	3 10.6	31.1	328 05.5	56.0	Sabik	102 25.7	S15 43.3
Y 11	5 26.7	27 11.1	00.8	337 51.1	03.1	18 12.5	30.9	343 07.6	56.1			
12	20 29.2	42 10.9	S 9 00.0	352 51.8	N11 03.8	33 14.5	S 6 30.7	358 09.8	N 6 56.2	Schedar	349 54.4	N56 31.5
13	35 31.7	57 10.8	8 59.2	7 52.5	04.5	48 16.4	30.5	13 12.0	56.3	Shaula	96 37.6	S37 06.0
14	50 34.1	72 10.7	58.5	22 53.1	05.2	63 18.4	30.3	28 14.1	56.4	Sirius	258 44.1	S16 43.1
15	65 36.6	87 10.6	.. 57.7	37 53.8	.. 05.8	78 20.4	.. 30.1	43 16.3	.. 56.5	Spica	158 43.2	S11 09.2
16	80 39.0	102 10.4	56.9	52 54.5	06.5	93 22.3	29.9	58 18.5	56.7	Suhail	223 00.8	S43 25.9
17	95 41.5	117 10.3	56.1	67 55.2	07.2	108 24.3	29.7	73 20.6	56.8			
18	110 44.0	132 10.2	S 8 55.3	82 55.9	N11 07.9	123 26.2	S 6 29.5	88 22.8	N 6 56.9	Vega	80 46.8	N38 46.8
19	125 46.4	147 10.1	54.5	97 56.6	08.6	138 28.2	29.3	103 25.0	57.0	Zuben'ubi	137 18.0	S16 02.0
20	140 48.9	162 09.9	53.7	112 57.3	09.3	153 30.2	29.1	118 27.1	57.1		SHA	Mer. Pass.
21	155 51.4	177 09.8	.. 52.9	127 58.0	.. 09.9	168 32.1	.. 28.9	133 29.3	.. 57.2	Venus	23 14.8	9 11
22	170 53.8	192 09.7	52.2	142 58.7	10.6	183 34.1	28.7	148 31.5	57.4	Mars	333 26.3	12 30
23	185 56.3	207 09.6	51.4	157 59.4	11.3	198 36.0	28.5	163 33.6	57.5	Jupiter	13 03.5	9 50
Mer. Pass. 10 42.2		v −0.1	d 0.8	v 0.7	d 0.7	v 2.0	d 0.2	v 2.2	d 0.1	Saturn	337 51.3	12 11

UT	SUN GHA	SUN Dec	MOON GHA	v	Dec	d	HP
d h	° ′	° ′	° ′	′	° ′	′	′
10 00	179 37.6	N 7 47.5	18 17.4	15.4	N 1 56.8	9.4	54.0
01	194 37.7	48.4	32 51.8	15.4	1 47.4	9.5	54.0
02	209 37.9	49.3	47 26.2	15.4	1 37.9	9.4	54.0
03	224 38.1	.. 50.2	62 00.6	15.4	1 28.5	9.4	54.0
04	239 38.3	51.2	76 35.0	15.4	1 19.1	9.5	54.0
05	254 38.4	52.1	91 09.4	15.4	1 09.6	9.4	54.0
06	269 38.6	N 7 53.0	105 43.8	15.4	N 1 00.2	9.5	54.0
07	284 38.8	53.9	120 18.2	15.4	0 50.7	9.5	54.0
F 08	299 38.9	54.9	134 52.6	15.4	0 41.2	9.4	54.0
R 09	314 39.1	.. 55.8	149 27.0	15.4	0 31.8	9.5	54.0
I 10	329 39.3	56.7	164 01.4	15.4	0 22.3	9.4	54.0
D 11	344 39.4	57.6	178 35.8	15.5	0 12.9	9.5	54.0
A 12	359 39.6	N 7 58.6	193 10.3	15.4	N 0 03.4	9.5	54.0
Y 13	14 39.8	7 59.5	207 44.7	15.4	S 0 06.1	9.4	54.0
14	29 39.9	8 00.4	222 19.1	15.4	0 15.5	9.5	54.0
15	44 40.1	.. 01.3	236 53.5	15.5	0 25.0	9.5	54.0
16	59 40.3	02.2	251 28.0	15.4	0 34.5	9.4	54.0
17	74 40.4	03.2	266 02.4	15.4	0 43.9	9.5	54.0
18	89 40.6	N 8 04.1	280 36.8	15.4	S 0 53.4	9.4	54.0
19	104 40.8	05.0	295 11.2	15.5	1 02.8	9.5	54.0
20	119 41.0	05.9	309 45.7	15.4	1 12.3	9.4	54.0
21	134 41.1	.. 06.9	324 20.1	15.4	1 21.7	9.5	54.0
22	149 41.3	07.8	338 54.5	15.4	1 31.2	9.4	54.0
23	164 41.5	08.7	353 28.9	15.4	1 40.6	9.4	54.0
11 00	179 41.6	N 8 09.6	8 03.3	15.4	S 1 50.0	9.4	54.0
01	194 41.8	10.5	22 37.7	15.4	1 59.4	9.5	54.0
02	209 42.0	11.5	37 12.1	15.4	2 08.9	9.4	54.0
03	224 42.1	.. 12.4	51 46.5	15.4	2 18.3	9.4	54.0
04	239 42.3	13.3	66 20.9	15.3	2 27.7	9.3	54.0
05	254 42.5	14.2	80 55.2	15.4	2 37.0	9.4	54.0
06	269 42.6	N 8 15.1	95 29.6	15.3	S 2 46.4	9.4	54.0
S 07	284 42.8	16.1	110 03.9	15.4	2 55.8	9.3	54.0
A 08	299 43.0	17.0	124 38.3	15.3	3 05.1	9.4	54.0
T 09	314 43.1	.. 17.9	139 12.6	15.3	3 14.5	9.3	54.0
U 10	329 43.3	18.8	153 46.9	15.4	3 23.8	9.3	54.0
R 11	344 43.5	19.7	168 21.3	15.3	3 33.1	9.3	54.0
D 12	359 43.6	N 8 20.7	182 55.6	15.3	S 3 42.4	9.3	54.0
A 13	14 43.8	21.6	197 29.9	15.2	3 51.7	9.3	54.0
Y 14	29 43.9	22.5	212 04.1	15.3	4 01.0	9.3	54.0
15	44 44.1	.. 23.4	226 38.4	15.2	4 10.3	9.2	54.0
16	59 44.3	24.3	241 12.6	15.3	4 19.5	9.3	54.0
17	74 44.4	25.2	255 46.9	15.2	4 28.8	9.2	54.0
18	89 44.6	N 8 26.2	270 21.1	15.2	S 4 38.0	9.2	54.0
19	104 44.8	27.1	284 55.3	15.2	4 47.2	9.1	54.0
20	119 44.9	28.0	299 29.5	15.2	4 56.3	9.2	54.0
21	134 45.1	.. 28.9	314 03.7	15.2	5 05.5	9.1	54.0
22	149 45.3	29.8	328 37.9	15.1	5 14.6	9.2	54.0
23	164 45.4	30.7	343 12.0	15.1	5 23.8	9.1	54.0
12 00	179 45.6	N 8 31.6	357 46.1	15.1	S 5 32.9	9.0	54.0
01	194 45.8	32.6	12 20.2	15.1	5 41.9	9.1	54.0
02	209 45.9	33.5	26 54.3	15.1	5 51.0	9.0	54.0
03	224 46.1	.. 34.4	41 28.4	15.1	6 00.0	9.0	54.0
04	239 46.3	35.3	56 02.5	15.0	6 09.0	9.0	54.0
05	254 46.4	36.2	70 36.5	15.0	6 18.0	9.0	54.0
06	269 46.6	N 8 37.1	85 10.5	15.0	S 6 27.0	8.9	54.0
07	284 46.7	38.0	99 44.5	15.0	6 35.9	9.0	54.0
S 08	299 46.9	39.0	114 18.5	14.9	6 44.9	8.8	54.0
U 09	314 47.1	.. 39.9	128 52.4	14.9	6 53.7	8.9	54.0
N 10	329 47.2	40.8	143 26.3	14.9	7 02.6	8.8	54.0
11	344 47.4	41.7	158 00.2	14.9	7 11.4	8.8	54.0
D 12	359 47.6	N 8 42.6	172 34.1	14.9	S 7 20.2	8.8	54.1
A 13	14 47.7	43.5	187 08.0	14.8	7 29.0	8.8	54.1
Y 14	29 47.9	44.4	201 41.8	14.8	7 37.8	8.7	54.1
15	44 48.0	.. 45.3	216 15.6	14.8	7 46.5	8.7	54.1
16	59 48.2	46.3	230 49.4	14.8	7 55.2	8.6	54.1
17	74 48.4	47.2	245 23.2	14.7	8 03.8	8.7	54.1
18	89 48.5	N 8 48.1	259 56.9	14.7	S 8 12.5	8.6	54.1
19	104 48.7	49.0	274 30.6	14.7	8 21.1	8.5	54.1
20	119 48.9	49.9	289 04.3	14.6	8 29.6	8.6	54.1
21	134 49.0	.. 50.8	303 37.9	14.6	8 38.2	8.5	54.1
22	149 49.2	51.7	318 11.5	14.6	8 46.7	8.4	54.1
23	164 49.3	52.6	332 45.1	14.6	S 8 55.1	8.5	54.1
	SD 16.0	d 0.9	SD 14.7		14.7		14.7

Lat.	Twilight Naut.	Twilight Civil	Sunrise	Moonrise 10	11	12	13
°	h m	h m	h m	h m	h m	h m	h m
N 72	////	02 34	04 03	17 28	19 01	20 37	22 16
N 70	////	03 02	04 17	17 26	18 54	20 23	21 54
68	01 35	03 22	04 28	17 25	18 48	20 12	21 37
66	02 13	03 38	04 37	17 25	18 44	20 03	21 23
64	02 38	03 51	04 44	17 24	18 40	19 56	21 12
62	02 57	04 02	04 51	17 23	18 36	19 49	21 02
60	03 13	04 11	04 56	17 23	18 33	19 44	20 54
N 58	03 26	04 19	05 01	17 22	18 30	19 39	20 47
56	03 37	04 26	05 05	17 22	18 28	19 34	20 41
54	03 46	04 32	05 09	17 21	18 26	19 31	20 35
52	03 54	04 38	05 13	17 21	18 24	19 27	20 30
50	04 01	04 42	05 16	17 21	18 22	19 24	20 25
45	04 16	04 53	05 23	17 20	18 18	19 17	20 16
N 40	04 28	05 01	05 29	17 19	18 15	19 11	20 07
35	04 37	05 08	05 34	17 19	18 12	19 06	20 00
30	04 45	05 14	05 38	17 19	18 10	19 02	19 54
20	04 57	05 23	05 45	17 18	18 06	18 54	19 44
N 10	05 06	05 31	05 52	17 17	18 02	18 48	19 34
0	05 13	05 37	05 58	17 17	17 59	18 42	19 26
S 10	05 18	05 42	06 04	17 16	17 56	18 36	19 17
20	05 22	05 47	06 10	17 16	17 52	18 29	19 08
30	05 25	05 52	06 16	17 15	17 48	18 22	18 58
35	05 25	05 55	06 20	17 15	17 46	18 18	18 52
40	05 26	05 57	06 25	17 14	17 43	18 13	18 45
45	05 26	06 00	06 30	17 14	17 40	18 08	18 37
S 50	05 26	06 03	06 36	17 13	17 37	18 01	18 28
52	05 25	06 04	06 38	17 13	17 35	17 58	18 24
54	05 25	06 06	06 41	17 13	17 33	17 55	18 19
56	05 24	06 07	06 45	17 12	17 31	17 51	18 14
58	05 23	06 09	06 48	17 12	17 29	17 47	18 08
S 60	05 22	06 10	06 53	17 12	17 27	17 43	18 01

Lat.	Sunset	Twilight Civil	Twilight Naut.	Moonset 10	11	12	13
°	h m	h m	h m	h m	h m	h m	h m
N 72	20 03	21 34	////	05 08	05 03	04 57	04 50
N 70	19 48	21 05	////	05 07	05 06	05 06	05 06
68	19 37	20 43	22 36	05 06	05 09	05 13	05 18
66	19 28	20 27	21 55	05 05	05 12	05 19	05 28
64	19 20	20 13	21 28	05 04	05 14	05 25	05 36
62	19 13	20 02	21 08	05 03	05 16	05 29	05 44
60	19 06	19 53	20 52	05 03	05 18	05 33	05 50
N 58	19 03	19 45	20 39	05 02	05 19	05 37	05 56
56	18 58	19 38	20 28	05 02	05 20	05 40	06 01
54	18 54	19 31	20 18	05 01	05 22	05 43	06 05
52	18 50	19 26	20 10	05 01	05 23	05 45	06 09
50	18 47	19 21	20 02	05 00	05 24	05 48	06 13
45	18 40	19 10	19 47	05 00	05 26	05 53	06 21
N 40	18 34	19 02	19 35	04 59	05 28	05 57	06 28
35	18 29	18 55	19 26	04 58	05 29	06 01	06 34
30	18 25	18 49	19 18	04 58	05 31	06 04	06 39
20	18 17	18 39	19 06	04 57	05 33	06 10	06 48
N 10	18 10	18 32	18 56	04 56	05 35	06 15	06 55
0	18 04	18 25	18 50	04 55	05 37	06 20	07 03
S 10	17 58	18 20	18 44	04 54	05 39	06 24	07 10
20	17 52	18 14	18 40	04 53	05 41	06 29	07 18
30	17 45	18 09	18 37	04 52	05 43	06 35	07 27
35	17 41	18 07	18 36	04 51	05 45	06 38	07 32
40	17 37	18 04	18 35	04 51	05 46	06 42	07 38
45	17 32	18 01	18 35	04 50	05 48	06 47	07 45
S 50	17 26	17 58	18 36	04 49	05 50	06 52	07 53
52	17 23	17 57	18 36	04 48	05 51	06 54	07 57
54	17 20	17 55	18 36	04 48	05 52	06 57	08 01
56	17 16	17 54	18 37	04 47	05 54	07 00	08 06
58	17 13	17 52	18 38	04 47	05 55	07 03	08 11
S 60	17 09	17 51	18 39	04 46	05 56	07 07	08 17

Day	SUN Eqn. of Time 00h	SUN Eqn. of Time 12h	Mer. Pass.	MOON Mer. Pass. Upper	MOON Mer. Pass. Lower	Age	Phase
d	m s	m s	h m	h m	h m	d	%
10	01 30	01 22	12 01	23 27	11 06	13	98
11	01 14	01 06	12 01	24 09	11 48	14	100
12	00 58	00 50	12 01	00 09	12 31	15	100

1998 APRIL 13, 14, 15 (MON., TUES., WED.)

UT	ARIES GHA	VENUS −4.2 GHA Dec	MARS +1.3 GHA Dec	JUPITER −2.1 GHA Dec	SATURN +0.5 GHA Dec	STARS Name	SHA	Dec
d h 13 00	200 58.8	222 09.5 S 8 50.6	173 00.1 N11 12.0	213 38.0 S 6 28.3	178 35.8 N 6 57.6	Acamar	315 27.6	S40 19.0
01	216 01.2	237 09.3 49.8	188 00.8 12.7	228 40.0 28.1	193 38.0 57.7	Achernar	335 36.1	S57 14.9
02	231 03.7	252 09.2 49.0	203 01.4 13.3	243 41.9 27.9	208 40.1 57.8	Acrux	173 21.6	S63 05.5
03	246 06.2	267 09.1 .. 48.2	218 02.1 .. 14.0	258 43.9 .. 27.7	223 42.3 .. 57.9	Adhara	255 21.8	S28 58.6
04	261 08.6	282 09.0 47.4	233 02.8 14.7	273 45.8 27.5	238 44.5 58.1	Aldebaran	291 03.0	N16 30.2
05	276 11.1	297 08.8 46.6	248 03.5 15.4	288 47.8 27.3	253 46.7 58.2			
M 06	291 13.5	312 08.7 S 8 45.8	263 04.2 N11 16.1	303 49.8 S 6 27.1	268 48.8 N 6 58.3	Alioth	166 30.4	N55 58.2
O 07	306 16.0	327 08.6 45.0	278 04.9 16.8	318 51.7 26.9	283 51.0 58.4	Alkaid	153 07.7	N49 19.3
N 08	321 18.5	342 08.5 44.2	293 05.6 17.4	333 53.7 26.7	298 53.2 58.5	Al Na'ir	27 58.7	S46 58.0
D 09	336 20.9	357 08.3 .. 43.4	308 06.3 .. 18.1	348 55.7 .. 26.4	313 55.3 .. 58.6	Alnilam	275 58.4	S 1 12.4
A 10	351 23.4	12 08.2 42.6	323 07.0 18.8	3 57.6 26.2	328 57.5 58.8	Alphard	218 07.4	S 8 39.3
Y 11	6 25.9	27 08.1 41.8	338 07.7 19.5	18 59.6 26.0	343 59.7 58.9			
12	21 28.3	42 08.0 S 8 41.0	353 08.4 N11 20.1	34 01.5 S 6 25.8	359 01.8 N 6 59.0	Alphecca	126 20.6	N26 43.2
13	36 30.8	57 07.8 40.2	8 09.0 20.8	49 03.5 25.6	14 04.0 59.1	Alpheratz	357 55.9	N29 04.6
14	51 33.3	72 07.7 39.4	23 09.7 21.5	64 05.5 25.4	29 06.2 59.2	Altair	62 19.6	N 8 51.8
15	66 35.7	87 07.6 .. 38.6	38 10.4 .. 22.2	79 07.4 .. 25.2	44 08.3 .. 59.3	Ankaa	353 27.7	S42 19.0
16	81 38.2	102 07.5 37.8	53 11.1 22.9	94 09.4 25.0	59 10.5 59.5	Antares	112 40.3	S26 25.6
17	96 40.6	117 07.3 37.0	68 11.8 23.5	109 11.4 24.8	74 12.7 59.6			
18	111 43.1	132 07.2 S 8 36.2	83 12.5 N11 24.2	124 13.3 S 6 24.6	89 14.8 N 6 59.7	Arcturus	146 06.1	N19 11.5
19	126 45.6	147 07.1 35.4	98 13.2 24.9	139 15.3 24.4	104 17.0 59.8	Atria	107 52.3	S69 01.2
20	141 48.0	162 07.0 34.6	113 13.9 25.6	154 17.3 24.2	119 19.2 6 59.9	Avior	234 22.6	S59 30.7
21	156 50.5	177 06.8 .. 33.8	128 14.6 .. 26.2	169 19.2 .. 24.0	134 21.3 7 00.0	Bellatrix	278 44.7	N 6 20.7
22	171 53.0	192 06.7 33.0	143 15.3 26.9	184 21.2 23.8	149 23.5 00.2	Betelgeuse	271 14.1	N 7 24.2
23	186 55.4	207 06.6 32.2	158 15.9 27.6	199 23.1 23.6	164 25.7 00.3			
14 00	201 57.9	222 06.4 S 8 31.4	173 16.6 N11 28.3	214 25.1 S 6 23.4	179 27.8 N 7 00.4	Canopus	264 01.5	S52 42.1
01	217 00.4	237 06.3 30.6	188 17.3 29.0	229 27.1 23.2	194 30.0 00.5	Capella	280 52.0	N45 59.7
02	232 02.8	252 06.2 29.8	203 18.0 29.6	244 29.0 23.0	209 32.2 00.6	Deneb	49 39.5	N45 16.2
03	247 05.3	267 06.1 .. 29.0	218 18.7 .. 30.3	259 31.0 .. 22.8	224 34.3 .. 00.7	Denebola	182 45.3	N14 34.9
04	262 07.8	282 05.9 28.2	233 19.4 31.0	274 33.0 22.6	239 36.5 00.9	Diphda	349 07.9	S17 59.9
05	277 10.2	297 05.8 27.4	248 20.1 31.7	289 34.9 22.4	254 38.7 01.0			
T 06	292 12.7	312 05.7 S 8 26.5	263 20.8 N11 32.3	304 36.9 S 6 22.2	269 40.8 N 7 01.1	Dubhe	194 05.5	N61 45.7
U 07	307 15.1	327 05.6 25.7	278 21.5 33.0	319 38.9 22.0	284 43.0 01.2	Elnath	278 27.6	N28 36.2
E 08	322 17.6	342 05.4 24.9	293 22.1 33.7	334 40.8 21.8	299 45.2 01.3	Eltanin	90 51.3	N51 29.2
S 09	337 20.1	357 05.3 .. 24.1	308 22.8 .. 34.4	349 42.8 .. 21.6	314 47.3 .. 01.4	Enif	33 58.7	N 9 51.9
D 10	352 22.5	12 05.2 23.3	323 23.5 35.0	4 44.8 21.4	329 49.5 01.6	Fomalhaut	15 37.1	S29 37.9
A 11	7 25.0	27 05.0 22.5	338 24.2 35.7	19 46.7 21.2	344 51.7 01.7			
Y 12	22 27.5	42 04.9 S 8 21.7	353 24.9 N11 36.4	34 48.7 S 6 21.0	359 53.9 N 7 01.8	Gacrux	172 13.3	S57 06.3
13	37 29.9	57 04.8 20.9	8 25.6 37.1	49 50.6 20.8	14 56.0 01.9	Gienah	176 04.0	S17 32.0
14	52 32.4	72 04.7 20.0	23 26.3 37.7	64 52.6 20.6	29 58.2 02.0	Hadar	149 03.8	S60 21.8
15	67 34.9	87 04.5 .. 19.2	38 27.0 .. 38.4	79 54.6 .. 20.4	45 00.4 .. 02.1	Hamal	328 14.3	N23 27.0
16	82 37.3	102 04.4 18.4	53 27.7 39.1	94 56.5 20.2	60 02.5 02.2	Kaus Aust.	83 59.2	S34 22.9
17	97 39.8	117 04.3 17.6	68 28.3 39.7	109 58.5 20.0	75 04.7 02.4			
18	112 42.3	132 04.1 S 8 16.8	83 29.0 N11 40.4	125 00.5 S 6 19.8	90 06.9 N 7 02.5	Kochab	137 18.6	N74 09.8
19	127 44.7	147 04.0 16.0	98 29.7 41.1	140 02.4 19.6	105 09.0 02.6	Markab	13 50.2	N15 11.6
20	142 47.2	162 03.9 15.1	113 30.4 41.8	155 04.4 19.4	120 11.2 02.7	Menkar	314 27.6	N 4 04.7
21	157 49.6	177 03.8 .. 14.3	128 31.1 .. 42.4	170 06.4 .. 19.2	135 13.4 .. 02.8	Menkent	148 20.9	S36 21.7
22	172 52.1	192 03.6 13.5	143 31.8 43.1	185 08.3 19.0	150 15.5 02.9	Miaplacidus	221 41.7	S69 43.0
23	187 54.6	207 03.5 12.7	158 32.5 43.8	200 10.3 18.7	165 17.7 03.1			
15 00	202 57.0	222 03.4 S 8 11.9	173 33.2 N11 44.5	215 12.3 S 6 18.5	180 19.9 N 7 03.2	Mirfak	308 57.5	N49 51.2
01	217 59.5	237 03.2 11.1	188 33.9 45.1	230 14.2 18.3	195 22.0 03.3	Nunki	76 12.7	S26 17.8
02	233 02.0	252 03.1 10.2	203 34.5 45.8	245 16.2 18.1	210 24.2 03.4	Peacock	53 37.7	S56 44.2
03	248 04.4	267 03.0 .. 09.4	218 35.2 .. 46.5	260 18.2 .. 17.9	225 26.4 .. 03.5	Pollux	243 42.0	N28 01.8
04	263 06.9	282 02.9 08.6	233 35.9 47.1	275 20.1 17.7	240 28.5 03.6	Procyon	245 12.0	N 5 13.6
05	278 09.4	297 02.7 07.8	248 36.6 47.8	290 22.1 17.5	255 30.7 03.8			
W 06	293 11.8	312 02.6 S 8 06.9	263 37.3 N11 48.5	305 24.1 S 6 17.3	270 32.9 N 7 03.9	Rasalhague	96 17.1	N12 33.6
E 07	308 14.3	327 02.5 06.1	278 38.0 49.1	320 26.0 17.1	285 35.0 04.0	Regulus	207 55.7	N11 58.4
D 08	323 16.7	342 02.3 05.3	293 38.7 49.8	335 28.0 16.9	300 37.2 04.1	Rigel	281 23.5	S 8 12.5
N 09	338 19.2	357 02.2 .. 04.5	308 39.4 .. 50.5	350 30.0 .. 16.7	315 39.4 .. 04.2	Rigil Kent.	140 07.1	S60 49.5
E 10	353 21.7	12 02.1 03.6	323 40.0 51.2	5 31.9 16.5	330 41.5 04.3	Sabik	102 25.7	S15 43.3
S 11	8 24.1	27 01.9 02.8	338 40.7 51.8	20 33.9 16.3	345 43.7 04.5			
D 12	23 26.6	42 01.8 S 8 02.0	353 41.4 N11 52.5	35 35.9 S 6 16.1	0 45.9 N 7 04.6	Schedar	349 54.3	N56 31.5
A 13	38 29.1	57 01.7 01.2	8 42.1 53.2	50 37.9 15.9	15 48.0 04.7	Shaula	96 37.5	S37 06.0
Y 14	53 31.5	72 01.5 8 00.3	23 42.8 53.8	65 39.8 15.7	30 50.2 04.8	Sirius	258 44.1	S16 43.1
15	68 34.0	87 01.4 7 59.5	38 43.5 .. 54.5	80 41.8 .. 15.5	45 52.4 .. 04.9	Spica	158 43.2	S11 09.2
16	83 36.5	102 01.3 58.7	53 44.2 55.2	95 43.8 15.3	60 54.5 05.0	Suhail	223 00.8	S43 25.9
17	98 38.9	117 01.2 57.8	68 44.9 55.8	110 45.7 15.1	75 56.7 05.2			
18	113 41.4	132 01.0 S 7 57.0	83 45.5 N11 56.5	125 47.7 S 6 14.9	90 58.9 N 7 05.3	Vega	80 46.7	N38 46.8
19	128 43.9	147 00.9 56.2	98 46.2 57.2	140 49.7 14.7	106 01.0 05.4	Zuben'ubi	137 18.0	S16 02.0
20	143 46.3	162 00.8 55.4	113 46.9 57.8	155 51.6 14.5	121 03.2 05.5		SHA	Mer. Pass.
21	158 48.8	177 00.6 .. 54.5	128 47.6 .. 58.5	170 53.6 .. 14.3	136 05.4 .. 05.6	Venus	20 08.5	9 12
22	173 51.2	192 00.5 53.7	143 48.3 59.2	185 55.6 14.1	151 07.5 05.7	Mars	331 18.7	12 26
23	188 53.7	207 00.4 52.9	158 49.0 59.8	200 57.5 13.9	166 09.7 05.9	Jupiter	12 27.2	9 41
Mer. Pass. 10 30.4		v −0.1 d 0.8	v 0.7 d 0.7	v 2.0 d 0.2	v 2.2 d 0.1	Saturn	337 29.9	12 00

SUN / MOON

UT (d h)	SUN GHA	SUN Dec	MOON GHA	v	MOON Dec	d	HP
13 00	179 49.5	N 8 53.5	347 18.7	14.5	S 9 03.6	8.3	54.1
01	194 49.7	54.4	1 52.2	14.6	9 11.9	8.4	54.1
02	209 49.8	55.4	16 25.8	14.4	9 20.3	8.3	54.1
03	224 50.0	.. 56.3	30 59.2	14.5	9 28.6	8.3	54.1
04	239 50.1	57.2	45 32.7	14.4	9 36.9	8.2	54.1
05	254 50.3	58.1	60 06.1	14.4	9 45.1	8.2	54.2
06	269 50.5	N 8 59.0	74 39.5	14.4	S 9 53.3	8.2	54.2
07	284 50.6	8 59.9	89 12.9	14.3	10 01.5	8.1	54.2
M 08	299 50.8	9 00.8	103 46.2	14.3	10 09.6	8.1	54.2
O 09	314 50.9	.. 01.7	118 19.5	14.2	10 17.7	8.1	54.2
N 10	329 51.1	02.6	132 52.7	14.3	10 25.8	8.0	54.2
D 11	344 51.3	03.5	147 26.0	14.2	10 33.8	7.9	54.2
A 12	359 51.4	N 9 04.4	161 59.2	14.1	S10 41.7	7.9	54.2
Y 13	14 51.6	05.3	176 32.3	14.1	10 49.6	7.9	54.2
14	29 51.7	06.2	191 05.4	14.1	10 57.5	7.8	54.2
15	44 51.9	.. 07.1	205 38.5	14.1	11 05.3	7.8	54.2
16	59 52.1	08.0	220 11.6	14.0	11 13.1	7.8	54.3
17	74 52.2	08.9	234 44.6	14.0	11 20.9	7.6	54.3
18	89 52.4	N 9 09.9	249 17.6	14.0	S11 28.5	7.7	54.3
19	104 52.5	10.8	263 50.6	13.9	11 36.2	7.6	54.3
20	119 52.7	11.7	278 23.5	13.9	11 43.8	7.5	54.3
21	134 52.9	.. 12.6	292 56.4	13.8	11 51.3	7.5	54.3
22	149 53.0	13.5	307 29.2	13.8	11 58.8	7.5	54.3
23	164 53.2	14.4	322 02.0	13.8	12 06.3	7.4	54.3
14 00	179 53.3	N 9 15.3	336 34.8	13.7	S12 13.7	7.3	54.3
01	194 53.5	16.2	351 07.5	13.7	12 21.0	7.3	54.3
02	209 53.6	17.1	5 40.2	13.7	12 28.3	7.3	54.4
03	224 53.8	.. 18.0	20 12.9	13.6	12 35.6	7.2	54.4
04	239 54.0	18.9	34 45.5	13.6	12 42.8	7.1	54.4
05	254 54.1	19.8	49 18.1	13.5	12 49.9	7.1	54.4
06	269 54.3	N 9 20.7	63 50.6	13.5	S12 57.0	7.1	54.4
07	284 54.4	21.6	78 23.1	13.5	13 04.1	6.9	54.4
T 08	299 54.6	22.5	92 55.6	13.4	13 11.0	7.0	54.4
U 09	314 54.7	.. 23.4	107 28.0	13.4	13 18.0	6.8	54.4
E 10	329 54.9	24.3	122 00.4	13.4	13 24.8	6.8	54.4
S 11	344 55.0	25.2	136 32.8	13.3	13 31.6	6.8	54.5
D 12	359 55.2	N 9 26.1	151 05.1	13.2	S13 38.4	6.7	54.5
A 13	14 55.4	27.0	165 37.3	13.3	13 45.1	6.6	54.5
Y 14	29 55.5	27.9	180 09.6	13.1	13 51.7	6.6	54.5
15	44 55.7	.. 28.8	194 41.7	13.2	13 58.3	6.5	54.5
16	59 55.8	29.7	209 13.9	13.1	14 04.8	6.5	54.5
17	74 56.0	30.6	223 46.0	13.0	14 11.3	6.4	54.5
18	89 56.1	N 9 31.5	238 18.0	13.1	S14 17.7	6.3	54.6
19	104 56.3	32.4	252 50.1	12.9	14 24.0	6.3	54.6
20	119 56.4	33.3	267 22.0	13.0	14 30.3	6.2	54.6
21	134 56.6	.. 34.2	281 54.0	12.9	14 36.5	6.2	54.6
22	149 56.8	35.1	296 25.9	12.8	14 42.7	6.1	54.6
23	164 56.9	36.0	310 57.7	12.8	14 48.8	6.0	54.6
15 00	179 57.1	N 9 36.9	325 29.5	12.8	S14 54.8	5.9	54.6
01	194 57.2	37.8	340 01.3	12.7	15 00.7	5.9	54.6
02	209 57.4	38.7	354 33.0	12.7	15 06.6	5.9	54.7
03	224 57.5	.. 39.5	9 04.7	12.7	15 12.5	5.7	54.7
04	239 57.7	40.4	23 36.4	12.6	15 18.2	5.7	54.7
05	254 57.8	41.3	38 08.0	12.5	15 23.9	5.6	54.7
06	269 58.0	N 9 42.2	52 39.5	12.5	S15 29.5	5.6	54.7
W 07	284 58.1	43.1	67 11.0	12.5	15 35.1	5.5	54.7
E 08	299 58.3	44.0	81 42.5	12.4	15 40.6	5.4	54.8
D 09	314 58.4	.. 44.9	96 13.9	12.4	15 46.0	5.3	54.8
N 10	329 58.6	45.8	110 45.3	12.3	15 51.3	5.3	54.8
E 11	344 58.7	46.7	125 16.6	12.3	15 56.6	5.2	54.8
S 12	359 58.9	N 9 47.6	139 47.9	12.3	S16 01.8	5.1	54.8
D 13	14 59.1	48.5	154 19.2	12.2	16 06.9	5.1	54.8
A 14	29 59.2	49.4	168 50.4	12.2	16 12.0	5.0	54.9
Y 15	44 59.4	.. 50.3	183 21.6	12.1	16 17.0	4.9	54.9
16	59 59.5	51.2	197 52.7	12.1	16 21.9	4.8	54.9
17	74 59.7	52.1	212 23.8	12.0	16 26.7	4.8	54.9
18	89 59.8	N 9 52.9	226 54.8	12.0	S16 31.5	4.7	54.9
19	105 00.0	53.8	241 25.8	12.0	16 36.2	4.6	54.9
20	120 00.1	54.7	255 56.8	11.9	16 40.8	4.5	55.0
21	135 00.3	.. 55.6	270 27.7	11.8	16 45.3	4.5	55.0
22	150 00.4	56.5	284 58.5	11.8	16 49.8	4.3	55.0
23	165 00.6	57.4	299 29.3	11.8	S16 54.1	4.3	55.0
	SD 16.0	d 0.9	SD 14.8		14.8		14.9

Twilight / Sunrise / Moonrise

Lat.	Twilight Naut.	Civil	Sunrise	Moonrise 13	14	15	16
N 72	////	02 09	03 46	22 16	24 01	00 01	01 58
N 70	////	02 43	04 02	21 54	23 26	24 58	00 58
68	01 00	03 06	04 15	21 37	23 01	24 23	00 23
66	01 51	03 25	04 25	21 23	22 42	23 58	25 08
64	02 22	03 39	04 33	21 12	22 27	23 39	24 45
62	02 44	03 51	04 41	21 02	22 14	23 24	24 28
60	03 01	04 01	04 47	20 54	22 04	23 11	24 13
N 58	03 15	04 10	04 53	20 47	21 54	22 59	24 00
56	03 27	04 18	04 58	20 41	21 46	22 50	23 49
54	03 37	04 25	05 02	20 35	21 39	22 41	23 40
52	03 46	04 31	05 06	20 30	21 32	22 33	23 31
50	03 54	04 36	05 10	20 25	21 27	22 26	23 24
45	04 10	04 47	05 18	20 16	21 14	22 12	23 07
N 40	04 23	04 56	05 24	20 07	21 04	21 59	22 54
35	04 33	05 04	05 30	20 00	20 55	21 49	22 43
30	04 41	05 10	05 35	19 54	20 47	21 40	22 33
20	04 54	05 21	05 43	19 44	20 34	21 25	22 16
N 10	05 04	05 29	05 50	19 34	20 22	21 11	22 01
0	05 12	05 36	05 57	19 26	20 11	20 59	21 48
S 10	05 18	05 42	06 04	19 17	20 00	20 46	21 34
20	05 23	05 48	06 11	19 08	19 49	20 33	21 19
30	05 26	05 54	06 18	18 58	19 36	20 17	21 02
35	05 28	05 57	06 23	18 52	19 28	20 09	20 53
40	05 29	06 00	06 28	18 45	19 20	19 58	20 42
45	05 30	06 04	06 33	18 37	19 10	19 47	20 29
S 50	05 30	06 08	06 40	18 28	18 58	19 33	20 13
52	05 30	06 09	06 43	18 24	18 52	19 26	20 05
54	05 30	06 11	06 47	18 19	18 46	19 19	19 57
56	05 30	06 13	06 51	18 14	18 40	19 10	19 48
58	05 29	06 15	06 55	18 08	18 32	19 01	19 38
S 60	05 29	06 17	07 00	18 01	18 23	18 51	19 26

Sunset / Twilight / Moonset

Lat.	Sunset	Twilight Civil	Naut.	Moonset 13	14	15	16
N 72	20 18	21 59	////	04 50	04 43	04 33	04 15
N 70	20 02	21 23	////	05 06	05 06	05 09	05 16
68	19 49	20 58	23 16	05 18	05 24	05 34	05 51
66	19 38	20 39	22 16	05 28	05 39	05 54	06 16
64	19 29	20 24	21 43	05 36	05 51	06 10	06 36
62	19 22	20 12	21 20	05 44	06 01	06 23	06 52
60	19 15	20 01	21 02	05 50	06 10	06 34	07 05
N 58	19 09	19 52	20 48	05 56	06 18	06 44	07 16
56	19 04	19 44	20 36	06 01	06 25	06 53	07 26
54	19 00	19 38	20 25	06 05	06 31	07 00	07 35
52	18 56	19 31	20 16	06 09	06 36	07 07	07 43
50	18 52	19 26	20 08	06 13	06 41	07 13	07 50
45	18 44	19 14	19 51	06 21	06 52	07 27	08 06
N 40	18 37	19 05	19 39	06 28	07 01	07 38	08 18
35	18 31	18 57	19 28	06 34	07 09	07 47	08 29
30	18 26	18 51	19 20	06 39	07 16	07 55	08 38
20	18 18	18 40	19 07	06 48	07 27	08 10	08 55
N 10	18 10	18 32	18 57	06 55	07 38	08 22	09 09
0	18 04	18 25	18 49	07 03	07 48	08 34	09 22
S 10	17 57	18 18	18 43	07 10	07 57	08 46	09 35
20	17 50	18 12	18 38	07 18	08 08	08 58	09 49
30	17 42	18 06	18 34	07 27	08 20	09 13	10 06
35	17 37	18 03	18 32	07 32	08 27	09 21	10 15
40	17 32	18 00	18 31	07 38	08 34	09 30	10 26
45	17 27	17 56	18 30	07 45	08 44	09 42	10 38
S 50	17 20	17 52	18 30	07 53	08 55	09 55	10 54
52	17 16	17 51	18 30	07 57	09 00	10 02	11 01
54	17 13	17 49	18 30	08 01	09 06	10 09	11 09
56	17 09	17 47	18 30	08 06	09 12	10 16	11 18
58	17 05	17 45	18 30	08 11	09 19	10 25	11 28
S 60	17 00	17 42	18 30	08 17	09 27	10 35	11 40

SUN / MOON

Day	Eqn. of Time 00h	12h	Mer. Pass.	Mer. Pass. Upper	Lower	Age	Phase
13	00 42	00 35	12 01	00 52	13 14	16	98
14	00 27	00 19	12 00	01 37	13 59	17	94
15	00 12	00 05	12 00	02 23	14 46	18	89

1998 APRIL 16, 17, 18 (THURS., FRI., SAT.)

UT	ARIES GHA	VENUS GHA	VENUS Dec	MARS GHA	MARS Dec	JUPITER GHA	JUPITER Dec	SATURN GHA	SATURN Dec	Name (STARS)	SHA	Dec
16 00	203 56.2	222 00.2	S 7 52.0	173 49.7	N12 00.5	215 59.5	S 6 13.7	181 11.9	N 7 06.0	Acamar	315 27.6	S40 19.0
01	218 58.6	237 00.1	51.2	188 50.3	01.2	231 01.5	13.5	196 14.1	06.1	Achernar	335 36.1	S57 14.9
02	234 01.1	252 00.0	50.4	203 51.0	01.8	246 03.4	13.3	211 16.2	06.2	Acrux	173 21.6	S63 05.5
03	249 03.6	266 59.8	.. 49.5	218 51.7	.. 02.5	261 05.4	.. 13.1	226 18.4	.. 06.3	Adhara	255 21.8	S28 58.6
04	264 06.0	281 59.7	48.7	233 52.4	03.2	276 07.4	12.9	241 20.6	06.4	Aldebaran	291 03.0	N16 30.2
05	279 08.5	296 59.6	47.9	248 53.1	03.8	291 09.4	12.7	256 22.7	06.5			
06	294 11.0	311 59.4	S 7 47.0	263 53.8	N12 04.5	306 11.3	S 6 12.5	271 24.9	N 7 06.7	Alioth	166 30.4	N55 58.2
07	309 13.4	326 59.3	46.2	278 54.5	05.2	321 13.3	12.3	286 27.1	06.8	Alkaid	153 07.6	N49 19.3
08	324 15.9	341 59.2	45.3	293 55.2	05.8	336 15.3	12.1	301 29.2	06.9	Al Na'ir	27 58.6	S46 58.0
09	339 18.4	356 59.0	.. 44.5	308 55.8	.. 06.5	351 17.2	.. 11.9	316 31.4	.. 07.0	Alnilam	275 58.4	S 1 12.4
10	354 20.8	11 58.9	43.7	323 56.5	07.1	6 19.2	11.7	331 33.6	07.1	Alphard	218 07.4	S 8 39.3
11	9 23.3	26 58.8	42.8	338 57.2	07.8	21 21.2	11.5	346 35.7	07.2			
12	24 25.7	41 58.6	S 7 42.0	353 57.9	N12 08.5	36 23.1	S 6 11.3	1 37.9	N 7 07.4	Alphecca	126 20.6	N26 43.2
13	39 28.2	56 58.5	41.1	8 58.6	09.1	51 25.1	11.1	16 40.1	07.5	Alpheratz	357 55.9	N29 04.6
14	54 30.7	71 58.4	40.3	23 59.3	09.8	66 27.1	10.9	31 42.2	07.6	Altair	62 19.6	N 8 51.8
15	69 33.1	86 58.2	.. 39.5	39 00.0	.. 10.5	81 29.1	.. 10.7	46 44.4	.. 07.7	Ankaa	353 27.6	S42 19.0
16	84 35.6	101 58.1	38.6	54 00.6	11.1	96 31.0	10.5	61 46.6	07.8	Antares	112 40.3	S26 25.6
17	99 38.1	116 58.0	37.8	69 01.3	11.8	111 33.0	10.3	76 48.7	07.9			
18	114 40.5	131 57.8	S 7 36.9	84 02.0	N12 12.5	126 35.0	S 6 10.1	91 50.9	N 7 08.1	Arcturus	146 06.1	N19 11.5
19	129 43.0	146 57.7	36.1	99 02.7	13.1	141 36.9	09.9	106 53.1	08.2	Atria	107 52.3	S69 01.2
20	144 45.5	161 57.6	35.2	114 03.4	13.8	156 38.9	09.7	121 55.2	08.3	Avior	234 22.6	S59 30.7
21	159 47.9	176 57.4	.. 34.4	129 04.1	.. 14.4	171 40.9	.. 09.5	136 57.4	.. 08.4	Bellatrix	278 44.7	N 6 20.7
22	174 50.4	191 57.3	33.6	144 04.8	15.1	186 42.9	09.3	151 59.6	08.5	Betelgeuse	271 14.1	N 7 24.2
23	189 52.9	206 57.2	32.7	159 05.4	15.8	201 44.8	09.1	167 01.7	08.6			
17 00	204 55.3	221 57.0	S 7 31.9	174 06.1	N12 16.4	216 46.8	S 6 08.9	182 03.9	N 7 08.7	Canopus	264 01.5	S52 42.1
01	219 57.8	236 56.9	31.0	189 06.8	17.1	231 48.8	08.7	197 06.1	08.9	Capella	280 52.0	N45 59.7
02	235 00.2	251 56.8	30.2	204 07.5	17.7	246 50.7	08.5	212 08.2	09.0	Deneb	49 39.5	N45 16.2
03	250 02.7	266 56.6	.. 29.3	219 08.2	.. 18.4	261 52.7	.. 08.3	227 10.4	.. 09.1	Denebola	182 45.3	N14 34.9
04	265 05.2	281 56.5	28.5	234 08.9	19.1	276 54.7	08.1	242 12.6	09.2	Diphda	349 07.9	S17 59.9
05	280 07.6	296 56.4	27.6	249 09.5	19.7	291 56.7	07.9	257 14.7	09.3			
06	295 10.1	311 56.2	S 7 26.8	264 10.2	N12 20.4	306 58.6	S 6 07.7	272 16.9	N 7 09.4	Dubhe	194 05.5	N61 45.8
07	310 12.6	326 56.1	25.9	279 10.9	21.0	322 00.6	07.5	287 19.1	09.6	Elnath	278 27.6	N28 36.2
08	325 15.0	341 56.0	25.1	294 11.6	21.7	337 02.6	07.3	302 21.2	09.7	Eltanin	90 51.3	N51 29.2
09	340 17.5	356 55.8	.. 24.2	309 12.3	.. 22.4	352 04.6	.. 07.1	317 23.4	.. 09.8	Enif	33 58.7	N 9 51.9
10	355 20.0	11 55.7	23.4	324 13.0	23.0	7 06.5	06.9	332 25.6	09.9	Fomalhaut	15 37.1	S29 37.9
11	10 22.4	26 55.6	22.5	339 13.7	23.7	22 08.5	06.7	347 27.7	10.0			
12	25 24.9	41 55.4	S 7 21.7	354 14.3	N12 24.3	37 10.5	S 6 06.5	2 29.9	N 7 10.1	Gacrux	172 13.3	S57 06.3
13	40 27.3	56 55.3	20.8	9 15.0	25.0	52 12.4	06.3	17 32.1	10.3	Gienah	176 04.0	S17 32.0
14	55 29.8	71 55.1	20.0	24 15.7	25.6	67 14.4	06.1	32 34.2	10.4	Hadar	149 03.8	S60 21.8
15	70 32.3	86 55.0	.. 19.1	39 16.4	.. 26.3	82 16.4	.. 05.9	47 36.4	.. 10.5	Hamal	328 14.3	N23 27.0
16	85 34.7	101 54.9	18.2	54 17.1	27.0	97 18.4	05.7	62 38.6	10.6	Kaus Aust.	83 59.1	S34 22.9
17	100 37.2	116 54.7	17.4	69 17.8	27.6	112 20.3	05.5	77 40.8	10.7			
18	115 39.7	131 54.6	S 7 16.5	84 18.4	N12 28.3	127 22.3	S 6 05.3	92 42.9	N 7 10.8	Kochab	137 18.6	N74 09.8
19	130 42.1	146 54.5	15.7	99 19.1	28.9	142 24.3	05.1	107 45.1	10.9	Markab	13 50.2	N15 11.6
20	145 44.6	161 54.3	14.8	114 19.8	29.6	157 26.3	04.9	122 47.3	11.1	Menkar	314 27.6	N 4 04.7
21	160 47.1	176 54.2	.. 14.0	129 20.5	.. 30.2	172 28.2	.. 04.7	137 49.4	.. 11.2	Menkent	148 20.9	S36 21.7
22	175 49.5	191 54.1	13.1	144 21.2	30.9	187 30.2	04.5	152 51.6	11.3	Miaplacidus	221 41.7	S69 43.0
23	190 52.0	206 53.9	12.2	159 21.9	31.6	202 32.2	04.3	167 53.8	11.4			
18 00	205 54.5	221 53.8	S 7 11.4	174 22.5	N12 32.2	217 34.2	S 6 04.1	182 55.9	N 7 11.5	Mirfak	308 57.5	N49 51.2
01	220 56.9	236 53.7	10.5	189 23.2	32.9	232 36.1	03.9	197 58.1	11.6	Nunki	76 12.7	S26 17.8
02	235 59.4	251 53.5	09.7	204 23.9	33.5	247 38.1	03.7	213 00.3	11.8	Peacock	53 37.7	S56 44.2
03	251 01.8	266 53.4	.. 08.8	219 24.6	.. 34.2	262 40.1	.. 03.5	228 02.4	.. 11.9	Pollux	243 42.0	N28 01.8
04	266 04.3	281 53.2	07.9	234 25.3	34.8	277 42.1	03.3	243 04.6	12.0	Procyon	245 12.0	N 5 13.6
05	281 06.8	296 53.1	07.1	249 26.0	35.5	292 44.0	03.1	258 06.8	12.1			
06	296 09.2	311 53.0	S 7 06.2	264 26.6	N12 36.1	307 46.0	S 6 02.9	273 08.9	N 7 12.2	Rasalhague	96 17.1	N12 33.6
07	311 11.7	326 52.8	05.4	279 27.3	36.8	322 48.0	02.7	288 11.1	12.3	Regulus	207 55.8	N11 58.5
08	326 14.2	341 52.7	04.5	294 28.0	37.4	337 50.0	02.5	303 13.3	12.4	Rigel	281 23.5	S 8 12.5
09	341 16.6	356 52.6	.. 03.6	309 28.7	.. 38.1	352 51.9	.. 02.3	318 15.4	.. 12.6	Rigil Kent.	140 07.0	S60 49.5
10	356 19.1	11 52.4	02.8	324 29.4	38.8	7 53.9	02.1	333 17.6	12.7	Sabik	102 25.7	S15 43.3
11	11 21.6	26 52.3	01.9	339 30.1	39.4	22 55.9	01.9	348 19.8	12.8			
12	26 24.0	41 52.1	S 7 01.0	354 30.7	N12 40.1	37 57.9	S 6 01.7	3 21.9	N 7 12.9	Schedar	349 54.3	N56 31.5
13	41 26.5	56 52.0	7 00.2	9 31.4	40.7	52 59.8	01.5	18 24.1	13.0	Shaula	96 37.5	S37 06.0
14	56 29.0	71 51.9	6 59.3	24 32.1	41.4	68 01.8	01.3	33 26.3	13.1	Sirius	258 44.1	S16 43.1
15	71 31.4	86 51.7	.. 58.4	39 32.8	.. 42.0	83 03.8	.. 01.1	48 28.4	.. 13.3	Spica	158 43.2	S11 09.2
16	86 33.9	101 51.6	57.6	54 33.5	42.7	98 05.8	01.0	63 30.6	13.4	Suhail	223 00.8	S43 25.9
17	101 36.3	116 51.5	56.7	69 34.1	43.3	113 07.7	00.8	78 32.8	13.5			
18	116 38.8	131 51.3	S 6 55.8	84 34.8	N12 44.0	128 09.7	S 6 00.6	93 34.9	N 7 13.6	Vega	80 46.7	N38 46.8
19	131 41.3	146 51.2	55.0	99 35.5	44.6	143 11.7	00.4	108 37.1	13.7	Zuben'ubi	137 18.0	S16 02.0
20	146 43.7	161 51.0	54.1	114 36.2	45.3	158 13.7	00.2	123 39.3	13.8			
21	161 46.2	176 50.9	.. 53.2	129 36.9	.. 45.9	173 15.7	6 00.0	138 41.4	.. 13.9			
22	176 48.7	191 50.8	52.3	144 37.6	46.6	188 17.6	5 59.8	153 43.6	14.1			
23	191 51.1	206 50.6	51.5	159 38.2	47.2	203 19.6	S 5 59.6	168 45.8	14.2			

	SHA	Mer. Pass.
	° '	h m
Venus	17 01.7	9 12
Mars	329 10.8	12 23
Jupiter	11 51.5	9 32
Saturn	337 08.6	11 50

Mer. Pass. 10 18.6 | v −0.1 d 0.9 | v 0.7 d 0.7 | v 2.0 d 0.2 | v 2.2 d 0.1

SUN / MOON

UT	SUN GHA	SUN Dec	MOON GHA	v	MOON Dec	d	HP
16 00	180 00.7	N 9 58.3	314 00.1	11.8	S16 58.4	4.3	55.0
01	195 00.9	9 59.2	328 30.9	11.7	17 02.7	4.1	55.1
02	210 01.0	10 00.1	343 01.6	11.6	17 06.8	4.1	55.1
03	225 01.2	.. 01.0	357 32.2	11.6	17 10.9	3.9	55.1
04	240 01.3	01.8	12 02.8	11.6	17 14.8	3.9	55.1
05	255 01.5	02.7	26 33.4	11.5	17 18.7	3.9	55.1
T 06	270 01.6	N10 03.6	41 03.9	11.5	S17 22.6	3.7	55.2
H 07	285 01.8	04.5	55 34.4	11.4	17 26.3	3.6	55.2
U 08	300 01.9	05.4	70 04.8	11.4	17 29.9	3.6	55.2
R 09	315 02.1	.. 06.3	84 35.2	11.4	17 33.5	3.5	55.2
S 10	330 02.2	07.2	99 05.6	11.3	17 37.0	3.4	55.2
D 11	345 02.4	08.1	113 35.9	11.3	17 40.4	3.3	55.3
A 12	0 02.5	N10 08.9	128 06.2	11.2	S17 43.7	3.2	55.3
Y 13	15 02.7	09.8	142 36.4	11.2	17 46.9	3.2	55.3
14	30 02.8	10.7	157 06.6	11.1	17 50.1	3.1	55.3
15	45 02.9	.. 11.6	171 36.7	11.1	17 53.2	2.9	55.3
16	60 03.1	12.5	186 06.8	11.1	17 56.1	2.9	55.4
17	75 03.2	13.4	200 36.9	11.0	17 59.0	2.8	55.4
18	90 03.4	N10 14.2	215 06.9	11.0	S18 01.8	2.7	55.4
19	105 03.5	15.1	229 36.9	10.9	18 04.5	2.6	55.4
20	120 03.7	16.0	244 06.8	11.0	18 07.1	2.6	55.4
21	135 03.8	.. 16.9	258 36.8	10.8	18 09.7	2.4	55.5
22	150 04.0	17.8	273 06.6	10.8	18 12.1	2.4	55.5
23	165 04.1	18.7	287 36.4	10.8	18 14.5	2.2	55.5
17 00	180 04.3	N10 19.5	302 06.2	10.8	S18 16.7	2.2	55.5
01	195 04.4	20.4	316 36.0	10.7	18 18.9	2.1	55.6
02	210 04.6	21.3	331 05.7	10.7	18 21.0	2.0	55.6
03	225 04.7	.. 22.2	345 35.4	10.6	18 23.0	1.9	55.6
04	240 04.9	23.1	0 05.0	10.6	18 24.9	1.8	55.6
05	255 05.0	24.0	14 34.6	10.5	18 26.7	1.7	55.7
06	270 05.1	N10 24.8	29 04.1	10.6	S18 28.4	1.6	55.7
07	285 05.3	25.7	43 33.7	10.5	18 30.0	1.6	55.7
08	300 05.4	26.6	58 03.2	10.4	18 31.6	1.4	55.7
F 09	315 05.6	.. 27.5	72 32.6	10.4	18 33.0	1.3	55.8
R 10	330 05.7	28.4	87 02.0	10.4	18 34.3	1.3	55.8
I 11	345 05.9	29.2	101 31.4	10.3	18 35.6	1.1	55.8
D 12	0 06.0	N10 30.1	116 00.7	10.3	S18 36.7	1.1	55.8
A 13	15 06.2	31.0	130 30.0	10.3	18 37.8	0.9	55.9
Y 14	30 06.3	31.9	144 59.3	10.2	18 38.7	0.9	55.9
15	45 06.4	.. 32.8	159 28.5	10.2	18 39.6	0.8	55.9
16	60 06.6	33.6	173 57.7	10.2	18 40.4	0.6	55.9
17	75 06.7	34.5	188 26.9	10.2	18 41.0	0.6	56.0
18	90 06.9	N10 35.4	202 56.1	10.1	S18 41.6	0.5	56.0
19	105 07.0	36.3	217 25.2	10.0	18 42.1	0.3	56.0
20	120 07.2	37.1	231 54.2	10.1	18 42.4	0.3	56.0
21	135 07.3	.. 38.0	246 23.3	10.0	18 42.7	0.2	56.1
22	150 07.4	38.9	260 52.3	9.9	18 42.9	0.1	56.1
23	165 07.6	39.8	275 21.2	10.0	18 43.0	0.1	56.1
18 00	180 07.7	N10 40.6	289 50.2	9.9	S18 42.9	0.1	56.2
01	195 07.9	41.5	304 19.1	9.9	18 42.8	0.2	56.2
02	210 08.0	42.4	318 48.0	9.8	18 42.6	0.3	56.2
03	225 08.2	.. 43.3	333 16.8	9.8	18 42.3	0.4	56.2
04	240 08.3	44.1	347 45.6	9.8	18 41.9	0.6	56.3
05	255 08.4	45.0	2 14.4	9.8	18 41.3	0.6	56.3
06	270 08.6	N10 45.9	16 43.2	9.7	S18 40.7	0.7	56.3
07	285 08.7	46.8	31 11.9	9.7	18 40.0	0.8	56.3
S 08	300 08.9	47.6	45 40.6	9.7	18 39.2	0.9	56.4
A 09	315 09.0	.. 48.5	60 09.3	9.7	18 38.3	1.1	56.4
T 10	330 09.1	49.4	74 38.0	9.6	18 37.2	1.1	56.4
U 11	345 09.3	50.2	89 06.6	9.6	18 36.1	1.2	56.5
R 12	0 09.4	N10 51.1	103 35.2	9.6	S18 34.9	1.4	56.5
D 13	15 09.6	52.0	118 03.8	9.5	18 33.5	1.4	56.5
A 14	30 09.7	52.9	132 32.3	9.6	18 32.1	1.5	56.6
Y 15	45 09.8	.. 53.7	147 00.9	9.5	18 30.6	1.7	56.6
16	60 10.0	54.6	161 29.4	9.4	18 28.9	1.7	56.6
17	75 10.1	55.5	175 57.8	9.5	18 27.2	1.8	56.6
18	90 10.3	N10 56.3	190 26.3	9.4	S18 25.4	2.0	56.7
19	105 10.4	57.2	204 54.7	9.4	18 23.4	2.0	56.7
20	120 10.5	58.1	219 23.1	9.4	18 21.4	2.2	56.7
21	135 10.7	.. 59.0	233 51.5	9.4	18 19.2	2.2	56.8
22	150 10.8	10 59.8	248 19.9	9.3	18 17.0	2.4	56.8
23	165 10.9	N11 00.7	262 48.2	9.3	S18 14.6	2.5	56.8
	SD 16.0	d 0.9	SD 15.1		15.2		15.4

Twilight / Sunrise / Moonrise

Lat.	Twilight Naut.	Twilight Civil	Sunrise	Moonrise 16	17	18	19
N 72	////	01 40	03 29	01 58	■■	■■	■■
N 70	////	02 22	03 47	00 58	02 23	03 26	03 56
68	////	02 50	04 02	00 23	01 37	02 36	03 15
66	01 26	03 11	04 13	25 08	01 08	02 05	02 47
64	02 04	03 27	04 23	24 45	00 45	01 42	02 26
62	02 29	03 40	04 31	24 28	00 28	01 23	02 09
60	02 49	03 52	04 38	24 13	00 13	01 08	01 54
N 58	03 05	04 01	04 45	24 00	00 00	00 55	01 42
56	03 18	04 10	04 50	23 49	24 44	00 44	01 31
54	03 29	04 17	04 55	23 40	24 34	00 34	01 22
52	03 39	04 24	05 00	23 31	24 25	00 25	01 14
50	03 47	04 29	05 04	23 24	24 17	00 17	01 06
45	04 04	04 42	05 12	23 07	24 01	00 01	00 50
N 40	04 18	04 52	05 20	22 54	23 47	24 37	00 37
35	04 29	05 00	05 26	22 43	23 35	24 26	00 26
30	04 38	05 07	05 31	22 33	23 25	24 16	00 16
20	04 52	05 18	05 41	22 16	23 08	23 59	24 49
N 10	05 02	05 27	05 49	22 01	22 53	23 44	24 36
0	05 11	05 35	05 56	21 48	22 38	23 30	24 23
S 10	05 18	05 42	06 04	21 34	22 24	23 17	24 11
20	05 23	05 49	06 11	21 19	22 09	23 02	23 57
30	05 28	05 56	06 20	21 02	21 52	22 45	23 42
35	05 30	05 59	06 25	20 53	21 42	22 35	23 33
40	05 32	06 03	06 31	20 42	21 30	22 24	23 23
45	05 33	06 07	06 37	20 29	21 17	22 11	23 11
S 50	05 34	06 12	06 45	20 13	21 00	21 55	22 56
52	05 35	06 14	06 48	20 05	20 52	21 47	22 49
54	05 35	06 16	06 52	19 57	20 44	21 39	22 42
56	05 35	06 19	06 57	19 48	20 34	21 29	22 33
58	05 36	06 21	07 02	19 38	20 23	21 18	22 24
S 60	05 36	06 24	07 07	19 26	20 10	21 06	22 13

Sunset / Twilight / Moonset

Lat.	Sunset	Twilight Civil	Twilight Naut.	Moonset 16	17	18	19
N 72	20 34	22 29	////	04 15	■■	■■	■■
N 70	20 15	21 43	////	05 16	05 33	06 17	07 35
68	20 00	21 13	////	05 51	06 19	07 06	08 16
66	19 48	20 52	22 42	06 16	06 49	07 37	08 43
64	19 38	20 35	22 00	06 36	07 11	08 01	09 04
62	19 30	20 21	21 34	06 52	07 29	08 19	09 21
60	19 22	20 10	21 13	07 05	07 44	08 34	09 36
N 58	19 16	20 00	20 57	07 16	07 57	08 47	09 48
56	19 10	19 51	20 44	07 26	08 08	08 58	09 58
54	19 05	19 44	20 32	07 35	08 18	09 08	10 07
52	19 01	19 37	20 22	07 43	08 26	09 17	10 15
50	18 57	19 31	20 14	07 50	08 34	09 25	10 23
45	18 48	19 18	19 56	08 06	08 50	09 41	10 38
N 40	18 40	19 08	19 42	08 18	09 04	09 55	10 51
35	18 34	19 00	19 31	08 29	09 15	10 06	11 02
30	18 28	18 53	19 22	08 38	09 25	10 17	11 12
20	18 19	18 41	19 08	08 55	09 43	10 34	11 28
N 10	18 11	18 32	18 57	09 09	09 58	10 49	11 42
0	18 03	18 24	18 48	09 22	10 12	11 03	11 56
S 10	17 55	18 17	18 41	09 35	10 26	11 17	12 09
20	17 48	18 10	18 36	09 49	10 41	11 32	12 23
30	17 39	18 03	18 31	10 06	10 58	11 49	12 39
35	17 34	17 59	18 29	10 15	11 08	11 59	12 48
40	17 28	17 55	18 27	10 26	11 20	12 11	12 59
45	17 21	17 51	18 25	10 38	11 33	12 24	13 12
S 50	17 14	17 46	18 24	10 54	11 49	12 41	13 27
52	17 10	17 44	18 23	11 01	11 57	12 48	13 34
54	17 06	17 42	18 23	11 09	12 06	12 57	13 42
56	17 02	17 40	18 23	11 18	12 15	13 06	13 50
58	16 57	17 37	18 22	11 28	12 26	13 17	14 00
S 60	16 51	17 34	18 22	11 40	12 39	13 30	14 12

SUN / MOON

Day	SUN Eqn. of Time 00h	12h	Mer. Pass.	MOON Mer. Pass. Upper	Lower	Age	Phase
	m s	m s	h m	h m	h m	d	%
16	00 03	00 10	12 00	03 10	15 35	19	82
17	00 17	00 24	12 00	04 00	16 25	20	73
18	00 31	00 37	11 59	04 51	17 17	21	64

UT	ARIES	VENUS −4.2		MARS +1.3		JUPITER −2.1		SATURN +0.5		STARS		
	GHA	GHA	Dec	GHA	Dec	GHA	Dec	GHA	Dec	Name	SHA	Dec
d h	° ′	° ′	° ′	° ′	° ′	° ′	° ′	° ′	° ′		° ′	° ′
19 00	206 53.6	221 50.5	S 6 50.6	174 38.9	N12 47.9	218 21.6	S 5 59.4	183 47.9	N 7 14.3	Acamar	315 27.6	S40 19.0
01	221 56.1	236 50.3	49.7	189 39.6	48.5	233 23.6	59.2	198 50.1	14.4	Achernar	335 36.1	S57 14.9
02	236 58.5	251 50.2	48.9	204 40.3	49.2	248 25.5	59.0	213 52.3	14.5	Acrux	173 21.6	S63 05.5
03	252 01.0	266 50.1	.. 48.0	219 41.0	.. 49.8	263 27.5	.. 58.8	228 54.4	.. 14.6	Adhara	255 21.8	S28 58.6
04	267 03.4	281 49.9	47.1	234 41.6	50.5	278 29.5	58.6	243 56.6	14.7	Aldebaran	291 03.0	N16 30.2
05	282 05.9	296 49.8	46.2	249 42.3	51.1	293 31.5	58.4	258 58.8	14.9			
06	297 08.4	311 49.7	S 6 45.4	264 43.0	N12 51.8	308 33.5	S 5 58.2	274 01.0	N 7 15.0	Alioth	166 30.4	N55 58.2
07	312 10.8	326 49.5	44.5	279 43.7	52.4	323 35.4	58.0	289 03.1	15.1	Alkaid	153 07.6	N49 19.4
08	327 13.3	341 49.4	43.6	294 44.4	53.1	338 37.4	57.8	304 05.3	15.2	Al Na'ir	27 58.6	S46 58.0
S 09	342 15.8	356 49.2	.. 42.7	309 45.0	.. 53.7	353 39.4	.. 57.6	319 07.5	.. 15.3	Alnilam	275 58.4	S 1 12.4
U 10	357 18.2	11 49.1	41.9	324 45.7	54.4	8 41.4	57.4	334 09.6	15.4	Alphard	218 07.4	S 8 39.3
N 11	12 20.7	26 49.0	41.0	339 46.4	55.0	23 43.3	57.2	349 11.8	15.6			
D 12	27 23.2	41 48.8	S 6 40.1	354 47.1	N12 55.7	38 45.3	S 5 57.0	4 14.0	N 7 15.7	Alphecca	126 20.6	N26 43.2
A 13	42 25.6	56 48.7	39.2	9 47.8	56.3	53 47.3	56.8	19 16.1	15.8	Alpheratz	357 55.9	N29 04.6
Y 14	57 28.1	71 48.5	38.4	24 48.5	56.9	68 49.3	56.6	34 18.3	15.9	Altair	62 19.6	N 8 51.8
15	72 30.6	86 48.4	.. 37.5	39 49.1	.. 57.6	83 51.3	.. 56.4	49 20.5	.. 16.0	Ankaa	353 27.6	S42 19.0
16	87 33.0	101 48.3	36.6	54 49.8	58.2	98 53.2	56.2	64 22.6	16.1	Antares	112 40.3	S26 25.6
17	102 35.5	116 48.1	35.7	69 50.5	58.9	113 55.2	56.0	79 24.8	16.2			
18	117 37.9	131 48.0	S 6 34.8	84 51.2	N12 59.5	128 57.2	S 5 55.8	94 27.0	N 7 16.4	Arcturus	146 06.1	N19 11.5
19	132 40.4	146 47.8	34.0	99 51.9	13 00.2	143 59.2	55.6	109 29.1	16.5	Atria	107 52.2	S69 01.2
20	147 42.9	161 47.7	33.1	114 52.5	00.8	159 01.2	55.4	124 31.3	16.6	Avior	234 22.7	S59 30.7
21	162 45.3	176 47.6	.. 32.2	129 53.2	.. 01.5	174 03.1	.. 55.2	139 33.5	.. 16.7	Bellatrix	278 44.7	N 6 20.7
22	177 47.8	191 47.4	31.3	144 53.9	02.1	189 05.1	55.0	154 35.6	16.8	Betelgeuse	271 14.1	N 7 24.2
23	192 50.3	206 47.3	30.4	159 54.6	02.8	204 07.1	54.8	169 37.8	16.9			
20 00	207 52.7	221 47.1	S 6 29.5	174 55.3	N13 03.4	219 09.1	S 5 54.6	184 40.0	N 7 17.0	Canopus	264 01.5	S52 42.1
01	222 55.2	236 47.0	28.7	189 55.9	04.0	234 11.1	54.4	199 42.1	17.2	Capella	280 52.0	N45 59.7
02	237 57.7	251 46.9	27.8	204 56.6	04.7	249 13.0	54.2	214 44.3	17.3	Deneb	49 39.4	N45 16.2
03	253 00.1	266 46.7	.. 26.9	219 57.3	.. 05.3	264 15.0	.. 54.0	229 46.5	.. 17.4	Denebola	182 45.3	N14 34.9
04	268 02.6	281 46.6	26.0	234 58.0	06.0	279 17.0	53.9	244 48.6	17.5	Diphda	349 07.9	S17 59.9
05	283 05.1	296 46.4	25.1	249 58.6	06.6	294 19.0	53.7	259 50.8	17.6			
06	298 07.5	311 46.3	S 6 24.2	264 59.3	N13 07.3	309 21.0	S 5 53.5	274 53.0	N 7 17.7	Dubhe	194 05.6	N61 45.8
07	313 10.0	326 46.1	23.3	280 00.0	07.9	324 22.9	53.3	289 55.1	17.9	Elnath	278 27.6	N28 36.2
08	328 12.4	341 46.0	22.5	295 00.7	08.5	339 24.9	53.1	304 57.3	18.0	Eltanin	90 51.3	N51 29.2
M 09	343 14.9	356 45.9	.. 21.6	310 01.4	.. 09.2	354 26.9	.. 52.9	319 59.5	.. 18.1	Enif	33 58.7	N 9 51.9
O 10	358 17.4	11 45.7	20.7	325 02.0	09.8	9 28.9	52.7	335 01.6	18.2	Fomalhaut	15 37.1	S29 37.9
N 11	13 19.8	26 45.6	19.8	340 02.7	10.5	24 30.9	52.5	350 03.8	18.3			
D 12	28 22.3	41 45.4	S 6 18.9	355 03.4	N13 11.1	39 32.8	S 5 52.3	5 06.0	N 7 18.4	Gacrux	172 13.3	S57 06.3
A 13	43 24.8	56 45.3	18.0	10 04.1	11.8	54 34.8	52.1	20 08.1	18.5	Gienah	176 04.0	S17 32.0
Y 14	58 27.2	71 45.2	17.1	25 04.8	12.4	69 36.8	51.9	35 10.3	18.7	Hadar	149 03.8	S60 21.8
15	73 29.7	86 45.0	.. 16.2	40 05.4	.. 13.0	84 38.8	.. 51.7	50 12.5	.. 18.8	Hamal	328 14.3	N23 27.0
16	88 32.2	101 44.9	15.3	55 06.1	13.7	99 40.8	51.5	65 14.7	18.9	Kaus Aust.	83 59.1	S34 22.9
17	103 34.6	116 44.7	14.4	70 06.8	14.3	114 42.8	51.3	80 16.8	19.0			
18	118 37.1	131 44.6	S 6 13.6	85 07.5	N13 15.0	129 44.7	S 5 51.1	95 19.0	N 7 19.1	Kochab	137 18.6	N74 09.8
19	133 39.6	146 44.4	12.7	100 08.2	15.6	144 46.7	50.9	110 21.2	19.2	Markab	13 50.2	N15 11.6
20	148 42.0	161 44.3	11.8	115 08.8	16.2	159 48.7	50.7	125 23.3	19.3	Menkar	314 27.6	N 4 04.7
21	163 44.5	176 44.2	.. 10.9	130 09.5	.. 16.9	174 50.7	.. 50.5	140 25.5	.. 19.5	Menkent	148 20.9	S36 21.7
22	178 46.9	191 44.0	10.0	145 10.2	17.5	189 52.7	50.3	155 27.7	19.6	Miaplacidus	221 41.8	S69 43.0
23	193 49.4	206 43.9	09.1	160 10.9	18.2	204 54.7	50.1	170 29.8	19.7			
21 00	208 51.9	221 43.7	S 6 08.2	175 11.5	N13 18.8	219 56.6	S 5 49.9	185 32.0	N 7 19.8	Mirfak	308 57.5	N49 51.2
01	223 54.3	236 43.6	07.3	190 12.2	19.4	234 58.6	49.7	200 34.2	19.9	Nunki	76 12.6	S26 17.8
02	238 56.8	251 43.5	06.4	205 12.9	20.1	250 00.6	49.5	215 36.3	20.0	Peacock	53 37.6	S56 44.2
03	253 59.3	266 43.3	.. 05.5	220 13.6	.. 20.7	265 02.6	.. 49.3	230 38.5	.. 20.1	Pollux	243 42.0	N28 01.8
04	269 01.7	281 43.2	04.6	235 14.3	21.3	280 04.6	49.2	245 40.7	20.3	Procyon	245 12.0	N 5 13.6
05	284 04.2	296 43.0	03.7	250 14.9	22.0	295 06.6	49.0	260 42.8	20.4			
06	299 06.7	311 42.9	S 6 02.8	265 15.6	N13 22.6	310 08.5	S 5 48.8	275 45.0	N 7 20.5	Rasalhague	96 17.1	N12 33.6
07	314 09.1	326 42.7	01.9	280 16.3	23.3	325 10.5	48.6	290 47.2	20.6	Regulus	207 55.8	N11 58.5
08	329 11.6	341 42.6	01.0	295 17.0	23.9	340 12.5	48.4	305 49.3	20.7	Rigel	281 23.5	S 8 12.5
T 09	344 14.0	356 42.4	6 00.1	310 17.6	.. 24.5	355 14.5	.. 48.2	320 51.5	.. 20.8	Rigil Kent.	140 07.0	S60 49.6
U 10	359 16.5	11 42.3	5 59.2	325 18.3	25.2	10 16.5	48.0	335 53.7	20.9	Sabik	102 25.7	S15 43.3
E 11	14 19.0	26 42.2	58.3	340 19.0	25.8	25 18.5	47.8	350 55.8	21.1			
S 12	29 21.4	41 42.0	S 5 57.4	355 19.7	N13 26.4	40 20.4	S 5 47.6	5 58.0	N 7 21.2	Schedar	349 54.3	N56 31.4
D 13	44 23.9	56 41.9	56.5	10 20.3	27.1	55 22.4	47.4	21 00.2	21.3	Shaula	96 37.5	S37 06.0
A 14	59 26.4	71 41.7	55.6	25 21.0	27.7	70 24.4	47.2	36 02.3	21.4	Sirius	258 44.1	S16 43.1
Y 15	74 28.8	86 41.6	.. 54.7	40 21.7	.. 28.3	85 26.4	.. 47.0	51 04.5	.. 21.5	Spica	158 43.2	S11 09.2
16	89 31.3	101 41.4	53.8	55 22.4	29.0	100 28.4	46.8	66 06.7	21.6	Suhail	223 00.9	S43 25.9
17	104 33.8	116 41.3	52.9	70 23.1	29.6	115 30.4	46.6	81 08.8	21.7			
18	119 36.2	131 41.2	S 5 52.0	85 23.7	N13 30.2	130 32.4	S 5 46.4	96 11.0	N 7 21.9	Vega	80 46.7	N38 46.8
19	134 38.7	146 41.0	51.1	100 24.4	30.9	145 34.3	46.2	111 13.2	22.0	Zuben'ubi	137 18.0	S16 02.1
20	149 41.2	161 40.9	50.2	115 25.1	31.5	160 36.3	46.0	126 15.4	22.1		SHA	Mer. Pass.
21	164 43.6	176 40.7	.. 49.3	130 25.8	.. 32.1	175 38.3	.. 45.8	141 17.5	.. 22.2		° ′	h m
22	179 46.1	191 40.6	48.4	145 26.4	32.8	190 40.3	45.6	156 19.7	22.3	Venus	13 54.4	9 13
23	194 48.5	206 40.4	47.5	160 27.1	33.4	205 42.3	45.4	171 21.9	22.4	Mars	327 02.5	12 20
	h m									Jupiter	11 16.3	9 22
Mer. Pass. 10 06.8		v −0.1	d 0.9	v 0.7	d 0.6	v 2.0	d 0.2	v 2.2	d 0.1	Saturn	336 47.2	11 40

SUN / MOON

UT (d h)	SUN GHA	SUN Dec	MOON GHA	v	MOON Dec	d	HP
19 00	180 11.1	N11 01.6	277 16.5	9.3	S18 12.1	2.5	56.9
01	195 11.2	02.4	291 44.8	9.3	18 09.6	2.7	56.9
02	210 11.4	03.3	306 13.1	9.2	18 06.9	2.8	56.9
03	225 11.5 ..	04.2	320 41.3	9.3	18 04.1	2.8	57.0
04	240 11.6	05.0	335 09.6	9.2	18 01.3	3.0	57.0
05	255 11.8	05.9	349 37.8	9.2	17 58.3	3.1	57.0
06	270 11.9	N11 06.8	4 06.0	9.2	S17 55.2	3.2	57.1
S 07	285 12.0	07.6	18 34.2	9.1	17 52.0	3.3	57.1
U 08	300 12.2	08.5	33 02.3	9.2	17 48.7	3.4	57.1
N 09	315 12.3 ..	09.4	47 30.5	9.1	17 45.3	3.5	57.2
D 10	330 12.5	10.2	61 58.6	9.1	17 41.8	3.6	57.2
A 11	345 12.6	11.1	76 26.7	9.1	17 38.2	3.7	57.2
Y 12	0 12.7	N11 12.0	90 54.8	9.1	S17 34.5	3.8	57.3
13	15 12.9	12.8	105 22.9	9.1	17 30.7	3.9	57.3
14	30 13.0	13.7	119 51.0	9.0	17 26.8	4.0	57.3
15	45 13.1 ..	14.5	134 19.0	9.0	17 22.8	4.1	57.4
16	60 13.3	15.4	148 47.0	9.1	17 18.7	4.2	57.4
17	75 13.4	16.3	163 15.1	9.0	17 14.5	4.4	57.4
18	90 13.5	N11 17.1	177 43.1	9.0	S17 10.1	4.4	57.5
19	105 13.7	18.0	192 11.1	8.9	17 05.7	4.5	57.5
20	120 13.8	18.9	206 39.0	9.0	17 01.2	4.6	57.5
21	135 13.9 ..	19.7	221 07.0	9.0	16 56.6	4.8	57.6
22	150 14.1	20.6	235 35.0	8.9	16 51.8	4.8	57.6
23	165 14.2	21.4	250 02.9	8.9	16 47.0	4.9	57.6
20 00	180 14.3	N11 22.3	264 30.8	8.9	S16 42.1	5.1	57.7
01	195 14.5	23.2	278 58.7	8.9	16 37.0	5.1	57.7
02	210 14.6	24.0	293 26.6	8.9	16 31.9	5.3	57.7
03	225 14.7 ..	24.9	307 54.5	8.9	16 26.6	5.3	57.8
04	240 14.9	25.7	322 22.4	8.9	16 21.3	5.4	57.8
05	255 15.0	26.6	336 50.3	8.8	16 15.9	5.6	57.8
06	270 15.1	N11 27.5	351 18.1	8.9	S16 10.3	5.6	57.9
M 07	285 15.3	28.3	5 46.0	8.8	16 04.7	5.8	57.9
O 08	300 15.4	29.2	20 13.8	8.9	15 58.9	5.8	58.0
N 09	315 15.5 ..	30.0	34 41.7	8.8	15 53.1	5.9	58.0
D 10	330 15.7	30.9	49 09.5	8.8	15 47.2	6.1	58.0
A 11	345 15.8	31.7	63 37.3	8.8	15 41.1	6.1	58.1
Y 12	0 15.9	N11 32.6	78 05.1	8.8	S15 35.0	6.3	58.1
13	15 16.1	33.5	92 32.9	8.8	15 28.7	6.3	58.1
14	30 16.2	34.3	107 00.7	8.8	15 22.4	6.4	58.2
15	45 16.3 ..	35.2	121 28.5	8.7	15 16.0	6.6	58.2
16	60 16.4	36.0	135 56.2	8.8	15 09.4	6.6	58.2
17	75 16.6	36.9	150 24.0	8.7	15 02.8	6.7	58.3
18	90 16.7	N11 37.7	164 51.7	8.8	S14 56.1	6.8	58.3
19	105 16.8	38.6	179 19.5	8.7	14 49.3	6.9	58.3
20	120 17.0	39.4	193 47.2	8.8	14 42.4	7.0	58.4
21	135 17.1 ..	40.3	208 15.0	8.7	14 35.4	7.1	58.4
22	150 17.2	41.1	222 42.7	8.7	14 28.3	7.2	58.5
23	165 17.4	42.0	237 10.4	8.7	14 21.1	7.3	58.5
21 00	180 17.5	N11 42.8	251 38.1	8.7	S14 13.8	7.4	58.5
01	195 17.6	43.7	266 05.8	8.7	14 06.4	7.5	58.6
02	210 17.7	44.6	280 33.5	8.7	13 58.9	7.6	58.6
03	225 17.9 ..	45.4	295 01.2	8.7	13 51.3	7.6	58.6
04	240 18.0	46.3	309 28.9	8.7	13 43.7	7.8	58.7
05	255 18.1	47.1	323 56.6	8.7	13 35.9	7.8	58.7
06	270 18.2	N11 48.0	338 24.3	8.7	S13 28.1	7.9	58.7
T 07	285 18.4	48.8	352 52.0	8.6	13 20.2	8.1	58.8
U 08	300 18.5	49.7	7 19.6	8.7	13 12.1	8.1	58.8
E 09	315 18.6 ..	50.5	21 47.3	8.6	13 04.0	8.2	58.8
S 10	330 18.8	51.4	36 14.9	8.7	12 55.8	8.2	58.9
D 11	345 18.9	52.2	50 42.6	8.6	12 47.6	8.4	58.9
A 12	0 19.0	N11 53.1	65 10.2	8.7	S12 39.2	8.5	59.0
Y 13	15 19.1	53.9	79 37.9	8.6	12 30.7	8.5	59.0
14	30 19.3	54.7	94 05.5	8.6	12 22.2	8.6	59.0
15	45 19.4 ..	55.6	108 33.1	8.6	12 13.6	8.7	59.1
16	60 19.5	56.4	123 00.7	8.7	12 04.9	8.8	59.1
17	75 19.6	57.3	137 28.4	8.6	11 56.1	8.9	59.1
18	90 19.8	N11 58.1	151 56.0	8.6	S11 47.2	9.0	59.2
19	105 19.9	59.0	166 23.6	8.6	11 38.2	9.0	59.2
20	120 20.0	11 59.8	180 51.2	8.6	11 29.2	9.1	59.2
21	135 20.1	12 00.7	195 18.8	8.6	11 20.1	9.2	59.3
22	150 20.3	01.5	209 46.4	8.5	11 10.9	9.3	59.3
23	165 20.4	02.4	224 13.9	8.6	S11 01.6	9.4	59.3
	SD 15.9	d 0.9	SD 15.6		15.8		16.1

Twilight / Sunrise / Moonrise

Lat.	Naut.	Civil	Sunrise	Moonrise 19	20	21	22
°	h m	h m	h m	h m	h m	h m	h m
N 72	////	01 00	03 11	■■■	04 52	04 40	04 31
N 70	////	01 59	03 32	03 56	04 08	04 13	04 15
68	////	02 32	03 48	03 15	03 39	03 53	04 02
66	00 52	02 56	04 02	02 47	03 17	03 37	03 51
64	01 44	03 15	04 12	02 26	02 59	03 23	03 42
62	02 14	03 29	04 22	02 09	02 44	03 12	03 34
60	02 36	03 42	04 30	01 54	02 32	03 02	03 28
N 58	02 54	03 52	04 37	01 42	02 21	02 54	03 22
56	03 08	04 02	04 43	01 31	02 12	02 47	03 16
54	03 20	04 10	04 48	01 22	02 04	02 40	03 12
52	03 31	04 17	04 53	01 14	01 56	02 34	03 07
50	03 40	04 23	04 58	01 06	01 50	02 28	03 03
45	03 58	04 36	05 07	00 50	01 35	02 17	02 55
N 40	04 13	04 47	05 15	00 37	01 24	02 07	02 48
35	04 25	04 56	05 22	00 26	01 13	01 59	02 41
30	04 34	05 03	05 28	00 16	01 05	01 51	02 36
20	04 49	05 16	05 38	24 49	00 49	01 38	02 27
N 10	05 01	05 26	05 47	24 36	00 36	01 27	02 18
0	05 10	05 35	05 56	24 23	00 23	01 17	02 11
S 10	05 18	05 42	06 04	24 11	00 11	01 06	02 03
20	05 24	05 50	06 12	23 57	24 55	00 55	01 54
30	05 30	05 58	06 22	23 42	24 42	00 42	01 45
35	05 32	06 02	06 27	23 33	24 35	00 35	01 39
40	05 34	06 06	06 34	23 23	24 26	00 26	01 33
45	05 37	06 11	06 41	23 11	24 16	00 16	01 26
S 50	05 39	06 16	06 50	22 56	24 04	00 04	01 17
52	05 39	06 19	06 53	22 49	23 59	25 13	01 13
54	05 40	06 21	06 58	22 42	23 52	25 09	01 09
56	05 41	06 24	07 03	22 33	23 46	25 04	01 04
58	05 42	06 27	07 08	22 24	23 38	24 58	00 58
S 60	05 42	06 31	07 14	22 13	23 29	24 52	00 52

Sunset / Twilight / Moonset

Lat.	Sunset	Civil	Naut.	Moonset 19	20	21	22
°	h m	h m	h m	h m	h m	h m	h m
N 72	20 51	23 17	////	■■■	08 30	10 35	12 36
N 70	20 29	22 06	////	07 35	09 14	11 01	12 51
68	20 12	21 30	////	08 16	09 43	11 20	13 02
66	19 59	21 05	23 22	08 43	10 04	11 35	13 12
64	19 48	20 46	22 20	09 04	10 21	11 47	13 20
62	19 38	20 31	21 48	09 21	10 35	11 58	13 26
60	19 30	20 18	21 25	09 36	10 47	12 07	13 32
N 58	19 23	20 07	21 07	09 48	10 57	12 15	13 37
56	19 16	19 58	20 52	09 58	11 06	12 21	13 42
54	19 11	19 50	20 40	10 07	11 14	12 27	13 46
52	19 06	19 43	20 29	10 15	11 21	12 33	13 50
50	19 01	19 36	20 20	10 23	11 28	12 38	13 53
45	18 51	19 23	20 01	10 38	11 41	12 49	14 00
N 40	18 43	19 12	19 46	10 51	11 52	12 58	14 06
35	18 36	19 03	19 34	11 02	12 02	13 05	14 11
30	18 30	18 55	19 24	11 12	12 10	13 12	14 16
20	18 20	18 42	19 09	11 28	12 25	13 23	14 23
N 10	18 11	18 32	18 57	11 42	12 37	13 33	14 30
0	18 02	18 23	18 48	11 56	12 49	13 42	14 36
S 10	17 54	18 15	18 40	12 09	13 00	13 52	14 42
20	17 45	18 08	18 34	12 23	13 13	14 01	14 49
30	17 36	18 00	18 28	12 39	13 27	14 12	14 56
35	17 30	17 56	18 25	12 48	13 35	14 19	15 01
40	17 24	17 51	18 23	12 59	13 44	14 26	15 05
45	17 16	17 46	18 21	13 12	13 55	14 34	15 11
S 50	17 08	17 41	18 18	13 27	14 08	14 44	15 18
52	17 04	17 38	18 18	13 34	14 14	14 49	15 21
54	16 59	17 36	18 17	13 42	14 20	14 54	15 24
56	16 54	17 33	18 16	13 50	14 28	15 00	15 28
58	16 49	17 30	18 15	14 00	14 36	15 06	15 32
S 60	16 43	17 26	18 14	14 12	14 45	15 13	15 36

SUN / MOON

Day	Eqn. of Time 00h	12h	Mer. Pass.	Mer. Pass. Upper	Lower	Age	Phase
d	m s	m s	h m	h m	h m	d	%
19	00 44	00 51	11 59	05 43	18 09	22	54
20	00 57	01 03	11 59	06 36	19 03	23	43
21	01 10	01 16	11 59	07 30	19 56	24	32

1998 APRIL 22, 23, 24 (WED., THURS., FRI.)

UT	ARIES GHA	VENUS −4.2 GHA	Dec	MARS +1.3 GHA	Dec	JUPITER −2.1 GHA	Dec	SATURN +0.5 GHA	Dec	STARS Name	SHA	Dec
22 00	209 51.0	221 40.3	S 5 46.6	175 27.8	N13 34.0	220 44.3	S 5 45.3	186 24.0	N 7 22.5	Acamar	315 27.6	S40 18.9
01	224 53.5	236 40.1	45.7	190 28.5	34.7	235 46.3	45.1	201 26.2	22.7	Achernar	335 36.1	S57 14.8
02	239 55.9	251 40.0	44.8	205 29.1	35.3	250 48.2	44.9	216 28.4	22.8	Acrux	173 21.6	S63 05.5
03	254 58.4	266 39.9	.. 43.8	220 29.8	.. 35.9	265 50.2	.. 44.7	231 30.5	.. 22.9	Adhara	255 21.8	S28 58.5
04	270 00.9	281 39.7	42.9	235 30.5	36.6	280 52.2	44.5	246 32.7	23.0	Aldebaran	291 03.0	N16 30.2
05	285 03.3	296 39.6	42.0	250 31.2	37.2	295 54.2	44.3	261 34.9	23.1			
W 06	300 05.8	311 39.4	S 5 41.1	265 31.8	N13 37.8	310 56.2	S 5 44.1	276 37.0	N 7 23.2	Alioth	166 30.4	N55 58.3
E 07	315 08.3	326 39.3	40.2	280 32.5	38.5	325 58.2	43.9	291 39.2	23.3	Alkaid	153 07.6	N49 19.4
D 08	330 10.7	341 39.1	39.3	295 33.2	39.1	341 00.2	43.7	306 41.4	23.5	Al Na'ir	27 58.6	S46 58.0
N 09	345 13.2	356 39.0	.. 38.4	310 33.9	.. 39.7	356 02.1	.. 43.5	321 43.5	.. 23.6	Alnilam	275 58.4	S 1 12.4
E 10	0 15.6	11 38.8	37.5	325 34.5	40.4	11 04.1	43.3	336 45.7	23.7	Alphard	218 07.4	S 8 39.3
S 11	15 18.1	26 38.7	36.6	340 35.2	41.0	26 06.1	43.1	351 47.9	23.8			
D 12	30 20.6	41 38.6	S 5 35.7	355 35.9	N13 41.6	41 08.1	S 5 42.9	6 50.0	N 7 23.9	Alphecca	126 20.6	N26 43.2
A 13	45 23.0	56 38.4	34.7	10 36.6	42.3	56 10.1	42.7	21 52.2	24.0	Alpheratz	357 55.9	N29 04.6
Y 14	60 25.5	71 38.3	33.8	25 37.2	42.9	71 12.1	42.5	36 54.4	24.1	Altair	62 19.5	N 8 51.8
15	75 28.0	86 38.1	.. 32.9	40 37.9	.. 43.5	86 14.1	.. 42.3	51 56.5	.. 24.3	Ankaa	353 27.6	S42 19.0
16	90 30.4	101 38.0	32.0	55 38.6	44.1	101 16.1	42.2	66 58.7	24.4	Antares	112 40.3	S26 25.6
17	105 32.9	116 37.8	31.1	70 39.3	44.8	116 18.0	42.0	82 00.9	24.5			
18	120 35.4	131 37.7	S 5 30.2	85 39.9	N13 45.4	131 20.0	S 5 41.8	97 03.1	N 7 24.6	Arcturus	146 06.0	N19 11.5
19	135 37.8	146 37.5	29.3	100 40.6	46.0	146 22.0	41.6	112 05.2	24.7	Atria	107 52.2	S69 01.2
20	150 40.3	161 37.4	28.3	115 41.3	46.7	161 24.0	41.4	127 07.4	24.8	Avior	234 22.7	S59 30.7
21	165 42.8	176 37.2	.. 27.4	130 42.0	.. 47.3	176 26.0	.. 41.2	142 09.6	.. 24.9	Bellatrix	278 44.7	N 6 20.7
22	180 45.2	191 37.1	26.5	145 42.6	47.9	191 28.0	41.0	157 11.7	25.1	Betelgeuse	271 14.1	N 7 24.2
23	195 47.7	206 37.0	25.6	160 43.3	48.5	206 30.0	40.8	172 13.9	25.2			
23 00	210 50.1	221 36.8	S 5 24.7	175 44.0	N13 49.2	221 32.0	S 5 40.6	187 16.1	N 7 25.3	Canopus	264 01.6	S52 42.1
01	225 52.6	236 36.7	23.8	190 44.7	49.8	236 34.0	40.4	202 18.2	25.4	Capella	280 52.0	N45 59.7
02	240 55.1	251 36.5	22.8	205 45.3	50.4	251 35.9	40.2	217 20.4	25.5	Deneb	49 39.4	N45 16.2
03	255 57.5	266 36.4	.. 21.9	220 46.0	.. 51.0	266 37.9	.. 40.0	232 22.6	.. 25.6	Denebola	182 45.3	N14 34.9
04	271 00.0	281 36.2	21.0	235 46.7	51.7	281 39.9	39.8	247 24.7	25.7	Diphda	349 07.9	S17 59.9
05	286 02.5	296 36.1	20.1	250 47.4	52.3	296 41.9	39.6	262 26.9	25.8			
T 06	301 04.9	311 35.9	S 5 19.2	265 48.0	N13 52.9	311 43.9	S 5 39.4	277 29.1	N 7 26.0	Dubhe	194 05.6	N61 45.8
H 07	316 07.4	326 35.8	18.2	280 48.7	53.5	326 45.9	39.3	292 31.2	26.1	Elnath	278 27.6	N28 36.2
U 08	331 09.9	341 35.6	17.3	295 49.4	54.2	341 47.9	39.1	307 33.4	26.2	Eltanin	90 51.3	N51 29.2
R 09	346 12.3	356 35.5	.. 16.4	310 50.1	.. 54.8	356 49.9	.. 38.9	322 35.6	.. 26.3	Enif	33 58.7	N 9 51.9
S 10	1 14.8	11 35.3	15.5	325 50.7	55.4	11 51.9	38.7	337 37.7	26.4	Fomalhaut	15 37.1	S29 37.9
D 11	16 17.3	26 35.2	14.6	340 51.4	56.0	26 53.9	38.5	352 39.9	26.5			
A 12	31 19.7	41 35.1	S 5 13.6	355 52.1	N13 56.7	41 55.8	S 5 38.3	7 42.1	N 7 26.6	Gacrux	172 13.3	S57 06.3
Y 13	46 22.2	56 34.9	12.7	10 52.8	57.3	56 57.8	38.1	22 44.3	26.8	Gienah	176 04.0	S17 32.1
14	61 24.6	71 34.8	11.8	25 53.4	57.9	71 59.8	37.9	37 46.4	26.9	Hadar	149 03.8	S60 21.9
15	76 27.1	86 34.6	.. 10.9	40 54.1	.. 58.5	87 01.8	.. 37.7	52 48.6	.. 27.0	Hamal	328 14.3	N23 27.0
16	91 29.6	101 34.5	09.9	55 54.8	59.2	102 03.8	37.5	67 50.8	27.1	Kaus Aust.	83 59.1	S34 22.9
17	106 32.0	116 34.3	09.0	70 55.4	13 59.8	117 05.8	37.3	82 52.9	27.2			
18	121 34.5	131 34.2	S 5 08.1	85 56.1	N14 00.4	132 07.8	S 5 37.1	97 55.1	N 7 27.3	Kochab	137 18.6	N74 09.8
19	136 37.0	146 34.0	07.2	100 56.8	01.0	147 09.8	36.9	112 57.3	27.4	Markab	13 50.1	N15 11.6
20	151 39.4	161 33.9	06.2	115 57.5	01.6	162 11.8	36.8	127 59.4	27.6	Menkar	314 27.6	N 4 04.7
21	166 41.9	176 33.7	.. 05.3	130 58.1	.. 02.3	177 13.8	.. 36.6	143 01.6	.. 27.7	Menkent	148 20.9	S36 21.7
22	181 44.4	191 33.6	04.4	145 58.8	02.9	192 15.8	36.4	158 03.8	27.8	Miaplacidus	221 41.8	S69 43.0
23	196 46.8	206 33.4	03.5	160 59.5	03.5	207 17.8	36.2	173 05.9	27.9			
24 00	211 49.3	221 33.3	S 5 02.5	176 00.1	N14 04.1	222 19.7	S 5 36.0	188 08.1	N 7 28.0	Mirfak	308 57.5	N49 51.2
01	226 51.7	236 33.1	01.6	191 00.8	04.8	237 21.7	35.8	203 10.3	28.1	Nunki	76 12.6	S26 17.8
02	241 54.2	251 33.0	5 00.7	206 01.5	05.4	252 23.7	35.6	218 12.4	28.2	Peacock	53 37.6	S56 44.2
03	256 56.7	266 32.8	4 59.8	221 02.2	.. 06.0	267 25.7	.. 35.4	233 14.6	.. 28.3	Pollux	243 42.0	N28 01.8
04	271 59.1	281 32.7	58.8	236 02.8	06.6	282 27.7	35.2	248 16.8	28.5	Procyon	245 12.0	N 5 13.6
05	287 01.6	296 32.6	57.9	251 03.5	07.2	297 29.7	35.0	263 18.9	28.6			
F 06	302 04.1	311 32.4	S 4 57.0	266 04.2	N14 07.9	312 31.7	S 5 34.8	278 21.1	N 7 28.7	Rasalhague	96 17.1	N12 33.6
R 07	317 06.5	326 32.3	56.0	281 04.9	08.5	327 33.7	34.6	293 23.3	28.8	Regulus	207 55.8	N11 58.5
I 08	332 09.0	341 32.1	55.1	296 05.5	09.1	342 35.7	34.5	308 25.5	28.9	Rigel	281 23.5	S 8 12.5
D 09	347 11.5	356 32.0	.. 54.2	311 06.2	.. 09.7	357 37.7	.. 34.3	323 27.6	.. 29.0	Rigil Kent.	140 07.0	S60 49.6
A 10	2 13.9	11 31.8	53.2	326 06.9	10.3	12 39.7	34.1	338 29.8	29.1	Sabik	102 25.7	S15 43.3
Y 11	17 16.4	26 31.7	52.3	341 07.5	10.9	27 41.7	33.9	353 32.0	29.3			
12	32 18.9	41 31.5	S 4 51.4	356 08.2	N14 11.6	42 43.7	S 5 33.7	8 34.1	N 7 29.4	Schedar	349 54.3	N56 31.4
13	47 21.3	56 31.4	50.5	11 08.9	12.2	57 45.7	33.5	23 36.3	29.5	Shaula	96 37.5	S37 06.0
14	62 23.8	71 31.2	49.5	26 09.6	12.8	72 47.6	33.3	38 38.5	29.6	Sirius	258 44.1	S16 43.1
15	77 26.2	86 31.1	.. 48.6	41 10.2	.. 13.4	87 49.6	.. 33.1	53 40.6	.. 29.7	Spica	158 43.2	S11 09.2
16	92 28.7	101 30.9	47.7	56 10.9	14.0	102 51.6	32.9	68 42.8	29.8	Suhail	223 00.9	S43 25.9
17	107 31.2	116 30.8	46.7	71 11.6	14.7	117 53.6	32.7	83 45.0	29.9			
18	122 33.6	131 30.6	S 4 45.8	86 12.2	N14 15.3	132 55.6	S 5 32.5	98 47.1	N 7 30.0	Vega	80 46.7	N38 46.8
19	137 36.1	146 30.5	44.8	101 12.9	15.9	147 57.6	32.3	113 49.3	30.2	Zuben'ubi	137 18.0	S16 02.1
20	152 38.6	161 30.3	43.9	116 13.6	16.5	162 59.6	32.2	128 51.5	30.3		SHA	Mer. Pass.
21	167 41.0	176 30.2	.. 43.0	131 14.2	.. 17.1	178 01.6	.. 32.0	143 53.6	.. 30.4	Venus	10 46.7	9 14
22	182 43.5	191 30.0	42.0	146 14.9	17.7	193 03.6	31.8	158 55.8	30.5	Mars	324 53.8	12 17
23	197 46.0	206 29.9	41.1	161 15.6	18.3	208 05.6	31.6	173 58.0	30.6	Jupiter	10 41.8	9 13
Mer. Pass. 9 55.0		v −0.1	d 0.9	v 0.7	d 0.6	v 2.0	d 0.2	v 2.2	d 0.1	Saturn	336 25.9	11 29

UT	SUN GHA	Dec	MOON GHA	v	Dec	d	HP
22 00	180 20.5	N12 03.2	238 41.5	8.6	S10 52.2	9.4	59.4
01	195 20.6	04.1	253 09.1	8.5	10 42.8	9.5	59.4
02	210 20.8	04.9	267 36.6	8.6	10 33.3	9.6	59.4
03	225 20.9	.. 05.7	282 04.2	8.5	10 23.7	9.7	59.5
04	240 21.0	06.6	296 31.7	8.6	10 14.0	9.7	59.5
05	255 21.1	07.4	310 59.3	8.5	10 04.3	9.8	59.5
W 06	270 21.2	N12 08.3	325 26.8	8.6	S 9 54.5	9.9	59.6
E 07	285 21.4	09.1	339 54.4	8.5	9 44.6	9.9	59.6
D 08	300 21.5	10.0	354 21.9	8.5	9 34.7	10.1	59.6
N 09	315 21.6	.. 10.8	8 49.4	8.5	9 24.6	10.1	59.7
E 10	330 21.7	11.6	23 16.9	8.5	9 14.5	10.1	59.7
S 11	345 21.9	12.5	37 44.4	8.5	9 04.4	10.2	59.7
D 12	0 22.0	N12 13.3	52 11.9	8.5	S 8 54.2	10.3	59.8
A 13	15 22.1	14.2	66 39.4	8.5	8 43.9	10.4	59.8
Y 14	30 22.2	15.0	81 06.9	8.4	8 33.5	10.4	59.8
15	45 22.3	.. 15.8	95 34.3	8.5	8 23.1	10.5	59.9
16	60 22.5	16.7	110 01.8	8.5	8 12.6	10.6	59.9
17	75 22.6	17.5	124 29.3	8.4	8 02.0	10.6	59.9
18	90 22.7	N12 18.3	138 56.7	8.4	S 7 51.4	10.6	60.0
19	105 22.8	19.2	153 24.1	8.5	7 40.8	10.8	60.0
20	120 22.9	20.0	167 51.6	8.4	7 30.0	10.8	60.0
21	135 23.1	.. 20.9	182 19.0	8.4	7 19.2	10.8	60.1
22	150 23.2	21.7	196 46.4	8.4	7 08.4	10.9	60.1
23	165 23.3	22.5	211 13.8	8.4	6 57.5	11.0	60.1
23 00	180 23.4	N12 23.4	225 41.2	8.3	S 6 46.5	11.0	60.2
01	195 23.5	24.2	240 08.5	8.4	6 35.5	11.0	60.2
02	210 23.7	25.0	254 35.9	8.4	6 24.5	11.2	60.2
03	225 23.8	.. 25.9	269 03.3	8.3	6 13.3	11.1	60.2
04	240 23.9	26.7	283 30.6	8.3	6 02.2	11.2	60.3
05	255 24.0	27.5	297 57.9	8.4	5 51.0	11.3	60.3
T 06	270 24.1	N12 28.4	312 25.3	8.3	S 5 39.7	11.3	60.3
H 07	285 24.3	29.2	326 52.6	8.3	5 28.4	11.4	60.4
U 08	300 24.4	30.0	341 19.9	8.3	5 17.0	11.4	60.4
R 09	315 24.5	.. 30.9	355 47.2	8.2	5 05.6	11.4	60.4
S 10	330 24.6	31.7	10 14.4	8.3	4 54.2	11.5	60.4
D 11	345 24.7	32.5	24 41.7	8.2	4 42.7	11.5	60.5
A 12	0 24.8	N12 33.4	39 08.9	8.3	S 4 31.2	11.6	60.5
Y 13	15 25.0	34.2	53 36.2	8.2	4 19.6	11.6	60.5
14	30 25.1	35.0	68 03.4	8.2	4 08.0	11.6	60.5
15	45 25.2	.. 35.9	82 30.6	8.2	3 56.4	11.7	60.6
16	60 25.3	36.7	96 57.8	8.1	3 44.7	11.7	60.6
17	75 25.4	37.5	111 24.9	8.2	3 33.0	11.7	60.6
18	90 25.5	N12 38.4	125 52.1	8.1	S 3 21.3	11.8	60.6
19	105 25.6	39.2	140 19.2	8.1	3 09.5	11.8	60.7
20	120 25.8	40.0	154 46.3	8.1	2 57.7	11.8	60.7
21	135 25.9	.. 40.8	169 13.4	8.1	2 45.9	11.9	60.7
22	150 26.0	41.7	183 40.5	8.1	2 34.0	11.9	60.7
23	165 26.1	42.5	198 07.6	8.1	2 22.1	11.9	60.8
24 00	180 26.2	N12 43.3	212 34.7	8.0	S 2 10.2	11.9	60.8
01	195 26.3	44.2	227 01.7	8.0	1 58.3	12.0	60.8
02	210 26.4	45.0	241 28.7	8.0	1 46.3	12.0	60.8
03	225 26.6	.. 45.8	255 55.7	8.0	1 34.3	12.0	60.8
04	240 26.7	46.6	270 22.7	8.0	1 22.3	12.0	60.9
05	255 26.8	47.5	284 49.7	7.9	1 10.3	12.0	60.9
F 06	270 26.9	N12 48.3	299 16.6	7.9	S 0 58.3	12.0	60.9
R 07	285 27.0	49.1	313 43.5	7.9	0 46.3	12.1	60.9
I 08	300 27.1	49.9	328 10.4	7.9	0 34.2	12.1	60.9
D 09	315 27.2	.. 50.8	342 37.3	7.9	0 22.1	12.0	60.9
A 10	330 27.3	51.6	357 04.2	7.8	S 0 10.1	12.1	61.0
Y 11	345 27.5	52.4	11 31.0	7.8	N 0 02.0	12.1	61.0
12	0 27.6	N12 53.2	25 57.8	7.8	N 0 14.1	12.1	61.0
13	15 27.7	54.1	40 24.6	7.8	0 26.2	12.1	61.0
14	30 27.8	54.9	54 51.4	7.8	0 38.3	12.1	61.0
15	45 27.9	.. 55.7	69 18.2	7.7	0 50.4	12.1	61.0
16	60 28.0	56.5	83 44.9	7.7	1 02.5	12.2	61.1
17	75 28.1	57.3	98 11.6	7.7	1 14.7	12.1	61.1
18	90 28.2	N12 58.2	112 38.3	7.7	N 1 26.8	12.1	61.1
19	105 28.3	59.0	127 05.0	7.6	1 38.9	12.1	61.1
20	120 28.5	12 59.8	141 31.6	7.6	1 51.0	12.1	61.1
21	135 28.6	13 00.6	155 58.2	7.6	2 03.1	12.0	61.1
22	150 28.7	01.4	170 24.8	7.6	2 15.1	12.1	61.1
23	165 28.8	02.3	184 51.4	7.5	N 2 27.2	12.1	61.1
	SD 15.9	d 0.8	SD 16.3		16.5		16.6

Lat.	Twilight Naut.	Civil	Sunrise	Moonrise 22	23	24	25
N 72	////	////	02 52	04 31	04 24	04 18	04 11
N 70	////	01 32	03 16	04 15	04 16	04 16	04 16
68	////	02 14	03 35	04 02	04 09	04 14	04 19
66	////	02 41	03 50	03 51	04 03	04 13	04 22
64	01 20	03 02	04 02	03 42	03 58	04 11	04 25
62	01 58	03 18	04 12	03 34	03 53	04 10	04 27
60	02 23	03 32	04 21	03 28	03 49	04 09	04 30
N 58	02 43	03 44	04 29	03 22	03 46	04 09	04 31
56	02 58	03 53	04 35	03 16	03 43	04 08	04 33
54	03 11	04 02	04 41	03 12	03 40	04 07	04 35
52	03 23	04 10	04 47	03 07	03 38	04 07	04 36
50	03 33	04 17	04 52	03 03	03 35	04 06	04 37
45	03 53	04 31	05 02	02 55	03 30	04 05	04 40
N 40	04 08	04 43	05 11	02 48	03 26	04 04	04 42
35	04 20	04 52	05 19	02 41	03 23	04 03	04 44
30	04 31	05 00	05 25	02 36	03 20	04 03	04 46
20	04 47	05 14	05 36	02 27	03 14	04 01	04 49
N 10	04 59	05 24	05 46	02 18	03 09	04 00	04 52
0	05 09	05 34	05 55	02 11	03 05	03 59	04 54
S 10	05 18	05 42	06 04	02 03	03 00	03 58	04 57
20	05 25	05 51	06 13	01 54	02 55	03 57	05 00
30	05 31	05 59	06 24	01 45	02 50	03 56	05 03
35	05 34	06 04	06 30	01 39	02 46	03 55	05 05
40	05 37	06 09	06 37	01 33	02 43	03 55	05 08
45	05 40	06 14	06 45	01 26	02 39	03 54	05 10
S 50	05 43	06 21	06 54	01 17	02 34	03 53	05 13
52	05 44	06 24	06 58	01 13	02 31	03 52	05 15
54	05 45	06 27	07 03	01 09	02 29	03 52	05 16
56	05 46	06 30	07 09	01 04	02 26	03 51	05 18
58	05 47	06 33	07 15	00 58	02 23	03 50	05 20
S 60	05 49	06 37	07 21	00 52	02 19	03 50	05 22

Lat.	Sunset	Twilight Civil	Naut.	Moonset 22	23	24	25
N 72	21 10	////	////	12 36	14 37	16 39	18 43
N 70	20 44	22 34	////	12 51	14 43	16 37	18 33
68	20 25	21 48	////	13 02	14 48	16 36	18 26
66	20 09	21 19	////	13 12	14 52	16 35	18 19
64	19 57	20 58	22 44	13 20	14 56	16 34	18 14
62	19 46	20 41	22 04	13 26	14 59	16 33	18 10
60	19 37	20 27	21 37	13 32	15 01	16 33	18 06
N 58	19 30	20 15	21 17	13 37	15 04	16 32	18 02
56	19 25	20 05	21 01	13 42	15 06	16 32	17 59
54	19 16	19 56	20 47	13 46	15 07	16 31	17 56
52	19 11	19 48	20 36	13 50	15 09	16 31	17 54
50	19 06	19 41	20 26	13 53	15 11	16 31	17 51
45	18 55	19 27	20 05	14 00	15 14	16 30	17 47
N 40	18 46	19 15	19 50	14 06	15 17	16 29	17 42
35	18 39	19 05	19 37	14 11	15 19	16 28	17 39
30	18 32	18 57	19 27	14 16	15 21	16 28	17 36
20	18 21	18 43	19 10	14 23	15 25	16 27	17 30
N 10	18 11	18 32	18 58	14 30	15 28	16 26	17 26
0	18 02	18 23	18 48	14 36	15 30	16 25	17 21
S 10	17 53	18 14	18 39	14 42	15 33	16 24	17 17
20	17 43	18 06	18 32	14 49	15 36	16 24	17 12
30	17 33	17 57	18 25	14 56	15 39	16 22	17 06
35	17 26	17 52	18 22	15 01	15 41	16 22	17 03
40	17 19	17 47	18 19	15 05	15 43	16 21	17 00
45	17 11	17 42	18 16	15 11	15 46	16 20	16 56
S 50	17 02	17 35	18 13	15 18	15 49	16 19	16 51
52	16 57	17 32	18 12	15 21	15 50	16 19	16 49
54	16 53	17 29	18 11	15 24	15 52	16 18	16 46
56	16 47	17 26	18 09	15 28	15 53	16 18	16 43
58	16 41	17 22	18 08	15 32	15 55	16 17	16 40
S 60	16 34	17 18	18 07	15 36	15 57	16 17	16 37

Day	SUN Eqn. of Time 00h	12h	Mer. Pass.	MOON Mer. Pass. Upper	Lower	Age	Phase
	m s	m s	h m	h m	h m	d	%
22	01 22	01 28	11 59	08 23	20 50	25	22
23	01 33	01 39	11 58	09 17	21 45	26	13
24	01 45	01 50	11 58	10 12	22 40	27	6

1998 APRIL 25, 26, 27 (SAT., SUN., MON.)

UT	ARIES GHA	VENUS −4.2 GHA	VENUS Dec	MARS +1.3 GHA	MARS Dec	JUPITER −2.1 GHA	JUPITER Dec	SATURN +0.5 GHA	SATURN Dec	STARS Name	SHA	Dec
25 00	212 48.4	221 29.7	S 4 40.2	176 16.3	N14 19.0	223 07.6	S 5 31.4	189 00.2	N 7 30.7	Acamar	315 27.6	S40 18.9
01	227 50.9	236 29.6	39.2	191 16.9	19.6	238 09.6	31.2	204 02.3	30.8	Achernar	335 36.1	S57 14.8
02	242 53.3	251 29.4	38.3	206 17.6	20.2	253 11.6	31.0	219 04.5	30.9	Acrux	173 21.7	S63 05.5
03	257 55.8	266 29.3	.. 37.4	221 18.3	.. 20.8	268 13.6	.. 30.8	234 06.7	.. 31.1	Adhara	255 21.8	S28 58.5
04	272 58.3	281 29.1	36.4	236 18.9	21.4	283 15.6	30.6	249 08.8	31.2	Aldebaran	291 03.1	N16 30.2
05	288 00.7	296 29.0	35.5	251 19.6	22.0	298 17.6	30.4	264 11.0	31.3			
06	303 03.2	311 28.8	S 4 34.5	266 20.3	N14 22.6	313 19.6	S 5 30.3	279 13.2	N 7 31.4	Alioth	166 30.4	N55 58.3
07	318 05.7	326 28.7	33.6	281 21.0	23.3	328 21.6	30.1	294 15.3	31.5	Alkaid	153 07.6	N49 19.4
S 08	333 08.1	341 28.5	32.7	296 21.6	23.9	343 23.6	29.9	309 17.5	31.6	Al Na'ir	27 58.6	S46 58.0
A 09	348 10.6	356 28.4	.. 31.7	311 22.3	.. 24.5	358 25.6	.. 29.7	324 19.7	.. 31.7	Alnilam	275 58.4	S 1 12.4
T 10	3 13.1	11 28.2	30.8	326 23.0	25.1	13 27.6	29.5	339 21.8	31.8	Alphard	218 07.4	S 8 39.3
U 11	18 15.5	26 28.1	29.8	341 23.6	25.7	28 29.5	29.3	354 24.0	32.0			
R 12	33 18.0	41 27.9	S 4 28.9	356 24.3	N14 26.3	43 31.5	S 5 29.1	9 26.2	N 7 32.1	Alphecca	126 20.5	N26 43.2
D 13	48 20.5	56 27.8	27.9	11 25.0	26.9	58 33.5	28.9	24 28.4	32.2	Alpheratz	357 55.8	N29 04.6
A 14	63 22.9	71 27.6	27.0	26 25.6	27.5	73 35.5	28.7	39 30.5	32.3	Altair	62 19.5	N 8 51.8
Y 15	78 25.4	86 27.5	.. 26.1	41 26.3	.. 28.2	88 37.5	.. 28.5	54 32.7	.. 32.4	Ankaa	353 27.6	S42 18.9
16	93 27.8	101 27.3	25.1	56 27.0	28.8	103 39.5	28.4	69 34.9	32.5	Antares	112 40.3	S26 25.6
17	108 30.3	116 27.2	24.2	71 27.6	29.4	118 41.5	28.2	84 37.0	32.6			
18	123 32.8	131 27.0	S 4 23.2	86 28.3	N14 30.0	133 43.5	S 5 28.0	99 39.2	N 7 32.8	Arcturus	146 06.0	N19 11.5
19	138 35.2	146 26.9	22.3	101 29.0	30.6	148 45.5	27.8	114 41.4	32.9	Atria	107 52.1	S69 01.2
20	153 37.7	161 26.7	21.3	116 29.7	31.2	163 47.5	27.6	129 43.5	33.0	Avior	234 22.7	S59 30.7
21	168 40.2	176 26.6	.. 20.4	131 30.3	.. 31.8	178 49.5	.. 27.4	144 45.7	.. 33.1	Bellatrix	278 44.8	N 6 20.7
22	183 42.6	191 26.4	19.4	146 31.0	32.4	193 51.5	27.2	159 47.9	33.2	Betelgeuse	271 14.1	N 7 24.2
23	198 45.1	206 26.3	18.5	161 31.7	33.0	208 53.5	27.0	174 50.0	33.3			
26 00	213 47.6	221 26.1	S 4 17.6	176 32.3	N14 33.6	223 55.5	S 5 26.8	189 52.2	N 7 33.4	Canopus	264 01.6	S52 42.1
01	228 50.0	236 26.0	16.6	191 33.0	34.2	238 57.5	26.6	204 54.4	33.5	Capella	280 52.0	N45 59.7
02	243 52.5	251 25.8	15.7	206 33.7	34.9	253 59.5	26.5	219 56.6	33.7	Deneb	49 39.4	N45 16.2
03	258 55.0	266 25.7	.. 14.7	221 34.3	.. 35.5	269 01.5	.. 26.3	234 58.7	.. 33.8	Denebola	182 45.3	N14 34.9
04	273 57.4	281 25.5	13.8	236 35.0	36.1	284 03.5	26.1	250 00.9	33.9	Diphda	349 07.9	S17 59.9
05	288 59.9	296 25.4	12.8	251 35.7	36.7	299 05.5	25.9	265 03.1	34.0			
06	304 02.3	311 25.2	S 4 11.9	266 36.3	N14 37.3	314 07.5	S 5 25.7	280 05.2	N 7 34.1	Dubhe	194 05.6	N61 45.8
07	319 04.8	326 25.1	10.9	281 37.0	37.9	329 09.5	25.5	295 07.4	34.2	Elnath	278 27.6	N28 36.2
08	334 07.3	341 24.9	10.0	296 37.7	38.5	344 11.5	25.3	310 09.6	34.3	Eltanin	90 51.2	N51 29.2
S 09	349 09.7	356 24.8	.. 09.0	311 38.3	.. 39.1	359 13.5	.. 25.1	325 11.7	.. 34.4	Enif	33 58.7	N 9 51.9
U 10	4 12.2	11 24.6	08.1	326 39.0	39.7	14 15.5	24.9	340 13.9	34.6	Fomalhaut	15 37.1	S29 37.9
N 11	19 14.7	26 24.5	07.1	341 39.7	40.3	29 17.5	24.8	355 16.1	34.7			
D 12	34 17.1	41 24.3	S 4 06.2	356 40.3	N14 40.9	44 19.5	S 5 24.6	10 18.2	N 7 34.8	Gacrux	172 13.3	S57 06.3
A 13	49 19.6	56 24.2	05.2	11 41.0	41.5	59 21.5	24.4	25 20.4	34.9	Gienah	176 04.0	S17 32.1
Y 14	64 22.1	71 24.0	04.3	26 41.7	42.1	74 23.5	24.2	40 22.6	35.0	Hadar	149 03.8	S60 21.9
15	79 24.5	86 23.9	.. 03.3	41 42.3	.. 42.7	89 25.5	.. 24.0	55 24.8	.. 35.1	Hamal	328 14.3	N23 27.0
16	94 27.0	101 23.7	02.4	56 43.0	43.3	104 27.5	23.8	70 26.9	35.2	Kaus Aust.	83 59.1	S34 22.9
17	109 29.4	116 23.6	01.4	71 43.7	43.9	119 29.5	23.6	85 29.1	35.3			
18	124 31.9	131 23.4	S 4 00.5	86 44.3	N14 44.6	134 31.5	S 5 23.4	100 31.3	N 7 35.4	Kochab	137 18.5	N74 09.8
19	139 34.4	146 23.3	3 59.5	101 45.0	45.2	149 33.5	23.2	115 33.4	35.6	Markab	13 50.1	N15 11.6
20	154 36.8	161 23.1	58.5	116 45.7	45.8	164 35.5	23.1	130 35.6	35.7	Menkar	314 27.6	N 4 04.7
21	169 39.3	176 23.0	.. 57.6	131 46.3	.. 46.4	179 37.5	.. 22.9	145 37.8	.. 35.8	Menkent	148 20.9	S36 21.7
22	184 41.8	191 22.8	56.6	146 47.0	47.0	194 39.5	22.7	160 39.9	35.9	Miaplacidus	221 41.9	S69 43.0
23	199 44.2	206 22.7	55.7	161 47.7	47.6	209 41.5	22.5	175 42.1	36.0			
27 00	214 46.7	221 22.5	S 3 54.7	176 48.3	N14 48.2	224 43.5	S 5 22.3	190 44.3	N 7 36.1	Mirfak	308 57.5	N49 51.2
01	229 49.2	236 22.3	53.8	191 49.0	48.8	239 45.5	22.1	205 46.5	36.2	Nunki	76 12.6	S26 17.8
02	244 51.6	251 22.2	52.8	206 49.7	49.4	254 47.5	21.9	220 48.6	36.3	Peacock	53 37.6	S56 44.1
03	259 54.1	266 22.0	.. 51.9	221 50.3	.. 50.0	269 49.5	.. 21.7	235 50.8	.. 36.5	Pollux	243 42.1	N28 01.8
04	274 56.6	281 21.9	50.9	236 51.0	50.6	284 51.5	21.6	250 53.0	36.6	Procyon	245 12.0	N 5 13.6
05	289 59.0	296 21.7	49.9	251 51.7	51.2	299 53.5	21.4	265 55.1	36.7			
06	305 01.5	311 21.6	S 3 49.0	266 52.3	N14 51.8	314 55.5	S 5 21.2	280 57.3	N 7 36.8	Rasalhague	96 17.0	N12 33.7
07	320 03.9	326 21.4	48.0	281 53.0	52.4	329 57.5	21.0	295 59.5	36.9	Regulus	207 55.8	N11 58.5
08	335 06.4	341 21.3	47.1	296 53.7	53.0	344 59.5	20.8	311 01.6	37.0	Rigel	281 23.5	S 8 12.5
M 09	350 08.9	356 21.1	.. 46.1	311 54.3	.. 53.6	0 01.5	.. 20.6	326 03.8	.. 37.1	Rigil Kent.	140 07.0	S60 49.6
O 10	5 11.3	11 21.0	45.2	326 55.0	54.2	15 03.5	20.4	341 06.0	37.2	Sabik	102 25.7	S15 43.3
N 11	20 13.8	26 20.8	44.2	341 55.7	54.8	30 05.5	20.2	356 08.1	37.4			
D 12	35 16.3	41 20.7	S 3 43.2	356 56.3	N14 55.4	45 07.6	S 5 20.1	11 10.3	N 7 37.5	Schedar	349 54.3	N56 31.4
A 13	50 18.7	56 20.5	42.3	11 57.0	56.0	60 09.6	19.9	26 12.5	37.6	Shaula	96 37.5	S37 06.0
Y 14	65 21.2	71 20.4	41.3	26 57.7	56.6	75 11.6	19.7	41 14.7	37.7	Sirius	258 44.2	S16 43.1
15	80 23.7	86 20.2	.. 40.4	41 58.3	.. 57.2	90 13.6	.. 19.5	56 16.8	.. 37.8	Spica	158 43.2	S11 09.2
16	95 26.1	101 20.1	39.4	56 59.0	57.8	105 15.6	19.3	71 19.0	37.9	Suhail	223 00.9	S43 25.9
17	110 28.6	116 19.9	38.4	71 59.7	58.4	120 17.6	19.1	86 21.2	38.0			
18	125 31.1	131 19.7	S 3 37.5	87 00.3	N14 59.0	135 19.6	S 5 18.9	101 23.3	N 7 38.1	Vega	80 46.6	N38 46.8
19	140 33.5	146 19.6	36.5	102 01.0	14 59.6	150 21.6	18.7	116 25.5	38.2	Zuben'ubi	137 18.0	S16 02.1
20	155 36.0	161 19.4	35.5	117 01.7	15 00.2	165 23.6	18.6	131 27.7	38.4			
21	170 38.4	176 19.3	.. 34.6	132 02.3	.. 00.8	180 25.6	.. 18.4	146 29.9	.. 38.5		SHA	Mer. Pass.
22	185 40.9	191 19.1	33.6	147 03.0	01.4	195 27.6	18.2	161 32.0	38.6	Venus	7 38.6	9 14
23	200 43.4	206 19.0	32.7	162 03.7	02.0	210 29.6	18.0	176 34.2	38.7	Mars	322 44.8	12 13
Mer. Pass.	h m 9 43.2	v −0.2	d 1.0	v 0.7	d 0.6	v 2.0	d 0.2	v 2.2	d 0.1	Jupiter	10 08.0	9 03
										Saturn	336 04.7	11 19

UT	SUN GHA	SUN Dec	MOON GHA	v	MOON Dec	d	HP
25 00	180 28.9	N13 03.1	199 17.9	7.6	N 2 39.3	12.0	61.2
01	195 29.0	03.9	213 44.5	7.4	2 51.3	12.1	61.2
02	210 29.1	04.7	228 10.9	7.5	3 03.4	12.0	61.2
03	225 29.2 ..	05.5	242 37.4	7.5	3 15.4	12.0	61.2
04	240 29.3	06.4	257 03.9	7.4	3 27.4	12.0	61.2
05	255 29.4	07.2	271 30.3	7.4	3 39.4	11.9	61.2
06	270 29.5	N13 08.0	285 56.7	7.3	N 3 51.3	12.0	61.2
S 07	285 29.7	08.8	300 23.0	7.4	4 03.3	11.9	61.2
A 08	300 29.8	09.6	314 49.4	7.3	4 15.2	11.9	61.2
T 09	315 29.9 ..	10.4	329 15.7	7.3	4 27.1	11.9	61.2
U 10	330 30.0	11.3	343 42.0	7.2	4 39.0	11.8	61.2
R 11	345 30.1	12.1	358 08.2	7.3	4 50.8	11.8	61.2
D 12	0 30.2	N13 12.9	12 34.5	7.2	N 5 02.6	11.8	61.2
A 13	15 30.3	13.7	27 00.7	7.2	5 14.4	11.7	61.2
Y 14	30 30.4	14.5	41 26.9	7.1	5 26.1	11.7	61.2
15	45 30.5 ..	15.3	55 53.0	7.1	5 37.8	11.7	61.2
16	60 30.6	16.1	70 19.1	7.1	5 49.5	11.6	61.2
17	75 30.7	16.9	84 45.2	7.1	6 01.1	11.6	61.2
18	90 30.8	N13 17.8	99 11.3	7.0	N 6 12.7	11.6	61.2
19	105 30.9	18.6	113 37.3	7.1	6 24.3	11.5	61.2
20	120 31.0	19.4	128 03.4	6.9	6 35.8	11.5	61.2
21	135 31.1 ..	20.2	142 29.3	7.0	6 47.3	11.4	61.2
22	150 31.2	21.0	156 55.3	6.9	6 58.7	11.4	61.2
23	165 31.3	21.8	171 21.2	6.9	7 10.1	11.3	61.2
26 00	180 31.4	N13 22.6	185 47.1	6.9	N 7 21.4	11.3	61.2
01	195 31.6	23.4	200 13.0	6.9	7 32.7	11.2	61.2
02	210 31.7	24.2	214 38.9	6.8	7 43.9	11.2	61.2
03	225 31.8 ..	25.0	229 04.7	6.8	7 55.1	11.1	61.2
04	240 31.9	25.9	243 30.5	6.8	8 06.2	11.1	61.2
05	255 32.0	26.7	257 56.3	6.7	8 17.3	11.0	61.2
06	270 32.1	N13 27.5	272 22.0	6.7	N 8 28.3	10.9	61.2
S 07	285 32.2	28.3	286 47.7	6.7	8 39.2	10.9	61.2
U 08	300 32.3	29.1	301 13.4	6.7	8 50.1	10.8	61.2
N 09	315 32.4 ..	29.9	315 39.1	6.6	9 00.9	10.8	61.2
D 10	330 32.5	30.7	330 04.7	6.6	9 11.7	10.7	61.2
A 11	345 32.6	31.5	344 30.3	6.6	9 22.4	10.6	61.2
Y 12	0 32.7	N13 32.3	358 55.9	6.6	N 9 33.0	10.5	61.2
13	15 32.8	33.1	13 21.5	6.5	9 43.5	10.5	61.1
14	30 32.9	33.9	27 47.0	6.5	9 54.0	10.5	61.1
15	45 33.0 ..	34.7	42 12.5	6.5	10 04.5	10.3	61.1
16	60 33.1	35.5	56 38.0	6.5	10 14.8	10.3	61.1
17	75 33.2	36.3	71 03.5	6.4	10 25.1	10.2	61.1
18	90 33.3	N13 37.1	85 28.9	6.4	N10 35.3	10.1	61.1
19	105 33.4	37.9	99 54.3	6.4	10 45.4	10.0	61.1
20	120 33.5	38.7	114 19.7	6.3	10 55.4	10.0	61.1
21	135 33.6 ..	39.5	128 45.0	6.4	11 05.4	9.9	61.0
22	150 33.7	40.3	143 10.4	6.3	11 15.3	9.8	61.0
23	165 33.8	41.1	157 35.7	6.3	11 25.1	9.7	61.0
27 00	180 33.9	N13 41.9	172 01.0	6.3	N11 34.8	9.6	61.0
01	195 34.0	42.7	186 26.3	6.2	11 44.4	9.6	61.0
02	210 34.1	43.5	200 51.5	6.2	11 54.0	9.5	61.0
03	225 34.2 ..	44.3	215 16.7	6.2	12 03.5	9.3	60.9
04	240 34.3	45.1	229 41.9	6.2	12 12.8	9.3	60.9
05	255 34.4	45.9	244 07.1	6.2	12 22.1	9.2	60.9
06	270 34.5	N13 46.7	258 32.3	6.2	N12 31.3	9.1	60.9
M 07	285 34.6	47.5	272 57.5	6.1	12 40.4	9.0	60.9
O 08	300 34.7	48.3	287 22.6	6.1	12 49.4	9.0	60.8
N 09	315 34.8 ..	49.1	301 47.7	6.1	12 58.4	8.8	60.8
D 10	330 34.9	49.9	316 12.8	6.1	13 07.2	8.7	60.8
A 11	345 35.0	50.7	330 37.9	6.0	13 15.9	8.6	60.8
Y 12	0 35.0	N13 51.5	345 02.9	6.1	N13 24.5	8.6	60.8
13	15 35.1	52.3	359 28.0	6.0	13 33.1	8.4	60.7
14	30 35.2	53.1	13 53.0	6.0	13 41.5	8.3	60.7
15	45 35.3 ..	53.9	28 18.0	6.1	13 49.8	8.3	60.7
16	60 35.4	54.7	42 43.1	5.9	13 58.1	8.1	60.7
17	75 35.5	55.5	57 08.0	6.0	14 06.2	8.0	60.6
18	90 35.6	N13 56.3	71 33.0	6.0	N14 14.2	8.0	60.6
19	105 35.7	57.1	85 58.0	6.0	14 22.2	7.8	60.6
20	120 35.8	57.9	100 23.0	5.9	14 30.0	7.7	60.6
21	135 35.9 ..	58.7	114 47.9	6.0	14 37.7	7.6	60.5
22	150 36.0	13 59.5	129 12.9	5.9	14 45.3	7.5	60.5
23	165 36.1	N14 00.2	143 37.8	5.9	N14 52.8	7.4	60.5
	SD 15.9	d 0.8	SD 16.7		16.7		16.6

Moonrise / Twilight

Lat.	Naut.	Civil	Sunrise	25	26	27	28
N 72	////	////	02 32	04 11	04 05	03 57	03 48
N 70	////	00 56	03 00	04 16	04 16	04 18	04 22
68	////	01 53	03 22	04 19	04 26	04 34	04 47
66	////	02 25	03 38	04 22	04 33	04 47	05 07
64	00 49	02 49	03 52	04 25	04 40	04 58	05 22
62	01 40	03 07	04 03	04 27	04 46	05 08	05 35
60	02 09	03 22	04 13	04 30	04 51	05 16	05 47
N 58	02 31	03 35	04 21	04 31	04 56	05 23	05 56
56	02 48	03 45	04 28	04 33	05 00	05 30	06 05
54	03 03	03 55	04 35	04 35	05 03	05 35	06 12
52	03 15	04 03	04 41	04 36	05 07	05 41	06 19
50	03 26	04 10	04 46	04 37	05 10	05 45	06 25
45	03 47	04 26	04 58	04 40	05 16	05 55	06 39
N 40	04 03	04 38	05 07	04 42	05 22	06 04	06 50
35	04 16	04 48	05 15	04 44	05 26	06 11	06 59
30	04 27	04 57	05 22	04 46	05 31	06 18	07 08
20	04 45	05 11	05 34	04 49	05 38	06 29	07 22
N 10	04 58	05 23	05 45	04 52	05 45	06 39	07 35
0	05 08	05 33	05 54	04 54	05 51	06 48	07 47
S 10	05 18	05 43	06 04	04 57	05 57	06 58	07 59
20	05 26	05 52	06 14	05 00	06 04	07 08	08 12
30	05 33	06 01	06 26	05 03	06 12	07 20	08 26
35	05 36	06 06	06 32	05 05	06 16	07 26	08 35
40	05 40	06 12	06 40	05 08	06 21	07 34	08 45
45	05 43	06 18	06 48	05 10	06 27	07 43	08 56
S 50	05 47	06 25	06 59	05 13	06 34	07 54	09 11
52	05 48	06 28	07 03	05 15	06 38	07 59	09 17
54	05 50	06 32	07 09	05 16	06 41	08 05	09 25
56	05 52	06 35	07 15	05 18	06 45	08 11	09 33
58	05 53	06 39	07 21	05 20	06 50	08 18	09 42
S 60	05 55	06 44	07 28	05 22	06 55	08 26	09 53

Moonset / Twilight

Lat.	Sunset	Civil	Naut.	25	26	27	28
N 72	21 29	////	////	18 43	20 50	23 01	25 23
N 70	20 59	23 16	////	18 33	20 31	22 28	24 19
68	20 37	22 09	////	18 26	20 16	22 04	23 43
66	20 20	21 34	////	18 19	20 04	21 45	23 23
64	20 06	21 10	23 22	18 14	19 54	21 30	22 58
62	19 55	20 51	22 22	18 10	19 45	21 18	22 43
60	19 45	20 36	21 50	18 06	19 38	21 07	22 29
N 58	19 36	20 23	21 28	18 02	19 32	20 58	22 18
56	19 29	20 12	21 10	17 59	19 26	20 50	22 08
54	19 22	20 02	20 55	17 56	19 21	20 43	21 59
52	19 16	19 54	20 43	17 54	19 16	20 36	21 51
50	19 11	19 46	20 32	17 51	19 12	20 31	21 44
45	18 59	19 31	20 10	17 47	19 03	20 18	21 29
N 40	18 49	19 18	19 53	17 42	18 56	20 08	21 17
35	18 41	19 08	19 40	17 39	18 50	19 59	21 06
30	18 34	18 59	19 29	17 36	18 44	19 52	20 57
20	18 22	18 45	19 11	17 30	18 34	19 38	20 41
N 10	18 11	18 33	18 58	17 26	18 26	19 27	20 27
0	18 01	18 22	18 47	17 21	18 18	19 16	20 14
S 10	17 51	18 13	18 38	17 17	18 10	19 05	20 02
20	17 41	18 04	18 30	17 12	18 02	18 54	19 48
30	17 30	17 54	18 22	17 06	17 52	18 41	19 32
35	17 23	17 49	18 19	17 03	17 47	18 33	19 23
40	17 15	17 43	18 15	17 00	17 40	18 24	19 12
45	17 07	17 37	18 12	16 56	17 33	18 14	19 00
S 50	16 56	17 30	18 08	16 51	17 25	18 02	18 45
52	16 51	17 27	18 06	16 49	17 21	17 57	18 38
54	16 46	17 23	18 05	16 46	17 16	17 51	18 31
56	16 40	17 19	18 03	16 43	17 11	17 44	18 22
58	16 34	17 15	18 01	16 40	17 06	17 36	18 12
S 60	16 26	17 11	18 00	16 37	17 00	17 27	18 01

Day	SUN Eqn. of Time 00h	SUN Eqn. of Time 12h	Mer. Pass.	MOON Mer. Pass. Upper	MOON Mer. Pass. Lower	Age	Phase
	m s	m s	h m	h m	h m	d	%
25	01 55	02 01	11 58	11 08	23 36	28	2
26	02 06	02 11	11 58	12 04	24 33	00	0
27	02 15	02 20	11 58	13 02	00 33	01	2

1998 APRIL 28, 29, 30 (TUES., WED., THURS.)

UT	ARIES GHA	VENUS −4.1 GHA	Dec	MARS +1.3 GHA	Dec	JUPITER −2.1 GHA	Dec	SATURN +0.5 GHA	Dec	Name	SHA	Dec
28 00	215 45.8	221 18.8	S 3 31.7	177 04.3	N15 02.6	225 31.6	S 5 17.8	191 36.4	N 7 38.8	Acamar	315 27.6	S40 18.9
01	230 48.3	236 18.7	30.7	192 05.0	03.1	240 33.6	17.6	206 38.5	38.9	Achernar	335 36.1	S57 14.8
02	245 50.8	251 18.5	29.8	207 05.6	03.7	255 35.6	17.4	221 40.7	39.0	Acrux	173 21.7	S63 05.5
03	260 53.2	266 18.4	.. 28.8	222 06.3	.. 04.3	270 37.6	.. 17.2	236 42.9	.. 39.1	Adhara	255 21.8	S28 58.5
04	275 55.7	281 18.2	27.8	237 07.0	04.9	285 39.6	17.1	251 45.0	39.3	Aldebaran	291 03.1	N16 30.2
05	290 58.2	296 18.1	26.9	252 07.6	05.5	300 41.6	16.9	266 47.2	39.4			
06	306 00.6	311 17.9	S 3 25.9	267 08.3	N15 06.1	315 43.6	S 5 16.7	281 49.4	N 7 39.5	Alioth	166 30.5	N55 58.3
07	321 03.1	326 17.8	24.9	282 09.0	06.7	330 45.6	16.5	296 51.6	39.6	Alkaid	153 07.6	N49 19.4
T 08	336 05.5	341 17.6	24.0	297 09.6	07.3	345 47.6	16.3	311 53.7	39.7	Al Na'ir	27 58.5	S46 58.0
U 09	351 08.0	356 17.4	.. 23.0	312 10.3	.. 07.9	0 49.7	.. 16.1	326 55.9	.. 39.8	Alnilam	275 58.4	S 1 12.4
E 10	6 10.5	11 17.3	22.0	327 11.0	08.5	15 51.7	15.9	341 58.1	39.9	Alphard	218 07.5	S 8 39.3
S 11	21 12.9	26 17.1	21.1	342 11.6	09.1	30 53.7	15.8	357 00.2	40.0			
D 12	36 15.4	41 17.0	S 3 20.1	357 12.3	N15 09.7	45 55.7	S 5 15.6	12 02.4	N 7 40.1	Alphecca	126 20.5	N26 43.2
A 13	51 17.9	56 16.8	19.1	12 13.0	10.3	60 57.7	15.4	27 04.6	40.3	Alpheratz	357 55.8	N29 04.6
Y 14	66 20.3	71 16.7	18.2	27 13.6	10.9	75 59.7	15.2	42 06.7	40.4	Altair	62 19.5	N 8 51.8
15	81 22.8	86 16.5	.. 17.2	42 14.3	.. 11.5	91 01.7	.. 15.0	57 08.9	.. 40.5	Ankaa	353 27.6	S42 18.9
16	96 25.3	101 16.4	16.2	57 14.9	12.1	106 03.7	14.8	72 11.1	40.6	Antares	112 40.2	S26 25.6
17	111 27.7	116 16.2	15.2	72 15.6	12.6	121 05.7	14.6	87 13.3	40.7			
18	126 30.2	131 16.1	S 3 14.3	87 16.3	N15 13.2	136 07.7	S 5 14.5	102 15.4	N 7 40.8	Arcturus	146 06.0	N19 11.5
19	141 32.7	146 15.9	13.3	102 16.9	13.8	151 09.7	14.3	117 17.6	40.9	Atria	107 52.1	S69 01.2
20	156 35.1	161 15.7	12.3	117 17.6	14.4	166 11.7	14.1	132 19.8	41.0	Avior	234 22.8	S59 30.7
21	171 37.6	176 15.6	.. 11.4	132 18.3	.. 15.0	181 13.7	.. 13.9	147 21.9	.. 41.1	Bellatrix	278 44.8	N 6 20.7
22	186 40.0	191 15.4	10.4	147 18.9	15.6	196 15.7	13.7	162 24.1	41.3	Betelgeuse	271 14.1	N 7 24.2
23	201 42.5	206 15.3	09.4	162 19.6	16.2	211 17.8	13.5	177 26.3	41.4			
29 00	216 45.0	221 15.1	S 3 08.4	177 20.2	N15 16.8	226 19.8	S 5 13.3	192 28.5	N 7 41.5	Canopus	264 01.6	S52 42.1
01	231 47.4	236 15.0	07.5	192 20.9	17.4	241 21.8	13.2	207 30.6	41.6	Capella	280 52.0	N45 59.7
02	246 49.9	251 14.8	06.5	207 21.6	18.0	256 23.8	13.0	222 32.8	41.7	Deneb	49 39.4	N45 16.2
03	261 52.4	266 14.7	.. 05.5	222 22.2	.. 18.5	271 25.8	.. 12.8	237 35.0	.. 41.8	Denebola	182 45.3	N14 34.9
04	276 54.8	281 14.5	04.6	237 22.9	19.1	286 27.8	12.6	252 37.1	41.9	Diphda	349 07.9	S17 59.9
05	291 57.3	296 14.3	03.6	252 23.6	19.7	301 29.8	12.4	267 39.3	42.0			
06	306 59.8	311 14.2	S 3 02.6	267 24.2	N15 20.3	316 31.8	S 5 12.2	282 41.5	N 7 42.1	Dubhe	194 05.6	N61 45.8
W 07	322 02.2	326 14.0	01.6	282 24.9	20.9	331 33.8	12.0	297 43.6	42.3	Elnath	278 27.6	N28 36.2
E 08	337 04.7	341 13.9	3 00.7	297 25.5	21.5	346 35.8	11.9	312 45.8	42.4	Eltanin	90 51.2	N51 29.2
D 09	352 07.2	356 13.7	2 59.7	312 26.2	.. 22.1	1 37.8	.. 11.7	327 48.0	.. 42.5	Enif	33 58.6	N 9 51.9
N 10	7 09.6	11 13.6	58.7	327 26.9	22.7	16 39.7	11.5	342 50.2	42.6	Fomalhaut	15 37.0	S29 37.8
E 11	22 12.1	26 13.4	57.7	342 27.5	23.2	31 41.9	11.3	357 52.3	42.7			
S 12	37 14.5	41 13.3	S 2 56.8	357 28.2	N15 23.8	46 43.9	S 5 11.1	12 54.5	N 7 42.8	Gacrux	172 13.3	S57 06.4
D 13	52 17.0	56 13.1	55.8	12 28.9	24.4	61 45.9	10.9	27 56.7	42.9	Gienah	176 04.0	S17 32.1
A 14	67 19.5	71 12.9	54.8	27 29.5	25.0	76 47.9	10.8	42 58.8	43.0	Hadar	149 03.8	S60 21.9
Y 15	82 21.9	86 12.8	.. 53.8	42 30.2	.. 25.6	91 49.9	.. 10.6	58 01.0	.. 43.1	Hamal	328 14.2	N23 27.0
16	97 24.4	101 12.6	52.8	57 30.8	26.2	106 51.9	10.4	73 03.2	43.3	Kaus Aust.	83 59.0	S34 22.9
17	112 26.9	116 12.5	51.9	72 31.5	26.8	121 53.9	10.2	88 05.4	43.4			
18	127 29.3	131 12.3	S 2 50.9	87 32.2	N15 27.3	136 55.9	S 5 10.0	103 07.5	N 7 43.5	Kochab	137 18.5	N74 09.8
19	142 31.8	146 12.2	49.9	102 32.8	27.9	151 57.9	09.8	118 09.7	43.6	Markab	13 50.1	N15 11.6
20	157 34.3	161 12.0	48.9	117 33.5	28.5	167 00.0	09.6	133 11.9	43.7	Menkar	314 27.6	N 4 04.7
21	172 36.7	176 11.8	.. 48.0	132 34.1	.. 29.1	182 02.0	.. 09.5	148 14.0	.. 43.8	Menkent	148 20.9	S36 21.7
22	187 39.2	191 11.7	47.0	147 34.8	29.7	197 04.0	09.3	163 16.2	43.9	Miaplacidus	221 41.9	S69 43.0
23	202 41.7	206 11.5	46.0	162 35.5	30.3	212 06.0	09.1	178 18.4	44.0			
30 00	217 44.1	221 11.4	S 2 45.0	177 36.1	N15 30.8	227 08.0	S 5 08.9	193 20.6	N 7 44.1	Mirfak	308 57.5	N49 51.2
01	232 46.6	236 11.2	44.0	192 36.8	31.4	242 10.0	08.7	208 22.7	44.3	Nunki	76 12.6	S26 17.8
02	247 49.0	251 11.1	43.1	207 37.4	32.0	257 12.0	08.5	223 24.9	44.4	Peacock	53 37.5	S56 44.1
03	262 51.5	266 10.9	.. 42.1	222 38.1	.. 32.6	272 14.0	.. 08.4	238 27.1	.. 44.5	Pollux	243 42.1	N28 01.8
04	277 54.0	281 10.7	41.1	237 38.8	33.2	287 16.1	08.2	253 29.2	44.6	Procyon	245 12.0	N 5 13.6
05	292 56.4	296 10.6	40.1	252 39.4	33.8	302 18.1	08.0	268 31.4	44.7			
06	307 58.9	311 10.4	S 2 39.1	267 40.1	N15 34.3	317 20.1	S 5 07.8	283 33.6	N 7 44.8	Rasalhague	96 17.0	N12 33.7
07	323 01.4	326 10.3	38.1	282 40.8	34.9	332 22.1	07.6	298 35.8	44.9	Regulus	207 55.8	N11 58.5
T 08	338 03.8	341 10.1	37.2	297 41.4	35.5	347 24.1	07.4	313 37.9	45.0	Rigel	281 23.5	S 8 12.5
H 09	353 06.3	356 10.0	.. 36.2	312 42.1	.. 36.1	2 26.1	.. 07.3	328 40.1	.. 45.1	Rigil Kent.	140 07.0	S60 49.6
U 10	8 08.8	11 09.8	35.2	327 42.7	36.7	17 28.1	07.1	343 42.3	45.3	Sabik	102 25.6	S15 43.3
R 11	23 11.2	26 09.6	34.2	342 43.4	37.2	32 30.1	06.9	358 44.4	45.4			
S 12	38 13.7	41 09.5	S 2 33.2	357 44.1	N15 37.8	47 32.2	S 5 06.7	13 46.6	N 7 45.5	Schedar	349 54.3	N56 31.4
D 13	53 16.1	56 09.3	32.3	12 44.7	38.4	62 34.2	06.5	28 48.8	45.6	Shaula	96 37.4	S37 06.0
A 14	68 18.6	71 09.2	31.3	27 45.4	39.0	77 36.2	06.3	43 51.0	45.7	Sirius	258 44.2	S16 43.1
Y 15	83 21.1	86 09.0	.. 30.3	42 46.0	.. 39.6	92 38.2	.. 06.2	58 53.1	.. 45.8	Spica	158 43.2	S11 09.2
16	98 23.5	101 08.9	29.3	57 46.7	40.1	107 40.2	06.0	73 55.3	45.9	Suhail	223 00.9	S43 25.9
17	113 26.0	116 08.7	28.3	72 47.3	40.7	122 42.2	05.8	88 57.5	46.0			
18	128 28.5	131 08.5	S 2 27.3	87 48.0	N15 41.3	137 44.2	S 5 05.6	103 59.6	N 7 46.1	Vega	80 46.6	N38 46.8
19	143 30.9	146 08.4	26.3	102 48.7	41.9	152 46.3	05.4	119 01.8	46.2	Zuben'ubi	137 18.0	S16 02.1
20	158 33.4	161 08.2	25.4	117 49.3	42.5	167 48.3	05.2	134 04.0	46.4			
21	173 35.9	176 08.1	.. 24.4	132 50.0	.. 43.0	182 50.3	.. 05.1	149 06.2	.. 46.5			
22	188 38.3	191 07.9	23.4	147 50.6	43.6	197 52.3	04.9	164 08.3	46.6			
23	203 40.8	206 07.7	22.4	162 51.3	44.2	212 54.3	04.7	179 10.5	46.7			

											SHA	Mer. Pass.
										Venus	4 30.1	9 15
										Mars	320 35.3	12 10
										Jupiter	9 34.8	8 53
Mer. Pass. 9 31.4	v −0.2 d 1.0	v 0.7 d 0.6	v 2.0 d 0.2	v 2.2 d 0.1						Saturn	335 43.5	11 08

UT	SUN GHA	SUN Dec	MOON GHA	v	MOON Dec	d	HP
d h	° ′	° ′	° ′	′	° ′	′	′
28 00	180 36.2	N14 01.0	158 02.7	5.9	N15 00.2	7.3	60.5
01	195 36.3	01.8	172 27.6	5.9	15 07.5	7.1	60.4
02	210 36.4	02.6	186 52.5	5.9	15 14.6	7.1	60.4
03	225 36.5	.. 03.4	201 17.4	6.0	15 21.7	6.9	60.4
04	240 36.6	04.2	215 42.4	5.9	15 28.6	6.9	60.3
05	255 36.7	05.0	230 07.3	5.9	15 35.5	6.7	60.3
06	270 36.7	N14 05.8	244 32.2	5.9	N15 42.2	6.6	60.3
07	285 36.8	06.6	258 57.1	5.8	15 48.8	6.5	60.2
T 08	300 36.9	07.3	273 21.9	5.9	15 55.3	6.4	60.2
U 09	315 37.0	.. 08.1	287 46.8	5.9	16 01.7	6.2	60.2
E 10	330 37.1	08.9	302 11.7	5.9	16 07.9	6.2	60.2
S 11	345 37.2	09.7	316 36.6	5.9	16 14.1	6.0	60.1
D 12	0 37.3	N14 10.5	331 01.5	6.0	N16 20.1	5.9	60.1
A 13	15 37.4	11.3	345 26.5	5.9	16 26.0	5.8	60.1
Y 14	30 37.5	12.1	359 51.4	5.9	16 31.8	5.7	60.0
15	45 37.6	.. 12.9	14 16.3	5.9	16 37.5	5.6	60.0
16	60 37.7	13.6	28 41.2	6.0	16 43.1	5.4	60.0
17	75 37.7	14.4	43 06.2	5.9	16 48.5	5.3	59.9
18	90 37.8	N14 15.2	57 31.1	6.0	N16 53.8	5.2	59.9
19	105 37.9	16.0	71 56.1	5.9	16 59.0	5.1	59.9
20	120 38.0	16.8	86 21.0	6.0	17 04.1	5.0	59.8
21	135 38.1	.. 17.6	100 46.0	6.0	17 09.1	4.8	59.8
22	150 38.2	18.3	115 11.0	6.0	17 13.9	4.7	59.8
23	165 38.3	19.1	129 36.0	6.0	17 18.6	4.6	59.7
29 00	180 38.4	N14 19.9	144 01.0	6.0	N17 23.2	4.5	59.7
01	195 38.5	20.7	158 26.0	6.1	17 27.7	4.4	59.6
02	210 38.5	21.5	172 51.1	6.1	17 32.1	4.2	59.6
03	225 38.6	.. 22.2	187 16.2	6.1	17 36.3	4.1	59.6
04	240 38.7	23.0	201 41.3	6.1	17 40.4	4.0	59.5
05	255 38.8	23.8	216 06.4	6.1	17 44.4	3.9	59.5
06	270 38.9	N14 24.6	230 31.5	6.2	N17 48.3	3.7	59.5
W 07	285 39.0	25.4	244 56.7	6.1	17 52.0	3.6	59.4
E 08	300 39.1	26.1	259 21.8	6.2	17 55.6	3.5	59.4
D 09	315 39.2	.. 26.9	273 47.0	6.3	17 59.1	3.4	59.4
N 10	330 39.2	27.7	288 12.3	6.2	18 02.5	3.3	59.3
E 11	345 39.3	28.5	302 37.5	6.3	18 05.8	3.1	59.3
S 12	0 39.4	N14 29.2	317 02.8	6.3	N18 08.9	3.0	59.2
D 13	15 39.5	30.0	331 28.1	6.4	18 11.9	2.9	59.2
A 14	30 39.6	30.8	345 53.5	6.3	18 14.8	2.7	59.2
Y 15	45 39.7	.. 31.6	0 18.8	6.4	18 17.5	2.7	59.1
16	60 39.8	32.3	14 44.2	6.5	18 20.2	2.5	59.1
17	75 39.8	33.1	29 09.7	6.4	18 22.7	2.4	59.0
18	90 39.9	N14 33.9	43 35.1	6.5	N18 25.1	2.3	59.0
19	105 40.0	34.7	58 00.6	6.6	18 27.4	2.1	59.0
20	120 40.1	35.4	72 26.2	6.6	18 29.5	2.0	58.9
21	135 40.2	.. 36.2	86 51.8	6.6	18 31.5	1.9	58.9
22	150 40.3	37.0	101 17.4	6.6	18 33.4	1.8	58.9
23	165 40.3	37.8	115 43.0	6.7	18 35.2	1.7	58.8
30 00	180 40.4	N14 38.5	130 08.7	6.7	N18 36.9	1.5	58.8
01	195 40.5	39.3	144 34.4	6.8	18 38.4	1.4	58.7
02	210 40.6	40.1	159 00.2	6.8	18 39.8	1.3	58.7
03	225 40.7	.. 40.8	173 26.0	6.9	18 41.1	1.2	58.7
04	240 40.8	41.6	187 51.9	6.9	18 42.3	1.1	58.6
05	255 40.8	42.4	202 17.8	7.0	18 43.4	0.9	58.6
06	270 40.9	N14 43.1	216 43.8	6.9	N18 44.3	0.9	58.5
07	285 41.0	43.9	231 09.7	7.1	18 45.2	0.7	58.5
T 08	300 41.1	44.7	245 35.8	7.1	18 45.9	0.5	58.5
H 09	315 41.2	.. 45.4	260 01.9	7.1	18 46.4	0.5	58.4
U 10	330 41.2	46.2	274 28.0	7.2	18 46.9	0.4	58.4
R 11	345 41.3	47.0	288 54.2	7.2	18 47.3	0.2	58.3
S 12	0 41.4	N14 47.7	303 20.4	7.3	N18 47.5	0.1	58.3
D 13	15 41.5	48.5	317 46.7	7.4	18 47.6	0.0	58.3
A 14	30 41.6	49.3	332 13.1	7.4	18 47.6	0.1	58.2
Y 15	45 41.6	.. 50.0	346 39.5	7.4	18 47.5	0.2	58.2
16	60 41.7	50.8	1 05.9	7.5	18 47.3	0.3	58.2
17	75 41.8	51.6	15 32.4	7.6	18 46.9	0.4	58.1
18	90 41.9	N14 52.3	29 59.0	7.6	N18 46.5	0.6	58.1
19	105 42.0	53.1	44 25.6	7.6	18 45.9	0.7	58.0
20	120 42.0	53.9	58 52.2	7.8	18 45.2	0.8	58.0
21	135 42.1	.. 54.6	73 19.0	7.7	18 44.5	0.9	57.9
22	150 42.2	55.4	87 45.7	7.9	18 43.6	1.1	57.9
23	165 42.3	56.1	102 12.6	7.9	N18 42.5	1.1	57.9
	SD 15.9	d 0.8	SD 16.4		16.1		15.9

Twilight / Sunrise / Moonrise

Lat.	Naut.	Civil	Sunrise	Moonrise 28	29	30	1
°	h m	h m	h m	h m	h m	h m	h m
N 72	////	////	02 10	03 48	03 29	□	□
N 70	////	////	02 44	04 22	04 33	05 01	06 02
68	////	01 29	03 08	04 47	05 09	05 47	06 47
66	////	02 09	03 26	05 07	05 35	06 17	07 16
64	////	02 36	03 41	05 22	05 55	06 40	07 39
62	01 18	02 56	03 54	05 35	06 11	06 58	07 57
60	01 54	03 12	04 04	05 47	06 25	07 13	08 11
N 58	02 19	03 26	04 13	05 56	06 37	07 26	08 24
56	02 38	03 38	04 21	06 05	06 47	07 37	08 35
54	02 54	03 48	04 28	06 12	06 56	07 47	08 44
52	03 07	03 56	04 35	06 19	07 04	07 55	08 53
50	03 18	04 04	04 41	06 25	07 11	08 03	09 00
45	03 41	04 21	04 53	06 39	07 27	08 20	09 17
N 40	03 59	04 34	05 03	06 50	07 40	08 33	09 30
35	04 13	04 45	05 12	06 59	07 51	08 45	09 41
30	04 24	04 54	05 19	07 08	08 00	08 55	09 51
20	04 42	05 09	05 32	07 22	08 17	09 13	10 08
N 10	04 56	05 22	05 44	07 35	08 31	09 28	10 23
0	05 08	05 33	05 54	07 47	08 45	09 42	10 37
S 10	05 18	05 43	06 04	07 59	08 59	09 56	10 51
20	05 26	05 53	06 15	08 12	09 13	10 12	11 06
30	05 35	06 03	06 28	08 26	09 30	10 29	11 23
35	05 39	06 08	06 35	08 35	09 40	10 39	11 33
40	05 43	06 15	06 43	08 45	09 51	10 51	11 44
45	05 47	06 21	06 52	08 56	10 04	11 05	11 58
S 50	05 51	06 29	07 03	09 11	10 21	11 22	12 14
52	05 53	06 33	07 08	09 17	10 28	11 30	12 22
54	05 55	06 37	07 14	09 25	10 37	11 39	12 30
56	05 57	06 41	07 20	09 33	10 46	11 49	12 39
58	05 59	06 45	07 28	09 42	10 57	12 00	12 50
S 60	06 01	06 50	07 36	09 53	11 09	12 13	13 03

Sunset / Twilight / Moonset

Lat.	Sunset	Civil	Naut.	Moonset 28	29	30	1
°	h m	h m	h m	h m	h m	h m	h m
N 72	21 51	////	////	25 23	01 23	□	□
N 70	21 15	////	////	24 19	00 19	01 51	02 47
68	20 50	22 34	////	23 43	25 05	01 05	02 02
66	20 31	21 51	////	23 18	24 35	00 35	01 32
64	20 16	21 23	////	22 58	24 13	00 13	01 09
62	20 03	21 02	22 43	22 43	23 55	24 52	00 52
60	19 52	20 45	22 05	22 29	23 40	24 37	00 37
N 58	19 43	20 31	21 39	22 18	23 27	24 24	00 24
56	19 35	20 19	21 19	22 08	23 16	24 13	00 13
54	19 28	20 09	21 03	21 59	23 06	24 03	00 03
52	19 21	20 00	20 49	21 51	22 58	23 55	24 42
50	19 15	19 52	20 38	21 44	22 50	23 47	24 35
45	19 03	19 35	20 15	21 29	22 34	23 31	24 19
N 40	18 52	19 22	19 57	21 17	22 20	23 17	24 07
35	18 44	19 11	19 43	21 06	22 09	23 06	23 56
30	18 36	19 01	19 31	20 57	21 59	22 55	23 47
20	18 23	18 46	19 13	20 41	21 42	22 38	23 30
N 10	18 11	18 33	18 59	20 27	21 26	22 23	23 16
0	18 01	18 22	18 47	20 14	21 12	22 09	23 03
S 10	17 50	18 12	18 37	20 02	20 58	21 54	22 49
20	17 39	18 02	18 28	19 48	20 43	21 39	22 35
30	17 27	17 52	18 20	19 32	20 26	21 22	22 18
35	17 20	17 46	18 16	19 23	20 16	21 11	22 08
40	17 12	17 40	18 12	19 12	20 04	21 00	21 57
45	17 02	17 33	18 07	19 00	19 51	20 46	21 44
S 50	16 51	17 25	18 03	18 45	19 34	20 29	21 29
52	16 46	17 21	18 01	18 38	19 27	20 21	21 21
54	16 40	17 17	17 59	18 31	19 18	20 12	21 13
56	16 33	17 13	17 57	18 22	19 08	20 03	21 03
58	16 26	17 08	17 55	18 12	18 57	19 51	20 53
S 60	16 18	17 03	17 53	18 01	18 45	19 38	20 41

SUN / MOON

Day	Eqn. of Time 00ʰ	12ʰ	Mer. Pass.	Mer. Pass. Upper	Lower	Age	Phase
d	m s	m s	h m	h m	h m	d	%
28	02 25	02 29	11 58	14 01	01 31	02	6
29	02 33	02 37	11 57	14 59	02 30	03	13
30	02 42	02 45	11 57	15 55	03 27	04	21

1998 MAY 1, 2, 3 (FRI., SAT., SUN.)

UT	ARIES GHA	VENUS −4.1 GHA	Dec	MARS +1.3 GHA	Dec	JUPITER −2.1 GHA	Dec	SATURN +0.5 GHA	Dec
1 00	218 43.3	221 07.6	S 2 21.4	177 52.0	N15 44.8	227 56.3	S 5 04.5	194 12.7	N 7 46.8
01	233 45.7	236 07.4	20.4	192 52.6	45.3	242 58.3	04.3	209 14.8	46.9
02	248 48.2	251 07.3	19.4	207 53.3	45.9	258 00.4	04.1	224 17.0	47.0
03	263 50.6	266 07.1	.. 18.4	222 53.9	.. 46.5	273 02.4	.. 04.0	239 19.2	.. 47.1
04	278 53.1	281 06.9	17.5	237 54.6	47.1	288 04.4	03.8	254 21.4	47.2
05	293 55.6	296 06.8	16.5	252 55.3	47.6	303 06.4	03.6	269 23.5	47.3
06	308 58.0	311 06.6	S 2 15.5	267 55.9	N15 48.2	318 08.4	S 5 03.4	284 25.7	N 7 47.5
07	324 00.5	326 06.5	14.5	282 56.6	48.8	333 10.4	03.2	299 27.9	47.6
08	339 03.0	341 06.3	13.5	297 57.2	49.4	348 12.5	03.1	314 30.1	47.7
F 09	354 05.4	356 06.2	.. 12.5	312 57.9	.. 49.9	3 14.5	.. 02.9	329 32.2	.. 47.8
R 10	9 07.9	11 06.0	11.5	327 58.5	50.5	18 16.5	02.7	344 34.4	47.9
I 11	24 10.4	26 05.8	10.5	342 59.2	51.1	33 18.5	02.5	359 36.6	48.0
D 12	39 12.8	41 05.7	S 2 09.5	357 59.9	N15 51.7	48 20.5	S 5 02.3	14 38.7	N 7 48.1
A 13	54 15.3	56 05.5	08.5	13 00.5	52.2	63 22.5	02.1	29 40.9	48.2
Y 14	69 17.8	71 05.4	07.6	28 01.2	52.8	78 24.6	02.0	44 43.1	48.3
15	84 20.2	86 05.2	.. 06.6	43 01.8	.. 53.4	93 26.6	.. 01.8	59 45.3	.. 48.4
16	99 22.7	101 05.0	05.6	58 02.5	54.0	108 28.6	01.6	74 47.4	48.6
17	114 25.1	116 04.9	04.6	73 03.2	54.5	123 30.6	01.4	89 49.6	48.7
18	129 27.6	131 04.7	S 2 03.6	88 03.8	N15 55.1	138 32.6	S 5 01.2	104 51.8	N 7 48.8
19	144 30.1	146 04.6	02.6	103 04.5	55.7	153 34.6	01.1	119 53.9	48.9
20	159 32.5	161 04.4	01.6	118 05.1	56.2	168 36.7	00.9	134 56.1	49.0
21	174 35.0	176 04.2	2 00.6	133 05.8	.. 56.8	183 38.7	.. 00.7	149 58.3	.. 49.1
22	189 37.5	191 04.1	1 59.6	148 06.4	57.4	198 40.7	00.5	165 00.5	49.2
23	204 39.9	206 03.9	58.6	163 07.1	57.9	213 42.7	00.3	180 02.6	49.3
2 00	219 42.4	221 03.8	S 1 57.6	178 07.8	N15 58.5	228 44.7	S 5 00.1	195 04.8	N 7 49.4
01	234 44.9	236 03.6	56.6	193 08.4	59.1	243 46.8	5 00.0	210 07.0	49.5
02	249 47.3	251 03.4	55.6	208 09.1	15 59.7	258 48.8	4 59.8	225 09.2	49.7
03	264 49.8	266 03.3	.. 54.6	223 09.7	16 00.2	273 50.8	.. 59.6	240 11.3	.. 49.8
04	279 52.3	281 03.1	53.7	238 10.4	00.8	288 52.8	59.4	255 13.5	49.9
05	294 54.7	296 02.9	52.7	253 11.0	01.4	303 54.8	59.2	270 15.7	50.0
06	309 57.2	311 02.8	S 1 51.7	268 11.7	N16 01.9	318 56.9	S 4 59.1	285 17.8	N 7 50.1
07	324 59.6	326 02.6	50.7	283 12.4	02.5	333 58.9	58.9	300 20.0	50.2
S 08	340 02.1	341 02.5	49.7	298 13.0	03.1	349 00.9	58.7	315 22.2	50.3
A 09	355 04.6	356 02.3	.. 48.7	313 13.7	.. 03.6	4 02.9	.. 58.5	330 24.4	.. 50.4
T 10	10 07.0	11 02.1	47.7	328 14.3	04.2	19 04.9	58.3	345 26.5	50.5
U 11	25 09.5	26 02.0	46.7	343 15.0	04.8	34 07.0	58.2	0 28.7	50.6
R 12	40 12.0	41 01.8	S 1 45.7	358 15.6	N16 05.3	49 09.0	S 4 58.0	15 30.9	N 7 50.7
D 13	55 14.4	56 01.7	44.7	13 16.3	05.9	64 11.0	57.8	30 33.1	50.9
A 14	70 16.9	71 01.5	43.7	28 16.9	06.5	79 13.0	57.6	45 35.2	51.0
Y 15	85 19.4	86 01.3	.. 42.7	43 17.6	.. 07.0	94 15.0	.. 57.4	60 37.4	.. 51.1
16	100 21.8	101 01.2	41.7	58 18.3	07.6	109 17.1	57.3	75 39.6	51.2
17	115 24.3	116 01.0	40.7	73 18.9	08.2	124 19.1	57.1	90 41.7	51.3
18	130 26.7	131 00.8	S 1 39.7	88 19.6	N16 08.7	139 21.1	S 4 56.9	105 43.9	N 7 51.4
19	145 29.2	146 00.7	38.7	103 20.2	09.3	154 23.1	56.7	120 46.1	51.5
20	160 31.7	161 00.5	37.7	118 20.9	09.9	169 25.1	56.5	135 48.3	51.6
21	175 34.1	176 00.4	.. 36.7	133 21.5	.. 10.4	184 27.2	.. 56.4	150 50.4	.. 51.7
22	190 36.6	191 00.2	35.7	148 22.2	11.0	199 29.2	56.2	165 52.6	51.8
23	205 39.1	206 00.0	34.7	163 22.8	11.5	214 31.2	56.0	180 54.8	52.0
3 00	220 41.5	220 59.9	S 1 33.7	178 23.5	N16 12.1	229 33.2	S 4 55.8	195 57.0	N 7 52.1
01	235 44.0	235 59.7	32.7	193 24.2	12.7	244 35.3	55.6	210 59.1	52.2
02	250 46.5	250 59.5	31.7	208 24.8	13.2	259 37.3	55.5	226 01.3	52.3
03	265 48.9	265 59.4	.. 30.7	223 25.5	.. 13.8	274 39.3	.. 55.3	241 03.5	.. 52.4
04	280 51.4	280 59.2	29.7	238 26.1	14.4	289 41.3	55.1	256 05.7	52.5
05	295 53.9	295 59.1	28.7	253 26.8	14.9	304 43.3	54.9	271 07.8	52.6
06	310 56.3	310 58.9	S 1 27.7	268 27.4	N16 15.5	319 45.4	S 4 54.7	286 10.0	N 7 52.7
07	325 58.8	325 58.7	26.7	283 28.1	16.0	334 47.4	54.6	301 12.2	52.8
08	341 01.2	340 58.6	25.7	298 28.7	16.6	349 49.4	54.4	316 14.3	52.9
S 09	356 03.7	355 58.4	.. 24.7	313 29.4	.. 17.2	4 51.4	.. 54.2	331 16.5	.. 53.0
U 10	11 06.2	10 58.2	23.7	328 30.0	17.7	19 53.5	54.0	346 18.7	53.2
N 11	26 08.6	25 58.1	22.7	343 30.7	18.3	34 55.5	53.8	1 20.9	53.3
D 12	41 11.1	40 57.9	S 1 21.7	358 31.4	N16 18.8	49 57.5	S 4 53.7	16 23.0	N 7 53.4
A 13	56 13.6	55 57.7	20.7	13 32.0	19.4	64 59.5	53.5	31 25.2	53.5
Y 14	71 16.0	70 57.6	19.7	28 32.7	20.0	80 01.6	53.3	46 27.4	53.6
15	86 18.5	85 57.4	.. 18.7	43 33.3	.. 20.5	95 03.6	.. 53.1	61 29.6	.. 53.7
16	101 21.0	100 57.3	17.7	58 34.0	21.1	110 05.6	53.0	76 31.7	53.8
17	116 23.4	115 57.1	16.7	73 34.6	21.6	125 07.6	52.8	91 33.9	53.9
18	131 25.9	130 56.9	S 1 15.6	88 35.3	N16 22.2	140 09.7	S 4 52.6	106 36.1	N 7 54.0
19	146 28.4	145 56.8	14.6	103 35.9	22.8	155 11.7	52.4	121 38.3	54.1
20	161 30.8	160 56.6	13.6	118 36.6	23.3	170 13.7	52.2	136 40.4	54.2
21	176 33.3	175 56.4	.. 12.6	133 37.2	.. 23.9	185 15.7	.. 52.1	151 42.6	.. 54.3
22	191 35.7	190 56.3	11.6	148 37.9	24.4	200 17.8	51.9	166 44.8	54.5
23	206 38.2	205 56.1	10.6	163 38.5	25.0	215 19.8	51.7	181 47.0	54.6
Mer.Pass.	h m 9 19.6	v −0.2	d 1.0	v 0.7	d 0.6	v 2.0	d 0.2	v 2.2	d 0.1

STARS

Name	SHA	Dec
Acamar	315 27.6	S40 18.9
Achernar	335 36.1	S57 14.8
Acrux	173 21.7	S63 05.6
Adhara	255 21.8	S28 58.5
Aldebaran	291 03.1	N16 30.2
Alioth	166 30.5	N55 58.3
Alkaid	153 07.6	N49 19.4
Al Na'ir	27 58.5	S46 58.0
Alnilam	275 58.4	S 1 12.4
Alphard	218 07.5	S 8 39.3
Alphecca	126 20.5	N26 43.2
Alpheratz	357 55.8	N29 04.6
Altair	62 19.5	N 8 51.8
Ankaa	353 27.6	S42 18.9
Antares	112 40.2	S26 25.6
Arcturus	146 06.0	N19 11.5
Atria	107 52.0	S69 01.2
Avior	234 22.8	S59 30.7
Bellatrix	278 44.8	N 6 20.7
Betelgeuse	271 14.1	N 7 24.2
Canopus	264 01.6	S52 42.1
Capella	280 52.0	N45 59.7
Deneb	49 39.3	N45 16.2
Denebola	182 45.3	N14 34.9
Diphda	349 07.9	S17 59.9
Dubhe	194 05.6	N61 45.8
Elnath	278 27.6	N28 36.2
Eltanin	90 51.2	N51 29.2
Enif	33 58.6	N 9 51.9
Fomalhaut	15 37.0	S29 37.8
Gacrux	172 13.3	S57 06.4
Gienah	176 04.0	S17 32.1
Hadar	149 03.7	S60 21.9
Hamal	328 14.2	N23 27.0
Kaus Aust.	83 59.0	S34 22.9
Kochab	137 18.5	N74 09.9
Markab	13 50.1	N15 11.6
Menkar	314 27.6	N 4 04.8
Menkent	148 20.9	S36 21.7
Miaplacidus	221 41.9	S69 43.0
Mirfak	308 57.5	N49 51.1
Nunki	76 12.5	S26 17.8
Peacock	53 37.5	S56 44.1
Pollux	243 42.1	N28 01.8
Procyon	245 12.0	N 5 13.6
Rasalhague	96 17.0	N12 33.7
Regulus	207 55.8	N11 58.5
Rigel	281 23.5	S 8 12.5
Rigil Kent.	140 07.0	S60 49.6
Sabik	102 25.6	S15 43.3
Schedar	349 54.2	N56 31.4
Shaula	96 37.4	S37 06.0
Sirius	258 44.2	S16 43.1
Spica	158 43.2	S11 09.2
Suhail	223 00.9	S43 25.9
Vega	80 46.6	N38 46.8
Zuben'ubi	137 18.0	S16 02.1

	SHA	Mer.Pass.
Venus	1 21.4	9 16
Mars	318 25.4	12 07
Jupiter	9 02.3	8 44
Saturn	335 22.4	10 58

UT	SUN GHA	SUN Dec	MOON GHA	v	MOON Dec	d	HP
d h	° '	° '	° '	'	° '	'	'
1 00	180 42.4	N14 56.9	116 39.5	7.9	N18 41.4	1.2	57.8
01	195 42.4	57.7	131 06.4	8.0	18 40.2	1.4	57.8
02	210 42.5	58.4	145 33.4	8.1	18 38.8	1.4	57.7
03	225 42.6 ..	59.2	160 00.5	8.2	18 37.4	1.6	57.7
04	240 42.7	14 59.9	174 27.7	8.2	18 35.8	1.6	57.7
05	255 42.7	15 00.7	188 54.9	8.2	18 34.2	1.8	57.6
06	270 42.8	N15 01.5	203 22.1	8.3	N18 32.4	1.9	57.6
07	285 42.9	02.2	217 49.4	8.4	18 30.5	2.0	57.5
08	300 43.0	03.0	232 16.8	8.5	18 28.5	2.1	57.5
F 09	315 43.0 ..	03.7	246 44.3	8.5	18 26.4	2.2	57.5
R 10	330 43.1	04.5	261 11.8	8.6	18 24.2	2.3	57.4
I 11	345 43.2	05.3	275 39.4	8.6	18 21.9	2.4	57.4
D 12	0 43.3	N15 06.0	290 07.0	8.7	N18 19.5	2.4	57.3
A 13	15 43.3	06.8	304 34.7	8.8	18 17.1	2.6	57.3
Y 14	30 43.4	07.5	319 02.5	8.9	18 14.5	2.7	57.3
15	45 43.5 ..	08.3	333 30.4	8.9	18 11.8	2.8	57.2
16	60 43.6	09.0	347 58.3	8.9	18 09.0	2.9	57.2
17	75 43.6	09.8	2 26.2	9.1	18 06.1	3.0	57.2
18	90 43.7	N15 10.5	16 54.3	9.1	N18 03.1	3.1	57.1
19	105 43.8	11.3	31 22.4	9.2	18 00.0	3.2	57.1
20	120 43.9	12.0	45 50.6	9.2	17 56.8	3.3	57.0
21	135 43.9 ..	12.8	60 18.8	9.3	17 53.5	3.4	57.0
22	150 44.0	13.5	74 47.1	9.4	17 50.1	3.4	57.0
23	165 44.1	14.3	89 15.5	9.4	17 46.7	3.6	56.9
2 00	180 44.2	N15 15.0	103 43.9	9.6	N17 43.1	3.7	56.9
01	195 44.2	15.8	118 12.5	9.5	17 39.4	3.7	56.9
02	210 44.3	16.5	132 41.0	9.7	17 35.7	3.9	56.8
03	225 44.4 ..	17.3	147 09.7	9.7	17 31.8	3.9	56.8
04	240 44.4	18.0	161 38.4	9.8	17 27.9	4.0	56.7
05	255 44.5	18.8	176 07.2	9.9	17 23.9	4.1	56.7
06	270 44.6	N15 19.5	190 36.1	9.9	N17 19.8	4.2	56.7
S 07	285 44.7	20.3	205 05.0	10.0	17 15.6	4.3	56.6
A 08	300 44.7	21.0	219 34.0	10.1	17 11.3	4.4	56.6
T 09	315 44.8 ..	21.8	234 03.1	10.1	17 06.9	4.4	56.6
U 10	330 44.9	22.5	248 32.2	10.2	17 02.5	4.6	56.5
R 11	345 44.9	23.3	263 01.4	10.3	16 57.9	4.6	56.5
D 12	0 45.0	N15 24.0	277 30.7	10.3	N16 53.3	4.7	56.5
A 13	15 45.1	24.8	292 00.0	10.5	16 48.6	4.8	56.4
Y 14	30 45.1	25.5	306 29.5	10.4	16 43.8	4.8	56.4
15	45 45.2 ..	26.2	320 58.9	10.6	16 39.0	5.0	56.4
16	60 45.3	27.0	335 28.5	10.6	16 34.0	5.0	56.3
17	75 45.3	27.7	349 58.1	10.7	16 29.0	5.1	56.3
18	90 45.4	N15 28.5	4 27.8	10.8	N16 23.9	5.2	56.3
19	105 45.5	29.2	18 57.6	10.8	16 18.7	5.3	56.2
20	120 45.6	30.0	33 27.4	10.9	16 13.4	5.3	56.2
21	135 45.6 ..	30.7	47 57.3	10.9	16 08.1	5.4	56.2
22	150 45.7	31.4	62 27.2	11.1	16 02.7	5.5	56.1
23	165 45.8	32.2	76 57.3	11.1	15 57.2	5.6	56.1
3 00	180 45.8	N15 32.9	91 27.4	11.1	N15 51.6	5.6	56.1
01	195 45.9	33.7	105 57.5	11.3	15 46.0	5.7	56.0
02	210 46.0	34.4	120 27.8	11.3	15 40.3	5.8	56.0
03	225 46.0 ..	35.1	134 58.1	11.4	15 34.5	5.9	56.0
04	240 46.1	35.9	149 28.5	11.4	15 28.6	5.9	55.9
05	255 46.2	36.6	163 58.9	11.5	15 22.7	6.0	55.9
06	270 46.2	N15 37.4	178 29.4	11.6	N15 16.7	6.0	55.9
07	285 46.3	38.1	193 00.0	11.6	15 10.7	6.2	55.8
08	300 46.3	38.8	207 30.6	11.7	15 04.5	6.2	55.8
S 09	315 46.4 ..	39.6	222 01.3	11.8	14 58.3	6.2	55.8
U 10	330 46.5	40.3	236 32.1	11.8	14 52.1	6.4	55.7
N 11	345 46.5	41.0	251 02.9	11.9	14 45.7	6.4	55.7
D 12	0 46.6	N15 41.8	265 33.8	11.9	N14 39.3	6.4	55.7
A 13	15 46.7	42.5	280 04.7	12.1	14 32.9	6.5	55.7
Y 14	30 46.7	43.2	294 35.8	12.1	14 26.4	6.6	55.6
15	45 46.8 ..	44.0	309 06.9	12.1	14 19.8	6.7	55.6
16	60 46.9	44.7	323 38.0	12.2	14 13.1	6.7	55.6
17	75 46.9	45.4	338 09.2	12.3	14 06.4	6.7	55.5
18	90 47.0	N15 46.2	352 40.5	12.3	N13 59.7	6.9	55.5
19	105 47.0	46.9	7 11.8	12.4	13 52.8	6.8	55.5
20	120 47.1	47.6	21 43.2	12.5	13 46.0	7.0	55.5
21	135 47.2 ..	48.4	36 14.7	12.5	13 39.0	7.0	55.4
22	150 47.2	49.1	50 46.2	12.6	13 32.0	7.0	55.4
23	165 47.3	49.8	65 17.8	12.6	N13 25.0	7.1	55.4
	SD 15.9	d 0.7	SD 15.6		15.4		15.2

Lat.	Naut.	Civil	Sunrise	Moonrise 1	2	3	4
°	h m	h m	h m	h m	h m	h m	h m
N 72	////	////	01 46	▭	06 31	08 31	10 18
N 70	////	////	02 27	06 02	07 30	09 05	10 39
68	////	00 58	02 54	06 47	08 04	09 29	10 56
66	////	01 51	03 15	07 16	08 29	09 48	11 10
64	////	02 22	03 31	07 39	08 48	10 03	11 21
62	00 51	02 44	03 45	07 57	09 04	10 16	11 30
60	01 38	03 02	03 56	08 11	09 17	10 27	11 38
N 58	02 07	03 17	04 06	08 24	09 28	10 36	11 45
56	02 28	03 30	04 14	08 35	09 38	10 44	11 51
54	02 45	03 41	04 22	08 44	09 46	10 51	11 57
52	02 59	03 50	04 29	08 53	09 54	10 58	12 02
50	03 11	03 58	04 35	09 00	10 01	11 03	12 06
45	03 36	04 16	04 48	09 17	10 16	11 16	12 16
N 40	03 54	04 30	04 59	09 30	10 28	11 26	12 24
35	04 09	04 41	05 09	09 41	10 38	11 35	12 31
30	04 21	04 51	05 17	09 51	10 47	11 43	12 37
20	04 40	05 07	05 30	10 08	11 03	11 56	12 48
N 10	04 55	05 21	05 42	10 23	11 17	12 08	12 57
0	05 07	05 32	05 54	10 37	11 29	12 19	13 06
S 10	05 18	05 43	06 05	10 51	11 42	12 30	13 14
20	05 27	05 55	06 16	11 06	11 56	12 41	13 23
30	05 36	06 05	06 29	11 23	12 11	12 55	13 34
35	05 41	06 11	06 37	11 33	12 20	13 02	13 40
40	05 45	06 17	06 46	11 44	12 31	13 11	13 46
45	05 50	06 25	06 56	11 58	12 43	13 21	13 54
S 50	05 55	06 33	07 08	12 14	12 57	13 33	14 04
52	05 57	06 37	07 13	12 22	13 04	13 39	14 08
54	05 59	06 42	07 19	12 30	13 11	13 45	14 13
56	06 02	06 46	07 26	12 39	13 20	13 52	14 18
58	06 04	06 51	07 34	12 50	13 29	14 00	14 24
S 60	06 07	06 57	07 43	13 03	13 40	14 09	14 31

Lat.	Sunset	Civil	Naut.	Moonset 1	2	3	4
°	h m	h m	h m	h m	h m	h m	h m
N 72	22 16	////	////	▭	04 09	03 53	03 44
N 70	21 32	////	////	02 47	03 09	03 18	03 21
68	21 04	23 09	////	02 02	02 34	02 53	03 04
66	20 42	22 09	////	01 32	02 09	02 33	02 49
64	20 25	21 36	////	01 09	01 50	02 18	02 37
62	20 11	21 12	23 14	00 52	01 34	02 04	02 27
60	20 00	20 54	22 21	00 37	01 20	01 53	02 19
N 58	19 50	20 39	21 51	00 24	01 09	01 44	02 11
56	19 41	20 26	21 29	00 13	00 59	01 35	02 04
54	19 33	20 15	21 11	00 03	00 50	01 27	01 58
52	19 26	20 05	20 56	24 42	00 42	01 21	01 53
50	19 20	19 57	20 44	24 35	00 35	01 14	01 48
45	19 06	19 39	20 19	24 19	00 19	01 01	01 37
N 40	18 55	19 25	20 01	24 07	00 07	00 50	01 28
35	18 46	19 13	19 46	23 56	24 41	00 41	01 20
30	18 38	19 03	19 34	23 47	24 32	00 32	01 14
20	18 24	18 47	19 14	23 30	24 18	00 18	01 02
N 10	18 12	18 34	18 59	23 16	24 05	00 05	00 51
0	18 00	18 22	18 47	23 03	23 53	24 42	00 42
S 10	17 49	18 11	18 36	22 49	23 42	24 32	00 32
20	17 38	18 00	18 27	22 35	23 29	24 21	00 21
30	17 24	17 49	18 17	22 18	23 14	24 09	00 09
35	17 17	17 43	18 13	22 08	23 06	24 02	00 02
40	17 08	17 36	18 08	21 57	22 56	23 54	24 52
45	16 58	17 29	18 03	21 44	22 44	23 45	24 45
S 50	16 46	17 20	17 58	21 29	22 30	23 33	24 36
52	16 40	17 16	17 56	21 21	22 24	23 28	24 32
54	16 34	17 12	17 54	21 13	22 17	23 22	24 28
56	16 27	17 07	17 51	21 03	22 09	23 16	24 23
58	16 19	17 02	17 49	20 53	22 00	23 09	24 18
S 60	16 10	16 56	17 46	20 41	21 49	23 00	24 12

Day	Eqn. of Time 00h	Eqn. of Time 12h	Mer. Pass.	Mer. Pass. Upper	Mer. Pass. Lower	Age	Phase
d	m s	m s	h m	h m	h m	d	%
1	02 49	02 53	11 57	16 50	04 23	05	31
2	02 56	03 00	11 57	17 42	05 16	06	41
3	03 03	03 06	11 57	18 30	06 06	07	51

92   1998 MAY 4, 5, 6 (MON., TUES., WED.)

UT	ARIES	VENUS −4.1		MARS +1.3		JUPITER −2.2		SATURN +0.5		STARS		
	GHA	GHA	Dec	GHA	Dec	GHA	Dec	GHA	Dec	Name	SHA	Dec
d h	° ′	° ′	° ′	° ′	° ′	° ′	° ′	° ′	° ′		° ′	° ′
4 00	221 40.7	220 55.9	S 1 09.6	178 39.2	N16 25.5	230 21.8	S 4 51.5	196 49.1	N 7 54.7	Acamar	315 27.6	S40 18.9
01	236 43.1	235 55.8	08.6	193 39.9	26.1	245 23.8	51.3	211 51.3	54.8	Achernar	335 36.1	S57 14.8
02	251 45.6	250 55.6	07.6	208 40.5	26.7	260 25.9	51.2	226 53.5	54.9	Acrux	173 21.7	S63 05.6
03	266 48.1	265 55.4	.. 06.6	223 41.2	.. 27.2	275 27.9	.. 51.0	241 55.7	.. 55.0	Adhara	255 21.9	S28 58.5
04	281 50.5	280 55.3	05.6	238 41.8	27.8	290 29.9	50.8	256 57.8	55.1	Aldebaran	291 03.1	N16 30.2
05	296 53.0	295 55.1	04.6	253 42.5	28.3	305 31.9	50.6	272 00.0	55.2			
06	311 55.5	310 54.9	S 1 03.6	268 43.1	N16 28.9	320 34.0	S 4 50.5	287 02.2	N 7 55.3	Alioth	166 30.5	N55 58.3
07	326 57.9	325 54.8	02.6	283 43.8	29.4	335 36.0	50.3	302 04.3	55.4	Alkaid	153 07.6	N49 19.4
08	342 00.4	340 54.6	01.6	298 44.4	30.0	350 38.0	50.1	317 06.5	55.5	Al Na'ir	27 58.5	S46 58.0
M 09	357 02.8	355 54.4	1 00.5	313 45.1	.. 30.5	5 40.1	.. 49.9	332 08.7	.. 55.7	Alnilam	275 58.4	S 1 12.4
O 10	12 05.3	10 54.3	0 59.5	328 45.7	31.1	20 42.1	49.7	347 10.9	55.8	Alphard	218 07.5	S 8 39.3
N 11	27 07.8	25 54.1	58.5	343 46.4	31.6	35 44.1	49.6	2 13.0	55.9			
D 12	42 10.2	40 53.9	S 0 57.5	358 47.0	N16 32.2	50 46.1	S 4 49.4	17 15.2	N 7 56.0	Alphecca	126 20.5	N26 43.3
A 13	57 12.7	55 53.8	56.5	13 47.7	32.7	65 48.2	49.2	32 17.4	56.1	Alpheratz	357 55.8	N29 04.6
Y 14	72 15.2	70 53.6	55.5	28 48.3	33.3	80 50.2	49.0	47 19.6	56.2	Altair	62 19.5	N 8 51.8
15	87 17.6	85 53.5	.. 54.5	43 49.0	.. 33.8	95 52.2	.. 48.9	62 21.7	.. 56.3	Ankaa	353 27.5	S42 18.9
16	102 20.1	100 53.3	53.5	58 49.6	34.4	110 54.2	48.7	77 23.9	56.4	Antares	112 40.2	S26 25.6
17	117 22.6	115 53.1	52.5	73 50.3	35.0	125 56.3	48.5	92 26.1	56.5			
18	132 25.0	130 53.0	S 0 51.5	88 50.9	N16 35.5	140 58.3	S 4 48.3	107 28.3	N 7 56.6	Arcturus	146 06.0	N19 11.5
19	147 27.5	145 52.8	50.4	103 51.6	36.1	156 00.3	48.2	122 30.4	56.7	Atria	107 52.0	S69 01.2
20	162 30.0	160 52.6	49.4	118 52.2	36.6	171 02.4	48.0	137 32.6	56.8	Avior	234 22.8	S59 30.7
21	177 32.4	175 52.4	.. 48.4	133 52.9	.. 37.2	186 04.4	.. 47.8	152 34.8	.. 56.9	Bellatrix	278 44.8	N 6 20.7
22	192 34.9	190 52.3	47.4	148 53.5	37.7	201 06.4	47.6	167 37.0	57.1	Betelgeuse	271 14.1	N 7 24.2
23	207 37.3	205 52.1	46.4	163 54.2	38.3	216 08.4	47.4	182 39.1	57.2			
5 00	222 39.8	220 51.9	S 0 45.4	178 54.8	N16 38.8	231 10.5	S 4 47.3	197 41.3	N 7 57.3	Canopus	264 01.7	S52 42.1
01	237 42.3	235 51.8	44.4	193 55.5	39.4	246 12.5	47.1	212 43.5	57.4	Capella	280 52.0	N45 59.7
02	252 44.7	250 51.6	43.4	208 56.2	39.9	261 14.5	46.9	227 45.7	57.5	Deneb	49 39.3	N45 16.2
03	267 47.2	265 51.4	.. 42.4	223 56.8	.. 40.4	276 16.6	.. 46.7	242 47.8	.. 57.6	Denebola	182 45.3	N14 34.9
04	282 49.7	280 51.3	41.3	238 57.5	41.0	291 18.6	46.6	257 50.0	57.7	Diphda	349 07.8	S17 59.9
05	297 52.1	295 51.1	40.3	253 58.1	41.5	306 20.6	46.4	272 52.2	57.8			
06	312 54.6	310 50.9	S 0 39.3	268 58.8	N16 42.1	321 22.7	S 4 46.2	287 54.4	N 7 57.9	Dubhe	194 05.7	N61 45.8
07	327 57.1	325 50.8	38.3	283 59.4	42.6	336 24.7	46.0	302 56.5	58.0	Elnath	278 27.6	N28 36.2
T 08	342 59.5	340 50.6	37.3	299 00.1	43.2	351 26.7	45.9	317 58.7	58.1	Eltanin	90 51.2	N51 29.3
U 09	358 02.0	355 50.4	.. 36.3	314 00.7	.. 43.7	6 28.8	.. 45.7	333 00.9	.. 58.2	Enif	33 58.6	N 9 51.9
E 10	13 04.5	10 50.3	35.3	329 01.4	44.3	21 30.8	45.5	348 03.1	58.4	Fomalhaut	15 37.0	S29 37.8
S 11	28 06.9	25 50.1	34.2	344 02.0	44.8	36 32.8	45.3	3 05.2	58.5			
D 12	43 09.4	40 49.9	S 0 33.2	359 02.7	N16 45.4	51 34.8	S 4 45.2	18 07.4	N 7 58.6	Gacrux	172 13.4	S57 06.4
A 13	58 11.8	55 49.8	32.2	14 03.3	45.9	66 36.9	45.0	33 09.6	58.7	Gienah	176 04.0	S17 32.1
Y 14	73 14.3	70 49.6	31.2	29 04.0	46.5	81 38.9	44.8	48 11.8	58.8	Hadar	149 03.7	S60 21.9
15	88 16.8	85 49.4	.. 30.2	44 04.6	.. 47.0	96 40.9	.. 44.6	63 13.9	.. 58.9	Hamal	328 14.2	N23 27.0
16	103 19.2	100 49.3	29.2	59 05.3	47.6	111 43.0	44.5	78 16.1	59.0	Kaus Aust.	83 59.0	S34 22.9
17	118 21.7	115 49.1	28.2	74 05.9	48.1	126 45.0	44.3	93 18.3	59.1			
18	133 24.2	130 48.9	S 0 27.1	89 06.6	N16 48.6	141 47.0	S 4 44.1	108 20.5	N 7 59.2	Kochab	137 18.5	N74 09.9
19	148 26.6	145 48.7	26.1	104 07.2	49.2	156 49.1	43.9	123 22.6	59.3	Markab	13 50.1	N15 11.6
20	163 29.1	160 48.6	25.1	119 07.9	49.7	171 51.1	43.8	138 24.8	59.4	Menkar	314 27.6	N 4 04.8
21	178 31.6	175 48.4	.. 24.1	134 08.5	.. 50.3	186 53.1	.. 43.6	153 27.0	.. 59.5	Menkent	148 20.9	S36 21.7
22	193 34.0	190 48.2	23.1	149 09.2	50.8	201 55.2	43.4	168 29.2	59.6	Miaplacidus	221 42.0	S69 43.0
23	208 36.5	205 48.1	22.1	164 09.8	51.4	216 57.2	43.2	183 31.3	59.8			
6 00	223 38.9	220 47.9	S 0 21.0	179 10.5	N16 51.9	231 59.2	S 4 43.1	198 33.5	N 7 59.9	Mirfak	308 57.5	N49 51.1
01	238 41.4	235 47.7	20.0	194 11.1	52.4	247 01.3	42.9	213 35.7	8 00.0	Nunki	76 12.5	S26 17.8
02	253 43.9	250 47.6	19.0	209 11.8	53.0	262 03.3	42.7	228 37.9	00.1	Peacock	53 37.5	S56 44.1
03	268 46.3	265 47.4	.. 18.0	224 12.4	.. 53.5	277 05.3	.. 42.5	243 40.0	.. 00.2	Pollux	243 42.1	N28 01.8
04	283 48.8	280 47.2	17.0	239 13.1	54.1	292 07.4	42.4	258 42.2	00.3	Procyon	245 12.0	N 5 13.6
05	298 51.3	295 47.0	16.0	254 13.7	54.6	307 09.4	42.2	273 44.4	00.4			
06	313 53.7	310 46.9	S 0 14.9	269 14.3	N16 55.1	322 11.4	S 4 42.0	288 46.6	N 8 00.5	Rasalhague	96 17.0	N12 33.7
W 07	328 56.2	325 46.7	13.9	284 15.0	55.7	337 13.5	41.8	303 48.8	00.6	Regulus	207 55.8	N11 58.5
E 08	343 58.7	340 46.5	12.9	299 15.6	56.2	352 15.5	41.7	318 50.9	00.7	Rigel	281 23.5	S 8 12.5
D 09	359 01.1	355 46.4	.. 11.9	314 16.3	.. 56.8	7 17.5	.. 41.5	333 53.1	.. 00.8	Rigil Kent.	140 07.0	S60 49.6
N 10	14 03.6	10 46.2	10.9	329 16.9	57.3	22 19.6	41.3	348 55.3	00.9	Sabik	102 25.6	S15 43.3
E 11	29 06.1	25 46.0	09.8	344 17.6	57.8	37 21.6	41.1	3 57.5	01.0			
S 12	44 08.5	40 45.8	S 0 08.8	359 18.2	N16 58.4	52 23.6	S 4 41.0	18 59.6	N 8 01.1	Schedar	349 54.2	N56 31.4
D 13	59 11.0	55 45.7	07.8	14 18.9	58.9	67 25.7	40.8	34 01.8	01.3	Shaula	96 37.4	S37 06.0
A 14	74 13.4	70 45.5	06.8	29 19.5	16 59.5	82 27.7	40.6	49 04.0	01.4	Sirius	258 44.2	S16 43.1
Y 15	89 15.9	85 45.3	.. 05.8	44 20.2	17 00.0	97 29.8	.. 40.4	64 06.2	.. 01.5	Spica	158 43.2	S11 09.2
16	104 18.4	100 45.2	04.7	59 20.8	00.5	112 31.8	40.3	79 08.3	01.6	Suhail	223 01.0	S43 25.9
17	119 20.8	115 45.0	03.7	74 21.5	01.1	127 33.8	40.1	94 10.5	01.7			
18	134 23.3	130 44.8	S 0 02.7	89 22.1	N17 01.6	142 35.9	S 4 39.9	109 12.7	N 8 01.8	Vega	80 46.6	N38 46.8
19	149 25.8	145 44.6	01.7	104 22.8	02.1	157 37.9	39.7	124 14.9	01.9	Zuben'ubi	137 18.0	S16 02.1
20	164 28.2	160 44.5	S 00.7	119 23.4	02.7	172 39.9	39.6	139 17.0	02.0		SHA	Mer. Pass.
21	179 30.7	175 44.3	N 00.4	134 24.1	.. 03.2	187 42.0	.. 39.4	154 19.2	.. 02.1		° ′	h m
22	194 33.2	190 44.1	01.4	149 24.7	03.8	202 44.0	39.2	169 21.4	02.2	Venus	358 12.1	9 17
23	209 35.6	205 44.0	02.4	164 25.4	04.3	217 46.0	39.0	184 23.6	02.3	Mars	316 15.0	12 04
	h m									Jupiter	8 30.7	8 34
Mer. Pass.	9 07.8	v −0.2	d 1.0	v 0.7	d 0.5	v 2.0	d 0.2	v 2.2	d 0.1	Saturn	335 01.5	10 48

UT	SUN GHA	SUN Dec	MOON GHA	v	MOON Dec	d	HP
d h	° ′	° ′	° ′	′	° ′	′	′
4 00	180 47.4	N15 50.5	79 49.4	12.7	N13 17.9	7.2	55.4
01	195 47.4	51.3	94 21.1	12.8	13 10.7	7.2	55.3
02	210 47.5	52.0	108 52.9	12.8	13 03.5	7.3	55.3
03	225 47.5 ..	52.7	123 24.7	12.8	12 56.2	7.3	55.3
04	240 47.6	53.5	137 56.5	13.0	12 48.9	7.4	55.2
05	255 47.7	54.2	152 28.5	13.0	12 41.5	7.5	55.2
06	270 47.7	N15 54.9	167 00.5	13.0	N12 34.0	7.4	55.2
07	285 47.8	55.6	181 32.5	13.1	12 26.6	7.6	55.2
M 08	300 47.8	56.4	196 04.6	13.1	12 19.0	7.6	55.1
O 09	315 47.9 ..	57.1	210 36.7	13.2	12 11.4	7.6	55.1
N 10	330 48.0	57.8	225 08.9	13.3	12 03.8	7.7	55.1
D 11	345 48.0	58.5	239 41.2	13.3	11 56.1	7.7	55.1
A 12	0 48.1	N15 59.3	254 13.5	13.4	N11 48.4	7.8	55.1
Y 13	15 48.1	16 00.0	268 45.9	13.4	11 40.6	7.8	55.0
14	30 48.2	00.7	283 18.3	13.5	11 32.8	7.8	55.0
15	45 48.3 ..	01.4	297 50.8	13.5	11 25.0	8.0	55.0
16	60 48.3	02.2	312 23.3	13.6	11 17.0	7.9	55.0
17	75 48.4	02.9	326 55.9	13.6	11 09.1	8.0	54.9
18	90 48.4	N16 03.6	341 28.5	13.6	N11 01.1	8.0	54.9
19	105 48.5	04.3	356 01.1	13.8	10 53.1	8.1	54.9
20	120 48.5	05.0	10 33.9	13.7	10 45.0	8.2	54.9
21	135 48.6 ..	05.8	25 06.6	13.8	10 36.8	8.1	54.9
22	150 48.7	06.5	39 39.4	13.9	10 28.7	8.2	54.8
23	165 48.7	07.2	54 12.3	13.9	10 20.5	8.3	54.8
5 00	180 48.8	N16 07.9	68 45.2	14.0	N10 12.2	8.2	54.8
01	195 48.8	08.6	83 18.2	14.0	10 04.0	8.4	54.8
02	210 48.9	09.3	97 51.2	14.0	9 55.6	8.3	54.8
03	225 48.9 ..	10.1	112 24.2	14.1	9 47.3	8.4	54.7
04	240 49.0	10.8	126 57.3	14.1	9 38.9	8.4	54.7
05	255 49.0	11.5	141 30.4	14.2	9 30.5	8.5	54.7
06	270 49.1	N16 12.2	156 03.6	14.2	N 9 22.0	8.5	54.7
07	285 49.1	12.9	170 36.8	14.2	9 13.5	8.6	54.7
T 08	300 49.2	13.6	185 10.0	14.3	9 04.9	8.5	54.6
U 09	315 49.3 ..	14.3	199 43.3	14.4	8 56.4	8.6	54.6
E 10	330 49.3	15.1	214 16.7	14.3	8 47.8	8.7	54.6
S 11	345 49.4	15.8	228 50.0	14.5	8 39.1	8.6	54.6
D 12	0 49.4	N16 16.5	243 23.5	14.4	N 8 30.5	8.7	54.6
A 13	15 49.5	17.2	257 56.9	14.5	8 21.8	8.8	54.6
Y 14	30 49.5	17.9	272 30.4	14.5	8 13.0	8.7	54.5
15	45 49.6 ..	18.6	287 03.9	14.6	8 04.3	8.8	54.5
16	60 49.6	19.3	301 37.5	14.6	7 55.5	8.8	54.5
17	75 49.7	20.0	316 11.1	14.6	7 46.7	8.9	54.5
18	90 49.7	N16 20.7	330 44.7	14.7	N 7 37.8	8.8	54.5
19	105 49.8	21.5	345 18.4	14.7	7 29.0	8.9	54.5
20	120 49.8	22.2	359 52.1	14.7	7 20.1	9.0	54.5
21	135 49.9 ..	22.9	14 25.8	14.8	7 11.1	8.9	54.4
22	150 49.9	23.6	28 59.6	14.7	7 02.2	9.0	54.4
23	165 50.0	24.3	43 33.3	14.9	6 53.2	9.0	54.4
6 00	180 50.0	N16 25.0	58 07.2	14.8	N 6 44.2	9.0	54.4
01	195 50.1	25.7	72 41.0	14.9	6 35.2	9.1	54.4
02	210 50.1	26.4	87 14.9	14.9	6 26.1	9.0	54.4
03	225 50.2 ..	27.1	101 48.8	15.0	6 17.1	9.1	54.4
04	240 50.2	27.8	116 22.8	14.9	6 08.0	9.2	54.3
05	255 50.3	28.5	130 56.7	15.0	5 58.8	9.1	54.3
06	270 50.3	N16 29.2	145 30.7	15.0	N 5 49.7	9.1	54.3
W 07	285 50.4	29.9	160 04.7	15.1	5 40.6	9.2	54.3
E 08	300 50.4	30.6	174 38.8	15.0	5 31.4	9.2	54.3
D 09	315 50.5 ..	31.3	189 12.8	15.1	5 22.2	9.2	54.3
N 10	330 50.5	32.0	203 46.9	15.2	5 13.0	9.3	54.3
E 11	345 50.6	32.7	218 21.1	15.1	5 03.7	9.2	54.3
S 12	0 50.6	N16 33.4	232 55.2	15.2	N 4 54.5	9.3	54.3
D 13	15 50.7	34.1	247 29.4	15.1	4 45.2	9.3	54.2
A 14	30 50.7	34.8	262 03.5	15.2	4 35.9	9.3	54.2
Y 15	45 50.8 ..	35.5	276 37.7	15.3	4 26.6	9.3	54.2
16	60 50.8	36.2	291 12.0	15.2	4 17.3	9.4	54.2
17	75 50.8	36.9	305 46.2	15.3	4 07.9	9.3	54.2
18	90 50.9	N16 37.6	320 20.5	15.2	N 3 58.6	9.4	54.2
19	105 50.9	38.3	334 54.7	15.3	3 49.2	9.3	54.2
20	120 51.0	39.0	349 29.0	15.3	3 39.9	9.4	54.2
21	135 51.0 ..	39.7	4 03.3	15.4	3 30.5	9.4	54.2
22	150 51.1	40.4	18 37.7	15.3	3 21.1	9.5	54.2
23	165 51.1	41.1	33 12.0	15.3	N 3 11.6	9.4	54.2
	SD 15.9	d 0.7	SD 15.0		14.9		14.8

Lat.	Twilight Naut.	Twilight Civil	Sunrise	Moonrise 4	Moonrise 5	Moonrise 6	Moonrise 7
°	h m	h m	h m	h m	h m	h m	h m
N 72	////	////	01 17	10 18	11 58	13 34	15 08
N 70	////	////	02 08	10 39	12 11	13 41	15 09
68	////	////	02 40	10 56	12 22	13 46	15 09
66	////	01 30	03 03	11 10	12 31	13 51	15 10
64	////	02 07	03 21	11 21	12 38	13 54	15 10
62	////	02 33	03 36	11 30	12 44	13 58	15 11
60	01 20	02 53	03 48	11 38	12 49	14 00	15 11
N 58	01 54	03 09	03 59	11 45	12 54	14 03	15 11
56	02 18	03 22	04 08	11 51	12 58	14 05	15 12
54	02 36	03 34	04 16	11 57	13 02	14 07	15 12
52	02 52	03 44	04 23	12 02	13 06	14 09	15 12
50	03 05	03 53	04 30	12 06	13 09	14 11	15 12
45	03 30	04 11	04 44	12 16	13 16	14 14	15 13
N 40	03 50	04 26	04 56	12 24	13 21	14 17	15 13
35	04 05	04 38	05 05	12 31	13 26	14 20	15 14
30	04 18	04 49	05 14	12 37	13 30	14 22	15 14
20	04 38	05 06	05 29	12 48	13 38	14 26	15 14
N 10	04 54	05 19	05 41	12 57	13 44	14 30	15 15
0	05 07	05 32	05 53	13 06	13 50	14 33	15 15
S 10	05 18	05 43	06 05	13 14	13 56	14 36	15 16
20	05 28	05 54	06 17	13 23	14 02	14 40	15 16
30	05 38	06 06	06 31	13 34	14 10	14 44	15 17
35	05 43	06 13	06 39	13 40	14 14	14 46	15 17
40	05 48	06 20	06 49	13 46	14 19	14 49	15 18
45	05 53	06 28	06 59	13 54	14 24	14 52	15 18
S 50	05 59	06 38	07 12	14 04	14 31	14 55	15 18
52	06 01	06 42	07 18	14 08	14 34	14 57	15 19
54	06 04	06 46	07 25	14 13	14 37	14 59	15 19
56	06 07	06 51	07 32	14 18	14 41	15 01	15 19
58	06 10	06 57	07 40	14 24	14 45	15 03	15 20
S 60	06 13	07 03	07 50	14 31	14 49	15 05	15 20

Lat.	Sunset	Twilight Civil	Twilight Naut.	Moonset 4	Moonset 5	Moonset 6	Moonset 7
°	h m	h m	h m	h m	h m	h m	h m
N 72	22 48	////	////	03 44	03 37	03 31	03 24
N 70	21 51	////	////	03 21	03 22	03 22	03 21
68	21 17	////	////	03 04	03 10	03 15	03 19
66	20 54	22 30	////	02 49	03 01	03 09	03 16
64	20 35	21 50	////	02 37	02 52	03 04	03 14
62	20 20	21 24	////	02 27	02 45	03 00	03 13
60	20 07	21 03	22 39	02 19	02 39	02 56	03 11
N 58	19 56	20 47	22 03	02 11	02 33	02 53	03 10
56	19 47	20 33	21 38	02 04	02 28	02 50	03 09
54	19 39	20 21	21 19	01 58	02 24	02 47	03 08
52	19 31	20 11	21 04	01 53	02 20	02 44	03 07
50	19 24	20 02	20 50	01 48	02 16	02 42	03 06
45	19 10	19 43	20 24	01 37	02 09	02 37	03 04
N 40	18 58	19 28	20 05	01 28	02 02	02 33	03 02
35	18 48	19 16	19 49	01 20	01 56	02 29	03 01
30	18 40	19 05	19 36	01 14	01 51	02 25	03 00
20	18 25	18 48	19 16	01 02	01 42	02 21	02 57
N 10	18 12	18 34	19 00	00 51	01 35	02 16	02 55
0	18 00	18 22	18 47	00 42	01 27	02 11	02 53
S 10	17 48	18 10	18 35	00 32	01 20	02 06	02 52
20	17 36	17 59	18 25	00 21	01 12	02 01	02 50
30	17 22	17 47	18 15	00 09	01 03	01 55	02 47
35	17 14	17 40	18 10	00 02	00 58	01 52	02 46
40	17 04	17 33	18 05	24 52	00 52	01 48	02 44
45	16 54	17 25	18 00	24 45	00 45	01 44	02 43
S 50	16 41	17 15	17 54	24 36	00 36	01 39	02 40
52	16 35	17 11	17 51	24 32	00 32	01 36	02 39
54	16 28	17 06	17 49	24 28	00 28	01 33	02 38
56	16 21	17 01	17 46	24 23	00 23	01 31	02 37
58	16 12	16 56	17 43	24 18	00 18	01 27	02 36
S 60	16 03	16 49	17 40	24 12	00 12	01 23	02 34

	SUN			MOON			
Day	Eqn. of Time 00ʰ	Eqn. of Time 12ʰ	Mer. Pass.	Mer. Pass. Upper	Mer. Pass. Lower	Age	Phase
d	m s	m s	h m	h m	h m	d	%
4	03 09	03 12	11 57	19 16	06 54	08	61
5	03 15	03 18	11 57	20 01	07 39	09	70
6	03 20	03 22	11 57	20 43	08 22	10	78

1998 MAY 7, 8, 9 (THURS., FRI., SAT.)

UT	ARIES	VENUS −4.1		MARS +1.3		JUPITER −2.2		SATURN +0.5	
	GHA	GHA	Dec	GHA	Dec	GHA	Dec	GHA	Dec
d h	° ′	° ′	° ′	° ′	° ′	° ′	° ′	° ′	° ′
7 00	224 38.1	220 43.8	N 0 03.4	179 26.0	N17 04.8	232 48.1	S 4 38.9	199 25.7	N 8 02.4
01	239 40.5	235 43.6	04.4	194 26.7	05.4	247 50.1	38.7	214 27.9	02.5
02	254 43.0	250 43.4	05.5	209 27.3	05.9	262 52.2	38.5	229 30.1	02.6
03	269 45.5	265 43.3	.. 06.5	224 28.0	.. 06.4	277 54.2	.. 38.4	244 32.3	.. 02.7
04	284 47.9	280 43.1	07.5	239 28.6	07.0	292 56.2	38.2	259 34.5	02.9
05	299 50.4	295 42.9	08.5	254 29.3	07.5	307 58.3	38.0	274 36.6	03.0
T 06	314 52.9	310 42.7	N 0 09.6	269 29.9	N17 08.0	323 00.3	S 4 37.8	289 38.8	N 8 03.1
H 07	329 55.3	325 42.6	10.6	284 30.5	08.6	338 02.3	37.7	304 41.0	03.2
U 08	344 57.8	340 42.4	11.6	299 31.2	09.1	353 04.4	37.5	319 43.2	03.3
R 09	0 00.3	355 42.2	.. 12.6	314 31.8	.. 09.6	8 06.4	.. 37.3	334 45.3	.. 03.4
S 10	15 02.7	10 42.0	13.7	329 32.5	10.2	23 08.5	37.1	349 47.5	03.5
D 11	30 05.2	25 41.9	14.7	344 33.1	10.7	38 10.5	37.0	4 49.7	03.6
A 12	45 07.7	40 41.7	N 0 15.7	359 33.8	N17 11.2	53 12.5	S 4 36.8	19 51.9	N 8 03.7
Y 13	60 10.1	55 41.5	16.7	14 34.4	11.8	68 14.6	36.6	34 54.0	03.8
14	75 12.6	70 41.4	17.7	29 35.1	12.3	83 16.6	36.4	49 56.2	03.9
15	90 15.0	85 41.2	.. 18.8	44 35.7	.. 12.8	98 18.7	.. 36.3	64 58.4	.. 04.0
16	105 17.5	100 41.0	19.8	59 36.4	13.3	113 20.7	36.1	80 00.6	04.1
17	120 20.0	115 40.8	20.8	74 37.0	13.9	128 22.7	35.9	95 02.8	04.2
18	135 22.4	130 40.7	N 0 21.8	89 37.7	N17 14.4	143 24.8	S 4 35.8	110 04.9	N 8 04.3
19	150 24.9	145 40.5	22.9	104 38.3	14.9	158 26.8	35.6	125 07.1	04.4
20	165 27.4	160 40.3	23.9	119 38.9	15.5	173 28.9	35.4	140 09.3	04.6
21	180 29.8	175 40.1	.. 24.9	134 39.6	.. 16.0	188 30.9	.. 35.2	155 11.5	.. 04.7
22	195 32.3	190 39.9	25.9	149 40.2	16.5	203 32.9	35.1	170 13.6	04.8
23	210 34.8	205 39.8	27.0	164 40.9	17.0	218 35.0	34.9	185 15.8	04.9
8 00	225 37.2	220 39.6	N 0 28.0	179 41.5	N17 17.6	233 37.0	S 4 34.7	200 18.0	N 8 05.0
01	240 39.7	235 39.4	29.0	194 42.2	18.1	248 39.1	34.6	215 20.2	05.1
02	255 42.2	250 39.2	30.1	209 42.8	18.6	263 41.1	34.4	230 22.3	05.2
03	270 44.6	265 39.1	.. 31.1	224 43.5	.. 19.2	278 43.1	.. 34.2	245 24.5	.. 05.3
04	285 47.1	280 38.9	32.1	239 44.1	19.7	293 45.2	34.0	260 26.7	05.4
05	300 49.5	295 38.7	33.1	254 44.8	20.2	308 47.2	33.9	275 28.9	05.5
F 06	315 52.0	310 38.5	N 0 34.2	269 45.4	N17 20.7	323 49.3	S 4 33.7	290 31.1	N 8 05.6
R 07	330 54.5	325 38.4	35.2	284 46.0	21.3	338 51.3	33.5	305 33.2	05.7
I 08	345 56.9	340 38.2	36.2	299 46.7	21.8	353 53.3	33.4	320 35.4	05.8
D 09	0 59.4	355 38.0	.. 37.2	314 47.3	.. 22.3	8 55.4	.. 33.2	335 37.6	.. 05.9
A 10	16 01.9	10 37.8	38.3	329 48.0	22.8	23 57.4	33.0	350 39.8	06.0
Y 11	31 04.3	25 37.7	39.3	344 48.6	23.4	38 59.5	32.8	5 41.9	06.2
12	46 06.8	40 37.5	N 0 40.3	359 49.3	N17 23.9	54 01.5	S 4 32.7	20 44.1	N 8 06.3
13	61 09.3	55 37.3	41.4	14 49.9	24.4	69 03.6	32.5	35 46.3	06.4
14	76 11.7	70 37.1	42.4	29 50.6	24.9	84 05.6	32.3	50 48.5	06.5
15	91 14.2	85 36.9	.. 43.4	44 51.2	.. 25.5	99 07.6	.. 32.2	65 50.7	.. 06.6
16	106 16.6	100 36.8	44.4	59 51.8	26.0	114 09.7	32.0	80 52.8	06.7
17	121 19.1	115 36.6	45.5	74 52.5	26.5	129 11.7	31.8	95 55.0	06.8
18	136 21.6	130 36.4	N 0 46.5	89 53.1	N17 27.0	144 13.8	S 4 31.6	110 57.2	N 8 06.9
19	151 24.0	145 36.2	47.5	104 53.8	27.6	159 15.8	31.5	125 59.4	07.0
20	166 26.5	160 36.1	48.6	119 54.4	28.1	174 17.9	31.3	141 01.5	07.1
21	181 29.0	175 35.9	.. 49.6	134 55.1	.. 28.6	189 19.9	.. 31.1	156 03.7	.. 07.2
22	196 31.4	190 35.7	50.6	149 55.7	29.1	204 22.0	31.0	171 05.9	07.3
23	211 33.9	205 35.5	51.6	164 56.3	29.6	219 24.0	30.8	186 08.1	07.4
9 00	226 36.4	220 35.3	N 0 52.7	179 57.0	N17 30.2	234 26.0	S 4 30.6	201 10.3	N 8 07.5
01	241 38.8	235 35.2	53.7	194 57.6	30.7	249 28.1	30.4	216 12.4	07.6
02	256 41.3	250 35.0	54.7	209 58.3	31.2	264 30.1	30.3	231 14.6	07.7
03	271 43.8	265 34.8	.. 55.8	224 58.9	.. 31.7	279 32.2	.. 30.1	246 16.8	.. 07.8
04	286 46.2	280 34.6	56.8	239 59.6	32.2	294 34.2	29.9	261 19.0	07.9
05	301 48.7	295 34.4	57.8	255 00.2	32.8	309 36.3	29.8	276 21.2	08.1
S 06	316 51.1	310 34.3	N 0 58.9	270 00.9	N17 33.3	324 38.3	S 4 29.6	291 23.3	N 8 08.2
A 07	331 53.6	325 34.1	0 59.9	285 01.5	33.8	339 40.4	29.4	306 25.5	08.3
T 08	346 56.1	340 33.9	1 00.9	300 02.1	34.3	354 42.4	29.3	321 27.7	08.4
U 09	1 58.5	355 33.7	.. 01.9	315 02.8	.. 34.8	9 44.4	.. 29.1	336 29.9	.. 08.5
R 10	17 01.0	10 33.5	03.0	330 03.4	35.3	24 46.5	28.9	351 32.0	08.6
D 11	32 03.5	25 33.4	04.0	345 04.1	35.9	39 48.5	28.8	6 34.2	08.7
A 12	47 05.9	40 33.2	N 1 05.0	0 04.7	N17 36.4	54 50.6	S 4 28.6	21 36.4	N 8 08.8
Y 13	62 08.4	55 33.0	06.1	15 05.3	36.9	69 52.6	28.4	36 38.6	08.9
14	77 10.9	70 32.8	07.1	30 06.0	37.4	84 54.7	28.2	51 40.8	09.0
15	92 13.3	85 32.6	.. 08.1	45 06.6	.. 37.9	99 56.7	.. 28.1	66 42.9	.. 09.1
16	107 15.8	100 32.5	09.2	60 07.3	38.4	114 58.8	27.9	81 45.1	09.2
17	122 18.2	115 32.3	10.2	75 07.9	39.0	130 00.8	27.7	96 47.3	09.3
18	137 20.7	130 32.1	N 1 11.2	90 08.6	N17 39.5	145 02.9	S 4 27.6	111 49.5	N 8 09.4
19	152 23.2	145 31.9	12.3	105 09.2	40.0	160 04.9	27.4	126 51.7	09.5
20	167 25.6	160 31.7	13.3	120 09.8	40.5	175 07.0	27.2	141 53.8	09.6
21	182 28.1	175 31.5	.. 14.3	135 10.5	.. 41.0	190 09.0	.. 27.1	156 56.0	.. 09.7
22	197 30.6	190 31.4	15.4	150 11.1	41.5	205 11.1	26.9	171 58.2	09.8
23	212 33.0	205 31.2	16.4	165 11.8	42.1	220 13.1	26.7	187 00.4	09.9
Mer. Pass.	h m 8 56.1	v −0.2	d 1.0	v 0.6	d 0.5	v 2.0	d 0.2	v 2.2	d 0.1

STARS

Name	SHA	Dec
Acamar	315 27.6	S40 18.9
Achernar	335 36.0	S57 14.7
Acrux	173 21.7	S63 05.6
Adhara	255 21.9	S28 58.5
Aldebaran	291 03.1	N16 30.2
Alioth	166 30.5	N55 58.3
Alkaid	153 07.6	N49 19.4
Al Na'ir	27 58.4	S46 58.0
Alnilam	275 58.4	S 1 12.4
Alphard	218 07.5	S 8 39.3
Alphecca	126 20.5	N26 43.3
Alpheratz	357 55.8	N29 04.6
Altair	62 19.4	N 8 51.8
Ankaa	353 27.5	S42 18.9
Antares	112 40.2	S26 25.6
Arcturus	146 06.0	N19 11.5
Atria	107 52.0	S69 01.3
Avior	234 22.9	S59 30.7
Bellatrix	278 44.8	N 6 20.7
Betelgeuse	271 14.1	N 7 24.2
Canopus	264 01.7	S52 42.1
Capella	280 52.1	N45 59.7
Deneb	49 39.3	N45 16.2
Denebola	182 45.3	N14 34.9
Diphda	349 07.8	S17 59.8
Dubhe	194 05.7	N61 45.8
Elnath	278 27.6	N28 36.2
Eltanin	90 51.1	N51 29.3
Enif	33 58.6	N 9 51.9
Fomalhaut	15 37.0	S29 37.8
Gacrux	172 13.4	S57 06.4
Gienah	176 04.0	S17 32.1
Hadar	149 03.7	S60 21.9
Hamal	328 14.2	N23 27.0
Kaus Aust.	83 59.0	S34 22.9
Kochab	137 18.5	N74 09.9
Markab	13 50.1	N15 11.6
Menkar	314 27.5	N 4 04.8
Menkent	148 20.9	S36 21.7
Miaplacidus	221 42.0	S69 43.0
Mirfak	308 57.5	N49 51.1
Nunki	76 12.5	S26 17.8
Peacock	53 37.4	S56 44.1
Pollux	243 42.1	N28 01.8
Procyon	245 12.0	N 5 13.6
Rasalhague	96 17.0	N12 33.7
Regulus	207 55.8	N11 58.5
Rigel	281 23.5	S 8 12.5
Rigil Kent.	140 07.0	S60 49.6
Sabik	102 25.6	S15 43.3
Schedar	349 54.2	N56 31.4
Shaula	96 37.4	S37 06.0
Sirius	258 44.2	S16 43.1
Spica	158 43.2	S11 09.2
Suhail	223 01.0	S43 25.9
Vega	80 46.5	N38 46.8
Zuben'ubi	137 18.0	S16 02.1

	SHA	Mer. Pass.
	° ′	h m
Venus	355 02.4	9 17
Mars	314 04.3	12 01
Jupiter	7 59.8	8 24
Saturn	334 40.8	10 37

UT	SUN		MOON				
d h	GHA	Dec	GHA	v	Dec	d	HP
	° ′	° ′	° ′	′	° ′	′	′
7 00	180 51.2	N16 41.8	47 46.3	15.4	N 3 02.2	9.4	54.1
01	195 51.2	42.5	62 20.7	15.4	2 52.8	9.5	54.1
02	210 51.3	43.2	76 55.1	15.4	2 43.3	9.4	54.1
03	225 51.3	.. 43.9	91 29.5	15.4	2 33.9	9.5	54.1
04	240 51.3	44.6	106 03.9	15.4	2 24.4	9.4	54.1
05	255 51.4	45.3	120 38.3	15.4	2 15.0	9.5	54.1
06	270 51.4	N16 46.0	135 12.7	15.5	N 2 05.5	9.5	54.1
07	285 51.5	46.7	149 47.2	15.4	1 56.0	9.5	54.1
T 08	300 51.5	47.4	164 21.6	15.5	1 46.5	9.5	54.1
H 09	315 51.6	.. 48.0	178 56.1	15.4	1 37.0	9.5	54.1
U 10	330 51.6	48.7	193 30.5	15.5	1 27.5	9.5	54.1
R 11	345 51.6	49.4	208 05.0	15.5	1 18.0	9.5	54.1
S 12	0 51.7	N16 50.1	222 39.5	15.4	N 1 08.5	9.6	54.1
D 13	15 51.7	50.8	237 13.9	15.5	0 58.9	9.5	54.1
A 14	30 51.8	51.5	251 48.4	15.5	0 49.4	9.5	54.1
Y 15	45 51.8	.. 52.2	266 22.9	15.5	0 39.9	9.5	54.1
16	60 51.8	52.9	280 57.4	15.5	0 30.4	9.6	54.1
17	75 51.9	53.6	295 31.9	15.5	0 20.8	9.5	54.1
18	90 51.9	N16 54.2	310 06.4	15.5	N 0 11.3	9.5	54.1
19	105 52.0	54.9	324 40.9	15.5	N 0 01.8	9.6	54.0
20	120 52.0	55.6	339 15.4	15.5	S 0 07.8	9.5	54.0
21	135 52.0	.. 56.3	353 49.9	15.5	0 17.3	9.5	54.0
22	150 52.1	57.0	8 24.4	15.5	0 26.8	9.6	54.0
23	165 52.1	57.7	22 58.9	15.4	0 36.4	9.5	54.0
8 00	180 52.2	N16 58.3	37 33.3	15.5	S 0 45.9	9.5	54.0
01	195 52.2	59.0	52 07.8	15.5	0 55.4	9.6	54.0
02	210 52.2	16 59.7	66 42.3	15.5	1 05.0	9.5	54.0
03	225 52.3	17 00.4	81 16.8	15.5	1 14.5	9.5	54.0
04	240 52.3	01.1	95 51.3	15.5	1 24.0	9.5	54.0
05	255 52.3	01.8	110 25.8	15.4	1 33.5	9.5	54.0
06	270 52.4	N17 02.4	125 00.2	15.5	S 1 43.0	9.5	54.0
07	285 52.4	03.1	139 34.7	15.5	1 52.5	9.5	54.0
F 08	300 52.5	03.8	154 09.2	15.4	2 02.0	9.5	54.0
R 09	315 52.5	.. 04.5	168 43.6	15.4	2 11.5	9.5	54.0
I 10	330 52.5	05.2	183 18.0	15.5	2 21.0	9.4	54.0
D 11	345 52.6	05.8	197 52.5	15.4	2 30.4	9.5	54.0
A 12	0 52.6	N17 06.5	212 26.9	15.4	S 2 39.9	9.4	54.0
Y 13	15 52.6	07.2	227 01.3	15.4	2 49.3	9.5	54.0
14	30 52.7	07.9	241 35.7	15.4	2 58.8	9.4	54.0
15	45 52.7	.. 08.5	256 10.1	15.4	3 08.2	9.4	54.0
16	60 52.7	09.2	270 44.5	15.4	3 17.6	9.4	54.0
17	75 52.8	09.9	285 18.9	15.3	3 27.0	9.4	54.0
18	90 52.8	N17 10.6	299 53.2	15.4	S 3 36.4	9.4	54.0
19	105 52.8	11.2	314 27.6	15.3	3 45.8	9.4	54.0
20	120 52.9	11.9	329 01.9	15.3	3 55.2	9.3	54.0
21	135 52.9	.. 12.6	343 36.2	15.3	4 04.5	9.4	54.0
22	150 52.9	13.3	358 10.5	15.3	4 13.9	9.3	54.0
23	165 53.0	13.9	12 44.8	15.2	4 23.2	9.3	54.0
9 00	180 53.0	N17 14.6	27 19.0	15.3	S 4 32.5	9.3	54.0
01	195 53.0	15.3	41 53.3	15.2	4 41.8	9.3	54.1
02	210 53.1	15.9	56 27.5	15.2	4 51.1	9.3	54.1
03	225 53.1	.. 16.6	71 01.7	15.2	5 00.4	9.2	54.1
04	240 53.1	17.3	85 35.9	15.2	5 09.6	9.2	54.1
05	255 53.2	18.0	100 10.1	15.1	5 18.8	9.2	54.1
06	270 53.2	N17 18.6	114 44.2	15.2	S 5 28.0	9.2	54.1
07	285 53.2	19.3	129 18.4	15.1	5 37.2	9.2	54.1
S 08	300 53.3	20.0	143 52.5	15.1	5 46.4	9.1	54.1
A 09	315 53.3	.. 20.6	158 26.6	15.0	5 55.5	9.1	54.1
T 10	330 53.3	21.3	173 00.6	15.1	6 04.6	9.1	54.1
U 11	345 53.4	22.0	187 34.7	15.0	6 13.7	9.1	54.1
R 12	0 53.4	N17 22.6	202 08.7	15.0	S 6 22.8	9.1	54.1
D 13	15 53.4	23.3	216 42.7	15.0	6 31.9	9.0	54.1
A 14	30 53.4	23.9	231 16.7	14.9	6 40.9	9.0	54.1
Y 15	45 53.5	.. 24.6	245 50.6	14.9	6 49.9	9.0	54.1
16	60 53.5	25.3	260 24.5	14.9	6 58.9	8.9	54.1
17	75 53.5	26.0	274 58.4	14.9	7 07.8	9.0	54.1
18	90 53.6	N17 26.6	289 32.3	14.9	S 7 16.8	8.9	54.1
19	105 53.6	27.3	304 06.2	14.8	7 25.7	8.8	54.1
20	120 53.6	27.9	318 40.0	14.8	7 34.5	8.9	54.1
21	135 53.6	.. 28.6	333 13.8	14.7	7 43.4	8.8	54.2
22	150 53.7	29.2	347 47.5	14.7	7 52.2	8.8	54.2
23	165 53.7	29.9	2 21.2	14.7	S 8 01.0	8.7	54.2
	SD 15.9	d 0.7	SD 14.7		14.7		14.7

Moonrise

Lat.	Twilight Naut.	Twilight Civil	Sunrise	7	8	9	10
°	h m	h m	h m	h m	h m	h m	h m
N 72	////	////	00 31	15 08	16 42	18 17	19 55
N 70	////	////	01 48	15 09	16 37	18 05	19 36
68	////	////	02 25	15 09	16 33	17 56	19 21
66	////	01 06	02 51	15 10	16 29	17 49	19 09
64	////	01 52	03 11	15 10	16 26	17 43	19 00
62	////	02 21	03 27	15 11	16 24	17 37	18 51
60	00 58	02 43	03 40	15 11	16 22	17 32	18 43
N 58	01 40	03 00	03 52	15 11	16 20	17 28	18 37
56	02 07	03 15	04 02	15 12	16 18	17 25	18 31
54	02 28	03 27	04 10	15 12	16 17	17 21	18 26
52	02 44	03 38	04 18	15 12	16 15	17 18	18 22
50	02 58	03 47	04 25	15 12	16 14	17 16	18 18
45	03 25	04 07	04 40	15 13	16 11	17 10	18 09
N 40	03 46	04 22	04 52	15 13	16 09	17 05	18 01
35	04 02	04 35	05 03	15 14	16 07	17 01	17 55
30	04 15	04 46	05 12	15 14	16 05	16 57	17 49
20	04 36	05 04	05 27	15 14	16 02	16 51	17 40
N 10	04 53	05 19	05 41	15 15	16 00	16 45	17 31
0	05 06	05 31	05 53	15 15	15 57	16 40	17 24
S 10	05 18	05 43	06 05	15 16	15 55	16 35	17 16
20	05 29	05 55	06 18	15 16	15 52	16 29	17 08
30	05 39	06 08	06 33	15 17	15 50	16 23	16 58
35	05 45	06 15	06 42	15 17	15 48	16 20	16 53
40	05 50	06 23	06 51	15 18	15 46	16 16	16 47
45	05 56	06 31	07 03	15 18	15 44	16 11	16 40
S 50	06 03	06 42	07 16	15 18	15 42	16 05	16 31
52	06 05	06 46	07 23	15 19	15 40	16 03	16 27
54	06 08	06 51	07 30	15 19	15 39	16 00	16 23
56	06 12	06 57	07 38	15 19	15 38	15 57	16 18
58	06 15	07 03	07 47	15 20	15 38	15 54	16 13
S 60	06 19	07 09	07 57	15 20	15 35	15 50	16 07

Moonset

Lat.	Sunset	Twilight Civil	Twilight Naut.	7	8	9	10
°	h m	h m	h m	h m	h m	h m	h m
N 72	▭	▭	▭	03 24	03 18	03 12	03 05
N 70	22 11	////	////	03 21	03 20	03 19	03 18
68	21 32	////	////	03 19	03 22	03 25	03 29
66	21 05	22 56	////	03 16	03 23	03 30	03 37
64	20 45	22 06	////	03 14	03 24	03 34	03 45
62	20 28	21 35	////	03 13	03 25	03 38	03 51
60	20 15	21 13	23 03	03 11	03 26	03 41	03 57
N 58	20 03	20 55	22 17	03 10	03 27	03 43	04 02
56	19 53	20 40	21 49	03 09	03 27	03 46	04 06
54	19 44	20 28	21 28	03 08	03 28	03 48	04 10
52	19 36	20 17	21 11	03 07	03 28	03 50	04 14
50	19 29	20 07	20 57	03 06	03 29	03 52	04 17
45	19 14	19 47	20 29	03 04	03 30	03 56	04 24
N 40	19 01	19 31	20 08	03 02	03 31	04 00	04 30
35	18 51	19 18	19 52	03 01	03 32	04 03	04 35
30	18 42	19 07	19 38	03 00	03 32	04 05	04 40
20	18 26	18 49	19 17	02 57	03 34	04 10	04 47
N 10	18 13	18 35	19 00	02 55	03 35	04 14	04 54
0	18 00	18 22	18 47	02 53	03 36	04 18	05 01
S 10	17 48	18 09	18 35	02 52	03 36	04 22	05 07
20	17 34	17 57	18 24	02 50	03 38	04 26	05 14
30	17 19	17 45	18 13	02 47	03 39	04 30	05 22
35	17 11	17 37	18 08	02 46	03 39	04 33	05 27
40	17 01	17 30	18 02	02 44	03 40	04 36	05 32
45	16 50	17 21	17 56	02 43	03 41	04 39	05 38
S 50	16 36	17 11	17 50	02 40	03 42	04 44	05 45
52	16 30	17 06	17 47	02 39	03 42	04 45	05 49
54	16 22	17 01	17 44	02 38	03 43	04 48	05 52
56	16 14	16 56	17 41	02 37	03 44	04 50	05 57
58	16 06	16 50	17 37	02 36	03 44	04 53	06 01
S 60	15 55	16 43	17 34	02 34	03 45	04 55	06 06

	SUN			MOON			
Day	Eqn. of Time 00h	Eqn. of Time 12h	Mer. Pass.	Mer. Pass. Upper	Mer. Pass. Lower	Age	Phase
d	m s	m s	h m	h m	h m	d	%
7	03 25	03 27	11 57	21 25	09 04	11	85
8	03 29	03 30	11 56	22 08	09 46	12	91
9	03 32	03 33	11 56	22 50	10 29	13	96

1998 MAY 10, 11, 12 (SUN., MON., TUES.)

UT	ARIES	VENUS −4.1		MARS +1.3		JUPITER −2.2		SATURN +0.5		STARS		
d h	GHA	GHA	Dec	GHA	Dec	GHA	Dec	GHA	Dec	Name	SHA	Dec
10 00	227 35.5	220 31.0 N 1 17.4		180 12.4 N17 42.6		235 15.2 S 4 26.6		202 02.6 N 8 10.1		Acamar	315 27.6	S40 18.9
01	242 38.0	235 30.8	18.5	195 13.0	43.1	250 17.2	26.4	217 04.7	10.2	Achernar	335 36.0	S57 14.7
02	257 40.4	250 30.6	19.5	210 13.7	43.6	265 19.3	26.2	232 06.9	10.3	Acrux	173 21.7	S63 05.6
03	272 42.9	265 30.4 ..	20.5	225 14.3 ..	44.1	280 21.3 ..	26.1	247 09.1 ..	10.4	Adhara	255 21.9	S28 58.5
04	287 45.4	280 30.3	21.6	240 15.0	44.6	295 23.4	25.9	262 11.3	10.5	Aldebaran	291 03.1	N16 30.2
05	302 47.8	295 30.1	22.6	255 15.6	45.1	310 25.4	25.7	277 13.5	10.6			
06	317 50.3	310 29.9 N 1 23.6		270 16.3 N17 45.6		325 27.5 S 4 25.6		292 15.6 N 8 10.7		Alioth	166 30.5	N55 58.3
07	332 52.7	325 29.7	24.7	285 16.9	46.1	340 29.5	25.4	307 17.8	10.8	Alkaid	153 07.7	N49 19.5
08	347 55.2	340 29.5	25.7	300 17.5	46.7	355 31.6	25.2	322 20.0	10.9	Al Na'ir	27 58.4	S46 57.9
S 09	2 57.7	355 29.3 ..	26.7	315 18.2 ..	47.2	10 33.6 ..	25.0	337 22.2 ..	11.0	Alnilam	275 58.5	S 1 12.4
U 10	18 00.1	10 29.2	27.8	330 18.8	47.7	25 35.7	24.9	352 24.3	11.1	Alphard	218 07.5	S 8 39.3
N 11	33 02.6	25 29.0	28.8	345 19.5	48.2	40 37.7	24.7	7 26.5	11.2			
D 12	48 05.1	40 28.8 N 1 29.8		0 20.1 N17 48.7		55 39.8 S 4 24.5		22 28.7 N 8 11.3		Alphecca	126 20.5	N26 43.3
A 13	63 07.5	55 28.6	30.9	15 20.7	49.2	70 41.8	24.4	37 30.9	11.4	Alpheratz	357 55.7	N29 04.6
Y 14	78 10.0	70 28.4	31.9	30 21.4	49.7	85 43.9	24.2	52 33.1	11.5	Altair	62 19.4	N 8 51.8
15	93 12.5	85 28.2 ..	32.9	45 22.0 ..	50.2	100 45.9 ..	24.0	67 35.2 ..	11.6	Ankaa	353 27.5	S42 18.9
16	108 14.9	100 28.1	34.0	60 22.7	50.7	115 48.0	23.9	82 37.4	11.7	Antares	112 40.2	S26 25.6
17	123 17.4	115 27.9	35.0	75 23.3	51.2	130 50.0	23.7	97 39.6	11.8			
18	138 19.9	130 27.7 N 1 36.0		90 23.9 N17 51.8		145 52.1 S 4 23.5		112 41.8 N 8 11.9		Arcturus	146 06.0	N19 11.5
19	153 22.3	145 27.5	37.1	105 24.6	52.3	160 54.1	23.4	127 44.0	12.0	Atria	107 51.9	S69 01.3
20	168 24.8	160 27.3	38.1	120 25.2	52.8	175 56.2	23.2	142 46.1	12.1	Avior	234 22.9	S59 30.7
21	183 27.2	175 27.1 ..	39.2	135 25.9 ..	53.3	190 58.2 ..	23.0	157 48.3 ..	12.3	Bellatrix	278 44.8	N 6 20.7
22	198 29.7	190 26.9	40.2	150 26.5	53.8	206 00.3	22.9	172 50.5	12.4	Betelgeuse	271 14.2	N 7 24.2
23	213 32.2	205 26.8	41.2	165 27.1	54.3	221 02.3	22.7	187 52.7	12.5			
11 00	228 34.6	220 26.6 N 1 42.3		180 27.8 N17 54.8		236 04.4 S 4 22.5		202 54.9 N 8 12.6		Canopus	264 01.7	S52 42.1
01	243 37.1	235 26.4	43.3	195 28.4	55.3	251 06.4	22.4	217 57.0	12.7	Capella	280 52.1	N45 59.7
02	258 39.6	250 26.2	44.3	210 29.1	55.8	266 08.5	22.2	232 59.2	12.8	Deneb	49 39.3	N45 16.3
03	273 42.0	265 26.0 ..	45.4	225 29.7 ..	56.3	281 10.5 ..	22.0	248 01.4 ..	12.9	Denebola	182 45.3	N14 34.9
04	288 44.5	280 25.8	46.4	240 30.3	56.8	296 12.6	21.9	263 03.6	13.0	Diphda	349 07.8	S17 59.8
05	303 47.0	295 25.6	47.4	255 31.0	57.3	311 14.6	21.7	278 05.8	13.1			
06	318 49.4	310 25.4 N 1 48.5		270 31.6 N17 57.8		326 16.7 S 4 21.5		293 08.0 N 8 13.2		Dubhe	194 05.7	N61 45.8
07	333 51.9	325 25.3	49.5	285 32.2	58.3	341 18.7	21.4	308 10.1	13.3	Elnath	278 27.7	N28 36.2
08	348 54.3	340 25.1	50.6	300 32.9	58.8	356 20.8	21.2	323 12.3	13.4	Eltanin	90 51.1	N51 29.3
M 09	3 56.8	355 24.9 ..	51.6	315 33.5 ..	59.3	11 22.9 ..	21.0	338 14.5 ..	13.5	Enif	33 58.5	N 9 51.9
O 10	18 59.3	10 24.7	52.6	330 34.2 17 59.8		26 24.9	20.9	353 16.7	13.6	Fomalhaut	15 36.9	S29 37.8
N 11	34 01.7	25 24.5	53.7	345 34.8 18 00.3		41 27.0	20.7	8 18.9	13.7			
D 12	49 04.2	40 24.3 N 1 54.7		0 35.4 N18 00.8		56 29.0 S 4 20.5		23 21.0 N 8 13.8		Gacrux	172 13.4	S57 06.4
A 13	64 06.7	55 24.1	55.7	15 36.1	01.3	71 31.1	20.4	38 23.2	13.9	Gienah	176 04.0	S17 32.1
Y 14	79 09.1	70 23.9	56.8	30 36.7	01.8	86 33.1	20.2	53 25.4	14.0	Hadar	149 03.7	S60 21.9
15	94 11.6	85 23.8 ..	57.8	45 37.4 ..	02.4	101 35.2 ..	20.0	68 27.6 ..	14.1	Hamal	328 14.2	N23 27.0
16	109 14.1	100 23.6	58.8	60 38.0	02.9	116 37.2	19.9	83 29.8	14.2	Kaus Aust.	83 58.9	S34 22.9
17	124 16.5	115 23.4 1 59.9		75 38.6	03.4	131 39.3	19.7	98 31.9	14.3			
18	139 19.0	130 23.2 N 2 00.9		90 39.3 N18 03.9		146 41.3 S 4 19.5		113 34.1 N 8 14.4		Kochab	137 18.6	N74 09.9
19	154 21.5	145 23.0	02.0	105 39.9	04.4	161 43.4	19.4	128 36.3	14.5	Markab	13 50.0	N15 11.6
20	169 23.9	160 22.8	03.0	120 40.5	04.9	176 45.5	19.2	143 38.5	14.6	Menkar	314 27.5	N 4 04.8
21	184 26.4	175 22.6 ..	04.0	135 41.2 ..	05.4	191 47.5 ..	19.0	158 40.7 ..	14.7	Menkent	148 20.9	S36 21.7
22	199 28.8	190 22.4	05.1	150 41.8	05.9	206 49.6	18.9	173 42.8	14.9	Miaplacidus	221 42.1	S69 43.0
23	214 31.3	205 22.2	06.1	165 42.5	06.4	221 51.6	18.7	188 45.0	15.0			
12 00	229 33.8	220 22.1 N 2 07.2		180 43.1 N18 06.9		236 53.7 S 4 18.6		203 47.2 N 8 15.1		Mirfak	308 57.5	N49 51.1
01	244 36.2	235 21.9	08.2	195 43.7	07.4	251 55.7	18.4	218 49.4	15.2	Nunki	76 12.5	S26 17.8
02	259 38.7	250 21.7	09.2	210 44.4	07.9	266 57.8	18.2	233 51.6	15.3	Peacock	53 37.4	S56 44.1
03	274 41.2	265 21.5 ..	10.3	225 45.0 ..	08.4	281 59.9 ..	18.1	248 53.8 ..	15.4	Pollux	243 42.1	N28 01.8
04	289 43.6	280 21.3	11.3	240 45.6	08.8	297 01.9	17.9	263 55.9	15.5	Procyon	245 12.1	N 5 13.6
05	304 46.1	295 21.1	12.3	255 46.3	09.3	312 04.0	17.7	278 58.1	15.6			
06	319 48.6	310 20.9 N 2 13.4		270 46.9 N18 09.8		327 06.0 S 4 17.6		294 00.3 N 8 15.7		Rasalhague	96 17.0	N12 33.7
07	334 51.0	325 20.7	14.4	285 47.5	10.3	342 08.1	17.4	309 02.5	15.8	Regulus	207 55.8	N11 58.5
08	349 53.5	340 20.5	15.5	300 48.2	10.8	357 10.1	17.2	324 04.7	15.9	Rigel	281 23.5	S 8 12.5
T 09	4 56.0	355 20.3 ..	16.5	315 48.8 ..	11.3	12 12.2 ..	17.1	339 06.8 ..	16.0	Rigil Kent.	140 07.0	S60 49.7
U 10	19 58.4	10 20.1	17.5	330 49.4	11.8	27 14.3	16.9	354 09.0	16.1	Sabik	102 25.6	S15 43.3
E 11	35 00.9	25 19.9	18.6	345 50.1	12.3	42 16.3	16.7	9 11.2	16.2			
S 12	50 03.3	40 19.8 N 2 19.6		0 50.7 N18 12.8		57 18.4 S 4 16.6		24 13.4 N 8 16.3		Schedar	349 54.2	N56 31.4
D 13	65 05.8	55 19.6	20.7	15 51.3	13.3	72 20.4	16.4	39 15.6	16.4	Shaula	96 37.3	S37 06.0
A 14	80 08.3	70 19.4	21.7	30 52.0	13.8	87 22.5	16.2	54 17.8	16.5	Sirius	258 44.2	S16 43.1
Y 15	95 10.7	85 19.2 ..	22.7	45 52.6 ..	14.3	102 24.6 ..	16.1	69 19.9 ..	16.6	Spica	158 43.2	S11 09.2
16	110 13.2	100 19.0	23.8	60 53.3	14.9	117 26.6	15.9	84 22.1	16.7	Suhail	223 01.0	S43 25.9
17	125 15.7	115 18.8	24.8	75 53.9	15.4	132 28.7	15.8	99 24.3	16.8			
18	140 18.1	130 18.6 N 2 25.9		90 54.6 N18 15.9		147 30.7 S 4 15.6		114 26.5 N 8 16.9		Vega	80 46.5	N38 46.8
19	155 20.6	145 18.4	26.9	105 55.2	16.4	162 32.8	15.4	129 28.7	17.0	Zuben'ubi	137 18.0	S16 02.1
20	170 23.1	160 18.2	27.9	120 55.9	16.8	177 34.8	15.3	144 30.8	17.1		SHA	Mer. Pass.
21	185 25.5	175 18.0 ..	29.0	135 56.5 ..	17.3	192 36.9 ..	15.1	159 33.0 ..	17.2		° ′	h m
22	200 28.0	190 17.8	30.0	150 57.2	17.8	207 39.0	14.9	174 35.2	17.3	Venus	351 51.9	9 18
23	215 30.4	205 17.6	31.1	165 57.8	18.3	222 41.0	14.8	189 37.4	17.4	Mars	311 53.1	11 58
	h m									Jupiter	7 29.7	8 15
Mer. Pass. 8 44.3		v −0.2	d 1.0	v 0.6	d 0.5	v 2.1	d 0.2	v 2.2	d 0.1	Saturn	334 20.2	10 27

UT		SUN GHA	Dec	MOON GHA	v	Dec	d	HP
d	h	° ′	° ′	° ′	′	° ′	′	′
10	00	180 53.7	N17 30.6	16 54.9	14.7	S 8 09.7	8.8	54.2
	01	195 53.8	31.2	31 28.6	14.7	8 18.5	8.7	54.2
	02	210 53.8	31.9	46 02.3	14.6	8 27.2	8.6	54.2
	03	225 53.8 ..	32.5	60 35.9	14.6	8 35.8	8.6	54.2
	04	240 53.8	33.2	75 09.5	14.5	8 44.4	8.6	54.2
	05	255 53.9	33.9	89 43.0	14.5	8 53.0	8.6	54.2
	06	270 53.9	N17 34.5	104 16.5	14.5	S 9 01.6	8.5	54.2
	07	285 53.9	35.2	118 50.0	14.5	9 10.1	8.5	54.2
S	08	300 53.9	35.8	133 23.5	14.4	9 18.6	8.5	54.2
U	09	315 54.0 ..	36.5	147 56.9	14.4	9 27.1	8.4	54.2
N	10	330 54.0	37.1	162 30.3	14.3	9 35.5	8.4	54.2
D	11	345 54.0	37.8	177 03.6	14.3	9 43.9	8.3	54.3
A	12	0 54.0	N17 38.4	191 36.9	14.3	S 9 52.2	8.3	54.3
Y	13	15 54.1	39.1	206 10.2	14.2	10 00.5	8.3	54.3
	14	30 54.1	39.7	220 43.4	14.3	10 08.8	8.2	54.3
	15	45 54.1 ..	40.4	235 16.7	14.1	10 17.0	8.2	54.3
	16	60 54.1	41.0	249 49.8	14.2	10 25.2	8.2	54.3
	17	75 54.1	41.7	264 23.0	14.1	10 33.4	8.1	54.3
	18	90 54.2	N17 42.3	278 56.1	14.0	S10 41.5	8.0	54.3
	19	105 54.2	43.0	293 29.1	14.0	10 49.5	8.1	54.3
	20	120 54.2	43.6	308 02.1	14.0	10 57.6	7.9	54.3
	21	135 54.2 ..	44.3	322 35.1	14.0	11 05.5	8.0	54.4
	22	150 54.3	44.9	337 08.1	13.9	11 13.5	7.9	54.4
	23	165 54.3	45.6	351 41.0	13.9	11 21.4	7.8	54.4
11	00	180 54.3	N17 46.2	6 13.9	13.8	S11 29.2	7.8	54.4
	01	195 54.3	46.9	20 46.7	13.8	11 37.0	7.7	54.4
	02	210 54.3	47.5	35 19.5	13.7	11 44.7	7.7	54.4
	03	225 54.4 ..	48.2	49 52.2	13.7	11 52.4	7.7	54.4
	04	240 54.4	48.8	64 24.9	13.7	12 00.1	7.6	54.4
	05	255 54.4	49.5	78 57.6	13.6	12 07.7	7.6	54.4
	06	270 54.4	N17 50.1	93 30.2	13.6	S12 15.3	7.5	54.4
	07	285 54.4	50.8	108 02.8	13.6	12 22.8	7.4	54.5
M	08	300 54.5	51.4	122 35.4	13.5	12 30.2	7.4	54.5
O	09	315 54.5 ..	52.0	137 07.9	13.4	12 37.6	7.4	54.5
N	10	330 54.5	52.7	151 40.3	13.5	12 45.0	7.3	54.5
D	11	345 54.5	53.3	166 12.8	13.3	12 52.3	7.2	54.5
A	12	0 54.5	N17 54.0	180 45.1	13.4	S12 59.5	7.2	54.5
Y	13	15 54.6	54.6	195 17.5	13.3	13 06.7	7.1	54.5
	14	30 54.6	55.2	209 49.8	13.2	13 13.8	7.1	54.5
	15	45 54.6 ..	55.9	224 22.0	13.2	13 20.9	7.0	54.6
	16	60 54.6	56.5	238 54.2	13.2	13 27.9	7.0	54.6
	17	75 54.6	57.2	253 26.4	13.1	13 34.9	6.9	54.6
	18	90 54.6	N17 57.8	267 58.5	13.1	S13 41.8	6.9	54.6
	19	105 54.6	58.4	282 30.6	13.0	13 48.7	6.8	54.6
	20	120 54.7	59.1	297 02.6	13.0	13 55.5	6.7	54.6
	21	135 54.7	17 59.7	311 34.6	13.0	14 02.2	6.7	54.6
	22	150 54.7	18 00.3	326 06.6	12.9	14 08.9	6.6	54.6
	23	165 54.7	01.0	340 38.5	12.8	14 15.5	6.5	54.7
12	00	180 54.7	N18 01.6	355 10.3	12.8	S14 22.0	6.5	54.7
	01	195 54.7	02.2	9 42.1	12.8	14 28.5	6.4	54.7
	02	210 54.8	02.9	24 13.9	12.7	14 34.9	6.4	54.7
	03	225 54.8 ..	03.5	38 45.6	12.7	14 41.3	6.3	54.7
	04	240 54.8	04.1	53 17.3	12.6	14 47.6	6.2	54.7
	05	255 54.8	04.8	67 48.9	12.6	14 53.8	6.2	54.7
	06	270 54.8	N18 05.4	82 20.5	12.6	S15 00.0	6.1	54.8
	07	285 54.8	06.0	96 52.1	12.5	15 06.1	6.1	54.8
T	08	300 54.8	06.7	111 23.6	12.4	15 12.2	5.9	54.8
U	09	315 54.9 ..	07.3	125 55.0	12.4	15 18.1	5.9	54.8
E	10	330 54.9	07.9	140 26.4	12.4	15 24.0	5.9	54.8
S	11	345 54.9	08.6	154 57.8	12.3	15 29.9	5.7	54.8
D	12	0 54.9	N18 09.2	169 29.1	12.3	S15 35.6	5.7	54.8
A	13	15 54.9	09.8	184 00.4	12.2	15 41.3	5.6	54.9
Y	14	30 54.9	10.4	198 31.6	12.2	15 46.9	5.6	54.9
	15	45 54.9 ..	11.1	213 02.8	12.2	15 52.5	5.5	54.9
	16	60 54.9	11.7	227 34.0	12.1	15 58.0	5.4	54.9
	17	75 54.9	12.3	242 05.1	12.0	16 03.4	5.3	54.9
	18	90 55.0	N18 12.9	256 36.1	12.0	S16 08.7	5.3	54.9
	19	105 55.0	13.6	271 07.1	12.0	16 14.0	5.2	55.0
	20	120 55.0	14.2	285 38.1	11.9	16 19.2	5.1	55.0
	21	135 55.0 ..	14.8	300 09.0	11.9	16 24.3	5.0	55.0
	22	150 55.0	15.4	314 39.9	11.8	16 29.3	5.0	55.0
	23	165 55.0	16.1	329 10.7	11.8	S16 34.3	4.9	55.0
		SD 15.9	d 0.6	SD 14.8		14.9		14.9

Lat.	Twilight Naut.	Twilight Civil	Sunrise	Moonrise 10	11	12	13
°	h m	h m	h m	h m	h m	h m	h m
N 72	▭	▭	▭	19 55	21 40	23 35	■
N 70	////	////	01 25	19 36	21 09	22 44	24 15
68	////	////	02 10	19 21	20 47	22 12	23 31
66	////	00 29	02 39	19 09	20 30	21 49	23 02
64	////	01 35	03 01	18 59	20 16	21 31	22 40
62	////	02 08	03 18	18 51	20 04	21 16	22 23
60	00 25	02 33	03 33	18 43	19 54	21 03	22 08
N 58	01 25	02 52	03 45	18 37	19 45	20 52	21 56
56	01 56	03 07	03 55	18 31	19 38	20 43	21 45
54	02 19	03 21	04 05	18 26	19 31	20 35	21 36
52	02 37	03 32	04 13	18 22	19 25	20 27	21 27
50	02 51	03 42	04 20	18 18	19 20	20 21	21 20
45	03 20	04 03	04 36	18 09	19 08	20 06	21 04
N 40	03 42	04 19	04 49	18 01	18 58	19 55	20 51
35	03 59	04 32	05 00	17 55	18 50	19 45	20 39
30	04 13	04 44	05 09	17 49	18 42	19 36	20 30
20	04 35	05 02	05 26	17 40	18 30	19 21	20 13
N 10	04 52	05 18	05 40	17 31	18 19	19 08	19 58
0	05 06	05 31	05 53	17 24	18 09	18 56	19 45
S 10	05 18	05 44	06 06	17 16	17 59	18 44	19 31
20	05 30	05 56	06 20	17 08	17 48	18 31	19 17
30	05 41	06 10	06 35	16 58	17 36	18 16	19 00
35	05 47	06 17	06 44	16 53	17 28	18 08	18 51
40	05 53	06 25	06 54	16 47	17 20	17 58	18 40
45	05 59	06 35	07 06	16 40	17 11	17 47	18 27
S 50	06 06	06 46	07 21	16 31	17 00	17 33	18 11
52	06 09	06 50	07 28	16 27	16 55	17 27	18 04
54	06 13	06 56	07 35	16 23	16 49	17 20	17 56
56	06 16	07 02	07 43	16 18	16 43	17 12	17 47
58	06 20	07 08	07 53	16 13	16 36	17 03	17 37
S 60	06 24	07 15	08 04	16 07	16 28	16 53	17 25

Lat.	Sunset	Twilight Civil	Twilight Naut.	Moonset 10	11	12	13
°	h m	h m	h m	h m	h m	h m	h m
N 72	▭	▭	▭	03 05	02 58	02 48	02 31
N 70	22 35	////	////	03 18	03 18	03 19	03 23
68	21 47	////	////	03 29	03 34	03 42	03 55
66	21 17	////	////	03 37	03 47	04 00	04 19
64	20 54	22 22	////	03 45	03 58	04 15	04 38
62	20 37	21 47	////	03 51	04 07	04 27	04 53
60	20 22	21 22	////	03 57	04 15	04 38	05 06
N 58	20 09	21 03	22 32	04 02	04 22	04 47	05 17
56	19 59	20 47	22 00	04 06	04 29	04 55	05 27
54	19 49	20 34	21 36	04 10	04 34	05 02	05 35
52	19 41	20 22	21 18	04 14	04 39	05 09	05 43
50	19 33	20 12	21 03	04 17	04 44	05 14	05 50
45	19 17	19 51	20 34	04 24	04 54	05 27	06 05
N 40	19 04	19 35	20 12	04 30	05 02	05 38	06 17
35	18 53	19 21	19 55	04 35	05 09	05 47	06 27
30	18 44	19 10	19 41	04 40	05 16	05 54	06 37
20	18 27	18 51	19 18	04 47	05 27	06 08	06 52
N 10	18 13	18 35	19 01	04 54	05 36	06 20	07 06
0	18 00	18 22	18 47	05 01	05 45	06 31	07 19
S 10	17 47	18 09	18 34	05 07	05 54	06 42	07 32
20	17 33	17 56	18 23	05 14	06 04	06 54	07 46
30	17 17	17 43	18 11	05 22	06 15	07 08	08 02
35	17 08	17 35	18 05	05 27	06 21	07 16	08 11
40	16 58	17 27	17 59	05 32	06 29	07 25	08 22
45	16 46	17 18	17 53	05 38	06 37	07 36	08 34
S 50	16 31	17 07	17 46	05 45	06 47	07 49	08 49
52	16 25	17 02	17 43	05 49	06 52	07 55	08 56
54	16 17	16 56	17 39	05 52	06 57	08 02	09 04
56	16 09	16 50	17 36	05 57	07 03	08 09	09 13
58	15 59	16 44	17 32	06 01	07 10	08 17	09 23
S 60	15 48	16 37	17 28	06 06	07 17	08 27	09 34

Day	SUN Eqn. of Time 00h	12h	Mer. Pass.	MOON Mer. Pass. Upper	Lower	Age	Phase
d	m s	m s	h m	h m	h m	d	%
10	03 35	03 36	11 56	23 34	11 12	14	99
11	03 37	03 38	11 56	24 20	11 57	15	100
12	03 39	03 40	11 56	00 20	12 43	16	99

1998 MAY 13, 14, 15 (WED., THURS., FRI.)

UT	ARIES GHA	VENUS −4.1 GHA	Dec	MARS +1.3 GHA	Dec	JUPITER −2.2 GHA	Dec	SATURN +0.6 GHA	Dec
d h	° ′	° ′	° ′	° ′	° ′	° ′	° ′	° ′	° ′
13 00	230 32.9	220 17.4 N 2	32.1	180 58.5 N18	18.8	237 43.1 S 4	14.6	204 39.6 N 8	17.5
01	245 35.4	235 17.2	33.1	195 59.1	19.2	252 45.2	14.4	219 41.8	17.6
02	260 37.8	250 17.0	34.2	210 59.7	19.7	267 47.2	14.3	234 43.9	17.7
03	275 40.3	265 16.9 ..	35.2	226 00.3 ..	20.2	282 49.3 ..	14.1	249 46.1 ..	17.8
04	290 42.8	280 16.7	36.3	241 01.0	20.7	297 51.3	14.0	264 48.3	18.0
05	305 45.2	295 16.5	37.3	256 01.6	21.2	312 53.4	13.8	279 50.5	18.1
W 06	320 47.7	310 16.3 N 2	38.4	271 02.2 N18	21.7	327 55.5 S 4	13.6	294 52.7 N 8	18.2
E 07	335 50.2	325 16.1	39.4	286 02.9	22.2	342 57.5	13.5	309 54.9	18.3
D 08	350 52.6	340 15.9	40.4	301 03.5	22.6	357 59.6	13.3	324 57.0	18.4
N 09	5 55.1	355 15.7 ..	41.5	316 04.1 ..	23.1	13 01.6 ..	13.1	339 59.2 ..	18.5
E 10	20 57.6	10 15.5	42.5	331 04.8	23.6	28 03.7	13.0	355 01.4	18.6
S 11	36 00.0	25 15.3	43.6	346 05.4	24.1	43 05.8	12.8	10 03.6	18.7
D 12	51 02.5	40 15.1 N 2	44.6	1 06.0 N18	24.6	58 07.8 S 4	12.7	25 05.8 N 8	18.8
A 13	66 04.9	55 14.9	45.6	16 06.7	25.1	73 09.9	12.5	40 08.0	18.9
Y 14	81 07.4	70 14.7	46.7	31 07.3	25.6	88 12.0	12.3	55 10.1	19.0
15	96 09.9	85 14.5 ..	47.7	46 07.9 ..	26.0	103 14.0 ..	12.2	70 12.3 ..	19.1
16	111 12.3	100 14.3	48.8	61 08.6	26.5	118 16.1	12.0	85 14.5	19.2
17	126 14.8	115 14.1	49.8	76 09.2	27.0	133 18.1	11.8	100 16.7	19.3
18	141 17.3	130 13.9 N 2	50.9	91 09.8 N18	27.5	148 20.2 S 4	11.7	115 18.9 N 8	19.4
19	156 19.7	145 13.7	51.9	106 10.5	28.0	163 22.3	11.5	130 21.0	19.5
20	171 22.2	160 13.5	52.9	121 11.1	28.5	178 24.3	11.4	145 23.2	19.6
21	186 24.7	175 13.3 ..	54.0	136 11.7 ..	29.0	193 26.4 ..	11.2	160 25.4 ..	19.7
22	201 27.1	190 13.1	55.0	151 12.4	29.4	208 28.5	11.0	175 27.6	19.8
23	216 29.6	205 12.9	56.1	166 13.0	29.9	223 30.5	10.9	190 29.8	19.9
14 00	231 32.1	220 12.7 N 2	57.1	181 13.6 N18	30.4	238 32.6 S 4	10.7	205 32.0 N 8	20.0
01	246 34.5	235 12.5	58.1	196 14.3	30.9	253 34.7	10.6	220 34.2	20.1
02	261 37.0	250 12.3 2	59.2	211 14.9	31.4	268 36.7	10.4	235 36.3	20.2
03	276 39.4	265 12.1 3	00.2	226 15.5 ..	31.9	283 38.8 ..	10.2	250 38.5 ..	20.3
04	291 41.9	280 11.9	01.3	241 16.2	32.3	298 40.9	10.1	265 40.7	20.4
05	306 44.4	295 11.7	02.3	256 16.8	32.8	313 42.9	09.9	280 42.9	20.5
T 06	321 46.8	310 11.5 N 3	03.4	271 17.4 N18	33.3	328 45.0 S 4	09.7	295 45.1 N 8	20.6
H 07	336 49.3	325 11.3	04.4	286 18.1	33.8	343 47.1	09.6	310 47.3	20.7
U 08	351 51.8	340 11.1	05.4	301 18.7	34.3	358 49.1	09.4	325 49.4	20.8
R 09	6 54.2	355 10.9 ..	06.5	316 19.3 ..	34.7	13 51.2 ..	09.3	340 51.6 ..	20.9
S 10	21 56.7	10 10.7	07.5	331 20.0	35.2	28 53.3	09.1	355 53.8	21.0
D 11	36 59.2	25 10.5	08.6	346 20.6	35.7	43 55.3	08.9	10 56.0	21.1
A 12	52 01.6	40 10.3 N 3	09.6	1 21.2 N18	36.2	58 57.4 S 4	08.8	25 58.2 N 8	21.2
Y 13	67 04.1	55 10.1	10.7	16 21.9	36.7	73 59.5	08.6	41 00.4	21.3
14	82 06.6	70 09.9	11.7	31 22.5	37.1	89 01.5	08.5	56 02.5	21.4
15	97 09.0	85 09.7 ..	12.7	46 23.1 ..	37.6	104 03.6 ..	08.3	71 04.7 ..	21.5
16	112 11.5	100 09.5	13.8	61 23.8	38.1	119 05.7	08.1	86 06.9	21.6
17	127 13.9	115 09.3	14.8	76 24.4	38.6	134 07.7	08.0	101 09.1	21.7
18	142 16.4	130 09.1 N 3	15.9	91 25.0 N18	39.1	149 09.8 S 4	07.8	116 11.3 N 8	21.8
19	157 18.9	145 08.9	16.9	106 25.6	39.5	164 11.9	07.7	131 13.5	21.9
20	172 21.3	160 08.7	18.0	121 26.3	40.0	179 13.9	07.5	146 15.6	22.0
21	187 23.8	175 08.5 ..	19.0	136 26.9 ..	40.5	194 16.0 ..	07.3	161 17.8 ..	22.1
22	202 26.3	190 08.3	20.0	151 27.5	41.0	209 18.1	07.2	176 20.0	22.2
23	217 28.7	205 08.1	21.1	166 28.2	41.4	224 20.1	07.0	191 22.2	22.3
15 00	232 31.2	220 07.9 N 3	22.1	181 28.8 N18	41.9	239 22.2 S 4	06.9	206 24.4 N 8	22.4
01	247 33.7	235 07.7	23.2	196 29.4	42.4	254 24.3	06.7	221 26.6	22.6
02	262 36.1	250 07.5	24.2	211 30.1	42.9	269 26.3	06.5	236 28.8	22.7
03	277 38.6	265 07.3 ..	25.3	226 30.7 ..	43.3	284 28.4 ..	06.4	251 30.9 ..	22.8
04	292 41.0	280 07.1	26.3	241 31.3	43.8	299 30.5	06.2	266 33.1	22.9
05	307 43.5	295 06.9	27.3	256 32.0	44.3	314 32.5	06.1	281 35.3	23.0
06	322 46.0	310 06.7 N 3	28.4	271 32.6 N18	44.8	329 34.6 S 4	05.9	296 37.5 N 8	23.1
07	337 48.4	325 06.5	29.4	286 33.2	45.2	344 36.7	05.7	311 39.7	23.2
08	352 50.9	340 06.3	30.5	301 33.9	45.7	359 38.8	05.6	326 41.9	23.3
F 09	7 53.4	355 06.1 ..	31.5	316 34.5 ..	46.2	14 40.8 ..	05.4	341 44.1 ..	23.4
R 10	22 55.8	10 05.8	32.6	331 35.1	46.7	29 42.9	05.3	356 46.2	23.5
I 11	37 58.3	25 05.6	33.6	346 35.7	47.1	44 45.0	05.1	11 48.4	23.6
D 12	53 00.8	40 05.4 N 3	34.7	1 36.4 N18	47.6	59 47.0 S 4	04.9	26 50.6 N 8	23.7
A 13	68 03.2	55 05.2	35.7	16 37.0	48.1	74 49.1	04.8	41 52.8	23.8
Y 14	83 05.7	70 05.0	36.7	31 37.6	48.5	89 51.2	04.6	56 55.0	23.9
15	98 08.2	85 04.8 ..	37.8	46 38.3 ..	49.0	104 53.3 ..	04.5	71 57.2 ..	24.0
16	113 10.6	100 04.6	38.8	61 38.9	49.5	119 55.3	04.3	86 59.3	24.1
17	128 13.1	115 04.4	39.9	76 39.5	50.0	134 57.4	04.2	102 01.5	24.2
18	143 15.5	130 04.2 N 3	40.9	91 40.2 N18	50.4	149 59.5 S 4	04.0	117 03.7 N 8	24.3
19	158 18.0	145 04.0	42.0	106 40.8	50.9	165 01.5	03.8	132 05.9	24.4
20	173 20.5	160 03.8	43.0	121 41.4	51.4	180 03.6	03.7	147 08.1	24.5
21	188 22.9	175 03.6 ..	44.1	136 42.0 ..	51.8	195 05.7 ..	03.5	162 10.3 ..	24.6
22	203 25.4	190 03.4	45.1	151 42.7	52.3	210 07.8	03.4	177 12.5	24.7
23	218 27.9	205 03.2	46.1	166 43.3	52.8	225 09.8	03.2	192 14.6	24.8
Mer. Pass.	8 32.5	v −0.2 d 1.0		v 0.6 d 0.5		v 2.1 d 0.2		v 2.2 d 0.1	

STARS

Name	SHA ° ′	Dec ° ′
Acamar	315 27.6	S40 18.8
Achernar	335 36.0	S57 14.7
Acrux	173 21.7	S63 05.6
Adhara	255 21.9	S28 58.5
Aldebaran	291 03.1	N16 30.2
Alioth	166 30.5	N55 58.3
Alkaid	153 07.7	N49 19.5
Al Na'ir	27 58.4	S46 57.9
Alnilam	275 58.5	S 1 12.4
Alphard	218 07.5	S 8 39.3
Alphecca	126 20.5	N26 43.3
Alpheratz	357 55.7	N29 04.6
Altair	62 19.4	N 8 51.8
Ankaa	353 27.5	S42 18.9
Antares	112 40.2	S26 25.6
Arcturus	146 06.0	N19 11.5
Atria	107 51.9	S69 01.3
Avior	234 22.9	S59 30.7
Bellatrix	278 44.8	N 6 20.7
Betelgeuse	271 14.2	N 7 24.2
Canopus	264 01.7	S52 42.0
Capella	280 52.1	N45 59.7
Deneb	49 39.2	N45 16.3
Denebola	182 45.3	N14 34.9
Diphda	349 07.8	S17 59.8
Dubhe	194 05.7	N61 45.8
Elnath	278 27.7	N28 36.2
Eltanin	90 51.1	N51 29.3
Enif	33 58.5	N 9 51.9
Fomalhaut	15 36.9	S29 37.8
Gacrux	172 13.4	S57 06.4
Gienah	176 04.0	S17 32.1
Hadar	149 03.7	S60 21.9
Hamal	328 14.2	N23 27.0
Kaus Aust.	83 58.9	S34 22.9
Kochab	137 18.6	N74 09.9
Markab	13 50.0	N15 11.6
Menkar	314 27.5	N 4 04.8
Menkent	148 20.9	S36 21.7
Miaplacidus	221 42.1	S69 43.0
Mirfak	308 57.5	N49 51.1
Nunki	76 12.5	S26 17.8
Peacock	53 37.3	S56 44.1
Pollux	243 42.1	N28 01.8
Procyon	245 12.1	N 5 13.6
Rasalhague	96 16.9	N12 33.7
Regulus	207 55.8	N11 58.5
Rigel	281 23.5	S 8 12.5
Rigil Kent.	140 07.0	S60 49.7
Sabik	102 25.6	S15 43.3
Schedar	349 54.1	N56 31.4
Shaula	96 37.3	S37 06.0
Sirius	258 44.2	S16 43.1
Spica	158 43.2	S11 09.2
Suhail	223 01.0	S43 25.9
Vega	80 46.5	N38 46.9
Zuben'ubi	137 18.0	S16 02.1

	SHA	Mer. Pass.
	° ′	h m
Venus	348 40.7	9 19
Mars	309 41.6	11 55
Jupiter	7 00.5	8 05
Saturn	333 59.9	10 16

UT	SUN GHA	Dec	MOON GHA	v	Dec	d	HP
13 00	180 55.0	N18 16.7	343 41.5	11.7	S16 39.2	4.8	55.0
01	195 55.0	17.3	358 12.2	11.7	16 44.0	4.7	55.0
02	210 55.0	17.9	12 42.9	11.7	16 48.7	4.7	55.1
03	225 55.0 ..	18.5	27 13.6	11.6	16 53.4	4.5	55.1
04	240 55.0	19.2	41 44.2	11.6	16 57.9	4.5	55.1
05	255 55.1	19.8	56 14.8	11.5	17 02.4	4.4	55.1
W 06	270 55.1	N18 20.4	70 45.3	11.5	S17 06.8	4.4	55.1
E 07	285 55.1	21.0	85 15.8	11.4	17 11.2	4.2	55.1
D 08	300 55.1	21.6	99 46.2	11.4	17 15.4	4.2	55.2
N 09	315 55.1 ..	22.2	114 16.6	11.3	17 19.6	4.1	55.2
E 10	330 55.1	22.9	128 46.9	11.4	17 23.7	4.0	55.2
S 11	345 55.1	23.5	143 17.3	11.2	17 27.7	3.9	55.2
D 12	0 55.1	N18 24.1	157 47.5	11.3	S17 31.6	3.8	55.2
A 13	15 55.1	24.7	172 17.8	11.1	17 35.4	3.8	55.2
Y 14	30 55.1	25.3	186 47.9	11.2	17 39.2	3.6	55.3
15	45 55.1 ..	25.9	201 18.1	11.1	17 42.8	3.6	55.3
16	60 55.1	26.6	215 48.2	11.1	17 46.4	3.5	55.3
17	75 55.1	27.2	230 18.3	11.0	17 49.9	3.4	55.3
18	90 55.1	N18 27.8	244 48.3	11.0	S17 53.3	3.3	55.3
19	105 55.1	28.4	259 18.3	10.9	17 56.6	3.3	55.4
20	120 55.1	29.0	273 48.2	10.9	17 59.9	3.1	55.4
21	135 55.1 ..	29.6	288 18.1	10.9	18 03.0	3.1	55.4
22	150 55.1	30.2	302 48.0	10.8	18 06.1	2.9	55.4
23	165 55.2	30.8	317 17.8	10.8	18 09.0	2.9	55.4
14 00	180 55.2	N18 31.4	331 47.6	10.8	S18 11.9	2.8	55.4
01	195 55.2	32.0	346 17.4	10.7	18 14.7	2.7	55.5
02	210 55.2	32.7	0 47.1	10.7	18 17.4	2.6	55.5
03	225 55.2 ..	33.3	15 16.8	10.6	18 20.0	2.5	55.5
04	240 55.2	33.9	29 46.4	10.6	18 22.5	2.4	55.5
05	255 55.2	34.5	44 16.0	10.6	18 24.9	2.4	55.5
T 06	270 55.2	N18 35.1	58 45.6	10.5	S18 27.3	2.2	55.6
H 07	285 55.2	35.7	73 15.1	10.5	18 29.5	2.2	55.6
U 08	300 55.2	36.3	87 44.6	10.5	18 31.7	2.0	55.6
R 09	315 55.2 ..	36.9	102 14.1	10.5	18 33.7	2.0	55.6
S 10	330 55.2	37.5	116 43.6	10.4	18 35.7	1.8	55.6
D 11	345 55.2	38.1	131 13.0	10.3	18 37.5	1.8	55.7
A 12	0 55.2	N18 38.7	145 42.3	10.4	S18 39.3	1.7	55.7
Y 13	15 55.2	39.3	160 11.7	10.3	18 41.0	1.5	55.7
14	30 55.2	39.9	174 41.0	10.2	18 42.5	1.5	55.7
15	45 55.2 ..	40.5	189 10.2	10.3	18 44.0	1.4	55.7
16	60 55.2	41.1	203 39.5	10.2	18 45.4	1.3	55.8
17	75 55.2	41.7	218 08.7	10.2	18 46.7	1.2	55.8
18	90 55.2	N18 42.3	232 37.9	10.1	S18 47.9	1.1	55.8
19	105 55.2	42.9	247 07.0	10.2	18 49.0	1.0	55.8
20	120 55.2	43.5	261 36.2	10.1	18 50.0	0.9	55.8
21	135 55.2 ..	44.1	276 05.3	10.0	18 50.9	0.8	55.9
22	150 55.2	44.7	290 34.3	10.1	18 51.7	0.7	55.9
23	165 55.2	45.3	305 03.4	10.0	18 52.4	0.6	55.9
15 00	180 55.2	N18 45.9	319 32.4	10.0	S18 53.0	0.5	55.9
01	195 55.1	46.5	334 01.4	9.9	18 53.5	0.4	55.9
02	210 55.1	47.1	348 30.3	10.0	18 53.9	0.3	56.0
03	225 55.1 ..	47.7	2 59.3	9.9	18 54.2	0.2	56.0
04	240 55.1	48.3	17 28.2	9.9	18 54.4	0.1	56.0
05	255 55.1	48.9	31 57.1	9.9	18 54.5	0.0	56.0
F 06	270 55.1	N18 49.5	46 26.0	9.8	S18 54.5	0.1	56.1
R 07	285 55.1	50.0	60 54.8	9.8	18 54.4	0.2	56.1
I 08	300 55.1	50.6	75 23.6	9.8	18 54.2	0.3	56.1
D 09	315 55.1 ..	51.2	89 52.4	9.8	18 53.9	0.4	56.1
A 10	330 55.1	51.8	104 21.2	9.7	18 53.5	0.5	56.1
Y 11	345 55.1	52.4	118 49.9	9.8	18 53.0	0.6	56.2
12	0 55.1	N18 53.0	133 18.7	9.7	S18 52.4	0.7	56.2
13	15 55.1	53.6	147 47.4	9.7	18 51.7	0.8	56.2
14	30 55.1	54.2	162 16.1	9.7	18 50.9	0.9	56.2
15	45 55.1 ..	54.8	176 44.8	9.6	18 50.0	1.0	56.3
16	60 55.1	55.3	191 13.4	9.7	18 49.0	1.1	56.3
17	75 55.1	55.9	205 42.1	9.6	18 47.9	1.2	56.3
18	90 55.1	N18 56.5	220 10.7	9.6	S18 46.7	1.4	56.3
19	105 55.0	57.1	234 39.3	9.6	18 45.3	1.4	56.3
20	120 55.0	57.7	249 07.9	9.6	18 43.9	1.5	56.4
21	135 55.0 ..	58.3	263 36.5	9.5	18 42.4	1.6	56.4
22	150 55.0	58.9	278 05.0	9.6	18 40.8	1.7	56.4
23	165 55.0	59.4	292 33.6	9.5	S18 39.1	1.9	56.4
	SD 15.8 d 0.6		SD 15.0		15.2		15.3

Lat.	Twilight Naut.	Civil	Sunrise	Moonrise 13	14	15	16
N 72	□	□	□	■	■	■	■
N 70	////	////	00 56	24 15	00 15	01 30	02 09
68	////	////	01 54	23 31	24 37	00 37	01 22
66	////	////	02 27	23 02	24 05	00 05	00 52
64	////	01 17	02 51	22 40	23 41	24 29	00 52
62	////	01 56	03 10	22 23	23 22	24 11	00 11
60	////	02 23	03 25	22 08	23 06	23 56	24 36
N 58	01 09	02 44	03 38	21 56	22 53	23 43	24 24
56	01 45	03 00	03 50	21 45	22 42	23 32	24 14
54	02 10	03 14	03 59	21 36	22 32	23 22	24 06
52	02 29	03 26	04 08	21 27	22 23	23 13	23 58
50	02 45	03 37	04 16	21 20	22 15	23 06	23 51
45	03 15	03 58	04 32	21 04	21 58	22 49	23 35
N 40	03 38	04 16	04 46	20 51	21 44	22 35	23 23
35	03 56	04 29	04 57	20 39	21 33	22 24	23 12
30	04 10	04 41	05 07	20 30	21 22	22 14	23 03
20	04 33	05 01	05 24	20 13	21 05	21 56	22 47
N 10	04 51	05 17	05 39	19 58	20 50	21 41	22 33
0	05 06	05 31	05 53	19 45	20 35	21 27	22 19
S 10	05 19	05 44	06 06	19 31	20 21	21 13	22 06
20	05 31	05 57	06 21	19 17	20 06	20 58	21 52
30	05 43	06 12	06 37	19 00	19 48	20 40	21 36
35	05 49	06 19	06 46	18 51	19 38	20 30	21 26
40	05 55	06 28	06 57	18 40	19 27	20 19	21 16
45	06 02	06 38	07 10	18 27	19 13	20 05	21 03
S 50	06 10	06 49	07 25	18 11	18 56	19 49	20 48
52	06 13	06 55	07 32	18 04	18 49	19 41	20 40
54	06 17	07 00	07 40	17 56	18 40	19 32	20 32
56	06 21	07 07	07 49	17 47	18 30	19 22	20 23
58	06 25	07 13	07 59	17 37	18 19	19 11	20 13
S 60	06 29	07 21	08 10	17 25	18 06	18 58	20 01

Lat.	Sunset	Twilight Civil	Naut.	Moonset 13	14	15	16
N 72	□	□	□	02 31	■	■	■
N 70	23 08	////	////	03 23	03 34	04 05	05 14
68	22 03	////	////	03 55	04 18	04 58	06 00
66	21 29	////	////	04 19	04 47	05 30	06 31
64	21 04	22 42	////	04 38	05 10	05 54	06 53
62	20 45	22 00	////	04 53	05 27	06 13	07 11
60	20 29	21 32	////	05 06	05 42	06 28	07 26
N 58	20 16	21 11	22 49	05 17	05 55	06 42	07 39
56	20 04	20 54	22 11	05 27	06 06	06 53	07 50
54	19 54	20 40	21 45	05 35	06 15	07 03	07 59
52	19 46	20 28	21 25	05 43	06 24	07 12	08 08
50	19 38	20 17	21 09	05 50	06 31	07 20	08 15
45	19 21	19 55	20 38	06 05	06 48	07 37	08 32
N 40	19 07	19 38	20 16	06 17	07 01	07 51	08 45
35	18 56	19 24	19 58	06 27	07 13	08 02	08 57
30	18 46	19 12	19 43	06 37	07 23	08 13	09 06
20	18 28	18 52	19 20	06 52	07 40	08 30	09 23
N 10	18 14	18 36	19 02	07 06	07 55	08 46	09 38
0	18 00	18 22	18 47	07 19	08 09	09 00	09 52
S 10	17 46	18 08	18 34	07 32	08 23	09 14	10 06
20	17 32	17 55	18 22	07 46	08 38	09 30	10 21
30	17 15	17 41	18 10	08 02	08 55	09 47	10 37
35	17 06	17 33	18 03	08 11	09 05	09 57	10 47
40	16 55	17 24	17 57	08 22	09 16	10 09	10 58
45	16 42	17 14	17 50	08 34	09 30	10 23	11 11
S 50	16 27	17 03	17 42	08 49	09 46	10 39	11 27
52	16 20	16 57	17 39	08 56	09 54	10 47	11 34
54	16 12	16 52	17 35	09 04	10 03	10 56	11 43
56	16 03	16 45	17 31	09 13	10 12	11 06	11 52
58	15 53	16 38	17 27	09 23	10 23	11 17	12 02
S 60	15 41	16 31	17 22	09 34	10 36	11 30	12 14

Day	SUN Eqn. of Time 00h	12h	Mer. Pass.	MOON Mer. Pass. Upper	Lower	Age	Phase
	m s	m s	h m	h m	h m	d	%
13	03 40	03 40	11 56	01 07	13 32	17	96
14	03 41	03 41	11 56	01 57	14 22	18	92
15	03 41	03 40	11 56	02 48	15 13	19	85

1998 MAY 16, 17, 18 (SAT., SUN., MON.)

UT	ARIES GHA	VENUS −4.0 GHA	Dec	MARS +1.3 GHA	Dec	JUPITER −2.2 GHA	Dec	SATURN +0.6 GHA	Dec
16 00	233 30.3	220 02.9	N 3 47.2	181 43.9	N18 53.2	240 11.9	S 4 03.0	207 16.8	N 8 24.9
01	248 32.8	235 02.7	48.2	196 44.6	53.7	255 14.0	02.9	222 19.0	25.0
02	263 35.3	250 02.5	49.3	211 45.2	54.2	270 16.0	02.7	237 21.2	25.1
03	278 37.7	265 02.3 ..	50.3	226 45.8 ..	54.6	285 18.1 ..	02.6	252 23.4 ..	25.2
04	293 40.2	280 02.1	51.4	241 46.5	55.1	300 20.2	02.4	267 25.6	25.3
05	308 42.7	295 01.9	52.4	256 47.1	55.6	315 22.3	02.3	282 27.8	25.4
06	323 45.1	310 01.7	N 3 53.4	271 47.7	N18 56.0	330 24.3	S 4 02.1	297 30.0	N 8 25.5
07	338 47.6	325 01.5	54.5	286 48.3	56.5	345 26.4	01.9	312 32.1	25.6
S 08	353 50.0	340 01.3	55.5	301 49.0	57.0	0 28.5	01.8	327 34.3	25.7
A 09	8 52.5	355 01.1 ..	56.6	316 49.6 ..	57.4	15 30.6 ..	01.6	342 36.5 ..	25.8
T 10	23 55.0	10 00.9	57.6	331 50.2	57.9	30 32.6	01.5	357 38.7	25.9
U 11	38 57.4	25 00.6	58.7	346 50.9	58.4	45 34.7	01.3	12 40.9	26.0
R 12	53 59.9	40 00.4	N 3 59.7	1 51.5	N18 58.8	60 36.8	S 4 01.2	27 43.1	N 8 26.1
D 13	69 02.4	55 00.2	4 00.8	16 52.1	59.3	75 38.9	01.0	42 45.3	26.2
A 14	84 04.8	70 00.0	01.8	31 52.7	18 59.8	90 40.9	00.8	57 47.4	26.3
Y 15	99 07.3	84 59.8 ..	02.8	46 53.4	19 00.2	105 43.0 ..	00.7	72 49.6 ..	26.4
16	114 09.8	99 59.6	03.9	61 54.0	00.7	120 45.1	00.5	87 51.8	26.5
17	129 12.2	114 59.4	04.9	76 54.6	01.2	135 47.2	00.4	102 54.0	26.6
18	144 14.7	129 59.2	N 4 06.0	91 55.3	N19 01.6	150 49.2	S 4 00.2	117 56.2	N 8 26.7
19	159 17.2	144 58.9	07.0	106 55.9	02.1	165 51.3	4 00.1	132 58.4	26.8
20	174 19.6	159 58.7	08.1	121 56.5	02.5	180 53.4	3 59.9	148 00.6	26.9
21	189 22.1	174 58.5 ..	09.1	136 57.1 ..	03.0	195 55.5 ..	59.7	163 02.8 ..	27.0
22	204 24.5	189 58.3	10.2	151 57.8	03.5	210 57.6	59.6	178 04.9	27.1
23	219 27.0	204 58.1	11.2	166 58.4	03.9	225 59.6	59.4	193 07.1	27.2
17 00	234 29.5	219 57.9	N 4 12.2	181 59.0	N19 04.4	241 01.7	S 3 59.3	208 09.3	N 8 27.3
01	249 31.9	234 57.7	13.3	196 59.7	04.8	256 03.8	59.1	223 11.5	27.4
02	264 34.4	249 57.5	14.3	212 00.3	05.3	271 05.9	59.0	238 13.7	27.5
03	279 36.9	264 57.2 ..	15.4	227 00.9 ..	05.8	286 07.9 ..	58.8	253 15.9 ..	27.6
04	294 39.3	279 57.0	16.4	242 01.5	06.2	301 10.0	58.7	268 18.1	27.7
05	309 41.8	294 56.8	17.5	257 02.2	06.7	316 12.1	58.5	283 20.2	27.8
06	324 44.3	309 56.6	N 4 18.5	272 02.8	N19 07.1	331 14.2	S 3 58.3	298 22.4	N 8 27.9
07	339 46.7	324 56.4	19.6	287 03.4	07.6	346 16.3	58.2	313 24.6	28.0
08	354 49.2	339 56.2	20.6	302 04.0	08.1	1 18.3	58.0	328 26.8	28.1
S 09	9 51.7	354 56.0 ..	21.6	317 04.7 ..	08.5	16 20.4 ..	57.9	343 29.0 ..	28.2
U 10	24 54.1	9 55.7	22.7	332 05.3	09.0	31 22.5	57.7	358 31.2	28.3
N 11	39 56.6	24 55.5	23.7	347 05.9	09.4	46 24.6	57.6	13 33.4	28.4
D 12	54 59.0	39 55.3	N 4 24.8	2 06.5	N19 09.9	61 26.7	S 3 57.4	28 35.6	N 8 28.5
A 13	70 01.5	54 55.1	25.8	17 07.2	10.3	76 28.7	57.3	43 37.7	28.6
Y 14	85 04.0	69 54.9	26.9	32 07.8	10.8	91 30.8	57.1	58 39.9	28.7
15	100 06.4	84 54.7 ..	27.9	47 08.4 ..	11.3	106 32.9 ..	56.9	73 42.1 ..	28.8
16	115 08.9	99 54.4	29.0	62 09.1	11.7	121 35.0	56.8	88 44.3	28.9
17	130 11.4	114 54.2	30.0	77 09.7	12.2	136 37.1	56.6	103 46.5	29.0
18	145 13.8	129 54.0	N 4 31.0	92 10.3	N19 12.6	151 39.1	S 3 56.5	118 48.7	N 8 29.1
19	160 16.3	144 53.8	32.1	107 10.9	13.1	166 41.2	56.3	133 50.9	29.2
20	175 18.8	159 53.6	33.1	122 11.6	13.5	181 43.3	56.2	148 53.1	29.3
21	190 21.2	174 53.4 ..	34.2	137 12.2 ..	14.0	196 45.4 ..	56.0	163 55.3 ..	29.4
22	205 23.7	189 53.1	35.2	152 12.8	14.4	211 47.5	55.9	178 57.4	29.5
23	220 26.1	204 52.9	36.3	167 13.4	14.9	226 49.5	55.7	193 59.6	29.6
18 00	235 28.6	219 52.7	N 4 37.3	182 14.1	N19 15.3	241 51.6	S 3 55.6	209 01.8	N 8 29.7
01	250 31.1	234 52.5	38.4	197 14.7	15.8	256 53.7	55.4	224 04.0	29.8
02	265 33.5	249 52.3	39.4	212 15.3	16.2	271 55.8	55.2	239 06.2	29.9
03	280 36.0	264 52.0 ..	40.4	227 15.9 ..	16.7	286 57.9 ..	55.1	254 08.4 ..	30.0
04	295 38.5	279 51.8	41.5	242 16.6	17.1	301 59.9	54.9	269 10.6	30.1
05	310 40.9	294 51.6	42.5	257 17.2	17.6	317 02.0	54.8	284 12.8	30.2
06	325 43.4	309 51.4	N 4 43.6	272 17.8	N19 18.0	332 04.1	S 3 54.6	299 15.0	N 8 30.3
07	340 45.9	324 51.2	44.6	287 18.4	18.5	347 06.2	54.5	314 17.1	30.4
08	355 48.3	339 50.9	45.7	302 19.1	18.9	2 08.3	54.3	329 19.3	30.5
M 09	10 50.8	354 50.7 ..	46.7	317 19.7 ..	19.4	17 10.4 ..	54.2	344 21.5 ..	30.6
O 10	25 53.3	9 50.5	47.7	332 20.3	19.8	32 12.4	54.0	359 23.7	30.7
N 11	40 55.7	24 50.3	48.8	347 20.9	20.3	47 14.5	53.9	14 25.9	30.8
D 12	55 58.2	39 50.1	N 4 49.8	2 21.6	N19 20.7	62 16.6	S 3 53.7	29 28.1	N 8 30.9
A 13	71 00.6	54 49.8	50.9	17 22.2	21.2	77 18.7	53.6	44 30.3	31.0
Y 14	86 03.1	69 49.6	51.9	32 22.8	21.6	92 20.8	53.4	59 32.5	31.1
15	101 05.6	84 49.4 ..	53.0	47 23.4 ..	22.1	107 22.9 ..	53.2	74 34.7 ..	31.2
16	116 08.0	99 49.2	54.0	62 24.1	22.5	122 25.0	53.1	89 36.8	31.3
17	131 10.5	114 48.9	55.1	77 24.7	23.0	137 27.0	52.9	104 39.0	31.4
18	146 13.0	129 48.7	N 4 56.1	92 25.3	N19 23.4	152 29.1	S 3 52.8	119 41.2	N 8 31.5
19	161 15.4	144 48.5	57.1	107 25.9	23.9	167 31.2	52.6	134 43.4	31.6
20	176 17.9	159 48.3	58.2	122 26.6	24.3	182 33.3	52.5	149 45.6	31.7
21	191 20.4	174 48.0	4 59.2	137 27.2 ..	24.8	197 35.4 ..	52.3	164 47.8 ..	31.8
22	206 22.8	189 47.8	5 00.3	152 27.8	25.2	212 37.5	52.2	179 50.0	31.9
23	221 25.3	204 47.6	N 5 01.3	167 28.4	25.7	227 39.5	52.0	194 52.2	32.0
Mer. Pass.	h m 8 20.7	v −0.2	d 1.0	v 0.6	d 0.5	v 2.1	d 0.2	v 2.2	d 0.1

STARS

Name	SHA	Dec
Acamar	315 27.6	S40 18.8
Achernar	335 36.0	S57 14.7
Acrux	173 21.7	S63 05.6
Adhara	255 21.9	S28 58.5
Aldebaran	291 03.1	N16 30.2
Alioth	166 30.5	N55 58.4
Alkaid	153 07.7	N49 19.5
Al Na'ir	27 58.3	S46 57.9
Alnilam	275 58.5	S 1 12.4
Alphard	218 07.5	S 8 39.3
Alphecca	126 20.5	N26 43.3
Alpheratz	357 55.7	N29 04.6
Altair	62 19.4	N 8 51.8
Ankaa	353 27.5	S42 18.8
Antares	112 40.1	S26 25.6
Arcturus	146 06.0	N19 11.5
Atria	107 51.9	S69 01.3
Avior	234 22.9	S59 30.6
Bellatrix	278 44.8	N 6 20.7
Betelgeuse	271 14.2	N 7 24.2
Canopus	264 01.7	S52 42.0
Capella	280 52.1	N45 59.7
Deneb	49 39.2	N45 16.3
Denebola	182 45.3	N14 34.9
Diphda	349 07.8	S17 59.8
Dubhe	194 05.8	N61 45.8
Elnath	278 27.6	N28 36.2
Eltanin	90 51.1	N51 29.3
Enif	33 58.5	N 9 52.0
Fomalhaut	15 36.9	S29 37.8
Gacrux	172 13.4	S57 06.4
Gienah	176 04.0	S17 32.1
Hadar	149 03.7	S60 22.0
Hamal	328 14.2	N23 27.0
Kaus Aust.	83 58.9	S34 22.9
Kochab	137 18.6	N74 09.9
Markab	13 50.0	N15 11.6
Menkar	314 27.5	N 4 04.8
Menkent	148 20.9	S36 21.7
Miaplacidus	221 42.2	S69 43.0
Mirfak	308 57.5	N49 51.1
Nunki	76 12.4	S26 17.8
Peacock	53 37.3	S56 44.1
Pollux	243 42.1	N28 01.8
Procyon	245 12.1	N 5 13.6
Rasalhague	96 16.9	N12 33.7
Regulus	207 55.8	N11 58.5
Rigel	281 23.5	S 8 12.5
Rigil Kent.	140 07.0	S60 49.7
Sabik	102 25.5	S15 43.3
Schedar	349 54.1	N56 31.4
Shaula	96 37.3	S37 06.0
Sirius	258 44.0	S16 43.1
Spica	158 43.2	S11 09.2
Suhail	223 01.0	S43 25.9
Vega	80 46.5	N38 46.9
Zuben'ubi	137 17.9	S16 02.1

	SHA	Mer. Pass.
	° ′	h m
Venus	345 28.4	9 20
Mars	307 29.6	11 52
Jupiter	6 32.2	7 55
Saturn	333 39.8	10 06

SUN and MOON

UT	SUN GHA	SUN Dec	MOON GHA	v	MOON Dec	d	HP
16 00	180 55.0	N19 00.0	307 02.1	9.6	S18 37.2	1.9	56.5
01	195 55.0	00.6	321 30.7	9.5	18 35.3	2.0	56.5
02	210 55.0	01.2	335 59.2	9.5	18 33.3	2.2	56.5
03	225 55.0 ..	01.8	350 27.7	9.4	18 31.1	2.2	56.5
04	240 55.0	02.3	4 56.1	9.4	18 28.9	2.3	56.6
05	255 55.0	02.9	19 24.6	9.5	18 26.6	2.5	56.6
S 06	270 54.9	N19 03.5	33 53.1	9.4	S18 24.1	2.5	56.6
A 07	285 54.9	04.1	48 21.5	9.5	18 21.6	2.7	56.6
T 08	300 54.9	04.7	62 50.0	9.4	18 18.9	2.7	56.7
U 09	315 54.9 ..	05.2	77 18.4	9.5	18 16.2	2.9	56.7
R 10	330 54.9	05.8	91 46.9	9.4	18 13.3	2.9	56.7
D 11	345 54.9	06.4	106 15.3	9.4	18 10.4	3.1	56.7
A 12	0 54.9	N19 07.0	120 43.7	9.4	S18 07.3	3.1	56.7
Y 13	15 54.9	07.5	135 12.1	9.4	18 04.2	3.3	56.8
14	30 54.8	08.1	149 40.5	9.4	18 00.9	3.3	56.8
15	45 54.8 ..	08.7	164 08.9	9.4	17 57.6	3.5	56.8
16	60 54.8	09.3	178 37.3	9.4	17 54.1	3.6	56.8
17	75 54.8	09.8	193 05.7	9.3	17 50.5	3.6	56.9
18	90 54.8	N19 10.4	207 34.0	9.4	S17 46.9	3.8	56.9
19	105 54.8	11.0	222 02.4	9.4	17 43.1	3.9	56.9
20	120 54.8	11.6	236 30.8	9.4	17 39.2	3.9	56.9
21	135 54.8 ..	12.1	250 59.2	9.3	17 35.3	4.1	57.0
22	150 54.7	12.7	265 27.5	9.4	17 31.2	4.2	57.0
23	165 54.7	13.3	279 55.9	9.3	17 27.0	4.2	57.0
17 00	180 54.7	N19 13.8	294 24.2	9.4	S17 22.8	4.4	57.0
01	195 54.7	14.4	308 52.6	9.4	17 18.4	4.5	57.1
02	210 54.7	15.0	323 21.0	9.3	17 13.9	4.5	57.1
03	225 54.7 ..	15.5	337 49.3	9.4	17 09.4	4.7	57.1
04	240 54.6	16.1	352 17.7	9.3	17 04.7	4.8	57.2
05	255 54.6	16.7	6 46.0	9.4	16 59.9	4.8	57.2
S 06	270 54.6	N19 17.2	21 14.4	9.3	S16 55.1	5.0	57.2
U 07	285 54.6	17.8	35 42.7	9.4	16 50.1	5.1	57.2
N 08	300 54.6	18.4	50 11.1	9.3	16 45.0	5.1	57.3
D 09	315 54.6 ..	18.9	64 39.4	9.4	16 39.9	5.3	57.3
A 10	330 54.5	19.5	79 07.8	9.4	16 34.6	5.3	57.3
Y 11	345 54.5	20.1	93 36.2	9.3	16 29.3	5.5	57.3
12	0 54.5	N19 20.6	108 04.5	9.4	S16 23.8	5.5	57.4
13	15 54.5	21.2	122 32.9	9.4	16 18.3	5.7	57.4
14	30 54.5	21.7	137 01.3	9.3	16 12.6	5.7	57.4
15	45 54.5 ..	22.3	151 29.6	9.4	16 06.9	5.8	57.4
16	60 54.4	22.9	165 58.0	9.4	16 01.1	5.9	57.5
17	75 54.4	23.4	180 26.4	9.4	15 55.2	6.1	57.5
18	90 54.4	N19 24.0	194 54.8	9.3	S15 49.1	6.1	57.5
19	105 54.4	24.5	209 23.1	9.4	15 43.0	6.2	57.6
20	120 54.4	25.1	223 51.5	9.4	15 36.8	6.3	57.6
21	135 54.3 ..	25.7	238 19.9	9.4	15 30.5	6.4	57.6
22	150 54.3	26.2	252 48.3	9.4	15 24.1	6.5	57.6
23	165 54.3	26.8	267 16.7	9.4	15 17.6	6.5	57.7
18 00	180 54.3	N19 27.3	281 45.1	9.4	S15 11.1	6.7	57.7
01	195 54.3	27.9	296 13.5	9.4	15 04.4	6.8	57.7
02	210 54.2	28.4	310 41.9	9.5	14 57.6	6.8	57.7
03	225 54.2 ..	29.0	325 10.4	9.4	14 50.8	7.0	57.8
04	240 54.2	29.5	339 38.8	9.4	14 43.8	7.0	57.8
05	255 54.2	30.1	354 07.2	9.4	14 36.8	7.1	57.8
M 06	270 54.1	N19 30.6	8 35.6	9.5	S14 29.7	7.2	57.9
O 07	285 54.1	31.2	23 04.1	9.4	14 22.5	7.3	57.9
N 08	300 54.1	31.7	37 32.5	9.5	14 15.2	7.4	57.9
D 09	315 54.1 ..	32.3	52 01.0	9.4	14 07.8	7.4	57.9
A 10	330 54.0	32.8	66 29.4	9.5	14 00.4	7.6	58.0
Y 11	345 54.0	33.4	80 57.9	9.5	13 52.8	7.6	58.0
12	0 54.0	N19 33.9	95 26.4	9.4	S13 45.2	7.8	58.0
13	15 54.0	34.5	109 54.8	9.5	13 37.4	7.8	58.0
14	30 54.0	35.0	124 23.3	9.5	13 29.6	7.9	58.1
15	45 53.9 ..	35.6	138 51.8	9.5	13 21.7	7.9	58.1
16	60 53.9	36.1	153 20.3	9.5	13 13.8	8.1	58.1
17	75 53.9	36.7	167 48.8	9.5	13 05.7	8.1	58.2
18	90 53.9	N19 37.2	182 17.3	9.5	S12 57.6	8.2	58.2
19	105 53.8	37.8	196 45.8	9.5	12 49.4	8.3	58.2
20	120 53.8	38.3	211 14.3	9.5	12 41.1	8.4	58.2
21	135 53.8 ..	38.9	225 42.8	9.5	12 32.7	8.5	58.3
22	150 53.7	39.4	240 11.3	9.5	12 24.2	8.5	58.3
23	165 53.7	39.9	254 39.8	9.5	S12 15.7	8.6	58.3
	SD 15.8	d 0.6	SD 15.5		15.6		15.8

Twilight / Sunrise / Moonrise

Lat.	Naut.	Civil	Sunrise	Moonrise 16	17	18	19
N 72	□	□	□	■	03 21	03 01	02 50
N 70	□	□	□	02 09	02 24	02 29	02 30
68	////	////	01 37	01 22	01 49	02 05	02 14
66	////	////	02 15	00 52	01 24	01 46	02 01
64	////	00 54	02 42	00 29	01 05	01 31	01 51
62	////	01 44	03 02	00 11	00 49	01 19	01 41
60	////	02 13	03 18	24 36	00 36	01 08	01 34
N 58	00 49	02 36	03 32	24 24	00 24	00 58	01 27
56	01 34	02 53	03 44	24 14	00 14	00 50	01 20
54	02 01	03 08	03 54	24 06	00 06	00 43	01 15
52	02 22	03 21	04 04	23 58	24 36	00 36	01 10
50	02 39	03 32	04 12	23 51	24 30	00 30	01 05
45	03 11	03 55	04 29	23 35	24 17	00 17	00 55
N 40	03 34	04 12	04 43	23 23	24 07	00 07	00 47
35	03 53	04 27	04 55	23 12	23 57	24 40	00 40
30	04 08	04 39	05 06	23 03	23 49	24 34	00 34
20	04 32	05 00	05 23	22 47	23 35	24 23	00 23
N 10	04 50	05 16	05 39	22 33	23 23	24 13	00 13
0	05 05	05 31	05 53	22 19	23 12	24 04	00 04
S 10	05 19	05 45	06 07	22 06	23 00	23 55	24 50
20	05 32	05 59	06 22	21 52	22 48	23 46	24 44
30	05 44	06 13	06 39	21 36	22 34	23 35	24 37
35	05 51	06 22	06 49	21 26	22 26	23 28	24 31
40	05 58	06 31	07 00	21 16	22 17	23 21	24 28
45	06 05	06 41	07 13	21 03	22 06	23 13	24 22
S 50	06 13	06 53	07 29	20 48	21 53	23 02	24 15
52	06 17	06 59	07 37	20 40	21 47	22 58	24 12
54	06 21	07 05	07 45	20 32	21 40	22 52	24 09
56	06 25	07 11	07 54	20 23	21 32	22 47	24 05
58	06 30	07 19	08 05	20 13	21 24	22 40	24 01
S 60	06 35	07 27	08 17	20 01	21 14	22 33	23 56

Sunset / Twilight / Moonset

Lat.	Sunset	Civil	Naut.	Moonset 16	17	18	19
N 72	□	□	□	■	05 52	08 01	10 01
N 70	□	□	□	05 14	06 49	08 33	10 20
68	22 21	////	////	06 00	07 22	08 56	10 34
66	21 41	////	////	06 31	07 47	09 13	10 46
64	21 14	23 07	////	06 53	08 06	09 28	10 56
62	20 53	22 13	////	07 11	08 21	09 40	11 04
60	20 36	21 42	////	07 26	08 34	09 50	11 11
N 58	20 22	21 19	23 12	07 39	08 45	09 59	11 18
56	20 10	21 01	22 23	07 50	08 55	10 06	11 23
54	19 59	20 46	21 54	07 59	09 03	10 13	11 28
52	19 50	20 33	21 33	08 08	09 11	10 20	11 32
50	19 42	20 22	21 15	08 15	09 18	10 25	11 37
45	19 24	19 59	20 43	08 32	09 32	10 37	11 45
N 40	19 10	19 41	20 19	08 45	09 44	10 47	11 52
35	18 58	19 26	20 01	08 57	09 54	10 55	11 59
30	18 47	19 14	19 45	09 06	10 03	11 03	12 04
20	18 30	18 53	19 21	09 23	10 19	11 15	12 13
N 10	18 14	18 37	19 03	09 38	10 32	11 27	12 21
0	18 00	18 22	18 47	09 52	10 45	11 37	12 29
S 10	17 46	18 08	18 34	10 06	10 57	11 47	12 37
20	17 31	17 54	18 21	10 21	11 10	11 58	12 45
30	17 14	17 39	18 08	10 37	11 25	12 10	12 54
35	17 04	17 31	18 02	10 47	11 34	12 18	12 59
40	16 52	17 22	17 55	10 58	11 44	12 26	13 05
45	16 39	17 11	17 47	11 11	11 55	12 35	13 11
S 50	16 23	16 59	17 39	11 27	12 09	12 46	13 19
52	16 16	16 54	17 35	11 34	12 16	12 51	13 23
54	16 07	16 48	17 31	11 42	12 24	12 57	13 27
56	15 58	16 41	17 27	11 52	12 31	13 04	13 32
58	15 47	16 33	17 22	12 02	12 40	13 11	13 37
S 60	15 35	16 25	17 17	12 14	12 50	13 19	13 42

SUN and MOON (lower summary)

Day	SUN Eqn. of Time 00h	12h	Mer. Pass.	MOON Mer. Pass. Upper	Lower	Age	Phase
16	03 40	03 40	11 56	03 40	16 06	20	78
17	03 39	03 38	11 56	04 32	16 58	21	68
18	03 37	03 36	11 56	05 24	17 51	22	58

1998 MAY 19, 20, 21 (TUES., WED., THURS.)

UT	ARIES GHA	VENUS −4.0 GHA	Dec	MARS +1.3 GHA	Dec	JUPITER −2.2 GHA	Dec	SATURN +0.6 GHA	Dec	STARS Name	SHA	Dec
d h	° ′	° ′	° ′	° ′	° ′	° ′	° ′	° ′	° ′		° ′	° ′
19 00	236 27.8	219 47.4 N 5	02.4	182 29.1 N19	26.1	242 41.6 S 3	51.9	209 54.4 N 8	32.1	Acamar	315 27.6	S40 18.8
01	251 30.2	234 47.2	03.4	197 29.7	26.5	257 43.7	51.7	224 56.5	32.2	Achernar	335 36.0	S57 14.7
02	266 32.7	249 46.9	04.4	212 30.3	27.0	272 45.8	51.6	239 58.7	32.3	Acrux	173 21.8	S63 05.6
03	281 35.1	264 46.7 ..	05.5	227 30.9 ..	27.4	287 47.9 ..	51.4	255 00.9 ..	32.4	Adhara	255 21.9	S28 58.5
04	296 37.6	279 46.5	06.5	242 31.6	27.9	302 50.0	51.3	270 03.1	32.5	Aldebaran	291 03.0	N16 30.2
05	311 40.1	294 46.3	07.6	257 32.2	28.3	317 52.1	51.1	285 05.3	32.6			
06	326 42.5	309 46.0 N 5	08.6	272 32.8 N19	28.8	332 54.2 S 3	51.0	300 07.5 N 8	32.7	Alioth	166 30.5	N55 58.4
07	341 45.0	324 45.8	09.7	287 33.4	29.2	347 56.2	50.8	315 09.7	32.8	Alkaid	153 07.7	N49 19.5
T 08	356 47.5	339 45.6	10.7	302 34.0	29.6	2 58.3	50.7	330 11.9	32.9	Al Na'ir	27 58.3	S46 57.9
U 09	11 49.9	354 45.3 ..	11.7	317 34.7 ..	30.1	18 00.4 ..	50.5	345 14.1 ..	33.0	Alnilam	275 58.5	S 1 12.4
E 10	26 52.4	9 45.1	12.8	332 35.3	30.5	33 02.5	50.4	0 16.3	33.0	Alphard	218 07.5	S 8 39.3
S 11	41 54.9	24 44.9	13.8	347 35.9	31.0	48 04.6	50.2	15 18.4	33.1			
D 12	56 57.3	39 44.7 N 5	14.9	2 36.5 N19	31.4	63 06.7 S 3	50.1	30 20.6 N 8	33.2	Alphecca	126 20.5	N26 43.3
A 13	71 59.8	54 44.4	15.9	17 37.2	31.9	78 08.8	49.9	45 22.8	33.3	Alpheratz	357 55.7	N29 04.6
Y 14	87 02.2	69 44.2	17.0	32 37.8	32.3	93 10.9	49.8	60 25.0	33.4	Altair	62 19.4	N 8 51.8
15	102 04.7	84 44.0 ..	18.0	47 38.4 ..	32.7	108 12.9 ..	49.6	75 27.2 ..	33.5	Ankaa	353 27.4	S42 18.8
16	117 07.2	99 43.8	19.0	62 39.0	33.2	123 15.0	49.5	90 29.4	33.6	Antares	112 40.1	S26 25.6
17	132 09.6	114 43.5	20.1	77 39.7	33.6	138 17.1	49.3	105 31.6	33.7			
18	147 12.1	129 43.3 N 5	21.1	92 40.3 N19	34.1	153 19.2 S 3	49.1	120 33.8 N 8	33.8	Arcturus	146 06.0	N19 11.6
19	162 14.6	144 43.1	22.2	107 40.9	34.5	168 21.3	49.0	135 36.0	33.9	Atria	107 51.9	S69 01.3
20	177 17.0	159 42.8	23.2	122 41.5	34.9	183 23.4	48.8	150 38.2	34.0	Avior	234 23.0	S59 30.6
21	192 19.5	174 42.6 ..	24.3	137 42.1 ..	35.4	198 25.5 ..	48.7	165 40.4 ..	34.1	Bellatrix	278 44.8	N 6 20.7
22	207 22.0	189 42.4	25.3	152 42.8	35.8	213 27.6	48.5	180 42.5	34.2	Betelgeuse	271 14.2	N 7 24.2
23	222 24.4	204 42.1	26.3	167 43.4	36.2	228 29.7	48.4	195 44.7	34.3			
20 00	237 26.9	219 41.9 N 5	27.4	182 44.0 N19	36.7	243 31.8 S 3	48.2	210 46.9 N 8	34.4	Canopus	264 01.7	S52 42.0
01	252 29.4	234 41.7	28.4	197 44.6	37.1	258 33.8	48.1	225 49.1	34.5	Capella	280 52.1	N45 59.7
02	267 31.8	249 41.5	29.5	212 45.3	37.5	273 35.9	47.9	240 51.3	34.6	Deneb	49 39.2	N45 16.3
03	282 34.3	264 41.2 ..	30.5	227 45.9 ..	38.0	288 38.0 ..	47.8	255 53.5 ..	34.7	Denebola	182 45.4	N14 34.9
04	297 36.7	279 41.0	31.6	242 46.5	38.4	303 40.1	47.6	270 55.7	34.8	Diphda	349 07.8	S17 59.8
05	312 39.2	294 40.8	32.6	257 47.1	38.9	318 42.2	47.5	285 57.9	34.9			
06	327 41.7	309 40.5 N 5	33.6	272 47.7 N19	39.3	333 44.3 S 3	47.3	301 00.1 N 8	35.0	Dubhe	194 05.8	N61 45.8
W 07	342 44.1	324 40.3	34.7	287 48.4	39.7	348 46.4	47.2	316 02.3	35.1	Elnath	278 27.7	N28 36.2
E 08	357 46.6	339 40.1	35.7	302 49.0	40.2	3 48.5	47.0	331 04.5	35.2	Eltanin	90 51.1	N51 29.3
D 09	12 49.1	354 39.8 ..	36.8	317 49.6 ..	40.6	18 50.6 ..	46.9	346 06.7 ..	35.3	Enif	33 58.5	N 9 52.0
N 10	27 51.5	9 39.6	37.8	332 50.2	41.0	33 52.7	46.7	1 08.8	35.4	Fomalhaut	15 36.9	S29 37.8
E 11	42 54.0	24 39.4	38.8	347 50.8	41.5	48 54.8	46.6	16 11.0	35.5			
S 12	57 56.5	39 39.1 N 5	39.9	2 51.5 N19	41.9	63 56.9 S 3	46.4	31 13.2 N 8	35.6	Gacrux	172 13.4	S57 06.4
D 13	72 58.9	54 38.9	40.9	17 52.1	42.3	78 59.0	46.3	46 15.4	35.7	Gienah	176 04.0	S17 32.1
A 14	88 01.4	69 38.7	42.0	32 52.7	42.8	94 01.0	46.1	61 17.6	35.8	Hadar	149 03.8	S60 22.0
Y 15	103 03.9	84 38.4 ..	43.0	47 53.3 ..	43.2	109 03.1 ..	46.0	76 19.8 ..	35.9	Hamal	328 14.2	N23 27.0
16	118 06.3	99 38.2	44.1	62 53.9	43.6	124 05.2	45.9	91 22.0	36.0	Kaus Aust.	83 58.9	S34 22.9
17	133 08.8	114 38.0	45.1	77 54.6	44.1	139 07.3	45.7	106 24.2	36.1			
18	148 11.2	129 37.7 N 5	46.1	92 55.2 N19	44.5	154 09.4 S 3	45.6	121 26.4 N 8	36.2	Kochab	137 18.6	N74 09.9
19	163 13.7	144 37.5	47.2	107 55.8	44.9	169 11.5	45.4	136 28.6	36.3	Markab	13 50.0	N15 11.6
20	178 16.2	159 37.3	48.2	122 56.4	45.3	184 13.6	45.3	151 30.8	36.4	Menkar	314 27.5	N 4 04.8
21	193 18.6	174 37.0 ..	49.3	137 57.1 ..	45.8	199 15.7 ..	45.1	166 33.0 ..	36.5	Menkent	148 20.9	S36 21.8
22	208 21.1	189 36.8	50.3	152 57.7	46.2	214 17.8	45.0	181 35.1	36.6	Miaplacidus	221 42.2	S69 43.0
23	223 23.6	204 36.6	51.3	167 58.3	46.6	229 19.9	44.8	196 37.3	36.7			
21 00	238 26.0	219 36.3 N 5	52.4	182 58.9 N19	47.1	244 22.0 S 3	44.7	211 39.5 N 8	36.8	Mirfak	308 57.5	N49 51.1
01	253 28.5	234 36.1	53.4	197 59.5	47.5	259 24.1	44.5	226 41.7	36.9	Nunki	76 12.4	S26 17.8
02	268 31.0	249 35.8	54.5	213 00.2	47.9	274 26.2	44.4	241 43.9	37.0	Peacock	53 37.3	S56 44.1
03	283 33.4	264 35.6 ..	55.5	228 00.8 ..	48.3	289 28.3 ..	44.2	256 46.1 ..	37.1	Pollux	243 42.1	N28 01.8
04	298 35.9	279 35.4	56.5	243 01.4	48.8	304 30.4	44.1	271 48.3	37.1	Procyon	245 12.1	N 5 13.6
05	313 38.3	294 35.1	57.6	258 02.0	49.2	319 32.5	43.9	286 50.5	37.2			
06	328 40.8	309 34.9 N 5	58.6	273 02.6 N19	49.6	334 34.6 S 3	43.8	301 52.7 N 8	37.3	Rasalhague	96 16.9	N12 33.7
07	343 43.3	324 34.7	5 59.7	288 03.2	50.1	349 36.7	43.6	316 54.9	37.4	Regulus	207 55.9	N11 58.5
T 08	358 45.7	339 34.4	6 00.7	303 03.9	50.5	4 38.8	43.5	331 57.1	37.5	Rigel	281 23.5	S 8 12.4
H 09	13 48.2	354 34.2 ..	01.7	318 04.5 ..	50.9	19 40.9 ..	43.3	346 59.3 ..	37.6	Rigil Kent.	140 07.0	S60 49.7
U 10	28 50.7	9 33.9	02.8	333 05.1	51.3	34 42.9	43.2	2 01.5	37.7	Sabik	102 25.5	S15 43.3
R 11	43 53.1	24 33.7	03.8	348 05.7	51.8	49 45.0	43.0	17 03.7	37.8			
S 12	58 55.6	39 33.5 N 6	04.9	3 06.3 N19	52.2	64 47.1 S 3	42.9	32 05.8 N 8	37.9	Schedar	349 54.1	N56 31.4
D 13	73 58.1	54 33.2	05.9	18 07.0	52.6	79 49.2	42.7	47 08.0	38.0	Shaula	96 37.3	S37 06.0
A 14	89 00.5	69 33.0	06.9	33 07.6	53.0	94 51.3	42.6	62 10.2	38.1	Sirius	258 44.2	S16 43.1
Y 15	104 03.0	84 32.7 ..	08.0	48 08.2 ..	53.5	109 53.4 ..	42.4	77 12.4 ..	38.2	Spica	158 43.2	S11 09.2
16	119 05.5	99 32.5	09.0	63 08.8	53.9	124 55.5	42.3	92 14.6	38.3	Suhail	223 01.0	S43 25.9
17	134 07.9	114 32.3	10.1	78 09.4	54.3	139 57.6	42.2	107 16.8	38.4			
18	149 10.4	129 32.0 N 6	11.1	93 10.1 N19	54.7	154 59.7 S 3	42.0	122 19.0 N 8	38.5	Vega	80 46.5	N38 46.9
19	164 12.8	144 31.8	12.1	108 10.7	55.1	170 01.8	41.9	137 21.2	38.6	Zuben'ubi	137 17.9	S16 02.1
20	179 15.3	159 31.5	13.2	123 11.3	55.6	185 03.9	41.7	152 23.4	38.7		SHA	Mer. Pass.
21	194 17.8	174 31.3 ..	14.2	138 11.9 ..	56.0	200 06.0 ..	41.6	167 25.6 ..	38.8		° ′	h m
22	209 20.2	189 31.1	15.3	153 12.5	56.4	215 08.1	41.4	182 27.8	38.9	Venus	342 15.0	9 21
23	224 22.7	204 30.8	16.3	168 13.2	56.8	230 10.2	41.3	197 30.0	39.0	Mars	305 17.1	11 49
	h m									Jupiter	6 04.9	7 45
Mer. Pass. 8 08.9	v −0.2	d 1.0		v 0.6	d 0.4	v 2.1	d 0.1	v 2.2	d 0.1	Saturn	333 20.0	9 55

UT	SUN GHA	SUN Dec	MOON GHA	v	Dec	d	HP
d h	° ′	° ′	° ′	′	° ′	′	′
19 00	180 53.7	N19 40.5	269 08.3	9.6	S12 07.1	8.7	58.4
01	195 53.7	41.0	283 36.9	9.5	11 58.4	8.8	58.4
02	210 53.6	41.6	298 05.4	9.6	11 49.6	8.8	58.4
03	225 53.6 ..	42.1	312 33.9	9.6	11 40.8	8.9	58.4
04	240 53.6	42.6	327 02.5	9.5	11 31.9	9.0	58.5
05	255 53.6	43.2	341 31.0	9.6	11 22.9	9.1	58.5
06	270 53.5	N19 43.7	355 59.6	9.5	S11 13.8	9.1	58.5
07	285 53.5	44.3	10 28.1	9.6	11 04.7	9.2	58.5
08	300 53.5	44.8	24 56.7	9.6	10 55.5	9.3	58.6
T 09	315 53.4 ..	45.3	39 25.2	9.6	10 46.2	9.4	58.6
U 10	330 53.4	45.9	53 53.8	9.6	10 36.8	9.4	58.6
E 11	345 53.4	46.4	68 22.4	9.5	10 27.4	9.5	58.7
S 12	0 53.3	N19 46.9	82 50.9	9.6	S10 17.9	9.5	58.7
D 13	15 53.3	47.5	97 19.5	9.6	10 08.4	9.6	58.7
A 14	30 53.3	48.0	111 48.1	9.5	9 58.8	9.7	58.7
Y 15	45 53.3 ..	48.5	126 16.6	9.6	9 49.1	9.8	58.8
16	60 53.2	49.1	140 45.2	9.6	9 39.3	9.8	58.8
17	75 53.2	49.6	155 13.8	9.5	9 29.5	9.9	58.8
18	90 53.2	N19 50.1	169 42.3	9.6	S 9 19.6	9.9	58.9
19	105 53.1	50.7	184 10.9	9.6	9 09.7	10.0	58.9
20	120 53.1	51.2	198 39.5	9.5	8 59.7	10.1	58.9
21	135 53.1 ..	51.7	213 08.0	9.6	8 49.6	10.1	58.9
22	150 53.0	52.3	227 36.6	9.6	8 39.5	10.2	59.0
23	165 53.0	52.8	242 05.2	9.5	8 29.3	10.3	59.0
20 00	180 53.0	N19 53.3	256 33.7	9.6	S 8 19.0	10.3	59.0
01	195 52.9	53.8	271 02.3	9.5	8 08.7	10.4	59.0
02	210 52.9	54.4	285 30.8	9.6	7 58.3	10.4	59.1
03	225 52.9 ..	54.9	299 59.4	9.6	7 47.9	10.4	59.1
04	240 52.8	55.4	314 27.9	9.6	7 37.5	10.6	59.1
05	255 52.8	55.9	328 56.5	9.5	7 26.9	10.6	59.2
06	270 52.8	N19 56.5	343 25.0	9.6	S 7 16.3	10.6	59.2
W 07	285 52.7	57.0	357 53.6	9.5	7 05.7	10.7	59.2
E 08	300 52.7	57.5	12 22.1	9.5	6 55.0	10.7	59.2
D 09	315 52.7 ..	58.0	26 50.6	9.5	6 44.3	10.8	59.3
N 10	330 52.6	58.6	41 19.1	9.6	6 33.5	10.8	59.3
E 11	345 52.6	59.1	55 47.7	9.5	6 22.7	10.9	59.3
S 12	0 52.6	N19 59.6	70 16.2	9.5	S 6 11.8	10.9	59.3
D 13	15 52.5	20 00.1	84 44.7	9.4	6 00.9	11.0	59.4
A 14	30 52.5	00.6	99 13.1	9.5	5 49.9	11.0	59.4
Y 15	45 52.4 ..	01.2	113 41.6	9.5	5 38.9	11.1	59.4
16	60 52.4	01.7	128 10.1	9.5	5 27.8	11.1	59.4
17	75 52.4	02.2	142 38.6	9.4	5 16.7	11.1	59.5
18	90 52.3	N20 02.7	157 07.0	9.5	S 5 05.6	11.2	59.5
19	105 52.3	03.2	171 35.5	9.4	4 54.4	11.2	59.5
20	120 52.3	03.7	186 03.9	9.4	4 43.2	11.2	59.5
21	135 52.2 ..	04.3	200 32.3	9.4	4 32.0	11.3	59.6
22	150 52.2	04.8	215 00.7	9.4	4 20.7	11.4	59.6
23	165 52.1	05.3	229 29.1	9.4	4 09.3	11.3	59.6
21 00	180 52.1	N20 05.8	243 57.5	9.4	S 3 58.0	11.4	59.6
01	195 52.1	06.3	258 25.9	9.3	3 46.6	11.5	59.7
02	210 52.0	06.8	272 54.2	9.3	3 35.1	11.4	59.7
03	225 52.0 ..	07.3	287 22.5	9.4	3 23.7	11.5	59.7
04	240 51.9	07.8	301 50.9	9.3	3 12.2	11.5	59.7
05	255 51.9	08.4	316 19.2	9.3	3 00.7	11.6	59.8
06	270 51.9	N20 08.9	330 47.5	9.2	S 2 49.1	11.5	59.8
07	285 51.8	09.4	345 15.7	9.3	2 37.6	11.6	59.8
T 08	300 51.8	09.9	359 44.0	9.2	2 26.0	11.6	59.8
H 09	315 51.7 ..	10.4	14 12.2	9.3	2 14.4	11.7	59.8
U 10	330 51.7	10.9	28 40.5	9.2	2 02.7	11.7	59.9
R 11	345 51.7	11.4	43 08.7	9.1	1 51.0	11.6	59.9
S 12	0 51.6	N20 11.9	57 36.8	9.2	S 1 39.4	11.7	59.9
D 13	15 51.6	12.4	72 05.0	9.2	1 27.7	11.8	59.9
A 14	30 51.5	12.9	86 33.2	9.1	1 15.9	11.7	60.0
Y 15	45 51.5 ..	13.4	101 01.3	9.1	1 04.2	11.8	60.0
16	60 51.5	13.9	115 29.4	9.1	0 52.4	11.7	60.0
17	75 51.4	14.4	129 57.5	9.0	0 40.7	11.8	60.0
18	90 51.4	N20 14.9	144 25.5	9.0	S 0 28.9	11.8	60.0
19	105 51.3	15.4	158 53.5	9.0	0 17.1	11.8	60.1
20	120 51.3	15.9	173 21.5	9.0	S 0 05.3	11.8	60.1
21	135 51.2 ..	16.4	187 49.5	9.0	N 0 06.5	11.8	60.1
22	150 51.2	16.9	202 17.5	8.9	0 18.3	11.9	60.1
23	165 51.1	17.4	216 45.4	8.9	N 0 30.2	11.8	60.1
SD 15.8	d 0.5	SD	16.0		16.2		16.3

Lat.	Twilight Naut.	Twilight Civil	Sunrise	Moonrise 19	Moonrise 20	Moonrise 21	Moonrise 22
°	h m	h m	h m	h m	h m	h m	h m
N 72	▭	▭	▭	02 50	02 42	02 35	02 28
N 70	▭	▭		02 30	02 30	02 30	02 29
68	////	////	01 18	02 14	02 21	02 26	02 30
66	////	////	02 03	02 01	02 13	02 22	02 31
64	////	00 18	02 32	01 51	02 06	02 20	02 32
62	////	01 30	02 54	01 41	02 00	02 17	02 33
60	////	02 04	03 12	01 34	01 55	02 15	02 34
N 58	00 16	02 28	03 26	01 27	01 51	02 13	02 34
56	01 21	02 47	03 39	01 20	01 47	02 11	02 35
54	01 52	03 03	03 50	01 15	01 43	02 09	02 35
52	02 15	03 16	03 59	01 10	01 40	02 08	02 35
50	02 33	03 28	04 08	01 05	01 37	02 07	02 36
45	03 07	03 51	04 26	00 55	01 30	02 04	02 37
N 40	03 31	04 10	04 41	00 47	01 25	02 01	02 37
35	03 50	04 25	04 53	00 40	01 20	01 59	02 38
30	04 06	04 38	05 04	00 34	01 16	01 57	02 39
20	04 30	04 58	05 22	00 23	01 09	01 54	02 40
N 10	04 49	05 16	05 38	00 13	01 02	01 51	02 41
0	05 05	05 31	05 53	00 04	00 56	01 49	02 42
S 10	05 20	05 45	06 08	24 50	00 50	01 46	02 42
20	05 33	06 00	06 23	24 44	00 44	01 43	02 43
30	05 46	06 15	06 41	24 37	00 37	01 40	02 45
35	05 53	06 24	06 51	24 32	00 32	01 38	02 45
40	06 00	06 33	07 03	24 28	00 28	01 36	02 46
45	06 08	06 44	07 16	24 22	00 22	01 34	02 47
S 50	06 17	06 57	07 33	24 15	00 15	01 31	02 48
52	06 21	07 02	07 41	24 12	00 12	01 29	02 49
54	06 25	07 09	07 50	24 09	00 09	01 28	02 49
56	06 29	07 16	07 59	24 05	00 05	01 26	02 50
58	06 34	07 24	08 11	24 01	00 01	01 25	02 51
S 60	06 39	07 32	08 23	23 56	25 23	01 23	02 51

Lat.	Sunset	Twilight Civil	Twilight Naut.	Moonset 19	Moonset 20	Moonset 21	Moonset 22
°	h m	h m	h m	h m	h m	h m	h m
N 72	▭	▭	▭	10 01	11 59	13 56	15 55
N 70	▭	▭	▭	10 20	12 08	13 58	15 49
68	22 41	////	////	10 34	12 16	14 00	15 45
66	21 53	////	////	10 46	12 22	14 01	15 42
64	21 23	////	////	10 56	12 28	14 02	15 38
62	21 01	22 27	////	11 04	12 32	14 03	15 36
60	20 43	21 52	////	11 11	12 36	14 04	15 33
N 58	20 28	21 27	////	11 18	12 40	14 05	15 31
56	20 15	21 08	22 35	11 23	12 43	14 05	15 30
54	20 04	20 52	22 03	11 28	12 46	14 06	15 28
52	19 55	20 38	21 40	11 32	12 48	14 07	15 27
50	19 46	20 26	21 22	11 37	12 51	14 07	15 25
45	19 28	20 03	20 47	11 45	12 56	14 08	15 22
N 40	19 13	19 44	20 23	11 52	13 00	14 09	15 20
35	19 00	19 29	20 03	11 59	13 03	14 10	15 18
30	18 49	19 16	19 48	12 04	13 07	14 11	15 16
20	18 31	18 55	19 23	12 13	13 12	14 12	15 13
N 10	18 15	18 37	19 04	12 21	13 17	14 13	15 10
0	18 00	18 22	18 48	12 29	13 21	14 14	15 07
S 10	17 45	18 08	18 33	12 37	13 25	14 15	15 04
20	17 30	17 53	18 20	12 45	13 30	14 15	15 01
30	17 12	17 38	18 07	12 54	13 35	14 16	14 58
35	17 02	17 29	18 00	12 59	13 38	14 17	14 56
40	16 50	17 19	17 53	13 05	13 42	14 18	14 54
45	16 36	17 09	17 45	13 11	13 45	14 18	14 52
S 50	16 20	16 56	17 36	13 19	13 50	14 19	14 49
52	16 12	16 50	17 32	13 23	13 52	14 20	14 47
54	16 03	16 44	17 28	13 27	13 54	14 20	14 46
56	15 53	16 37	17 23	13 32	13 57	14 20	14 44
58	15 42	16 29	17 18	13 37	14 00	14 21	14 42
S 60	15 29	16 20	17 13	13 42	14 03	14 22	14 40

	SUN Eqn. of Time 00ʰ	SUN Eqn. of Time 12ʰ	SUN Mer. Pass.	MOON Mer. Pass. Upper	MOON Mer. Pass. Lower	MOON Age	MOON Phase
Day	m s	m s	h m	h m	h m	d	%
d							
19	03 35	03 33	11 56	06 17	18 43	23	47
20	03 32	03 30	11 56	07 09	19 35	24	36
21	03 28	03 27	11 57	08 01	20 28	25	25

1998 MAY 22, 23, 24 (FRI., SAT., SUN.)

UT	ARIES GHA	VENUS −4.0 GHA	Dec	MARS +1.4 GHA	Dec	JUPITER −2.2 GHA	Dec	SATURN +0.6 GHA	Dec
22 00	239 25.2	219 30.6 N 6 17.3		183 13.8 N19 57.3		245 12.3 S 3 41.1		212 32.2 N 8 39.1	
01	254 27.6	234 30.3	18.4	198 14.4	57.7	260 14.4	41.0	227 34.4	39.2
02	269 30.1	249 30.1	19.4	213 15.0	58.1	275 16.5	40.8	242 36.6	39.3
03	284 32.6	264 29.8 ..	20.4	228 15.6 ..	58.5	290 18.6 ..	40.7	257 38.8 ..	39.4
04	299 35.0	279 29.6	21.5	243 16.2	58.9	305 20.7	40.5	272 41.0	39.5
05	314 37.5	294 29.4	22.5	258 16.9	59.3	320 22.8	40.4	287 43.1	39.6
06	329 39.9	309 29.1 N 6 23.6		273 17.5 N19 59.8		335 24.9 S 3 40.3		302 45.3 N 8 39.7	
F 07	344 42.4	324 28.9	24.6	288 18.1 20 00.2		350 27.0	40.1	317 47.5	39.8
R 08	359 44.9	339 28.6	25.6	303 18.7	00.6	5 29.1	40.0	332 49.7	39.9
I 09	14 47.3	354 28.4 ..	26.7	318 19.3 ..	01.0	20 31.2 ..	39.8	347 51.9 ..	39.9
D 10	29 49.8	9 28.1	27.7	333 20.0	01.4	35 33.3	39.7	2 54.1	40.0
A 11	44 52.3	24 27.9	28.7	348 20.6	01.9	50 35.4	39.5	17 56.3	40.1
Y 12	59 54.7	39 27.6 N 6 29.8		3 21.2 N20 02.3		65 37.5 S 3 39.4		32 58.5 N 8 40.2	
13	74 57.2	54 27.4	30.8	18 21.8	02.7	80 39.6	39.2	48 00.7	40.3
14	89 59.7	69 27.1	31.9	33 22.4	03.1	95 41.7	39.1	63 02.9	40.4
15	105 02.1	84 26.9 ..	32.9	48 23.0 ..	03.5	110 43.8 ..	38.9	78 05.1 ..	40.5
16	120 04.6	99 26.7	33.9	63 23.7	03.9	125 45.9	38.8	93 07.3	40.6
17	135 07.1	114 26.4	35.0	78 24.3	04.4	140 48.1	38.7	108 09.5	40.7
18	150 09.5	129 26.2 N 6 36.0		93 24.9 N20 04.8		155 50.2 S 3 38.5		123 11.7 N 8 40.8	
19	165 12.0	144 25.9	37.0	108 25.5	05.2	170 52.3	38.4	138 13.9	40.9
20	180 14.4	159 25.7	38.1	123 26.1	05.6	185 54.4	38.2	153 16.1	41.0
21	195 16.9	174 25.4 ..	39.1	138 26.7 ..	06.0	200 56.5 ..	38.1	168 18.3 ..	41.1
22	210 19.4	189 25.2	40.2	153 27.4	06.4	215 58.6	37.9	183 20.5	41.2
23	225 21.8	204 24.9	41.2	168 28.0	06.8	231 00.7	37.8	198 22.7	41.3
23 00	240 24.3	219 24.7 N 6 42.2		183 28.6 N20 07.2		246 02.8 S 3 37.6		213 24.8 N 8 41.4	
01	255 26.8	234 24.4	43.3	198 29.2	07.7	261 04.9	37.5	228 27.0	41.5
02	270 29.2	249 24.2	44.3	213 29.8	08.1	276 07.0	37.4	243 29.2	41.6
03	285 31.7	264 23.9 ..	45.3	228 30.4 ..	08.5	291 09.1 ..	37.2	258 31.4 ..	41.7
04	300 34.2	279 23.7	46.4	243 31.1	08.9	306 11.2	37.1	273 33.6	41.8
05	315 36.6	294 23.4	47.4	258 31.7	09.3	321 13.3	36.9	288 35.8	41.9
06	330 39.1	309 23.2 N 6 48.4		273 32.3 N20 09.7		336 15.4 S 3 36.8		303 38.0 N 8 42.0	
S 07	345 41.6	324 22.9	49.5	288 32.9	10.1	351 17.5	36.6	318 40.2	42.0
A 08	0 44.0	339 22.7	50.5	303 33.5	10.5	6 19.6	36.5	333 42.4	42.1
T 09	15 46.5	354 22.4 ..	51.5	318 34.1 ..	10.9	21 21.7 ..	36.3	348 44.6 ..	42.2
U 10	30 48.9	9 22.2	52.6	333 34.8	11.4	36 23.8	36.2	3 46.8	42.3
R 11	45 51.4	24 21.9	53.6	348 35.4	11.8	51 25.9	36.1	18 49.0	42.4
D 12	60 53.9	39 21.7 N 6 54.6		3 36.0 N20 12.2		66 28.0 S 3 35.9		33 51.2 N 8 42.5	
A 13	75 56.3	54 21.4	55.7	18 36.6	12.6	81 30.2	35.8	48 53.4	42.6
Y 14	90 58.8	69 21.2	56.7	33 37.2	13.0	96 32.3	35.6	63 55.6	42.7
15	106 01.3	84 20.9 ..	57.7	48 37.8 ..	13.4	111 34.4 ..	35.5	78 57.8 ..	42.8
16	121 03.7	99 20.6	58.8	63 38.4	13.8	126 36.5	35.3	94 00.0	42.9
17	136 06.2	114 20.4 6 59.8		78 39.1	14.2	141 38.6	35.2	109 02.2	43.0
18	151 08.7	129 20.1 N 7 00.9		93 39.7 N20 14.6		156 40.7 S 3 35.1		124 04.4 N 8 43.1	
19	166 11.1	144 19.9	01.9	108 40.3	15.0	171 42.8	34.9	139 06.6	43.2
20	181 13.6	159 19.6	02.9	123 40.9	15.4	186 44.9	34.8	154 08.8	43.3
21	196 16.0	174 19.4 ..	04.0	138 41.5 ..	15.8	201 47.0 ..	34.6	169 11.0 ..	43.4
22	211 18.5	189 19.1	05.0	153 42.1	16.2	216 49.1	34.5	184 13.2	43.5
23	226 21.0	204 18.9	06.0	168 42.8	16.6	231 51.2	34.3	199 15.4	43.6
24 00	241 23.4	219 18.6 N 7 07.1		183 43.4 N20 17.0		246 53.3 S 3 34.2		214 17.6 N 8 43.7	
01	256 25.9	234 18.4	08.1	198 44.0	17.5	261 55.5	34.1	229 19.8	43.8
02	271 28.4	249 18.1	09.1	213 44.6	17.9	276 57.6	33.9	244 22.0	43.9
03	286 30.8	264 17.8 ..	10.1	228 45.2 ..	18.3	291 59.7 ..	33.8	259 24.2 ..	43.9
04	301 33.3	279 17.6	11.2	243 45.8	18.7	307 01.8	33.6	274 26.3	44.0
05	316 35.8	294 17.3	12.2	258 46.4	19.1	322 03.9	33.5	289 28.5	44.1
06	331 38.2	309 17.1 N 7 13.2		273 47.1 N20 19.5		337 06.0 S 3 33.4		304 30.7 N 8 44.2	
S 07	346 40.7	324 16.8	14.3	288 47.7	19.9	352 08.1	33.2	319 32.9	44.3
U 08	1 43.2	339 16.6	15.3	303 48.3	20.3	7 10.2	33.1	334 35.1	44.4
N 09	16 45.6	354 16.3 ..	16.3	318 48.9 ..	20.7	22 12.3 ..	32.9	349 37.3 ..	44.5
D 10	31 48.1	9 16.0	17.4	333 49.5	21.1	37 14.4	32.8	4 39.5	44.6
A 11	46 50.5	24 15.8	18.4	348 50.1	21.5	52 16.6	32.6	19 41.7	44.7
Y 12	61 53.0	39 15.5 N 7 19.4		3 50.7 N20 21.9		67 18.7 S 3 32.5		34 43.9 N 8 44.8	
13	76 55.5	54 15.3	20.5	18 51.4	22.3	82 20.8	32.4	49 46.1	44.9
14	91 57.9	69 15.0	21.5	33 52.0	22.7	97 22.9	32.2	64 48.3	45.0
15	107 00.4	84 14.7 ..	22.5	48 52.6 ..	23.1	112 25.0 ..	32.1	79 50.5 ..	45.1
16	122 02.9	99 14.5	23.6	63 53.2	23.5	127 27.1	31.9	94 52.7	45.2
17	137 05.3	114 14.2	24.6	78 53.8	23.9	142 29.2	31.8	109 54.9	45.3
18	152 07.8	129 14.0 N 7 25.6		93 54.4 N20 24.3		157 31.3 S 3 31.7		124 57.1 N 8 45.4	
19	167 10.3	144 13.7	26.6	108 55.0	24.7	172 33.5	31.5	139 59.3	45.5
20	182 12.7	159 13.4	27.7	123 55.6	25.1	187 35.6	31.4	155 01.5	45.5
21	197 15.2	174 13.2 ..	28.7	138 56.3 ..	25.5	202 37.7 ..	31.2	170 03.7 ..	45.6
22	212 17.7	189 12.9	29.7	153 56.9	25.9	217 39.8	31.1	185 05.9	45.7
23	227 20.1	204 12.6	30.8	168 57.5	26.3	232 41.9	31.0	200 08.1	45.8
Mer. Pass. 7 57.1		v −0.3 d 1.0		v 0.6 d 0.4		v 2.1 d 0.1		v 2.2 d 0.1	

STARS

Name	SHA	Dec
Acamar	315 27.6	S40 18.8
Achernar	335 36.0	S57 14.7
Acrux	173 21.8	S63 05.6
Adhara	255 21.9	S28 58.5
Aldebaran	291 03.0	N16 30.2
Alioth	166 30.6	N55 58.4
Alkaid	153 07.7	N49 19.5
Al Na'ir	27 58.3	S46 57.9
Alnilam	275 58.5	S 1 12.4
Alphard	218 07.5	S 8 39.3
Alphecca	126 20.5	N26 43.3
Alpheratz	357 55.7	N29 04.6
Altair	62 19.3	N 8 51.8
Ankaa	353 27.4	S42 18.8
Antares	112 40.1	S26 25.6
Arcturus	146 06.0	N19 11.6
Atria	107 51.8	S69 01.3
Avior	234 23.0	S59 30.6
Bellatrix	278 44.8	N 6 20.7
Betelgeuse	271 14.2	N 7 24.2
Canopus	264 01.8	S52 42.0
Capella	280 52.1	N45 59.7
Deneb	49 39.1	N45 16.3
Denebola	182 45.4	N14 34.9
Diphda	349 07.7	S17 59.8
Dubhe	194 05.8	N61 45.8
Elnath	278 27.7	N28 36.2
Eltanin	90 51.1	N51 29.3
Enif	33 58.5	N 9 52.0
Fomalhaut	15 36.9	S29 37.8
Gacrux	172 13.4	S57 06.4
Gienah	176 04.1	S17 32.1
Hadar	149 03.8	S60 22.0
Hamal	328 14.2	N23 27.0
Kaus Aust.	83 58.9	S34 22.9
Kochab	137 18.6	N74 10.0
Markab	13 49.9	N15 11.6
Menkar	314 27.5	N 4 04.8
Menkent	148 20.9	S36 21.8
Miaplacidus	221 42.3	S69 43.0
Mirfak	308 57.5	N49 51.1
Nunki	76 12.4	S26 17.8
Peacock	53 37.2	S56 44.1
Pollux	243 42.2	N28 01.8
Procyon	245 12.1	N 5 13.6
Rasalhague	96 16.9	N12 33.7
Regulus	207 55.9	N11 58.5
Rigel	281 23.5	S 8 12.4
Rigil Kent.	140 07.3	S60 49.7
Sabik	102 25.5	S15 43.3
Schedar	349 54.1	N56 31.4
Shaula	96 37.3	S37 06.0
Sirius	258 44.2	S16 43.1
Spica	158 43.2	S11 09.2
Suhail	223 01.1	S43 25.9
Vega	80 46.5	N38 46.9
Zuben'ubi	137 17.9	S16 02.1

	SHA	Mer. Pass.
	° ′	h m
Venus	339 00.4	9 23
Mars	303 04.3	11 46
Jupiter	5 38.5	7 35
Saturn	333 00.5	9 45

SUN and MOON

UT	SUN GHA	SUN Dec	MOON GHA	v	MOON Dec	d	HP
d h	° '	° '	° '	'	° '	'	'
22 00	180 51.1	N20 17.9	231 13.3	8.9	N 0 42.0	11.8	60.2
01	195 51.1	18.4	245 41.2	8.9	0 53.8	11.9	60.2
02	210 51.0	18.9	260 09.1	8.8	1 05.7	11.8	60.2
03	225 51.0 ..	19.4	274 36.9	8.8	1 17.5	11.9	60.2
04	240 50.9	19.9	289 04.7	8.8	1 29.4	11.8	60.2
05	255 50.9	20.4	303 32.5	8.7	1 41.2	11.9	60.2
06	270 50.8	N20 20.9	318 00.2	8.7	N 1 53.1	11.8	60.3
F 07	285 50.8	21.4	332 27.9	8.7	2 04.9	11.8	60.3
R 08	300 50.7	21.9	346 55.6	8.6	2 16.7	11.8	60.3
I 09	315 50.7 ..	22.4	1 23.2	8.7	2 28.5	11.9	60.3
D 10	330 50.6	22.9	15 50.9	8.6	2 40.4	11.8	60.3
A 11	345 50.6	23.4	30 18.5	8.5	2 52.2	11.8	60.3
Y 12	0 50.5	N20 23.9	44 46.0	8.5	N 3 04.0	11.7	60.4
13	15 50.5	24.4	59 13.5	8.5	3 15.7	11.8	60.4
14	30 50.5	24.9	73 41.0	8.5	3 27.5	11.8	60.4
15	45 50.4 ..	25.4	88 08.5	8.4	3 39.3	11.7	60.4
16	60 50.4	25.8	102 35.9	8.4	3 51.0	11.7	60.4
17	75 50.3	26.3	117 03.3	8.4	4 02.7	11.7	60.4
18	90 50.3	N20 26.8	131 30.7	8.3	N 4 14.4	11.7	60.4
19	105 50.2	27.3	145 58.0	8.3	4 26.1	11.7	60.4
20	120 50.2	27.8	160 25.3	8.3	4 37.8	11.6	60.5
21	135 50.1 ..	28.3	174 52.6	8.2	4 49.4	11.6	60.5
22	150 50.1	28.8	189 19.8	8.2	5 01.0	11.6	60.5
23	165 50.0	29.2	203 47.0	8.1	5 12.6	11.5	60.5
23 00	180 50.0	N20 29.7	218 14.1	8.1	N 5 24.1	11.6	60.5
01	195 49.9	30.2	232 41.2	8.1	5 35.7	11.5	60.5
02	210 49.9	30.7	247 08.3	8.0	5 47.2	11.4	60.5
03	225 49.8 ..	31.2	261 35.3	8.0	5 58.6	11.5	60.5
04	240 49.8	31.7	276 02.3	8.0	6 10.1	11.4	60.5
05	255 49.7	32.1	290 29.3	7.9	6 21.5	11.3	60.6
06	270 49.7	N20 32.6	304 56.2	7.9	N 6 32.8	11.3	60.6
S 07	285 49.6	33.1	319 23.1	7.8	6 44.1	11.3	60.6
A 08	300 49.6	33.6	333 49.9	7.9	6 55.4	11.3	60.6
T 09	315 49.5 ..	34.1	348 16.8	7.7	7 06.7	11.2	60.6
U 10	330 49.4	34.5	2 43.5	7.7	7 17.9	11.1	60.6
R 11	345 49.4	35.0	17 10.2	7.7	7 29.0	11.1	60.6
D 12	0 49.3	N20 35.5	31 36.9	7.7	N 7 40.1	11.1	60.6
A 13	15 49.3	36.0	46 03.6	7.6	7 51.2	11.0	60.6
Y 14	30 49.2	36.5	60 30.2	7.6	8 02.2	11.0	60.6
15	45 49.2 ..	36.9	74 56.8	7.5	8 13.2	10.9	60.6
16	60 49.1	37.4	89 23.3	7.5	8 24.1	10.9	60.6
17	75 49.1	37.9	103 49.8	7.4	8 35.0	10.8	60.6
18	90 49.0	N20 38.4	118 16.2	7.5	N 8 45.8	10.8	60.6
19	105 49.0	38.8	132 42.7	7.3	8 56.6	10.7	60.6
20	120 48.9	39.3	147 09.0	7.4	9 07.3	10.6	60.6
21	135 48.9 ..	39.8	161 35.4	7.3	9 17.9	10.6	60.6
22	150 48.8	40.2	176 01.7	7.2	9 28.5	10.5	60.6
23	165 48.8	40.7	190 27.9	7.2	9 39.0	10.5	60.6
24 00	180 48.7	N20 41.2	204 54.1	7.2	N 9 49.5	10.4	60.6
01	195 48.6	41.6	219 20.3	7.1	9 59.9	10.3	60.6
02	210 48.6	42.1	233 46.4	7.1	10 10.2	10.3	60.6
03	225 48.5 ..	42.6	248 12.5	7.1	10 20.5	10.2	60.6
04	240 48.5	43.1	262 38.6	7.0	10 30.7	10.1	60.6
05	255 48.4	43.5	277 04.6	7.0	10 40.8	10.1	60.6
06	270 48.3	N20 44.0	291 30.6	6.9	N10 50.9	10.0	60.6
07	285 48.3	44.5	305 56.5	6.9	11 00.9	9.9	60.6
08	300 48.2	44.9	320 22.4	6.8	11 10.8	9.8	60.6
S 09	315 48.2 ..	45.4	334 48.2	6.9	11 20.6	9.8	60.6
U 10	330 48.1	45.8	349 14.1	6.7	11 30.4	9.7	60.6
N 11	345 48.1	46.3	3 39.8	6.8	11 40.1	9.6	60.6
D 12	0 48.0	N20 46.8	18 05.6	6.7	N11 49.7	9.5	60.6
A 13	15 47.9	47.2	32 31.3	6.7	11 59.2	9.5	60.6
Y 14	30 47.9	47.7	46 57.0	6.6	12 08.7	9.3	60.6
15	45 47.8 ..	48.2	61 22.6	6.6	12 18.0	9.3	60.6
16	60 47.8	48.6	75 48.2	6.5	12 27.3	9.2	60.6
17	75 47.7	49.1	90 13.7	6.6	12 36.5	9.1	60.6
18	90 47.6	N20 49.5	104 39.3	6.5	N12 45.6	9.1	60.6
19	105 47.6	50.0	119 04.8	6.4	12 54.7	8.9	60.5
20	120 47.5	50.4	133 30.2	6.3	13 03.6	8.8	60.5
21	135 47.5 ..	50.9	147 55.6	6.4	13 12.4	8.8	60.5
22	150 47.4	51.4	162 21.0	6.4	13 21.2	8.7	60.5
23	165 47.3	51.8	176 46.4	6.3	N13 29.9	8.5	60.5
	SD 15.8	d 0.5	SD 16.4		16.5		16.5

Moonrise

Lat.	Twilight Naut.	Twilight Civil	Sunrise	Moonrise 22	23	24	25
°	h m	h m	h m	h m	h m	h m	h m
N 72	☐	☐	☐	02 28	02 21	02 13	02 04
N 70	☐	☐	☐	02 29	02 29	02 29	02 31
68	////	////	00 56	02 30	02 35	02 42	02 51
66	////	////	01 51	02 31	02 41	02 52	03 07
64	////	////	02 23	02 32	02 45	03 01	03 21
62	////	01 16	02 47	02 33	02 50	03 09	03 32
60	////	01 54	03 06	02 34	02 53	03 15	03 42
N 58	////	02 21	03 21	02 34	02 56	03 21	03 50
56	01 08	02 41	03 34	02 35	02 59	03 26	03 58
54	01 44	02 57	03 45	02 35	03 02	03 31	04 04
52	02 08	03 11	03 55	02 36	03 04	03 35	04 10
50	02 27	03 23	04 04	02 36	03 06	03 39	04 16
45	03 03	03 48	04 23	02 37	03 11	03 47	04 28
N 40	03 28	04 07	04 39	02 37	03 15	03 54	04 38
35	03 48	04 23	04 51	02 38	03 18	04 00	04 46
30	04 04	04 36	05 03	02 39	03 21	04 06	04 53
20	04 29	04 58	05 22	02 40	03 26	04 15	05 06
N 10	04 49	05 15	05 38	02 41	03 31	04 23	05 18
0	05 05	05 31	05 53	02 42	03 36	04 31	05 29
S 10	05 20	05 46	06 08	02 42	03 40	04 39	05 39
20	05 34	06 01	06 24	02 43	03 45	04 48	05 51
30	05 48	06 17	06 42	02 45	03 51	04 57	06 04
35	05 55	06 26	06 53	02 45	03 54	05 03	06 12
40	06 02	06 36	07 05	02 46	03 57	05 09	06 21
45	06 11	06 47	07 19	02 47	04 02	05 17	06 31
S 50	06 20	07 00	07 37	02 48	04 07	05 26	06 44
52	06 24	07 06	07 45	02 49	04 09	05 30	06 50
54	06 28	07 13	07 54	02 49	04 12	05 35	06 57
56	06 33	07 20	08 04	02 50	04 15	05 40	07 04
58	06 38	07 28	08 16	02 51	04 18	05 46	07 12
S 60	06 44	07 38	08 30	02 51	04 22	05 53	07 21

Moonset

Lat.	Sunset	Twilight Civil	Twilight Naut.	Moonset 22	23	24	25
°	h m	h m	h m	h m	h m	h m	h m
N 72	☐	☐	☐	15 55	17 58	20 06	22 22
N 70	☐	☐	☐	15 49	17 44	19 40	21 36
68	23 06	////	////	15 45	17 33	19 21	21 07
66	22 06	////	////	15 42	17 24	19 06	20 45
64	21 33	////	////	15 38	17 16	18 54	20 27
62	21 08	22 42	////	15 36	17 10	18 43	20 13
60	20 49	22 02	////	15 33	17 04	18 34	20 01
N 58	20 34	21 35	////	15 31	16 59	18 26	19 50
56	20 20	21 14	22 49	15 30	16 55	18 20	19 41
54	20 09	20 57	22 12	15 28	16 51	18 14	19 33
52	19 59	20 43	21 47	15 27	16 47	18 08	19 26
50	19 50	20 31	21 27	15 25	16 44	18 03	19 19
45	19 31	20 06	20 52	15 22	16 37	17 52	19 06
N 40	19 15	19 47	20 26	15 20	16 32	17 44	18 54
35	19 02	19 31	20 06	15 18	16 27	17 36	18 44
30	18 51	19 18	19 50	15 16	16 22	17 29	18 36
20	18 32	18 56	19 24	15 13	16 15	17 18	18 21
N 10	18 16	18 38	19 05	15 10	16 08	17 08	18 08
0	18 00	18 22	18 48	15 07	16 02	16 58	17 56
S 10	17 45	18 08	18 33	15 04	15 56	16 49	17 44
20	17 29	17 53	18 20	15 01	15 49	16 39	17 32
30	17 11	17 36	18 06	14 58	15 42	16 28	17 17
35	17 00	17 27	17 58	14 56	15 37	16 21	17 08
40	16 48	17 18	17 51	14 54	15 32	16 14	16 59
45	16 34	17 06	17 42	14 52	15 27	16 05	16 48
S 50	16 16	16 53	17 33	14 49	15 20	15 54	16 34
52	16 08	16 47	17 29	14 47	15 17	15 50	16 28
54	15 59	16 40	17 24	14 46	15 13	15 44	16 21
56	15 49	16 33	17 20	14 44	15 10	15 39	16 13
58	15 37	16 25	17 14	14 42	15 06	15 32	16 04
S 60	15 23	16 15	17 09	14 40	15 01	15 25	15 54

SUN and MOON

Day	SUN Eqn. of Time 00h	SUN Eqn. of Time 12h	SUN Mer. Pass.	MOON Mer. Pass. Upper	MOON Mer. Pass. Lower	Age	Phase
d	m s	m s	h m	h m	h m	d	%
22	03 24	03 22	11 57	08 54	21 21	26	15
23	03 20	03 17	11 57	09 49	22 17	27	8
24	03 15	03 12	11 57	10 45	23 13	28	3

1998 MAY 25, 26, 27 (MON., TUES., WED.)

UT	ARIES GHA	VENUS −4.0 GHA	Dec	MARS +1.4 GHA	Dec	JUPITER −2.3 GHA	Dec	SATURN +0.6 GHA	Dec	STARS Name	SHA	Dec
d h	° ′	° ′	° ′	° ′	° ′	° ′	° ′	° ′	° ′		° ′	° ′
25 00	242 22.6	219 12.4	N 7 31.8	183 58.1	N20 26.7	247 44.0	S 3 30.8	215 10.3	N 8 45.9	Acamar	315 27.6	S40 18.8
01	257 25.0	234 12.1	32.8	198 58.7	27.0	262 46.1	30.7	230 12.5	46.0	Achernar	335 35.9	S57 14.6
02	272 27.5	249 11.9	33.9	213 59.3	27.4	277 48.3	30.5	245 14.7	46.1	Acrux	173 21.8	S63 05.6
03	287 30.0	264 11.6	.. 34.9	228 59.9	.. 27.8	292 50.4	.. 30.4	260 16.9	.. 46.2	Adhara	255 21.9	S28 58.5
04	302 32.4	279 11.3	35.9	244 00.6	28.2	307 52.5	30.3	275 19.1	46.3	Aldebaran	291 03.0	N16 30.2
05	317 34.9	294 11.1	36.9	259 01.2	28.6	322 54.6	30.1	290 21.3	46.4			
06	332 37.4	309 10.8	N 7 38.0	274 01.8	N20 29.0	337 56.7	S 3 30.0	305 23.5	N 8 46.5	Alioth	166 30.6	N55 58.4
07	347 39.8	324 10.5	39.0	289 02.4	29.4	352 58.8	29.8	320 25.7	46.6	Alkaid	153 07.7	N49 19.5
M 08	2 42.3	339 10.3	40.0	304 03.0	29.8	8 00.9	29.7	335 27.9	46.7	Al Na'ir	27 58.3	S46 57.9
O 09	17 44.8	354 10.0	.. 41.1	319 03.6	.. 30.2	23 03.1	.. 29.6	350 30.1	.. 46.8	Alnilam	275 58.5	S 1 12.4
N 10	32 47.2	9 09.7	42.1	334 04.2	30.6	38 05.2	29.4	5 32.3	46.9	Alphard	218 07.6	S 8 39.3
11	47 49.7	24 09.5	43.1	349 04.8	31.0	53 07.3	29.3	20 34.5	47.0			
D 12	62 52.1	39 09.2	N 7 44.1	4 05.5	N20 31.4	68 09.4	S 3 29.1	35 36.7	N 8 47.0	Alphecca	126 20.5	N26 43.3
A 13	77 54.6	54 08.9	45.2	19 06.1	31.8	83 11.5	29.0	50 38.9	47.1	Alpheratz	357 55.6	N29 04.6
Y 14	92 57.1	69 08.7	46.2	34 06.7	32.2	98 13.6	28.9	65 41.1	47.2	Altair	62 19.3	N 8 51.9
15	107 59.5	84 08.4	.. 47.2	49 07.3	.. 32.6	113 15.8	.. 28.7	80 43.3	.. 47.3	Ankaa	353 27.4	S42 18.8
16	123 02.0	99 08.1	48.2	64 07.9	32.9	128 17.9	28.6	95 45.5	47.4	Antares	112 40.1	S26 25.6
17	138 04.5	114 07.9	49.3	79 08.5	33.3	143 20.0	28.5	110 47.7	47.5			
18	153 06.9	129 07.6	N 7 50.3	94 09.1	N20 33.7	158 22.1	S 3 28.3	125 49.9	N 8 47.6	Arcturus	146 06.0	N19 11.6
19	168 09.4	144 07.3	51.3	109 09.7	34.1	173 24.2	28.2	140 52.1	47.7	Atria	107 51.8	S69 01.3
20	183 11.9	159 07.1	52.4	124 10.4	34.5	188 26.3	28.0	155 54.3	47.8	Avior	234 23.0	S59 30.6
21	198 14.3	174 06.8	.. 53.4	139 11.0	.. 34.9	203 28.5	.. 27.9	170 56.5	.. 47.9	Bellatrix	278 44.8	N 6 20.7
22	213 16.8	189 06.5	54.4	154 11.6	35.3	218 30.6	27.8	185 58.7	48.0	Betelgeuse	271 14.2	N 7 24.2
23	228 19.3	204 06.3	55.4	169 12.2	35.7	233 32.7	27.6	201 00.9	48.1			
26 00	243 21.7	219 06.0	N 7 56.5	184 12.8	N20 36.1	248 34.8	S 3 27.5	216 03.1	N 8 48.2	Canopus	264 01.8	S52 42.0
01	258 24.2	234 05.7	57.5	199 13.4	36.4	263 36.9	27.3	231 05.3	48.3	Capella	280 52.1	N45 59.6
02	273 26.6	249 05.5	58.5	214 14.0	36.8	278 39.1	27.2	246 07.5	48.4	Deneb	49 39.1	N45 16.3
03	288 29.1	264 05.2	7 59.5	229 14.6	.. 37.2	293 41.2	.. 27.1	261 09.7	.. 48.4	Denebola	182 45.4	N14 34.9
04	303 31.6	279 04.9	8 00.6	244 15.2	37.6	308 43.3	26.9	276 11.9	48.5	Diphda	349 07.7	S17 59.8
05	318 34.0	294 04.6	01.6	259 15.9	38.0	323 45.4	26.8	291 14.1	48.6			
06	333 36.5	309 04.4	N 8 02.6	274 16.5	N20 38.4	338 47.5	S 3 26.7	306 16.3	N 8 48.7	Dubhe	194 05.9	N61 45.9
07	348 39.0	324 04.1	03.6	289 17.1	38.8	353 49.7	26.5	321 18.5	48.8	Elnath	278 27.6	N28 36.2
T 08	3 41.4	339 03.8	04.7	304 17.7	39.1	8 51.8	26.4	336 20.7	48.9	Eltanin	90 51.0	N51 29.4
U 09	18 43.9	354 03.6	.. 05.7	319 18.3	.. 39.5	23 53.9	.. 26.2	351 22.9	.. 49.0	Enif	33 58.4	N 9 52.0
E 10	33 46.4	9 03.3	06.7	334 18.9	39.9	38 56.0	26.1	6 25.1	49.1	Fomalhaut	15 36.8	S29 37.7
S 11	48 48.8	24 03.0	07.7	349 19.5	40.3	53 58.1	26.0	21 27.3	49.2			
D 12	63 51.3	39 02.7	N 8 08.7	4 20.1	N20 40.7	69 00.3	S 3 25.8	36 29.5	N 8 49.3	Gacrux	172 13.5	S57 06.5
A 13	78 53.8	54 02.5	09.8	19 20.7	41.1	84 02.4	25.7	51 31.7	49.4	Gienah	176 04.1	S17 32.1
Y 14	93 56.2	69 02.2	10.8	34 21.4	41.5	99 04.5	25.6	66 33.9	49.5	Hadar	149 03.8	S60 22.0
15	108 58.7	84 01.9	.. 11.8	49 22.0	.. 41.8	114 06.6	.. 25.4	81 36.1	.. 49.6	Hamal	328 14.1	N23 27.0
16	124 01.1	99 01.6	12.8	64 22.6	42.2	129 08.8	25.3	96 38.3	49.6	Kaus Aust.	83 58.8	S34 22.9
17	139 03.6	114 01.4	13.9	79 23.2	42.6	144 10.9	25.2	111 40.5	49.7			
18	154 06.1	129 01.1	N 8 14.9	94 23.8	N20 43.0	159 13.0	S 3 25.0	126 42.7	N 8 49.8	Kochab	137 18.6	N74 10.0
19	169 08.5	144 00.8	15.9	109 24.4	43.4	174 15.1	24.9	141 44.9	49.9	Markab	13 49.9	N15 11.6
20	184 11.0	159 00.5	16.9	124 25.0	43.7	189 17.2	24.7	156 47.1	50.0	Menkar	314 27.5	N 4 04.8
21	199 13.5	174 00.3	.. 17.9	139 25.6	.. 44.1	204 19.4	.. 24.6	171 49.3	.. 50.1	Menkent	148 20.9	S36 21.8
22	214 15.9	189 00.0	19.0	154 26.2	44.5	219 21.5	24.5	186 51.5	50.2	Miaplacidus	221 42.3	S69 43.0
23	229 18.4	203 59.7	20.0	169 26.9	44.9	234 23.6	24.3	201 53.7	50.3			
27 00	244 20.9	218 59.4	N 8 21.0	184 27.5	N20 45.3	249 25.7	S 3 24.2	216 55.9	N 8 50.4	Mirfak	308 57.4	N49 51.1
01	259 23.3	233 59.2	22.0	199 28.1	45.6	264 27.9	24.1	231 58.1	50.5	Nunki	76 12.4	S26 17.8
02	274 25.8	248 58.9	23.0	214 28.7	46.0	279 30.0	23.9	247 00.3	50.6	Peacock	53 37.2	S56 44.1
03	289 28.3	263 58.6	.. 24.1	229 29.3	.. 46.4	294 32.1	.. 23.8	262 02.5	.. 50.7	Pollux	243 42.2	N28 01.8
04	304 30.7	278 58.3	25.1	244 29.9	46.8	309 34.2	23.7	277 04.7	50.8	Procyon	245 12.1	N 5 13.6
05	319 33.2	293 58.0	26.1	259 30.5	47.2	324 36.4	23.5	292 06.9	50.8			
06	334 35.6	308 57.8	N 8 27.1	274 31.1	N20 47.5	339 38.5	S 3 23.4	307 09.1	N 8 50.9	Rasalhague	96 16.9	N12 33.7
W 07	349 38.1	323 57.5	28.1	289 31.7	47.9	354 40.6	23.3	322 11.3	51.0	Regulus	207 55.9	N11 58.5
E 08	4 40.6	338 57.2	29.2	304 32.3	48.3	9 42.7	23.1	337 13.5	51.1	Rigel	281 23.5	S 8 12.4
D 09	19 43.0	353 56.9	.. 30.2	319 33.0	.. 48.7	24 44.9	.. 23.0	352 15.7	.. 51.2	Rigil Kent.	140 07.0	S60 49.7
N 10	34 45.5	8 56.6	31.2	334 33.6	49.0	39 47.0	22.8	7 18.0	51.3	Sabik	102 25.5	S15 43.3
E 11	49 48.0	23 56.4	32.2	349 34.2	49.4	54 49.1	22.7	22 20.2	51.4			
S 12	64 50.4	38 56.1	N 8 33.2	4 34.8	N20 49.8	69 51.2	S 3 22.6	37 22.4	N 8 51.5	Schedar	349 54.0	N56 31.4
D 13	79 52.9	53 55.8	34.3	19 35.4	50.2	84 53.4	22.4	52 24.6	51.6	Shaula	96 37.3	S37 06.0
A 14	94 55.4	68 55.5	35.3	34 36.0	50.5	99 55.5	22.3	67 26.8	51.7	Sirius	258 44.2	S16 43.1
Y 15	109 57.8	83 55.2	.. 36.3	49 36.6	.. 50.9	114 57.6	.. 22.2	82 29.0	.. 51.8	Spica	158 43.2	S11 09.2
16	125 00.3	98 55.0	37.3	64 37.2	51.3	129 59.7	22.0	97 31.2	51.9	Suhail	223 01.1	S43 25.9
17	140 02.8	113 54.7	38.3	79 37.8	51.7	145 01.9	21.9	112 33.4	51.9			
18	155 05.2	128 54.4	N 8 39.3	94 38.4	N20 52.0	160 04.0	S 3 21.8	127 35.6	N 8 52.0	Vega	80 46.4	N38 46.9
19	170 07.7	143 54.1	40.4	109 39.0	52.4	175 06.1	21.6	142 37.8	52.1	Zuben'ubi	137 17.9	S16 02.1
20	185 10.1	158 53.8	41.4	124 39.7	52.8	190 08.3	21.5	157 40.0	52.2			
21	200 12.6	173 53.5	.. 42.4	139 40.3	.. 53.2	205 10.4	.. 21.4	172 42.2	.. 52.3		SHA	Mer. Pass.
22	215 15.1	188 53.3	43.4	154 40.9	53.5	220 12.5	21.2	187 44.4	52.4	Venus	335 44.3	9 24
23	230 17.5	203 53.0	44.4	169 41.5	53.9	235 14.6	21.1	202 46.6	52.5	Mars	300 51.1	11 43
Mer. Pass. 7 45.3		v −0.3 d 1.0		v 0.6 d 0.4		v 2.1 d 0.1		v 2.2 d 0.1		Jupiter	5 13.1	7 25
										Saturn	332 41.4	9 34

UT	SUN GHA	SUN Dec	MOON GHA	v	Dec	d	HP
	° ′	° ′	° ′	′	° ′	′	′
25 00	180 47.3	N20 52.3	191 11.7	6.3	N13 38.4	8.5	60.5
01	195 47.2	52.7	205 37.0	6.2	13 46.9	8.4	60.5
02	210 47.2	53.2	220 02.2	6.2	13 55.3	8.3	60.5
03	225 47.1	.. 53.6	234 27.4	6.2	14 03.6	8.2	60.5
04	240 47.0	54.1	248 52.6	6.2	14 11.8	8.1	60.5
05	255 47.0	54.5	263 17.8	6.1	14 19.9	7.9	60.4
06	270 46.9	N20 55.0	277 42.9	6.2	N14 27.8	7.9	60.4
07	285 46.9	55.4	292 08.1	6.0	14 35.7	7.8	60.4
M 08	300 46.8	55.9	306 33.1	6.1	14 43.5	7.7	60.4
O 09	315 46.7	.. 56.3	320 58.2	6.0	14 51.2	7.6	60.4
N 10	330 46.7	56.8	335 23.2	6.0	14 58.8	7.4	60.4
D 11	345 46.6	57.2	349 48.2	6.0	15 06.2	7.4	60.3
A 12	0 46.5	N20 57.7	4 13.2	6.0	N15 13.6	7.3	60.3
Y 13	15 46.5	58.1	18 38.2	5.9	15 20.9	7.1	60.3
14	30 46.4	58.6	33 03.1	5.9	15 28.0	7.1	60.3
15	45 46.3	.. 59.0	47 28.0	5.9	15 35.1	6.9	60.3
16	60 46.3	59.5	61 52.9	5.9	15 42.0	6.8	60.3
17	75 46.2	20 59.9	76 17.8	5.8	15 48.8	6.7	60.2
18	90 46.1	N21 00.4	90 42.6	5.8	N15 55.5	6.6	60.2
19	105 46.1	00.8	105 07.4	5.9	16 02.1	6.5	60.2
20	120 46.0	01.2	119 32.3	5.8	16 08.6	6.4	60.2
21	135 46.0	.. 01.7	133 57.1	5.8	16 15.0	6.2	60.2
22	150 45.9	02.1	148 21.9	5.7	16 21.2	6.2	60.1
23	165 45.8	02.6	162 46.6	5.8	16 27.4	6.0	60.1
26 00	180 45.8	N21 03.0	177 11.4	5.7	N16 33.4	5.9	60.1
01	195 45.7	03.4	191 36.1	5.8	16 39.3	5.8	60.1
02	210 45.6	03.9	206 00.9	5.7	16 45.1	5.7	60.1
03	225 45.6	.. 04.3	220 25.6	5.7	16 50.8	5.5	60.0
04	240 45.5	04.8	234 50.3	5.7	16 56.3	5.4	60.0
05	255 45.4	05.2	249 15.0	5.7	17 01.7	5.4	60.0
06	270 45.4	N21 05.6	263 39.7	5.7	N17 07.1	5.1	60.0
07	285 45.3	06.1	278 04.4	5.7	17 12.2	5.1	59.9
T 08	300 45.2	06.5	292 29.1	5.7	17 17.3	5.0	59.9
U 09	315 45.2	.. 06.9	306 53.8	5.7	17 22.3	4.8	59.9
E 10	330 45.1	07.4	321 18.5	5.7	17 27.1	4.7	59.9
S 11	345 45.0	07.8	335 43.2	5.7	17 31.8	4.6	59.8
D 12	0 44.9	N21 08.2	350 07.9	5.7	N17 36.4	4.4	59.8
A 13	15 44.9	08.7	4 32.6	5.7	17 40.8	4.4	59.8
Y 14	30 44.8	09.1	18 57.3	5.7	17 45.2	4.2	59.8
15	45 44.7	.. 09.5	33 22.0	5.7	17 49.4	4.1	59.7
16	60 44.7	10.0	47 46.7	5.7	17 53.5	3.9	59.7
17	75 44.6	10.4	62 11.4	5.7	17 57.4	3.9	59.7
18	90 44.5	N21 10.8	76 36.1	5.8	N18 01.3	3.7	59.7
19	105 44.5	11.2	91 00.9	5.7	18 05.0	3.6	59.6
20	120 44.4	11.7	105 25.6	5.8	18 08.6	3.4	59.6
21	135 44.3	.. 12.1	119 50.4	5.7	18 12.0	3.4	59.6
22	150 44.2	12.5	134 15.1	5.8	18 15.4	3.2	59.5
23	165 44.2	12.9	148 39.9	5.8	18 18.6	3.1	59.5
27 00	180 44.1	N21 13.4	163 04.7	5.8	N18 21.7	2.9	59.5
01	195 44.0	13.8	177 29.5	5.9	18 24.6	2.9	59.5
02	210 44.0	14.2	191 54.4	5.8	18 27.5	2.7	59.4
03	225 43.9	.. 14.6	206 19.2	5.9	18 30.2	2.6	59.4
04	240 43.8	15.1	220 44.1	5.9	18 32.8	2.4	59.4
05	255 43.7	15.5	235 09.0	5.9	18 35.2	2.3	59.3
06	270 43.7	N21 15.9	249 33.9	5.9	N18 37.5	2.2	59.3
07	285 43.6	16.3	263 58.8	6.0	18 39.7	2.1	59.3
W 08	300 43.5	16.7	278 23.8	6.0	18 41.8	2.0	59.2
E 09	315 43.5	.. 17.2	292 48.8	6.0	18 43.8	1.8	59.2
D 10	330 43.4	17.6	307 13.8	6.1	18 45.6	1.7	59.2
N 11	345 43.3	18.0	321 38.9	6.1	18 47.3	1.6	59.1
E 12	0 43.2	N21 18.4	336 04.0	6.1	N18 48.9	1.4	59.1
S 13	15 43.2	18.8	350 29.1	6.1	18 50.3	1.3	59.1
D 14	30 43.1	19.3	4 54.2	6.2	18 51.6	1.2	59.0
A 15	45 43.0	.. 19.7	19 19.4	6.2	18 52.8	1.1	59.0
Y 16	60 42.9	20.1	33 44.6	6.3	18 53.9	1.0	59.0
17	75 42.9	20.5	48 09.9	6.3	18 54.9	0.8	58.9
18	90 42.8	N21 20.9	62 35.2	6.3	N18 55.7	0.7	58.9
19	105 42.7	21.3	77 00.5	6.4	18 56.4	0.6	58.9
20	120 42.6	21.7	91 25.9	6.4	18 57.0	0.4	58.8
21	135 42.6	.. 22.1	105 51.3	6.5	18 57.4	0.3	58.8
22	150 42.5	22.6	120 16.8	6.5	18 57.8	0.2	58.8
23	165 42.4	23.0	134 42.3	6.5	N18 58.0	0.1	58.7
	SD 15.8	d 0.4	SD 16.4		16.3		16.1

Twilight / Sunrise / Moonrise

Lat.	Naut.	Civil	Sunrise	Moonrise 25	26	27	28
°	h m	h m	h m	h m	h m	h m	h m
N 72	□	□	□	02 04	01 50	□	□
N 70	□	□	□	02 31	02 36	02 52	□
68	////	////	00 21	02 51	03 07	03 35	03 34
66	////	////	01 39	03 07	03 30	04 04	04 24
64	////	////	02 14	03 21	03 48	04 26	04 55
62	////	01 00	02 40	03 32	04 03	04 44	05 19
60	////	01 45	03 00	03 42	04 15	04 58	05 53
N 58	////	02 13	03 16	03 50	04 26	05 11	06 06
56	00 54	02 35	03 30	03 58	04 36	05 22	06 17
54	01 35	02 52	03 42	04 04	04 44	05 31	06 27
52	02 02	03 07	03 52	04 10	04 52	05 40	06 35
50	02 22	03 20	04 01	04 16	04 58	05 48	06 43
45	02 59	03 45	04 21	04 28	05 13	06 04	07 00
N 40	03 25	04 05	04 37	04 38	05 25	06 18	07 14
35	03 46	04 21	04 50	04 46	05 36	06 29	07 26
30	04 02	04 35	05 01	04 53	05 45	06 39	07 36
20	04 28	04 57	05 21	05 06	06 00	06 56	07 53
N 10	04 49	05 15	05 38	05 18	06 14	07 12	08 09
0	05 05	05 31	05 53	05 29	06 27	07 26	08 23
S 10	05 21	05 46	06 09	05 39	06 40	07 40	08 38
20	05 35	06 02	06 25	05 51	06 54	07 55	08 53
30	05 49	06 18	06 44	06 04	07 10	08 13	09 11
35	05 57	06 28	06 55	06 12	07 19	08 23	09 21
40	06 04	06 38	07 08	06 21	07 30	08 35	09 33
45	06 13	06 50	07 22	06 31	07 43	08 49	09 47
S 50	06 23	07 03	07 40	06 44	07 58	09 05	10 04
52	06 27	07 10	07 49	06 50	08 05	09 13	10 12
54	06 32	07 17	07 58	06 57	08 13	09 22	10 20
56	06 37	07 24	08 09	07 04	08 22	09 32	10 30
58	06 42	07 33	08 21	07 12	08 33	09 44	10 42
S 60	06 48	07 43	08 35	07 21	08 44	09 57	10 55

Sunset / Twilight / Moonset

Lat.	Sunset	Civil	Naut.	Moonset 25	26	27	28
°	h m	h m	h m	h m	h m	h m	h m
N 72	□	□	□	22 22	□	□	□
N 70	□	□	□	21 36	23 24	24 43	00 43
68	□	□	□	21 07	22 41	23 53	24 38
66	22 19	////	////	20 45	22 13	23 22	24 09
64	21 42	////	////	20 27	21 51	22 58	23 48
62	21 16	23 00	////	20 13	21 33	22 40	23 30
60	20 56	22 12	////	20 01	21 19	22 24	23 16
N 58	20 39	21 42	////	19 50	21 06	22 11	23 03
56	20 25	21 21	23 05	19 41	20 56	22 00	22 53
54	20 13	21 03	22 21	19 33	20 46	21 50	22 43
52	20 03	20 48	21 54	19 26	20 38	21 41	22 35
50	19 53	20 35	21 33	19 19	20 30	21 33	22 27
45	19 34	20 10	20 56	19 06	20 14	21 17	22 11
N 40	19 18	19 50	20 29	18 54	20 01	21 03	21 57
35	19 05	19 33	20 09	18 44	19 50	20 51	21 46
30	18 53	19 20	19 52	18 36	19 40	20 41	21 36
20	18 33	18 57	19 26	18 21	19 23	20 23	21 19
N 10	18 16	18 39	19 06	18 08	19 09	20 08	21 04
0	18 01	18 23	18 49	17 56	18 55	19 53	20 50
S 10	17 45	18 07	18 33	17 44	18 41	19 39	20 36
20	17 28	17 52	18 19	17 32	18 27	19 23	20 21
30	17 10	17 35	18 05	17 17	18 10	19 06	20 03
35	16 59	17 26	17 57	17 08	18 00	18 55	19 53
40	16 46	17 16	17 49	16 59	17 49	18 44	19 42
45	16 31	17 04	17 40	16 48	17 36	18 30	19 28
S 50	16 13	16 50	17 31	16 34	17 20	18 13	19 11
52	16 05	16 44	17 26	16 28	17 12	18 05	19 03
54	15 55	16 37	17 22	16 21	17 04	17 56	18 55
56	15 45	16 29	17 17	16 13	16 55	17 46	18 45
58	15 32	16 21	17 11	16 04	16 44	17 34	18 34
S 60	15 18	16 11	17 05	15 54	16 32	17 21	18 21

SUN / MOON

Day	Eqn. of Time 00h	12h	Mer. Pass.	Mer. Pass. Upper	Lower	Age	Phase
d	m s	m s	h m	h m	h m	d	%
25	03 09	03 06	11 57	11 42	24 12	29	0
26	03 03	03 00	11 57	12 41	00 12	01	1
27	02 57	02 53	11 57	13 40	01 10	02	4

UT	ARIES GHA	VENUS −4.0 GHA	VENUS Dec	MARS +1.4 GHA	MARS Dec	JUPITER −2.3 GHA	JUPITER Dec	SATURN +0.6 GHA	SATURN Dec	STARS Name	SHA	Dec
28 00	245 20.0	218 52.7	N 8 45.4	184 42.1	N20 54.3	250 16.8	S 3 21.0	217 48.8	N 8 52.6	Acamar	315 27.6	S40 18.8
01	260 22.5	233 52.4	46.5	199 42.7	54.6	265 18.9	20.8	232 51.0	52.7	Achernar	335 35.9	S57 14.6
02	275 24.9	248 52.1	47.5	214 43.3	55.0	280 21.0	20.7	247 53.2	52.8	Acrux	173 21.8	S63 05.6
03	290 27.4	263 51.8	.. 48.5	229 43.9	.. 55.4	295 23.2	.. 20.6	262 55.4	.. 52.9	Adhara	255 21.9	S28 58.5
04	305 29.9	278 51.5	49.5	244 44.5	55.8	310 25.3	20.4	277 57.6	52.9	Aldebaran	291 03.0	N16 30.2
05	320 32.3	293 51.3	50.5	259 45.1	56.1	325 27.4	20.3	292 59.8	53.0			
06	335 34.8	308 51.0	N 8 51.5	274 45.7	N20 56.5	340 29.6	S 3 20.2	308 02.0	N 8 53.1	Alioth	166 30.6	N55 58.4
07	350 37.3	323 50.7	52.6	289 46.3	56.9	355 31.7	20.0	323 04.2	53.2	Alkaid	153 07.7	N49 19.5
T 08	5 39.7	338 50.4	53.6	304 47.0	57.2	10 33.8	19.9	338 06.4	53.3	Al Na'ir	27 58.2	S46 57.9
H 09	20 42.2	353 50.1	.. 54.6	319 47.6	.. 57.6	25 35.9	.. 19.8	353 08.6	.. 53.4	Alnilam	275 58.5	S 1 12.4
U 10	35 44.6	8 49.8	55.6	334 48.2	58.0	40 38.1	19.6	8 10.8	53.5	Alphard	218 07.6	S 8 39.3
R 11	50 47.1	23 49.5	56.6	349 48.8	58.3	55 40.2	19.5	23 13.0	53.6			
S 12	65 49.6	38 49.2	N 8 57.6	4 49.4	N20 58.7	70 42.3	S 3 19.4	38 15.2	N 8 53.7	Alphecca	126 20.5	N26 43.3
D 13	80 52.0	53 49.0	58.6	19 50.0	59.1	85 44.5	19.2	53 17.5	53.8	Alpheratz	357 55.6	N29 04.6
A 14	95 54.5	68 48.7	8 59.6	34 50.6	59.4	100 46.6	19.1	68 19.7	53.9	Altair	62 19.3	N 8 51.9
Y 15	110 57.0	83 48.4	9 00.7	49 51.2	20 59.8	115 48.7	.. 19.0	83 21.9	.. 53.9	Ankaa	353 27.4	S42 18.8
16	125 59.4	98 48.1	01.7	64 51.8	21 00.2	130 50.9	18.8	98 24.1	54.0	Antares	112 40.1	S26 25.6
17	141 01.9	113 47.8	02.7	79 52.4	00.5	145 53.0	18.7	113 26.3	54.1			
18	156 04.4	128 47.5	N 9 03.7	94 53.0	N21 00.9	160 55.1	S 3 18.6	128 28.5	N 8 54.2	Arcturus	146 06.0	N19 11.6
19	171 06.8	143 47.2	04.7	109 53.6	01.3	175 57.3	18.5	143 30.7	54.3	Atria	107 51.8	S69 01.3
20	186 09.3	158 46.9	05.7	124 54.3	01.6	190 59.4	18.3	158 32.9	54.4	Avior	234 23.0	S59 30.6
21	201 11.7	173 46.6	.. 06.7	139 54.9	.. 02.0	206 01.5	.. 18.2	173 35.1	.. 54.5	Bellatrix	278 44.8	N 6 20.7
22	216 14.2	188 46.3	07.7	154 55.5	02.4	221 03.7	18.1	188 37.3	54.6	Betelgeuse	271 14.2	N 7 24.2
23	231 16.7	203 46.0	08.8	169 56.1	02.7	236 05.8	17.9	203 39.5	54.7			
29 00	246 19.1	218 45.8	N 9 09.8	184 56.7	N21 03.1	251 07.9	S 3 17.8	218 41.7	N 8 54.8	Canopus	264 01.8	S52 42.0
01	261 21.6	233 45.5	10.8	199 57.3	03.4	266 10.1	17.7	233 43.9	54.9	Capella	280 52.0	N45 59.6
02	276 24.1	248 45.2	11.8	214 57.9	03.8	281 12.2	17.5	248 46.1	54.9	Deneb	49 39.1	N45 16.3
03	291 26.5	263 44.9	.. 12.8	229 58.5	.. 04.2	296 14.3	.. 17.4	263 48.3	.. 55.0	Denebola	182 45.4	N14 34.9
04	306 29.0	278 44.6	13.8	244 59.1	04.5	311 16.5	17.3	278 50.5	55.1	Diphda	349 07.7	S17 59.8
05	321 31.5	293 44.3	14.8	259 59.7	04.9	326 18.6	17.1	293 52.7	55.2			
06	336 33.9	308 44.0	N 9 15.8	275 00.3	N21 05.2	341 20.7	S 3 17.0	308 54.9	N 8 55.3	Dubhe	194 05.9	N61 45.9
07	351 36.4	323 43.7	16.8	290 00.9	05.6	356 22.9	16.9	323 57.2	55.4	Elnath	278 27.6	N28 36.2
08	6 38.9	338 43.4	17.8	305 01.5	06.0	11 25.0	16.7	338 59.4	55.5	Eltanin	90 51.0	N51 29.4
F 09	21 41.3	353 43.1	.. 18.8	320 02.1	.. 06.3	26 27.1	.. 16.6	354 01.6	.. 55.6	Enif	33 58.4	N 9 52.0
R 10	36 43.8	8 42.8	19.9	335 02.8	06.7	41 29.3	16.5	9 03.8	55.7	Fomalhaut	15 36.8	S29 37.7
I 11	51 46.2	23 42.5	20.9	350 03.4	07.0	56 31.4	16.4	24 06.0	55.8			
D 12	66 48.7	38 42.2	N 9 21.9	5 04.0	N21 07.4	71 33.6	S 3 16.2	39 08.2	N 8 55.8	Gacrux	172 13.5	S57 06.5
A 13	81 51.2	53 41.9	22.9	20 04.6	07.8	86 35.7	16.1	54 10.4	55.9	Gienah	176 04.1	S17 32.1
Y 14	96 53.6	68 41.6	23.9	35 05.2	08.1	101 37.8	16.0	69 12.6	56.0	Hadar	149 03.8	S60 22.0
15	111 56.1	83 41.3	.. 24.9	50 05.8	.. 08.5	116 40.0	.. 15.8	84 14.8	.. 56.1	Hamal	328 14.1	N23 27.0
16	126 58.6	98 41.0	25.9	65 06.4	08.8	131 42.1	15.7	99 17.0	56.2	Kaus Aust.	83 58.8	S34 22.9
17	142 01.0	113 40.7	26.9	80 07.0	09.2	146 44.2	15.6	114 19.2	56.3			
18	157 03.5	128 40.4	N 9 27.9	95 07.6	N21 09.5	161 46.4	S 3 15.4	129 21.4	N 8 56.4	Kochab	137 18.7	N74 10.0
19	172 06.0	143 40.1	28.9	110 08.2	09.9	176 48.5	15.3	144 23.6	56.5	Markab	13 49.9	N15 11.6
20	187 08.4	158 39.8	29.9	125 08.8	10.3	191 50.7	15.2	159 25.8	56.6	Menkar	314 27.5	N 4 04.8
21	202 10.9	173 39.5	.. 30.9	140 09.4	.. 10.6	206 52.8	.. 15.1	174 28.0	.. 56.6	Menkent	148 20.9	S36 21.8
22	217 13.4	188 39.2	31.9	155 10.0	11.0	221 54.9	14.9	189 30.3	56.7	Miaplacidus	221 42.3	S69 43.0
23	232 15.8	203 38.9	32.9	170 10.6	11.3	236 57.1	14.8	204 32.5	56.8			
30 00	247 18.3	218 38.6	N 9 33.9	185 11.2	N21 11.7	251 59.2	S 3 14.7	219 34.7	N 8 56.9	Mirfak	308 57.4	N49 51.1
01	262 20.7	233 38.3	34.9	200 11.9	12.0	267 01.3	14.5	234 36.9	57.0	Nunki	76 12.4	S26 17.8
02	277 23.2	248 38.0	35.9	215 12.5	12.4	282 03.5	14.4	249 39.1	57.1	Peacock	53 37.2	S56 44.1
03	292 25.7	263 37.7	.. 37.0	230 13.1	.. 12.7	297 05.6	.. 14.3	264 41.3	.. 57.2	Pollux	243 42.2	N28 01.8
04	307 28.1	278 37.4	38.0	245 13.7	13.1	312 07.8	14.2	279 43.5	57.3	Procyon	245 12.1	N 5 13.6
05	322 30.6	293 37.1	39.0	260 14.3	13.4	327 09.9	14.0	294 45.7	57.4			
06	337 33.1	308 36.8	N 9 40.0	275 14.9	N21 13.8	342 12.0	S 3 13.9	309 47.9	N 8 57.5	Rasalhague	96 16.9	N12 33.8
07	352 35.5	323 36.5	41.0	290 15.5	14.2	357 14.2	13.8	324 50.1	57.5	Regulus	207 55.9	N11 58.5
S 08	7 38.0	338 36.2	42.0	305 16.1	14.5	12 16.3	13.6	339 52.3	57.6	Rigel	281 23.5	S 8 12.4
A 09	22 40.5	353 35.9	.. 43.0	320 16.7	.. 14.9	27 18.5	.. 13.5	354 54.5	.. 57.7	Rigil Kent.	140 07.0	S60 49.7
T 10	37 42.9	8 35.6	44.0	335 17.3	15.2	42 20.6	13.4	9 56.7	57.8	Sabik	102 25.5	S15 43.3
U 11	52 45.4	23 35.3	45.0	350 17.9	15.6	57 22.8	13.3	24 59.0	57.9			
R 12	67 47.9	38 35.0	N 9 46.0	5 18.5	N21 15.9	72 24.9	S 3 13.1	40 01.2	N 8 58.0	Schedar	349 54.0	N56 31.4
D 13	82 50.3	53 34.7	47.0	20 19.1	16.3	87 27.0	13.0	55 03.4	58.1	Shaula	96 37.2	S37 06.0
A 14	97 52.8	68 34.4	48.0	35 19.7	16.6	102 29.2	12.9	70 05.6	58.2	Sirius	258 44.2	S16 43.0
Y 15	112 55.2	83 34.1	.. 49.0	50 20.3	.. 17.0	117 31.3	.. 12.7	85 07.8	.. 58.3	Spica	158 43.2	S11 09.2
16	127 57.7	98 33.8	50.0	65 20.9	17.3	132 33.5	12.6	100 10.0	58.3	Suhail	223 01.1	S43 25.9
17	143 00.2	113 33.5	51.0	80 21.5	17.7	147 35.6	12.5	115 12.2	58.4			
18	158 02.6	128 33.2	N 9 52.0	95 22.1	N21 18.0	162 37.7	S 3 12.4	130 14.4	N 8 58.5	Vega	80 46.4	N38 46.9
19	173 05.1	143 32.9	53.0	110 22.8	18.3	177 39.9	12.2	145 16.6	58.6	Zuben'ubi	137 17.9	S16 02.1
20	188 07.6	158 32.6	54.0	125 23.4	18.7	192 42.0	12.1	160 18.8	58.7			
21	203 10.0	173 32.2	.. 55.0	140 24.0	.. 19.0	207 44.2	.. 12.0	175 21.0	.. 58.8			
22	218 12.5	188 31.9	56.0	155 24.6	19.4	222 46.3	11.9	190 23.2	58.9			
23	233 15.0	203 31.6	57.0	170 25.2	19.7	237 48.5	11.7	205 25.5	59.0			

	SHA	Mer. Pass.
	° ′	h m
Venus	332 26.6	9 25
Mars	298 37.5	11 40
Jupiter	4 48.8	7 14
Saturn	332 22.6	9 24

	h m								
Mer. Pass.	7 33.5	*v* −0.3	*d* 1.0	*v* 0.6	*d* 0.4	*v* 2.1	*d* 0.1	*v* 2.2	*d* 0.1

UT	SUN GHA	Dec	MOON GHA	v	Dec	d	HP
d h	° ′	° ′	° ′	′	° ′	′	′
28 00	180 42.3	N21 23.4	149 07.8	6.6	N18 58.1	0.1	58.7
01	195 42.3	23.8	163 33.4	6.6	18 58.0	0.1	58.7
02	210 42.2	24.2	177 59.0	6.7	18 57.9	0.3	58.6
03	225 42.1 ..	24.6	192 24.7	6.8	18 57.6	0.4	58.6
04	240 42.0	25.0	206 50.5	6.8	18 57.2	0.5	58.6
05	255 42.0	25.4	221 16.3	6.8	18 56.7	0.6	58.5
06	270 41.9 N21	25.8	235 42.1	6.9	N18 56.1	0.8	58.5
T 07	285 41.8	26.2	250 08.0	6.9	18 55.3	0.9	58.5
H 08	300 41.7	26.6	264 33.9	7.0	18 54.4	0.9	58.4
U 09	315 41.6 ..	27.0	278 59.9	7.1	18 53.5	1.1	58.4
R 10	330 41.6	27.4	293 26.0	7.1	18 52.4	1.3	58.4
S 11	345 41.5	27.8	307 52.1	7.2	18 51.1	1.3	58.3
D 12	0 41.4 N21	28.2	322 18.3	7.2	N18 49.8	1.4	58.3
A 13	15 41.3	28.6	336 44.5	7.3	18 48.4	1.6	58.2
Y 14	30 41.3	29.0	351 10.8	7.3	18 46.8	1.7	58.2
15	45 41.2 ..	29.4	5 37.1	7.4	18 45.1	1.8	58.2
16	60 41.1	29.8	20 03.5	7.5	18 43.3	1.9	58.1
17	75 41.0	30.2	34 30.0	7.5	18 41.4	2.0	58.1
18	90 40.9 N21	30.6	48 56.5	7.6	N18 39.4	2.1	58.1
19	105 40.9	31.0	63 23.1	7.6	18 37.3	2.2	58.0
20	120 40.8	31.4	77 49.7	7.7	18 35.1	2.4	58.0
21	135 40.7 ..	31.8	92 16.4	7.8	18 32.7	2.4	58.0
22	150 40.6	32.2	106 43.2	7.8	18 30.3	2.6	57.9
23	165 40.5	32.6	121 10.0	8.0	18 27.7	2.7	57.9
29 00	180 40.5 N21	33.0	135 37.0	7.9	N18 25.0	2.7	57.8
01	195 40.4	33.4	150 03.9	8.1	18 22.3	2.9	57.8
02	210 40.3	33.8	164 31.0	8.1	18 19.4	3.0	57.8
03	225 40.2 ..	34.2	178 58.1	8.2	18 16.4	3.1	57.7
04	240 40.1	34.6	193 25.3	8.2	18 13.3	3.1	57.7
05	255 40.1	35.0	207 52.5	8.3	18 10.2	3.3	57.7
06	270 40.0 N21	35.4	222 19.8	8.4	N18 06.9	3.4	57.6
F 07	285 39.9	35.8	236 47.2	8.5	18 03.5	3.5	57.6
R 08	300 39.8	36.1	251 14.7	8.5	18 00.0	3.6	57.5
I 09	315 39.7 ..	36.5	265 42.2	8.6	17 56.4	3.7	57.5
D 10	330 39.6	36.9	280 09.8	8.7	17 52.7	3.8	57.5
A 11	345 39.6	37.3	294 37.5	8.7	17 48.9	3.9	57.4
Y 12	0 39.5 N21	37.7	309 05.2	8.8	N17 45.0	3.9	57.4
13	15 39.4	38.1	323 33.0	8.9	17 41.1	4.1	57.4
14	30 39.3	38.5	338 00.9	9.0	17 37.0	4.2	57.3
15	45 39.2 ..	38.8	352 28.9	9.0	17 32.8	4.2	57.3
16	60 39.1	39.2	6 56.9	9.1	17 28.6	4.4	57.3
17	75 39.1	39.6	21 25.0	9.2	17 24.2	4.4	57.2
18	90 39.0 N21	40.0	35 53.2	9.2	N17 19.8	4.6	57.2
19	105 38.9	40.4	50 21.4	9.4	17 15.2	4.6	57.1
20	120 38.8	40.8	64 49.8	9.4	17 10.6	4.7	57.1
21	135 38.7 ..	41.1	79 18.2	9.4	17 05.9	4.8	57.1
22	150 38.6	41.5	93 46.6	9.6	17 01.1	4.9	57.0
23	165 38.6	41.9	108 15.2	9.6	16 56.2	5.0	57.0
30 00	180 38.5 N21	42.3	122 43.8	9.7	N16 51.2	5.0	57.0
01	195 38.4	42.6	137 12.5	9.8	16 46.2	5.2	56.9
02	210 38.3	43.0	151 41.3	9.8	16 41.0	5.2	56.9
03	225 38.2 ..	43.4	166 10.1	10.0	16 35.8	5.3	56.9
04	240 38.1	43.8	180 39.1	10.0	16 30.5	5.4	56.8
05	255 38.0	44.2	195 08.1	10.0	16 25.1	5.5	56.8
06	270 38.0 N21	44.5	209 37.1	10.2	N16 19.6	5.5	56.8
S 07	285 37.9	44.9	224 06.3	10.2	16 14.1	5.7	56.7
A 08	300 37.8	45.3	238 35.5	10.3	16 08.4	5.7	56.7
T 09	315 37.7 ..	45.6	253 04.8	10.4	16 02.7	5.8	56.6
U 10	330 37.6	46.0	267 34.2	10.4	15 56.9	5.8	56.6
R 11	345 37.5	46.4	282 03.6	10.5	15 51.1	6.0	56.6
D 12	0 37.4 N21	46.8	296 33.1	10.6	N15 45.1	6.0	56.5
A 13	15 37.3	47.1	311 02.7	10.7	15 39.1	6.1	56.5
Y 14	30 37.3	47.5	325 32.4	10.8	15 33.0	6.1	56.5
15	45 37.2 ..	47.9	340 02.2	10.8	15 26.9	6.3	56.4
16	60 37.1	48.2	354 32.0	10.9	15 20.6	6.3	56.4
17	75 37.0	48.6	9 01.9	10.9	15 14.3	6.4	56.4
18	90 36.9 N21	49.0	23 31.8	11.1	N15 07.9	6.4	56.3
19	105 36.8	49.3	38 01.9	11.1	15 01.5	6.5	56.3
20	120 36.7	49.7	52 32.0	11.2	14 55.0	6.6	56.3
21	135 36.6 ..	50.0	67 02.2	11.2	14 48.4	6.6	56.2
22	150 36.6	50.4	81 32.4	11.3	14 41.8	6.7	56.2
23	165 36.5	50.8	96 02.7	11.4	N14 35.1	6.8	56.2
	SD 15.8	d 0.4	SD 15.9		15.6		15.4

Twilight / Moonrise

Lat.	Naut.	Civil	Sunrise	28	29	30	31
°	h m	h m	h m	h m	h m	h m	h m
N 72	□	□	□	□	□	05 47	07 43
N 70	□	□	□	03 34	04 54	06 32	08 10
68	□	□	□	04 24	05 36	07 01	08 30
66	////	////	01 26	04 55	06 04	07 23	08 46
64	////	////	02 06	05 19	06 25	07 40	08 59
62	////	00 40	02 33	05 37	06 42	07 55	09 10
60	////	01 36	02 54	05 53	06 57	08 07	09 20
N 58	////	02 07	03 11	06 06	07 09	08 17	09 28
56	00 37	02 30	03 26	06 17	07 19	08 26	09 35
54	01 27	02 48	03 38	06 27	07 28	08 34	09 41
52	01 56	03 03	03 49	06 35	07 37	08 41	09 47
50	02 17	03 16	03 58	06 43	07 44	08 48	09 52
45	02 56	03 43	04 19	07 00	08 00	09 02	10 03
N 40	03 23	04 03	04 35	07 14	08 13	09 13	10 13
35	03 44	04 19	04 48	07 26	08 24	09 23	10 21
30	04 01	04 33	05 00	07 36	08 34	09 31	10 27
20	04 28	04 56	05 20	07 53	08 50	09 46	10 39
N 10	04 48	05 15	05 38	08 09	09 05	09 59	10 50
0	05 06	05 32	05 54	08 23	09 18	10 11	11 00
S 10	05 21	05 47	06 10	08 38	09 32	10 23	11 09
20	05 36	06 03	06 27	08 53	09 46	10 35	11 20
30	05 51	06 20	06 46	09 11	10 03	10 50	11 32
35	05 58	06 29	06 57	09 21	10 13	10 58	11 38
40	06 07	06 40	07 10	09 33	10 24	11 08	11 46
45	06 16	06 52	07 25	09 47	10 37	11 19	11 55
S 50	06 26	07 06	07 44	10 04	10 52	11 32	12 06
52	06 30	07 13	07 52	10 12	11 00	11 39	12 11
54	06 35	07 20	08 02	10 20	11 08	11 46	12 17
56	06 40	07 28	08 13	10 30	11 17	11 53	12 23
58	06 46	07 37	08 26	10 42	11 27	12 02	12 30
S 60	06 52	07 47	08 41	10 55	11 39	12 12	12 37

Sunset / Twilight / Moonset

Lat.	Sunset	Civil	Naut.	28	29	30	31
°	h m	h m	h m	h m	h m	h m	h m
N 72	□	□	□	□	□	02 18	02 05
N 70	□	□	□	00 43	01 19	01 32	01 37
68	□	□	□	24 38	00 38	01 02	01 15
66	22 32	////	////	24 09	00 09	00 39	00 59
64	21 51	////	////	23 48	24 22	00 22	00 45
62	21 23	23 22	////	23 30	24 07	00 07	00 33
60	21 02	22 22	////	23 16	23 54	24 23	00 23
N 58	20 44	21 50	////	23 03	23 44	24 15	00 15
56	20 30	21 26	23 25	22 53	23 34	24 07	00 07
54	20 17	21 08	22 30	22 43	23 26	24 00	00 00
52	20 07	20 52	22 00	22 35	23 18	23 54	24 24
50	19 57	20 39	21 39	22 27	23 12	23 48	24 19
45	19 37	20 13	21 00	22 11	22 57	23 36	24 10
N 40	19 20	19 52	20 32	21 57	22 45	23 26	24 03
35	19 07	19 36	20 11	21 46	22 35	23 18	23 56
30	18 55	19 22	19 54	21 36	22 26	23 10	23 50
20	18 35	18 59	19 27	21 19	22 10	22 57	23 40
N 10	18 17	18 40	19 07	21 04	21 57	22 45	23 31
0	18 01	18 23	18 49	20 50	21 44	22 35	23 22
S 10	17 45	18 08	18 34	20 36	21 31	22 24	23 14
20	17 28	17 52	18 19	20 21	21 17	22 12	23 04
30	17 09	17 35	18 04	20 03	21 01	21 58	22 54
35	16 57	17 25	17 56	19 53	20 52	21 51	22 48
40	16 44	17 14	17 48	19 42	20 42	21 42	22 41
45	16 29	17 02	17 39	19 28	20 29	21 31	22 33
S 50	16 11	16 48	17 29	19 11	20 14	21 18	22 23
52	16 02	16 41	17 24	19 03	20 07	21 12	22 18
54	15 52	16 34	17 19	18 55	19 59	21 06	22 13
56	15 41	16 26	17 14	18 45	19 50	20 59	22 08
58	15 28	16 17	17 08	18 34	19 40	20 50	22 02
S 60	15 13	16 07	17 02	18 21	19 28	20 41	21 54

SUN / MOON

Day	Eqn. of Time 00ʰ	Eqn. of Time 12ʰ	Mer. Pass.	Mer. Pass. Upper	Mer. Pass. Lower	Age	Phase
d	m s	m s	h m	h m	h m	d	%
28	02 49	02 46	11 57	14 37	02 08	03	10
29	02 42	02 38	11 57	15 31	03 04	04	17
30	02 34	02 30	11 57	16 23	03 57	05	26

1998 MAY 31, JUNE 1, 2 (SUN., MON., TUES.)

UT	ARIES GHA	VENUS −4.0 GHA	Dec	MARS +1.4 GHA	Dec	JUPITER −2.3 GHA	Dec	SATURN +0.5 GHA	Dec
31 SUNDAY									
00	248 17.4	218 31.3	N 9 58.0	185 25.8	N21 20.1	252 50.6	S 3 11.6	220 27.7	N 8 59.1
01	263 19.9	233 31.0	9 59.0	200 26.4	20.4	267 52.8	11.5	235 29.9	59.1
02	278 22.3	248 30.7	10 00.0	215 27.0	20.8	282 54.9	11.4	250 32.1	59.2
03	293 24.8	263 30.4	.. 01.0	230 27.6	.. 21.1	297 57.0	.. 11.2	265 34.3	.. 59.3
04	308 27.3	278 30.1	02.0	245 28.2	21.5	312 59.2	11.1	280 36.5	59.4
05	323 29.7	293 29.8	03.0	260 28.8	21.8	328 01.3	11.0	295 38.7	59.5
06	338 32.2	308 29.5	N10 04.0	275 29.4	N21 22.1	343 03.5	S 3 10.8	310 40.9	N 8 59.6
07	353 34.7	323 29.1	05.0	290 30.0	22.5	358 05.6	10.7	325 43.1	59.7
08	8 37.1	338 28.8	05.9	305 30.6	22.8	13 07.8	10.6	340 45.3	59.8
09	23 39.6	353 28.5	.. 06.9	320 31.2	.. 23.2	28 09.9	.. 10.5	355 47.6	.. 59.8
10	38 42.1	8 28.2	07.9	335 31.8	23.5	43 12.1	10.3	10 49.8	8 59.9
11	53 44.5	23 27.9	08.9	350 32.4	23.9	58 14.2	10.2	25 52.0	9 00.0
12	68 47.0	38 27.6	N10 09.9	5 33.0	N21 24.2	73 16.4	S 3 10.1	40 54.2	N 9 00.1
13	83 49.5	53 27.3	10.9	20 33.6	24.5	88 18.5	10.0	55 56.4	00.2
14	98 51.9	68 27.0	11.9	35 34.2	24.9	103 20.7	09.8	70 58.6	00.3
15	113 54.4	83 26.6	.. 12.9	50 34.8	.. 25.2	118 22.8	.. 09.7	86 00.8	.. 00.4
16	128 56.8	98 26.3	13.9	65 35.5	25.6	133 25.0	09.6	101 03.0	00.5
17	143 59.3	113 26.0	14.9	80 36.1	25.9	148 27.1	09.5	116 05.2	00.5
18	159 01.8	128 25.7	N10 15.9	95 36.7	N21 26.2	163 29.2	S 3 09.3	131 07.4	N 9 00.6
19	174 04.2	143 25.4	16.9	110 37.3	26.6	178 31.4	09.2	146 09.7	00.7
20	189 06.7	158 25.1	17.9	125 37.9	26.9	193 33.5	09.1	161 11.9	00.8
21	204 09.2	173 24.7	.. 18.9	140 38.5	.. 27.3	208 35.7	.. 09.0	176 14.1	.. 00.9
22	219 11.6	188 24.4	19.9	155 39.1	27.6	223 37.8	08.8	191 16.3	01.0
23	234 14.1	203 24.1	20.9	170 39.7	27.9	238 40.0	08.7	206 18.5	01.1
1 MONDAY									
00	249 16.6	218 23.8	N10 21.8	185 40.3	N21 28.3	253 42.1	S 3 08.6	221 20.7	N 9 01.2
01	264 19.0	233 23.5	22.8	200 40.9	28.6	268 44.3	08.5	236 22.9	01.2
02	279 21.5	248 23.2	23.8	215 41.5	28.9	283 46.4	08.3	251 25.1	01.3
03	294 24.0	263 22.8	.. 24.8	230 42.1	.. 29.3	298 48.6	.. 08.2	266 27.3	.. 01.4
04	309 26.4	278 22.5	25.8	245 42.7	29.6	313 50.7	08.1	281 29.6	01.5
05	324 28.9	293 22.2	26.8	260 43.3	30.0	328 52.9	08.0	296 31.8	01.6
06	339 31.3	308 21.9	N10 27.8	275 43.9	N21 30.3	343 55.0	S 3 07.8	311 34.0	N 9 01.7
07	354 33.8	323 21.6	28.8	290 44.5	30.6	358 57.2	07.7	326 36.2	01.8
08	9 36.3	338 21.2	29.8	305 45.1	31.0	13 59.3	07.6	341 38.4	01.9
09	24 38.7	353 20.9	.. 30.7	320 45.7	.. 31.3	29 01.5	.. 07.5	356 40.6	.. 01.9
10	39 41.2	8 20.6	31.7	335 46.3	31.6	44 03.6	07.4	11 42.8	02.0
11	54 43.7	23 20.3	32.7	350 46.9	32.0	59 05.8	07.2	26 45.0	02.1
12	69 46.1	38 20.0	N10 33.7	5 47.5	N21 32.3	74 08.0	S 3 07.1	41 47.3	N 9 02.2
13	84 48.6	53 19.6	34.7	20 48.1	32.6	89 10.1	07.0	56 49.5	02.3
14	99 51.1	68 19.3	35.7	35 48.7	33.0	104 12.3	06.9	71 51.7	02.4
15	114 53.5	83 19.0	.. 36.7	50 49.3	.. 33.3	119 14.4	.. 06.7	86 53.9	.. 02.5
16	129 56.0	98 18.7	37.7	65 49.9	33.6	134 16.6	06.6	101 56.1	02.6
17	144 58.4	113 18.3	38.6	80 50.5	34.0	149 18.7	06.5	116 58.3	02.6
18	160 00.9	128 18.0	N10 39.6	95 51.1	N21 34.3	164 20.9	S 3 06.4	132 00.5	N 9 02.7
19	175 03.4	143 17.7	40.6	110 51.8	34.6	179 23.0	06.2	147 02.7	02.8
20	190 05.8	158 17.4	41.6	125 52.4	34.9	194 25.2	06.1	162 05.0	02.9
21	205 08.3	173 17.0	.. 42.6	140 53.0	.. 35.3	209 27.3	.. 06.0	177 07.2	.. 03.0
22	220 10.8	188 16.7	43.6	155 53.6	35.6	224 29.5	05.9	192 09.4	03.1
23	235 13.2	203 16.4	44.6	170 54.2	35.9	239 31.6	05.8	207 11.6	03.2
2 TUESDAY									
00	250 15.7	218 16.1	N10 45.5	185 54.8	N21 36.3	254 33.8	S 3 05.6	222 13.8	N 9 03.2
01	265 18.2	233 15.7	46.5	200 55.4	36.6	269 36.0	05.5	237 16.0	03.3
02	280 20.6	248 15.4	47.5	215 56.0	36.9	284 38.1	05.4	252 18.2	03.4
03	295 23.1	263 15.1	.. 48.5	230 56.6	.. 37.2	299 40.3	.. 05.3	267 20.4	.. 03.5
04	310 25.6	278 14.8	49.5	245 57.2	37.6	314 42.4	05.1	282 22.7	03.6
05	325 28.0	293 14.4	50.5	260 57.8	37.9	329 44.6	05.0	297 24.9	03.7
06	340 30.5	308 14.1	N10 51.4	275 58.4	N21 38.2	344 46.7	S 3 04.9	312 27.1	N 9 03.8
07	355 32.9	323 13.8	52.4	290 59.0	38.5	359 48.9	04.8	327 29.3	03.9
08	10 35.4	338 13.4	53.4	305 59.6	38.9	14 51.0	04.7	342 31.5	03.9
09	25 37.9	353 13.1	.. 54.4	321 00.2	.. 39.2	29 53.2	.. 04.5	357 33.7	.. 04.0
10	40 40.3	8 12.8	55.4	336 00.8	39.5	44 55.4	04.4	12 35.9	04.1
11	55 42.8	23 12.5	56.3	351 01.4	39.9	59 57.5	04.3	27 38.2	04.2
12	70 45.3	38 12.1	N10 57.3	6 02.0	N21 40.2	74 59.7	S 3 04.2	42 40.4	N 9 04.3
13	85 47.7	53 11.8	58.3	21 02.6	40.5	90 01.8	04.1	57 42.6	04.4
14	100 50.2	68 11.5	10 59.3	36 03.2	40.8	105 04.0	03.9	72 44.8	04.5
15	115 52.7	83 11.1	11 00.3	51 03.8	.. 41.1	120 06.1	.. 03.8	87 47.0	.. 04.5
16	130 55.1	98 10.8	01.2	66 04.4	41.5	135 08.3	03.7	102 49.2	04.6
17	145 57.6	113 10.5	02.2	81 05.0	41.8	150 10.5	03.6	117 51.4	04.7
18	161 00.1	128 10.1	N11 03.2	96 05.6	N21 42.1	165 12.6	S 3 03.5	132 53.7	N 9 04.8
19	176 02.5	143 09.8	04.2	111 06.2	42.4	180 14.8	03.3	147 55.9	04.9
20	191 05.0	158 09.5	05.2	126 06.8	42.8	195 16.9	03.2	162 58.1	05.0
21	206 07.4	173 09.1	.. 06.1	141 07.4	.. 43.1	210 19.1	.. 03.1	178 00.3	.. 05.1
22	221 09.9	188 08.8	07.1	156 08.0	43.4	225 21.3	03.0	193 02.5	05.1
23	236 12.4	203 08.5	08.1	171 08.6	43.7	240 23.4	02.9	208 04.7	05.2
Mer. Pass. 7 21.7	v −0.3 d 1.0			v 0.6 d 0.3		v 2.2 d 0.1		v 2.2 d 0.1	

STARS

Name	SHA	Dec
Acamar	315 27.5	S40 18.7
Achernar	335 35.9	S57 14.6
Acrux	173 21.8	S63 05.6
Adhara	255 21.9	S28 58.4
Aldebaran	291 03.0	N16 30.2
Alioth	166 30.6	N55 58.4
Alkaid	153 07.7	N49 19.5
Al Na'ir	27 58.2	S46 57.9
Alnilam	275 58.4	S 1 12.4
Alphard	218 07.6	S 8 39.2
Alphecca	126 20.5	N26 43.4
Alpheratz	357 55.6	N29 04.7
Altair	62 19.3	N 8 51.9
Ankaa	353 27.3	S42 18.8
Antares	112 40.1	S26 25.6
Arcturus	146 06.0	N19 11.6
Atria	107 51.8	S69 01.4
Avior	234 23.1	S59 30.6
Bellatrix	278 44.8	N 6 20.7
Betelgeuse	271 14.2	N 7 24.2
Canopus	264 01.8	S52 42.0
Capella	280 52.0	N45 59.6
Deneb	49 39.1	N45 16.3
Denebola	182 45.4	N14 34.9
Diphda	349 07.7	S17 59.7
Dubhe	194 05.9	N61 45.9
Elnath	278 27.6	N28 36.2
Eltanin	90 51.0	N51 29.4
Enif	33 58.4	N 9 52.0
Fomalhaut	15 36.8	S29 37.7
Gacrux	172 13.5	S57 06.5
Gienah	176 04.1	S17 32.1
Hadar	149 03.8	S60 22.0
Hamal	328 14.1	N23 27.0
Kaus Aust.	83 58.8	S34 22.9
Kochab	137 18.7	N74 10.0
Markab	13 49.9	N15 11.7
Menkar	314 27.5	N 4 04.8
Menkent	148 20.9	S36 21.8
Miaplacidus	221 42.4	S69 43.0
Mirfak	308 57.4	N49 51.1
Nunki	76 12.3	S26 17.8
Peacock	53 37.1	S56 44.1
Pollux	243 42.2	N28 01.8
Procyon	245 12.1	N 5 13.6
Rasalhague	96 16.9	N12 33.8
Regulus	207 55.9	N11 58.5
Rigel	281 23.5	S 8 12.4
Rigil Kent.	140 07.0	S60 49.7
Sabik	102 25.5	S15 43.3
Schedar	349 53.9	N56 31.4
Shaula	96 37.2	S37 06.0
Sirius	258 44.2	S16 43.0
Spica	158 43.2	S11 09.2
Suhail	223 01.1	S43 25.9
Vega	80 46.4	N38 46.9
Zuben'ubi	137 17.9	S16 02.1

	SHA	Mer. Pass.
Venus	329 07.2	9 27
Mars	296 23.7	11 37
Jupiter	4 25.6	7 04
Saturn	332 04.2	9 13

SUN and MOON

UT (d h)	SUN GHA	SUN Dec	MOON GHA	v	Dec	d	HP
31 00	180 36.4	N21 51.2	110 33.1	11.5	N14 28.3	6.9	56.1
01	195 36.3	51.5	125 03.6	11.6	14 21.4	6.9	56.1
02	210 36.2	51.9	139 34.2	11.6	14 14.5	6.9	56.1
03	225 36.1	.. 52.2	154 04.8	11.6	14 07.6	7.0	56.1
04	240 36.0	52.6	168 35.4	11.8	14 00.6	7.1	56.0
05	255 35.9	53.0	183 06.2	11.8	13 53.5	7.2	56.0
06	270 35.8	N21 53.3	197 37.0	11.9	N13 46.3	7.2	56.0
07	285 35.7	53.7	212 07.9	12.0	13 39.1	7.2	55.9
08	300 35.7	54.0	226 38.9	12.0	13 31.9	7.4	55.9
S 09	315 35.6	.. 54.4	241 09.9	12.1	13 24.5	7.3	55.9
U 10	330 35.5	54.7	255 41.0	12.1	13 17.2	7.5	55.8
N 11	345 35.4	55.1	270 12.1	12.3	13 09.7	7.4	55.8
D 12	0 35.3	N21 55.5	284 43.4	12.3	N13 02.3	7.6	55.8
A 13	15 35.2	55.8	299 14.7	12.3	12 54.7	7.6	55.7
Y 14	30 35.1	56.2	313 46.0	12.4	12 47.1	7.6	55.7
15	45 35.0	.. 56.5	328 17.4	12.5	12 39.5	7.7	55.7
16	60 34.9	56.9	342 48.9	12.6	12 31.8	7.8	55.7
17	75 34.8	57.2	357 20.5	12.6	12 24.0	7.8	55.6
18	90 34.7	N21 57.6	11 52.1	12.7	N12 16.2	7.8	55.6
19	105 34.6	57.9	26 23.8	12.7	12 08.4	7.9	55.6
20	120 34.6	58.3	40 55.5	12.8	12 00.5	7.9	55.5
21	135 34.5	.. 58.6	55 27.3	12.9	11 52.6	8.0	55.5
22	150 34.4	59.0	69 59.2	12.9	11 44.6	8.0	55.5
23	165 34.3	59.3	84 31.1	13.0	11 36.6	8.1	55.5
1 00	180 34.2	N21 59.7	99 03.1	13.0	N11 28.5	8.1	55.4
01	195 34.1	22 00.0	113 35.1	13.2	11 20.4	8.2	55.4
02	210 34.0	00.3	128 07.3	13.1	11 12.2	8.2	55.4
03	225 33.9	.. 00.7	142 39.4	13.2	11 04.0	8.3	55.4
04	240 33.8	01.0	157 11.6	13.3	10 55.7	8.3	55.3
05	255 33.7	01.4	171 43.9	13.3	10 47.4	8.3	55.3
06	270 33.6	N22 01.7	186 16.2	13.4	N10 39.1	8.4	55.3
07	285 33.5	02.1	200 48.6	13.5	10 30.7	8.4	55.3
08	300 33.4	02.4	215 21.1	13.5	10 22.3	8.4	55.2
M 09	315 33.3	.. 02.7	229 53.6	13.5	10 13.9	8.5	55.2
O 10	330 33.2	03.1	244 26.1	13.6	10 05.4	8.5	55.2
N 11	345 33.1	03.4	258 58.7	13.7	9 56.9	8.6	55.2
D 12	0 33.0	N22 03.8	273 31.4	13.7	N 9 48.3	8.6	55.1
A 13	15 32.9	04.1	288 04.1	13.8	9 39.7	8.6	55.1
Y 14	30 32.9	04.4	302 36.9	13.8	9 31.1	8.7	55.1
15	45 32.8	.. 04.8	317 09.7	13.8	9 22.4	8.7	55.1
16	60 32.7	05.1	331 42.5	13.9	9 13.7	8.7	55.0
17	75 32.6	05.4	346 15.4	14.0	9 05.0	8.8	55.0
18	90 32.5	N22 05.8	0 48.4	14.0	N 8 56.2	8.8	55.0
19	105 32.4	06.1	15 21.4	14.1	8 47.4	8.8	55.0
20	120 32.3	06.4	29 54.5	14.1	8 38.6	8.9	55.0
21	135 32.2	.. 06.8	44 27.6	14.1	8 29.7	8.9	54.9
22	150 32.1	07.1	59 00.7	14.2	8 20.8	8.9	54.9
23	165 32.0	07.4	73 33.9	14.2	8 11.9	8.9	54.9
2 00	180 31.9	N22 07.8	88 07.1	14.3	N 8 03.0	9.0	54.9
01	195 31.8	08.1	102 40.4	14.3	7 54.0	9.0	54.9
02	210 31.7	08.4	117 13.7	14.4	7 45.0	9.1	54.8
03	225 31.6	.. 08.8	131 47.1	14.4	7 35.9	9.1	54.8
04	240 31.5	09.1	146 20.5	14.4	7 26.9	9.1	54.8
05	255 31.4	09.4	160 53.9	14.5	7 17.8	9.1	54.8
06	270 31.3	N22 09.7	175 27.4	14.5	N 7 08.7	9.1	54.8
07	285 31.2	10.1	190 00.9	14.6	6 59.6	9.2	54.7
08	300 31.1	10.4	204 34.5	14.6	6 50.4	9.1	54.7
T 09	315 31.0	.. 10.7	219 08.1	14.6	6 41.3	9.2	54.7
U 10	330 30.9	11.0	233 41.7	14.7	6 32.1	9.2	54.7
E 11	345 30.8	11.4	248 15.4	14.7	6 22.9	9.3	54.7
S 12	0 30.7	N22 11.7	262 49.1	14.7	N 6 13.6	9.2	54.6
D 13	15 30.6	12.0	277 22.8	14.8	6 04.4	9.3	54.6
A 14	30 30.5	12.3	291 56.6	14.8	5 55.1	9.3	54.6
Y 15	45 30.4	.. 12.7	306 30.4	14.8	5 45.8	9.3	54.6
16	60 30.3	13.0	321 04.2	14.9	5 36.5	9.4	54.6
17	75 30.2	13.3	335 38.1	14.9	5 27.1	9.4	54.6
18	90 30.1	N22 13.6	350 12.0	14.9	N 5 17.8	9.4	54.6
19	105 30.0	13.9	4 45.9	15.0	5 08.4	9.4	54.5
20	120 29.9	14.2	19 19.9	15.0	4 59.0	9.4	54.5
21	135 29.8	.. 14.6	33 53.9	15.0	4 49.6	9.4	54.5
22	150 29.7	14.9	48 27.9	15.0	4 40.2	9.4	54.5
23	165 29.6	15.2	63 01.9	15.1	N 4 30.8	9.5	54.5
	SD 15.8	d 0.3	SD 15.2		15.0		14.9

Twilight, Sunrise, Moonrise

Lat.	Twilight Naut.	Twilight Civil	Sunrise	Moonrise 31	1	2	3
N 72	□	□	□	07 43	09 29	11 08	12 44
N 70	□	□	□	08 10	09 46	11 18	12 47
68	□	□	□	08 30	09 59	11 25	12 50
66	////	////	01 13	08 46	10 10	11 31	12 54
64	////	////	01 58	08 59	10 19	11 37	12 54
62	////	00 03	02 28	09 10	10 26	11 41	12 55
60	////	01 27	02 50	09 20	10 33	11 45	12 55
N 58	////	02 01	03 07	09 28	10 39	11 49	12 58
56	00 02	02 25	03 22	09 35	10 44	11 52	12 59
54	01 19	02 44	03 35	09 41	10 48	11 55	13 00
52	01 50	03 00	03 46	09 47	10 53	11 57	13 01
50	02 13	03 13	03 56	09 52	10 56	11 59	13 02
45	02 53	03 40	04 17	10 03	11 04	12 04	13 03
N 40	03 21	04 01	04 33	10 13	11 11	12 09	13 05
35	03 42	04 18	04 47	10 21	11 17	12 12	13 06
30	04 00	04 32	04 59	10 27	11 22	12 15	13 07
20	04 27	04 56	05 20	10 39	11 31	12 21	13 09
N 10	04 48	05 15	05 38	10 50	11 39	12 26	13 11
0	05 06	05 32	05 54	11 00	11 46	12 30	13 13
S 10	05 22	05 48	06 10	11 09	11 53	12 34	13 14
20	05 37	06 04	06 28	11 20	12 01	12 39	13 16
30	05 52	06 21	06 48	11 32	12 10	12 45	13 18
35	06 00	06 31	06 59	11 38	12 15	12 48	13 19
40	06 08	06 42	07 12	11 46	12 20	12 51	13 21
45	06 18	06 55	07 28	11 55	12 27	12 55	13 22
S 50	06 28	07 09	07 47	12 06	12 35	13 00	13 24
52	06 33	07 16	07 56	12 11	12 38	13 02	13 25
54	06 38	07 23	08 06	12 17	12 42	13 05	13 26
56	06 44	07 32	08 17	12 23	12 47	13 08	13 27
58	06 50	07 41	08 31	12 30	12 52	13 11	13 28
S 60	06 56	07 51	08 46	12 37	12 57	13 14	13 29

Sunset, Twilight, Moonset

Lat.	Sunset	Twilight Civil	Twilight Naut.	Moonset 31	1	2	3
N 72	□	□	□	02 05	01 56	01 49	01 42
N 70	□	□	□	01 37	01 38	01 38	01 37
68	□	□	□	01 15	01 23	01 29	01 32
66	22 46	////	////	00 59	01 12	01 21	01 28
64	21 59	////	////	00 45	01 02	01 15	01 25
62	21 30	////	////	00 33	00 53	01 09	01 22
60	21 07	22 31	////	00 23	00 46	01 04	01 20
N 58	20 49	21 57	////	00 15	00 39	01 00	01 18
56	20 34	21 32	////	00 07	00 34	00 56	01 16
54	20 21	21 13	22 39	00 00	00 28	00 53	01 14
52	20 10	20 57	22 07	24 24	00 24	00 49	01 12
50	20 00	20 43	21 44	24 19	00 19	00 47	01 11
45	19 39	20 16	21 03	24 10	00 10	00 40	01 08
N 40	19 23	19 55	20 35	24 03	00 03	00 35	01 05
35	19 09	19 38	20 14	23 56	24 30	00 30	01 03
30	18 56	19 23	19 56	23 50	24 26	00 26	01 01
20	18 36	19 00	19 29	23 40	24 19	00 19	00 57
N 10	18 18	18 41	19 08	23 31	24 13	00 13	00 54
0	18 01	18 24	18 50	23 22	24 07	00 07	00 51
S 10	17 45	18 08	18 34	23 14	24 01	00 01	00 48
20	17 28	17 51	18 19	23 04	23 55	24 44	00 44
30	17 08	17 34	18 03	22 54	23 48	24 40	00 40
35	16 56	17 24	17 55	22 48	23 44	24 38	00 38
40	16 43	17 13	17 47	22 41	23 39	24 36	00 36
45	16 28	17 01	17 37	22 33	23 33	24 33	00 33
S 50	16 08	16 46	17 27	22 23	23 27	24 29	00 29
52	15 59	16 39	17 22	22 18	23 24	24 28	00 28
54	15 49	16 32	17 17	22 13	23 20	24 26	00 26
56	15 38	16 24	17 12	22 08	23 16	24 24	00 24
58	15 25	16 14	17 06	22 02	23 12	24 22	00 22
S 60	15 09	16 04	16 59	21 54	23 07	24 19	00 19

SUN and MOON

Day	SUN Eqn. of Time 00h	12h	Mer. Pass.	MOON Mer. Pass. Upper	Lower	Age	Phase
31	02 26	02 21	11 58	17 11	04 47	06	35
1	02 17	02 12	11 58	17 57	05 34	07	45
2	02 08	02 03	11 58	18 40	06 19	08	54

1998 JUNE 3, 4, 5 (WED., THURS., FRI.)

UT	ARIES GHA	VENUS −4.0 GHA	VENUS Dec	MARS +1.4 GHA	MARS Dec	JUPITER −2.3 GHA	JUPITER Dec	SATURN +0.5 GHA	SATURN Dec
d h	° '	° '	° '	° '	° '	° '	° '	° '	° '
3 00	251 14.8	218 08.1	N11 09.1	186 09.2	N21 44.0	255 25.6	S 3 02.7	223 06.9	N 9 05.3
01	266 17.3	233 07.8	10.0	201 09.8	44.4	270 27.7	02.6	238 09.2	05.4
02	281 19.8	248 07.4	11.0	216 10.4	44.7	285 29.9	02.5	253 11.4	05.5
03	296 22.2	263 07.1 ..	12.0	231 11.0 ..	45.0	300 32.1 ..	02.4	268 13.6 ..	05.6
04	311 24.7	278 06.8	13.0	246 11.6	45.3	315 34.2	02.3	283 15.8	05.7
05	326 27.2	293 06.4	13.9	261 12.2	45.6	330 36.4	02.1	298 18.0	05.7
W 06	341 29.6	308 06.1	N11 14.9	276 12.8	N21 46.0	345 38.5	S 3 02.0	313 20.2	N 9 05.8
E 07	356 32.1	323 05.8	15.9	291 13.4	46.3	0 40.7	01.9	328 22.4	05.9
D 08	11 34.5	338 05.4	16.9	306 14.0	46.6	15 42.9	01.8	343 24.7	06.0
N 09	26 37.0	353 05.1 ..	17.8	321 14.6 ..	46.9	30 45.0 ..	01.7	358 26.9 ..	06.1
E 10	41 39.5	8 04.7	18.8	336 15.2	47.2	45 47.2	01.5	13 29.1	06.2
S 11	56 41.9	23 04.4	19.8	351 15.8	47.5	60 49.4	01.4	28 31.3	06.3
D 12	71 44.4	38 04.1	N11 20.7	6 16.4	N21 47.9	75 51.5	S 3 01.3	43 33.5	N 9 06.3
A 13	86 46.9	53 03.7	21.7	21 17.0	48.2	90 53.7	01.2	58 35.7	06.4
Y 14	101 49.3	68 03.4	22.7	36 17.6	48.5	105 55.8	01.1	73 38.0	06.5
15	116 51.8	83 03.0 ..	23.7	51 18.2 ..	48.8	120 58.0 ..	01.0	88 40.2 ..	06.6
16	131 54.3	98 02.7	24.6	66 18.8	49.1	136 00.2	00.8	103 42.4	06.7
17	146 56.7	113 02.4	25.6	81 19.4	49.4	151 02.3	00.7	118 44.6	06.8
18	161 59.2	128 02.0	N11 26.6	96 20.0	N21 49.7	166 04.5	S 3 00.6	133 46.8	N 9 06.8
19	177 01.7	143 01.7	27.5	111 20.6	50.1	181 06.7	00.5	148 49.0	06.9
20	192 04.1	158 01.3	28.5	126 21.2	50.4	196 08.8	00.4	163 51.3	07.0
21	207 06.6	173 01.0 ..	29.5	141 21.8 ..	50.7	211 11.0 ..	00.2	178 53.5 ..	07.1
22	222 09.0	188 00.6	30.4	156 22.4	51.0	226 13.2	00.1	193 55.7	07.2
23	237 11.5	203 00.3	31.4	171 23.0	51.3	241 15.3	3 00.0	208 57.9	07.3
4 00	252 14.0	218 00.0	N11 32.4	186 23.6	N21 51.6	256 17.5	S 2 59.9	224 00.1	N 9 07.4
01	267 16.4	232 59.6	33.3	201 24.2	51.9	271 19.7	59.8	239 02.3	07.4
02	282 18.9	247 59.3	34.3	216 24.9	52.2	286 21.8	59.7	254 04.6	07.5
03	297 21.4	262 58.9 ..	35.3	231 25.5 ..	52.5	301 24.0 ..	59.5	269 06.8 ..	07.6
04	312 23.8	277 58.6	36.2	246 26.1	52.9	316 26.2	59.4	284 09.0	07.7
05	327 26.3	292 58.2	37.2	261 26.7	53.2	331 28.3	59.3	299 11.2	07.8
T 06	342 28.8	307 57.9	N11 38.2	276 27.3	N21 53.5	346 30.5	S 2 59.2	314 13.4	N 9 07.9
H 07	357 31.2	322 57.5	39.1	291 27.9	53.8	1 32.7	59.1	329 15.7	07.9
U 08	12 33.7	337 57.2	40.1	306 28.5	54.1	16 34.8	59.0	344 17.9	08.0
R 09	27 36.2	352 56.8 ..	41.1	321 29.1 ..	54.4	31 37.0 ..	58.8	359 20.1 ..	08.1
S 10	42 38.6	7 56.5	42.0	336 29.7	54.7	46 39.2	58.7	14 22.3	08.2
D 11	57 41.1	22 56.1	43.0	351 30.3	55.0	61 41.3	58.6	29 24.5	08.3
A 12	72 43.5	37 55.8	N11 44.0	6 30.9	N21 55.3	76 43.5	S 2 58.5	44 26.7	N 9 08.4
Y 13	87 46.0	52 55.4	44.9	21 31.5	55.6	91 45.7	58.4	59 29.0	08.5
14	102 48.5	67 55.1	45.9	36 32.1	55.9	106 47.8	58.3	74 31.2	08.5
15	117 50.9	82 54.7 ..	46.9	51 32.7 ..	56.2	121 50.0 ..	58.1	89 33.4 ..	08.6
16	132 53.4	97 54.4	47.8	66 33.3	56.5	136 52.2	58.0	104 35.6	08.7
17	147 55.9	112 54.0	48.8	81 33.9	56.8	151 54.3	57.9	119 37.8	08.8
18	162 58.3	127 53.7	N11 49.7	96 34.5	N21 57.2	166 56.5	S 2 57.8	134 40.1	N 9 08.9
19	178 00.8	142 53.3	50.7	111 35.1	57.5	181 58.7	57.7	149 42.3	09.0
20	193 03.3	157 53.0	51.7	126 35.7	57.8	197 00.9	57.6	164 44.5	09.0
21	208 05.7	172 52.6 ..	52.6	141 36.3 ..	58.1	212 03.0 ..	57.5	179 46.7 ..	09.1
22	223 08.2	187 52.3	53.6	156 36.9	58.4	227 05.2	57.3	194 48.9	09.2
23	238 10.6	202 51.9	54.5	171 37.5	58.7	242 07.4	57.2	209 51.1	09.3
5 00	253 13.1	217 51.6	N11 55.5	186 38.1	N21 59.0	257 09.5	S 2 57.1	224 53.4	N 9 09.4
01	268 15.6	232 51.2	56.5	201 38.7	59.3	272 11.7	57.0	239 55.6	09.5
02	283 18.0	247 50.8	57.4	216 39.3	59.6	287 13.9	56.9	254 57.8	09.5
03	298 20.5	262 50.5 ..	58.4	231 39.9	21 59.9	302 16.1 ..	56.8	270 00.0 ..	09.6
04	313 23.0	277 50.1	11 59.3	246 40.5	22 00.2	317 18.2	56.7	285 02.2	09.7
05	328 25.4	292 49.8	12 00.3	261 41.1	00.5	332 20.4	56.5	300 04.5	09.8
F 06	343 27.9	307 49.4	N12 01.2	276 41.7	N22 00.8	347 22.6	S 2 56.4	315 06.7	N 9 09.9
R 07	358 30.4	322 49.1	02.2	291 42.3	01.1	2 24.7	56.3	330 08.9	10.0
I 08	13 32.8	337 48.7	03.2	306 42.9	01.4	17 26.9	56.2	345 11.1	10.0
D 09	28 35.3	352 48.3 ..	04.1	321 43.5 ..	01.7	32 29.1 ..	56.1	0 13.3 ..	10.1
A 10	43 37.8	7 48.0	05.1	336 44.1	02.0	47 31.3	56.0	15 15.6	10.2
Y 11	58 40.2	22 47.6	06.0	351 44.7	02.3	62 33.4	55.9	30 17.8	10.3
12	73 42.7	37 47.3	N12 07.0	6 45.3	N22 02.6	77 35.6	S 2 55.7	45 20.0	N 9 10.4
13	88 45.1	52 46.9	07.9	21 45.9	02.9	92 37.8	55.6	60 22.2	10.5
14	103 47.6	67 46.6	08.9	36 46.5	03.2	107 40.0	55.5	75 24.4	10.5
15	118 50.1	82 46.2 ..	09.8	51 47.1 ..	03.5	122 42.1 ..	55.4	90 26.7 ..	10.6
16	133 52.5	97 45.8	10.8	66 47.7	03.8	137 44.3	55.3	105 28.9	10.7
17	148 55.0	112 45.5	11.7	81 48.2	04.1	152 46.5	55.2	120 31.1	10.8
18	163 57.5	127 45.1	N12 12.7	96 48.8	N22 04.4	167 48.7	S 2 55.1	135 33.3	N 9 10.9
19	178 59.9	142 44.7	13.6	111 49.4	04.7	182 50.8	54.9	150 35.5	11.0
20	194 02.4	157 44.4	14.6	126 50.0	04.9	197 53.0	54.8	165 37.8	11.0
21	209 04.9	172 44.0 ..	15.6	141 50.6 ..	05.2	212 55.2 ..	54.7	180 40.0 ..	11.1
22	224 07.3	187 43.7	16.5	156 51.2	05.5	227 57.4	54.6	195 42.2	11.2
23	239 09.8	202 43.3	17.5	171 51.8	05.8	242 59.5	54.5	210 44.4	11.3
Mer. Pass.	h m 7 09.9	v −0.3	d 1.0	v 0.6	d 0.3	v 2.2	d 0.1	v 2.2	d 0.1

STARS

Name	SHA	Dec
Acamar	315 27.5	S40 18.7
Achernar	335 35.9	S57 14.6
Acrux	173 21.9	S63 05.7
Adhara	255 21.9	S28 58.4
Aldebaran	291 03.0	N16 30.2
Alioth	166 30.6	N55 58.4
Alkaid	153 07.7	N49 19.5
Al Na'ir	27 58.2	S46 57.9
Alnilam	275 58.4	S 1 12.4
Alphard	218 07.6	S 8 39.2
Alphecca	126 20.5	N26 43.4
Alpheratz	357 55.6	N29 04.7
Altair	62 19.3	N 8 51.9
Ankaa	353 27.3	S42 18.8
Antares	112 40.1	S26 25.6
Arcturus	146 06.0	N19 11.6
Atria	107 51.8	S69 01.4
Avior	234 23.1	S59 30.6
Bellatrix	278 44.8	N 6 20.7
Betelgeuse	271 14.2	N 7 24.2
Canopus	264 01.8	S52 41.9
Capella	280 52.0	N45 59.6
Deneb	49 39.0	N45 16.3
Denebola	182 45.4	N14 34.9
Diphda	349 07.7	S17 59.7
Dubhe	194 05.9	N61 45.9
Elnath	278 27.6	N28 36.2
Eltanin	90 51.0	N51 29.4
Enif	33 58.4	N 9 52.0
Fomalhaut	15 36.7	S29 37.7
Gacrux	172 13.5	S57 06.5
Gienah	176 04.1	S17 32.1
Hadar	149 03.8	S60 22.0
Hamal	328 14.1	N23 27.1
Kaus Aust.	83 58.8	S34 23.0
Kochab	137 18.7	N74 10.0
Markab	13 49.9	N15 11.7
Menkar	314 27.5	N 4 04.8
Menkent	148 20.9	S36 21.8
Miaplacidus	221 42.4	S69 43.0
Mirfak	308 57.4	N49 51.1
Nunki	76 12.3	S26 17.8
Peacock	53 37.1	S56 44.1
Pollux	243 42.2	N28 01.8
Procyon	245 12.1	N 5 13.6
Rasalhague	96 16.9	N12 33.8
Regulus	207 55.9	N11 58.5
Rigel	281 23.5	S 8 12.4
Rigil Kent.	140 07.0	S60 49.8
Sabik	102 25.5	S15 43.3
Schedar	349 53.9	N56 31.4
Shaula	96 37.2	S37 06.0
Sirius	258 44.2	S16 43.0
Spica	158 43.2	S11 09.2
Suhail	223 01.1	S43 25.9
Vega	80 46.4	N38 46.9
Zuben'ubi	137 17.9	S16 02.1

	SHA	Mer. Pass.
	° '	h m
Venus	325 46.0	9 28
Mars	294 09.7	11 34
Jupiter	4 03.5	6 54
Saturn	331 46.2	9 03

UT	SUN GHA	SUN Dec	MOON GHA	v	MOON Dec	d	HP
d h	° '	° '	° '	'	° '	'	'
3 00	180 29.5	N22 15.5	77 36.0	15.1	N 4 21.3	9.4	54.5
01	195 29.4	15.8	92 10.1	15.1	4 11.9	9.5	54.5
02	210 29.3	16.1	106 44.2	15.1	4 02.4	9.5	54.4
03	225 29.2	.. 16.4	121 18.3	15.2	3 52.9	9.5	54.4
04	240 29.1	16.8	135 52.5	15.2	3 43.4	9.5	54.4
05	255 29.0	17.1	150 26.7	15.2	3 33.9	9.5	54.4
06	270 28.9	N22 17.4	165 00.9	15.2	N 3 24.4	9.5	54.4
W 07	285 28.8	17.7	179 35.1	15.2	3 14.9	9.5	54.4
E 08	300 28.7	18.0	194 09.3	15.3	3 05.4	9.6	54.4
D 09	315 28.6	.. 18.3	208 43.6	15.3	2 55.8	9.5	54.4
N 10	330 28.5	18.6	223 17.9	15.3	2 46.3	9.6	54.3
E 11	345 28.4	18.9	237 52.2	15.3	2 36.7	9.6	54.3
S 12	0 28.3	N22 19.2	252 26.5	15.3	N 2 27.1	9.6	54.3
D 13	15 28.2	19.5	267 00.8	15.4	2 17.5	9.5	54.3
A 14	30 28.1	19.8	281 35.2	15.3	2 08.0	9.6	54.3
Y 15	45 28.0	.. 20.1	296 09.5	15.4	1 58.4	9.6	54.3
16	60 27.9	20.4	310 43.9	15.4	1 48.8	9.6	54.3
17	75 27.8	20.7	325 18.3	15.4	1 39.2	9.6	54.3
18	90 27.7	N22 21.0	339 52.7	15.4	N 1 29.6	9.6	54.3
19	105 27.6	21.3	354 27.1	15.4	1 20.0	9.6	54.3
20	120 27.5	21.6	9 01.5	15.4	1 10.4	9.6	54.3
21	135 27.3	.. 21.9	23 35.9	15.5	1 00.8	9.7	54.2
22	150 27.2	22.2	38 10.4	15.4	0 51.1	9.6	54.2
23	165 27.1	22.5	52 44.8	15.5	0 41.5	9.6	54.2
4 00	180 27.0	N22 22.8	67 19.3	15.4	N 0 31.9	9.6	54.2
01	195 26.9	23.1	81 53.7	15.5	0 22.3	9.6	54.2
02	210 26.8	23.4	96 28.2	15.5	0 12.7	9.6	54.2
03	225 26.7	.. 23.7	111 02.7	15.4	N 0 03.1	9.7	54.2
04	240 26.6	24.0	125 37.1	15.5	S 0 06.6	9.6	54.2
05	255 26.5	24.3	140 11.6	15.5	0 16.2	9.6	54.2
06	270 26.4	N22 24.6	154 46.1	15.5	S 0 25.8	9.6	54.2
T 07	285 26.3	24.9	169 20.6	15.5	0 35.4	9.6	54.2
H 08	300 26.2	25.2	183 55.1	15.4	0 45.0	9.6	54.2
U 09	315 26.1	.. 25.5	198 29.5	15.5	0 54.6	9.6	54.2
R 10	330 26.0	25.8	213 04.0	15.5	1 04.2	9.6	54.2
S 11	345 25.9	26.1	227 38.5	15.5	1 13.8	9.6	54.2
D 12	0 25.8	N22 26.4	242 13.0	15.5	S 1 23.4	9.6	54.2
A 13	15 25.7	26.7	256 47.5	15.4	1 33.0	9.5	54.2
Y 14	30 25.6	26.9	271 21.9	15.5	1 42.5	9.6	54.2
15	45 25.4	.. 27.2	285 56.4	15.5	1 52.1	9.6	54.2
16	60 25.3	27.5	300 30.9	15.4	2 01.7	9.5	54.2
17	75 25.2	27.8	315 05.3	15.5	2 11.2	9.6	54.2
18	90 25.1	N22 28.1	329 39.8	15.4	S 2 20.8	9.5	54.2
19	105 25.0	28.4	344 14.2	15.5	2 30.3	9.5	54.2
20	120 24.9	28.7	358 48.7	15.4	2 39.8	9.5	54.2
21	135 24.8	.. 28.9	13 23.1	15.4	2 49.3	9.5	54.2
22	150 24.7	29.2	27 57.5	15.5	2 58.8	9.5	54.2
23	165 24.6	29.5	42 32.0	15.4	3 08.3	9.5	54.2
5 00	180 24.5	N22 29.8	57 06.4	15.4	S 3 17.8	9.5	54.2
01	195 24.4	30.1	71 40.8	15.3	3 27.3	9.5	54.2
02	210 24.3	30.4	86 15.1	15.4	3 36.8	9.4	54.2
03	225 24.2	.. 30.6	100 49.5	15.4	3 46.2	9.4	54.2
04	240 24.0	30.9	115 23.9	15.3	3 55.6	9.5	54.2
05	255 23.9	31.2	129 58.2	15.3	4 05.1	9.4	54.2
06	270 23.8	N22 31.5	144 32.5	15.4	S 4 14.5	9.3	54.2
F 07	285 23.7	31.7	159 06.9	15.3	4 23.8	9.4	54.2
R 08	300 23.6	32.0	173 41.2	15.2	4 33.2	9.4	54.2
I 09	315 23.5	.. 32.3	188 15.4	15.3	4 42.6	9.3	54.2
D 10	330 23.4	32.6	202 49.7	15.2	4 51.9	9.3	54.2
A 11	345 23.3	32.8	217 23.9	15.3	5 01.2	9.3	54.2
Y 12	0 23.2	N22 33.1	231 58.2	15.2	S 5 10.5	9.3	54.2
13	15 23.1	33.4	246 32.4	15.2	5 19.8	9.3	54.2
14	30 23.0	33.7	261 06.6	15.1	5 29.1	9.2	54.2
15	45 22.8	.. 33.9	275 40.7	15.2	5 38.3	9.2	54.2
16	60 22.7	34.2	290 14.9	15.1	5 47.5	9.2	54.2
17	75 22.6	34.5	304 49.0	15.1	5 56.7	9.2	54.2
18	90 22.5	N22 34.7	319 23.1	15.1	S 6 05.9	9.2	54.2
19	105 22.4	35.0	333 57.2	15.0	6 15.1	9.1	54.2
20	120 22.3	35.3	348 31.2	15.1	6 24.2	9.1	54.2
21	135 22.2	.. 35.6	3 05.3	15.0	6 33.3	9.1	54.2
22	150 22.1	35.8	17 39.3	14.9	6 42.4	9.1	54.2
23	165 22.0	36.1	32 13.2	15.0	S 6 51.5	9.0	54.2
	SD 15.8	d 0.3	SD 14.8		14.8		14.8

Twilight / Sunrise / Moonrise

Lat.	Twilight Naut.	Civil	Sunrise	Moonrise 3	4	5	6
°	h m	h m	h m	h m	h m	h m	h m
N 72	☐	☐	☐	12 44	14 18	15 53	17 30
N 70	☐	☐	☐	12 47	14 15	15 44	17 14
68	☐	☐	☐	12 50	14 13	15 37	17 02
66	////	////	01 00	12 52	14 11	15 31	16 51
64	////	////	01 51	12 54	14 10	15 26	16 43
62	////	////	02 22	12 55	14 09	15 22	16 36
60	////	01 18	02 46	12 57	14 07	15 18	16 29
N 58	////	01 55	03 04	12 58	14 06	15 15	16 24
56	////	02 21	03 19	12 59	14 06	15 12	16 19
54	01 11	02 41	03 32	13 00	14 05	15 10	16 15
52	01 45	02 57	03 44	13 01	14 04	15 08	16 11
50	02 09	03 11	03 54	13 02	14 03	15 05	16 07
45	02 51	03 39	04 15	13 03	14 02	15 00	15 59
N 40	03 19	04 00	04 32	13 05	14 01	14 57	15 53
35	03 41	04 17	04 47	13 06	14 00	14 53	15 47
30	03 59	04 32	04 59	13 07	13 59	14 50	15 43
20	04 27	04 55	05 20	13 09	13 57	14 45	15 34
N 10	04 48	05 15	05 38	13 11	13 56	14 41	15 27
0	05 06	05 32	05 55	13 13	13 55	14 37	15 20
S 10	05 22	05 49	06 11	13 14	13 54	14 33	15 13
20	05 38	06 05	06 29	13 16	13 52	14 29	15 06
30	05 53	06 23	06 49	13 18	13 51	14 24	14 58
35	06 01	06 33	07 01	13 19	13 50	14 21	14 53
40	06 10	06 44	07 14	13 21	13 49	14 18	14 48
45	06 20	06 57	07 30	13 22	13 48	14 14	14 42
S 50	06 31	07 12	07 50	13 24	13 47	14 10	14 35
52	06 36	07 19	07 59	13 25	13 46	14 08	14 31
54	06 41	07 26	08 09	13 26	13 46	14 06	14 28
56	06 47	07 35	08 21	13 27	13 45	14 04	14 24
58	06 53	07 44	08 35	13 28	13 44	14 01	14 19
S 60	07 00	07 55	08 51	13 29	13 43	13 58	14 14

Sunset / Twilight / Moonset

Lat.	Sunset	Twilight Civil	Naut.	Moonset 3	4	5	6
°	h m	h m	h m	h m	h m	h m	h m
N 72	☐	☐	☐	01 42	01 36	01 29	01 22
N 70	☐	☐	☐	01 37	01 35	01 34	01 33
68	☐	☐	☐	01 32	01 35	01 38	01 41
66	23 01	////	////	01 28	01 35	01 41	01 48
64	22 07	////	////	01 25	01 35	01 44	01 54
62	21 36	////	////	01 22	01 35	01 47	02 00
60	21 12	22 41	////	01 20	01 35	01 49	02 04
N 58	20 53	22 03	////	01 18	01 35	01 51	02 08
56	20 38	21 37	////	01 16	01 34	01 53	02 12
54	20 25	21 17	22 48	01 14	01 34	01 54	02 15
52	20 13	21 00	22 13	01 12	01 34	01 56	02 18
50	20 03	20 46	21 48	01 11	01 34	01 57	02 21
45	19 42	20 18	21 07	01 08	01 34	02 00	02 27
N 40	19 25	19 57	20 38	01 05	01 34	02 02	02 32
35	19 10	19 40	20 16	01 03	01 34	02 05	02 36
30	18 58	19 25	19 58	01 01	01 34	02 06	02 40
20	18 37	19 01	19 30	00 57	01 33	02 10	02 46
N 10	18 19	18 42	19 08	00 54	01 33	02 12	02 52
0	18 02	18 24	18 50	00 51	01 33	02 15	02 58
S 10	17 45	18 08	18 34	00 48	01 33	02 18	03 03
20	17 28	17 51	18 19	00 44	01 33	02 21	03 09
30	17 07	17 33	18 03	00 40	01 32	02 24	03 16
35	16 56	17 23	17 55	00 38	01 32	02 26	03 19
40	16 42	17 12	17 46	00 36	01 32	02 28	03 24
45	16 26	17 00	17 36	00 33	01 32	02 30	03 29
S 50	16 07	16 44	17 25	00 29	01 31	02 33	03 35
52	15 57	16 37	17 21	00 28	01 31	02 34	03 38
54	15 47	16 30	17 15	00 26	01 31	02 36	03 41
56	15 35	16 21	17 10	00 24	01 31	02 37	03 44
58	15 22	16 12	17 03	00 22	01 31	02 39	03 48
S 60	15 06	16 01	16 57	00 19	01 31	02 41	03 52

SUN / MOON

Day	Eqn. of Time 00h	12h	Mer. Pass.	Mer. Pass. Upper	Lower	Age	Phase
d	m s	m s	h m	h m	h m	d	%
3	01 58	01 53	11 58	19 23	07 02	09	64
4	01 48	01 43	11 58	20 05	07 44	10	72
5	01 38	01 33	11 58	20 47	08 26	11	80

1998 JUNE 6, 7, 8 (SAT., SUN., MON.)

UT	ARIES	VENUS −4.0		MARS +1.4		JUPITER −2.3		SATURN +0.5		STARS		
	GHA	GHA	Dec	GHA	Dec	GHA	Dec	GHA	Dec	Name	SHA	Dec
d h	° ′	° ′	° ′	° ′	° ′	° ′	° ′	° ′	° ′		° ′	° ′
6 00	254 12.2	217 42.9	N12 18.4	186 52.4	N22 06.1	258 01.7	S 2 54.4	225 46.7	N 9 11.4	Acamar	315 27.5	S40 18.7
01	269 14.7	232 42.6	19.4	201 53.0	06.4	273 03.9	54.3	240 48.9	11.4	Achernar	335 35.8	S57 14.6
02	284 17.2	247 42.2	20.3	216 53.6	06.7	288 06.1	54.2	255 51.1	11.5	Acrux	173 21.9	S63 05.7
03	299 19.6	262 41.8	.. 21.2	231 54.2	.. 07.0	303 08.2	.. 54.0	270 53.3	.. 11.6	Adhara	255 22.0	S28 58.4
04	314 22.1	277 41.5	22.2	246 54.8	07.3	318 10.4	53.9	285 55.5	11.7	Aldebaran	291 03.0	N16 30.2
05	329 24.6	292 41.1	23.1	261 55.4	07.6	333 12.6	53.8	300 57.8	11.8			
06	344 27.0	307 40.7	N12 24.1	276 56.0	N22 07.9	348 14.8	S 2 53.7	316 00.0	N 9 11.9	Alioth	166 30.6	N55 58.4
07	359 29.5	322 40.4	25.0	291 56.6	08.2	3 17.0	53.6	331 02.2	11.9	Alkaid	153 07.7	N49 19.6
S 08	14 32.0	337 40.0	26.0	306 57.2	08.5	18 19.1	53.5	346 04.4	12.0	Al Na'ir	27 58.1	S46 57.9
A 09	29 34.4	352 39.6	.. 26.9	321 57.8	.. 08.8	33 21.3	.. 53.4	1 06.6	.. 12.1	Alnilam	275 58.4	S 1 12.4
T 10	44 36.9	7 39.3	27.9	336 58.4	09.0	48 23.5	53.3	16 08.9	12.2	Alphard	218 07.6	S 8 39.2
U 11	59 39.4	22 38.9	28.8	351 59.0	09.3	63 25.7	53.2	31 11.1	12.3			
R 12	74 41.8	37 38.5	N12 29.8	6 59.6	N22 09.6	78 27.8	S 2 53.0	46 13.3	N 9 12.4	Alphecca	126 20.5	N26 43.4
D 13	89 44.3	52 38.2	30.7	22 00.2	09.9	93 30.0	52.9	61 15.5	12.4	Alpheratz	357 55.5	N29 04.7
A 14	104 46.7	67 37.8	31.7	37 00.8	10.2	108 32.2	52.8	76 17.8	12.5	Altair	62 19.2	N 8 51.9
Y 15	119 49.2	82 37.4	.. 32.6	52 01.4	.. 10.5	123 34.4	.. 52.7	91 20.0	.. 12.6	Ankaa	353 27.3	S42 18.8
16	134 51.7	97 37.0	33.5	67 02.0	10.8	138 36.6	52.6	106 22.2	12.7	Antares	112 40.1	S26 25.6
17	149 54.1	112 36.7	34.5	82 02.6	11.1	153 38.7	52.5	121 24.4	12.8			
18	164 56.6	127 36.3	N12 35.4	97 03.2	N22 11.4	168 40.9	S 2 52.4	136 26.6	N 9 12.8	Arcturus	146 06.1	N19 11.6
19	179 59.1	142 35.9	36.4	112 03.8	11.6	183 43.1	52.3	151 28.9	12.9	Atria	107 51.7	S69 01.4
20	195 01.5	157 35.6	37.3	127 04.4	11.9	198 45.3	52.2	166 31.1	13.0	Avior	234 23.1	S59 30.6
21	210 04.0	172 35.2	.. 38.3	142 05.0	.. 12.2	213 47.5	.. 52.0	181 33.3	.. 13.1	Bellatrix	278 44.8	N 6 20.7
22	225 06.5	187 34.8	39.2	157 05.6	12.5	228 49.7	51.9	196 35.5	13.2	Betelgeuse	271 14.2	N 7 24.2
23	240 08.9	202 34.4	40.1	172 06.2	12.8	243 51.8	51.8	211 37.8	13.3			
7 00	255 11.4	217 34.1	N12 41.1	187 06.8	N22 13.1	258 54.0	S 2 51.7	226 40.0	N 9 13.3	Canopus	264 01.8	S52 41.9
01	270 13.9	232 33.7	42.0	202 07.4	13.4	273 56.2	51.6	241 42.2	13.4	Capella	280 52.0	N45 59.6
02	285 16.3	247 33.3	43.0	217 08.0	13.6	288 58.4	51.5	256 44.4	13.5	Deneb	49 39.0	N45 16.3
03	300 18.8	262 32.9	.. 43.9	232 08.6	.. 13.9	304 00.6	.. 51.4	271 46.7	.. 13.6	Denebola	182 45.4	N14 34.9
04	315 21.2	277 32.6	44.8	247 09.2	14.2	319 02.8	51.3	286 48.9	13.7	Diphda	349 07.6	S17 59.7
05	330 23.7	292 32.2	45.8	262 09.8	14.5	334 04.9	51.2	301 51.1	13.7			
06	345 26.2	307 31.8	N12 46.7	277 10.4	N22 14.8	349 07.1	S 2 51.1	316 53.3	N 9 13.8	Dubhe	194 06.0	N61 45.9
07	0 28.6	322 31.4	47.6	292 11.0	15.1	4 09.3	51.0	331 55.6	13.9	Elnath	278 27.6	N28 36.2
08	15 31.1	337 31.1	48.6	307 11.6	15.3	19 11.5	50.8	346 57.8	14.0	Eltanin	90 51.0	N51 29.4
S 09	30 33.6	352 30.7	.. 49.5	322 12.2	.. 15.6	34 13.7	.. 50.7	2 00.0	.. 14.1	Enif	33 58.3	N 9 52.0
U 10	45 36.0	7 30.3	50.5	337 12.8	15.9	49 15.9	50.6	17 02.2	14.1	Fomalhaut	15 36.7	S29 37.7
N 11	60 38.5	22 29.9	51.4	352 13.4	16.2	64 18.0	50.5	32 04.5	14.2			
D 12	75 41.0	37 29.5	N12 52.3	7 14.0	N22 16.5	79 20.2	S 2 50.4	47 06.7	N 9 14.3	Gacrux	172 13.5	S57 06.5
A 13	90 43.4	52 29.2	53.3	22 14.6	16.7	94 22.4	50.3	62 08.9	14.4	Gienah	176 04.1	S17 32.1
Y 14	105 45.9	67 28.8	54.2	37 15.2	17.0	109 24.6	50.2	77 11.1	14.5	Hadar	149 03.8	S60 22.0
15	120 48.3	82 28.4	.. 55.1	52 15.8	.. 17.3	124 26.8	.. 50.1	92 13.3	.. 14.6	Hamal	328 14.1	N23 27.1
16	135 50.8	97 28.0	56.1	67 16.4	17.6	139 29.0	50.0	107 15.6	14.6	Kaus Aust.	83 58.8	S34 23.0
17	150 53.3	112 27.6	57.0	82 17.0	17.9	154 31.2	49.9	122 17.8	14.7			
18	165 55.7	127 27.2	N12 57.9	97 17.6	N22 18.1	169 33.3	S 2 49.8	137 20.0	N 9 14.8	Kochab	137 18.7	N74 10.0
19	180 58.2	142 26.9	58.9	112 18.2	18.4	184 35.5	49.6	152 22.2	14.9	Markab	13 49.8	N15 11.7
20	196 00.7	157 26.5	12 59.8	127 18.8	18.7	199 37.7	49.5	167 24.5	15.0	Menkar	314 27.4	N 4 04.8
21	211 03.1	172 26.1	13 00.7	142 19.4	.. 19.0	214 39.9	.. 49.4	182 26.7	.. 15.0	Menkent	148 20.9	S36 21.8
22	226 05.6	187 25.7	01.7	157 20.0	19.2	229 42.1	49.3	197 28.9	15.1	Miaplacidus	221 42.5	S69 43.0
23	241 08.1	202 25.3	02.6	172 20.6	19.5	244 44.3	49.2	212 31.2	15.2			
8 00	256 10.5	217 24.9	N13 03.5	187 21.2	N22 19.8	259 46.5	S 2 49.1	227 33.4	N 9 15.3	Mirfak	308 57.4	N49 51.1
01	271 13.0	232 24.6	04.4	202 21.8	20.1	274 48.6	49.0	242 35.6	15.4	Nunki	76 12.3	S26 17.8
02	286 15.5	247 24.2	05.4	217 22.4	20.4	289 50.8	48.9	257 37.8	15.4	Peacock	53 37.1	S56 44.1
03	301 17.9	262 23.8	.. 06.3	232 23.0	.. 20.6	304 53.0	.. 48.8	272 40.1	.. 15.5	Pollux	243 42.2	N28 01.8
04	316 20.4	277 23.4	07.2	247 23.5	20.9	319 55.2	48.7	287 42.3	15.6	Procyon	245 12.1	N 5 13.6
05	331 22.8	292 23.0	08.2	262 24.1	21.2	334 57.4	48.6	302 44.5	15.7			
06	346 25.3	307 22.6	N13 09.1	277 24.7	N22 21.5	349 59.6	S 2 48.5	317 46.7	N 9 15.8	Rasalhague	96 16.8	N12 33.8
07	1 27.8	322 22.2	10.0	292 25.3	21.7	5 01.8	48.4	332 49.0	15.8	Regulus	207 55.9	N11 58.5
08	16 30.2	337 21.9	10.9	307 25.9	22.0	20 04.0	48.3	347 51.2	15.9	Rigel	281 23.5	S 8 12.4
M 09	31 32.7	352 21.5	.. 11.9	322 26.5	.. 22.3	35 06.2	.. 48.1	2 53.4	.. 16.0	Rigil Kent.	140 07.0	S60 49.8
O 10	46 35.2	7 21.1	12.8	337 27.1	22.5	50 08.4	48.0	17 55.6	16.1	Sabik	102 25.5	S15 43.3
N 11	61 37.6	22 20.7	13.7	352 27.7	22.8	65 10.5	47.9	32 57.9	16.2			
D 12	76 40.1	37 20.3	N13 14.6	7 28.3	N22 23.1	80 12.7	S 2 47.8	48 00.1	N 9 16.2	Schedar	349 53.9	N56 31.4
A 13	91 42.6	52 19.9	15.6	22 28.9	23.4	95 14.9	47.7	63 02.3	16.3	Shaula	96 37.2	S37 06.0
Y 14	106 45.0	67 19.5	16.5	37 29.5	23.6	110 17.1	47.6	78 04.5	16.4	Sirius	258 44.2	S16 43.0
15	121 47.5	82 19.1	.. 17.4	52 30.1	.. 23.9	125 19.3	.. 47.5	93 06.8	.. 16.5	Spica	158 43.3	S11 09.2
16	136 50.0	97 18.7	18.3	67 30.7	24.2	140 21.5	47.4	108 09.0	16.6	Suhail	223 01.1	S43 25.9
17	151 52.4	112 18.3	19.3	82 31.3	24.4	155 23.7	47.3	123 11.2	16.6			
18	166 54.9	127 17.9	N13 20.2	97 31.9	N22 24.7	170 25.9	S 2 47.2	138 13.5	N 9 16.7	Vega	80 46.4	N38 47.0
19	181 57.3	142 17.6	21.1	112 32.5	25.0	185 28.1	47.1	153 15.7	16.8	Zuben'ubi	137 17.9	S16 02.1
20	196 59.8	157 17.2	22.0	127 33.1	25.3	200 30.3	47.0	168 17.9	16.9		SHA	Mer. Pass.
21	212 02.3	172 16.8	.. 22.9	142 33.7	.. 25.5	215 32.5	.. 46.9	183 20.1	.. 17.0		° ′	h m
22	227 04.7	187 16.4	23.9	157 34.3	25.8	230 34.7	46.8	198 22.4	17.0	Venus	322 22.7	9 30
23	242 07.2	202 16.0	24.8	172 34.9	26.1	245 36.8	46.7	213 24.6	17.1	Mars	291 55.4	11 31
	h m									Jupiter	3 42.6	6 43
Mer. Pass.	6 58.1	v −0.4	d 0.9	v 0.6	d 0.3	v 2.2	d 0.1	v 2.2	d 0.1	Saturn	331 28.6	8 52

UT	SUN GHA	Dec	MOON GHA	v	Dec	d	HP
d h	° ′	° ′	° ′	′	° ′	′	′
6 00	180 21.8	N22 36.3	46 47.2	14.9	S 7 00.5	9.0	54.2
01	195 21.7	36.6	61 21.1	14.9	7 09.5	9.0	54.2
02	210 21.6	36.9	75 55.0	14.9	7 18.5	9.0	54.2
03	225 21.5 ..	37.1	90 28.9	14.8	7 27.5	8.9	54.2
04	240 21.4	37.4	105 02.7	14.8	7 36.4	8.9	54.3
05	255 21.3	37.7	119 36.5	14.8	7 45.3	8.9	54.3
S 06	270 21.2	N22 37.9	134 10.3	14.8	S 7 54.2	8.8	54.3
A 07	285 21.1	38.2	148 44.1	14.7	8 03.0	8.8	54.3
T 08	300 21.0	38.4	163 17.8	14.7	8 11.8	8.8	54.3
U 09	315 20.8 ..	38.7	177 51.5	14.6	8 20.6	8.8	54.3
R 10	330 20.7	39.0	192 25.1	14.6	8 29.4	8.7	54.3
D 11	345 20.6	39.2	206 58.7	14.6	8 38.1	8.7	54.3
A 12	0 20.5	N22 39.5	221 32.3	14.6	S 8 46.8	8.7	54.3
Y 13	15 20.4	39.7	236 05.9	14.5	8 55.5	8.6	54.3
14	30 20.3	40.0	250 39.4	14.5	9 04.1	8.6	54.3
15	45 20.2 ..	40.2	265 12.9	14.4	9 12.7	8.5	54.3
16	60 20.1	40.5	279 46.3	14.5	9 21.2	8.6	54.3
17	75 19.9	40.8	294 19.8	14.3	9 29.8	8.4	54.4
18	90 19.8	N22 41.0	308 53.1	14.4	S 9 38.2	8.5	54.4
19	105 19.7	41.3	323 26.5	14.3	9 46.7	8.4	54.4
20	120 19.6	41.5	337 59.8	14.2	9 55.1	8.4	54.4
21	135 19.5 ..	41.8	352 33.0	14.3	10 03.5	8.3	54.4
22	150 19.4	42.0	7 06.3	14.1	10 11.8	8.3	54.4
23	165 19.3	42.3	21 39.4	14.2	10 20.1	8.3	54.4
7 00	180 19.1	N22 42.5	36 12.6	14.1	S10 28.4	8.2	54.4
01	195 19.0	42.8	50 45.7	14.1	10 36.6	8.2	54.4
02	210 18.9	43.0	65 18.8	14.0	10 44.8	8.1	54.5
03	225 18.8 ..	43.2	79 51.8	14.0	10 52.9	8.1	54.5
04	240 18.7	43.5	94 24.8	13.9	11 01.0	8.0	54.5
05	255 18.6	43.7	108 57.7	13.9	11 09.0	8.0	54.5
S 06	270 18.5	N22 44.0	123 30.6	13.9	S11 17.0	8.0	54.5
U 07	285 18.3	44.3	138 03.5	13.8	11 25.0	7.9	54.5
N 08	300 18.2	44.5	152 36.3	13.8	11 32.9	7.9	54.5
D 09	315 18.1 ..	44.7	167 09.1	13.7	11 40.8	7.8	54.5
A 10	330 18.0	45.0	181 41.8	13.7	11 48.6	7.8	54.5
Y 11	345 17.9	45.2	196 14.5	13.6	11 56.4	7.7	54.6
12	0 17.8	N22 45.4	210 47.1	13.6	S12 04.1	7.7	54.6
13	15 17.6	45.7	225 19.7	13.6	12 11.8	7.6	54.6
14	30 17.5	45.9	239 52.3	13.5	12 19.4	7.6	54.6
15	45 17.4 ..	46.2	254 24.8	13.4	12 27.0	7.5	54.6
16	60 17.3	46.4	268 57.2	13.4	12 34.5	7.5	54.6
17	75 17.2	46.6	283 29.6	13.4	12 42.0	7.4	54.6
18	90 17.1	N22 46.9	298 02.0	13.3	S12 49.4	7.4	54.7
19	105 16.9	47.1	312 34.3	13.3	12 56.8	7.3	54.7
20	120 16.8	47.3	327 06.6	13.2	13 04.1	7.3	54.7
21	135 16.7 ..	47.6	341 38.8	13.2	13 11.4	7.2	54.7
22	150 16.6	47.8	356 11.0	13.1	13 18.6	7.1	54.7
23	165 16.5	48.0	10 43.1	13.1	13 25.7	7.1	54.7
8 00	180 16.4	N22 48.3	25 15.2	13.0	S13 32.8	7.1	54.7
01	195 16.2	48.5	39 47.2	13.0	13 39.9	7.0	54.8
02	210 16.1	48.7	54 19.2	12.9	13 46.9	6.9	54.8
03	225 16.0 ..	49.0	68 51.1	12.9	13 53.8	6.9	54.8
04	240 15.9	49.2	83 23.0	12.8	14 00.7	6.8	54.8
05	255 15.8	49.4	97 54.8	12.8	14 07.5	6.7	54.8
M 06	270 15.7	N22 49.6	112 26.6	12.7	S14 14.2	6.7	54.8
O 07	285 15.5	49.9	126 58.3	12.7	14 20.9	6.6	54.9
N 08	300 15.4	50.1	141 30.0	12.6	14 27.6	6.5	54.9
D 09	315 15.3 ..	50.3	156 01.6	12.6	14 34.1	6.5	54.9
A 10	330 15.2	50.5	170 33.2	12.6	14 40.6	6.5	54.9
Y 11	345 15.1	50.8	185 04.8	12.4	14 47.1	6.4	54.9
12	0 14.9	N22 51.0	199 36.2	12.5	S14 53.5	6.3	54.9
13	15 14.8	51.2	214 07.7	12.4	14 59.8	6.2	54.9
14	30 14.7	51.4	228 39.1	12.3	15 06.0	6.2	55.0
15	45 14.6 ..	51.7	243 10.4	12.3	15 12.2	6.1	55.0
16	60 14.5	51.9	257 41.7	12.2	15 18.3	6.1	55.0
17	75 14.4	52.1	272 12.9	12.2	15 24.4	5.9	55.0
18	90 14.2	N22 52.3	286 44.1	12.1	S15 30.3	5.9	55.0
19	105 14.1	52.5	301 15.2	12.1	15 36.2	5.9	55.1
20	120 14.0	52.8	315 46.3	12.0	15 42.1	5.7	55.1
21	135 13.9 ..	53.0	330 17.3	12.0	15 47.8	5.7	55.1
22	150 13.8	53.2	344 48.3	11.9	15 53.5	5.7	55.1
23	165 13.6	53.4	359 19.2	11.9	S15 59.2	5.5	55.1
	SD 15.8	d 0.2	SD 14.8		14.9		15.0

Moonrise

Lat.	Twilight Naut.	Civil	Sunrise	6	7	8	9
°	h m	h m	h m	h m	h m	h m	h m
N 72	□	□	□	17 30	19 13	21 04	■
N 70	□	□	□	17 14	18 47	20 23	21 58
68	□	□	□	17 02	18 28	19 55	21 18
66	////	////	00 46	16 51	18 13	19 33	20 51
64	////	////	01 45	16 43	18 00	19 17	20 30
62	////	////	02 18	16 36	17 50	19 03	20 13
60	////	01 10	02 42	16 29	17 41	18 51	19 59
N 58	////	01 50	03 01	16 24	17 33	18 41	19 47
56	////	02 17	03 17	16 19	17 26	18 32	19 37
54	01 04	02 38	03 30	16 15	17 20	18 25	19 28
52	01 41	02 55	03 42	16 11	17 14	18 18	19 20
50	02 06	03 09	03 52	16 07	17 09	18 11	19 12
45	02 49	03 37	04 14	15 59	16 59	17 58	18 57
N 40	03 18	03 59	04 31	15 53	16 50	17 47	18 44
35	03 40	04 16	04 46	15 47	16 42	17 38	18 33
30	03 58	04 31	04 58	15 43	16 36	17 29	18 24
20	04 26	04 55	05 20	15 34	16 24	17 15	18 07
N 10	04 48	05 15	05 38	15 27	16 14	17 03	17 53
0	05 07	05 33	05 55	15 20	16 05	16 51	17 40
S 10	05 23	05 49	06 12	15 13	15 55	16 40	17 27
20	05 39	06 06	06 30	15 06	15 46	16 28	17 13
30	05 55	06 24	06 50	14 58	15 34	16 14	16 57
35	06 03	06 34	07 02	14 53	15 28	16 06	16 47
40	06 12	06 46	07 16	14 48	15 21	15 57	16 37
45	06 22	06 59	07 32	14 42	15 12	15 46	16 25
S 50	06 33	07 14	07 52	14 35	15 02	15 33	16 10
52	06 38	07 21	08 02	14 31	14 57	15 27	16 03
54	06 43	07 29	08 12	14 28	14 52	15 21	15 55
56	06 49	07 38	08 24	14 24	14 46	15 13	15 46
58	06 56	07 48	08 38	14 19	14 40	15 05	15 36
S 60	07 03	07 59	08 55	14 14	14 33	14 56	15 25

Moonset

Lat.	Sunset	Twilight Civil	Naut.	6	7	8	9
°	h m	h m	h m	h m	h m	h m	h m
N 72	□	□	□	01 22	01 15	01 05	00 51
N 70	□	□	□	01 33	01 32	01 32	01 33
68	□	□	□	01 41	01 45	01 52	02 02
66	23 17	////	////	01 48	01 57	02 08	02 24
64	22 14	////	////	01 54	02 06	02 21	02 41
62	21 41	////	////	02 00	02 14	02 32	02 55
60	21 17	22 50	////	02 04	02 21	02 42	03 08
N 58	20 57	22 09	////	02 08	02 28	02 50	03 18
56	20 41	21 42	////	02 12	02 33	02 58	03 27
54	20 28	21 21	22 56	02 15	02 38	03 04	03 35
52	20 16	21 04	22 18	02 18	02 43	03 10	03 43
50	20 06	20 49	21 52	02 21	02 47	03 16	03 49
45	19 44	20 21	21 10	02 27	02 56	03 27	04 03
N 40	19 26	19 59	20 40	02 32	03 03	03 37	04 15
35	19 12	19 41	20 18	02 36	03 09	03 45	04 25
30	18 59	19 27	20 00	02 40	03 15	03 53	04 34
20	18 38	19 02	19 31	02 46	03 25	04 05	04 49
N 10	18 20	18 42	19 09	02 52	03 33	04 16	05 02
0	18 02	18 25	18 51	02 58	03 41	04 27	05 14
S 10	17 46	18 08	18 35	03 03	03 49	04 37	05 27
20	17 28	17 51	18 19	03 09	03 58	04 49	05 40
30	17 07	17 33	18 03	03 16	04 08	05 01	05 55
35	16 55	17 23	17 55	03 19	04 14	05 09	06 04
40	16 41	17 12	17 46	03 24	04 20	05 17	06 14
45	16 25	16 59	17 36	03 29	04 28	05 27	06 26
S 50	16 05	16 43	17 24	03 35	04 37	05 39	06 41
52	15 56	16 36	17 19	03 38	04 41	05 45	06 47
54	15 45	16 28	17 14	03 41	04 46	05 51	06 55
56	15 33	16 20	17 08	03 44	04 51	05 58	07 03
58	15 19	16 10	17 02	03 48	04 57	06 05	07 13
S 60	15 03	15 59	16 55	03 52	05 03	06 14	07 24

	SUN Eqn. of Time 00h	12h	Mer. Pass.	MOON Mer. Pass. Upper	Lower	Age	Phase
Day	m s	m s	h m	h m	h m	d	%
6	01 28	01 22	11 59	21 31	09 09	12	87
7	01 17	01 11	11 59	22 16	09 53	13	93
8	01 06	01 00	11 59	23 03	10 39	14	97

1998 JUNE 9, 10, 11 (TUES., WED., THURS.)

UT	ARIES GHA	VENUS −4.0 GHA	VENUS Dec	MARS +1.5 GHA	MARS Dec	JUPITER −2.4 GHA	JUPITER Dec	SATURN +0.5 GHA	SATURN Dec
d h	° ′	° ′	° ′	° ′	° ′	° ′	° ′	° ′	° ′
9 00	257 09.7	217 15.6	N13 25.7	187 35.5	N22 26.3	260 39.0	S 2 46.6	228 26.8	N 9 17.2
01	272 12.1	232 15.2	26.6	202 36.1	26.6	275 41.2	46.5	243 29.0	17.3
02	287 14.6	247 14.8	27.5	217 36.7	26.9	290 43.4	46.4	258 31.3	17.4
03	302 17.1	262 14.4	.. 28.5	232 37.3	.. 27.1	305 45.6	.. 46.2	273 33.5	.. 17.4
04	317 19.5	277 14.0	29.4	247 37.9	27.4	320 47.8	46.1	288 35.7	17.5
05	332 22.0	292 13.6	30.3	262 38.5	27.7	335 50.0	46.0	303 38.0	17.6
06	347 24.5	307 13.2	N13 31.2	277 39.1	N22 27.9	350 52.2	S 2 45.9	318 40.2	N 9 17.7
07	2 26.9	322 12.8	32.1	292 39.7	28.2	5 54.4	45.8	333 42.4	17.8
08	17 29.4	337 12.4	33.0	307 40.3	28.5	20 56.6	45.7	348 44.6	17.8
09	32 31.8	352 12.0	.. 34.0	322 40.9	.. 28.7	35 58.8	.. 45.6	3 46.9	.. 17.9
10	47 34.3	7 11.6	34.9	337 41.5	29.0	51 01.0	45.5	18 49.1	18.0
11	62 36.8	22 11.2	35.8	352 42.1	29.2	66 03.2	45.4	33 51.3	18.1
12	77 39.2	37 10.8	N13 36.7	7 42.7	N22 29.5	81 05.4	S 2 45.3	48 53.6	N 9 18.2
13	92 41.7	52 10.4	37.6	22 43.2	29.8	96 07.6	45.2	63 55.8	18.2
14	107 44.2	67 10.0	38.5	37 43.8	30.0	111 09.8	45.1	78 58.0	18.3
15	122 46.6	82 09.6	.. 39.4	52 44.4	.. 30.3	126 12.0	.. 45.0	94 00.2	.. 18.4
16	137 49.1	97 09.2	40.3	67 45.0	30.6	141 14.2	44.9	109 02.5	18.5
17	152 51.6	112 08.8	41.3	82 45.6	30.8	156 16.4	44.8	124 04.7	18.5
18	167 54.0	127 08.4	N13 42.2	97 46.2	N22 31.1	171 18.6	S 2 44.7	139 06.9	N 9 18.6
19	182 56.5	142 08.0	43.1	112 46.8	31.3	186 20.8	44.6	154 09.2	18.7
20	197 58.9	157 07.6	44.0	127 47.4	31.6	201 23.0	44.5	169 11.4	18.8
21	213 01.4	172 07.2	.. 44.9	142 48.0	.. 31.9	216 25.2	.. 44.4	184 13.6	.. 18.9
22	228 03.9	187 06.8	45.8	157 48.6	32.1	231 27.4	44.3	199 15.9	18.9
23	243 06.3	202 06.4	46.7	172 49.2	32.4	246 29.6	44.2	214 18.1	19.0
10 00	258 08.8	217 06.0	N13 47.6	187 49.8	N22 32.6	261 31.8	S 2 44.1	229 20.3	N 9 19.1
01	273 11.3	232 05.6	48.5	202 50.4	32.9	276 34.0	44.0	244 22.5	19.2
02	288 13.7	247 05.2	49.4	217 51.0	33.1	291 36.2	43.9	259 24.8	19.3
03	303 16.2	262 04.7	.. 50.3	232 51.6	.. 33.4	306 38.4	.. 43.8	274 27.0	.. 19.3
04	318 18.7	277 04.3	51.2	247 52.2	33.7	321 40.6	43.7	289 29.2	19.4
05	333 21.1	292 03.9	52.2	262 52.8	33.9	336 42.8	43.6	304 31.5	19.5
06	348 23.6	307 03.5	N13 53.1	277 53.4	N22 34.2	351 45.0	S 2 43.5	319 33.7	N 9 19.6
07	3 26.1	322 03.1	54.0	292 54.0	34.4	6 47.2	43.4	334 35.9	19.6
08	18 28.5	337 02.7	54.9	307 54.6	34.7	21 49.4	43.3	349 38.2	19.7
09	33 31.0	352 02.3	.. 55.8	322 55.2	.. 34.9	36 51.6	.. 43.2	4 40.4	.. 19.8
10	48 33.4	7 01.9	56.7	337 55.8	35.2	51 53.8	43.1	19 42.6	19.9
11	63 35.9	22 01.5	57.6	352 56.4	35.5	66 56.0	43.0	34 44.8	20.0
12	78 38.4	37 01.1	N13 58.5	7 57.0	N22 35.7	81 58.2	S 2 42.9	49 47.1	N 9 20.0
13	93 40.8	52 00.7	13 59.4	22 57.6	36.0	97 00.4	42.8	64 49.3	20.1
14	108 43.3	67 00.2	14 00.2	37 58.2	36.2	112 02.6	42.7	79 51.5	20.2
15	123 45.8	81 59.8	.. 01.2	52 58.8	.. 36.5	127 04.8	.. 42.6	94 53.8	.. 20.3
16	138 48.2	96 59.4	02.1	67 59.3	36.7	142 07.0	42.5	109 56.0	20.3
17	153 50.7	111 59.0	03.0	82 59.9	37.0	157 09.2	42.4	124 58.2	20.4
18	168 53.2	126 58.6	N14 03.9	98 00.5	N22 37.2	172 11.4	S 2 42.3	140 00.5	N 9 20.5
19	183 55.6	141 58.2	04.8	113 01.1	37.5	187 13.6	42.2	155 02.7	20.6
20	198 58.1	156 57.8	05.7	128 01.7	37.7	202 15.8	42.1	170 04.9	20.7
21	214 00.6	171 57.3	.. 06.6	143 02.3	.. 38.0	217 18.0	.. 42.0	185 07.2	.. 20.7
22	229 03.0	186 56.9	07.5	158 02.9	38.2	232 20.2	41.9	200 09.4	20.8
23	244 05.5	201 56.5	08.4	173 03.5	38.5	247 22.4	41.8	215 11.6	20.9
11 00	259 07.9	216 56.1	N14 09.3	188 04.1	N22 38.7	262 24.6	S 2 41.7	230 13.9	N 9 21.0
01	274 10.4	231 55.7	10.2	203 04.7	39.0	277 26.8	41.6	245 16.1	21.0
02	289 12.9	246 55.3	11.1	218 05.3	39.2	292 29.0	41.5	260 18.3	21.1
03	304 15.3	261 54.8	.. 12.0	233 05.9	.. 39.5	307 31.2	.. 41.4	275 20.6	.. 21.2
04	319 17.8	276 54.4	12.9	248 06.5	39.7	322 33.4	41.3	290 22.8	21.3
05	334 20.3	291 54.0	13.7	263 07.1	40.0	337 35.6	41.2	305 25.0	21.4
06	349 22.7	306 53.6	N14 14.6	278 07.7	N22 40.2	352 37.8	S 2 41.1	320 27.3	N 9 21.4
07	4 25.2	321 53.2	15.5	293 08.3	40.5	7 40.1	41.0	335 29.5	21.5
08	19 27.7	336 52.7	16.4	308 08.9	40.7	22 42.3	40.9	350 31.7	21.6
09	34 30.1	351 52.3	.. 17.3	323 09.5	.. 41.0	37 44.5	.. 40.8	5 34.0	.. 21.7
10	49 32.6	6 51.9	18.2	338 10.1	41.2	52 46.7	40.7	20 36.2	21.7
11	64 35.1	21 51.5	19.1	353 10.7	41.5	67 48.9	40.6	35 38.4	21.8
12	79 37.5	36 51.1	N14 20.0	8 11.3	N22 41.7	82 51.1	S 2 40.5	50 40.7	N 9 21.9
13	94 40.0	51 50.6	20.9	23 11.9	41.9	97 53.3	40.4	65 42.9	22.0
14	109 42.4	66 50.2	21.8	38 12.5	42.2	112 55.5	40.3	80 45.1	22.1
15	124 44.9	81 49.8	.. 22.7	53 13.0	.. 42.4	127 57.7	.. 40.2	95 47.4	.. 22.1
16	139 47.4	96 49.4	23.5	68 13.6	42.7	142 59.9	40.1	110 49.6	22.2
17	154 49.8	111 48.9	24.4	83 14.2	42.9	158 02.1	40.0	125 51.8	22.3
18	169 52.3	126 48.5	N14 25.3	98 14.8	N22 43.2	173 04.3	S 2 39.9	140 54.1	N 9 22.4
19	184 54.8	141 48.1	26.2	113 15.4	43.4	188 06.6	39.8	155 56.3	22.4
20	199 57.2	156 47.7	27.1	128 16.0	43.6	203 08.8	39.7	170 58.5	22.5
21	214 59.7	171 47.2	.. 28.0	143 16.6	.. 43.9	218 11.0	.. 39.6	186 00.8	.. 22.6
22	230 02.2	186 46.8	28.9	158 17.2	44.1	233 13.2	39.5	201 03.0	22.7
23	245 04.6	201 46.4	29.8	173 17.8	44.4	248 15.4	39.4	216 05.2	22.7
Mer. Pass.	h m 6 46.3	v −0.4	d 0.9	v 0.6	d 0.3	v 2.2	d 0.1	v 2.2	d 0.1

STARS

Name	SHA	Dec
Acamar	315 27.5	S40 18.7
Achernar	335 35.8	S57 14.6
Acrux	173 21.9	S63 05.7
Adhara	255 22.0	S28 58.4
Aldebaran	291 03.0	N16 30.2
Alioth	166 30.7	N55 58.4
Alkaid	153 07.8	N49 19.6
Al Na'ir	27 58.1	S46 57.9
Alnilam	275 58.4	S 1 12.4
Alphard	218 07.6	S 8 39.2
Alphecca	126 20.5	N26 43.4
Alpheratz	357 55.5	N29 04.7
Altair	62 19.2	N 8 51.9
Ankaa	353 27.3	S42 18.7
Antares	112 40.1	S26 25.6
Arcturus	146 06.1	N19 11.6
Atria	107 51.7	S69 01.4
Avior	234 23.1	S59 30.6
Bellatrix	278 44.8	N 6 20.7
Betelgeuse	271 14.2	N 7 24.2
Canopus	264 01.8	S52 41.9
Capella	280 52.0	N45 59.6
Deneb	49 39.0	N45 16.4
Denebola	182 45.4	N14 35.0
Diphda	349 07.6	S17 59.7
Dubhe	194 06.0	N61 45.9
Elnath	278 27.6	N28 36.2
Eltanin	90 51.0	N51 29.4
Enif	33 58.3	N 9 52.0
Fomalhaut	15 36.7	S29 37.7
Gacrux	172 13.5	S57 06.5
Gienah	176 04.1	S17 32.1
Hadar	149 03.8	S60 22.0
Hamal	328 14.0	N23 27.1
Kaus Aust.	83 58.7	S34 23.0
Kochab	137 18.8	N74 10.0
Markab	13 49.8	N15 11.7
Menkar	314 27.4	N 4 04.8
Menkent	148 20.9	S36 21.8
Miaplacidus	221 42.5	S69 43.0
Mirfak	308 57.4	N49 51.1
Nunki	76 12.3	S26 17.8
Peacock	53 37.0	S56 44.1
Pollux	243 42.2	N28 01.8
Procyon	245 12.1	N 5 13.6
Rasalhague	96 16.8	N12 33.8
Regulus	207 55.9	N11 58.5
Rigel	281 23.5	S 8 12.4
Rigil Kent.	140 07.0	S60 49.8
Sabik	102 25.4	S15 43.3
Schedar	349 53.8	N56 31.4
Shaula	96 37.2	S37 06.0
Sirius	258 44.2	S16 43.0
Spica	158 43.3	S11 09.2
Suhail	223 01.2	S43 25.9
Vega	80 46.4	N38 47.0
Zuben'ubi	137 17.9	S16 02.1

	SHA	Mer. Pass.
	° ′	h m
Venus	318 57.2	9 32
Mars	289 41.0	11 28
Jupiter	3 22.9	6 33
Saturn	331 11.5	8 41

SUN / MOON

UT		SUN GHA	Dec	MOON GHA	v	Dec	d	HP
d	h	° ′	° ′	° ′	′	° ′	′	′
9	00	180 13.5	N22 53.6	13 50.1	11.8	S16 04.7	5.5	55.1
	01	195 13.4	53.8	28 20.9	11.8	16 10.2	5.4	55.2
	02	210 13.3	54.1	42 51.7	11.7	16 15.6	5.3	55.2
	03	225 13.2	.. 54.3	57 22.4	11.6	16 20.9	5.3	55.2
	04	240 13.0	54.5	71 53.0	11.7	16 26.2	5.2	55.2
	05	255 12.9	54.7	86 23.7	11.5	16 31.4	5.1	55.2
T	06	270 12.8	N22 54.9	100 54.2	11.5	S16 36.5	5.0	55.2
U	07	285 12.7	55.1	115 24.7	11.5	16 41.5	4.9	55.3
E	08	300 12.6	55.3	129 55.2	11.4	16 46.4	4.9	55.3
S	09	315 12.4	.. 55.5	144 25.6	11.4	16 51.3	4.8	55.3
D	10	330 12.3	55.7	158 56.0	11.3	16 56.1	4.7	55.3
A	11	345 12.2	55.9	173 26.3	11.3	17 00.8	4.6	55.3
Y	12	0 12.1	N22 56.2	187 56.6	11.2	S17 05.4	4.6	55.4
	13	15 11.9	56.4	202 26.8	11.2	17 10.0	4.5	55.4
	14	30 11.8	56.6	216 57.0	11.1	17 14.5	4.3	55.4
	15	45 11.7	.. 56.8	231 27.1	11.1	17 18.8	4.3	55.4
	16	60 11.6	57.0	245 57.2	11.0	17 23.1	4.3	55.4
	17	75 11.5	57.2	260 27.2	11.0	17 27.4	4.1	55.5
	18	90 11.3	N22 57.4	274 57.2	10.9	S17 31.5	4.0	55.5
	19	105 11.2	57.6	289 27.1	10.9	17 35.5	4.0	55.5
	20	120 11.1	57.8	303 57.0	10.8	17 39.5	3.9	55.5
	21	135 11.0	.. 58.0	318 26.8	10.8	17 43.4	3.8	55.5
	22	150 10.8	58.2	332 56.6	10.8	17 47.2	3.7	55.6
	23	165 10.7	58.4	347 26.4	10.7	17 50.9	3.6	55.6
10	00	180 10.6	N22 58.6	1 56.1	10.6	S17 54.5	3.5	55.6
	01	195 10.5	58.8	16 25.7	10.6	17 58.0	3.4	55.6
	02	210 10.4	59.0	30 55.3	10.6	18 01.4	3.4	55.6
	03	225 10.2	.. 59.2	45 24.9	10.5	18 04.8	3.2	55.7
	04	240 10.1	59.4	59 54.4	10.5	18 08.0	3.2	55.7
	05	255 10.0	59.6	74 23.9	10.4	18 11.2	3.1	55.7
W	06	270 09.9	N22 59.8	88 53.3	10.4	S18 14.3	3.0	55.7
E	07	285 09.7	22 59.9	103 22.7	10.4	18 17.3	2.9	55.7
D	08	300 09.6	23 00.1	117 52.1	10.3	18 20.2	2.8	55.8
N	09	315 09.5	.. 00.3	132 21.4	10.2	18 23.0	2.7	55.8
E	10	330 09.4	00.5	146 50.6	10.3	18 25.7	2.6	55.8
S	11	345 09.3	00.7	161 19.9	10.1	18 28.3	2.5	55.8
D	12	0 09.1	N23 00.9	175 49.0	10.2	S18 30.8	2.4	55.8
A	13	15 09.0	01.1	190 18.2	10.1	18 33.2	2.3	55.9
Y	14	30 08.9	01.3	204 47.3	10.1	18 35.5	2.3	55.9
	15	45 08.8	.. 01.5	219 16.4	10.0	18 37.8	2.1	55.9
	16	60 08.6	01.7	233 45.4	10.0	18 39.9	2.0	55.9
	17	75 08.5	01.8	248 14.4	9.9	18 41.9	2.0	55.9
	18	90 08.4	N23 02.0	262 43.3	9.9	S18 43.9	1.8	56.0
	19	105 08.3	02.2	277 12.2	9.9	18 45.7	1.8	56.0
	20	120 08.1	02.4	291 41.1	9.9	18 47.5	1.6	56.0
	21	135 08.0	.. 02.6	306 10.0	9.8	18 49.1	1.5	56.0
	22	150 07.9	02.8	320 38.8	9.8	18 50.6	1.5	56.0
	23	165 07.8	03.0	335 07.6	9.7	18 52.1	1.3	56.1
11	00	180 07.6	N23 03.1	349 36.3	9.7	S18 53.4	1.3	56.1
	01	195 07.5	03.3	4 05.0	9.7	18 54.7	1.1	56.1
	02	210 07.4	03.5	18 33.7	9.6	18 55.8	1.1	56.1
	03	225 07.3	.. 03.7	33 02.3	9.6	18 56.9	0.9	56.1
	04	240 07.1	03.9	47 30.9	9.6	18 57.8	0.9	56.2
	05	255 07.0	04.0	61 59.5	9.6	18 58.7	0.7	56.2
T	06	270 06.9	N23 04.2	76 28.1	9.5	S18 59.4	0.6	56.2
H	07	285 06.8	04.4	90 56.6	9.5	19 00.0	0.6	56.2
U	08	300 06.6	04.6	105 25.1	9.5	19 00.6	0.4	56.2
R	09	315 06.5	.. 04.7	119 53.6	9.4	19 01.0	0.3	56.3
S	10	330 06.4	04.9	134 22.0	9.4	19 01.3	0.3	56.3
D	11	345 06.3	05.1	148 50.4	9.4	19 01.6	0.1	56.3
A	12	0 06.1	N23 05.3	163 18.8	9.4	S19 01.7	0.0	56.3
Y	13	15 06.0	05.4	177 47.2	9.4	19 01.7	0.1	56.3
	14	30 05.9	05.6	192 15.6	9.3	19 01.6	0.2	56.4
	15	45 05.8	.. 05.8	206 43.9	9.3	19 01.4	0.3	56.4
	16	60 05.6	05.9	221 12.2	9.3	19 01.1	0.4	56.4
	17	75 05.5	06.1	235 40.5	9.2	19 00.7	0.5	56.4
	18	90 05.4	N23 06.3	250 08.7	9.3	S19 00.2	0.6	56.5
	19	105 05.2	06.4	264 37.0	9.2	18 59.6	0.7	56.5
	20	120 05.1	06.6	279 05.2	9.2	18 58.9	0.8	56.5
	21	135 05.0	.. 06.8	293 33.4	9.2	18 58.1	0.9	56.5
	22	150 04.9	06.9	308 01.6	9.1	18 57.2	1.0	56.5
	23	165 04.7	07.1	322 29.7	9.2	S18 56.2	1.2	56.6
		SD 15.8	d 0.2	SD 15.1		15.2		15.3

Moonrise

Lat.	Twilight Naut.	Twilight Civil	Sunrise	9	10	11	12
°	h m	h m	h m	h m	h m	h m	h m
N 72	□	□	□	■	■	■	■
N 70	□	□	□	21 58	23 24	24 18	00 18
68	□	□	□	21 18	22 32	23 26	23 59
66	////	////	00 30	20 51	21 59	22 53	23 31
64	////	////	01 40	20 30	21 36	22 29	23 10
62	////	////	02 14	20 13	21 17	22 11	22 53
60	////	01 03	02 39	19 59	21 01	21 55	22 39
N 58	////	01 46	02 59	19 47	20 48	21 42	22 27
56	////	02 14	03 15	19 37	20 37	21 30	22 16
54	00 58	02 35	03 29	19 28	20 27	21 20	22 07
52	01 37	02 53	03 41	19 20	20 18	21 11	21 59
50	02 03	03 07	03 51	19 12	20 10	21 03	21 51
45	02 47	03 36	04 13	18 57	19 53	20 46	21 35
N 40	03 17	03 58	04 31	18 44	19 40	20 33	21 22
35	03 40	04 16	04 46	18 33	19 28	20 21	21 11
30	03 58	04 31	04 58	18 24	19 18	20 10	21 01
20	04 26	04 55	05 20	18 07	19 00	19 53	20 44
N 10	04 48	05 15	05 38	17 53	18 45	19 37	20 30
0	05 07	05 33	05 56	17 40	18 31	19 23	20 16
S 10	05 24	05 50	06 13	17 27	18 16	19 08	20 02
20	05 40	06 07	06 31	17 13	18 01	18 53	19 47
30	05 56	06 26	06 52	16 57	17 44	18 35	19 31
35	06 04	06 36	07 04	16 47	17 34	18 25	19 21
40	06 13	06 47	07 18	16 37	17 22	18 13	19 10
45	06 23	07 01	07 34	16 25	17 09	18 00	18 56
S 50	06 35	07 16	07 54	16 10	16 52	17 43	18 40
52	06 40	07 23	08 04	16 03	16 45	17 35	18 33
54	06 45	07 31	08 15	15 55	16 36	17 26	18 24
56	06 51	07 40	08 27	15 46	16 27	17 16	18 15
58	06 58	07 50	08 41	15 36	16 16	17 05	18 04
S 60	07 05	08 02	08 58	15 25	16 03	16 52	17 52

Moonset

Lat.	Sunset	Twilight Civil	Twilight Naut.	9	10	11	12
°	h m	h m	h m	h m	h m	h m	h m
N 72	□	□	□	00 51	■	■	■
N 70	□	□	□	01 33	01 40	02 00	02 55
68	□	□	□	02 02	02 20	02 52	03 47
66	23 36	////	////	02 24	02 48	03 25	04 20
64	22 21	////	////	02 41	03 09	03 49	04 43
62	21 45	////	////	02 55	03 26	04 08	05 02
60	21 20	22 58	////	03 08	03 41	04 23	05 18
N 58	21 01	22 14	////	03 18	03 53	04 37	05 31
56	20 44	21 45	////	03 27	04 03	04 48	05 42
54	20 30	21 24	23 03	03 35	04 13	04 58	05 52
52	20 18	21 07	22 22	03 43	04 21	05 07	06 01
50	20 08	20 52	21 56	03 49	04 29	05 15	06 09
45	19 46	20 23	21 12	04 03	04 45	05 32	06 26
N 40	19 28	20 01	20 42	04 15	04 58	05 46	06 39
35	19 13	19 43	20 19	04 25	05 09	05 58	06 51
30	19 01	19 28	20 01	04 34	05 19	06 08	07 01
20	18 39	19 03	19 33	04 49	05 36	06 26	07 19
N 10	18 20	18 43	19 10	05 02	05 50	06 41	07 34
0	18 03	18 26	18 52	05 14	06 04	06 56	07 48
S 10	17 46	18 09	18 35	05 27	06 18	07 10	08 02
20	17 28	17 52	18 19	05 40	06 33	07 25	08 18
30	17 07	17 33	18 03	05 55	06 50	07 43	08 35
35	16 55	17 23	17 54	06 04	06 59	07 53	08 45
40	16 41	17 11	17 45	06 14	07 11	08 05	08 56
45	16 24	16 58	17 35	06 26	07 24	08 19	09 10
S 50	16 04	16 42	17 24	06 41	07 40	08 36	09 26
52	15 55	16 35	17 19	06 47	07 48	08 44	09 34
54	15 44	16 27	17 13	06 55	07 57	08 52	09 42
56	15 31	16 18	17 07	07 03	08 06	09 02	09 52
58	15 17	16 08	17 01	07 13	08 16	09 14	10 03
S 60	15 00	15 57	16 53	07 24	08 29	09 27	10 16

SUN / MOON

	SUN			MOON			
Day	Eqn. of Time 00h	Eqn. of Time 12h	Mer. Pass.	Mer. Pass. Upper	Mer. Pass. Lower	Age	Phase
d	m s	m s	h m	h m	h m	d	%
9	00 54	00 49	11 59	23 52	11 27	15	99
10	00 43	00 37	11 59	24 43	12 17	16	100
11	00 31	00 25	12 00	00 43	13 09	17	98	○

1998 JUNE 12, 13, 14 (FRI., SAT., SUN.)

UT	ARIES GHA	VENUS −4.0 GHA	VENUS Dec	MARS +1.5 GHA	MARS Dec	JUPITER −2.4 GHA	JUPITER Dec	SATURN +0.5 GHA	SATURN Dec
12 FRIDAY									
00	260 07.1	216 46.0	N14 30.6	188 18.4	N22 44.6	263 17.6	S 2 39.3	231 07.5	N 9 22.8
01	275 09.6	231 45.5	31.5	203 19.0	44.9	278 19.8	39.2	246 09.7	22.9
02	290 12.0	246 45.1	32.4	218 19.6	45.1	293 22.0	39.1	261 11.9	23.0
03	305 14.5	261 44.7 ..	33.3	233 20.2 ..	45.3	308 24.2 ..	39.0	276 14.2 ..	23.0
04	320 16.9	276 44.2	34.2	248 20.8	45.6	323 26.5	38.9	291 16.4	23.1
05	335 19.4	291 43.8	35.1	263 21.4	45.8	338 28.7	38.8	306 18.6	23.2
06	350 21.9	306 43.4	N14 35.9	278 22.0	N22 46.1	353 30.9	S 2 38.7	321 20.9	N 9 23.3
07	5 24.3	321 42.9	36.8	293 22.6	46.3	8 33.1	38.6	336 23.1	23.4
08	20 26.8	336 42.5	37.7	308 23.2	46.5	23 35.3	38.5	351 25.3	23.4
09	35 29.3	351 42.1 ..	38.6	323 23.8 ..	46.8	38 37.5 ..	38.4	6 27.6 ..	23.5
10	50 31.7	6 41.6	39.5	338 24.4	47.0	53 39.7	38.3	21 29.8	23.6
11	65 34.2	21 41.2	40.3	353 25.0	47.2	68 41.9	38.2	36 32.1	23.7
12	80 36.7	36 40.8	N14 41.2	8 25.5	N22 47.4	83 44.2	S 2 38.1	51 34.3	N 9 23.7
13	95 39.1	51 40.3	42.1	23 26.1	47.7	98 46.4	38.0	66 36.5	23.8
14	110 41.6	66 39.9	43.0	38 26.7	47.9	113 48.6	37.9	81 38.8	23.9
15	125 44.1	81 39.5 ..	43.8	53 27.3 ..	48.2	128 50.8 ..	37.8	96 41.0 ..	24.0
16	140 46.5	96 39.0	44.7	68 27.9	48.4	143 53.0	37.7	111 43.2	24.0
17	155 49.0	111 38.6	45.6	83 28.5	48.7	158 55.2	37.7	126 45.5	24.1
18	170 51.4	126 38.2	N14 46.5	98 29.1	N22 48.9	173 57.4	S 2 37.6	141 47.7	N 9 24.2
19	185 53.9	141 37.7	47.3	113 29.7	49.1	188 59.7	37.5	156 49.9	24.3
20	200 56.4	156 37.3	48.2	128 30.3	49.4	204 01.9	37.4	171 52.2	24.3
21	215 58.8	171 36.9 ..	49.1	143 30.9 ..	49.6	219 04.1 ..	37.3	186 54.4 ..	24.4
22	231 01.3	186 36.4	50.0	158 31.5	49.8	234 06.3	37.2	201 56.7	24.5
23	246 03.8	201 36.0	50.8	173 32.1	50.1	249 08.5	37.1	216 58.9	24.6
13 SATURDAY									
00	261 06.2	216 35.5	N14 51.7	188 32.7	N22 50.3	264 10.7	S 2 37.0	232 01.1	N 9 24.6
01	276 08.7	231 35.1	52.6	203 33.3	50.5	279 13.0	36.9	247 03.4	24.7
02	291 11.2	246 34.7	53.5	218 33.9	50.7	294 15.2	36.8	262 05.6	24.8
03	306 13.6	261 34.2 ..	54.3	233 34.5 ..	51.0	309 17.4 ..	36.7	277 07.8 ..	24.9
04	321 16.1	276 33.8	55.2	248 35.1	51.2	324 19.6	36.6	292 10.1	24.9
05	336 18.5	291 33.3	56.1	263 35.7	51.4	339 21.8	36.5	307 12.3	25.0
06	351 21.0	306 32.9	N14 56.9	278 36.3	N22 51.7	354 24.0	S 2 36.4	322 14.6	N 9 25.1
07	6 23.5	321 32.5	57.8	293 36.8	51.9	9 26.3	36.3	337 16.8	25.2
08	21 25.9	336 32.0	58.7	308 37.4	52.1	24 28.5	36.2	352 19.0	25.2
09	36 28.4	351 31.6	14 59.5	323 38.0 ..	52.4	39 30.7 ..	36.1	7 21.3 ..	25.3
10	51 30.9	6 31.1	15 00.4	338 38.6	52.6	54 32.9	36.0	22 23.5	25.4
11	66 33.3	21 30.7	01.3	353 39.2	52.8	69 35.1	36.0	37 25.7	25.5
12	81 35.8	36 30.2	N15 02.1	8 39.8	N22 53.0	84 37.4	S 2 35.9	52 28.0	N 9 25.5
13	96 38.3	51 29.8	03.0	23 40.4	53.3	99 39.6	35.8	67 30.2	25.6
14	111 40.7	66 29.3	03.9	38 41.0	53.5	114 41.8	35.7	82 32.5	25.7
15	126 43.2	81 28.9 ..	04.7	53 41.6 ..	53.7	129 44.0 ..	35.6	97 34.7 ..	25.8
16	141 45.7	96 28.5	05.6	68 42.2	53.9	144 46.2	35.5	112 36.9	25.8
17	156 48.1	111 28.0	06.4	83 42.8	54.2	159 48.5	35.4	127 39.2	25.9
18	171 50.6	126 27.6	N15 07.3	98 43.4	N22 54.4	174 50.7	S 2 35.3	142 41.4	N 9 26.0
19	186 53.0	141 27.1	08.2	113 44.0	54.6	189 52.9	35.2	157 43.6	26.1
20	201 55.5	156 26.7	09.0	128 44.6	54.8	204 55.1	35.1	172 45.9	26.1
21	216 58.0	171 26.2 ..	09.9	143 45.2 ..	55.1	219 57.4 ..	35.0	187 48.1 ..	26.2
22	232 00.4	186 25.8	10.8	158 45.8	55.3	234 59.6	34.9	202 50.4	26.3
23	247 02.9	201 25.3	11.6	173 46.4	55.5	250 01.8	34.8	217 52.6	26.4
14 SUNDAY									
00	262 05.4	216 24.9	N15 12.5	188 47.0	N22 55.7	265 04.0	S 2 34.8	232 54.8	N 9 26.4
01	277 07.8	231 24.4	13.3	203 47.6	56.0	280 06.2	34.7	247 57.1	26.5
02	292 10.3	246 24.0	14.2	218 48.1	56.2	295 08.5	34.6	262 59.3	26.6
03	307 12.8	261 23.5 ..	15.0	233 48.7 ..	56.4	310 10.7 ..	34.5	278 01.6 ..	26.7
04	322 15.2	276 23.1	15.9	248 49.3	56.6	325 12.9	34.4	293 03.8	26.7
05	337 17.7	291 22.6	16.8	263 49.9	56.9	340 15.1	34.3	308 06.0	26.8
06	352 20.2	306 22.1	N15 17.6	278 50.5	N22 57.1	355 17.4	S 2 34.2	323 08.3	N 9 26.9
07	7 22.6	321 21.7	18.5	293 51.1	57.3	10 19.6	34.1	338 10.5	27.0
08	22 25.1	336 21.2	19.3	308 51.7	57.5	25 21.8	34.0	353 12.8	27.0
09	37 27.5	351 20.8 ..	20.2	323 52.3 ..	57.7	40 24.0 ..	33.9	8 15.0 ..	27.1
10	52 30.0	6 20.3	21.0	338 52.9	58.0	55 26.3	33.8	23 17.2	27.2
11	67 32.5	21 19.9	21.9	353 53.5	58.2	70 28.5	33.7	38 19.5	27.2
12	82 34.9	36 19.4	N15 22.7	8 54.1	N22 58.4	85 30.7	S 2 33.7	53 21.7	N 9 27.3
13	97 37.4	51 19.0	23.6	23 54.7	58.6	100 32.9	33.6	68 24.0	27.4
14	112 39.9	66 18.5	24.4	38 55.3	58.8	115 35.2	33.5	83 26.2	27.5
15	127 42.3	81 18.0 ..	25.3	53 55.9 ..	59.0	130 37.4 ..	33.4	98 28.4 ..	27.5
16	142 44.8	96 17.6	26.1	68 56.5	59.3	145 39.6	33.3	113 30.7	27.6
17	157 47.3	111 17.1	27.0	83 57.1	59.5	160 41.8	33.2	128 32.9	27.7
18	172 49.7	126 16.7	N15 27.8	98 57.7	N22 59.7	175 44.1	S 2 33.1	143 35.2	N 9 27.8
19	187 52.2	141 16.2	28.7	113 58.3	22 59.9	190 46.3	33.0	158 37.4	27.8
20	202 54.7	156 15.8	29.5	128 58.8	23 00.1	205 48.5	32.9	173 39.7	27.9
21	217 57.1	171 15.3 ..	30.4	143 59.4 ..	00.3	220 50.8 ..	32.8	188 41.9 ..	28.0
22	232 59.6	186 14.8	31.2	159 00.0	00.6	235 53.0	32.8	203 44.1	28.1
23	248 02.0	201 14.4	32.1	174 00.6	00.8	250 55.2	32.7	218 46.4	28.1
Mer. Pass.	6 34.5 h m	v −0.4 d 0.9		v 0.6 d 0.2		v 2.2 d 0.1		v 2.2 d 0.1	

STARS

Name	SHA	Dec
Acamar	315 27.5	S40 18.7
Achernar	335 35.8	S57 14.6
Acrux	173 21.9	S63 05.7
Adhara	255 22.0	S28 58.4
Aldebaran	291 03.0	N16 30.2
Alioth	166 30.7	N55 58.4
Alkaid	153 07.8	N49 19.6
Al Na'ir	27 58.1	S46 57.9
Alnilam	275 58.4	S 1 12.3
Alphard	218 07.6	S 8 39.2
Alphecca	126 20.5	N26 43.4
Alpheratz	357 55.5	N29 04.7
Altair	62 19.2	N 8 51.9
Ankaa	353 27.2	S42 18.7
Antares	112 40.1	S26 25.6
Arcturus	146 06.1	N19 11.6
Atria	107 51.7	S69 01.4
Avior	234 23.1	S59 30.6
Bellatrix	278 44.7	N 6 20.7
Betelgeuse	271 14.1	N 7 24.2
Canopus	264 01.8	S52 41.9
Capella	280 52.0	N45 59.6
Deneb	49 39.0	N45 16.4
Denebola	182 45.4	N14 35.0
Diphda	349 07.6	S17 59.7
Dubhe	194 06.0	N61 45.9
Elnath	278 27.6	N28 36.2
Eltanin	90 51.0	N51 29.5
Enif	33 58.3	N 9 52.0
Fomalhaut	15 36.7	S29 37.7
Gacrux	172 13.6	S57 06.5
Gienah	176 04.1	S17 32.1
Hadar	149 03.8	S60 22.1
Hamal	328 14.0	N23 27.1
Kaus Aust.	83 58.7	S34 23.0
Kochab	137 18.8	N74 10.1
Markab	13 49.8	N15 11.7
Menkar	314 27.4	N 4 04.8
Menkent	148 20.9	S36 21.8
Miaplacidus	221 42.5	S69 43.0
Mirfak	308 57.3	N49 51.1
Nunki	76 12.3	S26 17.8
Peacock	53 37.0	S56 44.1
Pollux	243 42.2	N28 01.8
Procyon	245 12.1	N 5 13.6
Rasalhague	96 16.8	N12 33.8
Regulus	207 55.9	N11 58.5
Rigel	281 23.5	S 8 12.4
Rigil Kent.	140 07.0	S60 49.8
Sabik	102 25.4	S15 43.3
Schedar	349 53.8	N56 31.4
Shaula	96 37.2	S37 06.0
Sirius	258 44.2	S16 43.0
Spica	158 43.3	S11 09.2
Suhail	223 01.2	S43 25.8
Vega	80 46.3	N38 47.0
Zuben'ubi	137 17.9	S16 02.1

	SHA	Mer. Pass.
	° ′	h m
Venus	315 29.3	9 34
Mars	287 26.5	11 25
Jupiter	3 04.5	6 22
Saturn	330 54.9	8 31

UT	SUN GHA	Dec	MOON GHA	v	Dec	d	HP
d h	° ′	° ′	° ′	′	° ′	′	′
12 00	180 04.6	N23 07.3	336 57.9	9.1	S18 55.0	1.2	56.6
01	195 04.5	07.4	351 26.0	9.1	18 53.8	1.3	56.6
02	210 04.4	07.6	5 54.1	9.1	18 52.5	1.5	56.6
03	225 04.2 ..	07.8	20 22.2	9.1	18 51.0	1.5	56.6
04	240 04.1	07.9	34 50.3	9.1	18 49.5	1.7	56.7
05	255 04.0	08.1	49 18.4	9.1	18 47.8	1.8	56.7
06	270 03.9	N23 08.3	63 46.5	9.0	S18 46.0	1.8	56.7
07	285 03.7	08.4	78 14.5	9.1	18 44.2	2.0	56.7
08	300 03.6	08.6	92 42.6	9.0	18 42.2	2.1	56.7
F 09	315 03.5 ..	08.7	107 10.6	9.0	18 40.1	2.2	56.8
R 10	330 03.3	08.9	121 38.6	9.1	18 37.9	2.3	56.8
I 11	345 03.2	09.0	136 06.7	9.0	18 35.6	2.4	56.8
D 12	0 03.1	N23 09.2	150 34.7	9.0	S18 33.2	2.5	56.8
A 13	15 03.0	09.4	165 02.7	9.0	18 30.7	2.6	56.9
Y 14	30 02.8	09.5	179 30.7	9.0	18 28.1	2.7	56.9
15	45 02.7 ..	09.7	193 58.7	8.9	18 25.4	2.8	56.9
16	60 02.6	09.8	208 26.6	9.0	18 22.6	3.0	56.9
17	75 02.4	10.0	222 54.6	9.0	18 19.6	3.0	56.9
18	90 02.3	N23 10.1	237 22.6	9.0	S18 16.6	3.1	57.0
19	105 02.2	10.3	251 50.6	9.0	18 13.5	3.3	57.0
20	120 02.1	10.4	266 18.6	8.9	18 10.2	3.3	57.0
21	135 01.9 ..	10.6	280 46.5	9.0	18 06.9	3.5	57.0
22	150 01.8	10.7	295 14.5	9.0	18 03.4	3.5	57.0
23	165 01.7	10.9	309 42.5	9.0	17 59.9	3.7	57.1
13 00	180 01.5	N23 11.0	324 10.5	8.9	S17 56.2	3.7	57.1
01	195 01.4	11.2	338 38.4	9.0	17 52.5	3.9	57.1
02	210 01.3	11.3	353 06.4	9.0	17 48.6	4.0	57.1
03	225 01.2 ..	11.5	7 34.4	9.0	17 44.6	4.0	57.1
04	240 01.0	11.6	22 02.4	9.0	17 40.6	4.2	57.2
05	255 00.9	11.7	36 30.4	9.0	17 36.4	4.3	57.2
06	270 00.8	N23 11.9	50 58.4	9.0	S17 32.1	4.4	57.2
S 07	285 00.6	12.0	65 26.4	9.0	17 27.7	4.4	57.2
A 08	300 00.5	12.2	79 54.4	9.0	17 23.3	4.6	57.2
T 09	315 00.4 ..	12.3	94 22.4	9.0	17 18.7	4.7	57.3
U 10	330 00.3	12.5	108 50.4	9.0	17 14.0	4.8	57.3
R 11	345 00.1	12.6	123 18.4	9.0	17 09.2	4.8	57.3
D 12	0 00.0	N23 12.7	137 46.4	9.1	S17 04.4	5.0	57.3
A 13	14 59.9	12.9	152 14.5	9.0	16 59.4	5.1	57.4
Y 14	29 59.7	13.0	166 42.5	9.1	16 54.3	5.2	57.4
15	44 59.6 ..	13.1	181 10.6	9.0	16 49.1	5.3	57.4
16	59 59.5	13.3	195 38.6	9.1	16 43.8	5.3	57.4
17	74 59.3	13.4	210 06.7	9.1	16 38.5	5.5	57.4
18	89 59.2	N23 13.6	224 34.8	9.1	S16 33.0	5.6	57.5
19	104 59.1	13.7	239 02.9	9.1	16 27.4	5.6	57.5
20	119 59.0	13.8	253 31.0	9.1	16 21.8	5.8	57.5
21	134 58.8 ..	14.0	267 59.1	9.2	16 16.0	5.8	57.5
22	149 58.7	14.1	282 27.3	9.1	16 10.2	6.0	57.5
23	164 58.6	14.2	296 55.4	9.2	16 04.2	6.0	57.6
14 00	179 58.3	N23 14.3	311 23.6	9.1	S15 58.2	6.2	57.6
01	194 58.3	14.5	325 51.7	9.2	15 52.0	6.2	57.6
02	209 58.2	14.6	340 19.9	9.2	15 45.8	6.3	57.6
03	224 58.0 ..	14.7	354 48.1	9.2	15 39.5	6.5	57.6
04	239 57.9	14.9	9 16.3	9.2	15 33.0	6.5	57.7
05	254 57.8	15.0	23 44.5	9.3	15 26.5	6.6	57.7
06	269 57.6	N23 15.1	38 12.8	9.2	S15 19.9	6.7	57.7
07	284 57.5	15.2	52 41.0	9.3	15 13.2	6.7	57.7
08	299 57.4	15.4	67 09.3	9.3	15 06.5	6.9	57.7
S 09	314 57.3 ..	15.5	81 37.6	9.3	14 59.6	7.0	57.8
U 10	329 57.1	15.6	96 05.9	9.3	14 52.6	7.0	57.8
N 11	344 57.0	15.7	110 34.2	9.3	14 45.6	7.2	57.8
D 12	359 56.9	N23 15.9	125 02.5	9.4	S14 38.4	7.2	57.8
A 13	14 56.7	16.0	139 30.9	9.4	14 31.2	7.3	57.8
Y 14	29 56.6	16.1	153 59.3	9.3	14 23.9	7.4	57.9
15	44 56.5 ..	16.2	168 27.6	9.4	14 16.5	7.5	57.9
16	59 56.3	16.3	182 56.0	9.4	14 09.0	7.5	57.9
17	74 56.2	16.5	197 24.4	9.5	14 01.5	7.7	57.9
18	89 56.1	N23 16.6	211 52.9	9.4	S13 53.8	7.7	57.9
19	104 55.9	16.7	226 21.3	9.5	13 46.1	7.8	58.0
20	119 55.8	16.8	240 49.8	9.5	13 38.3	7.9	58.0
21	134 55.7 ..	16.9	255 18.3	9.5	13 30.4	8.0	58.0
22	149 55.5	17.0	269 46.8	9.5	13 22.4	8.1	58.0
23	164 55.4	17.2	284 15.3	9.5	S13 14.3	8.1	58.0
	SD 15.8	d 0.1	SD 15.5		15.6		15.8

Lat.	Twilight Naut.	Civil	Sunrise	Moonrise 12	13	14	15
°	h m	h m	h m	h m	h m	h m	h m
N 72	☐	☐	☐	■	02 07	01 24	01 10
N 70	☐	☐	☐	00 18	00 38	00 44	00 46
68	☐	☐	☐	23 59	24 17	00 17	00 28
66	☐	☐	☐	23 31	23 56	24 13	00 13
64	////	////	01 35	23 10	23 39	24 01	00 01
62	////	////	02 12	22 53	23 25	23 50	24 10
60	////	00 57	02 37	22 39	23 14	23 41	24 04
N 58	////	01 43	02 57	22 27	23 03	23 33	23 58
56	////	02 12	03 14	22 16	22 54	23 26	23 53
54	00 52	02 34	03 28	22 07	22 46	23 20	23 49
52	01 35	02 51	03 40	21 59	22 39	23 14	23 45
50	02 02	03 06	03 50	21 51	22 33	23 09	23 42
45	02 46	03 36	04 13	21 35	22 19	22 58	23 34
N 40	03 16	03 58	04 31	21 22	22 07	22 49	23 27
35	03 39	04 16	04 45	21 11	21 58	22 41	23 22
30	03 58	04 31	04 58	21 01	21 49	22 34	23 17
20	04 26	04 56	05 20	20 44	21 34	22 22	23 08
N 10	04 49	05 16	05 39	20 30	21 21	22 11	23 01
0	05 08	05 34	05 56	20 16	21 09	22 01	22 53
S 10	05 24	05 51	06 14	20 02	20 57	21 52	22 46
20	05 40	06 08	06 32	19 47	20 44	21 41	22 39
30	05 57	06 27	06 53	19 31	20 29	21 29	22 30
35	06 05	06 37	07 05	19 21	20 20	21 22	22 25
40	06 15	06 49	07 19	19 10	20 10	21 14	22 19
45	06 25	07 02	07 36	18 56	19 58	21 04	22 13
S 50	06 37	07 18	07 56	18 40	19 44	20 53	22 05
52	06 42	07 25	08 06	18 33	19 38	20 48	22 01
54	06 47	07 33	08 17	18 24	19 30	20 42	21 57
56	06 53	07 42	08 30	18 15	19 22	20 35	21 52
58	07 00	07 52	08 44	18 04	19 13	20 28	21 47
S 60	07 07	08 04	09 01	17 52	19 02	20 20	21 41

Lat.	Sunset	Twilight Civil	Naut.	Moonset 12	13	14	15
°	h m	h m	h m	h m	h m	h m	h m
N 72	☐	☐	☐	■	02 58	05 31	07 34
N 70	☐	☐	☐	02 55	04 26	06 10	07 57
68	☐	☐	☐	03 47	05 05	06 36	08 14
66	☐	☐	☐	04 20	05 32	06 57	08 28
64	22 26	////	////	04 43	05 53	07 13	08 40
62	21 49	////	////	05 02	06 09	07 26	08 49
60	21 23	23 05	////	05 18	06 23	07 37	08 57
N 58	21 03	22 18	////	05 31	06 35	07 47	09 05
56	20 47	21 49	////	05 42	06 45	07 56	09 11
54	20 33	21 27	23 10	05 52	06 54	08 03	09 17
52	20 21	21 09	22 26	06 01	07 02	08 10	09 22
50	20 10	20 54	21 59	06 09	07 10	08 16	09 26
45	19 47	20 25	21 14	06 26	07 25	08 29	09 36
N 40	19 30	20 02	20 44	06 39	07 38	08 40	09 45
35	19 15	19 44	20 21	06 51	07 49	08 49	09 52
30	19 02	19 29	20 02	07 01	07 58	08 57	09 58
20	18 40	19 04	19 34	07 19	08 14	09 11	10 08
N 10	18 21	18 44	19 11	07 34	08 28	09 23	10 18
0	18 04	18 26	18 52	07 48	08 41	09 34	10 26
S 10	17 46	18 09	18 36	08 02	08 54	09 45	10 35
20	17 28	17 52	18 19	08 18	09 08	09 57	10 44
30	17 07	17 33	18 03	08 35	09 24	10 11	10 54
35	16 55	17 23	17 54	08 45	09 33	10 18	11 00
40	16 41	17 11	17 45	08 56	09 44	10 27	11 07
45	16 24	16 58	17 35	09 10	09 56	10 37	11 15
S 50	16 04	16 42	17 23	09 26	10 11	10 50	11 24
52	15 54	16 35	17 18	09 34	10 18	10 55	11 28
54	15 43	16 26	17 13	09 42	10 25	11 02	11 33
56	15 30	16 17	17 06	09 52	10 34	11 09	11 38
58	15 16	16 07	17 00	10 03	10 44	11 17	11 44
S 60	14 59	15 56	16 52	10 16	10 55	11 26	11 50

	SUN			MOON			
Day	Eqn. of Time 00h	12h	Mer. Pass.	Mer. Pass. Upper	Lower	Age	Phase
d	m s	m s	h m	h m	h m	d	%
12	00 19	00 13	12 00	01 36	14 02	18	94
13	00 06	00 00	12 00	02 29	14 55	19	88
14	00 06	00 12	12 00	03 22	15 48	20	81

1998 JUNE 15, 16, 17 (MON., TUES., WED.)

UT	ARIES GHA	VENUS −3.9 GHA	Dec	MARS +1.5 GHA	Dec	JUPITER −2.4 GHA	Dec	SATURN +0.5 GHA	Dec
15 00	263 04.5	216 13.9	N15 32.9	189 01.2	N23 01.0	265 57.4	S 2 32.6	233 48.6	N 9 28.2
01	278 07.0	231 13.4	33.8	204 01.8	01.2	280 59.7	32.5	248 50.9	28.3
02	293 09.4	246 13.0	34.6	219 02.4	01.4	296 01.9	32.4	263 53.1	28.3
03	308 11.9	261 12.5 ..	35.4	234 03.0 ..	01.6	311 04.1 ..	32.3	278 55.3 ..	28.4
04	323 14.4	276 12.1	36.3	249 03.6	01.8	326 06.4	32.2	293 57.6	28.5
05	338 16.8	291 11.6	37.1	264 04.2	02.1	341 08.6	32.1	308 59.8	28.6
06	353 19.3	306 11.1	N15 38.0	279 04.8	N23 02.3	356 10.8	S 2 32.0	324 02.1	N 9 28.6
07	8 21.8	321 10.7	38.8	294 05.4	02.5	11 13.1	32.0	339 04.3	28.7
08	23 24.2	336 10.2	39.7	309 06.0	02.7	26 15.3	31.9	354 06.6	28.8
M 09	38 26.7	351 09.7 ..	40.5	324 06.6 ..	02.9	41 17.5 ..	31.8	9 08.8 ..	28.9
O 10	53 29.1	6 09.3	41.3	339 07.2	03.1	56 19.7	31.7	24 11.0	28.9
N 11	68 31.6	21 08.8	42.2	354 07.8	03.3	71 22.0	31.6	39 13.3	29.0
D 12	83 34.1	36 08.3	N15 43.0	9 08.4	N23 03.5	86 24.2	S 2 31.5	54 15.5	N 9 29.1
A 13	98 36.5	51 07.9	43.8	24 08.9	03.7	101 26.4	31.4	69 17.8	29.1
Y 14	113 39.0	66 07.4	44.7	39 09.5	03.9	116 28.7	31.3	84 20.0	29.2
15	128 41.5	81 06.9 ..	45.5	54 10.1 ..	04.2	131 30.9 ..	31.3	99 22.3 ..	29.3
16	143 43.9	96 06.4	46.4	69 10.7	04.4	146 33.1	31.2	114 24.5	29.4
17	158 46.4	111 06.0	47.2	84 11.3	04.6	161 35.4	31.1	129 26.7	29.4
18	173 48.9	126 05.5	N15 48.0	99 11.9	N23 04.8	176 37.6	S 2 31.0	144 29.0	N 9 29.5
19	188 51.3	141 05.0	48.9	114 12.5	05.0	191 39.8	30.9	159 31.2	29.6
20	203 53.8	156 04.6	49.7	129 13.1	05.2	206 42.1	30.8	174 33.5	29.7
21	218 56.3	171 04.1 ..	50.5	144 13.7 ..	05.4	221 44.3 ..	30.7	189 35.7 ..	29.7
22	233 58.7	186 03.6	51.4	159 14.3	05.6	236 46.5	30.6	204 38.0	29.8
23	249 01.2	201 03.1	52.2	174 14.9	05.8	251 48.8	30.6	219 40.2	29.9
16 00	264 03.6	216 02.7	N15 53.0	189 15.5	N23 06.0	266 51.0	S 2 30.5	234 42.5	N 9 29.9
01	279 06.1	231 02.2	53.9	204 16.1	06.2	281 53.2	30.4	249 44.7	30.0
02	294 08.6	246 01.7	54.7	219 16.7	06.4	296 55.5	30.3	264 46.9	30.1
03	309 11.0	261 01.2 ..	55.5	234 17.3 ..	06.6	311 57.7 ..	30.2	279 49.2 ..	30.2
04	324 13.5	276 00.8	56.3	249 17.9	06.8	327 00.0	30.1	294 51.4	30.2
05	339 16.0	291 00.3	57.2	264 18.5	07.0	342 02.2	30.0	309 53.7	30.3
06	354 18.4	305 59.8	N15 58.0	279 19.0	N23 07.2	357 04.4	S 2 30.0	324 55.9	N 9 30.4
07	9 20.9	320 59.3	58.8	294 19.6	07.4	12 06.7	29.9	339 58.2	30.4
T 08	24 23.4	335 58.9	15 59.7	309 20.2	07.6	27 08.9	29.8	355 00.4	30.5
U 09	39 25.8	350 58.4	16 00.5	324 20.8 ..	07.8	42 11.1 ..	29.7	10 02.7 ..	30.6
E 10	54 28.3	5 57.9	01.3	339 21.4	08.0	57 13.4	29.6	25 04.9	30.6
S 11	69 30.8	20 57.4	02.1	354 22.0	08.2	72 15.6	29.5	40 07.1	30.7
D 12	84 33.2	35 57.0	N16 03.0	9 22.6	N23 08.4	87 17.9	S 2 29.4	55 09.4	N 9 30.8
A 13	99 35.7	50 56.5	03.8	24 23.2	08.7	102 20.1	29.4	70 11.6	30.9
Y 14	114 38.1	65 56.0	04.6	39 23.8	08.9	117 22.3	29.3	85 13.9	30.9
15	129 40.6	80 55.5 ..	05.4	54 24.4 ..	09.1	132 24.6 ..	29.2	100 16.1 ..	31.0
16	144 43.1	95 55.0	06.2	69 25.0	09.2	147 26.8	29.1	115 18.4	31.1
17	159 45.5	110 54.5	07.1	84 25.6	09.4	162 29.0	29.0	130 20.6	31.2
18	174 48.0	125 54.1	N16 07.9	99 26.2	N23 09.6	177 31.3	S 2 28.9	145 22.9	N 9 31.2
19	189 50.5	140 53.6	08.7	114 26.8	09.8	192 33.5	28.9	160 25.1	31.3
20	204 52.9	155 53.1	09.5	129 27.4	10.0	207 35.8	28.8	175 27.4	31.4
21	219 55.4	170 52.6 ..	10.3	144 28.0 ..	10.2	222 38.0 ..	28.7	190 29.6 ..	31.4
22	234 57.9	185 52.1	11.2	159 28.5	10.4	237 40.2	28.6	205 31.9	31.5
23	250 00.3	200 51.6	12.0	174 29.1	10.6	252 42.5	28.5	220 34.1	31.6
17 00	265 02.8	215 51.2	N16 12.8	189 29.7	N23 10.8	267 44.7	S 2 28.4	235 36.3	N 9 31.7
01	280 05.2	230 50.7	13.6	204 30.3	11.0	282 47.0	28.3	250 38.6	31.7
02	295 07.7	245 50.2	14.4	219 30.9	11.2	297 49.2	28.3	265 40.8	31.8
03	310 10.2	260 49.7 ..	15.2	234 31.5 ..	11.4	312 51.5 ..	28.2	280 43.1 ..	31.9
04	325 12.6	275 49.2	16.1	249 32.1	11.6	327 53.7	28.1	295 45.3	31.9
05	340 15.1	290 48.7	16.9	264 32.7	11.8	342 55.9	28.0	310 47.6	32.0
06	355 17.6	305 48.2	N16 17.7	279 33.3	N23 12.0	357 58.2	S 2 27.9	325 49.8	N 9 32.1
W 07	10 20.0	320 47.7	18.5	294 33.9	12.2	13 00.4	27.8	340 52.1	32.2
E 08	25 22.5	335 47.3	19.3	309 34.5	12.4	28 02.7	27.8	355 54.3	32.2
D 09	40 25.0	350 46.8 ..	20.1	324 35.1 ..	12.6	43 04.9 ..	27.7	10 56.6 ..	32.3
N 10	55 27.4	5 46.3	20.9	339 35.7	12.8	58 07.2	27.6	25 58.8	32.4
E 11	70 29.9	20 45.8	21.7	354 36.3	13.0	73 09.4	27.5	41 01.1	32.4
S 12	85 32.4	35 45.3	N16 22.5	9 36.9	N23 13.2	88 11.6	S 2 27.4	56 03.3	N 9 32.5
D 13	100 34.8	50 44.8	23.4	24 37.5	13.3	103 13.9	27.4	71 05.6	32.6
A 14	115 37.3	65 44.3	24.2	39 38.1	13.5	118 16.1	27.3	86 07.8	32.6
Y 15	130 39.7	80 43.8 ..	25.0	54 38.6 ..	13.7	133 18.4 ..	27.2	101 10.1 ..	32.7
16	145 42.2	95 43.3	25.8	69 39.2	13.9	148 20.6	27.1	116 12.3	32.8
17	160 44.7	110 42.8	26.6	84 39.8	14.1	163 22.9	27.0	131 14.6	32.9
18	175 47.1	125 42.3	N16 27.4	99 40.4	N23 14.3	178 25.1	S 2 26.9	146 16.8	N 9 32.9
19	190 49.6	140 41.8	28.2	114 41.0	14.5	193 27.4	26.9	161 19.1	33.0
20	205 52.1	155 41.4	29.0	129 41.6	14.7	208 29.6	26.8	176 21.3	33.1
21	220 54.5	170 40.9 ..	29.8	144 42.2 ..	14.9	223 31.9 ..	26.7	191 23.6 ..	33.1
22	235 57.0	185 40.4	30.6	159 42.8	15.1	238 34.1	26.6	206 25.8	33.2
23	250 59.5	200 39.9	31.4	174 43.4	15.2	253 36.4	26.5	221 28.1	33.3
Mer. Pass. 6 22.7		v −0.5 d 0.8		v 0.6 d 0.2		v 2.2 d 0.1		v 2.2 d 0.1	

STARS

Name	SHA	Dec
Acamar	315 27.5	S40 18.7
Achernar	335 35.7	S57 14.5
Acrux	173 22.0	S63 05.7
Adhara	255 22.0	S28 58.4
Aldebaran	291 03.0	N16 30.2
Alioth	166 30.7	N55 58.4
Alkaid	153 07.8	N49 19.6
Al Na'ir	27 58.0	S46 57.9
Alnilam	275 58.4	S 1 12.3
Alphard	218 07.6	S 8 39.2
Alphecca	126 20.5	N26 43.4
Alpheratz	357 55.5	N29 04.7
Altair	62 19.2	N 8 51.9
Ankaa	353 27.2	S42 18.7
Antares	112 40.1	S26 25.6
Arcturus	146 06.1	N19 11.6
Atria	107 51.7	S69 01.4
Avior	234 23.2	S59 30.6
Bellatrix	278 44.7	N 6 20.7
Betelgeuse	271 14.1	N 7 24.3
Canopus	264 01.8	S52 41.9
Capella	280 52.0	N45 59.6
Deneb	49 39.0	N45 16.4
Denebola	182 45.4	N14 35.0
Diphda	349 07.6	S17 59.7
Dubhe	194 06.0	N61 45.9
Elnath	278 27.6	N28 36.2
Eltanin	90 51.0	N51 29.5
Enif	33 58.3	N 9 52.1
Fomalhaut	15 36.6	S29 37.7
Gacrux	172 13.6	S57 06.5
Gienah	176 04.1	S17 32.1
Hadar	149 03.8	S60 22.1
Hamal	328 14.0	N23 27.1
Kaus Aust.	83 58.7	S34 23.0
Kochab	137 18.8	N74 10.1
Markab	13 49.8	N15 11.7
Menkar	314 27.4	N 4 04.9
Menkent	148 20.9	S36 21.8
Miaplacidus	221 42.6	S69 43.0
Mirfak	308 57.3	N49 51.1
Nunki	76 12.3	S26 17.8
Peacock	53 37.0	S56 44.1
Pollux	243 42.2	N28 01.8
Procyon	245 12.1	N 5 13.6
Rasalhague	96 16.8	N12 33.8
Regulus	207 55.9	N11 58.5
Rigel	281 23.5	S 8 12.4
Rigil Kent.	140 07.0	S60 49.8
Sabik	102 25.4	S15 43.3
Schedar	349 53.8	N56 31.4
Shaula	96 37.2	S37 06.0
Sirius	258 44.2	S16 43.0
Spica	158 43.3	S11 09.2
Suhail	223 01.2	S43 25.8
Vega	80 46.3	N38 47.0
Zuben'ubi	137 17.9	S16 02.1

	SHA	Mer. Pass.
Venus	311 59.0	9 36
Mars	285 11.8	11 23
Jupiter	2 47.4	6 12
Saturn	330 38.8	8 20

UT	SUN GHA	SUN Dec	MOON GHA	v	MOON Dec	d	HP
d h	° '	° '	° '	'	° '	'	'
15 00	179 55.3	N23 17.3	298 43.8	9.6	S13 06.2	8.2	58.1
01	194 55.1	17.4	313 12.4	9.5	12 58.0	8.3	58.1
02	209 55.0	17.5	327 40.9	9.6	12 49.7	8.4	58.1
03	224 54.9 ..	17.6	342 09.5	9.6	12 41.3	8.4	58.1
04	239 54.7	17.7	356 38.1	9.6	12 32.9	8.6	58.1
05	254 54.6	17.8	11 06.7	9.7	12 24.3	8.6	58.2
06	269 54.5	N23 17.9	25 35.4	9.6	S12 15.7	8.6	58.2
07	284 54.3	18.0	40 04.0	9.7	12 07.1	8.8	58.2
08	299 54.2	18.2	54 32.7	9.7	11 58.3	8.8	58.2
M 09	314 54.1 ..	18.3	69 01.4	9.7	11 49.5	8.9	58.2
O 10	329 53.9	18.4	83 30.1	9.7	11 40.6	8.9	58.2
N 11	344 53.8	18.5	97 58.8	9.7	11 31.7	9.1	58.3
D 12	359 53.7	N23 18.6	112 27.5	9.7	S11 22.6	9.1	58.3
A 13	14 53.5	18.7	126 56.2	9.8	11 13.5	9.1	58.3
Y 14	29 53.4	18.8	141 25.0	9.8	11 04.4	9.3	58.3
15	44 53.3 ..	18.9	155 53.8	9.8	10 55.1	9.3	58.3
16	59 53.2	19.0	170 22.6	9.8	10 45.8	9.3	58.4
17	74 53.0	19.1	184 51.4	9.8	10 36.5	9.5	58.4
18	89 52.9	N23 19.2	199 20.2	9.8	S10 27.0	9.5	58.4
19	104 52.8	19.3	213 49.0	9.9	10 17.5	9.5	58.4
20	119 52.6	19.4	228 17.9	9.8	10 08.0	9.6	58.4
21	134 52.5 ..	19.5	242 46.7	9.9	9 58.4	9.7	58.5
22	149 52.4	19.6	257 15.6	9.9	9 48.7	9.8	58.5
23	164 52.2	19.7	271 44.5	9.9	9 38.9	9.8	58.5
16 00	179 52.1	N23 19.8	286 13.4	9.9	S 9 29.1	9.8	58.5
01	194 51.9	19.9	300 42.3	10.0	9 19.3	10.0	58.5
02	209 51.8	20.0	315 11.3	9.9	9 09.3	9.9	58.5
03	224 51.7 ..	20.1	329 40.2	9.9	8 59.4	10.1	58.6
04	239 51.5	20.2	344 09.1	10.0	8 49.3	10.1	58.6
05	254 51.4	20.2	358 38.1	10.0	8 39.2	10.1	58.6
06	269 51.3	N23 20.3	13 07.1	10.0	S 8 29.1	10.2	58.6
07	284 51.1	20.4	27 36.1	9.9	8 18.9	10.2	58.6
T 08	299 51.0	20.5	42 05.0	10.0	8 08.7	10.3	58.7
U 09	314 50.9 ..	20.6	56 34.0	10.1	7 58.4	10.4	58.7
E 10	329 50.7	20.7	71 03.1	10.0	7 48.0	10.4	58.7
S 11	344 50.6	20.8	85 32.1	10.0	7 37.6	10.5	58.7
D 12	359 50.5	N23 20.9	100 01.1	10.0	S 7 27.1	10.4	58.7
A 13	14 50.3	21.0	114 30.1	10.1	7 16.7	10.6	58.7
Y 14	29 50.2	21.1	128 59.2	10.0	7 06.1	10.6	58.8
15	44 50.1 ..	21.1	143 28.2	10.1	6 55.5	10.6	58.8
16	59 49.9	21.2	157 57.3	10.0	6 44.9	10.7	58.8
17	74 49.8	21.3	172 26.3	10.1	6 34.2	10.7	58.8
18	89 49.7	N23 21.4	186 55.4	10.1	S 6 23.5	10.8	58.8
19	104 49.5	21.5	201 24.5	10.0	6 12.7	10.8	58.8
20	119 49.4	21.6	215 53.5	10.1	6 01.9	10.8	58.9
21	134 49.3 ..	21.6	230 22.6	10.1	5 51.1	10.9	58.9
22	149 49.1	21.7	244 51.7	10.1	5 40.2	10.9	58.9
23	164 49.0	21.8	259 20.8	10.1	5 29.3	11.0	58.9
17 00	179 48.9	N23 21.9	273 49.9	10.0	S 5 18.3	11.0	58.9
01	194 48.7	22.0	288 18.9	10.1	5 07.3	11.0	58.9
02	209 48.6	22.0	302 48.0	10.1	4 56.3	11.1	59.0
03	224 48.5 ..	22.1	317 17.1	10.1	4 45.2	11.1	59.0
04	239 48.3	22.2	331 46.2	10.1	4 34.1	11.1	59.0
05	254 48.2	22.3	346 15.3	10.1	4 23.0	11.2	59.0
06	269 48.1	N23 22.3	0 44.4	10.0	S 4 11.8	11.1	59.0
W 07	284 47.9	22.4	15 13.4	10.1	4 00.7	11.3	59.0
E 08	299 47.8	22.5	29 42.5	10.1	3 49.4	11.2	59.1
D 09	314 47.6 ..	22.6	44 11.6	10.1	3 38.2	11.3	59.1
N 10	329 47.5	22.6	58 40.7	10.0	3 26.9	11.3	59.1
E 11	344 47.4	22.7	73 09.7	10.1	3 15.6	11.3	59.1
S 12	359 47.2	N23 22.8	87 38.8	10.1	S 3 04.3	11.3	59.1
D 13	14 47.1	22.8	102 07.9	10.0	2 53.0	11.4	59.1
A 14	29 47.0	22.9	116 36.9	10.1	2 41.6	11.4	59.2
Y 15	44 46.8 ..	23.0	131 06.0	10.0	2 30.2	11.4	59.2
16	59 46.7	23.0	145 35.0	10.0	2 18.8	11.4	59.2
17	74 46.6	23.1	160 04.0	10.0	2 07.4	11.4	59.2
18	89 46.4	N23 23.2	174 33.0	10.0	S 1 56.0	11.5	59.2
19	104 46.3	23.2	189 02.0	10.0	1 44.5	11.5	59.2
20	119 46.2	23.3	203 31.0	10.0	1 33.0	11.5	59.2
21	134 46.0 ..	23.4	218 00.0	10.0	1 21.5	11.5	59.3
22	149 45.9	23.4	232 29.0	10.0	1 10.0	11.5	59.3
23	164 45.8	23.5	246 58.0	9.9	S 0 58.5	11.5	59.3
	SD 15.8	d 0.1	SD 15.9		16.0		16.1

Lat.	Twilight Naut.	Twilight Civil	Sunrise	Moonrise 15	16	17	18
°	h m	h m	h m	h m	h m	h m	h m
N 72	☐	☐	☐	01 10	01 01	00 53	00 46
N 70	☐	☐	☐	00 46	00 46	00 46	00 45
68	☐	☐	☐	00 28	00 35	00 40	00 44
66	☐	☐	☐	00 13	00 25	00 35	00 44
64	////	////	01 33	00 01	00 17	00 30	00 42
62	////	////	02 10	24 10	00 10	00 26	00 42
60	////	00 52	02 36	24 04	00 04	00 23	00 41
N 58	////	01 41	02 56	23 58	24 20	00 20	00 41
56	////	02 11	03 13	23 53	24 18	00 18	00 41
54	00 48	02 33	03 27	23 49	24 15	00 15	00 40
52	01 33	02 51	03 39	23 45	24 13	00 13	00 40
50	02 00	03 06	03 50	23 42	24 11	00 11	00 40
45	02 46	03 35	04 13	23 34	24 07	00 07	00 39
N 40	03 16	03 58	04 31	23 27	24 03	00 03	00 39
35	03 39	04 16	04 46	23 22	24 00	00 00	00 38
30	03 58	04 31	04 59	23 17	23 58	24 38	00 38
20	04 27	04 56	05 21	23 08	23 53	24 37	00 37
N 10	04 49	05 16	05 39	23 01	23 49	24 37	00 37
0	05 08	05 34	05 57	22 53	23 45	24 36	00 36
S 10	05 25	05 52	06 14	22 46	23 41	24 36	00 36
20	05 41	06 09	06 33	22 39	23 37	24 35	00 35
30	05 58	06 28	06 54	22 30	23 32	24 35	00 35
35	06 06	06 38	07 06	22 25	23 29	24 34	00 34
40	06 16	06 50	07 20	22 19	23 26	24 34	00 34
45	06 26	07 03	07 37	22 13	23 22	24 34	00 34
S 50	06 38	07 19	07 58	22 05	23 18	24 33	00 33
52	06 43	07 27	08 08	22 01	23 16	24 33	00 33
54	06 49	07 35	08 19	21 57	23 14	24 33	00 33
56	06 55	07 44	08 31	21 52	23 12	24 32	00 32
58	07 02	07 54	08 46	21 47	23 09	24 32	00 32
S 60	07 09	08 06	09 03	21 41	23 06	24 32	00 32

Lat.	Sunset	Twilight Civil	Twilight Naut.	Moonset 15	16	17	18
°	h m	h m	h m	h m	h m	h m	h m
N 72	☐	☐	☐	07 34	09 32	11 27	13 22
N 70	☐	☐	☐	07 57	09 44	11 32	13 20
68	☐	☐	☐	08 14	09 55	11 36	13 18
66	☐	☐	☐	08 28	10 03	11 39	13 16
64	22 29	////	////	08 40	10 10	11 42	13 15
62	21 52	////	////	08 49	10 16	11 44	13 14
60	21 26	23 10	////	08 57	10 21	11 46	13 13
N 58	21 05	22 21	////	09 05	10 25	11 48	13 12
56	20 49	21 51	////	09 11	10 29	11 50	13 11
54	20 34	21 29	23 15	09 17	10 33	11 51	13 10
52	20 22	21 11	22 29	09 22	10 36	11 52	13 10
50	20 11	20 56	22 01	09 26	10 39	11 53	13 09
45	19 49	20 26	21 16	09 36	10 46	11 56	13 08
N 40	19 31	20 04	20 45	09 45	10 51	11 58	13 06
35	19 16	19 45	20 22	09 52	10 55	12 00	13 06
30	19 03	19 30	20 04	09 58	10 59	12 02	13 05
20	18 41	19 05	19 35	10 08	11 06	12 04	13 03
N 10	18 22	18 45	19 12	10 18	11 12	12 07	13 02
0	18 04	18 27	18 53	10 26	11 18	12 09	13 01
S 10	17 47	18 10	18 36	10 35	11 23	12 11	12 59
20	17 28	17 52	18 20	10 44	11 29	12 14	12 58
30	17 07	17 34	18 03	10 54	11 36	12 16	12 56
35	16 55	17 23	17 55	11 00	11 40	12 18	12 56
40	16 41	17 11	17 45	11 07	11 44	12 19	12 54
45	16 24	16 58	17 35	11 15	11 49	12 21	12 53
S 50	16 03	16 42	17 23	11 24	11 55	12 24	12 52
52	15 53	16 34	17 18	11 28	11 57	12 25	12 51
54	15 42	16 26	17 12	11 33	12 00	12 26	12 51
56	15 30	16 17	17 06	11 38	12 04	12 27	12 50
58	15 15	16 07	17 00	11 44	12 07	12 28	12 49
S 60	14 58	15 55	16 52	11 50	12 11	12 30	12 48

	SUN Eqn. of Time 00h	SUN Eqn. of Time 12h	SUN Mer. Pass.	MOON Mer. Pass. Upper	MOON Mer. Pass. Lower	Age	Phase
Day							
d	m s	m s	h m	h m	h m	d	%
15	00 19	00 25	12 00	04 14	16 40	21	71
16	00 31	00 38	12 01	05 06	17 31	22	61
17	00 44	00 51	12 01	05 57	18 23	23	49

UT	ARIES	VENUS −3.9		MARS +1.5		JUPITER −2.4		SATURN +0.5		STARS		
	GHA	GHA	Dec	GHA	Dec	GHA	Dec	GHA	Dec	Name	SHA	Dec
d h	° ′	° ′	° ′	° ′	° ′	° ′	° ′	° ′	° ′		° ′	° ′
18 00	266 01.9	215 39.4	N16 32.2	189 44.0	N23 15.4	268 38.6	S 2 26.5	236 30.3	N 9 33.3	Acamar	315 27.4	S40 18.6
01	281 04.4	230 38.9	33.0	204 44.6	15.6	283 40.8	26.4	251 32.6	33.4	Achernar	335 35.7	S57 14.5
02	296 06.9	245 38.4	33.8	219 45.2	15.8	298 43.1	26.3	266 34.8	33.5	Acrux	173 22.0	S63 05.7
03	311 09.3	260 37.9	.. 34.6	234 45.8	.. 16.0	313 45.3	.. 26.2	281 37.1	.. 33.6	Adhara	255 22.0	S28 58.4
04	326 11.8	275 37.4	35.4	249 46.4	16.2	328 47.6	26.1	296 39.3	33.6	Aldebaran	291 03.0	N16 30.2
05	341 14.2	290 36.9	36.2	264 47.0	16.4	343 49.8	26.1	311 41.6	33.7			
06	356 16.7	305 36.4	N16 37.0	279 47.6	N23 16.5	358 52.1	S 2 26.0	326 43.8	N 9 33.8	Alioth	166 30.7	N55 58.4
07	11 19.2	320 35.9	37.8	294 48.1	16.7	13 54.3	25.9	341 46.1	33.8	Alkaid	153 07.8	N49 19.6
T 08	26 21.6	335 35.4	38.6	309 48.7	16.9	28 56.6	25.8	356 48.3	33.9	Al Na'ir	27 58.0	S46 57.9
H 09	41 24.1	350 34.9	.. 39.4	324 49.3	.. 17.1	43 58.8	.. 25.7	11 50.6	.. 34.0	Alnilam	275 58.4	S 1 12.3
U 10	56 26.6	5 34.4	40.2	339 49.9	17.3	59 01.1	25.7	26 52.8	34.0	Alphard	218 07.6	S 8 39.2
R 11	71 29.0	20 33.9	41.0	354 50.5	17.5	74 03.3	25.6	41 55.1	34.1			
S 12	86 31.5	35 33.4	N16 41.8	9 51.1	N23 17.6	89 05.6	S 2 25.5	56 57.3	N 9 34.2	Alphecca	126 20.5	N26 43.4
D 13	101 34.0	50 32.9	42.6	24 51.7	17.8	104 07.8	25.4	71 59.6	34.2	Alpheratz	357 55.4	N29 04.7
A 14	116 36.4	65 32.4	43.4	39 52.3	18.0	119 10.1	25.3	87 01.8	34.3	Altair	62 19.2	N 8 51.9
Y 15	131 38.9	80 31.8	.. 44.2	54 52.9	.. 18.2	134 12.3	.. 25.3	102 04.1	.. 34.4	Ankaa	353 27.2	S42 18.7
16	146 41.3	95 31.3	45.0	69 53.5	18.4	149 14.6	25.2	117 06.3	34.5	Antares	112 40.1	S26 25.6
17	161 43.8	110 30.8	45.7	84 54.1	18.6	164 16.8	25.1	132 08.6	34.5			
18	176 46.3	125 30.3	N16 46.5	99 54.7	N23 18.7	179 19.1	S 2 25.0	147 10.8	N 9 34.6	Arcturus	146 06.1	N19 11.6
19	191 48.7	140 29.8	47.3	114 55.3	18.9	194 21.4	24.9	162 13.1	34.7	Atria	107 51.7	S69 01.4
20	206 51.2	155 29.3	48.1	129 55.9	19.1	209 23.6	24.9	177 15.3	34.7	Avior	234 23.2	S59 30.5
21	221 53.7	170 28.8	.. 48.9	144 56.5	.. 19.3	224 25.9	.. 24.8	192 17.6	.. 34.8	Bellatrix	278 44.7	N 6 20.7
22	236 56.1	185 28.3	49.7	159 57.1	19.5	239 28.1	24.7	207 19.8	34.9	Betelgeuse	271 14.1	N 7 24.3
23	251 58.6	200 27.8	50.5	174 57.6	19.6	254 30.4	24.6	222 22.1	34.9			
19 00	267 01.1	215 27.3	N16 51.3	189 58.2	N23 19.8	269 32.6	S 2 24.5	237 24.3	N 9 35.0	Canopus	264 01.8	S52 41.9
01	282 03.5	230 26.8	52.0	204 58.8	20.0	284 34.9	24.5	252 26.6	35.1	Capella	280 52.0	N45 59.6
02	297 06.0	245 26.3	52.8	219 59.4	20.2	299 37.1	24.4	267 28.8	35.1	Deneb	49 38.9	N45 16.4
03	312 08.5	260 25.8	.. 53.6	235 00.0	.. 20.3	314 39.4	.. 24.3	282 31.1	.. 35.2	Denebola	182 45.4	N14 35.0
04	327 10.9	275 25.2	54.4	250 00.6	20.5	329 41.6	24.2	297 33.3	35.3	Diphda	349 07.6	S17 59.7
05	342 13.4	290 24.7	55.2	265 01.2	20.7	344 43.9	24.2	312 35.6	35.3			
06	357 15.8	305 24.2	N16 56.0	280 01.8	N23 20.9	359 46.1	S 2 24.1	327 37.8	N 9 35.4	Dubhe	194 06.0	N61 45.9
07	12 18.3	320 23.7	56.7	295 02.4	21.1	14 48.4	24.0	342 40.1	35.5	Elnath	278 27.6	N28 36.2
F 08	27 20.8	335 23.2	57.5	310 03.0	21.2	29 50.7	23.9	357 42.3	35.6	Eltanin	90 51.0	N51 29.5
R 09	42 23.2	350 22.7	.. 58.3	325 03.6	.. 21.4	44 52.9	.. 23.9	12 44.6	.. 35.6	Enif	33 58.3	N 9 52.1
I 10	57 25.7	5 22.2	59.1	340 04.2	21.6	59 55.2	23.8	27 46.9	35.7	Fomalhaut	15 36.6	S29 37.7
D 11	72 28.2	20 21.7	16 59.9	355 04.8	21.8	74 57.4	23.7	42 49.1	35.8			
A 12	87 30.6	35 21.1	N17 00.6	10 05.4	N23 21.9	89 59.7	S 2 23.6	57 51.4	N 9 35.8	Gacrux	172 13.6	S57 06.5
Y 13	102 33.1	50 20.6	01.4	25 06.0	22.1	105 01.9	23.5	72 53.6	35.9	Gienah	176 04.1	S17 32.0
14	117 35.6	65 20.1	02.2	40 06.6	22.3	120 04.2	23.5	87 55.9	36.0	Hadar	149 03.9	S60 22.1
15	132 38.0	80 19.6	.. 03.0	55 07.2	.. 22.4	135 06.5	.. 23.4	102 58.1	.. 36.0	Hamal	328 14.0	N23 27.1
16	147 40.5	95 19.1	03.8	70 07.7	22.6	150 08.7	23.3	118 00.4	36.1	Kaus Aust.	83 58.7	S34 23.0
17	162 42.9	110 18.6	04.5	85 08.3	22.8	165 11.0	23.2	133 02.6	36.2			
18	177 45.4	125 18.0	N17 05.3	100 08.9	N23 23.0	180 13.2	S 2 23.2	148 04.9	N 9 36.2	Kochab	137 18.9	N74 10.1
19	192 47.9	140 17.5	06.1	115 09.5	23.1	195 15.5	23.1	163 07.1	36.3	Markab	13 49.7	N15 11.7
20	207 50.3	155 17.0	06.8	130 10.1	23.3	210 17.8	23.0	178 09.4	36.4	Menkar	314 27.4	N 4 04.9
21	222 52.8	170 16.5	.. 07.6	145 10.7	.. 23.5	225 20.0	.. 22.9	193 11.6	.. 36.4	Menkent	148 20.9	S36 21.8
22	237 55.3	185 16.0	08.4	160 11.3	23.6	240 22.3	22.9	208 13.9	36.5	Miaplacidus	221 42.6	S69 43.0
23	252 57.7	200 15.4	09.2	175 11.9	23.8	255 24.5	22.8	223 16.2	36.6			
20 00	268 00.2	215 14.9	N17 09.9	190 12.5	N23 24.0	270 26.8	S 2 22.7	238 18.4	N 9 36.6	Mirfak	308 57.3	N49 51.0
01	283 02.7	230 14.4	10.7	205 13.1	24.2	285 29.1	22.6	253 20.7	36.7	Nunki	76 12.2	S26 17.8
02	298 05.1	245 13.9	11.5	220 13.7	24.3	300 31.3	22.6	268 22.9	36.8	Peacock	53 36.9	S56 44.1
03	313 07.6	260 13.4	.. 12.2	235 14.3	.. 24.5	315 33.6	.. 22.5	283 25.2	.. 36.8	Pollux	243 42.2	N28 01.8
04	328 10.1	275 12.8	13.0	250 14.9	24.7	330 35.8	22.4	298 27.4	36.9	Procyon	245 12.1	N 5 13.6
05	343 12.5	290 12.3	13.8	265 15.5	24.8	345 38.1	22.3	313 29.7	37.0			
06	358 15.0	305 11.8	N17 14.5	280 16.1	N23 25.0	0 40.4	S 2 22.3	328 31.9	N 9 37.0	Rasalhague	96 16.8	N12 33.8
07	13 17.4	320 11.3	15.3	295 16.7	25.2	15 42.6	22.2	343 34.2	37.1	Regulus	207 55.9	N11 58.5
S 08	28 19.9	335 10.7	16.1	310 17.3	25.3	30 44.9	22.1	358 36.4	37.2	Rigel	281 23.5	S 8 12.4
A 09	43 22.4	350 10.2	.. 16.8	325 17.8	.. 25.5	45 47.1	.. 22.0	13 38.7	.. 37.2	Rigil Kent.	140 07.1	S60 49.8
T 10	58 24.8	5 09.7	17.6	340 18.4	25.7	60 49.4	22.0	28 41.0	37.3	Sabik	102 25.4	S15 43.3
U 11	73 27.3	20 09.2	18.4	355 19.0	25.8	75 51.7	21.9	43 43.2	37.4			
R 12	88 29.8	35 08.6	N17 19.1	10 19.6	N23 26.0	90 53.9	S 2 21.8	58 45.5	N 9 37.4	Schedar	349 53.7	N56 31.4
D 13	103 32.2	50 08.1	19.9	25 20.2	26.2	105 56.2	21.7	73 47.7	37.5	Shaula	96 37.2	S37 06.0
A 14	118 34.7	65 07.6	20.6	40 20.8	26.3	120 58.5	21.7	88 50.0	37.6	Sirius	258 44.2	S16 43.0
Y 15	133 37.2	80 07.1	.. 21.4	55 21.4	.. 26.5	136 00.7	.. 21.6	103 52.2	.. 37.6	Spica	158 43.3	S11 09.2
16	148 39.6	95 06.5	22.2	70 22.0	26.6	151 03.0	21.5	118 54.5	37.7	Suhail	223 01.2	S43 25.8
17	163 42.1	110 06.0	22.9	85 22.6	26.8	166 05.3	21.4	133 56.8	37.8			
18	178 44.6	125 05.5	N17 23.7	100 23.2	N23 27.0	181 07.5	S 2 21.4	148 59.0	N 9 37.8	Vega	80 46.3	N38 47.0
19	193 47.0	140 04.9	24.4	115 23.8	27.1	196 09.8	21.3	164 01.3	37.9	Zuben'ubi	137 18.0	S16 02.1
20	208 49.5	155 04.4	25.2	130 24.4	27.3	211 12.1	21.2	179 03.5	38.0		SHA	Mer. Pass.
21	223 51.9	170 03.9	.. 26.0	145 25.0	.. 27.5	226 14.3	.. 21.2	194 05.8	.. 38.0		° ′	h m
22	238 54.4	185 03.3	26.7	160 25.6	27.6	241 16.6	21.1	209 08.0	38.1	Venus	308 26.2	9 39
23	253 56.9	200 02.8	27.5	175 26.2	27.8	256 18.8	21.0	224 10.3	38.2	Mars	282 57.2	11 20
	h m									Jupiter	2 31.6	6 01
Mer. Pass.	6 10.9	*v* −0.5	*d* 0.8	*v* 0.6	*d* 0.2	*v* 2.3	*d* 0.1	*v* 2.3	*d* 0.1	Saturn	330 23.3	8 09

UT	SUN GHA	Dec	MOON GHA	v	Dec	d	HP
d h	° ′	° ′	° ′	′	° ′	′	′
18 00	179 45.6	N23 23.6	261 26.9	10.0	S 0 47.0	11.6	59.3
01	194 45.5	23.6	275 55.9	9.9	0 35.4	11.5	59.3
02	209 45.3	23.7	290 24.8	9.9	0 23.9	11.5	59.3
03	224 45.2 ..	23.7	304 53.7	9.9	0 12.3	11.5	59.3
04	239 45.1	23.8	319 22.6	9.9	S 0 00.8	11.6	59.4
05	254 44.9	23.9	333 51.5	9.8	N 0 10.8	11.6	59.4
06	269 44.8	N23 23.9	348 20.3	9.9	N 0 22.4	11.5	59.4
T 07	284 44.7	24.0	2 49.2	9.8	0 33.9	11.6	59.4
H 08	299 44.5	24.0	17 18.0	9.8	0 45.5	11.6	59.4
U 09	314 44.4 ..	24.1	31 46.8	9.8	0 57.1	11.6	59.4
R 10	329 44.3	24.1	46 15.6	9.8	1 08.7	11.5	59.4
S 11	344 44.1	24.2	60 44.4	9.7	1 20.2	11.6	59.4
D 12	359 44.0	N23 24.3	75 13.1	9.8	N 1 31.8	11.6	59.5
A 13	14 43.9	24.3	89 41.9	9.7	1 43.4	11.6	59.5
Y 14	29 43.7	24.4	104 10.6	9.7	1 55.0	11.5	59.5
15	44 43.6 ..	24.4	118 39.3	9.6	2 06.5	11.6	59.5
16	59 43.4	24.5	133 07.9	9.7	2 18.1	11.5	59.5
17	74 43.3	24.5	147 36.6	9.6	2 29.6	11.5	59.5
18	89 43.2	N23 24.6	162 05.2	9.6	N 2 41.1	11.6	59.5
19	104 43.0	24.6	176 33.8	9.5	2 52.7	11.5	59.5
20	119 42.9	24.7	191 02.3	9.6	3 04.2	11.5	59.6
21	134 42.8 ..	24.7	205 30.9	9.5	3 15.7	11.5	59.6
22	149 42.6	24.7	219 59.4	9.5	3 27.2	11.4	59.6
23	164 42.5	24.8	234 27.9	9.4	3 38.6	11.5	59.6
19 00	179 42.4	N23 24.8	248 56.3	9.5	N 3 50.1	11.4	59.6
01	194 42.2	24.9	263 24.8	9.4	4 01.5	11.4	59.6
02	209 42.1	24.9	277 53.2	9.3	4 12.9	11.4	59.6
03	224 41.9 ..	25.0	292 21.5	9.4	4 24.3	11.4	59.6
04	239 41.8	25.0	306 49.9	9.3	4 35.7	11.4	59.6
05	254 41.7	25.1	321 18.2	9.3	4 47.1	11.3	59.6
06	269 41.5	N23 25.1	335 46.5	9.2	N 4 58.4	11.3	59.6
F 07	284 41.4	25.1	350 14.7	9.2	5 09.7	11.3	59.7
R 08	299 41.3	25.2	4 42.9	9.2	5 21.0	11.2	59.7
I 09	314 41.1 ..	25.2	19 11.1	9.2	5 32.2	11.2	59.7
D 10	329 41.0	25.2	33 39.3	9.1	5 43.4	11.2	59.7
A 11	344 40.9	25.3	48 07.4	9.1	5 54.6	11.2	59.7
Y 12	359 40.7	N23 25.4	62 35.5	9.0	N 6 05.8	11.1	59.7
13	14 40.6	25.4	77 03.5	9.0	6 16.9	11.1	59.7
14	29 40.4	25.4	91 31.5	9.0	6 28.0	11.0	59.7
15	44 40.3 ..	25.5	105 59.5	9.0	6 39.0	11.1	59.7
16	59 40.2	25.5	120 27.5	8.9	6 50.1	10.9	59.7
17	74 40.0	25.5	134 55.4	8.8	7 01.0	11.0	59.7
18	89 39.9	N23 25.5	149 23.2	8.9	N 7 12.0	10.9	59.7
19	104 39.8	25.6	163 51.1	8.8	7 22.9	10.8	59.7
20	119 39.6	25.6	178 18.9	8.7	7 33.7	10.9	59.8
21	134 39.5 ..	25.6	192 46.6	8.8	7 44.6	10.7	59.8
22	149 39.3	25.6	207 14.4	8.6	7 55.3	10.8	59.8
23	164 39.2	25.7	221 42.0	8.7	8 06.1	10.6	59.8
20 00	179 39.1	N23 25.7	236 09.7	8.6	N 8 16.7	10.7	59.8
01	194 38.9	25.7	250 37.3	8.5	8 27.4	10.6	59.8
02	209 38.8	25.8	265 04.8	8.6	8 38.0	10.5	59.8
03	224 38.7 ..	25.8	279 32.4	8.5	8 48.5	10.5	59.8
04	239 38.5	25.8	293 59.9	8.4	8 59.0	10.4	59.8
05	254 38.4	25.8	308 27.3	8.4	9 09.4	10.4	59.8
06	269 38.3	N23 25.8	322 54.7	8.4	N 9 19.8	10.3	59.8
S 07	284 38.1	25.9	337 22.1	8.3	9 30.1	10.3	59.8
A 08	299 38.0	25.9	351 49.4	8.3	9 40.4	10.2	59.8
T 09	314 37.8 ..	25.9	6 16.7	8.2	9 50.6	10.1	59.8
U 10	329 37.7	25.9	20 43.9	8.2	10 00.7	10.1	59.8
R 11	344 37.6	26.0	35 11.1	8.2	10 10.8	10.0	59.8
D 12	359 37.4	N23 26.0	49 38.3	8.1	N10 20.8	10.0	59.8
A 13	14 37.3	26.0	64 05.4	8.1	10 30.8	9.9	59.8
Y 14	29 37.2	26.0	78 32.5	8.0	10 40.7	9.8	59.8
15	44 37.0 ..	26.0	92 59.5	8.0	10 50.5	9.8	59.8
16	59 36.9	26.0	107 26.5	7.9	11 00.3	9.7	59.8
17	74 36.8	26.1	121 53.4	8.0	11 10.0	9.6	59.8
18	89 36.6	N23 26.1	136 20.4	7.8	N11 19.6	9.6	59.8
19	104 36.5	26.1	150 47.2	7.8	11 29.2	9.5	59.8
20	119 36.3	26.1	165 14.0	7.8	11 38.7	9.4	59.8
21	134 36.2 ..	26.1	179 40.8	7.8	11 48.1	9.3	59.8
22	149 36.1	26.1	194 07.6	7.7	11 57.4	9.3	59.8
23	164 35.9	26.1	208 34.3	7.6	N12 06.7	9.2	59.8
	SD 15.8	d 0.0	SD 16.2		16.3		16.3

Moonrise

Lat.	Twilight Naut.	Civil	Sunrise	18	19	20	21
°	h m	h m	h m	h m	h m	h m	h m
N 72	☐	☐	☐	00 46	00 39	00 31	00 23
N 70	☐	☐	☐	00 45	00 44	00 43	00 44
68	☐	☐	☐	00 44	00 48	00 53	01 00
66	☐	☐	☐	00 43	00 51	01 01	01 14
64	////	////	01 31	00 42	00 55	01 08	01 25
62	////	////	02 09	00 42	00 57	01 14	01 34
60	////	00 49	02 35	00 41	01 00	01 19	01 43
N 58	////	01 40	02 56	00 41	01 02	01 24	01 50
56	////	02 10	03 13	00 41	01 04	01 28	01 56
54	00 45	02 33	03 27	00 40	01 05	01 32	02 02
52	01 32	02 50	03 39	00 40	01 07	01 35	02 07
50	02 00	03 06	03 50	00 40	01 08	01 39	02 12
45	02 46	03 35	04 13	00 39	01 11	01 45	02 22
N 40	03 16	03 58	04 31	00 39	01 14	01 51	02 31
35	03 39	04 16	04 46	00 38	01 16	01 56	02 38
30	03 58	04 31	04 59	00 38	01 18	02 00	02 45
20	04 27	04 56	05 21	00 37	01 22	02 08	02 56
N 10	04 50	05 17	05 40	00 37	01 25	02 15	03 06
0	05 09	05 35	05 58	00 36	01 28	02 21	03 16
S 10	05 26	05 52	06 15	00 36	01 31	02 28	03 25
20	05 42	06 10	06 34	00 35	01 34	02 34	03 36
30	05 59	06 28	06 55	00 35	01 38	02 42	03 47
35	06 07	06 39	07 07	00 34	01 40	02 47	03 54
40	06 17	06 51	07 21	00 34	01 43	02 52	04 02
45	06 27	07 04	07 38	00 34	01 46	02 58	04 11
S 50	06 39	07 20	07 59	00 33	01 49	03 06	04 22
52	06 44	07 28	08 10	00 33	01 51	03 09	04 27
54	06 50	07 36	08 20	00 33	01 52	03 13	04 33
56	06 56	07 45	08 33	00 32	01 54	03 17	04 39
58	07 03	07 56	08 47	00 32	01 57	03 22	04 46
S 60	07 10	08 07	09 05	00 32	01 59	03 27	04 55

Moonset

Lat.	Sunset	Twilight Civil	Naut.	18	19	20	21
°	h m	h m	h m	h m	h m	h m	h m
N 72	☐	☐	☐	13 22	15 19	17 21	19 30
N 70	☐	☐	☐	13 20	15 09	17 02	18 56
68	☐	☐	☐	13 18	15 01	16 47	18 31
66	☐	☐	☐	13 16	14 55	16 34	18 13
64	22 32	////	////	13 15	14 49	16 24	17 58
62	21 54	////	////	13 14	14 44	16 15	17 45
60	21 27	23 14	////	13 13	14 40	16 08	17 34
N 58	21 07	22 23	////	13 12	14 37	16 02	17 25
56	20 50	21 53	////	13 11	14 33	15 56	17 17
54	20 36	21 30	23 18	13 10	14 30	15 51	17 10
52	20 23	21 12	22 31	13 10	14 28	15 46	17 03
50	20 12	20 57	22 03	13 09	14 25	15 42	16 58
45	19 50	20 27	21 17	13 08	14 20	15 33	16 45
N 40	19 32	20 05	20 46	13 06	14 16	15 25	16 35
35	19 17	19 46	20 23	13 06	14 12	15 19	16 26
30	19 04	19 31	20 04	13 05	14 09	15 13	16 18
20	18 42	19 06	19 35	13 03	14 03	15 03	16 05
N 10	18 23	18 46	19 13	13 02	13 58	14 55	15 53
0	18 05	18 28	18 54	13 01	13 53	14 47	15 43
S 10	17 48	18 10	18 37	12 59	13 48	14 39	15 32
20	17 29	17 53	18 20	12 58	13 43	14 30	15 20
30	17 08	17 34	18 04	12 56	13 38	14 21	15 07
35	16 55	17 24	17 55	12 56	13 34	14 15	14 59
40	16 41	17 12	17 46	12 54	13 30	14 09	14 51
45	16 24	16 58	17 35	12 53	13 26	14 01	14 41
S 50	16 04	16 42	17 24	12 52	13 21	13 53	14 28
52	15 54	16 35	17 18	12 51	13 19	13 49	14 23
54	15 42	16 26	17 13	12 51	13 16	13 44	14 17
56	15 30	16 17	17 06	12 50	13 13	13 39	14 10
58	15 15	16 07	17 00	12 49	13 10	13 34	14 02
S 60	14 58	15 55	16 52	12 48	13 07	13 28	13 53

Day	SUN Eqn. of Time 00ʰ	12ʰ	Mer. Pass.	MOON Mer. Pass. Upper	Lower	Age	Phase
d	m s	m s	h m	h m	h m	d	%
18	00 57	01 04	12 01	06 48	19 14	24	38
19	01 10	01 17	12 01	07 40	20 07	25	27
20	01 23	01 30	12 01	08 34	21 01	26	17

1998 JUNE 21, 22, 23 (SUN., MON., TUES.)

UT	ARIES GHA	VENUS −3.9 GHA	Dec	MARS +1.5 GHA	Dec	JUPITER −2.4 GHA	Dec	SATURN +0.5 GHA	Dec
d h	° ′	° ′	° ′	° ′	° ′	° ′	° ′	° ′	° ′
21 00	268 59.3	215 02.3	N17 28.2	190 26.8	N23 27.9	271 21.1	S 2 20.9	239 12.6	N 9 38.2
01	284 01.8	230 01.7	29.0	205 27.4	28.1	286 23.4	20.9	254 14.8	38.3
02	299 04.3	245 01.2	29.7	220 27.9	28.3	301 25.7	20.8	269 17.1	38.4
03	314 06.7	260 00.7 ..	30.5	235 28.5 ..	28.4	316 27.9 ..	20.7	284 19.3 ..	38.4
04	329 09.2	275 00.1	31.2	250 29.1	28.6	331 30.2	20.6	299 21.6	38.5
05	344 11.7	289 59.6	32.0	265 29.7	28.7	346 32.5	20.6	314 23.8	38.6
06	359 14.1	304 59.1	N17 32.7	280 30.3	N23 28.9	1 34.7	S 2 20.5	329 26.1	N 9 38.6
07	14 16.6	319 58.5	33.5	295 30.9	29.0	16 37.0	20.4	344 28.4	38.7
08	29 19.1	334 58.0	34.2	310 31.5	29.2	31 39.3	20.4	359 30.6	38.8
S 09	44 21.5	349 57.5 ..	35.0	325 32.1 ..	29.4	46 41.5 ..	20.3	14 32.9 ..	38.8
U 10	59 24.0	4 56.9	35.7	340 32.7	29.5	61 43.8	20.2	29 35.1	38.9
N 11	74 26.4	19 56.4	36.5	355 33.3	29.7	76 46.1	20.1	44 37.4	39.0
D 12	89 28.9	34 55.8	N17 37.2	10 33.9	N23 29.8	91 48.3	S 2 20.1	59 39.7	N 9 39.0
A 13	104 31.4	49 55.3	38.0	25 34.5	30.0	106 50.6	20.0	74 41.9	39.1
Y 14	119 33.8	64 54.8	38.7	40 35.1	30.1	121 52.9	19.9	89 44.2	39.2
15	134 36.3	79 54.2 ..	39.4	55 35.7 ..	30.3	136 55.2 ..	19.9	104 46.4 ..	39.2
16	149 38.8	94 53.7	40.2	70 36.3	30.5	151 57.4	19.8	119 48.7	39.3
17	164 41.2	109 53.1	40.9	85 36.9	30.6	166 59.7	19.7	134 50.9	39.4
18	179 43.7	124 52.6	N17 41.7	100 37.5	N23 30.8	182 02.0	S 2 19.7	149 53.2	N 9 39.4
19	194 46.2	139 52.1	42.4	115 38.1	30.9	197 04.2	19.6	164 55.5	39.5
20	209 48.6	154 51.5	43.1	130 38.6	31.1	212 06.5	19.5	179 57.7	39.6
21	224 51.1	169 51.0 ..	43.9	145 39.2 ..	31.2	227 08.8 ..	19.4	195 00.0 ..	39.6
22	239 53.5	184 50.4	44.6	160 39.8	31.4	242 11.1	19.4	210 02.2	39.7
23	254 56.0	199 49.9	45.4	175 40.4	31.5	257 13.3	19.3	225 04.5	39.8
22 00	269 58.5	214 49.3	N17 46.1	190 41.0	N23 31.7	272 15.6	S 2 19.2	240 06.8	N 9 39.8
01	285 00.9	229 48.8	46.8	205 41.6	31.8	287 17.9	19.2	255 09.0	39.9
02	300 03.4	244 48.3	47.6	220 42.2	32.0	302 20.1	19.1	270 11.3	40.0
03	315 05.9	259 47.7 ..	48.3	235 42.8 ..	32.1	317 22.4 ..	19.0	285 13.5 ..	40.0
04	330 08.3	274 47.2	49.0	250 43.4	32.3	332 24.7	19.0	300 15.8	40.1
05	345 10.8	289 46.6	49.8	265 44.0	32.4	347 27.0	18.9	315 18.1	40.1
06	0 13.3	304 46.1	N17 50.5	280 44.6	N23 32.6	2 29.2	S 2 18.8	330 20.3	N 9 40.2
07	15 15.7	319 45.5	51.2	295 45.2	32.7	17 31.5	18.8	345 22.6	40.3
08	30 18.2	334 45.0	52.0	310 45.8	32.9	32 33.8	18.7	0 24.8	40.3
M 09	45 20.7	349 44.4 ..	52.7	325 46.4 ..	33.0	47 36.1 ..	18.6	15 27.1 ..	40.4
O 10	60 23.1	4 43.9	53.4	340 47.0	33.2	62 38.3	18.5	30 29.4	40.5
N 11	75 25.6	19 43.3	54.2	355 47.6	33.3	77 40.6	18.5	45 31.6	40.5
D 12	90 28.0	34 42.8	N17 54.9	10 48.2	N23 33.5	92 42.9	S 2 18.4	60 33.9	N 9 40.6
A 13	105 30.5	49 42.2	55.6	25 48.8	33.6	107 45.2	18.3	75 36.2	40.7
Y 14	120 33.0	64 41.7	56.3	40 49.4	33.8	122 47.5	18.3	90 38.4	40.7
15	135 35.4	79 41.1 ..	57.1	55 50.0 ..	33.9	137 49.7 ..	18.2	105 40.7 ..	40.8
16	150 37.9	94 40.6	57.8	70 50.5	34.0	152 52.0	18.1	120 42.9	40.9
17	165 40.4	109 40.0	58.5	85 51.1	34.2	167 54.3	18.1	135 45.2	40.9
18	180 42.8	124 39.5	N17 59.2	100 51.7	N23 34.3	182 56.6	S 2 18.0	150 47.5	N 9 41.0
19	195 45.3	139 38.9	18 00.0	115 52.3	34.5	197 58.8	17.9	165 49.7	41.1
20	210 47.8	154 38.4	00.7	130 52.9	34.6	213 01.1	17.9	180 52.0	41.1
21	225 50.2	169 37.8 ..	01.4	145 53.5 ..	34.8	228 03.4 ..	17.8	195 54.3 ..	41.2
22	240 52.7	184 37.2	02.1	160 54.1	34.9	243 05.7	17.7	210 56.5	41.2
23	255 55.2	199 36.7	02.9	175 54.7	35.1	258 08.0	17.7	225 58.8	41.3
23 00	270 57.6	214 36.1	N18 03.6	190 55.3	N23 35.2	273 10.2	S 2 17.6	241 01.0	N 9 41.4
01	286 00.1	229 35.6	04.3	205 55.9	35.3	288 12.5	17.5	256 03.3	41.4
02	301 02.5	244 35.0	05.0	220 56.5	35.5	303 14.8	17.5	271 05.6	41.5
03	316 05.0	259 34.5 ..	05.7	235 57.1 ..	35.6	318 17.1 ..	17.4	286 07.8 ..	41.6
04	331 07.5	274 33.9	06.4	250 57.7	35.8	333 19.4	17.3	301 10.1	41.6
05	346 09.9	289 33.3	07.2	265 58.3	35.9	348 21.6	17.3	316 12.4	41.7
06	1 12.4	304 32.8	N18 07.9	280 58.9	N23 36.0	3 23.9	S 2 17.2	331 14.6	N 9 41.8
07	16 14.9	319 32.2	08.6	295 59.5	36.2	18 26.2	17.1	346 16.9	41.8
T 08	31 17.3	334 31.7	09.3	311 00.1	36.3	33 28.5	17.1	1 19.2	41.9
U 09	46 19.8	349 31.1 ..	10.0	326 00.7 ..	36.5	48 30.8 ..	17.0	16 21.4 ..	41.9
E 10	61 22.3	4 30.5	10.7	341 01.3	36.6	63 33.1	16.9	31 23.7	42.0
S 11	76 24.7	19 30.0	11.4	356 01.9	36.7	78 35.3	16.9	46 25.9	42.1
D 12	91 27.2	34 29.4	N18 12.2	11 02.5	N23 36.9	93 37.6	S 2 16.8	61 28.2	N 9 42.1
A 13	106 29.7	49 28.9	12.9	26 03.0	37.0	108 39.9	16.7	76 30.5	42.2
Y 14	121 32.1	64 28.3	13.6	41 03.6	37.2	123 42.2	16.7	91 32.7	42.3
15	136 34.6	79 27.7 ..	14.3	56 04.2 ..	37.3	138 44.5 ..	16.6	106 35.0 ..	42.3
16	151 37.0	94 27.2	15.0	71 04.8	37.4	153 46.8	16.5	121 37.3	42.4
17	166 39.5	109 26.6	15.7	86 05.4	37.6	168 49.0	16.5	136 39.5	42.5
18	181 42.0	124 26.0	N18 16.4	101 06.0	N23 37.7	183 51.3	S 2 16.4	151 41.8	N 9 42.5
19	196 44.4	139 25.5	17.1	116 06.6	37.8	198 53.6	16.4	166 44.1	42.6
20	211 46.9	154 24.9	17.8	131 07.2	38.0	213 55.9	16.3	181 46.3	42.6
21	226 49.4	169 24.3 ..	18.5	146 07.8 ..	38.1	228 58.2 ..	16.2	196 48.6 ..	42.7
22	241 51.8	184 23.8	19.2	161 08.4	38.2	244 00.5	16.2	211 50.9	42.8
23	256 54.3	199 23.2	19.9	176 09.0	38.4	259 02.8	16.1	226 53.1	42.8
	h m								
Mer. Pass.	5 59.1	v −0.6	d 0.7	v 0.6	d 0.1	v 2.3	d 0.1	v 2.3	d 0.1

STARS

Name	SHA	Dec
	° ′	° ′
Acamar	315 27.4	S40 18.6
Achernar	335 35.7	S57 14.5
Acrux	173 22.0	S63 05.7
Adhara	255 22.0	S28 58.4
Aldebaran	291 02.9	N16 30.2
Alioth	166 30.7	N55 58.4
Alkaid	153 07.8	N49 19.6
Al Na'ir	27 58.0	S46 57.9
Alnilam	275 58.4	S 1 12.3
Alphard	218 07.6	S 8 39.2
Alphecca	126 20.5	N26 43.4
Alpheratz	357 55.4	N29 04.7
Altair	62 19.2	N 8 51.9
Ankaa	353 27.2	S42 18.7
Antares	112 40.1	S26 25.6
Arcturus	146 06.1	N19 11.6
Atria	107 51.7	S69 01.4
Avior	234 23.2	S59 30.5
Bellatrix	278 44.7	N 6 20.7
Betelgeuse	271 14.1	N 7 24.3
Canopus	264 01.8	S52 41.9
Capella	280 52.0	N45 59.6
Deneb	49 38.9	N45 16.4
Denebola	182 45.4	N14 35.0
Diphda	349 07.5	S17 59.7
Dubhe	194 06.1	N61 45.8
Elnath	278 27.6	N28 36.2
Eltanin	90 51.0	N51 29.5
Enif	33 58.2	N 9 52.1
Fomalhaut	15 36.6	S29 37.7
Gacrux	172 13.6	S57 06.5
Gienah	176 04.1	S17 32.0
Hadar	149 03.9	S60 22.1
Hamal	328 14.0	N23 27.1
Kaus Aust.	83 58.7	S34 23.0
Kochab	137 18.9	N74 10.1
Markab	13 49.7	N15 11.7
Menkar	314 27.4	N 4 04.9
Menkent	148 20.9	S36 21.8
Miaplacidus	221 42.6	S69 42.9
Mirfak	308 57.3	N49 51.0
Nunki	76 12.2	S26 17.8
Peacock	53 36.9	S56 44.2
Pollux	243 42.2	N28 01.8
Procyon	245 12.1	N 5 13.6
Rasalhague	96 16.8	N12 33.8
Regulus	207 56.0	N11 58.5
Rigel	281 23.5	S 8 12.3
Rigil Kent.	140 07.1	S60 49.8
Sabik	102 25.4	S15 43.2
Schedar	349 53.7	N56 31.4
Shaula	96 37.1	S37 06.0
Sirius	258 44.2	S16 43.0
Spica	158 43.3	S11 09.2
Suhail	223 01.2	S43 25.8
Vega	80 46.3	N38 47.0
Zuben'ubi	137 18.0	S16 02.1

	SHA	Mer. Pass.
	° ′	h m
Venus	304 50.9	9 41
Mars	280 42.6	11 17
Jupiter	2 17.1	5 50
Saturn	330 08.3	7 58

UT	SUN GHA	Dec	MOON GHA	v	Dec	d	HP
d h	° '	° '	° '	'	° '	'	'
21 00	179 35.8	N23 26.1	223 00.9	7.6	N12 15.9	9.1	59.8
01	194 35.7	26.2	237 27.5	7.6	12 25.0	9.1	59.8
02	209 35.5	26.2	251 54.1	7.5	12 34.1	8.9	59.8
03	224 35.4 ..	26.2	266 20.6	7.5	12 43.0	8.9	59.8
04	239 35.2	26.2	280 47.1	7.5	12 51.9	8.8	59.8
05	254 35.1	26.2	295 13.6	7.4	13 00.7	8.7	59.8
06	269 35.0	N23 26.2	309 40.0	7.4	N13 09.4	8.6	59.8
07	284 34.8	26.2	324 06.4	7.3	13 18.0	8.6	59.8
S 08	299 34.7	26.2	338 32.7	7.3	13 26.6	8.4	59.8
U 09	314 34.6 ..	26.2	352 59.0	7.2	13 35.0	8.4	59.8
N 10	329 34.4	26.2	7 25.2	7.2	13 43.4	8.3	59.8
D 11	344 34.3	26.2	21 51.4	7.2	13 51.7	8.2	59.8
A 12	359 34.2	N23 26.2	36 17.6	7.1	N13 59.9	8.1	59.8
Y 13	14 34.0	26.2	50 43.7	7.1	14 08.0	8.0	59.8
14	29 33.9	26.2	65 09.8	7.1	14 16.0	7.9	59.8
15	44 33.7 ..	26.2	79 35.9	7.0	14 23.9	7.8	59.7
16	59 33.6	26.2	94 01.9	7.0	14 31.7	7.7	59.7
17	74 33.5	26.2	108 27.9	7.0	14 39.4	7.7	59.7
18	89 33.3	N23 26.2	122 53.9	6.9	N14 47.1	7.5	59.7
19	104 33.2	26.2	137 19.8	6.8	14 54.6	7.4	59.7
20	119 33.1	26.2	151 45.6	6.9	15 02.0	7.4	59.7
21	134 32.9 ..	26.2	166 11.5	6.8	15 09.4	7.2	59.7
22	149 32.8	26.2	180 37.3	6.8	15 16.6	7.2	59.7
23	164 32.6	26.2	195 03.1	6.7	15 23.8	7.0	59.7
22 00	179 32.5	N23 26.2	209 28.8	6.7	N15 30.8	6.9	59.7
01	194 32.4	26.2	223 54.5	6.7	15 37.7	6.9	59.7
02	209 32.2	26.2	238 20.2	6.7	15 44.6	6.7	59.7
03	224 32.1 ..	26.2	252 45.9	6.6	15 51.3	6.6	59.6
04	239 32.0	26.1	267 11.5	6.6	15 57.9	6.5	59.6
05	254 31.8	26.1	281 37.1	6.5	16 04.4	6.4	59.6
06	269 31.7	N23 26.1	296 02.6	6.6	N16 10.8	6.3	59.6
07	284 31.6	26.1	310 28.2	6.5	16 17.1	6.2	59.6
M 08	299 31.4	26.1	324 53.7	6.5	16 23.3	6.1	59.6
O 09	314 31.3 ..	26.1	339 19.2	6.4	16 29.4	6.0	59.6
N 10	329 31.1	26.1	353 44.6	6.5	16 35.4	5.8	59.6
D 11	344 31.0	26.1	8 10.1	6.4	16 41.2	5.8	59.5
A 12	359 30.9	N23 26.0	22 35.5	6.3	N16 47.0	5.6	59.5
Y 13	14 30.7	26.0	37 00.8	6.4	16 52.6	5.5	59.5
14	29 30.6	26.0	51 26.2	6.4	16 58.1	5.4	59.5
15	44 30.5 ..	26.0	65 51.6	6.3	17 03.5	5.3	59.5
16	59 30.3	26.0	80 16.9	6.3	17 08.8	5.2	59.5
17	74 30.2	26.0	94 42.2	6.3	17 14.0	5.1	59.5
18	89 30.1	N23 25.9	109 07.5	6.3	N17 19.1	4.9	59.5
19	104 29.9	25.9	123 32.8	6.2	17 24.0	4.9	59.4
20	119 29.8	25.9	137 58.0	6.3	17 28.9	4.7	59.4
21	134 29.6 ..	25.9	152 23.3	6.2	17 33.6	4.6	59.4
22	149 29.5	25.8	166 48.5	6.2	17 38.2	4.4	59.4
23	164 29.4	25.8	181 13.7	6.2	17 42.6	4.4	59.4
23 00	179 29.2	N23 25.8	195 38.9	6.2	N17 47.0	4.2	59.4
01	194 29.1	25.8	210 04.1	6.2	17 51.2	4.2	59.3
02	209 29.0	25.8	224 29.3	6.2	17 55.4	4.0	59.3
03	224 28.8 ..	25.7	238 54.5	6.2	17 59.4	3.8	59.3
04	239 28.7	25.7	253 19.7	6.2	18 03.2	3.8	59.3
05	254 28.6	25.7	267 44.9	6.1	18 07.0	3.6	59.3
06	269 28.4	N23 25.6	282 10.0	6.2	N18 10.6	3.5	59.2
07	284 28.3	25.6	296 35.2	6.1	18 14.1	3.4	59.2
T 08	299 28.1	25.6	311 00.3	6.2	18 17.5	3.3	59.2
U 09	314 28.0 ..	25.6	325 25.5	6.2	18 20.8	3.1	59.2
E 10	329 27.9	25.5	339 50.7	6.1	18 23.9	3.1	59.2
S 11	344 27.7	25.5	354 15.8	6.2	18 27.0	2.9	59.2
D 12	359 27.6	N23 25.5	8 41.0	6.2	N18 29.9	2.7	59.1
A 13	14 27.5	25.4	23 06.2	6.1	18 32.6	2.7	59.1
Y 14	29 27.3	25.4	37 31.3	6.2	18 35.3	2.5	59.1
15	44 27.2 ..	25.4	51 56.5	6.2	18 37.8	2.4	59.1
16	59 27.1	25.3	66 21.7	6.2	18 40.2	2.3	59.0
17	74 26.9	25.3	80 46.9	6.2	18 42.5	2.2	59.0
18	89 26.8	N23 25.2	95 12.1	6.3	N18 44.7	2.0	59.0
19	104 26.7	25.2	109 37.4	6.2	18 46.7	1.9	59.0
20	119 26.5	25.2	124 02.6	6.3	18 48.6	1.8	59.0
21	134 26.4 ..	25.1	138 27.9	6.2	18 50.4	1.7	58.9
22	149 26.2	25.1	152 53.1	6.3	18 52.1	1.5	58.9
23	164 26.1	25.1	167 18.4	6.3	N18 53.6	1.4	58.9
SD	15.8	d 0.0	SD 16.3		16.2		16.1

Lat.	Twilight Naut.	Civil	Sunrise	Moonrise 21	22	23	24
°	h m	h m	h m	h m	h m	h m	h m
N 72	□	□	□	00 23	{00 11 / 23 46}	□	□
N 70	□	□	□	00 44	00 46	00 55	01 19
68	□	□	□	01 00	01 12	01 32	02 08
66	□	□	□	01 14	01 31	01 58	02 39
64	////	////	01 31	01 25	01 47	02 18	03 03
62	////	////	02 09	01 34	02 00	02 35	03 21
60	////	00 49	02 36	01 43	02 11	02 49	03 37
N 58	////	01 41	02 56	01 50	02 21	03 01	03 50
56	////	02 11	03 13	01 56	02 30	03 11	04 01
54	00 45	02 33	03 28	02 02	02 38	03 20	04 11
52	01 32	02 51	03 40	02 07	02 44	03 28	04 20
50	02 00	03 06	03 51	02 12	02 51	03 36	04 28
45	02 46	03 36	04 13	02 22	03 04	03 51	04 44
N 40	03 17	03 59	04 31	02 31	03 15	04 04	04 58
35	03 40	04 17	04 47	02 38	03 25	04 15	05 10
30	03 59	04 32	05 00	02 45	03 33	04 25	05 20
20	04 28	04 57	05 22	02 56	03 48	04 42	05 38
N 10	04 50	05 18	05 41	03 06	04 00	04 55	05 53
0	05 09	05 36	05 58	03 16	04 12	05 10	06 08
S 10	05 27	05 53	06 16	03 25	04 24	05 24	06 22
20	05 43	06 10	06 34	03 36	04 37	05 39	06 38
30	05 59	06 29	06 56	03 47	04 52	05 56	06 56
35	06 08	06 40	07 08	03 54	05 01	06 06	07 06
40	06 17	06 52	07 22	04 02	05 11	06 17	07 18
45	06 28	07 05	07 39	04 11	05 23	06 30	07 32
S 50	06 40	07 21	08 00	04 22	05 37	06 47	07 49
52	06 45	07 29	08 10	04 27	05 43	06 54	07 57
54	06 51	07 37	08 21	04 33	05 51	07 03	08 06
56	06 57	07 46	08 34	04 39	05 59	07 12	08 16
58	07 04	07 56	08 48	04 46	06 08	07 23	08 28
S 60	07 11	08 08	09 06	04 55	06 19	07 36	08 41

Lat.	Sunset	Twilight Civil	Naut.	Moonset 21	22	23	24
°	h m	h m	h m	h m	h m	h m	h m
N 72	□	□	□	19 30	21 55	□	□
N 70	□	□	□	18 56	20 47	22 24	23 22
68	□	□	□	18 31	20 11	21 35	22 34
66	□	□	□	18 13	19 45	21 04	22 03
64	22 33	////	////	17 58	19 25	20 41	21 40
62	21 54	////	////	17 45	19 09	20 22	21 21
60	21 28	23 14	////	17 34	18 56	20 07	21 06
N 58	21 07	22 23	////	17 25	18 44	19 54	20 53
56	20 51	21 53	////	17 17	18 34	19 43	20 42
54	20 36	21 31	23 18	17 10	18 25	19 33	20 32
52	20 24	21 13	22 31	17 03	18 17	19 25	20 23
50	20 13	20 58	22 03	16 58	18 10	19 17	20 15
45	19 50	20 28	21 18	16 45	17 55	19 00	19 58
N 40	19 32	20 05	20 47	16 35	17 43	18 46	19 45
35	19 17	19 47	20 24	16 26	17 32	18 35	19 33
30	19 04	19 32	20 05	16 18	17 23	18 25	19 23
20	18 42	19 07	19 36	16 05	17 07	18 07	19 05
N 10	18 23	18 46	19 13	15 53	16 53	17 52	18 50
0	18 06	18 28	18 54	15 43	16 40	17 38	18 35
S 10	17 48	18 11	18 37	15 32	16 27	17 23	18 21
20	17 30	17 54	18 21	15 20	16 13	17 08	18 05
30	17 08	17 35	18 05	15 07	15 57	16 51	17 47
35	16 56	17 24	17 56	14 59	15 48	16 40	17 37
40	16 42	17 12	17 46	14 51	15 37	16 29	17 25
45	16 25	16 59	17 36	14 41	15 25	16 15	17 11
S 50	16 04	16 43	17 24	14 28	15 10	15 58	16 54
52	15 54	16 35	17 19	14 23	15 03	15 51	16 46
54	15 43	16 27	17 13	14 17	14 55	15 42	16 37
56	15 30	16 18	17 07	14 10	14 47	15 32	16 27
58	15 16	16 08	17 00	14 02	14 37	15 21	16 15
S 60	14 58	15 56	16 53	13 53	14 26	15 08	16 02

Day	SUN Eqn. of Time 00h	12h	Mer. Pass.	MOON Mer. Pass. Upper	Lower	Age	Phase
d	m s	m s	h m	h m	h m	d	%
21	01 37	01 43	12 02	09 29	21 57	27	10
22	01 50	01 56	12 02	10 26	22 55	28	4
23	02 03	02 09	12 02	11 24	23 53	29	1

1998 JUNE 24, 25, 26 (WED., THURS., FRI.)

UT	ARIES GHA	VENUS −3.9 GHA	Dec	MARS +1.5 GHA	Dec	JUPITER −2.5 GHA	Dec	SATURN +0.5 GHA	Dec	STARS Name	SHA	Dec
24 00	271 56.8	214 22.6	N18 20.6	191 09.6	N23 38.5	274 05.0	S 2 16.0	241 55.4	N 9 42.9	Acamar	315 27.4	S40 18.6
01	286 59.2	229 22.1	21.3	206 10.2	38.6	289 07.3	16.0	256 57.7	43.0	Achernar	335 35.6	S57 14.5
02	302 01.7	244 21.5	22.0	221 10.8	38.8	304 09.6	15.9	271 59.9	43.0	Acrux	173 22.0	S63 05.7
03	317 04.2	259 20.9 ..	22.7	236 11.4 ..	38.9	319 11.9 ..	15.8	287 02.2 ..	43.1	Adhara	255 21.9	S28 58.4
04	332 06.6	274 20.4	23.4	251 12.0	39.0	334 14.2	15.8	302 04.5	43.1	Aldebaran	291 02.9	N16 30.2
05	347 09.1	289 19.8	24.1	266 12.6	39.2	349 16.5	15.7	317 06.7	43.2			
06	2 11.5	304 19.2	N18 24.8	281 13.2	N23 39.3	4 18.8	S 2 15.7	332 09.0	N 9 43.3	Alioth	166 30.8	N55 58.4
W 07	17 14.0	319 18.7	25.5	296 13.8	39.4	19 21.1	15.6	347 11.3	43.3	Alkaid	153 07.8	N49 19.6
E 08	32 16.5	334 18.1	26.2	311 14.4	39.6	34 23.3	15.5	2 13.5	43.4	Al Na'ir	27 57.9	S46 57.9
D 09	47 18.9	349 17.5 ..	26.9	326 15.0 ..	39.7	49 25.6 ..	15.5	17 15.8 ..	43.5	Alnilam	275 58.4	S 1 12.3
N 10	62 21.4	4 16.9	27.6	341 15.6	39.8	64 27.9	15.4	32 18.1	43.5	Alphard	218 07.6	S 8 39.2
E 11	77 23.9	19 16.4	28.3	356 16.2	39.9	79 30.2	15.3	47 20.3	43.6			
S 12	92 26.3	34 15.8	N18 29.0	11 16.8	N23 40.1	94 32.5	S 2 15.3	62 22.6	N 9 43.6	Alphecca	126 20.5	N26 43.4
D 13	107 28.8	49 15.2	29.7	26 17.3	40.2	109 34.8	15.2	77 24.9	43.7	Alpheratz	357 55.4	N29 04.7
A 14	122 31.3	64 14.6	30.4	41 17.9	40.3	124 37.1	15.2	92 27.1	43.8	Altair	62 19.1	N 8 52.0
Y 15	137 33.7	79 14.1 ..	31.1	56 18.5 ..	40.5	139 39.4 ..	15.1	107 29.4 ..	43.8	Ankaa	353 27.1	S42 18.7
16	152 36.2	94 13.5	31.7	71 19.1	40.6	154 41.7	15.0	122 31.7	43.9	Antares	112 40.1	S26 25.6
17	167 38.6	109 12.9	32.4	86 19.7	40.7	169 44.0	15.0	137 33.9	44.0			
18	182 41.1	124 12.3	N18 33.1	101 20.3	N23 40.8	184 46.2	S 2 14.9	152 36.2	N 9 44.0	Arcturus	146 06.1	N19 11.6
19	197 43.6	139 11.8	33.8	116 20.9	41.0	199 48.5	14.8	167 38.5	44.1	Atria	107 51.7	S69 01.5
20	212 46.0	154 11.2	34.5	131 21.5	41.1	214 50.8	14.8	182 40.7	44.1	Avior	234 23.2	S59 30.5
21	227 48.5	169 10.6 ..	35.2	146 22.1 ..	41.2	229 53.1 ..	14.7	197 43.0 ..	44.2	Bellatrix	278 44.7	N 6 20.7
22	242 51.0	184 10.0	35.9	161 22.7	41.3	244 55.4	14.7	212 45.3	44.3	Betelgeuse	271 14.1	N 7 24.3
23	257 53.4	199 09.4	36.6	176 23.3	41.5	259 57.7	14.6	227 47.5	44.3			
25 00	272 55.9	214 08.9	N18 37.2	191 23.9	N23 41.6	275 00.0	S 2 14.5	242 49.8	N 9 44.4	Canopus	264 01.8	S52 41.8
01	287 58.4	229 08.3	37.9	206 24.5	41.7	290 02.3	14.5	257 52.1	44.4	Capella	280 51.9	N45 59.6
02	303 00.8	244 07.7	38.6	221 25.1	41.8	305 04.6	14.4	272 54.3	44.5	Deneb	49 38.9	N45 16.4
03	318 03.3	259 07.1 ..	39.3	236 25.7 ..	42.0	320 06.9 ..	14.4	287 56.6 ..	44.6	Denebola	182 45.5	N14 35.0
04	333 05.8	274 06.5	40.0	251 26.3	42.1	335 09.2	14.3	302 58.9	44.6	Diphda	349 07.5	S17 59.7
05	348 08.2	289 06.0	40.6	266 26.9	42.2	350 11.5	14.2	318 01.1	44.7			
06	3 10.7	304 05.4	N18 41.3	281 27.5	N23 42.3	5 13.8	S 2 14.2	333 03.4	N 9 44.8	Dubhe	194 06.1	N61 45.8
T 07	18 13.1	319 04.8	42.0	296 28.1	42.5	20 16.1	14.1	348 05.7	44.8	Elnath	278 27.6	N28 36.2
H 08	33 15.6	334 04.2	42.7	311 28.7	42.6	35 18.4	14.1	3 08.0	44.9	Eltanin	90 51.0	N51 29.5
U 09	48 18.1	349 03.6 ..	43.4	326 29.3 ..	42.7	50 20.7 ..	14.0	18 10.2 ..	44.9	Enif	33 58.2	N 9 52.1
R 10	63 20.5	4 03.0	44.0	341 29.9	42.8	65 22.9	13.9	33 12.5	45.0	Fomalhaut	15 36.6	S29 37.7
S 11	78 23.0	19 02.5	44.7	356 30.5	42.9	80 25.2	13.9	48 14.8	45.1			
D 12	93 25.5	34 01.9	N18 45.4	11 31.1	N23 43.1	95 27.5	S 2 13.8	63 17.0	N 9 45.1	Gacrux	172 13.6	S57 06.5
A 13	108 27.9	49 01.3	46.1	26 31.7	43.2	110 29.8	13.8	78 19.3	45.2	Gienah	176 04.1	S17 32.0
A 14	123 30.4	64 00.7	46.7	41 32.3	43.3	125 32.1	13.7	93 21.6	45.2	Hadar	149 03.9	S60 22.1
Y 15	138 32.9	79 00.1 ..	47.4	56 32.9 ..	43.4	140 34.4 ..	13.6	108 23.8 ..	45.3	Hamal	328 13.9	N23 27.1
16	153 35.3	93 59.5	48.1	71 33.5	43.5	155 36.7	13.6	123 26.1	45.4	Kaus Aust.	83 58.7	S34 23.0
17	168 37.8	108 58.9	48.7	86 34.1	43.7	170 39.0	13.5	138 28.4	45.4			
18	183 40.3	123 58.4	N18 49.4	101 34.7	N23 43.8	185 41.3	S 2 13.5	153 30.7	N 9 45.5	Kochab	137 19.0	N74 10.1
19	198 42.7	138 57.8	50.1	116 35.2	43.9	200 43.6	13.4	168 32.9	45.5	Markab	13 49.7	N15 11.7
20	213 45.2	153 57.2	50.7	131 35.8	44.0	215 45.9	13.4	183 35.2	45.6	Menkar	314 27.3	N 4 04.9
21	228 47.6	168 56.6 ..	51.4	146 36.4 ..	44.1	230 48.2 ..	13.3	198 37.5 ..	45.7	Menkent	148 20.9	S36 21.8
22	243 50.1	183 56.0	52.1	161 37.0	44.2	245 50.5	13.2	213 39.7	45.7	Miaplacidus	221 42.7	S69 42.9
23	258 52.6	198 55.4	52.7	176 37.6	44.4	260 52.8	13.2	228 42.0	45.8			
26 00	273 55.0	213 54.8	N18 53.4	191 38.2	N23 44.5	275 55.1	S 2 13.1	243 44.3	N 9 45.8	Mirfak	308 57.2	N49 51.0
01	288 57.5	228 54.2	54.1	206 38.8	44.6	290 57.4	13.1	258 46.6	45.9	Nunki	76 12.2	S26 17.8
02	304 00.0	243 53.6	54.7	221 39.4	44.7	305 59.7	13.0	273 48.8	46.0	Peacock	53 36.9	S56 44.2
03	319 02.4	258 53.0 ..	55.4	236 40.0 ..	44.8	321 02.0 ..	12.9	288 51.1 ..	46.0	Pollux	243 42.2	N28 01.8
04	334 04.9	273 52.5	56.1	251 40.6	44.9	336 04.3	12.9	303 53.4	46.1	Procyon	245 12.1	N 5 13.6
05	349 07.4	288 51.9	56.7	266 41.2	45.0	351 06.6	12.8	318 55.6	46.1			
06	4 09.8	303 51.3	N18 57.4	281 41.8	N23 45.2	6 08.9	S 2 12.8	333 57.9	N 9 46.2	Rasalhague	96 16.8	N12 33.8
07	19 12.3	318 50.7	58.0	296 42.4	45.3	21 11.2	12.7	349 00.2	46.3	Regulus	207 56.0	N11 58.5
08	34 14.8	333 50.1	58.7	311 43.0	45.4	36 13.5	12.7	4 02.5	46.3	Rigel	281 23.5	S 8 12.3
F 09	49 17.2	348 49.5	18 59.4	326 43.6 ..	45.5	51 15.8 ..	12.6	19 04.7 ..	46.4	Rigil Kent.	140 07.1	S60 49.8
R 10	64 19.7	3 48.9	19 00.0	341 44.2	45.6	66 18.1	12.5	34 07.0	46.4	Sabik	102 25.4	S15 43.2
I 11	79 22.1	18 48.3	00.7	356 44.8	45.7	81 20.4	12.5	49 09.3	46.5			
D 12	94 24.6	33 47.7	N19 01.3	11 45.4	N23 45.8	96 22.7	S 2 12.4	64 11.6	N 9 46.6	Schedar	349 53.7	N56 31.4
A 13	109 27.1	48 47.1	02.0	26 46.0	45.9	111 25.0	12.4	79 13.8	46.6	Shaula	96 37.1	S37 06.0
Y 14	124 29.5	63 46.5	02.6	41 46.6	46.1	126 27.4	12.3	94 16.1	46.7	Sirius	258 44.2	S16 43.0
15	139 32.0	78 45.9 ..	03.3	56 47.2 ..	46.2	141 29.7 ..	12.3	109 18.4 ..	46.7	Spica	158 43.3	S11 09.2
16	154 34.5	93 45.3	03.9	71 47.8	46.3	156 32.0	12.2	124 20.7	46.8	Suhail	223 01.2	S43 25.8
17	169 36.9	108 44.7	04.6	86 48.4	46.4	171 34.3	12.2	139 22.9	46.9			
18	184 39.4	123 44.1	N19 05.2	101 49.0	N23 46.5	186 36.6	S 2 12.1	154 25.2	N 9 46.9	Vega	80 46.3	N38 47.1
19	199 41.9	138 43.5	05.9	116 49.6	46.6	201 38.9	12.0	169 27.5	47.0	Zuben'ubi	137 18.0	S16 02.1
20	214 44.3	153 42.9	06.5	131 50.2	46.7	216 41.2	12.0	184 29.7	47.0			
21	229 46.8	168 42.3 ..	07.2	146 50.8 ..	46.8	231 43.5 ..	11.9	199 32.0 ..	47.1		SHA	Mer. Pass.
22	244 49.3	183 41.7	07.8	161 51.4	46.9	246 45.8	11.9	214 34.3	47.2	Venus	301 13.0	9 44
23	259 51.7	198 41.1	08.5	176 52.0	47.0	261 48.1	11.8	229 36.6	47.2	Mars	278 28.0	11 14
										Jupiter	2 04.1	5 39
Mer. Pass. 5 47.3		v −0.6	d 0.7	v 0.6	d 0.1	v 2.3	d 0.1	v 2.3	d 0.1	Saturn	329 53.9	7 48

UT	SUN GHA	SUN Dec	MOON GHA	v	Dec	d	HP
d h	° ′	° ′	° ′	′	° ′	′	′
24 00	179 26.0	N23 25.0	181 43.7	6.3	N18 55.0	1.3	58.9
01	194 25.8	25.0	196 09.0	6.4	18 56.3	1.2	58.8
02	209 25.7	24.9	210 34.4	6.3	18 57.5	1.0	58.8
03	224 25.6	.. 24.9	224 59.7	6.4	18 58.5	1.0	58.8
04	239 25.4	24.8	239 25.1	6.4	18 59.5	0.8	58.8
05	254 25.3	24.8	253 50.5	6.5	19 00.3	0.6	58.7
06	269 25.2	N23 24.7	268 16.0	6.4	N19 00.9	0.6	58.7
W 07	284 25.0	24.7	282 41.4	6.5	19 01.5	0.4	58.7
E 08	299 24.9	24.7	297 06.9	6.6	19 01.9	0.3	58.7
D 09	314 24.8	.. 24.6	311 32.5	6.5	19 02.2	0.2	58.6
N 10	329 24.6	24.6	325 58.0	6.6	19 02.4	0.1	58.6
E 11	344 24.5	24.5	340 23.6	6.6	19 02.5	0.1	58.6
S 12	359 24.4	N23 24.5	354 49.2	6.7	N19 02.4	0.2	58.6
D 13	14 24.2	24.4	9 14.9	6.7	19 02.2	0.3	58.5
A 14	29 24.1	24.4	23 40.6	6.7	19 01.9	0.4	58.5
Y 15	44 23.9	.. 24.3	38 06.3	6.8	19 01.5	0.5	58.5
16	59 23.8	24.3	52 32.1	6.8	19 01.0	0.7	58.5
17	74 23.7	24.2	66 57.9	6.8	19 00.3	0.8	58.4
18	89 23.5	N23 24.1	81 23.7	6.9	N18 59.5	0.9	58.4
19	104 23.4	24.1	95 49.6	6.9	18 58.6	1.0	58.4
20	119 23.3	24.0	110 15.5	7.0	18 57.6	1.1	58.4
21	134 23.1	.. 24.0	124 41.5	7.0	18 56.5	1.3	58.3
22	149 23.0	23.9	139 07.5	7.1	18 55.2	1.3	58.3
23	164 22.9	23.9	153 33.6	7.1	18 53.9	1.5	58.3
25 00	179 22.7	N23 23.8	167 59.7	7.1	N18 52.4	1.6	58.2
01	194 22.6	23.7	182 25.8	7.2	18 50.8	1.7	58.2
02	209 22.5	23.7	196 52.0	7.3	18 49.1	1.9	58.2
03	224 22.3	.. 23.6	211 18.3	7.3	18 47.2	1.9	58.2
04	239 22.2	23.6	225 44.6	7.3	18 45.3	2.1	58.1
05	254 22.1	23.5	240 10.9	7.4	18 43.2	2.1	58.1
06	269 21.9	N23 23.4	254 37.3	7.5	N18 41.1	2.3	58.1
T 07	284 21.8	23.4	269 03.8	7.5	18 38.8	2.4	58.0
H 08	299 21.7	23.3	283 30.3	7.6	18 36.4	2.5	58.0
U 09	314 21.5	.. 23.2	297 56.9	7.6	18 33.9	2.6	58.0
R 10	329 21.4	23.2	312 23.5	7.7	18 31.3	2.8	57.9
S 11	344 21.3	23.1	326 50.2	7.7	18 28.5	2.8	57.9
D 12	359 21.1	N23 23.1	341 16.9	7.8	N18 25.7	2.9	57.9
A 13	14 21.0	23.0	355 43.7	7.9	18 22.8	3.1	57.9
Y 14	29 20.9	22.9	10 10.6	7.9	18 19.7	3.1	57.8
15	44 20.7	.. 22.8	24 37.5	8.0	18 16.6	3.3	57.8
16	59 20.6	22.8	39 04.5	8.0	18 13.3	3.3	57.8
17	74 20.5	22.7	53 31.5	8.1	18 10.0	3.5	57.7
18	89 20.3	N23 22.6	67 58.6	8.2	N18 06.5	3.6	57.7
19	104 20.2	22.6	82 25.8	8.2	18 02.9	3.6	57.7
20	119 20.1	22.5	96 53.0	8.3	17 59.3	3.8	57.6
21	134 19.9	.. 22.4	111 20.3	8.3	17 55.5	3.9	57.6
22	149 19.8	22.3	125 47.6	8.5	17 51.6	4.0	57.6
23	164 19.7	22.3	140 15.1	8.4	17 47.6	4.0	57.6
26 00	179 19.5	N23 22.2	154 42.5	8.6	N17 43.6	4.2	57.5
01	194 19.4	22.1	169 10.1	8.6	17 39.4	4.3	57.5
02	209 19.3	22.0	183 37.7	8.7	17 35.1	4.3	57.5
03	224 19.1	.. 22.0	198 05.4	8.8	17 30.8	4.5	57.4
04	239 19.0	21.9	212 33.2	8.8	17 26.3	4.5	57.4
05	254 18.9	21.8	227 01.0	8.9	17 21.8	4.7	57.4
06	269 18.7	N23 21.6	241 28.9	8.9	N17 17.1	4.7	57.3
07	284 18.6	21.6	255 56.8	9.1	17 12.4	4.9	57.3
08	299 18.5	21.6	270 24.9	9.1	17 07.5	4.9	57.3
F 09	314 18.3	.. 21.5	284 53.0	9.1	17 02.6	5.0	57.2
R 10	329 18.2	21.4	299 21.1	9.3	16 57.6	5.1	57.2
I 11	344 18.1	21.3	313 49.4	9.3	16 52.5	5.2	57.2
D 12	359 17.9	N23 21.2	328 17.7	9.4	N16 47.3	5.3	57.1
A 13	14 17.8	21.1	342 46.1	9.4	16 42.0	5.3	57.1
Y 14	29 17.7	21.1	357 14.5	9.5	16 36.7	5.5	57.1
15	44 17.5	.. 21.0	11 43.0	9.6	16 31.2	5.5	57.1
16	59 17.4	20.9	26 11.6	9.7	16 25.7	5.6	57.0
17	74 17.3	20.8	40 40.3	9.7	16 20.1	5.7	57.0
18	89 17.1	N23 20.7	55 09.0	9.8	N16 14.4	5.8	57.0
19	104 17.0	20.6	69 37.8	9.9	16 08.6	5.8	56.9
20	119 16.9	20.5	84 06.7	10.0	16 02.8	6.0	56.9
21	134 16.8	.. 20.4	98 35.7	10.0	15 56.8	6.0	56.9
22	149 16.6	20.4	113 04.7	10.1	15 50.8	6.1	56.8
23	164 16.5	20.3	127 33.8	10.2	N15 44.7	6.2	56.8
	SD 15.8	d 0.1	SD 16.0		15.8		15.6

Lat.	Twilight Naut.	Twilight Civil	Sunrise	Moonrise 24	25	26	27
°	h m	h m	h m	h m	h m	h m	h m
N 72	□	□	□	□	□	02 49	05 00
N 70	□	□	□	01 19	02 21	03 54	05 35
68	□	□	□	02 08	03 09	04 30	06 00
66	□	□	□	02 39	03 40	04 55	06 19
64	////	////	01 33	03 03	04 03	05 15	06 34
62	////	////	02 11	03 21	04 21	05 31	06 47
60	////	00 52	02 37	03 37	04 36	05 44	06 57
N 58	////	01 42	02 58	03 50	04 49	05 56	07 07
56	////	02 12	03 14	04 01	05 00	06 06	07 15
54	00 47	02 34	03 29	04 11	05 10	06 14	07 22
52	01 34	02 52	03 41	04 20	05 18	06 22	07 29
50	02 02	03 07	03 52	04 28	05 26	06 29	07 34
45	02 47	03 37	04 14	04 44	05 43	06 44	07 47
N 40	03 18	03 59	04 32	04 58	05 56	06 57	07 57
35	03 41	04 18	04 47	05 10	06 08	07 07	08 06
30	04 00	04 33	05 00	05 20	06 18	07 16	08 14
20	04 29	04 58	05 22	05 38	06 35	07 32	08 28
N 10	04 51	05 18	05 41	05 53	06 50	07 46	08 39
0	05 10	05 36	05 59	06 08	07 04	07 59	08 50
S 10	05 27	05 54	06 16	06 22	07 19	08 12	09 01
20	05 43	06 11	06 35	06 38	07 34	08 26	09 13
30	06 00	06 30	06 56	06 56	07 51	08 41	09 26
35	06 09	06 40	07 08	07 06	08 01	08 50	09 34
40	06 18	06 52	07 23	07 18	08 13	09 01	09 43
45	06 28	07 06	07 39	07 32	08 26	09 13	09 53
S 50	06 40	07 22	08 00	07 49	08 43	09 28	10 05
52	06 45	07 29	08 10	07 57	08 51	09 35	10 11
54	06 51	07 37	08 21	08 06	08 59	09 42	10 17
56	06 57	07 46	08 34	08 16	09 09	09 51	10 24
58	07 04	07 57	08 48	08 28	09 20	10 01	10 32
S 60	07 11	08 08	09 06	08 41	09 33	10 12	10 41

Lat.	Sunset	Twilight Civil	Twilight Naut.	Moonset 24	25	26	27
°	h m	h m	h m	h m	h m	h m	h m
N 72	□	□	□			00 49	00 26
N 70	□	□	□	23 22	23 43	23 51	23 53
68	□	□	□	22 34	23 07	23 25	23 35
66	□	□	□	22 03	22 41	23 05	23 21
64	22 32	////	////	21 40	22 21	22 49	23 09
62	21 54	////	////	21 21	22 05	22 36	22 59
60	21 28	23 13	////	21 06	21 51	22 25	22 51
N 58	21 07	22 23	////	20 53	21 39	22 15	22 43
56	20 51	21 53	////	20 42	21 29	22 06	22 36
54	20 36	21 31	23 17	20 32	21 20	21 59	22 30
52	20 24	21 13	22 31	20 23	21 12	21 52	22 25
50	20 13	20 58	22 02	20 15	21 05	21 46	22 20
45	19 51	20 28	21 18	19 58	20 49	21 32	22 09
N 40	19 33	20 06	20 47	19 45	20 36	21 21	22 00
35	19 18	19 47	20 24	19 33	20 25	21 11	21 53
30	19 05	19 32	20 06	19 23	20 16	21 03	21 46
20	18 43	19 07	19 37	19 05	19 59	20 49	21 34
N 10	18 24	18 47	19 14	18 50	19 44	20 36	21 24
0	18 06	18 29	18 55	18 35	19 31	20 24	21 14
S 10	17 49	18 12	18 38	18 21	19 17	20 12	21 04
20	17 30	17 54	18 22	18 05	19 03	19 59	20 54
30	17 09	17 36	18 05	17 47	18 46	19 44	20 41
35	16 57	17 25	17 57	17 37	18 36	19 35	20 34
40	16 43	17 13	17 47	17 25	18 25	19 26	20 26
45	16 26	17 00	17 37	17 11	18 11	19 14	20 17
S 50	16 05	16 44	17 25	16 54	17 55	19 00	20 06
52	15 55	16 36	17 20	16 46	17 48	18 53	20 00
54	15 44	16 28	17 14	16 37	17 39	18 46	19 55
56	15 32	16 19	17 08	16 27	17 30	18 38	19 48
58	15 17	16 09	17 01	16 15	17 19	18 29	19 41
S 60	14 59	15 57	16 54	16 02	17 06	18 18	19 33

Day	SUN Eqn. of Time 00h	12h	SUN Mer. Pass.	MOON Mer. Pass. Upper	Lower	Age	Phase
d	m s	m s	h m	h m	h m	d	%
24	02 16	02 22	12 02	12 22	24 50	00	0
25	02 29	02 35	12 03	13 18	00 50	01	2
26	02 42	02 48	12 03	14 11	01 45	02	7

1998 JUNE 27, 28, 29 (SAT., SUN., MON.)

UT	ARIES	VENUS −3.9		MARS +1.5		JUPITER −2.5		SATURN +0.5	
	GHA	GHA	Dec	GHA	Dec	GHA	Dec	GHA	Dec
d h	° ′	° ′	° ′	° ′	° ′	° ′	° ′	° ′	° ′
27 00	274 54.2	213 40.5	N19 09.1	191 52.6	N23 47.1	276 50.4	S 2 11.8	244 38.8	N 9 47.3
01	289 56.6	228 39.9	09.8	206 53.2	47.2	291 52.7	11.7	259 41.1	47.3
02	304 59.1	243 39.3	10.4	221 53.8	47.3	306 55.0	11.7	274 43.4	47.4
03	320 01.6	258 38.7 ..	11.1	236 54.4 ..	47.5	321 57.3 ..	11.6	289 45.7 ..	47.5
04	335 04.0	273 38.1	11.7	251 55.0	47.6	336 59.6	11.6	304 47.9	47.5
05	350 06.5	288 37.5	12.3	266 55.6	47.7	352 01.9	11.5	319 50.2	47.6
S 06	5 09.0	303 36.9	N19 13.0	281 56.2	N23 47.8	7 04.3	S 2 11.4	334 52.5	N 9 47.6
A 07	20 11.4	318 36.3	13.6	296 56.8	47.9	22 06.6	11.4	349 54.8	47.7
T 08	35 13.9	333 35.7	14.3	311 57.4	48.0	37 08.9	11.3	4 57.0	47.7
U 09	50 16.4	348 35.1 ..	14.9	326 58.0 ..	48.1	52 11.2 ..	11.3	19 59.3 ..	47.8
R 10	65 18.8	3 34.4	15.5	341 58.6	48.2	67 13.5	11.2	35 01.6	47.9
D 11	80 21.3	18 33.8	16.2	356 59.2	48.3	82 15.8	11.2	50 03.9	47.9
A 12	95 23.7	33 33.2	N19 16.8	11 59.8	N23 48.4	97 18.1	S 2 11.1	65 06.1	N 9 48.0
Y 13	110 26.2	48 32.6	17.4	27 00.4	48.5	112 20.4	11.1	80 08.4	48.0
14	125 28.7	63 32.0	18.1	42 01.0	48.6	127 22.7	11.0	95 10.7	48.1
15	140 31.1	78 31.4 ..	18.7	57 01.6 ..	48.7	142 25.0 ..	11.0	110 13.0 ..	48.2
16	155 33.6	93 30.8	19.3	72 02.2	48.8	157 27.4	10.9	125 15.3	48.2
17	170 36.1	108 30.2	20.0	87 02.8	48.9	172 29.7	10.9	140 17.5	48.3
18	185 38.5	123 29.6	N19 20.6	102 03.4	N23 49.0	187 32.0	S 2 10.8	155 19.8	N 9 48.3
19	200 41.0	138 29.0	21.2	117 04.0	49.1	202 34.3	10.8	170 22.1	48.4
20	215 43.5	153 28.3	21.9	132 04.6	49.2	217 36.6	10.7	185 24.4	48.5
21	230 45.9	168 27.7 ..	22.5	147 05.2 ..	49.3	232 38.9 ..	10.6	200 26.6 ..	48.5
22	245 48.4	183 27.1	23.1	162 05.8	49.4	247 41.2	10.6	215 28.9	48.6
23	260 50.9	198 26.5	23.8	177 06.4	49.5	262 43.5	10.5	230 31.2	48.6
28 00	275 53.3	213 25.9	N19 24.4	192 07.0	N23 49.6	277 45.9	S 2 10.5	245 33.5	N 9 48.7
01	290 55.8	228 25.3	25.0	207 07.6	49.7	292 48.2	10.4	260 35.8	48.7
02	305 58.2	243 24.7	25.6	222 08.2	49.8	307 50.5	10.4	275 38.0	48.8
03	321 00.7	258 24.1 ..	26.3	237 08.8 ..	49.9	322 52.8 ..	10.3	290 40.3 ..	48.9
04	336 03.2	273 23.4	26.9	252 09.4	50.0	337 55.1	10.3	305 42.6	48.9
05	351 05.6	288 22.8	27.5	267 10.0	50.1	352 57.4	10.2	320 44.9	49.0
S 06	6 08.1	303 22.2	N19 28.1	282 10.6	N23 50.2	7 59.7	S 2 10.2	335 47.1	N 9 49.0
U 07	21 10.6	318 21.6	28.7	297 11.2	50.3	23 02.1	10.1	350 49.4	49.1
N 08	36 13.0	333 21.0	29.4	312 11.8	50.4	38 04.4	10.1	5 51.7	49.1
D 09	51 15.5	348 20.4 ..	30.0	327 12.4 ..	50.4	53 06.7 ..	10.0	20 54.0 ..	49.2
A 10	66 18.0	3 19.7	30.6	342 13.0	50.5	68 09.0	10.0	35 56.3	49.3
Y 11	81 20.4	18 19.1	31.2	357 13.6	50.6	83 11.3	09.9	50 58.5	49.3
12	96 22.9	33 18.5	N19 31.8	12 14.2	N23 50.7	98 13.6	S 2 09.9	66 00.8	N 9 49.4
13	111 25.4	48 17.9	32.4	27 14.8	50.8	113 16.0	09.8	81 03.1	49.4
14	126 27.8	63 17.3	33.1	42 15.4	50.9	128 18.3	09.8	96 05.4	49.5
15	141 30.3	78 16.6 ..	33.7	57 16.0 ..	51.0	143 20.6 ..	09.7	111 07.7 ..	49.5
16	156 32.7	93 16.0	34.3	72 16.6	51.1	158 22.9	09.7	126 09.9	49.6
17	171 35.2	108 15.4	34.9	87 17.2	51.2	173 25.2	09.6	141 12.2	49.7
18	186 37.7	123 14.8	N19 35.5	102 17.8	N23 51.3	188 27.6	S 2 09.6	156 14.5	N 9 49.7
19	201 40.1	138 14.1	36.1	117 18.4	51.4	203 29.9	09.5	171 16.8	49.8
20	216 42.6	153 13.5	36.7	132 19.0	51.5	218 32.2	09.5	186 19.1	49.8
21	231 45.1	168 12.9 ..	37.3	147 19.6 ..	51.5	233 34.5 ..	09.4	201 21.3 ..	49.9
22	246 47.5	183 12.3	37.9	162 20.2	51.6	248 36.8	09.4	216 23.6	49.9
23	261 50.0	198 11.7	38.5	177 20.8	51.7	263 39.2	09.3	231 25.9	50.0
29 00	276 52.5	213 11.0	N19 39.2	192 21.4	N23 51.8	278 41.5	S 2 09.3	246 28.2	N 9 50.1
01	291 54.9	228 10.4	39.8	207 22.0	51.9	293 43.8	09.2	261 30.5	50.1
02	306 57.4	243 09.8	40.4	222 22.6	52.0	308 46.1	09.2	276 32.7	50.2
03	321 59.8	258 09.2 ..	41.0	237 23.2 ..	52.1	323 48.4 ..	09.1	291 35.0 ..	50.2
04	337 02.3	273 08.5	41.6	252 23.8	52.2	338 50.8	09.1	306 37.3	50.3
05	352 04.8	288 07.9	42.3	267 24.4	52.3	353 53.1	09.0	321 39.6	50.3
M 06	7 07.2	303 07.3	N19 42.8	282 25.0	N23 52.3	8 55.4	S 2 09.0	336 41.9	N 9 50.4
O 07	22 09.7	318 06.6	43.4	297 25.6	52.4	23 57.7	09.0	351 44.1	50.4
N 08	37 12.2	333 06.0	44.0	312 26.2	52.5	39 00.0	08.9	6 46.4	50.5
D 09	52 14.6	348 05.4 ..	44.6	327 26.8 ..	52.6	54 02.4 ..	08.9	21 48.7 ..	50.6
A 10	67 17.1	3 04.8	45.2	342 27.4	52.7	69 04.7	08.8	36 51.0	50.6
Y 11	82 19.6	18 04.1	45.8	357 28.0	52.8	84 07.0	08.8	51 53.3	50.7
12	97 22.0	33 03.5	N19 46.4	12 28.6	N23 52.9	99 09.3	S 2 08.7	66 55.5	N 9 50.7
13	112 24.5	48 02.9	47.0	27 29.2	52.9	114 11.7	08.7	81 57.8	50.8
14	127 27.0	63 02.2	47.5	42 29.8	53.0	129 14.0	08.6	97 00.1	50.8
15	142 29.4	78 01.6 ..	48.1	57 30.4 ..	53.1	144 16.3 ..	08.6	112 02.4 ..	50.9
16	157 31.9	93 01.0	48.7	72 31.0	53.2	159 18.6	08.5	127 04.7	51.0
17	172 34.3	108 00.3	49.3	87 31.6	53.3	174 21.0	08.5	142 07.0	51.0
18	187 36.8	122 59.7	N19 49.9	102 32.2	N23 53.3	189 23.3	S 2 08.4	157 09.2	N 9 51.1
19	202 39.3	137 59.1	50.5	117 32.8	53.4	204 25.6	08.4	172 11.5	51.1
20	217 41.7	152 58.4	51.1	132 33.4	53.5	219 27.9	08.3	187 13.8	51.2
21	232 44.2	167 57.8 ..	51.7	147 34.0 ..	53.6	234 30.3 ..	08.3	202 16.1 ..	51.2
22	247 46.7	182 57.2	52.3	162 34.6	53.7	249 32.6	08.2	217 18.4	51.3
23	262 49.1	197 56.5	52.9	177 35.2	53.8	264 34.9	08.2	232 20.7	51.3
Mer.Pass.	h m 5 35.5	v −0.6	d 0.6	v 0.6	d 0.1	v 2.3	d 0.1	v 2.3	d 0.1

STARS

Name	SHA	Dec
Acamar	315 27.4	S40 18.6
Achernar	335 35.6	S57 14.5
Acrux	173 22.1	S63 05.7
Adhara	255 21.9	S28 58.3
Aldebaran	291 02.9	N16 30.2
Alioth	166 30.8	N55 58.4
Alkaid	153 07.8	N49 19.6
Al Na'ir	27 57.9	S46 57.9
Alnilam	275 58.4	S 1 12.3
Alphard	218 07.6	S 8 39.2
Alphecca	126 20.5	N26 43.4
Alpheratz	357 55.4	N29 04.7
Altair	62 19.1	N 8 52.0
Ankaa	353 27.1	S42 18.7
Antares	112 40.1	S26 25.6
Arcturus	146 06.1	N19 11.6
Atria	107 51.7	S69 01.5
Avior	234 23.2	S59 30.5
Bellatrix	278 44.7	N 6 20.8
Betelgeuse	271 14.1	N 7 24.3
Canopus	264 01.8	S52 41.8
Capella	280 51.9	N45 59.6
Deneb	49 38.9	N45 16.5
Denebola	182 45.5	N14 35.0
Diphda	349 07.5	S17 59.6
Dubhe	194 06.1	N61 45.8
Elnath	278 27.5	N28 36.2
Eltanin	90 51.0	N51 29.5
Enif	33 58.2	N 9 52.1
Fomalhaut	15 36.5	S29 37.7
Gacrux	172 13.7	S57 06.5
Gienah	176 04.2	S17 32.0
Hadar	149 03.9	S60 22.1
Hamal	328 13.9	N23 27.1
Kaus Aust.	83 58.7	S34 23.0
Kochab	137 19.0	N74 10.1
Markab	13 49.7	N15 11.7
Menkar	314 27.3	N 4 04.9
Menkent	148 21.0	S36 21.8
Miaplacidus	221 42.7	S69 42.9
Mirfak	308 57.2	N49 51.0
Nunki	76 12.2	S26 17.8
Peacock	53 36.8	S56 44.2
Pollux	243 42.2	N28 01.8
Procyon	245 12.1	N 5 13.7
Rasalhague	96 16.8	N12 33.8
Regulus	207 56.0	N11 58.5
Rigel	281 23.5	S 8 12.3
Rigil Kent.	140 07.1	S60 49.8
Sabik	102 25.4	S15 43.2
Schedar	349 53.6	N56 31.4
Shaula	96 37.1	S37 06.0
Sirius	258 44.2	S16 42.9
Spica	158 43.3	S11 09.2
Suhail	223 01.2	S43 25.8
Vega	80 46.3	N38 47.1
Zuben'ubi	137 18.0	S16 02.1

	SHA	Mer.Pass.
	° ′	h m
Venus	297 32.6	9 47
Mars	276 13.6	11 11
Jupiter	1 52.5	5 28
Saturn	329 40.2	7 37

SUN / MOON

UT	SUN GHA	SUN Dec	MOON GHA	v	MOON Dec	d	HP
27 00	179 16.4	N23 20.2	142 03.0	10.2	N15 38.5	6.2	56.8
01	194 16.2	20.1	156 32.2	10.3	15 32.3	6.4	56.7
02	209 16.1	20.0	171 01.5	10.4	15 25.9	6.3	56.7
03	224 16.0	.. 19.9	185 30.9	10.5	15 19.6	6.5	56.7
04	239 15.8	19.8	200 00.4	10.5	15 13.1	6.6	56.6
05	254 15.7	19.7	214 29.9	10.6	15 06.5	6.6	56.6
S 06	269 15.6	N23 19.6	228 59.5	10.7	N14 59.9	6.7	56.6
A 07	284 15.4	19.5	243 29.2	10.8	14 53.2	6.7	56.6
T 08	299 15.3	19.4	257 59.0	10.8	14 46.5	6.8	56.5
U 09	314 15.2	.. 19.3	272 28.8	10.9	14 39.7	6.9	56.5
R 10	329 15.0	19.2	286 58.7	10.9	14 32.8	7.0	56.5
D 11	344 14.9	19.1	301 28.6	11.1	14 25.8	7.0	56.4
A 12	359 14.8	N23 19.0	315 58.7	11.1	N14 18.8	7.1	56.4
Y 13	14 14.7	18.9	330 28.8	11.2	14 11.7	7.2	56.4
14	29 14.5	18.8	344 59.0	11.2	14 04.5	7.2	56.3
15	44 14.4	.. 18.7	359 29.2	11.3	13 57.3	7.2	56.3
16	59 14.3	18.6	13 59.5	11.4	13 50.1	7.4	56.3
17	74 14.1	18.5	28 29.9	11.5	13 42.7	7.4	56.3
18	89 14.0	N23 18.4	43 00.4	11.5	N13 35.3	7.4	56.2
19	104 13.9	18.3	57 30.9	11.6	13 27.9	7.6	56.2
20	119 13.7	18.2	72 01.5	11.7	13 20.3	7.5	56.2
21	134 13.6	.. 18.1	86 32.2	11.7	13 12.8	7.7	56.1
22	149 13.5	17.9	101 02.9	11.8	13 05.1	7.6	56.1
23	164 13.4	17.8	115 33.7	11.9	12 57.5	7.8	56.1
28 00	179 13.2	N23 17.7	130 04.6	11.9	N12 49.7	7.8	56.0
01	194 13.1	17.6	144 35.5	12.0	12 41.9	7.8	56.0
02	209 13.0	17.5	159 06.5	12.1	12 34.1	7.9	56.0
03	224 12.8	.. 17.4	173 37.6	12.1	12 26.2	8.0	56.0
04	239 12.7	17.3	188 08.7	12.2	12 18.2	8.0	55.9
05	254 12.6	17.2	202 39.9	12.3	12 10.2	8.0	55.9
S 06	269 12.5	N23 17.1	217 11.2	12.3	N12 02.2	8.1	55.9
U 07	284 12.3	16.9	231 42.5	12.4	11 54.1	8.2	55.8
N 08	299 12.2	16.8	246 13.9	12.5	11 45.9	8.2	55.8
D 09	314 12.1	.. 16.7	260 45.4	12.5	11 37.7	8.2	55.8
A 10	329 11.9	16.6	275 16.9	12.6	11 29.5	8.3	55.8
Y 11	344 11.8	16.5	289 48.5	12.7	11 21.2	8.3	55.7
12	359 11.7	N23 16.4	304 20.2	12.7	N11 12.9	8.4	55.7
13	14 11.6	16.2	318 51.9	12.7	11 04.5	8.4	55.7
14	29 11.4	16.1	333 23.6	12.9	10 56.1	8.5	55.7
15	44 11.3	.. 16.0	347 55.5	12.9	10 47.6	8.5	55.6
16	59 11.2	15.9	2 27.4	12.9	10 39.1	8.5	55.6
17	74 11.0	15.8	16 59.3	13.0	10 30.6	8.6	55.6
18	89 10.9	N23 15.6	31 31.3	13.1	N10 22.0	8.6	55.6
19	104 10.8	15.5	46 03.4	13.1	10 13.4	8.7	55.5
20	119 10.7	15.4	60 35.5	13.2	10 04.7	8.7	55.5
21	134 10.5	.. 15.3	75 07.7	13.3	9 56.0	8.7	55.5
22	149 10.4	15.1	89 40.0	13.3	9 47.3	8.7	55.5
23	164 10.3	15.0	104 12.3	13.3	9 38.6	8.9	55.4
29 00	179 10.1	N23 14.9	118 44.6	13.4	N 9 29.7	8.8	55.4
01	194 10.0	14.8	133 17.0	13.5	9 20.9	8.9	55.4
02	209 09.9	14.6	147 49.5	13.5	9 12.0	8.9	55.4
03	224 09.8	.. 14.5	162 22.0	13.6	9 03.1	8.9	55.3
04	239 09.6	14.4	176 54.6	13.6	8 54.2	9.0	55.3
05	254 09.5	14.2	191 27.2	13.6	8 45.2	8.9	55.3
M 06	269 09.4	N23 14.1	205 59.8	13.8	N 8 36.3	9.1	55.3
O 07	284 09.3	14.0	220 32.6	13.7	8 27.2	9.0	55.2
N 08	299 09.1	13.8	235 05.3	13.9	8 18.2	9.1	55.2
D 09	314 09.0	.. 13.7	249 38.2	13.8	8 09.1	9.1	55.2
A 10	329 08.9	13.6	264 11.0	13.9	8 00.0	9.2	55.2
Y 11	344 08.8	13.4	278 43.9	14.0	7 50.8	9.1	55.1
12	359 08.6	N23 13.3	293 16.9	14.0	N 7 41.7	9.2	55.1
13	14 08.5	13.2	307 49.9	14.1	7 32.5	9.2	55.1
14	29 08.4	13.0	322 23.0	14.1	7 23.3	9.3	55.1
15	44 08.3	.. 12.9	336 56.1	14.1	7 14.0	9.2	55.1
16	59 08.1	12.8	351 29.2	14.2	7 04.8	9.3	55.0
17	74 08.0	12.6	6 02.4	14.2	6 55.5	9.3	55.0
18	89 07.9	N23 12.5	20 35.6	14.3	N 6 46.2	9.3	55.0
19	104 07.7	12.3	35 08.9	14.3	6 36.9	9.4	55.0
20	119 07.6	12.2	49 42.2	14.4	6 27.5	9.3	55.0
21	134 07.5	.. 12.1	64 15.6	14.4	6 18.2	9.4	54.9
22	149 07.4	11.9	78 49.0	14.4	6 08.8	9.4	54.9
23	164 07.2	11.8	93 22.4	14.5	N 5 59.4	9.5	54.9
SD	15.8	d 0.1	SD 15.4		15.2		15.0

Moonrise

Lat.	Twilight Naut.	Twilight Civil	Sunrise	Moonrise 27	28	29	30
N 72	□	□	□	05 00	06 53	08 37	10 15
N 70	□	□	□	05 35	07 14	08 49	10 21
68	□	□	□	06 00	07 31	09 00	10 26
66	□	□	□	06 19	07 44	09 08	10 30
64	////	////	01 36	06 34	07 55	09 15	10 33
62	////	////	02 13	06 47	08 04	09 21	10 36
60	////	00 56	02 39	06 57	08 12	09 26	10 39
N 58	////	01 45	02 59	07 07	08 19	09 31	10 41
56	////	02 14	03 16	07 15	08 25	09 35	10 43
54	00 52	02 36	03 30	07 22	08 31	09 38	10 45
52	01 36	02 54	03 42	07 29	08 36	09 42	10 47
50	02 04	03 09	03 53	07 34	08 40	09 45	10 48
45	02 49	03 38	04 15	07 47	08 50	09 51	10 51
N 40	03 19	04 01	04 33	07 57	08 58	09 57	10 54
35	03 42	04 19	04 48	08 06	09 05	10 01	10 56
30	04 01	04 34	05 01	08 14	09 11	10 05	10 59
20	04 29	04 59	05 23	08 28	09 21	10 12	11 02
N 10	04 52	05 19	05 42	08 39	09 30	10 19	11 05
0	05 11	05 37	06 00	08 50	09 39	10 24	11 08
S 10	05 28	05 54	06 17	09 01	09 47	10 30	11 11
20	05 44	06 11	06 35	09 13	09 56	10 36	11 14
30	06 00	06 30	06 56	09 26	10 07	10 43	11 18
35	06 09	06 41	07 09	09 34	10 12	10 47	11 20
40	06 18	06 52	07 23	09 43	10 19	10 52	11 22
45	06 28	07 06	07 39	09 53	10 27	10 57	11 25
S 50	06 40	07 22	08 00	10 05	10 36	11 04	11 28
52	06 45	07 29	08 10	10 11	10 41	11 07	11 30
54	06 51	07 37	08 21	10 17	10 45	11 10	11 31
56	06 57	07 46	08 33	10 24	10 51	11 13	11 33
58	07 04	07 56	08 48	10 32	10 57	11 17	11 35
S 60	07 11	08 08	09 05	10 41	11 03	11 21	11 37

Moonset

Lat.	Sunset	Twilight Civil	Twilight Naut.	Moonset 27	28	29	30
N 72	□	□	□	00 26	00 15	00 07	{00 00 / 23 54}
N 70	□	□	□	23 53	23 53	23 52	23 51
68	□	□	□	23 35	23 42	23 46	23 49
66	□	□	□	23 21	23 32	23 40	23 47
64	22 29	////	////	23 09	23 24	23 36	23 46
62	21 53	////	////	22 59	23 17	23 31	23 44
60	21 27	23 09	////	22 51	23 11	23 28	23 43
N 58	21 07	22 21	////	22 43	23 06	23 25	23 42
56	20 50	21 52	////	22 36	23 01	23 22	23 41
54	20 36	21 30	23 13	22 30	22 56	23 19	23 40
52	20 24	21 12	22 30	22 25	22 53	23 17	23 39
50	20 13	20 58	22 02	22 20	22 49	23 15	23 39
45	19 51	20 28	21 18	22 09	22 41	23 10	23 37
N 40	19 33	20 06	20 47	22 00	22 35	23 06	23 36
35	19 18	19 48	20 24	21 53	22 29	23 03	23 35
30	19 05	19 32	20 06	21 46	22 22	23 00	23 34
20	18 43	19 08	19 37	21 34	22 16	22 55	23 32
N 10	18 24	18 47	19 14	21 24	22 08	22 50	23 30
0	18 07	18 29	18 56	21 14	22 01	22 46	23 29
S 10	17 50	18 12	18 39	21 04	21 54	22 41	23 27
20	17 31	17 55	18 23	20 54	21 46	22 37	23 26
30	17 10	17 36	18 06	20 41	21 37	22 31	23 24
35	16 58	17 26	17 58	20 34	21 32	22 28	23 23
40	16 44	17 14	17 48	20 26	21 26	22 24	23 21
45	16 27	17 01	17 38	20 17	21 19	22 20	23 20
S 50	16 06	16 45	17 26	20 06	21 11	22 15	23 18
52	15 57	16 38	17 21	20 00	21 07	22 13	23 17
54	15 46	16 29	17 16	19 55	21 03	22 10	23 17
56	15 33	16 20	17 09	19 48	20 58	22 08	23 16
58	15 19	16 10	17 03	19 41	20 53	22 04	23 14
S 60	15 01	15 59	16 55	19 33	20 47	22 01	23 13

SUN / MOON

Day	Eqn. of Time 00h	Eqn. of Time 12h	Mer. Pass.	Mer. Pass. Upper	Mer. Pass. Lower	Age	Phase %
	m s	m s	h m	h m	h m	d	%
27	02 54	03 01	12 03	15 02	02 37	03	13
28	03 07	03 13	12 03	15 50	03 26	04	20
29	03 19	03 25	12 03	16 35	04 13	05	29

1998 JUNE 30, JULY 1, 2 (TUES., WED., THURS.)

UT	ARIES GHA	VENUS −3.9 GHA	VENUS Dec	MARS +1.6 GHA	MARS Dec	JUPITER −2.5 GHA	JUPITER Dec	SATURN +0.5 GHA	SATURN Dec	STARS Name	SHA	Dec
30 00	277 51.6	212 55.9	N19 53.4	192 35.8	N23 53.8	279 37.3	S 2 08.2	247 22.9	N 9 51.4	Acamar	315 27.4	S40 18.6
01	292 54.1	227 55.3	54.0	207 36.4	53.9	294 39.6	08.1	262 25.2	51.5	Achernar	335 35.6	S57 14.5
02	307 56.5	242 54.6	54.6	222 37.0	54.0	309 41.9	08.1	277 27.5	51.5	Acrux	173 22.1	S63 05.7
03	322 59.0	257 54.0	.. 55.2	237 37.6	.. 54.1	324 44.2	.. 08.0	292 29.8	.. 51.6	Adhara	255 21.9	S28 58.3
04	338 01.5	272 53.3	55.8	252 38.2	54.1	339 46.6	08.0	307 32.1	51.6	Aldebaran	291 02.9	N16 30.2
05	353 03.9	287 52.7	56.4	267 38.8	54.2	354 48.9	07.9	322 34.4	51.7			
06	8 06.4	302 52.1	N19 56.9	282 39.4	N23 54.3	9 51.2	S 2 07.9	337 36.7	N 9 51.7	Alioth	166 30.8	N55 58.4
07	23 08.8	317 51.4	57.5	297 40.0	54.4	24 53.6	07.8	352 38.9	51.8	Alkaid	153 07.9	N49 19.6
T 08	38 11.3	332 50.8	58.1	312 40.6	54.5	39 55.9	07.8	7 41.2	51.8	Al Na'ir	27 57.9	S46 57.9
U 09	53 13.8	347 50.2	.. 58.7	327 41.2	.. 54.5	54 58.2	.. 07.8	22 43.5	.. 51.9	Alnilam	275 58.4	S 1 12.3
E 10	68 16.2	2 49.5	59.2	342 41.8	54.6	70 00.5	07.7	37 45.8	51.9	Alphard	218 07.6	S 8 39.2
S 11	83 18.7	17 48.9	19 59.8	357 42.4	54.7	85 02.9	07.7	52 48.1	52.0			
D 12	98 21.2	32 48.2	N20 00.4	12 43.0	N23 54.8	100 05.2	S 2 07.6	67 50.4	N 9 52.1	Alphecca	126 20.5	N26 43.5
A 13	113 23.6	47 47.6	01.0	27 43.6	54.8	115 07.5	07.6	82 52.6	52.1	Alpheratz	357 55.3	N29 04.7
Y 14	128 26.1	62 46.9	01.5	42 44.2	54.9	130 09.9	07.5	97 54.9	52.2	Altair	62 19.1	N 8 52.0
15	143 28.6	77 46.3	.. 02.1	57 44.8	.. 55.0	145 12.2	.. 07.5	112 57.2	.. 52.2	Ankaa	353 27.1	S42 18.7
16	158 31.0	92 45.7	02.7	72 45.4	55.1	160 14.5	07.4	127 59.5	52.3	Antares	112 40.1	S26 25.6
17	173 33.5	107 45.0	03.3	87 46.0	55.1	175 16.9	07.4	143 01.8	52.3			
18	188 35.9	122 44.4	N20 03.8	102 46.6	N23 55.2	190 19.2	S 2 07.4	158 04.1	N 9 52.4	Arcturus	146 06.1	N19 11.7
19	203 38.4	137 43.7	04.4	117 47.2	55.3	205 21.5	07.3	173 06.4	52.4	Atria	107 51.7	S69 01.5
20	218 40.9	152 43.1	05.0	132 47.8	55.4	220 23.9	07.3	188 08.6	52.5	Avior	234 23.2	S59 30.5
21	233 43.3	167 42.4	.. 05.5	147 48.4	.. 55.4	235 26.2	.. 07.2	203 10.9	.. 52.5	Bellatrix	278 44.7	N 6 20.8
22	248 45.8	182 41.8	06.1	162 49.0	55.5	250 28.5	07.2	218 13.2	52.6	Betelgeuse	271 14.1	N 7 24.3
23	263 48.3	197 41.1	06.7	177 49.6	55.5	265 30.9	07.1	233 15.5	52.7			
1 00	278 50.7	212 40.5	N20 07.2	192 50.2	N23 55.6	280 33.2	S 2 07.1	248 17.8	N 9 52.7	Canopus	264 01.8	S52 41.8
01	293 53.2	227 39.9	07.8	207 50.9	55.7	295 35.5	07.1	263 20.1	52.8	Capella	280 51.9	N45 59.6
02	308 55.7	242 39.2	08.4	222 51.5	55.8	310 37.9	07.0	278 22.4	52.8	Deneb	49 38.9	N45 16.5
03	323 58.1	257 38.6	.. 08.9	237 52.1	.. 55.8	325 40.2	.. 07.0	293 24.7	.. 52.9	Denebola	182 45.5	N14 35.0
04	339 00.6	272 37.9	09.5	252 52.7	55.9	340 42.5	06.9	308 26.9	52.9	Diphda	349 07.5	S17 59.6
05	354 03.1	287 37.3	10.0	267 53.3	56.0	355 44.9	06.9	323 29.2	53.0			
06	9 05.5	302 36.6	N20 10.6	282 53.9	N23 56.1	10 47.2	S 2 06.8	338 31.5	N 9 53.0	Dubhe	194 06.1	N61 45.8
W 07	24 08.0	317 36.0	11.2	297 54.5	56.1	25 49.6	06.8	353 33.8	53.1	Elnath	278 27.5	N28 36.2
E 08	39 10.4	332 35.3	11.7	312 55.1	56.2	40 51.9	06.8	8 36.1	53.1	Eltanin	90 51.0	N51 29.5
D 09	54 12.9	347 34.7	.. 12.3	327 55.7	.. 56.3	55 54.2	.. 06.7	23 38.4	.. 53.2	Enif	33 58.2	N 9 52.1
N 10	69 15.4	2 34.0	12.8	342 56.3	56.3	70 56.6	06.7	38 40.7	53.2	Fomalhaut	15 36.5	S29 37.7
N 11	84 17.8	17 33.4	13.4	357 56.9	56.4	85 58.9	06.6	53 43.0	53.3			
S 12	99 20.3	32 32.7	N20 13.9	12 57.5	N23 56.5	101 01.2	S 2 06.6	68 45.2	N 9 53.4	Gacrux	172 13.7	S57 06.5
D 13	114 22.8	47 32.1	14.5	27 58.1	56.5	116 03.6	06.6	83 47.5	53.4	Gienah	176 04.2	S17 32.0
A 14	129 25.2	62 31.4	15.0	42 58.7	56.6	131 05.9	06.5	98 49.8	53.5	Hadar	149 03.9	S60 22.1
Y 15	144 27.7	77 30.7	.. 15.6	57 59.3	.. 56.7	146 08.3	.. 06.5	113 52.1	.. 53.5	Hamal	328 13.9	N23 27.1
16	159 30.2	92 30.1	16.1	72 59.9	56.7	161 10.6	06.4	128 54.4	53.6	Kaus Aust.	83 58.7	S34 23.0
17	174 32.6	107 29.4	16.7	88 00.5	56.8	176 12.9	06.4	143 56.7	53.6			
18	189 35.1	122 28.8	N20 17.2	103 01.1	N23 56.8	191 15.3	S 2 06.4	158 59.0	N 9 53.7	Kochab	137 19.1	N74 10.1
19	204 37.6	137 28.1	17.8	118 01.7	56.9	206 17.6	06.3	174 01.3	53.7	Markab	13 49.6	N15 11.8
20	219 40.0	152 27.5	18.3	133 02.3	57.0	221 20.0	06.3	189 03.6	53.8	Menkar	314 27.3	N 4 04.9
21	234 42.5	167 26.8	.. 18.9	148 02.9	.. 57.0	236 22.3	.. 06.2	204 05.8	.. 53.8	Menkent	148 21.0	S36 21.8
22	249 44.9	182 26.2	19.4	163 03.5	57.1	251 24.6	06.2	219 08.1	53.9	Miaplacidus	221 42.7	S69 42.9
23	264 47.4	197 25.5	20.0	178 04.1	57.2	266 27.0	06.2	234 10.4	53.9			
2 00	279 49.9	212 24.8	N20 20.5	193 04.7	N23 57.2	281 29.3	S 2 06.1	249 12.7	N 9 54.0	Mirfak	308 57.2	N49 51.0
01	294 52.3	227 24.2	21.1	208 05.3	57.3	296 31.7	06.1	264 15.0	54.0	Nunki	76 12.2	S26 17.8
02	309 54.8	242 23.5	21.6	223 06.0	57.3	311 34.0	06.0	279 17.3	54.1	Peacock	53 36.8	S56 44.2
03	324 57.3	257 22.9	.. 22.1	238 06.6	.. 57.4	326 36.3	.. 06.0	294 19.6	.. 54.1	Pollux	243 42.2	N28 01.8
04	339 59.7	272 22.2	22.7	253 07.2	57.5	341 38.7	06.0	309 21.9	54.2	Procyon	245 12.1	N 5 13.7
05	355 02.2	287 21.5	23.2	268 07.8	57.5	356 41.0	05.9	324 24.2	54.3			
06	10 04.7	302 20.9	N20 23.8	283 08.4	N23 57.6	11 43.4	S 2 05.9	339 26.5	N 9 54.3	Rasalhague	96 16.8	N12 33.9
07	25 07.1	317 20.2	24.3	298 09.0	57.6	26 45.7	05.8	354 28.7	54.4	Regulus	207 56.0	N11 58.5
T 08	40 09.6	332 19.6	24.8	313 09.6	57.7	41 48.1	05.8	9 31.0	54.4	Rigel	281 23.4	S 8 12.3
H 09	55 12.0	347 18.9	.. 25.4	328 10.2	.. 57.8	56 50.4	.. 05.8	24 33.3	.. 54.5	Rigil Kent.	140 07.1	S60 49.8
U 10	70 14.5	2 18.2	25.9	343 10.8	57.8	71 52.7	05.7	39 35.6	54.5	Sabik	102 25.4	S15 43.2
R 11	85 17.0	17 17.6	26.4	358 11.4	57.9	86 55.1	05.7	54 37.9	54.6			
S 12	100 19.4	32 16.9	N20 27.0	13 12.0	N23 57.9	101 57.4	S 2 05.7	69 40.2	N 9 54.6	Schedar	349 53.6	N56 31.4
D 13	115 21.9	47 16.3	27.5	28 12.6	58.0	116 59.8	05.6	84 42.5	54.7	Shaula	96 37.1	S37 06.1
A 14	130 24.4	62 15.6	28.0	43 13.2	58.1	132 02.1	05.6	99 44.8	54.7	Sirius	258 44.2	S16 42.9
Y 15	145 26.8	77 14.9	.. 28.6	58 13.8	.. 58.1	147 04.5	.. 05.5	114 47.1	.. 54.8	Spica	158 43.3	S11 09.2
16	160 29.3	92 14.3	29.1	73 14.4	58.2	162 06.8	05.5	129 49.4	54.8	Suhail	223 01.2	S43 25.8
17	175 31.8	107 13.6	29.6	88 15.0	58.2	177 09.2	05.5	144 51.7	54.9			
18	190 34.2	122 12.9	N20 30.1	103 15.6	N23 58.3	192 11.5	S 2 05.4	159 54.0	N 9 54.9	Vega	80 46.3	N38 47.1
19	205 36.7	137 12.3	30.7	118 16.2	58.3	207 13.9	05.4	174 56.2	55.0	Zuben'ubi	137 18.0	S16 02.1
20	220 39.2	152 11.6	31.2	133 16.8	58.4	222 16.2	05.4	189 58.5	55.0		SHA	Mer. Pass.
21	235 41.6	167 10.9	.. 31.7	148 17.5	.. 58.4	237 18.6	.. 05.3	205 00.8	.. 55.1	Venus	293 49.8	9 50
22	250 44.1	182 10.3	32.2	163 18.1	58.5	252 20.9	05.3	220 03.1	55.1	Mars	273 59.5	11 08
23	265 46.5	197 09.6	32.8	178 18.7	58.5	267 23.3	05.2	235 05.4	55.2	Jupiter	1 42.5	5 17
Mer. Pass. 5h 23.7m	v −0.7 d 0.6			v 0.6 d 0.1		v 2.3 d 0.0		v 2.3 d 0.1		Saturn	329 27.1	7 26

UT		SUN		MOON					Lat.	Twilight		Sunrise	Moonrise			
		GHA	Dec	GHA	v	Dec	d	HP		Naut.	Civil		30	1	2	3
d	h	° ′	° ′	° ′	′	° ′	′	′	°	h m	h m	h m	h m	h m	h m	h m
30	00	179 07.1	N23 11.6	107 55.9	14.5	N 5 49.9	9.4	54.9	N 72	☐	☐	☐	10 15	11 51	13 25	15 01
	01	194 07.0	11.5	122 29.4	14.6	5 40.5	9.5	54.9	N 70	☐	☐	☐	10 21	11 51	13 19	14 49
	02	209 06.9	11.3	137 03.0	14.5	5 31.0	9.4	54.8	68	☐	☐	☐	10 26	11 51	13 14	14 39
	03	224 06.8	.. 11.2	151 36.5	14.7	5 21.6	9.5	54.8	66	////	////	00 17	10 30	11 51	13 10	14 30
	04	239 06.6	11.1	166 10.2	14.6	5 12.1	9.5	54.8	64	////	////	01 41	10 33	11 51	13 07	14 23
	05	254 06.5	10.9	180 43.8	14.7	5 02.6	9.6	54.8	62	////	////	02 16	10 36	11 51	13 04	14 18
	06	269 06.4	N23 10.8	195 17.5	14.7	N 4 53.0	9.5	54.8	60	////	01 03	02 42	10 39	11 51	13 02	14 12
	07	284 06.3	10.6	209 51.2	14.8	4 43.5	9.5	54.8	N 58	////	01 48	03 02	10 41	11 51	12 59	14 08
	08	299 06.1	10.5	224 25.0	14.8	4 34.0	9.6	54.7	56	////	02 17	03 18	10 43	11 51	12 57	14 04
T	09	314 06.0	.. 10.3	238 58.8	14.8	4 24.4	9.6	54.7	54	00 58	02 38	03 32	10 45	11 51	12 56	14 00
U	10	329 05.9	10.2	253 32.6	14.8	4 14.8	9.5	54.7	52	01 40	02 56	03 44	10 47	11 51	12 54	13 57
E	11	344 05.8	10.0	268 06.4	14.9	4 05.3	9.6	54.7	50	02 06	03 11	03 55	10 48	11 51	12 52	13 54
S	12	359 05.6	N23 09.9	282 40.3	14.9	N 3 55.7	9.6	54.7	45	02 50	03 40	04 17	10 51	11 51	12 49	13 48
D	13	14 05.5	09.7	297 14.2	14.9	3 46.1	9.7	54.7	N 40	03 21	04 02	04 35	10 54	11 51	12 47	13 43
A	14	29 05.4	09.5	311 48.1	15.0	3 36.4	9.6	54.6	35	03 43	04 20	04 50	10 56	11 51	12 44	13 38
Y	15	44 05.3	.. 09.4	326 22.1	15.0	3 26.8	9.6	54.6	30	04 02	04 35	05 02	10 59	11 51	12 42	13 34
	16	59 05.1	09.2	340 56.1	15.0	3 17.2	9.7	54.6	20	04 30	05 00	05 24	11 02	11 51	12 39	13 27
	17	74 05.0	09.1	355 30.1	15.0	3 07.5	9.6	54.6	N 10	04 53	05 20	05 43	11 05	11 51	12 36	13 21
	18	89 04.9	N23 08.9	10 04.1	15.1	N 2 57.9	9.7	54.6	0	05 11	05 38	06 00	11 08	11 51	12 33	13 16
	19	104 04.8	08.8	24 38.2	15.1	2 48.2	9.6	54.6	S 10	05 28	05 55	06 17	11 11	11 51	12 30	13 10
	20	119 04.7	08.6	39 12.3	15.1	2 38.6	9.7	54.6	20	05 44	06 12	06 36	11 14	11 51	12 27	13 04
	21	134 04.5	.. 08.5	53 46.4	15.1	2 28.9	9.7	54.5	30	06 00	06 30	06 57	11 18	11 51	12 24	12 57
	22	149 04.4	08.3	68 20.5	15.2	2 19.2	9.6	54.5	35	06 09	06 41	07 09	11 20	11 51	12 22	12 54
	23	164 04.3	08.1	82 54.7	15.1	2 09.6	9.7	54.5	40	06 18	06 52	07 23	11 22	11 51	12 20	12 49
1	00	179 04.2	N23 08.0	97 28.8	15.2	N 1 59.9	9.7	54.5	45	06 28	07 06	07 39	11 25	11 51	12 17	12 44
	01	194 04.0	07.8	112 03.0	15.2	1 50.2	9.7	54.5	S 50	06 40	07 21	08 00	11 28	11 51	12 14	12 38
	02	209 03.9	07.7	126 37.2	15.2	1 40.5	9.7	54.5	52	06 45	07 29	08 09	11 30	11 51	12 13	12 36
	03	224 03.8	.. 07.5	141 11.4	15.3	1 30.8	9.7	54.5	54	06 51	07 37	08 20	11 31	11 52	12 12	12 33
	04	239 03.7	07.3	155 45.7	15.2	1 21.1	9.6	54.5	56	06 57	07 46	08 33	11 33	11 52	12 10	12 29
	05	254 03.6	07.2	170 19.9	15.3	1 11.5	9.7	54.4	58	07 03	07 56	08 47	11 35	11 52	12 08	12 26
	06	269 03.4	N23 07.0	184 54.2	15.3	N 1 01.8	9.7	54.4	S 60	07 11	08 07	09 04	11 37	11 52	12 06	12 21

UT		SUN		MOON					Lat.	Sunset	Twilight		Moonset			
		GHA	Dec	GHA	v	Dec	d	HP			Civil	Naut.	30	1	2	3
d	h	° ′	° ′	° ′	′	° ′	′	′	°	h m	h m	h m	h m	h m	h m	h m
W	07	284 03.3	06.8	199 28.5	15.3	0 52.1	9.7	54.4	N 72	☐	☐	☐	{00 00 / 23 54}	23 47	23 40	23 33
E	08	299 03.2	06.7	214 02.8	15.3	0 42.4	9.7	54.4	N 70	☐	☐	☐	23 51	23 49	23 48	23 47
D	09	314 03.1	.. 06.5	228 37.1	15.3	0 32.7	9.7	54.4	68	☐	☐	☐	23 49	23 52	23 55	23 58
N	10	329 02.9	06.3	243 11.4	15.3	0 23.0	9.7	54.4	66	23 42	////	////	23 47	23 53	24 00	00 00
E	11	344 02.8	06.2	257 45.7	15.4	0 13.3	9.7	54.4	64	22 26	////	////	23 46	23 54	24 05	00 05
S	12	359 02.7	N23 06.0	272 20.1	15.3	N 0 03.6	9.7	54.4	62	21 50	////	////	23 44	23 56	24 09	00 09
D	13	14 02.6	05.8	286 54.4	15.4	S 0 06.1	9.6	54.4	60	21 25	23 03	////	23 43	23 57	24 12	00 12
A	14	29 02.5	05.7	301 28.8	15.3	0 15.7	9.7	54.4	N 58	21 05	22 18	////	23 42	23 58	24 15	00 15
Y	15	44 02.3	.. 05.5	316 03.1	15.4	0 25.4	9.7	54.3	56	20 49	21 50	////	23 41	23 59	24 18	00 18
	16	59 02.2	05.3	330 37.5	15.4	0 35.1	9.6	54.3	54	20 35	21 29	23 08	23 40	24 00	00 00	00 21
	17	74 02.1	05.1	345 11.9	15.4	0 44.7	9.7	54.3	52	20 23	21 11	22 27	23 39	24 01	00 01	00 23
	18	89 02.0	N23 04.9	359 46.3	15.4	S 0 54.4	9.7	54.3	50	20 13	20 57	22 01	23 39	24 02	00 02	00 25
	19	104 01.9	04.8	14 20.7	15.4	1 04.1	9.6	54.3	45	19 50	20 28	21 17	23 37	24 03	00 03	00 30
	20	119 01.7	04.6	28 55.1	15.4	1 13.7	9.6	54.3	N 40	19 33	20 05	20 47	23 36	24 04	00 04	00 33
	21	134 01.6	.. 04.4	43 29.5	15.4	1 23.3	9.7	54.3	35	19 18	19 48	20 24	23 35	24 06	00 06	00 37
	22	149 01.5	04.3	58 03.9	15.4	1 33.0	9.6	54.3	30	19 05	19 33	20 06	23 34	24 07	00 07	00 40
	23	164 01.4	04.1	72 38.3	15.4	1 42.6	9.6	54.3	20	18 43	19 08	19 37	23 32	24 08	00 08	00 45
2	00	179 01.3	N23 03.9	87 12.7	15.4	S 1 52.2	9.6	54.3	N 10	18 25	18 48	19 15	23 30	24 10	00 10	00 49
	01	194 01.1	03.7	101 47.1	15.4	2 01.8	9.6	54.3	0	18 07	18 30	18 56	23 29	24 11	00 11	00 53
	02	209 01.0	03.6	116 21.5	15.4	2 11.4	9.6	54.3	S 10	17 50	18 13	18 39	23 27	24 12	00 12	00 58
	03	224 00.9	.. 03.4	130 55.9	15.4	2 21.0	9.6	54.3	20	17 32	17 56	18 23	23 26	24 14	00 14	01 02
	04	239 00.8	03.2	145 30.3	15.3	2 30.6	9.5	54.3	30	17 11	17 37	18 07	23 24	24 16	00 16	01 07
	05	254 00.7	03.0	160 04.7	15.4	2 40.1	9.6	54.3	35	16 59	17 27	17 59	23 23	24 17	00 17	01 10
	06	269 00.6	N23 02.8	174 39.1	15.4	S 2 49.7	9.5	54.3	40	16 45	17 16	17 50	23 23	24 18	00 18	01 14
T	07	284 00.4	02.7	189 13.5	15.4	2 59.2	9.6	54.2	45	16 29	17 02	17 39	23 20	24 19	00 19	01 17
H	08	299 00.3	02.5	203 47.9	15.3	3 08.8	9.5	54.2	S 50	16 08	16 46	17 28	23 18	24 20	00 20	01 22
U	09	314 00.2	.. 02.3	218 22.2	15.4	3 18.3	9.5	54.2	52	15 59	16 39	17 23	23 17	24 21	00 21	01 24
R	10	329 00.1	02.1	232 56.6	15.4	3 27.8	9.5	54.2	54	15 48	16 31	17 17	23 17	24 22	00 22	01 27
S	11	344 00.0	01.9	247 31.0	15.3	3 37.3	9.4	54.2	56	15 35	16 22	17 11	23 16	24 23	00 23	01 29
D	12	358 59.8	N23 01.7	262 05.3	15.4	S 3 46.7	9.5	54.2	58	15 21	16 12	17 05	23 14	24 24	00 24	01 32
A	13	13 59.7	01.5	276 39.7	15.3	3 56.2	9.4	54.2	S 60	15 04	16 01	16 57	23 13	24 25	00 25	01 35
Y	14	28 59.6	01.4	291 14.0	15.3	4 05.6	9.4	54.2								
	15	43 59.5	.. 01.2	305 48.3	15.3	4 15.0	9.4	54.2								
	16	58 59.4	01.0	320 22.6	15.3	4 24.4	9.4	54.2								
	17	73 59.3	00.8	334 56.9	15.3	4 33.8	9.4	54.2								

		SUN		MOON		
Day	Eqn. of Time		Mer.	Mer. Pass.		Age Phase
	00ʰ	12ʰ	Pass.	Upper	Lower	
d	m s	m s	h m	h m	h m	d %
30	03 31	03 37	12 04	17 19	04 57	06 38
1	03 43	03 49	12 04	18 01	05 40	07 47
2	03 55	04 00	12 04	18 43	06 22	08 57

Additional SUN rows (continued, Thursday):

		GHA	Dec
2	18	88 59.1	N23 00.6
	19	103 59.0	00.4
	20	118 58.9	00.2
	21	133 58.8	23 00.0
	22	148 58.7	22 59.8
	23	163 58.6	N22 59.6

Additional MOON rows (continued):

GHA	v	Dec	d	HP
349 31.2	15.3	S 4 43.2	9.4	54.2
4 05.5	15.2	4 52.6	9.3	54.2
18 39.7	15.3	5 01.9	9.3	54.2
33 14.0	15.2	5 11.2	9.3	54.2
47 48.2	15.2	5 20.5	9.3	54.2
62 22.4	15.2	S 5 29.6	9.2	54.3

SD 15.8	d 0.2	SD 14.9	14.8	14.8

1998 JULY 3, 4, 5 (FRI., SAT., SUN.)

UT	ARIES GHA	VENUS −3.9 GHA	Dec	MARS +1.6 GHA	Dec	JUPITER −2.5 GHA	Dec	SATURN +0.5 GHA	Dec	STARS Name	SHA	Dec
3 00	280 49.0	212 08.9 N20	33.3	193 19.3 N23	58.6	282 25.6 S 2	05.2	250 07.7 N 9	55.2	Acamar	315 27.3	S40 18.6
01	295 51.5	227 08.3	33.8	208 19.9	58.7	297 28.0	05.2	265 10.0	55.3	Achernar	335 35.5	S57 14.5
02	310 53.9	242 07.6	34.3	223 20.5	58.7	312 30.3	05.1	280 12.3	55.3	Acrux	173 22.1	S63 05.7
03	325 56.4	257 06.9 ..	34.8	238 21.1 ..	58.8	327 32.6 ..	05.1	295 14.6 ..	55.4	Adhara	255 21.9	S28 58.3
04	340 58.9	272 06.3	35.4	253 21.7	58.8	342 35.0	05.1	310 16.9	55.4	Aldebaran	291 02.9	N16 30.2
05	356 01.3	287 05.6	35.9	268 22.3	58.9	357 37.3	05.0	325 19.2	55.5			
06	11 03.8	302 04.9 N20	36.4	283 22.9 N23	58.9	12 39.7 S 2	05.0	340 21.5 N 9	55.5	Alioth	166 30.8	N55 58.5
07	26 06.3	317 04.2	36.9	298 23.5	59.0	27 42.0	05.0	355 23.8	55.6	Alkaid	153 07.9	N49 19.6
F 08	41 08.7	332 03.6	37.4	313 24.1	59.0	42 44.4	04.9	10 26.1	55.7	Al Na'ir	27 57.9	S46 57.9
R 09	56 11.2	347 02.9 ..	37.9	328 24.7 ..	59.1	57 46.8 ..	04.9	25 28.4 ..	55.7	Alnilam	275 58.4	S 1 12.3
I 10	71 13.7	2 02.2	38.4	343 25.3	59.1	72 49.1	04.8	40 30.7	55.8	Alphard	218 07.6	S 8 39.2
D 11	86 16.1	17 01.6	39.0	358 25.9	59.2	87 51.5	04.8	55 32.9	55.8			
A 12	101 18.6	32 00.9 N20	39.5	13 26.6 N23	59.2	102 53.8 S 2	04.8	70 35.2 N 9	55.9	Alphecca	126 20.5	N26 43.5
Y 13	116 21.0	47 00.2	40.0	28 27.2	59.3	117 56.2	04.7	85 37.5	55.9	Alpheratz	357 55.3	N29 04.7
14	131 23.5	61 59.5	40.5	43 27.8	59.3	132 58.5	04.7	100 39.8	56.0	Altair	62 19.1	N 8 52.0
15	146 26.0	76 58.9 ..	41.0	58 28.4 ..	59.4	148 00.9 ..	04.7	115 42.1 ..	56.0	Ankaa	353 27.0	S42 18.7
16	161 28.4	91 58.2	41.5	73 29.0	59.4	163 03.2	04.6	130 44.4	56.1	Antares	112 40.1	S26 25.6
17	176 30.9	106 57.5	42.0	88 29.6	59.4	178 05.6	04.6	145 46.7	56.1			
18	191 33.4	121 56.8 N20	42.5	103 30.2 N23	59.5	193 07.9 S 2	04.6	160 49.0 N 9	56.2	Arcturus	146 06.1	N19 11.7
19	206 35.8	136 56.2	43.0	118 30.8	59.5	208 10.3	04.5	175 51.3	56.2	Atria	107 51.7	S69 01.5
20	221 38.3	151 55.5	43.5	133 31.4	59.6	223 12.6	04.5	190 53.6	56.3	Avior	234 23.3	S59 30.5
21	236 40.8	166 54.8 ..	44.0	148 32.0 ..	59.6	238 15.0 ..	04.5	205 55.9 ..	56.3	Bellatrix	278 44.7	N 6 20.8
22	251 43.2	181 54.1	44.5	163 32.6	59.7	253 17.3	04.4	220 58.2	56.4	Betelgeuse	271 14.1	N 7 24.3
23	266 45.7	196 53.4	45.0	178 33.2	59.7	268 19.7	04.4	236 00.5	56.4			
4 00	281 48.1	211 52.8 N20	45.5	193 33.8 N23	59.8	283 22.1 S 2	04.4	251 02.8 N 9	56.5	Canopus	264 01.8	S52 41.8
01	296 50.6	226 52.1	46.0	208 34.4	59.8	298 24.4	04.3	266 05.1	56.5	Capella	280 51.9	N45 59.6
02	311 53.1	241 51.4	46.5	223 35.1	59.9	313 26.8	04.3	281 07.4	56.6	Deneb	49 38.8	N45 16.5
03	326 55.5	256 50.7 ..	47.0	238 35.7 ..	59.9	328 29.1 ..	04.3	296 09.7 ..	56.6	Denebola	182 45.5	N14 35.0
04	341 58.0	271 50.1	47.5	253 36.3 23	59.9	343 31.5	04.2	311 12.0	56.7	Diphda	349 07.4	S17 59.6
05	357 00.5	286 49.4	48.0	268 36.9 24	00.0	358 33.8	04.2	326 14.3	56.7			
06	12 02.9	301 48.7 N20	48.5	283 37.5 N24	00.0	13 36.2 S 2	04.2	341 16.6 N 9	56.8	Dubhe	194 06.2	N61 45.8
07	27 05.4	316 48.0	49.0	298 38.1	00.1	28 38.6	04.1	356 18.9	56.8	Elnath	278 27.5	N28 36.2
S 08	42 07.9	331 47.3	49.5	313 38.7	00.1	43 40.9	04.1	11 21.2	56.9	Eltanin	90 51.0	N51 29.6
A 09	57 10.3	346 46.6 ..	50.0	328 39.3 ..	00.1	58 43.3 ..	04.1	26 23.5 ..	56.9	Enif	33 58.2	N 9 52.1
T 10	72 12.8	1 46.0	50.5	343 39.9	00.2	73 45.6	04.0	41 25.8	57.0	Fomalhaut	15 36.5	S29 37.6
U 11	87 15.3	16 45.3	50.9	358 40.5	00.2	88 48.0	04.0	56 28.1	57.0			
R 12	102 17.7	31 44.6 N20	51.4	13 41.1 N24	00.3	103 50.3 S 2	04.0	71 30.4 N 9	57.1	Gacrux	172 13.7	S57 06.5
D 13	117 20.2	46 43.9	51.9	28 41.7	00.3	118 52.7	03.9	86 32.7	57.1	Gienah	176 04.2	S17 32.0
A 14	132 22.6	61 43.2	52.4	43 42.4	00.3	133 55.1	03.9	101 35.0	57.2	Hadar	149 04.0	S60 22.1
Y 15	147 25.1	76 42.5 ..	52.9	58 43.0 ..	00.4	148 57.4 ..	03.9	116 37.2 ..	57.2	Hamal	328 13.9	N23 27.1
16	162 27.6	91 41.9	53.4	73 43.6	00.4	163 59.8	03.9	131 39.5	57.3	Kaus Aust.	83 58.7	S34 23.0
17	177 30.0	106 41.2	53.9	88 44.2	00.5	179 02.1	03.8	146 41.8	57.3			
18	192 32.5	121 40.5 N20	54.3	103 44.8 N24	00.5	194 04.5 S 2	03.8	161 44.1 N 9	57.4	Kochab	137 19.1	N74 10.1
19	207 35.0	136 39.8	54.8	118 45.4	00.5	209 06.9	03.8	176 46.4	57.4	Markab	13 49.6	N15 11.8
20	222 37.4	151 39.1	55.3	133 46.0	00.6	224 09.2	03.7	191 48.7	57.5	Menkar	314 27.3	N 4 04.9
21	237 39.9	166 38.4 ..	55.8	148 46.6 ..	00.6	239 11.6 ..	03.7	206 51.0 ..	57.5	Menkent	148 21.0	S36 21.8
22	252 42.4	181 37.7	56.3	163 47.2	00.6	254 14.0	03.7	221 53.3	57.6	Miaplacidus	221 42.7	S69 42.9
23	267 44.8	196 37.1	56.7	178 47.8	00.7	269 16.3	03.6	236 55.6	57.6			
5 00	282 47.3	211 36.4 N20	57.2	193 48.4 N24	00.7	284 18.7 S 2	03.6	251 57.9 N 9	57.6	Mirfak	308 57.1	N49 51.0
01	297 49.8	226 35.7	57.7	208 49.1	00.8	299 21.0	03.6	267 00.2	57.7	Nunki	76 12.2	S26 17.8
02	312 52.2	241 35.0	58.2	223 49.7	00.8	314 23.4	03.5	282 02.5	57.7	Peacock	53 36.8	S56 44.2
03	327 54.7	256 34.3 ..	58.6	238 50.3 ..	00.8	329 25.8 ..	03.5	297 04.8 ..	57.8	Pollux	243 42.2	N28 01.7
04	342 57.1	271 33.6	59.1	253 50.9	00.9	344 28.1	03.5	312 07.1	57.8	Procyon	245 12.1	N 5 13.7
05	357 59.6	286 32.9 20	59.6	268 51.5	00.9	359 30.5	03.5	327 09.4	57.9			
06	13 02.1	301 32.2 N21	00.1	283 52.1 N24	00.9	14 32.9 S 2	03.4	342 11.7 N 9	57.9	Rasalhague	96 16.8	N12 33.9
07	28 04.5	316 31.5	00.5	298 52.7	01.0	29 35.2	03.4	357 14.0	58.0	Regulus	207 56.0	N11 58.5
S 08	43 07.0	331 30.8	01.0	313 53.3	01.0	44 37.6	03.4	12 16.3	58.0	Rigel	281 23.4	S 8 12.3
U 09	58 09.5	346 30.2 ..	01.5	328 53.9 ..	01.0	59 40.0 ..	03.3	27 18.6 ..	58.1	Rigil Kent.	140 07.1	S60 49.8
N 10	73 11.9	1 29.5	01.9	343 54.5	01.0	74 42.3	03.3	42 20.9	58.1	Sabik	102 25.4	S15 43.2
11	88 14.4	16 28.8	02.4	358 55.2	01.1	89 44.7	03.3	57 23.2	58.2			
D 12	103 16.9	31 28.1 N21	02.9	13 55.8 N24	01.1	104 47.1 S 2	03.3	72 25.5 N 9	58.2	Schedar	349 53.6	N56 31.4
A 13	118 19.3	46 27.4	03.3	28 56.4	01.1	119 49.4	03.2	87 27.8	58.3	Shaula	96 37.1	S37 06.1
Y 14	133 21.8	61 26.7	03.8	43 57.0	01.2	134 51.8	03.2	102 30.1	58.3	Sirius	258 44.2	S16 42.9
15	148 24.2	76 26.0 ..	04.3	58 57.6 ..	01.2	149 54.2 ..	03.2	117 32.5 ..	58.4	Spica	158 43.3	S11 09.1
16	163 26.7	91 25.3	04.7	73 58.2	01.2	164 56.5	03.1	132 34.8	58.4	Suhail	223 01.2	S43 25.8
17	178 29.2	106 24.6	05.2	88 58.8	01.3	179 58.9	03.1	147 37.1	58.5			
18	193 31.6	121 23.9 N21	05.6	103 59.4 N24	01.3	195 01.3 S 2	03.1	162 39.4 N 9	58.5	Vega	80 46.3	N38 47.1
19	208 34.1	136 23.2	06.1	119 00.0	01.3	210 03.6	03.1	177 41.7	58.6	Zuben'ubi	137 18.0	S16 02.1
20	223 36.6	151 22.5	06.6	134 00.7	01.3	225 06.0	03.0	192 44.0	58.6		SHA	Mer.Pass.
21	238 39.0	166 21.8 ..	07.0	149 01.3 ..	01.4	240 08.4 ..	03.0	207 46.3 ..	58.7	Venus	290 04.6	9 53
22	253 41.5	181 21.1	07.5	164 01.9	01.4	255 10.7	03.0	222 48.6	58.7	Mars	271 45.7	11 05
23	268 44.0	196 20.4	07.9	179 02.5	01.4	270 13.1	02.9	237 50.9	58.8	Jupiter	1 33.9	5 06
Mer. Pass. 5 11.9		v −0.7 d 0.5		v 0.6 d 0.0		v 2.4 d 0.0		v 2.3 d 0.0		Saturn	329 14.6	7 15

1998 JULY 3, 4, 5 (FRI., SAT., SUN.)

UT	SUN GHA	SUN Dec	MOON GHA	v	Dec	d	HP
d h	° ′	° ′	° ′	′	° ′	′	′
3 00	178 58.4	N22 59.4	76 56.6	15.2	S 5 39.0	9.2	54.3
01	193 58.3	59.3	91 30.8	15.2	5 48.2	9.2	54.3
02	208 58.2	59.1	106 05.0	15.1	5 57.4	9.2	54.3
03	223 58.1	.. 58.9	120 39.1	15.1	6 06.6	9.2	54.3
04	238 58.0	58.7	135 13.2	15.1	6 15.8	9.1	54.3
05	253 57.9	58.5	149 47.3	15.1	6 24.9	9.1	54.3
06	268 57.7	N22 58.3	164 21.4	15.0	S 6 34.0	9.1	54.3
07	283 57.6	58.1	178 55.4	15.1	6 43.1	9.1	54.3
08	298 57.5	57.9	193 29.5	15.0	6 52.2	9.0	54.3
F 09	313 57.4	.. 57.7	208 03.5	15.0	7 01.2	9.0	54.3
R 10	328 57.3	57.5	222 37.5	14.9	7 10.2	9.0	54.3
I 11	343 57.2	57.3	237 11.4	14.9	7 19.2	9.0	54.3
D 12	358 57.1	N22 57.1	251 45.3	14.9	S 7 28.2	8.9	54.3
A 13	13 56.9	56.9	266 19.2	14.9	7 37.1	8.9	54.3
Y 14	28 56.8	56.7	280 53.1	14.9	7 46.0	8.9	54.3
15	43 56.7	.. 56.5	295 27.0	14.8	7 54.9	8.8	54.3
16	58 56.6	56.2	310 00.8	14.8	8 03.7	8.8	54.3
17	73 56.5	56.0	324 34.6	14.7	8 12.5	8.8	54.3
18	88 56.4	N22 55.8	339 08.3	14.8	S 8 21.3	8.7	54.3
19	103 56.3	55.6	353 42.1	14.7	8 30.0	8.7	54.3
20	118 56.1	55.4	8 15.8	14.6	8 38.7	8.7	54.3
21	133 56.0	.. 55.2	22 49.4	14.7	8 47.4	8.7	54.4
22	148 55.9	55.0	37 23.1	14.6	8 56.1	8.6	54.4
23	163 55.8	54.8	51 56.7	14.5	9 04.7	8.6	54.4
4 00	178 55.7	N22 54.6	66 30.2	14.6	S 9 13.3	8.5	54.4
01	193 55.6	54.4	81 03.8	14.5	9 21.8	8.5	54.4
02	208 55.5	54.2	95 37.3	14.4	9 30.3	8.5	54.4
03	223 55.4	.. 53.9	110 10.7	14.5	9 38.8	8.4	54.4
04	238 55.2	53.7	124 44.2	14.3	9 47.2	8.4	54.4
05	253 55.1	53.5	139 17.5	14.4	9 55.6	8.4	54.4
06	268 55.0	N22 53.3	153 50.9	14.3	S10 04.0	8.3	54.4
S 07	283 54.9	53.1	168 24.2	14.3	10 12.3	8.3	54.4
A 08	298 54.8	52.9	182 57.5	14.2	10 20.6	8.2	54.5
T 09	313 54.7	.. 52.6	197 30.7	14.2	10 28.8	8.2	54.5
U 10	328 54.6	52.4	212 03.9	14.2	10 37.0	8.2	54.5
R 11	343 54.5	52.2	226 37.1	14.1	10 45.2	8.1	54.5
D 12	358 54.4	N22 52.0	241 10.2	14.1	S10 53.3	8.1	54.5
A 13	13 54.2	51.8	255 43.3	14.0	11 01.4	8.0	54.5
Y 14	28 54.1	51.6	270 16.3	14.0	11 09.4	8.0	54.5
15	43 54.0	.. 51.3	284 49.3	13.9	11 17.4	7.9	54.5
16	58 53.9	51.1	299 22.2	13.9	11 25.3	7.9	54.6
17	73 53.8	50.9	313 55.1	13.9	11 33.2	7.9	54.6
18	88 53.7	N22 50.7	328 28.0	13.8	S11 41.1	7.8	54.6
19	103 53.6	50.4	343 00.8	13.8	11 48.9	7.8	54.6
20	118 53.5	50.2	357 33.6	13.7	11 56.7	7.7	54.6
21	133 53.4	.. 50.0	12 06.3	13.7	12 04.4	7.7	54.6
22	148 53.2	49.8	26 39.0	13.6	12 12.1	7.6	54.6
23	163 53.1	49.5	41 11.6	13.6	12 19.7	7.5	54.6
5 00	178 53.0	N22 49.3	55 44.2	13.5	S12 27.2	7.6	54.7
01	193 52.9	49.1	70 16.7	13.5	12 34.8	7.4	54.7
02	208 52.8	48.9	84 49.2	13.5	12 42.2	7.4	54.7
03	223 52.7	.. 48.6	99 21.7	13.4	12 49.6	7.4	54.7
04	238 52.6	48.4	113 54.1	13.3	12 57.0	7.3	54.7
05	253 52.5	48.2	128 26.4	13.3	13 04.3	7.3	54.7
06	268 52.4	N22 47.9	142 58.7	13.3	S13 11.6	7.2	54.8
07	283 52.3	47.7	157 31.0	13.1	13 18.8	7.1	54.8
08	298 52.2	47.5	172 03.1	13.2	13 25.9	7.1	54.8
S 09	313 52.1	.. 47.2	186 35.3	13.1	13 33.0	7.1	54.8
U 10	328 51.9	47.0	201 07.4	13.0	13 40.1	7.0	54.8
N 11	343 51.8	46.8	215 39.4	13.0	13 47.1	6.9	54.8
D 12	358 51.7	N22 46.5	230 11.4	13.0	S13 54.0	6.9	54.9
A 13	13 51.6	46.3	244 43.4	12.8	14 00.9	6.8	54.9
Y 14	28 51.5	46.1	259 15.2	12.9	14 07.7	6.7	54.9
15	43 51.4	.. 45.8	273 47.1	12.8	14 14.4	6.7	54.9
16	58 51.3	45.6	288 18.9	12.7	14 21.1	6.6	54.9
17	73 51.2	45.3	302 50.6	12.7	14 27.7	6.6	54.9
18	88 51.1	N22 45.1	317 22.3	12.6	S14 34.3	6.5	55.0
19	103 51.0	44.9	331 53.9	12.5	14 40.8	6.5	55.0
20	118 50.9	44.6	346 25.4	12.6	14 47.3	6.3	55.0
21	133 50.8	.. 44.4	0 57.0	12.4	14 53.6	6.3	55.0
22	148 50.7	44.1	15 28.4	12.4	14 59.9	6.3	55.0
23	163 50.6	43.9	29 59.8	12.4	S15 06.2	6.2	55.1
	SD 15.8	d 0.2	SD 14.8		14.9		14.9

Moonrise

Lat.	Naut.	Civil	Sunrise	3	4	5	6
°	h m	h m	h m	h m	h m	h m	h m
N 72	□	□	□	15 01	16 41	18 27	20 30
N 70	□	□	□	14 49	16 20	17 54	19 30
68	□	□	□	14 39	16 04	17 30	18 56
66	////	////	00 40	14 30	15 51	17 12	18 32
64	////	////	01 46	14 23	15 40	16 57	18 13
62	////	////	02 21	14 18	15 31	16 45	17 57
60	////	01 10	02 45	14 12	15 24	16 34	17 44
N 58	////	01 53	03 05	14 08	15 17	16 25	17 33
56	////	02 20	03 21	14 04	15 11	16 17	17 23
54	01 05	02 41	03 34	14 00	15 05	16 10	17 14
52	01 44	02 58	03 46	13 57	15 01	16 04	17 07
50	02 09	03 13	03 57	13 54	14 56	15 58	17 00
45	02 53	03 42	04 19	13 48	14 47	15 46	16 45
N 40	03 22	04 04	04 36	13 43	14 39	15 36	16 33
35	03 45	04 21	04 51	13 38	14 32	15 27	16 23
30	04 03	04 36	05 04	13 34	14 26	15 20	16 14
20	04 32	05 01	05 25	13 27	14 16	15 07	15 58
N 10	04 54	05 21	05 44	13 21	14 08	14 55	15 45
0	05 12	05 38	06 01	13 16	13 59	14 45	15 33
S 10	05 29	05 55	06 18	13 10	13 51	14 34	15 20
20	05 44	06 12	06 36	13 04	13 42	14 23	15 07
30	06 00	06 30	06 56	12 57	13 33	14 10	14 52
35	06 09	06 40	07 08	12 54	13 27	14 03	14 43
40	06 18	06 52	07 22	12 49	13 20	13 55	14 33
45	06 28	07 05	07 39	12 44	13 13	13 45	14 21
S 50	06 39	07 21	07 59	12 38	13 04	13 33	14 07
52	06 44	07 28	08 08	12 36	13 00	13 28	14 01
54	06 50	07 36	08 19	12 33	12 56	13 22	13 53
56	06 56	07 44	08 31	12 29	12 51	13 15	13 45
58	07 02	07 54	08 45	12 26	12 45	13 08	13 36
S 60	07 09	08 05	09 02	12 21	12 39	13 00	13 26

Moonset

Lat.	Sunset	Civil	Naut.	3	4	5	6
°	h m	h m	h m	h m	h m	h m	h m
N 72	□	□	□	23 33	23 24	23 13	22 51
N 70	□	□	□	23 47	23 46	23 47	23 50
68	□	□	□	23 58	24 03	00 03	00 11
66	23 23	////	////	00 00	00 08	00 17	00 30
64	22 21	////	////	00 05	00 15	00 29	00 46
62	21 47	////	////	00 09	00 22	00 38	00 59
60	21 23	22 56	////	00 12	00 28	00 47	01 10
N 58	21 03	22 15	////	00 15	00 33	00 54	01 19
56	20 47	21 48	////	00 18	00 38	01 01	01 28
54	20 34	21 27	23 02	00 21	00 42	01 07	01 35
52	20 22	21 10	22 24	00 23	00 46	01 12	01 42
50	20 11	20 55	21 58	00 25	00 50	01 17	01 48
45	19 50	20 27	21 15	00 30	00 57	01 27	02 01
N 40	19 32	20 05	20 46	00 33	01 04	01 36	02 12
35	19 18	19 47	20 24	00 37	01 09	01 44	02 21
30	19 05	19 32	20 05	00 40	01 14	01 50	02 29
20	18 44	19 08	19 37	00 45	01 22	02 02	02 44
N 10	18 25	18 48	19 15	00 49	01 30	02 12	02 56
0	18 08	18 30	18 57	00 53	01 36	02 21	03 07
S 10	17 51	18 14	18 40	00 58	01 43	02 30	03 19
20	17 33	17 57	18 24	01 02	01 51	02 40	03 32
30	17 12	17 39	18 08	01 07	01 59	02 52	03 46
35	17 00	17 28	18 00	01 10	02 04	02 59	03 54
40	16 47	17 17	17 51	01 14	02 10	03 06	04 03
45	16 30	17 04	17 41	01 17	02 16	03 15	04 14
S 50	16 10	16 48	17 30	01 22	02 24	03 26	04 28
52	16 01	16 41	17 25	01 24	02 28	03 31	04 34
54	15 50	16 33	17 19	01 27	02 32	03 36	04 41
56	15 38	16 25	17 13	01 29	02 36	03 43	04 49
58	15 24	16 15	17 07	01 32	02 41	03 49	04 57
S 60	15 07	16 04	17 00	01 35	02 46	03 57	05 07

	SUN			MOON			
Day	Eqn. of Time 00h	12h	Mer. Pass.	Mer. Pass. Upper	Lower	Age	Phase
d	m s	m s	h m	h m	h m	d	%
3	04 06	04 12	12 04	19 26	07 04	09	66
4	04 17	04 22	12 04	20 10	07 48	10	75
5	04 28	04 33	12 05	20 56	08 33	11	83

1998 JULY 6, 7, 8 (MON., TUES., WED.)

UT	ARIES GHA	VENUS −3.9 GHA	Dec	MARS +1.6 GHA	Dec	JUPITER −2.6 GHA	Dec	SATURN +0.5 GHA	Dec
6 00	283 46.4	211 19.7	N21 08.4	194 03.1	N24 01.5	285 15.5	S 2 02.9	252 53.2	N 9 58.8
01	298 48.9	226 19.0	08.8	209 03.7	01.5	300 17.8	02.9	267 55.5	58.9
02	313 51.4	241 18.3	09.3	224 04.3	01.5	315 20.2	02.9	282 57.8	58.9
03	328 53.8	256 17.6 ..	09.7	239 04.9 ..	01.5	330 22.6 ..	02.8	298 00.1 ..	58.9
04	343 56.3	271 16.9	10.2	254 05.5	01.6	345 25.0	02.8	313 02.4	59.0
05	358 58.7	286 16.2	10.6	269 06.2	01.6	0 27.3	02.8	328 04.7	59.0
06	14 01.2	301 15.5	N21 11.1	284 06.8	N24 01.6	15 29.7	S 2 02.8	343 07.0	N 9 59.1
07	29 03.7	316 14.8	11.5	299 07.4	01.6	30 32.1	02.7	358 09.3	59.1
M 08	44 06.1	331 14.1	12.0	314 08.0	01.7	45 34.4	02.7	13 11.6	59.2
O 09	59 08.6	346 13.4 ..	12.4	329 08.6 ..	01.7	60 36.8 ..	02.7	28 13.9 ..	59.2
N 10	74 11.1	1 12.7	12.9	344 09.2	01.7	75 39.2	02.7	43 16.2	59.3
D 11	89 13.5	16 12.0	13.3	359 09.8	01.7	90 41.6	02.6	58 18.5	59.3
A 12	104 16.0	31 11.3	N21 13.8	14 10.4	N24 01.7	105 43.9	S 2 02.6	73 20.8	N 9 59.4
Y 13	119 18.5	46 10.6	14.2	29 11.1	01.8	120 46.3	02.6	88 23.1	59.4
14	134 20.9	61 09.9	14.6	44 11.7	01.8	135 48.7	02.6	103 25.4	59.5
15	149 23.4	76 09.2 ..	15.1	59 12.3 ..	01.8	150 51.1 ..	02.5	118 27.7 ..	59.5
16	164 25.9	91 08.5	15.5	74 12.9	01.8	165 53.4	02.5	133 30.0	59.6
17	179 28.3	106 07.8	16.0	89 13.5	01.8	180 55.8	02.5	148 32.3	59.6
18	194 30.8	121 07.1	N21 16.4	104 14.1	N24 01.9	195 58.2	S 2 02.5	163 34.6	N 9 59.7
19	209 33.2	136 06.4	16.8	119 14.7	01.9	211 00.6	02.4	178 37.0	59.7
20	224 35.7	151 05.7	17.3	134 15.3	01.9	226 02.9	02.4	193 39.3	59.7
21	239 38.2	166 05.0 ..	17.7	149 16.0 ..	01.9	241 05.3 ..	02.4	208 41.6 ..	59.8
22	254 40.6	181 04.3	18.1	164 16.6	01.9	256 07.7	02.4	223 43.9	59.8
23	269 43.1	196 03.6	18.6	179 17.2	02.0	271 10.1	02.3	238 46.2	59.9
7 00	284 45.6	211 02.8	N21 19.0	194 17.8	N24 02.0	286 12.4	S 2 02.3	253 48.5	N 9 59.9
01	299 48.0	226 02.1	19.4	209 18.4	02.0	301 14.8	02.3	268 50.8	10 00.0
02	314 50.5	241 01.4	19.8	224 19.0	02.0	316 17.2	02.3	283 53.1	00.0
03	329 53.0	256 00.7 ..	20.3	239 19.6 ..	02.0	331 19.6 ..	02.2	298 55.4 ..	00.1
04	344 55.4	271 00.0	20.7	254 20.2	02.0	346 22.0	02.2	313 57.7	00.1
05	359 57.9	285 59.3	21.1	269 20.9	02.1	1 24.3	02.2	329 00.0	00.2
06	15 00.4	300 58.6	N21 21.6	284 21.5	N24 02.1	16 26.7	S 2 02.2	344 02.3	N10 00.2
07	30 02.8	315 57.9	22.0	299 22.1	02.1	31 29.1	02.1	359 04.6	00.3
T 08	45 05.3	330 57.2	22.4	314 22.7	02.1	46 31.5	02.1	14 06.9	00.3
U 09	60 07.7	345 56.5 ..	22.8	329 23.3 ..	02.1	61 33.9 ..	02.1	29 09.2 ..	00.3
E 10	75 10.2	0 55.7	23.2	344 23.9	02.1	76 36.2	02.1	44 11.5	00.4
S 11	90 12.7	15 55.0	23.7	359 24.5	02.2	91 38.6	02.1	59 13.9	00.4
D 12	105 15.1	30 54.3	N21 24.1	14 25.2	N24 02.2	106 41.0	S 2 02.0	74 16.2	N10 00.5
A 13	120 17.6	45 53.6	24.5	29 25.8	02.2	121 43.4	02.0	89 18.5	00.5
Y 14	135 20.1	60 52.9	24.9	44 26.4	02.2	136 45.8	02.0	104 20.8	00.6
15	150 22.5	75 52.2 ..	25.3	59 27.0 ..	02.2	151 48.1 ..	02.0	119 23.1 ..	00.6
16	165 25.0	90 51.5	25.8	74 27.6	02.2	166 50.5	01.9	134 25.4	00.7
17	180 27.5	105 50.8	26.2	89 28.2	02.2	181 52.9	01.9	149 27.7	00.7
18	195 29.9	120 50.0	N21 26.6	104 28.8	N24 02.2	196 55.3	S 2 01.9	164 30.0	N10 00.8
19	210 32.4	135 49.3	27.0	119 29.5	02.3	211 57.7	01.9	179 32.3	00.8
20	225 34.8	150 48.6	27.4	134 30.1	02.3	227 00.1	01.9	194 34.6	00.8
21	240 37.3	165 47.9 ..	27.8	149 30.7 ..	02.3	242 02.4 ..	01.8	209 36.9 ..	00.9
22	255 39.8	180 47.2	28.2	164 31.3	02.3	257 04.8	01.8	224 39.2	00.9
23	270 42.2	195 46.5	28.6	179 31.9	02.3	272 07.2	01.8	239 41.6	01.0
8 00	285 44.7	210 45.7	N21 29.0	194 32.5	N24 02.3	287 09.6	S 2 01.8	254 43.9	N10 01.0
01	300 47.2	225 45.0	29.4	209 33.1	02.3	302 12.0	01.8	269 46.2	01.1
02	315 49.6	240 44.3	29.8	224 33.8	02.3	317 14.4	01.7	284 48.5	01.1
03	330 52.1	255 43.6 ..	30.3	239 34.4 ..	02.3	332 16.7 ..	01.7	299 50.8 ..	01.2
04	345 54.6	270 42.9	30.7	254 35.0	02.3	347 19.1	01.7	314 53.1	01.2
05	0 57.0	285 42.1	31.1	269 35.6	02.3	2 21.5	01.7	329 55.4	01.3
06	15 59.5	300 41.4	N21 31.5	284 36.2	N24 02.3	17 23.9	S 2 01.7	344 57.7	N10 01.3
W 07	31 02.0	315 40.7	31.9	299 36.8	02.4	32 26.3	01.6	0 00.0	01.3
E 08	46 04.4	330 40.0	32.3	314 37.5	02.4	47 28.7	01.6	15 02.3	01.4
D 09	61 06.9	345 39.3 ..	32.7	329 38.1 ..	02.4	62 31.1 ..	01.6	30 04.7 ..	01.4
N 10	76 09.3	0 38.5	33.1	344 38.7	02.4	77 33.4	01.6	45 07.0	01.5
E 11	91 11.8	15 37.8	33.5	359 39.3	02.4	92 35.8	01.6	60 09.3	01.5
S 12	106 14.3	30 37.1	N21 33.8	14 39.9	N24 02.4	107 38.2	S 2 01.5	75 11.6	N10 01.6
D 13	121 16.7	45 36.4	34.2	29 40.5	02.4	122 40.6	01.5	90 13.9	01.6
A 14	136 19.2	60 35.7	34.6	44 41.1	02.4	137 43.0	01.5	105 16.2	01.7
Y 15	151 21.7	75 34.9 ..	35.0	59 41.8 ..	02.4	152 45.4 ..	01.5	120 18.5 ..	01.7
16	166 24.1	90 34.2	35.4	74 42.4	02.4	167 47.8	01.5	135 20.8	01.7
17	181 26.6	105 33.5	35.8	89 43.0	02.4	182 50.2	01.4	150 23.2	01.8
18	196 29.1	120 32.8	N21 36.2	104 43.6	N24 02.4	197 52.6	S 2 01.4	165 25.5	N10 01.8
19	211 31.5	135 32.0	36.6	119 44.2	02.4	212 54.9	01.4	180 27.8	01.9
20	226 34.0	150 31.3	37.0	134 44.8	02.4	227 57.3	01.4	195 30.1	01.9
21	241 36.5	165 30.6 ..	37.4	149 45.5 ..	02.4	242 59.7 ..	01.4	210 32.4 ..	02.0
22	256 38.9	180 29.9	37.7	164 46.1	02.4	258 02.1	01.3	225 34.7	02.0
23	271 41.4	195 29.1	38.1	179 46.7	02.4	273 04.5	01.3	240 37.0	02.0
Mer. Pass. 5 00.1		v −0.7 d 0.4		v 0.6 d 0.0		v 2.4 d 0.0		v 2.3 d 0.0	

STARS

Name	SHA	Dec
Acamar	315 27.3	S40 18.6
Achernar	335 35.5	S57 14.5
Acrux	173 22.2	S63 05.7
Adhara	255 21.9	S28 58.3
Aldebaran	291 02.9	N16 30.2
Alioth	166 30.9	N55 58.5
Alkaid	153 07.9	N49 19.6
Al Na'ir	27 57.8	S46 57.9
Alnilam	275 58.4	S 1 12.3
Alphard	218 07.6	S 8 39.2
Alphecca	126 20.5	N26 43.5
Alpheratz	357 55.3	N29 04.8
Altair	62 19.1	N 8 52.0
Ankaa	353 27.0	S42 18.7
Antares	112 40.1	S26 25.6
Arcturus	146 06.1	N19 11.7
Atria	107 51.7	S69 01.5
Avior	234 23.3	S59 30.5
Bellatrix	278 44.7	N 6 20.8
Betelgeuse	271 14.1	N 7 24.3
Canopus	264 01.8	S52 41.8
Capella	280 51.9	N45 59.6
Deneb	49 38.8	N45 16.5
Denebola	182 45.5	N14 35.0
Diphda	349 07.4	S17 59.6
Dubhe	194 06.2	N61 45.8
Elnath	278 27.5	N28 36.2
Eltanin	90 51.0	N51 29.6
Enif	33 58.1	N 9 52.1
Fomalhaut	15 36.5	S29 37.6
Gacrux	172 13.7	S57 06.5
Gienah	176 04.2	S17 32.0
Hadar	149 04.0	S60 22.1
Hamal	328 13.8	N23 27.1
Kaus Aust.	83 58.7	S34 23.0
Kochab	137 19.2	N74 10.1
Markab	13 49.6	N15 11.8
Menkar	314 27.3	N 4 04.9
Menkent	148 21.0	S36 21.8
Miaplacidus	221 42.8	S69 42.9
Mirfak	308 57.1	N49 51.0
Nunki	76 12.2	S26 17.8
Peacock	53 36.8	S56 44.2
Pollux	243 42.2	N28 01.7
Procyon	245 12.1	N 5 13.7
Rasalhague	96 16.8	N12 33.9
Regulus	207 56.0	N11 58.5
Rigel	281 23.4	S 8 12.3
Rigil Kent.	140 07.2	S60 49.8
Sabik	102 25.4	S15 43.2
Schedar	349 53.5	N56 31.4
Shaula	96 37.1	S37 06.1
Sirius	258 44.2	S16 42.9
Spica	158 43.3	S11 09.1
Suhail	223 01.2	S43 25.8
Vega	80 46.3	N38 47.1
Zuben'ubi	137 18.0	S16 02.1

	SHA	Mer. Pass.
Venus	286 17.3	9 56
Mars	269 32.2	11 02
Jupiter	1 26.9	4 54
Saturn	329 02.9	7 04

UT	SUN GHA	SUN Dec	MOON GHA	v	MOON Dec	d	HP
6 MONDAY							
00	178 50.5	N22 43.6	44 31.2	12.3	S15 12.4	6.1	55.1
01	193 50.3	43.4	59 02.5	12.2	15 18.5	6.1	55.1
02	208 50.2	43.2	73 33.7	12.2	15 24.6	5.9	55.1
03	223 50.1	.. 42.9	88 04.9	12.1	15 30.5	6.0	55.1
04	238 50.0	42.7	102 36.0	12.1	15 36.5	5.8	55.2
05	253 49.9	42.4	117 07.1	12.0	15 42.3	5.8	55.2
06	268 49.8	N22 42.2	131 38.1	12.0	S15 48.1	5.7	55.2
07	283 49.7	41.9	146 09.1	11.9	15 53.8	5.6	55.2
08	298 49.6	41.7	160 40.0	11.8	15 59.4	5.6	55.2
09	313 49.5	.. 41.4	175 10.8	11.8	16 05.0	5.5	55.3
10	328 49.4	41.2	189 41.6	11.7	16 10.5	5.4	55.3
11	343 49.3	40.9	204 12.3	11.7	16 15.9	5.3	55.3
12	358 49.2	N22 40.7	218 43.0	11.6	S16 21.2	5.3	55.3
13	13 49.1	40.4	233 13.6	11.6	16 26.5	5.2	55.3
14	28 49.0	40.2	247 44.2	11.5	16 31.7	5.1	55.4
15	43 48.9	.. 39.9	262 14.7	11.5	16 36.8	5.1	55.4
16	58 48.8	39.7	276 45.2	11.4	16 41.9	4.9	55.4
17	73 48.7	39.4	291 15.6	11.3	16 46.8	4.9	55.4
18	88 48.6	N22 39.1	305 45.9	11.3	S16 51.7	4.8	55.5
19	103 48.5	38.9	320 16.2	11.3	16 56.5	4.7	55.5
20	118 48.4	38.6	334 46.5	11.1	17 01.2	4.7	55.5
21	133 48.3	.. 38.4	349 16.6	11.2	17 05.9	4.5	55.5
22	148 48.2	38.1	3 46.8	11.0	17 10.4	4.5	55.5
23	163 48.1	37.8	18 16.8	11.0	17 14.9	4.4	55.6
7 TUESDAY							
00	178 48.0	N22 37.6	32 46.8	11.0	S17 19.3	4.3	55.6
01	193 47.9	37.3	47 16.8	10.9	17 23.6	4.3	55.6
02	208 47.8	37.1	61 46.7	10.9	17 27.9	4.1	55.6
03	223 47.7	.. 36.8	76 16.6	10.7	17 32.0	4.1	55.7
04	238 47.6	36.5	90 46.3	10.8	17 36.1	4.0	55.7
05	253 47.5	36.3	105 16.1	10.7	17 40.1	3.9	55.7
06	268 47.4	N22 36.0	119 45.8	10.6	S17 44.0	3.8	55.7
07	283 47.3	35.7	134 15.4	10.6	17 47.8	3.7	55.8
08	298 47.2	35.5	148 45.0	10.5	17 51.5	3.6	55.8
09	313 47.1	.. 35.2	163 14.5	10.5	17 55.1	3.6	55.8
10	328 47.0	34.9	177 44.0	10.4	17 58.7	3.4	55.8
11	343 46.9	34.7	192 13.4	10.4	18 02.1	3.4	55.9
12	358 46.8	N22 34.4	206 42.8	10.3	S18 05.5	3.2	55.9
13	13 46.7	34.1	221 12.1	10.3	18 08.7	3.2	55.9
14	28 46.6	33.9	235 41.4	10.2	18 11.9	3.1	55.9
15	43 46.5	.. 33.6	250 10.6	10.1	18 15.0	3.0	55.9
16	58 46.4	33.3	264 39.7	10.2	18 18.0	2.9	56.0
17	73 46.3	33.1	279 08.9	10.0	18 20.9	2.8	56.0
18	88 46.2	N22 32.8	293 37.9	10.0	S18 23.7	2.7	56.0
19	103 46.1	32.5	308 06.9	10.0	18 26.4	2.6	56.0
20	118 46.0	32.2	322 35.9	9.9	18 29.0	2.6	56.1
21	133 45.9	.. 32.0	337 04.8	9.9	18 31.6	2.4	56.1
22	148 45.8	31.7	351 33.7	9.8	18 34.0	2.3	56.1
23	163 45.7	31.4	6 02.5	9.8	18 36.3	2.2	56.1
8 WEDNESDAY							
00	178 45.6	N22 31.1	20 31.3	9.7	S18 38.5	2.2	56.2
01	193 45.5	30.9	35 00.0	9.7	18 40.7	2.0	56.2
02	208 45.4	30.6	49 28.7	9.6	18 42.7	1.9	56.2
03	223 45.3	.. 30.3	63 57.3	9.6	18 44.6	1.9	56.2
04	238 45.2	30.0	78 25.9	9.6	18 46.5	1.7	56.3
05	253 45.1	29.7	92 54.5	9.5	18 48.2	1.7	56.3
06	268 45.0	N22 29.5	107 23.0	9.5	S18 49.9	1.5	56.3
07	283 44.9	29.2	121 51.5	9.4	18 51.4	1.4	56.3
08	298 44.8	28.9	136 19.9	9.4	18 52.8	1.4	56.4
09	313 44.7	.. 28.6	150 48.3	9.3	18 54.2	1.2	56.4
10	328 44.6	28.3	165 16.6	9.3	18 55.4	1.1	56.4
11	343 44.5	28.1	179 44.9	9.3	18 56.5	1.1	56.4
12	358 44.4	N22 27.8	194 13.2	9.2	S18 57.6	0.9	56.5
13	13 44.3	27.5	208 41.4	9.2	18 58.5	0.8	56.5
14	28 44.2	27.2	223 09.6	9.1	18 59.3	0.7	56.5
15	43 44.1	.. 26.9	237 37.7	9.2	19 00.0	0.6	56.6
16	58 44.0	26.6	252 05.9	9.0	19 00.6	0.5	56.6
17	73 43.9	26.3	266 33.9	9.1	19 01.1	0.4	56.6
18	88 43.8	N22 26.0	281 02.0	9.0	S19 01.5	0.3	56.6
19	103 43.7	25.8	295 30.0	9.0	19 01.8	0.2	56.7
20	118 43.7	25.5	309 58.0	8.9	19 02.0	0.1	56.7
21	133 43.6	.. 25.2	324 25.9	8.9	19 02.1	0.2	56.7
22	148 43.5	24.9	338 53.8	8.9	19 02.1	0.2	56.7
23	163 43.4	24.6	353 21.7	8.8	S19 01.9	0.2	56.8
SD	15.8	d 0.3	SD 15.1		15.2		15.4

Lat.	Twilight Naut.	Twilight Civil	Sunrise	Moonrise 6	7	8	9
N 72	☐	☐	☐	20 30	▬	▬	▬
N 70	☐	☐	☐	19 30	21 04	22 16	22 48
68	☐	☐	☐	18 56	20 16	21 20	22 02
66	////	////	00 56	18 32	19 45	20 47	21 32
64	////	////	01 53	18 13	19 23	20 23	21 10
62	////	////	02 26	17 57	19 04	20 04	20 52
60	////	01 19	02 49	17 44	18 49	19 48	20 37
N 58	////	01 58	03 08	17 33	18 37	19 34	20 24
56	////	02 24	03 24	17 23	18 26	19 23	20 13
54	01 12	02 45	03 37	17 14	18 16	19 13	20 03
52	01 49	03 02	03 49	17 07	18 07	19 04	19 55
50	02 13	03 16	03 59	17 00	18 00	18 56	19 47
45	02 56	03 44	04 21	16 45	17 43	18 39	19 30
N 40	03 25	04 06	04 38	16 33	17 30	18 25	19 17
35	03 47	04 23	04 52	16 23	17 18	18 13	19 05
30	04 05	04 38	05 05	16 14	17 08	18 02	18 55
20	04 33	05 02	05 26	15 58	16 51	17 45	18 38
N 10	04 54	05 21	05 44	15 45	16 36	17 29	18 23
0	05 13	05 39	06 01	15 33	16 23	17 15	18 08
S 10	05 29	05 55	06 18	15 20	16 09	17 00	17 54
20	05 45	06 12	06 36	15 07	15 54	16 45	17 39
30	06 00	06 30	06 56	14 52	15 37	16 27	17 22
35	06 09	06 40	07 08	14 43	15 27	16 17	17 12
40	06 18	06 51	07 22	14 33	15 16	16 05	17 00
45	06 27	07 04	07 38	14 21	15 03	15 51	16 47
S 50	06 38	07 19	07 57	14 07	14 47	15 35	16 30
52	06 43	07 26	08 07	14 01	14 40	15 27	16 22
54	06 49	07 34	08 17	13 53	14 31	15 18	16 14
56	06 54	07 43	08 29	13 45	14 22	15 08	16 04
58	07 01	07 52	08 43	13 36	14 12	14 57	15 53
S 60	07 08	08 03	08 59	13 26	14 00	14 44	15 40

Lat.	Sunset	Twilight Civil	Twilight Naut.	Moonset 6	7	8	9
N 72	☐	☐	☐	22 51	▬	▬	▬
N 70	☐	☐	☐	23 50	24 02	00 02	00 40
68	☐	☐	☐	00 11	00 25	00 50	01 35
66	23 09	////	////	00 30	00 50	01 21	02 08
64	22 15	////	////	00 46	01 10	01 44	02 33
62	21 43	////	////	00 59	01 26	02 03	02 52
60	21 19	22 48	////	01 10	01 39	02 18	03 08
N 58	21 01	22 10	////	01 19	01 51	02 31	03 21
56	20 45	21 44	////	01 28	02 01	02 42	03 32
54	20 32	21 24	22 55	01 35	02 10	02 52	03 43
52	20 20	21 07	22 20	01 42	02 18	03 00	03 51
50	20 10	20 53	21 55	01 48	02 25	03 08	04 00
45	19 49	20 25	21 14	02 01	02 40	03 25	04 17
N 40	19 31	20 04	20 45	02 12	02 53	03 39	04 31
35	19 17	19 46	20 23	02 21	03 03	03 50	04 42
30	19 05	19 32	20 05	02 29	03 13	04 00	04 53
20	18 44	19 08	19 37	02 44	03 29	04 18	05 10
N 10	18 25	18 48	19 15	02 56	03 43	04 33	05 26
0	18 09	18 31	18 57	03 07	03 56	04 47	05 40
S 10	17 52	18 14	18 41	03 19	04 10	05 02	05 55
20	17 34	17 58	18 25	03 32	04 24	05 17	06 10
30	17 14	17 40	18 10	03 46	04 40	05 34	06 28
35	17 02	17 30	18 01	03 54	04 50	05 45	06 38
40	16 48	17 19	17 52	04 03	05 00	05 56	06 50
45	16 32	17 06	17 43	04 14	05 13	06 10	07 03
S 50	16 13	16 51	17 32	04 28	05 29	06 27	07 20
52	16 03	16 44	17 27	04 34	05 36	06 34	07 28
54	15 53	16 36	17 21	04 41	05 44	06 43	07 37
56	15 41	16 27	17 16	04 49	05 53	06 53	07 47
58	15 27	16 18	17 09	04 57	06 03	07 04	07 58
S 60	15 11	16 07	17 02	05 07	06 15	07 17	08 11

Day	SUN Eqn. of Time 00ʰ	SUN Eqn. of Time 12ʰ	SUN Mer. Pass.	MOON Mer. Pass. Upper	MOON Mer. Pass. Lower	Age	Phase %
6	04 38	04 43	12 05	21 44	09 20	12	89
7	04 48	04 53	12 05	22 35	10 09	13	95
8	04 58	05 02	12 05	23 28	11 01	14	98

1998 JULY 9, 10, 11 (THURS., FRI., SAT.)

UT	ARIES GHA	VENUS −3.9 GHA	Dec	MARS +1.6 GHA	Dec	JUPITER −2.6 GHA	Dec	SATURN +0.5 GHA	Dec	STARS Name	SHA	Dec
9 00	286 43.8	210 28.4	N21 38.5	194 47.3	N24 02.4	288 06.9	S 2 01.3	255 39.3	N10 02.1	Acamar	315 27.3	S40 18.5
01	301 46.3	225 27.7	38.9	209 47.9	02.4	303 09.3	01.3	270 41.7	02.1	Achernar	335 35.5	S57 14.5
02	316 48.8	240 27.0	39.3	224 48.6	02.4	318 11.7	01.3	285 44.0	02.2	Acrux	173 22.2	S63 05.7
03	331 51.2	255 26.2 ..	39.7	239 49.2 ..	02.4	333 14.1 ..	01.3	300 46.3 ..	02.2	Adhara	255 21.9	S28 58.3
04	346 53.7	270 25.5	40.0	254 49.8	02.4	348 16.5	01.2	315 48.6	02.3	Aldebaran	291 02.8	N16 30.2
05	1 56.2	285 24.8	40.4	269 50.4	02.4	3 18.9	01.2	330 50.9	02.3			
06	16 58.6	300 24.1	N21 40.8	284 51.0	N24 02.4	18 21.3	S 2 01.2	345 53.2	N10 02.4	Alioth	166 30.9	N55 58.4
07	32 01.1	315 23.3	41.2	299 51.6	02.4	33 23.7	01.2	0 55.5	02.4	Alkaid	153 07.9	N49 19.6
T 08	47 03.6	330 22.6	41.5	314 52.3	02.4	48 26.0	01.2	15 57.8	02.4	Al Na'ir	27 57.8	S46 57.9
H 09	62 06.0	345 21.9 ..	41.9	329 52.9 ..	02.4	63 28.4 ..	01.2	31 00.2 ..	02.5	Alnilam	275 58.3	S 1 12.3
U 10	77 08.5	0 21.1	42.3	344 53.5	02.4	78 30.8	01.1	46 02.5	02.5	Alphard	218 07.6	S 8 39.2
R 11	92 11.0	15 20.4	42.7	359 54.1	02.4	93 33.2	01.1	61 04.8	02.6			
S 12	107 13.4	30 19.7	N21 43.0	14 54.7	N24 02.4	108 35.6	S 2 01.1	76 07.1	N10 02.6	Alphecca	126 20.5	N26 43.5
D 13	122 15.9	45 18.9	43.4	29 55.3	02.4	123 38.0	01.1	91 09.4	02.7	Alpheratz	357 55.3	N29 04.8
A 14	137 18.3	60 18.2	43.8	44 56.0	02.4	138 40.4	01.1	106 11.7	02.7	Altair	62 19.1	N 8 52.0
Y 15	152 20.8	75 17.5 ..	44.1	59 56.6 ..	02.4	153 42.8 ..	01.1	121 14.0 ..	02.7	Ankaa	353 27.0	S42 18.7
16	167 23.3	90 16.8	44.5	74 57.2	02.4	168 45.2	01.0	136 16.4	02.8	Antares	112 40.1	S26 25.6
17	182 25.7	105 16.0	44.9	89 57.8	02.4	183 47.6	01.0	151 18.7	02.8			
18	197 28.2	120 15.3	N21 45.2	104 58.4	N24 02.4	198 50.0	S 2 01.0	166 21.0	N10 02.9	Arcturus	146 06.1	N19 11.7
19	212 30.7	135 14.6	45.6	119 59.1	02.3	213 52.4	01.0	181 23.3	02.9	Atria	107 51.7	S69 01.5
20	227 33.1	150 13.8	46.0	134 59.7	02.3	228 54.8	01.0	196 25.6	03.0	Avior	234 23.3	S59 30.5
21	242 35.6	165 13.1 ..	46.3	150 00.3 ..	02.3	243 57.2 ..	01.0	211 27.9 ..	03.0	Bellatrix	278 44.6	N 6 20.8
22	257 38.1	180 12.4	46.7	165 00.9	02.3	258 59.6	01.0	226 30.3	03.0	Betelgeuse	271 14.1	N 7 24.3
23	272 40.5	195 11.6	47.1	180 01.5	02.3	274 02.0	00.9	241 32.6	03.1			
10 00	287 43.0	210 10.9	N21 47.4	195 02.2	N24 02.3	289 04.4	S 2 00.9	256 34.9	N10 03.1	Canopus	264 01.8	S52 41.8
01	302 45.5	225 10.1	47.8	210 02.8	02.3	304 06.8	00.9	271 37.2	03.2	Capella	280 51.9	N45 59.6
02	317 47.9	240 09.4	48.1	225 03.4	02.3	319 09.2	00.9	286 39.5	03.2	Deneb	49 38.8	N45 16.5
03	332 50.4	255 08.7 ..	48.5	240 04.0 ..	02.3	334 11.6 ..	00.9	301 41.8 ..	03.2	Denebola	182 45.5	N14 35.0
04	347 52.8	270 07.9	48.8	255 04.6	02.3	349 14.0	00.9	316 44.2	03.3	Diphda	349 07.4	S17 59.6
05	2 55.3	285 07.2	49.2	270 05.3	02.3	4 16.4	00.9	331 46.5	03.3			
06	17 57.8	300 06.5	N21 49.6	285 05.9	N24 02.2	19 18.8	S 2 00.8	346 48.8	N10 03.4	Dubhe	194 06.2	N61 45.8
07	33 00.2	315 05.7	49.9	300 06.5	02.2	34 21.2	00.8	1 51.1	03.4	Elnath	278 27.5	N28 36.2
F 08	48 02.7	330 05.0	50.3	315 07.1	02.2	49 23.6	00.8	16 53.4	03.5	Eltanin	90 51.0	N51 29.6
R 09	63 05.2	345 04.3 ..	50.6	330 07.7 ..	02.2	64 26.0 ..	00.8	31 55.7 ..	03.5	Enif	33 58.1	N 9 52.1
I 10	78 07.6	0 03.5	51.0	345 08.4	02.2	79 28.4	00.8	46 58.1	03.5	Fomalhaut	15 36.4	S29 37.6
D 11	93 10.1	15 02.8	51.3	0 09.0	02.2	94 30.8	00.8	62 00.4	03.6			
A 12	108 12.6	30 02.0	N21 51.6	15 09.6	N24 02.2	109 33.2	S 2 00.8	77 02.7	N10 03.6	Gacrux	172 13.7	S57 06.5
Y 13	123 15.0	45 01.3	52.0	30 10.2	02.2	124 35.6	00.8	92 05.0	03.7	Gienah	176 04.2	S17 32.0
14	138 17.5	60 00.6	52.3	45 10.8	02.2	139 38.0	00.7	107 07.3	03.7	Hadar	149 04.0	S60 22.1
15	153 19.9	74 59.8 ..	52.7	60 11.5 ..	02.1	154 40.4 ..	00.7	122 09.6 ..	03.8	Hamal	328 13.8	N23 27.1
16	168 22.4	89 59.1	53.0	75 12.1	02.1	169 42.8	00.7	137 12.0	03.8	Kaus Aust.	83 58.6	S34 23.0
17	183 24.9	104 58.3	53.4	90 12.7	02.1	184 45.2	00.7	152 14.3	03.8			
18	198 27.3	119 57.6	N21 53.7	105 13.3	N24 02.1	199 47.6	S 2 00.7	167 16.6	N10 03.9	Kochab	137 19.2	N74 10.1
19	213 29.8	134 56.9	54.0	120 13.9	02.1	214 50.0	00.7	182 18.9	03.9	Markab	13 49.6	N15 11.8
20	228 32.3	149 56.1	54.4	135 14.6	02.1	229 52.4	00.7	197 21.2	04.0	Menkar	314 27.2	N 4 04.9
21	243 34.7	164 55.4 ..	54.7	150 15.2 ..	02.0	244 54.8 ..	00.7	212 23.6 ..	04.0	Menkent	148 21.0	S36 21.8
22	258 37.2	179 54.6	55.1	165 15.8	02.0	259 57.2	00.6	227 25.9	04.0	Miaplacidus	221 42.8	S69 42.9
23	273 39.7	194 53.9	55.4	180 16.4	02.0	274 59.6	00.6	242 28.2	04.1			
11 00	288 42.1	209 53.1	N21 55.7	195 17.0	N24 02.0	290 02.1	S 2 00.6	257 30.5	N10 04.1	Mirfak	308 57.1	N49 51.0
01	303 44.6	224 52.4	56.1	210 17.7	02.0	305 04.5	00.6	272 32.8	04.2	Nunki	76 12.2	S26 17.8
02	318 47.1	239 51.7	56.4	225 18.3	02.0	320 06.9	00.6	287 35.2	04.2	Peacock	53 36.8	S56 44.2
03	333 49.5	254 50.9 ..	56.7	240 18.9 ..	01.9	335 09.3 ..	00.6	302 37.5 ..	04.2	Pollux	243 42.2	N28 01.7
04	348 52.0	269 50.2	57.1	255 19.5	01.9	350 11.7	00.6	317 39.8	04.3	Procyon	245 12.1	N 5 13.7
05	3 54.4	284 49.4	57.4	270 20.1	01.9	5 14.1	00.6	332 42.1	04.3			
06	18 56.9	299 48.7	N21 57.7	285 20.8	N24 01.9	20 16.5	S 2 00.5	347 44.4	N10 04.4	Rasalhague	96 16.8	N12 33.9
07	33 59.4	314 47.9	58.0	300 21.4	01.9	35 18.9	00.5	2 46.8	04.4	Regulus	207 56.0	N11 58.5
S 08	49 01.8	329 47.2	58.4	315 22.0	01.9	50 21.3	00.5	17 49.1	04.4	Rigel	281 23.4	S 8 12.3
A 09	64 04.3	344 46.4 ..	58.7	330 22.6 ..	01.8	65 23.7 ..	00.5	32 51.4 ..	04.5	Rigil Kent.	140 07.2	S60 49.8
T 10	79 06.8	359 45.7	59.0	345 23.3	01.8	80 26.1	00.5	47 53.7	04.5	Sabik	102 25.4	S15 43.2
U 11	94 09.2	14 45.0	59.3	0 23.9	01.8	95 28.5	00.5	62 56.0	04.6			
R 12	109 11.7	29 44.2	N21 59.7	15 24.5	N24 01.8	110 31.0	S 2 00.5	77 58.4	N10 04.6	Schedar	349 53.5	N56 31.4
D 13	124 14.2	44 43.5	22 00.0	30 25.1	01.7	125 33.4	00.5	93 00.7	04.6	Shaula	96 37.1	S37 06.1
A 14	139 16.6	59 42.7	00.3	45 25.7	01.7	140 35.8	00.5	108 03.0	04.7	Sirius	258 44.2	S16 42.9
Y 15	154 19.1	74 42.0 ..	00.6	60 26.4 ..	01.7	155 38.2 ..	00.5	123 05.3 ..	04.7	Spica	158 43.3	S11 09.1
16	169 21.6	89 41.2	00.9	75 27.0	01.7	170 40.6	00.5	138 07.7	04.8	Suhail	223 01.2	S43 25.7
17	184 24.0	104 40.5	01.3	90 27.6	01.7	185 43.0	00.5	153 10.0	04.8			
18	199 26.5	119 39.7	N22 01.6	105 28.2	N24 01.6	200 45.4	S 2 00.4	168 12.3	N10 04.8	Vega	80 46.3	N38 47.1
19	214 28.9	134 39.0	01.9	120 28.9	01.6	215 47.8	00.4	183 14.6	04.9	Zuben'ubi	137 18.0	S16 02.1
20	229 31.4	149 38.2	02.2	135 29.5	01.6	230 50.2	00.4	198 16.9	04.9			
21	244 33.9	164 37.5 ..	02.5	150 30.1 ..	01.6	245 52.7 ..	00.4	213 19.3 ..	05.0			
22	259 36.3	179 36.7	02.8	165 30.7	01.5	260 55.1	00.4	228 21.6	05.0			
23	274 38.8	194 36.0	03.1	180 31.4	01.5	275 57.5	00.4	243 23.9	05.0			

		SHA	Mer. Pass.
Venus		282 27.9	10 00
Mars		267 19.2	10 59
Jupiter		1 21.4	4 43
Saturn		328 51.9	6 53

Mer. Pass.	4 48.3	v −0.7 d 0.3	v 0.6 d 0.0	v 2.4 d 0.0	v 2.3 d 0.0

UT	SUN GHA	SUN Dec	MOON GHA	v	Dec	d	HP
d h	° ′	° ′	° ′	′	° ′	′	′
9 00	178 43.3	N22 24.3	7 49.5	8.9	S19 01.7	0.4	56.8
01	193 43.2	24.0	22 17.4	8.7	19 01.3	0.4	56.8
02	208 43.1	23.7	36 45.1	8.8	19 00.9	0.6	56.8
03	223 43.0	.. 23.4	51 12.9	8.7	19 00.3	0.6	56.9
04	238 42.9	23.1	65 40.6	8.8	18 59.7	0.8	56.9
05	253 42.8	22.8	80 08.4	8.6	18 58.9	0.9	56.9
06	268 42.7	N22 22.5	94 36.0	8.7	S18 58.0	1.0	56.9
T 07	283 42.6	22.2	109 03.7	8.6	18 57.0	1.1	57.0
H 08	298 42.5	21.9	123 31.3	8.6	18 55.9	1.2	57.0
U 09	313 42.4	.. 21.6	137 58.9	8.6	18 54.7	1.3	57.0
R 10	328 42.4	21.3	152 26.5	8.6	18 53.4	1.5	57.0
S 11	343 42.3	21.0	166 54.1	8.6	18 51.9	1.5	57.1
D 12	358 42.2	N22 20.7	181 21.7	8.5	S18 50.4	1.7	57.1
A 13	13 42.1	20.4	195 49.2	8.5	18 48.7	1.7	57.1
Y 14	28 42.0	20.1	210 16.7	8.5	18 47.0	1.9	57.1
15	43 41.9	.. 19.8	224 44.2	8.5	18 45.1	2.0	57.2
16	58 41.8	19.5	239 11.7	8.4	18 43.1	2.1	57.2
17	73 41.7	19.2	253 39.1	8.5	18 41.0	2.2	57.2
18	88 41.6	N22 18.9	268 06.6	8.4	S18 38.8	2.3	57.2
19	103 41.5	18.6	282 34.0	8.4	18 36.5	2.4	57.2
20	118 41.4	18.3	297 01.4	8.4	18 34.1	2.5	57.3
21	133 41.4	.. 18.0	311 28.8	8.4	18 31.6	2.7	57.3
22	148 41.3	17.7	325 56.2	8.4	18 28.9	2.7	57.3
23	163 41.2	17.4	340 23.6	8.4	18 26.2	2.9	57.3
10 00	178 41.1	N22 17.1	354 51.0	8.3	S18 23.3	2.9	57.4
01	193 41.0	16.8	9 18.3	8.4	18 20.4	3.1	57.4
02	208 40.9	16.5	23 45.7	8.3	18 17.3	3.2	57.4
03	223 40.8	.. 16.1	38 13.0	8.4	18 14.1	3.3	57.4
04	238 40.7	15.8	52 40.4	8.3	18 10.8	3.4	57.5
05	253 40.6	15.5	67 07.7	8.4	18 07.4	3.5	57.5
06	268 40.6	N22 15.2	81 35.1	8.3	S18 03.9	3.6	57.5
F 07	283 40.5	14.9	96 02.4	8.3	18 00.3	3.8	57.5
R 08	298 40.4	14.6	110 29.7	8.3	17 56.5	3.8	57.6
I 09	313 40.3	.. 14.3	124 57.0	8.4	17 52.7	3.9	57.6
D 10	328 40.2	14.0	139 24.4	8.3	17 48.8	4.1	57.6
A 11	343 40.1	13.6	153 51.7	8.3	17 44.7	4.2	57.6
Y 12	358 40.0	N22 13.3	168 19.0	8.3	S17 40.5	4.2	57.6
13	13 39.9	13.0	182 46.3	8.3	17 36.3	4.4	57.7
14	28 39.9	12.7	197 13.6	8.4	17 31.9	4.5	57.7
15	43 39.8	.. 12.4	211 41.0	8.3	17 27.4	4.6	57.7
16	58 39.7	12.0	226 08.3	8.4	17 22.8	4.6	57.7
17	73 39.6	11.7	240 35.6	8.3	17 18.2	4.8	57.8
18	88 39.5	N22 11.4	255 02.9	8.4	S17 13.4	4.9	57.8
19	103 39.4	11.1	269 30.3	8.3	17 08.5	5.0	57.8
20	118 39.3	10.8	283 57.6	8.4	17 03.5	5.1	57.8
21	133 39.3	.. 10.4	298 25.0	8.3	16 58.4	5.3	57.8
22	148 39.2	10.1	312 52.3	8.4	16 53.1	5.3	57.9
23	163 39.1	09.8	327 19.7	8.4	16 47.8	5.4	57.9
11 00	178 39.0	N22 09.5	341 47.1	8.4	S16 42.4	5.5	57.9
01	193 38.9	09.1	356 14.5	8.4	16 36.9	5.6	57.9
02	208 38.8	08.8	10 41.9	8.4	16 31.3	5.7	58.0
03	223 38.7	.. 08.5	25 09.3	8.4	16 25.6	5.9	58.0
04	238 38.7	08.2	39 36.7	8.4	16 19.7	5.9	58.0
05	253 38.6	07.8	54 04.1	8.5	16 13.8	6.0	58.0
06	268 38.5	N22 07.5	68 31.6	8.4	S16 07.8	6.1	58.0
S 07	283 38.4	07.2	82 59.0	8.5	16 01.7	6.2	58.1
A 08	298 38.3	06.8	97 26.5	8.5	15 55.5	6.4	58.1
T 09	313 38.2	.. 06.5	111 54.0	8.5	15 49.1	6.4	58.1
U 10	328 38.2	06.2	126 21.5	8.5	15 42.7	6.5	58.1
R 11	343 38.1	05.9	140 49.0	8.5	15 36.2	6.6	58.1
D 12	358 38.0	N22 05.5	155 16.5	8.5	S15 29.6	6.7	58.2
A 13	13 37.9	05.2	169 44.0	8.6	15 22.9	6.8	58.2
Y 14	28 37.8	04.9	184 11.6	8.6	15 16.1	6.9	58.2
15	43 37.8	.. 04.5	198 39.2	8.6	15 09.2	7.0	58.2
16	58 37.7	04.2	213 06.8	8.6	15 02.2	7.0	58.2
17	73 37.6	03.8	227 34.4	8.6	14 55.2	7.2	58.3
18	88 37.5	N22 03.5	242 02.0	8.6	S14 48.0	7.3	58.3
19	103 37.4	03.2	256 29.6	8.7	14 40.7	7.3	58.3
20	118 37.3	02.8	270 57.3	8.7	14 33.4	7.4	58.3
21	133 37.3	.. 02.5	285 25.0	8.7	14 26.0	7.6	58.3
22	148 37.2	02.2	299 52.7	8.7	14 18.4	7.6	58.3
23	163 37.1	01.8	314 20.4	8.8	S14 10.8	7.7	58.4
SD	15.8	d 0.3	SD 15.6		15.7		15.8

Lat.	Twilight Naut.	Twilight Civil	Sunrise	Moonrise 9	10	11	12
°	h m	h m	h m	h m	h m	h m	h m
N 72	☐	☐	☐	■■	23 48	23 30	23 19
N 70	☐	☐	☐	22 48	22 58	23 01	23 01
68	☐	☐	☐	22 02	22 26	22 39	22 47
66	////	////	01 11	21 32	22 02	22 22	22 36
64	////	////	02 01	21 10	21 44	22 08	22 26
62	////	////	02 31	20 52	21 29	21 56	22 18
60	////	01 28	02 54	20 37	21 16	21 46	22 11
N 58	////	02 04	03 12	20 24	21 05	21 38	22 05
56	////	02 29	03 28	20 13	20 55	21 30	21 59
54	01 21	02 49	03 41	20 03	20 46	21 23	21 54
52	01 54	03 05	03 52	19 55	20 39	21 17	21 50
50	02 18	03 19	04 02	19 47	20 32	21 11	21 45
45	02 59	03 47	04 23	19 30	20 17	20 59	21 36
N 40	03 27	04 08	04 40	19 17	20 05	20 49	21 29
35	03 49	04 25	04 54	19 05	19 54	20 40	21 22
30	04 06	04 39	05 06	18 55	19 45	20 32	21 17
20	04 34	05 03	05 27	18 38	19 29	20 19	21 07
N 10	04 55	05 22	05 45	18 23	19 16	20 08	20 58
0	05 13	05 39	06 02	18 08	19 03	19 57	20 50
S 10	05 29	05 56	06 18	17 54	18 50	19 46	20 42
20	05 45	06 12	06 36	17 39	18 36	19 34	20 33
30	06 00	06 30	06 56	17 22	18 20	19 21	20 23
35	06 08	06 39	07 07	17 12	18 11	19 13	20 17
40	06 17	06 50	07 21	17 00	18 00	19 04	20 10
45	06 26	07 03	07 36	16 47	17 48	18 54	20 03
S 50	06 37	07 18	07 56	16 30	17 33	18 41	19 54
52	06 42	07 25	08 05	16 22	17 26	18 35	19 49
54	06 47	07 32	08 15	16 14	17 18	18 29	19 44
56	06 53	07 41	08 27	16 04	17 09	18 22	19 39
58	06 59	07 50	08 40	15 53	16 59	18 14	19 33
S 60	07 05	08 01	08 56	15 40	16 48	18 04	19 27

Lat.	Sunset	Twilight Civil	Twilight Naut.	Moonset 9	10	11	12
°	h m	h m	h m	h m	h m	h m	h m
N 72	☐	☐	☐	■■	■■	02 54	05 05
N 70	☐	☐	☐	00 40	02 00	03 43	05 32
68	☐	☐	☐	01 35	02 45	04 14	05 53
66	22 55	////	////	02 08	03 15	04 37	06 09
64	22 08	////	////	02 33	03 37	04 55	06 22
62	21 38	////	////	02 52	03 55	05 10	06 33
60	21 15	22 40	////	03 08	04 10	05 22	06 42
N 58	20 57	22 05	////	03 21	04 22	05 33	06 51
56	20 42	21 40	////	03 32	04 33	05 42	06 58
54	20 29	21 21	22 47	03 43	04 43	05 51	07 04
52	20 18	21 05	22 15	03 51	04 51	05 58	07 10
50	20 08	20 51	21 52	04 00	04 59	06 05	07 15
45	19 47	20 24	21 11	04 17	05 15	06 19	07 26
N 40	19 30	20 03	20 43	04 31	05 28	06 30	07 36
35	19 16	19 46	20 21	04 42	05 39	06 40	07 43
30	19 04	19 31	20 04	04 53	05 49	06 49	07 50
20	18 43	19 08	19 37	05 10	06 06	07 04	08 02
N 10	18 26	18 48	19 15	05 26	06 21	07 16	08 13
0	18 09	18 31	18 57	05 40	06 34	07 29	08 22
S 10	17 53	18 15	18 41	05 55	06 48	07 41	08 32
20	17 35	17 59	18 26	06 10	07 03	07 53	08 42
30	17 15	17 41	18 11	06 28	07 19	08 08	08 53
35	17 04	17 31	18 03	06 38	07 29	08 16	09 00
40	16 50	17 20	17 54	06 50	07 40	08 26	09 08
45	16 35	17 08	17 45	07 03	07 53	08 37	09 16
S 50	16 15	16 53	17 34	07 20	08 08	08 50	09 27
52	16 06	16 46	17 29	07 28	08 15	08 56	09 32
54	15 56	16 39	17 24	07 37	08 24	09 03	09 37
56	15 44	16 30	17 18	07 47	08 33	09 11	09 43
58	15 31	16 21	17 12	07 58	08 43	09 19	09 49
S 60	15 15	16 10	17 06	08 11	08 55	09 29	09 57

Day	SUN Eqn. of Time 00h	12h	Mer. Pass.	MOON Mer. Pass. Upper	Lower	Age	Phase
d	m s	m s	h m	h m	h m	d	%
9	05 07	05 11	12 05	24 21	11 54	15	100
10	05 15	05 20	12 05	00 21	12 48	16	99
11	05 24	05 28	12 05	01 16	13 43	17	96

1998 JULY 12, 13, 14 (SUN., MON., TUES.)

UT	ARIES GHA	VENUS −3.9 GHA	Dec	MARS +1.6 GHA	Dec	JUPITER −2.6 GHA	Dec	SATURN +0.5 GHA	Dec
12 00	289 41.3	209 35.2	N22 03.5	195 32.0	N24 01.5	290 59.9	S 2 00.4	258 26.2	N10 05.1
01	304 43.7	224 34.5	03.8	210 32.6	01.5	306 02.3	00.4	273 28.6	05.1
02	319 46.2	239 33.7	04.1	225 33.2	01.4	321 04.7	00.4	288 30.9	05.2
03	334 48.7	254 33.0 ..	04.4	240 33.9 ..	01.4	336 07.1 ..	00.4	303 33.2 ..	05.2
04	349 51.1	269 32.2	04.7	255 34.5	01.4	351 09.6	00.4	318 35.5	05.2
05	4 53.6	284 31.5	05.0	270 35.1	01.3	6 12.0	00.4	333 37.9	05.3
06	19 56.1	299 30.7	N22 05.3	285 35.7	N24 01.3	21 14.4	S 2 00.3	348 40.2	N10 05.3
07	34 58.5	314 29.9	05.6	300 36.3	01.3	36 16.8	00.3	3 42.5	05.4
S 08	50 01.0	329 29.2	05.9	315 37.0	01.3	51 19.2	00.3	18 44.8	05.4
U 09	65 03.4	344 28.4 ..	06.2	330 37.6 ..	01.2	66 21.6 ..	00.3	33 47.2 ..	05.4
N 10	80 05.9	359 27.7	06.5	345 38.2	01.2	81 24.0	00.3	48 49.5	05.5
D 11	95 08.4	14 26.9	06.8	0 38.8	01.2	96 26.5	00.3	63 51.8	05.5
A 12	110 10.8	29 26.2	N22 07.1	15 39.5	N24 01.1	111 28.9	S 2 00.3	78 54.1	N10 05.6
Y 13	125 13.3	44 25.4	07.4	30 40.1	01.1	126 31.3	00.3	93 56.5	05.6
14	140 15.8	59 24.7	07.7	45 40.7	01.1	141 33.7	00.3	108 58.8	05.6
15	155 18.2	74 23.9 ..	08.0	60 41.3 ..	01.1	156 36.1 ..	00.3	124 01.1 ..	05.7
16	170 20.7	89 23.2	08.3	75 42.0	01.0	171 38.6	00.3	139 03.4	05.7
17	185 23.2	104 22.4	08.6	90 42.6	01.0	186 41.0	00.3	154 05.8	05.8
18	200 25.6	119 21.6	N22 08.9	105 43.2	N24 01.0	201 43.4	S 2 00.3	169 08.1	N10 05.8
19	215 28.1	134 20.9	09.1	120 43.8	00.9	216 45.8	00.3	184 10.4	05.8
20	230 30.5	149 20.1	09.4	135 44.5	00.9	231 48.2	00.3	199 12.7	05.9
21	245 33.0	164 19.4 ..	09.7	150 45.1 ..	00.9	246 50.7 ..	00.3	214 15.1 ..	05.9
22	260 35.5	179 18.6	10.0	165 45.7	00.8	261 53.1	00.2	229 17.4	05.9
23	275 37.9	194 17.9	10.3	180 46.4	00.8	276 55.5	00.2	244 19.7	06.0
13 00	290 40.4	209 17.1	N22 10.6	195 47.0	N24 00.8	291 57.9	S 2 00.2	259 22.0	N10 06.0
01	305 42.9	224 16.3	10.9	210 47.6	00.7	307 00.3	00.2	274 24.4	06.1
02	320 45.3	239 15.6	11.2	225 48.2	00.7	322 02.8	00.2	289 26.7	06.1
03	335 47.8	254 14.8 ..	11.4	240 48.9 ..	00.7	337 05.2 ..	00.2	304 29.0 ..	06.1
04	350 50.3	269 14.1	11.7	255 49.5	00.6	352 07.6	00.2	319 31.3	06.2
05	5 52.7	284 13.3	12.0	270 50.1	00.6	7 10.0	00.2	334 33.7	06.2
06	20 55.2	299 12.5	N22 12.3	285 50.7	N24 00.5	22 12.4	S 2 00.2	349 36.0	N10 06.3
07	35 57.7	314 11.8	12.5	300 51.4	00.5	37 14.9	00.2	4 38.3	06.3
M 08	51 00.1	329 11.0	12.8	315 52.0	00.5	52 17.3	00.2	19 40.7	06.3
O 09	66 02.6	344 10.3 ..	13.1	330 52.6 ..	00.4	67 19.7 ..	00.2	34 43.0 ..	06.4
N 10	81 05.0	359 09.5	13.4	345 53.2	00.4	82 22.1	00.2	49 45.3	06.4
D 11	96 07.5	14 08.7	13.6	0 53.9	00.4	97 24.6	00.2	64 47.6	06.4
A 12	111 10.0	29 08.0	N22 13.9	15 54.5	N24 00.3	112 27.0	S 2 00.2	79 50.0	N10 06.5
Y 13	126 12.4	44 07.2	14.2	30 55.1	00.3	127 29.4	00.2	94 52.3	06.5
14	141 14.9	59 06.4	14.5	45 55.8	00.2	142 31.8	00.2	109 54.6	06.6
15	156 17.4	74 05.7 ..	14.7	60 56.4 ..	00.2	157 34.3 ..	00.2	124 57.0 ..	06.6
16	171 19.8	89 04.9	15.0	75 57.0	00.2	172 36.7	00.2	139 59.3	06.6
17	186 22.3	104 04.2	15.3	90 57.6	00.1	187 39.1	00.2	155 01.6	06.7
18	201 24.8	119 03.4	N22 15.5	105 58.3	N24 00.1	202 41.5	S 2 00.2	170 03.9	N10 06.7
19	216 27.2	134 02.6	15.8	120 58.9	00.0	217 44.0	00.2	185 06.3	06.7
20	231 29.7	149 01.9	16.1	135 59.5	00.0	232 46.4	00.2	200 08.6	06.8
21	246 32.2	164 01.1 ..	16.3	151 00.1	24 00.0	247 48.8 ..	00.2	215 10.9 ..	06.8
22	261 34.6	179 00.3	16.6	166 00.8	23 59.9	262 51.3	00.2	230 13.3	06.9
23	276 37.1	193 59.6	16.8	181 01.4	59.9	277 53.7	00.2	245 15.6	06.9
14 00	291 39.5	208 58.8	N22 17.1	196 02.0	N23 59.8	292 56.1	S 2 00.2	260 17.9	N10 07.0
01	306 42.0	223 58.0	17.4	211 02.7	59.8	307 58.5	00.2	275 20.2	07.0
02	321 44.5	238 57.3	17.6	226 03.3	59.7	323 01.0	00.2	290 22.6	07.0
03	336 46.9	253 56.5 ..	17.9	241 03.9 ..	59.7	338 03.4 ..	00.2	305 24.9 ..	07.0
04	351 49.4	268 55.7	18.1	256 04.5	59.7	353 05.8	00.2	320 27.2	07.1
05	6 51.9	283 55.0	18.4	271 05.2	59.6	8 08.3	00.2	335 29.6	07.1
06	21 54.3	298 54.2	N22 18.6	286 05.8	N23 59.6	23 10.7	S 2 00.2	350 31.9	N10 07.1
07	36 56.8	313 53.4	18.9	301 06.4	59.5	38 13.1	00.2	5 34.2	07.2
T 08	51 59.3	328 52.7	19.1	316 07.1	59.5	53 15.5	00.2	20 36.6	07.2
U 09	67 01.7	343 51.9 ..	19.4	331 07.7 ..	59.4	68 18.0 ..	00.2	35 38.9 ..	07.3
E 10	82 04.2	358 51.1	19.6	346 08.3	59.4	83 20.4	00.2	50 41.2	07.3
S 11	97 06.6	13 50.4	19.9	1 08.9	59.3	98 22.8	00.2	65 43.6	07.3
D 12	112 09.1	28 49.6	N22 20.1	16 09.6	N23 59.3	113 25.3	S 2 00.2	80 45.9	N10 07.4
A 13	127 11.6	43 48.8	20.4	31 10.2	59.2	128 27.7	00.2	95 48.2	07.4
Y 14	142 14.0	58 48.0	20.6	46 10.8	59.2	143 30.1	00.2	110 50.6	07.4
15	157 16.5	73 47.3 ..	20.9	61 11.5 ..	59.2	158 32.6 ..	00.2	125 52.9 ..	07.5
16	172 19.0	88 46.5	21.1	76 12.1	59.1	173 35.0	00.2	140 55.2	07.5
17	187 21.4	103 45.7	21.4	91 12.7	59.1	188 37.4	00.2	155 57.6	07.5
18	202 23.9	118 45.0	N22 21.6	106 13.4	N23 59.0	203 39.9	S 2 00.2	170 59.9	N10 07.6
19	217 26.4	133 44.2	21.8	121 14.0	59.0	218 42.3	00.2	186 02.2	07.6
20	232 28.8	148 43.4	22.1	136 14.6	58.9	233 44.7	00.2	201 04.6	07.7
21	247 31.3	163 42.7 ..	22.3	151 15.2 ..	58.9	248 47.2 ..	00.2	216 06.9 ..	07.7
22	262 33.8	178 41.9	22.5	166 15.9	58.8	263 49.6	00.2	231 09.2	07.7
23	277 36.2	193 41.1	22.8	181 16.5	58.8	278 52.0	00.2	246 11.6	07.8
Mer.Pass.	h m 4 36.5	v −0.8	d 0.3	v 0.6	d 0.0	v 2.4	d 0.0	v 2.3	d 0.0

STARS

Name	SHA	Dec
Acamar	315 27.3	S40 18.5
Achernar	335 35.4	S57 14.4
Acrux	173 22.2	S63 05.7
Adhara	255 21.9	S28 58.3
Aldebaran	291 02.8	N16 30.2
Alioth	166 30.9	N55 58.4
Alkaid	153 07.9	N49 19.6
Al Na'ir	27 57.8	S46 57.9
Alnilam	275 58.3	S 1 12.3
Alphard	218 07.6	S 8 39.2
Alphecca	126 20.5	N26 43.5
Alpheratz	357 55.2	N29 04.8
Altair	62 19.1	N 8 52.0
Ankaa	353 27.0	S42 18.6
Antares	112 40.1	S26 25.6
Arcturus	146 06.1	N19 11.7
Atria	107 51.7	S69 01.5
Avior	234 23.3	S59 30.4
Bellatrix	278 44.6	N 6 20.8
Betelgeuse	271 14.0	N 7 24.3
Canopus	264 01.8	S52 41.7
Capella	280 51.8	N45 59.6
Deneb	49 38.8	N45 16.5
Denebola	182 45.5	N14 35.0
Diphda	349 07.4	S17 59.6
Dubhe	194 06.2	N61 45.8
Elnath	278 27.5	N28 36.2
Eltanin	90 51.0	N51 29.6
Enif	33 58.1	N 9 52.1
Fomalhaut	15 36.4	S29 37.6
Gacrux	172 13.8	S57 06.5
Gienah	176 04.2	S17 32.0
Hadar	149 04.0	S60 22.1
Hamal	328 13.8	N23 27.1
Kaus Aust.	83 58.6	S34 23.0
Kochab	137 19.3	N74 10.1
Markab	13 49.6	N15 11.8
Menkar	314 27.2	N 4 04.9
Menkent	148 21.0	S36 21.8
Miaplacidus	221 42.8	S69 42.9
Mirfak	308 57.1	N49 51.0
Nunki	76 12.2	S26 17.8
Peacock	53 36.7	S56 44.2
Pollux	243 42.1	N28 01.7
Procyon	245 12.1	N 5 13.7
Rasalhague	96 16.8	N12 33.9
Regulus	207 56.0	N11 58.5
Rigel	281 23.4	S 8 12.3
Rigil Kent.	140 07.2	S60 49.8
Sabik	102 25.4	S15 43.2
Schedar	349 53.5	N56 31.4
Shaula	96 37.1	S37 06.1
Sirius	258 44.2	S16 42.9
Spica	158 43.3	S11 09.1
Suhail	223 01.2	S43 25.7
Vega	80 46.3	N38 47.1
Zuben'ubi	137 18.0	S16 02.1

	SHA	Mer.Pass.
Venus	278 36.7	h m 10 03
Mars	265 06.6	10 56
Jupiter	1 17.5	4 31
Saturn	328 41.6	6 41

SUN and MOON

UT	SUN GHA	SUN Dec	MOON GHA	v	MOON Dec	d	HP
12 00	178 37.0	N22 01.5	328 48.2	8.7	S14 03.1	7.8	58.4
01	193 36.9	01.1	343 15.9	8.8	13 55.3	7.8	58.4
02	208 36.9	00.8	357 43.7	8.8	13 47.5	8.0	58.4
03	223 36.8 . .	00.4	12 11.5	8.8	13 39.5	8.0	58.4
04	238 36.7	22 00.1	26 39.3	8.9	13 31.5	8.2	58.4
05	253 36.6	21 59.8	41 07.2	8.9	13 23.3	8.2	58.5
06	268 36.5	N21 59.4	55 35.1	8.9	S13 15.1	8.3	58.5
07	283 36.5	59.1	70 03.0	8.9	13 06.8	8.3	58.5
S 08	298 36.4	58.7	84 30.9	8.9	12 58.5	8.5	58.5
U 09	313 36.3 . .	58.4	98 58.8	9.0	12 50.0	8.5	58.5
N 10	328 36.2	58.0	113 26.8	9.0	12 41.5	8.6	58.5
D 11	343 36.2	57.7	127 54.7	9.0	12 32.9	8.7	58.6
A 12	358 36.1	N21 57.3	142 22.7	9.1	S12 24.2	8.8	58.6
Y 13	13 36.0	57.0	156 50.8	9.0	12 15.4	8.8	58.6
14	28 35.9	56.6	171 18.8	9.1	12 06.6	8.9	58.6
15	43 35.8 . .	56.3	185 46.9	9.1	11 57.7	9.0	58.6
16	58 35.8	55.9	200 15.0	9.1	11 48.7	9.0	58.6
17	73 35.7	55.6	214 43.1	9.1	11 39.7	9.2	58.7
18	88 35.6	N21 55.2	229 11.2	9.2	S11 30.5	9.2	58.7
19	103 35.5	54.9	243 39.4	9.2	11 21.3	9.2	58.7
20	118 35.5	54.5	258 07.6	9.2	11 12.1	9.4	58.7
21	133 35.4 . .	54.2	272 35.8	9.2	11 02.7	9.4	58.7
22	148 35.3	53.8	287 04.0	9.2	10 53.3	9.4	58.7
23	163 35.2	53.5	301 32.2	9.3	10 43.9	9.6	58.7
13 00	178 35.2	N21 53.1	316 00.5	9.3	S10 34.3	9.6	58.8
01	193 35.1	52.8	330 28.8	9.3	10 24.7	9.6	58.8
02	208 35.0	52.4	344 57.1	9.4	10 15.1	9.8	58.8
03	223 34.9 . .	52.0	359 25.5	9.3	10 05.3	9.8	58.8
04	238 34.9	51.7	13 53.8	9.4	9 55.5	9.8	58.8
05	253 34.8	51.3	28 22.2	9.4	9 45.7	9.9	58.8
06	268 34.7	N21 51.0	42 50.6	9.4	S 9 35.8	10.0	58.8
07	283 34.6	50.6	57 19.0	9.5	9 25.8	10.0	58.8
08	298 34.6	50.2	71 47.5	9.4	9 15.8	10.1	58.9
M 09	313 34.5 . .	49.9	86 15.9	9.5	9 05.7	10.1	58.9
O 10	328 34.4	49.5	100 44.4	9.5	8 55.6	10.2	58.9
N 11	343 34.3	49.1	115 12.9	9.5	8 45.4	10.3	58.9
D 12	358 34.3	N21 48.8	129 41.4	9.6	S 8 35.1	10.3	58.9
A 13	13 34.2	48.4	144 10.0	9.6	8 24.8	10.3	58.9
Y 14	28 34.1	48.1	158 38.6	9.5	8 14.5	10.4	58.9
15	43 34.0 . .	47.7	173 07.1	9.6	8 04.1	10.5	58.9
16	58 34.0	47.3	187 35.7	9.7	7 53.6	10.5	58.9
17	73 33.9	47.0	202 04.4	9.6	7 43.1	10.5	59.0
18	88 33.8	N21 46.6	216 33.0	9.7	S 7 32.6	10.6	59.0
19	103 33.8	46.2	231 01.7	9.6	7 22.0	10.7	59.0
20	118 33.7	45.8	245 30.3	9.7	7 11.3	10.6	59.0
21	133 33.6 . .	45.5	259 59.0	9.7	7 00.7	10.8	59.0
22	148 33.5	45.1	274 27.7	9.8	6 49.9	10.7	59.0
23	163 33.5	44.7	288 56.5	9.7	6 39.2	10.8	59.0
14 00	178 33.4	N21 44.4	303 25.2	9.8	S 6 28.4	10.9	59.0
01	193 33.3	44.0	317 54.0	9.7	6 17.5	10.9	59.0
02	208 33.3	43.6	332 22.7	9.8	6 06.6	10.9	59.0
03	223 33.2 . .	43.2	346 51.5	9.8	5 55.7	11.0	59.1
04	238 33.1	42.9	1 20.3	9.8	5 44.7	11.0	59.1
05	253 33.1	42.5	15 49.1	9.9	5 33.7	11.0	59.1
06	268 33.0	N21 42.1	30 18.0	9.8	S 5 22.7	11.0	59.1
07	283 32.9	41.7	44 46.8	9.9	5 11.7	11.1	59.1
T 08	298 32.8	41.4	59 15.7	9.8	5 00.6	11.2	59.1
U 09	313 32.8 . .	41.0	73 44.5	9.9	4 49.4	11.1	59.1
E 10	328 32.7	40.6	88 13.4	9.9	4 38.3	11.2	59.1
S 11	343 32.6	40.2	102 42.3	9.9	4 27.1	11.2	59.1
D 12	358 32.6	N21 39.9	117 11.2	9.9	S 4 15.9	11.2	59.1
A 13	13 32.5	39.5	131 40.1	9.9	4 04.7	11.3	59.1
Y 14	28 32.4	39.1	146 09.0	9.9	3 53.4	11.3	59.1
15	43 32.4 . .	38.7	160 37.9	10.0	3 42.1	11.3	59.2
16	58 32.3	38.3	175 06.9	9.9	3 30.8	11.3	59.2
17	73 32.2	37.9	189 35.8	9.9	3 19.5	11.4	59.2
18	88 32.2	N21 37.6	204 04.7	10.0	S 3 08.1	11.4	59.2
19	103 32.1	37.2	218 33.7	10.0	2 56.7	11.3	59.2
20	118 32.0	36.8	233 02.7	9.9	2 45.4	11.4	59.2
21	133 32.0 . .	36.4	247 31.6	10.0	2 34.0	11.5	59.2
22	148 31.9	36.0	262 00.6	10.0	2 22.5	11.4	59.2
23	163 31.8	35.6	276 29.6	9.9	S 2 11.1	11.5	59.2
	SD 15.8	d 0.4	SD 16.0		16.0		16.1

Twilight, Sunrise, Moonrise

Lat.	Twilight Naut.	Twilight Civil	Sunrise	Moonrise 12	Moonrise 13	Moonrise 14	Moonrise 15
N 72	□	□	□	23 19	23 10	23 03	22 56
N 70	□	□	□	23 01	23 01	23 00	22 59
68	□	□	□	22 47	22 53	22 57	23 01
66	////	////	01 25	22 36	22 46	22 55	23 03
64	////	////	02 09	22 26	22 41	22 53	23 05
62	////	00 28	02 38	22 18	22 36	22 52	23 07
60	////	01 38	02 59	22 11	22 32	22 50	23 08
N 58	////	02 11	03 17	22 05	22 28	22 49	23 09
56	00 26	02 34	03 32	21 59	22 25	22 48	23 11
54	01 30	02 53	03 44	21 54	22 22	22 47	23 12
52	02 00	03 09	03 55	21 50	22 19	22 46	23 13
50	02 23	03 22	04 05	21 45	22 16	22 45	23 13
45	03 02	03 49	04 25	21 36	22 11	22 43	23 15
N 40	03 30	04 10	04 42	21 29	22 06	22 42	23 17
35	03 51	04 27	04 56	21 22	22 02	22 40	23 18
30	04 08	04 41	05 08	21 17	21 59	22 39	23 19
20	04 35	05 04	05 28	21 07	21 53	22 37	23 21
N 10	04 56	05 23	05 46	20 58	21 47	22 35	23 23
0	05 14	05 40	06 02	20 50	21 42	22 34	23 25
S 10	05 30	05 56	06 18	20 42	21 37	22 32	23 27
20	05 44	06 12	06 35	20 33	21 32	22 30	23 29
30	05 59	06 29	06 55	20 23	21 26	22 28	23 31
35	06 07	06 39	07 06	20 17	21 22	22 27	23 33
40	06 16	06 49	07 19	20 10	21 18	22 26	23 34
45	06 25	07 02	07 35	20 03	21 13	22 24	23 36
S 50	06 35	07 16	07 53	19 54	21 08	22 23	23 38
52	06 40	07 23	08 02	19 49	21 05	22 22	23 39
54	06 45	07 30	08 12	19 44	21 02	22 21	23 40
56	06 50	07 38	08 24	19 39	20 59	22 20	23 41
58	06 56	07 47	08 37	19 33	20 56	22 19	23 43
S 60	07 03	07 58	08 52	19 27	20 52	22 18	23 44

Sunset, Twilight, Moonset

Lat.	Sunset	Twilight Civil	Twilight Naut.	Moonset 12	Moonset 13	Moonset 14	Moonset 15
N 72	□	□	□	05 05	07 06	09 03	10 58
N 70	□	□	□	05 32	07 22	09 10	10 58
68	□	□	□	05 53	07 34	09 16	10 58
66	22 42	////	////	06 09	07 44	09 21	10 58
64	22 00	////	////	06 22	07 53	09 25	10 58
62	21 32	23 33	////	06 33	08 00	09 29	10 58
60	21 11	22 31	////	06 42	08 06	09 32	10 58
N 58	20 54	21 59	////	06 51	08 12	09 35	10 58
56	20 39	21 36	23 36	06 58	08 17	09 37	10 58
54	20 26	21 17	22 39	07 04	08 21	09 39	10 58
52	20 16	21 01	22 10	07 10	08 25	09 41	10 58
50	20 06	20 48	21 48	07 15	08 28	09 43	10 58
45	19 45	20 22	21 09	07 26	08 36	09 47	10 58
N 40	19 29	20 01	20 41	07 36	08 42	09 50	10 58
35	19 15	19 44	20 20	07 43	08 48	09 53	10 58
30	19 03	19 30	20 03	07 50	08 53	09 55	10 58
20	18 43	19 07	19 36	08 02	09 01	10 00	10 58
N 10	18 26	18 48	19 15	08 13	09 08	10 03	10 58
0	18 09	18 32	18 57	08 22	09 15	10 07	10 58
S 10	17 53	18 16	18 42	08 32	09 22	10 10	10 58
20	17 36	18 00	18 27	08 42	09 29	10 14	10 58
30	17 17	17 43	18 12	08 53	09 36	10 18	10 58
35	17 05	17 33	18 04	09 00	09 41	10 20	10 58
40	16 52	17 22	17 56	09 08	09 46	10 22	10 58
45	16 37	17 10	17 47	09 16	09 52	10 25	10 57
S 50	16 18	16 56	17 36	09 27	09 59	10 29	10 57
52	16 09	16 49	17 32	09 32	10 02	10 31	10 57
54	15 59	16 42	17 27	09 37	10 06	10 32	10 57
56	15 48	16 34	17 21	09 43	10 10	10 34	10 57
58	15 35	16 25	17 16	09 49	10 14	10 36	10 57
S 60	15 20	16 14	17 09	09 57	10 19	10 39	10 57

SUN and MOON data

Day	SUN Eqn. of Time 00h	SUN Eqn. of Time 12h	SUN Mer. Pass.	MOON Mer. Pass. Upper	MOON Mer. Pass. Lower	Age	Phase
	m s	m s	h m	h m	h m	d	%
12	05 32	05 36	12 06	02 09	14 36	18	91
13	05 39	05 43	12 06	03 02	15 29	19	83
14	05 46	05 50	12 06	03 54	16 20	20	74

UT	ARIES GHA	VENUS GHA	VENUS Dec	MARS GHA	MARS Dec	JUPITER GHA	JUPITER Dec	SATURN GHA	SATURN Dec
d h	° ′	° ′	° ′	° ′	° ′	° ′	° ′	° ′	° ′
15 00	292 38.7	208 40.3	N22 23.0	196 17.1	N23 58.7	293 54.5	S 2 00.2	261 13.9	N10 07.8
01	307 41.1	223 39.6	23.2	211 17.8	58.7	308 56.9	00.2	276 16.2	07.8
02	322 43.6	238 38.8	23.5	226 18.4	58.6	323 59.4	00.2	291 18.6	07.9
03	337 46.1	253 38.0 ..	23.7	241 19.0 ..	58.5	339 01.8 ..	00.2	306 20.9 ..	07.9
04	352 48.5	268 37.2	23.9	256 19.7	58.5	354 04.2	00.2	321 23.2	07.9
05	7 51.0	283 36.5	24.2	271 20.3	58.4	9 06.7	00.2	336 25.6	08.0
W 06	22 53.5	298 35.7	N22 24.4	286 20.9	N23 58.4	24 09.1	S 2 00.2	351 27.9	N10 08.0
E 07	37 55.9	313 34.9	24.6	301 21.6	58.3	39 11.5	00.2	6 30.2	08.0
D 08	52 58.4	328 34.1	24.6	316 22.2	58.3	54 14.0	00.2	21 32.6	08.1
N 09	68 00.9	343 33.4 ..	25.1	331 22.8 ..	58.2	69 16.4 ..	00.2	36 34.9 ..	08.1
E 10	83 03.3	358 32.6	25.3	346 23.5	58.2	84 18.9	00.2	51 37.2	08.1
S 11	98 05.8	13 31.8	25.5	1 24.1	58.1	99 21.3	00.2	66 39.6	08.2
D 12	113 08.2	28 31.0	N22 25.7	16 24.7	N23 58.1	114 23.7	S 2 00.2	81 41.9	N10 08.2
A 13	128 10.7	43 30.3	26.0	31 25.3	58.0	129 26.2	00.2	96 44.2	08.2
Y 14	143 13.2	58 29.5	26.2	46 26.0	57.9	144 28.6	00.2	111 46.6	08.3
15	158 15.6	73 28.7 ..	26.4	61 26.6 ..	57.9	159 31.1 ..	00.2	126 48.9 ..	08.3
16	173 18.1	88 27.9	26.6	76 27.2	57.8	174 33.5	00.2	141 51.2	08.4
17	188 20.6	103 27.2	26.8	91 27.9	57.8	189 35.9	00.2	156 53.6	08.4
18	203 23.0	118 26.4	N22 27.0	106 28.5	N23 57.7	204 38.4	S 2 00.2	171 55.9	N10 08.4
19	218 25.5	133 25.6	27.3	121 29.1	57.7	219 40.8	00.2	186 58.3	08.5
20	233 28.0	148 24.8	27.5	136 29.8	57.6	234 43.3	00.2	202 00.6	08.5
21	248 30.4	163 24.1 ..	27.7	151 30.4 ..	57.5	249 45.7 ..	00.2	217 02.9 ..	08.5
22	263 32.9	178 23.3	27.9	166 31.0	57.5	264 48.2	00.2	232 05.3	08.6
23	278 35.4	193 22.5	28.1	181 31.7	57.4	279 50.6	00.2	247 07.6	08.6
16 00	293 37.8	208 21.7	N22 28.3	196 32.3	N23 57.4	294 53.0	S 2 00.2	262 09.9	N10 08.6
01	308 40.3	223 20.9	28.5	211 32.9	57.3	309 55.5	00.2	277 12.3	08.7
02	323 42.7	238 20.2	28.7	226 33.6	57.2	324 57.9	00.3	292 14.6	08.7
03	338 45.2	253 19.4 ..	28.9	241 34.2 ..	57.2	340 00.4 ..	00.3	307 17.0 ..	08.7
04	353 47.7	268 18.6	29.1	256 34.8	57.1	355 02.8	00.3	322 19.3	08.8
05	8 50.1	283 17.8	29.3	271 35.5	57.1	10 05.3	00.3	337 21.6	08.8
T 06	23 52.6	298 17.0	N22 29.5	286 36.1	N23 57.0	25 07.7	S 2 00.3	352 24.0	N10 08.8
H 07	38 55.1	313 16.3	29.7	301 36.7	56.9	40 10.1	00.3	7 26.3	08.9
U 08	53 57.5	328 15.5	29.9	316 37.4	56.9	55 12.6	00.3	22 28.6	08.9
R 09	69 00.0	343 14.7 ..	30.1	331 38.0 ..	56.8	70 15.0 ..	00.3	37 31.0 ..	08.9
S 10	84 02.5	358 13.9	30.3	346 38.6	56.7	85 17.5	00.3	52 33.3	09.0
11	99 04.9	13 13.1	30.5	1 39.3	56.7	100 19.9	00.3	67 35.7	09.0
D 12	114 07.4	28 12.4	N22 30.7	16 39.9	N23 56.6	115 22.4	S 2 00.3	82 38.0	N10 09.0
A 13	129 09.9	43 11.6	30.9	31 40.6	56.6	130 24.8	00.3	97 40.3	09.1
Y 14	144 12.3	58 10.8	31.1	46 41.2	56.5	145 27.3	00.3	112 42.7	09.1
15	159 14.8	73 10.0 ..	31.3	61 41.8 ..	56.4	160 29.7 ..	00.3	127 45.0 ..	09.1
16	174 17.2	88 09.2	31.5	76 42.5	56.4	175 32.2	00.3	142 47.4	09.2
17	189 19.7	103 08.4	31.7	91 43.1	56.3	190 34.6	00.3	157 49.7	09.2
18	204 22.2	118 07.7	N22 31.9	106 43.7	N23 56.2	205 37.1	S 2 00.4	172 52.0	N10 09.2
19	219 24.6	133 06.9	32.1	121 44.4	56.2	220 39.5	00.4	187 54.4	09.3
20	234 27.1	148 06.1	32.2	136 45.0	56.1	235 42.0	00.4	202 56.7	09.3
21	249 29.6	163 05.3 ..	32.4	151 45.6 ..	56.0	250 44.4 ..	00.4	217 59.1 ..	09.3
22	264 32.0	178 04.5	32.6	166 46.3	56.0	265 46.9	00.4	233 01.4	09.4
23	279 34.5	193 03.7	32.8	181 46.9	55.9	280 49.3	00.4	248 03.7	09.4
17 00	294 37.0	208 03.0	N22 33.0	196 47.5	N23 55.8	295 51.8	S 2 00.4	263 06.1	N10 09.4
01	309 39.4	223 02.2	33.2	211 48.2	55.8	310 54.2	00.4	278 08.4	09.5
02	324 41.9	238 01.4	33.3	226 48.8	55.7	325 56.7	00.4	293 10.8	09.5
03	339 44.3	253 00.6 ..	33.5	241 49.4 ..	55.6	340 59.1 ..	00.4	308 13.1 ..	09.5
04	354 46.8	267 59.8	33.7	256 50.1	55.6	356 01.6	00.4	323 15.4	09.6
05	9 49.3	282 59.0	33.9	271 50.7	55.5	11 04.0	00.4	338 17.8	09.6
F 06	24 51.7	297 58.2	N22 34.0	286 51.4	N23 55.4	26 06.5	S 2 00.4	353 20.1	N10 09.6
R 07	39 54.2	312 57.5	34.2	301 52.0	55.3	41 08.9	00.5	8 22.5	09.7
I 08	54 56.7	327 56.7	34.4	316 52.6	55.3	56 11.4	00.5	23 24.8	09.7
09	69 59.1	342 55.9 ..	34.6	331 53.3 ..	55.2	71 13.8 ..	00.5	38 27.2 ..	09.7
10	85 01.6	357 55.1	34.7	346 53.9	55.1	86 16.3	00.5	53 29.5	09.8
11	100 04.1	12 54.3	34.9	1 54.5	55.1	101 18.8	00.5	68 31.8	09.8
D 12	115 06.5	27 53.5	N22 35.1	16 55.2	N23 55.0	116 21.2	S 2 00.5	83 34.2	N10 09.8
A 13	130 09.0	42 52.7	35.2	31 55.8	54.9	131 23.7	00.5	98 36.5	09.9
Y 14	145 11.5	57 52.0	35.4	46 56.5	54.8	146 26.1	00.5	113 38.9	09.9
15	160 13.9	72 51.2 ..	35.6	61 57.1 ..	54.8	161 28.6 ..	00.5	128 41.2 ..	09.9
16	175 16.4	87 50.4	35.7	76 57.7	54.7	176 31.0	00.5	143 43.6	09.9
17	190 18.8	102 49.6	35.9	91 58.4	54.6	191 33.5	00.5	158 45.9	10.0
18	205 21.3	117 48.8	N22 36.1	106 59.0	N23 54.5	206 35.9	S 2 00.6	173 48.2	N10 10.0
19	220 23.8	132 48.0	36.2	121 59.6	54.5	221 38.4	00.6	188 50.6	10.0
20	235 26.2	147 47.2	36.4	137 00.3	54.4	236 40.9	00.6	203 52.9	10.1
21	250 28.7	162 46.4 ..	36.5	152 00.9 ..	54.3	251 43.3 ..	00.6	218 55.3 ..	10.1
22	265 31.1	177 45.6	36.7	167 01.6	54.2	266 45.8	00.6	233 57.6	10.1
23	280 33.6	192 44.9	36.9	182 02.2	54.2	281 48.2	00.6	249 00.0	10.2
Mer. Pass.	h m 4 24.8	v −0.8	d 0.2	v 0.6	d 0.1	v 2.4	d 0.0	v 2.3	d 0.0

STARS

Name	SHA	Dec
Acamar	315 27.2	S40 18.5
Achernar	335 35.4	S57 14.4
Acrux	173 22.2	S63 05.7
Adhara	255 21.9	S28 58.3
Aldebaran	291 02.8	N16 30.2
Alioth	166 30.9	N55 58.4
Alkaid	153 07.9	N49 19.6
Al Na'ir	27 57.8	S46 57.9
Alnilam	275 58.3	S 1 12.3
Alphard	218 07.6	S 8 39.2
Alphecca	126 20.5	N26 43.5
Alpheratz	357 55.2	N29 04.8
Altair	62 19.1	N 8 52.0
Ankaa	353 26.9	S42 18.6
Antares	112 40.1	S26 25.6
Arcturus	146 06.2	N19 11.7
Atria	107 51.8	S69 01.5
Avior	234 23.3	S59 30.4
Bellatrix	278 44.6	N 6 20.8
Betelgeuse	271 14.0	N 7 24.3
Canopus	264 01.8	S52 41.7
Capella	280 51.8	N45 59.6
Deneb	49 38.8	N45 16.5
Denebola	182 45.5	N14 35.0
Diphda	349 07.3	S17 59.6
Dubhe	194 06.2	N61 45.8
Elnath	278 27.5	N28 36.2
Eltanin	90 51.0	N51 29.6
Enif	33 58.1	N 9 52.2
Fomalhaut	15 36.4	S29 37.6
Gacrux	172 13.8	S57 06.5
Gienah	176 04.2	S17 32.0
Hadar	149 04.0	S60 22.1
Hamal	328 13.8	N23 27.1
Kaus Aust.	83 58.6	S34 23.0
Kochab	137 19.3	N74 10.1
Markab	13 49.5	N15 11.8
Menkar	314 27.2	N 4 04.9
Menkent	148 21.0	S36 21.8
Miaplacidus	221 42.8	S69 42.8
Mirfak	308 57.0	N49 51.0
Nunki	76 12.2	S26 17.8
Peacock	53 36.7	S56 44.2
Pollux	243 42.1	N28 01.7
Procyon	245 12.1	N 5 13.7
Rasalhague	96 16.8	N12 33.9
Regulus	207 56.0	N11 58.5
Rigel	281 23.4	S 8 12.3
Rigil Kent.	140 07.2	S60 49.8
Sabik	102 25.4	S15 43.2
Schedar	349 53.4	N56 31.4
Shaula	96 37.1	S37 06.1
Sirius	258 44.2	S16 42.9
Spica	158 43.4	S11 09.1
Suhail	223 01.3	S43 25.7
Vega	80 46.3	N38 47.2
Zuben'ubi	137 18.0	S16 02.0

	SHA	Mer. Pass.
	° ′	h m
Venus	274 43.9	10 07
Mars	262 54.5	10 53
Jupiter	1 15.2	4 20
Saturn	328 32.1	6 30

UT	SUN GHA	SUN Dec	MOON GHA	v	Dec	d	HP
d h	° ′	° ′	° ′	′	° ′	′	′
15 00	178 31.8	N21 35.2	290 58.5	10.0	S 1 59.6	11.4	59.2
01	193 31.7	34.9	305 27.5	10.0	1 48.2	11.5	59.2
02	208 31.6	34.5	319 56.5	10.0	1 36.7	11.5	59.2
03	223 31.6 ..	34.1	334 25.5	10.0	1 25.2	11.5	59.2
04	238 31.5	33.7	348 54.5	10.0	1 13.7	11.5	59.2
05	253 31.4	33.3	3 23.5	10.0	1 02.2	11.5	59.2
06	268 31.4	N21 32.9	17 52.5	9.9	S 0 50.7	11.5	59.2
W 07	283 31.3	32.5	32 21.4	10.0	0 39.2	11.5	59.2
E 08	298 31.3	32.1	46 50.4	10.0	0 27.7	11.6	59.2
D 09	313 31.2 ..	31.7	61 19.4	10.0	0 16.1	11.5	59.3
N 10	328 31.1	31.3	75 48.4	10.0	S 0 04.6	11.5	59.3
E 11	343 31.1	30.9	90 17.4	9.9	N 0 06.9	11.6	59.3
S 12	358 31.0	N21 30.6	104 46.3	10.0	N 0 18.5	11.5	59.3
D 13	13 30.9	30.2	119 15.3	10.0	0 30.0	11.5	59.3
A 14	28 30.9	29.8	133 44.3	9.9	0 41.5	11.5	59.3
Y 15	43 30.8 ..	29.4	148 13.2	10.0	0 53.0	11.6	59.3
16	58 30.7	29.0	162 42.2	9.9	1 04.6	11.5	59.3
17	73 30.7	28.6	177 11.1	10.0	1 16.1	11.5	59.3
18	88 30.6	N21 28.2	191 40.1	9.9	N 1 27.6	11.5	59.3
19	103 30.6	27.8	206 09.0	9.9	1 39.1	11.5	59.3
20	118 30.5	27.4	220 37.9	9.9	1 50.6	11.5	59.3
21	133 30.4 ..	27.0	235 06.8	9.9	2 02.1	11.4	59.3
22	148 30.4	26.6	249 35.7	9.9	2 13.5	11.5	59.3
23	163 30.3	26.2	264 04.6	9.9	2 25.0	11.5	59.3
16 00	178 30.3	N21 25.8	278 33.5	9.9	N 2 36.5	11.4	59.3
01	193 30.2	25.4	293 02.4	9.8	2 47.9	11.4	59.3
02	208 30.1	25.0	307 31.2	9.9	2 59.3	11.4	59.3
03	223 30.1 ..	24.6	322 00.1	9.8	3 10.7	11.4	59.3
04	238 30.0	24.1	336 28.9	9.8	3 22.1	11.3	59.3
05	253 30.0	23.7	350 57.7	9.8	3 33.4	11.4	59.3
06	268 29.9	N21 23.3	5 26.5	9.8	N 3 44.8	11.3	59.3
T 07	283 29.8	22.9	19 55.3	9.8	3 56.1	11.3	59.3
H 08	298 29.8	22.5	34 24.1	9.7	4 07.4	11.3	59.3
U 09	313 29.7 ..	22.1	48 52.8	9.8	4 18.7	11.2	59.3
R 10	328 29.7	21.7	63 21.6	9.7	4 29.9	11.3	59.3
S 11	343 29.6	21.3	77 50.3	9.7	4 41.2	11.2	59.3
D 12	358 29.5	N21 20.9	92 19.0	9.7	N 4 52.4	11.1	59.3
A 13	13 29.5	20.5	106 47.7	9.6	5 03.5	11.2	59.3
Y 14	28 29.4	20.1	121 16.3	9.7	5 14.7	11.1	59.3
15	43 29.4 ..	19.7	135 45.0	9.6	5 25.8	11.1	59.3
16	58 29.3	19.2	150 13.6	9.6	5 36.9	11.0	59.3
17	73 29.3	18.8	164 42.2	9.6	5 47.9	11.0	59.3
18	88 29.2	N21 18.4	179 10.8	9.5	N 5 58.9	11.0	59.3
19	103 29.1	18.0	193 39.3	9.6	6 09.9	11.0	59.3
20	118 29.1	17.6	208 07.9	9.5	6 20.9	10.9	59.3
21	133 29.0 ..	17.2	222 36.4	9.5	6 31.8	10.8	59.3
22	148 29.0	16.8	237 04.9	9.4	6 42.6	10.9	59.3
23	163 28.9	16.3	251 33.3	9.5	6 53.5	10.7	59.3
17 00	178 28.9	N21 15.9	266 01.8	9.4	N 7 04.2	10.8	59.3
01	193 28.8	15.5	280 30.2	9.4	7 15.0	10.7	59.3
02	208 28.7	15.1	294 58.6	9.3	7 25.7	10.7	59.3
03	223 28.7 ..	14.7	309 26.9	9.4	7 36.4	10.6	59.3
04	238 28.6	14.2	323 55.3	9.3	7 47.0	10.5	59.3
05	253 28.6	13.8	338 23.6	9.3	7 57.5	10.6	59.3
06	268 28.5	N21 13.4	352 51.9	9.2	N 8 08.1	10.4	59.3
F 07	283 28.5	13.0	7 20.1	9.2	8 18.5	10.5	59.3
R 08	298 28.4	12.6	21 48.3	9.2	8 29.0	10.3	59.3
I 09	313 28.4 ..	12.1	36 16.5	9.2	8 39.3	10.4	59.3
D 10	328 28.3	11.7	50 44.7	9.1	8 49.7	10.2	59.3
A 11	343 28.3	11.3	65 12.8	9.1	8 59.9	10.2	59.3
Y 12	358 28.2	N21 10.9	79 40.9	9.1	N 9 10.1	10.2	59.3
13	13 28.2	10.4	94 09.0	9.1	9 20.3	10.1	59.3
14	28 28.1	10.0	108 37.1	9.0	9 30.4	10.1	59.3
15	43 28.1 ..	09.6	123 05.1	9.0	9 40.5	9.9	59.3
16	58 28.0	09.1	137 33.1	8.9	9 50.4	10.0	59.3
17	73 27.9	08.7	152 01.0	8.9	10 00.4	9.8	59.3
18	88 27.9	N21 08.3	166 28.9	8.9	N10 10.2	9.9	59.3
19	103 27.8	07.9	180 56.8	8.9	10 20.1	9.7	59.3
20	118 27.8	07.4	195 24.7	8.8	10 29.8	9.7	59.2
21	133 27.7 ..	07.0	209 52.5	8.8	10 39.5	9.6	59.2
22	148 27.7	06.6	224 20.3	8.8	10 49.1	9.6	59.2
23	163 27.6	06.1	238 48.1	8.7	N10 58.7	9.4	59.2
	SD 15.8 d 0.4		SD 16.1	16.2		16.2	

Twilight / Sunrise / Moonrise

Lat.	Twilight Naut.	Twilight Civil	Sunrise	Moonrise 15	16	17	18
°	h m	h m	h m	h m	h m	h m	h m
N 72	▢	▢	▢	22 56	22 49	22 41	22 31
N 70	▢	▢	▢	22 59	22 58	22 58	23 00
68	▢	▢	▢	23 01	23 06	23 12	23 21
66	////	////	01 39	23 03	23 12	23 23	23 38
64	////	////	02 18	23 05	23 18	23 33	23 52
62	////	00 55	02 44	23 07	23 23	23 41	24 04
60	////	01 47	03 05	23 08	23 27	23 48	24 14
N 58	////	02 18	03 22	23 09	23 31	23 55	24 23
56	00 51	02 40	03 36	23 11	23 34	24 00	00 00
54	01 38	02 58	03 48	23 12	23 37	24 05	00 05
52	02 07	03 13	03 59	23 13	23 40	24 10	00 10
50	02 28	03 26	04 08	23 13	23 42	24 14	00 14
45	03 06	03 52	04 28	23 15	23 48	24 23	00 23
N 40	03 33	04 12	04 44	23 17	23 53	24 31	00 31
35	03 53	04 29	04 58	23 18	23 57	24 37	00 37
30	04 10	04 43	05 09	23 19	24 00	00 00	00 43
20	04 37	05 05	05 29	23 21	24 06	00 06	00 53
N 10	04 57	05 24	05 47	23 23	24 12	00 12	01 02
0	05 14	05 40	06 02	23 25	24 17	00 17	01 10
S 10	05 30	05 56	06 18	23 27	24 22	00 22	01 19
20	05 44	06 11	06 35	23 29	24 28	00 28	01 28
30	05 59	06 28	06 54	23 31	24 34	00 34	01 38
35	06 06	06 38	07 05	23 33	24 38	00 38	01 44
40	06 15	06 48	07 18	23 34	24 43	00 43	01 51
45	06 23	07 00	07 33	23 36	24 48	00 48	01 59
S 50	06 33	07 14	07 51	23 38	24 54	00 54	02 09
52	06 38	07 20	08 00	23 39	24 56	00 56	02 13
54	06 43	07 27	08 09	23 40	24 59	00 59	02 18
56	06 48	07 35	08 20	23 41	25 03	01 03	02 24
58	06 53	07 44	08 33	23 43	25 07	01 07	02 30
S 60	07 00	07 54	08 47	23 44	25 11	01 11	02 37

Sunset / Twilight / Moonset

Lat.	Sunset	Twilight Civil	Twilight Naut.	Moonset 15	16	17	18
°	h m	h m	h m	h m	h m	h m	h m
N 72	▢	▢	▢	10 58	12 54	14 51	16 54
N 70	▢	▢	▢	10 58	12 46	14 36	16 26
68	▢	▢	▢	10 58	12 40	14 23	16 06
66	22 30	////	////	10 58	12 36	14 13	15 50
64	21 52	////	////	10 58	12 31	14 05	15 37
62	21 26	23 11	////	10 58	12 28	13 57	15 26
60	21 06	22 22	////	10 58	12 25	13 51	15 16
N 58	20 49	21 53	////	10 58	12 22	13 46	15 08
56	20 35	21 30	23 16	10 58	12 20	13 41	15 01
54	20 23	21 13	22 31	10 58	12 18	13 37	14 54
52	20 13	20 58	22 04	10 58	12 16	13 33	14 49
50	20 03	20 45	21 43	10 58	12 14	13 29	14 43
45	19 43	20 19	21 05	10 58	12 10	13 21	14 32
N 40	19 27	19 59	20 39	10 58	12 07	13 15	14 23
35	19 14	19 43	20 18	10 58	12 04	13 10	14 15
30	19 02	19 29	20 02	10 58	12 01	13 05	14 08
20	18 43	19 07	19 35	10 58	11 57	12 56	13 56
N 10	18 25	18 48	19 15	10 58	11 53	12 49	13 46
0	18 10	18 32	18 58	10 58	11 50	12 42	13 36
S 10	17 54	18 16	18 42	10 58	11 46	12 35	13 26
20	17 37	18 01	18 28	10 58	11 42	12 28	13 16
30	17 18	17 44	18 14	10 58	11 38	12 20	13 04
35	17 07	17 35	18 06	10 58	11 36	12 15	12 57
40	16 55	17 24	17 58	10 58	11 33	12 10	12 49
45	16 40	17 13	17 49	10 57	11 30	12 03	12 40
S 50	16 21	16 59	17 39	10 57	11 26	11 56	12 29
52	16 13	16 52	17 35	10 57	11 24	11 52	12 24
54	16 03	16 45	17 30	10 57	11 22	11 49	12 18
56	15 52	16 37	17 25	10 57	11 20	11 44	12 12
58	15 40	16 29	17 19	10 57	11 18	11 40	12 05
S 60	15 25	16 19	17 13	10 57	11 15	11 35	11 58

SUN / MOON

Day	SUN Eqn. of Time 00h	SUN Eqn. of Time 12h	Mer. Pass.	MOON Mer. Pass. Upper	MOON Mer. Pass. Lower	Age	Phase
d	m s	m s	h m	h m	h m	d	%
15	05 53	05 56	12 06	04 46	17 12	21	63
16	05 59	06 02	12 06	05 37	18 03	22	52
17	06 04	06 07	12 06	06 30	18 56	23	40

1998 JULY 18, 19, 20 (SAT., SUN., MON.)

UT	ARIES GHA	VENUS −3.9 GHA	Dec	MARS +1.6 GHA	Dec	JUPITER −2.6 GHA	Dec	SATURN +0.4 GHA	Dec
18 00	295 36.1	207 44.1	N22 37.0	197 02.8	N23 54.1	296 50.7	S 2 00.6	264 02.3	N10 10.2
01	310 38.6	222 43.3	37.2	212 03.5	54.0	311 53.1	00.6	279 04.7	10.2
02	325 41.0	237 42.5	37.3	227 04.1	53.9	326 55.6	00.7	294 07.0	10.3
03	340 43.5	252 41.7 ..	37.5	242 04.8 ..	53.9	341 58.1 ..	00.7	309 09.3 ..	10.3
04	355 46.0	267 40.9	37.6	257 05.4	53.8	357 00.5	00.7	324 11.7	10.3
05	10 48.4	282 40.1	37.8	272 06.0	53.7	12 03.0	00.7	339 14.0	10.4
06	25 50.9	297 39.3	N22 37.9	287 06.7	N23 53.6	27 05.4	S 2 00.7	354 16.4	N10 10.4
S 07	40 53.3	312 38.5	38.1	302 07.3	53.5	42 07.9	00.7	9 18.7	10.4
A 08	55 55.8	327 37.7	38.2	317 08.0	53.5	57 10.4	00.7	24 21.1	10.4
T 09	70 58.3	342 36.9 ..	38.4	332 08.6 ..	53.4	72 12.8 ..	00.7	39 23.4 ..	10.5
U 10	86 00.7	357 36.2	38.5	347 09.2	53.3	87 15.3	00.8	54 25.8	10.5
R 11	101 03.2	12 35.4	38.7	2 09.9	53.2	102 17.7	00.8	69 28.1	10.5
D 12	116 05.7	27 34.6	N22 38.8	17 10.5	N23 53.1	117 20.2	S 2 00.8	84 30.5	N10 10.6
A 13	131 08.1	42 33.8	38.9	32 11.2	53.1	132 22.7	00.8	99 32.8	10.6
Y 14	146 10.6	57 33.0	39.1	47 11.8	53.0	147 25.1	00.8	114 35.2	10.6
15	161 13.1	72 32.2 ..	39.2	62 12.4 ..	52.9	162 27.6 ..	00.8	129 37.5 ..	10.7
16	176 15.5	87 31.4	39.4	77 13.1	52.8	177 30.1	00.8	144 39.8	10.7
17	191 18.0	102 30.6	39.5	92 13.7	52.7	192 32.5	00.8	159 42.2	10.7
18	206 20.4	117 29.8	N22 39.6	107 14.4	N23 52.7	207 35.0	S 2 00.9	174 44.5	N10 10.8
19	221 22.9	132 29.0	39.8	122 15.0	52.6	222 37.5	00.9	189 46.9	10.8
20	236 25.4	147 28.2	39.9	137 15.6	52.5	237 39.9	00.9	204 49.2	10.8
21	251 27.8	162 27.4 ..	40.0	152 16.3 ..	52.4	252 42.4 ..	00.9	219 51.6 ..	10.8
22	266 30.3	177 26.6	40.2	167 16.9	52.3	267 44.8	00.9	234 53.9	10.9
23	281 32.8	192 25.8	40.3	182 17.6	52.2	282 47.3	00.9	249 56.3	10.9
19 00	296 35.2	207 25.0	N22 40.4	197 18.2	N23 52.1	297 49.8	S 2 00.9	264 58.6	N10 10.9
01	311 37.7	222 24.3	40.6	212 18.8	52.1	312 52.2	01.0	280 01.0	11.0
02	326 40.2	237 23.5	40.7	227 19.5	52.0	327 54.7	01.0	295 03.3	11.0
03	341 42.6	252 22.7 ..	40.8	242 20.1 ..	51.9	342 57.2 ..	01.0	310 05.7 ..	11.0
04	356 45.1	267 21.9	40.9	257 20.8	51.8	357 59.6	01.0	325 08.0	11.1
05	11 47.6	282 21.1	41.1	272 21.4	51.7	13 02.1	01.0	340 10.4	11.1
06	26 50.0	297 20.3	N22 41.2	287 22.1	N23 51.6	28 04.6	S 2 01.0	355 12.7	N10 11.1
07	41 52.5	312 19.5	41.3	302 22.7	51.5	43 07.1	01.0	10 15.1	11.1
08	56 54.9	327 18.7	41.4	317 23.3	51.5	58 09.5	01.1	25 17.4	11.2
S 09	71 57.4	342 17.9 ..	41.5	332 24.0 ..	51.4	73 12.0 ..	01.1	40 19.8 ..	11.2
U 10	86 59.9	357 17.1	41.7	347 24.6	51.3	88 14.5	01.1	55 22.1	11.2
N 11	102 02.3	12 16.3	41.8	2 25.3	51.2	103 16.9	01.1	70 24.5	11.3
D 12	117 04.8	27 15.5	N22 41.9	17 25.9	N23 51.1	118 19.4	S 2 01.1	85 26.8	N10 11.3
A 13	132 07.3	42 14.7	42.0	32 26.5	51.0	133 21.9	01.1	100 29.2	11.3
Y 14	147 09.7	57 13.9	42.1	47 27.2	50.9	148 24.3	01.2	115 31.5	11.3
15	162 12.2	72 13.1 ..	42.2	62 27.8 ..	50.8	163 26.8 ..	01.2	130 33.9 ..	11.4
16	177 14.7	87 12.3	42.3	77 28.5	50.7	178 29.3	01.2	145 36.2	11.4
17	192 17.1	102 11.5	42.5	92 29.1	50.7	193 31.7	01.2	160 38.6	11.4
18	207 19.6	117 10.7	N22 42.6	107 29.8	N23 50.6	208 34.2	S 2 01.2	175 40.9	N10 11.5
19	222 22.1	132 09.9	42.7	122 30.4	50.5	223 36.7	01.2	190 43.3	11.5
20	237 24.5	147 09.1	42.8	137 31.1	50.4	238 39.2	01.3	205 45.6	11.5
21	252 27.0	162 08.3 ..	42.9	152 31.7 ..	50.3	253 41.6 ..	01.3	220 48.0 ..	11.5
22	267 29.4	177 07.5	43.0	167 32.3	50.2	268 44.1	01.3	235 50.3	11.6
23	282 31.9	192 06.7	43.1	182 33.0	50.1	283 46.6	01.3	250 52.7	11.6
20 00	297 34.4	207 05.9	N22 43.2	197 33.6	N23 50.0	298 49.1	S 2 01.3	265 55.0	N10 11.7
01	312 36.8	222 05.1	43.3	212 34.3	49.9	313 51.5	01.3	280 57.4	11.7
02	327 39.3	237 04.3	43.4	227 34.9	49.8	328 54.0	01.4	295 59.7	11.7
03	342 41.8	252 03.5 ..	43.5	242 35.6 ..	49.7	343 56.5 ..	01.4	311 02.1 ..	11.7
04	357 44.2	267 02.7	43.6	257 36.2	49.6	358 59.0	01.4	326 04.5	11.7
05	12 46.7	282 01.9	43.7	272 36.9	49.5	14 01.4	01.4	341 06.8	11.8
06	27 49.2	297 01.1	N22 43.8	287 37.5	N23 49.4	29 03.9	S 2 01.4	356 09.2	N10 11.8
07	42 51.6	312 00.3	43.9	302 38.1	49.3	44 06.4	01.5	11 11.5	11.8
08	57 54.1	326 59.5	44.0	317 38.8	49.2	59 08.9	01.5	26 13.9	11.9
M 09	72 56.6	341 58.7 ..	44.1	332 39.4 ..	49.2	74 11.3 ..	01.5	41 16.2 ..	11.9
O 10	87 59.0	356 57.9	44.2	347 40.1	49.1	89 13.8	01.5	56 18.6	11.9
N 11	103 01.5	11 57.1	44.3	2 40.7	49.0	104 16.3	01.5	71 20.9	11.9
D 12	118 03.9	26 56.3	N22 44.3	17 41.4	N23 48.9	119 18.8	S 2 01.6	86 23.3	N10 12.0
A 13	133 06.4	41 55.5	44.4	32 42.0	48.8	134 21.2	01.6	101 25.6	12.0
Y 14	148 08.9	56 54.7	44.5	47 42.7	48.7	149 23.7	01.6	116 28.0	12.0
15	163 11.3	71 53.9 ..	44.6	62 43.3 ..	48.6	164 26.2 ..	01.6	131 30.3 ..	12.1
16	178 13.8	86 53.1	44.7	77 44.0	48.5	179 28.7	01.6	146 32.7	12.1
17	193 16.3	101 52.3	44.8	92 44.6	48.4	194 31.2	01.7	161 35.1	12.1
18	208 18.7	116 51.5	N22 44.9	107 45.3	N23 48.3	209 33.6	S 2 01.7	176 37.4	N10 12.1
19	223 21.2	131 50.7	44.9	122 45.9	48.2	224 36.1	01.7	191 39.8	12.2
20	238 23.7	146 49.9	45.0	137 46.5	48.1	239 38.6	01.7	206 42.1	12.2
21	253 26.1	161 49.1 ..	45.1	152 47.2 ..	48.0	254 41.1 ..	01.7	221 44.5 ..	12.2
22	268 28.6	176 48.3	45.2	167 47.8	47.9	269 43.6	01.8	236 46.8	12.2
23	283 31.1	191 47.5	45.3	182 48.5	47.8	284 46.0	01.8	251 49.2	12.3
Mer. Pass.	h m 4 13.0	v −0.8	d 0.1	v 0.6	d 0.1	v 2.5	d 0.0	v 2.4	d 0.0

STARS

Name	SHA	Dec
Acamar	315 27.2	S40 18.5
Achernar	335 35.4	S57 14.4
Acrux	173 22.3	S63 05.7
Adhara	255 21.9	S28 58.2
Aldebaran	291 02.8	N16 30.2
Alioth	166 30.9	N55 58.4
Alkaid	153 08.0	N49 19.6
Al Na'ir	27 57.7	S46 57.9
Alnilam	275 58.3	S 1 12.3
Alphard	218 07.6	S 8 39.2
Alphecca	126 20.6	N26 43.5
Alpheratz	357 55.2	N29 04.8
Altair	62 19.1	N 8 52.0
Ankaa	353 26.9	S42 18.6
Antares	112 40.1	S26 25.6
Arcturus	146 06.2	N19 11.7
Atria	107 51.8	S69 01.6
Avior	234 23.3	S59 30.4
Bellatrix	278 44.6	N 6 20.8
Betelgeuse	271 14.0	N 7 24.3
Canopus	264 01.7	S52 41.7
Capella	280 51.8	N45 59.6
Deneb	49 38.8	N45 16.6
Denebola	182 45.5	N14 35.0
Diphda	349 07.3	S17 59.6
Dubhe	194 06.2	N61 45.8
Elnath	278 27.4	N28 36.2
Eltanin	90 51.0	N51 29.6
Enif	33 58.1	N 9 52.2
Fomalhaut	15 36.4	S29 37.6
Gacrux	172 13.8	S57 06.5
Gienah	176 04.2	S17 32.0
Hadar	149 04.1	S60 22.1
Hamal	328 13.7	N23 27.1
Kaus Aust.	83 58.6	S34 23.0
Kochab	137 19.4	N74 10.1
Markab	13 49.5	N15 11.8
Menkar	314 27.2	N 4 04.9
Menkent	148 21.0	S36 21.8
Miaplacidus	221 42.9	S69 42.8
Mirfak	308 57.0	N49 51.0
Nunki	76 12.2	S26 17.8
Peacock	53 36.7	S56 44.2
Pollux	243 42.1	N28 01.7
Procyon	245 12.1	N 5 13.7
Rasalhague	96 16.8	N12 33.9
Regulus	207 56.0	N11 58.5
Rigel	281 23.4	S 8 12.3
Rigil Kent.	140 07.3	S60 49.8
Sabik	102 25.4	S15 43.2
Schedar	349 53.4	N56 31.5
Shaula	96 37.1	S37 06.1
Sirius	258 44.2	S16 42.9
Spica	158 43.4	S11 09.1
Suhail	223 01.3	S43 25.7
Vega	80 46.3	N38 47.2
Zuben'ubi	137 18.0	S16 02.0

	SHA	Mer. Pass.
	o '	h m
Venus	270 49.8	10 11
Mars	260 43.0	10 50
Jupiter	1 14.5	4 08
Saturn	328 23.4	6 19

UT	SUN GHA	SUN Dec	MOON GHA	v	MOON Dec	d	HP
d h	° ′	° ′	° ′	′	° ′	′	′
18 00	178 27.6	N21 05.7	253 15.8	8.7	N11 08.1	9.5	59.2
01	193 27.5	05.3	267 43.5	8.6	11 17.6	9.3	59.2
02	208 27.5	04.8	282 11.1	8.7	11 26.9	9.3	59.2
03	223 27.4	.. 04.4	296 38.8	8.6	11 36.2	9.2	59.2
04	238 27.4	04.0	311 06.4	8.5	11 45.4	9.1	59.2
05	253 27.3	03.5	325 33.9	8.6	11 54.5	9.1	59.2
06	268 27.3	N21 03.1	340 01.5	8.4	N12 03.6	9.0	59.2
S 07	283 27.2	02.7	354 28.9	8.5	12 12.6	8.9	59.2
A 08	298 27.2	02.2	8 56.4	8.4	12 21.5	8.9	59.2
T 09	313 27.1	.. 01.8	23 23.8	8.4	12 30.4	8.7	59.2
U 10	328 27.1	01.3	37 51.2	8.4	12 39.1	8.7	59.2
R 11	343 27.1	00.9	52 18.6	8.3	12 47.8	8.6	59.2
D 12	358 27.0	N21 00.5	66 45.9	8.3	N12 56.4	8.5	59.2
A 13	13 27.0	21 00.0	81 13.2	8.2	13 04.9	8.5	59.1
Y 14	28 26.9	20 59.6	95 40.4	8.3	13 13.4	8.4	59.1
15	43 26.9	.. 59.1	110 07.7	8.1	13 21.8	8.2	59.1
16	58 26.8	58.7	124 34.8	8.2	13 30.0	8.2	59.1
17	73 26.8	58.3	139 02.0	8.1	13 38.2	8.1	59.1
18	88 26.7	N20 57.8	153 29.1	8.1	N13 46.3	8.1	59.1
19	103 26.7	57.4	167 56.2	8.0	13 54.4	7.9	59.1
20	118 26.6	56.9	182 23.2	8.1	14 02.3	7.9	59.1
21	133 26.6	.. 56.5	196 50.3	7.9	14 10.2	7.7	59.1
22	148 26.5	56.0	211 17.2	8.0	14 17.9	7.7	59.1
23	163 26.5	55.6	225 44.2	7.9	14 25.6	7.6	59.1
19 00	178 26.4	N20 55.1	240 11.1	7.9	N14 33.2	7.5	59.1
01	193 26.4	54.7	254 38.0	7.9	14 40.7	7.4	59.1
02	208 26.4	54.2	269 04.9	7.8	14 48.1	7.3	59.0
03	223 26.3	.. 53.8	283 31.7	7.8	14 55.4	7.2	59.0
04	238 26.3	53.3	297 58.5	7.7	15 02.6	7.1	59.0
05	253 26.2	52.9	312 25.2	7.8	15 09.7	7.1	59.0
06	268 26.2	N20 52.4	326 52.0	7.7	N15 16.8	6.9	59.0
S 07	283 26.1	52.0	341 18.7	7.6	15 23.7	6.8	59.0
U 08	298 26.1	51.5	355 45.3	7.7	15 30.5	6.8	59.0
N 09	313 26.1	.. 51.1	10 12.0	7.6	15 37.3	6.6	59.0
D 10	328 26.0	50.6	24 38.6	7.6	15 43.9	6.5	59.0
A 11	343 26.0	50.2	39 05.2	7.5	15 50.4	6.5	59.0
Y 12	358 25.9	N20 49.7	53 31.7	7.6	N15 56.9	6.3	58.9
13	13 25.9	49.3	67 58.3	7.5	16 03.2	6.2	58.9
14	28 25.8	48.8	82 24.8	7.4	16 09.4	6.2	58.9
15	43 25.8	.. 48.4	96 51.2	7.5	16 15.6	6.0	58.9
16	58 25.8	47.9	111 17.7	7.4	16 21.6	5.9	58.9
17	73 25.7	47.4	125 44.1	7.4	16 27.5	5.9	58.9
18	88 25.7	N20 47.0	140 10.5	7.4	N16 33.4	5.7	58.9
19	103 25.6	46.5	154 36.9	7.3	16 39.1	5.6	58.9
20	118 25.6	46.1	169 03.2	7.4	16 44.7	5.5	58.9
21	133 25.6	.. 45.6	183 29.6	7.3	16 50.2	5.4	58.8
22	148 25.5	45.1	197 55.9	7.3	16 55.6	5.3	58.8
23	163 25.5	44.7	212 22.2	7.2	17 00.9	5.1	58.8
20 00	178 25.4	N20 44.2	226 48.4	7.3	N17 06.0	5.1	58.8
01	193 25.4	43.8	241 14.7	7.2	17 11.1	5.0	58.8
02	208 25.4	43.3	255 40.9	7.2	17 16.1	4.8	58.8
03	223 25.3	.. 42.8	270 07.1	7.2	17 20.9	4.8	58.8
04	238 25.3	42.4	284 33.3	7.1	17 25.7	4.6	58.8
05	253 25.2	41.9	298 59.4	7.2	17 30.3	4.5	58.7
06	268 25.2	N20 41.4	313 25.6	7.1	N17 34.8	4.4	58.7
07	283 25.2	41.0	327 51.7	7.2	17 39.2	4.3	58.7
08	298 25.1	40.5	342 17.9	7.1	17 43.5	4.2	58.7
M 09	313 25.1	.. 40.0	356 44.0	7.1	17 47.7	4.1	58.7
O 10	328 25.1	39.6	11 10.1	7.0	17 51.8	3.9	58.7
N 11	343 25.0	39.1	25 36.1	7.1	17 55.7	3.9	58.7
D 12	358 25.0	N20 38.6	40 02.2	7.1	N17 59.6	3.7	58.6
A 13	13 24.9	38.2	54 28.3	7.0	18 03.3	3.6	58.6
Y 14	28 24.9	37.7	68 54.3	7.1	18 06.9	3.5	58.6
15	43 24.9	.. 37.2	83 20.4	7.0	18 10.4	3.4	58.6
16	58 24.8	36.7	97 46.4	7.0	18 13.8	3.2	58.6
17	73 24.8	36.3	112 12.4	7.1	18 17.0	3.2	58.6
18	88 24.8	N20 35.8	126 38.5	7.0	N18 20.2	3.0	58.6
19	103 24.7	35.3	141 04.5	7.0	18 23.2	2.9	58.5
20	118 24.7	34.9	155 30.5	7.0	18 26.1	2.8	58.5
21	133 24.7	.. 34.4	169 56.5	7.0	18 28.9	2.7	58.5
22	148 24.6	33.9	184 22.5	7.0	18 31.6	2.6	58.5
23	163 24.6	33.4	198 48.5	7.0	N18 34.2	2.4	58.5
	SD 15.8	d 0.5	SD 16.1		16.1		16.0

Lat.	Naut. Twilight	Civil Twilight	Sunrise	Moonrise 18	Moonrise 19	Moonrise 20	Moonrise 21
°	h m	h m	h m	h m	h m	h m	h m
N 72	□	□	□	22 31	22 16	□	□
N 70	□	□	□	23 00	23 05	23 20	24 02
68	////	////	00 43	23 21	23 37	24 04	00 04
66	////	////	01 52	23 38	24 00	00 00	00 34
64	////	////	02 27	23 52	24 19	00 19	00 56
62	////	01 14	02 51	24 04	00 04	00 34	01 14
60	////	01 57	03 11	24 14	00 14	00 47	01 29
N 58	////	02 25	03 27	24 23	00 23	00 58	01 41
56	01 08	02 46	03 41	00 00	00 31	01 07	01 52
54	01 47	03 03	03 52	00 05	00 38	01 16	02 02
52	02 13	03 18	04 03	00 10	00 44	01 24	02 10
50	02 33	03 30	04 12	00 14	00 49	01 30	02 18
45	03 10	03 56	04 31	00 23	01 02	01 45	02 35
N 40	03 36	04 15	04 47	00 31	01 12	01 58	02 48
35	03 56	04 31	05 00	00 37	01 21	02 08	03 00
30	04 12	04 44	05 11	00 43	01 28	02 17	03 10
20	04 38	05 06	05 30	00 53	01 42	02 33	03 27
N 10	04 58	05 25	05 47	01 02	01 53	02 42	03 42
0	05 15	05 41	06 03	01 10	02 04	03 00	03 57
S 10	05 30	05 56	06 18	01 19	02 16	03 13	04 11
20	05 44	06 11	06 34	01 28	02 28	03 27	04 26
30	05 58	06 27	06 53	01 38	02 41	03 44	04 44
35	06 05	06 36	07 04	01 44	02 49	03 53	04 54
40	06 13	06 46	07 16	01 51	02 58	04 04	05 06
45	06 22	06 58	07 30	01 59	03 09	04 17	05 20
S 50	06 31	07 11	07 48	02 09	03 22	04 32	05 36
52	06 35	07 18	07 56	02 13	03 28	04 39	05 44
54	06 40	07 24	08 06	02 18	03 35	04 48	05 53
56	06 45	07 32	08 16	02 24	03 42	04 57	06 03
58	06 50	07 40	08 28	02 30	03 51	05 07	06 15
S 60	06 56	07 50	08 42	02 37	04 01	05 19	06 28

Lat.	Sunset	Civil Twilight	Naut. Twilight	Moonset 18	Moonset 19	Moonset 20	Moonset 21
°	h m	h m	h m	h m	h m	h m	h m
N 72	□	□	□	16 54	19 05		
N 70	□	□	□	16 26	18 16	19 59	21 15
68	23 19	////	////	16 06	17 45	19 15	20 24
66	22 17	////	////	15 50	17 23	18 46	19 53
64	21 44	////	////	15 37	17 05	18 24	19 29
62	21 19	22 54	////	15 26	16 50	18 06	19 10
60	21 00	22 13	////	15 16	16 38	17 52	18 55
N 58	20 44	21 46	////	15 08	16 27	17 39	18 42
56	20 31	21 25	23 01	15 01	16 18	17 29	18 31
54	20 19	21 08	22 23	14 54	16 09	17 19	18 21
52	20 09	20 54	21 57	14 49	16 02	17 11	18 12
50	20 00	20 41	21 38	14 43	15 56	17 03	18 04
45	19 41	20 16	21 02	14 32	15 41	16 47	17 47
N 40	19 25	19 57	20 36	14 23	15 30	16 34	17 33
35	19 12	19 41	20 16	14 15	15 20	16 22	17 21
30	19 01	19 28	20 00	14 09	15 11	16 13	17 11
20	18 42	19 06	19 34	13 56	14 56	15 56	16 53
N 10	18 25	18 48	19 14	13 46	14 43	15 41	16 38
0	18 10	18 32	18 58	13 36	14 31	15 27	16 24
S 10	17 55	18 17	18 43	13 26	14 19	15 13	16 09
20	17 38	18 02	18 29	13 16	14 06	14 58	15 54
30	17 20	17 46	18 15	13 04	13 51	14 42	15 36
35	17 09	17 37	18 08	12 57	13 42	14 32	15 25
40	16 57	17 27	18 00	12 49	13 32	14 21	15 14
45	16 42	17 15	17 51	12 40	13 21	14 07	15 00
S 50	16 25	17 02	17 42	12 29	13 07	13 51	14 43
52	16 17	16 55	17 38	12 24	13 01	13 44	14 34
54	16 07	16 49	17 33	12 18	12 53	13 35	14 26
56	15 57	16 41	17 28	12 12	12 46	13 26	14 15
58	15 45	16 33	17 23	12 05	12 37	13 15	14 04
S 60	15 31	16 23	17 17	11 58	12 26	13 03	13 51

Day	SUN Eqn. of Time 00h	SUN Eqn. of Time 12h	SUN Mer. Pass.	MOON Mer. Pass. Upper	MOON Mer. Pass. Lower	Age	Phase
d	m s	m s	h m	h m	h m	d	%
18	06 10	06 12	12 06	07 23	19 50	24	29
19	06 14	06 16	12 06	08 18	20 45	25	19
20	06 18	06 20	12 06	09 14	21 42	26	11

1998 JULY 21, 22, 23 (TUES., WED., THURS.)

UT	ARIES GHA	VENUS −3.9 GHA	Dec	MARS +1.6 GHA	Dec	JUPITER −2.7 GHA	Dec	SATURN +0.4 GHA	Dec
21 00	298 33.5	206 46.7	N22 45.3	197 49.1	N23 47.7	299 48.5	S 2 01.8	266 51.5	N10 12.3
01	313 36.0	221 45.9	45.4	212 49.8	47.6	314 51.0	01.8	281 53.9	12.3
02	328 38.4	236 45.1	45.5	227 50.4	47.5	329 53.5	01.8	296 56.3	12.4
03	343 40.9	251 44.3 ..	45.6	242 51.1 ..	47.4	344 56.0 ..	01.9	311 58.6 ..	12.4
04	358 43.4	266 43.5	45.6	257 51.7	47.3	359 58.4	01.9	327 01.0	12.4
05	13 45.8	281 42.7	45.7	272 52.4	47.2	15 00.9	01.9	342 03.3	12.4
T 06	28 48.3	296 41.9	N22 45.8	287 53.0	N23 47.1	30 03.4	S 2 01.9	357 05.7	N10 12.5
U 07	43 50.8	311 41.1	45.8	302 53.7	47.0	45 05.9	01.9	12 08.0	12.5
E 08	58 53.2	326 40.3	45.9	317 54.3	46.8	60 08.4	02.0	27 10.4	12.5
S 09	73 55.7	341 39.5 ..	46.0	332 55.0 ..	46.7	75 10.9 ..	02.0	42 12.7 ..	12.5
D 10	88 58.2	356 38.7	46.0	347 55.6	46.6	90 13.3	02.0	57 15.1	12.6
A 11	104 00.6	11 37.9	46.1	2 56.3	46.5	105 15.8	02.0	72 17.5	12.6
Y 12	119 03.1	26 37.1	N22 46.2	17 56.9	N23 46.4	120 18.3	S 2 02.1	87 19.8	N10 12.6
13	134 05.5	41 36.3	46.2	32 57.6	46.3	135 20.8	02.1	102 22.2	12.7
14	149 08.0	56 35.5	46.3	47 58.2	46.2	150 23.3	02.1	117 24.5	12.7
15	164 10.5	71 34.6 ..	46.3	62 58.9 ..	46.1	165 25.8 ..	02.1	132 26.9 ..	12.7
16	179 12.9	86 33.8	46.4	77 59.5	46.0	180 28.3	02.2	147 29.3	12.7
17	194 15.4	101 33.0	46.5	93 00.2	45.9	195 30.7	02.2	162 31.6	12.8
18	209 17.9	116 32.2	N22 46.5	108 00.8	N23 45.8	210 33.2	S 2 02.2	177 34.0	N10 12.8
19	224 20.3	131 31.4	46.6	123 01.5	45.7	225 35.7	02.2	192 36.3	12.8
20	239 22.8	146 30.6	46.6	138 02.1	45.6	240 38.2	02.2	207 38.7	12.8
21	254 25.3	161 29.8 ..	46.7	153 02.8 ..	45.5	255 40.7 ..	02.3	222 41.0 ..	12.9
22	269 27.7	176 29.0	46.7	168 03.4	45.4	270 43.2	02.3	237 43.4	12.9
23	284 30.2	191 28.2	46.8	183 04.1	45.2	285 45.7	02.3	252 45.8	12.9
22 00	299 32.7	206 27.4	N22 46.8	198 04.7	N23 45.1	300 48.2	S 2 02.3	267 48.1	N10 12.9
01	314 35.1	221 26.6	46.9	213 05.4	45.0	315 50.6	02.4	282 50.5	13.0
02	329 37.6	236 25.8	46.9	228 06.0	44.9	330 53.1	02.4	297 52.8	13.0
03	344 40.0	251 25.0 ..	47.0	243 06.7 ..	44.8	345 55.6 ..	02.4	312 55.2 ..	13.0
04	359 42.5	266 24.2	47.0	258 07.3	44.7	0 58.1	02.4	327 57.6	13.0
05	14 45.0	281 23.4	47.0	273 08.0	44.6	16 00.6	02.5	342 59.9	13.1
W 06	29 47.4	296 22.6	N22 47.1	288 08.6	N23 44.5	31 03.1	S 2 02.5	358 02.3	N10 13.1
E 07	44 49.9	311 21.8	47.1	303 09.3	44.4	46 05.6	02.5	13 04.7	13.1
D 08	59 52.4	326 20.9	47.2	318 09.9	44.2	61 08.1	02.5	28 07.0	13.1
N 09	74 54.8	341 20.1 ..	47.2	333 10.6 ..	44.1	76 10.6 ..	02.6	43 09.4 ..	13.2
E 10	89 57.3	356 19.3	47.2	348 11.2	44.0	91 13.1	02.6	58 11.7	13.2
S 11	104 59.8	11 18.5	47.3	3 11.9	43.9	106 15.6	02.6	73 14.1	13.2
D 12	120 02.2	26 17.7	N22 47.3	18 12.5	N23 43.8	121 18.0	S 2 02.6	88 16.5	N10 13.2
A 13	135 04.7	41 16.9	47.4	33 13.2	43.7	136 20.5	02.7	103 18.8	13.3
Y 14	150 07.2	56 16.1	47.4	48 13.8	43.6	151 23.0	02.7	118 21.2	13.3
15	165 09.6	71 15.3 ..	47.4	63 14.5 ..	43.5	166 25.5 ..	02.7	133 23.5 ..	13.3
16	180 12.1	86 14.5	47.5	78 15.1	43.3	181 28.0	02.7	148 25.9	13.3
17	195 14.5	101 13.7	47.5	93 15.8	43.2	196 30.5	02.8	163 28.3	13.4
18	210 17.0	116 12.9	N22 47.5	108 16.4	N23 43.1	211 33.0	S 2 02.8	178 30.6	N10 13.4
19	225 19.5	131 12.1	47.5	123 17.1	43.0	226 35.5	02.8	193 33.0	13.4
20	240 21.9	146 11.3	47.6	138 17.7	42.9	241 38.0	02.9	208 35.4	13.4
21	255 24.4	161 10.5 ..	47.6	153 18.4 ..	42.8	256 40.5 ..	02.9	223 37.7 ..	13.5
22	270 26.9	176 09.6	47.6	168 19.1	42.6	271 43.0	02.9	238 40.1	13.5
23	285 29.3	191 08.8	47.6	183 19.7	42.5	286 45.5	02.9	253 42.4	13.5
23 00	300 31.8	206 08.0	N22 47.7	198 20.4	N23 42.4	301 48.0	S 2 03.0	268 44.8	N10 13.5
01	315 34.3	221 07.2	47.7	213 21.0	42.3	316 50.5	03.0	283 47.2	13.6
02	330 36.7	236 06.4	47.7	228 21.7	42.2	331 53.0	03.0	298 49.5	13.6
03	345 39.2	251 05.6 ..	47.7	243 22.3 ..	42.1	346 55.5 ..	03.0	313 51.9 ..	13.6
04	0 41.7	266 04.8	47.7	258 23.0	41.9	1 58.0	03.1	328 54.3	13.6
05	15 44.1	281 04.0	47.8	273 23.6	41.8	17 00.5	03.1	343 56.6	13.7
T 06	30 46.6	296 03.2	N22 47.8	288 24.3	N23 41.7	32 03.0	S 2 03.1	358 59.0	N10 13.7
H 07	45 49.0	311 02.4	47.8	303 24.9	41.6	47 05.5	03.2	14 01.4	13.7
U 08	60 51.5	326 01.6	47.8	318 25.6	41.5	62 08.0	03.2	29 03.7	13.7
R 09	75 54.0	341 00.7 ..	47.8	333 26.3 ..	41.3	77 10.5 ..	03.2	44 06.1 ..	13.8
S 10	90 56.4	355 59.9	47.8	348 26.9	41.2	92 13.0	03.2	59 08.5	13.8
D 11	105 58.9	10 59.1	47.8	3 27.6	41.1	107 15.5	03.3	74 10.8	13.8
A 12	121 01.4	25 58.3	N22 47.8	18 28.2	N23 41.0	122 18.0	S 2 03.3	89 13.2	N10 13.8
Y 13	136 03.8	40 57.5	47.8	33 28.9	40.8	137 20.5	03.3	104 15.6	13.8
14	151 06.3	55 56.7	47.9	48 29.5	40.7	152 23.0	03.4	119 17.9	13.9
15	166 08.8	70 55.9 ..	47.9	63 30.2 ..	40.6	167 25.5 ..	03.4	134 20.3 ..	13.9
16	181 11.2	85 55.1	47.9	78 30.8	40.5	182 28.0	03.4	149 22.7	13.9
17	196 13.7	100 54.3	47.9	93 31.5	40.4	197 30.5	03.5	164 25.0	13.9
18	211 16.1	115 53.5	N22 47.9	108 32.2	N23 40.2	212 33.0	S 2 03.5	179 27.4	N10 14.0
19	226 18.6	130 52.7	47.9	123 32.8	40.1	227 35.5	03.5	194 29.7	14.0
20	241 21.1	145 51.8	47.9	138 33.5	40.0	242 38.0	03.5	209 32.1	14.0
21	256 23.5	160 51.0 ..	47.9	153 34.1 ..	39.9	257 40.5 ..	03.6	224 34.5 ..	14.0
22	271 26.0	175 50.2	47.9	168 34.8	39.7	272 43.0	03.6	239 36.9	14.1
23	286 28.5	190 49.4	47.9	183 35.4	39.6	287 45.5	03.6	254 39.2	14.1
Mer.Pass.	4 01.2ʰ ᵐ	v −0.8 d 0.0		v 0.7 d 0.1		v 2.5 d 0.0		v 2.4 d 0.0	

STARS

Name	SHA	Dec
Acamar	315 27.2	S40 18.5
Achernar	335 35.3	S57 14.4
Acrux	173 22.3	S63 05.7
Adhara	255 21.9	S28 58.2
Aldebaran	291 02.8	N16 30.2
Alioth	166 31.0	N55 58.4
Alkaid	153 08.0	N49 19.6
Al Na'ir	27 57.7	S46 57.9
Alnilam	275 58.3	S 1 12.2
Alphard	218 07.6	S 8 39.1
Alphecca	126 20.6	N26 43.5
Alpheratz	357 55.2	N29 04.8
Altair	62 19.1	N 8 52.0
Ankaa	353 26.9	S42 18.6
Antares	112 40.1	S26 25.6
Arcturus	146 06.2	N19 11.7
Atria	107 51.8	S69 01.6
Avior	234 23.3	S59 30.4
Bellatrix	278 44.6	N 6 20.8
Betelgeuse	271 14.0	N 7 24.3
Canopus	264 01.7	S52 41.7
Capella	280 51.8	N45 59.5
Deneb	49 38.8	N45 16.6
Denebola	182 45.5	N14 35.0
Diphda	349 07.3	S17 59.6
Dubhe	194 06.3	N61 45.8
Elnath	278 27.4	N28 36.2
Eltanin	90 51.0	N51 29.7
Enif	33 58.1	N 9 52.2
Fomalhaut	15 36.4	S29 37.6
Gacrux	172 13.8	S57 06.5
Gienah	176 04.2	S17 32.0
Hadar	149 04.1	S60 22.1
Hamal	328 13.7	N23 27.2
Kaus Aust.	83 58.6	S34 23.0
Kochab	137 19.4	N74 10.1
Markab	13 49.5	N15 11.8
Menkar	314 27.1	N 4 05.0
Menkent	148 21.0	S36 21.8
Miaplacidus	221 42.9	S69 42.8
Mirfak	308 57.0	N49 51.0
Nunki	76 12.1	S26 17.8
Peacock	53 36.7	S56 44.2
Pollux	243 42.1	N28 01.7
Procyon	245 12.0	N 5 13.7
Rasalhague	96 16.8	N12 33.9
Regulus	207 56.0	N11 58.5
Rigel	281 23.3	S 8 12.2
Rigil Kent.	140 07.3	S60 49.8
Sabik	102 25.4	S15 43.2
Schedar	349 53.4	N56 31.5
Shaula	96 37.1	S37 06.1
Sirius	258 44.2	S16 42.9
Spica	158 43.4	S11 09.1
Suhail	223 01.3	S43 25.7
Vega	80 46.3	N38 47.2
Zuben'ubi	137 18.0	S16 02.0

	SHA	Mer.Pass.
Venus	266 54.7	10 15
Mars	258 32.1	10 47
Jupiter	1 15.5	3 56
Saturn	328 15.5	6 08

UT	SUN GHA	SUN Dec	MOON GHA	v	MOON Dec	d	HP
d h	° ′	° ′	° ′	′	° ′	′	′
21 00	178 24.6	N20 32.9	213 14.5	7.0	N18 36.6	2.3	58.5
01	193 24.5	32.5	227 40.5	7.0	18 38.9	2.3	58.4
02	208 24.5	.. 32.0	242 06.5	7.1	18 41.2	2.0	58.4
03	223 24.5	31.5	256 32.6	7.0	18 43.2	2.0	58.4
04	238 24.4	31.0	270 58.6	7.0	18 45.2	1.9	58.4
05	253 24.4	30.6	285 24.6	7.0	18 47.1	1.7	58.4
06	268 24.4	N20 30.1	299 50.6	7.1	N18 48.8	1.6	58.4
07	283 24.3	29.6	314 16.7	7.0	18 50.4	1.5	58.3
T 08	298 24.3	29.1	328 42.7	7.1	18 51.9	1.4	58.3
U 09	313 24.3	.. 28.6	343 08.8	7.1	18 53.3	1.3	58.3
E 10	328 24.2	28.1	357 34.9	7.0	18 54.6	1.1	58.3
S 11	343 24.2	27.7	12 00.9	7.1	18 55.7	1.0	58.3
D 12	358 24.2	N20 27.2	26 27.0	7.2	N18 56.7	0.9	58.2
A 13	13 24.1	26.7	40 53.2	7.1	18 57.6	0.8	58.2
Y 14	28 24.1	26.2	55 19.3	7.1	18 58.4	0.7	58.2
15	43 24.1	.. 25.7	69 45.4	7.2	18 59.1	0.5	58.2
16	58 24.1	25.2	84 11.6	7.1	18 59.6	0.5	58.2
17	73 24.0	24.8	98 37.7	7.2	19 00.1	0.3	58.2
18	88 24.0	N20 24.3	113 03.9	7.3	N19 00.4	0.2	58.1
19	103 24.0	23.8	127 30.2	7.2	19 00.6	0.1	58.1
20	118 23.9	23.3	141 56.4	7.3	19 00.7	0.1	58.1
21	133 23.9	.. 22.8	156 22.7	7.2	19 00.6	0.1	58.1
22	148 23.9	22.3	170 48.9	7.3	19 00.5	0.3	58.1
23	163 23.9	21.8	185 15.2	7.4	19 00.2	0.4	58.0
22 00	178 23.8	N20 21.3	199 41.6	7.3	N18 59.8	0.5	58.0
01	193 23.8	20.8	214 07.9	7.4	18 59.3	0.6	58.0
02	208 23.8	20.3	228 34.3	7.4	18 58.7	0.8	58.0
03	223 23.7	.. 19.9	243 00.7	7.5	18 57.9	0.8	57.9
04	238 23.7	19.4	257 27.2	7.5	18 57.1	1.0	57.9
05	253 23.7	18.9	271 53.7	7.5	18 56.1	1.1	57.9
06	268 23.7	N20 18.4	286 20.2	7.5	N18 55.0	1.2	57.9
W 07	283 23.6	17.9	300 46.7	7.6	18 53.8	1.3	57.9
E 08	298 23.6	17.4	315 13.3	7.6	18 52.5	1.4	57.8
D 09	313 23.6	.. 16.9	329 39.9	7.7	18 51.1	1.5	57.8
N 10	328 23.6	16.4	344 06.6	7.6	18 49.6	1.7	57.8
E 11	343 23.5	15.9	358 33.2	7.8	18 47.9	1.7	57.8
S 12	358 23.5	N20 15.4	13 00.0	7.7	N18 46.2	1.9	57.8
D 13	13 23.5	14.9	27 26.7	7.8	18 44.3	2.0	57.7
A 14	28 23.5	14.4	41 53.5	7.9	18 42.3	2.1	57.7
Y 15	43 23.4	.. 13.9	56 20.4	7.9	18 40.2	2.2	57.7
16	58 23.4	13.4	70 47.3	7.9	18 38.0	2.3	57.7
17	73 23.4	12.9	85 14.2	7.9	18 35.7	2.4	57.6
18	88 23.4	N20 12.4	99 41.1	8.1	N18 33.3	2.5	57.6
19	103 23.3	11.9	114 08.2	8.0	18 30.8	2.7	57.6
20	118 23.3	11.4	128 35.2	8.1	18 28.1	2.7	57.6
21	133 23.3	.. 10.9	143 02.3	8.2	18 25.4	2.9	57.6
22	148 23.3	10.4	157 29.5	8.2	18 22.5	2.9	57.5
23	163 23.3	09.8	171 56.7	8.2	18 19.6	3.1	57.5
23 00	178 23.2	N20 09.4	186 23.9	8.3	N18 16.5	3.2	57.5
01	193 23.2	08.9	200 51.2	8.3	18 13.3	3.2	57.5
02	208 23.2	08.4	215 18.5	8.4	18 10.1	3.4	57.4
03	223 23.2	.. 07.9	229 45.9	8.5	18 06.7	3.5	57.4
04	238 23.1	07.4	244 13.4	8.5	18 03.2	3.6	57.4
05	253 23.1	06.8	258 40.9	8.5	17 59.6	3.6	57.4
06	268 23.1	N20 06.3	273 08.4	8.6	N17 56.0	3.8	57.3
07	283 23.1	05.8	287 36.0	8.7	17 52.2	3.9	57.3
T 08	298 23.1	05.3	302 03.7	8.7	17 48.3	4.0	57.3
H 09	313 23.0	.. 04.8	316 31.4	8.8	17 44.3	4.1	57.3
U 10	328 23.0	04.3	330 59.2	8.8	17 40.2	4.1	57.2
R 11	343 23.0	03.8	345 27.0	8.8	17 36.1	4.3	57.2
S 12	358 23.0	N20 03.3	359 54.8	9.0	N17 31.8	4.4	57.2
D 13	13 23.0	02.8	14 22.8	9.0	17 27.4	4.4	57.2
A 14	28 22.9	02.2	28 50.8	9.0	17 23.0	4.6	57.2
Y 15	43 22.9	.. 01.7	43 18.8	9.1	17 18.4	4.6	57.1
16	58 22.9	01.2	57 46.9	9.2	17 13.8	4.8	57.1
17	73 22.9	00.7	72 15.1	9.2	17 09.0	4.8	57.1
18	88 22.9	N20 00.2	86 43.3	9.3	N17 04.2	4.9	57.1
19	103 22.9	19 59.7	101 11.6	9.3	16 59.3	5.1	57.0
20	118 22.8	59.2	115 39.9	9.4	16 54.2	5.1	57.0
21	133 22.8	.. 58.6	130 08.3	9.5	16 49.1	5.2	57.0
22	148 22.8	58.1	144 36.8	9.5	16 43.9	5.2	57.0
23	163 22.8	57.6	159 05.3	9.6	N16 38.7	5.4	56.9
	SD 15.8	d 0.5	SD 15.9		15.7		15.6

Twilight / Sunrise / Moonrise

Lat.	Naut.	Civil	Sunrise	21	22	23	24
°	h m	h m	h m	h m	h m	h m	h m
N 72	□	□	□	□	□	□	02 16
N 70	□	□	□	24 02	00 02	01 23	03 01
68	////	////	01 12	00 04	00 52	02 04	03 31
66	////	////	02 04	00 34	01 24	02 32	03 53
64	////	////	02 36	00 56	01 48	02 54	04 10
62	////	01 30	02 59	01 14	02 06	03 11	04 24
60	////	02 07	03 17	01 29	02 22	03 25	04 36
N 58	////	02 33	03 33	01 41	02 35	03 38	04 47
56	01 22	02 53	03 45	01 52	02 46	03 48	04 56
54	01 56	03 09	03 57	02 02	02 56	03 57	05 04
52	02 20	03 23	04 07	02 10	03 05	04 06	05 11
50	02 39	03 35	04 15	02 18	03 13	04 13	05 17
45	03 14	03 59	04 34	02 35	03 30	04 29	05 31
N 40	03 39	04 18	04 49	02 48	03 43	04 42	05 43
35	03 58	04 33	05 02	03 00	03 55	04 53	05 52
30	04 14	04 46	05 13	03 10	04 05	05 03	06 01
20	04 39	05 08	05 32	03 27	04 23	05 19	06 15
N 10	04 59	05 25	05 48	03 42	04 38	05 34	06 28
0	05 15	05 41	06 03	03 57	04 53	05 48	06 40
S 10	05 30	05 55	06 18	04 11	05 07	06 01	06 52
20	05 43	06 10	06 34	04 26	05 23	06 16	07 05
30	05 57	06 26	06 52	04 44	05 40	06 32	07 19
35	06 04	06 35	07 02	04 54	05 51	06 42	07 28
40	06 11	06 44	07 14	05 06	06 02	06 53	07 37
45	06 19	06 55	07 28	05 20	06 16	07 06	07 48
S 50	06 28	07 08	07 45	05 36	06 33	07 22	08 02
52	06 32	07 14	07 53	05 44	06 41	07 29	08 08
54	06 37	07 21	08 02	05 53	06 50	07 37	08 15
56	06 41	07 28	08 12	06 03	07 00	07 46	08 23
58	06 46	07 36	08 23	06 15	07 11	07 57	08 32
S 60	06 52	07 45	08 37	06 28	07 25	08 08	08 42

Sunset / Twilight / Moonset

Lat.	Sunset	Civil	Naut.	21	22	23	24
°	h m	h m	h m	h m	h m	h m	h m
N 72	□	□	□	□	□	22 47	22 34
N 70	□	□	□	21 15	21 50	22 02	22 06
68	22 54	////	////	20 24	21 08	21 31	21 45
66	22 05	////	////	19 53	20 39	21 09	21 28
64	21 35	////	////	19 29	20 17	20 51	21 14
62	21 12	22 39	////	19 10	20 00	20 36	21 03
60	20 54	22 03	////	18 55	19 45	20 24	20 53
N 58	20 39	21 38	////	18 42	19 33	20 13	20 44
56	20 26	21 19	22 47	18 31	19 22	20 03	20 37
54	20 15	21 03	22 14	18 21	19 13	19 55	20 30
52	20 05	20 49	21 51	18 12	19 04	19 48	20 24
50	19 57	20 37	21 32	18 04	18 56	19 41	20 18
45	19 38	20 13	20 58	17 47	18 40	19 26	20 06
N 40	19 23	19 54	20 34	17 33	18 27	19 14	19 56
35	19 11	19 39	20 14	17 21	18 15	19 04	19 47
30	19 00	19 26	19 58	17 11	18 05	18 55	19 40
20	18 41	19 05	19 33	16 53	17 48	18 40	19 27
N 10	18 25	18 47	19 14	16 38	17 33	18 26	19 15
0	18 10	18 32	18 58	16 24	17 19	18 13	19 04
S 10	17 55	18 17	18 43	16 09	17 05	18 00	18 53
20	17 39	18 03	18 30	15 54	16 50	17 46	18 42
30	17 22	17 47	18 16	15 36	16 33	17 31	18 28
35	17 11	17 38	18 09	15 25	16 22	17 21	18 21
40	16 59	17 29	18 02	15 14	16 11	17 11	18 12
45	16 45	17 18	17 54	15 00	15 57	16 58	18 01
S 50	16 28	17 05	17 45	14 43	15 40	16 43	17 48
52	16 20	16 59	17 41	14 34	15 32	16 36	17 43
54	16 12	16 52	17 37	14 26	15 24	16 28	17 36
56	16 02	16 45	17 32	14 15	15 14	16 19	17 29
58	15 50	16 37	17 27	14 04	15 03	16 09	17 20
S 60	15 37	16 28	17 22	13 51	14 50	15 58	17 11

	SUN			MOON			
Day	Eqn. of Time 00h	12h	Mer. Pass.	Mer. Pass. Upper	Lower	Age	Phase
d	m s	m s	h m	h m	h m	d	%
21	06 22	06 23	12 06	10 10	22 38	27	5
22	06 25	06 26	12 06	11 06	23 33	28	1
23	06 27	06 28	12 06	12 00	24 27	29	0

1998 JULY 24, 25, 26 (FRI., SAT., SUN.)

UT	ARIES GHA	VENUS −3.9 GHA	Dec	MARS +1.6 GHA	Dec	JUPITER −2.7 GHA	Dec	SATURN +0.4 GHA	Dec
24 00	301 30.9	205 48.6	N22 47.9	198 36.1	N23 39.5	302 48.0	S 2 03.7	269 41.6	N10 14.1
01	316 33.4	220 47.8	47.8	213 36.8	39.4	317 50.5	03.7	284 44.0	14.1
02	331 35.9	235 47.0	47.8	228 37.4	39.2	332 53.0	03.7	299 46.3	14.2
03	346 38.3	250 46.2 ..	47.8	243 38.1 ..	39.1	347 55.5 ..	03.8	314 48.7 ..	14.2
04	1 40.8	265 45.4	47.8	258 38.7	39.0	2 58.0	03.8	329 51.1	14.2
05	16 43.3	280 44.5	47.8	273 39.4	38.9	18 00.5	03.8	344 53.4	14.2
06	31 45.7	295 43.7	N22 47.8	288 40.0	N23 38.7	33 03.0	S 2 03.9	359 55.8	N10 14.2
07	46 48.2	310 42.9	47.8	303 40.7	38.6	48 05.5	03.9	14 58.2	14.3
08	61 50.6	325 42.1	47.8	318 41.4	38.5	63 08.0	03.9	30 00.5	14.3
F 09	76 53.1	340 41.3 ..	47.8	333 42.0 ..	38.3	78 10.5 ..	03.9	45 02.9 ..	14.3
R 10	91 55.6	355 40.5	47.7	348 42.7	38.2	93 13.0	04.0	60 05.3	14.3
I 11	106 58.0	10 39.7	47.7	3 43.3	38.1	108 15.5	04.0	75 07.6	14.4
D 12	122 00.5	25 38.9	N22 47.7	18 44.0	N23 38.0	123 18.1	S 2 04.0	90 10.0	N10 14.4
A 13	137 03.0	40 38.1	47.7	33 44.6	37.8	138 20.6	04.1	105 12.4	14.4
Y 14	152 05.4	55 37.2	47.7	48 45.3	37.7	153 23.1	04.1	120 14.7	14.4
15	167 07.9	70 36.4 ..	47.6	63 46.0 ..	37.6	168 25.6 ..	04.1	135 17.1 ..	14.4
16	182 10.4	85 35.6	47.6	78 46.6	37.4	183 28.1	04.2	150 19.5	14.5
17	197 12.8	100 34.8	47.6	93 47.3	37.3	198 30.6	04.2	165 21.9	14.5
18	212 15.3	115 34.0	N22 47.6	108 47.9	N23 37.2	213 33.1	S 2 04.2	180 24.2	N10 14.5
19	227 17.8	130 33.2	47.5	123 48.6	37.0	228 35.6	04.3	195 26.6	14.5
20	242 20.2	145 32.4	47.5	138 49.3	36.9	243 38.1	04.3	210 29.0	14.5
21	257 22.7	160 31.6 ..	47.5	153 49.9 ..	36.8	258 40.6 ..	04.3	225 31.3 ..	14.6
22	272 25.1	175 30.8	47.5	168 50.6	36.6	273 43.1	04.4	240 33.7	14.6
23	287 27.6	190 29.9	47.4	183 51.2	36.5	288 45.7	04.4	255 36.1	14.6
25 00	302 30.1	205 29.1	N22 47.4	198 51.9	N23 36.4	303 48.2	S 2 04.4	270 38.5	N10 14.6
01	317 32.5	220 28.3	47.4	213 52.6	36.2	318 50.7	04.5	285 40.8	14.7
02	332 35.0	235 27.5	47.3	228 53.2	36.1	333 53.2	04.5	300 43.2	14.7
03	347 37.5	250 26.7 ..	47.3	243 53.9 ..	36.0	348 55.7 ..	04.5	315 45.6 ..	14.7
04	2 39.9	265 25.9	47.2	258 54.5	35.8	3 58.2	04.6	330 47.9	14.7
05	17 42.4	280 25.1	47.2	273 55.2	35.7	19 00.7	04.6	345 50.3	14.7
06	32 44.9	295 24.3	N22 47.2	288 55.9	N23 35.6	34 03.2	S 2 04.6	0 52.7	N10 14.8
07	47 47.3	310 23.4	47.1	303 56.5	35.4	49 05.8	04.7	15 55.1	14.8
S 08	62 49.8	325 22.6	47.1	318 57.2	35.3	64 08.3	04.7	30 57.4	14.8
A 09	77 52.3	340 21.8 ..	47.1	333 57.9 ..	35.2	79 10.8 ..	04.8	45 59.8 ..	14.8
T 10	92 54.7	355 21.0	47.0	348 58.5	35.0	94 13.3	04.8	61 02.2	14.8
U 11	107 57.2	10 20.2	47.0	3 59.2	34.9	109 15.8	04.8	76 04.5	14.9
R 12	122 59.6	25 19.4	N22 46.9	18 59.8	N23 34.7	124 18.3	S 2 04.9	91 06.9	N10 14.9
D 13	138 02.1	40 18.6	46.9	34 00.5	34.6	139 20.8	04.9	106 09.3	14.9
A 14	153 04.6	55 17.8	46.8	49 01.2	34.5	154 23.4	04.9	121 11.7	14.9
Y 15	168 07.0	70 16.9 ..	46.8	64 01.8 ..	34.3	169 25.9 ..	05.0	136 14.0 ..	14.9
16	183 09.5	85 16.1	46.7	79 02.5	34.2	184 28.4	05.0	151 16.4	15.0
17	198 12.0	100 15.3	46.7	94 03.2	34.0	199 30.9	05.0	166 18.8	15.0
18	213 14.4	115 14.5	N22 46.6	109 03.8	N23 33.9	214 33.4	S 2 05.1	181 21.2	N10 15.0
19	228 16.9	130 13.7	46.6	124 04.5	33.8	229 35.9	05.1	196 23.5	15.0
20	243 19.4	145 12.9	46.5	139 05.1	33.6	244 38.5	05.1	211 25.9	15.1
21	258 21.8	160 12.1 ..	46.5	154 05.8 ..	33.5	259 41.0 ..	05.2	226 28.3 ..	15.1
22	273 24.3	175 11.3	46.4	169 06.5	33.3	274 43.5	05.2	241 30.7	15.1
23	288 26.7	190 10.4	46.3	184 07.1	33.2	289 46.0	05.3	256 33.0	15.1
26 00	303 29.2	205 09.6	N22 46.3	199 07.8	N23 33.1	304 48.5	S 2 05.3	271 35.4	N10 15.1
01	318 31.7	220 08.8	46.2	214 08.5	32.9	319 51.0	05.3	286 37.8	15.2
02	333 34.1	235 08.0	46.1	229 09.1	32.8	334 53.6	05.4	301 40.2	15.2
03	348 36.6	250 07.2 ..	46.1	244 09.8 ..	32.6	349 56.1 ..	05.4	316 42.5 ..	15.2
04	3 39.1	265 06.4	46.0	259 10.5	32.5	4 58.6	05.4	331 44.9	15.2
05	18 41.5	280 05.6	46.0	274 11.1	32.3	20 01.1	05.5	346 47.3	15.2
06	33 44.0	295 04.8	N22 45.9	289 11.8	N23 32.2	35 03.6	S 2 05.5	1 49.7	N10 15.3
07	48 46.5	310 03.9	45.8	304 12.5	32.1	50 06.2	05.6	16 52.0	15.3
08	63 48.9	325 03.1	45.8	319 13.1	31.9	65 08.7	05.6	31 54.4	15.3
S 09	78 51.4	340 02.3 ..	45.7	334 13.8 ..	31.8	80 11.2 ..	05.6	46 56.8 ..	15.3
U 10	93 53.9	355 01.5	45.6	349 14.4	31.6	95 13.7	05.7	61 59.2	15.3
N 11	108 56.3	10 00.7	45.6	4 15.1	31.5	110 16.3	05.7	77 01.6	15.3
D 12	123 58.8	24 59.9	N22 45.5	19 15.8	N23 31.3	125 18.8	S 2 05.7	92 03.9	N10 15.4
A 13	139 01.2	39 59.1	45.4	34 16.4	31.2	140 21.3	05.8	107 06.3	15.4
Y 14	154 03.7	54 58.3	45.3	49 17.1	31.0	155 23.8	05.8	122 08.7	15.4
15	169 06.2	69 57.4 ..	45.3	64 17.8 ..	30.9	170 26.3 ..	05.9	137 11.1 ..	15.4
16	184 08.6	84 56.6	45.2	79 18.4	30.8	185 28.9	05.9	152 13.4	15.4
17	199 11.1	99 55.8	45.1	94 19.1	30.6	200 31.4	05.9	167 15.8	15.5
18	214 13.6	114 55.0	N22 45.0	109 19.8	N23 30.5	215 33.9	S 2 06.0	182 18.2	N10 15.5
19	229 16.0	129 54.2	44.9	124 20.4	30.3	230 36.4	06.0	197 20.6	15.5
20	244 18.5	144 53.4	44.9	139 21.1	30.2	245 39.0	06.1	212 23.0	15.5
21	259 21.0	159 52.6 ..	44.8	154 21.8 ..	30.0	260 41.5 ..	06.1	227 25.3 ..	15.5
22	274 23.4	174 51.7	44.7	169 22.4	29.9	275 44.0	06.1	242 27.7	15.6
23	289 25.9	189 50.9	44.6	184 23.1	29.7	290 46.5	06.2	257 30.1	15.6
Mer. Pass.	h m 3 49.4	v −0.8	d 0.0	v 0.7	d 0.1	v 2.5	d 0.0	v 2.4	d 0.0

STARS

Name	SHA	Dec
Acamar	315 27.2	S40 18.5
Achernar	335 35.3	S57 14.4
Acrux	173 22.3	S63 05.7
Adhara	255 21.9	S28 58.2
Aldebaran	291 02.7	N16 30.2
Alioth	166 31.0	N55 58.4
Alkaid	153 08.0	N49 19.6
Al Na'ir	27 57.7	S46 57.9
Alnilam	275 58.3	S 1 12.2
Alphard	218 07.6	S 8 39.1
Alphecca	126 20.6	N26 43.5
Alpheratz	357 55.1	N29 04.8
Altair	62 19.1	N 8 52.0
Ankaa	353 26.8	S42 18.6
Antares	112 40.1	S26 25.6
Arcturus	146 06.2	N19 11.7
Atria	107 51.8	S69 01.6
Avior	234 23.3	S59 30.4
Bellatrix	278 44.6	N 6 20.8
Betelgeuse	271 14.0	N 7 24.3
Canopus	264 01.7	S52 41.7
Capella	280 51.7	N45 59.5
Deneb	49 38.8	N45 16.6
Denebola	182 45.5	N14 35.0
Diphda	349 07.3	S17 59.6
Dubhe	194 06.3	N61 45.8
Elnath	278 27.4	N28 36.2
Eltanin	90 51.0	N51 29.7
Enif	33 58.0	N 9 52.2
Fomalhaut	15 36.3	S29 37.6
Gacrux	172 13.9	S57 06.5
Gienah	176 04.2	S17 32.0
Hadar	149 04.1	S60 22.1
Hamal	328 13.7	N23 27.2
Kaus Aust.	83 58.6	S34 23.0
Kochab	137 19.5	N74 10.1
Markab	13 49.5	N15 11.8
Menkar	314 27.1	N 4 05.0
Menkent	148 21.1	S36 21.8
Miaplacidus	221 42.9	S69 42.8
Mirfak	308 56.9	N49 51.0
Nunki	76 12.1	S26 17.8
Peacock	53 36.7	S56 44.2
Pollux	243 42.1	N28 01.7
Procyon	245 12.0	N 5 13.7
Rasalhague	96 16.8	N12 33.9
Regulus	207 56.0	N11 58.5
Rigel	281 23.3	S 8 12.2
Rigil Kent.	140 07.3	S60 49.9
Sabik	102 25.4	S15 43.2
Schedar	349 53.3	N56 31.5
Shaula	96 37.1	S37 06.1
Sirius	258 44.1	S16 42.9
Spica	158 43.4	S11 09.1
Suhail	223 01.3	S43 25.7
Vega	80 46.3	N38 47.2
Zuben'ubi	137 18.0	S16 02.0

	SHA	Mer. Pass.
Venus	262 59.1	h m 10 19
Mars	256 21.8	10 44
Jupiter	1 18.1	3 44
Saturn	328 08.4	5 56

UT	SUN		MOON					Lat.	Twilight		Sunrise	Moonrise			
	GHA	Dec	GHA	v	Dec	d	HP		Naut.	Civil		24	25	26	27
d h	° ′	° ′	° ′	′	° ′	′	′	°	h m	h m	h m	h m	h m	h m	h m
24 00	178 22.8	N19 57.1	173 33.9	9.7	N16 33.3	5.5	56.9	N 72	▭	▭	▭	02 16	04 15	06 03	07 45
01	193 22.8	56.6	188 02.6	9.7	16 27.8	5.5	56.9	N 70	▭	▭	▭	03 01	04 42	06 20	07 54
02	208 22.7	56.0	202 31.3	9.7	16 22.3	5.6	56.9	68	////	////	01 34	03 31	05 02	06 32	08 01
03	223 22.7 ..	55.5	217 00.0	9.9	16 16.7	5.7	56.8	66	////	////	02 17	03 53	05 18	06 43	08 07
04	238 22.7	55.0	231 28.9	9.9	16 11.0	5.8	56.8	64	////	00 41	02 45	04 10	05 31	06 52	08 12
05	253 22.7	54.5	245 57.8	9.9	16 05.2	5.9	56.8	62	////	01 44	03 07	04 24	05 41	06 59	08 16
06	268 22.7	N19 54.0	260 26.7	10.1	N15 59.3	5.9	56.8	60	////	02 17	03 24	04 36	05 51	07 06	08 20
07	283 22.7	53.4	274 55.8	10.0	15 53.4	6.1	56.7	N 58	00 37	02 40	03 38	04 47	05 59	07 11	08 23
08	298 22.7	52.9	289 24.8	10.2	15 47.3	6.1	56.7	56	01 35	02 59	03 51	04 56	06 06	07 16	08 26
F 09	313 22.6 ..	52.4	303 54.0	10.2	15 41.2	6.2	56.7	54	02 05	03 14	04 01	05 04	06 12	07 21	08 29
R 10	328 22.6	51.9	318 23.2	10.3	15 35.0	6.2	56.7	52	02 27	03 28	04 11	05 11	06 18	07 25	08 31
I 11	343 22.6	51.3	332 52.5	10.3	15 28.8	6.4	56.6	50	02 45	03 39	04 19	05 17	06 23	07 29	08 33
D 12	358 22.6	N19 50.8	347 21.8	10.4	N15 22.4	6.4	56.6	45	03 18	04 02	04 37	05 31	06 34	07 37	08 38
A 13	13 22.6	50.3	1 51.2	10.5	15 16.0	6.5	56.6	N 40	03 42	04 21	04 52	05 43	06 43	07 43	08 42
Y 14	28 22.6	49.8	16 20.7	10.5	15 09.5	6.5	56.6	35	04 01	04 36	05 04	05 52	06 51	07 49	08 45
15	43 22.6 ..	49.2	30 50.2	10.6	15 03.0	6.7	56.5	30	04 17	04 48	05 15	06 01	06 58	07 54	08 48
16	58 22.6	48.7	45 19.8	10.7	14 56.3	6.7	56.5	20	04 41	05 09	05 33	06 15	07 10	08 03	08 53
17	73 22.5	48.2	59 49.5	10.7	14 49.6	6.8	56.5	N 10	05 00	05 26	05 48	06 28	07 20	08 10	08 58
18	88 22.5	N19 47.6	74 19.2	10.8	N14 42.8	6.8	56.5	0	05 15	05 41	06 03	06 40	07 30	08 17	09 02
19	103 22.5	47.1	88 49.0	10.9	14 36.0	6.9	56.4	S 10	05 29	05 55	06 17	06 52	07 40	08 24	09 07
20	118 22.5	46.6	103 18.9	10.9	14 29.1	7.0	56.4	20	05 42	06 09	06 33	07 05	07 50	08 32	09 11
21	133 22.5 ..	46.1	117 48.8	11.0	14 22.1	7.1	56.4	30	05 55	06 24	06 50	07 19	08 02	08 40	09 16
22	148 22.5	45.5	132 18.8	11.1	14 15.0	7.1	56.4	35	06 02	06 33	07 00	07 28	08 09	08 45	09 19
23	163 22.5	45.0	146 48.9	11.1	14 07.9	7.2	56.3	40	06 09	06 42	07 12	07 37	08 16	08 51	09 23
25 00	178 22.5	N19 44.5	161 19.0	11.1	N14 00.7	7.2	56.3	45	06 17	06 53	07 25	07 48	08 25	08 57	09 26
01	193 22.5	43.9	175 49.1	11.3	13 53.5	7.4	56.3	S 50	06 26	07 05	07 41	08 02	08 36	09 05	09 31
02	208 22.5	43.4	190 19.4	11.3	13 46.1	7.3	56.3	52	06 29	07 11	07 49	08 08	08 41	09 09	09 33
03	223 22.4 ..	42.9	204 49.7	11.4	13 38.8	7.5	56.2	54	06 33	07 17	07 58	08 15	08 46	09 13	09 36
04	238 22.4	42.3	219 20.1	11.4	13 31.3	7.5	56.2	56	06 38	07 24	08 07	08 23	08 53	09 17	09 38
05	253 22.4	41.8	233 50.5	11.5	13 23.8	7.5	56.2	58	06 42	07 32	08 18	08 32	08 59	09 22	09 41
06	268 22.4	N19 41.2	248 21.0	11.6	N13 16.3	7.6	56.2	S 60	06 47	07 40	08 31	08 42	09 07	09 27	09 44
07	283 22.4	40.7	262 51.6	11.6	13 08.7	7.7	56.1								
S 08	298 22.4	40.2	277 22.2	11.7	13 01.0	7.7	56.1	Lat.	Sunset	Twilight		Moonset			
A 09	313 22.4 ..	39.6	291 52.9	11.7	12 53.3	7.8	56.1			Civil	Naut.	24	25	26	27
T 10	328 22.4	39.1	306 23.6	11.8	12 45.5	7.9	56.1								
U 11	343 22.4	38.6	320 54.4	11.9	12 37.6	7.9	56.0	°	h m	h m	h m	h m	h m	h m	h m
R 12	358 22.4	N19 38.0	335 25.3	11.9	N12 29.7	7.9	56.0	N 72	▭	▭	▭	22 34	22 25	22 17	22 10
D 13	13 22.4	37.5	349 56.2	12.0	12 21.8	8.0	56.0	N 70	▭	▭	▭	22 06	22 07	22 07	22 06
A 14	28 22.4	36.9	4 27.2	12.1	12 13.8	8.1	56.0	68	22 34	////	////	21 45	21 53	21 58	22 01
Y 15	43 22.4 ..	36.4	18 58.3	12.1	12 05.7	8.1	55.9	66	22 13	////	////	21 28	21 41	21 51	21 58
16	58 22.3	35.9	33 29.4	12.2	11 57.6	8.2	55.9	64	21 26	23 21	////	21 14	21 31	21 44	21 55
17	73 22.3	35.3	48 00.6	12.2	11 49.4	8.2	55.9	62	21 04	22 25	////	21 03	21 23	21 39	21 53
18	88 22.3	N19 34.8	62 31.8	12.3	N11 41.2	8.3	55.9	60	20 47	21 54	////	20 53	21 16	21 34	21 50
19	103 22.3	34.2	77 03.1	12.4	11 33.0	8.3	55.8	N 58	20 33	21 31	23 25	20 44	21 09	21 30	21 48
20	118 22.3	33.7	91 34.5	12.4	11 24.7	8.4	55.8	56	20 21	21 12	22 35	20 37	21 04	21 26	21 46
21	133 22.3 ..	33.1	106 05.9	12.5	11 16.3	8.4	55.8	54	20 11	20 57	22 06	20 30	20 58	21 23	21 45
22	148 22.3	32.6	120 37.4	12.5	11 07.9	8.4	55.8	52	20 01	20 44	21 44	20 24	20 54	21 20	21 43
23	163 22.3	32.1	135 08.9	12.6	10 59.5	8.5	55.7	50	19 53	20 33	21 27	20 18	20 50	21 17	21 42
26 00	178 22.3	N19 31.5	149 40.5	12.7	N10 51.0	8.5	55.7	45	19 35	20 10	20 54	20 06	20 41	21 11	21 39
01	193 22.3	31.0	164 12.2	12.7	10 42.5	8.6	55.7	N 40	19 21	19 52	20 30	19 56	20 33	21 06	21 37
02	208 22.3	30.4	178 43.9	12.7	10 33.9	8.6	55.7	35	19 09	19 37	20 11	19 47	20 26	21 02	21 34
03	223 22.3 ..	29.9	193 15.6	12.9	10 25.3	8.6	55.6	30	18 58	19 24	19 56	19 40	20 20	20 58	21 33
04	238 22.3	29.3	207 47.5	12.8	10 16.7	8.7	55.6	20	18 40	19 04	19 32	19 27	20 10	20 51	21 29
05	253 22.3	28.8	222 19.3	13.0	10 08.0	8.8	55.6	N 10	18 24	18 47	19 13	19 15	20 01	20 45	21 26
06	268 22.3	N19 28.2	236 51.3	13.0	N 9 59.2	8.7	55.6	0	18 10	18 32	18 58	19 04	19 53	20 39	21 23
07	283 22.3	27.7	251 23.3	13.0	9 50.5	8.8	55.5	S 10	17 56	18 18	18 44	18 53	19 45	20 33	21 21
08	298 22.3	27.1	265 55.3	13.1	9 41.7	8.9	55.5	20	17 41	18 04	18 31	18 42	19 36	20 27	21 17
S 09	313 22.3 ..	26.6	280 27.4	13.1	9 32.8	8.9	55.5	30	17 23	17 49	18 18	18 28	19 25	20 20	21 14
U 10	328 22.3	26.0	294 59.5	13.2	9 23.9	8.9	55.5	35	17 13	17 40	18 11	18 21	19 19	20 16	21 12
N 11	343 22.3	25.5	309 31.7	13.3	9 15.0	8.9	55.5	40	17 02	17 31	18 04	18 12	19 12	20 12	21 10
D 12	358 22.3	N19 24.9	324 04.0	13.3	N 9 06.1	9.0	55.4	45	16 48	17 21	17 57	18 01	19 04	20 06	21 07
A 13	13 22.3	24.4	338 36.3	13.4	8 57.1	9.0	55.4	S 50	16 32	17 08	17 48	17 48	18 54	20 00	21 04
Y 14	28 22.3	23.8	353 08.7	13.4	8 48.1	9.1	55.4	52	16 25	17 03	17 44	17 43	18 50	19 57	21 02
15	43 22.3 ..	23.2	7 41.1	13.4	8 39.0	9.0	55.4	54	16 16	16 56	17 40	17 36	18 45	19 53	21 01
16	58 22.3	22.7	22 13.5	13.5	8 30.0	9.1	55.3	56	16 06	16 50	17 36	17 29	18 39	19 50	20 59
17	73 22.3	22.1	36 46.0	13.6	8 20.9	9.2	55.3	58	15 56	16 42	17 31	17 20	18 33	19 46	20 57
18	88 22.3	N19 21.6	51 18.6	13.6	N 8 11.7	9.1	55.3	S 60	15 43	16 34	17 26	17 11	18 26	19 41	20 55
19	103 22.3	21.0	65 51.2	13.6	8 02.6	9.2	55.3								
20	118 22.3	20.5	80 23.8	13.7	7 53.4	9.2	55.3			SUN			MOON		
21	133 22.3 ..	19.9	94 56.5	13.8	7 44.2	9.3	55.2	Day	Eqn. of Time		Mer.	Mer. Pass.		Age	Phase
22	148 22.3	19.3	109 29.3	13.8	7 34.9	9.3	55.2		00ʰ	12ʰ	Pass.	Upper	Lower		
23	163 22.3	18.8	124 02.1	13.8	N 7 25.6	9.3	55.2	d	m s	m s	h m	h m	h m	d	%
								24	06 29	06 30	12 06	12 52	00 27	01	1
SD 15.8	d 0.5		SD 15.4		15.3		15.1	25	06 30	06 30	12 07	13 42	01 17	02	4
								26	06 31	06 31	12 07	14 28	02 05	03	9

UT	ARIES	VENUS −3.9		MARS +1.7		JUPITER −2.7		SATURN +0.4		STARS		
	GHA	GHA	Dec	GHA	Dec	GHA	Dec	GHA	Dec	Name	SHA	Dec
d h	° ′	° ′	° ′	° ′	° ′	° ′	° ′	° ′	° ′		° ′	° ′
27 00	304 28.4	204 50.1	N22 44.5	199 23.8	N23 29.6	305 49.1	S 2 06.2	272 32.5	N10 15.6	Acamar	315 27.1	S40 18.5
01	319 30.8	219 49.3	44.4	214 24.4	29.4	320 51.6	06.3	287 34.9	15.6	Achernar	335 35.3	S57 14.4
02	334 33.3	234 48.5	44.3	229 25.1	29.3	335 54.1	06.3	302 37.2	15.6	Acrux	173 22.4	S63 05.6
03	349 35.7	249 47.7	.. 44.3	244 25.8	.. 29.1	350 56.7	.. 06.3	317 39.6	.. 15.6	Adhara	255 21.9	S28 58.2
04	4 38.2	264 46.9	44.2	259 26.5	29.0	5 59.2	06.4	332 42.0	15.7	Aldebaran	291 02.7	N16 30.2
05	19 40.7	279 46.1	44.1	274 27.1	28.8	21 01.7	06.4	347 44.4	15.7			
06	34 43.1	294 45.2	N22 44.0	289 27.8	N23 28.7	36 04.2	S 2 06.5	2 46.8	N10 15.7	Alioth	166 31.0	N55 58.4
07	49 45.6	309 44.4	43.9	304 28.5	28.5	51 06.8	06.5	17 49.1	15.7	Alkaid	153 08.0	N49 19.6
08	64 48.1	324 43.6	43.8	319 29.1	28.4	66 09.3	06.6	32 51.5	15.7	Al Na'ir	27 57.7	S46 57.9
M 09	79 50.5	339 42.8	.. 43.7	334 29.8	.. 28.2	81 11.8	.. 06.6	47 53.9	.. 15.8	Alnilam	275 58.2	S 1 12.2
O 10	94 53.0	354 42.0	43.6	349 30.5	28.1	96 14.4	06.6	62 56.3	15.8	Alphard	218 07.6	S 8 39.1
N 11	109 55.5	9 41.2	43.5	4 31.1	27.9	111 16.9	06.7	77 58.7	15.8			
D 12	124 57.9	24 40.4	N22 43.4	19 31.8	N23 27.7	126 19.4	S 2 06.7	93 01.0	N10 15.8	Alphecca	126 20.6	N26 43.5
A 13	140 00.4	39 39.5	43.3	34 32.5	27.6	141 21.9	06.8	108 03.4	15.8	Alpheratz	357 55.1	N29 04.8
Y 14	155 02.8	54 38.7	43.2	49 33.1	27.4	156 24.5	06.8	123 05.8	15.8	Altair	62 19.1	N 8 52.1
15	170 05.3	69 37.9	.. 43.1	64 33.8	.. 27.3	171 27.0	.. 06.8	138 08.2	.. 15.9	Ankaa	353 26.8	S42 18.6
16	185 07.8	84 37.1	43.0	79 34.5	27.1	186 29.5	06.9	153 10.6	15.9	Antares	112 40.1	S26 25.6
17	200 10.2	99 36.3	42.9	94 35.2	27.0	201 32.1	06.9	168 12.9	15.9			
18	215 12.7	114 35.5	N22 42.8	109 35.8	N23 26.8	216 34.6	S 2 07.0	183 15.3	N10 15.9	Arcturus	146 06.2	N19 11.7
19	230 15.2	129 34.7	42.7	124 36.5	26.7	231 37.1	07.0	198 17.7	15.9	Atria	107 51.9	S69 01.6
20	245 17.6	144 33.9	42.6	139 37.2	26.5	246 39.7	07.1	213 20.1	16.0	Avior	234 23.3	S59 30.4
21	260 20.1	159 33.0	.. 42.4	154 37.8	.. 26.3	261 42.2	.. 07.1	228 22.5	.. 16.0	Bellatrix	278 44.5	N 6 20.8
22	275 22.6	174 32.2	42.3	169 38.5	26.2	276 44.7	07.1	243 24.9	16.0	Betelgeuse	271 14.0	N 7 24.3
23	290 25.0	189 31.4	42.2	184 39.2	26.0	291 47.3	07.2	258 27.2	16.0			
28 00	305 27.5	204 30.6	N22 42.1	199 39.8	N23 25.9	306 49.8	S 2 07.2	273 29.6	N10 16.0	Canopus	264 01.7	S52 41.7
01	320 30.0	219 29.8	41.9	214 40.5	25.7	321 52.3	07.3	288 32.0	16.0	Capella	280 51.7	N45 59.5
02	335 32.4	234 29.0	41.9	229 41.2	25.6	336 54.9	07.3	303 34.4	16.1	Deneb	49 38.8	N45 16.6
03	350 34.9	249 28.2	.. 41.8	244 41.9	.. 25.4	351 57.4	.. 07.4	318 36.8	.. 16.1	Denebola	182 45.5	N14 35.0
04	5 37.3	264 27.4	41.6	259 42.5	25.2	6 59.9	07.4	333 39.2	16.1	Diphda	349 07.2	S17 59.6
05	20 39.8	279 26.5	41.5	274 43.2	25.1	22 02.5	07.5	348 41.5	16.1			
06	35 42.3	294 25.7	N22 41.4	289 43.9	N23 24.9	37 05.0	S 2 07.5	3 43.9	N10 16.1	Dubhe	194 06.3	N61 45.8
07	50 44.7	309 24.9	41.3	304 44.5	24.8	52 07.5	07.5	18 46.3	16.1	Elnath	278 27.4	N28 36.2
T 08	65 47.2	324 24.1	41.2	319 45.2	24.6	67 10.1	07.6	33 48.7	16.2	Eltanin	90 51.0	N51 29.7
U 09	80 49.7	339 23.3	.. 41.0	334 45.9	.. 24.4	82 12.6	.. 07.6	48 51.1	.. 16.2	Enif	33 58.0	N 9 52.2
E 10	95 52.1	354 22.5	40.9	349 46.6	24.3	97 15.2	07.7	63 53.5	16.2	Fomalhaut	15 36.3	S29 37.6
S 11	110 54.6	9 21.7	40.8	4 47.2	24.1	112 17.7	07.7	78 55.8	16.2			
D 12	125 57.1	24 20.8	N22 40.7	19 47.9	N23 24.0	127 20.2	S 2 07.8	93 58.2	N10 16.2	Gacrux	172 13.9	S57 06.5
A 13	140 59.5	39 20.0	40.5	34 48.6	23.8	142 22.8	07.8	109 00.6	16.2	Gienah	176 04.2	S17 32.0
Y 14	156 02.0	54 19.2	40.4	49 49.3	23.6	157 25.3	07.9	124 03.0	16.3	Hadar	149 04.1	S60 22.1
15	171 04.4	69 18.4	.. 40.3	64 49.9	.. 23.5	172 27.8	.. 07.9	139 05.4	.. 16.3	Hamal	328 13.7	N23 27.2
16	186 06.9	84 17.6	40.1	79 50.6	23.3	187 30.4	07.9	154 07.8	16.3	Kaus Aust.	83 58.6	S34 23.0
17	201 09.4	99 16.8	40.0	94 51.3	23.2	202 32.9	08.0	169 10.2	16.3			
18	216 11.8	114 16.0	N22 39.9	109 52.0	N23 23.0	217 35.5	S 2 08.0	184 12.5	N10 16.3	Kochab	137 19.5	N74 10.1
19	231 14.3	129 15.2	39.7	124 52.6	22.8	232 38.0	08.1	199 14.9	16.3	Markab	13 49.5	N15 11.9
20	246 16.8	144 14.3	39.6	139 53.3	22.7	247 40.5	08.1	214 17.3	16.3	Menkar	314 27.1	N 4 05.0
21	261 19.2	159 13.5	.. 39.5	154 54.0	.. 22.5	262 43.1	.. 08.2	229 19.7	.. 16.4	Menkent	148 21.1	S36 21.8
22	276 21.7	174 12.7	39.3	169 54.7	22.3	277 45.6	08.2	244 22.1	16.4	Miaplacidus	221 42.9	S69 42.8
23	291 24.2	189 11.9	39.2	184 55.3	22.2	292 48.2	08.3	259 24.5	16.4			
29 00	306 26.6	204 11.1	N22 39.0	199 56.0	N23 22.0	307 50.7	S 2 08.3	274 26.9	N10 16.4	Mirfak	308 56.9	N49 51.0
01	321 29.1	219 10.3	38.9	214 56.7	21.8	322 53.2	08.4	289 29.3	16.4	Nunki	76 12.1	S26 17.8
02	336 31.6	234 09.5	38.7	229 57.4	21.7	337 55.8	08.4	304 31.6	16.4	Peacock	53 36.7	S56 44.2
03	351 34.0	249 08.7	.. 38.6	244 58.0	.. 21.5	352 58.3	.. 08.5	319 34.0	.. 16.5	Pollux	243 42.0	N28 01.7
04	6 36.5	264 07.8	38.5	259 58.7	21.3	8 00.9	08.5	334 36.4	16.5	Procyon	245 12.0	N 5 13.7
05	21 38.9	279 07.0	38.3	274 59.4	21.2	23 03.4	08.5	349 38.8	16.5			
06	36 41.4	294 06.2	N22 38.2	290 00.1	N23 21.0	38 06.0	S 2 08.6	4 41.2	N10 16.5	Rasalhague	96 16.8	N12 33.9
W 07	51 43.9	309 05.4	38.0	305 00.7	20.8	53 08.5	08.6	19 43.6	16.5	Regulus	207 56.0	N11 58.5
E 08	66 46.3	324 04.6	37.9	320 01.4	20.7	68 11.0	08.7	34 46.0	16.5	Rigel	281 23.3	S 8 12.2
D 09	81 48.8	339 03.8	.. 37.7	335 02.1	.. 20.5	83 13.6	.. 08.7	49 48.4	.. 16.5	Rigil Kent.	140 07.3	S60 49.9
N 10	96 51.3	354 03.0	37.6	350 02.8	20.3	98 16.1	08.8	64 50.8	16.6	Sabik	102 25.4	S15 43.2
E 11	111 53.7	9 02.2	37.4	5 03.4	20.2	113 18.7	08.8	79 53.1	16.6			
S 12	126 56.2	24 01.4	N22 37.3	20 04.1	N23 20.0	128 21.2	S 2 08.9	94 55.5	N10 16.6	Schedar	349 53.3	N56 31.5
D 13	141 58.7	39 00.5	37.1	35 04.8	19.8	143 23.8	08.9	109 57.9	16.6	Shaula	96 37.1	S37 06.1
A 14	157 01.1	53 59.7	36.9	50 05.5	19.7	158 26.3	09.0	125 00.3	16.6	Sirius	258 44.1	S16 42.8
Y 15	172 03.6	68 58.9	.. 36.8	65 06.1	.. 19.5	173 28.9	.. 09.0	140 02.7	.. 16.6	Spica	158 43.4	S11 09.1
16	187 06.1	83 58.1	36.6	80 06.8	19.3	188 31.4	09.1	155 05.1	16.7	Suhail	223 01.3	S43 25.7
17	202 08.5	98 57.3	36.5	95 07.5	19.1	203 33.9	09.1	170 07.5	16.7			
18	217 11.0	113 56.5	N22 36.3	110 08.2	N23 19.0	218 36.5	S 2 09.2	185 09.9	N10 16.7	Vega	80 46.3	N38 47.2
19	232 13.4	128 55.7	36.1	125 08.9	18.8	233 39.0	09.2	200 12.3	16.7	Zuben'ubi	137 18.0	S16 02.0
20	247 15.9	143 54.9	36.0	140 09.5	18.6	248 41.6	09.3	215 14.6	16.7		SHA	Mer. Pass.
21	262 18.4	158 54.0	.. 35.8	155 10.2	.. 18.5	263 44.1	.. 09.3	230 17.0	.. 16.7		° ′	h m
22	277 20.8	173 53.2	35.6	170 10.9	18.3	278 46.7	09.4	245 19.4	16.7	Venus	259 03.1	10 23
23	292 23.3	188 52.4	35.5	185 11.6	18.1	293 49.2	09.4	260 21.8	16.8	Mars	254 12.4	10 41
	h m									Jupiter	1 22.3	3 32
Mer. Pass. 3 37.6		v −0.8	d 0.1	v 0.7	d 0.2	v 2.5	d 0.0	v 2.4	d 0.0	Saturn	328 02.1	5 45

UT		SUN GHA	SUN Dec	MOON GHA	v	Dec	d	HP
d h		° '	° '	° '	'	° '	'	'
27	00	178 22.3	N19 18.2	138 34.9	13.9	N 7 16.3	9.3	55.2
	01	193 22.3	17.7	153 07.8	13.9	7 07.0	9.3	55.2
	02	208 22.3	17.1	167 40.7	14.0	6 57.7	9.4	55.1
	03	223 22.3	.. 16.5	182 13.7	14.0	6 48.3	9.4	55.1
	04	238 22.3	16.0	196 46.7	14.1	6 38.9	9.4	55.1
	05	253 22.3	15.4	211 19.8	14.0	6 29.5	9.4	55.1
	06	268 22.3	N19 14.9	225 52.8	14.2	N 6 20.1	9.5	55.1
	07	283 22.3	14.3	240 26.0	14.2	6 10.6	9.4	55.0
M	08	298 22.3	13.7	254 59.2	14.2	6 01.2	9.5	55.0
O	09	313 22.3	.. 13.2	269 32.4	14.2	5 51.7	9.5	55.0
N	10	328 22.4	12.6	284 05.6	14.3	5 42.2	9.5	55.0
D	11	343 22.4	12.0	298 38.9	14.4	5 32.7	9.6	55.0
A	12	358 22.4	N19 11.5	313 12.3	14.3	N 5 23.1	9.5	54.9
Y	13	13 22.4	10.9	327 45.6	14.4	5 13.6	9.6	54.9
	14	28 22.4	10.3	342 19.0	14.5	5 04.0	9.6	54.9
	15	43 22.4	.. 09.8	356 52.5	14.5	4 54.4	9.6	54.9
	16	58 22.4	09.2	11 26.0	14.5	4 44.8	9.6	54.9
	17	73 22.4	08.6	25 59.5	14.5	4 35.2	9.6	54.9
	18	88 22.4	N19 08.1	40 33.0	14.6	N 4 25.6	9.7	54.8
	19	103 22.4	07.5	55 06.6	14.6	4 15.9	9.6	54.8
	20	118 22.4	06.9	69 40.2	14.7	4 06.3	9.7	54.8
	21	133 22.4	.. 06.4	84 13.9	14.6	3 56.6	9.6	54.8
	22	148 22.4	05.8	98 47.5	14.7	3 47.0	9.7	54.8
	23	163 22.4	05.2	113 21.2	14.8	3 37.3	9.7	54.8
28	00	178 22.5	N19 04.6	127 55.0	14.7	N 3 27.6	9.7	54.7
	01	193 22.5	04.1	142 28.7	14.8	3 17.9	9.7	54.7
	02	208 22.5	03.5	157 02.5	14.8	3 08.2	9.7	54.7
	03	223 22.5	.. 02.9	171 36.3	14.9	2 58.5	9.7	54.7
	04	238 22.5	02.3	186 10.2	14.8	2 48.8	9.7	54.7
	05	253 22.5	01.8	200 44.0	14.9	2 39.1	9.7	54.7
	06	268 22.5	N19 01.2	215 17.9	15.0	N 2 29.4	9.8	54.6
	07	283 22.5	00.6	229 51.9	14.9	2 19.6	9.7	54.6
T	08	298 22.5	19 00.0	244 25.8	15.0	2 09.9	9.7	54.6
U	09	313 22.6	18 59.5	258 59.8	15.0	2 00.2	9.8	54.6
E	10	328 22.6	58.9	273 33.8	15.0	1 50.4	9.7	54.6
S	11	343 22.6	58.3	288 07.8	15.0	1 40.7	9.7	54.6
D	12	358 22.6	N18 57.7	302 41.8	15.1	N 1 31.0	9.8	54.6
A	13	13 22.6	57.1	317 15.9	15.0	1 21.2	9.7	54.6
Y	14	28 22.6	56.6	331 49.9	15.1	1 11.5	9.7	54.5
	15	43 22.6	.. 56.0	346 24.0	15.1	1 01.8	9.8	54.5
	16	58 22.6	55.4	0 58.1	15.1	0 52.0	9.7	54.5
	17	73 22.7	54.8	15 32.2	15.2	0 42.3	9.7	54.5
	18	88 22.7	N18 54.2	30 06.4	15.1	N 0 32.6	9.8	54.5
	19	103 22.7	53.6	44 40.5	15.2	0 22.8	9.7	54.5
	20	118 22.7	53.1	59 14.7	15.2	0 13.1	9.7	54.5
	21	133 22.7	.. 52.5	73 48.9	15.2	N 0 03.4	9.7	54.5
	22	148 22.7	51.9	88 23.1	15.2	S 0 06.3	9.8	54.4
	23	163 22.7	51.3	102 57.3	15.2	0 16.1	9.7	54.4
29	00	178 22.8	N18 50.7	117 31.5	15.3	S 0 25.8	9.7	54.4
	01	193 22.8	50.1	132 05.8	15.2	0 35.5	9.7	54.4
	02	208 22.8	49.6	146 40.0	15.3	0 45.2	9.7	54.4
	03	223 22.8	.. 49.0	161 14.3	15.2	0 54.9	9.6	54.4
	04	238 22.8	48.4	175 48.5	15.3	1 04.5	9.7	54.4
	05	253 22.8	47.8	190 22.8	15.3	1 14.2	9.7	54.4
	06	268 22.9	N18 47.2	204 57.1	15.3	S 1 23.9	9.6	54.4
W	07	283 22.9	46.6	219 31.4	15.3	1 33.5	9.7	54.4
E	08	298 22.9	46.0	234 05.7	15.3	1 43.2	9.6	54.3
D	09	313 22.9	.. 45.4	248 40.0	15.3	1 52.8	9.6	54.3
N	10	328 22.9	44.8	263 14.3	15.3	2 02.4	9.6	54.3
E	11	343 22.9	44.2	277 48.6	15.3	2 12.0	9.6	54.3
S	12	358 23.0	N18 43.7	292 22.9	15.3	S 2 21.6	9.6	54.3
D	13	13 23.0	43.1	306 57.2	15.3	2 31.2	9.6	54.3
A	14	28 23.0	42.5	321 31.5	15.3	2 40.8	9.6	54.3
Y	15	43 23.0	.. 41.9	336 05.8	15.4	2 50.4	9.5	54.3
	16	58 23.0	41.3	350 40.2	15.3	2 59.9	9.5	54.3
	17	73 23.1	40.7	5 14.5	15.3	3 09.4	9.5	54.3
	18	88 23.1	N18 40.1	19 48.8	15.3	S 3 18.9	9.5	54.3
	19	103 23.1	39.5	34 23.1	15.3	3 28.4	9.5	54.3
	20	118 23.1	38.9	48 57.4	15.3	3 37.9	9.5	54.3
	21	133 23.1	.. 38.3	63 31.7	15.3	3 47.4	9.4	54.3
	22	148 23.2	37.7	78 06.0	15.3	3 56.8	9.5	54.3
	23	163 23.2	37.1	92 40.3	15.3	S 4 06.3	9.4	54.3
		SD 15.8	d 0.6	SD 15.0		14.9		14.8

Moonrise

Lat.	Twilight Naut.	Twilight Civil	Sunrise	27	28	29	30
°	h m	h m	h m	h m	h m	h m	h m
N 72	□	□	□	07 45	09 22	10 57	12 32
N 70	////	////	00 30	07 54	09 25	10 54	12 23
68	////	////	01 52	08 01	09 27	10 51	12 15
66	////	////	02 29	08 07	09 28	10 49	12 09
64	////	01 11	02 55	08 12	09 30	10 47	12 03
62	////	01 57	03 14	08 16	09 31	10 45	11 59
60	////	02 26	03 31	08 20	09 32	10 44	11 55
N 58	01 04	02 48	03 44	08 23	09 33	10 43	11 51
56	01 47	03 06	03 56	08 26	09 34	10 42	11 48
54	02 14	03 20	04 06	08 29	09 35	10 41	11 46
52	02 34	03 33	04 15	08 31	09 36	10 40	11 43
50	02 51	03 44	04 23	08 33	09 37	10 39	11 41
45	03 22	04 06	04 40	08 38	09 38	10 37	11 36
N 40	03 46	04 24	04 54	08 42	09 39	10 36	11 32
35	04 04	04 38	05 06	08 45	09 40	10 34	11 28
30	04 19	04 50	05 16	08 48	09 41	10 33	11 25
20	04 42	05 10	05 34	08 53	09 43	10 31	11 20
N 10	05 00	05 27	05 49	08 58	09 44	10 30	11 15
0	05 16	05 41	06 03	09 02	09 46	10 28	11 11
S 10	05 29	05 55	06 17	09 07	09 47	10 27	11 06
20	05 42	06 08	06 32	09 11	09 49	10 25	11 02
30	05 54	06 23	06 48	09 16	09 50	10 23	10 56
35	06 00	06 31	06 58	09 19	09 51	10 22	10 53
40	06 07	06 40	07 09	09 23	09 52	10 21	10 50
45	06 14	06 50	07 22	09 26	09 54	10 20	10 46
S 50	06 22	07 02	07 37	09 31	09 55	10 18	10 42
52	06 26	07 07	07 45	09 33	09 56	10 17	10 39
54	06 30	07 13	07 53	09 36	09 56	10 17	10 37
56	06 34	07 19	08 02	09 38	09 57	10 16	10 35
58	06 38	07 27	08 12	09 41	09 58	10 15	10 32
S 60	06 43	07 35	08 24	09 44	09 59	10 14	10 29

Moonset

Lat.	Sunset	Twilight Civil	Twilight Naut.	27	28	29	30
°	h m	h m	h m	h m	h m	h m	h m
N 72	□	□	□	22 10	22 04	21 57	21 50
N 70	23 24	////	////	22 06	22 04	22 03	22 02
68	22 16	////	////	22 01	22 04	22 07	22 11
66	21 41	////	////	21 58	22 05	22 11	22 18
64	21 16	22 55	////	21 55	22 05	22 14	22 25
62	20 57	22 12	////	21 53	22 05	22 17	22 30
60	20 41	21 44	////	21 50	22 05	22 20	22 35
N 58	20 27	21 23	23 02	21 48	22 05	22 22	22 39
56	20 16	21 06	22 23	21 46	22 05	22 24	22 43
54	20 06	20 51	21 57	21 45	22 05	22 26	22 47
52	19 57	20 39	21 37	21 43	22 05	22 27	22 50
50	19 49	20 28	21 21	21 42	22 05	22 29	22 53
45	19 32	20 06	20 49	21 39	22 06	22 32	22 59
N 40	19 18	19 49	20 27	21 37	22 06	22 35	23 04
35	19 06	19 34	20 09	21 34	22 06	22 37	23 08
30	18 56	19 22	19 54	21 33	22 06	22 39	23 12
20	18 39	19 03	19 30	21 29	22 06	22 43	23 19
N 10	18 24	18 46	19 12	21 26	22 06	22 46	23 25
0	18 10	18 32	18 57	21 23	22 06	22 49	23 31
S 10	17 56	18 18	18 44	21 21	22 06	22 52	23 37
20	17 42	18 05	18 32	21 17	22 06	22 55	23 43
30	17 25	17 50	18 19	21 14	22 07	22 58	23 50
35	17 15	17 42	18 13	21 12	22 07	23 00	23 54
40	17 04	17 34	18 06	21 10	22 07	23 03	23 59
45	16 52	17 24	17 59	21 07	22 07	23 05	24 04
S 50	16 36	17 12	17 51	21 04	22 07	23 09	24 10
52	16 29	17 06	17 48	21 02	22 07	23 10	24 13
54	16 21	17 01	17 44	21 01	22 07	23 12	24 17
56	16 12	16 54	17 40	20 59	22 07	23 14	24 20
58	16 01	16 47	17 36	20 57	22 07	23 16	24 24
S 60	15 49	16 39	17 31	20 55	22 07	23 18	24 29

Day	SUN Eqn. of Time 00h	SUN Eqn. of Time 12h	SUN Mer. Pass.	MOON Mer. Pass. Upper	MOON Mer. Pass. Lower	Age	Phase
d	m s	m s	h m	h m	h m	d	%
27	06 31	06 31	12 07	15 13	02 51	04	15
28	06 30	06 30	12 06	15 56	03 35	05	23
29	06 29	06 28	12 06	16 38	04 17	06	32

1998 JULY 30, 31, AUG. 1 (THURS., FRI., SAT.)

UT	ARIES GHA	VENUS −3.9 GHA	Dec	MARS +1.7 GHA	Dec	JUPITER −2.7 GHA	Dec	SATURN +0.4 GHA	Dec	STARS Name	SHA	Dec
30 00	307 25.8	203 51.6	N22 35.3	200 12.2	N23 17.9	308 51.8	S 2 09.5	275 24.2	N10 16.8	Acamar	315 27.1	S40 18.5
01	322 28.2	218 50.8	35.1	215 12.9	17.8	323 54.3	09.5	290 26.6	16.8	Achernar	335 35.2	S57 14.4
02	337 30.7	233 50.0	35.0	230 13.6	17.6	338 56.9	09.6	305 29.0	16.8	Acrux	173 22.4	S63 05.6
03	352 33.2	248 49.2	.. 34.8	245 14.3	.. 17.4	353 59.4	.. 09.6	320 31.4	.. 16.8	Adhara	255 21.9	S28 58.2
04	7 35.6	263 48.4	34.6	260 15.0	17.3	9 02.0	09.7	335 33.8	16.8	Aldebaran	291 02.7	N16 30.2
05	22 38.1	278 47.6	34.5	275 15.6	17.1	24 04.5	09.7	350 36.2	16.8			
06	37 40.5	293 46.7	N22 34.3	290 16.3	N23 16.9	39 07.1	S 2 09.8	5 38.6	N10 16.9	Alioth	166 31.0	N55 58.4
07	52 43.0	308 45.9	34.1	305 17.0	16.7	54 09.6	09.8	20 41.0	16.9	Alkaid	153 08.0	N49 19.6
T 08	67 45.5	323 45.1	33.9	320 17.7	16.6	69 12.2	09.9	35 43.3	16.9	Al Na'ir	27 57.7	S46 57.9
H 09	82 47.9	338 44.3	.. 33.7	335 18.4	.. 16.4	84 14.7	.. 09.9	50 45.7	.. 16.9	Alnilam	275 58.2	S 1 12.2
U 10	97 50.4	353 43.5	33.6	350 19.0	16.2	99 17.3	10.0	65 48.1	16.9	Alphard	218 07.6	S 8 39.1
R 11	112 52.9	8 42.7	33.4	5 19.7	16.0	114 19.8	10.0	80 50.5	16.9			
S 12	127 55.3	23 41.9	N22 33.2	20 20.4	N23 15.8	129 22.4	S 2 10.1	95 52.9	N10 16.9	Alphecca	126 20.6	N26 43.5
D 13	142 57.8	38 41.1	33.0	35 21.1	15.7	144 24.9	10.1	110 55.3	16.9	Alpheratz	357 55.1	N29 04.8
A 14	158 00.3	53 40.3	32.8	50 21.8	15.5	159 27.5	10.2	125 57.7	17.0	Altair	62 19.1	N 8 52.1
Y 15	173 02.7	68 39.5	.. 32.6	65 22.4	.. 15.3	174 30.0	.. 10.2	141 00.1	.. 17.0	Ankaa	353 26.8	S42 18.6
16	188 05.2	83 38.6	32.5	80 23.1	15.1	189 32.6	10.3	156 02.5	17.0	Antares	112 40.1	S26 25.6
17	203 07.7	98 37.8	32.3	95 23.8	15.0	204 35.1	10.3	171 04.9	17.0			
18	218 10.1	113 37.0	N22 32.1	110 24.5	N23 14.8	219 37.7	S 2 10.4	186 07.3	N10 17.0	Arcturus	146 06.2	N19 11.7
19	233 12.6	128 36.2	31.9	125 25.2	14.6	234 40.3	10.4	201 09.7	17.0	Atria	107 51.9	S69 01.6
20	248 15.0	143 35.4	31.7	140 25.9	14.4	249 42.8	10.5	216 12.1	17.0	Avior	234 23.3	S59 30.3
21	263 17.5	158 34.6	.. 31.5	155 26.5	.. 14.2	264 45.4	.. 10.5	231 14.5	.. 17.1	Bellatrix	278 44.5	N 6 20.8
22	278 20.0	173 33.8	31.3	170 27.2	14.1	279 47.9	10.6	246 16.9	17.1	Betelgeuse	271 14.0	N 7 24.3
23	293 22.4	188 33.0	31.1	185 27.9	13.9	294 50.5	10.6	261 19.3	17.1			
31 00	308 24.9	203 32.2	N22 30.9	200 28.6	N23 13.7	309 53.0	S 2 10.7	276 21.7	N10 17.1	Canopus	264 01.7	S52 41.6
01	323 27.4	218 31.4	30.7	215 29.3	13.5	324 55.6	10.8	291 24.0	17.1	Capella	280 51.7	N45 59.5
02	338 29.8	233 30.5	30.5	230 29.9	13.3	339 58.1	10.8	306 26.4	17.1	Deneb	49 38.8	N45 16.6
03	353 32.3	248 29.7	.. 30.3	245 30.6	.. 13.2	355 00.7	.. 10.9	321 28.8	.. 17.1	Denebola	182 45.5	N14 35.0
04	8 34.8	263 28.9	30.1	260 31.3	13.0	10 03.3	10.9	336 31.2	17.1	Diphda	349 07.2	S17 59.6
05	23 37.2	278 28.1	29.9	275 32.0	12.8	25 05.8	11.0	351 33.6	17.2			
06	38 39.7	293 27.3	N22 29.7	290 32.7	N23 12.6	40 08.4	S 2 11.0	6 36.0	N10 17.2	Dubhe	194 06.3	N61 45.8
07	53 42.1	308 26.5	29.5	305 33.4	12.4	55 10.9	11.1	21 38.4	17.2	Elnath	278 27.4	N28 36.2
F 08	68 44.6	323 25.7	29.3	320 34.0	12.2	70 13.5	11.1	36 40.8	17.2	Eltanin	90 51.0	N51 29.7
R 09	83 47.1	338 24.9	.. 29.1	335 34.7	.. 12.1	85 16.1	.. 11.2	51 43.2	.. 17.2	Enif	33 58.0	N 9 52.2
I 10	98 49.5	353 24.1	28.9	350 35.4	11.9	100 18.6	11.2	66 45.6	17.2	Fomalhaut	15 36.3	S29 37.6
D 11	113 52.0	8 23.3	28.7	5 36.1	11.7	115 21.2	11.3	81 48.0	17.2			
A 12	128 54.5	23 22.5	N22 28.5	20 36.8	N23 11.5	130 23.7	S 2 11.3	96 50.4	N10 17.2	Gacrux	172 13.9	S57 06.4
Y 13	143 56.9	38 21.7	28.3	35 37.5	11.3	145 26.3	11.4	111 52.8	17.2	Gienah	176 04.3	S17 32.0
14	158 59.4	53 20.8	28.1	50 38.2	11.1	160 28.8	11.5	126 55.2	17.3	Hadar	149 04.2	S60 22.1
15	174 01.9	68 20.0	.. 27.9	65 38.8	.. 11.0	175 31.4	.. 11.5	141 57.6	.. 17.3	Hamal	328 13.6	N23 27.2
16	189 04.3	83 19.2	27.6	80 39.5	10.8	190 34.0	11.6	157 00.0	17.3	Kaus Aust.	83 58.6	S34 23.0
17	204 06.8	98 18.4	27.4	95 40.2	10.6	205 36.5	11.6	172 02.4	17.3			
18	219 09.3	113 17.6	N22 27.2	110 40.9	N23 10.4	220 39.1	S 2 11.7	187 04.8	N10 17.3	Kochab	137 19.6	N74 10.1
19	234 11.7	128 16.8	27.0	125 41.6	10.2	235 41.7	11.7	202 07.2	17.3	Markab	13 45.9	N15 11.9
20	249 14.2	143 16.0	26.8	140 42.3	10.0	250 44.2	11.8	217 09.6	17.3	Menkar	314 27.1	N 4 05.0
21	264 16.6	158 15.2	.. 26.6	155 43.0	.. 09.8	265 46.8	.. 11.8	232 12.0	.. 17.3	Menkent	148 21.1	S36 21.8
22	279 19.1	173 14.4	26.3	170 43.6	09.6	280 49.3	11.9	247 14.4	17.4	Miaplacidus	221 42.9	S69 42.8
23	294 21.6	188 13.6	26.1	185 44.3	09.5	295 51.9	11.9	262 16.8	17.4			
1 00	309 24.0	203 12.8	N22 25.9	200 45.0	N23 09.3	310 54.5	S 2 12.0	277 19.2	N10 17.4	Mirfak	308 56.9	N49 51.1
01	324 26.5	218 12.0	25.7	215 45.7	09.1	325 57.0	12.1	292 21.6	17.4	Nunki	76 12.2	S26 17.8
02	339 29.0	233 11.2	25.5	230 46.4	08.9	340 59.6	12.1	307 24.0	17.4	Peacock	53 36.7	S56 44.3
03	354 31.4	248 10.3	.. 25.2	245 47.1	.. 08.7	356 02.2	.. 12.2	322 26.4	.. 17.4	Pollux	243 42.1	N28 01.7
04	9 33.9	263 09.5	25.0	260 47.8	08.5	11 04.7	12.2	337 28.8	17.4	Procyon	245 12.0	N 5 13.7
05	24 36.4	278 08.7	24.8	275 48.4	08.3	26 07.3	12.3	352 31.2	17.4			
06	39 38.8	293 07.9	N22 24.5	290 49.1	N23 08.1	41 09.8	S 2 12.3	7 33.6	N10 17.4	Rasalhague	96 16.8	N12 33.9
07	54 41.3	308 07.1	24.3	305 49.8	07.9	56 12.4	12.4	22 36.0	17.5	Regulus	207 56.0	N11 58.5
S 08	69 43.8	323 06.3	24.1	320 50.5	07.8	71 15.0	12.5	37 38.4	17.5	Rigel	281 23.3	S 8 12.2
A 09	84 46.2	338 05.5	.. 23.8	335 51.2	.. 07.6	86 17.5	.. 12.5	52 40.8	.. 17.5	Rigil Kent.	140 07.4	S60 49.8
T 10	99 48.7	353 04.7	23.6	350 51.9	07.4	101 20.1	12.6	67 43.2	17.5	Sabik	102 25.4	S15 43.2
U 11	114 51.1	8 03.9	23.4	5 52.6	07.2	116 22.7	12.6	82 45.6	17.5			
R 12	129 53.6	23 03.1	N22 23.1	20 53.3	N23 07.0	131 25.2	S 2 12.7	97 48.0	N10 17.5	Schedar	349 53.3	N56 31.5
D 13	144 56.1	38 02.3	22.9	35 54.0	06.8	146 27.8	12.7	112 50.4	17.5	Shaula	96 37.1	S37 06.1
A 14	159 58.5	53 01.5	22.7	50 54.6	06.6	161 30.4	12.8	127 52.8	17.5	Sirius	258 44.1	S16 42.8
Y 15	175 01.0	68 00.7	.. 22.4	65 55.3	.. 06.4	176 32.9	.. 12.8	142 55.2	.. 17.5	Spica	158 43.4	S11 09.1
16	190 03.5	82 59.9	22.2	80 56.0	06.2	191 35.5	12.9	157 57.6	17.5	Suhail	223 01.3	S43 25.7
17	205 05.9	97 59.1	21.9	95 56.7	06.0	206 38.1	13.0	173 00.0	17.6			
18	220 08.4	112 58.3	N22 21.7	110 57.4	N23 05.8	221 40.7	S 2 13.0	188 02.4	N10 17.6	Vega	80 46.3	N38 47.2
19	235 10.9	127 57.5	21.5	125 58.1	05.6	236 43.2	13.1	203 04.8	17.6	Zuben'ubi	137 18.1	S16 02.0
20	250 13.3	142 56.6	21.2	140 58.8	05.4	251 45.8	13.1	218 07.2	17.6		SHA	Mer. Pass.
21	265 15.8	157 55.8	.. 21.0	155 59.5	.. 05.3	266 48.4	.. 13.2	233 09.6	.. 17.6		° ′	h m
22	280 18.2	172 55.0	20.7	171 00.2	05.1	281 50.9	13.3	248 12.0	17.6	Venus	255 07.3	10 26
23	295 20.7	187 54.2	20.5	186 00.8	04.9	296 53.5	13.3	263 14.4	17.6	Mars	252 03.7	10 38
Mer. Pass.	h m 3 25.8	*v* −0.8	*d* 0.2	*v* 0.7	*d* 0.2	*v* 2.6	*d* 0.1	*v* 2.4	*d* 0.0	Jupiter	1 28.1	3 20
										Saturn	327 56.8	5 34

UT	SUN GHA	SUN Dec	MOON GHA	v	MOON Dec	d	HP
30	° '	° '	° '	'	° '	'	'
00	178 23.2	N18 36.5	107 14.6	15.3	S 4 15.7	9.4	54.3
01	193 23.2	35.9	121 48.9	15.3	4 25.1	9.4	54.3
02	208 23.3	35.3	136 23.2	15.2	4 34.5	9.3	54.3
03	223 23.3	.. 34.7	150 57.4	15.3	4 43.8	9.3	54.2
04	238 23.3	34.1	165 31.7	15.2	4 53.1	9.4	54.2
05	253 23.3	33.5	180 05.9	15.3	5 02.5	9.3	54.2
06	268 23.4	N18 32.9	194 40.2	15.2	S 5 11.8	9.2	54.2
T 07	283 23.4	32.3	209 14.4	15.2	5 21.0	9.3	54.2
H 08	298 23.4	31.7	223 48.6	15.2	5 30.3	9.2	54.2
U 09	313 23.4	.. 31.1	238 22.8	15.2	5 39.5	9.2	54.2
R 10	328 23.5	30.5	252 57.0	15.2	5 48.7	9.2	54.2
S 11	343 23.5	29.9	267 31.2	15.2	5 57.9	9.1	54.2
D 12	358 23.5	N18 29.3	282 05.4	15.1	S 6 07.0	9.2	54.2
A 13	13 23.5	28.7	296 39.5	15.1	6 16.2	9.1	54.2
Y 14	28 23.6	28.1	311 13.6	15.2	6 25.3	9.0	54.2
15	43 23.6	.. 27.5	325 47.8	15.1	6 34.3	9.1	54.2
16	58 23.6	26.9	340 21.9	15.0	6 43.4	9.0	54.2
17	73 23.6	26.2	354 55.9	15.1	6 52.4	9.0	54.2
18	88 23.7	N18 25.6	9 30.0	15.0	S 7 01.4	9.0	54.2
19	103 23.7	25.0	24 04.0	15.1	7 10.4	8.9	54.2
20	118 23.7	24.4	38 38.1	15.0	7 19.3	8.9	54.2
21	133 23.7	.. 23.8	53 12.1	14.9	7 28.2	8.9	54.2
22	148 23.8	23.2	67 46.0	15.0	7 37.1	8.9	54.2
23	163 23.8	22.6	82 20.0	14.9	7 46.0	8.8	54.3
31 00	178 23.8	N18 22.0	96 53.9	15.0	S 7 54.8	8.8	54.3
01	193 23.9	21.4	111 27.9	14.8	8 03.6	8.7	54.3
02	208 23.9	20.8	126 01.7	14.9	8 12.3	8.7	54.3
03	223 23.9	.. 20.1	140 35.6	14.8	8 21.0	8.7	54.3
04	238 23.9	19.5	155 09.4	14.8	8 29.7	8.7	54.3
05	253 24.0	18.9	169 43.2	14.8	8 38.4	8.6	54.3
06	268 24.0	N18 18.3	184 17.0	14.8	S 8 47.0	8.6	54.3
07	283 24.0	17.7	198 50.8	14.7	8 55.6	8.5	54.3
F 08	298 24.1	17.1	213 24.5	14.7	9 04.1	8.6	54.3
R 09	313 24.1	.. 16.5	227 58.2	14.7	9 12.7	8.4	54.3
I 10	328 24.1	15.8	242 31.9	14.6	9 21.1	8.5	54.3
D 11	343 24.2	15.2	257 05.5	14.6	9 29.6	8.4	54.3
A 12	358 24.2	N18 14.6	271 39.1	14.6	S 9 38.0	8.4	54.3
Y 13	13 24.2	14.0	286 12.7	14.5	9 46.4	8.3	54.3
14	28 24.3	13.4	300 46.2	14.6	9 54.7	8.3	54.3
15	43 24.3	.. 12.8	315 19.8	14.4	10 03.0	8.2	54.3
16	58 24.3	12.1	329 53.2	14.5	10 11.2	8.3	54.4
17	73 24.4	11.5	344 26.7	14.4	10 19.5	8.1	54.4
18	88 24.4	N18 10.9	359 00.1	14.4	S10 27.6	8.2	54.4
19	103 24.4	10.3	13 33.5	14.3	10 35.8	8.1	54.4
20	118 24.5	09.6	28 06.8	14.3	10 43.9	8.0	54.4
21	133 24.5	.. 09.0	42 40.1	14.3	10 51.9	8.0	54.4
22	148 24.5	08.4	57 13.4	14.2	10 59.9	8.0	54.4
23	163 24.6	07.8	71 46.6	14.2	11 07.9	7.9	54.4
1 00	178 24.6	N18 07.2	86 19.8	14.1	S11 15.8	7.9	54.4
01	193 24.6	06.5	100 52.9	14.1	11 23.7	7.8	54.4
02	208 24.7	05.9	115 26.0	14.1	11 31.5	7.8	54.4
03	223 24.7	.. 05.3	129 59.1	14.0	11 39.3	7.7	54.5
04	238 24.7	04.7	144 32.1	14.0	11 47.0	7.7	54.5
05	253 24.8	04.0	159 05.1	14.0	11 54.7	7.6	54.5
06	268 24.8	N18 03.4	173 38.1	13.9	S12 02.3	7.6	54.5
S 07	283 24.9	02.8	188 11.0	13.9	12 09.9	7.6	54.5
A 08	298 24.9	02.1	202 43.9	13.8	12 17.5	7.5	54.5
T 09	313 24.9	.. 01.5	217 16.7	13.7	12 25.0	7.4	54.5
U 10	328 25.0	00.9	231 49.4	13.8	12 32.4	7.4	54.5
R 11	343 25.0	18 00.3	246 22.2	13.7	12 39.8	7.3	54.6
D 12	358 25.0	N17 59.6	260 54.9	13.6	S12 47.1	7.3	54.6
A 13	13 25.1	59.0	275 28.5	13.6	12 54.4	7.3	54.6
Y 14	28 25.1	58.4	290 00.1	13.6	13 01.7	7.2	54.6
15	43 25.2	.. 57.7	304 32.7	13.5	13 08.9	7.1	54.6
16	58 25.2	57.1	319 05.2	13.4	13 16.0	7.1	54.6
17	73 25.2	56.5	333 37.6	13.4	13 23.1	7.0	54.7
18	88 25.3	N17 55.8	348 10.0	13.4	S13 30.1	6.9	54.7
19	103 25.3	55.2	2 42.4	13.3	13 37.1	6.9	54.7
20	118 25.4	54.6	17 14.7	13.3	13 44.0	6.8	54.7
21	133 25.4	.. 53.9	31 47.0	13.3	13 50.8	6.8	54.7
22	148 25.4	53.3	46 19.2	13.1	13 57.6	6.8	54.7
23	163 25.5	52.7	60 51.3	13.1	S14 04.4	6.7	54.8
	SD 15.8	d 0.6	SD 14.8		14.8		14.9

Lat.	Twilight Naut.	Twilight Civil	Sunrise	Moonrise 30	31	1	2
°	h m	h m	h m	h m	h m	h m	h m
N 72	▭	▭	▭	12 32	14 09	15 51	17 42
N 70	////	////	01 14	12 23	13 52	15 25	16 59
68	////	////	02 08	12 15	13 39	15 05	16 30
66	////	////	02 41	12 09	13 29	14 49	16 09
64	////	01 32	03 04	12 03	13 20	14 36	15 52
62	////	02 10	03 22	11 59	13 12	14 25	15 38
60	////	02 36	03 38	11 55	13 06	14 16	15 26
N 58	01 23	02 56	03 50	11 51	13 00	14 08	15 15
56	01 58	03 12	04 01	11 48	12 55	14 01	15 07
54	02 22	03 26	04 11	11 46	12 50	13 55	14 59
52	02 41	03 38	04 20	11 43	12 46	13 49	14 52
50	02 57	03 48	04 27	11 41	12 42	13 44	14 45
45	03 27	04 10	04 44	11 36	12 34	13 33	14 32
N 40	03 49	04 27	04 57	11 32	12 28	13 24	14 20
35	04 07	04 40	05 08	11 28	12 22	13 16	14 11
30	04 21	04 52	05 18	11 25	12 17	13 09	14 03
20	04 44	05 11	05 35	11 20	12 08	12 58	13 48
N 10	05 01	05 27	05 49	11 15	12 01	12 47	13 36
0	05 16	05 41	06 03	11 11	11 54	12 38	13 24
S 10	05 29	05 54	06 16	11 06	11 47	12 28	13 12
20	05 41	06 07	06 30	11 02	11 39	12 18	13 00
30	05 52	06 21	06 46	10 56	11 31	12 07	12 46
35	05 58	06 29	06 56	10 53	11 26	12 00	12 38
40	06 05	06 37	07 06	10 50	11 20	11 53	12 29
45	06 11	06 47	07 18	10 46	11 14	11 44	12 18
S 50	06 19	06 58	07 33	10 42	11 06	11 34	12 05
52	06 22	07 03	07 40	10 39	11 03	11 29	11 59
54	06 25	07 09	07 48	10 37	10 59	11 24	11 52
56	06 29	07 15	07 57	10 35	10 55	11 18	11 45
58	06 33	07 21	08 06	10 32	10 50	11 11	11 37
S 60	06 37	07 29	08 18	10 29	10 45	11 04	11 27

Lat.	Sunset	Twilight Civil	Twilight Naut.	Moonset 30	31	1	2
°	h m	h m	h m	h m	h m	h m	h m
N 72	▭	▭	▭	21 50	21 43	21 33	21 19
N 70	22 50	////	////	22 02	22 01	22 01	22 03
68	22 00	////	////	22 11	22 15	22 22	22 32
66	21 29	////	////	22 18	22 27	22 38	22 54
64	21 06	22 35	////	22 25	22 37	22 52	23 12
62	20 48	21 59	////	22 30	22 45	23 03	23 27
60	20 33	21 34	////	22 35	22 52	23 13	23 39
N 58	20 21	21 15	22 45	22 39	22 59	23 22	23 49
56	20 10	20 59	22 11	22 43	23 04	23 29	23 59
54	20 00	20 45	21 48	22 47	23 10	23 36	24 07
52	19 52	20 33	21 30	22 50	23 14	23 42	24 14
50	19 44	20 23	21 14	22 53	23 18	23 47	24 21
45	19 28	20 02	20 45	22 59	23 28	23 59	24 35
N 40	19 15	19 45	20 23	23 04	23 35	24 09	00 09
35	19 04	19 32	20 06	23 08	23 42	24 18	00 18
30	18 54	19 20	19 51	23 12	23 48	24 25	00 25
20	18 38	19 01	19 29	23 19	23 58	24 38	00 38
N 10	18 23	18 45	19 11	23 25	24 06	00 06	00 49
0	18 10	18 32	18 57	23 31	24 15	00 15	01 00
S 10	17 57	18 19	18 44	23 37	24 23	00 23	01 11
20	17 43	18 06	18 32	23 43	24 32	00 32	01 22
30	17 27	17 52	18 21	23 50	24 42	00 42	01 35
35	17 18	17 44	18 15	23 54	24 48	00 48	01 42
40	17 07	17 36	18 09	23 59	24 55	00 55	01 51
45	16 55	17 27	18 02	24 04	00 04	01 03	02 01
S 50	16 40	17 15	17 55	24 10	00 10	01 12	02 13
52	16 33	17 10	17 51	24 13	00 13	01 16	02 19
54	16 25	17 05	17 48	24 17	00 17	01 21	02 25
56	16 17	16 59	17 44	24 20	00 20	01 26	02 32
58	16 07	16 52	17 40	24 24	00 24	01 32	02 40
S 60	15 56	16 45	17 36	24 29	00 29	01 39	02 49

Day	SUN Eqn. of Time 00h	12h	Mer. Pass.	MOON Mer. Pass. Upper	Lower	Age	Phase
d	m s	m s	h m	h m	h m	d	%
30	06 27	06 26	12 06	17 21	05 00	07	41
31	06 25	06 23	12 06	18 04	05 42	08	50
1	06 22	06 20	12 06	18 49	06 26	09	60

1998 AUGUST 2, 3, 4 (SUN., MON., TUES.)

UT	ARIES GHA	VENUS −3.9 GHA	Dec	MARS +1.7 GHA	Dec	JUPITER −2.7 GHA	Dec	SATURN +0.4 GHA	Dec
d h	° ′	° ′	° ′	° ′	° ′	° ′	° ′	° ′	° ′
2 00	310 23.2	202 53.4	N22 20.2	201 01.5	N23 04.7	311 56.1	S 2 13.4	278 16.8	N10 17.6
01	325 25.6	217 52.6	20.0	216 02.2	04.5	326 58.6	13.4	293 19.2	17.6
02	340 28.1	232 51.8	19.7	231 02.9	04.3	342 01.2	13.5	308 21.6	17.6
03	355 30.6	247 51.0 ..	19.5	246 03.6 ..	04.1	357 03.8 ..	13.6	323 24.0 ..	17.7
04	10 33.0	262 50.2	19.2	261 04.3	03.9	12 06.4	13.6	338 26.4	17.7
05	25 35.5	277 49.4	19.0	276 05.0	03.7	27 08.9	13.7	353 28.8	17.7
06	40 38.0	292 48.6	N22 18.7	291 05.7	N23 03.5	42 11.5	S 2 13.7	8 31.2	N10 17.7
07	55 40.4	307 47.8	18.4	306 06.4	03.3	57 14.1	13.8	23 33.6	17.7
08	70 42.9	322 47.0	18.2	321 07.1	03.1	72 16.6	13.9	38 36.1	17.7
S 09	85 45.4	337 46.2 ..	17.9	336 07.8 ..	02.9	87 19.2 ..	13.9	53 38.5 ..	17.7
U 10	100 47.8	352 45.4	17.7	351 08.5	02.7	102 21.8	14.0	68 40.9	17.7
N 11	115 50.3	7 44.6	17.4	6 09.1	02.5	117 24.4	14.0	83 43.3	17.7
D 12	130 52.7	22 43.8	N22 17.1	21 09.8	N23 02.3	132 26.9	S 2 14.1	98 45.7	N10 17.7
A 13	145 55.2	37 43.0	16.9	36 10.5	02.1	147 29.5	14.2	113 48.1	17.8
Y 14	160 57.7	52 42.2	16.6	51 11.2	01.9	162 32.1	14.2	128 50.5	17.8
15	176 00.1	67 41.4 ..	16.3	66 11.9 ..	01.7	177 34.7 ..	14.3	143 52.9 ..	17.8
16	191 02.6	82 40.6	16.1	81 12.6	01.5	192 37.2	14.3	158 55.3	17.8
17	206 05.1	97 39.8	15.8	96 13.3	01.3	207 39.8	14.4	173 57.7	17.8
18	221 07.5	112 39.0	N22 15.5	111 14.0	N23 01.1	222 42.4	S 2 14.5	189 00.1	N10 17.8
19	236 10.0	127 38.2	15.3	126 14.7	00.9	237 45.0	14.5	204 02.5	17.8
20	251 12.5	142 37.4	15.0	141 15.4	00.7	252 47.5	14.6	219 04.9	17.8
21	266 14.9	157 36.6 ..	14.7	156 16.1 ..	00.5	267 50.1 ..	14.6	234 07.3 ..	17.8
22	281 17.4	172 35.8	14.4	171 16.8	00.3	282 52.7	14.7	249 09.7	17.8
23	296 19.9	187 35.0	14.2	186 17.5	23 00.1	297 55.3	14.8	264 12.1	17.8
3 00	311 22.3	202 34.2	N22 13.9	201 18.2	N22 59.9	312 57.8	S 2 14.8	279 14.5	N10 17.8
01	326 24.8	217 33.4	13.6	216 18.9	59.7	328 00.4	14.9	294 17.0	17.9
02	341 27.2	232 32.6	13.3	231 19.5	59.5	343 03.0	15.0	309 19.4	17.9
03	356 29.7	247 31.8 ..	13.1	246 20.2 ..	59.3	358 05.6 ..	15.0	324 21.8 ..	17.9
04	11 32.2	262 31.0	12.8	261 20.9	59.1	13 08.2	15.1	339 24.2	17.9
05	26 34.6	277 30.2	12.5	276 21.6	58.9	28 10.7	15.1	354 26.6	17.9
06	41 37.1	292 29.4	N22 12.2	291 22.3	N22 58.7	43 13.3	S 2 15.2	9 29.0	N10 17.9
07	56 39.6	307 28.6	11.9	306 23.0	58.4	58 15.9	15.3	24 31.4	17.9
08	71 42.0	322 27.8	11.6	321 23.7	58.2	73 18.5	15.3	39 33.8	17.9
M 09	86 44.5	337 27.0 ..	11.4	336 24.4 ..	58.0	88 21.1 ..	15.4	54 36.2 ..	17.9
O 10	101 47.0	352 26.2	11.1	351 25.1	57.8	103 23.6	15.5	69 38.6	17.9
N 11	116 49.4	7 25.4	10.8	6 25.8	57.6	118 26.2	15.5	84 41.0	17.9
D 12	131 51.9	22 24.6	N22 10.5	21 26.5	N22 57.4	133 28.8	S 2 15.6	99 43.5	N10 17.9
A 13	146 54.4	37 23.8	10.2	36 27.2	57.2	148 31.4	15.6	114 45.9	17.9
Y 14	161 56.8	52 23.0	09.9	51 27.9	57.0	163 34.0	15.7	129 48.3	18.0
15	176 59.3	67 22.2 ..	09.6	66 28.6 ..	56.8	178 36.5 ..	15.8	144 50.7 ..	18.0
16	192 01.7	82 21.4	09.3	81 29.3	56.6	193 39.1	15.8	159 53.1	18.0
17	207 04.2	97 20.6	09.0	96 30.0	56.4	208 41.7	15.9	174 55.5	18.0
18	222 06.7	112 19.8	N22 08.7	111 30.7	N22 56.2	223 44.3	S 2 16.0	189 57.9	N10 18.0
19	237 09.1	127 19.0	08.4	126 31.4	56.0	238 46.9	16.0	205 00.3	18.0
20	252 11.6	142 18.2	08.1	141 32.1	55.7	253 49.5	16.1	220 02.7	18.0
21	267 14.1	157 17.4 ..	07.8	156 32.8 ..	55.5	268 52.0 ..	16.2	235 05.1 ..	18.0
22	282 16.5	172 16.6	07.5	171 33.5	55.3	283 54.6	16.2	250 07.6	18.0
23	297 19.0	187 15.8	07.2	186 34.2	55.1	298 57.2	16.3	265 10.0	18.0
4 00	312 21.5	202 15.0	N22 06.9	201 34.9	N22 54.9	313 59.8	S 2 16.3	280 12.4	N10 18.0
01	327 23.9	217 14.2	06.6	216 35.6	54.7	329 02.4	16.4	295 14.8	18.0
02	342 26.4	232 13.4	06.3	231 36.3	54.5	344 05.0	16.5	310 17.2	18.0
03	357 28.8	247 12.6 ..	06.0	246 37.0 ..	54.3	359 07.5 ..	16.5	325 19.6 ..	18.0
04	12 31.3	262 11.8	05.7	261 37.7	54.1	14 10.1	16.6	340 22.0	18.1
05	27 33.8	277 11.0	05.4	276 38.4	53.9	29 12.7	16.7	355 24.4	18.1
06	42 36.2	292 10.2	N22 05.1	291 39.1	N22 53.6	44 15.3	S 2 16.7	10 26.8	N10 18.1
07	57 38.7	307 09.4	04.8	306 39.8	53.4	59 17.9	16.8	25 29.3	18.1
08	72 41.2	322 08.6	04.5	321 40.5	53.2	74 20.5	16.9	40 31.7	18.1
T 09	87 43.6	337 07.8 ..	04.1	336 41.2 ..	53.0	89 23.1 ..	16.9	55 34.1 ..	18.1
U 10	102 46.1	352 07.0	03.8	351 41.9	52.8	104 25.7	17.0	70 36.5	18.1
E 11	117 48.6	7 06.2	03.5	6 42.6	52.6	119 28.2	17.1	85 38.9	18.1
S 12	132 51.0	22 05.4	N22 03.2	21 43.3	N22 52.4	134 30.8	S 2 17.1	100 41.3	N10 18.1
D 13	147 53.5	37 04.6	02.9	36 44.0	52.1	149 33.4	17.2	115 43.7	18.1
A 14	162 56.0	52 03.8	02.6	51 44.7	51.9	164 36.0	17.3	130 46.2	18.1
Y 15	177 58.4	67 03.0 ..	02.2	66 45.4 ..	51.7	179 38.6 ..	17.3	145 48.6 ..	18.1
16	193 00.9	82 02.2	01.9	81 46.1	51.5	194 41.2	17.4	160 51.0	18.1
17	208 03.3	97 01.4	01.6	96 46.8	51.3	209 43.8	17.5	175 53.4	18.1
18	223 05.8	112 00.6	N22 01.3	111 47.5	N22 51.1	224 46.4	S 2 17.5	190 55.8	N10 18.1
19	238 08.3	126 59.8	00.9	126 48.2	50.9	239 49.0	17.6	205 58.2	18.1
20	253 10.7	141 59.1	00.6	141 48.9	50.6	254 51.5	17.7	221 00.6	18.2
21	268 13.2	156 58.3 ..	00.3	156 49.6 ..	50.4	269 54.1 ..	17.7	236 03.1 ..	18.2
22	283 15.7	171 57.5	22 00.0	171 50.3	50.2	284 56.7	17.8	251 05.5	18.2
23	298 18.1	186 56.7	N21 59.6	186 51.0	50.0	299 59.3	17.9	266 07.9	18.2
Mer. Pass. 3 14.0		v −0.8 d 0.3		v 0.7 d 0.2		v 2.6 d 0.1		v 2.4 d 0.0	

STARS

Name	SHA	Dec
Acamar	315 27.1	S40 18.5
Achernar	335 35.2	S57 14.4
Acrux	173 22.4	S63 05.6
Adhara	255 21.8	S28 58.2
Aldebaran	291 02.7	N16 30.3
Alioth	166 31.0	N55 58.4
Alkaid	153 08.1	N49 19.6
Al Na'ir	27 57.7	S46 57.9
Alnilam	275 58.2	S 1 12.2
Alphard	218 07.6	S 8 39.1
Alphecca	126 20.6	N26 43.5
Alpheratz	357 55.1	N29 04.9
Altair	62 19.1	N 8 52.1
Ankaa	353 26.8	S42 18.6
Antares	112 40.1	S26 25.6
Arcturus	146 06.2	N19 11.7
Atria	107 51.9	S69 01.6
Avior	234 23.3	S59 30.3
Bellatrix	278 44.5	N 6 20.8
Betelgeuse	271 13.9	N 7 24.3
Canopus	264 01.7	S52 41.6
Capella	280 51.7	N45 59.5
Deneb	49 38.8	N45 16.6
Denebola	182 45.6	N14 35.0
Diphda	349 07.2	S17 59.6
Dubhe	194 06.3	N61 45.7
Elnath	278 27.3	N28 36.2
Eltanin	90 51.1	N51 29.7
Enif	33 58.0	N 9 52.2
Fomalhaut	15 36.3	S29 37.6
Gacrux	172 13.9	S57 06.4
Gienah	176 04.3	S17 32.0
Hadar	149 04.2	S60 22.1
Hamal	328 13.6	N23 27.2
Kaus Aust.	83 58.6	S34 23.0
Kochab	137 19.7	N74 10.1
Markab	13 49.4	N15 11.9
Menkar	314 27.1	N 4 05.0
Menkent	148 21.1	S36 21.8
Miaplacidus	221 42.9	S69 42.8
Mirfak	308 56.8	N49 51.1
Nunki	76 12.2	S26 17.8
Peacock	53 36.7	S56 44.3
Pollux	243 42.1	N28 01.7
Procyon	245 12.0	N 5 13.7
Rasalhague	96 16.8	N12 33.9
Regulus	207 56.0	N11 58.5
Rigel	281 23.3	S 8 12.2
Rigil Kent.	140 07.4	S60 49.8
Sabik	102 25.4	S15 43.2
Schedar	349 53.2	N56 31.5
Shaula	96 37.2	S37 06.1
Sirius	258 44.1	S16 42.8
Spica	158 43.4	S11 09.1
Suhail	223 01.3	S43 25.6
Vega	80 46.3	N38 47.2
Zuben'ubi	137 18.1	S16 02.0

	SHA	Mer. Pass.
	° ′	h m
Venus	251 11.8	10 30
Mars	249 55.8	10 34
Jupiter	1 35.5	3 08
Saturn	327 52.2	5 22

UT	SUN GHA	Dec	MOON GHA	v	Dec	d	HP
d h	° ′	° ′	° ′	′	° ′	′	′
2 00	178 25.5	N17 52.0	75 23.4	13.1	S14 11.1	6.6	54.8
01	193 25.6	51.4	89 55.5	13.0	14 17.7	6.5	54.8
02	208 25.6	50.8	104 27.5	13.0	14 24.2	6.5	54.8
03	223 25.6	.. 50.1	118 59.5	12.9	14 30.7	6.5	54.8
04	238 25.7	49.5	133 31.4	12.8	14 37.2	6.4	54.8
05	253 25.7	48.9	148 03.2	12.9	14 43.6	6.3	54.9
06	268 25.8	N17 48.2	162 35.1	12.7	S14 49.9	6.2	54.9
07	283 25.8	47.6	177 06.8	12.7	14 56.1	6.2	54.9
S 08	298 25.9	46.9	191 38.5	12.7	15 02.3	6.2	54.9
U 09	313 25.9	.. 46.3	206 10.2	12.5	15 08.5	6.0	54.9
N 10	328 26.0	45.7	220 41.7	12.6	15 14.5	6.0	55.0
D 11	343 26.0	45.0	235 13.3	12.5	15 20.5	5.9	55.0
A 12	358 26.0	N17 44.4	249 44.8	12.4	S15 26.4	5.9	55.0
Y 13	13 26.1	43.7	264 16.2	12.4	15 32.3	5.8	55.0
14	28 26.1	43.1	278 47.6	12.3	15 38.1	5.7	55.0
15	43 26.2	.. 42.4	293 18.9	12.3	15 43.8	5.7	55.1
16	58 26.2	41.8	307 50.2	12.2	15 49.5	5.6	55.1
17	73 26.3	41.2	322 21.4	12.1	15 55.1	5.5	55.1
18	88 26.3	N17 40.5	336 52.5	12.1	S16 00.6	5.4	55.1
19	103 26.4	39.9	351 23.6	12.1	16 06.0	5.4	55.1
20	118 26.4	39.2	5 54.7	11.9	16 11.4	5.3	55.2
21	133 26.5	.. 38.6	20 25.6	12.0	16 16.7	5.3	55.2
22	148 26.5	37.9	34 56.6	11.8	16 22.0	5.1	55.2
23	163 26.6	37.3	49 27.4	11.9	16 27.1	5.1	55.2
3 00	178 26.6	N17 36.6	63 58.3	11.7	S16 32.2	5.0	55.3
01	193 26.7	36.0	78 29.0	11.7	16 37.2	4.9	55.3
02	208 26.7	35.3	92 59.7	11.7	16 42.1	4.9	55.3
03	223 26.8	.. 34.7	107 30.4	11.6	16 47.0	4.8	55.3
04	238 26.8	34.0	122 01.0	11.5	16 51.8	4.7	55.4
05	253 26.8	33.4	136 31.5	11.5	16 56.5	4.6	55.4
06	268 26.9	N17 32.7	151 02.0	11.4	S17 01.1	4.5	55.4
07	283 26.9	32.1	165 32.4	11.4	17 05.6	4.5	55.4
08	298 27.0	31.4	180 02.8	11.3	17 10.1	4.4	55.5
M 09	313 27.1	.. 30.8	194 33.1	11.2	17 14.5	4.3	55.5
O 10	328 27.1	30.1	209 03.3	11.2	17 18.8	4.2	55.5
N 11	343 27.2	29.5	223 33.5	11.1	17 23.0	4.2	55.5
D 12	358 27.2	N17 28.8	238 03.6	11.1	S17 27.2	4.0	55.6
A 13	13 27.3	28.2	252 33.7	11.0	17 31.2	4.0	55.6
Y 14	28 27.3	27.5	267 03.7	11.0	17 35.2	3.9	55.6
15	43 27.4	.. 26.9	281 33.7	10.9	17 39.1	3.8	55.6
16	58 27.4	26.2	296 03.6	10.8	17 42.9	3.7	55.7
17	73 27.5	25.5	310 33.4	10.8	17 46.6	3.6	55.7
18	88 27.5	N17 24.9	325 03.2	10.8	S17 50.2	3.6	55.7
19	103 27.6	24.2	339 33.0	10.6	17 53.8	3.4	55.7
20	118 27.6	23.6	354 02.6	10.7	17 57.2	3.4	55.8
21	133 27.7	.. 22.9	8 32.3	10.5	18 00.6	3.3	55.8
22	148 27.7	22.3	23 01.8	10.5	18 03.9	3.2	55.8
23	163 27.8	21.6	37 31.4	10.4	18 07.1	3.1	55.9
4 00	178 27.8	N17 20.9	52 00.8	10.4	S18 10.2	3.0	55.9
01	193 27.9	20.3	66 30.2	10.4	18 13.2	2.9	55.9
02	208 28.0	19.6	80 59.6	10.2	18 16.1	2.8	55.9
03	223 28.0	.. 19.0	95 28.8	10.3	18 18.9	2.8	56.0
04	238 28.1	18.3	109 58.1	10.2	18 21.7	2.6	56.0
05	253 28.1	17.6	124 27.3	10.1	18 24.3	2.6	56.0
06	268 28.2	N17 17.0	138 56.4	10.1	S18 26.9	2.4	56.1
07	283 28.2	16.3	153 25.5	10.0	18 29.3	2.4	56.1
08	298 28.3	15.6	167 54.5	10.0	18 31.7	2.2	56.1
T 09	313 28.3	.. 15.0	182 23.5	9.9	18 33.9	2.2	56.1
U 10	328 28.4	14.3	196 52.4	9.9	18 36.1	2.1	56.2
E 11	343 28.5	13.6	211 21.3	9.8	18 38.2	2.0	56.2
S 12	358 28.5	N17 13.0	225 50.1	9.7	S18 40.2	1.8	56.2
D 13	13 28.6	12.3	240 18.8	9.8	18 42.0	1.8	56.3
A 14	28 28.6	11.6	254 47.6	9.6	18 43.8	1.7	56.3
Y 15	43 28.7	.. 11.0	269 16.2	9.6	18 45.5	1.6	56.3
16	58 28.7	10.3	283 44.8	9.6	18 47.1	1.5	56.4
17	73 28.8	09.6	298 13.4	9.5	18 48.6	1.3	56.4
18	88 28.9	N17 09.0	312 41.9	9.5	S18 49.9	1.3	56.4
19	103 28.9	08.3	327 10.4	9.4	18 51.2	1.2	56.4
20	118 29.0	07.6	341 38.8	9.4	18 52.4	1.1	56.5
21	133 29.1	.. 07.0	356 07.2	9.3	18 53.5	0.9	56.5
22	148 29.1	06.3	10 35.5	9.3	18 54.4	0.9	56.5
23	163 29.2	05.6	25 03.8	9.2	S18 55.3	0.8	56.6
	SD 15.8	d 0.7	SD 15.0		15.1		15.3

Lat.	Twilight Naut.	Civil	Sunrise	Moonrise 2	3	4	5
°	h m	h m	h m	h m	h m	h m	h m
N 72	□	□	□	17 42	■	■	■
N 70	////	////	01 41	16 59	18 33	19 57	20 48
68	////	////	02 24	16 30	17 53	19 05	19 57
66	////	00 50	02 52	16 09	17 25	18 32	19 26
64	////	01 49	03 14	15 52	17 04	18 09	19 02
62	////	02 22	03 31	15 38	16 47	17 50	18 43
60	00 45	02 45	03 45	15 26	16 33	17 34	18 28
N 58	01 39	03 04	03 57	15 15	16 21	17 21	18 15
56	02 09	03 19	04 07	15 07	16 10	17 10	18 03
54	02 31	03 32	04 16	14 59	16 01	17 00	17 53
52	02 48	03 43	04 24	14 52	15 53	16 51	17 45
50	03 03	03 53	04 32	14 45	15 45	16 43	17 37
45	03 31	04 14	04 47	14 32	15 30	16 26	17 20
N 40	03 53	04 30	05 00	14 20	15 17	16 12	17 06
35	04 09	04 43	05 11	14 11	15 06	16 01	16 54
30	04 23	04 54	05 20	14 03	14 57	15 51	16 44
20	04 45	05 13	05 36	13 48	14 40	15 33	16 26
N 10	05 02	05 28	05 50	13 36	14 26	15 18	16 11
0	05 16	05 41	06 03	13 24	14 13	15 04	15 57
S 10	05 28	05 53	06 15	13 12	13 59	14 49	15 42
20	05 39	06 06	06 29	13 00	13 45	14 34	15 27
30	05 50	06 19	06 44	12 46	13 29	14 17	15 09
35	05 56	06 26	06 53	12 38	13 20	14 07	14 59
40	06 02	06 34	07 03	12 29	13 09	13 55	14 47
45	06 08	06 43	07 15	12 18	12 57	13 42	14 34
S 50	06 15	06 54	07 29	12 05	12 42	13 25	14 17
52	06 18	06 59	07 35	11 59	12 35	13 17	14 09
54	06 21	07 04	07 43	11 52	12 27	13 09	14 00
56	06 24	07 09	07 51	11 45	12 18	12 59	13 50
58	06 28	07 16	08 00	11 37	12 08	12 48	13 39
S 60	06 32	07 23	08 11	11 27	11 57	12 36	13 26

Lat.	Sunset	Twilight Civil	Naut.	Moonset 2	3	4	5
°	h m	h m	h m	h m	h m	h m	h m
N 72	□	□	□	21 19	■	■	■
N 70	22 25	////	////	22 03	22 10	22 33	23 34
68	21 45	////	////	22 32	22 51	23 26	24 24
66	21 17	23 11	////	22 54	23 19	23 58	24 56
64	20 56	22 18	////	23 12	23 41	24 22	00 22
62	20 40	21 47	////	23 27	23 58	24 41	00 41
60	20 26	21 24	23 17	23 39	24 13	00 13	00 57
N 58	20 14	21 06	22 29	23 49	24 25	00 25	01 10
56	20 04	20 51	22 00	23 59	24 36	00 36	01 21
54	19 55	20 39	21 39	24 07	00 07	00 45	01 31
52	19 47	20 28	21 22	24 14	00 14	00 53	01 40
50	19 40	20 18	21 08	24 21	00 21	01 01	01 48
45	19 24	19 58	20 40	24 35	00 35	01 17	02 05
N 40	19 12	19 42	20 19	00 09	00 47	01 30	02 19
35	19 01	19 29	20 02	00 18	00 57	01 41	02 31
30	18 52	19 18	19 49	00 25	01 06	01 51	02 41
20	18 36	19 00	19 27	00 38	01 21	02 08	02 59
N 10	18 22	18 45	19 10	00 49	01 35	02 23	03 14
0	18 10	18 31	18 57	01 00	01 47	02 37	03 29
S 10	17 57	18 19	18 44	01 11	02 00	02 51	03 43
20	17 44	18 07	18 33	01 22	02 13	03 05	03 59
30	17 29	17 54	18 22	01 35	02 28	03 22	04 16
35	17 20	17 46	18 17	01 42	02 37	03 32	04 26
40	17 10	17 39	18 11	01 51	02 48	03 44	04 38
45	16 58	17 30	18 05	02 01	03 00	03 57	04 52
S 50	16 44	17 19	17 58	02 13	03 14	04 13	05 09
52	16 38	17 14	17 55	02 19	03 21	04 21	05 17
54	16 30	17 09	17 52	02 25	03 28	04 29	05 25
56	16 22	17 04	17 49	02 32	03 37	04 39	05 35
58	16 13	16 57	17 45	02 40	03 46	04 49	05 47
S 60	16 03	16 50	17 41	02 49	03 57	05 02	06 00

	SUN			MOON			
Day	Eqn. of Time 00h	12h	Mer. Pass.	Mer. Pass. Upper	Lower	Age	Phase
d	m s	m s	h m	h m	h m	d	%
2	06 18	06 16	12 06	19 36	07 12	10	69
3	06 14	06 11	12 06	20 25	08 00	11	78
4	06 09	06 06	12 06	21 16	08 50	12	85

UT	ARIES GHA	VENUS −3.9 GHA	Dec	MARS +1.7 GHA	Dec	JUPITER −2.8 GHA	Dec	SATURN +0.4 GHA	Dec	STARS Name	SHA	Dec
d h	° '	° '	° '	° '	° '	° '	° '	° '	° '		° '	° '
5 00	313 20.6	201 55.9	N21 59.3	201 51.7	N22 49.8	315 01.9	S 2 17.9	281 10.3	N10 18.2	Acamar	315 27.1	S40 18.5
01	328 23.1	216 55.1	59.0	216 52.4	49.5	330 04.5	18.0	296 12.7	18.2	Achernar	335 35.2	S57 14.4
02	343 25.5	231 54.3	58.6	231 53.1	49.3	345 07.1	18.1	311 15.1	18.2	Acrux	173 22.4	S63 05.6
03	358 28.0	246 53.5	.. 58.3	246 53.8	.. 49.1	0 09.7	.. 18.1	326 17.5	.. 18.2	Adhara	255 21.8	S28 58.2
04	13 30.5	261 52.7	58.0	261 54.5	48.9	15 12.3	18.2	341 20.0	18.2	Aldebaran	291 02.7	N16 30.3
05	28 32.9	276 51.9	57.6	276 55.2	48.7	30 14.9	18.3	356 22.4	18.2			
06	43 35.4	291 51.1	N21 57.3	291 55.9	N22 48.5	45 17.5	S 2 18.3	11 24.8	N10 18.2	Alioth	166 31.0	N55 58.4
W 07	58 37.8	306 50.3	57.0	306 56.6	48.2	60 20.1	18.4	26 27.2	18.2	Alkaid	153 08.1	N49 19.6
E 08	73 40.3	321 49.5	56.6	321 57.3	48.0	75 22.7	18.5	41 29.6	18.2	Al Na'ir	27 57.6	S46 57.9
D 09	88 42.8	336 48.7	.. 56.3	336 58.0	.. 47.8	90 25.2	.. 18.6	56 32.0	.. 18.2	Alnilam	275 58.2	S 1 12.2
N 10	103 45.2	351 48.0	55.9	351 58.7	47.6	105 27.8	18.6	71 34.5	18.2	Alphard	218 07.6	S 8 39.1
E 11	118 47.7	6 47.2	55.6	6 59.4	47.4	120 30.4	18.7	86 36.9	18.2			
S 12	133 50.2	21 46.4	N21 55.3	22 00.1	N22 47.1	135 33.0	S 2 18.8	101 39.3	N10 18.2	Alphecca	126 20.6	N26 43.5
D 13	148 52.6	36 45.6	54.9	37 00.8	46.9	150 35.6	18.8	116 41.7	18.2	Alpheratz	357 55.1	N29 04.9
A 14	163 55.1	51 44.8	54.6	52 01.5	46.7	165 38.2	18.9	131 44.1	18.2	Altair	62 19.0	N 8 52.1
Y 15	178 57.6	66 44.0	.. 54.2	67 02.2	.. 46.5	180 40.8	.. 19.0	146 46.6	.. 18.3	Ankaa	353 26.8	S42 18.6
16	194 00.0	81 43.2	53.9	82 03.0	46.2	195 43.4	19.0	161 49.0	18.3	Antares	112 40.1	S26 25.6
17	209 02.5	96 42.4	53.5	97 03.7	46.0	210 46.0	19.1	176 51.4	18.3			
18	224 05.0	111 41.6	N21 53.2	112 04.4	N22 45.8	225 48.6	S 2 19.2	191 53.8	N10 18.3	Arcturus	146 06.2	N19 11.7
19	239 07.4	126 40.8	52.8	127 05.1	45.6	240 51.2	19.2	206 56.2	18.3	Atria	107 51.9	S69 01.6
20	254 09.9	141 40.0	52.4	142 05.8	45.3	255 53.8	19.3	221 58.6	18.3	Avior	234 23.3	S59 30.3
21	269 12.3	156 39.3	.. 52.1	157 06.5	.. 45.1	270 56.4	.. 19.4	237 01.1	.. 18.3	Bellatrix	278 44.5	N 6 20.8
22	284 14.8	171 38.5	51.8	172 07.2	44.9	285 59.0	19.5	252 03.5	18.3	Betelgeuse	271 13.9	N 7 24.3
23	299 17.3	186 37.7	51.4	187 07.9	44.7	301 01.6	19.5	267 05.9	18.3			
6 00	314 19.7	201 36.9	N21 51.1	202 08.6	N22 44.4	316 04.2	S 2 19.6	282 08.3	N10 18.3	Canopus	264 01.6	S52 41.6
01	329 22.2	216 36.1	50.7	217 09.3	44.2	331 06.8	19.7	297 10.7	18.3	Capella	280 51.6	N45 59.5
02	344 24.7	231 35.3	50.3	232 10.0	44.0	346 09.4	19.7	312 13.2	18.3	Deneb	49 38.8	N45 16.7
03	359 27.1	246 34.5	.. 50.0	247 10.7	.. 43.8	1 12.0	.. 19.8	327 15.6	.. 18.3	Denebola	182 45.6	N14 35.0
04	14 29.6	261 33.7	49.6	262 11.4	43.5	16 14.6	19.9	342 18.0	18.3	Diphda	349 07.2	S17 59.6
05	29 32.1	276 32.9	49.3	277 12.1	43.3	31 17.2	20.0	357 20.4	18.3			
06	44 34.5	291 32.2	N21 48.9	292 12.8	N22 43.1	46 19.8	S 2 20.0	12 22.8	N10 18.3	Dubhe	194 06.3	N61 45.7
T 07	59 37.0	306 31.4	48.5	307 13.6	42.9	61 22.4	20.1	27 25.3	18.3	Elnath	278 27.3	N28 36.2
H 08	74 39.4	321 30.6	48.2	322 14.3	42.6	76 25.0	20.2	42 27.7	18.3	Eltanin	90 51.1	N51 29.7
U 09	89 41.9	336 29.8	.. 47.8	337 15.0	.. 42.4	91 27.6	.. 20.2	57 30.1	.. 18.3	Enif	33 58.0	N 9 52.2
R 10	104 44.3	351 29.0	47.4	352 15.7	42.2	106 30.2	20.3	72 32.5	18.3	Fomalhaut	15 36.3	S29 37.6
S 11	119 46.8	6 28.2	47.1	7 16.4	42.0	121 32.8	20.4	87 35.0	18.3			
D 12	134 49.3	21 27.4	N21 46.7	22 17.1	N22 41.7	136 35.4	S 2 20.5	102 37.4	N10 18.3	Gacrux	172 13.9	S57 06.4
A 13	149 51.8	36 26.7	46.3	37 17.8	41.5	151 38.0	20.5	117 39.8	18.3	Gienah	176 04.3	S17 32.0
Y 14	164 54.2	51 25.9	46.0	52 18.5	41.3	166 40.6	20.6	132 42.2	18.3	Hadar	149 04.2	S60 22.1
15	179 56.7	66 25.1	.. 45.6	67 19.2	.. 41.0	181 43.2	.. 20.7	147 44.6	.. 18.3	Hamal	328 13.6	N23 27.2
16	194 59.2	81 24.3	45.2	82 19.9	40.8	196 45.8	20.7	162 47.1	18.3	Kaus Aust.	83 58.6	S34 23.0
17	210 01.6	96 23.5	44.8	97 20.6	40.6	211 48.4	20.8	177 49.5	18.3			
18	225 04.1	111 22.7	N21 44.5	112 21.4	N22 40.3	226 51.0	S 2 20.9	192 51.9	N10 18.4	Kochab	137 19.7	N74 10.1
19	240 06.6	126 21.9	44.1	127 22.1	40.1	241 53.6	21.0	207 54.3	18.4	Markab	13 49.4	N15 11.9
20	255 09.0	141 21.2	43.7	142 22.8	39.9	256 56.2	21.0	222 56.8	18.4	Menkar	314 27.0	N 4 05.0
21	270 11.5	156 20.4	.. 43.3	157 23.5	.. 39.7	271 58.8	.. 21.1	237 59.2	.. 18.4	Menkent	148 21.1	S36 21.8
22	285 13.9	171 19.6	42.9	172 24.2	39.4	287 01.4	21.2	253 01.6	18.4	Miaplacidus	221 42.9	S69 42.7
23	300 16.4	186 18.8	42.6	187 24.9	39.2	302 04.0	21.3	268 04.0	18.4			
7 00	315 18.9	201 18.0	N21 42.2	202 25.6	N22 39.0	317 06.6	S 2 21.3	283 06.4	N10 18.4	Mirfak	308 56.8	N49 51.1
01	330 21.3	216 17.2	41.8	217 26.3	38.7	332 09.3	21.4	298 08.9	18.4	Nunki	76 12.1	S26 17.8
02	345 23.8	231 16.4	41.4	232 27.0	38.5	347 11.9	21.5	313 11.3	18.4	Peacock	53 36.6	S56 44.3
03	0 26.3	246 15.7	.. 41.0	247 27.7	.. 38.3	2 14.5	.. 21.6	328 13.7	.. 18.4	Pollux	243 42.1	N28 01.7
04	15 28.7	261 14.9	40.6	262 28.5	38.0	17 17.1	21.6	343 16.1	18.4	Procyon	245 12.0	N 5 13.7
05	30 31.2	276 14.1	40.2	277 29.2	37.8	32 19.7	21.7	358 18.6	18.4			
06	45 33.7	291 13.3	N21 39.9	292 29.9	N22 37.6	47 22.3	S 2 21.8	13 21.0	N10 18.4	Rasalhague	96 16.8	N12 33.9
07	60 36.1	306 12.5	39.5	307 30.6	37.3	62 24.9	21.8	28 23.4	18.4	Regulus	207 56.0	N11 58.5
08	75 38.6	321 11.8	39.1	322 31.3	37.1	77 27.5	21.9	43 25.8	18.4	Rigel	281 23.2	S 8 12.2
F 09	90 41.1	336 11.0	.. 38.7	337 32.0	.. 36.9	92 30.1	.. 22.0	58 28.3	.. 18.4	Rigil Kent.	140 07.4	S60 49.8
R 10	105 43.5	351 10.2	38.3	352 32.7	36.6	107 32.7	22.1	73 30.7	18.4	Sabik	102 25.5	S15 43.2
I 11	120 46.0	6 09.4	37.9	7 33.4	36.4	122 35.3	22.1	88 33.1	18.4			
D 12	135 48.4	21 08.6	N21 37.5	22 34.2	N22 36.1	137 37.9	S 2 22.2	103 35.5	N10 18.4	Schedar	349 53.2	N56 31.5
A 13	150 50.9	36 07.8	37.1	37 34.9	35.9	152 40.5	22.3	118 38.0	18.4	Shaula	96 37.2	S37 06.1
Y 14	165 53.4	51 07.1	36.7	52 35.6	35.7	167 43.2	22.4	133 40.4	18.4	Sirius	258 44.1	S16 42.8
15	180 55.8	66 06.3	.. 36.3	67 36.3	.. 35.4	182 45.8	.. 22.4	148 42.8	.. 18.4	Spica	158 43.4	S11 09.1
16	195 58.3	81 05.5	35.9	82 37.0	35.2	197 48.4	22.5	163 45.3	18.4	Suhail	223 01.3	S43 25.6
17	211 00.8	96 04.7	35.5	97 37.7	35.0	212 51.0	22.6	178 47.7	18.4			
18	226 03.2	111 03.9	N21 35.1	112 38.4	N22 34.7	227 53.6	S 2 22.7	193 50.1	N10 18.4	Vega	80 46.3	N38 47.2
19	241 05.7	126 03.2	34.7	127 39.2	34.5	242 56.2	22.7	208 52.5	18.4	Zuben'ubi	137 18.1	S16 02.0
20	256 08.2	141 02.4	34.3	142 39.9	34.2	257 58.8	22.8	223 55.0	18.4		SHA	Mer. Pass.
21	271 10.6	156 01.6	.. 33.9	157 40.6	.. 34.0	273 01.4	.. 22.9	238 57.4	.. 18.4		° '	h m
22	286 13.1	171 00.8	33.5	172 41.3	33.8	288 04.0	23.0	253 59.8	18.4	Venus	247 17.2	10 34
23	301 15.6	186 00.0	33.1	187 42.0	33.5	303 06.7	23.1	269 02.2	18.4	Mars	247 48.9	10 31
	h m									Jupiter	1 44.5	2 55
Mer. Pass.	3 02.2	v −0.8	d 0.4	v 0.7	d 0.2	v 2.6	d 0.1	v 2.4	d 0.0	Saturn	327 48.6	5 11

UT	SUN GHA	SUN Dec	MOON GHA	v	MOON Dec	d	HP
d h	° ′	° ′	° ′	′	° ′	′	′
5 00	178 29.2	N17 05.0	39 32.0	9.2	S18 56.1	0.6	56.6
01	193 29.3	04.3	54 00.2	9.1	18 56.7	0.6	56.6
02	208 29.4	03.6	68 28.3	9.1	18 57.3	0.5	56.7
03	223 29.4	.. 02.9	82 56.4	9.1	18 57.8	0.3	56.7
04	238 29.5	02.3	97 24.5	9.0	18 58.1	0.2	56.7
05	253 29.5	01.6	111 52.5	9.0	18 58.3	0.2	56.8
06	268 29.6	N17 00.9	126 20.5	8.9	S18 58.5	0.0	56.8
W 07	283 29.7	17 00.2	140 48.4	8.9	18 58.5	0.1	56.8
E 08	298 29.7	16 59.6	155 16.3	8.9	18 58.4	0.1	56.8
D 09	313 29.8	.. 58.9	169 44.2	8.8	18 58.3	0.3	56.9
N 10	328 29.9	58.2	184 12.0	8.8	18 58.0	0.4	56.9
E 11	343 29.9	57.5	198 39.8	8.7	18 57.6	0.5	56.9
S 12	358 30.0	N16 56.9	213 07.5	8.7	S18 57.1	0.6	57.0
D 13	13 30.1	56.2	227 35.2	8.7	18 56.5	0.8	57.0
A 14	28 30.1	55.5	242 02.9	8.6	18 55.7	0.8	57.0
Y 15	43 30.2	.. 54.8	256 30.5	8.6	18 54.9	0.9	57.1
16	58 30.2	54.2	270 58.1	8.5	18 54.0	1.1	57.1
17	73 30.3	53.5	285 25.6	8.6	18 52.9	1.2	57.1
18	88 30.4	N16 52.8	299 53.2	8.5	S18 51.7	1.2	57.2
19	103 30.4	52.1	314 20.7	8.4	18 50.5	1.4	57.2
20	118 30.5	51.4	328 48.1	8.5	18 49.1	1.5	57.2
21	133 30.6	.. 50.8	343 15.6	8.4	18 47.6	1.6	57.3
22	148 30.6	50.1	357 43.0	8.3	18 46.0	1.7	57.3
23	163 30.7	49.4	12 10.3	8.4	18 44.3	1.8	57.3
6 00	178 30.8	N16 48.7	26 37.7	8.3	S18 42.5	2.0	57.4
01	193 30.8	48.0	41 05.0	8.3	18 40.5	2.0	57.4
02	208 30.9	47.3	55 32.3	8.2	18 38.5	2.2	57.4
03	223 31.0	.. 46.7	69 59.5	8.3	18 36.3	2.2	57.4
04	238 31.1	46.0	84 26.8	8.2	18 34.1	2.4	57.5
05	253 31.1	45.3	98 54.0	8.2	18 31.7	2.5	57.5
06	268 31.2	N16 44.6	113 21.2	8.1	S18 29.2	2.6	57.5
T 07	283 31.3	43.9	127 48.3	8.2	18 26.6	2.7	57.6
H 08	298 31.3	43.2	142 15.5	8.1	18 23.9	2.9	57.6
U 09	313 31.4	.. 42.5	156 42.6	8.1	18 21.0	2.9	57.6
R 10	328 31.5	41.9	171 09.7	8.1	18 18.1	3.0	57.7
S 11	343 31.5	41.2	185 36.8	8.1	18 15.1	3.2	57.7
D 12	358 31.6	N16 40.5	200 03.9	8.0	S18 11.9	3.3	57.7
A 13	13 31.7	39.8	214 30.9	8.0	18 08.6	3.4	57.8
Y 14	28 31.7	39.1	228 57.9	8.1	18 05.2	3.5	57.8
15	43 31.8	.. 38.4	243 25.0	8.0	18 01.7	3.6	57.8
16	58 31.9	37.7	257 52.0	7.9	17 58.1	3.7	57.8
17	73 32.0	37.0	272 18.9	8.0	17 54.4	3.8	57.9
18	88 32.0	N16 36.3	286 45.9	8.0	S17 50.6	4.0	57.9
19	103 32.1	35.6	301 12.9	7.9	17 46.6	4.0	57.9
20	118 32.2	35.0	315 39.8	8.0	17 42.6	4.2	58.0
21	133 32.3	.. 34.3	330 06.8	7.9	17 38.4	4.2	58.0
22	148 32.3	33.6	344 33.7	7.9	17 34.2	4.4	58.0
23	163 32.4	32.9	359 00.6	7.9	17 29.8	4.5	58.1
7 00	178 32.5	N16 32.2	13 27.5	7.9	S17 25.3	4.6	58.1
01	193 32.5	31.5	27 54.4	7.9	17 20.7	4.7	58.1
02	208 32.6	30.8	42 21.3	7.9	17 16.0	4.9	58.1
03	223 32.7	.. 30.1	56 48.2	7.9	17 11.1	4.9	58.2
04	238 32.8	29.4	71 15.1	7.8	17 06.2	5.0	58.2
05	253 32.8	28.7	85 41.9	7.9	17 01.2	5.2	58.2
06	268 32.9	N16 28.0	100 08.8	7.9	S16 56.0	5.2	58.3
07	283 33.0	27.3	114 35.7	7.8	16 50.8	5.4	58.3
08	298 33.1	26.6	129 02.5	7.9	16 45.4	5.5	58.3
F 09	313 33.1	.. 25.9	143 29.4	7.9	16 39.9	5.5	58.3
R 10	328 33.2	25.2	157 56.3	7.8	16 34.4	5.7	58.4
I 11	343 33.3	24.5	172 23.1	7.9	16 28.7	5.8	58.4
D 12	358 33.4	N16 23.8	186 50.0	7.9	S16 22.9	5.9	58.4
A 13	13 33.5	23.1	201 16.9	7.8	16 17.0	6.0	58.5
Y 14	28 33.5	22.4	215 43.7	7.9	16 11.0	6.1	58.5
15	43 33.6	.. 21.7	230 10.6	7.9	16 04.9	6.2	58.5
16	58 33.7	21.0	244 37.5	7.9	15 58.7	6.3	58.5
17	73 33.8	20.3	259 04.4	7.8	15 52.4	6.4	58.6
18	88 33.8	N16 19.6	273 31.2	7.9	S15 46.0	6.6	58.6
19	103 33.9	18.9	287 58.1	7.9	15 39.4	6.6	58.6
20	118 34.0	18.2	302 25.0	7.9	15 32.8	6.7	58.6
21	133 34.1	.. 17.5	316 51.9	7.9	15 26.1	6.8	58.7
22	148 34.2	16.8	331 18.8	7.9	15 19.3	6.9	58.7
23	163 34.2	16.1	345 45.7	8.0	S15 12.4	7.0	58.7
	SD 15.8	d 0.7	SD 15.5		15.7		15.9

Twilight / Sunrise / Moonrise

Lat.	Twilight Naut.	Twilight Civil	Sunrise	Moonrise 5	6	7	8
°	h m	h m	h m	h m	h m	h m	h m
N 72	////	////	00 54	■■	22 18	21 48	21 36
N 70	////	////	02 03	20 48	21 07	21 13	21 14
68	////	////	02 38	19 57	20 29	20 47	20 58
66	////	01 21	03 04	19 26	20 03	20 27	20 44
64	////	02 05	03 23	19 02	19 42	20 11	20 32
62	////	02 34	03 39	18 43	19 26	19 58	20 23
60	01 13	02 55	03 52	18 28	19 12	19 47	20 14
N 58	01 53	03 12	04 03	18 15	19 00	19 37	20 07
56	02 19	03 26	04 13	18 03	18 50	19 28	20 01
54	02 39	03 38	04 21	17 53	18 40	19 21	19 55
52	02 55	03 49	04 29	17 45	18 32	19 14	19 49
50	03 09	03 58	04 36	17 37	18 25	19 07	19 45
45	03 36	04 18	04 51	17 20	18 09	18 54	19 34
N 40	03 56	04 33	05 03	17 06	17 56	18 43	19 26
35	04 12	04 45	05 13	16 54	17 45	18 33	19 18
30	04 25	04 56	05 22	16 44	17 36	18 25	19 12
20	04 46	05 14	05 37	16 26	17 19	18 11	19 00
N 10	05 02	05 28	05 50	16 11	17 05	17 58	18 50
0	05 16	05 41	06 02	15 57	16 51	17 46	18 41
S 10	05 27	05 53	06 15	15 42	16 37	17 34	18 31
20	05 38	06 04	06 27	15 27	16 23	17 21	18 21
30	05 48	06 17	06 42	15 09	16 06	17 07	18 10
35	05 53	06 24	06 50	14 59	15 57	16 58	18 03
40	05 59	06 31	07 00	14 47	15 46	16 49	17 56
45	06 04	06 40	07 11	14 34	15 33	16 38	17 47
S 50	06 11	06 49	07 24	14 17	15 17	16 24	17 36
52	06 13	06 54	07 30	14 09	15 09	16 17	17 31
54	06 16	06 59	07 37	14 00	15 01	16 10	17 26
56	06 19	07 04	07 45	13 50	14 52	16 02	17 20
58	06 22	07 10	07 53	13 39	14 41	15 53	17 13
S 60	06 26	07 16	08 03	13 26	14 29	15 43	17 05

Sunset / Twilight / Moonset

Lat.	Sunset	Twilight Civil	Twilight Naut.	Moonset 5	6	7	8
°	h m	h m	h m	h m	h m	h m	h m
N 72	23 04	////	////	■■	23 58	26 23	02 22
N 70	22 04	////	////	23 34	25 09	01 09	02 58
68	21 30	////	////	24 24	00 24	01 46	03 23
66	21 05	22 43	////	24 56	00 56	02 12	03 42
64	20 46	22 02	////	00 22	01 19	02 32	03 57
62	20 31	21 35	////	00 41	01 38	02 48	04 10
60	20 18	21 15	22 52	00 57	01 53	03 02	04 20
N 58	20 07	20 58	22 15	01 10	02 06	03 14	04 30
56	19 58	20 44	21 50	01 21	02 17	03 24	04 38
54	19 49	20 32	21 30	01 31	02 27	03 32	04 45
52	19 42	20 22	21 14	01 40	02 36	03 40	04 52
50	19 35	20 12	21 01	01 48	02 44	03 47	04 57
45	19 20	19 53	20 35	02 05	03 01	04 03	05 10
N 40	19 08	19 38	20 15	02 19	03 14	04 15	05 20
35	18 58	19 26	19 59	02 31	03 26	04 26	05 29
30	18 50	19 15	19 46	02 41	03 36	04 35	05 37
20	18 35	18 58	19 25	02 59	03 53	04 51	05 50
N 10	18 21	18 43	19 09	03 14	04 08	05 04	06 01
0	18 09	18 31	18 56	03 29	04 22	05 17	06 12
S 10	17 57	18 19	18 44	03 43	04 37	05 30	06 23
20	17 45	18 08	18 34	03 59	04 52	05 44	06 34
30	17 30	17 55	18 24	04 16	05 09	05 59	06 47
35	17 22	17 49	18 19	04 26	05 19	06 08	06 55
40	17 13	17 41	18 13	04 38	05 30	06 18	07 03
45	17 02	17 33	18 08	04 52	05 43	06 30	07 13
S 50	16 48	17 23	18 02	05 09	06 00	06 45	07 25
52	16 42	17 19	17 59	05 17	06 07	06 52	07 30
54	16 35	17 14	17 56	05 25	06 16	06 59	07 36
56	16 28	17 09	17 53	05 35	06 25	07 07	07 43
58	16 19	17 03	17 50	05 47	06 36	07 17	07 50
S 60	16 09	16 56	17 47	06 00	06 48	07 27	07 59

SUN / MOON

Day	SUN Eqn. of Time 00h	SUN Eqn. of Time 12h	Mer. Pass.	MOON Mer. Pass. Upper	MOON Mer. Pass. Lower	Age	Phase
d	m s	m s	h m	h m	h m	d	%
5	06 03	06 00	12 06	22 09	09 43	13	92
6	05 57	05 54	12 06	23 04	10 37	14	97
7	05 50	05 47	12 06	23 59	11 32	15	100

UT	ARIES GHA	VENUS GHA	VENUS Dec	MARS GHA	MARS Dec	JUPITER GHA	JUPITER Dec	SATURN GHA	SATURN Dec
SATURDAY									
8 00	316 18.0	200 59.3	N21 32.7	202 42.7	N22 33.3	318 09.3	S 2 23.1	284 04.7	N10 18.4
01	331 20.5	215 58.5	32.2	217 43.4	33.1	333 11.9	23.2	299 07.1	18.4
02	346 22.9	230 57.7	31.8	232 44.2	32.8	348 14.5	23.3	314 09.5	18.4
03	1 25.4	245 56.9 ..	31.4	247 44.9 ..	32.6	3 17.1 ..	23.4	329 12.0 ..	18.4
04	16 27.9	260 56.2	31.0	262 45.6	32.3	18 19.7	23.4	344 14.4	18.4
05	31 30.3	275 55.4	30.6	277 46.3	32.1	33 22.3	23.5	359 16.8	18.4
06	46 32.8	290 54.6	N21 30.2	292 47.0	N22 31.9	48 24.9	S 2 23.6	14 19.2	N10 18.4
07	61 35.3	305 53.8	29.8	307 47.7	31.6	63 27.6	23.7	29 21.7	18.4
08	76 37.7	320 53.0	29.3	322 48.5	31.4	78 30.2	23.7	44 24.1	18.4
09	91 40.2	335 52.3 ..	28.9	337 49.2 ..	31.1	93 32.8 ..	23.8	59 26.5 ..	18.4
10	106 42.7	350 51.5	28.5	352 49.9	30.9	108 35.4	23.9	74 29.0	18.4
11	121 45.1	5 50.7	28.1	7 50.6	30.6	123 38.0	24.0	89 31.4	18.4
12	136 47.6	20 49.9	N21 27.7	22 51.3	N22 30.4	138 40.6	S 2 24.1	104 33.8	N10 18.4
13	151 50.0	35 49.2	27.2	37 52.0	30.2	153 43.3	24.1	119 36.2	18.4
14	166 52.5	50 48.4	26.8	52 52.8	29.9	168 45.9	24.2	134 38.7	18.4
15	181 55.0	65 47.6 ..	26.4	67 53.5 ..	29.7	183 48.5 ..	24.3	149 41.1 ..	18.4
16	196 57.4	80 46.8	26.0	82 54.2	29.4	198 51.1	24.4	164 43.5	18.4
17	211 59.9	95 46.1	25.5	97 54.9	29.2	213 53.7	24.4	179 46.0	18.4
18	227 02.4	110 45.3	N21 25.1	112 55.6	N22 28.9	228 56.3	S 2 24.5	194 48.4	N10 18.4
19	242 04.8	125 44.5	24.7	127 56.3	28.7	243 59.0	24.6	209 50.8	18.4
20	257 07.3	140 43.7	24.2	142 57.1	28.4	259 01.6	24.7	224 53.3	18.4
21	272 09.8	155 43.0 ..	23.8	157 57.8 ..	28.2	274 04.2 ..	24.8	239 55.7 ..	18.4
22	287 12.2	170 42.2	23.4	172 58.5	28.0	289 06.8	24.8	254 58.1	18.4
23	302 14.7	185 41.4	23.0	187 59.2	27.7	304 09.4	24.9	270 00.6	18.4
SUNDAY									
9 00	317 17.2	200 40.7	N21 22.5	202 59.9	N22 27.5	319 12.0	S 2 25.0	285 03.0	N10 18.4
01	332 19.6	215 39.9	22.1	218 00.7	27.2	334 14.7	25.1	300 05.4	18.4
02	347 22.1	230 39.1	21.6	233 01.4	27.0	349 17.3	25.2	315 07.9	18.4
03	2 24.5	245 38.3 ..	21.2	248 02.1 ..	26.7	4 19.9 ..	25.2	330 10.3 ..	18.4
04	17 27.0	260 37.6	20.8	263 02.8	26.5	19 22.5	25.3	345 12.7	18.4
05	32 29.5	275 36.8	20.3	278 03.5	26.2	34 25.1	25.4	0 15.1	18.4
06	47 31.9	290 36.0	N21 19.9	293 04.3	N22 26.0	49 27.8	S 2 25.5	15 17.6	N10 18.4
07	62 34.4	305 35.2	19.4	308 05.0	25.7	64 30.4	25.6	30 20.0	18.4
08	77 36.9	320 34.4	19.0	323 05.7	25.5	79 33.0	25.6	45 22.4	18.4
09	92 39.3	335 33.7 ..	18.6	338 06.4 ..	25.2	94 35.6 ..	25.7	60 24.9 ..	18.4
10	107 41.8	350 32.9	18.1	353 07.1	25.0	109 38.3	25.8	75 27.3	18.4
11	122 44.3	5 32.2	17.7	8 07.9	24.7	124 40.9	25.9	90 29.8	18.4
12	137 46.7	20 31.4	N21 17.2	23 08.6	N22 24.5	139 43.5	S 2 26.0	105 32.2	N10 18.4
13	152 49.2	35 30.6	16.8	38 09.3	24.2	154 46.1	26.0	120 34.6	18.4
14	167 51.7	50 29.9	16.3	53 10.0	24.0	169 48.7	26.1	135 37.1	18.4
15	182 54.1	65 29.1 ..	15.9	68 10.7 ..	23.7	184 51.4 ..	26.2	150 39.5 ..	18.4
16	197 56.6	80 28.3	15.4	83 11.5	23.5	199 54.0	26.3	165 41.9	18.4
17	212 59.0	95 27.6	15.0	98 12.2	23.2	214 56.6	26.4	180 44.4	18.4
18	228 01.5	110 26.8	N21 14.5	113 12.9	N22 23.0	229 59.2	S 2 26.4	195 46.8	N10 18.4
19	243 04.0	125 26.0	14.0	128 13.6	22.7	245 01.9	26.5	210 49.2	18.4
20	258 06.4	140 25.2	13.6	143 14.4	22.5	260 04.5	26.6	225 51.7	18.4
21	273 08.9	155 24.5 ..	13.1	158 15.1 ..	22.2	275 07.1 ..	26.7	240 54.1 ..	18.4
22	288 11.4	170 23.7	12.7	173 15.8	22.0	290 09.7	26.8	255 56.5	18.4
23	303 13.8	185 22.9	12.2	188 16.5	21.7	305 12.4	26.8	270 59.0	18.4
MONDAY									
10 00	318 16.3	200 22.2	N21 11.8	203 17.3	N22 21.5	320 15.0	S 2 26.9	286 01.4	N10 18.4
01	333 18.8	215 21.4	11.3	218 18.0	21.2	335 17.6	27.0	301 03.8	18.4
02	348 21.2	230 20.6	10.8	233 18.7	21.0	350 20.2	27.1	316 06.3	18.4
03	3 23.7	245 19.9 ..	10.4	248 19.4 ..	20.7	5 22.9 ..	27.2	331 08.7 ..	18.4
04	18 26.1	260 19.1	09.9	263 20.1	20.4	20 25.5	27.3	346 11.2	18.4
05	33 28.6	275 18.4	09.4	278 20.9	20.2	35 28.1	27.3	1 13.6	18.4
06	48 31.1	290 17.6	N21 09.0	293 21.6	N22 19.9	50 30.7	S 2 27.4	16 16.0	N10 18.4
07	63 33.5	305 16.8	08.5	308 22.3	19.7	65 33.4	27.5	31 18.5	18.4
08	78 36.0	320 16.1	08.0	323 23.0	19.4	80 36.0	27.6	46 20.9	18.4
09	93 38.5	335 15.3 ..	07.6	338 23.8 ..	19.2	95 38.6 ..	27.7	61 23.3 ..	18.4
10	108 40.9	350 14.5	07.1	353 24.5	18.9	110 41.3	27.7	76 25.8	18.4
11	123 43.4	5 13.8	06.6	8 25.2	18.7	125 43.9	27.8	91 28.2	18.4
12	138 45.9	20 13.0	N21 06.1	23 25.9	N22 18.4	140 46.5	S 2 27.9	106 30.7	N10 18.4
13	153 48.3	35 12.2	05.7	38 26.7	18.1	155 49.1	28.0	121 33.1	18.4
14	168 50.8	50 11.5	05.2	53 27.4	17.9	170 51.8	28.1	136 35.5	18.4
15	183 53.3	65 10.7 ..	04.7	68 28.1 ..	17.6	185 54.4 ..	28.2	151 38.0 ..	18.4
16	198 55.7	80 10.0	04.2	83 28.9	17.4	200 57.0	28.2	166 40.4	18.4
17	213 58.2	95 09.2	03.8	98 29.6	17.1	215 59.7	28.3	181 42.8	18.4
18	229 00.6	110 08.4	N21 03.3	113 30.3	N22 16.8	231 02.3	S 2 28.4	196 45.3	N10 18.4
19	244 03.1	125 07.7	02.8	128 31.0	16.6	246 04.9	28.5	211 47.7	18.3
20	259 05.6	140 06.9	02.3	143 31.8	16.3	261 07.6	28.6	226 50.2	18.3
21	274 08.0	155 06.1 ..	01.8	158 32.5 ..	16.1	276 10.2 ..	28.7	241 52.6 ..	18.3
22	289 10.5	170 05.4	01.4	173 33.2	15.8	291 12.8	28.8	256 55.0	18.3
23	304 13.0	185 04.6	00.9	188 33.9	15.6	306 15.5	28.8	271 57.5	18.3
Mer. Pass. 2 50.4		v −0.8	d 0.4	v 0.7	d 0.3	v 2.6	d 0.1	v 2.4	d 0.0

STARS

Name	SHA	Dec
Acamar	315 27.0	S40 18.5
Achernar	335 35.1	S57 14.4
Acrux	173 22.4	S63 05.6
Adhara	255 21.8	S28 58.2
Aldebaran	291 02.6	N16 30.3
Alioth	166 31.1	N55 58.4
Alkaid	153 08.1	N49 19.6
Al Na'ir	27 57.6	S46 57.9
Alnilam	275 58.2	S 1 12.2
Alphard	218 07.6	S 8 39.1
Alphecca	126 20.6	N26 43.5
Alpheratz	357 55.1	N29 04.9
Altair	62 19.0	N 8 52.1
Ankaa	353 26.7	S42 18.6
Antares	112 40.1	S26 25.6
Arcturus	146 06.2	N19 11.7
Atria	107 52.0	S69 01.6
Avior	234 23.3	S59 30.3
Bellatrix	278 44.5	N 6 20.8
Betelgeuse	271 13.9	N 7 24.3
Canopus	264 01.6	S52 41.6
Capella	280 51.6	N45 59.5
Deneb	49 38.8	N45 16.7
Denebola	182 45.6	N14 35.0
Diphda	349 07.2	S17 59.5
Dubhe	194 06.3	N61 45.7
Elnath	278 27.3	N28 36.2
Eltanin	90 51.1	N51 29.7
Enif	33 58.0	N 9 52.2
Fomalhaut	15 36.3	S29 37.6
Gacrux	172 14.0	S57 06.4
Gienah	176 04.3	S17 32.0
Hadar	149 04.2	S60 22.1
Hamal	328 13.6	N23 27.2
Kaus Aust.	83 58.6	S34 23.0
Kochab	137 19.8	N74 10.1
Markab	13 49.4	N15 11.9
Menkar	314 27.0	N 4 05.0
Menkent	148 21.1	S36 21.8
Miaplacidus	221 42.9	S69 42.7
Mirfak	308 56.8	N49 51.1
Nunki	76 12.2	S26 17.8
Peacock	53 36.6	S56 44.3
Pollux	243 42.0	N28 01.7
Procyon	245 12.0	N 5 13.7
Rasalhague	96 16.8	N12 33.9
Regulus	207 56.0	N11 58.5
Rigel	281 23.2	S 8 12.2
Rigil Kent.	140 07.4	S60 49.8
Sabik	102 25.5	S15 43.2
Schedar	349 53.2	N56 31.5
Shaula	96 37.2	S37 06.1
Sirius	258 44.1	S16 42.8
Spica	158 43.4	S11 09.1
Suhail	223 01.3	S43 25.6
Vega	80 46.3	N38 47.3
Zuben'ubi	137 18.1	S16 02.0

	SHA	Mer. Pass.
Venus	243 23.5	10 38
Mars	245 42.8	10 28
Jupiter	1 54.9	2 43
Saturn	327 45.8	4 59

UT	SUN GHA	SUN Dec	MOON GHA	MOON v	MOON Dec	MOON d	MOON HP
d h	° ′	° ′	° ′	′	° ′	′	′
8 00	178 34.3	N16 15.4	0 12.7	7.9	S15 05.4	7.2	58.7
01	193 34.4	14.7	14 39.6	7.9	14 58.2	7.2	58.8
02	208 34.5	14.0	29 06.5	8.0	14 51.0	7.3	58.8
03	223 34.6 ..	13.3	43 33.5	7.9	14 43.7	7.4	58.8
04	238 34.6	12.6	58 00.4	8.0	14 36.3	7.5	58.8
05	253 34.7	11.9	72 27.4	8.0	14 28.8	7.6	58.9
S 06	268 34.8	N16 11.2	86 54.4	8.0	S14 21.2	7.7	58.9
A 07	283 34.9	10.4	101 21.4	8.0	14 13.5	7.7	58.9
T 08	298 35.0	09.7	115 48.4	8.0	14 05.8	7.9	58.9
U 09	313 35.1 ..	09.0	130 15.4	8.1	13 57.9	8.0	59.0
R 10	328 35.1	08.3	144 42.5	8.0	13 49.9	8.0	59.0
D 11	343 35.2	07.6	159 09.5	8.1	13 41.9	8.2	59.0
A 12	358 35.3	N16 06.9	173 36.6	8.1	S13 33.7	8.2	59.0
Y 13	13 35.4	06.2	188 03.7	8.0	13 25.5	8.3	59.1
14	28 35.5	05.5	202 30.7	8.2	13 17.2	8.4	59.1
15	43 35.5 ..	04.8	216 57.9	8.1	13 08.8	8.5	59.1
16	58 35.6	04.1	231 25.0	8.1	13 00.3	8.6	59.1
17	73 35.7	03.3	245 52.1	8.2	12 51.7	8.7	59.1
18	88 35.8	N16 02.6	260 19.3	8.1	S12 43.0	8.7	59.2
19	103 35.9	01.9	274 46.4	8.2	12 34.3	8.9	59.2
20	118 36.0	01.2	289 13.6	8.2	12 25.4	8.9	59.2
21	133 36.1	16 00.5	303 40.8	8.3	12 16.5	9.0	59.2
22	148 36.1	15 59.8	318 08.1	8.2	12 07.5	9.1	59.2
23	163 36.2	59.1	332 35.3	8.3	11 58.4	9.1	59.3
9 00	178 36.3	N15 58.3	347 02.6	8.2	S11 49.3	9.2	59.3
01	193 36.4	57.6	1 29.8	8.3	11 40.1	9.4	59.3
02	208 36.5	56.9	15 57.1	8.4	11 30.7	9.3	59.3
03	223 36.6 ..	56.2	30 24.5	8.3	11 21.4	9.5	59.3
04	238 36.7	55.5	44 51.8	8.3	11 11.9	9.5	59.4
05	253 36.7	54.8	59 19.1	8.4	11 02.4	9.6	59.4
S 06	268 36.8	N15 54.0	73 46.5	8.4	S10 52.8	9.7	59.4
U 07	283 36.9	53.3	88 13.9	8.4	10 43.1	9.7	59.4
N 08	298 37.0	52.6	102 41.3	8.4	10 33.4	9.9	59.4
D 09	313 37.1 ..	51.9	117 08.7	8.5	10 23.5	9.8	59.5
A 10	328 37.2	51.2	131 36.2	8.4	10 13.7	10.0	59.5
Y 11	343 37.3	50.4	146 03.6	8.5	10 03.7	10.0	59.5
12	358 37.4	N15 49.7	160 31.1	8.5	S 9 53.7	10.1	59.5
13	13 37.5	49.0	174 58.6	8.6	9 43.6	10.1	59.5
14	28 37.5	48.3	189 26.2	8.5	9 33.5	10.2	59.5
15	43 37.6 ..	47.6	203 53.7	8.6	9 23.3	10.3	59.5
16	58 37.7	46.8	218 21.3	8.5	9 13.0	10.3	59.5
17	73 37.8	46.1	232 48.8	8.7	9 02.7	10.4	59.6
18	88 37.9	N15 45.4	247 16.5	8.6	S 8 52.3	10.5	59.6
19	103 38.0	44.7	261 44.1	8.6	8 41.8	10.5	59.6
20	118 38.1	43.9	276 11.7	8.7	8 31.3	10.5	59.6
21	133 38.2 ..	43.2	290 39.4	8.6	8 20.8	10.6	59.6
22	148 38.3	42.5	305 07.0	8.7	8 10.2	10.7	59.6
23	163 38.4	41.8	319 34.7	8.8	7 59.5	10.7	59.6
10 00	178 38.5	N15 41.0	334 02.5	8.7	S 7 48.8	10.8	59.6
01	193 38.5	40.3	348 30.2	8.7	7 38.0	10.8	59.7
02	208 38.6	39.6	2 57.9	8.8	7 27.2	10.9	59.7
03	223 38.7 ..	38.9	17 25.7	8.8	7 16.3	10.9	59.7
04	238 38.8	38.1	31 53.5	8.8	7 05.4	10.9	59.7
05	253 38.9	37.4	46 21.3	8.8	6 54.5	11.0	59.7
M 06	268 39.0	N15 36.7	60 49.1	8.9	S 6 43.5	11.1	59.7
O 07	283 39.1	35.9	75 17.0	8.8	6 32.4	11.1	59.7
N 08	298 39.2	35.2	89 44.8	8.9	6 21.3	11.1	59.7
D 09	313 39.3 ..	34.5	104 12.7	8.9	6 10.2	11.2	59.7
A 10	328 39.4	33.8	118 40.6	8.9	5 59.0	11.2	59.7
Y 11	343 39.5	33.0	133 08.5	8.9	5 47.8	11.2	59.7
12	358 39.6	N15 32.3	147 36.4	9.0	S 5 36.6	11.3	59.8
13	13 39.7	31.6	162 04.4	8.9	5 25.3	11.3	59.8
14	28 39.8	30.8	176 32.3	9.0	5 14.0	11.4	59.8
15	43 39.9 ..	30.1	191 00.3	9.0	5 02.6	11.3	59.8
16	58 40.0	29.4	205 28.3	9.0	4 51.3	11.4	59.8
17	73 40.1	28.6	219 56.3	9.0	4 39.9	11.5	59.8
18	88 40.1	N15 27.9	234 24.3	9.0	S 4 28.4	11.5	59.8
19	103 40.2	27.2	248 52.3	9.1	4 16.9	11.4	59.8
20	118 40.3	26.4	263 20.4	9.0	4 05.5	11.6	59.8
21	133 40.4 ..	25.7	277 48.4	9.1	3 53.9	11.5	59.8
22	148 40.5	25.0	292 16.5	9.1	3 42.4	11.6	59.8
23	163 40.6	24.2	306 44.6	9.1	S 3 30.8	11.6	59.8
	SD 15.8	d 0.7	SD 16.1		16.2		16.3

Lat.	Twilight Naut.	Twilight Civil	Sunrise	Moonrise 8	Moonrise 9	Moonrise 10	Moonrise 11
°	h m	h m	h m	h m	h m	h m	h m
N 72	////	////	01 33	21 36	21 27	21 19	21 12
N 70	////	////	02 22	21 14	21 14	21 14	21 13
68	////	00 18	02 52	20 58	21 04	21 09	21 14
66	////	01 44	03 15	20 44	20 56	21 06	21 14
64	////	02 20	03 32	20 32	20 49	21 02	21 15
62	00 17	02 45	03 47	20 23	20 43	21 00	21 15
60	01 33	03 04	03 59	20 14	20 37	20 57	21 16
N 58	02 06	03 20	04 09	20 07	20 32	20 55	21 16
56	02 29	03 33	04 19	20 01	20 28	20 53	21 16
54	02 47	03 44	04 27	19 55	20 24	20 51	21 17
52	03 02	03 54	04 34	19 49	20 21	20 50	21 17
50	03 15	04 03	04 40	19 45	20 18	20 48	21 17
45	03 41	04 21	04 54	19 34	20 11	20 45	21 18
N 40	04 00	04 36	05 06	19 26	20 05	20 42	21 18
35	04 15	04 48	05 15	19 18	20 00	20 40	21 19
30	04 28	04 58	05 24	19 12	19 56	20 38	21 19
20	04 48	05 15	05 38	19 00	19 48	20 34	21 20
N 10	05 03	05 29	05 51	18 50	19 41	20 31	21 20
0	05 16	05 41	06 02	18 41	19 35	20 28	21 21
S 10	05 27	05 52	06 14	18 31	19 29	20 25	21 22
20	05 37	06 03	06 26	18 21	19 22	20 22	21 22
30	05 46	06 14	06 39	18 10	19 14	20 19	21 23
35	05 51	06 21	06 47	18 03	19 10	20 17	21 24
40	05 56	06 28	06 56	17 56	19 05	20 14	21 24
45	06 01	06 36	07 07	17 47	18 59	20 12	21 25
S 50	06 06	06 45	07 19	17 36	18 52	20 09	21 26
52	06 09	06 49	07 25	17 31	18 48	20 07	21 26
54	06 11	06 53	07 31	17 26	18 45	20 05	21 26
56	06 14	06 58	07 39	17 20	18 41	20 04	21 27
58	06 17	07 04	07 47	17 13	18 36	20 02	21 27
S 60	06 20	07 10	07 56	17 05	18 31	19 59	21 28

Lat.	Sunset	Twilight Civil	Twilight Naut.	Moonset 8	Moonset 9	Moonset 10	Moonset 11
°	h m	h m	h m	h m	h m	h m	h m
N 72	22 30	////	////	02 23	04 30	06 32	08 31
N 70	21 45	////	////	02 58	04 50	06 42	08 33
68	21 15	23 26	////	03 23	05 06	06 51	08 35
66	20 54	22 21	////	03 42	05 18	06 57	08 37
64	20 36	21 48	////	03 57	05 29	07 03	08 39
62	20 22	21 23	23 30	04 10	05 37	07 08	08 40
60	20 10	21 05	22 32	04 20	05 45	07 12	08 41
N 58	20 00	20 49	22 02	04 30	05 52	07 16	08 42
56	19 51	20 36	21 39	04 38	05 57	07 20	08 43
54	19 43	20 25	21 21	04 45	06 03	07 23	08 44
52	19 36	20 15	21 07	04 52	06 07	07 25	08 44
50	19 30	20 07	20 54	04 57	06 12	07 28	08 45
45	19 16	19 49	20 29	05 10	06 21	07 33	08 46
N 40	19 05	19 34	20 10	05 20	06 28	07 38	08 48
35	18 55	19 23	19 55	05 29	06 35	07 42	08 49
30	18 47	19 12	19 43	05 37	06 40	07 45	08 49
20	18 33	18 56	19 23	05 50	06 50	07 51	08 51
N 10	18 20	18 42	19 08	06 01	06 59	07 56	08 52
0	18 09	18 30	18 54	06 12	07 07	08 00	08 53
S 10	17 58	18 19	18 45	06 23	07 15	08 05	08 55
20	17 46	18 08	18 35	06 34	07 23	08 10	08 56
30	17 32	17 57	18 25	06 47	07 32	08 15	08 57
35	17 24	17 51	18 21	06 55	07 38	08 19	08 58
40	17 15	17 44	18 16	07 03	07 44	08 22	08 59
45	17 05	17 36	18 11	07 13	07 51	08 26	09 00
S 50	16 53	17 27	18 05	07 25	08 00	08 31	09 01
52	16 47	17 23	18 03	07 30	08 03	08 33	09 01
54	16 40	17 18	18 01	07 36	08 08	08 36	09 02
56	16 33	17 14	17 58	07 43	08 12	08 39	09 02
58	16 25	17 08	17 55	07 50	08 18	08 42	09 03
S 60	16 16	17 02	17 52	07 59	08 24	08 45	09 04

Day	SUN Eqn. of Time 00ʰ	SUN Eqn. of Time 12ʰ	SUN Mer. Pass.	MOON Mer. Pass. Upper	MOON Mer. Pass. Lower	MOON Age	MOON Phase
d	m s	m s	h m	h m	h m	d	%
8	05 43	05 39	12 06	24 54	12 27	16	100
9	05 35	05 31	12 06	00 54	13 21	17	97
10	05 26	05 22	12 05	01 48	14 14	18	92

158 1998 AUGUST 11, 12, 13 (TUES., WED., THURS.)

UT	ARIES GHA	VENUS −3.9 GHA	VENUS Dec	MARS +1.7 GHA	MARS Dec	JUPITER −2.8 GHA	JUPITER Dec	SATURN +0.3 GHA	SATURN Dec	STARS Name	SHA	Dec
11 00	319 15.4	200 03.9	N21 00.4	203 34.7	N22 15.3	321 18.1	S 2 28.9	286 59.9	N10 18.3	Acamar	315 27.0	S40 18.5
01	334 17.9	215 03.1	20 59.9	218 35.4	15.0	336 20.7	29.0	302 02.4	18.3	Achernar	335 35.1	S57 14.4
02	349 20.4	230 02.3	59.4	233 36.1	14.8	351 23.4	29.1	317 04.8	18.3	Acrux	173 22.5	S63 05.6
03	4 22.8	245 01.6	.. 58.9	248 36.9	.. 14.5	6 26.0	.. 29.2	332 07.2	.. 18.3	Adhara	255 21.8	S28 58.2
04	19 25.3	260 00.8	58.4	263 37.6	14.2	21 28.6	29.3	347 09.7	18.3	Aldebaran	291 02.6	N16 30.3
05	34 27.7	275 00.1	57.9	278 38.3	14.0	36 31.3	29.3	2 12.1	18.3			
06	49 30.2	289 59.3	N20 57.4	293 39.0	N22 13.7	51 33.9	S 2 29.4	17 14.6	N10 18.3	Alioth	166 31.1	N55 58.4
07	64 32.7	304 58.5	56.9	308 39.8	13.5	66 36.5	29.5	32 17.0	18.3	Alkaid	153 08.1	N49 19.6
T 08	79 35.1	319 57.8	56.5	323 40.5	13.2	81 39.2	29.6	47 19.5	18.3	Al Na'ir	27 57.6	S46 57.9
U 09	94 37.6	334 57.0	.. 56.0	338 41.2	.. 12.9	96 41.8	.. 29.7	62 21.9	.. 18.3	Alnilam	275 58.2	S 1 12.2
E 10	109 40.1	349 56.3	55.5	353 42.0	12.7	111 44.4	29.8	77 24.3	18.3	Alphard	218 07.6	S 8 39.1
S 11	124 42.5	4 55.5	55.0	8 42.7	12.4	126 47.1	29.9	92 26.8	18.3			
D 12	139 45.0	19 54.8	N20 54.5	23 43.4	N22 12.1	141 49.7	S 2 29.9	107 29.2	N10 18.3	Alphecca	126 20.7	N26 43.5
A 13	154 47.5	34 54.0	54.0	38 44.1	11.9	156 52.3	30.0	122 31.7	18.3	Alpheratz	357 55.0	N29 04.9
Y 14	169 49.9	49 53.2	53.5	53 44.9	11.6	171 55.0	30.1	137 34.1	18.3	Altair	62 19.1	N 8 52.1
15	184 52.4	64 52.5	.. 53.0	68 45.6	.. 11.4	186 57.6	.. 30.2	152 36.5	.. 18.3	Ankaa	353 26.7	S42 18.6
16	199 54.9	79 51.7	52.5	83 46.3	11.1	202 00.2	30.3	167 39.0	18.3	Antares	112 40.2	S26 25.6
17	214 57.3	94 51.0	52.0	98 47.1	10.8	217 02.9	30.4	182 41.4	18.3			
18	229 59.8	109 50.2	N20 51.5	113 47.8	N22 10.6	232 05.5	S 2 30.5	197 43.9	N10 18.3	Arcturus	146 06.3	N19 11.7
19	245 02.2	124 49.5	50.9	128 48.5	10.3	247 08.2	30.5	212 46.3	18.3	Atria	107 52.0	S69 01.6
20	260 04.7	139 48.7	50.4	143 49.3	10.0	262 10.8	30.6	227 48.8	18.3	Avior	234 23.2	S59 30.3
21	275 07.2	154 48.0	.. 49.9	158 50.0	.. 09.8	277 13.4	.. 30.7	242 51.2	.. 18.3	Bellatrix	278 44.4	N 6 20.8
22	290 09.6	169 47.2	49.4	173 50.7	09.5	292 16.1	30.8	257 53.7	18.2	Betelgeuse	271 13.9	N 7 24.3
23	305 12.1	184 46.5	48.9	188 51.5	09.2	307 18.7	30.9	272 56.1	18.2			
12 00	320 14.6	199 45.7	N20 48.4	203 52.2	N22 09.0	322 21.3	S 2 31.0	287 58.5	N10 18.2	Canopus	264 01.6	S52 41.6
01	335 17.0	214 44.9	47.9	218 52.9	08.7	337 24.0	31.1	303 01.0	18.2	Capella	280 51.6	N45 59.5
02	350 19.5	229 44.2	47.4	233 53.6	08.4	352 26.6	31.2	318 03.4	18.2	Deneb	49 38.8	N45 16.7
03	5 22.0	244 43.4	.. 46.9	248 54.4	.. 08.2	7 29.3	.. 31.2	333 05.9	.. 18.2	Denebola	182 45.6	N14 35.0
04	20 24.4	259 42.7	46.3	263 55.1	07.9	22 31.9	31.3	348 08.3	18.2	Diphda	349 07.1	S17 59.5
05	35 26.9	274 41.9	45.8	278 55.8	07.6	37 34.5	31.4	3 10.8	18.2			
06	50 29.4	289 41.2	N20 45.3	293 56.6	N22 07.3	52 37.2	S 2 31.5	18 13.2	N10 18.2	Dubhe	194 06.3	N61 45.7
W 07	65 31.8	304 40.4	44.8	308 57.3	07.1	67 39.8	31.6	33 15.7	18.2	Elnath	278 27.3	N28 36.2
E 08	80 34.3	319 39.7	44.3	323 58.0	06.8	82 42.5	31.7	48 18.1	18.2	Eltanin	90 51.1	N51 29.7
D 09	95 36.7	334 38.9	.. 43.7	338 58.8	.. 06.5	97 45.1	.. 31.8	63 20.6	.. 18.2	Enif	33 58.0	N 9 52.2
N 10	110 39.2	349 38.2	43.2	353 59.5	06.3	112 47.8	31.9	78 23.0	18.2	Fomalhaut	15 36.2	S29 37.6
E 11	125 41.7	4 37.4	42.7	9 00.2	06.0	127 50.4	32.0	93 25.4	18.2			
S 12	140 44.1	19 36.7	N20 42.2	24 01.0	N22 05.7	142 53.0	S 2 32.0	108 27.9	N10 18.2	Gacrux	172 14.0	S57 06.4
D 13	155 46.6	34 35.9	41.7	39 01.7	05.5	157 55.7	32.1	123 30.3	18.2	Gienah	176 04.3	S17 32.0
A 14	170 49.1	49 35.2	41.1	54 02.5	05.2	172 58.3	32.2	138 32.8	18.2	Hadar	149 04.3	S60 22.1
Y 15	185 51.5	64 34.4	.. 40.6	69 03.2	.. 04.9	188 01.0	.. 32.3	153 35.2	.. 18.2	Hamal	328 13.6	N23 27.2
16	200 54.0	79 33.7	40.1	84 03.9	04.6	203 03.6	32.4	168 37.7	18.2	Kaus Aust.	83 58.7	S34 23.0
17	215 56.5	94 32.9	39.5	99 04.7	04.4	218 06.3	32.5	183 40.1	18.1			
18	230 58.9	109 32.2	N20 39.0	114 05.4	N22 04.1	233 08.9	S 2 32.6	198 42.6	N10 18.1	Kochab	137 19.8	N74 10.1
19	246 01.4	124 31.4	38.5	129 06.1	03.8	248 11.5	32.7	213 45.0	18.1	Markab	13 49.4	N15 11.9
20	261 03.8	139 30.7	38.0	144 06.9	03.6	263 14.2	32.8	228 47.5	18.1	Menkar	314 27.0	N 4 05.0
21	276 06.3	154 29.9	.. 37.4	159 07.6	.. 03.3	278 16.8	.. 32.8	243 49.9	.. 18.1	Menkent	148 21.1	S36 21.8
22	291 08.8	169 29.2	36.9	174 08.3	03.0	293 19.5	32.9	258 52.4	18.1	Miaplacidus	221 42.9	S69 42.7
23	306 11.2	184 28.5	36.3	189 09.1	02.7	308 22.1	33.0	273 54.8	18.1			
13 00	321 13.7	199 27.7	N20 35.8	204 09.8	N22 02.5	323 24.8	S 2 33.1	288 57.3	N10 18.1	Mirfak	308 56.7	N49 51.1
01	336 16.2	214 27.0	35.3	219 10.5	02.2	338 27.4	33.2	303 59.7	18.1	Nunki	76 12.2	S26 17.8
02	351 18.6	229 26.2	34.7	234 11.3	01.9	353 30.1	33.3	319 02.2	18.1	Peacock	53 36.6	S56 44.3
03	6 21.1	244 25.5	.. 34.2	249 12.0	.. 01.6	8 32.7	.. 33.4	334 04.6	.. 18.1	Pollux	243 42.0	N28 01.7
04	21 23.6	259 24.7	33.7	264 12.8	01.4	23 35.3	33.5	349 07.1	18.1	Procyon	245 12.0	N 5 13.7
05	36 26.0	274 24.0	33.1	279 13.5	01.1	38 38.0	33.6	4 09.5	18.1			
06	51 28.5	289 23.2	N20 32.6	294 14.2	N22 00.8	53 40.6	S 2 33.7	19 12.0	N10 18.1	Rasalhague	96 16.9	N12 34.0
07	66 31.0	304 22.5	32.0	309 15.0	00.5	68 43.3	33.7	34 14.4	18.1	Regulus	207 56.0	N11 58.5
T 08	81 33.4	319 21.8	31.5	324 15.7	00.3	83 45.9	33.8	49 16.9	18.1	Rigel	281 23.2	S 8 12.2
H 09	96 35.9	334 21.0	.. 30.9	339 16.4	22 00.0	98 48.6	.. 33.9	64 19.3	.. 18.0	Rigil Kent.	140 07.5	S60 49.8
U 10	111 38.3	349 20.3	30.4	354 17.2	21 59.7	113 51.2	34.0	79 21.8	18.0	Sabik	102 25.5	S15 43.2
R 11	126 40.8	4 19.5	29.8	9 17.9	59.4	128 53.9	34.1	94 24.2	18.0			
S 12	141 43.3	19 18.8	N20 29.3	24 18.7	N21 59.2	143 56.5	S 2 34.2	109 26.7	N10 18.0	Schedar	349 53.1	N56 31.6
D 13	156 45.7	34 18.0	28.7	39 19.4	58.9	158 59.2	34.3	124 29.1	18.0	Shaula	96 37.2	S37 06.1
A 14	171 48.2	49 17.3	28.2	54 20.1	58.6	174 01.8	34.4	139 31.6	18.0	Sirius	258 44.1	S16 42.8
Y 15	186 50.7	64 16.6	.. 27.6	69 20.9	.. 58.3	189 04.5	.. 34.5	154 34.0	.. 18.0	Spica	158 43.4	S11 09.1
16	201 53.1	79 15.8	27.1	84 21.6	58.0	204 07.1	34.6	169 36.5	18.0	Suhail	223 01.3	S43 25.6
17	216 55.6	94 15.1	26.5	99 22.4	57.8	219 09.8	34.7	184 38.9	18.0			
18	231 58.1	109 14.3	N20 26.0	114 23.1	N21 57.5	234 12.4	S 2 34.7	199 41.4	N10 18.0	Vega	80 46.4	N38 47.3
19	247 00.5	124 13.6	25.4	129 23.8	57.2	249 15.1	34.8	214 43.8	18.0	Zuben'ubi	137 18.1	S16 02.0
20	262 03.0	139 12.9	24.9	144 24.6	56.9	264 17.7	34.9	229 46.3	18.0		SHA	Mer.Pass.
21	277 05.4	154 12.1	.. 24.3	159 25.3	.. 56.6	279 20.4	.. 35.0	244 48.7	.. 18.0	Venus	239 31.1	10 41
22	292 07.9	169 11.4	23.8	174 26.1	56.4	294 23.0	35.1	259 51.2	18.0	Mars	243 37.6	10 24
23	307 10.4	184 10.6	23.2	189 26.8	56.1	309 25.7	35.2	274 53.6	17.9	Jupiter	2 06.8	2 30
Mer. Pass. 2 38.6		v −0.7	d 0.5	v 0.7	d 0.3	v 2.6	d 0.1	v 2.4	d 0.0	Saturn	327 44.0	4 47

UT	SUN GHA	Dec	MOON GHA	v	Dec	d	HP
d h	° '	° '	° '	'	° '	'	'
11 00	178 40.7	N15 23.5	321 12.7	9.1	S 3 19.2	11.6	59.8
01	193 40.8	22.7	335 40.8	9.1	3 07.6	11.6	59.8
02	208 40.9	22.0	350 08.9	9.1	2 56.0	11.6	59.8
03	223 41.0 ..	21.3	4 37.0	9.2	2 44.4	11.7	59.8
04	238 41.1	20.5	19 05.2	9.1	2 32.7	11.7	59.8
05	253 41.2	19.8	33 33.3	9.2	2 21.0	11.7	59.8
06	268 41.3	N15 19.1	48 01.5	9.1	S 2 09.3	11.7	59.8
T 07	283 41.4	18.3	62 29.6	9.2	1 57.6	11.7	59.8
U 08	298 41.5	17.6	76 57.8	9.2	1 45.9	11.7	59.8
E 09	313 41.6 ..	16.8	91 26.0	9.2	1 34.2	11.7	59.8
S 10	328 41.7	16.1	105 54.2	9.2	1 22.5	11.8	59.8
D 11	343 41.8	15.4	120 22.4	9.2	1 10.7	11.7	59.8
A 12	358 41.9	N15 14.6	134 50.6	9.2	S 0 59.0	11.8	59.8
Y 13	13 42.0	13.9	149 18.8	9.2	0 47.2	11.7	59.8
14	28 42.1	13.1	163 47.0	9.2	0 35.5	11.8	59.8
15	43 42.2 ..	12.4	178 15.2	9.3	0 23.7	11.8	59.8
16	58 42.3	11.6	192 43.5	9.2	0 11.9	11.7	59.8
17	73 42.4	10.9	207 11.7	9.2	S 0 00.2	11.8	59.8
18	88 42.5	N15 10.2	221 39.9	9.3	N 0 11.6	11.7	59.8
19	103 42.6	09.4	236 08.2	9.2	0 23.3	11.8	59.8
20	118 42.7	08.7	250 36.4	9.3	0 35.1	11.7	59.8
21	133 42.8 ..	07.9	265 04.7	9.2	0 46.8	11.8	59.8
22	148 42.9	07.2	279 32.9	9.3	0 58.6	11.7	59.8
23	163 43.0	06.4	294 01.2	9.2	1 10.3	11.7	59.8
12 00	178 43.1	N15 05.7	308 29.4	9.3	N 1 22.0	11.7	59.8
01	193 43.3	04.9	322 57.7	9.2	1 33.7	11.7	59.8
02	208 43.4	04.2	337 25.9	9.3	1 45.4	11.7	59.8
03	223 43.5 ..	03.4	351 54.2	9.2	1 57.1	11.7	59.8
04	238 43.6	02.7	6 22.4	9.3	2 08.8	11.6	59.8
05	253 43.7	01.9	20 50.7	9.2	2 20.4	11.7	59.8
06	268 43.8	N15 01.2	35 18.9	9.3	N 2 32.1	11.6	59.8
W 07	283 43.9	15 00.4	49 47.2	9.2	2 43.7	11.6	59.8
E 08	298 44.0	14 59.7	64 15.4	9.2	2 55.3	11.6	59.8
D 09	313 44.1 ..	58.9	78 43.6	9.3	3 06.9	11.6	59.8
N 10	328 44.2	58.2	93 11.9	9.2	3 18.5	11.5	59.8
E 11	343 44.3	57.4	107 40.1	9.2	3 30.0	11.5	59.8
S 12	358 44.4	N14 56.7	122 08.3	9.3	N 3 41.5	11.5	59.8
D 13	13 44.5	55.9	136 36.6	9.2	3 53.0	11.5	59.8
A 14	28 44.6	55.2	151 04.8	9.2	4 04.5	11.4	59.7
Y 15	43 44.7 ..	54.4	165 33.0	9.2	4 15.9	11.4	59.7
16	58 44.8	53.7	180 01.2	9.2	4 27.3	11.4	59.7
17	73 44.9	52.9	194 29.4	9.2	4 38.7	11.3	59.7
18	88 45.1	N14 52.2	208 57.6	9.1	N 4 50.0	11.4	59.7
19	103 45.2	51.4	223 25.7	9.2	5 01.4	11.2	59.7
20	118 45.3	50.7	237 53.9	9.2	5 12.6	11.3	59.7
21	133 45.4 ..	49.9	252 22.1	9.1	5 23.9	11.2	59.7
22	148 45.5	49.1	266 50.2	9.2	5 35.1	11.2	59.7
23	163 45.6	48.4	281 18.4	9.1	5 46.3	11.1	59.7
13 00	178 45.7	N14 47.6	295 46.5	9.1	N 5 57.4	11.1	59.7
01	193 45.8	46.9	310 14.6	9.1	6 08.5	11.1	59.7
02	208 45.9	46.1	324 42.7	9.1	6 19.6	11.0	59.6
03	223 46.0 ..	45.4	339 10.8	9.1	6 30.6	10.9	59.6
04	238 46.1	44.6	353 38.9	9.1	6 41.5	11.0	59.6
05	253 46.3	43.8	8 07.0	9.1	6 52.5	10.8	59.6
06	268 46.4	N14 43.1	22 35.1	9.0	N 7 03.3	10.9	59.6
T 07	283 46.5	42.3	37 03.1	9.1	7 14.2	10.8	59.6
H 08	298 46.6	41.6	51 31.2	9.0	7 25.0	10.7	59.6
U 09	313 46.7 ..	40.8	65 59.2	9.0	7 35.7	10.7	59.6
R 10	328 46.8	40.0	80 27.2	9.0	7 46.4	10.6	59.6
S 11	343 46.9	39.3	94 55.2	9.0	7 57.0	10.6	59.5
D 12	358 47.0	N14 38.5	109 23.2	8.9	N 8 07.6	10.5	59.5
A 13	13 47.1	37.8	123 51.1	9.0	8 18.1	10.5	59.5
Y 14	28 47.3	37.0	138 19.1	8.9	8 28.6	10.4	59.5
15	43 47.4 ..	36.2	152 47.0	8.9	8 39.0	10.4	59.5
16	58 47.5	35.5	167 14.9	8.9	8 49.4	10.3	59.5
17	73 47.6	34.7	181 42.8	8.9	8 59.7	10.3	59.5
18	88 47.7	N14 33.9	196 10.7	8.8	N 9 10.0	10.2	59.5
19	103 47.8	33.2	210 38.5	8.9	9 20.2	10.1	59.5
20	118 47.9	32.4	225 06.4	8.8	9 30.3	10.1	59.4
21	133 48.0 ..	31.7	239 34.2	8.8	9 40.4	10.0	59.4
22	148 48.2	30.9	254 02.0	8.8	9 50.4	9.9	59.4
23	163 48.3	30.1	268 29.8	8.8	N10 00.3	9.9	59.4
SD	15.8	d 0.8	SD 16.3		16.3		16.2

Lat.	Naut.	Civil	Sunrise	Moonrise 11	12	13	14
°	h m	h m	h m	h m	h m	h m	h m
N 72	////	////	02 00	21 12	21 05	20 58	20 49
N 70	////	////	02 39	21 13	21 12	21 12	21 13
68	////	01 13	03 05	21 14	21 18	21 24	21 32
66	////	02 03	03 26	21 14	21 23	21 34	21 47
64	////	02 33	03 42	21 15	21 28	21 42	22 00
62	01 06	02 55	03 55	21 15	21 31	21 49	22 10
60	01 50	03 13	04 06	21 16	21 35	21 55	22 19
N 58	02 18	03 28	04 16	21 16	21 38	22 01	22 27
56	02 39	03 40	04 24	21 16	21 40	22 06	22 34
54	02 55	03 50	04 32	21 17	21 42	22 10	22 41
52	03 09	04 00	04 39	21 17	21 45	22 14	22 46
50	03 21	04 08	04 45	21 17	21 47	22 18	22 52
45	03 45	04 25	04 58	21 18	21 51	22 25	23 03
N 40	04 03	04 39	05 08	21 18	21 55	22 32	23 12
35	04 18	04 50	05 17	21 19	21 58	22 38	23 20
30	04 30	05 00	05 25	21 19	22 00	22 43	23 27
20	04 49	05 16	05 39	21 20	22 05	22 52	23 40
N 10	05 03	05 29	05 51	21 20	22 09	23 00	23 51
0	05 15	05 40	06 02	21 21	22 14	23 07	24 01
S 10	05 26	05 51	06 12	21 22	22 18	23 14	24 11
20	05 35	06 01	06 24	21 22	22 22	23 22	24 22
30	05 44	06 12	06 37	21 23	22 27	23 31	24 35
35	05 48	06 18	06 44	21 24	22 30	23 37	24 42
40	05 52	06 24	06 52	21 24	22 34	23 43	24 51
45	05 57	06 32	07 02	21 25	22 38	23 50	25 00
S 50	06 02	06 40	07 14	21 26	22 43	23 58	25 12
52	06 04	06 44	07 19	21 26	22 45	24 02	00 02
54	06 06	06 48	07 25	21 26	22 47	24 07	00 07
56	06 08	06 52	07 32	21 27	22 50	24 12	00 12
58	06 10	06 57	07 39	21 27	22 53	24 17	00 17
S 60	06 13	07 02	07 48	21 28	22 56	24 23	00 23

Lat.	Sunset	Civil	Naut.	Moonset 11	12	13	14
°	h m	h m	h m	h m	h m	h m	h m
N 72	22 04	////	////	08 31	10 28	12 27	14 27
N 70	21 27	////	////	08 33	10 24	12 14	14 04
68	21 01	22 47	////	08 35	10 20	12 04	13 47
66	20 42	22 02	////	08 37	10 17	11 55	13 33
64	20 26	21 33	////	08 39	10 14	11 48	13 21
62	20 13	21 12	22 56	08 40	10 11	11 42	13 11
60	20 02	20 55	22 16	08 41	10 09	11 37	13 03
N 58	19 53	20 41	21 49	08 42	10 08	11 33	12 56
56	19 44	20 29	21 29	08 43	10 06	11 29	12 49
54	19 37	20 18	21 13	08 44	10 05	11 25	12 44
52	19 30	20 09	20 59	08 44	10 03	11 22	12 38
50	19 24	20 01	20 47	08 45	10 02	11 19	12 34
45	19 12	19 44	20 24	08 46	10 00	11 12	12 24
N 40	19 01	19 30	20 06	08 48	09 57	11 07	12 15
35	18 52	19 19	19 51	08 49	09 56	11 02	12 08
30	18 44	19 10	19 40	08 49	09 54	10 58	12 02
20	18 31	18 54	19 21	08 51	09 51	10 51	11 51
N 10	18 19	18 41	19 07	08 52	09 48	10 45	11 41
0	18 08	18 30	18 55	08 53	09 46	10 39	11 32
S 10	17 58	18 19	18 45	08 54	09 44	10 33	11 24
20	17 47	18 09	18 35	08 56	09 41	10 27	11 14
30	17 34	17 58	18 27	08 57	09 38	10 20	11 03
35	17 26	17 53	18 23	08 58	09 37	10 16	10 57
40	17 18	17 46	18 18	08 59	09 35	10 11	10 50
45	17 08	17 39	18 14	09 00	09 32	10 06	10 42
S 50	16 57	17 31	18 09	09 01	09 30	10 00	10 32
52	16 51	17 27	18 07	09 01	09 29	09 57	10 27
54	16 45	17 23	18 05	09 02	09 27	09 54	10 23
56	16 39	17 19	18 03	09 02	09 26	09 50	10 17
58	16 31	17 14	18 01	09 03	09 24	09 46	10 11
S 60	16 23	17 09	17 58	09 04	09 22	09 42	10 04

Day	SUN Eqn. of Time 00h	12h	Mer. Pass.	MOON Mer. Pass. Upper	Lower	Age	Phase
d	m s	m s	h m	h m	h m	d	%
11	05 17	05 13	12 05	02 41	15 07	19	85
12	05 08	05 03	12 05	03 34	16 00	20	76
13	04 57	04 52	12 05	04 26	16 53	21	65

1998 AUGUST 14, 15, 16 (FRI., SAT., SUN.)

UT	ARIES GHA	VENUS −3.9 GHA	Dec	MARS +1.7 GHA	Dec	JUPITER −2.8 GHA	Dec	SATURN +0.3 GHA	Dec	STARS Name	SHA	Dec
14 00	322 12.8	199 09.9	N20 22.6	204 27.5	N21 55.8	324 28.3	S 2 35.3	289 56.1	N10 17.9	Acamar	315 27.0	S40 18.5
01	337 15.3	214 09.2	22.1	219 28.3	55.5	339 31.0	35.4	304 58.5	17.9	Achernar	335 35.1	S57 14.4
02	352 17.8	229 08.4	21.5	234 29.0	55.2	354 33.6	35.5	320 01.0	17.9	Acrux	173 22.5	S63 05.6
03	7 20.2	244 07.7	.. 20.9	249 29.8	.. 55.0	9 36.3	.. 35.6	335 03.4	.. 17.9	Adhara	255 21.8	S28 58.1
04	22 22.7	259 06.9	20.4	264 30.5	54.7	24 38.9	35.7	350 05.9	17.9	Aldebaran	291 02.6	N16 30.3
05	37 25.2	274 06.2	19.8	279 31.2	54.4	39 41.6	35.8	5 08.3	17.9			
06	52 27.6	289 05.5	N20 19.2	294 32.0	N21 54.1	54 44.3	S 2 35.9	20 10.8	N10 17.9	Alioth	166 31.1	N55 58.4
07	67 30.1	304 04.7	18.7	309 32.7	53.8	69 46.9	35.9	35 13.3	17.9	Alkaid	153 08.1	N49 19.6
F 08	82 32.6	319 04.0	18.1	324 33.5	53.5	84 49.6	36.0	50 15.7	17.9	Al Na'ir	27 57.6	S46 57.9
R 09	97 35.0	334 03.3	.. 17.5	339 34.2	.. 53.3	99 52.2	.. 36.1	65 18.2	.. 17.9	Alnilam	275 58.1	S 1 12.2
I 10	112 37.5	349 02.5	17.0	354 34.9	53.0	114 54.9	36.2	80 20.6	17.9	Alphard	218 07.6	S 8 39.1
D 11	127 39.9	4 01.8	16.4	9 35.7	52.7	129 57.5	36.3	95 23.1	17.8			
A 12	142 42.4	19 01.1	N20 15.8	24 36.4	N21 52.4	145 00.2	S 2 36.4	110 25.5	N10 17.8	Alphecca	126 20.7	N26 43.5
Y 13	157 44.9	34 00.3	15.2	39 37.2	52.1	160 02.8	36.5	125 28.0	17.8	Alpheratz	357 55.0	N29 04.9
14	172 47.3	48 59.6	14.7	54 37.9	51.8	175 05.5	36.6	140 30.4	17.8	Altair	62 19.1	N 8 52.1
15	187 49.8	63 58.9	.. 14.1	69 38.7	.. 51.6	190 08.1	.. 36.7	155 32.9	.. 17.8	Ankaa	353 26.7	S42 18.6
16	202 52.3	78 58.1	13.5	84 39.4	51.3	205 10.8	36.8	170 35.3	17.8	Antares	112 40.2	S26 25.6
17	217 54.7	93 57.4	12.9	99 40.1	51.0	220 13.5	36.9	185 37.8	17.8			
18	232 57.2	108 56.7	N20 12.4	114 40.9	N21 50.7	235 16.1	S 2 37.0	200 40.3	N10 17.8	Arcturus	146 06.3	N19 11.7
19	247 59.7	123 55.9	11.8	129 41.6	50.4	250 18.8	37.1	215 42.7	17.8	Atria	107 52.0	S69 01.6
20	263 02.1	138 55.2	11.2	144 42.4	50.1	265 21.4	37.2	230 45.2	17.8	Avior	234 23.2	S59 30.3
21	278 04.6	153 54.5	.. 10.6	159 43.1	.. 49.8	280 24.1	.. 37.3	245 47.6	.. 17.8	Bellatrix	278 44.4	N 6 20.8
22	293 07.1	168 53.7	10.0	174 43.9	49.6	295 26.7	37.4	260 50.1	17.8	Betelgeuse	271 13.9	N 7 24.4
23	308 09.5	183 53.0	09.4	189 44.6	49.3	310 29.4	37.4	275 52.5	17.7			
15 00	323 12.0	198 52.3	N20 08.9	204 45.4	N21 49.0	325 32.1	S 2 37.5	290 55.0	N10 17.7	Canopus	264 01.6	S52 41.6
01	338 14.4	213 51.5	08.3	219 46.1	48.7	340 34.7	37.6	305 57.5	17.7	Capella	280 51.5	N45 59.5
02	353 16.9	228 50.8	07.7	234 46.9	48.4	355 37.4	37.7	320 59.9	17.7	Deneb	49 38.8	N45 16.7
03	8 19.4	243 50.1	.. 07.1	249 47.6	.. 48.1	10 40.0	.. 37.8	336 02.4	.. 17.7	Denebola	182 45.6	N14 35.0
04	23 21.8	258 49.3	06.5	264 48.3	47.8	25 42.7	37.9	351 04.8	17.7	Diphda	349 07.1	S17 59.5
05	38 24.3	273 48.6	05.9	279 49.1	47.5	40 45.3	38.0	6 07.3	17.7			
06	53 26.8	288 47.9	N20 05.3	294 49.8	N21 47.3	55 48.0	S 2 38.1	21 09.7	N10 17.7	Dubhe	194 06.3	N61 45.7
S 07	68 29.2	303 47.2	04.7	309 50.6	47.0	70 50.7	38.2	36 12.2	17.7	Elnath	278 27.2	N28 36.2
A 08	83 31.7	318 46.4	04.1	324 51.3	46.7	85 53.3	38.3	51 14.7	17.7	Eltanin	90 51.1	N51 29.7
T 09	98 34.2	333 45.7	.. 03.5	339 52.1	.. 46.4	100 56.0	.. 38.4	66 17.1	.. 17.7	Enif	33 58.0	N 9 52.3
U 10	113 36.6	348 45.0	02.9	354 52.8	46.1	115 58.6	38.5	81 19.6	17.6	Fomalhaut	15 36.2	S29 37.6
R 11	128 39.1	3 44.2	02.4	9 53.6	45.8	131 01.3	38.6	96 22.0	17.6			
D 12	143 41.5	18 43.5	N20 01.8	24 54.3	N21 45.5	146 04.0	S 2 38.7	111 24.5	N10 17.6	Gacrux	172 14.0	S57 06.4
A 13	158 44.0	33 42.8	01.2	39 55.1	45.2	161 06.6	38.8	126 27.0	17.6	Gienah	176 04.3	S17 32.0
Y 14	173 46.5	48 42.1	00.6	54 55.8	44.9	176 09.3	38.9	141 29.4	17.6	Hadar	149 04.3	S60 22.1
15	188 48.9	63 41.3	20 00.0	69 56.6	.. 44.6	191 12.0	.. 39.0	156 31.9	.. 17.6	Hamal	328 13.5	N23 27.2
16	203 51.4	78 40.6	19 59.4	84 57.3	44.3	206 14.6	39.1	171 34.3	17.6	Kaus Aust.	83 58.7	S34 23.0
17	218 53.9	93 39.9	58.8	99 58.1	44.1	221 17.3	39.2	186 36.8	17.6			
18	233 56.3	108 39.2	N19 58.1	114 58.8	N21 43.8	236 19.9	S 2 39.3	201 39.3	N10 17.6	Kochab	137 19.9	N74 10.1
19	248 58.8	123 38.4	57.5	129 59.6	43.5	251 22.6	39.4	216 41.7	17.6	Markab	13 49.4	N15 11.9
20	264 01.3	138 37.7	56.9	145 00.3	43.2	266 25.3	39.5	231 44.2	17.5	Menkar	314 27.0	N 4 05.0
21	279 03.7	153 37.0	.. 56.3	160 01.0	.. 42.9	281 27.9	.. 39.6	246 46.6	.. 17.5	Menkent	148 21.2	S36 21.8
22	294 06.2	168 36.3	55.7	175 01.8	42.6	296 30.6	39.7	261 49.1	17.5	Miaplacidus	221 42.9	S69 42.7
23	309 08.7	183 35.5	55.1	190 02.5	42.3	311 33.3	39.8	276 51.6	17.5			
16 00	324 11.1	198 34.8	N19 54.5	205 03.3	N21 42.0	326 35.9	S 2 39.8	291 54.0	N10 17.5	Mirfak	308 56.7	N49 51.1
01	339 13.6	213 34.1	53.9	220 04.0	41.7	341 38.6	39.9	306 56.5	17.5	Nunki	76 12.2	S26 17.8
02	354 16.0	228 33.4	53.3	235 04.8	41.4	356 41.3	40.0	321 58.9	17.5	Peacock	53 36.6	S56 44.3
03	9 18.5	243 32.7	.. 52.7	250 05.5	.. 41.1	11 43.9	.. 40.1	337 01.4	.. 17.5	Pollux	243 42.0	N28 01.7
04	24 21.0	258 31.9	52.1	265 06.3	40.8	26 46.6	40.2	352 03.9	17.5	Procyon	245 12.0	N 5 13.7
05	39 23.4	273 31.2	51.4	280 07.0	40.5	41 49.2	40.3	7 06.3	17.4			
06	54 25.9	288 30.5	N19 50.8	295 07.8	N21 40.2	56 51.9	S 2 40.4	22 08.8	N10 17.4	Rasalhague	96 16.9	N12 34.0
07	69 28.4	303 29.8	50.2	310 08.5	39.9	71 54.6	40.5	37 11.3	17.4	Regulus	207 56.0	N11 58.5
08	84 30.8	318 29.1	49.6	325 09.3	39.6	86 57.2	40.6	52 13.7	17.4	Rigel	281 23.2	S 8 12.2
S 09	99 33.3	333 28.3	.. 49.0	340 10.0	.. 39.4	101 59.9	.. 40.7	67 16.2	.. 17.4	Rigil Kent.	140 07.5	S60 49.8
U 10	114 35.8	348 27.6	48.4	355 10.8	39.1	117 02.6	40.8	82 18.6	17.4	Sabik	102 25.5	S15 43.2
N 11	129 38.2	3 26.9	47.7	10 11.6	38.8	132 05.2	40.9	97 21.1	17.4			
D 12	144 40.7	18 26.2	N19 47.1	25 12.3	N21 38.5	147 07.9	S 2 41.0	112 23.6	N10 17.4	Schedar	349 53.1	N56 31.6
A 13	159 43.2	33 25.5	46.5	40 13.1	38.2	162 10.6	41.1	127 26.0	17.4	Shaula	96 37.2	S37 06.1
Y 14	174 45.6	48 24.7	45.9	55 13.8	37.9	177 13.3	41.2	142 28.5	17.3	Sirius	258 44.0	S16 42.8
15	189 48.1	63 24.0	.. 45.2	70 14.6	.. 37.6	192 15.9	.. 41.3	157 31.0	.. 17.3	Spica	158 43.4	S11 09.1
16	204 50.5	78 23.3	44.6	85 15.3	37.3	207 18.6	41.4	172 33.4	17.3	Suhail	223 01.2	S43 25.6
17	219 53.0	93 22.6	44.0	100 16.1	37.0	222 21.3	41.5	187 35.9	17.3			
18	234 55.5	108 21.9	N19 43.4	115 16.8	N21 36.7	237 23.9	S 2 41.6	202 38.3	N10 17.3	Vega	80 46.4	N38 47.3
19	249 57.9	123 21.2	42.7	130 17.6	36.4	252 26.6	41.7	217 40.8	17.3	Zuben'ubi	137 18.1	S16 02.0
20	265 00.4	138 20.4	42.1	145 18.3	36.1	267 29.3	41.8	232 43.3	17.3			
21	280 02.9	153 19.7	.. 41.5	160 19.1	.. 35.8	282 31.9	.. 41.9	247 45.7	.. 17.3		SHA	Mer. Pass.
22	295 05.3	168 19.0	40.8	175 19.8	35.5	297 34.6	42.0	262 48.2	17.3	Venus	235 40.3	10 45
23	310 07.8	183 18.3	40.2	190 20.6	35.2	312 37.3	42.1	277 50.7	17.2	Mars	241 33.4	10 20
	h m									Jupiter	2 20.1	2 17
Mer. Pass. 2 26.8	v −0.7	d 0.6		v 0.7	d 0.3	v 2.7	d 0.1	v 2.5	d 0.0	Saturn	327 43.0	4 36

UT	SUN GHA	SUN Dec	MOON GHA	v	MOON Dec	d	HP
d h	° '	° '	° '	'	° '	'	'
14 00	178 48.4	N14 29.4	282 57.6	8.7	N10 10.2	9.8	59.4
01	193 48.5	28.6	297 25.3	8.8	10 20.0	9.8	59.4
02	208 48.6	27.8	311 53.1	8.7	10 29.8	9.6	59.4
03	223 48.7 ..	27.1	326 20.8	8.7	10 39.4	9.7	59.3
04	238 48.8	26.3	340 48.5	8.6	10 49.1	9.5	59.3
05	253 49.0	25.5	355 16.1	8.7	10 58.6	9.5	59.3
06	268 49.1	N14 24.7	9 43.8	8.6	N11 08.1	9.4	59.3
07	283 49.2	24.0	24 11.4	8.6	11 17.5	9.3	59.3
08	298 49.3	23.2	38 39.0	8.6	11 26.8	9.2	59.3
F 09	313 49.4 ..	22.4	53 06.6	8.6	11 36.0	9.2	59.3
R 10	328 49.5	21.7	67 34.2	8.6	11 45.2	9.1	59.2
I 11	343 49.7	20.9	82 01.8	8.5	11 54.3	9.1	59.2
D 12	358 49.8	N14 20.1	96 29.3	8.5	N12 03.4	8.9	59.2
A 13	13 49.9	19.4	110 56.8	8.5	12 12.3	8.9	59.2
Y 14	28 50.0	18.6	125 24.3	8.5	12 21.2	8.8	59.2
15	43 50.1 ..	17.8	139 51.8	8.4	12 30.0	8.7	59.2
16	58 50.3	17.0	154 19.2	8.4	12 38.7	8.6	59.2
17	73 50.4	16.3	168 46.6	8.4	12 47.3	8.6	59.1
18	88 50.5	N14 15.5	183 14.0	8.4	N12 55.9	8.4	59.1
19	103 50.6	14.7	197 41.4	8.4	13 04.3	8.4	59.1
20	118 50.7	13.9	212 08.8	8.3	13 12.7	8.3	59.1
21	133 50.8 ..	13.2	226 36.1	8.4	13 21.0	8.2	59.1
22	148 51.0	12.4	241 03.5	8.3	13 29.2	8.2	59.1
23	163 51.1	11.6	255 30.8	8.2	13 37.4	8.0	59.0
15 00	178 51.2	N14 10.8	269 58.0	8.3	N13 45.4	7.9	59.0
01	193 51.3	10.1	284 25.3	8.2	13 53.3	7.9	59.0
02	208 51.4	09.3	298 52.5	8.3	14 01.2	7.8	59.0
03	223 51.6 ..	08.5	313 19.8	8.2	14 09.0	7.7	59.0
04	238 51.7	07.7	327 47.0	8.1	14 16.7	7.6	59.0
05	253 51.8	07.0	342 14.1	8.2	14 24.3	7.5	58.9
06	268 51.9	N14 06.2	356 41.3	8.1	N14 31.8	7.4	58.9
S 07	283 52.1	05.4	11 08.4	8.1	14 39.2	7.3	58.9
A 08	298 52.2	04.6	25 35.5	8.1	14 46.5	7.2	58.9
T 09	313 52.3 ..	03.8	40 02.6	8.1	14 53.7	7.2	58.9
U 10	328 52.4	03.1	54 29.7	8.1	15 00.9	7.0	58.9
R 11	343 52.5	02.3	68 56.8	8.0	15 07.9	6.9	58.8
D 12	358 52.7	N14 01.5	83 23.8	8.1	N15 14.8	6.9	58.8
A 13	13 52.8	14 00.7	97 50.9	8.0	15 21.7	6.7	58.8
Y 14	28 52.9	13 59.9	112 17.9	7.9	15 28.4	6.7	58.8
15	43 53.0 ..	59.2	126 44.8	8.0	15 35.1	6.5	58.8
16	58 53.2	58.4	141 11.8	8.0	15 41.6	6.5	58.8
17	73 53.3	57.6	155 38.8	7.9	15 48.1	6.3	58.7
18	88 53.4	N13 56.8	170 05.7	7.9	N15 54.4	6.3	58.7
19	103 53.5	56.0	184 32.6	7.9	16 00.7	6.1	58.7
20	118 53.7	55.2	198 59.5	7.9	16 06.8	6.1	58.7
21	133 53.8 ..	54.5	213 26.4	7.9	16 12.9	5.9	58.7
22	148 53.9	53.7	227 53.3	7.8	16 18.8	5.9	58.6
23	163 54.0	52.9	242 20.1	7.9	16 24.7	5.7	58.6
16 00	178 54.2	N13 52.1	256 47.0	7.8	N16 30.4	5.6	58.6
01	193 54.3	51.3	271 13.8	7.8	16 36.0	5.6	58.6
02	208 54.4	50.5	285 40.6	7.8	16 41.6	5.4	58.6
03	223 54.5 ..	49.7	300 07.4	7.8	16 47.0	5.3	58.6
04	238 54.6	49.0	314 34.2	7.8	16 52.3	5.2	58.5
05	253 54.8	48.2	329 01.0	7.8	16 57.5	5.1	58.5
06	268 54.9	N13 47.4	343 27.8	7.7	N17 02.6	5.0	58.5
07	283 55.0	46.6	357 54.5	7.8	17 07.6	4.9	58.5
08	298 55.2	45.8	12 21.3	7.7	17 12.5	4.8	58.5
S 09	313 55.3 ..	45.0	26 48.0	7.7	17 17.3	4.7	58.4
U 10	328 55.4	44.2	41 14.7	7.7	17 22.0	4.5	58.4
N 11	343 55.5	43.4	55 41.4	7.7	17 26.5	4.5	58.4
D 12	358 55.7	N13 42.6	70 08.1	7.7	N17 31.0	4.4	58.4
A 13	13 55.8	41.9	84 34.8	7.7	17 35.4	4.2	58.4
Y 14	28 55.9	41.1	99 01.5	7.7	17 39.6	4.1	58.3
15	43 56.1 ..	40.3	113 28.2	7.7	17 43.7	4.0	58.3
16	58 56.2	39.5	127 54.9	7.7	17 47.7	4.0	58.3
17	73 56.3	38.7	142 21.6	7.7	17 51.7	3.8	58.3
18	88 56.4	N13 37.9	156 48.3	7.6	N17 55.5	3.6	58.3
19	103 56.6	37.1	171 14.9	7.7	17 59.1	3.6	58.2
20	118 56.7	36.3	185 41.6	7.7	18 02.7	3.5	58.2
21	133 56.8 ..	35.5	200 08.3	7.6	18 06.2	3.3	58.2
22	148 57.0	34.7	214 34.9	7.7	18 09.5	3.3	58.2
23	163 57.1	33.9	229 01.6	7.6	N18 12.8	3.1	58.2
	SD 15.8	d 0.8	SD 16.1		16.0		15.9

Lat.	Twilight Naut.	Twilight Civil	Sunrise	Moonrise 14	Moonrise 15	Moonrise 16	Moonrise 17
°	h m	h m	h m	h m	h m	h m	h m
N 72	////	////	02 22	20 49	20 37	20 01	◻
N 70	////	////	02 55	21 13	21 17	21 28	21 58
68	////	01 41	03 18	21 32	21 45	22 07	22 47
66	////	02 20	03 36	21 47	22 06	22 35	23 18
64	////	02 46	03 51	22 00	22 23	22 56	23 42
62	01 30	03 06	04 03	22 10	22 37	23 13	24 00
60	02 05	03 22	04 13	22 19	22 49	23 27	24 15
N 58	02 29	03 35	04 22	22 27	22 59	23 39	24 28
56	02 48	03 47	04 30	22 34	23 09	23 50	24 40
54	03 03	03 57	04 37	22 41	23 17	23 59	24 49
52	03 16	04 05	04 44	22 46	23 24	24 07	00 07
50	03 27	04 13	04 49	22 52	23 30	24 15	00 15
45	03 50	04 29	05 01	23 03	23 44	24 31	00 31
N 40	04 07	04 42	05 11	23 12	23 56	24 44	00 44
35	04 21	04 53	05 20	23 20	24 06	00 06	00 55
30	04 32	05 02	05 27	23 27	24 18	00 18	01 05
20	04 50	05 17	05 40	23 40	24 30	00 30	01 22
N 10	05 04	05 29	05 51	23 51	24 43	00 43	01 37
0	05 15	05 40	06 01	24 01	00 01	00 56	01 51
S 10	05 25	05 50	06 11	24 11	00 11	01 08	02 05
20	05 33	05 59	06 22	24 22	00 22	01 21	02 20
30	05 41	06 09	06 34	24 35	00 35	01 37	02 37
35	05 45	06 15	06 41	24 42	00 42	01 46	02 47
40	05 49	06 21	06 49	24 51	00 51	01 56	02 58
45	05 53	06 27	06 58	25 00	01 00	02 08	03 12
S 50	05 57	06 35	07 09	25 12	01 12	02 23	03 28
52	05 58	06 38	07 14	00 02	01 18	02 30	03 36
54	06 00	06 42	07 19	00 07	01 24	02 38	03 45
56	06 02	06 46	07 25	00 12	01 31	02 46	03 55
58	06 04	06 50	07 32	00 17	01 39	02 56	04 06
S 60	06 06	06 55	07 40	00 23	01 48	03 07	04 18

Lat.	Sunset	Twilight Civil	Twilight Naut.	Moonset 14	Moonset 15	Moonset 16	Moonset 17
°	h m	h m	h m	h m	h m	h m	h m
N 72	21 41	////	////	14 27	16 34	19 05	◻
N 70	21 10	23 38	////	14 04	15 54	17 39	19 04
68	20 47	22 21	////	13 47	15 27	17 00	18 15
66	20 30	21 45	////	13 33	15 07	16 32	17 44
64	20 16	21 20	23 41	13 21	14 50	16 12	17 21
62	20 04	21 01	22 33	13 11	14 37	15 55	17 02
60	19 54	20 45	22 00	13 03	14 25	15 41	16 47
N 58	19 45	20 32	21 37	12 56	14 16	15 29	16 34
56	19 37	20 21	21 19	12 49	14 07	15 19	16 23
54	19 31	20 11	21 04	12 44	13 59	15 10	16 13
52	19 24	20 02	20 51	12 38	13 52	15 02	16 04
50	19 19	19 55	20 40	12 34	13 46	14 54	15 57
45	19 07	19 39	20 18	12 24	13 33	14 39	15 40
N 40	18 57	19 26	20 01	12 15	13 22	14 26	15 26
35	18 49	19 16	19 48	12 08	13 13	14 15	15 14
30	18 41	19 07	19 36	12 02	13 05	14 06	15 04
20	18 29	18 52	19 19	11 51	12 51	13 49	14 47
N 10	18 18	18 40	19 05	11 41	12 38	13 35	14 32
0	18 08	18 29	18 54	11 32	12 27	13 22	14 17
S 10	17 58	18 19	18 44	11 24	12 15	13 09	14 03
20	17 47	18 10	18 36	11 14	12 03	12 54	13 48
30	17 35	18 00	18 28	11 03	11 49	12 38	13 30
35	17 29	17 55	18 25	10 57	11 41	12 29	13 20
40	17 21	17 49	18 21	10 50	11 32	12 18	13 08
45	17 12	17 42	18 17	10 42	11 21	12 05	12 54
S 50	17 01	17 35	18 13	10 32	11 08	11 50	12 38
52	16 56	17 32	18 11	10 27	11 02	11 43	12 30
54	16 51	17 28	18 10	10 23	10 56	11 35	12 21
56	16 45	17 24	18 08	10 17	10 48	11 26	12 11
58	16 38	17 20	18 06	10 11	10 40	11 16	12 00
S 60	16 30	17 15	18 04	10 04	10 31	11 04	11 47

	SUN			MOON			
Day	Eqn. of Time 00h	Eqn. of Time 12h	Mer. Pass.	Mer. Pass. Upper	Mer. Pass. Lower	Age	Phase
d	m s	m s	h m	h m	h m	d	%
14	04 47	04 41	12 05	05 20	17 47	22	54
15	04 35	04 30	12 04	06 14	18 41	23	42
16	04 24	04 18	12 04	07 09	19 36	24	32

UT	ARIES	VENUS −3.9		MARS +1.7		JUPITER −2.8		SATURN +0.3	
d h	GHA	GHA	Dec	GHA	Dec	GHA	Dec	GHA	Dec
17 00	325 10.3	198 17.6 N19 39.6		205 21.3 N21 34.9		327 39.9 S 2 42.2		292 53.1 N10 17.2	
01	340 12.7	213 16.9	38.9	220 22.1	34.6	342 42.6	42.3	307 55.6	17.2
02	355 15.2	228 16.1	38.3	235 22.8	34.3	357 45.3	42.4	322 58.1	17.2
03	10 17.7	243 15.4 .. 37.7		250 23.6 .. 34.0		12 48.0 .. 42.5		338 00.5 .. 17.2	
04	25 20.1	258 14.7	37.0	265 24.4	33.7	27 50.6	42.6	353 03.0	17.2
05	40 22.6	273 14.0	36.4	280 25.1	33.4	42 53.3	42.7	8 05.5	17.2
06	55 25.0	288 13.3 N19 35.8		295 25.9 N21 33.1		57 56.0 S 2 42.8		23 07.9 N10 17.2	
07	70 27.5	303 12.6	35.1	310 26.6	32.8	72 58.6	42.9	38 10.4	17.1
08	85 30.0	318 11.9	34.5	325 27.4	32.5	88 01.3	43.0	53 12.9	17.1
M 09	100 32.4	333 11.2 .. 33.8		340 28.1 .. 32.2		103 04.0 .. 43.1		68 15.3 .. 17.1	
O 10	115 34.9	348 10.5	33.2	355 28.9	31.9	118 06.7	43.2	83 17.8	17.1
N 11	130 37.4	3 09.7	32.6	10 29.6	31.6	133 09.3	43.3	98 20.3	17.1
D 12	145 39.8	18 09.0 N19 31.9		25 30.4 N21 31.2		148 12.0 S 2 43.4		113 22.7 N10 17.1	
A 13	160 42.3	33 08.3	31.3	40 31.2	30.9	163 14.7	43.5	128 25.2	17.1
Y 14	175 44.8	48 07.6	30.6	55 31.9	30.6	178 17.4	43.6	143 27.7	17.1
15	190 47.2	63 06.9 .. 30.0		70 32.7 .. 30.3		193 20.0 .. 43.7		158 30.1 .. 17.0	
16	205 49.7	78 06.2	29.3	85 33.4	30.0	208 22.7	43.8	173 32.6	17.0
17	220 52.1	93 05.5	28.7	100 34.2	29.7	223 25.4	43.9	188 35.1	17.0
18	235 54.6	108 04.8 N19 28.0		115 34.9 N21 29.4		238 28.0 S 2 44.0		203 37.5 N10 17.0	
19	250 57.1	123 04.1	27.4	130 35.7	29.1	253 30.7	44.1	218 40.0	17.0
20	265 59.5	138 03.4	26.7	145 36.5	28.8	268 33.4	44.2	233 42.5	17.0
21	281 02.0	153 02.7 .. 26.1		160 37.2 .. 28.5		283 36.1 .. 44.3		248 45.0 .. 17.0	
22	296 04.5	168 01.9	25.4	175 38.0	28.2	298 38.7	44.4	263 47.4	17.0
23	311 06.9	183 01.2	24.7	190 38.7	27.9	313 41.4	44.5	278 49.9	16.9
18 00	326 09.4	198 00.5 N19 24.1		205 39.5 N21 27.6		328 44.1 S 2 44.6		293 52.4 N10 16.9	
01	341 11.9	212 59.8	23.4	220 40.2	27.3	343 46.8	44.7	308 54.8	16.9
02	356 14.3	227 59.1	22.8	235 41.0	27.0	358 49.5	44.8	323 57.3	16.9
03	11 16.8	242 58.4 .. 22.1		250 41.8 .. 26.7		13 52.1 .. 44.9		338 59.8 .. 16.9	
04	26 19.3	257 57.7	21.5	265 42.5	26.4	28 54.8	45.0	354 02.2	16.9
05	41 21.7	272 57.0	20.8	280 43.3	26.1	43 57.5	45.1	9 04.7	16.9
06	56 24.2	287 56.3 N19 20.1		295 44.0 N21 25.7		59 00.2 S 2 45.2		24 07.2 N10 16.8	
07	71 26.6	302 55.6	19.5	310 44.8	25.4	74 02.8	45.4	39 09.7	16.8
T 08	86 29.1	317 54.9	18.8	325 45.6	25.1	89 05.5	45.5	54 12.1	16.8
U 09	101 31.6	332 54.2 .. 18.1		340 46.3 .. 24.8		104 08.2 .. 45.6		69 14.6 .. 16.8	
E 10	116 34.0	347 53.5	17.5	355 47.1	24.5	119 10.9	45.7	84 17.1	16.8
S 11	131 36.5	2 52.8	16.8	10 47.8	24.2	134 13.6	45.8	99 19.5	16.8
D 12	146 39.0	17 52.1 N19 16.1		25 48.6 N21 23.9		149 16.2 S 2 45.9		114 22.0 N10 16.8	
A 13	161 41.4	32 51.4	15.5	40 49.4	23.6	164 18.9	46.0	129 24.5	16.7
Y 14	176 43.9	47 50.7	14.8	55 50.1	23.3	179 21.6	46.1	144 27.0	16.7
15	191 46.4	62 50.0 .. 14.1		70 50.9 .. 23.0		194 24.3 .. 46.2		159 29.4 .. 16.7	
16	206 48.8	77 49.3	13.5	85 51.6	22.6	209 27.0	46.3	174 31.9	16.7
17	221 51.3	92 48.6	12.8	100 52.4	22.3	224 29.6	46.4	189 34.4	16.7
18	236 53.8	107 47.9 N19 12.1		115 53.2 N21 22.0		239 32.3 S 2 46.5		204 36.8 N10 16.7	
19	251 56.2	122 47.2	11.4	130 53.9	21.7	254 35.0	46.6	219 39.3	16.7
20	266 58.7	137 46.5	10.8	145 54.7	21.4	269 37.7	46.7	234 41.8	16.6
21	282 01.1	152 45.8 .. 10.1		160 55.5 .. 21.1		284 40.4 .. 46.8		249 44.3 .. 16.6	
22	297 03.6	167 45.1	09.4	175 56.2	20.8	299 43.0	46.9	264 46.7	16.6
23	312 06.1	182 44.4	08.7	190 57.0	20.5	314 45.7	47.0	279 49.2	16.6
19 00	327 08.5	197 43.7 N19 08.0		205 57.7 N21 20.1		329 48.4 S 2 47.1		294 51.7 N10 16.6	
01	342 11.0	212 43.0	07.4	220 58.5	19.8	344 51.1	47.2	309 54.2	16.6
02	357 13.5	227 42.3	06.7	235 59.3	19.5	359 53.8	47.3	324 56.6	16.6
03	12 15.9	242 41.6 .. 06.0		251 00.0 .. 19.2		14 56.4 .. 47.4		339 59.1 .. 16.5	
04	27 18.4	257 40.9	05.3	266 00.8	18.9	29 59.1	47.5	355 01.6	16.5
05	42 20.9	272 40.2	04.6	281 01.6	18.6	45 01.8	47.6	10 04.1	16.5
06	57 23.3	287 39.5 N19 03.9		296 02.3 N21 18.3		60 04.5 S 2 47.7		25 06.5 N10 16.5	
W 07	72 25.8	302 38.8	03.3	311 03.1	18.0	75 07.2	47.8	40 09.0	16.5
E 08	87 28.3	317 38.1	02.6	326 03.9	17.6	90 09.9	47.9	55 11.5	16.5
D 09	102 30.7	332 37.4 .. 01.9		341 04.6 .. 17.3		105 12.5 .. 48.1		70 14.0 .. 16.5	
N 10	117 33.2	347 36.8	01.2	356 05.4	17.0	120 15.2	48.2	85 16.4	16.4
E 11	132 35.6	2 36.1 19 00.5		11 06.2	16.7	135 17.9	48.3	100 18.9	16.4
S 12	147 38.1	17 35.4 N18 59.8		26 06.9 N21 16.4		150 20.6 S 2 48.4		115 21.4 N10 16.4	
D 13	162 40.6	32 34.7	59.1	41 07.7	16.1	165 23.3	48.5	130 23.9	16.4
A 14	177 43.0	47 34.0	58.4	56 08.4	15.7	180 26.0	48.6	145 26.3	16.4
Y 15	192 45.5	62 33.3 .. 57.7		71 09.2 .. 15.4		195 28.7 .. 48.7		160 28.8 .. 16.4	
16	207 48.0	77 32.6	57.0	86 10.0	15.1	210 31.3	48.8	175 31.3	16.3
17	222 50.4	92 31.9	56.3	101 10.7	14.8	225 34.0	48.9	190 33.8	16.3
18	237 52.9	107 31.2 N18 55.6		116 11.5 N21 14.5		240 36.7 S 2 49.0		205 36.2 N10 16.3	
19	252 55.4	122 30.5	54.9	131 12.3	14.2	255 39.4	49.1	220 38.7	16.3
20	267 57.8	137 29.8	54.3	146 13.0	13.8	270 42.1	49.2	235 41.2	16.3
21	283 00.3	152 29.2 .. 53.6		161 13.8 .. 13.5		285 44.8 .. 49.3		250 43.7 .. 16.3	
22	298 02.7	167 28.5	52.9	176 14.6	13.2	300 47.5	49.4	265 46.2	16.2
23	313 05.2	182 27.8	52.1	191 15.3	12.9	315 50.2	49.5	280 48.6	16.2
Mer. Pass.	h m 2 15.0	v −0.7 d 0.7		v 0.8 d 0.3		v 2.7 d 0.1		v 2.5 d 0.0	

STARS

Name	SHA	Dec
Acamar	315 27.0	S40 18.4
Achernar	335 35.0	S57 14.4
Acrux	173 22.5	S63 05.6
Adhara	255 21.8	S28 58.1
Aldebaran	291 02.6	N16 30.3
Alioth	166 31.1	N55 58.4
Alkaid	153 08.1	N49 19.6
Al Na'ir	27 57.6	S46 57.9
Alnilam	275 58.1	S 1 12.2
Alphard	218 07.6	S 8 39.1
Alphecca	126 20.7	N26 43.5
Alpheratz	357 55.0	N29 04.9
Altair	62 19.1	N 8 52.1
Ankaa	353 26.7	S42 18.6
Antares	112 40.2	S26 25.6
Arcturus	146 06.3	N19 11.7
Atria	107 52.1	S69 01.6
Avior	234 23.2	S59 30.2
Bellatrix	278 44.4	N 6 20.9
Betelgeuse	271 13.8	N 7 24.4
Canopus	264 01.6	S52 41.6
Capella	280 51.5	N45 59.5
Deneb	49 38.8	N45 16.7
Denebola	182 45.6	N14 35.0
Diphda	349 07.1	S17 59.5
Dubhe	194 06.3	N61 45.7
Elnath	278 27.2	N28 36.2
Eltanin	90 51.1	N51 29.8
Enif	33 58.0	N 9 52.3
Fomalhaut	15 36.2	S29 37.6
Gacrux	172 14.0	S57 06.4
Gienah	176 04.3	S17 32.0
Hadar	149 04.3	S60 22.1
Hamal	328 13.5	N23 27.2
Kaus Aust.	83 58.7	S34 23.0
Kochab	137 19.9	N74 10.1
Markab	13 49.4	N15 11.9
Menkar	314 26.9	N 4 05.0
Menkent	148 21.2	S36 21.8
Miaplacidus	221 42.9	S69 42.7
Mirfak	308 56.7	N49 51.1
Nunki	76 12.2	S26 17.8
Peacock	53 36.6	S56 44.3
Pollux	243 42.0	N28 01.7
Procyon	245 11.9	N 5 13.7
Rasalhague	96 16.9	N12 34.0
Regulus	207 56.0	N11 58.5
Rigel	281 23.2	S 8 12.2
Rigil Kent.	140 07.5	S60 49.8
Sabik	102 25.5	S15 43.2
Schedar	349 53.1	N56 31.6
Shaula	96 37.2	S37 06.1
Sirius	258 44.0	S16 42.8
Spica	158 43.5	S11 09.1
Suhail	223 01.2	S43 25.6
Vega	80 46.4	N38 47.3
Zuben'ubi	137 18.1	S16 02.0

	SHA	Mer. Pass.
Venus	231 51.1	10 48
Mars	239 30.1	10 17
Jupiter	2 34.7	2 05
Saturn	327 43.0	4 24

SUN and MOON — GHA / Dec

UT	SUN GHA	SUN Dec	MOON GHA	v	MOON Dec	d	HP
17	° ′	° ′	° ′	′	° ′	′	′
00	178 57.2	N13 33.1	243 28.2	7.7	N18 15.9	3.0	58.2
01	193 57.4	32.3	257 54.9	7.7	18 18.9	2.9	58.1
02	208 57.5	31.5	272 21.6	7.6	18 21.8	2.8	58.1
03	223 57.6	.. 30.7	286 48.2	7.7	18 24.6	2.7	58.1
04	238 57.8	30.0	301 14.9	7.7	18 27.3	2.5	58.1
05	253 57.9	29.2	315 41.6	7.7	18 29.8	2.5	58.1
06	268 58.0	N13 28.4	330 08.3	7.6	N18 32.3	2.3	58.0
M 07	283 58.1	27.6	344 34.9	7.7	18 34.6	2.2	58.0
O 08	298 58.3	26.8	359 01.6	7.7	18 36.8	2.1	58.0
N 09	313 58.4	.. 26.0	13 28.3	7.7	18 38.9	2.0	58.0
D 10	328 58.5	25.2	27 55.0	7.7	18 40.9	1.9	58.0
A 11	343 58.7	24.4	42 21.7	7.8	18 42.8	1.8	57.9
Y 12	358 58.8	N13 23.6	56 48.5	7.7	N18 44.6	1.6	57.9
13	13 58.9	22.8	71 15.2	7.7	18 46.2	1.5	57.9
14	28 59.1	22.0	85 41.9	7.8	18 47.7	1.5	57.9
15	43 59.2	.. 21.2	100 08.7	7.7	18 49.2	1.3	57.9
16	58 59.3	20.4	114 35.4	7.8	18 50.5	1.2	57.8
17	73 59.5	19.6	129 02.2	7.8	18 51.7	1.0	57.8
18	88 59.6	N13 18.8	143 29.0	7.8	N18 52.7	1.0	57.8
19	103 59.7	18.0	157 55.8	7.8	18 53.7	0.9	57.8
20	118 59.9	17.2	172 22.6	7.9	18 54.6	0.7	57.8
21	134 00.0	.. 16.4	186 49.5	7.8	18 55.3	0.6	57.7
22	149 00.2	15.6	201 16.3	7.9	18 55.9	0.5	57.7
23	164 00.3	14.8	215 43.2	7.9	18 56.4	0.4	57.7
18 00	179 00.4	N13 14.0	230 10.1	7.9	N18 56.8	0.3	57.7
01	194 00.6	13.2	244 37.0	8.0	18 57.1	0.2	57.7
02	209 00.7	12.3	259 04.0	7.9	18 57.3	0.0	57.6
03	224 00.8	.. 11.5	273 30.9	8.0	18 57.3	0.0	57.6
04	239 01.0	10.7	287 57.9	8.0	18 57.3	0.3	57.6
05	254 01.1	09.9	302 24.9	8.0	18 57.1	0.3	57.6
06	269 01.2	N13 09.1	316 51.9	8.1	N18 56.8	0.4	57.6
T 07	284 01.4	08.3	331 19.0	8.1	18 56.4	0.5	57.5
U 08	299 01.5	07.5	345 46.0	8.1	18 55.9	0.6	57.5
E 09	314 01.7	.. 06.7	0 13.1	8.2	18 55.3	0.7	57.5
S 10	329 01.8	05.9	14 40.3	8.1	18 54.6	0.8	57.5
D 11	344 01.9	05.1	29 07.4	8.2	18 53.8	1.0	57.5
A 12	359 02.1	N13 04.3	43 34.6	8.2	N18 52.8	1.0	57.4
Y 13	14 02.2	03.5	58 01.8	8.3	18 51.8	1.2	57.4
14	29 02.3	02.7	72 29.1	8.3	18 50.6	1.3	57.4
15	44 02.5	.. 01.9	86 56.4	8.3	18 49.3	1.4	57.4
16	59 02.6	01.1	101 23.7	8.3	18 47.9	1.4	57.3
17	74 02.8	13 00.2	115 51.0	8.4	18 46.5	1.6	57.3
18	89 02.9	N12 59.4	130 18.4	8.4	N18 44.9	1.8	57.3
19	104 03.0	58.6	144 45.8	8.4	18 43.1	1.8	57.3
20	119 03.2	57.8	159 13.2	8.5	18 41.3	1.9	57.3
21	134 03.3	.. 57.0	173 40.7	8.5	18 39.4	2.0	57.2
22	149 03.5	56.2	188 08.2	8.6	18 37.4	2.2	57.2
23	164 03.6	55.4	202 35.8	8.6	18 35.2	2.2	57.2
19 00	179 03.7	N12 54.6	217 03.4	8.6	N18 33.0	2.3	57.2
01	194 03.9	53.8	231 31.0	8.7	18 30.7	2.5	57.2
02	209 04.0	52.9	245 58.7	8.7	18 28.2	2.5	57.1
03	224 04.2	.. 52.1	260 26.4	8.7	18 25.7	2.7	57.1
04	239 04.3	51.3	274 54.1	8.8	18 23.0	2.8	57.1
05	254 04.5	50.5	289 21.9	8.8	18 20.2	2.8	57.1
06	269 04.6	N12 49.7	303 49.7	8.9	N18 17.4	3.0	57.1
W 07	284 04.7	48.9	318 17.6	8.9	18 14.4	3.1	57.0
E 08	299 04.9	48.1	332 45.5	9.0	18 11.3	3.1	57.0
D 09	314 05.0	.. 47.2	347 13.5	9.0	18 08.2	3.3	57.0
N 10	329 05.2	46.4	1 41.5	9.0	18 04.9	3.4	57.0
E 11	344 05.3	45.6	16 09.5	9.1	18 01.5	3.4	57.0
S 12	359 05.4	N12 44.8	30 37.6	9.2	N17 58.1	3.6	56.9
D 13	14 05.6	44.0	45 05.8	9.1	17 54.5	3.7	56.9
A 14	29 05.7	43.2	59 33.9	9.3	17 50.8	3.7	56.9
Y 15	44 05.9	.. 42.3	74 02.2	9.2	17 47.1	3.9	56.9
16	59 06.0	41.5	88 30.4	9.4	17 43.2	3.9	56.8
17	74 06.2	40.7	102 58.8	9.3	17 39.3	4.1	56.8
18	89 06.3	N12 39.9	117 27.1	9.5	N17 35.2	4.1	56.8
19	104 06.5	39.1	131 55.6	9.4	17 31.1	4.2	56.8
20	119 06.6	38.3	146 24.0	9.6	17 26.9	4.4	56.8
21	134 06.7	.. 37.4	160 52.6	9.5	17 22.5	4.4	56.7
22	149 06.9	36.6	175 21.1	9.7	17 18.1	4.5	56.7
23	164 07.0	35.8	189 49.8	9.6	N17 13.6	4.6	56.7
SD	15.8	d 0.8	SD 15.8		15.6		15.5

Twilight, Sunrise and Moonrise

Lat.	Twilight Naut.	Twilight Civil	Sunrise	Moonrise 17	18	19	20
°	h m	h m	h m	h m	h m	h m	h m
N 72	////	////	02 43	☐	☐	23 39	25 43
N 70	////	01 10	03 10	21 58	23 04	24 36	00 36
68	////	02 03	03 31	22 47	23 49	25 10	01 10
66	////	02 35	03 47	23 18	24 19	00 19	01 34
64	01 01	02 58	04 00	23 42	24 41	00 41	01 53
62	01 49	03 16	04 11	24 00	00 00	00 59	02 08
60	02 18	03 31	04 21	24 15	00 15	01 14	02 21
N 58	02 40	03 43	04 29	24 28	00 28	01 27	02 32
56	02 57	03 53	04 36	24 40	00 40	01 38	02 42
54	03 11	04 03	04 43	24 49	00 49	01 47	02 51
52	03 23	04 11	04 48	00 07	00 58	01 56	02 58
50	03 33	04 18	04 54	00 15	01 06	02 03	03 05
45	03 54	04 33	05 05	00 31	01 23	02 20	03 20
N 40	04 10	04 45	05 14	00 44	01 37	02 33	03 32
35	04 23	04 55	05 22	00 55	01 48	02 44	03 42
30	04 34	05 04	05 29	01 05	01 59	02 54	03 51
20	04 51	05 18	05 41	01 22	02 16	03 11	04 07
N 10	05 04	05 29	05 51	01 37	02 32	03 26	04 20
0	05 15	05 39	06 01	01 51	02 46	03 40	04 33
S 10	05 23	05 48	06 10	02 05	03 00	03 54	04 45
20	05 31	05 57	06 20	02 20	03 16	04 09	04 59
30	05 38	06 06	06 31	02 37	03 34	04 26	05 14
35	05 42	06 11	06 37	02 47	03 44	04 36	05 23
40	05 45	06 17	06 45	02 58	03 56	04 47	05 33
45	05 48	06 23	06 53	03 12	04 10	05 01	05 45
S 50	05 52	06 29	07 03	03 28	04 27	05 17	06 00
52	05 53	06 33	07 08	03 36	04 35	05 25	06 06
54	05 54	06 36	07 13	03 45	04 44	05 33	06 14
56	05 56	06 39	07 18	03 55	04 54	05 43	06 22
58	05 57	06 43	07 25	04 06	05 05	05 53	06 31
S 60	05 59	06 48	07 32	04 18	05 18	06 06	06 42

Sunset, Twilight and Moonset

Lat.	Sunset	Twilight Civil	Twilight Naut.	Moonset 17	18	19	20
°	h m	h m	h m	h m	h m	h m	h m
N 72	21 20	////	////	☐	☐	21 08	20 50
N 70	20 54	22 47	////	19 04	19 52	20 10	20 17
68	20 34	21 59	////	18 15	19 07	19 36	19 53
66	20 18	21 29	////	17 44	18 36	19 11	19 34
64	20 05	21 07	22 56	17 21	18 14	18 52	19 19
62	19 55	20 49	22 13	17 02	17 56	18 36	19 06
60	19 45	20 35	21 46	16 47	17 41	18 23	18 55
N 58	19 37	20 23	21 25	16 34	17 28	18 11	18 45
56	19 30	20 13	21 09	16 23	17 17	18 01	18 37
54	19 24	20 04	20 55	16 13	17 07	17 52	18 29
52	19 18	19 56	20 43	16 04	16 59	17 45	18 23
50	19 13	19 49	20 33	15 57	16 51	17 37	18 17
45	19 02	19 34	20 12	15 40	16 34	17 22	18 04
N 40	18 53	19 22	19 56	15 26	16 21	17 10	17 53
35	18 45	19 12	19 44	15 14	16 09	16 59	17 44
30	18 38	19 03	19 33	15 04	15 59	16 49	17 35
20	18 27	18 49	19 16	14 47	15 42	16 33	17 21
N 10	18 17	18 38	19 03	14 32	15 26	16 19	17 09
0	18 07	18 28	18 53	14 17	15 12	16 06	16 57
S 10	17 58	18 19	18 44	14 03	14 58	15 52	16 45
20	17 48	18 11	18 37	13 48	14 43	15 38	16 33
30	17 37	18 02	18 30	13 30	14 25	15 21	16 18
35	17 31	17 57	18 27	13 20	14 15	15 12	16 10
40	17 24	17 52	18 23	13 08	14 03	15 01	16 00
45	17 15	17 46	18 20	12 54	13 49	14 48	15 49
S 50	17 05	17 39	18 17	12 38	13 32	14 32	15 35
52	17 01	17 36	18 16	12 30	13 24	14 24	15 29
54	16 56	17 33	18 14	12 21	13 15	14 16	15 22
56	16 50	17 29	18 13	12 11	13 05	14 07	15 14
58	16 44	17 25	18 12	12 00	12 54	13 56	15 05
S 60	16 37	17 21	18 10	11 47	12 41	13 44	14 55

SUN and MOON

Day	SUN Eqn. of Time 00ʰ	SUN Eqn. of Time 12ʰ	SUN Mer. Pass.	MOON Mer. Pass. Upper	MOON Mer. Pass. Lower	Age	Phase
d	m s	m s	h m	h m	h m	d	%
17	04 11	04 05	12 04	08 04	20 32	25	22
18	03 59	03 52	12 04	08 59	21 26	26	14
19	03 45	03 38	12 04	09 53	22 19	27	7

1998 AUGUST 20, 21, 22 (THURS., FRI., SAT.)

UT	ARIES GHA	VENUS −3.9 GHA	Dec	MARS +1.7 GHA	Dec	JUPITER −2.8 GHA	Dec	SATURN +0.3 GHA	Dec	STARS Name	SHA	Dec
20 00	328 07.7	197 27.1	N18 51.4	206 16.1	N21 12.6	330 52.8	S 2 49.6	295 51.1	N10 16.2	Acamar	315 26.9	S40 18.4
01	343 10.1	212 26.4	50.7	221 16.9	12.2	345 55.5	49.7	310 53.6	16.2	Achernar	335 35.0	S57 14.4
02	358 12.6	227 25.7	50.0	236 17.7	11.9	0 58.2	49.9	325 56.1	16.2	Acrux	173 22.5	S63 05.6
03	13 15.1	242 25.0	.. 49.3	251 18.4	.. 11.6	16 00.9	.. 50.0	340 58.5	.. 16.2	Adhara	255 21.8	S28 58.1
04	28 17.5	257 24.3	48.6	266 19.2	11.3	31 03.6	50.1	356 01.0	16.1	Aldebaran	291 02.5	N16 30.3
05	43 20.0	272 23.7	47.9	281 20.0	11.0	46 06.3	50.2	11 03.5	16.1			
06	58 22.5	287 23.0	N18 47.2	296 20.7	N21 10.6	61 09.0	S 2 50.3	26 06.0	N10 16.1	Alioth	166 31.1	N55 58.4
07	73 24.9	302 22.3	46.5	311 21.5	10.3	76 11.7	50.4	41 08.5	16.1	Alkaid	153 08.2	N49 19.6
T 08	88 27.4	317 21.6	45.8	326 22.3	10.0	91 14.3	50.5	56 10.9	16.1	Al Na'ir	27 57.6	S46 57.9
H 09	103 29.9	332 20.9	.. 45.1	341 23.0	.. 09.7	106 17.0	.. 50.6	71 13.4	.. 16.1	Alnilam	275 58.1	S 1 12.2
U 10	118 32.3	347 20.2	44.4	356 23.8	09.4	121 19.7	50.7	86 15.9	16.0	Alphard	218 07.6	S 8 39.1
R 11	133 34.8	2 19.5	43.7	11 24.6	09.0	136 22.4	50.8	101 18.4	16.0			
S 12	148 37.2	17 18.9	N18 42.9	26 25.3	N21 08.7	151 25.1	S 2 50.9	116 20.9	N10 16.0	Alphecca	126 20.7	N26 43.5
D 13	163 39.7	32 18.2	42.2	41 26.1	08.4	166 27.8	51.0	131 23.3	16.0	Alpheratz	357 55.0	N29 04.9
A 14	178 42.2	47 17.5	41.5	56 26.9	08.1	181 30.5	51.1	146 25.8	16.0	Altair	62 19.1	N 8 52.1
Y 15	193 44.6	62 16.8	.. 40.8	71 27.7	.. 07.7	196 33.2	.. 51.2	161 28.3	.. 16.0	Ankaa	353 26.6	S42 18.6
16	208 47.1	77 16.1	40.1	86 28.4	07.4	211 35.9	51.4	176 30.8	15.9	Antares	112 40.2	S26 25.6
17	223 49.6	92 15.5	39.4	101 29.2	07.1	226 38.6	51.5	191 33.3	15.9			
18	238 52.0	107 14.8	N18 38.6	116 30.0	N21 06.8	241 41.3	S 2 51.6	206 35.7	N10 15.9	Arcturus	146 06.3	N19 11.7
19	253 54.5	122 14.1	37.9	131 30.7	06.4	256 43.9	51.7	221 38.2	15.9	Atria	107 52.1	S69 01.6
20	268 57.0	137 13.4	37.2	146 31.5	06.1	271 46.6	51.8	236 40.7	15.9	Avior	234 23.2	S59 30.2
21	283 59.4	152 12.7	.. 36.5	161 32.3	.. 05.8	286 49.3	.. 51.9	251 43.2	.. 15.9	Bellatrix	278 44.4	N 6 20.9
22	299 01.9	167 12.1	35.8	176 33.1	05.5	301 52.0	52.0	266 45.7	15.8	Betelgeuse	271 13.8	N 7 24.4
23	314 04.4	182 11.4	35.0	191 33.8	05.1	316 54.7	52.1	281 48.2	15.8			
21 00	329 06.8	197 10.7	N18 34.3	206 34.6	N21 04.8	331 57.4	S 2 52.2	296 50.6	N10 15.8	Canopus	264 01.5	S52 41.6
01	344 09.3	212 10.0	33.6	221 35.4	04.5	347 00.1	52.3	311 53.1	15.8	Capella	280 51.5	N45 59.5
02	359 11.7	227 09.3	32.9	236 36.1	04.2	2 02.8	52.4	326 55.6	15.8	Deneb	49 38.8	N45 16.7
03	14 14.2	242 08.7	.. 32.1	251 36.9	.. 03.8	17 05.5	.. 52.5	341 58.1	.. 15.8	Denebola	182 45.6	N14 35.0
04	29 16.7	257 08.0	31.4	266 37.7	03.5	32 08.2	52.7	357 00.6	15.7	Diphda	349 07.1	S17 59.5
05	44 19.1	272 07.3	30.7	281 38.5	03.2	47 10.9	52.8	12 03.0	15.7			
06	59 21.6	287 06.6	N18 29.9	296 39.2	N21 02.9	62 13.6	S 2 52.9	27 05.5	N10 15.7	Dubhe	194 06.3	N61 45.7
07	74 24.1	302 06.0	29.2	311 40.0	02.5	77 16.3	53.0	42 08.0	15.7	Elnath	278 27.2	N28 36.2
F 08	89 26.5	317 05.3	28.5	326 40.8	02.2	92 19.0	53.1	57 10.5	15.7	Eltanin	90 51.2	N51 29.8
R 09	104 29.0	332 04.6	.. 27.7	341 41.6	.. 01.9	107 21.7	.. 53.2	72 13.0	.. 15.6	Enif	33 58.0	N 9 52.3
I 10	119 31.5	347 03.9	27.0	356 42.3	01.6	122 24.4	53.3	87 15.5	15.6	Fomalhaut	15 36.2	S29 37.6
D 11	134 33.9	2 03.3	26.3	11 43.1	01.2	137 27.0	53.4	102 18.0	15.6			
A 12	149 36.4	17 02.6	N18 25.5	26 43.9	N21 00.9	152 29.7	S 2 53.5	117 20.4	N10 15.6	Gacrux	172 14.0	S57 06.4
Y 13	164 38.8	32 01.9	24.8	41 44.7	00.6	167 32.4	53.6	132 22.9	15.6	Gienah	176 04.3	S17 32.0
14	179 41.3	47 01.2	24.1	56 45.4	21 00.2	182 35.1	53.8	147 25.4	15.6	Hadar	149 04.3	S60 22.1
15	194 43.8	62 00.6	.. 23.3	71 46.2	20 59.9	197 37.8	.. 53.9	162 27.9	.. 15.5	Hamal	328 13.5	N23 27.2
16	209 46.2	76 59.9	22.6	86 47.0	59.6	212 40.5	54.0	177 30.4	15.5	Kaus Aust.	83 58.7	S34 23.0
17	224 48.7	91 59.2	21.9	101 47.8	59.3	227 43.2	54.1	192 32.9	15.5			
18	239 51.2	106 58.5	N18 21.1	116 48.5	N20 58.9	242 45.9	S 2 54.2	207 35.3	N10 15.5	Kochab	137 20.0	N74 10.1
19	254 53.6	121 57.9	20.4	131 49.3	58.6	257 48.6	54.3	222 37.8	15.5	Markab	13 49.4	N15 11.9
20	269 56.1	136 57.2	19.6	146 50.1	58.3	272 51.3	54.4	237 40.3	15.4	Menkar	314 26.9	N 4 05.0
21	284 58.6	151 56.5	.. 18.9	161 50.9	.. 57.9	287 54.0	.. 54.5	252 42.8	.. 15.4	Menkent	148 21.2	S36 21.8
22	300 01.0	166 55.9	18.1	176 51.6	57.6	302 56.7	54.6	267 45.3	15.4	Miaplacidus	221 42.9	S69 42.7
23	315 03.5	181 55.2	17.4	191 52.4	57.3	317 59.4	54.7	282 47.8	15.4			
22 00	330 06.0	196 54.5	N18 16.6	206 53.2	N20 56.9	333 02.1	S 2 54.9	297 50.3	N10 15.4	Mirfak	308 56.6	N49 51.1
01	345 08.4	211 53.9	15.9	221 54.0	56.6	348 04.8	55.0	312 52.7	15.3	Nunki	76 12.2	S26 17.8
02	0 10.9	226 53.2	15.2	236 54.7	56.3	3 07.5	55.1	327 55.2	15.3	Peacock	53 36.6	S56 44.3
03	15 13.3	241 52.5	.. 14.4	251 55.5	.. 55.9	18 10.2	.. 55.2	342 57.7	.. 15.3	Pollux	243 42.0	N28 01.7
04	30 15.8	256 51.9	13.7	266 56.3	55.6	33 12.9	55.3	358 00.2	15.3	Procyon	245 11.9	N 5 13.7
05	45 18.3	271 51.2	12.9	281 57.1	55.3	48 15.6	55.4	13 02.7	15.3			
06	60 20.7	286 50.5	N18 12.1	296 57.9	N20 54.9	63 18.3	S 2 55.5	28 05.2	N10 15.2	Rasalhague	96 16.9	N12 34.0
07	75 23.2	301 49.9	11.4	311 58.6	54.6	78 21.0	55.6	43 07.7	15.2	Regulus	207 56.0	N11 58.5
S 08	90 25.7	316 49.2	10.6	326 59.4	54.3	93 23.7	55.7	58 10.2	15.2	Rigel	281 23.1	S 8 12.2
A 09	105 28.1	331 48.5	.. 09.9	342 00.2	.. 53.9	108 26.4	.. 55.9	73 12.6	.. 15.2	Rigil Kent.	140 07.5	S60 49.8
T 10	120 30.6	346 47.9	09.1	357 01.0	53.6	123 29.1	56.0	88 15.1	15.2	Sabik	102 25.5	S15 43.2
U 11	135 33.1	1 47.2	08.4	12 01.8	53.3	138 31.8	56.1	103 17.6	15.2			
R 12	150 35.5	16 46.5	N18 07.6	27 02.5	N20 52.9	153 34.5	S 2 56.2	118 20.1	N10 15.1	Schedar	349 53.1	N56 31.6
D 13	165 38.0	31 45.9	06.9	42 03.3	52.6	168 37.2	56.3	133 22.6	15.1	Shaula	96 37.2	S37 06.1
A 14	180 40.5	46 45.2	06.1	57 04.1	52.3	183 39.9	56.4	148 25.1	15.1	Sirius	258 44.0	S16 42.8
Y 15	195 42.9	61 44.5	.. 05.3	72 04.9	.. 51.9	198 42.6	.. 56.5	163 27.6	.. 15.1	Spica	158 43.5	S11 09.1
16	210 45.4	76 43.9	04.6	87 05.6	51.6	213 45.3	56.6	178 30.1	15.1	Suhail	223 01.2	S43 25.6
17	225 47.8	91 43.2	03.8	102 06.4	51.3	228 48.0	56.8	193 32.6	15.0			
18	240 50.3	106 42.5	N18 03.1	117 07.2	N20 50.9	243 50.7	S 2 56.9	208 35.0	N10 15.0	Vega	80 46.4	N38 47.3
19	255 52.8	121 41.9	02.3	132 08.0	50.6	258 53.4	57.0	223 37.5	15.0	Zuben'ubi	137 18.1	S16 02.0
20	270 55.2	136 41.2	01.5	147 08.8	50.3	273 56.1	57.1	238 40.0	15.0		SHA	Mer. Pass.
21	285 57.7	151 40.6	.. 00.8	162 09.6	.. 49.9	288 58.8	.. 57.2	253 42.5	.. 14.9			h m
22	301 00.2	166 39.9	18 00.0	177 10.3	49.6	304 01.5	57.3	268 45.0	14.9	Venus	228 03.9	10 52
23	316 02.6	181 39.2	N17 59.2	192 11.1	49.3	319 04.2	57.4	283 47.5	14.9	Mars	237 27.8	10 13
Mer. Pass. 2 03.2		v −0.7	d 0.7	v 0.8	d 0.3	v 2.7	d 0.1	v 2.5	d 0.0	Jupiter	2 50.6	1 52
										Saturn	327 43.8	4 12

UT	SUN GHA	SUN Dec	MOON GHA	v	MOON Dec	d	HP
d h	° ′	° ′	° ′	′	° ′	′	′
20 00	179 07.2	N12 35.0	204 18.4	9.7	N17 09.0	4.7	56.7
01	194 07.3	34.2	218 47.1	9.8	17 04.3	4.8	56.7
02	209 07.5	33.3	233 15.9	9.8	16 59.5	4.9	56.6
03	224 07.6 ..	32.5	247 44.7	9.9	16 54.6	4.9	56.6
04	239 07.8	31.7	262 13.6	10.0	16 49.7	5.1	56.6
05	254 07.9	30.9	276 42.6	9.9	16 44.6	5.1	56.6
06	269 08.1	N12 30.0	291 11.5	10.1	N16 39.5	5.2	56.6
T 07	284 08.2	29.2	305 40.6	10.1	16 34.3	5.3	56.5
H 08	299 08.4	28.4	320 09.7	10.1	16 29.0	5.4	56.5
U 09	314 08.5 ..	27.6	334 38.8	10.2	16 23.6	5.5	56.5
R 10	329 08.7	26.8	349 08.0	10.3	16 18.1	5.5	56.5
S 11	344 08.8	25.9	3 37.3	10.3	16 12.6	5.7	56.5
D 12	359 08.9	N12 25.1	18 06.6	10.3	N16 06.9	5.7	56.4
A 13	14 09.1	24.3	32 35.9	10.4	16 01.2	5.8	56.4
Y 14	29 09.2	23.5	47 05.3	10.5	15 55.4	5.9	56.4
15	44 09.4 ..	22.6	61 34.8	10.5	15 49.5	5.9	56.4
16	59 09.5	21.8	76 04.3	10.6	15 43.6	6.1	56.3
17	74 09.7	21.0	90 33.9	10.7	15 37.5	6.1	56.3
18	89 09.8	N12 20.1	105 03.6	10.7	N15 31.4	6.2	56.3
19	104 10.0	19.3	119 33.3	10.7	15 25.2	6.2	56.3
20	119 10.1	18.5	134 03.0	10.8	15 19.0	6.4	56.3
21	134 10.3 ..	17.7	148 32.8	10.9	15 12.6	6.4	56.2
22	149 10.4	16.8	163 02.7	10.9	15 06.2	6.5	56.2
23	164 10.6	16.0	177 32.6	11.0	14 59.7	6.5	56.2
21 00	179 10.7	N12 15.2	192 02.6	11.0	N14 53.2	6.6	56.2
01	194 10.9	14.4	206 32.6	11.1	14 46.6	6.7	56.2
02	209 11.0	13.5	221 02.7	11.1	14 39.9	6.8	56.1
03	224 11.2 ..	12.7	235 32.8	11.2	14 33.1	6.8	56.1
04	239 11.3	11.9	250 03.0	11.3	14 26.3	7.0	56.1
05	254 11.5	11.0	264 33.3	11.3	14 19.3	6.9	56.1
06	269 11.7	N12 10.2	279 03.6	11.3	N14 12.4	7.1	56.1
07	284 11.8	09.4	293 33.9	11.5	14 05.3	7.1	56.0
F 08	299 12.0	08.5	308 04.4	11.4	13 58.2	7.1	56.0
R 09	314 12.1 ..	07.7	322 34.8	11.6	13 51.1	7.3	56.0
I 10	329 12.3	06.9	337 05.4	11.6	13 43.8	7.3	56.0
D 11	344 12.4	06.0	351 36.0	11.6	13 36.5	7.3	56.0
A 12	359 12.6	N12 05.2	6 06.6	11.7	N13 29.2	7.5	55.9
Y 13	14 12.7	04.4	20 37.3	11.8	13 21.7	7.4	55.9
14	29 12.9	03.5	35 08.1	11.8	13 14.3	7.6	55.9
15	44 13.0 ..	02.7	49 38.9	11.8	13 06.7	7.6	55.9
16	59 13.2	01.9	64 09.7	11.9	12 59.1	7.6	55.8
17	74 13.3	01.0	78 40.6	12.0	12 51.5	7.8	55.8
18	89 13.5	N12 00.2	93 11.6	12.0	N12 43.7	7.7	55.8
19	104 13.6	11 59.4	107 42.6	12.1	12 36.0	7.9	55.8
20	119 13.8	58.5	122 13.7	12.2	12 28.1	7.8	55.8
21	134 14.0 ..	57.7	136 44.9	12.2	12 20.3	8.0	55.7
22	149 14.1	56.9	151 16.1	12.2	12 12.3	8.0	55.7
23	164 14.3	56.0	165 47.3	12.3	N12 04.3	8.0	55.7
22 00	179 14.4	N11 55.2					
01	194 14.6	54.4					
02	209 14.7	53.5	An annular eclipse of				
03	224 14.9 ..	52.7	the Sun occurs on this				
04	239 15.0	51.8	date. See page 5.				
05	254 15.2	51.0					
06	269 15.4	N11 50.2	267 27.5	12.7	N11 07.0	8.4	55.6
07	284 15.5	49.3	281 59.2	12.7	10 58.6	8.4	55.5
S 08	299 15.7	48.5	296 30.9	12.8	10 50.2	8.5	55.5
A 09	314 15.8 ..	47.7	311 02.7	12.8	10 41.7	8.5	55.5
T 10	329 16.0	46.8	325 34.5	12.8	10 33.2	8.5	55.5
U 11	344 16.1	46.0	340 06.3	13.0	10 24.7	8.6	55.5
R 12	359 16.3	N11 45.1	354 38.3	12.9	N10 16.1	8.7	55.4
D 13	14 16.4	44.3	9 10.2	13.0	10 07.4	8.7	55.4
A 14	29 16.6	43.4	23 42.2	13.1	9 58.7	8.7	55.4
Y 15	44 16.8 ..	42.6	38 14.3	13.1	9 50.0	8.8	55.4
16	59 16.9	41.8	52 46.4	13.2	9 41.2	8.8	55.4
17	74 17.1	40.9	67 18.6	13.2	9 32.4	8.8	55.3
18	89 17.2	N11 40.1	81 50.8	13.2	N 9 23.6	8.9	55.3
19	104 17.4	39.2	96 23.0	13.3	9 14.7	8.9	55.3
20	119 17.6	38.4	110 55.3	13.4	9 05.8	8.9	55.3
21	134 17.7 ..	37.6	125 27.1	13.4	8 56.9	9.0	55.3
22	149 17.9	36.7	140 00.1	13.4	8 47.9	9.0	55.2
23	164 18.0	35.9	154 32.5	13.5	N 8 38.9	9.1	55.2
	SD 15.8	d 0.8	SD 15.4		15.2		15.1

Lat.	Naut.	Civil	Sunrise	Moonrise 20	21	22	23
°	h m	h m	h m	h m	h m	h m	h m
N 72	////	////	03 01	25 43	01 43	03 33	05 16
N 70	////	01 42	03 25	00 36	02 15	03 53	05 28
68	////	02 22	03 43	01 10	02 38	04 09	05 38
66	////	02 49	03 57	01 34	02 56	04 21	05 45
64	01 30	03 09	04 09	01 53	03 11	04 32	05 52
62	02 06	03 26	04 19	02 08	03 23	04 40	05 57
60	02 31	03 39	04 28	02 21	03 34	04 48	06 02
N 58	02 50	03 50	04 35	02 32	03 43	04 55	06 07
56	03 05	04 00	04 42	02 42	03 51	05 00	06 10
54	03 18	04 09	04 48	02 51	03 58	05 06	06 14
52	03 29	04 16	04 53	02 58	04 04	05 10	06 17
50	03 39	04 23	04 58	03 05	04 10	05 15	06 20
45	03 59	04 37	05 08	03 20	04 22	05 24	06 26
N 40	04 14	04 48	05 17	03 32	04 32	05 32	06 31
35	04 26	04 58	05 24	03 42	04 40	05 38	06 35
30	04 36	05 06	05 31	03 51	04 48	05 44	06 39
20	04 52	05 19	05 42	04 07	05 01	05 54	06 45
N 10	05 04	05 30	05 51	04 20	05 12	06 03	06 51
0	05 14	05 39	06 00	04 33	05 23	06 11	06 57
S 10	05 22	05 47	06 08	04 45	05 34	06 19	07 02
20	05 29	05 55	06 18	04 59	05 45	06 28	07 08
30	05 35	06 03	06 28	05 14	05 58	06 38	07 15
35	05 38	06 08	06 34	05 23	06 05	06 43	07 18
40	05 41	06 13	06 40	05 33	06 14	06 50	07 23
45	05 44	06 18	06 48	05 45	06 24	06 57	07 28
S 50	05 46	06 24	06 57	06 00	06 36	07 06	07 33
52	05 47	06 27	07 01	06 06	06 41	07 10	07 36
54	05 48	06 30	07 06	06 14	06 47	07 15	07 39
56	05 49	06 33	07 11	06 22	06 54	07 20	07 42
58	05 50	06 36	07 17	06 31	07 01	07 26	07 46
S 60	05 51	06 40	07 23	06 42	07 10	07 32	07 50

Lat.	Sunset	Twilight Civil	Naut.	Moonset 20	21	22	23
°	h m	h m	h m	h m	h m	h m	h m
N 72	21 01	23 42	////	20 50	20 40	20 33	20 26
N 70	20 38	22 16	////	20 17	20 19	20 19	20 19
68	20 21	21 39	////	19 53	20 02	20 09	20 13
66	20 07	21 13	23 44	19 34	19 49	20 00	20 08
64	19 55	20 54	22 30	19 19	19 38	19 52	20 03
62	19 45	20 38	21 56	19 06	19 28	19 45	20 00
60	19 37	20 25	21 32	18 55	19 20	19 40	19 57
N 58	19 29	20 14	21 14	18 45	19 12	19 35	19 54
56	19 23	20 05	20 59	18 37	19 06	19 30	19 51
54	19 17	19 56	20 46	18 29	19 00	19 26	19 49
52	19 12	19 49	20 35	18 23	18 55	19 22	19 47
50	19 07	19 42	20 26	18 17	18 50	19 19	19 45
45	18 57	19 28	20 06	18 04	18 40	19 12	19 40
N 40	18 49	19 17	19 51	17 53	18 31	19 05	19 37
35	18 42	19 08	19 39	17 44	18 24	19 00	19 34
30	18 35	19 00	19 29	17 35	18 17	18 55	19 31
20	18 24	18 47	19 14	17 21	18 06	18 47	19 26
N 10	18 15	18 37	19 02	17 09	17 56	18 40	19 22
0	18 07	18 28	18 52	16 57	17 46	18 33	19 18
S 10	17 58	18 19	18 44	16 45	17 37	18 26	19 14
20	17 49	18 11	18 37	16 33	17 27	18 19	19 10
30	17 39	18 03	18 31	16 18	17 15	18 11	19 05
35	17 33	17 59	18 29	16 10	17 08	18 06	19 02
40	17 26	17 54	18 26	16 00	17 01	18 00	18 59
45	17 19	17 49	18 23	15 49	16 52	17 54	18 55
S 50	17 10	17 43	18 21	15 35	16 41	17 46	18 50
52	17 06	17 40	18 20	15 29	16 35	17 42	18 48
54	17 01	17 38	18 19	15 22	16 30	17 38	18 46
56	16 56	17 35	18 18	15 14	16 24	17 34	18 43
58	16 50	17 31	18 17	15 05	16 17	17 29	18 41
S 60	16 44	17 27	18 16	14 55	16 09	17 23	18 37

	SUN			MOON			
Day	Eqn. of Time 00h	12h	Mer. Pass.	Mer. Pass. Upper	Lower	Age	Phase
d	m s	m s	h m	h m	h m	d	%
20	03 32	03 24	12 03	10 45	23 10	28	3
21	03 17	03 10	12 03	11 35	23 59	29	0
22	03 03	02 55	12 03	12 22	24 45	00	0

1998 AUGUST 23, 24, 25 (SUN., MON., TUES.)

UT	ARIES GHA	VENUS −3.9 GHA	VENUS Dec	MARS +1.7 GHA	MARS Dec	JUPITER −2.9 GHA	JUPITER Dec	SATURN +0.3 GHA	SATURN Dec	STARS Name	SHA	Dec
d h	° ′	° ′	° ′	° ′	° ′	° ′	° ′	° ′	° ′		° ′	° ′
23 00	331 05.1	196 38.6	N17 58.5	207 11.9	N20 48.9	334 06.9	S 2 57.5	298 50.0	N10 14.9	Acamar	315 26.9	S40 18.4
01	346 07.6	211 37.9	57.7	222 12.7	48.6	349 09.6	57.7	313 52.5	14.9	Achernar	335 35.0	S57 14.4
02	1 10.0	226 37.3	56.9	237 13.5	48.2	4 12.3	57.8	328 55.0	14.8	Acrux	173 22.6	S63 05.6
03	16 12.5	241 36.6	.. 56.2	252 14.2	.. 47.9	19 15.0	.. 57.9	343 57.5	.. 14.8	Adhara	255 21.7	S28 58.1
04	31 14.9	256 35.9	55.4	267 15.0	47.6	34 17.8	58.0	359 00.0	14.8	Aldebaran	291 02.5	N16 30.3
05	46 17.4	271 35.3	54.6	282 15.8	47.2	49 20.5	58.1	14 02.4	14.8			
06	61 19.9	286 34.6	N17 53.8	297 16.6	N20 46.9	64 23.2	S 2 58.2	29 04.9	N10 14.8	Alioth	166 31.1	N55 58.4
07	76 22.3	301 34.0	53.1	312 17.4	46.5	79 25.9	58.3	44 07.4	14.7	Alkaid	153 08.2	N49 19.6
08	91 24.8	316 33.3	52.3	327 18.2	46.2	94 28.6	58.4	59 09.9	14.7	Al Na'ir	27 57.6	S46 57.9
S 09	106 27.3	331 32.7	.. 51.5	342 18.9	.. 45.9	109 31.3	.. 58.6	74 12.4	.. 14.7	Alnilam	275 58.1	S 1 12.2
U 10	121 29.7	346 32.0	50.7	357 19.7	45.5	124 34.0	58.7	89 14.9	14.7	Alphard	218 07.6	S 8 39.1
N 11	136 32.2	1 31.4	50.0	12 20.5	45.2	139 36.7	58.8	104 17.4	14.7			
D 12	151 34.7	16 30.7	N17 49.2	27 21.3	N20 44.8	154 39.4	S 2 58.9	119 19.9	N10 14.6	Alphecca	126 20.7	N26 43.5
A 13	166 37.1	31 30.0	48.4	42 22.1	44.5	169 42.1	59.0	134 22.4	14.6	Alpheratz	357 55.0	N29 04.9
Y 14	181 39.6	46 29.4	47.6	57 22.9	44.2	184 44.8	59.1	149 24.9	14.6	Altair	62 19.1	N 8 52.1
15	196 42.1	61 28.7	.. 46.8	72 23.6	.. 43.8	199 47.5	.. 59.2	164 27.4	.. 14.6	Ankaa	353 26.6	S42 18.7
16	211 44.5	76 28.1	46.1	87 24.4	43.5	214 50.2	59.4	179 29.9	14.6	Antares	112 40.2	S26 25.6
17	226 47.0	91 27.4	45.3	102 25.2	43.1	229 52.9	59.5	194 32.4	14.5			
18	241 49.4	106 26.8	N17 44.5	117 26.0	N20 42.8	244 55.6	S 2 59.6	209 34.8	N10 14.5	Arcturus	146 06.3	N19 11.7
19	256 51.9	121 26.1	43.7	132 26.8	42.5	259 58.3	59.7	224 37.3	14.5	Atria	107 52.1	S69 01.6
20	271 54.4	136 25.5	42.9	147 27.6	42.1	275 01.1	59.8	239 39.8	14.5	Avior	234 23.2	S59 30.2
21	286 56.8	151 24.8	.. 42.1	162 28.4	.. 41.8	290 03.8	2 59.9	254 42.3	.. 14.4	Bellatrix	278 44.4	N 6 20.9
22	301 59.3	166 24.2	41.3	177 29.1	41.4	305 06.5	3 00.0	269 44.8	14.4	Betelgeuse	271 13.8	N 7 24.4
23	317 01.8	181 23.5	40.6	192 29.9	41.1	320 09.2	00.2	284 47.3	14.4			
24 00	332 04.2	196 22.9	N17 39.8	207 30.7	N20 40.7	335 11.9	S 3 00.3	299 49.8	N10 14.4	Canopus	264 01.5	S52 41.5
01	347 06.7	211 22.2	39.0	222 31.5	40.4	350 14.6	00.4	314 52.3	14.4	Capella	280 51.5	N45 59.4
02	2 09.2	226 21.6	38.2	237 32.3	40.1	5 17.3	00.5	329 54.8	14.3	Deneb	49 38.8	N45 16.8
03	17 11.6	241 20.9	.. 37.4	252 33.1	.. 39.7	20 20.0	.. 00.6	344 57.3	.. 14.3	Denebola	182 45.6	N14 35.0
04	32 14.1	256 20.3	36.6	267 33.9	39.4	35 22.7	00.7	359 59.8	14.3	Diphda	349 07.1	S17 59.5
05	47 16.5	271 19.6	35.8	282 34.7	39.0	50 25.4	00.8	15 02.3	14.3			
06	62 19.0	286 19.0	N17 35.0	297 35.4	N20 38.7	65 28.1	S 3 01.0	30 04.8	N10 14.2	Dubhe	194 06.3	N61 45.6
07	77 21.5	301 18.3	34.2	312 36.2	38.3	80 30.9	01.1	45 07.3	14.2	Elnath	278 27.2	N28 36.2
08	92 23.9	316 17.7	33.4	327 37.0	38.0	95 33.6	01.2	60 09.8	14.2	Eltanin	90 51.2	N51 29.8
M 09	107 26.4	331 17.0	.. 32.6	342 37.8	.. 37.6	110 36.3	.. 01.3	75 12.3	.. 14.2	Enif	33 58.0	N 9 52.3
O 10	122 28.9	346 16.4	31.8	357 38.6	37.3	125 39.0	01.4	90 14.8	14.2	Fomalhaut	15 36.2	S29 37.6
N 11	137 31.3	1 15.7	31.0	12 39.4	37.0	140 41.7	01.5	105 17.3	14.1			
D 12	152 33.8	16 15.1	N17 30.2	27 40.2	N20 36.6	155 44.4	S 3 01.6	120 19.8	N10 14.1	Gacrux	172 14.0	S57 06.4
A 13	167 36.3	31 14.5	29.4	42 41.0	36.3	170 47.1	01.8	135 22.3	14.1	Gienah	176 04.3	S17 31.9
Y 14	182 38.7	46 13.8	28.6	57 41.7	35.9	185 49.8	01.9	150 24.8	14.1	Hadar	149 04.4	S60 22.1
15	197 41.2	61 13.2	.. 27.8	72 42.5	.. 35.6	200 52.5	.. 02.0	165 27.3	.. 14.0	Hamal	328 13.5	N23 27.3
16	212 43.7	76 12.5	27.0	87 43.3	35.2	215 55.2	02.1	180 29.8	14.0	Kaus Aust.	83 58.7	S34 23.0
17	227 46.1	91 11.9	26.2	102 44.1	34.9	230 58.0	02.2	195 32.3	14.0			
18	242 48.6	106 11.2	N17 25.4	117 44.9	N20 34.5	246 00.7	S 3 02.3	210 34.8	N10 14.0	Kochab	137 20.0	N74 10.1
19	257 51.0	121 10.6	24.6	132 45.7	34.2	261 03.4	02.5	225 37.2	13.9	Markab	13 49.4	N15 12.0
20	272 53.5	136 10.0	23.8	147 46.5	33.8	276 06.1	02.6	240 39.7	13.9	Menkar	314 26.9	N 4 05.0
21	287 56.0	151 09.3	.. 23.0	162 47.3	.. 33.5	291 08.8	.. 02.7	255 42.2	.. 13.9	Menkent	148 21.2	S36 21.8
22	302 58.4	166 08.7	22.2	177 48.1	33.1	306 11.5	02.8	270 44.7	13.9	Miaplacidus	221 42.9	S69 42.6
23	318 00.9	181 08.0	21.4	192 48.9	32.8	321 14.2	02.9	285 47.2	13.9			
25 00	333 03.4	196 07.4	N17 20.6	207 49.7	N20 32.4	336 17.0	S 3 03.0	300 49.7	N10 13.8	Mirfak	308 56.6	N49 51.1
01	348 05.8	211 06.8	19.8	222 50.4	32.1	351 19.7	03.2	315 52.2	13.8	Nunki	76 12.2	S26 17.8
02	3 08.3	226 06.1	18.9	237 51.2	31.7	6 22.4	03.3	330 54.7	13.8	Peacock	53 36.6	S56 44.3
03	18 10.8	241 05.5	.. 18.1	252 52.0	.. 31.4	21 25.1	.. 03.4	345 57.2	.. 13.8	Pollux	243 42.0	N28 01.7
04	33 13.2	256 04.8	17.3	267 52.8	31.0	36 27.8	03.5	0 59.7	13.7	Procyon	245 11.9	N 5 13.7
05	48 15.7	271 04.2	16.5	282 53.6	30.7	51 30.5	03.6	16 02.2	13.7			
06	63 18.1	286 03.6	N17 15.7	297 54.4	N20 30.3	66 33.2	S 3 03.7	31 04.7	N10 13.7	Rasalhague	96 16.9	N12 34.0
07	78 20.6	301 02.9	14.9	312 55.2	30.0	81 36.0	03.9	46 07.2	13.7	Regulus	207 56.0	N11 58.5
08	93 23.1	316 02.3	14.1	327 56.0	29.6	96 38.7	04.0	61 09.7	13.6	Rigel	281 23.1	S 8 12.2
T 09	108 25.5	331 01.7	.. 13.2	342 56.8	.. 29.3	111 41.4	.. 04.1	76 12.2	.. 13.6	Rigil Kent.	140 07.6	S60 49.8
U 10	123 28.0	346 01.0	12.4	357 57.6	28.9	126 44.1	04.2	91 14.7	13.6	Sabik	102 25.5	S15 43.2
E 11	138 30.5	1 00.4	11.6	12 58.4	28.6	141 46.8	04.3	106 17.2	13.6			
S 12	153 32.9	15 59.7	N17 10.8	27 59.2	N20 28.2	156 49.5	S 3 04.4	121 19.7	N10 13.5	Schedar	349 53.0	N56 31.6
D 13	168 35.4	30 59.1	10.0	43 00.0	27.9	171 52.2	04.6	136 22.2	13.5	Shaula	96 37.2	S37 06.1
A 14	183 37.9	45 58.5	09.1	58 00.7	27.5	186 55.0	04.7	151 24.7	13.5	Sirius	258 44.0	S16 42.8
Y 15	198 40.3	60 57.8	.. 08.3	73 01.5	.. 27.2	201 57.7	.. 04.8	166 27.2	.. 13.5	Spica	158 43.5	S11 09.1
16	213 42.8	75 57.2	07.5	88 02.3	26.8	217 00.4	04.9	181 29.7	13.5	Suhail	223 01.2	S43 25.5
17	228 45.3	90 56.6	06.7	103 03.1	26.5	232 03.1	05.0	196 32.3	13.4			
18	243 47.7	105 55.9	N17 05.8	118 03.9	N20 26.1	247 05.8	S 3 05.1	211 34.8	N10 13.4	Vega	80 46.4	N38 47.3
19	258 50.2	120 55.3	05.0	133 04.7	25.8	262 08.5	05.3	226 37.3	13.4	Zuben'ubi	137 18.1	S16 02.0
20	273 52.6	135 54.7	04.2	148 05.5	25.4	277 11.3	05.4	241 39.8	13.4		SHA	Mer. Pass.
21	288 55.1	150 54.0	.. 03.4	163 06.3	.. 25.0	292 14.0	.. 05.5	256 42.3	.. 13.3		° ′	h m
22	303 57.6	165 53.4	02.5	178 07.1	24.7	307 16.7	05.6	271 44.8	13.3	Venus	224 18.6	10 55
23	319 00.0	180 52.8	01.7	193 07.9	24.3	322 19.4	05.7	286 47.3	13.3	Mars	235 26.5	10 09
	h m									Jupiter	3 07.7	1 39
Mer. Pass. 1 51.4	v −0.6 d 0.8		v 0.8 d 0.3		v 2.7 d 0.1		v 2.5 d 0.0		Saturn	327 45.6	4 00

UT	SUN GHA	SUN Dec	MOON GHA	v	Dec	d	HP
d h	° ′	° ′	° ′	′	° ′	′	′
23 00	179 18.2	N11 35.0	169 05.0	13.5	N 8 29.8	9.0	55.2
01	194 18.4	34.2	183 37.5	13.6	8 20.8	9.1	55.2
02	209 18.5	33.3	198 10.1	13.6	8 11.7	9.2	55.2
03	224 18.7 ..	32.5	212 42.7	13.7	8 02.5	9.1	55.2
04	239 18.9	31.6	227 15.4	13.7	7 53.4	9.2	55.1
05	254 19.0	30.8	241 48.1	13.7	7 44.2	9.3	55.1
06	269 19.2	N11 29.9	256 20.8	13.8	N 7 34.9	9.2	55.1
07	284 19.3	29.1	270 53.6	13.9	7 25.7	9.3	55.1
S 08	299 19.5	28.3	285 26.5	13.8	7 16.4	9.3	55.1
U 09	314 19.7 ..	27.4	299 59.3	13.9	7 07.1	9.3	55.1
N 10	329 19.8	26.6	314 32.2	14.0	6 57.8	9.4	55.1
D 11	344 20.0	25.7	329 05.2	14.0	6 48.4	9.3	55.0
A 12	359 20.1	N11 24.9	343 38.2	14.0	N 6 39.1	9.4	55.0
Y 13	14 20.3	24.0	358 11.2	14.1	6 29.7	9.5	55.0
14	29 20.5	23.2	12 44.3	14.1	6 20.2	9.4	55.0
15	44 20.6 ..	22.3	27 17.4	14.1	6 10.8	9.5	55.0
16	59 20.8	21.5	41 50.5	14.2	6 01.3	9.5	54.9
17	74 21.0	20.6	56 23.7	14.2	5 51.8	9.5	54.9
18	89 21.1	N11 19.8	70 56.9	14.2	N 5 42.3	9.5	54.9
19	104 21.3	18.9	85 30.1	14.3	5 32.8	9.5	54.9
20	119 21.5	18.1	100 03.4	14.3	5 23.3	9.6	54.9
21	134 21.6 ..	17.2	114 36.7	14.4	5 13.7	9.5	54.9
22	149 21.8	16.4	129 10.1	14.3	5 04.2	9.6	54.8
23	164 21.9	15.5	143 43.4	14.5	4 54.6	9.6	54.8
24 00	179 22.1	N11 14.7	158 16.9	14.4	N 4 45.0	9.6	54.8
01	194 22.3	13.8	172 50.3	14.5	4 35.4	9.7	54.8
02	209 22.4	13.0	187 23.8	14.5	4 25.7	9.6	54.8
03	224 22.6 ..	12.1	201 57.3	14.5	4 16.1	9.7	54.8
04	239 22.8	11.3	216 30.8	14.6	4 06.4	9.6	54.7
05	254 22.9	10.4	231 04.4	14.6	3 56.8	9.7	54.7
06	269 23.1	N11 09.5	245 38.0	14.6	N 3 47.1	9.7	54.7
07	284 23.3	08.7	260 11.6	14.7	3 37.4	9.7	54.7
M 08	299 23.4	07.8	274 45.3	14.6	3 27.7	9.7	54.7
O 09	314 23.6 ..	07.0	289 18.9	14.8	3 18.0	9.7	54.7
N 10	329 23.8	06.1	303 52.7	14.7	3 08.3	9.7	54.7
D 11	344 23.9	05.3	318 26.4	14.7	2 58.6	9.7	54.6
A 12	359 24.1	N11 04.4	333 00.1	14.8	N 2 48.9	9.8	54.6
Y 13	14 24.3	03.6	347 33.9	14.8	2 39.1	9.7	54.6
14	29 24.4	02.7	2 07.7	14.9	2 29.4	9.7	54.6
15	44 24.6 ..	01.8	16 41.6	14.8	2 19.7	9.8	54.6
16	59 24.8	01.0	31 15.4	14.9	2 09.9	9.7	54.6
17	74 24.9	11 00.1	45 49.3	14.9	2 00.2	9.8	54.6
18	89 25.1	N10 59.3	60 23.2	14.9	N 1 50.4	9.7	54.5
19	104 25.3	58.4	74 57.1	14.9	1 40.7	9.8	54.5
20	119 25.4	57.6	89 31.0	15.0	1 30.9	9.8	54.5
21	134 25.6 ..	56.7	104 05.0	15.0	1 21.1	9.7	54.5
22	149 25.8	55.8	118 39.0	15.0	1 11.4	9.8	54.5
23	164 25.9	55.0	133 13.0	15.0	1 01.6	9.7	54.5
25 00	179 26.1	N10 54.1	147 47.0	15.0	N 0 51.9	9.8	54.5
01	194 26.3	53.3	162 21.0	15.1	0 42.1	9.7	54.5
02	209 26.5	52.4	176 55.1	15.0	0 32.4	9.8	54.4
03	224 26.6 ..	51.5	191 29.1	15.1	0 22.6	9.7	54.4
04	239 26.8	50.7	206 03.2	15.1	0 12.9	9.8	54.4
05	254 27.0	49.8	220 37.3	15.1	N 0 03.1	9.7	54.4
06	269 27.1	N10 49.0	235 11.4	15.1	S 0 06.6	9.7	54.4
07	284 27.3	48.1	249 45.5	15.2	0 16.3	9.8	54.4
T 08	299 27.5	47.2	264 19.7	15.1	0 26.1	9.7	54.4
U 09	314 27.6 ..	46.4	278 53.8	15.2	0 35.8	9.7	54.4
E 10	329 27.8	45.5	293 28.0	15.2	0 45.5	9.7	54.4
S 11	344 28.0	44.7	308 02.2	15.1	0 55.2	9.7	54.4
D 12	359 28.2	N10 43.8	322 36.3	15.2	S 1 04.9	9.7	54.3
A 13	14 28.3	42.9	337 10.5	15.2	1 14.6	9.7	54.3
Y 14	29 28.5	42.1	351 44.7	15.2	1 24.3	9.6	54.3
15	44 28.7 ..	41.2	6 18.9	15.2	1 33.9	9.7	54.3
16	59 28.8	40.3	20 53.1	15.3	1 43.6	9.6	54.3
17	74 29.0	39.5	35 27.4	15.2	1 53.2	9.7	54.3
18	89 29.2	N10 38.6	50 01.6	15.2	S 2 02.9	9.6	54.3
19	104 29.4	37.7	64 35.8	15.3	2 12.5	9.6	54.3
20	119 29.5	36.9	79 10.1	15.2	2 22.1	9.6	54.3
21	134 29.7 ..	36.0	93 44.3	15.3	2 31.7	9.6	54.2
22	149 29.9	35.1	108 18.6	15.2	2 41.3	9.6	54.2
23	164 30.1	34.3	122 52.8	15.3	S 2 50.9	9.5	54.2
	SD 15.8	d 0.9	SD 15.0		14.9		14.8

Lat.	Twilight Naut.	Twilight Civil	Sunrise	Moonrise 23	24	25	26
°	h m	h m	h m	h m	h m	h m	h m
N 72	////	01 11	03 18	05 16	06 55	08 31	10 06
N 70	////	02 07	03 39	05 28	07 00	08 30	09 59
68	////	02 39	03 55	05 38	07 04	08 30	09 54
66	01 01	03 03	04 08	05 45	07 08	08 29	09 49
64	01 51	03 21	04 18	05 52	07 11	08 29	09 45
62	02 20	03 35	04 27	05 57	07 13	08 28	09 42
60	02 42	03 47	04 35	06 02	07 16	08 28	09 39
N 58	02 59	03 58	04 42	06 07	07 18	08 27	09 36
56	03 13	04 07	04 48	06 10	07 19	08 27	09 34
54	03 25	04 15	04 53	06 14	07 21	08 27	09 32
52	03 36	04 22	04 58	06 17	07 22	08 27	09 30
50	03 45	04 28	05 03	06 20	07 24	08 26	09 29
45	04 03	04 41	05 12	06 26	07 26	08 26	09 25
N 40	04 17	04 51	05 20	06 31	07 29	08 26	09 22
35	04 29	05 00	05 26	06 35	07 31	08 25	09 19
30	04 38	05 08	05 32	06 39	07 32	08 25	09 17
20	04 53	05 20	05 42	06 45	07 36	08 25	09 13
N 10	05 05	05 30	05 51	06 51	07 38	08 24	09 09
0	05 14	05 38	05 59	06 57	07 41	08 24	09 06
S 10	05 21	05 46	06 07	07 02	07 43	08 23	09 03
20	05 27	05 53	06 15	07 08	07 46	08 23	09 00
30	05 32	06 00	06 25	07 15	07 49	08 23	08 56
35	05 35	06 04	06 30	07 18	07 51	08 22	08 54
40	05 37	06 09	06 36	07 23	07 53	08 22	08 51
45	05 39	06 13	06 43	07 28	07 55	08 22	08 48
S 50	05 41	06 18	06 51	07 33	07 58	08 22	08 45
52	05 41	06 21	06 55	07 36	07 59	08 21	08 43
54	05 42	06 23	06 59	07 39	08 01	08 21	08 42
56	05 43	06 26	07 04	07 42	08 02	08 21	08 40
58	05 43	06 29	07 09	07 46	08 04	08 21	08 38
S 60	05 44	06 32	07 15	07 50	08 06	08 21	08 35

Lat.	Sunset	Twilight Civil	Twilight Naut.	Moonset 23	24	25	26
°	h m	h m	h m	h m	h m	h m	h m
N 72	20 42	22 41	////	20 26	20 19	20 13	20 06
N 70	20 23	21 52	////	20 19	20 17	20 16	20 15
68	20 07	21 21	////	20 13	20 16	20 19	20 22
66	19 55	20 59	22 52	20 08	20 15	20 21	20 28
64	19 44	20 41	22 08	20 03	20 14	20 23	20 33
62	19 36	20 27	21 40	20 00	20 13	20 25	20 38
60	19 28	20 15	21 20	19 57	20 12	20 26	20 41
N 58	19 21	20 05	21 03	19 54	20 11	20 28	20 45
56	19 15	19 56	20 49	19 51	20 10	20 29	20 48
54	19 10	19 49	20 37	19 49	20 10	20 30	20 51
52	19 06	19 42	20 27	19 47	20 09	20 31	20 53
50	19 01	19 36	20 19	19 45	20 09	20 32	20 56
45	18 52	19 23	20 00	19 40	20 08	20 34	21 01
N 40	18 44	19 13	19 46	19 37	20 07	20 36	21 05
35	18 38	19 04	19 35	19 34	20 06	20 37	21 08
30	18 32	18 57	19 26	19 31	20 05	20 38	21 12
20	18 22	18 45	19 11	19 26	20 04	20 40	21 17
N 10	18 14	18 35	19 00	19 22	20 03	20 42	21 22
0	18 06	18 27	18 51	19 18	20 02	20 44	21 27
S 10	17 58	18 19	18 44	19 14	20 00	20 46	21 31
20	17 50	18 12	18 38	19 10	19 59	20 48	21 36
30	17 41	18 05	18 33	19 05	19 58	20 50	21 42
35	17 35	18 01	18 31	19 02	19 57	20 51	21 45
40	17 29	17 57	18 28	18 59	19 56	20 53	21 49
45	17 22	17 52	18 27	18 55	19 55	20 54	21 53
S 50	17 14	17 47	18 25	18 50	19 54	20 56	21 58
52	17 10	17 45	18 24	18 48	19 53	20 57	22 01
54	17 06	17 43	18 24	18 46	19 53	20 58	22 03
56	17 02	17 40	18 23	18 43	19 52	20 59	22 06
58	16 57	17 37	18 23	18 41	19 51	21 01	22 09
S 60	16 51	17 34	18 22	18 37	19 50	21 02	22 13

Day	SUN Eqn. of Time 00h	12h	SUN Mer. Pass.	MOON Mer. Pass. Upper	Lower	Age	Phase
d	m s	m s	h m	h m	h m	d	%
23	02 47	02 40	12 03	13 07	00 45	01	2
24	02 32	02 24	12 02	13 51	01 30	02	6
25	02 16	02 08	12 02	14 34	02 13	03	11

1998 AUGUST 26, 27, 28 (WED., THURS., FRI.)

UT	ARIES GHA	VENUS −3.9 GHA	Dec	MARS +1.7 GHA	Dec	JUPITER −2.9 GHA	Dec	SATURN +0.3 GHA	Dec
26 00	334 02.5	195 52.2	N17 00.9	208 08.7	N20 24.0	337 22.1	S 3 05.9	301 49.8	N10 13.3
01	349 05.0	210 51.5	17 00.0	223 09.5	23.6	352 24.8	06.0	316 52.3	13.2
02	4 07.4	225 50.9	16 59.2	238 10.3	23.3	7 27.6	06.1	331 54.8	13.2
03	19 09.9	240 50.3	.. 58.4	253 11.1	.. 22.9	22 30.3	.. 06.2	346 57.3	.. 13.2
04	34 12.4	255 49.6	57.5	268 11.9	22.6	37 33.0	06.3	1 59.8	13.2
05	49 14.8	270 49.0	56.7	283 12.7	22.2	52 35.7	06.4	17 02.3	13.1
06	64 17.3	285 48.4	N16 55.9	298 13.5	N20 21.9	67 38.4	S 3 06.6	32 04.8	N10 13.1
W 07	79 19.8	300 47.8	55.0	313 14.3	21.5	82 41.2	06.7	47 07.3	13.1
E 08	94 22.2	315 47.1	54.2	328 15.1	21.1	97 43.9	06.8	62 09.8	13.1
D 09	109 24.7	330 46.5	.. 53.4	343 15.9	.. 20.8	112 46.6	.. 06.9	77 12.3	.. 13.0
N 10	124 27.1	345 45.9	52.5	358 16.7	20.4	127 49.3	07.0	92 14.8	13.0
E 11	139 29.6	0 45.2	51.7	13 17.5	20.1	142 52.0	07.2	107 17.3	13.0
S 12	154 32.1	15 44.6	N16 50.8	28 18.3	N20 19.7	157 54.8	S 3 07.3	122 19.8	N10 13.0
D 13	169 34.5	30 44.0	50.0	43 19.1	19.4	172 57.5	07.4	137 22.3	12.9
A 14	184 37.0	45 43.4	49.2	58 19.9	19.0	188 00.2	07.5	152 24.8	12.9
Y 15	199 39.5	60 42.7	.. 48.3	73 20.7	.. 18.6	203 02.9	.. 07.6	167 27.3	.. 12.9
16	214 41.9	75 42.1	47.5	88 21.5	18.3	218 05.6	07.8	182 29.8	12.9
17	229 44.4	90 41.5	46.6	103 22.3	17.9	233 08.4	07.9	197 32.3	12.8
18	244 46.9	105 40.9	N16 45.8	118 23.1	N20 17.6	248 11.1	S 3 08.0	212 34.9	N10 12.8
19	259 49.3	120 40.3	44.9	133 23.9	17.2	263 13.8	08.1	227 37.4	12.8
20	274 51.8	135 39.6	44.1	148 24.7	16.8	278 16.5	08.2	242 39.9	12.7
21	289 54.2	150 39.0	.. 43.2	163 25.5	.. 16.5	293 19.3	.. 08.4	257 42.4	.. 12.7
22	304 56.7	165 38.4	42.4	178 26.3	16.1	308 22.0	08.5	272 44.9	12.7
23	319 59.2	180 37.8	41.5	193 27.1	15.8	323 24.7	08.6	287 47.4	12.7
27 00	335 01.6	195 37.2	N16 40.7	208 27.9	N20 15.4	338 27.4	S 3 08.7	302 49.9	N10 12.6
01	350 04.1	210 36.5	39.8	223 28.7	15.0	353 30.1	08.8	317 52.4	12.6
02	5 06.6	225 35.9	39.0	238 29.5	14.7	8 32.9	08.9	332 54.9	12.6
03	20 09.0	240 35.3	.. 38.1	253 30.3	.. 14.3	23 35.6	.. 09.1	347 57.4	.. 12.6
04	35 11.5	255 34.7	37.3	268 31.1	14.0	38 38.3	09.2	2 59.9	12.5
05	50 14.0	270 34.1	36.4	283 31.9	13.6	53 41.0	09.3	18 02.4	12.5
06	65 16.4	285 33.4	N16 35.6	298 32.7	N20 13.2	68 43.8	S 3 09.4	33 04.9	N10 12.5
T 07	80 18.9	300 32.8	34.7	313 33.5	12.9	83 46.5	09.6	48 07.4	12.5
H 08	95 21.4	315 32.2	33.9	328 34.3	12.5	98 49.2	09.7	63 10.0	12.4
U 09	110 23.8	330 31.6	.. 33.0	343 35.1	.. 12.1	113 51.9	.. 09.8	78 12.5	.. 12.4
R 10	125 26.3	345 31.0	32.2	358 35.9	11.8	128 54.7	09.9	93 15.0	12.4
S 11	140 28.7	0 30.4	31.3	13 36.7	11.4	143 57.4	10.0	108 17.5	12.4
D 12	155 31.2	15 29.7	N16 30.4	28 37.5	N20 11.1	159 00.1	S 3 10.2	123 20.0	N10 12.3
A 13	170 33.7	30 29.1	29.6	43 38.3	10.7	174 02.8	10.3	138 22.5	12.3
Y 14	185 36.1	45 28.5	28.7	58 39.1	10.3	189 05.6	10.4	153 25.0	12.3
15	200 38.6	60 27.9	.. 27.9	73 39.9	.. 10.0	204 08.3	.. 10.5	168 27.5	.. 12.2
16	215 41.1	75 27.3	27.0	88 40.7	09.6	219 11.0	10.6	183 30.0	12.2
17	230 43.5	90 26.7	26.1	103 41.5	09.2	234 13.7	10.8	198 32.5	12.2
18	245 46.0	105 26.1	N16 25.3	118 42.3	N20 08.9	249 16.5	S 3 10.9	213 35.1	N10 12.2
19	260 48.5	120 25.4	24.4	133 43.1	08.5	264 19.2	11.0	228 37.6	12.1
20	275 50.9	135 24.8	23.5	148 43.9	08.1	279 21.9	11.1	243 40.1	12.1
21	290 53.4	150 24.2	.. 22.7	163 44.7	.. 07.8	294 24.6	.. 11.2	258 42.6	.. 12.1
22	305 55.8	165 23.6	21.8	178 45.5	07.4	309 27.4	11.4	273 45.1	12.1
23	320 58.3	180 23.0	20.9	193 46.3	07.0	324 30.1	11.5	288 47.6	12.0
28 00	336 00.8	195 22.4	N16 20.1	208 47.1	N20 06.7	339 32.8	S 3 11.6	303 50.1	N10 12.0
01	351 03.2	210 21.8	19.2	223 48.0	06.3	354 35.5	11.7	318 52.6	12.0
02	6 05.7	225 21.2	18.3	238 48.8	05.9	9 38.3	11.8	333 55.1	11.9
03	21 08.2	240 20.6	.. 17.4	253 49.6	.. 05.6	24 41.0	.. 12.0	348 57.7	.. 11.9
04	36 10.6	255 19.9	16.6	268 50.4	05.2	39 43.7	12.1	4 00.2	11.9
05	51 13.1	270 19.3	15.7	283 51.2	04.8	54 46.5	12.2	19 02.7	11.9
06	66 15.6	285 18.7	N16 14.8	298 52.0	N20 04.5	69 49.2	S 3 12.3	34 05.2	N10 11.8
07	81 18.0	300 18.1	13.9	313 52.8	04.1	84 51.9	12.5	49 07.7	11.8
08	96 20.5	315 17.5	13.1	328 53.6	03.7	99 54.6	12.6	64 10.2	11.8
F 09	111 23.0	330 16.9	.. 12.2	343 54.4	.. 03.4	114 57.4	.. 12.7	79 12.7	.. 11.8
R 10	126 25.4	345 16.3	11.3	358 55.2	03.0	130 00.1	12.8	94 15.2	11.7
I 11	141 27.9	0 15.7	10.4	13 56.0	02.6	145 02.8	12.9	109 17.8	11.7
D 12	156 30.3	15 15.1	N16 09.6	28 56.8	N20 02.3	160 05.5	S 3 13.1	124 20.3	N10 11.7
A 13	171 32.8	30 14.5	08.7	43 57.6	01.9	175 08.3	13.2	139 22.8	11.6
Y 14	186 35.3	45 13.9	07.8	58 58.4	01.5	190 11.0	13.3	154 25.3	11.6
15	201 37.7	60 13.3	.. 06.9	73 59.3	.. 01.2	205 13.7	.. 13.4	169 27.8	.. 11.6
16	216 40.2	75 12.7	06.0	89 00.1	00.8	220 16.5	13.6	184 30.3	11.6
17	231 42.7	90 12.1	05.2	104 00.9	00.4	235 19.2	13.7	199 32.8	11.5
18	246 45.1	105 11.5	N16 04.3	119 01.7	N20 00.0	250 21.9	S 3 13.8	214 35.4	N10 11.5
19	261 47.6	120 10.9	03.4	134 02.5	19 59.7	265 24.7	13.9	229 37.9	11.5
20	276 50.1	135 10.3	02.5	149 03.3	59.3	280 27.4	14.0	244 40.4	11.4
21	291 52.5	150 09.7	.. 01.6	164 04.1	.. 58.9	295 30.1	.. 14.2	259 42.9	.. 11.4
22	306 55.0	165 09.1	16 00.7	179 04.9	58.6	310 32.8	14.3	274 45.4	11.4
23	321 57.5	180 08.5	N15 59.8	194 05.7	58.2	325 35.6	14.4	289 47.9	11.4
Mer. Pass.	h m 1 39.6	v −0.6 d 0.9		v 0.8 d 0.4		v 2.7 d 0.1		v 2.5 d 0.0	

STARS

Name	SHA	Dec
Acamar	315 26.9	S40 18.4
Achernar	335 34.9	S57 14.4
Acrux	173 22.6	S63 05.5
Adhara	255 21.7	S28 58.1
Aldebaran	291 02.5	N16 30.3
Alioth	166 31.2	N55 58.3
Alkaid	153 08.2	N49 19.6
Al Na'ir	27 57.6	S46 58.0
Alnilam	275 58.1	S 1 12.2
Alphard	218 07.6	S 8 39.1
Alphecca	126 20.7	N26 43.5
Alpheratz	357 55.0	N29 05.0
Altair	62 19.1	N 8 52.1
Ankaa	353 26.6	S42 18.7
Antares	112 40.2	S26 25.6
Arcturus	146 06.3	N19 11.7
Atria	107 52.2	S69 01.6
Avior	234 23.2	S59 30.2
Bellatrix	278 44.3	N 6 20.9
Betelgeuse	271 13.8	N 7 24.4
Canopus	264 01.5	S52 41.5
Capella	280 51.4	N45 59.5
Deneb	49 38.8	N45 16.8
Denebola	182 45.6	N14 35.0
Diphda	349 07.1	S17 59.5
Dubhe	194 06.3	N61 45.6
Elnath	278 27.2	N28 36.2
Eltanin	90 51.2	N51 29.8
Enif	33 58.0	N 9 52.3
Fomalhaut	15 36.2	S29 37.6
Gacrux	172 14.1	S57 06.4
Gienah	176 04.3	S17 31.9
Hadar	149 04.4	S60 22.1
Hamal	328 13.4	N23 27.3
Kaus Aust.	83 58.7	S34 23.0
Kochab	137 20.1	N74 10.1
Markab	13 49.4	N15 12.0
Menkar	314 26.9	N 4 05.0
Menkent	148 21.2	S36 21.8
Miaplacidus	221 42.9	S69 42.6
Mirfak	308 56.6	N49 51.1
Nunki	76 12.2	S26 17.8
Peacock	53 36.7	S56 44.3
Pollux	243 42.0	N28 01.7
Procyon	245 11.9	N 5 13.7
Rasalhague	96 16.9	N12 34.0
Regulus	207 56.0	N11 58.5
Rigel	281 23.1	S 8 12.2
Rigil Kent.	140 07.6	S60 49.8
Sabik	102 25.5	S15 43.2
Schedar	349 53.0	N56 31.6
Shaula	96 37.2	S37 06.1
Sirius	258 44.0	S16 42.8
Spica	158 43.5	S11 09.1
Suhail	223 01.2	S43 25.5
Vega	80 46.4	N38 47.3
Zuben'ubi	137 18.2	S16 02.0

	SHA	Mer. Pass.
	° ′	h m
Venus	220 35.5	10 58
Mars	233 26.2	10 06
Jupiter	3 25.8	1 26
Saturn	327 48.3	3 48

UT	SUN		MOON					Lat.	Twilight		Sunrise	Moonrise			
	GHA	Dec	GHA	v	Dec	d	HP		Naut.	Civil		26	27	28	29
d h	° ′	° ′	° ′	′	° ′	′	′	°	h m	h m	h m	h m	h m	h m	h m
26 00	179 30.2	N10 33.4	137 27.1	15.2	S 3 00.4	9.6	54.2	N 72	////	01 47	03 34	10 06	11 42	13 21	15 06
01	194 30.4	32.6	152 01.3	15.3	3 10.0	9.5	54.2	N 70	////	02 28	03 52	09 59	11 28	12 59	14 31
02	209 30.6	31.7	166 35.6	15.2	3 19.5	9.5	54.2	68	////	02 55	04 06	09 54	11 18	12 42	14 07
03	224 30.8	.. 30.8	181 09.8	15.3	3 29.0	9.5	54.2	66	01 32	03 15	04 18	09 49	11 09	12 29	13 48
04	239 30.9	29.9	195 44.1	15.2	3 38.5	9.5	54.2	64	02 09	03 31	04 27	09 45	11 02	12 17	13 32
05	254 31.1	29.1	210 18.3	15.3	3 48.0	9.4	54.2	62	02 34	03 45	04 35	09 42	10 55	12 08	13 20
06	269 31.3	N10 28.2	224 52.6	15.2	S 3 57.4	9.4	54.2	60	02 53	03 56	04 42	09 39	10 50	12 00	13 09
W 07	284 31.5	27.3	239 26.8	15.3	4 06.8	9.5	54.2	N 58	03 09	04 05	04 48	09 36	10 45	11 53	13 00
E 08	299 31.6	26.5	254 01.1	15.2	4 16.3	9.4	54.2	56	03 21	04 13	04 54	09 34	10 41	11 47	12 52
D 09	314 31.8	.. 25.6	268 35.3	15.3	4 25.7	9.3	54.2	54	03 32	04 21	04 59	09 32	10 37	11 41	12 44
N 10	329 32.0	24.7	283 09.6	15.2	4 35.0	9.4	54.2	52	03 42	04 27	05 03	09 30	10 33	11 36	12 38
E 11	344 32.2	23.9	297 43.8	15.3	4 44.4	9.3	54.2	50	03 50	04 33	05 07	09 29	10 30	11 31	12 32
S 12	359 32.3	N10 23.0	312 18.1	15.2	S 4 53.7	9.3	54.2	45	04 08	04 45	05 16	09 25	10 23	11 22	12 20
D 13	14 32.5	22.1	326 52.3	15.2	5 03.0	9.3	54.2	N 40	04 21	04 55	05 23	09 22	10 18	11 13	12 09
A 14	29 32.7	21.3	341 26.5	15.2	5 12.3	9.3	54.2	35	04 32	05 03	05 29	09 19	10 13	11 07	12 00
Y 15	44 32.9	.. 20.4	356 00.7	15.2	5 21.6	9.2	54.2	30	04 40	05 09	05 34	09 17	10 09	11 00	11 53
16	59 33.0	19.5	10 34.9	15.2	5 30.8	9.2	54.1	20	04 54	05 21	05 43	09 13	10 01	10 50	11 39
17	74 33.2	18.6	25 09.1	15.2	5 40.0	9.2	54.1	N 10	05 05	05 30	05 51	09 09	09 55	10 41	11 28
18	89 33.4	N10 17.8	39 43.3	15.2	S 5 49.2	9.2	54.1	0	05 13	05 37	05 58	09 06	09 49	10 32	11 17
19	104 33.6	16.9	54 17.5	15.1	5 58.4	9.1	54.1	S 10	05 19	05 44	06 05	09 03	09 43	10 24	11 06
20	119 33.7	16.0	68 51.6	15.2	6 07.5	9.1	54.1	20	05 25	05 51	06 13	09 00	09 37	10 15	10 55
21	134 33.9	.. 15.2	83 25.8	15.1	6 16.6	9.1	54.1	30	05 29	05 57	06 21	08 56	09 29	10 04	10 42
22	149 34.1	14.3	97 59.9	15.2	6 25.7	9.1	54.1	35	05 31	06 01	06 26	08 54	09 25	09 59	10 34
23	164 34.3	13.4	112 34.1	15.1	6 34.8	9.0	54.1	40	05 33	06 04	06 32	08 51	09 21	09 52	10 26
27 00	179 34.5	N10 12.5	127 08.2	15.1	S 6 43.8	9.0	54.1	45	05 34	06 08	06 38	08 48	09 15	09 44	10 16
01	194 34.6	11.7	141 42.3	15.1	6 52.8	9.0	54.1	S 50	05 35	06 13	06 45	08 45	09 09	09 35	10 04
02	209 34.8	10.8	156 16.4	15.1	7 01.8	8.9	54.1	52	05 35	06 14	06 49	08 43	09 06	09 31	09 58
03	224 35.0	.. 09.9	170 50.5	15.0	7 10.7	9.0	54.1	54	05 36	06 17	06 53	08 42	09 03	09 26	09 52
04	239 35.2	09.0	185 24.5	15.1	7 19.7	8.8	54.1	56	05 36	06 19	06 57	08 40	08 59	09 21	09 46
05	254 35.3	08.2	199 58.6	15.0	7 28.5	8.9	54.1	58	05 36	06 21	07 01	08 38	08 55	09 15	09 38
06	269 35.5	N10 07.3	214 32.6	15.0	S 7 37.4	8.8	54.1	S 60	05 36	06 24	07 06	08 35	08 51	09 08	09 29

UT	SUN		MOON					Lat.	Sunset	Twilight		Moonset			
	GHA	Dec	GHA	v	Dec	d	HP			Civil	Naut.	26	27	28	29
								°	h m	h m	h m	h m	h m	h m	h m
07	284 35.7	06.4	229 06.6	15.0	7 46.2	8.8	54.1	N 72	20 25	22 08	////	20 06	19 59	19 50	19 39
08	299 35.9	05.5	243 40.6	15.0	7 55.0	8.8	54.1	N 70	20 08	21 30	////	20 15	20 14	20 14	20 15
T 09	314 36.1	.. 04.7	258 14.6	15.0	8 03.8	8.7	54.1	68	19 54	21 04	23 36	20 22	20 26	20 31	20 40
H 10	329 36.2	03.8	272 48.6	14.9	8 12.5	8.7	54.1	66	19 43	20 45	22 24	20 28	20 36	20 46	21 00
U 11	344 36.4	02.9	287 22.5	14.9	8 21.2	8.6	54.1	64	19 34	20 29	21 50	20 33	20 44	20 58	21 15
R 12	359 36.6	N10 02.0	301 56.4	14.9	S 8 29.8	8.6	54.1	62	19 26	20 16	21 26	20 38	20 51	21 08	21 29
S 13	14 36.8	01.2	316 30.3	14.9	8 38.4	8.6	54.1	60	19 19	20 05	21 07	20 41	20 58	21 17	21 40
D 14	29 37.0	10 00.3	331 04.2	14.8	8 47.0	8.6	54.1	N 58	19 13	19 56	20 52	20 45	21 03	21 24	21 50
A 15	44 37.1	9 59.4	345 38.0	14.9	8 55.6	8.5	54.1	56	19 08	19 48	20 40	20 48	21 08	21 31	21 58
Y 16	59 37.3	58.5	0 11.9	14.8	9 04.1	8.5	54.1	54	19 03	19 41	20 29	20 51	21 13	21 37	22 06
17	74 37.5	57.7	14 45.7	14.8	9 12.6	8.4	54.1	52	18 59	19 35	20 19	20 53	21 17	21 43	22 13
18	89 37.7	N 9 56.8	29 19.5	14.7	S 9 21.0	8.4	54.1	50	18 55	19 29	20 11	20 56	21 20	21 48	22 19
19	104 37.9	55.9	43 53.2	14.7	9 29.4	8.4	54.1	45	18 47	19 17	19 54	21 01	21 28	21 58	22 32
20	119 38.0	55.0	58 26.9	14.8	9 37.8	8.3	54.1	N 40	18 40	19 08	19 41	21 05	21 35	22 07	22 43
21	134 38.2	.. 54.1	73 00.7	14.6	9 46.1	8.3	54.2	35	18 34	19 00	19 31	21 08	21 41	22 15	22 53
22	149 38.4	53.3	87 34.3	14.7	9 54.4	8.2	54.2	30	18 29	18 53	19 22	21 12	21 46	22 22	23 01
23	164 38.6	52.4	102 08.0	14.6	10 02.6	8.2	54.2	20	18 20	18 42	19 09	21 17	21 55	22 34	23 15
28 00	179 38.8	N 9 51.5	116 41.6	14.6	S10 10.8	8.2	54.2	N 10	18 12	18 33	18 58	21 22	22 02	22 44	23 28
01	194 39.0	50.6	131 15.2	14.6	10 19.0	8.1	54.2	0	18 05	18 26	18 50	21 27	22 10	22 54	23 39
02	209 39.1	49.7	145 48.8	14.5	10 27.1	8.1	54.2	S 10	17 58	18 19	18 44	21 31	22 17	23 03	23 51
03	224 39.3	.. 48.9	160 22.3	14.5	10 35.2	8.0	54.2	20	17 50	18 13	18 39	21 36	22 25	23 14	24 04
04	239 39.5	48.0	174 55.8	14.5	10 43.2	8.0	54.2	30	17 42	18 06	18 34	21 42	22 34	23 26	24 18
05	254 39.7	47.1	189 29.3	14.5	10 51.2	7.9	54.2	35	17 37	18 03	18 33	21 45	22 39	23 33	24 26
06	269 39.9	N 9 46.2	204 02.8	14.4	S10 59.1	7.9	54.2	40	17 32	18 00	18 31	21 49	22 45	23 40	24 36
07	284 40.0	45.3	218 36.2	14.4	11 07.0	7.9	54.2	45	17 26	17 56	18 30	21 53	22 51	23 50	24 47
08	299 40.2	44.5	233 09.6	14.3	11 14.9	7.8	54.2	S 50	17 19	17 51	18 29	21 58	23 00	24 01	00 01
F 09	314 40.4	.. 43.6	247 42.9	14.4	11 22.7	7.8	54.2	52	17 15	17 50	18 29	22 01	23 03	24 06	00 06
R 10	329 40.6	42.7	262 16.3	14.3	11 30.5	7.7	54.2	54	17 11	17 48	18 29	22 03	23 08	24 11	00 11
I 11	344 40.8	41.8	276 49.6	14.2	11 38.2	7.7	54.2	56	17 07	17 45	18 29	22 06	23 12	24 18	00 18
D 12	359 41.0	N 9 40.9	291 22.8	14.2	S11 45.9	7.6	54.2	58	17 03	17 43	18 29	22 09	23 17	24 25	00 25
A 13	14 41.2	40.0	305 56.0	14.2	11 53.5	7.6	54.2	S 60	16 58	17 40	18 29	22 13	23 23	24 33	00 33
Y 14	29 41.3	39.2	320 29.2	14.2	12 01.1	7.5	54.3								
15	44 41.5	.. 38.3	335 02.4	14.1	12 08.6	7.5	54.3								
16	59 41.7	37.4	349 35.5	14.1	12 16.1	7.4	54.3								
17	74 41.9	36.5	4 08.6	14.0	12 23.5	7.4	54.3								
18	89 42.1	N 9 35.6	18 41.6	14.0	S12 30.9	7.3	54.3								
19	104 42.3	34.7	33 14.6	14.0	12 38.2	7.3	54.3								
20	119 42.4	33.9	47 47.6	13.9	12 45.5	7.2	54.3								
21	134 42.6	.. 33.0	62 20.5	13.9	12 52.7	7.2	54.3								
22	149 42.8	32.1	76 53.4	13.9	12 59.9	7.1	54.3								
23	164 43.0	31.2	91 26.3	13.8	S13 07.0	7.1	54.3								

							Day	SUN		Mer.	MOON		Age	Phase
								Eqn. of Time		Mer.	Mer. Pass.			
								00h	12h	Pass.	Upper	Lower		
							d	m s	m s	h m	h m	h m	d	%
SD 15.9	d 0.9		SD	14.8	14.7	14.8	26	01 59	01 51	12 02	15 16	02 55	04	18
							27	01 43	01 34	12 02	15 59	03 38	05	25
							28	01 25	01 17	12 01	16 43	04 21	06	34

1998 AUGUST 29, 30, 31 (SAT., SUN., MON.)

UT	ARIES GHA	VENUS −3.9 GHA	Dec	MARS +1.7 GHA	Dec	JUPITER −2.9 GHA	Dec	SATURN +0.2 GHA	Dec	STARS Name	SHA	Dec
29 00	336 59.9	195 07.9	N15 59.0	209 06.5	N19 57.8	340 38.3	S 3 14.5	304 50.4	N10 11.3	Acamar	315 26.9	S40 18.4
01	352 02.4	210 07.3	58.1	224 07.4	57.5	355 41.0	14.7	319 53.0	11.3	Achernar	335 34.9	S57 14.4
02	7 04.8	225 06.7	57.2	239 08.2	57.1	10 43.8	14.8	334 55.5	11.3	Acrux	173 22.6	S63 05.5
03	22 07.3	240 06.1 ..	56.3	254 09.0 ..	56.7	25 46.5 ..	14.9	349 58.0 ..	11.2	Adhara	255 21.7	S28 58.1
04	37 09.8	255 05.5	55.4	269 09.8	56.3	40 49.2	15.0	5 00.5	11.2	Aldebaran	291 02.5	N16 30.3
05	52 12.2	270 04.9	54.5	284 10.6	56.0	55 52.0	15.1	20 03.0	11.2			
06	67 14.7	285 04.3	N15 53.6	299 11.4	N19 55.6	70 54.7	S 3 15.3	35 05.5	N10 11.2	Alioth	166 31.2	N55 58.3
07	82 17.2	300 03.7	52.7	314 12.2	55.2	85 57.4	15.4	50 08.1	11.1	Alkaid	153 08.2	N49 19.6
S 08	97 19.6	315 03.1	51.8	329 13.0	54.8	101 00.2	15.5	65 10.6	11.1	Al Na'ir	27 57.6	S46 58.0
A 09	112 22.1	330 02.5 ..	50.9	344 13.8 ..	54.5	116 02.9 ..	15.6	80 13.1 ..	11.1	Alnilam	275 58.0	S 1 12.2
T 10	127 24.6	345 01.9	50.0	359 14.7	54.1	131 05.6	15.8	95 15.6	11.0	Alphard	218 07.6	S 8 39.1
U 11	142 27.0	0 01.3	49.1	14 15.5	53.7	146 08.4	15.9	110 18.1	11.0			
R 12	157 29.5	15 00.7	N15 48.2	29 16.3	N19 53.3	161 11.1	S 3 16.0	125 20.6	N10 11.0	Alphecca	126 20.7	N26 43.5
D 13	172 31.9	30 00.1	47.3	44 17.1	53.0	176 13.8	16.1	140 23.2	10.9	Alpheratz	357 55.0	N29 05.0
A 14	187 34.4	44 59.5	46.4	59 17.9	52.6	191 16.6	16.3	155 25.7	10.9	Altair	62 19.1	N 8 52.1
Y 15	202 36.9	59 58.9 ..	45.5	74 18.7 ..	52.2	206 19.3 ..	16.4	170 28.2 ..	10.9	Ankaa	353 26.6	S42 18.7
16	217 39.3	74 58.3	44.6	89 19.5	51.8	221 22.0	16.5	185 30.7	10.9	Antares	112 40.2	S26 25.6
17	232 41.8	89 57.7	43.7	104 20.4	51.5	236 24.8	16.6	200 33.2	10.8			
18	247 44.3	104 57.1	N15 42.8	119 21.2	N19 51.1	251 27.5	S 3 16.8	215 35.7	N10 10.8	Arcturus	146 06.3	N19 11.7
19	262 46.7	119 56.5	41.9	134 22.0	50.7	266 30.2	16.9	230 38.3	10.8	Atria	107 52.2	S69 01.6
20	277 49.2	134 55.9	41.0	149 22.8	50.3	281 33.0	17.0	245 40.8	10.7	Avior	234 23.2	S59 30.2
21	292 51.7	149 55.3 ..	40.1	164 23.6 ..	50.0	296 35.7 ..	17.1	260 43.3 ..	10.7	Bellatrix	278 44.3	N 6 20.9
22	307 54.1	164 54.7	39.2	179 24.4	49.6	311 38.4	17.3	275 45.8	10.7	Betelgeuse	271 13.8	N 7 24.4
23	322 56.6	179 54.2	38.3	194 25.2	49.2	326 41.2	17.4	290 48.3	10.6			
30 00	337 59.1	194 53.6	N15 37.4	209 26.1	N19 48.8	341 43.9	S 3 17.5	305 50.9	N10 10.6	Canopus	264 01.5	S52 41.5
01	353 01.5	209 53.0	36.5	224 26.9	48.5	356 46.6	17.6	320 53.4	10.6	Capella	280 51.4	N45 59.5
02	8 04.0	224 52.4	35.6	239 27.7	48.1	11 49.4	17.7	335 55.9	10.6	Deneb	49 38.8	N45 16.8
03	23 06.4	239 51.8 ..	34.7	254 28.5 ..	47.7	26 52.1 ..	17.9	350 58.4 ..	10.5	Denebola	182 45.6	N14 35.0
04	38 08.9	254 51.2	33.8	269 29.3	47.3	41 54.8	18.0	6 00.9	10.5	Diphda	349 07.1	S17 59.5
05	53 11.4	269 50.6	32.9	284 30.1	47.0	56 57.6	18.1	21 03.5	10.5			
06	68 13.8	284 50.0	N15 31.9	299 31.0	N19 46.6	72 00.3	S 3 18.2	36 06.0	N10 10.4	Dubhe	194 06.3	N61 45.6
07	83 16.3	299 49.4	31.0	314 31.8	46.2	87 03.0	18.4	51 08.5	10.4	Elnath	278 27.1	N28 36.2
08	98 18.8	314 48.8	30.1	329 32.6	45.8	102 05.8	18.5	66 11.0	10.4	Eltanin	90 51.2	N51 29.8
S 09	113 21.2	329 48.3 ..	29.2	344 33.4 ..	45.4	117 08.5 ..	18.6	81 13.5 ..	10.3	Enif	33 58.0	N 9 52.3
U 10	128 23.7	344 47.7	28.3	359 34.2	45.1	132 11.3	18.7	96 16.1	10.3	Fomalhaut	15 36.2	S29 37.6
N 11	143 26.2	359 47.1	27.4	14 35.0	44.7	147 14.0	18.9	111 18.6	10.3			
D 12	158 28.6	14 46.5	N15 26.5	29 35.9	N19 44.3	162 16.7	S 3 19.0	126 21.1	N10 10.2	Gacrux	172 14.1	S57 06.3
A 13	173 31.1	29 45.9	25.5	44 36.7	43.9	177 19.5	19.1	141 23.6	10.2	Gienah	176 04.3	S17 31.9
Y 14	188 33.6	44 45.3	24.6	59 37.5	43.5	192 22.2	19.2	156 26.1	10.2	Hadar	149 04.4	S60 22.0
15	203 36.0	59 44.7 ..	23.7	74 38.3 ..	43.2	207 24.9 ..	19.4	171 28.7 ..	10.2	Hamal	328 13.4	N23 27.3
16	218 38.5	74 44.2	22.8	89 39.1	42.8	222 27.7	19.5	186 31.2	10.1	Kaus Aust.	83 58.7	S34 23.1
17	233 40.9	89 43.6	21.9	104 40.0	42.4	237 30.4	19.6	201 33.7	10.1			
18	248 43.4	104 43.0	N15 20.9	119 40.8	N19 42.0	252 33.2	S 3 19.7	216 36.2	N10 10.1	Kochab	137 20.2	N74 10.1
19	263 45.9	119 42.4	20.0	134 41.6	41.6	267 35.9	19.9	231 38.8	10.0	Markab	13 49.3	N15 12.0
20	278 48.3	134 41.8	19.1	149 42.4	41.3	282 38.6	20.0	246 41.3	10.0	Menkar	314 26.9	N 4 05.0
21	293 50.8	149 41.2 ..	18.2	164 43.2 ..	40.9	297 41.4 ..	20.1	261 43.8 ..	10.0	Menkent	148 21.2	S36 21.7
22	308 53.3	164 40.7	17.3	179 44.0	40.5	312 44.1	20.2	276 46.3	09.9	Miaplacidus	221 42.9	S69 42.6
23	323 55.7	179 40.1	16.3	194 44.9	40.1	327 46.8	20.4	291 48.9	09.9			
31 00	338 58.2	194 39.5	N15 15.4	209 45.7	N19 39.7	342 49.6	S 3 20.5	306 51.4	N10 09.9	Mirfak	308 56.5	N49 51.1
01	354 00.7	209 38.9	14.5	224 46.5	39.3	357 52.3	20.6	321 53.9	09.8	Nunki	76 12.2	S26 17.8
02	9 03.1	224 38.3	13.6	239 47.3	39.0	12 55.1	20.7	336 56.4	09.8	Peacock	53 36.7	S56 44.4
03	24 05.6	239 37.8 ..	12.6	254 48.1 ..	38.6	27 57.8 ..	20.9	351 58.9 ..	09.8	Pollux	243 41.9	N28 01.7
04	39 08.0	254 37.2	11.7	269 49.0	38.2	43 00.5	21.0	7 01.5	09.7	Procyon	245 11.9	N 5 13.7
05	54 10.5	269 36.6	10.8	284 49.8	37.8	58 03.3	21.1	22 04.0	09.7			
06	69 13.0	284 36.0	N15 09.8	299 50.6	N19 37.4	73 06.0	S 3 21.2	37 06.5	N10 09.7	Rasalhague	96 16.9	N12 34.0
07	84 15.4	299 35.4	08.9	314 51.4	37.0	88 08.8	21.4	52 09.0	09.7	Regulus	207 56.0	N11 58.5
08	99 17.9	314 34.9	08.0	329 52.3	36.7	103 11.5	21.5	67 11.6	09.6	Rigel	281 23.1	S 8 12.2
M 09	114 20.4	329 34.3 ..	07.0	344 53.1 ..	36.3	118 14.2 ..	21.6	82 14.1 ..	09.6	Rigil Kent.	140 07.6	S60 49.8
O 10	129 22.8	344 33.7	06.1	359 53.9	35.9	133 17.0	21.8	97 16.6	09.6	Sabik	102 25.5	S15 43.2
N 11	144 25.3	359 33.1	05.2	14 54.7	35.5	148 19.7	21.9	112 19.1	09.5			
D 12	159 27.8	14 32.6	N15 04.2	29 55.5	N19 35.1	163 22.5	S 3 22.0	127 21.7	N10 09.5	Schedar	349 53.0	N56 31.6
A 13	174 30.2	29 32.0	03.3	44 56.4	34.7	178 25.2	22.1	142 24.2	09.5	Shaula	96 37.3	S37 06.1
Y 14	189 32.7	44 31.4	02.4	59 57.2	34.3	193 27.9	22.3	157 26.7	09.4	Sirius	258 44.0	S16 42.8
15	204 35.2	59 30.8 ..	01.4	74 58.0 ..	34.0	208 30.7 ..	22.4	172 29.2 ..	09.4	Spica	158 43.5	S11 09.1
16	219 37.6	74 30.3	15 00.5	89 58.8	33.6	223 33.4	22.5	187 31.8	09.4	Suhail	223 01.2	S43 25.5
17	234 40.1	89 29.7	14 59.6	104 59.7	33.2	238 36.2	22.6	202 34.3	09.3			
18	249 42.5	104 29.1	N14 58.6	120 00.5	N19 32.8	253 38.9	S 3 22.8	217 36.8	N10 09.3	Vega	80 46.4	N38 47.3
19	264 45.0	119 28.5	57.7	135 01.3	32.4	268 41.6	22.9	232 39.4	09.3	Zuben'ubi	137 18.2	S16 02.0
20	279 47.5	134 28.0	56.7	150 02.1	32.0	283 44.4	23.0	247 41.9	09.2			
21	294 49.9	149 27.4 ..	55.8	165 03.0 ..	31.6	298 47.1 ..	23.1	262 44.4 ..	09.2			
22	309 52.4	164 26.8	54.9	180 03.8	31.3	313 49.9	23.3	277 46.9	09.2	Venus	216 54.5	11 01
23	324 54.9	179 26.3	53.9	195 04.6	30.9	328 52.6	23.4	292 49.5	09.1	Mars	231 27.0	10 02

	SHA	Mer. Pass.
	° ′	h m
Venus	216 54.5	11 01
Mars	231 27.0	10 02
Jupiter	3 44.8	1 13
Saturn	327 51.8	3 36

Mer. Pass. 1 27.8 | *v* −0.6 *d* 0.9 | *v* 0.8 *d* 0.4 | *v* 2.7 *d* 0.1 | *v* 2.5 *d* 0.0

UT	SUN GHA	Dec	MOON GHA	v	Dec	d	HP
d h	° ′	° ′	° ′	′	° ′	′	′
29 00	179 43.2	N 9 30.3	105 59.1	13.7	S13 14.1	7.0	54.4
01	194 43.4	29.4	120 31.8	13.8	13 21.1	6.9	54.4
02	209 43.6	28.5	135 04.6	13.7	13 28.0	6.9	54.4
03	224 43.7	.. 27.7	149 37.3	13.6	13 34.9	6.9	54.4
04	239 43.9	26.8	164 09.9	13.6	13 41.8	6.8	54.4
05	254 44.1	25.9	178 42.5	13.6	13 48.6	6.7	54.4
06	269 44.3	N 9 25.0	193 15.1	13.5	S13 55.3	6.7	54.4
S 07	284 44.5	24.1	207 47.6	13.5	14 02.0	6.6	54.5
A 08	299 44.7	23.2	222 20.1	13.4	14 08.6	6.5	54.5
T 09	314 44.9	.. 22.3	236 52.5	13.4	14 15.1	6.5	54.5
U 10	329 45.1	21.4	251 24.9	13.4	14 21.6	6.5	54.5
R 11	344 45.2	20.6	265 57.3	13.3	14 28.1	6.3	54.5
D 12	359 45.4	N 9 19.7	280 29.6	13.2	S14 34.4	6.3	54.5
A 13	14 45.6	18.8	295 01.8	13.2	14 40.7	6.3	54.5
Y 14	29 45.8	17.9	309 34.0	13.2	14 47.0	6.2	54.6
15	44 46.0	.. 17.0	324 06.2	13.1	14 53.2	6.1	54.6
16	59 46.2	16.1	338 38.3	13.1	14 59.3	6.1	54.6
17	74 46.4	15.2	353 10.4	13.0	15 05.4	6.0	54.6
18	89 46.6	N 9 14.3	7 42.4	13.0	S15 11.4	5.9	54.6
19	104 46.7	13.4	22 14.4	12.9	15 17.3	5.9	54.6
20	119 46.9	12.5	36 46.3	12.9	15 23.2	5.8	54.7
21	134 47.1	.. 11.6	51 18.2	12.8	15 29.0	5.7	54.7
22	149 47.3	10.8	65 50.0	12.8	15 34.7	5.7	54.7
23	164 47.5	09.9	80 21.8	12.7	15 40.4	5.6	54.7
30 00	179 47.7	N 9 09.0	94 53.5	12.7	S15 46.0	5.5	54.7
01	194 47.9	08.1	109 25.2	12.7	15 51.5	5.5	54.7
02	209 48.1	07.2	123 56.9	12.6	15 57.0	5.4	54.8
03	224 48.3	.. 06.3	138 28.5	12.5	16 02.4	5.3	54.8
04	239 48.5	05.4	153 00.0	12.5	16 07.7	5.3	54.8
05	254 48.6	04.5	167 31.5	12.4	16 13.0	5.2	54.8
06	269 48.8	N 9 03.6	182 02.9	12.4	S16 18.2	5.1	54.8
S 07	284 49.0	02.7	196 34.3	12.3	16 23.3	5.0	54.9
U 08	299 49.2	01.8	211 05.6	12.3	16 28.3	5.0	54.9
N 09	314 49.4	.. 00.9	225 36.9	12.3	16 33.3	4.9	54.9
D 10	329 49.6	9 00.0	240 08.2	12.2	16 38.2	4.8	54.9
A 11	344 49.8	8 59.1	254 39.4	12.1	16 43.0	4.7	55.0
Y 12	359 50.0	N 8 58.2	269 10.5	12.1	S16 47.7	4.7	55.0
13	14 50.2	57.4	283 41.6	12.0	16 52.4	4.6	55.0
14	29 50.4	56.5	298 12.6	12.0	16 57.0	4.5	55.0
15	44 50.6	.. 55.6	312 43.6	11.9	17 01.5	4.5	55.0
16	59 50.7	54.7	327 14.5	11.9	17 06.0	4.3	55.1
17	74 50.9	53.8	341 45.4	11.8	17 10.3	4.3	55.1
18	89 51.1	N 8 52.9	356 16.2	11.8	S17 14.6	4.2	55.1
19	104 51.3	52.0	10 47.0	11.7	17 18.8	4.1	55.1
20	119 51.5	51.1	25 17.7	11.6	17 22.9	4.1	55.2
21	134 51.7	.. 50.2	39 48.3	11.7	17 27.0	4.0	55.2
22	149 51.9	49.3	54 19.0	11.5	17 31.0	3.8	55.2
23	164 52.1	48.4	68 49.5	11.5	17 34.8	3.8	55.2
31 00	179 52.3	N 8 47.5	83 20.0	11.5	S17 38.6	3.8	55.3
01	194 52.5	46.6	97 50.5	11.4	17 42.4	3.6	55.3
02	209 52.7	45.7	112 20.9	11.3	17 46.0	3.6	55.3
03	224 52.9	.. 44.8	126 51.2	11.3	17 49.6	3.4	55.3
04	239 53.1	43.9	141 21.5	11.3	17 53.0	3.4	55.4
05	254 53.3	43.0	155 51.8	11.2	17 56.4	3.3	55.4
06	269 53.5	N 8 42.1	170 22.0	11.1	S17 59.7	3.2	55.4
07	284 53.6	41.2	184 52.1	11.1	18 02.9	3.2	55.4
08	299 53.8	40.3	199 22.2	11.0	18 06.1	3.0	55.5
M 09	314 54.0	.. 39.4	213 52.2	11.0	18 09.1	2.9	55.5
O 10	329 54.2	38.5	228 22.2	10.9	18 12.0	2.9	55.5
N 11	344 54.4	37.6	242 52.1	10.9	18 14.9	2.8	55.6
D 12	359 54.6	N 8 36.7	257 22.0	10.8	S18 17.7	2.7	55.6
A 13	14 54.8	35.8	271 51.8	10.8	18 20.4	2.6	55.6
Y 14	29 55.0	34.9	286 21.6	10.7	18 23.0	2.5	55.6
15	44 55.2	.. 34.0	300 51.3	10.7	18 25.5	2.4	55.7
16	59 55.4	33.1	315 21.0	10.6	18 27.9	2.3	55.7
17	74 55.6	32.2	329 50.6	10.6	18 30.2	2.2	55.7
18	89 55.8	N 8 31.3	344 20.2	10.5	S18 32.4	2.2	55.8
19	104 56.0	30.4	358 49.7	10.5	18 34.6	2.0	55.8
20	119 56.2	29.5	13 19.2	10.4	18 36.6	2.0	55.8
21	134 56.4	.. 28.6	27 48.6	10.4	18 38.6	1.8	55.8
22	149 56.6	27.7	42 18.0	10.3	18 40.4	1.8	55.9
23	164 56.8	26.8	56 47.3	10.3	S18 42.2	1.7	55.9
	SD 15.9	d 0.9	SD 14.9	15.0	15.1		

Twilight — Sunrise — Moonrise

Lat.	Naut.	Civil	Sunrise	29	30	31	1
°	h m	h m	h m	h m	h m	h m	h m
N 72	////	02 14	03 50	15 06	17 06	■	■
N 70	////	02 46	04 05	14 31	16 04	17 33	18 41
68	01 06	03 10	04 18	14 07	15 30	16 46	17 47
66	01 55	03 28	04 28	13 48	15 05	16 15	17 15
64	02 25	03 42	04 36	13 32	14 45	15 53	16 51
62	02 46	03 54	04 43	13 20	14 29	15 34	16 32
60	03 03	04 04	04 49	13 09	14 16	15 19	16 16
N 58	03 17	04 12	04 55	13 00	14 05	15 07	16 03
56	03 29	04 20	05 00	12 52	13 55	14 56	15 51
54	03 39	04 26	05 04	12 44	13 47	14 46	15 41
52	03 48	04 32	05 08	12 38	13 39	14 37	15 32
50	03 56	04 38	05 12	12 32	13 32	14 30	15 24
45	04 12	04 49	05 19	12 20	13 17	14 13	15 07
N 40	04 24	04 58	05 26	12 09	13 05	14 00	14 53
35	04 34	05 05	05 31	12 00	12 55	13 49	14 42
30	04 42	05 11	05 36	11 53	12 46	13 39	14 31
20	04 55	05 21	05 44	11 39	12 30	13 21	14 14
N 10	05 05	05 30	05 51	11 28	12 16	13 07	13 58
0	05 12	05 36	05 57	11 17	12 04	12 53	13 44
S 10	05 18	05 43	06 04	11 06	11 51	12 39	13 29
20	05 23	05 48	06 10	10 55	11 38	12 24	13 14
30	05 26	05 54	06 18	10 42	11 23	12 07	12 57
35	05 27	05 57	06 22	10 34	11 14	11 57	12 46
40	05 28	06 00	06 27	10 26	11 04	11 46	12 35
45	05 29	06 03	06 33	10 16	10 52	11 33	12 21
S 50	05 29	06 07	06 39	10 04	10 38	11 17	12 04
52	05 29	06 08	06 42	09 58	10 31	11 10	11 56
54	05 29	06 10	06 46	09 52	10 24	11 01	11 48
56	05 29	06 12	06 49	09 46	10 15	10 52	11 38
58	05 28	06 13	06 53	09 38	10 06	10 42	11 26
S 60	05 27	06 16	06 58	09 29	09 56	10 30	11 14

Sunset — Twilight — Moonset

Lat.	Sunset	Civil	Naut.	29	30	31	1
°	h m	h m	h m	h m	h m	h m	h m
N 72	20 08	21 41	////	19 39	19 17	■	■
N 70	19 53	21 10	////	20 15	20 19	20 33	21 13
68	19 41	20 48	22 44	20 40	20 55	21 20	22 06
66	19 31	20 31	22 00	21 00	21 20	21 51	22 39
64	19 23	20 17	21 32	21 15	21 40	22 14	23 03
62	19 16	20 05	21 12	21 29	21 56	22 33	23 22
60	19 10	19 56	20 55	21 40	22 09	22 48	23 37
N 58	19 05	19 47	20 42	21 50	22 21	23 01	23 51
56	19 00	19 40	20 30	21 58	22 31	23 12	24 02
54	18 56	19 34	20 20	22 06	22 40	23 22	24 12
52	18 52	19 28	20 12	22 13	22 48	23 30	24 21
50	18 49	19 23	20 04	22 19	22 55	23 38	24 29
45	18 41	19 12	19 48	22 32	23 10	23 55	24 46
N 40	18 35	19 03	19 36	22 43	23 23	24 08	00 08
35	18 30	18 56	19 26	22 53	23 34	24 20	00 20
30	18 25	18 50	19 18	23 01	23 43	24 30	00 30
20	18 17	18 40	19 06	23 15	24 00	00 00	00 48
N 10	18 10	18 32	18 56	23 28	24 14	00 14	01 03
0	18 04	18 25	18 49	23 39	24 27	00 27	01 17
S 10	17 58	18 19	18 44	23 51	24 40	00 40	01 31
20	17 51	18 13	18 39	24 04	00 04	00 55	01 46
30	17 44	18 08	18 36	24 18	00 18	01 11	02 04
35	17 40	18 05	18 35	24 26	00 26	01 20	02 14
40	17 35	18 02	18 34	24 36	00 36	01 31	02 26
45	17 29	17 59	18 33	24 47	00 47	01 44	02 39
S 50	17 23	17 56	18 33	00 01	01 01	02 00	02 56
52	17 20	17 54	18 33	00 06	01 07	02 07	03 04
54	17 17	17 53	18 34	00 11	01 14	02 15	03 12
56	17 13	17 51	18 34	00 18	01 22	02 24	03 22
58	17 09	17 49	18 34	00 25	01 31	02 34	03 33
S 60	17 05	17 47	18 35	00 33	01 41	02 46	03 46

SUN — MOON

Day	Eqn. of Time 00h	12h	Mer. Pass.	Mer. Pass. Upper	Lower	Age	Phase
d	m s	m s	h m	h m	h m	d	%
29	01 08	00 59	12 01	17 28	05 05	07	43
30	00 50	00 40	12 01	18 15	05 52	08	53
31	00 31	00 22	12 00	19 05	06 40	09	63

1998 SEPTEMBER 1, 2, 3 (TUES., WED., THURS.)

UT	ARIES GHA	VENUS −3.9 GHA	Dec	MARS +1.7 GHA	Dec	JUPITER −2.9 GHA	Dec	SATURN +0.2 GHA	Dec	STARS Name	SHA	Dec
1 00	339 57.3	194 25.7	N14 53.0	210 05.4	N19 30.5	343 55.3	S 3 23.5	307 52.0	N10 09.1	Acamar	315 26.8	S40 18.4
01	354 59.8	209 25.1	52.0	225 06.3	30.1	358 58.1	23.6	322 54.5	09.1	Achernar	335 34.9	S57 14.5
02	10 02.3	224 24.5	51.1	240 07.1	29.7	14 00.8	23.8	337 57.0	09.0	Acrux	173 22.6	S63 05.5
03	25 04.7	239 24.0	.. 50.1	255 07.9	.. 29.3	29 03.6	.. 23.9	352 59.6	.. 09.0	Adhara	255 21.7	S28 58.1
04	40 07.2	254 23.4	49.2	270 08.7	28.9	44 06.3	24.0	8 02.1	09.0	Aldebaran	291 02.5	N16 30.3
05	55 09.7	269 22.8	48.3	285 09.6	28.5	59 09.1	24.2	23 04.6	08.9			
06	70 12.1	284 22.3	N14 47.3	300 10.4	N19 28.2	74 11.8	S 3 24.3	38 07.2	N10 08.9	Alioth	166 31.2	N55 58.3
07	85 14.6	299 21.7	46.4	315 11.2	27.8	89 14.5	24.4	53 09.7	08.9	Alkaid	153 08.2	N49 19.6
T 08	100 17.0	314 21.1	45.4	330 12.0	27.4	104 17.3	24.5	68 12.2	08.8	Al Na'ir	27 57.6	S46 58.0
U 09	115 19.5	329 20.6	.. 44.5	345 12.9	.. 27.0	119 20.0	.. 24.7	83 14.7	.. 08.8	Alnilam	275 58.0	S 1 12.2
E 10	130 22.0	344 20.0	43.5	0 13.7	26.6	134 22.8	24.8	98 17.3	08.8	Alphard	218 07.6	S 8 39.1
S 11	145 24.4	359 19.4	42.6	15 14.5	26.2	149 25.5	24.9	113 19.8	08.7			
D 12	160 26.9	14 18.9	N14 41.6	30 15.3	N19 25.8	164 28.3	S 3 25.0	128 22.3	N10 08.7	Alphecca	126 20.8	N26 43.5
A 13	175 29.4	29 18.3	40.7	45 16.2	25.4	179 31.0	25.2	143 24.9	08.7	Alpheratz	357 54.9	N29 05.0
Y 14	190 31.8	44 17.7	39.7	60 17.0	25.0	194 33.8	25.3	158 27.4	08.6	Altair	62 19.1	N 8 52.1
15	205 34.3	59 17.2	.. 38.8	75 17.8	.. 24.6	209 36.5	.. 25.4	173 29.9	.. 08.6	Ankaa	353 26.6	S42 18.7
16	220 36.8	74 16.6	37.8	90 18.7	24.3	224 39.2	25.6	188 32.4	08.6	Antares	112 40.2	S26 25.6
17	235 39.2	89 16.0	36.8	105 19.5	23.9	239 42.0	25.7	203 35.0	08.5			
18	250 41.7	104 15.5	N14 35.9	120 20.3	N19 23.5	254 44.7	S 3 25.8	218 37.5	N10 08.5	Arcturus	146 06.3	N19 11.7
19	265 44.1	119 14.9	34.9	135 21.1	23.1	269 47.5	25.9	233 40.0	08.5	Atria	107 52.3	S69 01.6
20	280 46.6	134 14.3	34.0	150 22.0	22.7	284 50.2	26.1	248 42.6	08.4	Avior	234 23.1	S59 30.2
21	295 49.1	149 13.8	.. 33.0	165 22.8	.. 22.3	299 53.0	.. 26.2	263 45.1	.. 08.4	Bellatrix	278 44.3	N 6 20.9
22	310 51.5	164 13.2	32.1	180 23.6	21.9	314 55.7	26.3	278 47.6	08.4	Betelgeuse	271 13.7	N 7 24.4
23	325 54.0	179 12.7	31.1	195 24.5	21.5	329 58.5	26.4	293 50.2	08.3			
2 00	340 56.5	194 12.1	N14 30.1	210 25.3	N19 21.1	345 01.2	S 3 26.6	308 52.7	N10 08.3	Canopus	264 01.4	S52 41.5
01	355 58.9	209 11.5	29.2	225 26.1	20.7	0 03.9	26.7	323 55.2	08.3	Capella	280 51.4	N45 59.5
02	11 01.4	224 11.0	28.2	240 27.0	20.3	15 06.7	26.8	338 57.8	08.2	Deneb	49 38.8	N45 16.8
03	26 03.9	239 10.4	.. 27.3	255 27.8	.. 19.9	30 09.4	.. 27.0	354 00.3	.. 08.2	Denebola	182 45.6	N14 35.0
04	41 06.3	254 09.8	26.3	270 28.6	19.5	45 12.2	27.1	9 02.8	08.2	Diphda	349 07.0	S17 59.5
05	56 08.8	269 09.3	25.3	285 29.4	19.2	60 14.9	27.2	24 05.4	08.1			
06	71 11.3	284 08.7	N14 24.4	300 30.3	N19 18.8	75 17.7	S 3 27.3	39 07.9	N10 08.1	Dubhe	194 06.3	N61 45.6
W 07	86 13.7	299 08.2	23.4	315 31.1	18.4	90 20.4	27.5	54 10.4	08.1	Elnath	278 27.1	N28 36.2
E 08	101 16.2	314 07.6	22.4	330 31.9	18.0	105 23.2	27.6	69 12.9	08.0	Eltanin	90 51.3	N51 29.8
D 09	116 18.6	329 07.1	.. 21.5	345 32.8	.. 17.6	120 25.9	.. 27.7	84 15.5	.. 08.0	Enif	33 58.0	N 9 52.3
N 10	131 21.1	344 06.5	20.5	0 33.6	17.2	135 28.7	27.9	99 18.0	08.0	Fomalhaut	15 36.2	S29 37.7
E 11	146 23.6	359 05.9	19.5	15 34.4	16.8	150 31.4	28.0	114 20.5	07.9			
S 12	161 26.0	14 05.4	N14 18.6	30 35.3	N19 16.4	165 34.2	S 3 28.1	129 23.1	N10 07.9	Gacrux	172 14.1	S57 06.3
D 13	176 28.5	29 04.8	17.6	45 36.1	16.0	180 36.9	28.2	144 25.6	07.9	Gienah	176 04.3	S17 31.9
A 14	191 31.0	44 04.3	16.6	60 36.9	15.6	195 39.6	28.4	159 28.1	07.8	Hadar	149 04.4	S60 22.0
Y 15	206 33.4	59 03.7	.. 15.7	75 37.8	.. 15.2	210 42.4	.. 28.5	174 30.7	.. 07.8	Hamal	328 13.4	N23 27.3
16	221 35.9	74 03.2	14.7	90 38.6	14.8	225 45.1	28.6	189 33.2	07.7	Kaus Aust.	83 58.7	S34 23.1
17	236 38.4	89 02.6	13.7	105 39.4	14.4	240 47.9	28.8	204 35.8	07.7			
18	251 40.8	104 02.1	N14 12.7	120 40.3	N19 14.0	255 50.6	S 3 28.9	219 38.3	N10 07.7	Kochab	137 20.2	N74 10.1
19	266 43.3	119 01.5	11.8	135 41.1	13.6	270 53.4	29.0	234 40.8	07.6	Markab	13 49.3	N15 12.0
20	281 45.8	134 00.9	10.8	150 41.9	13.2	285 56.1	29.1	249 43.4	07.6	Menkar	314 26.8	N 4 05.1
21	296 48.2	149 00.4	.. 09.8	165 42.8	.. 12.8	300 58.9	.. 29.3	264 45.9	.. 07.6	Menkent	148 21.2	S36 21.7
22	311 50.7	163 59.8	08.8	180 43.6	12.4	316 01.6	29.4	279 48.4	07.5	Miaplacidus	221 42.8	S69 42.6
23	326 53.1	178 59.3	07.9	195 44.4	12.0	331 04.4	29.5	294 51.0	07.5			
3 00	341 55.6	193 58.7	N14 06.9	210 45.3	N19 11.6	346 07.1	S 3 29.7	309 53.5	N10 07.5	Mirfak	308 56.5	N49 51.1
01	356 58.1	208 58.2	05.9	225 46.1	11.2	1 09.9	29.8	324 56.0	07.4	Nunki	76 12.2	S26 17.8
02	12 00.5	223 57.6	04.9	240 46.9	10.8	16 12.6	29.9	339 58.6	07.4	Peacock	53 36.7	S56 44.4
03	27 03.0	238 57.1	.. 04.0	255 47.8	.. 10.4	31 15.4	.. 30.0	355 01.1	.. 07.4	Pollux	243 41.9	N28 01.7
04	42 05.5	253 56.5	03.0	270 48.6	10.0	46 18.1	30.2	10 03.6	07.3	Procyon	245 11.9	N 5 13.7
05	57 07.9	268 56.0	02.0	285 49.4	09.6	61 20.9	30.3	25 06.2	07.3			
06	72 10.4	283 55.4	N14 01.0	300 50.3	N19 09.2	76 23.6	S 3 30.4	40 08.7	N10 07.3	Rasalhague	96 16.9	N12 34.0
07	87 12.9	298 54.9	14 00.0	315 51.1	08.8	91 26.4	30.6	55 11.2	07.2	Regulus	207 56.0	N11 58.5
T 08	102 15.3	313 54.3	13 59.1	330 52.0	08.4	106 29.1	30.7	70 13.8	07.2	Rigel	281 23.1	S 8 12.2
H 09	117 17.8	328 53.8	.. 58.1	345 52.8	.. 08.0	121 31.9	.. 30.8	85 16.3	.. 07.1	Rigil Kent.	140 07.7	S60 49.8
U 10	132 20.3	343 53.2	57.1	0 53.6	07.6	136 34.6	30.9	100 18.9	07.1	Sabik	102 25.6	S15 43.2
R 11	147 22.7	358 52.7	56.1	15 54.5	07.2	151 37.4	31.1	115 21.4	07.1			
S 12	162 25.2	13 52.1	N13 55.1	30 55.3	N19 06.8	166 40.1	S 3 31.2	130 23.9	N10 07.0	Schedar	349 53.0	N56 31.7
D 13	177 27.6	28 51.6	54.1	45 56.1	06.4	181 42.9	31.3	145 26.5	07.0	Shaula	96 37.3	S37 06.1
A 14	192 30.1	43 51.0	53.2	60 57.0	06.0	196 45.6	31.5	160 29.0	07.0	Sirius	258 43.9	S16 42.7
Y 15	207 32.6	58 50.5	.. 52.2	75 57.8	.. 05.6	211 48.4	.. 31.6	175 31.5	.. 06.9	Spica	158 43.5	S11 09.1
16	222 35.0	73 49.9	51.2	90 58.7	05.2	226 51.1	31.7	190 34.1	06.9	Suhail	223 01.2	S43 25.5
17	237 37.5	88 49.4	50.2	105 59.5	04.8	241 53.9	31.9	205 36.6	06.9			
18	252 40.0	103 48.9	N13 49.2	121 00.3	N19 04.4	256 56.6	S 3 32.0	220 39.2	N10 06.8	Vega	80 46.5	N38 47.3
19	267 42.4	118 48.3	48.2	136 01.2	04.0	271 59.4	32.1	235 41.7	06.8	Zuben'ubi	137 18.2	S16 02.0
20	282 44.9	133 47.8	47.2	151 02.0	03.6	287 02.1	32.2	250 44.2	06.8		SHA	Mer. Pass.
21	297 47.4	148 47.2	.. 46.2	166 02.8	.. 03.2	302 04.9	.. 32.4	265 46.8	.. 06.7		° '	h m
22	312 49.8	163 46.7	45.2	181 03.7	02.8	317 07.6	32.5	280 49.3	06.7	Venus	213 15.6	11 04
23	327 52.3	178 46.1	44.2	196 04.5	02.4	332 10.4	32.6	295 51.8	06.6	Mars	229 28.8	9 58
Mer. Pass.	h m 1 16.0	v −0.6	d 1.0	v 0.8	d 0.4	v 2.7	d 0.1	v 2.5	d 0.0	Jupiter	4 04.7	1 00
										Saturn	327 56.2	3 24

UT	SUN GHA	SUN Dec	MOON GHA	v	Dec	d	HP
d h	° ′	° ′	° ′	′	° ′	′	′
1 00	179 57.0	N 8 25.9	71 16.6	10.2	S18 43.9	1.5	55.9
01	194 57.2	25.0	85 45.8	10.1	18 45.4	1.5	56.0
02	209 57.4	24.1	100 14.9	10.2	18 46.9	1.4	56.0
03	224 57.6	.. 23.1	114 44.1	10.0	18 48.3	1.3	56.0
04	239 57.8	22.2	129 13.1	10.1	18 49.6	1.2	56.1
05	254 58.0	21.3	143 42.2	9.9	18 50.8	1.1	56.1
06	269 58.2	N 8 20.4	158 11.1	10.0	S18 51.9	0.9	56.1
07	284 58.4	19.5	172 40.1	9.8	18 52.8	0.9	56.2
08	299 58.5	18.6	187 08.9	9.9	18 53.7	0.8	56.2
09	314 58.7	.. 17.7	201 37.8	9.8	18 54.5	0.7	56.2
10	329 58.9	16.8	216 06.6	9.7	18 55.2	0.6	56.3
11	344 59.1	15.9	230 35.3	9.7	18 55.8	0.5	56.3
12	359 59.3	N 8 15.0	245 04.0	9.6	S18 56.3	0.4	56.3
13	14 59.5	14.1	259 32.6	9.6	18 56.7	0.3	56.4
14	29 59.7	13.2	274 01.2	9.6	18 57.0	0.1	56.4
15	44 59.9	.. 12.3	288 29.8	9.5	18 57.1	0.1	56.4
16	60 00.1	11.4	302 58.3	9.5	18 57.2	0.0	56.5
17	75 00.3	10.5	317 26.8	9.4	18 57.2	0.1	56.5
18	90 00.5	N 8 09.6	331 55.2	9.4	S18 57.1	0.3	56.5
19	105 00.7	08.6	346 23.6	9.3	18 56.8	0.3	56.6
20	120 00.9	07.7	0 51.9	9.3	18 56.5	0.4	56.6
21	135 01.1	.. 06.8	15 20.2	9.3	18 56.1	0.6	56.6
22	150 01.3	05.9	29 48.5	9.2	18 55.5	0.6	56.7
23	165 01.5	05.0	44 16.7	9.2	18 54.9	0.8	56.7
2 00	180 01.7	N 8 04.1	58 44.9	9.1	S18 54.1	0.9	56.7
01	195 01.9	03.2	73 13.0	9.1	18 53.2	0.9	56.8
02	210 02.1	02.3	87 41.1	9.1	18 52.3	1.1	56.8
03	225 02.3	.. 01.4	102 09.2	9.0	18 51.2	1.2	56.9
04	240 02.5	8 00.5	116 37.2	9.0	18 50.0	1.3	56.9
05	255 02.7	7 59.5	131 05.2	9.0	18 48.7	1.4	56.9
06	270 02.9	N 7 58.6	145 33.2	8.9	S18 47.3	1.5	57.0
07	285 03.1	57.7	160 01.1	8.9	18 45.8	1.6	57.0
08	300 03.3	56.8	174 29.0	8.8	18 44.2	1.7	57.0
09	315 03.5	.. 55.9	188 56.8	8.8	18 42.5	1.8	57.1
10	330 03.7	55.0	203 24.6	8.8	18 40.7	2.0	57.1
11	345 03.9	54.1	217 52.4	8.7	18 38.7	2.0	57.1
12	0 04.1	N 7 53.2	232 20.1	8.7	S18 36.7	2.2	57.2
13	15 04.3	52.3	246 47.8	8.7	18 34.5	2.3	57.2
14	30 04.5	51.3	261 15.5	8.6	18 32.2	2.3	57.3
15	45 04.7	.. 50.4	275 43.1	8.6	18 29.9	2.5	57.3
16	60 04.9	49.5	290 10.7	8.6	18 27.4	2.6	57.3
17	75 05.2	48.6	304 38.3	8.6	18 24.8	2.7	57.4
18	90 05.4	N 7 47.7	319 05.9	8.5	S18 22.1	2.8	57.4
19	105 05.6	46.8	333 33.4	8.5	18 19.3	3.0	57.4
20	120 05.8	45.9	348 00.9	8.5	18 16.3	3.0	57.5
21	135 06.0	.. 45.0	2 28.4	8.4	18 13.3	3.1	57.5
22	150 06.2	44.0	16 55.8	8.4	18 10.2	3.3	57.5
23	165 06.4	43.1	31 23.2	8.4	18 06.9	3.4	57.6
3 00	180 06.6	N 7 42.2	45 50.6	8.4	S18 03.5	3.4	57.6
01	195 06.8	41.3	60 18.0	8.3	18 00.1	3.6	57.7
02	210 07.0	40.4	74 45.3	8.3	17 56.5	3.7	57.7
03	225 07.2	.. 39.5	89 12.6	8.3	17 52.8	3.8	57.7
04	240 07.4	38.5	103 39.9	8.3	17 49.0	3.9	57.8
05	255 07.6	37.6	118 07.2	8.2	17 45.1	4.1	57.8
06	270 07.8	N 7 36.7	132 34.4	8.2	S17 41.0	4.1	57.8
07	285 08.0	35.8	147 01.6	8.2	17 36.9	4.3	57.9
08	300 08.2	34.9	161 28.8	8.2	17 32.6	4.3	57.9
09	315 08.4	.. 34.0	175 56.0	8.2	17 28.3	4.5	57.9
10	330 08.6	33.1	190 23.2	8.1	17 23.8	4.6	58.0
11	345 08.8	32.1	204 50.3	8.1	17 19.2	4.6	58.0
12	0 09.0	N 7 31.2	219 17.4	8.1	S17 14.6	4.8	58.1
13	15 09.2	30.3	233 44.5	8.1	17 09.8	4.9	58.1
14	30 09.4	29.4	248 11.6	8.1	17 04.9	5.1	58.1
15	45 09.6	.. 28.5	262 38.7	8.1	16 59.8	5.1	58.2
16	60 09.8	27.5	277 05.8	8.0	16 54.7	5.2	58.2
17	75 10.0	26.6	291 32.8	8.0	16 49.5	5.3	58.3
18	90 10.2	N 7 25.7	305 59.8	8.0	S16 44.2	5.5	58.3
19	105 10.4	24.8	320 26.8	8.0	16 38.7	5.5	58.3
20	120 10.7	23.9	334 53.8	8.0	16 33.2	5.7	58.4
21	135 10.9	.. 23.0	349 20.8	8.0	16 27.5	5.8	58.4
22	150 11.1	22.0	3 47.8	8.0	16 21.7	5.9	58.4
23	165 11.3	21.1	18 14.8	7.9	S16 15.8	5.9	58.5
	SD 15.9	d 0.9	SD 15.3		15.6		15.8

Row side labels: 1 TUESDAY, 2 WEDNESDAY, 3 THURSDAY

Lat.	Twilight Naut.	Twilight Civil	Sunrise	Moonrise 1	2	3	4
°	h m	h m	h m	h m	h m	h m	h m
N 72	////	02 36	04 05	■■	■■	20 10	19 53
N 70	////	03 03	04 18	18 41	19 12	19 22	19 26
68	01 38	03 24	04 29	17 47	18 28	18 51	19 05
66	02 14	03 39	04 37	17 15	17 58	18 28	18 48
64	02 39	03 52	04 45	16 51	17 36	18 10	18 35
62	02 58	04 03	04 51	16 32	17 19	17 55	18 23
60	03 13	04 12	04 57	16 16	17 04	17 43	18 14
N 58	03 26	04 19	05 01	16 03	16 51	17 32	18 05
56	03 37	04 26	05 06	15 51	16 40	17 22	17 57
54	03 46	04 32	05 09	15 41	16 31	17 14	17 51
52	03 54	04 38	05 13	15 32	16 22	17 06	17 45
50	04 01	04 42	05 16	15 24	16 14	16 59	17 39
45	04 16	04 52	05 23	15 07	15 58	16 45	17 27
N 40	04 28	05 01	05 28	14 53	15 45	16 33	17 17
35	04 37	05 07	05 33	14 42	15 33	16 22	17 09
30	04 44	05 13	05 37	14 31	15 23	16 13	17 01
20	04 56	05 23	05 45	14 14	15 06	15 58	16 48
N 10	05 05	05 29	05 51	13 58	14 51	15 44	16 37
0	05 11	05 36	05 56	13 44	14 37	15 31	16 26
S 10	05 16	05 41	06 02	13 29	14 23	15 18	16 15
20	05 20	05 46	06 08	13 14	14 08	15 05	16 04
30	05 23	05 50	06 14	12 57	13 51	14 49	15 51
35	05 23	05 53	06 18	12 46	13 41	14 40	15 43
40	05 24	05 55	06 22	12 35	13 29	14 29	15 35
45	05 24	05 58	06 27	12 21	13 16	14 17	15 25
S 50	05 23	06 00	06 33	12 04	12 59	14 02	15 12
52	05 23	06 02	06 36	11 56	12 52	13 55	15 07
54	05 22	06 03	06 38	11 48	12 43	13 48	15 00
56	05 21	06 04	06 42	11 38	12 33	13 39	14 53
58	05 20	06 06	06 45	11 26	12 22	13 29	14 45
S 60	05 19	06 07	06 49	11 14	12 10	13 18	14 36

Lat.	Sunset	Twilight Civil	Twilight Naut.	Moonset 1	2	3	4
°	h m	h m	h m	h m	h m	h m	h m
N 72	19 51	21 17	////	■■	■■	23 30	25 42
N 70	19 38	20 52	23 21	21 13	22 33	24 16	00 16
68	19 28	20 33	22 14	22 06	23 17	24 46	00 46
66	19 20	20 17	21 40	22 39	23 46	25 09	01 09
64	19 13	20 05	21 17	23 03	24 07	00 07	01 26
62	19 07	19 55	20 58	23 22	24 25	00 25	01 41
60	19 01	19 46	20 43	23 37	24 40	00 40	01 53
N 58	18 57	19 38	20 31	23 51	24 52	00 52	02 04
56	18 53	19 32	20 21	24 02	00 02	01 03	02 13
54	18 49	19 26	20 12	24 12	00 12	01 12	02 21
52	18 46	19 21	20 04	24 21	00 21	01 20	02 28
50	18 42	19 16	19 57	24 29	00 29	01 28	02 35
45	18 36	19 06	19 42	24 46	00 46	01 44	02 48
N 40	18 30	18 58	19 31	00 08	01 00	01 57	03 00
35	18 26	18 52	19 22	00 20	01 11	02 08	03 10
30	18 22	18 46	19 15	00 30	01 22	02 18	03 18
20	18 15	18 37	19 03	00 48	01 39	02 35	03 33
N 10	18 09	18 30	18 55	01 03	01 55	02 49	03 45
0	18 03	18 24	18 48	01 17	02 09	03 03	03 57
S 10	17 58	18 19	18 43	01 31	02 23	03 16	04 09
20	17 52	18 14	18 40	01 46	02 39	03 31	04 22
30	17 45	18 09	18 37	02 04	02 56	03 47	04 36
35	17 42	18 07	18 37	02 14	03 06	03 57	04 44
40	17 38	18 05	18 36	02 26	03 18	04 07	04 54
45	17 33	18 02	18 36	02 39	03 32	04 20	05 05
S 50	17 27	18 00	18 37	02 56	03 48	04 36	05 18
52	17 25	17 59	18 38	03 04	03 56	04 43	05 24
54	17 22	17 58	18 39	03 12	04 05	04 51	05 31
56	17 19	17 56	18 39	03 22	04 14	05 00	05 38
58	17 15	17 55	18 40	03 33	04 24	05 10	05 46
S 60	17 12	17 54	18 42	03 46	04 38	05 21	05 56

Day	SUN Eqn. of Time 00h	12h	Mer. Pass.	MOON Mer. Pass. Upper	Lower	Age	Phase
d	m s	m s	h m	h m	h m	d	%
1	00 13	00 03	12 00	19 56	07 30	10	72
2	00 07	00 16	12 00	20 50	08 23	11	81
3	00 26	00 36	11 59	21 44	09 17	12	89

1998 SEPTEMBER 4, 5, 6 (FRI., SAT., SUN.)

UT	ARIES GHA	VENUS −3.9 GHA	VENUS Dec	MARS +1.7 GHA	MARS Dec	JUPITER −2.9 GHA	JUPITER Dec	SATURN +0.2 GHA	SATURN Dec	STARS Name	SHA	Dec
FRIDAY 4 00	342 54.7	193 45.6	N13 43.3	211 05.4	N19 02.0	347 13.1	S 3 32.8	310 54.4	N10 06.6	Acamar	315 26.8	S40 18.4
01	357 57.2	208 45.1	42.3	226 06.2	01.6	2 15.9	32.9	325 56.9	06.6	Achernar	335 34.9	S57 14.5
02	12 59.7	223 44.5	41.3	241 07.0	01.2	17 18.6	33.0	340 59.5	06.5	Acrux	173 22.6	S63 05.5
03	28 02.1	238 44.0 ..	40.3	256 07.9 ..	00.8	32 21.4 ..	33.1	356 02.0 ..	06.5	Adhara	255 21.7	S28 58.1
04	43 04.6	253 43.4	39.3	271 08.7	00.4	47 24.1	33.3	11 04.5	06.5	Aldebaran	291 02.4	N16 30.3
05	58 07.1	268 42.9	38.3	286 09.6	19 00.0	62 26.9	33.4	26 07.1	06.4			
06	73 09.5	283 42.3	N13 37.3	301 10.4	N18 59.6	77 29.6	S 3 33.5	41 09.6	N10 06.4	Alioth	166 31.2	N55 58.3
07	88 12.0	298 41.8	36.3	316 11.2	59.2	92 32.4	33.7	56 12.2	06.3	Alkaid	153 08.2	N49 19.5
08	103 14.5	313 41.3	35.3	331 12.1	58.8	107 35.1	33.8	71 14.7	06.3	Al Na'ir	27 57.6	S46 58.0
09	118 16.9	328 40.7 ..	34.3	346 12.9 ..	58.4	122 37.9 ..	33.9	86 17.2 ..	06.3	Alnilam	275 58.0	S 1 12.2
10	133 19.4	343 40.2	33.3	1 13.8	58.0	137 40.6	34.1	101 19.8	06.2	Alphard	218 07.6	S 8 39.1
11	148 21.9	358 39.7	32.3	16 14.6	57.6	152 43.4	34.2	116 22.3	06.2			
12	163 24.3	13 39.1	N13 31.3	31 15.5	N18 57.2	167 46.2	S 3 34.3	131 24.9	N10 06.2	Alphecca	126 20.8	N26 43.5
13	178 26.8	28 38.6	30.3	46 16.3	56.8	182 48.9	34.4	146 27.4	06.1	Alpheratz	357 54.9	N29 05.0
14	193 29.2	43 38.0	29.3	61 17.1	56.4	197 51.7	34.6	161 30.0	06.1	Altair	62 19.1	N 8 52.1
15	208 31.7	58 37.5 ..	28.3	76 18.0 ..	56.0	212 54.4 ..	34.7	176 32.5 ..	06.0	Ankaa	353 26.6	S42 18.7
16	223 34.2	73 37.0	27.3	91 18.8	55.5	227 57.2	34.8	191 35.0	06.0	Antares	112 40.3	S26 25.6
17	238 36.6	88 36.4	26.3	106 19.7	55.1	242 59.9	35.0	206 37.6	06.0			
18	253 39.1	103 35.9	N13 25.3	121 20.5	N18 54.7	258 02.7	S 3 35.1	221 40.1	N10 05.9	Arcturus	146 06.3	N19 11.7
19	268 41.6	118 35.4	24.3	136 21.4	54.3	273 05.4	35.2	236 42.7	05.9	Atria	107 52.3	S69 01.6
20	283 44.0	133 34.8	23.3	151 22.2	53.9	288 08.2	35.4	251 45.2	05.9	Avior	234 23.1	S59 30.2
21	298 46.5	148 34.3 ..	22.3	166 23.0 ..	53.5	303 10.9 ..	35.5	266 47.7 ..	05.8	Bellatrix	278 44.3	N 6 20.9
22	313 49.0	163 33.8	21.2	181 23.9	53.1	318 13.7	35.6	281 50.3	05.8	Betelgeuse	271 13.7	N 7 24.4
23	328 51.4	178 33.2	20.2	196 24.7	52.7	333 16.4	35.8	296 52.8	05.7			
SATURDAY 5 00	343 53.9	193 32.7	N13 19.2	211 25.6	N18 52.3	348 19.2	S 3 35.9	311 55.4	N10 05.7	Canopus	264 01.4	S52 41.5
01	358 56.4	208 32.2	18.2	226 26.4	51.9	3 21.9	36.0	326 57.9	05.7	Capella	280 51.3	N45 59.5
02	13 58.8	223 31.6	17.2	241 27.3	51.5	18 24.7	36.1	342 00.5	05.6	Deneb	49 38.8	N45 16.8
03	29 01.3	238 31.1 ..	16.2	256 28.1 ..	51.1	33 27.5 ..	36.3	357 03.0 ..	05.6	Denebola	182 45.6	N14 35.0
04	44 03.7	253 30.6	15.2	271 29.0	50.7	48 30.2	36.4	12 05.6	05.6	Diphda	349 07.0	S17 59.5
05	59 06.2	268 30.0	14.2	286 29.8	50.2	63 33.0	36.5	27 08.1	05.5			
06	74 08.7	283 29.5	N13 13.2	301 30.6	N18 49.8	78 35.7	S 3 36.7	42 10.6	N10 05.5	Dubhe	194 06.3	N61 45.6
07	89 11.1	298 29.0	12.2	316 31.5	49.4	93 38.5	36.8	57 13.2	05.4	Elnath	278 27.1	N28 36.2
08	104 13.6	313 28.4	11.1	331 32.3	49.0	108 41.2	36.9	72 15.7	05.4	Eltanin	90 51.3	N51 29.8
09	119 16.1	328 27.9 ..	10.1	346 33.2 ..	48.6	123 44.0 ..	37.1	87 18.3 ..	05.4	Enif	33 58.0	N 9 52.3
10	134 18.5	343 27.4	09.1	1 34.0	48.2	138 46.7	37.2	102 20.8	05.3	Fomalhaut	15 36.2	S29 37.7
11	149 21.0	358 26.8	08.1	16 34.9	47.8	153 49.5	37.3	117 23.4	05.3			
12	164 23.5	13 26.3	N13 07.1	31 35.7	N18 47.4	168 52.2	S 3 37.5	132 25.9	N10 05.3	Gacrux	172 14.1	S57 06.3
13	179 25.9	28 25.8	06.1	46 36.6	47.0	183 55.0	37.6	147 28.5	05.2	Gienah	176 04.3	S17 31.9
14	194 28.4	43 25.3	05.0	61 37.4	46.6	198 57.8	37.7	162 31.0	05.2	Hadar	149 04.5	S60 22.0
15	209 30.8	58 24.7 ..	04.0	76 38.3 ..	46.1	214 00.5 ..	37.8	177 33.5 ..	05.1	Hamal	328 13.4	N23 27.3
16	224 33.3	73 24.2	03.0	91 39.1	45.7	229 03.3	38.0	192 36.1	05.1	Kaus Aust.	83 58.7	S34 23.1
17	239 35.8	88 23.7	02.0	106 40.0	45.3	244 06.0	38.1	207 38.6	05.1			
18	254 38.2	103 23.1	N13 01.0	121 40.8	N18 44.9	259 08.8	S 3 38.2	222 41.2	N10 05.0	Kochab	137 20.3	N74 10.1
19	269 40.7	118 22.6	12 59.9	136 41.6	44.5	274 11.5	38.4	237 43.7	05.0	Markab	13 49.3	N15 12.0
20	284 43.2	133 22.1	58.9	151 42.5	44.1	289 14.3	38.5	252 46.3	04.9	Menkar	314 26.8	N 4 05.1
21	299 45.6	148 21.6 ..	57.9	166 43.3 ..	43.7	304 17.1 ..	38.6	267 48.8 ..	04.9	Menkent	148 21.3	S36 21.7
22	314 48.1	163 21.0	56.9	181 44.2	43.3	319 19.8	38.8	282 51.4	04.9	Miaplacidus	221 42.8	S69 42.6
23	329 50.6	178 20.5	55.9	196 45.0	42.9	334 22.6	38.9	297 53.9	04.8			
SUNDAY 6 00	344 53.0	193 20.0	N12 54.8	211 45.9	N18 42.4	349 25.3	S 3 39.0	312 56.5	N10 04.8	Mirfak	308 56.5	N49 51.1
01	359 55.5	208 19.5	53.8	226 46.7	42.0	4 28.1	39.2	327 59.0	04.7	Nunki	76 12.2	S26 17.8
02	14 58.0	223 18.9	52.8	241 47.6	41.6	19 30.8	39.3	343 01.6	04.7	Peacock	53 36.7	S56 44.4
03	30 00.4	238 18.4 ..	51.8	256 48.4 ..	41.2	34 33.6 ..	39.4	358 04.1 ..	04.7	Pollux	243 41.9	N28 01.7
04	45 02.9	253 17.9	50.7	271 49.3	40.8	49 36.3	39.6	13 06.6	04.6	Procyon	245 11.8	N 5 13.7
05	60 05.3	268 17.4	49.7	286 50.1	40.4	64 39.1	39.7	28 09.2	04.6			
06	75 07.8	283 16.8	N12 48.7	301 51.0	N18 40.0	79 41.9	S 3 39.8	43 11.7	N10 04.5	Rasalhague	96 17.0	N12 34.0
07	90 10.3	298 16.3	47.7	316 51.8	39.5	94 44.6	39.9	58 14.3	04.5	Regulus	207 55.9	N11 58.5
08	105 12.7	313 15.8	46.6	331 52.7	39.1	109 47.4	40.1	73 16.8	04.5	Rigel	281 23.0	S 8 12.1
09	120 15.2	328 15.3 ..	45.6	346 53.5 ..	38.7	124 50.1 ..	40.2	88 19.4 ..	04.4	Rigil Kent.	140 07.7	S60 49.8
10	135 17.7	343 14.8	44.6	1 54.4	38.3	139 52.9	40.3	103 21.9	04.4	Sabik	102 25.6	S15 43.2
11	150 20.1	358 14.2	43.5	16 55.2	37.9	154 55.7	40.5	118 24.5	04.3			
12	165 22.6	13 13.7	N12 42.5	31 56.1	N18 37.5	169 58.4	S 3 40.6	133 27.0	N10 04.3	Schedar	349 53.0	N56 31.7
13	180 25.1	28 13.2	41.5	46 56.9	37.1	185 01.2	40.7	148 29.6	04.3	Shaula	96 37.3	S37 06.1
14	195 27.5	43 12.7	40.4	61 57.8	36.6	200 03.9	40.9	163 32.1	04.2	Sirius	258 43.9	S16 42.7
15	210 30.0	58 12.2 ..	39.4	76 58.7 ..	36.2	215 06.7 ..	41.0	178 34.7 ..	04.2	Spica	158 43.5	S11 09.1
16	225 32.4	73 11.6	38.4	91 59.5	35.8	230 09.4	41.1	193 37.2	04.2	Suhail	223 01.2	S43 25.5
17	240 34.9	88 11.1	37.3	107 00.4	35.4	245 12.2	41.3	208 39.8	04.1			
18	255 37.4	103 10.6	N12 36.3	122 01.2	N18 35.0	260 15.0	S 3 41.4	223 42.3	N10 04.1	Vega	80 46.5	N38 47.3
19	270 39.8	118 10.1	35.3	137 02.1	34.6	275 17.7	41.5	238 44.9	04.0	Zuben'ubi	137 18.2	S16 02.0
20	285 42.3	133 09.6	34.2	152 02.9	34.1	290 20.5	41.7	253 47.4	04.0			
21	300 44.8	148 09.1 ..	33.2	167 03.8 ..	33.7	305 23.2 ..	41.8	268 50.0 ..	04.0		SHA	Mer. Pass.
22	315 47.2	163 08.5	32.2	182 04.6	33.3	320 26.0	41.9	283 52.5	03.9	Venus	209 38.8	11 06
23	330 49.7	178 08.0	31.1	197 05.5	32.9	335 28.7	42.1	298 55.1	03.9	Mars	227 31.7	9 54
Mer. Pass. 1 04.2		v −0.5 d 1.0		v 0.8 d 0.4		v 2.8 d 0.1		v 2.5 d 0.0		Jupiter	4 25.3	0 47
										Saturn	328 01.5	3 12

UT	SUN GHA	SUN Dec	MOON GHA	v	MOON Dec	d	HP
4 FRIDAY	° ′	° ′	° ′	′	° ′	′	′
00	180 11.5	N 7 20.2	32 41.7	8.0	S16 09.9	6.1	58.5
01	195 11.7	19.3	47 08.7	7.9	16 03.8	6.2	58.5
02	210 11.9	18.4	61 35.6	7.9	15 57.6	6.3	58.6
03	225 12.1 ..	17.4	76 02.5	8.0	15 51.3	6.4	58.6
04	240 12.3	16.5	90 29.5	7.9	15 44.9	6.5	58.7
05	255 12.5	15.6	104 56.4	7.9	15 38.4	6.7	58.7
06	270 12.7	N 7 14.7	119 23.3	7.9	S15 31.7	6.7	58.7
07	285 12.9	13.8	133 50.2	7.9	15 25.0	6.8	58.8
08	300 13.1	12.8	148 17.1	7.8	15 18.2	6.9	58.8
09	315 13.3 ..	11.9	162 43.9	7.9	15 11.3	7.0	58.8
10	330 13.5	11.0	177 10.8	7.9	15 04.3	7.2	58.8
11	345 13.7	10.1	191 37.7	7.9	14 57.1	7.2	58.9
12	0 14.0	N 7 09.1	206 04.6	7.8	S14 49.9	7.3	58.9
13	15 14.2	08.2	220 31.4	7.9	14 42.6	7.4	59.0
14	30 14.4	07.3	234 58.3	7.8	14 35.2	7.6	59.0
15	45 14.6 ..	06.4	249 25.1	7.9	14 27.6	7.6	59.0
16	60 14.8	05.5	263 52.0	7.9	14 20.0	7.7	59.1
17	75 15.0	04.5	278 18.9	7.8	14 12.3	7.8	59.1
18	90 15.2	N 7 03.6	292 45.7	7.9	S14 04.5	7.9	59.1
19	105 15.4	02.7	307 12.6	7.8	13 56.6	8.0	59.2
20	120 15.6	01.8	321 39.4	7.9	13 48.6	8.1	59.2
21	135 15.8	7 00.8	336 06.3	7.8	13 40.5	8.2	59.2
22	150 16.0	6 59.9	350 33.1	7.9	13 32.3	8.3	59.3
23	165 16.2	59.0	5 00.0	7.8	13 24.0	8.4	59.3
5 SATURDAY							
00	180 16.5	N 6 58.1	19 26.8	7.9	S13 15.6	8.5	59.3
01	195 16.7	57.1	33 53.7	7.9	13 07.1	8.6	59.4
02	210 16.9	56.2	48 20.6	7.8	12 58.5	8.6	59.4
03	225 17.1 ..	55.3	62 47.4	7.9	12 49.9	8.8	59.4
04	240 17.3	54.4	77 14.3	7.9	12 41.1	8.8	59.5
05	255 17.5	53.4	91 41.2	7.8	12 32.3	8.9	59.5
06	270 17.7	N 6 52.5	106 08.0	7.9	S12 23.4	9.0	59.5
07	285 17.9	51.6	120 34.9	7.9	12 14.4	9.1	59.5
08	300 18.1	50.7	135 01.8	7.9	12 05.3	9.2	59.6
09	315 18.3 ..	49.7	149 28.7	7.8	11 56.1	9.3	59.6
10	330 18.5	48.8	163 55.5	7.9	11 46.8	9.3	59.6
11	345 18.8	47.9	178 22.4	7.9	11 37.5	9.5	59.7
12	0 19.0	N 6 47.0	192 49.3	7.9	S11 28.0	9.5	59.7
13	15 19.2	46.0	207 16.2	7.9	11 18.5	9.6	59.7
14	30 19.4	45.1	221 43.1	7.9	11 08.9	9.6	59.7
15	45 19.6 ..	44.2	236 10.0	8.0	10 59.3	9.8	59.8
16	60 19.8	43.3	250 37.0	7.9	10 49.5	9.8	59.8
17	75 20.0	42.3	265 03.9	7.9	10 39.7	9.9	59.8
18	90 20.2	N 6 41.4	279 30.8	7.9	S10 29.8	10.0	59.9
19	105 20.4	40.5	293 57.7	8.0	10 19.8	10.0	59.9
20	120 20.6	39.5	308 24.7	7.9	10 09.8	10.1	59.9
21	135 20.9	38.6	322 51.6	8.0	9 59.7	10.2	59.9
22	150 21.1	37.7	337 18.6	7.9	9 49.5	10.3	60.0
23	165 21.3	36.8	351 45.5	8.0	9 39.2	10.3	60.0
6 SUNDAY							
00	180 21.5	N 6 35.8	6 12.5	8.0	S 9 28.9	10.4	60.0
01	195 21.7	34.9	20 39.5	7.9	9 18.5	10.5	60.0
02	210 21.9	34.0	35 06.4	8.0	9 08.0	10.5	60.1
03	225 22.1 ..	33.0	49 33.4	8.0	8 57.5	10.6	60.1
04	240 22.3	32.1	64 00.4	8.0	8 46.9	10.7	60.1
05	255 22.5	31.2	78 27.4	8.0	8 36.2	10.7	60.1
06	270 22.8	N 6 30.2	92 54.4	8.1	S 8 25.5	10.8	60.1
07	285 23.0	29.3	107 21.5	8.0	8 14.7	10.8	60.2
08	300 23.2	28.4	121 48.5	8.0	8 03.9	10.9	60.2
09	315 23.4 ..	27.5	136 15.5	8.0	7 53.0	11.0	60.2
10	330 23.6	26.5	150 42.5	8.1	7 42.0	11.0	60.2
11	345 23.8	25.6	165 09.6	8.0	7 31.0	11.1	60.2
12	0 24.0	N 6 24.7	179 36.6	8.1	S 7 19.9	11.1	60.3
13	15 24.2	23.7	194 03.7	8.0	7 08.8	11.2	60.3
14	30 24.4	22.8	208 30.7	8.1	6 57.6	11.2	60.3
15	45 24.7 ..	21.9	222 57.8	8.1	6 46.4	11.3	60.3
16	60 24.9	20.9	237 24.9	8.1	6 35.1	11.3	60.3
17	75 25.1	20.0	251 52.0	8.1	6 23.8	11.4	60.4
18	90 25.3	N 6 19.1	266 19.1	8.1	S 6 12.4	11.4	60.4
19	105 25.5	18.1	280 46.2	8.1	6 01.0	11.5	60.4
20	120 25.7	17.2	295 13.3	8.1	5 49.5	11.5	60.4
21	135 25.9 ..	16.3	309 40.4	8.1	5 38.0	11.5	60.4
22	150 26.1	15.3	324 07.5	8.1	5 26.5	11.6	60.4
23	165 26.4	14.4	338 34.6	8.1	S 5 14.9	11.6	60.5
	SD 15.9	d 0.9	SD 16.1		16.3		16.4

Lat.	Twilight Naut.	Twilight Civil	Sunrise	Moonrise 4	5	6	7
°	h m	h m	h m	h m	h m	h m	h m
N 72	////	02 56	04 20	19 53	19 43	19 35	19 27
N 70	01 14	03 19	04 31	19 26	19 27	19 26	19 26
68	02 02	03 37	04 40	19 05	19 13	19 19	19 24
66	02 31	03 51	04 47	18 48	19 03	19 13	19 23
64	02 52	04 02	04 54	18 35	18 53	19 08	19 22
62	03 09	04 11	04 59	18 23	18 46	19 04	19 21
60	03 23	04 19	05 04	18 14	18 39	19 00	19 20
N 58	03 34	04 26	05 08	18 05	18 33	18 57	19 19
56	03 44	04 33	05 11	17 57	18 27	18 54	19 19
54	03 53	04 38	05 15	17 51	18 23	18 51	19 18
52	04 00	04 43	05 18	17 45	18 18	18 49	19 17
50	04 07	04 47	05 20	17 39	18 14	18 47	19 17
45	04 20	04 56	05 26	17 27	18 06	18 42	19 16
N 40	04 31	05 04	05 31	17 17	17 59	18 38	19 15
35	04 39	05 10	05 35	17 09	17 52	18 34	19 14
30	04 46	05 15	05 39	17 01	17 47	18 31	19 14
20	04 57	05 23	05 45	16 48	17 37	18 25	19 12
N 10	05 05	05 29	05 51	16 37	17 29	18 21	19 11
0	05 10	05 35	05 55	16 26	17 21	18 16	19 11
S 10	05 15	05 39	06 00	16 15	17 13	18 11	19 10
20	05 18	05 43	06 05	16 04	17 05	18 07	19 09
30	05 19	05 47	06 11	15 51	16 55	18 01	19 08
35	05 19	05 49	06 14	15 43	16 50	17 58	19 07
40	05 19	05 51	06 18	15 35	16 43	17 54	19 06
45	05 18	05 52	06 22	15 25	16 36	17 50	19 06
S 50	05 17	05 54	06 27	15 12	16 27	17 45	19 05
52	05 16	05 55	06 29	15 07	16 23	17 43	19 04
54	05 15	05 56	06 31	15 00	16 19	17 40	19 04
56	05 14	05 57	06 34	14 53	16 14	17 38	19 03
58	05 12	05 58	06 37	14 45	16 08	17 34	19 03
S 60	05 10	05 58	06 40	14 36	16 02	17 31	19 02

Lat.	Sunset	Twilight Civil	Twilight Naut.	Moonset 4	5	6	7
°	h m	h m	h m	h m	h m	h m	h m
N 72	19 35	20 56	////	25 42	01 42	03 47	05 50
N 70	19 24	20 34	22 33	00 16	02 08	04 02	05 56
68	19 15	20 18	21 50	00 46	02 27	04 13	06 01
66	19 08	20 04	21 22	01 09	02 43	04 23	06 05
64	19 02	19 53	21 02	01 26	02 56	04 30	06 08
62	18 57	19 44	20 45	01 41	03 06	04 37	06 11
60	18 52	19 36	20 32	01 53	03 15	04 43	06 14
N 58	18 48	19 29	20 21	02 04	03 23	04 48	06 16
56	18 45	19 23	20 11	02 13	03 30	04 53	06 18
54	18 42	19 18	20 03	02 21	03 37	04 57	06 20
52	18 39	19 13	19 56	02 28	03 42	05 00	06 21
50	18 36	19 09	19 49	02 35	03 47	05 04	06 23
45	18 30	19 00	19 36	02 48	03 58	05 11	06 26
N 40	18 26	18 53	19 18	03 00	04 07	05 17	06 28
35	18 22	18 47	19 13	03 10	04 15	05 22	06 31
30	18 18	18 42	19 11	03 18	04 21	05 27	06 33
20	18 12	18 34	19 00	03 33	04 33	05 34	06 36
N 10	18 07	18 28	18 53	03 45	04 43	05 41	06 39
0	18 02	18 23	18 47	03 57	04 52	05 47	06 42
S 10	17 57	18 18	18 43	04 09	05 02	05 53	06 45
20	17 52	18 15	18 40	04 22	05 12	06 00	06 49
30	17 47	18 11	18 39	04 36	05 23	06 07	06 51
35	17 44	18 09	18 39	04 44	05 29	06 12	06 53
40	17 40	18 08	18 39	04 54	05 36	06 16	06 55
45	17 36	18 06	18 40	05 05	05 45	06 22	06 57
S 50	17 32	18 04	18 42	05 18	05 55	06 29	07 00
52	17 30	18 03	18 42	05 24	06 00	06 32	07 01
54	17 27	18 03	18 44	05 31	06 05	06 35	07 02
56	17 25	18 02	18 47	05 38	06 11	06 39	07 04
58	17 22	18 01	18 47	05 46	06 17	06 43	07 06
S 60	17 19	18 00	18 49	05 56	06 24	06 47	07 08

	SUN Eqn. of Time 00ʰ	SUN Eqn. of Time 12ʰ	SUN Mer. Pass.	MOON Mer. Pass. Upper	MOON Mer. Pass. Lower	Age	Phase %
Day	m s	m s	h m	h m	h m	d	%
4	00 45	00 55	11 59	22 39	10 12	13	95
5	01 05	01 15	11 59	23 34	11 07	14	99
6	01 26	01 36	11 58	24 29	12 02	15	100

1998 SEPTEMBER 7, 8, 9 (MON., TUES., WED.)

UT	ARIES GHA	VENUS −3.9 GHA	Dec	MARS +1.7 GHA	Dec	JUPITER −2.9 GHA	Dec	SATURN +0.2 GHA	Dec	STARS Name	SHA	Dec
d h	° ′	° ′	° ′	° ′	° ′	° ′	° ′	° ′	° ′		° ′	° ′
7 00	345 52.2	193 07.5	N12 30.1	212 06.3	N18 32.5	350 31.5	S 3 42.2	313 57.6	N10 03.8	Acamar	315 26.8	S40 18.4
01	0 54.6	208 07.0	29.0	227 07.2	32.1	5 34.3	42.3	329 00.2	03.8	Achernar	335 34.8	S57 14.5
02	15 57.1	223 06.5	28.0	242 08.0	31.6	20 37.0	42.5	344 02.7	03.7	Acrux	173 22.6	S63 05.5
03	30 59.6	238 06.0 ..	27.0	257 08.9 ..	31.2	35 39.8 ..	42.6	359 05.3 ..	03.7	Adhara	255 21.6	S28 58.1
04	46 02.0	253 05.4	25.9	272 09.7	30.8	50 42.5	42.7	14 07.8	03.7	Aldebaran	291 02.4	N16 30.3
05	61 04.5	268 04.9	24.9	287 10.6	30.4	65 45.3	42.8	29 10.4	03.6			
06	76 06.9	283 04.4	N12 23.8	302 11.5	N18 30.0	80 48.1	S 3 43.0	44 12.9	N10 03.6	Alioth	166 31.2	N55 58.3
07	91 09.4	298 03.9	22.8	317 12.3	29.6	95 50.8	43.1	59 15.5	03.5	Alkaid	153 08.3	N49 19.5
08	106 11.9	313 03.4	21.7	332 13.2	29.1	110 53.6	43.2	74 18.0	03.5	Al Na'ir	27 57.6	S46 58.0
M 09	121 14.3	328 02.9 ..	20.7	347 14.0 ..	28.7	125 56.3 ..	43.4	89 20.6 ..	03.5	Alnilam	275 58.0	S 1 12.2
O 10	136 16.8	343 02.4	19.7	2 14.9	28.3	140 59.1	43.5	104 23.1	03.4	Alphard	218 07.6	S 8 39.1
N 11	151 19.3	358 01.9	18.6	17 15.7	27.9	156 01.9	43.6	119 25.7	03.4			
D 12	166 21.7	13 01.3	N12 17.6	32 16.6	N18 27.5	171 04.6	S 3 43.8	134 28.2	N10 03.3	Alphecca	126 20.8	N26 43.5
A 13	181 24.2	28 00.8	16.5	47 17.4	27.0	186 07.4	43.9	149 30.8	03.3	Alpheratz	357 54.9	N29 05.0
Y 14	196 26.7	43 00.3	15.5	62 18.3	26.6	201 10.1	44.0	164 33.4	03.3	Altair	62 19.1	N 8 52.1
15	211 29.1	57 59.8 ..	14.4	77 19.2 ..	26.2	216 12.9 ..	44.2	179 35.9 ..	03.2	Ankaa	353 26.6	S42 18.7
16	226 31.6	72 59.3	13.4	92 20.0	25.8	231 15.7	44.3	194 38.5	03.2	Antares	112 40.3	S26 25.6
17	241 34.1	87 58.8	12.3	107 20.9	25.4	246 18.4	44.4	209 41.0	03.1			
18	256 36.5	102 58.3	N12 11.3	122 21.7	N18 24.9	261 21.2	S 3 44.6	224 43.6	N10 03.1	Arcturus	146 06.4	N19 11.7
19	271 39.0	117 57.8	10.2	137 22.6	24.5	276 23.9	44.7	239 46.1	03.1	Atria	107 52.3	S69 01.6
20	286 41.4	132 57.3	09.2	152 23.4	24.1	291 26.7	44.8	254 48.7	03.0	Avior	234 23.1	S59 30.2
21	301 43.9	147 56.8 ..	08.1	167 24.3 ..	23.7	306 29.5 ..	45.0	269 51.2 ..	03.0	Bellatrix	278 44.3	N 6 20.9
22	316 46.4	162 56.2	07.1	182 25.2	23.2	321 32.2	45.1	284 53.8	02.9	Betelgeuse	271 13.7	N 7 24.4
23	331 48.8	177 55.7	06.0	197 26.0	22.8	336 35.0	45.2	299 56.3	02.9			
8 00	346 51.3	192 55.2	N12 05.0	212 26.9	N18 22.4	351 37.7	S 3 45.4	314 58.9	N10 02.8	Canopus	264 01.4	S52 41.5
01	1 53.8	207 54.7	03.9	227 27.7	22.0	6 40.5	45.5	330 01.4	02.8	Capella	280 51.3	N45 59.5
02	16 56.2	222 54.2	02.9	242 28.6	21.6	21 43.3	45.6	345 04.0	02.8	Deneb	49 38.8	N45 16.8
03	31 58.7	237 53.7 ..	01.8	257 29.4 ..	21.1	36 46.0 ..	45.8	0 06.6 ..	02.7	Denebola	182 45.6	N14 35.0
04	47 01.2	252 53.2	12 00.8	272 30.3	20.7	51 48.8	45.9	15 09.1	02.7	Diphda	349 07.0	S17 59.5
05	62 03.6	267 52.7	11 59.7	287 31.2	20.3	66 51.6	46.0	30 11.7	02.6			
06	77 06.1	282 52.2	N11 58.6	302 32.0	N18 19.9	81 54.3	S 3 46.2	45 14.2	N10 02.6	Dubhe	194 06.3	N61 45.6
07	92 08.5	297 51.7	57.6	317 32.9	19.4	96 57.1	46.3	60 16.8	02.6	Elnath	278 27.1	N28 36.2
T 08	107 11.0	312 51.2	56.5	332 33.7	19.0	111 59.8	46.4	75 19.3	02.5	Eltanin	90 51.3	N51 29.8
U 09	122 13.5	327 50.7 ..	55.5	347 34.6 ..	18.6	127 02.6 ..	46.6	90 21.9 ..	02.5	Enif	33 58.0	N 9 52.3
E 10	137 15.9	342 50.2	54.4	2 35.5	18.2	142 05.4	46.7	105 24.4	02.4	Fomalhaut	15 36.2	S29 37.7
S 11	152 18.4	357 49.7	53.4	17 36.3	17.7	157 08.1	46.8	120 27.0	02.4			
D 12	167 20.9	12 49.2	N11 52.3	32 37.2	N18 17.3	172 10.9	S 3 47.0	135 29.6	N10 02.3	Gacrux	172 14.1	S57 06.3
A 13	182 23.3	27 48.7	51.2	47 38.0	16.9	187 13.6	47.1	150 32.1	02.3	Gienah	176 04.3	S17 31.9
Y 14	197 25.8	42 48.2	50.2	62 38.9	16.5	202 16.4	47.2	165 34.7	02.3	Hadar	149 04.5	S60 22.0
15	212 28.3	57 47.7 ..	49.1	77 39.8 ..	16.1	217 19.2 ..	47.4	180 37.2 ..	02.2	Hamal	328 13.4	N23 27.3
16	227 30.7	72 47.2	48.0	92 40.6	15.6	232 21.9	47.5	195 39.8	02.2	Kaus Aust.	83 58.8	S34 23.1
17	242 33.2	87 46.7	47.0	107 41.5	15.2	247 24.7	47.6	210 42.3	02.1			
18	257 35.7	102 46.2	N11 45.9	122 42.4	N18 14.8	262 27.5	S 3 47.8	225 44.9	N10 02.1	Kochab	137 20.3	N74 10.1
19	272 38.1	117 45.7	44.9	137 43.2	14.3	277 30.2	47.9	240 47.5	02.0	Markab	13 49.3	N15 12.0
20	287 40.6	132 45.2	43.8	152 44.1	13.9	292 33.0	48.0	255 50.0	02.0	Menkar	314 26.8	N 4 05.1
21	302 43.0	147 44.7 ..	42.7	167 44.9 ..	13.5	307 35.7 ..	48.2	270 52.6 ..	02.0	Menkent	148 21.3	S36 21.7
22	317 45.5	162 44.2	41.7	182 45.8	13.1	322 38.5	48.3	285 55.1	01.9	Miaplacidus	221 42.8	S69 42.6
23	332 48.0	177 43.7	40.6	197 46.7	12.6	337 41.3	48.4	300 57.7	01.9			
9 00	347 50.4	192 43.2	N11 39.5	212 47.5	N18 12.2	352 44.0	S 3 48.5	316 00.2	N10 01.8	Mirfak	308 56.5	N49 51.1
01	2 52.9	207 42.7	38.5	227 48.4	11.8	7 46.8	48.7	331 02.8	01.8	Nunki	76 12.2	S26 17.8
02	17 55.4	222 42.2	37.4	242 49.3	11.4	22 49.6	48.8	346 05.4	01.7	Peacock	53 36.7	S56 44.4
03	32 57.8	237 41.7 ..	36.3	257 50.1 ..	10.9	37 52.3 ..	48.9	1 07.9 ..	01.7	Pollux	243 41.9	N28 01.7
04	48 00.3	252 41.2	35.3	272 51.0	10.5	52 55.1	49.1	16 10.5	01.7	Procyon	245 11.8	N 5 13.7
05	63 02.8	267 40.7	34.2	287 51.8	10.1	67 57.9	49.2	31 13.0	01.6			
06	78 05.2	282 40.2	N11 33.1	302 52.7	N18 09.7	83 00.6	S 3 49.3	46 15.6	N10 01.6	Rasalhague	96 17.0	N12 34.0
07	93 07.7	297 39.7	32.0	317 53.6	09.2	98 03.4	49.5	61 18.2	01.5	Regulus	207 55.9	N11 58.5
W 08	108 10.1	312 39.2	31.0	332 54.4	08.8	113 06.1	49.6	76 20.7	01.5	Rigel	281 23.0	S 8 12.1
E 09	123 12.6	327 38.7 ..	29.9	347 55.3 ..	08.4	128 08.9 ..	49.7	91 23.3 ..	01.4	Rigil Kent.	140 07.7	S60 49.8
D 10	138 15.1	342 38.2	28.8	2 56.2	07.9	143 11.7	49.9	106 25.8	01.4	Sabik	102 25.6	S15 43.2
N 11	153 17.5	357 37.7	27.8	17 57.0	07.5	158 14.4	50.0	121 28.4	01.4			
E 12	168 20.0	12 37.2	N11 26.7	32 57.9	N18 07.1	173 17.2	S 3 50.1	136 30.9	N10 01.3	Schedar	349 53.0	N56 31.7
S 13	183 22.5	27 36.7	25.6	47 58.8	06.7	188 20.0	50.3	151 33.5	01.3	Shaula	96 37.3	S37 06.1
D 14	198 24.9	42 36.2	24.5	62 59.6	06.2	203 22.7	50.4	166 36.1	01.2	Sirius	258 43.9	S16 42.7
A 15	213 27.4	57 35.7 ..	23.5	78 00.5 ..	05.8	218 25.5 ..	50.5	181 38.6 ..	01.2	Spica	158 43.5	S11 09.1
Y 16	228 29.9	72 35.2	22.4	93 01.4	05.4	233 28.3	50.7	196 41.2	01.1	Suhail	223 01.2	S43 25.5
17	243 32.3	87 34.7	21.3	108 02.2	04.9	248 31.0	50.8	211 43.8	01.1			
18	258 34.8	102 34.2	N11 20.2	123 03.1	N18 04.5	263 33.8	S 3 50.9	226 46.3	N10 01.1	Vega	80 46.5	N38 47.3
19	273 37.3	117 33.7	19.1	138 04.0	04.1	278 36.5	51.1	241 48.9	01.0	Zuben'ubi	137 18.2	S16 02.0
20	288 39.7	132 33.3	18.0	153 04.8	03.6	293 39.3	51.2	256 51.4	01.0		SHA	Mer. Pass.
21	303 42.2	147 32.8 ..	17.0	168 05.7 ..	03.2	308 42.1 ..	51.3	271 54.0 ..	00.9	Venus	206 03.9	11 09
22	318 44.6	162 32.3	15.9	183 06.6	02.8	323 44.8	51.5	286 56.6	00.9	Mars	225 35.6	9 50
23	333 47.1	177 31.8	14.8	198 07.4	02.3	338 47.6	51.6	301 59.1	00.8	Jupiter	4 46.4	0 33
	h m									Saturn	328 07.6	3 00
Mer. Pass. 0 52.4		v −0.5	d 1.1	v 0.9	d 0.4	v 2.8	d 0.1	v 2.6	d 0.0			

UT	SUN GHA	SUN Dec	MOON GHA	v	MOON Dec	d	HP
d h	° ′	° ′	° ′	′	° ′	′	′
7 00	180 26.6	N 6 13.5	353 01.7	8.2	S 5 03.3	11.7	60.5
01	195 26.8	12.5	7 28.9	8.1	4 51.6	11.7	60.5
02	210 27.0	11.6	21 56.0	8.1	4 39.9	11.7	60.5
03	225 27.2	.. 10.7	36 23.1	8.2	4 28.2	11.8	60.5
04	240 27.4	09.7	50 50.3	8.1	4 16.4	11.8	60.5
05	255 27.6	08.8	65 17.4	8.2	4 04.6	11.8	60.5
06	270 27.9	N 6 07.9	79 44.6	8.2	S 3 52.8	11.8	60.5
M 07	285 28.1	06.9	94 11.8	8.1	3 41.0	11.9	60.6
O 08	300 28.3	06.0	108 38.9	8.2	3 29.1	11.9	60.6
N 09	315 28.5	.. 05.1	123 06.1	8.2	3 17.2	11.9	60.6
D 10	330 28.7	04.1	137 33.3	8.1	3 05.3	12.0	60.6
A 11	345 28.9	03.2	152 00.4	8.2	2 53.3	12.0	60.6
Y 12	0 29.1	N 6 02.3	166 27.6	8.2	S 2 41.3	12.0	60.6
13	15 29.4	01.3	180 54.8	8.2	2 29.3	12.0	60.6
14	30 29.6	6 00.4	195 22.0	8.2	2 17.3	12.0	60.6
15	45 29.8	5 59.5	209 49.2	8.2	2 05.3	12.1	60.6
16	60 30.0	58.5	224 16.4	8.1	1 53.2	12.0	60.6
17	75 30.2	57.6	238 43.5	8.2	1 41.2	12.1	60.6
18	90 30.4	N 5 56.6	253 10.7	8.2	S 1 29.1	12.1	60.6
19	105 30.6	55.7	267 37.9	8.2	1 17.0	12.1	60.7
20	120 30.9	54.8	282 05.1	8.2	1 04.9	12.1	60.7
21	135 31.1	.. 53.8	296 32.3	8.2	0 52.8	12.1	60.7
22	150 31.3	52.9	310 59.5	8.2	0 40.7	12.1	60.7
23	165 31.5	52.0	325 26.7	8.2	0 28.6	12.1	60.7
8 00	180 31.7	N 5 51.0	339 53.9	8.2	S 0 16.5	12.1	60.7
01	195 31.9	50.1	354 21.1	8.2	S 0 04.4	12.2	60.7
02	210 32.1	49.1	8 48.3	8.2	N 0 07.8	12.1	60.7
03	225 32.4	.. 48.2	23 15.5	8.2	0 19.9	12.1	60.7
04	240 32.6	47.3	37 42.7	8.2	0 32.0	12.1	60.7
05	255 32.8	46.3	52 09.9	8.2	0 44.1	12.1	60.7
06	270 33.0	N 5 45.4	66 37.1	8.2	N 0 56.2	12.1	60.7
T 07	285 33.2	44.5	81 04.3	8.1	1 08.3	12.1	60.7
U 08	300 33.4	43.5	95 31.4	8.2	1 20.4	12.1	60.7
E 09	315 33.7	.. 42.6	109 58.6	8.2	1 32.5	12.1	60.7
S 10	330 33.9	41.6	124 25.8	8.2	1 44.6	12.0	60.7
D 11	345 34.1	40.7	138 53.0	8.1	1 56.6	12.1	60.7
A 12	0 34.3	N 5 39.8	153 20.1	8.2	N 2 08.7	12.0	60.7
Y 13	15 34.5	38.8	167 47.3	8.2	2 20.7	12.0	60.7
14	30 34.7	37.9	182 14.5	8.1	2 32.7	12.0	60.7
15	45 34.9	.. 36.9	196 41.6	8.2	2 44.7	12.0	60.6
16	60 35.2	36.0	211 08.8	8.1	2 56.7	12.0	60.6
17	75 35.4	35.1	225 35.9	8.2	3 08.7	11.9	60.6
18	90 35.6	N 5 34.1	240 03.1	8.1	N 3 20.6	11.9	60.6
19	105 35.8	33.2	254 30.2	8.2	3 32.5	11.9	60.6
20	120 36.0	32.2	268 57.4	8.1	3 44.4	11.9	60.6
21	135 36.2	.. 31.3	283 24.5	8.1	3 56.3	11.8	60.6
22	150 36.5	30.4	297 51.6	8.1	4 08.1	11.8	60.6
23	165 36.7	29.4	312 18.7	8.1	4 19.9	11.8	60.6
9 00	180 36.9	N 5 28.5	326 45.8	8.1	N 4 31.7	11.7	60.6
01	195 37.1	27.5	341 12.9	8.1	4 43.4	11.7	60.6
02	210 37.3	26.6	355 40.0	8.1	4 55.1	11.7	60.6
03	225 37.5	.. 25.6	10 07.1	8.1	5 06.8	11.6	60.6
04	240 37.8	24.7	24 34.2	8.0	5 18.4	11.6	60.6
05	255 38.0	23.8	39 01.2	8.1	5 30.0	11.6	60.6
06	270 38.2	N 5 22.8	53 28.3	8.0	N 5 41.6	11.5	60.6
W 07	285 38.4	21.9	67 55.3	8.1	5 53.1	11.5	60.5
E 08	300 38.6	20.9	82 22.4	8.0	6 04.6	11.4	60.5
D 09	315 38.8	.. 20.0	96 49.4	8.0	6 16.0	11.4	60.5
N 10	330 39.1	19.0	111 16.4	8.0	6 27.4	11.3	60.5
E 11	345 39.3	18.1	125 43.4	8.0	6 38.7	11.3	60.5
S 12	0 39.5	N 5 17.2	140 10.4	8.0	N 6 50.0	11.3	60.5
D 13	15 39.7	16.2	154 37.4	8.0	7 01.3	11.2	60.5
A 14	30 39.9	15.3	169 04.4	7.9	7 12.5	11.1	60.5
Y 15	45 40.2	.. 14.3	183 31.3	8.0	7 23.6	11.1	60.4
16	60 40.4	13.4	197 58.3	7.9	7 34.7	11.0	60.4
17	75 40.6	12.4	212 25.2	8.0	7 45.7	11.0	60.4
18	90 40.8	N 5 11.5	226 52.2	7.9	N 7 56.7	10.9	60.4
19	105 41.0	10.6	241 19.1	7.9	8 07.6	10.9	60.4
20	120 41.2	09.6	255 46.0	7.9	8 18.5	10.8	60.4
21	135 41.5	.. 08.7	270 12.9	7.9	8 29.3	10.7	60.4
22	150 41.7	07.7	284 39.8	7.9	8 40.0	10.7	60.3
23	165 41.9	06.8	299 06.7	7.8	N 8 50.7	10.7	60.3
	SD 15.9	d 0.9	SD 16.5		16.5		16.5

Lat.	Twilight Naut.	Twilight Civil	Sunrise	Moonrise 7	8	9	10
°	h m	h m	h m	h m	h m	h m	h m
N 72	00 28	03 15	04 34	19 27	19 20	19 13	19 04
N 70	01 46	03 34	04 43	19 26	19 25	19 24	19 25
68	02 22	03 49	04 51	19 24	19 29	19 34	19 41
66	02 46	04 02	04 57	19 23	19 32	19 42	19 54
64	03 05	04 12	05 02	19 22	19 35	19 49	20 06
62	03 20	04 20	05 07	19 21	19 37	19 55	20 15
60	03 32	04 27	05 11	19 20	19 39	20 00	20 23
N 58	03 43	04 33	05 14	19 19	19 41	20 04	20 30
56	03 51	04 39	05 17	19 19	19 43	20 08	20 37
54	03 59	04 44	05 20	19 18	19 44	20 12	20 42
52	04 06	04 48	05 23	19 17	19 46	20 15	20 47
50	04 12	04 52	05 25	19 17	19 47	20 18	20 52
45	04 24	05 00	05 30	19 16	19 50	20 25	21 02
N 40	04 34	05 07	05 34	19 15	19 52	20 30	21 11
35	04 42	05 12	05 38	19 14	19 54	20 35	21 18
30	04 48	05 16	05 41	19 14	19 56	20 40	21 25
20	04 58	05 24	05 46	19 12	19 59	20 47	21 36
N 10	05 05	05 29	05 50	19 11	20 02	20 54	21 46
0	05 10	05 34	05 54	19 11	20 05	21 00	21 55
S 10	05 13	05 37	05 58	19 10	20 08	21 06	22 04
20	05 15	05 41	06 03	19 09	20 11	21 13	22 15
30	05 16	05 43	06 07	19 08	20 14	21 21	22 26
35	05 15	05 45	06 10	19 07	20 16	21 25	22 33
40	05 14	05 46	06 13	19 06	20 19	21 30	22 41
45	05 13	05 47	06 16	19 06	20 21	21 36	22 50
S 50	05 11	05 48	06 20	19 05	20 24	21 43	23 01
52	05 09	05 48	06 22	19 04	20 26	21 47	23 06
54	05 08	05 49	06 24	19 04	20 27	21 50	23 11
56	05 06	05 49	06 26	19 03	20 29	21 55	23 17
58	05 04	05 49	06 28	19 03	20 31	21 59	23 25
S 60	05 02	05 50	06 31	19 02	20 34	22 04	23 32

Lat.	Sunset	Twilight Civil	Twilight Naut.	Moonset 7	8	9	10
°	h m	h m	h m	h m	h m	h m	h m
N 72	19 19	20 36	23 03	05 50	07 52	09 54	11 57
N 70	19 10	20 17	22 02	05 56	07 50	09 44	11 38
68	19 03	20 03	21 29	06 01	07 49	09 36	11 24
66	18 57	19 51	21 05	06 05	07 47	09 30	11 12
64	18 51	19 42	20 47	06 08	07 46	09 25	11 02
62	18 47	19 33	20 33	06 11	07 46	09 20	10 53
60	18 43	19 27	20 21	06 14	07 45	09 16	10 46
N 58	18 40	19 20	20 11	06 16	07 44	09 13	10 39
56	18 37	19 15	20 02	06 18	07 44	09 09	10 34
54	18 34	19 10	19 55	06 20	07 43	09 07	10 29
52	18 32	19 06	19 48	06 21	07 43	09 04	10 24
50	18 30	19 02	19 42	06 23	07 42	09 02	10 20
45	18 25	18 55	19 30	06 26	07 41	08 57	10 11
N 40	18 21	18 48	19 21	06 28	07 41	08 53	10 04
35	18 17	18 43	19 13	06 31	07 40	08 49	09 57
30	18 14	18 39	19 07	06 33	07 39	08 46	09 52
20	18 09	18 31	18 57	06 36	07 38	08 40	09 42
N 10	18 05	18 26	18 51	06 39	07 37	08 36	09 34
0	18 01	18 22	18 46	06 42	07 36	08 31	09 26
S 10	17 57	18 18	18 43	06 45	07 35	08 26	09 18
20	17 53	18 15	18 41	06 47	07 34	08 22	09 10
30	17 49	18 13	18 40	06 51	07 33	08 16	09 00
35	17 46	18 11	18 41	06 53	07 33	08 13	08 55
40	17 43	18 10	18 42	06 55	07 32	08 09	08 48
45	17 40	18 09	18 43	06 57	07 31	08 05	08 41
S 50	17 36	18 08	18 46	07 00	07 30	08 00	08 33
52	17 34	18 08	18 47	07 01	07 29	07 58	08 29
54	17 32	18 08	18 49	07 02	07 29	07 56	08 24
56	17 30	18 08	18 51	07 04	07 28	07 53	08 19
58	17 28	18 07	18 53	07 06	07 28	07 50	08 14
S 60	17 25	18 07	18 55	07 08	07 27	07 46	08 08

	SUN			MOON			
Day	Eqn. of Time 00h	12h	Mer. Pass.	Mer. Pass. Upper	Lower	Age	Phase
d	m s	m s	h m	h m	h m	d	%
7	01 46	01 56	11 58	00 29	12 56	16	98
8	02 06	02 17	11 58	01 23	13 51	17	94
9	02 27	02 38	11 57	02 18	14 45	18	87

1998 SEPTEMBER 10, 11, 12 (THURS., FRI., SAT.)

UT	ARIES GHA	VENUS −3.9 GHA	Dec	MARS +1.7 GHA	Dec	JUPITER −2.9 GHA	Dec	SATURN +0.2 GHA	Dec
d h	° ′	° ′	° ′	° ′	° ′	° ′	° ′	° ′	° ′
10 00	348 49.6	192 31.3	N11 13.8	213 08.3	N18 01.9	353 50.4	S 3 51.7	317 01.7	N10 00.8
01	3 52.0	207 30.8	12.7	228 09.2	01.5	8 53.1	51.9	332 04.2	00.7
02	18 54.5	222 30.3	11.6	243 10.0	01.1	23 55.9	52.0	347 06.8	00.7
03	33 57.0	237 29.8	.. 10.5	258 10.9	.. 00.6	38 58.7	.. 52.1	2 09.4	.. 00.7
04	48 59.4	252 29.3	09.4	273 11.8	18 00.2	54 01.4	52.3	17 11.9	00.6
05	64 01.9	267 28.8	08.3	288 12.6	17 59.8	69 04.2	52.4	32 14.5	00.6
T 06	79 04.4	282 28.3	N11 07.3	303 13.5	N17 59.3	84 07.0	S 3 52.5	47 17.1	N10 00.5
H 07	94 06.8	297 27.9	06.2	318 14.4	58.9	99 09.7	52.7	62 19.6	00.5
U 08	109 09.3	312 27.4	05.1	333 15.3	58.5	114 12.5	52.8	77 22.2	00.4
R 09	124 11.7	327 26.9	.. 04.0	348 16.1	.. 58.0	129 15.3	.. 52.9	92 24.7	.. 00.4
S 10	139 14.2	342 26.4	02.9	3 17.0	57.6	144 18.0	53.1	107 27.3	00.3
D 11	154 16.7	357 25.9	01.8	18 17.9	57.2	159 20.8	53.2	122 29.9	00.3
A 12	169 19.1	12 25.4	N11 00.7	33 18.7	N17 56.7	174 23.5	S 3 53.3	137 32.4	N10 00.3
Y 13	184 21.6	27 24.9	10 59.7	48 19.6	56.3	189 26.3	53.5	152 35.0	00.2
14	199 24.1	42 24.4	58.6	63 20.5	55.9	204 29.1	53.6	167 37.6	00.2
15	214 26.5	57 24.0	.. 57.5	78 21.3	.. 55.4	219 31.8	.. 53.7	182 40.1	.. 00.1
16	229 29.0	72 23.5	56.4	93 22.2	55.0	234 34.6	53.9	197 42.7	00.1
17	244 31.5	87 23.0	55.3	108 23.1	54.6	249 37.4	54.0	212 45.3	00.0
18	259 33.9	102 22.5	N10 54.2	123 24.0	N17 54.1	264 40.1	S 3 54.1	227 47.8	N10 00.0
19	274 36.4	117 22.0	53.1	138 24.8	53.7	279 42.9	54.3	242 50.4	9 59.9
20	289 38.9	132 21.5	52.0	153 25.7	53.3	294 45.7	54.4	257 52.9	59.9
21	304 41.3	147 21.0	.. 50.9	168 26.6	.. 52.8	309 48.4	.. 54.5	272 55.5	.. 59.8
22	319 43.8	162 20.6	49.8	183 27.4	52.4	324 51.2	54.7	287 58.1	59.8
23	334 46.2	177 20.1	48.8	198 28.3	51.9	339 54.0	54.8	303 00.6	59.8
11 00	349 48.7	192 19.6	N10 47.7	213 29.2	N17 51.5	354 56.7	S 3 55.0	318 03.2	N 9 59.7
01	4 51.2	207 19.1	46.6	228 30.1	51.1	9 59.5	55.1	333 05.8	59.7
02	19 53.6	222 18.6	45.5	243 30.9	50.6	25 02.3	55.2	348 08.3	59.6
03	34 56.1	237 18.2	.. 44.4	258 31.8	.. 50.2	40 05.0	.. 55.4	3 10.9	.. 59.6
04	49 58.6	252 17.7	43.3	273 32.7	49.8	55 07.8	55.5	18 13.5	59.5
05	65 01.0	267 17.2	42.2	288 33.5	49.3	70 10.6	55.6	33 16.0	59.5
06	80 03.5	282 16.7	N10 41.1	303 34.4	N17 48.9	85 13.3	S 3 55.8	48 18.6	N 9 59.4
F 07	95 06.0	297 16.2	40.0	318 35.3	48.5	100 16.1	55.9	63 21.2	59.4
R 08	110 08.4	312 15.7	38.9	333 36.2	48.0	115 18.9	56.0	78 23.7	59.4
I 09	125 10.9	327 15.3	.. 37.8	348 37.0	.. 47.6	130 21.6	.. 56.2	93 26.3	.. 59.3
D 10	140 13.4	342 14.8	36.7	3 37.9	47.1	145 24.4	56.3	108 28.9	59.3
A 11	155 15.8	357 14.3	35.6	18 38.8	46.7	160 27.2	56.4	123 31.4	59.2
Y 12	170 18.3	12 13.8	N10 34.5	33 39.7	N17 46.3	175 29.9	S 3 56.6	138 34.0	N 9 59.2
13	185 20.7	27 13.3	33.4	48 40.5	45.8	190 32.7	56.7	153 36.6	59.1
14	200 23.2	42 12.9	32.3	63 41.4	45.4	205 35.5	56.8	168 39.1	59.1
15	215 25.7	57 12.4	.. 31.2	78 42.3	.. 45.0	220 38.2	.. 57.0	183 41.7	.. 59.0
16	230 28.1	72 11.9	30.1	93 43.2	44.5	235 41.0	57.1	198 44.3	59.0
17	245 30.6	87 11.4	29.0	108 44.0	44.1	250 43.8	57.2	213 46.8	58.9
18	260 33.1	102 11.0	N10 27.9	123 44.9	N17 43.6	265 46.5	S 3 57.4	228 49.4	N 9 58.9
19	275 35.5	117 10.5	26.8	138 45.8	43.2	280 49.3	57.5	243 52.0	58.9
20	290 38.0	132 10.0	25.7	153 46.7	42.8	295 52.1	57.6	258 54.5	58.8
21	305 40.5	147 09.5	.. 24.6	168 47.5	.. 42.3	310 54.8	.. 57.8	273 57.1	.. 58.8
22	320 42.9	162 09.0	23.5	183 48.4	41.9	325 57.6	57.9	288 59.7	58.7
23	335 45.4	177 08.6	22.4	198 49.3	41.4	341 00.4	58.0	304 02.3	58.7
12 00	350 47.8	192 08.1	N10 21.3	213 50.2	N17 41.0	356 03.1	S 3 58.2	319 04.8	N 9 58.6
01	5 50.3	207 07.6	20.2	228 51.1	40.6	11 05.9	58.3	334 07.4	58.6
02	20 52.8	222 07.1	19.1	243 51.9	40.1	26 08.7	58.4	349 10.0	58.5
03	35 55.2	237 06.7	.. 18.0	258 52.8	.. 39.7	41 11.4	.. 58.6	4 12.5	.. 58.5
04	50 57.7	252 06.2	16.8	273 53.7	39.2	56 14.2	58.7	19 15.1	58.4
05	66 00.2	267 05.7	15.7	288 54.6	38.8	71 17.0	58.8	34 17.7	58.4
06	81 02.6	282 05.3	N10 14.6	303 55.4	N17 38.4	86 19.7	S 3 59.0	49 20.2	N 9 58.3
S 07	96 05.1	297 04.8	13.5	318 56.3	37.9	101 22.5	59.1	64 22.8	58.3
A 08	111 07.6	312 04.3	12.4	333 57.2	37.5	116 25.3	59.2	79 25.4	58.3
T 09	126 10.0	327 03.8	.. 11.3	348 58.1	.. 37.0	131 28.0	.. 59.4	94 28.0	.. 58.2
U 10	141 12.5	342 03.4	10.2	3 59.0	36.6	146 30.8	59.5	109 30.5	58.2
R 11	156 15.0	357 02.9	09.1	18 59.8	36.1	161 33.6	59.6	124 33.1	58.1
D 12	171 17.4	12 02.4	N10 08.0	34 00.7	N17 35.7	176 36.4	S 3 59.8	139 35.7	N 9 58.1
A 13	186 19.9	27 01.9	06.9	49 01.6	35.3	191 39.1	59.9	154 38.2	58.0
Y 14	201 22.3	42 01.5	05.7	64 02.5	34.8	206 41.9	4 00.0	169 40.8	58.0
15	216 24.8	57 01.0	.. 04.6	79 03.4	.. 34.4	221 44.7	.. 00.2	184 43.4	.. 57.9
16	231 27.3	72 00.5	03.5	94 04.2	33.9	236 47.4	00.3	199 45.9	57.9
17	246 29.7	87 00.0	02.4	109 05.1	33.5	251 50.2	00.4	214 48.5	57.8
18	261 32.2	101 59.6	N10 01.3	124 06.0	N17 33.0	266 53.0	S 4 00.6	229 51.1	N 9 57.8
19	276 34.7	116 59.1	10 00.2	139 06.9	32.6	281 55.7	00.7	244 53.7	57.7
20	291 37.1	131 58.6	9 59.1	154 07.8	32.2	296 58.5	00.8	259 56.2	57.7
21	306 39.6	146 58.2	.. 57.9	169 08.6	.. 31.7	312 01.3	.. 01.0	274 58.8	.. 57.6
22	321 42.1	161 57.7	56.8	184 09.5	31.3	327 04.0	01.1	290 01.4	57.6
23	336 44.5	176 57.2	55.7	199 10.4	30.8	342 06.8	01.2	305 04.0	57.5
Mer.Pass.	h m 0 40.6	v −0.5	d 1.1	v 0.9	d 0.4	v 2.8	d 0.1	v 2.6	d 0.0

STARS

Name	SHA	Dec
Acamar	315 26.8	S40 18.5
Achernar	335 34.8	S57 14.5
Acrux	173 22.6	S63 05.5
Adhara	255 21.6	S28 58.1
Aldebaran	291 02.4	N16 30.3
Alioth	166 31.2	N55 58.3
Alkaid	153 08.3	N49 19.5
Al Na'ir	27 57.6	S46 58.0
Alnilam	275 58.0	S 1 12.2
Alphard	218 07.5	S 8 39.1
Alphecca	126 20.8	N26 43.5
Alpheratz	357 54.9	N29 05.0
Altair	62 19.1	N 8 52.1
Ankaa	353 26.5	S42 18.7
Antares	112 40.3	S26 25.6
Arcturus	146 06.4	N19 11.7
Atria	107 52.4	S69 01.6
Avior	234 23.1	S59 30.1
Bellatrix	278 44.2	N 6 20.9
Betelgeuse	271 13.7	N 7 24.4
Canopus	264 01.3	S52 41.6
Capella	280 51.3	N45 59.5
Deneb	49 38.8	N45 16.8
Denebola	182 45.6	N14 35.0
Diphda	349 07.0	S17 59.5
Dubhe	194 06.3	N61 45.6
Elnath	278 27.0	N28 36.2
Eltanin	90 51.3	N51 29.8
Enif	33 58.0	N 9 52.3
Fomalhaut	15 36.2	S29 37.7
Gacrux	172 14.1	S57 06.3
Gienah	176 04.3	S17 31.9
Hadar	149 04.5	S60 22.0
Hamal	328 13.4	N23 27.3
Kaus Aust.	83 58.8	S34 23.1
Kochab	137 20.4	N74 10.1
Markab	13 49.3	N15 12.0
Menkar	314 26.8	N 4 05.1
Menkent	148 21.3	S36 21.7
Miaplacidus	221 42.8	S69 42.6
Mirfak	308 56.4	N49 51.1
Nunki	76 12.3	S26 17.8
Peacock	53 36.7	S56 44.4
Pollux	243 41.9	N28 01.7
Procyon	245 11.8	N 5 13.7
Rasalhague	96 17.0	N12 34.0
Regulus	207 55.9	N11 58.5
Rigel	281 23.0	S 8 12.1
Rigil Kent.	140 07.7	S60 49.8
Sabik	102 25.6	S15 43.2
Schedar	349 52.9	N56 31.7
Shaula	96 37.3	S37 06.1
Sirius	258 43.9	S16 42.7
Spica	158 43.5	S11 09.1
Suhail	223 01.2	S43 25.5
Vega	80 46.5	N38 47.3
Zuben'ubi	137 18.2	S16 02.0

	SHA	Mer.Pass.
	° ′	h m
Venus	202 30.9	11 11
Mars	223 40.5	9 45
Jupiter	5 08.0	0 20
Saturn	328 14.5	2 47

UT	SUN GHA	SUN Dec	MOON GHA	v	Dec	d	HP
d h	° '	° '	° '	'	° '	'	'
10 00	180 42.1	N 5 05.8	313 33.5	7.9	N 9 01.4	10.5	60.3
01	195 42.3	04.9	328 00.4	7.8	9 11.9	10.5	60.3
02	210 42.6	03.9	342 27.2	7.8	9 22.4	10.4	60.3
03	225 42.8	. . 03.0	356 54.0	7.8	9 32.8	10.4	60.2
04	240 43.0	02.0	11 20.8	7.8	9 43.2	10.3	60.2
05	255 43.2	01.1	25 47.6	7.8	9 53.5	10.2	60.2
06	270 43.4	N 5 00.2	40 14.4	7.8	N10 03.7	10.2	60.2
T 07	285 43.6	4 59.2	54 41.2	7.7	10 13.9	10.1	60.2
H 08	300 43.9	58.3	69 07.9	7.8	10 24.0	10.0	60.2
U 09	315 44.1	. . 57.3	83 34.7	7.7	10 34.0	9.9	60.1
R 10	330 44.3	56.4	98 01.4	7.8	10 43.9	9.9	60.1
S 11	345 44.5	55.4	112 28.2	7.7	10 53.8	9.7	60.1
D 12	0 44.7	N 4 54.5	126 54.9	7.7	N11 03.5	9.7	60.1
A 13	15 45.0	53.5	141 21.6	7.6	11 13.2	9.7	60.1
Y 14	30 45.2	52.6	155 48.2	7.7	11 22.9	9.5	60.0
15	45 45.4	. . 51.6	170 14.9	7.7	11 32.4	9.5	60.0
16	60 45.6	50.7	184 41.6	7.6	11 41.9	9.4	60.0
17	75 45.8	49.7	199 08.2	7.6	11 51.3	9.3	60.0
18	90 46.1	N 4 48.8	213 34.8	7.7	N12 00.6	9.2	59.9
19	105 46.3	47.8	228 01.5	7.6	12 09.8	9.1	59.9
20	120 46.5	46.9	242 28.1	7.6	12 18.9	9.1	59.9
21	135 46.7	. . 45.9	256 54.7	7.5	12 28.0	8.9	59.9
22	150 46.9	45.0	271 21.2	7.6	12 36.9	8.9	59.9
23	165 47.1	44.0	285 47.8	7.6	12 45.8	8.8	59.8
11 00	180 47.4	N 4 43.1	300 14.4	7.5	N12 54.6	8.7	59.8
01	195 47.6	42.1	314 40.9	7.6	13 03.3	8.6	59.8
02	210 47.8	41.2	329 07.5	7.5	13 11.9	8.5	59.8
03	225 48.0	. . 40.3	343 34.0	7.5	13 20.4	8.5	59.7
04	240 48.2	39.3	358 00.5	7.5	13 28.9	8.3	59.7
05	255 48.5	38.4	12 27.0	7.5	13 37.2	8.3	59.7
06	270 48.7	N 4 37.4	26 53.5	7.5	N13 45.5	8.1	59.7
F 07	285 48.9	36.5	41 20.0	7.4	13 53.6	8.1	59.6
R 08	300 49.1	35.5	55 46.4	7.5	14 01.7	7.9	59.6
I 09	315 49.3	. . 34.6	70 12.9	7.5	14 09.6	7.9	59.6
D 10	330 49.6	33.6	84 39.4	7.4	14 17.5	7.7	59.6
A 11	345 49.8	32.7	99 05.8	7.4	14 25.2	7.7	59.5
Y 12	0 50.0	N 4 31.7	113 32.2	7.5	N14 32.9	7.6	59.5
13	15 50.2	30.8	127 58.7	7.4	14 40.5	7.4	59.5
14	30 50.4	29.8	142 25.1	7.4	14 47.9	7.4	59.5
15	45 50.7	. . 28.8	156 51.5	7.4	14 55.3	7.3	59.4
16	60 50.9	27.9	171 17.9	7.4	15 02.6	7.1	59.4
17	75 51.1	26.9	185 44.3	7.4	15 09.7	7.1	59.4
18	90 51.3	N 4 26.0	200 10.7	7.3	N15 16.8	7.0	59.4
19	105 51.5	25.0	214 37.0	7.4	15 23.8	6.8	59.3
20	120 51.8	24.1	229 03.4	7.4	15 30.6	6.8	59.3
21	135 52.0	. . 23.1	243 29.8	7.3	15 37.4	6.6	59.3
22	150 52.2	22.2	257 56.1	7.4	15 44.0	6.6	59.3
23	165 52.4	21.2	272 22.5	7.4	15 50.6	6.4	59.2
12 00	180 52.6	N 4 20.3	286 48.9	7.3	N15 57.0	6.3	59.2
01	195 52.9	19.3	301 15.2	7.4	16 03.3	6.3	59.2
02	210 53.1	18.4	315 41.6	7.3	16 09.6	6.1	59.1
03	225 53.3	. . 17.4	330 07.9	7.3	16 15.7	6.0	59.1
04	240 53.5	16.5	344 34.2	7.4	16 21.7	5.9	59.1
05	255 53.7	15.5	359 00.6	7.3	16 27.6	5.8	59.1
06	270 54.0	N 4 14.6	13 26.9	7.3	N16 33.4	5.7	59.0
S 07	285 54.2	13.6	27 53.2	7.4	16 39.1	5.5	59.0
A 08	300 54.4	12.7	42 19.6	7.3	16 44.6	5.5	59.0
T 09	315 54.6	. . 11.7	56 45.9	7.3	16 50.1	5.4	59.0
U 10	330 54.8	10.8	71 12.2	7.4	16 55.5	5.2	58.9
R 11	345 55.1	09.8	85 38.6	7.3	17 00.7	5.1	58.9
D 12	0 55.3	N 4 08.8	100 04.9	7.4	N17 05.8	5.0	58.9
A 13	15 55.5	07.9	114 31.3	7.3	17 10.8	5.0	58.8
Y 14	30 55.7	06.9	128 57.6	7.4	17 15.8	4.7	58.8
15	45 56.0	. . 06.0	143 24.0	7.3	17 20.5	4.7	58.8
16	60 56.2	05.0	157 50.3	7.4	17 25.2	4.6	58.8
17	75 56.4	04.1	172 16.7	7.3	17 29.8	4.5	58.7
18	90 56.6	N 4 03.1	186 43.0	7.4	N17 34.3	4.3	58.7
19	105 56.8	02.2	201 09.4	7.4	17 38.6	4.2	58.7
20	120 57.1	01.2	215 35.8	7.3	17 42.8	4.1	58.6
21	135 57.3	4 00.3	230 02.1	7.4	17 46.9	4.0	58.6
22	150 57.5	3 59.3	244 28.5	7.4	17 50.9	3.9	58.6
23	165 57.7	N 3 58.3	258 54.9	7.4	N17 54.8	3.8	58.6
SD 15.9	d 1.0		SD 16.4		16.2		16.0

Lat.	Twilight Naut.	Twilight Civil	Sunrise	Moonrise 10	Moonrise 11	Moonrise 12	Moonrise 13
°	h m	h m	h m	h m	h m	h m	h m
N 72	01 24	03 32	04 48	19 04	18 54	18 32	☐
N 70	02 10	03 49	04 55	19 25	19 28	19 36	19 58
68	02 39	04 02	05 01	19 41	19 52	20 11	20 45
66	03 00	04 12	05 06	19 54	20 12	20 37	21 16
64	03 17	04 21	05 11	20 06	20 27	20 57	21 39
62	03 30	04 28	05 14	20 15	20 40	21 13	21 57
60	03 41	04 35	05 18	20 23	20 51	21 27	22 12
N 58	03 50	04 40	05 21	20 30	21 01	21 39	22 25
56	03 58	04 45	05 23	20 37	21 10	21 49	22 36
54	04 05	04 49	05 25	20 42	21 17	21 58	22 46
52	04 12	04 53	05 27	20 47	21 24	22 06	22 55
50	04 17	04 57	05 29	20 52	21 30	22 13	23 03
45	04 28	05 04	05 33	21 02	21 43	22 29	23 19
N 40	04 37	05 09	05 37	21 11	21 54	22 42	23 33
35	04 44	05 14	05 40	21 18	22 04	22 53	23 45
30	04 50	05 18	05 42	21 25	22 12	23 02	23 55
20	04 59	05 24	05 46	21 36	22 26	23 19	24 12
N 10	05 05	05 29	05 50	21 46	22 39	23 33	24 28
0	05 09	05 33	05 53	21 55	22 51	23 47	24 42
S 10	05 11	05 36	05 57	22 04	23 03	24 00	00 00
20	05 12	05 38	06 00	22 15	23 16	24 15	00 15
30	05 12	05 40	06 04	22 26	23 30	24 32	00 32
35	05 11	05 40	06 06	22 33	23 39	24 42	00 42
40	05 10	05 41	06 08	22 41	23 49	24 53	00 53
45	05 07	05 41	06 11	22 50	24 00	00 00	01 06
S 50	05 04	05 41	06 14	23 01	24 14	00 14	01 22
52	05 02	05 41	06 15	23 06	24 21	00 21	01 30
54	05 00	05 41	06 17	23 11	24 28	00 28	01 38
56	04 58	05 41	06 18	23 17	24 36	00 36	01 48
58	04 56	05 41	06 20	23 25	24 45	00 45	01 58
S 60	04 53	05 41	06 22	23 32	24 56	00 56	02 11

Lat.	Sunset	Twilight Civil	Twilight Naut.	Moonset 10	Moonset 11	Moonset 12	Moonset 13
°	h m	h m	h m	h m	h m	h m	h m
N 72	19 03	20 17	22 18	11 57	14 05	16 24	☐
N 70	18 56	20 01	21 37	11 38	13 32	15 21	16 55
68	18 50	19 49	21 10	11 24	13 08	14 45	16 08
66	18 45	19 39	20 49	11 12	12 50	14 20	15 37
64	18 41	19 30	20 34	11 02	12 35	14 01	15 15
62	18 37	19 23	20 21	10 53	12 22	13 45	14 56
60	18 34	19 17	20 10	10 46	12 12	13 32	14 41
N 58	18 31	19 12	20 01	10 39	12 03	13 20	14 29
56	18 29	19 07	19 53	10 34	11 55	13 10	14 18
54	18 27	19 03	19 46	10 29	11 48	13 02	14 08
52	18 25	18 59	19 40	10 24	11 41	12 54	13 59
50	18 23	18 56	19 34	10 20	11 36	12 47	13 51
45	18 19	18 49	19 24	10 11	11 23	12 32	13 35
N 40	18 16	18 43	19 15	10 04	11 13	12 19	13 21
35	18 13	18 39	19 08	09 57	11 05	12 09	13 10
30	18 11	18 35	19 03	09 52	10 57	12 00	13 00
20	18 07	18 29	18 54	09 42	10 44	11 44	12 42
N 10	18 03	18 24	18 49	09 34	10 32	11 30	12 27
0	18 00	18 21	18 45	09 26	10 22	11 17	12 13
S 10	17 57	18 18	18 42	09 18	10 11	11 05	11 59
20	17 54	18 16	18 41	09 10	09 59	10 51	11 44
30	17 50	18 14	18 42	09 00	09 46	10 35	11 27
35	17 48	18 14	18 43	08 55	09 39	10 26	11 17
40	17 46	18 13	18 44	08 48	09 30	10 16	11 05
45	17 44	18 13	18 47	08 41	09 20	10 04	10 52
S 50	17 41	18 13	18 50	08 33	09 08	09 49	10 35
52	17 39	18 13	18 52	08 29	09 03	09 42	10 27
54	17 38	18 13	18 54	08 24	08 57	09 34	10 19
56	17 36	18 13	18 56	08 19	08 50	09 26	10 09
58	17 34	18 14	18 59	08 14	08 42	09 16	09 58
S 60	17 32	18 14	19 02	08 08	08 34	09 05	09 45

Day	SUN Eqn. of Time 00h	SUN Eqn. of Time 12h	SUN Mer. Pass.	MOON Mer. Pass. Upper	MOON Mer. Pass. Lower	Age	Phase
d	m s	m s	h m	h m	h m	d	%
10	02 48	02 59	11 57	03 13	15 41	19	78
11	03 09	03 20	11 57	04 08	16 36	20	68
12	03 30	03 41	11 56	05 04	17 32	21	57

1998 SEPTEMBER 13, 14, 15 (SUN., MON., TUES.)

UT	ARIES GHA	VENUS −3.9 GHA	Dec	MARS +1.7 GHA	Dec	JUPITER −2.9 GHA	Dec	SATURN +0.2 GHA	Dec
13 00	351 47.0	191 56.8	N 9 54.6	214 11.3	N17 30.4	357 09.6	S 4 01.4	320 06.5	N 9 57.5
01	6 49.5	206 56.3	53.5	229 12.2	29.9	12 12.3	01.5	335 09.1	57.5
02	21 51.9	221 55.8	52.3	244 13.0	29.5	27 15.1	01.6	350 11.7	57.4
03	36 54.4	236 55.4 ..	51.2	259 13.9 ..	29.0	42 17.9 ..	01.8	5 14.2 ..	57.4
04	51 56.8	251 54.9	50.1	274 14.8	28.6	57 20.6	01.9	20 16.8	57.3
05	66 59.3	266 54.4	49.0	289 15.7	28.2	72 23.4	02.0	35 19.4	57.3
06	82 01.8	281 54.0	N 9 47.9	304 16.6	N17 27.7	87 26.2	S 4 02.2	50 22.0	N 9 57.2
07	97 04.2	296 53.5	46.7	319 17.5	27.3	102 28.9	02.3	65 24.5	57.2
08	112 06.7	311 53.0	45.6	334 18.3	26.8	117 31.7	02.4	80 27.1	57.1
S 09	127 09.2	326 52.6 ..	44.5	349 19.2 ..	26.4	132 34.5 ..	02.6	95 29.7 ..	57.1
U 10	142 11.6	341 52.1	43.4	4 20.1	25.9	147 37.3	02.7	110 32.3	57.0
N 11	157 14.1	356 51.6	42.3	19 21.0	25.5	162 40.0	02.8	125 34.8	57.0
D 12	172 16.6	11 51.2	N 9 41.1	34 21.9	N17 25.0	177 42.8	S 4 03.0	140 37.4	N 9 56.9
A 13	187 19.0	26 50.7	40.0	49 22.8	24.6	192 45.6	03.1	155 40.0	56.9
Y 14	202 21.5	41 50.2	38.9	64 23.6	24.1	207 48.3	03.2	170 42.6	56.8
15	217 24.0	56 49.8 ..	37.8	79 24.5 ..	23.7	222 51.1 ..	03.4	185 45.1 ..	56.8
16	232 26.4	71 49.3	36.6	94 25.4	23.2	237 53.9	03.5	200 47.7	56.7
17	247 28.9	86 48.9	35.5	109 26.3	22.8	252 56.6	03.6	215 50.3	56.7
18	262 31.3	101 48.4	N 9 34.4	124 27.2	N17 22.4	267 59.4	S 4 03.8	230 52.9	N 9 56.6
19	277 33.8	116 47.9	33.3	139 28.1	21.9	283 02.2	03.9	245 55.4	56.6
20	292 36.3	131 47.5	32.1	154 28.9	21.5	298 04.9	04.0	260 58.0	56.5
21	307 38.7	146 47.0 ..	31.0	169 29.8 ..	21.0	313 07.7 ..	04.2	276 00.6 ..	56.5
22	322 41.2	161 46.5	29.9	184 30.7	20.6	328 10.5	04.3	291 03.2	56.4
23	337 43.7	176 46.1	28.7	199 31.6	20.1	343 13.3	04.4	306 05.7	56.4
14 00	352 46.1	191 45.6	N 9 27.6	214 32.5	N17 19.7	358 16.0	S 4 04.6	321 08.3	N 9 56.3
01	7 48.6	206 45.2	26.5	229 33.4	19.2	13 18.8	04.7	336 10.9	56.3
02	22 51.1	221 44.7	25.4	244 34.3	18.8	28 21.6	04.9	351 13.5	56.3
03	37 53.5	236 44.2 ..	24.2	259 35.1 ..	18.3	43 24.3 ..	05.0	6 16.0 ..	56.2
04	52 56.0	251 43.8	23.1	274 36.0	17.9	58 27.1	05.1	21 18.6	56.2
05	67 58.4	266 43.3	22.0	289 36.9	17.4	73 29.9	05.3	36 21.2	56.1
06	83 00.9	281 42.9	N 9 20.8	304 37.8	N17 17.0	88 32.6	S 4 05.4	51 23.8	N 9 56.1
07	98 03.4	296 42.4	19.7	319 38.7	16.5	103 35.4	05.5	66 26.3	56.0
08	113 05.8	311 41.9	18.6	334 39.6	16.1	118 38.2	05.7	81 28.9	56.0
M 09	128 08.3	326 41.5 ..	17.4	349 40.5 ..	15.6	133 40.9 ..	05.8	96 31.5 ..	55.9
O 10	143 10.8	341 41.0	16.3	4 41.4	15.2	148 43.7	05.9	111 34.1	55.9
N 11	158 13.2	356 40.6	15.2	19 42.2	14.7	163 46.5	06.1	126 36.7	55.8
D 12	173 15.7	11 40.1	N 9 14.0	34 43.1	N17 14.3	178 49.3	S 4 06.2	141 39.2	N 9 55.8
A 13	188 18.2	26 39.6	12.9	49 44.0	13.8	193 52.0	06.3	156 41.8	55.7
Y 14	203 20.6	41 39.2	11.8	64 44.9	13.4	208 54.8	06.5	171 44.4	55.7
15	218 23.1	56 38.7 ..	10.6	79 45.8 ..	12.9	223 57.6 ..	06.6	186 47.0 ..	55.6
16	233 25.6	71 38.3	09.5	94 46.7	12.5	239 00.3	06.7	201 49.5	55.6
17	248 28.0	86 37.8	08.4	109 47.6	12.0	254 03.1	06.9	216 52.1	55.5
18	263 30.5	101 37.4	N 9 07.2	124 48.5	N17 11.6	269 05.9	S 4 07.0	231 54.7	N 9 55.4
19	278 32.9	116 36.9	06.1	139 49.4	11.1	284 08.6	07.1	246 57.3	55.4
20	293 35.4	131 36.5	04.9	154 50.3	10.7	299 11.4	07.3	261 59.9	55.4
21	308 37.9	146 36.0 ..	03.8	169 51.1 ..	10.2	314 14.2 ..	07.4	277 02.4 ..	55.3
22	323 40.3	161 35.5	02.7	184 52.0	09.7	329 16.9	07.5	292 05.0	55.3
23	338 42.8	176 35.1	01.5	199 52.9	09.3	344 19.7	07.7	307 07.6	55.2
15 00	353 45.3	191 34.6	N 9 00.4	214 53.8	N17 08.8	359 22.5	S 4 07.8	322 10.2	N 9 55.2
01	8 47.7	206 34.2	8 59.2	229 54.7	08.4	14 25.3	07.9	337 12.8	55.1
02	23 50.2	221 33.7	58.1	244 55.6	07.9	29 28.0	08.1	352 15.3	55.1
03	38 52.7	236 33.3 ..	57.0	259 56.5 ..	07.5	44 30.8 ..	08.2	7 17.9 ..	55.0
04	53 55.1	251 32.8	55.8	274 57.4	07.0	59 33.6	08.3	22 20.5	55.0
05	68 57.6	266 32.4	54.7	289 58.3	06.6	74 36.3	08.5	37 23.1	54.9
06	84 00.1	281 31.9	N 8 53.5	304 59.2	N17 06.1	89 39.1	S 4 08.6	52 25.7	N 9 54.9
07	99 02.5	296 31.5	52.4	320 00.1	05.7	104 41.9	08.7	67 28.2	54.8
08	114 05.0	311 31.0	51.2	335 00.9	05.2	119 44.6	08.9	82 30.8	54.8
T 09	129 07.4	326 30.6 ..	50.1	350 01.8 ..	04.8	134 47.4 ..	09.0	97 33.4 ..	54.7
U 10	144 09.9	341 30.1	49.0	5 02.7	04.3	149 50.2	09.1	112 36.0	54.7
E 11	159 12.4	356 29.7	47.8	20 03.6	03.9	164 53.0	09.3	127 38.6	54.6
S 12	174 14.8	11 29.2	N 8 46.7	35 04.5	N17 03.4	179 55.7	S 4 09.4	142 41.1	N 9 54.6
D 13	189 17.3	26 28.7	45.5	50 05.4	02.9	194 58.5	09.5	157 43.7	54.5
A 14	204 19.8	41 28.3	44.4	65 06.3	02.5	210 01.3	09.7	172 46.3	54.5
Y 15	219 22.2	56 27.8 ..	43.2	80 07.2 ..	02.0	225 04.0 ..	09.8	187 48.9 ..	54.4
16	234 24.7	71 27.4	42.1	95 08.1	01.6	240 06.8	09.9	202 51.5	54.4
17	249 27.2	86 26.9	40.9	110 09.0	01.1	255 09.6	10.1	217 54.1	54.3
18	264 29.6	101 26.5	N 8 39.8	125 09.9	N17 00.7	270 12.3	S 4 10.2	232 56.6	N 9 54.3
19	279 32.1	116 26.0	38.6	140 10.8	17 00.2	285 15.1	10.3	247 59.2	54.2
20	294 34.5	131 25.6	37.5	155 11.7	16 59.8	300 17.9	10.5	263 01.8	54.2
21	309 37.0	146 25.1 ..	36.3	170 12.6 ..	59.3	315 20.7 ..	10.6	278 04.4 ..	54.1
22	324 39.5	161 24.7	35.2	185 13.5	58.8	330 23.4	10.7	293 07.0	54.1
23	339 41.9	176 24.3	34.0	200 14.4	58.4	345 26.2	10.9	308 09.6	54.0
Mer. Pass. h m 0 28.8		v −0.5	d 1.1	v 0.9	d 0.5	v 2.8	d 0.1	v 2.6	d 0.0

STARS

Name	SHA	Dec
Acamar	315 26.7	S40 18.5
Achernar	335 34.8	S57 14.5
Acrux	173 22.6	S63 05.5
Adhara	255 21.6	S28 58.1
Aldebaran	291 02.4	N16 30.3
Alioth	166 31.2	N55 58.3
Alkaid	153 08.3	N49 19.5
Al Na'ir	27 57.6	S46 58.0
Alnilam	275 57.9	S 1 12.2
Alphard	218 07.5	S 8 39.1
Alphecca	126 20.8	N26 43.5
Alpheratz	357 54.9	N29 05.0
Altair	62 19.1	N 8 52.1
Ankaa	353 26.5	S42 18.7
Antares	112 40.3	S26 25.6
Arcturus	146 06.4	N19 11.7
Atria	107 52.4	S69 01.6
Avior	234 23.0	S59 30.1
Bellatrix	278 44.2	N 6 20.9
Betelgeuse	271 13.7	N 7 24.4
Canopus	264 01.3	S52 41.5
Capella	280 51.2	N45 59.5
Deneb	49 38.9	N45 16.8
Denebola	182 45.6	N14 34.9
Diphda	349 07.0	S17 59.5
Dubhe	194 06.3	N61 45.5
Elnath	278 27.0	N28 36.2
Eltanin	90 51.4	N51 29.8
Enif	33 58.0	N 9 52.3
Fomalhaut	15 36.2	S29 37.7
Gacrux	172 14.1	S57 06.3
Gienah	176 04.3	S17 31.9
Hadar	149 04.5	S60 22.0
Hamal	328 13.3	N23 27.3
Kaus Aust.	83 58.8	S34 23.1
Kochab	137 20.4	N74 10.0
Markab	13 49.3	N15 12.0
Menkar	314 26.8	N 4 05.1
Menkent	148 21.3	S36 21.7
Miaplacidus	221 42.7	S69 42.5
Mirfak	308 56.4	N49 51.1
Nunki	76 12.3	S26 17.8
Peacock	53 36.7	S56 44.4
Pollux	243 41.8	N28 01.7
Procyon	245 11.8	N 5 13.7
Rasalhague	96 17.0	N12 34.0
Regulus	207 55.9	N11 58.5
Rigel	281 23.0	S 8 12.1
Rigil Kent.	140 07.8	S60 49.8
Sabik	102 25.6	S15 43.2
Schedar	349 52.9	N56 31.7
Shaula	96 37.3	S37 06.1
Sirius	258 43.9	S16 42.7
Spica	158 43.5	S11 09.1
Suhail	223 01.1	S43 25.5
Vega	80 46.5	N38 47.4
Zuben'ubi	137 18.2	S16 02.0

	SHA	Mer. Pass.
	° '	h m
Venus	198 59.5	11 13
Mars	221 46.4	9 41
Jupiter	5 29.9	0 07
Saturn	328 22.2	2 35

UT	SUN GHA	SUN Dec	MOON GHA	v	MOON Dec	d	HP
d h	° ′	° ′	° ′	′	° ′	′	′
13 00	180 57.9	N 3 57.4	273 21.3	7.4	N17 58.6	3.7	58.5
01	195 58.2	56.4	287 47.7	7.5	18 02.3	3.5	58.5
02	210 58.4	55.5	302 14.2	7.4	18 05.8	3.4	58.5
03	225 58.6	.. 54.5	316 40.6	7.5	18 09.2	3.4	58.4
04	240 58.8	53.6	331 07.1	7.4	18 12.6	3.2	58.4
05	255 59.0	52.6	345 33.5	7.5	18 15.8	3.0	58.4
06	270 59.3	N 3 51.7	0 00.0	7.5	N18 18.8	3.0	58.4
07	285 59.5	50.7	14 26.5	7.5	18 21.8	2.9	58.3
08	300 59.7	49.7	28 53.0	7.5	18 24.7	2.7	58.3
S 09	315 59.9	.. 48.8	43 19.5	7.5	18 27.4	2.6	58.3
U 10	331 00.2	47.8	57 46.0	7.6	18 30.0	2.5	58.2
N 11	346 00.4	46.9	72 12.6	7.6	18 32.5	2.4	58.2
D 12	1 00.6	N 3 45.9	86 39.2	7.6	N18 34.9	2.3	58.2
A 13	16 00.8	45.0	101 05.8	7.6	18 37.2	2.2	58.2
Y 14	31 01.0	44.0	115 32.4	7.6	18 39.4	2.0	58.1
15	46 01.3	.. 43.0	129 59.0	7.6	18 41.4	1.9	58.1
16	61 01.5	42.1	144 25.6	7.7	18 43.3	1.8	58.1
17	76 01.7	41.1	158 52.3	7.7	18 45.1	1.7	58.0
18	91 01.9	N 3 40.2	173 19.0	7.7	N18 46.8	1.6	58.0
19	106 02.1	39.2	187 45.7	7.7	18 48.4	1.5	58.0
20	121 02.4	38.3	202 12.4	7.8	18 49.9	1.3	58.0
21	136 02.6	.. 37.3	216 39.2	7.8	18 51.2	1.3	57.9
22	151 02.8	36.3	231 06.0	7.8	18 52.5	1.1	57.9
23	166 03.0	35.4	245 32.8	7.8	18 53.6	1.0	57.9
14 00	181 03.3	N 3 34.4	259 59.6	7.9	N18 54.6	0.9	57.9
01	196 03.5	33.5	274 26.5	7.9	18 55.5	0.8	57.8
02	211 03.7	32.5	288 53.4	7.9	18 56.3	0.7	57.8
03	226 03.9	.. 31.6	303 20.3	7.9	18 57.0	0.5	57.8
04	241 04.1	30.6	317 47.2	8.0	18 57.5	0.5	57.7
05	256 04.4	29.6	332 14.2	8.0	18 58.0	0.3	57.7
06	271 04.6	N 3 28.7	346 41.2	8.0	N18 58.3	0.2	57.7
07	286 04.8	27.7	1 08.2	8.1	18 58.5	0.1	57.7
08	301 05.0	26.8	15 35.3	8.1	18 58.6	0.0	57.6
M 09	316 05.3	.. 25.8	30 02.4	8.1	18 58.6	0.1	57.6
O 10	331 05.5	24.8	44 29.5	8.2	18 58.5	0.2	57.6
N 11	346 05.7	23.9	58 56.7	8.2	18 58.3	0.4	57.5
D 12	1 05.9	N 3 22.9	73 23.9	8.2	N18 57.9	0.4	57.5
A 13	16 06.1	22.0	87 51.1	8.3	18 57.5	0.6	57.5
Y 14	31 06.4	21.0	102 18.4	8.3	18 56.9	0.6	57.5
15	46 06.6	.. 20.0	116 45.7	8.4	18 56.3	0.8	57.4
16	61 06.8	19.1	131 13.1	8.3	18 55.5	0.9	57.4
17	76 07.0	18.1	145 40.4	8.5	18 54.6	1.0	57.4
18	91 07.2	N 3 17.2	160 07.9	8.4	N18 53.6	1.1	57.4
19	106 07.5	16.2	174 35.3	8.5	18 52.5	1.2	57.3
20	121 07.7	15.2	189 02.8	8.5	18 51.3	1.4	57.3
21	136 07.9	.. 14.3	203 30.3	8.6	18 49.9	1.4	57.3
22	151 08.1	13.3	217 57.9	8.6	18 48.5	1.5	57.2
23	166 08.4	12.4	232 25.5	8.7	18 47.0	1.7	57.2
15 00	181 08.6	N 3 11.4	246 53.2	8.7	N18 45.3	1.7	57.2
01	196 08.8	10.4	261 20.9	8.7	18 43.6	1.9	57.2
02	211 09.0	09.5	275 48.6	8.8	18 41.7	1.9	57.1
03	226 09.2	.. 08.5	290 16.4	8.8	18 39.8	2.1	57.1
04	241 09.5	07.6	304 44.2	8.9	18 37.7	2.2	57.1
05	256 09.7	06.6	319 12.1	8.9	18 35.5	2.3	57.1
06	271 09.9	N 3 05.6	333 40.0	9.0	N18 33.2	2.3	57.0
07	286 10.1	04.7	348 08.0	9.0	18 30.9	2.5	57.0
08	301 10.4	03.7	2 36.0	9.0	18 28.4	2.6	57.0
T 09	316 10.6	.. 02.7	17 04.0	9.1	18 25.8	2.7	57.0
U 10	331 10.8	01.8	31 32.1	9.1	18 23.1	2.7	56.9
E 11	346 11.0	3 00.8	46 00.2	9.2	18 20.4	2.9	56.9
S 12	1 11.2	N 2 59.9	60 28.4	9.3	N18 17.5	3.0	56.9
D 13	16 11.5	58.9	74 56.7	9.3	18 14.5	3.1	56.9
A 14	31 11.7	57.9	89 25.0	9.3	18 11.4	3.2	56.8
Y 15	46 11.9	.. 57.0	103 53.3	9.4	18 08.2	3.2	56.8
16	61 12.1	56.0	118 21.7	9.4	18 05.0	3.4	56.8
17	76 12.4	55.1	132 50.1	9.5	18 01.6	3.5	56.8
18	91 12.6	N 2 54.1	147 18.6	9.5	N17 58.1	3.5	56.7
19	106 12.8	53.1	161 47.1	9.6	17 54.6	3.7	56.7
20	121 13.0	52.2	176 15.7	9.6	17 50.9	3.7	56.7
21	136 13.2	.. 51.2	190 44.3	9.7	17 47.2	3.9	56.7
22	151 13.5	50.2	205 13.0	9.7	17 43.3	3.9	56.6
23	166 13.7	49.3	219 41.7	9.8	N17 39.4	4.0	56.6
	SD 15.9	d 1.0	SD 15.9		15.7		15.5

Lat.	Twilight Naut.	Twilight Civil	Sunrise	Moonrise 13	14	15	16
°	h m	h m	h m	h m	h m	h m	h m
N 72	01 56	03 48	05 01	▭	▭	21 03	23 16
N 70	02 31	04 02	05 07	19 58	20 52	22 18	23 55
68	02 55	04 14	05 12	20 45	21 40	22 55	24 21
66	03 13	04 23	05 16	21 16	22 11	23 22	24 41
64	03 28	04 30	05 19	21 39	22 34	23 42	24 57
62	03 40	04 37	05 22	21 57	22 52	23 58	25 11
60	03 50	04 42	05 25	22 12	23 08	24 12	00 12
N 58	03 58	04 47	05 27	22 25	23 20	24 23	00 23
56	04 05	04 51	05 29	22 36	23 32	24 34	00 34
54	04 12	04 55	05 31	22 46	23 41	24 43	00 43
52	04 17	04 58	05 32	22 55	23 50	24 51	00 51
50	04 22	05 01	05 34	23 03	23 58	24 58	00 58
45	04 32	05 07	05 37	23 19	24 14	00 14	01 13
N 40	04 40	05 12	05 40	23 33	24 28	00 28	01 26
35	04 47	05 16	05 42	23 45	24 40	00 40	01 36
30	04 52	05 20	05 44	23 55	24 50	00 50	01 46
20	04 59	05 25	05 47	24 12	00 12	01 07	02 02
N 10	05 04	05 29	05 50	24 28	00 28	01 22	02 16
0	05 08	05 32	05 52	24 42	00 42	01 36	02 29
S 10	05 09	05 34	05 55	00 00	00 57	01 51	02 42
20	05 10	05 35	05 57	00 15	01 12	02 06	02 56
30	05 08	05 36	06 00	00 32	01 30	02 23	03 12
35	05 07	05 36	06 01	00 42	01 40	02 33	03 21
40	05 05	05 36	06 03	00 53	01 52	02 45	03 32
45	05 02	05 36	06 05	01 06	02 06	02 59	03 44
S 50	04 58	05 35	06 07	01 22	02 23	03 15	04 00
52	04 55	05 35	06 08	01 30	02 31	03 23	04 07
54	04 53	05 34	06 09	01 38	02 40	03 32	04 14
56	04 50	05 33	06 10	01 48	02 50	03 42	04 23
58	04 47	05 33	06 12	01 58	03 01	03 53	04 33
S 60	04 43	05 32	06 13	02 11	03 15	04 05	04 44

Lat.	Sunset	Twilight Civil	Twilight Naut.	Moonset 13	14	15	16
°	h m	h m	h m	h m	h m	h m	h m
N 72	18 47	19 59	21 47	▭	▭	19 35	19 08
N 70	18 42	19 46	21 15	16 55	17 55	18 20	18 28
68	18 37	19 35	20 52	16 08	17 07	17 42	18 01
66	18 33	19 26	20 34	15 37	16 36	17 15	17 40
64	18 30	19 19	20 21	15 15	16 12	16 54	17 24
62	18 27	19 13	20 09	14 56	15 54	16 38	17 10
60	18 25	19 07	19 59	14 41	15 39	16 24	16 58
N 58	18 23	19 03	19 51	14 29	15 26	16 12	16 48
56	18 21	18 59	19 44	14 18	15 15	16 01	16 39
54	18 19	18 55	19 38	14 08	15 05	15 52	16 31
52	18 18	18 52	19 33	13 59	14 56	15 44	16 24
50	18 16	18 49	19 28	13 51	14 48	15 37	16 17
45	18 13	18 43	19 18	13 35	14 31	15 21	16 04
N 40	18 11	18 38	19 10	13 21	14 18	15 08	15 52
35	18 09	18 34	19 04	13 10	14 06	14 57	15 42
30	18 07	18 31	18 59	13 00	13 56	14 47	15 34
20	18 04	18 26	18 52	12 42	13 38	14 30	15 19
N 10	18 01	18 22	18 47	12 27	13 23	14 16	15 06
0	17 59	18 20	18 44	12 13	13 08	14 02	14 53
S 10	17 57	18 18	18 42	11 59	12 54	13 48	14 41
20	17 54	18 16	18 42	11 44	12 38	13 33	14 28
30	17 52	18 16	18 43	11 27	12 21	13 16	14 13
35	17 50	18 16	18 45	11 17	12 10	13 06	14 04
40	17 49	18 16	18 47	11 05	11 59	12 55	13 54
45	17 47	18 16	18 49	10 52	11 45	12 42	13 42
S 50	17 45	18 17	18 55	10 35	11 28	12 25	13 27
52	17 44	18 18	18 57	10 27	11 20	12 18	13 20
54	17 43	18 18	19 00	10 19	11 11	12 09	13 13
56	17 42	18 19	19 02	10 09	11 01	12 00	13 04
58	17 41	18 20	19 06	09 58	10 49	11 49	12 55
S 60	17 39	18 21	19 10	09 45	10 36	11 36	12 44

Day	SUN Eqn. of Time 00h	12h	Mer. Pass.	MOON Mer. Pass. Upper	Lower	Age	Phase
d	m s	m s	h m	h m	h m	d	%
13	03 51	04 02	11 56	06 00	18 28	22	45
14	04 13	04 23	11 56	06 55	19 22	23	35
15	04 34	04 45	11 55	07 49	20 15	24	25

1998 SEPTEMBER 16, 17, 18 (WED., THURS., FRI.)

UT	ARIES GHA	VENUS −3.9 GHA	Dec	MARS +1.7 GHA	Dec	JUPITER −2.9 GHA	Dec	SATURN +0.1 GHA	Dec	Name	SHA	Dec
d h	° ′	° ′	° ′	° ′	° ′	° ′	° ′	° ′	° ′		° ′	° ′
16 00	354 44.4	191 23.8	N 8 32.9	215 15.2	N16 57.9	0 29.0	S 4 11.0	323 12.1	N 9 54.0	Acamar	315 26.7	S40 18.5
01	9 46.9	206 23.4	31.7	230 16.1	57.5	15 31.7	11.1	338 14.7	53.9	Achernar	335 34.8	S57 14.5
02	24 49.3	221 22.9	30.6	245 17.0	57.0	30 34.5	11.3	353 17.3	53.9	Acrux	173 22.6	S63 05.5
03	39 51.8	236 22.5	.. 29.4	260 17.9	.. 56.6	45 37.3	.. 11.4	8 19.9	.. 53.8	Adhara	255 21.6	S28 58.1
04	54 54.3	251 22.0	28.3	275 18.8	56.1	60 40.0	11.5	23 22.5	53.8	Aldebaran	291 02.3	N16 30.3
05	69 56.7	266 21.6	27.1	290 19.7	55.6	75 42.8	11.7	38 25.1	53.7			
06	84 59.2	281 21.1	N 8 26.0	305 20.6	N16 55.2	90 45.6	S 4 11.8	53 27.6	N 9 53.7	Alioth	166 31.2	N55 58.3
W 07	100 01.7	296 20.7	24.8	320 21.5	54.7	105 48.4	11.9	68 30.2	53.6	Alkaid	153 08.3	N49 19.5
E 08	115 04.1	311 20.2	23.7	335 22.4	54.3	120 51.1	12.1	83 32.8	53.6	Al Na'ir	27 57.6	S46 58.0
D 09	130 06.6	326 19.8	.. 22.5	350 23.3	.. 53.8	135 53.9	.. 12.2	98 35.4	.. 53.5	Alnilam	275 57.9	S 1 12.1
N 10	145 09.0	341 19.3	21.4	5 24.2	53.3	150 56.7	12.3	113 38.0	53.5	Alphard	218 07.5	S 8 39.1
E 11	160 11.5	356 18.9	20.2	20 25.1	52.9	165 59.4	12.5	128 40.6	53.4			
S 12	175 14.0	11 18.4	N 8 19.1	35 26.0	N16 52.4	181 02.2	S 4 12.6	143 43.1	N 9 53.4	Alphecca	126 20.8	N26 43.5
D 13	190 16.4	26 18.0	17.9	50 26.9	52.0	196 05.0	12.7	158 45.7	53.3	Alpheratz	357 54.9	N29 05.0
A 14	205 18.9	41 17.6	16.7	65 27.8	51.5	211 07.8	12.9	173 48.3	53.3	Altair	62 19.1	N 8 52.1
Y 15	220 21.4	56 17.1	.. 15.6	80 28.7	.. 51.1	226 10.5	.. 13.0	188 50.9	.. 53.2	Ankaa	353 26.5	S42 18.7
16	235 23.8	71 16.7	14.4	95 29.6	50.6	241 13.3	13.1	203 53.5	53.2	Antares	112 40.3	S26 25.6
17	250 26.3	86 16.2	13.3	110 30.5	50.1	256 16.1	13.3	218 56.1	53.1			
18	265 28.8	101 15.8	N 8 12.1	125 31.4	N16 49.7	271 18.8	S 4 13.4	233 58.7	N 9 53.1	Arcturus	146 06.4	N19 11.7
19	280 31.2	116 15.3	11.0	140 32.3	49.2	286 21.6	13.5	249 01.2	53.0	Atria	107 52.5	S69 01.6
20	295 33.7	131 14.9	09.8	155 33.2	48.8	301 24.4	13.7	264 03.8	52.9	Avior	234 23.0	S59 30.1
21	310 36.2	146 14.5	.. 08.6	170 34.1	.. 48.3	316 27.1	.. 13.8	279 06.4	.. 52.9	Bellatrix	278 44.2	N 6 20.9
22	325 38.6	161 14.0	07.5	185 35.0	47.8	331 29.9	13.9	294 09.0	52.8	Betelgeuse	271 13.6	N 7 24.4
23	340 41.1	176 13.6	06.3	200 35.9	47.4	346 32.7	14.1	309 11.6	52.8			
17 00	355 43.5	191 13.1	N 8 05.2	215 36.8	N16 46.9	1 35.5	S 4 14.2	324 14.2	N 9 52.7	Canopus	264 01.3	S52 41.5
01	10 46.0	206 12.7	04.0	230 37.7	46.5	16 38.2	14.3	339 16.8	52.7	Capella	280 51.2	N45 59.5
02	25 48.5	221 12.2	02.8	245 38.6	46.0	31 41.0	14.5	354 19.3	52.6	Deneb	49 38.9	N45 16.9
03	40 50.9	236 11.8	.. 01.7	260 39.5	.. 45.5	46 43.8	.. 14.6	9 21.9	.. 52.6	Denebola	182 45.6	N14 34.9
04	55 53.4	251 11.4	8 00.5	275 40.4	45.1	61 46.5	14.7	24 24.5	52.5	Diphda	349 07.0	S17 59.5
05	70 55.9	266 10.9	7 59.4	290 41.3	44.6	76 49.3	14.9	39 27.1	52.5			
06	85 58.3	281 10.5	N 7 58.2	305 42.2	N16 44.1	91 52.1	S 4 15.0	54 29.7	N 9 52.4	Dubhe	194 06.3	N61 45.5
T 07	101 00.8	296 10.0	57.0	320 43.1	43.7	106 54.8	15.1	69 32.3	52.4	Elnath	278 27.0	N28 36.2
H 08	116 03.3	311 09.6	55.9	335 44.0	43.2	121 57.6	15.3	84 34.9	52.3	Eltanin	90 51.4	N51 29.8
U 09	131 05.7	326 09.2	.. 54.7	350 44.9	.. 42.8	137 00.4	.. 15.4	99 37.5	.. 52.3	Enif	33 58.0	N 9 52.3
R 10	146 08.2	341 08.7	53.5	5 45.8	42.3	152 03.2	15.5	114 40.0	52.2	Fomalhaut	15 36.2	S29 37.7
S 11	161 10.6	356 08.3	52.4	20 46.7	41.8	167 05.9	15.7	129 42.6	52.2			
D 12	176 13.1	11 07.8	N 7 51.2	35 47.6	N16 41.4	182 08.7	S 4 15.8	144 45.2	N 9 52.1	Gacrux	172 14.1	S57 06.3
A 13	191 15.6	26 07.4	50.0	50 48.5	40.9	197 11.5	15.9	159 47.8	52.1	Gienah	176 04.3	S17 31.9
Y 14	206 18.0	41 07.0	48.9	65 49.4	40.4	212 14.2	16.1	174 50.4	52.0	Hadar	149 04.5	S60 22.0
15	221 20.5	56 06.5	.. 47.7	80 50.3	.. 40.0	227 17.0	.. 16.2	189 53.0	.. 52.0	Hamal	328 13.3	N23 27.3
16	236 23.0	71 06.1	46.5	95 51.2	39.5	242 19.8	16.3	204 55.6	51.9	Kaus Aust.	83 58.8	S34 23.1
17	251 25.4	86 05.6	45.4	110 52.1	39.1	257 22.5	16.5	219 58.2	51.9			
18	266 27.9	101 05.2	N 7 44.2	125 53.0	N16 38.6	272 25.3	S 4 16.6	235 00.7	N 9 51.8	Kochab	137 20.5	N74 10.0
19	281 30.4	116 04.8	43.0	140 53.9	38.1	287 28.1	16.7	250 03.3	51.8	Markab	13 49.3	N15 12.0
20	296 32.8	131 04.3	41.9	155 54.8	37.7	302 30.9	16.9	265 05.9	51.7	Menkar	314 26.7	N 4 05.1
21	311 35.3	146 03.9	.. 40.7	170 55.7	.. 37.2	317 33.6	.. 17.0	280 08.5	.. 51.6	Menkent	148 21.3	S36 21.7
22	326 37.8	161 03.5	39.5	185 56.6	36.7	332 36.4	17.1	295 11.1	51.6	Miaplacidus	221 42.7	S69 42.5
23	341 40.2	176 03.0	38.4	200 57.5	36.3	347 39.2	17.3	310 13.7	51.5			
18 00	356 42.7	191 02.6	N 7 37.2	215 58.4	N16 35.8	2 41.9	S 4 17.4	325 16.3	N 9 51.5	Mirfak	308 56.4	N49 51.2
01	11 45.1	206 02.2	36.0	230 59.4	35.3	17 44.7	17.5	340 18.9	51.4	Nunki	76 12.3	S26 17.8
02	26 47.6	221 01.7	34.9	246 00.3	34.9	32 47.5	17.7	355 21.5	51.4	Peacock	53 36.7	S56 44.4
03	41 50.1	236 01.3	.. 33.7	261 01.2	.. 34.4	47 50.2	.. 17.8	10 24.1	.. 51.3	Pollux	243 41.8	N28 01.7
04	56 52.5	251 00.8	32.5	276 02.1	33.9	62 53.0	17.9	25 26.6	51.3	Procyon	245 11.8	N 5 13.7
05	71 55.0	266 00.4	31.3	291 03.0	33.5	77 55.8	18.1	40 29.2	51.2			
06	86 57.5	281 00.0	N 7 30.2	306 03.9	N16 33.0	92 58.6	S 4 18.2	55 31.8	N 9 51.2	Rasalhague	96 17.0	N12 34.0
07	101 59.9	295 59.5	29.0	321 04.8	32.5	108 01.3	18.3	70 34.4	51.1	Regulus	207 55.9	N11 58.5
08	117 02.4	310 59.1	27.8	336 05.7	32.1	123 04.1	18.5	85 37.0	51.1	Rigel	281 23.0	S 8 12.1
F 09	132 04.9	325 58.7	.. 26.6	351 06.6	.. 31.6	138 06.9	.. 18.6	100 39.6	.. 51.0	Rigil Kent.	140 07.8	S60 49.8
R 10	147 07.3	340 58.2	25.5	6 07.5	31.2	153 09.6	18.7	115 42.2	51.0	Sabik	102 25.6	S15 43.2
I 11	162 09.8	355 57.8	24.3	21 08.4	30.7	168 12.4	18.8	130 44.8	50.9			
D 12	177 12.3	10 57.4	N 7 23.1	36 09.3	N16 30.2	183 15.2	S 4 19.0	145 47.4	N 9 50.9	Schedar	349 52.9	N56 31.7
A 13	192 14.7	25 56.9	22.0	51 10.2	29.8	198 17.9	19.1	160 50.0	50.8	Shaula	96 37.4	S37 06.1
Y 14	207 17.2	40 56.5	20.8	66 11.1	29.3	213 20.7	19.2	175 52.6	50.7	Sirius	258 43.8	S16 42.7
15	222 19.6	55 56.1	.. 19.6	81 12.0	.. 28.8	228 23.5	.. 19.4	190 55.1	.. 50.7	Spica	158 43.5	S11 09.1
16	237 22.1	70 55.6	18.4	96 12.9	28.4	243 26.3	19.5	205 57.7	50.6	Suhail	223 01.1	S43 25.5
17	252 24.6	85 55.2	17.3	111 13.9	27.9	258 29.0	19.6	221 00.3	50.6			
18	267 27.0	100 54.8	N 7 16.1	126 14.8	N16 27.4	273 31.8	S 4 19.8	236 02.9	N 9 50.5	Vega	80 46.5	N38 47.4
19	282 29.5	115 54.3	14.9	141 15.7	26.9	288 34.6	19.9	251 05.5	50.5	Zuben'ubi	137 18.2	S16 02.0
20	297 32.0	130 53.9	13.7	156 16.6	26.5	303 37.3	20.0	266 08.1	50.4			
21	312 34.4	145 53.5	.. 12.5	171 17.5	.. 26.0	318 40.1	.. 20.2	281 10.7	.. 50.4			
22	327 36.9	160 53.0	11.4	186 18.4	25.5	333 42.9	20.3	296 13.3	50.3			
23	342 39.4	175 52.6	10.2	201 19.3	25.1	348 45.6	20.4	311 15.9	50.3			

	SHA	Mer.Pass.
	° ′	h m
Venus	195 29.6	11 15
Mars	219 53.2	9 37
Jupiter	5 51.9	23 49
Saturn	328 30.6	2 23

Mer. Pass. 0 17.1   *v* −0.4 *d* 1.2   *v* 0.9 *d* 0.5   *v* 2.8 *d* 0.1   *v* 2.6 *d* 0.1

UT	SUN GHA	SUN Dec	MOON GHA	v	MOON Dec	d	HP
d h	° ′	° ′	° ′	′	° ′	′	′
16 00	181 13.9	N 2 48.3	234 10.5	9.9	N17 35.4	4.2	56.6
01	196 14.1	47.3	248 39.4	9.8	17 31.2	4.2	56.6
02	211 14.4	46.4	263 08.2	10.0	17 27.0	4.3	56.5
03	226 14.6	.. 45.4	277 37.2	10.0	17 22.7	4.4	56.5
04	241 14.8	44.5	292 06.2	10.0	17 18.3	4.5	56.5
05	256 15.0	43.5	306 35.2	10.1	17 13.8	4.5	56.5
06	271 15.2	N 2 42.5	321 04.3	10.2	N17 09.3	4.7	56.4
W 07	286 15.5	41.6	335 33.5	10.2	17 04.6	4.7	56.4
E 08	301 15.7	40.6	350 02.7	10.2	16 59.9	4.9	56.4
D 09	316 15.9	.. 39.6	4 31.9	10.3	16 55.0	4.9	56.4
N 10	331 16.1	38.7	19 01.2	10.4	16 50.1	5.0	56.3
E 11	346 16.4	37.7	33 30.6	10.4	16 45.1	5.1	56.3
S 12	1 16.6	N 2 36.7	48 00.0	10.5	N16 40.0	5.2	56.3
D 13	16 16.8	35.8	62 29.5	10.5	16 34.8	5.2	56.3
A 14	31 17.0	34.8	76 59.0	10.6	16 29.6	5.4	56.3
Y 15	46 17.2	.. 33.9	91 28.6	10.6	16 24.2	5.4	56.2
16	61 17.5	32.9	105 58.2	10.7	16 18.8	5.5	56.2
17	76 17.7	31.9	120 27.9	10.8	16 13.3	5.5	56.2
18	91 17.9	N 2 31.0	134 57.7	10.7	N16 07.8	5.7	56.2
19	106 18.1	30.0	149 27.4	10.9	16 02.1	5.7	56.1
20	121 18.4	29.0	163 57.3	10.9	15 56.4	5.9	56.1
21	136 18.6	.. 28.1	178 27.2	11.0	15 50.5	5.8	56.1
22	151 18.8	27.1	192 57.2	11.0	15 44.7	6.0	56.1
23	166 19.0	26.1	207 27.2	11.0	15 38.7	6.1	56.1
17 00	181 19.3	N 2 25.2	221 57.2	11.2	N15 32.6	6.1	56.0
01	196 19.5	24.2	236 27.4	11.1	15 26.5	6.2	56.0
02	211 19.7	23.2	250 57.5	11.3	15 20.3	6.2	56.0
03	226 19.9	.. 22.3	265 27.8	11.3	15 14.1	6.4	56.0
04	241 20.1	21.3	279 58.1	11.3	15 07.7	6.4	55.9
05	256 20.4	20.3	294 28.4	11.4	15 01.3	6.4	55.9
06	271 20.6	N 2 19.4	308 58.8	11.4	N14 54.9	6.6	55.9
T 07	286 20.8	18.4	323 29.2	11.5	14 48.3	6.6	55.9
H 08	301 21.0	17.4	337 59.7	11.6	14 41.7	6.7	55.9
U 09	316 21.3	.. 16.5	352 30.3	11.6	14 35.0	6.7	55.8
R 10	331 21.5	15.5	7 00.9	11.7	14 28.3	6.9	55.8
S 11	346 21.7	14.5	21 31.6	11.7	14 21.4	6.8	55.8
D 12	1 21.9	N 2 13.6	36 02.3	11.8	N14 14.6	7.0	55.8
A 13	16 22.1	12.6	50 33.1	11.8	14 07.6	7.0	55.8
Y 14	31 22.4	11.6	65 03.9	11.9	14 00.6	7.1	55.7
15	46 22.6	.. 10.7	79 34.8	11.9	13 53.5	7.1	55.7
16	61 22.8	09.7	94 05.7	12.0	13 46.4	7.2	55.7
17	76 23.0	08.7	108 36.7	12.0	13 39.2	7.3	55.7
18	91 23.3	N 2 07.8	123 07.7	12.1	N13 31.9	7.3	55.7
19	106 23.5	06.8	137 38.8	12.1	13 24.6	7.4	55.6
20	121 23.7	05.8	152 09.9	12.2	13 17.2	7.5	55.6
21	136 23.9	.. 04.9	166 41.1	12.3	13 09.7	7.5	55.6
22	151 24.1	03.9	181 12.4	12.3	13 02.2	7.5	55.6
23	166 24.4	02.9	195 43.7	12.3	12 54.7	7.7	55.6
18 00	181 24.6	N 2 02.0	210 15.0	12.4	N12 47.0	7.6	55.5
01	196 24.8	01.0	224 46.4	12.4	12 39.4	7.8	55.5
02	211 25.0	2 00.0	239 17.8	12.5	12 31.6	7.7	55.5
03	226 25.3	1 59.1	253 49.3	12.6	12 23.9	7.9	55.5
04	241 25.5	58.1	268 20.9	12.6	12 16.0	7.9	55.5
05	256 25.7	57.1	282 52.5	12.6	12 08.1	7.9	55.4
06	271 25.9	N 1 56.2	297 24.1	12.7	N12 00.2	8.0	55.4
F 07	286 26.1	55.2	311 55.8	12.8	11 52.2	8.1	55.4
R 08	301 26.4	54.2	326 27.6	12.8	11 44.1	8.1	55.4
I 09	316 26.6	.. 53.3	340 59.4	12.8	11 36.0	8.1	55.4
D 10	331 26.8	52.3	355 31.2	12.9	11 27.9	8.2	55.4
A 11	346 27.0	51.3	10 03.1	12.9	11 19.7	8.2	55.3
Y 12	1 27.3	N 1 50.4	24 35.0	13.0	N11 11.5	8.3	55.3
13	16 27.5	49.4	39 07.0	13.0	11 03.2	8.4	55.3
14	31 27.7	48.4	53 39.0	13.1	10 54.8	8.3	55.3
15	46 27.9	.. 47.5	68 11.1	13.1	10 46.5	8.5	55.3
16	61 28.1	46.5	82 43.2	13.2	10 38.0	8.4	55.2
17	76 28.4	45.5	97 15.4	13.2	10 29.6	8.5	55.2
18	91 28.6	N 1 44.6	111 47.6	13.3	N10 21.1	8.6	55.2
19	106 28.8	43.6	126 19.9	13.3	10 12.5	8.6	55.2
20	121 29.0	42.6	140 52.2	13.3	10 03.9	8.6	55.2
21	136 29.2	.. 41.6	155 24.5	13.4	9 55.3	8.7	55.2
22	151 29.5	40.7	169 56.9	13.4	9 46.6	8.7	55.1
23	166 29.7	39.7	184 29.3	13.5	N 9 37.9	8.8	55.1
	SD 15.9	d 1.0	SD 15.3		15.2		15.1

Lat.	Twilight Naut.	Twilight Civil	Sunrise	Moonrise 16	Moonrise 17	Moonrise 18	Moonrise 19
°	h m	h m	h m	h m	h m	h m	h m
N 72	02 22	04 04	05 15	23 16	25 09	01 09	02 53
N 70	02 50	04 16	05 19	23 55	25 32	01 32	03 07
68	03 10	04 25	05 23	24 21	00 21	01 50	03 19
66	03 26	04 33	05 25	24 41	00 41	02 05	03 28
64	03 39	04 39	05 28	24 57	00 57	02 16	03 36
62	03 49	04 45	05 30	25 11	01 11	02 26	03 43
60	03 58	04 50	05 32	00 12	01 22	02 35	03 49
N 58	04 06	04 54	05 33	00 23	01 32	02 42	03 54
56	04 12	04 57	05 35	00 34	01 40	02 49	03 58
54	04 18	05 00	05 36	00 43	01 48	02 55	04 02
52	04 23	05 03	05 37	00 51	01 55	03 00	04 06
50	04 27	05 06	05 38	00 58	02 01	03 05	04 09
45	04 36	05 11	05 41	01 13	02 14	03 15	04 16
N 40	04 43	05 15	05 42	01 26	02 25	03 24	04 22
35	04 49	05 19	05 44	01 36	02 34	03 31	04 27
30	04 53	05 21	05 45	01 46	02 42	03 37	04 32
20	05 00	05 26	05 48	02 02	02 56	03 49	04 40
N 10	05 04	05 29	05 50	02 16	03 08	03 58	04 47
0	05 07	05 31	05 51	02 29	03 19	04 07	04 53
S 10	05 08	05 32	05 53	02 42	03 31	04 16	05 00
20	05 07	05 32	05 54	02 56	03 43	04 26	05 06
30	05 05	05 32	05 56	03 12	03 56	04 37	05 14
35	05 03	05 32	05 57	03 21	04 04	04 43	05 19
40	05 00	05 31	05 58	03 32	04 14	04 50	05 24
45	04 56	05 30	05 59	03 44	04 24	04 59	05 30
S 50	04 51	05 28	06 01	04 00	04 37	05 09	05 37
52	04 48	05 28	06 01	04 07	04 43	05 13	05 40
54	04 45	05 27	06 02	04 14	04 49	05 18	05 43
56	04 42	05 26	06 02	04 23	04 57	05 24	05 47
58	04 38	05 24	06 03	04 33	05 05	05 30	05 52
S 60	04 34	05 23	06 04	04 44	05 14	05 37	05 56

Lat.	Sunset	Twilight Civil	Twilight Naut.	Moonset 16	Moonset 17	Moonset 18	Moonset 19
°	h m	h m	h m	h m	h m	h m	h m
N 72	18 32	19 42	21 21	19 08	18 56	18 48	18 41
N 70	18 28	19 30	20 55	18 28	18 31	18 32	18 31
68	18 24	19 21	20 35	18 01	18 12	18 19	18 24
66	18 22	19 14	20 20	17 40	17 57	18 09	18 17
64	18 20	19 08	20 08	17 24	17 44	18 00	18 12
62	18 18	19 02	19 58	17 10	17 34	17 52	18 07
60	18 16	18 58	19 49	16 58	17 24	17 46	18 03
N 58	18 14	18 54	19 42	16 48	17 16	17 40	17 59
56	18 13	18 50	19 35	16 39	17 09	17 34	17 56
54	18 12	18 47	19 30	16 31	17 03	17 30	17 53
52	18 11	18 45	19 25	16 24	16 57	17 26	17 51
50	18 10	18 42	19 21	16 17	16 52	17 22	17 48
45	18 08	18 37	19 12	16 04	16 41	17 13	17 43
N 40	18 06	18 33	19 05	15 52	16 31	17 06	17 38
35	18 04	18 30	18 59	15 42	16 23	17 00	17 34
30	18 03	18 27	18 55	15 34	16 16	16 55	17 31
20	18 01	18 23	18 49	15 19	16 04	16 46	17 25
N 10	17 59	18 20	18 45	15 06	15 53	16 37	17 20
0	17 58	18 18	18 42	14 53	15 43	16 30	17 15
S 10	17 56	18 17	18 42	14 41	15 32	16 22	17 10
20	17 55	18 17	18 42	14 28	15 21	16 13	17 04
30	17 53	18 17	18 45	14 13	15 09	16 04	16 58
35	17 53	18 18	18 47	14 04	15 01	15 58	16 54
40	17 52	18 19	18 50	13 54	14 53	15 52	16 50
45	17 51	18 20	18 54	13 42	14 43	15 45	16 46
S 50	17 49	18 22	18 59	13 27	14 31	15 36	16 40
52	17 49	18 23	19 02	13 20	14 26	15 32	16 37
54	17 48	18 24	19 05	13 13	14 19	15 27	16 34
56	17 48	18 25	19 08	13 04	14 13	15 22	16 31
58	17 47	18 26	19 12	12 55	14 05	15 16	16 28
S 60	17 46	18 28	19 17	12 44	13 56	15 10	16 24

Day	SUN Eqn. of Time 00h	SUN Eqn. of Time 12h	SUN Mer. Pass.	MOON Mer. Pass. Upper	MOON Mer. Pass. Lower	Age	Phase
d	m s	m s	h m	h m	h m	d	%
16	04 55	05 06	11 55	08 41	21 06	25	17
17	05 17	05 27	11 55	09 31	21 55	26	10
18	05 38	05 49	11 54	10 18	22 41	27	5

UT	ARIES GHA	VENUS −3.9 GHA	Dec	MARS +1.7 GHA	Dec	JUPITER −2.9 GHA	Dec	SATURN +0.1 GHA	Dec
19 00	357 41.8	190 52.2 N 7 09.0		216 20.2 N16 24.6		3 48.4 S 4 20.6		326 18.5 N 9 50.2	
01	12 44.3	205 51.8	07.8	231 21.1	24.1	18 51.2	20.7	341 21.1	50.2
02	27 46.7	220 51.3	06.6	246 22.0	23.7	33 54.0	20.8	356 23.7	50.1
03	42 49.2	235 50.9 ..	05.5	261 22.9 ..	23.2	48 56.7 ..	21.0	11 26.3 ..	50.0
04	57 51.7	250 50.5	04.3	276 23.9	22.7	63 59.5	21.1	26 28.8	50.0
05	72 54.1	265 50.0	03.1	291 24.8	22.3	79 02.3	21.2	41 31.4	49.9
S 06	87 56.6	280 49.6 N 7 01.9		306 25.7 N16 21.8		94 05.0 S 4 21.4		56 34.0 N 9 49.9	
A 07	102 59.1	295 49.2 7 00.7		321 26.6	21.3	109 07.8	21.5	71 36.6	49.8
T 08	118 01.5	310 48.8 6 59.6		336 27.5	20.9	124 10.6	21.6	86 39.2	49.8
U 09	133 04.0	325 48.3 ..	58.4	351 28.4 ..	20.4	139 13.3 ..	21.8	101 41.8 ..	49.7
R 10	148 06.5	340 47.9	57.2	6 29.3	19.9	154 16.1	21.9	116 44.4	49.7
D 11	163 08.9	355 47.5	56.0	21 30.2	19.4	169 18.9	22.0	131 47.0	49.6
A 12	178 11.4	10 47.0 N 6 54.8		36 31.1 N16 19.0		184 21.6 S 4 22.2		146 49.6 N 9 49.6	
Y 13	193 13.9	25 46.6	53.7	51 32.1	18.5	199 24.4	22.3	161 52.2	49.5
14	208 16.3	40 46.2	52.5	66 33.0	18.0	214 27.2	22.4	176 54.8	49.5
15	223 18.8	55 45.8 ..	51.3	81 33.9 ..	17.6	229 30.0 ..	22.5	191 57.4 ..	49.4
16	238 21.2	70 45.3	50.1	96 34.8	17.1	244 32.7	22.7	207 00.0	49.3
17	253 23.7	85 44.9	48.9	111 35.7	16.6	259 35.5	22.8	222 02.6	49.3
18	268 26.2	100 44.5 N 6 47.7		126 36.6 N16 16.2		274 38.3 S 4 22.9		237 05.2 N 9 49.2	
19	283 28.6	115 44.0	46.5	141 37.5	15.7	289 41.0	23.1	252 07.8	49.2
20	298 31.1	130 43.6	45.4	156 38.4	15.2	304 43.8	23.2	267 10.4	49.1
21	313 33.6	145 43.2 ..	44.2	171 39.4 ..	14.7	319 46.6 ..	23.3	282 13.0 ..	49.1
22	328 36.0	160 42.8	43.0	186 40.3	14.3	334 49.3	23.5	297 15.6	49.0
23	343 38.5	175 42.3	41.8	201 41.2	13.8	349 52.1	23.6	312 18.1	49.0
20 00	358 41.0	190 41.9 N 6 40.6		216 42.1 N16 13.3		4 54.9 S 4 23.7		327 20.7 N 9 48.9	
01	13 43.4	205 41.5	39.4	231 43.0	12.8	19 57.6	23.9	342 23.3	48.9
02	28 45.9	220 41.1	38.2	246 43.9	12.4	35 00.4	24.0	357 25.9	48.8
03	43 48.3	235 40.6 ..	37.1	261 44.8 ..	11.9	50 03.2 ..	24.1	12 28.5 ..	48.7
04	58 50.8	250 40.2	35.9	276 45.7	11.4	65 06.0	24.3	27 31.1	48.7
05	73 53.3	265 39.8	34.7	291 46.7	11.0	80 08.7	24.4	42 33.7	48.6
S 06	88 55.7	280 39.4 N 6 33.5		306 47.6 N16 10.5		95 11.5 S 4 24.5		57 36.3 N 9 48.6	
U 07	103 58.2	295 38.9	32.3	321 48.5	10.0	110 14.3	24.7	72 38.9	48.5
N 08	119 00.7	310 38.5	31.1	336 49.4	09.5	125 17.0	24.8	87 41.5	48.5
D 09	134 03.1	325 38.1 ..	29.9	351 50.3 ..	09.1	140 19.8 ..	24.9	102 44.1 ..	48.4
A 10	149 05.6	340 37.7	28.7	6 51.2	08.6	155 22.6	25.0	117 46.7	48.4
Y 11	164 08.1	355 37.3	27.5	21 52.2	08.1	170 25.3	25.2	132 49.3	48.3
12	179 10.5	10 36.8 N 6 26.3		36 53.1 N16 07.6		185 28.1 S 4 25.3		147 51.9 N 9 48.3	
13	194 13.0	25 36.4	25.2	51 54.0	07.2	200 30.9	25.4	162 54.5	48.2
14	209 15.5	40 36.0	24.0	66 54.9	06.7	215 33.6	25.6	177 57.1	48.1
15	224 17.9	55 35.6 ..	22.8	81 55.8 ..	06.2	230 36.4 ..	25.7	192 59.7 ..	48.1
16	239 20.4	70 35.1	21.6	96 56.7	05.7	245 39.2	25.8	208 02.3	48.0
17	254 22.8	85 34.7	20.4	111 57.7	05.3	260 41.9	26.0	223 04.9	48.0
18	269 25.3	100 34.3 N 6 19.2		126 58.6 N16 04.8		275 44.7 S 4 26.1		238 07.5 N 9 47.9	
19	284 27.8	115 33.9	18.0	141 59.5	04.3	290 47.5	26.2	253 10.1	47.9
20	299 30.2	130 33.5	16.8	157 00.4	03.8	305 50.3	26.4	268 12.7	47.8
21	314 32.7	145 33.0 ..	15.6	172 01.3 ..	03.4	320 53.0 ..	26.5	283 15.3 ..	47.8
22	329 35.2	160 32.6	14.4	187 02.2	02.9	335 55.8	26.6	298 17.9	47.7
23	344 37.6	175 32.2	13.2	202 03.2	02.4	350 58.6	26.7	313 20.5	47.6
21 00	359 40.1	190 31.8 N 6 12.0		217 04.1 N16 01.9		6 01.3 S 4 26.9		328 23.1 N 9 47.6	
01	14 42.6	205 31.3	10.8	232 05.0	01.5	21 04.1	27.0	343 25.7	47.5
02	29 45.0	220 30.9	09.6	247 05.9	01.0	36 06.9	27.1	358 28.3	47.5
03	44 47.5	235 30.5 ..	08.4	262 06.8 ..	00.5	51 09.6 ..	27.3	13 30.9 ..	47.4
04	59 49.9	250 30.1	07.2	277 07.8 16 00.0		66 12.4	27.4	28 33.5	47.4
05	74 52.4	265 29.7	06.0	292 08.7 15 59.6		81 15.2	27.5	43 36.1	47.3
M 06	89 54.9	280 29.2 N 6 04.8		307 09.6 N15 59.1		96 17.9 S 4 27.7		58 38.7 N 9 47.3	
O 07	104 57.3	295 28.8	03.7	322 10.5	58.6	111 20.7	27.8	73 41.3	47.2
N 08	119 59.8	310 28.4	02.5	337 11.4	58.1	126 23.5	27.9	88 43.9	47.1
D 09	135 02.3	325 28.0 ..	01.3	352 12.4 ..	57.7	141 26.2 ..	28.1	103 46.5 ..	47.1
A 10	150 04.7	340 27.6 6 00.1		7 13.3	57.2	156 29.0	28.2	118 49.1	47.0
Y 11	165 07.2	355 27.2 5 58.9		22 14.2	56.7	171 31.8	28.3	133 51.7	47.0
12	180 09.7	10 26.7 N 5 57.7		37 15.1 N15 56.2		186 34.5 S 4 28.4		148 54.3 N 9 46.9	
13	195 12.1	25 26.3	56.5	52 16.0	55.7	201 37.3	28.6	163 56.9	46.9
14	210 14.6	40 25.9	55.3	67 17.0	55.3	216 40.1	28.7	178 59.5	46.8
15	225 17.1	55 25.5 ..	54.1	82 17.9 ..	54.8	231 42.8 ..	28.8	194 02.1 ..	46.7
16	240 19.5	70 25.1	52.9	97 18.8	54.3	246 45.6	29.0	209 04.7	46.7
17	255 22.0	85 24.6	51.7	112 19.7	53.8	261 48.4	29.1	224 07.3	46.6
18	270 24.4	100 24.2 N 5 50.5		127 20.6 N15 53.4		276 51.1 S 4 29.2		239 09.9 N 9 46.6	
19	285 26.9	115 23.8	49.3	142 21.6	52.9	291 53.9	29.4	254 12.5	46.5
20	300 29.4	130 23.4	48.1	157 22.5	52.4	306 56.7	29.5	269 15.1	46.5
21	315 31.8	145 23.0 ..	46.9	172 23.4 ..	51.9	321 59.4 ..	29.6	284 17.7 ..	46.4
22	330 34.3	160 22.6	45.7	187 24.3	51.4	337 02.2	29.7	299 20.3	46.4
23	345 36.8	175 22.1	44.5	202 25.3	51.0	352 05.0	29.9	314 22.9	46.3
Mer.Pass. 0 05.3		v −0.4 d 1.2		v 0.9 d 0.5		v 2.8 d 0.1		v 2.6 d 0.1	

STARS

Name	SHA	Dec
Acamar	315 26.7	S40 18.5
Achernar	335 34.8	S57 14.5
Acrux	173 22.7	S63 05.4
Adhara	255 21.6	S28 58.1
Aldebaran	291 02.3	N16 30.3
Alioth	166 31.2	N55 58.2
Alkaid	153 08.3	N49 19.5
Al Na'ir	27 57.6	S46 58.0
Alnilam	275 57.9	S 1 12.1
Alphard	218 07.5	S 8 39.1
Alphecca	126 20.8	N26 43.5
Alpheratz	357 54.9	N29 05.0
Altair	62 19.2	N 8 52.2
Ankaa	353 26.5	S42 18.7
Antares	112 40.3	S26 25.6
Arcturus	146 06.4	N19 11.6
Atria	107 52.5	S69 01.6
Avior	234 23.0	S59 30.1
Bellatrix	278 44.2	N 6 20.9
Betelgeuse	271 13.6	N 7 24.4
Canopus	264 01.3	S52 41.5
Capella	280 51.2	N45 59.6
Deneb	49 38.9	N45 16.9
Denebola	182 45.6	N14 34.9
Diphda	349 07.0	S17 59.5
Dubhe	194 06.3	N61 45.5
Elnath	278 27.0	N28 36.2
Eltanin	90 51.4	N51 29.8
Enif	33 58.0	N 9 52.3
Fomalhaut	15 36.2	S29 37.7
Gacrux	172 14.1	S57 06.3
Gienah	176 04.3	S17 31.9
Hadar	149 04.6	S60 22.0
Hamal	328 13.3	N23 27.3
Kaus Aust.	83 58.8	S34 23.1
Kochab	137 20.5	N74 10.0
Markab	13 49.3	N15 12.0
Menkar	314 26.7	N 4 05.1
Menkent	148 21.3	S36 21.7
Miaplacidus	221 42.7	S69 42.5
Mirfak	308 56.3	N49 51.2
Nunki	76 12.3	S26 17.8
Peacock	53 36.8	S56 44.4
Pollux	243 41.8	N28 01.7
Procyon	245 11.8	N 5 13.7
Rasalhague	96 17.0	N12 34.0
Regulus	207 55.9	N11 58.5
Rigel	281 22.9	S 8 12.1
Rigil Kent.	140 07.8	S60 49.8
Sabik	102 25.6	S15 43.2
Schedar	349 52.9	N56 31.8
Shaula	96 37.4	S37 06.1
Sirius	258 43.8	S16 42.7
Spica	158 43.5	S11 09.1
Suhail	223 01.1	S43 25.4
Vega	80 46.6	N38 47.4
Zuben'ubi	137 18.2	S16 02.0

	SHA	Mer. Pass.
	° ′	h m
Venus	192 01.0	11 18
Mars	218 01.1	9 33
Jupiter	6 13.9	23 36
Saturn	328 39.8	2 10

SUN / MOON

UT	SUN GHA	SUN Dec	MOON GHA	v	MOON Dec	d	HP
d h	° ′	° ′	° ′	′	° ′	′	′
19 00	181 29.9	N 1 38.7	199 01.8	13.5	N 9 29.1	8.8	55.1
01	196 30.1	37.8	213 34.3	13.6	9 20.3	8.8	55.1
02	211 30.4	36.8	228 06.9	13.6	9 11.5	8.9	55.1
03	226 30.6	.. 35.8	242 39.5	13.6	9 02.7	8.9	55.1
04	241 30.8	34.9	257 12.1	13.7	8 53.8	9.0	55.0
05	256 31.0	33.9	271 44.8	13.8	8 44.8	8.9	55.0
06	271 31.2	N 1 32.9	286 17.6	13.7	N 8 35.9	9.0	55.0
S 07	286 31.5	32.0	300 50.3	13.8	8 26.9	9.1	55.0
A 08	301 31.7	31.0	315 23.1	13.9	8 17.8	9.0	55.0
T 09	316 31.9	.. 30.0	329 56.0	13.8	8 08.8	9.1	55.0
U 10	331 32.1	29.0	344 28.8	14.0	7 59.7	9.2	54.9
R 11	346 32.4	28.1	359 01.8	13.9	7 50.5	9.1	54.9
D 12	1 32.6	N 1 27.1	13 34.7	14.0	N 7 41.4	9.2	54.9
A 13	16 32.8	26.1	28 07.7	14.0	7 32.2	9.2	54.9
Y 14	31 33.0	25.2	42 40.7	14.1	7 23.0	9.3	54.9
15	46 33.2	.. 24.2	57 13.8	14.1	7 13.7	9.2	54.9
16	61 33.5	23.2	71 46.9	14.1	7 04.5	9.3	54.8
17	76 33.7	22.3	86 20.0	14.2	6 55.2	9.3	54.8
18	91 33.9	N 1 21.3	100 53.2	14.2	N 6 45.9	9.4	54.8
19	106 34.1	20.3	115 26.4	14.2	6 36.5	9.4	54.8
20	121 34.3	19.3	129 59.6	14.3	6 27.1	9.3	54.8
21	136 34.6	.. 18.4	144 32.9	14.3	6 17.8	9.5	54.8
22	151 34.8	17.4	159 06.2	14.4	6 08.3	9.4	54.8
23	166 35.0	16.4	173 39.6	14.3	5 58.9	9.4	54.7
20 00	181 35.2	N 1 15.5	188 12.9	14.4	N 5 49.5	9.5	54.7
01	196 35.5	14.5	202 46.3	14.4	5 40.0	9.5	54.7
02	211 35.7	13.5	217 19.7	14.5	5 30.5	9.5	54.7
03	226 35.9	.. 12.6	231 53.2	14.5	5 21.0	9.6	54.7
04	241 36.1	11.6	246 26.7	14.5	5 11.4	9.5	54.7
05	256 36.3	10.6	261 00.2	14.6	5 01.9	9.6	54.7
06	271 36.6	N 1 09.6	275 33.8	14.5	N 4 52.3	9.6	54.6
S 07	286 36.8	08.7	290 07.3	14.6	4 42.7	9.6	54.6
U 08	301 37.0	07.7	304 40.9	14.6	4 33.1	9.6	54.6
N 09	316 37.2	.. 06.7	319 14.5	14.7	4 23.5	9.6	54.6
D 10	331 37.4	05.8	333 48.2	14.7	4 13.9	9.6	54.6
A 11	346 37.7	04.8	348 21.9	14.7	4 04.3	9.7	54.6
Y 12	1 37.9	N 1 03.8	2 55.6	14.7	N 3 54.6	9.7	54.6
13	16 38.1	02.8	17 29.3	14.8	3 44.9	9.6	54.6
14	31 38.3	01.9	32 03.1	14.7	3 35.3	9.7	54.5
15	46 38.6	1 00.9	46 36.8	14.8	3 25.6	9.7	54.5
16	61 38.8	0 59.9	61 10.6	14.9	3 15.9	9.7	54.5
17	76 39.0	59.0	75 44.5	14.8	3 06.2	9.7	54.5
18	91 39.2	N 0 58.0	90 18.3	14.9	N 2 56.5	9.8	54.5
19	106 39.4	57.0	104 52.2	14.8	2 46.7	9.7	54.5
20	121 39.7	56.0	119 26.0	14.9	2 37.0	9.7	54.5
21	136 39.9	.. 55.1	133 59.9	15.0	2 27.3	9.8	54.5
22	151 40.1	54.1	148 33.9	14.9	2 17.5	9.7	54.4
23	166 40.3	53.1	163 07.8	15.0	2 07.8	9.8	54.4
21 00	181 40.5	N 0 52.2	177 41.8	15.0	N 1 58.0	9.7	54.4
01	196 40.8	51.2	192 15.8	14.9	1 48.3	9.8	54.4
02	211 41.0	50.2	206 49.7	15.1	1 38.5	9.8	54.4
03	226 41.2	.. 49.2	221 23.8	15.0	1 28.7	9.7	54.4
04	241 41.4	48.3	235 57.8	15.0	1 19.0	9.8	54.4
05	256 41.6	47.3	250 31.8	15.1	1 09.2	9.8	54.4
06	271 41.9	N 0 46.3	265 05.9	15.1	N 0 59.4	9.7	54.4
07	286 42.1	45.4	279 40.0	15.1	0 49.7	9.8	54.3
08	301 42.3	44.4	294 14.1	15.1	0 39.9	9.8	54.3
M 09	316 42.5	.. 43.4	308 48.2	15.1	0 30.1	9.7	54.3
O 10	331 42.7	42.4	323 22.3	15.1	0 20.4	9.8	54.3
N 11	346 43.0	41.5	337 56.4	15.1	0 10.6	9.8	54.3
D 12	1 43.2	N 0 40.5	352 30.5	15.2	N 0 00.8	9.7	54.3
A 13	16 43.4	39.5	7 04.7	15.2	S 0 08.9	9.8	54.3
Y 14	31 43.6	38.6	21 38.9	15.1	0 18.7	9.7	54.3
15	46 43.8	.. 37.6	36 13.0	15.2	0 28.4	9.8	54.3
16	61 44.1	36.6	50 47.2	15.2	0 38.2	9.7	54.3
17	76 44.3	35.6	65 21.4	15.2	0 47.9	9.7	54.2
18	91 44.5	N 0 34.7	79 55.6	15.2	S 0 57.6	9.7	54.2
19	106 44.7	33.7	94 29.8	15.2	1 07.3	9.8	54.2
20	121 44.9	32.7	109 04.0	15.2	1 17.1	9.7	54.2
21	136 45.2	.. 31.7	123 38.2	15.3	1 26.8	9.7	54.2
22	151 45.4	30.8	138 12.5	15.2	1 36.5	9.7	54.2
23	166 45.6	29.8	152 46.7	15.2	S 1 46.2	9.6	54.2
	SD 16.0	d 1.0	SD 15.0		14.9		14.8

Twilight / Moonrise

Lat.	Twilight Naut.	Twilight Civil	Sunrise	Moonrise 19	20	21	22
°	h m	h m	h m	h m	h m	h m	h m
N 72	02 43	04 19	05 28	02 53	04 33	06 09	07 44
N 70	03 07	04 29	05 31	03 07	04 40	06 10	07 40
68	03 24	04 36	05 33	03 19	04 46	06 11	07 36
66	03 38	04 43	05 35	03 28	04 51	06 12	07 33
64	03 49	04 48	05 36	03 36	04 55	06 13	07 30
62	03 58	04 53	05 38	03 43	04 59	06 14	07 28
60	04 06	04 57	05 39	03 49	05 02	06 14	07 26
N 58	04 13	05 00	05 40	03 54	05 05	06 15	07 24
56	04 19	05 03	05 41	03 58	05 07	06 15	07 22
54	04 24	05 06	05 41	04 02	05 09	06 15	07 21
52	04 28	05 08	05 42	04 06	05 11	06 16	07 20
50	04 32	05 10	05 43	04 09	05 13	06 16	07 19
45	04 40	05 15	05 44	04 16	05 17	06 17	07 16
N 40	04 46	05 18	05 45	04 22	05 20	06 17	07 14
35	04 51	05 21	05 46	04 28	05 23	06 18	07 12
30	04 55	05 23	05 47	04 32	05 26	06 18	07 10
20	05 01	05 26	05 48	04 40	05 30	06 19	07 08
N 10	05 04	05 28	05 49	04 47	05 34	06 20	07 05
0	05 06	05 30	05 50	04 53	05 37	06 21	07 03
S 10	05 06	05 30	05 51	05 00	05 41	06 21	07 01
20	05 04	05 30	05 52	05 06	05 45	06 22	06 59
30	05 01	05 29	05 52	05 14	05 49	06 23	06 56
35	04 58	05 28	05 53	05 19	05 52	06 23	06 54
40	04 55	05 26	05 53	05 24	05 55	06 24	06 53
45	04 50	05 24	05 54	05 30	05 58	06 25	06 51
S 50	04 44	05 22	05 54	05 37	06 02	06 25	06 48
52	04 41	05 21	05 54	05 40	06 03	06 26	06 47
54	04 38	05 19	05 54	05 43	06 05	06 26	06 46
56	04 34	05 18	05 55	05 47	06 08	06 26	06 45
58	04 29	05 16	05 55	05 52	06 10	06 27	06 44
S 60	04 24	05 14	05 55	05 56	06 13	06 27	06 42

Moonset

Lat.	Sunset	Twilight Civil	Twilight Naut.	Moonset 19	20	21	22
°	h m	h m	h m	h m	h m	h m	h m
N 72	18 16	19 25	20 58	18 41	18 34	18 27	18 20
N 70	18 14	19 16	20 36	18 31	18 30	18 29	18 27
68	18 12	19 08	20 19	18 24	18 27	18 30	18 33
66	18 10	19 02	20 06	18 17	18 24	18 31	18 37
64	18 09	18 57	19 55	18 12	18 22	18 32	18 41
62	18 08	18 52	19 46	18 07	18 20	18 33	18 45
60	18 07	18 48	19 39	18 03	18 19	18 33	18 48
N 58	18 06	18 45	19 32	17 59	18 17	18 34	18 51
56	18 05	18 42	19 27	17 56	18 16	18 34	18 53
54	18 04	18 40	19 22	17 53	18 15	18 35	18 55
52	18 04	18 37	19 17	17 51	18 13	18 35	18 57
50	18 03	18 35	19 14	17 48	18 12	18 36	18 59
45	18 02	18 31	19 06	17 43	18 10	18 37	19 03
N 40	18 01	18 28	19 00	17 38	18 08	18 37	19 06
35	18 00	18 25	18 55	17 34	18 07	18 38	19 09
30	17 59	18 23	18 51	17 31	18 05	18 39	19 12
20	17 58	18 20	18 46	17 25	18 03	18 39	19 16
N 10	17 57	18 18	18 43	17 20	18 00	18 40	19 20
0	17 57	18 17	18 41	17 15	17 58	18 41	19 24
S 10	17 56	18 17	18 41	17 10	17 56	18 42	19 27
20	17 56	18 18	18 43	17 04	17 54	18 43	19 31
30	17 55	18 19	18 47	16 58	17 51	18 44	19 36
35	17 55	18 20	18 49	16 54	17 50	18 44	19 38
40	17 54	18 22	18 53	16 50	17 48	18 45	19 41
45	17 54	18 24	18 58	16 46	17 46	18 45	19 44
S 50	17 54	18 26	19 04	16 40	17 43	18 46	19 48
52	17 54	18 28	19 07	16 37	17 42	18 47	19 50
54	17 54	18 29	19 11	16 34	17 41	18 47	19 52
56	17 54	18 31	19 15	16 31	17 40	18 47	19 54
58	17 53	18 33	19 19	16 28	17 38	18 48	19 57
S 60	17 53	18 35	19 25	16 24	17 36	18 49	20 00

SUN / MOON

Day	SUN Eqn. of Time 00h	SUN Eqn. of Time 12h	SUN Mer. Pass.	MOON Mer. Pass. Upper	MOON Mer. Pass. Lower	Age	Phase
d	m s	m s	h m	h m	h m	d	%
19	05 59	06 10	11 54	11 04	23 26	28	1
20	06 20	06 31	11 53	11 48	24 09	29	0
21	06 42	06 52	11 53	12 31	00 09	01	1

1998 SEPTEMBER 22, 23, 24 (TUES., WED., THURS.)

UT	ARIES	VENUS −3.9		MARS +1.7		JUPITER −2.9		SATURN +0.1		STARS		
	GHA	GHA	Dec	GHA	Dec	GHA	Dec	GHA	Dec	Name	SHA	Dec
d h	° ′	° ′	° ′	° ′	° ′	° ′	° ′	° ′	° ′		° ′	° ′
22 00	0 39.2	190 21.7	N 5 43.3	217 26.2	N15 50.5	7 07.7	S 4 30.0	329 25.5	N 9 46.2	Acamar	315 26.7	S40 18.5
01	15 41.7	205 21.3	42.0	232 27.1	50.0	22 10.5	30.1	344 28.1	46.2	Achernar	335 34.7	S57 14.5
02	30 44.2	220 20.9	40.8	247 28.0	49.5	37 13.3	30.3	359 30.7	46.1	Acrux	173 22.7	S63 05.4
03	45 46.6	235 20.5	.. 39.6	262 28.9	.. 49.0	52 16.0	.. 30.4	14 33.3	.. 46.1	Adhara	255 21.5	S28 58.1
04	60 49.1	250 20.1	38.4	277 29.9	48.6	67 18.8	30.5	29 35.9	46.0	Aldebaran	291 02.3	N16 30.3
05	75 51.5	265 19.7	37.2	292 30.8	48.1	82 21.6	30.7	44 38.5	46.0			
06	90 54.0	280 19.2	N 5 36.0	307 31.7	N15 47.6	97 24.3	S 4 30.8	59 41.1	N 9 45.9	Alioth	166 31.2	N55 58.2
07	105 56.5	295 18.8	34.8	322 32.6	47.1	112 27.1	30.9	74 43.7	45.8	Alkaid	153 08.3	N49 19.5
T 08	120 58.9	310 18.4	33.6	337 33.6	46.6	127 29.9	31.0	89 46.3	45.8	Al Na'ir	27 57.6	S46 58.0
U 09	136 01.4	325 18.0	.. 32.4	352 34.5	.. 46.2	142 32.6	.. 31.2	104 48.9	.. 45.7	Alnilam	275 57.9	S 1 12.1
E 10	151 03.9	340 17.6	31.2	7 35.4	45.7	157 35.4	31.3	119 51.5	45.7	Alphard	218 07.5	S 8 39.0
S 11	166 06.3	355 17.2	30.0	22 36.3	45.2	172 38.2	31.4	134 54.1	45.6			
D 12	181 08.8	10 16.7	N 5 28.8	37 37.3	N15 44.7	187 40.9	S 4 31.6	149 56.7	N 9 45.6	Alphecca	126 20.9	N26 43.5
A 13	196 11.3	25 16.3	27.6	52 38.2	44.2	202 43.7	31.7	164 59.3	45.5	Alpheratz	357 54.9	N29 05.1
Y 14	211 13.7	40 15.9	26.4	67 39.1	43.8	217 46.5	31.8	180 02.0	45.4	Altair	62 19.2	N 8 52.2
15	226 16.2	55 15.5	.. 25.2	82 40.0	.. 43.3	232 49.2	.. 32.0	195 04.6	.. 45.4	Ankaa	353 26.5	S42 18.7
16	241 18.7	70 15.1	24.0	97 41.0	42.8	247 52.0	32.1	210 07.2	45.3	Antares	112 40.4	S26 25.6
17	256 21.1	85 14.7	22.8	112 41.9	42.3	262 54.8	32.2	225 09.8	45.3			
18	271 23.6	100 14.3	N 5 21.6	127 42.8	N15 41.8	277 57.5	S 4 32.3	240 12.4	N 9 45.2	Arcturus	146 06.4	N19 11.6
19	286 26.0	115 13.9	20.4	142 43.8	41.4	293 00.3	32.5	255 15.0	45.2	Atria	107 52.5	S69 01.6
20	301 28.5	130 13.4	19.1	157 44.7	40.9	308 03.1	32.6	270 17.6	45.1	Avior	234 23.0	S59 30.1
21	316 31.0	145 13.0	.. 17.9	172 45.6	.. 40.4	323 05.8	.. 32.7	285 20.2	.. 45.0	Bellatrix	278 44.2	N 6 20.9
22	331 33.4	160 12.6	16.7	187 46.5	39.9	338 08.6	32.9	300 22.8	45.0	Betelgeuse	271 13.6	N 7 24.4
23	346 35.9	175 12.2	15.5	202 47.5	39.4	353 11.4	33.0	315 25.4	44.9			
23 00	1 38.4	190 11.8	N 5 14.3	217 48.4	N15 38.9	8 14.1	S 4 33.1	330 28.0	N 9 44.9	Canopus	264 01.2	S52 41.5
01	16 40.8	205 11.4	13.1	232 49.3	38.5	23 16.9	33.2	345 30.6	44.8	Capella	280 51.2	N45 59.6
02	31 43.3	220 11.0	11.9	247 50.2	38.0	38 19.7	33.4	0 33.2	44.8	Deneb	49 38.9	N45 16.9
03	46 45.8	235 10.6	.. 10.7	262 51.2	.. 37.5	53 22.4	.. 33.5	15 35.8	.. 44.7	Denebola	182 45.6	N14 34.9
04	61 48.2	250 10.1	09.5	277 52.1	37.0	68 25.2	33.6	30 38.4	44.6	Diphda	349 07.0	S17 59.6
05	76 50.7	265 09.7	08.3	292 53.0	36.5	83 28.0	33.8	45 41.0	44.6			
06	91 53.2	280 09.3	N 5 07.0	307 54.0	N15 36.0	98 30.7	S 4 33.9	60 43.6	N 9 44.5	Dubhe	194 06.3	N61 45.5
W 07	106 55.6	295 08.9	05.8	322 54.9	35.6	113 33.5	34.0	75 46.2	44.5	Elnath	278 26.9	N28 36.2
E 08	121 58.1	310 08.5	04.6	337 55.8	35.1	128 36.3	34.1	90 48.8	44.4	Eltanin	90 51.4	N51 29.8
D 09	137 00.5	325 08.1	.. 03.4	352 56.7	.. 34.6	143 39.0	.. 34.3	105 51.4	.. 44.4	Enif	33 58.0	N 9 52.3
N 10	152 03.0	340 07.7	02.2	7 57.7	34.1	158 41.8	34.4	120 54.1	44.3	Fomalhaut	15 36.2	S29 37.7
E 11	167 05.5	355 07.3	5 01.0	22 58.6	33.6	173 44.6	34.5	135 56.7	44.2			
S 12	182 07.9	10 06.9	N 4 59.8	37 59.5	N15 33.1	188 47.3	S 4 34.7	150 59.3	N 9 44.2	Gacrux	172 14.1	S57 06.2
D 13	197 10.4	25 06.5	58.6	53 00.5	32.7	203 50.1	34.8	166 01.9	44.1	Gienah	176 04.3	S17 31.9
A 14	212 12.9	40 06.0	57.3	68 01.4	32.2	218 52.9	34.9	181 04.5	44.1	Hadar	149 04.6	S60 22.0
Y 15	227 15.3	55 05.6	.. 56.1	83 02.3	.. 31.7	233 55.6	.. 35.0	196 07.1	.. 44.0	Hamal	328 13.3	N23 27.3
16	242 17.8	70 05.2	54.9	98 03.3	31.2	248 58.4	35.2	211 09.7	43.9	Kaus Aust.	83 58.8	S34 23.1
17	257 20.3	85 04.8	53.7	113 04.2	30.7	264 01.1	35.3	226 12.3	43.9			
18	272 22.7	100 04.4	N 4 52.5	128 05.1	N15 30.2	279 03.9	S 4 35.4	241 14.9	N 9 43.8	Kochab	137 20.5	N74 10.0
19	287 25.2	115 04.0	51.3	143 06.0	29.7	294 06.7	35.6	256 17.5	43.8	Markab	13 49.3	N15 12.0
20	302 27.6	130 03.6	50.1	158 07.0	29.3	309 09.4	35.7	271 20.1	43.7	Menkar	314 26.7	N 4 05.1
21	317 30.1	145 03.2	.. 48.8	173 07.9	.. 28.8	324 12.2	.. 35.8	286 22.7	.. 43.7	Menkent	148 21.3	S36 21.7
22	332 32.6	160 02.8	47.6	188 08.8	28.3	339 15.0	35.9	301 25.3	43.6	Miaplacidus	221 42.7	S69 42.5
23	347 35.0	175 02.4	46.4	203 09.8	27.8	354 17.7	36.1	316 27.9	43.5			
24 00	2 37.5	190 02.0	N 4 45.2	218 10.7	N15 27.3	9 20.5	S 4 36.2	331 30.6	N 9 43.5	Mirfak	308 56.3	N49 51.2
01	17 40.0	205 01.5	44.0	233 11.6	26.8	24 23.3	36.3	346 33.2	43.4	Nunki	76 12.3	S26 17.8
02	32 42.4	220 01.1	42.8	248 12.6	26.3	39 26.0	36.5	1 35.8	43.4	Peacock	53 36.8	S56 44.4
03	47 44.9	235 00.7	.. 41.6	263 13.5	.. 25.9	54 28.8	.. 36.6	16 38.4	.. 43.3	Pollux	243 41.8	N28 01.7
04	62 47.4	250 00.3	40.3	278 14.4	25.4	69 31.6	36.7	31 41.0	43.2	Procyon	245 11.7	N 5 13.7
05	77 49.8	264 59.9	39.1	293 15.4	24.9	84 34.3	36.8	46 43.6	43.2			
06	92 52.3	279 59.5	N 4 37.9	308 16.3	N15 24.4	99 37.1	S 4 37.0	61 46.2	N 9 43.1	Rasalhague	96 17.0	N12 34.0
07	107 54.8	294 59.1	36.7	323 17.2	23.9	114 39.8	37.1	76 48.8	43.1	Regulus	207 55.9	N11 58.5
T 08	122 57.2	309 58.7	35.5	338 18.2	23.4	129 42.6	37.2	91 51.4	43.0	Rigel	281 22.9	S 8 12.1
H 09	137 59.7	324 58.3	.. 34.2	353 19.1	.. 22.9	144 45.4	.. 37.3	106 54.0	.. 43.0	Rigil Kent.	140 07.8	S60 49.7
U 10	153 02.1	339 57.9	33.0	8 20.0	22.4	159 48.1	37.5	121 56.6	42.9	Sabik	102 25.7	S15 43.2
R 11	168 04.6	354 57.5	31.8	23 21.0	22.0	174 50.9	37.6	136 59.3	42.8			
S 12	183 07.1	9 57.1	N 4 30.6	38 21.9	N15 21.5	189 53.7	S 4 37.7	152 01.9	N 9 42.8	Schedar	349 52.9	N56 31.8
D 13	198 09.5	24 56.7	29.4	53 22.8	21.0	204 56.4	37.9	167 04.5	42.7	Shaula	96 37.4	S37 06.1
A 14	213 12.0	39 56.3	28.2	68 23.8	20.5	219 59.2	38.0	182 07.1	42.7	Sirius	258 43.8	S16 42.7
Y 15	228 14.5	54 55.9	.. 26.9	83 24.7	.. 20.0	235 02.0	.. 38.1	197 09.7	.. 42.6	Spica	158 43.5	S11 09.1
16	243 16.9	69 55.4	25.7	98 25.6	19.5	250 04.7	38.2	212 12.3	42.5	Suhail	223 01.1	S43 25.4
17	258 19.4	84 55.0	24.5	113 26.6	19.0	265 07.5	38.4	227 14.9	42.5			
18	273 21.9	99 54.6	N 4 23.3	128 27.5	N15 18.5	280 10.2	S 4 38.5	242 17.5	N 9 42.4	Vega	80 46.6	N38 47.4
19	288 24.3	114 54.2	22.1	143 28.5	18.1	295 13.0	38.6	257 20.1	42.4	Zuben'ubi	137 18.3	S16 02.0
20	303 26.8	129 53.8	20.8	158 29.4	17.6	310 15.8	38.7	272 22.7	42.3		SHA	Mer. Pass.
21	318 29.2	144 53.4	.. 19.6	173 30.3	.. 17.1	325 18.5	.. 38.9	287 25.4	.. 42.2		° ′	h m
22	333 31.7	159 53.0	18.4	188 31.3	16.6	340 21.3	39.0	302 28.0	42.2	Venus	188 33.4	11 20
23	348 34.2	174 52.6	17.2	203 32.2	16.1	355 24.1	39.1	317 30.6	42.1	Mars	216 10.0	9 28
	h m									Jupiter	6 35.8	23 23
Mer. Pass. 23 49.5		v −0.4 d 1.2		v 0.9 d 0.5		v 2.8 d 0.1		v 2.6 d 0.1		Saturn	328 49.6	1 58

UT	SUN GHA	SUN Dec	MOON GHA	v	MOON Dec	d	HP
d h	° ′	° ′	° ′	′	° ′	′	′
22 00	181 45.8	N 0 28.8	167 20.9	15.3	S 1 55.8	9.7	54.2
01	196 46.0	27.9	181 55.2	15.2	2 05.5	9.7	54.2
02	211 46.3	26.9	196 29.4	15.3	2 15.2	9.6	54.2
03	226 46.5 ..	25.9	211 03.7	15.2	2 24.8	9.6	54.2
04	241 46.7	24.9	225 37.9	15.2	2 34.4	9.6	54.2
05	256 46.9	24.0	240 12.2	15.2	2 44.0	9.6	54.1
T 06	271 47.1	N 0 23.0	254 46.4	15.3	S 2 53.6	9.6	54.1
U 07	286 47.4	22.0	269 20.7	15.2	3 03.2	9.6	54.1
E 08	301 47.6	21.0	283 54.9	15.3	3 12.8	9.6	54.1
S 09	316 47.8 ..	20.1	298 29.2	15.2	3 22.4	9.5	54.1
D 10	331 48.0	19.1	313 03.4	15.3	3 31.9	9.5	54.1
A 11	346 48.2	18.1	327 37.7	15.2	3 41.4	9.5	54.1
Y 12	1 48.5	N 0 17.2	342 11.9	15.3	S 3 50.9	9.5	54.1
13	16 48.7	16.2	356 46.2	15.2	4 00.4	9.5	54.1
14	31 48.9	15.2	11 20.4	15.3	4 09.9	9.4	54.1
15	46 49.1 ..	14.2	25 54.7	15.2	4 19.3	9.5	54.1
16	61 49.3	13.3	40 28.9	15.2	4 28.8	9.4	54.1
17	76 49.6	12.3	55 03.1	15.3	4 38.2	9.4	54.1
18	91 49.8	N 0 11.3	69 37.4	15.2	S 4 47.6	9.3	54.1
19	106 50.0	10.3	84 11.6	15.2	4 56.9	9.4	54.1
20	121 50.2	09.4	98 45.8	15.2	5 06.3	9.3	54.1
21	136 50.4 ..	08.4	113 20.0	15.2	5 15.6	9.3	54.0
22	151 50.7	07.4	127 54.2	15.2	5 24.9	9.3	54.0
23	166 50.9	06.5	142 28.4	15.2	5 34.2	9.2	54.0
23 00	181 51.1	N 0 05.5	157 02.6	15.2	S 5 43.4	9.3	54.0
01	196 51.3	04.5	171 36.8	15.2	5 52.7	9.2	54.0
02	211 51.5	03.5	186 11.0	15.1	6 01.9	9.1	54.0
03	226 51.7 ..	02.6	200 45.1	15.2	6 11.0	9.2	54.0
04	241 52.0	01.6	215 19.3	15.1	6 20.2	9.1	54.0
05	256 52.2	N 00.6	229 53.4	15.1	6 29.3	9.1	54.0
06	271 52.4	S 0 00.4	244 27.5	15.2	S 6 38.4	9.1	54.0
W 07	286 52.6	01.3	259 01.7	15.1	6 47.5	9.0	54.0
E 08	301 52.8	02.3	273 35.8	15.0	6 56.5	9.0	54.0
D 09	316 53.1 ..	03.3	288 09.8	15.1	7 05.5	9.0	54.0
N 10	331 53.3	04.3	302 43.9	15.1	7 14.5	8.9	54.0
E 11	346 53.5	05.2	317 18.0	15.0	7 23.4	9.0	54.0
S 12	1 53.7	S 0 06.2	331 52.0	15.1	S 7 32.4	8.8	54.0
D 13	16 53.9	07.2	346 26.1	15.0	7 41.2	8.9	54.0
A 14	31 54.1	08.1	1 00.1	15.0	7 50.1	8.8	54.0
Y 15	46 54.4 ..	09.1	15 34.1	15.0	7 58.9	8.8	54.0
16	61 54.6	10.1	30 08.1	14.9	8 07.7	8.7	54.0
17	76 54.8	11.1	44 42.0	15.0	8 16.4	8.8	54.0
18	91 55.0	S 0 12.0	59 16.0	14.9	S 8 25.2	8.6	54.0
19	106 55.2	13.0	73 49.9	14.9	8 33.8	8.7	54.0
20	121 55.5	14.0	88 23.8	14.9	8 42.5	8.6	54.0
21	136 55.7 ..	15.0	102 57.7	14.9	8 51.1	8.6	54.0
22	151 55.9	15.9	117 31.6	14.8	8 59.7	8.5	54.0
23	166 56.1	16.9	132 05.4	14.9	9 08.2	8.5	54.0
24 00	181 56.3	S 0 17.9	146 39.3	14.8	S 9 16.7	8.5	54.0
01	196 56.5	18.9	161 13.1	14.8	9 25.2	8.4	54.0
02	211 56.8	19.8	175 46.9	14.7	9 33.6	8.4	54.0
03	226 57.0 ..	20.8	190 20.6	14.8	9 42.0	8.3	54.0
04	241 57.2	21.8	204 54.4	14.7	9 50.3	8.3	54.0
05	256 57.4	22.8	219 28.1	14.7	9 58.6	8.3	54.0
T 06	271 57.6	S 0 23.7	234 01.8	14.7	S10 06.9	8.2	54.0
H 07	286 57.8	24.7	248 35.5	14.6	10 15.1	8.1	54.0
U 08	301 58.1	25.7	263 09.1	14.7	10 23.2	8.2	54.0
R 09	316 58.3 ..	26.6	277 42.8	14.6	10 31.4	8.1	54.0
S 10	331 58.5	27.6	292 16.4	14.5	10 39.5	8.0	54.0
D 11	346 58.7	28.6	306 49.9	14.6	10 47.5	8.0	54.0
A 12	1 58.9	S 0 29.6	321 23.5	14.5	S10 55.5	8.0	54.0
Y 13	16 59.1	30.5	335 57.0	14.5	11 03.5	7.9	54.0
14	31 59.4	31.5	350 30.5	14.5	11 11.4	7.8	54.0
15	46 59.6 ..	32.5	5 04.0	14.4	11 19.2	7.9	54.0
16	61 59.8	33.5	19 37.4	14.4	11 27.1	7.7	54.0
17	77 00.0	34.4	34 10.8	14.4	11 34.8	7.8	54.0
18	92 00.2	S 0 35.4	48 44.2	14.4	S11 42.6	7.6	54.0
19	107 00.4	36.4	63 17.6	14.3	11 50.2	7.7	54.0
20	122 00.7	37.4	77 50.9	14.3	11 57.9	7.5	54.0
21	137 00.9 ..	38.3	92 24.2	14.3	12 05.4	7.6	54.0
22	152 01.1	39.3	106 57.5	14.2	12 13.0	7.5	54.0
23	167 01.3	40.3	121 30.7	14.2	S12 20.4	7.5	54.1
	SD 16.0	d 1.0	SD 14.7		14.7		14.7

Twilight / Sunrise / Moonrise

Lat.	Naut.	Civil	Sunrise	22	23	24	25
°	h m	h m	h m	h m	h m	h m	h m
N 72	03 03	04 33	05 42	07 44	09 20	10 58	12 40
N 70	03 22	04 41	05 43	07 40	09 09	10 39	12 11
68	03 37	04 48	05 44	07 36	09 00	10 24	11 49
66	03 49	04 53	05 44	07 33	08 53	10 13	11 32
64	03 59	04 57	05 45	07 30	08 47	10 03	11 18
62	04 07	05 01	05 45	07 28	08 41	09 54	11 06
60	04 14	05 04	05 46	07 26	08 37	09 47	10 57
N 58	04 20	05 07	05 46	07 24	08 33	09 41	10 48
56	04 25	05 09	05 47	07 22	08 29	09 35	10 41
54	04 30	05 12	05 47	07 21	08 26	09 30	10 34
52	04 33	05 13	05 47	07 20	08 23	09 26	10 28
50	04 37	05 15	05 47	07 19	08 20	09 22	10 23
45	04 44	05 18	05 48	07 16	08 15	09 13	10 11
N 40	04 49	05 21	05 48	07 14	08 10	09 06	10 01
35	04 54	05 23	05 48	07 12	08 06	08 59	09 53
30	04 57	05 25	05 49	07 10	08 02	08 54	09 46
20	05 01	05 27	05 49	07 08	07 56	08 45	09 34
N 10	05 04	05 28	05 49	07 05	07 51	08 36	09 23
0	05 05	05 29	05 49	07 03	07 46	08 29	09 13
S 10	05 04	05 28	05 49	07 01	07 41	08 21	09 03
20	05 01	05 27	05 49	06 59	07 35	08 13	08 52
30	04 57	05 25	05 49	06 56	07 29	08 04	08 40
35	04 54	05 23	05 48	06 54	07 26	07 58	08 33
40	04 50	05 21	05 48	06 53	07 22	07 52	08 25
45	04 44	05 19	05 48	06 51	07 17	07 45	08 16
S 50	04 37	05 15	05 47	06 48	07 12	07 37	08 05
52	04 34	05 13	05 47	06 47	07 10	07 33	08 00
54	04 30	05 12	05 47	06 46	07 07	07 29	07 54
56	04 25	05 09	05 47	06 45	07 04	07 24	07 48
58	04 20	05 07	05 46	06 44	07 01	07 19	07 41
S 60	04 14	05 04	05 46	06 42	06 57	07 14	07 33

Sunset / Twilight / Moonset

Lat.	Sunset	Civil	Naut.	22	23	24	25
°	h m	h m	h m	h m	h m	h m	h m
N 72	18 01	19 09	20 37	18 20	18 13	18 05	17 55
N 70	18 00	19 01	20 19	18 27	18 26	18 25	18 25
68	17 59	18 55	20 04	18 33	18 36	18 41	18 48
66	17 59	18 50	19 53	18 37	18 45	18 53	19 05
64	17 58	18 46	19 43	18 41	18 52	19 04	19 20
62	17 58	18 42	19 36	18 45	18 58	19 13	19 32
60	17 58	18 39	19 29	18 48	19 03	19 21	19 42
N 58	17 57	18 36	19 23	18 51	19 08	19 28	19 51
56	17 57	18 34	19 18	18 53	19 12	19 34	19 59
54	17 57	18 32	19 14	18 55	19 16	19 40	20 06
52	17 57	18 30	19 10	18 57	19 20	19 45	20 13
50	17 57	18 29	19 07	18 59	19 23	19 49	20 19
45	17 56	18 25	19 00	19 03	19 30	19 59	20 31
N 40	17 56	18 23	18 55	19 06	19 36	20 07	20 41
35	17 56	18 21	18 51	19 09	19 41	20 14	20 50
30	17 56	18 20	18 47	19 12	19 45	20 21	20 58
20	17 56	18 18	18 43	19 16	19 53	20 31	21 12
N 10	17 56	18 17	18 41	19 20	20 00	20 41	21 23
0	17 56	18 16	18 40	19 24	20 06	20 50	21 34
S 10	17 56	18 17	18 41	19 27	20 13	20 59	21 46
20	17 56	18 18	18 44	19 31	20 19	21 08	21 57
30	17 57	18 21	18 48	19 36	20 27	21 19	22 11
35	17 57	18 22	18 52	19 38	20 32	21 25	22 19
40	17 57	18 24	18 56	19 41	20 37	21 33	22 28
45	17 58	18 27	19 02	19 44	20 43	21 41	22 38
S 50	17 58	18 31	19 09	19 48	20 50	21 51	22 51
52	17 59	18 33	19 12	19 50	20 53	21 56	22 57
54	17 59	18 34	19 16	19 52	20 57	22 01	23 04
56	17 59	18 37	19 21	19 54	21 01	22 07	23 11
58	18 00	18 39	19 26	19 57	21 05	22 13	23 20
S 60	18 00	18 42	19 32	20 00	21 10	22 20	23 29

SUN / MOON

Day	Eqn. of Time 00h	Eqn. of Time 12h	Mer. Pass.	Mer. Pass. Upper	Mer. Pass. Lower	Age	Phase
d	m s	m s	h m	h m	h m	d	%
22	07 03	07 13	11 53	13 13	00 52	02	3
23	07 24	07 34	11 52	13 56	01 35	03	7
24	07 45	07 55	11 52	14 39	02 17	04	13

1998 SEPTEMBER 25, 26, 27 (FRI., SAT., SUN.)

UT	ARIES	VENUS −3.9		MARS +1.7		JUPITER −2.9		SATURN +0.1		STARS		
	GHA	GHA	Dec	GHA	Dec	GHA	Dec	GHA	Dec	Name	SHA	Dec
d h	° ′	° ′	° ′	° ′	° ′	° ′	° ′	° ′	° ′		° ′	° ′
25 00	3 36.6	189 52.2	N 4 16.0	218 33.1	N15 15.6	10 26.8	S 4 39.3	332 33.2	N 9 42.1	Acamar	315 26.7	S40 18.5
01	18 39.1	204 51.8	14.7	233 34.1	15.1	25 29.6	39.4	347 35.8	42.0	Achernar	335 34.7	S57 14.5
02	33 41.6	219 51.4	13.5	248 35.0	14.6	40 32.3	39.5	2 38.4	42.0	Acrux	173 22.7	S63 05.4
03	48 44.0	234 51.0 ..	12.3	263 35.9 ..	14.1	55 35.1 ..	39.6	17 41.0 ..	41.9	Adhara	255 21.5	S28 58.1
04	63 46.5	249 50.6	11.1	278 36.9	13.6	70 37.9	39.8	32 43.6	41.8	Aldebaran	291 02.3	N16 30.3
05	78 49.0	264 50.2	09.8	293 37.8	13.2	85 40.6	39.9	47 46.2	41.8			
06	93 51.4	279 49.8	N 4 08.6	308 38.8	N15 12.7	100 43.4	S 4 40.0	62 48.9	N 9 41.7	Alioth	166 31.2	N55 58.2
07	108 53.9	294 49.4	07.4	323 39.7	12.2	115 46.1	40.1	77 51.5	41.7	Alkaid	153 08.3	N49 19.5
08	123 56.4	309 49.0	06.2	338 40.6	11.7	130 48.9	40.3	92 54.1	41.6	Al Na'ir	27 57.6	S46 58.1
F 09	138 58.8	324 48.6 ..	04.9	353 41.6 ..	11.2	145 51.7 ..	40.4	107 56.7 ..	41.5	Alnilam	275 57.8	S 1 12.1
R 10	154 01.3	339 48.2	03.7	8 42.5	10.7	160 54.4	40.5	122 59.3	41.5	Alphard	218 07.5	S 8 39.0
I 11	169 03.7	354 47.8	02.5	23 43.4	10.2	175 57.2	40.6	138 01.9	41.4			
D 12	184 06.2	9 47.4	N 4 01.3	38 44.4	N15 09.7	191 00.0	S 4 40.8	153 04.5	N 9 41.4	Alphecca	126 20.9	N26 43.5
A 13	199 08.7	24 47.0	4 00.0	53 45.3	09.2	206 02.7	40.9	168 07.1	41.3	Alpheratz	357 54.9	N29 05.1
Y 14	214 11.1	39 46.6	3 58.8	68 46.3	08.7	221 05.5	41.0	183 09.7	41.2	Altair	62 19.2	N 8 52.2
15	229 13.6	54 46.1 ..	57.6	83 47.2 ..	08.3	236 08.2 ..	41.1	198 12.4 ..	41.2	Ankaa	353 26.5	S42 18.7
16	244 16.1	69 45.7	56.4	98 48.1	07.8	251 11.0	41.3	213 15.0	41.1	Antares	112 40.4	S26 25.6
17	259 18.5	84 45.3	55.1	113 49.1	07.3	266 13.8	41.4	228 17.6	41.1			
18	274 21.0	99 44.9	N 3 53.9	128 50.0	N15 06.8	281 16.5	S 4 41.5	243 20.2	N 9 41.0	Arcturus	146 06.4	N19 11.6
19	289 23.5	114 44.5	52.7	143 51.0	06.3	296 19.3	41.6	258 22.8	40.9	Atria	107 52.6	S69 01.6
20	304 25.9	129 44.1	51.5	158 51.9	05.8	311 22.0	41.8	273 25.4	40.9	Avior	234 22.9	S59 30.1
21	319 28.4	144 43.7 ..	50.2	173 52.8 ..	05.3	326 24.8 ..	41.9	288 28.0 ..	40.8	Bellatrix	278 44.1	N 6 20.9
22	334 30.9	159 43.3	49.0	188 53.8	04.8	341 27.6	42.0	303 30.7	40.8	Betelgeuse	271 13.6	N 7 24.4
23	349 33.3	174 42.9	47.8	203 54.7	04.3	356 30.3	42.2	318 33.3	40.7			
26 00	4 35.8	189 42.5	N 3 46.6	218 55.7	N15 03.8	11 33.1	S 4 42.3	333 35.9	N 9 40.6	Canopus	264 01.2	S52 41.5
01	19 38.2	204 42.1	45.3	233 56.6	03.3	26 35.8	42.4	348 38.5	40.6	Capella	280 51.1	N45 59.6
02	34 40.7	219 41.7	44.1	248 57.6	02.8	41 38.6	42.5	3 41.1	40.5	Deneb	49 38.9	N45 16.9
03	49 43.2	234 41.3 ..	42.9	263 58.5 ..	02.3	56 41.4 ..	42.7	18 43.7 ..	40.5	Denebola	182 45.6	N14 34.9
04	64 45.6	249 40.9	41.7	278 59.4	01.9	71 44.1	42.8	33 46.3	40.4	Diphda	349 07.0	S17 59.6
05	79 48.1	264 40.5	40.4	294 00.4	01.4	86 46.9	42.9	48 48.9	40.3			
06	94 50.6	279 40.1	N 3 39.2	309 01.3	N15 00.9	101 49.6	S 4 43.0	63 51.6	N 9 40.3	Dubhe	194 06.3	N61 45.5
07	109 53.0	294 39.7	38.0	324 02.3	15 00.4	116 52.4	43.2	78 54.2	40.2	Elnath	278 26.9	N28 36.2
S 08	124 55.5	309 39.3	36.7	339 03.2	14 59.9	131 55.2	43.3	93 56.8	40.2	Eltanin	90 51.5	N51 29.8
A 09	139 58.0	324 38.9 ..	35.5	354 04.1 ..	59.4	146 57.9 ..	43.4	108 59.4 ..	40.1	Enif	33 58.0	N 9 52.3
T 10	155 00.4	339 38.5	34.3	9 05.1	58.9	162 00.7	43.5	124 02.0	40.0	Fomalhaut	15 36.2	S29 37.7
U 11	170 02.9	354 38.1	33.1	24 06.0	58.4	177 03.4	43.7	139 04.6	40.0			
R 12	185 05.3	9 37.7	N 3 31.8	39 07.0	N14 57.9	192 06.2	S 4 43.8	154 07.2	N 9 39.9	Gacrux	172 14.1	S57 06.2
D 13	200 07.8	24 37.3	30.6	54 07.9	57.4	207 09.0	43.9	169 09.9	39.9	Gienah	176 04.3	S17 31.9
A 14	215 10.3	39 36.9	29.4	69 08.9	56.9	222 11.7	44.0	184 12.5	39.8	Hadar	149 04.6	S60 22.0
Y 15	230 12.7	54 36.5 ..	28.1	84 09.8 ..	56.4	237 14.5 ..	44.2	199 15.1 ..	39.7	Hamal	328 13.3	N23 27.3
16	245 15.2	69 36.1	26.9	99 10.8	55.9	252 17.2	44.3	214 17.7	39.7	Kaus Aust.	83 58.9	S34 23.1
17	260 17.7	84 35.7	25.7	114 11.7	55.4	267 20.0	44.4	229 20.3	39.6			
18	275 20.1	99 35.3	N 3 24.4	129 12.6	N14 54.9	282 22.7	S 4 44.5	244 22.9	N 9 39.6	Kochab	137 20.6	N74 10.0
19	290 22.6	114 34.9	23.2	144 13.6	54.4	297 25.5	44.6	259 25.6	39.5	Markab	13 49.3	N15 12.0
20	305 25.1	129 34.5	22.0	159 14.5	53.9	312 28.3	44.8	274 28.2	39.4	Menkar	314 26.7	N 4 05.1
21	320 27.5	144 34.1 ..	20.8	174 15.5 ..	53.4	327 31.0 ..	44.9	289 30.8 ..	39.4	Menkent	148 21.3	S36 21.7
22	335 30.0	159 33.7	19.5	189 16.4	53.0	342 33.8	45.0	304 33.4	39.3	Miaplacidus	221 42.6	S69 42.5
23	350 32.5	174 33.3	18.3	204 17.4	52.5	357 36.5	45.1	319 36.0	39.3			
27 00	5 34.9	189 32.9	N 3 17.1	219 18.3	N14 52.0	12 39.3	S 4 45.3	334 38.6	N 9 39.2	Mirfak	308 56.3	N49 51.2
01	20 37.4	204 32.5	15.8	234 19.3	51.5	27 42.1	45.4	349 41.3	39.1	Nunki	76 12.3	S26 17.8
02	35 39.8	219 32.1	14.6	249 20.2	51.0	42 44.8	45.5	4 43.9	39.1	Peacock	53 36.8	S56 44.4
03	50 42.3	234 31.7 ..	13.4	264 21.2 ..	50.5	57 47.6 ..	45.6	19 46.5 ..	39.0	Pollux	243 41.8	N28 01.7
04	65 44.8	249 31.3	12.1	279 22.1	50.0	72 50.3	45.8	34 49.1	38.9	Procyon	245 11.7	N 5 13.7
05	80 47.2	264 30.9	10.9	294 23.0	49.5	87 53.1	45.9	49 51.7	38.9			
06	95 49.7	279 30.5	N 3 09.7	309 24.0	N14 49.0	102 55.8	S 4 46.0	64 54.3	N 9 38.8	Rasalhague	96 17.1	N12 34.0
07	110 52.2	294 30.1	08.4	324 24.9	48.5	117 58.6	46.1	79 57.0	38.8	Regulus	207 55.9	N11 58.5
08	125 54.6	309 29.7	07.2	339 25.9	48.0	133 01.4	46.3	94 59.6	38.7	Rigel	281 22.9	S 8 12.1
S 09	140 57.1	324 29.3 ..	06.0	354 26.8 ..	47.5	148 04.1 ..	46.4	110 02.2 ..	38.6	Rigil Kent.	140 07.8	S60 49.7
U 10	155 59.6	339 28.9	04.7	9 27.8	47.0	163 06.9	46.5	125 04.8	38.6	Sabik	102 25.7	S15 43.2
N 11	171 02.0	354 28.5	03.5	24 28.7	46.5	178 09.6	46.6	140 07.4	38.5			
D 12	186 04.5	9 28.1	N 3 02.3	39 29.7	N14 46.0	193 12.4	S 4 46.8	155 10.0	N 9 38.5	Schedar	349 52.9	N56 31.8
A 13	201 07.0	24 27.7	3 01.0	54 30.6	45.5	208 15.1	46.9	170 12.7	38.4	Shaula	96 37.4	S37 06.1
Y 14	216 09.4	39 27.3	2 59.8	69 31.6	45.0	223 17.9	47.0	185 15.3	38.3	Sirius	258 43.8	S16 42.7
15	231 11.9	54 26.9 ..	58.6	84 32.5 ..	44.5	238 20.7 ..	47.1	200 17.9 ..	38.3	Spica	158 43.5	S11 09.1
16	246 14.3	69 26.6	57.3	99 33.5	44.0	253 23.4	47.2	215 20.5	38.2	Suhail	223 01.1	S43 25.4
17	261 16.8	84 26.2	56.1	114 34.4	43.5	268 26.2	47.4	230 23.1	38.2			
18	276 19.3	99 25.8	N 2 54.9	129 35.4	N14 43.0	283 28.9	S 4 47.5	245 25.7	N 9 38.1	Vega	80 46.6	N38 47.4
19	291 21.7	114 25.4	53.6	144 36.3	42.5	298 31.7	47.6	260 28.4	38.0	Zuben'ubi	137 18.3	S16 02.0
20	306 24.2	129 25.0	52.4	159 37.3	42.0	313 34.4	47.7	275 31.0	38.0		SHA	Mer. Pass.
21	321 26.7	144 24.6 ..	51.1	174 38.2 ..	41.5	328 37.2 ..	47.9	290 33.6 ..	37.9		° ′	h m
22	336 29.1	159 24.2	49.9	189 39.2	41.0	343 39.9	48.0	305 36.2	37.8	Venus	185 06.8	11 21
23	351 31.6	174 23.8	48.7	204 40.1	40.5	358 42.7	48.1	320 38.8	37.8	Mars	214 19.9	9 24
	h m									Jupiter	6 57.3	23 10
Mer. Pass. 23 37.7		v −0.4	d 1.2	v 0.9	d 0.5	v 2.8	d 0.1	v 2.6	d 0.1	Saturn	329 00.1	1 45

SUN and MOON

UT	SUN GHA	SUN Dec	MOON GHA	v	MOON Dec	d	HP
25 00	182 01.5	S 0 41.3	136 03.9	14.2	S12 27.9	7.3	54.1
01	197 01.7	42.2	150 37.1	14.1	12 35.2	7.4	54.1
02	212 02.0	43.2	165 10.2	14.1	12 42.6	7.2	54.1
03	227 02.2	.. 44.2	179 43.3	14.1	12 49.8	7.2	54.1
04	242 02.4	45.2	194 16.4	14.0	12 57.0	7.2	54.1
05	257 02.6	46.1	208 49.4	14.0	13 04.2	7.1	54.1
06	272 02.8	S 0 47.1	223 22.4	14.0	S13 11.3	7.1	54.1
07	287 03.0	48.1	237 55.4	13.9	13 18.4	7.0	54.1
08	302 03.3	49.0	252 28.3	13.9	13 25.4	6.9	54.1
F 09	317 03.5	.. 50.0	267 01.2	13.9	13 32.3	6.9	54.1
R 10	332 03.7	51.0	281 34.1	13.9	13 39.2	6.8	54.1
I 11	347 03.9	52.0	296 07.0	13.8	13 46.0	6.8	54.1
D 12	2 04.1	S 0 52.9	310 39.8	13.7	S13 52.8	6.7	54.1
A 13	17 04.3	53.9	325 12.5	13.8	13 59.5	6.7	54.2
Y 14	32 04.6	54.9	339 45.3	13.6	14 06.2	6.6	54.2
15	47 04.8	.. 55.9	354 17.9	13.7	14 12.8	6.5	54.2
16	62 05.0	56.8	8 50.6	13.6	14 19.3	6.5	54.2
17	77 05.2	57.8	23 23.2	13.6	14 25.8	6.4	54.2
18	92 05.4	S 0 58.8	37 55.8	13.5	S14 32.2	6.4	54.2
19	107 05.6	0 59.8	52 28.3	13.5	14 38.6	6.3	54.2
20	122 05.8	1 00.7	67 00.8	13.5	14 44.9	6.2	54.2
21	137 06.1	.. 01.7	81 33.3	13.4	14 51.1	6.2	54.2
22	152 06.3	02.7	96 05.7	13.4	14 57.3	6.1	54.2
23	167 06.5	03.7	110 38.1	13.4	15 03.4	6.0	54.3
26 00	182 06.7	S 1 04.6	125 10.5	13.3	S15 09.4	6.0	54.3
01	197 06.9	05.6	139 42.8	13.3	15 15.4	5.9	54.3
02	212 07.1	06.6	154 15.1	13.2	15 21.3	5.8	54.3
03	227 07.3	.. 07.6	168 47.3	13.2	15 27.1	5.8	54.3
04	242 07.6	08.5	183 19.5	13.1	15 32.9	5.7	54.3
05	257 07.8	09.5	197 51.6	13.2	15 38.6	5.7	54.3
06	272 08.0	S 1 10.5	212 23.8	13.0	S15 44.3	5.6	54.3
S 07	287 08.2	11.4	226 55.8	13.1	15 49.9	5.5	54.4
A 08	302 08.4	12.4	241 27.9	13.0	15 55.4	5.4	54.4
T 09	317 08.6	.. 13.4	255 59.9	12.9	16 00.8	5.4	54.4
U 10	332 08.8	14.4	270 31.8	12.9	16 06.2	5.3	54.4
R 11	347 09.1	15.3	285 03.7	12.9	16 11.5	5.2	54.4
D 12	2 09.3	S 1 16.3	299 35.6	12.8	S16 16.7	5.2	54.4
A 13	17 09.5	17.3	314 07.4	12.8	16 21.9	5.1	54.4
Y 14	32 09.7	18.3	328 39.2	12.7	16 27.0	5.0	54.5
15	47 09.9	.. 19.2	343 10.9	12.7	16 32.0	5.0	54.5
16	62 10.1	20.2	357 42.6	12.7	16 37.0	4.8	54.5
17	77 10.3	21.2	12 14.3	12.6	16 41.8	4.8	54.5
18	92 10.6	S 1 22.2	26 45.9	12.6	S16 46.6	4.8	54.5
19	107 10.8	23.1	41 17.5	12.5	16 51.4	4.6	54.5
20	122 11.0	24.1	55 49.0	12.5	16 56.0	4.6	54.6
21	137 11.2	.. 25.1	70 20.5	12.4	17 00.6	4.5	54.6
22	152 11.4	26.1	84 51.9	12.4	17 05.1	4.4	54.6
23	167 11.6	27.0	99 23.3	12.4	17 09.5	4.4	54.6
27 00	182 11.8	S 1 28.0	113 54.7	12.3	S17 13.9	4.2	54.6
01	197 12.0	29.0	128 26.0	12.2	17 18.1	4.2	54.6
02	212 12.3	30.0	142 57.2	12.3	17 22.3	4.2	54.7
03	227 12.5	.. 30.9	157 28.5	12.2	17 26.5	4.0	54.7
04	242 12.7	31.9	171 59.7	12.1	17 30.5	4.0	54.7
05	257 12.9	32.9	186 30.8	12.1	17 34.5	3.8	54.7
06	272 13.1	S 1 33.8	201 01.9	12.0	S17 38.3	3.8	54.7
07	287 13.3	34.8	215 32.9	12.1	17 42.1	3.8	54.8
08	302 13.5	35.8	230 04.0	11.9	17 45.9	3.6	54.8
S 09	317 13.7	.. 36.8	244 34.9	11.9	17 49.5	3.6	54.8
U 10	332 14.0	37.7	259 05.8	11.9	17 53.1	3.4	54.8
N 11	347 14.2	38.7	273 36.7	11.8	17 56.5	3.4	54.8
D 12	2 14.4	S 1 39.7	288 07.5	11.8	S17 59.9	3.3	54.9
A 13	17 14.6	40.7	302 38.3	11.8	18 03.2	3.3	54.9
Y 14	32 14.8	41.6	317 09.1	11.7	18 06.5	3.1	54.9
15	47 15.0	.. 42.6	331 39.8	11.6	18 09.6	3.1	54.9
16	62 15.2	43.6	346 10.4	11.7	18 12.7	2.9	55.0
17	77 15.4	44.6	0 41.1	11.5	18 15.6	2.9	55.0
18	92 15.6	S 1 45.5	15 11.6	11.6	S18 18.5	2.8	55.0
19	107 15.9	46.5	29 42.2	11.5	18 21.3	2.7	55.0
20	122 16.1	47.5	44 12.7	11.4	18 24.0	2.7	55.1
21	137 16.3	.. 48.4	58 43.1	11.4	18 26.7	2.5	55.1
22	152 16.5	49.4	73 13.5	11.4	18 29.2	2.4	55.1
23	167 16.7	50.4	87 43.9	11.3	S18 31.6	2.4	55.1
	SD 16.0	d 1.0	SD 14.8		14.8		15.0

Twilight, Sunrise and Moonrise

Lat.	Twilight Naut.	Twilight Civil	Sunrise	Moonrise 25	Moonrise 26	Moonrise 27	Moonrise 28
°	h m	h m	h m	h m	h m	h m	h m
N 72	03 21	04 47	05 55	12 40	14 32	■■	■■
N 70	03 37	04 54	05 55	12 11	13 43	15 14	16 33
68	03 50	04 59	05 54	11 49	13 12	14 31	15 38
66	04 00	05 03	05 54	11 32	12 49	14 02	15 05
64	04 09	05 06	05 53	11 18	12 31	13 40	14 41
62	04 16	05 09	05 53	11 06	12 17	13 23	14 22
60	04 22	05 11	05 53	10 57	12 04	13 08	14 06
N 58	04 27	05 14	05 53	10 48	11 54	12 56	13 53
56	04 31	05 15	05 52	10 41	11 44	12 45	13 42
54	04 35	05 17	05 52	10 34	11 36	12 36	13 32
52	04 39	05 18	05 52	10 28	11 29	12 28	13 23
50	04 42	05 20	05 52	10 23	11 22	12 20	13 15
45	04 48	05 22	05 51	10 11	11 08	12 04	12 58
N 40	04 53	05 24	05 51	10 01	10 57	11 51	12 44
35	04 56	05 25	05 51	09 53	10 47	11 40	12 32
30	04 59	05 26	05 50	09 46	10 38	11 30	12 22
20	05 02	05 28	05 50	09 34	10 23	11 14	12 04
N 10	05 04	05 28	05 49	09 23	10 10	10 59	11 49
0	05 03	05 27	05 48	09 13	09 58	10 45	11 35
S 10	05 02	05 26	05 47	09 03	09 46	10 32	11 20
20	04 59	05 24	05 46	08 52	09 33	10 18	11 05
30	04 53	05 21	05 45	08 40	09 19	10 01	10 48
35	04 49	05 19	05 44	08 33	09 10	09 52	10 37
40	04 44	05 16	05 43	08 25	09 01	09 41	10 26
45	04 38	05 13	05 42	08 16	08 50	09 28	10 12
S 50	04 30	05 08	05 41	08 05	08 36	09 13	09 55
52	04 26	05 06	05 40	08 00	08 30	09 05	09 48
54	04 22	05 04	05 39	07 54	08 23	08 57	09 39
56	04 17	05 01	05 39	07 48	08 15	08 48	09 29
58	04 11	04 58	05 38	07 41	08 06	08 38	09 18
S 60	04 04	04 55	05 37	07 33	07 57	08 27	09 05

Sunset, Twilight and Moonset

Lat.	Sunset	Twilight Civil	Twilight Naut.	Moonset 25	Moonset 26	Moonset 27	Moonset 28
°	h m	h m	h m	h m	h m	h m	h m
N 72	17 45	18 53	20 18	17 55	17 37	■■	■■
N 70	17 46	18 47	20 02	18 25	18 27	18 35	19 00
68	17 47	18 42	19 50	18 48	18 59	19 19	19 54
66	17 47	18 38	19 40	19 05	19 22	19 48	20 27
64	17 48	18 35	19 32	19 20	19 41	20 10	20 51
62	17 48	18 32	19 25	19 32	19 56	20 28	21 10
60	17 49	18 30	19 19	19 42	20 09	20 42	21 26
N 58	17 49	18 28	19 14	19 51	20 20	20 55	21 39
56	17 49	18 26	19 10	19 59	20 29	21 06	21 51
54	17 49	18 25	19 06	20 06	20 38	21 15	22 01
52	17 50	18 23	19 03	20 13	20 45	21 24	22 10
50	17 50	18 22	19 00	20 19	20 52	21 32	22 18
45	17 51	18 20	18 54	20 31	21 07	21 48	22 35
N 40	17 51	18 18	18 49	20 41	21 19	22 01	22 49
35	17 52	18 17	18 46	20 50	21 29	22 13	23 01
30	17 52	18 16	18 44	20 58	21 39	22 23	23 11
20	17 53	18 15	18 40	21 12	21 54	22 40	23 29
N 10	17 54	18 15	18 39	21 23	22 08	22 55	23 44
0	17 55	18 15	18 39	21 34	22 21	23 09	23 59
S 10	17 56	18 17	18 41	21 46	22 34	23 23	24 13
20	17 57	18 19	18 45	21 57	22 47	23 38	24 29
30	17 58	18 22	18 50	22 11	23 03	23 55	24 46
35	17 59	18 25	18 54	22 19	23 12	24 05	00 05
40	18 00	18 27	18 59	22 28	23 23	24 17	00 17
45	18 02	18 31	19 06	22 38	23 35	24 30	00 30
S 50	18 03	18 35	19 14	22 51	23 50	24 46	00 46
52	18 04	18 38	19 18	22 57	23 57	24 54	00 54
54	18 05	18 40	19 22	23 04	24 05	00 05	01 03
56	18 05	18 43	19 28	23 11	24 14	00 14	01 12
58	18 06	18 46	19 34	23 20	24 23	00 23	01 23
S 60	18 07	18 49	19 40	23 29	24 35	00 35	01 36

SUN and MOON

Day	SUN Eqn. of Time 00h	SUN Eqn. of Time 12h	SUN Mer. Pass.	MOON Mer. Pass. Upper	MOON Mer. Pass. Lower	Age	Phase
d	m s	m s	h m	h m	h m	d	%
25	08 06	08 16	11 52	15 24	03 01	05	20
26	08 26	08 37	11 51	16 09	03 46	06	28
27	08 47	08 57	11 51	16 57	04 33	07	37

1998 SEPTEMBER 28, 29, 30 (MON., TUES., WED.)

UT	ARIES GHA	VENUS −3.9 GHA	VENUS Dec	MARS +1.7 GHA	MARS Dec	JUPITER −2.9 GHA	JUPITER Dec	SATURN +0.1 GHA	SATURN Dec
28 00	6 34.1	189 23.4	N 2 47.4	219 41.1	N14 40.0	13 45.5	S 4 48.2	335 41.5	N 9 37.7
01	21 36.5	204 23.0	46.2	234 42.0	39.5	28 48.2	48.4	350 44.1	37.7
02	36 39.0	219 22.6	45.0	249 43.0	39.0	43 51.0	48.5	5 46.7	37.6
03	51 41.4	234 22.2 ..	43.7	264 43.9 ..	38.5	58 53.7 ..	48.6	20 49.3 ..	37.5
04	66 43.9	249 21.8	42.5	279 44.9	38.0	73 56.5	48.7	35 51.9	37.5
05	81 46.4	264 21.4	41.3	294 45.8	37.5	88 59.2	48.8	50 54.6	37.4
06	96 48.8	279 21.0	N 2 40.0	309 46.8	N14 37.0	104 02.0	S 4 49.0	65 57.2 N 9 37.4
07	111 51.3	294 20.6	38.8	324 47.7	36.5	119 04.7	49.1	80 59.8	37.3
08	126 53.8	309 20.2	37.5	339 48.7	36.0	134 07.5	49.2	96 02.4	37.2
09	141 56.2	324 19.8 ..	36.3	354 49.6 ..	35.5	149 10.2 ..	49.3	111 05.0 ..	37.1
10	156 58.7	339 19.4	35.1	9 50.6	35.0	164 13.0	49.5	126 07.7	37.1
11	172 01.2	354 19.0	33.8	24 51.5	34.5	179 15.8	49.6	141 10.3	37.0
12	187 03.6	9 18.6	N 2 32.6	39 52.5	N14 34.0	194 18.5	S 4 49.7	156 12.9 N 9 37.0
13	202 06.1	24 18.2	31.3	54 53.4	33.5	209 21.3	49.8	171 15.5	36.9
14	217 08.6	39 17.8	30.1	69 54.4	33.0	224 24.0	49.9	186 18.1	36.9
15	232 11.0	54 17.4 ..	28.9	84 55.3 ..	32.5	239 26.8 ..	50.1	201 20.8 ..	36.8
16	247 13.5	69 17.0	27.6	99 56.3	32.0	254 29.5	50.2	216 23.4	36.7
17	262 15.9	84 16.6	26.4	114 57.2	31.5	269 32.3	50.3	231 26.0	36.7
18	277 18.4	99 16.2	N 2 25.2	129 58.2	N14 31.0	284 35.0	S 4 50.4	246 28.6 N 9 36.6
19	292 20.9	114 15.9	23.9	144 59.2	30.5	299 37.8	50.5	261 31.2	36.6
20	307 23.3	129 15.5	22.7	160 00.1	30.0	314 40.5	50.7	276 33.9	36.5
21	322 25.8	144 15.1 ..	21.4	175 01.1 ..	29.5	329 43.3 ..	50.8	291 36.5 ..	36.4
22	337 28.3	159 14.7	20.2	190 02.0	29.0	344 46.0	50.9	306 39.1	36.4
23	352 30.7	174 14.3	19.0	205 03.0	28.5	359 48.8	51.0	321 41.7	36.3
29 00	7 33.2	189 13.9	N 2 17.7	220 03.9	N14 28.0	14 51.5	S 4 51.2	336 44.3 N 9 36.2
01	22 35.7	204 13.5	16.5	235 04.9	27.5	29 54.3	51.3	351 47.0	36.2
02	37 38.1	219 13.1	15.2	250 05.8	27.0	44 57.1	51.4	6 49.6	36.1
03	52 40.6	234 12.7 ..	14.0	265 06.8 ..	26.5	59 59.8 ..	51.5	21 52.2 ..	36.1
04	67 43.1	249 12.3	12.8	280 07.7	26.0	75 02.6	51.6	36 54.8	36.0
05	82 45.5	264 11.9	11.5	295 08.7	25.5	90 05.3	51.8	51 57.4	35.9
06	97 48.0	279 11.5	N 2 10.3	310 09.7	N14 25.0	105 08.1	S 4 51.9	67 00.1 N 9 35.9
07	112 50.4	294 11.1	09.0	325 10.6	24.5	120 10.8	52.0	82 02.7	35.8
08	127 52.9	309 10.7	07.8	340 11.6	24.0	135 13.6	52.1	97 05.3	35.7
09	142 55.4	324 10.3 ..	06.6	355 12.5 ..	23.5	150 16.3 ..	52.2	112 07.9 ..	35.7
10	157 57.8	339 09.9	05.3	10 13.5	23.0	165 19.1	52.4	127 10.6	35.6
11	173 00.3	354 09.5	04.1	25 14.4	22.5	180 21.8	52.5	142 13.2	35.6
12	188 02.8	9 09.1	N 2 02.8	40 15.4	N14 22.0	195 24.6	S 4 52.6	157 15.8 N 9 35.5
13	203 05.2	24 08.8	01.6	55 16.4	21.5	210 27.3	52.7	172 18.4	35.4
14	218 07.7	39 08.4	2 00.3	70 17.3	21.0	225 30.1	52.8	187 21.0	35.4
15	233 10.2	54 08.0	1 59.1	85 18.3 ..	20.5	240 32.8 ..	53.0	202 23.7 ..	35.3
16	248 12.6	69 07.6	57.9	100 19.2	20.0	255 35.6	53.1	217 26.3	35.2
17	263 15.1	84 07.2	56.6	115 20.2	19.5	270 38.3	53.2	232 28.9	35.2
18	278 17.5	99 06.8	N 1 55.4	130 21.1	N14 19.0	285 41.1	S 4 53.3	247 31.5 N 9 35.1
19	293 20.0	114 06.4	54.1	145 22.1	18.5	300 43.8	53.4	262 34.2	35.1
20	308 22.5	129 06.0	52.9	160 23.1	18.0	315 46.6	53.6	277 36.8	35.0
21	323 24.9	144 05.6 ..	51.6	175 24.0 ..	17.5	330 49.3 ..	53.7	292 39.4 ..	34.9
22	338 27.4	159 05.2	50.4	190 25.0	17.0	345 52.1	53.8	307 42.0	34.9
23	353 29.9	174 04.8	49.2	205 25.9	16.5	0 54.8	53.9	322 44.7	34.8
30 00	8 32.3	189 04.4	N 1 47.9	220 26.9	N14 15.9	15 57.6	S 4 54.0	337 47.3 N 9 34.7
01	23 34.8	204 04.0	46.7	235 27.9	15.4	31 00.3	54.2	352 49.9	34.7
02	38 37.3	219 03.6	45.4	250 28.8	14.9	46 03.1	54.3	7 52.5	34.6
03	53 39.7	234 03.2 ..	44.2	265 29.8 ..	14.4	61 05.8 ..	54.4	22 55.2 ..	34.6
04	68 42.2	249 02.8	42.9	280 30.7	13.9	76 08.6	54.5	37 57.8	34.5
05	83 44.7	264 02.5	41.7	295 31.7	13.4	91 11.3	54.6	53 00.4	34.4
06	98 47.1	279 02.1	N 1 40.5	310 32.7	N14 12.9	106 14.1	S 4 54.8	68 03.0 N 9 34.4
07	113 49.6	294 01.7	39.2	325 33.6	12.4	121 16.8	54.9	83 05.6	34.3
08	128 52.0	309 01.3	38.0	340 34.6	11.9	136 19.6	55.0	98 08.3	34.2
09	143 54.5	324 00.9 ..	36.7	355 35.5 ..	11.4	151 22.3 ..	55.1	113 10.9 ..	34.2
10	158 57.0	339 00.5	35.5	10 36.5	10.9	166 25.1	55.2	128 13.5	34.1
11	173 59.4	354 00.1	34.2	25 37.5	10.4	181 27.8	55.3	143 16.1	34.0
12	189 01.9	8 59.7	N 1 33.0	40 38.4	N14 09.9	196 30.6	S 4 55.5	158 18.8 N 9 34.0
13	204 04.4	23 59.3	31.7	55 39.4	09.4	211 33.3	55.6	173 21.4	33.9
14	219 06.8	38 58.9	30.5	70 40.3	08.9	226 36.1	55.7	188 24.0	33.9
15	234 09.3	53 58.5 ..	29.2	85 41.3 ..	08.4	241 38.8 ..	55.8	203 26.6 ..	33.8
16	249 11.8	68 58.1	28.0	100 42.3	07.9	256 41.6	55.9	218 29.3	33.7
17	264 14.2	83 57.7	26.8	115 43.2	07.4	271 44.3	56.1	233 31.9	33.7
18	279 16.7	98 57.3	N 1 25.5	130 44.2	N14 06.8	286 47.0	S 4 56.2	248 34.5 N 9 33.6
19	294 19.2	113 57.0	24.3	145 45.1	06.3	301 49.8	56.3	263 37.1	33.5
20	309 21.6	128 56.6	23.0	160 46.1	05.8	316 52.5	56.4	278 39.8	33.5
21	324 24.1	143 56.2 ..	21.8	175 47.1 ..	05.3	331 55.3 ..	56.5	293 42.4 ..	33.4
22	339 26.5	158 55.8	20.5	190 48.0	04.8	346 58.0	56.7	308 45.0	33.4
23	354 29.0	173 55.4	19.3	205 49.0	04.3	2 00.8	56.8	323 47.7	33.3
Mer. Pass.	h m 23 25.9	v −0.4 d 1.2		v 1.0 d 0.5		v 2.8 d 0.1		v 2.6 d 0.1	

Left margin day labels: MONDAY (28), TUESDAY (29), WEDNESDAY (30)

STARS

Name	SHA	Dec
Acamar	315 26.6	S40 18.5
Achernar	335 34.7	S57 14.5
Acrux	173 22.6	S63 05.4
Adhara	255 21.5	S28 58.1
Aldebaran	291 02.3	N16 30.3
Alioth	166 31.2	N55 58.2
Alkaid	153 08.3	N49 19.5
Al Na'ir	27 57.6	S46 58.1
Alnilam	275 57.8	S 1 12.2
Alphard	218 07.5	S 8 39.1
Alphecca	126 20.9	N26 43.5
Alpheratz	357 54.9	N29 05.1
Altair	62 19.2	N 8 52.2
Ankaa	353 26.5	S42 18.8
Antares	112 40.4	S26 25.6
Arcturus	146 06.4	N19 11.6
Atria	107 52.6	S69 01.6
Avior	234 22.9	S59 30.1
Bellatrix	278 44.1	N 6 20.9
Betelgeuse	271 13.5	N 7 24.4
Canopus	264 01.2	S52 41.5
Capella	280 51.1	N45 59.6
Deneb	49 38.9	N45 16.9
Denebola	182 45.6	N14 34.9
Diphda	349 07.0	S17 59.6
Dubhe	194 06.2	N61 45.5
Elnath	278 26.9	N28 36.2
Eltanin	90 51.5	N51 29.8
Enif	33 58.0	N 9 52.3
Fomalhaut	15 36.2	S29 37.7
Gacrux	172 14.1	S57 06.2
Gienah	176 04.3	S17 31.9
Hadar	149 04.6	S60 21.9
Hamal	328 13.3	N23 27.3
Kaus Aust.	83 58.9	S34 23.1
Kochab	137 20.6	N74 10.0
Markab	13 49.3	N15 12.0
Menkar	314 26.7	N 4 05.1
Menkent	148 21.3	S36 21.7
Miaplacidus	221 42.6	S69 42.5
Mirfak	308 56.3	N49 51.2
Nunki	76 12.3	S26 17.8
Peacock	53 36.8	S56 44.4
Pollux	243 41.7	N28 01.7
Procyon	245 11.7	N 5 13.7
Rasalhague	96 17.1	N12 34.0
Regulus	207 55.9	N11 58.5
Rigel	281 22.9	S 8 12.1
Rigil Kent.	140 07.8	S60 49.7
Sabik	102 25.7	S15 43.2
Schedar	349 52.9	N56 31.8
Shaula	96 37.4	S37 06.1
Sirius	258 43.7	S16 42.7
Spica	158 43.5	S11 09.1
Suhail	223 01.0	S43 25.4
Vega	80 46.6	N38 47.4
Zuben'ubi	137 18.3	S16 02.0

	SHA	Mer. Pass.
	° ′	h m
Venus	181 40.7	11 23
Mars	212 30.7	9 19
Jupiter	7 18.4	22 56
Saturn	329 11.1	1 33

UT	SUN GHA	SUN Dec	MOON GHA	MOON v	MOON Dec	MOON d	MOON HP
d h	° '	° '	° '	'	° '	'	'
28 00	182 16.9	S 1 51.4	102 14.2	11.2	S18 34.0	2.3	55.1
01	197 17.1	52.3	116 44.4	11.3	18 36.3	2.2	55.2
02	212 17.3	53.3	131 14.7	11.2	18 38.5	2.1	55.2
03	227 17.5 ..	54.3	145 44.9	11.1	18 40.6	2.0	55.2
04	242 17.8	55.3	160 15.0	11.1	18 42.6	1.9	55.3
05	257 18.0	56.2	174 45.1	11.1	18 44.5	1.8	55.3
06	272 18.2	S 1 57.2	189 15.2	11.0	S18 46.3	1.7	55.3
M 07	287 18.4	58.2	203 45.2	11.0	18 48.0	1.7	55.3
O 08	302 18.6	1 59.1	218 15.2	10.9	18 49.7	1.5	55.4
N 09	317 18.8	2 00.1	232 45.1	10.9	18 51.2	1.5	55.4
D 10	332 19.0	01.1	247 15.0	10.9	18 52.7	1.4	55.4
A 11	347 19.2	02.1	261 44.9	10.8	18 54.1	1.2	55.4
Y 12	2 19.4	S 2 03.0	276 14.7	10.7	S18 55.3	1.2	55.5
13	17 19.6	04.0	290 44.4	10.8	18 56.5	1.1	55.5
14	32 19.9	05.0	305 14.2	10.7	18 57.6	1.0	55.5
15	47 20.1 ..	06.0	319 43.9	10.6	18 58.6	0.8	55.5
16	62 20.3	06.9	334 13.5	10.6	18 59.4	0.8	55.6
17	77 20.5	07.9	348 43.1	10.6	19 00.2	0.7	55.6
18	92 20.7	S 2 08.9	3 12.7	10.5	S19 00.9	0.6	55.6
19	107 20.9	09.8	17 42.2	10.5	19 01.5	0.6	55.7
20	122 21.1	10.8	32 11.7	10.5	19 02.1	0.4	55.7
21	137 21.3 ..	11.8	46 41.2	10.4	19 02.5	0.3	55.7
22	152 21.5	12.8	61 10.6	10.4	19 02.8	0.2	55.8
23	167 21.7	13.7	75 40.0	10.3	19 03.0	0.1	55.8
29 00	182 21.9	S 2 14.7	90 09.3	10.3	S19 03.1	0.0	55.8
01	197 22.2	15.7	104 38.6	10.3	19 03.1	0.1	55.9
02	212 22.4	16.7	119 07.9	10.2	19 03.0	0.1	55.9
03	227 22.6 ..	17.6	133 37.1	10.2	19 02.9	0.3	55.9
04	242 22.8	18.6	148 06.3	10.2	19 02.6	0.4	55.9
05	257 23.0	19.6	162 35.5	10.1	19 02.2	0.5	56.0
06	272 23.2	S 2 20.5	177 04.6	10.1	S19 01.7	0.6	56.0
T 07	287 23.4	21.5	191 33.7	10.1	19 01.1	0.7	56.0
U 08	302 23.6	22.5	206 02.8	10.0	19 00.4	0.7	56.1
E 09	317 23.8 ..	23.5	220 31.8	9.9	18 59.7	0.9	56.1
S 10	332 24.0	24.4	235 00.7	10.0	18 58.8	1.0	56.1
D 11	347 24.2	25.4	249 29.7	9.9	18 57.8	1.1	56.2
A 12	2 24.4	S 2 26.4	263 58.6	9.9	S18 56.7	1.2	56.2
Y 13	17 24.6	27.3	278 27.5	9.8	18 55.5	1.3	56.2
14	32 24.9	28.3	292 56.3	9.8	18 54.2	1.4	56.3
15	47 25.1 ..	29.3	307 25.1	9.8	18 52.8	1.5	56.3
16	62 25.3	30.3	321 53.9	9.8	18 51.3	1.6	56.3
17	77 25.5	31.2	336 22.7	9.7	18 49.7	1.7	56.4
18	92 25.7	S 2 32.2	350 51.4	9.7	S18 48.0	1.8	56.4
19	107 25.9	33.2	5 20.1	9.6	18 46.2	1.9	56.5
20	122 26.1	34.2	19 48.7	9.7	18 44.3	2.0	56.5
21	137 26.3 ..	35.1	34 17.4	9.6	18 42.3	2.1	56.5
22	152 26.5	36.1	48 46.0	9.5	18 40.2	2.2	56.6
23	167 26.7	37.1	63 14.5	9.5	18 38.0	2.4	56.6
30 00	182 26.9	S 2 38.0	77 43.0	9.6	S18 35.6	2.4	56.6
01	197 27.1	39.0	92 11.6	9.4	18 33.2	2.5	56.7
02	212 27.3	40.0	106 40.0	9.5	18 30.7	2.7	56.7
03	227 27.5 ..	40.9	121 08.5	9.4	18 28.0	2.7	56.7
04	242 27.7	41.9	135 36.9	9.4	18 25.3	2.9	56.8
05	257 28.0	42.9	150 05.3	9.4	18 22.4	2.9	56.8
06	272 28.2	S 2 43.9	164 33.7	9.3	S18 19.5	3.1	56.9
W 07	287 28.4	44.8	179 02.0	9.3	18 16.4	3.2	56.9
E 08	302 28.6	45.8	193 30.3	9.3	18 13.2	3.2	56.9
D 09	317 28.8 ..	46.8	207 58.6	9.3	18 10.0	3.4	57.0
N 10	332 29.0	47.7	222 26.9	9.2	18 06.6	3.5	57.0
E 11	347 29.2	48.7	236 55.1	9.2	18 03.1	3.6	57.0
S 12	2 29.4	S 2 49.7	251 23.3	9.2	S17 59.5	3.7	57.1
D 13	17 29.6	50.7	265 51.5	9.2	17 55.8	3.8	57.1
A 14	32 29.8	51.6	280 19.7	9.1	17 52.0	3.9	57.2
Y 15	47 30.0 ..	52.6	294 47.8	9.1	17 48.1	4.0	57.2
16	62 30.2	53.6	309 15.9	9.1	17 44.1	4.1	57.2
17	77 30.4	54.5	323 44.0	9.1	17 40.0	4.2	57.3
18	92 30.6	S 2 55.5	338 12.1	9.1	S17 35.8	4.4	57.3
19	107 30.8	56.5	352 40.2	9.0	17 31.4	4.4	57.4
20	122 31.0	57.5	7 08.2	9.0	17 27.0	4.5	57.4
21	137 31.2 ..	58.4	21 36.2	9.0	17 22.5	4.7	57.4
22	152 31.4	2 59.4	36 04.2	9.0	17 17.8	4.7	57.5
23	167 31.6	S 3 00.4	50 32.2	8.9	S17 13.1	4.9	57.5
	SD 16.0	d 1.0	SD 15.1	15.3		15.6	

Lat.	Twilight Naut.	Twilight Civil	Sunrise	Moonrise 28	Moonrise 29	Moonrise 30	Moonrise 1
°	h m	h m	h m	h m	h m	h m	h m
N 72	03 37	05 01	06 09	■■■	■■■	18 44	18 14
N 70	03 51	05 06	06 06	16 33	17 17	17 33	17 38
68	04 02	05 09	06 05	15 38	16 26	16 55	17 12
66	04 11	05 12	06 03	15 05	15 55	16 29	16 52
64	04 18	05 15	06 02	14 41	15 31	16 09	16 36
62	04 24	05 17	06 01	14 22	15 12	15 52	16 23
60	04 29	05 19	06 00	14 06	14 57	15 38	16 12
N 58	04 34	05 20	05 59	13 53	14 44	15 26	16 02
56	04 38	05 21	05 58	13 42	14 32	15 16	15 53
54	04 41	05 22	05 58	13 32	14 22	15 07	15 46
52	04 44	05 23	05 57	13 23	14 14	14 59	15 39
50	04 47	05 24	05 56	13 15	14 06	14 51	15 32
45	04 52	05 26	05 55	12 58	13 49	14 36	15 19
N 40	04 56	05 27	05 54	12 44	13 35	14 23	15 08
35	04 58	05 28	05 53	12 32	13 23	14 12	14 58
30	05 00	05 28	05 52	12 22	13 13	14 02	14 50
20	05 03	05 28	05 50	12 04	12 55	13 46	14 36
N 10	05 03	05 28	05 49	11 49	12 40	13 31	14 23
0	05 02	05 26	05 47	11 35	12 25	13 18	14 11
S 10	05 00	05 24	05 45	11 20	12 11	13 04	13 59
20	04 56	05 21	05 44	11 05	11 56	12 50	13 46
30	04 49	05 17	05 41	10 48	11 38	12 33	13 32
35	04 45	05 15	05 40	10 37	11 28	12 23	13 23
40	04 39	05 11	05 38	10 26	11 16	12 12	13 14
45	04 32	05 07	05 36	10 12	11 02	11 59	13 03
S 50	04 23	05 02	05 34	09 55	10 46	11 44	12 49
52	04 19	04 59	05 33	09 48	10 38	11 36	12 42
54	04 14	04 56	05 32	09 39	10 29	11 28	12 35
56	04 08	04 53	05 31	09 29	10 19	11 19	12 27
58	04 02	04 50	05 29	09 18	10 08	11 08	12 18
S 60	03 54	04 46	05 28	09 05	09 55	10 56	12 08

Lat.	Sunset	Twilight Civil	Twilight Naut.	Moonset 28	Moonset 29	Moonset 30	Moonset 1
°	h m	h m	h m	h m	h m	h m	h m
N 72	17 30	18 37	20 00	■■■	■■■	20 26	22 47
N 70	17 32	18 33	19 46	19 00	20 02	21 36	23 22
68	17 34	18 29	19 36	19 54	20 52	22 12	23 47
66	17 36	18 27	19 27	20 27	21 24	22 38	24 06
64	17 37	18 24	19 20	20 51	21 47	22 58	24 21
62	17 38	18 22	19 15	21 10	22 06	23 15	24 34
60	17 39	18 21	19 10	21 26	22 21	23 28	24 45
N 58	17 40	18 19	19 05	21 39	22 34	23 40	24 54
56	17 41	18 18	19 02	21 51	22 45	23 50	25 02
54	17 42	18 17	18 58	22 01	22 55	23 58	25 09
52	17 43	18 16	18 56	22 10	23 04	24 06	00 06
50	17 43	18 16	18 53	22 18	23 12	24 13	00 13
45	17 45	18 14	18 48	22 35	23 29	24 29	00 29
N 40	17 46	18 13	18 44	22 49	23 42	24 41	00 41
35	17 47	18 12	18 42	23 01	23 54	24 51	00 51
30	17 48	18 12	18 40	23 11	24 04	00 04	01 01
20	17 50	18 12	18 38	23 29	24 21	00 21	01 17
N 10	17 52	18 13	18 37	23 44	24 36	00 36	01 30
0	17 54	18 14	18 38	23 59	24 50	00 50	01 43
S 10	17 55	18 16	18 41	24 13	00 13	01 04	01 56
20	17 58	18 20	18 45	24 29	00 29	01 19	02 10
30	18 00	18 24	18 52	24 46	00 46	01 37	02 25
35	18 01	18 27	18 57	00 05	00 57	01 46	02 34
40	18 03	18 30	19 02	00 17	01 08	01 58	02 44
45	18 05	18 35	19 10	00 30	01 22	02 11	02 56
S 50	18 08	18 40	19 19	00 46	01 39	02 27	03 11
52	18 09	18 43	19 23	00 54	01 47	02 35	03 17
54	18 10	18 46	19 29	01 03	01 56	02 43	03 25
56	18 11	18 49	19 34	01 12	02 06	02 53	03 33
58	18 13	18 53	19 41	01 23	02 17	03 04	03 42
S 60	18 15	18 57	19 49	01 36	02 30	03 16	03 53

	SUN Eqn. of Time 00h	SUN Eqn. of Time 12h	SUN Mer. Pass.	MOON Mer. Pass. Upper	MOON Mer. Pass. Lower	MOON Age	MOON Phase
Day							
d	m s	m s	h m	h m	h m	d	%
28	09 07	09 17	11 51	17 47	05 22	08	46
29	09 27	09 37	11 50	18 38	06 12	09	56
30	09 47	09 57	11 50	19 30	07 04	10	66

1998 OCTOBER 1, 2, 3 (THURS., FRI., SAT.)

UT	ARIES GHA	VENUS −3.9 GHA	VENUS Dec	MARS +1.7 GHA	MARS Dec	JUPITER −2.9 GHA	JUPITER Dec	SATURN +0.1 GHA	SATURN Dec
d h	° ′	° ′	° ′	° ′	° ′	° ′	° ′	° ′	° ′
1 00	9 31.5	188 55.0	N 1 18.0	220 50.0	N14 03.8	17 03.5	S 4 56.9	338 50.3	N 9 33.2
01	24 33.9	203 54.6	16.8	235 50.9	03.3	32 06.3	57.0	353 52.9	33.2
02	39 36.4	218 54.2	15.5	250 51.9	02.8	47 09.0	57.1	8 55.5	33.1
03	54 38.9	233 53.8 ..	14.3	265 52.9 ..	02.3	62 11.8 ..	57.2	23 58.2 ..	33.0
04	69 41.3	248 53.4	13.1	280 53.8	01.8	77 14.5	57.4	39 00.8	33.0
05	84 43.8	263 53.0	11.8	295 54.8	01.3	92 17.3	57.5	54 03.4	32.9
T 06	99 46.3	278 52.6 N 1 10.6		310 55.7 N14 00.8		107 20.0 S 4 57.6		69 06.0 N 9 32.8	
H 07	114 48.7	293 52.2	09.3	325 56.7	14 00.2	122 22.7	57.7	84 08.7	32.8
U 08	129 51.2	308 51.9	08.1	340 57.7	13 59.7	137 25.5	57.8	99 11.3	32.7
R 09	144 53.7	323 51.5 ..	06.8	355 58.6 ..	59.2	152 28.2 ..	57.9	114 13.9 ..	32.7
S 10	159 56.1	338 51.1	05.6	10 59.6	58.7	167 31.0	58.1	129 16.5	32.6
D 11	174 58.6	353 50.7	04.3	26 00.6	58.2	182 33.7	58.2	144 19.2	32.5
A 12	190 01.0	8 50.3 N 1 03.1		41 01.5 N13 57.7		197 36.5 S 4 58.3		159 21.8 N 9 32.5	
Y 13	205 03.5	23 49.9	01.8	56 02.5	57.2	212 39.2	58.4	174 24.4	32.4
14	220 06.0	38 49.5	1 00.6	71 03.5	56.7	227 42.0	58.5	189 27.0	32.3
15	235 08.4	53 49.1	0 59.3	86 04.4 ..	56.2	242 44.7 ..	58.6	204 29.7 ..	32.3
16	250 10.9	68 48.7	58.1	101 05.4	55.7	257 47.5	58.8	219 32.3	32.2
17	265 13.4	83 48.3	56.8	116 06.4	55.2	272 50.2	58.9	234 34.9	32.1
18	280 15.8	98 47.9 N 0 55.6		131 07.3 N13 54.6		287 52.9 S 4 59.0		249 37.6 N 9 32.1	
19	295 18.3	113 47.5	54.3	146 08.3	54.1	302 55.7	59.1	264 40.2	32.0
20	310 20.8	128 47.2	53.1	161 09.3	53.6	317 58.4	59.2	279 42.8	32.0
21	325 23.2	143 46.8 ..	51.8	176 10.2 ..	53.1	333 01.2 ..	59.3	294 45.4 ..	31.9
22	340 25.7	158 46.4	50.6	191 11.2	52.6	348 03.9	59.5	309 48.1	31.8
23	355 28.1	173 46.0	49.3	206 12.2	52.1	3 06.7	59.6	324 50.7	31.8
2 00	10 30.6	188 45.6 N 0 48.1		221 13.1 N13 51.6		18 09.4 S 4 59.7		339 53.3 N 9 31.7	
01	25 33.1	203 45.2	46.8	236 14.1	51.1	33 12.1	59.8	354 56.0	31.6
02	40 35.5	218 44.8	45.6	251 15.1	50.6	48 14.9	4 59.9	9 58.6	31.6
03	55 38.0	233 44.4 ..	44.4	266 16.0 ..	50.1	63 17.6	5 00.0	25 01.2 ..	31.5
04	70 40.5	248 44.0	43.1	281 17.0	49.5	78 20.4	00.2	40 03.8	31.4
05	85 42.9	263 43.6	41.9	296 18.0	49.0	93 23.1	00.3	55 06.5	31.4
06	100 45.4	278 43.2 N 0 40.6		311 18.9 N13 48.5		108 25.9 S 5 00.4		70 09.1 N 9 31.3	
F 07	115 47.9	293 42.8	39.4	326 19.9	48.0	123 28.6	00.5	85 11.7	31.2
R 08	130 50.3	308 42.5	38.1	341 20.9	47.5	138 31.3	00.6	100 14.4	31.2
I 09	145 52.8	323 42.1 ..	36.9	356 21.9 ..	47.0	153 34.1 ..	00.7	115 17.0 ..	31.1
D 10	160 55.3	338 41.7	35.6	11 22.8	46.5	168 36.8	00.8	130 19.6	31.1
A 11	175 57.7	353 41.3	34.4	26 23.8	46.0	183 39.6	01.0	145 22.2	31.0
Y 12	191 00.2	8 40.9 N 0 33.1		41 24.8 N13 45.5		198 42.3 S 5 01.1		160 24.9 N 9 30.9	
13	206 02.6	23 40.5	31.9	56 25.7	44.9	213 45.0	01.2	175 27.5	30.9
14	221 05.1	38 40.1	30.6	71 26.7	44.4	228 47.8	01.3	190 30.1	30.8
15	236 07.6	53 39.7 ..	29.4	86 27.7 ..	43.9	243 50.5 ..	01.4	205 32.8 ..	30.7
16	251 10.0	68 39.3	28.1	101 28.6	43.4	258 53.3	01.5	220 35.4	30.7
17	266 12.5	83 38.9	26.9	116 29.6	42.9	273 56.0	01.6	235 38.0	30.6
18	281 15.0	98 38.5 N 0 25.6		131 30.6 N13 42.4		288 58.7 S 5 01.8		250 40.6 N 9 30.5	
19	296 17.4	113 38.1	24.4	146 31.6	41.9	304 01.5	01.9	265 43.3	30.5
20	311 19.9	128 37.8	23.1	161 32.5	41.4	319 04.2	02.0	280 45.9	30.4
21	326 22.4	143 37.4 ..	21.9	176 33.5 ..	40.9	334 07.0 ..	02.1	295 48.5 ..	30.3
22	341 24.8	158 37.0	20.6	191 34.5	40.3	349 09.7	02.2	310 51.2	30.3
23	356 27.3	173 36.6	19.4	206 35.4	39.8	4 12.4	02.3	325 53.8	30.2
3 00	11 29.7	188 36.2 N 0 18.1		221 36.4 N13 39.3		19 15.2 S 5 02.5		340 56.4 N 9 30.2	
01	26 32.2	203 35.8	16.9	236 37.4	38.8	34 17.9	02.6	355 59.1	30.1
02	41 34.7	218 35.4	15.6	251 38.4	38.3	49 20.7	02.7	11 01.7	30.0
03	56 37.1	233 35.0 ..	14.4	266 39.3 ..	37.8	64 23.4 ..	02.8	26 04.3 ..	30.0
04	71 39.6	248 34.6	13.1	281 40.3	37.3	79 26.1	02.9	41 07.0	29.9
05	86 42.1	263 34.2	11.9	296 41.3	36.8	94 28.9	03.0	56 09.6	29.8
06	101 44.5	278 33.8 N 0 10.6		311 42.3 N13 36.2		109 31.6 S 5 03.1		71 12.2 N 9 29.8	
S 07	116 47.0	293 33.4	09.4	326 43.2	35.7	124 34.4	03.2	86 14.8	29.7
A 08	131 49.5	308 33.1	08.1	341 44.2	35.2	139 37.1	03.4	101 17.5	29.6
T 09	146 51.9	323 32.7 ..	06.9	356 45.2 ..	34.7	154 39.8 ..	03.5	116 20.1 ..	29.6
U 10	161 54.4	338 32.3	05.6	11 46.1	34.2	169 42.6	03.6	131 22.7	29.5
R 11	176 56.9	353 31.9	04.4	26 47.1	33.7	184 45.3	03.7	146 25.4	29.4
D 12	191 59.3	8 31.5 N 0 03.1		41 48.1 N13 33.2		199 48.0 S 5 03.8		161 28.0 N 9 29.4	
A 13	207 01.8	23 31.1	01.8	56 49.1	32.6	214 50.8	03.9	176 30.6	29.3
Y 14	222 04.2	38 30.7 N	00.6	71 50.0	32.1	229 53.5	04.0	191 33.3	29.2
15	237 06.7	53 30.3 S	00.7	86 51.0 ..	31.6	244 56.3 ..	04.2	206 35.9 ..	29.2
16	252 09.2	68 29.9	01.9	101 52.0	31.1	259 59.0	04.3	221 38.5	29.1
17	267 11.6	83 29.5	03.2	116 53.0	30.6	275 01.7	04.4	236 41.2	29.0
18	282 14.1	98 29.1 S 0 04.4		131 53.9 N13 30.1		290 04.5 S 5 04.5		251 43.8 N 9 29.0	
19	297 16.6	113 28.7	05.7	146 54.9	29.6	305 07.2	04.6	266 46.4	28.9
20	312 19.0	128 28.4	06.9	161 55.9	29.0	320 09.9	04.7	281 49.1	28.9
21	327 21.5	143 28.0 ..	08.2	176 56.9 ..	28.5	335 12.7 ..	04.8	296 51.7 ..	28.8
22	342 24.0	158 27.6	09.4	191 57.8	28.0	350 15.4	04.9	311 54.3	28.7
23	357 26.4	173 27.2	10.7	206 58.8	27.5	5 18.2	05.1	326 57.0	28.7
Mer.Pass. 23 14.1		v −0.4 d 1.2		v 1.0 d 0.5		v 2.7 d 0.1		v 2.6 d 0.1	

STARS

Name	SHA	Dec
Acamar	315 26.6	S40 18.5
Achernar	335 34.7	S57 14.6
Acrux	173 22.6	S63 05.4
Adhara	255 21.5	S28 58.1
Aldebaran	291 02.2	N16 30.3
Alioth	166 31.2	N55 58.2
Alkaid	153 08.3	N49 19.4
Al Na'ir	27 57.6	S46 58.1
Alnilam	275 57.8	S 1 12.1
Alphard	218 07.4	S 8 39.1
Alphecca	126 20.9	N26 43.5
Alpheratz	357 54.9	N29 05.1
Altair	62 19.2	N 8 52.2
Ankaa	353 26.5	S42 18.8
Antares	112 40.4	S26 25.6
Arcturus	146 06.4	N19 11.6
Atria	107 52.7	S69 01.6
Avior	234 22.9	S59 30.1
Bellatrix	278 44.1	N 6 20.9
Betelgeuse	271 13.5	N 7 24.4
Canopus	264 01.1	S52 41.5
Capella	280 51.1	N45 59.6
Deneb	49 39.0	N45 16.9
Denebola	182 45.6	N14 34.9
Diphda	349 06.9	S17 59.6
Dubhe	194 06.2	N61 45.4
Elnath	278 26.9	N28 36.2
Eltanin	90 51.5	N51 29.8
Enif	33 58.0	N 9 52.3
Fomalhaut	15 36.2	S29 37.7
Gacrux	172 14.1	S57 06.2
Gienah	176 04.3	S17 31.9
Hadar	149 04.6	S60 21.9
Hamal	328 13.3	N23 27.4
Kaus Aust.	83 58.9	S34 23.1
Kochab	137 20.7	N74 10.0
Markab	13 49.3	N15 12.0
Menkar	314 26.7	N 4 05.1
Menkent	148 21.3	S36 21.7
Miaplacidus	221 42.5	S69 42.5
Mirfak	308 56.2	N49 51.2
Nunki	76 12.4	S26 17.8
Peacock	53 36.8	S56 44.5
Pollux	243 41.7	N28 01.6
Procyon	245 11.7	N 5 13.7
Rasalhague	96 17.1	N12 34.0
Regulus	207 55.8	N11 58.5
Rigel	281 22.9	S 8 12.1
Rigil Kent.	140 07.9	S60 49.7
Sabik	102 25.7	S15 43.2
Schedar	349 52.9	N56 31.8
Shaula	96 37.4	S37 06.1
Sirius	258 43.7	S16 42.7
Spica	158 43.5	S11 09.1
Suhail	223 01.0	S43 25.4
Vega	80 46.6	N38 47.4
Zuben'ubi	137 18.3	S16 02.0

	SHA	Mer. Pass.
	° ′	h m
Venus	178 15.0	11 25
Mars	210 42.5	9 15
Jupiter	7 38.8	22 43
Saturn	329 22.7	1 20

UT	SUN GHA	SUN Dec	MOON GHA	v	Dec	d	HP
1 00	182 31.8	S 3 01.3	65 00.1	8.9	S17 08.2	4.9	57.6
01	197 32.0	02.3	79 28.0	8.9	17 03.3	5.1	57.6
02	212 32.2	03.3	93 55.9	8.9	16 58.2	5.2	57.6
03	227 32.4	.. 04.2	108 23.8	8.9	16 53.0	5.2	57.7
04	242 32.7	05.2	122 51.7	8.9	16 47.8	5.4	57.7
05	257 32.9	06.2	137 19.6	8.8	16 42.4	5.5	57.8
06	272 33.1	S 3 07.2	151 47.4	8.8	S16 36.9	5.6	57.8
T 07	287 33.3	08.1	166 15.2	8.8	16 31.3	5.6	57.8
H 08	302 33.5	09.1	180 43.0	8.8	16 25.7	5.8	57.9
U 09	317 33.7	.. 10.1	195 10.8	8.8	16 19.9	5.9	57.9
R 10	332 33.9	11.0	209 38.6	8.8	16 14.0	6.0	58.0
S 11	347 34.1	12.0	224 06.4	8.7	16 08.0	6.1	58.0
D 12	2 34.3	S 3 13.0	238 34.1	8.7	S16 01.9	6.2	58.0
A 13	17 34.5	13.9	253 01.8	8.7	15 55.7	6.3	58.1
Y 14	32 34.7	14.9	267 29.5	8.7	15 49.4	6.4	58.1
15	47 34.9	.. 15.9	281 57.2	8.7	15 43.0	6.5	58.2
16	62 35.1	16.8	296 24.9	8.7	15 36.5	6.6	58.2
17	77 35.3	17.8	310 52.6	8.6	15 29.9	6.7	58.2
18	92 35.5	S 3 18.8	325 20.2	8.7	S15 23.2	6.9	58.3
19	107 35.7	19.8	339 47.9	8.6	15 16.3	6.9	58.3
20	122 35.9	20.7	354 15.5	8.6	15 09.4	7.0	58.4
21	137 36.1	.. 21.7	8 43.1	8.6	15 02.4	7.1	58.4
22	152 36.3	22.7	23 10.7	8.6	14 55.3	7.2	58.4
23	167 36.5	23.6	37 38.3	8.6	14 48.1	7.3	58.5
2 00	182 36.7	S 3 24.6	52 05.9	8.5	S14 40.8	7.4	58.5
01	197 36.9	25.6	66 33.4	8.6	14 33.4	7.5	58.6
02	212 37.1	26.5	81 01.0	8.5	14 25.9	7.6	58.6
03	227 37.3	.. 27.5	95 28.5	8.6	14 18.3	7.7	58.6
04	242 37.5	28.5	109 56.1	8.5	14 10.6	7.7	58.7
05	257 37.7	29.4	124 23.6	8.5	14 02.9	7.9	58.7
06	272 37.9	S 3 30.4	138 51.1	8.5	S13 55.0	8.0	58.8
F 07	287 38.1	31.4	153 18.6	8.5	13 47.0	8.1	58.8
R 08	302 38.3	32.3	167 46.1	8.5	13 38.9	8.1	58.8
I 09	317 38.5	.. 33.3	182 13.6	8.4	13 30.8	8.3	58.9
D 10	332 38.7	34.3	196 41.0	8.5	13 22.5	8.3	58.9
A 11	347 38.9	35.2	211 08.5	8.5	13 14.2	8.5	59.0
Y 12	2 39.1	S 3 36.2	225 36.0	8.4	S13 05.7	8.5	59.0
13	17 39.3	37.2	240 03.4	8.4	12 57.2	8.6	59.0
14	32 39.5	38.1	254 30.8	8.5	12 48.6	8.7	59.1
15	47 39.7	.. 39.1	268 58.3	8.4	12 39.9	8.8	59.1
16	62 39.9	40.1	283 25.7	8.4	12 31.1	8.9	59.2
17	77 40.1	41.1	297 53.1	8.4	12 22.2	9.0	59.2
18	92 40.3	S 3 42.0	312 20.5	8.4	S12 13.2	9.1	59.2
19	107 40.5	43.0	326 47.9	8.3	12 04.1	9.1	59.3
20	122 40.7	44.0	341 15.2	8.4	11 55.0	9.3	59.3
21	137 40.9	.. 44.9	355 42.6	8.4	11 45.7	9.3	59.4
22	152 41.1	45.9	10 10.0	8.3	11 36.4	9.4	59.4
23	167 41.3	46.9	24 37.3	8.4	11 27.0	9.5	59.4
3 00	182 41.5	S 3 47.8	39 04.7	8.3	S11 17.5	9.5	59.5
01	197 41.7	48.8	53 32.0	8.3	11 08.0	9.7	59.5
02	212 41.9	49.8	67 59.3	8.4	10 58.3	9.7	59.5
03	227 42.1	.. 50.7	82 26.7	8.3	10 48.6	9.8	59.6
04	242 42.3	51.7	96 54.0	8.3	10 38.8	9.9	59.6
05	257 42.5	52.7	111 21.3	8.3	10 28.9	9.9	59.7
06	272 42.7	S 3 53.6	125 48.6	8.3	S10 19.0	10.1	59.7
S 07	287 42.8	54.6	140 15.9	8.3	10 08.9	10.1	59.7
A 08	302 43.0	55.6	154 43.2	8.2	9 58.8	10.2	59.8
T 09	317 43.2	.. 56.5	169 10.4	8.3	9 48.6	10.3	59.8
U 10	332 43.4	57.5	183 37.7	8.3	9 38.3	10.3	59.8
R 11	347 43.6	58.5	198 05.0	8.2	9 28.0	10.4	59.9
D 12	2 43.8	S 3 59.4	212 32.2	8.3	S 9 17.6	10.5	59.9
A 13	17 44.0	4 00.4	226 59.5	8.2	9 07.1	10.5	59.9
Y 14	32 44.2	01.3	241 26.7	8.2	8 56.6	10.6	60.0
15	47 44.4	.. 02.3	255 53.9	8.2	8 46.0	10.7	60.0
16	62 44.6	03.3	270 21.1	8.2	8 35.3	10.8	60.0
17	77 44.8	04.2	284 48.3	8.2	8 24.5	10.8	60.1
18	92 45.0	S 4 05.2	299 15.5	8.2	S 8 13.7	10.9	60.1
19	107 45.2	06.2	313 42.7	8.2	8 02.8	10.9	60.1
20	122 45.4	07.1	328 09.9	8.2	7 51.9	11.0	60.2
21	137 45.6	.. 08.1	342 37.1	8.2	7 40.9	11.1	60.2
22	152 45.8	09.1	357 04.3	8.1	7 29.8	11.1	60.2
23	167 46.0	10.0	11 31.4	8.2	S 7 18.7	11.2	60.3
	SD 16.0	d 1.0	SD 15.8		16.1		16.3

Lat.	Naut.	Civil	Sunrise	Moonrise 1	2	3	4
N 72	03 53	05 15	06 22	18 14	18 01	17 51	17 43
N 70	04 05	05 18	06 18	17 38	17 39	17 39	17 38
68	04 14	05 18	06 15	17 12	17 22	17 29	17 34
66	04 21	05 22	06 13	16 52	17 08	17 20	17 30
64	04 27	05 23	06 11	16 36	16 57	17 13	17 27
62	04 33	05 25	06 09	16 23	16 47	17 07	17 24
60	04 37	05 26	06 07	16 12	16 39	17 01	17 22
N 58	04 41	05 27	06 06	16 02	16 31	16 57	17 20
56	04 44	05 27	06 04	15 53	16 25	16 52	17 18
54	04 47	05 28	06 03	15 46	16 19	16 49	17 16
52	04 49	05 28	06 02	15 39	16 14	16 45	17 14
50	04 51	05 29	06 01	15 32	16 09	16 42	17 13
45	04 55	05 30	05 59	15 19	15 58	16 35	17 10
N 40	04 58	05 30	05 57	15 08	15 50	16 29	17 07
35	05 01	05 30	05 55	14 58	15 42	16 24	17 05
30	05 02	05 30	05 54	14 50	15 36	16 20	17 03
20	05 03	05 29	05 51	14 36	15 24	16 12	17 00
N 10	05 03	05 28	05 49	14 23	15 14	16 06	16 57
0	05 01	05 25	05 46	14 11	15 05	15 59	16 54
S 10	04 58	05 23	05 44	13 59	14 55	15 53	16 51
20	04 53	05 19	05 41	13 46	14 45	15 46	16 48
30	04 45	05 14	05 38	13 32	14 34	15 38	16 45
35	04 40	05 10	05 36	13 23	14 27	15 34	16 43
40	04 34	05 06	05 33	13 14	14 20	15 29	16 41
45	04 26	05 01	05 31	13 03	14 11	15 23	16 38
S 50	04 16	04 55	05 28	12 49	14 00	15 16	16 35
52	04 11	04 52	05 26	12 42	13 55	15 13	16 33
54	04 06	04 49	05 24	12 35	13 50	15 09	16 32
56	03 59	04 45	05 23	12 27	13 43	15 05	16 30
58	03 52	04 41	05 21	12 18	13 37	15 01	16 28
S 60	03 44	04 36	05 18	12 08	13 29	14 56	16 26

Lat.	Sunset	Civil	Naut.	Moonset 1	2	3	4
N 72	17 15	18 21	19 42	22 47	24 54	00 54	02 57
N 70	17 18	18 19	19 31	23 22	25 14	01 14	03 07
68	17 22	18 17	19 22	23 47	25 29	01 29	03 16
66	17 24	18 15	19 15	24 06	00 06	01 42	03 23
64	17 27	18 14	19 09	24 21	00 21	01 52	03 28
62	17 29	18 13	19 04	24 34	00 34	02 01	03 33
60	17 30	18 12	19 00	24 45	00 45	02 09	03 38
N 58	17 32	18 11	18 57	24 54	00 54	02 15	03 41
56	17 33	18 10	18 53	25 02	01 02	02 21	03 45
54	17 35	18 10	18 51	25 09	01 09	02 26	03 48
52	17 36	18 09	18 48	00 06	01 16	02 31	03 50
50	17 37	18 09	18 46	00 10	01 22	02 35	03 53
45	17 39	18 08	18 42	00 29	01 34	02 45	03 58
N 40	17 41	18 08	18 40	00 41	01 45	02 52	04 03
35	17 43	18 08	18 38	00 51	01 53	02 59	04 06
30	17 45	18 08	18 36	01 01	02 01	03 04	04 10
20	17 47	18 09	18 35	01 17	02 14	03 14	04 15
N 10	17 50	18 11	18 35	01 30	02 26	03 23	04 20
0	17 53	18 13	18 37	01 43	02 37	03 31	04 25
S 10	17 55	18 16	18 41	01 56	02 47	03 39	04 30
20	17 58	18 20	18 46	02 10	02 59	03 47	04 35
30	18 02	18 26	18 54	02 25	03 12	03 56	04 40
35	18 04	18 29	18 59	02 34	03 19	04 02	04 43
40	18 06	18 33	19 06	02 44	03 27	04 08	04 47
45	18 09	18 39	19 14	02 56	03 37	04 15	04 51
S 50	18 12	18 45	19 24	03 11	03 49	04 24	04 56
52	18 14	18 48	19 29	03 17	03 55	04 28	04 58
54	18 15	18 51	19 35	03 25	04 01	04 32	05 00
56	18 17	18 55	19 41	03 33	04 07	04 37	05 03
58	18 19	19 00	19 49	03 42	04 15	04 42	05 06
S 60	18 22	19 04	19 57	03 53	04 23	04 48	05 09

Day	SUN Eqn. of Time 00h	12h	Mer. Pass.	MOON Mer. Pass. Upper	Lower	Age	Phase
	m s	m s	h m	h m	h m	d	%
1	10 07	10 17	11 50	20 24	07 57	11	76
2	10 26	10 36	11 49	21 18	08 51	12	85
3	10 45	10 55	11 49	22 12	09 45	13	92

1998 OCTOBER 4, 5, 6 (SUN., MON., TUES.)

UT	ARIES GHA	VENUS −3.9 GHA	Dec	MARS +1.7 GHA	Dec	JUPITER −2.9 GHA	Dec	SATURN +0.0 GHA	Dec	STARS Name	SHA	Dec
4 00	12 28.9	188 26.8	S 0 11.9	221 59.8	N13 27.0	20 20.9	S 5 05.2	341 59.6	N 9 28.6	Acamar	315 26.6	S40 18.5
01	27 31.4	203 26.4	13.2	237 00.8	26.5	35 23.6	05.3	357 02.2	28.5	Achernar	335 34.7	S57 14.6
02	42 33.8	218 26.0	14.4	252 01.8	25.9	50 26.4	05.4	12 04.8	28.5	Acrux	173 22.6	S63 05.4
03	57 36.3	233 25.6 . .	15.7	267 02.7 . .	25.4	65 29.1 . .	05.5	27 07.5 . .	28.4	Adhara	255 21.4	S28 58.1
04	72 38.7	248 25.2	16.9	282 03.7	24.9	80 31.8	05.6	42 10.1	28.3	Aldebaran	291 02.2	N16 30.3
05	87 41.2	263 24.8	18.2	297 04.7	24.4	95 34.6	05.7	57 12.7	28.3			
S 06	102 43.7	278 24.4	S 0 19.4	312 05.7	N13 23.9	110 37.3	S 5 05.8	72 15.4	N 9 28.2	Alioth	166 31.2	N55 58.2
U 07	117 46.1	293 24.0	20.7	327 06.6	23.4	125 40.0	05.9	87 18.0	28.1	Alkaid	153 08.3	N49 19.4
N 08	132 48.6	308 23.6	21.9	342 07.6	22.9	140 42.8	06.1	102 20.6	28.1	Al Na'ir	27 57.6	S46 58.1
D 09	147 51.1	323 23.3 . .	23.2	357 08.6 . .	22.3	155 45.5 . .	06.2	117 23.3 . .	28.0	Alnilam	275 57.8	S 1 12.1
A 10	162 53.5	338 22.9	24.4	12 09.6	21.8	170 48.2	06.3	132 25.9	27.9	Alphard	218 07.4	S 8 39.0
Y 11	177 56.0	353 22.5	25.7	27 10.5	21.3	185 51.0	06.4	147 28.5	27.9			
12	192 58.5	8 22.1	S 0 26.9	42 11.5	N13 20.8	200 53.7	S 5 06.5	162 31.2	N 9 27.8	Alphecca	126 20.9	N26 43.5
13	208 00.9	23 21.7	28.2	57 12.5	20.3	215 56.4	06.6	177 33.8	27.7	Alpheratz	357 54.9	N29 05.1
14	223 03.4	38 21.3	29.4	72 13.5	19.8	230 59.2	06.7	192 36.4	27.7	Altair	62 19.2	N 8 52.2
15	238 05.8	53 20.9 . .	30.7	87 14.5 . .	19.2	246 01.9 . .	06.8	207 39.1 . .	27.6	Ankaa	353 26.5	S42 18.8
16	253 08.3	68 20.5	31.9	102 15.4	18.7	261 04.6	06.9	222 41.7	27.5	Antares	112 40.4	S26 25.6
17	268 10.8	83 20.1	33.2	117 16.4	18.2	276 07.4	07.1	237 44.3	27.5			
18	283 13.2	98 19.7	S 0 34.5	132 17.4	N13 17.7	291 10.1	S 5 07.2	252 47.0	N 9 27.4	Arcturus	146 06.4	N19 11.6
19	298 15.7	113 19.3	35.7	147 18.4	17.2	306 12.8	07.3	267 49.6	27.3	Atria	107 52.7	S69 01.6
20	313 18.2	128 18.9	37.0	162 19.4	16.6	321 15.6	07.4	282 52.2	27.3	Avior	234 22.8	S59 30.1
21	328 20.6	143 18.5 . .	38.2	177 20.3 . .	16.1	336 18.3 . .	07.5	297 54.9 . .	27.2	Bellatrix	278 44.1	N 6 20.9
22	343 23.1	158 18.2	39.5	192 21.3	15.6	351 21.0	07.6	312 57.5	27.2	Betelgeuse	271 13.5	N 7 24.4
23	358 25.6	173 17.8	40.7	207 22.3	15.1	6 23.8	07.7	328 00.2	27.1			
5 00	13 28.0	188 17.4	S 0 42.0	222 23.3	N13 14.6	21 26.5	S 5 07.8	343 02.8	N 9 27.0	Canopus	264 01.1	S52 41.5
01	28 30.5	203 17.0	43.2	237 24.3	14.1	36 29.2	07.9	358 05.4	27.0	Capella	280 51.0	N45 59.6
02	43 33.0	218 16.6	44.5	252 25.2	13.5	51 32.0	08.1	13 08.1	26.9	Deneb	49 39.0	N45 16.9
03	58 35.4	233 16.2 . .	45.7	267 26.2 . .	13.0	66 34.7 . .	08.2	28 10.7 . .	26.8	Denebola	182 45.5	N14 34.9
04	73 37.9	248 15.8	47.0	282 27.2	12.5	81 37.4	08.3	43 13.3	26.8	Diphda	349 06.9	S17 59.6
05	88 40.3	263 15.4	48.2	297 28.2	12.0	96 40.2	08.4	58 16.0	26.7			
M 06	103 42.8	278 15.0	S 0 49.5	312 29.2	N13 11.5	111 42.9	S 5 08.5	73 18.6	N 9 26.6	Dubhe	194 06.2	N61 45.4
O 07	118 45.3	293 14.6	50.7	327 30.1	11.0	126 45.6	08.6	88 21.2	26.6	Elnath	278 26.8	N28 36.2
N 08	133 47.7	308 14.2	52.0	342 31.1	10.4	141 48.3	08.7	103 23.9	26.5	Eltanin	90 51.5	N51 29.8
D 09	148 50.2	323 13.8 . .	53.2	357 32.1 . .	09.9	156 51.1 . .	08.8	118 26.5 . .	26.4	Enif	33 58.0	N 9 52.3
A 10	163 52.7	338 13.4	54.5	12 33.1	09.4	171 53.8	08.9	133 29.1	26.4	Fomalhaut	15 36.2	S29 37.7
Y 11	178 55.1	353 13.1	55.7	27 34.1	08.9	186 56.5	09.0	148 31.8	26.3			
12	193 57.6	8 12.7	S 0 57.0	42 35.1	N13 08.4	201 59.3	S 5 09.1	163 34.4	N 9 26.2	Gacrux	172 14.1	S57 06.2
13	209 00.1	23 12.3	58.2	57 36.0	07.8	217 02.0	09.3	178 37.0	26.2	Gienah	176 04.3	S17 31.9
14	224 02.5	38 11.9	0 59.5	72 37.0	07.3	232 04.7	09.4	193 39.7	26.1	Hadar	149 04.6	S60 21.9
15	239 05.0	53 11.5	1 00.8	87 38.0 . .	06.8	247 07.5 . .	09.5	208 42.3 . .	26.0	Hamal	328 13.2	N23 27.4
16	254 07.4	68 11.1	02.0	102 39.0	06.3	262 10.2	09.6	223 44.9	26.0	Kaus Aust.	83 58.9	S34 23.1
17	269 09.9	83 10.7	03.3	117 40.0	05.8	277 12.9	09.7	238 47.6	25.9			
18	284 12.4	98 10.3	S 1 04.5	132 41.0	N13 05.2	292 15.6	S 5 09.8	253 50.2	N 9 25.8	Kochab	137 20.7	N74 09.9
19	299 14.8	113 09.9	05.8	147 41.9	04.7	307 18.4	09.9	268 52.9	25.8	Markab	13 49.3	N15 12.0
20	314 17.3	128 09.5	07.0	162 42.9	04.2	322 21.1	10.0	283 55.5	25.7	Menkar	314 26.7	N 4 05.1
21	329 19.8	143 09.1 . .	08.3	177 43.9 . .	03.7	337 23.8 . .	10.1	298 58.1 . .	25.6	Menkent	148 21.3	S36 21.7
22	344 22.2	158 08.7	09.5	192 44.9	03.2	352 26.6	10.2	314 00.8	25.6	Miaplacidus	221 42.5	S69 42.5
23	359 24.7	173 08.3	10.8	207 45.9	02.6	7 29.3	10.3	329 03.4	25.5			
6 00	14 27.2	188 07.9	S 1 12.0	222 46.9	N13 02.1	22 32.0	S 5 10.4	344 06.0	N 9 25.4	Mirfak	308 56.2	N49 51.2
01	29 29.6	203 07.5	13.3	237 47.8	01.6	37 34.7	10.6	359 08.7	25.4	Nunki	76 12.4	S26 17.8
02	44 32.1	218 07.1	14.5	252 48.8	01.1	52 37.5	10.7	14 11.3	25.3	Peacock	53 36.9	S56 44.5
03	59 34.6	233 06.8 . .	15.8	267 49.8 . .	00.6	67 40.2 . .	10.8	29 13.9 . .	25.2	Pollux	243 41.7	N28 01.6
04	74 37.0	248 06.4	17.0	282 50.8	13 00.0	82 42.9	10.9	44 16.6	25.2	Procyon	245 11.7	N 5 13.7
05	89 39.5	263 06.0	18.3	297 51.8	12 59.5	97 45.6	11.0	59 19.2	25.1			
T 06	104 41.9	278 05.6	S 1 19.5	312 52.8	N12 59.0	112 48.4	S 5 11.1	74 21.9	N 9 25.0	Rasalhague	96 17.1	N12 34.0
U 07	119 44.4	293 05.2	20.8	327 53.8	58.5	127 51.1	11.2	89 24.5	25.0	Regulus	207 55.8	N11 58.5
E 08	134 46.9	308 04.8	22.1	342 54.7	58.0	142 53.8	11.3	104 27.1	24.9	Rigel	281 22.8	S 8 12.1
S 09	149 49.3	323 04.4 . .	23.3	357 55.7 . .	57.4	157 56.5 . .	11.4	119 29.8 . .	24.8	Rigil Kent.	140 07.9	S60 49.7
D 10	164 51.8	338 04.0	24.6	12 56.7	56.9	172 59.3	11.5	134 32.4	24.8	Sabik	102 25.7	S15 43.2
A 11	179 54.3	353 03.6	25.8	27 57.7	56.4	188 02.0	11.6	149 35.0	24.7			
Y 12	194 56.7	8 03.2	S 1 27.1	42 58.7	N12 55.9	203 04.7	S 5 11.7	164 37.7	N 9 24.6	Schedar	349 52.9	N56 31.8
13	209 59.2	23 02.8	28.3	57 59.7	55.3	218 07.5	11.8	179 40.3	24.6	Shaula	96 37.5	S37 06.1
14	225 01.7	38 02.4	29.6	73 00.7	54.8	233 10.2	12.0	194 42.9	24.5	Sirius	258 43.7	S16 42.7
15	240 04.1	53 02.0 . .	30.8	88 01.6 . .	54.3	248 12.9 . .	12.1	209 45.6 . .	24.4	Spica	158 43.5	S11 09.1
16	255 06.6	68 01.6	32.1	103 02.6	53.8	263 15.6	12.2	224 48.2	24.4	Suhail	223 01.0	S43 25.4
17	270 09.0	83 01.2	33.3	118 03.6	53.3	278 18.4	12.3	239 50.9	24.3			
18	285 11.5	98 00.8	S 1 34.6	133 04.6	N12 52.7	293 21.1	S 5 12.4	254 53.5	N 9 24.2	Vega	80 46.7	N38 47.4
19	300 14.0	113 00.4	35.8	148 05.6	52.2	308 23.8	12.5	269 56.1	24.2	Zuben'ubi	137 18.3	S16 02.0
20	315 16.4	128 00.0	37.1	163 06.6	51.7	323 26.5	12.6	284 58.8	24.1		SHA	Mer. Pass.
21	330 18.9	142 59.7 . .	38.3	178 07.6 . .	51.2	338 29.2 . .	12.7	300 01.4 . .	24.0		° ′	h m
22	345 21.4	157 59.3	39.6	193 08.6	50.6	353 32.0	12.8	315 04.1	24.0	Venus	174 49.3	11 27
23	0 23.8	172 58.9	40.8	208 09.5	50.1	8 34.7	12.9	330 06.7	23.9	Mars	208 55.3	9 10
										Jupiter	7 58.5	22 30
Mer. Pass. 23 02.3	v −0.4 d 1.3			v 1.0 d 0.5		v 2.7 d 0.1		v 2.6 d 0.1		Saturn	329 34.8	1 08

UT	SUN GHA	SUN Dec	MOON GHA	v	MOON Dec	d	HP
d h	° ′	° ′	° ′	′	° ′	′	′
4 00	182 46.2	S 4 11.0	25 58.6	8.1	S 7 07.5	11.2	60.3
01	197 46.4	12.0	40 25.7	8.1	6 56.3	11.3	60.3
02	212 46.6	12.9	54 52.8	8.1	6 45.0	11.3	60.4
03	227 46.8	. . 13.9	69 19.9	8.2	6 33.7	11.4	60.4
04	242 46.9	14.9	83 47.1	8.1	6 22.3	11.5	60.4
05	257 47.1	15.8	98 14.2	8.0	6 10.8	11.5	60.5
06	272 47.3	S 4 16.8	112 41.2	8.1	S 5 59.3	11.6	60.5
07	287 47.5	17.8	127 08.3	8.1	5 47.7	11.6	60.5
S 08	302 47.7	18.7	141 35.4	8.0	5 36.1	11.6	60.5
U 09	317 47.9	. . 19.7	156 02.4	8.1	5 24.5	11.7	60.6
N 10	332 48.1	20.6	170 29.5	8.0	5 12.8	11.7	60.6
D 11	347 48.3	21.6	184 56.5	8.1	5 01.1	11.8	60.6
A 12	2 48.5	S 4 22.6	199 23.6	8.0	S 4 49.3	11.9	60.6
Y 13	17 48.7	23.5	213 50.6	8.0	4 37.4	11.8	60.7
14	32 48.9	24.5	228 17.6	8.0	4 25.6	11.9	60.7
15	47 49.1	. . 25.5	242 44.6	7.9	4 13.7	12.0	60.7
16	62 49.3	26.4	257 11.5	8.0	4 01.7	11.9	60.7
17	77 49.5	27.4	271 38.5	7.9	3 49.8	12.0	60.8
18	92 49.6	S 4 28.4	286 05.4	8.0	S 3 37.8	12.1	60.8
19	107 49.8	29.3	300 32.4	7.9	3 25.7	12.1	60.8
20	122 50.0	30.3	314 59.3	7.9	3 13.6	12.1	60.8
21	137 50.2	. . 31.2	329 26.2	7.9	3 01.5	12.1	60.9
22	152 50.4	32.2	343 53.1	7.9	2 49.4	12.2	60.9
23	167 50.6	33.2	358 20.0	7.9	2 37.2	12.2	60.9
5 00	182 50.8	S 4 34.1	12 46.9	7.8	S 2 25.0	12.2	60.9
01	197 51.0	35.1	27 13.7	7.9	2 12.8	12.2	60.9
02	212 51.2	36.1	41 40.6	7.8	2 00.6	12.3	61.0
03	227 51.4	. . 37.0	56 07.4	7.8	1 48.3	12.2	61.0
04	242 51.6	38.0	70 34.2	7.8	1 36.1	12.3	61.0
05	257 51.7	38.9	85 01.0	7.8	1 23.8	12.3	61.0
06	272 51.9	S 4 39.9	99 27.8	7.8	S 1 11.5	12.4	61.0
07	287 52.1	40.9	113 54.6	7.7	0 59.1	12.3	61.1
M 08	302 52.3	41.8	128 21.3	7.8	0 46.8	12.4	61.1
O 09	317 52.5	. . 42.8	142 48.1	7.7	0 34.4	12.3	61.1
N 10	332 52.7	43.7	157 14.8	7.7	0 22.1	12.4	61.1
D 11	347 52.9	44.7	171 41.5	7.7	S 0 09.7	12.4	61.1
A 12	2 53.1	S 4 45.7	186 08.2	7.7	N 0 02.7	12.4	61.1
Y 13	17 53.3	46.6	200 34.9	7.6	0 15.1	12.3	61.1
14	32 53.4	47.6	215 01.5	7.6	0 27.4	12.4	61.2
15	47 53.6	. . 48.6	229 28.1	7.7	0 39.8	12.4	61.2
16	62 53.8	49.5	243 54.8	7.6	0 52.2	12.4	61.2
17	77 54.0	50.5	258 21.4	7.5	1 04.6	12.4	61.2
18	92 54.2	S 4 51.4	272 47.9	7.6	N 1 17.0	12.4	61.2
19	107 54.4	52.4	287 14.5	7.5	1 29.4	12.4	61.2
20	122 54.6	53.4	301 41.0	7.6	1 41.8	12.4	61.2
21	137 54.8	. . 54.3	316 07.6	7.5	1 54.2	12.3	61.2
22	152 55.0	55.3	330 34.1	7.4	2 06.5	12.4	61.2
23	167 55.1	56.2	345 00.5	7.5	2 18.9	12.3	61.3
6 00	182 55.3	S 4 57.2	359 27.0	7.5	N 2 31.2	12.4	61.3
01	197 55.5	58.2	13 53.5	7.4	2 43.6	12.3	61.3
02	212 55.7	4 59.1	28 19.9	7.4	2 55.9	12.3	61.3
03	227 55.9	5 00.1	42 46.3	7.4	3 08.2	12.3	61.3
04	242 56.1	01.0	57 12.7	7.3	3 20.5	12.2	61.3
05	257 56.3	02.0	71 39.0	7.4	3 32.7	12.3	61.3
06	272 56.4	S 5 03.0	86 05.4	7.3	N 3 45.0	12.2	61.3
07	287 56.6	03.9	100 31.7	7.3	3 57.2	12.2	61.3
T 08	302 56.8	04.9	114 58.0	7.3	4 09.4	12.1	61.3
U 09	317 57.0	. . 05.8	129 24.3	7.2	4 21.5	12.2	61.3
E 10	332 57.2	06.8	143 50.5	7.3	4 33.7	12.1	61.3
S 11	347 57.4	07.8	158 16.8	7.2	4 45.8	12.1	61.3
D 12	2 57.6	S 5 08.7	172 43.0	7.2	N 4 57.9	12.0	61.3
A 13	17 57.7	09.7	187 09.2	7.2	5 09.9	12.0	61.3
Y 14	32 57.9	10.6	201 35.4	7.1	5 21.9	12.0	61.3
15	47 58.1	. . 11.6	216 01.5	7.1	5 33.9	11.9	61.3
16	62 58.3	12.5	230 27.6	7.1	5 45.8	11.9	61.3
17	77 58.5	13.5	244 53.7	7.1	5 57.7	11.9	61.3
18	92 58.7	S 5 14.5	259 19.8	7.1	N 6 09.6	11.8	61.3
19	107 58.8	15.4	273 45.9	7.0	6 21.4	11.8	61.3
20	122 59.0	16.4	288 11.9	7.0	6 33.2	11.7	61.3
21	137 59.2	. . 17.3	302 37.9	7.0	6 44.9	11.7	61.3
22	152 59.4	18.3	317 03.9	7.0	6 56.6	11.6	61.3
23	167 59.6	19.3	331 29.9	6.9	N 7 08.2	11.6	61.3
	SD 16.0	d 1.0	SD 16.5		16.7		16.7

Lat.	Twilight Naut.	Twilight Civil	Sunrise	Moonrise 4	5	6	7
°	h m	h m	h m	h m	h m	h m	h m
N 72	04 08	05 28	06 36	17 43	17 35	17 28	17 19
N 70	04 17	05 30	06 30	17 38	17 37	17 36	17 36
68	04 25	05 31	06 26	17 34	17 38	17 43	17 49
66	04 31	05 31	06 22	17 30	17 39	17 48	18 00
64	04 36	05 32	06 19	17 27	17 40	17 53	18 09
62	04 41	05 32	06 17	17 24	17 40	17 57	18 17
60	04 44	05 33	06 14	17 22	17 41	18 01	18 23
N 58	04 47	05 33	06 12	17 20	17 42	18 04	18 29
56	04 50	05 33	06 10	17 18	17 42	18 07	18 35
54	04 52	05 33	06 09	17 16	17 43	18 10	18 40
52	04 54	05 33	06 07	17 14	17 43	18 12	18 44
50	04 56	05 33	06 06	17 13	17 43	18 15	18 48
45	04 59	05 33	06 03	17 10	17 44	18 19	18 57
N 40	05 01	05 33	06 00	17 07	17 45	18 23	19 04
35	05 03	05 32	05 58	17 05	17 46	18 27	19 10
30	05 04	05 32	05 55	17 03	17 46	18 30	19 16
20	05 04	05 30	05 52	17 00	17 47	18 36	19 25
N 10	05 03	05 27	05 48	16 57	17 48	18 40	19 34
0	05 00	05 24	05 45	16 54	17 49	18 45	19 42
S 10	04 56	05 21	05 42	16 51	17 50	18 50	19 50
20	04 50	05 16	05 38	16 48	17 51	18 55	19 59
30	04 42	05 10	05 34	16 45	17 52	19 01	20 09
35	04 36	05 06	05 31	16 43	17 53	19 04	20 15
40	04 29	05 01	05 28	16 41	17 54	19 08	20 21
45	04 20	04 55	05 25	16 38	17 55	19 12	20 29
S 50	04 09	04 48	05 21	16 35	17 56	19 18	20 39
52	04 03	04 45	05 19	16 33	17 56	19 20	20 43
54	03 57	04 41	05 17	16 32	17 57	19 23	20 48
56	03 50	04 37	05 15	16 30	17 58	19 26	20 53
58	03 42	04 32	05 12	16 28	17 58	19 29	20 59
S 60	03 33	04 27	05 09	16 26	17 59	19 33	21 06

Lat.	Sunset	Twilight Civil	Twilight Naut.	Moonset 4	5	6	7
°	h m	h m	h m	h m	h m	h m	h m
N 72	16 59	18 06	19 26	02 57	05 01	07 05	09 12
N 70	17 05	18 05	19 17	03 07	05 03	07 00	08 58
68	17 09	18 04	19 09	03 16	05 05	06 55	08 46
66	17 13	18 04	19 03	03 23	05 06	06 51	08 37
64	17 16	18 03	18 58	03 28	05 07	06 48	08 29
62	17 19	18 03	18 54	03 33	05 08	06 45	08 22
60	17 21	18 03	18 51	03 38	05 09	06 43	08 17
N 58	17 24	18 03	18 48	03 41	05 10	06 41	08 11
56	17 26	18 03	18 46	03 45	05 11	06 39	08 07
54	17 27	18 02	18 43	03 48	05 12	06 37	08 03
52	17 29	18 03	18 42	03 50	05 12	06 36	07 59
50	17 30	18 03	18 40	03 53	05 13	06 34	07 56
45	17 34	18 03	18 37	03 58	05 14	06 31	07 49
N 40	17 36	18 03	18 35	04 03	05 15	06 29	07 43
35	17 39	18 04	18 33	04 06	05 16	06 27	07 38
30	17 41	18 05	18 32	04 10	05 17	06 25	07 33
20	17 45	18 07	18 32	04 15	05 18	06 21	07 25
N 10	17 48	18 09	18 34	04 20	05 19	06 18	07 18
0	17 52	18 12	18 37	04 25	05 20	06 16	07 12
S 10	17 55	18 16	18 41	04 30	05 21	06 13	07 06
20	17 59	18 21	18 47	04 35	05 22	06 10	06 59
30	18 04	18 28	18 56	04 40	05 23	06 06	06 51
35	18 06	18 32	19 02	04 43	05 24	06 04	06 46
40	18 09	18 36	19 09	04 47	05 24	06 02	06 41
45	18 13	18 42	19 18	04 51	05 25	06 00	06 36
S 50	18 17	18 50	19 29	04 56	05 26	05 57	06 29
52	18 19	18 53	19 35	04 58	05 26	05 55	06 25
54	18 21	18 57	19 41	05 00	05 27	05 54	06 22
56	18 23	19 02	19 48	05 03	05 27	05 52	06 18
58	18 26	19 07	19 57	05 06	05 28	05 50	06 14
S 60	18 29	19 12	20 06	05 09	05 29	05 48	06 09

	SUN			MOON			
Day	Eqn. of Time 00h	Eqn. of Time 12h	Mer. Pass.	Mer. Pass. Upper	Mer. Pass. Lower	Age	Phase
d	m s	m s	h m	h m	h m	d	%
4	11 04	11 14	11 49	23 07	10 39	14	97
5	11 23	11 32	11 48	24 02	11 35	15	100
6	11 41	11 50	11 48	00 02	12 30	16	99

1998 OCTOBER 7, 8, 9 (WED., THURS., FRI.)

UT	ARIES GHA	VENUS −3.9 GHA	VENUS Dec	MARS +1.7 GHA	MARS Dec	JUPITER −2.9 GHA	JUPITER Dec	SATURN +0.0 GHA	SATURN Dec
7 00	15 26.3	187 58.5	S 1 42.1	223 10.5	N12 49.6	23 37.4	S 5 13.0	345 09.3	N 9 23.8
01	30 28.8	202 58.1	43.3	238 11.5	49.1	38 40.1	13.1	0 12.0	23.8
02	45 31.2	217 57.7	44.6	253 12.5	48.6	53 42.9	13.2	15 14.6	23.7
03	60 33.7	232 57.3 ..	45.9	268 13.5 ..	48.0	68 45.6 ..	13.3	30 17.2 ..	23.6
04	75 36.2	247 56.9	47.1	283 14.5	47.5	83 48.3	13.4	45 19.9	23.6
05	90 38.6	262 56.5	48.4	298 15.5	47.0	98 51.0	13.5	60 22.5	23.5
W 06	105 41.1	277 56.1	S 1 49.6	313 16.5	N12 46.5	113 53.8	S 5 13.6	75 25.2	N 9 23.4
E 07	120 43.5	292 55.7	50.9	328 17.5	45.9	128 56.5	13.8	90 27.8	23.4
D 08	135 46.0	307 55.3	52.1	343 18.5	45.4	143 59.2	13.9	105 30.4	23.3
N 09	150 48.5	322 54.9 ..	53.4	358 19.4 ..	44.9	159 01.9 ..	14.0	120 33.1 ..	23.2
E 10	165 50.9	337 54.5	54.6	13 20.4	44.4	174 04.6	14.1	135 35.7	23.2
S 11	180 53.4	352 54.1	55.9	28 21.4	43.8	189 07.4	14.2	150 38.3	23.1
D 12	195 55.9	7 53.7	S 1 57.1	43 22.4	N12 43.3	204 10.1	S 5 14.3	165 41.0	N 9 23.0
A 13	210 58.3	22 53.3	58.4	58 23.4	42.8	219 12.8	14.4	180 43.6	23.0
Y 14	226 00.8	37 52.9	1 59.6	73 24.4	42.3	234 15.5	14.5	195 46.3	22.9
15	241 03.3	52 52.5	2 00.9	88 25.4 ..	41.8	249 18.2 ..	14.6	210 48.9 ..	22.8
16	256 05.7	67 52.1	02.1	103 26.4	41.2	264 21.0	14.7	225 51.5	22.8
17	271 08.2	82 51.7	03.4	118 27.4	40.7	279 23.7	14.8	240 54.2	22.7
18	286 10.7	97 51.3	S 2 04.6	133 28.4	N12 40.2	294 26.4	S 5 14.9	255 56.8	N 9 22.6
19	301 13.1	112 50.9	05.9	148 29.4	39.7	309 29.1	15.0	270 59.5	22.6
20	316 15.6	127 50.5	07.1	163 30.3	39.1	324 31.8	15.1	286 02.1	22.5
21	331 18.0	142 50.1 ..	08.4	178 31.3 ..	38.6	339 34.6 ..	15.2	301 04.7 ..	22.4
22	346 20.5	157 49.7	09.6	193 32.3	38.1	354 37.3	15.3	316 07.4	22.4
23	1 23.0	172 49.4	10.9	208 33.3	37.6	9 40.0	15.4	331 10.0	22.3
8 00	16 25.4	187 49.0	S 2 12.1	223 34.3	N12 37.0	24 42.7	S 5 15.5	346 12.7	N 9 22.2
01	31 27.9	202 48.6	13.4	238 35.3	36.5	39 45.4	15.6	1 15.3	22.2
02	46 30.4	217 48.2	14.6	253 36.3	36.0	54 48.2	15.7	16 17.9	22.1
03	61 32.8	232 47.8 ..	15.9	268 37.3 ..	35.5	69 50.9 ..	15.8	31 20.6 ..	22.0
04	76 35.3	247 47.4	17.2	283 38.3	34.9	84 53.6	15.9	46 23.2	22.0
05	91 37.8	262 47.0	18.4	298 39.3	34.4	99 56.3	16.0	61 25.9	21.9
T 06	106 40.2	277 46.6	S 2 19.7	313 40.3	N12 33.9	114 59.0	S 5 16.1	76 28.5	N 9 21.8
H 07	121 42.7	292 46.2	20.9	328 41.3	33.4	130 01.7	16.3	91 31.1	21.8
U 08	136 45.1	307 45.8	22.2	343 42.3	32.8	145 04.5	16.4	106 33.8	21.7
R 09	151 47.6	322 45.4 ..	23.4	358 43.3 ..	32.3	160 07.2 ..	16.5	121 36.4 ..	21.6
S 10	166 50.1	337 45.0	24.7	13 44.2	31.8	175 09.9	16.6	136 39.1	21.6
D 11	181 52.5	352 44.6	25.9	28 45.2	31.2	190 12.6	16.7	151 41.7	21.5
A 12	196 55.0	7 44.2	S 2 27.2	43 46.2	N12 30.7	205 15.3	S 5 16.8	166 44.3	N 9 21.4
Y 13	211 57.5	22 43.8	28.4	58 47.2	30.2	220 18.0	16.9	181 47.0	21.4
14	226 59.9	37 43.4	29.7	73 48.2	29.7	235 20.8	17.0	196 49.6	21.3
15	242 02.4	52 43.0 ..	30.9	88 49.2 ..	29.1	250 23.5 ..	17.1	211 52.3 ..	21.2
16	257 04.9	67 42.6	32.2	103 50.2	28.6	265 26.2	17.2	226 54.9	21.2
17	272 07.3	82 42.2	33.4	118 51.2	28.1	280 28.9	17.3	241 57.5	21.1
18	287 09.8	97 41.8	S 2 34.7	133 52.2	N12 27.6	295 31.6	S 5 17.4	257 00.2	N 9 21.0
19	302 12.3	112 41.4	35.9	148 53.2	27.0	310 34.3	17.5	272 02.8	21.0
20	317 14.7	127 41.0	37.2	163 54.2	26.5	325 37.1	17.6	287 05.5	20.9
21	332 17.2	142 40.6 ..	38.4	178 55.2 ..	26.0	340 39.8 ..	17.7	302 08.1 ..	20.8
22	347 19.6	157 40.2	39.7	193 56.2	25.5	355 42.5	17.8	317 10.8	20.8
23	2 22.1	172 39.8	40.9	208 57.2	24.9	10 45.2	17.9	332 13.4	20.7
9 00	17 24.6	187 39.4	S 2 42.2	223 58.2	N12 24.4	25 47.9	S 5 18.0	347 16.0	N 9 20.6
01	32 27.0	202 39.0	43.4	238 59.2	23.9	40 50.6	18.1	2 18.7	20.5
02	47 29.5	217 38.6	44.7	254 00.2	23.3	55 53.3	18.2	17 21.3	20.5
03	62 32.0	232 38.2 ..	45.9	269 01.2 ..	22.8	70 56.1 ..	18.3	32 24.0 ..	20.4
04	77 34.4	247 37.8	47.2	284 02.2	22.3	85 58.8	18.4	47 26.6	20.3
05	92 36.9	262 37.4	48.4	299 03.2	21.8	101 01.5	18.5	62 29.2	20.3
F 06	107 39.4	277 37.0	S 2 49.7	314 04.2	N12 21.2	116 04.2	S 5 18.6	77 31.9	N 9 20.2
R 07	122 41.8	292 36.6	50.9	329 05.2	20.7	131 06.9	18.7	92 34.5	20.1
I 08	137 44.3	307 36.2	52.2	344 06.2	20.2	146 09.6	18.8	107 37.2	20.1
D 09	152 46.8	322 35.8 ..	53.4	359 07.2 ..	19.7	161 12.3 ..	18.9	122 39.8 ..	20.0
A 10	167 49.2	337 35.4	54.7	14 08.2	19.1	176 15.0	19.0	137 42.5	19.9
Y 11	182 51.7	352 35.0	55.9	29 09.2	18.6	191 17.8	19.1	152 45.1	19.9
12	197 54.1	7 34.6	S 2 57.2	44 10.2	N12 18.1	206 20.5	S 5 19.2	167 47.7	N 9 19.8
13	212 56.6	22 34.2	58.4	59 11.2	17.5	221 23.2	19.3	182 50.4	19.7
14	227 59.1	37 33.8	2 59.7	74 12.2	17.0	236 25.9	19.4	197 53.0	19.7
15	243 01.5	52 33.4	3 00.9	89 13.2 ..	16.5	251 28.6 ..	19.5	212 55.7 ..	19.6
16	258 04.0	67 33.0	02.2	104 14.2	16.0	266 31.3	19.6	227 58.3	19.5
17	273 06.5	82 32.6	03.4	119 15.1	15.4	281 34.0	19.7	243 00.9	19.5
18	288 08.9	97 32.2	S 3 04.7	134 16.1	N12 14.9	296 36.7	S 5 19.8	258 03.6	N 9 19.4
19	303 11.4	112 31.8	05.9	149 17.1	14.4	311 39.4	19.9	273 06.2	19.3
20	318 13.9	127 31.4	07.2	164 18.1	13.8	326 42.2	20.0	288 08.9	19.3
21	333 16.3	142 31.0 ..	08.4	179 19.1 ..	13.3	341 44.9 ..	20.1	303 11.5 ..	19.2
22	348 18.8	157 30.6	09.7	194 20.1	12.8	356 47.6	20.2	318 14.2	19.1
23	3 21.2	172 30.2	10.9	209 21.1	12.3	11 50.3	20.3	333 16.8	19.1
Mer. Pass.	h m 22 50.6	v −0.4	d 1.3	v 1.0	d 0.5	v 2.7	d 0.1	v 2.6	d 0.1

STARS

Name	SHA	Dec
Acamar	315 26.6	S40 18.5
Achernar	335 34.7	S57 14.6
Acrux	173 22.6	S63 05.4
Adhara	255 21.4	S28 58.1
Aldebaran	291 02.2	N16 30.3
Alioth	166 31.2	N55 58.1
Alkaid	153 08.3	N49 19.4
Al Na'ir	27 57.6	S46 58.1
Alnilam	275 57.8	S 1 12.2
Alphard	218 07.4	S 8 39.1
Alphecca	126 20.9	N26 43.5
Alpheratz	357 54.9	N29 05.1
Altair	62 19.2	N 8 52.2
Ankaa	353 26.5	S42 18.8
Antares	112 40.4	S26 25.6
Arcturus	146 06.4	N19 11.6
Atria	107 52.7	S69 01.6
Avior	234 22.8	S59 30.1
Bellatrix	278 44.0	N 6 20.9
Betelgeuse	271 13.5	N 7 24.4
Canopus	264 01.1	S52 41.5
Capella	280 51.0	N45 59.6
Deneb	49 39.0	N45 16.9
Denebola	182 45.5	N14 34.9
Diphda	349 06.9	S17 59.6
Dubhe	194 06.2	N61 45.4
Elnath	278 26.8	N28 36.2
Eltanin	90 51.6	N51 29.8
Enif	33 58.0	N 9 52.3
Fomalhaut	15 36.2	S29 37.7
Gacrux	172 14.1	S57 06.2
Gienah	176 04.3	S17 31.9
Hadar	149 04.6	S60 21.9
Hamal	328 13.2	N23 27.4
Kaus Aust.	83 58.9	S34 23.1
Kochab	137 20.7	N74 09.9
Markab	13 49.3	N15 12.0
Menkar	314 26.6	N 4 05.1
Menkent	148 21.3	S36 21.7
Miaplacidus	221 42.5	S69 42.5
Mirfak	308 56.2	N49 51.2
Nunki	76 12.4	S26 17.8
Peacock	53 36.9	S56 44.5
Pollux	243 41.7	N28 01.6
Procyon	245 11.6	N 5 13.7
Rasalhague	96 17.1	N12 34.0
Regulus	207 55.8	N11 58.5
Rigel	281 22.8	S 8 12.1
Rigil Kent.	140 07.9	S60 49.7
Sabik	102 25.7	S15 43.2
Schedar	349 52.9	N56 31.9
Shaula	96 37.5	S37 06.1
Sirius	258 43.7	S16 42.7
Spica	158 43.5	S11 09.1
Suhail	223 01.0	S43 25.4
Vega	80 46.7	N38 47.4
Zuben'ubi	137 18.3	S16 02.0

	SHA	Mer. Pass.
	° '	h m
Venus	171 23.5	11 29
Mars	207 08.9	9 05
Jupiter	8 17.3	22 17
Saturn	329 47.2	0 55

UT	SUN GHA	SUN Dec	MOON GHA	v	MOON Dec	d	HP
d h	° '	° '	° '	'	° '	'	'
7 00	182 59.8	S 5 20.2	345 55.8	7.0	N 7 19.8	11.5	61.3
01	197 59.9	21.2	0 21.8	6.9	7 31.3	11.5	61.3
02	213 00.1	22.1	14 47.7	6.8	7 42.8	11.4	61.3
03	228 00.3	.. 23.1	29 13.5	6.9	7 54.2	11.4	61.3
04	243 00.5	24.0	43 39.4	6.8	8 05.6	11.3	61.2
05	258 00.7	25.0	58 05.2	6.8	8 16.9	11.3	61.2
W 06	273 00.9	S 5 26.0	72 31.0	6.8	N 8 28.2	11.1	61.2
E 07	288 01.0	26.9	86 56.8	6.8	8 39.3	11.2	61.2
D 08	303 01.2	27.9	101 22.6	6.7	8 50.5	11.0	61.2
N 09	318 01.4	.. 28.8	115 48.3	6.8	9 01.5	11.0	61.2
E 10	333 01.6	29.8	130 14.1	6.6	9 12.5	11.0	61.2
S 11	348 01.8	30.7	144 39.7	6.7	9 23.5	10.8	61.2
D 12	3 01.9	S 5 31.7	159 05.4	6.7	N 9 34.3	10.8	61.2
A 13	18 02.1	32.6	173 31.1	6.6	9 45.1	10.7	61.2
Y 14	33 02.3	33.6	187 56.7	6.6	9 55.8	10.7	61.1
15	48 02.5	.. 34.6	202 22.3	6.6	10 06.5	10.6	61.1
16	63 02.7	35.5	216 47.9	6.6	10 17.1	10.5	61.1
17	78 02.8	36.5	231 13.5	6.5	10 27.6	10.4	61.1
18	93 03.0	S 5 37.4	245 39.0	6.6	N10 38.0	10.3	61.1
19	108 03.2	38.4	260 04.6	6.5	10 48.3	10.3	61.0
20	123 03.4	39.3	274 30.1	6.5	10 58.6	10.2	61.0
21	138 03.6	.. 40.3	288 55.6	6.4	11 08.8	10.1	61.0
22	153 03.7	41.2	303 21.0	6.5	11 18.9	10.0	61.0
23	168 03.9	42.2	317 46.5	6.4	11 28.9	10.0	61.0
8 00	183 04.1	S 5 43.2	332 11.9	6.4	N11 38.9	9.8	61.0
01	198 04.3	44.1	346 37.3	6.4	11 48.7	9.8	61.0
02	213 04.5	45.1	1 02.7	6.4	11 58.5	9.7	60.9
03	228 04.6	.. 46.0	15 28.1	6.3	12 08.2	9.6	60.9
04	243 04.8	47.0	29 53.4	6.4	12 17.8	9.5	60.9
05	258 05.0	47.9	44 18.8	6.3	12 27.3	9.4	60.9
T 06	273 05.2	S 5 48.9	58 44.1	6.3	N12 36.7	9.3	60.9
H 07	288 05.3	49.8	73 09.4	6.3	12 46.0	9.2	60.8
U 08	303 05.5	50.8	87 34.7	6.3	12 55.2	9.1	60.8
R 09	318 05.7	.. 51.7	102 00.0	6.2	13 04.3	9.1	60.8
S 10	333 05.9	52.7	116 25.2	6.3	13 13.4	8.9	60.8
D 11	348 06.0	53.6	130 50.5	6.2	13 22.3	8.9	60.7
A 12	3 06.2	S 5 54.6	145 15.7	6.2	N13 31.2	8.7	60.7
Y 13	18 06.4	55.6	159 40.9	6.2	13 39.9	8.6	60.7
14	33 06.6	56.5	174 06.1	6.2	13 48.5	8.6	60.7
15	48 06.8	.. 57.5	188 31.3	6.2	13 57.1	8.4	60.6
16	63 06.9	58.4	202 56.5	6.1	14 05.5	8.3	60.6
17	78 07.1	5 59.4	217 21.6	6.2	14 13.8	8.3	60.6
18	93 07.3	S 6 00.3	231 46.8	6.1	N14 22.1	8.1	60.6
19	108 07.5	01.3	246 11.9	6.2	14 30.2	8.0	60.5
20	123 07.6	02.2	260 37.1	6.1	14 38.2	7.9	60.5
21	138 07.8	.. 03.2	275 02.2	6.1	14 46.1	7.8	60.5
22	153 08.0	04.1	289 27.3	6.1	14 53.9	7.6	60.5
23	168 08.1	05.1	303 52.4	6.1	15 01.6	7.6	60.4
9 00	183 08.3	S 6 06.0	318 17.5	6.1	N15 09.2	7.5	60.4
01	198 08.5	07.0	332 42.6	6.1	15 16.7	7.4	60.4
02	213 08.7	07.9	347 07.7	6.1	15 24.1	7.2	60.3
03	228 08.8	.. 08.9	1 32.8	6.0	15 31.3	7.2	60.3
04	243 09.0	09.8	15 57.8	6.1	15 38.5	7.0	60.3
05	258 09.2	10.8	30 22.9	6.1	15 45.5	6.9	60.3
F 06	273 09.4	S 6 11.7	44 48.0	6.0	N15 52.4	6.8	60.2
R 07	288 09.5	12.7	59 13.0	6.1	15 59.2	6.7	60.2
I 08	303 09.7	13.6	73 38.1	6.1	16 05.9	6.6	60.2
D 09	318 09.9	.. 14.6	88 03.2	6.0	16 12.5	6.4	60.1
A 10	333 10.1	15.5	102 28.2	6.1	16 18.9	6.4	60.1
Y 11	348 10.2	16.5	116 53.3	6.1	16 25.3	6.2	60.1
12	3 10.4	S 6 17.4	131 18.4	6.0	N16 31.5	6.1	60.0
13	18 10.6	18.4	145 43.4	6.1	16 37.6	6.0	60.0
14	33 10.7	19.3	160 08.5	6.1	16 43.6	5.9	60.0
15	48 10.9	.. 20.3	174 33.6	6.1	16 49.5	5.7	59.9
16	63 11.1	21.2	188 58.7	6.1	16 55.2	5.7	59.9
17	78 11.2	22.2	203 23.8	6.1	17 00.9	5.5	59.9
18	93 11.4	S 6 23.1	217 48.9	6.1	N17 06.4	5.4	59.9
19	108 11.6	24.1	232 14.0	6.1	17 11.8	5.2	59.8
20	123 11.8	25.0	246 39.1	6.1	17 17.0	5.2	59.8
21	138 11.9	.. 26.0	261 04.2	6.1	17 22.2	5.0	59.8
22	153 12.1	26.9	275 29.3	6.2	17 27.2	4.9	59.7
23	168 12.3	27.9	289 54.5	6.1	N17 32.1	4.8	59.7
	SD 16.0	d 1.0	SD 16.7		16.5		16.4

Lat.	Twilight Naut.	Twilight Civil	Sunrise	Moonrise 7	8	9	10
°	h m	h m	h m	h m	h m	h m	h m
N 72	04 22	05 42	06 50	17 19	17 09	16 51	▢
N 70	04 30	05 41	06 43	17 36	17 37	17 41	17 55
68	04 36	05 41	06 37	17 49	17 58	18 13	18 41
66	04 41	05 41	06 32	18 00	18 15	18 37	19 11
64	04 45	05 41	06 28	18 09	18 28	18 55	19 33
62	04 49	05 40	06 25	18 17	18 40	19 11	19 51
60	04 52	05 40	06 22	18 23	18 50	19 24	20 06
N 58	04 54	05 40	06 19	18 29	18 59	19 35	20 19
56	04 56	05 39	06 16	18 35	19 06	19 44	20 30
54	04 58	05 39	06 14	18 40	19 13	19 53	20 40
52	04 59	05 38	06 12	18 44	19 20	20 01	20 49
50	05 01	05 38	06 10	18 48	19 25	20 08	20 56
45	05 03	05 37	06 06	18 57	19 37	20 23	21 13
N 40	05 04	05 36	06 03	19 04	19 47	20 35	21 27
35	05 05	05 35	06 00	19 10	19 56	20 45	21 38
30	05 06	05 33	05 57	19 16	20 04	20 55	21 48
20	05 05	05 31	05 53	19 25	20 17	21 11	22 06
N 10	05 03	05 27	05 48	19 34	20 29	21 25	22 21
0	04 59	05 24	05 44	19 42	20 40	21 38	22 35
S 10	04 54	05 19	05 40	19 50	20 51	21 51	22 50
20	04 48	05 13	05 36	19 59	21 03	22 05	23 05
30	04 38	05 06	05 30	20 09	21 16	22 22	23 23
35	04 32	05 02	05 27	20 15	21 24	22 31	23 33
40	04 24	04 56	05 24	20 21	21 33	22 42	23 45
45	04 14	04 50	05 20	20 29	21 44	22 55	23 59
S 50	04 02	04 42	05 15	20 39	21 57	23 10	24 16
52	03 56	04 38	05 12	20 43	22 03	23 18	24 24
54	03 49	04 33	05 10	20 48	22 10	23 26	24 33
56	03 41	04 29	05 07	20 53	22 17	23 35	24 43
58	03 33	04 23	05 04	20 59	22 26	23 46	24 55
S 60	03 22	04 17	05 00	21 06	22 36	23 58	25 08

Lat.	Sunset	Twilight Civil	Twilight Naut.	Moonset 7	8	9	10
°	h m	h m	h m	h m	h m	h m	h m
N 72	16 44	17 51	19 10	09 12	11 23	13 43	▢
N 70	16 51	17 52	19 03	08 58	10 57	12 54	14 41
68	16 57	17 52	18 57	08 46	10 37	12 23	13 56
66	17 01	17 53	18 52	08 37	10 21	12 00	13 26
64	17 06	17 53	18 48	08 29	10 08	11 42	13 04
62	17 09	17 53	18 45	08 22	09 57	11 27	12 46
60	17 12	17 54	18 42	08 17	09 48	11 14	12 31
N 58	17 15	17 54	18 40	08 11	09 40	11 04	12 19
56	17 18	17 55	18 38	08 07	09 33	10 54	12 08
54	17 20	17 55	18 36	08 03	09 27	10 46	11 58
52	17 22	17 56	18 35	07 59	09 21	10 39	11 50
50	17 24	17 56	18 34	07 56	09 16	10 32	11 42
45	17 28	17 57	18 31	07 49	09 05	10 18	11 26
N 40	17 32	17 59	18 30	07 43	08 56	10 06	11 12
35	17 35	18 00	18 29	07 38	08 48	09 56	11 01
30	17 37	18 01	18 29	07 33	08 41	09 48	10 51
20	17 42	18 04	18 30	07 25	08 29	09 33	10 34
N 10	17 47	18 08	18 32	07 18	08 19	09 20	10 19
0	17 51	18 12	18 36	07 12	08 09	09 07	10 05
S 10	17 55	18 16	18 41	07 06	08 00	08 55	09 52
20	18 00	18 22	18 48	06 59	07 49	08 42	09 37
30	18 05	18 30	18 58	06 51	07 38	08 27	09 20
35	18 09	18 34	19 04	06 46	07 31	08 19	09 10
40	18 12	18 40	19 12	06 41	07 23	08 09	08 58
45	18 16	18 46	19 22	06 36	07 14	07 57	08 45
S 50	18 22	18 55	19 35	06 29	07 04	07 43	08 29
52	18 24	18 59	19 41	06 25	06 59	07 37	08 21
54	18 27	19 03	19 48	06 21	06 53	07 30	08 13
56	18 30	19 08	19 56	06 18	06 47	07 22	08 03
58	18 33	19 14	20 05	06 14	06 41	07 13	07 53
S 60	18 36	19 20	20 15	06 09	06 33	07 03	07 40

	SUN Eqn. of Time 00h	SUN Eqn. of Time 12h	SUN Mer. Pass.	MOON Mer. Pass. Upper	MOON Mer. Pass. Lower	Age	Phase
Day							
d	m s	m s	h m	h m	h m	d	%
7	11 59	12 07	11 48	00 58	13 27	17	96
8	12 16	12 25	11 48	01 56	14 25	18	89
9	12 33	12 41	11 47	02 54	15 23	19	81

1998 OCTOBER 10, 11, 12 (SAT., SUN., MON.)

UT	ARIES GHA	VENUS −3.9 GHA	VENUS Dec	MARS +1.7 GHA	MARS Dec	JUPITER −2.8 GHA	JUPITER Dec	SATURN +0.0 GHA	SATURN Dec
10 00	18 23.7	187 29.7	S 3 12.2	224 22.1	N12 11.7	26 53.0	S 5 20.4	348 19.4	N 9 19.0
01	33 26.2	202 29.3	13.4	239 23.2	11.2	41 55.7	20.5	3 22.1	18.9
02	48 28.6	217 28.9	14.7	254 24.2	10.7	56 58.4	20.6	18 24.7	18.9
03	63 31.1	232 28.5 ..	15.9	269 25.2 ..	10.1	72 01.1 ..	20.7	33 27.4 ..	18.8
04	78 33.6	247 28.1	17.2	284 26.2	09.6	87 03.8	20.8	48 30.0	18.7
05	93 36.0	262 27.7	18.4	299 27.2	09.1	102 06.5	20.9	63 32.7	18.7
06	108 38.5	277 27.3	S 3 19.7	314 28.2	N12 08.5	117 09.2	S 5 21.0	78 35.3	N 9 18.6
07	123 41.0	292 26.9	20.9	329 29.2	08.0	132 12.0	21.1	93 38.0	18.5
08	138 43.4	307 26.5	22.2	344 30.2	07.5	147 14.7	21.2	108 40.6	18.4
09	153 45.9	322 26.1 ..	23.4	359 31.2 ..	07.0	162 17.4 ..	21.3	123 43.2 ..	18.4
10	168 48.4	337 25.7	24.7	14 32.2	06.4	177 20.1	21.4	138 45.9	18.3
11	183 50.8	352 25.3	25.9	29 33.2	05.9	192 22.8	21.5	153 48.5	18.2
12	198 53.3	7 24.9	S 3 27.1	44 34.2	N12 05.4	207 25.5	S 5 21.6	168 51.2	N 9 18.2
13	213 55.7	22 24.5	28.4	59 35.2	04.8	222 28.2	21.7	183 53.8	18.1
14	228 58.2	37 24.1	29.6	74 36.2	04.3	237 30.9	21.8	198 56.5	18.0
15	244 00.7	52 23.7 ..	30.9	89 37.2 ..	03.8	252 33.6 ..	21.9	213 59.1 ..	18.0
16	259 03.1	67 23.3	32.1	104 38.2	03.2	267 36.3	22.0	229 01.7	17.9
17	274 05.6	82 22.9	33.4	119 39.2	02.7	282 39.0	22.1	244 04.4	17.8
18	289 08.1	97 22.5	S 3 34.6	134 40.2	N12 02.2	297 41.7	S 5 22.1	259 07.0	N 9 17.8
19	304 10.5	112 22.1	35.9	149 41.2	01.6	312 44.4	22.2	274 09.7	17.7
20	319 13.0	127 21.7	37.1	164 42.2	01.1	327 47.1	22.3	289 12.3	17.6
21	334 15.5	142 21.3 ..	38.4	179 43.2 ..	00.6	342 49.8 ..	22.4	304 15.0 ..	17.6
22	349 17.9	157 20.9	39.6	194 44.2	12 00.1	357 52.5	22.5	319 17.6	17.5
23	4 20.4	172 20.4	40.9	209 45.2	11 59.5	12 55.3	22.6	334 20.3	17.4
11 00	19 22.9	187 20.0	S 3 42.1	224 46.2	N11 59.0	27 58.0	S 5 22.7	349 22.9	N 9 17.4
01	34 25.3	202 19.6	43.4	239 47.2	58.5	43 00.7	22.8	4 25.5	17.3
02	49 27.8	217 19.2	44.6	254 48.2	57.9	58 03.4	22.9	19 28.2	17.2
03	64 30.2	232 18.8 ..	45.9	269 49.2 ..	57.4	73 06.1 ..	23.0	34 30.8 ..	17.2
04	79 32.7	247 18.4	47.1	284 50.2	56.9	88 08.8	23.1	49 33.5	17.1
05	94 35.2	262 18.0	48.3	299 51.2	56.3	103 11.5	23.2	64 36.1	17.0
06	109 37.6	277 17.6	S 3 49.6	314 52.2	N11 55.8	118 14.2	S 5 23.3	79 38.8	N 9 17.0
07	124 40.1	292 17.2	50.8	329 53.2	55.3	133 16.9	23.4	94 41.4	16.9
08	139 42.6	307 16.8	52.1	344 54.3	54.7	148 19.6	23.5	109 44.1	16.8
09	154 45.0	322 16.4 ..	53.3	359 55.3 ..	54.2	163 22.3 ..	23.6	124 46.7 ..	16.7
10	169 47.5	337 16.0	54.6	14 56.3	53.7	178 25.0	23.7	139 49.3	16.7
11	184 50.0	352 15.6	55.8	29 57.3	53.1	193 27.7	23.8	154 52.0	16.6
12	199 52.4	7 15.1	S 3 57.1	44 58.3	N11 52.6	208 30.4	S 5 23.9	169 54.6	N 9 16.5
13	214 54.9	22 14.7	58.3	59 59.3	52.1	223 33.1	24.0	184 57.3	16.5
14	229 57.4	37 14.3	3 59.6	75 00.3	51.5	238 35.8	24.1	199 59.9	16.4
15	244 59.8	52 13.9	4 00.8	90 01.3 ..	51.0	253 38.5 ..	24.2	215 02.6 ..	16.3
16	260 02.3	67 13.5	02.0	105 02.3	50.5	268 41.2	24.3	230 05.2	16.3
17	275 04.7	82 13.1	03.3	120 03.3	49.9	283 43.9	24.4	245 07.9	16.2
18	290 07.2	97 12.7	S 4 04.5	135 04.3	N11 49.4	298 46.6	S 5 24.4	260 10.5	N 9 16.1
19	305 09.7	112 12.3	05.8	150 05.3	48.9	313 49.3	24.5	275 13.2	16.1
20	320 12.1	127 11.9	07.0	165 06.3	48.3	328 52.0	24.6	290 15.8	16.0
21	335 14.6	142 11.5 ..	08.3	180 07.3 ..	47.8	343 54.7 ..	24.7	305 18.4 ..	15.9
22	350 17.1	157 11.1	09.5	195 08.3	47.3	358 57.4	24.8	320 21.1	15.9
23	5 19.5	172 10.6	10.8	210 09.4	46.7	14 00.1	24.9	335 23.7	15.8
12 00	20 22.0	187 10.2	S 4 12.0	225 10.4	N11 46.2	29 02.8	S 5 25.0	350 26.4	N 9 15.7
01	35 24.5	202 09.8	13.2	240 11.4	45.7	44 05.5	25.1	5 29.0	15.7
02	50 26.9	217 09.4	14.5	255 12.4	45.1	59 08.2	25.2	20 31.7	15.6
03	65 29.4	232 09.0 ..	15.7	270 13.4 ..	44.6	74 10.9 ..	25.3	35 34.3 ..	15.5
04	80 31.8	247 08.6	17.0	285 14.4	44.1	89 13.6	25.4	50 37.0	15.4
05	95 34.3	262 08.2	18.2	300 15.4	43.5	104 16.3	25.5	65 39.6	15.4
06	110 36.8	277 07.8	S 4 19.5	315 16.4	N11 43.0	119 19.0	S 5 25.6	80 42.3	N 9 15.3
07	125 39.2	292 07.4	20.7	330 17.4	42.5	134 21.7	25.7	95 44.9	15.2
08	140 41.7	307 07.0	21.9	345 18.4	41.9	149 24.4	25.8	110 47.5	15.2
09	155 44.2	322 06.5 ..	23.2	0 19.4 ..	41.4	164 27.1 ..	25.9	125 50.2 ..	15.1
10	170 46.6	337 06.1	24.4	15 20.5	40.9	179 29.8	25.9	140 52.8	15.0
11	185 49.1	352 05.7	25.7	30 21.5	40.3	194 32.5	26.0	155 55.5	15.0
12	200 51.6	7 05.3	S 4 26.9	45 22.5	N11 39.8	209 35.2	S 5 26.1	170 58.1	N 9 14.9
13	215 54.0	22 04.9	28.2	60 23.5	39.3	224 37.9	26.2	186 00.8	14.8
14	230 56.5	37 04.5	29.4	75 24.5	38.7	239 40.6	26.3	201 03.4	14.8
15	245 59.0	52 04.1 ..	30.6	90 25.5 ..	38.2	254 43.3 ..	26.4	216 06.1 ..	14.7
16	261 01.4	67 03.7	31.9	105 26.5	37.7	269 46.0	26.5	231 08.7	14.6
17	276 03.9	82 03.2	33.1	120 27.5	37.1	284 48.6	26.6	246 11.4	14.6
18	291 06.3	97 02.8	S 4 34.4	135 28.5	N11 36.6	299 51.3	S 5 26.7	261 14.0	N 9 14.5
19	306 08.8	112 02.4	35.6	150 29.6	36.1	314 54.0	26.8	276 16.7	14.4
20	321 11.3	127 02.0	36.9	165 30.6	35.5	329 56.7	26.9	291 19.3	14.4
21	336 13.7	142 01.6 ..	38.1	180 31.6 ..	35.0	344 59.4 ..	27.0	306 22.0 ..	14.3
22	351 16.2	157 01.2	39.3	195 32.6	34.5	0 02.1	27.1	321 24.6	14.2
23	6 18.7	172 00.8	40.6	210 33.6	33.9	15 04.8	27.1	336 27.2	14.1
Mer. Pass. 22 38.8		v −0.4	d 1.2	v 1.0	d 0.5	v 2.7	d 0.1	v 2.6	d 0.1

STARS

Name	SHA	Dec
Acamar	315 26.6	S40 18.5
Achernar	335 34.7	S57 14.6
Acrux	173 22.6	S63 05.3
Adhara	255 21.4	S28 58.1
Aldebaran	291 02.2	N16 30.3
Alioth	166 31.2	N55 58.1
Alkaid	153 08.3	N49 19.4
Al Na'ir	27 57.6	S46 58.1
Alnilam	275 57.7	S 1 12.2
Alphard	218 07.4	S 8 39.1
Alphecca	126 20.9	N26 43.5
Alpheratz	357 54.9	N29 05.1
Altair	62 19.2	N 8 52.2
Ankaa	353 26.5	S42 18.8
Antares	112 40.4	S26 25.6
Arcturus	146 06.4	N19 11.6
Atria	107 52.8	S69 01.6
Avior	234 22.8	S59 30.1
Bellatrix	278 44.0	N 6 20.9
Betelgeuse	271 13.5	N 7 24.4
Canopus	264 01.0	S52 41.5
Capella	280 51.0	N45 59.6
Deneb	49 39.0	N45 16.9
Denebola	182 45.5	N14 34.9
Diphda	349 06.9	S17 59.6
Dubhe	194 06.2	N61 45.4
Elnath	278 26.8	N28 36.2
Eltanin	90 51.6	N51 29.8
Enif	33 58.1	N 9 52.3
Fomalhaut	15 36.2	S29 37.7
Gacrux	172 14.1	S57 06.2
Gienah	176 04.3	S17 31.9
Hadar	149 04.6	S60 21.9
Hamal	328 13.2	N23 27.4
Kaus Aust.	83 58.9	S34 23.1
Kochab	137 20.8	N74 09.9
Markab	13 49.3	N15 12.1
Menkar	314 26.6	N 4 05.1
Menkent	148 21.3	S36 21.7
Miaplacidus	221 42.4	S69 42.5
Mirfak	308 56.2	N49 51.2
Nunki	76 12.4	S26 17.8
Peacock	53 36.9	S56 44.5
Pollux	243 41.6	N28 01.6
Procyon	245 11.6	N 5 13.7
Rasalhague	96 17.1	N12 34.0
Regulus	207 55.8	N11 58.5
Rigel	281 22.8	S 8 12.2
Rigil Kent.	140 07.9	S60 49.7
Sabik	102 25.7	S15 43.2
Schedar	349 52.9	N56 31.9
Shaula	96 37.5	S37 06.1
Sirius	258 43.7	S16 42.7
Spica	158 43.5	S11 09.1
Suhail	223 01.0	S43 25.4
Vega	80 46.7	N38 47.4
Zuben'ubi	137 18.3	S16 02.0

	SHA	Mer. Pass.
Venus	167 57.2	11 31
Mars	205 23.4	9 00
Jupiter	8 35.1	22 04
Saturn	330 00.0	0 42

UT	SUN GHA	SUN Dec	MOON GHA	v	MOON Dec	d	HP
	° ′	° ′	° ′	′	° ′	′	′
10 SATURDAY							
00	183 12.4	S 6 28.8	304 19.6	6.2	N17 36.9	4.7	59.6
01	198 12.6	29.8	318 44.8	6.1	17 41.6	4.5	59.6
02	213 12.8	30.7	333 09.9	6.2	17 46.1	4.4	59.6
03	228 12.9	.. 31.7	347 35.1	6.2	17 50.5	4.3	59.5
04	243 13.1	32.6	2 00.3	6.3	17 54.8	4.2	59.5
05	258 13.3	33.6	16 25.6	6.2	17 59.0	4.0	59.5
06	273 13.4	S 6 34.5	30 50.8	6.3	N18 03.0	3.9	59.4
07	288 13.6	35.4	45 16.1	6.2	18 06.9	3.8	59.4
08	303 13.8	36.4	59 41.3	6.3	18 10.7	3.7	59.4
09	318 13.9	.. 37.3	74 06.6	6.4	18 14.4	3.6	59.3
10	333 14.1	38.3	88 32.0	6.3	18 18.0	3.4	59.3
11	348 14.3	39.2	102 57.3	6.4	18 21.4	3.3	59.3
12	3 14.4	S 6 40.2	117 22.7	6.4	N18 24.7	3.2	59.2
13	18 14.6	41.1	131 48.1	6.4	18 27.9	3.0	59.2
14	33 14.8	42.1	146 13.5	6.4	18 30.9	3.0	59.2
15	48 14.9	.. 43.0	160 38.9	6.5	18 33.9	2.8	59.1
16	63 15.1	44.0	175 04.4	6.5	18 36.7	2.7	59.1
17	78 15.3	44.9	189 29.9	6.5	18 39.4	2.5	59.0
18	93 15.4	S 6 45.9	203 55.4	6.5	N18 41.9	2.5	59.0
19	108 15.6	46.8	218 20.9	6.6	18 44.4	2.3	59.0
20	123 15.8	47.7	232 46.5	6.6	18 46.7	2.2	58.9
21	138 15.9	.. 48.7	247 12.1	6.7	18 48.9	2.0	58.9
22	153 16.1	49.6	261 37.8	6.6	18 50.9	2.0	58.9
23	168 16.3	50.6	276 03.4	6.7	18 52.9	1.8	58.8
11 SUNDAY							
00	183 16.4	S 6 51.5	290 29.1	6.8	N18 54.7	1.7	58.8
01	198 16.6	52.5	304 54.9	6.8	18 56.4	1.6	58.8
02	213 16.8	53.4	319 20.7	6.8	18 58.0	1.4	58.7
03	228 16.9	.. 54.4	333 46.5	6.8	18 59.4	1.4	58.7
04	243 17.1	55.3	348 12.3	6.9	19 00.8	1.2	58.6
05	258 17.2	56.2	2 38.2	6.9	19 02.0	1.1	58.6
06	273 17.4	S 6 57.2	17 04.1	7.0	N19 03.1	0.9	58.6
07	288 17.6	58.1	31 30.1	7.0	19 04.0	0.9	58.5
08	303 17.7	6 59.1	45 56.1	7.1	19 04.9	0.7	58.5
09	318 17.9	7 00.0	60 22.2	7.0	19 05.6	0.6	58.5
10	333 18.1	01.0	74 48.2	7.2	19 06.2	0.5	58.4
11	348 18.2	01.9	89 14.4	7.2	19 06.7	0.4	58.4
12	3 18.4	S 7 02.8	103 40.6	7.2	N19 07.1	0.2	58.4
13	18 18.5	03.8	118 06.8	7.2	19 07.3	0.2	58.3
14	33 18.7	04.7	132 33.0	7.4	19 07.5	0.0	58.3
15	48 18.9	.. 05.7	146 59.4	7.3	19 07.5	0.1	58.2
16	63 19.0	06.6	161 25.7	7.4	19 07.4	0.2	58.2
17	78 19.2	07.6	175 52.1	7.5	19 07.2	0.4	58.2
18	93 19.3	S 7 08.5	190 18.6	7.5	N19 06.8	0.6	58.1
19	108 19.5	09.4	204 45.1	7.5	19 06.4	0.6	58.1
20	123 19.7	10.4	219 11.6	7.6	19 05.8	0.7	58.1
21	138 19.8	.. 11.3	233 38.2	7.7	19 05.1	0.8	58.0
22	153 20.0	12.3	248 04.9	7.7	19 04.3	0.9	58.0
23	168 20.1	13.2	262 31.6	7.8	19 03.4	1.0	58.0
12 MONDAY							
00	183 20.3	S 7 14.1	276 58.4	7.8	N19 02.4	1.1	57.9
01	198 20.5	15.1	291 25.2	7.8	19 01.3	1.3	57.9
02	213 20.6	16.0	305 52.0	7.9	19 00.0	1.3	57.8
03	228 20.8	.. 17.0	320 18.9	8.0	18 58.7	1.5	57.8
04	243 20.9	17.9	334 45.9	8.0	18 57.2	1.6	57.8
05	258 21.1	18.8	349 12.9	8.1	18 55.6	1.7	57.7
06	273 21.2	S 7 19.8	3 40.0	8.2	N18 53.9	1.8	57.7
07	288 21.4	20.7	18 07.2	8.2	18 52.1	1.9	57.7
08	303 21.6	21.7	32 34.4	8.2	18 50.2	2.0	57.6
09	318 21.7	.. 22.6	47 01.6	8.3	18 48.2	2.1	57.6
10	333 21.9	23.5	61 28.9	8.4	18 46.1	2.3	57.6
11	348 22.0	24.5	75 56.3	8.4	18 43.8	2.3	57.5
12	3 22.2	S 7 25.4	90 23.7	8.5	N18 41.5	2.4	57.5
13	18 22.3	26.4	104 51.2	8.5	18 39.1	2.6	57.5
14	33 22.5	27.3	119 18.7	8.6	18 36.5	2.6	57.4
15	48 22.7	.. 28.2	133 46.3	8.7	18 33.9	2.8	57.4
16	63 22.8	29.2	148 14.0	8.7	18 31.1	2.9	57.4
17	78 23.0	30.1	162 41.7	8.8	18 28.2	2.9	57.3
18	93 23.1	S 7 31.0	177 09.5	8.9	N18 25.3	3.1	57.3
19	108 23.3	32.0	191 37.4	8.9	18 22.2	3.1	57.3
20	123 23.4	32.9	206 05.3	8.9	18 19.1	3.3	57.2
21	138 23.6	.. 33.9	220 33.2	9.1	18 15.8	3.4	57.2
22	153 23.7	34.8	235 01.3	9.1	18 12.4	3.4	57.1
23	168 23.9	35.7	249 29.4	9.1	N18 09.0	3.6	57.1
SD 16.0	d 0.9		SD 16.1	15.9	15.7		

Lat.	Naut.	Civil	Sunrise	Moonrise 10	11	12	13
°	h m	h m	h m	h m	h m	h m	h m
N 72	04 36	05 55	07 04	▭	▭	▭	20 48
N 70	04 42	05 53	06 55	17 55	18 37	19 57	21 35
68	04 47	05 52	06 48	18 41	19 29	20 40	22 05
66	04 51	05 50	06 42	19 11	20 01	21 09	22 27
64	04 54	05 49	06 37	19 33	20 25	21 31	22 45
62	04 57	05 48	06 33	19 51	20 44	21 48	22 59
60	04 59	05 47	06 29	20 06	21 00	22 02	23 12
N 58	05 01	05 46	06 25	20 19	21 13	22 15	23 22
56	05 02	05 45	06 22	20 30	21 24	22 25	23 31
54	05 03	05 44	06 20	20 40	21 34	22 35	23 39
52	05 04	05 43	06 17	20 49	21 43	22 43	23 47
50	05 05	05 43	06 15	20 56	21 51	22 51	23 53
45	05 07	05 41	06 10	21 13	22 08	23 07	24 07
N 40	05 07	05 39	06 06	21 27	22 22	23 20	24 19
35	05 08	05 37	06 02	21 38	22 34	23 31	24 29
30	05 07	05 35	05 59	21 48	22 44	23 41	24 37
20	05 06	05 31	05 53	22 06	23 02	23 57	24 52
N 10	05 03	05 27	05 48	22 21	23 17	24 12	00 12
0	04 58	05 23	05 43	22 35	23 32	24 26	00 26
S 10	04 53	05 17	05 39	22 50	23 46	24 39	00 39
20	04 45	05 11	05 33	23 05	24 02	00 02	00 54
30	04 34	05 03	05 27	23 23	24 19	00 19	01 11
35	04 27	04 57	05 23	23 33	24 30	00 30	01 20
40	04 19	04 51	05 19	23 45	24 42	00 42	01 31
45	04 08	04 44	05 14	23 59	24 56	00 56	01 44
S 50	03 55	04 35	05 08	24 16	00 16	01 13	02 00
52	03 48	04 31	05 05	24 24	00 24	01 21	02 08
54	03 41	04 26	05 02	24 33	00 33	01 30	02 16
56	03 32	04 20	04 59	24 43	00 43	01 40	02 25
58	03 22	04 14	04 55	24 55	00 55	01 51	02 36
S 60	03 11	04 07	04 51	25 08	01 08	02 05	02 48

Lat.	Sunset	Civil	Naut.	Moonset 10	11	12	13
°	h m	h m	h m	h m	h m	h m	h m
N 72	16 28	17 37	18 55	▭	▭	▭	17 30
N 70	16 37	17 38	18 49	14 41	15 58	16 31	16 42
68	16 44	17 40	18 44	13 56	15 06	15 48	16 11
66	16 50	17 42	18 41	13 26	14 33	15 19	15 48
64	16 55	17 43	18 38	13 04	14 09	14 57	15 30
62	17 00	17 44	18 35	12 46	13 51	14 40	15 15
60	17 04	17 45	18 33	12 31	13 35	14 25	15 02
N 58	17 07	17 46	18 32	12 19	13 22	14 12	14 52
56	17 10	17 47	18 30	12 08	13 10	14 01	14 42
54	17 13	17 48	18 29	11 58	13 00	13 52	14 34
52	17 15	17 49	18 28	11 50	12 52	13 43	14 26
50	17 18	17 50	18 27	11 42	12 44	13 36	14 19
45	17 23	17 52	18 26	11 26	12 27	13 19	14 04
N 40	17 27	17 54	18 25	11 12	12 13	13 06	13 52
35	17 31	17 56	18 25	11 01	12 01	12 54	13 42
30	17 34	17 58	18 26	10 51	11 50	12 44	13 33
20	17 40	18 02	18 28	10 34	11 33	12 27	13 17
N 10	17 45	18 06	18 31	10 19	11 17	12 12	13 03
0	17 50	18 11	18 35	10 05	11 03	11 58	12 51
S 10	17 55	18 16	18 41	09 52	10 48	11 44	12 38
20	18 01	18 23	18 49	09 37	10 32	11 28	12 24
30	18 07	18 31	19 00	09 20	10 14	11 11	12 08
35	18 11	18 37	19 07	09 10	10 04	11 01	11 58
40	18 15	18 43	19 16	08 58	09 52	10 49	11 48
45	18 20	18 50	19 27	08 45	09 38	10 35	11 35
S 50	18 26	19 00	19 40	08 29	09 21	10 18	11 20
52	18 29	19 04	19 47	08 21	09 13	10 10	11 13
54	18 32	19 09	19 55	08 13	09 04	10 02	11 05
56	18 36	19 15	20 03	08 03	08 54	09 52	10 56
58	18 40	19 21	20 13	07 53	08 42	09 40	10 46
S 60	18 44	19 28	20 25	07 40	08 29	09 27	10 34

Day	SUN Eqn. of Time 00h	12h	Mer. Pass.	MOON Mer. Pass. Upper	Lower	Age	Phase
d	m s	m s	h m	h m	h m	d	%
10	12 49	12 57	11 47	03 52	16 20	20	71
11	13 05	13 13	11 47	04 49	17 17	21	61
12	13 21	13 28	11 47	05 45	18 12	22	50

UT	ARIES GHA	VENUS −3.9 GHA	Dec	MARS +1.7 GHA	Dec	JUPITER −2.8 GHA	Dec	SATURN +0.0 GHA	Dec
13 00	21 21.1	187 00.3	S 4 41.8	225 34.6	N11 33.4	30 07.5	S 5 27.2	351 29.9	N 9 14.1
01	36 23.6	201 59.9	43.1	240 35.6	32.9	45 10.2	27.3	6 32.5	14.0
02	51 26.1	216 59.5	44.3	255 36.6	32.3	60 12.9	27.4	21 35.2	13.9
03	66 28.5	231 59.1 ..	45.5	270 37.7 ..	31.8	75 15.6 ..	27.5	36 37.8 ..	13.9
04	81 31.0	246 58.7	46.8	285 38.7	31.2	90 18.3	27.6	51 40.5	13.8
05	96 33.5	261 58.3	48.0	300 39.7	30.7	105 21.0	27.7	66 43.1	13.7
T 06	111 35.9	276 57.9	S 4 49.3	315 40.7	N11 30.2	120 23.7	S 5 27.8	81 45.8	N 9 13.7
U 07	126 38.4	291 57.4	50.5	330 41.7	29.6	135 26.4	27.9	96 48.4	13.6
E 08	141 40.8	306 57.0	51.7	345 42.7	29.1	150 29.1	28.0	111 51.1	13.5
S 09	156 43.3	321 56.6 ..	53.0	0 43.7 ..	28.6	165 31.7 ..	28.1	126 53.7 ..	13.5
D 10	171 45.8	336 56.2	54.2	15 44.7	28.0	180 34.4	28.1	141 56.4	13.4
A 11	186 48.2	351 55.8	55.4	30 45.8	27.5	195 37.1	28.2	156 59.0	13.3
Y 12	201 50.7	6 55.4	S 4 56.7	45 46.8	N11 27.0	210 39.8	S 5 28.3	172 01.7	N 9 13.3
13	216 53.2	21 54.9	57.9	60 47.8	26.4	225 42.5	28.4	187 04.3	13.2
14	231 55.6	36 54.5	4 59.2	75 48.8	25.9	240 45.2	28.5	202 07.0	13.1
15	246 58.1	51 54.1	5 00.4	90 49.8 ..	25.3	255 47.9 ..	28.6	217 09.6 ..	13.0
16	262 00.6	66 53.7	01.6	105 50.8	24.8	270 50.6	28.7	232 12.3	13.0
17	277 03.0	81 53.3	02.9	120 51.8	24.3	285 53.3	28.8	247 14.9	12.9
18	292 05.5	96 52.8	S 5 04.1	135 52.9	N11 23.7	300 56.0	S 5 28.9	262 17.5	N 9 12.8
19	307 07.9	111 52.4	05.4	150 53.9	23.2	315 58.6	29.0	277 20.2	12.8
20	322 10.4	126 52.0	06.6	165 54.9	22.7	331 01.3	29.0	292 22.8	12.7
21	337 12.9	141 51.6 ..	07.8	180 55.9 ..	22.1	346 04.0 ..	29.1	307 25.5 ..	12.6
22	352 15.3	156 51.2	09.1	195 56.9	21.6	1 06.7	29.2	322 28.1	12.6
23	7 17.8	171 50.8	10.3	210 57.9	21.1	16 09.4	29.3	337 30.8	12.5
14 00	22 20.3	186 50.3	S 5 11.5	225 59.0	N11 20.5	31 12.1	S 5 29.4	352 33.4	N 9 12.4
01	37 22.7	201 49.9	12.8	241 00.0	20.0	46 14.8	29.5	7 36.1	12.4
02	52 25.2	216 49.5	14.0	256 01.0	19.4	61 17.5	29.6	22 38.7	12.3
03	67 27.7	231 49.1 ..	15.2	271 02.0 ..	18.9	76 20.2 ..	29.7	37 41.4 ..	12.2
04	82 30.1	246 48.7	16.5	286 03.0	18.4	91 22.8	29.7	52 44.0	12.2
05	97 32.6	261 48.2	17.7	301 04.0	17.8	106 25.5	29.8	67 46.7	12.1
W 06	112 35.1	276 47.8	S 5 19.0	316 05.1	N11 17.3	121 28.2	S 5 29.9	82 49.3	N 9 12.0
E 07	127 37.5	291 47.4	20.2	331 06.1	16.8	136 30.9	30.0	97 52.0	11.9
D 08	142 40.0	306 47.0	21.4	346 07.1	16.2	151 33.6	30.1	112 54.6	11.9
N 09	157 42.4	321 46.6 ..	22.7	1 08.1 ..	15.7	166 36.3 ..	30.2	127 57.3 ..	11.8
E 10	172 44.9	336 46.1	23.9	16 09.1	15.1	181 39.0	30.3	142 59.9	11.7
S 11	187 47.4	351 45.7	25.1	31 10.1	14.6	196 41.6	30.4	158 02.6	11.7
D 12	202 49.8	6 45.3	S 5 26.4	46 11.2	N11 14.1	211 44.3	S 5 30.5	173 05.2	N 9 11.6
A 13	217 52.3	21 44.9	27.6	61 12.2	13.5	226 47.0	30.5	188 07.9	11.5
Y 14	232 54.8	36 44.4	28.8	76 13.2	13.0	241 49.7	30.6	203 10.5	11.5
15	247 57.2	51 44.0 ..	30.1	91 14.2 ..	12.5	256 52.4 ..	30.7	218 13.2 ..	11.4
16	262 59.7	66 43.6	31.3	106 15.2	11.9	271 55.1	30.8	233 15.8	11.3
17	278 02.2	81 43.2	32.5	121 16.3	11.4	286 57.8	30.9	248 18.5	11.3
18	293 04.6	96 42.8	S 5 33.8	136 17.3	N11 10.8	302 00.4	S 5 31.0	263 21.1	N 9 11.2
19	308 07.1	111 42.3	35.0	151 18.3	10.3	317 03.1	31.1	278 23.8	11.1
20	323 09.6	126 41.9	36.2	166 19.3	09.8	332 05.8	31.1	293 26.4	11.0
21	338 12.0	141 41.5 ..	37.5	181 20.3 ..	09.2	347 08.5 ..	31.2	308 29.1 ..	11.0
22	353 14.5	156 41.1	38.7	196 21.4	08.7	2 11.2	31.3	323 31.7	10.9
23	8 16.9	171 40.6	39.9	211 22.4	08.1	17 13.9	31.4	338 34.4	10.8
15 00	23 19.4	186 40.2	S 5 41.2	226 23.4	N11 07.6	32 16.5	S 5 31.5	353 37.0	N 9 10.8
01	38 21.9	201 39.8	42.4	241 24.4	07.0	47 19.2	31.6	8 39.7	10.7
02	53 24.3	216 39.4	43.6	256 25.4	06.5	62 21.9	31.7	23 42.3	10.6
03	68 26.8	231 38.9 ..	44.9	271 26.5 ..	06.0	77 24.6 ..	31.7	38 45.0 ..	10.6
04	83 29.3	246 38.5	46.1	286 27.5	05.5	92 27.3	31.8	53 47.6	10.5
05	98 31.7	261 38.1	47.3	301 28.5	04.9	107 29.9	31.9	68 50.3	10.4
T 06	113 34.2	276 37.7	S 5 48.6	316 29.5	N11 04.4	122 32.6	S 5 32.0	83 52.9	N 9 10.4
H 07	128 36.7	291 37.2	49.8	331 30.5	03.8	137 35.3	32.1	98 55.6	10.3
U 08	143 39.1	306 36.8	51.0	346 31.6	03.3	152 38.0	32.2	113 58.2	10.2
R 09	158 41.6	321 36.4 ..	52.2	1 32.6 ..	02.8	167 40.7 ..	32.3	129 00.9 ..	10.1
S 10	173 44.0	336 36.0	53.5	16 33.6	02.2	182 43.3	32.3	144 03.5	10.1
D 11	188 46.5	351 35.5	54.7	31 34.6	01.7	197 46.0	32.4	159 06.1	10.0
A 12	203 49.0	6 35.1	S 5 55.9	46 35.6	N11 01.1	212 48.7	S 5 32.5	174 08.8	N 9 09.9
Y 13	218 51.4	21 34.7	57.2	61 36.7	00.6	227 51.4	32.6	189 11.4	09.9
14	233 53.9	36 34.3	58.4	76 37.7	11 00.1	242 54.1	32.7	204 14.1	09.8
15	248 56.4	51 33.8	5 59.6	91 38.7	10 59.5	257 56.7 ..	32.8	219 16.7 ..	09.7
16	263 58.8	66 33.4	6 00.9	106 39.7	59.0	272 59.4	32.9	234 19.4	09.7
17	279 01.3	81 33.0	02.1	121 40.8	58.4	288 02.1	32.9	249 22.0	09.6
18	294 03.8	96 32.5	S 6 03.3	136 41.8	N10 57.9	303 04.8	S 5 33.0	264 24.7	N 9 09.5
19	309 06.2	111 32.1	04.5	151 42.8	57.4	318 07.5	33.1	279 27.3	09.5
20	324 08.7	126 31.7	05.8	166 43.8	56.8	333 10.1	33.2	294 30.0	09.4
21	339 11.2	141 31.3 ..	07.0	181 44.8 ..	56.3	348 12.8 ..	33.3	309 32.6 ..	09.3
22	354 13.6	156 30.8	08.2	196 45.9	55.7	3 15.5	33.4	324 35.3	09.2
23	9 16.1	171 30.4	09.5	211 46.9	55.2	18 18.2	33.4	339 37.9	09.2
Mer. Pass. 22 27.0		v −0.4	d 1.2	v 1.0	d 0.5	v 2.7	d 0.1	v 2.6	d 0.1

STARS

Name	SHA	Dec
Acamar	315 26.6	S40 18.5
Achernar	335 34.7	S57 14.6
Acrux	173 22.6	S63 05.3
Adhara	255 21.4	S28 58.1
Aldebaran	291 02.2	N16 30.3
Alioth	166 31.2	N55 58.1
Alkaid	153 08.3	N49 19.4
Al Na'ir	27 57.7	S46 58.1
Alnilam	275 57.7	S 1 12.2
Alphard	218 07.4	S 8 39.1
Alphecca	126 20.9	N26 43.4
Alpheratz	357 54.9	N29 05.1
Altair	62 19.3	N 8 52.2
Ankaa	353 26.5	S42 18.8
Antares	112 40.4	S26 25.6
Arcturus	146 06.4	N19 11.6
Atria	107 52.8	S69 01.6
Avior	234 22.7	S59 30.1
Bellatrix	278 44.0	N 6 20.9
Betelgeuse	271 13.4	N 7 24.4
Canopus	264 01.0	S52 41.5
Capella	280 50.9	N45 59.6
Deneb	49 39.0	N45 16.9
Denebola	182 45.5	N14 34.9
Diphda	349 06.9	S17 59.6
Dubhe	194 06.1	N61 45.4
Elnath	278 26.8	N28 36.2
Eltanin	90 51.6	N51 29.8
Enif	33 58.1	N 9 52.3
Fomalhaut	15 36.2	S29 37.7
Gacrux	172 14.1	S57 06.2
Gienah	176 04.3	S17 31.9
Hadar	149 04.6	S60 21.9
Hamal	328 13.2	N23 27.4
Kaus Aust.	83 59.0	S34 23.1
Kochab	137 20.8	N74 09.9
Markab	13 49.3	N15 12.1
Menkar	314 26.6	N 4 05.1
Menkent	148 21.3	S36 21.6
Miaplacidus	221 42.4	S69 42.4
Mirfak	308 56.2	N49 51.2
Nunki	76 12.4	S26 17.8
Peacock	53 36.9	S56 44.5
Pollux	243 41.6	N28 01.6
Procyon	245 11.6	N 5 13.7
Rasalhague	96 17.1	N12 34.0
Regulus	207 55.8	N11 58.5
Rigel	281 22.8	S 8 12.2
Rigil Kent.	140 07.9	S60 49.7
Sabik	102 25.7	S15 43.2
Schedar	349 52.9	N56 31.9
Shaula	96 37.5	S37 06.1
Sirius	258 43.6	S16 42.7
Spica	158 43.5	S11 09.1
Suhail	223 00.9	S43 25.4
Vega	80 46.7	N38 47.4
Zuben'ubi	137 18.3	S16 02.0

	SHA	Mer. Pass.
	° ′	h m
Venus	164 30.1	11 33
Mars	203 38.7	8 55
Jupiter	8 51.8	21 51
Saturn	330 13.2	0 30

UT	SUN GHA	SUN Dec	MOON GHA	v	Dec	d	HP
d h	° ′	° ′	° ′	′	° ′	′	′
13 00	183 24.0	S 7 36.7	263 57.5	9.2	N18 05.4	3.6	57.1
01	198 24.2	37.6	278 25.7	9.3	18 01.8	3.8	57.0
02	213 24.3	38.5	292 54.0	9.4	17 58.0	3.8	57.0
03	228 24.5 ..	39.5	307 22.4	9.4	17 54.2	4.0	57.0
04	243 24.7	40.4	321 50.8	9.4	17 50.2	4.0	57.0
05	258 24.8	41.3	336 19.2	9.6	17 46.2	4.1	56.9
06	273 25.0 S 7	42.3	350 47.8	9.6	N17 42.1	4.2	56.9
T 07	288 25.1	43.2	5 16.4	9.6	17 37.9	4.3	56.9
U 08	303 25.3	44.1	19 45.0	9.8	17 33.6	4.4	56.8
E 09	318 25.4 ..	45.1	34 13.8	9.8	17 29.2	4.5	56.8
S 10	333 25.6	46.0	48 42.6	9.8	17 24.7	4.6	56.8
D 11	348 25.7	46.9	63 11.4	9.9	17 20.1	4.7	56.7
A 12	3 25.9 S 7 47.9		77 40.3	10.0	N17 15.4	4.7	56.7
Y 13	18 26.0	48.8	92 09.3	10.1	17 10.7	4.9	56.7
14	33 26.2	49.7	106 38.4	10.1	17 05.8	4.9	56.6
15	48 26.3 ..	50.7	121 07.5	10.2	17 00.9	5.0	56.6
16	63 26.5	51.6	135 36.7	10.2	16 55.9	5.1	56.6
17	78 26.6	52.5	150 05.9	10.3	16 50.8	5.2	56.5
18	93 26.8 S 7 53.5		164 35.2	10.4	N16 45.6	5.2	56.5
19	108 26.9	54.4	179 04.6	10.4	16 40.4	5.4	56.5
20	123 27.1	55.3	193 34.0	10.5	16 35.0	5.4	56.4
21	138 27.2 ..	56.3	208 03.5	10.6	16 29.6	5.5	56.4
22	153 27.4	57.2	222 33.1	10.6	16 24.1	5.6	56.4
23	168 27.5	58.1	237 02.7	10.7	16 18.5	5.7	56.4
14 00	183 27.6 S 7 59.1		251 32.4	10.8	N16 12.8	5.7	56.3
01	198 27.8 8 00.0		266 02.2	10.8	16 07.1	5.8	56.3
02	213 27.9	00.9	280 32.0	10.9	16 01.3	5.9	56.3
03	228 28.1 ..	01.9	295 01.9	11.0	15 55.4	6.0	56.2
04	243 28.2	02.8	309 31.9	11.0	15 49.4	6.0	56.2
05	258 28.4	03.7	324 01.9	11.0	15 43.4	6.2	56.2
06	273 28.5 S 8 04.7		338 31.9	11.2	N15 37.2	6.2	56.2
W 07	288 28.7	05.6	353 02.1	11.2	15 31.0	6.2	56.1
E 08	303 28.8	06.5	7 32.3	11.3	15 24.8	6.4	56.1
D 09	318 29.0 ..	07.5	22 02.6	11.3	15 18.4	6.4	56.1
N 10	333 29.1	08.4	36 32.9	11.3	15 12.0	6.4	56.0
E 11	348 29.3	09.3	51 03.3	11.4	15 05.6	6.6	56.0
S 12	3 29.4 S 8 10.2		65 33.7	11.6	N14 59.0	6.6	56.0
D 13	18 29.5	11.2	80 04.3	11.5	14 52.4	6.7	56.0
A 14	33 29.7	12.1	94 34.8	11.7	14 45.7	6.8	55.9
Y 15	48 29.8 ..	13.0	109 05.5	11.7	14 38.9	6.8	55.9
16	63 30.0	14.0	123 36.2	11.7	14 32.1	6.9	55.9
17	78 30.1	14.9	138 06.9	11.9	14 25.2	6.9	55.9
18	93 30.3 S 8 15.8		152 37.8	11.8	N14 18.3	7.0	55.8
19	108 30.4	16.7	167 08.6	12.0	14 11.3	7.1	55.8
20	123 30.6	17.7	181 39.6	12.0	14 04.2	7.1	55.8
21	138 30.7 ..	18.6	196 10.6	12.0	13 57.1	7.2	55.8
22	153 30.8	19.5	210 41.6	12.2	13 49.9	7.3	55.7
23	168 31.0	20.5	225 12.8	12.1	13 42.6	7.3	55.7
15 00	183 31.1 S 8 21.4		239 43.9	12.3	N13 35.3	7.4	55.7
01	198 31.3	22.3	254 15.2	12.3	13 27.9	7.4	55.7
02	213 31.4	23.2	268 46.5	12.3	13 20.5	7.5	55.6
03	228 31.5 ..	24.2	283 17.8	12.4	13 13.0	7.5	55.6
04	243 31.7	25.1	297 49.2	12.5	13 05.5	7.7	55.6
05	258 31.8	26.0	312 20.7	12.5	12 57.8	7.6	55.6
06	273 32.0 S 8 26.9		326 52.2	12.6	N12 50.2	7.7	55.5
T 07	288 32.1	27.9	341 23.8	12.6	12 42.5	7.8	55.5
H 08	303 32.2	28.8	355 55.4	12.7	12 34.7	7.8	55.5
U 09	318 32.4 ..	29.7	10 27.1	12.7	12 26.9	7.9	55.5
R 10	333 32.5	30.6	24 58.8	12.8	12 19.0	7.9	55.4
S 11	348 32.7	31.6	39 30.6	12.9	12 11.1	8.0	55.4
D 12	3 32.8 S 8 32.5		54 02.5	12.9	N12 03.1	8.0	55.4
A 13	18 32.9	33.4	68 34.4	12.9	11 55.1	8.1	55.4
Y 14	33 33.1	34.3	83 06.3	13.0	11 47.0	8.1	55.3
15	48 33.2 ..	35.3	97 38.3	13.1	11 38.9	8.2	55.3
16	63 33.4	36.2	112 10.4	13.1	11 30.7	8.2	55.3
17	78 33.5	37.1	126 42.5	13.1	11 22.5	8.3	55.3
18	93 33.6 S 8 38.0		141 14.6	13.2	N11 14.2	8.3	55.3
19	108 33.8	39.0	155 46.8	13.3	11 05.9	8.3	55.2
20	123 33.9	39.9	170 19.1	13.3	10 57.6	8.4	55.2
21	138 34.0 ..	40.8	184 51.4	13.3	10 49.2	8.5	55.2
22	153 34.2	41.7	199 23.7	13.4	10 40.7	8.5	55.2
23	168 34.3	42.7	213 56.1	13.5	N10 32.2	8.5	55.2
	SD 16.1	d 0.9	SD 15.4		15.3		15.1

Lat.	Twilight Naut.	Twilight Civil	Sunrise	Moonrise 13	14	15	16
°	h m	h m	h m	h m	h m	h m	h m
N 72	04 50	06 08	07 18	20 48	22 46	24 32	00 32
N 70	04 54	06 05	07 08	21 35	23 13	24 49	00 49
68	04 58	06 02	06 59	22 05	23 34	25 03	01 03
66	05 00	06 00	06 52	22 27	23 50	25 14	01 14
64	05 03	05 58	06 46	22 45	24 04	00 04	01 23
62	05 04	05 56	06 41	22 59	24 15	00 15	01 31
60	05 06	05 54	06 36	23 12	24 24	00 24	01 38
N 58	05 07	05 52	06 32	23 22	24 32	00 32	01 43
56	05 08	05 51	06 29	23 31	24 40	00 40	01 49
54	05 09	05 50	06 25	23 39	24 46	00 46	01 53
52	05 09	05 48	06 22	23 47	24 52	00 52	01 58
50	05 10	05 47	06 20	23 53	24 57	00 57	02 01
45	05 10	05 44	06 14	24 07	00 07	01 09	02 10
N 40	05 10	05 42	06 09	24 19	00 19	01 18	02 17
35	05 10	05 39	06 05	24 29	00 29	01 26	02 22
30	05 09	05 37	06 01	24 37	00 37	01 33	02 28
20	05 07	05 32	05 54	24 52	00 52	01 45	02 37
N 10	05 03	05 27	05 48	00 12	01 05	01 56	02 44
0	04 58	05 22	05 43	00 26	01 17	02 06	02 52
S 10	04 51	05 16	05 37	00 39	01 29	02 15	02 59
20	04 42	05 08	05 31	00 54	01 42	02 26	03 07
30	04 30	04 59	05 23	01 11	01 56	02 38	03 16
35	04 23	04 53	05 19	01 20	02 05	02 45	03 21
40	04 14	04 47	05 14	01 31	02 15	02 53	03 27
45	04 02	04 38	05 09	01 44	02 26	03 02	03 33
S 50	03 47	04 28	05 02	02 00	02 40	03 13	03 41
52	03 40	04 24	04 59	02 08	02 46	03 18	03 45
54	03 32	04 18	04 55	02 16	02 53	03 23	03 49
56	03 23	04 12	04 51	02 25	03 01	03 30	03 54
58	03 12	04 05	04 47	02 36	03 10	03 37	03 59
S 60	03 00	03 58	04 42	02 48	03 20	03 44	04 04

Lat.	Sunset	Twilight Civil	Twilight Naut.	Moonset 13	14	15	16
°	h m	h m	h m	h m	h m	h m	h m
N 72	16 12	17 22	18 40	17 30	17 14	17 04	16 56
N 70	16 23	17 25	18 36	16 42	16 45	16 46	16 45
68	16 32	17 28	18 32	16 11	16 24	16 31	16 36
66	16 39	17 31	18 30	15 48	16 07	16 19	16 28
64	16 45	17 33	18 28	15 30	15 53	16 09	16 22
62	16 50	17 35	18 26	15 15	15 41	16 00	16 16
60	16 55	17 37	18 25	15 02	15 31	15 53	16 11
N 58	16 59	17 38	18 24	14 52	15 22	15 46	16 07
56	17 03	17 40	18 23	14 42	15 14	15 40	16 03
54	17 06	17 41	18 22	14 34	15 07	15 35	15 59
52	17 09	17 43	18 22	14 26	15 01	15 30	15 56
50	17 11	17 44	18 21	14 19	14 55	15 26	15 53
45	17 17	17 47	18 21	14 04	14 43	15 17	15 47
N 40	17 22	17 50	18 21	13 52	14 33	15 09	15 41
35	17 27	17 52	18 22	13 42	14 24	15 02	15 37
30	17 31	17 55	18 23	13 33	14 17	14 56	15 33
20	17 37	18 00	18 25	13 17	14 03	14 46	15 26
N 10	17 44	18 05	18 29	13 03	13 52	14 37	15 19
0	17 49	18 10	18 34	12 51	13 41	14 28	15 13
S 10	17 55	18 17	18 41	12 38	13 30	14 19	15 07
20	18 02	18 24	18 50	12 24	13 18	14 10	15 01
30	18 09	18 34	19 02	12 08	13 04	13 59	14 54
35	18 13	18 39	19 10	11 58	12 56	13 53	14 49
40	18 18	18 46	19 19	11 48	12 47	13 46	14 44
45	18 24	18 55	19 31	11 35	12 36	13 38	14 39
S 50	18 31	19 05	19 46	11 20	12 24	13 28	14 32
52	18 34	19 10	19 53	11 13	12 18	13 23	14 29
54	18 38	19 15	20 02	11 05	12 11	13 18	14 25
56	18 42	19 21	20 11	10 56	12 03	13 13	14 22
58	18 46	19 28	20 22	10 46	11 55	13 06	14 17
S 60	18 51	19 36	20 35	10 34	11 46	12 59	14 12

	SUN			MOON			
Day	Eqn. of Time 00h	12h	Mer. Pass.	Mer. Pass. Upper	Lower	Age	Phase
d	m s	m s	h m	h m	h m	d	%
13	13 36	13 43	11 46	06 38	19 04	23	39
14	13 50	13 57	11 46	07 29	19 53	24	30
15	14 04	14 11	11 46	08 17	20 40	25	21

1998 OCTOBER 16, 17, 18 (FRI., SAT., SUN.)

UT	ARIES GHA	VENUS −3.9 GHA	Dec	MARS +1.6 GHA	Dec	JUPITER −2.8 GHA	Dec	SATURN +0.0 GHA	Dec
16 00	24 18.5	186 30.0	S 6 10.7	226 47.9	N10 54.7	33 20.8	S 5 33.5	354 40.6	N 9 09.1
01	39 21.0	201 29.5	11.9	241 48.9	54.1	48 23.5	33.6	9 43.2	09.0
02	54 23.5	216 29.1	13.1	256 50.0	53.6	63 26.2	33.7	24 45.9	09.0
03	69 25.9	231 28.7 ..	14.4	271 51.0 ..	53.0	78 28.9 ..	33.8	39 48.6 ..	08.9
04	84 28.4	246 28.2	15.6	286 52.0	52.5	93 31.5	33.9	54 51.2	08.8
05	99 30.9	261 27.8	16.8	301 53.0	52.0	108 34.2	33.9	69 53.9	08.8
F 06	114 33.3	276 27.4	S 6 18.0	316 54.1	N10 51.4	123 36.9	S 5 34.0	84 56.5	N 9 08.7
R 07	129 35.8	291 27.0	19.3	331 55.1	50.9	138 39.6	34.1	99 59.2	08.6
I 08	144 38.3	306 26.5	20.5	346 56.1	50.3	153 42.2	34.2	115 01.8	08.6
D 09	159 40.7	321 26.1 ..	21.7	1 57.1 ..	49.8	168 44.9 ..	34.3	130 04.5 ..	08.5
A 10	174 43.2	336 25.7	22.9	16 58.2	49.2	183 47.6	34.3	145 07.1	08.4
Y 11	189 45.7	351 25.2	24.2	31 59.2	48.7	198 50.3	34.4	160 09.8	08.3
12	204 48.1	6 24.8	S 6 25.4	47 00.2	N10 48.2	213 52.9	S 5 34.5	175 12.4	N 9 08.3
13	219 50.6	21 24.4	26.6	62 01.2	47.6	228 55.6	34.6	190 15.1	08.2
14	234 53.0	36 23.9	27.8	77 02.3	47.1	243 58.3	34.7	205 17.7	08.1
15	249 55.5	51 23.5 ..	29.1	92 03.3 ..	46.5	259 01.0 ..	34.8	220 20.4 ..	08.1
16	264 58.0	66 23.1	30.3	107 04.3	46.0	274 03.6	34.8	235 23.0	08.0
17	280 00.4	81 22.6	31.5	122 05.3	45.5	289 06.3	34.9	250 25.7	07.9
18	295 02.9	96 22.2	S 6 32.7	137 06.4	N10 44.9	304 09.0	S 5 35.0	265 28.3	N 9 07.8
19	310 05.4	111 21.8	34.0	152 07.4	44.4	319 11.7	35.1	280 31.0	07.7
20	325 07.8	126 21.3	35.2	167 08.4	43.8	334 14.3	35.2	295 33.6	07.7
21	340 10.3	141 20.9 ..	36.4	182 09.5 ..	43.3	349 17.0 ..	35.2	310 36.3 ..	07.6
22	355 12.8	156 20.5	37.6	197 10.5	42.7	4 19.7	35.3	325 38.9	07.6
23	10 15.2	171 20.0	38.9	212 11.5	42.2	19 22.3	35.4	340 41.6	07.5
17 00	25 17.7	186 19.6	S 6 40.1	227 12.5	N10 41.7	34 25.0	S 5 35.5	355 44.2	N 9 07.4
01	40 20.1	201 19.1	41.3	242 13.6	41.1	49 27.7	35.6	10 46.9	07.4
02	55 22.6	216 18.7	42.5	257 14.6	40.6	64 30.4	35.6	25 49.5	07.3
03	70 25.1	231 18.3 ..	43.7	272 15.6 ..	40.0	79 33.0 ..	35.7	40 52.2 ..	07.2
04	85 27.5	246 17.8	45.0	287 16.7	39.5	94 35.7	35.8	55 54.8	07.2
05	100 30.0	261 17.4	46.2	302 17.7	39.0	109 38.4	35.9	70 57.5	07.1
S 06	115 32.5	276 17.0	S 6 47.4	317 18.7	N10 38.4	124 41.0	S 5 36.0	86 00.1	N 9 07.0
A 07	130 34.9	291 16.5	48.6	332 19.7	37.9	139 43.7	36.0	101 02.8	07.0
T 08	145 37.4	306 16.1	49.8	347 20.8	37.3	154 46.4	36.1	116 05.4	06.9
U 09	160 39.9	321 15.7 ..	51.1	2 21.8 ..	36.8	169 49.0 ..	36.2	131 08.1 ..	06.8
R 10	175 42.3	336 15.2	52.3	17 22.8	36.2	184 51.7	36.3	146 10.7	06.8
D 11	190 44.8	351 14.8	53.5	32 23.9	35.7	199 54.4	36.4	161 13.4	06.7
A 12	205 47.3	6 14.3	S 6 54.7	47 24.9	N10 35.2	214 57.0	S 5 36.4	176 16.0	N 9 06.6
Y 13	220 49.7	21 13.9	55.9	62 25.9	34.6	229 59.7	36.5	191 18.7	06.5
14	235 52.2	36 13.5	57.2	77 26.9	34.1	245 02.4	36.6	206 21.3	06.5
15	250 54.6	51 13.0 ..	58.4	92 28.0 ..	33.5	260 05.0 ..	36.7	221 24.0 ..	06.4
16	265 57.1	66 12.6	6 59.6	107 29.0	33.0	275 07.7	36.8	236 26.6	06.3
17	280 59.6	81 12.1	7 00.8	122 30.0	32.4	290 10.4	36.8	251 29.3	06.3
18	296 02.0	96 11.7	S 7 02.0	137 31.1	N10 31.9	305 13.0	S 5 36.9	266 31.9	N 9 06.2
19	311 04.5	111 11.3	03.2	152 32.1	31.4	320 15.7	37.0	281 34.6	06.1
20	326 07.0	126 10.8	04.5	167 33.1	30.8	335 18.4	37.1	296 37.2	06.1
21	341 09.4	141 10.4 ..	05.7	182 34.2 ..	30.3	350 21.0 ..	37.2	311 39.9 ..	06.0
22	356 11.9	156 09.9	06.9	197 35.2	29.7	5 23.7	37.2	326 42.5	05.9
23	11 14.4	171 09.5	08.1	212 36.2	29.2	20 26.4	37.3	341 45.2	05.9
18 00	26 16.8	186 09.1	S 7 09.3	227 37.3	N10 28.6	35 29.0	S 5 37.4	356 47.8	N 9 05.8
01	41 19.3	201 08.6	10.5	242 38.3	28.1	50 31.7	37.5	11 50.5	05.7
02	56 21.7	216 08.2	11.8	257 39.3	27.5	65 34.4	37.5	26 53.2	05.6
03	71 24.2	231 07.7 ..	13.0	272 40.3 ..	27.0	80 37.0 ..	37.6	41 55.8 ..	05.6
04	86 26.7	246 07.3	14.2	287 41.4	26.5	95 39.7	37.7	56 58.5	05.5
05	101 29.1	261 06.8	15.4	302 42.4	25.9	110 42.3	37.8	72 01.1	05.4
S 06	116 31.6	276 06.4	S 7 16.6	317 43.4	N10 25.4	125 45.0	S 5 37.8	87 03.8	N 9 05.4
U 07	131 34.1	291 06.0	17.8	332 44.5	24.8	140 47.7	37.9	102 06.4	05.3
N 08	146 36.5	306 05.5	19.0	347 45.5	24.3	155 50.3	38.0	117 09.1	05.2
D 09	161 39.0	321 05.1 ..	20.3	2 46.5 ..	23.7	170 53.0 ..	38.1	132 11.7 ..	05.2
A 10	176 41.5	336 04.6	21.5	17 47.6	23.2	185 55.7	38.2	147 14.4	05.1
Y 11	191 43.9	351 04.2	22.7	32 48.6	22.7	200 58.3	38.2	162 17.0	05.0
12	206 46.4	6 03.7	S 7 23.9	47 49.6	N10 22.1	216 01.0	S 5 38.3	177 19.7	N 9 05.0
13	221 48.9	21 03.3	25.1	62 50.7	21.6	231 03.6	38.4	192 22.3	04.9
14	236 51.3	36 02.8	26.3	77 51.7	21.0	246 06.3	38.5	207 25.0	04.8
15	251 53.8	51 02.4 ..	27.5	92 52.7 ..	20.5	261 09.0 ..	38.5	222 27.6 ..	04.7
16	266 56.2	66 02.0	28.7	107 53.8	19.9	276 11.6	38.6	237 30.3	04.7
17	281 58.7	81 01.5	30.0	122 54.8	19.4	291 14.3	38.7	252 32.9	04.6
18	297 01.2	96 01.1	S 7 31.2	137 55.8	N10 18.8	306 16.9	S 5 38.8	267 35.6	N 9 04.5
19	312 03.6	111 00.6	32.4	152 56.9	18.3	321 19.6	38.8	282 38.2	04.5
20	327 06.1	126 00.2	33.6	167 57.9	17.8	336 22.3	38.9	297 40.9	04.4
21	342 08.6	140 59.7 ..	34.8	182 59.0 ..	17.2	351 24.9 ..	39.0	312 43.5 ..	04.3
22	357 11.0	155 59.3	36.0	198 00.0	16.7	6 27.6	39.1	327 46.2	04.3
23	12 13.5	170 58.8	37.2	213 01.0	16.1	21 30.2	39.1	342 48.8	04.2
Mer. Pass. 22 15.2		v −0.4	d 1.2	v 1.0	d 0.5	v 2.7	d 0.1	v 2.7	d 0.1

STARS

Name	SHA	Dec
Acamar	315 26.6	S40 18.5
Achernar	335 34.7	S57 14.6
Acrux	173 22.6	S63 05.3
Adhara	255 21.4	S28 58.1
Aldebaran	291 02.1	N16 30.3
Alioth	166 31.2	N55 58.1
Alkaid	153 08.3	N49 19.4
Al Na'ir	27 57.7	S46 58.1
Alnilam	275 57.7	S 1 12.2
Alphard	218 07.4	S 8 39.1
Alphecca	126 20.9	N26 43.4
Alpheratz	357 54.9	N29 05.1
Altair	62 19.3	N 8 52.2
Ankaa	353 26.5	S42 18.8
Antares	112 40.4	S26 25.6
Arcturus	146 06.4	N19 11.6
Atria	107 52.8	S69 01.6
Avior	234 22.7	S59 30.1
Bellatrix	278 44.0	N 6 20.9
Betelgeuse	271 13.4	N 7 24.4
Canopus	264 01.0	S52 41.5
Capella	280 50.9	N45 59.6
Deneb	49 39.1	N45 16.9
Denebola	182 45.5	N14 34.9
Diphda	349 06.9	S17 59.6
Dubhe	194 06.1	N61 45.4
Elnath	278 26.7	N28 36.2
Eltanin	90 51.6	N51 29.8
Enif	33 58.1	N 9 52.3
Fomalhaut	15 36.2	S29 37.8
Gacrux	172 14.1	S57 06.1
Gienah	176 04.3	S17 31.9
Hadar	149 04.6	S60 21.9
Hamal	328 13.2	N23 27.4
Kaus Aust.	83 59.0	S34 23.1
Kochab	137 20.8	N74 09.9
Markab	13 49.4	N15 12.1
Menkar	314 26.6	N 4 05.1
Menkent	148 21.3	S36 21.6
Miaplacidus	221 42.3	S69 42.4
Mirfak	308 56.1	N49 51.3
Nunki	76 12.4	S26 17.8
Peacock	53 37.0	S56 44.5
Pollux	243 41.6	N28 01.6
Procyon	245 11.6	N 5 13.7
Rasalhague	96 17.1	N12 34.0
Regulus	207 55.8	N11 58.5
Rigel	281 22.8	S 8 12.2
Rigil Kent.	140 07.9	S60 49.6
Sabik	102 25.7	S15 43.2
Schedar	349 52.9	N56 31.9
Shaula	96 37.5	S37 06.1
Sirius	258 43.6	S16 42.7
Spica	158 43.5	S11 09.1
Suhail	223 00.9	S43 25.4
Vega	80 46.7	N38 47.3
Zuben'ubi	137 18.3	S16 02.0

	SHA	Mer. Pass.
Venus	161 01.9	11 35
Mars	201 54.9	8 51
Jupiter	9 07.3	21 38
Saturn	330 26.5	0 17

UT	SUN GHA	SUN Dec	MOON GHA	v	MOON Dec	d	HP
d h	° ′	° ′	° ′	′	° ′	′	′
16 00	183 34.4	S 8 43.6	228 28.6	13.5	N10 23.7	8.6	55.1
01	198 34.6	44.5	243 01.1	13.5	10 15.1	8.6	55.1
02	213 34.7	45.4	257 33.6	13.6	10 06.5	8.6	55.1
03	228 34.9 ..	46.3	272 06.2	13.6	9 57.9	8.7	55.1
04	243 35.0	47.3	286 38.8	13.7	9 49.2	8.7	55.1
05	258 35.1	48.2	301 11.5	13.7	9 40.5	8.8	55.0
06	273 35.3	S 8 49.1	315 44.2	13.8	N 9 31.7	8.8	55.0
F 07	288 35.4	50.0	330 17.0	13.8	9 22.9	8.8	55.0
R 08	303 35.5	50.9	344 49.8	13.8	9 14.1	8.9	55.0
I 09	318 35.7 ..	51.9	359 22.6	13.9	9 05.2	8.9	55.0
D 10	333 35.8	52.8	13 55.5	14.0	8 56.3	9.0	54.9
A 11	348 35.9	53.7	28 28.5	13.9	8 47.3	8.9	54.9
Y 12	3 36.1	S 8 54.6	43 01.4	14.1	N 8 38.4	9.0	54.9
13	18 36.2	55.5	57 34.5	14.0	8 29.4	9.1	54.9
14	33 36.3	56.5	72 07.5	14.1	8 20.3	9.0	54.9
15	48 36.5 ..	57.4	86 40.6	14.1	8 11.3	9.1	54.9
16	63 36.6	58.3	101 13.7	14.2	8 02.2	9.2	54.8
17	78 36.7	8 59.2	115 46.9	14.2	7 53.0	9.1	54.8
18	93 36.9	S 9 00.1	130 20.1	14.3	N 7 43.9	9.2	54.8
19	108 37.0	01.1	144 53.4	14.2	7 34.7	9.2	54.8
20	123 37.1	02.0	159 26.6	14.4	7 25.5	9.3	54.8
21	138 37.2 ..	02.9	174 00.0	14.3	7 16.2	9.2	54.8
22	153 37.4	03.8	188 33.3	14.4	7 07.0	9.3	54.7
23	168 37.5	04.7	203 06.7	14.4	6 57.7	9.4	54.7
17 00	183 37.6	S 9 05.6	217 40.1	14.5	N 6 48.3	9.3	54.7
01	198 37.8	06.6	232 13.6	14.5	6 39.0	9.4	54.7
02	213 37.9	07.5	246 47.1	14.5	6 29.6	9.4	54.7
03	228 38.0 ..	08.4	261 20.6	14.5	6 20.2	9.4	54.7
04	243 38.1	09.3	275 54.1	14.6	6 10.8	9.4	54.6
05	258 38.3	10.2	290 27.7	14.6	6 01.4	9.5	54.6
06	273 38.4	S 9 11.1	305 01.3	14.7	N 5 51.9	9.5	54.6
S 07	288 38.5	12.1	319 35.0	14.6	5 42.4	9.5	54.6
A 08	303 38.7	13.0	334 08.6	14.7	5 32.9	9.5	54.6
T 09	318 38.8 ..	13.9	348 42.3	14.7	5 23.4	9.5	54.6
U 10	333 38.9	14.8	3 16.0	14.8	5 13.9	9.6	54.6
R 11	348 39.0	15.7	17 49.8	14.8	5 04.3	9.5	54.5
D 12	3 39.2	S 9 16.6	32 23.6	14.8	N 4 54.8	9.6	54.5
A 13	18 39.3	17.5	46 57.4	14.8	4 45.2	9.6	54.5
Y 14	33 39.4	18.5	61 31.2	14.9	4 35.6	9.6	54.5
15	48 39.5 ..	19.4	76 05.1	14.8	4 26.0	9.7	54.5
16	63 39.7	20.3	90 38.9	15.0	4 16.3	9.6	54.5
17	78 39.8	21.2	105 12.9	14.9	4 06.7	9.7	54.5
18	93 39.9	S 9 22.1	119 46.8	14.9	N 3 57.0	9.7	54.4
19	108 40.0	23.0	134 20.7	15.0	3 47.3	9.6	54.4
20	123 40.2	23.9	148 54.7	15.0	3 37.7	9.7	54.4
21	138 40.3 ..	24.8	163 28.7	15.0	3 28.0	9.7	54.4
22	153 40.4	25.8	178 02.7	15.0	3 18.3	9.8	54.4
23	168 40.5	26.7	192 36.7	15.1	3 08.5	9.7	54.4
18 00	183 40.7	S 9 27.6	207 10.8	15.1	N 2 58.8	9.7	54.4
01	198 40.8	28.5	221 44.9	15.0	2 49.1	9.8	54.4
02	213 40.9	29.4	236 18.9	15.2	2 39.3	9.7	54.3
03	228 41.0 ..	30.3	250 53.1	15.1	2 29.6	9.8	54.3
04	243 41.2	31.2	265 27.2	15.1	2 19.8	9.8	54.3
05	258 41.3	32.1	280 01.3	15.2	2 10.0	9.7	54.3
06	273 41.4	S 9 33.0	294 35.5	15.1	N 2 00.3	9.8	54.3
S 07	288 41.5	34.0	309 09.6	15.2	1 50.5	9.8	54.3
U 08	303 41.6	34.9	323 43.8	15.2	1 40.7	9.8	54.3
N 09	318 41.8 ..	35.8	338 18.0	15.2	1 30.9	9.8	54.3
D 10	333 41.9	36.7	352 52.2	15.3	1 21.1	9.8	54.3
A 11	348 42.0	37.6	7 26.5	15.2	1 11.3	9.8	54.3
Y 12	3 42.1	S 9 38.5	22 00.7	15.3	N 1 01.5	9.7	54.2
13	18 42.2	39.4	36 35.0	15.2	0 51.8	9.8	54.2
14	33 42.4	40.3	51 09.2	15.3	0 42.0	9.8	54.2
15	48 42.5 ..	41.2	65 43.5	15.3	0 32.2	9.8	54.2
16	63 42.6	42.1	80 17.8	15.3	0 22.4	9.8	54.2
17	78 42.7	43.0	94 52.1	15.3	0 12.6	9.8	54.2
18	93 42.8	S 9 43.9	109 26.4	15.3	N 0 02.8	9.8	54.2
19	108 43.0	44.9	124 00.7	15.3	S 0 07.0	9.8	54.2
20	123 43.1	45.8	138 35.0	15.3	0 16.8	9.8	54.2
21	138 43.2 ..	46.7	153 09.3	15.3	0 26.6	9.8	54.2
22	153 43.3	47.6	167 43.6	15.4	0 36.4	9.8	54.1
23	168 43.4	48.5	182 18.0	15.3	S 0 46.2	9.7	54.1
	SD 16.1	d 0.9	SD 15.0		14.9		14.8

Lat.	Twilight Naut.	Twilight Civil	Sunrise	Moonrise 16	17	18	19
°	h m	h m	h m	h m	h m	h m	h m
N 72	05 03	06 21	07 33	00 32	02 13	03 50	05 25
N 70	05 06	06 16	07 20	00 49	02 23	03 53	05 23
68	05 08	06 12	07 10	01 03	02 30	03 56	05 20
66	05 10	06 09	07 02	01 14	02 37	03 58	05 19
64	05 11	06 06	06 55	01 23	02 42	04 00	05 17
62	05 12	06 03	06 49	01 31	02 47	04 01	05 16
60	05 13	06 01	06 44	01 38	02 51	04 03	05 15
N 58	05 13	05 59	06 39	01 43	02 54	04 03	05 14
56	05 14	05 57	06 35	01 49	02 57	04 05	05 13
54	05 14	05 55	06 31	01 53	03 00	04 06	05 12
52	05 14	05 53	06 28	01 58	03 03	04 07	05 11
50	05 14	05 52	06 25	02 01	03 05	04 08	05 10
45	05 14	05 48	06 18	02 10	03 10	04 10	05 09
N 40	05 13	05 45	06 12	02 17	03 14	04 11	05 08
35	05 12	05 42	06 07	02 22	03 18	04 13	05 07
30	05 11	05 39	06 03	02 28	03 21	04 14	05 06
20	05 07	05 33	05 55	02 37	03 27	04 16	05 04
N 10	05 03	05 27	05 49	02 44	03 32	04 17	05 03
0	04 57	05 21	05 42	02 52	03 36	04 19	05 01
S 10	04 49	05 14	05 35	02 59	03 41	04 21	05 00
20	04 40	05 06	05 28	03 07	03 45	04 23	04 59
30	04 27	04 56	05 20	03 16	03 51	04 25	04 57
35	04 19	04 49	05 15	03 21	03 54	04 26	04 57
40	04 09	04 42	05 10	03 27	03 58	04 27	04 56
45	03 56	04 33	05 03	03 33	04 02	04 29	04 54
S 50	03 40	04 22	04 56	03 41	04 07	04 30	04 53
52	03 32	04 17	04 52	03 45	04 09	04 31	04 53
54	03 24	04 11	04 48	03 49	04 12	04 32	04 52
56	03 14	04 04	04 44	03 54	04 14	04 33	04 51
58	03 02	03 57	04 39	03 59	04 17	04 34	04 50
S 60	02 48	03 48	04 33	04 04	04 21	04 35	04 50

Lat.	Sunset	Twilight Civil	Twilight Naut.	Moonset 16	17	18	19
°	h m	h m	h m	h m	h m	h m	h m
N 72	15 56	17 08	18 25	16 56	16 49	16 42	16 35
N 70	16 09	17 12	18 23	16 45	16 44	16 42	16 40
68	16 19	17 17	18 21	16 36	16 39	16 42	16 44
66	16 27	17 20	18 19	16 28	16 35	16 41	16 47
64	16 35	17 23	18 18	16 22	16 32	16 41	16 50
62	16 41	17 26	18 17	16 16	16 29	16 41	16 53
60	16 46	17 28	18 17	16 11	16 27	16 41	16 55
N 58	16 51	17 31	18 16	16 07	16 24	16 41	16 57
56	16 55	17 33	18 16	16 03	16 22	16 41	16 59
54	16 59	17 35	18 16	15 59	16 21	16 41	17 01
52	17 02	17 36	18 15	15 56	16 19	16 41	17 02
50	17 05	17 38	18 15	15 53	16 18	16 41	17 04
45	17 12	17 42	18 16	15 47	16 14	16 41	17 06
N 40	17 18	17 45	18 17	15 41	16 12	16 40	17 09
35	17 23	17 49	18 18	15 37	16 09	16 40	17 11
30	17 27	17 52	18 19	15 33	16 07	16 40	17 13
20	17 35	17 57	18 23	15 26	16 03	16 40	17 16
N 10	17 42	18 03	18 28	15 19	16 00	16 40	17 19
0	17 49	18 10	18 34	15 13	15 57	16 40	17 22
S 10	17 55	18 17	18 42	15 07	15 54	16 39	17 25
20	18 03	18 25	18 51	15 01	15 50	16 39	17 27
30	18 11	18 36	19 05	14 54	15 47	16 39	17 31
35	18 16	18 42	19 13	14 49	15 44	16 39	17 33
40	18 22	18 50	19 23	14 44	15 42	16 39	17 35
45	18 28	18 59	19 36	14 39	15 39	16 38	17 37
S 50	18 36	19 10	19 52	14 32	15 35	16 38	17 40
52	18 40	19 16	20 00	14 29	15 34	16 38	17 42
54	18 44	19 22	20 09	14 25	15 32	16 38	17 43
56	18 48	19 28	20 19	14 22	15 30	16 38	17 45
58	18 53	19 36	20 31	14 17	15 28	16 37	17 47
S 60	18 59	19 45	20 46	14 12	15 25	16 37	17 49

Day	SUN Eqn. of Time 00h	12h	SUN Mer. Pass.	MOON Mer. Pass. Upper	Lower	Age	Phase
d	m s	m s	h m	h m	h m	d	%
16	14 18	14 24	11 46	09 03	21 25	26	14
17	14 30	14 36	11 45	09 47	22 08	27	8
18	14 42	14 48	11 45	10 29	22 51	28	3

1998 OCTOBER 19, 20, 21 (MON., TUES., WED.)

UT	ARIES GHA	VENUS −3.9 GHA	Dec	MARS +1.6 GHA	Dec	JUPITER −2.8 GHA	Dec	SATURN +0.0 GHA	Dec
19 00	27 16.0	185 58.4	S 7 38.4	228 02.1	N10 15.6	36 32.9	S 5 39.2	357 51.5	N 9 04.1
01	42 18.4	200 57.9	39.6	243 03.1	15.0	51 35.6	39.3	12 54.2	04.0
02	57 20.9	215 57.5	40.8	258 04.1	14.5	66 38.2	39.4	27 56.8	04.0
03	72 23.3	230 57.0 ..	42.1	273 05.2 ..	13.9	81 40.9 ..	39.4	42 59.5 ..	03.9
04	87 25.8	245 56.6	43.3	288 06.2	13.4	96 43.5	39.5	58 02.1	03.8
05	102 28.3	260 56.1	44.5	303 07.2	12.8	111 46.2	39.6	73 04.8	03.8
06	117 30.7	275 55.7	S 7 45.7	318 08.3	N10 12.3	126 48.8	S 5 39.7	88 07.4	N 9 03.7
07	132 33.2	290 55.2	46.9	333 09.3	11.8	141 51.5	39.7	103 10.1	03.6
08	147 35.7	305 54.8	48.1	348 10.3	11.2	156 54.2	39.8	118 12.7	03.6
09	162 38.1	320 54.3 ..	49.3	3 11.4 ..	10.7	171 56.8 ..	39.9	133 15.4 ..	03.5
10	177 40.6	335 53.9	50.5	18 12.4	10.1	186 59.5	39.9	148 18.0	03.4
11	192 43.1	350 53.4	51.7	33 13.5	09.6	202 02.1	40.0	163 20.7	03.4
12	207 45.5	5 53.0	S 7 52.9	48 14.5	N10 09.0	217 04.8	S 5 40.1	178 23.3	N 9 03.3
13	222 48.0	20 52.5	54.1	63 15.5	08.5	232 07.4	40.2	193 26.0	03.2
14	237 50.5	35 52.1	55.3	78 16.6	07.9	247 10.1	40.2	208 28.6	03.1
15	252 52.9	50 51.6 ..	56.5	93 17.6 ..	07.4	262 12.7 ..	40.3	223 31.3 ..	03.1
16	267 55.4	65 51.2	57.7	108 18.6	06.8	277 15.4	40.4	238 33.9	03.0
17	282 57.8	80 50.7	7 58.9	123 19.7	06.3	292 18.0	40.5	253 36.6	02.9
18	298 00.3	95 50.3	S 8 00.1	138 20.7	N10 05.8	307 20.7	S 5 40.5	268 39.2	N 9 02.9
19	313 02.8	110 49.8	01.4	153 21.8	05.2	322 23.4	40.6	283 41.9	02.8
20	328 05.2	125 49.3	02.6	168 22.8	04.7	337 26.0	40.7	298 44.6	02.7
21	343 07.7	140 48.9 ..	03.8	183 23.8 ..	04.1	352 28.7 ..	40.7	313 47.2 ..	02.7
22	358 10.2	155 48.4	05.0	198 24.9	03.6	7 31.3	40.8	328 49.9	02.6
23	13 12.6	170 48.0	06.2	213 25.9	03.0	22 34.0	40.9	343 52.5	02.5
20 00	28 15.1	185 47.5	S 8 07.4	228 27.0	N10 02.5	37 36.6	S 5 41.0	358 55.2	N 9 02.5
01	43 17.6	200 47.1	08.6	243 28.0	01.9	52 39.3	41.0	13 57.8	02.4
02	58 20.0	215 46.6	09.8	258 29.0	01.3	67 41.9	41.1	29 00.5	02.3
03	73 22.5	230 46.2 ..	11.0	273 30.1 ..	00.8	82 44.6 ..	41.2	44 03.1 ..	02.2
04	88 24.9	245 45.7	12.2	288 31.1	10 00.3	97 47.2	41.3	59 05.8	02.2
05	103 27.4	260 45.2	13.4	303 32.2	9 59.7	112 49.9	41.3	74 08.4	02.1
06	118 29.9	275 44.8	S 8 14.6	318 33.2	N 9 59.2	127 52.5	S 5 41.4	89 11.1	N 9 02.0
07	133 32.3	290 44.3	15.8	333 34.2	58.7	142 55.2	41.5	104 13.7	02.0
08	148 34.8	305 43.9	17.0	348 35.3	58.1	157 57.8	41.5	119 16.4	01.9
09	163 37.3	320 43.4 ..	18.2	3 36.3 ..	57.6	173 00.5 ..	41.6	134 19.0 ..	01.8
10	178 39.7	335 43.0	19.4	18 37.4	57.0	188 03.1	41.7	149 21.7	01.8
11	193 42.2	350 42.5	20.6	33 38.4	56.5	203 05.8	41.7	164 24.3	01.7
12	208 44.7	5 42.0	S 8 21.8	48 39.4	N 9 55.9	218 08.4	S 5 41.8	179 27.0	N 9 01.6
13	223 47.1	20 41.6	23.0	63 40.5	55.4	233 11.1	41.9	194 29.7	01.6
14	238 49.6	35 41.1	24.2	78 41.5	54.8	248 13.7	42.0	209 32.3	01.5
15	253 52.1	50 40.7 ..	25.4	93 42.6 ..	54.3	263 16.4 ..	42.0	224 35.0 ..	01.4
16	268 54.5	65 40.2	26.6	108 43.6	53.7	278 19.0	42.1	239 37.6	01.3
17	283 57.0	80 39.7	27.8	123 44.7	53.2	293 21.7	42.2	254 40.3	01.3
18	298 59.4	95 39.3	S 8 29.0	138 45.7	N 9 52.6	308 24.3	S 5 42.2	269 42.9	N 9 01.2
19	314 01.9	110 38.8	30.2	153 46.7	52.1	323 26.9	42.3	284 45.6	01.1
20	329 04.4	125 38.4	31.3	168 47.8	51.5	338 29.6	42.4	299 48.2	01.1
21	344 06.8	140 37.9 ..	32.5	183 48.8 ..	51.0	353 32.2 ..	42.4	314 50.9 ..	01.0
22	359 09.3	155 37.4	33.7	198 49.9	50.4	8 34.9	42.5	329 53.5	00.9
23	14 11.8	170 37.0	34.9	213 50.9	49.9	23 37.5	42.6	344 56.2	00.9
21 00	29 14.2	185 36.5	S 8 36.1	228 51.9	N 9 49.3	38 40.2	S 5 42.6	359 58.8	N 9 00.8
01	44 16.7	200 36.0	37.3	243 53.0	48.8	53 42.8	42.7	15 01.5	00.7
02	59 19.2	215 35.6	38.5	258 54.0	48.3	68 45.5	42.8	30 04.1	00.6
03	74 21.6	230 35.1 ..	39.7	273 55.1 ..	47.7	83 48.1 ..	42.9	45 06.8 ..	00.6
04	89 24.1	245 34.7	40.9	288 56.1	47.2	98 50.8	42.9	60 09.5	00.5
05	104 26.6	260 34.2	42.1	303 57.2	46.6	113 53.4	43.0	75 12.1	00.4
06	119 29.0	275 33.7	S 8 43.3	318 58.2	N 9 46.1	128 56.0	S 5 43.1	90 14.8	N 9 00.4
07	134 31.5	290 33.3	44.5	333 59.3	45.5	143 58.7	43.1	105 17.4	00.3
08	149 33.9	305 32.8	45.7	349 00.3	45.0	159 01.3	43.2	120 20.1	00.2
09	164 36.4	320 32.3 ..	46.9	4 01.3 ..	44.4	174 04.0 ..	43.3	135 22.7 ..	00.2
10	179 38.9	335 31.9	48.1	19 02.4	43.9	189 06.6	43.3	150 25.4	00.1
11	194 41.3	350 31.4	49.3	34 03.4	43.3	204 09.3	43.4	165 28.0	00.0
12	209 43.8	5 30.9	S 8 50.4	49 04.5	N 9 42.8	219 11.9	S 5 43.5	180 30.7	N 9 00.0
13	224 46.3	20 30.5	51.6	64 05.5	42.2	234 14.5	43.5	195 33.3	8 59.9
14	239 48.7	35 30.0	52.8	79 06.6	41.7	249 17.2	43.6	210 36.0	59.8
15	254 51.2	50 29.5 ..	54.0	94 07.6 ..	41.1	264 19.8 ..	43.7	225 38.6 ..	59.7
16	269 53.7	65 29.1	55.2	109 08.7	40.6	279 22.5	43.7	240 41.3	59.7
17	284 56.1	80 28.6	56.4	124 09.7	40.0	294 25.1	43.8	255 43.9	59.6
18	299 58.6	95 28.1	S 8 57.6	139 10.8	N 9 39.5	309 27.7	S 5 43.9	270 46.6	N 8 59.5
19	315 01.0	110 27.6	8 58.8	154 11.8	38.9	324 30.4	43.9	285 49.3	59.5
20	330 03.5	125 27.2	9 00.0	169 12.8	38.4	339 33.0	44.0	300 51.9	59.4
21	345 06.0	140 26.7 ..	01.1	184 13.9 ..	37.8	354 35.7 ..	44.1	315 54.6 ..	59.3
22	0 08.4	155 26.2	02.3	199 14.9	37.3	9 38.3	44.1	330 57.2	59.3
23	15 10.9	170 25.8	03.5	214 16.0	36.7	24 40.9	44.2	345 59.9	59.2
Mer. Pass.	h m 22 03.4	v −0.5	d 1.2	v 1.0	d 0.5	v 2.6	d 0.1	v 2.7	d 0.1

STARS

Name	SHA	Dec
Acamar	315 26.5	S40 18.6
Achernar	335 34.7	S57 14.6
Acrux	173 22.6	S63 05.3
Adhara	255 21.3	S28 58.1
Aldebaran	291 02.1	N16 30.3
Alioth	166 31.2	N55 58.1
Alkaid	153 08.3	N49 19.3
Al Na'ir	27 57.7	S46 58.1
Alnilam	275 57.7	S 1 12.2
Alphard	218 07.3	S 8 39.1
Alphecca	126 20.9	N26 43.4
Alpheratz	357 54.9	N29 05.1
Altair	62 19.3	N 8 52.2
Ankaa	353 26.5	S42 18.8
Antares	112 40.5	S26 25.6
Arcturus	146 06.4	N19 11.6
Atria	107 52.9	S69 01.6
Avior	234 22.7	S59 30.1
Bellatrix	278 44.0	N 6 20.9
Betelgeuse	271 13.4	N 7 24.4
Canopus	264 01.0	S52 41.5
Capella	280 50.9	N45 59.6
Deneb	49 39.1	N45 16.9
Denebola	182 45.5	N14 34.9
Diphda	349 06.9	S17 59.6
Dubhe	194 06.1	N61 45.3
Elnath	278 26.7	N28 36.2
Eltanin	90 51.7	N51 29.8
Enif	33 58.1	N 9 52.3
Fomalhaut	15 36.2	S29 37.8
Gacrux	172 14.1	S57 06.1
Gienah	176 04.3	S17 31.9
Hadar	149 04.6	S60 21.9
Hamal	328 13.2	N23 27.4
Kaus Aust.	83 59.0	S34 23.1
Kochab	137 20.8	N74 09.9
Markab	13 49.4	N15 12.1
Menkar	314 26.6	N 4 05.1
Menkent	148 21.3	S36 21.6
Miaplacidus	221 42.3	S69 42.4
Mirfak	308 56.1	N49 51.3
Nunki	76 12.4	S26 17.8
Peacock	53 37.0	S56 44.5
Pollux	243 41.6	N28 01.6
Procyon	245 11.6	N 5 13.7
Rasalhague	96 17.2	N12 34.0
Regulus	207 55.7	N11 58.5
Rigel	281 22.7	S 8 12.2
Rigil Kent.	140 07.9	S60 49.6
Sabik	102 25.8	S15 43.2
Schedar	349 52.9	N56 31.9
Shaula	96 37.5	S37 06.1
Sirius	258 43.6	S16 42.8
Spica	158 43.5	S11 09.1
Suhail	223 00.9	S43 25.4
Vega	80 46.8	N38 47.3
Zuben'ubi	137 18.3	S16 02.0

	SHA	Mer. Pass.
	° ′	h m
Venus	157 32.4	11 37
Mars	200 11.9	8 46
Jupiter	9 21.5	21 26
Saturn	330 40.1	0 04

UT	SUN GHA	SUN Dec	MOON GHA	v	MOON Dec	d	HP
	° ′	° ′	° ′	′	° ′	′	′
19 00	183 43.5	S 9 49.4	196 52.3	15.3	S 0 55.9	9.8	54.1
01	198 43.7	50.3	211 26.6	15.3	1 05.7	9.8	54.1
02	213 43.8	51.2	226 01.0	15.3	1 15.5	9.7	54.1
03	228 43.9	.. 52.1	240 35.3	15.4	1 25.2	9.8	54.1
04	243 44.0	53.0	255 09.7	15.3	1 35.0	9.7	54.1
05	258 44.1	53.9	269 44.0	15.4	1 44.7	9.7	54.1
06	273 44.2	S 9 54.8	284 18.4	15.4	S 1 54.4	9.7	54.1
07	288 44.4	55.7	298 52.8	15.3	2 04.1	9.7	54.1
08	303 44.5	56.6	313 27.1	15.4	2 13.8	9.7	54.1
M 09	318 44.6	.. 57.5	328 01.5	15.3	2 23.5	9.7	54.1
O 10	333 44.7	58.4	342 35.8	15.4	2 33.2	9.7	54.1
N 11	348 44.8	9 59.3	357 10.2	15.4	2 42.9	9.7	54.1
D 12	3 44.9	S10 00.2	11 44.6	15.3	S 2 52.6	9.6	54.0
A 13	18 45.0	01.1	26 18.9	15.4	3 02.2	9.7	54.0
Y 14	33 45.2	02.0	40 53.3	15.3	3 11.9	9.6	54.0
15	48 45.3	.. 02.9	55 27.6	15.4	3 21.5	9.6	54.0
16	63 45.4	03.8	70 02.0	15.3	3 31.1	9.6	54.0
17	78 45.5	04.7	84 36.3	15.3	3 40.7	9.5	54.0
18	93 45.6	S10 05.6	99 10.6	15.4	S 3 50.2	9.6	54.0
19	108 45.7	06.5	113 45.0	15.3	3 59.8	9.5	54.0
20	123 45.8	07.4	128 19.3	15.3	4 09.3	9.6	54.0
21	138 45.9	.. 08.3	142 53.6	15.3	4 18.9	9.5	54.0
22	153 46.1	09.2	157 27.9	15.3	4 28.4	9.5	54.0
23	168 46.2	10.1	172 02.2	15.3	4 37.9	9.4	54.0
20 00	183 46.3	S10 11.0	186 36.5	15.3	S 4 47.3	9.5	54.0
01	198 46.4	11.9	201 10.8	15.3	4 56.8	9.4	54.0
02	213 46.5	12.8	215 45.1	15.2	5 06.2	9.4	54.0
03	228 46.6	.. 13.7	230 19.3	15.3	5 15.6	9.4	54.0
04	243 46.7	14.6	244 53.6	15.2	5 25.0	9.3	54.0
05	258 46.8	15.5	259 27.8	15.3	5 34.3	9.4	54.0
06	273 46.9	S10 16.4	274 02.1	15.2	S 5 43.7	9.3	54.0
07	288 47.0	17.3	288 36.3	15.2	5 53.0	9.3	54.0
T 08	303 47.1	18.2	303 10.5	15.2	6 02.3	9.2	54.0
U 09	318 47.3	.. 19.1	317 44.7	15.2	6 11.5	9.3	54.0
E 10	333 47.4	20.0	332 18.9	15.1	6 20.8	9.2	53.9
S 11	348 47.5	20.9	346 53.0	15.2	6 30.0	9.2	53.9
D 12	3 47.6	S10 21.8	1 27.2	15.1	S 6 39.2	9.1	53.9
A 13	18 47.7	22.7	16 01.3	15.2	6 48.3	9.1	53.9
Y 14	33 47.8	23.6	30 35.5	15.1	6 57.4	9.1	53.9
15	48 47.9	.. 24.5	45 09.6	15.1	7 06.5	9.1	53.9
16	63 48.0	25.4	59 43.7	15.1	7 15.6	9.1	53.9
17	78 48.1	26.3	74 17.8	15.0	7 24.7	9.0	53.9
18	93 48.2	S10 27.2	88 51.8	15.1	S 7 33.7	8.9	53.9
19	108 48.3	28.1	103 25.9	15.0	7 42.6	9.0	53.9
20	123 48.4	29.0	117 59.9	15.0	7 51.6	8.9	53.9
21	138 48.5	.. 29.9	132 33.9	15.0	8 00.5	8.9	53.9
22	153 48.6	30.8	147 07.9	14.9	8 09.4	8.8	53.9
23	168 48.7	31.7	161 41.8	15.0	8 18.2	8.9	53.9
21 00	183 48.8	S10 32.6	176 15.8	14.9	S 8 27.1	8.7	53.9
01	198 48.9	33.5	190 49.7	14.9	8 35.8	8.8	53.9
02	213 49.0	34.3	205 23.6	14.9	8 44.6	8.7	53.9
03	228 49.2	.. 35.2	219 57.5	14.9	8 53.3	8.7	53.9
04	243 49.3	36.1	234 31.4	14.8	9 02.0	8.6	53.9
05	258 49.4	37.0	249 05.2	14.9	9 10.6	8.6	53.9
06	273 49.5	S10 37.9	263 39.1	14.8	S 9 19.2	8.6	53.9
W 07	288 49.6	38.8	278 12.9	14.7	9 27.8	8.5	53.9
E 08	303 49.7	39.7	292 46.6	14.8	9 36.3	8.5	53.9
D 09	318 49.8	.. 40.6	307 20.4	14.7	9 44.8	8.4	53.9
N 10	333 49.9	41.5	321 54.1	14.7	9 53.2	8.4	53.9
E 11	348 50.0	42.4	336 27.8	14.7	10 01.6	8.4	53.9
S 12	3 50.1	S10 43.3	351 01.5	14.7	S10 10.0	8.3	53.9
D 13	18 50.2	44.1	5 35.2	14.6	10 18.3	8.3	53.9
A 14	33 50.3	45.0	20 08.8	14.6	10 26.6	8.2	53.9
Y 15	48 50.4	.. 45.9	34 42.4	14.6	10 34.8	8.2	53.9
16	63 50.5	46.8	49 16.0	14.5	10 43.0	8.2	53.9
17	78 50.6	47.7	63 49.5	14.6	10 51.2	8.1	53.9
18	93 50.7	S10 48.6	78 23.1	14.4	S10 59.3	8.0	53.9
19	108 50.8	49.5	92 56.5	14.5	11 07.3	8.1	53.9
20	123 50.9	50.4	107 30.0	14.5	11 15.4	7.9	53.9
21	138 51.0	.. 51.3	122 03.5	14.4	11 23.3	8.0	53.9
22	153 51.1	52.1	136 36.9	14.4	11 31.3	7.8	53.9
23	168 51.1	53.0	151 10.3	14.3	S11 39.1	7.9	53.9
	SD 16.1	d 0.9	SD 14.7		14.7		14.7

Lat.	Twilight Naut.	Twilight Civil	Sunrise	Moonrise 19	Moonrise 20	Moonrise 21	Moonrise 22
°	h m	h m	h m	h m	h m	h m	h m
N 72	05 16	06 34	07 48	05 25	07 01	08 38	10 20
N 70	05 17	06 28	07 33	05 23	06 52	08 22	09 54
68	05 18	06 23	07 22	05 20	06 45	08 10	09 35
66	05 19	06 18	07 12	05 19	06 39	07 59	09 19
64	05 19	06 15	07 04	05 17	06 34	07 50	09 07
62	05 20	06 11	06 57	05 16	06 30	07 43	08 56
60	05 20	06 08	06 51	05 15	06 26	07 37	08 47
N 58	05 20	06 05	06 46	05 14	06 23	07 31	08 39
56	05 20	06 03	06 41	05 13	06 20	07 26	08 32
54	05 20	06 01	06 37	05 12	06 17	07 22	08 26
52	05 19	05 59	06 33	05 11	06 15	07 18	08 20
50	05 19	05 57	06 29	05 10	06 12	07 14	08 15
45	05 18	05 52	06 22	05 09	06 08	07 06	08 05
N 40	05 16	05 48	06 15	05 08	06 04	07 00	07 56
35	05 15	05 44	06 10	05 07	06 00	06 54	07 48
30	05 13	05 41	06 05	05 06	05 57	06 49	07 41
20	05 08	05 34	05 56	05 04	05 52	06 41	07 30
N 10	05 03	05 27	05 49	05 03	05 48	06 33	07 20
0	04 56	05 20	05 41	05 01	05 44	06 27	07 10
S 10	04 48	05 13	05 34	05 00	05 40	06 20	07 01
20	04 37	05 04	05 26	04 59	05 35	06 12	06 51
30	04 23	04 52	05 17	04 57	05 30	06 04	06 40
35	04 14	04 45	05 12	04 57	05 27	05 59	06 33
40	04 04	04 37	05 05	04 56	05 24	05 54	06 26
45	03 50	04 28	04 58	04 54	05 21	05 48	06 17
S 50	03 33	04 15	04 50	04 53	05 16	05 41	06 07
52	03 25	04 10	04 46	04 53	05 14	05 37	06 02
54	03 15	04 03	04 41	04 52	05 12	05 33	05 57
56	03 04	03 56	04 36	04 51	05 10	05 29	05 51
58	02 51	03 48	04 31	04 50	05 07	05 25	05 45
S 60	02 36	03 38	04 25	04 50	05 04	05 20	05 38

Lat.	Sunset	Twilight Civil	Twilight Naut.	Moonset 19	Moonset 20	Moonset 21	Moonset 22
°	h m	h m	h m	h m	h m	h m	h m
N 72	15 40	16 53	18 11	16 35	16 27	16 19	16 08
N 70	15 55	17 00	18 10	16 40	16 38	16 36	16 35
68	16 06	17 05	18 10	16 44	16 47	16 50	16 56
66	16 16	17 10	18 09	16 47	16 54	17 02	17 12
64	16 24	17 14	18 09	16 50	17 00	17 11	17 25
62	16 31	17 17	18 08	16 53	17 05	17 20	17 37
60	16 38	17 20	18 08	16 55	17 10	17 27	17 46
N 58	16 43	17 23	18 09	16 57	17 14	17 33	17 55
56	16 48	17 26	18 09	16 59	17 18	17 39	18 02
54	16 52	17 28	18 09	17 01	17 21	17 44	18 09
52	16 56	17 30	18 09	17 02	17 24	17 48	18 15
50	16 59	17 32	18 10	17 04	17 27	17 52	18 20
45	17 07	17 37	18 11	17 06	17 33	18 01	18 32
N 40	17 14	17 41	18 13	17 09	17 38	18 09	18 42
35	17 19	17 45	18 14	17 11	17 42	18 15	18 50
30	17 24	17 49	18 16	17 13	17 46	18 21	18 57
20	17 33	17 55	18 21	17 16	17 53	18 31	19 10
N 10	17 41	18 02	18 27	17 19	17 59	18 39	19 21
0	17 48	18 09	18 34	17 22	18 04	18 48	19 32
S 10	17 56	18 17	18 42	17 25	18 10	18 56	19 42
20	18 04	18 26	18 53	17 27	18 16	19 04	19 54
30	18 13	18 38	19 07	17 31	18 22	19 14	20 06
35	18 19	18 45	19 16	17 33	18 26	19 20	20 14
40	18 25	18 53	19 27	17 35	18 31	19 27	20 22
45	18 32	19 03	19 41	17 37	18 36	19 34	20 32
S 50	18 41	19 16	19 58	17 40	18 42	19 44	20 44
52	18 45	19 21	20 07	17 42	18 45	19 48	20 50
54	18 50	19 28	20 16	17 43	18 48	19 53	20 56
56	18 55	19 35	20 28	17 45	18 52	19 58	21 03
58	19 00	19 44	20 41	17 47	18 55	20 04	21 11
S 60	19 07	19 54	20 57	17 49	19 00	20 10	21 20

	SUN			MOON			
Day	Eqn. of Time 00ʰ	Eqn. of Time 12ʰ	Mer. Pass.	Mer. Pass. Upper	Mer. Pass. Lower	Age	Phase
d	m s	m s	h m	h m	h m	d	%
19	14 54	15 00	11 45	11 12	23 33	29	1
20	15 05	15 10	11 45	11 54	24 15	00	0
21	15 15	15 20	11 45	12 37	00 15	01	1

1998 OCTOBER 22, 23, 24 (THURS., FRI., SAT.)

UT	ARIES	VENUS −3.9		MARS +1.6		JUPITER −2.8		SATURN +0.0	
	GHA	GHA	Dec	GHA	Dec	GHA	Dec	GHA	Dec
d h	° ′	° ′	° ′	° ′	° ′	° ′	° ′	° ′	° ′
22 00	30 13.4	185 25.3	S 9 04.7	229 17.0	N 9 36.2	39 43.6	S 5 44.3	1 02.5	N 8 59.1
01	45 15.8	200 24.8	05.9	244 18.1	35.6	54 46.2	44.3	16 05.2	59.1
02	60 18.3	215 24.4	07.1	259 19.1	35.1	69 48.8	44.4	31 07.8	59.0
03	75 20.8	230 23.9	.. 08.3	274 20.2	.. 34.5	84 51.5	.. 44.5	46 10.5	.. 58.9
04	90 23.2	245 23.4	09.4	289 21.2	34.0	99 54.1	44.5	61 13.1	58.8
05	105 25.7	260 22.9	10.6	304 22.3	33.4	114 56.8	44.6	76 15.8	58.8
06	120 28.2	275 22.5	S 9 11.8	319 23.3	N 9 32.9	129 59.4	S 5 44.6	91 18.4	N 8 58.7
T 07	135 30.6	290 22.0	13.0	334 24.4	32.3	145 02.0	44.7	106 21.1	58.6
H 08	150 33.1	305 21.5	14.2	349 25.4	31.8	160 04.7	44.8	121 23.8	58.6
U 09	165 35.5	320 21.0	.. 15.4	4 26.5	.. 31.3	175 07.3	.. 44.8	136 26.4	.. 58.5
R 10	180 38.0	335 20.6	16.6	19 27.5	30.7	190 09.9	44.9	151 29.1	58.4
S 11	195 40.5	350 20.1	17.7	34 28.6	30.2	205 12.6	45.0	166 31.7	58.4
D 12	210 42.9	5 19.6	S 9 18.9	49 29.6	N 9 29.6	220 15.2	S 5 45.0	181 34.4	N 8 58.3
A 13	225 45.4	20 19.2	20.1	64 30.7	29.1	235 17.8	45.1	196 37.0	58.2
Y 14	240 47.9	35 18.7	21.3	79 31.7	28.5	250 20.5	45.2	211 39.7	58.2
15	255 50.3	50 18.2	.. 22.5	94 32.8	.. 28.0	265 23.1	.. 45.2	226 42.3	.. 58.1
16	270 52.8	65 17.7	23.6	109 33.8	27.4	280 25.8	45.3	241 45.0	58.0
17	285 55.3	80 17.2	24.8	124 34.9	26.9	295 28.4	45.4	256 47.6	57.9
18	300 57.7	95 16.8	S 9 26.0	139 35.9	N 9 26.3	310 31.0	S 5 45.4	271 50.3	N 8 57.9
19	316 00.2	110 16.3	27.2	154 36.9	25.8	325 33.7	45.5	286 52.9	57.8
20	331 02.7	125 15.8	28.4	169 38.0	25.2	340 36.3	45.5	301 55.6	57.7
21	346 05.1	140 15.3	.. 29.5	184 39.1	.. 24.7	355 38.9	.. 45.6	316 58.2	.. 57.7
22	1 07.6	155 14.9	30.7	199 40.1	24.1	10 41.5	45.7	332 00.9	57.6
23	16 10.0	170 14.4	31.9	214 41.2	23.6	25 44.2	45.7	347 03.6	57.5
23 00	31 12.5	185 13.9	S 9 33.1	229 42.2	N 9 23.0	40 46.8	S 5 45.8	2 06.2	N 8 57.5
01	46 15.0	200 13.4	34.3	244 43.3	22.5	55 49.4	45.9	17 08.9	57.4
02	61 17.4	215 12.9	35.4	259 44.3	21.9	70 52.1	45.9	32 11.5	57.3
03	76 19.9	230 12.5	.. 36.6	274 45.4	.. 21.4	85 54.7	.. 46.0	47 14.2	.. 57.3
04	91 22.4	245 12.0	37.8	289 46.4	20.8	100 57.3	46.0	62 16.8	57.2
05	106 24.8	260 11.5	39.0	304 47.5	20.3	116 00.0	46.1	77 19.5	57.1
06	121 27.3	275 11.0	S 9 40.1	319 48.5	N 9 19.7	131 02.6	S 5 46.2	92 22.1	N 8 57.1
07	136 29.8	290 10.5	41.3	334 49.6	19.2	146 05.2	46.2	107 24.8	57.0
F 08	151 32.2	305 10.1	42.5	349 50.6	18.6	161 07.9	46.3	122 27.4	56.9
R 09	166 34.7	320 09.6	.. 43.7	4 51.7	.. 18.1	176 10.5	.. 46.4	137 30.1	.. 56.8
I 10	181 37.1	335 09.1	44.8	19 52.7	17.5	191 13.1	46.4	152 32.7	56.8
11	196 39.6	350 08.6	46.0	34 53.8	17.0	206 15.7	46.5	167 35.4	56.7
D 12	211 42.1	5 08.1	S 9 47.2	49 54.8	N 9 16.4	221 18.4	S 5 46.5	182 38.1	N 8 56.6
A 13	226 44.5	20 07.7	48.4	64 55.9	15.9	236 21.0	46.6	197 40.7	56.6
Y 14	241 47.0	35 07.2	49.5	79 56.9	15.3	251 23.6	46.7	212 43.4	56.5
15	256 49.5	50 06.7	.. 50.7	94 58.0	.. 14.8	266 26.3	.. 46.7	227 46.0	.. 56.4
16	271 51.9	65 06.2	51.9	109 59.0	14.2	281 28.9	46.8	242 48.7	56.4
17	286 54.4	80 05.7	53.0	125 00.1	13.7	296 31.5	46.8	257 51.3	56.3
18	301 56.9	95 05.2	S 9 54.2	140 01.1	N 9 13.1	311 34.1	S 5 46.9	272 54.0	N 8 56.2
19	316 59.3	110 04.7	55.4	155 02.2	12.6	326 36.8	47.0	287 56.6	56.2
20	332 01.8	125 04.3	56.6	170 03.2	12.0	341 39.4	47.0	302 59.3	56.1
21	347 04.3	140 03.8	.. 57.7	185 04.3	.. 11.5	356 42.0	.. 47.1	318 01.9	.. 56.0
22	2 06.7	155 03.3	9 58.9	200 05.4	10.9	11 44.6	47.1	333 04.6	55.9
23	17 09.2	170 02.8	10 00.1	215 06.4	10.3	26 47.3	47.2	348 07.3	55.9
24 00	32 11.6	185 02.3	S10 01.2	230 07.5	N 9 09.8	41 49.9	S 5 47.3	3 09.9	N 8 55.8
01	47 14.1	200 01.8	02.4	245 08.5	09.2	56 52.5	47.3	18 12.6	55.7
02	62 16.6	215 01.3	03.6	260 09.6	08.7	71 55.1	47.4	33 15.2	55.7
03	77 19.0	230 00.8	.. 04.7	275 10.6	.. 08.1	86 57.8	.. 47.4	48 17.9	.. 55.6
04	92 21.5	245 00.4	05.9	290 11.7	07.6	102 00.4	47.5	63 20.5	55.5
05	107 24.0	259 59.9	07.1	305 12.7	07.0	117 03.0	47.5	78 23.2	55.5
06	122 26.4	274 59.4	S10 08.2	320 13.8	N 9 06.5	132 05.6	S 5 47.6	93 25.8	N 8 55.4
07	137 28.9	289 58.9	09.4	335 14.8	05.9	147 08.2	47.7	108 28.5	55.3
S 08	152 31.4	304 58.4	10.6	350 15.9	05.4	162 10.9	47.7	123 31.1	55.3
A 09	167 33.8	319 57.9	.. 11.7	5 17.0	.. 04.8	177 13.5	.. 47.8	138 33.8	.. 55.2
T 10	182 36.3	334 57.4	12.9	20 18.0	04.3	192 16.1	47.8	153 36.4	55.1
U 11	197 38.8	349 56.9	14.1	35 19.1	03.7	207 18.7	47.9	168 39.1	55.1
R 12	212 41.2	4 56.4	S10 15.2	50 20.1	N 9 03.2	222 21.4	S 5 48.0	183 41.8	N 8 55.0
D 13	227 43.7	19 55.9	16.4	65 21.2	02.6	237 24.0	48.0	198 44.4	54.9
A 14	242 46.1	34 55.5	17.5	80 22.2	02.1	252 26.6	48.1	213 47.1	54.8
Y 15	257 48.6	49 55.0	.. 18.7	95 23.3	.. 01.5	267 29.2	.. 48.1	228 49.7	.. 54.8
16	272 51.1	64 54.5	19.9	110 24.4	01.0	282 31.8	48.2	243 52.4	54.7
17	287 53.5	79 54.0	21.0	125 25.4	9 00.4	297 34.5	48.2	258 55.0	54.6
18	302 56.0	94 53.5	S10 22.2	140 26.5	N 8 59.9	312 37.1	S 5 48.3	273 57.7	N 8 54.6
19	317 58.5	109 53.0	23.4	155 27.5	59.3	327 39.7	48.4	289 00.3	54.5
20	333 00.9	124 52.5	24.5	170 28.6	58.8	342 42.3	48.4	304 03.0	54.4
21	348 03.4	139 52.0	.. 25.7	185 29.6	.. 58.2	357 44.9	.. 48.5	319 05.6	.. 54.4
22	3 05.9	154 51.5	26.8	200 30.7	57.7	12 47.6	48.5	334 08.3	54.3
23	18 08.3	169 51.0	28.0	215 31.8	57.1	27 50.2	48.6	349 10.9	54.2
Mer. Pass.	21 51.6	v −0.5	d 1.2	v 1.1	d 0.6	v 2.6	d 0.1	v 2.7	d 0.1

STARS

Name	SHA	Dec
	° ′	° ′
Acamar	315 26.5	S40 18.6
Achernar	335 34.7	S57 14.7
Acrux	173 22.6	S63 05.3
Adhara	255 21.3	S28 58.1
Aldebaran	291 02.1	N16 30.3
Alioth	166 31.2	N55 58.1
Alkaid	153 08.3	N49 19.3
Al Na'ir	27 57.7	S46 58.1
Alnilam	275 57.7	S 1 12.2
Alphard	218 07.3	S 8 39.1
Alphecca	126 20.9	N26 43.4
Alpheratz	357 54.9	N29 05.1
Altair	62 19.3	N 8 52.2
Ankaa	353 26.5	S42 18.8
Antares	112 40.5	S26 25.6
Arcturus	146 06.4	N19 11.6
Atria	107 52.9	S69 01.5
Avior	234 22.6	S59 30.1
Bellatrix	278 43.9	N 6 20.9
Betelgeuse	271 13.4	N 7 24.4
Canopus	264 00.9	S52 41.5
Capella	280 50.9	N45 59.6
Deneb	49 39.1	N45 16.9
Denebola	182 45.5	N14 34.9
Diphda	349 06.9	S17 59.6
Dubhe	194 06.1	N61 45.3
Elnath	278 26.7	N28 36.2
Eltanin	90 51.7	N51 29.8
Enif	33 58.1	N 9 52.3
Fomalhaut	15 36.2	S29 37.8
Gacrux	172 14.0	S57 06.1
Gienah	176 04.3	S17 31.9
Hadar	149 04.6	S60 21.8
Hamal	328 13.2	N23 27.4
Kaus Aust.	83 59.0	S34 23.1
Kochab	137 20.9	N74 09.8
Markab	13 49.4	N15 12.1
Menkar	314 26.6	N 4 05.1
Menkent	148 21.3	S36 21.6
Miaplacidus	221 42.2	S69 42.4
Mirfak	308 56.1	N49 51.3
Nunki	76 12.5	S26 17.8
Peacock	53 37.0	S56 44.5
Pollux	243 41.5	N28 01.6
Procyon	245 11.5	N 5 13.7
Rasalhague	96 17.2	N12 34.0
Regulus	207 55.7	N11 58.4
Rigel	281 22.7	S 8 12.2
Rigil Kent.	140 07.9	S60 49.6
Sabik	102 25.8	S15 43.2
Schedar	349 52.9	N56 31.9
Shaula	96 37.5	S37 06.1
Sirius	258 43.6	S16 42.8
Spica	158 43.5	S11 09.1
Suhail	223 00.9	S43 25.4
Vega	80 46.8	N38 47.3
Zuben'ubi	137 18.3	S16 02.0

	SHA	Mer. Pass.
	° ′	h m
Venus	154 01.4	11 39
Mars	198 29.7	8 41
Jupiter	9 34.3	21 13
Saturn	330 53.7	23 47

UT	SUN		MOON				Lat.	Twilight		Sunrise	Moonrise			
								Naut.	Civil		22	23	24	25
	GHA	Dec	GHA	v	Dec	d	HP	°	h m	h m	h m	h m	h m	h m
d h	° '	° '	° '	'	° '	'	'	N 72	05 28	06 48	08 04	10 20	12 10	▬▬	▬▬
22 00	183 51.2	S10 53.9	165 43.6	14.3	S11 47.0	7.7	53.9	N 70	05 28	06 40	07 47	09 54	11 28	13 01	14 29
01	198 51.3	54.8	180 16.9	14.3	11 54.7	7.8	54.0	68	05 28	06 33	07 33	09 35	10 59	12 21	13 33
02	213 51.4	55.7	194 50.2	14.3	12 02.5	7.6	54.0	66	05 28	06 28	07 22	09 19	10 38	11 53	13 00
03	228 51.5	. . 56.6	209 23.5	14.2	12 10.1	7.7	54.0	64	05 28	06 23	07 13	09 07	10 21	11 32	12 36
04	243 51.6	57.5	223 56.7	14.3	12 17.8	7.6	54.0	62	05 27	06 19	07 05	08 56	10 07	11 15	12 17
05	258 51.7	58.3	238 30.0	14.1	12 25.4	7.5	54.0	60	05 27	06 15	06 59	08 47	09 56	11 01	12 02
06	273 51.8	S10 59.2	253 03.1	14.2	S12 32.9	7.5	54.0	N 58	05 26	06 12	06 53	08 39	09 46	10 49	11 48
07	288 51.9	11 00.1	267 36.3	14.1	12 40.4	7.4	54.0	56	05 26	06 09	06 47	08 32	09 37	10 39	11 37
T 08	303 52.0	01.0	282 09.4	14.1	12 47.8	7.3	54.0	54	05 25	06 06	06 43	08 26	09 29	10 30	11 27
H 09	318 52.1	. . 01.9	296 42.5	14.0	12 55.1	7.4	54.0	52	05 24	06 04	06 38	08 20	09 22	10 22	11 18
U 10	333 52.2	02.8	311 15.5	14.0	13 02.5	7.2	54.0	50	05 24	06 01	06 34	08 15	09 16	10 14	11 10
R 11	348 52.3	03.7	325 48.5	14.0	13 09.7	7.2	54.0	45	05 22	05 56	06 26	08 05	09 02	09 59	10 53
S 12	3 52.4	S11 04.5	340 21.5	14.0	S13 16.9	7.2	54.0	N 40	05 19	05 51	06 19	07 56	08 51	09 46	10 39
D 13	18 52.5	05.4	354 54.5	13.9	13 24.1	7.1	54.0	35	05 17	05 47	06 13	07 48	08 42	09 35	10 27
A 14	33 52.6	06.3	9 27.4	13.9	13 31.2	7.0	54.0	30	05 15	05 43	06 07	07 41	08 33	09 26	10 17
Y 15	48 52.7	. . 07.2	24 00.3	13.9	13 38.2	7.0	54.0	20	05 09	05 35	05 58	07 30	08 19	09 09	10 00
16	63 52.8	08.1	38 33.2	13.8	13 45.2	6.9	54.0	N 10	05 03	05 28	05 49	07 20	08 07	08 55	09 44
17	78 52.8	08.9	53 06.0	13.8	13 52.1	6.9	54.0	0	04 55	05 20	05 41	07 10	07 55	08 42	09 30
18	93 52.9	S11 09.8	67 38.8	13.7	S13 59.0	6.8	54.0	S 10	04 46	05 11	05 33	07 01	07 44	08 29	09 16
19	108 53.0	10.7	82 11.5	13.7	14 05.8	6.8	54.0	20	04 35	05 01	05 24	06 51	07 32	08 15	09 00
20	123 53.1	11.6	96 44.2	13.7	14 12.6	6.6	54.0	30	04 20	04 49	05 14	06 40	07 18	07 59	08 43
21	138 53.2	. . 12.5	111 16.9	13.7	14 19.2	6.7	54.0	35	04 10	04 42	05 08	06 33	07 09	07 49	08 33
22	153 53.3	13.4	125 49.6	13.6	14 25.9	6.5	54.1	40	03 59	04 33	05 01	06 26	07 00	07 39	08 21
23	168 53.4	14.2	140 22.2	13.6	14 32.4	6.5	54.1	45	03 44	04 22	04 53	06 17	06 50	07 26	08 08
23 00	183 53.5	S11 15.1	154 54.8	13.5	S14 38.9	6.5	54.1	S 50	03 26	04 09	04 44	06 07	06 37	07 11	07 51
01	198 53.6	16.0	169 27.3	13.5	14 45.4	6.4	54.1	52	03 17	04 03	04 39	06 02	06 31	07 04	07 43
02	213 53.7	16.9	183 59.8	13.5	14 51.8	6.3	54.1	54	03 07	03 56	04 34	05 57	06 24	06 56	07 35
03	228 53.8	. . 17.7	198 32.3	13.4	14 58.1	6.2	54.1	56	02 55	03 48	04 29	05 51	06 17	06 48	07 25
04	243 53.8	18.6	213 04.7	13.5	15 04.3	6.2	54.1	58	02 40	03 39	04 23	05 45	06 09	06 38	07 14
05	258 53.9	19.5	227 37.2	13.3	15 10.5	6.2	54.1	S 60	02 23	03 28	04 16	05 38	05 59	06 26	07 01

UT	SUN		MOON				Lat.	Sunset	Twilight		Moonset			
									Civil	Naut.	22	23	24	25
	GHA	Dec	GHA	v	Dec	d	HP	°	h m	h m	h m	h m	h m	h m
06	273 54.0	S11 20.4	242 09.5	13.4	S15 16.7	6.0	54.1	N 72	15 23	16 39	17 58	16 08	15 53	▬▬	▬▬
07	288 54.1	21.3	256 41.9	13.3	15 22.7	6.0	54.1	N 70	15 40	16 47	17 58	16 35	16 36	16 39	16 52
08	303 54.2	22.1	271 14.2	13.2	15 28.7	5.9	54.1	68	15 54	16 54	17 59	16 56	17 04	17 20	17 48
F 09	318 54.3	. . 23.0	285 46.4	13.2	15 34.6	5.9	54.1	66	16 05	16 59	17 59	17 12	17 26	17 48	18 21
R 10	333 54.4	23.9	300 18.6	13.2	15 40.5	5.8	54.1	64	16 14	17 04	18 00	17 25	17 44	18 09	18 45
I 11	348 54.5	24.8	314 50.8	13.2	15 46.3	5.7	54.2	62	16 22	17 09	18 00	17 37	17 58	18 26	19 04
D 12	3 54.5	S11 25.6	329 23.0	13.1	S15 52.0	5.7	54.2	60	16 29	17 12	18 01	17 46	18 10	18 41	19 20
A 13	18 54.6	26.5	343 55.1	13.1	15 57.7	5.6	54.2	N 58	16 35	17 16	18 01	17 55	18 21	18 53	19 33
Y 14	33 54.7	27.4	358 27.2	13.0	16 03.3	5.5	54.2	56	16 40	17 19	18 02	18 02	18 30	19 04	19 45
15	48 54.8	. . 28.3	12 59.2	13.0	16 08.8	5.4	54.2	54	16 45	17 22	18 03	18 09	18 38	19 13	19 55
16	63 54.9	29.1	27 31.2	13.0	16 14.2	5.4	54.2	52	16 50	17 24	18 04	18 15	18 45	19 21	20 04
17	78 55.0	30.0	42 03.2	12.9	16 19.6	5.3	54.2	50	16 54	17 27	18 04	18 20	18 52	19 29	20 12
18	93 55.1	S11 30.9	56 35.1	12.9	S16 24.9	5.3	54.2	45	17 02	17 32	18 06	18 32	19 06	19 45	20 29
19	108 55.1	31.8	71 07.0	12.9	16 30.2	5.1	54.2	N 40	17 10	17 37	18 09	18 42	19 18	19 58	20 43
20	123 55.2	32.6	85 38.9	12.8	16 35.3	5.1	54.2	35	17 16	17 41	18 11	18 50	19 28	20 09	20 55
21	138 55.3	. . 33.5	100 10.7	12.8	16 40.4	5.0	54.3	30	17 21	17 46	18 14	18 57	19 37	20 19	21 06
22	153 55.4	34.4	114 42.5	12.7	16 45.4	5.0	54.3	20	17 31	17 53	18 19	19 10	19 52	20 36	21 24
23	168 55.5	35.3	129 14.2	12.7	16 50.4	4.8	54.3	N 10	17 40	18 01	18 26	19 21	20 05	20 51	21 39
24 00	183 55.6	S11 36.1	143 45.9	12.7	S16 55.2	4.8	54.3	0	17 48	18 09	18 33	19 32	20 18	21 05	21 54
01	198 55.6	37.0	158 17.6	12.6	17 00.0	4.8	54.3	S 10	17 56	18 18	18 43	19 42	20 30	21 19	22 08
02	213 55.7	37.9	172 49.2	12.6	17 04.8	4.6	54.3	20	18 05	18 28	18 54	19 54	20 43	21 33	22 24
03	228 55.8	. . 38.8	187 20.8	12.6	17 09.4	4.6	54.3	30	18 15	18 40	19 09	20 06	20 59	21 50	22 42
04	243 55.9	39.6	201 52.4	12.5	17 14.0	4.5	54.3	35	18 21	18 48	19 19	20 14	21 07	22 00	22 52
05	258 56.0	40.5	216 23.9	12.5	17 18.5	4.4	54.4	40	18 28	18 57	19 31	20 22	21 18	22 12	23 04
06	273 56.0	S11 41.4	230 55.4	12.5	S17 22.9	4.3	54.4	45	18 36	19 07	19 45	20 32	21 30	22 25	23 18
07	288 56.1	42.2	245 26.9	12.4	17 27.2	4.3	54.4	S 50	18 46	19 21	20 04	20 44	21 44	22 41	23 35
S 08	303 56.2	43.1	259 58.3	12.4	17 31.5	4.1	54.4	52	18 51	19 27	20 14	20 50	21 51	22 49	23 43
A 09	318 56.3	. . 44.0	274 29.7	12.3	17 35.6	4.1	54.4	54	18 56	19 35	20 24	20 56	21 58	22 57	23 52
T 10	333 56.4	44.8	289 01.0	12.3	17 39.7	4.1	54.4	56	19 01	19 43	20 36	21 03	22 07	23 07	24 02
U 11	348 56.4	45.7	303 32.3	12.3	17 43.8	3.9	54.4	58	19 07	19 52	20 51	21 11	22 16	23 18	24 13
R 12	3 56.5	S11 46.6	318 03.6	12.2	S17 47.7	3.9	54.5	S 60	19 14	20 03	21 09	21 20	22 27	23 30	24 27
D 13	18 56.6	47.5	332 34.8	12.2	17 51.6	3.7	54.5								
A 14	33 56.7	48.3	347 06.0	12.2	17 55.3	3.7	54.5		SUN			MOON			
Y 15	48 56.8	. . 49.2	1 37.2	12.1	17 59.0	3.7	54.5	Day	Eqn. of Time		Mer.	Mer. Pass.		Age	Phase
16	63 56.8	50.1	16 08.3	12.1	18 02.7	3.5	54.5		00ʰ	12ʰ	Pass.	Upper	Lower		
17	78 56.9	50.9	30 39.4	12.0	18 06.2	3.4	54.5	d	m s	m s	h m	h m	h m	d	%
18	93 57.0	S11 51.8	45 10.4	12.1	S18 09.6	3.4	54.5	22	15 25	15 29	11 45	13 21	00 59	02	4
19	108 57.1	52.7	59 41.5	11.9	18 13.0	3.3	54.6	23	15 34	15 38	11 44	14 06	01 44	03	8
20	123 57.1	53.5	74 12.4	12.0	18 16.3	3.2	54.6	24	15 42	15 46	11 44	14 53	02 30	04	15
21	138 57.2	. . 54.4	88 43.4	11.9	18 19.5	3.1	54.6								
22	153 57.3	55.3	103 14.3	11.9	18 22.6	3.0	54.6								
23	168 57.4	56.1	117 45.2	11.8	S18 25.6	3.0	54.6								
	SD 16.1	d 0.9	SD	14.7	14.8		14.8								

1998 OCTOBER 25, 26, 27 (SUN., MON., TUES.)

UT	ARIES GHA	VENUS −3.9 GHA	Dec	MARS +1.6 GHA	Dec	JUPITER −2.8 GHA	Dec	SATURN +0.0 GHA	Dec	STARS Name	SHA	Dec
25 00	33 10.8	184 50.5	S10 29.2	230 32.8	N 8 56.6	42 52.8	S 5 48.6	4 13.6	N 8 54.2	Acamar	315 26.5	S40 18.6
01	48 13.2	199 50.0	30.3	245 33.9	56.0	57 55.4	48.7	19 16.3	54.1	Achernar	335 34.7	S57 14.7
02	63 15.7	214 49.5	31.5	260 34.9	55.5	72 58.0	48.8	34 18.9	54.0	Acrux	173 22.5	S63 05.3
03	78 18.2	229 49.0	.. 32.6	275 36.0	.. 54.9	88 00.6	.. 48.8	49 21.6	.. 54.0	Adhara	255 21.3	S28 58.1
04	93 20.6	244 48.5	33.8	290 37.1	54.4	103 03.3	48.9	64 24.2	53.9	Aldebaran	291 02.1	N16 30.3
05	108 23.1	259 48.0	34.9	305 38.1	53.8	118 05.9	48.9	79 26.9	53.8			
06	123 25.6	274 47.5	S10 36.1	320 39.2	N 8 53.3	133 08.5	S 5 49.0	94 29.5	N 8 53.7	Alioth	166 31.2	N55 58.0
07	138 28.0	289 47.0	37.3	335 40.2	52.7	148 11.1	49.0	109 32.2	53.7	Alkaid	153 08.3	N49 19.3
08	153 30.5	304 46.5	38.4	350 41.3	52.1	163 13.7	49.1	124 34.8	53.6	Al Na'ir	27 57.7	S46 58.1
S 09	168 33.0	319 46.0	.. 39.6	5 42.3	.. 51.6	178 16.3	.. 49.1	139 37.5	.. 53.5	Alnilam	275 57.6	S 1 12.2
U 10	183 35.4	334 45.5	40.7	20 43.4	51.0	193 18.9	49.2	154 40.1	53.5	Alphard	218 07.3	S 8 39.1
N 11	198 37.9	349 45.0	41.9	35 44.5	50.5	208 21.6	49.2	169 42.8	53.4			
D 12	213 40.4	4 44.5	S10 43.0	50 45.5	N 8 49.9	223 24.2	S 5 49.3	184 45.4	N 8 53.3	Alphecca	126 21.0	N26 43.4
A 13	228 42.8	19 44.0	44.2	65 46.6	49.4	238 26.8	49.4	199 48.1	53.3	Alpheratz	357 54.9	N29 05.1
Y 14	243 45.3	34 43.5	45.3	80 47.6	48.8	253 29.4	49.4	214 50.8	53.2	Altair	62 19.3	N 8 52.2
15	258 47.7	49 43.0	.. 46.5	95 48.7	.. 48.3	268 32.0	.. 49.5	229 53.4	.. 53.1	Ankaa	353 26.5	S42 18.9
16	273 50.2	64 42.5	47.6	110 49.8	47.7	283 34.6	49.5	244 56.1	53.1	Antares	112 40.5	S26 25.6
17	288 52.7	79 42.0	48.8	125 50.8	47.2	298 37.2	49.6	259 58.7	53.0			
18	303 55.1	94 41.5	S10 49.9	140 51.9	N 8 46.6	313 39.9	S 5 49.6	275 01.4	N 8 52.9	Arcturus	146 06.4	N19 11.5
19	318 57.6	109 41.0	51.1	155 53.0	46.1	328 42.5	49.7	290 04.0	52.9	Atria	107 52.9	S69 01.5
20	334 00.1	124 40.5	52.2	170 54.0	45.5	343 45.1	49.7	305 06.7	52.8	Avior	234 22.6	S59 30.1
21	349 02.5	139 40.0	.. 53.4	185 55.1	.. 45.0	358 47.7	.. 49.8	320 09.3	.. 52.7	Bellatrix	278 43.9	N 6 20.9
22	4 05.0	154 39.5	54.5	200 56.1	44.4	13 50.3	49.8	335 12.0	52.7	Betelgeuse	271 13.4	N 7 24.4
23	19 07.5	169 39.0	55.7	215 57.2	43.9	28 52.9	49.9	350 14.6	52.6			
26 00	34 09.9	184 38.5	S10 56.8	230 58.3	N 8 43.3	43 55.5	S 5 49.9	5 17.3	N 8 52.5	Canopus	264 00.9	S52 41.5
01	49 12.4	199 38.0	58.0	245 59.3	42.8	58 58.1	50.0	20 19.9	52.4	Capella	280 50.8	N45 59.6
02	64 14.9	214 37.5	10 59.1	261 00.4	42.2	74 00.7	50.1	35 22.6	52.4	Deneb	49 39.1	N45 16.9
03	79 17.3	229 37.0	11 00.3	276 01.4	.. 41.7	89 03.4	.. 50.1	50 25.2	.. 52.3	Denebola	182 45.5	N14 34.8
04	94 19.8	244 36.5	01.4	291 02.5	41.1	104 06.0	50.2	65 27.9	52.2	Diphda	349 06.9	S17 59.6
05	109 22.2	259 36.0	02.6	306 03.6	40.5	119 08.6	50.2	80 30.6	52.2			
06	124 24.7	274 35.5	S11 03.7	321 04.6	N 8 40.0	134 11.2	S 5 50.3	95 33.2	N 8 52.1	Dubhe	194 06.0	N61 45.3
07	139 27.2	289 35.0	04.9	336 05.7	39.4	149 13.8	50.3	110 35.9	52.0	Elnath	278 26.7	N28 36.2
08	154 29.6	304 34.4	06.0	351 06.8	38.9	164 16.4	50.4	125 38.5	52.0	Eltanin	90 51.7	N51 29.7
M 09	169 32.1	319 33.9	.. 07.2	6 07.8	.. 38.3	179 19.0	.. 50.4	140 41.2	.. 51.9	Enif	33 58.1	N 9 52.3
O 10	184 34.6	334 33.4	08.3	21 08.9	37.8	194 21.6	50.5	155 43.8	51.8	Fomalhaut	15 36.2	S29 37.8
N 11	199 37.0	349 32.9	09.4	36 09.9	37.2	209 24.2	50.5	170 46.5	51.8			
D 12	214 39.5	4 32.4	S11 10.6	51 11.0	N 8 36.7	224 26.8	S 5 50.6	185 49.1	N 8 51.7	Gacrux	172 14.0	S57 06.1
A 13	229 42.0	19 31.9	11.7	66 12.1	36.1	239 29.4	50.6	200 51.8	51.6	Gienah	176 04.2	S17 31.9
Y 14	244 44.4	34 31.4	12.9	81 13.1	35.6	254 32.0	50.7	215 54.4	51.6	Hadar	149 04.6	S60 21.8
15	259 46.9	49 30.9	.. 14.0	96 14.2	.. 35.0	269 34.6	.. 50.7	230 57.1	.. 51.5	Hamal	328 13.2	N23 27.4
16	274 49.4	64 30.4	15.2	111 15.3	34.5	284 37.2	50.8	245 59.7	51.4	Kaus Aust.	83 59.0	S34 23.1
17	289 51.8	79 29.9	16.3	126 16.3	33.9	299 39.9	50.8	261 02.4	51.4			
18	304 54.3	94 29.3	S11 17.4	141 17.4	N 8 33.4	314 42.5	S 5 50.9	276 05.1	N 8 51.3	Kochab	137 20.9	N74 09.8
19	319 56.7	109 28.8	18.6	156 18.5	32.8	329 45.1	50.9	291 07.7	51.2	Markab	13 49.4	N15 12.1
20	334 59.2	124 28.3	19.7	171 19.5	32.2	344 47.7	51.0	306 10.4	51.2	Menkar	314 26.6	N 4 05.1
21	350 01.7	139 27.8	.. 20.9	186 20.6	.. 31.7	359 50.3	.. 51.0	321 13.0	.. 51.1	Menkent	148 21.3	S36 21.6
22	5 04.1	154 27.3	22.0	201 21.7	31.1	14 52.9	51.1	336 15.7	51.0	Miaplacidus	221 42.2	S69 42.4
23	20 06.6	169 26.8	23.1	216 22.7	30.6	29 55.5	51.1	351 18.3	50.9			
27 00	35 09.1	184 26.3	S11 24.3	231 23.8	N 8 30.0	44 58.1	S 5 51.2	6 21.0	N 8 50.9	Mirfak	308 56.1	N49 51.3
01	50 11.5	199 25.7	25.4	246 24.9	29.5	60 00.7	51.2	21 23.6	50.8	Nunki	76 12.5	S26 17.8
02	65 14.0	214 25.2	26.5	261 25.9	28.9	75 03.3	51.3	36 26.3	50.7	Peacock	53 37.0	S56 44.5
03	80 16.5	229 24.7	.. 27.7	276 27.0	.. 28.4	90 05.9	.. 51.3	51 28.9	.. 50.7	Pollux	243 41.5	N28 01.6
04	95 18.9	244 24.2	28.8	291 28.1	27.8	105 08.5	51.4	66 31.6	50.6	Procyon	245 11.5	N 5 13.7
05	110 21.4	259 23.7	29.9	306 29.1	27.3	120 11.1	51.4	81 34.2	50.5			
06	125 23.8	274 23.2	S11 31.1	321 30.2	N 8 26.7	135 13.7	S 5 51.5	96 36.9	N 8 50.5	Rasalhague	96 17.2	N12 33.9
07	140 26.3	289 22.6	32.2	336 31.3	26.2	150 16.3	51.5	111 39.5	50.4	Regulus	207 55.7	N11 58.4
08	155 28.8	304 22.1	33.3	351 32.3	25.6	165 18.9	51.6	126 42.2	50.3	Rigel	281 22.7	S 8 12.2
T 09	170 31.2	319 21.6	.. 34.5	6 33.4	.. 25.1	180 21.5	.. 51.6	141 44.9	.. 50.3	Rigil Kent.	140 07.9	S60 49.6
U 10	185 33.7	334 21.1	35.6	21 34.5	24.5	195 24.1	51.7	156 47.5	50.2	Sabik	102 25.8	S15 43.2
E 11	200 36.2	349 20.6	36.7	36 35.5	23.9	210 26.7	51.7	171 50.2	50.1			
S 12	215 38.6	4 20.1	S11 37.9	51 36.6	N 8 23.4	225 29.3	S 5 51.8	186 52.8	N 8 50.1	Schedar	349 52.9	N56 31.9
D 13	230 41.1	19 19.5	39.0	66 37.7	22.8	240 31.9	51.8	201 55.5	50.0	Shaula	96 37.6	S37 06.1
A 14	245 43.6	34 19.0	40.1	81 38.7	22.3	255 34.5	51.9	216 58.1	49.9	Sirius	258 43.6	S16 42.8
Y 15	260 46.0	49 18.5	.. 41.3	96 39.8	.. 21.7	270 37.1	.. 51.9	232 00.8	.. 49.9	Spica	158 43.5	S11 09.1
16	275 48.5	64 18.0	42.4	111 40.9	21.2	285 39.7	52.0	247 03.4	49.8	Suhail	223 00.8	S43 25.4
17	290 51.0	79 17.5	43.5	126 41.9	20.6	300 42.3	52.0	262 06.1	49.7			
18	305 53.4	94 16.9	S11 44.7	141 43.0	N 8 20.1	315 44.9	S 5 52.0	277 08.7	N 8 49.7	Vega	80 46.8	N38 47.3
19	320 55.9	109 16.4	45.8	156 44.1	19.5	330 47.5	52.1	292 11.4	49.6	Zuben'ubi	137 18.3	S16 02.0
20	335 58.3	124 15.9	46.9	171 45.1	19.0	345 50.1	52.1	307 14.0	49.5		SHA	Mer. Pass.
21	351 00.8	139 15.4	.. 48.0	186 46.2	.. 18.4	0 52.7	.. 52.2	322 16.7	.. 49.5	Venus	150 28.6	11 42
22	6 03.3	154 14.8	49.2	201 47.3	17.9	15 55.3	52.2	337 19.3	49.4	Mars	196 48.3	8 36
23	21 05.7	169 14.3	50.3	216 48.3	17.3	30 57.9	52.3	352 22.0	49.3	Jupiter	9 45.6	21 01
Mer. Pass. 21 39.8		v −0.5	d 1.1	v 1.1	d 0.6	v 2.6	d 0.1	v 2.7	d 0.1	Saturn	331 07.4	23 35

UT		SUN		MOON					Lat.	Twilight Naut.	Twilight Civil	Sunrise	Moonrise 25	26	27	28
		GHA	Dec	GHA	v	Dec	d	HP								
d h		° ′	° ′	° ′	′	° ′	′	′	°	h m	h m	h m	h m	h m	h m	h m
25 00		183 57.5	S11 57.0	132 16.0	11.8	S18 28.6	2.8	54.6	N 72	05 41	07 01	08 20	■	■	■	16 41
01		198 57.5	57.9	146 46.8	11.8	18 31.4	2.8	54.7	N 70	05 40	06 52	08 01	14 29	15 27	15 47	15 53
02		213 57.6	58.7	161 17.6	11.7	18 34.2	2.7	54.7	68	05 38	06 44	07 45	13 33	14 28	15 03	15 22
03		228 57.7	11 59.6	175 48.3	11.8	18 36.9	2.6	54.7	66	05 37	06 37	07 33	13 00	13 54	14 33	14 59
04		243 57.7	12 00.4	190 19.1	11.6	18 39.5	2.5	54.7	64	05 36	06 32	07 23	12 36	13 29	14 10	14 40
05		258 57.8	01.3	204 49.7	11.7	18 42.0	2.4	54.7	62	05 35	06 27	07 14	12 17	13 10	13 53	14 25
06		273 57.9	S12 02.2	219 20.4	11.6	S18 44.4	2.4	54.7	60	05 34	06 22	07 06	12 02	12 54	13 38	14 13
07		288 58.0	03.0	233 51.0	11.5	18 46.8	2.2	54.8	N 58	05 32	06 18	07 00	11 48	12 41	13 25	14 02
08		303 58.0	03.9	248 21.5	11.6	18 49.0	2.2	54.8	56	05 31	06 15	06 54	11 37	12 29	13 14	13 52
S 09		318 58.1	04.8	262 52.1	11.5	18 51.2	2.1	54.8	54	05 30	06 12	06 48	11 27	12 19	13 04	13 44
U 10		333 58.2	05.6	277 22.6	11.4	18 53.3	1.9	54.8	52	05 29	06 09	06 44	11 18	12 10	12 56	13 36
N 11		348 58.3	06.5	291 53.0	11.5	18 55.2	1.9	54.8	50	05 28	06 06	06 39	11 10	12 01	12 48	13 29
D 12		3 58.3	S12 07.3	306 23.5	11.4	S18 57.1	1.8	54.9	45	05 25	06 00	06 30	10 53	11 44	12 32	13 15
A 13		18 58.4	08.2	320 53.9	11.4	18 58.9	1.7	54.9	N 40	05 22	05 54	06 22	10 39	11 30	12 18	13 03
Y 14		33 58.5	09.1	335 24.3	11.3	19 00.6	1.7	54.9	35	05 20	05 49	06 15	10 27	11 18	12 07	12 52
15		48 58.5	09.9	349 54.6	11.3	19 02.3	1.5	54.9	30	05 17	05 45	06 09	10 17	11 08	11 56	12 43
16		63 58.6	10.8	4 24.9	11.3	19 03.8	1.4	54.9	20	05 10	05 36	05 59	10 00	10 50	11 39	12 28
17		78 58.7	11.6	18 55.2	11.3	19 05.2	1.4	55.0	N 10	05 03	05 28	05 49	09 44	10 34	11 24	12 14
18		93 58.8	S12 12.5	33 25.5	11.2	S19 06.6	1.2	55.0	0	04 55	05 19	05 41	09 30	10 19	11 10	12 01
19		108 58.8	13.4	47 55.7	11.2	19 07.8	1.2	55.0	S 10	04 45	05 10	05 32	09 16	10 05	10 56	11 48
20		123 58.9	14.2	62 25.9	11.1	19 09.0	1.0	55.0	20	04 33	04 59	05 22	09 00	09 49	10 41	11 35
21		138 59.0	15.1	76 56.0	11.2	19 10.0	1.0	55.0	30	04 17	04 46	05 11	08 43	09 31	10 23	11 19
22		153 59.0	15.9	91 26.2	11.1	19 11.0	0.9	55.1	35	04 06	04 38	05 04	08 33	09 21	10 13	11 10
23		168 59.1	16.8	105 56.3	11.0	19 11.9	0.8	55.1	40	03 54	04 29	04 57	08 21	09 09	10 02	10 59
26 00		183 59.2	S12 17.7	120 26.3	11.1	S19 12.7	0.7	55.1	45	03 39	04 17	04 48	08 08	08 55	09 48	10 47
01		198 59.2	18.5	134 56.4	11.0	19 13.4	0.6	55.1	S 50	03 19	04 03	04 38	07 51	08 38	09 32	10 32
02		213 59.3	19.4	149 26.4	11.0	19 14.0	0.5	55.2	52	03 09	03 56	04 33	07 43	08 30	09 24	10 25
03		228 59.4	20.2	163 56.4	10.9	19 14.5	0.4	55.2	54	02 58	03 48	04 28	07 35	08 21	09 15	10 18
04		243 59.4	21.1	178 26.3	11.0	19 14.9	0.3	55.2	56	02 45	03 40	04 22	07 25	08 11	09 05	10 09
05		258 59.5	21.9	192 56.3	10.9	19 15.2	0.2	55.2	58	02 29	03 30	04 15	07 14	07 59	08 54	09 59
06		273 59.6	S12 22.8	207 26.2	10.8	S19 15.4	0.1	55.2	S 60	02 10	03 19	04 07	07 01	07 46	08 42	09 48
07		288 59.6	23.6	221 56.0	10.9	19 15.5	0.0	55.3								
08		303 59.7	24.5	236 25.9	10.8	19 15.5	0.2	55.3	Lat.	Sunset	Twilight Civil	Twilight Naut.	Moonset 25	26	27	28
M 09		318 59.8	25.4	250 55.7	10.8	19 15.5	0.2	55.3								
O 10		333 59.8	26.2	265 25.5	10.8	19 15.3	0.3	55.3	°	h m	h m	h m	h m	h m	h m	h m
N 11		348 59.9	27.1	279 55.3	10.7	19 15.0	0.4	55.4	N 72	15 06	16 25	17 45	■	■	■	19 57
D 12		4 00.0	S12 27.9	294 25.0	10.7	S19 14.7	0.5	55.4	N 70	15 26	16 35	17 47	16 52	17 38	19 03	20 44
A 13		19 00.0	28.8	308 54.7	10.7	19 14.2	0.5	55.4	68	15 41	16 43	17 48	17 48	18 36	19 47	21 15
Y 14		34 00.1	29.6	323 24.4	10.7	19 13.7	0.7	55.4	66	15 54	16 49	17 49	18 21	19 10	20 17	21 37
15		49 00.2	30.5	337 54.1	10.7	19 13.0	0.7	55.5	64	16 04	16 55	17 51	18 45	19 35	20 39	21 55
16		64 00.2	31.3	352 23.8	10.6	19 12.3	0.9	55.5	62	16 13	17 00	17 52	19 04	19 54	20 56	22 09
17		79 00.3	32.2	6 53.4	10.6	19 11.4	0.9	55.5	60	16 21	17 05	17 53	19 20	20 10	21 11	22 22
18		94 00.3	S12 33.0	21 23.0	10.6	S19 10.5	1.1	55.6	N 58	16 28	17 09	17 54	19 33	20 23	21 23	22 32
19		109 00.4	33.9	35 52.6	10.5	19 09.4	1.1	55.6	56	16 33	17 12	17 56	19 45	20 35	21 34	22 41
20		124 00.5	34.7	50 22.1	10.6	19 08.3	1.3	55.6	54	16 39	17 15	17 57	19 55	20 45	21 43	22 49
21		139 00.5	35.6	64 51.7	10.5	19 07.0	1.3	55.6	52	16 44	17 19	17 58	20 04	20 54	21 52	22 57
22		154 00.6	36.4	79 21.2	10.5	19 05.7	1.5	55.7	50	16 48	17 21	17 59	20 12	21 02	21 59	23 03
23		169 00.7	37.3	93 50.7	10.4	19 04.2	1.5	55.7	45	16 58	17 28	18 02	20 29	21 19	22 16	23 17
27 00		184 00.7	S12 38.1	108 20.1	10.5	S19 02.7	1.6	55.7	N 40	17 05	17 33	18 05	20 43	21 33	22 29	23 28
01		199 00.8	39.0	122 49.6	10.4	19 01.1	1.8	55.7	35	17 12	17 38	18 08	20 55	21 45	22 40	23 38
02		214 00.8	39.8	137 19.0	10.4	18 59.3	1.8	55.7	30	17 18	17 43	18 11	21 06	21 56	22 50	23 47
03		229 00.9	40.7	151 48.4	10.4	18 57.5	2.0	55.8	20	17 29	17 52	18 17	21 24	22 14	23 06	24 01
04		244 01.0	41.5	166 17.8	10.4	18 55.5	2.0	55.8	N 10	17 38	18 00	18 25	21 39	22 29	23 21	24 14
05		259 01.0	42.4	180 47.2	10.3	18 53.5	2.1	55.9	0	17 47	18 09	18 33	21 54	22 44	23 35	24 26
06		274 01.1	S12 43.2	195 16.5	10.3	S18 51.4	2.3	55.9	S 10	17 56	18 18	18 43	22 08	22 58	23 48	24 38
07		289 01.1	44.1	209 45.8	10.3	18 49.1	2.3	55.9	20	18 06	18 29	18 56	22 24	23 13	24 02	00 02
08		304 01.2	44.9	224 15.1	10.3	18 46.8	2.4	55.9	30	18 18	18 42	19 12	22 42	23 31	24 19	00 19
T 09		319 01.3	45.8	238 44.4	10.3	18 44.4	2.6	56.0	35	18 24	18 51	19 22	22 52	23 41	24 29	00 29
U 10		334 01.3	46.6	253 13.7	10.3	18 41.8	2.6	56.0	40	18 32	19 00	19 35	23 04	23 53	24 39	00 39
E 11		349 01.4	47.4	267 43.0	10.2	18 39.2	2.7	56.0	45	18 40	19 12	19 50	23 18	24 07	00 07	00 52
S 12		4 01.4	S12 48.3	282 12.2	10.2	S18 36.5	2.9	56.1	S 50	18 51	19 26	20 11	23 35	24 24	00 24	01 08
D 13		19 01.5	49.1	296 41.4	10.2	18 33.6	2.9	56.1	52	18 56	19 33	20 21	23 43	24 32	00 32	01 15
A 14		34 01.5	50.0	311 10.6	10.2	18 30.7	3.1	56.1	54	19 02	19 41	20 32	23 52	24 40	00 40	01 23
Y 15		49 01.6	50.8	325 39.8	10.2	18 27.6	3.1	56.2	56	19 08	19 50	20 46	24 02	00 02	00 50	01 32
16		64 01.6	51.7	340 09.0	10.1	18 24.5	3.2	56.2	58	19 15	20 00	21 02	24 13	00 13	01 02	01 42
17		79 01.7	52.5	354 38.1	10.2	18 21.3	3.4	56.2	S 60	19 22	20 12	21 22	24 27	00 27	01 15	01 54
18		94 01.8	S12 53.4	9 07.3	10.1	S18 17.9	3.4	56.3								
19		109 01.8	54.2	23 36.4	10.1	18 14.5	3.5	56.3			SUN			MOON		
20		124 01.9	55.0	38 05.5	10.1	18 11.0	3.7	56.3	Day	Eqn. of Time 00h	12h	Mer. Pass.	Mer. Pass. Upper	Lower	Age	Phase
21		139 01.9	55.9	52 34.6	10.1	18 07.3	3.7	56.4								
22		154 02.0	56.7	67 03.7	10.0	18 03.6	3.8	56.4	d	m s	m s	h m	h m	h m	d %	
23		169 02.0	57.6	81 32.7	10.1	S17 59.8	4.0	56.4	25	15 50	15 53	11 44	15 42	03 17	05 22	
									26	15 57	16 00	11 44	16 31	04 06	06 30	
		SD 16.1	d 0.9	SD 14.9		15.1		15.3	27	16 03	16 06	11 44	17 22	04 57	07 40	

1998 OCTOBER 28, 29, 30 (WED., THURS., FRI.)

UT	ARIES GHA	VENUS −3.9 GHA	Dec	MARS +1.6 GHA	Dec	JUPITER −2.7 GHA	Dec	SATURN +0.0 GHA	Dec	STARS Name	SHA	Dec
28 00	36 08.2	184 13.8	S11 51.4	231 49.4	N 8 16.7	46 00.5	S 5 52.3	7 24.7	N 8 49.3	Acamar	315 26.5	S40 18.6
01	51 10.7	199 13.3	52.5	246 50.5	16.2	61 03.1	52.4	22 27.3	49.2	Achernar	335 34.7	S57 14.7
02	66 13.1	214 12.7	53.7	261 51.5	15.6	76 05.7	52.4	37 30.0	49.1	Acrux	173 22.5	S63 05.3
03	81 15.6	229 12.2	.. 54.8	276 52.6	.. 15.1	91 08.3	.. 52.5	52 32.6	.. 49.0	Adhara	255 21.3	S28 58.1
04	96 18.1	244 11.7	55.9	291 53.7	14.5	106 10.8	52.5	67 35.3	49.0	Aldebaran	291 02.1	N16 30.3
05	111 20.5	259 11.2	57.0	306 54.8	14.0	121 13.4	52.6	82 37.9	48.9			
06	126 23.0	274 10.6	S11 58.2	321 55.8	N 8 13.4	136 16.0	S 5 52.6	97 40.6	N 8 48.8	Alioth	166 31.2	N55 58.0
W 07	141 25.5	289 10.1	11 59.3	336 56.9	12.9	151 18.6	52.7	112 43.2	48.8	Alkaid	153 08.3	N49 19.3
E 08	156 27.9	304 09.6	12 00.4	351 58.0	12.3	166 21.2	52.7	127 45.9	48.7	Al Na'ir	27 57.7	S46 58.2
D 09	171 30.4	319 09.1	.. 01.5	6 59.0	.. 11.8	181 23.8	.. 52.7	142 48.5	.. 48.6	Alnilam	275 57.6	S 1 12.2
N 10	186 32.8	334 08.5	02.7	22 00.1	11.2	196 26.4	52.8	157 51.2	48.6	Alphard	218 07.3	S 8 39.1
E 11	201 35.3	349 08.0	03.8	37 01.2	10.6	211 29.0	52.8	172 53.8	48.5			
S 12	216 37.8	4 07.5	S12 04.9	52 02.2	N 8 10.1	226 31.6	S 5 52.9	187 56.5	N 8 48.4	Alphecca	126 21.0	N26 43.4
D 13	231 40.2	19 06.9	06.0	67 03.3	09.5	241 34.2	52.9	202 59.1	48.4	Alpheratz	357 54.9	N29 05.2
A 14	246 42.7	34 06.4	07.1	82 04.4	09.0	256 36.8	53.0	218 01.8	48.3	Altair	62 19.3	N 8 52.1
Y 15	261 45.2	49 05.9	.. 08.3	97 05.5	.. 08.4	271 39.4	.. 53.0	233 04.4	.. 48.2	Ankaa	353 26.5	S42 18.9
16	276 47.6	64 05.3	09.4	112 06.5	07.9	286 42.0	53.1	248 07.1	48.2	Antares	112 40.5	S26 25.6
17	291 50.1	79 04.8	10.5	127 07.6	07.3	301 44.6	53.1	263 09.8	48.1			
18	306 52.6	94 04.3	S12 11.6	142 08.7	N 8 06.8	316 47.1	S 5 53.1	278 12.4	N 8 48.0	Arcturus	146 06.4	N19 11.5
19	321 55.0	109 03.8	12.7	157 09.7	06.2	331 49.7	53.2	293 15.1	48.0	Atria	107 52.9	S69 01.5
20	336 57.5	124 03.2	13.8	172 10.8	05.7	346 52.3	53.2	308 17.7	47.9	Avior	234 22.6	S59 30.1
21	351 59.9	139 02.7	.. 15.0	187 11.9	.. 05.1	1 54.9	.. 53.3	323 20.4	.. 47.8	Bellatrix	278 43.9	N 6 20.9
22	7 02.4	154 02.2	16.1	202 13.0	04.5	16 57.5	53.3	338 23.0	47.8	Betelgeuse	271 13.3	N 7 24.4
23	22 04.9	169 01.6	17.2	217 14.0	04.0	32 00.1	53.4	353 25.7	47.7			
29 00	37 07.3	184 01.1	S12 18.3	232 15.1	N 8 03.4	47 02.7	S 5 53.4	8 28.3	N 8 47.6	Canopus	264 00.9	S52 41.5
01	52 09.8	199 00.5	19.4	247 16.2	02.9	62 05.3	53.4	23 31.0	47.6	Capella	280 50.8	N45 59.6
02	67 12.3	214 00.0	20.5	262 17.3	02.3	77 07.9	53.5	38 33.6	47.5	Deneb	49 39.1	N45 16.9
03	82 14.7	228 59.5	.. 21.6	277 18.3	.. 01.8	92 10.4	.. 53.5	53 36.3	.. 47.4	Denebola	182 45.4	N14 34.8
04	97 17.2	243 58.9	22.7	292 19.4	01.2	107 13.0	53.6	68 38.9	47.4	Diphda	349 06.9	S17 59.6
05	112 19.7	258 58.4	23.9	307 20.5	00.7	122 15.6	53.6	83 41.6	47.3			
06	127 22.1	273 57.9	S12 25.0	322 21.6	N 8 00.1	137 18.2	S 5 53.7	98 44.2	N 8 47.2	Dubhe	194 06.0	N61 45.3
07	142 24.6	288 57.3	26.1	337 22.6	7 59.5	152 20.8	53.7	113 46.9	47.2	Elnath	278 26.7	N28 36.2
T 08	157 27.1	303 56.8	27.2	352 23.7	59.0	167 23.4	53.7	128 49.5	47.1	Eltanin	90 51.7	N51 29.7
H 09	172 29.5	318 56.3	.. 28.3	7 24.8	.. 58.4	182 26.0	.. 53.8	143 52.2	.. 47.0	Enif	33 58.1	N 9 52.3
U 10	187 32.0	333 55.7	29.4	22 25.8	57.9	197 28.5	53.8	158 54.8	47.0	Fomalhaut	15 36.2	S29 37.8
R 11	202 34.4	348 55.2	30.5	37 26.9	57.3	212 31.1	53.9	173 57.5	46.9			
S 12	217 36.9	3 54.6	S12 31.6	52 28.0	N 7 56.8	227 33.7	S 5 53.9	189 00.2	N 8 46.8	Gacrux	172 14.0	S57 06.1
D 13	232 39.4	18 54.1	32.7	67 29.1	56.2	242 36.3	54.0	204 02.8	46.8	Gienah	176 04.2	S17 31.9
A 14	247 41.8	33 53.6	33.8	82 30.1	55.7	257 38.9	54.0	219 05.5	46.7	Hadar	149 04.6	S60 21.8
Y 15	262 44.3	48 53.0	.. 34.9	97 31.2	.. 55.1	272 41.5	.. 54.0	234 08.1	.. 46.6	Hamal	328 13.2	N23 27.4
16	277 46.8	63 52.5	36.1	112 32.3	54.6	287 44.1	54.1	249 10.8	46.6	Kaus Aust.	83 59.0	S34 23.1
17	292 49.2	78 51.9	37.2	127 33.4	54.0	302 46.6	54.1	264 13.4	46.5			
18	307 51.7	93 51.4	S12 38.3	142 34.4	N 7 53.4	317 49.2	S 5 54.2	279 16.1	N 8 46.4	Kochab	137 20.9	N74 09.8
19	322 54.2	108 50.9	39.4	157 35.5	52.9	332 51.8	54.2	294 18.7	46.4	Markab	13 49.4	N15 12.1
20	337 56.6	123 50.3	40.5	172 36.6	52.3	347 54.4	54.2	309 21.4	46.3	Menkar	314 26.6	N 4 05.1
21	352 59.1	138 49.8	.. 41.6	187 37.7	.. 51.8	2 57.0	.. 54.3	324 24.0	.. 46.2	Menkent	148 21.3	S36 21.6
22	8 01.6	153 49.2	42.7	202 38.7	51.2	17 59.5	54.3	339 26.7	46.2	Miaplacidus	221 42.1	S69 42.4
23	23 04.0	168 48.7	43.8	217 39.8	50.7	33 02.1	54.4	354 29.3	46.1			
30 00	38 06.5	183 48.1	S12 44.9	232 40.9	N 7 50.1	48 04.7	S 5 54.4	9 32.0	N 8 46.0	Mirfak	308 56.1	N49 51.3
01	53 08.9	198 47.6	46.0	247 42.0	49.6	63 07.3	54.4	24 34.6	46.0	Nunki	76 12.5	S26 17.8
02	68 11.4	213 47.0	47.1	262 43.1	49.0	78 09.9	54.5	39 37.3	45.9	Peacock	53 37.1	S56 44.5
03	83 13.9	228 46.5	.. 48.2	277 44.1	.. 48.4	93 12.5	.. 54.5	54 39.9	.. 45.8	Pollux	243 41.5	N28 01.6
04	98 16.3	243 45.9	49.3	292 45.2	47.9	108 15.0	54.6	69 42.6	45.8	Procyon	245 11.5	N 5 13.7
05	113 18.8	258 45.4	50.4	307 46.3	47.3	123 17.6	54.6	84 45.2	45.7			
06	128 21.3	273 44.9	S12 51.5	322 47.4	N 7 46.8	138 20.2	S 5 54.6	99 47.9	N 8 45.6	Rasalhague	96 17.2	N12 33.9
07	143 23.7	288 44.3	52.6	337 48.4	46.2	153 22.8	54.7	114 50.5	45.6	Regulus	207 55.7	N11 58.4
08	158 26.2	303 43.8	53.7	352 49.5	45.7	168 25.3	54.7	129 53.2	45.5	Rigel	281 22.7	S 8 12.2
F 09	173 28.7	318 43.2	.. 54.8	7 50.6	.. 45.1	183 27.9	.. 54.8	144 55.8	.. 45.4	Rigil Kent.	140 07.9	S60 49.6
R 10	188 31.1	333 42.7	55.9	22 51.7	44.6	198 30.5	54.8	159 58.5	45.4	Sabik	102 25.8	S15 43.2
I 11	203 33.6	348 42.1	57.0	37 52.8	44.0	213 33.1	54.8	175 01.2	45.3			
D 12	218 36.0	3 41.6	S12 58.1	52 53.8	N 7 43.4	228 35.7	S 5 54.9	190 03.8	N 8 45.2	Schedar	349 52.9	N56 32.0
A 13	233 38.5	18 41.0	12 59.1	67 54.9	42.9	243 38.2	54.9	205 06.5	45.2	Shaula	96 37.6	S37 06.1
Y 14	248 41.0	33 40.5	13 00.2	82 56.0	42.3	258 40.8	54.9	220 09.1	45.1	Sirius	258 43.5	S16 42.8
15	263 43.4	48 39.9	.. 01.3	97 57.1	.. 41.8	273 43.4	.. 55.0	235 11.8	.. 45.0	Spica	158 43.5	S11 09.1
16	278 45.9	63 39.4	02.4	112 58.1	41.2	288 46.0	55.0	250 14.4	45.0	Suhail	223 00.8	S43 25.4
17	293 48.4	78 38.8	03.5	127 59.2	40.7	303 48.5	55.1	265 17.1	44.9			
18	308 50.8	93 38.3	S13 04.6	143 00.3	N 7 40.1	318 51.1	S 5 55.1	280 19.7	N 8 44.8	Vega	80 46.8	N38 47.3
19	323 53.3	108 37.7	05.7	158 01.4	39.5	333 53.7	55.1	295 22.4	44.8	Zuben'ubi	137 18.3	S16 02.0
20	338 55.8	123 37.1	06.8	173 02.5	39.0	348 56.3	55.2	310 25.0	44.7		SHA	Mer. Pass.
21	353 58.2	138 36.6	.. 07.9	188 03.5	.. 38.4	3 58.8	.. 55.2	325 27.7	.. 44.6	Venus	146 53.7	11 44
22	9 00.7	153 36.0	09.0	203 04.6	37.9	19 01.4	55.2	340 30.3	44.6	Mars	195 07.8	8 30
23	24 03.2	168 35.5	10.1	218 05.7	37.3	34 04.0	55.3	355 33.0	44.5	Jupiter	9 55.3	20 48
Mer. Pass. 21 28.0		v −0.5 d 1.1		v 1.1 d 0.6		v 2.6 d 0.0		v 2.7 d 0.1		Saturn	331 21.0	23 22

UT	SUN GHA	SUN Dec	MOON GHA	v	MOON Dec	d	HP
d h	° ′	° ′	° ′	′	° ′	′	′
28 00	184 02.1	S12 58.4	96 01.8	10.0	S17 55.8	4.0	56.5
01	199 02.1	12 59.3	110 30.8	10.1	17 51.8	4.1	56.5
02	214 02.2	13 00.1	124 59.9	10.0	17 47.7	4.3	56.5
03	229 02.2	.. 00.9	139 28.9	10.0	17 43.4	4.3	56.6
04	244 02.3	01.8	153 57.9	10.0	17 39.1	4.4	56.6
05	259 02.3	02.6	168 26.9	9.9	17 34.7	4.6	56.6
W 06	274 02.4	S13 03.4	182 55.8	10.0	S17 30.1	4.6	56.7
E 07	289 02.4	04.3	197 24.8	9.9	17 25.5	4.7	56.7
D 08	304 02.5	05.1	211 53.7	9.9	17 20.8	4.9	56.7
N 09	319 02.5	.. 06.0	226 22.7	9.9	17 15.9	4.9	56.8
E 10	334 02.6	06.8	240 51.6	9.9	17 11.0	5.0	56.8
S 11	349 02.6	07.6	255 20.5	9.9	17 06.0	5.2	56.8
D 12	4 02.7	S13 08.5	269 49.4	9.9	S17 00.8	5.2	56.9
A 13	19 02.7	09.3	284 18.3	9.9	16 55.6	5.3	56.9
Y 14	34 02.8	10.1	298 47.2	9.9	16 50.3	5.4	56.9
15	49 02.8	.. 11.0	313 16.1	9.8	16 44.9	5.5	57.0
16	64 02.9	11.8	327 44.9	9.9	16 39.4	5.7	57.0
17	79 02.9	12.7	342 13.8	9.8	16 33.7	5.7	57.0
18	94 03.0	S13 13.5	356 42.6	9.9	S16 28.0	5.8	57.1
19	109 03.0	14.3	11 11.5	9.8	16 22.2	5.9	57.1
20	124 03.1	15.2	25 40.3	9.8	16 16.3	6.0	57.2
21	139 03.1	.. 16.0	40 09.1	9.8	16 10.3	6.1	57.2
22	154 03.2	16.8	54 37.9	9.8	16 04.2	6.2	57.2
23	169 03.2	17.7	69 06.7	9.8	15 58.0	6.3	57.3
29 00	184 03.3	S13 18.5	83 35.5	9.7	S15 51.7	6.3	57.3
01	199 03.3	19.3	98 04.2	9.8	15 45.4	6.5	57.3
02	214 03.4	20.2	112 33.0	9.8	15 38.9	6.4	57.4
03	229 03.4	.. 21.0	127 01.8	9.7	15 32.3	6.7	57.4
04	244 03.4	21.8	141 30.5	9.7	15 25.6	6.7	57.5
05	259 03.5	22.6	155 59.2	9.8	15 18.9	6.9	57.5
T 06	274 03.5	S13 23.5	170 28.0	9.7	S15 12.0	6.9	57.5
H 07	289 03.6	24.3	184 56.7	9.7	15 05.1	7.1	57.6
U 08	304 03.6	25.1	199 25.4	9.7	14 58.0	7.1	57.6
R 09	319 03.7	.. 26.0	213 54.1	9.7	14 50.9	7.2	57.6
S 10	334 03.7	26.8	228 22.8	9.7	14 43.7	7.3	57.7
D 11	349 03.7	27.6	242 51.5	9.7	14 36.4	7.4	57.7
A 12	4 03.8	S13 28.4	257 20.2	9.6	S14 29.0	7.5	57.8
Y 13	19 03.8	29.3	271 48.8	9.7	14 21.5	7.6	57.8
14	34 03.9	30.1	286 17.5	9.6	14 13.9	7.6	57.8
15	49 03.9	.. 30.9	300 46.1	9.7	14 06.2	7.8	57.9
16	64 04.0	31.8	315 14.8	9.6	13 58.4	7.8	57.9
17	79 04.0	32.6	329 43.4	9.6	13 50.6	8.0	58.0
18	94 04.0	S13 33.4	344 12.0	9.7	S13 42.6	8.0	58.0
19	109 04.1	34.2	358 40.7	9.6	13 34.6	8.1	58.0
20	124 04.1	35.1	13 09.3	9.6	13 26.5	8.2	58.1
21	139 04.1	.. 35.9	27 37.9	9.6	13 18.3	8.3	58.1
22	154 04.2	36.7	42 06.5	9.5	13 10.0	8.4	58.1
23	169 04.2	37.5	56 35.0	9.6	13 01.6	8.5	58.2
30 00	184 04.3	S13 38.4	71 03.6	9.6	S12 53.1	8.5	58.2
01	199 04.3	39.2	85 32.2	9.5	12 44.6	8.7	58.3
02	214 04.3	40.0	100 00.7	9.6	12 35.9	8.7	58.3
03	229 04.4	.. 40.8	114 29.3	9.5	12 27.2	8.8	58.3
04	244 04.4	41.6	128 57.8	9.5	12 18.4	8.9	58.4
05	259 04.4	42.5	143 26.3	9.5	12 09.5	8.9	58.4
F 06	274 04.5	S13 43.3	157 54.8	9.5	S12 00.6	9.1	58.5
R 07	289 04.5	44.1	172 23.3	9.5	11 51.5	9.1	58.5
I 08	304 04.6	44.9	186 51.8	9.5	11 42.4	9.2	58.5
D 09	319 04.6	.. 45.7	201 20.3	9.5	11 33.2	9.3	58.6
A 10	334 04.6	46.6	215 48.8	9.5	11 23.9	9.4	58.6
Y 11	349 04.7	47.4	230 17.3	9.4	11 14.5	9.4	58.7
12	4 04.7	S13 48.2	244 45.7	9.5	S11 05.1	9.5	58.7
13	19 04.7	49.0	259 14.2	9.4	10 55.6	9.6	58.7
14	34 04.8	49.8	273 42.6	9.4	10 46.0	9.7	58.8
15	49 04.8	.. 50.7	288 11.0	9.4	10 36.3	9.7	58.8
16	64 04.8	51.5	302 39.4	9.4	10 26.6	9.9	58.8
17	79 04.9	52.3	317 07.8	9.4	10 16.7	9.9	58.9
18	94 04.9	S13 53.1	331 36.2	9.3	S10 06.8	9.9	58.9
19	109 04.9	53.9	346 04.5	9.4	9 56.9	10.1	59.0
20	124 05.0	54.7	0 32.9	9.3	9 46.8	10.1	59.0
21	139 05.0	.. 55.6	15 01.2	9.3	9 36.7	10.1	59.1
22	154 05.0	56.4	29 29.6	9.3	9 26.6	10.3	59.1
23	169 05.0	57.2	43 57.9	9.3	S 9 16.3	10.3	59.1
SD	16.1	d 0.8	SD 15.5		15.7		16.0

Lat.	Twilight Naut.	Twilight Civil	Sunrise	Moonrise 28	29	30	31
°	h m	h m	h m	h m	h m	h m	h m
N 72	05 53	07 15	08 37	16 41	16 22	16 10	16 01
N 70	05 50	07 03	08 15	15 53	15 54	15 53	15 52
68	05 48	06 54	07 57	15 22	15 33	15 40	15 44
66	05 46	06 47	07 43	14 59	15 16	15 29	15 38
64	05 44	06 40	07 32	14 40	15 02	15 19	15 33
62	05 42	06 34	07 22	14 25	14 51	15 11	15 28
60	05 40	06 29	07 14	14 13	14 41	15 04	15 24
N 58	05 39	06 25	07 06	14 02	14 32	14 58	15 21
56	05 37	06 21	07 00	13 52	14 25	14 53	15 18
54	05 35	06 17	06 54	13 44	14 18	14 48	15 15
52	05 34	06 14	06 49	13 36	14 12	14 43	15 12
50	05 33	06 11	06 44	13 29	14 06	14 39	15 10
45	05 29	06 04	06 34	13 15	13 54	14 31	15 05
N 40	05 26	05 57	06 25	13 03	13 44	14 23	15 01
35	05 22	05 52	06 18	12 52	13 36	14 17	14 57
30	05 19	05 47	06 11	12 43	13 28	14 11	14 53
20	05 11	05 37	06 00	12 28	13 15	14 02	14 48
N 10	05 03	05 28	05 50	12 14	13 04	13 53	14 43
0	04 54	05 19	05 40	12 01	12 53	13 45	14 38
S 10	04 44	05 09	05 31	11 48	12 42	13 37	14 33
20	04 31	04 57	05 20	11 35	12 31	13 29	14 28
30	04 13	04 43	05 08	11 19	12 18	13 19	14 22
35	04 03	04 35	05 01	11 10	12 10	13 13	14 19
40	03 49	04 24	04 53	10 59	12 01	13 07	14 15
45	03 33	04 12	04 44	10 47	11 51	12 59	14 11
S 50	03 12	03 57	04 32	10 32	11 39	12 50	14 06
52	03 01	03 49	04 27	10 25	11 33	12 46	14 03
54	02 49	03 41	04 21	10 18	11 27	12 42	14 01
56	02 35	03 32	04 15	10 09	11 20	12 36	13 58
58	02 18	03 21	04 07	09 59	11 12	12 31	13 54
S 60	01 56	03 09	03 59	09 48	11 03	12 24	13 51

Lat.	Sunset	Twilight Civil	Twilight Naut.	Moonset 28	29	30	31
°	h m	h m	h m	h m	h m	h m	h m
N 72	14 49	16 11	17 32	19 57	22 05	24 06	00 06
N 70	15 11	16 23	17 35	20 44	22 32	24 21	00 21
68	15 29	16 32	17 38	21 15	22 52	24 33	00 33
66	15 43	16 40	17 40	21 37	23 07	24 43	00 43
64	15 54	16 46	17 42	21 55	23 20	24 51	00 51
62	16 04	16 52	17 44	22 09	23 31	24 58	00 58
60	16 13	16 57	17 46	22 22	23 40	25 04	01 04
N 58	16 20	17 02	17 48	22 32	23 48	25 09	01 09
56	16 27	17 06	17 49	22 41	23 55	25 14	01 14
54	16 32	17 10	17 51	22 49	24 01	00 01	01 18
52	16 38	17 13	17 53	22 57	24 07	00 07	01 22
50	16 43	17 16	17 54	23 03	24 12	00 12	01 26
45	16 53	17 23	17 58	23 17	24 23	00 23	01 33
N 40	17 02	17 30	18 01	23 28	24 32	00 32	01 39
35	17 09	17 35	18 05	23 38	24 40	00 40	01 44
30	17 16	17 40	18 08	23 47	24 47	00 47	01 49
20	17 27	17 50	18 16	24 01	00 01	00 58	01 57
N 10	17 37	17 59	18 24	24 14	00 14	01 08	02 04
0	17 47	18 08	18 33	24 26	00 26	01 18	02 10
S 10	17 57	18 19	18 44	24 38	00 38	01 27	02 17
20	18 08	18 30	18 57	00 02	00 50	01 37	02 23
30	18 20	18 45	19 15	00 19	01 05	01 49	02 31
35	18 27	18 54	19 26	00 29	01 13	01 55	02 35
40	18 35	19 04	19 39	00 39	01 22	02 02	02 40
45	18 45	19 16	19 56	00 52	01 33	02 11	02 46
S 50	18 56	19 32	20 17	01 08	01 47	02 21	02 53
52	19 02	19 39	20 28	01 15	01 53	02 26	02 56
54	19 08	19 48	20 40	01 23	02 00	02 31	02 59
56	19 14	19 57	20 55	01 32	02 07	02 37	03 03
58	19 22	20 08	21 13	01 42	02 16	02 43	03 07
S 60	19 30	20 21	21 35	01 54	02 25	02 51	03 12

Day	SUN Eqn. of Time 00h	12h	Mer. Pass.	MOON Mer. Pass. Upper	Lower	Age	Phase
d	m s	m s	h m	h m	h m	d	%
28	16 08	16 11	11 44	18 14	05 48	08	50
29	16 13	16 15	11 44	19 05	06 40	09	61
30	16 17	16 19	11 44	19 58	07 32	10	71

1998 OCT. 31, NOV. 1, 2 (SAT., SUN., MON.)

UT	ARIES GHA	VENUS −3.9 GHA	VENUS Dec	MARS +1.6 GHA	MARS Dec	JUPITER −2.7 GHA	JUPITER Dec	SATURN +0.0 GHA	SATURN Dec
31 00	39 05.6	183 34.9	S13 11.1	233 06.8	N 7 36.8	49 06.6	S 5 55.3	10 35.6	N 8 44.4
01	54 08.1	198 34.4	12.2	248 07.9	36.2	64 09.1	55.4	25 38.3	44.4
02	69 10.5	213 33.8	13.3	263 08.9	35.7	79 11.7	55.4	40 40.9	44.3
03	84 13.0	228 33.3 ..	14.4	278 10.0 ..	35.1	94 14.3 ..	55.4	55 43.6 ..	44.2
04	99 15.5	243 32.7	15.5	293 11.1	34.5	109 16.9	55.5	70 46.2	44.2
05	114 17.9	258 32.1	16.6	308 12.2	34.0	124 19.4	55.5	85 48.9	44.1
S 06	129 20.4	273 31.6	S13 17.7	323 13.3	N 7 33.4	139 22.0	S 5 55.5	100 51.5	N 8 44.0
A 07	144 22.9	288 31.0	18.7	338 14.3	32.9	154 24.6	55.6	115 54.2	44.0
T 08	159 25.3	303 30.5	19.8	353 15.4	32.3	169 27.1	55.6	130 56.8	43.9
U 09	174 27.8	318 29.9 ..	20.9	8 16.5 ..	31.8	184 29.7 ..	55.6	145 59.5 ..	43.8
R 10	189 30.3	333 29.3	22.0	23 17.6	31.2	199 32.3	55.7	161 02.1	43.8
D 11	204 32.7	348 28.8	23.1	38 18.7	30.7	214 34.9	55.7	176 04.8	43.7
A 12	219 35.2	3 28.2	S13 24.2	53 19.7	N 7 30.1	229 37.4	S 5 55.7	191 07.4	N 8 43.6
Y 13	234 37.7	18 27.7	25.2	68 20.8	29.5	244 40.0	55.8	206 10.1	43.6
14	249 40.1	33 27.1	26.3	83 21.9	29.0	259 42.6	55.8	221 12.7	43.5
15	264 42.6	48 26.5 ..	27.4	98 23.0 ..	28.4	274 45.1 ..	55.8	236 15.4 ..	43.4
16	279 45.0	63 26.0	28.5	113 24.1	27.9	289 47.7	55.9	251 18.0	43.4
17	294 47.5	78 25.4	29.6	128 25.2	27.3	304 50.3	55.9	266 20.7	43.3
18	309 50.0	93 24.9	S13 30.6	143 26.2	N 7 26.8	319 52.8	S 5 55.9	281 23.3	N 8 43.2
19	324 52.4	108 24.3	31.7	158 27.3	26.2	334 55.4	56.0	296 26.0	43.2
20	339 54.9	123 23.7	32.8	173 28.4	25.6	349 58.0	56.0	311 28.6	43.1
21	354 57.4	138 23.2 ..	33.9	188 29.5 ..	25.1	5 00.5 ..	56.1	326 31.3 ..	43.0
22	9 59.8	153 22.6	34.9	203 30.6	24.5	20 03.1	56.1	341 33.9	43.0
23	25 02.3	168 22.0	36.0	218 31.7	24.0	35 05.7	56.1	356 36.6	42.9
1 00	40 04.8	183 21.5	S13 37.1	233 32.7	N 7 23.4	50 08.2	S 5 56.2	11 39.2	N 8 42.8
01	55 07.2	198 20.9	38.2	248 33.8	22.9	65 10.8	56.2	26 41.9	42.8
02	70 09.7	213 20.3	39.2	263 34.9	22.3	80 13.4	56.2	41 44.5	42.7
03	85 12.1	228 19.8 ..	40.3	278 36.0 ..	21.8	95 15.9 ..	56.3	56 47.2 ..	42.6
04	100 14.6	243 19.2	41.4	293 37.1	21.2	110 18.5	56.3	71 49.9	42.6
05	115 17.1	258 18.6	42.5	308 38.2	20.6	125 21.1	56.3	86 52.5	42.5
S 06	130 19.5	273 18.1	S13 43.5	323 39.2	N 7 20.1	140 23.6	S 5 56.3	101 55.2	N 8 42.4
U 07	145 22.0	288 17.5	44.6	338 40.3	19.5	155 26.2	56.4	116 57.8	42.4
N 08	160 24.5	303 16.9	45.7	353 41.4	19.0	170 28.8	56.4	132 00.5	42.3
D 09	175 26.9	318 16.3 ..	46.7	8 42.5 ..	18.4	185 31.3 ..	56.4	147 03.1 ..	42.2
A 10	190 29.4	333 15.8	47.8	23 43.6	17.9	200 33.9	56.5	162 05.8	42.2
Y 11	205 31.9	348 15.2	48.9	38 44.7	17.3	215 36.4	56.5	177 08.4	42.1
12	220 34.3	3 14.6	S13 49.9	53 45.7	N 7 16.7	230 39.0	S 5 56.5	192 11.1	N 8 42.0
13	235 36.8	18 14.1	51.0	68 46.8	16.2	245 41.6	56.6	207 13.7	42.0
14	250 39.3	33 13.5	52.1	83 47.9	15.6	260 44.1	56.6	222 16.4	41.9
15	265 41.7	48 12.9 ..	53.1	98 49.0 ..	15.1	275 46.7 ..	56.6	237 19.0 ..	41.8
16	280 44.2	63 12.3	54.2	113 50.1	14.5	290 49.2	56.7	252 21.7	41.8
17	295 46.6	78 11.8	55.3	128 51.2	14.0	305 51.8	56.7	267 24.3	41.7
18	310 49.1	93 11.2	S13 56.3	143 52.3	N 7 13.4	320 54.4	S 5 56.7	282 27.0	N 8 41.7
19	325 51.6	108 10.6	57.4	158 53.3	12.8	335 56.9	56.8	297 29.6	41.6
20	340 54.0	123 10.0	58.5	173 54.4	12.3	350 59.5	56.8	312 32.3	41.5
21	355 56.5	138 09.5	13 59.5	188 55.5 ..	11.7	6 02.0 ..	56.8	327 34.9 ..	41.5
22	10 59.0	153 08.9	14 00.6	203 56.6	11.2	21 04.6	56.8	342 37.6	41.4
23	26 01.4	168 08.3	01.6	218 57.7	10.6	36 07.2	56.9	357 40.2	41.3
2 00	41 03.9	183 07.7	S14 02.7	233 58.8	N 7 10.1	51 09.7	S 5 56.9	12 42.9	N 8 41.3
01	56 06.4	198 07.2	03.8	248 59.9	09.5	66 12.3	56.9	27 45.5	41.2
02	71 08.8	213 06.6	04.8	264 01.0	08.9	81 14.8	57.0	42 48.2	41.1
03	86 11.3	228 06.0 ..	05.9	279 02.0 ..	08.4	96 17.4 ..	57.0	57 50.8 ..	41.1
04	101 13.7	243 05.4	06.9	294 03.1	07.8	111 20.0	57.0	72 53.5	41.0
05	116 16.2	258 04.8	08.0	309 04.2	07.3	126 22.5	57.1	87 56.1	40.9
M 06	131 18.7	273 04.3	S14 09.1	324 05.3	N 7 06.7	141 25.1	S 5 57.1	102 58.8	N 8 40.9
O 07	146 21.1	288 03.7	10.1	339 06.4	06.2	156 27.6	57.1	118 01.4	40.8
N 08	161 23.6	303 03.1	11.2	354 07.5	05.6	171 30.2	57.1	133 04.0	40.7
D 09	176 26.1	318 02.5 ..	12.2	9 08.6 ..	05.0	186 32.7 ..	57.2	148 06.7 ..	40.7
A 10	191 28.5	333 01.9	13.3	24 09.7	04.5	201 35.3	57.2	163 09.3	40.6
Y 11	206 31.0	348 01.4	14.3	39 10.7	03.9	216 37.8	57.2	178 12.0	40.5
12	221 33.5	3 00.8	S14 15.4	54 11.8	N 7 03.4	231 40.4	S 5 57.3	193 14.6	N 8 40.5
13	236 35.9	18 00.2	16.4	69 12.9	02.8	246 43.0	57.3	208 17.3	40.4
14	251 38.4	32 59.6	17.5	84 14.0	02.3	261 45.5	57.3	223 19.9	40.3
15	266 40.9	47 59.0 ..	18.5	99 15.1 ..	01.7	276 48.1 ..	57.3	238 22.6 ..	40.3
16	281 43.3	62 58.4	19.6	114 16.2	01.1	291 50.6	57.4	253 25.2	40.2
17	296 45.8	77 57.8	20.6	129 17.3	00.6	306 53.2	57.4	268 27.9	40.1
18	311 48.2	92 57.3	S14 21.7	144 18.4	N 7 00.0	321 55.7	S 5 57.4	283 30.5	N 8 40.1
19	326 50.7	107 56.7	22.7	159 19.5	6 59.5	336 58.3	57.4	298 33.2	40.0
20	341 53.2	122 56.1	23.8	174 20.5	58.9	352 00.8	57.5	313 35.8	40.0
21	356 55.6	137 55.5 ..	24.8	189 21.6 ..	58.4	7 03.4 ..	57.5	328 38.5 ..	39.9
22	11 58.1	152 54.9	25.9	204 22.7	57.8	22 05.9	57.5	343 41.1	39.8
23	27 00.6	167 54.3	26.9	219 23.8	57.3	37 08.5	57.6	358 43.8	39.8
Mer. Pass.	h m 21 16.2	v −0.6	d 1.1	v 1.1	d 0.6	v 2.6	d 0.0	v 2.7	d 0.1

STARS

Name	SHA	Dec
Acamar	315 26.5	S40 18.6
Achernar	335 34.7	S57 14.7
Acrux	173 22.5	S63 05.3
Adhara	255 21.2	S28 58.1
Aldebaran	291 02.1	N16 30.3
Alioth	166 31.2	N55 58.0
Alkaid	153 08.3	N49 19.3
Al Na'ir	27 57.8	S46 58.2
Alnilam	275 57.6	S 1 12.2
Alphard	218 07.2	S 8 39.1
Alphecca	126 21.0	N26 43.4
Alpheratz	357 54.9	N29 05.2
Altair	62 19.3	N 8 52.1
Ankaa	353 26.5	S42 18.9
Antares	112 40.5	S26 25.6
Arcturus	146 06.4	N19 11.5
Atria	107 52.9	S69 01.5
Avior	234 22.5	S59 30.1
Bellatrix	278 43.9	N 6 20.9
Betelgeuse	271 13.3	N 7 24.4
Canopus	264 00.8	S52 41.5
Capella	280 50.8	N45 59.6
Deneb	49 39.2	N45 16.9
Denebola	182 45.4	N14 34.8
Diphda	349 06.9	S17 59.6
Dubhe	194 06.0	N61 45.3
Elnath	278 26.6	N28 36.2
Eltanin	90 51.8	N51 29.7
Enif	33 58.1	N 9 52.3
Fomalhaut	15 36.2	S29 37.8
Gacrux	172 14.0	S57 06.1
Gienah	176 04.2	S17 31.9
Hadar	149 04.6	S60 21.8
Hamal	328 13.2	N23 27.4
Kaus Aust.	83 59.0	S34 23.1
Kochab	137 20.9	N74 09.8
Markab	13 49.4	N15 12.1
Menkar	314 26.6	N 4 05.1
Menkent	148 21.3	S36 21.6
Miaplacidus	221 42.1	S69 42.4
Mirfak	308 56.1	N49 51.3
Nunki	76 12.5	S26 17.8
Peacock	53 37.1	S56 44.5
Pollux	243 41.5	N28 01.6
Procyon	245 11.5	N 5 13.7
Rasalhague	96 17.2	N12 33.9
Regulus	207 55.7	N11 58.4
Rigel	281 22.7	S 8 12.2
Rigil Kent.	140 07.9	S60 49.6
Sabik	102 25.8	S15 43.2
Schedar	349 52.9	N56 32.0
Shaula	96 37.6	S37 06.1
Sirius	258 43.5	S16 42.8
Spica	158 43.5	S11 09.1
Suhail	223 00.8	S43 25.4
Vega	80 46.8	N38 47.3
Zuben'ubi	137 18.3	S16 02.0

	SHA	Mer. Pass.
	° '	h m
Venus	143 16.7	11 47
Mars	193 28.0	8 25
Jupiter	10 03.5	20 36
Saturn	331 34.5	23 09

UT	SUN GHA	SUN Dec	MOON GHA	v	Dec	d	HP
d h	° ′	° ′	° ′	′	° ′	′	′
31 00	184 05.1	S13 58.0	58 26.2	9.3	S 9 06.0	10.4	59.2
01	199 05.1	58.8	72 54.5	9.2	8 55.6	10.4	59.2
02	214 05.1	13 59.6	87 22.7	9.3	8 45.2	10.6	59.3
03	229 05.2	14 00.4	101 51.0	9.2	8 34.6	10.5	59.3
04	244 05.2	01.3	116 19.2	9.3	8 24.1	10.7	59.3
05	259 05.2	02.1	130 47.5	9.2	8 13.4	10.7	59.4
S 06	274 05.2	S14 02.9	145 15.7	9.2	S 8 02.7	10.8	59.4
07	289 05.3	03.7	159 43.9	9.2	7 51.9	10.8	59.4
A 08	304 05.3	04.5	174 12.1	9.1	7 41.1	10.9	59.5
T 09	319 05.3 ..	05.3	188 40.2	9.2	7 30.2	10.9	59.5
U 10	334 05.4	06.1	203 08.4	9.1	7 19.3	11.1	59.6
R 11	349 05.4	06.9	217 36.5	9.1	7 08.2	11.0	59.6
D 12	4 05.4	S14 07.7	232 04.6	9.1	S 6 57.2	11.1	59.6
A 13	19 05.4	08.6	246 32.7	9.1	6 46.1	11.2	59.7
Y 14	34 05.5	09.4	261 00.8	9.0	6 34.9	11.3	59.7
15	49 05.5 ..	10.2	275 28.8	9.0	6 23.6	11.2	59.7
16	64 05.5	11.0	289 56.8	9.0	6 12.4	11.4	59.8
17	79 05.5	11.8	304 24.8	9.0	6 01.0	11.4	59.8
18	94 05.6	S14 12.6	318 52.8	9.0	S 5 49.6	11.4	59.9
19	109 05.6	13.4	333 20.8	9.0	5 38.2	11.5	59.9
20	124 05.6	14.2	347 48.8	8.9	5 26.7	11.5	59.9
21	139 05.6 ..	15.0	2 16.7	8.9	5 15.2	11.6	60.0
22	154 05.6	15.8	16 44.6	8.9	5 03.6	11.6	60.0
23	169 05.6	16.6	31 12.5	8.8	4 52.0	11.7	60.0
1 00	184 05.7	S14 17.4	45 40.3	8.9	S 4 40.3	11.7	60.1
01	199 05.7	18.2	60 08.2	8.8	4 28.6	11.8	60.1
02	214 05.7	19.0	74 36.0	8.8	4 16.8	11.8	60.1
03	229 05.8 ..	19.8	89 03.8	8.8	4 05.0	11.8	60.2
04	244 05.8	20.6	103 31.6	8.7	3 53.2	11.9	60.2
05	259 05.8	21.4	117 59.3	8.7	3 41.3	11.9	60.2
06	274 05.8	S14 22.2	132 27.0	8.7	S 3 29.4	11.9	60.3
07	289 05.9	23.0	146 54.7	8.7	3 17.5	12.0	60.3
S 08	304 05.9	23.8	161 22.4	8.6	3 05.5	12.0	60.3
U 09	319 05.9 ..	24.7	175 50.0	8.6	2 53.5	12.1	60.4
N 10	334 05.9	25.5	190 17.6	8.6	2 41.4	12.1	60.4
D 11	349 05.9	26.3	204 45.2	8.6	2 29.3	12.1	60.4
A 12	4 05.9	S14 27.1	219 12.8	8.5	S 2 17.2	12.1	60.5
Y 13	19 05.9	27.9	233 40.3	8.5	2 05.1	12.1	60.5
14	34 06.0	28.7	248 07.8	8.4	1 53.0	12.2	60.5
15	49 06.0 ..	29.4	262 35.2	8.5	1 40.8	12.2	60.6
16	64 06.0	30.2	277 02.7	8.4	1 28.6	12.3	60.6
17	79 06.0	31.0	291 30.1	8.4	1 16.3	12.2	60.6
18	94 06.0	S14 31.8	305 57.5	8.3	S 1 04.1	12.3	60.6
19	109 06.0	32.6	320 24.8	8.3	0 51.8	12.3	60.7
20	124 06.1	33.4	334 52.1	8.3	0 39.5	12.3	60.7
21	139 06.1 ..	34.2	349 19.4	8.3	0 27.2	12.3	60.7
22	154 06.1	35.0	3 46.7	8.2	0 14.9	12.3	60.8
23	169 06.1	35.8	18 13.9	8.2	S 0 02.6	12.4	60.8
2 00	184 06.1	S14 36.6	32 41.1	8.1	N 0 09.8	12.3	60.8
01	199 06.1	37.4	47 08.2	8.1	0 22.1	12.4	60.8
02	214 06.1	38.2	61 35.3	8.1	0 34.5	12.4	60.9
03	229 06.2 ..	39.0	76 02.4	8.1	0 46.9	12.4	60.9
04	244 06.2	39.8	90 29.5	8.0	0 59.3	12.3	60.9
05	259 06.2	40.6	104 56.5	8.0	1 11.6	12.4	60.9
06	274 06.2	S14 41.4	119 23.5	7.9	N 1 24.0	12.4	61.0
07	289 06.2	42.2	133 50.4	7.9	1 36.4	12.4	61.0
08	304 06.2	43.0	148 17.3	7.9	1 48.8	12.4	61.0
M 09	319 06.2 ..	43.8	162 44.2	7.8	2 01.2	12.4	61.0
O 10	334 06.2	44.5	177 11.0	7.8	2 13.6	12.4	61.1
N 11	349 06.2	45.3	191 37.8	7.8	2 26.0	12.4	61.1
D 12	4 06.2	S14 46.1	206 04.6	7.7	N 2 38.4	12.3	61.1
A 13	19 06.3	46.9	220 31.3	7.7	2 50.7	12.4	61.1
Y 14	34 06.3	47.7	234 58.0	7.6	3 03.1	12.3	61.1
15	49 06.3 ..	48.5	249 24.6	7.6	3 15.4	12.4	61.2
16	64 06.3	49.3	263 51.2	7.6	3 27.8	12.3	61.2
17	79 06.3	50.1	278 17.8	7.5	3 40.1	12.3	61.2
18	94 06.3	S14 50.9	292 44.3	7.5	N 3 52.4	12.3	61.2
19	109 06.3	51.6	307 10.8	7.5	4 04.7	12.3	61.2
20	124 06.3	52.4	321 37.3	7.4	4 17.0	12.2	61.2
21	139 06.3 ..	53.2	336 03.7	7.3	4 29.2	12.2	61.3
22	154 06.3	54.0	350 30.0	7.3	4 41.4	12.2	61.3
23	169 06.3	54.8	4 56.3	7.3	N 4 53.6	12.2	61.3
	SD 16.1	d 0.8	SD 16.2		16.5		16.6

Lat.	Twilight Naut.	Twilight Civil	Sunrise	Moonrise 31	1	2	3
°	h m	h m	h m	h m	h m	h m	h m
N 72	06 05	07 28	08 55	16 01	15 52	15 44	15 35
N 70	06 01	07 15	08 29	15 52	15 50	15 48	15 47
68	05 58	07 05	08 10	15 44	15 48	15 52	15 57
66	05 55	06 56	07 54	15 38	15 47	15 55	16 05
64	05 52	06 49	07 41	15 33	15 45	15 58	16 11
62	05 49	06 42	07 31	15 28	15 44	16 00	16 17
60	05 47	06 36	07 22	15 24	15 43	16 02	16 22
N 58	05 45	06 31	07 13	15 21	15 42	16 04	16 27
56	05 43	06 27	07 06	15 18	15 41	16 05	16 31
54	05 41	06 23	07 00	15 15	15 41	16 07	16 35
52	05 39	06 19	06 54	15 12	15 40	16 08	16 38
50	05 37	06 15	06 49	15 10	15 39	16 09	16 41
45	05 33	06 07	06 38	15 05	15 38	16 12	16 48
N 40	05 29	06 01	06 29	15 01	15 37	16 14	16 53
35	05 25	05 55	06 21	14 57	15 36	16 16	16 58
30	05 21	05 49	06 14	14 53	15 35	16 18	17 02
20	05 13	05 39	06 01	14 48	15 34	16 21	17 10
N 10	05 04	05 29	05 50	14 43	15 33	16 24	17 16
0	04 54	05 19	05 40	14 38	15 31	16 26	17 23
S 10	04 43	05 08	05 30	14 33	15 30	16 29	17 29
20	04 29	04 56	05 19	14 28	15 29	16 32	17 36
30	04 10	04 40	05 06	14 22	15 28	16 35	17 44
35	03 59	04 31	04 58	14 19	15 27	16 37	17 48
40	03 45	04 20	04 49	14 15	15 26	16 39	17 53
45	03 28	04 07	04 39	14 11	15 25	16 42	17 59
S 50	03 05	03 51	04 27	14 06	15 24	16 45	18 07
52	02 54	03 43	04 21	14 03	15 23	16 46	18 10
54	02 41	03 34	04 15	14 01	15 23	16 48	18 14
56	02 25	03 24	04 08	13 58	15 22	16 49	18 18
58	02 06	03 13	04 00	13 54	15 21	16 51	18 23
S 60	01 41	02 59	03 51	13 51	15 21	16 53	18 28

Lat.	Sunset	Twilight Civil	Twilight Naut.	Moonset 31	1	2	3
°	h m	h m	h m	h m	h m	h m	h m
N 72	14 30	15 57	17 20	00 06	02 07	04 09	06 14
N 70	14 57	16 10	17 24	00 21	02 13	04 07	06 05
68	15 16	16 21	17 28	00 33	02 18	04 06	05 57
66	15 32	16 30	17 31	00 43	02 23	04 06	05 51
64	15 45	16 38	17 34	00 51	02 26	04 05	05 46
62	15 55	16 44	17 37	00 58	02 30	04 04	05 41
60	16 05	16 50	17 39	01 04	02 32	04 04	05 37
N 58	16 13	16 55	17 41	01 09	02 35	04 03	05 34
56	16 20	17 00	17 44	01 14	02 37	04 03	05 31
54	16 26	17 04	17 46	01 18	02 39	04 02	05 28
52	16 32	17 08	17 48	01 22	02 41	04 02	05 25
50	16 37	17 11	17 49	01 26	02 42	04 02	05 23
45	16 49	17 19	17 54	01 33	02 46	04 01	05 18
N 40	16 58	17 26	17 58	01 39	02 49	04 00	05 14
35	17 06	17 32	18 02	01 44	02 51	04 00	05 11
30	17 13	17 38	18 06	01 49	02 53	03 59	05 07
20	17 26	17 48	18 14	01 57	02 57	03 59	05 02
N 10	17 37	17 58	18 23	02 04	03 00	03 58	04 57
0	17 47	18 08	18 33	02 10	03 03	03 57	04 53
S 10	17 58	18 19	18 45	02 17	03 06	03 57	04 48
20	18 09	18 32	18 59	02 23	03 09	03 56	04 44
30	18 22	18 47	19 17	02 31	03 13	03 55	04 38
35	18 30	18 57	19 29	02 35	03 15	03 54	04 35
40	18 38	19 08	19 43	02 40	03 17	03 54	04 31
45	18 49	19 21	20 01	02 46	03 20	03 53	04 27
S 50	19 01	19 38	20 24	02 53	03 23	03 52	04 23
52	19 07	19 46	20 35	02 56	03 24	03 52	04 20
54	19 14	19 55	20 49	02 59	03 26	03 51	04 18
56	19 21	20 05	21 05	03 03	03 27	03 51	04 15
58	19 29	20 17	21 25	03 07	03 29	03 50	04 12
S 60	19 38	20 31	21 50	03 12	03 31	03 50	04 09

Day	SUN Eqn. of Time 00h	SUN Eqn. of Time 12h	SUN Mer. Pass.	MOON Mer. Pass. Upper	MOON Mer. Pass. Lower	Age	Phase
d	m s	m s	h m	h m	h m	d	%
31	16 20	16 22	11 44	20 51	08 24	11	81
1	16 23	16 24	11 44	21 44	09 17	12	89
2	16 24	16 25	11 44	22 39	10 12	13	95

1998 NOVEMBER 3, 4, 5 (TUES., WED., THURS.)

UT	ARIES GHA	VENUS −3.9 GHA	Dec	MARS +1.6 GHA	Dec	JUPITER −2.7 GHA	Dec	SATURN +0.0 GHA	Dec	STARS Name	SHA	Dec
3 00	42 03.0	182 53.7	S14 28.0	234 24.9	N 6 56.7	52 11.0	S 5 57.6	13 46.4	N 8 39.7	Acamar	315 26.5	S40 18.6
01	57 05.5	197 53.1	29.0	249 26.0	56.1	67 13.6	57.6	28 49.1	39.6	Achernar	335 34.7	S57 14.7
02	72 08.0	212 52.6	30.1	264 27.1	55.6	82 16.1	57.6	43 51.7	39.6	Acrux	173 22.5	S63 05.3
03	87 10.4	227 52.0	.. 31.1	279 28.2	.. 55.0	97 18.7	.. 57.7	58 54.4	.. 39.5	Adhara	255 21.2	S28 58.1
04	102 12.9	242 51.4	32.1	294 29.3	54.5	112 21.2	57.7	73 57.0	39.4	Aldebaran	291 02.1	N16 30.3
05	117 15.4	257 50.8	33.2	309 30.4	53.9	127 23.8	57.7	88 59.7	39.4			
06	132 17.8	272 50.2	S14 34.2	324 31.5	N 6 53.3	142 26.3	S 5 57.7	104 02.3	N 8 39.3	Alioth	166 31.2	N55 58.0
07	147 20.3	287 49.6	35.3	339 32.5	52.8	157 28.9	57.8	119 05.0	39.2	Alkaid	153 08.3	N49 19.3
08	162 22.7	302 49.0	36.3	354 33.6	52.2	172 31.4	57.8	134 07.6	39.2	Al Na'ir	27 57.8	S46 58.2
09	177 25.2	317 48.4	.. 37.3	9 34.7	.. 51.7	187 34.0	.. 57.8	149 10.3	.. 39.1	Alnilam	275 57.6	S 1 12.2
10	192 27.7	332 47.8	38.4	24 35.8	51.1	202 36.5	57.8	164 12.9	39.0	Alphard	218 07.2	S 8 39.1
11	207 30.1	347 47.2	39.4	39 36.9	50.6	217 39.1	57.9	179 15.6	39.0			
12	222 32.6	2 46.6	S14 40.5	54 38.0	N 6 50.0	232 41.6	S 5 57.9	194 18.2	N 8 38.9	Alphecca	126 21.0	N26 43.4
13	237 35.1	17 46.0	41.5	69 39.1	49.4	247 44.2	57.9	209 20.9	38.9	Alpheratz	357 54.9	N29 05.2
14	252 37.5	32 45.4	42.5	84 40.2	48.9	262 46.7	57.9	224 23.5	38.8	Altair	62 19.4	N 8 52.1
15	267 40.0	47 44.8	.. 43.6	99 41.3	.. 48.3	277 49.3	.. 58.0	239 26.2	.. 38.7	Ankaa	353 26.5	S42 18.9
16	282 42.5	62 44.2	44.6	114 42.4	47.8	292 51.8	58.0	254 28.8	38.7	Antares	112 40.5	S26 25.6
17	297 44.9	77 43.6	45.6	129 43.5	47.2	307 54.3	58.0	269 31.5	38.6			
18	312 47.4	92 43.1	S14 46.7	144 44.6	N 6 46.7	322 56.9	S 5 58.0	284 34.1	N 8 38.5	Arcturus	146 06.4	N19 11.5
19	327 49.8	107 42.5	47.7	159 45.6	46.1	337 59.4	58.1	299 36.8	38.5	Atria	107 53.0	S69 01.5
20	342 52.3	122 41.9	48.7	174 46.7	45.5	353 02.0	58.1	314 39.4	38.4	Avior	234 22.5	S59 30.1
21	357 54.8	137 41.3	.. 49.8	189 47.8	.. 45.0	8 04.5	.. 58.1	329 42.1	.. 38.3	Bellatrix	278 43.9	N 6 20.9
22	12 57.2	152 40.7	50.8	204 48.9	44.4	23 07.1	58.1	344 44.7	38.3	Betelgeuse	271 13.3	N 7 24.4
23	27 59.7	167 40.1	51.8	219 50.0	43.9	38 09.6	58.2	359 47.3	38.2			
4 00	43 02.2	182 39.5	S14 52.9	234 51.1	N 6 43.3	53 12.2	S 5 58.2	14 50.0	N 8 38.1	Canopus	264 00.8	S52 41.6
01	58 04.6	197 38.9	53.9	249 52.2	42.8	68 14.7	58.2	29 52.6	38.1	Capella	280 50.8	N45 59.6
02	73 07.1	212 38.3	54.9	264 53.3	42.2	83 17.2	58.2	44 55.3	38.0	Deneb	49 39.2	N45 16.9
03	88 09.6	227 37.7	.. 56.0	279 54.4	.. 41.6	98 19.8	.. 58.2	59 57.9	.. 38.0	Denebola	182 45.4	N14 34.8
04	103 12.0	242 37.0	57.0	294 55.5	41.1	113 22.3	58.3	75 00.6	37.9	Diphda	349 06.9	S17 59.6
05	118 14.5	257 36.4	58.0	309 56.6	40.5	128 24.9	58.3	90 03.2	37.8			
06	133 17.0	272 35.8	S14 59.0	324 57.7	N 6 40.0	143 27.4	S 5 58.3	105 05.9	N 8 37.8	Dubhe	194 05.9	N61 45.3
07	148 19.4	287 35.2	15 00.1	339 58.8	39.4	158 29.9	58.3	120 08.5	37.7	Elnath	278 26.6	N28 36.2
08	163 21.9	302 34.6	01.1	354 59.9	38.9	173 32.5	58.4	135 11.2	37.6	Eltanin	90 51.8	N51 29.7
09	178 24.3	317 34.0	.. 02.1	10 01.0	.. 38.3	188 35.0	.. 58.4	150 13.8	.. 37.6	Enif	33 58.1	N 9 52.3
10	193 26.8	332 33.4	03.1	25 02.1	37.7	203 37.6	58.4	165 16.5	37.5	Fomalhaut	15 36.3	S29 37.8
11	208 29.3	347 32.8	04.2	40 03.2	37.2	218 40.1	58.4	180 19.1	37.4			
12	223 31.7	2 32.2	S15 05.2	55 04.2	N 6 36.6	233 42.6	S 5 58.4	195 21.8	N 8 37.4	Gacrux	172 14.0	S57 06.1
13	238 34.2	17 31.6	06.2	70 05.3	36.1	248 45.2	58.5	210 24.4	37.3	Gienah	176 04.2	S17 31.9
14	253 36.7	32 31.0	07.2	85 06.4	35.5	263 47.7	58.5	225 27.1	37.2	Hadar	149 04.6	S60 21.8
15	268 39.1	47 30.4	.. 08.2	100 07.5	.. 35.0	278 50.3	.. 58.5	240 29.7	.. 37.2	Hamal	328 13.2	N23 27.4
16	283 41.6	62 29.8	09.3	115 08.6	34.4	293 52.8	58.5	255 32.4	37.1	Kaus Aust.	83 59.1	S34 23.1
17	298 44.1	77 29.2	10.3	130 09.7	33.8	308 55.3	58.5	270 35.0	37.1			
18	313 46.5	92 28.6	S15 11.3	145 10.8	N 6 33.3	323 57.9	S 5 58.6	285 37.6	N 8 37.0	Kochab	137 20.9	N74 09.8
19	328 49.0	107 27.9	12.3	160 11.9	32.7	339 00.4	58.6	300 40.3	36.9	Markab	13 49.4	N15 12.1
20	343 51.4	122 27.3	13.3	175 13.0	32.2	354 02.9	58.6	315 42.9	36.9	Menkar	314 26.6	N 4 05.1
21	358 53.9	137 26.7	.. 14.3	190 14.1	.. 31.6	9 05.5	.. 58.6	330 45.6	.. 36.8	Menkent	148 21.3	S36 21.6
22	13 56.4	152 26.1	15.4	205 15.2	31.1	24 08.0	58.6	345 48.2	36.7	Miaplacidus	221 42.0	S69 42.4
23	28 58.8	167 25.5	16.4	220 16.3	30.5	39 10.6	58.7	0 50.9	36.7			
5 00	44 01.3	182 24.9	S15 17.4	235 17.4	N 6 29.9	54 13.1	S 5 58.7	15 53.5	N 8 36.6	Mirfak	308 56.0	N49 51.3
01	59 03.8	197 24.3	18.4	250 18.5	29.4	69 15.6	58.7	30 56.2	36.5	Nunki	76 12.5	S26 17.8
02	74 06.2	212 23.7	19.4	265 19.6	28.8	84 18.2	58.7	45 58.8	36.5	Peacock	53 37.1	S56 44.5
03	89 08.7	227 23.1	.. 20.4	280 20.7	.. 28.3	99 20.7	.. 58.7	61 01.5	.. 36.4	Pollux	243 41.4	N28 01.6
04	104 11.2	242 22.4	21.4	295 21.8	27.7	114 23.2	58.8	76 04.1	36.4	Procyon	245 11.4	N 5 13.7
05	119 13.6	257 21.8	22.5	310 22.9	27.1	129 25.8	58.8	91 06.8	36.3			
06	134 16.1	272 21.2	S15 23.5	325 24.0	N 6 26.6	144 28.3	S 5 58.8	106 09.4	N 8 36.2	Rasalhague	96 17.2	N12 33.9
07	149 18.6	287 20.6	24.5	340 25.1	26.0	159 30.8	58.8	121 12.1	36.2	Regulus	207 55.6	N11 58.4
08	164 21.0	302 20.0	25.5	355 26.2	25.5	174 33.4	58.8	136 14.7	36.1	Rigel	281 22.7	S 8 12.2
09	179 23.5	317 19.4	.. 26.5	10 27.3	.. 24.9	189 35.9	.. 58.9	151 17.3	.. 36.0	Rigil Kent.	140 07.9	S60 49.6
10	194 25.9	332 18.7	27.5	25 28.4	24.4	204 38.4	58.9	166 20.0	36.0	Sabik	102 25.8	S15 43.2
11	209 28.4	347 18.1	28.5	40 29.5	23.8	219 41.0	58.9	181 22.6	35.9			
12	224 30.9	2 17.5	S15 29.5	55 30.6	N 6 23.2	234 43.5	S 5 58.9	196 25.3	N 8 35.9	Schedar	349 52.9	N56 32.0
13	239 33.3	17 16.9	30.5	70 31.7	22.7	249 46.0	58.9	211 27.9	35.8	Shaula	96 37.6	S37 06.1
14	254 35.8	32 16.3	31.5	85 32.8	22.1	264 48.6	59.0	226 30.6	35.7	Sirius	258 43.5	S16 42.8
15	269 38.3	47 15.6	.. 32.5	100 33.9	.. 21.6	279 51.1	.. 59.0	241 33.2	.. 35.7	Spica	158 43.5	S11 09.1
16	284 40.7	62 15.0	33.5	115 35.0	21.0	294 53.6	59.0	256 35.9	35.6	Suhail	223 00.7	S43 25.4
17	299 43.2	77 14.4	34.5	130 36.1	20.5	309 56.1	59.0	271 38.5	35.5			
18	314 45.7	92 13.8	S15 35.5	145 37.2	N 6 19.9	324 58.7	S 5 59.0	286 41.2	N 8 35.5	Vega	80 46.9	N38 47.3
19	329 48.1	107 13.2	36.5	160 38.3	19.3	340 01.2	59.0	301 43.8	35.4	Zuben'ubi	137 18.3	S16 02.0
20	344 50.6	122 12.5	37.5	175 39.4	18.8	355 03.7	59.1	316 46.4	35.3			
21	359 53.1	137 11.9	.. 38.5	190 40.5	.. 18.2	10 06.3	.. 59.1	331 49.1	.. 35.3			
22	14 55.5	152 11.3	39.5	205 41.6	17.7	25 08.8	59.1	346 51.7	35.2			
23	29 58.0	167 10.7	40.5	220 42.7	17.1	40 11.3	59.1	1 54.4	35.2			
Mer. Pass.	21 04.4	v −0.6	d 1.0	v 1.1	d 0.6	v 2.5	d 0.0	v 2.6	d 0.1			

Tuesday — day 3; Wednesday — day 4; Thursday — day 5

	SHA	Mer. Pass.
	° '	h m
Venus	139 37.3	11 50
Mars	191 48.9	8 20
Jupiter	10 10.0	20 24
Saturn	331 47.8	22 57

UT	SUN GHA	SUN Dec	MOON GHA	v	Dec	d	HP
3 d h	° ′	° ′	° ′	′	° ′	′	′
00	184 06.3	S14 55.6	19 22.6	7.3	N 5 05.8	12.2	61.3
01	199 06.3	56.4	33 48.9	7.2	5 18.0	12.1	61.3
02	214 06.3	57.1	48 15.1	7.1	5 30.1	12.1	61.3
03	229 06.4 ..	57.9	62 41.2	7.1	5 42.2	12.0	61.3
04	244 06.4	58.7	77 07.3	7.1	5 54.2	12.1	61.4
05	259 06.4	14 59.5	91 33.4	7.1	6 06.3	12.0	61.4
T 06	274 06.4	S15 00.3	105 59.5	6.9	N 6 18.3	11.9	61.4
U 07	289 06.4	01.1	120 25.4	7.0	6 30.2	11.9	61.4
E 08	304 06.4	01.8	134 51.4	6.9	6 42.1	11.9	61.4
S 09	319 06.4 ..	02.6	149 17.3	6.9	6 54.0	11.8	61.4
D 10	334 06.4	03.4	163 43.2	6.8	7 05.8	11.8	61.4
A 11	349 06.4	04.2	178 09.0	6.8	7 17.6	11.8	61.4
Y 12	4 06.4	S15 05.0	192 34.8	6.7	N 7 29.4	11.7	61.4
13	19 06.4	05.7	207 00.5	6.7	7 41.1	11.6	61.4
14	34 06.4	06.5	221 26.2	6.6	7 52.7	11.6	61.5
15	49 06.4 ..	07.3	235 51.8	6.6	8 04.3	11.6	61.5
16	64 06.4	08.1	250 17.4	6.6	8 15.9	11.5	61.5
17	79 06.4	08.9	264 43.0	6.5	8 27.4	11.4	61.5
18	94 06.4	S15 09.6	279 08.5	6.5	N 8 38.8	11.4	61.5
19	109 06.4	10.4	293 34.0	6.4	8 50.2	11.3	61.5
20	124 06.4	11.2	307 59.4	6.4	9 01.5	11.3	61.5
21	139 06.4 ..	12.0	322 24.8	6.4	9 12.8	11.2	61.5
22	154 06.4	12.7	336 50.2	6.3	9 24.0	11.1	61.5
23	169 06.4	13.5	351 15.5	6.3	9 35.1	11.1	61.5
4 00	184 06.4	S15 14.3	5 40.8	6.2	N 9 46.2	11.0	61.5
01	199 06.4	15.1	20 06.0	6.2	9 57.2	10.9	61.5
02	214 06.4	15.8	34 31.2	6.1	10 08.1	10.9	61.5
03	229 06.3 ..	16.6	48 56.3	6.1	10 19.0	10.8	61.5
04	244 06.3	17.4	63 21.4	6.1	10 29.8	10.7	61.5
05	259 06.3	18.2	77 46.5	6.0	10 40.5	10.6	61.5
W 06	274 06.3	S15 18.9	92 11.5	6.0	N10 51.1	10.6	61.5
E 07	289 06.3	19.7	106 36.5	5.9	11 01.7	10.5	61.5
D 08	304 06.3	20.5	121 01.4	5.9	11 12.2	10.4	61.5
N 09	319 06.3 ..	21.2	135 26.3	5.9	11 22.6	10.3	61.5
E 10	334 06.3	22.0	149 51.2	5.8	11 32.9	10.3	61.5
S 11	349 06.3	22.8	164 16.0	5.8	11 43.2	10.2	61.5
D 12	4 06.3	S15 23.6	178 40.8	5.7	N11 53.4	10.0	61.4
A 13	19 06.3	24.3	193 05.5	5.7	12 03.4	10.0	61.4
Y 14	34 06.3	25.1	207 30.2	5.7	12 13.4	9.9	61.4
15	49 06.3 ..	25.9	221 54.9	5.6	12 23.3	9.9	61.4
16	64 06.3	26.6	236 19.5	5.6	12 33.2	9.7	61.4
17	79 06.3	27.4	250 44.1	5.6	12 42.9	9.6	61.4
18	94 06.2	S15 28.2	265 08.7	5.5	N12 52.5	9.6	61.4
19	109 06.2	28.9	279 33.2	5.5	13 02.1	9.4	61.4
20	124 06.2	29.7	293 57.7	5.5	13 11.5	9.3	61.4
21	139 06.2 ..	30.5	308 22.2	5.4	13 20.8	9.3	61.4
22	154 06.2	31.2	322 46.6	5.4	13 30.1	9.1	61.4
23	169 06.2	32.0	337 11.0	5.4	13 39.2	9.1	61.3
5 00	184 06.2	S15 32.8	351 35.4	5.3	N13 48.3	8.9	61.3
01	199 06.2	33.5	5 59.7	5.3	13 57.2	8.9	61.3
02	214 06.2	34.3	20 24.0	5.3	14 06.1	8.7	61.3
03	229 06.1 ..	35.0	34 48.3	5.3	14 14.8	8.7	61.3
04	244 06.1	35.8	49 12.6	5.2	14 23.5	8.5	61.3
05	259 06.1	36.6	63 36.8	5.2	14 32.0	8.4	61.2
T 06	274 06.1	S15 37.3	78 01.0	5.1	N14 40.4	8.3	61.2
H 07	289 06.1	38.1	92 25.1	5.2	14 48.7	8.2	61.2
U 08	304 06.1	38.9	106 49.3	5.1	14 56.9	8.1	61.2
R 09	319 06.1 ..	39.6	121 13.4	5.1	15 05.0	8.0	61.2
S 10	334 06.0	40.4	135 37.5	5.1	15 13.0	7.9	61.2
D 11	349 06.0	41.1	150 01.6	5.0	15 20.9	7.7	61.1
A 12	4 06.0	S15 41.9	164 25.6	5.1	N15 28.6	7.6	61.1
Y 13	19 06.0	42.6	178 49.7	5.0	15 36.2	7.6	61.1
14	34 06.0	43.4	193 13.7	5.0	15 43.8	7.4	61.1
15	49 06.0 ..	44.2	207 37.7	4.9	15 51.2	7.2	61.1
16	64 05.9	44.9	222 01.6	5.0	15 58.4	7.2	61.0
17	79 05.9	45.7	236 25.6	5.0	16 05.6	7.0	61.0
18	94 05.9	S15 46.4	250 49.6	4.9	N16 12.6	7.0	61.0
19	109 05.9	47.2	265 13.5	4.9	16 19.6	6.8	61.0
20	124 05.9	47.9	279 37.4	4.9	16 26.4	6.6	60.9
21	139 05.8 ..	48.7	294 01.3	4.9	16 33.0	6.6	60.9
22	154 05.8	49.5	308 25.2	4.9	16 39.6	6.4	60.9
23	169 05.8	50.2	322 49.1	4.9	N16 46.0	6.3	60.9
	SD 16.2	d 0.8	SD 16.7		16.7		16.7

Lat.	Twilight Naut.	Twilight Civil	Sunrise	Moonrise 3	4	5	6
°	h m	h m	h m	h m	h m	h m	h m
N 72	06 17	07 42	09 15	15 35	15 25	15 10	14 18
N 70	06 12	07 27	08 45	15 47	15 47	15 48	15 54
68	06 07	07 15	08 22	15 57	16 03	16 14	16 34
66	06 03	07 05	08 05	16 05	16 17	16 35	17 02
64	06 00	06 57	07 51	16 11	16 28	16 51	17 24
62	05 56	06 50	07 39	16 17	16 38	17 05	17 41
60	05 53	06 43	07 29	16 22	16 46	17 16	17 55
N 58	05 51	06 38	07 20	16 27	16 54	17 26	18 08
56	05 48	06 33	07 13	16 31	17 00	17 35	18 18
54	05 46	06 28	07 06	16 35	17 06	17 43	18 28
52	05 44	06 24	07 00	16 38	17 11	17 50	18 36
50	05 41	06 20	06 54	16 41	17 16	17 57	18 44
45	05 36	06 11	06 42	16 48	17 27	18 10	19 00
N 40	05 32	06 04	06 32	16 53	17 35	18 22	19 13
35	05 27	05 57	06 24	16 58	17 43	18 32	19 25
30	05 23	05 51	06 16	17 02	17 50	18 40	19 35
20	05 14	05 40	06 03	17 10	18 01	18 55	19 52
N 10	05 04	05 29	05 51	17 16	18 11	19 08	20 07
0	04 54	05 19	05 40	17 23	18 21	19 21	20 21
S 10	04 42	05 07	05 29	17 29	18 31	19 33	20 35
20	04 27	04 54	05 17	17 36	18 41	19 46	20 50
30	04 08	04 38	05 03	17 44	18 53	20 02	21 08
35	03 55	04 28	04 55	17 48	19 00	20 11	21 18
40	03 41	04 16	04 46	17 53	19 08	20 21	21 29
45	03 22	04 03	04 35	17 59	19 17	20 33	21 43
S 50	02 58	03 45	04 22	18 07	19 29	20 47	22 00
52	02 46	03 37	04 16	18 10	19 34	20 54	22 08
54	02 32	03 27	04 09	18 14	19 40	21 02	22 17
56	02 15	03 16	04 01	18 18	19 46	21 10	22 27
58	01 54	03 04	03 53	18 23	19 53	21 20	22 38
S 60	01 25	02 49	03 43	18 28	20 02	21 31	22 51

Lat.	Sunset	Twilight Civil	Twilight Naut.	Moonset 3	4	5	6
°	h m	h m	h m	h m	h m	h m	h m
N 72	14 11	15 44	17 08	06 14	08 25	10 45	13 44
N 70	14 41	15 59	17 14	06 05	08 06	10 08	12 08
68	15 04	16 11	17 19	05 57	07 50	09 43	11 28
66	15 21	16 21	17 23	05 51	07 38	09 23	11 01
64	15 35	16 29	17 26	05 46	07 28	09 08	10 40
62	15 47	16 36	17 30	05 41	07 19	08 55	10 23
60	15 57	16 43	17 33	05 37	07 12	08 44	10 09
N 58	16 06	16 49	17 35	05 34	07 05	08 34	09 57
56	16 14	16 54	17 38	05 31	06 59	08 26	09 47
54	16 20	16 58	17 40	05 28	06 54	08 19	09 38
52	16 27	17 03	17 43	05 25	06 50	08 12	09 30
50	16 32	17 07	17 45	05 23	06 45	08 06	09 22
45	16 44	17 15	17 50	05 18	06 36	07 53	09 07
N 40	16 55	17 23	17 55	05 14	06 29	07 43	08 54
35	17 03	17 30	18 00	05 11	06 22	07 34	08 43
30	17 11	17 36	18 04	05 07	06 17	07 26	08 33
20	17 24	17 47	18 13	05 02	06 07	07 12	08 17
N 10	17 36	17 58	18 23	04 57	05 58	07 00	08 03
0	17 47	18 08	18 33	04 53	05 50	06 49	07 49
S 10	17 58	18 20	18 46	04 48	05 42	06 38	07 36
20	18 10	18 34	19 01	04 44	05 34	06 26	07 22
30	18 24	18 50	19 20	04 38	05 24	06 13	07 05
35	18 33	19 00	19 32	04 35	05 18	06 05	06 56
40	18 42	19 11	19 47	04 31	05 12	05 56	06 45
45	18 53	19 26	20 06	04 27	05 05	05 46	06 32
S 50	19 06	19 43	20 31	04 23	04 56	05 33	06 17
52	19 13	19 52	20 43	04 20	04 52	05 27	06 10
54	19 20	20 02	20 58	04 18	04 47	05 21	06 02
56	19 27	20 13	21 15	04 15	04 42	05 14	05 53
58	19 36	20 25	21 37	04 12	04 37	05 06	05 43
S 60	19 46	20 40	22 07	04 09	04 31	04 57	05 31

	SUN			MOON			
Day	Eqn. of Time 00h	12h	Mer. Pass.	Mer. Pass. Upper	Lower	Age	Phase
d	m s	m s	h m	h m	h m	d	%
3	16 25	16 25	11 44	23 36	11 08	14	99
4	16 25	16 25	11 44	24 35	12 05	15	100
5	16 25	16 24	11 44	00 35	13 05	16	97

1998 NOVEMBER 6, 7, 8 (FRI., SAT., SUN.)

UT	ARIES GHA	VENUS −3.9 GHA	Dec	MARS +1.5 GHA	Dec	JUPITER −2.7 GHA	Dec	SATURN +0.0 GHA	Dec
6 00	45 00.4	182 10.0	S15 41.5	235 43.8	N 6 16.5	55 13.8	S 5 59.1	16 57.0	N 8 35.1
01	60 02.9	197 09.4	42.5	250 44.9	16.0	70 16.4	59.1	31 59.7	35.0
02	75 05.4	212 08.8	43.5	265 46.0	15.4	85 18.9	59.2	47 02.3	35.0
03	90 07.8	227 08.2 ..	44.5	280 47.1 ..	14.9	100 21.4 ..	59.2	62 05.0 ..	34.9
04	105 10.3	242 07.5	45.5	295 48.2	14.3	115 24.0	59.2	77 07.6	34.8
05	120 12.8	257 06.9	46.5	310 49.3	13.8	130 26.5	59.2	92 10.3	34.8
06	135 15.2	272 06.3	S15 47.5	325 50.4	N 6 13.2	145 29.0	S 5 59.2	107 12.9	N 8 34.7
07	150 17.7	287 05.7	48.5	340 51.5	12.6	160 31.5	59.2	122 15.5	34.7
08	165 20.2	302 05.0	49.5	355 52.6	12.1	175 34.1	59.2	137 18.2	34.6
F 09	180 22.6	317 04.4 ..	50.5	10 53.7 ..	11.5	190 36.6 ..	59.3	152 20.8 ..	34.5
R 10	195 25.1	332 03.8	51.5	25 54.8	11.0	205 39.1	59.3	167 23.5	34.5
I 11	210 27.6	347 03.1	52.5	40 55.9	10.4	220 41.6	59.3	182 26.1	34.4
D 12	225 30.0	2 02.5	S15 53.4	55 57.0	N 6 09.9	235 44.2	S 5 59.3	197 28.8	N 8 34.3
A 13	240 32.5	17 01.9	54.4	70 58.1	09.3	250 46.7	59.3	212 31.4	34.3
Y 14	255 34.9	32 01.2	55.4	85 59.2	08.7	265 49.2	59.3	227 34.1	34.2
15	270 37.4	47 00.6 ..	56.4	101 00.3 ..	08.2	280 51.7 ..	59.3	242 36.7 ..	34.2
16	285 39.9	62 00.0	57.4	116 01.4	07.6	295 54.2	59.4	257 39.3	34.1
17	300 42.3	76 59.3	58.4	131 02.5	07.1	310 56.8	59.4	272 42.0	34.0
18	315 44.8	91 58.7	S15 59.4	146 03.6	N 6 06.5	325 59.3	S 5 59.4	287 44.6	N 8 34.0
19	330 47.3	106 58.1	16 00.3	161 04.7	06.0	341 01.8	59.4	302 47.3	33.9
20	345 49.7	121 57.4	01.3	176 05.8	05.4	356 04.3	59.4	317 49.9	33.8
21	0 52.2	136 56.8 ..	02.3	191 06.9 ..	04.8	11 06.9 ..	59.4	332 52.6 ..	33.8
22	15 54.7	151 56.2	03.3	206 08.0	04.3	26 09.4	59.4	347 55.2	33.7
23	30 57.1	166 55.5	04.3	221 09.1	03.7	41 11.9	59.5	2 57.9	33.7
7 00	45 59.6	181 54.9	S16 05.3	236 10.2	N 6 03.2	56 14.4	S 5 59.5	18 00.5	N 8 33.6
01	61 02.1	196 54.3	06.2	251 11.3	02.6	71 16.9	59.5	33 03.1	33.5
02	76 04.5	211 53.6	07.2	266 12.4	02.0	86 19.4	59.5	48 05.8	33.5
03	91 07.0	226 53.0 ..	08.2	281 13.5 ..	01.5	101 22.0 ..	59.5	63 08.4 ..	33.4
04	106 09.4	241 52.3	09.2	296 14.6	00.9	116 24.5	59.5	78 11.1	33.3
05	121 11.9	256 51.7	10.1	311 15.7	6 00.4	131 27.0	59.5	93 13.7	33.3
06	136 14.4	271 51.1	S16 11.1	326 16.8	N 5 59.8	146 29.5	S 5 59.5	108 16.4	N 8 33.2
07	151 16.8	286 50.4	12.1	341 17.9	59.3	161 32.0	59.6	123 19.0	33.2
S 08	166 19.3	301 49.8	13.1	356 19.0	58.7	176 34.6	59.6	138 21.6	33.1
A 09	181 21.8	316 49.1 ..	14.0	11 20.1 ..	58.1	191 37.1 ..	59.6	153 24.3 ..	33.0
T 10	196 24.2	331 48.5	15.0	26 21.2	57.6	206 39.6	59.6	168 26.9	33.0
U 11	211 26.7	346 47.8	16.0	41 22.3	57.0	221 42.1	59.6	183 29.6	32.9
R 12	226 29.2	1 47.2	S16 17.0	56 23.5	N 5 56.5	236 44.6	S 5 59.6	198 32.2	N 8 32.8
D 13	241 31.6	16 46.6	17.9	71 24.6	55.9	251 47.1	59.6	213 34.9	32.8
A 14	256 34.1	31 45.9	18.9	86 25.7	55.4	266 49.7	59.6	228 37.5	32.7
Y 15	271 36.5	46 45.3 ..	19.9	101 26.8 ..	54.8	281 52.2 ..	59.6	243 40.2 ..	32.7
16	286 39.0	61 44.6	20.8	116 27.9	54.2	296 54.7	59.7	258 42.8	32.6
17	301 41.5	76 44.0	21.8	131 29.0	53.7	311 57.2	59.7	273 45.4	32.5
18	316 43.9	91 43.3	S16 22.8	146 30.1	N 5 53.1	326 59.7	S 5 59.7	288 48.1	N 8 32.5
19	331 46.4	106 42.7	23.8	161 31.2	52.6	342 02.2	59.7	303 50.7	32.4
20	346 48.9	121 42.0	24.7	176 32.3	52.0	357 04.7	59.7	318 53.4	32.4
21	1 51.3	136 41.4 ..	25.7	191 33.4 ..	51.4	12 07.3 ..	59.7	333 56.0 ..	32.3
22	16 53.8	151 40.7	26.6	206 34.5	50.9	27 09.8	59.7	348 58.7	32.2
23	31 56.3	166 40.1	27.6	221 35.6	50.3	42 12.3	59.7	4 01.3	32.2
8 00	46 58.7	181 39.4	S16 28.6	236 36.7	N 5 49.8	57 14.8	S 5 59.7	19 03.9	N 8 32.1
01	62 01.2	196 38.8	29.5	251 37.8	49.2	72 17.3	59.7	34 06.6	32.0
02	77 03.7	211 38.1	30.5	266 38.9	48.7	87 19.8	59.8	49 09.2	32.0
03	92 06.1	226 37.5 ..	31.5	281 40.0 ..	48.1	102 22.3 ..	59.8	64 11.9 ..	31.9
04	107 08.6	241 36.8	32.4	296 41.2	47.5	117 24.8	59.8	79 14.5	31.9
05	122 11.0	256 36.2	33.4	311 42.3	47.0	132 27.3	59.8	94 17.1	31.8
06	137 13.5	271 35.5	S16 34.3	326 43.4	N 5 46.4	147 29.9	S 5 59.8	109 19.8	N 8 31.7
07	152 16.0	286 34.9	35.3	341 44.5	45.9	162 32.4	59.8	124 22.4	31.7
08	167 18.4	301 34.2	36.3	356 45.6	45.3	177 34.9	59.8	139 25.1	31.6
S 09	182 20.9	316 33.6 ..	37.2	11 46.7 ..	44.8	192 37.4 ..	59.8	154 27.7 ..	31.6
U 10	197 23.4	331 32.9	38.2	26 47.8	44.2	207 39.9	59.8	169 30.4	31.5
N 11	212 25.8	346 32.3	39.1	41 48.9	43.6	222 42.4	59.8	184 33.0	31.4
D 12	227 28.3	1 31.6	S16 40.1	56 50.0	N 5 43.1	237 44.9	S 5 59.8	199 35.6	N 8 31.4
A 13	242 30.8	16 31.0	41.0	71 51.1	42.5	252 47.4	59.8	214 38.3	31.3
Y 14	257 33.2	31 30.3	42.0	86 52.2	42.0	267 49.9	59.9	229 40.9	31.3
15	272 35.7	46 29.6 ..	42.9	101 53.3 ..	41.4	282 52.4 ..	59.9	244 43.6 ..	31.2
16	287 38.2	61 29.0	43.9	116 54.4	40.8	297 54.9	59.9	259 46.2	31.1
17	302 40.6	76 28.3	44.8	131 55.5	40.3	312 57.4	59.9	274 48.9	31.1
18	317 43.1	91 27.7	S16 45.8	146 56.7	N 5 39.7	328 00.0	S 5 59.9	289 51.5	N 8 31.0
19	332 45.5	106 27.0	46.7	161 57.8	39.2	343 02.5	59.9	304 54.1	30.9
20	347 48.0	121 26.3	47.7	176 58.9	38.6	358 05.0	59.9	319 56.8	30.9
21	2 50.5	136 25.7 ..	48.6	192 00.0 ..	38.1	13 07.5 ..	59.9	334 59.4 ..	30.8
22	17 52.9	151 25.0	49.6	207 01.1	37.5	28 10.0	59.9	350 02.1	30.8
23	32 55.4	166 24.4	50.5	222 02.2	36.9	43 12.5	59.9	5 04.7	30.7
	h m Mer. Pass. 20 52.6	v −0.6	d 1.0	v 1.1	d 0.6	v 2.5	d 0.0	v 2.6	d 0.1

STARS

Name	SHA	Dec
Acamar	315 26.5	S40 18.6
Achernar	335 34.7	S57 14.7
Acrux	173 25.2	S63 05.2
Adhara	255 21.2	S28 58.1
Aldebaran	291 02.0	N16 30.3
Alioth	166 31.1	N55 58.0
Alkaid	153 08.3	N49 19.2
Al Na'ir	27 57.8	S46 58.2
Alnilam	275 57.6	S 1 12.2
Alphard	218 07.2	S 8 39.1
Alphecca	126 21.0	N26 43.4
Alpheratz	357 54.9	N29 05.2
Altair	62 19.4	N 8 52.1
Ankaa	353 26.5	S42 18.9
Antares	112 40.5	S26 25.6
Arcturus	146 06.4	N19 11.5
Atria	107 53.0	S69 01.5
Avior	234 22.4	S59 30.1
Bellatrix	278 43.9	N 6 20.9
Betelgeuse	271 13.3	N 7 24.3
Canopus	264 00.8	S52 41.6
Capella	280 50.7	N45 59.6
Deneb	49 39.2	N45 16.9
Denebola	182 45.4	N14 34.8
Diphda	349 07.0	S17 59.6
Dubhe	194 05.9	N61 45.3
Elnath	278 26.6	N28 36.2
Eltanin	90 51.8	N51 29.7
Enif	33 58.2	N 9 52.3
Fomalhaut	15 36.3	S29 37.8
Gacrux	172 13.9	S57 06.1
Gienah	176 04.2	S17 31.9
Hadar	149 04.6	S60 21.8
Hamal	328 13.2	N23 27.4
Kaus Aust.	83 59.1	S34 23.1
Kochab	137 20.9	N74 09.7
Markab	13 49.4	N15 12.1
Menkar	314 26.5	N 4 05.1
Menkent	148 21.3	S36 21.6
Miaplacidus	221 42.0	S69 42.4
Mirfak	308 56.0	N49 51.3
Nunki	76 12.5	S26 17.8
Peacock	53 37.1	S56 44.5
Pollux	243 41.4	N28 01.6
Procyon	245 11.4	N 5 13.7
Rasalhague	96 17.2	N12 33.9
Regulus	207 55.6	N11 58.4
Rigel	281 22.6	S 8 12.2
Rigil Kent.	140 07.9	S60 49.6
Sabik	102 25.8	S15 43.2
Schedar	349 52.9	N56 32.0
Shaula	96 37.6	S37 06.1
Sirius	258 43.5	S16 42.8
Spica	158 43.5	S11 09.1
Suhail	223 00.7	S43 25.4
Vega	80 46.9	N38 47.3
Zuben'ubi	137 18.3	S16 02.0

	SHA	Mer. Pass.
	° '	h m
Venus	135 55.3	11 53
Mars	190 10.6	8 15
Jupiter	10 14.8	20 12
Saturn	332 00.9	22 44

UT	SUN GHA	SUN Dec	MOON GHA	v	MOON Dec	d	HP
d h	° ′	° ′	° ′	′	° ′	′	′
6 00	184 05.8	S15 51.0	337 13.0	4.9	N16 52.3	6.2	60.8
01	199 05.8	51.7	351 36.9	4.8	16 58.5	6.1	60.8
02	214 05.7	52.5	6 00.7	4.9	17 04.6	5.9	60.8
03	229 05.7	.. 53.2	20 24.6	4.9	17 10.5	5.8	60.8
04	244 05.7	54.0	34 48.5	4.8	17 16.3	5.7	60.7
05	259 05.7	54.7	49 12.3	4.9	17 22.0	5.5	60.7
06	274 05.7	S15 55.5	63 36.2	4.8	N17 27.5	5.4	60.7
07	289 05.6	56.2	78 00.0	4.9	17 32.9	5.3	60.6
08	304 05.6	57.0	92 23.9	4.8	17 38.2	5.1	60.6
F 09	319 05.6	.. 57.7	106 47.7	4.9	17 43.3	5.1	60.6
R 10	334 05.6	58.5	121 11.6	4.9	17 48.4	4.9	60.6
I 11	349 05.5	15 59.2	135 35.5	4.9	17 53.3	4.7	60.5
D 12	4 05.5	S16 00.0	149 59.4	4.8	N17 58.0	4.6	60.5
A 13	19 05.5	00.7	164 23.2	4.9	18 02.6	4.5	60.5
Y 14	34 05.5	01.5	178 47.1	4.9	18 07.1	4.4	60.4
15	49 05.4	.. 02.2	193 11.0	5.0	18 11.5	4.2	60.4
16	64 05.4	03.0	207 35.0	4.9	18 15.7	4.1	60.4
17	79 05.4	03.7	221 58.9	4.9	18 19.8	4.0	60.3
18	94 05.4	S16 04.4	236 22.8	5.0	N18 23.8	3.8	60.3
19	109 05.3	05.2	250 46.8	5.0	18 27.6	3.7	60.3
20	124 05.3	05.9	265 10.8	5.0	18 31.3	3.6	60.2
21	139 05.3	.. 06.7	279 34.8	5.0	18 34.9	3.4	60.2
22	154 05.2	07.4	293 58.8	5.0	18 38.3	3.3	60.2
23	169 05.2	08.2	308 22.8	5.1	18 41.6	3.2	60.1
7 00	184 05.2	S16 08.9	322 46.9	5.1	N18 44.8	3.0	60.1
01	199 05.2	09.6	337 11.0	5.1	18 47.8	2.9	60.1
02	214 05.1	10.4	351 35.1	5.1	18 50.7	2.8	60.0
03	229 05.1	.. 11.1	5 59.2	5.2	18 53.5	2.6	60.0
04	244 05.1	11.9	20 23.4	5.2	18 56.1	2.5	60.0
05	259 05.0	12.6	34 47.6	5.2	18 58.6	2.3	59.9
06	274 05.0	S16 13.4	49 11.8	5.3	N19 00.9	2.3	59.9
S 07	289 05.0	14.1	63 36.1	5.3	19 03.2	2.0	59.8
A 08	304 04.9	14.8	78 00.4	5.3	19 05.2	2.0	59.8
T 09	319 04.9	.. 15.6	92 24.7	5.4	19 07.2	1.8	59.8
U 10	334 04.9	16.3	106 49.1	5.4	19 09.0	1.7	59.7
R 11	349 04.8	17.0	121 13.5	5.4	19 10.7	1.6	59.7
D 12	4 04.8	S16 17.8	135 37.9	5.5	N19 12.3	1.4	59.7
A 13	19 04.8	18.5	150 02.4	5.5	19 13.7	1.3	59.6
Y 14	34 04.7	19.3	164 26.9	5.6	19 15.0	1.2	59.6
15	49 04.7	.. 20.0	178 51.5	5.6	19 16.2	1.0	59.5
16	64 04.7	20.7	193 16.1	5.6	19 17.2	0.9	59.5
17	79 04.6	21.5	207 40.7	5.7	19 18.1	0.8	59.5
18	94 04.6	S16 22.2	222 05.4	5.8	N19 18.9	0.6	59.4
19	109 04.6	22.9	236 30.2	5.8	19 19.5	0.5	59.4
20	124 04.5	23.7	250 55.0	5.8	19 20.0	0.4	59.3
21	139 04.5	.. 24.4	265 19.8	5.9	19 20.4	0.3	59.3
22	154 04.5	25.1	279 44.7	5.9	19 20.7	0.1	59.3
23	169 04.4	25.9	294 09.6	6.0	19 20.8	0.0	59.2
8 00	184 04.4	S16 26.6	308 34.6	6.1	N19 20.8	0.1	59.2
01	199 04.3	27.3	322 59.7	6.1	19 20.7	0.3	59.1
02	214 04.3	28.0	337 24.8	6.1	19 20.4	0.4	59.1
03	229 04.3	.. 28.8	351 49.9	6.3	19 20.0	0.5	59.1
04	244 04.2	29.5	6 15.2	6.2	19 19.5	0.6	59.0
05	259 04.2	30.2	20 40.4	6.4	19 18.9	0.8	59.0
06	274 04.1	S16 31.0	35 05.8	6.4	N19 18.1	0.8	58.9
07	289 04.1	31.7	49 31.2	6.4	19 17.3	1.0	58.9
08	304 04.1	32.4	63 56.6	6.5	19 16.3	1.2	58.9
S 09	319 04.0	.. 33.1	78 22.1	6.6	19 15.1	1.2	58.8
U 10	334 04.0	33.9	92 47.7	6.6	19 13.9	1.4	58.8
N 11	349 03.9	34.6	107 13.3	6.7	19 12.5	1.5	58.7
D 12	4 03.9	S16 35.3	121 39.0	6.8	N19 11.0	1.6	58.7
A 13	19 03.8	36.0	136 04.8	6.8	19 09.4	1.7	58.7
Y 14	34 03.8	36.8	150 30.6	6.9	19 07.7	1.8	58.6
15	49 03.8	.. 37.5	164 56.5	7.0	19 05.9	2.0	58.6
16	64 03.7	38.2	179 22.5	7.0	19 03.9	2.1	58.5
17	79 03.7	38.9	193 48.5	7.1	19 01.8	2.1	58.5
18	94 03.6	S16 39.7	208 14.6	7.2	N18 59.7	2.4	58.5
19	109 03.6	40.4	222 40.8	7.3	18 57.3	2.4	58.4
20	124 03.5	41.1	237 07.1	7.3	18 54.9	2.5	58.4
21	139 03.5	.. 41.8	251 33.4	7.4	18 52.4	2.6	58.3
22	154 03.4	42.5	265 59.8	7.4	18 49.8	2.8	58.3
23	169 03.4	43.3	280 26.2	7.6	N18 47.0	2.9	58.3
	SD 16.2	d 0.7	SD 16.5		16.3		16.0

Moonrise

Lat.	Twilight Naut.	Twilight Civil	Sunrise	Moonrise 6	7	8	9
°	h m	h m	h m	h m	h m	h m	h m
N 72	06 29	07 56	09 36	14 18	□	□	17 58
N 70	06 22	07 39	09 00	15 54	16 18	17 26	19 05
68	06 16	07 26	08 35	16 34	17 13	18 17	19 42
66	06 12	07 15	08 16	17 02	17 46	18 49	20 07
64	06 07	07 05	08 01	17 24	18 10	19 13	20 27
62	06 03	06 57	07 48	17 41	18 30	19 31	20 43
60	06 00	06 50	07 37	17 55	18 45	19 47	20 56
N 58	05 57	06 44	07 27	18 08	18 59	20 00	21 08
56	05 54	06 38	07 19	18 18	19 10	20 11	21 18
54	05 51	06 33	07 12	18 28	19 21	20 21	21 27
52	05 48	06 29	07 05	18 36	19 30	20 30	21 34
50	05 46	06 25	06 59	18 44	19 38	20 38	21 42
45	05 40	06 15	06 46	19 00	19 55	20 54	21 57
N 40	05 35	06 07	06 36	19 13	20 09	21 08	22 09
35	05 30	06 00	06 26	19 25	20 21	21 20	22 20
30	05 25	05 53	06 18	19 35	20 32	21 30	22 29
20	05 15	05 41	06 04	19 52	20 50	21 48	22 45
N 10	05 05	05 30	05 52	20 07	21 05	22 03	22 58
0	04 54	05 19	05 40	20 21	21 20	22 17	23 11
S 10	04 41	05 06	05 28	20 35	21 35	22 32	23 24
20	04 25	04 52	05 16	20 50	21 51	22 47	23 38
30	04 05	04 35	05 01	21 08	22 09	23 04	23 54
35	03 52	04 25	04 52	21 18	22 20	23 15	24 03
40	03 37	04 13	04 43	21 29	22 32	23 26	24 13
45	03 17	03 58	04 31	21 43	22 46	23 40	24 26
S 50	02 51	03 40	04 17	22 00	23 04	23 57	24 40
52	02 38	03 31	04 10	22 08	23 12	24 05	00 05
54	02 23	03 21	04 03	22 17	23 21	24 13	00 13
56	02 04	03 09	03 55	22 27	23 31	24 23	00 23
58	01 41	02 55	03 45	22 38	23 43	24 34	00 34
S 60	01 07	02 39	03 35	22 51	23 57	24 47	00 47

Moonset

Lat.	Sunset	Twilight Civil	Twilight Naut.	Moonset 6	7	8	9
°	h m	h m	h m	h m	h m	h m	h m
N 72	13 50	15 30	16 57	13 44	□	□	16 05
N 70	14 26	15 47	17 04	12 08	13 50	14 43	14 58
68	14 51	16 01	17 10	11 28	12 55	13 51	14 21
66	15 10	16 12	17 15	11 01	12 22	13 19	13 55
64	15 26	16 21	17 19	10 40	11 58	12 55	13 35
62	15 39	16 29	17 23	10 23	11 38	12 36	13 18
60	15 50	16 36	17 26	10 09	11 23	12 21	13 04
N 58	15 59	16 42	17 30	09 57	11 09	12 08	12 53
56	16 07	16 48	17 33	09 47	10 58	11 56	12 42
54	16 15	16 53	17 36	09 38	10 48	11 46	12 33
52	16 22	16 58	17 38	09 30	10 39	11 37	12 25
50	16 28	17 02	17 41	09 22	10 31	11 29	12 18
45	16 41	17 12	17 47	09 07	10 14	11 12	12 02
N 40	16 51	17 20	17 52	08 54	10 00	10 58	11 49
35	17 01	17 27	17 57	08 43	09 48	10 46	11 38
30	17 09	17 34	18 02	08 33	09 37	10 36	11 28
20	17 23	17 46	18 12	08 17	09 19	10 18	11 12
N 10	17 35	17 57	18 22	08 03	09 04	10 02	10 57
0	17 47	18 09	18 34	07 49	08 49	09 48	10 44
S 10	17 59	18 21	18 47	07 36	08 35	09 33	10 30
20	18 12	18 35	19 03	07 22	08 19	09 17	10 15
30	18 27	18 52	19 23	07 05	08 01	08 59	09 58
35	18 36	19 03	19 36	06 56	07 51	08 49	09 48
40	18 45	19 15	19 52	06 45	07 39	08 37	09 37
45	18 57	19 30	20 11	06 32	07 25	08 22	09 24
S 50	19 12	19 49	20 38	06 17	07 08	08 05	09 07
52	19 18	19 58	20 51	06 10	06 59	07 57	09 00
54	19 26	20 08	21 07	06 02	06 50	07 47	08 51
56	19 34	20 20	21 26	05 53	06 40	07 37	08 41
58	19 44	20 34	21 51	05 43	06 29	07 25	08 31
S 60	19 54	20 51	22 27	05 31	06 15	07 12	08 18

	SUN Eqn. of Time 00h	12h	SUN Mer. Pass.	MOON Mer. Pass. Upper	Lower	Age	Phase
Day	m s	m s	h m	h m	h m	d	%
6	16 23	16 22	11 44	01 35	14 05	17	92
7	16 21	16 19	11 44	02 35	15 05	18	85
8	16 18	16 16	11 44	03 34	16 03	19	76

1998 NOVEMBER 9, 10, 11 (MON., TUES., WED.)

UT	ARIES GHA	VENUS −3.9 GHA	Dec	MARS +1.5 GHA	Dec	JUPITER −2.7 GHA	Dec	SATURN +0.1 GHA	Dec
d h	° ′	° ′	° ′	° ′	° ′	° ′	° ′	° ′	° ′
9 00	47 57.9	181 23.7	S16 51.5	237 03.3	N 5 36.4	58 15.0	S 5 59.9	20 07.3	N 8 30.6
01	63 00.3	196 23.0	52.4	252 04.4	35.8	73 17.5	59.9	35 10.0	30.6
02	78 02.8	211 22.4	53.3	267 05.5	35.3	88 20.0	59.9	50 12.6	30.5
03	93 05.3	226 21.7 ..	54.3	282 06.7 ..	34.7	103 22.5 ..	59.9	65 15.3 ..	30.5
04	108 07.7	241 21.0	55.2	297 07.8	34.2	118 25.0	5 59.9	80 17.9	30.4
05	123 10.2	256 20.4	56.2	312 08.9	33.6	133 27.5	6 00.0	95 20.5	30.3
M 06	138 12.7	271 19.7	S16 57.1	327 10.0	N 5 33.0	148 30.0	S 6 00.0	110 23.2	N 8 30.3
O 07	153 15.1	286 19.0	58.1	342 11.1	32.5	163 32.5	00.0	125 25.8	30.2
N 08	168 17.6	301 18.4	59.0	357 12.2	31.9	178 35.0	00.0	140 28.5	30.2
D 09	183 20.0	316 17.7	16 59.9	12 13.3 ..	31.4	193 37.5 ..	00.0	155 31.1 ..	30.1
A 10	198 22.5	331 17.0	17 00.9	27 14.4	30.8	208 40.0	00.0	170 33.7	30.0
Y 11	213 25.0	346 16.4	01.8	42 15.5	30.2	223 42.5	00.0	185 36.4	30.0
12	228 27.4	1 15.7	S17 02.7	57 16.7	N 5 29.7	238 45.0	S 6 00.0	200 39.0	N 8 29.9
13	243 29.9	16 15.0	03.7	72 17.8	29.1	253 47.5	00.0	215 41.7	29.9
14	258 32.4	31 14.4	04.6	87 18.9	28.6	268 50.0	00.0	230 44.3	29.8
15	273 34.8	46 13.7 ..	05.5	102 20.0 ..	28.0	283 52.5 ..	00.0	245 46.9 ..	29.7
16	288 37.3	61 13.0	06.5	117 21.1	27.5	298 55.0	00.0	260 49.6	29.7
17	303 39.8	76 12.4	07.4	132 22.2	26.9	313 57.5	00.0	275 52.2	29.6
18	318 42.2	91 11.7	S17 08.3	147 23.3	N 5 26.3	329 00.0	S 6 00.0	290 54.9	N 8 29.6
19	333 44.7	106 11.0	09.3	162 24.4	25.8	344 02.5	00.0	305 57.5	29.5
20	348 47.1	121 10.3	10.2	177 25.5	25.2	359 05.0	00.0	321 00.1	29.4
21	3 49.6	136 09.7 ..	11.1	192 26.7 ..	24.7	14 07.5 ..	00.0	336 02.8 ..	29.4
22	18 52.1	151 09.0	12.1	207 27.8	24.1	29 10.0	00.0	351 05.4	29.3
23	33 54.5	166 08.3	13.0	222 28.9	23.6	44 12.5	00.0	6 08.1	29.3
10 00	48 57.0	181 07.6	S17 13.9	237 30.0	N 5 23.0	59 15.0	S 6 00.0	21 10.7	N 8 29.2
01	63 59.5	196 07.0	14.8	252 31.1	22.4	74 17.5	00.0	36 13.3	29.1
02	79 01.9	211 06.3	15.8	267 32.2	21.9	89 20.0	00.0	51 16.0	29.1
03	94 04.4	226 05.6 ..	16.7	282 33.3 ..	21.3	104 22.5 ..	00.0	66 18.6 ..	29.0
04	109 06.9	241 04.9	17.6	297 34.5	20.8	119 25.0	00.0	81 21.3	29.0
05	124 09.3	256 04.3	18.5	312 35.6	20.2	134 27.5	00.0	96 23.9	28.9
T 06	139 11.8	271 03.6	S17 19.4	327 36.7	N 5 19.7	149 30.0	S 6 00.0	111 26.5	N 8 28.8
U 07	154 14.3	286 02.9	20.4	342 37.8	19.1	164 32.4	00.0	126 29.2	28.8
E 08	169 16.7	301 02.2	21.3	357 38.9	18.5	179 34.9	00.0	141 31.8	28.7
S 09	184 19.2	316 01.5 ..	22.2	12 40.0 ..	18.0	194 37.4 ..	00.0	156 34.5 ..	28.7
D 10	199 21.6	331 00.9	23.1	27 41.1	17.4	209 39.9	00.0	171 37.1	28.6
A 11	214 24.1	346 00.2	24.0	42 42.3	16.9	224 42.4	00.0	186 39.7	28.5
Y 12	229 26.6	0 59.5	S17 25.0	57 43.4	N 5 16.3	239 44.9	S 6 00.1	201 42.4	N 8 28.5
13	244 29.0	15 58.8	25.9	72 44.5	15.8	254 47.4	00.1	216 45.0	28.4
14	259 31.5	30 58.1	26.8	87 45.6	15.2	269 49.9	00.1	231 47.6	28.4
15	274 34.0	45 57.5 ..	27.7	102 46.7 ..	14.6	284 52.4 ..	00.1	246 50.3 ..	28.3
16	289 36.4	60 56.8	28.6	117 47.8	14.1	299 54.9	00.1	261 52.9	28.2
17	304 38.9	75 56.1	29.5	132 48.9	13.5	314 57.4	00.1	276 55.6	28.2
18	319 41.4	90 55.4	S17 30.4	147 50.1	N 5 13.0	329 59.9	S 6 00.1	291 58.2	N 8 28.1
19	334 43.8	105 54.7	31.4	162 51.2	12.4	345 02.4	00.1	307 00.8	28.1
20	349 46.3	120 54.0	32.3	177 52.3	11.8	0 04.8	00.1	322 03.5	28.0
21	4 48.8	135 53.3 ..	33.2	192 53.4 ..	11.3	15 07.3 ..	00.1	337 06.1 ..	28.0
22	19 51.2	150 52.7	34.1	207 54.5	10.7	30 09.8	00.1	352 08.8	27.9
23	34 53.7	165 52.0	35.0	222 55.6	10.2	45 12.3	00.1	7 11.4	27.8
11 00	49 56.1	180 51.3	S17 35.9	237 56.8	N 5 09.6	60 14.8	S 6 00.1	22 14.0	N 8 27.8
01	64 58.6	195 50.6	36.8	252 57.9	09.1	75 17.3	00.1	37 16.7	27.7
02	80 01.1	210 49.9	37.7	267 59.0	08.5	90 19.8	00.1	52 19.3	27.7
03	95 03.5	225 49.2 ..	38.6	283 00.1 ..	07.9	105 22.3 ..	00.0	67 21.9 ..	27.6
04	110 06.0	240 48.5	39.5	298 01.2	07.4	120 24.7	00.0	82 24.6	27.5
05	125 08.5	255 47.8	40.4	313 02.3	06.8	135 27.2	00.0	97 27.2	27.5
W 06	140 10.9	270 47.1	S17 41.3	328 03.5	N 5 06.3	150 29.7	S 6 00.0	112 29.9	N 8 27.4
E 07	155 13.4	285 46.5	42.2	343 04.6	05.7	165 32.2	00.0	127 32.5	27.4
D 08	170 15.9	300 45.8	43.1	358 05.7	05.2	180 34.7	00.0	142 35.1	27.3
N 09	185 18.3	315 45.1 ..	44.0	13 06.8 ..	04.6	195 37.2 ..	00.0	157 37.8 ..	27.2
E 10	200 20.8	330 44.4	44.9	28 07.9	04.0	210 39.7	00.0	172 40.4	27.2
S 11	215 23.3	345 43.7	45.8	43 09.0	03.5	225 42.1	00.0	187 43.0	27.1
D 12	230 25.7	0 43.0	S17 46.7	58 10.2	N 5 02.9	240 44.6	S 6 00.0	202 45.7	N 8 27.1
A 13	245 28.2	15 42.3	47.6	73 11.3	02.4	255 47.1	00.0	217 48.3	27.0
Y 14	260 30.6	30 41.6	48.5	88 12.4	01.8	270 49.6	00.0	232 50.9	27.0
15	275 33.1	45 40.9 ..	49.4	103 13.5 ..	01.3	285 52.1 ..	00.0	247 53.6 ..	26.9
16	290 35.6	60 40.2	50.3	118 14.6	00.7	300 54.6	00.0	262 56.2	26.8
17	305 38.0	75 39.5	51.2	133 15.7	5 00.1	315 57.1	00.0	277 58.9	26.8
18	320 40.5	90 38.8	S17 52.1	148 16.9	N 4 59.6	330 59.5	S 6 00.0	293 01.5	N 8 26.7
19	335 43.0	105 38.1	53.0	163 18.0	59.0	346 02.0	00.0	308 04.1	26.7
20	350 45.4	120 37.4	53.9	178 19.1	58.5	1 04.5	00.0	323 06.8	26.6
21	5 47.9	135 36.7 ..	54.8	193 20.2 ..	57.9	16 07.0 ..	00.0	338 09.4 ..	26.5
22	20 50.4	150 36.0	55.6	208 21.3	57.4	31 09.5	00.0	353 12.0	26.5
23	35 52.8	165 35.3	56.5	223 22.5	56.8	46 11.9	00.0	8 14.7	26.4
Mer. Pass.	20 40.8	*v* −0.7	*d* 0.9	*v* 1.1	*d* 0.6	*v* 2.5	*d* 0.0	*v* 2.6	*d* 0.1

STARS

Name	SHA	Dec
Acamar	315 26.5	S40 18.7
Achernar	335 34.7	S57 14.7
Acrux	173 22.4	S63 05.2
Adhara	255 21.2	S28 58.1
Aldebaran	291 02.0	N16 30.3
Alioth	166 31.1	N55 57.9
Alkaid	153 08.3	N49 19.2
Al Na'ir	27 57.8	S46 58.2
Alnilam	275 57.6	S 1 12.2
Alphard	218 07.2	S 8 39.1
Alphecca	126 21.0	N26 43.3
Alpheratz	357 54.9	N29 05.2
Altair	62 19.4	N 8 52.1
Ankaa	353 26.5	S42 18.9
Antares	112 40.5	S26 25.6
Arcturus	146 06.4	N19 11.5
Atria	107 53.0	S69 01.5
Avior	234 22.4	S59 30.1
Bellatrix	278 43.8	N 6 20.9
Betelgeuse	271 13.3	N 7 24.3
Canopus	264 00.8	S52 41.6
Capella	280 50.7	N45 59.6
Deneb	49 39.2	N45 16.9
Denebola	182 45.4	N14 34.8
Diphda	349 07.0	S17 59.6
Dubhe	194 05.9	N61 45.2
Elnath	278 26.6	N28 36.3
Eltanin	90 51.8	N51 29.7
Enif	33 58.2	N 9 52.3
Fomalhaut	15 36.3	S29 37.8
Gacrux	172 13.9	S57 06.1
Gienah	176 04.2	S17 31.9
Hadar	149 04.5	S60 21.8
Hamal	328 13.2	N23 27.4
Kaus Aust.	83 59.1	S34 23.1
Kochab	137 20.9	N74 09.7
Markab	13 49.4	N15 12.1
Menkar	314 26.5	N 4 05.1
Menkent	148 21.3	S36 21.6
Miaplacidus	221 41.9	S69 42.4
Mirfak	308 56.0	N49 51.3
Nunki	76 12.5	S26 17.8
Peacock	53 37.2	S56 44.5
Pollux	243 41.4	N28 01.6
Procyon	245 11.4	N 5 13.7
Rasalhague	96 17.2	N12 33.9
Regulus	207 55.6	N11 58.4
Rigel	281 22.6	S 8 12.2
Rigil Kent.	140 07.9	S60 49.6
Sabik	102 25.8	S15 43.2
Schedar	349 52.9	N56 32.0
Shaula	96 37.6	S37 06.1
Sirius	258 43.4	S16 42.8
Spica	158 43.4	S11 09.1
Suhail	223 00.7	S43 25.4
Vega	80 46.9	N38 47.3
Zuben'ubi	137 18.3	S16 02.0

	SHA	Mer. Pass.
	° ′	h m
Venus	132 10.6	11 56
Mars	188 33.0	8 09
Jupiter	10 18.0	20 00
Saturn	332 13.7	22 31

UT	SUN GHA	SUN Dec	MOON GHA	v	MOON Dec	d	HP
9 MONDAY							
00	184 03.3	S16 44.0	294 52.8	7.6	N18 44.1	2.9	58.2
01	199 03.3	44.7	309 19.4	7.6	18 41.2	3.1	58.2
02	214 03.3	45.4	323 46.0	7.8	18 38.1	3.2	58.1
03	229 03.2	.. 46.1	338 12.8	7.8	18 34.9	3.3	58.1
04	244 03.2	46.9	352 39.6	7.9	18 31.6	3.4	58.1
05	259 03.1	47.6	7 06.5	8.0	18 28.2	3.5	58.0
06	274 03.1	S16 48.3	21 33.5	8.0	N18 24.7	3.6	58.0
07	289 03.0	49.0	36 00.5	8.2	18 21.1	3.8	57.9
08	304 03.0	49.7	50 27.7	8.2	18 17.3	3.8	57.9
09	319 02.9	.. 50.4	64 54.9	8.2	18 13.5	3.9	57.9
10	334 02.9	51.2	79 22.1	8.4	18 09.6	4.0	57.8
11	349 02.8	51.9	93 49.5	8.4	18 05.6	4.1	57.8
12	4 02.8	S16 52.6	108 16.9	8.5	N18 01.5	4.3	57.8
13	19 02.7	53.3	122 44.4	8.6	17 57.2	4.3	57.7
14	34 02.7	54.0	137 12.0	8.7	17 52.9	4.4	57.7
15	49 02.6	.. 54.7	151 39.7	8.7	17 48.5	4.5	57.6
16	64 02.5	55.4	166 07.4	8.9	17 44.0	4.6	57.6
17	79 02.5	56.1	180 35.3	8.9	17 39.4	4.7	57.5
18	94 02.4	S16 56.9	195 03.2	9.0	N17 34.7	4.8	57.5
19	109 02.4	57.6	209 31.2	9.0	17 29.9	4.9	57.5
20	124 02.3	58.3	223 59.2	9.2	17 25.0	5.0	57.4
21	139 02.3	.. 59.0	238 27.4	9.2	17 20.0	5.0	57.4
22	154 02.2	16 59.7	252 55.6	9.3	17 15.0	5.2	57.3
23	169 02.2	17 00.4	267 23.9	9.4	17 09.8	5.2	57.3
10 TUESDAY							
00	184 02.1	S17 01.1	281 52.3	9.4	N17 04.6	5.3	57.3
01	199 02.1	01.8	296 20.7	9.6	16 59.3	5.5	57.2
02	214 02.0	02.5	310 49.3	9.6	16 53.8	5.5	57.2
03	229 01.9	.. 03.2	325 17.9	9.7	16 48.3	5.6	57.1
04	244 01.9	03.9	339 46.6	9.7	16 42.7	5.6	57.1
05	259 01.8	04.6	354 15.3	9.9	16 37.1	5.8	57.1
06	274 01.8	S17 05.3	8 44.2	9.9	N16 31.3	5.8	57.0
07	289 01.7	06.0	23 13.1	10.0	16 25.5	5.9	57.0
08	304 01.6	06.7	37 42.1	10.1	16 19.6	6.0	57.0
09	319 01.6	.. 07.4	52 11.2	10.2	16 13.6	6.1	56.9
10	334 01.5	08.2	66 40.4	10.2	16 07.5	6.1	56.9
11	349 01.5	08.9	81 09.6	10.3	16 01.4	6.3	56.8
12	4 01.4	S17 09.6	95 38.9	10.4	N15 55.1	6.3	56.8
13	19 01.3	10.3	110 08.3	10.5	15 48.8	6.3	56.8
14	34 01.3	11.0	124 37.8	10.6	15 42.5	6.5	56.7
15	49 01.2	.. 11.7	139 07.4	10.6	15 36.0	6.5	56.7
16	64 01.2	12.4	153 37.0	10.7	15 29.5	6.6	56.7
17	79 01.1	13.1	168 06.7	10.8	15 22.9	6.7	56.6
18	94 01.0	S17 13.8	182 36.5	10.8	N15 16.2	6.7	56.6
19	109 01.0	14.5	197 06.3	11.0	15 09.5	6.8	56.6
20	124 00.9	15.2	211 36.3	11.0	15 02.7	6.9	56.5
21	139 00.8	.. 15.8	226 06.3	11.1	14 55.8	7.0	56.5
22	154 00.8	16.5	240 36.4	11.1	14 48.8	7.0	56.5
23	169 00.7	17.2	255 06.5	11.3	14 41.8	7.0	56.4
11 WEDNESDAY							
00	184 00.6	S17 17.9	269 36.8	11.3	N14 34.8	7.2	56.4
01	199 00.6	18.6	284 07.1	11.3	14 27.6	7.2	56.4
02	214 00.5	19.3	298 37.4	11.5	14 20.4	7.3	56.3
03	229 00.5	.. 20.0	313 07.9	11.5	14 13.1	7.3	56.3
04	244 00.4	20.7	327 38.4	11.6	14 05.8	7.4	56.2
05	259 00.3	21.4	342 09.0	11.7	13 58.4	7.4	56.2
06	274 00.2	S17 22.1	356 39.7	11.7	N13 51.0	7.5	56.2
07	289 00.2	22.8	11 10.4	11.8	13 43.5	7.6	56.2
08	304 00.1	23.5	25 41.2	11.9	13 35.9	7.6	56.1
09	319 00.0	.. 24.2	40 12.1	12.0	13 28.3	7.7	56.1
10	334 00.0	24.9	54 43.1	12.0	13 20.6	7.8	56.1
11	348 59.9	25.6	69 14.1	12.1	13 12.8	7.8	56.0
12	3 59.8	S17 26.2	83 45.2	12.1	N13 05.0	7.8	56.0
13	18 59.8	26.9	98 16.3	12.2	12 57.2	7.9	56.0
14	33 59.7	27.6	112 47.5	12.3	12 49.3	8.0	55.9
15	48 59.6	.. 28.3	127 18.8	12.4	12 41.3	8.0	55.9
16	63 59.6	29.0	141 50.2	12.4	12 33.3	8.0	55.9
17	78 59.5	29.7	156 21.6	12.5	12 25.3	8.1	55.8
18	93 59.4	S17 30.4	170 53.1	12.6	N12 17.2	8.2	55.8
19	108 59.3	31.1	185 24.7	12.6	12 09.0	8.2	55.8
20	123 59.2	31.7	199 56.3	12.6	12 00.8	8.3	55.7
21	138 59.2	.. 32.4	214 27.9	12.8	11 52.5	8.3	55.7
22	153 59.1	33.1	228 59.7	12.8	11 44.2	8.3	55.7
23	168 59.0	33.8	243 31.5	12.8	N11 35.9	8.4	55.7
	SD 16.2	d 0.7	SD 15.7		15.5		15.3

Moonrise / Twilight

Lat.	Naut.	Civil	Sunrise	9	10	11	12
N 72	06 40	08 10	10 00	17 58	20 14	22 07	23 51
N 70	06 32	07 51	09 17	19 05	20 49	22 28	24 03
68	06 25	07 36	08 49	19 42	21 13	22 44	24 13
66	06 20	07 24	08 27	20 07	21 32	22 57	24 21
64	06 15	07 14	08 10	20 27	21 47	23 08	24 28
62	06 10	07 05	07 56	20 43	21 59	23 17	24 34
60	06 06	06 57	07 45	20 56	22 10	23 25	24 39
N 58	06 03	06 50	07 34	21 08	22 19	23 32	24 43
56	05 59	06 44	07 26	21 18	22 27	23 38	24 47
54	05 56	06 39	07 18	21 27	22 35	23 43	24 51
52	05 53	06 34	07 11	21 34	22 41	23 48	24 54
50	05 50	06 29	07 04	21 42	22 47	23 52	24 57
45	05 44	06 19	06 50	21 57	22 59	24 02	00 02
N 40	05 38	06 10	06 39	22 09	23 10	24 10	00 10
35	05 32	06 03	06 29	22 20	23 19	24 16	00 16
30	05 27	05 56	06 21	22 29	23 26	24 22	00 22
20	05 16	05 43	06 06	22 45	23 40	24 32	00 32
N 10	05 05	05 31	05 53	22 58	23 51	24 41	00 41
0	04 54	05 19	05 40	23 11	24 02	00 02	00 50
S 10	04 40	05 06	05 28	23 24	24 13	00 13	00 58
20	04 24	04 51	05 14	23 38	24 25	00 25	01 07
30	04 02	04 33	04 59	23 54	24 38	00 38	01 17
35	03 49	04 22	04 50	24 03	00 03	00 45	01 23
40	03 33	04 09	04 39	24 13	00 13	00 54	01 30
45	03 12	03 54	04 27	24 26	00 26	01 04	01 37
S 50	02 45	03 34	04 12	24 40	00 40	01 16	01 46
52	02 31	03 25	04 05	00 05	00 47	01 22	01 51
54	02 14	03 14	03 57	00 13	00 55	01 28	01 55
56	01 54	03 02	03 49	00 23	01 03	01 35	02 00
58	01 27	02 47	03 39	00 34	01 13	01 43	02 06
S 60	00 44	02 30	03 27	00 47	01 24	01 52	02 13

Moonset / Twilight

Lat.	Sunset	Civil	Naut.	9	10	11	12
N 72	13 26	15 17	16 46	16 05	15 37	15 24	15 15
N 70	14 09	15 36	16 54	14 58	15 02	15 02	15 01
68	14 38	15 51	17 01	14 21	14 36	14 45	14 50
66	14 59	16 03	17 07	13 55	14 17	14 31	14 40
64	15 16	16 13	17 12	13 35	14 01	14 19	14 32
62	15 31	16 22	17 17	13 18	13 48	14 09	14 26
60	15 42	16 30	17 21	13 04	13 36	14 01	14 20
N 58	15 53	16 37	17 24	12 53	13 27	13 53	14 15
56	16 02	16 43	17 28	12 42	13 18	13 47	14 10
54	16 10	16 48	17 31	12 33	13 11	13 41	14 06
52	16 17	16 53	17 34	12 25	13 04	13 35	14 02
50	16 23	16 58	17 37	12 18	12 57	13 30	13 59
45	16 37	17 08	17 43	12 02	12 44	13 20	13 51
N 40	16 48	17 17	17 49	11 49	12 33	13 11	13 45
35	16 58	17 25	17 55	11 38	12 24	13 04	13 39
30	17 07	17 32	18 01	11 28	12 15	12 57	13 35
20	17 22	17 45	18 11	11 12	12 01	12 45	13 26
N 10	17 35	17 57	18 22	10 57	11 48	12 35	13 19
0	17 47	18 09	18 34	10 44	11 36	12 25	13 12
S 10	18 00	18 22	18 48	10 30	11 24	12 16	13 05
20	18 14	18 37	19 05	10 15	11 11	12 05	12 57
30	18 29	18 55	19 26	09 58	10 56	11 53	12 49
35	18 38	19 06	19 40	09 48	10 48	11 46	12 44
40	18 49	19 19	19 56	09 37	10 38	11 39	12 38
45	19 01	19 35	20 17	09 24	10 26	11 29	12 31
S 50	19 17	19 55	20 45	09 07	10 12	11 18	12 23
52	19 24	20 04	20 59	09 00	10 06	11 13	12 19
54	19 32	20 15	21 16	08 51	09 58	11 07	12 15
56	19 41	20 28	21 37	08 41	09 50	11 01	12 11
58	19 51	20 43	22 05	08 31	09 41	10 54	12 06
S 60	20 02	21 01	22 52	08 18	09 30	10 45	12 00

SUN / MOON

Day	Eqn. of Time 00h	12h	Mer. Pass.	Mer. Pass. Upper	Lower	Age	Phase
	m s	m s	h m	h m	h m	d	%
9	16 13	16 11	11 44	04 30	16 58	20	66
10	16 09	16 06	11 44	05 24	17 49	21	55
11	16 03	15 59	11 44	06 14	18 38	22	45

UT	ARIES GHA	VENUS −3.9 GHA	Dec	MARS +1.5 GHA	Dec	JUPITER −2.6 GHA	Dec	SATURN +0.1 GHA	Dec	STARS Name	SHA	Dec
d h	° ′	° ′	° ′	° ′	° ′	° ′	° ′	° ′	° ′		° ′	° ′
12 00	50 55.3	180 34.6	S17 57.4	238 23.6	N 4 56.2	61 14.4	S 6 00.0	23 17.3	N 8 26.4	Acamar	315 26.5	S40 18.7
01	65 57.7	195 33.9	58.3	253 24.7	55.7	76 16.9	00.0	38 19.9	26.3	Achernar	335 34.7	S57 14.8
02	81 00.2	210 33.2	17 59.2	268 25.8	55.1	91 19.4	00.0	53 22.6	26.3	Acrux	173 22.4	S63 05.2
03	96 02.7	225 32.5	18 00.1	283 26.9 ..	54.6	106 21.9 ..	00.0	68 25.2 ..	26.2	Adhara	255 21.1	S28 58.1
04	111 05.1	240 31.8	01.0	298 28.1	54.0	121 24.3	00.0	83 27.9	26.1	Aldebaran	291 02.0	N16 30.3
05	126 07.6	255 31.1	01.8	313 29.2	53.5	136 26.8	00.0	98 30.5	26.1			
06	141 10.1	270 30.4	S18 02.7	328 30.3	N 4 52.9	151 29.3	S 6 00.0	113 33.1	N 8 26.0	Alioth	166 31.1	N55 57.9
07	156 12.5	285 29.7	03.6	343 31.4	52.3	166 31.8	00.0	128 35.8	26.0	Alkaid	153 08.3	N49 19.2
T 08	171 15.0	300 29.0	04.5	358 32.5	51.8	181 34.3	00.0	143 38.4	25.9	Al Na'ir	27 57.8	S46 58.2
H 09	186 17.5	315 28.3 ..	05.4	13 33.7 ..	51.2	196 36.7	6 00.0	158 41.0 ..	25.8	Alnilam	275 57.5	S 1 12.2
U 10	201 19.9	330 27.6	06.3	28 34.8	50.7	211 39.2	5 59.9	173 43.7	25.8	Alphard	218 07.2	S 8 39.1
R 11	216 22.4	345 26.9	07.1	43 35.9	50.1	226 41.7	59.9	188 46.3	25.7			
S 12	231 24.9	0 26.2	S18 08.0	58 37.0	N 4 49.6	241 44.2	S 5 59.9	203 48.9	N 8 25.7	Alphecca	126 21.0	N26 43.3
D 13	246 27.3	15 25.5	08.9	73 38.1	49.0	256 46.6	59.9	218 51.6	25.6	Alpheratz	357 54.9	N29 05.2
A 14	261 29.8	30 24.7	09.8	88 39.3	48.5	271 49.1	59.9	233 54.2	25.6	Altair	62 19.4	N 8 52.1
Y 15	276 32.2	45 24.0 ..	10.6	103 40.4 ..	47.9	286 51.6 ..	59.9	248 56.8 ..	25.5	Ankaa	353 26.6	S42 18.9
16	291 34.7	60 23.3	11.5	118 41.5	47.3	301 54.1	59.9	263 59.5	25.4	Antares	112 40.5	S26 25.6
17	306 37.2	75 22.6	12.4	133 42.6	46.8	316 56.5	59.9	279 02.1	25.4			
18	321 39.6	90 21.9	S18 13.3	148 43.8	N 4 46.2	331 59.0	S 5 59.9	294 04.7	N 8 25.3	Arcturus	146 06.4	N19 11.5
19	336 42.1	105 21.2	14.1	163 44.9	45.7	347 01.5	59.9	309 07.4	25.3	Atria	107 53.0	S69 01.5
20	351 44.6	120 20.5	15.0	178 46.0	45.1	2 04.0	59.9	324 10.0	25.2	Avior	234 22.4	S59 30.1
21	6 47.0	135 19.8 ..	15.9	193 47.1 ..	44.6	17 06.4 ..	59.9	339 12.6 ..	25.2	Bellatrix	278 43.8	N 6 20.8
22	21 49.5	150 19.1	16.7	208 48.2	44.0	32 08.9	59.9	354 15.3	25.1	Betelgeuse	271 13.2	N 7 24.3
23	36 52.0	165 18.3	17.6	223 49.4	43.4	47 11.4	59.9	9 17.9	25.0			
13 00	51 54.4	180 17.6	S18 18.5	238 50.5	N 4 42.9	62 13.9	S 5 59.9	24 20.5	N 8 25.0	Canopus	264 00.7	S52 41.6
01	66 56.9	195 16.9	19.3	253 51.6	42.3	77 16.3	59.9	39 23.2	24.9	Capella	280 50.7	N45 59.6
02	81 59.3	210 16.2	20.2	268 52.7	41.8	92 18.8	59.8	54 25.8	24.9	Deneb	49 39.2	N45 16.9
03	97 01.8	225 15.5 ..	21.1	283 53.9 ..	41.2	107 21.3 ..	59.8	69 28.4 ..	24.8	Denebola	182 45.4	N14 34.8
04	112 04.3	240 14.8	21.9	298 55.0	40.7	122 23.7	59.8	84 31.1	24.8	Diphda	349 07.0	S17 59.7
05	127 06.7	255 14.1	22.8	313 56.1	40.1	137 26.2	59.8	99 33.7	24.7			
06	142 09.2	270 13.3	S18 23.7	328 57.2	N 4 39.5	152 28.7	S 5 59.8	114 36.3	N 8 24.6	Dubhe	194 05.8	N61 45.2
07	157 11.7	285 12.6	24.5	343 58.4	39.0	167 31.2	59.8	129 39.0	24.6	Elnath	278 26.6	N28 36.3
08	172 14.1	300 11.9	25.4	358 59.5	38.4	182 33.6	59.8	144 41.6	24.5	Eltanin	90 51.8	N51 29.7
F 09	187 16.6	315 11.2 ..	26.2	14 00.6 ..	37.9	197 36.1 ..	59.8	159 44.2 ..	24.5	Enif	33 58.2	N 9 52.3
R 10	202 19.1	330 10.5	27.1	29 01.7	37.3	212 38.6	59.8	174 46.9	24.4	Fomalhaut	15 36.3	S29 37.8
I 11	217 21.5	345 09.7	27.9	44 02.9	36.8	227 41.0	59.8	189 49.5	24.4			
D 12	232 24.0	0 09.0	S18 28.8	59 04.0	N 4 36.2	242 43.5	S 5 59.8	204 52.1	N 8 24.3	Gacrux	172 13.9	S57 06.1
A 13	247 26.5	15 08.3	29.7	74 05.1	35.6	257 46.0	59.7	219 54.8	24.2	Gienah	176 04.1	S17 31.9
Y 14	262 28.9	30 07.6	30.5	89 06.2	35.1	272 48.4	59.7	234 57.4	24.2	Hadar	149 04.5	S60 21.8
15	277 31.4	45 06.9 ..	31.4	104 07.3 ..	34.5	287 50.9 ..	59.7	250 00.0 ..	24.1	Hamal	328 13.2	N23 27.4
16	292 33.8	60 06.1	32.2	119 08.5	34.0	302 53.4	59.7	265 02.7	24.1	Kaus Aust.	83 59.1	S34 23.1
17	307 36.3	75 05.4	33.1	134 09.6	33.4	317 55.8	59.7	280 05.3	24.0			
18	322 38.8	90 04.7	S18 33.9	149 10.7	N 4 32.9	332 58.3	S 5 59.7	295 07.9	N 8 24.0	Kochab	137 20.9	N74 09.7
19	337 41.2	105 04.0	34.8	164 11.9	32.3	348 00.8	59.7	310 10.6	23.9	Markab	13 49.4	N15 12.1
20	352 43.7	120 03.2	35.6	179 13.0	31.8	3 03.2	59.7	325 13.2	23.9	Menkar	314 26.5	N 4 05.1
21	7 46.2	135 02.5 ..	36.5	194 14.1 ..	31.2	18 05.7 ..	59.7	340 15.8 ..	23.8	Menkent	148 21.3	S36 21.6
22	22 48.6	150 01.8	37.3	209 15.2	30.6	33 08.2	59.7	355 18.5	23.7	Miaplacidus	221 41.9	S69 42.4
23	37 51.1	165 01.1	38.2	224 16.4	30.1	48 10.6	59.6	10 21.1	23.7			
14 00	52 53.6	180 00.3	S18 39.0	239 17.5	N 4 29.5	63 13.1	S 5 59.6	25 23.7	N 8 23.6	Mirfak	308 56.0	N49 51.4
01	67 56.0	194 59.6	39.9	254 18.6	29.0	78 15.6	59.6	40 26.4	23.6	Nunki	76 12.5	S26 17.8
02	82 58.5	209 58.9	40.7	269 19.7	28.4	93 18.0	59.6	55 29.0	23.5	Peacock	53 37.2	S56 44.5
03	98 01.0	224 58.2 ..	41.5	284 20.9 ..	27.9	108 20.5 ..	59.6	70 31.6 ..	23.5	Pollux	243 41.4	N28 01.6
04	113 03.4	239 57.4	42.4	299 22.0	27.3	123 23.0	59.6	85 34.3	23.4	Procyon	245 11.4	N 5 13.6
05	128 05.9	254 56.7	43.2	314 23.1	26.7	138 25.4	59.6	100 36.9	23.3			
06	143 08.3	269 56.0	S18 44.1	329 24.2	N 4 26.2	153 27.9	S 5 59.6	115 39.5	N 8 23.3	Rasalhague	96 17.2	N12 33.9
07	158 10.8	284 55.2	44.9	344 25.4	25.6	168 30.3	59.6	130 42.2	23.2	Regulus	207 55.6	N11 58.4
S 08	173 13.3	299 54.5	45.8	359 26.5	25.1	183 32.8	59.5	145 44.8	23.2	Rigel	281 22.6	S 8 12.2
A 09	188 15.7	314 53.8 ..	46.6	14 27.6 ..	24.5	198 35.3 ..	59.5	160 47.4 ..	23.1	Rigil Kent.	140 07.8	S60 49.5
T 10	203 18.2	329 53.0	47.4	29 28.7	24.0	213 37.7	59.5	175 50.0	23.1	Sabik	102 25.8	S15 43.2
U 11	218 20.7	344 52.3	48.3	44 29.9	23.4	228 40.2	59.5	190 52.7	23.0			
R 12	233 23.1	359 51.6	S18 49.1	59 31.0	N 4 22.9	243 42.6	S 5 59.5	205 55.3	N 8 23.0	Schedar	349 52.9	N56 32.0
D 13	248 25.6	14 50.8	49.9	74 32.1	22.3	258 45.1	59.5	220 57.9	22.9	Shaula	96 37.6	S37 06.1
A 14	263 28.1	29 50.1	50.8	89 33.3	21.7	273 47.6	59.5	236 00.6	22.8	Sirius	258 43.4	S16 42.8
Y 15	278 30.5	44 49.4 ..	51.6	104 34.4 ..	21.2	288 50.0 ..	59.5	251 03.2 ..	22.8	Spica	158 43.4	S11 09.1
16	293 33.0	59 48.6	52.4	119 35.5	20.6	303 52.5	59.4	266 05.8	22.7	Suhail	223 00.7	S43 25.4
17	308 35.4	74 47.9	53.3	134 36.6	20.1	318 54.9	59.4	281 08.5	22.7			
18	323 37.9	89 47.2	S18 54.1	149 37.8	N 4 19.5	333 57.4	S 5 59.4	296 11.1	N 8 22.6	Vega	80 46.9	N38 47.3
19	338 40.4	104 46.4	54.9	164 38.9	19.0	348 59.9	59.4	311 13.7	22.6	Zuben'ubi	137 18.3	S16 02.0
20	353 42.8	119 45.7	55.8	179 40.0	18.4	4 02.3	59.4	326 16.4	22.5			
21	8 45.3	134 45.0 ..	56.6	194 41.2 ..	17.9	19 04.8 ..	59.4	341 19.0 ..	22.5		SHA	Mer. Pass.
22	23 47.8	149 44.2	57.4	209 42.3	17.3	34 07.2	59.4	356 21.6	22.4		° ′	h m
23	38 50.2	164 43.5	58.2	224 43.4	16.7	49 09.7	59.3	11 24.2	22.3	Venus	128 23.2	11 59
	h m									Mars	186 56.1	8 04
Mer. Pass. 20 29.0		v −0.7 d 0.9		v 1.1 d 0.6		v 2.5 d 0.0		v 2.6 d 0.1		Jupiter	10 19.4	19 48
										Saturn	332 26.1	22 19

UT	SUN GHA	SUN Dec	MOON GHA	v	MOON Dec	d	HP
d h	° ′	° ′	° ′	′	° ′	′	′
12 00	183 59.0	S17 34.5	258 03.3	13.0	N11 27.5	8.4	55.6
01	198 58.9	35.2	272 35.3	13.0	11 19.1	8.5	55.6
02	213 58.8	35.8	287 07.3	13.0	11 10.6	8.6	55.6
03	228 58.7 ..	36.5	301 39.3	13.1	11 02.0	8.5	55.5
04	243 58.7	37.2	316 11.4	13.2	10 53.5	8.6	55.5
05	258 58.6	37.9	330 43.6	13.2	10 44.9	8.7	55.5
06	273 58.5	S17 38.6	345 15.8	13.3	N10 36.2	8.7	55.5
T 07	288 58.4	39.2	359 48.1	13.3	10 27.5	8.7	55.4
H 08	303 58.4	39.9	14 20.4	13.4	10 18.8	8.8	55.4
U 09	318 58.3 ..	40.6	28 52.8	13.4	10 10.0	8.8	55.4
R 10	333 58.2	41.3	43 25.2	13.5	10 01.2	8.8	55.4
S 11	348 58.1	41.9	57 57.7	13.6	9 52.4	8.9	55.3
D 12	3 58.1	S17 42.6	72 30.3	13.6	N 9 43.5	8.9	55.3
A 13	18 58.0	43.3	87 02.9	13.6	9 34.6	9.0	55.3
Y 14	33 57.9	44.0	101 35.5	13.7	9 25.6	8.9	55.3
15	48 57.8 ..	44.7	116 08.2	13.8	9 16.7	9.1	55.2
16	63 57.7	45.3	130 41.0	13.8	9 07.6	9.0	55.2
17	78 57.7	46.0	145 13.8	13.8	8 58.6	9.1	55.2
18	93 57.6	S17 46.7	159 46.6	13.9	N 8 49.5	9.1	55.2
19	108 57.5	47.3	174 19.5	14.0	8 40.4	9.2	55.1
20	123 57.4	48.0	188 52.5	14.0	8 31.2	9.1	55.1
21	138 57.3 ..	48.7	203 25.5	14.0	8 22.1	9.2	55.1
22	153 57.2	49.4	217 58.5	14.1	8 12.9	9.3	55.1
23	168 57.2	50.0	232 31.6	14.2	8 03.6	9.2	55.0
13 00	183 57.1	S17 50.7	247 04.8	14.1	N 7 54.4	9.3	55.0
01	198 57.0	51.4	261 37.9	14.3	7 45.1	9.3	55.0
02	213 56.9	52.0	276 11.2	14.2	7 35.8	9.4	55.0
03	228 56.8 ..	52.7	290 44.4	14.3	7 26.4	9.3	54.9
04	243 56.7	53.4	305 17.7	14.4	7 17.1	9.4	54.9
05	258 56.7	54.0	319 51.1	14.4	7 07.7	9.4	54.9
06	273 56.6	S17 54.7	334 24.5	14.4	N 6 58.3	9.5	54.9
07	288 56.5	55.4	348 57.9	14.5	6 48.8	9.4	54.9
F 08	303 56.4	56.0	3 31.4	14.5	6 39.4	9.5	54.8
R 09	318 56.3 ..	56.7	18 04.9	14.5	6 29.9	9.5	54.8
I 10	333 56.2	57.4	32 38.4	14.6	6 20.4	9.5	54.8
D 11	348 56.1	58.0	47 12.0	14.6	6 10.9	9.6	54.8
A 12	3 56.1	S17 58.7	61 45.6	14.7	N 6 01.3	9.5	54.8
Y 13	18 56.0	17 59.4	76 19.3	14.7	5 51.8	9.6	54.7
14	33 55.9	18 00.0	90 53.0	14.7	5 42.2	9.6	54.7
15	48 55.8 ..	00.7	105 26.7	14.7	5 32.6	9.6	54.7
16	63 55.7	01.4	120 00.4	14.8	5 23.0	9.7	54.7
17	78 55.6	02.0	134 34.2	14.8	5 13.3	9.6	54.7
18	93 55.5	S18 02.7	149 08.0	14.9	N 5 03.7	9.7	54.7
19	108 55.4	03.3	163 41.9	14.9	4 54.0	9.7	54.6
20	123 55.3	04.0	178 15.8	14.9	4 44.3	9.7	54.6
21	138 55.3 ..	04.7	192 49.7	14.9	4 34.6	9.7	54.6
22	153 55.2	05.3	207 23.6	15.0	4 24.9	9.7	54.6
23	168 55.1	06.0	221 57.6	15.0	4 15.2	9.7	54.6
14 00	183 55.0	S18 06.6	236 31.6	15.0	N 4 05.5	9.8	54.5
01	198 54.9	07.3	251 05.6	15.0	3 55.7	9.7	54.5
02	213 54.8	07.9	265 39.6	15.1	3 46.0	9.8	54.5
03	228 54.7 ..	08.6	280 13.7	15.1	3 36.2	9.8	54.5
04	243 54.6	09.3	294 47.8	15.1	3 26.4	9.8	54.5
05	258 54.5	09.9	309 21.9	15.2	3 16.6	9.8	54.5
06	273 54.4	S18 10.6	323 56.1	15.1	N 3 06.8	9.8	54.5
07	288 54.3	11.2	338 30.2	15.2	2 57.0	9.8	54.4
S 08	303 54.2	11.9	353 04.4	15.2	2 47.2	9.8	54.4
A 09	318 54.1 ..	12.5	7 38.6	15.2	2 37.4	9.8	54.4
T 10	333 54.0	13.2	22 12.8	15.3	2 27.6	9.8	54.4
U 11	348 53.9	13.8	36 47.1	15.2	2 17.8	9.9	54.4
R 12	3 53.8	S18 14.5	51 21.3	15.3	N 2 07.9	9.8	54.4
D 13	18 53.7	15.1	65 55.6	15.3	1 58.1	9.9	54.4
A 14	33 53.6	15.8	80 29.9	15.3	1 48.2	9.8	54.3
Y 15	48 53.5 ..	16.4	95 04.2	15.3	1 38.4	9.9	54.3
16	63 53.4	17.1	109 38.5	15.4	1 28.5	9.8	54.3
17	78 53.3	17.7	124 12.9	15.3	1 18.7	9.9	54.3
18	93 53.2	S18 18.4	138 47.2	15.4	N 1 08.8	9.9	54.3
19	108 53.1	19.0	153 21.6	15.4	0 58.9	9.8	54.3
20	123 53.0	19.7	167 56.0	15.4	0 49.1	9.9	54.3
21	138 52.9 ..	20.3	182 30.4	15.4	0 39.2	9.8	54.3
22	153 52.8	21.0	197 04.8	15.4	0 29.4	9.9	54.2
23	168 52.7	21.6	211 39.2	15.4	N 0 19.5	9.9	54.2
	SD 16.2 d 0.7		SD 15.1		14.9		14.8

Lat.	Twilight Naut.	Twilight Civil	Sunrise	Moonrise 12	Moonrise 13	Moonrise 14	Moonrise 15
°	h m	h m	h m	h m	h m	h m	h m
N 72	06 51	08 24	10 30	23 51	25 30	01 30	03 06
N 70	06 42	08 03	09 35	24 03	00 03	01 35	03 05
68	06 34	07 46	09 02	24 13	00 13	01 40	03 05
66	06 28	07 33	08 39	24 21	00 21	01 43	03 04
64	06 22	07 22	08 20	24 28	00 28	01 47	03 04
62	06 17	07 12	08 05	24 34	00 34	01 49	03 04
60	06 12	07 04	07 52	24 39	00 39	01 52	03 03
N 58	06 08	06 57	07 41	24 43	00 43	01 54	03 03
56	06 05	06 50	07 32	24 47	00 47	01 55	03 03
54	06 01	06 44	07 23	24 51	00 51	01 57	03 03
52	05 58	06 39	07 16	24 54	00 54	01 59	03 03
50	05 55	06 34	07 09	24 57	00 57	02 00	03 02
45	05 47	06 23	06 54	00 02	01 03	02 03	03 02
N 40	05 41	06 14	06 42	00 10	01 08	02 05	03 02
35	05 35	06 05	06 32	00 16	01 13	02 08	03 02
30	05 29	05 58	06 23	00 22	01 16	02 09	03 01
20	05 18	05 44	06 08	00 32	01 23	02 13	03 01
N 10	05 06	05 32	05 54	00 41	01 29	02 16	03 01
0	04 54	05 19	05 41	00 50	01 35	02 18	03 01
S 10	04 39	05 05	05 28	00 58	01 41	02 21	03 00
20	04 22	04 50	05 13	01 07	01 46	02 24	03 00
30	04 00	04 31	04 57	01 17	01 53	02 27	03 00
35	03 46	04 20	04 48	01 23	01 57	02 29	03 00
40	03 29	04 06	04 37	01 30	02 02	02 31	03 00
45	03 08	03 50	04 24	01 37	02 07	02 34	02 59
S 50	02 39	03 29	04 08	01 46	02 13	02 37	02 59
52	02 24	03 19	04 00	01 51	02 15	02 38	02 59
54	02 06	03 08	03 52	01 55	02 19	02 39	02 59
56	01 43	02 55	03 43	02 00	02 22	02 41	02 59
58	01 12	02 39	03 32	02 06	02 26	02 43	02 59
S 60	00 08	02 20	03 20	02 13	02 30	02 45	02 59

Lat.	Sunset	Twilight Civil	Twilight Naut.	Moonset 12	Moonset 13	Moonset 14	Moonset 15
°	h m	h m	h m	h m	h m	h m	h m
N 72	12 57	15 04	16 36	15 15	15 06	14 59	14 51
N 70	13 52	15 25	16 45	15 01	14 59	14 57	14 54
68	14 25	15 41	16 53	14 50	14 53	14 55	14 57
66	14 49	15 55	17 00	14 40	14 48	14 54	14 59
64	15 08	16 06	17 05	14 32	14 43	14 52	15 01
62	15 23	16 15	17 11	14 26	14 39	14 51	15 03
60	15 35	16 24	17 15	14 20	14 36	14 50	15 04
N 58	15 46	16 31	17 19	14 15	14 33	14 49	15 05
56	15 56	16 38	17 23	14 10	14 30	14 49	15 06
54	16 04	16 44	17 27	14 06	14 28	14 48	15 07
52	16 12	16 49	17 30	14 02	14 26	14 47	15 08
50	16 19	16 54	17 33	13 59	14 24	14 47	15 09
45	16 34	17 05	17 41	13 51	14 19	14 46	15 11
N 40	16 46	17 14	17 47	13 45	14 15	14 44	15 13
35	16 56	17 23	17 53	13 39	14 12	14 44	15 14
30	17 05	17 30	17 59	13 35	14 09	14 43	15 15
20	17 21	17 44	18 11	13 26	14 05	14 41	15 17
N 10	17 35	17 57	18 22	13 19	14 00	14 40	15 19
0	17 48	18 10	18 35	13 12	13 56	14 39	15 21
S 10	18 01	18 23	18 49	13 05	13 52	14 38	15 23
20	18 15	18 39	19 07	12 57	13 47	14 36	15 24
30	18 32	18 58	19 29	12 49	13 42	14 35	15 26
35	18 41	19 09	19 43	12 44	13 39	14 34	15 28
40	18 53	19 23	20 00	12 38	13 36	14 33	15 29
45	19 06	19 39	20 22	12 31	13 32	14 32	15 30
S 50	19 22	20 00	20 52	12 23	13 27	14 30	15 32
52	19 29	20 11	21 07	12 19	13 25	14 29	15 33
54	19 38	20 22	21 25	12 15	13 23	14 29	15 34
56	19 47	20 36	21 49	12 11	13 20	14 28	15 35
58	19 58	20 52	22 22	12 06	13 17	14 27	15 36
S 60	20 11	21 11	////	12 00	13 14	14 26	15 38

	SUN			MOON			
Day	Eqn. of Time 00h	Eqn. of Time 12h	Mer. Pass.	Mer. Pass. Upper	Mer. Pass. Lower	Age	Phase
d	m s	m s	h m	h m	h m	d %	
12	15 56	15 52	11 44	07 01	19 23	23 36	
13	15 48	15 44	11 44	07 45	20 07	24 27	
14	15 40	15 36	11 44	08 29	20 50	25 19	

UT d h	ARIES GHA	VENUS −3.9 GHA	VENUS Dec	MARS +1.5 GHA	MARS Dec	JUPITER −2.6 GHA	JUPITER Dec	SATURN +0.1 GHA	SATURN Dec	STARS Name	SHA	Dec
15 00	53 52.7	179 42.7	S18 59.1	239 44.5	N 4 16.2	64 12.1	S 5 59.3	26 26.9	N 8 22.3	Acamar	315 26.5	S40 18.7
01	68 55.2	194 42.0	18 59.9	254 45.7	15.6	79 14.6	59.3	41 29.5	22.2	Achernar	335 34.7	S57 14.8
02	83 57.6	209 41.3	19 00.7	269 46.8	15.1	94 17.1	59.3	56 32.1	22.2	Acrux	173 22.3	S63 05.2
03	99 00.1	224 40.5	.. 01.5	284 47.9	.. 14.5	109 19.5	.. 59.3	71 34.8	.. 22.1	Adhara	255 21.1	S28 58.1
04	114 02.6	239 39.8	02.3	299 49.1	14.0	124 22.0	59.3	86 37.4	22.1	Aldebaran	291 02.0	N16 30.3
05	129 05.0	254 39.0	03.2	314 50.2	13.4	139 24.4	59.3	101 40.0	22.0			
06	144 07.5	269 38.3	S19 04.0	329 51.3	N 4 12.9	154 26.9	S 5 59.2	116 42.6	N 8 22.0	Alioth	166 31.1	N55 57.9
07	159 09.9	284 37.5	04.8	344 52.5	12.3	169 29.3	59.2	131 45.3	21.9	Alkaid	153 08.3	N49 19.2
08	174 12.4	299 36.8	05.6	359 53.6	11.7	184 31.8	59.2	146 47.9	21.8	Al Na'ir	27 57.8	S46 58.2
S 09	189 14.9	314 36.1	.. 06.4	14 54.7	.. 11.2	199 34.2	.. 59.2	161 50.5	.. 21.8	Alnilam	275 55.5	S 1 12.2
U 10	204 17.3	329 35.3	07.3	29 55.8	10.6	214 36.7	59.2	176 53.2	21.7	Alphard	218 07.1	S 8 39.1
N 11	219 19.8	344 34.6	08.1	44 57.0	10.1	229 39.1	59.2	191 55.8	21.7			
D 12	234 22.3	359 33.8	S19 08.9	59 58.1	N 4 09.5	244 41.6	S 5 59.1	206 58.4	N 8 21.6	Alphecca	126 21.0	N26 43.3
A 13	249 24.7	14 33.1	09.7	74 59.2	09.0	259 44.1	59.1	222 01.0	21.6	Alpheratz	357 54.9	N29 05.2
Y 14	264 27.2	29 32.3	10.5	90 00.4	08.4	274 46.5	59.1	237 03.7	21.5	Altair	62 19.4	N 8 52.1
15	279 29.7	44 31.6	.. 11.3	105 01.5	.. 07.9	289 49.0	.. 59.1	252 06.3	.. 21.5	Ankaa	353 26.6	S42 18.9
16	294 32.1	59 30.8	12.1	120 02.6	07.3	304 51.4	59.1	267 08.9	21.4	Antares	112 40.5	S26 25.6
17	309 34.6	74 30.1	12.9	135 03.8	06.7	319 53.9	59.1	282 11.6	21.4			
18	324 37.1	89 29.3	S19 13.7	150 04.9	N 4 06.2	334 56.3	S 5 59.0	297 14.2	N 8 21.3	Arcturus	146 06.4	N19 11.5
19	339 39.5	104 28.6	14.6	165 06.0	05.6	349 58.8	59.0	312 16.8	21.2	Atria	107 53.0	S69 01.5
20	354 42.0	119 27.8	15.4	180 07.2	05.1	5 01.2	59.0	327 19.4	21.2	Avior	234 22.3	S59 30.1
21	9 44.4	134 27.1	.. 16.2	195 08.3	.. 04.5	20 03.7	.. 59.0	342 22.1	.. 21.1	Bellatrix	278 43.8	N 6 20.8
22	24 46.9	149 26.3	17.0	210 09.4	04.0	35 06.1	59.0	357 24.7	21.1	Betelgeuse	271 13.2	N 7 24.3
23	39 49.4	164 25.6	17.8	225 10.6	03.4	50 08.6	59.0	12 27.3	21.0			
16 00	54 51.8	179 24.8	S19 18.6	240 11.7	N 4 02.9	65 11.0	S 5 58.9	27 30.0	N 8 21.0	Canopus	264 00.7	S52 41.6
01	69 54.3	194 24.1	19.4	255 12.8	02.3	80 13.5	58.9	42 32.6	20.9	Capella	280 50.7	N45 59.6
02	84 56.8	209 23.3	20.2	270 14.0	01.8	95 15.9	58.9	57 35.2	20.9	Deneb	49 39.3	N45 16.9
03	99 59.2	224 22.6	.. 21.0	285 15.1	.. 01.2	110 18.4	.. 58.9	72 37.8	.. 20.8	Denebola	182 45.3	N14 34.8
04	115 01.7	239 21.8	21.8	300 16.2	00.6	125 20.8	58.9	87 40.5	20.8	Diphda	349 07.0	S17 59.7
05	130 04.2	254 21.1	22.6	315 17.4	4 00.1	140 23.2	58.9	102 43.1	20.7			
06	145 06.6	269 20.3	S19 23.4	330 18.5	N 3 59.5	155 25.7	S 5 58.8	117 45.7	N 8 20.7	Dubhe	194 05.8	N61 45.2
07	160 09.1	284 19.6	24.2	345 19.6	59.0	170 28.1	58.8	132 48.3	20.6	Elnath	278 26.5	N28 36.3
08	175 11.5	299 18.8	25.0	0 20.8	58.4	185 30.6	58.8	147 51.0	20.5	Eltanin	90 51.8	N51 29.7
M 09	190 14.0	314 18.0	.. 25.8	15 21.9	.. 57.9	200 33.0	.. 58.8	162 53.6	.. 20.5	Enif	33 58.2	N 9 52.3
O 10	205 16.5	329 17.3	26.6	30 23.0	57.3	215 35.5	58.8	177 56.2	20.4	Fomalhaut	15 36.3	S29 37.8
N 11	220 18.9	344 16.5	27.3	45 24.2	56.8	230 37.9	58.7	192 58.9	20.4			
D 12	235 21.4	359 15.8	S19 28.1	60 25.3	N 3 56.2	245 40.4	S 5 58.7	208 01.5	N 8 20.3	Gacrux	172 13.9	S57 06.1
A 13	250 23.9	14 15.0	28.9	75 26.4	55.6	260 42.8	58.7	223 04.1	20.3	Gienah	176 04.1	S17 31.9
Y 14	265 26.3	29 14.2	29.7	90 27.6	55.1	275 45.3	58.7	238 06.7	20.2	Hadar	149 04.5	S60 21.8
15	280 28.8	44 13.5	.. 30.5	105 28.7	.. 54.5	290 47.7	.. 58.7	253 09.4	.. 20.2	Hamal	328 13.2	N23 27.4
16	295 31.3	59 12.7	31.3	120 29.8	54.0	305 50.1	58.6	268 12.0	20.1	Kaus Aust.	83 59.1	S34 23.0
17	310 33.7	74 12.0	32.1	135 31.0	53.4	320 52.6	58.6	283 14.6	20.1			
18	325 36.2	89 11.2	S19 32.9	150 32.1	N 3 52.9	335 55.0	S 5 58.6	298 17.2	N 8 20.0	Kochab	137 20.9	N74 09.7
19	340 38.7	104 10.4	33.7	165 33.2	52.3	350 57.5	58.6	313 19.9	20.0	Markab	13 49.4	N15 12.1
20	355 41.1	119 09.7	34.4	180 34.4	51.8	5 59.9	58.6	328 22.5	19.9	Menkar	314 26.5	N 4 05.1
21	10 43.6	134 08.9	.. 35.2	195 35.5	.. 51.2	21 02.4	.. 58.5	343 25.1	.. 19.8	Menkent	148 21.3	S36 21.6
22	25 46.0	149 08.1	36.0	210 36.6	50.7	36 04.8	58.5	358 27.7	19.8	Miaplacidus	221 41.8	S69 42.4
23	40 48.5	164 07.4	36.8	225 37.8	50.1	51 07.2	58.5	13 30.4	19.7			
17 00	55 51.0	179 06.6	S19 37.6	240 38.9	N 3 49.5	66 09.7	S 5 58.5	28 33.0	N 8 19.7	Mirfak	308 56.0	N49 51.4
01	70 53.4	194 05.9	38.3	255 40.1	49.0	81 12.1	58.5	43 35.6	19.6	Nunki	76 12.5	S26 17.8
02	85 55.9	209 05.1	39.1	270 41.2	48.4	96 14.6	58.4	58 38.2	19.6	Peacock	53 37.2	S56 44.5
03	100 58.4	224 04.3	.. 39.9	285 42.3	.. 47.9	111 17.0	.. 58.4	73 40.9	.. 19.5	Pollux	243 41.3	N28 01.6
04	116 00.8	239 03.6	40.7	300 43.5	47.3	126 19.4	58.4	88 43.5	19.5	Procyon	245 11.4	N 5 13.6
05	131 03.3	254 02.8	41.5	315 44.6	46.8	141 21.9	58.4	103 46.1	19.4			
06	146 05.8	269 02.0	S19 42.2	330 45.7	N 3 46.2	156 24.3	S 5 58.3	118 48.7	N 8 19.4	Rasalhague	96 17.2	N12 33.9
07	161 08.2	284 01.2	43.0	345 46.9	45.7	171 26.8	58.3	133 51.4	19.3	Regulus	207 55.6	N11 58.4
08	176 10.7	299 00.5	43.8	0 48.0	45.1	186 29.2	58.3	148 54.0	19.3	Rigel	281 22.6	S 8 12.2
T 09	191 13.1	313 59.7	.. 44.5	15 49.1	.. 44.6	201 31.6	.. 58.3	163 56.6	.. 19.2	Rigil Kent.	140 07.8	S60 49.5
U 10	206 15.6	328 58.9	45.3	30 50.3	44.0	216 34.1	58.3	178 59.2	19.2	Sabik	102 25.8	S15 43.2
E 11	221 18.1	343 58.2	46.1	45 51.4	43.5	231 36.5	58.2	194 01.9	19.1			
S 12	236 20.5	358 57.4	S19 46.9	60 52.6	N 3 42.9	246 38.9	S 5 58.2	209 04.5	N 8 19.1	Schedar	349 52.9	N56 32.0
D 13	251 23.0	13 56.6	47.6	75 53.7	42.3	261 41.4	58.2	224 07.1	19.0	Shaula	96 37.6	S37 06.1
A 14	266 25.5	28 55.9	48.4	90 54.8	41.8	276 43.8	58.2	239 09.7	18.9	Sirius	258 43.4	S16 42.8
Y 15	281 27.9	43 55.1	.. 49.2	105 56.0	.. 41.2	291 46.2	.. 58.1	254 12.4	.. 18.9	Spica	158 43.4	S11 09.1
16	296 30.4	58 54.3	49.9	120 57.1	40.7	306 48.7	58.1	269 15.0	18.8	Suhail	223 00.6	S43 25.4
17	311 32.9	73 53.5	50.7	135 58.2	40.1	321 51.1	58.1	284 17.6	18.8			
18	326 35.3	88 52.8	S19 51.5	150 59.4	N 3 39.6	336 53.6	S 5 58.1	299 20.2	N 8 18.7	Vega	80 46.9	N38 47.3
19	341 37.8	103 52.0	52.2	166 00.5	39.0	351 56.0	58.0	314 22.8	18.7	Zuben'ubi	137 18.3	S16 02.0
20	356 40.3	118 51.2	53.0	181 01.7	38.5	6 58.4	58.0	329 25.5	18.6			
21	11 42.7	133 50.4	.. 53.7	196 02.8	.. 37.9	22 00.9	.. 58.0	344 28.1	.. 18.6		SHA	Mer. Pass.
22	26 45.2	148 49.7	54.5	211 03.9	37.4	37 03.3	58.0	359 30.7	18.5	Venus	124 33.0	12 03
23	41 47.6	163 48.9	55.3	226 05.1	36.8	52 05.7	58.0	14 33.3	18.5	Mars	185 19.9	7 59
	h m									Jupiter	10 19.2	19 36
Mer. Pass. 20 17.2	v −0.8 d 0.8			v 1.1 d 0.6		v 2.4 d 0.0		v 2.6 d 0.1		Saturn	332 38.1	22 06

UT	SUN GHA	SUN Dec	MOON GHA	v	MOON Dec	d	HP
15 00	183 52.6	S18 22.2	226 13.6	15.5	N 0 09.6	9.8	54.2
01	198 52.5	22.9	240 48.1	15.4	S 0 00.2	9.9	54.2
02	213 52.4	23.5	255 22.5	15.4	0 10.1	9.8	54.2
03	228 52.3	.. 24.2	269 56.9	15.5	0 19.9	9.9	54.2
04	243 52.2	24.8	284 31.4	15.5	0 29.8	9.8	54.2
05	258 52.1	25.5	299 05.9	15.4	0 39.6	9.8	54.2
06	273 52.0	S18 26.1	313 40.3	15.5	S 0 49.4	9.9	54.2
07	288 51.9	26.7	328 14.8	15.5	0 59.3	9.8	54.2
08	303 51.8	27.4	342 49.3	15.5	1 09.1	9.8	54.1
S 09	318 51.7	.. 28.0	357 23.8	15.5	1 18.9	9.8	54.1
U 10	333 51.6	28.6	11 58.3	15.4	1 28.7	9.8	54.1
N 11	348 51.5	29.3	26 32.7	15.5	1 38.5	9.8	54.1
D 12	3 51.4	S18 29.9	41 07.2	15.5	S 1 48.3	9.8	54.1
A 13	18 51.3	30.6	55 41.7	15.5	1 58.1	9.8	54.1
Y 14	33 51.2	31.2	70 16.2	15.5	2 07.9	9.8	54.1
15	48 51.1	.. 31.8	84 50.7	15.5	2 17.7	9.7	54.1
16	63 51.0	32.5	99 25.2	15.5	2 27.4	9.8	54.1
17	78 50.9	33.1	113 59.7	15.5	2 37.2	9.7	54.1
18	93 50.8	S18 33.7	128 34.2	15.5	S 2 46.9	9.7	54.1
19	108 50.7	34.4	143 08.7	15.4	2 56.6	9.7	54.1
20	123 50.5	35.0	157 43.1	15.5	3 06.3	9.7	54.1
21	138 50.4	.. 35.6	172 17.6	15.5	3 16.0	9.7	54.0
22	153 50.3	36.3	186 52.1	15.4	3 25.7	9.7	54.0
23	168 50.2	36.9	201 26.5	15.5	3 35.4	9.6	54.0
16 00	183 50.1	S18 37.5	216 01.0	15.5	S 3 45.0	9.7	54.0
01	198 50.0	38.2	230 35.5	15.4	3 54.7	9.6	54.0
02	213 49.9	38.8	245 09.9	15.4	4 04.3	9.6	54.0
03	228 49.8	.. 39.4	259 44.3	15.5	4 13.9	9.6	54.0
04	243 49.7	40.0	274 18.8	15.4	4 23.5	9.5	54.0
05	258 49.5	40.7	288 53.2	15.4	4 33.0	9.6	54.0
06	273 49.4	S18 41.3	303 27.6	15.4	S 4 42.6	9.5	54.0
07	288 49.3	41.9	318 02.0	15.4	4 52.1	9.5	54.0
08	303 49.2	42.5	332 36.4	15.3	5 01.6	9.5	54.0
M 09	318 49.1	.. 43.2	347 10.7	15.4	5 11.1	9.5	54.0
O 10	333 49.0	43.8	1 45.1	15.4	5 20.6	9.4	54.0
N 11	348 48.9	44.4	16 19.5	15.3	5 30.0	9.5	54.0
D 12	3 48.8	S18 45.0	30 53.8	15.3	S 5 39.5	9.4	54.0
A 13	18 48.6	45.7	45 28.1	15.3	5 48.9	9.3	54.0
Y 14	33 48.5	46.3	60 02.4	15.3	5 58.2	9.4	54.0
15	48 48.4	.. 46.9	74 36.7	15.3	6 07.6	9.3	54.0
16	63 48.3	47.5	89 11.0	15.3	6 16.9	9.3	54.0
17	78 48.2	48.2	103 45.3	15.2	6 26.2	9.3	54.0
18	93 48.1	S18 48.8	118 19.5	15.2	S 6 35.5	9.3	54.0
19	108 47.9	49.4	132 53.7	15.2	6 44.8	9.2	54.0
20	123 47.8	50.0	147 27.9	15.2	6 54.0	9.2	54.0
21	138 47.7	.. 50.6	162 02.1	15.2	7 03.2	9.2	54.0
22	153 47.6	51.3	176 36.3	15.2	7 12.4	9.1	53.9
23	168 47.5	51.9	191 10.5	15.1	7 21.5	9.1	53.9
17 00	183 47.4	S18 52.5	205 44.6	15.1	S 7 30.6	9.1	53.9
01	198 47.2	53.1	220 18.7	15.1	7 39.7	9.1	53.9
02	213 47.1	53.7	234 52.8	15.1	7 48.8	9.0	53.9
03	228 47.0	.. 54.3	249 26.9	15.0	7 57.8	9.0	53.9
04	243 46.9	54.9	264 00.9	15.1	8 06.8	9.0	53.9
05	258 46.8	55.6	278 35.0	15.0	8 15.8	8.9	53.9
06	273 46.6	S18 56.2	293 09.0	14.9	S 8 24.7	8.9	53.9
07	288 46.5	56.8	307 42.9	15.0	8 33.6	8.8	53.9
08	303 46.4	57.4	322 16.9	14.9	8 42.4	8.9	53.9
T 09	318 46.3	.. 58.0	336 50.8	14.9	8 51.3	8.8	53.9
U 10	333 46.1	58.6	351 24.7	14.9	9 00.1	8.7	53.9
E 11	348 46.0	59.2	5 58.6	14.9	9 08.8	8.7	53.9
S 12	3 45.9	S18 59.8	20 32.5	14.8	S 9 17.5	8.7	53.9
D 13	18 45.8	19 00.4	35 06.3	14.8	9 26.2	8.7	53.9
A 14	33 45.6	01.1	49 40.1	14.8	9 34.9	8.6	53.9
Y 15	48 45.5	.. 01.7	64 13.9	14.7	9 43.5	8.6	53.9
16	63 45.4	02.3	78 47.6	14.8	9 52.1	8.5	54.0
17	78 45.3	02.9	93 21.4	14.7	10 00.6	8.5	54.0
18	93 45.1	S19 03.5	107 55.1	14.6	S10 09.1	8.4	54.0
19	108 45.0	04.1	122 28.7	14.7	10 17.5	8.4	54.0
20	123 44.9	04.7	137 02.4	14.6	10 25.9	8.4	54.0
21	138 44.8	.. 05.3	151 36.0	14.5	10 34.3	8.3	54.0
22	153 44.6	05.9	166 09.5	14.6	10 42.6	8.3	54.0
23	168 44.5	06.5	180 43.1	14.5	S10 50.9	8.3	54.0
	SD 16.2	d 0.6	SD 14.7		14.7		14.7

Lat.	Twilight Naut.	Twilight Civil	Sunrise	Moonrise 15	16	17	18
N 72	07 02	08 38	11 21	03 06	04 41	06 18	07 59
N 70	06 52	08 14	09 55	03 05	04 35	06 05	07 37
68	06 43	07 56	09 16	03 05	04 29	05 54	07 20
66	06 36	07 42	08 50	03 04	04 25	05 45	07 06
64	06 29	07 30	08 30	03 04	04 21	05 38	06 55
62	06 24	07 19	08 13	03 04	04 18	05 31	06 45
60	06 19	07 10	08 00	03 03	04 15	05 26	06 37
N 58	06 14	07 03	07 48	03 03	04 12	05 21	06 30
56	06 10	06 56	07 38	03 03	04 10	05 17	06 23
54	06 06	06 49	07 29	03 03	04 08	05 13	06 18
52	06 02	06 44	07 21	03 03	04 06	05 09	06 13
50	05 59	06 38	07 14	03 02	04 04	05 06	06 08
45	05 51	06 27	06 59	03 02	04 01	05 00	05 58
N 40	05 44	06 17	06 46	03 02	03 58	04 54	05 50
35	05 38	06 08	06 35	03 02	03 55	04 49	05 43
30	05 31	06 00	06 26	03 01	03 53	04 45	05 37
20	05 19	05 46	06 09	03 01	03 49	04 37	05 26
N 10	05 07	05 33	05 55	03 01	03 46	04 31	05 17
0	04 54	05 19	05 41	03 01	03 43	04 25	05 08
S 10	04 39	05 05	05 27	03 00	03 39	04 19	05 00
20	04 21	04 49	05 13	03 00	03 36	04 13	04 51
30	03 58	04 29	04 56	03 00	03 32	04 06	04 40
35	03 44	04 18	04 45	03 00	03 30	04 01	04 34
40	03 26	04 03	04 34	03 00	03 28	03 57	04 28
45	03 03	03 46	04 20	02 59	03 25	03 51	04 20
S 50	02 32	03 25	04 04	02 59	03 22	03 45	04 10
52	02 17	03 14	03 56	02 59	03 20	03 42	04 06
54	01 57	03 02	03 47	02 59	03 19	03 39	04 01
56	01 32	02 48	03 37	02 59	03 17	03 35	03 56
58	00 55	02 31	03 26	02 59	03 15	03 32	03 50
S 60	////	02 10	03 13	02 59	03 12	03 27	03 44

Lat.	Sunset	Twilight Civil	Twilight Naut.	Moonset 15	16	17	18
N 72	12 07	14 50	16 26	14 51	14 43	14 35	14 24
N 70	13 34	15 14	16 37	14 54	14 52	14 50	14 48
68	14 12	15 32	16 45	14 57	14 59	15 02	15 06
66	14 39	15 47	16 53	14 59	15 05	15 12	15 20
64	14 59	15 59	16 59	15 01	15 10	15 20	15 32
62	15 15	16 09	17 05	15 03	15 14	15 27	15 43
60	15 29	16 18	17 10	15 04	15 18	15 34	15 52
N 58	15 41	16 26	17 15	15 05	15 22	15 39	15 59
56	15 51	16 33	17 19	15 06	15 25	15 44	16 06
54	16 00	16 40	17 23	15 07	15 27	15 49	16 12
52	16 08	16 45	17 27	15 08	15 30	15 53	16 18
50	16 15	16 51	17 30	15 09	15 32	15 56	16 23
45	16 30	17 02	17 38	15 11	15 37	16 04	16 34
N 40	16 43	17 12	17 45	15 13	15 41	16 11	16 43
35	16 54	17 21	17 52	15 14	15 45	16 17	16 51
30	17 04	17 29	17 58	15 15	15 48	16 22	16 58
20	17 20	17 43	18 10	15 17	15 54	16 31	17 10
N 10	17 35	17 57	18 22	15 19	15 58	16 38	17 20
0	17 48	18 10	18 36	15 21	16 03	16 46	17 30
S 10	18 02	18 24	18 51	15 23	16 08	16 53	17 40
20	18 17	18 41	19 09	15 24	16 12	17 01	17 50
30	18 34	19 00	19 32	15 26	16 18	17 10	18 02
35	18 44	19 13	19 47	15 28	16 21	17 15	18 09
40	18 56	19 27	20 05	15 29	16 25	17 21	18 17
45	19 10	19 44	20 28	15 30	16 29	17 28	18 26
S 50	19 27	20 06	20 59	15 32	16 34	17 36	18 38
52	19 35	20 17	21 15	15 33	16 37	17 40	18 43
54	19 44	20 29	21 35	15 34	16 39	17 44	18 49
56	19 54	20 44	22 01	15 35	16 42	17 49	18 55
58	20 05	21 01	22 41	15 36	16 45	17 54	19 02
S 60	20 19	21 22	////	15 38	16 49	18 00	19 10

Day	SUN Eqn. of Time 00h	12h	Mer. Pass.	MOON Mer. Pass. Upper	Lower	Age	Phase
	m s	m s	h m	h m	h m	d	%
15	15 31	15 26	11 45	09 11	21 32	26	12
16	15 21	15 15	11 45	09 53	22 14	27	6
17	15 10	15 04	11 45	10 35	22 57	28	3

1998 NOVEMBER 18, 19, 20 (WED., THURS., FRI.)

UT	ARIES GHA	VENUS −3.9 GHA	Dec	MARS +1.5 GHA	Dec	JUPITER −2.6 GHA	Dec	SATURN +0.1 GHA	Dec
18 00	56 50.1	178 48.1	S19 56.0	241 06.2	N 3 36.3	67 08.2	S 5 57.9	29 36.0	N 8 18.4
01	71 52.6	193 47.3	56.8	256 07.4	35.7	82 10.6	57.9	44 38.6	18.4
02	86 55.0	208 46.5	57.5	271 08.5	35.1	97 13.0	57.9	59 41.2	18.3
03	101 57.5	223 45.8	.. 58.3	286 09.6	.. 34.6	112 15.4	.. 57.9	74 43.8	.. 18.3
04	117 00.0	238 45.0	59.0	301 10.8	34.0	127 17.9	57.8	89 46.5	18.2
05	132 02.4	253 44.2	19 59.8	316 11.9	33.5	142 20.3	57.8	104 49.1	18.2
W 06	147 04.9	268 43.4	S20 00.5	331 13.1	N 3 32.9	157 22.7	S 5 57.8	119 51.7	N 8 18.1
E 07	162 07.4	283 42.6	01.3	346 14.2	32.4	172 25.2	57.8	134 54.3	18.1
D 08	177 09.8	298 41.9	02.0	1 15.3	31.8	187 27.6	57.7	149 56.9	18.0
N 09	192 12.3	313 41.1	.. 02.8	16 16.5	.. 31.3	202 30.0	.. 57.7	164 59.6	.. 18.0
E 10	207 14.8	328 40.3	03.5	31 17.6	30.7	217 32.5	57.7	180 02.2	17.9
S 11	222 17.2	343 39.5	04.3	46 18.8	30.2	232 34.9	57.6	195 04.8	17.9
D 12	237 19.7	358 38.7	S20 05.0	61 19.9	N 3 29.6	247 37.3	S 5 57.6	210 07.4	N 8 17.8
A 13	252 22.1	13 37.9	05.8	76 21.0	29.1	262 39.7	57.6	225 10.0	17.8
Y 14	267 24.6	28 37.2	06.5	91 22.2	28.5	277 42.2	57.6	240 12.7	17.7
15	282 27.1	43 36.4	.. 07.3	106 23.3	.. 28.0	292 44.6	.. 57.5	255 15.3	.. 17.6
16	297 29.5	58 35.6	08.0	121 24.5	27.4	307 47.0	57.5	270 17.9	17.6
17	312 32.0	73 34.8	08.7	136 25.6	26.8	322 49.5	57.5	285 20.5	17.5
18	327 34.5	88 34.0	S20 09.5	151 26.7	N 3 26.3	337 51.9	S 5 57.5	300 23.2	N 8 17.5
19	342 36.9	103 33.2	10.2	166 27.9	25.7	352 54.3	57.4	315 25.8	17.4
20	357 39.4	118 32.4	11.0	181 29.0	25.2	7 56.7	57.4	330 28.4	17.4
21	12 41.9	133 31.6	.. 11.7	196 30.2	.. 24.6	22 59.2	.. 57.4	345 31.0	.. 17.3
22	27 44.3	148 30.9	12.4	211 31.3	24.1	38 01.6	57.4	0 33.6	17.3
23	42 46.8	163 30.1	13.2	226 32.5	23.5	53 04.0	57.3	15 36.3	17.2
19 00	57 49.2	178 29.3	S20 13.9	241 33.6	N 3 23.0	68 06.4	S 5 57.3	30 38.9	N 8 17.2
01	72 51.7	193 28.5	14.6	256 34.7	22.4	83 08.9	57.3	45 41.5	17.1
02	87 54.2	208 27.7	15.4	271 35.9	21.9	98 11.3	57.2	60 44.1	17.1
03	102 56.6	223 26.9	.. 16.1	286 37.0	.. 21.3	113 13.7	.. 57.2	75 46.7	.. 17.0
04	117 59.1	238 26.1	16.8	301 38.2	20.8	128 16.1	57.2	90 49.4	17.0
05	133 01.6	253 25.3	17.6	316 39.3	20.2	143 18.6	57.2	105 52.0	16.9
T 06	148 04.0	268 24.5	S20 18.3	331 40.5	N 3 19.7	158 21.0	S 5 57.1	120 54.6	N 8 16.9
H 07	163 06.5	283 23.7	19.0	346 41.6	19.1	173 23.4	57.1	135 57.2	16.8
U 08	178 09.0	298 22.9	19.7	1 42.7	18.6	188 25.8	57.1	150 59.8	16.8
R 09	193 11.4	313 22.1	.. 20.5	16 43.9	.. 18.0	203 28.2	.. 57.0	166 02.5	.. 16.7
S 10	208 13.9	328 21.3	21.2	31 45.0	17.5	218 30.7	57.0	181 05.1	16.7
D 11	223 16.4	343 20.6	21.9	46 46.2	16.9	233 33.1	57.0	196 07.7	16.6
A 12	238 18.8	358 19.8	S20 22.6	61 47.3	N 3 16.3	248 35.5	S 5 56.9	211 10.3	N 8 16.6
Y 13	253 21.3	13 19.0	23.4	76 48.5	15.8	263 37.9	56.9	226 12.9	16.5
14	268 23.7	28 18.2	24.1	91 49.6	15.2	278 40.4	56.9	241 15.6	16.5
15	283 26.2	43 17.4	.. 24.8	106 50.8	.. 14.7	293 42.8	.. 56.9	256 18.2	.. 16.4
16	298 28.7	58 16.6	25.5	121 51.9	14.1	308 45.2	56.8	271 20.8	16.4
17	313 31.1	73 15.8	26.2	136 53.0	13.6	323 47.6	56.8	286 23.4	16.3
18	328 33.6	88 15.0	S20 26.9	151 54.2	N 3 13.0	338 50.0	S 5 56.8	301 26.0	N 8 16.3
19	343 36.1	103 14.2	27.7	166 55.3	12.5	353 52.4	56.7	316 28.6	16.2
20	358 38.5	118 13.4	28.4	181 56.5	11.9	8 54.9	56.7	331 31.3	16.2
21	13 41.0	133 12.6	.. 29.1	196 57.6	.. 11.4	23 57.3	.. 56.7	346 33.9	.. 16.1
22	28 43.5	148 11.8	29.8	211 58.8	10.8	38 59.7	56.6	1 36.5	16.1
23	43 45.9	163 11.0	30.5	226 59.9	10.3	54 02.1	56.6	16 39.1	16.0
20 00	58 48.4	178 10.2	S20 31.2	242 01.1	N 3 09.7	69 04.5	S 5 56.6	31 41.7	N 8 16.0
01	73 50.9	193 09.4	31.9	257 02.2	09.2	84 07.0	56.6	46 44.4	15.9
02	88 53.3	208 08.6	32.6	272 03.4	08.6	99 09.4	56.5	61 47.0	15.9
03	103 55.8	223 07.7	.. 33.3	287 04.5	.. 08.1	114 11.8	.. 56.5	76 49.6	.. 15.8
04	118 58.2	238 06.9	34.0	302 05.7	07.5	129 14.2	56.5	91 52.2	15.8
05	134 00.7	253 06.1	34.8	317 06.8	07.0	144 16.6	56.4	106 54.8	15.7
06	149 03.2	268 05.3	S20 35.5	332 07.9	N 3 06.4	159 19.0	S 5 56.4	121 57.4	N 8 15.7
07	164 05.6	283 04.5	36.2	347 09.1	05.9	174 21.4	56.4	137 00.1	15.6
08	179 08.1	298 03.7	36.9	2 10.2	05.3	189 23.9	56.3	152 02.7	15.6
F 09	194 10.6	313 02.9	.. 37.6	17 11.4	.. 04.8	204 26.3	.. 56.3	167 05.3	.. 15.5
R 10	209 13.0	328 02.1	38.3	32 12.5	04.2	219 28.7	56.3	182 07.9	15.5
I 11	224 15.5	343 01.3	39.0	47 13.7	03.7	234 31.1	56.2	197 10.5	15.4
D 12	239 18.0	358 00.5	S20 39.7	62 14.8	N 3 03.1	249 33.5	S 5 56.2	212 13.1	N 8 15.4
A 13	254 20.4	12 59.7	40.4	77 16.0	02.6	264 35.9	56.2	227 15.8	15.3
Y 14	269 22.9	27 58.9	41.1	92 17.1	02.0	279 38.3	56.1	242 18.4	15.3
15	284 25.4	42 58.1	.. 41.7	107 18.3	.. 01.5	294 40.7	.. 56.1	257 21.0	.. 15.2
16	299 27.8	57 57.3	42.4	122 19.4	00.9	309 43.2	56.1	272 23.6	15.2
17	314 30.3	72 56.4	43.1	137 20.6	3 00.3	324 45.6	56.0	287 26.2	15.1
18	329 32.7	87 55.6	S20 43.8	152 21.7	N 2 59.8	339 48.0	S 5 56.0	302 28.8	N 8 15.1
19	344 35.2	102 54.8	44.5	167 22.9	59.2	354 50.4	56.0	317 31.4	15.0
20	359 37.7	117 54.0	45.2	182 24.0	58.7	9 52.8	55.9	332 34.1	15.0
21	14 40.1	132 53.2	.. 45.9	197 25.2	.. 58.1	24 55.2	.. 55.9	347 36.7	.. 14.9
22	29 42.6	147 52.4	46.6	212 26.3	57.6	39 57.6	55.9	2 39.3	14.9
23	44 45.1	162 51.6	47.3	227 27.5	57.0	55 00.0	55.8	17 41.9	14.8
Mer. Pass.	h m 20 05.4	v −0.8	d 0.7	v 1.1	d 0.6	v 2.4	d 0.0	v 2.6	d 0.1

STARS

Name	SHA	Dec
Acamar	315 26.5	S40 18.7
Achernar	335 34.7	S57 14.8
Acrux	173 22.3	S63 05.2
Adhara	255 21.1	S28 58.2
Aldebaran	291 02.0	N16 30.3
Alioth	166 31.0	N55 57.9
Alkaid	153 08.3	N49 19.2
Al Na'ir	27 57.9	S46 58.2
Alnilam	275 57.5	S 1 12.2
Alphard	218 07.1	S 8 39.1
Alphecca	126 20.9	N26 43.3
Alpheratz	357 54.9	N29 05.2
Altair	62 19.4	N 8 52.1
Ankaa	353 26.6	S42 18.9
Antares	112 40.5	S26 25.6
Arcturus	146 06.4	N19 11.4
Atria	107 53.0	S69 01.4
Avior	234 22.3	S59 30.1
Bellatrix	278 43.8	N 6 20.8
Betelgeuse	271 13.2	N 7 24.3
Canopus	264 00.7	S52 41.6
Capella	280 50.7	N45 59.6
Deneb	49 39.3	N45 16.9
Denebola	182 45.3	N14 34.8
Diphda	349 07.0	S17 59.7
Dubhe	194 05.7	N61 45.2
Elnath	278 26.5	N28 36.3
Eltanin	90 51.9	N51 29.7
Enif	33 58.2	N 9 52.3
Fomalhaut	15 36.3	S29 37.8
Gacrux	172 13.8	S57 06.1
Gienah	176 04.1	S17 31.9
Hadar	149 04.5	S60 21.7
Hamal	328 13.2	N23 27.4
Kaus Aust.	83 59.1	S34 23.0
Kochab	137 20.9	N74 09.7
Markab	13 49.5	N15 12.1
Menkar	314 26.5	N 4 05.1
Menkent	148 21.2	S36 21.6
Miaplacidus	221 41.8	S69 42.5
Mirfak	308 56.0	N49 51.4
Nunki	76 12.6	S26 17.8
Peacock	53 37.2	S56 44.5
Pollux	243 41.3	N28 01.6
Procyon	245 11.3	N 5 13.6
Rasalhague	96 17.2	N12 33.9
Regulus	207 55.5	N11 58.4
Rigel	281 22.6	S 8 12.2
Rigil Kent.	140 07.8	S60 49.5
Sabik	102 25.8	S15 43.2
Schedar	349 52.9	N56 32.0
Shaula	96 37.6	S37 06.1
Sirius	258 43.4	S16 42.8
Spica	158 43.4	S11 09.1
Suhail	223 00.6	S43 25.4
Vega	80 46.9	N38 47.3
Zuben'ubi	137 18.2	S16 02.0

	SHA	Mer. Pass.
	° '	h m
Venus	120 40.0	12 07
Mars	183 44.4	7 53
Jupiter	10 17.2	19 24
Saturn	332 49.6	21 54

UT		SUN GHA	Dec	MOON GHA	v	Dec	d	HP
d	h	° ′	° ′	° ′	′	° ′	′	′
18	00	183 44.4	S19 07.1	195 16.6	14.5	S10 59.2	8.2	54.0
	01	198 44.3	07.7	209 50.1	14.4	11 07.4	8.1	54.0
	02	213 44.1	08.3	224 23.5	14.5	11 15.5	8.1	54.0
	03	228 44.0	.. 08.9	238 57.0	14.3	11 23.6	8.1	54.0
	04	243 43.9	09.5	253 30.3	14.4	11 31.7	8.0	54.0
	05	258 43.7	10.1	268 03.7	14.3	11 39.7	8.0	54.0
	06	273 43.6	S19 10.7	282 37.0	14.3	S11 47.7	7.9	54.0
W	07	288 43.5	11.3	297 10.3	14.3	11 55.6	7.8	54.0
E	08	303 43.4	11.9	311 43.6	14.2	12 03.4	7.9	54.0
D	09	318 43.2	.. 12.5	326 16.8	14.2	12 11.3	7.7	54.0
N	10	333 43.1	13.1	340 50.0	14.1	12 19.0	7.8	54.0
E	11	348 43.0	13.7	355 23.1	14.2	12 26.8	7.6	54.0
S	12	3 42.8	S19 14.3	9 56.3	14.1	S12 34.4	7.7	54.0
D	13	18 42.7	14.9	24 29.4	14.0	12 42.1	7.5	54.0
A	14	33 42.6	15.5	39 02.4	14.0	12 49.6	7.5	54.0
Y	15	48 42.4	.. 16.1	53 35.4	14.0	12 57.1	7.5	54.0
	16	63 42.3	16.7	68 08.4	13.9	13 04.6	7.4	54.0
	17	78 42.2	17.3	82 41.3	14.0	13 12.0	7.4	54.0
	18	93 42.0	S19 17.9	97 14.3	13.8	S13 19.4	7.3	54.0
	19	108 41.9	18.4	111 47.1	13.9	13 26.7	7.2	54.0
	20	123 41.8	19.0	126 20.0	13.8	13 33.9	7.2	54.0
	21	138 41.6	.. 19.6	140 52.8	13.7	13 41.1	7.1	54.1
	22	153 41.5	20.2	155 25.5	13.8	13 48.2	7.1	54.1
	23	168 41.4	20.8	169 58.3	13.6	13 55.3	7.0	54.1
19	00	183 41.2	S19 21.4	184 30.9	13.7	S14 02.3	7.0	54.1
	01	198 41.1	22.0	199 03.6	13.6	14 09.3	6.9	54.1
	02	213 40.9	22.6	213 36.2	13.6	14 16.2	6.8	54.1
	03	228 40.8	.. 23.2	228 08.8	13.5	14 23.0	6.8	54.1
	04	243 40.7	23.7	242 41.3	13.5	14 29.8	6.7	54.1
	05	258 40.5	24.3	257 13.8	13.5	14 36.5	6.7	54.1
	06	273 40.4	S19 24.9	271 46.3	13.4	S14 43.2	6.6	54.1
T	07	288 40.2	25.5	286 18.7	13.4	14 49.8	6.5	54.1
H	08	303 40.1	26.1	300 51.1	13.3	14 56.3	6.5	54.1
U	09	318 40.0	.. 26.7	315 23.4	13.4	15 02.8	6.4	54.1
R	10	333 39.8	27.2	329 55.8	13.2	15 09.2	6.3	54.2
S	11	348 39.7	27.8	344 28.0	13.3	15 15.5	6.3	54.2
D	12	3 39.6	S19 28.4	359 00.3	13.1	S15 21.8	6.2	54.2
A	13	18 39.4	29.0	13 32.4	13.2	15 28.0	6.1	54.2
Y	14	33 39.3	29.6	28 04.6	13.1	15 34.1	6.1	54.2
	15	48 39.1	.. 30.1	42 36.7	13.1	15 40.2	6.0	54.2
	16	63 39.0	30.7	57 08.8	13.0	15 46.2	5.9	54.2
	17	78 38.8	31.3	71 40.8	13.0	15 52.1	5.9	54.2
	18	93 38.7	S19 31.9	86 12.8	13.0	S15 58.0	5.8	54.2
	19	108 38.6	32.5	100 44.8	12.9	16 03.8	5.7	54.2
	20	123 38.4	33.0	115 16.7	12.9	16 09.5	5.7	54.2
	21	138 38.3	.. 33.6	129 48.6	12.8	16 15.2	5.6	54.2
	22	153 38.1	34.2	144 20.4	12.8	16 20.8	5.5	54.3
	23	168 38.0	34.8	158 52.2	12.8	16 26.3	5.5	54.3
20	00	183 37.8	S19 35.3	173 24.0	12.7	S16 31.8	5.3	54.3
	01	198 37.7	35.9	187 55.7	12.7	16 37.1	5.3	54.3
	02	213 37.5	36.5	202 27.4	12.7	16 42.4	5.3	54.3
	03	228 37.4	.. 37.0	216 59.1	12.6	16 47.7	5.1	54.3
	04	243 37.3	37.6	231 30.7	12.5	16 52.8	5.1	54.3
	05	258 37.1	38.2	246 02.2	12.6	16 57.9	5.0	54.3
	06	273 37.0	S19 38.8	260 33.8	12.5	S17 02.9	4.9	54.3
F	07	288 36.8	39.3	275 05.3	12.4	17 07.8	4.9	54.3
R	08	303 36.7	39.9	289 36.7	12.5	17 12.7	4.7	54.4
I	09	318 36.5	.. 40.5	304 08.2	12.3	17 17.4	4.7	54.4
D	10	333 36.4	41.0	318 39.5	12.4	17 22.1	4.7	54.4
A	11	348 36.2	41.6	333 10.9	12.3	17 26.8	4.5	54.4
Y	12	3 36.1	S19 42.2	347 42.2	12.3	S17 31.3	4.4	54.4
	13	18 35.9	42.7	2 13.5	12.2	17 35.7	4.4	54.4
	14	33 35.8	43.3	16 44.7	12.2	17 40.1	4.3	54.4
	15	48 35.6	.. 43.9	31 15.9	12.2	17 44.4	4.2	54.4
	16	63 35.5	44.4	45 47.1	12.1	17 48.6	4.2	54.4
	17	78 35.3	45.0	60 18.2	12.1	17 52.8	4.0	54.5
	18	93 35.2	S19 45.5	74 49.3	12.0	S17 56.8	4.0	54.5
	19	108 35.0	46.1	89 20.3	12.0	18 00.8	3.9	54.5
	20	123 34.9	46.7	103 51.3	12.0	18 04.7	3.8	54.5
	21	138 34.7	.. 47.2	118 22.3	11.9	18 08.5	3.7	54.5
	22	153 34.6	47.8	132 53.2	12.0	18 12.2	3.7	54.5
	23	168 34.4	48.4	147 24.2	11.8	S18 15.9	3.5	54.5
		SD 16.2	d 0.6	SD 14.7		14.8		14.8

Moonrise

Lat.	Twilight Naut.	Civil	Sunrise	18	19	20	21
°	h m	h m	h m	h m	h m	h m	h m
N 72	07 13	08 52	■	07 59	09 47	12 00	■
N 70	07 01	08 26	10 16	07 37	09 11	10 48	12 22
68	06 51	08 06	09 31	07 20	08 46	10 10	11 29
66	06 43	07 51	09 01	07 06	08 26	09 44	10 56
64	06 36	07 38	08 39	06 55	08 11	09 24	10 32
62	06 30	07 26	08 22	06 45	07 58	09 08	10 13
60	06 24	07 17	08 07	06 37	07 47	08 55	09 58
N 58	06 19	07 09	07 55	06 30	07 37	08 43	09 45
56	06 15	07 01	07 44	06 23	07 29	08 33	09 33
54	06 11	06 54	07 35	06 18	07 22	08 24	09 23
52	06 07	06 48	07 26	06 13	07 15	08 16	09 14
50	06 03	06 43	07 19	06 08	07 09	08 09	09 07
45	05 55	06 30	07 03	05 58	06 57	07 54	08 50
N 40	05 47	06 20	06 49	05 50	06 46	07 42	08 36
35	05 40	06 11	06 38	05 43	06 37	07 31	08 24
30	05 34	06 03	06 28	05 37	06 29	07 22	08 14
20	05 21	05 48	06 11	05 26	06 16	07 06	07 57
N 10	05 08	05 34	05 56	05 17	06 04	06 52	07 41
0	04 54	05 20	05 42	05 08	05 53	06 39	07 27
S 10	04 39	05 05	05 27	05 00	05 42	06 27	07 13
20	04 20	04 48	05 12	04 51	05 31	06 13	06 58
30	03 56	04 28	04 54	04 40	05 17	05 57	06 41
35	03 41	04 15	04 44	04 34	05 10	05 48	06 31
40	03 23	04 01	04 32	04 28	05 01	05 38	06 19
45	02 59	03 43	04 17	04 20	04 51	05 26	06 06
S 50	02 27	03 20	04 00	04 10	04 39	05 11	05 50
52	02 10	03 09	03 52	04 06	04 33	05 05	05 42
54	01 48	02 56	03 42	04 01	04 27	04 57	05 33
56	01 20	02 41	03 32	03 56	04 20	04 49	05 24
58	00 33	02 23	03 20	03 50	04 12	04 39	05 13
S 60	////	02 01	03 06	03 44	04 04	04 29	05 00

Moonset

Lat.	Sunset	Twilight Civil	Naut.	18	19	20	21
°	h m	h m	h m	h m	h m	h m	h m
N 72	■	14 38	16 17	14 24	14 09	13 33	■
N 70	13 13	15 04	16 29	14 48	14 46	14 46	14 52
68	13 59	15 23	16 38	15 06	15 12	15 24	15 46
66	14 28	15 39	16 47	15 20	15 32	15 51	16 19
64	14 51	15 52	16 54	15 32	15 49	16 11	16 43
62	15 08	16 03	17 00	15 43	16 02	16 28	17 02
60	15 23	16 13	17 06	15 52	16 14	16 42	17 18
N 58	15 35	16 22	17 11	15 59	16 24	16 53	17 31
56	15 46	16 29	17 15	16 06	16 32	17 04	17 42
54	15 55	16 36	17 20	16 12	16 40	17 13	17 52
52	16 04	16 42	17 23	16 18	16 47	17 21	18 01
50	16 11	16 48	17 27	16 23	16 53	17 28	18 09
45	16 28	17 00	17 36	16 34	17 07	17 44	18 27
N 40	16 41	17 10	17 43	16 43	17 18	17 57	18 41
35	16 52	17 19	17 50	16 51	17 28	18 08	18 53
30	17 02	17 28	17 57	16 58	17 36	18 18	19 03
20	17 20	17 43	18 10	17 10	17 51	18 34	19 21
N 10	17 35	17 57	18 23	17 20	18 03	18 49	19 36
0	17 49	18 11	18 36	17 30	18 15	19 02	19 51
S 10	18 03	18 26	18 52	17 40	18 27	19 16	20 05
20	18 19	18 43	19 11	17 50	18 40	19 30	20 21
30	18 37	19 03	19 35	18 02	18 55	19 47	20 39
35	18 47	19 16	19 50	18 09	19 03	19 57	20 49
40	19 00	19 31	20 09	18 17	19 13	20 08	21 01
45	19 14	19 49	20 33	18 26	19 24	20 21	21 15
S 50	19 32	20 12	21 06	18 38	19 38	20 37	21 32
52	19 40	20 23	21 22	18 43	19 45	20 44	21 40
54	19 49	20 36	21 45	18 49	19 52	20 53	21 49
56	20 00	20 51	22 14	18 55	20 00	21 02	21 59
58	20 12	21 10	23 07	19 02	20 09	21 13	22 11
S 60	20 26	21 33	////	19 10	20 19	21 25	22 24

	SUN Eqn. of Time 00ʰ	12ʰ	Mer. Pass.	MOON Mer. Pass. Upper	Lower	Age	Phase
Day	m s	m s	h m	h m	h m	d	%
d							
18	14 58	14 52	11 45	11 19	23 41	29	1
19	14 45	14 38	11 45	12 04	24 27	00	0
20	14 32	14 25	11 46	12 51	00 27	01	2

1998 NOVEMBER 21, 22, 23 (SAT., SUN., MON.)

UT	ARIES GHA	VENUS −3.9 GHA	VENUS Dec	MARS +1.4 GHA	MARS Dec	JUPITER −2.6 GHA	JUPITER Dec	SATURN +0.1 GHA	SATURN Dec	STARS Name	SHA	Dec
21 00	59 47.5	177 50.8	S20 48.0	242 28.6	N 2 56.5	70 02.4	S 5 55.8	32 44.5	N 8 14.8	Acamar	315 26.5	S40 18.7
01	74 50.0	192 49.9	48.6	257 29.8	55.9	85 04.8	55.8	47 47.1	14.8	Achernar	335 34.7	S57 14.8
02	89 52.5	207 49.1	49.3	272 30.9	55.4	100 07.3	55.7	62 49.8	14.7	Acrux	173 22.3	S63 05.2
03	104 54.9	222 48.3 ..	50.0	287 32.1 ..	54.8	115 09.7 ..	55.7	77 52.4 ..	14.7	Adhara	255 21.1	S28 58.2
04	119 57.4	237 47.5	50.7	302 33.2	54.3	130 12.1	55.6	92 55.0	14.6	Aldebaran	291 02.0	N16 30.3
05	134 59.8	252 46.7	51.4	317 34.4	53.7	145 14.5	55.6	107 57.6	14.6			
06	150 02.3	267 45.9	S20 52.0	332 35.5	N 2 53.2	160 16.9	S 5 55.6	123 00.2 N 8 14.5		Alioth	166 31.0	N55 57.9
07	165 04.8	282 45.0	52.7	347 36.7	52.6	175 19.3	55.5	138 02.8	14.5	Alkaid	153 08.2	N49 19.1
08	180 07.2	297 44.2	53.4	2 37.8	52.1	190 21.7	55.5	153 05.4	14.4	Al Na'ir	27 57.9	S46 58.2
09	195 09.7	312 43.4 ..	54.1	17 39.0 ..	51.5	205 24.1 ..	55.5	168 08.1 ..	14.4	Alnilam	275 57.5	S 1 12.2
10	210 12.2	327 42.6	54.8	32 40.1	51.0	220 26.5	55.4	183 10.7	14.3	Alphard	218 07.1	S 8 39.1
11	225 14.6	342 41.8	55.4	47 41.3	50.4	235 28.9	55.4	198 13.3	14.3			
12	240 17.1	357 40.9	S20 56.1	62 42.4	N 2 49.9	250 31.3	S 5 55.4	213 15.9	N 8 14.2	Alphecca	126 20.9	N26 43.3
13	255 19.6	12 40.1	56.8	77 43.6	49.3	265 33.7	55.3	228 18.5	14.2	Alpheratz	357 55.0	N29 05.2
14	270 22.0	27 39.3	57.4	92 44.7	48.8	280 36.1	55.3	243 21.1	14.1	Altair	62 19.4	N 8 52.1
15	285 24.5	42 38.5 ..	58.1	107 45.9 ..	48.2	295 38.5 ..	55.3	258 23.7 ..	14.1	Ankaa	353 26.6	S42 19.0
16	300 27.0	57 37.7	58.8	122 47.0	47.7	310 40.9	55.2	273 26.3	14.0	Antares	112 40.5	S26 25.6
17	315 29.4	72 36.8	20 59.5	137 48.2	47.1	325 43.3	55.2	288 29.0	14.0			
18	330 31.9	87 36.0	S21 00.1	152 49.3	N 2 46.6	340 45.7	S 5 55.1	303 31.6	N 8 13.9	Arcturus	146 06.3	N19 11.4
19	345 34.3	102 35.2	00.8	167 50.5	46.0	355 48.1	55.1	318 34.2	13.9	Atria	107 53.0	S69 01.4
20	0 36.8	117 34.4	01.5	182 51.6	45.5	10 50.5	55.1	333 36.8	13.8	Avior	234 22.3	S59 30.1
21	15 39.3	132 33.5 ..	02.1	197 52.8 ..	44.9	25 53.0 ..	55.0	348 39.4 ..	13.8	Bellatrix	278 43.8	N 6 20.8
22	30 41.7	147 32.7	02.8	212 53.9	44.4	40 55.4	55.0	3 42.0	13.7	Betelgeuse	271 13.2	N 7 24.3
23	45 44.2	162 31.9	03.4	227 55.1	43.8	55 57.8	55.0	18 44.6	13.7			
22 00	60 46.7	177 31.1	S21 04.1	242 56.2	N 2 43.3	71 00.2	S 5 54.9	33 47.3	N 8 13.6	Canopus	264 00.7	S52 41.6
01	75 49.1	192 30.2	04.8	257 57.4	42.7	86 02.6	54.9	48 49.9	13.6	Capella	280 50.6	N45 59.7
02	90 51.6	207 29.4	05.4	272 58.5	42.2	101 05.0	54.8	63 52.5	13.6	Deneb	49 39.3	N45 16.9
03	105 54.1	222 28.6 ..	06.1	287 59.7 ..	41.6	116 07.4 ..	54.8	78 55.1 ..	13.5	Denebola	182 45.3	N14 34.7
04	120 56.5	237 27.7	06.7	303 00.8	41.1	131 09.8	54.8	93 57.7	13.5	Diphda	349 07.0	S17 59.7
05	135 59.0	252 26.9	07.4	318 02.0	40.5	146 12.2	54.7	109 00.3	13.4			
06	151 01.5	267 26.1	S21 08.0	333 03.2	N 2 40.0	161 14.6	S 5 54.7	124 02.9	N 8 13.4	Dubhe	194 05.7	N61 45.2
07	166 03.9	282 25.3	08.7	348 04.3	39.4	176 17.0	54.6	139 05.5	13.3	Elnath	278 26.5	N28 36.3
08	181 06.4	297 24.4	09.4	3 05.5	38.9	191 19.4	54.6	154 08.1	13.3	Eltanin	90 51.9	N51 29.6
09	196 08.8	312 23.6 ..	10.0	18 06.6 ..	38.3	206 21.7 ..	54.6	169 10.8 ..	13.2	Enif	33 58.2	N 9 52.3
10	211 11.3	327 22.8	10.7	33 07.8	37.8	221 24.1	54.5	184 13.4	13.2	Fomalhaut	15 36.3	S29 37.8
11	226 13.8	342 21.9	11.3	48 08.9	37.2	236 26.5	54.5	199 16.0	13.1			
12	241 16.2	357 21.1	S21 11.9	63 10.1	N 2 36.7	251 28.9	S 5 54.4	214 18.6	N 8 13.1	Gacrux	172 13.8	S57 06.1
13	256 18.7	12 20.3	12.6	78 11.2	36.1	266 31.3	54.4	229 21.2	13.0	Gienah	176 04.1	S17 31.9
14	271 21.2	27 19.4	13.2	93 12.4	35.6	281 33.7	54.4	244 23.8	13.0	Hadar	149 04.4	S60 21.7
15	286 23.6	42 18.6 ..	13.9	108 13.5 ..	35.0	296 36.1 ..	54.3	259 26.4 ..	12.9	Hamal	328 13.2	N23 27.4
16	301 26.1	57 17.8	14.5	123 14.7	34.5	311 38.5	54.3	274 29.0	12.9	Kaus Aust.	83 59.1	S34 23.0
17	316 28.6	72 16.9	15.2	138 15.9	33.9	326 40.9	54.2	289 31.6	12.9			
18	331 31.0	87 16.1	S21 15.8	153 17.0	N 2 33.4	341 43.3	S 5 54.2	304 34.3	N 8 12.8	Kochab	137 20.9	N74 09.7
19	346 33.5	102 15.3	16.5	168 18.2	32.8	356 45.7	54.2	319 36.9	12.8	Markab	13 49.5	N15 12.1
20	1 36.0	117 14.4	17.1	183 19.3	32.3	11 48.1	54.1	334 39.5	12.7	Menkar	314 26.5	N 4 05.1
21	16 38.4	132 13.6 ..	17.7	198 20.5 ..	31.7	26 50.5 ..	54.1	349 42.1 ..	12.7	Menkent	148 21.2	S36 21.6
22	31 40.9	147 12.8	18.4	213 21.6	31.2	41 52.9	54.0	4 44.7	12.6	Miaplacidus	221 41.7	S69 42.5
23	46 43.3	162 11.9	19.0	228 22.8	30.7	56 55.3	54.0	19 47.3	12.6			
23 00	61 45.8	177 11.1	S21 19.6	243 23.9	N 2 30.1	71 57.7	S 5 54.0	34 49.9	N 8 12.5	Mirfak	308 56.0	N49 51.4
01	76 48.3	192 10.2	20.3	258 25.1	29.6	87 00.1	53.9	49 52.5	12.5	Nunki	76 12.6	S26 17.8
02	91 50.7	207 09.4	20.9	273 26.3	29.0	102 02.5	53.9	64 55.1	12.4	Peacock	53 37.2	S56 44.5
03	106 53.2	222 08.6 ..	21.5	288 27.4 ..	28.5	117 04.9 ..	53.8	79 57.7 ..	12.4	Pollux	243 41.3	N28 01.6
04	121 55.7	237 07.7	22.2	303 28.6	27.9	132 07.3	53.8	95 00.3	12.3	Procyon	245 11.3	N 5 13.6
05	136 58.1	252 06.9	22.8	318 29.7	27.4	147 09.6	53.8	110 03.0	12.3			
06	152 00.6	267 06.0	S21 23.4	333 30.9	N 2 26.8	162 12.0	S 5 53.7	125 05.6	N 8 12.3	Rasalhague	96 17.2	N12 33.9
07	167 03.1	282 05.2	24.1	348 32.0	26.3	177 14.4	53.7	140 08.2	12.2	Regulus	207 55.5	N11 58.4
08	182 05.5	297 04.4	24.7	3 33.2	25.7	192 16.8	53.6	155 10.8	12.2	Rigel	281 22.6	S 8 12.2
09	197 08.0	312 03.5 ..	25.3	18 34.3 ..	25.2	207 19.2 ..	53.6	170 13.4 ..	12.1	Rigil Kent.	140 07.8	S60 49.5
10	212 10.5	327 02.7	25.9	33 35.5	24.6	222 21.6	53.5	185 16.0	12.1	Sabik	102 25.8	S15 43.2
11	227 12.9	342 01.8	26.6	48 36.7	24.1	237 24.0	53.5	200 18.6	12.0			
12	242 15.4	357 01.0	S21 27.2	63 37.8	N 2 23.5	252 26.4	S 5 53.4	215 21.2	N 8 12.0	Schedar	349 53.0	N56 32.0
13	257 17.8	12 00.1	27.8	78 39.0	23.0	267 28.8	53.4	230 23.8	11.9	Shaula	96 37.6	S37 06.1
14	272 20.3	26 59.3	28.4	93 40.1	22.4	282 31.2	53.4	245 26.4	11.9	Sirius	258 43.4	S16 42.8
15	287 22.8	41 58.5 ..	29.0	108 41.3 ..	21.9	297 33.5 ..	53.3	260 29.0 ..	11.8	Spica	158 43.4	S11 09.1
16	302 25.2	56 57.6	29.7	123 42.5	21.3	312 35.9	53.3	275 31.6	11.8	Suhail	223 00.6	S43 25.5
17	317 27.7	71 56.8	30.3	138 43.6	20.8	327 38.3	53.2	290 34.3	11.8			
18	332 30.2	86 55.9	S21 30.9	153 44.8	N 2 20.2	342 40.7	S 5 53.2	305 36.9	N 8 11.7	Vega	80 46.9	N38 47.3
19	347 32.6	101 55.1	31.5	168 45.9	19.7	357 43.1	53.1	320 39.5	11.7	Zuben'ubi	137 18.2	S16 02.0
20	2 35.1	116 54.2	32.1	183 47.1	19.1	12 45.5	53.1	335 42.1	11.6			
21	17 37.6	131 53.4 ..	32.7	198 48.2 ..	18.6	27 47.9 ..	53.1	350 44.7 ..	11.6		SHA	Mer. Pass.
22	32 40.0	146 52.5	33.3	213 49.4	18.0	42 50.2	53.0	5 47.3	11.5	Venus	116 44.4	12 11
23	47 42.5	161 51.7	34.0	228 50.6	17.5	57 52.6	53.0	20 49.9	11.5	Mars	182 09.6	7 48
										Jupiter	10 13.5	19 13
Mer. Pass. 19 53.6		v −0.8 d 0.6		v 1.2 d 0.5		v 2.4 d 0.0		v 2.6 d 0.0		Saturn	333 00.6	21 41

Left margin day labels: SATURDAY, SUNDAY, MONDAY

UT	SUN GHA	SUN Dec	MOON GHA	v	Dec	d	HP
21 d h	° ′	° ′	° ′	′	° ′	′	′
00	183 34.3	S19 48.9	161 55.0	11.9	S18 19.4	3.5	54.5
01	198 34.1	49.5	176 25.9	11.8	18 22.9	3.4	54.6
02	213 33.9	50.0	190 56.7	11.8	18 26.3	3.3	54.6
03	228 33.8	.. 50.6	205 27.5	11.7	18 29.6	3.2	54.6
04	243 33.6	51.1	219 58.2	11.7	18 32.8	3.2	54.6
05	258 33.5	51.7	234 28.9	11.7	18 36.0	3.0	54.6
06	273 33.3	S19 52.2	248 59.6	11.6	S18 39.0	3.0	54.6
07	288 33.2	52.8	263 30.2	11.6	18 42.0	2.8	54.6
08	303 33.0	53.4	278 00.8	11.6	18 44.8	2.8	54.7
09	318 32.9	.. 53.9	292 31.4	11.5	18 47.6	2.7	54.7
10	333 32.7	54.5	307 01.9	11.5	18 50.3	2.6	54.7
11	348 32.5	55.0	321 32.4	11.5	18 52.9	2.5	54.7
12	3 32.4	S19 55.6	336 02.9	11.5	S18 55.4	2.4	54.7
13	18 32.2	56.1	350 33.4	11.4	18 57.8	2.4	54.7
14	33 32.1	56.7	5 03.8	11.4	19 00.2	2.2	54.7
15	48 31.9	.. 57.2	19 34.2	11.3	19 02.4	2.1	54.8
16	63 31.8	57.8	34 04.5	11.4	19 04.5	2.1	54.8
17	78 31.6	58.3	48 34.9	11.3	19 06.6	2.0	54.8
18	93 31.4	S19 58.9	63 05.2	11.3	S19 08.6	1.8	54.8
19	108 31.3	19 59.4	77 35.5	11.2	19 10.4	1.8	54.8
20	123 31.1	20 00.0	92 05.7	11.2	19 12.2	1.7	54.8
21	138 31.0	.. 00.5	106 35.9	11.2	19 13.9	1.6	54.9
22	153 30.8	01.0	121 06.1	11.2	19 15.5	1.5	54.9
23	168 30.6	01.6	135 36.3	11.1	19 17.0	1.4	54.9
22 00	183 30.5	S20 02.1	150 06.4	11.1	S19 18.4	1.4	54.9
01	198 30.3	02.7	164 36.5	11.1	19 19.8	1.2	54.9
02	213 30.1	03.2	179 06.6	11.1	19 21.0	1.1	54.9
03	228 30.0	.. 03.8	193 36.7	11.0	19 22.1	1.0	54.9
04	243 29.8	04.3	208 06.7	11.1	19 23.1	1.0	55.0
05	258 29.7	04.8	222 36.8	11.0	19 24.1	0.8	55.0
06	273 29.5	S20 05.4	237 06.8	10.9	S19 24.9	0.8	55.0
07	288 29.3	05.9	251 36.7	11.0	19 25.7	0.6	55.0
08	303 29.2	06.5	266 06.7	10.9	19 26.3	0.6	55.0
09	318 29.0	.. 07.0	280 36.6	10.9	19 26.9	0.4	55.0
10	333 28.8	07.5	295 06.5	10.9	19 27.3	0.4	55.0
11	348 28.7	08.1	309 36.4	10.9	19 27.7	0.3	55.1
12	3 28.5	S20 08.6	324 06.3	10.8	S19 28.0	0.2	55.1
13	18 28.3	09.1	338 36.1	10.8	19 28.2	0.0	55.1
14	33 28.2	09.7	353 05.9	10.8	19 28.2	0.0	55.1
15	48 28.0	.. 10.2	7 35.7	10.8	19 28.2	0.1	55.2
16	63 27.8	10.7	22 05.5	10.8	19 28.1	0.2	55.2
17	78 27.7	11.3	36 35.3	10.7	19 27.9	0.3	55.2
18	93 27.5	S20 11.8	51 05.0	10.8	S19 27.6	0.5	55.2
19	108 27.3	12.3	65 34.8	10.7	19 27.1	0.5	55.2
20	123 27.2	12.9	80 04.5	10.7	19 26.6	0.6	55.2
21	138 27.0	.. 13.4	94 34.2	10.7	19 26.0	0.7	55.3
22	153 26.8	13.9	109 03.9	10.6	19 25.3	0.8	55.3
23	168 26.7	14.5	123 33.5	10.7	19 24.5	0.9	55.3
23 00	183 26.5	S20 15.0	138 03.2	10.6	S19 23.6	1.0	55.3
01	198 26.3	15.5	152 32.8	10.7	19 22.6	1.1	55.3
02	213 26.1	16.0	167 02.5	10.6	19 21.5	1.2	55.4
03	228 26.0	.. 16.6	181 32.1	10.6	19 20.3	1.3	55.4
04	243 25.8	17.1	196 01.7	10.6	19 19.0	1.4	55.4
05	258 25.6	17.6	210 31.3	10.5	19 17.6	1.5	55.4
06	273 25.5	S20 18.1	225 00.8	10.6	S19 16.1	1.6	55.4
07	288 25.3	18.7	239 30.4	10.6	19 14.5	1.7	55.5
08	303 25.1	19.2	254 00.0	10.5	19 12.8	1.7	55.5
09	318 24.9	.. 19.7	268 29.5	10.5	19 11.1	1.9	55.5
10	333 24.8	20.2	282 59.0	10.6	19 09.2	2.0	55.5
11	348 24.6	20.8	297 28.6	10.5	19 07.2	2.1	55.6
12	3 24.4	S20 21.3	311 58.1	10.5	S19 05.1	2.2	55.6
13	18 24.3	21.8	326 27.6	10.5	19 02.9	2.3	55.6
14	33 24.1	22.3	340 57.1	10.5	19 00.6	2.4	55.6
15	48 23.9	.. 22.8	355 26.6	10.5	18 58.2	2.4	55.6
16	63 23.7	23.3	9 56.1	10.4	18 55.8	2.6	55.7
17	78 23.6	23.9	24 25.5	10.5	18 53.2	2.7	55.7
18	93 23.4	S20 24.4	38 55.0	10.5	S18 50.5	2.8	55.7
19	108 23.2	24.9	53 24.5	10.4	18 47.7	2.8	55.7
20	123 23.0	25.4	67 53.9	10.5	18 44.9	3.0	55.8
21	138 22.8	.. 25.9	82 23.4	10.4	18 41.9	3.1	55.8
22	153 22.7	26.4	96 52.8	10.5	18 38.8	3.2	55.8
23	168 22.5	27.0	111 22.3	10.4	S18 35.6	3.2	55.8
SD	16.2	d 0.5	SD 14.9		15.0		15.1

Days of week: SATURDAY (21), SUNDAY (22), MONDAY (23)

Moonrise

Lat.	Twilight Naut.	Twilight Civil	Sunrise	21	22	23	24
°	h m	h m	h m	h m	h m	h m	h m
N 72	07 23	09 06	■	■	■	■	15 19
N 70	07 10	08 38	10 43	12 22	13 37	14 05	14 11
68	06 59	08 16	09 46	11 29	12 31	13 12	13 34
66	06 50	07 59	09 13	10 56	11 55	12 39	13 08
64	06 43	07 45	08 49	10 32	11 30	12 15	12 48
62	06 36	07 33	08 30	10 13	11 10	11 56	12 31
60	06 30	07 23	08 14	09 58	10 53	11 40	12 17
N 58	06 25	07 14	08 01	09 45	10 40	11 27	12 06
56	06 20	07 06	07 50	09 33	10 28	11 15	11 55
54	06 15	06 59	07 40	09 23	10 17	11 05	11 46
52	06 11	06 53	07 31	09 14	10 08	10 56	11 38
50	06 07	06 47	07 23	09 07	10 00	10 48	11 31
45	05 58	06 34	07 07	08 50	09 42	10 31	11 15
N 40	05 50	06 23	06 53	08 36	09 28	10 17	11 02
35	05 43	06 14	06 41	08 24	09 16	10 05	10 51
30	05 36	06 05	06 31	08 14	09 05	09 55	10 42
20	05 22	05 49	06 13	07 57	08 47	09 37	10 25
N 10	05 09	05 35	05 57	07 41	08 31	09 21	10 11
0	04 55	05 20	05 42	07 27	08 17	09 07	09 57
S 10	04 39	05 05	05 28	07 13	08 02	08 52	09 44
20	04 19	04 48	05 12	06 58	07 46	08 37	09 29
30	03 55	04 27	04 53	06 41	07 28	08 19	09 13
35	03 39	04 14	04 42	06 31	07 17	08 08	09 03
40	03 20	03 59	04 30	06 19	07 05	07 56	08 52
45	02 55	03 40	04 15	06 06	06 51	07 42	08 39
S 50	02 21	03 16	03 57	05 50	06 34	07 25	08 23
52	02 03	03 04	03 48	05 42	06 26	07 17	08 16
54	01 40	02 51	03 38	05 33	06 17	07 08	08 08
56	01 08	02 35	03 27	05 24	06 07	06 58	07 58
58	////	02 16	03 15	05 13	05 55	06 47	07 48
S 60	////	01 51	03 00	05 00	05 42	06 33	07 36

Moonset

Lat.	Sunset	Twilight Civil	Twilight Naut.	21	22	23	24
°	h m	h m	h m	h m	h m	h m	h m
N 72	■	14 25	16 08	■	■	■	17 09
N 70	12 49	14 54	16 21	14 52	15 20	16 37	18 17
68	13 46	15 15	16 32	15 46	16 26	17 30	18 53
66	14 19	15 32	16 41	16 19	17 02	18 03	19 18
64	14 43	15 46	16 49	16 43	17 27	18 26	19 38
62	15 02	15 58	16 55	17 02	17 47	18 45	19 54
60	15 17	16 08	17 01	17 18	18 04	19 01	20 08
N 58	15 30	16 17	17 07	17 31	18 18	19 14	20 19
56	15 42	16 25	17 12	17 42	18 29	19 25	20 29
54	15 52	16 32	17 16	17 52	18 40	19 35	20 38
52	16 00	16 39	17 21	18 01	18 49	19 44	20 46
50	16 08	16 45	17 25	18 09	18 57	19 52	20 53
45	16 25	16 58	17 34	18 27	19 15	20 09	21 08
N 40	16 39	17 09	17 42	18 41	19 29	20 22	21 20
35	16 51	17 18	17 49	18 53	19 41	20 34	21 30
30	17 01	17 27	17 56	19 03	19 52	20 44	21 40
20	17 19	17 43	18 10	19 21	20 10	21 02	21 55
N 10	17 35	17 57	18 23	19 36	20 26	21 17	22 09
0	17 50	18 12	18 37	19 51	20 41	21 31	22 22
S 10	18 05	18 27	18 54	20 05	20 55	21 45	22 34
20	18 21	18 45	19 13	20 21	21 11	22 00	22 48
30	18 39	19 06	19 38	20 39	21 29	22 17	23 03
35	18 50	19 19	19 54	20 49	21 40	22 27	23 12
40	19 03	19 34	20 13	21 01	21 52	22 39	23 22
45	19 18	19 53	20 38	21 15	22 06	22 52	23 34
S 50	19 36	20 17	21 13	21 32	22 23	23 09	23 48
52	19 45	20 29	21 31	21 40	22 31	23 16	23 55
54	19 55	20 43	21 55	21 49	22 40	23 25	24 03
56	20 06	20 59	22 28	21 59	22 50	23 34	24 11
58	20 19	21 19	////	22 11	23 02	23 45	24 20
S 60	20 34	21 44	////	22 24	23 16	23 57	24 31

SUN and MOON

Day	SUN Eqn. of Time 00h	SUN Eqn. of Time 12h	SUN Mer. Pass.	MOON Mer. Pass. Upper	MOON Mer. Pass. Lower	Age	Phase
d	m s	m s	h m	h m	h m	d	%
21	14 17	14 10	11 46	13 39	01 15	02	5
22	14 02	13 54	11 46	14 29	02 04	03	10
23	13 46	13 38	11 46	15 19	02 54	04	17

UT	ARIES GHA	VENUS −3.9 GHA	VENUS Dec	MARS +1.4 GHA	MARS Dec	JUPITER −2.6 GHA	JUPITER Dec	SATURN +0.2 GHA	SATURN Dec	STARS Name	SHA	Dec
24 00	62 44.9	176 50.8	S21 34.6	243 51.7	N 2 17.0	72 55.0	S 5 52.9	35 52.5	N 8 11.4	Acamar	315 26.5	S40 18.7
01	77 47.4	191 50.0	35.2	258 52.9	16.4	87 57.4	52.9	50 55.1	11.4	Achernar	335 34.7	S57 14.8
02	92 49.9	206 49.1	35.8	273 54.0	15.9	102 59.8	52.8	65 57.7	11.3	Acrux	173 22.2	S63 05.2
03	107 52.3	221 48.3	. . 36.4	288 55.2	. . 15.3	118 02.2	. . 52.8	81 00.3	. . 11.3	Adhara	255 21.1	S28 58.2
04	122 54.8	236 47.4	37.0	303 56.4	14.8	133 04.6	52.7	96 02.9	11.3	Aldebaran	291 02.0	N16 30.3
05	137 57.3	251 46.6	37.6	318 57.5	14.2	148 06.9	52.7	111 05.5	11.2			
06	152 59.7	266 45.7	S21 38.2	333 58.7	N 2 13.7	163 09.3	S 5 52.7	126 08.1	N 8 11.2	Alioth	166 31.0	N55 57.9
07	168 02.2	281 44.9	38.8	348 59.8	13.1	178 11.7	52.6	141 10.7	11.1	Alkaid	153 08.2	N49 19.1
08	183 04.7	296 44.0	39.4	4 01.0	12.6	193 14.1	52.6	156 13.3	11.1	Al Na'ir	27 57.9	S46 58.2
T 09	198 07.1	311 43.2	.. 40.0	19 02.2	.. 12.0	208 16.5	.. 52.5	171 16.0	.. 11.0	Alnilam	275 57.5	S 1 12.2
U 10	213 09.6	326 42.3	40.6	34 03.3	11.5	223 18.9	52.5	186 18.6	11.0	Alphard	218 07.1	S 8 39.1
E 11	228 12.1	341 41.5	41.2	49 04.5	10.9	238 21.2	52.4	201 21.2	10.9			
S 12	243 14.5	356 40.6	S21 41.8	64 05.6	N 2 10.4	253 23.6	S 5 52.4	216 23.8	N 8 10.9	Alphecca	126 20.9	N26 43.3
D 13	258 17.0	11 39.7	42.4	79 06.8	09.8	268 26.0	52.3	231 26.4	10.9	Alpheratz	357 55.0	N29 05.2
A 14	273 19.4	26 38.9	43.0	94 08.0	09.3	283 28.4	52.3	246 29.0	10.8	Altair	62 19.4	N 8 52.1
Y 15	288 21.9	41 38.0	.. 43.6	109 09.1	.. 08.7	298 30.8	.. 52.2	261 31.6	.. 10.8	Ankaa	353 26.6	S42 19.0
16	303 24.4	56 37.2	44.2	124 10.3	08.2	313 33.1	52.2	276 34.2	10.7	Antares	112 40.5	S26 25.6
17	318 26.8	71 36.3	44.8	139 11.5	07.7	328 35.5	52.1	291 36.8	10.7			
18	333 29.3	86 35.5	S21 45.4	154 12.6	N 2 07.1	343 37.9	S 5 52.1	306 39.4	N 8 10.6	Arcturus	146 06.3	N19 11.4
19	348 31.8	101 34.6	45.9	169 13.8	06.6	358 40.3	52.1	321 42.0	10.6	Atria	107 53.0	S69 01.4
20	3 34.2	116 33.7	46.5	184 14.9	06.0	13 42.7	52.0	336 44.6	10.6	Avior	234 22.2	S59 30.1
21	18 36.7	131 32.9	.. 47.1	199 16.1	.. 05.5	28 45.0	.. 52.0	351 47.2	.. 10.5	Bellatrix	278 43.8	N 6 20.8
22	33 39.2	146 32.0	47.7	214 17.3	04.9	43 47.4	51.9	6 49.8	10.5	Betelgeuse	271 13.2	N 7 24.3
23	48 41.6	161 31.2	48.3	229 18.4	04.4	58 49.8	51.9	21 52.4	10.4			
25 00	63 44.1	176 30.3	S21 48.9	244 19.6	N 2 03.8	73 52.2	S 5 51.8	36 55.0	N 8 10.4	Canopus	264 00.7	S52 41.7
01	78 46.6	191 29.4	49.5	259 20.8	03.3	88 54.5	51.8	51 57.6	10.3	Capella	280 50.6	N45 59.7
02	93 49.0	206 28.6	50.0	274 21.9	02.7	103 56.9	51.7	67 00.2	10.3	Deneb	49 39.3	N45 16.9
03	108 51.5	221 27.7	.. 50.6	289 23.1	.. 02.2	118 59.3	.. 51.7	82 02.8	.. 10.2	Denebola	182 45.3	N14 34.7
04	123 53.9	236 26.9	51.2	304 24.2	01.6	134 01.7	51.6	97 05.4	10.2	Diphda	349 07.0	S17 59.7
05	138 56.4	251 26.0	51.8	319 25.4	01.1	149 04.1	51.6	112 08.0	10.2			
06	153 58.9	266 25.1	S21 52.4	334 26.6	N 2 00.6	164 06.4	S 5 51.5	127 10.6	N 8 10.1	Dubhe	194 05.7	N61 45.2
W 07	169 01.3	281 24.3	52.9	349 27.7	2 00.0	179 08.8	51.5	142 13.2	10.1	Elnath	278 26.5	N28 36.3
E 08	184 03.8	296 23.4	53.5	4 28.9	1 59.5	194 11.2	51.4	157 15.8	10.0	Eltanin	90 51.9	N51 29.6
D 09	199 06.3	311 22.5	.. 54.1	19 30.1	.. 58.9	209 13.6	.. 51.4	172 18.4	.. 10.0	Enif	33 58.2	N 9 52.3
N 10	214 08.7	326 21.7	54.6	34 31.2	58.4	224 15.9	51.3	187 21.0	09.9	Fomalhaut	15 36.3	S29 37.8
E 11	229 11.2	341 20.8	55.2	49 32.4	57.8	239 18.3	51.3	202 23.6	09.9			
S 12	244 13.7	356 20.0	S21 55.8	64 33.6	N 1 57.3	254 20.7	S 5 51.2	217 26.2	N 8 09.9	Gacrux	172 13.7	S57 06.1
D 13	259 16.1	11 19.1	56.4	79 34.7	56.7	269 23.0	51.2	232 28.9	09.8	Gienah	176 04.1	S17 31.9
A 14	274 18.6	26 18.2	56.9	94 35.9	56.2	284 25.4	51.1	247 31.5	09.8	Hadar	149 04.4	S60 21.7
Y 15	289 21.1	41 17.4	.. 57.5	109 37.0	.. 55.6	299 27.8	.. 51.1	262 34.1	.. 09.7	Hamal	328 13.2	N23 27.4
16	304 23.5	56 16.5	58.1	124 38.2	55.1	314 30.2	51.0	277 36.7	09.7	Kaus Aust.	83 59.1	S34 23.0
17	319 26.0	71 15.6	58.6	139 39.4	54.6	329 32.5	51.0	292 39.3	09.6			
18	334 28.4	86 14.8	S21 59.2	154 40.5	N 1 54.0	344 34.9	S 5 50.9	307 41.9	N 8 09.6	Kochab	137 20.9	N74 09.6
19	349 30.9	101 13.9	21 59.7	169 41.7	53.5	359 37.3	50.9	322 44.5	09.6	Markab	13 49.5	N15 12.1
20	4 33.4	116 13.0	22 00.3	184 42.9	52.9	14 39.7	50.8	337 47.1	09.5	Menkar	314 26.5	N 4 05.0
21	19 35.8	131 12.1	.. 00.9	199 44.0	.. 52.4	29 42.0	.. 50.8	352 49.7	.. 09.5	Menkent	148 21.2	S36 21.6
22	34 38.3	146 11.3	01.4	214 45.2	51.8	44 44.4	50.7	7 52.3	09.4	Miaplacidus	221 41.7	S69 42.5
23	49 40.8	161 10.4	02.0	229 46.4	51.3	59 46.8	50.7	22 54.9	09.4			
26 00	64 43.2	176 09.5	S22 02.5	244 47.5	N 1 50.7	74 49.1	S 5 50.6	37 57.5	N 8 09.3	Mirfak	308 56.0	N49 51.4
01	79 45.7	191 08.7	03.1	259 48.7	50.2	89 51.5	50.6	53 00.1	09.3	Nunki	76 12.6	S26 17.8
02	94 48.2	206 07.8	03.7	274 49.9	49.6	104 53.9	50.5	68 02.7	09.3	Peacock	53 37.3	S56 44.5
03	109 50.6	221 06.9	.. 04.2	289 51.0	.. 49.1	119 56.2	.. 50.5	83 05.3	.. 09.2	Pollux	243 41.3	N28 01.6
04	124 53.1	236 06.0	04.8	304 52.2	48.6	134 58.6	50.4	98 07.9	09.2	Procyon	245 11.3	N 5 13.6
05	139 55.5	251 05.2	05.3	319 53.4	48.0	150 01.0	50.4	113 10.5	09.1			
06	154 58.0	266 04.3	S22 05.9	334 54.5	N 1 47.5	165 03.3	S 5 50.3	128 13.1	N 8 09.1	Rasalhague	96 17.2	N12 33.9
07	170 00.5	281 03.4	06.4	349 55.7	46.9	180 05.7	50.3	143 15.7	09.1	Regulus	207 55.5	N11 58.3
T 08	185 02.9	296 02.6	07.0	4 56.9	46.4	195 08.1	50.2	158 18.3	09.0	Rigel	281 22.6	S 8 12.2
H 09	200 05.4	311 01.7	.. 07.5	19 58.0	.. 45.8	210 10.4	.. 50.2	173 20.9	.. 09.0	Rigil Kent.	140 07.8	S60 49.5
U 10	215 07.9	326 00.8	08.0	34 59.2	45.3	225 12.8	50.1	188 23.5	08.9	Sabik	102 25.8	S15 43.2
R 11	230 10.3	340 59.9	08.6	50 00.4	44.7	240 15.2	50.0	203 26.1	08.9			
S 12	245 12.8	355 59.1	S22 09.1	65 01.5	N 1 44.2	255 17.5	S 5 50.0	218 28.7	N 8 08.8	Schedar	349 53.0	N56 32.1
D 13	260 15.3	10 58.2	09.7	80 02.7	43.7	270 19.9	49.9	233 31.3	08.8	Shaula	96 37.6	S37 06.1
A 14	275 17.7	25 57.3	10.2	95 03.9	43.1	285 22.3	49.9	248 33.9	08.8	Sirius	258 43.4	S16 42.9
Y 15	290 20.2	40 56.4	.. 10.8	110 05.0	.. 42.6	300 24.6	.. 49.8	263 36.5	.. 08.7	Spica	158 43.4	S11 09.1
16	305 22.7	55 55.5	11.3	125 06.2	42.0	315 27.0	49.8	278 39.1	08.7	Suhail	223 00.5	S43 25.5
17	320 25.1	70 54.7	11.8	140 07.4	41.5	330 29.4	49.7	293 41.7	08.6			
18	335 27.6	85 53.8	S22 12.4	155 08.6	N 1 40.9	345 31.7	S 5 49.7	308 44.2	N 8 08.6	Vega	80 46.9	N38 47.2
19	350 30.0	100 52.9	12.9	170 09.7	40.4	0 34.1	49.6	323 46.8	08.6	Zuben'ubi	137 18.2	S16 02.0
20	5 32.5	115 52.0	13.4	185 10.9	39.9	15 36.5	49.6	338 49.4	08.5		SHA	Mer. Pass.
21	20 35.0	130 51.2	.. 14.0	200 12.1	.. 39.3	30 38.8	.. 49.5	353 52.0	.. 08.5	Venus	112 46.2	12 15
22	35 37.4	145 50.3	14.5	215 13.2	38.8	45 41.2	49.5	8 54.6	08.4	Mars	180 35.5	7 42
23	50 39.9	160 49.4	15.0	230 14.4	38.2	60 43.5	49.4	23 57.2	08.4	Jupiter	10 08.1	19 02
Mer. Pass. 19 41.8		v −0.9 d 0.6		v 1.2 d 0.5		v 2.4 d 0.0		v 2.6 d 0.0		Saturn	333 10.9	21 29

SUN / MOON

UT	SUN GHA	SUN Dec	MOON GHA	v	MOON Dec	d	HP
d h	° ′	° ′	° ′	′	° ′	′	′
24 00	183 22.3	S20 27.5	125 51.7	10.4	S18 32.4	3.4	55.8
01	198 22.1	28.0	140 21.1	10.5	18 29.0	3.5	55.9
02	213 22.0	28.5	154 50.6	10.4	18 25.5	3.5	55.9
03	228 21.8	.. 29.0	169 20.0	10.4	18 22.0	3.7	55.9
04	243 21.6	29.5	183 49.4	10.4	18 18.3	3.7	55.9
05	258 21.4	30.0	198 18.8	10.5	18 14.6	3.9	56.0
06	273 21.2	S20 30.5	212 48.3	10.4	S18 10.7	3.9	56.0
07	288 21.1	31.0	227 17.7	10.4	18 06.8	4.1	56.0
08	303 20.9	31.5	241 47.1	10.4	18 02.7	4.1	56.0
T 09	318 20.7	.. 32.0	256 16.5	10.4	17 58.6	4.3	56.1
U 10	333 20.5	32.6	270 45.9	10.4	17 54.3	4.3	56.1
E 11	348 20.3	33.1	285 15.3	10.4	17 50.0	4.4	56.1
S 12	3 20.2	S20 33.6	299 44.7	10.5	S17 45.6	4.6	56.1
D 13	18 20.0	34.1	314 14.2	10.4	17 41.0	4.6	56.2
A 14	33 19.8	34.6	328 43.6	10.4	17 36.4	4.7	56.2
Y 15	48 19.6	.. 35.1	343 13.0	10.4	17 31.7	4.8	56.2
16	63 19.4	35.6	357 42.4	10.4	17 26.9	4.9	56.2
17	78 19.2	36.1	12 11.8	10.4	17 22.0	5.0	56.3
18	93 19.1	S20 36.6	26 41.2	10.4	S17 17.0	5.1	56.3
19	108 18.9	37.1	41 10.6	10.5	17 11.9	5.2	56.3
20	123 18.7	37.6	55 40.1	10.4	17 06.7	5.3	56.4
21	138 18.5	.. 38.1	70 09.5	10.4	17 01.4	5.4	56.4
22	153 18.3	38.6	84 38.9	10.4	16 56.0	5.5	56.4
23	168 18.1	39.1	99 08.3	10.5	16 50.5	5.5	56.4
25 00	183 18.0	S20 39.6	113 37.8	10.4	S16 45.0	5.7	56.5
01	198 17.8	40.1	128 07.2	10.4	16 39.3	5.7	56.5
02	213 17.6	40.6	142 36.6	10.5	16 33.6	5.9	56.5
03	228 17.4	.. 41.1	157 06.1	10.4	16 27.7	5.9	56.5
04	243 17.2	41.5	171 35.5	10.4	16 21.8	6.0	56.6
05	258 17.0	42.0	186 04.9	10.5	16 15.8	6.1	56.6
06	273 16.8	S20 42.5	200 34.4	10.4	S16 09.7	6.2	56.6
W 07	288 16.6	43.0	215 03.8	10.5	16 03.5	6.3	56.7
E 08	303 16.5	43.5	229 33.3	10.4	15 57.2	6.4	56.7
D 09	318 16.3	.. 44.0	244 02.7	10.5	15 50.8	6.5	56.7
N 10	333 16.1	44.5	258 32.2	10.5	15 44.3	6.6	56.7
E 11	348 15.9	45.0	273 01.7	10.4	15 37.7	6.6	56.8
S 12	3 15.7	S20 45.5	287 31.1	10.5	S15 31.1	6.8	56.8
D 13	18 15.5	46.0	302 00.6	10.5	15 24.3	6.8	56.8
A 14	33 15.3	46.5	316 30.1	10.5	15 17.5	6.9	56.9
Y 15	48 15.1	.. 46.9	330 59.6	10.4	15 10.6	7.0	56.9
16	63 14.9	47.4	345 29.0	10.5	15 03.6	7.1	56.9
17	78 14.8	47.9	359 58.5	10.5	14 56.5	7.2	56.9
18	93 14.6	S20 48.4	14 28.0	10.5	S14 49.3	7.2	57.0
19	108 14.4	48.9	28 57.5	10.5	14 42.1	7.4	57.0
20	123 14.2	49.4	43 27.0	10.5	14 34.7	7.4	57.0
21	138 14.0	.. 49.8	57 56.5	10.5	14 27.3	7.5	57.1
22	153 13.8	50.3	72 26.0	10.5	14 19.8	7.6	57.1
23	168 13.6	50.8	86 55.5	10.5	14 12.2	7.7	57.1
26 00	183 13.4	S20 51.3	101 25.0	10.5	S14 04.5	7.7	57.2
01	198 13.2	51.8	115 54.5	10.5	13 56.8	7.9	57.2
02	213 13.0	52.2	130 24.0	10.5	13 48.9	7.9	57.2
03	228 12.8	.. 52.7	144 53.5	10.5	13 41.0	8.0	57.3
04	243 12.6	53.2	159 23.0	10.6	13 33.0	8.1	57.3
05	258 12.4	53.7	173 52.6	10.5	13 24.9	8.2	57.3
06	273 12.2	S20 54.2	188 22.1	10.5	S13 16.7	8.2	57.3
T 07	288 12.0	54.6	202 51.6	10.5	13 08.5	8.3	57.4
H 08	303 11.9	55.1	217 21.1	10.5	13 00.2	8.4	57.4
U 09	318 11.7	.. 55.6	231 50.6	10.6	12 51.8	8.5	57.4
R 10	333 11.5	56.1	246 20.2	10.5	12 43.3	8.6	57.5
S 11	348 11.3	56.5	260 49.7	10.5	12 34.7	8.6	57.5
D 12	3 11.1	S20 57.0	275 19.2	10.5	S12 26.1	8.7	57.5
A 13	18 10.9	57.5	289 48.7	10.6	12 17.4	8.8	57.6
Y 14	33 10.7	57.9	304 18.3	10.5	12 08.6	8.8	57.6
15	48 10.5	.. 58.4	318 47.8	10.5	11 59.8	9.0	57.6
16	63 10.3	58.9	333 17.3	10.6	11 50.8	9.0	57.7
17	78 10.1	59.4	347 46.9	10.5	11 41.8	9.1	57.7
18	93 09.9	S20 59.8	2 16.4	10.5	S11 32.7	9.1	57.7
19	108 09.7	21 00.3	16 45.9	10.5	11 23.6	9.2	57.8
20	123 09.5	00.8	31 15.4	10.6	11 14.4	9.3	57.8
21	138 09.3	.. 01.2	45 45.0	10.5	11 05.1	9.4	57.8
22	153 09.1	01.7	60 14.5	10.5	10 55.7	9.4	57.9
23	168 08.9	02.2	74 44.0	10.5	S10 46.3	9.5	57.9
	SD 16.2	d 0.5	SD 15.3		15.5		15.7

Twilight / Sunrise / Moonrise

Lat.	Naut.	Civil	Sunrise	24	25	26	27
°	h m	h m	h m	h m	h m	h m	h m
N 72	07 33	09 21	■■	15 19	14 46	14 31	14 20
N 70	07 19	08 49	11 25	14 11	14 11	14 10	14 08
68	07 07	08 26	10 01	13 34	13 46	13 53	13 58
66	06 57	08 07	09 24	13 08	13 27	13 40	13 49
64	06 49	07 52	08 58	12 48	13 11	13 29	13 42
62	06 42	07 40	08 38	12 31	12 58	13 19	13 36
60	06 35	07 29	08 21	12 17	12 47	13 11	13 31
N 58	06 30	07 20	08 08	12 06	12 37	13 04	13 26
56	06 24	07 12	07 56	11 55	12 29	12 57	13 22
54	06 20	07 04	07 45	11 46	12 21	12 51	13 18
52	06 15	06 57	07 36	11 38	12 14	12 46	13 15
50	06 11	06 51	07 28	11 31	12 08	12 41	13 12
45	06 01	06 38	07 10	11 15	11 55	12 31	13 05
N 40	05 53	06 26	06 56	11 02	11 44	12 23	12 59
35	05 45	06 16	06 44	10 51	11 35	12 15	12 54
30	05 38	06 07	06 33	10 42	11 26	12 09	12 50
20	05 24	05 51	06 15	10 25	11 12	11 58	12 42
N 10	05 10	05 36	05 59	10 11	11 00	11 48	12 35
0	04 55	05 21	05 43	09 57	10 48	11 39	12 29
S 10	04 39	05 05	05 28	09 44	10 36	11 29	12 23
20	04 19	04 47	05 11	09 29	10 24	11 19	12 16
30	03 53	04 26	04 52	09 13	10 09	11 08	12 08
35	03 37	04 12	04 41	09 03	10 01	11 01	12 04
40	03 17	03 57	04 28	08 52	09 51	10 54	11 59
45	02 52	03 37	04 13	08 39	09 40	10 45	11 53
S 50	02 16	03 12	03 54	08 23	09 27	10 35	11 46
52	01 57	03 00	03 45	08 16	09 20	10 30	11 43
54	01 32	02 46	03 34	08 08	09 13	10 24	11 39
56	00 55	02 29	03 23	07 58	09 06	10 18	11 35
58	////	02 09	03 08	07 48	08 57	10 12	11 31
S 60	////	01 42	02 54	07 36	08 47	10 04	11 26

Sunset / Twilight / Moonset

Lat.	Sunset	Civil	Naut.	24	25	26	27
°	h m	h m	h m	h m	h m	h m	h m
N 72	■■	14 12	16 00	17 09	19 28	21 29	23 25
N 70	12 08	14 44	16 14	18 17	20 02	21 48	23 36
68	13 32	15 08	16 26	18 53	20 26	22 03	23 44
66	14 09	15 26	16 36	19 18	20 44	22 16	23 51
64	14 35	15 41	16 51	19 38	20 59	22 26	23 56
62	14 55	15 53	16 51	19 54	21 12	22 35	24 01
60	15 12	16 04	16 58	20 08	21 22	22 42	24 06
N 58	15 26	16 14	17 04	20 19	21 31	22 48	24 09
56	15 38	16 22	17 09	20 29	21 39	22 54	24 13
54	15 48	16 29	17 14	20 38	21 46	22 59	24 16
52	15 57	16 36	17 18	20 46	21 53	23 04	24 18
50	16 06	16 42	17 23	20 53	21 58	23 08	24 21
45	16 23	16 56	17 32	21 08	22 11	23 17	24 26
N 40	16 38	17 07	17 40	21 20	22 21	23 25	24 31
35	16 50	17 17	17 48	21 30	22 30	23 31	24 34
30	17 01	17 26	17 56	21 40	22 37	23 37	24 38
20	17 19	17 43	18 10	21 55	22 50	23 46	24 44
N 10	17 35	17 58	18 24	22 09	23 02	23 55	24 49
0	17 51	18 13	18 38	22 22	23 12	24 03	00 03
S 10	18 06	18 29	18 55	22 34	23 23	24 10	00 10
20	18 23	18 47	19 15	22 48	23 34	24 19	00 19
30	18 42	19 09	19 41	23 03	23 47	24 28	00 28
35	18 53	19 22	19 57	23 12	23 54	24 33	00 33
40	19 06	19 38	20 17	23 22	24 02	00 02	00 40
45	19 22	19 57	20 43	23 34	24 12	00 12	00 47
S 50	19 41	20 22	21 19	23 48	24 24	00 24	00 55
52	19 50	20 35	21 39	23 55	24 29	00 29	00 59
54	20 00	20 49	22 05	24 03	00 03	00 35	01 03
56	20 12	21 06	22 44	24 11	00 11	00 41	01 08
58	20 26	21 27	////	24 20	00 20	00 49	01 13
S 60	20 42	21 55	////	24 31	00 31	00 57	01 19

SUN / MOON

Day	Eqn. of Time 00h	Eqn. of Time 12h	Mer. Pass.	Mer. Pass. Upper	Mer. Pass. Lower	Age	Phase
d	m s	m s	h m	h m	h m	d	%
24	13 30	13 21	11 47	16 09	03 44	05	25
25	13 12	13 03	11 47	17 00	04 35	06	34
26	12 54	12 45	11 47	17 51	05 25	07	45

1998 NOVEMBER 27, 28, 29 (FRI., SAT., SUN.)

UT	ARIES	VENUS −3.9		MARS +1.4		JUPITER −2.5		SATURN +0.2	
	GHA	GHA	Dec	GHA	Dec	GHA	Dec	GHA	Dec
d h	° ′	° ′	° ′	° ′	° ′	° ′	° ′	° ′	° ′
27 00	65 42.4	175 48.5	S22 15.6	245 15.6	N 1 37.7	75 45.9	S 5 49.3	38 59.8	N 8 08.4
01	80 44.8	190 47.6	16.1	260 16.7	37.1	90 48.3	49.3	54 02.4	08.3
02	95 47.3	205 46.7	16.6	275 17.9	36.6	105 50.6	49.2	69 05.0	08.3
03	110 49.8	220 45.9	.. 17.1	290 19.1	.. 36.0	120 53.0	.. 49.2	84 07.6	.. 08.2
04	125 52.2	235 45.0	17.7	305 20.2	35.5	135 55.4	49.1	99 10.2	08.2
05	140 54.7	250 44.1	18.2	320 21.4	35.0	150 57.7	49.1	114 12.8	08.1
F 06	155 57.2	265 43.2	S22 18.7	335 22.6	N 1 34.4	166 00.1	S 5 49.0	129 15.4	N 8 08.1
R 07	170 59.6	280 42.3	19.2	350 23.8	33.9	181 02.4	49.0	144 18.0	08.1
I 08	186 02.1	295 41.4	19.8	5 24.9	33.3	196 04.8	48.9	159 20.6	08.0
D 09	201 04.5	310 40.6	.. 20.3	20 26.1	.. 32.8	211 07.2	.. 48.9	174 23.2	.. 08.0
A 10	216 07.0	325 39.7	20.8	35 27.3	32.2	226 09.5	48.8	189 25.8	07.9
Y 11	231 09.5	340 38.8	21.3	50 28.4	31.7	241 11.9	48.7	204 28.4	07.9
12	246 11.9	355 37.9	S22 21.8	65 29.6	N 1 31.2	256 14.2	S 5 48.7	219 31.0	N 8 07.9
13	261 14.4	10 37.0	22.3	80 30.8	30.6	271 16.6	48.6	234 33.6	07.8
14	276 16.9	25 36.1	22.9	95 32.0	30.1	286 18.9	48.6	249 36.2	07.8
15	291 19.3	40 35.2	.. 23.4	110 33.1	.. 29.5	301 21.3	.. 48.5	264 38.8	.. 07.7
16	306 21.8	55 34.4	23.9	125 34.3	29.0	316 23.7	48.5	279 41.4	07.7
17	321 24.3	70 33.5	24.4	140 35.5	28.4	331 26.0	48.4	294 44.0	07.7
18	336 26.7	85 32.6	S22 24.9	155 36.6	N 1 27.9	346 28.4	S 5 48.3	309 46.6	N 8 07.6
19	351 29.2	100 31.7	25.4	170 37.8	27.4	1 30.7	48.3	324 49.2	07.6
20	6 31.6	115 30.8	25.9	185 39.0	26.8	16 33.1	48.2	339 51.8	07.5
21	21 34.1	130 29.9	.. 26.4	200 40.2	.. 26.3	31 35.4	.. 48.2	354 54.3	.. 07.5
22	36 36.6	145 29.0	26.9	215 41.3	25.7	46 37.8	48.1	9 56.9	07.5
23	51 39.0	160 28.1	27.4	230 42.5	25.2	61 40.1	48.1	24 59.5	07.4
28 00	66 41.5	175 27.2	S22 27.9	245 43.7	N 1 24.7	76 42.5	S 5 48.0	40 02.1	N 8 07.4
01	81 44.0	190 26.4	28.4	260 44.8	24.1	91 44.9	47.9	55 04.7	07.3
02	96 46.4	205 25.5	28.9	275 46.0	23.6	106 47.2	47.9	70 07.3	07.3
03	111 48.9	220 24.6	.. 29.4	290 47.2	.. 23.0	121 49.6	.. 47.8	85 09.9	.. 07.3
04	126 51.4	235 23.7	29.9	305 48.4	22.5	136 51.9	47.8	100 12.5	07.2
05	141 53.8	250 22.8	30.4	320 49.5	21.9	151 54.3	47.7	115 15.1	07.2
S 06	156 56.3	265 21.9	S22 30.9	335 50.7	N 1 21.4	166 56.6	S 5 47.7	130 17.7	N 8 07.1
A 07	171 58.8	280 21.0	31.4	350 51.9	20.9	181 59.0	47.6	145 20.3	07.1
T 08	187 01.2	295 20.1	31.9	5 53.1	20.3	197 01.3	47.5	160 22.9	07.1
U 09	202 03.7	310 19.2	.. 32.4	20 54.2	.. 19.8	212 03.7	.. 47.5	175 25.5	.. 07.0
R 10	217 06.1	325 18.3	32.9	35 55.4	19.2	227 06.0	47.4	190 28.1	07.0
D 11	232 08.6	340 17.4	33.4	50 56.6	18.7	242 08.4	47.4	205 30.7	07.0
A 12	247 11.1	355 16.5	S22 33.9	65 57.8	N 1 18.2	257 10.7	S 5 47.3	220 33.2	N 8 06.9
Y 13	262 13.5	10 15.6	34.3	80 58.9	17.6	272 13.1	47.2	235 35.8	06.9
14	277 16.0	25 14.7	34.8	96 00.1	17.1	287 15.4	47.2	250 38.4	06.8
15	292 18.5	40 13.8	.. 35.3	111 01.3	.. 16.5	302 17.8	.. 47.1	265 41.0	.. 06.8
16	307 20.9	55 12.9	35.8	126 02.5	16.0	317 20.1	47.1	280 43.6	06.8
17	322 23.4	70 12.0	36.3	141 03.6	15.4	332 22.5	47.0	295 46.2	06.7
18	337 25.9	85 11.1	S22 36.8	156 04.8	N 1 14.9	347 24.8	S 5 46.9	310 48.8	N 8 06.7
19	352 28.3	100 10.2	37.2	171 06.0	14.4	2 27.2	46.9	325 51.4	06.6
20	7 30.8	115 09.4	37.7	186 07.2	13.8	17 29.5	46.8	340 54.0	06.6
21	22 33.2	130 08.5	.. 38.2	201 08.3	.. 13.3	32 31.9	.. 46.8	355 56.6	.. 06.5
22	37 35.7	145 07.6	38.7	216 09.5	12.7	47 34.2	46.7	10 59.2	06.5
23	52 38.2	160 06.7	39.2	231 10.7	12.2	62 36.6	46.6	26 01.8	06.5
29 00	67 40.6	175 05.8	S22 39.6	246 11.9	N 1 11.7	77 38.9	S 5 46.6	41 04.3	N 8 06.5
01	82 43.1	190 04.9	40.1	261 13.0	11.1	92 41.3	46.5	56 06.9	06.4
02	97 45.6	205 04.0	40.6	276 14.2	10.6	107 43.6	46.5	71 09.5	06.4
03	112 48.0	220 03.0	.. 41.0	291 15.4	.. 10.0	122 46.0	.. 46.4	86 12.1	.. 06.3
04	127 50.5	235 02.1	41.5	306 16.6	09.5	137 48.3	46.3	101 14.7	06.3
05	142 53.0	250 01.2	42.0	321 17.7	09.0	152 50.6	46.3	116 17.3	06.3
S 06	157 55.4	265 00.3	S22 42.4	336 18.9	N 1 08.4	167 53.0	S 5 46.2	131 19.9	N 8 06.2
U 07	172 57.9	279 59.4	42.9	351 20.1	07.9	182 55.3	46.1	146 22.5	06.2
N 08	188 00.4	294 58.5	43.4	6 21.3	07.3	197 57.7	46.1	161 25.1	06.2
D 09	203 02.8	309 57.6	.. 43.8	21 22.5	.. 06.8	213 00.0	.. 46.0	176 27.7	.. 06.1
A 10	218 05.3	324 56.7	44.3	36 23.6	06.3	228 02.4	46.0	191 30.2	06.1
Y 11	233 07.7	339 55.8	44.8	51 24.8	05.7	243 04.7	45.9	206 32.8	06.0
12	248 10.2	354 54.9	S22 45.2	66 26.0	N 1 05.2	258 07.1	S 5 45.8	221 35.4	N 8 06.0
13	263 12.7	9 54.0	45.7	81 27.2	04.6	273 09.4	45.8	236 38.0	06.0
14	278 15.1	24 53.1	46.1	96 28.3	04.1	288 11.7	45.7	251 40.6	05.9
15	293 17.6	39 52.2	.. 46.6	111 29.5	.. 03.6	303 14.1	.. 45.6	266 43.2	.. 05.9
16	308 20.1	54 51.3	47.1	126 30.7	03.0	318 16.4	45.6	281 45.8	05.9
17	323 22.5	69 50.4	47.5	141 31.9	02.5	333 18.8	45.5	296 48.4	05.8
18	338 25.0	84 49.5	S22 48.0	156 33.1	N 1 01.9	348 21.1	S 5 45.5	311 51.0	N 8 05.8
19	353 27.5	99 48.6	48.4	171 34.2	01.4	3 23.4	45.4	326 53.5	05.7
20	8 29.9	114 47.7	48.9	186 35.4	00.9	18 25.8	45.3	341 56.1	05.7
21	23 32.4	129 46.8	.. 49.3	201 36.6	1 00.3	33 28.1	.. 45.3	356 58.7	.. 05.7
22	38 34.9	144 45.9	49.8	216 37.8	0 59.8	48 30.5	45.2	12 01.3	05.6
23	53 37.3	159 44.9	50.2	231 38.9	N 0 59.3	63 32.8	45.1	27 03.9	05.6
Mer. Pass. 19ʰ 30.0ᵐ		v −0.9	d 0.5	v 1.2	d 0.5	v 2.4	d 0.1	v 2.6	d 0.0

STARS

Name	SHA	Dec
Acamar	315 26.5	S40 18.7
Achernar	335 34.7	S57 14.8
Acrux	173 22.2	S63 05.2
Adhara	255 21.1	S28 58.2
Aldebaran	291 01.9	N16 30.3
Alioth	166 31.0	N55 57.8
Alkaid	153 08.2	N49 19.1
Al Na'ir	27 57.9	S46 58.2
Alnilam	275 57.5	S 1 12.2
Alphard	218 07.0	S 8 39.2
Alphecca	126 20.9	N26 43.3
Alpheratz	357 55.0	N29 05.2
Altair	62 19.4	N 8 52.1
Ankaa	353 26.6	S42 19.0
Antares	112 40.5	S26 25.6
Arcturus	146 06.3	N19 11.4
Atria	107 53.0	S69 01.4
Avior	234 22.2	S59 30.2
Bellatrix	278 43.8	N 6 20.8
Betelgeuse	271 13.2	N 7 24.3
Canopus	264 00.6	S52 41.7
Capella	280 50.6	N45 59.7
Deneb	49 39.3	N45 16.9
Denebola	182 45.3	N14 34.7
Diphda	349 07.0	S17 59.7
Dubhe	194 05.6	N61 45.2
Elnath	278 26.5	N28 36.3
Eltanin	90 51.9	N51 29.6
Enif	33 58.2	N 9 52.3
Fomalhaut	15 36.4	S29 37.8
Gacrux	172 13.7	S57 06.1
Gienah	176 04.0	S17 31.9
Hadar	149 04.4	S60 21.7
Hamal	328 13.2	N23 27.4
Kaus Aust.	83 59.1	S34 23.0
Kochab	137 20.8	N74 09.6
Markab	13 49.5	N15 12.1
Menkar	314 26.5	N 4 05.0
Menkent	148 21.2	S36 21.6
Miaplacidus	221 41.6	S69 42.5
Mirfak	308 56.0	N49 51.4
Nunki	76 12.6	S26 17.8
Peacock	53 37.3	S56 44.5
Pollux	243 41.2	N28 01.6
Procyon	245 11.3	N 5 13.6
Rasalhague	96 17.2	N12 33.9
Regulus	207 55.5	N11 58.3
Rigel	281 22.5	S 8 12.3
Rigil Kent.	140 07.7	S60 49.5
Sabik	102 25.8	S15 43.2
Schedar	349 53.0	N56 32.1
Shaula	96 37.6	S37 06.1
Sirius	258 43.3	S16 42.9
Spica	158 43.4	S11 09.1
Suhail	223 00.5	S43 25.5
Vega	80 46.9	N38 47.2
Zuben'ubi	137 18.2	S16 02.0

	SHA	Mer. Pass.
	° ′	h m
Venus	108 45.7	12 19
Mars	179 02.2	7 36
Jupiter	10 01.0	18 50
Saturn	333 20.6	21 16

UT	SUN GHA	SUN Dec	MOON GHA	v	MOON Dec	d	HP
27 FRIDAY	° ′	° ′	° ′	′	° ′	′	′
00	183 08.7	S21 02.6	89 13.5	10.5	S10 36.8	9.6	57.9
01	198 08.5	03.1	103 43.0	10.5	10 27.2	9.6	58.0
02	213 08.3	03.5	118 12.5	10.5	10 17.6	9.7	58.0
03	228 08.1	.. 04.0	132 42.0	10.5	10 07.9	9.8	58.0
04	243 07.9	04.5	147 11.5	10.5	9 58.1	9.8	58.1
05	258 07.7	04.9	161 41.0	10.5	9 48.3	9.9	58.1
06	273 07.5	S21 05.4	176 10.5	10.5	S 9 38.4	10.0	58.1
07	288 07.3	05.8	190 40.0	10.5	9 28.4	10.0	58.1
08	303 07.1	06.3	205 09.5	10.4	9 18.4	10.1	58.2
09	318 06.9	.. 06.8	219 38.9	10.5	9 08.3	10.1	58.2
10	333 06.7	07.2	234 08.4	10.5	8 58.2	10.3	58.3
11	348 06.5	07.7	248 37.9	10.4	8 47.9	10.2	58.3
12	3 06.2	S21 08.1	263 07.3	10.4	S 8 37.7	10.3	58.3
13	18 06.0	08.6	277 36.7	10.5	8 27.4	10.4	58.4
14	33 05.8	09.0	292 06.2	10.4	8 17.0	10.5	58.4
15	48 05.6	.. 09.5	306 35.6	10.4	8 06.5	10.5	58.4
16	63 05.4	09.9	321 05.0	10.4	7 56.0	10.5	58.5
17	78 05.2	10.4	335 34.4	10.4	7 45.5	10.6	58.5
18	93 05.0	S21 10.9	350 03.8	10.4	S 7 34.9	10.7	58.5
19	108 04.8	11.3	4 33.2	10.3	7 24.2	10.7	58.6
20	123 04.6	11.8	19 02.5	10.4	7 13.5	10.8	58.6
21	138 04.4	.. 12.2	33 31.9	10.3	7 02.7	10.8	58.6
22	153 04.2	12.7	48 01.2	10.3	6 51.9	10.9	58.7
23	168 04.0	13.1	62 30.5	10.3	6 41.0	10.9	58.7
28 SATURDAY							
00	183 03.8	S21 13.5	76 59.8	10.3	S 6 30.1	11.0	58.7
01	198 03.6	14.0	91 29.1	10.3	6 19.1	11.0	58.8
02	213 03.4	14.4	105 58.4	10.3	6 08.1	11.1	58.8
03	228 03.1	.. 14.9	120 27.7	10.2	5 57.0	11.1	58.8
04	243 02.9	15.3	134 56.9	10.3	5 45.9	11.2	58.9
05	258 02.7	15.8	149 26.2	10.2	5 34.7	11.2	58.9
06	273 02.5	S21 16.2	163 55.4	10.2	S 5 23.5	11.2	58.9
07	288 02.3	16.7	178 24.6	10.1	5 12.3	11.3	59.0
08	303 02.1	17.1	192 53.7	10.2	5 01.0	11.3	59.0
09	318 01.9	.. 17.5	207 22.9	10.1	4 49.7	11.4	59.0
10	333 01.7	18.0	221 52.0	10.1	4 38.3	11.4	59.1
11	348 01.5	18.4	236 21.1	10.1	4 26.9	11.5	59.1
12	3 01.3	S21 18.9	250 50.2	10.1	S 4 15.4	11.5	59.1
13	18 01.0	19.3	265 19.3	10.0	4 03.9	11.5	59.2
14	33 00.8	19.7	279 48.3	10.0	3 52.4	11.5	59.2
15	48 00.6	.. 20.2	294 17.3	10.0	3 40.9	11.6	59.2
16	63 00.4	20.6	308 46.3	10.0	3 29.3	11.7	59.3
17	78 00.2	21.0	323 15.3	10.0	3 17.6	11.6	59.3
18	93 00.0	S21 21.5	337 44.3	9.9	S 3 06.0	11.7	59.3
19	107 59.8	21.9	352 13.2	9.9	2 54.3	11.7	59.4
20	122 59.6	22.4	6 42.1	9.8	2 42.6	11.8	59.4
21	137 59.3	.. 22.8	21 10.9	9.9	2 30.8	11.8	59.4
22	152 59.1	23.2	35 39.8	9.8	2 19.0	11.8	59.5
23	167 58.9	23.6	50 08.6	9.8	2 07.2	11.8	59.5
29 SUNDAY							
00	182 58.7	S21 24.1	64 37.4	9.7	S 1 55.4	11.9	59.5
01	197 58.5	24.5	79 06.1	9.7	1 43.5	11.8	59.6
02	212 58.3	24.9	93 34.8	9.7	1 31.7	11.9	59.6
03	227 58.0	.. 25.4	108 03.5	9.7	1 19.8	12.0	59.6
04	242 57.8	25.8	122 32.2	9.6	1 07.8	11.9	59.6
05	257 57.6	26.2	137 00.8	9.6	0 55.9	12.0	59.7
06	272 57.4	S21 26.6	151 29.4	9.6	S 0 43.9	12.0	59.7
07	287 57.2	27.1	165 58.0	9.5	0 31.9	11.9	59.7
08	302 57.0	27.5	180 26.5	9.5	0 20.0	12.1	59.8
09	317 56.7	.. 27.9	194 55.0	9.4	S 0 07.9	12.0	59.8
10	332 56.5	28.3	209 23.4	9.4	N 0 04.1	12.0	59.8
11	347 56.3	28.8	223 51.8	9.4	0 16.1	12.1	59.9
12	2 56.1	S21 29.2	238 20.2	9.3	N 0 28.2	12.0	59.9
13	17 55.9	29.6	252 48.5	9.3	0 40.2	12.1	59.9
14	32 55.7	30.0	267 16.8	9.3	0 52.3	12.1	59.9
15	47 55.4	.. 30.5	281 45.1	9.2	1 04.4	12.0	60.0
16	62 55.2	30.9	296 13.3	9.2	1 16.4	12.1	60.0
17	77 55.0	31.3	310 41.5	9.2	1 28.5	12.1	60.0
18	92 54.8	S21 31.7	325 09.7	9.1	N 1 40.6	12.1	60.1
19	107 54.6	32.1	339 37.8	9.0	1 52.7	12.1	60.1
20	122 54.3	32.5	354 05.8	9.0	2 04.8	12.1	60.1
21	137 54.1	.. 33.0	8 33.8	9.0	2 16.9	12.1	60.1
22	152 53.9	33.4	23 01.8	8.9	2 29.0	12.1	60.2
23	167 53.7	33.8	37 29.7	8.9	N 2 41.1	12.1	60.2
	SD 16.2	d 0.4	SD 15.9		16.1		16.3

Lat.	Twilight Naut.	Twilight Civil	Sunrise	Moonrise 27	28	29	30
°	h m	h m	h m	h m	h m	h m	h m
N 72	07 42	09 35	■	14 20	14 11	14 02	13 53
N 70	07 27	09 00		14 08	14 05	14 03	14 01
68	07 14	08 35	10 18	13 58	14 01	14 04	14 07
66	07 04	08 15	09 35	13 49	13 57	14 05	14 13
64	06 55	07 59	09 07	13 42	13 54	14 05	14 17
62	06 47	07 46	08 45	13 36	13 51	14 06	14 21
60	06 41	07 35	08 28	13 31	13 49	14 06	14 24
N 58	06 34	07 25	08 14	13 26	13 47	14 07	14 28
56	06 29	07 16	08 01	13 22	13 45	14 07	14 30
54	06 24	07 09	07 50	13 18	13 43	14 07	14 33
52	06 19	07 02	07 41	13 15	13 41	14 08	14 35
50	06 15	06 55	07 32	13 12	13 40	14 08	14 37
45	06 05	06 41	07 14	13 05	13 37	14 08	14 42
N 40	05 56	06 29	06 59	12 59	13 34	14 09	14 45
35	05 48	06 19	06 47	12 54	13 32	14 09	14 49
30	05 40	06 10	06 36	12 50	13 30	14 10	14 52
20	05 26	05 53	06 17	12 42	13 26	14 11	14 57
N 10	05 11	05 37	06 01	12 35	13 23	14 11	15 01
0	04 56	05 22	05 44	12 29	13 20	14 12	15 06
S 10	04 39	05 06	05 28	12 23	13 17	14 13	15 10
20	04 19	04 47	05 11	12 16	13 14	14 13	15 15
30	03 52	04 25	04 52	12 08	13 10	14 14	15 20
35	03 36	04 11	04 40	12 04	13 08	14 15	15 23
40	03 15	03 55	04 27	11 59	13 06	14 15	15 27
45	02 49	03 35	04 11	11 53	13 03	14 16	15 31
S 50	02 11	03 09	03 51	11 46	13 00	14 17	15 36
52	01 51	02 56	03 42	11 43	12 59	14 17	15 38
54	01 23	02 42	03 31	11 39	12 57	14 18	15 41
56	00 39	02 24	03 19	11 35	12 55	14 18	15 43
58	////	02 02	03 05	11 31	12 53	14 19	15 47
S 60	////	01 32	02 49	11 26	12 51	14 19	15 50

Lat.	Sunset	Twilight Civil	Twilight Naut.	Moonset 27	28	29	30
°	h m	h m	h m	h m	h m	h m	h m
N 72	■	14 00	15 53	23 25	25 22	01 22	03 21
N 70	■	14 35	16 08	23 36	25 25	01 25	03 16
68	13 18	15 01	16 21	23 44	25 27	01 27	03 12
66	14 00	15 20	16 31	23 51	25 28	01 28	03 09
64	14 28	15 36	16 40	23 56	25 30	01 30	03 06
62	14 50	15 49	16 48	24 01	00 01	01 31	03 04
60	15 07	16 00	16 55	24 06	00 06	01 32	03 02
N 58	15 22	16 10	17 01	24 09	00 09	01 33	03 00
56	15 34	16 19	17 06	24 13	00 13	01 34	02 58
54	15 45	16 27	17 12	24 16	00 16	01 35	02 57
52	15 55	16 34	17 16	24 18	00 18	01 36	02 55
50	16 03	16 40	17 21	24 21	00 21	01 36	02 54
45	16 21	16 54	17 31	24 26	00 26	01 38	02 51
N 40	16 36	17 06	17 40	24 31	00 31	01 39	02 49
35	16 49	17 17	17 48	24 34	00 34	01 40	02 47
30	17 00	17 26	17 55	24 38	00 38	01 41	02 45
20	17 19	17 43	18 10	24 44	00 44	01 42	02 42
N 10	17 36	17 58	18 24	24 49	00 49	01 43	02 40
0	17 52	18 14	18 40	00 03	00 53	01 45	02 37
S 10	18 07	18 30	18 57	00 10	00 58	01 46	02 35
20	18 25	18 49	19 17	00 19	01 03	01 47	02 32
30	18 44	19 11	19 44	00 33	01 08	01 48	02 29
35	18 56	19 25	20 01	00 40	01 11	01 49	02 27
40	19 10	19 41	20 21	00 40	01 15	01 50	02 25
45	19 26	20 02	20 48	00 47	01 19	01 51	02 23
S 50	19 46	20 28	21 26	00 55	01 24	01 52	02 20
52	19 55	20 40	21 47	00 59	01 26	01 52	02 19
54	20 06	20 56	22 15	01 03	01 29	01 53	02 17
56	20 18	21 14	23 02	01 08	01 31	01 54	02 16
58	20 32	21 36	////	01 13	01 34	01 54	02 14
S 60	20 49	22 06	////	01 19	01 38	01 55	02 12

Day	SUN Eqn. of Time 00h	12h	SUN Mer. Pass.	MOON Mer. Pass. Upper	Lower	Age	Phase
d	m s	m s	h m	h m	h m	d	%
27	12 35	12 25	11 48	18 41	06 16	08	55
28	12 16	12 05	11 48	19 32	07 07	09	66
29	11 55	11 45	11 48	20 24	07 58	10	77

1998 NOV. 30, DEC. 1, 2 (MON., TUES., WED.)

UT	ARIES GHA	VENUS −3.9 GHA	Dec	MARS +1.4 GHA	Dec	JUPITER −2.5 GHA	Dec	SATURN +0.2 GHA	Dec
30 00	68 39.8	174 44.0	S22 50.7	246 40.1	N 0 58.7	78 35.1	S 5 45.1	42 06.5	N 8 05.6
01	83 42.2	189 43.1	51.1	261 41.3	58.2	93 37.5	45.0	57 09.1	05.5
02	98 44.7	204 42.2	51.5	276 42.5	57.6	108 39.8	44.9	72 11.7	05.5
03	113 47.2	219 41.3	. . 52.0	291 43.7	. . 57.1	123 42.2	. . 44.9	87 14.2	. . 05.4
04	128 49.6	234 40.4	52.4	306 44.8	56.6	138 44.5	44.8	102 16.8	05.4
05	143 52.1	249 39.5	52.9	321 46.0	56.0	153 46.8	44.8	117 19.4	05.4
M 06	158 54.6	264 38.6	S22 53.3	336 47.2	N 0 55.5	168 49.2	S 5 44.7	132 22.0	N 8 05.3
O 07	173 57.0	279 37.7	53.7	351 48.4	54.9	183 51.5	44.6	147 24.6	05.3
N 08	188 59.5	294 36.8	54.2	6 49.6	54.4	198 53.8	44.6	162 27.2	05.3
D 09	204 02.0	309 35.8	. . 54.6	21 50.7	. . 53.9	213 56.2	. . 44.5	177 29.8	. . 05.2
A 10	219 04.4	324 34.9	55.0	36 51.9	53.3	228 58.5	44.4	192 32.4	05.2
Y 11	234 06.9	339 34.0	55.5	51 53.1	52.8	244 00.9	44.4	207 34.9	05.2
12	249 09.3	354 33.1	S22 55.9	66 54.3	N 0 52.3	259 03.2	S 5 44.3	222 37.5	N 8 05.1
13	264 11.8	9 32.2	56.3	81 55.5	51.7	274 05.5	44.2	237 40.1	05.1
14	279 14.3	24 31.3	56.8	96 56.7	51.2	289 07.9	44.2	252 42.7	05.0
15	294 16.7	39 30.4	. . 57.2	111 57.8	. . 50.6	304 10.2	. . 44.1	267 45.3	. . 05.0
16	309 19.2	54 29.4	57.6	126 59.0	50.1	319 12.5	44.0	282 47.9	05.0
17	324 21.7	69 28.5	58.0	142 00.2	49.6	334 14.9	44.0	297 50.4	04.9
18	339 24.1	84 27.6	S22 58.5	157 01.4	N 0 49.0	349 17.2	S 5 43.9	312 53.0	N 8 04.9
19	354 26.6	99 26.7	58.9	172 02.6	48.5	4 19.5	43.8	327 55.6	04.9
20	9 29.1	114 25.8	59.3	187 03.7	47.9	19 21.9	43.8	342 58.2	04.8
21	24 31.5	129 24.9	22 59.7	202 04.9	. . 47.4	34 24.2	. . 43.7	358 00.8	. . 04.8
22	39 34.0	144 23.9	23 00.2	217 06.1	46.9	49 26.5	43.6	13 03.4	04.8
23	54 36.5	159 23.0	00.6	232 07.3	46.3	64 28.9	43.6	28 06.0	04.7
1 00	69 38.9	174 22.1	S23 01.0	247 08.5	N 0 45.8	79 31.2	S 5 43.5	43 08.5	N 8 04.7
01	84 41.4	189 21.2	01.4	262 09.7	45.3	94 33.5	43.4	58 11.1	04.7
02	99 43.8	204 20.3	01.8	277 10.8	44.7	109 35.9	43.4	73 13.7	04.6
03	114 46.3	219 19.4	. . 02.2	292 12.0	. . 44.2	124 38.2	. . 43.3	88 16.3	. . 04.6
04	129 48.8	234 18.4	02.6	307 13.2	43.7	139 40.5	43.2	103 18.9	04.6
05	144 51.2	249 17.5	03.1	322 14.4	43.1	154 42.8	43.2	118 21.5	04.5
T 06	159 53.7	264 16.6	S23 03.5	337 15.6	N 0 42.6	169 45.2	S 5 43.1	133 24.0	N 8 04.5
U 07	174 56.2	279 15.7	03.9	352 16.8	42.0	184 47.5	43.0	148 26.6	04.4
E 08	189 58.6	294 14.8	04.3	7 17.9	41.5	199 49.8	43.0	163 29.2	04.4
S 09	205 01.1	309 13.8	. . 04.7	22 19.1	. . 41.0	214 52.2	. . 42.9	178 31.8	. . 04.4
D 10	220 03.6	324 12.9	05.1	37 20.3	40.4	229 54.5	42.8	193 34.4	04.3
A 11	235 06.0	339 12.0	05.5	52 21.5	39.9	244 56.8	42.7	208 37.0	04.3
Y 12	250 08.5	354 11.1	S23 05.9	67 22.7	N 0 39.4	259 59.1	S 5 42.7	223 39.5	N 8 04.3
13	265 11.0	9 10.2	06.3	82 23.9	38.8	275 01.5	42.6	238 42.1	04.2
14	280 13.4	24 09.2	06.7	97 25.0	38.3	290 03.8	42.5	253 44.7	04.2
15	295 15.9	39 08.3	. . 07.1	112 26.2	. . 37.7	305 06.1	. . 42.5	268 47.3	. . 04.2
16	310 18.3	54 07.4	07.5	127 27.4	37.2	320 08.5	42.4	283 49.9	04.1
17	325 20.8	69 06.5	07.9	142 28.6	36.7	335 10.8	42.3	298 52.4	04.1
18	340 23.3	84 05.5	S23 08.3	157 29.8	N 0 36.1	350 13.1	S 5 42.3	313 55.0	N 8 04.1
19	355 25.7	99 04.6	08.7	172 31.0	35.6	5 15.4	42.2	328 57.6	04.0
20	10 28.2	114 03.7	09.1	187 32.2	35.1	20 17.8	42.1	344 00.2	04.0
21	25 30.7	129 02.8	. . 09.5	202 33.3	. . 34.5	35 20.1	. . 42.1	359 02.8	. . 04.0
22	40 33.1	144 01.8	09.9	217 34.5	34.0	50 22.4	42.0	14 05.4	03.9
23	55 35.6	159 00.9	10.3	232 35.7	33.5	65 24.7	41.9	29 07.9	03.9
2 00	70 38.1	174 00.0	S23 10.6	247 36.9	N 0 32.9	80 27.1	S 5 41.8	44 10.5	N 8 03.9
01	85 40.5	188 59.1	11.0	262 38.1	32.4	95 29.4	41.8	59 13.1	03.8
02	100 43.0	203 58.1	11.4	277 39.3	31.9	110 31.7	41.7	74 15.7	03.8
03	115 45.4	218 57.2	. . 11.8	292 40.5	. . 31.3	125 34.0	. . 41.6	89 18.3	. . 03.8
04	130 47.9	233 56.3	12.2	307 41.6	30.8	140 36.4	41.6	104 20.8	03.7
05	145 50.4	248 55.4	12.6	322 42.8	30.2	155 38.7	41.5	119 23.4	03.7
W 06	160 52.8	263 54.4	S23 12.9	337 44.0	N 0 29.7	170 41.0	S 5 41.4	134 26.0	N 8 03.7
E 07	175 55.3	278 53.5	13.3	352 45.2	29.2	185 43.3	41.3	149 28.6	03.6
D 08	190 57.8	293 52.6	13.7	7 46.4	28.6	200 45.6	41.3	164 31.2	03.6
N 09	206 00.2	308 51.6	. . 14.1	22 47.6	. . 28.1	215 48.0	. . 41.2	179 33.7	. . 03.6
E 10	221 02.7	323 50.7	14.5	37 48.8	27.6	230 50.3	41.1	194 36.3	03.5
S 11	236 05.2	338 49.8	14.8	52 50.0	27.0	245 52.6	41.1	209 38.9	03.5
D 12	251 07.6	353 48.9	S23 15.2	67 51.1	N 0 26.5	260 54.9	S 5 41.0	224 41.5	N 8 03.5
A 13	266 10.1	8 47.9	15.6	82 52.3	26.0	275 57.2	40.9	239 44.1	03.4
Y 14	281 12.6	23 47.0	16.0	97 53.5	25.4	290 59.6	40.8	254 46.6	03.4
15	296 15.0	38 46.1	. . 16.3	112 54.7	. . 24.9	306 01.9	. . 40.8	269 49.2	. . 03.4
16	311 17.5	53 45.1	16.7	127 55.9	24.4	321 04.2	40.7	284 51.8	03.3
17	326 19.9	68 44.2	17.1	142 57.1	23.8	336 06.5	40.6	299 54.4	03.3
18	341 22.4	83 43.3	S23 17.4	157 58.3	N 0 23.3	351 08.8	S 5 40.5	314 56.9	N 8 03.3
19	356 24.9	98 42.3	17.8	172 59.5	22.8	6 11.2	40.5	329 59.5	03.2
20	11 27.3	113 41.4	18.2	188 00.6	22.2	21 13.5	40.4	345 02.1	03.2
21	26 29.8	128 40.5	. . 18.5	203 01.8	. . 21.7	36 15.8	. . 40.3	0 04.7	. . 03.2
22	41 32.3	143 39.6	18.9	218 03.0	21.2	51 18.1	40.3	15 07.3	03.1
23	56 34.7	158 38.6	19.2	233 04.2	20.6	66 20.4	40.2	30 09.8	03.1
Mer. Pass. 19 18.2		v −0.9	d 0.4	v 1.2	d 0.5	v 2.3	d 0.1	v 2.6	d 0.0

STARS

Name	SHA	Dec
Acamar	315 26.5	S40 18.8
Achernar	335 34.8	S57 14.8
Acrux	173 22.1	S63 05.2
Adhara	255 21.0	S28 58.2
Aldebaran	291 01.9	N16 30.3
Alioth	166 30.9	N55 57.8
Alkaid	153 08.2	N49 19.1
Al Na'ir	27 57.9	S46 58.2
Alnilam	275 57.5	S 1 12.2
Alphard	218 07.0	S 8 39.2
Alphecca	126 20.9	N26 43.2
Alpheratz	357 55.0	N29 05.2
Altair	62 19.4	N 8 52.1
Ankaa	353 26.6	S42 19.0
Antares	112 40.4	S26 25.6
Arcturus	146 06.3	N19 11.4
Atria	107 53.0	S69 01.4
Avior	234 22.2	S59 30.2
Bellatrix	278 43.7	N 6 20.8
Betelgeuse	271 13.2	N 7 24.3
Canopus	264 00.6	S52 41.7
Capella	280 50.6	N45 59.7
Deneb	49 39.3	N45 16.9
Denebola	182 45.2	N14 34.7
Diphda	349 07.0	S17 59.7
Dubhe	194 05.6	N61 45.2
Elnath	278 26.5	N28 36.3
Eltanin	90 51.9	N51 29.6
Enif	33 58.2	N 9 52.3
Fomalhaut	15 36.4	S29 37.8
Gacrux	172 13.7	S57 06.1
Gienah	176 04.0	S17 31.9
Hadar	149 04.4	S60 21.7
Hamal	328 13.2	N23 27.4
Kaus Aust.	83 59.1	S34 23.0
Kochab	137 20.8	N74 09.6
Markab	13 49.5	N15 12.1
Menkar	314 26.5	N 4 05.0
Menkent	148 21.2	S36 21.6
Miaplacidus	221 41.6	S69 42.5
Mirfak	308 56.0	N49 51.4
Nunki	76 12.6	S26 17.8
Peacock	53 37.3	S56 44.5
Pollux	243 41.2	N28 01.6
Procyon	245 11.2	N 5 13.6
Rasalhague	96 17.2	N12 33.9
Regulus	207 55.4	N11 58.3
Rigel	281 22.5	S 8 12.3
Rigil Kent.	140 07.7	S60 49.5
Sabik	102 25.8	S15 43.2
Schedar	349 53.0	N56 32.1
Shaula	96 37.6	S37 06.0
Sirius	258 43.3	S16 42.9
Spica	158 43.3	S11 09.1
Suhail	223 00.5	S43 25.5
Vega	80 47.0	N38 47.2
Zuben'ubi	137 18.2	S16 02.0

	SHA	Mer. Pass.
Venus	104 43.2	12 23
Mars	177 29.6	7 31
Jupiter	9 52.3	18 39
Saturn	333 29.6	21 04

UT	SUN GHA	SUN Dec	MOON GHA	v	MOON Dec	d	HP
d h	° ′	° ′	° ′	′	° ′	′	′
30 00	182 53.4	S21 34.2	51 57.6	8.8	N 2 53.2	12.0	60.2
01	197 53.2	34.6	66 25.4	8.8	3 05.2	12.1	60.3
02	212 53.0	35.0	80 53.2	8.8	3 17.3	12.1	60.3
03	227 52.8	35.4	95 21.0	8.7	3 29.4	12.0	60.3
04	242 52.6	35.9	109 48.7	8.6	3 41.4	12.1	60.3
05	257 52.3	36.3	124 16.3	8.6	3 53.5	12.0	60.4
06	272 52.1	S21 36.7	138 43.9	8.6	N 4 05.5	12.0	60.4
M 07	287 51.9	37.1	153 11.5	8.5	4 17.5	12.0	60.4
O 08	302 51.7	37.5	167 39.0	8.4	4 29.5	12.0	60.4
N 09	317 51.4	37.9	182 06.4	8.5	4 41.5	12.0	60.5
D 10	332 51.2	38.3	196 33.9	8.3	4 53.5	11.9	60.5
A 11	347 51.0	38.7	211 01.2	8.3	5 05.4	11.9	60.5
Y 12	2 50.8	S21 39.1	225 28.5	8.3	N 5 17.3	11.9	60.5
13	17 50.5	39.5	239 55.8	8.2	5 29.2	11.9	60.5
14	32 50.3	39.9	254 23.0	8.1	5 41.1	11.8	60.6
15	47 50.1	40.3	268 50.1	8.1	5 52.9	11.9	60.6
16	62 49.9	40.7	283 17.2	8.1	6 04.8	11.8	60.6
17	77 49.7	41.1	297 44.3	7.9	6 16.6	11.7	60.6
18	92 49.4	S21 41.5	312 11.2	8.0	N 6 28.3	11.8	60.7
19	107 49.2	41.9	326 38.2	7.9	6 40.1	11.7	60.7
20	122 48.9	42.3	341 05.1	7.8	6 51.8	11.6	60.7
21	137 48.7	42.7	355 31.9	7.8	7 03.4	11.6	60.7
22	152 48.5	43.1	9 58.7	7.7	7 15.0	11.6	60.7
23	167 48.3	43.5	24 25.4	7.6	7 26.6	11.6	60.7
1 00	182 48.0	S21 43.9	38 52.0	7.6	N 7 38.2	11.5	60.8
01	197 47.8	44.3	53 18.6	7.6	7 49.7	11.4	60.8
02	212 47.6	44.7	67 45.2	7.5	8 01.1	11.5	60.8
03	227 47.3	45.1	82 11.7	7.4	8 12.6	11.3	60.8
04	242 47.1	45.5	96 38.1	7.4	8 23.9	11.4	60.8
05	257 46.9	45.9	111 04.5	7.3	8 35.3	11.2	60.9
06	272 46.7	S21 46.3	125 30.8	7.3	N 8 46.5	11.3	60.9
T 07	287 46.4	46.7	139 57.1	7.2	8 57.8	11.1	60.9
U 08	302 46.2	47.1	154 23.3	7.2	9 08.9	11.2	60.9
E 09	317 46.0	47.5	168 49.5	7.1	9 20.1	11.0	60.9
S 10	332 45.7	47.8	183 15.6	7.0	9 31.1	11.0	60.9
D 11	347 45.5	48.2	197 41.6	7.0	9 42.1	11.0	60.9
A 12	2 45.3	S21 48.6	212 07.6	6.9	N 9 53.1	10.9	61.0
Y 13	17 45.0	49.0	226 33.5	6.9	10 04.0	10.8	61.0
14	32 44.8	49.4	240 59.4	6.8	10 14.8	10.8	61.0
15	47 44.6	49.8	255 25.2	6.7	10 25.6	10.7	61.0
16	62 44.3	50.2	269 50.9	6.7	10 36.3	10.6	61.0
17	77 44.1	50.5	284 16.6	6.6	10 46.9	10.6	61.0
18	92 43.9	S21 50.9	298 42.2	6.6	N10 57.5	10.5	61.0
19	107 43.6	51.3	313 07.8	6.5	11 08.0	10.4	61.0
20	122 43.4	51.7	327 33.3	6.5	11 18.4	10.3	61.0
21	137 43.2	52.1	341 58.8	6.4	11 28.7	10.3	61.0
22	152 42.9	52.5	356 24.2	6.4	11 39.0	10.2	61.1
23	167 42.7	52.8	10 49.6	6.2	11 49.2	10.1	61.1
2 00	182 42.5	S21 53.2	25 14.8	6.3	N11 59.3	10.1	61.1
01	197 42.2	53.6	39 40.1	6.2	12 09.4	9.9	61.1
02	212 42.0	54.0	54 05.3	6.1	12 19.3	9.9	61.1
03	227 41.8	54.3	68 30.4	6.1	12 29.2	9.8	61.1
04	242 41.5	54.7	82 55.5	6.0	12 39.0	9.8	61.1
05	257 41.3	55.1	97 20.5	5.9	12 48.8	9.6	61.1
06	272 41.0	S21 55.5	111 45.4	5.9	N12 58.4	9.5	61.1
W 07	287 40.8	55.8	126 10.3	5.9	13 07.9	9.5	61.1
E 08	302 40.6	56.2	140 35.2	5.8	13 17.4	9.3	61.1
D 09	317 40.3	56.6	155 00.0	5.7	13 26.7	9.3	61.1
N 10	332 40.1	57.0	169 24.7	5.7	13 36.0	9.2	61.1
E 11	347 39.9	57.3	183 49.4	5.7	13 45.2	9.1	61.1
S 12	2 39.6	S21 57.7	198 14.1	5.6	N13 54.3	9.0	61.1
D 13	17 39.4	58.1	212 38.7	5.5	14 03.3	8.9	61.1
A 14	32 39.1	58.4	227 03.2	5.5	14 12.2	8.7	61.1
Y 15	47 38.9	58.8	241 27.7	5.5	14 20.9	8.7	61.1
16	62 38.7	59.2	255 52.2	5.4	14 29.6	8.6	61.1
17	77 38.4	59.5	270 16.6	5.3	14 38.2	8.5	61.1
18	92 38.2	S21 59.9	284 40.9	5.3	N14 46.7	8.4	61.1
19	107 37.9	22 00.3	299 05.2	5.3	14 55.1	8.3	61.1
20	122 37.7	00.6	313 29.5	5.2	15 03.4	8.1	61.1
21	137 37.5	01.0	327 53.7	5.1	15 11.5	8.1	61.1
22	152 37.2	01.4	342 17.8	5.2	15 19.6	7.9	61.1
23	167 37.0	01.7	356 42.0	5.0	N15 27.5	7.9	61.1
	SD 16.2	d 0.4	SD 16.5		16.6		16.6

Lat.	Twilight Naut.	Twilight Civil	Sunrise	Moonrise 30	1	2	3
°	h m	h m	h m	h m	h m	h m	h m
N 72	07 50	09 49	■■■	13 53	13 44	13 31	13 08
N 70	07 34	09 10	■■■	14 01	13 59	13 58	14 00
68	07 21	08 43	10 35	14 07	14 12	14 19	14 32
66	07 10	08 22	09 46	14 13	14 22	14 36	14 56
64	07 01	08 06	09 15	14 17	14 31	14 49	15 15
62	06 53	07 52	08 53	14 21	14 39	15 01	15 31
60	06 45	07 40	08 34	14 24	14 45	15 11	15 44
N 58	06 39	07 30	08 19	14 28	14 51	15 19	15 55
56	06 33	07 21	08 06	14 30	14 56	15 27	16 05
54	06 28	07 13	07 55	14 33	15 01	15 34	16 14
52	06 23	07 06	07 45	14 35	15 05	15 40	16 21
50	06 18	06 59	07 36	14 37	15 09	15 46	16 29
45	06 08	06 45	07 18	14 42	15 17	15 58	16 44
N 40	05 59	06 32	07 02	14 45	15 24	16 08	16 56
35	05 50	06 22	06 49	14 49	15 30	16 16	17 07
30	05 42	06 12	06 38	14 52	15 36	16 24	17 16
20	05 28	05 55	06 18	14 57	15 45	16 37	17 32
N 10	05 13	05 39	06 01	15 01	15 54	16 49	17 46
0	04 57	05 23	05 45	15 06	16 01	17 00	18 00
S 10	04 40	05 06	05 29	15 10	16 09	17 11	18 13
20	04 19	04 47	05 12	15 15	16 18	17 22	18 28
30	03 52	04 24	04 51	15 20	16 27	17 36	18 44
35	03 35	04 10	04 39	15 23	16 33	17 44	18 54
40	03 14	03 54	04 26	15 27	16 40	17 53	19 05
45	02 46	03 33	04 09	15 31	16 47	18 04	19 18
S 50	02 07	03 06	03 49	15 36	16 56	18 17	19 34
52	01 45	02 53	03 39	15 38	17 01	18 23	19 41
54	01 15	02 38	03 28	15 41	17 05	18 29	19 50
56	00 19	02 19	03 16	15 43	17 10	18 37	19 59
58	////	01 56	03 01	15 47	17 16	18 45	20 10
S 60	////	01 24	02 44	15 50	17 23	18 55	20 22

Lat.	Sunset	Twilight Civil	Twilight Naut.	Moonset 30	1	2	3
°	h m	h m	h m	h m	h m	h m	h m
N 72	■■■	13 48	15 47	03 21	05 26	07 39	10 07
N 70	■■■	14 27	16 03	03 16	05 12	07 13	09 16
68	13 03	14 54	16 16	03 12	05 02	06 54	08 45
66	13 51	15 15	16 27	03 09	04 53	06 38	08 22
64	14 22	15 32	16 37	03 06	04 45	06 26	08 03
62	14 45	15 45	16 45	03 04	04 39	06 15	07 48
60	15 03	15 57	16 52	03 02	04 33	06 06	07 36
N 58	15 18	16 07	16 59	03 00	04 28	05 58	07 25
56	15 31	16 17	17 04	02 58	04 24	05 51	07 16
54	15 42	16 25	17 10	02 57	04 20	05 45	07 07
52	15 52	16 32	17 15	02 55	04 17	05 39	07 00
50	16 01	16 39	17 19	02 54	04 14	05 34	06 53
45	16 20	16 53	17 30	02 51	04 07	05 23	06 39
N 40	16 35	17 05	17 39	02 49	04 01	05 14	06 27
35	16 48	17 16	17 47	02 47	03 56	05 07	06 17
30	17 00	17 26	17 55	02 45	03 52	05 00	06 09
20	17 19	17 43	18 10	02 42	03 44	04 48	05 53
N 10	17 36	17 59	18 25	02 40	03 38	04 38	05 40
0	17 53	18 15	18 41	02 37	03 32	04 29	05 28
S 10	18 09	18 32	18 58	02 35	03 26	04 19	05 16
20	18 27	18 51	19 20	02 32	03 19	04 09	05 03
30	18 47	19 14	19 47	02 29	03 12	03 58	04 48
35	18 59	19 28	20 04	02 27	03 07	03 51	04 39
40	19 13	19 45	20 25	02 25	03 03	03 43	04 29
45	19 29	20 05	20 53	02 23	02 57	03 35	04 18
S 50	19 50	20 32	21 32	02 20	02 50	03 24	04 04
52	20 00	20 46	21 55	02 17	02 47	03 19	03 57
54	20 11	21 01	22 25	02 17	02 44	03 14	03 50
56	20 23	21 20	23 29	02 16	02 40	03 08	03 42
58	20 38	21 44	////	02 14	02 36	03 02	03 33
S 60	20 55	22 17	////	02 12	02 32	02 54	03 23

	SUN Eqn. of Time 00h	SUN Eqn. of Time 12h	SUN Mer. Pass.	MOON Mer. Pass. Upper	MOON Mer. Pass. Lower	Age	Phase
Day							
d	m s	m s	h m	h m	h m	d	%
30	11 34	11 23	11 49	21 19	08 51	11	86
1	11 13	11 02	11 49	22 15	09 46	12	93
2	10 50	10 39	11 49	23 14	10 44	13	98

1998 DECEMBER 3, 4, 5 (THURS., FRI., SAT.)

UT	ARIES GHA	VENUS −3.9 GHA	Dec	MARS +1.3 GHA	Dec	JUPITER −2.5 GHA	Dec	SATURN +0.2 GHA	Dec
d h	° ′	° ′	° ′	° ′	° ′	° ′	° ′	° ′	° ′
3 00	71 37.2	173 37.7	S23 19.6	248 05.4	N 0 20.1	81 22.8	S 5 40.1	45 12.4	N 8 03.1
01	86 39.7	188 36.8	20.0	263 06.6	19.6	96 25.1	40.0	60 15.0	03.0
02	101 42.1	203 35.8	20.3	278 07.8	19.0	111 27.4	40.0	75 17.6	03.0
03	116 44.6	218 34.9 ..	20.7	293 09.0 ..	18.5	126 29.7 ..	39.9	90 20.1 ..	03.0
04	131 47.1	233 34.0	21.0	308 10.2	18.0	141 32.0	39.8	105 22.7	02.9
05	146 49.5	248 33.0	21.4	323 11.4	17.4	156 34.3	39.7	120 25.3	02.9
T 06	161 52.0	263 32.1	S23 21.7	338 12.5	N 0 16.9	171 36.6	S 5 39.7	135 27.9	N 8 02.9
H 07	176 54.4	278 31.1	22.1	353 13.7	16.4	186 39.0	39.6	150 30.4	02.8
U 08	191 56.9	293 30.2	22.4	8 14.9	15.8	201 41.3	39.5	165 33.0	02.8
R 09	206 59.4	308 29.3 ..	22.8	23 16.1 ..	15.3	216 43.6 ..	39.4	180 35.6 ..	02.8
S 10	222 01.8	323 28.3	23.1	38 17.3	14.8	231 45.9	39.4	195 38.2	02.7
D 11	237 04.3	338 27.4	23.5	53 18.5	14.2	246 48.2	39.3	210 40.7	02.7
A 12	252 06.8	353 26.5	S23 23.8	68 19.7	N 0 13.7	261 50.5	S 5 39.2	225 43.3	N 8 02.7
Y 13	267 09.2	8 25.5	24.2	83 20.9	13.2	276 52.8	39.1	240 45.9	02.7
14	282 11.7	23 24.6	24.5	98 22.1	12.6	291 55.2	39.1	255 48.5	02.6
15	297 14.2	38 23.7 ..	24.8	113 23.3 ..	12.1	306 57.5 ..	39.0	270 51.0 ..	02.6
16	312 16.6	53 22.7	25.2	128 24.5	11.6	321 59.8	38.9	285 53.6	02.6
17	327 19.1	68 21.8	25.5	143 25.6	11.0	337 02.1	38.8	300 56.2	02.5
18	342 21.6	83 20.8	S23 25.9	158 26.8	N 0 10.5	352 04.4	S 5 38.8	315 58.8	N 8 02.5
19	357 24.0	98 19.9	26.2	173 28.0	10.0	7 06.7	38.7	331 01.3	02.5
20	12 26.5	113 19.0	26.5	188 29.2	09.4	22 09.0	38.6	346 03.9	02.4
21	27 28.9	128 18.0 ..	26.9	203 30.4 ..	08.9	37 11.3 ..	38.5	1 06.5 ..	02.4
22	42 31.4	143 17.1	27.2	218 31.6	08.4	52 13.7	38.5	16 09.1	02.4
23	57 33.9	158 16.1	27.5	233 32.8	07.8	67 16.0	38.4	31 11.6	02.3
4 00	72 36.3	173 15.2	S23 27.8	248 34.0	N 0 07.3	82 18.3	S 5 38.3	46 14.2	N 8 02.3
01	87 38.8	188 14.3	28.2	263 35.2	06.8	97 20.6	38.2	61 16.8	02.3
02	102 41.3	203 13.3	28.5	278 36.4	06.2	112 22.9	38.2	76 19.4	02.2
03	117 43.7	218 12.4 ..	28.8	293 37.6 ..	05.7	127 25.2 ..	38.1	91 21.9 ..	02.2
04	132 46.2	233 11.4	29.2	308 38.8	05.2	142 27.5	38.0	106 24.5	02.2
05	147 48.7	248 10.5	29.5	323 40.0	04.6	157 29.8	37.9	121 27.1	02.2
F 06	162 51.1	263 09.6	S23 29.8	338 41.1	N 0 04.1	172 32.1	S 5 37.8	136 29.7	N 8 02.1
R 07	177 53.6	278 08.6	30.1	353 42.3	03.6	187 34.4	37.8	151 32.2	02.1
I 08	192 56.1	293 07.7	30.4	8 43.5	03.0	202 36.7	37.7	166 34.8	02.1
D 09	207 58.5	308 06.7 ..	30.8	23 44.7 ..	02.5	217 39.0 ..	37.6	181 37.4 ..	02.0
A 10	223 01.0	323 05.8	31.1	38 45.9	02.0	232 41.4	37.5	196 39.9	02.0
Y 11	238 03.4	338 04.8	31.4	53 47.1	01.5	247 43.7	37.5	211 42.5	02.0
12	253 05.9	353 03.9	S23 31.7	68 48.3	N 0 00.9	262 46.0	S 5 37.4	226 45.1	N 8 01.9
13	268 08.4	8 03.0	32.0	83 49.5	N 00.4	277 48.3	37.3	241 47.7	01.9
14	283 10.8	23 02.0	32.3	98 50.7	S 00.1	292 50.6	37.2	256 50.2	01.9
15	298 13.3	38 01.1 ..	32.6	113 51.9 ..	00.7	307 52.9 ..	37.1	271 52.8 ..	01.9
16	313 15.8	53 00.1	33.0	128 53.1	01.2	322 55.2	37.1	286 55.4	01.8
17	328 18.2	67 59.2	33.3	143 54.3	01.7	337 57.5	37.0	301 57.9	01.8
18	343 20.7	82 58.2	S23 33.6	158 55.5	S 0 02.3	352 59.8	S 5 36.9	317 00.5	N 8 01.8
19	358 23.2	97 57.3	33.9	173 56.7	02.8	8 02.1	36.8	332 03.1	01.7
20	13 25.6	112 56.3	34.2	188 57.9	03.3	23 04.4	36.7	347 05.7	01.7
21	28 28.1	127 55.4 ..	34.5	203 59.1 ..	03.9	38 06.7 ..	36.7	2 08.2 ..	01.7
22	43 30.6	142 54.5	34.8	219 00.3	04.4	53 09.0	36.6	17 10.8	01.6
23	58 33.0	157 53.5	35.1	234 01.5	04.9	68 11.3	36.5	32 13.4	01.6
5 00	73 35.5	172 52.6	S23 35.4	249 02.7	S 0 05.4	83 13.6	S 5 36.4	47 15.9	N 8 01.6
01	88 37.9	187 51.6	35.7	264 03.8	06.0	98 15.9	36.3	62 18.5	01.6
02	103 40.4	202 50.7	36.0	279 05.0	06.5	113 18.2	36.3	77 21.1	01.5
03	118 42.9	217 49.7 ..	36.3	294 06.2 ..	07.0	128 20.5 ..	36.2	92 23.6 ..	01.5
04	133 45.3	232 48.8	36.6	309 07.4	07.6	143 22.8	36.1	107 26.2	01.5
05	148 47.8	247 47.8	36.9	324 08.6	08.1	158 25.1	36.0	122 28.8	01.4
S 06	163 50.3	262 46.9	S23 37.2	339 09.8	S 0 08.6	173 27.4	S 5 35.9	137 31.4	N 8 01.4
A 07	178 52.7	277 45.9	37.5	354 11.0	09.2	188 29.7	35.9	152 33.9	01.4
T 08	193 55.2	292 45.0	37.7	9 12.2	09.7	203 32.0	35.8	167 36.5	01.4
U 09	208 57.7	307 44.0 ..	38.0	24 13.4 ..	10.2	218 34.3 ..	35.7	182 39.1 ..	01.3
R 10	224 00.1	322 43.1	38.3	39 14.6	10.7	233 36.6	35.6	197 41.6	01.3
D 11	239 02.6	337 42.1	38.6	54 15.8	11.3	248 38.9	35.5	212 44.2	01.3
A 12	254 05.0	352 41.2	S23 38.9	69 17.0	S 0 11.8	263 41.2	S 5 35.4	227 46.8	N 8 01.2
Y 13	269 07.5	7 40.2	39.2	84 18.2	12.3	278 43.5	35.4	242 49.3	01.2
14	284 10.0	22 39.3	39.5	99 19.4	12.9	293 45.8	35.3	257 51.9	01.2
15	299 12.4	37 38.3 ..	39.7	114 20.6 ..	13.4	308 48.1 ..	35.2	272 54.5 ..	01.2
16	314 14.9	52 37.4	40.0	129 21.8	13.9	323 50.4	35.1	287 57.0	01.1
17	329 17.4	67 36.4	40.3	144 23.0	14.4	338 52.7	35.1	302 59.6	01.1
18	344 19.8	82 35.5	S23 40.6	159 24.2	S 0 15.0	353 55.0	S 5 35.0	318 02.2	N 8 01.1
19	359 22.3	97 34.5	40.8	174 25.4	15.5	8 57.3	34.9	333 04.7	01.0
20	14 24.8	112 33.6	41.1	189 26.6	16.0	23 59.6	34.8	348 07.3	01.0
21	29 27.2	127 32.6 ..	41.4	204 27.8 ..	16.6	39 01.9 ..	34.7	3 09.9 ..	01.0
22	44 29.7	142 31.7	41.7	219 29.0	17.1	54 04.2	34.6	18 12.4	01.0
23	59 32.2	157 30.7	41.9	234 30.2	17.6	69 06.5	34.6	33 15.0	00.9
Mer. Pass. 19 06.4		v −0.9 d 0.3		v 1.2 d 0.5		v 2.3 d 0.1		v 2.6 d 0.0	

STARS

Name	SHA	Dec
Acamar	315 26.5	S40 18.8
Achernar	335 34.8	S57 14.8
Acrux	173 22.1	S63 05.2
Adhara	255 21.0	S28 58.2
Aldebaran	291 01.9	N16 30.3
Alioth	166 30.9	N55 57.8
Alkaid	153 08.2	N49 19.1
Al Na'ir	27 57.9	S46 58.2
Alnilam	275 57.4	S 1 12.3
Alphard	218 07.0	S 8 39.2
Alphecca	126 20.9	N26 43.2
Alpheratz	357 55.0	N29 05.2
Altair	62 19.4	N 8 52.1
Ankaa	353 26.6	S42 19.0
Antares	112 40.4	S26 25.6
Arcturus	146 06.3	N19 11.4
Atria	107 52.9	S69 01.4
Avior	234 22.1	S59 30.2
Bellatrix	278 43.7	N 6 20.8
Betelgeuse	271 13.1	N 7 24.3
Canopus	264 00.6	S52 41.7
Capella	280 50.6	N45 59.7
Deneb	49 39.4	N45 16.9
Denebola	182 45.2	N14 34.7
Diphda	349 07.0	S17 59.7
Dubhe	194 05.5	N61 45.2
Elnath	278 26.4	N28 36.3
Eltanin	90 51.9	N51 29.6
Enif	33 58.2	N 9 52.3
Fomalhaut	15 36.4	S29 37.8
Gacrux	172 13.6	S57 06.1
Gienah	176 04.0	S17 31.9
Hadar	149 04.3	S60 21.7
Hamal	328 13.2	N23 27.4
Kaus Aust.	83 59.1	S34 23.0
Kochab	137 20.8	N74 09.6
Markab	13 49.5	N15 12.1
Menkar	314 26.5	N 4 05.0
Menkent	148 21.1	S36 21.6
Miaplacidus	221 41.5	S69 42.5
Mirfak	308 56.0	N49 51.4
Nunki	76 12.6	S26 17.8
Peacock	53 37.3	S56 44.5
Pollux	243 41.2	N28 01.6
Procyon	245 11.2	N 5 13.6
Rasalhague	96 17.2	N12 33.8
Regulus	207 55.4	N11 58.3
Rigel	281 22.5	S 8 12.3
Rigil Kent.	140 07.7	S60 49.5
Sabik	102 25.8	S15 43.2
Schedar	349 53.0	N56 32.1
Shaula	96 37.6	S37 06.0
Sirius	258 43.3	S16 42.9
Spica	158 43.3	S11 09.1
Suhail	223 00.5	S43 25.5
Vega	80 47.0	N38 47.2
Zuben'ubi	137 18.2	S16 02.0

	SHA	Mer.Pass.
	° ′	h m
Venus	100 38.9	12 28
Mars	175 57.7	7 25
Jupiter	9 41.9	18 28
Saturn	333 37.9	20 51

UT	SUN GHA	SUN Dec	MOON GHA	v	MOON Dec	d	HP
d h	° '	° '	° '	'	° '	'	'
3 00	182 36.7	S22 02.1	11 06.0	5.1	N15 35.4	7.7	61.1
01	197 36.5	02.5	25 30.1	5.0	15 43.1	7.6	61.1
02	212 36.3	02.8	39 54.1	5.0	15 50.7	7.5	61.1
03	227 36.0	.. 03.2	54 18.1	4.9	15 58.2	7.4	61.0
04	242 35.8	03.5	68 42.0	4.9	16 05.6	7.3	61.0
05	257 35.5	03.9	83 05.9	4.8	16 12.9	7.1	61.0
06	272 35.3	S22 04.2	97 29.7	4.8	N16 20.0	7.0	61.0
T 07	287 35.0	04.6	111 53.5	4.8	16 27.0	6.9	61.0
H 08	302 34.8	05.0	126 17.3	4.8	16 33.9	6.8	61.0
U 09	317 34.5	.. 05.3	140 41.1	4.7	16 40.7	6.7	61.0
R 10	332 34.3	05.7	155 04.8	4.7	16 47.4	6.5	61.0
S 11	347 34.1	06.0	169 28.5	4.6	16 53.9	6.5	61.0
D 12	2 33.8	S22 06.4	183 52.1	4.7	N17 00.4	6.3	61.0
A 13	17 33.6	06.7	198 15.8	4.6	17 06.7	6.1	60.9
Y 14	32 33.3	07.1	212 39.4	4.5	17 12.8	6.1	60.9
15	47 33.1	.. 07.4	227 02.9	4.6	17 18.9	5.9	60.9
16	62 32.8	07.8	241 26.5	4.5	17 24.8	5.8	60.9
17	77 32.6	08.1	255 50.0	4.5	17 30.6	5.6	60.9
18	92 32.3	S22 08.5	270 13.5	4.5	N17 36.2	5.6	60.9
19	107 32.1	08.8	284 37.0	4.5	17 41.8	5.4	60.9
20	122 31.8	09.2	299 00.5	4.5	17 47.2	5.2	60.8
21	137 31.6	.. 09.5	313 24.0	4.4	17 52.4	5.2	60.8
22	152 31.4	09.9	327 47.4	4.4	17 57.6	5.0	60.8
23	167 31.1	10.2	342 10.8	4.5	18 02.6	4.8	60.8
4 00	182 30.9	S22 10.5	356 34.3	4.4	N18 07.4	4.8	60.8
01	197 30.6	10.9	10 57.7	4.4	18 12.2	4.6	60.7
02	212 30.4	11.2	25 21.1	4.4	18 16.8	4.4	60.7
03	227 30.1	.. 11.6	39 44.5	4.3	18 21.2	4.4	60.7
04	242 29.9	11.9	54 07.8	4.4	18 25.6	4.2	60.7
05	257 29.6	12.3	68 31.2	4.4	18 29.8	4.0	60.7
06	272 29.4	S22 12.6	82 54.6	4.4	N18 33.8	3.9	60.6
F 07	287 29.1	12.9	97 18.0	4.3	18 37.7	3.8	60.6
R 08	302 28.9	13.3	111 41.3	4.4	18 41.5	3.7	60.6
I 09	317 28.6	.. 13.6	126 04.7	4.4	18 45.2	3.5	60.6
D 10	332 28.4	13.9	140 28.1	4.4	18 48.7	3.3	60.5
A 11	347 28.1	14.3	154 51.5	4.4	18 52.0	3.3	60.5
Y 12	2 27.9	S22 14.6	169 14.9	4.4	N18 55.3	3.1	60.5
13	17 27.6	14.9	183 38.3	4.4	18 58.4	2.9	60.5
14	32 27.4	15.3	198 01.7	4.4	19 01.3	2.8	60.5
15	47 27.1	.. 15.6	212 25.1	4.4	19 04.1	2.7	60.4
16	62 26.9	15.9	226 48.5	4.5	19 06.8	2.6	60.4
17	77 26.6	16.3	241 12.0	4.5	19 09.4	2.4	60.4
18	92 26.3	S22 16.6	255 35.5	4.4	N19 11.8	2.2	60.3
19	107 26.1	16.9	269 58.9	4.5	19 14.0	2.2	60.3
20	122 25.8	17.3	284 22.4	4.6	19 16.2	1.9	60.3
21	137 25.6	.. 17.6	298 46.0	4.5	19 18.1	1.9	60.3
22	152 25.3	17.9	313 09.5	4.6	19 20.0	1.7	60.2
23	167 25.1	18.2	327 33.1	4.6	19 21.7	1.6	60.2
5 00	182 24.8	S22 18.6	341 56.7	4.6	N19 23.3	1.4	60.2
01	197 24.6	18.9	356 20.3	4.7	19 24.7	1.3	60.1
02	212 24.3	19.2	10 44.0	4.7	19 26.0	1.1	60.1
03	227 24.1	.. 19.5	25 07.7	4.7	19 27.1	1.0	60.1
04	242 23.8	19.9	39 31.4	4.7	19 28.1	0.9	60.1
05	257 23.6	20.2	53 55.1	4.8	19 29.0	0.8	60.0
06	272 23.3	S22 20.5	68 18.9	4.8	N19 29.8	0.6	60.0
S 07	287 23.0	20.8	82 42.7	4.9	19 30.4	0.4	60.0
A 08	302 22.8	21.1	97 06.6	4.9	19 30.8	0.4	59.9
T 09	317 22.5	.. 21.5	111 30.5	5.0	19 31.2	0.1	59.9
U 10	332 22.3	21.8	125 54.5	5.0	19 31.3	0.1	59.9
R 11	347 22.0	22.1	140 18.5	5.0	19 31.4	0.1	59.8
D 12	2 21.8	S22 22.4	154 42.5	5.1	N19 31.3	0.2	59.8
A 13	17 21.5	22.7	169 06.6	5.1	19 31.1	0.3	59.8
Y 14	32 21.2	23.1	183 30.7	5.2	19 30.8	0.5	59.7
15	47 21.0	.. 23.4	197 54.9	5.2	19 30.3	0.6	59.7
16	62 20.7	23.7	212 19.1	5.3	19 29.7	0.8	59.7
17	77 20.5	24.0	226 43.4	5.3	19 28.9	0.8	59.6
18	92 20.2	S22 24.3	241 07.7	5.4	N19 28.1	1.0	59.6
19	107 20.0	24.6	255 32.1	5.5	19 27.1	1.2	59.6
20	122 19.7	24.9	269 56.6	5.5	19 25.9	1.3	59.5
21	137 19.4	.. 25.2	284 21.1	5.5	19 24.6	1.4	59.5
22	152 19.2	25.5	298 45.6	5.7	19 23.2	1.5	59.4
23	167 18.9	25.9	313 10.3	5.7	N19 21.7	1.6	59.4
	SD 16.3	d 0.3	SD 16.6		16.5		16.3

Lat.	Naut. (Twilight)	Civil (Twilight)	Sunrise	Moonrise 3	4	5	6
°	h m	h m	h m	h m	h m	h m	h m
N 72	07 58	10 03	■■■	13 08	□	□	□
N 70	07 41	09 20	■■■	14 00	14 09	14 49	16 22
68	07 27	08 51	10 54	14 32	14 59	15 49	17 08
66	07 16	08 29	09 56	14 56	15 30	16 24	17 38
64	07 06	08 12	09 23	15 15	15 54	16 49	18 00
62	06 57	07 57	08 59	15 31	16 12	17 09	18 18
60	06 50	07 45	08 40	15 44	16 28	17 25	18 33
N 58	06 43	07 35	08 25	15 55	16 41	17 38	18 45
56	06 37	07 25	08 11	16 05	16 52	17 50	18 56
54	06 32	07 17	08 00	16 14	17 02	18 00	19 06
52	06 26	07 09	07 49	16 21	17 11	18 09	19 14
50	06 22	07 03	07 40	16 29	17 19	18 18	19 22
45	06 11	06 48	07 21	16 44	17 36	18 35	19 38
N 40	06 01	06 35	07 05	16 56	17 50	18 49	19 52
35	05 53	06 24	06 52	17 07	18 02	19 01	20 03
30	05 45	06 14	06 40	17 16	18 13	19 12	20 13
20	05 29	05 57	06 20	17 32	18 30	19 30	20 30
N 10	05 14	05 40	06 03	17 46	18 46	19 46	20 45
0	04 58	05 24	05 46	18 00	19 01	20 01	20 59
S 10	04 40	05 07	05 30	18 13	19 16	20 16	21 12
20	04 19	04 48	05 12	18 28	19 31	20 32	21 27
30	03 51	04 24	04 51	18 44	19 49	20 50	21 44
35	03 34	04 10	04 39	18 54	20 00	21 01	21 54
40	03 12	03 53	04 25	19 05	20 12	21 13	22 05
45	02 44	03 32	04 08	19 18	20 27	21 27	22 19
S 50	02 03	03 04	03 47	19 34	20 44	21 45	22 35
52	01 40	02 51	03 37	19 41	20 52	21 53	22 42
54	01 08	02 34	03 26	19 50	21 02	22 02	22 51
56	////	02 15	03 13	19 59	21 12	22 13	23 00
58	////	01 50	02 58	20 10	21 24	22 24	23 11
S 60	////	01 15	02 40	20 22	21 38	22 38	23 23

Lat.	Sunset	Civil (Twilight)	Naut. (Twilight)	Moonset 3	4	5	6
°	h m	h m	h m	h m	h m	h m	h m
N 72	■■■	13 37	15 41	10 07	□	□	□
N 70	■■■	14 20	15 59	09 16	11 15	12 42	13 12
68	12 46	14 49	16 12	08 45	10 26	11 42	12 26
66	13 43	15 11	16 24	08 22	09 55	11 08	11 55
64	14 16	15 28	16 34	08 03	09 32	10 42	11 32
62	14 41	15 42	16 42	07 48	09 13	10 23	11 14
60	15 00	15 55	16 50	07 36	08 58	10 06	10 59
N 58	15 15	16 05	16 57	07 25	08 45	09 53	10 46
56	15 29	16 15	17 03	07 16	08 34	09 41	10 35
54	15 40	16 23	17 08	07 07	08 24	09 31	10 25
52	15 51	16 31	17 14	07 00	08 15	09 22	10 17
50	16 00	16 37	17 18	06 53	08 07	09 13	10 09
45	16 19	16 52	17 29	06 39	07 51	08 56	09 52
N 40	16 35	17 05	17 39	06 27	07 37	08 42	09 38
35	16 48	17 16	17 47	06 17	07 26	08 29	09 27
30	17 00	17 26	17 55	06 09	07 16	08 19	09 16
20	17 20	17 44	18 11	05 53	06 58	08 00	08 59
N 10	17 37	18 00	18 26	05 40	06 43	07 45	08 43
0	17 54	18 16	18 42	05 28	06 29	07 30	08 29
S 10	18 10	18 33	19 00	05 16	06 15	07 15	08 14
20	18 28	18 53	19 22	05 03	05 59	06 59	07 59
30	18 49	19 16	19 49	04 48	05 42	06 40	07 41
35	19 02	19 31	20 07	04 39	05 32	06 30	07 30
40	19 16	19 48	20 29	04 29	05 20	06 17	07 18
45	19 33	20 09	20 57	04 18	05 07	06 03	07 04
S 50	19 54	20 37	21 38	04 04	04 50	05 45	06 47
52	20 04	20 51	22 02	03 57	04 43	05 37	06 39
54	20 15	21 07	22 35	03 50	04 34	05 27	06 30
56	20 28	21 27	////	03 42	04 24	05 17	06 19
58	20 43	21 52	////	03 33	04 13	05 05	06 08
S 60	21 02	22 28	////	03 23	04 01	04 51	05 54

Day	SUN Eqn. of Time 00h	12h	Mer. Pass.	MOON Mer. Pass. Upper	Lower	Age	Phase
d	m s	m s	h m	h m	h m	d	%
3	10 27	10 16	11 50	24 14	11 44	14	100
4	10 04	09 52	11 50	00 14	12 45	15	99
5	09 40	09 28	11 51	01 15	13 45	16	95

(Phase: ◯)

1998 DECEMBER 6, 7, 8 (SUN., MON., TUES.)

UT	ARIES GHA	VENUS −3.9 GHA	Dec	MARS +1.3 GHA	Dec	JUPITER −2.5 GHA	Dec	SATURN +0.2 GHA	Dec	Star Name	SHA	Dec
d h	o ′	o ′	o ′	o ′	o ′	o ′	o ′	o ′	o ′		o ′	o ′
6 00	74 34.6	172 29.8	S23 42.2	249 31.4	S 0 18.1	84 08.8	S 5 34.5	48 17.6	N 8 00.9	Acamar	315 26.5	S40 18.8
01	89 37.1	187 28.8	42.5	264 32.6	18.7	99 11.1	34.4	63 20.1	00.9	Achernar	335 34.8	S57 14.9
02	104 39.5	202 27.9	42.7	279 33.8	19.2	114 13.4	34.3	78 22.7	00.8	Acrux	173 22.0	S63 05.2
03	119 42.0	217 26.9	.. 43.0	294 35.0	.. 19.7	129 15.7	.. 34.2	93 25.3	.. 00.8	Adhara	255 21.0	S28 58.2
04	134 44.5	232 26.0	43.3	309 36.2	20.3	144 18.0	34.1	108 27.8	00.8	Aldebaran	291 01.9	N16 30.3
05	149 46.9	247 25.0	43.5	324 37.4	20.8	159 20.3	34.1	123 30.4	00.8			
06	164 49.4	262 24.1	S23 43.8	339 38.6	S 0 21.3	174 22.5	S 5 34.0	138 33.0	N 8 00.7	Alioth	166 30.9	N55 57.8
07	179 51.9	277 23.1	44.1	354 39.8	21.8	189 24.8	33.9	153 35.5	00.7	Alkaid	153 08.1	N49 19.1
08	194 54.3	292 22.1	44.3	9 41.0	22.4	204 27.1	33.8	168 38.1	00.7	Al Na'ir	27 58.0	S46 58.2
S 09	209 56.8	307 21.2	.. 44.6	24 42.2	.. 22.9	219 29.4	.. 33.7	183 40.7	.. 00.7	Alnilam	275 57.4	S 1 12.3
U 10	224 59.3	322 20.2	44.8	39 43.4	23.4	234 31.7	33.6	198 43.2	00.6	Alphard	218 07.0	S 8 39.2
N 11	240 01.7	337 19.3	45.1	54 44.6	24.0	249 34.0	33.6	213 45.8	00.6			
D 12	255 04.2	352 18.3	S23 45.3	69 45.8	S 0 24.5	264 36.3	S 5 33.5	228 48.4	N 8 00.6	Alphecca	126 20.9	N26 43.2
A 13	270 06.7	7 17.4	45.6	84 47.0	25.0	279 38.6	33.4	243 50.9	00.5	Alpheratz	357 55.0	N29 05.2
Y 14	285 09.1	22 16.4	45.9	99 48.2	25.5	294 40.9	33.3	258 53.5	00.5	Altair	62 19.4	N 8 52.1
15	300 11.6	37 15.5	.. 46.1	114 49.4	.. 26.1	309 43.2	.. 33.2	273 56.0	.. 00.5	Ankaa	353 26.6	S42 19.0
16	315 14.0	52 14.5	46.4	129 50.6	26.6	324 45.5	33.1	288 58.6	00.5	Antares	112 40.4	S26 25.6
17	330 16.5	67 13.5	46.6	144 51.8	27.1	339 47.8	33.0	304 01.2	00.4			
18	345 19.0	82 12.6	S23 46.8	159 53.0	S 0 27.6	354 50.0	S 5 33.0	319 03.7	N 8 00.4	Arcturus	146 06.3	N19 11.4
19	0 21.4	97 11.6	47.1	174 54.2	28.2	9 52.3	32.9	334 06.3	00.4	Atria	107 52.9	S69 01.4
20	15 23.9	112 10.7	47.3	189 55.4	28.7	24 54.6	32.8	349 08.9	00.4	Avior	234 22.1	S59 30.2
21	30 26.4	127 09.7	.. 47.6	204 56.6	.. 29.2	39 56.9	.. 32.7	4 11.4	.. 00.3	Bellatrix	278 43.7	N 6 20.8
22	45 28.8	142 08.8	47.8	219 57.8	29.7	54 59.2	32.6	19 14.0	00.3	Betelgeuse	271 13.1	N 7 24.3
23	60 31.3	157 07.8	48.1	234 59.0	30.3	70 01.5	32.5	34 16.6	00.3			
7 00	75 33.8	172 06.8	S23 48.3	250 00.2	S 0 30.8	85 03.8	S 5 32.5	49 19.1	N 8 00.3	Canopus	264 00.6	S52 41.7
01	90 36.2	187 05.9	48.5	265 01.4	31.3	100 06.1	32.4	64 21.7	00.2	Capella	280 50.6	N45 59.7
02	105 38.7	202 04.9	48.8	280 02.6	31.9	115 08.4	32.3	79 24.2	00.2	Deneb	49 39.4	N45 16.9
03	120 41.2	217 04.0	.. 49.0	295 03.8	.. 32.4	130 10.6	.. 32.2	94 26.8	.. 00.2	Denebola	182 45.2	N14 34.7
04	135 43.6	232 03.0	49.3	310 05.0	32.9	145 12.9	32.1	109 29.4	00.2	Diphda	349 07.0	S17 59.7
05	150 46.1	247 02.1	49.5	325 06.2	33.4	160 15.2	32.0	124 31.9	00.1			
06	165 48.5	262 01.1	S23 49.7	340 07.4	S 0 34.0	175 17.5	S 5 31.9	139 34.5	N 8 00.1	Dubhe	194 05.5	N61 45.1
07	180 51.0	277 00.1	49.9	355 08.6	34.5	190 19.8	31.8	154 37.1	00.1	Elnath	278 26.4	N28 36.3
08	195 53.5	291 59.2	50.2	10 09.8	35.0	205 22.1	31.8	169 39.6	00.0	Eltanin	90 51.9	N51 29.6
M 09	210 55.9	306 58.2	.. 50.4	25 11.0	.. 35.5	220 24.4	.. 31.7	184 42.2	.. 00.0	Enif	33 58.3	N 9 52.3
O 10	225 58.4	321 57.3	50.6	40 12.3	36.1	235 26.6	31.6	199 44.7	00.0	Fomalhaut	15 36.4	S29 37.8
N 11	241 00.9	336 56.3	50.9	55 13.5	36.6	250 28.9	31.5	214 47.3	8 00.0			
D 12	256 03.3	351 55.3	S23 51.1	70 14.7	S 0 37.1	265 31.2	S 5 31.4	229 49.9	N 7 59.9	Gacrux	172 13.6	S57 06.1
A 13	271 05.8	6 54.4	51.3	85 15.9	37.6	280 33.5	31.3	244 52.4	59.9	Gienah	176 04.0	S17 32.0
Y 14	286 08.3	21 53.4	51.5	100 17.1	38.2	295 35.8	31.2	259 55.0	59.9	Hadar	149 04.3	S60 21.7
15	301 10.7	36 52.5	.. 51.7	115 18.3	.. 38.7	310 38.1	.. 31.2	274 57.5	.. 59.9	Hamal	328 13.2	N23 27.4
16	316 13.2	51 51.5	52.0	130 19.5	39.2	325 40.3	31.1	290 00.1	59.8	Kaus Aust.	83 59.1	S34 23.0
17	331 15.7	66 50.5	52.2	145 20.7	39.7	340 42.6	31.0	305 02.7	59.8			
18	346 18.1	81 49.6	S23 52.4	160 21.9	S 0 40.3	355 44.9	S 5 30.9	320 05.2	N 7 59.8	Kochab	137 20.8	N74 09.6
19	1 20.6	96 48.6	52.6	175 23.1	40.8	10 47.2	30.8	335 07.8	59.8	Markab	13 49.5	N15 12.1
20	16 23.0	111 47.6	52.8	190 24.3	41.3	25 49.5	30.7	350 10.3	59.7	Menkar	314 26.5	N 4 05.0
21	31 25.5	126 46.7	.. 53.0	205 25.5	.. 41.8	40 51.8	.. 30.6	5 12.9	.. 59.7	Menkent	148 21.1	S36 21.6
22	46 28.0	141 45.7	53.3	220 26.7	42.4	55 54.0	30.5	20 15.5	59.7	Miaplacidus	221 41.5	S69 42.5
23	61 30.4	156 44.8	53.5	235 27.9	42.9	70 56.3	30.4	35 18.0	59.7			
8 00	76 32.9	171 43.8	S23 53.7	250 29.1	S 0 43.4	85 58.6	S 5 30.4	50 20.6	N 7 59.6	Mirfak	308 55.9	N49 51.4
01	91 35.4	186 42.8	53.9	265 30.3	43.9	101 00.9	30.3	65 23.1	59.6	Nunki	76 12.6	S26 17.8
02	106 37.8	201 41.9	54.1	280 31.5	44.5	116 03.2	30.2	80 25.7	59.6	Peacock	53 37.3	S56 44.4
03	121 40.3	216 40.9	.. 54.3	295 32.7	.. 45.0	131 05.4	.. 30.1	95 28.2	.. 59.6	Pollux	243 41.2	N28 01.6
04	136 42.8	231 39.9	54.5	310 34.0	45.5	146 07.7	30.0	110 30.8	59.5	Procyon	245 11.2	N 5 13.6
05	151 45.2	246 39.0	54.7	325 35.2	46.0	161 10.0	29.9	125 33.4	59.5			
06	166 47.7	261 38.0	S23 54.9	340 36.4	S 0 46.6	176 12.3	S 5 29.8	140 35.9	N 7 59.5	Rasalhague	96 17.2	N12 33.8
07	181 50.1	276 37.1	55.1	355 37.6	47.1	191 14.6	29.7	155 38.5	59.5	Regulus	207 55.4	N11 58.3
08	196 52.6	291 36.1	55.3	10 38.8	47.6	206 16.8	29.6	170 41.0	59.4	Rigel	281 22.5	S 8 12.3
T 09	211 55.1	306 35.1	.. 55.5	25 40.0	.. 48.1	221 19.1	.. 29.6	185 43.6	.. 59.4	Rigil Kent.	140 07.7	S60 49.5
U 10	226 57.5	321 34.2	55.7	40 41.2	48.6	236 21.4	29.5	200 46.2	59.4	Sabik	102 25.8	S15 43.2
E 11	242 00.0	336 33.2	55.9	55 42.4	49.2	251 23.7	29.4	215 48.7	59.4			
S 12	257 02.5	351 32.2	S23 56.1	70 43.6	S 0 49.7	266 26.0	S 5 29.3	230 51.3	N 7 59.4	Schedar	349 53.0	N56 32.1
D 13	272 04.9	6 31.3	56.3	85 44.8	50.2	281 28.2	29.2	245 53.8	59.3	Shaula	96 37.6	S37 06.0
A 14	287 07.4	21 30.3	56.5	100 46.0	50.7	296 30.5	29.1	260 56.4	59.3	Sirius	258 43.3	S16 42.9
Y 15	302 09.9	36 29.3	.. 56.7	115 47.2	.. 51.3	311 32.8	.. 29.0	275 58.9	.. 59.3	Spica	158 43.3	S11 09.1
16	317 12.3	51 28.4	56.8	130 48.4	51.8	326 35.1	28.9	291 01.5	59.3	Suhail	223 00.4	S43 25.5
17	332 14.8	66 27.4	57.0	145 49.6	52.3	341 37.3	28.8	306 04.1	59.2			
18	347 17.3	81 26.4	S23 57.2	160 50.9	S 0 52.8	356 39.6	S 5 28.7	321 06.6	N 7 59.2	Vega	80 47.0	N38 47.2
19	2 19.7	96 25.5	57.4	175 52.1	53.4	11 41.9	28.6	336 09.2	59.2	Zuben'ubi	137 18.2	S16 02.0
20	17 22.2	111 24.5	57.6	190 53.3	53.9	26 44.2	28.6	351 11.7	59.2		SHA	Mer. Pass.
21	32 24.6	126 23.5	.. 57.8	205 54.5	.. 54.4	41 46.4	.. 28.5	6 14.3	.. 59.1		o ′	h m
22	47 27.1	141 22.6	58.0	220 55.7	54.9	56 48.7	28.4	21 16.8	59.1	Venus	96 33.1	12 32
23	62 29.6	156 21.6	58.1	235 56.9	55.4	71 51.0	28.3	36 19.4	59.1	Mars	174 26.5	7 19
	h m									Jupiter	9 30.0	18 17
Mer. Pass.	18 54.6	v −1.0	d 0.2	v 1.2	d 0.5	v 2.3	d 0.1	v 2.6	d 0.0	Saturn	333 45.4	20 39

UT	SUN GHA	SUN Dec	MOON GHA	v	MOON Dec	d	HP
d h	° ′	° ′	° ′	′	° ′	′	′
6 00	182 18.7	S22 26.2	327 35.0	5.7	N19 20.1	1.8	59.4
01	197 18.4	26.5	341 59.7	5.8	19 18.3	1.9	59.3
02	212 18.1	26.8	356 24.5	5.9	19 16.4	2.0	59.3
03	227 17.9	27.1	10 49.4	5.9	19 14.4	2.2	59.3
04	242 17.6	27.4	25 14.3	6.0	19 12.2	2.3	59.2
05	257 17.4	27.7	39 39.3	6.1	19 09.9	2.4	59.2
06	272 17.1	S22 28.0	54 04.4	6.2	N19 07.5	2.5	59.2
07	287 16.8	28.3	68 29.6	6.2	19 05.0	2.7	59.1
08	302 16.6	28.6	82 54.8	6.2	19 02.3	2.7	59.1
S 09	317 16.3	28.9	97 20.0	6.4	18 59.6	2.9	59.0
U 10	332 16.0	29.2	111 45.4	6.4	18 56.7	3.0	59.0
N 11	347 15.8	29.5	126 10.8	6.5	18 53.7	3.1	59.0
D 12	2 15.5	S22 29.8	140 36.3	6.6	N18 50.6	3.3	58.9
A 13	17 15.3	30.1	155 01.9	6.7	18 47.3	3.3	58.9
Y 14	32 15.0	30.4	169 27.6	6.7	18 44.0	3.5	58.8
15	47 14.7	30.7	183 53.3	6.8	18 40.5	3.6	58.8
16	62 14.5	31.0	198 19.1	6.9	18 36.9	3.7	58.8
17	77 14.2	31.3	212 45.0	6.9	18 33.2	3.8	58.7
18	92 13.9	S22 31.6	227 10.9	7.1	N18 29.4	3.9	58.7
19	107 13.7	31.9	241 37.0	7.1	18 25.5	4.0	58.7
20	122 13.4	32.2	256 03.1	7.2	18 21.5	4.1	58.6
21	137 13.1	32.4	270 29.3	7.3	18 17.4	4.3	58.6
22	152 12.9	32.7	284 55.6	7.4	18 13.1	4.3	58.5
23	167 12.6	33.0	299 22.0	7.4	18 08.8	4.5	58.5
7 00	182 12.4	S22 33.3	313 48.4	7.5	N18 04.3	4.5	58.5
01	197 12.1	33.6	328 14.9	7.6	17 59.8	4.7	58.4
02	212 11.8	33.9	342 41.5	7.7	17 55.1	4.7	58.4
03	227 11.6	34.2	357 08.2	7.8	17 50.4	4.9	58.3
04	242 11.3	34.5	11 35.0	7.9	17 45.5	5.0	58.3
05	257 11.0	34.8	26 01.9	7.9	17 40.5	5.0	58.3
06	272 10.8	S22 35.0	40 28.8	8.1	N17 35.5	5.2	58.2
07	287 10.5	35.3	54 55.9	8.1	17 30.3	5.2	58.2
08	302 10.2	35.6	69 23.0	8.2	17 25.1	5.4	58.1
M 09	317 10.0	35.9	83 50.2	8.3	17 19.7	5.4	58.1
O 10	332 09.7	36.2	98 17.5	8.3	17 14.3	5.6	58.1
N 11	347 09.4	36.5	112 44.8	8.5	17 08.7	5.6	58.0
D 12	2 09.1	S22 36.7	127 12.3	8.5	N17 03.1	5.7	58.0
A 13	17 08.9	37.0	141 39.8	8.7	16 57.4	5.8	57.9
Y 14	32 08.6	37.3	156 07.5	8.7	16 51.6	5.9	57.9
15	47 08.3	37.6	170 35.2	8.8	16 45.7	6.0	57.9
16	62 08.1	37.8	185 03.0	8.9	16 39.7	6.1	57.8
17	77 07.8	38.1	199 30.9	9.0	16 33.6	6.2	57.8
18	92 07.5	S22 38.4	213 58.9	9.0	N16 27.4	6.2	57.7
19	107 07.3	38.7	228 26.9	9.2	16 21.2	6.3	57.7
20	122 07.0	38.9	242 55.1	9.2	16 14.9	6.5	57.7
21	137 06.7	39.2	257 23.3	9.4	16 08.4	6.5	57.6
22	152 06.5	39.5	271 51.7	9.4	16 01.9	6.5	57.6
23	167 06.2	39.8	286 20.1	9.5	15 55.4	6.7	57.5
8 00	182 05.9	S22 40.0	300 48.6	9.6	N15 48.7	6.7	57.5
01	197 05.6	40.3	315 17.2	9.6	15 42.0	6.8	57.5
02	212 05.4	40.6	329 45.8	9.8	15 35.2	6.9	57.4
03	227 05.1	40.8	344 14.6	9.9	15 28.3	6.9	57.4
04	242 04.8	41.1	358 43.5	9.9	15 21.4	7.1	57.3
05	257 04.6	41.4	13 12.4	10.0	15 14.3	7.1	57.3
06	272 04.3	S22 41.6	27 41.4	10.1	N15 07.2	7.2	57.2
07	287 04.0	41.9	42 10.5	10.2	15 00.0	7.2	57.2
08	302 03.7	42.2	56 39.7	10.3	14 52.8	7.3	57.2
T 09	317 03.5	42.4	71 09.0	10.3	14 45.5	7.4	57.1
U 10	332 03.2	42.7	85 38.3	10.5	14 38.1	7.4	57.1
E 11	347 02.9	43.0	100 07.8	10.5	14 30.7	7.5	57.1
S 12	2 02.6	S22 43.2	114 37.3	10.6	N14 23.2	7.6	57.0
D 13	17 02.4	43.5	129 06.9	10.7	14 15.6	7.7	57.0
A 14	32 02.1	43.7	143 36.6	10.8	14 07.9	7.7	57.0
Y 15	47 01.8	44.0	158 06.4	10.8	14 00.2	7.7	56.9
16	62 01.5	44.3	172 36.2	10.9	13 52.5	7.9	56.9
17	77 01.3	44.5	187 06.1	11.1	13 44.6	7.9	56.8
18	92 01.0	S22 44.8	201 36.2	11.1	N13 36.7	7.9	56.8
19	107 00.7	45.0	216 06.3	11.1	13 28.8	8.0	56.8
20	122 00.5	45.3	230 36.4	11.3	13 20.8	8.1	56.7
21	137 00.2	45.5	245 06.7	11.3	13 12.7	8.1	56.7
22	151 59.9	45.8	259 37.0	11.4	13 04.6	8.2	56.7
23	166 59.6	46.1	274 07.4	11.5	N12 56.4	8.2	56.6
	SD 16.3	d 0.3	SD 16.1		15.8		15.5

Lat.	Twilight Naut.	Twilight Civil	Sunrise	Moonrise 6	Moonrise 7	Moonrise 8	Moonrise 9
°	h m	h m	h m	h m	h m	h m	h m
N 72	08 05	10 16	■■■	■■■	17 24	19 30	21 21
N 70	07 47	09 29	▭	16 22	18 11	19 57	21 37
68	07 33	08 58	11 17	17 08	18 41	20 16	21 49
66	07 21	08 35	10 06	17 38	19 03	20 32	21 59
64	07 11	08 17	09 31	18 00	19 21	20 45	22 08
62	07 02	08 02	09 05	18 18	19 35	20 56	22 15
60	06 54	07 50	08 46	18 33	19 48	21 05	22 21
N 58	06 47	07 39	08 29	18 45	19 58	21 13	22 27
56	06 41	07 29	08 16	18 56	20 07	21 20	22 32
54	06 35	07 21	08 04	19 06	20 15	21 26	22 36
52	06 30	07 13	07 53	19 14	20 23	21 32	22 40
50	06 25	07 06	07 44	19 22	20 29	21 37	22 44
45	06 14	06 51	07 24	19 38	20 43	21 48	22 51
N 40	06 04	06 38	07 08	19 52	20 55	21 57	22 58
35	05 55	06 27	06 54	20 03	21 05	22 05	23 03
30	05 47	06 16	06 43	20 13	21 13	22 12	23 08
20	05 31	05 58	06 22	20 30	21 28	22 24	23 17
N 10	05 16	05 42	06 04	20 45	21 41	22 34	23 24
0	04 59	05 25	05 48	20 59	21 53	22 43	23 31
S 10	04 41	05 08	05 31	21 12	22 05	22 53	23 38
20	04 19	04 48	05 13	21 27	22 18	23 03	23 45
30	03 51	04 24	04 51	21 44	22 32	23 15	23 53
35	03 33	04 10	04 39	21 54	22 41	23 22	23 58
40	03 11	03 52	04 25	22 05	22 51	23 29	24 03
45	02 42	03 31	04 07	22 19	23 02	23 38	24 10
S 50	02 00	03 03	03 46	22 35	23 15	23 49	24 17
52	01 36	02 48	03 36	22 42	23 22	23 54	24 21
54	01 00	02 32	03 24	22 51	23 29	23 59	24 24
56	////	02 11	03 11	23 00	23 37	24 05	00 05
58	////	01 45	02 55	23 11	23 45	24 12	00 12
S 60	////	01 07	02 36	23 23	23 55	24 20	00 20

Lat.	Sunset	Twilight Civil	Twilight Naut.	Moonset 6	Moonset 7	Moonset 8	Moonset 9
°	h m	h m	h m	h m	h m	h m	h m
N 72	■■■	13 26	15 37	▭	14 06	13 46	13 35
N 70	■■■	14 14	15 55	13 12	13 18	13 19	13 17
68	12 25	14 44	16 09	12 26	12 47	12 58	13 04
66	13 36	15 07	16 21	11 55	12 24	12 41	12 52
64	14 12	15 25	16 32	11 32	12 05	12 27	12 43
62	14 37	15 40	16 41	11 14	11 50	12 16	12 35
60	14 57	15 53	16 48	10 59	11 38	12 06	12 27
N 58	15 13	16 04	16 55	10 46	11 27	11 57	12 21
56	15 27	16 13	17 02	10 35	11 17	11 50	12 16
54	15 39	16 22	17 07	10 25	11 09	11 43	12 11
52	15 49	16 30	17 13	10 17	11 01	11 37	12 06
50	15 59	16 37	17 18	10 09	10 54	11 31	12 02
45	16 18	16 52	17 29	09 52	10 39	11 19	11 53
N 40	16 35	17 05	17 39	09 38	10 27	11 09	11 46
35	16 48	17 16	17 48	09 27	10 17	11 01	11 39
30	17 00	17 26	17 56	09 16	10 08	10 53	11 33
20	17 20	17 44	18 12	08 59	09 52	10 40	11 24
N 10	17 38	18 01	18 27	08 43	09 38	10 28	11 15
0	17 55	18 18	18 44	08 29	09 25	10 18	11 07
S 10	18 12	18 35	19 02	08 14	09 12	10 07	10 58
20	18 30	18 55	19 24	07 59	08 58	09 55	10 49
30	18 52	19 19	19 52	07 41	08 42	09 41	10 39
35	19 04	19 33	20 10	07 30	08 32	09 34	10 33
40	19 19	19 51	20 32	07 18	08 22	09 25	10 26
45	19 36	20 13	21 01	07 04	08 09	09 14	10 18
S 50	19 57	20 41	21 44	06 47	07 53	09 01	10 09
52	20 08	20 55	22 08	06 39	07 46	08 55	10 05
54	20 19	21 12	22 45	06 30	07 38	08 49	10 00
56	20 33	21 32	////	06 19	07 29	08 42	09 54
58	20 48	21 59	////	06 08	07 19	08 33	09 48
S 60	21 07	22 39	////	05 54	07 07	08 24	09 41

	SUN Eqn. of Time 00h	SUN Eqn. of Time 12h	SUN Mer. Pass.	MOON Mer. Pass. Upper	MOON Mer. Pass. Lower	Age	Phase
Day							
d	m s	m s	h m	h m	h m	d	%
6	09 15	09 03	11 51	02 15	14 44	17	89
7	08 50	08 37	11 51	03 12	15 39	18	81
8	08 24	08 11	11 52	04 05	16 31	19	72

UT	ARIES GHA	VENUS GHA	VENUS Dec	MARS GHA	MARS Dec	JUPITER GHA	JUPITER Dec	SATURN GHA	SATURN Dec	Name	SHA	Dec
9 00	77 32.0	171 20.6	S23 58.3	250 58.1	S 0 56.0	86 53.3	S 5 28.2	51 21.9	N 7 59.1	Acamar	315 26.5	S40 18.8
01	92 34.5	186 19.7	58.5	265 59.3	56.5	101 55.5	28.1	66 24.5	59.0	Achernar	335 34.8	S57 14.9
02	107 37.0	201 18.7	58.7	281 00.5	57.0	116 57.8	28.0	81 27.0	59.0	Acrux	173 22.0	S63 05.2
03	122 39.4	216 17.7	.. 58.8	296 01.7	.. 57.5	132 00.1	.. 27.9	96 29.6	.. 59.0	Adhara	255 21.0	S28 58.3
04	137 41.9	231 16.8	59.0	311 03.0	58.0	147 02.4	27.8	111 32.2	59.0	Aldebaran	291 01.9	N16 30.3
05	152 44.4	246 15.8	59.2	326 04.2	58.6	162 04.6	27.7	126 34.7	59.0			
W 06	167 46.8	261 14.8	S23 59.4	341 05.4	S 0 59.1	177 06.9	S 5 27.6	141 37.3	N 7 58.9	Alioth	166 30.8	N55 57.8
E 07	182 49.3	276 13.9	59.5	356 06.6	0 59.6	192 09.2	27.5	156 39.8	58.9	Alkaid	153 08.1	N49 19.0
D 08	197 51.8	291 12.9	59.7	11 07.8	1 00.1	207 11.4	27.5	171 42.4	58.9	Al Na'ir	27 58.0	S46 58.2
N 09	212 54.2	306 11.9	23 59.9	26 09.0	.. 00.7	222 13.7	.. 27.4	186 44.9	.. 58.9	Alnilam	275 57.4	S 1 12.3
E 10	227 56.7	321 11.0	24 00.0	41 10.2	01.2	237 16.0	27.3	201 47.5	58.8	Alphard	218 06.9	S 8 39.2
S 11	242 59.1	336 10.0	00.2	56 11.4	01.7	252 18.3	27.2	216 50.0	58.8			
D 12	258 01.6	351 09.0	S24 00.4	71 12.6	S 1 02.2	267 20.5	S 5 27.1	231 52.6	N 7 58.8	Alphecca	126 20.9	N26 43.2
A 13	273 04.1	6 08.1	00.5	86 13.8	02.7	282 22.8	27.0	246 55.1	58.8	Alpheratz	357 55.0	N29 05.2
Y 14	288 06.5	21 07.1	00.7	101 15.1	03.3	297 25.1	26.9	261 57.7	58.7	Altair	62 19.4	N 8 52.1
15	303 09.0	36 06.1	.. 00.8	116 16.3	.. 03.8	312 27.3	.. 26.8	277 00.2	.. 58.7	Ankaa	353 26.7	S42 19.0
16	318 11.5	51 05.2	01.0	131 17.5	04.3	327 29.6	26.7	292 02.8	58.7	Antares	112 40.4	S26 25.6
17	333 13.9	66 04.2	01.2	146 18.7	04.8	342 31.9	26.6	307 05.4	58.7			
18	348 16.4	81 03.2	S24 01.3	161 19.9	S 1 05.3	357 34.1	S 5 26.5	322 07.9	N 7 58.7	Arcturus	146 06.2	N19 11.3
19	3 18.9	96 02.3	01.5	176 21.1	05.9	12 36.4	26.4	337 10.5	58.6	Atria	107 52.9	S69 01.4
20	18 21.3	111 01.3	01.6	191 22.3	06.4	27 38.7	26.3	352 13.0	58.6	Avior	234 22.1	S59 30.2
21	33 23.8	126 00.3	.. 01.8	206 23.5	.. 06.9	42 40.9	.. 26.2	7 15.6	.. 58.6	Bellatrix	278 43.7	N 6 20.8
22	48 26.2	140 59.3	01.9	221 24.8	07.4	57 43.2	26.1	22 18.1	58.6	Betelgeuse	271 13.1	N 7 24.3
23	63 28.7	155 58.4	02.1	236 26.0	07.9	72 45.5	26.0	37 20.7	58.6			
10 00	78 31.2	170 57.4	S24 02.2	251 27.2	S 1 08.5	87 47.7	S 5 26.0	52 23.2	N 7 58.5	Canopus	264 00.6	S52 41.7
01	93 33.6	185 56.4	02.4	266 28.4	09.0	102 50.0	25.9	67 25.8	58.5	Capella	280 50.5	N45 59.7
02	108 36.1	200 55.5	02.5	281 29.6	09.5	117 52.3	25.8	82 28.3	58.5	Deneb	49 39.4	N45 16.9
03	123 38.6	215 54.5	.. 02.7	296 30.8	.. 10.0	132 54.5	.. 25.7	97 30.9	.. 58.5	Denebola	182 45.2	N14 34.7
04	138 41.0	230 53.5	02.8	311 32.0	10.5	147 56.8	25.6	112 33.4	58.4	Diphda	349 07.0	S17 59.7
05	153 43.5	245 52.6	02.9	326 33.3	11.1	162 59.1	25.5	127 36.0	58.4			
T 06	168 46.0	260 51.6	S24 03.1	341 34.5	S 1 11.6	178 01.3	S 5 25.4	142 38.5	N 7 58.4	Dubhe	194 05.4	N61 45.1
H 07	183 48.4	275 50.6	03.2	356 35.7	12.1	193 03.6	25.3	157 41.1	58.4	Elnath	278 26.4	N28 36.3
U 08	198 50.9	290 49.6	03.4	11 36.9	12.6	208 05.9	25.2	172 43.6	58.4	Eltanin	90 51.9	N51 29.5
R 09	213 53.4	305 48.7	.. 03.5	26 38.1	.. 13.1	223 08.1	.. 25.1	187 46.2	.. 58.3	Enif	33 58.3	N 9 52.3
S 10	228 55.8	320 47.7	03.6	41 39.3	13.7	238 10.4	25.0	202 48.7	58.3	Fomalhaut	15 36.4	S29 37.8
D 11	243 58.3	335 46.7	03.8	56 40.5	14.2	253 12.7	24.9	217 51.3	58.3			
A 12	259 00.7	350 45.8	S24 03.9	71 41.8	S 1 14.7	268 14.9	S 5 24.8	232 53.8	N 7 58.3	Gacrux	172 13.6	S57 06.1
Y 13	274 03.2	5 44.8	04.0	86 43.0	15.2	283 17.2	24.7	247 56.4	58.3	Gienah	176 03.9	S17 32.0
14	289 05.7	20 43.8	04.1	101 44.2	15.7	298 19.5	24.6	262 58.9	58.2	Hadar	149 04.3	S60 21.7
15	304 08.1	35 42.8	.. 04.3	116 45.4	.. 16.3	313 21.7	.. 24.5	278 01.5	.. 58.2	Hamal	328 13.2	N23 27.4
16	319 10.6	50 41.9	04.4	131 46.6	16.8	328 24.0	24.4	293 04.0	58.2	Kaus Aust.	83 59.1	S34 23.0
17	334 13.1	65 40.9	04.5	146 47.8	17.3	343 26.2	24.3	308 06.6	58.2			
18	349 15.5	80 39.9	S24 04.7	161 49.0	S 1 17.8	358 28.5	S 5 24.2	323 09.1	N 7 58.2	Kochab	137 20.7	N74 09.5
19	4 18.0	95 38.9	04.8	176 50.3	18.3	13 30.8	24.1	338 11.7	58.1	Markab	13 49.5	N15 12.1
20	19 20.5	110 38.0	04.9	191 51.5	18.8	28 33.0	24.0	353 14.2	58.1	Menkar	314 26.5	N 4 05.0
21	34 22.9	125 37.0	.. 05.0	206 52.7	.. 19.4	43 35.3	.. 23.9	8 16.8	.. 58.1	Menkent	148 21.1	S36 21.6
22	49 25.4	140 36.0	05.1	221 53.9	19.9	58 37.5	23.8	23 19.3	58.1	Miaplacidus	221 41.4	S69 42.5
23	64 27.9	155 35.1	05.3	236 55.1	20.4	73 39.8	23.7	38 21.8	58.1			
11 00	79 30.3	170 34.1	S24 05.4	251 56.3	S 1 20.9	88 42.1	S 5 23.6	53 24.4	N 7 58.0	Mirfak	308 56.0	N49 51.4
01	94 32.8	185 33.1	05.5	266 57.6	21.4	103 44.3	23.5	68 26.9	58.0	Nunki	76 12.6	S26 17.8
02	109 35.2	200 32.1	05.6	281 58.8	21.9	118 46.6	23.4	83 29.5	58.0	Peacock	53 37.3	S56 44.4
03	124 37.7	215 31.2	.. 05.7	297 00.0	.. 22.5	133 48.8	.. 23.4	98 32.0	.. 58.0	Pollux	243 41.1	N28 01.6
04	139 40.2	230 30.2	05.8	312 01.2	23.0	148 51.1	23.3	113 34.6	58.0	Procyon	245 11.2	N 5 13.6
05	154 42.6	245 29.2	05.9	327 02.4	23.5	163 53.4	23.2	128 37.1	57.9			
F 06	169 45.1	260 28.2	S24 06.0	342 03.6	S 1 24.0	178 55.6	S 5 23.1	143 39.7	N 7 57.9	Rasalhague	96 17.2	N12 33.8
R 07	184 47.6	275 27.3	06.2	357 04.9	24.5	193 57.9	23.0	158 42.2	57.9	Regulus	207 55.3	N11 58.3
I 08	199 50.0	290 26.3	06.3	12 06.1	25.0	209 00.1	22.9	173 44.8	57.9	Rigel	281 22.5	S 8 12.3
D 09	214 52.5	305 25.3	.. 06.4	27 07.3	.. 25.6	224 02.4	.. 22.8	188 47.3	.. 57.9	Rigil Kent.	140 07.6	S60 49.5
A 10	229 55.0	320 24.4	06.5	42 08.5	26.1	239 04.7	22.7	203 49.9	57.8	Sabik	102 25.8	S15 43.2
Y 11	244 57.4	335 23.4	06.6	57 09.7	26.6	254 06.9	22.6	218 52.4	57.8			
12	259 59.9	350 22.4	S24 06.7	72 11.0	S 1 27.1	269 09.2	S 5 22.5	233 55.0	N 7 57.8	Schedar	349 53.1	N56 32.1
13	275 02.3	5 21.4	06.8	87 12.2	27.6	284 11.4	22.4	248 57.5	57.8	Shaula	96 37.6	S37 06.0
14	290 04.8	20 20.5	06.9	102 13.4	28.1	299 13.7	22.3	264 00.0	57.8	Sirius	258 43.3	S16 42.9
15	305 07.3	35 19.5	.. 07.0	117 14.6	.. 28.7	314 15.9	.. 22.2	279 02.6	.. 57.7	Spica	158 43.3	S11 09.2
16	320 09.7	50 18.5	07.1	132 15.8	29.2	329 18.2	22.1	294 05.1	57.7	Suhail	223 00.4	S43 25.5
17	335 12.2	65 17.5	07.2	147 17.0	29.7	344 20.4	22.0	309 07.7	57.7			
18	350 14.7	80 16.5	S24 07.3	162 18.3	S 1 30.2	359 22.7	S 5 21.9	324 10.2	N 7 57.7	Vega	80 47.0	N38 47.2
19	5 17.1	95 15.6	07.3	177 19.5	30.7	14 25.0	21.8	339 12.8	57.7	Zuben'ubi	137 18.1	S16 02.0
20	20 19.6	110 14.6	07.4	192 20.7	31.2	29 27.2	21.7	354 15.3	57.6			
21	35 22.1	125 13.6	.. 07.5	207 21.9	.. 31.8	44 29.5	.. 21.6	9 17.9	.. 57.6			
22	50 24.5	140 12.7	07.6	222 23.1	32.3	59 31.7	21.5	24 20.4	57.6			
23	65 27.0	155 11.7	07.7	237 24.4	32.8	74 34.0	21.4	39 22.9	57.6			

	SHA	Mer. Pass.
	° '	h m
Venus	92 26.2	12 37
Mars	172 56.0	7 14
Jupiter	9 16.6	18 06
Saturn	333 52.0	20 27

	ARIES	VENUS	MARS	JUPITER	SATURN
Mer. Pass. 18 42.8		v −1.0 d 0.1	v 1.2 d 0.5	v 2.3 d 0.1	v 2.5 d 0.0

UT	SUN GHA	SUN Dec	MOON GHA	v	MOON Dec	d	HP
9 00	181 59.3	S22 46.3	288 37.9	11.6	N12 48.2	8.3	56.6
01	196 59.1	46.6	303 08.5	11.6	12 39.9	8.3	56.5
02	211 58.8	46.8	317 39.1	11.8	12 31.6	8.4	56.5
03	226 58.5	.. 47.1	332 09.9	11.8	12 23.2	8.4	56.5
04	241 58.2	47.3	346 40.7	11.8	12 14.8	8.5	56.4
05	256 58.0	47.6	1 11.5	12.0	12 06.3	8.5	56.4
W 06	271 57.7	S22 47.8	15 42.5	12.0	N11 57.8	8.6	56.4
E 07	286 57.4	48.1	30 13.5	12.1	11 49.2	8.6	56.3
D 08	301 57.1	48.3	44 44.6	12.2	11 40.6	8.7	56.3
N 09	316 56.9	.. 48.5	59 15.8	12.2	11 31.9	8.7	56.3
E 10	331 56.6	48.8	73 47.0	12.3	11 23.2	8.8	56.2
S 11	346 56.3	49.0	88 18.3	12.4	11 14.4	8.8	56.2
D 12	1 56.0	S22 49.3	102 49.7	12.5	N11 05.6	8.8	56.2
A 13	16 55.7	49.5	117 21.2	12.5	10 56.8	8.9	56.1
Y 14	31 55.5	49.8	131 52.7	12.6	10 47.9	8.9	56.1
15	46 55.2	.. 50.0	146 24.3	12.6	10 39.0	9.0	56.1
16	61 54.9	50.2	160 55.9	12.8	10 30.0	9.0	56.0
17	76 54.6	50.5	175 27.7	12.7	10 21.0	9.0	56.0
18	91 54.3	S22 50.7	189 59.4	12.9	N10 12.0	9.1	56.0
19	106 54.1	51.0	204 31.3	12.9	10 02.9	9.1	55.9
20	121 53.8	51.2	219 03.2	13.0	9 53.8	9.1	55.9
21	136 53.5	.. 51.4	233 35.2	13.1	9 44.7	9.2	55.9
22	151 53.2	51.7	248 07.3	13.1	9 35.5	9.2	55.8
23	166 52.9	51.9	262 39.4	13.1	9 26.3	9.3	55.8
10 00	181 52.7	S22 52.1	277 11.5	13.3	N 9 17.0	9.2	55.8
01	196 52.4	52.4	291 43.8	13.3	9 07.8	9.4	55.7
02	211 52.1	52.6	306 16.1	13.3	8 58.4	9.3	55.7
03	226 51.8	.. 52.8	320 48.4	13.4	8 49.1	9.4	55.7
04	241 51.5	53.1	335 20.8	13.5	8 39.7	9.4	55.6
05	256 51.3	53.3	349 53.3	13.5	8 30.3	9.4	55.6
T 06	271 51.0	S22 53.5	4 25.8	13.6	N 8 20.9	9.4	55.6
H 07	286 50.7	53.7	18 58.4	13.7	8 11.5	9.5	55.6
U 08	301 50.4	54.0	33 31.1	13.7	8 02.0	9.5	55.5
R 09	316 50.1	.. 54.2	48 03.8	13.7	7 52.5	9.5	55.5
S 10	331 49.8	54.4	62 36.5	13.8	7 43.0	9.6	55.5
D 11	346 49.6	54.7	77 09.3	13.9	7 33.4	9.6	55.4
A 12	1 49.3	S22 54.9	91 42.2	13.9	N 7 23.8	9.6	55.4
Y 13	16 49.0	55.1	106 15.1	13.9	7 14.2	9.6	55.4
14	31 48.7	55.3	120 48.0	14.0	7 04.6	9.6	55.4
15	46 48.4	.. 55.5	135 21.0	14.1	6 55.0	9.7	55.3
16	61 48.1	55.8	149 54.1	14.1	6 45.3	9.7	55.3
17	76 47.9	56.0	164 27.2	14.1	6 35.6	9.7	55.3
18	91 47.6	S22 56.2	179 00.3	14.2	N 6 25.9	9.7	55.3
19	106 47.3	56.4	193 33.5	14.3	6 16.2	9.7	55.2
20	121 47.0	56.6	208 06.8	14.3	6 06.5	9.8	55.2
21	136 46.7	.. 56.9	222 40.1	14.3	5 56.7	9.8	55.2
22	151 46.4	57.1	237 13.4	14.4	5 46.9	9.8	55.2
23	166 46.2	57.3	251 46.8	14.4	5 37.1	9.8	55.1
11 00	181 45.9	S22 57.5	266 20.2	14.5	N 5 27.3	9.8	55.1
01	196 45.6	57.7	280 53.7	14.5	5 17.5	9.8	55.1
02	211 45.3	57.9	295 27.2	14.5	5 07.7	9.9	55.1
03	226 45.0	.. 58.1	310 00.7	14.6	4 57.8	9.8	55.0
04	241 44.7	58.4	324 34.3	14.6	4 48.0	9.9	55.0
05	256 44.4	58.6	339 07.9	14.6	4 38.1	9.9	55.0
F 06	271 44.2	S22 58.8	353 41.5	14.7	N 4 28.2	9.8	54.9
R 07	286 43.9	59.0	8 15.2	14.8	4 18.4	10.0	54.9
I 08	301 43.6	59.2	22 49.0	14.7	4 08.4	9.9	54.9
D 09	316 43.3	.. 59.4	37 22.7	14.8	3 58.5	9.9	54.9
A 10	331 43.0	59.6	51 56.5	14.8	3 48.6	9.9	54.9
Y 11	346 42.7	22 59.8	66 30.3	14.9	3 38.7	9.9	54.8
12	1 42.4	S23 00.0	81 04.2	14.9	N 3 28.8	10.0	54.8
13	16 42.1	00.2	95 38.1	14.9	3 18.8	9.9	54.8
14	31 41.9	00.4	110 12.0	15.0	3 08.9	10.0	54.8
15	46 41.6	.. 00.6	124 46.0	14.9	2 58.9	10.0	54.8
16	61 41.3	00.8	139 19.9	15.1	2 48.9	9.9	54.7
17	76 41.0	01.0	153 54.0	15.0	2 39.0	10.0	54.7
18	91 40.7	S23 01.2	168 28.0	15.0	N 2 29.0	10.0	54.7
19	106 40.4	01.4	183 02.0	15.1	2 19.0	9.9	54.7
20	121 40.1	01.6	197 36.1	15.1	2 09.1	10.0	54.7
21	136 39.8	.. 01.8	212 10.2	15.2	1 59.1	10.0	54.6
22	151 39.5	02.0	226 44.4	15.1	1 49.1	10.0	54.6
23	166 39.3	02.2	241 18.5	15.2	N 1 39.1	10.0	54.6
	SD 16.3	d 0.2	SD 15.3		15.1		14.9

Lat.	Twilight Naut.	Twilight Civil	Sunrise	Moonrise 9	10	11	12
N 72	08 12	10 29	■■	21 21	23 04	24 42	00 42
N 70	07 53	09 37	■■	21 37	23 12	24 44	00 44
68	07 38	09 05	■■	21 49	23 19	24 45	00 45
66	07 25	08 41	10 15	21 59	23 24	24 47	00 47
64	07 15	08 22	09 37	22 08	23 29	24 48	00 48
62	07 06	08 07	09 11	22 15	23 33	24 49	00 49
60	06 58	07 54	08 50	22 21	23 36	24 49	00 49
N 58	06 51	07 43	08 34	22 27	23 39	24 50	00 50
56	06 44	07 33	08 20	22 32	23 42	24 51	00 51
54	06 38	07 24	08 07	22 36	23 44	24 51	00 51
52	06 33	07 16	07 57	22 40	23 47	24 52	00 52
50	06 28	07 09	07 47	22 44	23 49	24 52	00 52
45	06 16	06 53	07 27	22 51	23 53	24 53	00 53
N 40	06 06	06 40	07 11	22 58	23 57	24 54	00 54
35	05 57	06 29	06 57	23 03	24 00	00 00	00 55
30	05 49	06 19	06 45	23 08	24 03	00 03	00 55
20	05 33	06 00	06 24	23 17	24 07	00 07	00 57
N 10	05 17	05 43	06 06	23 24	24 12	00 12	00 58
0	05 00	05 27	05 49	23 31	24 16	00 16	00 59
S 10	04 42	05 09	05 32	23 38	24 19	00 19	01 00
20	04 20	04 49	05 13	23 45	24 24	00 24	01 01
30	03 51	04 25	04 52	23 53	24 28	00 28	01 02
35	03 33	04 10	04 39	23 58	24 31	00 31	01 02
40	03 11	03 52	04 25	24 03	00 03	00 34	01 03
45	02 41	03 30	04 07	24 10	00 10	00 38	01 04
S 50	01 58	03 01	03 45	24 17	00 17	00 42	01 05
52	01 32	02 47	03 35	24 21	00 21	00 44	01 06
54	00 54	02 30	03 23	24 24	00 24	00 46	01 06
56	////	02 09	03 09	00 05	00 29	00 49	01 07
58	////	01 41	02 53	00 12	00 33	00 51	01 07
S 60	////	00 59	02 34	00 20	00 38	00 54	01 08

Lat.	Sunset	Twilight Civil	Twilight Naut.	Moonset 9	10	11	12
N 72	■■	13 16	15 33	13 35	13 25	13 17	13 09
N 70	■■	14 08	15 52	13 17	13 15	13 13	13 10
68	■■	14 41	16 07	13 04	13 07	13 09	13 11
66	13 30	15 04	16 20	12 52	13 00	13 06	13 12
64	14 08	15 23	16 30	12 43	12 54	13 04	13 12
62	14 34	15 38	16 39	12 35	12 49	13 02	13 13
60	14 55	15 51	16 47	12 27	12 45	13 00	13 13
N 58	15 11	16 03	16 55	12 21	12 41	12 58	13 14
56	15 26	16 12	17 01	12 16	12 37	12 56	13 14
54	15 38	16 21	17 07	12 11	12 34	12 55	13 15
52	15 49	16 29	17 13	12 06	12 31	12 54	13 15
50	15 58	16 36	17 18	12 02	12 29	12 53	13 15
45	16 18	16 52	17 29	11 53	12 23	12 50	13 16
N 40	16 35	17 05	17 39	11 46	12 18	12 48	13 16
35	16 48	17 17	17 48	11 39	12 14	12 46	13 17
30	17 01	17 27	17 57	11 33	12 10	12 44	13 17
20	17 21	17 45	18 13	11 24	12 04	12 41	13 18
N 10	17 39	18 02	18 28	11 15	11 58	12 39	13 18
0	17 56	18 19	18 45	11 07	11 53	12 36	13 19
S 10	18 14	18 37	19 04	10 58	11 47	12 34	13 19
20	18 32	18 57	19 26	10 49	11 41	12 31	13 20
30	18 54	19 21	19 54	10 39	11 35	12 28	13 21
35	19 06	19 36	20 12	10 33	11 31	12 26	13 21
40	19 21	19 54	20 35	10 26	11 26	12 24	13 21
45	19 39	20 16	21 05	10 18	11 21	12 22	13 22
S 50	20 01	20 45	21 48	10 09	11 15	12 19	13 22
52	20 11	20 59	22 14	10 05	11 12	12 18	13 23
54	20 23	21 16	22 54	10 00	11 09	12 17	13 23
56	20 37	21 38	////	09 54	11 05	12 15	13 23
58	20 53	22 05	////	09 48	11 02	12 13	13 23
S 60	21 12	22 48	////	09 41	10 57	12 11	13 24

Day	SUN Eqn. of Time 00ʰ	SUN Eqn. of Time 12ʰ	Mer. Pass.	MOON Mer. Pass. Upper	MOON Mer. Pass. Lower	Age	Phase
9	07 58	07 45	11 52	04 55	17 19	20	62
10	07 31	07 18	11 53	05 42	18 04	21	53
11	07 04	06 50	11 53	06 26	18 48	22	43

1998 DECEMBER 12, 13, 14 (SAT., SUN., MON.)

UT	ARIES GHA	VENUS −3.9 GHA	Dec	MARS +1.3 GHA	Dec	JUPITER −2.4 GHA	Dec	SATURN +0.3 GHA	Dec
12 00	80 29.5	170 10.7	S24 07.8	252 25.6	S 1 33.3	89 36.2	S 5 21.3	54 25.5	N 7 57.6
01	95 31.9	185 09.7	07.9	267 26.8	33.8	104 38.5	21.2	69 28.0	57.6
02	110 34.4	200 08.8	08.0	282 28.0	34.3	119 40.7	21.1	84 30.6	57.5
03	125 36.8	215 07.8 ..	08.0	297 29.2 ..	34.8	134 43.0 ..	21.0	99 33.1 ..	57.5
04	140 39.3	230 06.8	08.1	312 30.5	35.4	149 45.2	20.9	114 35.7	57.5
05	155 41.8	245 05.8	08.2	327 31.7	35.9	164 47.5	20.8	129 38.2	57.5
S 06	170 44.2	260 04.9	S24 08.3	342 32.9	S 1 36.4	179 49.7	S 5 20.7	144 40.7	N 7 57.5
A 07	185 46.7	275 03.9	08.4	357 34.1	36.9	194 52.0	20.6	159 43.3	57.4
T 08	200 49.2	290 02.9	08.4	12 35.4	37.4	209 54.2	20.5	174 45.8	57.4
U 09	215 51.6	305 01.9 ..	08.5	27 36.6 ..	37.9	224 56.5 ..	20.4	189 48.4 ..	57.4
R 10	230 54.1	320 01.0	08.6	42 37.8	38.5	239 58.7	20.3	204 50.9	57.4
D 11	245 56.6	335 00.0	08.7	57 39.0	39.0	255 01.0	20.2	219 53.4	57.4
A 12	260 59.0	349 59.0	S24 08.7	72 40.2	S 1 39.5	270 03.2	S 5 20.1	234 56.0	N 7 57.4
Y 13	276 01.5	4 58.0	08.8	87 41.5	40.0	285 05.5	20.0	249 58.5	57.3
14	291 04.0	19 57.0	08.9	102 42.7	40.5	300 07.7	19.8	265 01.1	57.3
15	306 06.4	34 56.1 ..	08.9	117 43.9 ..	41.0	315 10.0 ..	19.7	280 03.6 ..	57.3
16	321 08.9	49 55.1	09.0	132 45.1	41.5	330 12.2	19.6	295 06.2	57.3
17	336 11.3	64 54.1	09.1	147 46.4	42.0	345 14.5	19.5	310 08.7	57.3
18	351 13.8	79 53.1	S24 09.1	162 47.6	S 1 42.6	0 16.7	S 5 19.4	325 11.2	N 7 57.2
19	6 16.3	94 52.2	09.2	177 48.8	43.1	15 19.0	19.3	340 13.8	57.2
20	21 18.7	109 51.2	09.2	192 50.0	43.6	30 21.2	19.2	355 16.3	57.2
21	36 21.2	124 50.2 ..	09.3	207 51.2 ..	44.1	45 23.5 ..	19.1	10 18.9 ..	57.2
22	51 23.7	139 49.2	09.4	222 52.5	44.6	60 25.7	19.0	25 21.4	57.2
23	66 26.1	154 48.3	09.4	237 53.7	45.1	75 28.0	18.9	40 23.9	57.2
13 00	81 28.6	169 47.3	S24 09.5	252 54.9	S 1 45.6	90 30.2	S 5 18.8	55 26.5	N 7 57.1
01	96 31.1	184 46.3	09.5	267 56.1	46.2	105 32.5	18.7	70 29.0	57.1
02	111 33.5	199 45.3	09.6	282 57.4	46.7	120 34.7	18.6	85 31.6	57.1
03	126 36.0	214 44.4 ..	09.6	297 58.6 ..	47.2	135 36.9 ..	18.5	100 34.1 ..	57.1
04	141 38.4	229 43.4	09.7	312 59.8	47.7	150 39.2	18.4	115 36.6	57.1
05	156 40.9	244 42.4	09.7	328 01.0	48.2	165 41.4	18.3	130 39.2	57.1
S 06	171 43.4	259 41.4	S24 09.8	343 02.3	S 1 48.7	180 43.7	S 5 18.2	145 41.7	N 7 57.0
U 07	186 45.8	274 40.5	09.8	358 03.5	49.2	195 45.9	18.1	160 44.2	57.0
N 08	201 48.3	289 39.5	09.9	13 04.7	49.7	210 48.2	18.0	175 46.8	57.0
D 09	216 50.8	304 38.5 ..	09.9	28 05.9 ..	50.2	225 50.4 ..	17.9	190 49.3 ..	57.0
A 10	231 53.2	319 37.5	09.9	43 07.2	50.8	240 52.7	17.8	205 51.9	57.0
Y 11	246 55.7	334 36.5	10.0	58 08.4	51.3	255 54.9	17.7	220 54.4	57.0
12	261 58.2	349 35.6	S24 10.0	73 09.6	S 1 51.8	270 57.1	S 5 17.6	235 56.9	N 7 56.9
13	277 00.6	4 34.6	10.1	88 10.8	52.3	285 59.4	17.5	250 59.5	56.9
14	292 03.1	19 33.6	10.1	103 12.1	52.8	301 01.6	17.4	266 02.0	56.9
15	307 05.6	34 32.6 ..	10.1	118 13.3 ..	53.3	316 03.9 ..	17.3	281 04.5 ..	56.9
16	322 08.0	49 31.7	10.2	133 14.5	53.8	331 06.1	17.2	296 07.1	56.9
17	337 10.5	64 30.7	10.2	148 15.8	54.3	346 08.4	17.0	311 09.6	56.9
18	352 12.9	79 29.7	S24 10.2	163 17.0	S 1 54.9	1 10.6	S 5 16.9	326 12.2	N 7 56.9
19	7 15.4	94 28.7	10.3	178 18.2	55.4	16 12.8	16.8	341 14.7	56.8
20	22 17.9	109 27.7	10.3	193 19.4	55.9	31 15.1	16.7	356 17.2	56.8
21	37 20.3	124 26.8 ..	10.3	208 20.7 ..	56.4	46 17.3 ..	16.6	11 19.8 ..	56.8
22	52 22.8	139 25.8	10.3	223 21.9	56.9	61 19.6	16.5	26 22.3	56.8
23	67 25.3	154 24.8	10.4	238 23.1	57.4	76 21.8	16.4	41 24.8	56.8
14 00	82 27.7	169 23.8	S24 10.4	253 24.3	S 1 57.9	91 24.0	S 5 16.3	56 27.4	N 7 56.8
01	97 30.2	184 22.9	10.4	268 25.6	58.4	106 26.3	16.2	71 29.9	56.7
02	112 32.7	199 21.9	10.4	283 26.8	58.9	121 28.5	16.1	86 32.4	56.7
03	127 35.1	214 20.9 ..	10.5	298 28.0	1 59.4	136 30.8 ..	16.0	101 35.0 ..	56.7
04	142 37.6	229 19.9	10.5	313 29.3	2 00.0	151 33.0	15.9	116 37.5	56.7
05	157 40.1	244 19.0	10.5	328 30.5	00.5	166 35.2	15.8	131 40.0	56.7
M 06	172 42.5	259 18.0	S24 10.5	343 31.7	S 2 01.0	181 37.5	S 5 15.7	146 42.6	N 7 56.7
O 07	187 45.0	274 17.0	10.5	358 32.9	01.5	196 39.7	15.6	161 45.1	56.7
N 08	202 47.4	289 16.0	10.5	13 34.2	02.0	211 41.9	15.5	176 47.6	56.6
D 09	217 49.9	304 15.0 ..	10.5	28 35.4 ..	02.5	226 44.2 ..	15.3	191 50.2 ..	56.6
A 10	232 52.4	319 14.1	10.6	43 36.6	03.0	241 46.4	15.2	206 52.7	56.6
Y 11	247 54.8	334 13.1	10.6	58 37.9	03.5	256 48.7	15.1	221 55.2	56.6
12	262 57.3	349 12.1	S24 10.6	73 39.1	S 2 04.0	271 50.9	S 5 15.0	236 57.8	N 7 56.6
13	277 59.8	4 11.1	10.6	88 40.3	04.5	286 53.1	14.9	252 00.3	56.6
14	293 02.2	19 10.2	10.6	103 41.5	05.0	301 55.4	14.8	267 02.8	56.6
15	308 04.7	34 09.2 ..	10.6	118 42.8 ..	05.6	316 57.6 ..	14.7	282 05.4 ..	56.5
16	323 07.2	49 08.2	10.6	133 44.0	06.1	331 59.8	14.6	297 07.9	56.5
17	338 09.6	64 07.2	10.6	148 45.2	06.6	347 02.1	14.5	312 10.4	56.5
18	353 12.1	79 06.2	S24 10.6	163 46.5	S 2 07.1	2 04.3	S 5 14.4	327 13.0	N 7 56.5
19	8 14.5	94 05.3	10.6	178 47.7	07.6	17 06.5	14.3	342 15.5	56.5
20	23 17.0	109 04.3	10.6	193 48.9	08.1	32 08.8	14.2	357 18.0	56.5
21	38 19.5	124 03.3 ..	10.6	208 50.2 ..	08.6	47 11.0 ..	14.1	12 20.6 ..	56.5
22	53 21.9	139 02.3	10.6	223 51.4	09.1	62 13.2	13.9	27 23.1	56.4
23	68 24.4	154 01.4	10.6	238 52.6	09.6	77 15.5	13.8	42 25.6	56.4
Mer. Pass. 18 31.1		v −1.0 d 0.0		v 1.2 d 0.5		v 2.2 d 0.1		v 2.5 d 0.0	

STARS

Name	SHA	Dec
Acamar	315 26.5	S40 18.8
Achernar	335 34.8	S57 14.9
Acrux	173 22.0	S63 05.2
Adhara	255 21.0	S28 58.3
Aldebaran	291 01.9	N16 30.3
Alioth	166 30.8	N55 57.8
Alkaid	153 08.1	N49 19.0
Al Na'ir	27 58.0	S46 58.2
Alnilam	275 57.4	S 1 12.3
Alphard	218 06.9	S 8 39.2
Alphecca	126 20.9	N26 43.2
Alpheratz	357 55.0	N29 05.2
Altair	62 19.4	N 8 52.1
Ankaa	353 26.7	S42 19.0
Antares	112 40.4	S26 25.6
Arcturus	146 06.2	N19 11.3
Atria	107 52.9	S69 01.3
Avior	234 22.0	S59 30.2
Bellatrix	278 43.7	N 6 20.8
Betelgeuse	271 13.1	N 7 24.3
Canopus	264 00.6	S52 41.8
Capella	280 50.5	N45 59.7
Deneb	49 39.4	N45 16.8
Denebola	182 45.1	N14 34.7
Diphda	349 07.0	S17 59.7
Dubhe	194 05.4	N61 45.1
Elnath	278 26.4	N28 36.3
Eltanin	90 51.9	N51 29.5
Enif	33 58.3	N 9 52.3
Fomalhaut	15 36.4	S29 37.9
Gacrux	172 13.5	S57 06.1
Gienah	176 03.9	S17 32.0
Hadar	149 04.2	S60 21.7
Hamal	328 13.2	N23 27.4
Kaus Aust.	83 59.1	S34 23.0
Kochab	137 20.7	N74 09.5
Markab	13 49.5	N15 12.0
Menkar	314 26.5	N 4 05.0
Menkent	148 21.1	S36 21.6
Miaplacidus	221 41.4	S69 42.5
Mirfak	308 56.0	N49 51.5
Nunki	76 12.6	S26 17.8
Peacock	53 37.3	S56 44.4
Pollux	243 41.1	N28 01.6
Procyon	245 11.2	N 5 13.6
Rasalhague	96 17.2	N12 33.8
Regulus	207 55.3	N11 58.3
Rigel	281 22.5	S 8 12.3
Rigil Kent.	140 07.6	S60 49.5
Sabik	102 25.8	S15 43.2
Schedar	349 53.1	N56 32.1
Shaula	96 37.6	S37 06.0
Sirius	258 43.3	S16 42.9
Spica	158 43.2	S11 09.2
Suhail	223 00.4	S43 25.5
Vega	80 47.0	N38 47.2
Zuben'ubi	137 18.1	S16 02.0

	SHA	Mer. Pass.
Venus	88 18.7	12 42
Mars	171 26.3	7 08
Jupiter	9 01.6	17 55
Saturn	333 57.9	20 15

UT		SUN		MOON				
		GHA	Dec	GHA	v	Dec	d	HP
d	h	° '	° '	° '	'	° '	'	'
12	00	181 39.0	S23 02.4	255 52.7	15.2	N 1 29.1	9.9	54.6
	01	196 38.7	02.6	270 26.9	15.2	1 19.2	10.0	54.6
	02	211 38.4	02.8	285 01.1	15.3	1 09.2	10.0	54.6
	03	226 38.1	.. 03.0	299 35.4	15.3	0 59.2	10.0	54.5
	04	241 37.8	03.2	314 09.7	15.2	0 49.2	10.0	54.5
	05	256 37.5	03.4	328 43.9	15.3	0 39.2	9.9	54.5
	06	271 37.2	S23 03.6	343 18.2	15.3	N 0 29.3	10.0	54.5
S	07	286 36.9	03.8	357 52.5	15.4	0 19.3	10.0	54.5
A	08	301 36.6	04.0	12 26.9	15.3	N 0 09.3	9.9	54.5
T	09	316 36.4	.. 04.2	27 01.2	15.4	S 0 00.6	10.0	54.4
U	10	331 36.1	04.3	41 35.6	15.3	0 10.6	9.9	54.4
R	11	346 35.8	04.5	56 09.9	15.4	0 20.5	10.0	54.4
D	12	1 35.5	S23 04.7	70 44.3	15.4	S 0 30.5	9.9	54.4
A	13	16 35.2	04.9	85 18.7	15.4	0 40.4	10.0	54.4
Y	14	31 34.9	05.1	99 53.1	15.4	0 50.4	9.9	54.4
	15	46 34.6	.. 05.3	114 27.5	15.5	1 00.3	9.9	54.4
	16	61 34.3	05.5	129 02.0	15.4	1 10.2	9.9	54.3
	17	76 34.0	05.6	143 36.4	15.4	1 20.1	9.9	54.3
	18	91 33.7	S23 05.8	158 10.8	15.5	S 1 30.0	9.9	54.3
	19	106 33.4	06.0	172 45.3	15.4	1 39.9	9.9	54.3
	20	121 33.1	06.2	187 19.7	15.5	1 49.8	9.8	54.3
	21	136 32.9	.. 06.4	201 54.2	15.5	1 59.6	9.8	54.3
	22	151 32.6	06.5	216 28.7	15.4	2 09.5	9.8	54.3
	23	166 32.3	06.7	231 03.1	15.5	2 19.3	9.9	54.3
13	00	181 32.0	S23 06.9	245 37.6	15.5	S 2 29.2	9.8	54.3
	01	196 31.7	07.1	260 12.1	15.5	2 39.0	9.8	54.2
	02	211 31.4	07.2	274 46.6	15.4	2 48.8	9.8	54.2
	03	226 31.1	.. 07.4	289 21.0	15.5	2 58.6	9.8	54.2
	04	241 30.8	07.6	303 55.5	15.5	3 08.4	9.7	54.2
	05	256 30.5	07.8	318 30.0	15.4	3 18.1	9.8	54.2
	06	271 30.2	S23 07.9	333 04.4	15.5	S 3 27.9	9.7	54.2
S	07	286 29.9	08.1	347 38.9	15.5	3 37.6	9.7	54.2
U	08	301 29.6	08.3	2 13.4	15.4	3 47.3	9.7	54.2
N	09	316 29.3	.. 08.5	16 47.8	15.5	3 57.0	9.7	54.2
D	10	331 29.0	08.6	31 22.3	15.5	4 06.7	9.7	54.2
A	11	346 28.7	08.8	45 56.8	15.4	4 16.4	9.6	54.2
Y	12	1 28.4	S23 09.0	60 31.2	15.4	S 4 26.0	9.6	54.1
	13	16 28.1	09.1	75 05.6	15.4	4 35.6	9.6	54.1
	14	31 27.9	09.3	89 40.1	15.4	4 45.2	9.6	54.1
	15	46 27.6	.. 09.5	104 14.5	15.4	4 54.8	9.6	54.1
	16	61 27.3	09.6	118 48.9	15.4	5 04.4	9.5	54.1
	17	76 27.0	09.8	133 23.3	15.4	5 13.9	9.6	54.1
	18	91 26.7	S23 09.9	147 57.7	15.4	S 5 23.5	9.5	54.1
	19	106 26.4	10.1	162 32.1	15.4	5 33.0	9.4	54.1
	20	121 26.1	10.3	177 06.5	15.3	5 42.4	9.5	54.1
	21	136 25.8	.. 10.4	191 40.8	15.4	5 51.9	9.4	54.1
	22	151 25.5	10.6	206 15.2	15.3	6 01.3	9.4	54.1
	23	166 25.2	10.7	220 49.5	15.4	6 10.7	9.4	54.1
14	00	181 24.9	S23 10.9	235 23.9	15.3	S 6 20.1	9.3	54.1
	01	196 24.6	11.1	249 58.2	15.2	6 29.4	9.4	54.1
	02	211 24.3	11.2	264 32.4	15.3	6 38.8	9.3	54.1
	03	226 24.0	.. 11.4	279 06.7	15.3	6 48.1	9.2	54.1
	04	241 23.7	11.5	293 41.0	15.2	6 57.3	9.3	54.1
	05	256 23.4	11.7	308 15.2	15.2	7 06.6	9.2	54.1
	06	271 23.1	S23 11.8	322 49.4	15.2	S 7 15.8	9.2	54.1
M	07	286 22.8	12.0	337 23.6	15.2	7 25.0	9.1	54.1
O	08	301 22.5	12.1	351 57.8	15.2	7 34.1	9.2	54.1
N	09	316 22.2	.. 12.3	6 32.0	15.1	7 43.3	9.1	54.0
D	10	331 21.9	12.4	21 06.1	15.1	7 52.4	9.0	54.0
A	11	346 21.6	12.6	35 40.2	15.1	8 01.4	9.1	54.0
Y	12	1 21.3	S23 12.7	50 14.3	15.1	S 8 10.5	9.0	54.0
	13	16 21.0	12.9	64 48.4	15.1	8 19.5	8.9	54.0
	14	31 20.7	13.0	79 22.5	15.0	8 28.4	9.0	54.0
	15	46 20.4	.. 13.2	93 56.5	15.0	8 37.4	8.9	54.0
	16	61 20.1	13.3	108 30.5	15.0	8 46.3	8.8	54.0
	17	76 19.8	13.5	123 04.5	14.9	8 55.1	8.9	54.0
	18	91 19.5	S23 13.6	137 38.4	14.9	S 9 04.0	8.7	54.0
	19	106 19.2	13.7	152 12.3	14.9	9 12.7	8.8	54.0
	20	121 18.9	13.9	166 46.2	14.9	9 21.5	8.7	54.0
	21	136 18.6	.. 14.0	181 20.1	14.9	9 30.2	8.7	54.0
	22	151 18.3	14.2	195 54.0	14.8	9 38.9	8.7	54.0
	23	166 18.0	14.3	210 27.8	14.8	S 9 47.6	8.6	54.0
		SD 16.3	d 0.2	SD 14.8		14.8		14.7

Lat.	Twilight		Sunrise	Moonrise			
	Naut.	Civil		12	13	14	15
°	h m	h m	h m	h m	h m	h m	h m
N 72	08 17	10 40	■■■	00 42	02 18	03 54	05 34
N 70	07 57	09 43	■■■	00 44	02 14	03 44	05 15
68	07 42	09 10	■■■	00 45	02 10	03 35	05 01
66	07 29	08 46	10 22	00 47	02 08	03 28	04 49
64	07 18	08 26	09 43	00 48	02 05	03 22	04 39
62	07 09	08 11	09 16	00 49	02 03	03 17	04 31
60	07 01	07 57	08 55	00 49	02 01	03 13	04 24
N 58	06 54	07 46	08 37	00 50	02 00	03 09	04 17
56	06 47	07 36	08 23	00 51	01 58	03 05	04 12
54	06 41	07 27	08 11	00 51	01 57	03 02	04 07
52	06 35	07 19	08 00	00 52	01 56	02 59	04 03
50	06 30	07 12	07 50	00 52	01 55	02 57	03 58
45	06 19	06 56	07 30	00 53	01 52	02 51	03 50
N 40	06 09	06 43	07 13	00 54	01 50	02 46	03 43
35	05 59	06 31	06 59	00 55	01 49	02 43	03 36
30	05 51	06 20	06 47	00 55	01 47	02 39	03 31
20	05 34	06 02	06 26	00 57	01 45	02 33	03 22
N 10	05 18	05 45	06 08	00 58	01 43	02 28	03 13
0	05 02	05 28	05 50	00 59	01 41	02 23	03 06
S 10	04 43	05 10	05 33	01 00	01 39	02 18	02 58
20	04 21	04 50	05 14	01 01	01 37	02 13	02 50
30	03 52	04 25	04 53	01 02	01 34	02 07	02 41
35	03 34	04 10	04 40	01 02	01 33	02 04	02 36
40	03 11	03 52	04 25	01 03	01 31	02 00	02 30
45	02 41	03 30	04 07	01 04	01 30	01 56	02 23
S 50	01 56	03 01	03 45	01 05	01 28	01 50	02 15
52	01 30	02 46	03 34	01 06	01 27	01 48	02 11
54	00 48	02 28	03 22	01 06	01 26	01 45	02 07
56	////	02 07	03 08	01 07	01 24	01 42	02 02
58	////	01 38	02 52	01 07	01 23	01 39	01 57
S 60	////	00 53	02 32	01 08	01 22	01 36	01 51

Lat.	Sunset	Twilight		Moonset			
		Civil	Naut.	12	13	14	15
°	h m	h m	h m	h m	h m	h m	h m
N 72	■■■	13 08	15 31	13 09	13 01	12 52	12 42
N 70	■■■	14 04	15 50	13 10	13 08	13 05	13 02
68	■■■	14 38	16 06	13 11	13 13	13 15	13 18
66	13 25	15 02	16 19	13 12	13 17	13 23	13 31
64	14 05	15 22	16 29	13 12	13 21	13 30	13 41
62	14 32	15 37	16 39	13 13	13 24	13 36	13 50
60	14 53	15 51	16 47	13 13	13 27	13 42	13 58
N 58	15 11	16 02	16 54	13 14	13 30	13 46	14 05
56	15 25	16 12	17 01	13 14	13 32	13 51	14 11
54	15 37	16 21	17 07	13 15	13 34	13 54	14 17
52	15 48	16 29	17 13	13 15	13 36	13 58	14 22
50	15 58	16 36	17 18	13 15	13 38	14 01	14 26
45	16 19	16 52	17 29	13 16	13 41	14 08	14 36
N 40	16 35	17 06	17 40	13 16	13 45	14 14	14 44
35	16 49	17 17	17 49	13 17	13 47	14 19	14 51
30	17 01	17 28	17 57	13 17	13 50	14 23	14 58
20	17 22	17 46	18 14	13 18	13 54	14 30	15 08
N 10	17 41	18 03	18 30	13 18	13 58	14 37	15 18
0	17 58	18 20	18 47	13 19	14 01	14 43	15 27
S 10	18 15	18 38	19 05	13 19	14 04	14 50	15 36
20	18 34	18 58	19 28	13 20	14 08	14 56	15 45
30	18 56	19 23	19 56	13 21	14 12	15 04	15 56
35	19 09	19 38	20 15	13 21	14 15	15 08	16 02
40	19 24	19 56	20 38	13 21	14 18	15 14	16 10
45	19 41	20 19	21 08	13 22	14 21	15 19	16 18
S 50	20 04	20 48	21 52	13 22	14 25	15 26	16 28
52	20 14	21 03	22 19	13 23	14 26	15 30	16 33
54	20 26	21 20	23 02	13 23	14 28	15 33	16 38
56	20 40	21 42	////	13 23	14 30	15 37	16 44
58	20 57	22 11	////	13 23	14 33	15 42	16 50
S 60	21 17	22 57	////	13 24	14 35	15 47	16 58

	SUN			MOON			
Day	Eqn. of Time		Mer. Pass.	Mer. Pass.		Age	Phase
	00ʰ	12ʰ		Upper	Lower		
d	m s	m s	h m	h m	h m	d	%
12	06 36	06 23	11 54	07 09	19 30	23	34
13	06 08	05 54	11 54	07 51	20 12	24	25
14	05 40	05 26	11 55	08 33	20 54	25	17

1998 DECEMBER 15, 16, 17 (TUES., WED., THURS.)

UT	ARIES GHA	VENUS −3.9 GHA	Dec	MARS +1.2 GHA	Dec	JUPITER −2.4 GHA	Dec	SATURN +0.3 GHA	Dec
d h	° ′	° ′	° ′	° ′	° ′	° ′	° ′	° ′	° ′
15 00	83 26.9	169 00.4	S24 10.6	253 53.9	S 2 10.1	92 17.7	S 5 13.7	57 28.2	N 7 56.4
01	98 29.3	183 59.4	10.6	268 55.1	10.6	107 19.9	13.6	72 30.7	56.4
02	113 31.8	198 58.4	10.5	283 56.3	11.1	122 22.2	13.5	87 33.2	56.4
03	128 34.3	213 57.5	.. 10.5	298 57.6	.. 11.7	137 24.4	.. 13.4	102 35.8	.. 56.4
04	143 36.7	228 56.5	10.5	313 58.8	12.2	152 26.6	13.3	117 38.3	56.4
05	158 39.2	243 55.5	10.5	329 00.0	12.7	167 28.9	13.2	132 40.8	56.4
06	173 41.7	258 54.5	S24 10.5	344 01.3	S 2 13.2	182 31.1	S 5 13.1	147 43.4	N 7 56.3
T 07	188 44.1	273 53.5	10.5	359 02.5	13.7	197 33.3	13.0	162 45.9	56.3
U 08	203 46.6	288 52.6	10.5	14 03.7	14.2	212 35.6	12.8	177 48.4	56.3
E 09	218 49.0	303 51.6	.. 10.4	29 04.9	.. 14.7	227 37.8	.. 12.7	192 50.9	.. 56.3
S 10	233 51.5	318 50.6	10.4	44 06.2	15.2	242 40.0	12.6	207 53.5	56.3
D 11	248 54.0	333 49.6	10.4	59 07.4	15.7	257 42.2	12.5	222 56.0	56.3
A 12	263 56.4	348 48.7	S24 10.4	74 08.6	S 2 16.2	272 44.5	S 5 12.4	237 58.5	N 7 56.3
Y 13	278 58.9	3 47.7	10.3	89 09.9	16.7	287 46.7	12.3	253 01.1	56.2
14	294 01.4	18 46.7	10.3	104 11.1	17.2	302 48.9	12.2	268 03.6	56.2
15	309 03.8	33 45.7	.. 10.3	119 12.4	.. 17.7	317 51.2	.. 12.1	283 06.1	.. 56.2
16	324 06.3	48 44.7	10.3	134 13.6	18.2	332 53.4	12.0	298 08.6	56.2
17	339 08.8	63 43.8	10.2	149 14.8	18.7	347 55.6	11.9	313 11.2	56.2
18	354 11.2	78 42.8	S24 10.2	164 16.1	S 2 19.3	2 57.8	S 5 11.7	328 13.7	N 7 56.2
19	9 13.7	93 41.8	10.2	179 17.3	19.8	18 00.1	11.6	343 16.2	56.2
20	24 16.2	108 40.8	10.1	194 18.5	20.3	33 02.3	11.5	358 18.8	56.2
21	39 18.6	123 39.9	.. 10.1	209 19.8	.. 20.8	48 04.5	.. 11.4	13 21.3	.. 56.1
22	54 21.1	138 38.9	10.1	224 21.0	21.3	63 06.8	11.3	28 23.8	56.1
23	69 23.5	153 37.9	10.0	239 22.2	21.8	78 09.0	11.2	43 26.3	56.1
16 00	84 26.0	168 36.9	S24 10.0	254 23.5	S 2 22.3	93 11.2	S 5 11.1	58 28.9	N 7 56.1
01	99 28.5	183 36.0	10.0	269 24.7	22.8	108 13.4	11.0	73 31.4	56.1
02	114 30.9	198 35.0	09.9	284 25.9	23.3	123 15.7	10.8	88 33.9	56.1
03	129 33.4	213 34.0	.. 09.9	299 27.2	.. 23.8	138 17.9	.. 10.7	103 36.4	.. 56.1
04	144 35.9	228 33.0	09.8	314 28.4	24.3	153 20.1	10.6	118 39.0	56.1
05	159 38.3	243 32.1	09.8	329 29.6	24.8	168 22.3	10.5	133 41.5	56.1
06	174 40.8	258 31.1	S24 09.7	344 30.9	S 2 25.3	183 24.6	S 5 10.4	148 44.0	N 7 56.0
W 07	189 43.3	273 30.1	09.7	359 32.1	25.8	198 26.8	10.3	163 46.6	56.0
E 08	204 45.7	288 29.1	09.6	14 33.4	26.3	213 29.0	10.2	178 49.1	56.0
D 09	219 48.2	303 28.1	.. 09.6	29 34.6	.. 26.8	228 31.2	.. 10.1	193 51.6	.. 56.0
N 10	234 50.6	318 27.2	09.5	44 35.8	27.3	243 33.5	09.9	208 54.1	56.0
E 11	249 53.1	333 26.2	09.5	59 37.1	27.8	258 35.7	09.8	223 56.7	56.0
S 12	264 55.6	348 25.2	S24 09.4	74 38.3	S 2 28.3	273 37.9	S 5 09.7	238 59.2	N 7 56.0
D 13	279 58.0	3 24.2	09.4	89 39.5	28.8	288 40.1	09.6	254 01.7	56.0
A 14	295 00.5	18 23.3	09.3	104 40.8	29.3	303 42.3	09.5	269 04.2	56.0
Y 15	310 03.0	33 22.3	.. 09.3	119 42.0	.. 29.8	318 44.6	.. 09.4	284 06.8	.. 55.9
16	325 05.4	48 21.3	09.2	134 43.3	30.3	333 46.8	09.3	299 09.3	55.9
17	340 07.9	63 20.3	09.1	149 44.5	30.9	348 49.0	09.2	314 11.8	55.9
18	355 10.4	78 19.4	S24 09.1	164 45.7	S 2 31.4	3 51.2	S 5 09.0	329 14.3	N 7 55.9
19	10 12.8	93 18.4	09.0	179 47.0	31.9	18 53.5	08.9	344 16.9	55.9
20	25 15.3	108 17.4	08.9	194 48.2	32.4	33 55.7	08.8	359 19.4	55.9
21	40 17.8	123 16.4	.. 08.9	209 49.5	.. 32.9	48 57.9	.. 08.7	14 21.9	.. 55.9
22	55 20.2	138 15.5	08.8	224 50.7	33.4	64 00.1	08.6	29 24.4	55.9
23	70 22.7	153 14.5	08.7	239 51.9	33.9	79 02.3	08.5	44 26.9	55.9
17 00	85 25.1	168 13.5	S24 08.7	254 53.2	S 2 34.4	94 04.6	S 5 08.4	59 29.5	N 7 55.8
01	100 27.6	183 12.5	08.6	269 54.4	34.9	109 06.8	08.2	74 32.0	55.8
02	115 30.1	198 11.6	08.5	284 55.6	35.4	124 09.0	08.1	89 34.5	55.8
03	130 32.5	213 10.6	.. 08.5	299 56.9	.. 35.9	139 11.2	.. 08.0	104 37.0	.. 55.8
04	145 35.0	228 09.6	08.4	314 58.1	36.4	154 13.4	07.9	119 39.6	55.8
05	160 37.5	243 08.6	08.3	329 59.4	36.9	169 15.7	07.8	134 42.1	55.8
06	175 39.9	258 07.7	S24 08.2	345 00.6	S 2 37.4	184 17.9	S 5 07.7	149 44.6	N 7 55.8
T 07	190 42.4	273 06.7	08.1	0 01.9	37.9	199 20.1	07.6	164 47.1	55.8
H 08	205 44.9	288 05.7	08.1	15 03.1	38.4	214 22.3	07.4	179 49.6	55.8
U 09	220 47.3	303 04.7	.. 08.0	30 04.3	.. 38.9	229 24.5	.. 07.3	194 52.2	.. 55.8
R 10	235 49.8	318 03.8	07.9	45 05.6	39.4	244 26.7	07.2	209 54.7	55.8
S 11	250 52.3	333 02.8	07.8	60 06.8	39.9	259 29.0	07.1	224 57.2	55.7
D 12	265 54.7	348 01.8	S24 07.7	75 08.1	S 2 40.4	274 31.2	S 5 07.0	239 59.7	N 7 55.7
A 13	280 57.2	3 00.8	07.6	90 09.3	40.9	289 33.4	06.9	255 02.3	55.7
Y 14	295 59.6	17 59.9	07.6	105 10.5	41.4	304 35.6	06.7	270 04.8	55.7
15	311 02.1	32 58.9	.. 07.5	120 11.8	.. 41.9	319 37.8	.. 06.6	285 07.3	.. 55.7
16	326 04.6	47 57.9	07.4	135 13.0	42.4	334 40.0	06.5	300 09.8	55.7
17	341 07.0	62 56.9	07.3	150 14.3	42.9	349 42.3	06.4	315 12.3	55.7
18	356 09.5	77 56.0	S24 07.2	165 15.5	S 2 43.4	4 44.5	S 5 06.3	330 14.9	N 7 55.7
19	11 12.0	92 55.0	07.1	180 16.8	43.9	19 46.7	06.2	345 17.4	55.7
20	26 14.4	107 54.0	07.0	195 18.0	44.4	34 48.9	06.0	0 19.9	55.7
21	41 16.9	122 53.0	.. 06.9	210 19.2	.. 44.9	49 51.1	.. 05.9	15 22.4	.. 55.6
22	56 19.4	137 52.1	06.8	225 20.5	45.4	64 53.3	05.8	30 24.9	55.6
23	71 21.8	152 51.1	06.7	240 21.7	45.9	79 55.5	05.7	45 27.4	55.6
Mer. Pass. 18 19.3		v −1.0 d 0.1		v 1.2 d 0.5		v 2.2 d 0.1		v 2.5 d 0.0	

STARS

Name	SHA	Dec
	° ′	° ′
Acamar	315 26.5	S40 18.8
Achernar	335 34.9	S57 14.9
Acrux	173 21.9	S63 05.2
Adhara	255 21.0	S28 58.3
Aldebaran	291 01.9	N16 30.3
Alioth	166 30.8	N55 57.8
Alkaid	153 08.1	N49 19.0
Al Na'ir	27 58.0	S46 58.2
Alnilam	275 57.4	S 1 12.3
Alphard	218 06.9	S 8 39.2
Alphecca	126 20.8	N26 43.2
Alpheratz	357 55.0	N29 05.2
Altair	62 19.5	N 8 52.1
Ankaa	353 26.7	S42 19.0
Antares	112 40.4	S26 25.6
Arcturus	146 06.2	N19 11.3
Atria	107 52.9	S69 01.3
Avior	234 22.0	S59 30.2
Bellatrix	278 43.7	N 6 20.8
Betelgeuse	271 13.1	N 7 24.3
Canopus	264 00.6	S52 41.8
Capella	280 50.5	N45 59.7
Deneb	49 39.4	N45 16.8
Denebola	182 45.1	N14 34.7
Diphda	349 07.0	S17 59.7
Dubhe	194 05.4	N61 45.1
Elnath	278 26.4	N28 36.3
Eltanin	90 51.9	N51 29.5
Enif	33 58.3	N 9 52.3
Fomalhaut	15 36.4	S29 37.9
Gacrux	172 13.5	S57 06.1
Gienah	176 03.9	S17 32.0
Hadar	149 04.2	S60 21.7
Hamal	328 13.2	N23 27.4
Kaus Aust.	83 59.1	S34 23.0
Kochab	137 20.7	N74 09.5
Markab	13 49.5	N15 12.0
Menkar	314 26.5	N 4 05.0
Menkent	148 21.0	S36 21.6
Miaplacidus	221 41.3	S69 42.5
Mirfak	308 56.0	N49 51.5
Nunki	76 12.6	S26 17.8
Peacock	53 37.4	S56 44.4
Pollux	243 41.1	N28 01.6
Procyon	245 11.2	N 5 13.6
Rasalhague	96 17.2	N12 33.8
Regulus	207 55.3	N11 58.3
Rigel	281 22.5	S 8 12.3
Rigil Kent.	140 07.6	S60 49.5
Sabik	102 25.8	S15 43.2
Schedar	349 53.1	N56 32.1
Shaula	96 37.6	S37 06.0
Sirius	258 43.3	S16 42.9
Spica	158 43.2	S11 09.2
Suhail	223 00.4	S43 25.5
Vega	80 47.0	N38 47.1
Zuben'ubi	137 18.1	S16 02.0

	SHA	Mer. Pass.
	° ′	h m
Venus	84 10.9	12 46
Mars	169 57.5	7 02
Jupiter	8 45.2	17 45
Saturn	334 02.9	20 03

UT	SUN GHA	SUN Dec	MOON GHA	v	Dec	d	HP
d h	° ′	° ′	° ′	′	° ′	′	′
15 00	181 17.7	S23 14.4	225 01.6	14.7	S 9 56.2	8.5	54.0
01	196 17.4	14.6	239 35.3	14.8	10 04.7	8.6	54.0
02	211 17.1	14.7	254 09.1	14.7	10 13.3	8.5	54.0
03	226 16.8 ..	14.9	268 42.8	14.6	10 21.8	8.4	54.1
04	241 16.5	15.0	283 16.4	14.7	10 30.2	8.4	54.1
05	256 16.2	15.1	297 50.1	14.6	10 38.6	8.4	54.1
06	271 15.9 S23 15.3		312 23.7	14.5	S10 47.0	8.3	54.1
07	286 15.6	15.4	326 57.2	14.6	10 55.3	8.3	54.1
08	301 15.3	15.5	341 30.8	14.5	11 03.6	8.2	54.1
09	316 15.0 ..	15.7	356 04.3	14.4	11 11.8	8.2	54.1
10	331 14.7	15.8	10 37.7	14.5	11 20.0	8.2	54.1
11	346 14.4	15.9	25 11.2	14.4	11 28.2	8.1	54.1
12	1 14.1 S23 16.0		39 44.6	14.3	S11 36.3	8.0	54.1
13	16 13.8	16.2	54 17.9	14.4	11 44.3	8.0	54.1
14	31 13.5	16.3	68 51.3	14.3	11 52.3	8.0	54.1
15	46 13.2 ..	16.4	83 24.6	14.2	12 00.3	7.9	54.1
16	61 12.9	16.6	97 57.8	14.3	12 08.2	7.9	54.1
17	76 12.6	16.7	112 31.1	14.1	12 16.1	7.8	54.1
18	91 12.3 S23 16.8		127 04.2	14.2	S12 23.9	7.7	54.1
19	106 12.0	16.9	141 37.4	14.1	12 31.6	7.7	54.1
20	121 11.7	17.0	156 10.5	14.1	12 39.3	7.7	54.1
21	136 11.4 ..	17.2	170 43.6	14.0	12 47.0	7.6	54.1
22	151 11.1	17.3	185 16.6	14.0	12 54.6	7.6	54.1
23	166 10.8	17.4	199 49.6	13.9	13 02.2	7.5	54.1
16 00	181 10.5 S23 17.5		214 22.5	14.0	S13 09.7	7.4	54.1
01	196 10.2	17.6	228 55.5	13.8	13 17.1	7.4	54.2
02	211 09.9	17.8	243 28.3	13.9	13 24.5	7.4	54.2
03	226 09.6 ..	17.9	258 01.2	13.8	13 31.9	7.3	54.2
04	241 09.3	18.0	272 34.0	13.7	13 39.2	7.2	54.2
05	256 09.0	18.1	287 06.7	13.7	13 46.4	7.2	54.2
06	271 08.7 S23 18.2		301 39.4	13.7	S13 53.6	7.1	54.2
07	286 08.4	18.3	316 12.1	13.6	14 00.7	7.1	54.2
08	301 08.1	18.5	330 44.7	13.6	14 07.8	7.0	54.2
09	316 07.8 ..	18.6	345 17.3	13.6	14 14.8	6.9	54.2
10	331 07.5	18.7	359 49.9	13.5	14 21.7	6.9	54.2
11	346 07.2	18.8	14 22.4	13.4	14 28.6	6.8	54.2
12	1 06.8 S23 18.9		28 54.8	13.4	S14 35.4	6.8	54.2
13	16 06.5	19.0	43 27.2	13.4	14 42.2	6.7	54.2
14	31 06.2	19.1	57 59.6	13.3	14 48.9	6.6	54.3
15	46 05.9 ..	19.2	72 31.9	13.3	14 55.5	6.6	54.3
16	61 05.6	19.3	87 04.2	13.3	15 02.1	6.5	54.3
17	76 05.3	19.4	101 36.5	13.2	15 08.6	6.5	54.3
18	91 05.0 S23 19.5		116 08.7	13.1	S15 15.1	6.4	54.3
19	106 04.7	19.6	130 40.8	13.1	15 21.5	6.3	54.3
20	121 04.4	19.7	145 12.9	13.1	15 27.8	6.2	54.3
21	136 04.1 ..	19.8	159 45.0	13.0	15 34.0	6.2	54.3
22	151 03.8	19.9	174 17.0	13.0	15 40.2	6.2	54.3
23	166 03.5	20.1	188 49.0	13.0	15 46.4	6.0	54.3
17 00	181 03.2 S23 20.2		203 21.0	12.8	S15 52.4	6.0	54.4
01	196 02.9	20.2	217 52.8	12.9	15 58.4	5.9	54.4
02	211 02.6	20.3	232 24.7	12.8	16 04.3	5.9	54.4
03	226 02.3 ..	20.4	246 56.5	12.8	16 10.2	5.7	54.4
04	241 02.0	20.5	261 28.3	12.7	16 15.9	5.7	54.4
05	256 01.7	20.6	276 00.0	12.6	16 21.6	5.7	54.4
06	271 01.4 S23 20.7		290 31.6	12.7	S16 27.3	5.5	54.4
07	286 01.1	20.8	305 03.3	12.6	16 32.8	5.5	54.4
08	301 00.7	20.9	319 34.9	12.5	16 38.3	5.4	54.4
09	316 00.4 ..	21.0	334 06.4	12.5	16 43.7	5.4	54.5
10	331 00.1	21.1	348 37.9	12.4	16 49.1	5.3	54.5
11	345 59.8	21.2	3 09.3	12.4	16 54.4	5.2	54.5
12	0 59.5 S23 21.3		17 40.7	12.4	S16 59.6	5.1	54.5
13	15 59.2	21.4	32 12.1	12.3	17 04.7	5.0	54.5
14	30 58.9	21.5	46 43.4	12.3	17 09.7	5.0	54.5
15	45 58.6 ..	21.6	61 14.7	12.2	17 14.7	4.9	54.5
16	60 58.3	21.6	75 45.9	12.2	17 19.6	4.8	54.5
17	75 58.0	21.7	90 17.1	12.2	17 24.4	4.7	54.6
18	90 57.7 S23 21.8		104 48.3	12.1	S17 29.1	4.7	54.6
19	105 57.4	21.9	119 19.4	12.0	17 33.8	4.5	54.6
20	120 57.1	22.0	133 50.4	12.0	17 38.3	4.5	54.6
21	135 56.8 ..	22.1	148 21.4	12.0	17 42.8	4.4	54.6
22	150 56.5	22.1	162 52.4	11.9	17 47.2	4.4	54.6
23	165 56.1	22.2	177 23.3	11.9	S17 51.6	4.2	54.6
	SD 16.3 d 0.1		SD 14.7		14.8		14.8

(TUESDAY = Dec 15 block; WEDNESDAY = Dec 16 block; THURSDAY = Dec 17 block)

Twilight / Sunrise / Moonrise

Lat.	Twilight Naut.	Civil	Sunrise	Moonrise 15	16	17	18
°	h m	h m	h m	h m	h m	h m	h m
N 72	08 21	10 49	■	05 34	07 19	09 19	■
N 70	08 01	09 49	■	05 15	06 49	08 26	10 05
68	07 46	09 14	■	05 01	06 27	07 54	09 17
66	07 33	08 49	10 29	04 49	06 10	07 30	08 46
64	07 22	08 30	09 47	04 39	05 56	07 12	08 23
62	07 12	08 14	09 19	04 31	05 44	06 57	08 05
60	07 04	08 00	08 58	04 24	05 34	06 44	07 50
N 58	06 56	07 49	08 41	04 17	05 26	06 33	07 37
56	06 50	07 39	08 26	04 12	05 18	06 24	07 26
54	06 43	07 30	08 13	04 07	05 12	06 15	07 16
52	06 38	07 21	08 02	04 03	05 06	06 08	07 08
50	06 33	07 14	07 52	03 58	05 00	06 01	07 00
45	06 21	06 58	07 32	03 50	04 48	05 47	06 44
N 40	06 11	06 45	07 15	03 43	04 39	05 35	06 30
35	06 01	06 33	07 01	03 36	04 30	05 25	06 19
30	05 52	06 22	06 48	03 31	04 23	05 16	06 09
20	05 36	06 04	06 28	03 22	04 11	05 01	05 52
N 10	05 20	05 46	06 09	03 13	04 00	04 48	05 37
0	05 03	05 29	05 52	03 06	03 50	04 36	05 23
S 10	04 44	05 11	05 34	02 58	03 40	04 23	05 10
20	04 22	04 51	05 15	02 50	03 29	04 10	04 55
30	03 53	04 26	04 53	02 41	03 17	03 56	04 38
35	03 34	04 11	04 41	02 36	03 10	03 47	04 28
40	03 11	03 53	04 26	02 30	03 02	03 37	04 17
45	02 41	03 30	04 08	02 23	02 53	03 26	04 04
S 50	01 56	03 01	03 45	02 15	02 41	03 12	03 48
52	01 28	02 46	03 34	02 11	02 36	03 06	03 41
54	00 44	02 28	03 22	02 07	02 31	02 59	03 33
56	////	02 06	03 08	02 02	02 24	02 51	03 23
58	////	01 37	02 51	01 57	02 17	02 42	03 13
S 60	////	00 48	02 31	01 51	02 09	02 32	03 01

Sunset / Twilight / Moonset

Lat.	Sunset	Twilight Civil	Naut.	Moonset 15	16	17	18
°	h m	h m	h m	h m	h m	h m	h m
N 72	■	13 02	15 30	12 42	12 29	12 05	■
N 70	■	14 02	15 50	13 02	13 00	12 59	13 00
68	■	14 37	16 05	13 18	13 23	13 32	13 48
66	13 22	15 02	16 18	13 31	13 41	13 56	14 20
64	14 04	15 21	16 29	13 41	13 56	14 15	14 43
62	14 32	15 37	16 39	13 50	14 08	14 31	15 01
60	14 53	15 51	16 47	13 58	14 18	14 44	15 16
N 58	15 10	16 02	16 55	14 05	14 28	14 55	15 29
56	15 25	16 12	17 01	14 11	14 36	15 05	15 41
54	15 38	16 21	17 08	14 17	14 43	15 13	15 50
52	15 49	16 30	17 13	14 22	14 49	15 21	15 59
50	15 58	16 37	17 18	14 26	14 55	15 28	16 07
45	16 19	16 53	17 30	14 36	15 08	15 43	16 24
N 40	16 36	17 06	17 40	14 44	15 18	15 56	16 38
35	16 50	17 18	17 50	14 51	15 27	16 06	16 49
30	17 02	17 29	17 59	14 58	15 35	16 15	17 00
20	17 23	17 47	18 15	15 08	15 49	16 31	17 17
N 10	17 42	18 05	18 31	15 18	16 00	16 45	17 33
0	17 59	18 22	18 48	15 27	16 12	16 58	17 47
S 10	18 17	18 40	19 07	15 36	16 23	17 12	18 01
20	18 36	19 00	19 29	15 45	16 35	17 26	18 17
30	18 58	19 25	19 58	15 56	16 49	17 42	18 34
35	19 11	19 40	20 17	16 02	16 57	17 51	18 45
40	19 26	19 58	20 40	16 10	17 06	18 02	18 56
45	19 44	20 21	21 11	16 18	17 16	18 14	19 10
S 50	20 06	20 51	21 56	16 28	17 29	18 29	19 27
52	20 17	21 06	22 23	16 33	17 35	18 37	19 35
54	20 29	21 23	23 09	16 38	17 42	18 45	19 44
56	20 43	21 46	////	16 44	17 50	18 54	19 54
58	21 00	22 15	////	16 50	17 58	19 04	20 05
S 60	21 20	23 04	////	16 58	18 08	19 15	20 18

Day	SUN Eqn. of Time 00ʰ	12ʰ	Mer. Pass.	MOON Mer. Pass. Upper	Lower	Age	Phase
d	m s	m s	h m	h m	h m	d	%
15	05 12	04 57	11 55	09 16	21 38	26	11
16	04 43	04 28	11 56	10 01	22 24	27	6
17	04 13	03 59	11 56	10 47	23 11	28	2

UT (d h)	ARIES GHA	VENUS −3.9 GHA	Dec	MARS +1.2 GHA	Dec	JUPITER −2.4 GHA	Dec	SATURN +0.3 GHA	Dec	STARS Name	SHA	Dec
18 00	86 24.3	167 50.1	S24 06.6	255 23.0	S 2 46.4	94 57.7	S 5 05.6	60 30.0	N 7 55.6	Acamar	315 26.5	S40 18.8
01	101 26.8	182 49.2	06.5	270 24.2	46.9	110 00.0	05.5	75 32.5	55.6	Achernar	335 34.9	S57 14.9
02	116 29.2	197 48.2	06.4	285 25.5	47.4	125 02.2	05.3	90 35.0	55.6	Acrux	173 21.9	S63 05.2
03	131 31.7	212 47.2	.. 06.3	300 26.7	.. 47.9	140 04.4	.. 05.2	105 37.5	.. 55.6	Adhara	255 20.9	S28 58.3
04	146 34.1	227 46.2	06.2	315 27.9	48.4	155 06.6	05.1	120 40.0	55.6	Aldebaran	291 01.9	N16 30.3
05	161 36.6	242 45.3	06.1	330 29.2	48.9	170 08.8	05.0	135 42.6	55.6			
06	176 39.1	257 44.3	S24 06.0	345 30.4	S 2 49.4	185 11.0	S 5 04.9	150 45.1	N 7 55.6	Alioth	166 30.7	N55 57.8
07	191 41.5	272 43.3	05.9	0 31.7	49.9	200 13.2	04.7	165 47.6	55.6	Alkaid	153 08.0	N49 19.0
08	206 44.0	287 42.3	05.7	15 32.9	50.4	215 15.4	04.6	180 50.1	55.6	Al Na'ir	27 58.0	S46 58.2
F 09	221 46.5	302 41.4	.. 05.6	30 34.2	.. 50.9	230 17.7	.. 04.5	195 52.6	.. 55.6	Alnilam	275 57.4	S 1 12.3
R 10	236 48.9	317 40.4	05.5	45 35.4	51.4	245 19.9	04.4	210 55.1	55.5	Alphard	218 06.9	S 8 39.2
I 11	251 51.4	332 39.4	05.4	60 36.7	51.9	260 22.1	04.3	225 57.7	55.5			
D 12	266 53.9	347 38.5	S24 05.3	75 37.9	S 2 52.4	275 24.3	S 5 04.2	241 00.2	N 7 55.5	Alphecca	126 20.8	N26 43.2
A 13	281 56.3	2 37.5	05.2	90 39.2	52.9	290 26.5	04.0	256 02.7	55.5	Alpheratz	357 55.0	N29 05.2
Y 14	296 58.8	17 36.5	05.0	105 40.4	53.4	305 28.7	03.9	271 05.2	55.5	Altair	62 19.5	N 8 52.1
15	312 01.3	32 35.5	.. 04.9	120 41.6	.. 53.9	320 30.9	.. 03.8	286 07.7	.. 55.5	Ankaa	353 26.7	S42 19.0
16	327 03.7	47 34.6	04.8	135 42.9	54.4	335 33.1	03.7	301 10.2	55.5	Antares	112 40.4	S26 25.6
17	342 06.2	62 33.6	04.7	150 44.1	54.9	350 35.3	03.6	316 12.8	55.5			
18	357 08.6	77 32.6	S24 04.6	165 45.4	S 2 55.4	5 37.5	S 5 03.4	331 15.3	N 7 55.5	Arcturus	146 06.2	N19 11.3
19	12 11.1	92 31.7	04.4	180 46.6	55.9	20 39.7	03.3	346 17.8	55.5	Atria	107 52.8	S69 01.3
20	27 13.6	107 30.7	04.3	195 47.9	56.4	35 42.0	03.2	1 20.3	55.5	Avior	234 22.0	S59 30.3
21	42 16.0	122 29.7	.. 04.2	210 49.1	.. 56.9	50 44.2	.. 03.1	16 22.8	.. 55.5	Bellatrix	278 43.7	N 6 20.8
22	57 18.5	137 28.7	04.0	225 50.4	57.4	65 46.4	03.0	31 25.3	55.5	Betelgeuse	271 13.1	N 7 24.3
23	72 21.0	152 27.8	03.9	240 51.6	57.9	80 48.6	02.8	46 27.9	55.4			
19 00	87 23.4	167 26.8	S24 03.8	255 52.9	S 2 58.4	95 50.8	S 5 02.7	61 30.4	N 7 55.4	Canopus	264 00.5	S52 41.8
01	102 25.9	182 25.8	03.7	270 54.1	58.8	110 53.0	02.6	76 32.9	55.4	Capella	280 50.5	N45 59.7
02	117 28.4	197 24.9	03.5	285 55.4	59.3	125 55.2	02.5	91 35.4	55.4	Deneb	49 39.4	N45 16.8
03	132 30.8	212 23.9	.. 03.4	300 56.6	2 59.8	140 57.4	.. 02.4	106 37.9	.. 55.4	Denebola	182 45.1	N14 34.6
04	147 33.3	227 22.9	03.2	315 57.9	3 00.3	155 59.6	02.2	121 40.4	55.4	Diphda	349 07.1	S17 59.7
05	162 35.7	242 21.9	03.1	330 59.1	00.8	171 01.8	02.1	136 42.9	55.4			
06	177 38.2	257 21.0	S24 03.0	346 00.4	S 3 01.3	186 04.0	S 5 02.0	151 45.5	N 7 55.4	Dubhe	194 05.3	N61 45.1
07	192 40.7	272 20.0	02.8	1 01.6	01.8	201 06.2	01.9	166 48.0	55.4	Elnath	278 26.4	N28 36.3
S 08	207 43.1	287 19.0	02.7	16 02.9	02.3	216 08.4	01.8	181 50.5	55.4	Eltanin	90 51.9	N51 29.5
A 09	222 45.6	302 18.1	.. 02.5	31 04.1	.. 02.8	231 10.6	.. 01.6	196 53.0	.. 55.4	Enif	33 58.3	N 9 52.3
T 10	237 48.1	317 17.1	02.4	46 05.4	03.3	246 12.8	01.5	211 55.5	55.4	Fomalhaut	15 36.4	S29 37.9
U 11	252 50.5	332 16.1	02.2	61 06.6	03.8	261 15.0	01.4	226 58.0	55.4			
R 12	267 53.0	347 15.2	S24 02.1	76 07.9	S 3 04.3	276 17.2	S 5 01.3	242 00.5	N 7 55.4	Gacrux	172 13.5	S57 06.1
D 13	282 55.5	2 14.2	01.9	91 09.1	04.8	291 19.4	01.2	257 03.0	55.4	Gienah	176 03.9	S17 32.0
A 14	297 57.9	17 13.2	01.8	106 10.4	05.3	306 21.7	01.0	272 05.6	55.4	Hadar	149 04.1	S60 21.7
Y 15	313 00.4	32 12.3	.. 01.6	121 11.6	.. 05.8	321 23.9	.. 00.9	287 08.1	.. 55.3	Hamal	328 13.2	N23 27.4
16	328 02.9	47 11.3	01.5	136 12.9	06.3	336 26.1	00.8	302 10.6	55.3	Kaus Aust.	83 59.1	S34 23.0
17	343 05.3	62 10.3	01.3	151 14.1	06.8	351 28.3	00.7	317 13.1	55.3			
18	358 07.8	77 09.3	S24 01.2	166 15.4	S 3 07.3	6 30.5	S 5 00.5	332 15.6	N 7 55.3	Kochab	137 20.6	N74 09.5
19	13 10.2	92 08.4	01.0	181 16.6	07.8	21 32.7	00.4	347 18.1	55.3	Markab	13 49.5	N15 12.0
20	28 12.7	107 07.4	00.9	196 17.9	08.3	36 34.9	00.3	2 20.6	55.3	Menkar	314 26.5	N 4 05.0
21	43 15.2	122 06.4	.. 00.7	211 19.1	.. 08.8	51 37.1	.. 00.2	17 23.1	.. 55.3	Menkent	148 21.0	S36 21.6
22	58 17.6	137 05.5	00.5	226 20.4	09.2	66 39.3	5 00.1	32 25.6	55.3	Miaplacidus	221 41.3	S69 42.6
23	73 20.1	152 04.5	00.4	241 21.6	09.7	81 41.5	4 59.9	47 28.2	55.3			
20 00	88 22.6	167 03.5	S24 00.2	256 22.9	S 3 10.2	96 43.7	S 4 59.8	62 30.7	N 7 55.3	Mirfak	308 56.0	N49 51.5
01	103 25.0	182 02.6	24 00.1	271 24.1	10.7	111 45.9	59.7	77 33.2	55.3	Nunki	76 12.6	S26 17.8
02	118 27.5	197 01.6	23 59.9	286 25.4	11.2	126 48.1	59.6	92 35.7	55.3	Peacock	53 37.4	S56 44.4
03	133 30.0	212 00.6	.. 59.7	301 26.6	.. 11.7	141 50.3	.. 59.4	107 38.2	.. 55.3	Pollux	243 41.1	N28 01.6
04	148 32.4	226 59.7	59.6	316 27.9	12.2	156 52.5	59.3	122 40.7	55.3	Procyon	245 11.1	N 5 13.6
05	163 34.9	241 58.7	59.4	331 29.1	12.7	171 54.7	59.2	137 43.2	55.3			
06	178 37.4	256 57.7	S23 59.2	346 30.4	S 3 13.2	186 56.9	S 4 59.1	152 45.7	N 7 55.3	Rasalhague	96 17.2	N12 33.8
07	193 39.8	271 56.8	59.0	1 31.6	13.7	201 59.1	58.9	167 48.2	55.3	Regulus	207 55.3	N11 58.3
08	208 42.3	286 55.8	58.9	16 32.9	14.2	217 01.3	58.8	182 50.7	55.3	Rigel	281 22.5	S 8 12.3
S 09	223 44.7	301 54.8	.. 58.7	31 34.1	.. 14.7	232 03.5	.. 58.7	197 53.3	.. 55.3	Rigil Kent.	140 07.5	S60 49.5
U 10	238 47.2	316 53.9	58.5	46 35.4	15.2	247 05.7	58.6	212 55.8	55.2	Sabik	102 25.7	S15 43.2
N 11	253 49.7	331 52.9	58.3	61 36.7	15.7	262 07.9	58.5	227 58.3	55.2			
D 12	268 52.1	346 52.0	S23 58.2	76 37.9	S 3 16.2	277 10.1	S 4 58.3	243 00.8	N 7 55.2	Schedar	349 53.1	N56 32.1
A 13	283 54.6	1 51.0	58.0	91 39.2	16.6	292 12.2	58.2	258 03.3	55.2	Shaula	96 37.6	S37 06.0
Y 14	298 57.1	16 50.0	57.8	106 40.4	17.1	307 14.4	58.1	273 05.8	55.2	Sirius	258 43.2	S16 43.0
15	313 59.5	31 49.1	.. 57.6	121 41.7	.. 17.6	322 16.6	.. 58.0	288 08.3	.. 55.2	Spica	158 43.2	S11 09.2
16	329 02.0	46 48.1	57.4	136 42.9	18.1	337 18.8	57.8	303 10.8	55.2	Suhail	223 00.3	S43 25.6
17	344 04.5	61 47.1	57.3	151 44.2	18.6	352 21.0	57.7	318 13.3	55.2			
18	359 06.9	76 46.2	S23 57.1	166 45.4	S 3 19.1	7 23.2	S 4 57.6	333 15.8	N 7 55.2	Vega	80 47.0	N38 47.1
19	14 09.4	91 45.2	56.9	181 46.7	19.6	22 25.4	57.5	348 18.3	55.2	Zuben'ubi	137 18.1	S16 02.0
20	29 11.9	106 44.2	56.7	196 47.9	20.1	37 27.6	57.3	3 20.8	55.2		SHA	Mer. Pass.
21	44 14.3	121 43.3	.. 56.5	211 49.2	.. 20.6	52 29.8	.. 57.2	18 23.3	.. 55.2		° ′	h m
22	59 16.8	136 42.3	56.3	226 50.5	21.1	67 32.0	57.1	33 25.9	55.2	Venus	80 03.4	12 51
23	74 19.2	151 41.3	56.1	241 51.7	21.6	82 34.2	57.0	48 28.4	55.2	Mars	168 29.4	6 56
	h m									Jupiter	8 27.4	17 34
Mer. Pass. 18 07.5	v −1.0 d 0.1	v 1.3 d 0.5		v 2.2 d 0.1		v 2.5 d 0.0				Saturn	334 06.9	19 51

UT	SUN GHA	SUN Dec	MOON GHA	v	MOON Dec	d	HP
d h	° ′	° ′	° ′	′	° ′	′	′
18 00	180 55.8	S23 22.3	191 54.2	11.9	S17 55.8	4.2	54.6
01	195 55.5	22.4	206 25.1	11.8	18 00.0	4.1	54.7
02	210 55.2	22.5	220 55.9	11.7	18 04.1	4.0	54.7
03	225 54.9	22.5	235 26.6	11.7	18 08.1	3.9	54.7
04	240 54.6	22.6	249 57.3	11.7	18 12.0	3.8	54.7
05	255 54.3	22.7	264 28.0	11.7	18 15.8	3.8	54.7
06	270 54.0	S23 22.8	278 58.7	11.6	S18 19.6	3.6	54.7
07	285 53.7	22.8	293 29.3	11.5	18 23.2	3.6	54.7
08	300 53.4	22.9	307 59.8	11.5	18 26.8	3.5	54.8
F 09	315 53.1	23.0	322 30.3	11.5	18 30.3	3.4	54.8
R 10	330 52.8	23.1	337 00.8	11.5	18 33.7	3.3	54.8
I 11	345 52.5	23.1	351 31.3	11.4	18 37.0	3.2	54.8
D 12	0 52.1	S23 23.2	6 01.7	11.3	S18 40.2	3.2	54.8
A 13	15 51.8	23.3	20 32.0	11.4	18 43.4	3.0	54.8
Y 14	30 51.5	23.3	35 02.4	11.2	18 46.4	3.0	54.9
15	45 51.2	23.4	49 32.6	11.3	18 49.2	2.8	54.9
16	60 50.9	23.5	64 02.9	11.2	18 52.2	2.8	54.9
17	75 50.6	23.5	78 33.1	11.2	18 55.0	2.7	54.9
18	90 50.3	S23 23.6	93 03.3	11.1	S18 57.7	2.6	54.9
19	105 50.0	23.7	107 33.4	11.1	19 00.3	2.5	54.9
20	120 49.7	23.7	122 03.5	11.1	19 02.8	2.4	54.9
21	135 49.4	23.8	136 33.6	11.1	19 05.2	2.3	55.0
22	150 49.1	23.9	151 03.7	11.0	19 07.5	2.3	55.0
23	165 48.8	23.9	165 33.7	10.9	19 09.8	2.1	55.0
19 00	180 48.4	S23 24.0	180 03.6	11.0	S19 11.9	2.0	55.0
01	195 48.1	24.0	194 33.6	10.9	19 13.9	2.0	55.0
02	210 47.8	24.1	209 03.5	10.9	19 15.9	1.8	55.0
03	225 47.5	24.2	223 33.4	10.8	19 17.7	1.8	55.1
04	240 47.2	24.2	238 03.2	10.8	19 19.5	1.6	55.1
05	255 46.9	24.3	252 33.0	10.8	19 21.1	1.6	55.1
06	270 46.6	S23 24.3	267 02.8	10.7	S19 22.7	1.5	55.1
07	285 46.3	24.4	281 32.5	10.8	19 24.2	1.3	55.1
S 08	300 46.0	24.4	296 02.3	10.7	19 25.5	1.3	55.1
A 09	315 45.7	24.5	310 32.0	10.6	19 26.8	1.2	55.2
T 10	330 45.3	24.6	325 01.6	10.7	19 28.0	1.1	55.2
U 11	345 45.0	24.6	339 31.3	10.6	19 29.1	1.0	55.2
R 12	0 44.7	S23 24.7	354 00.9	10.6	S19 30.1	0.8	55.2
D 13	15 44.4	24.7	8 30.5	10.5	19 30.9	0.8	55.2
A 14	30 44.1	24.8	23 00.0	10.6	19 31.7	0.7	55.2
Y 15	45 43.8	24.8	37 29.6	10.5	19 32.4	0.6	55.3
16	60 43.5	24.9	51 59.1	10.5	19 33.0	0.5	55.3
17	75 43.2	24.9	66 28.6	10.5	19 33.5	0.4	55.3
18	90 42.9	S23 24.9	80 58.1	10.4	S19 33.9	0.3	55.3
19	105 42.6	25.0	95 27.5	10.4	19 34.2	0.2	55.3
20	120 42.3	25.0	109 56.9	10.4	19 34.4	0.1	55.3
21	135 41.9	25.1	124 26.3	10.4	19 34.5	0.0	55.4
22	150 41.6	25.1	138 55.7	10.4	19 34.5	0.1	55.4
23	165 41.3	25.2	153 25.1	10.3	19 34.4	0.2	55.4
20 00	180 41.0	S23 25.2	167 54.4	10.3	S19 34.2	0.4	55.4
01	195 40.7	25.2	182 23.7	10.3	19 33.8	0.4	55.4
02	210 40.4	25.3	196 53.0	10.3	19 33.4	0.5	55.5
03	225 40.1	25.3	211 22.3	10.3	19 32.9	0.6	55.5
04	240 39.8	25.4	225 51.6	10.3	19 32.3	0.7	55.5
05	255 39.5	25.4	240 20.9	10.2	19 31.6	0.8	55.5
06	270 39.1	S23 25.4	254 50.1	10.2	S19 30.8	0.9	55.5
07	285 38.8	25.5	269 19.3	10.3	19 29.9	1.0	55.5
08	300 38.5	25.5	283 48.6	10.2	19 28.9	1.2	55.6
S 09	315 38.2	25.5	298 17.8	10.1	19 27.7	1.2	55.6
U 10	330 37.9	25.6	312 46.9	10.2	19 26.5	1.3	55.6
N 11	345 37.6	25.6	327 16.1	10.2	19 25.2	1.4	55.6
D 12	0 37.3	S23 25.6	341 45.3	10.1	S19 23.8	1.5	55.6
A 13	15 37.0	25.7	356 14.4	10.2	19 22.3	1.7	55.7
Y 14	30 36.7	25.7	10 43.6	10.1	19 20.6	1.7	55.7
15	45 36.4	25.7	25 12.7	10.1	19 18.9	1.8	55.7
16	60 36.0	25.8	39 41.8	10.2	19 17.1	1.9	55.7
17	75 35.7	25.8	54 11.0	10.1	19 15.2	2.1	55.7
18	90 35.4	S23 25.8	68 40.1	10.1	S19 13.1	2.1	55.8
19	105 35.1	25.8	83 09.2	10.1	19 11.0	2.2	55.8
20	120 34.8	25.9	97 38.3	10.1	19 08.8	2.4	55.8
21	135 34.5	25.9	112 07.4	10.0	19 06.4	2.4	55.8
22	150 34.2	25.9	126 36.4	10.1	19 04.0	2.6	55.8
23	165 33.9	25.9	141 05.5	10.1	S19 01.4	2.6	55.8
SD	16.3	d 0.1	SD 14.9		15.0		15.2

Moonrise

Lat.	Twilight Naut.	Twilight Civil	Sunrise	18	19	20	21
°	h m	h m	h m	h m	h m	h m	h m
N 72	08 24	10 55	■	■	■	■	■
N 70	08 04	09 53	■	10 05	11 35	12 20	12 28
68	07 48	09 17	■	09 17	10 28	11 17	11 45
66	07 35	08 52	10 33	08 46	09 52	10 42	11 16
64	07 24	08 32	09 50	08 23	09 26	10 16	10 54
62	07 14	08 16	09 22	08 05	09 06	09 56	10 36
60	07 06	08 03	09 01	07 50	08 49	09 40	10 21
N 58	06 58	07 51	08 43	07 37	08 36	09 26	10 09
56	06 52	07 41	08 28	07 26	08 24	09 15	09 58
54	06 45	07 32	08 16	07 16	08 13	09 04	09 48
52	06 40	07 24	08 04	07 08	08 04	08 55	09 40
50	06 35	07 16	07 55	07 00	07 56	08 47	09 32
45	06 23	07 00	07 34	06 44	07 38	08 29	09 16
N 40	06 12	06 46	07 17	06 30	07 24	08 15	09 02
35	06 03	06 35	07 03	06 19	07 12	08 03	08 51
30	05 54	06 24	06 50	06 09	07 01	07 52	08 41
20	05 38	06 05	06 29	05 52	06 43	07 34	08 24
N 10	05 21	05 48	06 11	05 37	06 27	07 18	08 09
0	05 04	05 31	05 53	05 23	06 13	07 03	07 54
S 10	04 45	05 13	05 36	05 10	05 58	06 48	07 40
20	04 23	04 52	05 17	04 55	05 42	06 33	07 25
30	03 54	04 27	04 55	04 38	05 24	06 14	07 08
35	03 35	04 12	04 42	04 28	05 14	06 04	06 58
40	03 12	03 54	04 27	04 17	05 02	05 52	06 47
45	02 41	03 31	04 09	04 04	04 48	05 38	06 33
S 50	01 56	03 01	03 46	03 48	04 31	05 20	06 16
52	01 28	02 46	03 35	03 41	04 23	05 12	06 09
54	00 41	02 28	03 23	03 33	04 14	05 03	06 00
56	////	02 06	03 09	03 23	04 03	04 52	05 50
58	////	01 36	02 52	03 13	03 52	04 41	05 39
S 60	////	00 46	02 31	03 01	03 39	04 27	05 26

Moonset

Lat.	Sunset	Twilight Civil	Twilight Naut.	18	19	20	21
°	h m	h m	h m	h m	h m	h m	h m
N 72	■	12 59	15 30	■	■	■	■
N 70	■	14 01	15 50	13 00	13 13	14 14	15 53
68	■	14 36	16 06	13 48	14 21	15 17	16 35
66	13 21	15 02	16 19	14 20	14 57	15 52	17 04
64	14 03	15 21	16 30	14 43	15 23	16 17	17 26
62	14 32	15 38	16 40	15 01	15 43	16 37	17 43
60	14 53	15 51	16 48	15 16	15 59	16 53	17 57
N 58	15 11	16 03	16 56	15 29	16 13	17 07	18 10
56	15 26	16 13	17 02	15 41	16 25	17 18	18 20
54	15 38	16 22	17 08	15 50	16 35	17 29	18 30
52	15 49	16 30	17 14	15 59	16 45	17 38	18 38
50	15 59	16 38	17 19	16 07	16 53	17 46	18 45
45	16 20	16 54	17 31	16 24	17 11	18 03	19 01
N 40	16 37	17 08	17 42	16 38	17 25	18 17	19 14
35	16 51	17 19	17 51	16 49	17 37	18 29	19 25
30	17 04	17 30	18 00	17 00	17 48	18 40	19 35
20	17 25	17 49	18 16	17 17	18 06	18 58	19 51
N 10	17 43	18 06	18 33	17 33	18 22	19 13	20 06
0	18 01	18 23	18 50	17 47	18 37	19 28	20 19
S 10	18 18	18 41	19 09	18 01	18 52	19 42	20 32
20	18 37	19 02	19 31	18 17	19 08	19 58	20 47
30	18 59	19 27	20 00	18 34	19 26	20 16	21 03
35	19 12	19 42	20 19	18 45	19 37	20 26	21 12
40	19 28	20 00	20 42	18 56	19 49	20 38	21 23
45	19 46	20 23	21 13	19 10	20 03	20 52	21 36
S 50	20 08	20 53	21 58	19 27	20 20	21 09	21 51
52	20 19	21 08	22 26	19 35	20 29	21 16	21 58
54	20 31	21 26	23 13	19 44	20 38	21 25	22 06
56	20 46	21 48	////	19 54	20 48	21 35	22 15
58	21 02	22 18	////	20 05	21 00	21 46	22 24
S 60	21 23	23 09	////	20 18	21 14	21 59	22 36

Day	SUN Eqn. of Time 00h	SUN Eqn. of Time 12h	SUN Mer. Pass.	MOON Mer. Pass. Upper	MOON Mer. Pass. Lower	Age	Phase
d	m s	m s	h m	h m	h m	d	%
18	03 44	03 29	11 57	11 35	24 00	29	0
19	03 14	03 00	11 57	12 25	00 00	01	0
20	02 45	02 30	11 58	13 16	00 50	02	2

1998 DECEMBER 21, 22, 23 (MON., TUES., WED.)

UT	ARIES GHA	VENUS −3.9 GHA	Dec	MARS +1.2 GHA	Dec	JUPITER −2.4 GHA	Dec	SATURN +0.3 GHA	Dec	STARS Name	SHA	Dec
d h	° ′	° ′	° ′	° ′	° ′	° ′	° ′	° ′	° ′		° ′	° ′
21 00	89 21.7	166 40.4	S23 55.9	256 53.0	S 3 22.1	97 36.4	S 4 56.8	63 30.9	N 7 55.2	Acamar	315 26.5	S40 18.8
01	104 24.2	181 39.4	55.7	271 54.2	22.5	112 38.6	56.7	78 33.4	55.2	Achernar	335 34.9	S57 14.9
02	119 26.6	196 38.5	55.5	286 55.5	23.0	127 40.8	56.6	93 35.9	55.2	Acrux	173 21.8	S63 05.2
03	134 29.1	211 37.5	.. 55.3	301 56.7	.. 23.5	142 43.0	.. 56.5	108 38.4	.. 55.2	Adhara	255 20.9	S28 58.3
04	149 31.6	226 36.5	55.1	316 58.0	24.0	157 45.2	56.3	123 40.9	55.2	Aldebaran	291 01.9	N16 30.3
05	164 34.0	241 35.6	54.9	331 59.3	24.5	172 47.4	56.2	138 43.4	55.2			
06	179 36.5	256 34.6	S23 54.7	347 00.5	S 3 25.0	187 49.6	S 4 56.1	153 45.9	N 7 55.2	Alioth	166 30.7	N55 57.7
07	194 39.0	271 33.6	54.5	2 01.8	25.5	202 51.7	56.0	168 48.4	55.2	Alkaid	153 08.0	N49 19.0
M 08	209 41.4	286 32.7	54.3	17 03.0	26.0	217 53.9	55.8	183 50.9	55.2	Al Na'ir	27 58.0	S46 58.2
O 09	224 43.9	301 31.7	.. 54.1	32 04.3	.. 26.5	232 56.1	.. 55.7	198 53.4	.. 55.2	Alnilam	275 57.4	S 1 12.3
N 10	239 46.4	316 30.8	53.9	47 05.5	27.0	247 58.3	55.6	213 55.9	55.2	Alphard	218 06.9	S 8 39.2
D 11	254 48.8	331 29.8	53.7	62 06.8	27.4	263 00.5	55.4	228 58.4	55.2			
A 12	269 51.3	346 28.8	S23 53.5	77 08.1	S 3 27.9	278 02.7	S 4 55.3	244 00.9	N 7 55.2	Alphecca	126 20.8	N26 43.1
Y 13	284 53.7	1 27.9	53.3	92 09.3	28.4	293 04.9	55.2	259 03.4	55.2	Alpheratz	357 55.1	N29 05.2
14	299 56.2	16 26.9	53.1	107 10.6	28.9	308 07.1	55.1	274 05.9	55.2	Altair	62 19.4	N 8 52.0
15	314 58.7	31 26.0	.. 52.9	122 11.8	.. 29.4	323 09.3	.. 54.9	289 08.4	.. 55.2	Ankaa	353 26.7	S42 19.0
16	330 01.1	46 25.0	52.6	137 13.1	29.9	338 11.5	54.8	304 10.9	55.1	Antares	112 40.3	S26 25.6
17	345 03.6	61 24.0	52.4	152 14.4	30.4	353 13.7	54.7	319 13.4	55.1			
18	0 06.1	76 23.1	S23 52.2	167 15.6	S 3 30.9	8 15.8	S 4 54.6	334 15.9	N 7 55.1	Arcturus	146 06.2	N19 11.3
19	15 08.5	91 22.1	52.0	182 16.9	31.4	23 18.0	54.4	349 18.4	55.1	Atria	107 52.8	S69 01.3
20	30 11.0	106 21.2	51.8	197 18.1	31.8	38 20.2	54.3	4 20.9	55.1	Avior	234 22.0	S59 30.3
21	45 13.5	121 20.2	.. 51.5	212 19.4	.. 32.3	53 22.4	.. 54.2	19 23.5	.. 55.1	Bellatrix	278 43.7	N 6 20.8
22	60 15.9	136 19.2	51.3	227 20.7	32.8	68 24.6	54.0	34 26.0	55.1	Betelgeuse	271 13.1	N 7 24.3
23	75 18.4	151 18.3	51.1	242 21.9	33.3	83 26.8	53.9	49 28.5	55.1			
22 00	90 20.8	166 17.3	S23 50.9	257 23.2	S 3 33.8	98 29.0	S 4 53.8	64 31.0	N 7 55.1	Canopus	264 00.5	S52 41.8
01	105 23.3	181 16.4	50.7	272 24.4	34.3	113 31.2	53.7	79 33.5	55.1	Capella	280 50.5	N45 59.7
02	120 25.8	196 15.4	50.4	287 25.7	34.8	128 33.4	53.5	94 36.0	55.1	Deneb	49 39.4	N45 16.8
03	135 28.2	211 14.5	.. 50.2	302 27.0	.. 35.3	143 35.5	.. 53.4	109 38.5	.. 55.1	Denebola	182 45.1	N14 34.6
04	150 30.7	226 13.5	50.0	317 28.2	35.7	158 37.7	53.3	124 41.0	55.1	Diphda	349 07.1	S17 59.7
05	165 33.2	241 12.5	49.7	332 29.5	36.2	173 39.9	53.2	139 43.5	55.1			
06	180 35.6	256 11.6	S23 49.5	347 30.7	S 3 36.7	188 42.1	S 4 53.0	154 46.0	N 7 55.1	Dubhe	194 05.3	N61 45.1
07	195 38.1	271 10.6	49.3	2 32.0	37.2	203 44.3	52.9	169 48.5	55.1	Elnath	278 26.4	N28 36.3
T 08	210 40.6	286 09.7	49.0	17 33.3	37.7	218 46.5	52.8	184 51.0	55.1	Eltanin	90 51.9	N51 29.5
U 09	225 43.0	301 08.7	.. 48.8	32 34.5	.. 38.2	233 48.7	.. 52.6	199 53.5	.. 55.1	Enif	33 58.3	N 9 52.3
E 10	240 45.5	316 07.8	48.6	47 35.8	38.7	248 50.8	52.5	214 56.0	55.1	Fomalhaut	15 36.4	S29 37.9
S 11	255 48.0	331 06.8	48.3	62 37.1	39.2	263 53.0	52.4	229 58.5	55.1			
D 12	270 50.4	346 05.8	S23 48.1	77 38.3	S 3 39.6	278 55.2	S 4 52.2	245 01.0	N 7 55.1	Gacrux	172 13.4	S57 06.1
A 13	285 52.9	1 04.9	47.8	92 39.6	40.1	293 57.4	52.1	260 03.5	55.1	Gienah	176 03.8	S17 32.0
Y 14	300 55.3	16 03.9	47.6	107 40.9	40.6	308 59.6	52.0	275 06.0	55.1	Hadar	149 04.1	S60 21.7
15	315 57.8	31 03.0	.. 47.4	122 42.1	.. 41.1	324 01.8	.. 51.9	290 08.5	.. 55.1	Hamal	328 13.2	N23 27.4
16	331 00.3	46 02.0	47.1	137 43.4	41.6	339 03.9	51.7	305 11.0	55.1	Kaus Aust.	83 59.1	S34 23.0
17	346 02.7	61 01.1	46.9	152 44.6	42.1	354 06.1	51.6	320 13.5	55.1			
18	1 05.2	76 00.1	S23 46.6	167 45.9	S 3 42.6	9 08.3	S 4 51.5	335 16.0	N 7 55.1	Kochab	137 20.6	N74 09.5
19	16 07.7	90 59.2	46.4	182 47.2	43.0	24 10.5	51.3	350 18.5	55.1	Markab	13 49.6	N15 12.0
20	31 10.1	105 58.2	46.1	197 48.4	43.5	39 12.7	51.2	5 21.0	55.1	Menkar	314 26.5	N 4 05.0
21	46 12.6	120 57.3	.. 45.9	212 49.7	.. 44.0	54 14.9	.. 51.1	20 23.5	.. 55.1	Menkent	148 21.0	S36 21.6
22	61 15.1	135 56.3	45.6	227 51.0	44.5	69 17.0	51.0	35 26.0	55.1	Miaplacidus	221 41.3	S69 42.6
23	76 17.5	150 55.4	45.4	242 52.2	45.0	84 19.2	50.8	50 28.5	55.1			
23 00	91 20.0	165 54.4	S23 45.1	257 53.5	S 3 45.5	99 21.4	S 4 50.7	65 30.9	N 7 55.1	Mirfak	308 56.0	N49 51.5
01	106 22.5	180 53.4	44.8	272 54.8	45.9	114 23.6	50.6	80 33.4	55.1	Nunki	76 12.5	S26 17.8
02	121 24.9	195 52.5	44.6	287 56.0	46.4	129 25.8	50.4	95 35.9	55.1	Peacock	53 37.4	S56 44.4
03	136 27.4	210 51.5	.. 44.3	302 57.3	.. 46.9	144 28.0	.. 50.3	110 38.4	.. 55.1	Pollux	243 41.1	N28 01.6
04	151 29.8	225 50.6	44.1	317 58.6	47.4	159 30.1	50.2	125 40.9	55.1	Procyon	245 11.1	N 5 13.6
05	166 32.3	240 49.6	43.8	332 59.8	47.9	174 32.3	50.0	140 43.4	55.1			
06	181 34.8	255 48.7	S23 43.5	348 01.1	S 3 48.4	189 34.5	S 4 49.9	155 45.9	N 7 55.1	Rasalhague	96 17.2	N12 33.8
07	196 37.2	270 47.7	43.3	3 02.4	48.9	204 36.7	49.8	170 48.4	55.1	Regulus	207 55.3	N11 58.3
W 08	211 39.7	285 46.8	43.0	18 03.6	49.3	219 38.9	49.6	185 50.9	55.1	Rigel	281 22.5	S 8 12.3
E 09	226 42.2	300 45.8	.. 42.7	33 04.9	.. 49.8	234 41.0	.. 49.5	200 53.4	.. 55.1	Rigil Kent.	140 07.5	S60 49.5
D 10	241 44.6	315 44.9	42.5	48 06.2	50.3	249 43.2	49.4	215 55.9	55.1	Sabik	102 25.7	S15 43.3
N 11	256 47.1	330 43.9	42.2	63 07.4	50.8	264 45.4	49.2	230 58.4	55.1			
E 12	271 49.6	345 43.0	S23 41.9	78 08.7	S 3 51.3	279 47.6	S 4 49.1	246 00.9	N 7 55.1	Schedar	349 53.1	N56 32.1
S 13	286 52.0	0 42.0	41.7	93 10.0	51.8	294 49.8	49.0	261 03.4	55.1	Shaula	96 37.5	S37 06.0
D 14	301 54.5	15 41.1	41.4	108 11.2	52.2	309 51.9	48.9	276 05.9	55.1	Sirius	258 43.2	S16 43.0
A 15	316 57.0	30 40.1	.. 41.1	123 12.5	.. 52.7	324 54.1	.. 48.7	291 08.4	.. 55.1	Spica	158 43.2	S11 09.2
Y 16	331 59.4	45 39.2	40.8	138 13.8	53.2	339 56.3	48.6	306 10.9	55.1	Suhail	223 00.3	S43 25.6
17	347 01.9	60 38.2	40.6	153 15.0	53.7	354 58.5	48.5	321 13.4	55.1			
18	2 04.3	75 37.3	S23 40.3	168 16.3	S 3 54.2	10 00.6	S 4 48.3	336 15.9	N 7 55.1	Vega	80 47.0	N38 47.1
19	17 06.8	90 36.3	40.0	183 17.6	54.6	25 02.8	48.2	351 18.4	55.1	Zuben'ubi	137 18.1	S16 02.0
20	32 09.3	105 35.4	39.7	198 18.8	55.1	40 05.0	48.1	6 20.9	55.1		SHA	Mer. Pass.
21	47 11.7	120 34.4	.. 39.4	213 20.1	.. 55.6	55 07.2	.. 47.9	21 23.4	.. 55.1		° ′	h m
22	62 14.2	135 33.5	39.2	228 21.4	56.1	70 09.3	47.8	36 25.8	55.1	Venus	75 56.5	12 56
23	77 16.7	150 32.6	38.9	243 22.6	56.6	85 11.5	47.7	51 28.3	55.1	Mars	167 02.3	6 50
	h m									Jupiter	8 08.1	17 24
Mer. Pass. 17 55.7		v −1.0 d 0.2		v 1.3 d 0.5		v 2.2 d 0.1		v 2.5 d 0.0		Saturn	334 10.1	19 39

UT	SUN GHA	SUN Dec	MOON GHA	v	MOON Dec	d	HP
d h	° ′	° ′	° ′	′	° ′	′	′
21 00	180 33.6	S23 26.0	155 34.6	10.1	S18 58.8	2.7	55.9
01	195 33.5	26.0	170 03.7	10.0	18 56.1	2.9	55.9
02	210 32.9	26.0	184 32.7	10.1	18 53.2	2.9	55.9
03	225 32.6 ..	26.0	199 01.8	10.1	18 50.3	3.1	55.9
04	240 32.3	26.0	213 30.9	10.1	18 47.2	3.1	55.9
05	255 32.0	26.0	228 00.0	10.0	18 44.1	3.2	56.0
06	270 31.7	S23 26.1	242 29.0	10.1	S18 40.9	3.4	56.0
07	285 31.4	26.1	256 58.1	10.1	18 37.5	3.4	56.0
M 08	300 31.1	26.1	271 27.2	10.0	18 34.1	3.6	56.0
O 09	315 30.8 ..	26.1	285 56.2	10.1	18 30.5	3.6	56.0
N 10	330 30.4	26.1	300 25.3	10.1	18 26.9	3.8	56.1
D 11	345 30.1	26.1	314 54.4	10.1	18 23.1	3.8	56.1
A 12	0 29.8	S23 26.1	329 23.5	10.1	S18 19.3	3.9	56.1
Y 13	15 29.5	26.2	343 52.6	10.0	18 15.4	4.1	56.1
14	30 29.2	26.2	358 21.6	10.1	18 11.3	4.1	56.1
15	45 28.9 ..	26.2	12 50.7	10.1	18 07.2	4.3	56.2
16	60 28.6	26.2	27 19.8	10.1	18 02.9	4.3	56.2
17	75 28.3	26.2	41 48.9	10.1	17 58.6	4.4	56.2
18	90 27.9	S23 26.2	56 18.0	10.2	S17 54.2	4.5	56.2
19	105 27.6	26.2	70 47.2	10.1	17 49.7	4.7	56.2
20	120 27.3	26.2	85 16.3	10.1	17 45.0	4.7	56.3
21	135 27.0 ..	26.2	99 45.4	10.2	17 40.3	4.8	56.3
22	150 26.7	26.2	114 14.6	10.1	17 35.5	4.9	56.3
23	165 26.4	26.2	128 43.7	10.2	17 30.6	5.0	56.3
22 00	180 26.1	S23 26.2	143 12.9	10.1	S17 25.6	5.1	56.3
01	195 25.8	26.2	157 42.0	10.2	17 20.5	5.2	56.4
02	210 25.5	26.2	172 11.2	10.2	17 15.3	5.3	56.4
03	225 25.1 ..	26.2	186 40.4	10.2	17 10.0	5.4	56.4
04	240 24.8	26.2	201 09.6	10.2	17 04.6	5.5	56.4
05	255 24.5	26.2	215 38.8	10.2	16 59.1	5.6	56.5
06	270 24.2	S23 26.2	230 08.0	10.2	S16 53.5	5.6	56.5
T 07	285 23.9	26.2	244 37.2	10.3	16 47.9	5.8	56.5
U 08	300 23.6	26.2	259 06.5	10.2	16 42.1	5.8	56.5
E 09	315 23.3 ..	26.2	273 35.7	10.3	16 36.3	6.0	56.5
S 10	330 23.0	26.2	288 05.0	10.2	16 30.3	6.0	56.6
D 11	345 22.7	26.2	302 34.2	10.3	16 24.3	6.1	56.6
A 12	0 22.3	S23 26.2	317 03.5	10.3	S16 18.2	6.3	56.6
Y 13	15 22.0	26.2	331 32.8	10.3	16 11.9	6.3	56.6
14	30 21.7	26.2	346 02.1	10.4	16 05.6	6.4	56.6
15	45 21.4 ..	26.2	0 31.5	10.3	15 59.2	6.4	56.7
16	60 21.1	26.1	15 00.8	10.4	15 52.8	6.6	56.7
17	75 20.8	26.1	29 30.2	10.3	15 46.2	6.7	56.7
18	90 20.5	S23 26.1	43 59.5	10.4	S15 39.5	6.7	56.7
19	105 20.2	26.1	58 28.9	10.4	15 32.8	6.8	56.7
20	120 19.8	26.1	72 58.3	10.4	15 26.0	6.9	56.8
21	135 19.5 ..	26.1	87 27.7	10.4	15 19.1	7.0	56.8
22	150 19.2	26.1	101 57.1	10.5	15 12.1	7.1	56.8
23	165 18.9	26.0	116 26.6	10.4	15 05.0	7.2	56.8
23 00	180 18.6	S23 26.0	130 56.0	10.5	S14 57.8	7.3	56.9
01	195 18.3	26.0	145 25.5	10.5	14 50.5	7.3	56.9
02	210 18.0	26.0	159 55.0	10.5	14 43.2	7.4	56.9
03	225 17.7 ..	26.0	174 24.5	10.5	14 35.8	7.5	56.9
04	240 17.4	26.0	188 54.0	10.5	14 28.3	7.6	56.9
05	255 17.0	25.9	203 23.5	10.5	14 20.7	7.7	57.0
06	270 16.7	S23 25.9	217 53.0	10.6	S14 13.0	7.7	57.0
W 07	285 16.4	25.9	232 22.6	10.6	14 05.3	7.8	57.0
E 08	300 16.1	25.9	246 52.2	10.5	13 57.5	7.9	57.0
D 09	315 15.8 ..	25.8	261 21.7	10.6	13 49.6	8.0	57.1
N 10	330 15.5	25.8	275 51.3	10.7	13 41.6	8.1	57.1
E 11	345 15.2	25.8	290 21.0	10.6	13 33.5	8.1	57.1
S 12	0 14.9	S23 25.8	304 50.6	10.6	S13 25.4	8.2	57.1
D 13	15 14.6	25.7	319 20.2	10.7	13 17.2	8.3	57.1
A 14	30 14.2	25.7	333 49.9	10.6	13 08.9	8.4	57.2
Y 15	45 13.9 ..	25.7	348 19.5	10.7	13 00.5	8.4	57.2
16	60 13.6	25.6	2 49.2	10.7	12 52.1	8.5	57.2
17	75 13.3	25.6	17 18.9	10.7	12 43.6	8.6	57.2
18	90 13.0	S23 25.6	31 48.6	10.8	S12 35.0	8.6	57.3
19	105 12.7	25.5	46 18.4	10.7	12 26.4	8.8	57.3
20	120 12.4	25.5	60 48.1	10.7	12 17.6	8.8	57.3
21	135 12.1 ..	25.5	75 17.8	10.8	12 08.8	8.8	57.3
22	150 11.7	25.4	89 47.6	10.8	12 00.0	9.0	57.3
23	165 11.4	25.4	104 17.4	10.8	S11 51.0	9.0	57.4
	SD 16.3	d 0.0	SD 15.3		15.4		15.6

Lat.	Twilight Naut.	Twilight Civil	Sunrise	Moonrise 21	22	23	24
°	h m	h m	h m	h m	h m	h m	h m
N 72	08 26	10 58	▆	▆	13 11	12 52	12 39
N 70	08 06	09 55	▆	12 28	12 28	12 26	12 24
68	07 50	09 19	▆	11 45	11 59	12 07	12 12
66	07 37	08 54	10 35	11 16	11 37	11 52	12 02
64	07 26	08 34	09 52	10 54	11 20	11 39	11 53
62	07 16	08 18	09 24	10 36	11 06	11 28	11 46
60	07 08	08 05	09 02	10 21	10 53	11 19	11 40
N 58	07 00	07 53	08 45	10 09	10 43	11 11	11 34
56	06 53	07 43	08 30	09 58	10 34	11 03	11 29
54	06 47	07 33	08 17	09 48	10 25	10 57	11 24
52	06 41	07 25	08 06	09 40	10 18	10 51	11 20
50	06 36	07 18	07 56	09 32	10 11	10 46	11 17
45	06 24	07 02	07 36	09 16	09 57	10 35	11 08
N 40	06 14	06 48	07 19	09 02	09 46	10 25	11 02
35	06 05	06 36	07 04	08 51	09 35	10 17	10 56
30	05 56	06 26	06 52	08 41	09 27	10 10	10 51
20	05 39	06 07	06 31	08 24	09 11	09 57	10 42
N 10	05 23	05 49	06 12	08 09	08 58	09 47	10 34
0	05 06	05 32	05 55	07 54	08 46	09 36	10 26
S 10	04 47	05 14	05 37	07 40	08 33	09 26	10 19
20	04 24	04 54	05 18	07 25	08 20	09 15	10 11
30	03 55	04 29	04 56	07 08	08 04	09 03	10 02
35	03 36	04 13	04 43	06 58	07 56	08 55	09 57
40	03 13	03 55	04 28	06 47	07 45	08 47	09 51
45	02 43	03 32	04 10	06 33	07 33	08 37	09 44
S 50	01 57	03 03	03 47	06 16	07 19	08 26	09 35
52	01 29	02 47	03 36	06 09	07 12	08 20	09 32
54	00 42	02 29	03 24	06 00	07 04	08 14	09 27
56	////	02 07	03 10	05 50	06 56	08 07	09 23
58	////	01 37	02 53	05 39	06 46	08 00	09 17
S 60	////	00 46	02 32	05 26	06 35	07 51	09 11

Lat.	Sunset	Twilight Civil	Twilight Naut.	Moonset 21	22	23	24
°	h m	h m	h m	h m	h m	h m	h m
N 72	▆	12 59	15 31	▆	16 57	19 02	20 58
N 70	▆	14 02	15 51	15 53	17 39	19 25	21 11
68	▆	14 38	16 07	16 35	18 07	19 44	21 22
66	13 22	15 03	16 20	17 04	18 28	19 58	21 31
64	14 05	15 23	16 31	17 26	18 45	20 10	21 38
62	14 33	15 39	16 41	17 43	18 59	20 20	21 45
60	14 55	15 52	16 49	17 57	19 10	20 28	21 50
N 58	15 12	16 04	16 57	18 10	19 20	20 36	21 55
56	15 27	16 14	17 04	18 20	19 29	20 43	21 59
54	15 40	16 24	17 10	18 30	19 37	20 48	22 03
52	15 51	16 32	17 16	18 38	19 44	20 54	22 06
50	16 01	16 39	17 21	18 45	19 50	20 58	22 09
45	16 22	16 55	17 33	19 01	20 03	21 09	22 16
N 40	16 38	17 09	17 43	19 14	20 14	21 17	22 22
35	16 53	17 21	17 57	19 25	20 24	21 24	22 26
30	17 05	17 31	18 01	19 35	20 32	21 31	22 31
20	17 26	17 50	18 18	19 51	20 46	21 42	22 38
N 10	17 45	18 08	18 34	20 06	20 59	21 51	22 44
0	18 02	18 25	18 51	20 19	21 10	22 00	22 50
S 10	18 20	18 43	19 10	20 32	21 21	22 09	22 56
20	18 39	19 03	19 33	20 47	21 34	22 19	23 02
30	19 01	19 28	20 02	21 03	21 47	22 29	23 09
35	19 14	19 44	20 21	21 12	21 55	22 35	23 13
40	19 29	20 02	20 44	21 23	22 04	22 42	23 18
45	19 47	20 25	21 14	21 36	22 15	22 50	23 23
S 50	20 10	20 54	22 00	21 51	22 28	23 00	23 29
52	20 21	21 10	22 28	21 58	22 33	23 04	23 32
54	20 33	21 28	23 15	22 06	22 40	23 09	23 35
56	20 47	21 50	////	22 15	22 47	23 15	23 38
58	21 04	22 20	////	22 24	22 55	23 21	23 42
S 60	21 25	23 11	////	22 36	23 04	23 27	23 47

Day	SUN Eqn. of Time 00ʰ	12ʰ	Mer. Pass.	MOON Mer. Pass. Upper	Lower	Age	Phase
d	m s	m s	h m	h m	h m	d	%
21	02 15	02 00	11 58	14 07	01 41	03	6
22	01 45	01 30	11 58	14 58	02 32	04	12
23	01 15	01 00	11 59	15 48	03 23	05	20

1998 DECEMBER 24, 25, 26 (THURS., FRI., SAT.)

UT	ARIES GHA	VENUS −3.9 GHA	Dec	MARS +1.1 GHA	Dec	JUPITER −2.3 GHA	Dec	SATURN +0.4 GHA	Dec
24 00	92 19.1	165 31.6	S23 38.6	258 23.9	S 3 57.1	100 13.7	S 4 47.5	66 30.8	N 7 55.1
01	107 21.6	180 30.7	38.3	273 25.2	57.5	115 15.9	47.4	81 33.3	55.1
02	122 24.1	195 29.7	38.0	288 26.5	58.0	130 18.0	47.3	96 35.8	55.1
03	137 26.5	210 28.8 ..	37.7	303 27.7 ..	58.5	145 20.2 ..	47.1	111 38.3 ..	55.1
04	152 29.0	225 27.8	37.4	318 29.0	59.0	160 22.4	47.0	126 40.8	55.1
05	167 31.4	240 26.9	37.1	333 30.3	59.5	175 24.6	46.9	141 43.3	55.1
06	182 33.9	255 25.9	S23 36.8	348 31.5	S 3 59.9	190 26.7	S 4 46.7	156 45.8	N 7 55.1
07	197 36.4	270 25.0	36.6	3 32.8	4 00.4	205 28.9	46.6	171 48.3	55.1
T 08	212 38.8	285 24.0	36.3	18 34.1	00.9	220 31.1	46.5	186 50.8	55.1
H 09	227 41.3	300 23.1 ..	36.0	33 35.3 ..	01.4	235 33.3 ..	46.3	201 53.3 ..	55.1
U 10	242 43.8	315 22.2	35.7	48 36.6	01.9	250 35.4	46.2	216 55.8	55.2
R 11	257 46.2	330 21.2	35.4	63 37.9	02.3	265 37.6	46.1	231 58.2	55.2
S 12	272 48.7	345 20.3	S23 35.1	78 39.2	S 4 02.8	280 39.8	S 4 45.9	247 00.7	N 7 55.2
D 13	287 51.2	0 19.3	34.8	93 40.4	03.3	295 42.0	45.8	262 03.2	55.2
A 14	302 53.6	15 18.4	34.5	108 41.7	03.8	310 44.1	45.7	277 05.7	55.2
Y 15	317 56.1	30 17.4 ..	34.1	123 43.0 ..	04.3	325 46.3 ..	45.5	292 08.2 ..	55.2
16	332 58.6	45 16.5	33.8	138 44.3	04.7	340 48.5	45.4	307 10.7	55.2
17	348 01.0	60 15.6	33.5	153 45.5	05.2	355 50.6	45.2	322 13.2	55.2
18	3 03.5	75 14.6	S23 33.2	168 46.8	S 4 05.7	10 52.8	S 4 45.1	337 15.7	N 7 55.2
19	18 05.9	90 13.7	32.9	183 48.1	06.2	25 55.0	45.0	352 18.2	55.2
20	33 08.4	105 12.7	32.6	198 49.3	06.7	40 57.2	44.8	7 20.7	55.2
21	48 10.9	120 11.8 ..	32.3	213 50.6 ..	07.1	55 59.3 ..	44.7	22 23.2 ..	55.2
22	63 13.3	135 10.9	32.0	228 51.9	07.6	71 01.5	44.6	37 25.6	55.2
23	78 15.8	150 09.9	31.7	243 53.2	08.1	86 03.7	44.4	52 28.1	55.2
25 00	93 18.3	165 09.0	S23 31.4	258 54.4	S 4 08.6	101 05.8	S 4 44.3	67 30.6	N 7 55.2
01	108 20.7	180 08.0	31.0	273 55.7	09.1	116 08.0	44.2	82 33.1	55.2
02	123 23.2	195 07.1	30.7	288 57.0	09.5	131 10.2	44.0	97 35.6	55.2
03	138 25.7	210 06.2 ..	30.4	303 58.3 ..	10.0	146 12.3 ..	43.9	112 38.1 ..	55.2
04	153 28.1	225 05.2	30.1	318 59.5	10.5	161 14.5	43.8	127 40.6	55.2
05	168 30.6	240 04.3	29.8	334 00.8	11.0	176 16.7	43.6	142 43.1	55.2
06	183 33.1	255 03.3	S23 29.4	349 02.1	S 4 11.4	191 18.9	S 4 43.5	157 45.5	N 7 55.2
07	198 35.5	270 02.4	29.1	4 03.4	11.9	206 21.0	43.4	172 48.0	55.2
08	213 38.0	285 01.5	28.8	19 04.6	12.4	221 23.2	43.2	187 50.5	55.2
F 09	228 40.4	300 00.5 ..	28.4	34 05.9 ..	12.9	236 25.4 ..	43.1	202 53.0 ..	55.2
R 10	243 42.9	314 59.6	28.1	49 07.2	13.4	251 27.5	42.9	217 55.5	55.2
I 11	258 45.4	329 58.7	27.8	64 08.5	13.8	266 29.7	42.8	232 58.0	55.2
D 12	273 47.8	344 57.7	S23 27.5	79 09.7	S 4 14.3	281 31.9	S 4 42.7	248 00.5	N 7 55.2
A 13	288 50.3	359 56.8	27.1	94 11.0	14.8	296 34.0	42.5	263 03.0	55.2
Y 14	303 52.8	14 55.8	26.8	109 12.3	15.3	311 36.2	42.4	278 05.4	55.2
15	318 55.2	29 54.9 ..	26.5	124 13.6 ..	15.7	326 38.4 ..	42.3	293 07.9 ..	55.2
16	333 57.7	44 54.0	26.1	139 14.9	16.2	341 40.5	42.1	308 10.4	55.3
17	349 00.2	59 53.0	25.8	154 16.1	16.7	356 42.7	42.0	323 12.9	55.3
18	4 02.6	74 52.1	S23 25.4	169 17.4	S 4 17.2	11 44.9	S 4 41.8	338 15.4	N 7 55.3
19	19 05.1	89 51.2	25.1	184 18.7	17.6	26 47.0	41.7	353 17.9	55.3
20	34 07.5	104 50.2	24.8	199 20.0	18.1	41 49.2	41.6	8 20.4	55.3
21	49 10.0	119 49.3 ..	24.4	214 21.2 ..	18.6	56 51.3 ..	41.4	23 22.8 ..	55.3
22	64 12.5	134 48.4	24.1	229 22.5	19.1	71 53.5	41.3	38 25.3	55.3
23	79 14.9	149 47.4	23.7	244 23.8	19.5	86 55.7	41.2	53 27.8	55.3
26 00	94 17.4	164 46.5	S23 23.4	259 25.1	S 4 20.0	101 57.8	S 4 41.0	68 30.3	N 7 55.3
01	109 19.9	179 45.6	23.0	274 26.4	20.5	117 00.0	40.9	83 32.8	55.3
02	124 22.3	194 44.6	22.7	289 27.6	21.0	132 02.2	40.7	98 35.3	55.3
03	139 24.8	209 43.7 ..	22.3	304 28.9 ..	21.4	147 04.3 ..	40.6	113 37.7 ..	55.3
04	154 27.3	224 42.8	22.0	319 30.2	21.9	162 06.5	40.5	128 40.2	55.3
05	169 29.7	239 41.8	21.6	334 31.5	22.4	177 08.7	40.3	143 42.7	55.3
06	184 32.2	254 40.9	S23 21.3	349 32.8	S 4 22.9	192 10.8	S 4 40.2	158 45.2	N 7 55.3
07	199 34.7	269 40.0	20.9	4 34.0	23.3	207 13.0	40.0	173 47.7	55.3
S 08	214 37.1	284 39.1	20.6	19 35.3	23.8	222 15.1	39.9	188 50.2	55.3
A 09	229 39.6	299 38.1 ..	20.2	34 36.6 ..	24.3	237 17.3 ..	39.8	203 52.6 ..	55.3
T 10	244 42.0	314 37.2	19.9	49 37.9	24.8	252 19.5	39.6	218 55.1	55.3
U 11	259 44.5	329 36.3	19.5	64 39.2	25.2	267 21.6	39.5	233 57.6	55.4
R 12	274 47.0	344 35.3	S23 19.1	79 40.4	S 4 25.7	282 23.8	S 4 39.4	249 00.1	N 7 55.4
D 13	289 49.4	359 34.4	18.8	94 41.7	26.2	297 26.0	39.2	264 02.6	55.4
A 14	304 51.9	14 33.5	18.4	109 43.0	26.6	312 28.1	39.1	279 05.1	55.4
Y 15	319 54.4	29 32.6 ..	18.1	124 44.3 ..	27.1	327 30.3 ..	38.9	294 07.5 ..	55.4
16	334 56.8	44 31.6	17.7	139 45.6	27.6	342 32.4	38.8	309 10.0	55.4
17	349 59.3	59 30.7	17.3	154 46.9	28.1	357 34.6	38.7	324 12.5	55.4
18	5 01.8	74 29.8	S23 16.9	169 48.1	S 4 28.5	12 36.8	S 4 38.5	339 15.0	N 7 55.4
19	20 04.2	89 28.8	16.6	184 49.4	29.0	27 38.9	38.4	354 17.5	55.4
20	35 06.7	104 27.9	16.2	199 50.7	29.5	42 41.1	38.2	9 19.9	55.4
21	50 09.2	119 27.0 ..	15.8	214 52.0 ..	30.0	57 43.2 ..	38.1	24 22.4 ..	55.4
22	65 11.6	134 26.1	15.5	229 53.3	30.4	72 45.4	38.0	39 24.9	55.4
23	80 14.1	149 25.1	15.1	244 54.6	30.9	87 47.5	37.8	54 27.4	55.4
Mer. Pass. 17ʰ 43.9ᵐ		v −0.9 d 0.3		v 1.3 d 0.5		v 2.2 d 0.1		v 2.5 d 0.0	

STARS

Name	SHA	Dec
Acamar	315 26.6	S40 18.8
Achernar	335 34.9	S57 14.9
Acrux	173 21.8	S63 05.2
Adhara	255 20.9	S28 58.3
Aldebaran	291 01.9	N16 30.3
Alioth	166 30.7	N55 57.7
Alkaid	153 08.0	N49 19.0
Al Na'ir	27 58.0	S46 58.2
Alnilam	275 57.4	S 1 12.3
Alphard	218 06.8	S 8 39.3
Alphecca	126 20.8	N26 43.1
Alpheratz	357 55.1	N29 05.2
Altair	62 19.4	N 8 52.0
Ankaa	353 26.7	S42 19.0
Antares	112 40.3	S26 25.6
Arcturus	146 06.1	N19 11.3
Atria	107 52.8	S69 01.3
Avior	234 22.0	S59 30.3
Bellatrix	278 43.7	N 6 20.8
Betelgeuse	271 13.1	N 7 24.3
Canopus	264 00.5	S52 41.8
Capella	280 50.5	N45 59.7
Deneb	49 39.4	N45 16.8
Denebola	182 45.0	N14 34.6
Diphda	349 07.1	S17 59.7
Dubhe	194 05.2	N61 45.1
Elnath	278 26.4	N28 36.3
Eltanin	90 51.9	N51 29.5
Enif	33 58.3	N 9 52.3
Fomalhaut	15 36.4	S29 37.9
Gacrux	172 13.4	S57 06.1
Gienah	176 03.8	S17 32.0
Hadar	149 04.1	S60 21.7
Hamal	328 13.2	N23 27.4
Kaus Aust.	83 59.1	S34 23.0
Kochab	137 20.5	N74 09.5
Markab	13 49.6	N15 12.0
Menkar	314 26.5	N 4 05.0
Menkent	148 21.0	S36 21.6
Miaplacidus	221 41.2	S69 42.6
Mirfak	308 56.0	N49 51.5
Nunki	76 12.5	S26 17.8
Peacock	53 37.4	S56 44.4
Pollux	243 41.0	N28 01.6
Procyon	245 11.1	N 5 13.5
Rasalhague	96 17.2	N12 33.8
Regulus	207 55.2	N11 58.3
Rigel	281 22.5	S 8 12.3
Rigil Kent.	140 07.4	S60 49.5
Sabik	102 25.7	S15 43.3
Schedar	349 53.2	N56 32.1
Shaula	96 37.5	S37 06.0
Sirius	258 43.2	S16 43.0
Spica	158 43.2	S11 09.2
Suhail	223 00.3	S43 25.6
Vega	80 47.0	N38 47.1
Zuben'ubi	137 18.0	S16 02.1

	SHA	Mer. Pass.
Venus	71 50.7	13ʰ 00ᵐ
Mars	165 36.2	6 44
Jupiter	7 47.6	17 13
Saturn	334 12.4	19 27

UT	SUN GHA	Dec	MOON GHA	v	Dec	d	HP
d h	° ′	° ′	° ′	′	° ′	′	′
24 00	180 11.1	S23 25.4	118 47.2	10.8	S11 42.0	9.1	57.4
01	195 10.8	25.3	133 17.0	10.8	11 32.9	9.1	57.4
02	210 10.5	25.3	147 46.8	10.8	11 23.8	9.2	57.4
03	225 10.2	.. 25.3	162 16.6	10.8	11 14.6	9.3	57.5
04	240 09.9	25.2	176 46.4	10.9	11 05.3	9.3	57.5
05	255 09.6	25.2	191 16.3	10.8	10 56.0	9.4	57.5
06	270 09.3	S23 25.1	205 46.1	10.9	S10 46.6	9.5	57.5
T 07	285 08.9	25.1	220 16.0	10.9	10 37.1	9.5	57.5
H 08	300 08.6	25.0	234 45.9	10.9	10 27.6	9.6	57.6
U 09	315 08.3	.. 25.0	249 15.8	10.8	10 18.0	9.6	57.6
R 10	330 08.0	25.0	263 45.6	11.0	10 08.4	9.7	57.6
S 11	345 07.7	24.9	278 15.6	10.9	9 58.6	9.7	57.6
D 12	0 07.4	S23 24.9	292 45.5	10.9	S 9 48.9	9.9	57.7
A 13	15 07.1	24.8	307 15.4	10.9	9 39.0	9.9	57.7
Y 14	30 06.8	24.8	321 45.3	10.9	9 29.1	9.9	57.7
15	45 06.5	.. 24.7	336 15.2	11.0	9 19.2	10.0	57.7
16	60 06.1	24.7	350 45.2	10.9	9 09.2	10.1	57.8
17	75 05.8	24.6	5 15.1	11.0	8 59.1	10.1	57.8
18	90 05.5	S23 24.6	19 45.1	10.9	S 8 49.0	10.1	57.8
19	105 05.2	24.5	34 15.0	11.0	8 38.9	10.3	57.8
20	120 04.9	24.5	48 45.0	11.0	8 28.6	10.2	57.9
21	135 04.6	.. 24.4	63 15.0	10.9	8 18.4	10.4	57.9
22	150 04.3	24.3	77 44.9	11.0	8 08.0	10.3	57.9
23	165 04.0	24.3	92 14.9	11.0	7 57.7	10.5	57.9
25 00	180 03.7	S23 24.2	106 44.9	11.0	S 7 47.2	10.4	57.9
01	195 03.4	24.2	121 14.9	10.9	7 36.8	10.5	58.0
02	210 03.0	24.1	135 44.8	11.0	7 26.3	10.6	58.0
03	225 02.7	.. 24.1	150 14.8	11.0	7 15.7	10.6	58.0
04	240 02.4	24.0	164 44.8	11.0	7 05.1	10.7	58.0
05	255 02.1	23.9	179 14.8	11.0	6 54.4	10.7	58.1
06	270 01.8	S23 23.9	193 44.8	10.9	S 6 43.7	10.8	58.1
F 07	285 01.5	23.8	208 14.7	11.0	6 32.9	10.7	58.1
R 08	300 01.2	23.7	222 44.7	11.0	6 22.2	10.9	58.1
I 09	315 00.9	.. 23.7	237 14.7	11.0	6 11.3	10.9	58.2
D 10	330 00.6	23.6	251 44.7	10.9	6 00.4	10.9	58.2
A 11	345 00.2	23.6	266 14.6	11.0	5 49.5	10.9	58.2
Y 12	0 00.0	S23 23.5	280 44.6	11.0	S 5 38.6	11.0	58.2
13	14 59.6	23.4	295 14.6	10.9	5 27.6	11.1	58.3
14	29 59.3	23.3	309 44.5	11.0	5 16.5	11.0	58.3
15	44 59.0	.. 23.3	324 14.5	10.9	5 05.5	11.1	58.3
16	59 58.7	23.2	338 44.4	11.0	4 54.4	11.2	58.3
17	74 58.4	23.1	353 14.4	10.9	4 43.2	11.1	58.3
18	89 58.1	S23 23.1	7 44.3	10.9	S 4 32.1	11.2	58.4
19	104 57.8	23.0	22 14.2	10.9	4 20.9	11.3	58.4
20	119 57.5	22.9	36 44.1	10.9	4 09.6	11.3	58.4
21	134 57.2	.. 22.8	51 14.0	10.9	3 58.3	11.3	58.4
22	149 56.8	22.8	65 43.9	10.9	3 47.0	11.3	58.5
23	164 56.5	22.7	80 13.8	10.8	3 35.7	11.3	58.5
26 00	179 56.2	S23 22.6	94 43.6	10.9	S 3 24.4	11.4	58.5
01	194 55.9	22.5	109 13.5	10.9	3 13.0	11.4	58.5
02	209 55.6	22.5	123 43.3	10.9	3 01.6	11.5	58.6
03	224 55.3	.. 22.4	138 13.2	10.8	2 50.1	11.4	58.6
04	239 55.0	22.3	152 43.0	10.8	2 38.7	11.5	58.6
05	254 54.7	22.2	167 12.8	10.7	2 27.2	11.5	58.6
06	269 54.4	S23 22.1	181 42.5	10.8	S 2 15.7	11.5	58.7
S 07	284 54.1	22.1	196 12.3	10.7	2 04.2	11.6	58.7
A 08	299 53.7	22.0	210 42.0	10.8	1 52.6	11.6	58.7
T 09	314 53.4	.. 21.9	225 11.8	10.7	1 41.0	11.5	58.7
U 10	329 53.1	21.8	239 41.5	10.7	1 29.5	11.6	58.7
R 11	344 52.8	21.7	254 11.2	10.6	1 17.9	11.7	58.8
D 12	359 52.5	S23 21.6	268 40.8	10.7	S 1 06.2	11.6	58.8
A 13	14 52.2	21.6	283 10.5	10.6	0 54.6	11.6	58.8
Y 14	29 51.9	21.5	297 40.1	10.6	0 43.0	11.7	58.8
15	44 51.6	.. 21.4	312 09.7	10.6	0 31.3	11.7	58.9
16	59 51.3	21.3	326 39.3	10.5	0 19.6	11.7	58.9
17	74 51.0	21.2	341 08.8	10.5	S 0 07.9	11.7	58.9
18	89 50.7	S23 21.1	355 38.3	10.5	N 0 03.8	11.7	58.9
19	104 50.4	21.0	10 07.8	10.5	0 15.5	11.7	58.9
20	119 50.0	20.9	24 37.3	10.4	0 27.2	11.7	59.0
21	134 49.7	.. 20.8	39 06.7	10.5	0 38.9	11.7	59.0
22	149 49.4	20.7	53 36.2	10.4	0 50.6	11.7	59.0
23	164 49.1	20.6	68 05.6	10.3	N 1 02.3	11.8	59.0
SD	16.3	d 0.1	SD	15.7	15.9		16.0

Lat.	Twilight Naut.	Civil	Sunrise	Moonrise 24	25	26	27
°	h m	h m	h m	h m	h m	h m	h m
N 72	08 27	10 57	■	12 39	12 30	12 21	12 12
N 70	08 07	09 55	■	12 24	12 22	12 19	12 17
68	07 51	09 20	■	12 12	12 15	12 18	12 20
66	07 38	08 55	10 35	12 02	12 10	12 17	12 24
64	07 27	08 35	09 53	11 53	12 05	12 16	12 27
62	07 17	08 19	09 25	11 46	12 01	12 15	12 29
60	07 09	08 06	09 03	11 40	11 58	12 14	12 31
N 58	07 01	07 54	08 46	11 34	11 54	12 14	12 33
56	06 55	07 44	08 31	11 29	11 52	12 13	12 35
54	06 48	07 35	08 19	11 24	11 49	12 13	12 36
52	06 43	07 26	08 07	11 20	11 47	12 12	12 38
50	06 37	07 19	07 58	11 17	11 45	12 12	12 39
45	06 26	07 03	07 37	11 08	11 40	12 11	12 42
N 40	06 15	06 49	07 20	11 02	11 36	12 10	12 44
35	06 06	06 38	07 06	10 56	11 33	12 09	12 46
30	05 57	06 27	06 53	10 51	11 30	12 09	12 48
20	05 41	06 08	06 32	10 42	11 25	12 08	12 51
N 10	05 24	05 51	06 14	10 34	11 20	12 07	12 54
0	05 07	05 34	05 56	10 26	11 16	12 06	12 57
S 10	04 48	05 16	05 39	10 19	11 12	12 05	13 00
20	04 26	04 55	05 20	10 11	11 07	12 04	13 03
30	03 57	04 30	04 58	10 02	11 02	12 04	13 06
35	03 38	04 15	04 45	09 57	10 59	12 03	13 08
40	03 15	03 57	04 30	09 51	10 56	12 02	13 10
45	02 44	03 34	04 12	09 44	10 52	12 02	13 13
S 50	01 59	03 04	03 49	09 35	10 47	12 01	13 16
52	01 31	02 49	03 38	09 32	10 45	12 01	13 18
54	00 44	02 31	03 26	09 27	10 43	12 00	13 19
56	////	02 09	03 12	09 23	10 40	12 00	13 21
58	////	01 39	02 55	09 17	10 37	11 59	13 23
S 60	////	00 49	02 34	09 11	10 34	11 59	13 25

Lat.	Sunset	Twilight Civil	Naut.	Moonset 24	25	26	27
°	h m	h m	h m	h m	h m	h m	h m
N 72	■	13 03	15 33	20 58	22 52	24 46	00 46
N 70	■	14 05	15 53	21 11	22 57	24 45	00 45
68	■	14 40	16 09	21 22	23 02	24 43	00 43
66	13 25	15 05	16 22	21 31	23 06	24 42	00 42
64	14 07	15 25	16 33	21 38	23 09	24 41	00 41
62	14 35	15 41	16 43	21 45	23 11	24 40	00 40
60	14 57	15 54	16 51	21 50	23 14	24 39	00 39
N 58	15 14	16 06	16 59	21 55	23 16	24 39	00 39
56	15 29	16 16	17 05	21 59	23 18	24 38	00 38
54	15 42	16 25	17 12	22 03	23 19	24 37	00 37
52	15 53	16 34	17 17	22 06	23 21	24 37	00 37
50	16 03	16 41	17 23	22 09	23 22	24 36	00 36
45	16 23	16 57	17 34	22 16	23 25	24 35	00 35
N 40	16 40	17 11	17 45	22 22	23 27	24 35	00 35
35	16 54	17 22	17 54	22 26	23 30	24 34	00 34
30	17 07	17 33	18 03	22 31	23 31	24 33	00 33
20	17 28	17 52	18 19	22 38	23 35	24 32	00 32
N 10	17 46	18 09	18 36	22 44	23 37	24 31	00 31
0	18 04	18 26	18 53	22 50	23 40	24 30	00 30
S 10	18 21	18 44	19 11	22 56	23 42	24 29	00 29
20	18 40	19 05	19 34	23 02	23 45	24 28	00 28
30	19 02	19 30	20 03	23 09	23 48	24 27	00 27
35	19 15	19 45	20 22	23 13	23 50	24 26	00 26
40	19 30	20 03	20 45	23 18	23 52	24 25	00 25
45	19 48	20 26	21 15	23 23	23 54	24 24	00 24
S 50	20 11	20 55	22 01	23 29	23 56	24 23	00 23
52	20 22	21 11	22 29	23 32	23 58	24 23	00 23
54	20 34	21 28	23 15	23 35	23 59	24 22	00 22
56	20 48	21 51	////	23 38	24 00	00 00	00 22
58	21 05	22 20	////	23 42	24 02	00 02	00 21
S 60	21 25	23 10	////	23 47	24 04	00 04	00 20

Day	SUN Eqn. of Time 00h	12h	Mer. Pass.	MOON Mer. Pass. Upper	Lower	Age	Phase
d	m s	m s	h m	h m	h m	d	%
24	00 45	00 30	11 59	16 38	04 13	06	29
25	00 15	00 00	12 00	17 28	05 03	07	40
26	00 14	00 29	12 00	18 18	05 53	08	51

1998 DECEMBER 27, 28, 29 (SUN., MON., TUES.)

UT	ARIES GHA	VENUS −3.9 GHA	Dec	MARS +1.1 GHA	Dec	JUPITER −2.3 GHA	Dec	SATURN +0.4 GHA	Dec	STARS Name	SHA	Dec
27 00	95 16.5	164 24.2	S23 14.7	259 55.8	S 4 31.4	102 49.7	S 4 37.7	69 29.9	N 7 55.4	Acamar	315 26.6	S40 18.9
01	110 19.0	179 23.3	14.3	274 57.1	31.8	117 51.9	37.5	84 32.4	55.4	Achernar	335 35.0	S57 14.9
02	125 21.5	194 22.4	14.0	289 58.4	32.3	132 54.0	37.4	99 34.8	55.4	Acrux	173 21.7	S63 05.2
03	140 23.9	209 21.5 ..	13.6	304 59.7 ..	32.8	147 56.2 ..	37.3	114 37.3 ..	55.5	Adhara	255 20.9	S28 58.3
04	155 26.4	224 20.5	13.2	320 01.0	33.3	162 58.3	37.1	129 39.8	55.5	Aldebaran	291 01.9	N16 30.3
05	170 28.9	239 19.6	12.8	335 02.3	33.7	178 00.5	37.0	144 42.3	55.5			
06	185 31.3	254 18.7	S23 12.4	350 03.5	S 4 34.2	193 02.6	S 4 36.8	159 44.7	N 7 55.5	Alioth	166 30.6	N55 57.7
07	200 33.8	269 17.8	12.0	5 04.8	34.7	208 04.8	36.7	174 47.2	55.5	Alkaid	153 07.9	N49 19.0
08	215 36.3	284 16.8	11.7	20 06.1	35.1	223 07.0	36.5	189 49.7	55.5	Al Na'ir	27 58.0	S46 58.2
S 09	230 38.7	299 15.9 ..	11.3	35 07.4 ..	35.6	238 09.1 ..	36.4	204 52.2 ..	55.5	Alnilam	275 57.4	S 1 12.3
U 10	245 41.2	314 15.0	10.9	50 08.7	36.1	253 11.3	36.3	219 54.7	55.5	Alphard	218 06.8	S 8 39.3
N 11	260 43.6	329 14.1	10.5	65 10.0	36.6	268 13.4	36.1	234 57.1	55.5			
D 12	275 46.1	344 13.2	S23 10.1	80 11.3	S 4 37.0	283 15.6	S 4 36.0	249 59.6	N 7 55.5	Alphecca	126 20.8	N26 43.1
A 13	290 48.6	359 12.2	09.7	95 12.5	37.5	298 17.7	35.8	265 02.1	55.5	Alpheratz	357 55.1	N29 05.2
Y 14	305 51.0	14 11.3	09.3	110 13.8	38.0	313 19.9	35.7	280 04.6	55.5	Altair	62 19.5	N 8 52.0
15	320 53.5	29 10.4 ..	08.9	125 15.1 ..	38.4	328 22.0 ..	35.6	295 07.1 ..	55.5	Ankaa	353 26.8	S42 19.0
16	335 56.0	44 09.5	08.5	140 16.4	38.9	343 24.2	35.4	310 09.5	55.5	Antares	112 40.3	S26 25.6
17	350 58.4	59 08.6	08.1	155 17.7	39.4	358 26.4	35.3	325 12.0	55.6			
18	6 00.9	74 07.6	S23 07.7	170 19.0	S 4 39.8	13 28.5	S 4 35.1	340 14.5	N 7 55.6	Arcturus	146 06.1	N19 11.3
19	21 03.4	89 06.7	07.3	185 20.3	40.3	28 30.7	35.0	355 17.0	55.6	Atria	107 52.7	S69 01.3
20	36 05.8	104 05.8	06.9	200 21.5	40.8	43 32.8	34.8	10 19.4	55.6	Avior	234 21.9	S59 30.3
21	51 08.3	119 04.9 ..	06.5	215 22.8 ..	41.2	58 35.0 ..	34.7	25 21.9 ..	55.6	Bellatrix	278 43.7	N 6 20.8
22	66 10.8	134 04.0	06.1	230 24.1	41.7	73 37.1	34.6	40 24.4	55.6	Betelgeuse	271 13.1	N 7 24.3
23	81 13.2	149 03.1	05.7	245 25.4	42.2	88 39.3	34.4	55 26.9	55.6			
28 00	96 15.7	164 02.1	S23 05.3	260 26.7	S 4 42.7	103 41.4	S 4 34.3	70 29.3	N 7 55.6	Canopus	264 00.5	S52 41.8
01	111 18.1	179 01.2	04.9	275 28.0	43.1	118 43.6	34.1	85 31.8	55.6	Capella	280 50.5	N45 59.7
02	126 20.6	194 00.3	04.5	290 29.3	43.6	133 45.7	34.0	100 34.3	55.6	Deneb	49 39.5	N45 16.8
03	141 23.1	208 59.4 ..	04.1	305 30.6 ..	44.1	148 47.9 ..	33.8	115 36.8 ..	55.6	Denebola	182 45.0	N14 34.6
04	156 25.5	223 58.5	03.7	320 31.9	44.5	163 50.0	33.7	130 39.2	55.6	Diphda	349 07.1	S17 59.7
05	171 28.0	238 57.6	03.3	335 33.1	45.0	178 52.2	33.6	145 41.7	55.7			
06	186 30.5	253 56.7	S23 02.9	350 34.4	S 4 45.5	193 54.3	S 4 33.4	160 44.2	N 7 55.7	Dubhe	194 05.2	N61 45.1
07	201 32.9	268 55.7	02.5	5 35.7	45.9	208 56.5	33.3	175 46.7	55.7	Elnath	278 26.4	N28 36.3
08	216 35.4	283 54.8	02.0	20 37.0	46.4	223 58.6	33.1	190 49.1	55.7	Eltanin	90 51.9	N51 29.4
M 09	231 37.9	298 53.9 ..	01.6	35 38.3 ..	46.9	239 00.8 ..	33.0	205 51.6 ..	55.7	Enif	33 58.3	N 9 52.3
O 10	246 40.3	313 53.0	01.2	50 39.6	47.3	254 02.9	32.8	220 54.1	55.7	Fomalhaut	15 36.5	S29 37.9
N 11	261 42.8	328 52.1	00.8	65 40.9	47.8	269 05.1	32.7	235 56.6	55.7			
D 12	276 45.3	343 51.2	S23 00.4	80 42.2	S 4 48.3	284 07.2	S 4 32.6	250 59.0	N 7 55.7	Gacrux	172 13.3	S57 06.1
A 13	291 47.7	358 50.3	22 59.9	95 43.5	48.7	299 09.4	32.4	266 01.5	55.7	Gienah	176 03.8	S17 32.0
Y 14	306 50.2	13 49.4	59.5	110 44.8	49.2	314 11.5	32.3	281 04.0	55.7	Hadar	149 04.0	S60 21.7
15	321 52.6	28 48.4 ..	59.1	125 46.0 ..	49.7	329 13.7 ..	32.1	296 06.5 ..	55.7	Hamal	328 13.2	N23 27.4
16	336 55.1	43 47.5	58.7	140 47.3	50.1	344 15.8	32.0	311 08.9	55.8	Kaus Aust.	83 59.1	S34 23.0
17	351 57.6	58 46.6	58.3	155 48.6	50.6	359 18.0	31.8	326 11.4	55.8			
18	7 00.0	73 45.7	S22 57.8	170 49.9	S 4 51.1	14 20.1	S 4 31.7	341 13.9	N 7 55.8	Kochab	137 20.5	N74 09.4
19	22 02.5	88 44.8	57.4	185 51.2	51.5	29 22.3	31.5	356 16.3	55.8	Markab	13 49.6	N15 12.0
20	37 05.0	103 43.9	57.0	200 52.5	52.0	44 24.4	31.4	11 18.8	55.8	Menkar	314 26.5	N 4 05.0
21	52 07.4	118 43.0 ..	56.5	215 53.8 ..	52.5	59 26.6 ..	31.3	26 21.3 ..	55.8	Menkent	148 20.9	S36 21.6
22	67 09.9	133 42.1	56.1	230 55.1	52.9	74 28.7	31.1	41 23.8	55.8	Miaplacidus	221 41.2	S69 42.6
23	82 12.4	148 41.2	55.7	245 56.4	53.4	89 30.9	31.0	56 26.2	55.8			
29 00	97 14.8	163 40.3	S22 55.2	260 57.7	S 4 53.8	104 33.0	S 4 30.8	71 28.7	N 7 55.8	Mirfak	308 56.0	N49 51.5
01	112 17.3	178 39.4	54.8	275 59.0	54.3	119 35.2	30.7	86 31.2	55.8	Nunki	76 12.5	S26 17.8
02	127 19.7	193 38.5	54.4	291 00.3	54.8	134 37.3	30.5	101 33.6	55.9	Peacock	53 37.4	S56 44.4
03	142 22.2	208 37.5 ..	53.9	306 01.6 ..	55.2	149 39.5 ..	30.4	116 36.1 ..	55.9	Pollux	243 41.0	N28 01.6
04	157 24.7	223 36.6	53.5	321 02.9	55.7	164 41.6	30.2	131 38.6	55.9	Procyon	245 11.1	N 5 13.5
05	172 27.1	238 35.7	53.0	336 04.1	56.2	179 43.7	30.1	146 41.1	55.9			
06	187 29.6	253 34.8	S22 52.6	351 05.4	S 4 56.6	194 45.9	S 4 29.9	161 43.5	N 7 55.9	Rasalhague	96 17.2	N12 33.8
07	202 32.1	268 33.9	52.2	6 06.7	57.1	209 48.0	29.8	176 46.0	55.9	Regulus	207 55.2	N11 58.3
08	217 34.5	283 33.0	51.7	21 08.0	57.6	224 50.2	29.6	191 48.5	55.9	Rigel	281 22.5	S 8 12.3
T 09	232 37.0	298 32.1 ..	51.3	36 09.3 ..	58.0	239 52.3 ..	29.5	206 50.9 ..	55.9	Rigil Kent.	140 07.4	S60 49.4
U 10	247 39.5	313 31.2	50.8	51 10.6	58.5	254 54.5	29.4	221 53.4	55.9	Sabik	102 25.7	S15 43.3
E 11	262 41.9	328 30.3	50.4	66 11.9	59.0	269 56.6	29.2	236 55.9	56.0			
S 12	277 44.4	343 29.4	S22 49.9	81 13.2	S 4 59.4	284 58.8	S 4 29.1	251 58.4	N 7 56.0	Schedar	349 53.2	N56 32.1
D 13	292 46.9	358 28.5	49.5	96 14.5	4 59.9	300 00.9	28.9	267 00.8	56.0	Shaula	96 37.5	S37 06.0
A 14	307 49.3	13 27.6	49.0	111 15.8	5 00.3	315 03.0	28.8	282 03.3	56.0	Sirius	258 43.2	S16 42.4
Y 15	322 51.8	28 26.7 ..	48.6	126 17.1 ..	00.8	330 05.2 ..	28.6	297 05.8 ..	56.0	Spica	158 43.1	S11 09.2
16	337 54.2	43 25.8	48.1	141 18.4	01.3	345 07.3	28.5	312 08.2	56.0	Suhail	223 00.3	S43 25.6
17	352 56.7	58 24.9	47.7	156 19.7	01.7	0 09.5	28.3	327 10.7	56.0			
18	7 59.2	73 24.0	S22 47.2	171 21.0	S 5 02.2	15 11.6	S 4 28.2	342 13.2	N 7 56.0	Vega	80 47.0	N38 47.1
19	23 01.6	88 23.1	46.8	186 22.3	02.7	30 13.8	28.0	357 15.6	56.0	Zuben'ubi	137 18.0	S16 02.1
20	38 04.1	103 22.2	46.3	201 23.6	03.1	45 15.9	27.9	12 18.1	56.1		SHA	Mer.Pass.
21	53 06.6	118 21.3 ..	45.8	216 24.9 ..	03.6	60 18.0 ..	27.7	27 20.6 ..	56.1	Venus	67 46.5	13 05
22	68 09.0	133 20.4	45.4	231 26.2	04.0	75 20.2	27.6	42 23.0	56.1	Mars	164 11.0	6 38
23	83 11.5	148 19.5	44.9	246 27.5	04.5	90 22.3	27.4	57 25.5	56.1	Jupiter	7 25.7	17 03
Mer.Pass. 17 32.1		v −0.9	d 0.4	v 1.3	d 0.5	v 2.1	d 0.1	v 2.5	d 0.0	Saturn	334 13.7	19 15

UT	SUN GHA	SUN Dec	MOON GHA	v	Dec	d	HP
d h	° ′	° ′	° ′	′	° ′	′	′
27 00	179 48.8	S23 20.5	82 34.9	10.3	N 1 14.1	11.7	59.1
01	194 48.5	20.4	97 04.2	10.3	1 25.8	11.7	59.1
02	209 48.2	20.4	111 33.5	10.3	1 37.5	11.8	59.1
03	224 47.9	20.3	126 02.8	10.2	1 49.3	11.7	59.1
04	239 47.6	20.2	140 32.0	10.2	2 01.0	11.7	59.1
05	254 47.3	20.1	155 01.2	10.2	2 12.7	11.8	59.2
06	269 47.0	S23 20.0	169 30.4	10.1	N 2 24.5	11.7	59.2
07	284 46.7	19.9	183 59.5	10.1	2 36.2	11.7	59.2
08	299 46.3	19.7	198 28.6	10.1	2 47.9	11.7	59.2
S 09	314 46.0	19.6	212 57.7	10.0	2 59.6	11.7	59.3
U 10	329 45.7	19.5	227 26.7	10.0	3 11.3	11.7	59.3
N 11	344 45.4	19.4	241 55.7	9.9	3 23.0	11.7	59.3
D 12	359 45.1	S23 19.3	256 24.6	9.9	N 3 34.7	11.6	59.3
A 13	14 44.8	19.2	270 53.5	9.9	3 46.3	11.7	59.3
Y 14	29 44.5	19.1	285 22.4	9.8	3 58.0	11.6	59.4
15	44 44.2	19.0	299 51.2	9.8	4 09.6	11.7	59.4
16	59 43.9	18.9	314 20.0	9.7	4 21.3	11.6	59.4
17	74 43.6	18.8	328 48.7	9.7	4 32.9	11.6	59.4
18	89 43.3	S23 18.7	343 17.4	9.7	N 4 44.5	11.6	59.4
19	104 43.0	18.6	357 46.1	9.6	4 56.1	11.5	59.5
20	119 42.7	18.5	12 14.7	9.6	5 07.6	11.5	59.5
21	134 42.4	18.3	26 43.3	9.5	5 19.1	11.6	59.5
22	149 42.1	18.2	41 11.8	9.5	5 30.7	11.4	59.5
23	164 41.7	18.1	55 40.3	9.4	5 42.1	11.5	59.5
28 00	179 41.4	S23 18.0	70 08.7	9.4	N 5 53.6	11.4	59.6
01	194 41.1	17.9	84 37.1	9.3	6 05.0	11.4	59.6
02	209 40.8	17.8	99 05.4	9.3	6 16.4	11.4	59.6
03	224 40.5	17.6	113 33.7	9.2	6 27.8	11.4	59.6
04	239 40.2	17.5	128 01.9	9.2	6 39.2	11.3	59.6
05	254 39.9	17.4	142 30.1	9.2	6 50.5	11.3	59.7
06	269 39.6	S23 17.3	156 58.3	9.1	N 7 01.8	11.2	59.7
07	284 39.3	17.2	171 26.4	9.0	7 13.0	11.2	59.7
08	299 39.0	17.0	185 54.4	9.0	7 24.2	11.2	59.7
M 09	314 38.7	16.9	200 22.4	8.9	7 35.4	11.2	59.7
O 10	329 38.4	16.8	214 50.3	8.9	7 46.6	11.1	59.7
N 11	344 38.1	16.7	229 18.2	8.8	7 57.7	11.0	59.8
D 12	359 37.8	S23 16.6	243 46.0	8.8	N 8 08.7	11.0	59.8
A 13	14 37.5	16.4	258 13.8	8.7	8 19.7	11.0	59.8
Y 14	29 37.2	16.3	272 41.5	8.7	8 30.7	10.9	59.8
15	44 36.9	16.2	287 09.2	8.6	8 41.6	10.9	59.8
16	59 36.6	16.0	301 36.8	8.6	8 52.5	10.9	59.8
17	74 36.2	15.9	316 04.4	8.5	9 03.4	10.7	59.9
18	89 35.9	S23 15.8	330 31.9	8.4	N 9 14.1	10.8	59.9
19	104 35.6	15.7	344 59.3	8.4	9 24.9	10.7	59.9
20	119 35.3	15.5	359 26.7	8.3	9 35.6	10.6	59.9
21	134 35.0	15.4	13 54.0	8.3	9 46.2	10.6	59.9
22	149 34.7	15.3	28 21.3	8.2	9 56.8	10.5	59.9
23	164 34.4	15.1	42 48.5	8.2	10 07.3	10.5	60.0
29 00	179 34.1	S23 15.0	57 15.7	8.1	N10 17.8	10.4	60.0
01	194 33.8	14.8	71 42.8	8.0	10 28.2	10.3	60.0
02	209 33.5	14.7	86 09.8	8.0	10 38.5	10.3	60.0
03	224 33.2	14.6	100 36.8	7.9	10 48.8	10.3	60.0
04	239 32.9	14.4	115 03.7	7.9	10 59.1	10.1	60.0
05	254 32.6	14.3	129 30.6	7.8	11 09.2	10.1	60.0
06	269 32.3	S23 14.2	143 57.4	7.7	N11 19.3	10.1	60.0
07	284 32.0	14.0	158 24.1	7.7	11 29.4	9.9	60.1
08	299 31.7	13.9	172 50.8	7.7	11 39.3	9.9	60.1
T 09	314 31.4	13.7	187 17.5	7.5	11 49.2	9.8	60.1
U 10	329 31.1	13.6	201 44.0	7.6	11 59.0	9.8	60.1
E 11	344 30.8	13.4	216 10.6	7.4	12 08.8	9.7	60.1
S 12	359 30.5	S23 13.3	230 37.0	7.4	N12 18.5	9.6	60.1
D 13	14 30.2	13.2	245 03.4	7.3	12 28.1	9.5	60.1
A 14	29 29.9	13.0	259 29.7	7.3	12 37.6	9.5	60.1
Y 15	44 29.6	12.9	273 56.0	7.2	12 47.1	9.4	60.1
16	59 29.3	12.7	288 22.2	7.2	12 56.5	9.2	60.2
17	74 29.0	12.6	302 48.4	7.1	13 05.7	9.3	60.2
18	89 28.7	S23 12.4	317 14.5	7.0	N13 15.0	9.1	60.2
19	104 28.4	12.3	331 40.5	7.0	13 24.1	9.0	60.2
20	119 28.1	12.1	346 06.5	6.9	13 33.1	9.0	60.2
21	134 27.8	12.0	0 32.4	6.9	13 42.1	8.9	60.2
22	149 27.4	11.8	14 58.3	6.8	13 51.0	8.8	60.2
23	164 27.1	11.7	29 24.1	6.7	N13 59.8	8.7	60.2
	SD 16.3	d 0.1	SD 16.2		16.3		16.4

Lat.	Twilight Naut.	Twilight Civil	Sunrise	Moonrise 27	Moonrise 28	Moonrise 29	Moonrise 30
°	h m	h m	h m	h m	h m	h m	h m
N 72	08 27	10 53	■■	12 12	12 03	11 52	11 36
N 70	08 07	09 54	■■	12 17	12 14	12 13	12 12
68	07 51	09 20		12 20	12 24	12 29	12 38
66	07 38	08 55	10 34	12 24	12 32	12 42	12 58
64	07 27	08 35	09 53	12 27	12 39	12 54	13 14
62	07 18	08 20	09 25	12 29	12 44	13 03	13 28
60	07 09	08 06	09 04	12 31	12 50	13 11	13 39
N 58	07 02	07 55	08 46	12 33	12 54	13 19	13 49
56	06 55	07 44	08 32	12 35	12 58	13 25	13 58
54	06 49	07 35	08 19	12 36	13 02	13 31	14 05
52	06 44	07 27	08 08	12 38	13 05	13 36	14 12
50	06 38	07 20	07 58	12 39	13 08	13 41	14 19
45	06 27	07 04	07 38	12 42	13 15	13 51	14 32
N 40	06 16	06 50	07 21	12 44	13 20	14 00	14 44
35	06 07	06 39	07 07	12 46	13 25	14 07	14 53
30	05 58	06 28	06 55	12 48	13 29	14 13	15 02
20	05 42	06 10	06 34	12 51	13 37	14 25	15 17
N 10	05 26	05 52	06 15	12 54	13 43	14 35	15 29
0	05 09	05 35	05 58	12 57	13 49	14 44	15 42
S 10	04 50	05 17	05 40	13 00	13 56	14 54	15 54
20	04 28	04 57	05 21	13 03	14 03	15 04	16 07
30	03 59	04 32	04 59	13 06	14 10	15 16	16 22
35	03 40	04 17	04 46	13 08	14 15	15 23	16 31
40	03 17	03 59	04 32	13 10	14 20	15 30	16 41
45	02 47	03 36	04 14	13 13	14 26	15 40	16 53
S 50	02 02	03 07	03 51	13 16	14 33	15 51	17 07
52	01 34	02 52	03 40	13 18	14 36	15 56	17 14
54	00 50	02 34	03 28	13 19	14 40	16 02	17 22
56	////	02 12	03 14	13 21	14 44	16 08	17 30
58	////	01 43	02 57	13 23	14 49	16 15	17 40
S 60	////	00 54	02 37	13 25	14 54	16 23	17 51

Lat.	Sunset	Twilight Civil	Twilight Naut.	Moonset 27	Moonset 28	Moonset 29	Moonset 30
°	h m	h m	h m	h m	h m	h m	h m
N 72	■■	13 10	15 37	00 46	02 44	04 48	07 02
N 70	■■	14 09	15 56	00 45	02 35	04 29	06 27
68	■■	14 43	16 12	00 43	02 27	04 14	06 02
66	13 30	15 08	16 25	00 42	02 21	04 01	05 43
64	14 11	15 28	16 36	00 41	02 15	03 51	05 28
62	14 38	15 44	16 45	00 40	02 11	03 43	05 15
60	15 00	15 57	16 54	00 39	02 07	03 36	05 04
N 58	15 17	16 09	17 01	00 39	02 03	03 29	04 55
56	15 31	16 19	17 08	00 38	02 00	03 23	04 47
54	15 44	16 28	17 14	00 37	01 57	03 18	04 39
52	15 55	16 36	17 19	00 37	01 55	03 14	04 33
50	16 05	16 43	17 25	00 36	01 52	03 10	04 27
45	16 25	16 59	17 36	00 35	01 47	03 01	04 14
N 40	16 42	17 13	17 47	00 35	01 43	02 53	04 04
35	16 56	17 24	17 56	00 34	01 40	02 47	03 55
30	17 08	17 35	18 05	00 33	01 36	02 41	03 47
20	17 29	17 53	18 21	00 32	01 31	02 32	03 34
N 10	17 48	18 11	18 37	00 31	01 26	02 23	03 22
0	18 05	18 28	18 54	00 30	01 22	02 15	03 11
S 10	18 23	18 46	19 13	00 29	01 17	02 07	03 00
20	18 42	19 06	19 35	00 28	01 12	01 59	02 49
30	19 03	19 31	20 04	00 27	01 07	01 49	02 35
35	19 16	19 46	20 23	00 26	01 04	01 44	02 28
40	19 31	20 04	20 46	00 25	01 00	01 38	02 19
45	19 49	20 26	21 16	00 24	00 56	01 30	02 09
S 50	20 12	20 56	22 01	00 23	00 51	01 22	01 57
52	20 22	21 11	22 28	00 23	00 49	01 18	01 51
54	20 34	21 29	23 12	00 22	00 46	01 13	01 45
56	20 49	21 50	////	00 22	00 44	01 08	01 38
58	21 05	22 20	////	00 21	00 41	01 03	01 30
S 60	21 25	23 07	////	00 20	00 37	00 57	01 21

Day	SUN Eqn. of Time 00h	SUN Eqn. of Time 12h	SUN Mer. Pass.	MOON Mer. Pass. Upper	MOON Mer. Pass. Lower	Age	Phase
d	m s	m s	h m	h m	h m	d	%
27	00 44	00 59	12 01	19 09	06 43	09	62
28	01 14	01 28	12 01	20 02	07 36	10	73
29	01 43	01 58	12 02	20 58	08 30	11	82

1998 DEC. 30, 31, JAN. 1 (WED., THURS., FRI.)

UT	ARIES GHA	VENUS −3.9 GHA	Dec	MARS +1.0 GHA	Dec	JUPITER −2.3 GHA	Dec	SATURN +0.4 GHA	Dec	Name	SHA	Dec
30 00	98 14.0	163 18.6	S22 44.5	261 28.8	S 5 05.0	105 24.5	S 4 27.3	72 28.0	N 7 56.1	Acamar	315 26.6	S40 18.9
01	113 16.4	178 17.7	44.0	276 30.1	05.4	120 26.6	27.2	87 30.4	56.1	Achernar	335 35.0	S57 14.9
02	128 18.9	193 16.8	43.5	291 31.4	05.9	135 28.7	27.0	102 32.9	56.1	Acrux	173 21.7	S63 05.2
03	143 21.4	208 15.9 ..	43.1	306 32.7 ..	06.3	150 30.9 ..	26.9	117 35.4 ..	56.1	Adhara	255 20.9	S28 58.4
04	158 23.8	223 15.0	42.6	321 34.0	06.8	165 33.0	26.7	132 37.8	56.1	Aldebaran	291 01.9	N16 30.3
05	173 26.3	238 14.1	42.1	336 35.3	07.3	180 35.2	26.6	147 40.3	56.2			
W 06	188 28.7	253 13.2	S22 41.6	351 36.6	S 5 07.7	195 37.3	S 4 26.4	162 42.8	N 7 56.2	Alioth	166 30.6	N55 57.7
07	203 31.2	268 12.3	41.2	6 37.9	08.2	210 39.4	26.3	177 45.2	56.2	Alkaid	153 07.9	N49 19.0
E 08	218 33.7	283 11.5	40.7	21 39.2	08.6	225 41.6	26.1	192 47.7	56.2	Al Na'ir	27 58.1	S46 58.2
D 09	233 36.1	298 10.6 ..	40.2	36 40.5 ..	09.1	240 43.7 ..	26.0	207 50.2 ..	56.2	Alnilam	275 57.4	S 1 12.3
N 10	248 38.6	313 09.7	39.8	51 41.8	09.6	255 45.9	25.8	222 52.6	56.2	Alphard	218 06.8	S 8 39.3
E 11	263 41.1	328 08.8	39.3	66 43.1	10.0	270 48.0	25.7	237 55.1	56.2			
S 12	278 43.5	343 07.9	S22 38.8	81 44.4	S 5 10.5	285 50.1	S 4 25.5	252 57.6	N 7 56.2	Alphecca	126 20.8	N26 43.1
D 13	293 46.0	358 07.0	38.3	96 45.7	10.9	300 52.3	25.4	268 00.0	56.3	Alpheratz	357 55.1	N29 05.2
A 14	308 48.5	13 06.1	37.8	111 47.0	11.4	315 54.4	25.2	283 02.5	56.3	Altair	62 19.4	N 8 52.0
Y 15	323 50.9	28 05.2 ..	37.4	126 48.3 ..	11.9	330 56.6 ..	25.1	298 05.0 ..	56.3	Ankaa	353 26.8	S42 19.0
16	338 53.4	43 04.3	36.9	141 49.6	12.3	345 58.7	24.9	313 07.4	56.3	Antares	112 40.3	S26 25.6
17	353 55.9	58 03.4	36.4	156 50.9	12.8	1 00.8	24.8	328 09.9	56.3			
18	8 58.3	73 02.5	S22 35.9	171 52.2	S 5 13.2	16 03.0	S 4 24.6	343 12.3	N 7 56.3	Arcturus	146 06.1	N19 11.3
19	24 00.8	88 01.6	35.4	186 53.5	13.7	31 05.1	24.5	358 14.8	56.3	Atria	107 52.7	S69 01.3
20	39 03.2	103 00.8	34.9	201 54.8	14.2	46 07.2	24.3	13 17.3	56.4	Avior	234 21.9	S59 30.3
21	54 05.7	117 59.9 ..	34.4	216 56.1 ..	14.6	61 09.4 ..	24.2	28 19.7 ..	56.4	Bellatrix	278 43.7	N 6 20.8
22	69 08.2	132 59.0	34.0	231 57.4	15.1	76 11.5	24.0	43 22.2	56.4	Betelgeuse	271 13.1	N 7 24.3
23	84 10.6	147 58.1	33.5	246 58.7	15.5	91 13.7	23.9	58 24.7	56.4			
31 00	99 13.1	162 57.2	S22 33.0	262 00.0	S 5 16.0	106 15.8	S 4 23.7	73 27.1	N 7 56.4	Canopus	264 00.5	S52 41.9
01	114 15.6	177 56.3	32.5	277 01.3	16.5	121 17.9	23.6	88 29.6	56.4	Capella	280 50.5	N45 59.7
02	129 18.0	192 55.4	32.0	292 02.6	16.9	136 20.1	23.4	103 32.0	56.4	Deneb	49 39.5	N45 16.8
03	144 20.5	207 54.5 ..	31.5	307 03.9 ..	17.4	151 22.2 ..	23.3	118 34.5 ..	56.4	Denebola	182 45.0	N14 34.6
04	159 23.0	222 53.7	31.0	322 05.2	17.8	166 24.3	23.1	133 37.0	56.5	Diphda	349 07.1	S17 59.7
05	174 25.4	237 52.8	30.5	337 06.5	18.3	181 26.5	23.0	148 39.4	56.5			
T 06	189 27.9	252 51.9	S22 30.0	352 07.8	S 5 18.7	196 28.6	S 4 22.8	163 41.9	N 7 56.5	Dubhe	194 05.1	N61 45.1
07	204 30.3	267 51.0	29.5	7 09.1	19.2	211 30.7	22.7	178 44.4	56.5	Elnath	278 26.4	N28 36.3
H 08	219 32.8	282 50.1	29.0	22 10.4	19.7	226 32.9	22.5	193 46.8	56.5	Eltanin	90 51.9	N51 29.4
U 09	234 35.3	297 49.2 ..	28.5	37 11.7 ..	20.1	241 35.0 ..	22.4	208 49.3 ..	56.5	Enif	33 58.3	N 9 52.3
R 10	249 37.7	312 48.4	28.0	52 13.0	20.6	256 37.1	22.2	223 51.7	56.5	Fomalhaut	15 36.5	S29 37.9
S 11	264 40.2	327 47.5	27.5	67 14.3	21.0	271 39.3	22.1	238 54.2	56.6			
D 12	279 42.7	342 46.6	S22 27.0	82 15.6	S 5 21.5	286 41.4	S 4 21.9	253 56.7	N 7 56.6	Gacrux	172 13.3	S57 06.1
A 13	294 45.1	357 45.7	26.5	97 16.9	21.9	301 43.5	21.8	268 59.1	56.6	Gienah	176 03.8	S17 32.0
Y 14	309 47.6	12 44.8	26.0	112 18.3	22.4	316 45.7	21.6	284 01.6	56.6	Hadar	149 04.0	S60 21.7
15	324 50.1	27 43.9 ..	25.5	127 19.6 ..	22.8	331 47.8 ..	21.5	299 04.0 ..	56.6	Hamal	328 13.2	N23 27.4
16	339 52.5	42 43.1	24.9	142 20.9	23.3	346 49.9	21.3	314 06.5	56.6	Kaus Aust.	83 59.0	S34 23.0
17	354 55.0	57 42.2	24.4	157 22.2	23.8	1 52.1	21.2	329 09.0	56.6			
18	9 57.5	72 41.3	S22 23.9	172 23.5	S 5 24.2	16 54.2	S 4 21.0	344 11.4	N 7 56.7	Kochab	137 20.4	N74 09.4
19	24 59.9	87 40.4	23.4	187 24.8	24.7	31 56.3	20.9	359 13.9	56.7	Markab	13 49.6	N15 12.0
20	40 02.4	102 39.5	22.9	202 26.1	25.1	46 58.5	20.7	14 16.3	56.7	Menkar	314 26.5	N 4 05.0
21	55 04.8	117 38.7 ..	22.4	217 27.4 ..	25.6	62 00.6 ..	20.6	29 18.8 ..	56.7	Menkent	148 20.9	S36 21.6
22	70 07.3	132 37.8	21.9	232 28.7	26.0	77 02.7	20.4	44 21.3	56.7	Miaplacidus	221 41.2	S69 42.6
23	85 09.8	147 36.9	21.3	247 30.0	26.5	92 04.8	20.3	59 23.7	56.7			
1 00	100 12.2	162 36.0	S22 20.8	262 31.3	S 5 26.9	107 07.0	S 4 20.1	74 26.2	N 7 56.8	Mirfak	308 56.0	N49 51.5
01	115 14.7	177 35.2	20.3	277 32.6	27.4	122 09.1	19.9	89 28.6	56.8	Nunki	76 12.5	S26 17.8
02	130 17.2	192 34.3	19.8	292 33.9	27.8	137 11.2	19.8	104 31.1	56.8	Peacock	53 37.4	S56 44.4
03	145 19.6	207 33.4 ..	19.3	307 35.2 ..	28.3	152 13.4 ..	19.6	119 33.5 ..	56.8	Pollux	243 41.0	N28 01.6
04	160 22.1	222 32.5	18.7	322 36.6	28.8	167 15.5	19.5	134 36.0	56.8	Procyon	245 11.1	N 5 13.5
05	175 24.6	237 31.7	18.2	337 37.9	29.2	182 17.6	19.3	149 38.5	56.8			
06	190 27.0	252 30.8	S22 17.7	352 39.2	S 5 29.7	197 19.8	S 4 19.2	164 40.9	N 7 56.8	Rasalhague	96 17.2	N12 33.7
07	205 29.5	267 29.9	17.1	7 40.5	30.1	212 21.9	19.0	179 43.4	56.9	Regulus	207 55.2	N11 58.2
08	220 32.0	282 29.0	16.6	22 41.8	30.6	227 24.0	18.9	194 45.8	56.9	Rigel	281 22.5	S 8 12.4
F 09	235 34.4	297 28.2 ..	16.1	37 43.1 ..	31.0	242 26.1 ..	18.7	209 48.3 ..	56.9	Rigil Kent.	140 07.4	S60 49.4
R 10	250 36.9	312 27.3	15.6	52 44.4	31.5	257 28.3	18.6	224 50.7	56.9	Sabik	102 25.7	S15 43.3
I 11	265 39.3	327 26.4	15.0	67 45.7	31.9	272 30.4	18.4	239 53.2	56.9			
D 12	280 41.8	342 25.5	S22 14.5	82 47.0	S 5 32.4	287 32.5	S 4 18.3	254 55.7	N 7 56.9	Schedar	349 53.2	N56 32.1
A 13	295 44.3	357 24.7	14.0	97 48.3	32.8	302 34.7	18.1	269 58.1	57.0	Shaula	96 37.5	S37 06.0
Y 14	310 46.7	12 23.8	13.4	112 49.7	33.3	317 36.8	18.0	285 00.6	57.0	Sirius	258 43.2	S16 43.0
15	325 49.2	27 22.9 ..	12.9	127 51.0 ..	33.7	332 38.9 ..	17.8	300 03.0 ..	57.0	Spica	158 43.1	S11 09.2
16	340 51.7	42 22.1	12.3	142 52.3	34.2	347 41.0	17.7	315 05.5	57.0	Suhail	223 00.2	S43 25.6
17	355 54.1	57 21.2	11.8	157 53.6	34.6	2 43.2	17.5	330 07.9	57.0			
18	10 56.6	72 20.3	S22 11.3	172 54.9	S 5 35.1	17 45.3	S 4 17.3	345 10.4	N 7 57.0	Vega	80 47.0	N38 47.1
19	25 59.1	87 19.5	10.7	187 56.2	35.5	32 47.4	17.2	0 12.9	57.1	Zuben'ubi	137 18.0	S16 02.1
20	41 01.5	102 18.6	10.2	202 57.5	36.0	47 49.5	17.0	15 15.3	57.1		SHA	Mer. Pass.
21	56 04.0	117 17.7 ..	09.6	217 58.8 ..	36.4	62 51.7 ..	16.9	30 17.8 ..	57.1		° '	h m
22	71 06.5	132 16.8	09.1	233 00.2	36.9	77 53.8	16.7	45 20.2	57.1	Venus	63 44.1	13 09
23	86 08.9	147 16.0	08.5	248 01.5	37.3	92 55.9	16.6	60 22.7	57.1	Mars	162 46.9	6 31
Mer. Pass. 17 20.3		v −0.9	d 0.5	v 1.3	d 0.5	v 2.1	d 0.2	v 2.5	d 0.0	Jupiter	7 02.7	16 53
										Saturn	334 14.0	19 03

UT		SUN		MOON				
		GHA	Dec	GHA	v	Dec	d	HP
d	h	° ′	° ′	° ′	′	° ′	′	′
30	00	179 26.8	S23 11.5	43 49.8	6.7	N14 08.5	8.6	60.2
	01	194 26.5	11.3	58 15.5	6.6	14 17.1	8.5	60.2
	02	209 26.2	11.2	72 41.1	6.6	14 25.6	8.4	60.2
	03	224 25.9 ..	11.0	87 06.7	6.5	14 34.0	8.4	60.2
	04	239 25.6	10.9	101 32.2	6.5	14 42.4	8.2	60.2
	05	254 25.3	10.7	115 57.7	6.4	14 50.6	8.2	60.2
W	06	269 25.0	S23 10.6	130 23.1	6.4	N14 58.8	8.0	60.2
E	07	284 24.7	10.4	144 48.5	6.3	15 06.8	7.9	60.2
D	08	299 24.4	10.2	159 13.8	6.2	15 14.7	7.9	60.3
N	09	314 24.1 ..	10.1	173 39.0	6.2	15 22.6	7.7	60.3
E	10	329 23.8	09.9	188 04.2	6.1	15 30.3	7.7	60.3
S	11	344 23.5	09.8	202 29.3	6.1	15 38.0	7.5	60.3
D	12	359 23.2	S23 09.6	216 54.4	6.1	N15 45.5	7.4	60.3
A	13	14 22.9	09.4	231 19.5	6.0	15 52.9	7.4	60.3
Y	14	29 22.6	09.3	245 44.5	5.9	16 00.3	7.2	60.3
	15	44 22.3 ..	09.1	260 09.4	5.9	16 07.5	7.1	60.3
	16	59 22.0	08.9	274 34.3	5.8	16 14.6	7.0	60.3
	17	74 21.7	08.8	288 59.1	5.8	16 21.6	6.9	60.3
	18	89 21.4	S23 08.6	303 23.9	5.7	N16 28.5	6.7	60.3
	19	104 21.1	08.4	317 48.6	5.7	16 35.2	6.7	60.3
	20	119 20.8	08.2	332 13.3	5.7	16 41.9	6.6	60.3
	21	134 20.5 ..	08.1	346 38.0	5.6	16 48.5	6.4	60.3
	22	149 20.2	07.9	1 02.6	5.6	16 54.9	6.3	60.3
	23	164 19.9	07.7	15 27.2	5.5	17 01.2	6.2	60.3
31	00	179 19.6	S23 07.6	29 51.7	5.5	N17 07.4	6.1	60.3
	01	194 19.3	07.4	44 16.2	5.4	17 13.5	6.0	60.3
	02	209 19.0	07.2	58 40.6	5.4	17 19.5	5.8	60.3
	03	224 18.7 ..	07.0	73 05.0	5.4	17 25.3	5.7	60.3
	04	239 18.4	06.9	87 29.4	5.3	17 31.0	5.7	60.3
	05	254 18.1	06.7	101 53.7	5.3	17 36.7	5.4	60.3
T	06	269 17.8	S23 06.5	116 18.0	5.3	N17 42.1	5.4	60.2
H	07	284 17.5	06.3	130 42.3	5.2	17 47.5	5.2	60.2
U	08	299 17.3	06.1	145 06.5	5.2	17 52.7	5.1	60.2
R	09	314 17.0 ..	06.0	159 30.7	5.1	17 57.8	5.0	60.2
S	10	329 16.7	05.8	173 54.8	5.2	18 02.8	4.9	60.2
D	11	344 16.4	05.6	188 19.0	5.1	18 07.7	4.7	60.2
A	12	359 16.1	S23 05.4	202 43.1	5.1	N18 12.4	4.6	60.2
Y	13	14 15.8	05.2	217 07.2	5.0	18 17.0	4.5	60.2
	14	29 15.5	05.0	231 31.2	5.0	18 21.5	4.3	60.2
	15	44 15.2 ..	04.9	245 55.2	5.0	18 25.8	4.3	60.2
	16	59 14.9	04.7	260 19.2	5.0	18 30.1	4.1	60.2
	17	74 14.6	04.5	274 43.2	5.0	18 34.2	3.9	60.2
	18	89 14.3	S23 04.3	289 07.2	4.9	N18 38.1	3.8	60.2
	19	104 14.0	04.1	303 31.1	5.0	18 41.9	3.7	60.1
	20	119 13.7	03.9	317 55.1	4.9	18 45.6	3.6	60.1
	21	134 13.4 ..	03.7	332 19.0	4.9	18 49.2	3.4	60.1
	22	149 13.1	03.5	346 42.9	4.9	18 52.6	3.3	60.1
	23	164 12.8	03.3	1 06.8	4.8	18 55.9	3.2	60.1
1	00	179 12.5	S23 03.2	15 30.6	4.9	N18 59.1	3.0	60.1
	01	194 12.2	03.0	29 54.5	4.9	19 02.1	2.9	60.1
	02	209 11.9	02.8	44 18.4	4.8	19 05.0	2.7	60.1
	03	224 11.6 ..	02.6	58 42.2	4.9	19 07.7	2.7	60.1
	04	239 11.3	02.4	73 06.1	4.8	19 10.4	2.5	60.0
	05	254 11.0	02.2	87 29.9	4.8	19 12.9	2.3	60.0
	06	269 10.7	S23 02.0	101 53.7	4.9	N19 15.2	2.2	60.0
	07	284 10.4	01.8	116 17.6	4.8	19 17.4	2.1	60.0
F	08	299 10.1	01.6	130 41.4	4.9	19 19.5	1.9	60.0
R	09	314 09.8 ..	01.4	145 05.3	4.8	19 21.4	1.9	60.0
I	10	329 09.5	01.2	159 29.1	4.9	19 23.3	1.6	59.9
D	11	344 09.3	01.0	173 53.0	4.8	19 24.9	1.6	59.9
A	12	359 09.0	S23 00.8	188 16.8	4.9	N19 26.5	1.4	59.9
Y	13	14 08.7	00.6	202 40.7	4.9	19 27.9	1.2	59.9
	14	29 08.4	00.4	217 04.6	4.9	19 29.1	1.1	59.9
	15	44 08.1 ..	00.2	231 28.5	4.9	19 30.2	0.9	59.9
	16	59 07.8	23 00.0	245 52.4	4.9	19 31.2	0.9	59.8
	17	74 07.5	22 59.8	260 16.3	5.0	19 32.1	0.7	59.8
	18	89 07.2	S22 59.5	274 40.3	4.9	N19 32.8	0.6	59.8
	19	104 06.9	59.3	289 04.2	5.0	19 33.4	0.4	59.8
	20	119 06.6	59.1	303 28.2	5.0	19 33.8	0.3	59.8
	21	134 06.3 ..	58.9	317 52.2	5.1	19 34.1	0.2	59.7
	22	149 06.0	58.7	332 16.3	5.0	19 34.3	0.0	59.7
	23	164 05.7	58.5	346 40.3	5.1	N19 34.3	0.1	59.7
		SD 16.3	d 0.2	SD 16.4		16.4		16.3

Lat.	Twilight Naut.	Civil	Sunrise	Moonrise 30	31	1	2
°	h m	h m	h m	h m	h m	h m	h m
N 72	08 25	10 46	▬	11 36	10 45	☐	☐
N 70	08 06	09 51	▬	12 12	12 15	12 32	13 37
68	07 50	09 18	▬	12 38	12 55	13 30	14 33
66	07 38	08 54	10 30	12 58	13 23	14 04	15 07
64	07 27	08 35	09 51	13 14	13 44	14 29	15 32
62	07 18	08 19	09 24	13 28	14 01	14 49	15 51
60	07 10	08 06	09 03	13 39	14 16	15 05	16 07
N 58	07 02	07 55	08 46	13 49	14 28	15 18	16 20
56	06 56	07 45	08 32	13 58	14 39	15 30	16 32
54	06 50	07 36	08 19	14 05	14 48	15 40	16 42
52	06 44	07 28	08 08	14 12	14 56	15 49	16 51
50	06 39	07 20	07 59	14 19	15 04	15 58	16 59
45	06 27	07 05	07 38	14 32	15 20	16 15	17 16
N 40	06 17	06 51	07 22	14 44	15 33	16 29	17 30
35	06 08	06 40	07 08	14 53	15 45	16 41	17 42
30	05 59	06 29	06 56	15 02	15 55	16 52	17 53
20	05 43	06 11	06 35	15 17	16 12	17 10	18 10
N 10	05 27	05 54	06 16	15 29	16 27	17 26	18 26
0	05 11	05 37	05 59	15 42	16 41	17 41	18 41
S 10	04 52	05 19	05 42	15 54	16 55	17 56	18 55
20	04 29	04 59	05 23	16 07	17 10	18 12	19 11
30	04 01	04 34	05 01	16 22	17 28	18 31	19 29
35	03 42	04 19	04 49	16 31	17 38	18 41	19 39
40	03 20	04 01	04 34	16 41	17 50	18 54	19 51
45	02 50	03 39	04 16	16 53	18 04	19 08	20 05
S 50	02 05	03 10	03 54	17 07	18 20	19 26	20 22
52	01 39	02 55	03 43	17 14	18 28	19 35	20 30
54	00 57	02 37	03 31	17 22	18 37	19 44	20 39
56	////	02 16	03 17	17 30	18 47	19 54	20 49
58	////	01 47	03 01	17 40	18 59	20 07	21 01
S 60	////	01 02	02 41	17 51	19 12	20 21	21 14

Lat.	Sunset	Twilight Civil	Naut.	Moonset 30	31	1	2
°	h m	h m	h m	h m	h m	h m	h m
N 72	▬	13 20	15 41	07 02	09 57	☐	☐
N 70	▬	14 15	16 00	06 27	08 27	10 17	11 17
68	▬	14 48	16 16	06 02	07 48	09 19	10 20
66	13 36	15 12	16 28	05 43	07 20	08 45	09 46
64	14 15	15 31	16 39	05 28	07 00	08 20	09 21
62	14 42	15 47	16 48	05 15	06 43	08 00	09 02
60	15 03	16 00	16 57	05 04	06 29	07 44	08 46
N 58	15 20	16 12	17 04	04 55	06 17	07 31	08 32
56	15 35	16 22	17 12	04 47	06 07	07 19	08 21
54	15 47	16 30	17 16	04 39	05 58	07 09	08 10
52	15 58	16 38	17 22	04 33	05 50	07 00	08 01
50	16 07	16 46	17 27	04 27	05 42	06 52	07 53
45	16 28	17 01	17 39	04 14	05 27	06 35	07 36
N 40	16 44	17 15	17 49	04 04	05 14	06 21	07 21
35	16 58	17 26	17 58	03 55	05 03	06 09	07 09
30	17 10	17 37	18 06	03 47	04 54	05 58	06 59
20	17 31	17 55	18 23	03 34	04 37	05 40	06 41
N 10	17 49	18 12	18 39	03 22	04 23	05 24	06 25
0	18 07	18 29	18 55	03 11	04 10	05 10	06 10
S 10	18 24	18 47	19 14	03 00	03 56	04 55	05 55
20	18 43	19 07	19 36	02 49	03 42	04 39	05 39
30	19 04	19 32	20 05	02 35	03 26	04 21	05 20
35	19 17	19 47	20 23	02 28	03 17	04 11	05 10
40	19 32	20 05	20 46	02 19	03 06	03 59	04 57
45	19 50	20 27	21 16	02 09	02 53	03 45	04 43
S 50	20 12	20 56	22 00	01 57	02 38	03 27	04 25
52	20 22	21 10	22 26	01 51	02 31	03 19	04 17
54	20 34	21 28	23 07	01 45	02 23	03 10	04 07
56	20 48	21 49	////	01 38	02 14	03 00	03 57
58	21 04	22 17	////	01 30	02 04	02 48	03 45
S 60	21 24	23 02	////	01 21	01 53	02 35	03 31

Day	SUN Eqn. of Time 00h	12h	Mer. Pass.	MOON Mer. Pass. Upper	Lower	Age	Phase
d	m s	m s	h m	h m	h m	d	%
30	02 12	02 26	12 02	21 56	09 26	12	90
31	02 41	02 55	12 03	22 55	10 25	13	96
1	03 09	03 24	12 03	23 56	11 25	14	99 ◯

EXPLANATION

PRINCIPLE AND ARRANGEMENT

1. *Object.* The object of this Almanac is to provide, in a convenient form, the data required for the practice of astronomical navigation at sea.

2. *Principle.* The main contents of the Almanac consist of data from which the *Greenwich Hour Angle* (GHA) and the *Declination* (Dec) of all the bodies used for navigation can be obtained for any instant of *Universal Time* (UT), or *Greenwich Mean Time* (GMT). The *Local Hour Angle* (LHA) can then be obtained by means of the formula:

$$\text{LHA} = \text{GHA} \, {-\text{west} \atop +\text{east}} \, \text{longitude}$$

The remaining data consist of: times of rising and setting of the Sun and Moon, and times of twilight; miscellaneous calendarial and planning data and auxiliary tables, including a list of Standard Times; corrections to be applied to observed altitude.

For the Sun, Moon, and planets the GHA and Dec are tabulated directly for each hour of UT throughout the year. For the stars the *Sidereal Hour Angle* (SHA) is given, and the GHA is obtained from:

$$\text{GHA Star} = \text{GHA Aries} + \text{SHA Star}$$

The SHA and Dec of the stars change slowly and may be regarded as constant over periods of several days. GHA Aries, or the Greenwich Hour Angle of the first point of Aries (the Vernal Equinox), is tabulated for each hour. Permanent tables give the appropriate increments and corrections to the tabulated hourly values of GHA and Dec for the minutes and seconds of UT.

The six-volume series of *Sight Reduction Tables for Marine Navigation* (published in U.S.A. as Pub. No. 229 and in U.K. as N.P. 401) has been designed for the solution of the navigational triangle and is intended for use with *The Nautical Almanac.*

Two alternative procedures for sight reduction are described on pages 277–318. The first requires the use of programmable calculators or computers, while the second uses a set of concise tables that is given on pages 286–317.

The tabular accuracy is 0ʹ1 throughout. The time argument on the daily pages of this Almanac is $12^h +$ the Greenwich Hour Angle of the mean sun and is here denoted by UT, although it is also known as GMT. This scale may differ from the broadcast time signals (UTC) by an amount which, if ignored, will introduce an error of up to 0ʹ2 in longitude determined from astronomical observations. (The difference arises because the time argument depends on the variable rate of rotation of the Earth while the broadcast time signals are now based on an atomic time-scale.) Step adjustments of exactly one second are made to the time signals as required (normally at 24^h on December 31 and June 30) so that the difference between the time signals and UT, as used in this Almanac, may not exceed 0ˢ9. Those who require to reduce observations to a precision of better than 1^s must therefore obtain the correction (DUT1) to the time signals from coding in the signal, or from other sources; the required time is given by UT1=UTC+DUT1 to a precision of 0ˢ1. Alternatively, the longitude, when determined from astronomical observations, may be corrected by the corresponding amount shown in the following table:

Correction to time signals	Correction to longitude
−0ˢ9 to −0ˢ7	0ʹ2 to east
−0ˢ6 to −0ˢ3	0ʹ1 to east
−0ˢ2 to +0ˢ2	no correction
+0ˢ3 to +0ˢ6	0ʹ1 to west
+0ˢ7 to +0ˢ9	0ʹ2 to west

a_1, which is a function of LHA Aries and latitude, is the excess of the value of the second term over its mean value for latitude 50°, increased by a constant (0.́6) to make it always positive. a_2, which is a function of LHA Aries and date, is the correction to the first term for the variation of *Polaris* from its adopted mean position; it is increased by a constant (0.́6) to make it positive. The sum of the added constants is 1°, so that:

$$\text{Latitude} = \text{Apparent altitude (corrected for refraction)} - 1° + a_0 + a_1 + a_2$$

RISING AND SETTING PHENOMENA

10. *General.* On the right-hand daily pages are given the times of sunrise and sunset, of the beginning and end of civil and nautical twilights, and of moonrise and moonset for a range of latitudes from N 72° to S 60°. These times, which are given to the nearest minute, are strictly the UT of the phenomena on the Greenwich meridian; they are given for every day for moonrise and moonset, but only for the middle day of the three on each page for the solar phenomena.

They are approximately the Local Mean Times (LMT) of the corresponding phenomena on other meridians; they can be formally interpolated if desired. The UT of a phenomenon is obtained from the LMT by:

$$\text{UT} = \text{LMT} {\textstyle{+ \text{ west} \atop - \text{ east}}} \text{ longitude}$$

in which the longitude must first be converted to time by the table on page i or otherwise.

Interpolation for latitude can be done mentally or with the aid of Table I on page xxxii.

The following symbols are used to indicate the conditions under which, in high latitudes, some of the phenomena do not occur:

        ☐ Sun or Moon remains continuously above the horizon;

        ■ Sun or Moon remains continuously below the horizon;

        //// twilight lasts all night.

Basis of the tabulations. At sunrise and sunset 16' is allowed for semi-diameter and 34' for horizontal refraction, so that at the times given the Sun's upper limb is on the visible horizon; all times refer to phenomena as seen from sea level with a clear horizon.

At the times given for the beginning and end of twilight, the Sun's zenith distance is 96° for civil, and 102° for nautical twilight. The degree of illumination at the times given for civil twilight (in good conditions and in the absence of other illumination) is such that the brightest stars are visible and the horizon is clearly defined. At the times given for nautical twilight the horizon is in general not visible, and it is too dark for observation with a marine sextant.

Times corresponding to other depressions of the Sun may be obtained by interpolation or, for depressions of more than 12°, less reliably, by extrapolation; times so obtained will be subject to considerable uncertainty near extreme conditions.

At moonrise and moonset allowance is made for semi-diameter, parallax, and refraction (34'), so that at the times given the Moon's upper limb is on the visible horizon as seen from sea level.

11. *Sunrise, sunset, twilight.* The tabulated times may be regarded, without serious error, as the LMT of the phenomena on any of the three days on the page and in any longitude. Precise times may normally be obtained by interpolating the tabular values for latitude and to the correct day and longitude, the latter being expressed as a fraction of a day by dividing it by 360°, positive for west and negative for east longitudes. In the extreme conditions near ☐, ■ or //// interpolation may not be possible in one direction, but accurate times are of little value in these circumstances.

Examples. Required the UT of (a) the beginning of morning twilights and sunrise on 1998 January 22 for latitude S 48° 55', longitude E 75° 18'; (b) sunset and the end of evening twilights on 1998 January 24 for latitude N 67° 10', longitude W 168° 05'.

	(a)	Twilight Nautical	Twilight Civil	Sunrise	(b)	Sunset	Twilight Civil	Twilight Nautical
		d h m	d h m	d h m		d h m	d h m	d h m
From p. 25								
LMT for Lat	S 45°	22 03 25	22 04 09	22 04 43	N 66°	24 14 57	24 16 07	24 17 14
Corr. to	S 48° 55′	−27	−20	−12	N 67° 10′	−17	−9	−6
(p. xxxii, Table I)								
Long (p. i)	E 75° 18′	−5 01	−5 01	−5 01	W 168° 05′	+11 12	+11 12	+11 12
UT		21 21 57	21 22 48	21 23 30		25 01 52	25 03 10	25 04 20

The LMT are strictly for January 23 (middle date on page) and 0° longitude; for more precise times it is necessary to interpolate, but rounding errors may accumulate to about 2^m.

(a) to January $22^d - 75°/360° =$ Jan. $21^d\!\!.8$, i.e. $\frac{1}{3}(1\cdot2) = 0\cdot4$ backwards towards the data for the same latitude interpolated similarly from page 23; the corrections are -2^m to nautical twilight, -2^m to civil twilight and -3^m to sunrise.

(b) to January $24^d + 168°/360° =$ Jan. $24^d\!\!.5$, i.e. $\frac{1}{3}(1\cdot5) = 0\cdot5$ forwards towards the data for the same latitude interpolated similarly from page 27; the corrections are $+7^m$ to sunset, $+5^m$ to civil twilight, and $+6^m$ to nautical twilight.

12. *Moonrise, moonset.* Precise times of moonrise and moonset are rarely needed; a glance at the tables will generally give sufficient indication of whether the Moon is available for observation and of the hours of rising and setting. If needed, precise times may be obtained as follows. Interpolate for latitude, using Table I on page xxxii, on the day wanted and also on the preceding day in east longitudes or the following day in west longitudes; take the difference between these times and interpolate for longitude by applying to the time for the day wanted the correction from Table II on page xxxii, so that the resulting time is between the two times used. In extreme conditions near □ or ■ interpolation for latitude or longitude may be possible only in one direction; accurate times are of little value in these circumstances.

To facilitate this interpolation the times of moonrise and moonset are given for four days on each page; where no phenomenon occurs during a particular day (as happens once a month) the time of the phenomenon on the following day, increased by 24^h, is given; extra care must be taken when interpolating between two values, when one of those values exceeds 24^h. In practice it suffices to use the daily difference between the times for the nearest tabular latitude, and generally, to enter Table II with the nearest tabular arguments as in the examples below.

Examples. Required the UT of moonrise and moonset in latitude S 47° 10′, longitudes E 124° 00′ and W 78° 31′ on 1998 January 12.

	Longitude E 124° 00′ Moonrise	Longitude E 124° 00′ Moonset	Longitude W 78° 31′ Moonrise	Longitude W 78° 31′ Moonset
	d h m	d h m	d h m	d h m
LMT for Lat. S 45°	12 19 25	12 04 20	12 19 25	12 04 20
Lat correction (p. xxxii, Table I)	+07	−07	+07	−07
Long correction (p. xxxii, Table II)	−17	−20	+09	+13
Correct LMT	12 19 15	12 03 53	12 19 41	12 04 26
Longitude (p. i)	−8 16	−8 16	+5 14	+5 14
UT	12 10 59	11 19 37	13 00 55	12 09 40

ALTITUDE CORRECTION TABLES

13. *General.* In general two corrections are given for application to altitudes observed with a marine sextant; additional corrections are required for Venus and Mars and also for very low altitudes.

Tables of the correction for dip of the horizon, due to height of eye above sea level, are given on pages A2 and xxxiv. Strictly this correction should be applied first and subtracted from the sextant altitude to give apparent altitude, which is the correct argument for the other tables.

Separate tables are given of the second correction for the Sun, for stars and planets (on pages A2 and A3), and for the Moon (on pages xxxiv and xxxv). For the Sun, values are given for both lower and upper limbs, for two periods of the year. The star tables are used for the planets, but additional corrections for parallax (page A2) are required for Venus and Mars. The Moon tables are in two parts: the main correction is a function of apparent altitude only and is tabulated for the lower limb (30′ must be subtracted to obtain the correction for the upper limb); the other, which is given for both lower and upper limbs, depends also on the horizontal parallax, which has to be taken from the daily pages.

An additional correction, given on page A4, is required for the change in the refraction, due to variations of pressure and temperature from the adopted standard conditions; it may generally be ignored for altitudes greater than 10°, except possibly in extreme conditions. The correction tables for the Sun, stars, and planets are in two parts; only those for altitudes greater than 10° are reprinted on the bookmark.

14. *Critical tables.* Some of the altitude correction tables are arranged as critical tables. In these an interval of apparent altitude (or height of eye) corresponds to a single value of the correction; no interpolation is required. At a "critical" entry the upper of the two possible values of the correction is to be taken. For example, in the table of dip, a correction of −4′1 corresponds to all values of the height of eye from 5·3 to 5·5 metres (17·5 to 18·3 feet) inclusive.

15. *Examples.* The following examples illustrate the use of the altitude correction tables; the sextant altitudes given are assumed to be taken on 1998 March 3 with a marine sextant at height 5·4 metres (18 feet), temperature −3°C and pressure 982 mb, the Moon sights being taken at about 10^h UT.

	SUN lower limb	SUN upper limb	MOON lower limb	MOON upper limb	VENUS	Polaris
	° ′	° ′	° ′	° ′	° ′	° ′
Sextant altitude	21 19·7	3 20·2	33 27·6	26 06·7	4 32·6	49 36·5
Dip, height 5·4 metres (18 feet)	−4·1	−4·1	−4·1	−4·1	−4·1	−4·1
Main correction	+13·8	−29·6	+57·4	+60·5	−10·8	−0·8
−30′ for upper limb (Moon)	—	—	—	−30·0	—	—
L, U correction for Moon	—	—	+6·9	+4·7	—	—
Additional correction for Venus	—	—	—	—	+0·3	—
Additional refraction correction	−0·1	−0·3	0·0	−0·1	−0·3	0·0
Corrected sextant altitude	21 29·3	2 46·2	34 27·8	26 37·7	4 17·7	49 31·6

The main corrections have been taken out with apparent altitude (sextant altitude corrected for dip) as argument, interpolating where possible. These refinements are rarely necessary.

16. *Composition of the Corrections.* The table for the dip of the sea horizon is based on the formula:

$$\text{Correction for dip} = -1\!\cdot\!76\sqrt{(\text{height of eye in metres})} = -0\!\cdot\!97\sqrt{(\text{height of eye in feet})}$$

The correction table for the Sun includes the effects of semi-diameter, parallax and mean refraction.

The correction tables for the stars and planets allow for the effect of mean refraction.

The phase correction for Venus has been incorporated in the tabulations for GHA and Dec, and no correction for phase is required. The additional corrections for Venus and Mars allow for parallax. Alternatively, the correction for parallax may be calculated from $p \cos H$, where p is the parallax and H is the altitude. In 1998 the values for p are:

	Jan. 1	Feb. 5	Feb. 20	Mar. 15	May 5	Dec. 31
Venus	0′5	0′4	0′3	0′2	0′1	

	Jan. 1	Dec. 31
Mars	0′1	

The correction table for the Moon includes the effect of semi-diameter, parallax, augmentation and mean refraction.

Mean refraction is calculated for a temperature of 10°C (50°F) and a pressure of 1010 mb (29·83 inches).

17. *Bubble sextant observations.* When observing with a bubble sextant no correction is necessary for dip, semi-diameter, or augmentation. The altitude corrections for the stars and planets on page A2 and on the bookmark should be used for the Sun as well as for the stars and planets; for the Moon it is easiest to take the mean of the corrections for lower and upper limbs and subtract 15′ from the altitude; the correction for dip must not be applied.

AUXILIARY AND PLANNING DATA

18. *Sun and Moon.* On the daily pages are given: hourly values of the horizontal parallax of the Moon; the semi-diameters and the times of meridian passage of both Sun and Moon over the Greenwich meridian; the equation of time; the age of the Moon, the percent (%) illuminated and a symbol indicating the phase. The times of the phases of the Moon are given in UT on page 4. For the Moon, the semi-diameters for each of the three days are given at the foot of the column; for the Sun a single value is sufficient. Table II on page xxxii may be used for interpolating the time of the Moon's meridian passage for longitude. The equation of time is given daily at 00^h and 12^h UT. The sign is *positive* for unshaded values and *negative* for shaded values. To obtain apparent time add the equation of time to mean time when the sign is *positive*. Subtract the equation of time from mean time when the sign is *negative*. At 12^h UT, when the sign is *positive*, meridian passage of the Sun occurs *before* 12^h UT, otherwise is occurs *after* 12^h UT.

19. *Planets.* The magnitudes of the planets are given immediately following their names in the headings on the daily pages; also given, for the middle day of the three on the page, are their SHA at 00^h UT and their times of meridian passage.

The planet notes and diagram on pages 8 and 9 provide descriptive information as to the suitability of the planets for observation during the year, and of their positions and movements.

20. *Stars.* The time of meridian passage of the first point of Aries over the Greenwich meridian is given on the daily pages, for the middle day of the three on the page, to 0^m1. The interval between successive meridian passages is $23^h 56^m1$ (24^h less 3^m9) so that times for intermediate days and other meridians can readily be derived. If a precise time is required it may be obtained by finding the UT at which LHA Aries is zero.

The meridian passage of a star occurs when its LHA is zero, that is when LHA Aries + SHA = 360°. An approximate time can be obtained from the planet diagram on page 9.

The star charts on pages 266 and 267 are intended to assist identification. They show the relative positions of the stars in the sky as seen from the Earth and include all 173 stars used in the Almanac, together with a few others to complete the main constellation configurations. The local meridian at any time may be located on the chart by means of its SHA which is 360° − LHA Aries, or west longitude − GHA Aries.

21. *Star globe.* To set a star globe on which is printed a scale of LHA Aries, first set the globe for latitude and then rotate about the polar axis until the scale under the edge of the meridian circle reads LHA Aries.

To mark the positions of the Sun, Moon, and planets on the star globe, take the difference GHA Aries − GHA body and use this along the LHA Aries scale, in conjunction with the declination, to plot the position. GHA Aries − GHA body is most conveniently found by taking the difference when the GHA of the body is small (less than 15°), which happens once a day.

22. *Calendar.* On page 4 are given lists of ecclesiastical festivals, and of the principal anniversaries and holidays in the United Kingdom and the United States of America. The calendar on page 5 includes the day of the year as well as the day of the week.

Brief particulars are given, at the foot of page 5, of the solar and lunar eclipses occurring during the year; the times given are in UT. The principal features of the more important solar eclipses are shown on the maps on pages 6 and 7.

23. *Standard times.* The lists on pages 262–265 give the standard times used in most countries. In general no attempt is made to give details of the beginning and end of summer time, since they are liable to frequent changes at short notice. For the latest information consult Admiralty List of Radio Signals Volume 2 (NP 282) corrected by Section VI of the weekly edition of Admiralty Notices to Mariners.

The Date or Calendar Line is an arbitrary line, on either side of which the date differs by one day; when crossing this line on a westerly course, the date must be advanced one day; when crossing it on an easterly course, the date must be put back one day. The line is a modification of the line of the 180th meridian, and is drawn so as to include, as far as possible, islands of any one group, etc., on the same side of the line. It may be traced by starting at the South Pole and joining up to the following positions:

Lat	S 51·0	S 45·0	S 15·0	S 5·0	N 48·0	N 53·0	N 65·5
Long	180·0	W 172·5	W 172·5	180·0	180·0	E 170·0	W 169·0

thence through the middle of the Diomede Islands to Lat N 68°0, Long W 169°0, passing east of Ostrov Vrangelya (Wrangel Island) to Lat N 75°0, Long 180°0, and thence to the North Pole.

ACCURACY

24. *Main data.* The quantities tabulated in this Almanac are generally correct to the nearest 0′·1; the exception is the Sun's GHA which is deliberately adjusted by up to 0′·15 to reduce the error due to ignoring the v-correction. The GHA and Dec at intermediate times cannot be obtained to this precision, since at least two quantities must be added; moreover, the v- and d-corrections are based on mean values of v and d and are taken from tables for the whole minute only. The largest error that can occur in the GHA or Dec of any body other than the Sun or Moon is less than 0′·2; it may reach 0′·25 for the GHA of the Sun and 0′·3 for that of the Moon.

In practice it may be expected that only one third of the values of GHA and Dec taken out will have errors larger than 0′·05 and less than one tenth will have errors larger than 0′·1.

25. *Altitude corrections.* The errors in the altitude corrections are nominally of the same order as those in GHA and Dec, as they result from the addition of several quantities each correctly rounded off to 0′·1. But the actual values of the dip and of the refraction at low altitudes may, in extreme atmospheric conditions, differ considerably from the mean values used in the tables.

USE OF THIS ALMANAC IN 1999

This Almanac may be used for the Sun and stars in 1999 in the following manner.

For the Sun, take out the GHA and Dec for the same date but for a time $5^h 48^m 00^s$ *earlier* than the UT of observation; add 87° 00′ to the GHA so obtained. The error, mainly due to planetary perturbations of the Earth, is unlikely to exceed 0′·4.

For the stars, calculate the GHA and Dec for the same date and the same time, but *subtract* 15′·1 from the GHA so found. The error, due to incomplete correction for precession and nutation, is unlikely to exceed 0′·4. If preferred, the same result can be obtained by using a time $5^h 48^m 00^s$ earlier than the UT of observation (as for the Sun) and adding 86° 59′·2 to the GHA (or adding 87° as for the Sun and subtracting 0′·8, for precession, from the SHA of the star).

The Almanac cannot be so used for the Moon or planets.

LIST I — PLACES FAST ON UTC (mainly those EAST OF GREENWICH)

The times given } *added* to UTC to give Standard Time
below should be } *subtracted* from Standard Time to give UTC.

	h	m		h	m
Admiralty Islands	10		Egypt, Arab Republic of*	02	
Afghanistan	04	30	Equatorial Guinea, Republic of	01	
Albania*	01		Eritrea	03	
Algeria	01		Estonia*	02	
Amirante Islands	04		Ethiopia	03	
Andaman Islands	05	30			
Angola	01		Fiji	12	
Armenia	04		Finland*	02	
Australia			France*	01	
Australian Capital Territory*	10				
New South Wales[1]*	10		Gabon	01	
Northern Territory	09	30	Georgia*	04	
Queensland	10		Germany*	01	
South Australia*	09	30	Gibraltar*	01	
Tasmania*	10		Greece*	02	
Victoria*	10		Guam	10	
Western Australia	08				
Whitsunday Islands	10		Holland (The Netherlands)*	01	
Austria*	01		Hong Kong	08	
Azerbaijan*	04		Hungary*	01	
Bahrain	03		India	05	30
Balearic Islands*	01		Indonesia, Republic of		
Bangladesh	06		Bangka, Billiton, Java, West and		
Belarus*	02		Central Kalimantan, Madura, Sumatra	07	
Belgium*	01		Bali, Flores, South and East		
Benin	01		Kalimantan, Lombok, Sulawesi,		
Bosnia and Herzegovina*	01		Sumba, Sumbawa, Timor	08	
Botswana, Republic of	02		Aru, Irian Jaya, Kai, Moluccas,		
Brunei	08		Tanimbar	09	
Bulgaria*	02		Iran*	03	30
Burma (Myanmar)	06	30	Iraq*	03	
Burundi	02		Israel*	02	
			Italy*	01	
Cambodia	07				
Cameroon Republic	01		Jan Mayen Island*	01	
Caroline Islands[2]	10		Japan	09	
Central African Republic	01		Jordan*	02	
Chad	01				
Chagos Archipelago,	05		Kazakhstan		
Diego Garcia	06		Western (Aktau)*	04	
Chatham Islands*	12	45	Central (Atyrau)*	05	
China, People's Republic of	08		Eastern*	06	
Christmas Island, Indian Ocean	07		Kenya	03	
Cocos (Keeling) Islands	06	30	Kiribati Republic[3]	12	
Comoro Islands (Comoros)	03		Korea, North,	09	
Congo Republic	01		Republic of (South)	09	
Corsica*	01		Kuril Islands	11	
Crete*	02		Kuwait	03	
Croatia*	01		Kyrgyzstan*	05	
Cyprus: Ercan*, Larnaca*	02				
Czech Republic*	01		Laccadive Islands	05	30
			Laos	07	
Denmark*	01		Latvia*	02	
Djibouti	03		Lebanon*	02	
			Lesotho	02	
			Libya*	01	

* Summer time may be kept in these places.
[1] Except Broken Hill Area which keeps $09^h 30^m$.
[2] Except Pohnpei, Pingelap and Kosrae which keep 11^h.
[3] Kiritimati Island and those Line Islands that are not part of the Kiribati Republic keep 10^h slow on UTC.

LIST I — (continued)

	h	m
Liechtenstein*	01	
Lithuania*	02	
Lord Howe Island*	10	30
Luxembourg*	01	
Macau	08	
Macedonia*, former Yugoslav Republic	01	
Macias Nguema (Fernando Póo)	01	
Madagascar, Democratic Republic of	03	
Malawi	02	
Malaysia, Malaya, Sabah, Sarawak	08	
Maldives, Republic of The	05	
Malta*	01	
Mariana Islands	10	
Marshall Islands [1]	12	
Mauritius	04	
Moldova*	02	
Monaco*	01	
Mongolia*	08	
Mozambique	02	
Namibia*	01	
Nauru	12	
Nepal	05	45
Netherlands, The*	01	
New Caledonia	11	
New Zealand*	12	
Nicobar Islands	05	30
Niger	01	
Nigeria, Republic of	01	
Norfolk Island	11	30
Norway*	01	
Novaya Zemlya	03	
Okinawa	09	
Oman	04	
Pagalu (Annobon Islands)	01	
Pakistan	05	
Palau Islands	09	
Papua New Guinea	10	
Pescadores Islands	08	
Philippine Republic	08	
Poland*	01	
Qatar	03	
Reunion	04	
Romania*	02	
Russia [2]*		
Zone 1 Kaliningrad	02	
Zone 2 Moscow, St Petersburg, Arkhangelsk, Astrakhan	03	
Zone 3 Samara, Izhevsk	04	
Zone 4 Perm, Amderna, Novyy Port	05	
Zone 5 Omsk, Novosibirsk	06	
Zone 6 Norilsk, Kyzyl, Dikson	07	
Zone 7 Bratsk, Irkutsk, Ulan-Ude	08	
Zone 8 Yakutsk, Chita, Tiksi	09	

	h	m
Russia (continued)		
Zone 9 Vladivostok, Khabarovsk, Okhotsk	10	
Zone 10 Magadan, Yuzhno	11	
Zone 11 Petropavlovsk, Pevek	12	
Rwanda	02	
Ryukyu Islands	09	
Sakhalin Island*	11	
Santa Cruz Islands	11	
Sardinia*	01	
Saudi Arabia	03	
Schouten Islands	10	
Serbia*	01	
Seychelles	04	
Sicily*	01	
Singapore	08	
Slovakia*	01	
Slovenia*	01	
Socotra	03	
Solomon Islands	11	
Somalia Republic	03	
South Africa, Republic of	02	
Spain*	01	
Spanish Possessions in North Africa (Ceuta, Melilla)*	01	
Spitsbergen (Svalbard)*	01	
Sri Lanka	06	30
Sudan, Republic of	02	
Swaziland	02	
Sweden*	01	
Switzerland*	01	
Syria (Syrian Arab Republic)*	02	
Taiwan	08	
Tajikistan	05	
Tanzania	03	
Thailand	07	
Tonga	13	
Tunisia	01	
Turkey*	02	
Turkmenistan	05	
Tuvalu	12	
Uganda	03	
Ukraine*	02	
Simferopol*	03	
United Arab Emirates	04	
Uzbekistan	05	
Vanuatu, Republic of	11	
Vietnam, Socialist Republic of	07	
Yemen	03	
Yugoslavia*, Federal Republic of	01	
Zaire		
Kinshasa, Mbandaka	01	
Haut-Zaire, Kasai, Kivu, Shaba	02	
Zambia, Republic of	02	
Zimbabwe	02	

* Summer time may be kept in these places.
[1] Except the Ebon Atol which keeps time 24^h slow on that of the rest of the islands.
[2] The boundaries between the zones are irregular; listed are chief towns in each zone.

STANDARD TIMES (Corrected to October 1996)

LIST II — PLACES NORMALLY KEEPING UTC

Ascension Island	Ghana	Irish Republic*	Morocco	Sierra Leone
Burkina-Faso	Great Britain [1]	Ivory Coast	Portugal*	Togo Republic
Canary Islands*	Guinea-Bissau	Liberia	Principe	Tristan da Cunha
Channel Islands [1]	Guinea Republic	Madeira*	St. Helena	
Faeroes*, The	Iceland	Mali	São Tomé	
Gambia	Ireland, Northern [1]	Mauritania	Senegal	

* Summer time may be kept in these places.
[1] Summer time, one hour in advance of UTC, is kept from 1998 March 29^d 01^h to October 25^d 01^h UTC, subject to confirmation.

LIST III — PLACES SLOW ON UTC (WEST OF GREENWICH)

The times given ⎱ *subtracted* from UTC to give Standard Time
below should be ⎰ *added* to Standard Time to give UTC.

	h	m		h	m
Argentina	03		Canada (*continued*)		
Austral Islands [1]	10		Quebec, east of long. W. 63°[3]	04	
Azores*	01		west of long. W. 63°*	05	
			Saskatchewan [3]	06	
			Yukon*	08	
Bahamas*	05		Cape Verde Islands	01	
Barbados	04		Cayman Islands	05	
Belize	06		Chile*	04	
Bermuda*	04		Colombia	05	
Bolivia	04		Cook Islands	10	
Brazil			Costa Rica	06	
SE coastal states, Bahia, Goiás,			Cuba*	05	
Brasilia*	03		Curaçao Island	04	
NE coastal states, Eastern Para	03				
Mato Grosso, Mato Grosso do Sul*	04		Dominican Republic	04	
Amazones, NW States, Amapa,					
Western Para	04		Easter Island (I. de Pascua)*	06	
Territory of Acre	05		Ecuador	05	
British Antarctic Territory [2]	03		El Salvador	06	
			Falkland Islands*	04	
			Fanning Island	10	
Canada			Fernando de Noronha Island	02	
Alberta*	07		French Guiana	03	
British Columbia [3]*	08				
Labrador [3]*	04		Galápagos Islands	06	
Manitoba*	06		Greenland [4]		
New Brunswick*	04		General*	03	
Newfoundland*	03	30	Scoresby Sound*	01	
Northwest Territories [3]*			Thule area *	04	
east of long. W. 85°	05		Grenada	04	
long. W. 85° to W. 102°	06		Guadeloupe	04	
west of long. W. 102°	07		Guatemala	06	
Nova Scotia*	04		Guyana, Republic of	04	
Ontario, east of long. W. 90°*	05				
west of long. W. 90°*	06		Haiti*	05	
Prince Edward Island*	04		Honduras	06	

* Summer time may be kept in these places.
[1] This is the legal standard time, but local mean time is generally used.
[2] Most stations use UTC.
[3] Some areas may keep another time zone.
[4] Mesters Vig and Danmarkshavn keep UTC.

	h	m
Jamaica	05	
Johnston Island	10	
Juan Fernandez Islands*	04	
Leeward Islands	04	
Marquesas Islands	09	30
Martinique	04	
Mexico*[1]	06	
Midway Islands	11	
Nicaragua	06	
Niue	11	
Panama, Republic of	05	
Paraguay*	04	
Peru	05	
Puerto Rico	04	
St. Pierre and Miquelon*	03	
Samoa	11	
Society Islands	10	
South Georgia	02	
Suriname	03	
Trindade Island, South Atlantic ...	02	
Trinidad and Tobago	04	
Tuamotu Archipelago	10	
Tubuai Islands	10	
Turks and Caicos Islands*	05	
United States of America[2]		
Alabama	06	
Alaska, east of W. 169° 30'	09	
Aleutian Islands, west of W. 169° 30'	10	
Arizona[3]	07	
Arkansas	06	
California	08	
Colorado	07	
Connecticut	05	
Delaware	05	
District of Columbia	05	
Florida[4]	05	
Georgia	05	
Hawaii[3]	10	
Idaho[4]	07	
Illinois	06	
Indiana[3,4]	05	

	h	m
United States of America[2] (*continued*)		
Iowa	06	
Kansas[4]	06	
Kentucky, eastern part	05	
western part	06	
Louisiana	06	
Maine	05	
Maryland	05	
Massachusetts	05	
Michigan[4]	05	
Minnesota	06	
Mississippi	06	
Missouri	06	
Montana	07	
Nebraska, eastern part	06	
western part	07	
Nevada	08	
New Hampshire	05	
New Jersey	05	
New Mexico	07	
New York	05	
North Carolina	05	
North Dakota[4]	06	
Ohio	05	
Oklahoma	06	
Oregon[4]	08	
Pennsylvania	05	
Rhode Island	05	
South Carolina	05	
South Dakota, eastern part	06	
western part	07	
Tennessee, eastern part	05	
western part	06	
Texas[4]	06	
Utah	07	
Vermont	05	
Virginia	05	
Washington D.C.	05	
Washington	08	
West Virginia	05	
Wisconsin	06	
Wyoming	07	
Uruguay	03	
Venezuela	04	
Virgin Islands	04	
Windward Islands	04	

* Summer time may be kept in these places.
[1] Except the states of Sonora, Sinaloa, Nayarit and the Southern District of Lower California which keep 07[h], and the Northern District* of Lower California which keeps 08[h].
[2] Daylight-saving (Summer) time, one hour fast on the time given, is kept from the first Sunday in April to the last Sunday in October, changing at 02[h] 00[m] local clock time.
[3] Exempt from keeping daylight-saving time.
[4] A small portion of the state is in another time zone.

NORTHERN STARS

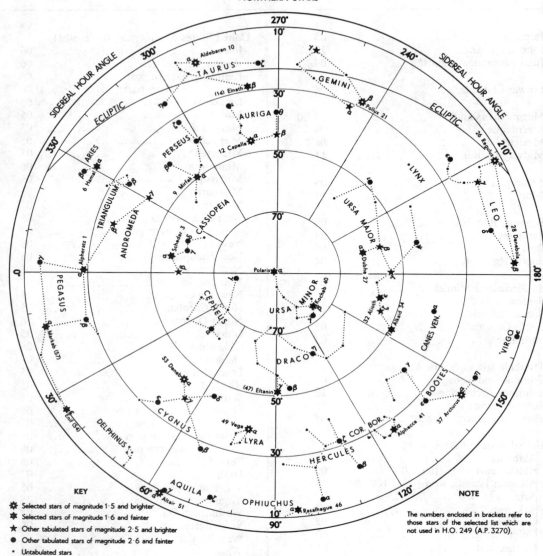

KEY

⚛ Selected stars of magnitude 1·5 and brighter
✦ Selected stars of magnitude 1·6 and fainter
★ Other tabulated stars of magnitude 2·5 and brighter
● Other tabulated stars of magnitude 2·6 and fainter
· Untabulated stars

NOTE

The numbers enclosed in brackets refer to those stars of the selected list which are not used in H.O. 249 (A.P. 3270).

EQUATORIAL STARS (S.H.A. 0° to 180°)

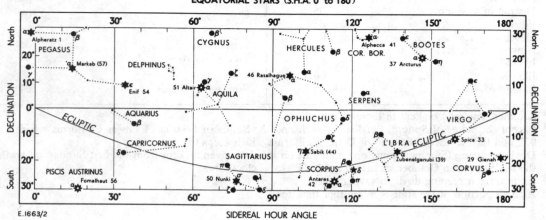

SIDEREAL HOUR ANGLE

E.1663/2

SOUTHERN STARS

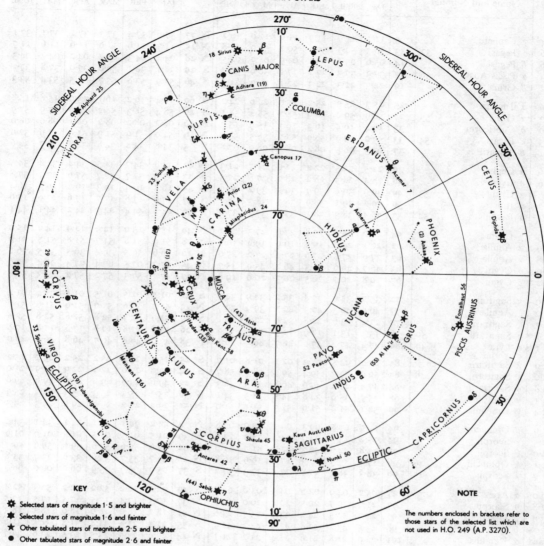

KEY

- ✹ Selected stars of magnitude 1·5 and brighter
- ★ Selected stars of magnitude 1·6 and fainter
- ★ Other tabulated stars of magnitude 2·5 and brighter
- ● Other tabulated stars of magnitude 2·6 and fainter
- · Untabulated stars

NOTE

The numbers enclosed in brackets refer to those stars of the selected list which are not used in H.O. 249 (A.P. 3270).

EQUATORIAL STARS (S.H.A. 180° to 360°)

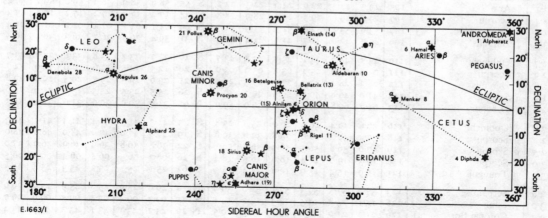

E.1663/1

SIDEREAL HOUR ANGLE

Mag.	Name and Number		SHA							Declination						
		°	JAN.	FEB.	MAR.	APR.	MAY	JUNE	°	JAN.	FEB.	MAR.	APR.	MAY	JUNE
3·4	γ Cephei	5	11·1	11·7	11·9	11·6	11·0	10·2	N 77	37·6	37·4	37·3	37·1	37·1	37·1
2·6	α Pegasi 57	13	50·3	50·3	50·3	50·2	50·0	49·8	N 15	11·7	11·6	11·6	11·6	11·6	11·7
2·6	β Pegasi	14	05·0	05·1	05·1	04·9	04·7	04·5	N 28	04·4	04·3	04·2	04·2	04·2	04·3
1·3	α Piscis Aust. 56	15	37·3	37·3	37·2	37·1	36·9	36·7	S 29	38·1	38·1	38·0	37·9	37·8	37·7
2·2	β Gruis	19	22·3	22·3	22·3	22·1	21·9	21·5	S 46	53·9	53·8	53·7	53·5	53·4	53·3
2·9	α Tucanæ	25	25·3	25·3	25·2	25·0	24·6	24·2	S 60	16·4	16·2	16·1	15·9	15·8	15·8
2·2	α Gruis 55	27	58·9	59·0	58·9	58·7	58·4	58·1	S 46	58·4	58·3	58·2	58·0	57·9	57·9
3·0	δ Capricorni	33	16·4	16·4	16·3	16·1	15·9	15·7	S 16	08·2	08·2	08·2	08·1	08·0	08·0
2·5	ε Pegasi 54	33	59·0	59·0	58·9	58·7	58·5	58·3	N 9	52·0	51·9	51·9	51·9	51·9	52·0
3·1	β Aquarii	37	08·5	08·5	08·4	08·2	08·0	07·8	S 5	34·8	34·8	34·8	34·8	34·7	34·6
2·6	α Cephei	40	22·5	22·5	22·4	22·0	21·7	21·3	N 62	34·8	34·7	34·5	34·4	34·4	34·5
2·6	ε Cygni	48	28·4	28·3	28·2	28·0	27·8	27·5	N 33	57·9	57·7	57·7	57·6	57·7	57·8
1·3	α Cygni 53	49	39·9	39·9	39·7	39·5	39·2	39·0	N 45	16·5	16·4	16·3	16·2	16·3	16·4
3·2	α Indi	50	39·2	39·1	38·9	38·7	38·3	38·0	S 47	17·9	17·8	17·7	17·6	17·6	17·6
2·1	α Pavonis 52	53	38·4	38·3	38·1	37·7	37·3	37·0	S 56	44·5	44·4	44·3	44·2	44·1	44·1
2·3	γ Cygni	54	28·0	28·0	27·8	27·6	27·3	27·1	N 40	15·1	15·0	14·9	14·9	14·9	15·0
0·9	α Aquilæ 51	62	20·1	20·0	19·8	19·6	19·4	19·2	N 8	51·9	51·8	51·7	51·8	51·8	51·9
2·8	γ Aquilæ	63	27·9	27·8	27·6	27·4	27·2	27·0	N 10	36·6	36·5	36·4	36·5	36·5	36·6
3·0	δ Cygni	63	46·7	46·6	46·5	46·2	45·9	45·7	N 45	07·7	07·5	07·4	07·4	07·4	07·6
3·2	β Cygni	67	20·8	20·7	20·5	20·3	20·0	19·8	N 27	57·4	57·3	57·2	57·2	57·3	57·4
3·0	π Sagittarii	72	35·7	35·6	35·4	35·2	34·9	34·7	S 21	01·5	01·5	01·5	01·5	01·4	01·4
3·0	ζ Aquilæ	73	40·6	40·5	40·3	40·1	39·9	39·7	N 13	51·7	51·6	51·6	51·6	51·6	51·7
2·7	ζ Sagittarii	74	23·2	23·0	22·8	22·6	22·3	22·1	S 29	52·9	52·9	52·8	52·8	52·8	52·8
2·1	σ Sagittarii 50	76	13·3	13·1	12·9	12·7	12·5	12·3	S 26	17·8	17·8	17·8	17·8	17·8	17·8
0·1	α Lyræ 49	80	47·4	47·2	47·0	46·7	46·5	46·3	N 38	47·0	46·8	46·7	46·8	46·9	47·0
2·9	λ Sagittarii	83	02·7	02·5	02·3	02·1	01·8	01·7	S 25	25·2	25·2	25·2	25·2	25·2	25·2
2·0	ε Sagittarii 48	83	59·9	59·7	59·4	59·2	58·9	58·7	S 34	23·0	23·0	22·9	22·9	22·9	23·0
2·8	δ Sagittarii	84	47·4	47·2	47·0	46·8	46·5	46·3	S 29	49·6	49·6	49·6	49·6	49·6	49·6
3·1	γ Sagittarii	88	35·2	35·0	34·8	34·5	34·3	34·1	S 30	25·3	25·3	25·3	25·3	25·3	25·3
2·4	γ Draconis 47	90	52·1	51·9	51·6	51·3	51·1	51·0	N 51	29·4	29·2	29·2	29·2	29·3	29·5
2·9	β Ophiuchi	94	09·7	09·5	09·3	09·1	09·0	08·8	N 4	34·2	34·1	34·1	34·1	34·1	34·2
2·5	κ Scorpii	94	25·2	24·9	24·7	24·4	24·2	24·0	S 39	01·5	01·5	01·5	01·5	01·6	01·6
2·0	θ Scorpii	95	42·8	42·6	42·3	42·0	41·7	41·6	S 42	59·6	59·6	59·6	59·6	59·6	59·7
2·1	α Ophiuchi 46	96	17·7	17·5	17·3	17·1	16·9	16·8	N 12	33·7	33·7	33·6	33·6	33·7	33·8
1·7	λ Scorpii 45	96	38·3	38·1	37·8	37·5	37·3	37·2	S 37	05·9	05·9	05·9	06·0	06·0	06·0
3·0	α Aræ	97	05·2	04·9	04·6	04·3	04·0	03·8	S 49	52·2	52·2	52·2	52·2	52·3	52·4
2·8	υ Scorpii	97	21·0	20·7	20·5	20·2	20·0	19·8	S 37	17·5	17·4	17·5	17·5	17·5	17·5
3·0	β Draconis	97	24·7	24·5	24·2	23·9	23·7	23·6	N 52	18·2	18·0	18·0	18·0	18·1	18·3
2·8	β Aræ	98	43·5	43·2	42·8	42·5	42·2	42·0	S 55	31·4	31·4	31·4	31·4	31·5	31·6
Var.‡	α Herculis	101	22·0	21·8	21·6	21·4	21·2	21·1	N 14	23·6	23·5	23·5	23·5	23·6	23·7
2·6	η Ophiuchi 44	102	26·4	26·2	25·9	25·7	25·5	25·4	S 15	43·2	43·2	43·3	43·3	43·3	43·3
3·1	ζ Aræ	105	23·7	23·3	22·9	22·6	22·3	22·2	S 55	58·9	58·9	58·9	59·0	59·1	59·2
2·4	ε Scorpii	107	29·8	29·6	29·3	29·1	28·9	28·8	S 34	17·2	17·2	17·2	17·2	17·3	17·3
1·9	α Triang. Aust. 43	107	53·9	53·4	52·8	52·3	51·9	51·7	S 69	01·1	01·1	01·1	01·2	01·3	01·4
3·0	ζ Herculis	109	42·2	42·0	41·7	41·5	41·4	41·3	N 31	36·4	36·3	36·2	36·3	36·4	36·5
2·7	ζ Ophiuchi	110	44·5	44·3	44·1	43·9	43·8	43·7	S 10	33·7	33·7	33·8	33·8	33·7	33·7
2·9	τ Scorpii	111	03·9	03·7	03·4	03·2	03·0	03·0	S 28	12·5	12·5	12·6	12·6	12·6	12·7
2·8	β Herculis	112	28·3	28·1	27·9	27·7	27·5	27·5	N 21	29·6	29·5	29·5	29·5	29·6	29·8
1·2	α Scorpii 42	112	41·0	40·8	40·5	40·3	40·2	40·1	S 26	25·5	25·5	25·5	25·6	25·6	25·6
2·9	η Draconis	113	61·2	60·8	60·4	60·1	59·9	59·9	N 61	31·0	30·9	30·9	31·0	31·1	31·3
3·0	δ Ophiuchi	116	26·6	26·4	26·2	26·0	25·9	25·8	S 3	41·3	41·3	41·4	41·4	41·3	41·3
2·8	β Scorpii	118	40·4	40·2	40·0	39·8	39·6	39·6	S 19	47·8	47·9	47·9	48·0	48·0	48·0
2·5	δ Scorpii	119	57·0	56·8	56·5	56·3	56·2	56·1	S 22	36·8	36·8	36·9	36·9	36·9	37·0
3·0	π Scorpii	120	19·3	19·0	18·8	18·6	18·5	18·4	S 26	06·3	06·3	06·4	06·4	06·5	06·5
3·0	β Trianguli Aust.	121	15·9	15·4	15·0	14·6	14·4	14·3	S 63	25·1	25·1	25·2	25·3	25·4	25·5
2·8	α Serpentis	123	57·7	57·5	57·3	57·1	57·0	56·9	N 6	26·0	25·9	25·8	25·9	25·9	26·0
3·0	γ Lupi	126	15·1	14·8	14·5	14·3	14·2	14·1	S 41	09·3	09·4	09·4	09·5	09·6	09·7
2·3	α Coronæ Bor. 41	126	21·2	21·0	20·8	20·6	20·5	20·5	N 26	43·2	43·1	43·1	43·2	43·3	43·4

‡ 3·0 — 3·7

Mag.	Name and Number		SHA							Declination					
		°	JULY	AUG.	SEPT.	OCT.	NOV.	DEC.	°	JULY	AUG.	SEPT.	OCT.	NOV.	DEC.
3·4	γ Cephei	5	09·5	08·9	08·7	08·8	09·3	09·9	N 77	37·1	37·3	37·5	37·7	37·8	37·9
2·6	*Markab* 57	13	49·6	49·4	49·3	49·3	49·4	49·5	N 15	11·8	11·9	12·0	12·1	12·1	12·0
2·6	*Scheat*	14	04·3	04·1	04·0	04·1	04·2	04·3	N 28	04·4	04·5	04·7	04·7	04·8	04·8
1·3	*Fomalhaut* 56	15	36·4	36·2	36·2	36·2	36·3	36·4	S 29	37·6	37·6	37·7	37·7	37·8	37·9
2·2	β Gruis	19	21·2	21·0	21·0	21·0	21·2	21·4	S 46	53·3	53·4	53·5	53·6	53·6	53·7
2·9	α Tucanæ	25	23·9	23·6	23·6	23·7	23·9	24·2	S 60	15·8	15·9	16·0	16·1	16·2	16·2
2·2	*Al Na'ir* 55	27	57·8	57·6	57·6	57·7	57·8	58·0	S 46	57·9	57·9	58·0	58·1	58·2	58·2
3·0	δ Capricorni	33	15·5	15·4	15·3	15·4	15·5	15·6	S 16	07·9	07·9	07·9	07·9	08·0	08·0
2·5	*Enif* 54	33	58·1	58·0	58·0	58·1	58·2	58·3	N 9	52·2	52·2	52·3	52·3	52·3	52·3
3·1	β Aquarii	37	07·6	07·5	07·5	07·6	07·7	07·8	S 5	34·5	34·5	34·5	34·5	34·5	34·5
2·6	*Alderamin*	40	21·0	21·0	21·1	21·3	21·7	22·0	N 62	34·7	34·9	35·1	35·2	35·2	35·2
2·6	ε Cygni	48	27·4	27·3	27·4	27·5	27·7	27·8	N 33	57·9	58·1	58·2	58·3	58·3	58·2
1·3	*Deneb* 53	49	38·8	38·8	38·9	39·0	39·2	39·4	N 45	16·5	16·7	16·8	16·9	16·9	16·8
3·2	α Indi	50	37·8	37·7	37·8	37·9	38·1	38·3	S 47	17·6	17·7	17·8	17·8	17·9	17·8
2·1	*Peacock* 52	53	36·7	36·6	36·7	36·9	37·2	37·4	S 56	44·2	44·3	44·4	44·5	44·5	44·4
2·3	γ Cygni	54	26·9	26·9	27·0	27·2	27·4	27·5	N 40	15·2	15·3	15·5	15·5	15·5	15·4
0·9	*Altair* 51	62	19·1	19·1	19·1	19·3	19·4	19·4	N 8	52·0	52·1	52·1	52·2	52·1	52·1
2·8	γ Aquilæ	63	26·9	26·9	27·0	27·1	27·2	27·3	N 10	36·7	36·8	36·9	36·9	36·8	36·8
3·0	δ Cygni	63	45·6	45·6	45·7	46·0	46·2	46·3	N 45	07·7	07·9	08·0	08·1	08·0	07·9
3·2	*Albireo*	67	19·7	19·8	19·9	20·0	20·2	20·3	N 27	57·5	57·7	57·8	57·8	57·7	57·6
3·0	π Sagittarii	72	34·6	34·6	34·7	34·9	35·0	35·0	S 21	01·4	01·4	01·4	01·4	01·4	01·4
3·0	ζ Aquilæ	73	39·6	39·6	39·7	39·9	40·0	40·0	N 13	51·9	51·9	52·0	52·0	51·9	51·9
2·7	ζ Sagittarii	74	22·0	22·0	22·1	22·2	22·4	22·4	S 29	52·8	52·8	52·9	52·9	52·9	52·9
2·1	*Nunki* 50	76	12·2	12·2	12·3	12·4	12·5	12·6	S 26	17·8	17·8	17·8	17·8	17·8	17·8
0·1	*Vega* 49	80	46·3	46·4	46·5	46·7	46·9	47·0	N 38	47·2	47·3	47·4	47·4	47·3	47·2
2·9	λ Sagittarii	83	01·6	01·6	01·7	01·9	02·0	02·0	S 25	25·2	25·2	25·2	25·2	25·2	25·2
2·0	*Kaus Australis* 48	83	58·6	58·7	58·8	59·0	59·1	59·1	S 34	23·0	23·0	23·1	23·1	23·0	23·0
2·8	δ Sagittarii	84	46·3	46·3	46·4	46·6	46·7	46·7	S 29	49·6	49·6	49·6	49·6	49·6	49·6
3·1	γ Sagittarii	88	34·1	34·1	34·2	34·4	34·5	34·5	S 30	25·3	25·3	25·4	25·4	25·4	25·3
2·4	*Eltanin* 47	90	51·0	51·1	51·4	51·6	51·8	51·9	N 51	29·6	29·7	29·8	29·8	29·7	29·5
2·9	β Ophiuchi	94	08·8	08·9	09·0	09·1	09·2	09·2	N 4	34·3	34·3	34·4	34·3	34·3	34·2
2·5	κ Scorpii	94	24·0	24·0	24·2	24·3	24·4	24·4	S 39	01·7	01·7	01·7	01·7	01·7	01·6
2·0	θ Scorpii	95	41·5	41·6	41·8	41·9	42·1	42·0	S 42	59·7	59·8	59·8	59·8	59·7	59·7
2·1	*Rasalhague* 46	96	16·8	16·9	17·0	17·1	17·2	17·2	N 12	33·9	34·0	34·0	34·0	33·9	33·8
1·7	*Shaula* 45	96	37·1	37·2	37·3	37·5	37·6	37·6	S 37	06·1	06·1	06·1	06·1	06·1	06·0
3·0	α Aræ	97	03·8	03·9	04·0	04·3	04·4	04·3	S 49	52·4	52·5	52·5	52·5	52·5	52·4
2·8	υ Scorpii	97	19·8	19·9	20·0	20·2	20·3	20·2	S 37	17·6	17·6	17·6	17·6	17·6	17·5
3·0	β Draconis	97	23·6	23·8	24·0	24·3	24·5	24·5	N 52	18·5	18·6	18·6	18·6	18·4	18·3
2·8	β Aræ	98	42·0	42·1	42·3	42·5	42·7	42·7	S 55	31·7	31·7	31·8	31·7	31·7	31·6
Var.‡	α Herculis	101	21·1	21·2	21·3	21·4	21·5	21·5	N 14	23·8	23·8	23·8	23·8	23·7	23·6
2·6	*Sabik* 44	102	25·4	25·5	25·6	25·7	25·8	25·8	S 15	43·2	43·2	43·2	43·2	43·2	43·2
3·1	ζ Aræ	105	22·2	22·3	22·5	22·8	22·9	22·8	S 55	59·3	59·3	59·3	59·3	59·2	59·1
2·4	ε Scorpii	107	28·8	28·9	29·0	29·2	29·2	29·2	S 34	17·4	17·4	17·4	17·4	17·3	17·3
1·9	*Atria* 43	107	51·8	52·0	52·4	52·8	53·0	52·9	S 69	01·5	01·6	01·6	01·6	01·5	01·3
3·0	ζ Herculis	109	41·3	41·5	41·6	41·8	41·9	41·8	N 31	36·6	36·7	36·7	36·7	36·6	36·4
2·7	ζ Ophiuchi	110	43·7	43·8	43·9	44·0	44·1	44·0	S 10	33·7	33·7	33·7	33·7	33·7	33·7
2·9	τ Scorpii	111	03·0	03·0	03·2	03·3	03·4	03·3	S 28	12·7	12·7	12·7	12·7	12·6	12·6
2·8	β Herculis	112	27·5	27·6	27·7	27·9	27·9	27·9	N 21	29·8	29·9	29·9	29·9	29·8	29·6
1·2	*Antares* 42	112	40·1	40·2	40·3	40·4	40·5	40·4	S 26	25·6	25·6	25·6	25·6	25·6	25·6
2·9	η Draconis	114	00·1	00·4	00·7	01·0	01·2	01·2	N 61	31·4	31·5	31·5	31·4	31·2	31·1
3·0	δ Ophiuchi	116	25·8	25·9	26·1	26·2	26·2	26·1	S 3	41·3	41·2	41·2	41·2	41·3	41·3
2·8	β Scorpii	118	39·6	39·7	39·8	39·9	40·0	39·9	S 19	48·0	48·0	48·0	47·9	47·9	47·9
2·5	*Dschubba*	119	56·2	56·3	56·4	56·5	56·5	56·4	S 22	37·0	37·0	36·9	36·9	36·9	36·9
3·0	π Scorpii	120	18·4	18·5	18·7	18·8	18·8	18·7	S 26	06·5	06·5	06·5	06·5	06·4	06·4
3·0	β Trianguli Aust.	121	14·4	14·7	15·0	15·2	15·3	15·1	S 63	25·6	25·7	25·7	25·6	25·5	25·4
2·8	α Serpentis	123	57·0	57·1	57·2	57·3	57·3	57·2	N 6	26·0	26·1	26·1	26·0	26·0	25·9
3·0	γ Lupi	126	14·2	14·3	14·5	14·6	14·6	14·5	S 41	09·7	09·7	09·7	09·6	09·6	09·5
2·3	*Alphecca* 41	126	20·5	20·7	20·8	20·9	21·0	20·9	N 26	43·5	43·5	43·5	43·4	43·3	43·2

‡ 3·0 — 3·7

Mag.	Name and Number	SHA °	JAN.	FEB.	MAR.	APR.	MAY	JUNE	Dec.	JAN.	FEB.	MAR.	APR.	MAY	JUNE
3·1	γ Ursæ Minoris	129	49·9	49·3	48·8	48·4	48·3	48·4	N 71	50·3	50·2	50·3	50·4	50·5	50·7
3·1	γ Trianguli Aust.	130	19·4	18·9	18·4	18·0	17·8	17·8	S 68	40·0	40·0	40·1	40·2	40·3	40·5
2·7	β Libræ	130	46·7	46·5	46·5	46·1	46·0	46·0	S 9	22·4	22·5	22·5	22·6	22·6	22·5
2·8	β Lupi	135	24·1	23·8	23·6	23·4	23·3	23·2	S 43	07·3	07·3	07·4	07·5	07·6	07·7
2·9	α Libræ 39	137	18·6	18·4	18·2	18·0	18·0	17·9	S 16	01·9	01·9	02·0	02·0	02·1	02·1
2·2	β Ursæ Minoris 40	137	20·2	19·5	19·0	18·6	18·6	18·8	N 74	09·6	09·6	09·6	09·8	09·9	10·1
2·6	ε Bootis	138	46·7	46·5	46·3	46·1	46·1	46·1	N 27	04·9	04·8	04·8	04·9	05·0	05·1
2·9	α Lupi	139	33·1	32·8	32·6	32·4	32·3	32·3	S 47	22·5	22·5	22·6	22·8	22·9	22·9
0·1	α Centauri 38	140	08·0	07·6	07·3	07·1	07·0	07·0	S 60	49·2	49·3	49·4	49·5	49·7	49·8
2·6	η Centauri	141	09·4	09·1	08·8	08·7	08·6	08·6	S 42	08·7	08·7	08·8	08·9	09·0	09·1
3·0	γ Bootis	141	60·3	60·0	59·8	59·6	59·6	59·7	N 38	18·9	18·8	18·8	18·9	19·1	19·2
0·2	α Bootis 37	146	06·6	06·4	06·2	06·1	06·0	06·1	N 19	11·5	11·4	11·4	11·5	11·5	11·6
2·3	θ Centauri 36	148	21·5	21·3	21·1	20·9	20·9	20·9	S 36	21·4	21·5	21·6	21·7	21·7	21·8
0·9	β Centauri 35	149	04·7	04·3	04·0	03·8	03·7	03·8	S 60	21·4	21·5	21·7	21·8	21·9	22·1
3·1	ζ Centauri	151	08·7	08·4	08·2	08·1	08·0	08·1	S 47	16·4	16·5	16·6	16·8	16·9	16·9
2·8	η Bootis	151	21·3	21·0	20·9	20·8	20·7	20·8	N 18	24·4	24·3	24·3	24·4	24·4	24·5
1·9	η Ursæ Majoris 34	153	08·3	08·0	07·8	07·6	07·7	07·8	N 49	19·2	19·2	19·2	19·3	19·5	19·6
2·6	ε Centauri	155	03·5	03·1	02·9	02·8	02·8	02·8	S 53	27·1	27·2	27·3	27·4	27·6	27·6
1·2	α Virginis 33	158	43·7	43·5	43·3	43·2	43·2	43·3	S 11	09·0	09·1	09·1	09·2	09·2	09·2
2·2	ζ Ursæ Majoris	159	02·5	02·2	02·0	01·9	01·9	02·1	N 54	55·9	55·9	56·0	56·1	56·2	56·3
2·9	ι Centauri	159	52·6	52·4	52·2	52·1	52·1	52·2	S 36	41·9	42·0	42·1	42·2	42·3	42·3
3·0	ε Virginis	164	28·9	28·6	28·5	28·5	28·5	28·5	N 10	58·1	58·1	58·0	58·1	58·1	58·2
2·9	α Canum Venat.	166	01·1	00·8	00·7	00·6	00·6	00·7	N 38	19·5	19·5	19·6	19·7	19·8	19·9
1·7	ε Ursæ Majoris 32	166	31·0	30·7	30·5	30·4	30·5	30·7	N 55	58·0	58·0	58·1	58·2	58·3	58·4
1·5	β Crucis	168	05·6	05·3	05·1	05·0	05·1	05·2	S 59	40·4	40·5	40·6	40·8	40·9	41·0
2·9	γ Virginis	169	36·6	36·4	36·3	36·2	36·2	36·3	S 1	26·3	26·4	26·4	26·4	26·4	26·4
2·4	γ Centauri	169	38·7	38·4	38·2	38·2	38·2	38·3	S 48	56·7	56·8	56·9	57·1	57·2	57·2
2·9	λ Muscæ	170	43·5	43·1	42·8	42·7	42·9	43·2	S 69	07·1	07·3	07·4	07·6	07·8	07·9
2·8	β Corvi	171	25·6	25·4	25·3	25·3	25·3	25·4	S 23	23·0	23·1	23·2	23·3	23·4	23·4
1·6	γ Crucis 31	172	13·8	13·5	13·4	13·3	13·4	13·6	S 57	05·8	06·0	06·1	06·3	06·4	06·5
1·1	α Crucis 30	173	22·2	21·8	21·7	21·6	21·7	22·0	S 63	05·0	05·1	05·3	05·5	05·6	05·7
2·8	γ Corvi 29	176	04·3	04·1	04·0	04·0	04·0	04·1	S 17	31·8	31·9	32·0	32·0	32·1	32·1
2·9	δ Centauri	177	55·8	55·6	55·5	55·5	55·5	55·7	S 50	42·4	42·6	42·7	42·9	43·0	43·0
2·5	γ Ursæ Majoris	181	34·1	33·8	33·7	33·7	33·8	34·0	N 53	42·1	42·1	42·2	42·3	42·5	42·5
2·2	β Leonis 28	182	45·6	45·4	45·3	45·3	45·3	45·4	N 14	34·9	34·8	34·8	34·9	34·9	35·0
2·6	δ Leonis	191	29·9	29·7	29·6	29·6	29·7	29·8	N 20	31·9	31·9	31·9	32·0	32·0	32·1
3·2	ψ Ursæ Majoris	192	36·6	36·4	36·3	36·4	36·5	36·7	N 44	30·3	30·3	30·4	30·5	30·6	30·7
2·0	α Ursæ Majoris 27	194	05·8	05·5	05·4	05·5	05·8	06·0	N 61	45·4	45·5	45·6	45·7	45·8	45·9
2·4	β Ursæ Majoris	194	34·1	33·8	33·7	33·8	34·0	34·2	N 56	23·3	23·4	23·5	23·6	23·7	23·7
2·8	μ Velorum	198	19·3	19·1	19·1	19·2	19·3	19·5	S 49	24·4	24·6	24·8	24·9	25·0	25·0
3·0	θ Carinæ	199	15·9	15·7	15·7	15·9	16·1	16·5	S 64	22·9	23·1	23·2	23·4	23·5	23·5
2·3	γ Leonis	205	01·9	01·7	01·7	01·8	01·9	02·0	N 19	50·9	50·9	50·9	51·0	51·0	51·0
1·3	α Leonis 26	207	55·8	55·7	55·7	55·7	55·8	55·9	N 11	58·5	58·4	58·4	58·4	58·5	58·5
3·1	ε Leonis	213	33·7	33·6	33·6	33·7	33·8	33·9	N 23	46·8	46·8	46·8	46·9	46·9	46·9
3·0	N Velorum	217	11·9	11·8	11·9	12·1	12·3	12·6	S 57	01·5	01·7	01·8	02·0	02·0	02·0
2·2	α Hydræ 25	218	07·4	07·3	07·3	07·4	07·5	07·6	S 8	39·1	39·2	39·3	39·3	39·3	39·2
2·6	κ Velorum	219	28·5	28·5	28·6	28·8	29·0	29·2	S 55	00·1	00·3	00·5	00·6	00·6	00·5
2·2	ι Carinæ	220	43·7	43·6	43·7	44·0	44·3	44·5	S 59	16·0	16·2	16·3	16·4	16·5	16·4
1·8	β Carinæ 24	221	41·1	41·1	41·3	41·7	42·1	42·5	S 69	42·5	42·7	42·9	43·0	43·0	43·0
2·2	λ Velorum 23	223	00·6	00·6	00·7	00·8	01·0	01·2	S 43	25·5	25·7	25·8	25·9	25·9	25·8
3·1	ι Ursæ Majoris	225	13·6	13·5	13·6	13·8	13·9	14·1	N 48	02·7	02·8	02·9	03·0	03·0	03·0
2·0	δ Velorum	228	49·6	49·6	49·7	49·9	50·2	50·4	S 54	42·1	42·3	42·4	42·5	42·5	42·4
1·7	ε Carinæ 22	234	22·1	22·1	22·3	22·6	22·9	23·2	S 59	30·3	30·4	30·6	30·7	30·7	30·6
1·9	γ Velorum	237	37·4	37·4	37·5	37·7	37·9	38·1	S 47	20·0	20·1	20·2	20·3	20·3	20·2
2·9	ρ Puppis	238	07·7	07·7	07·8	08·0	08·1	08·2	S 24	18·1	18·2	18·3	18·3	18·3	18·2
2·3	ζ Puppis	239	06·8	06·8	06·9	07·1	07·3	07·4	S 40	00·0	00·1	00·2	00·3	00·3	00·2
1·2	β Geminorum 21	243	41·8	41·8	41·9	42·0	42·1	42·2	N 28	01·7	01·7	01·7	01·8	01·8	01·8
0·5	α Canis Minoris 20	245	11·7	11·7	11·8	12·0	12·1	12·1	N 5	13·6	13·6	13·6	13·6	13·6	13·6

Mag.	Name and Number		SHA						Declination						
			JULY	AUG.	SEPT.	OCT.	NOV.	DEC.		JULY	AUG.	SEPT.	OCT.	NOV.	DEC.
		°	′	′	′	′	′	′	°	′	′	′	′	′	′
3·1	γ Ursæ Minoris	129	48·8	49·3	49·8	50·2	50·4	50·2	N 71	50·8	50·8	50·7	50·6	50·4	50·2
3·1	γ Trianguli Aust.	130	18·0	18·3	18·7	19·0	19·0	18·7	S 68	40·6	40·6	40·6	40·5	40·3	40·2
2·7	β Libræ	130	46·0	46·1	46·2	46·3	46·3	46·2	S 9	22·5	22·5	22·5	22·5	22·5	22·6
2·8	β Lupi	135	23·3	23·5	23·6	23·7	23·7	23·5	S 43	07·7	07·7	07·7	07·6	07·5	07·5
2·9	Zubenelgenubi 39	137	18·0	18·1	18·2	18·3	18·3	18·1	S 16	02·1	02·0	02·0	02·0	02·0	02·0
2·2	Kochab 40	137	19·3	19·9	20·4	20·8	20·9	20·7	N 74	10·1	10·1	10·0	09·9	09·7	09·5
2·6	ε Bootis	138	46·2	46·3	46·4	46·5	46·5	46·4	N 27	05·1	05·2	05·1	05·0	04·9	04·8
2·9	α Lupi	139	32·4	32·6	32·7	32·8	32·8	32·6	S 47	23·0	23·0	22·9	22·8	22·7	22·7
0·1	Rigil Kent. 38	140	07·2	07·5	07·8	07·9	07·8	07·6	S 60	49·8	49·8	49·8	49·7	49·5	49·5
2·6	η Centauri	141	08·7	08·8	09·0	09·1	09·0	08·8	S 42	09·1	09·1	09·1	09·0	08·9	08·9
3·0	γ Bootis	141	59·8	59·9	60·1	60·2	60·1	60·0	N 38	19·2	19·3	19·2	19·1	18·9	18·8
0·2	Arcturus 37	146	06·1	06·3	06·4	06·4	06·4	06·2	N 19	11·7	11·7	11·7	11·6	11·5	11·3
2·3	Menkent 36	148	21·0	21·2	21·3	21·3	21·3	21·1	S 36	21·8	21·8	21·7	21·6	21·6	21·6
0·9	Hadar 35	149	04·0	04·3	04·5	04·6	04·5	04·2	S 60	22·1	22·1	22·0	21·9	21·8	21·7
3·1	ζ Centauri	151	08·2	08·4	08·5	08·6	08·5	08·3	S 47	17·0	16·9	16·9	16·8	16·7	16·6
2·8	η Bootis	151	20·9	21·0	21·1	21·1	21·0	20·9	N 18	24·6	24·6	24·5	24·5	24·4	24·2
1·9	Alkaid 34	153	07·9	08·1	08·3	08·3	08·3	08·1	N 49	19·6	19·6	19·5	19·4	19·2	19·0
2·6	ε Centauri	155	03·0	03·2	03·4	03·4	03·3	03·0	S 53	27·7	27·6	27·5	27·4	27·3	27·3
1·2	Spica 33	158	43·3	43·4	43·5	43·5	43·4	43·2	S 11	09·1	09·1	09·1	09·1	09·1	09·2
2·2	Mizar	159	02·3	02·5	02·6	02·7	02·6	02·3	N 54	56·4	56·3	56·2	56·1	55·9	55·7
2·9	ι Centauri	159	52·3	52·4	52·5	52·6	52·4	52·2	S 36	42·3	42·3	42·2	42·1	42·1	42·1
3·0	ε Virginis	164	28·6	28·7	28·8	28·7	28·6	28·4	N 10	58·2	58·2	58·2	58·2	58·1	57·9
2·9	Cor Caroli	166	00·9	01·0	01·1	01·1	00·9	00·7	N 38	19·9	19·9	19·8	19·7	19·5	19·4
1·7	Alioth 32	166	30·9	31·1	31·2	31·2	31·1	30·8	N 55	58·4	58·4	58·3	58·1	57·9	57·8
1·5	Mimosa	168	05·5	05·7	05·9	05·8	05·6	05·3	S 59	41·0	41·0	40·8	40·7	40·6	40·6
2·9	γ Virginis	169	36·4	36·5	36·5	36·5	36·4	36·2	S 1	26·4	26·3	26·3	26·3	26·4	26·5
2·4	Muhlifain	169	38·5	38·7	38·8	38·8	38·6	38·3	S 48	57·2	57·2	57·1	57·0	56·9	56·9
2·9	α Muscæ	170	43·5	43·9	44·1	44·1	43·8	43·3	S 69	07·9	07·8	07·7	07·5	07·4	07·4
2·8	β Corvi	171	25·5	25·6	25·6	25·6	25·4	25·2	S 23	23·3	23·3	23·2	23·2	23·2	23·2
1·6	Gacrux 31	172	13·8	14·0	14·1	14·1	13·9	13·5	S 57	06·5	06·4	06·3	06·2	06·1	06·1
1·1	Acrux 30	173	22·2	22·5	22·6	22·6	22·4	21·9	S 63	05·7	05·6	05·5	05·3	05·2	05·2
2·8	Gienah 29	176	04·2	04·3	04·3	04·3	04·1	03·9	S 17	32·0	32·0	31·9	31·9	31·9	32·0
2·9	δ Centauri	177	55·9	56·0	56·1	56·0	55·8	55·5	S 50	43·0	42·9	42·8	42·7	42·6	42·6
2·5	Phecda	181	34·2	34·3	34·4	34·3	34·1	33·8	N 53	42·5	42·4	42·3	42·1	42·0	41·8
2·2	Denebola 28	182	45·5	45·6	45·6	45·5	45·4	45·1	N 14	35·0	35·0	34·9	34·9	34·8	34·7
2·6	δ Leonis	191	29·9	29·9	29·9	29·8	29·6	29·4	N 20	32·1	32·1	32·0	31·9	31·8	31·7
3·2	ψ Ursæ Majoris	192	36·8	36·9	36·8	36·7	36·5	36·2	N 44	30·6	30·6	30·4	30·3	30·2	30·1
2·0	Dubhe 27	194	06·2	06·3	06·3	06·1	05·8	05·4	N 61	45·8	45·7	45·5	45·4	45·2	45·1
2·4	Merak	194	34·4	34·5	34·4	34·3	34·0	33·6	N 56	23·7	23·6	23·4	23·3	23·1	23·0
2·8	μ Velorum	198	19·7	19·7	19·7	19·6	19·3	19·0	S 49	24·9	24·8	24·6	24·5	24·5	24·6
3·0	θ Carinæ	199	16·7	16·9	16·9	16·7	16·3	15·9	S 64	23·4	23·3	23·1	23·0	23·0	23·0
2·3	Algeiba	205	02·0	02·0	02·0	01·8	01·6	01·4	N 19	51·1	51·0	51·0	50·9	50·8	50·7
1·3	Regulus 26	207	56·0	56·0	55·9	55·8	55·6	55·3	N 11	58·5	58·5	58·5	58·5	58·4	58·3
3·1	ε Leonis	213	33·9	33·9	33·8	33·7	33·4	33·2	N 23	46·9	46·9	46·9	46·8	46·7	46·6
3·0	N Velorum	217	12·7	12·8	12·7	12·4	12·1	11·7	S 57	01·8	01·7	01·5	01·5	01·5	01·6
2·2	Alphard 25	218	07·6	07·6	07·5	07·4	07·1	06·9	S 8	39·2	39·1	39·1	39·1	39·1	39·2
2·6	κ Velorum	219	29·4	29·4	29·3	29·1	28·7	28·4	S 55	00·4	00·3	00·1	00·1	00·1	00·2
2·2	ι Carinæ	220	44·7	44·7	44·6	44·3	44·0	43·6	S 59	16·3	16·2	16·0	15·9	15·9	16·0
1·8	Miaplacidus 24	221	42·8	42·9	42·7	42·4	41·8	41·4	S 69	42·9	42·7	42·5	42·4	42·4	42·5
2·2	Suhail 23	223	01·3	01·2	01·1	00·9	00·6	00·4	S 43	25·7	25·6	25·5	25·4	25·4	25·5
3·1	ι Ursæ Majoris	225	14·1	14·0	13·9	13·6	13·3	13·0	N 48	02·9	02·8	02·7	02·6	02·5	02·5
2·0	δ Velorum	228	50·5	50·5	50·3	50·1	49·7	49·4	S 54	42·3	42·2	42·0	42·0	42·0	42·1
1·7	Avior 22	234	23·3	23·2	23·0	22·7	22·4	22·0	S 59	30·4	30·3	30·1	30·1	30·1	30·2
1·9	γ Velorum	237	38·1	38·1	37·9	37·7	37·4	37·1	S 47	20·0	19·9	19·8	19·7	19·8	19·9
2·9	ρ Puppis	238	08·2	08·1	08·0	07·8	07·5	07·3	S 24	18·1	18·0	17·9	17·9	17·9	18·0
2·3	ζ Puppis	239	07·4	07·4	07·2	07·0	06·7	06·5	S 39	60·0	59·9	59·8	59·8	59·8	59·9
1·2	Pollux 21	243	42·1	42·0	41·8	41·6	41·3	41·1	N 28	01·7	01·7	01·7	01·6	01·6	01·6
0·5	Procyon 20	245	12·1	12·0	11·8	11·6	11·4	11·2	N 5	13·7	13·7	13·7	13·7	13·6	13·6

Mag.	Name and Number		SHA							Declination						
		°	JAN.	FEB.	MAR.	APR.	MAY	JUNE	°		JAN.	FEB.	MAR.	APR.	MAY	JUNE
1·6	α Geminorum		246	22·6	22·6	22·7	22·8	23·0	23·0	N 31		53·4	53·4	53·5	53·5	53·5	53·5
3·3	σ Puppis		247	42·0	42·0	42·2	42·4	42·6	42·7	S 43		18·0	18·2	18·3	18·3	18·2	18·1
3·1	β Canis Minoris		248	14·1	14·1	14·1	14·3	14·4	14·4	N 8		17·4	17·4	17·4	17·4	17·4	17·4
2·4	η Canis Majoris		248	59·4	59·4	59·5	59·7	59·8	59·9	S 29		18·1	18·3	18·3	18·4	18·3	18·2
2·7	π Puppis		250	43·5	43·5	43·7	43·9	44·0	44·1	S 37		05·8	06·0	06·0	06·1	06·0	05·9
2·0	δ Canis Majoris		252	55·0	55·0	55·1	55·3	55·4	55·5	S 26		23·6	23·7	23·8	23·8	23·7	23·6
3·1	o Canis Majoris		254	15·5	15·6	15·7	15·8	16·0	16·0	S 23		50·0	50·1	50·2	50·2	50·2	50·1
1·6	ε Canis Majoris	19	255	21·4	21·5	21·6	21·8	21·9	22·0	S 28		58·4	58·5	58·6	58·6	58·5	58·4
2·8	τ Puppis		257	31·1	31·3	31·5	31·7	31·9	32·1	S 50		37·0	37·1	37·2	37·2	37·1	37·0
−1·6	α Canis Majoris	18	258	43·8	43·8	43·9	44·1	44·2	44·2	S 16		43·0	43·1	43·1	43·1	43·1	43·0
1·9	γ Geminorum		260	35·7	35·8	35·9	36·0	36·1	36·1	N 16		23·9	23·9	23·9	23·9	23·9	23·9
−0·9	α Carinæ	17	264	00·8	01·0	01·2	01·5	01·7	01·8	S 52		41·9	42·1	42·1	42·1	42·0	41·9
2·0	β Canis Majoris		264	20·5	20·6	20·7	20·9	21·0	21·0	S 17		57·5	57·6	57·7	57·6	57·6	57·5
2·7	θ Aurigæ		270	05·9	05·9	06·1	06·3	06·4	06·3	N 37		12·6	12·7	12·7	12·7	12·6	12·6
2·1	β Aurigæ		270	08·9	08·9	09·1	09·3	09·4	09·4	N 44		56·7	56·8	56·8	56·8	56·8	56·7
Var.‡	α Orionis	16	271	13·8	13·8	13·9	14·1	14·2	14·1	N 7		24·2	24·2	24·2	24·2	24·2	24·2
2·2	κ Orionis		273	04·8	04·9	05·0	05·1	05·2	05·2	S 9		40·5	40·5	40·5	40·5	40·5	40·4
1·9	ζ Orionis		274	49·9	49·9	50·1	50·2	50·3	50·2	S 1		56·8	56·9	56·9	56·9	56·8	56·8
2·8	α Columbæ		275	06·0	06·1	06·3	06·5	06·6	06·6	S 34		04·8	04·9	04·9	04·9	04·8	04·7
3·0	ζ Tauri		275	36·8	36·9	37·0	37·2	37·2	37·2	N 21		08·3	08·3	08·3	08·3	08·3	08·3
1·8	ε Orionis	15	275	58·1	58·1	58·2	58·4	58·5	58·4	S 1		12·4	12·4	12·5	12·4	12·4	12·3
2·9	ι Orionis		276	09·7	09·8	09·9	10·0	10·1	10·1	S 5		54·9	54·9	55·0	54·9	54·9	54·8
2·7	α Leporis		276	50·1	50·2	50·3	50·5	50·5	50·5	S 17		49·7	49·7	49·8	49·7	49·7	49·6
2·5	δ Orionis		277	01·2	01·2	01·4	01·5	01·6	01·5	S 0		18·2	18·3	18·3	18·3	18·2	18·2
3·0	β Leporis		277	57·3	57·4	57·6	57·7	57·8	57·8	S 20		45·9	46·0	46·0	46·0	45·9	45·8
1·8	β Tauri	14	278	27·2	27·3	27·4	27·6	27·7	27·6	N 28		36·2	36·3	36·3	36·2	36·2	36·2
1·7	γ Orionis	13	278	44·4	44·5	44·6	44·7	44·8	44·7	N 6		20·7	20·7	20·6	20·7	20·7	20·7
0·2	α Aurigæ	12	280	51·5	51·6	51·8	52·0	52·1	52·0	N 45		59·7	59·8	59·8	59·7	59·7	59·6
0·3	β Orionis	11	281	23·1	23·2	23·3	23·5	23·5	23·5	S 8		12·5	12·5	12·5	12·5	12·5	12·4
2·9	β Eridani		283	03·5	03·6	03·7	03·8	03·9	03·8	S 5		05·6	05·6	05·6	05·6	05·6	05·5
2·9	ι Aurigæ		285	46·7	46·8	47·0	47·1	47·2	47·1	N 33		09·7	09·7	09·7	09·7	09·7	09·6
1·1	α Tauri	10	291	02·7	02·8	02·9	03·0	03·1	03·0	N 16		30·2	30·2	30·2	30·2	30·2	30·2
3·2	γ Eridani		300	30·8	30·9	31·0	31·2	31·2	31·1	S 13		31·1	31·1	31·1	31·1	31·0	30·9
3·0	ε Persei		300	33·9	34·1	34·2	34·4	34·4	34·2	N 40		00·3	00·3	00·3	00·0	00·1	00·1
2·9	ζ Persei		301	29·7	29·8	29·9	30·1	30·1	29·9	N 31		52·6	52·6	52·6	52·6	52·5	52·5
3·0	η Tauri		303	09·3	09·4	09·6	09·7	09·7	09·5	N 24		05·9	05·9	05·8	05·8	05·8	05·8
1·9	α Persei	9	308	57·0	57·2	57·4	57·5	57·5	57·3	N 49		51·3	51·3	51·3	51·2	51·1	51·1
Var.§	β Persei		312	59·2	59·3	59·5	59·6	59·6	59·4	N 40		56·9	56·9	56·9	56·8	56·7	56·7
2·8	α Ceti	8	314	27·3	27·4	27·5	27·6	27·5	27·4	N 4		04·8	04·7	04·7	04·7	04·8	04·9
3·1	θ Eridani	7	315	27·2	27·3	27·5	27·6	27·6	27·5	S 40		19·1	19·1	19·1	19·0	18·8	18·7
2·1	α Ursæ Minoris		322	21·5	33·4	43·4	48·7	46·6	39·0	N 89		15·5	15·5	15·5	15·3	15·2	15·1
3·1	β Trianguli		327	38·5	38·7	38·8	38·8	38·7	38·5	N 34		58·7	58·7	58·6	58·6	58·5	58·5
2·2	α Arietis	6	328	14·0	14·1	14·2	14·3	14·2	14·0	N 23		27·2	27·1	27·1	27·0	27·0	27·1
2·2	γ Andromedæ		329	03·2	03·4	03·5	03·6	03·5	03·2	N 42		19·3	19·3	19·2	19·1	19·0	19·0
3·0	α Hydri		330	19·6	19·9	20·1	20·2	20·2	19·9	S 61		35·2	35·2	35·0	34·9	34·7	34·5
2·7	β Arietis		331	22·0	22·1	22·2	22·2	22·2	22·0	N 20		47·9	47·8	47·8	47·8	47·8	47·8
0·6	α Eridani	5	335	35·6	35·9	36·0	36·1	36·0	35·8	S 57		15·2	15·2	15·1	14·9	14·7	14·5
2·8	δ Cassiopeiæ		338	34·5	34·8	35·0	35·0	34·8	34·5	N 60		13·7	13·6	13·5	13·4	13·3	13·3
2·4	β Andromedæ		342	35·7	35·8	35·9	35·9	35·7	35·5	N 35		36·7	36·6	36·6	36·5	36·4	36·5
Var.‖	γ Cassiopeiæ		345	51·1	51·3	51·5	51·5	51·3	50·9	N 60		42·6	42·5	42·4	42·2	42·2	42·1
2·2	β Ceti	4	349	07·8	07·9	08·0	07·9	07·8	07·6	S 17		60·1	60·1	60·0	59·9	59·8	59·7
2·5	α Cassiopeiæ	3	349	54·0	54·3	54·4	54·3	54·1	53·8	N 56		31·8	31·7	31·6	31·5	31·4	31·4
2·4	α Phœnicis	2	353	27·5	27·6	27·7	27·6	27·5	27·2	S 42		19·3	19·3	19·2	19·0	18·9	18·7
2·9	β Hydri		353	36·4	36·9	37·2	37·1	36·7	36·0	S 77		16·3	16·2	16·0	15·8	15·7	15·5
2·9	γ Pegasi		356	43·1	43·2	43·2	43·2	43·0	42·8	N 15		10·4	10·3	10·3	10·2	10·3	10·3
2·4	β Cassiopeiæ		357	43·9	44·2	44·3	44·2	43·9	43·5	N 59		08·6	08·5	08·3	08·2	08·1	08·1
2·2	α Andromedæ	1	357	55·8	55·9	56·0	55·9	55·7	55·5	N 29		04·9	04·8	04·7	04·6	04·6	04·7

‡ 0·1 — 1·2   § 2·3 — 3·5   ‖ Irregular variable; 1996 mag. 2·3

Mag.	Name and Number		SHA						Declination						
		°	JULY	AUG.	SEPT.	OCT.	NOV.	DEC.	°	JULY	AUG.	SEPT.	OCT.	NOV.	DEC.
1·6	*Castor*	246	23·0	22·8	22·6	22·4	22·1	21·9	N 31	53·4	53·4	53·4	53·3	53·3	53·3
3·3	σ Puppis	247	42·7	42·6	42·4	42·2	41·9	41·7	S 43	18·0	17·8	17·7	17·7	17·8	17·9
3·1	β Canis Minoris	248	14·4	14·3	14·1	13·9	13·7	13·5	N 8	17·5	17·5	17·5	17·5	17·4	17·4
2·4	η Canis Majoris	248	59·9	59·8	59·6	59·4	59·1	58·9	S 29	18·1	18·0	17·9	17·9	17·9	18·1
2·7	π Puppis	250	44·1	44·0	43·8	43·5	43·3	43·1	S 37	05·8	05·6	05·5	05·5	05·6	05·7
2·0	*Wezen*	252	55·5	55·3	55·2	54·9	54·7	54·5	S 26	23·5	23·4	23·3	23·3	23·4	23·5
3·1	o Canis Majoris	254	16·0	15·9	15·7	15·5	15·2	15·1	S 23	49·9	49·8	49·7	49·7	49·8	49·9
1·6	*Adhara* 19	255	21·9	21·8	21·6	21·4	21·1	21·0	S 28	58·3	58·1	58·1	58·1	58·1	58·3
2·8	τ Puppis	257	32·0	31·9	31·6	31·3	31·1	30·9	S 50	36·8	36·7	36·6	36·6	36·7	36·8
−1·6	*Sirius* 18	258	44·2	44·0	43·8	43·6	43·4	43·3	S 16	42·9	42·8	42·7	42·7	42·8	42·9
1·9	*Alhena*	260	36·1	35·9	35·7	35·5	35·2	35·1	N 16	23·9	24·0	24·0	23·9	23·9	23·9
−0·9	*Canopus* 17	264	01·8	01·6	01·3	01·0	00·7	00·6	S 52	41·7	41·6	41·5	41·5	41·6	41·8
2·0	*Mirzam*	264	20·9	20·8	20·5	20·3	20·1	20·0	S 17	57·4	57·3	57·2	57·2	57·3	57·4
2·7	θ Aurigæ	270	06·2	06·0	05·7	05·5	05·2	05·0	N 37	12·6	12·5	12·5	12·5	12·6	12·6
2·1	*Menkalinan*	270	09·3	09·0	08·7	08·4	08·2	08·0	N 44	56·7	56·6	56·6	56·6	56·6	56·7
Var.‡	*Betelgeuse* 16	271	14·0	13·9	13·6	13·4	13·2	13·1	N 7	24·3	24·4	24·4	24·4	24·3	24·3
2·2	κ Orionis	273	05·1	04·9	04·7	04·5	04·3	04·2	S 9	40·3	40·2	40·1	40·2	40·2	40·3
1·9	*Alnitak*	274	50·1	50·0	49·7	49·5	49·4	49·2	S 1	56·7	56·6	56·6	56·6	56·6	56·7
2·8	*Phact*	275	06·5	06·3	06·1	05·9	05·7	05·5	S 34	04·5	04·4	04·3	04·4	04·5	04·6
3·0	ζ Tauri	275	37·1	36·9	36·7	36·4	36·2	36·1	N 21	08·4	08·4	08·4	08·4	08·4	08·4
1·8	*Alnilam* 15	275	58·3	58·1	57·9	57·7	57·5	57·4	S 1	12·3	12·2	12·2	12·2	12·2	12·3
2·9	ι Orionis	276	10·0	09·8	09·6	09·4	09·2	09·1	S 5	54·7	54·6	54·6	54·6	54·7	54·8
2·7	α Leporis	276	50·4	50·2	50·0	49·8	49·6	49·5	S 17	49·4	49·3	49·3	49·3	49·4	49·5
2·5	δ Orionis	277	01·4	01·2	01·0	00·8	00·6	00·5	S 0	18·1	18·0	18·0	18·0	18·0	18·1
3·0	β Leporis	277	57·7	57·5	57·3	57·1	56·9	56·8	S 20	45·7	45·6	45·5	45·5	45·6	45·7
1·8	*Elnath* 14	278	27·5	27·3	27·0	26·8	26·5	26·4	N 28	36·2	36·2	36·2	36·2	36·3	36·3
1·7	*Bellatrix* 13	278	44·6	44·4	44·2	44·0	43·8	43·7	N 6	20·8	20·8	20·9	20·9	20·8	20·8
0·2	*Capella* 12	280	51·8	51·6	51·2	50·9	50·7	50·5	N 45	59·6	59·5	59·5	59·6	59·6	59·7
0·3	*Rigel* 11	281	23·4	23·2	23·0	22·8	22·6	22·5	S 8	12·3	12·2	12·1	12·2	12·2	12·3
2·9	β Eridani	283	03·7	03·5	03·3	03·1	02·9	02·8	S 5	05·4	05·3	05·3	05·3	05·3	05·4
2·9	ι Aurigæ	285	47·0	46·7	46·5	46·2	46·0	45·9	N 33	09·6	09·6	09·7	09·7	09·7	09·8
1·1	*Aldebaran* 10	291	02·8	02·6	02·4	02·2	02·0	01·9	N 16	30·2	30·3	30·3	30·3	30·3	30·3
3·2	γ Eridani	300	30·9	30·7	30·5	30·3	30·2	30·1	S 13	30·8	30·7	30·7	30·7	30·8	30·9
3·0	ε Persei	300	34·0	33·7	33·5	33·2	33·1	33·0	N 40	00·1	00·1	00·2	00·2	00·3	00·4
2·9	ζ Persei	301	29·7	29·5	29·2	29·0	28·9	28·8	N 31	52·5	52·6	52·6	52·7	52·7	52·8
3·0	*Alcyone*	303	09·3	09·1	08·9	08·7	08·5	08·5	N 24	05·8	05·9	05·9	06·0	06·0	06·0
1·9	*Mirfak* 9	308	57·0	56·7	56·4	56·1	56·0	56·0	N 49	51·0	51·1	51·2	51·3	51·4	51·5
Var.§	*Algol*	312	59·2	58·9	58·6	58·4	58·3	58·3	N 40	56·7	56·8	56·8	56·9	57·0	57·1
2·8	*Menkar* 8	314	27·2	27·0	26·8	26·6	26·5	26·5	N 4	04·9	05·0	05·1	05·1	05·1	05·0
3·1	*Acamar* 7	315	27·3	27·0	26·7	26·6	26·5	26·5	S 40	18·5	18·5	18·5	18·5	18·7	18·8
2·1	*Polaris*	321	87·4	74·3	62·9	55·5	53·3	57·9	N 89	15·0	15·1	15·2	15·3	15·5	15·7
3·1	β Trianguli	327	38·3	38·0	37·8	37·7	37·6	37·6	N 34	58·6	58·6	58·8	58·9	58·9	59·0
2·2	*Hamal* 6	328	13·8	13·5	13·3	13·2	13·2	13·2	N 23	27·1	27·2	27·3	27·4	27·4	27·4
2·2	*Almak*	329	03·0	02·7	02·4	02·3	02·2	02·3	N 42	19·1	19·2	19·3	19·4	19·5	19·6
3·0	α Hydri	330	19·5	19·1	18·8	18·6	18·7	18·8	S 61	34·4	34·4	34·4	34·6	34·7	34·9
2·7	*Sheratan*	331	21·7	21·5	21·3	21·2	21·2	21·2	N 20	47·9	48·0	48·1	48·1	48·2	48·2
0·6	*Achernar* 5	335	35·4	35·1	34·8	34·7	34·7	34·9	S 57	14·4	14·4	14·5	14·6	14·8	14·9
2·8	*Ruchbah*	338	34·1	33·7	33·5	33·3	33·3	33·5	N 60	13·3	13·4	13·6	13·7	13·9	14·0
2·4	*Mirach*	342	35·2	35·0	34·8	34·7	34·7	34·8	N 35	36·5	36·6	36·8	36·9	37·0	37·0
Var.‖	γ Cassiopeiæ	345	50·5	50·1	49·9	49·8	49·9	50·0	N 60	42·2	42·3	42·5	42·6	42·8	42·9
2·2	*Diphda* 4	349	07·3	07·1	07·0	06·9	07·0	07·0	S 17	59·6	59·5	59·5	59·6	59·7	59·7
2·5	*Schedar* 3	349	53·4	53·1	52·9	52·9	52·9	53·1	N 56	31·4	31·6	31·7	31·9	32·0	32·1
2·4	*Ankaa* 2	353	26·9	26·7	26·5	26·5	26·6	26·7	S 42	18·6	18·6	18·7	18·8	18·9	19·0
2·9	β Hydri	353	35·2	34·5	34·0	34·0	34·3	34·9	S 77	15·5	15·5	15·7	15·8	16·0	16·0
2·9	*Algenib*	356	42·6	42·4	42·2	42·2	42·3	42·3	N 15	10·4	10·5	10·6	10·7	10·7	10·7
2·4	*Caph*	357	43·2	42·9	42·7	42·7	42·8	43·0	N 59	08·2	08·3	08·5	08·7	08·8	08·9
2·2	*Alpheratz* 1	357	55·2	55·0	54·9	54·9	54·9	55·0	N 29	04·8	04·9	05·0	05·1	05·2	05·2

‡ 0·1 — 1·2 § 2·3 — 3·5 ‖ Irregular variable; 1996 mag. 2·3

POLARIS (POLE STAR) TABLES, 1998
FOR DETERMINING LATITUDE FROM SEXTANT ALTITUDE AND FOR AZIMUTH

LHA ARIES	0° – 9°	10° – 19°	20° – 29°	30° – 39°	40° – 49°	50° – 59°	60° – 69°	70° – 79°	80° – 89°	90° – 99°	100° – 109°	110° – 119°
	a_0	a_0	a_0	a_0	a_0	a_0	a_0	a_0	a_0	a_0	a_0	a_0
0	0 23.6	0 19.4	0 16.3	0 14.6	0 14.2	0 15.2	0 17.6	0 21.2	0 26.0	0 31.8	0 38.4	0 45.6
1	23.1	19.0	16.1	14.5	14.3	15.4	17.9	21.6	26.5	32.4	39.1	46.3
2	22.7	18.7	15.9	14.4	14.3	15.6	18.2	22.1	27.1	33.0	39.8	47.1
3	22.2	18.3	15.7	14.3	14.4	15.8	18.6	22.5	27.6	33.7	40.5	47.8
4	21.8	18.0	15.5	14.3	14.5	16.0	18.9	23.0	28.2	34.3	41.2	48.6
5	0 21.3	0 17.7	0 15.3	0 14.2	0 14.6	0 16.3	0 19.3	0 23.5	0 28.8	0 35.0	0 41.9	0 49.4
6	20.9	17.4	15.1	14.2	14.7	16.5	19.6	24.0	29.4	35.6	42.6	50.1
7	20.5	17.1	15.0	14.2	14.8	16.8	20.0	24.5	29.9	36.3	43.4	50.9
8	20.1	16.8	14.8	14.2	14.9	17.0	20.4	25.0	30.5	37.0	44.1	51.7
9	19.7	16.6	14.7	14.2	15.1	17.3	20.8	25.5	31.2	37.7	44.8	52.4
10	0 19.4	0 16.3	0 14.6	0 14.2	0 15.2	0 17.6	0 21.2	0 26.0	0 31.8	0 38.4	0 45.6	0 53.2

Lat. °	a_1	a_1	a_1	a_1	a_1	a_1	a_1	a_1	a_1	a_1	a_1	a_1
0	0.5	0.5	0.6	0.6	0.6	0.6	0.5	0.5	0.4	0.4	0.3	0.3
10	.5	.6	.6	.6	.6	.6	.5	.5	.4	.4	.3	.3
20	.5	.6	.6	.6	.6	.6	.5	.5	.5	.4	.4	.4
30	.5	.6	.6	.6	.6	.6	.6	.5	.5	.5	.4	.4
40	0.6	0.6	0.6	0.6	0.6	0.6	0.6	0.6	0.5	0.5	0.5	0.5
45	.6	.6	.6	.6	.6	.6	.6	.6	.6	.6	.6	.5
50	.6	.6	.6	.6	.6	.6	.6	.6	.6	.6	.6	.6
55	.6	.6	.6	.6	.6	.6	.6	.6	.6	.6	.7	.7
60	.6	.6	.6	.6	.6	.6	.6	.7	.7	.7	.7	.7
62	0.7	0.6	0.6	0.6	0.6	0.6	0.6	0.7	0.7	0.7	0.8	0.8
64	.7	.6	.6	.6	.6	.6	.7	.7	.7	.8	.8	.8
66	.7	.6	.6	.6	.6	.6	.7	.7	.8	.8	.9	0.9
68	0.7	0.7	0.6	0.6	0.6	0.6	0.7	0.7	0.8	0.9	0.9	1.0

Month	a_2	a_2	a_2	a_2	a_2	a_2	a_2	a_2	a_2	a_2	a_2	a_2
Jan.	0.7	0.7	0.7	0.7	0.7	0.7	0.7	0.7	0.7	0.7	0.7	0.6
Feb.	.6	.7	.7	.7	.8	.8	.8	.8	.8	.8	.8	.8
Mar.	.5	.5	.6	.7	.7	.7	.8	.8	.9	.9	.9	.9
Apr.	0.3	0.4	0.4	0.5	0.6	0.6	0.7	0.8	0.8	0.9	0.9	0.9
May	.2	.3	.3	.4	.4	.5	.6	.6	.7	.8	.8	.9
June	.2	.2	.2	.3	.3	.4	.4	.5	.5	.6	.7	.8
July	0.2	0.2	0.2	0.2	0.2	0.3	0.3	0.4	0.4	0.5	0.5	0.6
Aug.	.4	.3	.3	.3	.3	.3	.3	.3	.3	.3	.4	.4
Sept.	.5	.5	.4	.4	.3	.3	.3	.3	.3	.3	.3	.3
Oct.	0.7	0.7	0.6	0.5	0.5	0.4	0.4	0.3	0.3	0.3	0.3	0.3
Nov.	0.9	0.8	.8	.7	.7	.6	.5	.5	.4	.4	.3	.3
Dec.	1.0	1.0	0.9	0.9	0.8	0.8	0.7	0.6	0.6	0.5	0.4	0.4

Lat. °	AZIMUTH											
0	0.4	0.3	0.2	0.0	359.9	359.8	359.7	359.5	359.5	359.4	359.3	359.3
20	0.4	0.3	0.2	0.0	359.9	359.8	359.6	359.5	359.4	359.3	359.3	359.2
40	0.5	0.4	0.2	0.0	359.9	359.7	359.5	359.4	359.3	359.2	359.1	359.1
50	0.6	0.5	0.3	0.1	359.8	359.7	359.5	359.3	359.1	359.0	358.9	358.9
55	0.7	0.5	0.3	0.1	359.8	359.6	359.4	359.2	359.0	358.9	358.8	358.7
60	0.8	0.6	0.3	0.1	359.8	359.5	359.3	359.1	358.9	358.7	358.6	358.5
65	1.0	0.7	0.4	0.1	359.8	359.5	359.2	358.9	358.7	358.5	358.4	358.3

Latitude = Apparent altitude (corrected for refraction) $-1° + a_0 + a_1 + a_2$

The table is entered with LHA Aries to determine the column to be used; each column refers to a range of 10°. a_0 is taken, with mental interpolation, from the upper table with the units of LHA Aries in degrees as argument; a_1, a_2 are taken, without interpolation, from the second and third tables with arguments latitude and month respectively. a_0, a_1, a_2, are always positive. The final table gives the azimuth of *Polaris*.

FOR DETERMINING LATITUDE FROM SEXTANT ALTITUDE AND FOR AZIMUTH

a_0

LHA ARIES	120°–129°	130°–139°	140°–149°	150°–159°	160°–169°	170°–179°	180°–189°	190°–199°	200°–209°	210°–219°	220°–229°	230°–239°
0	0 53·2	1 01·0	1 08·7	1 16·1	1 22·9	1 29·0	1 34·2	1 38·4	1 41·3	1 43·0	1 43·4	1 42·4
1	54·0	01·7	09·4	16·8	23·6	29·6	34·7	38·7	41·6	43·1	43·3	42·2
2	54·8	02·5	10·2	17·5	24·2	30·2	35·2	39·1	41·8	43·2	43·3	42·0
3	55·5	03·3	10·9	18·2	24·8	30·7	35·6	39·4	42·0	43·3	43·2	41·8
4	56·3	04·1	11·7	18·9	25·5	31·2	36·0	39·7	42·2	43·3	43·1	41·6
5	0 57·1	1 04·8	1 12·4	1 19·6	1 26·1	1 31·8	1 36·5	1 40·0	1 42·3	1 43·4	1 43·0	1 41·4
6	57·9	05·6	13·2	20·2	26·7	32·3	36·9	40·3	42·5	43·4	42·9	41·2
7	58·6	06·4	13·9	20·9	27·3	32·8	37·3	40·6	42·6	43·4	42·9	41·2
8	0 59·4	07·1	14·6	21·6	27·9	33·3	37·6	40·8	42·8	43·4	42·8	40·9
9	1 00·2	07·9	15·3	22·3	28·5	33·8	38·0	41·1	42·9	43·4	42·7	40·7
10	1 01·0	1 08·7	1 16·1	1 22·9	1 29·0	1 34·2	1 38·4	1 41·3	1 43·0	1 43·4	1 42·4	1 40·1

a_1

Lat.	120°–129°	130°–139°	140°–149°	150°–159°	160°–169°	170°–179°	180°–189°	190°–199°	200°–209°	210°–219°	220°–229°	230°–239°
0	0·3	0·3	0·3	0·3	0·4	0·4	0·5	0·5	0·6	0·6	0·6	0·6
10	·3	·3	·3	·4	·4	·5	·5	·6	·6	·6	·6	·6
20	·4	·4	·4	·4	·4	·5	·5	·6	·6	·6	·6	·6
30	·4	·4	·4	·5	·5	·5	·5	·6	·6	·6	·6	·6
40	0·5	0·5	0·5	0·5	0·5	0·6	0·6	0·6	0·6	0·6	0·6	0·6
45	·5	·5	·5	·6	·6	·6	·6	·6	·6	·6	·6	·6
50	·6	·6	·6	·6	·6	·6	·6	·6	·6	·6	·6	·6
55	·7	·7	·7	·7	·6	·6	·6	·6	·6	·6	·6	·6
60	·8	·8	·7	·7	·7	·7	·6	·6	·6	·6	·6	·6
62	0·8	0·8	0·8	0·8	0·7	0·7	0·7	0·6	0·6	0·6	0·6	0·6
64	·8	·8	·8	·8	·8	·7	·7	·6	·6	·6	·6	·6
66	0·9	0·9	·9	·8	·8	·7	·7	·6	·6	·6	·6	·6
68	1·0	1·0	0·9	0·9	0·8	0·8	0·7	0·7	0·6	0·6	0·6	0·6

a_2

Month	120°–129°	130°–139°	140°–149°	150°–159°	160°–169°	170°–179°	180°–189°	190°–199°	200°–209°	210°–219°	220°–229°	230°–239°
Jan.	0·6	0·6	0·6	0·5	0·5	0·5	0·5	0·5	0·5	0·5	0·5	0·5
Feb.	·8	·7	·7	·7	·6	·6	·6	·5	·5	·5	·4	·5
Mar.	0·9	0·9	0·9	0·8	·8	·7	·7	·7	·6	·5	·5	·4
Apr.	1·0	1·0	1·0	1·0	0·9	0·9	0·9	0·8	0·8	0·7	0·6	·5
May	0·9	1·0	1·0	1·0	1·0	1·0	1·0	0·9	0·9	·8	·8	0·6
June	·8	0·9	0·9	1·0	1·0	1·0	1·0	1·0	1·0	0·9	0·9	·7
July	0·7	0·7	0·8	0·8	0·9	0·9	1·0	1·0	1·0	1·0	1·0	·8
Aug.	·5	·6	·6	·7	·7	·8	0·8	0·9	0·9	0·9	1·0	0·9
Sept.	·4	·4	·4	·5	·6	·6	·7	·7	·8	·8	·9	·9
Oct.	0·3	0·3	0·3	0·3	0·4	0·4	0·5	0·5	0·6	0·7	0·7	0·8
Nov.	·2	·2	·2	·2	·2	·3	·3	·4	·4	·5	·5	·6
Dec.	0·3	0·3	0·2	0·2	0·2	0·2	0·2	0·2	0·3	0·3	0·4	0·4

AZIMUTH

Lat.	120°–129°	130°–139°	140°–149°	150°–159°	160°–169°	170°–179°	180°–189°	190°–199°	200°–209°	210°–219°	220°–229°	230°–239°
0	359·3	359·3	359·3	359·3	359·4	359·5	359·6	359·7	359·8	0·0	0·1	0·2
20	359·2	359·2	359·2	359·3	359·4	359·5	359·6	359·7	359·8	0·0	0·1	0·2
40	359·0	359·0	359·1	359·1	359·2	359·3	359·5	359·6	359·8	0·0	0·1	0·2
50	358·8	358·9	358·9	359·0	359·1	359·2	359·4	359·6	359·8	0·0	0·1	0·3
55	358·7	358·7	358·8	358·9	359·0	359·2	359·4	359·6	359·8	359·9	0·1	0·3
60	358·5	358·5	358·6	358·7	358·8	359·0	359·3	359·5	359·7	359·9	0·2	0·4
65	358·2	358·3	358·3	358·5	358·6	358·8	359·1	359·3	359·6	359·9	0·2	0·5

ILLUSTRATION

On 1998 April 21 at 23h 18m 56s UT in longitude W 37° 14' the apparent altitude (corrected for refraction), H_0, of Polaris was 49° 31'·6

From the daily pages:		
GHA Aries (23h)	194	48·5
Increment (18m 56s)	4	44·8
Longitude (west)	−37	14
LHA Aries	162	19

	°	'
H_0	49	31·6
a_0 (argument 162° 19')	1	24·4
a_1 (Lat 50° approx.)		0·6
a_2 (April)		0·9
Sum − 1° = Lat =	49	57·5

POLARIS (POLE STAR) TABLES, 1998
FOR DETERMINING LATITUDE FROM SEXTANT ALTITUDE AND FOR AZIMUTH

LHA ARIES	240°–249°	250°–259°	260°–269°	270°–279°	280°–289°	290°–299°	300°–309°	310°–319°	320°–329°	330°–339°	340°–349°	350°–359°
°	a_0	a_0	a_0	a_0	a_0	a_0	a_0	a_0	a_0	a_0	a_0	a_0
0	I 40·1	I 36·6	I 31·9	I 26·3	I 19·8	I 12·6	I 05·1	0 57·3	0 49·6	0 42·1	0 35·2	0 28·9
1	39·8	36·2	31·4	25·6	19·1	11·9	04·3	56·5	48·8	41·4	34·5	28·4
2	39·5	35·7	30·9	25·0	18·4	11·2	03·5	55·8	48·1	40·7	33·9	27·8
3	39·2	35·3	30·3	24·4	17·7	10·4	02·8	55·0	47·3	40·0	33·2	27·2
4	38·8	34·8	29·8	23·8	17·0	09·7	02·0	54·2	46·6	39·3	32·6	26·7
5	I 38·5	I 34·4	I 29·2	I 23·1	I 16·3	I 08·9	I 01·2	0 53·4	0 45·8	0 38·6	0 32·0	0 26·2
6	38·1	33·9	28·6	22·5	15·6	08·1	I 00·4	52·7	45·1	37·9	31·3	25·6
7	37·8	33·4	28·1	21·8	14·8	07·4	0 59·7	51·9	44·3	37·2	30·7	25·1
8	37·4	32·9	27·5	21·1	14·1	06·6	58·9	51·1	43·6	36·5	30·1	24·6
9	37·0	32·4	26·9	20·4	13·4	05·8	58·1	50·4	42·9	35·8	29·5	24·1
10	I 36·6	I 31·9	I 26·3	I 19·8	I 12·6	I 05·1	0 57·3	0 49·6	0 42·1	0 35·2	0 28·9	0 23·6

Lat. °	a_1	a_1	a_1	a_1	a_1	a_1	a_1	a_1	a_1	a_1	a_1	a_1
0	0·5	0·5	0·4	0·4	0·3	0·3	0·3	0·3	0·3	0·3	0·4	0·4
10	·5	·5	·4	·4	·3	·3	·3	·3	·3	·4	·4	·5
20	·5	·5	·5	·4	·4	·4	·4	·4	·4	·4	·4	·5
30	·6	·5	·5	·5	·4	·4	·4	·4	·4	·5	·5	·5
40	0·6	0·6	0·5	0·5	0·5	0·5	0·5	0·5	0·5	0·5	0·5	0·6
45	·6	·6	·6	·6	·6	·5	·5	·5	·5	·6	·6	·6
50	·6	·6	·6	·6	·6	·6	·6	·6	·6	·6	·6	·6
55	·6	·6	·6	·6	·7	·7	·7	·7	·7	·7	·6	·6
60	·6	·7	·7	·7	·7	·7	·8	·8	·7	·7	·7	·7
62	0·6	0·7	0·7	0·7	0·8	0·8	0·8	0·8	0·8	0·8	0·7	0·7
64	·7	·7	·7	·8	·8	·8	·8	·8	·8	·8	·8	·7
66	·7	·7	·8	·8	·9	0·9	0·9	0·9	·9	·8	·8	·7
68	0·7	0·7	0·8	0·9	0·9	I·0	I·0	I·0	0·9	0·9	0·8	0·8

Month	a_2	a_2	a_2	a_2	a_2	a_2	a_2	a_2	a_2	a_2	a_2	a_2
Jan.	0·5	0·5	0·5	0·5	0·5	0·6	0·6	0·6	0·6	0·7	0·7	0·7
Feb.	·4	·4	·4	·4	·4	·4	·4	·5	·5	·5	·6	·6
Mar.	·4	·4	·3	·3	·3	·3	·3	·3	·3	·4	·4	·5
Apr.	0·5	0·4	0·4	0·3	0·3	0·3	0·2	0·2	0·2	0·2	0·3	0·3
May	·6	·6	·5	·4	·4	·3	·3	·2	·2	·2	·2	·2
June	·8	·7	·7	·6	·5	·4	·4	·3	·3	·2	·2	·2
July	0·9	0·8	0·8	0·7	0·7	0·6	0·5	0·5	0·4	0·4	0·3	0·3
Aug.	·9	·9	·9	·9	·8	·8	·7	·6	·6	·5	·5	·4
Sept.	·9	·9	·9	·9	·9	·9	·8	·8	·8	·7	·6	·6
Oct.	0·8	0·9	0·9	0·9	0·9	0·9	0·9	0·9	0·9	0·9	0·8	0·8
Nov.	·7	·7	·8	·8	·9	·9	I·0	I·0	I·0	I·0	I·0	0·9
Dec.	0·5	0·6	0·6	0·7	0·8	0·8	0·9	0·9	I·0	I·0	I·0	I·0

Lat. °	AZIMUTH											
0	0·3	0·5	0·5	0·6	0·7	0·7	0·7	0·7	0·7	0·7	0·6	0·5
20	0·4	0·5	0·6	0·7	0·7	0·8	0·8	0·8	0·8	0·7	0·6	0·5
40	0·4	0·6	0·7	0·8	0·9	0·9	I·0	I·0	0·9	0·9	0·8	0·7
50	0·5	0·7	0·8	I·0	I·1	I·1	I·2	I·1	I·1	I·0	0·9	0·8
55	0·6	0·8	0·9	I·1	I·2	I·3	I·3	I·3	I·2	I·2	I·0	0·9
60	0·7	0·9	I·1	I·2	I·4	I·4	I·5	I·5	I·4	I·3	I·2	I·0
65	0·8	I·0	I·3	I·5	I·6	I·7	I·8	I·8	I·7	I·6	I·4	I·2

Latitude $=$ Apparent altitude (corrected for refraction) $-1° + a_0 + a_1 + a_2$

The table is entered with LHA Aries to determine the column to be used; each column refers to a range of 10°. a_0 is taken, with mental interpolation, from the upper table with the units of LHA Aries in degrees as argument; a_1, a_2 are taken, without interpolation, from the second and third tables with arguments latitude and month respectively. a_0, a_1, a_2, are always positive. The final table gives the azimuth of *Polaris*.

SIGHT REDUCTION PROCEDURES
METHODS AND FORMULAE FOR DIRECT COMPUTATION

1. *Introduction.* In this section formulae and methods are provided for *calculating* position at sea from observed altitudes taken with a marine sextant using a computer or programmable calculator.

The method uses analogous concepts and similar terminology as that used in *manual* methods of astro-navigation, where position is found by plotting position lines from their intercept and azimuth on a marine chart.

The algorithms are presented in standard algebra suitable for translating into the programming language of the user's computer. The basic ephemeris data may be taken directly from the main tabular pages of a current version of *The Nautical Almanac*. Formulae are given for calculating altitude and azimuth from the *GHA* and *Dec* of a body, and the estimated position of the observer. Formulae are also given for reducing sextant observations to observed altitudes by applying the corrections for dip, refraction, parallax and semi-diameter.

The intercept and azimuth obtained from each observation determines a position line, and the observer should lie on or close to each position line. The method of least squares is used to calculate the fix by finding the position where the sum of the squares of the distances from the position lines is a minimum. The use of least squares has other advantages. For example it is possible to improve the estimated position at the time of fix by repeating the calculation. It is also possible to include more observations in the solution and to reject doubtful ones.

2. *Notation.*

GHA = Greenwich hour angle. The range of GHA is from $0°$ to $360°$ starting at $0°$ on the Greenwich meridian increasing to the west, back to $360°$ on the Greenwich meridian.

SHA = sidereal hour angle. The range is $0°$ to $360°$.

Dec = declination. The sign convention for declination is north is positive, south is negative. The range is from $-90°$ at the south celestial pole to $+90°$ at the north celestial pole.

$Long$ = longitude. The sign convention is east is positive, west is negative. The range is $-180°$ to $+180°$.

Lat = latitude. The sign convention is north is positive, south is negative. The range is from $-90°$ to $+90°$.

LHA = $GHA + Long$ = local hour angle. The LHA increases to the west from $0°$ on the local meridian to $360°$.

H_C = calculated altitude. Above the horizon is positive, below the horizon is negative. The range is from $-90°$ in the nadir to $+90°$ in the zenith.

H_S = sextant altitude.

H = apparent altitude = sextant altitude corrected for instrumental error and dip.

H_O = observed altitude = apparent altitude corrected for refraction and, in appropriate cases, corrected for parallax and semi-diameter.

Z = Z_n = true azimuth. Z is measured from true north through east, south, west and back to north. The range is from $0°$ to $360°$.

I = sextant index error.

D = dip of horizon.

R = atmospheric refraction.

HP = horizontal parallax of the Sun, Moon, Venus or Mars.
PA = parallax in altitude of the Sun, Moon, Venus or Mars.
S = semi-diameter of the Sun or Moon.
p = intercept = $H_O - H_C$. Towards is positive, away is negative.
T = course or track, measured as for azimuth from the north.
V = speed in knots.

3. *Entering Basic Data.* When quantities such as *GHA* are entered, which in *The Nautical Almanac* are given in degrees and minutes, convert them to degrees and decimals of a degree by dividing the minutes by 60 and adding to the degrees; for example, if $GHA = 123° 45'6$, enter the two numbers 123 and 45·6 into the memory and set $GHA = 123 + 45·6/60 = 123°7600$. Although four decimal places of a degree are shown in the examples, it is assumed that full precision is maintained in the calculations.

When using a computer or programmable calculator, write a subroutine to convert degrees and minutes to degrees and decimals. Scientific calculators usually have a special key for this purpose. For quantities like *Dec* which require a minus sign for southern declination, change the sign from plus to minus after the value has been converted to degrees and decimals, *e.g.* $Dec = S 0° 12'3 = S 0°2050 = -0°2050$. Other quantities which require conversion are semi-diameter, horizontal parallax, longitude and latitude.

4. *Interpolation of GHA and Dec.* The *GHA* and *Dec* of the Sun, Moon and planets are interpolated to the time of observation by direct calculation as follows: If the universal time is $a^h b^m c^s$, form the interpolation factor $x = b/60 + c/3600$. Enter the tabular value GHA_0 for the preceeding hour (a) and the tabular value GHA_1 for the following hour ($a + 1$) then the interpolated value *GHA* is given by

$$GHA = GHA_0 + x(GHA_1 - GHA_0)$$

If the *GHA* passes through 360° between tabular values add 360° to GHA_1 before interpolation. If the interpolated value exceeds 360°, subtract 360° from *GHA*.

Similarly for declination, enter the tabular value Dec_0 for the preceeding hour (a) and the tabular value Dec_1 for the following hour ($a + 1$), then the interpolated value *Dec* is given by
$$Dec = Dec_0 + x(Dec_1 - Dec_0)$$

5. *Example.* (a) Find the *GHA* and *Dec* of the Sun on 1998 March 3 at $13^h 13^m 25^s$ UT.

The interpolation factor $x = 13/60 + 25/3600 = 0^h2236$

page 51 $13^h GHA_0 = 12° 00'5 = 12°0083$

$14^h GHA_1 = 27° 00'6 = 27°0100$

$13^h2236 GHA = 12·0083 + 0·2236(27·0100 - 12·0083) = 15°3629$

$13^h Dec_0 = S 6° 46'5 = -6°7750$

$14^h Dec_1 = S 6° 45'5 = -6°7583$

$13^h2236 Dec = -6·7750 + 0·2236(-6·7583 + 6·7750) = -6°7713$

GHA Aries is interpolated in the same way as *GHA* of a body. For a star the *SHA* and *Dec* are taken from the tabular page and do not require interpolation, then

$$GHA = GHA \text{ Aries} + SHA$$

where *GHA* Aries is interpolated to the time of observation.

(b) Find the *GHA* and *Dec* of *Vega* on 1998 March 3 at 13^h 13^m 25^s UT.
The interpolation factor $x = 0^h\!.2236$ as in the previous example

page 50 13^h GHA Aries$_0$ = 356° 06$'$.1 = 356°.1017
14^h GHA Aries$_1$ = 11° 08$'$.6 = 371°.1433   (360° added)
$13^h\!.2236$ GHA Aries = $356.1017 + 0.2236(371.1433 - 356.1017) = 359°.4652$
SHA = 80° 47$'$.1 = 80°.7850
GHA = GHA Aries + SHA = 80°.2502   (multiple of 360° removed)
Dec = N 38° 46$'$.8 = +38°.7800

6. *The calculated altitude and azimuth.* The calculated altitude H_C and true azimuth Z are determined from the *GHA* and *Dec* interpolated to the time of observation and from the *Long* and *Lat* estimated at the time of observation as follows:

Step 1.   Calculate the local hour angle

$$LHA = GHA + Long$$

Add or subtract multiples of 360° to set *LHA* in the range 0° to 360°.

Step 2.   Calculate S, C and the altitude H_C from

$$S = \sin Dec$$
$$C = \cos Dec \cos LHA$$
$$H_C = \sin^{-1}(S \sin Lat + C \cos Lat)$$

where $\sin^{-1}$ is the inverse function of sine.

Step 3.   Calculate X and A from

$$X = (S \cos Lat - C \sin Lat)/\cos H_C$$
If $X > +1$ set $X = +1$
If $X < -1$ set $X = -1$
$$A = \cos^{-1} X$$

where $\cos^{-1}$ is the inverse function of cosine.

Step 4.   Determine the azimuth Z

If $LHA > 180°$ then $Z = A$
Otherwise $Z = 360° - A$

7. *Example.*   Find the calculated altitude H_C and azimuth Z when

$GHA = 53°$ $Dec = S 15°$ $Lat = N 32°$ $Long = W 16°$
For the calculation
$GHA = 53°.0000$ $Dec = -15°.0000$ $Lat = +32°.0000$ $Long = -16°.0000$

Step 1. $LHA = 53.0000 - 16.0000 = 37.0000$
Step 2. $S = -0.2588$
$C = +0.9659 \times 0.7986 = 0.7714$
$\sin H_C = -0.2588 \times 0.5299 + 0.7714 \times 0.8480 = 0.5171$
$H_C = 31°.1346$

Step 3. $\qquad X = (-0.2588 \times 0.8480 - 0.7714 \times 0.5299)/0.8560 = -0.7340$

$\qquad\qquad\qquad A = 137°2239$

Step 4. $\qquad$ Since $\qquad LHA \leq 180°$ $\quad$ then $\quad Z = 360° - A = 222°7761$

8. *Reduction from sextant altitude to observed altitude.* The sextant altitude H_S is corrected for both dip and index error to produce the apparent altitude. The observed altitude H_O is calculated by applying a correction for refraction. For the Sun, Moon, Venus and Mars a correction for parallax is also applied to H, and for the Sun and Moon a further correction for semi-diameter is required. The corrections are calculated as follows:

Step 1. $\quad$ Calculate dip

$$D = 0°0293\sqrt{h}$$

where h is the height of eye above the horizon in metres.

Step 2. $\quad$ Calculate apparent altitude

$$H = H_S + I - D$$

where I is the sextant index error.

Step 3. $\quad$ Calculate refraction (R) at a standard temperature of 10° Celsius (C) and pressure of 1010 millibars (mb)

$$R_0 = 0°0167/\tan(H + 7.31/(H + 4.4))$$

If the temperature $T°$ C and pressure P mb are known calculate the refraction from

$$R = fR_0 \qquad \text{where} \qquad f = 0.28P/(T + 273)$$

$\qquad$ otherwise set $\quad R = R_0$

Step 4. $\quad$ Calculate the parallax in altitude (PA) from the horizontal parallax (HP) and the apparent altitude (H) for the Sun, Moon, Venus and Mars as follows:

$$PA = HP\cos H$$

For the Sun $HP = 0°0024$. This correction is very small and could be ignored.

For the Moon HP is taken for the nearest hour from the main tabular page and converted to degrees.

For Venus and Mars the HP is taken from the critical table at the bottom of page 259 and converted to degrees.

For the navigational stars and the remaining planets, Jupiter and Saturn set $PA = 0$.

If an error of $0°2$ is significant the expression for the parallax in altitude for the Moon should include a small correction OB for the oblateness of the Earth as follows:

$$PA = HP\cos H + OB$$

where $\quad OB = -0°0032\sin^2 Lat \cos H + 0°0032\sin(2Lat)\cos Z \sin H$

At mid-latitudes and for altitudes of the Moon below 60° a simple approximation to OB is

$$OB = -0°0017\cos H$$

Step 5. Calculate the semi-diameter for the Sun and Moon as follows:

Sun: S is taken from the main tabular page and converted to degrees.

Moon: $S = 0°2724HP$ where HP is taken for the nearest hour from the main tabular page and converted to degrees.

Step 6. Calculate the observed altitude

$$H_O = H - R + PA \pm S$$

where the plus sign is used if the lower limb of the Sun or Moon was observed and the minus sign if the upper limb was observed.

9. *Example.* The following example illustrates how to use a calculator to reduce the sextant altitude (H_S) to observed altitude (H_O); the sextant altitudes given are assumed to be taken on 1998 March 3 with a marine sextant, zero index error, at height 5·4 m, temperature $-3°$ C and pressure 982 mb, the Moon sights are assumed to be taken at 10^h UT.

Body limb	Sun lower	Sun upper	Moon lower	Moon upper	Venus —	*Polaris* —
Sextant altitude: H_S	21·3283	3·3367	33·4600	26·1117	4·5433	49·6083
Step 1. Dip: $D = 0·0293\sqrt{h}$	0·0681	0·0681	0·0681	0·0681	0·0681	0·0681
Step 2. Apparent altitude: $H = H_S + I - D$	21·2602	3·2686	33·3919	26·0436	4·4752	49·5402
Step 3. Refraction: R_0	0·0423	0·2262	0·0251	0·0338	0·1801	0·0142
f	1·0184	1·0184	1·0184	1·0184	1·0184	1·0184
$R = fR_0$	0·0431	0·2304	0·0256	0·0344	0·1834	0·0144
Step 4. Parallax: HP	0·0024	0·0024	(59″4) 0·9900	(59″4) 0·9900	(0″3) 0·0050	—
Parallax in altitude: $PA = HP\cos H$	0·0022	0·0024	0·8266	0·8895	0·0050	—
Step 5. Semi-diameter: Sun : $S = 16·2/60$	0·2700	0·2700	—	—	—	—
Moon : $S = 0·2724HP$	—	—	0·2697	0·2697	—	—
Step 6. Observed altitude: $H_O = H - R + PA \pm S$	21·4894	2·7706	34·4626	26·6290	4·2968	49·5258

Note that for the Moon the correction for the oblateness of the Earth of about $-0°0017\cos H$, which equals $-0°0014$ for the lower limb and $-0°0015$ for the upper limb, has been ignored in the above calculation.

10. *Position from intercept and azimuth using a chart.* An estimate is made of the position at the adopted time of fix. The position at the time of observation is then calculated by dead reckoning from the time of fix. For example if the course (track) T and the speed V (in knots) of the observer are constant then *Long* and *Lat* at the time of observation are calculated from

$$Long = L_F + t(V/60)\sin T / \cos B_F$$
$$Lat = B_F + t(V/60)\cos T$$

where L_F and B_F are the estimated longitude and latitude at the time of fix and t is the time interval in hours from the time of fix to the time of observation, t is positive if the time of observation is after the time of fix and negative if it was before.

The position line of an observation is plotted on a chart using the intercept

$$p = H_O - H_C$$

and azimuth Z with origin at the calculated position ($Long$, Lat) at the time of observation, where H_C and Z are calculated using the method in section 6, page 279. Starting from this calculated position a line is drawn on the chart along the direction of the azimuth to the body. Convert p to nautical miles by multiplying by 60. The position line is drawn at right angles to the azimuth line, distance p from ($Long$, Lat) towards the body if p is positive and distance p away from the body if p is negative. Provided there are no gross errors the navigator should be somewhere on or near the position line at the time of observation. Two or more position lines are required to determine a fix.

11. *Position from intercept and azimuth by calculation.* The position of the fix may be calculated from two or more sextant observations as follows.

If p_1, Z_1, are the intercept and azimuth of the first observation, p_2, Z_2, of the second observation and so on, form the summations

$$A = \cos^2 Z_1 + \cos^2 Z_2 + \cdots$$
$$B = \cos Z_1 \sin Z_1 + \cos Z_2 \sin Z_2 + \cdots$$
$$C = \sin^2 Z_1 + \sin^2 Z_2 + \cdots$$
$$D = p_1 \cos Z_1 + p_2 \cos Z_2 + \cdots$$
$$E = p_1 \sin Z_1 + p_2 \sin Z_2 + \cdots$$

where the number of terms in each summation is equal to the number of observations.

With $G = AC - B^2$, an improved estimate of the position at the time of fix (L_I, B_I) is given by

$$L_I = L_F + (AE - BD)/(G\cos B_F), \qquad B_I = B_F + (CD - BE)/G$$

Calculate the distance d between the initial estimated position (L_F, B_F) at the time of fix and the improved estimated position (L_I, B_I) in nautical miles from

$$d = 60\sqrt{((L_I - L_F)^2 \cos^2 B_F + (B_I - B_F)^2)}$$

If d exceeds about 20 nautical miles set $L_F = L_I$, $B_F = B_I$ and repeat the calculation until d, the distance between the position at the previous estimate and the improved estimate, is less than about 20 nautical miles.

12. *Example of direct computation.* Using the method described above, calculate the position of a ship on 1998 July 13 at $21^h\ 00^m\ 00^s$ UT from the marine sextant observations of the three stars *Regulus* (No. 26) at $20^h\ 39^m\ 23^s$ UT, *Antares* (No. 42) at $20^h\ 45^m\ 47^s$ UT and *Kochab* (No. 40) at $21^h\ 09^m\ 13^s$ UT, where the observed altitudes of the three stars corrected for the effects of refraction, dip and instrumental error, are $19°4909$, $29°1099$ and $47°1030$ respectively. The ship was travelling at a constant speed of 20 knots on a course of 325° during the period of observation, and the position of the ship at the time of fix $21^h\ 00^m\ 00^s$ UT is only known to the nearest whole degree W 15°, N 32°.

Intermediate values for the first iteration are shown in the table. *GHA* Aries was interpolated from the nearest tabular values on page 138. For the first iteration set $L_F = -15°0000$, $B_F = +32°0000$ at the time of fix at 21^h 00^m 00^s UT.

	First Iteration		
Body No.	Regulus 26	Antares 42	Kochab 40
time of observation	20^h 39^m 23^s	20^h 45^m 47^s	21^h 09^m 13^s
H_O	19·4909	29·1099	47·1030
interpolation factor	0·6564	0·7631	0·1536
GHA Aries	241·3682	242·9726	248·8470
SHA (page 138)	207·9333	112·6683	137·3217
GHA	89·3015	355·6410	26·1686
Dec (page 138)	+11·9750	−26·4267	+74·1683
t	−0·3436	−0·2369	+0·1536
Long	−14·9225	−14·9466	−15·0346
Lat	+31·9062	+31·9353	+32·0419
Z	272·2439	160·2532	355·5274
H_C	19·4678	28·8041	47·5031
p	+0·0231	+0·3058	−0·4001

$$A = 1·8813 \quad B = -0·4349 \quad C = 1·1187 \quad D = -0·6858 \quad E = 0·1114 \quad G = 1·9155$$
$$(A\,E - B\,D)/(G \cos B_F) = -0·0545, \qquad (C\,D - B\,E)/G = -0·3752$$

An improved estimate of the position at the time of fix is

$$L_I = L_F - 0·0545 = -15·0545 \quad \text{and} \quad B_I = B_F - 0·3752 = +31·6248$$

Since the distance between the previous estimated position and the improved estimate $d = 22·7$ nautical miles set $L_F = -15·0545$, and $B_F = +31·6248$ and repeat the calculation. The table shows the intermediate values of the calculation for the second iteration. In each iteration the quantities H_O, *GHA*, *Dec* and t do not change.

	Second Iteration		
Body No.	Regulus 26	Antares 42	Kochab 40
Long	−14·9774	−15·0013	−15·0890
Lat	+31·5310	+31·5601	+31·6667
Z	272·3484	160·1297	355·5802
H_C	19·4994	29·1414	47·1326
p	−0·0085	−0·0314	−0·0296

$$A = 1·8802 \quad B = -0·4374 \quad C = 1·1198 \quad D = -0·0003 \quad E = 0·0001 \quad G = 1·9141$$
$$(A\,E - B\,D)/(G \cos B_F) = +0·0000, \qquad (C\,D - B\,E)/G = -0·0001$$

An improved estimate of the position at the time of fix is

$$L_I = L_F + 0·0000 = -15·0545 \quad \text{and} \quad B_I = B_F - 0·0001 = +31·6246$$

The distance between the previous estimated position and the improved estimated position $d = 0·01$ nautical miles is so small that a third iteration would produce a negligible improvement to the estimate of the position.

USE OF CONCISE SIGHT REDUCTION TABLES

1. *Introduction.* The concise sight reduction tables given on pages 286 to 317 are intended for use when neither more extensive tables nor electronic computing aids are available. These "NAO sight reduction tables" provide for the reduction of the local hour angle and declination of a celestial object to azimuth and altitude, referred to an assumed position on the Earth, for use in the intercept method of celestial navigation which is now standard practice.

2. *Form of tables.* Entries in the reduction table are at a fixed interval of one degree for all latitudes and hour angles. A compact arrangement results from division of the navigational triangle into two right spherical triangles, so that the table has to be entered twice. Assumed latitude and local hour angle are the arguments for the first entry. The reduction table responds with the intermediate arguments A, B, and Z_1, where A is used as one of the arguments for the second entry to the table, B has to be incremented by the declination to produce the quantity F, and Z_1 is a component of the azimuth angle. The reduction table is then reentered with A and F and yields H, P, and Z_2 where H is the altitude, P is the complement of the parallactic angle, and Z_2 is the second component of the azimuth angle. It is usually necessary to adjust the tabular altitude for the fractional parts of the intermediate entering arguments to derive computed altitude, and an auxiliary table is provided for the purpose. Rules governing signs of the quantities which must be added or subtracted are given in the instructions and summarized on each tabular page. Azimuth angle is the sum of two components and is converted to true azimuth by familiar rules, repeated at the bottom of the tabular pages.

Tabular altitude and intermediate quantities are given to the nearest minute of arc, although errors of $2'$ in computed altitude may accrue during adjustment for the minutes parts of entering arguments. Components of azimuth angle are stated to $0°1$; for derived true azimuth, only whole degrees are warranted. Since objects near the zenith are difficult to observe with a marine sextant, they should be avoided; altitudes greater than about $80°$ are not suited to reduction by this method.

In many circumstances the accuracy provided by these tables is sufficient. However, to maintain the full accuracy $(0'1)$ of the ephemeral data in the almanac throughout their reduction to altitude and azimuth, more extensive tables or a calculator should be used.

3. *Use of Tables.*

Step 1. Determine the Greenwich hour angle (*GHA*) and Declination (*Dec*) of the body from the almanac. Select an assumed latitude (*Lat*) of integral degrees nearest to the estimated latitude. Choose an assumed longitude nearest to the estimated longitude such that the local hour angle

$$LHA = GHA \begin{array}{l} - \text{ west} \\ + \text{ east} \end{array} \text{ longitude}$$

has integral degrees.

Step 2. Enter the reduction table with *Lat* and *LHA* as arguments. Record the quantities A, B and Z_1. Apply the rules for the sign of B and Z_1: B is minus if $90° < LHA < 270°$: Z_1 has the same sign as B. Set $A° =$ nearest whole degree of A and $A' =$ minutes part of A. This step may be repeated for all reductions before leaving the latitude opening of the table.

Step 3. Record the declination *Dec*. Apply the rules for the sign of *Dec*: *Dec* is minus if the name of *Dec* (*i.e.* N or S) is contrary to latitude. Add B and *Dec* algebraically to produce F. If F is negative, the object is below the horizon (in sight reduction, this can occur when the objects are close to the horizon). Regard F as positive until step 7. Set $F° =$ nearest whole degree of F and $F' =$ minutes part of F.

Step 4. Enter the reduction table a second time with $A°$ and $F°$ as arguments and record H, P, and Z_2. Set $P° =$ nearest whole degree of P and $Z_2° =$ nearest whole degree of Z_2.

Step 5. Enter the auxiliary table with F' and $P°$ as arguments to obtain $corr_1$ to H for F'. Apply the rule for the sign of $corr_1$: $corr_1$ is minus if $F < 90°$ and $F' > 29'$ or if $F > 90°$ and $F' < 30'$, otherwise $corr_1$ is plus.

Step 6. Enter the auxiliary table with A' and $Z_2°$ as arguments to obtain $corr_2$ to H for A'. Apply the rule for the sign of $corr_2$: $corr_2$ is minus if $A' < 30'$, otherwise $corr_2$ is plus.

Step 7. Calculate the computed altitude H_C as the sum of H, $corr_1$ and $corr_2$. Apply the rule for the sign of H_C: H_C is minus if F is negative.

Step 8. Apply the rule for the sign of Z_2: Z_2 is minus if $F > 90°$. If F is negative, replace Z_2 by $180° - Z_2$. Set the azimuth angle Z equal to the algebraic sum of Z_1 and Z_2 and ignore the resulting sign. Obtain the true azimuth Z_n from the rules

For N latitude, if	$LHA > 180°$	$Z_n = Z$	
if	$LHA < 180°$	$Z_n = 360° - Z$	
For S latitude, if	$LHA > 180°$	$Z_n = 180° - Z$	
if	$LHA < 180°$	$Z_n = 180° + Z$	

Observed altitude H_O is compared with H_C to obtain the altitude difference, which, with Z_n, is used to plot the position line.

4. *Example.* (a) Required the altitude and azimuth of *Schedar* on 1998 February 5 at UT 06^h 28^m from the estimated position 5° east, 53° north.

1. Assumed latitude $Lat =$ 53° N
   From the almanac $GHA =$ 222° 07'
   Assumed longitude   4° 53' E
   Local hour angle $LHA =$ 227

2. Reduction table, 1st entry
   $(Lat, LHA) = (53, 227)$ $A =$ 26   07 $A° = 26, A' = 7$
                              $B = -27$ 12 $Z_1 = -49·4,$ $90° < LHA < 270°$
3. From the almanac $Dec = +56$ 32   *Lat* and *Dec* same
   Sum $= B + Dec$ $F = +29$ 20 $F° = 29, F' = 20$

4. Reduction table, 2nd entry
   $(A°, F°) = (26, 29)$ $H =$ 25   50 $P° = 61$
                                          $Z_2 = 76·3$

5. Auxiliary table, 1st entry
   $(F', P°) = (20, 61)$ $corr_1 =$ $+17$ $F < 90°, F' < 29'$
   Sum   26   07
6. Auxiliary table, 2nd entry
   $(A', Z_2°) = (7, 76)$ $corr_2 =$ -2 $A' < 30'$
7. Sum $=$ computed altitude $H_C = +26°$ 05' $F > 0°$

8. Azimuth,   first component $Z_1 = -49·4$ same sign as B
          second component $Z_2 = +76·3$ $F < 90°, F > 0°$
   Sum $=$ azimuth angle $Z =$ 26·9

   True azimuth $Z_n =$ 27°   N *Lat*, $LHA > 180°$

continued on page 318

B: (−) for 90°< LHA <270°
Dec: (−) for Lat. contrary name

Z₁: same sign as B
Z₂: (−) for F > 90°

SIGHT REDUCTION TABLE

Lat. / A	0°			1°			2°			3°			4°			5°			Lat. / A
LHA/F	A/H	B/P	Z₁/Z₂	A/H	B/P	Z₁/Z₂	A/H	B/P	Z₁/Z₂	A/H	B/P	Z₁/Z₂	A/H	B/P	Z₁/Z₂	A/H	B/P	Z₁/Z₂	LHA
0 / 180	0 00	90 00	90.0	0 00	89 00	90.0	0 00	88 00	90.0	0 00	87 00	90.0	0 00	86 00	90.0	0 00	85 00	90.0	180 / 360
1 / 179	1 00	90 00	90.0	1 00	89 00	90.0	1 00	88 00	90.0	1 00	87 00	89.9	1 00	86 00	89.9	1 00	85 00	89.9	181 / 359
2 / 178	2 00	90 00	90.0	2 00	89 00	90.0	2 00	88 00	89.9	2 00	87 00	89.9	2 00	86 00	89.9	2 00	85 00	89.8	182 / 358
3 / 177	3 00	90 00	90.0	3 00	89 00	90.0	3 00	88 00	89.9	3 00	87 00	89.9	3 00	86 00	89.9	2 59	85 00	89.7	183 / 357
4 / 176	4 00	90 00	90.0	4 00	89 00	89.9	4 00	88 00	89.9	4 00	87 00	89.8	3 59	85 59	89.8	3 59	84 59	89.6	184 / 356
5 / 175	5 00	90 00	90.0	5 00	89 00	89.9	5 00	88 00	89.8	5 00	86 59	89.7	4 59	85 59	89.7	4 59	84 59	89.5	185 / 355
6 / 174	6 00	90 00	90.0	6 00	89 00	89.9	6 00	87 59	89.8	6 00	86 59	89.7	5 59	85 59	89.7	5 59	84 58	89.4	186 / 354
7 / 173	7 00	90 00	90.0	7 00	89 00	89.9	7 00	87 59	89.8	6 59	86 59	89.6	6 59	85 58	89.6	6 58	84 58	89.3	187 / 353
8 / 172	8 00	90 00	90.0	8 00	88 59	89.9	8 00	87 59	89.7	7 59	86 58	89.6	7 59	85 58	89.6	7 58	84 56	89.2	188 / 352
9 / 171	9 00	90 00	90.0	9 00	88 59	89.9	9 00	87 59	89.7	8 59	86 58	89.5	8 59	85 57	89.5	8 58	84 55	89.1	189 / 351
10 / 170	10 00	90 00	90.0	10 00	88 59	89.8	10 00	87 58	89.7	9 59	86 57	89.5	9 59	85 56	89.4	9 58	84 54	89.0	190 / 350
11 / 169	11 00	90 00	90.0	11 00	88 59	89.8	11 00	87 58	89.6	10 59	86 57	89.4	10 58	85 56	89.4	10 57	84 53	88.9	191 / 349
12 / 168	12 00	90 00	90.0	12 00	88 58	89.8	12 00	87 57	89.6	11 59	86 56	89.4	11 58	85 55	89.3	11 57	84 52	88.8	192 / 348
13 / 167	13 00	90 00	90.0	13 00	88 58	89.8	13 00	87 57	89.6	12 59	86 55	89.3	12 58	85 54	89.2	12 57	84 51	88.7	193 / 347
14 / 166	14 00	90 00	90.0	14 00	88 58	89.8	13 59	87 56	89.6	13 59	86 54	89.3	13 58	85 53	89.2	13 57	84 49	88.6	194 / 346
15 / 165	15 00	90 00	90.0	15 00	88 58	89.7	14 59	87 56	89.5	14 59	86 54	89.3	14 58	85 52	89.1	14 56	84 48	88.5	195 / 345
16 / 164	16 00	90 00	90.0	16 00	88 58	89.7	15 59	87 55	89.5	15 59	86 53	89.2	15 58	85 50	89.0	15 56	84 46	88.4	196 / 344
17 / 163	17 00	90 00	90.0	17 00	88 57	89.7	16 59	87 55	89.4	16 59	86 52	89.1	16 57	85 49	88.9	16 56	84 44	88.3	197 / 343
18 / 162	18 00	90 00	90.0	18 00	88 57	89.7	17 59	87 54	89.4	17 58	86 51	89.1	17 57	85 48	88.9	17 56	84 43	88.2	198 / 342
19 / 161	19 00	90 00	90.0	19 00	88 57	89.7	18 59	87 53	89.4	18 58	86 50	89.0	18 57	85 46	88.8	18 55	84 41	88.1	199 / 341
20 / 160	20 00	90 00	90.0	20 00	88 56	89.6	19 59	87 52	89.3	19 58	86 48	89.0	19 57	85 45	88.7	19 55	84 39	88.0	200 / 340
21 / 159	21 00	90 00	90.0	21 00	88 56	89.6	20 59	87 51	89.3	20 58	86 47	88.9	20 57	85 43	88.6	20 55	84 37	87.9	201 / 339
22 / 158	22 00	90 00	90.0	22 00	88 55	89.6	21 59	87 51	89.3	21 58	86 46	88.9	21 57	85 41	88.5	21 55	84 34	87.8	202 / 338
23 / 157	23 00	90 00	90.0	23 00	88 55	89.6	22 59	87 50	89.2	22 58	86 44	88.8	22 56	85 39	88.5	22 54	84 32	87.7	203 / 337
24 / 156	24 00	90 00	90.0	24 00	88 54	89.6	23 59	87 49	89.2	23 58	86 43	88.7	23 56	85 37	88.4	23 54	84 29	87.6	204 / 336
25 / 155	25 00	90 00	90.0	25 00	88 54	89.5	24 59	87 48	89.2	24 58	86 41	88.7	24 56	85 35	88.3	24 54	84 26	87.5	205 / 335
26 / 154	26 00	90 00	90.0	26 00	88 53	89.5	25 59	87 46	89.1	25 58	86 40	88.6	25 56	85 33	88.2	25 54	84 24	87.4	206 / 334
27 / 153	27 00	90 00	90.0	27 00	88 53	89.5	26 59	87 45	89.1	26 58	86 38	88.5	26 56	85 31	88.1	26 53	84 20	87.3	207 / 333
28 / 152	28 00	90 00	90.0	28 00	88 52	89.5	27 59	87 44	89.0	27 57	86 36	88.5	27 56	85 28	88.1	27 53	84 17	87.2	208 / 332
29 / 151	29 00	90 00	90.0	29 00	88 51	89.4	28 59	87 43	89.0	28 57	86 34	88.4	28 55	85 26	88.0	28 53	84 14	87.1	209 / 331
30 / 150	30 00	90 00	90.0	30 00	88 51	89.4	29 59	87 41	89.0	29 57	86 32	88.3	29 55	85 23	87.9	29 52	84 10	87.0	210 / 330
31 / 149	31 00	90 00	90.0	31 00	88 50	89.4	30 59	87 40	89.0	30 57	86 30	88.3	30 55	85 20	87.8	30 52	84 07	86.9	211 / 329
32 / 148	32 00	90 00	90.0	32 00	88 49	89.4	31 59	87 39	88.9	31 57	86 28	88.2	31 55	85 17	87.7	31 52	84 03	86.8	212 / 328
33 / 147	33 00	90 00	90.0	33 00	88 48	89.4	32 59	87 37	88.8	32 57	86 25	88.1	32 54	85 14	87.6	32 52	83 59	86.7	213 / 327
34 / 146	34 00	90 00	90.0	34 00	88 48	89.3	33 59	87 35	88.8	33 57	86 23	88.1	33 54	85 11	87.5	33 51	83 54	86.6	214 / 326
35 / 145	35 00	90 00	90.0	35 00	88 47	89.3	34 59	87 34	88.7	34 57	86 20	88.0	34 54	85 07	87.4	34 51	83 50	86.5	215 / 325
36 / 144	36 00	90 00	90.0	36 00	88 46	89.3	35 58	87 32	88.7	35 57	86 18	87.9	35 54	85 04	87.1	35 51	83 45	86.4	216 / 324
37 / 143	37 00	90 00	90.0	37 00	88 45	89.2	36 58	87 30	88.6	36 56	86 15	87.8	36 54	85 00	87.0	36 50	83 40	86.2	217 / 323
38 / 142	38 00	90 00	90.0	38 00	88 44	89.2	37 58	87 28	88.5	37 56	86 12	87.7	37 53	84 56	86.9	37 50	83 35	86.1	218 / 322
39 / 141	39 00	90 00	90.0	39 00	88 43	89.2	38 58	87 26	88.5	38 56	86 09	87.7	38 53	84 52	86.8	38 49	83 29	86.0	219 / 321
40 / 140	40 00	90 00	90.0	40 00	88 42	89.2	39 58	87 23	88.4	39 56	86 05	87.6	39 53	84 47	86.7	39 49	83 23	85.8	220 / 320
41 / 139	41 00	90 00	90.0	41 00	88 40	89.1	40 58	87 21	88.4	40 56	86 02	87.5	40 53	84 42	86.5	40 49	83 17	85.7	221 / 319
42 / 138	42 00	90 00	90.0	42 00	88 39	89.1	41 58	87 19	88.3	41 56	85 58	87.4	41 52	84 37	86.4	41 48	83 11	85.5	222 / 318
43 / 137	43 00	90 00	90.0	43 00	88 38	89.1	42 58	87 16	88.2	42 56	85 54	87.3	42 52	84 32	86.3	42 48	83 04	85.4	223 / 317
44 / 136	44 00	90 00	90.0	44 00	88 37	89.0	43 58	87 13	88.1	43 55	85 50	87.2	43 52	84 27	86.1	43 47	82 57	85.2	224 / 316
45 / 135	45 00	90 00	90.0	44 59	88 35	89.0	44 58	87 10	88.0	44 55	85 46	87.0	44 52	84 21	86.0	44 47	82 57	85.0	225 / 315

Lat./A	LHA/F	0° A/H	0° B/P	0° Z₁/Z₂	1° A/H	1° B/P	1° Z₁/Z₂	2° A/H	2° B/P	2° Z₁/Z₂	3° A/H	3° B/P	3° Z₁/Z₂	4° A/H	4° B/P	4° Z₁/Z₂	5° A/H	5° B/P	5° Z₁/Z₂	Lat./A	LHA
45	135	45 00	90 00	90.0	44 59	88 35	89.0	44 58	87 10	88.0	44 55	85 46	87.0	44 52	84 21	86.0	44 47	82 57	85.0	315	225
46	134	46 00	90 00	90.0	45 59	88 34	89.0	45 58	87 07	87.9	45 55	85 41	86.9	45 51	84 15	85.9	45 46	82 49	84.8	314	226
47	133	47 00	90 00	90.0	46 59	88 33	88.9	46 58	87 04	87.9	46 55	85 36	86.8	46 51	84 09	85.7	46 46	82 41	84.7	313	227
48	132	48 00	90 00	90.0	47 59	88 32	88.9	47 58	87 01	87.8	47 55	85 31	86.7	47 51	84 02	85.6	47 46	82 33	84.5	312	228
49	131	49 00	90 00	90.0	48 59	88 30	88.9	48 58	86 57	87.8	48 55	85 26	86.6	48 50	83 55	85.4	48 45	82 24	84.3	311	229
50	130	50 00	90 00	90.0	49 59	88 29	88.8	49 58	86 53	87.7	49 54	85 20	86.3	49 50	83 47	85.2	49 44	82 15	84.1	310	230
51	129	51 00	90 00	90.0	50 59	88 27	88.8	50 57	86 49	87.5	50 54	85 14	86.2	50 50	83 40	85.1	50 44	82 05	83.9	309	231
52	128	52 00	90 00	90.0	51 59	88 25	88.8	51 57	86 45	87.4	51 54	85 08	86.0	51 49	83 31	84.9	51 43	81 55	83.6	308	232
53	127	53 00	90 00	90.0	52 59	88 23	88.7	52 57	86 41	87.3	52 54	85 01	85.9	52 49	83 22	84.7	52 43	81 44	83.4	307	233
54	126	54 00	90 00	90.0	53 59	88 20	88.7	53 57	86 36	87.2	53 54	84 54	85.7	53 49	83 13	84.5	53 42	81 32	83.2	306	234
55	125	55 00	90 00	90.0	54 59	88 18	88.6	54 57	86 31	87.1	54 54	84 47	85.6	54 48	83 03	84.3	54 41	81 20	82.9	305	235
56	124	56 00	90 00	90.0	55 59	88 15	88.6	55 57	86 26	87.0	55 53	84 39	85.4	55 48	82 52	84.1	55 41	81 06	82.6	304	236
57	123	57 00	90 00	90.0	56 59	88 13	88.5	56 57	86 20	86.9	56 53	84 30	85.2	56 47	82 41	83.9	56 40	80 52	82.4	303	237
58	122	58 00	90 00	90.0	57 59	88 10	88.5	57 57	86 14	86.8	57 52	84 21	85.0	57 47	82 29	83.6	57 39	80 38	82.1	302	238
59	121	59 00	90 00	90.0	58 59	88 07	88.4	58 57	86 07	86.7	58 52	84 11	84.8	58 46	82 16	83.4	58 38	80 22	81.7	301	239
60	120	60 00	90 00	90.0	59 59	88 04	88.3	59 56	86 00	86.6	59 52	84 01	84.6	59 46	82 02	83.1	59 37	80 04	81.4	300	240
61	119	61 00	90 00	90.0	60 59	88 00	88.3	60 56	85 53	86.4	60 52	83 50	84.4	60 45	81 48	82.8	60 37	79 46	81.1	299	241
62	118	62 00	90 00	90.0	61 59	87 56	88.2	61 56	85 45	86.2	61 51	83 38	84.1	61 44	81 32	82.5	61 36	79 27	80.7	298	242
63	117	63 00	90 00	90.0	62 59	87 52	88.1	62 56	85 36	86.1	62 51	83 25	83.9	62 44	81 15	82.2	62 35	79 06	80.3	297	243
64	116	64 00	90 00	90.0	63 59	87 48	88.0	63 56	85 27	85.9	63 50	83 11	83.6	63 43	80 56	81.9	63 33	78 43	79.9	296	244
65	115	65 00	90 00	90.0	64 59	87 43	88.0	64 56	85 17	85.7	64 50	82 56	83.3	64 42	80 36	81.5	64 32	78 18	79.4	295	245
66	114	66 00	90 00	90.0	65 59	87 39	87.9	65 55	85 06	85.5	65 49	82 39	83.0	65 41	80 15	81.1	65 31	77 52	78.9	294	246
67	113	67 00	90 00	90.0	66 59	87 33	87.8	66 55	84 54	85.3	66 49	82 22	82.6	66 40	79 51	80.7	66 29	77 23	78.4	293	247
68	112	68 00	90 00	90.0	67 59	87 26	87.8	67 55	84 40	85.1	67 48	82 02	82.2	67 39	79 26	80.2	67 28	76 51	77.8	292	248
69	111	69 00	90 00	90.0	68 59	87 20	87.6	68 55	84 26	84.8	68 48	81 41	81.8	68 38	78 57	79.7	68 26	76 17	77.2	291	249
70	110	70 00	90 00	90.0	69 59	87 13	87.5	69 54	84 10	84.5	69 47	81 17	81.4	69 37	78 27	79.2	69 25	75 39	76.5	290	250
71	109	71 00	90 00	90.0	70 58	87 05	87.4	70 54	83 53	84.2	70 46	80 51	80.8	70 36	77 53	78.5	70 23	74 57	75.8	289	251
72	108	72 00	90 00	90.0	71 58	86 56	87.3	71 54	83 33	83.9	71 46	80 22	80.3	71 35	77 15	77.9	71 20	74 12	75.0	288	252
73	107	73 00	90 00	90.0	72 58	86 46	87.1	72 53	83 11	83.5	72 45	79 50	79.7	72 33	76 33	77.1	72 18	73 20	74.1	287	253
74	106	74 00	90 00	90.0	73 58	86 35	86.9	73 53	82 47	83.1	73 44	79 14	78.9	73 31	75 46	76.3	73 20	72 23	73.1	286	254
75	105	75 00	90 00	90.0	74 58	86 22	86.7	74 52	82 19	82.6	74 43	78 33	78.1	74 29	74 53	75.4	74 12	71 19	72.0	285	255
76	104	76 00	90 00	90.0	75 58	86 08	86.5	75 52	81 47	82.0	75 41	77 47	77.2	75 27	73 53	74.4	75 09	70 07	70.7	284	256
77	103	77 00	90 00	90.0	76 58	85 52	86.3	76 51	81 11	81.4	76 40	76 53	76.2	76 25	72 44	73.2	76 05	68 45	69.3	283	257
78	102	78 00	90 00	90.0	77 58	85 34	86.0	77 50	80 28	80.7	77 38	75 51	74.9	77 22	71 25	71.8	77 01	67 11	67.7	282	258
79	101	79 00	90 00	90.0	78 57	85 12	85.7	78 49	79 38	79.8	78 36	74 38	73.5	78 18	69 52	70.3	77 56	65 22	65.8	281	259
80	100	80 00	90 00	90.0	79 57	84 46	85.3	79 48	78 38	78.8	79 34	73 12	71.7	79 14	68 04	68.4	78 50	63 16	63.7	280	260
81	99	81 00	90 00	90.0	80 57	84 16	84.9	80 47	77 25	77.6	80 31	71 29	69.6	80 09	65 55	66.2	79 43	60 47	61.2	279	261
82	98	82 00	90 00	90.0	81 56	83 38	84.3	81 45	75 55	76.1	81 28	69 22	66.9	81 04	63 19	63.6	80 34	57 51	58.2	278	262
83	97	83 00	90 00	90.0	82 56	82 51	83.7	82 43	74 01	74.1	82 23	66 44	63.5	81 57	60 09	60.4	81 24	54 20	54.6	277	263
84	96	84 00	90 00	90.0	83 55	81 51	82.9	83 41	74 01	74.1	83 18	66 44	66.9	82 48	60 09	60.4	82 12	54 20	54.6	276	264
85	95	85 00	90 00	90.0	84 54	80 31	80.6	84 37	71 32	71.6	84 10	63 22	63.5	83 36	56 13	56.4	82 56	50 04	50.3	275	265
86	94	86 00	90 00	90.0	85 53	78 40	78.7	85 32	68 10	68.3	85 00	58 59	59.1	84 21	51 16	51.4	83 36	44 53	45.1	274	266
87	93	87 00	90 00	90.0	86 50	75 57	76.0	86 24	63 24	63.5	85 45	53 05	53.2	85 00	44 56	45.1	84 10	38 34	38.7	273	267
88	92	88 00	90 00	90.0	87 46	71 33	71.6	87 10	56 17	56.3	86 24	44 58	45.0	85 32	36 49	36.9	84 37	30 53	31.0	272	268
89	91	89 00	90 00	90.0	88 35	63 26	63.4	87 46	44 59	45.0	86 50	33 40	33.7	85 53	26 31	26.6	84 54	21 45	21.8	271	269
90	90	90 00	90 00	90.0	89 00	45 00	45.0	88 00	26 33	26.6	87 00	18 25	18.4	86 00	14 01	14.0	85 00	11 17	11.3	270	270
			0 00	0.0		0 00	0.0		0 00	0.0		0 00	0.0		0 00	0.0		0 00	0.0		

N. Lat.: for LHA > 180° .... $Z_n = Z$ for LHA < 180° .... $Z_n = 360° − Z$

Lat. / A $Z_n = 180° − Z$ for LHA > 180° $Z_n = 180° + Z$ for LHA < 180°

S. Lat.: for LHA > 180° .... $Z_n = 180° − Z$ for LHA < 180° .... $Z_n = 180° + Z$

SIGHT REDUCTION TABLE

B: (−) for 90° < LHA < 270°
Dec: (−) for Lat. contrary name

Z₁: same sign as B
Z₂: (−) for F > 90°

LHA/F	6° A/H	6° B/P	6° Z₁/Z₂	7° A/H	7° B/P	7° Z₁/Z₂	8° A/H	8° B/P	8° Z₁/Z₂	9° A/H	9° B/P	9° Z₁/Z₂	10° A/H	10° B/P	10° Z₁/Z₂	11° A/H	11° B/P	11° Z₁/Z₂	Lat./A	LHA
0	0 00	84 00	90.0	0 00	83 00	90.0	0 00	82 00	90.0	0 00	81 00	90.0	0 00	80 00	90.0	0 00	79 00	90.0	360	180
1	1 00	84 00	89.9	1 00	83 00	89.9	0 59	82 00	89.9	0 59	81 00	89.8	0 59	80 00	89.8	0 59	79 00	89.8	359	181
2	1 59	84 00	89.9	1 59	83 00	89.8	1 59	82 00	89.7	1 59	81 00	89.7	1 58	80 00	89.7	1 58	79 00	89.6	358	182
3	2 59	84 00	89.7	2 59	82 59	89.7	2 58	81 59	89.6	2 58	80 59	89.6	2 57	79 59	89.5	2 57	78 59	89.4	357	183
4	3 59	83 59	89.6	3 58	82 59	89.6	3 58	81 59	89.5	3 57	80 59	89.4	3 56	79 59	89.3	3 56	78 58	89.2	356	184
5	4 58	83 59	89.5	4 58	82 58	89.4	4 57	81 58	89.3	4 56	80 58	89.2	4 55	79 58	89.1	4 54	78 58	89.0	355	185
6	5 58	83 58	89.4	5 57	82 58	89.3	5 56	81 57	89.2	5 56	80 57	89.1	5 55	79 57	89.0	5 53	78 56	88.9	354	186
7	6 58	83 57	89.3	6 57	82 57	89.1	6 56	81 56	89.0	6 55	80 56	88.9	6 54	79 56	88.8	6 52	78 55	88.7	353	187
8	7 57	83 56	89.2	7 56	82 56	89.0	7 55	81 55	88.9	7 54	80 55	88.7	7 53	79 54	88.6	7 51	78 54	88.5	352	188
9	8 57	83 56	89.1	8 56	82 55	88.9	8 55	81 54	88.7	8 53	80 53	88.6	8 52	79 53	88.4	8 50	78 52	88.3	351	189
10	9 57	83 54	88.9	9 55	82 54	88.8	9 54	81 53	88.6	9 53	80 52	88.4	9 51	79 51	88.2	9 49	78 50	88.1	350	190
11	10 56	83 53	88.8	10 55	82 52	88.6	10 53	81 51	88.5	10 52	80 50	88.3	10 50	79 49	88.1	10 48	78 48	87.9	349	191
12	11 56	83 52	88.7	11 55	82 51	88.5	11 53	81 49	88.3	11 51	80 48	88.1	11 49	79 47	87.9	11 47	78 46	87.7	348	192
13	12 56	83 51	88.6	12 54	82 49	88.4	12 52	81 48	88.2	12 50	80 46	87.9	12 48	79 45	87.7	12 45	78 43	87.5	347	193
14	13 55	83 49	88.5	13 54	82 47	88.3	13 52	81 46	88.0	13 49	80 44	87.8	13 47	79 42	87.5	13 44	78 40	87.3	346	194
15	14 55	83 47	88.4	14 53	82 45	88.1	14 51	81 43	87.9	14 49	80 41	87.6	14 46	79 39	87.3	14 43	78 37	87.1	345	195
16	15 55	83 46	88.3	15 53	82 43	88.0	15 50	81 41	87.7	15 48	80 39	87.4	15 45	79 36	87.1	15 42	78 34	86.9	344	196
17	16 54	83 44	88.2	16 52	82 41	87.9	16 50	81 38	87.6	16 47	80 36	87.3	16 44	79 33	87.0	16 41	78 31	86.7	343	197
18	17 54	83 42	88.1	17 52	82 39	87.7	17 49	81 36	87.4	17 46	80 33	87.1	17 43	79 30	86.8	17 39	78 27	86.5	342	198
19	18 54	83 39	87.9	18 51	82 36	87.6	18 48	81 33	87.3	18 45	80 29	86.9	18 42	79 26	86.6	18 38	78 23	86.2	341	199
20	19 53	83 37	87.8	19 51	82 33	87.5	19 48	81 30	87.1	19 45	80 26	86.7	19 41	79 22	86.4	19 37	78 19	86.0	340	200
21	20 53	83 35	87.7	20 50	82 30	87.3	20 47	81 26	86.9	20 44	80 22	86.6	20 40	79 18	86.2	20 36	78 14	85.8	339	201
22	21 52	83 32	87.6	21 50	82 27	87.2	21 46	81 23	86.8	21 43	80 18	86.4	21 39	79 14	86.0	21 35	78 10	85.6	338	202
23	22 52	83 29	87.5	22 49	82 24	87.0	22 46	81 19	86.6	22 42	80 14	86.2	22 38	79 09	85.8	22 33	78 05	85.4	337	203
24	23 52	83 26	87.3	23 49	82 21	86.9	23 45	81 15	86.5	23 41	80 10	86.0	23 37	79 05	85.6	23 32	77 59	85.1	336	204
25	24 51	83 23	87.2	24 48	82 17	86.7	24 44	81 11	86.3	24 40	80 05	85.8	24 36	78 59	85.4	24 31	77 54	84.9	335	205
26	25 51	83 20	87.1	25 48	82 13	86.6	25 44	81 07	86.1	25 39	80 00	85.6	25 35	78 54	85.2	25 29	77 48	84.7	334	206
27	26 50	83 16	87.0	26 47	82 09	86.4	26 43	81 02	85.9	26 38	79 55	85.4	26 33	78 48	84.9	26 28	77 42	84.4	333	207
28	27 50	83 13	86.8	27 46	82 05	86.3	27 42	80 57	85.8	27 38	79 50	85.2	27 32	78 42	84.7	27 27	77 35	84.2	332	208
29	28 50	83 09	86.7	28 46	82 01	86.1	28 41	80 52	85.6	28 37	79 44	85.0	28 31	78 36	84.5	28 25	77 28	84.0	331	209
30	29 49	83 05	86.5	29 45	81 56	86.0	29 41	80 47	85.4	29 36	79 38	84.8	29 30	78 29	84.3	29 24	77 21	83.7	330	210
31	30 49	83 01	86.4	30 44	81 51	85.8	30 40	80 41	85.2	30 35	79 32	84.6	30 29	78 23	84.0	30 22	77 13	83.5	329	211
32	31 48	82 56	86.3	31 44	81 46	85.6	31 39	80 35	85.0	31 34	79 25	84.4	31 27	78 15	83.8	31 21	77 05	83.2	328	212
33	32 48	82 51	86.1	32 43	81 40	85.5	32 38	80 29	84.8	32 33	79 18	84.2	32 26	78 08	83.6	32 19	76 57	82.9	327	213
34	33 47	82 46	86.0	33 43	81 35	85.3	33 37	80 23	84.6	33 32	79 11	84.0	33 25	78 00	83.3	33 18	76 48	82.7	326	214
35	34 47	82 41	85.8	34 42	81 29	85.1	34 37	80 16	84.4	34 30	79 03	83.7	34 24	77 51	83.1	34 16	76 39	82.4	325	215
36	35 46	82 36	85.7	35 41	81 22	84.9	35 36	80 09	84.2	35 29	78 55	83.5	35 22	77 42	82.8	35 14	76 29	82.1	324	216
37	36 46	82 30	85.5	36 41	81 16	84.8	36 35	80 01	84.0	36 28	78 47	83.3	36 21	77 33	82.5	36 13	76 19	81.8	323	217
38	37 45	82 24	85.3	37 40	81 09	84.6	37 34	79 53	83.8	37 27	78 38	83.0	37 19	77 23	82.3	37 11	76 09	81.5	322	218
39	38 45	82 18	85.2	38 39	81 01	84.4	38 33	79 45	83.6	38 26	78 29	82.8	38 18	77 13	82.0	38 09	75 57	81.2	321	219
40	39 44	82 11	85.0	39 39	80 54	84.2	39 32	79 36	83.3	39 25	78 19	82.5	39 16	77 02	81.7	39 07	75 46	80.9	320	220
41	40 44	82 04	84.8	40 38	80 46	84.0	40 31	79 27	83.1	40 23	78 09	82.3	40 15	76 51	81.4	40 05	75 33	80.6	319	221
42	41 43	81 57	84.6	41 37	80 37	83.7	41 30	79 17	82.9	41 22	77 58	82.0	41 13	76 39	81.1	41 04	75 21	80.3	318	222
43	42 42	81 49	84.4	42 36	80 28	83.5	42 29	79 07	82.6	42 21	77 47	81.7	42 12	76 27	80.8	42 02	75 07	79.9	317	223
44	43 42	81 41	84.2	43 35	80 19	83.3	43 28	78 57	82.3	43 19	77 35	81.4	43 10	76 14	80.5	43 00	74 53	79.6	316	224
45	44 41	81 33	84.0	44 34	80 09	83.1	44 27	78 46	82.1	44 18	77 22	81.1	44 08	76 00	80.1	43 57	74 38	79.2	315	225

Lat./F LHA/F		6° A/H	6° B/P	6° Z₁/Z₂	7° A/H	7° B/P	7° Z₁/Z₂	8° A/H	8° B/P	8° Z₁/Z₂	9° A/H	9° B/P	9° Z₁/Z₂	10° A/H	10° B/P	10° Z₁/Z₂	11° A/H	11° B/P	11° Z₁/Z₂	Lat./A LHA	
45	135	44 41	81 33	84.0	44 34	80 09	83.1	44 27	78 46	82.1	44 18	77 22	81.1	44 08	76 00	80.1	43 57	74 38	79.2	225	315
46	134	45 41	81 24	83.8	45 34	79 59	82.8	45 26	78 34	81.8	45 16	77 09	80.8	45 06	75 45	79.8	44 55	74 22	78.8	226	314
47	133	46 40	81 14	83.6	46 33	79 48	82.6	46 24	78 21	81.5	46 15	76 56	80.5	46 04	75 30	79.5	45 53	74 05	78.4	227	313
48	132	47 39	81 04	83.4	47 32	79 36	82.3	47 23	78 08	81.2	47 13	76 41	80.1	47 03	75 14	79.1	46 51	73 48	78.0	228	312
49	131	48 38	80 54	83.1	48 31	79 24	82.0	48 22	77 55	80.9	48 12	76 26	79.8	48 01	74 57	78.7	47 48	73 30	77.6	229	311
50	130	49 38	80 43	82.9	49 30	79 11	81.7	49 20	77 40	80.6	49 10	76 09	79.4	48 58	74 40	78.3	48 46	73 10	77.2	230	310
51	129	50 37	80 31	82.6	50 29	78 58	81.4	50 19	77 25	80.2	50 08	75 52	79.1	49 56	74 21	77.9	49 43	72 50	76.7	231	309
52	128	51 36	80 19	82.4	51 27	78 43	81.1	51 18	77 08	79.9	51 06	75 34	78.7	50 54	74 01	77.5	50 40	72 29	76.3	232	308
53	127	52 35	80 06	82.1	52 26	78 28	80.8	52 16	76 51	79.5	52 04	75 15	78.3	51 52	73 40	77.0	51 37	72 06	75.8	233	307
54	126	53 34	79 52	81.8	53 25	78 12	80.5	53 14	76 33	79.2	53 02	74 55	77.8	52 49	73 18	76.6	52 35	71 42	75.3	234	306
55	125	54 33	79 37	81.5	54 24	77 55	80.1	54 13	76 14	78.8	54 00	74 34	77.4	53 47	72 55	76.1	53 31	71 17	74.8	235	305
56	124	55 32	79 21	81.2	55 22	77 37	79.8	55 11	75 54	78.3	54 58	74 11	76.9	54 44	72 30	75.6	54 28	70 50	74.2	236	304
57	123	56 31	79 05	80.9	56 21	77 18	79.4	56 09	75 32	77.9	55 56	73 47	76.5	55 41	72 04	75.0	55 25	70 22	73.6	237	303
58	122	57 30	78 47	80.5	57 19	76 57	79.0	57 07	75 09	77.4	56 53	73 22	75.9	56 38	71 36	74.5	56 21	69 51	73.0	238	302
59	121	58 29	78 28	80.1	58 18	76 35	78.5	58 05	74 44	77.0	57 51	72 54	75.4	57 35	71 06	73.9	57 17	69 19	72.4	239	301
60	120	59 28	78 08	79.7	59 16	76 12	78.1	59 03	74 18	76.4	58 48	72 24	74.8	58 32	70 34	73.3	58 13	68 45	71.7	240	300
61	119	60 26	77 46	79.3	60 14	75 47	77.6	60 01	73 50	75.9	59 45	71 54	74.2	59 28	70 01	72.6	59 09	68 09	71.0	241	299
62	118	61 25	77 23	78.9	61 12	75 21	77.1	60 58	73 20	75.3	60 42	71 21	73.6	60 24	69 25	71.9	60 05	67 30	70.3	242	298
63	117	62 23	76 58	78.4	62 10	74 52	76.5	61 56	72 48	74.7	61 39	70 46	72.9	61 20	68 46	71.2	61 00	66 49	69.5	243	297
64	116	63 22	76 31	77.9	63 08	74 21	75.9	62 53	72 13	74.1	62 35	70 08	72.2	62 16	68 05	70.4	61 55	66 05	68.6	244	296
65	115	64 20	76 02	77.4	64 06	73 48	75.4	63 50	71 36	73.4	63 32	69 27	71.5	63 12	67 21	69.6	62 50	65 18	67.7	245	295
66	114	65 18	75 31	76.8	65 03	73 12	74.7	64 47	70 56	72.6	64 28	68 43	70.6	64 07	66 34	68.7	63 44	64 27	66.8	246	294
67	113	66 16	74 57	76.2	66 01	72 33	74.0	65 43	70 13	71.8	65 23	67 56	69.8	65 02	65 43	67.8	64 38	63 33	65.8	247	293
68	112	67 14	74 20	75.5	66 58	71 51	73.2	66 40	69 26	71.0	66 19	67 05	68.8	65 56	64 48	66.7	65 32	62 35	64.7	248	292
69	111	68 12	73 39	74.8	67 55	71 05	72.4	67 36	68 35	70.1	67 14	66 09	67.8	66 50	63 48	65.7	66 25	61 31	63.6	249	291
70	110	69 09	72 55	74.0	68 51	70 15	71.5	68 31	67 40	69.1	68 09	65 09	66.7	67 44	62 44	64.5	67 17	60 23	62.3	250	290
71	109	70 07	72 06	73.1	69 48	69 20	70.5	69 27	66 39	68.0	69 03	64 03	65.6	68 37	61 34	63.2	68 09	59 10	61.0	251	289
72	108	71 03	71 13	72.2	70 44	68 20	69.4	70 21	65 33	66.8	69 57	62 52	64.3	69 29	60 17	61.9	69 00	57 50	59.6	252	288
73	107	72 00	70 14	71.1	71 39	67 13	68.3	71 16	64 20	65.5	70 50	61 33	62.9	70 21	58 54	60.4	69 50	56 23	58.0	253	287
74	106	72 56	69 08	70.0	72 34	65 59	67.0	72 09	62 59	64.1	71 42	60 07	61.4	71 12	57 24	58.8	70 40	54 48	56.4	254	286
75	105	73 52	67 54	68.7	73 29	64 37	65.5	73 03	61 30	62.6	72 34	58 32	59.7	72 02	55 44	57.1	71 28	53 06	54.5	255	285
76	104	74 48	66 31	67.3	74 23	63 05	64.0	73 55	59 51	60.8	73 24	56 47	57.9	72 51	53 55	55.1	72 16	51 13	52.6	256	284
77	103	75 42	64 57	65.6	75 16	61 22	62.2	74 46	58 00	58.9	74 14	54 51	55.9	73 39	51 55	53.1	73 02	49 10	50.4	257	283
78	102	76 36	63 11	63.6	76 08	59 26	60.2	75 37	55 57	56.8	75 02	52 42	53.6	74 26	49 42	50.8	73 47	46 56	48.1	258	282
79	101	77 29	61 09	61.7	76 59	57 14	57.9	76 26	53 38	54.4	75 49	50 18	51.2	75 11	47 16	48.2	74 30	44 28	45.5	259	281
80	100	78 21	58 49	59.3	77 49	54 44	55.3	77 13	51 01	51.7	76 35	47 38	48.4	75 54	44 34	45.4	75 11	41 47	42.7	260	280
81	99	79 12	56 06	56.6	78 37	51 52	52.4	77 59	48 04	48.7	77 18	44 39	45.4	76 35	41 35	42.4	75 49	38 50	39.7	261	279
82	98	80 01	52 56	53.4	79 23	48 35	49.1	78 42	44 43	45.3	77 59	41 18	41.9	77 13	38 17	39.0	76 26	35 36	36.4	262	278
83	97	80 47	49 13	49.6	80 07	44 47	45.2	79 23	40 56	41.4	78 37	37 35	38.1	77 49	34 39	35.3	76 59	32 05	32.8	263	277
84	96	81 31	45 22	45.2	80 47	40 24	40.8	80 01	36 38	37.1	79 12	33 25	33.9	78 21	30 40	31.2	77 29	28 16	28.8	264	276
85	95	82 12	39 40	39.9	81 24	35 22	35.7	80 34	31 48	32.2	79 43	28 49	29.2	78 50	26 18	26.7	77 56	24 09	24.6	265	275
86	94	82 48	33 34	33.8	81 57	29 36	29.8	81 04	26 24	26.7	80 09	23 46	24.1	79 14	21 35	21.9	78 18	19 44	20.1	266	274
87	93	83 18	26 28	26.6	82 23	23 05	23.3	81 28	20 25	20.6	80 31	18 17	18.5	79 34	16 32	16.8	78 36	15 04	15.4	267	273
88	92	83 41	18 22	18.5	82 43	15 52	16.0	81 45	13 57	14.1	80 47	12 26	12.6	79 48	11 12	11.4	78 49	10 11	10.4	268	272
89	91	83 55	9 26	9.5	82 56	8 05	8.2	81 56	7 05	7.1	80 57	6 17	6.4	79 57	5 39	5.7	78 57	5 08	5.2	269	271
90	90	84 00	0 00	0.0	83 00	0 00	0.0	82 00	0 00	0.0	81 00	0 00	0.0	80 00	0 00	0.0	79 00	0 00	0.0	270	270

N. Lat.: for LHA > 180°.... Zn = Z
for LHA < 180°.... Zn = 360° − Z

S. Lat.: for LHA > 180°... Zn = 180° − Z
for LHA < 180°... Zn = 180° + Z

LATITUDE / A: 12° – 17°

SIGHT REDUCTION TABLE

B: (–) for 90°< LHA < 270°
Dec: (–) for Lat. contrary name

Z₁: same sign as B
Z₂: (–) for F > 90°

Lat./A	LHA/F	12° A/H	12° B/P	12° Z_1/Z_2	13° A/H	13° B/P	13° Z_1/Z_2	14° A/H	14° B/P	14° Z_1/Z_2	15° A/H	15° B/P	15° Z_1/Z_2	16° A/H	16° B/P	16° Z_1/Z_2	17° A/H	17° B/P	17° Z_1/Z_2	LHA	Lat./A
0	180	0 00	78 00	90.0	0 00	77 00	90.0	0 00	76 00	90.0	0 00	75 00	90.0	0 00	74 00	90.0	0 00	73 00	90.0	180	360
1	179	0 59	78 00	89.8	0 58	77 00	89.8	0 58	76 00	89.8	0 58	75 00	89.7	0 58	74 00	89.7	0 57	73 00	89.7	181	359
2	178	1 57	78 00	89.6	1 57	77 00	89.5	1 56	76 00	89.5	1 56	74 59	89.5	1 55	73 59	89.5	1 55	72 59	89.4	182	358
3	177	2 56	77 59	89.4	2 55	76 59	89.3	2 55	75 59	89.3	2 54	74 59	89.2	2 53	73 59	89.2	2 52	72 59	89.1	183	357
4	176	3 55	77 58	89.2	3 54	76 58	89.1	3 53	75 58	89.0	3 52	74 58	89.0	3 51	73 58	88.9	3 49	72 58	88.8	184	356
5	175	4 53	77 57	89.0	4 52	76 57	88.9	4 51	75 57	88.8	4 50	74 57	88.7	4 48	73 57	88.6	4 47	72 56	88.5	185	355
6	174	5 52	77 56	88.7	5 51	76 56	88.6	5 49	75 56	88.6	5 48	74 55	88.4	5 46	73 55	88.3	5 44	72 55	88.2	186	354
7	173	6 51	77 55	88.5	6 49	76 54	88.4	6 47	75 54	88.4	6 46	74 54	88.2	6 44	73 53	88.1	6 42	72 53	87.9	187	353
8	172	7 49	77 53	88.3	7 48	76 53	88.2	7 46	75 52	88.2	7 44	74 52	87.9	7 41	73 51	87.7	7 39	72 51	87.6	188	352
9	171	8 48	77 51	88.1	8 46	76 51	88.0	8 44	75 50	88.0	8 41	74 49	87.7	8 39	73 49	87.5	8 36	72 48	87.3	189	351
10	170	9 47	77 49	87.9	9 44	76 48	87.7	9 42	75 48	87.8	9 39	74 47	87.4	9 37	73 46	87.2	9 34	72 45	87.0	190	350
11	169	10 45	77 47	87.7	10 43	76 46	87.5	10 40	75 45	87.6	10 37	74 44	87.1	10 34	73 43	86.9	10 31	72 42	86.7	191	349
12	168	11 44	77 44	87.5	11 41	76 43	87.3	11 38	75 42	87.3	11 35	74 41	86.9	11 32	73 40	86.6	11 28	72 39	86.4	192	348
13	167	12 43	77 42	87.3	12 40	76 40	87.0	12 36	75 39	87.0	12 33	74 37	86.6	12 29	73 36	86.4	12 25	72 35	86.1	193	347
14	166	13 41	77 39	87.0	13 38	76 37	86.8	13 35	75 35	86.8	13 31	74 34	86.3	13 27	73 32	86.1	13 23	72 31	85.8	194	346
15	165	14 40	77 35	86.8	14 36	76 33	86.6	14 33	75 32	86.5	14 29	74 30	86.0	14 24	73 28	85.8	14 20	72 26	85.5	195	345
16	164	15 38	77 32	86.6	15 35	76 30	86.3	15 31	75 28	86.3	15 26	74 25	85.8	15 22	73 23	85.5	15 17	72 21	85.2	196	344
17	163	16 37	77 28	86.4	16 33	76 26	86.1	16 29	75 23	86.0	16 24	74 21	85.5	16 19	73 19	85.2	16 14	72 16	84.9	197	343
18	162	17 36	77 24	86.1	17 31	76 21	85.8	17 27	75 19	85.8	17 22	74 16	85.2	17 17	73 13	84.9	17 11	72 11	84.6	198	342
19	161	18 34	77 20	85.9	18 30	76 17	85.6	18 25	75 14	85.5	18 20	74 11	84.9	18 14	73 08	84.6	18 08	72 05	84.3	199	341
20	160	19 33	77 15	85.7	19 28	76 12	85.3	19 23	75 08	85.3	19 17	74 05	84.6	19 12	73 02	84.3	19 05	71 59	83.9	200	340
21	159	20 31	77 10	85.4	20 26	76 07	85.1	20 21	75 03	85.0	20 15	73 59	84.3	20 09	72 56	84.0	20 03	71 52	83.6	201	339
22	158	21 30	77 05	85.2	21 24	76 01	84.8	21 19	74 57	84.7	21 13	73 53	84.0	21 06	72 49	83.6	21 00	71 45	83.3	202	338
23	157	22 28	77 00	85.0	22 23	75 55	84.5	22 17	74 51	84.5	22 10	73 46	83.7	22 04	72 42	83.3	21 56	71 38	82.9	203	337
24	156	23 27	76 54	84.7	23 21	75 49	84.3	23 15	74 44	84.2	23 08	73 39	83.4	23 01	72 34	83.0	22 53	71 30	82.6	204	336
25	155	24 25	76 48	84.5	24 19	75 43	84.0	24 13	74 37	83.9	24 06	73 32	83.1	23 58	72 27	82.7	23 50	71 22	82.2	205	335
26	154	25 23	76 42	84.2	25 17	75 36	83.7	25 10	74 30	83.6	25 03	73 24	82.8	24 55	72 18	82.3	24 47	71 13	81.9	206	334
27	153	26 22	76 35	84.0	26 15	75 28	83.5	26 08	74 22	83.3	26 01	73 16	82.5	25 52	72 10	82.0	25 44	71 04	81.5	207	333
28	152	27 20	76 28	83.7	27 13	75 21	83.2	27 06	74 14	83.0	26 58	73 07	82.1	26 50	72 00	81.7	26 41	70 54	81.2	208	332
29	151	28 18	76 20	83.4	28 11	75 13	82.9	28 04	74 05	82.7	27 55	72 58	81.8	27 47	71 51	81.3	27 37	70 44	80.8	209	331
30	150	29 17	76 13	83.2	29 09	75 04	82.6	29 01	73 56	82.4	28 53	72 48	81.5	28 44	71 41	81.0	28 34	70 33	80.4	210	330
31	149	30 15	76 05	82.9	30 07	74 56	82.3	29 59	73 47	82.0	29 50	72 38	81.2	29 41	71 30	80.6	29 30	70 22	80.0	211	329
32	148	31 13	75 56	82.6	31 05	74 46	82.0	30 57	73 37	81.7	30 47	72 28	80.8	30 37	71 19	80.2	30 27	70 11	79.6	212	328
33	147	32 11	75 47	82.3	32 03	74 37	81.7	31 54	73 27	81.4	31 44	72 17	80.5	31 34	71 07	79.9	31 23	69 58	79.2	213	327
34	146	33 10	75 37	82.0	33 01	74 26	81.4	32 52	73 16	81.0	32 42	72 05	80.1	32 31	70 55	79.5	32 20	69 45	78.8	214	326
35	145	34 08	75 27	81.7	33 59	74 16	81.0	33 49	73 04	80.7	33 39	71 53	79.7	33 28	70 42	79.1	33 16	69 32	78.4	215	325
36	144	35 06	75 17	81.4	34 56	74 04	80.7	34 46	72 52	80.2	34 36	71 40	79.4	34 24	70 29	78.7	34 12	69 18	78.0	216	324
37	143	36 04	75 06	81.1	35 54	73 53	80.4	35 44	72 40	79.9	35 33	71 27	79.0	35 21	70 15	78.3	35 08	69 03	77.6	217	323
38	142	37 02	74 54	80.8	36 52	73 40	80.0	36 41	72 27	79.5	36 29	71 13	78.6	36 17	70 00	77.8	36 04	68 48	77.1	218	322
39	141	38 00	74 42	80.4	37 49	73 27	79.7	37 38	72 13	79.1	37 26	70 59	78.2	37 13	69 45	77.4	37 00	68 32	76.7	219	321
40	140	38 57	74 30	80.1	38 47	73 14	79.3	38 35	71 58	78.7	38 23	70 43	77.8	38 10	69 29	77.0	37 56	68 15	76.2	220	320
41	139	39 55	74 16	79.8	39 44	72 59	78.9	39 32	71 43	78.2	39 19	70 27	77.4	39 06	69 12	76.5	38 51	67 57	75.7	221	319
42	138	40 53	74 02	79.4	40 41	72 45	78.5	40 29	71 27	77.7	40 16	70 10	76.9	40 02	68 54	76.1	39 47	67 38	75.3	222	318
43	137	41 51	73 48	79.0	41 39	72 29	78.2	41 26	71 11	77.3	41 12	69 53	76.4	40 58	68 35	75.6	40 42	67 19	74.7	223	317
44	136	42 48	73 32	78.6	42 36	72 12	77.7	42 23	70 53	76.9	42 09	69 34	76.0	41 54	68 16	75.1	41 38	66 58	74.2	224	316
45	135	43 46	73 16	78.3	43 33	71 55	77.3	43 19	70 35	76.4	43 05	69 15	75.5	42 49	67 56	74.6	42 33	66 37	73.7	225	315

Lat/A	LHA/F	12° A/H	12° B/P	12° Z₁/Z₂	13° A/H	13° B/P	13° Z₁/Z₂	14° A/H	14° B/P	14° Z₁/Z₂	15° A/H	15° B/P	15° Z₁/Z₂	16° A/H	16° B/P	16° Z₁/Z₂	17° A/H	17° B/P	17° Z₁/Z₂	Lat/A	LHA
45	135	43 46	73 16	78.3	43 33	71 55	77.3	43 19	70 35	76.4	43 05	69 15	75.5	42 49	67 56	74.6	42 33	66 37	73.7	225	315
46	134	44 43	72 59	77.7	44 30	71 37	76.9	44 16	70 15	75.9	44 01	68 54	75.0	43 45	67 34	74.1	43 28	66 15	73.2	226	314
47	133	45 40	72 41	77.4	45 27	71 18	76.4	45 12	69 55	75.5	44 57	68 33	74.5	44 40	67 12	73.5	44 23	65 51	72.6	227	313
48	132	46 38	72 23	77.0	46 24	70 58	76.0	46 09	69 34	75.0	45 53	68 11	74.0	45 35	66 48	73.0	45 17	65 27	72.0	228	312
49	131	47 35	72 03	76.5	47 20	70 37	75.5	47 05	69 11	74.4	46 48	67 47	73.4	46 30	66 23	72.4	46 12	65 01	71.4	229	311
50	130	48 32	71 42	76.1	48 17	70 15	75.0	48 01	68 48	73.9	47 44	67 22	72.9	47 25	65 58	71.8	47 06	64 34	70.8	230	310
51	129	49 29	71 20	75.6	49 13	69 51	74.5	48 57	68 23	73.4	48 39	66 56	72.3	48 20	65 30	71.2	48 00	64 05	70.1	231	309
52	128	50 25	70 57	75.1	50 09	69 27	73.9	49 52	67 57	72.8	49 34	66 29	71.7	49 15	65 02	70.6	48 54	63 35	69.5	232	308
53	127	51 22	70 33	74.6	51 06	69 01	73.4	50 48	67 30	72.2	50 29	66 00	71.1	50 09	64 31	69.9	49 48	63 04	68.8	233	307
54	126	52 19	70 07	74.0	52 02	68 33	72.8	51 43	67 01	71.6	51 24	65 30	70.4	51 03	64 00	69.2	50 41	62 31	68.1	234	306
55	125	53 15	69 40	73.5	52 57	68 04	72.2	52 38	66 31	70.9	52 18	64 58	69.7	51 57	63 26	68.5	51 34	61 56	67.3	235	305
56	124	54 11	69 11	72.9	53 53	67 34	71.6	53 33	65 58	70.3	53 12	64 24	69.0	52 50	62 51	67.8	52 27	61 20	66.6	236	304
57	123	55 07	68 41	72.2	54 48	67 02	70.9	54 28	65 24	69.6	54 06	63 48	68.3	53 43	62 14	67.0	53 19	60 42	65.8	237	303
58	122	56 03	68 09	71.6	55 43	66 28	70.2	55 22	64 48	68.8	55 00	63 11	67.5	54 36	61 35	66.2	54 12	60 01	64.9	238	302
59	121	56 59	67 34	70.9	56 38	65 51	69.5	56 16	64 10	68.1	55 53	62 31	66.7	55 29	60 54	65.4	55 04	59 18	64.1	239	301
60	120	57 54	66 58	70.2	57 33	65 13	68.7	57 10	63 30	67.3	56 46	61 49	65.9	56 21	60 10	64.5	55 55	58 33	63.1	240	300
61	119	58 49	66 20	69.4	58 27	64 32	67.9	58 04	62 47	66.4	57 39	61 04	65.0	57 13	59 24	63.6	56 46	57 46	62.2	241	299
62	118	59 44	65 38	68.6	59 21	63 49	67.1	58 57	62 02	65.5	58 31	60 17	64.0	58 05	58 35	62.6	57 36	56 56	61.2	242	298
63	117	60 38	64 55	67.8	60 15	63 03	66.2	59 50	61 13	64.6	59 23	59 27	63.1	58 55	57 43	61.6	58 26	56 03	60.2	243	297
64	116	61 32	64 08	66.9	61 08	62 14	65.2	60 42	60 22	63.6	60 15	58 34	62.0	59 46	56 49	60.5	59 16	55 06	59.1	244	296
65	115	62 26	63 18	66.0	62 01	61 21	64.2	61 34	59 28	62.6	61 06	57 37	61.0	60 36	55 51	59.4	60 05	54 07	57.9	245	295
66	114	63 20	62 25	65.0	62 53	60 25	63.2	62 26	58 30	61.5	61 56	56 37	59.8	61 25	54 49	58.2	60 53	53 04	56.7	246	294
67	113	64 13	61 27	63.9	63 45	59 25	62.1	63 16	57 27	60.3	62 46	55 34	58.6	62 14	53 44	57.0	61 41	51 57	55.4	247	293
68	112	65 05	60 26	62.8	64 37	58 21	60.9	64 07	56 21	59.1	63 35	54 25	57.4	63 02	52 34	55.7	62 27	50 47	54.1	248	292
69	111	65 57	59 20	61.6	65 27	57 13	59.6	64 56	55 10	57.8	64 23	53 13	56.0	63 49	51 20	54.3	63 14	49 32	52.7	249	291
70	110	66 48	58 08	60.3	66 18	55 59	58.3	65 45	53 55	56.4	65 11	51 55	54.6	64 36	50 01	52.9	63 59	48 12	51.2	250	290
71	109	67 39	56 52	58.9	67 07	54 40	56.8	66 33	52 33	54.9	65 58	50 33	53.1	65 21	48 38	51.3	64 43	46 48	49.7	251	289
72	108	68 29	55 29	57.4	67 55	53 14	55.3	67 20	51 06	53.3	66 44	49 04	51.5	66 06	47 08	49.7	65 26	45 18	48.0	252	288
73	107	69 18	53 59	55.8	68 43	51 42	53.7	68 07	49 33	51.6	67 29	47 30	49.8	66 49	45 33	48.0	66 08	43 43	46.3	253	287
74	106	70 06	52 22	54.1	69 30	50 03	51.9	68 52	47 52	49.8	68 12	45 49	47.9	67 31	43 52	46.1	66 49	42 02	44.4	254	286
75	105	70 53	50 36	52.2	70 15	48 16	50.0	69 36	46 04	47.9	68 55	44 00	46.0	68 12	42 04	44.2	67 29	40 15	42.5	255	285
76	104	71 38	48 42	50.2	70 59	46 20	47.9	70 18	44 08	45.9	69 36	42 05	43.9	68 52	40 09	42.1	68 07	38 21	40.5	256	284
77	103	72 23	46 37	48.0	71 42	44 15	45.7	70 59	42 03	43.7	70 15	40 01	41.7	69 30	38 07	39.9	68 43	36 21	38.3	257	283
78	102	73 06	44 22	45.6	72 23	42 00	43.4	71 38	39 49	41.3	70 53	37 49	39.4	70 06	35 57	37.6	69 18	34 13	36.1	258	282
79	101	73 47	41 55	43.1	73 02	39 34	40.8	72 16	37 26	38.8	71 28	35 27	36.9	70 40	33 38	35.2	69 50	31 58	33.6	259	281
80	100	74 26	39 15	40.3	73 39	36 57	38.1	72 51	34 51	36.1	72 02	32 57	34.3	71 12	31 12	32.6	70 21	29 36	31.1	260	280
81	99	75 02	36 21	37.3	74 14	34 07	35.1	73 24	32 06	33.2	72 33	30 17	31.5	71 42	28 37	29.9	70 50	27 06	28.4	261	279
82	98	75 37	33 13	34.1	74 46	31 05	32.0	73 55	29 10	30.2	73 03	27 27	28.5	72 09	25 53	27.0	71 16	24 29	25.7	262	278
83	97	76 08	29 50	30.6	75 16	27 50	28.6	74 23	26 03	26.9	73 29	24 27	25.4	72 34	23 02	24.0	71 39	21 44	22.8	263	277
84	96	76 36	26 11	26.8	75 42	24 22	25.0	74 48	22 45	23.5	73 52	21 19	22.1	72 56	20 02	20.9	71 58	18 53	19.8	264	276
85	95	77 01	22 18	22.8	76 05	20 41	21.3	75 09	19 16	19.9	74 12	18 01	18.7	73 15	16 54	17.6	72 18	15 55	16.7	265	275
86	94	77 22	18 10	18.6	76 25	16 49	17.3	75 27	15 38	16.1	74 29	14 36	15.1	73 31	13 40	14.2	72 33	12 51	13.5	266	274
87	93	77 38	13 50	14.1	76 40	12 46	13.1	75 41	11 51	12.2	74 43	11 03	11.4	73 44	10 21	10.8	72 45	9 43	10.2	267	273
88	92	77 50	9 19	9.5	76 51	8 36	8.8	75 52	7 58	8.2	74 52	7 25	7.7	73 53	6 56	7.2	72 53	6 31	6.8	268	272
89	91	77 58	4 42	4.8	76 58	4 19	4.4	75 58	4 00	4.1	74 58	3 44	3.9	73 58	3 29	3.6	72 58	3 16	3.4	269	271
90	90	78 00	0 00	0.0	77 00	0 00	0.0	76 00	0 00	0.0	75 00	0 00	0.0	74 00	0 00	0.0	73 00	0 00	0.0	270	270

N. Lat.: for LHA > 180°.... Zn = Z
for LHA < 180°.... Zn = 360° − Z

S. Lat.: for LHA > 180°.... Zn = 180° − Z
for LHA < 180°.... Zn = 180° + Z

SIGHT REDUCTION TABLE

B: (−) for 90° < LHA < 270°
Dec: (−) for Lat. contrary name

Z_1: same sign as B
Z_2: (−) for F > 90°

Lat./A	LHA/F	18° A/H	18° B/P	18° Z_1/Z_2	19° A/H	19° B/P	19° Z_1/Z_2	20° A/H	20° B/P	20° Z_1/Z_2	21° A/H	21° B/P	21° Z_1/Z_2	22° A/H	22° B/P	22° Z_1/Z_2	23° A/H	23° B/P	23° Z_1/Z_2	Lat./A	LHA
180	0	0 00	72 00	90.0	0 00	71 00	90.0	0 00	70 00	90.0	0 00	69 00	90.0	0 00	68 00	90.0	0 00	67 00	90.0	360	180
179	1	0 57	72 00	89.7	0 57	71 00	89.7	0 56	70 00	89.7	0 56	69 00	89.6	0 56	68 00	89.6	0 55	67 00	89.6	359	181
178	2	1 54	71 59	89.4	1 53	70 59	89.3	1 53	69 59	89.3	1 52	68 59	89.3	1 51	67 59	89.3	1 50	66 59	89.2	358	182
177	3	2 51	71 59	89.1	2 50	70 59	89.0	2 49	69 58	89.0	2 48	68 58	88.9	2 47	67 58	88.9	2 46	66 58	88.8	357	183
176	4	3 48	71 58	88.8	3 47	70 57	88.7	3 46	69 57	88.6	3 44	68 57	88.6	3 42	67 57	88.5	3 41	66 57	88.4	356	184
175	5	4 45	71 56	88.5	4 44	70 56	88.4	4 42	69 56	88.3	4 40	68 56	88.2	4 38	67 55	88.1	4 36	66 55	88.0	355	185
174	6	5 42	71 54	88.1	5 40	70 54	88.1	5 38	69 54	87.9	5 36	68 54	87.8	5 34	67 53	87.7	5 31	66 53	87.6	354	186
173	7	6 39	71 52	87.8	6 37	70 52	87.7	6 35	69 52	87.6	6 32	68 51	87.5	6 29	67 51	87.4	6 26	66 51	87.3	353	187
172	8	7 36	71 50	87.5	7 34	70 50	87.4	7 31	69 49	87.2	7 28	68 49	87.1	7 25	67 48	87.0	7 22	66 48	86.9	352	188
171	9	8 33	71 47	87.2	8 30	70 47	87.0	8 27	69 46	86.9	8 24	68 46	86.8	8 20	67 45	86.6	8 17	66 45	86.5	351	189
170	10	9 30	71 44	86.9	9 27	70 44	86.7	9 23	69 43	86.5	9 20	68 42	86.4	9 16	67 42	86.2	9 12	66 41	86.1	350	190
169	11	10 27	71 41	86.6	10 24	70 40	86.4	10 20	69 39	86.2	10 16	68 39	86.0	10 11	67 38	85.8	10 07	66 37	85.7	349	191
168	12	11 24	71 37	86.2	11 20	70 36	86.0	11 16	69 35	85.8	11 12	68 34	85.6	11 07	67 33	85.4	11 02	66 32	85.3	348	192
167	13	12 21	71 33	85.9	12 17	70 32	85.7	12 12	69 31	85.5	12 07	68 30	85.3	12 02	67 29	85.1	11 57	66 28	84.8	347	193
166	14	13 18	71 29	85.6	13 13	70 28	85.4	13 08	69 26	85.1	13 03	68 25	84.9	12 58	67 24	84.7	12 52	66 22	84.4	346	194
165	15	14 15	71 24	85.3	14 10	70 23	85.0	14 05	69 21	84.8	13 59	68 20	84.5	13 53	67 18	84.3	13 47	66 17	84.0	345	195
164	16	15 12	71 19	84.9	15 06	70 18	84.7	15 01	69 16	84.4	14 55	68 14	84.1	14 48	67 12	83.9	14 42	66 10	83.6	344	196
163	17	16 09	71 14	84.6	16 03	70 12	84.3	15 57	69 10	84.0	15 50	68 08	83.7	15 44	67 06	83.5	15 37	66 04	83.2	343	197
162	18	17 05	71 08	84.3	16 59	70 06	84.0	16 53	69 03	83.7	16 46	68 01	83.4	16 39	66 59	83.1	16 32	65 57	82.8	342	198
161	19	18 02	71 02	83.9	17 56	69 59	83.6	17 49	68 57	83.3	17 42	67 54	83.0	17 34	66 52	82.7	17 26	65 49	82.3	341	199
160	20	18 59	70 56	83.6	18 52	69 53	83.2	18 45	68 50	82.9	18 37	67 47	82.6	18 29	66 44	82.2	18 21	65 41	81.9	340	200
159	21	19 56	70 49	83.2	19 48	69 45	82.9	19 41	68 42	82.5	19 33	67 39	82.2	19 24	66 36	81.8	19 16	65 33	81.5	339	201
158	22	20 52	70 41	82.9	20 45	69 38	82.5	20 37	68 34	82.1	20 28	67 31	81.8	20 19	66 27	81.4	20 10	65 24	81.0	338	202
157	23	21 49	70 33	82.5	21 41	69 29	82.1	21 32	68 26	81.7	21 24	67 22	81.4	21 14	66 18	81.0	21 05	65 15	80.6	337	203
156	24	22 45	70 25	82.2	22 37	69 21	81.8	22 28	68 17	81.3	22 19	67 12	80.9	22 09	66 09	80.5	21 59	65 05	80.1	336	204
155	25	23 42	70 17	81.8	23 33	69 12	81.4	23 24	68 07	80.9	23 14	67 03	80.5	23 04	65 58	80.1	22 54	64 54	79.7	335	205
154	26	24 38	70 07	81.4	24 29	69 02	81.0	24 20	67 57	80.5	24 09	66 52	80.1	23 59	65 48	79.6	23 48	64 43	79.2	334	206
153	27	25 35	69 58	81.1	25 25	68 52	80.6	25 15	67 47	80.1	25 05	66 42	79.7	24 54	65 36	79.2	24 42	64 32	78.7	333	207
152	28	26 31	69 48	80.7	26 21	68 42	80.2	26 11	67 36	79.7	26 00	66 30	79.2	25 48	65 25	78.7	25 36	64 19	78.3	332	208
151	29	27 27	69 37	80.3	27 17	68 31	79.8	27 06	67 24	79.3	26 55	66 18	78.8	26 43	65 12	78.3	26 30	64 07	77.8	331	209
150	30	28 24	69 26	79.9	28 13	68 19	79.4	28 01	67 12	78.8	27 50	66 06	78.3	27 37	64 59	77.8	27 24	63 53	77.3	330	210
149	31	29 20	69 14	79.5	29 09	68 07	79.0	28 57	67 00	78.4	28 44	65 53	77.8	28 31	64 46	77.3	28 18	63 39	76.8	329	211
148	32	30 16	69 02	79.1	30 04	67 54	78.5	29 52	66 46	77.9	29 39	65 39	77.4	29 26	64 32	76.8	29 12	63 25	76.3	328	212
147	33	31 12	68 49	78.7	31 00	67 41	78.1	30 47	66 32	77.5	30 34	65 24	76.9	30 20	64 17	76.3	30 05	63 09	75.8	327	213
146	34	32 08	68 36	78.2	31 55	67 27	77.6	31 42	66 18	77.0	31 28	65 09	76.4	31 14	64 01	75.8	30 59	62 53	75.2	326	214
145	35	33 04	68 22	77.8	32 51	67 12	77.2	32 37	66 03	76.5	32 23	64 54	75.9	32 08	63 45	75.3	31 52	62 36	74.7	325	215
144	36	33 59	68 07	77.3	33 46	66 57	76.7	33 32	65 47	76.0	33 17	64 37	75.4	33 01	63 28	74.8	32 45	62 19	74.2	324	216
143	37	34 55	67 52	76.9	34 41	66 41	76.2	34 26	65 30	75.5	34 11	64 20	74.9	33 55	63 10	74.2	33 38	62 01	73.6	323	217
142	38	35 50	67 36	76.4	35 36	66 24	75.7	35 21	65 13	75.0	35 05	64 02	74.4	34 48	62 51	73.7	34 31	61 41	73.0	322	218
141	39	36 46	67 19	76.0	36 31	66 06	75.2	36 15	64 54	74.5	35 59	63 43	73.8	35 42	62 32	73.1	35 24	61 21	72.4	321	219
140	40	37 41	67 01	75.5	37 26	65 48	74.7	37 10	64 35	74.0	36 53	63 23	73.3	36 35	62 12	72.6	36 17	61 01	71.8	320	220
139	41	38 36	66 42	75.0	38 20	65 29	74.2	38 04	64 15	73.4	37 46	63 02	72.7	37 28	61 50	72.0	37 09	60 39	71.2	319	221
138	42	39 31	66 23	74.5	39 15	65 08	73.7	38 58	63 54	72.9	38 40	62 41	72.1	38 21	61 28	71.4	38 01	60 16	70.6	318	222
137	43	40 26	66 03	73.9	40 09	64 47	73.1	39 51	63 33	72.3	39 33	62 18	71.5	39 13	61 05	70.7	38 53	59 52	70.0	317	223
136	44	41 21	65 42	73.4	41 03	64 25	72.5	40 45	63 10	71.7	40 26	61 55	70.9	40 06	60 41	70.1	39 45	59 27	69.3	316	224
135	45	42 16	65 19	72.8	41 57	64 02	72.0	41 38	62 46	71.1	41 19	61 30	70.3	40 58	60 15	69.5	40 37	59 01	68.7	315	225

Lat./A																			Lat./A
	18°			19°			20°			21°			22°			23°			
LHA/F	A/H	B/P	Z1/Z2	A/H	B/P	Z1/Z2	A/H	B/P	Z1/Z2	A/H	B/P	Z1/Z2	A/H	B/P	Z1/Z2	A/H	B/P	Z1/Z2	LHA
45	42 16	65 19	72.8	41 57	64 02	72.0	41 38	62 46	71.1	41 19	61 30	70.3	40 58	60 15	69.5	40 37	59 01	68.7	225 / 315
46	43 10	64 56	72.3	42 51	63 38	71.4	42 32	62 21	70.5	42 11	61 05	69.6	41 50	59 49	68.8	41 28	58 34	68.0	226 / 314
47	44 04	64 32	71.7	43 45	63 13	70.8	43 25	61 55	69.9	43 04	60 38	69.0	42 42	59 21	68.1	42 19	58 06	67.3	227 / 313
48	44 58	64 06	71.1	44 38	62 46	70.1	44 18	61 27	69.2	43 56	60 09	68.3	43 33	58 53	67.4	43 10	57 37	66.5	228 / 312
49	45 52	63 39	70.4	45 32	62 18	69.5	45 10	60 59	68.5	44 48	59 40	67.6	44 24	58 22	66.7	44 00	57 06	65.8	229 / 311
50	46 46	63 11	69.8	46 25	61 49	68.8	46 03	60 29	67.8	45 39	59 09	66.9	45 15	57 51	65.9	44 50	56 34	65.0	230 / 310
51	47 39	62 42	69.1	47 17	61 19	68.1	46 55	59 57	67.1	46 31	58 37	66.1	46 06	57 18	65.2	45 40	56 00	64.2	231 / 309
52	48 33	62 11	68.4	48 10	60 47	67.4	47 46	59 25	66.4	47 22	58 03	65.4	46 56	56 44	64.4	46 30	55 25	63.4	232 / 308
53	49 25	61 38	67.7	49 02	60 13	66.6	48 38	58 50	65.6	48 13	57 28	64.6	47 46	56 07	63.6	47 19	54 48	62.6	233 / 307
54	50 18	61 04	67.0	49 54	59 38	65.9	49 29	58 14	64.8	49 03	56 51	63.7	48 36	55 30	62.7	48 08	54 10	61.7	234 / 306
55	51 10	60 28	66.2	50 46	59 01	65.1	50 20	57 36	64.0	49 53	56 12	62.9	49 25	54 50	61.9	48 56	53 30	60.8	235 / 305
56	52 03	59 50	65.4	51 37	58 23	64.2	51 10	56 56	63.1	50 43	55 32	62.0	50 14	54 09	61.0	49 44	52 48	59.9	236 / 304
57	52 54	59 11	64.6	52 28	57 42	63.4	52 00	56 15	62.2	51 32	54 49	61.1	51 02	53 26	60.0	50 32	52 04	59.0	237 / 303
58	53 46	58 29	63.7	53 18	56 59	62.5	52 50	55 31	61.3	52 21	54 05	60.2	51 50	52 41	59.1	51 19	51 18	58.0	238 / 302
59	54 37	57 45	62.8	54 08	56 14	61.5	53 39	54 45	60.4	53 09	53 18	59.2	52 38	51 53	58.0	52 06	50 30	57.0	239 / 301
60	55 27	57 00	61.8	54 58	55 27	60.6	54 28	53 57	59.4	53 57	52 29	58.2	53 25	51 04	57.0	52 52	49 40	55.9	240 / 300
61	56 17	56 10	60.9	55 47	54 37	59.6	55 16	53 06	58.3	54 44	51 38	57.1	54 11	50 12	55.9	53 37	48 48	54.8	241 / 299
62	57 07	55 19	59.8	56 36	53 45	58.5	56 04	52 13	57.2	55 31	50 44	56.0	54 57	49 17	54.8	54 22	47 53	53.7	242 / 298
63	57 56	54 25	58.8	57 24	52 49	57.4	56 51	51 17	56.1	56 17	49 47	54.9	55 42	48 20	53.7	55 06	46 55	52.5	243 / 297
64	58 44	53 27	57.6	58 12	51 51	56.3	57 38	50 18	55.0	57 03	48 48	53.7	56 27	47 20	52.5	55 50	45 55	51.3	244 / 296
65	59 32	52 27	56.5	58 58	50 50	55.1	58 23	49 16	53.7	57 47	47 45	52.5	57 10	46 17	51.3	56 32	44 52	50.0	245 / 295
66	60 19	51 23	55.2	59 45	49 45	53.8	59 08	48 11	52.5	58 32	46 39	51.2	57 53	45 11	50.0	57 14	43 47	48.7	246 / 294
67	61 06	50 15	53.9	60 30	48 37	52.5	59 53	47 02	51.1	59 15	45 30	49.8	58 36	44 02	48.7	57 55	42 38	47.4	247 / 293
68	61 52	49 04	52.6	61 15	47 25	51.1	60 36	45 50	49.8	59 57	44 18	48.4	59 17	42 50	47.2	58 36	41 26	46.0	248 / 292
69	62 37	47 48	51.2	61 58	46 09	49.7	61 19	44 33	48.3	60 39	43 02	47.0	59 57	41 34	45.7	59 15	40 10	44.5	249 / 291
70	63 21	46 28	49.7	62 41	44 48	48.2	62 01	43 13	46.8	61 19	41 42	45.4	60 36	40 15	44.2	59 53	38 52	43.0	250 / 290
71	64 04	45 03	48.1	63 23	43 23	46.6	62 41	41 49	45.2	61 58	40 18	43.9	61 15	38 52	42.6	60 30	37 29	41.4	251 / 289
72	64 45	43 34	46.4	64 04	41 54	44.9	63 21	40 20	43.5	62 37	38 50	42.3	61 52	37 25	40.9	61 06	36 03	39.7	252 / 288
73	65 26	41 59	44.7	64 43	40 20	43.2	63 59	38 46	41.8	63 14	37 18	40.5	62 27	35 53	39.2	61 41	34 34	38.0	253 / 287
74	66 06	40 19	42.9	65 21	38 41	41.4	64 36	37 08	40.0	63 49	35 41	38.7	63 02	34 18	37.4	62 14	33 00	36.3	254 / 286
75	66 44	38 32	40.9	65 58	36 56	39.5	65 11	35 25	38.1	64 23	33 59	36.8	63 35	32 39	35.6	62 46	31 22	34.4	255 / 285
76	67 20	36 40	38.9	66 33	35 05	37.4	65 45	33 37	36.1	64 56	32 13	34.8	64 07	30 55	33.6	63 16	29 41	32.5	256 / 284
77	67 55	34 42	36.8	67 07	33 09	35.3	66 18	31 43	34.0	65 27	30 22	32.8	64 37	29 06	31.6	63 45	27 55	30.6	257 / 283
78	68 29	32 37	34.5	67 39	31 07	33.1	66 48	29 44	31.9	65 57	28 26	30.7	65 05	27 14	29.6	64 13	26 06	28.5	258 / 282
79	69 00	30 25	32.2	68 09	29 00	30.8	67 17	27 40	29.6	66 25	26 26	28.5	65 32	25 17	27.4	64 38	24 12	26.4	259 / 281
80	69 29	28 07	29.7	68 37	26 46	28.4	67 44	25 30	27.3	66 50	24 20	26.2	65 56	23 15	25.2	65 02	22 15	24.3	260 / 280
81	69 57	25 43	27.1	69 03	24 26	25.9	68 09	23 15	24.8	67 14	22 10	23.8	66 19	21 10	22.9	65 23	20 14	22.1	261 / 279
82	70 21	23 11	24.5	69 27	22 00	23.3	68 31	20 56	22.3	67 36	19 56	21.4	66 40	19 00	20.6	65 43	18 09	19.8	262 / 278
83	70 44	20 34	21.7	69 48	19 29	20.7	68 51	18 31	19.7	67 55	17 37	18.9	66 58	16 47	18.1	66 01	16 01	17.4	263 / 277
84	71 03	17 50	18.8	70 07	16 53	17.9	69 09	16 01	17.1	68 12	15 14	16.3	67 14	14 30	15.7	66 16	13 50	15.1	264 / 276
85	71 20	15 01	15.8	70 23	14 12	15.0	69 25	13 28	14.3	68 26	12 48	13.7	67 28	12 10	13.1	66 29	11 36	12.6	265 / 275
86	71 35	12 07	12.8	70 36	11 27	12.1	69 37	10 51	11.6	68 38	10 18	11.0	67 39	09 48	10.6	66 40	09 20	10.1	266 / 274
87	71 46	09 09	9.6	70 46	08 39	9.1	69 47	08 11	8.7	68 48	07 46	8.3	67 48	07 23	8.0	66 49	07 02	7.6	267 / 273
88	71 54	06 08	6.4	70 54	05 47	6.1	69 54	05 29	5.8	68 55	05 12	5.6	67 55	04 56	5.3	66 55	04 42	5.1	268 / 272
89	71 58	03 04	3.2	70 58	02 54	3.1	69 59	02 45	2.9	68 59	02 36	2.8	67 59	02 28	2.6	66 59	02 21	2.6	269 / 271
90	72 00	00 00	0.0	71 00	00 00	0.0	70 00	00 00	0.0	69 00	00 00	0.0	68 00	00 00	0.0	67 00	00 00	0.0	270 / 270

N. Lat.: for LHA > 180°.... Zn = Z
for LHA < 180°.... Zn = 360° − Z

S. Lat.: for LHA > 180°.... Zn = 180° − Z
for LHA < 180°.... Zn = 180° + Z

B: (−) for 90° < LHA < 270°
Dec: (−) for Lat. contrary name

Z_1: same sign as B
Z_2: (−) for F > 90°

SIGHT REDUCTION TABLE

Lat./A LHA/F	24° A/H	24° B/P	24° Z_1/Z_2	25° A/H	25° B/P	25° Z_1/Z_2	26° A/H	26° B/P	26° Z_1/Z_2	27° A/H	27° B/P	27° Z_1/Z_2	28° A/H	28° B/P	28° Z_1/Z_2	29° A/H	29° B/P	29° Z_1/Z_2	LHA	Lat./A LHA
0 / 180	0 00	66 00	90.0	0 00	65 00	90.0	0 00	64 00	90.0	0 00	63 00	90.0	0 00	62 00	90.0	0 00	61 00	90.0	180	360
1 / 179	0 55	66 00	89.6	0 54	65 00	89.6	0 54	64 00	89.6	0 53	63 00	89.5	0 53	62 00	89.5	0 52	61 00	89.5	181	359
2 / 178	1 50	65 59	89.2	1 49	64 59	89.2	1 48	63 59	89.1	1 47	62 59	89.1	1 46	61 59	89.1	1 45	60 59	89.0	182	358
3 / 177	2 44	65 58	88.8	2 43	64 58	88.7	2 42	63 58	88.7	2 40	62 58	88.6	2 39	61 58	88.6	2 37	60 58	88.5	183	357
4 / 176	3 39	65 57	88.4	3 37	64 57	88.3	3 36	63 57	88.2	3 34	62 57	88.2	3 32	61 57	88.1	3 30	60 56	88.1	184	356
5 / 175	4 34	65 55	88.0	4 32	64 55	87.9	4 30	63 55	87.8	4 27	62 55	87.7	4 25	61 55	87.6	4 22	60 54	87.6	185	355
6 / 174	5 29	65 53	87.5	5 26	64 53	87.5	5 23	63 53	87.4	5 21	62 52	87.3	5 18	61 52	87.2	5 15	60 52	87.1	186	354
7 / 173	6 24	65 50	87.1	6 20	64 50	87.0	6 17	63 50	86.9	6 14	62 50	86.8	6 11	61 49	86.7	6 07	60 49	86.6	187	353
8 / 172	7 18	65 47	86.7	7 15	64 47	86.6	7 11	63 47	86.5	7 07	62 46	86.3	7 04	61 46	86.2	6 59	60 46	86.1	188	352
9 / 171	8 13	65 44	86.3	8 09	64 44	86.2	8 05	63 43	86.0	8 00	62 43	85.9	7 56	61 42	85.7	7 52	60 42	85.6	189	351
10 / 170	9 08	65 40	85.9	9 03	64 40	85.7	8 59	63 39	85.6	8 54	62 39	85.4	8 49	61 38	85.3	8 44	60 38	85.1	190	350
11 / 169	10 02	65 36	85.5	9 57	64 35	85.3	9 52	63 35	85.1	9 47	62 34	85.0	9 42	61 33	84.8	9 36	60 33	84.6	191	349
12 / 168	10 57	65 32	85.1	10 52	64 31	84.9	10 46	63 30	84.7	10 41	62 29	84.5	10 35	61 28	84.3	10 29	60 28	84.1	192	348
13 / 167	11 52	65 27	84.6	11 46	64 26	84.4	11 40	63 25	84.2	11 34	62 24	84.0	11 27	61 23	83.8	11 21	60 22	83.6	193	347
14 / 166	12 46	65 21	84.2	12 40	64 20	84.0	12 34	63 20	83.8	12 27	62 18	83.5	12 20	61 17	83.3	12 13	60 16	83.1	194	346
15 / 165	13 41	65 15	83.8	13 34	64 14	83.5	13 27	63 13	83.3	13 20	62 11	83.1	13 13	61 10	82.8	13 05	60 09	82.6	195	345
16 / 164	14 35	65 09	83.3	14 28	64 07	83.1	14 21	63 06	82.8	14 13	62 04	82.6	14 05	61 03	82.3	13 57	60 02	82.1	196	344
17 / 163	15 29	65 02	82.9	15 22	64 00	82.6	15 14	62 59	82.4	15 06	61 57	82.1	14 58	60 56	81.8	14 49	59 54	81.6	197	343
18 / 162	16 24	64 55	82.5	16 16	63 53	82.2	16 08	62 51	81.9	15 59	61 49	81.6	15 50	60 47	81.3	15 41	59 46	81.0	198	342
19 / 161	17 18	64 47	82.0	17 10	63 45	81.7	17 01	62 43	81.4	16 52	61 41	81.1	16 42	60 39	80.8	16 33	59 37	80.5	199	341
20 / 160	18 12	64 39	81.6	18 03	63 36	81.3	17 54	62 34	80.9	17 45	61 32	80.6	17 35	60 30	80.3	17 24	59 28	80.0	200	340
21 / 159	19 07	64 30	81.1	18 57	63 28	80.8	18 47	62 25	80.4	18 37	61 23	80.1	18 27	60 20	79.8	18 16	59 18	79.5	201	339
22 / 158	20 01	64 21	80.7	19 51	63 18	80.3	19 41	62 15	80.0	19 30	61 13	79.6	19 19	60 10	79.3	19 08	59 08	78.9	202	338
23 / 157	20 55	64 11	80.2	20 44	63 08	79.8	20 34	62 05	79.5	20 22	61 02	79.1	20 11	59 59	78.7	19 59	58 57	78.4	203	337
24 / 156	21 49	64 01	79.7	21 38	62 58	79.3	21 27	61 54	79.0	21 15	60 51	78.6	21 03	59 48	78.2	20 50	58 45	77.8	204	336
25 / 155	22 43	63 50	79.3	22 31	62 46	78.9	22 19	61 43	78.4	22 07	60 39	78.0	21 55	59 36	77.7	21 42	58 33	77.3	205	335
26 / 154	23 36	63 39	78.8	23 25	62 35	78.4	23 12	61 31	77.9	22 59	60 27	77.5	22 46	59 24	77.1	22 33	58 20	76.7	206	334
27 / 153	24 30	63 27	78.3	24 18	62 22	77.8	24 05	61 18	77.4	23 52	60 14	77.0	23 38	59 10	76.5	23 24	58 07	76.1	207	333
28 / 152	25 24	63 14	77.8	25 11	62 10	77.3	24 57	61 05	76.9	24 44	60 01	76.4	24 29	58 57	76.0	24 15	57 53	75.5	208	332
29 / 151	26 17	63 01	77.3	26 04	61 56	76.8	25 50	60 51	76.3	25 36	59 47	75.9	25 21	58 42	75.4	25 05	57 38	75.0	209	331
30 / 150	27 11	62 48	76.8	26 57	61 42	76.3	26 42	60 37	75.8	26 27	59 32	75.3	26 12	58 27	74.8	25 56	57 23	74.4	210	330
31 / 149	28 04	62 33	76.3	27 50	61 27	75.8	27 35	60 22	75.2	27 19	59 16	74.7	27 03	58 11	74.2	26 46	57 07	73.8	211	329
32 / 148	28 57	62 18	75.7	28 42	61 12	75.2	28 27	60 06	74.7	28 10	59 00	74.2	27 54	57 55	73.6	27 37	56 50	73.1	212	328
33 / 147	29 50	62 02	75.2	29 35	60 56	74.7	29 19	59 49	74.1	29 02	58 43	73.6	28 45	57 38	73.0	28 27	56 32	72.5	213	327
34 / 146	30 43	61 46	74.7	30 27	60 39	74.1	30 10	59 32	73.5	29 53	58 26	73.0	29 35	57 20	72.4	29 17	56 14	71.9	214	326
35 / 145	31 36	61 28	74.1	31 19	60 21	73.5	31 02	59 14	72.9	30 44	58 07	72.4	30 26	57 01	71.8	30 07	55 55	71.2	215	325
36 / 144	32 29	61 10	73.5	32 11	60 02	72.9	31 53	58 55	72.3	31 35	57 48	71.7	31 16	56 41	71.2	30 56	55 35	70.6	216	324
37 / 143	33 21	60 52	73.0	33 03	59 43	72.3	32 45	58 35	71.7	32 26	57 28	71.1	32 06	56 21	70.5	31 46	55 14	69.9	217	323
38 / 142	34 13	60 32	72.4	33 55	59 23	71.7	33 36	58 15	71.1	33 16	57 07	70.5	32 56	56 00	69.9	32 35	54 53	69.3	218	322
39 / 141	35 06	60 11	71.8	34 47	59 02	71.1	34 27	57 53	70.5	34 06	56 45	69.8	33 45	55 37	69.2	33 24	54 30	68.6	219	321
40 / 140	35 58	59 50	71.2	35 38	58 40	70.5	35 17	57 31	69.8	34 56	56 22	69.1	34 35	55 14	68.5	34 12	54 07	67.9	220	320
41 / 139	36 49	59 28	70.5	36 29	58 17	69.8	36 08	57 08	69.1	35 46	55 59	68.5	35 24	54 50	67.8	35 01	53 42	67.1	221	319
42 / 138	37 41	59 04	69.9	37 20	57 54	69.2	36 58	56 43	68.5	36 36	55 34	67.8	36 13	54 25	67.1	35 49	53 17	66.4	222	318
43 / 137	38 32	58 40	69.2	38 11	57 29	68.5	37 48	56 18	67.8	37 25	55 08	67.1	37 02	53 59	66.4	36 37	52 50	65.7	223	317
44 / 136	39 23	58 15	68.6	39 01	57 03	67.8	38 38	55 52	67.1	38 14	54 41	66.3	37 50	53 32	65.6	37 25	52 23	64.9	224	316
45 / 135	40 14	57 48	67.9	39 51	56 36	67.1	39 28	55 24	66.3	39 03	54 13	65.6	38 38	53 04	64.9	38 12	51 54	64.1	225	315

LHA	F	A/H 24°	B/P 24°	Z₁/Z₂ 24°	A/H 25°	B/P 25°	Z₁/Z₂ 25°	A/H 26°	B/P 26°	Z₁/Z₂ 26°	A/H 27°	B/P 27°	Z₁/Z₂ 27°	A/H 28°	B/P 28°	Z₁/Z₂ 28°	A/H 29°	B/P 29°	Z₁/Z₂ 29°	LHA	A
45	135	40 14	57 48	67.9	39 51	56 36	67.1	39 28	55 24	66.3	39 03	54 13	65.6	38 38	53 04	64.9	38 12	51 54	64.1	225	315
46	134	41 05	57 21	67.2	40 41	56 08	66.6	40 17	54 56	65.6	39 52	53 44	65.0	39 26	52 34	64.1	38 59	51 25	63.3	226	314
47	133	41 55	56 52	66.4	41 31	55 38	65.6	41 06	54 26	64.8	40 40	53 14	64.0	40 13	52 04	63.3	39 46	50 54	62.5	227	313
48	132	42 45	56 22	65.7	42 20	55 08	64.9	41 54	53 55	64.0	41 28	52 43	63.2	41 00	51 32	62.5	40 32	50 22	61.7	228	312
49	131	43 35	55 50	64.9	43 09	54 36	64.1	42 43	53 22	63.2	42 15	52 10	62.4	41 47	50 59	61.6	41 18	49 48	60.9	229	311
50	130	44 25	55 17	64.1	43 58	54 02	63.3	43 31	52 49	62.4	43 03	51 36	61.6	42 34	50 24	60.8	42 04	49 14	60.0	230	310
51	129	45 14	54 43	63.3	44 47	53 28	62.4	44 18	52 13	61.6	43 49	51 00	60.7	43 20	49 48	59.9	42 49	48 38	59.1	231	309
52	128	46 03	54 08	62.5	45 35	52 52	61.6	45 06	51 37	60.7	44 36	50 23	59.8	44 05	49 11	59.0	43 34	48 00	58.2	232	308
53	127	46 51	53 30	61.6	46 22	52 14	60.7	45 52	50 59	59.8	45 22	49 45	58.9	44 51	48 32	58.1	44 18	47 21	57.2	233	307
54	126	47 39	52 51	60.8	47 09	51 34	59.8	46 39	50 19	58.9	46 07	49 05	58.0	45 35	47 52	57.1	45 02	46 41	56.3	234	306
55	125	48 27	52 11	59.8	47 56	50 53	58.9	47 25	49 37	58.0	46 53	48 23	57.0	46 19	47 10	56.2	45 46	45 59	55.3	235	305
56	124	49 14	51 28	58.9	48 43	50 11	57.9	48 10	48 54	57.0	47 37	47 40	56.1	47 03	46 27	55.2	46 29	45 15	54.3	236	304
57	123	50 01	50 44	57.9	49 28	49 26	56.9	48 55	48 09	56.0	48 21	46 54	55.0	47 46	45 41	54.1	47 11	44 30	53.3	237	303
58	122	50 47	49 58	56.9	50 14	48 39	55.9	49 40	47 22	54.9	49 05	46 07	54.0	48 29	44 54	53.1	47 53	43 43	52.2	238	302
59	121	51 33	49 09	55.8	50 58	47 51	54.9	50 23	46 34	53.9	49 48	45 18	52.9	49 11	44 05	52.0	48 34	42 54	51.1	239	301
60	120	52 18	48 19	54.8	51 43	47 00	53.8	51 07	45 43	52.8	50 30	44 28	51.8	49 53	43 14	50.9	49 14	42 03	50.0	240	300
61	119	53 02	47 26	53.7	52 26	46 07	52.7	51 49	44 50	51.7	51 12	43 35	50.7	50 33	42 22	49.7	49 54	41 10	48.8	241	299
62	118	53 46	46 31	52.6	53 09	45 12	51.5	52 31	43 54	50.5	51 53	42 39	49.5	51 13	41 27	48.6	50 33	40 16	47.6	242	298
63	117	54 29	45 33	51.4	53 51	44 14	50.3	53 13	42 57	49.3	52 33	41 42	48.3	51 53	40 30	47.3	51 12	39 19	46.4	243	297
64	116	55 11	44 33	50.2	54 33	43 14	49.1	53 53	41 57	48.1	53 13	40 42	47.1	52 31	39 30	46.1	51 49	38 20	45.2	244	296
65	115	55 53	43 30	48.9	55 13	42 11	47.8	54 33	40 55	46.8	53 51	39 40	45.8	53 09	38 29	44.8	52 26	37 19	43.9	245	295
66	114	56 34	42 25	47.6	55 53	41 06	46.5	55 12	39 50	45.4	54 29	38 36	44.4	53 46	37 25	43.5	53 02	36 16	42.6	246	294
67	113	57 14	41 16	46.2	56 32	39 58	45.1	55 50	38 42	44.1	55 06	37 29	43.1	54 22	36 19	42.1	53 37	35 11	41.2	247	293
68	112	57 53	40 05	44.8	57 10	38 47	43.7	56 27	37 32	42.7	55 42	36 19	41.7	54 57	35 10	40.7	54 11	34 03	39.8	248	292
69	111	58 32	38 50	43.3	57 47	37 33	42.2	57 03	36 18	41.2	56 17	35 07	40.2	55 31	33 59	39.3	54 44	32 53	38.4	249	291
70	110	59 09	37 32	41.8	58 23	36 16	40.7	57 38	35 02	39.7	56 51	33 52	38.7	56 04	32 45	37.8	55 16	31 41	36.9	250	290
71	109	59 45	36 11	40.2	58 58	34 55	39.2	58 12	33 43	38.1	57 24	32 35	37.2	56 36	31 29	36.3	55 47	30 26	35.4	251	289
72	108	60 19	34 46	38.6	59 32	33 32	37.6	58 44	32 21	36.5	57 56	31 14	35.6	57 07	30 10	34.7	56 17	29 08	33.8	252	288
73	107	60 53	33 18	36.9	60 05	32 05	35.9	59 16	30 56	34.9	58 26	29 51	34.0	57 36	28 48	33.1	56 46	27 49	32.2	253	287
74	106	61 25	31 46	35.2	60 36	30 35	34.2	59 46	29 28	33.2	58 55	28 25	32.3	58 05	27 24	31.4	57 13	26 26	30.6	254	286
75	105	61 56	30 10	33.4	61 06	29 02	32.4	60 15	27 57	31.4	59 23	26 56	30.5	58 31	25 57	29.7	57 39	25 02	28.9	255	285
76	104	62 26	28 31	31.5	61 34	27 25	30.5	60 42	26 23	29.6	59 50	25 24	28.8	58 57	24 28	28.0	58 04	23 35	27.2	256	284
77	103	62 53	26 48	29.6	62 01	25 45	28.6	61 08	24 46	27.8	60 15	23 49	27.0	59 21	22 56	26.2	58 27	22 05	25.5	257	283
78	102	63 20	25 02	27.6	62 26	24 02	26.7	61 32	23 05	25.9	60 38	22 12	25.1	59 44	21 21	24.4	58 49	20 34	23.7	258	282
79	101	63 44	23 12	25.5	62 50	22 15	24.7	61 55	21 20	23.9	61 00	20 32	23.2	60 05	19 44	22.5	59 09	19 00	21.8	259	281
80	100	64 07	21 18	23.4	63 12	20 25	22.6	62 16	19 36	21.9	61 20	18 49	21.2	60 24	18 05	20.6	59 28	17 24	20.0	260	280
81	99	64 28	19 22	21.3	63 32	18 33	20.5	62 35	17 47	19.8	61 39	17 04	19.2	60 42	16 24	18.6	59 45	15 46	18.1	261	279
82	98	64 47	17 22	19.1	63 50	16 37	18.4	62 53	15 56	17.8	61 56	15 17	17.2	60 58	14 40	16.7	60 01	14 06	16.2	262	278
83	97	65 03	15 18	16.8	64 06	14 39	16.2	63 08	14 02	15.6	62 10	13 27	15.1	61 12	12 55	14.7	60 14	12 24	14.2	263	277
84	96	65 18	13 13	14.5	64 20	12 38	14.0	63 22	12 06	13.5	62 23	11 36	13.0	61 25	11 07	12.6	60 26	10 41	12.2	264	276
85	95	65 31	11 05	12.1	64 32	10 35	11.7	63 33	10 08	11.3	62 35	9 42	10.9	61 36	9 19	10.6	60 37	8 56	10.2	265	275
86	94	65 41	8 54	9.8	64 42	8 30	9.4	63 43	8 08	9.1	62 44	7 48	8.8	61 44	7 28	8.5	60 45	7 10	8.2	266	274
87	93	65 49	6 42	7.3	64 50	6 24	7.1	63 50	6 07	6.8	62 51	5 52	6.6	61 51	5 37	6.4	60 52	5 24	6.2	267	273
88	92	65 55	4 29	4.9	64 56	4 17	4.7	63 56	4 06	4.6	62 56	3 55	4.4	61 56	3 45	4.3	60 56	3 36	4.1	268	272
89	91	65 59	2 15	2.5	64 59	2 09	2.4	63 59	2 03	2.3	62 59	1 58	2.2	61 59	1 53	2.1	60 59	1 48	2.1	269	271
90	90	66 00	0 00	0.0	65 00	0 00	0.0	64 00	0 00	0.0	63 00	0 00	0.0	62 00	0 00	0.0	61 00	0 00	0.0	270	270

N. Lat.: for LHA > 180° .... Zn = Z
         for LHA < 180° .... Zn = 360° − Z

S. Lat.: for LHA > 180° ... Zn = 180° − Z
         for LHA < 180° ... Zn = 180° + Z

SIGHT REDUCTION TABLE

B: (−) for 90° < LHA < 270°
Dec: (−) for Lat. contrary name

Z₁: same sign as B
Z₂: (−) for F > 90°

Lat./A LHA/F		30° A/H	30° B/P	30° Z₁/Z₂	31° A/H	31° B/P	31° Z₁/Z₂	32° A/H	32° B/P	32° Z₁/Z₂	33° A/H	33° B/P	33° Z₁/Z₂	34° A/H	34° B/P	34° Z₁/Z₂	35° A/H	35° B/P	35° Z₁/Z₂	Lat./A LHA	
0	180	0 00	60 00	90.0	0 00	59 00	90.0	0 00	58 00	90.0	0 00	57 00	90.0	0 00	56 00	90.0	0 00	55 00	90.0	180	360
1	179	0 52	60 00	89.5	0 51	59 00	89.5	0 51	58 00	89.5	0 50	57 00	89.5	0 50	56 00	89.4	0 49	55 00	89.4	181	359
2	178	1 44	59 59	89.0	1 43	58 59	89.0	1 42	57 59	89.0	1 41	56 59	88.9	1 39	55 59	88.9	1 38	54 59	88.9	182	358
3	177	2 36	59 58	88.5	2 34	58 58	88.5	2 33	57 59	88.4	2 31	56 58	88.4	2 29	55 58	88.3	2 27	54 58	88.3	183	357
4	176	3 28	59 56	88.0	3 26	58 56	87.9	3 23	57 56	87.9	3 21	56 56	87.8	3 19	55 56	87.8	3 17	54 56	87.7	184	356
5	175	4 20	59 54	87.5	4 17	58 54	87.4	4 14	57 54	87.3	4 12	56 54	87.3	4 09	55 54	87.2	4 06	54 54	87.1	185	355
6	174	5 12	59 52	87.0	5 08	58 52	86.9	5 05	57 52	86.8	5 02	56 51	86.7	4 58	55 51	86.6	4 55	54 51	86.6	186	354
7	173	6 04	59 49	86.5	6 00	58 49	86.4	5 56	57 48	86.4	5 52	56 48	86.2	5 48	55 48	86.1	5 44	54 48	86.0	187	353
8	172	6 55	59 45	86.0	6 51	58 45	85.9	6 47	57 45	85.7	6 42	56 45	85.6	6 38	55 44	85.5	6 33	54 44	85.4	188	352
9	171	7 47	59 42	85.5	7 42	58 41	85.3	7 37	57 41	85.2	7 32	56 40	85.1	7 27	55 40	84.9	7 22	54 40	84.8	189	351
10	170	8 39	59 37	85.0	8 34	58 37	84.8	8 28	57 36	84.7	8 22	56 36	84.5	8 17	55 36	84.4	8 11	54 35	84.2	190	350
11	169	9 31	59 32	84.4	9 25	58 32	84.3	9 19	57 31	84.1	9 13	56 31	84.0	9 06	55 30	83.8	9 00	54 30	83.6	191	349
12	168	10 22	59 27	83.9	10 16	58 26	83.8	10 09	57 26	83.6	10 03	56 25	83.4	9 56	55 25	83.2	9 48	54 24	83.0	192	348
13	167	11 14	59 21	83.4	11 07	58 20	83.2	11 00	57 20	83.2	10 52	56 19	82.8	10 45	55 18	82.6	10 37	54 18	82.5	193	347
14	166	12 06	59 15	82.9	11 58	58 14	82.7	11 50	57 13	82.5	11 42	56 12	82.3	11 34	55 12	82.1	11 26	54 11	81.9	194	346
15	165	12 57	59 08	82.4	12 49	58 07	82.1	12 41	57 06	81.9	12 32	56 05	81.7	12 23	55 04	81.5	12 14	54 04	81.3	195	345
16	164	13 49	59 01	81.8	13 40	57 59	81.6	13 31	56 58	81.4	13 22	55 57	81.1	13 13	54 57	80.9	13 03	53 56	80.7	196	344
17	163	14 40	58 53	81.3	14 31	57 51	81.1	14 21	56 50	80.8	14 12	55 49	80.5	14 02	54 48	80.3	13 51	53 47	80.1	197	343
18	162	15 31	58 44	80.8	15 22	57 43	80.5	15 12	56 42	80.2	15 01	55 40	80.0	14 51	54 39	79.7	14 40	53 38	79.4	198	342
19	161	16 23	58 35	80.2	16 12	57 34	79.9	16 02	56 32	79.7	15 51	55 31	79.4	15 40	54 30	79.1	15 28	53 29	78.8	199	341
20	160	17 14	58 26	79.7	17 03	57 24	79.4	16 52	56 23	79.1	16 40	55 21	78.8	16 28	54 20	78.5	16 16	53 19	78.2	200	340
21	159	18 05	58 16	79.1	17 53	57 14	78.8	17 42	56 12	78.5	17 29	55 11	78.2	17 17	54 09	77.9	17 04	53 08	77.6	201	339
22	158	18 56	58 05	78.6	18 44	57 03	78.2	18 31	56 01	77.9	18 19	55 00	77.6	18 06	53 58	77.3	17 52	52 56	77.0	202	338
23	157	19 47	57 54	78.0	19 34	56 52	77.7	19 21	55 50	77.3	19 08	54 48	77.0	18 54	53 46	76.6	18 40	52 44	76.3	203	337
24	156	20 37	57 42	77.4	20 24	56 40	77.1	20 11	55 38	76.7	19 57	54 36	76.4	19 42	53 34	76.0	19 28	52 32	75.7	204	336
25	155	21 28	57 30	76.9	21 14	56 27	76.5	21 00	55 25	76.1	20 46	54 23	75.7	20 31	53 21	75.4	20 15	52 19	75.0	205	335
26	154	22 19	57 17	76.3	22 04	56 14	75.9	21 49	55 12	75.5	21 34	54 09	75.1	21 19	53 07	74.7	21 03	52 05	74.4	206	334
27	153	23 09	57 03	75.7	22 54	56 00	75.3	22 39	54 57	74.9	22 23	53 55	74.5	22 07	52 52	74.1	21 50	51 50	73.7	207	333
28	152	23 59	56 49	75.1	23 44	55 46	74.7	23 28	54 43	74.3	23 11	53 40	73.8	22 54	52 37	73.4	22 37	51 35	73.0	208	332
29	151	24 50	56 34	74.5	24 33	55 31	74.1	24 17	54 27	73.6	23 59	53 24	73.2	23 42	52 22	72.8	23 24	51 19	72.4	209	331
30	150	25 40	56 19	73.9	25 23	55 15	73.4	25 05	54 11	73.0	24 48	53 08	72.5	24 29	52 05	72.1	24 11	51 03	71.7	210	330
31	149	26 29	56 02	73.3	26 12	54 58	72.8	25 54	53 54	72.3	25 35	52 51	71.9	25 17	51 48	71.4	24 57	50 45	71.0	211	329
32	148	27 19	55 45	72.6	27 01	54 41	72.2	26 42	53 37	71.7	26 23	52 33	71.2	26 04	51 30	70.7	25 44	50 27	70.3	212	328
33	147	28 09	55 27	72.0	27 50	54 23	71.5	27 31	53 19	71.0	27 11	52 15	70.5	26 50	51 12	70.0	26 30	50 08	69.6	213	327
34	146	28 58	55 09	71.4	28 38	54 04	70.8	28 19	53 00	70.3	27 58	51 56	69.8	27 37	50 52	69.3	27 16	49 49	68.8	214	326
35	145	29 47	54 49	70.7	29 27	53 44	70.2	29 06	52 40	69.6	28 45	51 36	69.1	28 24	50 32	68.6	28 01	49 29	68.1	215	325
36	144	30 36	54 29	70.0	30 15	53 24	69.5	29 54	52 19	68.9	29 32	51 15	68.4	29 10	50 11	67.9	28 47	49 07	67.4	216	324
37	143	31 25	54 08	69.4	31 03	53 03	68.8	30 41	51 58	68.2	30 19	50 53	67.7	29 56	49 49	67.2	29 32	48 45	66.6	217	323
38	142	32 13	53 46	68.7	31 51	52 40	68.1	31 28	51 35	67.5	31 05	50 30	66.9	30 41	49 26	66.4	30 17	48 23	65.9	218	322
39	141	33 02	53 23	68.0	32 39	52 17	67.4	32 15	51 12	66.8	31 51	50 07	66.2	31 27	49 03	65.6	31 02	47 59	65.1	219	321
40	140	33 50	53 00	67.2	33 26	51 53	66.6	33 02	50 48	66.0	32 37	49 43	65.4	32 12	48 38	64.9	31 46	47 34	64.3	220	320
41	139	34 37	52 35	66.5	34 13	51 29	65.9	33 48	50 23	65.3	33 23	49 17	64.7	32 57	48 13	64.1	32 30	47 09	63.5	221	319
42	138	35 25	52 09	65.8	35 00	51 03	65.1	34 34	49 56	64.5	34 08	48 51	63.9	33 42	47 46	63.3	33 14	46 42	62.7	222	318
43	137	36 12	51 43	65.0	35 46	50 36	64.3	35 20	49 29	63.7	34 53	48 24	63.1	34 26	47 19	62.5	33 58	46 15	61.9	223	317
44	136	36 59	51 15	64.2	36 33	50 08	63.6	36 06	49 01	62.9	35 38	47 55	62.3	35 10	46 51	61.6	34 41	45 46	61.0	224	316
45	135	37 46	50 46	63.4	37 19	49 39	62.7	36 51	48 32	62.1	36 22	47 26	61.4	35 53	46 21	60.8	35 24	45 17	60.2	225	315

Lat./A LHA/F		30° A/H	30° B/P	30° Z1/Z2	31° A/H	31° B/P	31° Z1/Z2	32° A/H	32° B/P	32° Z1/Z2	33° A/H	33° B/P	33° Z1/Z2	34° A/H	34° B/P	34° Z1/Z2	35° A/H	35° B/P	35° Z1/Z2	Lat./A LHA	
45	135	37 46	50 46	63.4	37 19	49 39	62.7	36 51	48 32	62.1	36 22	47 26	61.4	35 53	46 21	60.8	35 24	45 17	60.2	225	315
46	134	38 32	50 16	62.6	38 04	49 08	61.9	37 36	48 02	61.2	37 06	46 56	60.6	36 37	45 51	59.9	36 06	44 46	59.3	226	314
47	133	39 18	49 45	61.8	38 49	48 37	61.1	38 20	47 30	60.4	37 50	46 24	59.7	37 19	45 19	59.1	36 48	44 15	58.4	227	313
48	132	40 04	49 13	61.0	39 34	48 05	60.2	39 04	46 58	59.5	38 33	45 51	58.8	38 02	44 46	58.2	37 30	43 42	57.5	228	312
49	131	40 49	48 39	60.1	40 19	47 31	59.4	39 48	46 24	58.6	39 16	45 18	57.9	38 44	44 12	57.2	38 11	43 08	56.6	229	311
50	130	41 34	48 04	59.2	41 03	46 56	58.5	40 31	45 49	57.7	39 59	44 42	57.0	39 26	43 37	56.3	38 52	42 33	55.6	230	310
51	129	42 18	47 28	58.3	41 46	46 20	57.5	41 14	45 12	56.8	40 41	44 06	56.1	40 07	43 01	55.4	39 32	41 57	54.7	231	309
52	128	43 02	46 50	57.4	42 29	45 42	56.6	41 56	44 34	55.9	41 22	43 28	55.1	40 47	42 23	54.4	40 12	41 19	53.7	232	308
53	127	43 46	46 11	56.4	43 12	45 03	55.6	42 38	43 55	54.9	42 03	42 49	54.1	41 28	41 44	53.4	40 52	40 41	52.7	233	307
54	126	44 29	45 31	55.5	43 54	44 22	54.7	43 19	43 15	53.9	42 44	42 09	53.1	42 07	41 04	52.4	41 30	40 01	51.7	234	306
55	125	45 11	44 49	54.5	44 36	43 40	53.7	44 00	42 33	52.9	43 24	41 27	52.1	42 46	40 23	51.4	42 09	39 19	50.7	235	305
56	124	45 53	44 05	53.5	45 17	42 57	52.6	44 40	41 50	51.8	44 03	40 44	51.1	43 25	39 40	50.3	42 46	38 37	49.6	236	304
57	123	46 35	43 20	52.4	45 58	42 11	51.6	45 20	41 05	50.8	44 42	39 59	50.0	44 03	38 55	49.3	43 24	37 53	48.5	237	303
58	122	47 16	42 33	51.3	46 38	41 25	50.5	45 59	40 18	49.7	45 20	39 13	48.9	44 40	38 09	48.2	44 00	37 07	47.5	238	302
59	121	47 56	41 44	50.2	47 17	40 36	49.4	46 38	39 30	48.6	45 58	38 25	47.8	45 17	37 22	47.1	44 36	36 20	46.3	239	301
60	120	48 35	40 54	49.1	47 56	39 46	48.3	47 16	38 40	47.4	46 35	37 36	46.7	45 53	36 33	45.9	45 11	35 32	45.2	240	300
61	119	49 14	40 01	47.9	48 34	38 54	47.1	47 53	37 48	46.3	47 11	36 45	45.5	46 29	35 42	44.7	45 46	34 42	44.0	241	299
62	118	49 53	39 07	46.8	49 11	38 00	45.9	48 29	36 55	45.1	47 46	35 52	44.3	47 03	34 50	43.6	46 19	33 50	42.8	242	298
63	117	50 30	38 11	45.5	49 48	37 04	44.7	49 05	36 00	43.9	48 21	34 57	43.1	47 37	33 57	42.3	46 53	32 57	41.6	243	297
64	116	51 07	37 13	44.3	50 23	36 07	43.4	49 40	35 03	42.6	48 55	34 01	41.8	48 10	33 01	41.1	47 25	32 07	40.4	244	296
65	115	51 43	36 12	43.0	50 58	35 07	42.2	50 14	34 04	41.3	49 28	33 03	40.6	48 43	32 04	39.8	47 56	31 07	39.1	245	295
66	114	52 18	35 10	41.7	51 33	34 06	40.8	50 47	33 04	40.0	50 01	32 04	39.3	49 14	31 05	38.5	48 27	30 09	37.8	246	294
67	113	52 52	34 05	40.3	52 06	33 02	39.5	51 19	32 01	38.7	50 32	31 02	37.9	49 44	30 05	37.2	48 56	29 10	36.5	247	293
68	112	53 25	32 59	38.9	52 38	31 56	38.1	51 50	30 57	37.3	51 02	29 59	36.6	50 14	29 03	35.8	49 25	28 09	35.2	248	292
69	111	53 57	31 50	37.5	53 09	30 49	36.7	52 21	29 50	35.9	51 32	28 53	35.2	50 43	27 59	34.5	49 53	27 06	33.8	249	291
70	110	54 28	30 39	36.1	53 40	29 39	35.2	52 50	28 42	34.5	52 00	27 46	33.8	51 10	26 53	33.1	50 20	26 00	32.4	250	290
71	109	54 58	29 25	34.6	54 08	28 27	33.8	53 18	27 31	33.0	52 28	26 38	32.3	51 37	25 46	31.6	50 46	24 56	31.0	251	289
72	108	55 27	28 09	33.0	54 37	27 13	32.2	53 46	26 19	31.5	52 54	25 27	30.8	52 03	24 37	30.2	51 10	23 49	29.5	252	288
73	107	55 55	26 51	31.4	55 03	25 57	30.7	54 12	25 04	30.0	53 19	24 14	29.3	52 27	23 26	28.7	51 34	22 40	28.1	253	287
74	106	56 21	25 31	29.8	55 29	24 39	29.1	54 36	23 48	28.4	53 43	23 00	27.8	52 50	22 14	27.1	51 57	21 29	26.6	254	286
75	105	56 46	24 09	28.2	55 53	23 18	27.5	55 00	22 30	26.8	54 06	21 44	26.2	53 12	21 00	25.6	52 18	20 17	25.0	255	285
76	104	57 10	22 44	26.5	56 16	21 56	25.8	55 22	21 10	25.2	54 28	20 26	24.6	53 33	19 44	24.0	52 38	19 04	23.5	256	284
77	103	57 33	21 17	24.8	56 38	20 31	24.1	55 43	19 48	23.5	54 48	19 06	23.0	53 53	18 27	22.4	52 57	17 49	21.9	257	283
78	102	57 54	19 48	23.0	56 59	19 05	22.4	56 03	18 24	21.9	55 07	17 45	21.3	54 11	17 08	20.8	53 15	16 32	20.3	258	282
79	101	58 13	18 17	21.2	57 17	17 37	20.7	56 21	16 59	20.1	55 25	16 22	19.6	54 28	15 48	19.2	53 31	15 15	18.7	259	281
80	100	58 32	16 44	19.4	57 35	16 07	18.9	56 38	15 32	18.4	55 41	14 58	17.9	54 44	14 26	17.5	53 47	13 56	17.1	260	280
81	99	58 48	15 10	17.6	57 51	14 36	17.1	56 53	14 03	16.6	55 56	13 33	16.2	54 58	13 03	15.8	54 00	12 36	15.4	261	279
82	98	59 03	13 33	15.7	58 05	13 02	15.3	57 07	12 33	14.9	56 09	12 06	14.5	55 11	11 40	14.1	54 13	11 14	13.8	262	278
83	97	59 16	11 55	13.8	58 18	11 28	13.4	57 19	11 02	13.0	56 21	10 38	12.7	55 22	10 16	12.4	54 24	9 52	12.1	263	277
84	96	59 28	10 16	11.9	58 29	9 52	11.5	57 30	9 30	11.2	56 31	9 09	10.9	55 31	8 49	10.6	54 33	8 29	10.4	264	276
85	95	59 37	8 35	9.9	58 38	8 15	9.6	57 39	7 56	9.4	56 40	7 39	9.1	55 41	7 22	8.9	54 41	7 06	8.7	265	275
86	94	59 46	6 53	8.0	58 46	6 37	7.7	57 47	6 22	7.5	56 47	6 08	7.3	55 48	5 54	7.1	54 48	5 41	7.0	266	274
87	93	59 52	5 11	6.0	58 52	4 59	5.8	57 52	4 47	5.6	56 53	4 36	5.5	55 53	4 26	5.4	54 53	4 16	5.2	267	273
88	92	59 56	3 28	4.0	58 57	3 19	3.9	57 57	3 12	3.8	56 57	3 05	3.7	55 57	2 58	3.6	54 57	2 51	3.5	268	272
89	91	59 59	1 44	2.0	58 59	1 40	1.9	57 59	1 36	1.9	56 59	1 32	1.8	55 59	1 29	1.8	54 59	1 26	1.7	269	271
90	90	60 00	0 00	0.0	59 00	0 00	0.0	58 00	0 00	0.0	57 00	0 00	0.0	56 00	0 00	0.0	55 00	0 00	0.0	270	270

N. Lat.: for LHA > 180° .... Zn = Z
for LHA < 180° .... Zn = 360° − Z

S. Lat.: for LHA > 180° .... Zn = 180° − Z
for LHA < 180° .... Zn = 180° + Z

LATITUDE / A: 36° – 41°

SIGHT REDUCTION TABLE

B: (−) for 90° < LHA < 270°
Dec: (−) for Lat. contrary name

Z₁: same sign as B
Z₂: (−) for F > 90°

Lat. / A	36°			37°			38°			39°			40°			41°			Lat. / A
LHA/F	A/H	B/P	Z_1/Z_2	A/H	B/P	Z_1/Z_2	A/H	B/P	Z_1/Z_2	A/H	B/P	Z_1/Z_2	A/H	B/P	Z_1/Z_2	A/H	B/P	Z_1/Z_2	LHA
0	0 00	54 00	90.0	0 00	53 00	90.0	0 00	52 00	90.0	0 00	51 00	90.0	0 00	50 00	90.0	0 00	49 00	90.0	180
1	0 49	54 00	89.4	0 48	53 00	89.4	0 47	52 00	89.4	0 47	51 00	89.4	0 46	50 00	89.4	0 45	49 00	89.3	181
2	1 37	53 59	88.8	1 36	52 59	88.8	1 35	51 59	88.8	1 33	50 59	88.7	1 32	49 59	88.7	1 31	48 59	88.7	182
3	2 26	53 58	88.2	2 24	52 58	88.2	2 22	51 58	88.2	2 20	50 58	88.1	2 18	49 58	88.1	2 16	48 58	88.0	183
4	3 14	53 56	87.6	3 12	52 56	87.6	3 09	51 56	87.5	3 06	50 56	87.5	3 04	49 56	87.4	3 01	48 56	87.4	184
5	4 03	53 54	87.1	3 59	52 54	87.0	3 56	51 54	86.9	3 53	50 54	86.8	3 50	49 54	86.8	3 46	48 54	86.7	185
6	4 51	53 51	86.5	4 47	52 51	86.4	4 43	51 51	86.3	4 40	50 51	86.2	4 36	49 51	86.1	4 31	48 51	86.1	186
7	5 39	53 48	85.9	5 35	52 48	85.8	5 31	51 48	85.7	5 26	50 47	85.6	5 21	49 47	85.5	5 17	48 47	85.4	187
8	6 28	53 44	85.3	6 23	52 44	85.2	6 18	51 44	85.1	6 13	50 44	84.9	6 07	49 43	84.8	6 02	48 43	84.7	188
9	7 16	53 40	84.7	7 11	52 39	84.6	7 05	51 39	84.4	6 59	50 39	84.3	6 53	49 39	84.2	6 47	48 39	84.1	189
10	8 05	53 35	84.1	7 58	52 35	83.9	7 52	51 34	83.8	7 45	50 34	83.7	7 39	49 34	83.5	7 32	48 34	83.4	190
11	8 53	53 30	83.5	8 46	52 29	83.3	8 39	51 29	83.2	8 32	50 29	83.0	8 24	49 29	82.9	8 17	48 28	82.7	191
12	9 41	53 24	82.9	9 33	52 23	82.7	9 26	51 23	82.5	9 18	50 23	82.4	9 10	49 23	82.2	9 02	48 22	82.1	192
13	10 29	53 17	82.3	10 21	52 17	82.1	10 13	51 17	81.9	10 04	50 16	81.7	9 55	49 16	81.6	9 46	48 16	81.4	193
14	11 17	53 10	81.7	11 08	52 10	81.5	10 59	51 10	81.3	10 50	50 09	81.1	10 41	49 09	80.9	10 31	48 09	80.7	194
15	12 05	53 03	81.0	11 56	52 02	80.8	11 46	51 02	80.6	11 36	50 02	80.4	11 26	49 01	80.2	11 16	48 01	80.0	195
16	12 53	52 55	80.4	12 43	51 54	80.2	12 33	50 54	80.0	12 22	49 53	79.8	12 11	48 53	79.6	12 00	47 53	79.3	196
17	13 41	52 46	79.8	13 30	51 46	79.6	13 19	50 45	79.3	13 08	49 45	79.1	12 57	48 44	78.9	12 45	47 44	78.7	197
18	14 29	52 37	79.2	14 17	51 37	78.9	14 06	50 36	78.7	13 54	49 35	78.4	13 42	48 35	78.2	13 29	47 34	78.0	198
19	15 16	52 28	78.6	15 04	51 27	78.3	14 52	50 26	78.0	14 39	49 25	77.8	14 27	48 25	77.5	14 13	47 24	77.3	199
20	16 04	52 18	77.9	15 51	51 16	77.6	15 38	50 16	77.4	15 25	49 15	77.1	15 11	48 14	76.8	14 58	47 14	76.6	200
21	16 51	52 07	77.3	16 38	51 05	77.0	16 24	50 05	76.7	16 10	49 04	76.4	15 56	48 03	76.1	15 42	47 03	75.9	201
22	17 39	51 55	76.6	17 24	50 54	76.3	17 10	49 53	76.0	16 56	48 52	75.7	16 41	47 51	75.4	16 25	46 51	75.2	202
23	18 26	51 43	76.0	18 11	50 42	75.7	17 56	49 41	75.4	17 41	48 40	75.0	17 25	47 39	74.7	17 09	46 38	74.4	203
24	19 13	51 30	75.3	18 57	50 29	75.0	18 42	49 28	74.7	18 26	48 27	74.3	18 09	47 26	74.0	17 53	46 25	73.7	204
25	20 00	51 17	74.7	19 44	50 15	74.3	19 27	49 14	74.0	19 10	48 13	73.6	18 53	47 12	73.3	18 36	46 12	73.0	205
26	20 46	51 03	74.0	20 30	50 01	73.6	20 12	49 00	73.3	19 55	47 59	72.9	19 37	46 58	72.6	19 19	45 57	72.3	206
27	21 33	50 48	73.3	21 15	49 47	73.0	20 58	48 45	72.6	20 40	47 44	72.2	20 21	46 43	71.9	20 02	45 42	71.5	207
28	22 19	50 33	72.6	22 01	49 31	72.3	21 43	48 30	71.9	21 24	47 28	71.5	21 05	46 28	71.1	20 45	45 27	70.8	208
29	23 06	50 17	72.0	22 47	49 15	71.6	22 28	48 14	71.2	22 08	47 12	70.8	21 48	46 11	70.4	21 28	45 11	70.0	209
30	23 52	50 00	71.3	23 32	48 58	70.8	23 12	47 57	70.4	22 52	46 55	70.0	22 31	45 54	69.6	22 10	44 54	69.3	210
31	24 37	49 43	70.5	24 17	48 41	70.1	23 57	47 39	69.7	23 36	46 38	69.3	23 14	45 37	68.9	22 52	44 36	68.5	211
32	25 23	49 25	69.8	25 02	48 23	69.4	24 41	47 21	69.0	24 19	46 19	68.5	23 57	45 18	68.1	23 34	44 17	67.7	212
33	26 09	49 06	69.1	25 47	48 04	68.7	25 25	47 02	68.2	25 02	46 00	67.8	24 40	44 59	67.3	24 16	43 58	66.9	213
34	26 54	48 46	68.4	26 32	47 44	67.9	26 09	46 42	67.4	25 45	45 40	67.0	25 22	44 39	66.6	24 58	43 39	66.1	214
35	27 39	48 26	67.6	27 16	47 23	67.1	26 52	46 21	66.7	26 28	45 20	66.2	26 04	44 19	65.8	25 39	43 18	65.3	215
36	28 24	48 04	66.9	28 00	47 02	66.4	27 36	46 00	65.9	27 11	44 58	65.4	26 46	43 57	65.0	26 20	42 57	64.5	216
37	29 08	47 42	66.1	28 44	46 40	65.6	28 19	45 38	65.1	27 53	44 36	64.6	27 27	43 35	64.2	27 01	42 34	63.7	217
38	29 52	47 19	65.3	29 27	46 17	64.8	29 01	45 15	64.3	28 35	44 13	63.8	28 08	43 12	63.3	27 41	42 12	62.9	218
39	30 36	46 56	64.5	30 10	45 53	64.0	29 44	44 51	63.5	29 17	43 49	63.0	28 49	42 48	62.5	28 21	41 48	62.0	219
40	31 20	46 31	63.7	30 53	45 28	63.2	30 26	44 26	62.7	29 58	43 25	62.2	29 30	42 24	61.7	29 01	41 23	61.2	220
41	32 03	46 05	62.9	31 36	45 03	62.4	31 08	44 01	61.8	30 39	42 59	61.3	30 10	41 58	60.8	29 41	40 58	60.3	221
42	32 46	45 39	62.1	32 18	44 36	61.5	31 49	43 34	61.0	31 20	42 33	60.5	30 50	41 32	59.9	30 20	40 32	59.4	222
43	33 29	45 11	61.3	33 00	44 09	60.7	32 30	43 07	60.1	32 00	42 05	59.6	31 30	41 05	59.1	30 59	40 04	58.5	223
44	34 12	44 43	60.4	33 42	43 40	59.8	33 11	42 38	59.3	32 40	41 37	58.7	32 09	40 36	58.2	31 37	39 36	57.6	224
45	34 54	44 13	59.6	34 23	43 11	59.0	33 52	42 09	58.4	33 20	41 08	57.8	32 48	40 07	57.3	32 15	39 08	56.7	225

Lat./A LHA/F	36° A/H	36° B/P	36° Z₁/Z₂	37° A/H	37° B/P	37° Z₁/Z₂	38° A/H	38° B/P	38° Z₁/Z₂	39° A/H	39° B/P	39° Z₁/Z₂	40° A/H	40° B/P	40° Z₁/Z₂	41° A/H	41° B/P	41° Z₁/Z₂	Lat./A LHA
45 135	34 54	44 13	59.6	34 23	43 11	59.0	33 52	42 09	58.4	33 20	41 08	57.8	32 48	40 07	57.3	32 15	39 08	56.7	225 315
46 134	35 35	43 43	58.7	35 04	42 40	58.1	34 32	41 38	57.5	33 59	40 37	56.9	33 26	39 37	56.4	32 53	38 38	55.8	226 314
47 133	36 17	43 11	57.8	35 44	42 09	57.2	35 12	41 07	56.6	34 38	40 06	56.0	34 04	39 06	55.4	33 30	38 07	54.9	227 313
48 132	36 57	42 39	56.9	36 24	41 36	56.3	35 51	40 35	55.7	35 17	39 34	55.0	34 42	38 34	54.5	34 07	37 35	53.9	228 312
49 131	37 38	42 05	55.9	37 04	41 03	55.3	36 30	40 01	54.7	35 55	39 01	54.1	35 19	38 01	53.5	34 43	37 03	53.0	229 311
50 130	38 18	41 30	55.0	37 43	40 28	54.4	37 08	39 27	53.7	36 32	38 27	53.1	35 56	37 27	52.5	35 19	36 29	52.0	230 310
51 129	38 57	40 54	54.0	38 22	39 52	53.4	37 46	38 51	52.8	37 09	37 51	52.1	36 32	36 52	51.6	35 55	35 54	51.0	231 309
52 128	39 36	40 17	53.0	39 00	39 15	52.4	38 23	38 14	51.8	37 46	37 15	51.1	37 08	36 16	50.6	36 30	35 18	50.0	232 308
53 127	40 15	39 38	52.0	39 38	38 37	51.4	39 00	37 36	50.8	38 22	36 37	50.1	37 43	35 39	49.5	37 04	34 42	49.0	233 307
54 126	40 53	38 58	51.0	40 15	37 57	50.4	39 36	36 57	49.7	38 57	35 58	49.0	38 18	35 01	48.5	37 38	34 04	47.9	234 306
55 125	41 30	38 17	50.0	40 52	37 17	49.3	40 12	36 17	48.7	39 32	35 19	48.1	38 52	34 21	47.4	38 11	33 25	46.9	235 305
56 124	42 07	37 35	48.9	41 28	36 35	48.3	40 47	35 36	47.6	40 07	34 38	47.0	39 26	33 41	46.4	38 44	32 45	45.8	236 304
57 123	42 44	36 51	47.9	42 03	35 51	47.2	41 22	34 53	46.5	40 41	33 55	45.9	39 59	32 59	45.3	39 16	32 04	44.7	237 303
58 122	43 19	36 06	46.8	42 38	35 07	46.1	41 56	34 09	45.4	41 14	33 12	44.8	40 31	32 16	44.2	39 48	31 22	43.6	238 302
59 121	43 54	35 20	45.6	43 12	34 21	45.0	42 29	33 24	44.3	41 46	32 27	43.7	41 03	31 32	43.1	40 19	30 39	42.5	239 301
60 120	44 29	34 32	44.5	43 46	33 34	43.8	43 02	32 37	43.2	42 18	31 42	42.5	41 34	30 47	41.9	40 49	29 54	41.3	240 300
61 119	45 02	33 43	43.3	44 18	32 45	42.6	43 34	31 49	42.0	42 49	30 55	41.4	42 04	30 01	40.8	41 18	29 09	40.2	241 299
62 118	45 35	32 52	42.1	44 51	31 55	41.5	44 05	31 00	40.8	43 20	30 06	40.2	42 34	29 14	39.6	41 47	28 22	39.0	242 298
63 117	46 07	32 00	40.9	45 22	31 04	40.3	44 36	30 10	39.6	43 49	29 17	39.0	43 03	28 25	38.4	42 15	27 35	37.8	243 297
64 116	46 39	31 06	39.7	45 52	30 11	39.0	45 06	29 18	38.4	44 18	28 26	37.8	43 31	27 35	37.2	42 43	26 46	36.6	244 296
65 115	47 09	30 11	38.4	46 22	29 17	37.8	45 35	28 25	37.1	44 47	27 34	36.5	43 58	26 44	36.0	43 09	25 56	35.4	245 295
66 114	47 39	29 14	37.1	46 51	28 21	36.5	46 03	27 30	35.9	45 14	26 40	35.3	44 25	25 52	34.7	43 35	25 04	34.2	246 294
67 113	48 08	28 16	35.8	47 19	27 24	35.2	46 30	26 34	34.6	45 40	25 45	34.0	44 50	24 58	33.4	44 00	24 12	32.9	247 293
68 112	48 36	27 17	34.5	47 46	26 26	33.9	46 56	25 37	33.3	46 06	24 50	32.7	45 15	24 03	32.2	44 24	23 19	31.6	248 292
69 111	49 03	26 15	33.1	48 13	25 26	32.5	47 22	24 38	31.9	46 31	23 52	31.4	45 39	23 08	30.8	44 48	22 24	30.3	249 291
70 110	49 29	25 13	31.8	48 38	24 25	31.2	47 46	23 39	30.6	46 55	22 54	30.0	46 03	22 11	29.5	45 10	21 29	29.0	250 290
71 109	49 54	24 08	30.4	49 02	23 22	29.8	48 10	22 37	29.2	47 17	21 54	28.7	46 25	21 12	28.2	45 32	20 32	27.7	251 289
72 108	50 18	23 02	28.9	49 25	22 18	28.4	48 33	21 35	27.8	47 39	20 53	27.3	46 46	20 13	26.8	45 52	19 34	26.3	252 288
73 107	50 41	21 55	27.5	49 48	21 13	26.9	48 54	20 31	26.4	48 00	19 51	25.9	47 06	19 13	25.4	46 12	18 34	25.0	253 287
74 106	51 03	20 47	26.0	50 09	20 06	25.5	49 15	19 26	25.0	48 20	18 48	24.5	47 25	18 11	24.0	46 30	17 36	23.6	254 286
75 105	51 24	19 36	24.5	50 29	18 57	24.0	49 34	18 20	23.5	48 39	17 43	23.1	47 44	17 09	22.6	46 48	16 35	22.2	255 285
76 104	51 43	18 25	23.0	50 48	17 48	22.5	49 52	17 12	22.0	48 57	16 38	21.6	48 01	16 05	21.2	47 05	15 33	20.8	256 284
77 103	52 02	17 12	21.4	51 06	16 37	21.0	50 09	16 04	20.6	49 13	15 31	20.1	48 17	15 00	19.8	47 20	14 31	19.4	257 283
78 102	52 19	15 58	19.9	51 22	15 25	19.5	50 25	14 54	19.0	49 29	14 24	18.7	48 32	13 55	18.3	47 35	13 27	18.0	258 282
79 101	52 35	14 43	18.3	51 37	14 13	17.9	50 40	13 43	17.5	49 43	13 16	17.2	48 46	12 49	16.8	47 48	12 23	16.5	259 281
80 100	52 49	13 27	16.7	51 52	12 59	16.3	50 54	12 32	16.0	49 56	12 06	15.7	48 58	11 42	15.3	48 01	11 18	15.0	260 280
81 99	53 02	12 09	15.1	52 04	11 44	14.7	51 06	11 19	14.4	50 08	10 56	14.1	49 10	10 34	13.8	48 12	10 12	13.6	261 279
82 98	53 14	10 51	13.4	52 16	10 28	13.1	51 18	10 06	12.9	50 19	9 45	12.6	49 20	9 25	12.3	48 22	9 06	12.1	262 278
83 97	53 25	9 31	11.8	52 26	9 11	11.5	51 27	8 52	11.3	50 29	8 34	11.0	49 30	8 16	10.8	48 31	7 59	10.6	263 277
84 96	53 34	8 11	10.1	52 35	7 54	9.9	51 36	7 37	9.7	50 37	7 21	9.5	49 38	7 06	9.3	48 38	6 51	9.1	264 276
85 95	53 42	6 50	8.5	52 43	6 36	8.3	51 43	6 22	8.1	50 44	6 09	7.9	49 44	5 56	7.8	48 45	5 44	7.6	265 275
86 94	53 49	5 29	6.8	52 49	5 17	6.6	51 49	5 06	6.5	50 50	4 55	6.3	49 50	4 45	6.2	48 50	4 35	6.1	266 274
87 93	53 54	4 07	5.1	52 54	3 58	5.0	51 54	3 50	4.9	50 54	3 42	4.8	49 54	3 34	4.7	48 55	3 27	4.6	267 273
88 92	53 57	2 45	3.4	52 57	2 39	3.3	51 57	2 33	3.2	50 57	2 28	3.2	49 58	2 23	3.1	48 58	2 18	3.0	268 272
89 91	53 59	1 23	1.7	52 59	1 20	1.7	51 59	1 17	1.6	50 59	1 14	1.6	49 59	1 11	1.6	48 59	1 09	1.5	269 271
90 90	54 00	0 00	0.0	53 00	0 00	0.0	52 00	0 00	0.0	51 00	0 00	0.0	50 00	0 00	0.0	49 00	0 00	0.0	270 270

N. Lat.: for LHA > 180° ... Zn = Z
for LHA < 180° ... Zn = 360° − Z

S. Lat.: for LHA > 180° ... Zn = 180° − Z
for LHA < 180° ... Zn = 180° + Z

SIGHT REDUCTION TABLE

B: (−) for 90° < LHA < 270°
Dec: (−) for Lat. contrary name

Z₁: same sign as B
Z₂: (−) for F > 90°

Lat./A	LHA/F	42° A/H	42° B/P	42° Z₁/Z₂	43° A/H	43° B/P	43° Z₁/Z₂	44° A/H	44° B/P	44° Z₁/Z₂	45° A/H	45° B/P	45° Z₁/Z₂	46° A/H	46° B/P	46° Z₁/Z₂	47° A/H	47° B/P	47° Z₁/Z₂	LHA	Lat./A
0	180	0 00	48 00	90.0	0 00	47 00	90.0	0 00	46 00	90.0	0 00	45 00	90.0	0 00	44 00	90.0	0 00	43 00	90.0	180	360
1	179	0 45	48 00	89.3	0 44	47 00	89.3	0 43	46 00	89.3	0 42	45 00	89.3	0 42	44 00	89.3	0 41	43 00	89.3	181	359
2	178	1 29	47 59	88.7	1 28	46 59	88.6	1 26	45 59	88.6	1 25	44 59	88.6	1 23	43 59	88.6	1 22	42 59	88.5	182	358
3	177	2 14	47 58	88.0	2 12	46 58	88.0	2 09	45 58	87.9	2 07	44 58	87.9	2 05	43 58	87.8	2 03	42 58	87.8	183	357
4	176	2 58	47 56	87.3	2 55	46 56	87.3	2 53	45 56	87.2	2 50	44 56	87.2	2 47	43 56	87.1	2 44	42 56	87.1	184	356
5	175	3 43	47 53	86.6	3 39	46 53	86.6	3 36	45 53	86.5	3 32	44 53	86.5	3 28	43 53	86.4	3 24	42 53	86.3	185	355
6	174	4 27	47 51	86.0	4 23	46 51	85.9	4 19	45 51	85.8	4 14	44 51	85.7	4 10	43 51	85.7	4 05	42 51	85.6	186	354
7	173	5 12	47 47	85.3	5 07	46 47	85.2	5 02	45 47	85.1	4 57	44 47	85.0	4 51	43 47	85.0	4 46	42 47	84.9	187	353
8	172	5 56	47 43	84.6	5 51	46 43	84.5	5 45	45 43	84.4	5 39	44 43	84.3	5 33	43 43	84.2	5 27	42 43	84.1	188	352
9	171	6 41	47 39	84.0	6 34	46 39	83.8	6 28	45 39	83.7	6 21	44 39	83.6	6 14	43 39	83.5	6 07	42 39	83.4	189	351
10	170	7 25	47 34	83.3	7 18	46 34	83.1	7 11	45 34	83.0	7 03	44 34	82.9	6 56	43 34	82.8	6 48	42 34	82.7	190	350
11	169	8 09	47 28	82.6	8 01	46 28	82.4	7 53	45 28	82.3	7 45	44 28	82.2	7 37	43 28	82.0	7 29	42 28	81.9	191	349
12	168	8 53	47 22	81.9	8 45	46 22	81.8	8 36	45 22	81.6	8 27	44 22	81.5	8 18	43 22	81.3	8 09	42 22	81.2	192	348
13	167	9 37	47 16	81.2	9 28	46 15	81.1	9 19	45 15	80.9	9 09	44 15	80.7	8 59	43 15	80.6	8 49	42 16	80.4	193	347
14	166	10 21	47 08	80.5	10 11	46 08	80.3	10 01	45 08	80.2	9 51	44 08	80.0	9 40	43 08	79.8	9 30	42 08	79.7	194	346
15	165	11 05	47 01	79.8	10 55	46 00	79.6	10 44	45 00	79.5	10 33	44 00	79.3	10 21	43 00	79.1	10 10	42 01	78.9	195	345
16	164	11 49	46 52	79.1	11 38	45 52	78.9	11 26	44 52	78.7	11 14	43 52	78.5	11 02	42 52	78.3	10 50	41 52	78.2	196	344
17	163	12 33	46 43	78.4	12 21	45 43	78.2	12 08	44 43	78.0	11 56	43 43	77.8	11 43	42 43	77.6	11 30	41 44	77.4	197	343
18	162	13 17	46 34	77.7	13 04	45 34	77.5	12 51	44 34	77.3	12 37	43 34	77.1	12 24	42 34	76.8	12 10	41 34	76.6	198	342
19	161	14 00	46 24	77.0	13 46	45 24	76.8	13 33	44 24	76.5	13 19	43 24	76.3	13 04	42 24	76.1	12 50	41 24	75.9	199	341
20	160	14 43	46 13	76.3	14 29	45 13	76.1	14 15	44 13	75.8	14 00	43 13	75.6	13 45	42 13	75.3	13 29	41 14	75.1	200	340
21	159	15 27	46 02	75.6	15 12	45 02	75.3	14 56	44 02	75.1	14 41	43 02	74.8	14 25	42 02	74.6	14 09	41 03	74.3	201	339
22	158	16 10	45 50	74.9	15 54	44 50	74.6	15 38	43 50	74.3	15 22	42 50	74.1	15 05	41 50	73.8	14 48	40 51	73.5	202	338
23	157	16 53	45 38	74.1	16 36	44 38	73.9	16 19	43 38	73.6	16 02	42 38	73.3	15 45	41 38	73.0	15 27	40 39	72.8	203	337
24	156	17 36	45 25	73.4	17 18	44 25	73.1	17 01	43 25	72.8	16 43	42 25	72.5	16 25	41 25	72.2	16 06	40 26	72.0	204	336
25	155	18 18	45 11	72.7	18 00	44 11	72.4	17 42	43 11	72.1	17 23	42 11	71.8	17 04	41 12	71.5	16 45	40 12	71.2	205	335
26	154	19 01	44 57	71.9	18 42	43 57	71.6	18 23	42 57	71.3	18 03	41 57	71.0	17 44	40 57	70.7	17 24	39 58	70.4	206	334
27	153	19 43	44 42	71.2	19 24	43 42	70.8	19 04	42 42	70.5	18 43	41 42	70.2	18 23	40 43	69.9	18 02	39 43	69.6	207	333
28	152	20 25	44 26	70.4	20 05	43 26	70.1	19 44	42 26	69.7	19 23	41 27	69.4	19 02	40 27	69.1	18 40	39 28	68.8	208	332
29	151	21 07	44 10	69.6	20 46	43 10	69.3	20 25	42 10	68.9	20 03	41 10	68.6	19 41	40 11	68.3	19 18	39 12	67.9	209	331
30	150	21 49	43 53	68.9	21 27	42 53	68.5	21 05	41 53	68.1	20 42	40 54	67.8	20 19	39 54	67.4	19 56	38 55	67.1	210	330
31	149	22 30	43 35	68.1	22 08	42 35	67.7	21 45	41 36	67.3	21 21	40 36	67.0	20 58	39 37	66.6	20 34	38 38	66.3	211	329
32	148	23 11	43 17	67.3	22 48	42 17	66.9	22 24	41 17	66.5	22 00	40 18	66.2	21 36	39 19	65.8	21 11	38 20	65.4	212	328
33	147	23 53	42 58	66.5	23 28	41 58	66.1	23 04	40 58	65.7	22 39	39 59	65.3	22 14	39 00	65.0	21 48	38 02	64.6	213	327
34	146	24 33	42 38	65.7	24 08	41 38	65.3	23 43	40 39	64.9	23 17	39 40	64.5	22 51	38 41	64.1	22 25	37 42	63.7	214	326
35	145	25 14	42 18	64.9	24 48	41 18	64.5	24 22	40 18	64.1	23 56	39 19	63.7	23 29	38 21	63.3	23 02	37 23	62.9	215	325
36	144	25 54	41 56	64.1	25 28	40 57	63.6	25 01	39 57	63.2	24 34	38 58	62.8	24 06	38 00	62.4	23 38	37 02	62.0	216	324
37	143	26 34	41 34	63.2	26 07	40 35	62.8	25 39	39 35	62.4	25 11	38 37	61.9	24 43	37 38	61.5	24 14	36 41	61.1	217	323
38	142	27 14	41 11	62.4	26 46	40 12	61.9	26 17	39 13	61.5	25 48	38 14	61.1	25 19	37 16	60.7	24 50	36 19	60.3	218	322
39	141	27 53	40 48	61.5	27 24	39 48	61.1	26 55	38 50	60.6	26 25	37 51	60.2	25 55	36 53	59.8	25 25	35 56	59.4	219	321
40	140	28 32	40 23	60.7	28 02	39 24	60.2	27 32	38 25	59.8	27 02	37 27	59.3	26 31	36 30	58.9	26 00	35 32	58.5	220	320
41	139	29 11	39 58	59.8	28 40	38 59	59.3	28 10	38 01	58.9	27 38	37 03	58.4	27 07	36 05	58.0	26 35	35 08	57.6	221	319
42	138	29 49	39 32	58.9	29 18	38 33	58.4	28 46	37 35	58.0	28 14	36 37	57.5	27 42	35 40	57.1	27 09	34 43	56.6	222	318
43	137	30 27	39 05	58.0	29 55	38 06	57.5	29 23	37 08	57.1	28 50	36 11	56.6	28 17	35 14	56.1	27 43	34 18	55.7	223	317
44	136	31 05	38 37	57.1	30 32	37 39	56.6	29 59	36 41	56.1	29 25	35 44	55.7	28 51	34 47	55.2	28 17	33 51	54.8	224	316
45	135	31 42	38 09	56.2	31 08	37 10	55.7	30 34	36 13	55.2	30 00	35 16	54.7	29 25	34 20	54.3	28 50	33 24	53.8	225	315

Lat./A LHA/F		42° A/H	42° B/P	42° Z₁/Z₂	43° A/H	43° B/P	43° Z₁/Z₂	44° A/H	44° B/P	44° Z₁/Z₂	45° A/H	45° B/P	45° Z₁/Z₂	46° A/H	46° B/P	46° Z₁/Z₂	47° A/H	47° B/P	47° Z₁/Z₂	Lat./A LHA	
45	135	31 42	38 09	56.2	31 08	37 10	55.7	30 34	36 13	55.2	30 00	35 16	54.7	29 25	34 20	54.3	28 50	33 24	53.8	225	315
46	134	32 19	37 39	55.5	31 45	36 41	54.8	31 10	35 44	54.3	30 34	34 47	53.8	29 59	33 51	53.3	29 23	32 56	52.9	226	314
47	133	32 55	37 08	54.3	32 20	36 11	53.8	31 45	35 14	53.3	31 08	34 18	52.8	30 32	33 22	52.4	29 55	32 27	51.9	227	313
48	132	33 31	36 37	53.4	32 55	35 40	52.9	32 19	34 43	52.3	31 42	33 47	51.9	31 05	32 52	51.4	30 27	31 58	50.9	228	312
49	131	34 07	36 05	52.4	33 30	35 08	51.9	32 53	34 11	51.3	32 15	33 16	50.9	31 37	32 21	50.4	30 59	31 27	49.9	229	311
50	130	34 42	35 31	51.4	34 04	34 35	50.9	33 26	33 39	50.4	32 48	32 44	49.9	32 09	31 50	49.4	31 30	30 56	48.9	230	310
51	129	35 17	34 57	50.4	34 38	34 01	49.9	33 59	33 05	49.4	33 20	32 11	48.9	32 40	31 17	48.4	32 00	30 24	47.9	231	309
52	128	35 51	34 22	49.4	35 12	33 26	48.9	34 32	32 31	48.4	33 52	31 37	47.9	33 11	30 44	47.4	32 30	29 52	46.9	232	308
53	127	36 24	33 45	48.4	35 44	32 50	47.9	35 04	31 56	47.3	34 23	31 02	46.8	33 42	30 10	46.3	33 00	29 18	45.9	233	307
54	126	36 57	33 08	47.4	36 17	32 13	46.8	35 35	31 20	46.3	34 54	30 27	45.8	34 12	29 35	45.3	33 29	28 44	44.8	234	306
55	125	37 30	32 30	46.3	36 48	31 36	45.8	36 06	30 43	45.2	35 24	29 50	44.7	34 41	28 59	44.2	33 58	28 08	43.8	235	305
56	124	38 02	31 51	45.2	37 19	30 57	44.7	36 37	30 04	44.2	35 53	29 13	43.6	35 10	28 22	43.2	34 26	27 32	42.7	236	304
57	123	38 33	31 10	44.1	37 50	30 17	43.6	37 06	29 25	43.1	36 22	28 34	42.6	35 38	27 45	42.1	34 53	26 56	41.6	237	303
58	122	39 04	30 29	43.0	38 20	29 36	42.5	37 36	28 45	42.0	36 51	27 55	41.5	36 06	27 06	41.0	35 20	26 18	40.5	238	302
59	121	39 34	29 46	41.9	38 49	28 55	41.4	38 04	28 04	40.9	37 19	27 15	40.4	36 33	26 27	39.9	35 46	25 40	39.4	239	301
60	120	40 04	29 03	40.8	39 18	28 12	40.2	38 32	27 22	39.7	37 46	26 34	39.2	36 59	25 46	38.8	36 12	25 00	38.3	240	300
61	119	40 32	28 18	39.6	39 46	27 28	39.1	38 59	26 39	38.6	38 12	25 52	38.1	37 25	25 05	37.6	36 37	24 20	37.2	241	299
62	118	41 00	27 32	38.5	40 13	26 43	37.9	39 26	25 56	37.4	38 38	25 09	36.9	37 50	24 23	36.5	37 02	23 39	36.0	242	298
63	117	41 28	26 45	37.3	40 40	25 58	36.8	39 52	25 11	36.3	39 03	24 25	35.8	38 14	23 40	35.3	37 25	22 57	34.9	243	297
64	116	41 54	25 58	36.1	41 06	25 11	35.6	40 17	24 25	35.1	39 28	23 40	34.6	38 38	22 57	34.1	37 48	22 14	33.7	244	296
65	115	42 20	25 09	34.9	41 31	24 23	34.4	40 41	23 38	33.9	39 51	22 55	33.4	39 01	22 12	33.0	38 11	21 31	32.5	245	295
66	114	42 45	24 19	33.6	41 55	23 34	33.1	41 05	22 50	32.7	40 14	22 08	32.2	39 23	21 27	31.8	38 32	20 46	31.3	246	294
67	113	43 10	23 28	32.4	42 19	22 44	31.9	41 28	22 02	31.4	40 37	21 21	31.0	39 45	20 40	30.5	38 53	20 01	30.1	247	293
68	112	43 33	22 35	31.1	42 42	21 53	30.6	41 50	21 12	30.2	40 58	20 32	29.7	40 06	19 53	29.3	39 13	19 15	28.9	248	292
69	111	43 56	21 42	29.8	43 04	21 01	29.4	42 11	20 22	28.9	41 19	19 43	28.5	40 26	19 05	28.1	39 33	18 29	27.7	249	291
70	110	44 18	20 48	28.5	43 25	20 08	28.1	42 32	19 30	27.7	41 38	18 53	27.2	40 45	18 17	26.8	39 51	17 41	26.5	250	290
71	109	44 38	19 53	27.2	43 45	19 15	26.8	42 51	18 38	26.4	41 57	18 02	26.0	41 03	17 27	25.6	40 09	16 53	25.2	251	289
72	108	44 58	18 57	25.9	44 04	18 20	25.5	43 10	17 45	25.1	42 16	17 10	24.7	41 21	16 37	24.3	40 26	16 05	24.0	252	288
73	107	45 17	17 59	24.6	44 23	17 24	24.1	43 28	16 51	23.8	42 33	16 18	23.4	41 38	15 46	23.0	40 42	15 15	22.7	253	287
74	106	45 35	17 01	23.2	44 40	16 28	22.8	43 45	15 56	22.4	42 49	15 25	22.1	41 54	14 54	21.7	40 58	14 25	21.4	254	286
75	105	45 53	16 02	21.8	44 57	15 31	21.4	44 01	15 00	21.1	43 05	14 31	20.8	42 09	14 02	20.4	41 12	13 34	20.1	255	285
76	104	46 09	15 02	20.4	45 12	14 33	20.1	44 16	14 04	19.7	43 19	13 36	19.4	42 23	13 09	19.1	41 26	12 43	18.8	256	284
77	103	46 24	14 02	19.0	45 27	13 34	18.7	44 30	13 07	18.4	43 33	12 41	18.1	42 36	12 15	17.8	41 39	11 51	17.5	257	283
78	102	46 38	13 00	17.6	45 40	12 34	17.3	44 43	12 09	17.0	43 46	11 45	16.7	42 48	11 21	16.5	41 51	10 58	16.2	258	282
79	101	46 51	11 58	16.2	45 53	11 34	15.9	44 55	11 11	15.6	43 57	10 48	15.4	43 00	10 26	15.1	42 02	10 05	14.9	259	281
80	100	47 03	10 55	14.8	46 04	10 33	14.5	45 06	10 12	14.2	44 08	9 51	14.0	43 10	9 31	13.8	42 12	9 15	13.6	260	280
81	99	47 13	9 51	13.3	46 15	9 31	13.1	45 16	9 12	12.8	44 18	8 53	12.6	43 19	8 35	12.4	42 21	8 18	12.2	261	279
82	98	47 23	8 47	11.9	46 24	8 29	11.6	45 26	8 12	11.4	44 27	7 55	11.2	43 28	7 39	11.1	42 29	7 24	10.9	262	278
83	97	47 32	7 42	10.4	46 33	7 27	10.2	45 34	7 12	10.0	44 34	6 57	9.9	43 35	6 43	9.7	42 36	6 29	9.5	263	277
84	96	47 39	6 37	8.9	46 40	6 24	8.8	45 41	6 11	8.6	44 41	5 58	8.5	43 42	5 46	8.3	42 42	5 34	8.2	264	276
85	95	47 46	5 32	7.4	46 46	5 20	7.3	45 46	5 09	7.2	44 47	4 59	7.1	43 47	4 49	6.9	42 48	4 39	6.8	265	275
86	94	47 51	4 26	6.0	46 51	4 17	5.9	45 51	4 08	5.7	44 52	3 59	5.6	43 52	3 51	5.6	42 52	3 43	5.5	266	274
87	93	47 55	3 20	4.5	46 55	3 13	4.4	45 55	3 06	4.3	44 55	3 00	4.2	43 55	2 54	4.2	42 56	2 48	4.1	267	273
88	92	47 58	2 13	3.0	46 58	2 09	2.9	45 58	2 04	2.9	44 58	2 00	2.8	43 58	1 56	2.8	42 58	1 52	2.7	268	272
89	91	47 59	1 07	1.5	46 59	1 04	1.5	45 59	1 02	1.4	44 59	1 00	1.4	43 59	0 58	1.4	43 00	0 56	1.4	269	271
90	90	48 00	0 00	0.0	47 00	0 00	0.0	46 00	0 00	0.0	45 00	0 00	0.0	44 00	0 00	0.0	43 00	0 00	0.0	270	270

N. Lat.: for LHA > 180°..... Zn = Z
for LHA < 180°..... Zn = 360° − Z

S. Lat.: for LHA > 180°..... Zn = 180° − Z
for LHA < 180°..... Zn = 180° + Z

SIGHT REDUCTION TABLE

B: (–) for 90° < LHA < 270°
Dec: (–) for Lat. contrary name

Z1: same sign as B
Z2: (–) for F > 90°

LHA/F		48° A/H	48° B/P	48° Z1/Z2	49° A/H	49° B/P	49° Z1/Z2	50° A/H	50° B/P	50° Z1/Z2	51° A/H	51° B/P	51° Z1/Z2	52° A/H	52° B/P	52° Z1/Z2	53° A/H	53° B/P	53° Z1/Z2	LHA	
0	180	0 00	42 00	90.0	0 00	41 00	90.0	0 00	40 00	90.0	0 00	39 00	90.0	0 00	38 00	90.0	0 00	37 00	90.0	180	360
1	179	0 40	42 00	89.3	0 39	41 00	89.2	0 38	40 00	89.2	0 38	39 00	89.2	0 37	38 00	89.2	0 36	37 00	89.2	181	359
2	178	1 20	41 59	88.5	1 19	40 59	88.5	1 17	39 59	88.4	1 16	38 59	88.4	1 14	37 59	88.4	1 12	36 59	88.4	182	358
3	177	2 00	41 58	87.8	1 58	40 58	87.7	1 56	39 58	87.7	1 53	38 58	87.7	1 51	37 58	87.6	1 48	36 58	87.6	183	357
4	176	2 41	41 56	87.0	2 37	40 56	87.0	2 34	39 56	86.9	2 31	38 56	86.9	2 28	37 56	86.8	2 24	36 56	86.8	184	356
5	175	3 21	41 53	86.3	3 17	40 54	86.2	3 13	39 54	86.2	3 09	38 54	86.1	3 05	37 54	86.1	3 00	36 54	86.0	185	355
6	174	4 01	41 51	85.5	3 56	40 51	85.5	3 51	39 51	85.4	3 46	38 51	85.3	3 41	37 51	85.3	3 36	36 51	85.2	186	354
7	173	4 41	41 47	84.8	4 35	40 47	84.7	4 30	39 47	84.6	4 24	38 47	84.6	4 18	37 48	84.5	4 12	36 48	84.4	187	353
8	172	5 21	41 43	84.0	5 14	40 43	83.9	5 08	39 43	83.9	5 01	38 44	83.8	4 55	37 44	83.7	4 48	36 44	83.6	188	352
9	171	6 01	41 39	83.3	5 53	40 39	83.2	5 46	39 39	83.1	5 39	38 39	83.0	5 32	37 39	82.9	5 24	36 40	82.8	189	351
10	170	6 40	41 34	82.5	6 32	40 34	82.4	6 25	39 34	82.3	6 16	38 34	82.2	6 08	37 35	82.1	6 00	36 35	82.0	190	350
11	169	7 20	41 28	81.8	7 11	40 28	81.7	7 03	39 29	81.5	6 54	38 29	81.4	6 45	37 29	81.3	6 36	36 29	81.2	191	349
12	168	8 00	41 22	81.0	7 50	40 22	80.9	7 41	39 23	80.8	7 31	38 23	80.6	7 21	37 23	80.5	7 11	36 24	80.4	192	348
13	167	8 39	41 16	80.3	8 29	40 16	80.1	8 19	39 16	80.0	8 08	38 16	79.8	7 58	37 17	79.7	7 47	36 17	79.6	193	347
14	166	9 19	41 09	79.5	9 08	40 09	79.3	8 57	39 09	79.2	8 45	38 09	79.0	8 34	37 10	78.9	8 22	36 10	78.7	194	346
15	165	9 58	41 01	78.7	9 47	40 01	78.6	9 35	39 02	78.4	9 22	38 02	78.2	9 10	37 02	78.1	8 58	36 03	77.9	195	345
16	164	10 38	40 53	78.0	10 25	39 53	77.8	10 12	38 53	77.6	9 59	37 54	77.4	9 46	36 54	77.3	9 33	35 55	77.1	196	344
17	163	11 17	40 44	77.2	11 04	39 44	77.0	10 50	38 45	76.8	10 36	37 45	76.6	10 22	36 46	76.5	10 08	35 47	76.3	197	343
18	162	11 56	40 34	76.4	11 42	39 35	76.2	11 27	38 35	76.0	11 13	37 36	75.8	10 58	36 37	75.6	10 43	35 38	75.5	198	342
19	161	12 35	40 25	75.6	12 20	39 25	75.4	12 05	38 26	75.2	11 49	37 26	75.0	11 34	36 27	74.8	11 18	35 28	74.6	199	341
20	160	13 14	40 14	74.9	12 58	39 15	74.6	12 42	38 15	74.4	12 26	37 16	74.2	12 09	36 17	74.0	11 53	35 18	73.8	200	340
21	159	13 52	40 03	74.1	13 36	39 04	73.8	13 19	38 04	73.6	13 02	37 05	73.4	12 45	36 06	73.2	12 27	35 08	73.0	201	339
22	158	14 31	39 51	73.3	14 14	38 52	73.0	13 56	37 53	72.8	13 38	36 54	72.6	13 20	35 55	72.3	13 02	34 56	72.1	202	338
23	157	15 09	39 39	72.5	14 51	38 40	72.2	14 33	37 41	72.0	14 14	36 42	71.7	13 55	35 43	71.5	13 36	34 45	71.3	203	337
24	156	15 48	39 26	71.7	15 29	38 27	71.4	15 09	37 28	71.2	14 50	36 30	70.9	14 30	35 31	70.7	14 10	34 33	70.4	204	336
25	155	16 26	39 13	70.9	16 06	38 14	70.6	15 46	37 15	70.3	15 25	36 17	70.1	15 05	35 18	69.8	14 44	34 20	69.6	205	335
26	154	17 03	38 59	70.1	16 43	38 00	69.8	16 22	37 01	69.5	16 01	36 03	69.2	15 39	35 05	69.0	15 18	34 07	68.7	206	334
27	153	17 41	38 44	69.3	17 20	37 46	69.0	16 58	36 47	68.7	16 36	35 49	68.4	16 14	34 51	68.1	15 51	33 53	67.9	207	333
28	152	18 19	38 29	68.4	17 56	37 30	68.1	17 34	36 32	67.8	17 11	35 34	67.5	16 48	34 36	67.3	16 25	33 38	67.0	208	332
29	151	18 56	38 13	67.6	18 33	37 15	67.3	18 09	36 16	67.0	17 46	35 18	66.7	17 22	34 21	66.4	16 58	33 23	66.1	209	331
30	150	19 33	37 57	66.8	19 09	36 58	66.5	18 45	36 00	66.1	18 20	35 03	65.8	17 56	34 05	65.5	17 31	33 08	65.2	210	330
31	149	20 10	37 40	65.9	19 45	36 41	65.6	19 20	35 44	65.2	18 55	34 46	64.9	18 29	33 49	64.7	18 03	32 52	64.4	211	329
32	148	20 46	37 22	65.1	20 21	36 24	64.8	19 55	35 26	64.4	19 29	34 29	64.1	19 02	33 32	63.8	18 36	32 35	63.5	212	328
33	147	21 22	37 03	64.2	20 56	36 06	63.9	20 30	35 08	63.6	20 03	34 11	63.2	19 35	33 14	62.9	19 08	32 18	62.6	213	327
34	146	21 58	36 44	63.4	21 31	35 47	63.0	21 04	34 49	62.7	20 36	33 53	62.3	20 08	32 56	62.0	19 40	32 00	61.7	214	326
35	145	22 34	36 25	62.5	22 06	35 27	62.1	21 38	34 30	61.8	21 10	33 33	61.4	20 41	32 37	61.1	20 12	31 41	60.8	215	325
36	144	23 10	36 04	61.6	22 41	35 07	61.3	22 12	34 10	60.9	21 43	33 14	60.5	21 13	32 18	60.2	20 43	31 22	59.9	216	324
37	143	23 45	35 43	60.8	23 15	34 46	60.4	22 45	33 50	60.0	22 15	32 54	59.6	21 45	31 58	59.3	21 14	31 02	59.0	217	323
38	142	24 20	35 21	59.9	23 49	34 25	59.5	23 19	33 28	59.1	22 48	32 33	58.7	22 16	31 37	58.4	21 45	30 42	58.0	218	322
39	141	24 54	34 59	59.0	24 23	34 02	58.6	23 52	33 07	58.2	23 20	32 11	57.8	22 48	31 16	57.5	22 15	30 21	57.1	219	321
40	140	25 28	34 36	58.1	24 57	33 40	57.7	24 24	32 44	57.3	23 52	31 49	56.9	23 19	30 54	56.5	22 45	30 00	56.2	220	320
41	139	26 02	34 12	57.1	25 30	33 16	56.7	24 57	32 21	56.3	24 23	31 26	55.9	23 49	30 32	55.5	23 15	29 38	55.2	221	319
42	138	26 36	33 47	56.2	26 02	32 52	55.8	25 28	31 57	55.4	24 54	31 02	55.0	24 20	30 08	54.6	23 45	29 15	54.3	222	318
43	137	27 09	33 22	55.3	26 35	32 27	54.9	26 00	31 32	54.5	25 25	30 38	54.1	24 50	29 45	53.7	24 14	28 52	53.3	223	317
44	136	27 42	32 56	54.3	27 07	32 01	53.9	26 31	31 07	53.5	25 55	30 13	53.1	25 19	29 20	52.7	24 43	28 28	52.4	224	316
45	135	28 14	32 29	53.4	27 38	31 35	53.0	27 02	30 41	52.5	26 25	29 48	52.1	25 48	28 55	51.8	25 11	28 03	51.4	225	315

Lat./A		48°			49°			50°			51°			52°			53°			Lat./A	
LHA/F		A/H	B/P	Z_1/Z_2	A/H	B/P	Z_1/Z_2	A/H	B/P	Z_1/Z_2	A/H	B/P	Z_1/Z_2	A/H	B/P	Z_1/Z_2	A/H	B/P	Z_1/Z_2	LHA	
45	135	28 14	32 29	53.4	27 38	31 35	53.0	27 02	30 41	52.5	26 25	29 48	52.1	25 48	28 55	51.8	25 11	28 03	51.4	225	315
46	134	28 46	32 01	52.4	28 10	31 08	52.0	27 32	30 14	51.6	26 55	29 22	51.1	26 17	28 29	50.8	25 39	27 38	50.4	226	314
47	133	29 18	31 33	51.4	28 40	30 40	51.0	28 02	29 47	50.6	27 24	28 55	50.2	26 46	28 03	49.8	26 07	27 12	49.4	227	313
48	132	29 49	31 04	50.5	29 11	30 11	50.0	28 32	29 19	49.6	27 53	28 27	49.2	27 14	27 36	48.8	26 34	26 46	48.4	228	312
49	131	30 20	30 34	49.5	29 41	29 42	49.0	29 01	28 50	48.6	28 21	27 59	48.2	27 41	27 08	47.8	27 01	26 18	47.4	229	311
50	130	30 50	30 04	48.5	30 10	29 12	48.0	29 30	28 20	47.6	28 49	27 30	47.2	28 08	26 40	46.8	27 27	25 51	46.4	230	310
51	129	31 20	29 32	47.5	30 39	28 41	47.0	29 58	27 50	46.6	29 17	27 00	46.2	28 35	26 11	45.8	27 53	25 22	45.4	231	309
52	128	31 49	29 00	46.4	31 08	28 09	46.0	30 26	27 19	45.6	29 44	26 30	45.2	29 01	25 41	44.8	28 19	24 53	44.4	232	308
53	127	32 18	28 27	45.4	31 36	27 37	45.0	30 53	26 48	44.5	30 10	25 59	44.1	29 27	25 11	43.7	28 44	24 24	43.3	233	307
54	126	32 46	27 53	44.4	32 03	27 04	43.9	31 20	26 15	43.5	30 36	25 27	43.1	29 52	24 40	42.7	29 08	23 53	42.3	234	306
55	125	33 14	27 19	43.3	32 30	26 30	42.9	31 46	25 42	42.4	31 02	24 55	42.0	30 17	24 08	41.6	29 32	23 23	41.2	235	305
56	124	33 42	26 44	42.2	32 57	25 55	41.8	32 12	25 08	41.4	31 27	24 22	41.0	30 41	23 36	40.6	29 56	22 51	40.2	236	304
57	123	34 08	26 07	41.1	33 23	25 20	40.7	32 37	24 34	40.3	31 51	23 48	39.9	31 05	23 03	39.5	30 19	22 19	39.1	237	303
58	122	34 34	25 30	40.1	33 48	24 44	39.6	33 02	23 58	39.2	32 15	23 14	38.8	31 28	22 29	38.4	30 41	21 46	38.0	238	302
59	121	35 00	24 53	39.0	34 13	24 07	38.5	33 26	23 22	38.1	32 39	22 38	37.7	31 51	21 55	37.3	31 03	21 13	37.0	239	301
60	120	35 25	24 14	37.8	34 37	23 30	37.4	33 50	22 46	37.0	33 02	22 03	36.6	32 13	21 20	36.2	31 25	20 39	35.9	240	300
61	119	35 49	23 35	36.7	35 01	22 51	36.3	34 12	22 08	35.9	33 24	21 26	35.5	32 35	20 45	35.1	31 46	20 04	34.8	241	299
62	118	36 13	22 55	35.6	35 24	22 12	35.2	34 35	21 30	34.8	33 45	20 49	34.4	32 56	20 09	34.0	32 06	19 29	33.7	242	298
63	117	36 36	22 14	34.4	35 46	21 32	34.0	34 56	20 51	33.6	34 06	20 11	33.3	33 16	19 32	32.9	32 26	18 53	32.5	243	297
64	116	36 58	21 32	33.3	36 08	20 52	32.9	35 17	20 12	32.5	34 27	19 33	32.1	33 36	18 54	31.8	32 45	18 17	31.4	244	296
65	115	37 20	20 50	32.1	36 29	20 10	31.7	35 38	19 32	31.3	34 47	18 54	31.0	33 55	18 16	30.6	33 03	17 40	30.3	245	295
66	114	37 41	20 07	30.9	36 49	19 28	30.5	35 58	18 51	30.2	35 06	18 14	29.8	34 13	17 38	29.5	33 21	17 02	29.1	246	294
67	113	38 01	19 23	29.7	37 09	18 46	29.4	36 17	18 09	29.0	35 24	17 33	28.6	34 31	16 59	28.3	33 38	16 24	28.0	247	293
68	112	38 21	18 38	28.5	37 28	18 02	28.2	36 35	17 27	27.8	35 42	16 53	27.5	34 48	16 19	27.1	33 55	15 46	26.8	248	292
69	111	38 40	17 53	27.3	37 46	17 18	27.0	36 53	16 44	26.6	35 59	16 11	26.3	35 05	15 38	26.0	34 11	15 07	25.7	249	291
70	110	38 58	17 07	26.1	38 04	16 33	25.7	37 10	16 01	25.4	36 15	15 29	25.1	35 21	14 58	24.8	34 26	14 27	24.5	250	290
71	109	39 15	16 20	24.9	38 20	15 48	24.5	37 26	15 17	24.2	36 31	14 46	23.9	35 36	14 16	23.6	34 41	13 47	23.3	251	289
72	108	39 31	15 33	23.6	38 36	15 02	23.3	37 41	14 32	23.0	36 46	14 03	22.7	35 50	13 34	22.4	34 55	13 07	22.1	252	288
73	107	39 47	14 45	22.4	38 51	14 16	22.1	37 56	13 47	21.8	37 00	13 19	21.5	36 04	12 52	21.2	35 08	12 25	20.9	253	287
74	106	40 02	13 56	21.1	39 06	13 28	20.8	38 10	13 01	20.5	37 13	12 35	20.3	36 17	12 09	20.0	35 21	11 44	19.8	254	286
75	105	40 16	13 07	19.8	39 19	12 41	19.5	38 23	12 15	19.3	37 26	11 50	19.0	36 29	11 26	18.8	35 33	11 02	18.5	255	285
76	104	40 29	12 17	18.5	39 32	11 53	18.3	38 35	11 28	18.0	37 38	11 05	17.8	36 41	10 42	17.6	35 44	10 20	17.3	256	284
77	103	40 41	11 27	17.3	39 44	11 04	17.0	38 47	10 41	16.8	37 49	10 19	16.5	36 52	9 58	16.3	35 54	9 37	16.1	257	283
78	102	40 53	10 36	16.0	39 55	10 15	15.7	38 57	9 54	15.5	37 59	9 33	15.3	37 02	9 14	15.1	36 04	8 54	14.9	258	282
79	101	41 04	9 45	14.7	40 05	9 25	14.4	39 07	9 06	14.2	38 09	8 47	14.0	37 11	8 29	13.9	36 13	8 11	13.7	259	281
80	100	41 13	8 53	13.3	40 15	8 35	13.2	39 16	8 17	13.0	38 18	8 00	12.8	37 19	7 44	12.6	36 21	7 27	12.5	260	280
81	99	41 22	8 01	12.0	40 23	7 45	11.9	39 25	7 29	11.7	38 26	7 13	11.5	37 27	6 58	11.4	36 28	6 43	11.2	261	279
82	98	41 30	7 09	10.7	40 31	6 54	10.5	39 32	6 40	10.4	38 33	6 26	10.3	37 34	6 12	10.1	36 35	5 59	10.0	262	278
83	97	41 37	6 16	9.4	40 38	6 03	9.2	39 39	5 50	9.1	38 39	5 38	9.0	37 40	5 26	8.9	36 41	5 15	8.7	263	277
84	96	41 43	5 23	8.1	40 44	5 12	7.9	39 44	5 01	7.8	38 45	4 50	7.7	37 45	4 40	7.6	36 46	4 30	7.5	264	276
85	95	41 48	4 29	6.7	40 49	4 20	6.6	39 49	4 11	6.5	38 49	4 02	6.4	37 50	3 54	6.3	36 50	3 45	6.3	265	275
86	94	41 52	3 36	5.4	40 53	3 28	5.3	39 53	3 21	5.2	38 53	3 14	5.1	37 53	3 07	5.1	36 54	3 01	5.0	266	274
87	93	41 56	2 42	4.0	40 56	2 36	4.0	39 56	2 31	3.9	38 56	2 26	3.9	37 56	2 20	3.8	36 56	2 16	3.8	267	273
88	92	41 58	1 48	2.7	40 58	1 44	2.6	39 58	1 41	2.6	38 58	1 37	2.5	37 58	1 34	2.5	36 58	1 30	2.5	268	272
89	91	42 00	0 54	1.3	41 00	0 52	1.3	40 00	0 50	1.3	39 00	0 49	1.3	38 00	0 47	1.3	37 00	0 45	1.3	269	271
90	90	42 00	0 00	0.0	41 00	0 00	0.0	40 00	0 00	0.0	39 00	0 00	0.0	38 00	0 00	0.0	37 00	0 00	0.0	270	270

N. Lat.: for LHA > 180°... $Z_n = Z$
for LHA < 180°... $Z_n = 360° - Z$

S. Lat.: for LHA > 180°... $Z_n = 180° - Z$
for LHA < 180°... $Z_n = 180° + Z$

SIGHT REDUCTION TABLE

B: (–) for 90° < LHA < 270°
Dec: (–) for Lat. contrary name

Z₁: same sign as B
Z₂: (–) for F > 90°

Z_1: same sign as B
Z_2: (–) for F > 90°

Lat./A	LHA/F	54° A/H	54° B/P	54° Z_1/Z_2	55° A/H	55° B/P	55° Z_1/Z_2	56° A/H	56° B/P	56° Z_1/Z_2	57° A/H	57° B/P	57° Z_1/Z_2	58° A/H	58° B/P	58° Z_1/Z_2	59° A/H	59° B/P	59° Z_1/Z_2	Lat./A	LHA
0	180	0 00	36 00	90.0	0 00	35 00	90.0	0 00	34 00	90.0	0 00	33 00	90.0	0 00	32 00	90.0	0 00	31 00	90.0	0	180
1	179	0 35	36 00	89.2	0 34	35 00	89.2	0 34	34 00	89.2	0 33	33 00	89.2	0 32	32 00	89.2	0 31	31 00	89.1	1	181
2	178	1 11	35 59	88.4	1 09	34 59	88.4	1 07	33 59	88.3	1 05	32 59	88.3	1 04	31 59	88.3	1 02	30 59	88.3	2	182
3	177	1 46	35 58	87.6	1 43	34 58	87.5	1 41	33 58	87.5	1 38	32 58	87.5	1 35	31 58	87.5	1 33	30 58	87.4	3	183
4	176	2 21	35 56	86.8	2 18	34 56	86.7	2 14	33 56	86.7	2 11	32 56	86.6	2 07	31 56	86.6	2 04	30 56	86.6	4	184
5	175	2 56	35 54	86.0	2 52	34 54	85.9	2 48	33 54	85.9	2 43	32 54	85.8	2 39	31 54	85.8	2 34	30 54	85.7	5	185
6	174	3 31	35 51	85.1	3 26	34 51	85.1	3 21	33 51	85.0	3 16	32 51	85.0	3 11	31 52	84.9	3 05	30 52	84.9	6	186
7	173	4 06	35 48	84.3	4 00	34 48	84.3	3 54	33 48	84.2	3 48	32 48	84.1	3 42	31 48	84.1	3 36	30 49	84.0	7	187
8	172	4 42	35 44	83.5	4 35	34 44	83.4	4 28	33 44	83.4	4 21	32 45	83.3	4 14	31 45	83.2	4 07	30 45	83.1	8	188
9	171	5 17	35 40	82.7	5 09	34 40	82.6	5 01	33 40	82.5	4 53	32 41	82.4	4 45	31 41	82.3	4 37	30 41	82.3	9	189
10	170	5 51	35 35	81.9	5 43	34 35	81.8	5 34	33 36	81.7	5 26	32 36	81.6	5 17	31 36	81.5	5 08	30 37	81.4	10	190
11	169	6 26	35 30	81.1	6 17	34 30	81.0	6 08	33 31	80.8	5 58	32 31	80.7	5 48	31 31	80.6	5 38	30 32	80.5	11	191
12	168	7 01	35 24	80.2	6 51	34 24	80.1	6 41	33 25	80.0	6 30	32 25	79.9	6 20	31 26	79.8	6 09	30 27	79.7	12	192
13	167	7 36	35 18	79.4	7 25	34 18	79.3	7 14	33 19	79.2	7 02	32 19	79.0	6 51	31 20	78.9	6 39	30 21	78.8	13	193
14	166	8 11	35 11	78.6	7 59	34 12	78.5	7 46	33 12	78.3	7 34	32 13	78.2	7 22	31 14	78.1	7 09	30 15	77.9	14	194
15	165	8 45	35 04	77.8	8 32	34 04	77.6	8 19	33 05	77.5	8 06	32 06	77.3	7 53	31 07	77.2	7 40	30 08	77.1	15	195
16	164	9 19	34 56	76.9	9 06	33 57	76.8	8 52	32 58	76.6	8 38	31 59	76.5	8 24	31 00	76.3	8 10	30 01	76.2	16	196
17	163	9 54	34 47	76.1	9 39	33 48	75.9	9 25	32 49	75.8	9 10	31 50	75.6	8 55	30 52	75.5	8 40	29 53	75.3	17	197
18	162	10 28	34 39	75.3	10 13	33 40	75.1	9 57	32 41	74.9	9 41	31 42	74.8	9 25	30 43	74.6	9 09	29 45	74.4	18	198
19	161	11 02	34 29	74.4	10 46	33 31	74.2	10 29	32 32	74.1	10 13	31 33	73.9	9 56	30 35	73.7	9 39	29 36	73.6	19	199
20	160	11 36	34 19	73.6	11 19	33 21	73.4	11 02	32 22	73.2	10 44	31 24	73.0	10 27	30 25	72.8	10 08	29 27	72.7	20	200
21	159	12 10	34 09	72.7	11 52	33 10	72.5	11 34	32 12	72.3	11 15	31 14	72.2	10 57	30 15	72.0	10 38	29 17	71.8	21	201
22	158	12 43	33 58	71.9	12 24	33 00	71.7	12 06	32 01	71.5	11 46	31 03	71.3	11 27	30 05	71.1	11 07	29 07	70.9	22	202
23	157	13 17	33 46	71.0	12 57	32 48	70.8	12 37	31 50	70.6	12 17	30 52	70.4	11 57	29 54	70.2	11 37	28 57	70.0	23	203
24	156	13 50	33 34	70.2	13 29	32 36	70.0	13 09	31 38	69.7	12 48	30 41	69.5	12 27	29 43	69.3	12 06	28 46	69.1	24	204
25	155	14 23	33 22	69.3	14 02	32 24	69.1	13 40	31 26	68.9	13 18	30 29	68.6	12 56	29 31	68.4	12 34	28 34	68.2	25	205
26	154	14 56	33 09	68.5	14 34	32 11	68.2	14 11	31 14	68.0	13 49	30 16	67.8	13 26	29 19	67.5	13 03	28 22	67.3	26	206
27	153	15 29	32 55	67.6	15 06	31 58	67.3	14 42	31 01	67.1	14 19	30 03	66.9	13 55	29 06	66.6	13 31	28 10	66.4	27	207
28	152	16 01	32 41	66.7	15 37	31 44	66.5	15 13	30 47	66.2	14 49	29 50	66.0	14 24	28 53	65.7	14 00	27 57	65.5	28	208
29	151	16 33	32 26	65.8	16 09	31 29	65.6	15 44	30 32	65.3	15 19	29 36	65.1	14 53	28 39	64.8	14 28	27 43	64.6	29	209
30	150	17 05	32 11	65.0	16 40	31 14	64.7	16 14	30 17	64.4	15 48	29 21	64.2	15 22	28 25	63.9	14 55	27 29	63.7	30	210
31	149	17 37	31 55	64.1	17 11	30 58	63.8	16 44	30 02	63.5	16 17	29 06	63.3	15 50	28 10	63.0	15 23	27 15	62.7	31	211
32	148	18 09	31 38	63.2	17 42	30 42	62.9	17 14	29 46	62.6	16 47	28 51	62.3	16 19	27 55	62.1	15 50	27 00	61.8	32	212
33	147	18 40	31 21	62.3	18 12	30 25	62.0	17 44	29 30	61.7	17 15	28 34	61.4	16 47	27 39	61.2	16 17	26 45	60.9	33	213
34	146	19 11	31 04	61.4	18 42	30 08	61.1	18 13	29 13	60.8	17 44	28 18	60.5	17 14	27 23	60.2	16 44	26 29	60.0	34	214
35	145	19 42	30 46	60.5	19 12	29 50	60.2	18 42	28 55	59.9	18 12	28 01	59.6	17 42	27 06	59.3	17 11	26 12	59.0	35	215
36	144	20 13	30 27	59.6	19 42	29 32	59.2	19 11	28 37	58.9	18 40	27 43	58.6	18 09	26 49	58.4	17 37	25 55	58.1	36	216
37	143	20 43	30 07	58.6	20 12	29 13	58.3	19 40	28 19	58.0	19 08	27 25	57.7	18 36	26 31	57.4	18 03	25 38	57.1	37	217
38	142	21 13	29 48	57.7	20 41	28 53	57.4	20 08	27 59	57.1	19 35	27 06	56.8	19 02	26 13	56.5	18 29	25 20	56.2	38	218
39	141	21 43	29 27	56.8	21 10	28 33	56.4	20 36	27 40	56.1	20 03	26 47	55.8	19 29	25 54	55.5	18 55	25 02	55.2	39	219
40	140	22 12	29 06	55.8	21 38	28 13	55.5	21 04	27 20	55.2	20 30	26 27	54.9	19 55	25 35	54.6	19 20	24 43	54.3	40	220
41	139	22 41	28 44	54.9	22 06	27 51	54.5	21 31	26 59	54.2	20 56	26 07	53.9	20 21	25 15	53.6	19 45	24 24	53.3	41	221
42	138	23 10	28 22	53.9	22 34	27 29	53.6	21 58	26 37	53.3	21 22	25 46	52.9	20 46	24 55	52.6	20 10	24 04	52.3	42	222
43	137	23 38	27 59	53.0	23 02	27 07	52.6	22 25	26 15	52.3	21 48	25 24	52.0	21 11	24 34	51.7	20 34	23 43	51.4	43	223
44	136	24 06	27 36	52.0	23 29	26 44	51.7	22 51	25 53	51.3	22 14	25 02	51.0	21 36	24 12	50.7	20 58	23 23	50.4	44	224
45	135	24 34	27 11	51.0	23 56	26 20	50.7	23 17	25 30	50.3	22 39	24 40	50.0	22 00	23 50	49.7	21 21	23 01	49.4	45	225

Lat./F		54° A/H	54° B/P	54° Z1/Z2	55° A/H	55° B/P	55° Z1/Z2	56° A/H	56° B/P	56° Z1/Z2	57° A/H	57° B/P	57° Z1/Z2	58° A/H	58° B/P	58° Z1/Z2	59° A/H	59° B/P	59° Z1/Z2	Lat./A	
LHA/F °	°																			LHA °	°
45	135	24 34	27 11	51.0	23 56	26 20	50.7	23 17	25 30	50.3	22 39	24 40	50.0	22 00	23 50	49.7	21 21	23 01	49.4	225	315
46	134	25 01	26 47	50.0	24 22	25 56	49.7	23 43	25 06	49.4	23 04	24 17	49.0	22 24	23 28	48.7	21 45	22 39	48.4	226	314
47	133	25 28	26 22	49.1	24 48	25 32	48.7	24 08	24 42	48.4	23 28	23 53	48.0	22 48	23 05	47.7	22 08	22 17	47.4	227	313
48	132	25 54	25 56	48.1	25 14	25 06	47.7	24 33	24 17	47.4	23 53	23 29	47.0	23 11	22 41	46.7	22 30	21 54	46.4	228	312
49	131	26 20	25 29	47.1	25 39	24 40	46.7	24 58	23 52	46.4	24 16	23 05	46.0	23 34	22 17	45.7	22 52	21 31	45.4	229	311
50	130	26 46	25 02	46.0	26 04	24 14	45.7	25 22	23 26	45.3	24 40	22 39	45.0	23 57	21 53	44.7	23 14	21 07	44.4	230	310
51	129	27 11	24 34	45.0	26 28	23 47	44.7	25 45	23 00	44.3	25 02	22 14	44.0	24 19	21 28	43.7	23 36	20 43	43.4	231	309
52	128	27 36	24 06	44.0	26 52	23 19	43.6	26 09	22 33	43.3	25 25	21 48	43.0	24 41	21 03	42.7	23 57	20 18	42.3	232	308
53	127	28 00	23 37	43.0	27 16	22 51	42.6	26 32	22 06	42.3	25 47	21 21	41.9	25 02	20 37	41.6	24 17	19 53	41.3	233	307
54	126	28 24	23 07	41.9	27 39	22 22	41.6	26 54	21 38	41.2	26 09	20 54	40.9	25 23	20 10	40.6	24 37	19 27	40.3	234	306
55	125	28 47	22 37	40.9	28 01	21 53	40.5	27 16	21 09	40.2	26 30	20 26	39.9	25 44	19 43	39.5	24 57	19 01	39.2	235	305
56	124	29 10	22 07	39.8	28 24	21 23	39.5	27 37	20 40	39.1	26 50	19 57	38.8	26 04	19 16	38.5	25 17	18 34	38.2	236	304
57	123	29 32	21 35	38.8	28 45	20 52	38.4	27 58	20 10	38.1	27 11	19 29	37.8	26 23	18 48	37.4	25 35	18 07	37.1	237	303
58	122	29 54	21 03	37.7	29 06	20 21	37.3	28 19	19 40	37.0	27 31	18 59	36.7	26 42	18 19	36.4	25 54	17 40	36.1	238	302
59	121	30 15	20 31	36.6	29 27	19 50	36.3	28 38	19 09	35.9	27 50	18 30	35.6	27 01	17 50	35.3	26 12	17 12	35.0	239	301
60	120	30 36	19 58	35.5	29 47	19 18	35.2	28 58	18 38	34.9	28 09	17 59	34.5	27 19	17 21	34.2	26 29	16 43	33.9	240	300
61	119	30 56	19 24	34.4	30 07	18 45	34.1	29 17	18 06	33.8	28 27	17 29	33.5	27 37	16 51	33.2	26 46	16 14	32.9	241	299
62	118	31 16	18 50	33.3	30 26	18 12	33.0	29 35	17 34	32.7	28 45	16 57	32.4	27 54	16 21	32.1	27 03	15 45	31.8	242	298
63	117	31 35	18 15	32.2	30 44	17 38	31.9	29 53	17 02	31.6	29 02	16 26	31.3	28 10	15 50	31.0	27 19	15 15	30.7	243	297
64	116	31 53	17 40	31.1	31 02	17 04	30.8	30 10	16 28	30.5	29 19	15 53	30.2	28 27	15 19	29.9	27 35	14 45	29.6	244	296
65	115	32 11	17 04	30.0	31 19	16 29	29.7	30 27	15 55	29.4	29 35	15 21	29.1	28 42	14 48	28.8	27 50	14 15	28.5	245	295
66	114	32 29	16 28	28.8	31 36	15 54	28.5	30 43	15 20	28.2	29 50	14 48	28.0	28 57	14 16	27.7	28 04	13 44	27.4	246	294
67	113	32 45	15 51	27.7	31 52	15 18	27.4	30 59	14 46	27.1	30 05	14 14	26.8	29 12	13 43	26.6	28 18	13 13	26.3	247	293
68	112	33 01	15 14	26.5	32 08	14 42	26.3	31 14	14 11	26.0	30 20	13 40	25.7	29 26	13 10	25.5	28 31	12 41	25.2	248	292
69	111	33 17	14 36	25.4	32 23	14 05	25.1	31 28	13 35	24.8	30 34	13 06	24.6	29 39	12 37	24.4	28 44	12 09	24.1	249	291
70	110	33 32	13 57	24.2	32 38	13 28	24.0	31 42	12 59	23.7	30 47	12 31	23.5	29 52	12 04	23.2	28 57	11 37	23.0	250	290
71	109	33 46	13 18	23.1	32 51	12 51	22.8	31 55	12 23	22.6	31 00	11 56	22.3	30 04	11 30	22.1	29 09	11 04	21.9	251	289
72	108	33 59	12 39	21.9	33 04	12 13	21.6	32 08	11 46	21.4	31 12	11 21	21.2	30 16	10 56	21.0	29 20	10 31	20.8	252	288
73	107	34 12	12 00	20.7	33 16	11 34	20.5	32 20	11 09	20.3	31 23	10 45	20.0	30 27	10 21	19.8	29 30	9 58	19.6	253	287
74	106	34 24	11 19	19.5	33 28	10 55	19.3	32 31	10 32	19.1	31 34	10 09	18.9	30 37	9 46	18.7	29 41	9 24	18.5	254	286
75	105	34 36	10 39	18.3	33 39	10 16	18.1	32 42	9 54	17.9	31 44	9 32	17.7	30 47	9 11	17.5	29 50	8 50	17.4	255	285
76	104	34 46	9 58	17.1	33 49	9 37	16.9	32 52	9 16	16.7	31 54	8 56	16.6	30 57	8 36	16.4	29 59	8 16	16.2	256	284
77	103	34 56	9 17	15.9	33 59	8 57	15.7	33 01	8 38	15.6	32 03	8 19	15.4	31 05	8 00	15.2	30 07	7 42	15.1	257	283
78	102	35 06	8 35	14.7	34 08	8 17	14.5	33 10	7 59	14.4	32 11	7 41	14.2	31 13	7 24	14.1	30 15	7 07	13.9	258	282
79	101	35 14	7 54	13.5	34 16	7 37	13.3	33 18	7 20	13.2	32 19	7 04	13.0	31 21	6 48	12.9	30 22	6 32	12.8	259	281
80	100	35 22	7 11	12.3	34 24	6 56	12.1	33 25	6 41	12.0	32 26	6 26	11.9	31 27	6 12	11.7	30 29	5 57	11.6	260	280
81	99	35 29	6 29	11.1	34 30	6 15	10.9	33 32	6 01	10.8	32 33	5 48	10.7	31 34	5 35	10.6	30 35	5 22	10.5	261	279
82	98	35 36	5 46	9.9	34 37	5 34	9.7	33 37	5 22	9.6	32 38	5 10	9.5	31 39	4 58	9.4	30 40	4 47	9.3	262	278
83	97	35 41	5 04	8.6	34 42	4 53	8.5	33 43	4 42	8.4	32 43	4 32	8.3	31 44	4 21	8.2	30 45	4 11	8.2	263	277
84	96	35 46	4 21	7.4	34 47	4 11	7.3	33 47	4 02	7.2	32 48	3 53	7.1	31 48	3 44	7.1	30 49	3 36	7.0	264	276
85	95	35 51	3 37	6.2	34 51	3 30	6.1	33 51	3 22	6.0	32 52	3 14	6.0	31 52	3 07	5.9	30 52	3 00	5.8	265	275
86	94	35 54	2 54	4.9	34 54	2 48	4.9	33 54	2 42	4.8	32 55	2 36	4.8	31 55	2 30	4.7	30 55	2 24	4.7	266	274
87	93	35 57	2 11	3.7	34 57	2 06	3.7	33 57	2 01	3.6	32 57	1 57	3.6	31 57	1 52	3.5	30 57	1 48	3.5	267	273
88	92	35 58	1 27	2.5	34 59	1 24	2.4	33 59	1 21	2.4	32 59	1 18	2.4	31 59	1 15	2.4	30 59	1 12	2.3	268	272
89	91	36 00	0 44	1.2	35 00	0 42	1.2	34 00	0 40	1.2	33 00	0 39	1.2	32 00	0 37	1.2	31 00	0 36	1.2	269	271
90	90	36 00	0 00	0.0	35 00	0 00	0.0	34 00	0 00	0.0	33 00	0 00	0.0	32 00	0 00	0.0	31 00	0 00	0.0	270	270

N. Lat.: for LHA > 180° ... Zn = Z
for LHA < 180° ... Zn = 360° - Z

S. Lat.: for LHA > 180° ... Zn = 180° - Z
for LHA < 180° ... Zn = 180° + Z

B: (–) for 90°< LHA < 270°
Dec: (–) for Lat. contrary name

Z₁: same sign as B
Z₂: (–) for F > 90°

SIGHT REDUCTION TABLE

Lat./A LHA/F	F	60° A/H	60° B/P	60° Z₁/Z₂	61° A/H	61° B/P	61° Z₁/Z₂	62° A/H	62° B/P	62° Z₁/Z₂	63° A/H	63° B/P	63° Z₁/Z₂	64° A/H	64° B/P	64° Z₁/Z₂	65° A/H	65° B/P	65° Z₁/Z₂	Lat./A LHA	LHA
0	180	0 00	30 00	90.0	0 00	29 00	90.0	0 00	28 00	90.0	0 00	27 00	90.0	0 00	26 00	90.0	0 00	25 00	90.0	180	360
1	179	0 30	30 00	89.1	0 29	29 00	89.1	0 28	28 00	89.1	0 27	27 00	89.1	0 26	26 00	89.1	0 25	25 00	89.1	181	359
2	178	1 00	29 59	88.3	0 58	28 59	88.3	0 56	27 59	88.2	0 54	26 59	88.2	0 53	25 59	88.2	0 51	24 58	88.2	182	358
3	177	1 30	29 58	87.4	1 27	28 58	87.4	1 24	27 58	87.3	1 22	26 58	87.3	1 19	25 58	87.3	1 16	24 58	87.3	183	357
4	176	2 00	29 56	86.5	1 56	28 56	86.5	1 53	27 57	86.4	1 49	26 57	86.4	1 45	25 57	86.4	1 41	24 57	86.4	184	356
5	175	2 30	29 54	85.7	2 25	28 54	85.6	2 21	27 55	85.6	2 16	26 55	85.5	2 11	25 55	85.5	2 07	24 55	85.5	185	355
6	174	3 00	29 52	84.8	2 54	28 52	84.7	2 49	27 52	84.7	2 43	26 52	84.6	2 38	25 53	84.6	2 32	24 53	84.6	186	354
7	173	3 30	29 49	83.9	3 23	28 49	83.9	3 17	27 49	83.8	3 10	26 50	83.8	3 04	25 50	83.7	2 57	24 50	83.7	187	353
8	172	3 59	29 45	83.1	3 52	28 46	83.0	3 45	27 46	82.9	3 37	26 46	82.9	3 30	25 47	82.9	3 22	24 47	82.8	188	352
9	171	4 29	29 42	82.2	4 21	28 42	82.1	4 13	27 42	82.0	4 04	26 43	82.0	3 56	25 43	82.0	3 47	24 44	81.8	189	351
10	170	4 59	29 37	81.3	4 50	28 38	81.2	4 41	27 38	81.2	4 31	26 39	81.1	4 22	25 39	81.0	4 13	24 40	80.9	190	350
11	169	5 28	29 33	80.4	5 18	28 33	80.4	5 08	27 34	80.3	4 58	26 34	80.2	4 48	25 35	80.1	4 38	24 36	80.0	191	349
12	168	5 58	29 27	79.6	5 47	28 28	79.5	5 36	27 29	79.4	5 25	26 29	79.3	5 14	25 30	79.2	5 02	24 31	79.1	192	348
13	167	6 27	29 22	78.7	6 16	28 22	78.6	6 04	27 23	78.5	5 52	26 24	78.4	5 40	25 25	78.3	5 27	24 26	78.2	193	347
14	166	6 57	29 15	77.8	6 44	28 16	77.7	6 31	27 17	77.6	6 18	26 18	77.5	6 05	25 20	77.4	5 52	24 21	77.3	194	346
15	165	7 26	29 09	76.9	7 13	28 10	76.8	6 59	27 11	76.7	6 45	26 12	76.6	6 31	25 14	76.5	6 17	24 15	76.4	195	345
16	164	7 55	29 02	76.1	7 41	28 03	75.9	7 26	27 04	75.8	7 11	26 06	75.7	6 56	25 07	75.6	6 41	24 09	75.4	196	344
17	163	8 24	28 54	75.2	8 09	27 56	75.0	7 53	26 57	74.9	7 38	25 59	74.8	7 22	25 00	74.6	7 06	24 02	74.5	197	343
18	162	8 53	28 46	74.3	8 37	27 48	74.1	8 20	26 50	74.0	8 04	25 51	73.9	7 47	24 53	73.7	7 30	23 55	73.6	198	342
19	161	9 22	28 38	73.4	9 05	27 40	73.2	8 48	26 41	73.1	8 30	25 43	72.9	8 12	24 45	72.8	7 55	23 48	72.7	199	341
20	160	9 51	28 29	72.5	9 33	27 31	72.3	9 14	26 33	72.2	8 56	25 35	72.0	8 37	24 37	71.9	8 19	23 40	71.7	200	340
21	159	10 19	28 19	71.6	10 00	27 22	71.4	9 41	26 24	71.2	9 22	25 26	71.1	9 02	24 29	71.0	8 43	23 32	70.8	201	339
22	158	10 48	28 10	70.7	10 28	27 12	70.5	10 08	26 15	70.4	9 48	25 17	70.2	9 27	24 20	70.0	9 07	23 23	69.9	202	338
23	157	11 16	27 59	69.8	10 55	27 02	69.6	10 34	26 05	69.5	10 13	25 08	69.3	9 52	24 11	69.1	9 30	23 14	69.0	203	337
24	156	11 44	27 49	68.9	11 22	26 51	68.7	11 00	25 54	68.7	10 38	24 58	68.4	10 16	24 01	68.2	9 54	23 04	68.0	204	336
25	155	12 12	27 37	68.0	11 49	26 40	67.8	11 27	25 44	67.8	11 04	24 47	67.4	10 41	23 51	67.3	10 17	22 55	67.1	205	335
26	154	12 40	27 26	67.1	12 16	26 29	66.9	11 53	25 33	66.9	11 29	24 36	66.5	11 05	23 40	66.3	10 41	22 44	66.2	206	334
27	153	13 07	27 13	66.2	12 43	26 17	66.0	12 18	25 21	66.0	11 54	24 25	65.5	11 29	23 29	65.4	11 04	22 34	65.2	207	333
28	152	13 35	27 01	65.3	13 09	26 05	65.1	12 44	25 09	65.1	12 18	24 13	64.6	11 53	23 18	64.5	11 27	22 23	64.3	208	332
29	151	14 02	26 48	64.4	13 36	25 52	64.1	13 09	24 56	64.1	12 43	24 01	63.7	12 16	23 06	63.5	11 49	22 11	63.3	209	331
30	150	14 29	26 34	63.4	14 02	25 39	63.2	13 35	24 43	63.2	13 07	23 49	62.8	12 40	22 54	62.6	12 12	21 59	62.4	210	330
31	149	14 55	26 20	62.5	14 28	25 25	62.3	14 00	24 30	62.3	13 31	23 36	61.8	13 03	22 41	61.6	12 34	21 47	61.4	211	329
32	148	15 22	26 05	61.6	14 53	25 11	61.3	14 24	24 16	61.3	13 55	23 22	60.8	13 26	22 28	60.7	12 56	21 35	60.5	212	328
33	147	15 48	25 50	60.6	15 19	24 56	60.4	14 49	24 02	60.4	14 19	23 08	59.9	13 49	22 15	59.7	13 18	21 22	59.5	213	327
34	146	16 14	25 35	59.7	15 44	24 41	59.5	15 13	23 47	59.5	14 42	22 54	59.0	14 11	22 01	58.8	13 40	21 08	58.6	214	326
35	145	16 40	25 19	58.8	16 09	24 25	58.5	15 37	23 32	58.5	15 06	22 39	58.0	14 34	21 47	57.8	14 02	20 54	57.6	215	325
36	144	17 05	25 02	57.8	16 33	24 09	57.6	16 01	23 17	57.6	15 29	22 24	57.1	14 56	21 32	56.9	14 23	20 40	56.6	216	324
37	143	17 31	24 45	56.9	16 58	23 53	56.6	16 25	23 00	56.6	15 51	22 09	56.1	15 18	21 17	55.9	14 44	20 26	55.7	217	323
38	142	17 56	24 28	55.9	17 22	23 36	55.7	16 48	22 44	55.7	16 14	21 53	55.2	15 39	21 01	54.9	15 05	20 11	54.7	218	322
39	141	18 20	24 10	55.0	17 46	23 18	54.7	17 11	22 27	54.7	16 36	21 36	54.2	16 01	20 46	54.0	15 25	19 55	53.7	219	321
40	140	18 45	23 52	54.0	18 09	23 00	53.7	17 34	22 10	53.7	16 58	21 19	53.2	16 22	20 29	53.0	15 46	19 39	52.7	220	320
41	139	19 09	23 33	53.0	18 33	22 42	52.8	17 56	21 52	52.8	17 20	21 02	52.2	16 43	20 13	52.0	16 06	19 23	51.8	221	319
42	138	19 33	23 13	52.1	18 56	22 23	51.8	18 19	21 34	51.8	17 41	20 44	51.5	17 03	19 55	51.0	16 26	19 07	50.8	222	318
43	137	19 56	22 54	51.1	19 18	22 04	50.8	18 40	21 15	50.8	18 02	20 26	50.3	17 24	19 38	50.0	16 45	18 50	49.8	223	317
44	136	20 19	22 33	50.1	19 41	21 44	49.8	19 02	20 56	49.8	18 23	20 08	49.3	17 44	19 20	49.0	17 04	18 33	48.8	224	316
45	135	20 42	22 12	49.1	20 03	21 24	48.8	19 23	20 36	48.8	18 43	19 49	48.3	18 03	19 02	48.1	17 23	18 15	47.8	225	315

Lat./A	LHA/F	60° A/H	60° B/P	60° Z_1/Z_2	61° A/H	61° B/P	61° Z_1/Z_2	62° A/H	62° B/P	62° Z_1/Z_2	63° A/H	63° B/P	63° Z_1/Z_2	64° A/H	64° B/P	64° Z_1/Z_2	65° Z_1/Z_2	65° B/P	65° A/H	LHA	Lat./A
45	135	20 42	22 12	49.1	20 03	21 24	48.8	19 23	20 36	48.6	18 43	19 49	48.3	18 03	19 02	48.1	47.8	18 15	17 23	225	315
46	134	21 05	21 51	48.1	20 25	21 04	47.8	19 44	20 16	47.6	19 04	19 29	47.3	18 23	18 43	47.1	46.8	17 57	17 42	226	314
47	133	21 27	21 30	47.1	20 46	20 43	46.8	20 05	19 56	46.6	19 24	19 10	46.3	18 42	18 24	46.1	45.8	17 39	18 00	227	313
48	132	21 49	21 07	46.1	21 07	20 21	45.8	20 25	19 35	45.6	19 43	18 50	45.3	19 01	18 04	45.1	44.8	17 20	18 18	228	312
49	131	22 10	20 45	45.1	21 28	19 59	44.8	20 45	19 14	44.6	20 02	18 29	44.3	19 19	17 45	44.0	43.8	17 01	18 36	229	311
50	130	22 31	20 22	44.1	21 48	19 37	43.8	21 05	18 52	43.5	20 21	18 08	43.3	19 37	17 24	43.0	42.8	16 41	18 53	230	310
51	129	22 52	19 58	43.1	22 08	19 14	42.8	21 24	18 30	42.5	20 40	17 47	42.3	19 55	17 04	42.0	41.8	16 21	19 10	231	309
52	128	23 12	19 34	42.1	22 28	18 51	41.8	21 43	18 08	41.5	20 58	17 25	41.2	20 13	16 43	41.0	40.8	16 01	19 27	232	308
53	127	23 32	19 10	41.0	22 47	18 27	40.7	22 01	17 45	40.5	21 15	17 03	40.2	20 30	16 23	40.0	39.7	15 41	19 44	233	307
54	126	23 52	18 45	40.0	23 06	18 03	39.7	22 19	17 21	39.4	21 33	16 40	39.2	20 46	16 00	39.0	38.7	15 20	20 00	234	306
55	125	24 11	18 19	39.0	23 24	17 38	38.7	22 37	16 58	38.4	21 50	16 17	38.2	21 03	15 38	37.9	37.7	14 58	20 15	235	305
56	124	24 29	17 54	37.9	23 42	17 13	37.6	22 54	16 34	37.4	22 07	15 54	37.1	21 19	15 15	36.9	36.7	14 37	20 31	236	304
57	123	24 48	17 27	36.9	23 59	16 48	36.6	23 11	16 09	36.3	22 23	15 31	36.1	21 34	14 53	35.8	35.6	14 15	20 46	237	303
58	122	25 05	17 01	35.8	24 17	16 22	35.5	23 28	15 44	35.3	22 39	15 07	35.0	21 49	14 29	34.8	34.6	13 53	21 00	238	302
59	121	25 23	16 34	34.8	24 33	15 56	34.5	23 44	15 19	34.2	22 54	14 42	34.0	22 04	14 06	33.8	33.5	13 30	21 14	239	301
60	120	25 40	16 06	33.7	24 50	15 29	33.4	23 59	14 53	33.2	23 09	14 18	32.9	22 19	13 42	32.7	32.5	13 07	21 28	240	300
61	119	25 56	15 38	32.6	25 06	15 03	32.4	24 14	14 27	32.1	23 24	13 53	31.9	22 33	13 18	31.7	31.5	12 44	21 42	241	299
62	118	26 12	15 10	31.5	25 21	14 35	31.3	24 29	14 01	31.1	23 38	13 27	30.8	22 46	12 54	30.6	30.4	12 21	21 55	242	298
63	117	26 27	14 41	30.5	25 36	14 08	30.2	24 44	13 34	30.0	23 52	13 01	29.8	22 59	12 29	29.5	29.3	11 57	22 07	243	297
64	116	26 42	14 12	29.4	25 50	13 39	29.1	24 57	13 07	28.9	24 05	12 35	28.7	23 12	12 04	28.5	28.3	11 33	22 19	244	296
65	115	26 57	13 43	28.3	26 04	13 11	28.1	25 11	12 40	27.8	24 18	12 09	27.6	23 25	11 39	27.4	27.2	11 09	22 31	245	295
66	114	27 11	13 13	27.2	26 17	12 42	27.0	25 24	12 12	26.8	24 30	11 43	26.6	23 36	11 13	26.4	26.2	10 44	22 43	246	294
67	113	27 24	12 43	26.1	26 30	12 13	25.9	25 36	11 44	25.7	24 42	11 16	25.5	23 48	10 47	25.3	25.1	10 20	22 54	247	293
68	112	27 37	12 12	25.0	26 43	11 44	24.8	25 48	11 16	24.6	24 54	10 48	24.4	23 59	10 21	24.2	24.0	9 55	23 04	248	292
69	111	27 50	11 41	23.9	26 55	11 14	23.7	26 00	10 47	23.5	25 05	10 21	23.3	24 09	9 55	23.1	23.0	9 29	23 14	249	291
70	110	28 01	11 10	22.8	27 06	10 44	22.6	26 11	10 18	22.4	25 15	9 53	22.2	24 20	9 28	22.0	21.9	9 04	23 24	250	290
71	109	28 13	10 39	21.7	27 17	10 14	21.5	26 21	9 49	21.3	25 25	9 25	21.1	24 29	9 01	21.0	20.8	8 38	23 33	251	289
72	108	28 24	10 07	20.6	27 27	9 43	20.4	26 31	9 20	20.2	25 35	8 57	20.0	24 38	8 34	19.9	19.7	8 12	23 42	252	288
73	107	28 34	9 35	19.4	27 37	9 12	19.3	26 41	8 50	19.1	25 44	8 28	18.9	24 47	8 07	18.8	18.6	7 46	23 50	253	287
74	106	28 44	9 03	18.3	27 47	8 41	18.2	26 50	8 20	18.0	25 52	8 00	17.8	24 55	7 39	17.7	17.5	7 19	23 58	254	286
75	105	28 53	8 30	17.2	27 55	8 10	17.0	26 58	7 50	16.9	26 01	7 31	16.7	25 03	7 12	16.6	16.5	6 53	24 06	255	285
76	104	29 01	7 57	16.1	28 04	7 38	15.9	27 06	7 20	15.8	26 08	7 02	15.6	25 10	6 44	15.5	15.4	6 26	24 13	256	284
77	103	29 09	7 24	14.9	28 11	7 06	14.8	27 13	6 49	14.7	26 15	6 32	14.5	25 17	6 16	14.4	14.3	5 59	24 19	257	283
78	102	29 17	6 51	13.8	28 18	6 34	13.7	27 20	6 19	13.5	26 22	6 03	13.4	25 23	5 47	13.3	13.2	5 32	24 25	258	282
79	101	29 24	6 17	12.7	28 25	6 02	12.5	27 27	5 48	12.4	26 28	5 33	12.3	25 29	5 19	12.2	12.1	5 05	24 31	259	281
80	100	29 30	5 44	11.5	28 31	5 30	11.4	27 32	5 17	11.3	26 33	5 03	11.2	25 35	4 50	11.1	11.0	4 38	24 36	260	280
81	99	29 36	5 10	10.4	28 37	4 57	10.2	27 38	4 45	10.2	26 38	4 33	10.1	25 39	4 22	10.0	9.9	4 10	24 40	261	279
82	98	29 41	4 36	9.2	28 41	4 25	9.1	27 42	4 14	9.1	26 43	4 03	9.0	25 44	3 53	8.9	8.8	3 43	24 44	262	278
83	97	29 45	4 01	8.1	28 46	3 52	8.0	27 46	3 42	7.9	26 47	3 33	7.8	25 48	3 24	7.8	7.7	3 15	24 48	263	277
84	96	29 49	3 27	6.9	28 50	3 19	6.9	27 50	3 11	6.8	26 50	3 03	6.7	25 51	2 55	6.7	6.6	2 47	24 51	264	276
85	95	29 52	2 53	5.8	28 53	2 46	5.7	27 53	2 39	5.7	26 53	2 33	5.6	25 54	2 26	5.6	5.5	2 20	24 54	265	275
86	94	29 55	2 18	4.6	28 55	2 13	4.6	27 56	2 07	4.5	26 56	2 02	4.5	25 56	1 57	4.4	4.4	1 52	24 56	266	274
87	93	29 57	1 44	3.5	28 57	1 40	3.5	27 57	1 36	3.4	26 58	1 32	3.4	25 58	1 28	3.3	3.3	1 24	24 58	267	273
88	92	29 59	1 09	2.3	28 59	1 06	2.3	27 59	1 04	2.3	26 59	1 01	2.2	25 59	0 59	2.2	2.2	0 56	24 59	268	272
89	91	30 00	0 35	1.2	29 00	0 33	1.2	28 00	0 32	1.1	27 00	0 31	1.1	26 00	0 29	1.1	1.1	0 28	25 00	269	271
90	90	30 00	0 00	0.0	29 00	0 00	0.0	28 00	0 00	0.0	27 00	0 00	0.0	26 00	0 00	0.0	0.0	0 00	25 00	270	270

N. Lat.: for LHA > 180°.... $Z_n = Z$; for LHA < 180°.... $Z_n = 360° - Z$

S. Lat.: for LHA > 180°.... $Z_n = 180° - Z$; for LHA < 180°.... $Z_n = 180° + Z$

SIGHT REDUCTION TABLE

LATITUDE / A: 66° – 71°

B : (−) for 90° < LHA < 270°
Dec: (−) for Lat. contrary name

Z₁ : same sign as B
Z₂ : (−) for F > 90°

Lat./A LHA/F	66° A/H	66° B/P	66° Z₁/Z₂	67° A/H	67° B/P	67° Z₁/Z₂	68° A/H	68° B/P	68° Z₁/Z₂	69° A/H	69° B/P	69° Z₁/Z₂	70° A/H	70° B/P	70° Z₁/Z₂	71° A/H	71° B/P	71° Z₁/Z₂	Lat./A LHA
0 180	0 00	24 00	90.0	0 00	23 00	90.0	0 00	22 00	90.0	0 00	21 00	90.0	0 00	20 00	90.0	0 00	19 00	90.0	180 360
1 179	0 24	24 00	89.1	0 23	23 00	89.1	0 22	22 00	89.1	0 22	21 00	89.1	0 21	20 00	89.1	0 20	19 00	89.1	181 359
2 178	0 49	23 59	88.2	0 47	22 59	88.2	0 45	21 59	88.1	0 43	20 59	88.1	0 41	19 59	88.1	0 39	18 59	88.1	182 358
3 177	1 13	23 58	87.3	1 10	22 59	87.2	1 07	21 58	87.2	1 04	20 58	87.2	1 02	19 58	87.2	0 59	18 59	87.2	183 357
4 176	1 38	23 57	86.3	1 34	22 57	86.3	1 30	21 57	86.3	1 26	20 57	86.3	1 22	19 57	86.2	1 18	18 57	86.2	184 356
5 175	2 02	23 55	85.4	1 57	22 55	85.4	1 52	21 55	85.4	1 47	20 56	85.3	1 42	19 56	85.3	1 38	18 56	85.3	185 355
6 174	2 26	23 53	84.5	2 20	22 53	84.5	2 15	21 53	84.4	2 09	20 54	84.4	2 03	19 54	84.4	1 57	18 54	84.3	186 354
7 173	2 50	23 50	83.6	2 44	22 51	83.6	2 37	21 51	83.5	2 30	20 51	83.5	2 23	19 52	83.4	2 16	18 52	83.4	187 353
8 172	3 15	23 48	82.7	3 07	22 48	82.6	2 59	21 48	82.6	2 52	20 49	82.5	2 44	19 49	82.5	2 36	18 50	82.4	188 352
9 171	3 39	23 44	81.8	3 30	22 45	81.7	3 22	21 45	81.6	3 13	20 46	81.6	3 04	19 46	81.5	2 55	18 47	81.5	189 351
10 170	4 03	23 41	80.8	3 53	22 41	80.8	3 44	21 42	80.7	3 34	20 42	80.7	3 24	19 43	80.6	3 14	18 44	80.5	190 350
11 169	4 27	23 36	79.9	4 17	22 37	79.9	4 06	21 38	79.8	3 55	20 39	79.7	3 45	19 40	79.6	3 34	18 41	79.6	191 349
12 168	4 51	23 32	79.0	4 40	22 33	78.9	4 28	21 34	78.9	4 16	20 35	78.8	4 05	19 36	78.7	3 53	18 37	78.6	192 348
13 167	5 15	23 27	78.1	5 03	22 28	78.0	4 50	21 28	77.9	4 37	20 30	77.8	4 25	19 32	77.8	4 12	18 33	77.7	193 347
14 166	5 39	23 22	77.2	5 25	22 23	77.1	5 12	21 24	77.0	4 58	20 26	76.9	4 45	19 27	76.8	4 31	18 28	76.7	194 346
15 165	6 03	23 16	76.2	5 48	22 18	76.1	5 34	21 19	76.0	5 19	20 21	75.9	5 05	19 22	75.8	4 50	18 24	75.8	195 345
16 164	6 26	23 10	75.3	6 11	22 12	75.2	5 56	21 13	75.1	5 40	20 15	75.1	5 25	19 17	74.9	5 09	18 19	74.8	196 344
17 163	6 50	23 04	74.4	6 34	22 06	74.3	6 17	21 08	74.2	6 01	20 09	74.1	5 44	19 11	74.0	5 28	18 14	73.9	197 343
18 162	7 13	22 57	73.5	6 56	21 59	73.3	6 39	21 01	73.2	6 21	20 03	73.1	6 04	19 06	73.0	5 46	18 08	72.9	198 342
19 161	7 37	22 50	72.5	7 19	21 52	72.4	7 00	20 54	72.3	6 42	19 57	72.2	6 24	18 59	72.1	6 05	18 02	72.0	199 341
20 160	8 00	22 42	71.6	7 41	21 45	71.5	7 22	20 47	71.4	7 02	19 50	71.2	6 43	18 53	71.1	6 24	17 56	71.0	200 340
21 159	8 23	22 34	70.7	8 03	21 37	70.5	7 43	20 40	70.4	7 23	19 43	70.3	7 02	18 46	70.2	6 42	17 49	70.1	201 339
22 158	8 46	22 26	69.7	8 25	21 29	69.6	8 04	20 32	69.5	7 43	19 35	69.3	7 22	18 39	69.2	7 00	17 42	69.1	202 338
23 157	9 09	22 17	68.8	8 47	21 21	68.7	8 25	20 24	68.5	8 03	19 28	68.4	7 41	18 31	68.3	7 19	17 35	68.1	203 337
24 156	9 31	22 08	67.9	9 09	21 12	67.7	8 46	20 16	67.6	8 23	19 19	67.4	8 00	18 24	67.3	7 37	17 28	67.2	204 336
25 155	9 54	21 58	66.9	9 30	21 03	66.8	9 07	20 07	66.6	8 43	19 11	66.5	8 19	18 15	66.3	7 55	17 20	66.2	205 335
26 154	10 16	21 49	66.0	9 52	20 53	65.8	9 27	19 57	65.7	9 02	19 02	65.5	8 37	18 07	65.4	8 12	17 12	65.2	206 334
27 153	10 38	21 38	65.0	10 13	20 43	64.9	9 48	19 48	64.7	9 22	18 53	64.6	8 56	17 58	64.4	8 30	17 03	64.3	207 333
28 152	11 00	21 28	64.1	10 34	20 33	63.9	10 08	19 38	63.8	9 41	18 43	63.6	9 14	17 49	63.5	8 48	16 55	63.3	208 332
29 151	11 22	21 17	63.1	10 55	20 22	63.0	10 28	19 28	62.8	10 00	18 34	62.6	9 33	17 39	62.5	9 05	16 46	62.3	209 331
30 150	11 44	21 05	62.2	11 16	20 11	62.0	10 48	19 17	61.8	10 19	18 23	61.7	9 51	17 30	61.5	9 22	16 36	61.4	210 330
31 149	12 06	20 53	61.2	11 37	20 00	61.1	11 07	19 06	60.9	10 38	18 13	60.7	10 08	17 20	60.5	9 39	16 27	60.4	211 329
32 148	12 27	20 41	60.3	11 57	19 48	60.1	11 27	18 55	59.9	10 57	18 02	59.7	10 27	17 09	59.6	9 56	16 17	59.4	212 328
33 147	12 48	20 29	59.3	12 17	19 36	59.1	11 46	18 43	58.9	11 15	17 51	58.7	10 44	16 58	58.6	10 13	16 06	58.4	213 327
34 146	13 09	20 16	58.4	12 37	19 23	58.1	12 06	18 31	58.0	11 34	17 39	57.8	11 02	16 47	57.6	10 29	15 56	57.5	214 326
35 145	13 29	20 02	57.4	12 57	19 10	57.2	12 24	18 19	57.0	11 52	17 27	56.8	11 19	16 36	56.7	10 46	15 45	56.5	215 325
36 144	13 50	19 49	56.4	13 17	18 57	56.2	12 43	18 06	56.0	12 10	17 15	55.9	11 36	16 24	55.7	11 02	15 34	55.5	216 324
37 143	14 10	19 34	55.5	13 36	18 44	55.3	13 02	17 53	55.1	12 27	17 03	54.9	11 53	16 12	54.7	11 18	15 23	54.5	217 323
38 142	14 30	19 20	54.5	13 55	18 30	54.3	13 20	17 40	54.1	12 45	16 50	53.9	12 09	16 00	53.7	11 34	15 11	53.5	218 322
39 141	14 50	19 05	53.5	14 14	18 15	53.3	13 38	17 26	53.1	13 02	16 37	52.9	12 26	15 48	52.7	11 49	14 59	52.6	219 321
40 140	15 09	18 50	52.5	14 33	18 01	52.3	13 56	17 12	52.1	13 19	16 23	51.9	12 42	15 35	51.7	12 05	14 47	51.6	220 320
41 139	15 29	18 34	51.5	14 51	17 46	51.3	14 14	16 57	51.1	13 36	16 09	50.9	12 58	15 22	50.8	12 20	14 34	50.6	221 319
42 138	15 48	18 18	50.6	15 09	17 30	50.3	14 31	16 43	50.1	13 52	15 55	49.9	13 14	15 08	49.9	12 35	14 21	49.6	222 318
43 137	16 06	18 02	49.6	15 27	17 15	49.4	14 48	16 28	49.2	14 09	15 41	49.0	13 29	14 54	48.8	12 50	14 08	48.6	223 317
44 136	16 25	17 46	48.6	15 45	16 59	48.4	15 05	16 12	48.2	14 25	15 26	48.0	13 45	14 40	47.8	13 04	13 55	47.6	224 316
45 135	16 43	17 29	47.6	16 02	16 42	47.4	15 22	15 57	47.2	14 41	15 11	47.0	14 00	14 26	46.8	13 19	13 41	46.6	225 315

Lat./A	LHA/F	66° A/H	66° B/P	66° Z₁/Z₂	67° A/H	67° B/P	67° Z₁/Z₂	68° A/H	68° B/P	68° Z₁/Z₂	69° A/H	69° B/P	69° Z₁/Z₂	70° A/H	70° B/P	70° Z₁/Z₂	71° Z₁/Z₂	71° B/P	71° A/H	Lat./A	LHA
45	135	16 43	17 29	47.6	16 02	16 42	47.4	15 22	15 57	47.2	14 41	15 11	47.0	14 00	14 26	46.8	46.6	13 41	13 19	225	315
46	134	17 01	17 11	47.6	16 19	16 26	46.4	15 38	15 41	46.2	14 56	14 56	46.0	14 15	14 11	45.8	45.6	13 27	13 33	226	314
47	133	17 18	16 53	46.6	16 36	16 09	45.4	15 54	15 24	45.2	15 12	14 40	45.0	14 29	13 56	44.8	44.6	13 13	13 46	227	313
48	132	17 36	16 35	45.6	16 53	15 51	44.4	16 10	15 08	44.2	15 27	14 24	44.0	14 43	13 41	43.8	43.6	12 58	14 00	228	312
49	131	17 53	16 17	44.6	17 09	15 34	43.4	16 25	14 51	43.2	15 42	14 08	43.0	14 58	13 26	42.8	42.6	12 44	14 13	229	311
50	130	18 09	15 58	43.6	17 25	15 16	42.4	16 41	14 33	42.1	15 56	13 52	41.9	15 11	13 10	41.8	41.6	12 29	14 27	230	310
51	129	18 26	15 39	42.6	17 41	14 57	41.3	16 56	14 16	41.1	16 10	13 35	40.9	15 25	12 54	40.8	40.6	12 14	14 39	231	309
52	128	18 42	15 20	41.6	17 56	14 39	40.3	17 10	13 58	40.1	16 24	13 18	39.9	15 38	12 38	39.7	39.6	11 58	14 52	232	308
53	127	18 57	15 00	40.5	18 11	14 20	39.3	17 24	13 40	39.1	16 38	13 00	38.9	15 51	12 21	38.7	38.6	11 42	15 04	233	307
54	126	19 13	14 40	39.5	18 26	14 01	38.3	17 39	13 22	38.1	16 51	12 43	37.9	16 04	12 05	37.7	37.5	11 26	15 16	234	306
55	125	19 28	14 20	38.5	18 40	13 41	37.3	17 52	13 03	37.1	17 04	12 25	36.9	16 16	11 48	36.7	36.5	11 10	15 28	235	305
56	124	19 42	13 59	37.4	18 54	13 21	36.2	18 06	12 44	36.0	17 17	12 07	35.8	16 28	11 30	35.7	35.5	10 54	15 40	236	304
57	123	19 57	13 38	36.4	19 08	13 01	35.2	18 19	12 25	35.0	17 29	11 49	34.8	16 40	11 13	34.6	34.5	10 37	15 51	237	303
58	122	20 11	13 17	35.4	19 21	12 41	34.2	18 31	12 05	34.0	17 42	11 30	33.8	16 52	10 55	33.6	33.5	10 20	16 02	238	302
59	121	20 24	12 55	34.4	19 34	12 20	33.1	18 44	11 45	32.9	17 53	11 11	32.8	17 03	10 37	32.6	32.4	10 03	16 12	239	301
60	120	20 37	12 33	33.3	19 47	11 59	32.1	18 56	11 25	31.9	18 05	10 52	31.7	17 14	10 19	31.6	31.4	9 46	16 23	240	300
61	119	20 50	12 11	32.3	19 59	11 38	31.1	19 08	11 05	30.9	18 16	10 33	30.7	17 24	10 00	30.5	30.4	9 29	16 33	241	299
62	118	21 03	11 48	31.2	20 11	11 16	30.0	19 19	10 44	29.8	18 27	10 13	29.7	17 35	9 42	29.5	29.4	9 11	16 42	242	298
63	117	21 15	11 26	30.2	20 22	10 54	29.0	19 30	10 24	28.8	18 37	9 53	28.6	17 45	9 23	28.5	28.3	8 53	16 52	243	297
64	116	21 27	11 03	29.2	20 34	10 32	27.9	19 41	10 03	27.7	18 47	9 33	27.6	17 54	9 04	27.4	27.3	8 35	17 01	244	296
65	115	21 38	10 39	28.1	20 44	10 10	26.9	19 51	9 41	26.7	18 57	9 13	26.5	18 03	8 45	26.4	26.3	8 17	17 10	245	295
66	114	21 49	10 16	27.0	20 55	9 48	25.8	20 01	9 20	25.7	19 07	8 52	25.5	18 12	8 25	25.4	25.2	7 58	17 18	246	294
67	113	21 59	9 52	26.0	21 05	9 25	24.8	20 10	8 58	24.6	19 16	8 32	24.5	18 21	8 06	24.3	24.2	7 40	17 26	247	293
68	112	22 09	9 28	24.9	21 14	9 02	23.7	20 19	8 36	23.5	19 24	8 11	23.4	18 29	7 46	23.3	23.1	7 21	17 34	248	292
69	111	22 19	9 04	23.9	21 24	8 39	22.6	20 28	8 14	22.5	19 33	7 50	22.4	18 37	7 26	22.2	22.1	7 02	17 42	249	291
70	110	22 28	8 40	22.8	21 32	8 16	21.6	20 37	7 52	21.4	19 41	7 29	21.3	18 45	7 06	21.2	21.1	6 43	17 49	250	290
71	109	22 37	8 15	21.7	21 41	7 52	20.5	20 45	7 30	20.4	19 48	7 07	20.2	18 52	6 45	20.1	20.0	6 24	17 56	251	289
72	108	22 45	7 50	20.7	21 49	7 28	19.4	20 52	7 07	19.3	19 56	6 46	19.2	18 59	6 25	19.1	19.0	6 04	18 02	252	288
73	107	22 53	7 25	19.6	21 56	7 04	18.4	21 00	6 44	18.2	20 03	6 24	18.1	19 05	6 04	18.0	17.9	5 45	18 08	253	287
74	106	23 01	7 00	18.5	22 04	6 40	17.3	21 06	6 21	17.2	20 09	6 02	17.1	19 12	5 44	17.0	16.9	5 25	18 14	254	286
75	105	23 08	6 34	17.4	22 10	6 16	16.2	21 13	5 58	16.1	20 15	5 40	16.0	19 17	5 23	15.9	15.8	5 06	18 20	255	285
76	104	23 15	6 09	16.3	22 17	5 52	15.2	21 19	5 35	15.1	20 21	5 18	15.1	19 23	5 02	14.9	14.8	4 46	18 25	256	284
77	103	23 21	5 43	15.2	22 23	5 27	14.2	21 24	5 12	14.1	20 26	4 56	14.0	19 28	4 41	13.8	13.7	4 26	18 30	257	283
78	102	23 27	5 17	14.2	22 28	5 03	13.1	21 30	4 48	13.0	20 31	4 34	12.9	19 33	4 20	12.7	12.7	4 06	18 34	258	282
79	101	23 32	4 51	13.1	22 33	4 38	12.0	21 35	4 24	11.9	20 36	4 11	11.8	19 37	3 58	11.7	11.6	3 46	18 38	259	281
80	100	23 37	4 25	12.0	22 38	4 13	10.9	21 39	4 01	10.8	20 40	3 49	10.7	19 41	3 37	10.6	10.6	3 25	18 42	260	280
81	99	23 41	3 59	10.9	22 42	3 48	9.8	21 43	3 37	9.7	20 44	3 26	9.6	19 45	3 16	9.5	9.5	3 05	18 45	261	279
82	98	23 45	3 33	9.8	22 46	3 23	8.7	21 46	3 13	8.6	20 47	3 03	8.6	19 48	2 54	8.5	8.5	2 45	18 48	262	278
83	97	23 49	3 06	8.7	22 49	2 58	7.7	21 50	2 49	7.6	20 50	2 41	7.5	19 51	2 32	7.4	7.4	2 24	18 51	263	277
84	96	23 52	2 40	7.7	22 52	2 32	6.6	21 52	2 25	6.5	20 53	2 18	6.5	19 53	2 11	6.4	6.3	2 04	18 54	264	276
85	95	23 54	2 13	6.6	22 54	2 07	5.5	21 55	2 01	5.4	20 55	1 55	5.4	19 55	1 49	5.3	5.3	1 43	18 55	265	275
86	94	23 56	1 47	5.5	22 56	1 42	4.3	21 57	1 37	4.3	20 57	1 32	4.3	19 57	1 27	4.3	4.2	1 23	18 57	266	274
87	93	23 58	1 20	3.3	22 58	1 16	3.3	21 58	1 13	3.2	20 58	1 09	3.2	19 58	1 05	3.2	3.2	1 02	18 58	267	273
88	92	23 59	0 53	2.2	22 59	0 51	2.2	21 59	0 48	2.2	20 59	0 46	2.1	19 59	0 44	2.1	2.1	0 41	18 59	268	272
89	91	24 00	0 27	1.1	23 00	0 25	1.1	22 00	0 24	1.1	21 00	0 23	1.1	20 00	0 22	1.1	1.1	0 21	19 00	269	271
90	90	24 00	0 00	0.0	23 00	0 00	0.0	22 00	0 00	0.0	21 00	0 00	0.0	20 00	0 00	0.0	0.0	0 00	19 00	270	270

N. Lat.: for LHA > 180°... Zₙ = Z
for LHA < 180°... Zₙ = 360° − Z

S. Lat.: for LHA > 180°... Zₙ = 180° − Z
for LHA < 180°... Zₙ = 180° + Z

B: (−) for 90° < LHA < 270°
Dec: (−) for Lat. contrary name

Z_1 : same sign as B
Z_2 : (−) for F > 90°

Lat./A LHA/F	LHA/F	72° A/H	72° B/P	72° Z_1/Z_2	73° A/H	73° B/P	73° Z_1/Z_2	74° A/H	74° B/P	74° Z_1/Z_2	75° A/H	75° B/P	75° Z_1/Z_2	76° A/H	76° B/P	76° Z_1/Z_2	77° A/H	77° B/P	77° Z_1/Z_2	LHA	Lat./A LHA
0	180	0 00	18 00	90.0	0 00	17 00	90.0	0 00	16 00	90.0	0 00	15 00	90.0	0 00	14 00	90.0	0 00	13 00	90.0	180	360
1	179	0 19	18 00	89.0	0 18	17 00	89.0	0 17	16 00	89.0	0 16	15 00	89.0	0 15	14 00	89.0	0 13	13 00	89.0	181	359
2	178	0 37	17 59	88.1	0 35	16 59	88.1	0 33	15 59	88.1	0 31	14 59	88.1	0 29	14 00	88.1	0 27	13 00	88.1	182	358
3	177	0 56	17 59	87.1	0 53	16 59	87.1	0 50	15 59	87.1	0 47	14 59	87.1	0 44	13 59	87.1	0 40	12 59	87.1	183	357
4	176	1 14	17 58	86.2	1 10	16 58	86.2	1 06	15 58	86.2	1 02	14 58	86.1	0 58	13 58	86.1	0 54	12 58	86.1	184	356
5	175	1 33	17 56	85.2	1 28	16 56	85.2	1 23	15 57	85.2	1 18	14 57	85.2	1 12	13 57	85.1	1 07	12 57	85.1	185	355
6	174	1 51	17 54	84.3	1 45	16 55	84.3	1 39	15 55	84.2	1 33	14 55	84.2	1 27	13 56	84.2	1 21	12 56	84.2	186	354
7	173	2 09	17 52	83.3	2 03	16 53	83.3	1 56	15 53	83.3	1 48	14 54	83.2	1 41	13 54	83.2	1 34	12 54	83.2	187	353
8	172	2 28	17 50	82.4	2 20	16 51	82.3	2 12	15 51	82.3	2 04	14 52	82.3	1 56	13 52	82.2	1 48	12 53	82.2	188	352
9	171	2 46	17 48	81.4	2 37	16 48	81.4	2 28	15 49	81.3	2 19	14 49	81.3	2 10	13 50	81.3	2 01	12 51	81.2	189	351
10	170	3 05	17 45	80.5	2 55	16 45	80.4	2 45	15 46	80.4	2 35	14 47	80.3	2 24	13 48	80.3	2 14	12 49	80.3	190	350
11	169	3 23	17 41	79.5	3 12	16 42	79.5	3 01	15 43	79.4	2 50	14 44	79.4	2 39	13 45	79.3	2 28	12 46	79.3	191	349
12	168	3 41	17 38	78.6	3 29	16 39	78.5	3 17	15 40	78.5	3 05	14 41	78.4	2 53	13 42	78.3	2 41	12 44	78.3	192	348
13	167	3 59	17 34	77.6	3 46	16 35	77.5	3 33	15 37	77.5	3 20	14 38	77.4	3 07	13 39	77.4	2 54	12 41	77.3	193	347
14	166	4 17	17 30	76.7	4 03	16 31	76.6	3 49	15 33	76.5	3 35	14 34	76.5	3 21	13 36	76.4	3 07	12 38	76.4	194	346
15	165	4 35	17 25	75.7	4 20	16 27	75.6	4 05	15 29	75.6	3 50	14 31	75.5	3 35	13 32	75.4	3 20	12 34	75.4	195	345
16	164	4 53	17 21	74.7	4 37	16 23	74.7	4 21	15 25	74.6	4 05	14 27	74.5	3 49	13 29	74.4	3 33	12 31	74.4	196	344
17	163	5 11	17 16	73.8	4 54	16 18	73.7	4 37	15 20	73.6	4 20	14 22	73.5	4 03	13 25	73.5	3 46	12 27	73.4	197	343
18	162	5 29	17 10	72.8	5 11	16 13	72.7	4 53	15 15	72.7	4 35	14 18	72.6	4 17	13 20	72.5	3 59	12 23	72.4	198	342
19	161	5 46	17 05	71.9	5 28	16 07	71.8	5 09	15 10	71.7	4 50	14 13	71.6	4 31	13 16	71.5	4 12	12 19	71.5	199	341
20	160	6 04	16 59	70.9	5 44	16 02	70.8	5 25	15 05	70.7	5 05	14 08	70.6	4 45	13 11	70.5	4 25	12 14	70.5	200	340
21	159	6 21	16 52	69.9	6 01	15 56	69.8	5 40	14 59	69.7	5 19	14 03	69.7	4 58	13 06	69.6	4 37	12 10	69.5	201	339
22	158	6 39	16 46	69.0	6 17	15 50	68.9	5 56	14 53	68.8	5 34	13 57	68.7	5 12	13 01	68.6	4 50	12 05	68.5	202	338
23	157	6 56	16 39	68.0	6 34	15 43	67.9	6 11	14 47	67.8	5 48	13 51	67.7	5 25	12 56	67.6	5 03	12 00	67.5	203	337
24	156	7 13	16 32	67.1	6 50	15 36	66.9	6 26	14 41	66.8	6 03	13 45	66.7	5 39	12 50	66.6	5 15	11 55	66.5	204	336
25	155	7 30	16 25	66.1	7 06	15 29	66.0	6 41	14 34	65.9	6 17	13 39	65.8	5 52	12 44	65.7	5 27	11 49	65.6	205	335
26	154	7 47	16 17	65.1	7 22	15 22	65.0	6 56	14 27	64.9	6 31	13 32	64.8	6 05	12 38	64.7	5 40	11 43	64.6	206	334
27	153	8 04	16 09	64.1	7 38	15 14	64.0	7 11	14 20	63.9	6 45	13 26	63.8	6 18	12 32	63.7	5 52	11 37	63.6	207	333
28	152	8 20	16 00	63.2	7 53	15 06	63.0	7 26	14 12	62.9	6 59	13 19	62.8	6 31	12 25	62.7	6 04	11 31	62.6	208	332
29	151	8 37	15 52	62.2	8 09	14 58	62.1	7 41	14 05	61.9	7 13	13 11	61.8	6 44	12 18	61.7	6 16	11 25	61.6	209	331
30	150	8 53	15 43	61.2	8 24	14 50	61.1	7 55	13 57	61.0	7 26	13 04	60.9	6 57	12 11	60.7	6 27	11 18	60.6	210	330
31	149	9 09	15 34	60.3	8 40	14 41	60.1	8 10	13 49	60.0	7 40	12 56	59.9	7 09	12 04	59.8	6 39	11 12	59.7	211	329
32	148	9 25	15 24	59.3	8 55	14 32	59.1	8 24	13 40	59.0	7 53	12 48	58.9	7 22	11 56	58.8	6 51	11 05	58.7	212	328
33	147	9 41	15 15	58.3	9 10	14 23	58.2	8 38	13 31	58.0	8 06	12 40	57.9	7 34	11 49	57.8	7 02	10 57	57.7	213	327
34	146	9 57	15 05	57.3	9 25	14 13	57.2	8 52	13 22	57.0	8 19	12 31	56.9	7 46	11 41	56.8	7 14	10 50	56.7	214	326
35	145	10 13	14 54	56.3	9 39	14 04	56.2	9 06	13 13	56.1	8 32	12 23	55.9	7 59	11 33	55.8	7 25	10 43	55.7	215	325
36	144	10 28	14 44	55.4	9 54	13 54	55.2	9 19	13 04	55.1	8 45	12 14	54.9	8 11	11 24	54.8	7 36	10 35	54.7	216	324
37	143	10 43	14 33	54.4	10 08	13 43	54.2	9 33	12 54	54.1	8 58	12 05	53.9	8 22	11 16	53.8	7 47	10 27	53.7	217	323
38	142	10 58	14 22	53.4	10 22	13 33	53.2	9 46	12 44	53.1	9 10	11 55	53.0	8 34	11 07	52.8	7 58	10 19	52.7	218	322
39	141	11 13	14 10	52.4	10 36	13 22	52.3	9 59	12 34	52.1	9 22	11 46	52.0	8 45	10 58	51.8	8 08	10 10	51.7	219	321
40	140	11 27	13 59	51.4	10 50	13 11	51.3	10 12	12 23	51.1	9 35	11 36	51.0	8 57	10 49	50.8	8 19	10 02	50.7	220	320
41	139	11 42	13 47	50.4	11 04	13 00	50.3	10 25	12 13	50.1	9 47	11 26	50.0	9 08	10 39	49.9	8 29	9 53	49.7	221	319
42	138	11 56	13 34	49.4	11 17	12 48	49.3	10 38	12 02	49.1	9 58	11 16	49.0	9 19	10 30	48.9	8 39	9 44	48.7	222	318
43	137	12 10	13 22	48.4	11 30	12 36	48.3	10 50	11 51	48.1	10 10	11 05	48.0	9 30	10 20	47.9	8 49	9 35	47.7	223	317
44	136	12 24	13 09	47.4	11 43	12 24	47.3	11 02	11 39	47.1	10 21	10 55	47.0	9 40	10 10	46.9	8 59	9 26	46.7	224	316
45	135	12 37	12 56	46.4	11 56	12 12	46.3	11 14	11 28	46.1	10 33	10 44	46.0	9 51	10 00	45.9	9 09	9 16	45.7	225	315

Lat./A LHA/F	72° A/H	72° B/P	72° Z_1/Z_2	73° A/H	73° B/P	73° Z_1/Z_2	74° A/H	74° B/P	74° Z_1/Z_2	75° A/H	75° B/P	75° Z_1/Z_2	76° A/H	76° B/P	76° Z_1/Z_2	77° A/H	77° B/P	77° Z_1/Z_2	Lat./A LHA
45	12 37	12 56	46.4	11 56	12 12	46.3	11 14	11 28	46.1	10 33	10 44	46.0	9 51	10 00	45.9	9 09	9 16	45.7	225 315
46	12 51	12 43	45.4	12 08	11 59	45.3	11 26	11 16	45.1	10 44	10 33	45.0	10 01	9 50	44.9	9 19	9 07	44.7	226 314
47	13 04	12 30	44.4	12 21	11 47	44.3	11 38	11 04	44.1	10 55	10 21	44.0	10 11	9 39	43.9	9 28	8 57	43.7	227 313
48	13 17	12 16	43.4	12 33	11 34	43.3	11 49	10 52	43.1	11 05	10 10	43.0	10 21	9 28	42.9	9 37	8 47	42.7	228 312
49	13 29	12 02	42.4	12 45	11 21	42.3	12 00	10 39	42.1	11 16	9 58	42.0	10 31	9 17	41.9	9 46	8 37	41.7	229 311
50	13 42	11 48	41.4	12 57	11 07	41.3	12 11	10 27	41.1	11 26	9 46	41.0	10 41	9 06	40.9	9 55	8 26	40.7	230 310
51	13 54	11 33	40.4	13 08	10 53	40.3	12 22	10 14	40.1	11 36	9 34	40.0	10 50	8 55	39.8	10 04	8 16	39.7	231 309
52	14 06	11 19	39.4	13 19	10 40	39.2	12 33	10 01	39.1	11 46	9 22	39.0	10 59	8 44	38.8	10 13	8 05	38.7	232 308
53	14 17	11 04	38.4	13 30	10 26	38.2	12 43	9 47	38.1	11 56	9 10	37.9	11 08	8 32	37.8	10 21	7 55	37.7	233 307
54	14 29	10 49	37.4	13 41	10 11	37.2	12 53	9 34	37.1	12 05	8 57	36.9	11 17	8 20	36.8	10 29	7 44	36.7	234 306
55	14 40	10 33	36.4	13 51	9 57	36.2	13 03	9 20	36.1	12 14	8 44	35.9	11 26	8 08	35.8	10 37	7 33	35.7	235 305
56	14 51	10 18	35.3	14 02	9 42	35.2	13 13	9 07	35.1	12 23	8 31	34.9	11 34	7 56	34.8	10 45	7 21	34.7	236 304
57	15 01	10 02	34.3	14 12	9 27	34.2	13 22	8 53	34.0	12 32	8 18	33.9	11 42	7 44	33.8	10 52	7 10	33.7	237 303
58	15 12	9 46	33.3	14 21	9 12	33.2	13 31	8 38	33.0	12 41	8 05	32.9	11 50	7 32	32.8	11 00	6 59	32.7	238 302
59	15 22	9 30	32.3	14 31	8 57	32.1	13 40	8 24	32.0	12 49	7 51	31.9	11 58	7 19	31.8	11 07	6 47	31.7	239 301
60	15 31	9 14	31.3	14 40	8 41	31.1	13 49	8 10	31.0	12 57	7 38	30.8	12 06	7 06	30.8	11 14	6 35	30.6	240 300
61	15 41	8 57	30.2	14 49	8 26	30.1	13 57	7 55	30.0	13 05	7 24	29.7	12 13	6 54	29.7	11 21	6 23	29.6	241 299
62	15 50	8 40	29.2	14 58	8 10	29.1	14 05	7 40	28.9	13 13	7 10	28.8	12 20	6 41	28.7	11 27	6 11	28.6	242 298
63	15 59	8 23	28.2	15 06	7 54	28.0	14 13	7 25	27.9	13 20	6 56	27.8	12 27	6 27	27.7	11 34	5 59	27.6	243 297
64	16 08	8 06	27.2	15 14	7 38	27.0	14 21	7 10	26.9	13 27	6 42	26.8	12 34	6 14	26.7	11 40	5 47	26.6	244 296
65	16 16	7 49	26.1	15 22	7 22	26.0	14 28	6 55	25.9	13 34	6 28	25.7	12 40	6 01	25.7	11 46	5 34	25.6	245 295
66	16 24	7 32	25.1	15 29	7 05	25.0	14 35	6 39	24.9	13 41	6 13	24.7	12 46	5 47	24.6	11 52	5 22	24.6	246 294
67	16 32	7 14	24.1	15 37	6 49	23.9	14 42	6 24	23.8	13 47	5 59	23.7	12 52	5 34	23.6	11 57	5 09	23.5	247 293
68	16 39	6 56	23.0	15 44	6 32	22.9	14 48	6 08	22.8	13 53	5 44	22.7	12 58	5 20	22.6	12 02	4 57	22.5	248 292
69	16 46	6 39	22.0	15 50	6 15	21.9	14 55	5 52	21.8	13 59	5 29	21.7	13 03	5 06	21.6	12 07	4 44	21.5	249 291
70	16 53	6 20	20.9	15 57	5 58	20.8	15 01	5 36	20.7	14 05	5 14	20.6	13 08	4 52	20.6	12 12	4 31	20.5	250 290
71	16 59	6 02	19.9	16 03	5 41	19.8	15 06	5 20	19.7	14 10	4 59	19.6	13 13	4 38	19.5	12 17	4 18	19.5	251 289
72	17 05	5 44	18.9	16 09	5 24	18.8	15 12	5 04	18.7	14 15	4 44	18.6	13 18	4 24	18.5	12 21	4 05	18.4	252 288
73	17 11	5 26	17.8	16 14	5 06	17.7	15 17	4 48	17.6	14 20	4 29	17.6	13 23	4 10	17.5	12 25	3 52	17.4	253 287
74	17 17	5 07	16.8	16 19	4 49	16.7	15 22	4 31	16.6	14 24	4 13	16.5	13 27	3 56	16.5	12 29	3 38	16.4	254 286
75	17 22	4 48	15.7	16 24	4 31	15.7	15 26	4 15	15.6	14 29	3 58	15.5	13 31	3 42	15.4	12 33	3 25	15.4	255 285
76	17 27	4 30	14.7	16 29	4 14	14.6	15 31	3 58	14.6	14 33	3 43	14.5	13 35	3 27	14.4	12 36	3 12	14.4	256 284
77	17 31	4 11	13.6	16 33	3 56	13.6	15 35	3 41	13.5	14 36	3 27	13.4	13 38	3 13	13.4	12 40	2 58	13.3	257 283
78	17 36	3 52	12.6	16 37	3 38	12.5	15 38	3 25	12.5	14 40	3 11	12.4	13 41	2 58	12.4	12 43	2 45	12.3	258 282
79	17 39	3 33	11.6	16 41	3 20	11.5	15 42	3 08	11.4	14 43	2 56	11.4	13 44	2 43	11.3	12 45	2 31	11.3	259 281
80	17 43	3 14	10.5	16 44	3 02	10.4	15 45	2 51	10.4	14 46	2 40	10.3	13 47	2 29	10.3	12 48	2 18	10.3	260 280
81	17 46	2 55	9.5	16 47	2 44	9.4	15 48	2 34	9.4	14 49	2 24	9.3	13 49	2 14	9.3	12 50	2 04	9.2	261 279
82	17 49	2 35	8.4	16 50	2 26	8.4	15 50	2 17	8.3	14 51	2 08	8.3	13 52	1 59	8.2	12 52	1 50	8.2	262 278
83	17 52	2 16	7.4	16 52	2 08	7.3	15 53	2 00	7.3	14 53	1 52	7.2	13 54	1 44	7.2	12 54	1 37	7.2	263 277
84	17 54	1 57	6.3	16 54	1 50	6.3	15 55	1 43	6.2	14 55	1 36	6.2	13 55	1 30	6.2	12 56	1 23	6.2	264 276
85	17 56	1 37	5.3	16 56	1 32	5.2	15 56	1 26	5.2	14 56	1 20	5.2	13 57	1 15	5.2	12 57	1 09	5.1	265 275
86	17 57	1 18	4.2	16 57	1 13	4.2	15 58	1 09	4.2	14 58	1 04	4.1	13 58	1 00	4.1	12 58	0 55	4.1	266 274
87	17 58	0 58	3.2	16 59	0 55	3.1	15 59	0 52	3.1	14 59	0 48	3.1	13 59	0 45	3.1	12 59	0 42	3.1	267 273
88	17 59	0 39	2.1	16 59	0 37	2.1	15 59	0 34	2.1	14 59	0 32	2.1	13 59	0 30	2.1	13 00	0 28	2.1	268 272
89	18 00	0 19	1.1	17 00	0 18	1.1	16 00	0 17	1.0	15 00	0 16	1.0	14 00	0 15	1.0	13 00	0 14	1.0	269 271
90	18 00	0 00	0.0	17 00	0 00	0.0	16 00	0 00	0.0	15 00	0 00	0.0	14 00	0 00	0.0	13 00	0 00	0.0	270 270

N. Lat.: for LHA > 180° ..... $Z_n = Z$
for LHA < 180° ..... $Z_n = 360° − Z$

S. Lat.: for LHA > 180° ..... $Z_n = 180° − Z$
for LHA < 180° ..... $Z_n = 180° + Z$

SIGHT REDUCTION TABLE

B: (−) for 90° < LHA < 270°
Dec: (−) for Lat. contrary name

Z₁ : same sign as B
Z₂ : (−) for F > 90°

LHA/F	F	78° A/H	78° B/P	78° Z₁/Z₂	79° A/H	79° B/P	79° Z₁/Z₂	80° A/H	80° B/P	80° Z₁/Z₂	81° A/H	81° B/P	81° Z₁/Z₂	82° A/H	82° B/P	82° Z₁/Z₂	83° A/H	83° B/P	83° Z₁/Z₂	LHA	Lat./A LHA
0	180	0 00	12 00	90.0	0 00	11 00	90.0	0 00	10 00	90.0	0 00	9 00	90.0	0 00	8 00	90.0	0 00	7 00	90.0	180	360
1	179	0 12	12 00	89.0	0 11	11 00	89.0	0 10	10 00	89.0	0 09	9 00	89.0	0 08	8 00	89.0	0 07	7 00	89.0	181	359
2	178	0 25	12 00	88.0	0 23	11 00	88.0	0 21	10 00	88.0	0 19	9 00	88.0	0 17	8 00	88.0	0 15	7 00	88.0	182	358
3	177	0 37	11 59	87.1	0 34	10 59	87.1	0 31	9 59	87.0	0 28	8 59	87.0	0 25	7 59	87.0	0 22	6 59	87.0	183	357
4	176	0 50	11 58	86.1	0 46	10 58	86.1	0 42	9 59	86.1	0 38	8 59	86.0	0 33	7 59	86.0	0 29	6 59	86.0	184	356
5	175	1 02	11 57	85.1	0 57	10 58	85.1	0 52	9 58	85.1	0 47	8 58	85.1	0 42	7 58	85.0	0 37	6 58	85.0	185	355
6	174	1 15	11 56	84.1	1 09	10 56	84.1	1 02	9 57	84.1	0 56	8 57	84.1	0 50	7 57	84.1	0 44	6 58	84.0	186	354
7	173	1 27	11 55	83.2	1 20	10 55	83.1	1 13	9 56	83.1	1 06	8 56	83.1	0 58	7 56	83.1	0 51	6 57	83.1	187	353
8	172	1 39	11 53	82.2	1 31	10 54	82.1	1 23	9 54	82.1	1 15	8 55	82.1	1 07	7 55	82.1	0 58	6 56	82.1	188	352
9	171	1 52	11 51	81.2	1 43	10 52	81.2	1 33	9 53	81.1	1 24	8 53	81.1	1 15	7 54	81.1	1 06	6 55	81.1	189	351
10	170	2 04	11 49	80.2	1 54	10 50	80.2	1 44	9 51	80.1	1 33	8 52	80.1	1 23	7 53	80.1	1 13	6 54	80.1	190	350
11	169	2 16	11 47	79.2	2 05	10 48	79.2	1 54	9 49	79.2	1 43	8 50	79.1	1 31	7 51	79.1	1 20	6 52	79.1	191	349
12	168	2 29	11 45	78.3	2 16	10 46	78.2	2 04	9 47	78.2	1 52	8 48	78.1	1 39	7 50	78.1	1 27	6 51	78.1	192	348
13	167	2 41	11 42	77.3	2 28	10 43	77.2	2 14	9 45	77.2	2 01	8 46	77.2	1 48	7 48	77.1	1 34	6 49	77.1	193	347
14	166	2 53	11 39	76.3	2 39	10 41	76.2	2 24	9 43	76.2	2 10	8 44	76.2	1 56	7 46	76.1	1 41	6 48	76.1	194	346
15	165	3 05	11 36	75.3	2 50	10 38	75.3	2 35	9 41	75.3	2 19	8 42	75.2	2 04	7 44	75.1	1 48	6 46	75.1	195	345
16	164	3 17	11 33	74.3	3 01	10 35	74.3	2 45	9 37	74.2	2 28	8 39	74.2	2 12	7 42	74.1	1 56	6 44	74.1	196	344
17	163	3 29	11 29	73.4	3 12	10 32	73.3	2 55	9 34	73.2	2 37	8 37	73.2	2 20	7 39	73.2	2 03	6 42	73.1	197	343
18	162	3 41	11 26	72.4	3 23	10 28	72.3	3 05	9 31	72.3	2 46	8 34	72.2	2 28	7 37	72.2	2 09	6 40	72.1	198	342
19	161	3 53	11 22	71.4	3 34	10 25	71.3	3 14	9 28	71.3	2 55	8 31	71.2	2 36	7 34	71.2	2 16	6 37	71.1	199	341
20	160	4 05	11 18	70.4	3 45	10 21	70.3	3 24	9 24	70.3	3 04	8 28	70.2	2 44	7 31	70.2	2 23	6 35	70.1	200	340
21	159	4 16	11 13	69.4	3 55	10 17	69.4	3 34	9 21	69.3	3 13	8 25	69.2	2 52	7 28	69.2	2 30	6 32	69.1	201	339
22	158	4 28	11 09	68.4	4 06	10 13	68.4	3 44	9 17	68.3	3 22	8 21	68.2	2 59	7 25	68.2	2 37	6 30	68.1	202	338
23	157	4 40	11 04	67.5	4 17	10 09	67.4	3 53	9 13	67.3	3 30	8 18	67.3	3 07	7 22	67.2	2 44	6 27	67.2	203	337
24	156	4 51	10 59	66.5	4 27	10 04	66.4	4 03	9 09	66.3	3 39	8 14	66.3	3 15	7 19	66.2	2 50	6 24	66.2	204	336
25	155	5 02	10 54	65.5	4 38	9 59	65.4	4 13	9 05	65.3	3 47	8 10	65.3	3 22	7 16	65.2	2 57	6 21	65.2	205	335
26	154	5 14	10 48	64.5	4 48	9 55	64.4	4 22	9 00	64.3	3 56	8 06	64.3	3 30	7 12	64.2	3 04	6 18	64.2	206	334
27	153	5 25	10 43	63.5	4 58	9 50	63.5	4 31	8 56	63.3	4 04	8 02	63.3	3 37	7 08	63.2	3 10	6 15	63.2	207	333
28	152	5 36	10 38	62.5	5 08	9 44	62.5	4 41	8 51	62.4	4 13	7 58	62.3	3 45	7 04	62.2	3 17	6 11	62.2	208	332
29	151	5 47	10 32	61.5	5 18	9 39	61.4	4 50	8 46	61.4	4 21	7 53	61.3	3 52	7 00	61.2	3 23	6 08	61.2	209	331
30	150	5 58	10 26	60.5	5 28	9 33	60.5	4 59	8 41	60.4	4 29	7 49	60.3	3 59	6 56	60.2	3 30	6 04	60.2	210	330
31	149	6 09	10 20	59.6	5 38	9 28	59.5	5 08	8 36	59.4	4 37	7 44	59.3	4 07	6 52	59.2	3 36	6 00	59.2	211	329
32	148	6 20	10 13	58.6	5 48	9 22	58.5	5 17	8 30	58.4	4 45	7 39	58.3	4 14	6 48	58.3	3 42	5 57	58.2	212	328
33	147	6 30	10 06	57.6	5 58	9 16	57.5	5 26	8 25	57.4	4 53	7 34	57.3	4 21	6 43	57.3	3 48	5 53	57.2	213	327
34	146	6 41	10 00	56.6	6 08	9 09	56.5	5 34	8 19	56.4	5 01	7 29	56.3	4 28	6 39	56.3	3 54	5 49	56.2	214	326
35	145	6 51	9 53	55.6	6 17	9 03	55.5	5 43	8 13	55.4	5 09	7 24	55.3	4 35	6 34	55.3	4 00	5 45	55.2	215	325
36	144	7 01	9 45	54.6	6 26	8 56	54.5	5 51	8 07	54.4	5 17	7 18	54.3	4 42	6 29	54.3	4 06	5 40	54.2	216	324
37	143	7 11	9 38	53.6	6 36	8 49	53.5	6 00	8 01	53.4	5 24	7 13	53.3	4 48	6 24	53.3	4 12	5 36	53.2	217	323
38	142	7 21	9 31	52.6	6 45	8 43	52.5	6 08	7 55	52.4	5 32	7 07	52.3	4 55	6 19	52.3	4 18	5 32	52.2	218	322
39	141	7 31	9 23	51.6	6 54	8 35	51.5	6 16	7 48	51.4	5 39	7 01	51.3	5 01	6 14	51.3	4 24	5 27	51.2	219	321
40	140	7 41	9 15	50.6	7 03	8 28	50.5	6 25	7 42	50.4	5 46	6 55	50.3	5 08	6 09	50.3	4 30	5 22	50.2	220	320
41	139	7 50	9 07	49.6	7 11	8 21	49.5	6 32	7 35	49.4	5 53	6 49	49.4	5 14	6 03	49.3	4 35	5 18	49.2	221	319
42	138	8 00	8 59	48.6	7 20	8 13	48.5	6 40	7 28	48.4	6 01	6 43	48.4	5 21	5 58	48.3	4 41	5 13	48.2	222	318
43	137	8 09	8 50	47.6	7 29	8 05	47.5	6 48	7 21	47.4	6 07	6 36	47.4	5 27	5 52	47.3	4 46	5 08	47.2	223	317
44	136	8 18	8 42	46.6	7 37	7 58	46.5	6 56	7 14	46.4	6 14	6 30	46.4	5 33	5 46	46.3	4 51	5 03	46.2	224	316
45	135	8 27	8 33	45.6	7 45	7 50	45.5	7 03	7 06	45.4	6 21	6 23	45.4	5 39	5 41	45.3	4 57	4 58	45.2	225	315

Lat./A LHA/F	°	78° A/H	78° B/P	78° Z_1/Z_2	79° A/H	79° B/P	79° Z_1/Z_2	80° A/H	80° B/P	80° Z_1/Z_2	81° A/H	81° B/P	81° Z_1/Z_2	82° A/H	82° B/P	82° Z_1/Z_2	83° A/H	83° B/P	83° Z_1/Z_2	Lat./A LHA	LHA
135	45	8 27	8 33	45.6	7 45	7 50	45.5	7 03	7 06	45.4	6 21	6 23	45.4	5 39	5 41	45.3	4 57	4 58	45.2	225	315
134	46	8 36	8 24	44.6	7 53	7 41	44.5	7 11	6 59	44.4	6 28	6 17	44.4	5 45	5 35	44.3	5 02	4 53	44.2	226	314
133	47	8 45	8 15	43.6	8 01	7 33	43.5	7 18	6 51	43.4	6 34	6 10	43.4	5 51	5 29	43.3	5 07	4 47	43.2	227	313
132	48	8 53	8 06	42.6	8 09	7 25	42.5	7 25	6 44	42.4	6 41	6 03	42.4	5 56	5 22	42.3	5 12	4 42	42.2	228	312
131	49	9 02	7 56	41.6	8 17	7 16	41.5	7 32	6 36	41.4	6 47	5 56	41.4	6 02	5 16	41.3	5 17	4 36	41.2	229	311
130	50	9 10	7 47	40.6	8 24	7 07	40.5	7 39	6 28	40.4	6 53	5 49	40.3	6 07	5 10	40.3	5 21	4 31	40.2	230	310
129	51	9 18	7 37	39.6	8 32	6 58	39.5	7 45	6 20	39.4	6 59	5 42	39.3	6 13	5 03	39.3	5 26	4 25	39.2	231	309
128	52	9 26	7 27	38.6	8 39	6 49	38.5	7 52	6 12	38.4	7 05	5 34	38.3	6 18	4 57	38.3	5 31	4 19	38.2	232	308
127	53	9 33	7 17	37.6	8 46	6 40	37.5	7 58	6 03	37.4	7 11	5 27	37.3	6 23	4 50	37.3	5 35	4 14	37.2	233	307
126	54	9 41	7 07	36.6	8 53	6 31	36.5	8 05	5 55	36.4	7 16	5 19	36.3	6 28	4 43	36.3	5 39	4 08	36.2	234	306
125	55	9 48	6 57	35.6	9 00	6 22	35.5	8 11	5 47	35.4	7 22	5 11	35.3	6 33	4 37	35.3	5 44	4 02	35.2	235	305
124	56	9 56	6 47	34.6	9 06	6 12	34.5	8 17	5 38	34.4	7 27	5 04	34.3	6 38	4 30	34.3	5 48	3 56	34.2	236	304
123	57	10 03	6 36	33.6	9 13	6 03	33.5	8 22	5 29	33.4	7 32	4 56	33.3	6 42	4 23	33.3	5 52	3 50	33.2	237	303
122	58	10 09	6 26	32.6	9 19	5 53	32.5	8 28	5 20	32.4	7 37	4 48	32.3	6 47	4 16	32.3	5 56	3 43	32.2	238	302
121	59	10 16	6 15	31.6	9 25	5 43	31.5	8 34	5 11	31.4	7 42	4 40	31.3	6 51	4 08	31.2	6 00	3 37	31.2	239	301
120	60	10 22	6 04	30.6	9 31	5 33	30.5	8 39	5 02	30.4	7 47	4 32	30.3	6 55	4 01	30.2	6 04	3 31	30.2	240	300
119	61	10 29	5 53	29.5	9 36	5 23	29.5	8 44	4 53	29.4	7 52	4 23	29.3	6 59	3 54	29.2	6 07	3 24	29.2	241	299
118	62	10 35	5 42	28.5	9 42	5 13	28.4	8 49	4 44	28.4	7 56	4 15	28.3	7 04	3 46	28.2	6 11	3 18	28.2	242	298
117	63	10 41	5 31	27.5	9 47	5 03	27.4	8 54	4 35	27.4	8 01	4 07	27.3	7 07	3 39	27.2	6 14	3 11	27.2	243	297
116	64	10 46	5 19	26.5	9 52	4 52	26.4	8 59	4 25	26.3	8 05	3 58	26.3	7 11	3 32	26.2	6 17	3 05	26.2	244	296
115	65	10 52	5 08	25.5	9 57	4 42	25.4	9 03	4 16	25.3	8 09	3 50	25.3	7 15	3 24	25.2	6 20	2 58	25.2	245	295
114	66	10 57	4 56	24.5	10 02	4 31	24.4	9 08	4 06	24.3	8 13	3 41	24.3	7 18	3 16	24.2	6 24	2 52	24.2	246	294
113	67	11 02	4 45	23.5	10 07	4 21	23.4	9 12	3 56	23.3	8 17	3 32	23.3	7 22	3 09	23.2	6 26	2 45	23.2	247	293
112	68	11 07	4 33	22.4	10 11	4 10	22.4	9 16	3 47	22.3	8 20	3 24	22.2	7 25	3 01	22.2	6 29	2 38	22.1	248	292
111	69	11 12	4 21	21.4	10 16	3 59	21.4	9 20	3 37	21.3	8 24	3 15	21.2	7 28	2 53	21.2	6 32	2 31	21.1	249	291
110	70	11 16	4 09	20.4	10 20	3 48	20.3	9 23	3 27	20.3	8 27	3 06	20.2	7 31	2 45	20.2	6 35	2 24	20.1	250	290
109	71	11 20	3 58	19.4	10 24	3 37	19.3	9 27	3 17	19.3	8 30	2 57	19.2	7 34	2 37	19.2	6 37	2 17	19.1	251	289
108	72	11 24	3 45	18.4	10 27	3 26	18.3	9 30	3 07	18.3	8 33	2 48	18.2	7 36	2 29	18.2	6 39	2 10	18.1	252	288
107	73	11 28	3 33	17.4	10 31	3 15	17.3	9 34	2 57	17.2	8 36	2 39	17.2	7 39	2 21	17.2	6 41	2 03	17.1	253	287
106	74	11 32	3 21	16.3	10 34	3 04	16.3	9 37	2 47	16.2	8 39	2 30	16.2	7 41	2 13	16.1	6 44	1 56	16.1	254	286
105	75	11 35	3 09	15.3	10 37	2 53	15.3	9 39	2 37	15.2	8 41	2 21	15.2	7 44	2 05	15.1	6 46	1 49	15.1	255	285
104	76	11 38	2 57	14.3	10 40	2 42	14.3	9 42	2 27	14.2	8 44	2 12	14.2	7 46	1 57	14.1	6 47	1 42	14.1	256	284
103	77	11 41	2 44	13.3	10 43	2 30	13.2	9 44	2 16	13.2	8 46	2 02	13.2	7 48	1 49	13.1	6 49	1 35	13.1	257	283
102	78	11 44	2 32	12.3	10 45	2 19	12.2	9 47	2 06	12.2	8 48	1 53	12.1	7 49	1 40	12.1	6 51	1 28	12.1	258	282
101	79	11 47	2 19	11.2	10 48	2 07	11.2	9 49	1 56	11.2	8 50	1 44	11.1	7 51	1 32	11.1	6 52	1 21	11.1	259	281
100	80	11 49	2 07	10.2	10 50	1 56	10.2	9 51	1 45	10.2	8 52	1 35	10.1	7 53	1 24	10.1	6 54	1 13	10.1	260	280
99	81	11 51	1 54	9.2	10 52	1 45	9.1	9 53	1 35	9.1	8 53	1 25	9.1	7 54	1 16	9.1	6 55	1 06	9.1	261	279
98	82	11 53	1 42	8.2	10 53	1 33	8.1	9 54	1 24	8.1	8 55	1 16	8.1	7 55	1 07	8.1	6 56	0 59	8.1	262	278
97	83	11 55	1 29	7.2	10 55	1 21	7.1	9 55	1 14	7.1	8 56	1 06	7.1	7 56	0 59	7.1	6 57	0 51	7.1	263	277
96	84	11 56	1 16	6.1	10 56	1 10	6.1	9 57	1 03	6.1	8 57	0 57	6.1	7 57	0 50	6.1	6 58	0 44	6.0	264	276
95	85	11 57	1 04	5.1	10 57	0 58	5.1	9 58	0 53	5.1	8 58	0 47	5.1	7 58	0 42	5.0	6 58	0 37	5.0	265	275
94	86	11 58	0 51	4.1	10 58	0 47	4.1	9 59	0 42	4.1	8 59	0 38	4.0	7 59	0 34	4.0	6 59	0 29	4.0	266	274
93	87	11 59	0 38	3.1	10 59	0 35	3.1	9 59	0 32	3.0	8 59	0 28	3.0	7 59	0 25	3.0	6 59	0 22	3.0	267	273
92	88	12 00	0 26	2.0	11 00	0 23	2.0	10 00	0 21	2.0	9 00	0 19	2.0	8 00	0 17	2.0	7 00	0 15	2.0	268	272
91	89	12 00	0 13	1.0	11 00	0 12	1.0	10 00	0 11	1.0	9 00	0 10	1.0	8 00	0 08	1.0	7 00	0 07	1.0	269	271
90	90	12 00	0 00	0.0	11 00	0 00	0.0	10 00	0 00	0.0	9 00	0 00	0.0	8 00	0 00	0.0	7 00	0 00	0.0	270	270

N. Lat.: for LHA > 180°... $Z_n = Z$
for LHA < 180°... $Z_n = 360° - Z$

S. Lat.: for LHA > 180°... $Z_n = 180° - Z$
for LHA < 180°... $Z_n = 180° + Z$

Z₁: same sign as B
Z₂: (–) for F > 90°

B: (–) for 90° < LHA < 270°
Dec: (–) for Lat. contrary name

SIGHT REDUCTION TABLE

Lat./A LHA/F	84° A/H	84° B/P	84° Z₁/Z₂	85° A/H	85° B/P	85° Z₁/Z₂	86° A/H	86° B/P	86° Z₁/Z₂	87° A/H	87° B/P	87° Z₁/Z₂	88° A/H	88° B/P	88° Z₁/Z₂	89° A/H	89° B/P	89° Z₁/Z₂	Lat./A LHA	LHA
180	0 00	6 00	90.0	0 00	5 00	90.0	0 00	4 00	90.0	0 00	3 00	90.0	0 00	2 00	90.0	0 00	1 00	90.0	360	180
179	0 06	6 00	89.0	0 05	5 00	89.0	0 04	4 00	89.0	0 03	3 00	89.0	0 02	2 00	89.0	0 01	1 00	89.0	359	181
178	0 13	6 00	88.0	0 10	5 00	88.0	0 08	4 00	88.0	0 06	3 00	88.0	0 04	2 00	88.0	0 02	1 00	88.0	358	182
177	0 19	6 00	87.0	0 16	5 00	87.0	0 13	4 00	87.0	0 09	3 00	87.0	0 06	2 00	87.0	0 03	1 00	87.0	357	183
176	0 25	5 59	86.0	0 21	4 59	86.0	0 17	3 59	86.0	0 13	3 00	86.0	0 08	2 00	86.0	0 04	1 00	86.0	356	184
175	0 31	5 59	85.0	0 26	4 59	85.0	0 21	3 59	85.0	0 16	2 59	85.0	0 10	2 00	85.0	0 05	1 00	85.0	355	185
174	0 38	5 58	84.0	0 31	4 58	84.0	0 25	3 59	84.0	0 19	2 59	84.0	0 13	1 59	84.0	0 06	1 00	84.0	354	186
173	0 44	5 57	83.0	0 37	4 58	83.0	0 29	3 58	83.0	0 22	2 59	83.0	0 15	1 59	83.0	0 07	0 59	83.0	353	187
172	0 50	5 57	82.0	0 42	4 57	82.0	0 33	3 58	82.0	0 25	2 58	82.0	0 17	1 59	82.0	0 08	0 59	82.0	352	188
171	0 56	5 56	81.0	0 47	4 56	81.0	0 38	3 57	81.0	0 28	2 58	81.0	0 19	1 59	81.0	0 09	0 59	81.0	351	189
170	1 02	5 55	80.1	0 52	4 55	80.1	0 42	3 56	80.0	0 31	2 57	80.0	0 21	1 58	80.0	0 10	0 59	80.0	350	190
169	1 09	5 53	79.1	0 57	4 55	79.1	0 46	3 56	79.0	0 34	2 57	79.0	0 23	1 58	79.0	0 11	0 59	79.0	349	191
168	1 15	5 52	78.1	1 02	4 53	78.1	0 50	3 55	78.0	0 37	2 56	78.0	0 25	1 57	78.0	0 12	0 59	78.0	348	192
167	1 21	5 51	77.1	1 07	4 52	77.1	0 54	3 54	77.0	0 40	2 55	77.0	0 27	1 57	77.0	0 13	0 58	77.0	347	193
166	1 27	5 49	76.1	1 12	4 51	76.1	0 58	3 53	76.0	0 44	2 55	76.0	0 29	1 56	76.0	0 15	0 58	76.0	346	194
165	1 33	5 48	75.1	1 18	4 50	75.1	1 02	3 52	75.0	0 47	2 54	75.0	0 31	1 56	75.0	0 16	0 58	75.0	345	195
164	1 39	5 46	74.1	1 23	4 48	74.1	1 06	3 51	74.0	0 50	2 53	74.0	0 33	1 55	74.0	0 17	0 58	74.0	344	196
163	1 45	5 44	73.1	1 28	4 47	73.1	1 10	3 50	73.0	0 53	2 52	73.0	0 35	1 55	73.0	0 18	0 57	73.0	343	197
162	1 51	5 43	72.1	1 33	4 45	72.1	1 14	3 48	72.0	0 56	2 51	72.0	0 37	1 54	72.0	0 19	0 57	72.0	342	198
161	1 57	5 41	71.1	1 38	4 44	71.1	1 18	3 47	71.0	0 59	2 50	71.0	0 39	1 53	71.0	0 20	0 57	71.0	341	199
160	2 03	5 38	70.1	1 42	4 42	70.1	1 22	3 46	70.0	1 02	2 49	70.0	0 41	1 53	70.0	0 21	0 56	70.0	340	200
159	2 09	5 36	69.1	1 47	4 40	69.1	1 26	3 44	69.0	1 04	2 48	69.0	0 43	1 52	69.0	0 22	0 56	69.0	339	201
158	2 15	5 34	68.1	1 52	4 38	68.1	1 30	3 43	68.0	1 07	2 47	68.0	0 45	1 51	68.0	0 22	0 56	68.0	338	202
157	2 20	5 32	67.1	1 57	4 36	67.1	1 34	3 41	67.0	1 10	2 46	67.0	0 47	1 50	67.0	0 23	0 55	67.0	337	203
156	2 26	5 29	66.1	2 02	4 34	66.1	1 38	3 39	66.1	1 13	2 44	66.0	0 49	1 50	66.0	0 24	0 55	66.0	336	204
155	2 32	5 26	65.1	2 07	4 32	65.1	1 41	3 38	65.1	1 16	2 43	65.0	0 51	1 49	65.0	0 25	0 54	65.0	335	205
154	2 38	5 24	64.1	2 11	4 30	64.1	1 45	3 36	64.1	1 19	2 42	64.0	0 53	1 48	64.0	0 26	0 54	64.0	334	206
153	2 43	5 21	63.1	2 16	4 27	63.1	1 49	3 34	63.1	1 22	2 40	63.0	0 54	1 47	63.0	0 27	0 53	63.0	333	207
152	2 49	5 18	62.1	2 21	4 25	62.1	1 53	3 32	62.1	1 24	2 39	62.0	0 56	1 46	62.0	0 28	0 53	62.0	332	208
151	2 54	5 15	61.1	2 25	4 23	61.1	1 56	3 30	61.1	1 27	2 37	61.0	0 58	1 45	61.0	0 29	0 52	61.0	331	209
150	3 00	5 12	60.1	2 30	4 20	60.1	2 00	3 28	60.1	1 30	2 36	60.0	1 00	1 44	60.0	0 30	0 52	60.0	330	210
149	3 05	5 09	59.1	2 34	4 17	59.1	2 04	3 26	59.1	1 33	2 34	59.0	1 02	1 43	59.0	0 31	0 51	59.0	329	211
148	3 11	5 06	58.1	2 39	4 15	58.1	2 07	3 24	58.1	1 35	2 33	58.0	1 04	1 42	58.0	0 32	0 51	58.0	328	212
147	3 16	5 02	57.1	2 43	4 12	57.1	2 11	3 21	57.1	1 38	2 31	57.0	1 05	1 41	57.0	0 33	0 50	57.0	327	213
146	3 21	4 59	56.1	2 48	4 09	56.1	2 14	3 19	56.1	1 41	2 29	56.0	1 07	1 39	56.0	0 34	0 50	56.0	326	214
145	3 26	4 55	55.1	2 52	4 06	55.1	2 18	3 17	55.1	1 43	2 27	55.0	1 09	1 38	55.0	0 34	0 49	55.0	325	215
144	3 31	4 52	54.1	2 56	4 03	54.1	2 21	3 14	54.1	1 46	2 26	54.0	1 11	1 37	54.0	0 35	0 49	54.0	324	216
143	3 36	4 48	53.1	3 00	4 00	53.1	2 24	3 12	53.1	1 48	2 24	53.0	1 12	1 36	53.0	0 36	0 48	53.0	323	217
142	3 41	4 44	52.1	3 05	3 57	52.1	2 28	3 09	52.1	1 51	2 22	52.0	1 14	1 35	52.0	0 37	0 47	52.0	322	218
141	3 46	4 40	51.2	3 09	3 53	51.1	2 31	3 07	51.1	1 53	2 20	51.0	1 16	1 33	51.0	0 38	0 47	51.0	321	219
140	3 51	4 36	50.2	3 13	3 50	50.1	2 34	3 04	50.1	1 56	2 18	50.0	1 17	1 32	50.0	0 39	0 46	50.0	320	220
139	3 56	4 32	49.2	3 17	3 47	49.1	2 37	3 01	49.1	1 58	2 16	49.0	1 19	1 31	49.0	0 39	0 45	49.0	319	221
138	4 01	4 28	48.2	3 21	3 43	48.1	2 41	2 58	48.1	2 00	2 14	48.0	1 20	1 29	48.0	0 40	0 45	48.0	318	222
137	4 05	4 24	47.2	3 24	3 40	47.1	2 44	2 56	47.1	2 03	2 12	47.0	1 22	1 28	47.0	0 41	0 44	47.0	317	223
136	4 10	4 19	46.2	3 28	3 36	46.1	2 47	2 53	46.1	2 05	2 10	46.0	1 23	1 26	46.0	0 42	0 44	46.0	316	224
135	4 14	4 15	45.2	3 32	3 32	45.1	2 50	2 50	45.1	2 07	2 07	45.0	1 25	1 25	45.0	0 42	0 42	45.0	315	225

Lat./A LHA/F		84° A/H	84° B/P	84° Z₁/Z₂	85° A/H	85° B/P	85° Z₁/Z₂	86° A/H	86° B/P	86° Z₁/Z₂	87° A/H	87° B/P	87° Z₁/Z₂	88° A/H	88° B/P	88° Z₁/Z₂	89° A/H	89° B/P	89° Z₁/Z₂	Lat./A LHA	
45	135	4 14	4 15	45.2	3 32	3 32	45.1	2 50	2 50	45.1	2 07	2 07	45.0	1 25	1 25	45.0	0 42	0 42	45.0	225	315
46	134	4 19	4 11	44.2	3 36	3 29	44.1	2 53	2 47	44.1	2 09	2 05	44.0	1 26	1 23	44.0	0 43	0 42	44.0	226	314
47	133	4 23	4 06	43.2	3 39	3 25	43.1	2 55	2 44	43.1	2 12	2 03	43.0	1 28	1 22	43.0	0 44	0 41	43.0	227	313
48	132	4 27	4 01	42.2	3 43	3 21	42.2	2 58	2 41	42.1	2 14	2 01	42.0	1 29	1 20	42.0	0 45	0 40	42.0	228	312
49	131	4 31	3 57	41.2	3 46	3 17	41.1	3 01	2 38	41.1	2 16	1 58	41.0	1 31	1 19	41.0	0 45	0 39	41.0	229	311
50	130	4 36	3 52	40.2	3 50	3 13	40.1	3 04	2 34	40.1	2 18	1 56	40.0	1 32	1 17	40.0	0 46	0 39	40.0	230	310
51	129	4 40	3 47	39.2	3 53	3 09	39.1	3 06	2 31	39.1	2 20	1 53	39.0	1 33	1 16	39.0	0 47	0 38	39.0	231	309
52	128	4 43	3 42	38.2	3 56	3 05	38.1	3 09	2 28	38.1	2 22	1 51	38.0	1 35	1 14	38.0	0 47	0 37	38.0	232	308
53	127	4 47	3 37	37.2	3 59	3 01	37.1	3 12	2 25	37.1	2 24	1 48	37.0	1 36	1 12	37.0	0 48	0 36	37.0	233	307
54	126	4 51	3 32	36.2	4 03	2 57	36.1	3 14	2 21	36.1	2 26	1 46	36.0	1 37	1 11	36.0	0 49	0 35	36.0	234	306
55	125	4 55	3 27	35.1	4 06	2 52	35.1	3 17	2 18	35.1	2 27	1 43	35.0	1 38	1 09	35.0	0 49	0 34	35.0	235	305
56	124	4 58	3 22	34.1	4 09	2 48	34.1	3 19	2 14	34.1	2 29	1 41	34.0	1 39	1 07	34.0	0 50	0 34	34.0	236	304
57	123	5 02	3 17	33.1	4 12	2 44	33.1	3 21	2 11	33.1	2 31	1 38	33.0	1 41	1 05	33.0	0 50	0 33	33.0	237	303
58	122	5 05	3 11	32.1	4 14	2 39	32.1	3 23	2 07	32.1	2 33	1 35	32.0	1 42	1 04	32.0	0 51	0 32	32.0	238	302
59	121	5 08	3 06	31.1	4 17	2 35	31.1	3 26	2 04	31.1	2 34	1 33	31.0	1 43	1 02	31.0	0 51	0 31	31.0	239	301
60	120	5 12	3 00	30.1	4 20	2 30	30.1	3 28	2 00	30.1	2 36	1 30	30.0	1 44	1 00	30.0	0 52	0 30	30.0	240	300
61	119	5 15	2 55	29.1	4 22	2 26	29.1	3 30	1 56	29.1	2 37	1 27	29.0	1 45	0 58	29.0	0 52	0 29	29.0	241	299
62	118	5 18	2 49	28.1	4 25	2 21	28.1	3 32	1 53	28.1	2 39	1 25	28.0	1 46	0 56	28.0	0 53	0 28	28.0	242	298
63	117	5 21	2 44	27.1	4 27	2 16	27.1	3 34	1 49	27.1	2 40	1 22	27.0	1 47	0 54	27.0	0 53	0 27	27.0	243	297
64	116	5 23	2 38	26.1	4 30	2 12	26.1	3 36	1 45	26.1	2 42	1 19	26.0	1 48	0 53	26.0	0 54	0 26	26.0	244	296
65	115	5 26	2 33	25.1	4 32	2 07	25.1	3 37	1 42	25.1	2 43	1 16	25.0	1 49	0 51	25.0	0 54	0 25	25.0	245	295
66	114	5 29	2 27	24.1	4 34	2 02	24.1	3 39	1 38	24.1	2 44	1 13	24.0	1 50	0 49	24.0	0 55	0 24	24.0	246	294
67	113	5 31	2 21	23.1	4 36	1 57	23.1	3 41	1 34	23.1	2 46	1 10	23.0	1 50	0 47	23.0	0 55	0 23	23.0	247	293
68	112	5 34	2 15	22.1	4 38	1 53	22.1	3 42	1 30	22.1	2 47	1 07	22.0	1 51	0 45	22.0	0 56	0 22	22.0	248	292
69	111	5 36	2 09	21.1	4 40	1 48	21.1	3 44	1 26	21.1	2 48	1 05	21.0	1 52	0 43	21.0	0 56	0 22	21.0	249	291
70	110	5 38	2 04	20.1	4 42	1 43	20.1	3 46	1 22	20.1	2 49	1 02	20.0	1 53	0 41	20.0	0 56	0 21	20.0	250	290
71	109	5 40	1 58	19.1	4 44	1 38	19.1	3 47	1 18	19.1	2 50	0 59	19.0	1 53	0 39	19.0	0 57	0 20	19.0	251	289
72	108	5 42	1 52	18.1	4 45	1 33	18.1	3 48	1 14	18.1	2 51	0 56	18.0	1 54	0 37	18.0	0 57	0 19	18.0	252	288
73	107	5 44	1 46	17.1	4 47	1 28	17.1	3 49	1 10	17.1	2 52	0 53	17.0	1 55	0 35	17.0	0 57	0 18	17.0	253	287
74	106	5 46	1 40	16.1	4 48	1 23	16.1	3 51	1 06	16.1	2 53	0 50	16.0	1 55	0 33	16.0	0 58	0 17	16.0	254	286
75	105	5 48	1 33	15.1	4 50	1 18	15.1	3 52	1 02	15.0	2 54	0 47	15.0	1 56	0 31	15.0	0 58	0 16	15.0	255	285
76	104	5 49	1 27	14.1	4 51	1 13	14.1	3 53	0 58	14.0	2 55	0 44	14.0	1 56	0 29	14.0	0 58	0 15	14.0	256	284
77	103	5 51	1 21	13.1	4 52	1 08	13.0	3 54	0 54	13.0	2 55	0 41	13.0	1 57	0 27	13.0	0 59	0 13	13.0	257	283
78	102	5 52	1 15	12.1	4 53	1 03	12.0	3 55	0 50	12.0	2 56	0 37	12.0	1 57	0 25	12.0	0 59	0 12	12.0	258	282
79	101	5 53	1 09	11.1	4 54	0 57	11.0	3 56	0 46	11.0	2 57	0 34	11.0	1 58	0 23	11.0	0 59	0 11	11.0	259	281
80	100	5 55	1 03	10.1	4 55	0 52	10.0	3 56	0 42	10.0	2 57	0 31	10.0	1 58	0 21	10.0	0 59	0 10	10.0	260	280
81	99	5 56	0 57	9.0	4 56	0 47	9.0	3 57	0 38	9.0	2 58	0 28	9.0	1 59	0 19	9.0	0 59	0 09	9.0	261	279
82	98	5 56	0 50	8.0	4 57	0 42	8.0	3 58	0 33	8.0	2 58	0 25	8.0	1 59	0 17	8.0	0 59	0 08	8.0	262	278
83	97	5 57	0 44	7.0	4 57	0 37	7.0	3 58	0 29	7.0	2 59	0 22	7.0	1 59	0 15	7.0	1 00	0 07	7.0	263	277
84	96	5 58	0 38	6.0	4 58	0 31	6.0	3 58	0 25	6.0	2 59	0 19	6.0	1 59	0 13	6.0	1 00	0 06	6.0	264	276
85	95	5 59	0 31	5.0	4 59	0 26	5.0	3 59	0 21	5.0	2 59	0 16	5.0	2 00	0 10	5.0	1 00	0 05	5.0	265	275
86	94	5 59	0 25	4.0	4 59	0 21	4.0	3 59	0 17	4.0	3 00	0 13	4.0	2 00	0 08	4.0	1 00	0 04	4.0	266	274
87	93	6 00	0 19	3.0	5 00	0 16	3.0	4 00	0 13	3.0	3 00	0 09	3.0	2 00	0 06	3.0	1 00	0 03	3.0	267	273
88	92	6 00	0 13	2.0	5 00	0 10	2.0	4 00	0 08	2.0	3 00	0 06	2.0	2 00	0 04	2.0	1 00	0 02	2.0	268	272
89	91	6 00	0 06	1.0	5 00	0 05	1.0	4 00	0 04	1.0	3 00	0 03	1.0	2 00	0 02	1.0	1 00	0 01	1.0	269	271
90	90	6 00	0 00	0.0	5 00	0 00	0.0	4 00	0 00	0.0	3 00	0 00	0.0	2 00	0 00	0.0	1 00	0 00	0.0	270	270

N. Lat.: for LHA > 180°... Zn = Z
         for LHA < 180°... Zn = 360° − Z

S. Lat.: for LHA > 180°... Zn = 180° − Z
         for LHA < 180°... Zn = 180° + Z

AUXILIARY TABLE

Top-right note: F'/A' — A' < 30' : (−) corr.

Bottom-left note: F < 90° and F' > 29' : (−) corr. F > 90° and F' < 30' : (−) corr. — F'/A'

P°	Z₂°	30'/30	29'/31	28'/32	27'/33	26'/34	25'/35	24'/36	23'/37	22'/38	21'/39	20'/40	19'/41	18'/42	17'/43	16'/44	15'/45	14'/46	13'/47	12'/48	11'/49	10'/50	9'/51	8'/52	7'/53	6'/54	5'/55	4'/56	3'/57	2'/58	1'/59
1	89	1	1	0	0	0	0	0	0	0	0	0	0	0	0	0	0	0	0	0	0	0	0	0	0	0	0	0	0	0	0
2	88	1	1	1	1	1	1	1	1	1	1	1	1	1	1	1	1	0	0	0	0	0	0	0	0	0	0	0	0	0	0
3	87	2	2	1	1	1	1	1	1	1	1	1	1	1	1	1	1	1	1	1	1	1	0	0	0	0	0	0	0	0	0
4	86	2	2	2	2	2	2	2	2	2	1	1	1	1	1	1	1	1	1	1	1	1	1	1	0	0	0	0	0	0	0
5	85	3	3	2	2	2	2	2	2	2	2	2	2	2	1	1	1	1	1	1	1	1	1	1	1	1	0	0	0	0	0
6	84	3	3	3	3	3	3	2	2	2	2	2	2	2	2	2	2	1	1	1	1	1	1	1	1	1	1	0	0	0	0
7	83	4	4	3	3	3	3	3	3	3	3	2	2	2	2	2	2	2	2	1	1	1	1	1	1	1	1	0	0	0	0
8	82	4	4	4	4	4	3	3	3	3	3	3	3	3	2	2	2	2	2	2	2	2	1	1	1	1	1	1	0	0	0
9	81	5	5	4	4	4	4	4	4	3	3	3	3	3	3	3	2	2	2	2	2	2	1	1	1	1	1	1	0	0	0
10	80	5	5	5	5	5	4	4	4	4	4	3	3	3	3	3	3	2	2	2	2	2	2	2	1	1	1	1	1	0	0
11	79	6	6	5	5	5	5	4	4	4	4	4	4	3	3	3	3	3	2	2	2	2	2	2	1	1	1	1	1	0	0
12	78	6	6	6	6	5	5	5	5	4	4	4	4	4	4	3	3	3	3	2	2	2	2	2	2	1	1	1	1	0	0
13	77	7	7	6	6	6	6	5	5	5	5	4	4	4	4	4	3	3	3	3	3	2	2	2	2	2	1	1	1	0	0
14	76	7	7	7	7	6	6	6	6	5	5	5	5	4	4	4	4	3	3	3	3	3	2	2	2	2	1	1	1	0	0
15	75	8	8	7	7	7	6	6	6	6	5	5	5	5	4	4	4	4	3	3	3	3	2	2	2	2	2	1	1	1	0
16	74	8	8	8	7	7	7	7	6	6	6	6	5	5	5	4	4	4	4	3	3	3	3	2	2	2	2	1	1	1	0
17	73	9	8	8	8	8	7	7	7	7	6	6	6	5	5	5	5	4	4	4	4	3	3	3	2	2	2	1	1	1	0
18	72	9	9	9	8	8	8	7	7	7	6	6	6	6	5	5	5	4	4	4	4	3	3	3	2	2	2	1	1	1	0
19	71	10	9	9	9	9	8	8	7	8	7	7	6	6	6	5	5	5	4	4	4	3	3	3	2	2	2	1	1	1	0
20	70	10	10	10	9	9	9	8	8	8	7	7	6	6	6	5	5	5	4	4	4	4	3	3	3	2	2	2	1	1	0
21	69	11	10	10	10	9	9	9	8	8	8	7	7	6	6	6	5	5	5	4	4	4	3	3	3	2	2	2	1	1	0
22	68	11	11	11	10	10	9	9	8	9	8	7	7	7	6	6	6	5	5	4	4	4	4	3	3	2	2	2	1	1	0
23	67	12	11	11	11	10	10	9	9	9	8	8	8	7	7	6	6	5	5	5	5	4	4	3	3	3	2	2	1	1	0
24	66	12	12	12	11	11	10	10	9	9	9	8	8	7	7	7	6	6	5	5	5	4	4	4	3	3	2	2	1	1	0
25	65	13	12	12	11	11	11	10	9	10	9	8	8	8	7	7	6	6	6	5	5	5	4	4	3	3	2	2	1	1	0
26	64	13	13	12	12	11	11	10	10	10	9	9	8	8	7	7	7	6	6	5	5	5	4	4	3	3	2	2	1	1	0
27	63	14	13	13	12	12	11	11	10	10	10	9	9	8	8	7	7	6	6	6	5	5	4	4	3	3	3	2	1	1	0
28	62	14	14	13	13	12	12	11	11	11	10	9	9	9	8	8	7	7	6	6	5	5	4	4	3	3	3	2	1	1	0
29	61	14	14	14	13	13	12	11	11	11	10	10	9	9	8	8	7	7	6	6	6	5	5	4	4	3	3	2	1	1	0
30	60	15	14	14	13	13	12	12	11	11	10	10	9	9	8	8	8	7	7	6	6	5	5	4	4	3	3	2	1	1	0
31	59	15	15	15	14	13	13	12	11	12	11	10	10	9	9	8	8	7	7	6	6	5	5	5	4	3	3	2	1	1	0
32	58	16	15	15	14	14	13	12	12	12	11	11	10	10	9	9	8	8	7	7	6	6	5	5	4	3	3	2	1	1	0
33	57	16	16	16	15	14	14	13	12	12	11	11	11	10	9	9	8	8	7	7	6	6	5	5	4	4	3	2	1	1	0
34	56	17	16	16	15	15	14	14	13	13	12	11	11	10	10	9	9	8	8	7	6	6	6	5	4	4	3	2	1	1	1
35	55	17	17	16	15	15	14	14	13	13	12	12	11	11	10	10	9	9	8	7	7	6	6	5	4	4	3	2	2	1	1
36	54	18	17	16	16	15	15	14	14	13	13	12	12	11	10	10	9	9	8	7	7	6	6	5	4	4	3	2	2	1	1
37	53	18	17	17	16	16	15	14	14	14	13	12	12	11	11	10	9	9	8	7	7	6	6	5	4	4	3	2	2	1	1
38	52	18	18	17	17	16	16	15	14	14	13	12	12	12	11	10	9	9	8	8	7	6	6	5	4	4	3	3	2	1	1
39	51	19	18	18	17	16	16	15	14	14	13	13	12	12	11	10	9	9	8	8	7	6	6	5	4	4	3	3	2	1	1
40	50	19	19	18	17	17	16	15	15	14	13	13	12	12	11	10	10	9	8	8	7	6	6	5	4	4	3	3	2	1	1

F'/A' Z2°	30'/30	29'/31	28'/32	27'/33	26'/34	25'/35	24'/36	23'/37	22'/38	21'/39	20'/40	19'/41	18'/42	17'/43	16'/44	15'/45	14'/46	13'/47	12'/48	11'/49	10'/50	9'/51	8'/52	7'/53	6'/54	5'/55	4'/56	3'/57	2'/58	1'/59	P°
49	20	19	18	18	17	16	16	15	14	14	13	12	12	11	11	10	9	9	8	7	7	6	5	5	4	3	3	2	1	1	41
48	20	19	18	18	17	16	16	15	15	14	13	13	12	11	11	10	9	9	8	7	7	6	5	5	4	3	3	2	1	1	42
47	20	19	19	18	18	17	16	15	15	14	14	13	12	12	11	10	10	9	8	8	7	6	6	5	4	3	3	2	1	1	43
46	21	20	19	18	18	17	16	16	15	14	14	13	12	12	11	11	10	9	8	8	7	6	6	5	4	4	3	2	2	1	44
45	22	21	20	19	18	18	17	16	15	15	14	13	13	12	11	11	10	9	8	8	7	6	6	5	4	4	3	2	2	1	45
44	22	21	20	19	19	18	17	16	16	15	14	14	13	12	12	11	10	10	9	8	7	7	6	5	4	4	3	3	2	1	46
43	22	22	21	20	19	18	18	17	16	15	15	14	13	13	12	11	10	10	9	8	8	7	6	5	5	4	3	3	2	1	47
42	22	22	21	20	19	19	18	17	16	16	15	14	14	13	12	11	11	10	9	8	8	7	6	5	5	4	3	3	2	1	48
41	23	22	22	21	20	19	18	17	17	16	15	15	14	13	12	12	11	10	9	8	8	7	6	5	5	4	3	3	2	1	49
40	23	22	22	21	20	20	19	18	17	16	16	15	14	13	13	12	11	10	10	8	8	7	6	5	5	4	3	3	2	1	50
39	23	23	22	21	20	20	19	18	17	17	16	15	15	14	13	12	11	11	10	9	8	7	6	6	5	4	4	3	2	1	51
38	24	23	22	22	21	20	19	19	18	17	16	16	15	14	13	13	12	11	10	9	8	7	6	6	5	4	4	3	2	1	52
37	24	23	23	22	21	21	20	19	18	17	17	16	15	14	14	13	12	11	11	10	9	7	6	6	5	4	4	3	2	1	53
36	24	24	23	22	22	21	20	19	18	18	17	16	16	15	14	13	12	12	11	10	9	8	7	6	5	4	4	3	2	1	54
35	25	24	23	23	22	21	20	20	19	18	17	17	16	15	14	14	13	12	11	10	9	8	7	6	5	5	4	3	2	1	55
34	25	24	24	23	22	21	21	20	19	18	18	17	16	15	15	14	13	12	11	11	9	8	8	6	5	5	4	3	2	1	56
33	25	25	24	23	23	22	21	20	20	19	18	17	17	16	15	14	13	13	12	11	10	8	8	7	6	5	4	3	2	1	57
32	26	25	24	24	23	22	21	21	20	19	18	18	17	16	15	15	13	13	12	11	10	8	8	7	6	5	4	3	2	1	58
31	26	25	25	24	23	22	22	21	20	20	19	18	17	17	16	15	14	13	12	11	10	8	8	7	6	5	4	3	2	1	59
30	26	26	24	24	23	22	22	21	20	20	19	18	18	17	16	15	14	13	12	11	10	8	8	7	6	5	4	3	2	1	60
29	27	26	26	25	24	23	22	22	21	20	19	19	18	17	16	15	14	14	12	11	10	9	8	7	6	5	4	3	2	1	61
28	27	27	26	25	24	23	23	22	21	20	20	19	18	17	16	15	14	14	12	11	10	9	8	7	6	5	4	3	2	1	62
27	27	27	26	25	24	24	23	22	22	21	20	19	18	17	17	16	14	14	12	11	10	9	8	7	6	5	4	3	2	1	63
26	28	27	26	26	25	24	23	22	22	21	20	20	18	17	17	16	14	14	13	11	10	9	8	7	6	5	4	3	2	1	64
25	28	27	26	26	25	24	23	23	22	21	20	20	18	17	17	16	14	14	13	12	10	9	8	7	6	5	4	3	2	1	65
24	28	28	26	26	25	24	24	23	22	21	20	20	18	17	17	16	15	14	13	12	10	9	8	7	6	5	4	3	2	1	66
23	28	28	27	26	26	25	24	23	22	22	21	20	19	17	17	16	15	14	13	12	10	9	8	7	6	5	4	3	2	1	67
22	29	28	27	26	26	25	24	23	23	22	21	20	19	18	17	16	15	14	13	12	11	9	8	7	6	5	4	3	2	1	68
21	29	28	27	27	26	25	24	24	23	22	21	20	19	18	17	16	15	14	13	12	11	9	8	7	6	5	4	3	2	1	69
20	29	29	27	27	26	25	24	24	23	22	21	20	19	18	17	16	15	14	13	12	11	9	8	7	6	5	4	3	2	1	70
19	29	29	28	27	26	25	25	24	23	22	22	20	20	18	18	16	15	14	13	12	11	9	8	7	6	5	4	3	2	1	71
18	29	29	28	27	27	26	25	24	23	23	22	21	20	18	18	16	15	14	13	12	11	9	8	7	6	5	4	3	2	1	72
17	29	29	28	27	27	26	25	24	23	23	22	21	20	19	18	16	15	14	13	12	11	9	8	7	6	5	4	3	2	1	73
16	29	29	28	27	27	26	25	24	24	23	22	21	20	19	18	16	15	14	13	12	11	9	8	7	6	5	4	3	2	1	74
15	30	29	28	27	27	26	25	24	24	23	22	21	20	19	18	16	15	14	13	12	11	9	8	7	6	5	4	3	2	1	75
14	30	29	29	28	27	26	25	25	24	23	22	22	20	19	18	16	15	14	13	12	11	9	8	7	6	5	4	3	2	1	76
13	30	29	29	28	27	26	25	25	24	23	22	22	20	19	18	16	15	14	13	12	11	9	8	7	6	5	4	3	2	1	77
12	30	29	29	28	27	26	26	25	24	23	23	22	20	19	18	16	15	14	13	12	11	9	8	7	6	5	4	3	2	1	78
11	30	29	29	28	27	26	26	25	24	23	23	22	20	19	18	16	15	14	13	12	11	9	8	7	6	5	4	3	2	1	79
10	30	30	28	28	27	26	26	25	24	23	23	22	20	19	18	16	15	14	13	12	11	9	8	7	6	5	4	3	2	1	80

For Z2 < 10°, use 10°.

For P > 80°, use 80°.

USE OF CONCISE SIGHT REDUCTION TABLES (continued)

4. *Example.*   (b) Required the altitude and azimuth of *Vega* on 1998 July 29 at UT 04^h 51^m from the estimated position 152° west, 15° south.

1. Assumed latitude $Lat =$ 15° S
    From the almanac $GHA =$ 100° 10'
    Assumed longitude   152° 10' W
    Local hour angle $LHA =$ 308

2. Reduction table, 1st entry
    $(Lat, LHA) = (15, 308)$ $A =$ 49   34 $A° = 50, A' = 34$
                                  $B = +66$ 29 $Z_1 = +71.7,$ $LHA > 270°$
3. From the almanac $Dec = -38$ 47   *Lat* and *Dec* contrary
    $Sum = B + Dec$ $F = +27$ 42 $F° = 28, F' = 42$

4. Reduction table, 2nd entry
    $(A°, F°) = (50, 28)$ $H =$ 17   34 $P° = 37$
                                                         $Z_2 = 67.8$

5. Auxiliary table, 1st entry
    $(F', P°) = (42, 37)$ $corr_1 =$ _____ -11 $F < 90°, F' > 29'$
    Sum   17   23
6. Auxiliary table, 2nd entry
    $(A', Z_2°) = (34, 68)$ $corr_2 =$ _____ $+10$ $A' > 30'$
7. Sum = computed altitude $H_c = +17°$ 33' $F > 0°$

8. Azimuth, first component $Z_1 = +71.7$ same sign as B
           second component $Z_2 = +67.8$ $F < 90°, F > 0°$
    Sum = azimuth angle $Z =$ 139.5

    True azimuth $Z_n =$ 40°   S *Lat*, $LHA > 180°$

CONVERSION OF ARC TO TIME

°	h m	°	h m	°	h m	°	h m	°	h m	°	h m	′	0′.00 m s	0′.25 m s	0′.50 m s	0′.75 m s
0	0 00	60	4 00	120	8 00	180	12 00	240	16 00	300	20 00	0	0 00	0 01	0 02	0 03
1	0 04	61	4 04	121	8 04	181	12 04	241	16 04	301	20 04	1	0 04	0 05	0 06	0 07
2	0 08	62	4 08	122	8 08	182	12 08	242	16 08	302	20 08	2	0 08	0 09	0 10	0 11
3	0 12	63	4 12	123	8 12	183	12 12	243	16 12	303	20 12	3	0 12	0 13	0 14	0 15
4	0 16	64	4 16	124	8 16	184	12 16	244	16 16	304	20 16	4	0 16	0 17	0 18	0 19
5	0 20	65	4 20	125	8 20	185	12 20	245	16 20	305	20 20	5	0 20	0 21	0 22	0 23
6	0 24	66	4 24	126	8 24	186	12 24	246	16 24	306	20 24	6	0 24	0 25	0 26	0 27
7	0 28	67	4 28	127	8 28	187	12 28	247	16 28	307	20 28	7	0 28	0 29	0 30	0 31
8	0 32	68	4 32	128	8 32	188	12 32	248	16 32	308	20 32	8	0 32	0 33	0 34	0 35
9	0 36	69	4 36	129	8 36	189	12 36	249	16 36	309	20 36	9	0 36	0 37	0 38	0 39
10	0 40	70	4 40	130	8 40	190	12 40	250	16 40	310	20 40	10	0 40	0 41	0 42	0 43
11	0 44	71	4 44	131	8 44	191	12 44	251	16 44	311	20 44	11	0 44	0 45	0 46	0 47
12	0 48	72	4 48	132	8 48	192	12 48	252	16 48	312	20 48	12	0 48	0 49	0 50	0 51
13	0 52	73	4 52	133	8 52	193	12 52	253	16 52	313	20 52	13	0 52	0 53	0 54	0 55
14	0 56	74	4 56	134	8 56	194	12 56	254	16 56	314	20 56	14	0 56	0 57	0 58	0 59
15	1 00	75	5 00	135	9 00	195	13 00	255	17 00	315	21 00	15	1 00	1 01	1 02	1 03
16	1 04	76	5 04	136	9 04	196	13 04	256	17 04	316	21 04	16	1 04	1 05	1 06	1 07
17	1 08	77	5 08	137	9 08	197	13 08	257	17 08	317	21 08	17	1 08	1 09	1 10	1 11
18	1 12	78	5 12	138	9 12	198	13 12	258	17 12	318	21 12	18	1 12	1 13	1 14	1 15
19	1 16	79	5 16	139	9 16	199	13 16	259	17 16	319	21 16	19	1 16	1 17	1 18	1 19
20	1 20	80	5 20	140	9 20	200	13 20	260	17 20	320	21 20	20	1 20	1 21	1 22	1 23
21	1 24	81	5 24	141	9 24	201	13 24	261	17 24	321	21 24	21	1 24	1 25	1 26	1 27
22	1 28	82	5 28	142	9 28	202	13 28	262	17 28	322	21 28	22	1 28	1 29	1 30	1 31
23	1 32	83	5 32	143	9 32	203	13 32	263	17 32	323	21 32	23	1 32	1 33	1 34	1 35
24	1 36	84	5 36	144	9 36	204	13 36	264	17 36	324	21 36	24	1 36	1 37	1 38	1 39
25	1 40	85	5 40	145	9 40	205	13 40	265	17 40	325	21 40	25	1 40	1 41	1 42	1 43
26	1 44	86	5 44	146	9 44	206	13 44	266	17 44	326	21 44	26	1 44	1 45	1 46	1 47
27	1 48	87	5 48	147	9 48	207	13 48	267	17 48	327	21 48	27	1 48	1 49	1 50	1 51
28	1 52	88	5 52	148	9 52	208	13 52	268	17 52	328	21 52	28	1 52	1 53	1 54	1 55
29	1 56	89	5 56	149	9 56	209	13 56	269	17 56	329	21 56	29	1 56	1 57	1 58	1 59
30	2 00	90	6 00	150	10 00	210	14 00	270	18 00	330	22 00	30	2 00	2 01	2 02	2 03
31	2 04	91	6 04	151	10 04	211	14 04	271	18 04	331	22 04	31	2 04	2 05	2 06	2 07
32	2 08	92	6 08	152	10 08	212	14 08	272	18 08	332	22 08	32	2 08	2 09	2 10	2 11
33	2 12	93	6 12	153	10 12	213	14 12	273	18 12	333	22 12	33	2 12	2 13	2 14	2 15
34	2 16	94	6 16	154	10 16	214	14 16	274	18 16	334	22 16	34	2 16	2 17	2 18	2 19
35	2 20	95	6 20	155	10 20	215	14 20	275	18 20	335	22 20	35	2 20	2 21	2 22	2 23
36	2 24	96	6 24	156	10 24	216	14 24	276	18 24	336	22 24	36	2 24	2 25	2 26	2 27
37	2 28	97	6 28	157	10 28	217	14 28	277	18 28	337	22 28	37	2 28	2 29	2 30	2 31
38	2 32	98	6 32	158	10 32	218	14 32	278	18 32	338	22 32	38	2 32	2 33	2 34	2 35
39	2 36	99	6 36	159	10 36	219	14 36	279	18 36	339	22 36	39	2 36	2 37	2 38	2 39
40	2 40	100	6 40	160	10 40	220	14 40	280	18 40	340	22 40	40	2 40	2 41	2 42	2 43
41	2 44	101	6 44	161	10 44	221	14 44	281	18 44	341	22 44	41	2 44	2 45	2 46	2 47
42	2 48	102	6 48	162	10 48	222	14 48	282	18 48	342	22 48	42	2 48	2 49	2 50	2 51
43	2 52	103	6 52	163	10 52	223	14 52	283	18 52	343	22 52	43	2 52	2 53	2 54	2 55
44	2 56	104	6 56	164	10 56	224	14 56	284	18 56	344	22 56	44	2 56	2 57	2 58	2 59
45	3 00	105	7 00	165	11 00	225	15 00	285	19 00	345	23 00	45	3 00	3 01	3 02	3 03
46	3 04	106	7 04	166	11 04	226	15 04	286	19 04	346	23 04	46	3 04	3 05	3 06	3 07
47	3 08	107	7 08	167	11 08	227	15 08	287	19 08	347	23 08	47	3 08	3 09	3 10	3 11
48	3 12	108	7 12	168	11 12	228	15 12	288	19 12	348	23 12	48	3 12	3 13	3 14	3 15
49	3 16	109	7 16	169	11 16	229	15 16	289	19 16	349	23 16	49	3 16	3 17	3 18	3 19
50	3 20	110	7 20	170	11 20	230	15 20	290	19 20	350	23 20	50	3 20	3 21	3 22	3 23
51	3 24	111	7 24	171	11 24	231	15 24	291	19 24	351	23 24	51	3 24	3 25	3 26	3 27
52	3 28	112	7 28	172	11 28	232	15 28	292	19 28	352	23 28	52	3 28	3 29	3 30	3 31
53	3 32	113	7 32	173	11 32	233	15 32	293	19 32	353	23 32	53	3 32	3 33	3 34	3 35
54	3 36	114	7 36	174	11 36	234	15 36	294	19 36	354	23 36	54	3 36	3 37	3 38	3 39
55	3 40	115	7 40	175	11 40	235	15 40	295	19 40	355	23 40	55	3 40	3 41	3 42	3 43
56	3 44	116	7 44	176	11 44	236	15 44	296	19 44	356	23 44	56	3 44	3 45	3 46	3 47
57	3 48	117	7 48	177	11 48	237	15 48	297	19 48	357	23 48	57	3 48	3 49	3 50	3 51
58	3 52	118	7 52	178	11 52	238	15 52	298	19 52	358	23 52	58	3 52	3 53	3 54	3 55
59	3 56	119	7 56	179	11 56	239	15 56	299	19 56	359	23 56	59	3 56	3 57	3 58	3 59

The above table is for converting expressions in arc to their equivalent in time; its main use in this Almanac is for the conversion of longitude for application to L.M.T. (*added* if *west*, *subtracted* if *east*) to give UT or vice versa, particularly in the case of sunrise, sunset, etc.

0ᵐ	SUN PLANETS	ARIES	MOON	v or Corrⁿ d	v or Corrⁿ d	v or Corrⁿ d
s	° '	° '	° '	' '	' '	' '
00	0 00·0	0 00·0	0 00·0	0·0 0·0	6·0 0·1	12·0 0·1
01	0 00·3	0 00·3	0 00·2	0·1 0·0	6·1 0·1	12·1 0·1
02	0 00·5	0 00·5	0 00·5	0·2 0·0	6·2 0·1	12·2 0·1
03	0 00·8	0 00·8	0 00·7	0·3 0·0	6·3 0·1	12·3 0·1
04	0 01·0	0 01·0	0 01·0	0·4 0·0	6·4 0·1	12·4 0·1
05	0 01·3	0 01·3	0 01·2	0·5 0·0	6·5 0·1	12·5 0·1
06	0 01·5	0 01·5	0 01·4	0·6 0·0	6·6 0·1	12·6 0·1
07	0 01·8	0 01·8	0 01·7	0·7 0·0	6·7 0·1	12·7 0·1
08	0 02·0	0 02·0	0 01·9	0·8 0·0	6·8 0·1	12·8 0·1
09	0 02·3	0 02·3	0 02·1	0·9 0·0	6·9 0·1	12·9 0·1
10	0 02·5	0 02·5	0 02·4	1·0 0·0	7·0 0·1	13·0 0·1
11	0 02·8	0 02·8	0 02·6	1·1 0·0	7·1 0·1	13·1 0·1
12	0 03·0	0 03·0	0 02·9	1·2 0·0	7·2 0·1	13·2 0·1
13	0 03·3	0 03·3	0 03·1	1·3 0·0	7·3 0·1	13·3 0·1
14	0 03·5	0 03·5	0 03·3	1·4 0·0	7·4 0·1	13·4 0·1
15	0 03·8	0 03·8	0 03·6	1·5 0·0	7·5 0·1	13·5 0·1
16	0 04·0	0 04·0	0 03·8	1·6 0·0	7·6 0·1	13·6 0·1
17	0 04·3	0 04·3	0 04·1	1·7 0·0	7·7 0·1	13·7 0·1
18	0 04·5	0 04·5	0 04·3	1·8 0·0	7·8 0·1	13·8 0·1
19	0 04·8	0 04·8	0 04·5	1·9 0·0	7·9 0·1	13·9 0·1
20	0 05·0	0 05·0	0 04·8	2·0 0·0	8·0 0·1	14·0 0·1
21	0 05·3	0 05·3	0 05·0	2·1 0·0	8·1 0·1	14·1 0·1
22	0 05·5	0 05·5	0 05·2	2·2 0·0	8·2 0·1	14·2 0·1
23	0 05·8	0 05·8	0 05·5	2·3 0·0	8·3 0·1	14·3 0·1
24	0 06·0	0 06·0	0 05·7	2·4 0·0	8·4 0·1	14·4 0·1
25	0 06·3	0 06·3	0 06·0	2·5 0·0	8·5 0·1	14·5 0·1
26	0 06·5	0 06·5	0 06·2	2·6 0·0	8·6 0·1	14·6 0·1
27	0 06·8	0 06·8	0 06·4	2·7 0·0	8·7 0·1	14·7 0·1
28	0 07·0	0 07·0	0 06·7	2·8 0·0	8·8 0·1	14·8 0·1
29	0 07·3	0 07·3	0 06·9	2·9 0·0	8·9 0·1	14·9 0·1
30	0 07·5	0 07·5	0 07·2	3·0 0·0	9·0 0·1	15·0 0·1
31	0 07·8	0 07·8	0 07·4	3·1 0·0	9·1 0·1	15·1 0·1
32	0 08·0	0 08·0	0 07·6	3·2 0·0	9·2 0·1	15·2 0·1
33	0 08·3	0 08·3	0 07·9	3·3 0·0	9·3 0·1	15·3 0·1
34	0 08·5	0 08·5	0 08·1	3·4 0·0	9·4 0·1	15·4 0·1
35	0 08·8	0 08·8	0 08·4	3·5 0·0	9·5 0·1	15·5 0·1
36	0 09·0	0 09·0	0 08·6	3·6 0·0	9·6 0·1	15·6 0·1
37	0 09·3	0 09·3	0 08·8	3·7 0·0	9·7 0·1	15·7 0·1
38	0 09·5	0 09·5	0 09·1	3·8 0·0	9·8 0·1	15·8 0·1
39	0 09·8	0 09·8	0 09·3	3·9 0·0	9·9 0·1	15·9 0·1
40	0 10·0	0 10·0	0 09·5	4·0 0·0	10·0 0·1	16·0 0·1
41	0 10·3	0 10·3	0 09·8	4·1 0·0	10·1 0·1	16·1 0·1
42	0 10·5	0 10·5	0 10·0	4·2 0·0	10·2 0·1	16·2 0·1
43	0 10·8	0 10·8	0 10·3	4·3 0·0	10·3 0·1	16·3 0·1
44	0 11·0	0 11·0	0 10·5	4·4 0·0	10·4 0·1	16·4 0·1
45	0 11·3	0 11·3	0 10·7	4·5 0·0	10·5 0·1	16·5 0·1
46	0 11·5	0 11·5	0 11·0	4·6 0·0	10·6 0·1	16·6 0·1
47	0 11·8	0 11·8	0 11·2	4·7 0·0	10·7 0·1	16·7 0·1
48	0 12·0	0 12·0	0 11·5	4·8 0·0	10·8 0·1	16·8 0·1
49	0 12·3	0 12·3	0 11·7	4·9 0·0	10·9 0·1	16·9 0·1
50	0 12·5	0 12·5	0 11·9	5·0 0·0	11·0 0·1	17·0 0·1
51	0 12·8	0 12·8	0 12·2	5·1 0·0	11·1 0·1	17·1 0·1
52	0 13·0	0 13·0	0 12·4	5·2 0·0	11·2 0·1	17·2 0·1
53	0 13·3	0 13·3	0 12·6	5·3 0·0	11·3 0·1	17·3 0·1
54	0 13·5	0 13·5	0 12·9	5·4 0·0	11·4 0·1	17·4 0·1
55	0 13·8	0 13·8	0 13·1	5·5 0·0	11·5 0·1	17·5 0·1
56	0 14·0	0 14·0	0 13·4	5·6 0·0	11·6 0·1	17·6 0·1
57	0 14·3	0 14·3	0 13·6	5·7 0·0	11·7 0·1	17·7 0·1
58	0 14·5	0 14·5	0 13·8	5·8 0·0	11·8 0·1	17·8 0·1
59	0 14·8	0 14·8	0 14·1	5·9 0·0	11·9 0·1	17·9 0·1
60	0 15·0	0 15·0	0 14·3	6·0 0·1	12·0 0·1	18·0 0·2

1ᵐ	SUN PLANETS	ARIES	MOON	v or Corrⁿ d	v or Corrⁿ d	v or Corrⁿ d
s	° '	° '	° '	' '	' '	' '
00	0 15·0	0 15·0	0 14·3	0·0 0·0	6·0 0·2	12·0 0·3
01	0 15·3	0 15·3	0 14·6	0·1 0·0	6·1 0·2	12·1 0·3
02	0 15·5	0 15·5	0 14·8	0·2 0·0	6·2 0·2	12·2 0·3
03	0 15·8	0 15·8	0 15·0	0·3 0·0	6·3 0·2	12·3 0·3
04	0 16·0	0 16·0	0 15·3	0·4 0·0	6·4 0·2	12·4 0·3
05	0 16·3	0 16·3	0 15·5	0·5 0·0	6·5 0·2	12·5 0·3
06	0 16·5	0 16·5	0 15·7	0·6 0·0	6·6 0·2	12·6 0·3
07	0 16·8	0 16·8	0 16·0	0·7 0·0	6·7 0·2	12·7 0·3
08	0 17·0	0 17·0	0 16·2	0·8 0·0	6·8 0·2	12·8 0·3
09	0 17·3	0 17·3	0 16·5	0·9 0·0	6·9 0·2	12·9 0·3
10	0 17·5	0 17·5	0 16·7	1·0 0·0	7·0 0·2	13·0 0·3
11	0 17·8	0 17·8	0 16·9	1·1 0·0	7·1 0·2	13·1 0·3
12	0 18·0	0 18·0	0 17·2	1·2 0·0	7·2 0·2	13·2 0·3
13	0 18·3	0 18·3	0 17·4	1·3 0·0	7·3 0·2	13·3 0·3
14	0 18·5	0 18·6	0 17·7	1·4 0·0	7·4 0·2	13·4 0·3
15	0 18·8	0 18·8	0 17·9	1·5 0·0	7·5 0·2	13·5 0·3
16	0 19·0	0 19·1	0 18·1	1·6 0·0	7·6 0·2	13·6 0·3
17	0 19·3	0 19·3	0 18·4	1·7 0·0	7·7 0·2	13·7 0·3
18	0 19·5	0 19·6	0 18·6	1·8 0·0	7·8 0·2	13·8 0·3
19	0 19·8	0 19·8	0 18·9	1·9 0·0	7·9 0·2	13·9 0·3
20	0 20·0	0 20·1	0 19·1	2·0 0·1	8·0 0·2	14·0 0·4
21	0 20·3	0 20·3	0 19·3	2·1 0·1	8·1 0·2	14·1 0·4
22	0 20·5	0 20·6	0 19·6	2·2 0·1	8·2 0·2	14·2 0·4
23	0 20·8	0 20·8	0 19·8	2·3 0·1	8·3 0·2	14·3 0·4
24	0 21·0	0 21·1	0 20·0	2·4 0·1	8·4 0·2	14·4 0·4
25	0 21·3	0 21·3	0 20·3	2·5 0·1	8·5 0·2	14·5 0·4
26	0 21·5	0 21·6	0 20·5	2·6 0·1	8·6 0·2	14·6 0·4
27	0 21·8	0 21·8	0 20·8	2·7 0·1	8·7 0·2	14·7 0·4
28	0 22·0	0 22·1	0 21·0	2·8 0·1	8·8 0·2	14·8 0·4
29	0 22·3	0 22·3	0 21·2	2·9 0·1	8·9 0·2	14·9 0·4
30	0 22·5	0 22·6	0 21·5	3·0 0·1	9·0 0·2	15·0 0·4
31	0 22·8	0 22·8	0 21·7	3·1 0·1	9·1 0·2	15·1 0·4
32	0 23·0	0 23·1	0 22·0	3·2 0·1	9·2 0·2	15·2 0·4
33	0 23·3	0 23·3	0 22·2	3·3 0·1	9·3 0·2	15·3 0·4
34	0 23·5	0 23·6	0 22·4	3·4 0·1	9·4 0·2	15·4 0·4
35	0 23·8	0 23·8	0 22·7	3·5 0·1	9·5 0·2	15·5 0·4
36	0 24·0	0 24·1	0 22·9	3·6 0·1	9·6 0·2	15·6 0·4
37	0 24·3	0 24·3	0 23·1	3·7 0·1	9·7 0·2	15·7 0·4
38	0 24·5	0 24·6	0 23·4	3·8 0·1	9·8 0·2	15·8 0·4
39	0 24·8	0 24·8	0 23·6	3·9 0·1	9·9 0·2	15·9 0·4
40	0 25·0	0 25·1	0 23·9	4·0 0·1	10·0 0·3	16·0 0·4
41	0 25·3	0 25·3	0 24·1	4·1 0·1	10·1 0·3	16·1 0·4
42	0 25·5	0 25·6	0 24·3	4·2 0·1	10·2 0·3	16·2 0·4
43	0 25·8	0 25·8	0 24·6	4·3 0·1	10·3 0·3	16·3 0·4
44	0 26·0	0 26·1	0 24·8	4·4 0·1	10·4 0·3	16·4 0·4
45	0 26·3	0 26·3	0 25·1	4·5 0·1	10·5 0·3	16·5 0·4
46	0 26·5	0 26·6	0 25·3	4·6 0·1	10·6 0·3	16·6 0·4
47	0 26·8	0 26·8	0 25·5	4·7 0·1	10·7 0·3	16·7 0·4
48	0 27·0	0 27·1	0 25·8	4·8 0·1	10·8 0·3	16·8 0·4
49	0 27·3	0 27·3	0 26·0	4·9 0·1	10·9 0·3	16·9 0·4
50	0 27·5	0 27·6	0 26·2	5·0 0·1	11·0 0·3	17·0 0·4
51	0 27·8	0 27·8	0 26·5	5·1 0·1	11·1 0·3	17·1 0·4
52	0 28·0	0 28·1	0 26·7	5·2 0·1	11·2 0·3	17·2 0·4
53	0 28·3	0 28·3	0 27·0	5·3 0·1	11·3 0·3	17·3 0·4
54	0 28·5	0 28·6	0 27·2	5·4 0·1	11·4 0·3	17·4 0·4
55	0 28·8	0 28·8	0 27·4	5·5 0·1	11·5 0·3	17·5 0·4
56	0 29·0	0 29·1	0 27·7	5·6 0·1	11·6 0·3	17·6 0·4
57	0 29·3	0 29·3	0 27·9	5·7 0·1	11·7 0·3	17·7 0·4
58	0 29·5	0 29·6	0 28·2	5·8 0·1	11·8 0·3	17·8 0·4
59	0 29·8	0 29·8	0 28·4	5·9 0·1	11·9 0·3	17·9 0·4
60	0 30·0	0 30·1	0 28·6	6·0 0·2	12·0 0·3	18·0 0·5

2ᵐ

s	SUN PLANETS (° ′)	ARIES (° ′)	MOON (° ′)	v or Corrⁿ / d	v or Corrⁿ / d	v or Corrⁿ / d
00	0 30·0	0 30·1	0 28·6	0·0 0·0	6·0 0·3	12·0 0·5
01	0 30·3	0 30·3	0 28·9	0·1 0·0	6·1 0·3	12·1 0·5
02	0 30·5	0 30·6	0 29·1	0·2 0·0	6·2 0·3	12·2 0·5
03	0 30·8	0 30·8	0 29·3	0·3 0·0	6·3 0·3	12·3 0·5
04	0 31·0	0 31·1	0 29·6	0·4 0·0	6·4 0·3	12·4 0·5
05	0 31·3	0 31·3	0 29·8	0·5 0·0	6·5 0·3	12·5 0·5
06	0 31·5	0 31·6	0 30·1	0·6 0·0	6·6 0·3	12·6 0·5
07	0 31·8	0 31·8	0 30·3	0·7 0·0	6·7 0·3	12·7 0·5
08	0 32·0	0 32·1	0 30·5	0·8 0·0	6·8 0·3	12·8 0·5
09	0 32·3	0 32·3	0 30·8	0·9 0·0	6·9 0·3	12·9 0·5
10	0 32·5	0 32·6	0 31·0	1·0 0·0	7·0 0·3	13·0 0·5
11	0 32·8	0 32·8	0 31·3	1·1 0·0	7·1 0·3	13·1 0·5
12	0 33·0	0 33·1	0 31·5	1·2 0·1	7·2 0·3	13·2 0·6
13	0 33·3	0 33·3	0 31·7	1·3 0·1	7·3 0·3	13·3 0·6
14	0 33·5	0 33·6	0 32·0	1·4 0·1	7·4 0·3	13·4 0·6
15	0 33·8	0 33·8	0 32·2	1·5 0·1	7·5 0·3	13·5 0·6
16	0 34·0	0 34·1	0 32·5	1·6 0·1	7·6 0·3	13·6 0·6
17	0 34·3	0 34·3	0 32·7	1·7 0·1	7·7 0·3	13·7 0·6
18	0 34·5	0 34·6	0 32·9	1·8 0·1	7·8 0·3	13·8 0·6
19	0 34·8	0 34·8	0 33·2	1·9 0·1	7·9 0·3	13·9 0·6
20	0 35·0	0 35·1	0 33·4	2·0 0·1	8·0 0·3	14·0 0·6
21	0 35·3	0 35·3	0 33·6	2·1 0·1	8·1 0·3	14·1 0·6
22	0 35·5	0 35·6	0 33·9	2·2 0·1	8·2 0·3	14·2 0·6
23	0 35·8	0 35·8	0 34·1	2·3 0·1	8·3 0·3	14·3 0·6
24	0 36·0	0 36·1	0 34·4	2·4 0·1	8·4 0·4	14·4 0·6
25	0 36·3	0 36·3	0 34·6	2·5 0·1	8·5 0·4	14·5 0·6
26	0 36·5	0 36·6	0 34·8	2·6 0·1	8·6 0·4	14·6 0·6
27	0 36·8	0 36·9	0 35·1	2·7 0·1	8·7 0·4	14·7 0·6
28	0 37·0	0 37·1	0 35·3	2·8 0·1	8·8 0·4	14·8 0·6
29	0 37·3	0 37·4	0 35·6	2·9 0·1	8·9 0·4	14·9 0·6
30	0 37·5	0 37·6	0 35·8	3·0 0·1	9·0 0·4	15·0 0·6
31	0 37·8	0 37·9	0 36·0	3·1 0·1	9·1 0·4	15·1 0·6
32	0 38·0	0 38·1	0 36·3	3·2 0·1	9·2 0·4	15·2 0·6
33	0 38·3	0 38·4	0 36·5	3·3 0·1	9·3 0·4	15·3 0·6
34	0 38·5	0 38·6	0 36·7	3·4 0·1	9·4 0·4	15·4 0·6
35	0 38·8	0 38·9	0 37·0	3·5 0·1	9·5 0·4	15·5 0·6
36	0 39·0	0 39·1	0 37·2	3·6 0·2	9·6 0·4	15·6 0·7
37	0 39·3	0 39·4	0 37·5	3·7 0·2	9·7 0·4	15·7 0·7
38	0 39·5	0 39·6	0 37·7	3·8 0·2	9·8 0·4	15·8 0·7
39	0 39·8	0 39·9	0 37·9	3·9 0·2	9·9 0·4	15·9 0·7
40	0 40·0	0 40·1	0 38·2	4·0 0·2	10·0 0·4	16·0 0·7
41	0 40·3	0 40·4	0 38·4	4·1 0·2	10·1 0·4	16·1 0·7
42	0 40·5	0 40·6	0 38·7	4·2 0·2	10·2 0·4	16·2 0·7
43	0 40·8	0 40·9	0 38·9	4·3 0·2	10·3 0·4	16·3 0·7
44	0 41·0	0 41·1	0 39·1	4·4 0·2	10·4 0·4	16·4 0·7
45	0 41·3	0 41·4	0 39·4	4·5 0·2	10·5 0·4	16·5 0·7
46	0 41·5	0 41·6	0 39·6	4·6 0·2	10·6 0·4	16·6 0·7
47	0 41·8	0 41·9	0 39·8	4·7 0·2	10·7 0·4	16·7 0·7
48	0 42·0	0 42·1	0 40·1	4·8 0·2	10·8 0·5	16·8 0·7
49	0 42·3	0 42·4	0 40·3	4·9 0·2	10·9 0·5	16·9 0·7
50	0 42·5	0 42·6	0 40·6	5·0 0·2	11·0 0·5	17·0 0·7
51	0 42·8	0 42·9	0 40·8	5·1 0·2	11·1 0·5	17·1 0·7
52	0 43·0	0 43·1	0 41·0	5·2 0·2	11·2 0·5	17·2 0·7
53	0 43·3	0 43·4	0 41·3	5·3 0·2	11·3 0·5	17·3 0·7
54	0 43·5	0 43·6	0 41·5	5·4 0·2	11·4 0·5	17·4 0·7
55	0 43·8	0 43·9	0 41·8	5·5 0·2	11·5 0·5	17·5 0·7
56	0 44·0	0 44·1	0 42·0	5·6 0·2	11·6 0·5	17·6 0·7
57	0 44·3	0 44·4	0 42·2	5·7 0·2	11·7 0·5	17·7 0·7
58	0 44·5	0 44·6	0 42·5	5·8 0·2	11·8 0·5	17·8 0·7
59	0 44·8	0 44·9	0 42·7	5·9 0·2	11·9 0·5	17·9 0·7
60	0 45·0	0 45·1	0 43·0	6·0 0·3	12·0 0·5	18·0 0·8

3ᵐ

s	SUN PLANETS (° ′)	ARIES (° ′)	MOON (° ′)	v or Corrⁿ / d	v or Corrⁿ / d	v or Corrⁿ / d
00	0 45·0	0 45·1	0 43·0	0·0 0·0	6·0 0·4	12·0 0·7
01	0 45·3	0 45·4	0 43·2	0·1 0·0	6·1 0·4	12·1 0·7
02	0 45·5	0 45·6	0 43·4	0·2 0·0	6·2 0·4	12·2 0·7
03	0 45·8	0 45·9	0 43·7	0·3 0·0	6·3 0·4	12·3 0·7
04	0 46·0	0 46·1	0 43·9	0·4 0·0	6·4 0·4	12·4 0·7
05	0 46·3	0 46·4	0 44·1	0·5 0·0	6·5 0·4	12·5 0·7
06	0 46·5	0 46·6	0 44·4	0·6 0·0	6·6 0·4	12·6 0·7
07	0 46·8	0 46·9	0 44·6	0·7 0·0	6·7 0·4	12·7 0·7
08	0 47·0	0 47·1	0 44·9	0·8 0·0	6·8 0·4	12·8 0·7
09	0 47·3	0 47·4	0 45·1	0·9 0·1	6·9 0·4	12·9 0·8
10	0 47·5	0 47·6	0 45·3	1·0 0·1	7·0 0·4	13·0 0·8
11	0 47·8	0 47·9	0 45·6	1·1 0·1	7·1 0·4	13·1 0·8
12	0 48·0	0 48·1	0 45·8	1·2 0·1	7·2 0·4	13·2 0·8
13	0 48·3	0 48·4	0 46·1	1·3 0·1	7·3 0·4	13·3 0·8
14	0 48·5	0 48·6	0 46·3	1·4 0·1	7·4 0·4	13·4 0·8
15	0 48·8	0 48·9	0 46·5	1·5 0·1	7·5 0·4	13·5 0·8
16	0 49·0	0 49·1	0 46·8	1·6 0·1	7·6 0·4	13·6 0·8
17	0 49·3	0 49·4	0 47·0	1·7 0·1	7·7 0·4	13·7 0·8
18	0 49·5	0 49·6	0 47·2	1·8 0·1	7·8 0·5	13·8 0·8
19	0 49·8	0 49·9	0 47·5	1·9 0·1	7·9 0·5	13·9 0·8
20	0 50·0	0 50·1	0 47·7	2·0 0·1	8·0 0·5	14·0 0·8
21	0 50·3	0 50·4	0 48·0	2·1 0·1	8·1 0·5	14·1 0·8
22	0 50·5	0 50·6	0 48·2	2·2 0·1	8·2 0·5	14·2 0·8
23	0 50·8	0 50·9	0 48·4	2·3 0·1	8·3 0·5	14·3 0·8
24	0 51·0	0 51·1	0 48·7	2·4 0·1	8·4 0·5	14·4 0·8
25	0 51·3	0 51·4	0 48·9	2·5 0·1	8·5 0·5	14·5 0·8
26	0 51·5	0 51·6	0 49·2	2·6 0·2	8·6 0·5	14·6 0·9
27	0 51·8	0 51·9	0 49·4	2·7 0·2	8·7 0·5	14·7 0·9
28	0 52·0	0 52·1	0 49·6	2·8 0·2	8·8 0·5	14·8 0·9
29	0 52·3	0 52·4	0 49·9	2·9 0·2	8·9 0·5	14·9 0·9
30	0 52·5	0 52·6	0 50·1	3·0 0·2	9·0 0·5	15·0 0·9
31	0 52·8	0 52·9	0 50·3	3·1 0·2	9·1 0·5	15·1 0·9
32	0 53·0	0 53·1	0 50·6	3·2 0·2	9·2 0·5	15·2 0·9
33	0 53·3	0 53·4	0 50·8	3·3 0·2	9·3 0·5	15·3 0·9
34	0 53·5	0 53·6	0 51·1	3·4 0·2	9·4 0·5	15·4 0·9
35	0 53·8	0 53·9	0 51·3	3·5 0·2	9·5 0·6	15·5 0·9
36	0 54·0	0 54·1	0 51·5	3·6 0·2	9·6 0·6	15·6 0·9
37	0 54·3	0 54·4	0 51·8	3·7 0·2	9·7 0·6	15·7 0·9
38	0 54·5	0 54·6	0 52·0	3·8 0·2	9·8 0·6	15·8 0·9
39	0 54·8	0 54·9	0 52·3	3·9 0·2	9·9 0·6	15·9 0·9
40	0 55·0	0 55·2	0 52·5	4·0 0·2	10·0 0·6	16·0 0·9
41	0 55·3	0 55·4	0 52·7	4·1 0·2	10·1 0·6	16·1 0·9
42	0 55·5	0 55·7	0 53·0	4·2 0·2	10·2 0·6	16·2 0·9
43	0 55·8	0 55·9	0 53·2	4·3 0·3	10·3 0·6	16·3 1·0
44	0 56·0	0 56·2	0 53·4	4·4 0·3	10·4 0·6	16·4 1·0
45	0 56·3	0 56·4	0 53·7	4·5 0·3	10·5 0·6	16·5 1·0
46	0 56·5	0 56·7	0 53·9	4·6 0·3	10·6 0·6	16·6 1·0
47	0 56·8	0 56·9	0 54·2	4·7 0·3	10·7 0·6	16·7 1·0
48	0 57·0	0 57·2	0 54·4	4·8 0·3	10·8 0·6	16·8 1·0
49	0 57·3	0 57·4	0 54·6	4·9 0·3	10·9 0·6	16·9 1·0
50	0 57·5	0 57·7	0 54·9	5·0 0·3	11·0 0·6	17·0 1·0
51	0 57·8	0 57·9	0 55·1	5·1 0·3	11·1 0·6	17·1 1·0
52	0 58·0	0 58·2	0 55·4	5·2 0·3	11·2 0·7	17·2 1·0
53	0 58·3	0 58·4	0 55·6	5·3 0·3	11·3 0·7	17·3 1·0
54	0 58·5	0 58·7	0 55·8	5·4 0·3	11·4 0·7	17·4 1·0
55	0 58·8	0 58·9	0 56·1	5·5 0·3	11·5 0·7	17·5 1·0
56	0 59·0	0 59·2	0 56·3	5·6 0·3	11·6 0·7	17·6 1·0
57	0 59·3	0 59·4	0 56·6	5·7 0·3	11·7 0·7	17·7 1·0
58	0 59·5	0 59·7	0 56·8	5·8 0·3	11·8 0·7	17·8 1·0
59	0 59·8	0 59·9	0 57·0	5·9 0·3	11·9 0·7	17·9 1·0
60	1 00·0	1 00·2	0 57·3	6·0 0·4	12·0 0·7	18·0 1·1

4^m	SUN PLANETS	ARIES	MOON	v or Corr^n d		v or Corr^n d		v or Corr^n d	
s	° ′	° ′	° ′	′	′	′	′	′	′
00	1 00·0	1 00·2	0 57·3	0·0	0·0	6·0	0·5	12·0	0·9
01	1 00·3	1 00·4	0 57·5	0·1	0·0	6·1	0·5	12·1	0·9
02	1 00·5	1 00·7	0 57·7	0·2	0·0	6·2	0·5	12·2	0·9
03	1 00·8	1 00·9	0 58·0	0·3	0·0	6·3	0·5	12·3	0·9
04	1 01·0	1 01·2	0 58·2	0·4	0·0	6·4	0·5	12·4	0·9
05	1 01·3	1 01·4	0 58·5	0·5	0·0	6·5	0·5	12·5	0·9
06	1 01·5	1 01·7	0 58·7	0·6	0·0	6·6	0·5	12·6	0·9
07	1 01·8	1 01·9	0 58·9	0·7	0·1	6·7	0·5	12·7	1·0
08	1 02·0	1 02·2	0 59·2	0·8	0·1	6·8	0·5	12·8	1·0
09	1 02·3	1 02·4	0 59·4	0·9	0·1	6·9	0·5	12·9	1·0
10	1 02·5	1 02·7	0 59·7	1·0	0·1	7·0	0·5	13·0	1·0
11	1 02·8	1 02·9	0 59·9	1·1	0·1	7·1	0·5	13·1	1·0
12	1 03·0	1 03·2	1 00·1	1·2	0·1	7·2	0·5	13·2	1·0
13	1 03·3	1 03·4	1 00·4	1·3	0·1	7·3	0·5	13·3	1·0
14	1 03·5	1 03·7	1 00·6	1·4	0·1	7·4	0·6	13·4	1·0
15	1 03·8	1 03·9	1 00·8	1·5	0·1	7·5	0·6	13·5	1·0
16	1 04·0	1 04·2	1 01·1	1·6	0·1	7·6	0·6	13·6	1·0
17	1 04·3	1 04·4	1 01·3	1·7	0·1	7·7	0·6	13·7	1·0
18	1 04·5	1 04·7	1 01·6	1·8	0·1	7·8	0·6	13·8	1·0
19	1 04·8	1 04·9	1 01·8	1·9	0·1	7·9	0·6	13·9	1·0
20	1 05·0	1 05·2	1 02·0	2·0	0·1	8·0	0·6	14·0	1·1
21	1 05·3	1 05·4	1 02·3	2·1	0·2	8·1	0·6	14·1	1·1
22	1 05·5	1 05·7	1 02·5	2·2	0·2	8·2	0·6	14·2	1·1
23	1 05·8	1 05·9	1 02·8	2·3	0·2	8·3	0·6	14·3	1·1
24	1 06·0	1 06·2	1 03·0	2·4	0·2	8·4	0·6	14·4	1·1
25	1 06·3	1 06·4	1 03·2	2·5	0·2	8·5	0·6	14·5	1·1
26	1 06·5	1 06·7	1 03·5	2·6	0·2	8·6	0·6	14·6	1·1
27	1 06·8	1 06·9	1 03·7	2·7	0·2	8·7	0·7	14·7	1·1
28	1 07·0	1 07·2	1 03·9	2·8	0·2	8·8	0·7	14·8	1·1
29	1 07·3	1 07·4	1 04·2	2·9	0·2	8·9	0·7	14·9	1·1
30	1 07·5	1 07·7	1 04·4	3·0	0·2	9·0	0·7	15·0	1·1
31	1 07·8	1 07·9	1 04·7	3·1	0·2	9·1	0·7	15·1	1·1
32	1 08·0	1 08·2	1 04·9	3·2	0·2	9·2	0·7	15·2	1·1
33	1 08·3	1 08·4	1 05·1	3·3	0·2	9·3	0·7	15·3	1·1
34	1 08·5	1 08·7	1 05·4	3·4	0·3	9·4	0·7	15·4	1·2
35	1 08·8	1 08·9	1 05·6	3·5	0·3	9·5	0·7	15·5	1·2
36	1 09·0	1 09·2	1 05·9	3·6	0·3	9·6	0·7	15·6	1·2
37	1 09·3	1 09·4	1 06·1	3·7	0·3	9·7	0·7	15·7	1·2
38	1 09·5	1 09·7	1 06·3	3·8	0·3	9·8	0·7	15·8	1·2
39	1 09·8	1 09·9	1 06·6	3·9	0·3	9·9	0·7	15·9	1·2
40	1 10·0	1 10·2	1 06·8	4·0	0·3	10·0	0·8	16·0	1·2
41	1 10·3	1 10·4	1 07·0	4·1	0·3	10·1	0·8	16·1	1·2
42	1 10·5	1 10·7	1 07·3	4·2	0·3	10·2	0·8	16·2	1·2
43	1 10·8	1 10·9	1 07·5	4·3	0·3	10·3	0·8	16·3	1·2
44	1 11·0	1 11·2	1 07·8	4·4	0·3	10·4	0·8	16·4	1·2
45	1 11·3	1 11·4	1 08·0	4·5	0·3	10·5	0·8	16·5	1·2
46	1 11·5	1 11·7	1 08·2	4·6	0·3	10·6	0·8	16·6	1·2
47	1 11·8	1 11·9	1 08·5	4·7	0·4	10·7	0·8	16·7	1·3
48	1 12·0	1 12·2	1 08·7	4·8	0·4	10·8	0·8	16·8	1·3
49	1 12·3	1 12·4	1 09·0	4·9	0·4	10·9	0·8	16·9	1·3
50	1 12·5	1 12·7	1 09·2	5·0	0·4	11·0	0·8	17·0	1·3
51	1 12·8	1 12·9	1 09·4	5·1	0·4	11·1	0·8	17·1	1·3
52	1 13·0	1 13·2	1 09·7	5·2	0·4	11·2	0·8	17·2	1·3
53	1 13·3	1 13·5	1 09·9	5·3	0·4	11·3	0·8	17·3	1·3
54	1 13·5	1 13·7	1 10·2	5·4	0·4	11·4	0·9	17·4	1·3
55	1 13·8	1 14·0	1 10·4	5·5	0·4	11·5	0·9	17·5	1·3
56	1 14·0	1 14·2	1 10·6	5·6	0·4	11·6	0·9	17·6	1·3
57	1 14·3	1 14·5	1 10·9	5·7	0·4	11·7	0·9	17·7	1·3
58	1 14·5	1 14·7	1 11·1	5·8	0·4	11·8	0·9	17·8	1·3
59	1 14·8	1 15·0	1 11·3	5·9	0·4	11·9	0·9	17·9	1·3
60	1 15·0	1 15·2	1 11·6	6·0	0·5	12·0	0·9	18·0	1·4

5^m	SUN PLANETS	ARIES	MOON	v or Corr^n d		v or Corr^n d		v or Corr^n d	
s	° ′	° ′	° ′	′	′	′	′	′	′
00	1 15·0	1 15·2	1 11·6	0·0	0·0	6·0	0·6	12·0	1·1
01	1 15·3	1 15·5	1 11·8	0·1	0·0	6·1	0·6	12·1	1·1
02	1 15·5	1 15·7	1 12·1	0·2	0·0	6·2	0·6	12·2	1·1
03	1 15·8	1 16·0	1 12·3	0·3	0·0	6·3	0·6	12·3	1·1
04	1 16·0	1 16·2	1 12·5	0·4	0·0	6·4	0·6	12·4	1·1
05	1 16·3	1 16·5	1 12·8	0·5	0·0	6·5	0·6	12·5	1·1
06	1 16·5	1 16·7	1 13·0	0·6	0·1	6·6	0·6	12·6	1·2
07	1 16·8	1 17·0	1 13·3	0·7	0·1	6·7	0·6	12·7	1·2
08	1 17·0	1 17·2	1 13·5	0·8	0·1	6·8	0·6	12·8	1·2
09	1 17·3	1 17·5	1 13·7	0·9	0·1	6·9	0·6	12·9	1·2
10	1 17·5	1 17·7	1 14·0	1·0	0·1	7·0	0·6	13·0	1·2
11	1 17·8	1 18·0	1 14·2	1·1	0·1	7·1	0·7	13·1	1·2
12	1 18·0	1 18·2	1 14·4	1·2	0·1	7·2	0·7	13·2	1·2
13	1 18·3	1 18·5	1 14·7	1·3	0·1	7·3	0·7	13·3	1·2
14	1 18·5	1 18·7	1 14·9	1·4	0·1	7·4	0·7	13·4	1·2
15	1 18·8	1 19·0	1 15·2	1·5	0·1	7·5	0·7	13·5	1·2
16	1 19·0	1 19·2	1 15·4	1·6	0·1	7·6	0·7	13·6	1·2
17	1 19·3	1 19·5	1 15·6	1·7	0·2	7·7	0·7	13·7	1·3
18	1 19·5	1 19·7	1 15·9	1·8	0·2	7·8	0·7	13·8	1·3
19	1 19·8	1 20·0	1 16·1	1·9	0·2	7·9	0·7	13·9	1·3
20	1 20·0	1 20·2	1 16·4	2·0	0·2	8·0	0·7	14·0	1·3
21	1 20·3	1 20·5	1 16·6	2·1	0·2	8·1	0·7	14·1	1·3
22	1 20·5	1 20·7	1 16·8	2·2	0·2	8·2	0·8	14·2	1·3
23	1 20·8	1 21·0	1 17·1	2·3	0·2	8·3	0·8	14·3	1·3
24	1 21·0	1 21·2	1 17·3	2·4	0·2	8·4	0·8	14·4	1·3
25	1 21·3	1 21·5	1 17·5	2·5	0·2	8·5	0·8	14·5	1·3
26	1 21·5	1 21·7	1 17·8	2·6	0·2	8·6	0·8	14·6	1·3
27	1 21·8	1 22·0	1 18·0	2·7	0·2	8·7	0·8	14·7	1·3
28	1 22·0	1 22·2	1 18·3	2·8	0·3	8·8	0·8	14·8	1·4
29	1 22·3	1 22·5	1 18·5	2·9	0·3	8·9	0·8	14·9	1·4
30	1 22·5	1 22·7	1 18·7	3·0	0·3	9·0	0·8	15·0	1·4
31	1 22·8	1 23·0	1 19·0	3·1	0·3	9·1	0·8	15·1	1·4
32	1 23·0	1 23·2	1 19·2	3·2	0·3	9·2	0·8	15·2	1·4
33	1 23·3	1 23·5	1 19·5	3·3	0·3	9·3	0·9	15·3	1·4
34	1 23·5	1 23·7	1 19·7	3·4	0·3	9·4	0·9	15·4	1·4
35	1 23·8	1 24·0	1 19·9	3·5	0·3	9·5	0·9	15·5	1·4
36	1 24·0	1 24·2	1 20·2	3·6	0·3	9·6	0·9	15·6	1·4
37	1 24·3	1 24·5	1 20·4	3·7	0·3	9·7	0·9	15·7	1·4
38	1 24·5	1 24·7	1 20·7	3·8	0·3	9·8	0·9	15·8	1·4
39	1 24·8	1 25·0	1 20·9	3·9	0·4	9·9	0·9	15·9	1·5
40	1 25·0	1 25·2	1 21·1	4·0	0·4	10·0	0·9	16·0	1·5
41	1 25·3	1 25·5	1 21·4	4·1	0·4	10·1	0·9	16·1	1·5
42	1 25·5	1 25·7	1 21·6	4·2	0·4	10·2	0·9	16·2	1·5
43	1 25·8	1 26·0	1 21·8	4·3	0·4	10·3	0·9	16·3	1·5
44	1 26·0	1 26·2	1 22·1	4·4	0·4	10·4	1·0	16·4	1·5
45	1 26·3	1 26·5	1 22·3	4·5	0·4	10·5	1·0	16·5	1·5
46	1 26·5	1 26·7	1 22·6	4·6	0·4	10·6	1·0	16·6	1·5
47	1 26·8	1 27·0	1 22·8	4·7	0·4	10·7	1·0	16·7	1·5
48	1 27·0	1 27·2	1 23·0	4·8	0·4	10·8	1·0	16·8	1·5
49	1 27·3	1 27·5	1 23·3	4·9	0·4	10·9	1·0	16·9	1·5
50	1 27·5	1 27·7	1 23·5	5·0	0·5	11·0	1·0	17·0	1·6
51	1 27·8	1 28·0	1 23·8	5·1	0·5	11·1	1·0	17·1	1·6
52	1 28·0	1 28·2	1 24·0	5·2	0·5	11·2	1·0	17·2	1·6
53	1 28·3	1 28·5	1 24·2	5·3	0·5	11·3	1·0	17·3	1·6
54	1 28·5	1 28·7	1 24·5	5·4	0·5	11·4	1·0	17·4	1·6
55	1 28·8	1 29·0	1 24·7	5·5	0·5	11·5	1·1	17·5	1·6
56	1 29·0	1 29·2	1 24·9	5·6	0·5	11·6	1·1	17·6	1·6
57	1 29·3	1 29·5	1 25·2	5·7	0·5	11·7	1·1	17·7	1·6
58	1 29·5	1 29·7	1 25·4	5·8	0·5	11·8	1·1	17·8	1·6
59	1 29·8	1 30·0	1 25·7	5·9	0·5	11·9	1·1	17·9	1·6
60	1 30·0	1 30·2	1 25·9	6·0	0·6	12·0	1·1	18·0	1·7

6ᵐ

6	SUN PLANETS	ARIES	MOON	v or Corrⁿ d	v or Corrⁿ d	v or Corrⁿ d
s	° ′	° ′	° ′	′ ′	′ ′	′ ′
00	1 30·0	1 30·2	1 25·9	0·0 0·0	6·0 0·7	12·0 1·3
01	1 30·3	1 30·5	1 26·1	0·1 0·0	6·1 0·7	12·1 1·3
02	1 30·5	1 30·7	1 26·4	0·2 0·0	6·2 0·7	12·2 1·3
03	1 30·8	1 31·0	1 26·6	0·3 0·0	6·3 0·7	12·3 1·3
04	1 31·0	1 31·2	1 26·9	0·4 0·0	6·4 0·7	12·4 1·3
05	1 31·3	1 31·5	1 27·1	0·5 0·1	6·5 0·7	12·5 1·4
06	1 31·5	1 31·8	1 27·3	0·6 0·1	6·6 0·7	12·6 1·4
07	1 31·8	1 32·0	1 27·6	0·7 0·1	6·7 0·7	12·7 1·4
08	1 32·0	1 32·3	1 27·8	0·8 0·1	6·8 0·7	12·8 1·4
09	1 32·3	1 32·5	1 28·0	0·9 0·1	6·9 0·7	12·9 1·4
10	1 32·5	1 32·8	1 28·3	1·0 0·1	7·0 0·8	13·0 1·4
11	1 32·8	1 33·0	1 28·5	1·1 0·1	7·1 0·8	13·1 1·4
12	1 33·0	1 33·3	1 28·8	1·2 0·1	7·2 0·8	13·2 1·4
13	1 33·3	1 33·5	1 29·0	1·3 0·1	7·3 0·8	13·3 1·4
14	1 33·5	1 33·8	1 29·2	1·4 0·2	7·4 0·8	13·4 1·5
15	1 33·8	1 34·0	1 29·5	1·5 0·2	7·5 0·8	13·5 1·5
16	1 34·0	1 34·3	1 29·7	1·6 0·2	7·6 0·8	13·6 1·5
17	1 34·3	1 34·5	1 30·0	1·7 0·2	7·7 0·8	13·7 1·5
18	1 34·5	1 34·8	1 30·2	1·8 0·2	7·8 0·8	13·8 1·5
19	1 34·8	1 35·0	1 30·4	1·9 0·2	7·9 0·9	13·9 1·5
20	1 35·0	1 35·3	1 30·7	2·0 0·2	8·0 0·9	14·0 1·5
21	1 35·3	1 35·5	1 30·9	2·1 0·2	8·1 0·9	14·1 1·5
22	1 35·5	1 35·8	1 31·1	2·2 0·2	8·2 0·9	14·2 1·5
23	1 35·8	1 36·0	1 31·4	2·3 0·2	8·3 0·9	14·3 1·5
24	1 36·0	1 36·3	1 31·6	2·4 0·3	8·4 0·9	14·4 1·6
25	1 36·3	1 36·5	1 31·9	2·5 0·3	8·5 0·9	14·5 1·6
26	1 36·5	1 36·8	1 32·1	2·6 0·3	8·6 0·9	14·6 1·6
27	1 36·8	1 37·0	1 32·3	2·7 0·3	8·7 0·9	14·7 1·6
28	1 37·0	1 37·3	1 32·6	2·8 0·3	8·8 1·0	14·8 1·6
29	1 37·3	1 37·5	1 32·8	2·9 0·3	8·9 1·0	14·9 1·6
30	1 37·5	1 37·8	1 33·1	3·0 0·3	9·0 1·0	15·0 1·6
31	1 37·8	1 38·0	1 33·3	3·1 0·3	9·1 1·0	15·1 1·6
32	1 38·0	1 38·3	1 33·5	3·2 0·3	9·2 1·0	15·2 1·6
33	1 38·3	1 38·5	1 33·8	3·3 0·4	9·3 1·0	15·3 1·7
34	1 38·5	1 38·8	1 34·0	3·4 0·4	9·4 1·0	15·4 1·7
35	1 38·8	1 39·0	1 34·3	3·5 0·4	9·5 1·0	15·5 1·7
36	1 39·0	1 39·3	1 34·5	3·6 0·4	9·6 1·0	15·6 1·7
37	1 39·3	1 39·5	1 34·7	3·7 0·4	9·7 1·1	15·7 1·7
38	1 39·5	1 39·8	1 35·0	3·8 0·4	9·8 1·1	15·8 1·7
39	1 39·8	1 40·0	1 35·2	3·9 0·4	9·9 1·1	15·9 1·7
40	1 40·0	1 40·3	1 35·4	4·0 0·4	10·0 1·1	16·0 1·7
41	1 40·3	1 40·5	1 35·7	4·1 0·4	10·1 1·1	16·1 1·7
42	1 40·5	1 40·8	1 35·9	4·2 0·5	10·2 1·1	16·2 1·8
43	1 40·8	1 41·0	1 36·2	4·3 0·5	10·3 1·1	16·3 1·8
44	1 41·0	1 41·3	1 36·4	4·4 0·5	10·4 1·1	16·4 1·8
45	1 41·3	1 41·5	1 36·6	4·5 0·5	10·5 1·1	16·5 1·8
46	1 41·5	1 41·8	1 36·9	4·6 0·5	10·6 1·1	16·6 1·8
47	1 41·8	1 42·0	1 37·1	4·7 0·5	10·7 1·2	16·7 1·8
48	1 42·0	1 42·3	1 37·4	4·8 0·5	10·8 1·2	16·8 1·8
49	1 42·3	1 42·5	1 37·6	4·9 0·5	10·9 1·2	16·9 1·8
50	1 42·5	1 42·8	1 37·8	5·0 0·5	11·0 1·2	17·0 1·8
51	1 42·8	1 43·0	1 38·1	5·1 0·6	11·1 1·2	17·1 1·9
52	1 43·0	1 43·3	1 38·3	5·2 0·6	11·2 1·2	17·2 1·9
53	1 43·3	1 43·5	1 38·5	5·3 0·6	11·3 1·2	17·3 1·9
54	1 43·5	1 43·8	1 38·8	5·4 0·6	11·4 1·2	17·4 1·9
55	1 43·8	1 44·0	1 39·0	5·5 0·6	11·5 1·2	17·5 1·9
56	1 44·0	1 44·3	1 39·3	5·6 0·6	11·6 1·3	17·6 1·9
57	1 44·3	1 44·5	1 39·5	5·7 0·6	11·7 1·3	17·7 1·9
58	1 44·5	1 44·8	1 39·7	5·8 0·6	11·8 1·3	17·8 1·9
59	1 44·8	1 45·0	1 40·0	5·9 0·6	11·9 1·3	17·9 1·9
60	1 45·0	1 45·3	1 40·2	6·0 0·7	12·0 1·3	18·0 2·0

7ᵐ

7	SUN PLANETS	ARIES	MOON	v or Corrⁿ d	v or Corrⁿ d	v or Corrⁿ d
s	° ′	° ′	° ′	′ ′	′ ′	′ ′
00	1 45·0	1 45·3	1 40·2	0·0 0·0	6·0 0·8	12·0 1·5
01	1 45·3	1 45·5	1 40·5	0·1 0·0	6·1 0·8	12·1 1·5
02	1 45·5	1 45·8	1 40·7	0·2 0·0	6·2 0·8	12·2 1·5
03	1 45·8	1 46·0	1 40·9	0·3 0·0	6·3 0·8	12·3 1·5
04	1 46·0	1 46·3	1 41·2	0·4 0·1	6·4 0·8	12·4 1·6
05	1 46·3	1 46·5	1 41·4	0·5 0·1	6·5 0·8	12·5 1·6
06	1 46·5	1 46·8	1 41·6	0·6 0·1	6·6 0·8	12·6 1·6
07	1 46·8	1 47·0	1 41·9	0·7 0·1	6·7 0·8	12·7 1·6
08	1 47·0	1 47·3	1 42·1	0·8 0·1	6·8 0·9	12·8 1·6
09	1 47·3	1 47·5	1 42·4	0·9 0·1	6·9 0·9	12·9 1·6
10	1 47·5	1 47·8	1 42·6	1·0 0·1	7·0 0·9	13·0 1·6
11	1 47·8	1 48·0	1 42·8	1·1 0·1	7·1 0·9	13·1 1·6
12	1 48·0	1 48·3	1 43·1	1·2 0·2	7·2 0·9	13·2 1·7
13	1 48·3	1 48·5	1 43·3	1·3 0·2	7·3 0·9	13·3 1·7
14	1 48·5	1 48·8	1 43·6	1·4 0·2	7·4 0·9	13·4 1·7
15	1 48·8	1 49·0	1 43·8	1·5 0·2	7·5 0·9	13·5 1·7
16	1 49·0	1 49·3	1 44·0	1·6 0·2	7·6 1·0	13·6 1·7
17	1 49·3	1 49·5	1 44·3	1·7 0·2	7·7 1·0	13·7 1·7
18	1 49·5	1 49·8	1 44·5	1·8 0·2	7·8 1·0	13·8 1·7
19	1 49·8	1 50·1	1 44·8	1·9 0·2	7·9 1·0	13·9 1·7
20	1 50·0	1 50·3	1 45·0	2·0 0·3	8·0 1·0	14·0 1·8
21	1 50·3	1 50·6	1 45·2	2·1 0·3	8·1 1·0	14·1 1·8
22	1 50·5	1 50·8	1 45·5	2·2 0·3	8·2 1·0	14·2 1·8
23	1 50·8	1 51·1	1 45·7	2·3 0·3	8·3 1·0	14·3 1·8
24	1 51·0	1 51·3	1 45·9	2·4 0·3	8·4 1·1	14·4 1·8
25	1 51·3	1 51·6	1 46·2	2·5 0·3	8·5 1·1	14·5 1·8
26	1 51·5	1 51·8	1 46·4	2·6 0·3	8·6 1·1	14·6 1·8
27	1 51·8	1 52·1	1 46·7	2·7 0·3	8·7 1·1	14·7 1·8
28	1 52·0	1 52·3	1 46·9	2·8 0·4	8·8 1·1	14·8 1·9
29	1 52·3	1 52·6	1 47·1	2·9 0·4	8·9 1·1	14·9 1·9
30	1 52·5	1 52·8	1 47·4	3·0 0·4	9·0 1·1	15·0 1·9
31	1 52·8	1 53·1	1 47·6	3·1 0·4	9·1 1·1	15·1 1·9
32	1 53·0	1 53·3	1 47·9	3·2 0·4	9·2 1·2	15·2 1·9
33	1 53·3	1 53·6	1 48·1	3·3 0·4	9·3 1·2	15·3 1·9
34	1 53·5	1 53·8	1 48·3	3·4 0·4	9·4 1·2	15·4 1·9
35	1 53·8	1 54·1	1 48·6	3·5 0·4	9·5 1·2	15·5 1·9
36	1 54·0	1 54·3	1 48·8	3·6 0·5	9·6 1·2	15·6 2·0
37	1 54·3	1 54·6	1 49·0	3·7 0·5	9·7 1·2	15·7 2·0
38	1 54·5	1 54·8	1 49·3	3·8 0·5	9·8 1·2	15·8 2·0
39	1 54·8	1 55·1	1 49·5	3·9 0·5	9·9 1·2	15·9 2·0
40	1 55·0	1 55·3	1 49·8	4·0 0·5	10·0 1·3	16·0 2·0
41	1 55·3	1 55·6	1 50·0	4·1 0·5	10·1 1·3	16·1 2·0
42	1 55·5	1 55·8	1 50·2	4·2 0·5	10·2 1·3	16·2 2·0
43	1 55·8	1 56·1	1 50·5	4·3 0·5	10·3 1·3	16·3 2·0
44	1 56·0	1 56·3	1 50·7	4·4 0·6	10·4 1·3	16·4 2·1
45	1 56·3	1 56·6	1 51·0	4·5 0·6	10·5 1·3	16·5 2·1
46	1 56·5	1 56·8	1 51·2	4·6 0·6	10·6 1·3	16·6 2·1
47	1 56·8	1 57·1	1 51·4	4·7 0·6	10·7 1·3	16·7 2·1
48	1 57·0	1 57·3	1 51·7	4·8 0·6	10·8 1·4	16·8 2·1
49	1 57·3	1 57·6	1 51·9	4·9 0·6	10·9 1·4	16·9 2·1
50	1 57·5	1 57·8	1 52·1	5·0 0·6	11·0 1·4	17·0 2·1
51	1 57·8	1 58·1	1 52·4	5·1 0·6	11·1 1·4	17·1 2·1
52	1 58·0	1 58·3	1 52·6	5·2 0·7	11·2 1·4	17·2 2·2
53	1 58·3	1 58·6	1 52·9	5·3 0·7	11·3 1·4	17·3 2·2
54	1 58·5	1 58·8	1 53·1	5·4 0·7	11·4 1·4	17·4 2·2
55	1 58·8	1 59·1	1 53·3	5·5 0·7	11·5 1·4	17·5 2·2
56	1 59·0	1 59·3	1 53·6	5·6 0·7	11·6 1·5	17·6 2·2
57	1 59·3	1 59·6	1 53·8	5·7 0·7	11·7 1·5	17·7 2·2
58	1 59·5	1 59·8	1 54·1	5·8 0·7	11·8 1·5	17·8 2·2
59	1 59·8	2 00·1	1 54·3	5·9 0·7	11·9 1·5	17·9 2·2
60	2 00·0	2 00·3	1 54·5	6·0 0·8	12·0 1·5	18·0 2·3

v

8ᵐ

8	SUN PLANETS	ARIES	MOON	v or Corrⁿ d		v or Corrⁿ d		v or Corrⁿ d	
s	° ′	° ′	° ′	′	′	′	′	′	′
00	2 00·0	2 00·3	1 54·5	0·0	0·0	6·0	0·9	12·0	1·7
01	2 00·3	2 00·6	1 54·8	0·1	0·0	6·1	0·9	12·1	1·7
02	2 00·5	2 00·8	1 55·0	0·2	0·0	6·2	0·9	12·2	1·7
03	2 00·8	2 01·1	1 55·2	0·3	0·0	6·3	0·9	12·3	1·7
04	2 01·0	2 01·3	1 55·5	0·4	0·1	6·4	0·9	12·4	1·8
05	2 01·3	2 01·6	1 55·7	0·5	0·1	6·5	0·9	12·5	1·8
06	2 01·5	2 01·8	1 56·0	0·6	0·1	6·6	0·9	12·6	1·8
07	2 01·8	2 02·1	1 56·2	0·7	0·1	6·7	0·9	12·7	1·8
08	2 02·0	2 02·3	1 56·4	0·8	0·1	6·8	1·0	12·8	1·8
09	2 02·3	2 02·6	1 56·7	0·9	0·1	6·9	1·0	12·9	1·8
10	2 02·5	2 02·8	1 56·9	1·0	0·1	7·0	1·0	13·0	1·8
11	2 02·8	2 03·1	1 57·2	1·1	0·2	7·1	1·0	13·1	1·9
12	2 03·0	2 03·3	1 57·4	1·2	0·2	7·2	1·0	13·2	1·9
13	2 03·3	2 03·6	1 57·6	1·3	0·2	7·3	1·0	13·3	1·9
14	2 03·5	2 03·8	1 57·9	1·4	0·2	7·4	1·0	13·4	1·9
15	2 03·8	2 04·1	1 58·1	1·5	0·2	7·5	1·1	13·5	1·9
16	2 04·0	2 04·3	1 58·4	1·6	0·2	7·6	1·1	13·6	1·9
17	2 04·3	2 04·6	1 58·6	1·7	0·2	7·7	1·1	13·7	1·9
18	2 04·5	2 04·8	1 58·8	1·8	0·3	7·8	1·1	13·8	2·0
19	2 04·8	2 05·1	1 59·1	1·9	0·3	7·9	1·1	13·9	2·0
20	2 05·0	2 05·3	1 59·3	2·0	0·3	8·0	1·1	14·0	2·0
21	2 05·3	2 05·6	1 59·5	2·1	0·3	8·1	1·1	14·1	2·0
22	2 05·5	2 05·8	1 59·8	2·2	0·3	8·2	1·2	14·2	2·0
23	2 05·8	2 06·1	2 00·0	2·3	0·3	8·3	1·2	14·3	2·0
24	2 06·0	2 06·3	2 00·3	2·4	0·3	8·4	1·2	14·4	2·0
25	2 06·3	2 06·6	2 00·5	2·5	0·4	8·5	1·2	14·5	2·1
26	2 06·5	2 06·8	2 00·7	2·6	0·4	8·6	1·2	14·6	2·1
27	2 06·8	2 07·1	2 01·0	2·7	0·4	8·7	1·2	14·7	2·1
28	2 07·0	2 07·3	2 01·2	2·8	0·4	8·8	1·2	14·8	2·1
29	2 07·3	2 07·6	2 01·5	2·9	0·4	8·9	1·3	14·9	2·1
30	2 07·5	2 07·8	2 01·7	3·0	0·4	9·0	1·3	15·0	2·1
31	2 07·8	2 08·1	2 01·9	3·1	0·4	9·1	1·3	15·1	2·1
32	2 08·0	2 08·4	2 02·2	3·2	0·5	9·2	1·3	15·2	2·2
33	2 08·3	2 08·6	2 02·4	3·3	0·5	9·3	1·3	15·3	2·2
34	2 08·5	2 08·9	2 02·6	3·4	0·5	9·4	1·3	15·4	2·2
35	2 08·8	2 09·1	2 02·9	3·5	0·5	9·5	1·3	15·5	2·2
36	2 09·0	2 09·4	2 03·1	3·6	0·5	9·6	1·4	15·6	2·2
37	2 09·3	2 09·6	2 03·4	3·7	0·5	9·7	1·4	15·7	2·2
38	2 09·5	2 09·9	2 03·6	3·8	0·5	9·8	1·4	15·8	2·2
39	2 09·8	2 10·1	2 03·8	3·9	0·6	9·9	1·4	15·9	2·3
40	2 10·0	2 10·4	2 04·1	4·0	0·6	10·0	1·4	16·0	2·3
41	2 10·3	2 10·6	2 04·3	4·1	0·6	10·1	1·4	16·1	2·3
42	2 10·5	2 10·9	2 04·6	4·2	0·6	10·2	1·4	16·2	2·3
43	2 10·8	2 11·1	2 04·8	4·3	0·6	10·3	1·5	16·3	2·3
44	2 11·0	2 11·4	2 05·0	4·4	0·6	10·4	1·5	16·4	2·3
45	2 11·3	2 11·6	2 05·3	4·5	0·6	10·5	1·5	16·5	2·3
46	2 11·5	2 11·9	2 05·5	4·6	0·7	10·6	1·5	16·6	2·4
47	2 11·8	2 12·1	2 05·7	4·7	0·7	10·7	1·5	16·7	2·4
48	2 12·0	2 12·4	2 06·0	4·8	0·7	10·8	1·5	16·8	2·4
49	2 12·3	2 12·6	2 06·2	4·9	0·7	10·9	1·5	16·9	2·4
50	2 12·5	2 12·9	2 06·5	5·0	0·7	11·0	1·6	17·0	2·4
51	2 12·8	2 13·1	2 06·7	5·1	0·7	11·1	1·6	17·1	2·4
52	2 13·0	2 13·4	2 06·9	5·2	0·7	11·2	1·6	17·2	2·4
53	2 13·3	2 13·6	2 07·2	5·3	0·8	11·3	1·6	17·3	2·5
54	2 13·5	2 13·9	2 07·4	5·4	0·8	11·4	1·6	17·4	2·5
55	2 13·8	2 14·1	2 07·7	5·5	0·8	11·5	1·6	17·5	2·5
56	2 14·0	2 14·4	2 07·9	5·6	0·8	11·6	1·6	17·6	2·5
57	2 14·3	2 14·6	2 08·1	5·7	0·8	11·7	1·7	17·7	2·5
58	2 14·5	2 14·9	2 08·4	5·8	0·8	11·8	1·7	17·8	2·5
59	2 14·8	2 15·1	2 08·6	5·9	0·8	11·9	1·7	17·9	2·5
60	2 15·0	2 15·4	2 08·9	6·0	0·9	12·0	1·7	18·0	2·6

9ᵐ

9	SUN PLANETS	ARIES	MOON	v or Corrⁿ d		v or Corrⁿ d		v or Corrⁿ d	
s	° ′	° ′	° ′	′	′	′	′	′	′
00	2 15·0	2 15·4	2 08·9	0·0	0·0	6·0	1·0	12·0	1·9
01	2 15·3	2 15·6	2 09·1	0·1	0·0	6·1	1·0	12·1	1·9
02	2 15·5	2 15·9	2 09·3	0·2	0·0	6·2	1·0	12·2	1·9
03	2 15·8	2 16·1	2 09·6	0·3	0·0	6·3	1·0	12·3	1·9
04	2 16·0	2 16·4	2 09·8	0·4	0·1	6·4	1·0	12·4	2·0
05	2 16·3	2 16·6	2 10·0	0·5	0·1	6·5	1·0	12·5	2·0
06	2 16·5	2 16·9	2 10·3	0·6	0·1	6·6	1·0	12·6	2·0
07	2 16·8	2 17·1	2 10·5	0·7	0·1	6·7	1·1	12·7	2·0
08	2 17·0	2 17·4	2 10·8	0·8	0·1	6·8	1·1	12·8	2·0
09	2 17·3	2 17·6	2 11·0	0·9	0·1	6·9	1·1	12·9	2·0
10	2 17·5	2 17·9	2 11·2	1·0	0·2	7·0	1·1	13·0	2·1
11	2 17·8	2 18·1	2 11·5	1·1	0·2	7·1	1·1	13·1	2·1
12	2 18·0	2 18·4	2 11·7	1·2	0·2	7·2	1·1	13·2	2·1
13	2 18·3	2 18·6	2 12·0	1·3	0·2	7·3	1·2	13·3	2·1
14	2 18·5	2 18·9	2 12·2	1·4	0·2	7·4	1·2	13·4	2·1
15	2 18·8	2 19·1	2 12·4	1·5	0·2	7·5	1·2	13·5	2·1
16	2 19·0	2 19·4	2 12·7	1·6	0·3	7·6	1·2	13·6	2·2
17	2 19·3	2 19·6	2 12·9	1·7	0·3	7·7	1·2	13·7	2·2
18	2 19·5	2 19·9	2 13·1	1·8	0·3	7·8	1·2	13·8	2·2
19	2 19·8	2 20·1	2 13·4	1·9	0·3	7·9	1·3	13·9	2·2
20	2 20·0	2 20·4	2 13·6	2·0	0·3	8·0	1·3	14·0	2·2
21	2 20·3	2 20·6	2 13·9	2·1	0·3	8·1	1·3	14·1	2·2
22	2 20·5	2 20·9	2 14·1	2·2	0·3	8·2	1·3	14·2	2·2
23	2 20·8	2 21·1	2 14·3	2·3	0·4	8·3	1·3	14·3	2·3
24	2 21·0	2 21·4	2 14·6	2·4	0·4	8·4	1·3	14·4	2·3
25	2 21·3	2 21·6	2 14·8	2·5	0·4	8·5	1·3	14·5	2·3
26	2 21·5	2 21·9	2 15·1	2·6	0·4	8·6	1·4	14·6	2·3
27	2 21·8	2 22·1	2 15·3	2·7	0·4	8·7	1·4	14·7	2·3
28	2 22·0	2 22·4	2 15·5	2·8	0·4	8·8	1·4	14·8	2·3
29	2 22·3	2 22·6	2 15·8	2·9	0·5	8·9	1·4	14·9	2·4
30	2 22·5	2 22·9	2 16·0	3·0	0·5	9·0	1·4	15·0	2·4
31	2 22·8	2 23·1	2 16·2	3·1	0·5	9·1	1·4	15·1	2·4
32	2 23·0	2 23·4	2 16·5	3·2	0·5	9·2	1·5	15·2	2·4
33	2 23·3	2 23·6	2 16·7	3·3	0·5	9·3	1·5	15·3	2·4
34	2 23·5	2 23·9	2 17·0	3·4	0·5	9·4	1·5	15·4	2·4
35	2 23·8	2 24·1	2 17·2	3·5	0·6	9·5	1·5	15·5	2·5
36	2 24·0	2 24·4	2 17·4	3·6	0·6	9·6	1·5	15·6	2·5
37	2 24·3	2 24·6	2 17·7	3·7	0·6	9·7	1·5	15·7	2·5
38	2 24·5	2 24·9	2 17·9	3·8	0·6	9·8	1·6	15·8	2·5
39	2 24·8	2 25·1	2 18·2	3·9	0·6	9·9	1·6	15·9	2·5
40	2 25·0	2 25·4	2 18·4	4·0	0·6	10·0	1·6	16·0	2·5
41	2 25·3	2 25·6	2 18·6	4·1	0·6	10·1	1·6	16·1	2·5
42	2 25·5	2 25·9	2 18·9	4·2	0·7	10·2	1·6	16·2	2·6
43	2 25·8	2 26·1	2 19·1	4·3	0·7	10·3	1·6	16·3	2·6
44	2 26·0	2 26·4	2 19·3	4·4	0·7	10·4	1·6	16·4	2·6
45	2 26·3	2 26·7	2 19·6	4·5	0·7	10·5	1·7	16·5	2·6
46	2 26·5	2 26·9	2 19·8	4·6	0·7	10·6	1·7	16·6	2·6
47	2 26·8	2 27·2	2 20·1	4·7	0·7	10·7	1·7	16·7	2·6
48	2 27·0	2 27·4	2 20·3	4·8	0·8	10·8	1·7	16·8	2·7
49	2 27·3	2 27·7	2 20·5	4·9	0·8	10·9	1·7	16·9	2·7
50	2 27·5	2 27·9	2 20·8	5·0	0·8	11·0	1·7	17·0	2·7
51	2 27·8	2 28·2	2 21·0	5·1	0·8	11·1	1·8	17·1	2·7
52	2 28·0	2 28·4	2 21·3	5·2	0·8	11·2	1·8	17·2	2·7
53	2 28·3	2 28·7	2 21·5	5·3	0·8	11·3	1·8	17·3	2·7
54	2 28·5	2 28·9	2 21·7	5·4	0·9	11·4	1·8	17·4	2·8
55	2 28·8	2 29·2	2 22·0	5·5	0·9	11·5	1·8	17·5	2·8
56	2 29·0	2 29·4	2 22·2	5·6	0·9	11·6	1·8	17·6	2·8
57	2 29·3	2 29·7	2 22·5	5·7	0·9	11·7	1·9	17·7	2·8
58	2 29·5	2 29·9	2 22·7	5·8	0·9	11·8	1·9	17·8	2·8
59	2 29·8	2 30·2	2 22·9	5·9	0·9	11·9	1·9	17·9	2·8
60	2 30·0	2 30·4	2 23·2	6·0	1·0	12·0	1·9	18·0	2·9

10	SUN PLANETS	ARIES	MOON	v or Corrⁿ d		v or Corrⁿ d		v or Corrⁿ d	
s	° ′	° ′	° ′	′	′	′	′	′	′
00	2 30·0	2 30·4	2 23·2	0·0	0·0	6·0	1·1	12·0	2·1
01	2 30·3	2 30·7	2 23·4	0·1	0·0	6·1	1·1	12·1	2·1
02	2 30·5	2 30·9	2 23·6	0·2	0·0	6·2	1·1	12·2	2·1
03	2 30·8	2 31·2	2 23·9	0·3	0·1	6·3	1·1	12·3	2·2
04	2 31·0	2 31·4	2 24·1	0·4	0·1	6·4	1·1	12·4	2·2
05	2 31·3	2 31·7	2 24·4	0·5	0·1	6·5	1·1	12·5	2·2
06	2 31·5	2 31·9	2 24·6	0·6	0·1	6·6	1·2	12·6	2·2
07	2 31·8	2 32·2	2 24·8	0·7	0·1	6·7	1·2	12·7	2·2
08	2 32·0	2 32·4	2 25·1	0·8	0·1	6·8	1·2	12·8	2·2
09	2 32·3	2 32·7	2 25·3	0·9	0·2	6·9	1·2	12·9	2·3
10	2 32·5	2 32·9	2 25·6	1·0	0·2	7·0	1·2	13·0	2·3
11	2 32·8	2 33·2	2 25·8	1·1	0·2	7·1	1·2	13·1	2·3
12	2 33·0	2 33·4	2 26·0	1·2	0·2	7·2	1·3	13·2	2·3
13	2 33·3	2 33·7	2 26·3	1·3	0·2	7·3	1·3	13·3	2·3
14	2 33·5	2 33·9	2 26·5	1·4	0·2	7·4	1·3	13·4	2·3
15	2 33·8	2 34·2	2 26·7	1·5	0·3	7·5	1·3	13·5	2·4
16	2 34·0	2 34·4	2 27·0	1·6	0·3	7·6	1·3	13·6	2·4
17	2 34·3	2 34·7	2 27·2	1·7	0·3	7·7	1·3	13·7	2·4
18	2 34·5	2 34·9	2 27·5	1·8	0·3	7·8	1·4	13·8	2·4
19	2 34·8	2 35·2	2 27·7	1·9	0·3	7·9	1·4	13·9	2·4
20	2 35·0	2 35·4	2 27·9	2·0	0·4	8·0	1·4	14·0	2·5
21	2 35·3	2 35·7	2 28·2	2·1	0·4	8·1	1·4	14·1	2·5
22	2 35·5	2 35·9	2 28·4	2·2	0·4	8·2	1·4	14·2	2·5
23	2 35·8	2 36·2	2 28·7	2·3	0·4	8·3	1·5	14·3	2·5
24	2 36·0	2 36·4	2 28·9	2·4	0·4	8·4	1·5	14·4	2·5
25	2 36·3	2 36·7	2 29·1	2·5	0·4	8·5	1·5	14·5	2·5
26	2 36·5	2 36·9	2 29·4	2·6	0·5	8·6	1·5	14·6	2·6
27	2 36·8	2 37·2	2 29·6	2·7	0·5	8·7	1·5	14·7	2·6
28	2 37·0	2 37·4	2 29·8	2·8	0·5	8·8	1·5	14·8	2·6
29	2 37·3	2 37·7	2 30·1	2·9	0·5	8·9	1·6	14·9	2·6
30	2 37·5	2 37·9	2 30·3	3·0	0·5	9·0	1·6	15·0	2·6
31	2 37·8	2 38·2	2 30·6	3·1	0·5	9·1	1·6	15·1	2·6
32	2 38·0	2 38·4	2 30·8	3·2	0·6	9·2	1·6	15·2	2·7
33	2 38·3	2 38·7	2 31·0	3·3	0·6	9·3	1·6	15·3	2·7
34	2 38·5	2 38·9	2 31·3	3·4	0·6	9·4	1·6	15·4	2·7
35	2 38·8	2 39·2	2 31·5	3·5	0·6	9·5	1·7	15·5	2·7
36	2 39·0	2 39·4	2 31·8	3·6	0·6	9·6	1·7	15·6	2·7
37	2 39·3	2 39·7	2 32·0	3·7	0·6	9·7	1·7	15·7	2·7
38	2 39·5	2 39·9	2 32·2	3·8	0·7	9·8	1·7	15·8	2·8
39	2 39·8	2 40·2	2 32·5	3·9	0·7	9·9	1·7	15·9	2·8
40	2 40·0	2 40·4	2 32·7	4·0	0·7	10·0	1·8	16·0	2·8
41	2 40·3	2 40·7	2 32·9	4·1	0·7	10·1	1·8	16·1	2·8
42	2 40·5	2 40·9	2 33·2	4·2	0·7	10·2	1·8	16·2	2·8
43	2 40·8	2 41·2	2 33·4	4·3	0·8	10·3	1·8	16·3	2·9
44	2 41·0	2 41·4	2 33·7	4·4	0·8	10·4	1·8	16·4	2·9
45	2 41·3	2 41·7	2 33·9	4·5	0·8	10·5	1·8	16·5	2·9
46	2 41·5	2 41·9	2 34·1	4·6	0·8	10·6	1·9	16·6	2·9
47	2 41·8	2 42·2	2 34·4	4·7	0·8	10·7	1·9	16·7	2·9
48	2 42·0	2 42·4	2 34·6	4·8	0·8	10·8	1·9	16·8	2·9
49	2 42·3	2 42·7	2 34·9	4·9	0·9	10·9	1·9	16·9	3·0
50	2 42·5	2 42·9	2 35·1	5·0	0·9	11·0	1·9	17·0	3·0
51	2 42·8	2 43·2	2 35·3	5·1	0·9	11·1	1·9	17·1	3·0
52	2 43·0	2 43·4	2 35·6	5·2	0·9	11·2	2·0	17·2	3·0
53	2 43·3	2 43·7	2 35·8	5·3	0·9	11·3	2·0	17·3	3·0
54	2 43·5	2 43·9	2 36·1	5·4	0·9	11·4	2·0	17·4	3·0
55	2 43·8	2 44·2	2 36·3	5·5	1·0	11·5	2·0	17·5	3·1
56	2 44·0	2 44·4	2 36·5	5·6	1·0	11·6	2·0	17·6	3·1
57	2 44·3	2 44·7	2 36·8	5·7	1·0	11·7	2·0	17·7	3·1
58	2 44·5	2 45·0	2 37·0	5·8	1·0	11·8	2·1	17·8	3·1
59	2 44·8	2 45·2	2 37·2	5·9	1·0	11·9	2·1	17·9	3·1
60	2 45·0	2 45·5	2 37·5	6·0	1·1	12·0	2·1	18·0	3·2

11	SUN PLANETS	ARIES	MOON	v or Corrⁿ d		v or Corrⁿ d		v or Corrⁿ d	
s	° ′	° ′	° ′	′	′	′	′	′	′
00	2 45·0	2 45·5	2 37·5	0·0	0·0	6·0	1·2	12·0	2·3
01	2 45·3	2 45·7	2 37·7	0·1	0·0	6·1	1·2	12·1	2·3
02	2 45·5	2 46·0	2 38·0	0·2	0·0	6·2	1·2	12·2	2·3
03	2 45·8	2 46·2	2 38·2	0·3	0·1	6·3	1·2	12·3	2·4
04	2 46·0	2 46·5	2 38·4	0·4	0·1	6·4	1·2	12·4	2·4
05	2 46·3	2 46·7	2 38·7	0·5	0·1	6·5	1·2	12·5	2·4
06	2 46·5	2 47·0	2 38·9	0·6	0·1	6·6	1·3	12·6	2·4
07	2 46·8	2 47·2	2 39·2	0·7	0·1	6·7	1·3	12·7	2·4
08	2 47·0	2 47·5	2 39·4	0·8	0·2	6·8	1·3	12·8	2·5
09	2 47·3	2 47·7	2 39·6	0·9	0·2	6·9	1·3	12·9	2·5
10	2 47·5	2 48·0	2 39·9	1·0	0·2	7·0	1·3	13·0	2·5
11	2 47·8	2 48·2	2 40·1	1·1	0·2	7·1	1·4	13·1	2·5
12	2 48·0	2 48·5	2 40·3	1·2	0·2	7·2	1·4	13·2	2·5
13	2 48·3	2 48·7	2 40·6	1·3	0·2	7·3	1·4	13·3	2·5
14	2 48·5	2 49·0	2 40·8	1·4	0·3	7·4	1·4	13·4	2·6
15	2 48·8	2 49·2	2 41·1	1·5	0·3	7·5	1·4	13·5	2·6
16	2 49·0	2 49·5	2 41·3	1·6	0·3	7·6	1·5	13·6	2·6
17	2 49·3	2 49·7	2 41·5	1·7	0·3	7·7	1·5	13·7	2·6
18	2 49·5	2 50·0	2 41·8	1·8	0·3	7·8	1·5	13·8	2·6
19	2 49·8	2 50·2	2 42·0	1·9	0·4	7·9	1·5	13·9	2·7
20	2 50·0	2 50·5	2 42·3	2·0	0·4	8·0	1·5	14·0	2·7
21	2 50·3	2 50·7	2 42·5	2·1	0·4	8·1	1·6	14·1	2·7
22	2 50·5	2 51·0	2 42·7	2·2	0·4	8·2	1·6	14·2	2·7
23	2 50·8	2 51·2	2 43·0	2·3	0·4	8·3	1·6	14·3	2·7
24	2 51·0	2 51·5	2 43·2	2·4	0·5	8·4	1·6	14·4	2·8
25	2 51·3	2 51·7	2 43·4	2·5	0·5	8·5	1·6	14·5	2·8
26	2 51·5	2 52·0	2 43·7	2·6	0·5	8·6	1·6	14·6	2·8
27	2 51·8	2 52·2	2 43·9	2·7	0·5	8·7	1·7	14·7	2·8
28	2 52·0	2 52·5	2 44·2	2·8	0·5	8·8	1·7	14·8	2·8
29	2 52·3	2 52·7	2 44·4	2·9	0·6	8·9	1·7	14·9	2·9
30	2 52·5	2 53·0	2 44·6	3·0	0·6	9·0	1·7	15·0	2·9
31	2 52·8	2 53·2	2 44·9	3·1	0·6	9·1	1·7	15·1	2·9
32	2 53·0	2 53·5	2 45·1	3·2	0·6	9·2	1·8	15·2	2·9
33	2 53·3	2 53·7	2 45·4	3·3	0·6	9·3	1·8	15·3	2·9
34	2 53·5	2 54·0	2 45·6	3·4	0·7	9·4	1·8	15·4	3·0
35	2 53·8	2 54·2	2 45·8	3·5	0·7	9·5	1·8	15·5	3·0
36	2 54·0	2 54·5	2 46·1	3·6	0·7	9·6	1·8	15·6	3·0
37	2 54·3	2 54·7	2 46·3	3·7	0·7	9·7	1·9	15·7	3·0
38	2 54·5	2 55·0	2 46·6	3·8	0·7	9·8	1·9	15·8	3·0
39	2 54·8	2 55·2	2 46·8	3·9	0·7	9·9	1·9	15·9	3·0
40	2 55·0	2 55·5	2 47·0	4·0	0·8	10·0	1·9	16·0	3·1
41	2 55·3	2 55·7	2 47·3	4·1	0·8	10·1	1·9	16·1	3·1
42	2 55·5	2 56·0	2 47·5	4·2	0·8	10·2	2·0	16·2	3·1
43	2 55·8	2 56·2	2 47·7	4·3	0·8	10·3	2·0	16·3	3·1
44	2 56·0	2 56·5	2 48·0	4·4	0·8	10·4	2·0	16·4	3·1
45	2 56·3	2 56·7	2 48·2	4·5	0·9	10·5	2·0	16·5	3·2
46	2 56·5	2 57·0	2 48·5	4·6	0·9	10·6	2·0	16·6	3·2
47	2 56·8	2 57·2	2 48·7	4·7	0·9	10·7	2·1	16·7	3·2
48	2 57·0	2 57·5	2 48·9	4·8	0·9	10·8	2·1	16·8	3·2
49	2 57·3	2 57·7	2 49·2	4·9	0·9	10·9	2·1	16·9	3·2
50	2 57·5	2 58·0	2 49·4	5·0	1·0	11·0	2·1	17·0	3·3
51	2 57·8	2 58·2	2 49·7	5·1	1·0	11·1	2·1	17·1	3·3
52	2 58·0	2 58·5	2 49·9	5·2	1·0	11·2	2·1	17·2	3·3
53	2 58·3	2 58·7	2 50·1	5·3	1·0	11·3	2·2	17·3	3·3
54	2 58·5	2 59·0	2 50·4	5·4	1·0	11·4	2·2	17·4	3·3
55	2 58·8	2 59·2	2 50·6	5·5	1·1	11·5	2·2	17·5	3·4
56	2 59·0	2 59·5	2 50·8	5·6	1·1	11·6	2·2	17·6	3·4
57	2 59·3	2 59·7	2 51·1	5·7	1·1	11·7	2·2	17·7	3·4
58	2 59·5	3 00·0	2 51·3	5·8	1·1	11·8	2·3	17·8	3·4
59	2 59·8	3 00·2	2 51·6	5·9	1·1	11·9	2·3	17·9	3·4
60	3 00·0	3 00·5	2 51·8	6·0	1·2	12·0	2·3	18·0	3·5

12ᵐ	SUN PLANETS	ARIES	MOON	v or d	Corrⁿ	v or d	Corrⁿ	v or d	Corrⁿ
s	° ′	° ′	° ′	′	′	′	′	′	′
00	3 00.0	3 00.5	2 51.8	0.0	0.0	6.0	1.3	12.0	2.5
01	3 00.3	3 00.7	2 52.0	0.1	0.0	6.1	1.3	12.1	2.5
02	3 00.5	3 01.0	2 52.3	0.2	0.0	6.2	1.3	12.2	2.5
03	3 00.8	3 01.2	2 52.5	0.3	0.1	6.3	1.3	12.3	2.6
04	3 01.0	3 01.5	2 52.8	0.4	0.1	6.4	1.3	12.4	2.6
05	3 01.3	3 01.7	2 53.0	0.5	0.1	6.5	1.4	12.5	2.6
06	3 01.5	3 02.0	2 53.2	0.6	0.1	6.6	1.4	12.6	2.6
07	3 01.8	3 02.2	2 53.5	0.7	0.1	6.7	1.4	12.7	2.6
08	3 02.0	3 02.5	2 53.7	0.8	0.2	6.8	1.4	12.8	2.7
09	3 02.3	3 02.7	2 53.9	0.9	0.2	6.9	1.4	12.9	2.7
10	3 02.5	3 03.0	2 54.2	1.0	0.2	7.0	1.5	13.0	2.7
11	3 02.8	3 03.3	2 54.4	1.1	0.2	7.1	1.5	13.1	2.7
12	3 03.0	3 03.5	2 54.7	1.2	0.3	7.2	1.5	13.2	2.8
13	3 03.3	3 03.8	2 54.9	1.3	0.3	7.3	1.5	13.3	2.8
14	3 03.5	3 04.0	2 55.1	1.4	0.3	7.4	1.5	13.4	2.8
15	3 03.8	3 04.3	2 55.4	1.5	0.3	7.5	1.6	13.5	2.8
16	3 04.0	3 04.5	2 55.6	1.6	0.3	7.6	1.6	13.6	2.8
17	3 04.3	3 04.8	2 55.9	1.7	0.4	7.7	1.6	13.7	2.9
18	3 04.5	3 05.0	2 56.1	1.8	0.4	7.8	1.6	13.8	2.9
19	3 04.8	3 05.3	2 56.3	1.9	0.4	7.9	1.6	13.9	2.9
20	3 05.0	3 05.5	2 56.6	2.0	0.4	8.0	1.7	14.0	2.9
21	3 05.3	3 05.8	2 56.8	2.1	0.4	8.1	1.7	14.1	2.9
22	3 05.5	3 06.0	2 57.0	2.2	0.5	8.2	1.7	14.2	3.0
23	3 05.8	3 06.3	2 57.3	2.3	0.5	8.3	1.7	14.3	3.0
24	3 06.0	3 06.5	2 57.5	2.4	0.5	8.4	1.8	14.4	3.0
25	3 06.3	3 06.8	2 57.8	2.5	0.5	8.5	1.8	14.5	3.0
26	3 06.5	3 07.0	2 58.0	2.6	0.5	8.6	1.8	14.6	3.0
27	3 06.8	3 07.3	2 58.2	2.7	0.6	8.7	1.8	14.7	3.1
28	3 07.0	3 07.5	2 58.5	2.8	0.6	8.8	1.8	14.8	3.1
29	3 07.3	3 07.8	2 58.7	2.9	0.6	8.9	1.9	14.9	3.1
30	3 07.5	3 08.0	2 59.0	3.0	0.6	9.0	1.9	15.0	3.1
31	3 07.8	3 08.3	2 59.2	3.1	0.6	9.1	1.9	15.1	3.1
32	3 08.0	3 08.5	2 59.4	3.2	0.7	9.2	1.9	15.2	3.2
33	3 08.3	3 08.8	2 59.7	3.3	0.7	9.3	1.9	15.3	3.2
34	3 08.5	3 09.0	2 59.9	3.4	0.7	9.4	2.0	15.4	3.2
35	3 08.8	3 09.3	3 00.2	3.5	0.7	9.5	2.0	15.5	3.2
36	3 09.0	3 09.5	3 00.4	3.6	0.8	9.6	2.0	15.6	3.3
37	3 09.3	3 09.8	3 00.6	3.7	0.8	9.7	2.0	15.7	3.3
38	3 09.5	3 10.0	3 00.9	3.8	0.8	9.8	2.0	15.8	3.3
39	3 09.8	3 10.3	3 01.1	3.9	0.8	9.9	2.1	15.9	3.3
40	3 10.0	3 10.5	3 01.3	4.0	0.8	10.0	2.1	16.0	3.3
41	3 10.3	3 10.8	3 01.6	4.1	0.9	10.1	2.1	16.1	3.4
42	3 10.5	3 11.0	3 01.8	4.2	0.9	10.2	2.1	16.2	3.4
43	3 10.8	3 11.3	3 02.1	4.3	0.9	10.3	2.1	16.3	3.4
44	3 11.0	3 11.5	3 02.3	4.4	0.9	10.4	2.2	16.4	3.4
45	3 11.3	3 11.8	3 02.5	4.5	0.9	10.5	2.2	16.5	3.4
46	3 11.5	3 12.0	3 02.8	4.6	1.0	10.6	2.2	16.6	3.5
47	3 11.8	3 12.3	3 03.0	4.7	1.0	10.7	2.2	16.7	3.5
48	3 12.0	3 12.5	3 03.3	4.8	1.0	10.8	2.3	16.8	3.5
49	3 12.3	3 12.8	3 03.5	4.9	1.0	10.9	2.3	16.9	3.5
50	3 12.5	3 13.0	3 03.7	5.0	1.0	11.0	2.3	17.0	3.5
51	3 12.8	3 13.3	3 04.0	5.1	1.1	11.1	2.3	17.1	3.6
52	3 13.0	3 13.5	3 04.2	5.2	1.1	11.2	2.3	17.2	3.6
53	3 13.3	3 13.8	3 04.4	5.3	1.1	11.3	2.4	17.3	3.6
54	3 13.5	3 14.0	3 04.7	5.4	1.1	11.4	2.4	17.4	3.6
55	3 13.8	3 14.3	3 04.9	5.5	1.1	11.5	2.4	17.5	3.6
56	3 14.0	3 14.5	3 05.2	5.6	1.2	11.6	2.4	17.6	3.7
57	3 14.3	3 14.8	3 05.4	5.7	1.2	11.7	2.4	17.7	3.7
58	3 14.5	3 15.0	3 05.6	5.8	1.2	11.8	2.5	17.8	3.7
59	3 14.8	3 15.3	3 05.9	5.9	1.2	11.9	2.5	17.9	3.7
60	3 15.0	3 15.5	3 06.1	6.0	1.3	12.0	2.5	18.0	3.8

13ᵐ	SUN PLANETS	ARIES	MOON	v or d	Corrⁿ	v or d	Corrⁿ	v or d	Corrⁿ
s	° ′	° ′	° ′	′	′	′	′	′	′
00	3 15.0	3 15.5	3 06.1	0.0	0.0	6.0	1.4	12.0	2.7
01	3 15.3	3 15.8	3 06.4	0.1	0.0	6.1	1.4	12.1	2.7
02	3 15.5	3 16.0	3 06.6	0.2	0.0	6.2	1.4	12.2	2.7
03	3 15.8	3 16.3	3 06.8	0.3	0.1	6.3	1.4	12.3	2.8
04	3 16.0	3 16.5	3 07.1	0.4	0.1	6.4	1.4	12.4	2.8
05	3 16.3	3 16.8	3 07.3	0.5	0.1	6.5	1.5	12.5	2.8
06	3 16.5	3 17.0	3 07.5	0.6	0.1	6.6	1.5	12.6	2.8
07	3 16.8	3 17.3	3 07.8	0.7	0.2	6.7	1.5	12.7	2.9
08	3 17.0	3 17.5	3 08.0	0.8	0.2	6.8	1.5	12.8	2.9
09	3 17.3	3 17.8	3 08.3	0.9	0.2	6.9	1.6	12.9	2.9
10	3 17.5	3 18.0	3 08.5	1.0	0.2	7.0	1.6	13.0	2.9
11	3 17.8	3 18.3	3 08.7	1.1	0.2	7.1	1.6	13.1	2.9
12	3 18.0	3 18.5	3 09.0	1.2	0.3	7.2	1.6	13.2	3.0
13	3 18.3	3 18.8	3 09.2	1.3	0.3	7.3	1.6	13.3	3.0
14	3 18.5	3 19.0	3 09.5	1.4	0.3	7.4	1.7	13.4	3.0
15	3 18.8	3 19.3	3 09.7	1.5	0.3	7.5	1.7	13.5	3.0
16	3 19.0	3 19.5	3 09.9	1.6	0.4	7.6	1.7	13.6	3.1
17	3 19.3	3 19.8	3 10.2	1.7	0.4	7.7	1.7	13.7	3.1
18	3 19.5	3 20.0	3 10.4	1.8	0.4	7.8	1.8	13.8	3.1
19	3 19.8	3 20.3	3 10.7	1.9	0.4	7.9	1.8	13.9	3.1
20	3 20.0	3 20.5	3 10.9	2.0	0.5	8.0	1.8	14.0	3.2
21	3 20.3	3 20.8	3 11.1	2.1	0.5	8.1	1.8	14.1	3.2
22	3 20.5	3 21.0	3 11.4	2.2	0.5	8.2	1.8	14.2	3.2
23	3 20.8	3 21.3	3 11.6	2.3	0.5	8.3	1.9	14.3	3.2
24	3 21.0	3 21.6	3 11.8	2.4	0.5	8.4	1.9	14.4	3.2
25	3 21.3	3 21.8	3 12.1	2.5	0.6	8.5	1.9	14.5	3.3
26	3 21.5	3 22.1	3 12.3	2.6	0.6	8.6	1.9	14.6	3.3
27	3 21.8	3 22.3	3 12.6	2.7	0.6	8.7	2.0	14.7	3.3
28	3 22.0	3 22.6	3 12.8	2.8	0.6	8.8	2.0	14.8	3.3
29	3 22.3	3 22.8	3 13.0	2.9	0.7	8.9	2.0	14.9	3.4
30	3 22.5	3 23.1	3 13.3	3.0	0.7	9.0	2.0	15.0	3.4
31	3 22.8	3 23.3	3 13.5	3.1	0.7	9.1	2.0	15.1	3.4
32	3 23.0	3 23.6	3 13.8	3.2	0.7	9.2	2.1	15.2	3.4
33	3 23.3	3 23.8	3 14.0	3.3	0.7	9.3	2.1	15.3	3.4
34	3 23.5	3 24.1	3 14.2	3.4	0.8	9.4	2.1	15.4	3.5
35	3 23.8	3 24.3	3 14.5	3.5	0.8	9.5	2.1	15.5	3.5
36	3 24.0	3 24.6	3 14.7	3.6	0.8	9.6	2.2	15.6	3.5
37	3 24.3	3 24.8	3 14.9	3.7	0.8	9.7	2.2	15.7	3.5
38	3 24.5	3 25.1	3 15.2	3.8	0.9	9.8	2.2	15.8	3.6
39	3 24.8	3 25.3	3 15.4	3.9	0.9	9.9	2.2	15.9	3.6
40	3 25.0	3 25.6	3 15.7	4.0	0.9	10.0	2.3	16.0	3.6
41	3 25.3	3 25.8	3 15.9	4.1	0.9	10.1	2.3	16.1	3.6
42	3 25.5	3 26.1	3 16.1	4.2	0.9	10.2	2.3	16.2	3.6
43	3 25.8	3 26.3	3 16.4	4.3	1.0	10.3	2.3	16.3	3.7
44	3 26.0	3 26.6	3 16.6	4.4	1.0	10.4	2.3	16.4	3.7
45	3 26.3	3 26.8	3 16.9	4.5	1.0	10.5	2.4	16.5	3.7
46	3 26.5	3 27.1	3 17.1	4.6	1.0	10.6	2.4	16.6	3.7
47	3 26.8	3 27.3	3 17.3	4.7	1.1	10.7	2.4	16.7	3.8
48	3 27.0	3 27.6	3 17.6	4.8	1.1	10.8	2.4	16.8	3.8
49	3 27.3	3 27.8	3 17.8	4.9	1.1	10.9	2.5	16.9	3.8
50	3 27.5	3 28.1	3 18.0	5.0	1.1	11.0	2.5	17.0	3.8
51	3 27.8	3 28.3	3 18.3	5.1	1.1	11.1	2.5	17.1	3.9
52	3 28.0	3 28.6	3 18.5	5.2	1.2	11.2	2.5	17.2	3.9
53	3 28.3	3 28.8	3 18.8	5.3	1.2	11.3	2.5	17.3	3.9
54	3 28.5	3 29.1	3 19.0	5.4	1.2	11.4	2.6	17.4	3.9
55	3 28.8	3 29.3	3 19.2	5.5	1.2	11.5	2.6	17.5	3.9
56	3 29.0	3 29.6	3 19.5	5.6	1.3	11.6	2.6	17.6	4.0
57	3 29.3	3 29.8	3 19.7	5.7	1.3	11.7	2.6	17.7	4.0
58	3 29.5	3 30.1	3 20.0	5.8	1.3	11.8	2.7	17.8	4.0
59	3 29.8	3 30.3	3 20.2	5.9	1.3	11.9	2.7	17.9	4.0
60	3 30.0	3 30.6	3 20.4	6.0	1.4	12.0	2.7	18.0	4.1

14ᵐ

14 (s)	SUN PLANETS	ARIES	MOON	v or Corrⁿ d		v or Corrⁿ d		v or Corrⁿ d	
00	3 30.0	3 30.6	3 20.4	0.0	0.0	6.0	1.5	12.0	2.9
01	3 30.3	3 30.8	3 20.7	0.1	0.0	6.1	1.5	12.1	2.9
02	3 30.5	3 31.1	3 20.9	0.2	0.0	6.2	1.5	12.2	2.9
03	3 30.8	3 31.3	3 21.1	0.3	0.1	6.3	1.5	12.3	3.0
04	3 31.0	3 31.6	3 21.4	0.4	0.1	6.4	1.5	12.4	3.0
05	3 31.3	3 31.8	3 21.6	0.5	0.1	6.5	1.6	12.5	3.0
06	3 31.5	3 32.1	3 21.9	0.6	0.1	6.6	1.6	12.6	3.0
07	3 31.8	3 32.3	3 22.1	0.7	0.2	6.7	1.6	12.7	3.1
08	3 32.0	3 32.6	3 22.3	0.8	0.2	6.8	1.6	12.8	3.1
09	3 32.3	3 32.8	3 22.6	0.9	0.2	6.9	1.7	12.9	3.1
10	3 32.5	3 33.1	3 22.8	1.0	0.2	7.0	1.7	13.0	3.1
11	3 32.8	3 33.3	3 23.1	1.1	0.3	7.1	1.7	13.1	3.2
12	3 33.0	3 33.6	3 23.3	1.2	0.3	7.2	1.7	13.2	3.2
13	3 33.3	3 33.8	3 23.5	1.3	0.3	7.3	1.8	13.3	3.2
14	3 33.5	3 34.1	3 23.8	1.4	0.3	7.4	1.8	13.4	3.2
15	3 33.8	3 34.3	3 24.0	1.5	0.4	7.5	1.8	13.5	3.3
16	3 34.0	3 34.6	3 24.3	1.6	0.4	7.6	1.8	13.6	3.3
17	3 34.3	3 34.8	3 24.5	1.7	0.4	7.7	1.9	13.7	3.3
18	3 34.5	3 35.1	3 24.7	1.8	0.4	7.8	1.9	13.8	3.3
19	3 34.8	3 35.3	3 25.0	1.9	0.5	7.9	1.9	13.9	3.4
20	3 35.0	3 35.6	3 25.2	2.0	0.5	8.0	1.9	14.0	3.4
21	3 35.3	3 35.8	3 25.4	2.1	0.5	8.1	2.0	14.1	3.4
22	3 35.5	3 36.1	3 25.7	2.2	0.5	8.2	2.0	14.2	3.4
23	3 35.8	3 36.3	3 25.9	2.3	0.6	8.3	2.0	14.3	3.5
24	3 36.0	3 36.6	3 26.2	2.4	0.6	8.4	2.0	14.4	3.5
25	3 36.3	3 36.8	3 26.4	2.5	0.6	8.5	2.1	14.5	3.5
26	3 36.5	3 37.1	3 26.6	2.6	0.6	8.6	2.1	14.6	3.5
27	3 36.8	3 37.3	3 26.9	2.7	0.7	8.7	2.1	14.7	3.6
28	3 37.0	3 37.6	3 27.1	2.8	0.7	8.8	2.1	14.8	3.6
29	3 37.3	3 37.8	3 27.4	2.9	0.7	8.9	2.2	14.9	3.6
30	3 37.5	3 38.1	3 27.6	3.0	0.7	9.0	2.2	15.0	3.6
31	3 37.8	3 38.3	3 27.8	3.1	0.7	9.1	2.2	15.1	3.6
32	3 38.0	3 38.6	3 28.1	3.2	0.8	9.2	2.2	15.2	3.7
33	3 38.3	3 38.8	3 28.3	3.3	0.8	9.3	2.2	15.3	3.7
34	3 38.5	3 39.1	3 28.5	3.4	0.8	9.4	2.3	15.4	3.7
35	3 38.8	3 39.3	3 28.8	3.5	0.8	9.5	2.3	15.5	3.7
36	3 39.0	3 39.6	3 29.0	3.6	0.9	9.6	2.3	15.6	3.8
37	3 39.3	3 39.9	3 29.3	3.7	0.9	9.7	2.3	15.7	3.8
38	3 39.5	3 40.1	3 29.5	3.8	0.9	9.8	2.4	15.8	3.8
39	3 39.8	3 40.4	3 29.7	3.9	0.9	9.9	2.4	15.9	3.8
40	3 40.0	3 40.6	3 30.0	4.0	1.0	10.0	2.4	16.0	3.9
41	3 40.3	3 40.9	3 30.2	4.1	1.0	10.1	2.4	16.1	3.9
42	3 40.5	3 41.1	3 30.5	4.2	1.0	10.2	2.5	16.2	3.9
43	3 40.8	3 41.4	3 30.7	4.3	1.0	10.3	2.5	16.3	3.9
44	3 41.0	3 41.6	3 30.9	4.4	1.1	10.4	2.5	16.4	4.0
45	3 41.3	3 41.9	3 31.2	4.5	1.1	10.5	2.5	16.5	4.0
46	3 41.5	3 42.1	3 31.4	4.6	1.1	10.6	2.6	16.6	4.0
47	3 41.8	3 42.4	3 31.6	4.7	1.1	10.7	2.6	16.7	4.0
48	3 42.0	3 42.6	3 31.9	4.8	1.2	10.8	2.6	16.8	4.1
49	3 42.3	3 42.9	3 32.1	4.9	1.2	10.9	2.6	16.9	4.1
50	3 42.5	3 43.1	3 32.4	5.0	1.2	11.0	2.7	17.0	4.1
51	3 42.8	3 43.4	3 32.6	5.1	1.2	11.1	2.7	17.1	4.1
52	3 43.0	3 43.6	3 32.8	5.2	1.3	11.2	2.7	17.2	4.2
53	3 43.3	3 43.9	3 33.1	5.3	1.3	11.3	2.7	17.3	4.2
54	3 43.5	3 44.1	3 33.3	5.4	1.3	11.4	2.8	17.4	4.2
55	3 43.8	3 44.4	3 33.6	5.5	1.3	11.5	2.8	17.5	4.2
56	3 44.0	3 44.6	3 33.8	5.6	1.4	11.6	2.8	17.6	4.3
57	3 44.3	3 44.9	3 34.0	5.7	1.4	11.7	2.8	17.7	4.3
58	3 44.5	3 45.1	3 34.3	5.8	1.4	11.8	2.9	17.8	4.3
59	3 44.8	3 45.4	3 34.5	5.9	1.4	11.9	2.9	17.9	4.3
60	3 45.0	3 45.6	3 34.8	6.0	1.5	12.0	2.9	18.0	4.4

15ᵐ

15 (s)	SUN PLANETS	ARIES	MOON	v or Corrⁿ d		v or Corrⁿ d		v or Corrⁿ d	
00	3 45.0	3 45.6	3 34.8	0.0	0.0	6.0	1.6	12.0	3.1
01	3 45.3	3 45.9	3 35.0	0.1	0.0	6.1	1.6	12.1	3.1
02	3 45.5	3 46.1	3 35.2	0.2	0.1	6.2	1.6	12.2	3.2
03	3 45.8	3 46.4	3 35.5	0.3	0.1	6.3	1.6	12.3	3.2
04	3 46.0	3 46.6	3 35.7	0.4	0.1	6.4	1.7	12.4	3.2
05	3 46.3	3 46.9	3 35.9	0.5	0.1	6.5	1.7	12.5	3.2
06	3 46.5	3 47.1	3 36.2	0.6	0.2	6.6	1.7	12.6	3.3
07	3 46.8	3 47.4	3 36.4	0.7	0.2	6.7	1.7	12.7	3.3
08	3 47.0	3 47.6	3 36.7	0.8	0.2	6.8	1.8	12.8	3.3
09	3 47.3	3 47.9	3 36.9	0.9	0.2	6.9	1.8	12.9	3.3
10	3 47.5	3 48.1	3 37.1	1.0	0.3	7.0	1.8	13.0	3.4
11	3 47.8	3 48.4	3 37.4	1.1	0.3	7.1	1.8	13.1	3.4
12	3 48.0	3 48.6	3 37.6	1.2	0.3	7.2	1.9	13.2	3.4
13	3 48.3	3 48.9	3 37.9	1.3	0.3	7.3	1.9	13.3	3.4
14	3 48.5	3 49.1	3 38.1	1.4	0.4	7.4	1.9	13.4	3.5
15	3 48.8	3 49.4	3 38.3	1.5	0.4	7.5	1.9	13.5	3.5
16	3 49.0	3 49.6	3 38.6	1.6	0.4	7.6	2.0	13.6	3.5
17	3 49.3	3 49.9	3 38.8	1.7	0.4	7.7	2.0	13.7	3.5
18	3 49.5	3 50.1	3 39.0	1.8	0.5	7.8	2.0	13.8	3.6
19	3 49.8	3 50.4	3 39.3	1.9	0.5	7.9	2.0	13.9	3.6
20	3 50.0	3 50.6	3 39.5	2.0	0.5	8.0	2.1	14.0	3.6
21	3 50.3	3 50.9	3 39.8	2.1	0.5	8.1	2.1	14.1	3.6
22	3 50.5	3 51.1	3 40.0	2.2	0.6	8.2	2.1	14.2	3.7
23	3 50.8	3 51.4	3 40.2	2.3	0.6	8.3	2.1	14.3	3.7
24	3 51.0	3 51.6	3 40.5	2.4	0.6	8.4	2.2	14.4	3.7
25	3 51.3	3 51.9	3 40.7	2.5	0.6	8.5	2.2	14.5	3.7
26	3 51.5	3 52.1	3 41.0	2.6	0.7	8.6	2.2	14.6	3.8
27	3 51.8	3 52.4	3 41.2	2.7	0.7	8.7	2.2	14.7	3.8
28	3 52.0	3 52.6	3 41.4	2.8	0.7	8.8	2.3	14.8	3.8
29	3 52.3	3 52.9	3 41.7	2.9	0.7	8.9	2.3	14.9	3.8
30	3 52.5	3 53.1	3 41.9	3.0	0.8	9.0	2.3	15.0	3.9
31	3 52.8	3 53.4	3 42.1	3.1	0.8	9.1	2.4	15.1	3.9
32	3 53.0	3 53.6	3 42.4	3.2	0.8	9.2	2.4	15.2	3.9
33	3 53.3	3 53.9	3 42.6	3.3	0.9	9.3	2.4	15.3	4.0
34	3 53.5	3 54.1	3 42.9	3.4	0.9	9.4	2.4	15.4	4.0
35	3 53.8	3 54.4	3 43.1	3.5	0.9	9.5	2.5	15.5	4.0
36	3 54.0	3 54.6	3 43.3	3.6	0.9	9.6	2.5	15.6	4.0
37	3 54.3	3 54.9	3 43.6	3.7	1.0	9.7	2.5	15.7	4.1
38	3 54.5	3 55.1	3 43.8	3.8	1.0	9.8	2.5	15.8	4.1
39	3 54.8	3 55.4	3 44.1	3.9	1.0	9.9	2.6	15.9	4.1
40	3 55.0	3 55.6	3 44.3	4.0	1.0	10.0	2.6	16.0	4.1
41	3 55.3	3 55.9	3 44.5	4.1	1.1	10.1	2.6	16.1	4.2
42	3 55.5	3 56.1	3 44.8	4.2	1.1	10.2	2.6	16.2	4.2
43	3 55.8	3 56.4	3 45.0	4.3	1.1	10.3	2.7	16.3	4.2
44	3 56.0	3 56.6	3 45.2	4.4	1.1	10.4	2.7	16.4	4.2
45	3 56.3	3 56.9	3 45.5	4.5	1.2	10.5	2.7	16.5	4.3
46	3 56.5	3 57.1	3 45.7	4.6	1.2	10.6	2.7	16.6	4.3
47	3 56.8	3 57.4	3 46.0	4.7	1.2	10.7	2.8	16.7	4.3
48	3 57.0	3 57.6	3 46.2	4.8	1.2	10.8	2.8	16.8	4.3
49	3 57.3	3 57.9	3 46.4	4.9	1.3	10.9	2.8	16.9	4.4
50	3 57.5	3 58.2	3 46.7	5.0	1.3	11.0	2.8	17.0	4.4
51	3 57.8	3 58.4	3 46.9	5.1	1.3	11.1	2.9	17.1	4.4
52	3 58.0	3 58.7	3 47.2	5.2	1.3	11.2	2.9	17.2	4.4
53	3 58.3	3 58.9	3 47.4	5.3	1.4	11.3	2.9	17.3	4.5
54	3 58.5	3 59.2	3 47.6	5.4	1.4	11.4	2.9	17.4	4.5
55	3 58.8	3 59.4	3 47.9	5.5	1.4	11.5	3.0	17.5	4.5
56	3 59.0	3 59.7	3 48.1	5.6	1.4	11.6	3.0	17.6	4.5
57	3 59.3	3 59.9	3 48.4	5.7	1.5	11.7	3.0	17.7	4.6
58	3 59.5	4 00.2	3 48.6	5.8	1.5	11.8	3.0	17.8	4.6
59	3 59.8	4 00.4	3 48.8	5.9	1.5	11.9	3.1	17.9	4.6
60	4 00.0	4 00.7	3 49.1	6.0	1.6	12.0	3.1	18.0	4.7

16	SUN PLANETS	ARIES	MOON	v or Corrⁿ d	v or Corrⁿ d	v or Corrⁿ d	17	SUN PLANETS	ARIES	MOON	v or Corrⁿ d	v or Corrⁿ d	v or Corrⁿ d
s	° ′	° ′	° ′	′ ′	′ ′	′ ′	s	° ′	° ′	° ′	′ ′	′ ′	′ ′
00	4 00·0	4 00·7	3 49·1	0·0 0·0	6·0 1·7	12·0 3·3	00	4 15·0	4 15·7	4 03·4	0·0 0·0	6·0 1·8	12·0 3·5
01	4 00·3	4 00·9	3 49·3	0·1 0·0	6·1 1·7	12·1 3·3	01	4 15·3	4 15·9	4 03·6	0·1 0·0	6·1 1·8	12·1 3·5
02	4 00·5	4 01·2	3 49·5	0·2 0·1	6·2 1·7	12·2 3·4	02	4 15·5	4 16·2	4 03·9	0·2 0·1	6·2 1·8	12·2 3·6
03	4 00·8	4 01·4	3 49·8	0·3 0·1	6·3 1·7	12·3 3·4	03	4 15·8	4 16·5	4 04·1	0·3 0·1	6·3 1·8	12·3 3·6
04	4 01·0	4 01·7	3 50·0	0·4 0·1	6·4 1·8	12·4 3·4	04	4 16·0	4 16·7	4 04·3	0·4 0·1	6·4 1·9	12·4 3·6
05	4 01·3	4 01·9	3 50·3	0·5 0·1	6·5 1·8	12·5 3·4	05	4 16·3	4 17·0	4 04·6	0·5 0·1	6·5 1·9	12·5 3·6
06	4 01·5	4 02·2	3 50·5	0·6 0·2	6·6 1·8	12·6 3·5	06	4 16·5	4 17·2	4 04·8	0·6 0·2	6·6 1·9	12·6 3·7
07	4 01·8	4 02·4	3 50·7	0·7 0·2	6·7 1·8	12·7 3·5	07	4 16·8	4 17·5	4 05·1	0·7 0·2	6·7 2·0	12·7 3·7
08	4 02·0	4 02·7	3 51·0	0·8 0·2	6·8 1·9	12·8 3·5	08	4 17·0	4 17·7	4 05·3	0·8 0·2	6·8 2·0	12·8 3·7
09	4 02·3	4 02·9	3 51·2	0·9 0·2	6·9 1·9	12·9 3·5	09	4 17·3	4 18·0	4 05·5	0·9 0·3	6·9 2·0	12·9 3·8
10	4 02·5	4 03·2	3 51·5	1·0 0·3	7·0 1·9	13·0 3·6	10	4 17·5	4 18·2	4 05·8	1·0 0·3	7·0 2·0	13·0 3·8
11	4 02·8	4 03·4	3 51·7	1·1 0·3	7·1 2·0	13·1 3·6	11	4 17·8	4 18·5	4 06·0	1·1 0·3	7·1 2·1	13·1 3·8
12	4 03·0	4 03·7	3 51·9	1·2 0·3	7·2 2·0	13·2 3·6	12	4 18·0	4 18·7	4 06·2	1·2 0·4	7·2 2·1	13·2 3·9
13	4 03·3	4 03·9	3 52·2	1·3 0·4	7·3 2·0	13·3 3·7	13	4 18·3	4 19·0	4 06·5	1·3 0·4	7·3 2·1	13·3 3·9
14	4 03·5	4 04·2	3 52·4	1·4 0·4	7·4 2·0	13·4 3·7	14	4 18·5	4 19·2	4 06·7	1·4 0·4	7·4 2·2	13·4 3·9
15	4 03·8	4 04·4	3 52·6	1·5 0·4	7·5 2·1	13·5 3·7	15	4 18·8	4 19·5	4 07·0	1·5 0·4	7·5 2·2	13·5 3·9
16	4 04·0	4 04·7	3 52·9	1·6 0·4	7·6 2·1	13·6 3·7	16	4 19·0	4 19·7	4 07·2	1·6 0·5	7·6 2·2	13·6 4·0
17	4 04·3	4 04·9	3 53·1	1·7 0·5	7·7 2·1	13·7 3·8	17	4 19·3	4 20·0	4 07·4	1·7 0·5	7·7 2·2	13·7 4·0
18	4 04·5	4 05·2	3 53·4	1·8 0·5	7·8 2·1	13·8 3·8	18	4 19·5	4 20·2	4 07·7	1·8 0·5	7·8 2·3	13·8 4·0
19	4 04·8	4 05·4	3 53·6	1·9 0·5	7·9 2·2	13·9 3·8	19	4 19·8	4 20·5	4 07·9	1·9 0·6	7·9 2·3	13·9 4·1
20	4 05·0	4 05·7	3 53·8	2·0 0·6	8·0 2·2	14·0 3·9	20	4 20·0	4 20·7	4 08·2	2·0 0·6	8·0 2·3	14·0 4·1
21	4 05·3	4 05·9	3 54·1	2·1 0·6	8·1 2·2	14·1 3·9	21	4 20·3	4 21·0	4 08·4	2·1 0·6	8·1 2·4	14·1 4·1
22	4 05·5	4 06·2	3 54·3	2·2 0·6	8·2 2·3	14·2 3·9	22	4 20·5	4 21·2	4 08·6	2·2 0·6	8·2 2·4	14·2 4·1
23	4 05·8	4 06·4	3 54·6	2·3 0·6	8·3 2·3	14·3 3·9	23	4 20·8	4 21·5	4 08·9	2·3 0·7	8·3 2·4	14·3 4·2
24	4 06·0	4 06·7	3 54·8	2·4 0·7	8·4 2·3	14·4 4·0	24	4 21·0	4 21·7	4 09·1	2·4 0·7	8·4 2·5	14·4 4·2
25	4 06·3	4 06·9	3 55·0	2·5 0·7	8·5 2·3	14·5 4·0	25	4 21·3	4 22·0	4 09·3	2·5 0·7	8·5 2·5	14·5 4·2
26	4 06·5	4 07·2	3 55·3	2·6 0·7	8·6 2·4	14·6 4·0	26	4 21·5	4 22·2	4 09·6	2·6 0·8	8·6 2·5	14·6 4·3
27	4 06·8	4 07·4	3 55·5	2·7 0·7	8·7 2·4	14·7 4·0	27	4 21·8	4 22·5	4 09·8	2·7 0·8	8·7 2·5	14·7 4·3
28	4 07·0	4 07·7	3 55·7	2·8 0·8	8·8 2·4	14·8 4·1	28	4 22·0	4 22·7	4 10·1	2·8 0·8	8·8 2·6	14·8 4·3
29	4 07·3	4 07·9	3 56·0	2·9 0·8	8·9 2·4	14·9 4·1	29	4 22·3	4 23·0	4 10·3	2·9 0·8	8·9 2·6	14·9 4·3
30	4 07·5	4 08·2	3 56·2	3·0 0·8	9·0 2·5	15·0 4·1	30	4 22·5	4 23·2	4 10·5	3·0 0·9	9·0 2·6	15·0 4·4
31	4 07·8	4 08·4	3 56·5	3·1 0·9	9·1 2·5	15·1 4·2	31	4 22·8	4 23·5	4 10·8	3·1 0·9	9·1 2·7	15·1 4·4
32	4 08·0	4 08·7	3 56·7	3·2 0·9	9·2 2·5	15·2 4·2	32	4 23·0	4 23·7	4 11·0	3·2 0·9	9·2 2·7	15·2 4·4
33	4 08·3	4 08·9	3 56·9	3·3 0·9	9·3 2·6	15·3 4·2	33	4 23·3	4 24·0	4 11·3	3·3 1·0	9·3 2·7	15·3 4·5
34	4 08·5	4 09·2	3 57·2	3·4 0·9	9·4 2·6	15·4 4·2	34	4 23·5	4 24·2	4 11·5	3·4 1·0	9·4 2·7	15·4 4·5
35	4 08·8	4 09·4	3 57·4	3·5 1·0	9·5 2·6	15·5 4·3	35	4 23·8	4 24·5	4 11·7	3·5 1·0	9·5 2·8	15·5 4·5
36	4 09·0	4 09·7	3 57·7	3·6 1·0	9·6 2·6	15·6 4·3	36	4 24·0	4 24·7	4 12·0	3·6 1·1	9·6 2·8	15·6 4·6
37	4 09·3	4 09·9	3 57·9	3·7 1·0	9·7 2·7	15·7 4·3	37	4 24·3	4 25·0	4 12·2	3·7 1·1	9·7 2·8	15·7 4·6
38	4 09·5	4 10·2	3 58·1	3·8 1·0	9·8 2·7	15·8 4·3	38	4 24·5	4 25·2	4 12·5	3·8 1·1	9·8 2·9	15·8 4·6
39	4 09·8	4 10·4	3 58·4	3·9 1·1	9·9 2·7	15·9 4·4	39	4 24·8	4 25·5	4 12·7	3·9 1·1	9·9 2·9	15·9 4·6
40	4 10·0	4 10·7	3 58·6	4·0 1·1	10·0 2·8	16·0 4·4	40	4 25·0	4 25·7	4 12·9	4·0 1·2	10·0 2·9	16·0 4·7
41	4 10·3	4 10·9	3 58·8	4·1 1·1	10·1 2·8	16·1 4·4	41	4 25·3	4 26·0	4 13·2	4·1 1·2	10·1 2·9	16·1 4·7
42	4 10·5	4 11·2	3 59·1	4·2 1·2	10·2 2·8	16·2 4·5	42	4 25·5	4 26·2	4 13·4	4·2 1·2	10·2 3·0	16·2 4·7
43	4 10·8	4 11·4	3 59·3	4·3 1·2	10·3 2·8	16·3 4·5	43	4 25·8	4 26·5	4 13·6	4·3 1·3	10·3 3·0	16·3 4·8
44	4 11·0	4 11·7	3 59·6	4·4 1·2	10·4 2·9	16·4 4·5	44	4 26·0	4 26·7	4 13·9	4·4 1·3	10·4 3·0	16·4 4·8
45	4 11·3	4 11·9	3 59·8	4·5 1·2	10·5 2·9	16·5 4·5	45	4 26·3	4 27·0	4 14·1	4·5 1·3	10·5 3·1	16·5 4·8
46	4 11·5	4 12·2	4 00·0	4·6 1·3	10·6 2·9	16·6 4·6	46	4 26·5	4 27·2	4 14·4	4·6 1·3	10·6 3·1	16·6 4·8
47	4 11·8	4 12·4	4 00·3	4·7 1·3	10·7 2·9	16·7 4·6	47	4 26·8	4 27·5	4 14·6	4·7 1·4	10·7 3·1	16·7 4·9
48	4 12·0	4 12·7	4 00·5	4·8 1·3	10·8 3·0	16·8 4·6	48	4 27·0	4 27·7	4 14·8	4·8 1·4	10·8 3·2	16·8 4·9
49	4 12·3	4 12·9	4 00·8	4·9 1·3	10·9 3·0	16·9 4·6	49	4 27·3	4 28·0	4 15·1	4·9 1·4	10·9 3·2	16·9 4·9
50	4 12·5	4 13·2	4 01·0	5·0 1·4	11·0 3·0	17·0 4·7	50	4 27·5	4 28·2	4 15·3	5·0 1·5	11·0 3·2	17·0 5·0
51	4 12·8	4 13·4	4 01·2	5·1 1·4	11·1 3·1	17·1 4·7	51	4 27·8	4 28·5	4 15·6	5·1 1·5	11·1 3·2	17·1 5·0
52	4 13·0	4 13·7	4 01·5	5·2 1·4	11·2 3·1	17·2 4·7	52	4 28·0	4 28·7	4 15·8	5·2 1·5	11·2 3·3	17·2 5·0
53	4 13·3	4 13·9	4 01·7	5·3 1·5	11·3 3·1	17·3 4·8	53	4 28·3	4 29·0	4 16·0	5·3 1·5	11·3 3·3	17·3 5·0
54	4 13·5	4 14·2	4 02·0	5·4 1·5	11·4 3·1	17·4 4·8	54	4 28·5	4 29·2	4 16·3	5·4 1·6	11·4 3·3	17·4 5·1
55	4 13·8	4 14·4	4 02·2	5·5 1·5	11·5 3·2	17·5 4·8	55	4 28·8	4 29·5	4 16·5	5·5 1·6	11·5 3·4	17·5 5·1
56	4 14·0	4 14·7	4 02·4	5·6 1·5	11·6 3·2	17·6 4·8	56	4 29·0	4 29·7	4 16·7	5·6 1·6	11·6 3·4	17·6 5·1
57	4 14·3	4 14·9	4 02·7	5·7 1·6	11·7 3·2	17·7 4·9	57	4 29·3	4 30·0	4 17·0	5·7 1·7	11·7 3·4	17·7 5·2
58	4 14·5	4 15·2	4 02·9	5·8 1·6	11·8 3·2	17·8 4·9	58	4 29·5	4 30·2	4 17·2	5·8 1·7	11·8 3·4	17·8 5·2
59	4 14·8	4 15·4	4 03·1	5·9 1·6	11·9 3·3	17·9 4·9	59	4 29·8	4 30·5	4 17·5	5·9 1·7	11·9 3·5	17·9 5·2
60	4 15·0	4 15·7	4 03·4	6·0 1·7	12·0 3·3	18·0 5·0	60	4 30·0	4 30·7	4 17·7	6·0 1·8	12·0 3·5	18·0 5·3

18ᵐ	SUN PLANETS	ARIES	MOON	v or d Corrⁿ	v or d Corrⁿ	v or d Corrⁿ
s	° ′	° ′	° ′	′ ′	′ ′	′ ′
00	4 30·0	4 30·7	4 17·7	0·0 0·0	6·0 1·9	12·0 3·7
01	4 30·3	4 31·0	4 17·9	0·1 0·0	6·1 1·9	12·1 3·7
02	4 30·5	4 31·2	4 18·2	0·2 0·1	6·2 1·9	12·2 3·8
03	4 30·8	4 31·5	4 18·4	0·3 0·1	6·3 1·9	12·3 3·8
04	4 31·0	4 31·7	4 18·7	0·4 0·1	6·4 2·0	12·4 3·8
05	4 31·3	4 32·0	4 18·9	0·5 0·2	6·5 2·0	12·5 3·9
06	4 31·5	4 32·2	4 19·1	0·6 0·2	6·6 2·0	12·6 3·9
07	4 31·8	4 32·5	4 19·4	0·7 0·2	6·7 2·1	12·7 3·9
08	4 32·0	4 32·7	4 19·6	0·8 0·2	6·8 2·1	12·8 3·9
09	4 32·3	4 33·0	4 19·8	0·9 0·3	6·9 2·1	12·9 4·0
10	4 32·5	4 33·2	4 20·1	1·0 0·3	7·0 2·2	13·0 4·0
11	4 32·8	4 33·5	4 20·3	1·1 0·3	7·1 2·2	13·1 4·0
12	4 33·0	4 33·7	4 20·6	1·2 0·4	7·2 2·2	13·2 4·1
13	4 33·3	4 34·0	4 20·8	1·3 0·4	7·3 2·3	13·3 4·1
14	4 33·5	4 34·2	4 21·0	1·4 0·4	7·4 2·3	13·4 4·1
15	4 33·8	4 34·5	4 21·3	1·5 0·5	7·5 2·3	13·5 4·2
16	4 34·0	4 34·8	4 21·5	1·6 0·5	7·6 2·3	13·6 4·2
17	4 34·3	4 35·0	4 21·8	1·7 0·5	7·7 2·4	13·7 4·2
18	4 34·5	4 35·3	4 22·0	1·8 0·6	7·8 2·4	13·8 4·3
19	4 34·8	4 35·5	4 22·2	1·9 0·6	7·9 2·4	13·9 4·3
20	4 35·0	4 35·8	4 22·5	2·0 0·6	8·0 2·5	14·0 4·3
21	4 35·3	4 36·0	4 22·7	2·1 0·6	8·1 2·5	14·1 4·3
22	4 35·5	4 36·3	4 22·9	2·2 0·7	8·2 2·5	14·2 4·4
23	4 35·8	4 36·5	4 23·2	2·3 0·7	8·3 2·6	14·3 4·4
24	4 36·0	4 36·8	4 23·4	2·4 0·7	8·4 2·6	14·4 4·4
25	4 36·3	4 37·0	4 23·7	2·5 0·8	8·5 2·6	14·5 4·5
26	4 36·5	4 37·3	4 23·9	2·6 0·8	8·6 2·7	14·6 4·5
27	4 36·8	4 37·5	4 24·1	2·7 0·8	8·7 2·7	14·7 4·5
28	4 37·0	4 37·8	4 24·4	2·8 0·9	8·8 2·7	14·8 4·6
29	4 37·3	4 38·0	4 24·6	2·9 0·9	8·9 2·7	14·9 4·6
30	4 37·5	4 38·3	4 24·9	3·0 0·9	9·0 2·8	15·0 4·6
31	4 37·8	4 38·5	4 25·1	3·1 1·0	9·1 2·8	15·1 4·7
32	4 38·0	4 38·8	4 25·3	3·2 1·0	9·2 2·8	15·2 4·7
33	4 38·3	4 39·0	4 25·6	3·3 1·0	9·3 2·9	15·3 4·7
34	4 38·5	4 39·3	4 25·8	3·4 1·0	9·4 2·9	15·4 4·7
35	4 38·8	4 39·5	4 26·1	3·5 1·1	9·5 2·9	15·5 4·8
36	4 39·0	4 39·8	4 26·3	3·6 1·1	9·6 3·0	15·6 4·8
37	4 39·3	4 40·0	4 26·5	3·7 1·1	9·7 3·0	15·7 4·8
38	4 39·5	4 40·3	4 26·8	3·8 1·2	9·8 3·0	15·8 4·9
39	4 39·8	4 40·5	4 27·0	3·9 1·2	9·9 3·1	15·9 4·9
40	4 40·0	4 40·8	4 27·2	4·0 1·2	10·0 3·1	16·0 4·9
41	4 40·3	4 41·0	4 27·5	4·1 1·3	10·1 3·1	16·1 5·0
42	4 40·5	4 41·3	4 27·7	4·2 1·3	10·2 3·1	16·2 5·0
43	4 40·8	4 41·5	4 28·0	4·3 1·3	10·3 3·2	16·3 5·0
44	4 41·0	4 41·8	4 28·2	4·4 1·4	10·4 3·2	16·4 5·1
45	4 41·3	4 42·0	4 28·4	4·5 1·4	10·5 3·2	16·5 5·1
46	4 41·5	4 42·3	4 28·7	4·6 1·4	10·6 3·3	16·6 5·1
47	4 41·8	4 42·5	4 28·9	4·7 1·4	10·7 3·3	16·7 5·1
48	4 42·0	4 42·8	4 29·2	4·8 1·5	10·8 3·3	16·8 5·2
49	4 42·3	4 43·0	4 29·4	4·9 1·5	10·9 3·4	16·9 5·2
50	4 42·5	4 43·3	4 29·6	5·0 1·5	11·0 3·4	17·0 5·2
51	4 42·8	4 43·5	4 29·9	5·1 1·6	11·1 3·4	17·1 5·3
52	4 43·0	4 43·8	4 30·1	5·2 1·6	11·2 3·5	17·2 5·3
53	4 43·3	4 44·0	4 30·3	5·3 1·6	11·3 3·5	17·3 5·3
54	4 43·5	4 44·3	4 30·6	5·4 1·7	11·4 3·5	17·4 5·4
55	4 43·8	4 44·5	4 30·8	5·5 1·7	11·5 3·5	17·5 5·4
56	4 44·0	4 44·8	4 31·1	5·6 1·7	11·6 3·6	17·6 5·4
57	4 44·3	4 45·0	4 31·3	5·7 1·8	11·7 3·6	17·7 5·5
58	4 44·5	4 45·3	4 31·5	5·8 1·8	11·8 3·6	17·8 5·5
59	4 44·8	4 45·5	4 31·8	5·9 1·8	11·9 3·7	17·9 5·5
60	4 45·0	4 45·8	4 32·0	6·0 1·9	12·0 3·7	18·0 5·6

19ᵐ	SUN PLANETS	ARIES	MOON	v or d Corrⁿ	v or d Corrⁿ	v or d Corrⁿ
s	° ′	° ′	° ′	′ ′	′ ′	′ ′
00	4 45·0	4 45·8	4 32·0	0·0 0·0	6·0 2·0	12·0 3·9
01	4 45·3	4 46·0	4 32·3	0·1 0·0	6·1 2·0	12·1 3·9
02	4 45·5	4 46·3	4 32·5	0·2 0·1	6·2 2·0	12·2 4·0
03	4 45·8	4 46·5	4 32·7	0·3 0·1	6·3 2·0	12·3 4·0
04	4 46·0	4 46·8	4 33·0	0·4 0·1	6·4 2·1	12·4 4·0
05	4 46·3	4 47·0	4 33·2	0·5 0·2	6·5 2·1	12·5 4·1
06	4 46·5	4 47·3	4 33·4	0·6 0·2	6·6 2·1	12·6 4·1
07	4 46·8	4 47·5	4 33·7	0·7 0·2	6·7 2·2	12·7 4·1
08	4 47·0	4 47·8	4 33·9	0·8 0·3	6·8 2·2	12·8 4·2
09	4 47·3	4 48·0	4 34·2	0·9 0·3	6·9 2·2	12·9 4·2
10	4 47·5	4 48·3	4 34·4	1·0 0·3	7·0 2·3	13·0 4·2
11	4 47·8	4 48·5	4 34·6	1·1 0·4	7·1 2·3	13·1 4·3
12	4 48·0	4 48·8	4 34·9	1·2 0·4	7·2 2·3	13·2 4·3
13	4 48·3	4 49·0	4 35·1	1·3 0·4	7·3 2·4	13·3 4·3
14	4 48·5	4 49·3	4 35·4	1·4 0·5	7·4 2·4	13·4 4·4
15	4 48·8	4 49·5	4 35·6	1·5 0·5	7·5 2·4	13·5 4·4
16	4 49·0	4 49·8	4 35·8	1·6 0·5	7·6 2·5	13·6 4·4
17	4 49·3	4 50·0	4 36·1	1·7 0·6	7·7 2·5	13·7 4·5
18	4 49·5	4 50·3	4 36·3	1·8 0·6	7·8 2·5	13·8 4·5
19	4 49·8	4 50·5	4 36·6	1·9 0·6	7·9 2·6	13·9 4·5
20	4 50·0	4 50·8	4 36·8	2·0 0·7	8·0 2·6	14·0 4·6
21	4 50·3	4 51·0	4 37·0	2·1 0·7	8·1 2·6	14·1 4·6
22	4 50·5	4 51·3	4 37·3	2·2 0·7	8·2 2·7	14·2 4·6
23	4 50·8	4 51·5	4 37·5	2·3 0·7	8·3 2·7	14·3 4·6
24	4 51·0	4 51·8	4 37·7	2·4 0·8	8·4 2·7	14·4 4·7
25	4 51·3	4 52·0	4 38·0	2·5 0·8	8·5 2·8	14·5 4·7
26	4 51·5	4 52·3	4 38·2	2·6 0·8	8·6 2·8	14·6 4·7
27	4 51·8	4 52·5	4 38·5	2·7 0·9	8·7 2·8	14·7 4·8
28	4 52·0	4 52·8	4 38·7	2·8 0·9	8·8 2·9	14·8 4·8
29	4 52·3	4 53·1	4 38·9	2·9 0·9	8·9 2·9	14·9 4·8
30	4 52·5	4 53·3	4 39·2	3·0 1·0	9·0 2·9	15·0 4·9
31	4 52·8	4 53·6	4 39·4	3·1 1·0	9·1 3·0	15·1 4·9
32	4 53·0	4 53·8	4 39·7	3·2 1·0	9·2 3·0	15·2 4·9
33	4 53·3	4 54·1	4 39·9	3·3 1·1	9·3 3·0	15·3 5·0
34	4 53·5	4 54·3	4 40·1	3·4 1·1	9·4 3·1	15·4 5·0
35	4 53·8	4 54·6	4 40·4	3·5 1·1	9·5 3·1	15·5 5·0
36	4 54·0	4 54·8	4 40·6	3·6 1·2	9·6 3·1	15·6 5·1
37	4 54·3	4 55·1	4 40·8	3·7 1·2	9·7 3·2	15·7 5·1
38	4 54·5	4 55·3	4 41·1	3·8 1·2	9·8 3·2	15·8 5·1
39	4 54·8	4 55·6	4 41·3	3·9 1·3	9·9 3·2	15·9 5·2
40	4 55·0	4 55·8	4 41·6	4·0 1·3	10·0 3·3	16·0 5·2
41	4 55·3	4 56·1	4 41·8	4·1 1·3	10·1 3·3	16·1 5·2
42	4 55·5	4 56·3	4 42·0	4·2 1·4	10·2 3·3	16·2 5·3
43	4 55·8	4 56·6	4 42·3	4·3 1·4	10·3 3·3	16·3 5·3
44	4 56·0	4 56·8	4 42·5	4·4 1·4	10·4 3·4	16·4 5·3
45	4 56·3	4 57·1	4 42·8	4·5 1·5	10·5 3·4	16·5 5·4
46	4 56·5	4 57·3	4 43·0	4·6 1·5	10·6 3·4	16·6 5·4
47	4 56·8	4 57·6	4 43·2	4·7 1·5	10·7 3·5	16·7 5·4
48	4 57·0	4 57·8	4 43·5	4·8 1·6	10·8 3·5	16·8 5·5
49	4 57·3	4 58·1	4 43·7	4·9 1·6	10·9 3·5	16·9 5·5
50	4 57·5	4 58·3	4 43·9	5·0 1·6	11·0 3·6	17·0 5·5
51	4 57·8	4 58·6	4 44·2	5·1 1·7	11·1 3·6	17·1 5·6
52	4 58·0	4 58·8	4 44·4	5·2 1·7	11·2 3·6	17·2 5·6
53	4 58·3	4 59·1	4 44·7	5·3 1·7	11·3 3·7	17·3 5·6
54	4 58·5	4 59·3	4 44·9	5·4 1·8	11·4 3·7	17·4 5·7
55	4 58·8	4 59·6	4 45·1	5·5 1·8	11·5 3·7	17·5 5·7
56	4 59·0	4 59·8	4 45·4	5·6 1·8	11·6 3·8	17·6 5·7
57	4 59·3	5 00·1	4 45·6	5·7 1·9	11·7 3·8	17·7 5·8
58	4 59·5	5 00·3	4 45·9	5·8 1·9	11·8 3·8	17·8 5·8
59	4 59·8	5 00·6	4 46·1	5·9 1·9	11·9 3·9	17·9 5·8
60	5 00·0	5 00·8	4 46·3	6·0 2·0	12·0 3·9	18·0 5·9

20ᵐ	SUN PLANETS	ARIES	MOON	v or Corrn d		v or Corrn d		v or Corrn d	
s	° ′	° ′	° ′	′	′	′	′	′	′
00	5 00.0	5 00.8	4 46.3	0.0	0.0	6.0	2.1	12.0	4.1
01	5 00.3	5 01.1	4 46.6	0.1	0.0	6.1	2.1	12.1	4.1
02	5 00.5	5 01.3	4 46.8	0.2	0.1	6.2	2.1	12.2	4.2
03	5 00.8	5 01.6	4 47.0	0.3	0.1	6.3	2.2	12.3	4.2
04	5 01.0	5 01.8	4 47.3	0.4	0.1	6.4	2.2	12.4	4.2
05	5 01.3	5 02.1	4 47.5	0.5	0.2	6.5	2.2	12.5	4.3
06	5 01.5	5 02.3	4 47.8	0.6	0.2	6.6	2.3	12.6	4.3
07	5 01.8	5 02.6	4 48.0	0.7	0.2	6.7	2.3	12.7	4.3
08	5 02.0	5 02.8	4 48.2	0.8	0.3	6.8	2.3	12.8	4.4
09	5 02.3	5 03.1	4 48.5	0.9	0.3	6.9	2.4	12.9	4.4
10	5 02.5	5 03.3	4 48.7	1.0	0.3	7.0	2.4	13.0	4.4
11	5 02.8	5 03.6	4 49.0	1.1	0.4	7.1	2.4	13.1	4.5
12	5 03.0	5 03.8	4 49.2	1.2	0.4	7.2	2.5	13.2	4.5
13	5 03.3	5 04.1	4 49.4	1.3	0.4	7.3	2.5	13.3	4.5
14	5 03.5	5 04.3	4 49.7	1.4	0.5	7.4	2.5	13.4	4.6
15	5 03.8	5 04.6	4 49.9	1.5	0.5	7.5	2.6	13.5	4.6
16	5 04.0	5 04.8	4 50.2	1.6	0.5	7.6	2.6	13.6	4.6
17	5 04.3	5 05.1	4 50.4	1.7	0.6	7.7	2.6	13.7	4.7
18	5 04.5	5 05.3	4 50.6	1.8	0.6	7.8	2.7	13.8	4.7
19	5 04.8	5 05.6	4 50.9	1.9	0.6	7.9	2.7	13.9	4.7
20	5 05.0	5 05.8	4 51.1	2.0	0.7	8.0	2.7	14.0	4.8
21	5 05.3	5 06.1	4 51.3	2.1	0.7	8.1	2.8	14.1	4.8
22	5 05.5	5 06.3	4 51.6	2.2	0.8	8.2	2.8	14.2	4.9
23	5 05.8	5 06.6	4 51.8	2.3	0.8	8.3	2.8	14.3	4.9
24	5 06.0	5 06.8	4 52.1	2.4	0.8	8.4	2.9	14.4	4.9
25	5 06.3	5 07.1	4 52.3	2.5	0.9	8.5	2.9	14.5	5.0
26	5 06.5	5 07.3	4 52.5	2.6	0.9	8.6	2.9	14.6	5.0
27	5 06.8	5 07.6	4 52.8	2.7	0.9	8.7	3.0	14.7	5.0
28	5 07.0	5 07.8	4 53.0	2.8	1.0	8.8	3.0	14.8	5.1
29	5 07.3	5 08.1	4 53.3	2.9	1.0	8.9	3.0	14.9	5.1
30	5 07.5	5 08.3	4 53.5	3.0	1.0	9.0	3.1	15.0	5.1
31	5 07.8	5 08.6	4 53.7	3.1	1.1	9.1	3.1	15.1	5.2
32	5 08.0	5 08.8	4 54.0	3.2	1.1	9.2	3.1	15.2	5.2
33	5 08.3	5 09.1	4 54.2	3.3	1.1	9.3	3.2	15.3	5.2
34	5 08.5	5 09.3	4 54.4	3.4	1.2	9.4	3.2	15.4	5.3
35	5 08.8	5 09.6	4 54.7	3.5	1.2	9.5	3.2	15.5	5.3
36	5 09.0	5 09.8	4 54.9	3.6	1.2	9.6	3.3	15.6	5.3
37	5 09.3	5 10.1	4 55.2	3.7	1.3	9.7	3.3	15.7	5.4
38	5 09.5	5 10.3	4 55.4	3.8	1.3	9.8	3.3	15.8	5.4
39	5 09.8	5 10.6	4 55.6	3.9	1.3	9.9	3.4	15.9	5.4
40	5 10.0	5 10.8	4 55.9	4.0	1.4	10.0	3.4	16.0	5.5
41	5 10.3	5 11.1	4 56.1	4.1	1.4	10.1	3.5	16.1	5.5
42	5 10.5	5 11.4	4 56.4	4.2	1.4	10.2	3.5	16.2	5.5
43	5 10.8	5 11.6	4 56.6	4.3	1.5	10.3	3.5	16.3	5.6
44	5 11.0	5 11.9	4 56.8	4.4	1.5	10.4	3.6	16.4	5.6
45	5 11.3	5 12.1	4 57.1	4.5	1.5	10.5	3.6	16.5	5.6
46	5 11.5	5 12.4	4 57.3	4.6	1.6	10.6	3.6	16.6	5.7
47	5 11.8	5 12.6	4 57.5	4.7	1.6	10.7	3.7	16.7	5.7
48	5 12.0	5 12.9	4 57.8	4.8	1.6	10.8	3.7	16.8	5.7
49	5 12.3	5 13.1	4 58.0	4.9	1.7	10.9	3.7	16.9	5.8
50	5 12.5	5 13.4	4 58.3	5.0	1.7	11.0	3.8	17.0	5.8
51	5 12.8	5 13.6	4 58.5	5.1	1.7	11.1	3.8	17.1	5.8
52	5 13.0	5 13.9	4 58.7	5.2	1.8	11.2	3.8	17.2	5.9
53	5 13.3	5 14.1	4 59.0	5.3	1.8	11.3	3.9	17.3	5.9
54	5 13.5	5 14.4	4 59.2	5.4	1.8	11.4	3.9	17.4	5.9
55	5 13.8	5 14.6	4 59.5	5.5	1.9	11.5	3.9	17.5	6.0
56	5 14.0	5 14.9	4 59.7	5.6	1.9	11.6	4.0	17.6	6.0
57	5 14.3	5 15.1	4 59.9	5.7	1.9	11.7	4.0	17.7	6.0
58	5 14.5	5 15.4	5 00.2	5.8	2.0	11.8	4.0	17.8	6.1
59	5 14.8	5 15.6	5 00.4	5.9	2.0	11.9	4.1	17.9	6.1
60	5 15.0	5 15.9	5 00.7	6.0	2.1	12.0	4.1	18.0	6.2

21ᵐ	SUN PLANETS	ARIES	MOON	v or Corrn d		v or Corrn d		v or Corrn d	
s	° ′	° ′	° ′	′	′	′	′	′	′
00	5 15.0	5 15.9	5 00.7	0.0	0.0	6.0	2.2	12.0	4.3
01	5 15.3	5 16.1	5 00.9	0.1	0.0	6.1	2.2	12.1	4.3
02	5 15.5	5 16.4	5 01.1	0.2	0.1	6.2	2.2	12.2	4.4
03	5 15.8	5 16.6	5 01.4	0.3	0.1	6.3	2.3	12.3	4.4
04	5 16.0	5 16.9	5 01.6	0.4	0.1	6.4	2.3	12.4	4.4
05	5 16.3	5 17.1	5 01.8	0.5	0.2	6.5	2.3	12.5	4.5
06	5 16.5	5 17.4	5 02.1	0.6	0.2	6.6	2.4	12.6	4.5
07	5 16.8	5 17.6	5 02.3	0.7	0.3	6.7	2.4	12.7	4.6
08	5 17.0	5 17.9	5 02.6	0.8	0.3	6.8	2.4	12.8	4.6
09	5 17.3	5 18.1	5 02.8	0.9	0.3	6.9	2.5	12.9	4.6
10	5 17.5	5 18.4	5 03.0	1.0	0.4	7.0	2.5	13.0	4.7
11	5 17.8	5 18.6	5 03.3	1.1	0.4	7.1	2.5	13.1	4.7
12	5 18.0	5 18.9	5 03.5	1.2	0.4	7.2	2.6	13.2	4.7
13	5 18.3	5 19.1	5 03.8	1.3	0.5	7.3	2.6	13.3	4.8
14	5 18.5	5 19.4	5 04.0	1.4	0.5	7.4	2.7	13.4	4.8
15	5 18.8	5 19.6	5 04.2	1.5	0.5	7.5	2.7	13.5	4.8
16	5 19.0	5 19.9	5 04.5	1.6	0.6	7.6	2.7	13.6	4.9
17	5 19.3	5 20.1	5 04.7	1.7	0.6	7.7	2.8	13.7	4.9
18	5 19.5	5 20.4	5 04.9	1.8	0.6	7.8	2.8	13.8	4.9
19	5 19.8	5 20.6	5 05.2	1.9	0.7	7.9	2.8	13.9	5.0
20	5 20.0	5 20.9	5 05.4	2.0	0.7	8.0	2.9	14.0	5.0
21	5 20.3	5 21.1	5 05.7	2.1	0.8	8.1	2.9	14.1	5.1
22	5 20.5	5 21.4	5 05.9	2.2	0.8	8.2	2.9	14.2	5.1
23	5 20.8	5 21.6	5 06.1	2.3	0.8	8.3	3.0	14.3	5.1
24	5 21.0	5 21.9	5 06.4	2.4	0.9	8.4	3.0	14.4	5.2
25	5 21.3	5 22.1	5 06.6	2.5	0.9	8.5	3.0	14.5	5.2
26	5 21.5	5 22.4	5 06.9	2.6	0.9	8.6	3.1	14.6	5.2
27	5 21.8	5 22.6	5 07.1	2.7	1.0	8.7	3.1	14.7	5.3
28	5 22.0	5 22.9	5 07.3	2.8	1.0	8.8	3.2	14.8	5.3
29	5 22.3	5 23.1	5 07.6	2.9	1.0	8.9	3.2	14.9	5.3
30	5 22.5	5 23.4	5 07.8	3.0	1.1	9.0	3.2	15.0	5.4
31	5 22.8	5 23.6	5 08.0	3.1	1.1	9.1	3.3	15.1	5.4
32	5 23.0	5 23.9	5 08.3	3.2	1.1	9.2	3.3	15.2	5.4
33	5 23.3	5 24.1	5 08.5	3.3	1.2	9.3	3.3	15.3	5.5
34	5 23.5	5 24.4	5 08.8	3.4	1.2	9.4	3.4	15.4	5.5
35	5 23.8	5 24.6	5 09.0	3.5	1.3	9.5	3.4	15.5	5.6
36	5 24.0	5 24.9	5 09.2	3.6	1.3	9.6	3.4	15.6	5.6
37	5 24.3	5 25.1	5 09.5	3.7	1.3	9.7	3.5	15.7	5.6
38	5 24.5	5 25.4	5 09.7	3.8	1.4	9.8	3.5	15.8	5.7
39	5 24.8	5 25.6	5 10.0	3.9	1.4	9.9	3.5	15.9	5.7
40	5 25.0	5 25.9	5 10.2	4.0	1.4	10.0	3.6	16.0	5.7
41	5 25.3	5 26.1	5 10.4	4.1	1.5	10.1	3.6	16.1	5.8
42	5 25.5	5 26.4	5 10.7	4.2	1.5	10.2	3.7	16.2	5.8
43	5 25.8	5 26.6	5 10.9	4.3	1.5	10.3	3.7	16.3	5.8
44	5 26.0	5 26.9	5 11.1	4.4	1.6	10.4	3.7	16.4	5.9
45	5 26.3	5 27.1	5 11.4	4.5	1.6	10.5	3.8	16.5	5.9
46	5 26.5	5 27.4	5 11.6	4.6	1.6	10.6	3.8	16.6	5.9
47	5 26.8	5 27.6	5 11.9	4.7	1.7	10.7	3.8	16.7	6.0
48	5 27.0	5 27.9	5 12.1	4.8	1.7	10.8	3.9	16.8	6.0
49	5 27.3	5 28.1	5 12.3	4.9	1.8	10.9	3.9	16.9	6.1
50	5 27.5	5 28.4	5 12.6	5.0	1.8	11.0	3.9	17.0	6.1
51	5 27.8	5 28.6	5 12.8	5.1	1.8	11.1	4.0	17.1	6.1
52	5 28.0	5 28.9	5 13.1	5.2	1.9	11.2	4.0	17.2	6.2
53	5 28.3	5 29.1	5 13.3	5.3	1.9	11.3	4.0	17.3	6.2
54	5 28.5	5 29.4	5 13.5	5.4	1.9	11.4	4.1	17.4	6.2
55	5 28.8	5 29.7	5 13.8	5.5	2.0	11.5	4.1	17.5	6.3
56	5 29.0	5 29.9	5 14.0	5.6	2.0	11.6	4.2	17.6	6.3
57	5 29.3	5 30.2	5 14.3	5.7	2.0	11.7	4.2	17.7	6.3
58	5 29.5	5 30.4	5 14.5	5.8	2.1	11.8	4.2	17.8	6.4
59	5 29.8	5 30.7	5 14.7	5.9	2.1	11.9	4.3	17.9	6.4
60	5 30.0	5 30.9	5 15.0	6.0	2.2	12.0	4.3	18.0	6.5

22 (s)	SUN PLANETS	ARIES	MOON	v or d	Corr^n	v or d	Corr^n	v or d	Corr^n
00	5 30.0	5 30.9	5 15.0	0.0	0.0	6.0	2.3	12.0	4.5
01	5 30.3	5 31.2	5 15.2	0.1	0.0	6.1	2.3	12.1	4.5
02	5 30.5	5 31.4	5 15.4	0.2	0.1	6.2	2.3	12.2	4.6
03	5 30.8	5 31.7	5 15.7	0.3	0.1	6.3	2.4	12.3	4.6
04	5 31.0	5 31.9	5 15.9	0.4	0.2	6.4	2.4	12.4	4.7
05	5 31.3	5 32.2	5 16.2	0.5	0.2	6.5	2.4	12.5	4.7
06	5 31.5	5 32.4	5 16.4	0.6	0.2	6.6	2.5	12.6	4.7
07	5 31.8	5 32.7	5 16.6	0.7	0.3	6.7	2.5	12.7	4.8
08	5 32.0	5 32.9	5 16.9	0.8	0.3	6.8	2.6	12.8	4.8
09	5 32.3	5 33.2	5 17.1	0.9	0.3	6.9	2.6	12.9	4.8
10	5 32.5	5 33.4	5 17.4	1.0	0.4	7.0	2.6	13.0	4.9
11	5 32.8	5 33.7	5 17.6	1.1	0.4	7.1	2.7	13.1	4.9
12	5 33.0	5 33.9	5 17.8	1.2	0.5	7.2	2.7	13.2	5.0
13	5 33.3	5 34.2	5 18.1	1.3	0.5	7.3	2.7	13.3	5.0
14	5 33.5	5 34.4	5 18.3	1.4	0.5	7.4	2.8	13.4	5.0
15	5 33.8	5 34.7	5 18.5	1.5	0.6	7.5	2.8	13.5	5.1
16	5 34.0	5 34.9	5 18.8	1.6	0.6	7.6	2.9	13.6	5.1
17	5 34.3	5 35.2	5 19.0	1.7	0.6	7.7	2.9	13.7	5.1
18	5 34.5	5 35.4	5 19.3	1.8	0.7	7.8	2.9	13.8	5.2
19	5 34.8	5 35.7	5 19.5	1.9	0.7	7.9	3.0	13.9	5.2
20	5 35.0	5 35.9	5 19.7	2.0	0.8	8.0	3.0	14.0	5.3
21	5 35.3	5 36.2	5 20.0	2.1	0.8	8.1	3.0	14.1	5.3
22	5 35.5	5 36.4	5 20.2	2.2	0.8	8.2	3.1	14.2	5.3
23	5 35.8	5 36.7	5 20.5	2.3	0.9	8.3	3.1	14.3	5.4
24	5 36.0	5 36.9	5 20.7	2.4	0.9	8.4	3.2	14.4	5.4
25	5 36.3	5 37.2	5 20.9	2.5	0.9	8.5	3.2	14.5	5.4
26	5 36.5	5 37.4	5 21.2	2.6	1.0	8.6	3.2	14.6	5.5
27	5 36.8	5 37.7	5 21.4	2.7	1.0	8.7	3.3	14.7	5.5
28	5 37.0	5 37.9	5 21.6	2.8	1.1	8.8	3.3	14.8	5.6
29	5 37.3	5 38.2	5 21.9	2.9	1.1	8.9	3.3	14.9	5.6
30	5 37.5	5 38.4	5 22.1	3.0	1.1	9.0	3.4	15.0	5.6
31	5 37.8	5 38.7	5 22.4	3.1	1.2	9.1	3.4	15.1	5.7
32	5 38.0	5 38.9	5 22.6	3.2	1.2	9.2	3.5	15.2	5.7
33	5 38.3	5 39.2	5 22.8	3.3	1.2	9.3	3.5	15.3	5.7
34	5 38.5	5 39.4	5 23.1	3.4	1.3	9.4	3.5	15.4	5.8
35	5 38.8	5 39.7	5 23.3	3.5	1.3	9.5	3.6	15.5	5.8
36	5 39.0	5 39.9	5 23.6	3.6	1.4	9.6	3.6	15.6	5.9
37	5 39.3	5 40.2	5 23.8	3.7	1.4	9.7	3.6	15.7	5.9
38	5 39.5	5 40.4	5 24.0	3.8	1.4	9.8	3.7	15.8	5.9
39	5 39.8	5 40.7	5 24.3	3.9	1.5	9.9	3.7	15.9	6.0
40	5 40.0	5 40.9	5 24.5	4.0	1.5	10.0	3.8	16.0	6.0
41	5 40.3	5 41.2	5 24.7	4.1	1.5	10.1	3.8	16.1	6.0
42	5 40.5	5 41.4	5 25.0	4.2	1.6	10.2	3.8	16.2	6.1
43	5 40.8	5 41.7	5 25.2	4.3	1.6	10.3	3.9	16.3	6.1
44	5 41.0	5 41.9	5 25.5	4.4	1.7	10.4	3.9	16.4	6.2
45	5 41.3	5 42.2	5 25.7	4.5	1.7	10.5	3.9	16.5	6.2
46	5 41.5	5 42.4	5 25.9	4.6	1.7	10.6	4.0	16.6	6.2
47	5 41.8	5 42.7	5 26.2	4.7	1.8	10.7	4.0	16.7	6.3
48	5 42.0	5 42.9	5 26.4	4.8	1.8	10.8	4.1	16.8	6.3
49	5 42.3	5 43.2	5 26.7	4.9	1.8	10.9	4.1	16.9	6.3
50	5 42.5	5 43.4	5 26.9	5.0	1.9	11.0	4.1	17.0	6.4
51	5 42.8	5 43.7	5 27.1	5.1	1.9	11.1	4.2	17.1	6.4
52	5 43.0	5 43.9	5 27.4	5.2	2.0	11.2	4.2	17.2	6.5
53	5 43.3	5 44.2	5 27.6	5.3	2.0	11.3	4.2	17.3	6.5
54	5 43.5	5 44.4	5 27.9	5.4	2.0	11.4	4.3	17.4	6.5
55	5 43.8	5 44.7	5 28.1	5.5	2.1	11.5	4.3	17.5	6.6
56	5 44.0	5 44.9	5 28.3	5.6	2.1	11.6	4.4	17.6	6.6
57	5 44.3	5 45.2	5 28.6	5.7	2.1	11.7	4.4	17.7	6.6
58	5 44.5	5 45.4	5 28.8	5.8	2.2	11.8	4.4	17.8	6.7
59	5 44.8	5 45.7	5 29.0	5.9	2.2	11.9	4.5	17.9	6.7
60	5 45.0	5 45.9	5 29.3	6.0	2.3	12.0	4.5	18.0	6.8

23 (s)	SUN PLANETS	ARIES	MOON	v or d	Corr^n	v or d	Corr^n	v or d	Corr^n
00	5 45.0	5 45.9	5 29.3	0.0	0.0	6.0	2.4	12.0	4.7
01	5 45.3	5 46.2	5 29.5	0.1	0.0	6.1	2.4	12.1	4.7
02	5 45.5	5 46.4	5 29.8	0.2	0.1	6.2	2.4	12.2	4.8
03	5 45.8	5 46.7	5 30.0	0.3	0.1	6.3	2.5	12.3	4.8
04	5 46.0	5 46.9	5 30.2	0.4	0.2	6.4	2.5	12.4	4.9
05	5 46.3	5 47.2	5 30.5	0.5	0.2	6.5	2.5	12.5	4.9
06	5 46.5	5 47.4	5 30.7	0.6	0.2	6.6	2.6	12.6	4.9
07	5 46.8	5 47.7	5 31.0	0.7	0.3	6.7	2.6	12.7	5.0
08	5 47.0	5 48.0	5 31.2	0.8	0.3	6.8	2.7	12.8	5.0
09	5 47.3	5 48.2	5 31.4	0.9	0.4	6.9	2.7	12.9	5.1
10	5 47.5	5 48.5	5 31.7	1.0	0.4	7.0	2.7	13.0	5.1
11	5 47.8	5 48.7	5 31.9	1.1	0.4	7.1	2.8	13.1	5.1
12	5 48.0	5 49.0	5 32.1	1.2	0.5	7.2	2.8	13.2	5.2
13	5 48.3	5 49.2	5 32.4	1.3	0.5	7.3	2.9	13.3	5.2
14	5 48.5	5 49.5	5 32.6	1.4	0.5	7.4	2.9	13.4	5.2
15	5 48.8	5 49.7	5 32.9	1.5	0.6	7.5	2.9	13.5	5.3
16	5 49.0	5 50.0	5 33.1	1.6	0.6	7.6	3.0	13.6	5.3
17	5 49.3	5 50.2	5 33.3	1.7	0.7	7.7	3.0	13.7	5.4
18	5 49.5	5 50.5	5 33.6	1.8	0.7	7.8	3.1	13.8	5.4
19	5 49.8	5 50.7	5 33.8	1.9	0.7	7.9	3.1	13.9	5.4
20	5 50.0	5 51.0	5 34.1	2.0	0.8	8.0	3.1	14.0	5.5
21	5 50.3	5 51.2	5 34.3	2.1	0.8	8.1	3.2	14.1	5.5
22	5 50.5	5 51.5	5 34.5	2.2	0.9	8.2	3.2	14.2	5.6
23	5 50.8	5 51.7	5 34.8	2.3	0.9	8.3	3.3	14.3	5.6
24	5 51.0	5 52.0	5 35.0	2.4	0.9	8.4	3.3	14.4	5.6
25	5 51.3	5 52.2	5 35.2	2.5	1.0	8.5	3.3	14.5	5.7
26	5 51.5	5 52.5	5 35.5	2.6	1.0	8.6	3.4	14.6	5.7
27	5 51.8	5 52.7	5 35.7	2.7	1.1	8.7	3.4	14.7	5.8
28	5 52.0	5 53.0	5 36.0	2.8	1.1	8.8	3.4	14.8	5.8
29	5 52.3	5 53.2	5 36.2	2.9	1.1	8.9	3.5	14.9	5.8
30	5 52.5	5 53.5	5 36.4	3.0	1.2	9.0	3.5	15.0	5.9
31	5 52.8	5 53.7	5 36.7	3.1	1.2	9.1	3.6	15.1	5.9
32	5 53.0	5 54.0	5 36.9	3.2	1.3	9.2	3.6	15.2	6.0
33	5 53.3	5 54.2	5 37.2	3.3	1.3	9.3	3.6	15.3	6.0
34	5 53.5	5 54.5	5 37.4	3.4	1.3	9.4	3.7	15.4	6.0
35	5 53.8	5 54.7	5 37.6	3.5	1.4	9.5	3.7	15.5	6.1
36	5 54.0	5 55.0	5 37.9	3.6	1.4	9.6	3.8	15.6	6.1
37	5 54.3	5 55.2	5 38.1	3.7	1.4	9.7	3.8	15.7	6.1
38	5 54.5	5 55.5	5 38.4	3.8	1.5	9.8	3.8	15.8	6.2
39	5 54.8	5 55.7	5 38.6	3.9	1.5	9.9	3.9	15.9	6.2
40	5 55.0	5 56.0	5 38.8	4.0	1.6	10.0	3.9	16.0	6.3
41	5 55.3	5 56.2	5 39.1	4.1	1.6	10.1	4.0	16.1	6.3
42	5 55.5	5 56.5	5 39.3	4.2	1.6	10.2	4.0	16.2	6.3
43	5 55.8	5 56.7	5 39.5	4.3	1.7	10.3	4.0	16.3	6.4
44	5 56.0	5 57.0	5 39.8	4.4	1.7	10.4	4.1	16.4	6.4
45	5 56.3	5 57.2	5 40.0	4.5	1.8	10.5	4.1	16.5	6.5
46	5 56.5	5 57.5	5 40.3	4.6	1.8	10.6	4.2	16.6	6.5
47	5 56.8	5 57.7	5 40.5	4.7	1.8	10.7	4.2	16.7	6.5
48	5 57.0	5 58.0	5 40.7	4.8	1.9	10.8	4.2	16.8	6.6
49	5 57.3	5 58.2	5 41.0	4.9	1.9	10.9	4.3	16.9	6.6
50	5 57.5	5 58.5	5 41.2	5.0	2.0	11.0	4.3	17.0	6.7
51	5 57.8	5 58.7	5 41.5	5.1	2.0	11.1	4.3	17.1	6.7
52	5 58.0	5 59.0	5 41.7	5.2	2.0	11.2	4.4	17.2	6.7
53	5 58.3	5 59.2	5 41.9	5.3	2.1	11.3	4.4	17.3	6.8
54	5 58.5	5 59.5	5 42.2	5.4	2.1	11.4	4.5	17.4	6.8
55	5 58.8	5 59.7	5 42.4	5.5	2.2	11.5	4.5	17.5	6.9
56	5 59.0	6 00.0	5 42.6	5.6	2.2	11.6	4.5	17.6	6.9
57	5 59.3	6 00.2	5 42.9	5.7	2.2	11.7	4.6	17.7	6.9
58	5 59.5	6 00.5	5 43.1	5.8	2.3	11.8	4.6	17.8	7.0
59	5 59.8	6 00.7	5 43.4	5.9	2.3	11.9	4.7	17.9	7.0
60	6 00.0	6 01.0	5 43.6	6.0	2.4	12.0	4.7	18.0	7.1

24ᵐ s	SUN PLANETS ° ′	ARIES ° ′	MOON ° ′	v or d ′	Corrⁿ ′	v or d ′	Corrⁿ ′	v or d ′	Corrⁿ ′
00	6 00·0	6 01·0	5 43·6	0·0	0·0	6·0	2·5	12·0	4·9
01	6 00·3	6 01·2	5 43·8	0·1	0·0	6·1	2·5	12·1	4·9
02	6 00·5	6 01·5	5 44·1	0·2	0·1	6·2	2·5	12·2	5·0
03	6 00·8	6 01·7	5 44·3	0·3	0·1	6·3	2·6	12·3	5·0
04	6 01·0	6 02·0	5 44·6	0·4	0·2	6·4	2·6	12·4	5·1
05	6 01·3	6 02·2	5 44·8	0·5	0·2	6·5	2·7	12·5	5·1
06	6 01·5	6 02·5	5 45·0	0·6	0·2	6·6	2·7	12·6	5·1
07	6 01·8	6 02·7	5 45·3	0·7	0·3	6·7	2·7	12·7	5·2
08	6 02·0	6 03·0	5 45·5	0·8	0·3	6·8	2·8	12·8	5·2
09	6 02·3	6 03·2	5 45·7	0·9	0·4	6·9	2·8	12·9	5·3
10	6 02·5	6 03·5	5 46·0	1·0	0·4	7·0	2·9	13·0	5·3
11	6 02·8	6 03·7	5 46·2	1·1	0·4	7·1	2·9	13·1	5·3
12	6 03·0	6 04·0	5 46·5	1·2	0·5	7·2	2·9	13·2	5·4
13	6 03·3	6 04·2	5 46·7	1·3	0·5	7·3	3·0	13·3	5·4
14	6 03·5	6 04·5	5 46·9	1·4	0·6	7·4	3·0	13·4	5·5
15	6 03·8	6 04·7	5 47·2	1·5	0·6	7·5	3·1	13·5	5·5
16	6 04·0	6 05·0	5 47·4	1·6	0·7	7·6	3·1	13·6	5·6
17	6 04·3	6 05·2	5 47·7	1·7	0·7	7·7	3·1	13·7	5·6
18	6 04·5	6 05·5	5 47·9	1·8	0·7	7·8	3·2	13·8	5·6
19	6 04·8	6 05·7	5 48·1	1·9	0·8	7·9	3·2	13·9	5·7
20	6 05·0	6 06·0	5 48·4	2·0	0·8	8·0	3·3	14·0	5·7
21	6 05·3	6 06·3	5 48·6	2·1	0·9	8·1	3·3	14·1	5·8
22	6 05·5	6 06·5	5 48·8	2·2	0·9	8·2	3·3	14·2	5·8
23	6 05·8	6 06·8	5 49·1	2·3	0·9	8·3	3·4	14·3	5·8
24	6 06·0	6 07·0	5 49·3	2·4	1·0	8·4	3·4	14·4	5·9
25	6 06·3	6 07·3	5 49·6	2·5	1·0	8·5	3·5	14·5	5·9
26	6 06·5	6 07·5	5 49·8	2·6	1·1	8·6	3·5	14·6	6·0
27	6 06·8	6 07·8	5 50·0	2·7	1·1	8·7	3·6	14·7	6·0
28	6 07·0	6 08·0	5 50·3	2·8	1·1	8·8	3·6	14·8	6·0
29	6 07·3	6 08·3	5 50·5	2·9	1·2	8·9	3·6	14·9	6·1
30	6 07·5	6 08·5	5 50·8	3·0	1·2	9·0	3·7	15·0	6·1
31	6 07·8	6 08·8	5 51·0	3·1	1·3	9·1	3·7	15·1	6·2
32	6 08·0	6 09·0	5 51·2	3·2	1·3	9·2	3·8	15·2	6·2
33	6 08·3	6 09·3	5 51·5	3·3	1·3	9·3	3·8	15·3	6·2
34	6 08·5	6 09·5	5 51·7	3·4	1·4	9·4	3·8	15·4	6·3
35	6 08·8	6 09·8	5 52·0	3·5	1·4	9·5	3·9	15·5	6·3
36	6 09·0	6 10·0	5 52·2	3·6	1·5	9·6	3·9	15·6	6·4
37	6 09·3	6 10·3	5 52·4	3·7	1·5	9·7	4·0	15·7	6·4
38	6 09·5	6 10·5	5 52·7	3·8	1·6	9·8	4·0	15·8	6·5
39	6 09·8	6 10·8	5 52·9	3·9	1·6	9·9	4·0	15·9	6·5
40	6 10·0	6 11·0	5 53·1	4·0	1·6	10·0	4·1	16·0	6·5
41	6 10·3	6 11·3	5 53·4	4·1	1·7	10·1	4·1	16·1	6·6
42	6 10·5	6 11·5	5 53·6	4·2	1·7	10·2	4·2	16·2	6·6
43	6 10·8	6 11·8	5 53·9	4·3	1·8	10·3	4·2	16·3	6·7
44	6 11·0	6 12·0	5 54·1	4·4	1·8	10·4	4·2	16·4	6·7
45	6 11·3	6 12·3	5 54·3	4·5	1·8	10·5	4·3	16·5	6·7
46	6 11·5	6 12·5	5 54·6	4·6	1·9	10·6	4·3	16·6	6·8
47	6 11·8	6 12·8	5 54·8	4·7	1·9	10·7	4·4	16·7	6·8
48	6 12·0	6 13·0	5 55·1	4·8	2·0	10·8	4·4	16·8	6·9
49	6 12·3	6 13·3	5 55·3	4·9	2·0	10·9	4·5	16·9	6·9
50	6 12·5	6 13·5	5 55·5	5·0	2·0	11·0	4·5	17·0	6·9
51	6 12·8	6 13·8	5 55·8	5·1	2·1	11·1	4·5	17·1	7·0
52	6 13·0	6 14·0	5 56·0	5·2	2·1	11·2	4·6	17·2	7·0
53	6 13·3	6 14·3	5 56·2	5·3	2·2	11·3	4·6	17·3	7·1
54	6 13·5	6 14·5	5 56·5	5·4	2·2	11·4	4·7	17·4	7·1
55	6 13·8	6 14·8	5 56·7	5·5	2·2	11·5	4·7	17·5	7·1
56	6 14·0	6 15·0	5 57·0	5·6	2·3	11·6	4·7	17·6	7·2
57	6 14·3	6 15·3	5 57·2	5·7	2·3	11·7	4·8	17·7	7·2
58	6 14·5	6 15·5	5 57·4	5·8	2·4	11·8	4·8	17·8	7·3
59	6 14·8	6 15·8	5 57·7	5·9	2·4	11·9	4·9	17·9	7·3
60	6 15·0	6 16·0	5 57·9	6·0	2·5	12·0	4·9	18·0	7·4

25ᵐ s	SUN PLANETS ° ′	ARIES ° ′	MOON ° ′	v or d ′	Corrⁿ ′	v or d ′	Corrⁿ ′	v or d ′	Corrⁿ ′
00	6 15·0	6 16·0	5 57·9	0·0	0·0	6·0	2·6	12·0	5·1
01	6 15·3	6 16·3	5 58·2	0·1	0·0	6·1	2·6	12·1	5·1
02	6 15·5	6 16·5	5 58·4	0·2	0·1	6·2	2·6	12·2	5·2
03	6 15·8	6 16·8	5 58·6	0·3	0·1	6·3	2·7	12·3	5·2
04	6 16·0	6 17·0	5 58·9	0·4	0·2	6·4	2·7	12·4	5·3
05	6 16·3	6 17·3	5 59·1	0·5	0·2	6·5	2·8	12·5	5·3
06	6 16·5	6 17·5	5 59·3	0·6	0·3	6·6	2·8	12·6	5·4
07	6 16·8	6 17·8	5 59·6	0·7	0·3	6·7	2·8	12·7	5·4
08	6 17·0	6 18·0	5 59·8	0·8	0·3	6·8	2·9	12·8	5·4
09	6 17·3	6 18·3	6 00·1	0·9	0·4	6·9	2·9	12·9	5·5
10	6 17·5	6 18·5	6 00·3	1·0	0·4	7·0	3·0	13·0	5·5
11	6 17·8	6 18·8	6 00·5	1·1	0·5	7·1	3·0	13·1	5·6
12	6 18·0	6 19·0	6 00·8	1·2	0·5	7·2	3·1	13·2	5·6
13	6 18·3	6 19·3	6 01·0	1·3	0·6	7·3	3·1	13·3	5·7
14	6 18·5	6 19·5	6 01·3	1·4	0·6	7·4	3·1	13·4	5·7
15	6 18·8	6 19·8	6 01·5	1·5	0·6	7·5	3·2	13·5	5·7
16	6 19·0	6 20·0	6 01·7	1·6	0·7	7·6	3·2	13·6	5·8
17	6 19·3	6 20·3	6 02·0	1·7	0·7	7·7	3·3	13·7	5·8
18	6 19·5	6 20·5	6 02·2	1·8	0·8	7·8	3·3	13·8	5·9
19	6 19·8	6 20·8	6 02·5	1·9	0·8	7·9	3·4	13·9	5·9
20	6 20·0	6 21·0	6 02·7	2·0	0·9	8·0	3·4	14·0	6·0
21	6 20·3	6 21·3	6 02·9	2·1	0·9	8·1	3·4	14·1	6·0
22	6 20·5	6 21·5	6 03·2	2·2	0·9	8·2	3·5	14·2	6·0
23	6 20·8	6 21·8	6 03·4	2·3	1·0	8·3	3·5	14·3	6·1
24	6 21·0	6 22·0	6 03·6	2·4	1·0	8·4	3·6	14·4	6·1
25	6 21·3	6 22·3	6 03·9	2·5	1·1	8·5	3·6	14·5	6·2
26	6 21·5	6 22·5	6 04·1	2·6	1·1	8·6	3·7	14·6	6·2
27	6 21·8	6 22·8	6 04·4	2·7	1·1	8·7	3·7	14·7	6·2
28	6 22·0	6 23·0	6 04·6	2·8	1·2	8·8	3·7	14·8	6·3
29	6 22·3	6 23·3	6 04·8	2·9	1·2	8·9	3·8	14·9	6·3
30	6 22·5	6 23·5	6 05·1	3·0	1·3	9·0	3·8	15·0	6·4
31	6 22·8	6 23·8	6 05·3	3·1	1·3	9·1	3·9	15·1	6·4
32	6 23·0	6 24·0	6 05·6	3·2	1·4	9·2	3·9	15·2	6·5
33	6 23·3	6 24·3	6 05·8	3·3	1·4	9·3	4·0	15·3	6·5
34	6 23·5	6 24·5	6 06·0	3·4	1·4	9·4	4·0	15·4	6·5
35	6 23·8	6 24·8	6 06·3	3·5	1·5	9·5	4·0	15·5	6·6
36	6 24·0	6 25·1	6 06·5	3·6	1·5	9·6	4·1	15·6	6·6
37	6 24·3	6 25·3	6 06·7	3·7	1·6	9·7	4·1	15·7	6·7
38	6 24·5	6 25·6	6 07·0	3·8	1·6	9·8	4·2	15·8	6·7
39	6 24·8	6 25·8	6 07·2	3·9	1·7	9·9	4·2	15·9	6·8
40	6 25·0	6 26·1	6 07·5	4·0	1·7	10·0	4·3	16·0	6·8
41	6 25·3	6 26·3	6 07·7	4·1	1·7	10·1	4·3	16·1	6·8
42	6 25·5	6 26·6	6 07·9	4·2	1·8	10·2	4·3	16·2	6·9
43	6 25·8	6 26·8	6 08·2	4·3	1·8	10·3	4·4	16·3	6·9
44	6 26·0	6 27·1	6 08·4	4·4	1·9	10·4	4·4	16·4	7·0
45	6 26·3	6 27·3	6 08·7	4·5	1·9	10·5	4·5	16·5	7·0
46	6 26·5	6 27·6	6 08·9	4·6	2·0	10·6	4·5	16·6	7·1
47	6 26·8	6 27·8	6 09·1	4·7	2·0	10·7	4·5	16·7	7·1
48	6 27·0	6 28·1	6 09·4	4·8	2·0	10·8	4·6	16·8	7·1
49	6 27·3	6 28·3	6 09·6	4·9	2·1	10·9	4·6	16·9	7·2
50	6 27·5	6 28·6	6 09·8	5·0	2·1	11·0	4·7	17·0	7·2
51	6 27·8	6 28·8	6 10·1	5·1	2·2	11·1	4·7	17·1	7·3
52	6 28·0	6 29·1	6 10·3	5·2	2·2	11·2	4·8	17·2	7·3
53	6 28·3	6 29·3	6 10·6	5·3	2·3	11·3	4·8	17·3	7·4
54	6 28·5	6 29·6	6 10·8	5·4	2·3	11·4	4·8	17·4	7·4
55	6 28·8	6 29·8	6 11·0	5·5	2·3	11·5	4·9	17·5	7·4
56	6 29·0	6 30·1	6 11·3	5·6	2·4	11·6	4·9	17·6	7·5
57	6 29·3	6 30·3	6 11·5	5·7	2·4	11·7	5·0	17·7	7·5
58	6 29·5	6 30·6	6 11·8	5·8	2·5	11·8	5·0	17·8	7·6
59	6 29·8	6 30·8	6 12·0	5·9	2·5	11·9	5·1	17·9	7·6
60	6 30·0	6 31·1	6 12·2	6·0	2·6	12·0	5·1	18·0	7·7

26^m	SUN PLANETS	ARIES	MOON	v or Corrⁿ d	v or Corrⁿ d	v or Corrⁿ d
s	° ′	° ′	° ′	′ ′	′ ′	′ ′
00	6 30·0	6 31·1	6 12·2	0·0 0·0	6·0 2·7	12·0 5·3
01	6 30·3	6 31·3	6 12·5	0·1 0·0	6·1 2·7	12·1 5·3
02	6 30·5	6 31·6	6 12·7	0·2 0·1	6·2 2·7	12·2 5·4
03	6 30·8	6 31·8	6 12·9	0·3 0·1	6·3 2·8	12·3 5·4
04	6 31·0	6 32·1	6 13·2	0·4 0·2	6·4 2·8	12·4 5·5
05	6 31·3	6 32·3	6 13·4	0·5 0·2	6·5 2·9	12·5 5·5
06	6 31·5	6 32·6	6 13·7	0·6 0·3	6·6 2·9	12·6 5·6
07	6 31·8	6 32·8	6 13·9	0·7 0·3	6·7 3·0	12·7 5·6
08	6 32·0	6 33·1	6 14·1	0·8 0·4	6·8 3·0	12·8 5·7
09	6 32·3	6 33·3	6 14·4	0·9 0·4	6·9 3·0	12·9 5·7
10	6 32·5	6 33·6	6 14·6	1·0 0·4	7·0 3·1	13·0 5·7
11	6 32·8	6 33·8	6 14·9	1·1 0·5	7·1 3·1	13·1 5·8
12	6 33·0	6 34·1	6 15·1	1·2 0·5	7·2 3·2	13·2 5·8
13	6 33·3	6 34·3	6 15·3	1·3 0·6	7·3 3·2	13·3 5·9
14	6 33·5	6 34·6	6 15·6	1·4 0·6	7·4 3·3	13·4 5·9
15	6 33·8	6 34·8	6 15·8	1·5 0·7	7·5 3·3	13·5 6·0
16	6 34·0	6 35·1	6 16·1	1·6 0·7	7·6 3·4	13·6 6·0
17	6 34·3	6 35·3	6 16·3	1·7 0·8	7·7 3·4	13·7 6·1
18	6 34·5	6 35·6	6 16·5	1·8 0·8	7·8 3·4	13·8 6·1
19	6 34·8	6 35·8	6 16·8	1·9 0·8	7·9 3·5	13·9 6·1
20	6 35·0	6 36·1	6 17·0	2·0 0·9	8·0 3·5	14·0 6·2
21	6 35·3	6 36·3	6 17·2	2·1 0·9	8·1 3·6	14·1 6·2
22	6 35·5	6 36·6	6 17·5	2·2 1·0	8·2 3·6	14·2 6·3
23	6 35·8	6 36·8	6 17·7	2·3 1·0	8·3 3·7	14·3 6·3
24	6 36·0	6 37·1	6 18·0	2·4 1·1	8·4 3·7	14·4 6·4
25	6 36·3	6 37·3	6 18·2	2·5 1·1	8·5 3·8	14·5 6·4
26	6 36·5	6 37·6	6 18·4	2·6 1·1	8·6 3·8	14·6 6·4
27	6 36·8	6 37·8	6 18·7	2·7 1·2	8·7 3·8	14·7 6·5
28	6 37·0	6 38·1	6 18·9	2·8 1·2	8·8 3·9	14·8 6·5
29	6 37·3	6 38·3	6 19·2	2·9 1·3	8·9 3·9	14·9 6·6
30	6 37·5	6 38·6	6 19·4	3·0 1·3	9·0 4·0	15·0 6·6
31	6 37·8	6 38·8	6 19·6	3·1 1·4	9·1 4·0	15·1 6·7
32	6 38·0	6 39·1	6 19·9	3·2 1·4	9·2 4·1	15·2 6·7
33	6 38·3	6 39·3	6 20·1	3·3 1·5	9·3 4·1	15·3 6·8
34	6 38·5	6 39·6	6 20·3	3·4 1·5	9·4 4·2	15·4 6·8
35	6 38·8	6 39·8	6 20·6	3·5 1·5	9·5 4·2	15·5 6·8
36	6 39·0	6 40·1	6 20·8	3·6 1·6	9·6 4·2	15·6 6·9
37	6 39·3	6 40·3	6 21·1	3·7 1·6	9·7 4·3	15·7 6·9
38	6 39·5	6 40·6	6 21·3	3·8 1·7	9·8 4·3	15·8 7·0
39	6 39·8	6 40·8	6 21·5	3·9 1·7	9·9 4·4	15·9 7·0
40	6 40·0	6 41·1	6 21·8	4·0 1·8	10·0 4·4	16·0 7·1
41	6 40·3	6 41·3	6 22·0	4·1 1·8	10·1 4·5	16·1 7·1
42	6 40·5	6 41·6	6 22·3	4·2 1·9	10·2 4·5	16·2 7·2
43	6 40·8	6 41·8	6 22·5	4·3 1·9	10·3 4·5	16·3 7·2
44	6 41·0	6 42·1	6 22·7	4·4 1·9	10·4 4·6	16·4 7·2
45	6 41·3	6 42·3	6 23·0	4·5 2·0	10·5 4·6	16·5 7·3
46	6 41·5	6 42·6	6 23·2	4·6 2·0	10·6 4·7	16·6 7·3
47	6 41·8	6 42·8	6 23·4	4·7 2·1	10·7 4·7	16·7 7·4
48	6 42·0	6 43·1	6 23·7	4·8 2·1	10·8 4·8	16·8 7·4
49	6 42·3	6 43·4	6 23·9	4·9 2·2	10·9 4·8	16·9 7·5
50	6 42·5	6 43·6	6 24·2	5·0 2·2	11·0 4·9	17·0 7·5
51	6 42·8	6 43·9	6 24·4	5·1 2·3	11·1 4·9	17·1 7·6
52	6 43·0	6 44·1	6 24·6	5·2 2·3	11·2 4·9	17·2 7·6
53	6 43·3	6 44·4	6 24·9	5·3 2·3	11·3 5·0	17·3 7·6
54	6 43·5	6 44·6	6 25·1	5·4 2·4	11·4 5·0	17·4 7·7
55	6 43·8	6 44·9	6 25·4	5·5 2·4	11·5 5·1	17·5 7·7
56	6 44·0	6 45·1	6 25·6	5·6 2·5	11·6 5·1	17·6 7·8
57	6 44·3	6 45·4	6 25·8	5·7 2·5	11·7 5·2	17·7 7·8
58	6 44·5	6 45·6	6 26·1	5·8 2·6	11·8 5·2	17·8 7·9
59	6 44·8	6 45·9	6 26·3	5·9 2·6	11·9 5·3	17·9 7·9
60	6 45·0	6 46·1	6 26·6	6·0 2·7	12·0 5·3	18·0 8·0

27^m	SUN PLANETS	ARIES	MOON	v or Corrⁿ d	v or Corrⁿ d	v or Corrⁿ d
s	° ′	° ′	° ′	′ ′	′ ′	′ ′
00	6 45·0	6 46·1	6 26·6	0·0 0·0	6·0 2·8	12·0 5·5
01	6 45·3	6 46·4	6 26·8	0·1 0·0	6·1 2·8	12·1 5·5
02	6 45·5	6 46·6	6 27·0	0·2 0·1	6·2 2·8	12·2 5·6
03	6 45·8	6 46·9	6 27·3	0·3 0·1	6·3 2·9	12·3 5·6
04	6 46·0	6 47·1	6 27·5	0·4 0·2	6·4 2·9	12·4 5·7
05	6 46·3	6 47·4	6 27·7	0·5 0·2	6·5 3·0	12·5 5·7
06	6 46·5	6 47·6	6 28·0	0·6 0·3	6·6 3·0	12·6 5·8
07	6 46·8	6 47·9	6 28·2	0·7 0·3	6·7 3·1	12·7 5·8
08	6 47·0	6 48·1	6 28·5	0·8 0·4	6·8 3·1	12·8 5·9
09	6 47·3	6 48·4	6 28·7	0·9 0·4	6·9 3·2	12·9 5·9
10	6 47·5	6 48·6	6 28·9	1·0 0·5	7·0 3·2	13·0 6·0
11	6 47·8	6 48·9	6 29·2	1·1 0·5	7·1 3·3	13·1 6·0
12	6 48·0	6 49·1	6 29·4	1·2 0·6	7·2 3·3	13·2 6·1
13	6 48·3	6 49·4	6 29·7	1·3 0·6	7·3 3·3	13·3 6·1
14	6 48·5	6 49·6	6 29·9	1·4 0·6	7·4 3·4	13·4 6·1
15	6 48·8	6 49·9	6 30·1	1·5 0·7	7·5 3·4	13·5 6·2
16	6 49·0	6 50·1	6 30·4	1·6 0·7	7·6 3·5	13·6 6·2
17	6 49·3	6 50·4	6 30·6	1·7 0·8	7·7 3·5	13·7 6·3
18	6 49·5	6 50·6	6 30·8	1·8 0·8	7·8 3·6	13·8 6·3
19	6 49·8	6 50·9	6 31·1	1·9 0·9	7·9 3·6	13·9 6·4
20	6 50·0	6 51·1	6 31·3	2·0 0·9	8·0 3·7	14·0 6·4
21	6 50·3	6 51·4	6 31·6	2·1 1·0	8·1 3·7	14·1 6·5
22	6 50·5	6 51·6	6 31·8	2·2 1·0	8·2 3·8	14·2 6·5
23	6 50·8	6 51·9	6 32·0	2·3 1·1	8·3 3·8	14·3 6·6
24	6 51·0	6 52·1	6 32·3	2·4 1·1	8·4 3·9	14·4 6·6
25	6 51·3	6 52·4	6 32·5	2·5 1·1	8·5 3·9	14·5 6·6
26	6 51·5	6 52·6	6 32·8	2·6 1·2	8·6 3·9	14·6 6·7
27	6 51·8	6 52·9	6 33·0	2·7 1·2	8·7 4·0	14·7 6·7
28	6 52·0	6 53·1	6 33·2	2·8 1·3	8·8 4·0	14·8 6·8
29	6 52·3	6 53·4	6 33·5	2·9 1·3	8·9 4·1	14·9 6·8
30	6 52·5	6 53·6	6 33·7	3·0 1·4	9·0 4·1	15·0 6·9
31	6 52·8	6 53·9	6 33·9	3·1 1·4	9·1 4·2	15·1 6·9
32	6 53·0	6 54·1	6 34·2	3·2 1·5	9·2 4·2	15·2 7·0
33	6 53·3	6 54·4	6 34·4	3·3 1·5	9·3 4·3	15·3 7·0
34	6 53·5	6 54·6	6 34·7	3·4 1·6	9·4 4·3	15·4 7·1
35	6 53·8	6 54·9	6 34·9	3·5 1·6	9·5 4·4	15·5 7·1
36	6 54·0	6 55·1	6 35·1	3·6 1·7	9·6 4·4	15·6 7·2
37	6 54·3	6 55·4	6 35·4	3·7 1·7	9·7 4·4	15·7 7·2
38	6 54·5	6 55·6	6 35·6	3·8 1·7	9·8 4·5	15·8 7·2
39	6 54·8	6 55·9	6 35·9	3·9 1·8	9·9 4·5	15·9 7·3
40	6 55·0	6 56·1	6 36·1	4·0 1·8	10·0 4·6	16·0 7·3
41	6 55·3	6 56·4	6 36·3	4·1 1·9	10·1 4·6	16·1 7·4
42	6 55·5	6 56·6	6 36·6	4·2 1·9	10·2 4·7	16·2 7·4
43	6 55·8	6 56·9	6 36·8	4·3 2·0	10·3 4·7	16·3 7·5
44	6 56·0	6 57·1	6 37·0	4·4 2·0	10·4 4·8	16·4 7·5
45	6 56·3	6 57·4	6 37·3	4·5 2·1	10·5 4·8	16·5 7·6
46	6 56·5	6 57·6	6 37·5	4·6 2·1	10·6 4·9	16·6 7·6
47	6 56·8	6 57·9	6 37·8	4·7 2·2	10·7 4·9	16·7 7·7
48	6 57·0	6 58·1	6 38·0	4·8 2·2	10·8 5·0	16·8 7·7
49	6 57·3	6 58·4	6 38·2	4·9 2·2	10·9 5·0	16·9 7·7
50	6 57·5	6 58·6	6 38·5	5·0 2·3	11·0 5·0	17·0 7·8
51	6 57·8	6 58·9	6 38·7	5·1 2·3	11·1 5·1	17·1 7·8
52	6 58·0	6 59·1	6 39·0	5·2 2·4	11·2 5·1	17·2 7·9
53	6 58·3	6 59·4	6 39·2	5·3 2·4	11·3 5·2	17·3 7·9
54	6 58·5	6 59·6	6 39·4	5·4 2·5	11·4 5·2	17·4 8·0
55	6 58·8	6 59·9	6 39·7	5·5 2·5	11·5 5·3	17·5 8·0
56	6 59·0	7 00·1	6 39·9	5·6 2·6	11·6 5·3	17·6 8·1
57	6 59·3	7 00·4	6 40·2	5·7 2·6	11·7 5·4	17·7 8·1
58	6 59·5	7 00·6	6 40·4	5·8 2·7	11·8 5·4	17·8 8·2
59	6 59·8	7 00·9	6 40·6	5·9 2·7	11·9 5·5	17·9 8·2
60	7 00·0	7 01·1	6 40·9	6·0 2·8	12·0 5·5	18·0 8·3

28ᵐ	SUN PLANETS	ARIES	MOON	v or Corrⁿ d		v or Corrⁿ d		v or Corrⁿ d	
s	° ′	° ′	° ′	′	′	′	′	′	′
00	7 00·0	7 01·1	6 40·9	0·0	0·0	6·0	2·9	12·0	5·7
01	7 00·3	7 01·4	6 41·1	0·1	0·0	6·1	2·9	12·1	5·7
02	7 00·5	7 01·7	6 41·3	0·2	0·1	6·2	2·9	12·2	5·8
03	7 00·8	7 01·9	6 41·6	0·3	0·1	6·3	3·0	12·3	5·8
04	7 01·0	7 02·2	6 41·8	0·4	0·2	6·4	3·0	12·4	5·9
05	7 01·3	7 02·4	6 42·1	0·5	0·2	6·5	3·1	12·5	5·9
06	7 01·5	7 02·7	6 42·3	0·6	0·3	6·6	3·1	12·6	6·0
07	7 01·8	7 02·9	6 42·5	0·7	0·3	6·7	3·2	12·7	6·0
08	7 02·0	7 03·2	6 42·8	0·8	0·4	6·8	3·2	12·8	6·1
09	7 02·3	7 03·4	6 43·0	0·9	0·4	6·9	3·3	12·9	6·1
10	7 02·5	7 03·7	6 43·3	1·0	0·5	7·0	3·3	13·0	6·2
11	7 02·8	7 03·9	6 43·5	1·1	0·5	7·1	3·4	13·1	6·2
12	7 03·0	7 04·2	6 43·7	1·2	0·6	7·2	3·4	13·2	6·3
13	7 03·3	7 04·4	6 44·0	1·3	0·6	7·3	3·5	13·3	6·3
14	7 03·5	7 04·7	6 44·2	1·4	0·7	7·4	3·5	13·4	6·4
15	7 03·8	7 04·9	6 44·4	1·5	0·7	7·5	3·6	13·5	6·4
16	7 04·0	7 05·2	6 44·7	1·6	0·8	7·6	3·6	13·6	6·5
17	7 04·3	7 05·4	6 44·9	1·7	0·8	7·7	3·7	13·7	6·5
18	7 04·5	7 05·7	6 45·2	1·8	0·9	7·8	3·7	13·8	6·6
19	7 04·8	7 05·9	6 45·4	1·9	0·9	7·9	3·8	13·9	6·6
20	7 05·0	7 06·2	6 45·6	2·0	1·0	8·0	3·8	14·0	6·7
21	7 05·3	7 06·4	6 45·9	2·1	1·0	8·1	3·8	14·1	6·7
22	7 05·5	7 06·7	6 46·1	2·2	1·0	8·2	3·9	14·2	6·7
23	7 05·8	7 06·9	6 46·4	2·3	1·1	8·3	3·9	14·3	6·8
24	7 06·0	7 07·2	6 46·6	2·4	1·1	8·4	4·0	14·4	6·8
25	7 06·3	7 07·4	6 46·8	2·5	1·2	8·5	4·0	14·5	6·9
26	7 06·5	7 07·7	6 47·1	2·6	1·2	8·6	4·1	14·6	6·9
27	7 06·8	7 07·9	6 47·3	2·7	1·3	8·7	4·1	14·7	7·0
28	7 07·0	7 08·2	6 47·5	2·8	1·3	8·8	4·2	14·8	7·0
29	7 07·3	7 08·4	6 47·8	2·9	1·4	8·9	4·2	14·9	7·1
30	7 07·5	7 08·7	6 48·0	3·0	1·4	9·0	4·3	15·0	7·1
31	7 07·8	7 08·9	6 48·3	3·1	1·5	9·1	4·3	15·1	7·2
32	7 08·0	7 09·2	6 48·5	3·2	1·5	9·2	4·4	15·2	7·2
33	7 08·3	7 09·4	6 48·7	3·3	1·6	9·3	4·4	15·3	7·3
34	7 08·5	7 09·7	6 49·0	3·4	1·6	9·4	4·5	15·4	7·3
35	7 08·8	7 09·9	6 49·2	3·5	1·7	9·5	4·5	15·5	7·4
36	7 09·0	7 10·2	6 49·5	3·6	1·7	9·6	4·6	15·6	7·4
37	7 09·3	7 10·4	6 49·7	3·7	1·8	9·7	4·6	15·7	7·5
38	7 09·5	7 10·7	6 49·9	3·8	1·8	9·8	4·7	15·8	7·5
39	7 09·8	7 10·9	6 50·2	3·9	1·9	9·9	4·7	15·9	7·6
40	7 10·0	7 11·2	6 50·4	4·0	1·9	10·0	4·8	16·0	7·6
41	7 10·3	7 11·4	6 50·6	4·1	1·9	10·1	4·8	16·1	7·6
42	7 10·5	7 11·7	6 50·9	4·2	2·0	10·2	.4·8	16·2	7·7
43	7 10·8	7 11·9	6 51·1	4·3	2·0	10·3	4·9	16·3	7·7
44	7 11·0	7 12·2	6 51·4	4·4	2·1	10·4	4·9	16·4	7·8
45	7 11·3	7 12·4	6 51·6	4·5	2·1	10·5	5·0	16·5	7·8
46	7 11·5	7 12·7	6 51·8	4·6	2·2	10·6	5·0	16·6	7·9
47	7 11·8	7 12·9	6 52·1	4·7	2·2	10·7	5·1	16·7	7·9
48	7 12·0	7 13·2	6 52·3	4·8	2·3	10·8	5·1	16·8	8·0
49	7 12·3	7 13·4	6 52·6	4·9	2·3	10·9	5·2	16·9	8·0
50	7 12·5	7 13·7	6 52·8	5·0	2·4	11·0	5·2	17·0	8·1
51	7 12·8	7 13·9	6 53·0	5·1	2·4	11·1	5·3	17·1	8·1
52	7 13·0	7 14·2	6 53·3	5·2	2·5	11·2	5·3	17·2	8·2
53	7 13·3	7 14·4	6 53·5	5·3	2·5	11·3	5·4	17·3	8·2
54	7 13·5	7 14·7	6 53·8	5·4	2·6	11·4	5·4	17·4	8·3
55	7 13·8	7 14·9	6 54·0	5·5	2·6	11·5	5·5	17·5	8·3
56	7 14·0	7 15·2	6 54·2	5·6	2·7	11·6	5·5	17·6	8·4
57	7 14·3	7 15·4	6 54·5	5·7	2·7	11·7	5·6	17·7	8·4
58	7 14·5	7 15·7	6 54·7	5·8	2·8	11·8	5·6	17·8	8·5
59	7 14·8	7 15·9	6 54·9	5·9	2·8	11·9	5·7	17·9	8·5
60	7 15·0	7 16·2	6 55·2	6·0	2·9	12·0	5·7	18·0	8·6

29ᵐ	SUN PLANETS	ARIES	MOON	v or Corrⁿ d		v or Corrⁿ d		v or Corrⁿ d	
s	° ′	° ′	° ′	′	′	′	′	′	′
00	7 15·0	7 16·2	6 55·2	0·0	0·0	6·0	3·0	12·0	5·9
01	7 15·3	7 16·4	6 55·4	0·1	0·0	6·1	3·0	12·1	5·9
02	7 15·5	7 16·7	6 55·7	0·2	0·1	6·2	3·0	12·2	6·0
03	7 15·8	7 16·9	6 55·9	0·3	0·1	6·3	3·1	12·3	6·0
04	7 16·0	7 17·2	6 56·1	0·4	0·2	6·4	3·1	12·4	6·1
05	7 16·3	7 17·4	6 56·4	0·5	0·2	6·5	3·2	12·5	6·1
06	7 16·5	7 17·7	6 56·6	0·6	0·3	6·6	3·2	12·6	6·2
07	7 16·8	7 17·9	6 56·8	0·7	0·3	6·7	3·3	12·7	6·2
08	7 17·0	7 18·2	6 57·1	0·8	0·4	6·8	3·3	12·8	6·3
09	7 17·3	7 18·4	6 57·3	0·9	0·4	6·9	3·4	12·9	6·3
10	7 17·5	7 18·7	6 57·6	1·0	0·5	7·0	3·4	13·0	6·4
11	7 17·8	7 18·9	6 57·8	1·1	0·5	7·1	3·5	13·1	6·4
12	7 18·0	7 19·2	6 58·0	1·2	0·6	7·2	3·5	13·2	6·5
13	7 18·3	7 19·4	6 58·3	1·3	0·6	7·3	3·6	13·3	6·5
14	7 18·5	7 19·7	6 58·5	1·4	0·7	7·4	3·6	13·4	6·6
15	7 18·8	7 20·0	6 58·8	1·5	0·7	7·5	3·7	13·5	6·6
16	7 19·0	7 20·2	6 59·0	1·6	0·8	7·6	3·7	13·6	6·7
17	7 19·3	7 20·5	6 59·2	1·7	0·8	7·7	3·8	13·7	6·7
18	7 19·5	7 20·7	6 59·5	1·8	0·9	7·8	3·8	13·8	6·8
19	7 19·8	7 21·0	6 59·7	1·9	0·9	7·9	3·9	13·9	6·8
20	7 20·0	7 21·2	7 00·0	2·0	1·0	8·0	3·9	14·0	6·9
21	7 20·3	7 21·5	7 00·2	2·1	1·0	8·1	4·0	14·1	6·9
22	7 20·5	7 21·7	7 00·4	2·2	1·1	8·2	4·0	14·2	7·0
23	7 20·8	7 22·0	7 00·7	2·3	1·1	8·3	4·1	14·3	7·0
24	7 21·0	7 22·2	7 00·9	2·4	1·2	8·4	4·1	14·4	7·1
25	7 21·3	7 22·5	7 01·1	2·5	1·2	8·5	4·2	14·5	7·1
26	7 21·5	7 22·7	7 01·4	2·6	1·3	8·6	4·2	14·6	7·2
27	7 21·8	7 23·0	7 01·6	2·7	1·3	8·7	4·3	14·7	7·2
28	7 22·0	7 23·2	7 01·9	2·8	1·4	8·8	4·3	14·8	7·3
29	7 22·3	7 23·5	7 02·1	2·9	1·4	8·9	4·4	14·9	7·3
30	7 22·5	7 23·7	7 02·3	3·0	1·5	9·0	4·4	15·0	7·4
31	7 22·8	7 24·0	7 02·6	3·1	1·5	9·1	4·5	15·1	7·4
32	7 23·0	7 24·2	7 02·8	3·2	1·6	9·2	4·5	15·2	7·5
33	7 23·3	7 24·5	7 03·1	3·3	1·6	9·3	4·6	15·3	7·5
34	7 23·5	7 24·7	7 03·3	3·4	1·7	9·4	4·6	15·4	7·6
35	7 23·8	7 25·0	7 03·5	3·5	1·7	9·5	4·7	15·5	7·6
36	7 24·0	7 25·2	7 03·8	3·6	1·8	9·6	4·7	15·6	7·7
37	7 24·3	7 25·5	7 04·0	3·7	1·8	9·7	4·8	15·7	7·7
38	7 24·5	7 25·7	7 04·3	3·8	1·9	9·8	4·8	15·8	7·8
39	7 24·8	7 26·0	7 04·5	3·9	1·9	9·9	4·9	15·9	7·8
40	7 25·0	7 26·2	7 04·7	4·0	2·0	10·0	4·9	16·0	7·9
41	7 25·3	7 26·5	7 05·0	4·1	2·0	10·1	5·0	16·1	7·9
42	7 25·5	7 26·7	7 05·2	4·2	2·1	10·2	5·0	16·2	8·0
43	7 25·8	7 27·0	7 05·4	4·3	2·1	10·3	5·1	16·3	8·0
44	7 26·0	7 27·2	7 05·7	4·4	2·2	10·4	5·1	16·4	8·1
45	7 26·3	7 27·5	7 05·9	4·5	2·2	10·5	5·2	16·5	8·1
46	7 26·5	7 27·7	7 06·2	4·6	2·3	10·6	5·2	16·6	8·2
47	7 26·8	7 28·0	7 06·4	4·7	2·3	10·7	5·3	16·7	8·2
48	7 27·0	7 28·2	7 06·6	4·8	2·4	10·8	5·3	16·8	8·3
49	7 27·3	7 28·5	7 06·9	4·9	2·4	10·9	5·4	16·9	8·3
50	7 27·5	7 28·7	7 07·1	5·0	2·5	11·0	5·4	17·0	8·4
51	7 27·8	7 29·0	7 07·4	5·1	2·5	11·1	5·5	17·1	8·4
52	7 28·0	7 29·2	7 07·6	5·2	2·6	11·2	5·5	17·2	8·5
53	7 28·3	7 29·5	7 07·8	5·3	2·6	11·3	5·6	17·3	8·5
54	7 28·5	7 29·7	7 08·1	5·4	2·7	11·4	5·6	17·4	8·6
55	7 28·8	7 30·0	7 08·3	5·5	2·7	11·5	5·7	17·5	8·6
56	7 29·0	7 30·2	7 08·5	5·6	2·8	11·6	5·7	17·6	8·7
57	7 29·3	7 30·5	7 08·8	5·7	2·8	11·7	5·8	17·7	8·7
58	7 29·5	7 30·7	7 09·0	5·8	2·9	11·8	5·8	17·8	8·8
59	7 29·8	7 31·0	7 09·3	5·9	2·9	11·9	5·9	17·9	8·8
60	7 30·0	7 31·2	7 09·5	6·0	3·0	12·0	5·9	18·0	8·9

30ᵐ	SUN PLANETS	ARIES	MOON	v or Corrⁿ d	v or Corrⁿ d	v or Corrⁿ d
s	° ′	° ′	° ′	′ ′	′ ′	′ ′
00	7 30·0	7 31·2	7 09·5	0·0 0·0	6·0 3·1	12·0 6·1
01	7 30·3	7 31·5	7 09·7	0·1 0·1	6·1 3·1	12·1 6·2
02	7 30·5	7 31·7	7 10·0	0·2 0·1	6·2 3·2	12·2 6·2
03	7 30·8	7 32·0	7 10·2	0·3 0·2	6·3 3·2	12·3 6·3
04	7 31·0	7 32·2	7 10·5	0·4 0·2	6·4 3·3	12·4 6·3
05	7 31·3	7 32·5	7 10·7	0·5 0·3	6·5 3·3	12·5 6·4
06	7 31·5	7 32·7	7 10·9	0·6 0·3	6·6 3·4	12·6 6·4
07	7 31·8	7 33·0	7 11·2	0·7 0·4	6·7 3·4	12·7 6·5
08	7 32·0	7 33·2	7 11·4	0·8 0·4	6·8 3·5	12·8 6·5
09	7 32·3	7 33·5	7 11·6	0·9 0·5	6·9 3·5	12·9 6·6
10	7 32·5	7 33·7	7 11·9	1·0 0·5	7·0 3·6	13·0 6·6
11	7 32·8	7 34·0	7 12·1	1·1 0·6	7·1 3·6	13·1 6·7
12	7 33·0	7 34·2	7 12·4	1·2 0·6	7·2 3·7	13·2 6·7
13	7 33·3	7 34·5	7 12·6	1·3 0·7	7·3 3·7	13·3 6·8
14	7 33·5	7 34·7	7 12·8	1·4 0·7	7·4 3·8	13·4 6·8
15	7 33·8	7 35·0	7 13·1	1·5 0·8	7·5 3·8	13·5 6·9
16	7 34·0	7 35·2	7 13·3	1·6 0·8	7·6 3·9	13·6 6·9
17	7 34·3	7 35·5	7 13·6	1·7 0·9	7·7 3·9	13·7 7·0
18	7 34·5	7 35·7	7 13·8	1·8 0·9	7·8 4·0	13·8 7·0
19	7 34·8	7 36·0	7 14·0	1·9 1·0	7·9 4·0	13·9 7·1
20	7 35·0	7 36·2	7 14·3	2·0 1·0	8·0 4·1	14·0 7·1
21	7 35·3	7 36·5	7 14·5	2·1 1·1	8·1 4·1	14·1 7·2
22	7 35·5	7 36·7	7 14·7	2·2 1·1	8·2 4·2	14·2 7·2
23	7 35·8	7 37·0	7 15·0	2·3 1·2	8·3 4·2	14·3 7·3
24	7 36·0	7 37·2	7 15·2	2·4 1·2	8·4 4·3	14·4 7·3
25	7 36·3	7 37·5	7 15·5	2·5 1·3	8·5 4·3	14·5 7·4
26	7 36·5	7 37·7	7 15·7	2·6 1·3	8·6 4·4	14·6 7·4
27	7 36·8	7 38·0	7 15·9	2·7 1·4	8·7 4·4	14·7 7·5
28	7 37·0	7 38·3	7 16·2	2·8 1·4	8·8 4·5	14·8 7·5
29	7 37·3	7 38·5	7 16·4	2·9 1·5	8·9 4·5	14·9 7·6
30	7 37·5	7 38·8	7 16·7	3·0 1·5	9·0 4·6	15·0 7·6
31	7 37·8	7 39·0	7 16·9	3·1 1·6	9·1 4·6	15·1 7·7
32	7 38·0	7 39·3	7 17·1	3·2 1·6	9·2 4·7	15·2 7·7
33	7 38·3	7 39·5	7 17·4	3·3 1·7	9·3 4·7	15·3 7·8
34	7 38·5	7 39·8	7 17·6	3·4 1·7	9·4 4·8	15·4 7·8
35	7 38·8	7 40·0	7 17·9	3·5 1·8	9·5 4·8	15·5 7·9
36	7 39·0	7 40·3	7 18·1	3·6 1·8	9·6 4·9	15·6 7·9
37	7 39·3	7 40·5	7 18·3	3·7 1·9	9·7 4·9	15·7 8·0
38	7 39·5	7 40·8	7 18·6	3·8 1·9	9·8 5·0	15·8 8·0
39	7 39·8	7 41·0	7 18·8	3·9 2·0	9·9 5·0	15·9 8·1
40	7 40·0	7 41·3	7 19·0	4·0 2·0	10·0 5·1	16·0 8·1
41	7 40·3	7 41·5	7 19·3	4·1 2·1	10·1 5·1	16·1 8·2
42	7 40·5	7 41·8	7 19·5	4·2 2·1	10·2 5·2	16·2 8·2
43	7 40·8	7 42·0	7 19·8	4·3 2·2	10·3 5·2	16·3 8·3
44	7 41·0	7 42·3	7 20·0	4·4 2·2	10·4 5·3	16·4 8·3
45	7 41·3	7 42·5	7 20·2	4·5 2·3	10·5 5·3	16·5 8·4
46	7 41·5	7 42·8	7 20·5	4·6 2·3	10·6 5·4	16·6 8·4
47	7 41·8	7 43·0	7 20·7	4·7 2·4	10·7 5·4	16·7 8·5
48	7 42·0	7 43·3	7 21·0	4·8 2·4	10·8 5·5	16·8 8·5
49	7 42·3	7 43·5	7 21·2	4·9 2·5	10·9 5·5	16·9 8·6
50	7 42·5	7 43·8	7 21·4	5·0 2·5	11·0 5·6	17·0 8·6
51	7 42·8	7 44·0	7 21·7	5·1 2·6	11·1 5·6	17·1 8·7
52	7 43·0	7 44·3	7 21·9	5·2 2·6	11·2 5·7	17·2 8·7
53	7 43·3	7 44·5	7 22·1	5·3 2·7	11·3 5·7	17·3 8·8
54	7 43·5	7 44·8	7 22·4	5·4 2·7	11·4 5·8	17·4 8·8
55	7 43·8	7 45·0	7 22·6	5·5 2·8	11·5 5·8	17·5 8·9
56	7 44·0	7 45·3	7 22·9	5·6 2·8	11·6 5·9	17·6 8·9
57	7 44·3	7 45·5	7 23·1	5·7 2·9	11·7 5·9	17·7 9·0
58	7 44·5	7 45·8	7 23·3	5·8 2·9	11·8 6·0	17·8 9·0
59	7 44·8	7 46·0	7 23·6	5·9 3·0	11·9 6·0	17·9 9·1
60	7 45·0	7 46·3	7 23·8	6·0 3·1	12·0 6·1	18·0 9·2

31ᵐ	SUN PLANETS	ARIES	MOON	v or Corrⁿ d	v or Corrⁿ d	v or Corrⁿ d
s	° ′	° ′	° ′	′ ′	′ ′	′ ′
00	7 45·0	7 46·3	7 23·8	0·0 0·0	6·0 3·2	12·0 6·3
01	7 45·3	7 46·5	7 24·1	0·1 0·1	6·1 3·2	12·1 6·4
02	7 45·5	7 46·8	7 24·3	0·2 0·1	6·2 3·3	12·2 6·4
03	7 45·8	7 47·0	7 24·5	0·3 0·2	6·3 3·3	12·3 6·5
04	7 46·0	7 47·3	7 24·8	0·4 0·2	6·4 3·4	12·4 6·5
05	7 46·3	7 47·5	7 25·0	0·5 0·3	6·5 3·4	12·5 6·6
06	7 46·5	7 47·8	7 25·2	0·6 0·3	6·6 3·5	12·6 6·6
07	7 46·8	7 48·0	7 25·5	0·7 0·4	6·7 3·5	12·7 6·7
08	7 47·0	7 48·3	7 25·7	0·8 0·4	6·8 3·6	12·8 6·7
09	7 47·3	7 48·5	7 26·0	0·9 0·5	6·9 3·6	12·9 6·8
10	7 47·5	7 48·8	7 26·2	1·0 0·5	7·0 3·7	13·0 6·8
11	7 47·8	7 49·0	7 26·4	1·1 0·6	7·1 3·7	13·1 6·9
12	7 48·0	7 49·3	7 26·7	1·2 0·6	7·2 3·8	13·2 6·9
13	7 48·3	7 49·5	7 26·9	1·3 0·7	7·3 3·8	13·3 7·0
14	7 48·5	7 49·8	7 27·2	1·4 0·7	7·4 3·9	13·4 7·0
15	7 48·8	7 50·0	7 27·4	1·5 0·8	7·5 3·9	13·5 7·1
16	7 49·0	7 50·3	7 27·6	1·6 0·8	7·6 4·0	13·6 7·1
17	7 49·3	7 50·5	7 27·9	1·7 0·9	7·7 4·0	13·7 7·2
18	7 49·5	7 50·8	7 28·1	1·8 0·9	7·8 4·1	13·8 7·2
19	7 49·8	7 51·0	7 28·4	1·9 1·0	7·9 4·1	13·9 7·3
20	7 50·0	7 51·3	7 28·6	2·0 1·1	8·0 4·2	14·0 7·4
21	7 50·3	7 51·5	7 28·8	2·1 1·1	8·1 4·3	14·1 7·4
22	7 50·5	7 51·8	7 29·1	2·2 1·2	8·2 4·3	14·2 7·5
23	7 50·8	7 52·0	7 29·3	2·3 1·2	8·3 4·4	14·3 7·5
24	7 51·0	7 52·3	7 29·5	2·4 1·3	8·4 4·4	14·4 7·6
25	7 51·3	7 52·5	7 29·8	2·5 1·3	8·5 4·5	14·5 7·6
26	7 51·5	7 52·8	7 30·0	2·6 1·4	8·6 4·5	14·6 7·7
27	7 51·8	7 53·0	7 30·3	2·7 1·4	8·7 4·6	14·7 7·7
28	7 52·0	7 53·3	7 30·5	2·8 1·5	8·8 4·6	14·8 7·8
29	7 52·3	7 53·5	7 30·7	2·9 1·5	8·9 4·7	14·9 7·8
30	7 52·5	7 53·8	7 31·0	3·0 1·6	9·0 4·7	15·0 7·9
31	7 52·8	7 54·0	7 31·2	3·1 1·6	9·1 4·8	15·1 7·9
32	7 53·0	7 54·3	7 31·5	3·2 1·7	9·2 4·8	15·2 8·0
33	7 53·3	7 54·5	7 31·7	3·3 1·7	9·3 4·9	15·3 8·0
34	7 53·5	7 54·8	7 31·9	3·4 1·8	9·4 4·9	15·4 8·1
35	7 53·8	7 55·0	7 32·2	3·5 1·8	9·5 5·0	15·5 8·1
36	7 54·0	7 55·3	7 32·4	3·6 1·9	9·6 5·0	15·6 8·2
37	7 54·3	7 55·5	7 32·6	3·7 1·9	9·7 5·1	15·7 8·2
38	7 54·5	7 55·8	7 32·9	3·8 2·0	9·8 5·1	15·8 8·3
39	7 54·8	7 56·0	7 33·1	3·9 2·0	9·9 5·2	15·9 8·3
40	7 55·0	7 56·3	7 33·4	4·0 2·1	10·0 5·3	16·0 8·4
41	7 55·3	7 56·6	7 33·6	4·1 2·2	10·1 5·3	16·1 8·5
42	7 55·5	7 56·8	7 33·8	4·2 2·2	10·2 5·4	16·2 8·5
43	7 55·8	7 57·1	7 34·1	4·3 2·3	10·3 5·4	16·3 8·6
44	7 56·0	7 57·3	7 34·3	4·4 2·3	10·4 5·5	16·4 8·6
45	7 56·3	7 57·6	7 34·6	4·5 2·4	10·5 5·5	16·5 8·7
46	7 56·5	7 57·8	7 34·8	4·6 2·4	10·6 5·6	16·6 8·7
47	7 56·8	7 58·1	7 35·0	4·7 2·5	10·7 5·6	16·7 8·8
48	7 57·0	7 58·3	7 35·3	4·8 2·5	10·8 5·7	16·8 8·8
49	7 57·3	7 58·6	7 35·5	4·9 2·6	10·9 5·7	16·9 8·9
50	7 57·5	7 58·8	7 35·7	5·0 2·6	11·0 5·8	17·0 8·9
51	7 57·8	7 59·1	7 36·0	5·1 2·7	11·1 5·8	17·1 9·0
52	7 58·0	7 59·3	7 36·2	5·2 2·7	11·2 5·9	17·2 9·0
53	7 58·3	7 59·6	7 36·5	5·3 2·8	11·3 5·9	17·3 9·1
54	7 58·5	7 59·8	7 36·7	5·4 2·8	11·4 6·0	17·4 9·1
55	7 58·8	8 00·1	7 36·9	5·5 2·9	11·5 6·0	17·5 9·2
56	7 59·0	8 00·3	7 37·2	5·6 2·9	11·6 6·1	17·6 9·2
57	7 59·3	8 00·6	7 37·4	5·7 3·0	11·7 6·1	17·7 9·3
58	7 59·5	8 00·8	7 37·7	5·8 3·0	11·8 6·2	17·8 9·3
59	7 59·8	8 01·1	7 37·9	5·9 3·1	11·9 6·2	17·9 9·4
60	8 00·0	8 01·3	7 38·1	6·0 3·2	12·0 6·3	18·0 9·5

32ᵐ	SUN PLANETS	ARIES	MOON	v or Corrⁿ d	v or Corrⁿ d	v or Corrⁿ d
s	° ′	° ′	° ′	′ ′	′ ′	′ ′
00	8 00·0	8 01·3	7 38·1	0·0 0·0	6·0 3·3	12·0 6·5
01	8 00·3	8 01·6	7 38·4	0·1 0·1	6·1 3·3	12·1 6·6
02	8 00·5	8 01·8	7 38·6	0·2 0·1	6·2 3·4	12·2 6·6
03	8 00·8	8 02·1	7 38·8	0·3 0·2	6·3 3·4	12·3 6·7
04	8 01·0	8 02·3	7 39·1	0·4 0·2	6·4 3·5	12·4 6·7
05	8 01·3	8 02·6	7 39·3	0·5 0·3	6·5 3·5	12·5 6·8
06	8 01·5	8 02·8	7 39·6	0·6 0·3	6·6 3·6	12·6 6·8
07	8 01·8	8 03·1	7 39·8	0·7 0·4	6·7 3·6	12·7 6·9
08	8 02·0	8 03·3	7 40·0	0·8 0·4	6·8 3·7	12·8 6·9
09	8 02·3	8 03·6	7 40·3	0·9 0·5	6·9 3·7	12·9 7·0
10	8 02·5	8 03·8	7 40·5	1·0 0·5	7·0 3·8	13·0 7·0
11	8 02·8	8 04·1	7 40·8	1·1 0·6	7·1 3·8	13·1 7·1
12	8 03·0	8 04·3	7 41·0	1·2 0·7	7·2 3·9	13·2 7·2
13	8 03·3	8 04·6	7 41·2	1·3 0·7	7·3 4·0	13·3 7·2
14	8 03·5	8 04·8	7 41·5	1·4 0·8	7·4 4·0	13·4 7·3
15	8 03·8	8 05·1	7 41·7	1·5 0·8	7·5 4·1	13·5 7·3
16	8 04·0	8 05·3	7 42·0	1·6 0·9	7·6 4·1	13·6 7·4
17	8 04·3	8 05·6	7 42·2	1·7 0·9	7·7 4·2	13·7 7·4
18	8 04·5	8 05·8	7 42·4	1·8 1·0	7·8 4·2	13·8 7·5
19	8 04·8	8 06·1	7 42·7	1·9 1·0	7·9 4·3	13·9 7·5
20	8 05·0	8 06·3	7 42·9	2·0 1·1	8·0 4·3	14·0 7·6
21	8 05·3	8 06·6	7 43·1	2·1 1·1	8·1 4·4	14·1 7·6
22	8 05·5	8 06·8	7 43·4	2·2 1·2	8·2 4·4	14·2 7·7
23	8 05·8	8 07·1	7 43·6	2·3 1·2	8·3 4·5	14·3 7·7
24	8 06·0	8 07·3	7 43·9	2·4 1·3	8·4 4·6	14·4 7·8
25	8 06·3	8 07·6	7 44·1	2·5 1·4	8·5 4·6	14·5 7·9
26	8 06·5	8 07·8	7 44·3	2·6 1·4	8·6 4·7	14·6 7·9
27	8 06·8	8 08·1	7 44·6	2·7 1·5	8·7 4·7	14·7 8·0
28	8 07·0	8 08·3	7 44·8	2·8 1·5	8·8 4·8	14·8 8·0
29	8 07·3	8 08·6	7 45·1	2·9 1·6	8·9 4·8	14·9 8·1
30	8 07·5	8 08·8	7 45·3	3·0 1·6	9·0 4·9	15·0 8·1
31	8 07·8	8 09·1	7 45·5	3·1 1·7	9·1 4·9	15·1 8·2
32	8 08·0	8 09·3	7 45·8	3·2 1·7	9·2 5·0	15·2 8·2
33	8 08·3	8 09·6	7 46·0	3·3 1·8	9·3 5·0	15·3 8·3
34	8 08·5	8 09·8	7 46·2	3·4 1·8	9·4 5·1	15·4 8·3
35	8 08·8	8 10·1	7 46·5	3·5 1·9	9·5 5·1	15·5 8·4
36	8 09·0	8 10·3	7 46·7	3·6 2·0	9·6 5·2	15·6 8·5
37	8 09·3	8 10·6	7 47·0	3·7 2·0	9·7 5·3	15·7 8·5
38	8 09·5	8 10·8	7 47·2	3·8 2·1	9·8 5·3	15·8 8·6
39	8 09·8	8 11·1	7 47·4	3·9 2·1	9·9 5·4	15·9 8·6
40	8 10·0	8 11·3	7 47·7	4·0 2·2	10·0 5·4	16·0 8·7
41	8 10·3	8 11·6	7 47·9	4·1 2·2	10·1 5·5	16·1 8·7
42	8 10·5	8 11·8	7 48·2	4·2 2·3	10·2 5·5	16·2 8·8
43	8 10·8	8 12·1	7 48·4	4·3 2·3	10·3 5·6	16·3 8·8
44	8 11·0	8 12·3	7 48·6	4·4 2·4	10·4 5·6	16·4 8·9
45	8 11·3	8 12·6	7 48·9	4·5 2·4	10·5 5·7	16·5 8·9
46	8 11·5	8 12·8	7 49·1	4·6 2·5	10·6 5·7	16·6 9·0
47	8 11·8	8 13·1	7 49·3	4·7 2·5	10·7 5·8	16·7 9·0
48	8 12·0	8 13·3	7 49·6	4·8 2·6	10·8 5·9	16·8 9·1
49	8 12·3	8 13·6	7 49·8	4·9 2·7	10·9 5·9	16·9 9·2
50	8 12·5	8 13·8	7 50·1	5·0 2·7	11·0 6·0	17·0 9·2
51	8 12·8	8 14·1	7 50·3	5·1 2·8	11·1 6·0	17·1 9·3
52	8 13·0	8 14·3	7 50·5	5·2 2·8	11·2 6·1	17·2 9·3
53	8 13·3	8 14·6	7 50·8	5·3 2·9	11·3 6·1	17·3 9·4
54	8 13·5	8 14·9	7 51·0	5·4 2·9	11·4 6·2	17·4 9·4
55	8 13·8	8 15·1	7 51·3	5·5 3·0	11·5 6·2	17·5 9·5
56	8 14·0	8 15·4	7 51·5	5·6 3·0	11·6 6·3	17·6 9·5
57	8 14·3	8 15·6	7 51·7	5·7 3·1	11·7 6·3	17·7 9·6
58	8 14·5	8 15·9	7 52·0	5·8 3·1	11·8 6·4	17·8 9·6
59	8 14·8	8 16·1	7 52·2	5·9 3·2	11·9 6·4	17·9 9·7
60	8 15·0	8 16·4	7 52·5	6·0 3·3	12·0 6·5	18·0 9·8

33ᵐ	SUN PLANETS	ARIES	MOON	v or Corrⁿ d	v or Corrⁿ d	v or Corrⁿ d
s	° ′	° ′	° ′	′ ′	′ ′	′ ′
00	8 15·0	8 16·4	7 52·5	0·0 0·0	6·0 3·4	12·0 6·7
01	8 15·3	8 16·6	7 52·7	0·1 0·1	6·1 3·4	12·1 6·8
02	8 15·5	8 16·9	7 52·9	0·2 0·1	6·2 3·5	12·2 6·8
03	8 15·8	8 17·1	7 53·2	0·3 0·2	6·3 3·5	12·3 6·9
04	8 16·0	8 17·4	7 53·4	0·4 0·2	6·4 3·6	12·4 6·9
05	8 16·3	8 17·6	7 53·6	0·5 0·3	6·5 3·6	12·5 7·0
06	8 16·5	8 17·9	7 53·9	0·6 0·3	6·6 3·7	12·6 7·0
07	8 16·8	8 18·1	7 54·1	0·7 0·4	6·7 3·7	12·7 7·1
08	8 17·0	8 18·4	7 54·4	0·8 0·4	6·8 3·8	12·8 7·1
09	8 17·3	8 18·6	7 54·6	0·9 0·5	6·9 3·9	12·9 7·2
10	8 17·5	8 18·9	7 54·8	1·0 0·6	7·0 3·9	13·0 7·3
11	8 17·8	8 19·1	7 55·1	1·1 0·6	7·1 4·0	13·1 7·3
12	8 18·0	8 19·4	7 55·3	1·2 0·7	7·2 4·0	13·2 7·4
13	8 18·3	8 19·6	7 55·6	1·3 0·7	7·3 4·1	13·3 7·4
14	8 18·5	8 19·9	7 55·8	1·4 0·8	7·4 4·1	13·4 7·5
15	8 18·8	8 20·1	7 56·0	1·5 0·8	7·5 4·2	13·5 7·5
16	8 19·0	8 20·4	7 56·3	1·6 0·9	7·6 4·2	13·6 7·6
17	8 19·3	8 20·6	7 56·5	1·7 0·9	7·7 4·3	13·7 7·6
18	8 19·5	8 20·9	7 56·7	1·8 1·0	7·8 4·4	13·8 7·7
19	8 19·8	8 21·1	7 57·0	1·9 1·1	7·9 4·4	13·9 7·8
20	8 20·0	8 21·4	7 57·2	2·0 1·1	8·0 4·5	14·0 7·8
21	8 20·3	8 21·6	7 57·5	2·1 1·2	8·1 4·5	14·1 7·9
22	8 20·5	8 21·9	7 57·7	2·2 1·2	8·2 4·6	14·2 7·9
23	8 20·8	8 22·1	7 57·9	2·3 1·3	8·3 4·6	14·3 8·0
24	8 21·0	8 22·4	7 58·2	2·4 1·3	8·4 4·7	14·4 8·0
25	8 21·3	8 22·6	7 58·4	2·5 1·4	8·5 4·7	14·5 8·1
26	8 21·5	8 22·9	7 58·7	2·6 1·5	8·6 4·8	14·6 8·2
27	8 21·8	8 23·1	7 58·9	2·7 1·5	8·7 4·9	14·7 8·2
28	8 22·0	8 23·4	7 59·1	2·8 1·6	8·8 4·9	14·8 8·3
29	8 22·3	8 23·6	7 59·4	2·9 1·6	8·9 5·0	14·9 8·3
30	8 22·5	8 23·9	7 59·6	3·0 1·7	9·0 5·0	15·0 8·4
31	8 22·8	8 24·1	7 59·8	3·1 1·7	9·1 5·1	15·1 8·4
32	8 23·0	8 24·4	8 00·1	3·2 1·8	9·2 5·1	15·2 8·5
33	8 23·3	8 24·6	8 00·3	3·3 1·8	9·3 5·2	15·3 8·5
34	8 23·5	8 24·9	8 00·6	3·4 1·9	9·4 5·2	15·4 8·6
35	8 23·8	8 25·1	8 00·8	3·5 2·0	9·5 5·3	15·5 8·7
36	8 24·0	8 25·4	8 01·0	3·6 2·0	9·6 5·4	15·6 8·7
37	8 24·3	8 25·6	8 01·3	3·7 2·1	9·7 5·4	15·7 8·8
38	8 24·5	8 25·9	8 01·5	3·8 2·1	9·8 5·5	15·8 8·8
39	8 24·8	8 26·1	8 01·8	3·9 2·2	9·9 5·5	15·9 8·9
40	8 25·0	8 26·4	8 02·0	4·0 2·2	10·0 5·6	16·0 8·9
41	8 25·3	8 26·6	8 02·2	4·1 2·3	10·1 5·6	16·1 9·0
42	8 25·5	8 26·9	8 02·5	4·2 2·3	10·2 5·7	16·2 9·0
43	8 25·8	8 27·1	8 02·7	4·3 2·4	10·3 5·8	16·3 9·1
44	8 26·0	8 27·4	8 02·9	4·4 2·5	10·4 5·8	16·4 9·2
45	8 26·3	8 27·6	8 03·2	4·5 2·5	10·5 5·9	16·5 9·2
46	8 26·5	8 27·9	8 03·4	4·6 2·6	10·6 5·9	16·6 9·3
47	8 26·8	8 28·1	8 03·7	4·7 2·6	10·7 6·0	16·7 9·3
48	8 27·0	8 28·4	8 03·9	4·8 2·7	10·8 6·0	16·8 9·4
49	8 27·3	8 28·6	8 04·1	4·9 2·7	10·9 6·1	16·9 9·4
50	8 27·5	8 28·9	8 04·4	5·0 2·8	11·0 6·1	17·0 9·5
51	8 27·8	8 29·1	8 04·6	5·1 2·8	11·1 6·2	17·1 9·5
52	8 28·0	8 29·4	8 04·9	5·2 2·9	11·2 6·3	17·2 9·6
53	8 28·3	8 29·6	8 05·1	5·3 3·0	11·3 6·3	17·3 9·7
54	8 28·5	8 29·9	8 05·3	5·4 3·0	11·4 6·4	17·4 9·7
55	8 28·8	8 30·1	8 05·6	5·5 3·1	11·5 6·4	17·5 9·8
56	8 29·0	8 30·4	8 05·8	5·6 3·1	11·6 6·5	17·6 9·8
57	8 29·3	8 30·6	8 06·1	5·7 3·2	11·7 6·5	17·7 9·9
58	8 29·5	8 30·9	8 06·3	5·8 3·2	11·8 6·6	17·8 9·9
59	8 29·8	8 31·1	8 06·5	5·9 3·3	11·9 6·6	17·9 10·0
60	8 30·0	8 31·4	8 06·8	6·0 3·4	12·0 6·7	18·0 10·1

34ᵐ	SUN PLANETS	ARIES	MOON	v or Corrⁿ d		v or Corrⁿ d		v or Corrⁿ d	
s	° ′	° ′	° ′	′	′	′	′	′	′
00	8 30·0	8 31·4	8 06·8	0·0	0·0	6·0	3·5	12·0	6·9
01	8 30·3	8 31·6	8 07·0	0·1	0·1	6·1	3·5	12·1	7·0
02	8 30·5	8 31·9	8 07·2	0·2	0·1	6·2	3·6	12·2	7·0
03	8 30·8	8 32·1	8 07·5	0·3	0·2	6·3	3·6	12·3	7·1
04	8 31·0	8 32·4	8 07·7	0·4	0·2	6·4	3·7	12·4	7·1
05	8 31·3	8 32·6	8 08·0	0·5	0·3	6·5	3·7	12·5	7·2
06	8 31·5	8 32·9	8 08·2	0·6	0·3	6·6	3·8	12·6	7·2
07	8 31·8	8 33·2	8 08·4	0·7	0·4	6·7	3·9	12·7	7·3
08	8 32·0	8 33·4	8 08·7	0·8	0·5	6·8	3·9	12·8	7·4
09	8 32·3	8 33·7	8 08·9	0·9	0·5	6·9	4·0	12·9	7·4
10	8 32·5	8 33·9	8 09·2	1·0	0·6	7·0	4·0	13·0	7·5
11	8 32·8	8 34·2	8 09·4	1·1	0·6	7·1	4·1	13·1	7·5
12	8 33·0	8 34·4	8 09·6	1·2	0·7	7·2	4·1	13·2	7·6
13	8 33·3	8 34·7	8 09·9	1·3	0·7	7·3	4·2	13·3	7·6
14	8 33·5	8 34·9	8 10·1	1·4	0·8	7·4	4·3	13·4	7·7
15	8 33·8	8 35·2	8 10·3	1·5	0·9	7·5	4·3	13·5	7·8
16	8 34·0	8 35·4	8 10·6	1·6	0·9	7·6	4·4	13·6	7·8
17	8 34·3	8 35·7	8 10·8	1·7	1·0	7·7	4·4	13·7	7·9
18	8 34·5	8 35·9	8 11·1	1·8	1·0	7·8	4·5	13·8	7·9
19	8 34·8	8 36·2	8 11·3	1·9	1·1	7·9	4·5	13·9	8·0
20	8 35·0	8 36·4	8 11·5	2·0	1·2	8·0	4·6	14·0	8·1
21	8 35·3	8 36·7	8 11·8	2·1	1·2	8·1	4·7	14·1	8·1
22	8 35·5	8 36·9	8 12·0	2·2	1·3	8·2	4·7	14·2	8·2
23	8 35·8	8 37·2	8 12·3	2·3	1·3	8·3	4·8	14·3	8·2
24	8 36·0	8 37·4	8 12·5	2·4	1·4	8·4	4·8	14·4	8·3
25	8 36·3	8 37·7	8 12·7	2·5	1·4	8·5	4·9	14·5	8·3
26	8 36·5	8 37·9	8 13·0	2·6	1·5	8·6	4·9	14·6	8·4
27	8 36·8	8 38·2	8 13·2	2·7	1·6	8·7	5·0	14·7	8·5
28	8 37·0	8 38·4	8 13·4	2·8	1·6	8·8	5·1	14·8	8·5
29	8 37·3	8 38·7	8 13·7	2·9	1·7	8·9	5·1	14·9	8·6
30	8 37·5	8 38·9	8 13·9	3·0	1·7	9·0	5·2	15·0	8·6
31	8 37·8	8 39·2	8 14·2	3·1	1·8	9·1	5·2	15·1	8·7
32	8 38·0	8 39·4	8 14·4	3·2	1·8	9·2	5·3	15·2	8·7
33	8 38·3	8 39·7	8 14·6	3·3	1·9	9·3	5·3	15·3	8·8
34	8 38·5	8 39·9	8 14·9	3·4	2·0	9·4	5·4	15·4	8·9
35	8 38·8	8 40·2	8 15·1	3·5	2·0	9·5	5·5	15·5	8·9
36	8 39·0	8 40·4	8 15·4	3·6	2·1	9·6	5·5	15·6	9·0
37	8 39·3	8 40·7	8 15·6	3·7	2·1	9·7	5·6	15·7	9·0
38	8 39·5	8 40·9	8 15·8	3·8	2·2	9·8	5·6	15·8	9·1
39	8 39·8	8 41·2	8 16·1	3·9	2·2	9·9	5·7	15·9	9·1
40	8 40·0	8 41·4	8 16·3	4·0	2·3	10·0	5·8	16·0	9·2
41	8 40·3	8 41·7	8 16·5	4·1	2·4	10·1	5·8	16·1	9·3
42	8 40·5	8 41·9	8 16·8	4·2	2·4	10·2	5·9	16·2	9·3
43	8 40·8	8 42·2	8 17·0	4·3	2·5	10·3	5·9	16·3	9·4
44	8 41·0	8 42·4	8 17·3	4·4	2·5	10·4	6·0	16·4	9·4
45	8 41·3	8 42·7	8 17·5	4·5	2·6	10·5	6·0	16·5	9·5
46	8 41·5	8 42·9	8 17·7	4·6	2·6	10·6	6·1	16·6	9·5
47	8 41·8	8 43·2	8 18·0	4·7	2·7	10·7	6·2	16·7	9·6
48	8 42·0	8 43·4	8 18·2	4·8	2·8	10·8	6·2	16·8	9·7
49	8 42·3	8 43·7	8 18·5	4·9	2·8	10·9	6·3	16·9	9·7
50	8 42·5	8 43·9	8 18·7	5·0	2·9	11·0	6·3	17·0	9·8
51	8 42·8	8 44·2	8 18·9	5·1	2·9	11·1	6·4	17·1	9·8
52	8 43·0	8 44·4	8 19·2	5·2	3·0	11·2	6·4	17·2	9·9
53	8 43·3	8 44·7	8 19·4	5·3	3·0	11·3	6·5	17·3	9·9
54	8 43·5	8 44·9	8 19·7	5·4	3·1	11·4	6·6	17·4	10·0
55	8 43·8	8 45·2	8 19·9	5·5	3·2	11·5	6·6	17·5	10·1
56	8 44·0	8 45·4	8 20·1	5·6	3·2	11·6	6·7	17·6	10·1
57	8 44·3	8 45·7	8 20·4	5·7	3·3	11·7	6·7	17·7	10·2
58	8 44·5	8 45·9	8 20·6	5·8	3·3	11·8	6·8	17·8	10·2
59	8 44·8	8 46·2	8 20·8	5·9	3·4	11·9	6·8	17·9	10·3
60	8 45·0	8 46·4	8 21·1	6·0	3·5	12·0	6·9	18·0	10·4

35ᵐ	SUN PLANETS	ARIES	MOON	v or Corrⁿ d		v or Corrⁿ d		v or Corrⁿ d	
s	° ′	° ′	° ′	′	′	′	′	′	′
00	8 45·0	8 46·4	8 21·1	0·0	0·0	6·0	3·6	12·0	7·1
01	8 45·3	8 46·7	8 21·3	0·1	0·1	6·1	3·6	12·1	7·2
02	8 45·5	8 46·9	8 21·6	0·2	0·1	6·2	3·7	12·2	7·2
03	8 45·8	8 47·2	8 21·8	0·3	0·2	6·3	3·7	12·3	7·3
04	8 46·0	8 47·4	8 22·0	0·4	0·2	6·4	3·8	12·4	7·3
05	8 46·3	8 47·7	8 22·3	0·5	0·3	6·5	3·8	12·5	7·4
06	8 46·5	8 47·9	8 22·5	0·6	0·4	6·6	3·9	12·6	7·5
07	8 46·8	8 48·2	8 22·8	0·7	0·4	6·7	4·0	12·7	7·5
08	8 47·0	8 48·4	8 23·0	0·8	0·5	6·8	4·0	12·8	7·6
09	8 47·3	8 48·7	8 23·2	0·9	0·5	6·9	4·1	12·9	7·6
10	8 47·5	8 48·9	8 23·5	1·0	0·6	7·0	4·1	13·0	7·7
11	8 47·8	8 49·2	8 23·7	1·1	0·7	7·1	4·2	13·1	7·8
12	8 48·0	8 49·4	8 23·9	1·2	0·7	7·2	4·3	13·2	7·8
13	8 48·3	8 49·7	8 24·2	1·3	0·8	7·3	4·3	13·3	7·9
14	8 48·5	8 49·9	8 24·4	1·4	0·8	7·4	4·4	13·4	7·9
15	8 48·8	8 50·2	8 24·7	1·5	0·9	7·5	4·4	13·5	8·0
16	8 49·0	8 50·4	8 24·9	1·6	0·9	7·6	4·5	13·6	8·0
17	8 49·3	8 50·7	8 25·1	1·7	1·0	7·7	4·6	13·7	8·1
18	8 49·5	8 50·9	8 25·4	1·8	1·1	7·8	4·6	13·8	8·2
19	8 49·8	8 51·2	8 25·6	1·9	1·1	7·9	4·7	13·9	8·2
20	8 50·0	8 51·5	8 25·9	2·0	1·2	8·0	4·7	14·0	8·3
21	8 50·3	8 51·7	8 26·1	2·1	1·2	8·1	4·8	14·1	8·3
22	8 50·5	8 52·0	8 26·3	2·2	1·3	8·2	4·9	14·2	8·4
23	8 50·8	8 52·2	8 26·6	2·3	1·4	8·3	4·9	14·3	8·5
24	8 51·0	8 52·5	8 26·8	2·4	1·4	8·4	5·0	14·4	8·5
25	8 51·3	8 52·7	8 27·0	2·5	1·5	8·5	5·0	14·5	8·6
26	8 51·5	8 53·0	8 27·3	2·6	1·5	8·6	5·1	14·6	8·6
27	8 51·8	8 53·2	8 27·5	2·7	1·6	8·7	5·1	14·7	8·7
28	8 52·0	8 53·5	8 27·8	2·8	1·7	8·8	5·2	14·8	8·8
29	8 52·3	8 53·7	8 28·0	2·9	1·7	8·9	5·3	14·9	8·8
30	8 52·5	8 54·0	8 28·2	3·0	1·8	9·0	5·3	15·0	8·9
31	8 52·8	8 54·2	8 28·5	3·1	1·8	9·1	5·4	15·1	8·9
32	8 53·0	8 54·5	8 28·7	3·2	1·9	9·2	5·4	15·2	9·0
33	8 53·3	8 54·7	8 29·0	3·3	2·0	9·3	5·5	15·3	9·1
34	8 53·5	8 55·0	8 29·2	3·4	2·0	9·4	5·6	15·4	9·1
35	8 53·8	8 55·2	8 29·4	3·5	2·1	9·5	5·6	15·5	9·2
36	8 54·0	8 55·5	8 29·7	3·6	2·1	9·6	5·7	15·6	9·2
37	8 54·3	8 55·7	8 29·9	3·7	2·2	9·7	5·7	15·7	9·3
38	8 54·5	8 56·0	8 30·2	3·8	2·2	9·8	5·8	15·8	9·3
39	8 54·8	8 56·2	8 30·4	3·9	2·3	9·9	5·9	15·9	9·4
40	8 55·0	8 56·5	8 30·6	4·0	2·4	10·0	5·9	16·0	9·5
41	8 55·3	8 56·7	8 30·9	4·1	2·4	10·1	6·0	16·1	9·5
42	8 55·5	8 57·0	8 31·1	4·2	2·5	10·2	6·0	16·2	9·6
43	8 55·8	8 57·2	8 31·3	4·3	2·5	10·3	6·1	16·3	9·6
44	8 56·0	8 57·5	8 31·6	4·4	2·6	10·4	6·2	16·4	9·7
45	8 56·3	8 57·7	8 31·8	4·5	2·7	10·5	6·2	16·5	9·8
46	8 56·5	8 58·0	8 32·1	4·6	2·7	10·6	6·3	16·6	9·8
47	8 56·8	8 58·2	8 32·3	4·7	2·8	10·7	6·3	16·7	9·9
48	8 57·0	8 58·5	8 32·5	4·8	2·8	10·8	6·4	16·8	9·9
49	8 57·3	8 58·7	8 32·8	4·9	2·9	10·9	6·4	16·9	10·0
50	8 57·5	8 59·0	8 33·0	5·0	3·0	11·0	6·5	17·0	10·1
51	8 57·8	8 59·2	8 33·3	5·1	3·0	11·1	6·6	17·1	10·1
52	8 58·0	8 59·5	8 33·5	5·2	3·1	11·2	6·6	17·2	10·2
53	8 58·3	8 59·7	8 33·7	5·3	3·1	11·3	6·7	17·3	10·2
54	8 58·5	9 00·0	8 34·0	5·4	3·2	11·4	6·7	17·4	10·3
55	8 58·8	9 00·2	8 34·2	5·5	3·3	11·5	6·8	17·5	10·4
56	8 59·0	9 00·5	8 34·4	5·6	3·3	11·6	6·9	17·6	10·4
57	8 59·3	9 00·7	8 34·7	5·7	3·4	11·7	6·9	17·7	10·5
58	8 59·5	9 01·0	8 34·9	5·8	3·4	11·8	7·0	17·8	10·5
59	8 59·8	9 01·2	8 35·2	5·9	3·5	11·9	7·0	17·9	10·6
60	9 00·0	9 01·5	8 35·4	6·0	3·6	12·0	7·1	18·0	10·7

36ᵐ	SUN PLANETS	ARIES	MOON	v or Corrⁿ d	v or Corrⁿ d	v or Corrⁿ d
s	° ′	° ′	° ′	′ ′	′ ′	′ ′
00	9 00·0	9 01·5	8 35·4	0·0 0·0	6·0 3·7	12·0 7·3
01	9 00·3	9 01·7	8 35·6	0·1 0·1	6·1 3·7	12·1 7·4
02	9 00·5	9 02·0	8 35·9	0·2 0·1	6·2 3·8	12·2 7·4
03	9 00·8	9 02·2	8 36·1	0·3 0·2	6·3 3·8	12·3 7·5
04	9 01·0	9 02·5	8 36·4	0·4 0·2	6·4 3·9	12·4 7·5
05	9 01·3	9 02·7	8 36·6	0·5 0·3	6·5 4·0	12·5 7·6
06	9 01·5	9 03·0	8 36·8	0·6 0·4	6·6 4·0	12·6 7·7
07	9 01·8	9 03·2	8 37·1	0·7 0·4	6·7 4·1	12·7 7·7
08	9 02·0	9 03·5	8 37·3	0·8 0·5	6·8 4·1	12·8 7·8
09	9 02·3	9 03·7	8 37·5	0·9 0·5	6·9 4·2	12·9 7·8
10	9 02·5	9 04·0	8 37·8	1·0 0·6	7·0 4·3	13·0 7·9
11	9 02·8	9 04·2	8 38·0	1·1 0·7	7·1 4·3	13·1 8·0
12	9 03·0	9 04·5	8 38·3	1·2 0·7	7·2 4·4	13·2 8·0
13	9 03·3	9 04·7	8 38·5	1·3 0·8	7·3 4·4	13·3 8·1
14	9 03·5	9 05·0	8 38·7	1·4 0·9	7·4 4·5	13·4 8·2
15	9 03·8	9 05·2	8 39·0	1·5 0·9	7·5 4·6	13·5 8·2
16	9 04·0	9 05·5	8 39·2	1·6 1·0	7·6 4·6	13·6 8·3
17	9 04·3	9 05·7	8 39·5	1·7 1·0	7·7 4·7	13·7 8·3
18	9 04·5	9 06·0	8 39·7	1·8 1·1	7·8 4·7	13·8 8·4
19	9 04·8	9 06·2	8 39·9	1·9 1·2	7·9 4·8	13·9 8·5
20	9 05·0	9 06·5	8 40·2	2·0 1·2	8·0 4·9	14·0 8·5
21	9 05·3	9 06·7	8 40·4	2·1 1·3	8·1 4·9	14·1 8·6
22	9 05·5	9 07·0	8 40·6	2·2 1·3	8·2 5·0	14·2 8·6
23	9 05·8	9 07·2	8 40·9	2·3 1·4	8·3 5·0	14·3 8·7
24	9 06·0	9 07·5	8 41·1	2·4 1·5	8·4 5·1	14·4 8·8
25	9 06·3	9 07·7	8 41·4	2·5 1·5	8·5 5·2	14·5 8·8
26	9 06·5	9 08·0	8 41·6	2·6 1·6	8·6 5·2	14·6 8·9
27	9 06·8	9 08·2	8 41·8	2·7 1·6	8·7 5·3	14·7 8·9
28	9 07·0	9 08·5	8 42·1	2·8 1·7	8·8 5·4	14·8 9·0
29	9 07·3	9 08·7	8 42·3	2·9 1·8	8·9 5·4	14·9 9·1
30	9 07·5	9 09·0	8 42·6	3·0 1·8	9·0 5·5	15·0 9·1
31	9 07·8	9 09·2	8 42·8	3·1 1·9	9·1 5·5	15·1 9·2
32	9 08·0	9 09·5	8 43·0	3·2 1·9	9·2 5·6	15·2 9·2
33	9 08·3	9 09·8	8 43·3	3·3 2·0	9·3 5·7	15·3 9·3
34	9 08·5	9 10·0	8 43·5	3·4 2·1	9·4 5·7	15·4 9·4
35	9 08·8	9 10·3	8 43·8	3·5 2·1	9·5 5·8	15·5 9·4
36	9 09·0	9 10·5	8 44·0	3·6 2·2	9·6 5·8	15·6 9·5
37	9 09·3	9 10·8	8 44·2	3·7 2·3	9·7 5·9	15·7 9·6
38	9 09·5	9 11·0	8 44·5	3·8 2·3	9·8 6·0	15·8 9·6
39	9 09·8	9 11·3	8 44·7	3·9 2·4	9·9 6·0	15·9 9·7
40	9 10·0	9 11·5	8 44·9	4·0 2·4	10·0 6·1	16·0 9·7
41	9 10·3	9 11·8	8 45·2	4·1 2·5	10·1 6·1	16·1 9·8
42	9 10·5	9 12·0	8 45·4	4·2 2·6	10·2 6·2	16·2 9·9
43	9 10·8	9 12·3	8 45·7	4·3 2·6	10·3 6·3	16·3 9·9
44	9 11·0	9 12·5	8 45·9	4·4 2·7	10·4 6·3	16·4 10·0
45	9 11·3	9 12·8	8 46·1	4·5 2·7	10·5 6·4	16·5 10·0
46	9 11·5	9 13·0	8 46·4	4·6 2·8	10·6 6·4	16·6 10·1
47	9 11·8	9 13·3	8 46·6	4·7 2·9	10·7 6·5	16·7 10·2
48	9 12·0	9 13·5	8 46·9	4·8 2·9	10·8 6·6	16·8 10·2
49	9 12·3	9 13·8	8 47·1	4·9 3·0	10·9 6·6	16·9 10·3
50	9 12·5	9 14·0	8 47·3	5·0 3·0	11·0 6·7	17·0 10·3
51	9 12·8	9 14·3	8 47·6	5·1 3·1	11·1 6·8	17·1 10·4
52	9 13·0	9 14·5	8 47·8	5·2 3·2	11·2 6·8	17·2 10·5
53	9 13·3	9 14·8	8 48·0	5·3 3·2	11·3 6·9	17·3 10·5
54	9 13·5	9 15·0	8 48·3	5·4 3·3	11·4 6·9	17·4 10·6
55	9 13·8	9 15·3	8 48·5	5·5 3·3	11·5 7·0	17·5 10·6
56	9 14·0	9 15·5	8 48·8	5·6 3·4	11·6 7·1	17·6 10·7
57	9 14·3	9 15·8	8 49·0	5·7 3·5	11·7 7·1	17·7 10·8
58	9 14·5	9 16·0	8 49·2	5·8 3·5	11·8 7·2	17·8 10·8
59	9 14·8	9 16·3	8 49·5	5·9 3·6	11·9 7·2	17·9 10·9
60	9 15·0	9 16·5	8 49·7	6·0 3·7	12·0 7·3	18·0 11·0

37ᵐ	SUN PLANETS	ARIES	MOON	v or Corrⁿ d	v or Corrⁿ d	v or Corrⁿ d
s	° ′	° ′	° ′	′ ′	′ ′	′ ′
00	9 15·0	9 16·5	8 49·7	0·0 0·0	6·0 3·8	12·0 7·5
01	9 15·3	9 16·8	8 50·0	0·1 0·1	6·1 3·8	12·1 7·6
02	9 15·5	9 17·0	8 50·2	0·2 0·1	6·2 3·9	12·2 7·6
03	9 15·8	9 17·3	8 50·4	0·3 0·2	6·3 3·9	12·3 7·7
04	9 16·0	9 17·5	8 50·7	0·4 0·3	6·4 4·0	12·4 7·8
05	9 16·3	9 17·8	8 50·9	0·5 0·3	6·5 4·1	12·5 7·8
06	9 16·5	9 18·0	8 51·1	0·6 0·4	6·6 4·1	12·6 7·9
07	9 16·8	9 18·3	8 51·4	0·7 0·4	6·7 4·2	12·7 7·9
08	9 17·0	9 18·5	8 51·6	0·8 0·5	6·8 4·3	12·8 8·0
09	9 17·3	9 18·8	8 51·9	0·9 0·6	6·9 4·3	12·9 8·1
10	9 17·5	9 19·0	8 52·1	1·0 0·6	7·0 4·4	13·0 8·1
11	9 17·8	9 19·3	8 52·3	1·1 0·7	7·1 4·4	13·1 8·2
12	9 18·0	9 19·5	8 52·6	1·2 0·8	7·2 4·5	13·2 8·3
13	9 18·3	9 19·8	8 52·8	1·3 0·8	7·3 4·6	13·3 8·3
14	9 18·5	9 20·0	8 53·1	1·4 0·9	7·4 4·6	13·4 8·4
15	9 18·8	9 20·3	8 53·3	1·5 0·9	7·5 4·7	13·5 8·4
16	9 19·0	9 20·5	8 53·5	1·6 1·0	7·6 4·8	13·6 8·5
17	9 19·3	9 20·8	8 53·8	1·7 1·1	7·7 4·8	13·7 8·6
18	9 19·5	9 21·0	8 54·0	1·8 1·1	7·8 4·9	13·8 8·6
19	9 19·8	9 21·3	8 54·3	1·9 1·2	7·9 4·9	13·9 8·7
20	9 20·0	9 21·5	8 54·5	2·0 1·3	8·0 5·0	14·0 8·8
21	9 20·3	9 21·8	8 54·7	2·1 1·3	8·1 5·1	14·1 8·8
22	9 20·5	9 22·0	8 55·0	2·2 1·4	8·2 5·1	14·2 8·9
23	9 20·8	9 22·3	8 55·2	2·3 1·4	8·3 5·2	14·3 8·9
24	9 21·0	9 22·5	8 55·4	2·4 1·5	8·4 5·3	14·4 9·0
25	9 21·3	9 22·8	8 55·7	2·5 1·6	8·5 5·3	14·5 9·1
26	9 21·5	9 23·0	8 55·9	2·6 1·6	8·6 5·4	14·6 9·1
27	9 21·8	9 23·3	8 56·2	2·7 1·7	8·7 5·4	14·7 9·2
28	9 22·0	9 23·5	8 56·4	2·8 1·8	8·8 5·5	14·8 9·3
29	9 22·3	9 23·8	8 56·6	2·9 1·8	8·9 5·6	14·9 9·3
30	9 22·5	9 24·0	8 56·9	3·0 1·9	9·0 5·6	15·0 9·4
31	9 22·8	9 24·3	8 57·1	3·1 1·9	9·1 5·7	15·1 9·4
32	9 23·0	9 24·5	8 57·4	3·2 2·0	9·2 5·8	15·2 9·5
33	9 23·3	9 24·8	8 57·6	3·3 2·1	9·3 5·8	15·3 9·6
34	9 23·5	9 25·0	8 57·8	3·4 2·1	9·4 5·9	15·4 9·6
35	9 23·8	9 25·3	8 58·1	3·5 2·2	9·5 5·9	15·5 9·7
36	9 24·0	9 25·5	8 58·3	3·6 2·3	9·6 6·0	15·6 9·8
37	9 24·3	9 25·8	8 58·5	3·7 2·3	9·7 6·1	15·7 9·8
38	9 24·5	9 26·0	8 58·8	3·8 2·4	9·8 6·1	15·8 9·9
39	9 24·8	9 26·3	8 59·0	3·9 2·4	9·9 6·2	15·9 9·9
40	9 25·0	9 26·5	8 59·3	4·0 2·5	10·0 6·3	16·0 10·0
41	9 25·3	9 26·8	8 59·5	4·1 2·6	10·1 6·3	16·1 10·1
42	9 25·5	9 27·0	8 59·7	4·2 2·6	10·2 6·4	16·2 10·1
43	9 25·8	9 27·3	9 00·0	4·3 2·7	10·3 6·4	16·3 10·2
44	9 26·0	9 27·5	9 00·2	4·4 2·8	10·4 6·5	16·4 10·3
45	9 26·3	9 27·8	9 00·5	4·5 2·8	10·5 6·6	16·5 10·3
46	9 26·5	9 28·1	9 00·7	4·6 2·9	10·6 6·6	16·6 10·4
47	9 26·8	9 28·3	9 00·9	4·7 2·9	10·7 6·7	16·7 10·4
48	9 27·0	9 28·6	9 01·2	4·8 3·0	10·8 6·8	16·8 10·5
49	9 27·3	9 28·8	9 01·4	4·9 3·1	10·9 6·8	16·9 10·6
50	9 27·5	9 29·1	9 01·6	5·0 3·1	11·0 6·9	17·0 10·6
51	9 27·8	9 29·3	9 01·9	5·1 3·2	11·1 6·9	17·1 10·7
52	9 28·0	9 29·6	9 02·1	5·2 3·3	11·2 7·0	17·2 10·8
53	9 28·3	9 29·8	9 02·4	5·3 3·3	11·3 7·1	17·3 10·8
54	9 28·5	9 30·1	9 02·6	5·4 3·4	11·4 7·1	17·4 10·9
55	9 28·8	9 30·3	9 02·8	5·5 3·4	11·5 7·2	17·5 10·9
56	9 29·0	9 30·6	9 03·1	5·6 3·5	11·6 7·3	17·6 11·0
57	9 29·3	9 30·8	9 03·3	5·7 3·6	11·7 7·3	17·7 11·1
58	9 29·5	9 31·1	9 03·6	5·8 3·6	11·8 7·4	17·8 11·1
59	9 29·8	9 31·3	9 03·8	5·9 3·7	11·9 7·4	17·9 11·2
60	9 30·0	9 31·6	9 04·0	6·0 3·8	12·0 7·5	18·0 11·3

38ᵐ

38ᵐ s	SUN PLANETS	ARIES	MOON	v or Corrn d		v or Corrn d		v or Corrn d	
	° ′	° ′	° ′	′	′	′	′	′	′
00	9 30·0	9 31·6	9 04·0	0·0	0·0	6·0	3·9	12·0	7·7
01	9 30·3	9 31·8	9 04·3	0·1	0·1	6·1	3·9	12·1	7·8
02	9 30·5	9 32·1	9 04·5	0·2	0·1	6·2	4·0	12·2	7·8
03	9 30·8	9 32·3	9 04·7	0·3	0·2	6·3	4·0	12·3	7·9
04	9 31·0	9 32·6	9 05·0	0·4	0·3	6·4	4·1	12·4	8·0
05	9 31·3	9 32·8	9 05·2	0·5	0·3	6·5	4·2	12·5	8·0
06	9 31·5	9 33·1	9 05·5	0·6	0·4	6·6	4·2	12·6	8·1
07	9 31·8	9 33·3	9 05·7	0·7	0·4	6·7	4·3	12·7	8·1
08	9 32·0	9 33·6	9 05·9	0·8	0·5	6·8	4·4	12·8	8·2
09	9 32·3	9 33·8	9 06·2	0·9	0·6	6·9	4·4	12·9	8·3
10	9 32·5	9 34·1	9 06·4	1·0	0·6	7·0	4·5	13·0	8·3
11	9 32·8	9 34·3	9 06·7	1·1	0·7	7·1	4·6	13·1	8·4
12	9 33·0	9 34·6	9 06·9	1·2	0·8	7·2	4·6	13·2	8·5
13	9 33·3	9 34·8	9 07·1	1·3	0·8	7·3	4·7	13·3	8·5
14	9 33·5	9 35·1	9 07·4	1·4	0·9	7·4	4·7	13·4	8·6
15	9 33·8	9 35·3	9 07·6	1·5	1·0	7·5	4·8	13·5	8·7
16	9 34·0	9 35·6	9 07·9	1·6	1·0	7·6	4·9	13·6	8·7
17	9 34·3	9 35·8	9 08·1	1·7	1·1	7·7	4·9	13·7	8·8
18	9 34·5	9 36·1	9 08·3	1·8	1·2	7·8	5·0	13·8	8·9
19	9 34·8	9 36·3	9 08·6	1·9	1·2	7·9	5·1	13·9	8·9
20	9 35·0	9 36·6	9 08·8	2·0	1·3	8·0	5·1	14·0	9·0
21	9 35·3	9 36·8	9 09·0	2·1	1·3	8·1	5·2	14·1	9·0
22	9 35·5	9 37·1	9 09·3	2·2	1·4	8·2	5·3	14·2	9·1
23	9 35·8	9 37·3	9 09·5	2·3	1·5	8·3	5·3	14·3	9·2
24	9 36·0	9 37·6	9 09·8	2·4	1·5	8·4	5·4	14·4	9·2
25	9 36·3	9 37·8	9 10·0	2·5	1·6	8·5	5·5	14·5	9·3
26	9 36·5	9 38·1	9 10·2	2·6	1·7	8·6	5·5	14·6	9·4
27	9 36·8	9 38·3	9 10·5	2·7	1·7	8·7	5·6	14·7	9·4
28	9 37·0	9 38·6	9 10·7	2·8	1·8	8·8	5·6	14·8	9·5
29	9 37·3	9 38·8	9 11·0	2·9	1·9	8·9	5·7	14·9	9·6
30	9 37·5	9 39·1	9 11·2	3·0	1·9	9·0	5·8	15·0	9·6
31	9 37·8	9 39·3	9 11·4	3·1	2·0	9·1	5·8	15·1	9·7
32	9 38·0	9 39·6	9 11·7	3·2	2·1	9·2	5·9	15·2	9·8
33	9 38·3	9 39·8	9 11·9	3·3	2·1	9·3	6·0	15·3	9·8
34	9 38·5	9 40·1	9 12·1	3·4	2·2	9·4	6·0	15·4	9·9
35	9 38·8	9 40·3	9 12·4	3·5	2·2	9·5	6·1	15·5	9·9
36	9 39·0	9 40·6	9 12·6	3·6	2·3	9·6	6·2	15·6	10·0
37	9 39·3	9 40·8	9 12·9	3·7	2·4	9·7	6·2	15·7	10·1
38	9 39·5	9 41·1	9 13·1	3·8	2·4	9·8	6·3	15·8	10·1
39	9 39·8	9 41·3	9 13·3	3·9	2·5	9·9	6·4	15·9	10·2
40	9 40·0	9 41·6	9 13·6	4·0	2·6	10·0	6·4	16·0	10·3
41	9 40·3	9 41·8	9 13·8	4·1	2·6	10·1	6·5	16·1	10·3
42	9 40·5	9 42·1	9 14·1	4·2	2·7	10·2	6·5	16·2	10·4
43	9 40·8	9 42·3	9 14·3	4·3	2·8	10·3	6·6	16·3	10·5
44	9 41·0	9 42·6	9 14·5	4·4	2·8	10·4	6·7	16·4	10·5
45	9 41·3	9 42·8	9 14·8	4·5	2·9	10·5	6·7	16·5	10·6
46	9 41·5	9 43·1	9 15·0	4·6	3·0	10·6	6·8	16·6	10·7
47	9 41·8	9 43·3	9 15·2	4·7	3·0	10·7	6·9	16·7	10·7
48	9 42·0	9 43·6	9 15·5	4·8	3·1	10·8	6·9	16·8	10·8
49	9 42·3	9 43·8	9 15·7	4·9	3·1	10·9	7·0	16·9	10·8
50	9 42·5	9 44·1	9 16·0	5·0	3·2	11·0	7·1	17·0	10·9
51	9 42·8	9 44·3	9 16·2	5·1	3·3	11·1	7·1	17·1	11·0
52	9 43·0	9 44·6	9 16·4	5·2	3·3	11·2	7·2	17·2	11·0
53	9 43·3	9 44·8	9 16·7	5·3	3·4	11·3	7·3	17·3	11·1
54	9 43·5	9 45·1	9 16·9	5·4	3·5	11·4	7·3	17·4	11·2
55	9 43·8	9 45·3	9 17·2	5·5	3·5	11·5	7·4	17·5	11·2
56	9 44·0	9 45·6	9 17·4	5·6	3·6	11·6	7·4	17·6	11·3
57	9 44·3	9 45·8	9 17·6	5·7	3·7	11·7	7·5	17·7	11·4
58	9 44·5	9 46·1	9 17·9	5·8	3·7	11·8	7·6	17·8	11·4
59	9 44·8	9 46·4	9 18·1	5·9	3·8	11·9	7·6	17·9	11·5
60	9 45·0	9 46·6	9 18·4	6·0	3·9	12·0	7·7	18·0	11·6

39ᵐ

39ᵐ s	SUN PLANETS	ARIES	MOON	v or Corrn d		v or Corrn d		v or Corrn d	
	° ′	° ′	° ′	′	′	′	′	′	′
00	9 45·0	9 46·6	9 18·4	0·0	0·0	6·0	4·0	12·0	7·9
01	9 45·3	9 46·9	9 18·6	0·1	0·1	6·1	4·0	12·1	8·0
02	9 45·5	9 47·1	9 18·8	0·2	0·1	6·2	4·1	12·2	8·0
03	9 45·8	9 47·4	9 19·1	0·3	0·2	6·3	4·1	12·3	8·1
04	9 46·0	9 47·6	9 19·3	0·4	0·3	6·4	4·2	12·4	8·2
05	9 46·3	9 47·9	9 19·5	0·5	0·3	6·5	4·3	12·5	8·2
06	9 46·5	9 48·1	9 19·8	0·6	0·4	6·6	4·3	12·6	8·3
07	9 46·8	9 48·4	9 20·0	0·7	0·5	6·7	4·4	12·7	8·4
08	9 47·0	9 48·6	9 20·3	0·8	0·5	6·8	4·5	12·8	8·4
09	9 47·3	9 48·9	9 20·5	0·9	0·6	6·9	4·5	12·9	8·5
10	9 47·5	9 49·1	9 20·7	1·0	0·7	7·0	4·6	13·0	8·6
11	9 47·8	9 49·4	9 21·0	1·1	0·7	7·1	4·7	13·1	8·6
12	9 48·0	9 49·6	9 21·2	1·2	0·8	7·2	4·7	13·2	8·7
13	9 48·3	9 49·9	9 21·5	1·3	0·9	7·3	4·8	13·3	8·8
14	9 48·5	9 50·1	9 21·7	1·4	0·9	7·4	4·9	13·4	8·8
15	9 48·8	9 50·4	9 21·9	1·5	1·0	7·5	4·9	13·5	8·9
16	9 49·0	9 50·6	9 22·2	1·6	1·1	7·6	5·0	13·6	9·0
17	9 49·3	9 50·9	9 22·4	1·7	1·1	7·7	5·1	13·7	9·0
18	9 49·5	9 51·1	9 22·6	1·8	1·2	7·8	5·1	13·8	9·1
19	9 49·8	9 51·4	9 22·9	1·9	1·3	7·9	5·2	13·9	9·2
20	9 50·0	9 51·6	9 23·1	2·0	1·3	8·0	5·3	14·0	9·2
21	9 50·3	9 51·9	9 23·4	2·1	1·4	8·1	5·3	14·1	9·3
22	9 50·5	9 52·1	9 23·6	2·2	1·4	8·2	5·4	14·2	9·3
23	9 50·8	9 52·4	9 23·8	2·3	1·5	8·3	5·5	14·3	9·4
24	9 51·0	9 52·6	9 24·1	2·4	1·6	8·4	5·5	14·4	9·5
25	9 51·3	9 52·9	9 24·3	2·5	1·6	8·5	5·6	14·5	9·5
26	9 51·5	9 53·1	9 24·6	2·6	1·7	8·6	5·7	14·6	9·6
27	9 51·8	9 53·4	9 24·8	2·7	1·8	8·7	5·7	14·7	9·7
28	9 52·0	9 53·6	9 25·0	2·8	1·8	8·8	5·8	14·8	9·7
29	9 52·3	9 53·9	9 25·3	2·9	1·9	8·9	5·9	14·9	9·8
30	9 52·5	9 54·1	9 25·5	3·0	2·0	9·0	5·9	15·0	9·9
31	9 52·8	9 54·4	9 25·7	3·1	2·0	9·1	6·0	15·1	9·9
32	9 53·0	9 54·6	9 26·0	3·2	2·1	9·2	6·1	15·2	10·0
33	9 53·3	9 54·9	9 26·2	3·3	2·2	9·3	6·1	15·3	10·1
34	9 53·5	9 55·1	9 26·5	3·4	2·2	9·4	6·2	15·4	10·1
35	9 53·8	9 55·4	9 26·7	3·5	2·3	9·5	6·3	15·5	10·2
36	9 54·0	9 55·6	9 26·9	3·6	2·4	9·6	6·3	15·6	10·3
37	9 54·3	9 55·9	9 27·2	3·7	2·4	9·7	6·4	15·7	10·3
38	9 54·5	9 56·1	9 27·4	3·8	2·5	9·8	6·5	15·8	10·4
39	9 54·8	9 56·4	9 27·7	3·9	2·6	9·9	6·5	15·9	10·5
40	9 55·0	9 56·6	9 27·9	4·0	2·6	10·0	6·6	16·0	10·5
41	9 55·3	9 56·9	9 28·1	4·1	2·7	10·1	6·6	16·1	10·6
42	9 55·5	9 57·1	9 28·4	4·2	2·8	10·2	6·7	16·2	10·7
43	9 55·8	9 57·4	9 28·6	4·3	2·8	10·3	6·8	16·3	10·7
44	9 56·0	9 57·6	9 28·8	4·4	2·9	10·4	6·8	16·4	10·8
45	9 56·3	9 57·9	9 29·1	4·5	3·0	10·5	6·9	16·5	10·9
46	9 56·5	9 58·1	9 29·3	4·6	3·0	10·6	7·0	16·6	10·9
47	9 56·8	9 58·4	9 29·6	4·7	3·1	10·7	7·0	16·7	11·0
48	9 57·0	9 58·6	9 29·8	4·8	3·2	10·8	7·1	16·8	11·1
49	9 57·3	9 58·9	9 30·0	4·9	3·2	10·9	7·2	16·9	11·1
50	9 57·5	9 59·1	9 30·3	5·0	3·3	11·0	7·2	17·0	11·2
51	9 57·8	9 59·4	9 30·5	5·1	3·4	11·1	7·3	17·1	11·3
52	9 58·0	9 59·6	9 30·8	5·2	3·4	11·2	7·4	17·2	11·3
53	9 58·3	9 59·9	9 31·0	5·3	3·5	11·3	7·4	17·3	11·4
54	9 58·5	10 00·1	9 31·2	5·4	3·6	11·4	7·5	17·4	11·5
55	9 58·8	10 00·4	9 31·5	5·5	3·6	11·5	7·6	17·5	11·5
56	9 59·0	10 00·6	9 31·7	5·6	3·7	11·6	7·6	17·6	11·6
57	9 59·3	10 00·9	9 32·0	5·7	3·8	11·7	7·7	17·7	11·7
58	9 59·5	10 01·1	9 32·2	5·8	3·8	11·8	7·8	17·8	11·7
59	9 59·8	10 01·4	9 32·4	5·9	3·9	11·9	7·8	17·9	11·8
60	10 00·0	10 01·6	9 32·7	6·0	4·0	12·0	7·9	18·0	11·9

40^m	SUN PLANETS	ARIES	MOON	v or Corrⁿ d		v or Corrⁿ d		v or Corrⁿ d	
s	° ′	° ′	° ′	′	′	′	′	′	′
00	10 00·0	10 01·6	9 32·7	0·0	0·0	6·0	4·1	12·0	8·1
01	10 00·3	10 01·9	9 32·9	0·1	0·1	6·1	4·1	12·1	8·2
02	10 00·5	10 02·1	9 33·1	0·2	0·1	6·2	4·2	12·2	8·2
03	10 00·8	10 02·4	9 33·4	0·3	0·2	6·3	4·3	12·3	8·3
04	10 01·0	10 02·6	9 33·6	0·4	0·3	6·4	4·3	12·4	8·4
05	10 01·3	10 02·9	9 33·9	0·5	0·3	6·5	4·4	12·5	8·4
06	10 01·5	10 03·1	9 34·1	0·6	0·4	6·6	4·5	12·6	8·5
07	10 01·8	10 03·4	9 34·3	0·7	0·5	6·7	4·5	12·7	8·6
08	10 02·0	10 03·6	9 34·6	0·8	0·5	6·8	4·6	12·8	8·6
09	10 02·3	10 03·9	9 34·8	0·9	0·6	6·9	4·7	12·9	8·7
10	10 02·5	10 04·1	9 35·1	1·0	0·7	7·0	4·7	13·0	8·8
11	10 02·8	10 04·4	9 35·3	1·1	0·7	7·1	4·8	13·1	8·8
12	10 03·0	10 04·7	9 35·5	1·2	0·8	7·2	4·9	13·2	8·9
13	10 03·3	10 04·9	9 35·8	1·3	0·9	7·3	4·9	13·3	9·0
14	10 03·5	10 05·2	9 36·0	1·4	0·9	7·4	5·0	13·4	9·0
15	10 03·8	10 05·4	9 36·2	1·5	1·0	7·5	5·1	13·5	9·1
16	10 04·0	10 05·7	9 36·5	1·6	1·1	7·6	5·1	13·6	9·2
17	10 04·3	10 05·9	9 36·7	1·7	1·1	7·7	5·2	13·7	9·2
18	10 04·5	10 06·2	9 37·0	1·8	1·2	7·8	5·3	13·8	9·3
19	10 04·8	10 06·4	9 37·2	1·9	1·3	7·9	5·3	13·9	9·4
20	10 05·0	10 06·7	9 37·4	2·0	1·4	8·0	5·4	14·0	9·5
21	10 05·3	10 06·9	9 37·7	2·1	1·4	8·1	5·5	14·1	9·5
22	10 05·5	10 07·2	9 37·9	2·2	1·5	8·2	5·5	14·2	9·6
23	10 05·8	10 07·4	9 38·2	2·3	1·6	8·3	5·6	14·3	9·7
24	10 06·0	10 07·7	9 38·4	2·4	1·6	8·4	5·7	14·4	9·7
25	10 06·3	10 07·9	9 38·6	2·5	1·7	8·5	5·7	14·5	9·8
26	10 06·5	10 08·2	9 38·9	2·6	1·8	8·6	5·8	14·6	9·9
27	10 06·8	10 08·4	9 39·1	2·7	1·8	8·7	5·9	14·7	9·9
28	10 07·0	10 08·7	9 39·3	2·8	1·9	8·8	5·9	14·8	10·0
29	10 07·3	10 08·9	9 39·6	2·9	2·0	8·9	6·0	14·9	10·1
30	10 07·5	10 09·2	9 39·8	3·0	2·0	9·0	6·1	15·0	10·1
31	10 07·8	10 09·4	9 40·1	3·1	2·1	9·1	6·1	15·1	10·2
32	10 08·0	10 09·7	9 40·3	3·2	2·2	9·2	6·2	15·2	10·3
33	10 08·3	10 09·9	9 40·5	3·3	2·2	9·3	6·3	15·3	10·3
34	10 08·5	10 10·2	9 40·8	3·4	2·3	9·4	6·3	15·4	10·4
35	10 08·8	10 10·4	9 41·0	3·5	2·4	9·5	6·4	15·5	10·5
36	10 09·0	10 10·7	9 41·3	3·6	2·4	9·6	6·5	15·6	10·5
37	10 09·3	10 10·9	9 41·5	3·7	2·5	9·7	6·5	15·7	10·6
38	10 09·5	10 11·2	9 41·7	3·8	2·6	9·8	6·6	15·8	10·7
39	10 09·8	10 11·4	9 42·0	3·9	2·6	9·9	6·7	15·9	10·7
40	10 10·0	10 11·7	9 42·2	4·0	2·7	10·0	6·8	16·0	10·8
41	10 10·3	10 11·9	9 42·4	4·1	2·8	10·1	6·8	16·1	10·9
42	10 10·5	10 12·2	9 42·7	4·2	2·8	10·2	6·9	16·2	10·9
43	10 10·8	10 12·4	9 42·9	4·3	2·9	10·3	7·0	16·3	11·0
44	10 11·0	10 12·7	9 43·2	4·4	3·0	10·4	7·0	16·4	11·1
45	10 11·3	10 12·9	9 43·4	4·5	3·0	10·5	7·1	16·5	11·1
46	10 11·5	10 13·2	9 43·6	4·6	3·1	10·6	7·2	16·6	11·2
47	10 11·8	10 13·4	9 43·9	4·7	3·2	10·7	7·2	16·7	11·3
48	10 12·0	10 13·7	9 44·1	4·8	3·2	10·8	7·3	16·8	11·3
49	10 12·3	10 13·9	9 44·4	4·9	3·3	10·9	7·4	16·9	11·4
50	10 12·5	10 14·2	9 44·6	5·0	3·4	11·0	7·4	17·0	11·5
51	10 12·8	10 14·4	9 44·8	5·1	3·4	11·1	7·5	17·1	11·5
52	10 13·0	10 14·7	9 45·1	5·2	3·5	11·2	7·6	17·2	11·6
53	10 13·3	10 14·9	9 45·3	5·3	3·6	11·3	7·6	17·3	11·7
54	10 13·5	10 15·2	9 45·6	5·4	3·6	11·4	7·7	17·4	11·7
55	10 13·8	10 15·4	9 45·8	5·5	3·7	11·5	7·8	17·5	11·8
56	10 14·0	10 15·7	9 46·0	5·6	3·8	11·6	7·8	17·6	11·9
57	10 14·3	10 15·9	9 46·3	5·7	3·8	11·7	7·9	17·7	11·9
58	10 14·5	10 16·2	9 46·5	5·8	3·9	11·8	8·0	17·8	12·0
59	10 14·8	10 16·4	9 46·7	5·9	4·0	11·9	8·0	17·9	12·1
60	10 15·0	10 16·7	9 47·0	6·0	4·1	12·0	8·1	18·0	12·2

41^m	SUN PLANETS	ARIES	MOON	v or Corrⁿ d		v or Corrⁿ d		v or Corrⁿ d	
s	° ′	° ′	° ′	′	′	′	′	′	′
00	10 15·0	10 16·7	9 47·0	0·0	0·0	6·0	4·2	12·0	8·3
01	10 15·3	10 16·9	9 47·2	0·1	0·1	6·1	4·2	12·1	8·4
02	10 15·5	10 17·2	9 47·5	0·2	0·1	6·2	4·3	12·2	8·4
03	10 15·8	10 17·4	9 47·7	0·3	0·2	6·3	4·4	12·3	8·5
04	10 16·0	10 17·7	9 47·9	0·4	0·3	6·4	4·4	12·4	8·6
05	10 16·3	10 17·9	9 48·2	0·5	0·3	6·5	4·5	12·5	8·6
06	10 16·5	10 18·2	9 48·4	0·6	0·4	6·6	4·6	12·6	8·7
07	10 16·8	10 18·4	9 48·7	0·7	0·5	6·7	4·6	12·7	8·8
08	10 17·0	10 18·7	9 48·9	0·8	0·6	6·8	4·7	12·8	8·9
09	10 17·3	10 18·9	9 49·1	0·9	0·6	6·9	4·8	12·9	8·9
10	10 17·5	10 19·2	9 49·4	1·0	0·7	7·0	4·8	13·0	9·0
11	10 17·8	10 19·4	9 49·6	1·1	0·8	7·1	4·9	13·1	9·1
12	10 18·0	10 19·7	9 49·8	1·2	0·8	7·2	5·0	13·2	9·1
13	10 18·3	10 19·9	9 50·1	1·3	0·9	7·3	5·0	13·3	9·2
14	10 18·5	10 20·2	9 50·3	1·4	1·0	7·4	5·1	13·4	9·3
15	10 18·8	10 20·4	9 50·6	1·5	1·0	7·5	5·2	13·5	9·3
16	10 19·0	10 20·7	9 50·8	1·6	1·1	7·6	5·3	13·6	9·4
17	10 19·3	10 20·9	9 51·0	1·7	1·2	7·7	5·3	13·7	9·5
18	10 19·5	10 21·2	9 51·3	1·8	1·2	7·8	5·4	13·8	9·5
19	10 19·8	10 21·4	9 51·5	1·9	1·3	7·9	5·5	13·9	9·6
20	10 20·0	10 21·7	9 51·8	2·0	1·4	8·0	5·5	14·0	9·7
21	10 20·3	10 21·9	9 52·0	2·1	1·5	8·1	5·6	14·1	9·8
22	10 20·5	10 22·2	9 52·2	2·2	1·5	8·2	5·7	14·2	9·8
23	10 20·8	10 22·4	9 52·5	2·3	1·6	8·3	5·7	14·3	9·9
24	10 21·0	10 22·7	9 52·7	2·4	1·7	8·4	5·8	14·4	10·0
25	10 21·3	10 23·0	9 52·9	2·5	1·7	8·5	5·9	14·5	10·0
26	10 21·5	10 23·2	9 53·2	2·6	1·8	8·6	5·9	14·6	10·1
27	10 21·8	10 23·5	9 53·4	2·7	1·9	8·7	6·0	14·7	10·2
28	10 22·0	10 23·7	9 53·7	2·8	1·9	8·8	6·1	14·8	10·2
29	10 22·3	10 24·0	9 53·9	2·9	2·0	8·9	6·2	14·9	10·3
30	10 22·5	10 24·2	9 54·1	3·0	2·1	9·0	6·2	15·0	10·4
31	10 22·8	10 24·5	9 54·4	3·1	2·1	9·1	6·3	15·1	10·4
32	10 23·0	10 24·7	9 54·6	3·2	2·2	9·2	6·4	15·2	10·5
33	10 23·3	10 25·0	9 54·9	3·3	2·3	9·3	6·4	15·3	10·6
34	10 23·5	10 25·2	9 55·1	3·4	2·4	9·4	6·5	15·4	10·7
35	10 23·8	10 25·5	9 55·3	3·5	2·4	9·5	6·6	15·5	10·7
36	10 24·0	10 25·7	9 55·6	3·6	2·5	9·6	6·6	15·6	10·8
37	10 24·3	10 26·0	9 55·8	3·7	2·6	9·7	6·7	15·7	10·9
38	10 24·5	10 26·2	9 56·1	3·8	2·6	9·8	6·8	15·8	10·9
39	10 24·8	10 26·5	9 56·3	3·9	2·7	9·9	6·8	15·9	11·0
40	10 25·0	10 26·7	9 56·5	4·0	2·8	10·0	6·9	16·0	11·1
41	10 25·3	10 27·0	9 56·8	4·1	2·8	10·1	7·0	16·1	11·1
42	10 25·5	10 27·2	9 57·0	4·2	2·9	10·2	7·1	16·2	11·2
43	10 25·8	10 27·5	9 57·2	4·3	3·0	10·3	7·1	16·3	11·3
44	10 26·0	10 27·7	9 57·5	4·4	3·0	10·4	7·2	16·4	11·3
45	10 26·3	10 28·0	9 57·7	4·5	3·1	10·5	7·3	16·5	11·4
46	10 26·5	10 28·2	9 58·0	4·6	3·2	10·6	7·3	16·6	11·5
47	10 26·8	10 28·5	9 58·2	4·7	3·3	10·7	7·4	16·7	11·6
48	10 27·0	10 28·7	9 58·4	4·8	3·3	10·8	7·5	16·8	11·6
49	10 27·3	10 29·0	9 58·7	4·9	3·4	10·9	7·5	16·9	11·7
50	10 27·5	10 29·2	9 58·9	5·0	3·5	11·0	7·6	17·0	11·8
51	10 27·8	10 29·5	9 59·2	5·1	3·5	11·1	7·7	17·1	11·8
52	10 28·0	10 29·7	9 59·4	5·2	3·6	11·2	7·7	17·2	11·9
53	10 28·3	10 30·0	9 59·6	5·3	3·7	11·3	7·8	17·3	12·0
54	10 28·5	10 30·2	9 59·9	5·4	3·7	11·4	7·9	17·4	12·0
55	10 28·8	10 30·5	10 00·1	5·5	3·8	11·5	8·0	17·5	12·1
56	10 29·0	10 30·7	10 00·3	5·6	3·9	11·6	8·0	17·6	12·2
57	10 29·3	10 31·0	10 00·6	5·7	3·9	11·7	8·1	17·7	12·2
58	10 29·5	10 31·2	10 00·8	5·8	4·0	11·8	8·2	17·8	12·3
59	10 29·8	10 31·5	10 01·1	5·9	4·1	11·9	8·2	17·9	12·4
60	10 30·0	10 31·7	10 01·3	6·0	4·2	12·0	8·3	18·0	12·5

42ᵐ	SUN PLANETS	ARIES	MOON	v or Corrⁿ d		v or Corrⁿ d		v or Corrⁿ d	
s	° ′	° ′	° ′	′	′	′	′	′	′
00	10 30.0	10 31.7	10 01.3	0.0	0.0	6.0	4.3	12.0	8.5
01	10 30.3	10 32.0	10 01.5	0.1	0.1	6.1	4.3	12.1	8.6
02	10 30.5	10 32.2	10 01.8	0.2	0.1	6.2	4.4	12.2	8.6
03	10 30.8	10 32.5	10 02.0	0.3	0.2	6.3	4.5	12.3	8.7
04	10 31.0	10 32.7	10 02.3	0.4	0.3	6.4	4.5	12.4	8.8
05	10 31.3	10 33.0	10 02.5	0.5	0.4	6.5	4.6	12.5	8.9
06	10 31.5	10 33.2	10 02.7	0.6	0.4	6.6	4.7	12.6	8.9
07	10 31.8	10 33.5	10 03.0	0.7	0.5	6.7	4.7	12.7	9.0
08	10 32.0	10 33.7	10 03.2	0.8	0.6	6.8	4.8	12.8	9.1
09	10 32.3	10 34.0	10 03.4	0.9	0.6	6.9	4.9	12.9	9.1
10	10 32.5	10 34.2	10 03.7	1.0	0.7	7.0	5.0	13.0	9.2
11	10 32.8	10 34.5	10 03.9	1.1	0.8	7.1	5.0	13.1	9.3
12	10 33.0	10 34.7	10 04.2	1.2	0.9	7.2	5.1	13.2	9.4
13	10 33.3	10 35.0	10 04.4	1.3	0.9	7.3	5.2	13.3	9.4
14	10 33.5	10 35.2	10 04.6	1.4	1.0	7.4	5.2	13.4	9.5
15	10 33.8	10 35.5	10 04.9	1.5	1.1	7.5	5.3	13.5	9.6
16	10 34.0	10 35.7	10 05.1	1.6	1.1	7.6	5.4	13.6	9.6
17	10 34.3	10 36.0	10 05.4	1.7	1.2	7.7	5.5	13.7	9.7
18	10 34.5	10 36.2	10 05.6	1.8	1.3	7.8	5.5	13.8	9.8
19	10 34.8	10 36.5	10 05.8	1.9	1.3	7.9	5.6	13.9	9.8
20	10 35.0	10 36.7	10 06.1	2.0	1.4	8.0	5.7	14.0	9.9
21	10 35.3	10 37.0	10 06.3	2.1	1.5	8.1	5.7	14.1	10.0
22	10 35.5	10 37.2	10 06.5	2.2	1.6	8.2	5.8	14.2	10.1
23	10 35.8	10 37.5	10 06.8	2.3	1.6	8.3	5.9	14.3	10.1
24	10 36.0	10 37.7	10 07.0	2.4	1.7	8.4	6.0	14.4	10.2
25	10 36.3	10 38.0	10 07.3	2.5	1.8	8.5	6.0	14.5	10.3
26	10 36.5	10 38.2	10 07.5	2.6	1.8	8.6	6.1	14.6	10.3
27	10 36.8	10 38.5	10 07.7	2.7	1.9	8.7	6.2	14.7	10.4
28	10 37.0	10 38.7	10 08.0	2.8	2.0	8.8	6.2	14.8	10.5
29	10 37.3	10 39.0	10 08.2	2.9	2.1	8.9	6.3	14.9	10.6
30	10 37.5	10 39.2	10 08.5	3.0	2.1	9.0	6.4	15.0	10.6
31	10 37.8	10 39.5	10 08.7	3.1	2.2	9.1	6.4	15.1	10.7
32	10 38.0	10 39.7	10 08.9	3.2	2.3	9.2	6.5	15.2	10.8
33	10 38.3	10 40.0	10 09.2	3.3	2.3	9.3	6.6	15.3	10.8
34	10 38.5	10 40.2	10 09.4	3.4	2.4	9.4	6.7	15.4	10.9
35	10 38.8	10 40.5	10 09.7	3.5	2.5	9.5	6.7	15.5	11.0
36	10 39.0	10 40.7	10 09.9	3.6	2.6	9.6	6.8	15.6	11.1
37	10 39.3	10 41.0	10 10.1	3.7	2.6	9.7	6.9	15.7	11.1
38	10 39.5	10 41.3	10 10.4	3.8	2.7	9.8	6.9	15.8	11.2
39	10 39.8	10 41.5	10 10.6	3.9	2.8	9.9	7.0	15.9	11.3
40	10 40.0	10 41.8	10 10.8	4.0	2.8	10.0	7.1	16.0	11.3
41	10 40.3	10 42.0	10 11.1	4.1	2.9	10.1	7.2	16.1	11.4
42	10 40.5	10 42.3	10 11.3	4.2	3.0	10.2	7.2	16.2	11.5
43	10 40.8	10 42.5	10 11.6	4.3	3.0	10.3	7.3	16.3	11.5
44	10 41.0	10 42.8	10 11.8	4.4	3.1	10.4	7.4	16.4	11.6
45	10 41.3	10 43.0	10 12.0	4.5	3.2	10.5	7.4	16.5	11.7
46	10 41.5	10 43.3	10 12.3	4.6	3.3	10.6	7.5	16.6	11.8
47	10 41.8	10 43.5	10 12.5	4.7	3.3	10.7	7.6	16.7	11.8
48	10 42.0	10 43.8	10 12.8	4.8	3.4	10.8	7.7	16.8	11.9
49	10 42.3	10 44.0	10 13.0	4.9	3.5	10.9	7.7	16.9	12.0
50	10 42.5	10 44.3	10 13.2	5.0	3.5	11.0	7.8	17.0	12.0
51	10 42.8	10 44.5	10 13.5	5.1	3.6	11.1	7.9	17.1	12.1
52	10 43.0	10 44.8	10 13.7	5.2	3.7	11.2	7.9	17.2	12.2
53	10 43.3	10 45.0	10 13.9	5.3	3.8	11.3	8.0	17.3	12.3
54	10 43.5	10 45.3	10 14.2	5.4	3.8	11.4	8.1	17.4	12.3
55	10 43.8	10 45.5	10 14.4	5.5	3.9	11.5	8.1	17.5	12.4
56	10 44.0	10 45.8	10 14.7	5.6	4.0	11.6	8.2	17.6	12.5
57	10 44.3	10 46.0	10 14.9	5.7	4.0	11.7	8.3	17.7	12.5
58	10 44.5	10 46.3	10 15.1	5.8	4.1	11.8	8.4	17.8	12.6
59	10 44.8	10 46.5	10 15.4	5.9	4.2	11.9	8.4	17.9	12.7
60	10 45.0	10 46.8	10 15.6	6.0	4.3	12.0	8.5	18.0	12.8

43ᵐ	SUN PLANETS	ARIES	MOON	v or Corrⁿ d		v or Corrⁿ d		v or Corrⁿ d	
s	° ′	° ′	° ′	′	′	′	′	′	′
00	10 45.0	10 46.8	10 15.6	0.0	0.0	6.0	4.4	12.0	8.7
01	10 45.3	10 47.0	10 15.9	0.1	0.1	6.1	4.4	12.1	8.8
02	10 45.5	10 47.3	10 16.1	0.2	0.1	6.2	4.5	12.2	8.8
03	10 45.8	10 47.5	10 16.3	0.3	0.2	6.3	4.6	12.3	8.9
04	10 46.0	10 47.8	10 16.6	0.4	0.3	6.4	4.6	12.4	9.0
05	10 46.3	10 48.0	10 16.8	0.5	0.4	6.5	4.7	12.5	9.1
06	10 46.5	10 48.3	10 17.0	0.6	0.4	6.6	4.8	12.6	9.1
07	10 46.8	10 48.5	10 17.3	0.7	0.5	6.7	4.9	12.7	9.2
08	10 47.0	10 48.8	10 17.5	0.8	0.6	6.8	4.9	12.8	9.3
09	10 47.3	10 49.0	10 17.8	0.9	0.7	6.9	5.0	12.9	9.4
10	10 47.5	10 49.3	10 18.0	1.0	0.7	7.0	5.1	13.0	9.4
11	10 47.8	10 49.5	10 18.2	1.1	0.8	7.1	5.1	13.1	9.5
12	10 48.0	10 49.8	10 18.5	1.2	0.9	7.2	5.2	13.2	9.6
13	10 48.3	10 50.0	10 18.7	1.3	0.9	7.3	5.3	13.3	9.6
14	10 48.5	10 50.3	10 19.0	1.4	1.0	7.4	5.4	13.4	9.7
15	10 48.8	10 50.5	10 19.2	1.5	1.1	7.5	5.4	13.5	9.8
16	10 49.0	10 50.8	10 19.4	1.6	1.2	7.6	5.5	13.6	9.9
17	10 49.3	10 51.0	10 19.7	1.7	1.2	7.7	5.6	13.7	9.9
18	10 49.5	10 51.3	10 19.9	1.8	1.3	7.8	5.7	13.8	10.0
19	10 49.8	10 51.5	10 20.2	1.9	1.4	7.9	5.7	13.9	10.1
20	10 50.0	10 51.8	10 20.4	2.0	1.5	8.0	5.8	14.0	10.2
21	10 50.3	10 52.0	10 20.6	2.1	1.5	8.1	5.9	14.1	10.2
22	10 50.5	10 52.3	10 20.9	2.2	1.6	8.2	5.9	14.2	10.3
23	10 50.8	10 52.5	10 21.1	2.3	1.7	8.3	6.0	14.3	10.4
24	10 51.0	10 52.8	10 21.3	2.4	1.7	8.4	6.1	14.4	10.4
25	10 51.3	10 53.0	10 21.6	2.5	1.8	8.5	6.2	14.5	10.5
26	10 51.5	10 53.3	10 21.8	2.6	1.9	8.6	6.2	14.6	10.6
27	10 51.8	10 53.5	10 22.1	2.7	2.0	8.7	6.3	14.7	10.7
28	10 52.0	10 53.8	10 22.3	2.8	2.0	8.8	6.4	14.8	10.7
29	10 52.3	10 54.0	10 22.5	2.9	2.1	8.9	6.5	14.9	10.8
30	10 52.5	10 54.3	10 22.8	3.0	2.2	9.0	6.5	15.0	10.9
31	10 52.8	10 54.5	10 23.0	3.1	2.2	9.1	6.6	15.1	10.9
32	10 53.0	10 54.8	10 23.3	3.2	2.3	9.2	6.7	15.2	11.0
33	10 53.3	10 55.0	10 23.5	3.3	2.4	9.3	6.7	15.3	11.1
34	10 53.5	10 55.3	10 23.7	3.4	2.5	9.4	6.8	15.4	11.2
35	10 53.8	10 55.5	10 24.0	3.5	2.5	9.5	6.9	15.5	11.2
36	10 54.0	10 55.8	10 24.2	3.6	2.6	9.6	7.0	15.6	11.3
37	10 54.3	10 56.0	10 24.4	3.7	2.7	9.7	7.0	15.7	11.4
38	10 54.5	10 56.3	10 24.7	3.8	2.8	9.8	7.1	15.8	11.5
39	10 54.8	10 56.5	10 24.9	3.9	2.8	9.9	7.2	15.9	11.5
40	10 55.0	10 56.8	10 25.2	4.0	2.9	10.0	7.3	16.0	11.6
41	10 55.3	10 57.0	10 25.4	4.1	3.0	10.1	7.3	16.1	11.7
42	10 55.5	10 57.3	10 25.6	4.2	3.0	10.2	7.4	16.2	11.7
43	10 55.8	10 57.5	10 25.9	4.3	3.1	10.3	7.5	16.3	11.8
44	10 56.0	10 57.8	10 26.1	4.4	3.2	10.4	7.5	16.4	11.9
45	10 56.3	10 58.0	10 26.4	4.5	3.3	10.5	7.6	16.5	12.0
46	10 56.5	10 58.3	10 26.6	4.6	3.3	10.6	7.7	16.6	12.0
47	10 56.8	10 58.5	10 26.8	4.7	3.4	10.7	7.8	16.7	12.1
48	10 57.0	10 58.8	10 27.1	4.8	3.5	10.8	7.8	16.8	12.2
49	10 57.3	10 59.0	10 27.3	4.9	3.6	10.9	7.9	16.9	12.3
50	10 57.5	10 59.3	10 27.5	5.0	3.6	11.0	8.0	17.0	12.3
51	10 57.8	10 59.6	10 27.8	5.1	3.7	11.1	8.0	17.1	12.4
52	10 58.0	10 59.8	10 28.0	5.2	3.8	11.2	8.1	17.2	12.5
53	10 58.3	11 00.1	10 28.3	5.3	3.8	11.3	8.2	17.3	12.5
54	10 58.5	11 00.3	10 28.5	5.4	3.9	11.4	8.3	17.4	12.6
55	10 58.8	11 00.6	10 28.7	5.5	4.0	11.5	8.3	17.5	12.7
56	10 59.0	11 00.8	10 29.0	5.6	4.1	11.6	8.4	17.6	12.8
57	10 59.3	11 01.1	10 29.2	5.7	4.1	11.7	8.5	17.7	12.8
58	10 59.5	11 01.3	10 29.5	5.8	4.2	11.8	8.6	17.8	12.9
59	10 59.8	11 01.6	10 29.7	5.9	4.3	11.9	8.6	17.9	13.0
60	11 00.0	11 01.8	10 29.9	6.0	4.4	12.0	8.7	18.0	13.1

44ᵐ	SUN PLANETS	ARIES	MOON	v or d Corrⁿ	v or d Corrⁿ	v or d Corrⁿ	45ᵐ	SUN PLANETS	ARIES	MOON	v or d Corrⁿ	v or d Corrⁿ	v or d Corrⁿ
s	° ′	° ′	° ′	′ ′	′ ′	′ ′	s	° ′	° ′	° ′	′ ′	′ ′	′ ′
00	11 00·0	11 01·8	10 29·9	0·0 0·0	6·0 4·5	12·0 8·9	00	11 15·0	11 16·8	10 44·3	0·0 0·0	6·0 4·6	12·0 9·1
01	11 00·3	11 02·1	10 30·2	0·1 0·1	6·1 4·5	12·1 9·0	01	11 15·3	11 17·1	10 44·5	0·1 0·1	6·1 4·6	12·1 9·2
02	11 00·5	11 02·3	10 30·4	0·2 0·1	6·2 4·6	12·2 9·0	02	11 15·5	11 17·3	10 44·7	0·2 0·2	6·2 4·7	12·2 9·3
03	11 00·8	11 02·6	10 30·6	0·3 0·2	6·3 4·7	12·3 9·1	03	11 15·8	11 17·6	10 45·0	0·3 0·2	6·3 4·8	12·3 9·3
04	11 01·0	11 02·8	10 30·9	0·4 0·3	6·4 4·7	12·4 9·2	04	11 16·0	11 17·9	10 45·2	0·4 0·3	6·4 4·9	12·4 9·4
05	11 01·3	11 03·1	10 31·1	0·5 0·4	6·5 4·8	12·5 9·3	05	11 16·3	11 18·1	10 45·4	0·5 0·4	6·5 4·9	12·5 9·5
06	11 01·5	11 03·3	10 31·4	0·6 0·4	6·6 4·9	12·6 9·3	06	11 16·5	11 18·4	10 45·7	0·6 0·5	6·6 5·0	12·6 9·6
07	11 01·8	11 03·6	10 31·6	0·7 0·5	6·7 5·0	12·7 9·4	07	11 16·8	11 18·6	10 45·9	0·7 0·5	6·7 5·1	12·7 9·6
08	11 02·0	11 03·8	10 31·8	0·8 0·6	6·8 5·0	12·8 9·5	08	11 17·0	11 18·9	10 46·2	0·8 0·6	6·8 5·2	12·8 9·7
09	11 02·3	11 04·1	10 32·1	0·9 0·7	6·9 5·1	12·9 9·6	09	11 17·3	11 19·1	10 46·4	0·9 0·7	6·9 5·2	12·9 9·8
10	11 02·5	11 04·3	10 32·3	1·0 0·7	7·0 5·2	13·0 9·6	10	11 17·5	11 19·4	10 46·6	1·0 0·8	7·0 5·3	13·0 9·9
11	11 02·8	11 04·6	10 32·6	1·1 0·8	7·1 5·3	13·1 9·7	11	11 17·8	11 19·6	10 46·9	1·1 0·8	7·1 5·4	13·1 9·9
12	11 03·0	11 04·8	10 32·8	1·2 0·9	7·2 5·3	13·2 9·8	12	11 18·0	11 19·9	10 47·1	1·2 0·9	7·2 5·5	13·2 10·0
13	11 03·3	11 05·1	10 33·0	1·3 1·0	7·3 5·4	13·3 9·9	13	11 18·3	11 20·1	10 47·4	1·3 1·0	7·3 5·5	13·3 10·1
14	11 03·5	11 05·3	10 33·3	1·4 1·0	7·4 5·5	13·4 9·9	14	11 18·5	11 20·4	10 47·6	1·4 1·1	7·4 5·6	13·4 10·2
15	11 03·8	11 05·6	10 33·5	1·5 1·1	7·5 5·6	13·5 10·0	15	11 18·8	11 20·6	10 47·8	1·5 1·1	7·5 5·7	13·5 10·2
16	11 04·0	11 05·8	10 33·8	1·6 1·2	7·6 5·6	13·6 10·1	16	11 19·0	11 20·9	10 48·1	1·6 1·2	7·6 5·8	13·6 10·3
17	11 04·3	11 06·1	10 34·0	1·7 1·3	7·7 5·7	13·7 10·2	17	11 19·3	11 21·1	10 48·3	1·7 1·3	7·7 5·8	13·7 10·4
18	11 04·5	11 06·3	10 34·2	1·8 1·3	7·8 5·8	13·8 10·2	18	11 19·5	11 21·4	10 48·5	1·8 1·4	7·8 5·9	13·8 10·5
19	11 04·8	11 06·6	10 34·5	1·9 1·4	7·9 5·9	13·9 10·3	19	11 19·8	11 21·6	10 48·8	1·9 1·4	7·9 6·0	13·9 10·5
20	11 05·0	11 06·8	10 34·7	2·0 1·5	8·0 5·9	14·0 10·4	20	11 20·0	11 21·9	10 49·0	2·0 1·5	8·0 6·1	14·0 10·6
21	11 05·3	11 07·1	10 34·9	2·1 1·6	8·1 6·0	14·1 10·5	21	11 20·3	11 22·1	10 49·3	2·1 1·6	8·1 6·1	14·1 10·7
22	11 05·5	11 07·3	10 35·2	2·2 1·6	8·2 6·1	14·2 10·5	22	11 20·5	11 22·4	10 49·5	2·2 1·7	8·2 6·2	14·2 10·8
23	11 05·8	11 07·6	10 35·4	2·3 1·7	8·3 6·2	14·3 10·6	23	11 20·8	11 22·6	10 49·7	2·3 1·7	8·3 6·3	14·3 10·8
24	11 06·0	11 07·8	10 35·7	2·4 1·8	8·4 6·2	14·4 10·7	24	11 21·0	11 22·9	10 50·0	2·4 1·8	8·4 6·4	14·4 10·9
25	11 06·3	11 08·1	10 35·9	2·5 1·9	8·5 6·3	14·5 10·8	25	11 21·3	11 23·1	10 50·2	2·5 1·9	8·5 6·4	14·5 11·0
26	11 06·5	11 08·3	10 36·1	2·6 1·9	8·6 6·4	14·6 10·8	26	11 21·5	11 23·4	10 50·5	2·6 2·0	8·6 6·5	14·6 11·1
27	11 06·8	11 08·6	10 36·4	2·7 2·0	8·7 6·5	14·7 10·9	27	11 21·8	11 23·6	10 50·7	2·7 2·0	8·7 6·6	14·7 11·1
28	11 07·0	11 08·8	10 36·6	2·8 2·1	8·8 6·5	14·8 11·0	28	11 22·0	11 23·9	10 50·9	2·8 2·1	8·8 6·7	14·8 11·2
29	11 07·3	11 09·1	10 36·9	2·9 2·2	8·9 6·6	14·9 11·1	29	11 22·3	11 24·1	10 51·2	2·9 2·2	8·9 6·7	14·9 11·3
30	11 07·5	11 09·3	10 37·1	3·0 2·2	9·0 6·7	15·0 11·1	30	11 22·5	11 24·4	10 51·4	3·0 2·3	9·0 6·8	15·0 11·4
31	11 07·8	11 09·6	10 37·3	3·1 2·3	9·1 6·7	15·1 11·2	31	11 22·8	11 24·6	10 51·6	3·1 2·4	9·1 6·9	15·1 11·5
32	11 08·0	11 09·8	10 37·6	3·2 2·4	9·2 6·8	15·2 11·3	32	11 23·0	11 24·9	10 51·9	3·2 2·4	9·2 7·0	15·2 11·5
33	11 08·3	11 10·1	10 37·8	3·3 2·4	9·3 6·9	15·3 11·3	33	11 23·3	11 25·1	10 52·1	3·3 2·5	9·3 7·1	15·3 11·6
34	11 08·5	11 10·3	10 38·0	3·4 2·5	9·4 7·0	15·4 11·4	34	11 23·5	11 25·4	10 52·4	3·4 2·6	9·4 7·1	15·4 11·7
35	11 08·8	11 10·6	10 38·3	3·5 2·6	9·5 7·0	15·5 11·5	35	11 23·8	11 25·6	10 52·6	3·5 2·7	9·5 7·2	15·5 11·8
36	11 09·0	11 10·8	10 38·5	3·6 2·7	9·6 7·1	15·6 11·6	36	11 24·0	11 25·9	10 52·8	3·6 2·7	9·6 7·3	15·6 11·8
37	11 09·3	11 11·1	10 38·8	3·7 2·7	9·7 7·2	15·7 11·6	37	11 24·3	11 26·1	10 53·1	3·7 2·8	9·7 7·4	15·7 11·9
38	11 09·5	11 11·3	10 39·0	3·8 2·8	9·8 7·3	15·8 11·7	38	11 24·5	11 26·4	10 53·3	3·8 2·9	9·8 7·4	15·8 12·0
39	11 09·8	11 11·6	10 39·2	3·9 2·9	9·9 7·3	15·9 11·8	39	11 24·8	11 26·6	10 53·6	3·9 3·0	9·9 7·5	15·9 12·1
40	11 10·0	11 11·8	10 39·5	4·0 3·0	10·0 7·4	16·0 11·9	40	11 25·0	11 26·9	10 53·8	4·0 3·0	10·0 7·6	16·0 12·1
41	11 10·3	11 12·1	10 39·7	4·1 3·0	10·1 7·5	16·1 11·9	41	11 25·3	11 27·1	10 54·0	4·1 3·1	10·1 7·7	16·1 12·2
42	11 10·5	11 12·3	10 40·0	4·2 3·1	10·2 7·6	16·2 12·0	42	11 25·5	11 27·4	10 54·3	4·2 3·2	10·2 7·7	16·2 12·3
43	11 10·8	11 12·6	10 40·2	4·3 3·2	10·3 7·6	16·3 12·1	43	11 25·8	11 27·6	10 54·5	4·3 3·3	10·3 7·8	16·3 12·4
44	11 11·0	11 12·8	10 40·4	4·4 3·3	10·4 7·7	16·4 12·2	44	11 26·0	11 27·9	10 54·7	4·4 3·3	10·4 7·9	16·4 12·4
45	11 11·3	11 13·1	10 40·7	4·5 3·3	10·5 7·8	16·5 12·2	45	11 26·3	11 28·1	10 55·0	4·5 3·4	10·5 8·0	16·5 12·5
46	11 11·5	11 13·3	10 40·9	4·6 3·4	10·6 7·9	16·6 12·3	46	11 26·5	11 28·4	10 55·2	4·6 3·5	10·6 8·0	16·6 12·6
47	11 11·8	11 13·6	10 41·1	4·7 3·5	10·7 7·9	16·7 12·4	47	11 26·8	11 28·6	10 55·5	4·7 3·6	10·7 8·1	16·7 12·7
48	11 12·0	11 13·8	10 41·4	4·8 3·6	10·8 8·0	16·8 12·5	48	11 27·0	11 28·9	10 55·7	4·8 3·6	10·8 8·2	16·8 12·7
49	11 12·3	11 14·1	10 41·6	4·9 3·6	10·9 8·1	16·9 12·5	49	11 27·3	11 29·1	10 55·9	4·9 3·7	10·9 8·3	16·9 12·8
50	11 12·5	11 14·3	10 41·9	5·0 3·7	11·0 8·2	17·0 12·6	50	11 27·5	11 29·4	10 56·2	5·0 3·8	11·0 8·3	17·0 12·9
51	11 12·8	11 14·6	10 42·1	5·1 3·8	11·1 8·2	17·1 12·7	51	11 27·8	11 29·6	10 56·4	5·1 3·9	11·1 8·4	17·1 13·0
52	11 13·0	11 14·8	10 42·3	5·2 3·9	11·2 8·3	17·2 12·8	52	11 28·0	11 29·9	10 56·7	5·2 3·9	11·2 8·5	17·2 13·0
53	11 13·3	11 15·1	10 42·6	5·3 3·9	11·3 8·4	17·3 12·8	53	11 28·3	11 30·1	10 56·9	5·3 4·0	11·3 8·6	17·3 13·1
54	11 13·5	11 15·3	10 42·8	5·4 4·0	11·4 8·5	17·4 12·9	54	11 28·5	11 30·4	10 57·1	5·4 4·1	11·4 8·6	17·4 13·2
55	11 13·8	11 15·6	10 43·1	5·5 4·1	11·5 8·5	17·5 13·0	55	11 28·8	11 30·6	10 57·4	5·5 4·2	11·5 8·7	17·5 13·3
56	11 14·0	11 15·8	10 43·3	5·6 4·2	11·6 8·6	17·6 13·1	56	11 29·0	11 30·9	10 57·6	5·6 4·2	11·6 8·8	17·6 13·3
57	11 14·3	11 16·1	10 43·5	5·7 4·2	11·7 8·7	17·7 13·1	57	11 29·3	11 31·1	10 57·9	5·7 4·3	11·7 8·9	17·7 13·4
58	11 14·5	11 16·3	10 43·8	5·8 4·3	11·8 8·8	17·8 13·2	58	11 29·5	11 31·4	10 58·1	5·8 4·4	11·8 8·9	17·8 13·5
59	11 14·8	11 16·6	10 44·0	5·9 4·4	11·9 8·8	17·9 13·3	59	11 29·8	11 31·6	10 58·3	5·9 4·5	11·9 9·0	17·9 13·6
60	11 15·0	11 16·8	10 44·3	6·0 4·5	12·0 8·9	18·0 13·4	60	11 30·0	11 31·9	10 58·6	6·0 4·6	12·0 9·1	18·0 13·7

46ᵐ	SUN PLANETS	ARIES	MOON	v or d Corrⁿ	v or d Corrⁿ	v or d Corrⁿ
s	° ′	° ′	° ′	′ ′	′ ′	′ ′
00	11 30·0	11 31·9	10 58·6	0·0 0·0	6·0 4·7	12·0 9·3
01	11 30·3	11 32·1	10 58·8	0·1 0·1	6·1 4·7	12·1 9·4
02	11 30·5	11 32·4	10 59·0	0·2 0·2	6·2 4·8	12·2 9·5
03	11 30·8	11 32·6	10 59·3	0·3 0·2	6·3 4·9	12·3 9·5
04	11 31·0	11 32·9	10 59·5	0·4 0·3	6·4 5·0	12·4 9·6
05	11 31·3	11 33·1	10 59·8	0·5 0·4	6·5 5·0	12·5 9·7
06	11 31·5	11 33·4	11 00·0	0·6 0·5	6·6 5·1	12·6 9·8
07	11 31·8	11 33·6	11 00·2	0·7 0·5	6·7 5·2	12·7 9·8
08	11 32·0	11 33·9	11 00·5	0·8 0·6	6·8 5·3	12·8 9·9
09	11 32·3	11 34·1	11 00·7	0·9 0·7	6·9 5·3	12·9 10·0
10	11 32·5	11 34·4	11 01·0	1·0 0·8	7·0 5·4	13·0 10·1
11	11 32·8	11 34·6	11 01·2	1·1 0·9	7·1 5·5	13·1 10·2
12	11 33·0	11 34·9	11 01·4	1·2 0·9	7·2 5·6	13·2 10·2
13	11 33·3	11 35·1	11 01·7	1·3 1·0	7·3 5·7	13·3 10·3
14	11 33·5	11 35·4	11 01·9	1·4 1·1	7·4 5·7	13·4 10·4
15	11 33·8	11 35·6	11 02·1	1·5 1·2	7·5 5·8	13·5 10·5
16	11 34·0	11 35·9	11 02·4	1·6 1·2	7·6 5·9	13·6 10·5
17	11 34·3	11 36·2	11 02·6	1·7 1·3	7·7 6·0	13·7 10·6
18	11 34·5	11 36·4	11 02·9	1·8 1·4	7·8 6·0	13·8 10·7
19	11 34·8	11 36·7	11 03·1	1·9 1·5	7·9 6·1	13·9 10·8
20	11 35·0	11 36·9	11 03·3	2·0 1·6	8·0 6·2	14·0 10·9
21	11 35·3	11 37·2	11 03·6	2·1 1·6	8·1 6·3	14·1 10·9
22	11 35·5	11 37·4	11 03·8	2·2 1·7	8·2 6·4	14·2 11·0
23	11 35·8	11 37·7	11 04·1	2·3 1·8	8·3 6·4	14·3 11·1
24	11 36·0	11 37·9	11 04·3	2·4 1·9	8·4 6·5	14·4 11·2
25	11 36·3	11 38·2	11 04·5	2·5 1·9	8·5 6·6	14·5 11·2
26	11 36·5	11 38·4	11 04·8	2·6 2·0	8·6 6·7	14·6 11·3
27	11 36·8	11 38·7	11 05·0	2·7 2·1	8·7 6·7	14·7 11·4
28	11 37·0	11 38·9	11 05·2	2·8 2·2	8·8 6·8	14·8 11·5
29	11 37·3	11 39·2	11 05·5	2·9 2·2	8·9 6·9	14·9 11·5
30	11 37·5	11 39·4	11 05·7	3·0 2·3	9·0 7·0	15·0 11·6
31	11 37·8	11 39·7	11 06·0	3·1 2·4	9·1 7·1	15·1 11·7
32	11 38·0	11 39·9	11 06·2	3·2 2·5	9·2 7·1	15·2 11·8
33	11 38·3	11 40·2	11 06·4	3·3 2·6	9·3 7·2	15·3 11·9
34	11 38·5	11 40·4	11 06·7	3·4 2·6	9·4 7·3	15·4 11·9
35	11 38·8	11 40·7	11 06·9	3·5 2·7	9·5 7·4	15·5 12·0
36	11 39·0	11 40·9	11 07·2	3·6 2·8	9·6 7·4	15·6 12·1
37	11 39·3	11 41·2	11 07·4	3·7 2·9	9·7 7·5	15·7 12·2
38	11 39·5	11 41·4	11 07·6	3·8 2·9	9·8 7·6	15·8 12·2
39	11 39·8	11 41·7	11 07·9	3·9 3·0	9·9 7·7	15·9 12·3
40	11 40·0	11 41·9	11 08·1	4·0 3·1	10·0 7·8	16·0 12·4
41	11 40·3	11 42·2	11 08·3	4·1 3·2	10·1 7·8	16·1 12·5
42	11 40·5	11 42·4	11 08·6	4·2 3·3	10·2 7·9	16·2 12·6
43	11 40·8	11 42·7	11 08·8	4·3 3·3	10·3 8·0	16·3 12·6
44	11 41·0	11 42·9	11 09·1	4·4 3·4	10·4 8·1	16·4 12·7
45	11 41·3	11 43·2	11 09·3	4·5 3·5	10·5 8·1	16·5 12·8
46	11 41·5	11 43·4	11 09·5	4·6 3·6	10·6 8·2	16·6 12·9
47	11 41·8	11 43·7	11 09·8	4·7 3·6	10·7 8·3	16·7 12·9
48	11 42·0	11 43·9	11 10·0	4·8 3·7	10·8 8·4	16·8 13·0
49	11 42·3	11 44·2	11 10·3	4·9 3·8	10·9 8·4	16·9 13·1
50	11 42·5	11 44·4	11 10·5	5·0 3·9	11·0 8·5	17·0 13·2
51	11 42·8	11 44·7	11 10·7	5·1 4·0	11·1 8·6	17·1 13·3
52	11 43·0	11 44·9	11 11·0	5·2 4·0	11·2 8·7	17·2 13·3
53	11 43·3	11 45·2	11 11·2	5·3 4·1	11·3 8·8	17·3 13·4
54	11 43·5	11 45·4	11 11·5	5·4 4·2	11·4 8·8	17·4 13·5
55	11 43·8	11 45·7	11 11·7	5·5 4·3	11·5 8·9	17·5 13·6
56	11 44·0	11 45·9	11 11·9	5·6 4·3	11·6 9·0	17·6 13·6
57	11 44·3	11 46·2	11 12·2	5·7 4·4	11·7 9·1	17·7 13·7
58	11 44·5	11 46·4	11 12·4	5·8 4·5	11·8 9·1	17·8 13·8
59	11 44·8	11 46·7	11 12·6	5·9 4·6	11·9 9·2	17·9 13·9
60	11 45·0	11 46·9	11 12·9	6·0 4·7	12·0 9·3	18·0 14·0

47ᵐ	SUN PLANETS	ARIES	MOON	v or d Corrⁿ	v or d Corrⁿ	v or d Corrⁿ
s	° ′	° ′	° ′	′ ′	′ ′	′ ′
00	11 45·0	11 46·9	11 12·9	0·0 0·0	6·0 4·8	12·0 9·5
01	11 45·3	11 47·2	11 13·1	0·1 0·1	6·1 4·8	12·1 9·6
02	11 45·5	11 47·4	11 13·4	0·2 0·2	6·2 4·9	12·2 9·7
03	11 45·8	11 47·7	11 13·6	0·3 0·2	6·3 5·0	12·3 9·7
04	11 46·0	11 47·9	11 13·8	0·4 0·3	6·4 5·1	12·4 9·8
05	11 46·3	11 48·2	11 14·1	0·5 0·4	6·5 5·1	12·5 9·9
06	11 46·5	11 48·4	11 14·3	0·6 0·5	6·6 5·2	12·6 10·0
07	11 46·8	11 48·7	11 14·6	0·7 0·6	6·7 5·3	12·7 10·1
08	11 47·0	11 48·9	11 14·8	0·8 0·6	6·8 5·4	12·8 10·1
09	11 47·3	11 49·2	11 15·0	0·9 0·7	6·9 5·5	12·9 10·2
10	11 47·5	11 49·4	11 15·3	1·0 0·8	7·0 5·5	13·0 10·3
11	11 47·8	11 49·7	11 15·5	1·1 0·9	7·1 5·6	13·1 10·4
12	11 48·0	11 49·9	11 15·7	1·2 1·0	7·2 5·7	13·2 10·5
13	11 48·3	11 50·2	11 16·0	1·3 1·0	7·3 5·8	13·3 10·5
14	11 48·5	11 50·4	11 16·2	1·4 1·1	7·4 5·9	13·4 10·6
15	11 48·8	11 50·7	11 16·5	1·5 1·2	7·5 5·9	13·5 10·7
16	11 49·0	11 50·9	11 16·7	1·6 1·3	7·6 6·0	13·6 10·8
17	11 49·3	11 51·2	11 16·9	1·7 1·3	7·7 6·1	13·7 10·8
18	11 49·5	11 51·4	11 17·2	1·8 1·4	7·8 6·2	13·8 10·9
19	11 49·8	11 51·7	11 17·4	1·9 1·5	7·9 6·3	13·9 11·0
20	11 50·0	11 51·9	11 17·7	2·0 1·6	8·0 6·3	14·0 11·1
21	11 50·3	11 52·2	11 17·9	2·1 1·7	8·1 6·4	14·1 11·2
22	11 50·5	11 52·4	11 18·1	2·2 1·7	8·2 6·5	14·2 11·2
23	11 50·8	11 52·7	11 18·4	2·3 1·8	8·3 6·6	14·3 11·3
24	11 51·0	11 52·9	11 18·6	2·4 1·9	8·4 6·7	14·4 11·4
25	11 51·3	11 53·2	11 18·8	2·5 2·0	8·5 6·7	14·5 11·5
26	11 51·5	11 53·4	11 19·1	2·6 2·1	8·6 6·8	14·6 11·6
27	11 51·8	11 53·7	11 19·3	2·7 2·1	8·7 6·9	14·7 11·6
28	11 52·0	11 53·9	11 19·6	2·8 2·2	8·8 7·0	14·8 11·7
29	11 52·3	11 54·2	11 19·8	2·9 2·3	8·9 7·0	14·9 11·8
30	11 52·5	11 54·5	11 20·0	3·0 2·4	9·0 7·1	15·0 11·9
31	11 52·8	11 54·7	11 20·3	3·1 2·5	9·1 7·2	15·1 12·0
32	11 53·0	11 55·0	11 20·5	3·2 2·5	9·2 7·3	15·2 12·0
33	11 53·3	11 55·2	11 20·8	3·3 2·6	9·3 7·4	15·3 12·1
34	11 53·5	11 55·5	11 21·0	3·4 2·7	9·4 7·4	15·4 12·2
35	11 53·8	11 55·7	11 21·2	3·5 2·8	9·5 7·5	15·5 12·3
36	11 54·0	11 56·0	11 21·5	3·6 2·9	9·6 7·6	15·6 12·4
37	11 54·3	11 56·2	11 21·7	3·7 2·9	9·7 7·7	15·7 12·4
38	11 54·5	11 56·5	11 22·0	3·8 3·0	9·8 7·8	15·8 12·5
39	11 54·8	11 56·7	11 22·2	3·9 3·1	9·9 7·8	15·9 12·6
40	11 55·0	11 57·0	11 22·4	4·0 3·2	10·0 7·9	16·0 12·7
41	11 55·3	11 57·2	11 22·7	4·1 3·2	10·1 8·0	16·1 12·7
42	11 55·5	11 57·5	11 22·9	4·2 3·3	10·2 8·1	16·2 12·8
43	11 55·8	11 57·7	11 23·1	4·3 3·4	10·3 8·2	16·3 12·9
44	11 56·0	11 58·0	11 23·4	4·4 3·5	10·4 8·2	16·4 13·0
45	11 56·3	11 58·2	11 23·6	4·5 3·6	10·5 8·3	16·5 13·1
46	11 56·5	11 58·5	11 23·9	4·6 3·6	10·6 8·4	16·6 13·1
47	11 56·8	11 58·7	11 24·1	4·7 3·7	10·7 8·5	16·7 13·2
48	11 57·0	11 59·0	11 24·3	4·8 3·8	10·8 8·6	16·8 13·3
49	11 57·3	11 59·2	11 24·6	4·9 3·9	10·9 8·6	16·9 13·4
50	11 57·5	11 59·5	11 24·8	5·0 4·0	11·0 8·7	17·0 13·5
51	11 57·8	11 59·7	11 25·1	5·1 4·0	11·1 8·8	17·1 13·5
52	11 58·0	12 00·0	11 25·3	5·2 4·1	11·2 8·9	17·2 13·6
53	11 58·3	12 00·2	11 25·5	5·3 4·2	11·3 8·9	17·3 13·7
54	11 58·5	12 00·5	11 25·8	5·4 4·3	11·4 9·0	17·4 13·8
55	11 58·8	12 00·7	11 26·0	5·5 4·4	11·5 9·1	17·5 13·9
56	11 59·0	12 01·0	11 26·2	5·6 4·4	11·6 9·2	17·6 13·9
57	11 59·3	12 01·2	11 26·5	5·7 4·5	11·7 9·3	17·7 14·0
58	11 59·5	12 01·5	11 26·7	5·8 4·6	11·8 9·3	17·8 14·1
59	11 59·8	12 01·7	11 27·0	5·9 4·7	11·9 9·4	17·9 14·2
60	12 00·0	12 02·0	11 27·2	6·0 4·8	12·0 9·5	18·0 14·3

48ᵐ

s	SUN PLANETS	ARIES	MOON	v or Corrⁿ d	v or Corrⁿ d	v or Corrⁿ d
00	12 00·0	12 02·0	11 27·2	0·0 0·0	6·0 4·9	12·0 9·7
01	12 00·3	12 02·2	11 27·4	0·1 0·1	6·1 4·9	12·1 9·8
02	12 00·5	12 02·5	11 27·7	0·2 0·2	6·2 5·0	12·2 9·9
03	12 00·8	12 02·7	11 27·9	0·3 0·2	6·3 5·1	12·3 9·9
04	12 01·0	12 03·0	11 28·2	0·4 0·3	6·4 5·2	12·4 10·0
05	12 01·3	12 03·2	11 28·4	0·5 0·4	6·5 5·3	12·5 10·1
06	12 01·5	12 03·5	11 28·6	0·6 0·5	6·6 5·3	12·6 10·2
07	12 01·8	12 03·7	11 28·9	0·7 0·6	6·7 5·4	12·7 10·3
08	12 02·0	12 04·0	11 29·1	0·8 0·6	6·8 5·5	12·8 10·3
09	12 02·3	12 04·2	11 29·3	0·9 0·7	6·9 5·6	12·9 10·4
10	12 02·5	12 04·5	11 29·6	1·0 0·8	7·0 5·7	13·0 10·5
11	12 02·8	12 04·7	11 29·8	1·1 0·9	7·1 5·7	13·1 10·6
12	12 03·0	12 05·0	11 30·1	1·2 1·0	7·2 5·8	13·2 10·7
13	12 03·3	12 05·2	11 30·3	1·3 1·1	7·3 5·9	13·3 10·8
14	12 03·5	12 05·5	11 30·5	1·4 1·1	7·4 6·0	13·4 10·8
15	12 03·8	12 05·7	11 30·8	1·5 1·2	7·5 6·1	13·5 10·9
16	12 04·0	12 06·0	11 31·0	1·6 1·3	7·6 6·1	13·6 11·0
17	12 04·3	12 06·2	11 31·3	1·7 1·4	7·7 6·2	13·7 11·1
18	12 04·5	12 06·5	11 31·5	1·8 1·5	7·8 6·3	13·8 11·2
19	12 04·8	12 06·7	11 31·7	1·9 1·5	7·9 6·4	13·9 11·2
20	12 05·0	12 07·0	11 32·0	2·0 1·6	8·0 6·5	14·0 11·3
21	12 05·3	12 07·2	11 32·2	2·1 1·7	8·1 6·5	14·1 11·4
22	12 05·5	12 07·5	11 32·4	2·2 1·8	8·2 6·6	14·2 11·5
23	12 05·8	12 07·7	11 32·7	2·3 1·9	8·3 6·7	14·3 11·6
24	12 06·0	12 08·0	11 32·9	2·4 1·9	8·4 6·8	14·4 11·6
25	12 06·3	12 08·2	11 33·2	2·5 2·0	8·5 6·9	14·5 11·7
26	12 06·5	12 08·5	11 33·4	2·6 2·1	8·6 7·0	14·6 11·8
27	12 06·8	12 08·7	11 33·6	2·7 2·2	8·7 7·0	14·7 11·9
28	12 07·0	12 09·0	11 33·9	2·8 2·3	8·8 7·1	14·8 12·0
29	12 07·3	12 09·2	11 34·1	2·9 2·3	8·9 7·2	14·9 12·0
30	12 07·5	12 09·5	11 34·4	3·0 2·4	9·0 7·3	15·0 12·1
31	12 07·8	12 09·7	11 34·6	3·1 2·5	9·1 7·4	15·1 12·2
32	12 08·0	12 10·0	11 34·8	3·2 2·6	9·2 7·4	15·2 12·3
33	12 08·3	12 10·2	11 35·1	3·3 2·7	9·3 7·5	15·3 12·4
34	12 08·5	12 10·5	11 35·3	3·4 2·7	9·4 7·6	15·4 12·4
35	12 08·8	12 10·7	11 35·6	3·5 2·8	9·5 7·7	15·5 12·5
36	12 09·0	12 11·0	11 35·8	3·6 2·9	9·6 7·8	15·6 12·6
37	12 09·3	12 11·2	11 36·0	3·7 3·0	9·7 7·8	15·7 12·7
38	12 09·5	12 11·5	11 36·3	3·8 3·1	9·8 7·9	15·8 12·8
39	12 09·8	12 11·7	11 36·5	3·9 3·2	9·9 8·0	15·9 12·9
40	12 10·0	12 12·0	11 36·7	4·0 3·2	10·0 8·1	16·0 12·9
41	12 10·3	12 12·2	11 37·0	4·1 3·3	10·1 8·2	16·1 13·0
42	12 10·5	12 12·5	11 37·2	4·2 3·4	10·2 8·2	16·2 13·1
43	12 10·8	12 12·8	11 37·5	4·3 3·5	10·3 8·3	16·3 13·2
44	12 11·0	12 13·0	11 37·7	4·4 3·6	10·4 8·4	16·4 13·3
45	12 11·3	12 13·3	11 37·9	4·5 3·6	10·5 8·5	16·5 13·3
46	12 11·5	12 13·5	11 38·2	4·6 3·7	10·6 8·6	16·6 13·4
47	12 11·8	12 13·8	11 38·4	4·7 3·8	10·7 8·6	16·7 13·5
48	12 12·0	12 14·0	11 38·7	4·8 3·9	10·8 8·7	16·8 13·6
49	12 12·3	12 14·3	11 38·9	4·9 4·0	10·9 8·8	16·9 13·7
50	12 12·5	12 14·5	11 39·1	5·0 4·0	11·0 8·9	17·0 13·7
51	12 12·8	12 14·8	11 39·4	5·1 4·1	11·1 9·0	17·1 13·8
52	12 13·0	12 15·0	11 39·6	5·2 4·2	11·2 9·1	17·2 13·9
53	12 13·3	12 15·3	11 39·8	5·3 4·3	11·3 9·1	17·3 14·0
54	12 13·5	12 15·5	11 40·1	5·4 4·4	11·4 9·2	17·4 14·1
55	12 13·8	12 15·8	11 40·3	5·5 4·4	11·5 9·3	17·5 14·1
56	12 14·0	12 16·0	11 40·6	5·6 4·5	11·6 9·4	17·6 14·2
57	12 14·3	12 16·3	11 40·8	5·7 4·6	11·7 9·5	17·7 14·3
58	12 14·5	12 16·5	11 41·0	5·8 4·7	11·8 9·5	17·8 14·4
59	12 14·8	12 16·8	11 41·3	5·9 4·8	11·9 9·6	17·9 14·5
60	12 15·0	12 17·0	11 41·5	6·0 4·9	12·0 9·7	18·0 14·6

49ᵐ

s	SUN PLANETS	ARIES	MOON	v or Corrⁿ d	v or Corrⁿ d	v or Corrⁿ d
00	12 15·0	12 17·0	11 41·5	0·0 0·0	6·0 5·0	12·0 9·9
01	12 15·3	12 17·3	11 41·8	0·1 0·1	6·1 5·0	12·1 10·0
02	12 15·5	12 17·5	11 42·0	0·2 0·2	6·2 5·1	12·2 10·1
03	12 15·8	12 17·8	11 42·2	0·3 0·2	6·3 5·2	12·3 10·1
04	12 16·0	12 18·0	11 42·5	0·4 0·3	6·4 5·3	12·4 10·2
05	12 16·3	12 18·3	11 42·7	0·5 0·4	6·5 5·4	12·5 10·3
06	12 16·5	12 18·5	11 42·9	0·6 0·5	6·6 5·4	12·6 10·4
07	12 16·8	12 18·8	11 43·2	0·7 0·6	6·7 5·5	12·7 10·5
08	12 17·0	12 19·0	11 43·4	0·8 0·7	6·8 5·6	12·8 10·6
09	12 17·3	12 19·3	11 43·7	0·9 0·7	6·9 5·7	12·9 10·6
10	12 17·5	12 19·5	11 43·9	1·0 0·8	7·0 5·8	13·0 10·7
11	12 17·8	12 19·8	11 44·1	1·1 0·9	7·1 5·9	13·1 10·8
12	12 18·0	12 20·0	11 44·4	1·2 1·0	7·2 5·9	13·2 10·9
13	12 18·3	12 20·3	11 44·6	1·3 1·1	7·3 6·0	13·3 11·0
14	12 18·5	12 20·5	11 44·9	1·4 1·2	7·4 6·1	13·4 11·1
15	12 18·8	12 20·8	11 45·1	1·5 1·2	7·5 6·2	13·5 11·1
16	12 19·0	12 21·0	11 45·3	1·6 1·3	7·6 6·3	13·6 11·2
17	12 19·3	12 21·3	11 45·6	1·7 1·4	7·7 6·4	13·7 11·3
18	12 19·5	12 21·5	11 45·8	1·8 1·5	7·8 6·4	13·8 11·4
19	12 19·8	12 21·8	11 46·1	1·9 1·6	7·9 6·5	13·9 11·5
20	12 20·0	12 22·0	11 46·3	2·0 1·7	8·0 6·6	14·0 11·6
21	12 20·3	12 22·3	11 46·5	2·1 1·7	8·1 6·7	14·1 11·6
22	12 20·5	12 22·5	11 46·8	2·2 1·8	8·2 6·8	14·2 11·7
23	12 20·8	12 22·8	11 47·0	2·3 1·9	8·3 6·8	14·3 11·8
24	12 21·0	12 23·0	11 47·2	2·4 2·0	8·4 6·9	14·4 11·9
25	12 21·3	12 23·3	11 47·5	2·5 2·1	8·5 7·0	14·5 12·0
26	12 21·5	12 23·5	11 47·7	2·6 2·1	8·6 7·1	14·6 12·0
27	12 21·8	12 23·8	11 48·0	2·7 2·2	8·7 7·2	14·7 12·1
28	12 22·0	12 24·0	11 48·2	2·8 2·3	8·8 7·3	14·8 12·2
29	12 22·3	12 24·3	11 48·4	2·9 2·4	8·9 7·3	14·9 12·3
30	12 22·5	12 24·5	11 48·7	3·0 2·5	9·0 7·4	15·0 12·4
31	12 22·8	12 24·8	11 48·9	3·1 2·6	9·1 7·5	15·1 12·5
32	12 23·0	12 25·0	11 49·2	3·2 2·6	9·2 7·6	15·2 12·5
33	12 23·3	12 25·3	11 49·4	3·3 2·7	9·3 7·7	15·3 12·6
34	12 23·5	12 25·5	11 49·6	3·4 2·8	9·4 7·8	15·4 12·7
35	12 23·8	12 25·8	11 49·9	3·5 2·9	9·5 7·8	15·5 12·8
36	12 24·0	12 26·0	11 50·1	3·6 3·0	9·6 7·9	15·6 12·9
37	12 24·3	12 26·3	11 50·3	3·7 3·1	9·7 8·0	15·7 13·0
38	12 24·5	12 26·5	11 50·6	3·8 3·1	9·8 8·1	15·8 13·0
39	12 24·8	12 26·8	11 50·8	3·9 3·2	9·9 8·2	15·9 13·1
40	12 25·0	12 27·0	11 51·1	4·0 3·3	10·0 8·3	16·0 13·2
41	12 25·3	12 27·3	11 51·3	4·1 3·4	10·1 8·3	16·1 13·3
42	12 25·5	12 27·5	11 51·5	4·2 3·5	10·2 8·4	16·2 13·4
43	12 25·8	12 27·8	11 51·8	4·3 3·5	10·3 8·5	16·3 13·4
44	12 26·0	12 28·0	11 52·0	4·4 3·6	10·4 8·6	16·4 13·5
45	12 26·3	12 28·3	11 52·3	4·5 3·7	10·5 8·7	16·5 13·6
46	12 26·5	12 28·5	11 52·5	4·6 3·8	10·6 8·7	16·6 13·7
47	12 26·8	12 28·8	11 52·7	4·7 3·9	10·7 8·8	16·7 13·8
48	12 27·0	12 29·0	11 53·0	4·8 4·0	10·8 8·9	16·8 13·9
49	12 27·3	12 29·3	11 53·2	4·9 4·0	10·9 9·0	16·9 13·9
50	12 27·5	12 29·5	11 53·4	5·0 4·1	11·0 9·1	17·0 14·0
51	12 27·8	12 29·8	11 53·7	5·1 4·2	11·1 9·2	17·1 14·1
52	12 28·0	12 30·0	11 53·9	5·2 4·3	11·2 9·2	17·2 14·2
53	12 28·3	12 30·3	11 54·2	5·3 4·4	11·3 9·3	17·3 14·3
54	12 28·5	12 30·5	11 54·4	5·4 4·5	11·4 9·4	17·4 14·4
55	12 28·8	12 30·8	11 54·6	5·5 4·5	11·5 9·5	17·5 14·4
56	12 29·0	12 31·1	11 54·9	5·6 4·6	11·6 9·6	17·6 14·5
57	12 29·3	12 31·3	11 55·1	5·7 4·7	11·7 9·7	17·7 14·6
58	12 29·5	12 31·6	11 55·4	5·8 4·8	11·8 9·7	17·8 14·7
59	12 29·8	12 31·8	11 55·6	5·9 4·9	11·9 9·8	17·9 14·8
60	12 30·0	12 32·1	11 55·8	6·0 5·0	12·0 9·9	18·0 14·9

50ᵐ	SUN PLANETS	ARIES	MOON	v or Corrⁿ d	v or Corrⁿ d	v or Corrⁿ d
s	° ′	° ′	° ′	′ ′	′ ′	′ ′
00	12 30·0	12 32·1	11 55·8	0·0 0·0	6·0 5·1	12·0 10·1
01	12 30·3	12 32·3	11 56·1	0·1 0·1	6·1 5·1	12·1 10·2
02	12 30·5	12 32·6	11 56·3	0·2 0·2	6·2 5·2	12·2 10·3
03	12 30·8	12 32·8	11 56·5	0·3 0·3	6·3 5·3	12·3 10·4
04	12 31·0	12 33·1	11 56·8	0·4 0·4	6·4 5·4	12·4 10·4
05	12 31·3	12 33·3	11 57·0	0·5 0·4	6·5 5·5	12·5 10·5
06	12 31·5	12 33·6	11 57·3	0·6 0·5	6·6 5·6	12·6 10·6
07	12 31·8	12 33·8	11 57·5	0·7 0·6	6·7 5·6	12·7 10·7
08	12 32·0	12 34·1	11 57·7	0·8 0·7	6·8 5·7	12·8 10·8
09	12 32·3	12 34·3	11 58·0	0·9 0·8	6·9 5·8	12·9 10·9
10	12 32·5	12 34·6	11 58·2	1·0 0·8	7·0 5·9	13·0 10·9
11	12 32·8	12 34·8	11 58·5	1·1 0·9	7·1 6·0	13·1 11·0
12	12 33·0	12 35·1	11 58·7	1·2 1·0	7·2 6·1	13·2 11·1
13	12 33·3	12 35·3	11 58·9	1·3 1·1	7·3 6·1	13·3 11·2
14	12 33·5	12 35·6	11 59·2	1·4 1·2	7·4 6·2	13·4 11·3
15	12 33·8	12 35·8	11 59·4	1·5 1·3	7·5 6·3	13·5 11·4
16	12 34·0	12 36·1	11 59·7	1·6 1·3	7·6 6·4	13·6 11·4
17	12 34·3	12 36·3	11 59·9	1·7 1·4	7·7 6·5	13·7 11·5
18	12 34·5	12 36·6	12 00·1	1·8 1·5	7·8 6·6	13·8 11·6
19	12 34·8	12 36·8	12 00·4	1·9 1·6	7·9 6·6	13·9 11·7
20	12 35·0	12 37·1	12 00·6	2·0 1·7	8·0 6·7	14·0 11·8
21	12 35·3	12 37·3	12 00·8	2·1 1·8	8·1 6·8	14·1 11·9
22	12 35·5	12 37·6	12 01·1	2·2 1·9	8·2 6·9	14·2 12·0
23	12 35·8	12 37·8	12 01·3	2·3 1·9	8·3 7·0	14·3 12·0
24	12 36·0	12 38·1	12 01·6	2·4 2·0	8·4 7·1	14·4 12·1
25	12 36·3	12 38·3	12 01·8	2·5 2·1	8·5 7·2	14·5 12·2
26	12 36·5	12 38·6	12 02·0	2·6 2·2	8·6 7·2	14·6 12·3
27	12 36·8	12 38·8	12 02·3	2·7 2·3	8·7 7·3	14·7 12·4
28	12 37·0	12 39·1	12 02·5	2·8 2·3	8·8 7·4	14·8 12·5
29	12 37·3	12 39·3	12 02·8	2·9 2·4	8·9 7·5	14·9 12·5
30	12 37·5	12 39·6	12 03·0	3·0 2·5	9·0 7·6	15·0 12·6
31	12 37·8	12 39·8	12 03·2	3·1 2·6	9·1 7·7	15·1 12·7
32	12 38·0	12 40·1	12 03·5	3·2 2·7	9·2 7·7	15·2 12·8
33	12 38·3	12 40·3	12 03·7	3·3 2·8	9·3 7·8	15·3 12·9
34	12 38·5	12 40·6	12 03·9	3·4 2·9	9·4 7·9	15·4 13·0
35	12 38·8	12 40·8	12 04·2	3·5 2·9	9·5 8·0	15·5 13·0
36	12 39·0	12 41·1	12 04·4	3·6 3·0	9·6 8·1	15·6 13·1
37	12 39·3	12 41·3	12 04·7	3·7 3·1	9·7 8·2	15·7 13·2
38	12 39·5	12 41·6	12 04·9	3·8 3·2	9·8 8·2	15·8 13·3
39	12 39·8	12 41·8	12 05·1	3·9 3·3	9·9 8·3	15·9 13·4
40	12 40·0	12 42·1	12 05·4	4·0 3·4	10·0 8·4	16·0 13·5
41	12 40·3	12 42·3	12 05·6	4·1 3·5	10·1 8·5	16·1 13·6
42	12 40·5	12 42·6	12 05·9	4·2 3·5	10·2 8·6	16·2 13·6
43	12 40·8	12 42·8	12 06·1	4·3 3·6	10·3 8·7	16·3 13·7
44	12 41·0	12 43·1	12 06·3	4·4 3·7	10·4 8·8	16·4 13·8
45	12 41·3	12 43·3	12 06·6	4·5 3·8	10·5 8·8	16·5 13·9
46	12 41·5	12 43·6	12 06·8	4·6 3·9	10·6 8·9	16·6 14·0
47	12 41·8	12 43·8	12 07·0	4·7 4·0	10·7 9·0	16·7 14·1
48	12 42·0	12 44·1	12 07·3	4·8 4·0	10·8 9·1	16·8 14·1
49	12 42·3	12 44·3	12 07·5	4·9 4·1	10·9 9·2	16·9 14·2
50	12 42·5	12 44·6	12 07·8	5·0 4·2	11·0 9·3	17·0 14·3
51	12 42·8	12 44·8	12 08·0	5·1 4·3	11·1 9·3	17·1 14·4
52	12 43·0	12 45·1	12 08·2	5·2 4·4	11·2 9·4	17·2 14·5
53	12 43·3	12 45·3	12 08·5	5·3 4·5	11·3 9·5	17·3 14·6
54	12 43·5	12 45·6	12 08·7	5·4 4·5	11·4 9·6	17·4 14·6
55	12 43·8	12 45·8	12 09·0	5·5 4·6	11·5 9·7	17·5 14·7
56	12 44·0	12 46·1	12 09·2	5·6 4·7	11·6 9·8	17·6 14·8
57	12 44·3	12 46·3	12 09·4	5·7 4·8	11·7 9·8	17·7 14·9
58	12 44·5	12 46·6	12 09·7	5·8 4·9	11·8 9·9	17·8 15·0
59	12 44·8	12 46·8	12 09·9	5·9 5·0	11·9 10·0	17·9 15·1
60	12 45·0	12 47·1	12 10·2	6·0 5·1	12·0 10·1	18·0 15·2

51ᵐ	SUN PLANETS	ARIES	MOON	v or Corrⁿ d	v or Corrⁿ d	v or Corrⁿ d
s	° ′	° ′	° ′	′ ′	′ ′	′ ′
00	12 45·0	12 47·1	12 10·2	0·0 0·0	6·0 5·2	12·0 10·3
01	12 45·3	12 47·3	12 10·4	0·1 0·1	6·1 5·2	12·1 10·4
02	12 45·5	12 47·6	12 10·6	0·2 0·2	6·2 5·3	12·2 10·5
03	12 45·8	12 47·8	12 10·9	0·3 0·3	6·3 5·4	12·3 10·6
04	12 46·0	12 48·1	12 11·1	0·4 0·3	6·4 5·5	12·4 10·6
05	12 46·3	12 48·3	12 11·3	0·5 0·4	6·5 5·6	12·5 10·7
06	12 46·5	12 48·6	12 11·6	0·6 0·5	6·6 5·7	12·6 10·8
07	12 46·8	12 48·8	12 11·8	0·7 0·6	6·7 5·8	12·7 10·9
08	12 47·0	12 49·1	12 12·1	0·8 0·7	6·8 5·8	12·8 11·0
09	12 47·3	12 49·4	12 12·3	0·9 0·8	6·9 5·9	12·9 11·1
10	12 47·5	12 49·6	12 12·5	1·0 0·9	7·0 6·0	13·0 11·2
11	12 47·8	12 49·9	12 12·8	1·1 0·9	7·1 6·1	13·1 11·2
12	12 48·0	12 50·1	12 13·0	1·2 1·0	7·2 6·2	13·2 11·3
13	12 48·3	12 50·4	12 13·3	1·3 1·1	7·3 6·3	13·3 11·4
14	12 48·5	12 50·6	12 13·5	1·4 1·2	7·4 6·4	13·4 11·5
15	12 48·8	12 50·9	12 13·7	1·5 1·3	7·5 6·4	13·5 11·6
16	12 49·0	12 51·1	12 14·0	1·6 1·4	7·6 6·5	13·6 11·7
17	12 49·3	12 51·4	12 14·2	1·7 1·5	7·7 6·6	13·7 11·8
18	12 49·5	12 51·6	12 14·4	1·8 1·5	7·8 6·7	13·8 11·8
19	12 49·8	12 51·9	12 14·7	1·9 1·6	7·9 6·8	13·9 11·9
20	12 50·0	12 52·1	12 14·9	2·0 1·7	8·0 6·9	14·0 12·0
21	12 50·3	12 52·4	12 15·2	2·1 1·8	8·1 7·0	14·1 12·1
22	12 50·5	12 52·6	12 15·4	2·2 1·9	8·2 7·0	14·2 12·2
23	12 50·8	12 52·9	12 15·6	2·3 2·0	8·3 7·1	14·3 12·3
24	12 51·0	12 53·1	12 15·9	2·4 2·1	8·4 7·2	14·4 12·4
25	12 51·3	12 53·4	12 16·1	2·5 2·1	8·5 7·3	14·5 12·4
26	12 51·5	12 53·6	12 16·4	2·6 2·2	8·6 7·4	14·6 12·5
27	12 51·8	12 53·9	12 16·6	2·7 2·3	8·7 7·5	14·7 12·6
28	12 52·0	12 54·1	12 16·8	2·8 2·4	8·8 7·5	14·8 12·7
29	12 52·3	12 54·4	12 17·1	2·9 2·5	8·9 7·6	14·9 12·8
30	12 52·5	12 54·6	12 17·3	3·0 2·6	9·0 7·7	15·0 12·9
31	12 52·8	12 54·9	12 17·5	3·1 2·7	9·1 7·8	15·1 13·0
32	12 53·0	12 55·1	12 17·8	3·2 2·7	9·2 7·9	15·2 13·0
33	12 53·3	12 55·4	12 18·0	3·3 2·8	9·3 8·0	15·3 13·1
34	12 53·5	12 55·6	12 18·3	3·4 2·9	9·4 8·1	15·4 13·2
35	12 53·8	12 55·9	12 18·5	3·5 3·0	9·5 8·2	15·5 13·3
36	12 54·0	12 56·1	12 18·7	3·6 3·1	9·6 8·2	15·6 13·4
37	12 54·3	12 56·4	12 19·0	3·7 3·2	9·7 8·3	15·7 13·5
38	12 54·5	12 56·6	12 19·2	3·8 3·3	9·8 8·4	15·8 13·6
39	12 54·8	12 56·9	12 19·5	3·9 3·3	9·9 8·5	15·9 13·6
40	12 55·0	12 57·1	12 19·7	4·0 3·4	10·0 8·6	16·0 13·7
41	12 55·3	12 57·4	12 19·9	4·1 3·5	10·1 8·7	16·1 13·8
42	12 55·5	12 57·6	12 20·2	4·2 3·6	10·2 8·8	16·2 13·9
43	12 55·8	12 57·9	12 20·4	4·3 3·7	10·3 8·8	16·3 14·0
44	12 56·0	12 58·1	12 20·6	4·4 3·8	10·4 8·9	16·4 14·1
45	12 56·3	12 58·4	12 20·9	4·5 3·9	10·5 9·0	16·5 14·2
46	12 56·5	12 58·6	12 21·1	4·6 3·9	10·6 9·1	16·6 14·2
47	12 56·8	12 58·9	12 21·4	4·7 4·0	10·7 9·2	16·7 14·3
48	12 57·0	12 59·1	12 21·6	4·8 4·1	10·8 9·3	16·8 14·4
49	12 57·3	12 59·4	12 21·8	4·9 4·2	10·9 9·4	16·9 14·5
50	12 57·5	12 59·6	12 22·1	5·0 4·3	11·0 9·4	17·0 14·6
51	12 57·8	12 59·9	12 22·3	5·1 4·4	11·1 9·5	17·1 14·7
52	12 58·0	13 00·1	12 22·6	5·2 4·5	11·2 9·6	17·2 14·8
53	12 58·3	13 00·4	12 22·8	5·3 4·5	11·3 9·7	17·3 14·8
54	12 58·5	13 00·6	12 23·0	5·4 4·6	11·4 9·8	17·4 14·9
55	12 58·8	13 00·9	12 23·3	5·5 4·7	11·5 9·9	17·5 15·0
56	12 59·0	13 01·1	12 23·5	5·6 4·8	11·6 10·0	17·6 15·1
57	12 59·3	13 01·4	12 23·8	5·7 4·9	11·7 10·0	17·7 15·2
58	12 59·5	13 01·6	12 24·0	5·8 5·0	11·8 10·1	17·8 15·3
59	12 59·8	13 01·9	12 24·2	5·9 5·1	11·9 10·2	17·9 15·4
60	13 00·0	13 02·1	12 24·5	6·0 5·2	12·0 10·3	18·0 15·5

52ᵐ

52ᵐ	SUN PLANETS	ARIES	MOON	v or Corrn d	v or Corrn d	v or Corrn d
s	° ′	° ′	° ′	′ ′	′ ′	′ ′
00	13 00·0	13 02·1	12 24·5	0·0 0·0	6·0 5·3	12·0 10·5
01	13 00·3	13 02·4	12 24·7	0·1 0·1	6·1 5·3	12·1 10·6
02	13 00·5	13 02·6	12 24·9	0·2 0·2	6·2 5·4	12·2 10·7
03	13 00·8	13 02·9	12 25·2	0·3 0·3	6·3 5·5	12·3 10·8
04	13 01·0	13 03·1	12 25·4	0·4 0·4	6·4 5·6	12·4 10·9
05	13 01·3	13 03·4	12 25·7	0·5 0·4	6·5 5·7	12·5 10·9
06	13 01·5	13 03·6	12 25·9	0·6 0·5	6·6 5·8	12·6 11·0
07	13 01·8	13 03·9	12 26·1	0·7 0·6	6·7 5·9	12·7 11·1
08	13 02·0	13 04·1	12 26·4	0·8 0·7	6·8 6·0	12·8 11·2
09	13 02·3	13 04·4	12 26·6	0·9 0·8	6·9 6·0	12·9 11·3
10	13 02·5	13 04·6	12 26·9	1·0 0·9	7·0 6·1	13·0 11·4
11	13 02·8	13 04·9	12 27·1	1·1 1·0	7·1 6·2	13·1 11·5
12	13 03·0	13 05·1	12 27·3	1·2 1·1	7·2 6·3	13·2 11·6
13	13 03·3	13 05·4	12 27·6	1·3 1·1	7·3 6·4	13·3 11·6
14	13 03·5	13 05·6	12 27·8	1·4 1·2	7·4 6·5	13·4 11·7
15	13 03·8	13 05·9	12 28·0	1·5 1·3	7·5 6·6	13·5 11·8
16	13 04·0	13 06·1	12 28·3	1·6 1·4	7·6 6·7	13·6 11·9
17	13 04·3	13 06·4	12 28·5	1·7 1·5	7·7 6·7	13·7 12·0
18	13 04·5	13 06·6	12 28·8	1·8 1·6	7·8 6·8	13·8 12·1
19	13 04·8	13 06·9	12 29·0	1·9 1·7	7·9 6·9	13·9 12·2
20	13 05·0	13 07·1	12 29·2	2·0 1·8	8·0 7·0	14·0 12·3
21	13 05·3	13 07·4	12 29·5	2·1 1·8	8·1 7·1	14·1 12·3
22	13 05·5	13 07·7	12 29·7	2·2 1·9	8·2 7·2	14·2 12·4
23	13 05·8	13 07·9	12 30·0	2·3 2·0	8·3 7·3	14·3 12·5
24	13 06·0	13 08·2	12 30·2	2·4 2·1	8·4 7·4	14·4 12·6
25	13 06·3	13 08·4	12 30·4	2·5 2·2	8·5 7·4	14·5 12·7
26	13 06·5	13 08·7	12 30·7	2·6 2·3	8·6 7·5	14·6 12·8
27	13 06·8	13 08·9	12 30·9	2·7 2·4	8·7 7·6	14·7 12·9
28	13 07·0	13 09·2	12 31·1	2·8 2·5	8·8 7·7	14·8 13·0
29	13 07·3	13 09·4	12 31·4	2·9 2·5	8·9 7·8	14·9 13·0
30	13 07·5	13 09·7	12 31·6	3·0 2·6	9·0 7·9	15·0 13·1
31	13 07·8	13 09·9	12 31·9	3·1 2·7	9·1 8·0	15·1 13·2
32	13 08·0	13 10·2	12 32·1	3·2 2·8	9·2 8·1	15·2 13·3
33	13 08·3	13 10·4	12 32·3	3·3 2·9	9·3 8·1	15·3 13·4
34	13 08·5	13 10·7	12 32·6	3·4 3·0	9·4 8·2	15·4 13·5
35	13 08·8	13 10·9	12 32·8	3·5 3·1	9·5 8·3	15·5 13·6
36	13 09·0	13 11·2	12 33·1	3·6 3·2	9·6 8·4	15·6 13·7
37	13 09·3	13 11·4	12 33·3	3·7 3·2	9·7 8·5	15·7 13·7
38	13 09·5	13 11·7	12 33·5	3·8 3·3	9·8 8·6	15·8 13·8
39	13 09·8	13 11·9	12 33·8	3·9 3·4	9·9 8·7	15·9 13·9
40	13 10·0	13 12·2	12 34·0	4·0 3·5	10·0 8·8	16·0 14·0
41	13 10·3	13 12·4	12 34·2	4·1 3·6	10·1 8·8	16·1 14·1
42	13 10·5	13 12·7	12 34·5	4·2 3·7	10·2 8·9	16·2 14·2
43	13 10·8	13 12·9	12 34·7	4·3 3·8	10·3 9·0	16·3 14·3
44	13 11·0	13 13·2	12 35·0	4·4 3·9	10·4 9·1	16·4 14·4
45	13 11·3	13 13·4	12 35·2	4·5 3·9	10·5 9·2	16·5 14·4
46	13 11·5	13 13·7	12 35·4	4·6 4·0	10·6 9·3	16·6 14·5
47	13 11·8	13 13·9	12 35·7	4·7 4·1	10·7 9·4	16·7 14·6
48	13 12·0	13 14·2	12 35·9	4·8 4·2	10·8 9·5	16·8 14·7
49	13 12·3	13 14·4	12 36·2	4·9 4·3	10·9 9·5	16·9 14·8
50	13 12·5	13 14·7	12 36·4	5·0 4·4	11·0 9·6	17·0 14·9
51	13 12·8	13 14·9	12 36·6	5·1 4·5	11·1 9·7	17·1 15·0
52	13 13·0	13 15·2	12 36·9	5·2 4·6	11·2 9·8	17·2 15·1
53	13 13·3	13 15·4	12 37·1	5·3 4·6	11·3 9·9	17·3 15·1
54	13 13·5	13 15·7	12 37·4	5·4 4·7	11·4 10·0	17·4 15·2
55	13 13·8	13 15·9	12 37·6	5·5 4·8	11·5 10·1	17·5 15·3
56	13 14·0	13 16·2	12 37·8	5·6 4·9	11·6 10·2	17·6 15·4
57	13 14·3	13 16·4	12 38·1	5·7 5·0	11·7 10·2	17·7 15·5
58	13 14·5	13 16·7	12 38·3	5·8 5·1	11·8 10·3	17·8 15·6
59	13 14·8	13 16·9	12 38·5	5·9 5·2	11·9 10·4	17·9 15·7
60	13 15·0	13 17·2	12 38·8	6·0 5·3	12·0 10·5	18·0 15·8

53ᵐ

53ᵐ	SUN PLANETS	ARIES	MOON	v or Corrn d	v or Corrn d	v or Corrn d
s	° ′	° ′	° ′	′ ′	′ ′	′ ′
00	13 15·0	13 17·2	12 38·8	0·0 0·0	6·0 5·4	12·0 10·7
01	13 15·3	13 17·4	12 39·0	0·1 0·1	6·1 5·4	12·1 10·8
02	13 15·5	13 17·7	12 39·3	0·2 0·2	6·2 5·5	12·2 10·9
03	13 15·8	13 17·9	12 39·5	0·3 0·3	6·3 5·6	12·3 11·0
04	13 16·0	13 18·2	12 39·7	0·4 0·4	6·4 5·7	12·4 11·1
05	13 16·3	13 18·4	12 40·0	0·5 0·4	6·5 5·8	12·5 11·1
06	13 16·5	13 18·7	12 40·2	0·6 0·5	6·6 5·9	12·6 11·2
07	13 16·8	13 18·9	12 40·5	0·7 0·6	6·7 6·0	12·7 11·3
08	13 17·0	13 19·2	12 40·7	0·8 0·7	6·8 6·1	12·8 11·4
09	13 17·3	13 19·4	12 40·9	0·9 0·8	6·9 6·2	12·9 11·5
10	13 17·5	13 19·7	12 41·2	1·0 0·9	7·0 6·2	13·0 11·6
11	13 17·8	13 19·9	12 41·4	1·1 1·0	7·1 6·3	13·1 11·7
12	13 18·0	13 20·2	12 41·6	1·2 1·1	7·2 6·4	13·2 11·8
13	13 18·3	13 20·4	12 41·9	1·3 1·2	7·3 6·5	13·3 11·9
14	13 18·5	13 20·7	12 42·1	1·4 1·2	7·4 6·6	13·4 11·9
15	13 18·8	13 20·9	12 42·4	1·5 1·3	7·5 6·7	13·5 12·0
16	13 19·0	13 21·2	12 42·6	1·6 1·4	7·6 6·8	13·6 12·1
17	13 19·3	13 21·4	12 42·8	1·7 1·5	7·7 6·9	13·7 12·2
18	13 19·5	13 21·7	12 43·1	1·8 1·6	7·8 7·0	13·8 12·3
19	13 19·8	13 21·9	12 43·3	1·9 1·7	7·9 7·0	13·9 12·4
20	13 20·0	13 22·2	12 43·6	2·0 1·8	8·0 7·1	14·0 12·5
21	13 20·3	13 22·4	12 43·8	2·1 1·9	8·1 7·2	14·1 12·6
22	13 20·5	13 22·7	12 44·0	2·2 2·0	8·2 7·3	14·2 12·7
23	13 20·8	13 22·9	12 44·3	2·3 2·1	8·3 7·4	14·3 12·8
24	13 21·0	13 23·2	12 44·5	2·4 2·1	8·4 7·5	14·4 12·8
25	13 21·3	13 23·4	12 44·7	2·5 2·2	8·5 7·6	14·5 12·9
26	13 21·5	13 23·7	12 45·0	2·6 2·3	8·6 7·7	14·6 13·0
27	13 21·8	13 23·9	12 45·2	2·7 2·4	8·7 7·8	14·7 13·1
28	13 22·0	13 24·2	12 45·5	2·8 2·5	8·8 7·8	14·8 13·2
29	13 22·3	13 24·4	12 45·7	2·9 2·6	8·9 7·9	14·9 13·3
30	13 22·5	13 24·7	12 45·9	3·0 2·7	9·0 8·0	15·0 13·4
31	13 22·8	13 24·9	12 46·2	3·1 2·8	9·1 8·1	15·1 13·5
32	13 23·0	13 25·2	12 46·4	3·2 2·9	9·2 8·2	15·2 13·6
33	13 23·3	13 25·4	12 46·7	3·3 2·9	9·3 8·3	15·3 13·6
34	13 23·5	13 25·7	12 46·9	3·4 3·0	9·4 8·4	15·4 13·7
35	13 23·8	13 26·0	12 47·1	3·5 3·1	9·5 8·5	15·5 13·8
36	13 24·0	13 26·2	12 47·4	3·6 3·2	9·6 8·6	15·6 13·9
37	13 24·3	13 26·5	12 47·6	3·7 3·3	9·7 8·6	15·7 14·0
38	13 24·5	13 26·7	12 47·9	3·8 3·4	9·8 8·7	15·8 14·1
39	13 24·8	13 27·0	12 48·1	3·9 3·5	9·9 8·8	15·9 14·2
40	13 25·0	13 27·2	12 48·3	4·0 3·6	10·0 8·9	16·0 14·3
41	13 25·3	13 27·5	12 48·6	4·1 3·7	10·1 9·0	16·1 14·4
42	13 25·5	13 27·7	12 48·8	4·2 3·7	10·2 9·1	16·2 14·4
43	13 25·8	13 28·0	12 49·0	4·3 3·8	10·3 9·2	16·3 14·5
44	13 26·0	13 28·2	12 49·3	4·4 3·9	10·4 9·3	16·4 14·6
45	13 26·3	13 28·5	12 49·5	4·5 4·0	10·5 9·4	16·5 14·7
46	13 26·5	13 28·7	12 49·8	4·6 4·1	10·6 9·5	16·6 14·8
47	13 26·8	13 29·0	12 50·0	4·7 4·2	10·7 9·5	16·7 14·9
48	13 27·0	13 29·2	12 50·2	4·8 4·3	10·8 9·6	16·8 15·0
49	13 27·3	13 29·5	12 50·5	4·9 4·4	10·9 9·7	16·9 15·1
50	13 27·5	13 29·7	12 50·7	5·0 4·5	11·0 9·8	17·0 15·2
51	13 27·8	13 30·0	12 51·0	5·1 4·5	11·1 9·9	17·1 15·2
52	13 28·0	13 30·2	12 51·2	5·2 4·6	11·2 10·0	17·2 15·3
53	13 28·3	13 30·5	12 51·4	5·3 4·7	11·3 10·1	17·3 15·4
54	13 28·5	13 30·7	12 51·7	5·4 4·8	11·4 10·2	17·4 15·5
55	13 28·8	13 31·0	12 51·9	5·5 4·9	11·5 10·3	17·5 15·6
56	13 29·0	13 31·2	12 52·1	5·6 5·0	11·6 10·3	17·6 15·7
57	13 29·3	13 31·5	12 52·4	5·7 5·1	11·7 10·4	17·7 15·8
58	13 29·5	13 31·7	12 52·6	5·8 5·2	11·8 10·5	17·8 15·9
59	13 29·8	13 32·0	12 52·9	5·9 5·3	11·9 10·6	17·9 16·0
60	13 30·0	13 32·2	12 53·1	6·0 5·4	12·0 10·7	18·0 16·1

54ᵐ s	SUN PLANETS	ARIES	MOON	v or Corrⁿ d	v or Corrⁿ d	v or Corrⁿ d
00	13 30·0	13 32·2	12 53·1	0·0 0·0	6·0 5·5	12·0 10·9
01	13 30·3	13 32·5	12 53·3	0·1 0·1	6·1 5·5	12·1 11·0
02	13 30·5	13 32·7	12 53·6	0·2 0·2	6·2 5·6	12·2 11·1
03	13 30·8	13 33·0	12 53·8	0·3 0·3	6·3 5·7	12·3 11·2
04	13 31·0	13 33·2	12 54·1	0·4 0·4	6·4 5·8	12·4 11·3
05	13 31·3	13 33·5	12 54·3	0·5 0·5	6·5 5·9	12·5 11·4
06	13 31·5	13 33·7	12 54·5	0·6 0·5	6·6 6·0	12·6 11·4
07	13 31·8	13 34·0	12 54·8	0·7 0·6	6·7 6·1	12·7 11·5
08	13 32·0	13 34·2	12 55·0	0·8 0·7	6·8 6·2	12·8 11·6
09	13 32·3	13 34·5	12 55·2	0·9 0·8	6·9 6·3	12·9 11·7
10	13 32·5	13 34·7	12 55·5	1·0 0·9	7·0 6·4	13·0 11·8
11	13 32·8	13 35·0	12 55·7	1·1 1·0	7·1 6·4	13·1 11·9
12	13 33·0	13 35·2	12 56·0	1·2 1·1	7·2 6·5	13·2 12·0
13	13 33·3	13 35·5	12 56·2	1·3 1·2	7·3 6·6	13·3 12·1
14	13 33·5	13 35·7	12 56·4	1·4 1·3	7·4 6·7	13·4 12·2
15	13 33·8	13 36·0	12 56·7	1·5 1·4	7·5 6·8	13·5 12·3
16	13 34·0	13 36·2	12 56·9	1·6 1·5	7·6 6·9	13·6 12·4
17	13 34·3	13 36·5	12 57·2	1·7 1·5	7·7 7·0	13·7 12·4
18	13 34·5	13 36·7	12 57·4	1·8 1·6	7·8 7·1	13·8 12·5
19	13 34·8	13 37·0	12 57·6	1·9 1·7	7·9 7·2	13·9 12·6
20	13 35·0	13 37·2	12 57·9	2·0 1·8	8·0 7·3	14·0 12·7
21	13 35·3	13 37·5	12 58·1	2·1 1·9	8·1 7·4	14·1 12·8
22	13 35·5	13 37·7	12 58·3	2·2 2·0	8·2 7·4	14·2 12·9
23	13 35·8	13 38·0	12 58·6	2·3 2·1	8·3 7·5	14·3 13·0
24	13 36·0	13 38·2	12 58·8	2·4 2·2	8·4 7·6	14·4 13·1
25	13 36·3	13 38·5	12 59·1	2·5 2·3	8·5 7·7	14·5 13·2
26	13 36·5	13 38·7	12 59·3	2·6 2·4	8·6 7·8	14·6 13·3
27	13 36·8	13 39·0	12 59·5	2·7 2·5	8·7 7·9	14·7 13·4
28	13 37·0	13 39·2	12 59·8	2·8 2·5	8·8 8·0	14·8 13·4
29	13 37·3	13 39·5	13 00·0	2·9 2·6	8·9 8·1	14·9 13·5
30	13 37·5	13 39·7	13 00·3	3·0 2·7	9·0 8·2	15·0 13·6
31	13 37·8	13 40·0	13 00·5	3·1 2·8	9·1 8·3	15·1 13·7
32	13 38·0	13 40·2	13 00·7	3·2 2·9	9·2 8·4	15·2 13·8
33	13 38·3	13 40·5	13 01·0	3·3 3·0	9·3 8·4	15·3 13·9
34	13 38·5	13 40·7	13 01·2	3·4 3·1	9·4 8·5	15·4 14·0
35	13 38·8	13 41·0	13 01·5	3·5 3·2	9·5 8·6	15·5 14·1
36	13 39·0	13 41·2	13 01·7	3·6 3·3	9·6 8·7	15·6 14·2
37	13 39·3	13 41·5	13 01·9	3·7 3·4	9·7 8·8	15·7 14·3
38	13 39·5	13 41·7	13 02·2	3·8 3·5	9·8 8·9	15·8 14·4
39	13 39·8	13 42·0	13 02·4	3·9 3·5	9·9 9·0	15·9 14·4
40	13 40·0	13 42·2	13 02·6	4·0 3·6	10·0 9·1	16·0 14·5
41	13 40·3	13 42·5	13 02·9	4·1 3·7	10·1 9·2	16·1 14·6
42	13 40·5	13 42·7	13 03·1	4·2 3·8	10·2 9·3	16·2 14·7
43	13 40·8	13 43·0	13 03·4	4·3 3·9	10·3 9·4	16·3 14·8
44	13 41·0	13 43·2	13 03·6	4·4 4·0	10·4 9·4	16·4 14·9
45	13 41·3	13 43·5	13 03·8	4·5 4·1	10·5 9·5	16·5 15·0
46	13 41·5	13 43·7	13 04·1	4·6 4·2	10·6 9·6	16·6 15·1
47	13 41·8	13 44·0	13 04·3	4·7 4·3	10·7 9·7	16·7 15·2
48	13 42·0	13 44·2	13 04·6	4·8 4·4	10·8 9·8	16·8 15·3
49	13 42·3	13 44·5	13 04·8	4·9 4·5	10·9 9·9	16·9 15·4
50	13 42·5	13 44·8	13 05·0	5·0 4·5	11·0 10·0	17·0 15·4
51	13 42·8	13 45·0	13 05·3	5·1 4·6	11·1 10·1	17·1 15·5
52	13 43·0	13 45·3	13 05·5	5·2 4·7	11·2 10·2	17·2 15·6
53	13 43·3	13 45·5	13 05·7	5·3 4·8	11·3 10·3	17·3 15·7
54	13 43·5	13 45·8	13 06·0	5·4 4·9	11·4 10·4	17·4 15·8
55	13 43·8	13 46·0	13 06·2	5·5 5·0	11·5 10·4	17·5 15·9
56	13 44·0	13 46·3	13 06·5	5·6 5·1	11·6 10·5	17·6 16·0
57	13 44·3	13 46·5	13 06·7	5·7 5·2	11·7 10·6	17·7 16·1
58	13 44·5	13 46·8	13 06·9	5·8 5·3	11·8 10·7	17·8 16·2
59	13 44·8	13 47·0	13 07·2	5·9 5·4	11·9 10·8	17·9 16·3
60	13 45·0	13 47·3	13 07·4	6·0 5·5	12·0 10·9	18·0 16·4

55ᵐ s	SUN PLANETS	ARIES	MOON	v or Corrⁿ d	v or Corrⁿ d	v or Corrⁿ d
00	13 45·0	13 47·3	13 07·4	0·0 0·0	6·0 5·6	12·0 11·1
01	13 45·3	13 47·5	13 07·7	0·1 0·1	6·1 5·6	12·1 11·2
02	13 45·5	13 47·8	13 07·9	0·2 0·2	6·2 5·7	12·2 11·3
03	13 45·8	13 48·0	13 08·1	0·3 0·3	6·3 5·8	12·3 11·4
04	13 46·0	13 48·3	13 08·4	0·4 0·4	6·4 5·9	12·4 11·5
05	13 46·3	13 48·5	13 08·6	0·5 0·5	6·5 6·0	12·5 11·6
06	13 46·5	13 48·8	13 08·8	0·6 0·6	6·6 6·1	12·6 11·7
07	13 46·8	13 49·0	13 09·1	0·7 0·6	6·7 6·2	12·7 11·7
08	13 47·0	13 49·3	13 09·3	0·8 0·7	6·8 6·3	12·8 11·8
09	13 47·3	13 49·5	13 09·6	0·9 0·8	6·9 6·4	12·9 11·9
10	13 47·5	13 49·8	13 09·8	1·0 0·9	7·0 6·5	13·0 12·0
11	13 47·8	13 50·0	13 10·0	1·1 1·0	7·1 6·6	13·1 12·1
12	13 48·0	13 50·3	13 10·3	1·2 1·1	7·2 6·7	13·2 12·2
13	13 48·3	13 50·5	13 10·5	1·3 1·2	7·3 6·8	13·3 12·3
14	13 48·5	13 50·8	13 10·8	1·4 1·3	7·4 6·8	13·4 12·4
15	13 48·8	13 51·0	13 11·0	1·5 1·4	7·5 6·9	13·5 12·5
16	13 49·0	13 51·3	13 11·2	1·6 1·5	7·6 7·0	13·6 12·6
17	13 49·3	13 51·5	13 11·5	1·7 1·6	7·7 7·1	13·7 12·7
18	13 49·5	13 51·8	13 11·7	1·8 1·7	7·8 7·2	13·8 12·8
19	13 49·8	13 52·0	13 12·0	1·9 1·8	7·9 7·3	13·9 12·9
20	13 50·0	13 52·3	13 12·2	2·0 1·9	8·0 7·4	14·0 13·0
21	13 50·3	13 52·5	13 12·4	2·1 1·9	8·1 7·5	14·1 13·0
22	13 50·5	13 52·8	13 12·7	2·2 2·0	8·2 7·6	14·2 13·1
23	13 50·8	13 53·0	13 12·9	2·3 2·1	8·3 7·7	14·3 13·2
24	13 51·0	13 53·3	13 13·1	2·4 2·2	8·4 7·8	14·4 13·3
25	13 51·3	13 53·5	13 13·4	2·5 2·3	8·5 7·9	14·5 13·4
26	13 51·5	13 53·8	13 13·6	2·6 2·4	8·6 8·0	14·6 13·5
27	13 51·8	13 54·0	13 13·9	2·7 2·5	8·7 8·0	14·7 13·6
28	13 52·0	13 54·3	13 14·1	2·8 2·6	8·8 8·1	14·8 13·7
29	13 52·3	13 54·5	13 14·3	2·9 2·7	8·9 8·2	14·9 13·8
30	13 52·5	13 54·8	13 14·6	3·0 2·8	9·0 8·3	15·0 13·9
31	13 52·8	13 55·0	13 14·8	3·1 2·9	9·1 8·4	15·1 14·0
32	13 53·0	13 55·3	13 15·1	3·2 3·0	9·2 8·5	15·2 14·1
33	13 53·3	13 55·5	13 15·3	3·3 3·1	9·3 8·6	15·3 14·2
34	13 53·5	13 55·8	13 15·5	3·4 3·1	9·4 8·7	15·4 14·2
35	13 53·8	13 56·0	13 15·8	3·5 3·2	9·5 8·8	15·5 14·3
36	13 54·0	13 56·3	13 16·0	3·6 3·3	9·6 8·9	15·6 14·4
37	13 54·3	13 56·5	13 16·2	3·7 3·4	9·7 9·0	15·7 14·5
38	13 54·5	13 56·8	13 16·5	3·8 3·5	9·8 9·1	15·8 14·6
39	13 54·8	13 57·0	13 16·7	3·9 3·6	9·9 9·2	15·9 14·7
40	13 55·0	13 57·3	13 17·0	4·0 3·7	10·0 9·3	16·0 14·8
41	13 55·3	13 57·5	13 17·2	4·1 3·8	10·1 9·3	16·1 14·9
42	13 55·5	13 57·8	13 17·4	4·2 3·9	10·2 9·4	16·2 15·0
43	13 55·8	13 58·0	13 17·7	4·3 4·0	10·3 9·5	16·3 15·1
44	13 56·0	13 58·3	13 17·9	4·4 4·1	10·4 9·6	16·4 15·2
45	13 56·3	13 58·5	13 18·2	4·5 4·2	10·5 9·7	16·5 15·3
46	13 56·5	13 58·8	13 18·4	4·6 4·3	10·6 9·8	16·6 15·4
47	13 56·8	13 59·0	13 18·6	4·7 4·3	10·7 9·9	16·7 15·4
48	13 57·0	13 59·3	13 18·9	4·8 4·4	10·8 10·0	16·8 15·5
49	13 57·3	13 59·5	13 19·1	4·9 4·5	10·9 10·1	16·9 15·6
50	13 57·5	13 59·8	13 19·3	5·0 4·6	11·0 10·2	17·0 15·7
51	13 57·8	14 00·0	13 19·6	5·1 4·7	11·1 10·3	17·1 15·8
52	13 58·0	14 00·3	13 19·8	5·2 4·8	11·2 10·4	17·2 15·9
53	13 58·3	14 00·5	13 20·1	5·3 4·9	11·3 10·5	17·3 16·0
54	13 58·5	14 00·8	13 20·3	5·4 5·0	11·4 10·5	17·4 16·1
55	13 58·8	14 01·0	13 20·5	5·5 5·1	11·5 10·6	17·5 16·2
56	13 59·0	14 01·3	13 20·8	5·6 5·2	11·6 10·7	17·6 16·3
57	13 59·3	14 01·5	13 21·0	5·7 5·3	11·7 10·8	17·7 16·4
58	13 59·5	14 01·8	13 21·3	5·8 5·4	11·8 10·9	17·8 16·5
59	13 59·8	14 02·0	13 21·5	5·9 5·5	11·9 11·0	17·9 16·6
60	14 00·0	14 02·3	13 21·7	6·0 5·6	12·0 11·1	18·0 16·7

56ᵐ

56ᵐ s	SUN PLANETS	ARIES	MOON	v or Corrⁿ d		v or Corrⁿ d		v or Corrⁿ d	
00	14 00.0	14 02.3	13 21.7	0.0	0.0	6.0	5.7	12.0	11.3
01	14 00.3	14 02.6	13 22.0	0.1	0.1	6.1	5.7	12.1	11.4
02	14 00.5	14 02.8	13 22.2	0.2	0.2	6.2	5.8	12.2	11.5
03	14 00.8	14 03.1	13 22.4	0.3	0.3	6.3	5.9	12.3	11.6
04	14 01.0	14 03.3	13 22.7	0.4	0.4	6.4	6.0	12.4	11.7
05	14 01.3	14 03.6	13 22.9	0.5	0.5	6.5	6.1	12.5	11.8
06	14 01.5	14 03.8	13 23.2	0.6	0.6	6.6	6.2	12.6	11.9
07	14 01.8	14 04.1	13 23.4	0.7	0.7	6.7	6.3	12.7	12.0
08	14 02.0	14 04.3	13 23.7	0.8	0.8	6.8	6.4	12.8	12.1
09	14 02.3	14 04.6	13 23.9	0.9	0.8	6.9	6.5	12.9	12.1
10	14 02.5	14 04.8	13 24.1	1.0	0.9	7.0	6.6	13.0	12.2
11	14 02.8	14 05.1	13 24.4	1.1	1.0	7.1	6.7	13.1	12.3
12	14 03.0	14 05.3	13 24.6	1.2	1.1	7.2	6.8	13.2	12.4
13	14 03.3	14 05.6	13 24.8	1.3	1.2	7.3	6.9	13.3	12.5
14	14 03.5	14 05.8	13 25.1	1.4	1.3	7.4	7.0	13.4	12.6
15	14 03.8	14 06.1	13 25.3	1.5	1.4	7.5	7.1	13.5	12.7
16	14 04.0	14 06.3	13 25.6	1.6	1.5	7.6	7.2	13.6	12.8
17	14 04.3	14 06.6	13 25.8	1.7	1.6	7.7	7.3	13.7	12.9
18	14 04.5	14 06.8	13 26.0	1.8	1.7	7.8	7.3	13.8	13.0
19	14 04.8	14 07.1	13 26.3	1.9	1.8	7.9	7.4	13.9	13.1
20	14 05.0	14 07.3	13 26.5	2.0	1.9	8.0	7.5	14.0	13.2
21	14 05.3	14 07.6	13 26.7	2.1	2.0	8.1	7.6	14.1	13.3
22	14 05.5	14 07.8	13 27.0	2.2	2.1	8.2	7.7	14.2	13.4
23	14 05.8	14 08.1	13 27.2	2.3	2.2	8.3	7.8	14.3	13.5
24	14 06.0	14 08.3	13 27.5	2.4	2.3	8.4	7.9	14.4	13.6
25	14 06.3	14 08.6	13 27.7	2.5	2.4	8.5	8.0	14.5	13.7
26	14 06.5	14 08.8	13 27.9	2.6	2.4	8.6	8.1	14.6	13.7
27	14 06.8	14 09.1	13 28.2	2.7	2.5	8.7	8.2	14.7	13.8
28	14 07.0	14 09.3	13 28.4	2.8	2.6	8.8	8.3	14.8	13.9
29	14 07.3	14 09.6	13 28.7	2.9	2.7	8.9	8.4	14.9	14.0
30	14 07.5	14 09.8	13 28.9	3.0	2.8	9.0	8.5	15.0	14.1
31	14 07.8	14 10.1	13 29.1	3.1	2.9	9.1	8.6	15.1	14.2
32	14 08.0	14 10.3	13 29.4	3.2	3.0	9.2	8.7	15.2	14.3
33	14 08.3	14 10.6	13 29.6	3.3	3.1	9.3	8.8	15.3	14.4
34	14 08.5	14 10.8	13 29.8	3.4	3.2	9.4	8.9	15.4	14.5
35	14 08.8	14 11.1	13 30.1	3.5	3.3	9.5	8.9	15.5	14.6
36	14 09.0	14 11.3	13 30.3	3.6	3.4	9.6	9.0	15.6	14.7
37	14 09.3	14 11.6	13 30.6	3.7	3.5	9.7	9.1	15.7	14.8
38	14 09.5	14 11.8	13 30.8	3.8	3.6	9.8	9.2	15.8	14.9
39	14 09.8	14 12.1	13 31.0	3.9	3.7	9.9	9.3	15.9	15.0
40	14 10.0	14 12.3	13 31.3	4.0	3.8	10.0	9.4	16.0	15.1
41	14 10.3	14 12.6	13 31.5	4.1	3.9	10.1	9.5	16.1	15.2
42	14 10.5	14 12.8	13 31.8	4.2	4.0	10.2	9.6	16.2	15.3
43	14 10.8	14 13.1	13 32.0	4.3	4.0	10.3	9.7	16.3	15.3
44	14 11.0	14 13.3	13 32.2	4.4	4.1	10.4	9.8	16.4	15.4
45	14 11.3	14 13.6	13 32.5	4.5	4.2	10.5	9.9	16.5	15.5
46	14 11.5	14 13.8	13 32.7	4.6	4.3	10.6	10.0	16.6	15.6
47	14 11.8	14 14.1	13 32.9	4.7	4.4	10.7	10.1	16.7	15.7
48	14 12.0	14 14.3	13 33.2	4.8	4.5	10.8	10.2	16.8	15.8
49	14 12.3	14 14.6	13 33.4	4.9	4.6	10.9	10.3	16.9	15.9
50	14 12.5	14 14.8	13 33.7	5.0	4.7	11.0	10.4	17.0	16.0
51	14 12.8	14 15.1	13 33.9	5.1	4.8	11.1	10.5	17.1	16.1
52	14 13.0	14 15.3	13 34.1	5.2	4.9	11.2	10.5	17.2	16.2
53	14 13.3	14 15.6	13 34.4	5.3	5.0	11.3	10.6	17.3	16.3
54	14 13.5	14 15.8	13 34.6	5.4	5.1	11.4	10.7	17.4	16.4
55	14 13.8	14 16.1	13 34.9	5.5	5.2	11.5	10.8	17.5	16.5
56	14 14.0	14 16.3	13 35.1	5.6	5.3	11.6	10.9	17.6	16.6
57	14 14.3	14 16.6	13 35.3	5.7	5.4	11.7	11.0	17.7	16.7
58	14 14.5	14 16.8	13 35.6	5.8	5.5	11.8	11.1	17.8	16.8
59	14 14.8	14 17.1	13 35.8	5.9	5.6	11.9	11.2	17.9	16.9
60	14 15.0	14 17.3	13 36.1	6.0	5.7	12.0	11.3	18.0	17.0

57ᵐ

57ᵐ s	SUN PLANETS	ARIES	MOON	v or Corrⁿ d		v or Corrⁿ d		v or Corrⁿ d	
00	14 15.0	14 17.3	13 36.1	0.0	0.0	6.0	5.8	12.0	11.5
01	14 15.3	14 17.6	13 36.3	0.1	0.1	6.1	5.8	12.1	11.6
02	14 15.5	14 17.8	13 36.5	0.2	0.2	6.2	5.9	12.2	11.7
03	14 15.8	14 18.1	13 36.8	0.3	0.3	6.3	6.0	12.3	11.8
04	14 16.0	14 18.3	13 37.0	0.4	0.4	6.4	6.1	12.4	11.9
05	14 16.3	14 18.6	13 37.2	0.5	0.5	6.5	6.2	12.5	12.0
06	14 16.5	14 18.8	13 37.5	0.6	0.6	6.6	6.3	12.6	12.1
07	14 16.8	14 19.1	13 37.7	0.7	0.7	6.7	6.4	12.7	12.2
08	14 17.0	14 19.3	13 38.0	0.8	0.8	6.8	6.5	12.8	12.3
09	14 17.3	14 19.6	13 38.2	0.9	0.9	6.9	6.6	12.9	12.4
10	14 17.5	14 19.8	13 38.4	1.0	1.0	7.0	6.7	13.0	12.5
11	14 17.8	14 20.1	13 38.7	1.1	1.1	7.1	6.8	13.1	12.6
12	14 18.0	14 20.3	13 38.9	1.2	1.2	7.2	6.9	13.2	12.7
13	14 18.3	14 20.6	13 39.2	1.3	1.2	7.3	7.0	13.3	12.7
14	14 18.5	14 20.9	13 39.4	1.4	1.3	7.4	7.1	13.4	12.8
15	14 18.8	14 21.1	13 39.6	1.5	1.4	7.5	7.2	13.5	12.9
16	14 19.0	14 21.4	13 39.9	1.6	1.5	7.6	7.3	13.6	13.0
17	14 19.3	14 21.6	13 40.1	1.7	1.6	7.7	7.4	13.7	13.1
18	14 19.5	14 21.9	13 40.3	1.8	1.7	7.8	7.5	13.8	13.2
19	14 19.8	14 22.1	13 40.6	1.9	1.8	7.9	7.6	13.9	13.3
20	14 20.0	14 22.4	13 40.8	2.0	1.9	8.0	7.7	14.0	13.4
21	14 20.3	14 22.6	13 41.1	2.1	2.0	8.1	7.8	14.1	13.5
22	14 20.5	14 22.9	13 41.3	2.2	2.1	8.2	7.9	14.2	13.6
23	14 20.8	14 23.1	13 41.5	2.3	2.2	8.3	8.0	14.3	13.7
24	14 21.0	14 23.4	13 41.8	2.4	2.3	8.4	8.1	14.4	13.8
25	14 21.3	14 23.6	13 42.0	2.5	2.4	8.5	8.1	14.5	13.9
26	14 21.5	14 23.9	13 42.3	2.6	2.5	8.6	8.2	14.6	14.0
27	14 21.8	14 24.1	13 42.5	2.7	2.6	8.7	8.3	14.7	14.1
28	14 22.0	14 24.4	13 42.7	2.8	2.7	8.8	8.4	14.8	14.2
29	14 22.3	14 24.6	13 43.0	2.9	2.8	8.9	8.5	14.9	14.3
30	14 22.5	14 24.9	13 43.2	3.0	2.9	9.0	8.6	15.0	14.4
31	14 22.8	14 25.1	13 43.4	3.1	3.0	9.1	8.7	15.1	14.5
32	14 23.0	14 25.4	13 43.7	3.2	3.1	9.2	8.8	15.2	14.6
33	14 23.3	14 25.6	13 43.9	3.3	3.2	9.3	8.9	15.3	14.7
34	14 23.5	14 25.9	13 44.2	3.4	3.3	9.4	9.0	15.4	14.8
35	14 23.8	14 26.1	13 44.4	3.5	3.4	9.5	9.1	15.5	14.9
36	14 24.0	14 26.4	13 44.6	3.6	3.5	9.6	9.2	15.6	15.0
37	14 24.3	14 26.6	13 44.9	3.7	3.5	9.7	9.3	15.7	15.0
38	14 24.5	14 26.9	13 45.1	3.8	3.6	9.8	9.4	15.8	15.1
39	14 24.8	14 27.1	13 45.4	3.9	3.7	9.9	9.5	15.9	15.2
40	14 25.0	14 27.4	13 45.6	4.0	3.8	10.0	9.6	16.0	15.3
41	14 25.3	14 27.6	13 45.8	4.1	3.9	10.1	9.7	16.1	15.4
42	14 25.5	14 27.9	13 46.1	4.2	4.0	10.2	9.8	16.2	15.5
43	14 25.8	14 28.1	13 46.3	4.3	4.1	10.3	9.9	16.3	15.6
44	14 26.0	14 28.4	13 46.5	4.4	4.2	10.4	10.0	16.4	15.7
45	14 26.3	14 28.6	13 46.8	4.5	4.3	10.5	10.1	16.5	15.8
46	14 26.5	14 28.9	13 47.0	4.6	4.4	10.6	10.2	16.6	15.9
47	14 26.8	14 29.1	13 47.3	4.7	4.5	10.7	10.3	16.7	16.0
48	14 27.0	14 29.4	13 47.5	4.8	4.6	10.8	10.4	16.8	16.1
49	14 27.3	14 29.6	13 47.7	4.9	4.7	10.9	10.4	16.9	16.2
50	14 27.5	14 29.9	13 48.0	5.0	4.8	11.0	10.5	17.0	16.3
51	14 27.8	14 30.1	13 48.2	5.1	4.9	11.1	10.6	17.1	16.4
52	14 28.0	14 30.4	13 48.5	5.2	5.0	11.2	10.7	17.2	16.5
53	14 28.3	14 30.6	13 48.7	5.3	5.1	11.3	10.8	17.3	16.6
54	14 28.5	14 30.9	13 48.9	5.4	5.2	11.4	10.9	17.4	16.7
55	14 28.8	14 31.1	13 49.2	5.5	5.3	11.5	11.0	17.5	16.8
56	14 29.0	14 31.4	13 49.4	5.6	5.4	11.6	11.1	17.6	16.9
57	14 29.3	14 31.6	13 49.7	5.7	5.5	11.7	11.2	17.7	17.0
58	14 29.5	14 31.9	13 49.9	5.8	5.6	11.8	11.3	17.8	17.1
59	14 29.8	14 32.1	13 50.1	5.9	5.7	11.9	11.4	17.9	17.2
60	14 30.0	14 32.4	13 50.4	6.0	5.8	12.0	11.5	18.0	17.3

58ᵐ

58	SUN PLANETS	ARIES	MOON	v or Corrⁿ d		v or Corrⁿ d		v or Corrⁿ d	
s	° ′	° ′	° ′	′	′	′	′	′	′
00	14 30·0	14 32·4	13 50·4	0·0	0·0	6·0	5·9	12·0	11·7
01	14 30·3	14 32·6	13 50·6	0·1	0·1	6·1	5·9	12·1	11·8
02	14 30·5	14 32·9	13 50·8	0·2	0·2	6·2	6·0	12·2	11·9
03	14 30·8	14 33·1	13 51·1	0·3	0·3	6·3	6·1	12·3	12·0
04	14 31·0	14 33·4	13 51·3	0·4	0·4	6·4	6·2	12·4	12·1
05	14 31·3	14 33·6	13 51·6	0·5	0·5	6·5	6·3	12·5	12·2
06	14 31·5	14 33·9	13 51·8	0·6	0·6	6·6	6·4	12·6	12·3
07	14 31·8	14 34·1	13 52·0	0·7	0·7	6·7	6·5	12·7	12·4
08	14 32·0	14 34·4	13 52·3	0·8	0·8	6·8	6·6	12·8	12·5
09	14 32·3	14 34·6	13 52·5	0·9	0·9	6·9	6·7	12·9	12·6
10	14 32·5	14 34·9	13 52·8	1·0	1·0	7·0	6·8	13·0	12·7
11	14 32·8	14 35·1	13 53·0	1·1	1·1	7·1	6·9	13·1	12·8
12	14 33·0	14 35·4	13 53·2	1·2	1·2	7·2	7·0	13·2	12·9
13	14 33·3	14 35·6	13 53·5	1·3	1·3	7·3	7·1	13·3	13·0
14	14 33·5	14 35·9	13 53·7	1·4	1·4	7·4	7·2	13·4	13·1
15	14 33·8	14 36·1	13 53·9	1·5	1·5	7·5	7·3	13·5	13·2
16	14 34·0	14 36·4	13 54·2	1·6	1·6	7·6	7·4	13·6	13·3
17	14 34·3	14 36·6	13 54·4	1·7	1·7	7·7	7·5	13·7	13·4
18	14 34·5	14 36·9	13 54·6	1·8	1·8	7·8	7·6	13·8	13·5
19	14 34·8	14 37·1	13 54·9	1·9	1·9	7·9	7·7	13·9	13·6
20	14 35·0	14 37·4	13 55·1	2·0	2·0	8·0	7·8	14·0	13·7
21	14 35·3	14 37·6	13 55·4	2·1	2·0	8·1	7·9	14·1	13·7
22	14 35·5	14 37·9	13 55·6	2·2	2·1	8·2	8·0	14·2	13·8
23	14 35·8	14 38·1	13 55·9	2·3	2·2	8·3	8·1	14·3	13·9
24	14 36·0	14 38·4	13 56·1	2·4	2·3	8·4	8·2	14·4	14·0
25	14 36·3	14 38·6	13 56·3	2·5	2·4	8·5	8·3	14·5	14·1
26	14 36·5	14 38·9	13 56·6	2·6	2·5	8·6	8·4	14·6	14·2
27	14 36·8	14 39·2	13 56·8	2·7	2·6	8·7	8·5	14·7	14·3
28	14 37·0	14 39·4	13 57·0	2·8	2·7	8·8	8·6	14·8	14·4
29	14 37·3	14 39·7	13 57·3	2·9	2·8	8·9	8·7	14·9	14·5
30	14 37·5	14 39·9	13 57·5	3·0	2·9	9·0	8·8	15·0	14·6
31	14 37·8	14 40·2	13 57·8	3·1	3·0	9·1	8·9	15·1	14·7
32	14 38·0	14 40·4	13 58·0	3·2	3·1	9·2	9·0	15·2	14·8
33	14 38·3	14 40·7	13 58·2	3·3	3·2	9·3	9·1	15·3	14·9
34	14 38·5	14 40·9	13 58·5	3·4	3·3	9·4	9·2	15·4	15·0
35	14 38·8	14 41·2	13 58·7	3·5	3·4	9·5	9·3	15·5	15·1
36	14 39·0	14 41·4	13 59·0	3·6	3·5	9·6	9·4	15·6	15·2
37	14 39·3	14 41·7	13 59·2	3·7	3·6	9·7	9·5	15·7	15·3
38	14 39·5	14 41·9	13 59·4	3·8	3·7	9·8	9·6	15·8	15·4
39	14 39·8	14 42·2	13 59·7	3·9	3·8	9·9	9·7	15·9	15·5
40	14 40·0	14 42·4	13 59·9	4·0	3·9	10·0	9·8	16·0	15·6
41	14 40·3	14 42·7	14 00·1	4·1	4·0	10·1	9·8	16·1	15·7
42	14 40·5	14 42·9	14 00·4	4·2	4·1	10·2	9·9	16·2	15·8
43	14 40·8	14 43·2	14 00·6	4·3	4·2	10·3	10·0	16·3	15·9
44	14 41·0	14 43·4	14 00·9	4·4	4·3	10·4	10·1	16·4	16·0
45	14 41·3	14 43·7	14 01·1	4·5	4·4	10·5	10·2	16·5	16·1
46	14 41·5	14 43·9	14 01·3	4·6	4·5	10·6	10·3	16·6	16·2
47	14 41·8	14 44·2	14 01·6	4·7	4·6	10·7	10·4	16·7	16·3
48	14 42·0	14 44·4	14 01·8	4·8	4·7	10·8	10·5	16·8	16·4
49	14 42·3	14 44·7	14 02·1	4·9	4·8	10·9	10·6	16·9	16·5
50	14 42·5	14 44·9	14 02·3	5·0	4·9	11·0	10·7	17·0	16·6
51	14 42·8	14 45·2	14 02·5	5·1	5·0	11·1	10·8	17·1	16·7
52	14 43·0	14 45·4	14 02·8	5·2	5·1	11·2	10·9	17·2	16·8
53	14 43·3	14 45·7	14 03·0	5·3	5·2	11·3	11·0	17·3	16·9
54	14 43·5	14 45·9	14 03·3	5·4	5·3	11·4	11·1	17·4	17·0
55	14 43·8	14 46·2	14 03·5	5·5	5·4	11·5	11·2	17·5	17·1
56	14 44·0	14 46·4	14 03·7	5·6	5·5	11·6	11·3	17·6	17·2
57	14 44·3	14 46·7	14 04·0	5·7	5·6	11·7	11·4	17·7	17·3
58	14 44·5	14 46·9	14 04·2	5·8	5·7	11·8	11·5	17·8	17·4
59	14 44·8	14 47·2	14 04·4	5·9	5·8	11·9	11·6	17·9	17·5
60	14 45·0	14 47·4	14 04·7	6·0	5·9	12·0	11·7	18·0	17·6

59ᵐ

59	SUN PLANETS	ARIES	MOON	v or Corrⁿ d		v or Corrⁿ d		v or Corrⁿ d	
s	° ′	° ′	° ′	′	′	′	′	′	′
00	14 45·0	14 47·4	14 04·7	0·0	0·0	6·0	6·0	12·0	11·9
01	14 45·3	14 47·7	14 04·9	0·1	0·1	6·1	6·0	12·1	12·0
02	14 45·5	14 47·9	14 05·2	0·2	0·2	6·2	6·1	12·2	12·1
03	14 45·8	14 48·2	14 05·4	0·3	0·3	6·3	6·2	12·3	12·2
04	14 46·0	14 48·4	14 05·6	0·4	0·4	6·4	6·3	12·4	12·3
05	14 46·3	14 48·7	14 05·9	0·5	0·5	6·5	6·4	12·5	12·4
06	14 46·5	14 48·9	14 06·1	0·6	0·6	6·6	6·5	12·6	12·5
07	14 46·8	14 49·2	14 06·4	0·7	0·7	6·7	6·6	12·7	12·6
08	14 47·0	14 49·4	14 06·6	0·8	0·8	6·8	6·7	12·8	12·7
09	14 47·3	14 49·7	14 06·8	0·9	0·9	6·9	6·8	12·9	12·8
10	14 47·5	14 49·9	14 07·1	1·0	1·0	7·0	6·9	13·0	12·9
11	14 47·8	14 50·2	14 07·3	1·1	1·1	7·1	7·0	13·1	13·0
12	14 48·0	14 50·4	14 07·5	1·2	1·2	7·2	7·1	13·2	13·1
13	14 48·3	14 50·7	14 07·8	1·3	1·3	7·3	7·2	13·3	13·2
14	14 48·5	14 50·9	14 08·0	1·4	1·4	7·4	7·3	13·4	13·3
15	14 48·8	14 51·2	14 08·3	1·5	1·5	7·5	7·4	13·5	13·4
16	14 49·0	14 51·4	14 08·5	1·6	1·6	7·6	7·5	13·6	13·5
17	14 49·3	14 51·7	14 08·7	1·7	1·7	7·7	7·6	13·7	13·6
18	14 49·5	14 51·9	14 09·0	1·8	1·8	7·8	7·7	13·8	13·7
19	14 49·8	14 52·2	14 09·2	1·9	1·9	7·9	7·8	13·9	13·8
20	14 50·0	14 52·4	14 09·5	2·0	2·0	8·0	7·9	14·0	13·9
21	14 50·3	14 52·7	14 09·7	2·1	2·1	8·1	8·0	14·1	14·0
22	14 50·5	14 52·9	14 09·9	2·2	2·2	8·2	8·1	14·2	14·1
23	14 50·8	14 53·2	14 10·2	2·3	2·3	8·3	8·2	14·3	14·2
24	14 51·0	14 53·4	14 10·4	2·4	2·4	8·4	8·3	14·4	14·3
25	14 51·3	14 53·7	14 10·6	2·5	2·5	8·5	8·4	14·5	14·4
26	14 51·5	14 53·9	14 10·9	2·6	2·6	8·6	8·5	14·6	14·5
27	14 51·8	14 54·2	14 11·1	2·7	2·7	8·7	8·6	14·7	14·6
28	14 52·0	14 54·4	14 11·4	2·8	2·8	8·8	8·7	14·8	14·7
29	14 52·3	14 54·7	14 11·6	2·9	2·9	8·9	8·8	14·9	14·8
30	14 52·5	14 54·9	14 11·8	3·0	3·0	9·0	8·9	15·0	14·9
31	14 52·8	14 55·2	14 12·1	3·1	3·1	9·1	9·0	15·1	15·0
32	14 53·0	14 55·4	14 12·3	3·2	3·2	9·2	9·1	15·2	15·1
33	14 53·3	14 55·7	14 12·6	3·3	3·3	9·3	9·2	15·3	15·2
34	14 53·5	14 55·9	14 12·8	3·4	3·4	9·4	9·3	15·4	15·3
35	14 53·8	14 56·2	14 13·0	3·5	3·5	9·5	9·4	15·5	15·4
36	14 54·0	14 56·4	14 13·3	3·6	3·6	9·6	9·5	15·6	15·5
37	14 54·3	14 56·7	14 13·5	3·7	3·7	9·7	9·6	15·7	15·6
38	14 54·5	14 56·9	14 13·8	3·8	3·8	9·8	9·7	15·8	15·7
39	14 54·8	14 57·2	14 14·0	3·9	3·9	9·9	9·8	15·9	15·8
40	14 55·0	14 57·5	14 14·2	4·0	4·0	10·0	9·9	16·0	15·9
41	14 55·3	14 57·7	14 14·5	4·1	4·1	10·1	10·0	16·1	16·0
42	14 55·5	14 58·0	14 14·7	4·2	4·2	10·2	10·1	16·2	16·1
43	14 55·8	14 58·2	14 14·9	4·3	4·3	10·3	10·2	16·3	16·2
44	14 56·0	14 58·5	14 15·2	4·4	4·4	10·4	10·3	16·4	16·3
45	14 56·3	14 58·7	14 15·4	4·5	4·5	10·5	10·4	16·5	16·4
46	14 56·5	14 59·0	14 15·7	4·6	4·6	10·6	10·5	16·6	16·5
47	14 56·8	14 59·2	14 15·9	4·7	4·7	10·7	10·6	16·7	16·6
48	14 57·0	14 59·5	14 16·1	4·8	4·8	10·8	10·7	16·8	16·7
49	14 57·3	14 59·7	14 16·4	4·9	4·9	10·9	10·8	16·9	16·8
50	14 57·5	15 00·0	14 16·6	5·0	5·0	11·0	10·9	17·0	16·9
51	14 57·8	15 00·2	14 16·9	5·1	5·1	11·1	11·0	17·1	17·0
52	14 58·0	15 00·5	14 17·1	5·2	5·2	11·2	11·1	17·2	17·1
53	14 58·3	15 00·7	14 17·3	5·3	5·3	11·3	11·2	17·3	17·2
54	14 58·5	15 01·0	14 17·6	5·4	5·4	11·4	11·3	17·4	17·3
55	14 58·8	15 01·2	14 17·8	5·5	5·5	11·5	11·4	17·5	17·4
56	14 59·0	15 01·5	14 18·0	5·6	5·6	11·6	11·5	17·6	17·5
57	14 59·3	15 01·7	14 18·3	5·7	5·7	11·7	11·6	17·7	17·6
58	14 59·5	15 02·0	14 18·5	5·8	5·8	11·8	11·7	17·8	17·7
59	14 59·8	15 02·2	14 18·8	5·9	5·9	11·9	11·8	17·9	17·8
60	15 00·0	15 02·5	14 19·0	6·0	6·0	12·0	11·9	18·0	17·9

TABLES FOR INTERPOLATING SUNRISE, MOONRISE, ETC.

TABLE I—FOR LATITUDE

| Tabular Interval | | | Difference between the times for consecutive latitudes | | | | | | | | | | | | | | | |
10°	5°	2°	5^m	10^m	15^m	20^m	25^m	30^m	35^m	40^m	45^m	50^m	55^m	60^m	1^h 05^m	1^h 10^m	1^h 15^m	1^h 20^m
° ′	° ′	° ′	m	m	m	m	m	m	m	m	m	m	m	m	h m	h m	h m	h m
0 30	0 15	0 06	0	0	1	1	1	1	1	2	2	2	2	2	0 02	0 02	0 02	0 02
1 00	0 30	0 12	0	1	1	2	2	3	3	3	4	4	4	5	05	05	05	05
1 30	0 45	0 18	1	1	2	3	3	4	4	5	5	6	7	7	07	07	07	07
2 00	1 00	0 24	1	2	3	4	5	5	6	7	7	8	9	10	10	10	10	10
2 30	1 15	0 30	1	2	4	5	6	7	8	9	9	10	11	12	12	13	13	13
3 00	1 30	0 36	1	3	4	6	7	8	9	10	11	12	13	14	0 15	0 15	0 16	0 16
3 30	1 45	0 42	2	3	5	7	8	10	11	12	13	14	16	17	18	18	19	19
4 00	2 00	0 48	2	4	6	8	9	11	13	14	15	16	18	19	20	21	22	22
4 30	2 15	0 54	2	4	7	9	11	13	15	16	18	19	21	22	23	24	25	26
5 00	2 30	1 00	2	5	7	10	12	14	16	18	20	22	23	25	26	27	28	29
5 30	2 45	1 06	3	5	8	11	13	16	18	20	22	24	26	28	0 29	0 30	0 31	0 32
6 00	3 00	1 12	3	6	9	12	14	17	20	22	24	26	29	31	32	33	34	36
6 30	3 15	1 18	3	6	10	13	16	19	22	24	26	29	31	34	36	37	38	40
7 00	3 30	1 24	3	7	10	14	17	20	23	26	29	31	34	37	39	41	42	44
7 30	3 45	1 30	4	7	11	15	18	22	25	28	31	34	37	40	43	44	46	48
8 00	4 00	1 36	4	8	12	16	20	23	27	30	34	37	41	44	0 47	0 48	0 51	0 53
8 30	4 15	1 42	4	8	13	17	21	25	29	33	36	40	44	48	0 51	0 53	0 56	0 58
9 00	4 30	1 48	4	9	13	18	22	27	31	35	39	43	47	52	0 55	0 58	1 01	1 04
9 30	4 45	1 54	5	9	14	19	24	28	33	38	42	47	51	56	1 00	1 04	1 08	1 12
10 00	5 00	2 00	5	10	15	20	25	30	35	40	45	50	55	60	1 05	1 10	1 15	1 20

Table I is for interpolating the L.M.T. of sunrise, twilight, moonrise, etc., for latitude. It is to be entered, in the appropriate column on the left, with the difference between true latitude and the nearest tabular latitude which is *less* than the true latitude; and with the argument at the top which is the nearest value of the difference between the times for the tabular latitude and the next higher one; the correction so obtained is applied to the time for the tabular latitude; the sign of the correction can be seen by inspection. It is to be noted that the interpolation is not linear, so that when using this table it is essential to take out the tabular phenomenon for the latitude *less* than the true latitude.

TABLE II—FOR LONGITUDE

| Long. East or West | Difference between the times for given date and preceding date (for east longitude) or for given date and following date (for west longitude) | | | | | | | | | | | | | | | | | | |
	10^m	20^m	30^m	40^m	50^m	60^m	1^h+ 10^m	20^m	30^m	1^h+ 40^m	50^m	60^m	2^h 10^m	2^h 20^m	2^h 30^m	2^h 40^m	2^h 50^m	3^h 00^m
°	m	m	m	m	m	m	m	m	m	m	m	m	h m	h m	h m	h m	h m	h m
0	0	0	0	0	0	0	0	0	0	0	0	0	0 00	0 00	0 00	0 00	0 00	0 00
10	0	1	1	1	1	2	2	2	2	3	3	3	04	04	04	04	05	05
20	1	1	2	2	3	3	4	4	5	6	6	7	07	08	08	09	09	10
30	1	2	2	3	4	5	6	7	7	8	9	10	11	12	12	13	14	15
40	1	2	3	4	6	7	8	9	10	11	12	13	14	16	17	18	19	20
50	1	3	4	6	7	8	10	11	12	14	15	17	0 18	0 19	0 21	0 22	0 24	0 25
60	2	3	5	7	8	10	12	13	15	17	18	20	22	23	25	27	28	30
70	2	4	6	8	10	12	14	16	17	19	21	23	25	27	29	31	33	35
80	2	4	7	9	11	13	16	18	20	22	24	27	29	31	33	36	38	40
90	2	5	7	10	12	15	17	20	22	25	27	30	32	35	37	40	42	45
100	3	6	8	11	14	17	19	22	25	28	31	33	0 36	0 39	0 42	0 44	0 47	0 50
110	3	6	9	12	15	18	21	24	27	31	34	37	40	43	46	49	0 52	0 55
120	3	7	10	13	17	20	23	27	30	33	37	40	43	47	50	53	0 57	1 00
130	4	7	11	14	18	22	25	29	32	36	40	43	47	51	54	0 58	1 01	1 05
140	4	8	12	16	19	23	27	31	35	39	43	47	51	54	0 58	1 02	1 06	1 10
150	4	8	13	17	21	25	29	33	38	42	46	50	0 54	0 58	1 03	1 07	1 11	1 15
160	4	9	13	18	22	27	31	36	40	44	49	53	0 58	1 02	1 07	1 11	1 16	1 20
170	5	9	14	19	24	28	33	38	42	47	52	57	1 01	1 06	1 11	1 16	1 20	1 25
180	5	10	15	20	25	30	35	40	45	50	55	60	1 05	1 10	1 15	1 20	1 25	1 30

Table II is for interpolating the L.M.T. of moonrise, moonset and the Moon's meridian passage for longitude. It is entered with longitude and with the difference between the times for the given date and for the preceding date (in east longitudes) or following date (in west longitudes). The correction is normally *added* for west longitudes and *subtracted* for east longitudes, but if, as occasionally happens, the times become earlier each day instead of later, the signs of the corrections must be reversed.

INDEX TO SELECTED STARS, 1998

Name	No	Mag	SHA	Dec
Acamar	7	3·1	315	S 40
Achernar	5	0·6	336	S 57
Acrux	30	1·1	173	S 63
Adhara	19	1·6	255	S 29
Aldebaran	10	1·1	291	N 17
Alioth	32	1·7	167	N 56
Alkaid	34	1·9	153	N 49
Al Na'ir	55	2·2	28	S 47
Alnilam	15	1·8	276	S 1
Alphard	25	2·2	218	S 9
Alphecca	41	2·3	126	N 27
Alpheratz	1	2·2	358	N 29
Altair	51	0·9	62	N 9
Ankaa	2	2·4	353	S 42
Antares	42	1·2	113	S 26
Arcturus	37	0·2	146	N 19
Atria	43	1·9	108	S 69
Avior	22	1·7	234	S 60
Bellatrix	13	1·7	279	N 6
Betelgeuse	16	Var.*	271	N 7
Canopus	17	−0·9	264	S 53
Capella	12	0·2	281	N 46
Deneb	53	1·3	50	N 45
Denebola	28	2·2	183	N 15
Diphda	4	2·2	349	S 18
Dubhe	27	2·0	194	N 62
Elnath	14	1·8	278	N 29
Eltanin	47	2·4	91	N 51
Enif	54	2·5	34	N 10
Fomalhaut	56	1·3	16	S 30
Gacrux	31	1·6	172	S 57
Gienah	29	2·8	176	S 18
Hadar	35	0·9	149	S 60
Hamal	6	2·2	328	N 23
Kaus Australis	48	2·0	84	S 34
Kochab	40	2·2	137	N 74
Markab	57	2·6	14	N 15
Menkar	8	2·8	314	N 4
Menkent	36	2·3	148	S 36
Miaplacidus	24	1·8	222	S 70
Mirfak	9	1·9	309	N 50
Nunki	50	2·1	76	S 26
Peacock	52	2·1	54	S 57
Pollux	21	1·2	244	N 28
Procyon	20	0·5	245	N 5
Rasalhague	46	2·1	96	N 13
Regulus	26	1·3	208	N 12
Rigel	11	0·3	281	S 8
Rigil Kentaurus	38	0·1	140	S 61
Sabik	44	2·6	102	S 16
Schedar	3	2·5	350	N 57
Shaula	45	1·7	97	S 37
Sirius	18	−1·6	259	S 17
Spica	33	1·2	159	S 11
Suhail	23	2·2	223	S 43
Vega	49	0·1	81	N 39
Zubenelgenubi	39	2·9	137	S 16

No	Name	Mag	SHA	Dec
1	Alpheratz	2·2	358	N 29
2	Ankaa	2·4	353	S 42
3	Schedar	2·5	350	N 57
4	Diphda	2·2	349	S 18
5	Achernar	0·6	336	S 57
6	Hamal	2·2	328	N 23
7	Acamar	3·1	315	S 40
8	Menkar	2·8	314	N 4
9	Mirfak	1·9	309	N 50
10	Aldebaran	1·1	291	N 17
11	Rigel	0·3	281	S 8
12	Capella	0·2	281	N 46
13	Bellatrix	1·7	279	N 6
14	Elnath	1·8	278	N 29
15	Alnilam	1·8	276	S 1
16	Betelgeuse	Var.*	271	N 7
17	Canopus	−0·9	264	S 53
18	Sirius	−1·6	259	S 17
19	Adhara	1·6	255	S 29
20	Procyon	0·5	245	N 5
21	Pollux	1·2	244	N 28
22	Avior	1·7	234	S 60
23	Suhail	2·2	223	S 43
24	Miaplacidus	1·8	222	S 70
25	Alphard	2·2	218	S 9
26	Regulus	1·3	208	N 12
27	Dubhe	2·0	194	N 62
28	Denebola	2·2	183	N 15
29	Gienah	2·8	176	S 18
30	Acrux	1·1	173	S 63
31	Gacrux	1·6	172	S 57
32	Alioth	1·7	167	N 56
33	Spica	1·2	159	S 11
34	Alkaid	1·9	153	N 49
35	Hadar	0·9	149	S 60
36	Menkent	2·3	148	S 36
37	Arcturus	0·2	146	N 19
38	Rigil Kentaurus	0·1	140	S 61
39	Zubenelgenubi	2·9	137	S 16
40	Kochab	2·2	137	N 74
41	Alphecca	2·3	126	N 27
42	Antares	1·2	113	S 26
43	Atria	1·9	108	S 69
44	Sabik	2·6	102	S 16
45	Shaula	1·7	97	S 37
46	Rasalhague	2·1	96	N 13
47	Eltanin	2·4	91	N 51
48	Kaus Australis	2·0	84	S 34
49	Vega	0·1	81	N 39
50	Nunki	2·1	76	S 26
51	Altair	0·9	62	N 9
52	Peacock	2·1	54	S 57
53	Deneb	1·3	50	N 45
54	Enif	2·5	34	N 10
55	Al Na'ir	2·2	28	S 47
56	Fomalhaut	1·3	16	S 30
57	Markab	2·6	14	N 15

*0·1 — 1·2

ALTITUDE CORRECTION TABLES 0°–35°—MOON

App. Alt.	0°–4° Corrⁿ	5°–9° Corrⁿ	10°–14° Corrⁿ	15°–19° Corrⁿ	20°–24° Corrⁿ	25°–29° Corrⁿ	30°–34° Corrⁿ	App. Alt.
00	0 33.8	5 58.2	10 62.1	15 62.8	20 62.2	25 60.8	30 58.9	00
10	35.9	58.5	62.2	62.8	62.1	60.8	58.8	10
20	37.8	58.7	62.2	62.8	62.1	60.7	58.8	20
30	39.6	58.9	62.3	62.8	62.1	60.7	58.7	30
40	41.2	59.1	62.3	62.8	62.0	60.6	58.6	40
50	42.6	59.3	62.4	62.7	62.0	60.6	58.5	50
00	1 44.0	6 59.5	11 62.4	16 62.7	21 62.0	26 60.5	31 58.5	00
10	45.2	59.7	62.4	62.7	61.9	60.4	58.4	10
20	46.3	59.9	62.5	62.7	61.9	60.4	58.3	20
30	47.3	60.0	62.5	62.7	61.9	60.3	58.2	30
40	48.3	60.2	62.5	62.7	61.8	60.3	58.2	40
50	49.2	60.3	62.6	62.7	61.8	60.2	58.1	50
00	2 50.0	7 60.5	12 62.6	17 62.7	22 61.7	27 60.1	32 58.0	00
10	50.8	60.6	62.6	62.6	61.7	60.1	57.9	10
20	51.4	60.7	62.6	62.6	61.6	60.0	57.8	20
30	52.1	60.9	62.7	62.6	61.6	59.9	57.8	30
40	52.7	61.0	62.7	62.6	61.5	59.9	57.7	40
50	53.3	61.1	62.7	62.6	61.5	59.8	57.6	50
00	3 53.8	8 61.2	13 62.7	18 62.5	23 61.5	28 59.7	33 57.5	00
10	54.3	61.3	62.7	62.5	61.4	59.7	57.4	10
20	54.8	61.4	62.7	62.5	61.4	59.6	57.4	20
30	55.2	61.5	62.8	62.5	61.3	59.6	57.3	30
40	55.6	61.6	62.8	62.4	61.3	59.5	57.2	40
50	56.0	61.6	62.8	62.4	61.2	59.4	57.1	50
00	4 56.4	9 61.7	14 62.8	19 62.4	24 61.2	29 59.3	34 57.0	00
10	56.7	61.8	62.8	62.3	61.1	59.3	56.9	10
20	57.1	61.9	62.8	62.3	61.1	59.2	56.9	20
30	57.4	61.9	62.8	62.3	61.0	59.1	56.8	30
40	57.7	62.0	62.8	62.2	60.9	59.1	56.7	40
50	57.9	62.1	62.8	62.2	60.9	59.0	56.6	50

H.P.	L U	L U	L U	L U	L U	L U	L U	H.P.
54.0	0.3 0.9	0.3 0.9	0.4 1.0	0.5 1.1	0.6 1.2	0.7 1.3	0.9 1.5	54.0
54.3	0.7 1.1	0.7 1.2	0.7 1.2	0.8 1.3	0.9 1.4	1.1 1.5	1.2 1.7	54.3
54.6	1.1 1.4	1.1 1.4	1.1 1.4	1.2 1.5	1.3 1.6	1.4 1.7	1.5 1.8	54.6
54.9	1.4 1.6	1.5 1.6	1.5 1.6	1.6 1.7	1.6 1.8	1.8 1.9	1.9 2.0	54.9
55.2	1.8 1.8	1.8 1.8	1.9 1.9	1.9 1.9	2.0 2.0	2.1 2.1	2.2 2.2	55.2
55.5	2.2 2.0	2.2 2.0	2.3 2.1	2.3 2.1	2.4 2.2	2.4 2.3	2.5 2.4	55.5
55.8	2.6 2.2	2.6 2.2	2.6 2.3	2.7 2.3	2.7 2.4	2.8 2.4	2.9 2.5	55.8
56.1	3.0 2.4	3.0 2.5	3.0 2.5	3.0 2.5	3.1 2.6	3.1 2.6	3.2 2.7	56.1
56.4	3.4 2.7	3.4 2.7	3.4 2.7	3.4 2.7	3.4 2.8	3.5 2.8	3.5 2.9	56.4
56.7	3.7 2.9	3.7 2.9	3.8 2.9	3.8 2.9	3.8 3.0	3.8 3.0	3.9 3.0	56.7
57.0	4.1 3.1	4.1 3.1	4.1 3.1	4.1 3.1	4.2 3.1	4.2 3.2	4.2 3.2	57.0
57.3	4.5 3.3	4.5 3.3	4.5 3.3	4.5 3.3	4.5 3.3	4.5 3.4	4.6 3.4	57.3
57.6	4.9 3.5	4.9 3.5	4.9 3.5	4.9 3.5	4.9 3.5	4.9 3.5	4.9 3.6	57.6
57.9	5.3 3.8	5.3 3.8	5.2 3.8	5.2 3.7	5.2 3.7	5.2 3.7	5.2 3.7	57.9
58.2	5.6 4.0	5.6 4.0	5.6 4.0	5.6 4.0	5.6 3.9	5.6 3.9	5.6 3.9	58.2
58.5	6.0 4.2	6.0 4.2	6.0 4.2	6.0 4.2	6.0 4.1	5.9 4.1	5.9 4.1	58.5
58.8	6.4 4.4	6.4 4.4	6.4 4.4	6.3 4.4	6.3 4.3	6.3 4.3	6.2 4.2	58.8
59.1	6.8 4.6	6.8 4.6	6.7 4.6	6.7 4.6	6.7 4.5	6.6 4.5	6.6 4.4	59.1
59.4	7.2 4.8	7.1 4.8	7.1 4.8	7.1 4.8	7.0 4.7	7.0 4.7	6.9 4.6	59.4
59.7	7.5 5.1	7.5 5.0	7.5 5.0	7.5 5.0	7.4 4.9	7.3 4.8	7.2 4.7	59.7
60.0	7.9 5.3	7.9 5.3	7.9 5.2	7.8 5.2	7.8 5.1	7.7 5.0	7.6 4.9	60.0
60.3	8.3 5.5	8.3 5.5	8.2 5.4	8.2 5.4	8.1 5.3	8.0 5.2	7.9 5.1	60.3
60.6	8.7 5.7	8.7 5.7	8.6 5.6	8.6 5.6	8.5 5.5	8.4 5.4	8.2 5.3	60.6
60.9	9.1 5.9	9.0 5.9	9.0 5.9	8.9 5.8	8.8 5.7	8.7 5.6	8.6 5.4	60.9
61.2	9.5 6.2	9.4 6.1	9.4 6.1	9.3 6.0	9.2 5.9	9.1 5.8	8.9 5.6	61.2
61.5	9.8 6.4	9.8 6.3	9.7 6.3	9.7 6.2	9.5 6.1	9.4 5.9	9.2 5.8	61.5

DIP

Ht. of Eye (m)	Corrⁿ	Ht. of Eye (ft)	Ht. of Eye (m)	Corrⁿ	Ht. of Eye (ft)
m		ft	m		ft
2.4	−2.8	8.0	9.5	−5.5	31.5
2.6	−2.9	8.6	9.9	−5.6	32.7
2.8	−3.0	9.2	10.3	−5.7	33.9
3.0	−3.1	9.8	10.6	−5.8	35.1
3.2	−3.2	10.5	11.0	−5.9	36.3
3.4	−3.3	11.2	11.4	−6.0	37.6
3.6	−3.4	11.9	11.8	−6.1	38.9
3.8	−3.5	12.6	12.2	−6.2	40.1
4.0	−3.6	13.3	12.6	−6.3	41.5
4.3	−3.7	14.1	13.0	−6.4	42.8
4.5	−3.8	14.9	13.4	−6.5	44.2
4.7	−3.9	15.7	13.8	−6.6	45.5
5.0	−4.0	16.5	14.2	−6.7	46.9
5.2	−4.1	17.4	14.7	−6.8	48.4
5.5	−4.2	18.3	15.1	−6.9	49.8
5.8	−4.3	19.1	15.5	−7.0	51.3
6.1	−4.4	20.1	16.0	−7.1	52.8
6.3	−4.5	21.0	16.5	−7.2	54.3
6.6	−4.6	22.0	16.9	−7.3	55.8
6.9	−4.7	22.9	17.4	−7.4	57.4
7.2	−4.8	23.9	17.9	−7.5	58.9
7.5	−4.9	24.9	18.4	−7.6	60.5
7.9	−5.0	26.0	18.8	−7.7	62.1
8.2	−5.1	27.1	19.3	−7.8	63.8
8.5	−5.2	28.1	19.8	−7.9	65.4
8.8	−5.3	29.2	20.4	−8.0	67.1
9.2	−5.4	30.4	20.9	−8.1	68.8
9.5		31.5	21.4		70.5

MOON CORRECTION TABLE

The correction is in two parts; the first correction is taken from the upper part of the table with argument apparent altitude, and the second from the lower part, with argument H.P., in the same column as that from which the first correction was taken. Separate corrections are given in the lower part for lower (L) and upper (U) limbs. All corrections are to be **added** to apparent altitude, but 30′ is to be *subtracted from the altitude of the upper limb.*

For corrections for pressure and temperature see page A4.

For bubble sextant observations ignore dip, take the mean of upper and lower limb corrections and subtract 15′ from the altitude.

App. Alt. = Apparent altitude = Sextant altitude corrected for index error and dip.

App. Alt.	35°–39° Corrⁿ	40°–44° Corrⁿ	45°–49° Corrⁿ	50°–54° Corrⁿ	55°–59° Corrⁿ	60°–64° Corrⁿ	65°–69° Corrⁿ	70°–74° Corrⁿ	75°–79° Corrⁿ	80°–84° Corrⁿ	85°–89° Corrⁿ	App. Alt.
00	35 56.5	40 53.7	45 50.5	50 46.9	55 43.1	60 38.9	65 34.6	70 30.1	75 25.3	80 20.5	85 15.6	00
10	56.4	53.6	50.4	46.8	42.9	38.8	34.4	29.9	25.2	20.4	15.5	10
20	56.3	53.5	50.2	46.7	42.8	38.7	34.3	29.7	25.0	20.2	15.3	20
30	56.2	53.4	50.1	46.5	42.7	38.5	34.1	29.6	24.9	20.0	15.1	30
40	56.2	53.3	50.0	46.4	42.5	38.4	34.0	29.4	24.7	19.9	15.0	40
50	56.1	53.2	49.9	46.3	42.4	38.2	33.8	29.3	24.5	19.7	14.8	50
00	36 56.0	41 53.1	46 49.8	51 46.2	56 42.3	61 38.1	66 33.7	71 29.1	76 24.4	81 19.6	86 14.6	00
10	55.9	53.0	49.7	46.0	42.1	37.9	33.5	29.0	24.2	19.4	14.5	10
20	55.8	52.8	49.5	45.9	42.0	37.8	33.4	28.8	24.1	19.2	14.3	20
30	55.7	52.7	49.4	45.8	41.8	37.7	33.2	28.7	23.9	19.1	14.1	30
40	55.6	52.6	49.3	45.7	41.7	37.5	33.1	28.5	23.8	18.9	14.0	40
50	55.5	52.5	49.2	45.5	41.6	37.4	32.9	28.3	23.6	18.7	13.8	50
00	37 55.4	42 52.4	47 49.1	52 45.4	57 41.4	62 37.2	67 32.8	72 28.2	77 23.4	82 18.6	87 13.7	00
10	55.3	52.3	49.0	45.3	41.3	37.1	32.6	28.0	23.3	18.4	13.5	10
20	55.2	52.2	48.8	45.2	41.2	36.9	32.5	27.9	23.1	18.2	13.3	20
30	55.1	52.1	48.7	45.0	41.0	36.8	32.3	27.7	22.9	18.1	13.2	30
40	55.0	52.0	48.6	44.9	40.9	36.6	32.2	27.6	22.8	17.9	13.0	40
50	55.0	51.9	48.5	44.8	40.8	36.5	32.0	27.4	22.6	17.8	12.8	50
00	38 54.9	43 51.8	48 48.4	53 44.6	58 40.6	63 36.4	68 31.9	73 27.2	78 22.5	83 17.6	88 12.7	00
10	54.8	51.7	48.2	44.5	40.5	36.2	31.7	27.1	22.3	17.4	12.5	10
20	54.7	51.6	48.1	44.4	40.3	36.1	31.6	26.9	22.1	17.3	12.3	20
30	54.6	51.5	48.0	44.2	40.2	35.9	31.4	26.8	22.0	17.1	12.2	30
40	54.5	51.4	47.9	44.1	40.1	35.8	31.3	26.6	21.8	16.9	12.0	40
50	54.4	51.2	47.8	44.0	39.9	35.6	31.1	26.5	21.7	16.8	11.8	50
00	39 54.3	44 51.1	49 47.6	54 43.9	59 39.8	64 35.5	69 31.0	74 26.3	79 21.5	84 16.6	89 11.7	00
10	54.2	51.0	47.5	43.7	39.6	35.3	30.8	26.1	21.3	16.5	11.5	10
20	54.1	50.9	47.4	43.6	39.5	35.2	30.7	26.0	21.2	16.3	11.4	20
30	54.0	50.8	47.3	43.5	39.4	35.0	30.5	25.8	21.0	16.1	11.2	30
40	53.9	50.7	47.2	43.3	39.2	34.9	30.4	25.7	20.9	16.0	11.0	40
50	53.8	50.6	47.0	43.2	39.1	34.7	30.2	25.5	20.7	15.8	10.9	50

H.P.	L U	L U	L U	L U	L U	L U	L U	L U	L U	L U	L U	H.P.
54.0	1.1 1.7	1.3 1.9	1.5 2.1	1.7 2.4	2.0 2.6	2.3 2.9	2.6 3.2	2.9 3.5	3.2 3.8	3.5 4.1	3.8 4.5	54.0
54.3	1.4 1.8	1.6 2.0	1.8 2.2	2.0 2.5	2.3 2.7	2.5 3.0	2.8 3.2	3.0 3.5	3.3 3.8	3.6 4.1	3.9 4.4	54.3
54.6	1.7 2.0	1.9 2.2	2.1 2.4	2.3 2.6	2.5 2.8	2.7 3.0	3.0 3.3	3.2 3.5	3.5 3.8	3.7 4.1	4.0 4.3	54.6
54.9	2.0 2.2	2.2 2.3	2.3 2.5	2.5 2.7	2.7 2.9	2.9 3.1	3.2 3.3	3.4 3.5	3.6 3.8	3.9 4.0	4.1 4.3	54.9
55.2	2.3 2.3	2.5 2.4	2.6 2.6	2.8 2.8	3.0 2.9	3.2 3.1	3.4 3.3	3.6 3.5	3.8 3.7	4.0 4.0	4.2 4.2	55.2
55.5	2.7 2.5	2.8 2.6	2.9 2.7	3.1 2.9	3.2 3.0	3.4 3.2	3.6 3.4	3.7 3.5	3.9 3.7	4.1 3.9	4.3 4.1	55.5
55.8	3.0 2.6	3.1 2.7	3.2 2.8	3.3 3.0	3.5 3.1	3.6 3.3	3.8 3.4	3.9 3.6	4.1 3.7	4.2 3.9	4.4 4.0	55.8
56.1	3.3 2.8	3.4 2.9	3.5 3.0	3.6 3.1	3.7 3.2	3.8 3.3	4.0 3.4	4.1 3.6	4.2 3.7	4.3 3.8	4.5 4.0	56.1
56.4	3.6 2.9	3.7 3.0	3.8 3.1	3.9 3.2	3.9 3.3	4.0 3.4	4.1 3.5	4.3 3.6	4.4 3.7	4.5 3.8	4.6 3.9	56.4
56.7	3.9 3.1	4.0 3.1	4.1 3.2	4.1 3.3	4.2 3.3	4.3 3.4	4.3 3.5	4.4 3.6	4.5 3.7	4.6 3.8	4.7 3.8	56.7
57.0	4.3 3.2	4.3 3.3	4.3 3.3	4.4 3.4	4.4 3.4	4.5 3.5	4.5 3.5	4.6 3.6	4.7 3.6	4.7 3.7	4.8 3.8	57.0
57.3	4.6 3.4	4.6 3.4	4.6 3.4	4.6 3.5	4.7 3.5	4.7 3.5	4.7 3.6	4.8 3.6	4.8 3.6	4.8 3.7	4.9 3.7	57.3
57.6	4.9 3.6	4.9 3.6	4.9 3.6	4.9 3.6	4.9 3.6	4.9 3.6	4.9 3.6	4.9 3.6	5.0 3.6	5.0 3.6	5.0 3.6	57.6
57.9	5.2 3.7	5.2 3.7	5.2 3.7	5.2 3.7	5.2 3.7	5.1 3.6	5.1 3.6	5.1 3.6	5.1 3.6	5.1 3.6	5.1 3.6	57.9
58.2	5.5 3.9	5.5 3.8	5.5 3.8	5.4 3.8	5.4 3.7	5.4 3.7	5.3 3.7	5.3 3.6	5.2 3.6	5.2 3.5	5.2 3.5	58.2
58.5	5.9 4.0	5.8 4.0	5.8 3.9	5.7 3.9	5.6 3.8	5.6 3.8	5.5 3.7	5.5 3.6	5.4 3.6	5.3 3.5	5.3 3.4	58.5
58.8	6.2 4.2	6.1 4.1	6.0 4.1	6.0 4.0	5.9 3.9	5.8 3.8	5.7 3.7	5.6 3.6	5.5 3.5	5.4 3.5	5.3 3.4	58.8
59.1	6.5 4.3	6.4 4.3	6.3 4.2	6.2 4.1	6.1 4.0	6.0 3.9	5.9 3.8	5.8 3.6	5.7 3.5	5.6 3.4	5.4 3.3	59.1
59.4	6.8 4.5	6.7 4.4	6.6 4.3	6.5 4.2	6.4 4.1	6.2 3.9	6.1 3.8	6.0 3.7	5.8 3.5	5.7 3.4	5.5 3.2	59.4
59.7	7.1 4.6	7.0 4.5	6.9 4.4	6.8 4.3	6.6 4.1	6.5 4.0	6.3 3.8	6.2 3.7	6.0 3.5	5.8 3.3	5.6 3.2	59.7
60.0	7.5 4.8	7.3 4.7	7.2 4.5	7.0 4.4	6.9 4.2	6.7 4.0	6.5 3.9	6.3 3.7	6.1 3.5	5.9 3.3	5.7 3.1	60.0
60.3	7.8 5.0	7.6 4.8	7.5 4.7	7.3 4.5	7.1 4.3	6.9 4.1	6.7 3.9	6.5 3.7	6.3 3.5	6.0 3.2	5.8 3.0	60.3
60.6	8.1 5.1	7.9 5.0	7.7 4.8	7.6 4.6	7.3 4.4	7.1 4.2	6.9 3.9	6.7 3.7	6.4 3.4	6.2 3.2	5.9 2.9	60.6
60.9	8.4 5.3	8.2 5.1	8.0 4.9	7.8 4.7	7.6 4.5	7.3 4.2	7.1 4.0	6.8 3.7	6.6 3.4	6.3 3.2	6.0 2.9	60.9
61.2	8.7 5.4	8.5 5.2	8.3 5.0	8.1 4.8	7.8 4.5	7.6 4.3	7.3 4.0	7.0 3.7	6.7 3.4	6.4 3.1	6.1 2.8	61.2
61.5	9.1 5.6	8.8 5.4	8.6 5.1	8.3 4.9	8.1 4.6	7.8 4.3	7.5 4.0	7.2 3.7	6.9 3.4	6.5 3.1	6.2 2.7	61.5

LIST OF CONTENTS